Completely Revised Edition 2021-22

NEET
Exploring BIOLOGY

Volume 1

Sanjay Sharma
Sudhakar Banerjee
Supported by
Mahendra Singh Rawat

arihant
ARIHANT PRAKASHAN (Series), MEERUT

NEET
Exploring
BIOLOGY

arihant

ARIHANT PRAKASHAN (Series), MEERUT

卐 **Administrative & Production Offices**

Regd. Office

'Ramchhaya' 4577/15, Agarwal Road, Darya Ganj, New Delhi -110002
Tele: 011- 47630600, 43518550

卐 **Head Office**

Kalindi, TP Nagar, Meerut (UP) - 250002, Tel: 0121-7156203, 7156204

卐 **Sales & Support Offices**

Agra, Ahmedabad, Bengaluru, Bareilly, Chennai, Delhi, Guwahati, Hyderabad, Jaipur, Jhansi, Kolkata, Lucknow, Nagpur & Pune.

卐 **ISBN** 978-93-25793-09-5

卐 **PRICE** ₹850.00

PO No. : TXT-59-T042330-8-21

Published by Arihant Publications (India) Ltd.

PRODUCTION TEAM

Publishing Manager : Keshav Mohan
Project Head : Shiva Choudhary
Project Editor : Dr. Anshu Gupta, Sanubia, Kavita
Cover Designer : Bilal Hashmi
Inner Designer : Ankit Saini
Page Layouting : Deepak Bisht, Sachin, Naresh
Proof Readers : Nidhi, Nalini

For further information about the books published by Arihant, log on to www.arihantbooks.com or e-mail at info@arihantbooks.com

Follow us on

PREFACE

Mastery on Biology is important for students who are planning to make their career in medicine. To get this you require an objective approach in your studies. This Objective Approach can be defined as the approach through which a student is able to master the concepts and also the skills required to tackle the questions asked in different formats in entrances such as NEET (National Eligibility cum Entrance Test).

Although the number of medical colleges imparting quality education and training has significantly increased after independence in the country but due to the simultaneous increase in the number of serious aspirants, the competition is no longer easy for a seat in a prestigious medical college today. Present changes in NEET exam Pattern has increased this challenge many fold.

Exploring Biology (Vol.1 & 2) are borne out of our experience of teaching biology to medical aspirants. The content of these books is framed to provide helps in facing present day challenges by future medicos.

The plan of the presentation of the subject matter in the books is as follows

- The whole chapter has been divided under logical topic heads to cover the whole syllabi of NEET developing the concepts in an easy going manner, taking the help of suitable examples, diagrams, flow charts, table, etc.
- Important points of the topics have been highlighted in the text under notes, some extra points regarding the topics have been given in Notes to enrich the students.
- **Check Point** Exercises given in between the text of all chapter help the readers to remain linked with the text given as they provide them an opportunity to assess themselves while studying the text.
- At the end of the theory part **Exam Points** are given in every chapter. These points are framed from last 10 years (2011-2020) questions for providing fast track revision.
- Assignments at the end of the chapters have been divided into three parts:
  Part A- 'Topical Exercises' having simple NEET level MCQs arranged in accordance with content flow of the chapter. This exercise helps in building basics about the chapter.

  Part B- 'Cumulative Exercises' has advanced level objective questions of concerned chapter. The special point of this exercise is that questions have been arranged systematically with Assertion-Reason questions in the last. This exercise helps in gaining conceptual clarity of advanced level.

  Part C- 'Medical Entrances Gallery' covering all questions asked in last 10 years' (2020-2011) in NEET & other Medical Entrances.
- The Explanation of some selected question given in different exercises have been provided.
- Appendix given in the last is an additional feature to make the text more informative.
- Multi-Coloured pictures related to important topics of chapters are given at the end of book.

At the end we would like to say that suggestions from the respected Teachers & Students for the further improvement of the book will be welcomed open heartedly.

Huge efforts have been made from our side to keep this book error free, but inspite of that if any error or whatsoever is skipped in the book then that is purely incidental, apology for the same, please write to us about that so that it can be corrected in the further edition of the book.

Sanjay Sharma
Sudhakar Banerjee

SYLLABUS

UNIT I Diversity In Living World

What is living? ; Biodiversity, Need for classification, Three domains of life, Taxonomy and Systematics, Concept of species and taxonomical hierarchy, Binomial nomenclature, Tools for study of Taxonomy – Museums, Zoos, Herbaria, Botanical gardens. Five kingdom classification, salient features and classification of Monera, Protista and Fungi into major groups, Lichens, Viruses and Viroids. Salient features and classification of plants into major groups-Algae, Bryophytes, Pteridophytes, Gymnosperms and Angiosperms (three to five salient and distinguishing features and at least two examples of each category), Angiosperms-classification up to class, characteristic features and example.

Salient features and classification of animals-nonchordate upto phyla level and chordate up to classes level (three to five salient features and at least two examples).

UNIT II Structural Organisation in Animals and Plants

Morphology and modifications, Tissues, Anatomy and functions of different parts of flowering plants: Root, stem, leaf, inflorescence- cymose and recemose, flower, fruit and seed (To be dealt along with the relevant practical of the Practical Syllabus). Animal tissues; Morphology, anatomy and functions of different systems (digestive, circulatory, respiratory, nervous and reproductive) of an insect (cockroach). (Brief account only)

UNIT III Cell Structure and Function

Cell theory and cell as the basic unit of life; Structure of prokaryotic and eukaryotic cell, Plant cell and animal cell, Cell envelope, cell membrane, cell wall, Cell organelles-structure and function, Endomembrane system-endoplasmic reticulum, Golgi bodies, lysosomes, vacuoles, mitochondria, ribosomes, plastids, micro bodies, Cytoskeleton, cilia, flagella, centrioles (ultra structure and function), Nucleus-nuclear membrane, chromatin, nucleolus.

Chemical constituents of living cells: Biomolecules-structure and function of proteins, carbodydrates, lipids, nucleic acids, Enzymes-types, properties, enzyme action.

B Cell division: Cell cycle, mitosis, meiosis and their significance.

UNIT IV Plant Physiology

Transport in plants: Movement of water, gases and nutrients, Cell to cell transport-Diffusion, facilitated diffusion, active transport, Plant– water relations – Imbibition, water potential, osmosis, plasmolysis, Long distance transport of water – Absorption, apoplast, symplast, transpiration pull, root pressure and guttation, Transpiration-Opening and closing of stomata, Uptake and translocation of mineral nutrients-Transport of food, phloem transport, Mass flow hypothesis, Diffusion of gases (brief mention).

Mineral nutrition: Essential minerals, macro and micronutrients and their role, Deficiency symptoms, Mineral toxicity, Elementary idea of Hydroponics as a method to study mineral nutrition, **Nitrogen metabolism**-Nitrogen cycle, biological nitrogen fixation.

Photosynthesis: Photosynthes is as a means of Autotrophic nutrition, Site of photosynthesis take place, pigments involved in Photosynthesis (Elementary idea), Photochemical and biosynthetic phases of photosynthesis, Cyclic and non-cyclic and photophosphorylation, Chemiosmotic hypothesis, Photorespiration C_3 and C_4 pathways, Factors affecting photosynthesis.

Respiration: Exchange gases, Cellular respiration-glycolysis, fermentation (anaerobic), TCA cycle and electron transport system (aerobic), Energy relations–Number of ATP molecules generated; Amphibolic pathways, Respiratory quotient.

Plant growth and development: Seed germination, Phases of Plant growth and plant growth rate, Conditions of growth, Differentiation, dedifferentiation and redifferentiation, Sequence of developmental process in a plant cell, Growth regulators-auxin,gibberellin, cytokinin, ethylene, ABA; Seed dormancy, Vernalisation, Photoperiodism.

UNIT V Human Physiology

Digestion and absorption; Alimentary canal and digestive glands; Role of digestive enzymes and gastrointestinal hormones; Peristalsis, digestion, absorption and assimilation of proteins, carbohydrates and fats; Caloric value of proteins, carbohydrates and fats; Egestion; Nutritional and digestive disorders – PEM, indigestion, constipation, vomiting, jaundice, diarrhea.

Breathing and Respiration: Respiratory organs in animals (recall only), Respiratory system in humans, Mechanism of breathing and its regulation in humans-Exchange of gases, transport of gases and regulation of respiration. Respiratory volumes, Disorders related to respiration-Asthma, Emphysema, Occupational respiratory disorders.

Body fluids and circulation: Composition of blood, blood groups, coagulation of blood, Composition of lymph and its function, Human circulatory system-Structure of human heart and blood vessels; Cardiac cycle, cardiac output, ECG, Double circulation, Regulation of cardiac activity, Disorders of circulatory system-Hypertension, Coronary artery disease, Angina pectoris, Heart failure.

Excretory products and their elimination: Modes of excretion- Ammonotelism, ureotelism, uricotelism; Human excretory system-structure and function; Urine formation, Osmoregulation, Regulation of kidney function-Renin-angiotensin, Atrial Natriuretic Factor, ADH and Diabetes insipidus, Role of other organs in excretion, Disorders, Uraemia, Renal failure, Renal calculi, Nephritis, Dialysis and artificial kidney.

Locomotion and Movement: Types of movement- ciliary, fiagellar, muscular; Skeletal muscle- contractile proteins and muscle contraction; Skeletal system and its functions (To be dealt with the relevant practical of Practical syllabus), Joints, Disorders of muscular and skeletal system-Myasthenia gravis, Tetany, Muscular dystrophy, Arthritis, Osteoporosis, Gout.

Neural control and coordination: Neuron and nerves, Nervous system in humans- central nervous system, peripheral nervous system and visceral nervous system; Generation and conduction of nerve impulse, Reflex action, Sense organs, Elementary structure & function of eye and ear.

Chemical coordination and regulation: Endocrine glands and hormones, Human endocrine system-Hypothalamus, Pituitary, Pineal, Thyroid, Parathyroid, Adrenal, Pancreas, Gonads; Mechanism of hormone action (Elementary Idea), Role of hormones as messengers and regulators, Hypo-and hyperactivity and related disorders (Common disorders *e.g.*, Dwarfism, Acromegaly, Cretinism, goiter, exopthalmic goiter, diabetes, Addison's disease). (Important, Diseases and disorders mentioned above to be dealt in brief.)

CONTENTS

NEET Exploring BIOLOGY

CHAPTER 01

The Living World

Life can be defined as a unique aggregation of biomolecules. These biomolecules help in processes like growth, development, responsiveness, adaptations and reproduction. The forms that possess and express all these features are called **living beings** or **living organisms**.

These organisms live in almost all possible habitats like forests, mountains, deserts, oceans, freshwater bodies, hot springs, polar regions, etc. They are linked to one another by sharing the common genetic material.

Defining characteristics of living beings

Most important or defining characteristics of a living being are as follows

Metabolism

The term 'metabolism' is applied to all the constructive and destructive reactions by which living beings assimilate and utilise energy. This energy is used for various purposes like growth, movement, development, responsiveness, reproduction, etc. It is considered as the most important characteristic of living beings.

Metabolism is further categorised into anabolism and catabolism. **Anabolism** includes all constructive reactions whereas **catabolism** includes all destructive reactions.

Table 1.1 Differences between anabolism and catabolism

Anabolism	Catabolism
Anabolic reactions store energy, i.e. endergonic.	Catabolic reactions release energy, i.e. exergonic.
These reactions produce complex substances from simpler ones. e.g. Photosynthesis : $6CO_2 + 12H_2O \xrightarrow[\text{enzymes}]{\text{Chlorophyll, sunlight (energy)}} C_6H_{12}O_6 + 6H_2O + 6O_2 \uparrow$	These reactions form simple substances from complex ones.e.g. Respiration : $C_6H_{12}O_6 + 6O_2 \xrightarrow{\text{Enzymes}} 6CO_2 + 6H_2O + \text{Energy}$

Some metabolic reactions can also be carried out in cell-free systems or outside the cells such as *in vitro* conditions.

Consciousness

It is the characteristic which makes the living organisms aware of the surroundings environmental stimuli. These can be environment's physical, chemical or biological stimuli.

Inside

Animals sense these stimuli through their sensory organs, while plants have different ways of detecting these stimuli. Human being is the only organism who is aware of himself, i.e. has 'self consciousness'.

Responsiveness

All organisms (from the simplest prokaryotes to the most complex eukaryotes) respond to environmental signals or external stimuli. This property of living beings is also called **irritability**.

Cellular organisation

All organisms consist of one or more cells, so cellular organisation is a defining property of living beings. Unicellular organisms have a cell like body. On the other hand, in higher multicellular organisms cells work together in hierarchial manner to form tissues, organ and organ systems.

Non-defining characteristics of living beings

Following characteristics are considered as less important or less defining as these may easily be seen in non-livings as well.

Reproduction

It is the formation of new individuals of similar kind. Reproduction is of following two types

- **Asexual** (uniparental) reproduction
- **Sexual** (biparental) reproduction.

The process of reproduction is essential for maintaining the continuity of life on the earth.

Many organisms do not reproduce at all, e.g. mules, sterile worker bees, infertile human couple, etc.

Hence, reproduction cannot be an all inclusive defining characteristic of living organisms.

Reproductive strategies

Different living organisms have different reproductive strategies, e.g.

- For a unicellular organism reproduction is synonymous to growth in whole organism, while multicellular organisms may show growth in parts.
- The organisms can also be semelparous or iteroparous depending upon their reproductive strategies.
  Semelparous organisms reproduce only once before death, whereas, **iteroparous** organisms go through multiple reproductive cycles before death.

Metabolism *Vs* Reproduction

Although reproduction and metabolism both are most important features of living beings, but in comparative terms the importance of metabolism is higher. For example, we all know that viruses have power of reproduction, but they do not show metabolism. It is due to the lack of metabolism only we consider viruses as non-living.

Growth

This process is defined as 'An irreversible quantitative increase in the mass of an individual'.

Growth in living beings

All living beings show growth either by increase in number or by increase in size of cells. In living beings, the growth producing substances are of two types

(i) **Protoplasmic substances** These are components of living matter like cytoplasm and nucleoplasm.

(ii) **Apoplasmic substances** These are non-living materials formed by the cells, like cell wall, fibres of connective tissue, etc.

Growth through above both the substances is called **internal**, i.e. **intrinsic growth.**

Growth in non-living beings

Though growth is a common characteristic feature of all living beings, but it is not a defining property of living beings as non-living things also show growth, e.g. mountains boulders, sand dunes, etc. In them, growth occurs *via* accumulation of materials externally. This kind of growth is called **external** or **extrinsic growth.**

Nutrition

All living organisms require food as a source of energy for carrying out their life processes.

On the basis of nutritional habits the organism can be autotrophs or heterotrophs.

(i) **Autotrophs** are those organisms which have the ability to synthesise their own food, e.g. bacteria, algae and plants.

(ii) **Heterotrophs** are those organisms which are always dependent upon autotrophs for their nutritional needs, e.g. animals, humans and fungi.

Respiration

All living beings obtain energy *via* oxidation of nutrients taken. The process of nutrient oxidation is called respiration. The obtained energy is stored in an individual's body as ATP (Adenosine Triphosphate).

Locomotion and movement

Animals and some unicellular forms can move from one place to another on their own and this ability is called **locomotion.** Plants though lack locomotion, but fulfill their requirements *via* dispersal and certain very restricted slow movements.

All the living organisms usually move a body part or parts without change in their position. This process is called **movement**, e.g. pumping of heart, opening of flower buds into a flower.

Excretion

Every living cell, produces certain waste products that requires continuous removal as it might become poisonous. The process responsible for the elimination of these waste products is called **excretion.**

Carbon dioxide and water are the universal major waste products of living organisms. In animals, excretion is conducted through specialised organs, e.g. kidneys in most of the cases, while in plants excess water removal is done through stomata.

Homeostasis

(*Gr. Homois–alike; stasis–standing*)

Maintenance of a favourable consistantly dynamic internal environment in the body is called homeostasis. The term 'Homeostasis' was first used by **Cannon** (1932).

It is a property of all living organisms and is carried out by various regulatory mechanisms.
Homeostasis works even with the continuous changes in external environment. This process occurs at all levels, i.e. from cellular level to the ecosystem level. One of the examples of these processes is thermoregulation.

On the basis of thermoregulation there are two types of organisms

(i) **Ectothermic organisms** (Gr. *Ektos*–outside or external) The body temperature of these organisms vary with that of the surrounding environment, e.g. fishes, amphibians, reptiles, invertebrates, plants. They are also known as **poikilothermic** or **cold blooded** organisms.

(ii) **Endothermic organisms** (Gr. *Endon*–within or internal) The body temperature of these organisms remains constant and is independent of the environmental conditions, e.g. birds, mammals and few fishes. These organisms are also known as **homeotherms** or **warm blooded** organisms.

Adaptations

Inheritable traits that allow an organism to successfully meet specific requirements of its environment are called adaptations. These are responsible for the enhancement of their lifespan and ability to reproduce in a particular environment.

On the basis of the duration of adaptation in living beings, these are of following two types

(i) **Long term adaptations** These are permanent inheritable changes in the structure and function of an organism due to evolution. These are meant for better mode of living during favourable conditions. They do not disappear under changed conditions. For example, non-venomous scarlet snake superficially mimic the appearance of venomous coral snake to defer predators.

This process of mimicry in two different organisms with similar appearance, but opposite behaviour proves that mimicry is evolved as a long term adaptation.

(ii) **Short term adaptations** These are temporary changes that are developed in response to specific conditions. These are meant for smooth adjustment to unfavourable conditions. These adaptations disappear in the absence of stressfull conditions. Short term adaptations include **hibernation** or winter sleep and **aestivation** or summer sleep.

Ageing and death

Ageing is a progressive deterioration of structure and function of cells, tissues and vital organs. It leads to death.

Death is stoppage of all activities in an individual due to degeneration of body parts and increase in entropy.

The death can be **clinical**, i.e. when person's heart stops functioning. It may lead to **biological death** when along with heart, brain also stops functioning.

Death of an organism is essential for keeping a population under check and recycling of minerals.

After death, elements that constitute a body are returned to the ecosystem by nutrient recycling.

Lifespan

The period of time between the birth and natural death of an organism is known as lifespan. Some organisms possess a shorter lifespan like mayfly, whose adult lifespan can vary from just 30 minutes to one day. On the other hand, some others live for years, e.g. tortoises have a lifespan about 150 years.

Lifespans of some other common animals is given below

Animals	Lifespan	Animals	Lifespan
Elephant	60-90 years	Parrot	140 years
Horse	25-30 years	Eagle	55 years
Chimpanzee	40 years	Humming bird	8 years
Dog	25-30 years	Crow	15 years
Squirrel	16 years	Tortoise	100-150 years
Cow	20-25 years		

Animals	Lifespan	Plants	Lifespan
Catfish	60 years	Rose	5-7 years
Crocodile	60 years	Rice plant	3-7 months
Giant salamander	55 years	Banana tree	2-3 years
Toad	36 years	Banyan tree	200-1000 years
Ant (queen)	15 years		
Butterfly	1-2 weeks		

Biologically immortal

Some Jelly fishes (e.g. *Aurelia* sp., *Turritopsis dohrnii*) have the ability to revert their polyp stage rather than dying. They does this even after reproducing and the phenomenon is called **life cycle reversal.** Consequently these species are considered as biologically immortal and have maximum lifespan.

CHECK POINT 1.1

1. When green plants capture sunlight and utilise it to synthesise glucose, the reactions involved during this process are said to be
   (a) catabolic reactions
   (b) anabolic reactions
   (c) decomposition reactions
   (d) chain reactions
2. 'Viruses cannot die' as they do not show
   (a) reproduction (b) responsiveness
   (c) metabolism (d) cellular organisation
3. Consciousness is the defining property of living organisms because
   (a) photoperiod affects reproduction in seasonal breeders in both plants and animals
   (b) plants respond to external factors like temperature and light
   (c) human is aware of himself
   (d) All of the above
4. The differentiation between semelparous and iteroparous can be done on the basis of
   (a) cellular organisation (b) growth
   (c) nutritional cycles (d) reproductive cycle
5. One can observe growth in unicellular organism by
   (a) counting the mass of cultured cells
   (b) analysing the amount of nutrients absorbed by living organisms
   (c) simply counting the number of cells under microscope during *in vitro* culture
   (d) simply counting the number of cells under microscope during *in vivo* culture.
6. Choose the correct statement.
   (a) Long term adaptations are non-inheritable
   (b) Ectotherms are also called homeotherms
   (c) Short term adaptations are temporary
   (d) All the statements are correct

DIVERSITY IN THE LIVING WORLD

- Planet Earth exhibits a large variety of living organisms, like plants, animals, fungi, etc. These all constitute our 'Biodiversity'.
- In it each and every different kind of organism represents a species. The number of species that are known and described range between 1.7-1.8 million.
- These organisms inhabit distinct geographical locations.
- Thus, every distinct geographical location has its own flora (plants) and fauna (animals).
- Dense tropical rain forests and water reefs are the most common homes of living organisms.

Classification

- Due to the enormous diversity present in the living world, it is not possible to study each organism separately.
- So, for the sake of study, it is necessary to place them into groups on the basis of their characteristics.
- This arrangement of organisms into convenient categories or groups is called **classification.**
- It is done on the basis of their similarities and differences, in their certain easily observable fundamental characters.

Need for classification

Classification is necessary for organisms due to following reasons

- Convenience of study
- Knowledge of adaptations
- Knowledge of sequence of evolution
- Knowledge of phylogenetic relationships
- Discovery of new organisms

Taxonomy and systematics

The term taxonomy (Gk. *taxis*-arrangement; *nomos*-law) was given by **AP de Candolle** in 1813. It is the branch of science, that deals with principles and procedures of identification, nomenclature and classification of organisms.

Systematics is a branch of science that deals with identification, nomenclature, classification and evolutionary history of an organism. Thus, systematics includes evolutionary history along with taxonomic characteristics of an organism.

Branches of taxonomy

Various branches of taxonomy are as follows

(i) **Classical taxonomy** or α**-taxonomy** is based on the morphological traits.

(ii) **Morphotaxonomy** is based on morphological studies of organisms.

(iii) **Natural taxonomy** is based on the natural similarities amongst organisms.

(iv) **Phylogenetic taxonomy** is based on phylogeny or evolutionary history of a species.

(v) **Experimental taxonomy** is based on experimental determination of genetic inter-relationships.

(vi) **Chemotaxonomy** is based on the presence or absence of certain chemicals in cells or tissues.

(vii) **Numerical taxonomy** is based on the number of shared characters of various organisms.

(viii) **Cytotaxonomy** is based on cytological studies.

(ix) **Karyotaxonomy** is based on nuclear and chromosomal studies.

(x) **Practical taxonomy** is based on the utility of organisms.

Note ***Neosystematics*** *It is a concept of systematics that considers a species to be the product of evolution. This concept was developed by Julian Huxley in 1940. It takes into consideration all the known characteristics of organisms and all the known evidences from different fields of biology.*

Functional aspects of taxonomy

A normal taxonomic study involves following four steps

(i) **Characterisation** In this method, all the characteristics of an organism are noted and afterwards used for the description of organisms.

(ii) **Identification** To identify an organism, its similarities and dissimilarities are compared with already known organism.

(iii) **Classification** It is the method of arranging organisms into groups on the basis of their physical, cellular, morphological and other characteristics.

(iv) **Nomenclature** The organisms are given scientific names according to the internationaly by accepted rules.

Out of these steps, identification and classification deals with practical approach while nomenclature requires theoretical approach.

Nomenclature

It is the science of providing distinct and proper name to the organisms. These names are helpful in their recognition and differentiation from others. These names can be vernacular (common) and scientific (technical).

Vernacular names or common names

We know the plants and animals in our own area by their local names. These local names vary from place to place, even within a country.

Due to language variability, an organism can have many vernacular names. The vernacular names are easier to pronounce and are known to the residents of a particular area only.

Vernacular names *Vs* Taxonomy

Vernacular names are not suitable in taxonomy due to following reasons

- They cannot be used by biologists, because they may have no significance, e.g. widow's tears (*Tradescantia* sp.) and love-in-a-mist (*Nigella damascena*).
- Some common names may be misnomers, i.e. have incorrect meanings, e.g. silverfish, jellyfish, cuttlefish, starfish. All of these belong to different phyla and have no relationship with true fishes.
- Sometimes, a single plant is known by several vernacular names in same language, e.g. water lily has 81 Dutch names, 44 French names and 15 English names.
- A single name may often be used for two or more species, e.g. 'touch-me-not' is the name for both *Impatiens balsamifera* (balsam) and *Mimosa pudica* (chui-mui or sensitive plant).

Hence, from taxonomy prospective there is a need to standardise the naming of living organisms.

Scientific or Technical names

These are the names given to the organisms by biologists based on agreed principles and criteria. These are acceptable all over the world. To accomplish this, certain international codes have been established.

These codes are

- **ICN**, i.e. International Code of Nomenclature for algae, fungi and plants. Formely it was called as **ICBN** means International Code for Botanical Nomenclature.
- **ICZN**, i.e. International Code for Zoological Nomenclature
- **ICNB**, i.e. International Code for Nomenclature of Bacteria
- **ICNCP**, i.e. International Code of Nomenclature for Cultivated Plants
- **ICTV**, i.e. International Committee on Taxonomy of Viruses

Systems of scientific nomenclature

The systems of scientific ncmenclature are of following types

Binomial system of nomenclature

This system was proposed by **C Linnaeus** in his book *Species Plantarum* in 1753. The idea of binomial nomenclature was first introduced by **Gaspard Bauhin** in 1623.

As per this system, name of any organism consists of two parts or epithets, i.e. **generic epithet** and **specific epithet**. For example, in name *Homo sapiens,* word *Homo* is genus or generic epithet while word *sapiens* is species or specific epithet.

International rules of binomial nomenclature are as follows

- A scientific name consists of two words in Latin otherwise these are latinised if derived from any other language.
- The first word is genus and starts with a capital letter. The second name is species that starts with a small letter, e.g. *Homo sapiens.*
- Both names are printed in italics or are underlined separately if handwritten.
- The name of author is written in abbreviated form after the species name and it is printed in Roman and never underlined, e.g. *Mimosa pudica* Linn.
- All these three words (generic epithet, specific epithet and author citation) collectively form binomial epithet.
- If there are two or more words in species name, a hyphan is put between these words and constitute campound specific name, e.g. *Hibiscus rosa-sinensis.*
- **Rule of priority** It is the most important of all rules of ICBN. No names are recognised prior to those used by Linnaeus in 1758 in the 10th edition of *Systema Naturae.* So, according to the rule of priority in case two or more names are given the oldest, i.e. the name given first, is recognised as valid name and all other names are called synonyms.

Trinomial system of nomenclature

Sometimes, binomial nomenclature can also be extended to trinomial system of nomenclature. Here, the names of subspecies or varieties can also be incorporated.

Thus, making the name which comprise of three words, e.g. *Brassica oleracea botrytis* and *Homo sapiens sapiens.* The term 'subspecies' is used more commonly by zoologists, while variety is used commonly by botanists.

Polynomial system of nomenclature

This system was used before 1750. In this system biologists used descriptive names for organisms with each name made up of several Latin words. These words are also related to one or more major characteristics of organism.

For example, *Caryophyllum saxatilis folis gramineus umbellatus corymbia* means *Caryophyllum* growing on rocks having grass-like leaves an with umbellate corymb flower.

Nomenclature type and typification

A nomenclature type is that representative of a group, which acts as a source for the naming of a certain taxonomic group. Using a type for fixing the name of a taxon is called **typification.** The ICBN recognises following several types

(i) **Holotype** It is a single type specimen upon which the description and name of a new species is based.

(ii) **Isotype** It is the specimen which is a duplicate of the holotype. It is collected from the same place, same time and by the same person.

(iii) **Syntype** It is any one of the two or more specimens or biological types cited by the author when no holotype was designated.

In other words we can say it is the specimens having status equivalent to the holotype on which the description and name of a new species is based.

(iv) **Paratype** It is a specimen that is neither the holotype nor an isotype and nor one of the syntypes. However, it is used to define the scientific name of a species. Often, there is more than one paratypes.

(v) **Lectotype** A specimen or other element selected from the original material when no holotype was originally selected. It is usually needed in the condition when holotype no longer exists.

(vi) **Neotype** It is a specimen or illustration selected to serve as a nomenclatural type as long as all the materials (on which the name of the taxon was based) are missing.

(vii) **Epitype** It is a specimen selected to serve informational type when the actual holotype, lectotype or previously designated neotype are ambiguous. Thus, cannot be used for naming the taxon.

Tautonyms, Autonyms and Homonym

- **Tautonyms** These are the names with same generic and specific epithets, e.g. *Rattus rattus*. Tautonyms are not valid for plants though, these are commonly used for zoological names.
- **Autonyms** These are the names in which specific epithet and infraspecific epithet, i.e. subspecies are identical, e.g. *Homo sapiens sapiens*.
- **Homonym** This is the single name given for two different species. e.g. *Prunus dulcis* (for almond and plum).

Taxonomic categories or taxon

The term 'Taxon' was introduced for the first time by ICBN in 1956.

In 1964, **Mayr** defined taxon as a taxonomic group of any rank that is sufficiently distinct of being assigned a definite category.

Taxonomic hierarchy

- The system of arranging organisms in a definite sequence of various taxonomic categories is called taxonomic hierarchy.
- In this the arrangement is done in proper ascending or descending order.
- It is also called **Linnaean hierarchy** as it was first proposed by **Carolus Linnaeus.** That is why he is considered as the Father of Systematic Botany.
- The hierarchy includes following seven obligate categories

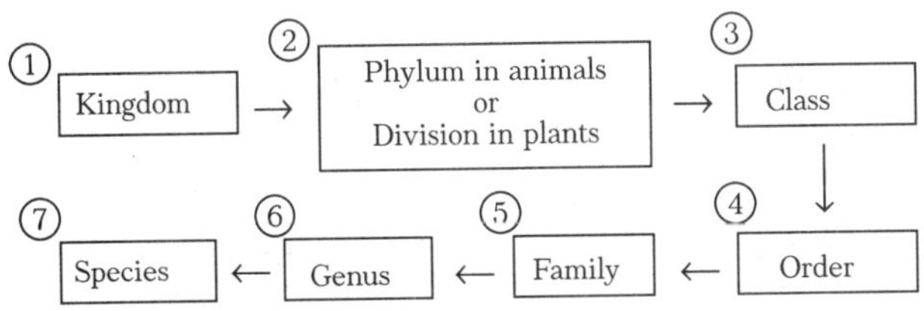

There connectivity can be described as

- One or more than one species can belong to a particular genus (pl. Genera). Many genera constitute a family.
- Families constitute order, then classes, division or phylum (pl. Phyla) and ultimately a kingdom is constituted by many phyla.

Detailed description of all these categories is given below

Species

- It is the fundamental or smallest unit of classification. The first defining concept of species was given by **John Ray**.
- According to him a species is a group of organisms that are morphologically similar irrespective of their breeding compatibility.
- In 1942, **Ernst Mayr** gave the most accepted definition of species. According to which a species can be defined as a group of organisms, who are capable of breeding freely in between themselves under natural conditions, on the other hand these are incapable of breeding with the members of any other species.
- Thus, **reproductive isolation** is an important boundary between different species.
- However, this definition also had some limitations like
  - It is restricted to only sexually reproducing organisms.
  - It cannot be applied to fossils.
  - It does not consider hybrid organisms.

Note
- *The concept of Mayr given above is called **Biological concept of species.***
- ***Static concept of species** is given by **Linnaeus**. It says that a species is fixed and immutable.*
- ***Dynamic concept of species** was given by **Lamarck**. According to this concept a species is mutable and dynamic.*
- ***Typological concept of species** was given by **Aristotle** and **Plato**. In this an ideal type of body for each species is fixed. Individuals of species is fixed match this type as closely as possible. It is considered as most simple and widely held species concept.*
- ***Nominalistic concept of species** was given by **Occam**. It says species have no existence, only the individuals does.*

Types of species

- **Morphospecies** Species decided on the basis of morphological characters only.
- **Taxonomic species** Species with a definite binomial name.
- **Sibling species** True species which do not interbreed but their separation is difficult on the basis of morphological characters also, e.g. *Drosophila pseudoobscura* and *D. persimilis.*
- **Allopatric species** Species belonging to distinct geographical areas.
- **Sympatric species** Species belonging to overlapping geographical areas.
- **Parapatric species** Species of adjacent geographic ranges meeting in a very narrow overlapping zone.
- **Polytypic species** Species with two or more subspecies.
- **Allochronic species** Species belonging to different time period.
- **Synchronic species** Species belonging to same fime period.

Genus

Term 'Genus' was coined by **Tournefort** in 1964. Genera are the aggregates of closely related species. For example, potato (*Solanum tuberosum*), tomato (*Solanum lycopersicum*) and brinjal (*Solanum nigrum*) are three different species, but all belong to the genus-*Solanum*.

Monotypic and Polytypic Genus

Genera having only one species are called **monotypic**, whereas those having a large number of closely related species are called **polytypic**.

Family

A family contains a group of related genera having some common features. It has less numbers of similarities as compared to the genus and species.

Families are characterised on the basis of both vegetative and reproductive features, e.g. among plants, three different genera *Solanum*, *Petunia* and *Datura* are placed in the family-Solanaceae.

Order

An order is an assemblage of families, which exhibit a few similar characters.

These similar characters are less in number as compared to different genera included in a family.

For example, plant families like Convolvulaceae and Solanaceae are included in the order–Polymoniales mainly based on the floral characters.

Class

Linnaeus gave the term class for a group of related orders. For example, order–Primata which comprises monkey, gorilla and gibbon is placed in class–Mammalia along with order–Carnivora that includes animals like tiger, cat and dog.

Phylum (for Animals) or division (for Plants)

The term 'Phylum' was coined by **Cuvier**.

Phylum is a group of various classes, e.g. Phylum–Chordata comprises of animals like fishes, amphibians, reptiles, birds along with mammals.

All of these have a dorsal notochord and nerve cord. Similarly in case of plants, classes with a few similar characters are assigned to a higher category called division. The term division was coined by **Eichler**.

Kingdom

It is the highest taxonomic category. All plants are included in kingdom–Plantae, while all animals in kingdom–Animalia. Some example showing taxonomic categorisation are given in the table below

Taxon	Wheat	Human	Housefly
Phylum/Division	Angiospermae	Chordata	Arthropoda
Class	Monocotyledonae	Mammalia	Insecta
Order	Poales	Primata	Diptera
Family	Poaceae	Hominidae	Muscidae
Genera	*Triticum*	*Homo*	*Musca*
Species	*aestivum*	*sapiens*	*domestica*

According to ICBN and ICZN, the names of different categories must end in the standard endings or suffixes. Details of these suffixes is given below

Table 1.2 Categories and corresponding suffixes

Plants

Category	Suffix	Category	Suffix
Division	phyta	Suborder	incae
Subdivision	phytina	Family	aceae
Class	phyceae or opsida or ae	Subfamily	oideae
Order	ales	Tribe	eae

Animals

Category	Suffix
Superfamily	oidea
Family	idea
Subfamily	inae
Tribe	ini

CHECK POINT 1.2

1. Taxonomy is connected with
   (a) collection of plants and animals
   (b) only classification of organisms
   (c) identification, nomenclature and classification of organisms
   (d) All of the above

2. In binomial nomenclature
   (a) Only genus is printed in Italics
   (b) Genus and species may be same
   (c) Both in genus and species the first letter is capital
   (d) Genus is written after the species

3. A single type specimen which is the basis of description and name of a new species is called
   (a) syntype (b) isotype
   (c) holotype (d) epitype

4. The closest category of family is
   (a) phylum
   (b) class
   (c) order
   (d) species

5. Connecting link between kingdom and class in plant hierarchy is
   (a) Phylum
   (b) order
   (c) family
   (d) division

6. If two plants belong to the same division but in different orders, they may belong to the same
   (a) Genus (b) Family
   (c) Class (d) Species

TAXONOMICAL AIDS

The correct identification is a primary task for classification of any organism. Taxonomical aids are the techniques, procedures and stored information useful in identification and classification of organisms. Some of these aids are described below

Herbarium (Pl. Herbaria)

- A herbarium is a collection of plants, which have been dried, pressed, mounted on herbarium sheets.
- Term 'Herbaria' can also refer to the building or room where the specimens are housed or to the scientific institute that not only stores but uses them for research.
- These specimens are identified and classified according to any universally accepted system of classification.
- A herbarium sheet (11.5 × 16.5 inches) carries label which provides information about date and place of collection, english, local and botanical names, family, collector's name, etc.

Tools and procedure used

- Tools for collection are digger, scissors, knife and a pole with hook.
- Moisture free environment is a must to keep the collections for longer duration.
- Dried specimens are poisoned by using chemicals, e.g. 0.1% solution of corrosive sublimate or $HgCl_2$.
- International size of herbarium is 41cm ×29cm.
- Preservance also require periodic spray of disinfectatnts or repellants like DDT powder or copper sulphate.

Function of herbarium

- To preserve specimen wealth including type material and palaeobotanical collection.
- To carry out research programmes for fundamental and applied values.
- To fulfill public needs by supplying materials and scientific information about specimens.

Herbaria types

Herbaria can be of following two types

(i) **Major** or **National** herbaria

(ii) **Minor** or **Regional** or Local herbaria

The local herbaria can be formed at college or university level. A few famous herbaria of the world are given below

Table 1.3 Some major herbaria of the world

Herbarium	Number of specimens (approx)
Museum of Natural History, Paris	10000000
Royal Botanical Garden, Kew	6000000
Komarov Botanical Institute, St Petersburg (Leningrad)	5700000
Conservatory and Botanical Garden, Geneva	5000000
British Museum of Natural History, London	4000000
Indian Botanical Garden, Kolkata (India)	1000000
Herbarium of Forest Research Institute, Dehradun (India)	300000
Herbaria of Agriculture College and Research Institute, Coimbatore (India)	200000
Herbarium of National Botanical Garden, Lucknow (India)	40000

Museums

They have collection of preserved plant and animal specimens for study and reference. Most educational institutes usually maintain museums. Museums provide information about the flora and fauna of different areas. In a museum, specimens are preserved as dry specimens, e.g.

- Larger animals like birds and mammals are usually stuffed and preserved.
- Skeletons of animals and humans are also preserved in museums.

Some important museums of the world are listed below

- Natural History Museum, London (England)
- United States National Museum, Washington (USA)
- Field Museum of Natural History, Chicago (USA)
- Zoology Museum, Amsterdam
- Prince of Wales Museum, Mumbai
- Indian Museum, Kolkata
- National Museum of Natural History (NMNH), Delhi
- Maharaja Sawai Man Singh (II) Museum, Jaipur

Botanical garden

A botanical garden is essentially a collection of living plants maintained for both pure and applied studies.

The botanical gardens play the following roles

- On-site teaching
- Integrated research projects
- Conservation
- Aesthetic appeal
- Material for botanical research
- Seed exchange
- Herbarium and library
- Public services

Thousands of botanical gardens are located worldwide. These are maintained by various institutes. Of these, nearly 800 important gardens are documented in the *International Directory of Botanical Gardens*, published by Henderson in 1983.

Table 1.4 Some major botanical gardens of the world

Name of the botanical garden	Location
New York Botanical Garden	New York, USA
Royal Botanical Garden	Kew, England
Pisa Botanical Garden	Pisa, Italy
Orto botanico or Padua Botanical Garden	Padua (PD), Italy
Berlin Botanical Garden and Museum	Berlin, Germany
Cambridge University Botanical Garden	Cambridge, UK
Indian Botanical Garden	Kolkata, India
National Botanical Garden (National Botanical Research Institute, Sikanderbagh)	Lucknow, India
Botanical Garden	Dehradun, India
Botanical Garden	Noida, India

Zoological parks

These are the places, where wild animals are kept in protected environments under human care. These enable us to learn about their food, habits and behaviour.

Some common zoos of India are listed below

- Mahendra Chaudhary Zoological Park, Chhatbir Punjab, established in 1977 in 200 hactare, exhibits about 1300 animals and has a lion safari
- National Zoological Park, Delhi established in 1959. It is one of the finest zoos of Asia, covers an area of 214 acres and displays about 22000 animals
- Nehru Zoological Park, Hyderabad
- Zoological Garden, Alipore, Kolkata
- Himalayan Zoological Park, Gangtok (Sikkim)
- Gandhi Zoological Park, Gwalior
- Padmaja Naidu Himalayan Zoological Park, Darjeling
- Kamala Nehru Zoological Garden, Ahemdabad
- Prince of Wales Zoological Park, Lucknow

Keys

- These are taxonomical aid used for the identification of plants and animals.
- The keys are based on the paired contrasting characters called as **couplet**. It represents two alternate forms of a given character.
- Keys are generally of two types
  (i) **Indented** or **Yoked keys** It has a sequence of forms of different characters.
  (ii) **Bracketed keys** It includes one pair of contrasting statements for a given character.
- Separate taxonomic keys are required for each taxonomic category such as family, genus and species for identification purposes.

Flora

- It is a book or other work detailing the plants of a particular region or habitat.
- Flora provide information about the plants taxonomy, nomenclature, habitat and distribution in particular area.
- It also include information on the biology and morphology of the taxa, as well as illustration keys and other notes.
- In general terminology 'Flora' refers to the plant life occurring in a particular region or time generally the naturally occurring. The corresponding term for animal life in 'Fauna'.

Monograph

- A monograph is a taxonomic aid that gives comprehensive information of a single taxon in biological taxonomic studies.
- The first ever monograph of a plant taxon was given in 1672 as *Plantarum Umbelliferarum Distribution Nova* by **Robert Morison**.

Manuals

- These are useful in providing information for identification of names of species found in a particular area.
- They also help in correct identification.
- They contain complete information about the occurrence in a particular area, their family, genus and species.

Catalogue

- It is a methodical indexed arrangement list of organisms found in a specific area.
- It consists of detailed descriptions about the organisms genera or family for identification.

CHECK POINT 1.3

1. Identification of organisms requires intensive
(a) field studies (b) laboratory studies
(c) habitat studies (d) Both (a) and (b)

2. In museums, insects are preserved by
(a) pinning without killing them and then preserving in containers
(b) killing, pinning and then displaying in special boxes
(c) mounting them alive in jars with various preservative solutions
(d) killing, pinning and then stuffing them

3. Which of the following taxonomical aid serve as a conservation and seed exchange?
(a) Key
(b) Botanical gardens
(c) Herbaria
(d) Museum

4. Which taxonomical aid contain information on any one taxon?
(a) Catalogues (b) Manuals
(c) Flora (d) Monograph

5. Which taxonomical aid help in correct identification?
(a) Monograph (b) Flora
(c) Manuals (d) All of these

6. Choose the correct statement.
(a) A university herbaria can be considered as national herbaria.
(b) In museums larger animals like birds and mammals are usually preserved.
(c) One can learn food habits and behaviour of animals from zoological parks.
(d) A catalogue is useful in providing the integration for identification of names of species of a particular area.

EXAM POINTS

- Biological organisation starts with cellular level where the cellular organelles interact with each other to form a cell. **[AFMC 2012]**
- The term 'taxonomy' was introduced by de Candolle. Classical taxonomy is based on the study of morphological trait of the organisms. It is also called alpha taxonomy or old systematics. **[J&K CET 2012, AIIMS 2017]**
- Scientific study of diversity of organism and their evolutionary relationship is called systematics. **[J&K CET 2012]**
- The term 'new systematics' was given by Julian Huxley in 1940. It is based on evolutionary, genetic and morphological traits. **[JCECE 2012]**
- The system of giving one name to an individual is termed as uninomial system. **[Chhattisgarh PMT 2015]**
- Binomial nomenclature was introduced by Carolus Linnaeus. **[Kerala CEE 2012]**
- Nomenclature is governed by certain universal rules like–the first word in a biological name represent genus name and the second is a specific epithet. Biological names originate from Latin language and printed in italics and when hand written should be underlined separately, i.e. *Panthera tigris*. **[NEET I 2016]**
- Taxonomic group of any rank is called taxon, it was introduced by ICBN.
- The stepwise arrangement of all categories for the classification of plants and animals is called taxonomic hierarchy. It was proposed by Carolus Linnaeus. **[AIIMS 2012]**
- The hierarchy includes seven obligate categories, i.e. kingdom → phylum/division → class → order → family → genus → species **[J&K CET 2015, Kerala CEE 2014]**
- As we go from kingdom to species in taxonomy hierarchy the number of common characteristics will increase. **[Kerala CEE 2012]**
- The word species was coined by John Ray. It is refers as the real basic unit of classification. **[J&K CET 2015]**
- Family is placed between order and genus. It is categorised on the basis of vegetative and reproductive features, e.g. cat belong to felidae and dog belong to Canidae family. **[OJEE 2013]**
- Order is the assemble age of genera which exhibit a few similar characters. **[Kerala CEE 2012]**
- Cohort is a formal taxonomic category in between infraclass and super-order and it contains the largest number of species. **[WB JEE 2012]**
- Herbarium is a place where dried and pressed plant specimens mounted on herbarium sheet. Which carry information about date and place of collection, botanical and local name, family, collector name, etc. **[NEET 2016, 2019]**
- The largest collection of herbarium in India is Central National Herbarium, Kolkata. **[MHT CET 2014]**
- Museum is a place having collection of preserved plants and animals specimens. It provide the information about different flora and fauna. **[NEET 2018]**
- Taxonomic key is one of the taxonomic tools in the identification and classification of plants and animals. It helps in preparing monographs and flora. **[JIPMER 2017]**
- The key is based on the paired contrasting characters called as couplet. **[NEET 2019]**
- Catalogue is a list that enumerates methodically all the species found in an area with brief description aiding identification. **[NEET 2018]**

Chapter Exercises

(A) Topical Exercise

Characteristics of Living Beings

1 Living beings are different from non-living in possessing
(a) organisation higher than molecular level
(b) response to stimulus
(c) reproduction
(d) All of the above

2 Which one of the following aspects is an exclusive characteristic of living things?
(a) Isolated metabolic reactions occur *in vitro*
(b) Increase in mass from inside only
(c) Perception of events happening in the environment and their memory
(d) Increase in mass by accumulation of material both on surface as well as internally

3 Consider the following statements.
I. Energy is required by cells for physiological activities only.
II. The energy yielding process that also requires input of energy is photosynthesis.

Choose the correct option.
(a) Statement I is true, but II is false
(b) Statement I is false, but II is true
(c) Both statements I and II are true
(d) Both statements I and II are false

4 Reproduction is synonymous with growth for which group of organisms?
(a) Multicellular filamentous organisms
(b) Colonial organisms
(c) Unicellular organisms
(d) All of the above

5 Reproduction cannot be an all-inclusive defining characteristic of living organisms because
(a) all living organisms do not show growth
(b) many organisms do not reproduce
(c) non-living things show reproduction
(d) all living organisms show small period of reproductive phase in their life

6 Consider the following statement.
I. Miniparity is the process of repeated reproduction of animals at regular intervals in the life cycle.
II. Two species can said to be reproductively isolated if they are not interfertile.

Choose the correct option.
(a) Statement I is true, but II is false
(b) Statement I is false, but II is true
(c) Both statements I and II are true
(d) Both statements I and II are false

7 The process of increase in size occurs due to synthesis of
(a) protoplasmic substances (b) apoplasmic substances
(c) Both (a) and (b) (d) None of these

8 Select the correct statement for growth as one of the characteristic of living organisms.
(a) Growth by increase in mass is a defining property of prokaryotic organisms only
(b) Non-living objects show growth by increase in mass of body
(c) Intrinsic growth is a characteristic of all living organisms
(d) Growth can be extrinsic or intrinsic for multicellular non-living organisms

9 Growth cannot be taken as a defining property of living organisms because
(a) all living organisms do not show growth
(b) non-living things also grow from inside
(c) non-living things also grow by increasing in mass
(d) some living organisms do not show process of reproduction

10 Consider the following statements.
I. Apoplasmic growth is through the formation of matrix and fibres of connective tissue only.
II. Growth involves only cell proliferation.

Choose the correct option.
(a) Statement I is true, but II is false
(b) Statement I is false, but II is true
(c) Both statements I and II are true
(d) Both statements I and II are false

11 Nothing lives forever, yet life continues. What does it mean?
(a) Death has nothing to do with the continuation of life
(b) Though older die, but new are produced due to reproduction
(c) Both (a) and (b)
(d) None of the above

12 What represents the steady state?
(a) Intake of food materials, oxygen and water
(b) Elimination of waste materials and energy
(c) Maintenance of constant internal environment in a living organism
(d) None of the above

13 Consider the following statement.

I. Homoestasis is the tendency to resist internal changes.
II. Sweating during exercise is an example of homoestasis.

Choose the correct option.

(a) Statement I is true, but II is false
(b) Statement I is false, but II is true
(c) Both statements I and II are true
(d) Both statements I and II are false

14 Read the statements with respect to homeostasis and choose the incorrect statement.

(a) It is the process of maintenance of a constant internal environment
(b) The term 'Homeostasis' was coined by Cannon for the first time
(c) Organisms adapt to changes in temperature, salinity, etc.
(d) Most fishes, amphibians, reptiles, plants are homeotherms

15 Consider the following statement.

I. In endothermic organisms, the body temperature is constant.
II. The organisms which can maintain a constant body temperature are birds and mammals.

Choose the correct option.

(a) Statement I is true, but II is false
(b) Statement I is false, but II is true
(c) Both statements I and II are true
(d) Both statements I and II are false

16 Which of the following characteristic is not shown by long term adaptation?

(a) Inheritable (b) Permanent
(c) Formed due to evolution (d) Can be reversed

17 Reason of diversity in living beings is

(a) varying environmental conditions
(b) long term evolutionary changes
(c) short term evolutionary changes
(d) gradual internal changes

18 Death is the most important regulatory process on the earth. It avoids

(a) pollution
(b) over crowding on earth
(c) passing of traits from one generation to the next
(d) reproduction

19 Clinical death is the condition in which

(a) brain and other body organs begin to degenerate due to non-availability of nutrients and oxygen
(b) all vital function are stopped
(c) deterioration of structure and functions of cells, tissues and vital organs
(d) elements that constitute a body are returned to the ecosystem

20 Consider these statements regarding 'characteristics of living beings' and choose the incorrect statement.

(a) Plants and animals grow throughout their life
(b) All living organisms are required to assimilate energy through metabolism
(c) Living beings are responsive to their environment
(d) Reproduction is a more important aspect of living beings as compared to metabolism

21 The two basic functions of life are

(a) energy transformation and synthesis of biomolecules
(b) homeostasis and energy flow
(c) Both (a) and (b)
(d) None of the above

22 All living organisms are linked to one another because

(a) they have common genetic material of the same type
(b) they share common genetic material, but to varying degrees
(c) All have common cellular organisation
(d) All of the above

Diversity in the Living World

23 Which of the following promotes similarity among living things?

(a) Classification of organisms
(b) Discovery of new organisms
(c) Inheritance from a common ancestor
(d) Homeostatic regulation

24 Most important criteria presently used for classification of living organisms is

(a) anatomical and physiological characteristics
(b) presence or absence of notochord
(c) resemblance in external features
(d) breeding habits

25 Taxonomy is the study of

(a) evolutionary relationships between organisms
(b) the classification of life forms by their similarities and differences
(c) genetics
(d) phylogeny

26 Neosystematics aims at

(a) the classification of organisms based on broad morphological characters
(b) delimiting various taxa of organisms and establishing their relationship
(c) the classification of organisms based on their evolutionary history and establishing their phylogeny on the totality of various parameters from all fields of studies
(d) identification and arrangement of organisms on the basis of their cytological characteristics

27 Karyotaxonomy is based on
(a) trinomial nomenclature
(b) organic evolution
(c) number of chromosomes
(d) utility of an organism

28 Consider the following statements.
I. Taxon includes only genus and species.
II. Taxonomy based on number of shared characteristics of various organisms is called numerical taxonomy.

Choose the correct option.
(a) Statement I is true, but II is false
(b) Statement I is false, but II is true
(c) Both statements I and II are true
(d) Both statements I and II are false

29 Match the following columns.

Column I	Column II
A. Numerical taxonomy	1. Nuclear and chromosomal studies
B. Classical taxonomy	2. Number of shared characters
C. Practical taxonomy	3. Utility of organism
D. Karyotaxonomy	4. α-taxonomy

Codes

	A	B	C	D		A	B	C	D
(a)	2	4	3	1	(b)	1	3	2	4
(c)	4	2	3	1	(d)	1	2	4	3

30 Taxonomy is connected with
(a) collection of plants and animals
(b) only classification of organisms
(c) identification, nomenclature and classification of organisms
(d) All of the above

31 Binomial nomenclature is
(a) two words in name of a species
(b) two names local and specific
(c) two names of a species
(d) two phases, asexual and sexual, in the life cycle of a species

32 According to binomial nomenclature, two words used for naming a plant or animal are
(a) species and genus
(b) genus and species
(c) species and family
(d) genus and family

33 Choose the wrong statement regarding universal rules of nomenclature.
(a) The first word in a biological name represents the genus
(b) The first word denoting the genus starts with a capital letter
(c) Both the words in biological name, when handwritten are separately underlined
(d) Biological names are generally in English and written in Italics

34 Scientific name of 'Tobacco Mosaic Virus' is standardised through rules given is
(a) ICBN
(b) ICZN
(c) ICTV
(d) ICNB

35 What do *A*, *B* and *C* represent in the given scientific name, respectively?

Solanum	*nigrum*	Linn
A	*B*	*C*

(a) Generic name, specific name and author's name
(b) Specific name, generic name and author's name
(c) Author's name, specific name and generic name
(d) Generic name, author's name and specific name

36 How many of the given features are associated with binomial nomenclature?
Descriptive names, Generic epithet, Specific epithet, Latin language name of subspecies/variety, should be Italicised or underlined
(a) 4
(b) 3
(c) 2
(d) 5

37 Third name in trinomial nomenclature indicate
(a) species
(b) subspecies
(c) subgenus
(d) ecotype of organism

38 If a botanist has to nomenclate a similar species, he will use
(a) mesotype
(b) isotype
(c) syntype
(d) neotype

39 The specimen selected to serve as holotype when the holotype, lectotype or neotypes are ambiguous
(a) neotype
(b) epitype
(c) paratype
(d) syntype

40 Consider the following statements.
I. Holotype is the type specimen used by the author in the original publication.
II. Isotype is the type specimen described along with holotype.

Choose the correct option.
(a) Statement I is true, but II is false
(b) Statement I is false, but II is true
(c) Both statements I and II are true
(d) Both statements I and II are false

41 A taxonomist found a new plant. He used lectotype for the identification of the new plant. Why did he use lectotype?
(a) Because lectotype is the single physical example or specimen of an organism
(b) Because holotype has the wrong information
(c) Because holotype was no longer exists
(d) Because is the duplicate of the holotype

42 Which one of the following represents a tautonym?
(a) *Rattus rattus*
(b) *Prunus dulcis*
(c) *Corvus splendens splendens*
(d) *Albugo candida*

43 *Corvus splendens splendens* is the scientific name of Indian crow, it represents
(a) binomial nomenclature (b) autonyms
(c) tautonyms (d) synonyms

44 The scientific name of a species 'X' was given by a scientist in 1960 as 'P' and another scientist gave the other name to the same species in 1990 as 'Q'. Which is considered as valid name?
(a) P (b) Q
(c) Both P nad Q (d) None of these

45 Consider the following statements.
I. First great taxonomist was Linnaeus.
II. *Species Plantarum* was written by John Ray.

Choose the correct option.
(a) Statement I is true, but II is false
(b) Statement I is false, but II is true
(c) Both statements I and II are true
(d) Both statements I and II are false

46 In the classification of organisms, which of the following is the correct hierarchy of taxonomic groups?

	Smallest	→			Largest
(a)	Family	Genus	Class	Order	Phylum
(b)	Family	Genus	Order	Phylum	Class
(c)	Genus	Family	Order	Class	Phylum
(d)	Genus	Phylum	Order	Family	Class

47 Consider the following statements.
I. Species is a specific unit of evolution.
II. It is consider as biological cohesive unit.

Choose the correct option.
(a) Statement I is true, but II is false
(b) Statement I is false, but II is true
(c) Both statements I and II are true
(d) Both statements I and II are false

48 Find out the incorrect statement from the following.
(a) Closely related species differ in morphological features
(b) Genus comprises a group of related species
(c) Taxonomic studies are useful in agricultural, forestry and industries
(d) Family includes related genera

49 Which one of the following has least similar characters?
(a) Family (b) Class (c) Genus (d) Species

50 Which taxonomic category of mango and wheat is similar?
(a) Order and Family
(b) Only Division
(c) Division and Class
(d) Division, Class and Order

51 The chart shows the classification of two species of crocodile.

Animalia
Chordata
Reptilia
Loricata
Crocodylidae
Crocodylus
Crocodylus niloticus (Nile crocodile)
Crocodylus porosus (Saltwater crocodile)

To which order do these crocodiles belong?
(a) Chordata (b) Crocodylidae
(c) Loricata (d) Reptilia

52 The members of which one of the following groups have many characteristics in common?
(a) Genus (b) Kingdom
(c) Phylum (d) Species

53 Which one of the following does not form basis of modern taxonomic studies?
(a) Structure of cell and morphology
(b) Development process
(c) Genetic basis
(d) Ecological information of organisms

54 As we go from species to kingdom in a taxonomic hierarchy, the number of common characteristics
(a) will decrease
(b) will increase
(c) remain same
(d) may increase or decrease

55 Match the following columns.

Column I		Column II	
A.	Family	1.	*tuberosum*
B.	Kingdom	2.	Polymoniales
C.	Order	3.	*Solanum*
D.	Species	4.	Plantae
E.	Genus	5.	Solanaceae

Codes

	A	B	C	D	E
(a)	5	4	2	1	3
(b)	4	3	5	2	1
(c)	4	3	5	1	2
(d)	4	3	2	5	1

Taxonomical Aids

56 Herbarium means
(a) storehouse/place where preserved identified dry/ liquid/ specimens are kept systematically
(b) a heavy card sheet carrying the dried and pressed specimen of plant
(c) a garden with all plants arranged systematically
(d) a system of arranging plants into different categories

57 In a plant herbarium, mostly to arrange plants, we make use of
(a) Hutchinson system of classification
(b) Bentham and Hooker's system
(c) Linnaeus system of classification
(d) Haeckel system of classification

58 The national herbarium is that which contains plants of
(a) a region (b) a country (c) a locality (d) world

59 By which chemical, specimens for herbaria are treated to prevent any fungal attack?
(a) 0.1% mercuric chloride
(b) 0.1% mercurous chloride
(c) Carbon disulphide
(d) Acetic acid

60 Over 6000000 specimens are preserved in which botanical garden or herbaria?
(a) Museum of National History, Paris
(b) Royal Botanical Garden, Kew
(c) Royal Botanical Garden, Edinburgh
(d) US National Herbarium, Washington DC

61 An institution or a place, where artistic and educational materials are exhibited to the public is called
(a) herbarium (b) sanctuary
(c) museum (d) parks

62 The headquarter of BSI (Botanical Survery of India) is at
(a) Howrah (Kolkata)
(b) New Delhi
(c) NBRI Lucknow (UP)
(d) FRI, Dehradun (Uttarakhand)

63 Consider the following statements regarding museums.
I. Museums are places/institutions where preserved plants, animals and educational materials are exhibited to public.
II. Museums are meant to conserve endangered animals and plants.

Choose the correct option.
(a) Statement I is true, but II is false
(b) Statement I is false, but II is true
(c) Both statements I and II are true
(d) Both statements I and II are false

64 Botanical gardens play which of the following roles?
I. On-site teaching
II. Deforestation sites
III. Aesthetic appeal
IV. Seed exchange

Codes
(a) I, II and III
(b) I and II
(c) III and IV
(d) I, III and IV

65 Taxonomic key is one of the taxonomic tools for the identification and classification of plants and animals. It is used in the preparation of
(a) monograph (b) flora
(c) Both (a) and (b) (d) None of these

66 Manuals contain information for
(a) habitat and distribution of animals of a given area
(b) habitat and distribution of all the plants found on earth
(c) identification of names of species found in an area
(d) habitat and distribution of some useful animals of a given area

67 Actual account of habitat and distribution of plants of a given area is provided by
(a) manuals
(b) catalogues
(c) monograph
(d) flora

68 Match the following columns.

Column I	Column II
A. Museum	1. Information on one taxon
B. Herbaria	2. Couplet
C. Botanical gardens	3. Arranged on universally accepted classification system
D. Taxonomic key	4. Educational institutes
	5. Records of local flora for monographic work

Codes

	A	B	C	D
(a)	4	3	5	2
(b)	3	5	1	2
(c)	4	1	3	5
(d)	5	3	4	2

(B) Cumulative Exercise

(Assorted Questions of the Chapter for Advanced Level Practice)

1 Read the given statements about taxon and select the code with correct statement.

I. Taxon is the taxonomic group of any rank.
II. The term 'Taxon' was introduced for the first time by ICBN.
III. Term 'Taxonomy' was given by Linnaeus.
IV. Padua botanical garden is present in Pune.

Codes

(a) I and III (b) I and II
(c) III and IV (d) II, III and IV

2 Go through the given statements regarding an order and choose the codes with all correct statements.

I. It is an assemblage of families.
II. It contains less similar characters than genera.
III. It is higher in systematic hierarchy than class.
IV. It is higher in systematic hierarchy than phylum.

Codes

(a) I, III and IV (b) I and II
(c) II and IV (d) II and III

3 A few statements regarding adaptations in organisms are given below. Go through these statements.

I. Mimicry is an example of short term adaptation.
II. Short term adaptations are though temporary changes, but are inherited in the next generation.
III. Aestivation is an example of long-term adaptation.
IV. Adaptations tend to enhance survival in the environment.

Select the alternative having all true statements.

(a) II and III (b) II and IV
(c) Only IV (d) I, II and III

4 Match the following columns.

Column I	Column II
A. Linnaeus	1. Living species of today
B. John Ray	2. Father of taxonomy
C. Synchronic species	3. Coined the term species
D. Neontological species	4. Species belonging to same time period

Codes

	A	B	C	D		A	B	C	D
(a)	2	3	4	1	(b)	1	3	2	4
(c)	4	3	2	1	(d)	1	2	4	3

5 Read the following statements about various taxonomic aids.

I. Herbaria serve as quick referral system in taxonomical studies.
II. In museums, large animal like birds and mammals are usually stuffed and preserved.
III. Museums have dry specimens of plants and animals as well as some specimens are preserved in solutions in jars or containers.
IV. Herbarium sheets are arranged according to universally accepted system of classification and these serve as a repository for future use.

Choose the correct option.

(a) I and II (b) III and IV
(c) II and IV (d) I, II, III and IV

6 Observe the gradation of taxonomic categories and identify the missing categories according to the statement given below.

Kingdom
↑
Division
↑
Y
↑
Order
↑
X
↑
Genus
↑
Species

I. X is a group of related species.
II. Y is a group of related divisions.
III. X is a group of related genera.
IV. Y is a group of related kingdoms.
V. Y is a group of related orders.

The correct options are

(a) II and IV (b) I and II
(c) I, II, IV and V (d) III and V

7 Identify the correct statement given below.

(a) Cellular organisation of the body is the defining feature of non-living forms
(b) Consciousness is the property shared by non-living organisms
(c) A patient with dead brain has no self-consciousness yet it is alive
(d) Human beings are the only organisms, who is aware of himself, i.e. self-conscious

8 Select true statements from the following and choose the right answer from the options given below.

I. Human's scientific name is *Homo sapiens.*
II. *Genera Plantarum* is written by John Ray.
III. Highest taxonomic category is division.
IV. Taxonomic group of any rank is taxon.
V. A group of closely related species of an organism represents genus.
VI. The term 'Systematics' was coined by de Candolle.

(a) II, III, IV and VI
(b) I, III, V and VI
(c) I, IV and V
(d) II, III and VI

9 Select true and false statements and choose the correct answer from the options given below.

I. Taxon is a group of related organisms.
II. Royal Botanical Garden is located in Bristol (England).
III. Dudhwa National Park is located at Lakhimpur in UP.

IV. There are about 89 Wildlife Sanctuaries and 492 National Parks in India.
V. When specific name is identical to generic name, it is an example of tautonym.

(a) True-I, II, IV False-III, V
(b) True-I, II False-III, IV, V
(c) True-III, V False-I, II, IV
(d) True-III, VI, V False-I, II

10 Match the following columns.

Column I	Column II
A. Genus	1. Eichler
B. Division	2. Cuvier
C. Phylum	3. Linnaeus
D. Class	4. Tournefort

Codes

	A	B	C	D		A	B	C	D
(a)	4	1	2	3	(b)	1	2	3	4
(c)	2	3	1	4	(d)	3	4	2	1

11 Go through the statements given below and choose the option with all incorrect statement(s).

I. Common names can be misleading.
II. Vernacular names are easier to pronounce.
III. Common name 'touch-me-not' is given to two species, i.e. *Nigella* and *Mimosa*.
IV. Vernacular names often have no significance.

Codes

(a) I and II (b) Only II
(c) I, II and III (d) Only III

12 Read the following statements and choose the option having all true statements.

I. The idea of binomial nomenclature was first introduced by Gaspard Bauhin.
II. In the name *Mangifera indica* L. Here L. is the symbol representing the publication where it was first published.
III. In polynomial system of nomenclature, descriptive names are used for organisms.
IV. Binomial system was proposed by Carolus Linnaeus.

Codes

(a) I, II and III (b) Only IV
(c) Only I (d) I, III and IV

13 Consider the following statements regarding 'scientific names'.

I. In the book *Pinax*, binomial system of nomenclature was proposed for the first time.
II. Scientific names are always in Greek.
III. The first word in a scientific name is generic name.
IV. Generic a .d specific epithets start with small letter.

Segregate the above statements as True (T) or False (F) by selecting the correct option.

	I	II	III	IV		I	II	III	IV
(a)	T	F	T	F	(b)	F	T	T	T
(c)	F	F	T	T	(d)	T	T	F	F

14 Read the following statements carefully and select the correct option.

I. In zoological parks, conditions similar to natural habitats are provided to animals.
II. Keys are generally analytical in nature.
III. In herbarium sheet, local names are not mentioned.
IV. Taxonomical aids are useful in knowing bioresources.

Codes

(a) I and II (b) I, III and IV
(c) I, II and IV (d) I, II, III and IV

15 Find the correct sequence of various steps of herbarium technique.

I. Drying II. Poisoning
III. Collection IV. Labelling
V. Mounting VI. Deposition
VII. Stitching

Codes

(a) III, I, II, V, VII, IV, VI (b) III, II, IV, V, VI, VII, I
(c) III, I, II, V, VII, VI, IV (d) III, I, II, VII, V, VI, IV

16 Select True (T) and False (F) statements and choose the correct answer from the options given below.

I. Polynomials represents a 'taxonomic category'.
II. Himalyan zoological park is located in Gangtok in Sikkim.
III. There are 40,000 species of plants are kept in herbarium of national botanical garden known in India.
IV. When single name gives for two different species it is an example of homonym.

	I	II	III	IV		I	II	III	IV
(a)	T	F	T	F	(b)	T	T	T	F
(c)	F	F	F	T	(d)	F	T	T	T

17 Match the following columns.

Column I (Taxonomical aids)	Column II (Features)
A. Monograph	1. Information for identification of name of species found in an area
B. Botanical garden	2. Living wild animals in their natural habitat
C. Zoological park	3. Information on any one taxon
D. Manual	4. Place having diversity of living plants

Codes

	A	B	C	D		A	B	C	D
(a)	3	4	2	1	(b)	1	4	2	3
(c)	4	3	1	2	(d)	4	1	2	3

18 Two different genera are classified in the same taxonomic category, family. Which of the following statement is correct about their classification?

(a) The same class, but different species
(b) A different class and different order
(c) The same phylum, but different class
(d) A different kingdom and different phylum

19 Match the following columns.

Column I	Column II
A. Introduced binomial nomenclature	1. Ernst Mayr
B. The Darwin of the 20th century	2. Carolus Linnaeus
C. Gave the concept of new systematics	3. John Ray
D. First described species as a unit of classification	4. Julian Huxley

Codes

	A	B	C	D
(a)	1	2	3	4
(b)	2	1	4	3
(c)	4	3	2	1
(d)	3	4	1	2

20 Which one of the following statement has organism scientifically correctly named, correctly printed according to the International Rules of Nomenclature and correctly described?

(a) *Musca domestica*, the common lizard is a reptile
(b) *Plasmodium falciparum* is a protozoan pathogen causing the most serious type of malaria
(c) *tigris* is the Indian tiger, well-protected in Gir forests
(d) *E. coli*, full name *Entamoeba coli*, is a commonly occurring bacterium in human intestine

21 Which of the following combinations is correct for wheat?

(a) Genus : *Triticum*, Family : Anacardiaceae, Order : Poales, Class : Monocotyledonae
(b) Genus : *Triticum*, Family : Poaceae, Order : Poales, Class : Dicotyledonae
(c) Genus : *Triticum*, Family : Poaceae, Order : Sapinadales, Class : Monocotyledonae
(d) Genus : *Triticum*, Family : Poaceae, Order : Poales, Class : Monocotyledonae

22 Which one of the following statement correctly define the term homonym?

(a) Identical name of two different taxon
(b) Two or more names belonging to the same taxon
(c) When species name repeats the genetic name
(d) Order name of a taxon given in a language other than the language of zoological/botanical nomenclature

23 Match the following columns.

Column I	Column II
A. *Ex situ* conservation	1. Central National Herbarium
B. Quick referral system	2. Museum
C. Preserved plants	3. Flora
D. Actual account of habitat and distribution of plants of a given area	4. Royal Botanical Garden, Kew

Codes

	A	B	C	D
(a)	2	3	4	1
(b)	1	4	2	3
(c)	4	1	3	2
(d)	4	1	2	3

Assertion and Reason

Direction (Q. Nos. 24-29) *In the following question, a statement of Assertion (A) is given followed by the corresponding statement of Reason (R). Of the statements, mark the correct answer as*

(a) Both A and R are true and R is the correct explanation of A
(b) Both A and R are true, but R is not the correct explanation of A
(c) A is true, but R is false
(d) A is false, but R is true
(e) Both A and R are false

24 **Assertion** (A) Scientific names are printed in italics.
Reason (R) The name of author is written in abbreviated form after the specific epithet and is printed in Roman.

25 **Assertion** (A) Cytotaxonomy involves the cytological information for classification purposes.
Reason (R) Chemotaxonomy is based on the chemical constituents of plants.

26 **Assertion** (A) Generic name is written in small letters.
Reason (R) Specific name is written in capital letters.

27 **Assertion** (A) Each taxonomic group can have two or more names.
Reason (R) Name should be short, precise and easy to pronounce.

28 **Assertion** (A) Binomial system of nomenclature was proposed by Theophrastus for the plants.
Reason (R) Binomial system of nomenclature was proposed by John Ray for animals.

29 **Assertion** (A) Carolus Linnaeus was a Swedish naturalist.
Reason (R) Linnaeus is called Father of Taxonomy.

(C) Medical Entrances' Gallery

(Collection of Questions Asked in NEET & in Various Medical Entrance Exams)

1 The contrasting characteristics generally in a pair used for identification of animals in taxonomic key are referred to as **[NEET 2019]**

(a) lead (b) couplet
(c) doublet (d) alternate

2 Match the items given in Column I with those in Column II and select the correct option given below. **[NEET 2018]**

	Column I		Column II
A.	Herbarium	1.	It is a place having a collection of preserved plants and animals.
B.	Key	2.	A list that enumerates methodically all the species found in an area with brief description aiding identification.
C.	Museum	3.	It is a place where dried and pressed plant specimens mounted on sheets are kept.
D.	Catalogue	4.	A booklet containing a list of characters and their alternates which are helpful in identification of various taxa.

Codes

	A	B	C	D		A	B	C	D
(a)	2	4	3	1	(b)	3	2	1	4
(c)	1	4	3	2	(d)	3	4	1	2

3 Taxonomic key is one of the taxonomic tools in the identification and classification of plants and animals. It is used in the preparation of **[JIPMER 2017]**

(a) monographs (b) flora
(c) Both (a) and (b) (d) None of these

4 Classical taxonomy is based on **[AIIMS 2017]**

(a) morphological traits
(b) habitat of organisms
(c) similarities and dissimilarities of behaviour
(d) phylogeny

5 The label of herbarium sheet does not carry information on **[NEET 2016]**

(a) date of collection (b) name of collector
(c) local names (d) height of the plant

6 Nomenclature is governed by certain universal rules. Which one of the following is contrary to the rules of nomenclature? **[NEET-I 2016]**

(a) The first word in a biological name represents the genus name and the second is a specific epithet
(b) The names are written in Latin and are Italicised
(c) When written by hand, the names are to be underlined
(d) Biological names can be written in any language

7 The system of giving one name to the individual is termed as **[Chhattisgarh PMT 2015]**

(a) quasi binomial system (b) semi-binomial system
(c) monomial system (d) uninomial system

8 Arrange the following in ascending order of Linnaean hierarchy. **[Kerala CEE 2014, UK PMT 2014]**

(a) Kingdom—Phylum—Class—Order—Family—Genus—Species
(b) Kingdom—Family—Genus—Species—Class—Phylum—Order
(c) Kingdom—Order—Species—Genus—Class—Family—Phylum
(d) Species—Genus—Family—Order—Class—Phylum—Kingdom

9 The word species was coined by **[J&K CET 2015]**

(a) Aristotle (b) Linnaeus
(c) John Ray (d) Engler

10 The largest collection of herbarium in India is

(a) Central National Herbarium, Kolkata **[MHT CET 2014]**
(b) Southern Circle Herbarium, Coimbatore
(c) Central Circle Herbarium, Allahabad
(d) Blatter Herbarium, Mumbai

11 Basis of life are **[AIIMS 2013]**

(a) nucleic acids (b) proteins
(c) nucleoproteins (d) amino acids

12 Taxonomic group of any rank is **[OJEE 2013]**

(a) taxon (b) tribe (c) race (d) variety

13 Family is placed between **[OJEE 2013]**

(a) genus and species (b) order and class
(c) class and genus (d) order and genus

14 Which one of the following is not a correct statement? **[NEET 2013]**

(a) Herbarium houses dried, pressed and preserved plant specimens
(b) Botanical gardens have collection of living plants for reference
(c) A museum has collection of photographs of plants and animals
(d) Key is a taxonomic aid for identification of specimens

15 Which one of the following animals is correctly matched with its particular named taxonomic category? **[CBSE AIPMT 2013]**

(a) Cuttle fish – Mollusca, a class
(b) Humans – Primata, the family
(c) Housefly – *Musca*, an order
(d) Tiger – *tigris*, the species

16 Read the statements given below and identify the incorrect statement. **[KCET 2012]**
(a) Scientific names favour multiple naming for the same kind of an organism
(b) Scientific names are used all over the world
(c) Scientific names are often descriptive and tell us some important characters of an organism
(d) Scientific names indicate relationship between species

17 Animals are classified into hierarchial group. In which one of the following, the largest number of species is found? **[WB JEE 2012]**
(a) Genus (b) Order (c) Family (d) Cohort

18 Taxonomic hierarchy refers to **[AIIMS 2012]**
(a) the stepwise arrangement of all categories for classification of plants and animals
(b) a group of senior taxonomists, who decide the nomenclature of plants and animals
(c) a list of botanists or zoologists, who have worked on taxonomy of a species or group
(d) classification of a species based on fossil records

19 Biological organisation starts with **[AFMC 2012]**
(a) sub-microscopic molecular level
(b) cellular level
(c) organismic level
(d) atomic level

20 Species are generally referred as **[BHU 2012]**
(a) real unit of classification
(b) real basic unit of classification
(c) artificial concept of human mind
(d) highest unit of classification

21 Term 'New systematics' was given by **[JCECE 2012]**
(a) Julian Huxley (b) Bateson
(c) Linnaeus (d) Darwin

22 Father of botany is **[OJEE 2012]**
(a) Aristotle (b) Robert Hooke
(c) Darwin (d) Theophrastus

23 Select the wrong statements.
I. Lower the taxon, more are the characteristics that the members within the taxon share.
II. Order is the assemblage of genera which exhibit a few similar characters.
III. Cat and dog are included in the same family–Felidae.
IV. Binomial nomenclature was introduced by Carolus Linnaeus. **[Kerala CEE 2012]**

Choose the correct option.
(a) I, II and III (b) II, III and IV
(c) I and IV (d) III and IV
(e) II and III

24 The framework system of classification in which various taxonomic categories are arranged in order of logical sequence is called **[J&K CET 2012]**
(a) systematics (b) classification
(c) hierarchy (d) taxon

25 The term 'Taxonomy' was introduced by **[J&K CET 2012]**
(a) de Candolle
(b) Bentham and Hooker
(c) Linnaeus
(d) Huxley

26 Scientific study of diversity of organisms and their evolutionary relationships is **[J&K CET 2012]**
(a) morphology
(b) anatomy
(c) taxonomy
(d) systematics

ANSWERS

Check Point 1.1

1. (b)	2. (c)	3. (d)	4. (d)	5. (d)	6. (c)

Check Point 1.2

1. (c)	2. (b)	3. (c)	4. (c)	5. (d)	6. (c)

Check Point 1.3

1. (d)	2. (b)	3. (b)	4. (d)	5. (c)	6. (c)

(A) Topical Exercise

1. (d)	2. (c)	3. (b)	4. (c)	5. (b)	6. (b)	7. (c)	8. (c)	9. (c)	10. (d)
11. (b)	12. (c)	13. (c)	14. (d)	15. (c)	16. (c)	17. (b)	18. (b)	19. (b)	20. (d)
21. (b)	22. (b)	23. (c)	24. (a)	25. (b)	26. (c)	27. (c)	28. (b)	29. (a)	30. (c)
31. (a)	32. (b)	33. (d)	34. (c)	35. (a)	36. (a)	37. (b)	38. (b)	39. (b)	40. (c)
41. (c)	42. (a)	43. (d)	44. (a)	45. (a)	46. (c)	47. (c)	48. (a)	49. (b)	50. (b)
51. (c)	52. (d)	53. (b)	54. (a)	55. (a)	56. (b)	57. (b)	58. (d)	59. (a)	60. (b)
61. (c)	62. (a)	63. (a)	64. (d)	65. (c)	66. (c)	67. (d)	68. (a)		

(B) Cumulative Exercise

1. (b)	2. (b)	3. (b)	4. (a)	5. (d)	6. (d)	7. (d)	8. (c)	9. (c)	10. (a)
11. (d)	12. (d)	13. (a)	14. (c)	15. (a)	16. (d)	17. (a)	18. (a)	19. (b)	20. (b)
21. (d)	22. (a)	23. (d)	24. (b)	25. (b)	26. (e)	27. (d)	28. (e)	29. (b)	

(C) Medical Entrances' Gallery

1. (b)	2. (d)	3. (c)	4. (a)	5. (d)	6. (d)	7. (d)	8. (d)	9. (c)	10. (a)
11. (a)	12. (a)	13. (d)	14. (c)	15. (d)	16. (d)	17. (d)	18. (a)	19. (b)	20. (b)
21. (a)	22. (d)	23. (e)	24. (c)	25. (a)	26. (d)				

Explanations

(A) Topical Exercise

3 (*b*) Statement II is true and I is false. It can be corrected as

Energy is required by both physiological and chemical activity. Cell need it for both synthesis and breakdown of molecules.

6 (*b*) Statement II is true and I is false, it can be corrected as

The repeated reproduction of animals at regular interval of time in their life time is known as iteroparity.

28 (*b*) Statement II is true and I is false. It can be corrected as

Taxon includes all the ranks of hierarchy, i.e. kingdom, phylum/division, class, order, family, genus and species.

41 (*c*) Lectotype is used because lectotype is the single physical example or specimen of an organisms and holotype has the wrong information as it no longer exists.

45 (*a*) Statement I is true and II is false. It can be corrected as

The book *Species Plantarum* was written by Carolus Linnaeus.

63 (*a*) Statement I is true and II is false. It can be corrected as

Museums are meant for the study of peserved plants and animals species. Endangered species are kept in wildlife sancturies.

(B) Cumulative Exercise

1 (*b*) Statements I and II are correct while statements III and IV are incorrect and can be corrected as

Term 'Taxonomy' was given by AP de Candolle. Padua botanical garden is present in Italy.

3 (*b*) Statements II and IV are correct while statements I and III are incorrect and can be corrected as

Mimicry is an example of long term adaptation and aestivation is an example of short term adaptation.

7 (*d*) The statement in option (d) is correct. Rest statements are incorrect and can be corrected as

- Cellular organisation of the body is the defining feature of living forms.
- All living organisms have the ability to sense their surroundings and respond to various stimuli and are thus aware of their surroundings, i.e. show consciousness.
- A patient with dead brain has no consciousness (as brain is the main controlling organ of the body) and hence is considered to be dead.

8 (*c*) Statements I, IV and V are correct. Statements II, III and VI are incorrect and can be corrected as

- *Genera Plantarum* was written by Bentham and Hooker.
- The highest taxonomic category is kingdom.
- The term 'Systematics' was first coined by Linnaeus.

9 (*c*) Statements III and V are true. Statements I, II and IV are false and can be corrected as

- Taxon is used to represent any rank in taxonomic hierarchy.
- Royal Botanical Garden is located at Kew (London).
- There are about 492 (approximately 504) Wildlife Sanctuaries and 89 National Parks in India.

11 (*d*) Statement III is incorrect. It can be corrected as

Common name 'touch me not' is gives to two species, i.e. *Impatiens* and *Mimosa*.

Rest statements are true.

12 (*d*) Statements I, III and IV are true while statement II is false and can be corrected as

In *Mangifera indica* L., 'L' is the name of the author who named it.

14 (*c*) Statements I, II and IV are correct while statement III is incorrect and can be corrected as

Herbarium are the store house of collected plant specimen, which are pressed, dried on a herbarium sheet on which their local name, botanical name, date and place are mentioned. Rest statements are correct.

18 (*a*) The statement in option (a) is correct.

In taxonomic categories, family occupies the position between class and species (lowest). It means two different families can have same class. But a family can accommodate different genera which in turn possess different species. Different species like dog, wolf, jackal belong to same genera *Cannis* and class-Carnivora. Thus, two different genera classified into the same family belong to the same class, but contain different species.

20 (*b*) The statement in option (b) is correct.

Rest statements are incorrect and can be corrected as

- *Musca domestica*, the housefly, belongs to the class-Insecta (Phylum-Arthropoda).
- *Panthera tigris* is the Indian tiger, well-protected in the Bandhavgarh National Park.
- *E. coli*, full name *Escherichia coli*, is a commonly occurring bacterium in the human intestine.

24 (*b*) Both A and R are true, but R is not the correct explanation of A.

Scientific names are printed in Italics or are separately underlined to indicate their Latin origin.

The name of the author is written in abbreviated form after the species name and is printed in Roman.

25 (*b*) Both A and R are true, but R is not the correct explanation of A.

Cytotaxonomy uses cytological informations of cell such as chromosome number, structure and behaviour of chromosomes during meiosis for classification purposes. On the other hand chemotaxonomy is based on the chemical constituents of plants, e.g. the fragrance and taste vary from species to species due to the presence of various types of chemicals.

26 (*e*) Both A and R are false and can be corrected as

As per ICBN rules, in botanical names first word denotes genus and second one represents species. Generic name starts with a capital letter and the specific name with a small letter, e.g. *Cocos nucifera.*

27 (*d*) A is false, but R is true because

According to binomial nomenclature, each taxonomic group can have only one correct name and the name should be short, precise and easy to pronounce to avoid confusions and make it simpler to study.

28 (*e*) Both A and R are false and can be correct as

Carolus Linnaeus proposed the system of nomenclature for plants and animals called Binomial System of Nomenclature. As per this system, scientific name of a species consists of two parts, i.e. genus name and species name.

Example – *Brassica* (genus) *campestris* (species)

John Ray is known for developing keys.

29 (*b*) Both A and R are true, but R is not the correct explanation of A.

Carolus Linnaeus (1707-1778) was a Swedish naturalist. He is considered as Father of Taxonomy. *Systema Naturae* (1758) and *Species Plantarum* (1753) are the famous publication of Linnaeus. He described about 4000 species of plants according to his system of classification based on sexual characters.

(C) Medical Entrances' Gallery

1 (*b*) Couplet is the contrasting characteristic generally in a pair used for identification of animals in taxonomic key. It represents the choice made between two opposite options, each half of a couplet is called lead. Separate taxonomic keys are required for each taxonomic category like family, genus, species.

3 (*c*) Taxonomic keys are the tools that help in the identification of organisms based on the description of specific characters. These are used in preparation of flora (a resource providing information on the taxonomy, nomenclature and descriptive data of newly discovered plants) and monographs (that contain information of any one taxon).

4 (*a*) Classical taxonomy is based on the study of morphological traits of the organisms. It is also called alpha taxonomy or old systematics.

5 (*d*) Herbarium is a collection of plants that usually have been dried, pressed and preserved on sheets and are arranged in accordance with any accepted system of classification providing information about the date and place of collection, family names, collector's name, etc., for future reference and study. However, it does not have information on height of the plant.

6 (*d*) Biological names originate from Latin language and are printed in Italics.

7 (*d*) Uninomial nomenclature is the designation of a taxon above species ranked by a scientific name consisting of a single word.

8 (*d*) Classification involves hierarchy of steps in which each step represents a rank or category. Since, the category is a part of overall taxonomic arrangement, it is called as the taxonomic category and all categories together constitute the taxonomic hierarchy. The correct hierarchial arrangement of taxonomic category of plants in are ascending order

Species → Genus → Family → Order → Class → Phylum → Kingdom

10 (*a*) Central National Herbarium is the largest herbarium in South-East Asia. It is situated within the Botanical Garden at Sibpur in Kolkata, West Bengal.

Presently, the Central National Herbarium possesses about 2.5 million of herbarium sheets belonging to nearly 350 families of plants, which are arranged according to Bentham and Hooker's system of classification.

11 (*a*) Nucleic acids (DNA and RNA) consisting carbon, hydrogen, oxygen, nitrogen and phosphorus are capable of replicating on their own therefore, considered as basis of life.

14 (*c*) Statement in option (c) is not correct because

Museums have collection of preserved plants and animal specimens for study and reference. Specimens are preserved in the containers in preservatives. Herbarium is a storehouse having collected dried, pressed and preserved plant specimens on sheets.

Botanical gardens have collections of living plants for reference. Key is a taxonomic aid for identification of plants and animals based on the similarities and dissimilarities.

15 (*d*) The correctly matched pairs is given by option (d). Tiger, the scientific name is *Panthera tigris* of which *tigris* is the species name.

The other can be corrected as

Cuttle fish-Class-Cephalopoda, phylum-Mollusca
Humans-Family-Hominidae, order-Primata
Housefly-Genus-*Musca*, order-Diptera

16 (*d*) Statement in option (d) is incorrect because Scientific name is unique to an organism, i.e. these names are used all over the world, to describe some important characters of organism, it also indicates the species and genus of the organism.

17 (*d*) Cohort is a formal taxonomic category in between infraclass and super-order and it contains the largest number of species.

Class → Subclass → Infraclass → Cohort → Super-order → Order

18 (*a*) Taxonomic hierarchy refers to the stepwise arrangement of all categories for the classification of plants and animals. It is also called Linnaean hierarchy as it was first proposed by Carolus Linnaeus. The hierarchy includes seven obligate categories namely; kingdom, division or phylum, class, order, family, genus and species.

19 (*b*) Biological organisation starts with cellular level where the cellular organelles interact with each other to form a structural and functional unit called the cell.

These interactions result in emergent properties at a higher level of organisation. This phenomenon is true in the hierarchy of organisational complexity at all levels.

20 (*b*) Species represents the real basic or lowest unit of classification. It is a group of organisms that can interbreed freely in their natural population and produces fertile offspring of their own kind.

21 (*a*) The term 'New systematics' was given by Julian Huxley (1940). New systematics or biosystematics is the concept of systematics, which defines taxonomic affinities on the basis of evolutionary, genetic and morphological traits.

22 (*d*) Father of Botany—Theophrastus

Father of Zoology/Biology— Aristotle

Father of Cytology — Robert Hooke

Father of Evolution —Darwin

23 (*e*) Statements II and III are wrong and can be corrected as

Order is the assemblage of families, which exhibit a few similar characters. The similar characters are less in number as compared to different genera included in a family.

Cat and dog belong to different families. Cat belongs to family-Felidae and dog belongs to family-Canidae.

Rest statements are correct.

CHAPTER 02

Biological Classification

Scientific arrangement of living organisms into groups and subgroups for convenience in study is known as **biological classification**. It is done on the basis of their similarities and dissimilarities.

The first pioneer work on classification was done by **Linnaeus**, for this contribution he is regarded as **Father of Taxonomy.** He classified living organisms into two kingdoms, i.e. Plantae and Animalia. The inadequacy and further refinements modified his classification many a times. This chapter contains a brief outline of all those modifications alongwith basics of biological classification.

Inside

SYSTEMS OF CLASSIFICATION

Following 3 scientific systems can be used to classify living organisms.

Artificial system

In artificial system of classification, living organisms are grouped together according to their habit, habitat and few morphological characters. For example, **Linnaeus** classified flowering plants on the basis of numerical relation of floral parts in his books *Genera Plantarum* (1737) and *Species Plantarum* (1753). This is known as **sexual system of classification**. Artificial system was preferential until the 18th century, after that, it was replaced by natural system.

Natural system

This system aims at classifying and arranging organisms according to their overall similarities in relation to phylogeny or evolutionary history. Through this system, the real kin relationships and taxonomic positions of living organisms can be revealed.

Phylogenetic system

This system of classification is based on the phylogenetic information in addition to many other taxonomic characters.

Table 2.1 Differences between artificial system, natural system and phylogenetic system

Artificial system	Natural system	Phylogenetic system
This was based on one or a few characters, which are not related to phylogenetic significance.	This was based on overall characters, which indicate natural evolutionary relationships.	This was based on evolutionary and genetic relationships of organisms.
This system allows living organisms to be used as identification keys in the various floras and faunas of the world.	This system brings out natural relationship with the help of all the available data, including fossils.	This system is also called **cladistics** as organisms belonging to same taxa are believed to have a common ancestor and may be represented as a family tree called **cladogram.**
First artificial system of plants was proposed by **Theophrastus,** who classified plants as herbs, shrubs and trees. Similarly, Aristotle classified animals into **Enaima** (with red blood) and **Anima** (without red blood). Enaima were further classified into **ovipera** (egg laying) and **vivipera** (giving birth to young ones).	First natural system was given by **AL de Jussieu.** The very first natural system for animal classification was given by **Linnaeus** in his book *Systema Naturae* (1758). The most pioneer natural plant classification was given by **Bentham** and **Hooker** (1863) in their *Genera Plantarum ad exemplaria.* It is the widely used classification system for practical purposes.	First phylogenetic system was proposed by **AW Eichler** in their book *'Syllabus der Vorlesungen'* (1892). His work was later modified by **Engler** and **Prantl** in their monograph *'Die Naturlichen Pflanzenfamilien'* (1887-1899). Afterwards **John Hutchinson** (1926) published more advanced phylogenetic classification in his book '*The families of flowering plants*'. The phylogenetic animal classification was given by **Dobzhansky** and **Mayr.**

Another system, i.e. **phenetic system of classification** is based on overall similarity of organisms regardless of their phylogeny or evolutionary relationship. This classification system did not have a strong impact on animal classification and thus, scientific interest in this approach is on decline.

CLASSIFICATION OF ORGANISMS

Kingdom is the most prominent category of taxonomical hierarchy. The earliest classification recognised only two kingdoms of living beings. However, the present number of kingdoms are six. The complete journey from two to six kingdoms can be explored as follows

Two kingdom classification

This system was given by **Carolus Linnaeus** in 1758. He divided all organisms of the world into following two kingdoms

- **Plantae** which includes photosynthetic autotrophic organisms.
- **Animalia** which includes non-photosynthetic heterotrophic organisms.

However, some organisms like *Chlamydomonas, Euglena* and the slime moulds, etc., share characteristics of both plants and animals.

Hence, it was suggested that a new kingdom should be established to accommodate such organisms. This led to the development of three kingdom system of classification.

Three kingdom classification

Ernst Haeckel, a German zoologist (1866), suggested a third kingdom in his book *'Generelle Morphologie der Organismen'.* Haeckel named this third kingdom as **Protista.**

The kingdom-Protista includes all single-celled eukaryotes. These are believed to have evolved from the prokaryotic organisms and are the precursors from which higher eukaryotic kingdoms-**Plantae, Fungi** and **Animalia** have evolved.

Four kingdom classification

It was given by **Copeland** (1956). According to this system, all organisms are divided into the following four kingdoms

- **Mycota** or **Monera** having prokaryotic organisms.
- **Protista** or **Protoctista** having true nucleate unicellular organisms.
- **Plantae** having photosynthetic and multicellular organisms.
- **Animalia** having non-photosynthetic and multicellular organisms.

Five kingdom classification

RH Whittaker (1969) divided living organisms into five kingdoms. These kingdoms are formed, on the basis of the following features

- Complexity of cell structure (i.e. prokaryote or eukaryote).
- Complexity of body structure (unicellular or multicellular).
- Mode of nutrition (autotrophic or heterotrophic)
- Ecological lifestyle (producers, consumers or decomposers)
- Phylogenetic relationships

The five kingdoms given by Whittaker are **Monera, Protista, Fungi, Plantae** and **Animalia**. The new kingdom-Fungi includes thalloid plants ranging from microscopic yeast to large mushrooms.

Six kingdom classification

The most recent classification of living world is six kingdom classification proposed by three microbiologists **Carl Woese, O Kandler** and **ML Wheelis** in 1990. It is based on recent nucleic acid hybridisation and DNA sequencing methods. After comparing base sequences of organisms, they proposed that kingdom-Monera should also be further divided into two kingdoms, named as Archaebacteria and Eubacteria.

Thus, the six kingdoms are **Archaebacteria, Eubacteria, Protista, Fungi, Plantae** and **Animalia**. The newly framed kingdom-Archaebacteria contain halobacteria, thermoacidophiles and methanogens. In kingdom-Eubacteria, true bacterial prokaryotes are placed.

Major characteristics of all these six kingdoms are given below

Table 2.2 Major characteristics of all six kingdom system

Characters	Monera		Protista	Fungi	Plantae	Animalia
	Eubacteria	Archaea				
Cell type	Prokaryotic	Prokaryotic	Eukaryotic	Eukaryotic	Eukaryotic	Eukaryotic
Multicellularity/ Cell specialisation	Absent	Absent	Absent in most, present in algae and water moulds	Present	Present	Present
Body organisation	Cellular	Cellular	Cellular	Multicellular/Loose tissue	Tissue/Organ	Tissue/Organ/Organ system
Cell wall	Non-cellulosic, peptidoglycan	Non-cellulosic, various types, different from Eubacteria	Absent or present, cellulosic or of various types	Chitinous or cellulosic	Cellulosic	Absent
Genetic material	DNA **not associated** with histone protein on chromosomes	DNA **associated** with histone protein on chromosomes	DNA **associated** with histone protein on chromosomes	DNA **associated** with histone protein on chromosomes	DNA **associated** with histone protein on chromosomes	DNA **associated** with histone protein on chromosomes
Nuclear envelope	Absent	Absent	Double or Single	Double	Double	Double
Chloroplasts	Absent	Absent	Present or Absent	Absent	Present	Absent
Other membrane-bound organelles	Absent	Absent	Present	Present	Present	Present
Flagella/Cilia	Solid (9 + 0) rotating	Solid (9 + 0) rotating	(9 + 2) microtubules	Absent in main body structures, but may be present in asexual reproductive bodies like spores	Absent or (9 + 2) microtubules in gametes of some groups	(9 + 2) microtubules

Characters	Monera		Protista	Fungi	Plantae	Animalia
	Eubacteria	Archaea				
Gene structure	Introns absent in most groups, present in some	Introns present	Introns present	Introns present	Introns present	Introns present
Means of genetic recombination	Conjugation, transduction, transformation	Conjugation, transformation, transduction	Fertilisation and meiosis in some groups only	Fertilisation and meiosis	Fertilisation and meiosis	Fertilisation and meiosis
Mode of nutrition	Autotrophic or heterotrophic by Absorption	Heterotrophic or chemoautotrophic (many are extremophiles)	Autotrophic or heterotrophic by absorption or phagocytosis	Heterotrophic by absorption	Autotrophic mostly, some parasite also	Heterotrophic by ingestion
Respiration	Aerobic or Anaerobic	Anaerobic or Aerobic	Aerobic	Aerobic or Anaerobic	Aerobic	Aerobic
Nervous system	Absent	Absent	Absent	Absent	Absent	Present, often complex
Life cycle	Haploid, fission	Haploid, fission	Mostly haploid, some forms diploid	Mostly haploid often two nuclei per cell, in some alternation of generation	Alternation of generations	Diploid except for gametes
Unicellular spore formation	Present in some groups	Absent	Present in some groups	Present in all groups	Present in some groups	Absent
Examples	Bacteria, cyanobacteria, rickettsias, etc.	Methanogens, halophiles and thermoacidophiles	Diatoms, dinoflagellates, slime moulds, etc.	*Rhizopus*, *Agaricus*, *Penicillium*, etc.	Algae, bryophytes, etc.	Tiger, cat, spider, earthworm, etc.

Note • *Based on 16S ribosomal RNA gene sequences,* ***Carl Woese*** *clustered the six kingdoms into three domains of life, i.e. Archae, Bacteria and Eukarya.*

• *Eichler (1883) divided the plant kingdom into two main groups viz flowerless or seedless plants called* ***Cryptogamia*** *and flowering or seed bearing plants called* ***Phanerogamia*** *or* ***Spermatophyta****.*

EXAM POINTS

> In Vedic times, the living organisms were classified into three classes, i.e. Jivaja (born from worm or viviparous), Andaja (born from egg or viviparous) and Udbhija (born from sprouts). **[Bihar CECE 2015]**

> Linnaeus system of classification is called **'sexual system' of classification**. **[NEET 2013]**

> Phenetic classification is based on observable characteristic of existing organisms. **[Manipal 2012]**

> Five kingdom system of classification suggested by **RH Whittaker** is not based on the presence or absence of a well-defined nucleus. **[CBSE AIPMT 2014]**

> According to five kingdom classification, bacteria belong to Monera. **[J&K CET 2011]**

> As per Whittaker's classification, an organism possessing eukaryotic cell structure multicellular organisation with a cell wall and nuclear membrane showing heterotrophic nutrition can be placed under the kingdom-Fungi. **[Kerala CEE 2014]**

> *Saccharomyces* or yeast is not a prokaryote. It is a eukaryote of kingdom–Fungi. **[NEET 2018]**

> In kingdom–Protista, green phytoplanktons are kept, under five kingdom classification. **[Chhattisgarh PMT 2015]**

> In five kingdom classification, single-celled eukaryotes are included in Protista. **[J&K CET 2015]**

Chapter Exercises

(A) Topical Exercise

Systems of Classification

1 What is correct about classification?
(a) It allows to confirm the already existing identifications
(b) It helps in new identifications
(c) It is a systematic arrangement of taxa
(d) All of the above

2 Which of the following classification is based on some morphological characters?
(a) Natural (b) Artificial
(c) Phylogenetic (d) Both (b) and (c)

3 Artificial system of classification of plants was first proposed by
(a) Engler and Prantl (b) Bentham and Hooker
(c) Sneath (d) Linnaeus

4 The artificial system of classification by Linnaeus was published in
(a) *Systema Naturae* (b) *Species Plantarum*
(c) *Genera Plantarum* (d) Both (b) and (c)

5 The system of classification proposed by Linnaeus was artificial because
(a) it considered physiological characters along with morphology
(b) it considered evolutionary relationships of plants
(c) it takes into account similarities and differences in few unifying characters
(d) None of the above

6 Floral characters are the basis of classification because
(a) they are more conserved
(b) they are variously pigmented
(c) they can be preserved for longer times
(d) All of the above

7 In his book '*Genera Plantarum*' Linnaeus classified flowering plants on the basis of
(a) numerical relation of floral parts
(b) presence or absence of floral parts
(c) position of ovary
(d) presence or absence of androecium

8 The natural system of classification differs from artificial system in
(a) taking only one floral trait
(b) employing the vegetative trait
(c) developing evolutionary trends
(d) bringing out similarities and dissimilarities between the organisms being studied

9 A system of classification, which takes into account, all noticeable characters in morphology in nature is
(a) phylogenetic system (b) natural system
(c) artificial system (d) cytotaxonomy

10 Phylogenetic system of classification is based on
(a) evolutionary relationships
(b) morphological features
(c) chemical constituents
(d) floral characters

11 Engler and Prantl published a phylogenetic system in the monograph.
(a) *Origin of Species*
(b) *Genera Plantarum*
(c) *Historia Plantarum*
(d) *Die Naturlichen Pflanzenfamilien*

12 The phylogenetic animal classification was given by
(a) Engler and Prantl (b) Bentham and Hooker
(c) Dobzansky and Mayr (d) Carolus Linnaeus

13 Advantage(s) of phylogenetic system over artificial system of classification is/are
(a) included overall similarities
(b) based on evolution and history
(c) included genetic relationships among organisms
(d) All of the above

Classification of Organisms

14 Two kingdom and three kingdom classification system was given by ...., respectively.
(a) Whittaker and Copeland (b) Linnaeus and Haeckel
(c) Copeland and Haeckel (d) None of these

15 What was the name of the book in which Ernst Haeckel gave the third kingdom called Protista?
(a) *Systema Animalia*
(b) *Systema Naturae*
(c) *Generelle Morphologie der Organismen*
(d) *Species Plantarum*

16 All eukaryotic unicellular organisms belong to **[NCERT Exemplar]**
(a) Monera (b) Protista (c) Fungi (d) Bacteria

17 In his four kingdom classification, Copeland divided organisms into
(a) Monera, Protista, Plantae and Animalia
(b) Bacteria, Fungi, Protozoa and Eukaryota
(c) Bacteria, Cyanobacteria, Protozoa and Eukaryota
(d) Archaebacteria, Eubacteria, Plantae and Animalia

18 Which of the following was given the status of kingdom in the classification system given by Copeland?
(a) Prokaryotes (b) Myxomycetes
(c) Eukaryotic algae (d) Protista

19 The kingdom-Monera includes
(a) unicellular photosynthetic organisms
(b) prokaryotic organism with incipient nucleus
(c) non-photosynthetic multicellular organisms
(d) true nucleate unicellular organisms

20 Consider the following statements.
I. Fungi possess DNA associated with histone protein which are absent in prokaryotes.
II. Nuclear envelope is present in Protista but not in Archaebacteria.
Choose the correct option.
(a) Statement I is true, but II is false
(b) Statement I is false, but II is true
(c) Both statements I and II are true
(d) Both statements I and II are false

21 The members of kingdom-Protista differ from members of kingdom-Monera mainly due to
(a) membrane bound nucleus
(b) the presence of cell wall
(c) DNA
(d) RNA

22 Who proposed a five kingdom system of classification and named kingdoms as Monera, Protista, Fungi, Plantae and Animalia? **[NCERT Exemplar]**
(a) Herbert Copeland (b) RH Whittaker
(c) Carl Woese (d) Carolus Linnaeus

23 In five kingdom system of classification how many kingdoms contain eukaryotes?
(a) Four kingdoms (b) One kingdom
(c) Two kingdoms (d) Three kingdoms

24 How many kingdom(s) in Whittaker's classification system strictly involves organisms with only heterotrophic nutrition?
(a) Two (b) Four
(c) Three (d) One

25 The unicellular eukaryotic wall-less organisms of Linnaeus's system should be placed in .... kingdom.
(a) Monera (b) Protista
(c) Animalia (d) Fungi

26 In which kingdom would you classify the Archaea and nitrogen-fixing organisms, if the five kingdom system of classification is used?
(a) Protista (b) Monera (c) Plantae (d) Fungi

27 A student viewed a drop of marine water under microscope and observed a unicellular organism with a distinct nucleus and a wall. In which of the following kingdom of Whittaker it should be placed?
(a) Monera (b) Plantae
(c) Protista (d) Animalia

28 Given below are certain characteristic features of members of kingdom-Protista. Pick out the one which is not true.
(a) The body is unicellular
(b) The cells are eukaryotic
(c) The mode of nutrition is always autotrophic
(d) Cell wall may or may not be present

29 The kingdom having both *Chlamydomonas* and *Amoeba* is
(a) Fungi (b) Protista
(c) Monera (d) Plantae

30 Which of the following have cellulosic cell wall?
(a) Monerans
(b) Plants
(c) Fungi
(d) Animals

31 The organisms which show the characteristics of both plants and animals are
(a) *Amoeba*, *Paramecium* and *Chlamydomonas*
(b) *Spirogyra*, *Paramecium* and *Amoeba*
(c) *Euglena*, *Amoeba* and *Spirogyra*
(d) Slime moulds, *Euglena* and *Chlamydomonas*

32 Naked cytoplasm, multinucleated and saprophytic nutrition are the characteristics of **[NCERT Exemplar]**
(a) Monera (b) Protista
(c) Fungi (d) Slime moulds

33 Choose the statement which is incorrect about members of kingdom-Animalia.
(a) It comprises of consumer organisms
(b) The members of kingdom-Animalia may be both unicellular or multicellular
(c) Their body has a definite shape and cellular, tissue and organ system
(d) Most of the members have a nervous system to coordinate their body actions and make rapid responses to external stimuli

34 Consider the given figure.

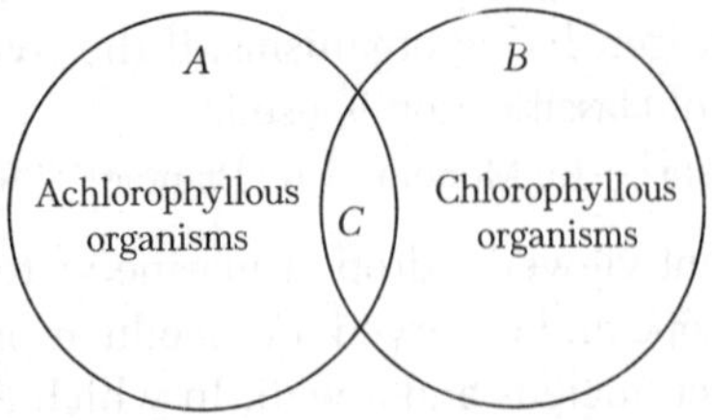

Which of the following option is correct for the labels *A* to *C*?

	A	*B*	*C*
(a)	Fungi	Plants	Animals
(b)	Monera	Fungi	Protista
(c)	Animals	Plants	Protista
(d)	Plants	Monera	Animals

35 (9+ 2) arrangement of microtubules is present in

(a) kingdom-Plantae (b) kingdom-Animalia
(c) kingdom-Protista (d) All of these

36 In six kingdom classification, Monera is sub-divided into

(a) Mycoplasma and Eubacteria
(b) Archaebacteria and Eubacteria
(c) Cyanobacteria and Bacteria
(d) Archaebacteria and Cyanobacteria

37 The newly framed kingdom-Archaebacteria consist of

(a) methanogens and bacterial prokaryotes
(b) thermoacidophiles and halobacteria
(c) halobacteria, thermoacidophiles and methanogens
(d) halobacteria and prokaryotic bacteria

38 Which of the following scientists proposed its classification system based on recent nucleic acid hybridisation and *r*RNA sequencing methods?

(a) Copeland
(b) Whittaker
(c) Haeckel
(d) Carl Woese

(B) Cumulative Exercise

(Assorted Questions of the Chapter for Advanced Level Practice)

1 What is true about classification?

I. It is the arrangement of living organisms in groups.
II. Aristotle was the first to classify organisms.
III. It can be artificial, natural and phylogenetic.
IV. Evolved by Robert Hooke.

Choose the correct option from the codes given below.

Codes

(a) I, II and III
(b) II and IV
(c) I and II
(d) I and III

2 Consider the following statements.

I. Biological classification is the scientific grouping of organisms in a hierarchical series of groups on the basis of their morphological and evolutionary relationships.
II. Whittaker classified organisms on the basis of autotrophic and heterotrophic mode of nutrition only.
III. In five kingdom system of classification, living organisms can be divided into prokaryotes and eukaryotes on the basis of their cell structure.

Which of the statements given above are correct?

(a) I and II
(b) I and III
(c) II and III
(d) I, II and III

3 Natural system of classification is

I. based on one or few characters which are not related to phylogenetic significance.
II. based on overall characters which indicate natural relationships.
III. based on evolutionary and genetic relationships of organisms only.
IV. given by AL de Jussieu.

Codes

(a) II and III are correct
(b) I, II and III are incorrect
(c) II and IV are correct
(d) Only IV is incorrect

4 Match the following columns.

Column I	Column II
A. Artificial system	1. AW Eichler
B. Phylogenetic system	2. Peter Sneath
C. Phenetic system	3. C Linnaeus
D. *Systema Naturae*	4. Theophrastus

Codes

	A	B	C	D
(a)	4	1	2	3
(b)	1	2	3	4
(c)	4	3	2	1
(d)	2	4	1	3

5 Choose the major characteristics of five kingdom system.

I. It was developed by RH Whittaker.
II. It is divided into five kingdom system based on cell structure, body structure, nutrition, phylogenetic relationship and lifestyle.
III. It is given by Copeland.
IV. It is divided into four divisions.

Codes

(a) I and II (b) III and IV (c) I and IV (d) II and IV

6 Select the correct combination of the statements (I-IV) regarding the characteristics of certain organisms.

I. Methanogens are Archaebacteria which produce methane in marshy areas.
II. *Nostoc* is a filamentous blue-green alga which fuses atmospheric nitrogen.
III. Chemosynthetic autotrophic bacteria synthesise cellulose from glucose.
IV. Mycoplasma lacks a cell wall and can survive without oxygen.

Codes

(a) II and III (b) I, II and III
(c) II, III and IV (d) I, II and IV

7 Consider the following statements.

I. All prokaryotic organisms were grouped together under kingdom-Monera.
II. The unicellular eukaryotic organisms were placed in kingdom-Protista.
III. *Chlorella* and *Chlamydomonas*, both have cell walls and are placed in kingdom-Protista.

Codes

(a) Only I and II are incorrect
(b) Only I is correct
(c) Only III is incorrect
(d) All statements are correct

8 Flagella of prokaryotic and eukaryotic cells differ in

(a) type of movement and placement in cell
(b) location in cell and mode of functioning
(c) microtubular organisation and type of movement
(d) macrotubular organisation and function

9 Which organism(s) is correctly assigned to its taxonomic group?

(a) *Paramecium* and *Plasmodium* belong to the same kingdom as that of *Penicillium*
(b) *Nostoc* and *Anabaena* are examples of Protista
(c) Desmids and *Euglena* belong to the same kingdom as that of *Trypanosoma*
(d) Lichen is a composite organism from the symbiotic association of an alga and a protist

10 Match the following columns.

Column I	Column II
A. Animalia	1. Unicellular prokaryotes
B. Monera	2. Multicellular and photosynthetic
C. Protista	3. Unicellular eukaryotes
D. Plantae	4. Multicellular ingestive heterotrophs

Codes

	A	B	C	D
(a)	4	3	1	2
(b)	4	1	3	2
(c)	1	2	4	3
(d)	2	3	1	4

Assertion and Reason

Direction (Q. Nos. 11-15) *In the following questions, a statement of Assertion (A) is given followed by corresponding statement of Reason (R). Of the statements, mark the correct answer as*

(a) Both A and R are true and R is the correct explanation of A
(b) Both A and R are true, but R is not the correct explanation of A
(c) A is true, but R is false
(d) A is false, but R is true
(e) Both A and R are false

11 **Assertion** (A) Two kingdom system was proposed by Linnaeus.

Reason (R) This includes kingdom-Prokarya and Eukarya.

12 **Assertion** (A) Monera includes all prokaryotes.

Reason (R) Nuclei of monerans are not organised with nuclear membrane, nucleoplasm, chromatic fibres and nucleolus.

13 **Assertion** (A) Artificial system of classification separated closely related species.

Reason (R) Artificial system gave equal weightage to vegetative and sexual characteristics.

14 **Assertion** (A) Kingdom-Monera is also known as kingdom-Prokaryota.

Reason (R) Monera cells lack membrane bound cell organelles.

15 **Assertion** (A) Algae were grouped as a separate kingdom by Whittaker.

Reason (R) There are red, blue and green algae only.

(C) Medical Entrances' Gallery

(Collection of questions Asked in NEET & Various Medical Entrance Exams)

1 Which among the following is not a prokaryote? **[NEET 2018]**

(a) *Nostoc* (b) *Mycobacterium*
(c) *Saccharomyces* (d) *Oscillatoria*

2 Green phytoplanktons are kept in which kingdom of five kingdom classification system? **[Chhattisgarh PMT 2015]**

(a) Kingdom-Monera (b) Kingdom-Protista
(c) Kingdom-Plantae (d) Kingdom-Fungi

3 In five kingdom classification, single-celled eukaryotes are included in **[J&K CET 2015]**

(a) Fungi (b) Protista
(c) Monera (d) Archaea

4 In Vedic times, living organisms were classified in the following number of classes. **[Bihar CECE 2015]**

(a) 2 (b) 3 (c) 4 (d) 5

5 Five kingdom system of classification suggested by RH Whittaker is not based on **[CBSE AIPMT 2014]**

(a) presence or absence of a well-defined nucleus
(b) mode of reproduction
(c) mode of nutrition
(d) complexity of body organisation

6 As per Whittaker's classification, an organism possessing eukaryotic cell structure, multicellular organisation with a cell wall and nuclear membrane showing heterotrophic nutrition can be placed under the kingdom **[Kerala CEE 2014]**

(a) Monera (b) Protista
(c) Plantae (d) Fungi

7 Which is called 'sexual system' of classification? **[NEET 2013]**

(a) Bentham and Hooker's (b) Tippo's
(c) Linnaeus (d) Takhtajan

8 Consider the following statements with respect to characteristic features of the kingdom. **[Kerala CEE 2012]**

I. In Animalia, the mode of nutrition is autotrophic.
II. In Monera, the nuclear membrane is present.
III. In Protista, the cell type is prokaryotic.
IV. In Plantae, the cell wall is present.

Which of the above statement(s) is/are correct?

(a) Only I (b) Only II
(c) Only III (d) Only IV
(e) I, II and III

9 Phenetic classification is based on **[Manipal 2012]**

(a) ancestral lineage of existing organisms
(b) observable characteristic of existing organisms
(c) dendrograms based on DNA characteristics
(d) sexual characteristics

10 Match the following columns.

	Column I (Kingdom)		Column II (Class)
A.	Plantae	1.	Archaebacteria
B.	Fungi	2.	Euglenoids
C.	Protista	3.	Phycomycetes
D.	Monera	4.	Algae

Codes **[Kerala CEE 2011]**

	A	B	C	D
(a)	4	3	2	1
(b)	1	2	3	4
(c)	3	4	2	1
(d)	4	2	3	1
(e)	2	3	4	1

11 According to five kingdom classification, bacteria belong to **[J&K CET 2011]**

(a) Protista (b) Monera
(c) Plantae (d) Archaea

12 **Assertion** (A) Bacteria, Protista do not have circulatory system.

Reason (R) These organisms live in moist and watery environment. **[AIIMS 2010]**

(a) Both A and R are true and R is the correct explanation of A
(b) Both A and R are true, but R is not the correct explanation of A
(c) A is true, but R is false
(d) A is false, but R is true
(e) Both A and R are false

ANSWERS

(A) Topical Exercise

1. (d)	2. (b)	3. (d)	4. (d)	5. (c)	6. (a)	7. (a)	8. (d)	9. (b)	10. (a)
11. (d)	12. (c)	13. (d)	14. (b)	15. (c)	16. (b)	17. (a)	18. (a)	19. (b)	20. (c)
21. (a)	22. (b)	23. (a)	24. (a)	25. (c)	26. (b)	27. (c)	28. (c)	29. (b)	30. (b)
31. (d)	32. (d)	33. (b)	34. (c)	35. (d)	36. (b)	37. (c)	38. (d)		

(B) Cumulative Exercise

1. (a)	2. (c)	3. (c)	4. (a)	5. (a)	6. (d)	7. (c)	8. (c)	9. (c)	10. (b)
11. (c)	12. (a)	13. (a)	14. (a)	15. (d)					

(C) Medical Entrances' Gallery

1. (c)	2. (b)	3. (b)	4. (b)	5. (a)	6. (d)	7. (c)	8. (d)	9. (b)	10. (a)
11. (b)	12. (a)								

Explanations

(A) Topical Exercise

2 (*b*) In artificial system of classification, only a few morphological characters of the plants are being considered. For example, the grouping of plants into shrubs and trees in the sexual system of Linnaeus was based on number of stamens and styles.

6 (*a*) Floral characters in a plant family usually are conserved, i.e. floral characters are less affected by environment than vegetative characters.

7 (*a*) In his book '*Genera Plantarum*' Linnaeus classified flowering plants on the basis of numerical relation of floral parts according to the number of stamens and pistil in the flowers.

8 (*d*) Artificial system of classification deals only with the morphological features of an organism, whereas natural classification involves similarities based upon cellular organisation, reproduction, mode of nutrition, etc.

16 (*b*) All single-celled eukaryotes are placed in kingdom-Protista. Phylogenetically, the kingdom-Protista acts as a connecting link between the prokaryotic Monera on one hand and the complex multicellular kingdom-Fungi, Plantae and Animalia on the other hand.

25 (*c*) Linnaeus placed all organisms without cell wall in kingdom—Animalia, but according to Whittaker's classification, eukaryotic, unicellular, wall-less organisms should be placed in Protista.

26 (*b*) Whittaker's classification includes five kingdom classification system. The five kingdoms include Monera, Protista, Fungi, Plantae and Animalia. Monera include non-nucleated unicellular organisms. They are prokaryotes and have a cell wall. They do not have membrane bound organelles and true nucleus. So, kingdom-monera classify the Archaea and nitrogen-fixing organisms.

27 (*c*) The organism should be placed in kingdom-Protista as it is unicellular and have well-defined nucleus, i.e. eukaryotic.

28 (*c*) Certain characteristic feature in option (c) is not true and can be corrected as

The mode of nutrition in Kingdom-Protista is either autotrophic or heterotrophic.

Rest other characteristic features of members of kingdom-protista are true.

32 (*d*) Slime moulds are saprophytic protists, moving along the dead leaves engulfing organic material. These are multinucleated and do not possess cell wall and have naked cytoplasm. Whereas, monerans are prokaryotes, which include all bacteria.

34 (*c*) The *venn* diagram depicts kingdom—Animalia (*A*) as achlorophyllous organisms, kingdom—Plantae (*B*) having chlorophyllous organisms and kingdom—Protista (*C*) which may have chlorophyllous as well as achlorophyllous organisms.

(B) Cumulative Exercise

1 (*a*) All the statements are true except (IV). Incorrect statement can be corrected as

Classification was evolved by Charles Darwin.

2 (*c*) All the statements are correct except (I). Incorrect statement can be corrected as

Biological classification is the process by which scientific group living organisms. Organisms are classified based on the basis of similarity among them. Historically, similarity was determined by examining the physical characteristics of an organism but modern classification uses a variety of techniques including genetic analysis.

3 (*c*) II and IV are correct. I and III are incorrect and can be corrected as

According to natural system of classification, the organisms are arranged according to their natural affinities through the use of all important permanent characteristics. The most important natural systems for classification of seed plant was developed by G Bentham and JD Hooker.

6 (*d*) All given statements are correct except (III). Incorrect statement can be corrected as

Chemosynthetic autotrophic bacteria oxidise various inorganic substances such as nitrates, nitrites and ammonia and use the released energy for their ATP production. They play a great role in recycling nutrients like nitrogen, phosphorus, iron and sulphur.

8 (*c*) Flagella of prokaryotic and eukaryotic species differ in microtubular organisations and type of movement. In eukaryotes, the tubular arrangement is (9+2) and specialised while in prokaryotes, the microtubular arrangement is (9+0) and is simple.

Movement is also different, prokaryotes have rotatory movement and eukaryotes have bending movement.

11 (*c*) A is true, but R is false because

As per two kingdom system of classification, given by Linnaeus living organisms are classified into kingdom-Plantae (includes photosynthetic autotrophic organisms). Animalia (includes non-photosynthetic heterotrophic organism).

12 (*a*) Both A and R are true and R is the correct explanation of A.

Kingdom–Monera includes prokaryotic organisms, i.e. primitive organisms which do not have a well-organised nucleus.

13 (*a*) Both A and R are true and R is the correct explanation of A.

Artificial system separate the closely related species as this system is based on a few vegetative and sexual characteristics.

14 (*a*) Both A and R true and R is correct explanation of A.

Kingdom–Monera is also known as kingdom–Prokaryota because it includes all prokaryotes. Monera cells lack membrane bound cell organelles.

15 (*d*) A is false, but R is true because

According to Whittaker's classification, algae are grouped into three kingdoms, i.e. Monera (having blue-green algae), Protista (having diatoms, etc.) and Plantae (having red, green and brown algae).

(C) Medical Entrances' Gallery

1 (*c*) Among the given options, *Saccharomyces* is not a prokaryote. It is a fungus, i.e. it is a eukaryote. It possesses a well-defined nucleus and other cell organelles.

Other options are explained as

Nostoc and *Oscillatoria* are cyanobacteria, while *Mycobacterium* is a true bacterium. Cyanobacteria and bacteria, both are prokaryotes as they lack a well-defined nucleus and other cell organelles.

2 (*b*) Green phytoplanktons are photosynthetic and have chloroplast for carrying out photosynthesis. Since, these are unicellular eukaryotic organisms,they are placed in kingdom–Protista, e.g. *Chlorella*, *Chlamydomonas*, etc.

4 (*b*) In ancient (Vedic) times, the living organisms were classified into three classes and these were

(i) Jivaja (born from worm or viviparous)
(ii) Andaja (born from egg or oviparous)
(iii) Udbhija (born from sprouts)

5 (*a*) RH Whittaker's classification is not based on the presence or absence of a well-defined nucleus. He gave the five kingdom classification system and used following criteria for delimiting kingdoms

(i) Complexity of cell structure
(ii) Complexity of body structure
(iii) Mode of nutrition
(iv) Ecological lifestyle including mode of reproduction
(v) Phylogenetic relationships

6 (*d*) As per Whittaker's classification, an organism possessing eukaryotic cell structure, multicellular organisation having a distinct cell wall and nuclear membrane and showing heterotrophic nutrition is placed under the kingdom-Fungi. Fungi are eukaryotic cells and have cell wall like plants but they cannot manufacture their own food.

They either derive it from living organisms (parasitic mode) or from dead organic matter (saprotrophic mode).

8 (*d*) Only statement IV is correct, while I, II and III are incorrect with respect to characteristic features of kingdom. Incorrect statements can be corrected as

- Mode of nutrition of kingdom-Animalia is heterotrophic.
- Nuclear envelope is absent in kingdom-Monera.
- Protists are eukaryotic. The only prokaryotic kingdom is Monera.

11 (*b*) Bacteria are prokaryotes and according to five kingdom system of classification, all bacteria are included in the kingdom-Monera.

12 (*a*) Both A and R are true and R is the correct explanation of A.

Bacteria, Protista are single-celled organisms. They do not have circulatory systems and live in moist and watery environments so that all their cells are close enough to the external environment and there is direct exchange of substances such as water and gases.

CHAPTER 03

Viruses

The term 'virus' means poisonous or venomous fluid in Latin language. The description of this term was given by **Louis Pasteur** for an unknown agent causing disease. **M Beijerinck** (1898) called this fluid as ***contagium vivum fluidum***, i.e. an infectious living fluid and later on named it 'virus'. The study of viruses is called **Virology**. It is a very important branch of Molecular Biology.

VIRUSES : THE EXCEPTIONS OF CELL THEORY

Viruses are very small infective particles. These are nucleoproteinaceous in nature. It simply means that viruses are made up of nucleic acids and proteins. These are considered as the connecting link between living and non-living beings. This is due to the following reasons

Viruses as living organisms

- Possess genetic material, i.e. DNA or RNA.
- Possess a protein coat.
- Can reproduce inside the hosts cell.

Viruses as non-living object

- Due to the absence of metabolism.
- Cannot reproduce by themselves.
- Can exist as inert crystalline structure.

Because of this doubtful biological status, viruses are considered as the exception to cells theory. Due to a very high degree of structural similarity with some eukaryotic genes, viruses are now considered as the detached fragments of the genomes of organisms also.

Origin of viruses

Today viruses co-exist with life wherever it occurs. However, their origin can be speculated through the following three theories

(i) **Regressive theory**, i.e. from small parasitic cells after losing their unused structures.

(ii) **Cellular theory**, i.e. from bits of nucleic acids that escaped from the genomes of higher organisms.

(iii) **Co-evolution theory**, i.e. simultaneous origin and evolution with other cells.

Inside

In today's perspective, the co-evolution theory is the most accepted one.

Discoveries related to viruses

Some important discoveries related to viruses are as follows

- **Edward Jenner** in 1796 developed a smallpox vaccine using milder cowpox viruses.
- **Louis Pasteur** in 1885 developed rabies vaccine, but the filtered form of the virus was still not discovered.
- In 1892 **Dmitri Ivanovsky** used bacterial filters to show that sap from a diseased tobacco plant remained infectious to healthy tobacco plants despite having being filtered.
- In 1898, **Martinus Beijerinck** called this filtered infectious substance in Dmitri experiment as 'virus'. Thus, TMV (Tobacco Mosaic Virus) is considered as the first discovered virus. M. Beijerinck is also called **father of virology**.
- In the same year, 1898, **Loeffler** and **Frosch** filtered the first animal virus causing foot and mouth disease in cattle.
- In 1900, **W. Reed** and **J. Carroll** filtered first human virus which causes yellow fever disease.
- **Landsteiner** and **Popper** discovered **polio virus of humans** in 1909.
- **FW Twort** gave the first description of bacterial viruses (bacteriophages) in 1915. **Felx d' Herelle** isolated bacteriophages in 1917. He also coined the term 'Bacteriophage'.
- **W Stanley** prepared the extract of TMV and purified its crystals in 1935. He discovered that viruses, are made up of nucleic acid and proteins.
- **Lwoff, Anderson** and **Jacob** introduced the term 'virion' for a complete infective form of a virus outside the host cell in 1959.
- **Howard Temin** and **D Baltimore** discovered the enzyme reverse transcriptase from DNA-RNA viruses, i.e. retroviruses in 1970. This enzyme is responsible for reverse transcription, i.e. DNA synthesis from RNA strand. This phenomenon is also called **teminism** after the name of its discoverer. Both the scientists got Nobel Prize in 1975 for their discovery.
- **Theodor Diener** discovered the disease causing RNA molecules, i.e. viroids in 1971.
- Small circular RNA molecules called **virusoids** were discovered by **JW Randles** in 1981. These molecules can causes disease in the presence of larger viral RNAs.
- **Stanley Prusiner** coined the term 'Prions' for infective particles of proteins in 1982.
- **Robert Gallo** discovered Human Immunodeficiency Virus (HIV) in 1984. HIV is an infectious agent that causes AIDS, i.e. Acquired Immuno Deficiency Syndrome.

General characteristics

All viruses show the following common general characteristics

- They are ultramicroscopic, disease producing obligate intracellular parasites, i.e. cannot reproduce outside the host cells.
- Being very small in size, they can easily pass through bacterial filters.
- They do not have a proper cellular organisation and their own metabolic machinery.
- Their basic structure is simple with nucleic acid wrapped in a protein coat.
- They multiply within the host cell by utilising its reproductive machinery. While doing so, they destabilise normal cell and cause disease.
- This multiplication in general, involves either virulent lytic phase or non-virulent lysogenic phase or both.
- They are completely incapable of growth and respond to their surroundings.
- They are very specific in action, i.e. the target organ or organisms of a virus is always fixed.
- They generally remain unaffected by antimicrobial compounds like antibiotics.
- They can be preserved as crystals outside the host cells. Their infertility is retained even in their crystalline form.
- They may undergo mutations *via* making changes in their nucleic acid structure.

Size

Viruses are extremely small in size, in general smaller than bacteria. Some viruses such as **poxviruses** (largest virus up to 450 nm long) are about the same size as some very small bacteria like mycoplasmas, rickettsias and chlamydias. The smallest viruses, **parvoviruses** are less than 20 nm in length.

However, among plant and animals, the smallest and the largest viruses are as follows

(i) **Smallest plant virus** Satellite Tobacco Necrosis virus, 17 nm.

(ii) **Largest plant virus** *Citrus tristeza virus*, 2000×12 nm.

(iii) **Smallest animal virus** Foot and mouth disease virus, 20 nm.

(iv) **Largest animal virus** Smallpox virus (*Variola*), $350 \times 250 \times 100$ nm.

Shape

Usually, the symmetry and the shape of viruses is same. However, viruses may also show variable shapes. For example,

- **spherical** herpes virus
- **rod-shaped** TMV
- **coiled** beet yellow virus
- **bullet-shaped** rabies virus
- **tadpole-like** bacteriophage.

Structure

A complete infective form of a virus is called **virion.** It is present outside the host cell.

It contain following parts

- core
- nucleoid protein capsid

Both core and capsid combine to form a nucleo-capsid. The infectivity of a virus is due to its nucleic acid, while host specificity is due to its protein coat. In some larger viruses, an external envelope may also be present. The generalised structure of a virus is shown below

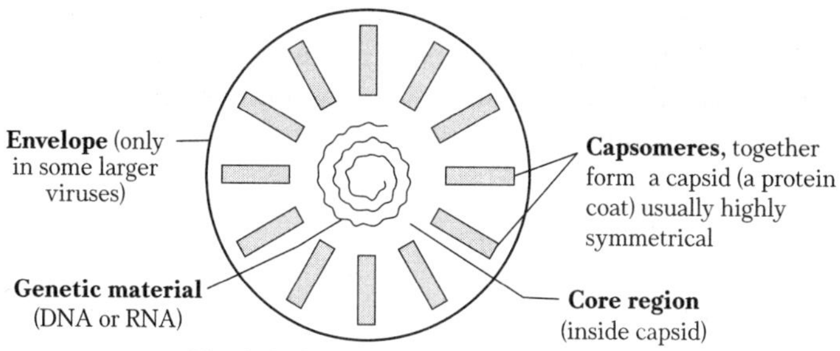

Fig. 3.1 Structure of a generalised virus

The components of the viral structure are further discussed as follows

Core or Nucleoid

It contains the nucleic acid, i.e. either DNA (e.g. $\phi \times 174$) or RNA (e.g. TMV) or both (e.g. retroviruses). The DNA or RNA present can be single stranded (*ss*) or double stranded (ds). This nature of genome is clearly indicated in **table 3.1**.

Function Nucleic acid is the only active part of a virus. Hence, viruses are sometimes called as **wandering genes** also. Nucleoid control all viral activities and reproduction.

Capsid

It is the outer protective coat of a virus. It is mostly made up of specific proteins. The identical subunits of capsid are called **capsomeres**. The shape and arrangement of capsomeres determines the shape or symmetry of a virus.

Function It protects the underlying nucleic acid from enzyme nuclease present freely in the environment.

It also serves as a vehicle of transmission for viral nucleic acid from one host to another.

There are enzymes present in the capsid which help in the production of viral particles, integration into host DNA and regulate their multiplication.

Envelope

Some viruses like HIV, herpes virus and influenza virus have an additional lipoprotein or glycoprotein layer around the capsid derived from the cell surface (membrane) of the host, this layer is called the **envelope**. Surface of envelope can be smooth or have outgrowth called **spikes**. The smaller subunits of envelope are called **peplomers**.

Function The spikes, help in binding the viruses at the host surface.

Symmetry

Viruses contain the following three types of symmetry, differentiated on the basis of their outline structures as seen in electron microscope.

(i) **Polyhedral** (icosahedral) **symmetry** Viruses with this symmetry show polyhedral shapes. Most common shape in this category is known as **icosahedron.** It is a polyhedron with 20 faces. For example,

- adenoviruses, polioviruses with naked capsid
- herpes simplex virus with enveloped capsid
- $\phi \times 174$ bacteriophage with enveloped capsid

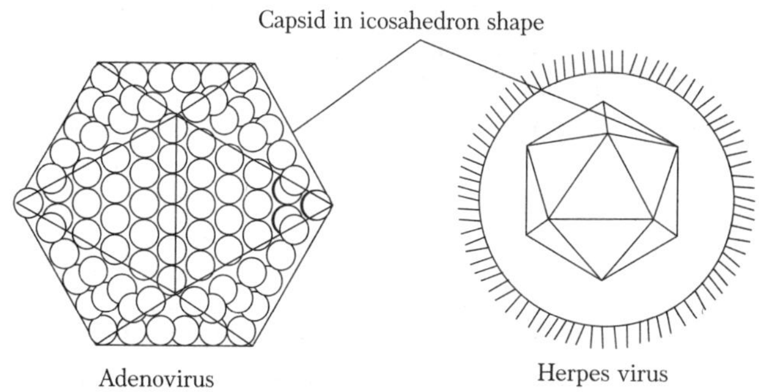

Fig. 3.2 Polyhedral viruses

(ii) **Helical** (Cylindrical) **symmetry** Viruses containing this symmetry resemble long rods. These rods may be rigid or flexible. Their nucleic acid is found within the hollow cylindrical capsid. This capsid has a superficial helical appearence.

For example,

- tobacco mosaic virus with naked capsid
- influenza virus with enveloped capsid
- mumps virus with enveloped capsid

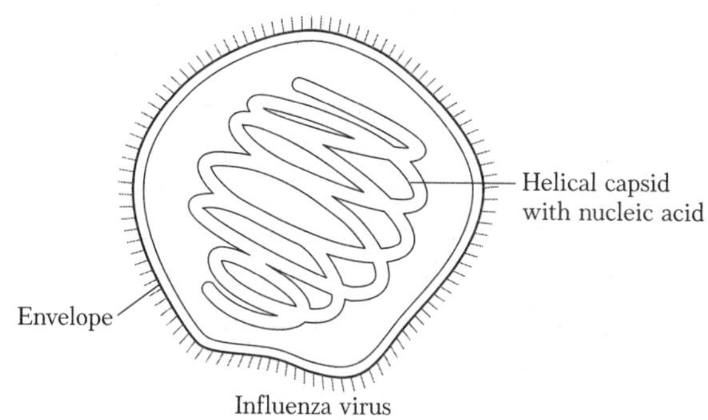

Fig. 3.3 Helical virus

(iii) **Complex symmetry** This symmetry is seen in those viruses which are neither purely helical nor purely icosahedral. Presence of additional structures like tail or comb is considered responsible for modification in symmetry.

For example, T_4-bacteriophage, *Vaccinia virus.*

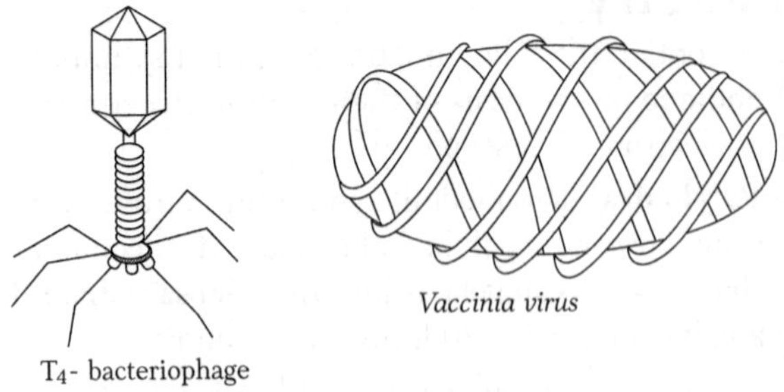

Fig. 3.4 *Complex viruses*

Classification

In taxonomy, the classification of viruses is difficult due to the lack of fossil records. Still, various classification schemes have been proposed from time to time to classify them.

Currently the main scheme used for the classification of viruses is Baltimore classification system, which is given in the table below

Table 3.1 Baltimore classification

Group names	Natures of genome	Replication strategy of genome	Virus family	Examples
Group I	*ds*DNA	*ds*DNA to *ds*DNA	Adenoviruses	Adenovirus, canine hepatitis virus, T_4-virus
			Herpes viruses	Herpes simplex virus, *Varicella zoster* virus, *Cytomegalovirus*, Epstein Barr virus
			Poxviruses	Smallpox virus, cowpox virus, sheeppox virus, orf virus, monkeypox virus, *Vaccinia virus*
Group II	*ss*DNA	*ss*DNA to *ds*DNA to *ss*DNA	Parvoviruses	Parvovirus B-19, *Canine parvovirus* 2, Coliphage $\phi \times 174$
Group III	*ds*RNA	*ds*RNA to *ss*RNA to *ds*RNA	Reoviruses	Reovirus, *Penicillium* virus, *Rotavirus*, wound tumour virus, rice dwarf virus
Group IV	(+) *ss*RNA	+RNA to -RNA to +RNA	Picornaviruses	*Enterovirus*, rhinovirus, *Hepatovirus, Cardiovirus, Aphthovirus*, poliovirus, *Parechovirus, Erbovirus, Kobuvirus, Teschovirus*, coxsackie virus
Group V	(-) *ss*RNA	–RNA to +RNA to –RNA	Orthomyxoviruses	Influenza virus-A, influenza virus-B, influenza virus-C, *isavirus, Thogotovirus*
			Togaviruses	*Rubella virus, Alphavirus*
Group VI	*ss*RNA RT	*ss*RNA to *ds*DNA to *ss*RNA	Retroviruses	HIV, *Rous sarcoma virus*
Group VII	*ds*DNA RT	*ds*DNA to + RNA to - DNA	Hepadnaviruses	Hepatitis-B virus, cauliflower mosaic virus

Nomenclature : Cryptograms

The binomial system of **Linnaeus** was not convenient enough for naming viruses. That is why in 1968, **Gibbs** and **Harrison** gave a new system of virus nomenclature. According to this, a virus name consists of four pairs. A representation of these pairs is known as cryptogram, e.g. cryptogram of Tobacco Mosaic Virus (TMV) is R/1 : 2/5 : E/E : S/O. This can be interpreted as follows

(i) **First pair** It describes type of nucleic acid, i.e. **D** for DNA and **R** for RNA (here, RNA) (R/1) → Describe number of strands in nucleic acid (here, 1 strand).

(ii) **Second pair** It describes the molecular weight of virus in millions (here, 2 million) (2/5) → Percentage of nucleic acid in virus (here 5%).

(iii) **Third pair** It describes the shape of virus (here, elongated) (E/E) → Describes shape of protein coat (here, elongated).

(iv) **Fourth pair** It describes the type of host infected (here, seeded plant) (S/S) → Describes type of vector used in its transmission (here, transfer without vector).

Likewise some other common cryptograms are as follows

- Poxvirus - D/2, 160/5-7.5, X / *, V/0 (Here, X = complex; * = Data not available)
- Poliovirus - R/1, 2.5/30, S/S, V/0 (Here, V = Vertebrate)
- T_4-bacteriophage - D/2, 130/40, X/X, B/0 (Here, X = Complex, B = Bacteria)

CHECK POINT 3.1

1. The non-living characteristic of virus can be attributed to its
   (a) ability to multiply only inside the host
   (b) ability to undergo mutation
   (c) ability to get crystalised
   (d) ability to cause diseases in the host
2. Which of the following is not concerned with viruses?
   (a) Viruses are made up of nucleic acid and proteins
   (b) Viruses multiply only in living host cell
   (c) Viruses cannot use oxygen for respiration
   (d) Viruses can be grown in sugary liquid
3. This virus for the first time was synthesised in the form of non-living crystals.
   (a) Poxvirus
   (b) Yellow fever virus
   (c) Tobacco mosaic virus
   (d) Foot and mouth disease virus
4. The viruses are known as
   (a) unicellular
   (b) acellular
   (c) multicellular
   (d) free genes
5. Which of the following is true for viruses?
   (a) They invariably contain DNA
   (b) They multiply only in host cell
   (c) They occur only inside bacteria
   (d) Their genetic material is RNA
6. In the given cryptogram D/2, 160/5, X/X, B/O the first pair represents
   (a) single stranded DNA
   (b) double stranded DNA
   (c) presence of two different DNA molecules
   (d) presence of both DNA and RNA

SOME IMPORTANT VIRUSES

'Viruses explore almost every field of life'. Following description of viruses is categorised with the same instinct.

Plant viruses

They are usually rod-shaped. Most of them contain nucleic acid in the form of RNA, e.g. Tobacco Mosaic Virus (TMV), Beet Yellow Virus (BYV), etc. However, Cauliflower Mosaic Virus (CMV) has double-stranded DNA as a genetic material.

Tobacco mosaic virus (TMV)

It has a rod-shape, like appearance. It is about 3000Å long and 180Å in diameter. It has 2130 capsomeres and each capsomere is made up of 158 amino acids. The capsomeres of TMV are elliptical and arranged helically around the central nucleic acid core.

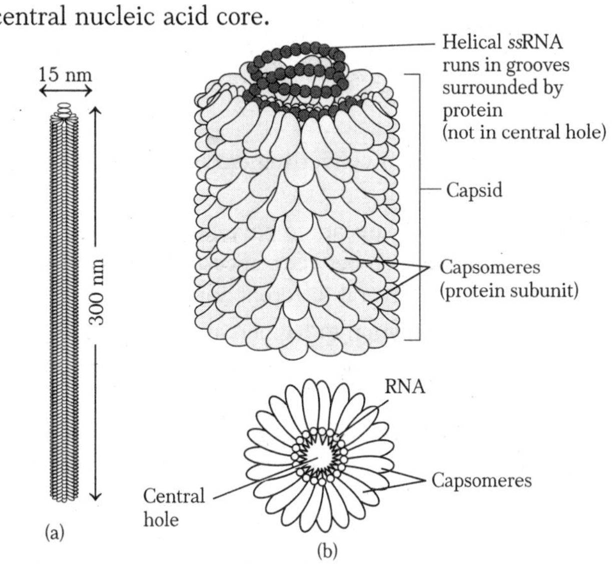

Fig. 3.5 TMV (a) External structure, (b) Internal structure

Hans Fraenkel Conrat *et. al.* demonstrated that the RNA is the genetic material in TMV. It has a molecular weight of 39.4 × 106 dalton.

The names of some important viruses causing different diseases in plants are given below

Table 3.2 Viral diseases of plants

Diseases	Casual viruses
Potato mild mosaic	Potato virus-X
Potato rugose mosaic	Potato virus-Y
Potato leaf roll	Potato leaf roll virus
Rosette of groundnut	Groundnut mosaic virus
Tristeza of Citrus	*Citrus tristeza virus*
Sugarcane mosaic	Sugarcane virus-I
Bunchy top of banana	Banana bunchy top virus
Tomato leaf curl	Tomato leaf curl virus

Animal viruses

They are generally polyhedral or spherical in shape. The capsid in some is surrounded by an envelope and the nucleic acid is either DNA or RNA. According to the type of tissue they infect, animal viruses can be classified as

(i) **Dermatotropic** Viruses infecting the skin, e.g. causing measles and chickenpox.

(ii) **Viscerotropic** Viruses infecting viscera (internal organs), e.g. causing yellow fever, jaundice and mumps.

(iii) **Neurotropic** Viruses infecting nerve cells, e.g. causing polio and meningitis.

Table 3.3 Some viral diseases of animals

Diseases	Pathogens	Reservoirs	Vectors/Epidemiology
AIDS	HIV (Human Immuno deficiency Virus)	Human and monkeys	Destroys immune defence, (T_4-lymphocytes) resulting in death by infection or cancer.
Chickenpox	Human herpes virus (*Varicella zoster*)	Human	Spread through contact with infected individuals. Rarely fatal. Vaccine approved in US in early 1995.
Ebola	Filoviruses	Unknown	Acute haemorrhagic fever, virus attacks connective tissue, leading to massive haemorrhage and death.
Hepatitis-B	Hepatitis-B Virus (HBV)	Human	Highly infectious through contact with infected body fluids. Vaccine available, can be fatal.
Herpes	Herpes Simplex Virus (HSV)	Human	Fever, blisters, spread primarily through direct contact with lesion and body fluids. Very prevalent worldwide.
Influenza	Influenza viruses	Human, ducks, chicken and pigs	Historically, a major killer (22 million died in 18 months in 1918-1919).
Measles and mumps	Paramyxoviruses	Human	Extremely contagious through contact with infected individuals. Vaccine available.
Pneumonia	Influenza virus	Human	Acute infection of the lungs, often fatal without treatment.
Polio	Poliovirus	Human	Acute viral infection of the CNS that can lead to paralysis and is often fatal.
Rabies	Rhabdovirus	Wild and domestic canids (dogs, foxes, wolves, coyotes), bats and raccoons	An acute viral encephalomyelitis transmitted by the bite of an infected animal. Fatal, if untreated.
Smallpox	*Variola* virus	Formerly humans, now only exists in research labs-may have been eliminated	Historically, a major killer, the last recorded case of smallpox was in 1977. A worldwide vaccination campaign wiped out the disease completely.
Yellow fever	*Flavivirus*	Human, mosquitoes	Spread by mosquito bites. If untreated, this disease has a peak mortality rate of 60%.
SARS (Severe Acute Respiratory Syndrome)	Paramyxovirus of corona virus family	Human	There is no vaccine for SARS, also antibiotics are ineffective against viruses. It is treated largely by supportive antipyretics.
Foot and mouth disease	*Aphtae epizooticae*	Domestic animals (cattle, goat, pig, sheep, etc.)	The virus causes high fever for 2-6 days, followed by blisters inside the mouth and on the feet which may rupture and cause lameness.
COVID-19 (corona virus disease)	SARS-CoV-2 of corona virus family	Human	Affect immune system and cause various respiratory ailments. Current fatality rate is 2%. Vaccine available.

CORONA VIRUSES

- Corona viruses are a group of related RNA viruses that causes disease in mammals and binds. In humans and birds they cause respiratory tract infections that can range from mild to lethal.
- Mild illness in humans include some cases of the common cold (which is also caused by other viruses predominantly rhinoviruses). More lethal varieties can cause SARS, MERS and COVID-19.
- SARS-COV-1 follows the replication strategy typical of the corona virus sub family. The primary human receptor of the virus is angiotensin-converting enzyme 2 (ACE2), first identified in 2003.
- Human SARS-COV-1 appears to have had a complex history of recombination between ancestral corona viruses that were hosted in several different animal group.
- In order for recombination to happen atleast two SARS-COV-1 genomes must be present in the same host cell.
- Recombination may occur during genome replication when the RNA polymerase switches from one template to another.

COVID-19 and its variants

- COVID-19 (Novel corona virus-disease 2019) is highly contagious. World wide pandemic is caused by a SARS-COV-2 of corona family.
- It was first reported in Wuhan city of China in early December, 2019. Till then several variants were detected in different continents.
- Virus constantly change through mutation and when a virus has one or more new mutations it is called a variant of the original virus. These variants includes
    - **Alpha** (α) = B.1.17 (in USA, Dec 2020)
    - **Beta** (β) = B.1.351 (in SA, Dec 2020)
    - **Gamma** (γ) = P.1 (In USA, Jan 2021)
    - **Delta** (Δ) = B.1.617.2 (in India, Dec 2020)
    - **Epsilon** = B.1.427 and B.1.429
- Out of these delta variant is a cause of great concern as it spread more easily. It reduces the effectiveness of some monoclonal antibody treatment.
- The **delta plus** or **Delta-AY.1** or **B.1.617.2.1** variant, a sub lineage of delta variant is a mutated version of B.1.617.2 variant. It is characterised by the k417N mutation in the spike protein of the SARS-COV-2 virus.

Retroviruses

These are one of the most dangerous category of viruses. These are also called RNA-DNA viruses, e.g. HIV (causes AIDS).

These possesses single-stranded RNA genome. However, they differ from other RNA viruses in working pattern. In them synthesis of DNA occurs *via* reverse transcription.

This produced DNA then transcribes *m*RNA in usual manner to produce proteins. Its RNA also has the capability to replicate itself.

HIV contains 8 proteins and is categorised among the most mutable virus.

A generalised diagram of HIV is shown below

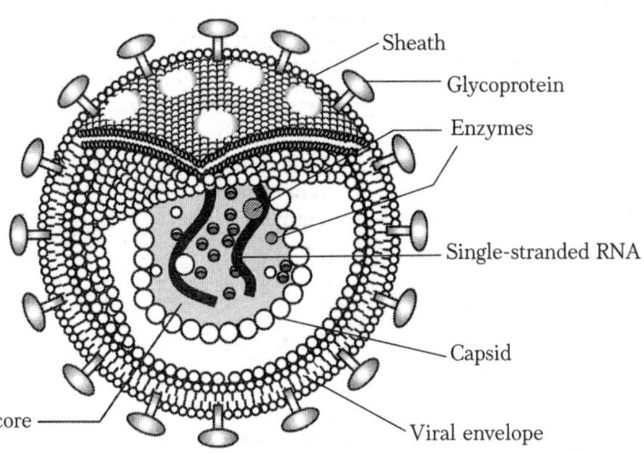

Fig. 3.6 Structure of HIV

Virus and Cancer

- Viruses cause tumour in animals and human beings. Some tumours are benign (harmless) and some are malignant (deadly).
- These malignant tumours are responsible for causing cancer. Thus, can be referred as carcinogenic. In fact these carcinogenic viruses are capable of triggering the expression of oncogenes.
- Oncogenes are cancer causing genes. The cancer causing viruses are also called **oncoviruses**.
- Oncoviruses are found within several families of DNA containing viruses. These groups include the Adenoviridae, Herpesviridae, Poxviridae, Papovaviridae and Hepadnaviridae.
- Retroviruses were the first viruses shown to cause cancer and have been studied most extensively for their carcinogenic characteristics.
- The relationship between the cancer and virus was first demonstrated in 1908 by **Wilhelm Ellerman** and **Olaf Bang**.
- Three years later, **Francis Rous** found that a chicken sarcoma (cancer of connective tissue) can be transmitted by the virus.
- In 1972, **Sarah Stewart** isolated the first oncovirus from a human sarcoma. It has been suggested that the virus contributes about 15% of all human cancer cases worldwide.

Phages or Microbial viruses

These viruses infect cells of microorganisms. On the basis of their host, these are further categoriesed into the following types

- Bacteriophage – Parasites on bacteria
- Actinophage – Parasites on Actinomycetes
- Zymophage – Parasites on yeast
- Cyanophage – Parasites on blue-green algae
- Phycophages – Parasites on algae
- Mycophages – Parasites on fungi

Bacteriophages

The name bacteriophage in itself signifies **'bacteria eaters'**. A typical bacteriophage has a tadpole-like structure.

Nucleic acid is present inside the head which is generally DNA in bacteriophages.

Types of bacteriophages

Bacteriophages are of two types

(i) **Virulent phages** The phages which cause lysis of bacterial cell, e.g. T_2, T_4-bacteriophages, etc.

(ii) **Non-virulent phages** The phages which do not cause lysis of bacterial cell, e.g. λ-phage, etc.

T_4-bacteriophage

The T_4-bacteriophage grows on colon bacterium *Escherichia coli*, hence commonly called **coliphage** also.

It comprises of the following parts

- **Head** It is polyhedral with about 2000 capsomeres. It contains a 53 μm long double-stranded DNA in it.

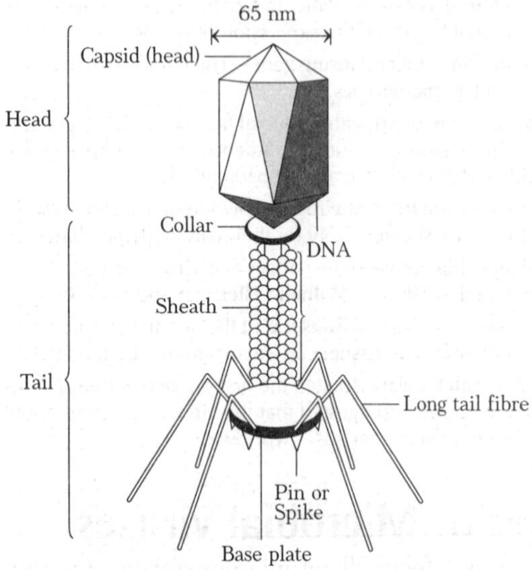

Fig. 3.7 Structure of T_4*-bacteriophage*

- **Tail** It is helically symmetrical with a core tube of 80 Å diameter. It is surrounded by a proteinaceous tail sheath of 144 subunits arranged in 24 rings of 6 subunits each.
- **Sheath** It is connected to a collar at its upper end and a hexagonal base plate at its lower end.
- **Base plate** It has tail fibres and pin or spikes. These are helpful in attachment of the virus to the host cell surface.

REPRODUCTION IN VIRUSES

The structure of viruses allows them to succeed in their main mission, i.e. reproduction. It is of two main types, i.e. phagic and pinocytic.

Reproduction in phages

In this reproduction, only the nucleoid enters inside the host cell. **Phagic reproduction** has two subtypes, i.e. lytic and lysogenic.

(i) **Lytic cycle** It is the reproductive cycle of virulent bacteriophages, e.g. T_4-bacteriophage. This cycle ends with the lysis and death of the host cell.

(ii) **Lysogenic cycle** Viruses which do not cause lysis and death of host cell are called **temperate** or **lysogenic phages**. In lysogenic cycle, the phage remains latent (inactive) and the viral DNA becomes a part of host (bacterial) DNA which is referred to as **prophage.** If the host is other than bacterial cell, i.e. plant or animal cell, it is called **provirus.**

Both of the cycles in diagrammatic form are given below

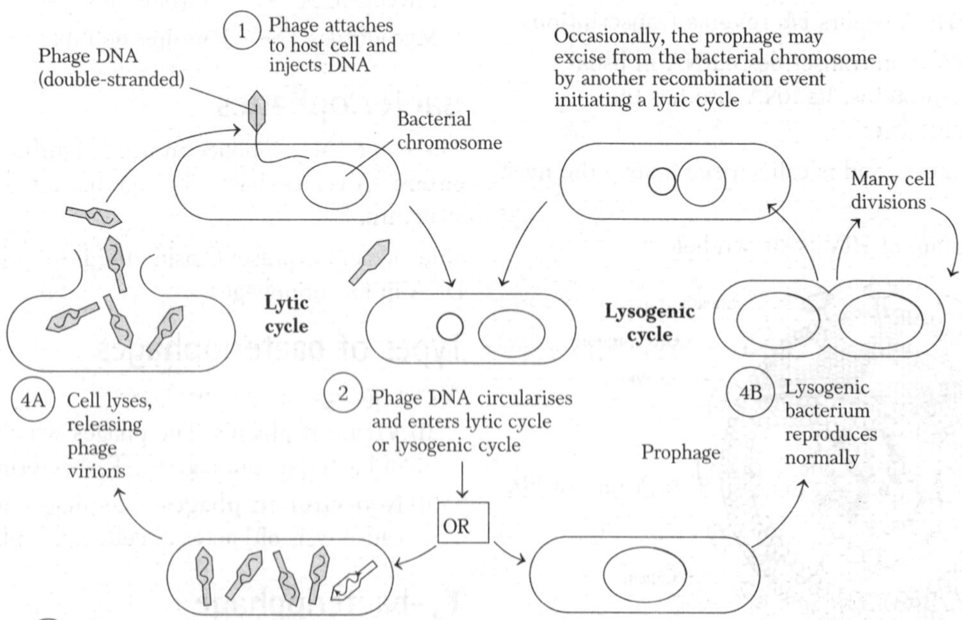

Fig. 3.8 Diagrammatic representation of lytic and lysogenic cycle

Reproduction in retrovirues

It is also known as **pinocytic reproduction**. In this, the whole virus (except envelope, if present) passes into host cell, e.g. in RNA-RNA and RNA-DNA viruses. Retroviruses are the best example to describe pinocytic reproduction.

Here, the transformation of RNA to DNA takes place by RNA-dependent DNA polymerase enzyme, commonly called as **reverse transcriptase**.

Following steps occurs in this type of reproduction process

- First of all, reverse transcriptase uses the *t*RNA as template to form a RNA-DNA hybrid.
- Then degradation of *t*RNA from RNA-DNA hybrid occurs.
- DNA produces double-stranded DNA called **proviral DNA**.
- This proviral DNA becomes integrated into host cell genome.
- From there, host RNA polymerase can direct the synthesis of *m*RNA and new *t*RNA genome.

Remember that during this process, genetic information is transferred from RNA to DNA rather than from DNA to RNA as in animal and plant cells. The representation of pinocytic reproduction in retrovirus is shown below

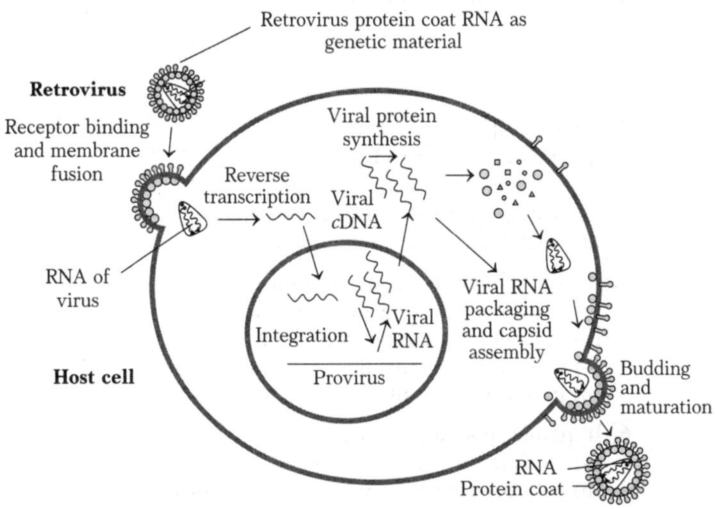

Fig. 3.9 Replication of retrovirus in host cell

SUBVIRAL AGENTS

Pathogenic structures or non-cellular organisms with size smaller than viruses come under this category. Structure wise also they may lack one of the essential features of a virus.

Subviral agents have the following four groups

Satellite viruses

- These were discovered by **Kassanis** in 1966.
- Satellite viruses or incomplete viruses are small *ss*RNA molecules.
- These are absolutely dependent on the presence of another virus for their multiplication, e.g. these are hyperparasites of viruses.
- Most satellites are associated with plant viruses, but a few are associated with bacteriophages or animal viruses also.
- Some viruses and their respective satellite viruses are given below

Viruses	Satellite Viruses
Barley yellow dwarf virus	Helper *Luteovirus*
Tobacco ring spot virus	Helper *Nepovirus*
Subterranean clover mottle virus	Helper *Sobemovirus*
Adenovirus	*Dependoparvovirus*

Viroids

- These were discovered by **TO Diener**, a plant pathologist in 1971.
- Viroids are **small, circular, single-stranded RNA molecules** that are considered as the smallest known pathogens.
- These range in size from 246-399 nucleotides and show a considerable degree of similarity in sequences, i.e. **sequence homology**. It suggests that they have common evolutionary roots.
- Viroids cause a number of important plant diseases and can have a severe agricultural impact.
- A few well-studied viroids include
  - coconut cadang-cadang viroid (246 nucleotides),
  - citrus exocortis viroid (375 nucleotides),
  - potato spindle tuber viroid (359 nucleotides).
- The viroid molecules enter the plant through a wound or an insect bite or other mechanical damage. Afterwards the viroid is replicated in the host cell nucleus by RNA polymerases.

Viroids differ from viruses in following aspects

- They lack virion like dormant phase.
- Genome is much smaller.
- They do not have capsid and envelope.
- They do not produce proteins.

The origin of viroids is attributed to the following two facts

- They may be degenerated viruses.
- They may be abnormal host RNAs.

Virusoids

- These were discovered by **JW Randle** and his associates in 1981.
- These are **small circular RNA** and are structurally similar to viroids.
- These are located inside the protein coat of a true virus and are always associated with a larger viral RNA.
- Some virusoids may be necessary for the replication of larger RNA.

Prions

- These are **proteinaceous infectious particles**. They have a distinct extracellular form, which is entirely a protein.
- The prion particle does not contain any nucleic acid. Thus, they are not destroyed by enzymes that digest DNA or RNA.
- However, they are infectious and known to cause several neurodegenerative diseases found in mammals as follows
    - Scrapie disease in sheep.
    - Bovine Spongiform Encephalopathy (BSE) or mad cow disease in cattle.
    - Chronic wasting disease in deer.
    - Creutzfeldt-Jakob Disease (CJD) in humans.
    - Kuru disease (Laughing death) in human.

No prion diseases of plants are known so far.

Beneficial uses of viruses in biotechnology

Despite their usual infectious (harmful) nature, viruses have been put to the following uses

- Viruses are used in gene therapy as gene delivery vehicles or vaccines and vaccine carriers.
- Viral genetic elements are used to construct eukaryotic expression plasmid vectors.
- Bacteriophages are used as antibacterial agents.
- In the production of interferons they are responsible for viral interferance. Interferons are low molecular weight proteins. These show non-specific antiviral activity by stimulating the host cells. Interferons were discovered by **Issacs** and **Lindemann** in 1957.

CHECK POINT 3.2

1. A viral genome inserted and integrated into the circular bacterial DNA or exists as an extra chromosomal plasmid is
(a) polyphage (b) cyanophage
(c) prophage (d) Both (a) and (b)

2. An specialised virus possessing reverse transcriptase enzyme which helps to build an RNA over its DNA.
(a) *Leukovirus* (b) Polio virus
(c) Cyanophage (d) Herpes virus

3. During lytic life cycle of a virulent DNA phage, which of the following does not occur?
(a) The host cell lyse
(b) New phages are released
(c) Host cell producers large number of virions
(d) The phage DNA integrated into the host chromosomes

4. A bacteriophage is
(a) a virus attacking a bacterium
(b) a bacterium attacking a virus
(c) a stage in the life cycle of bacterium
(d) a virus attacking another virus

5. Bacteriophage is similar to fungus
(a) in having DNA as genetic material
(b) in having RNA as genetic material
(c) in mode of reproduction
(d) in having cell wall

6. In phage (plaque) culture,
(a) only proton of virus enters and multiplies
(b) only DNA of virus enters the bacterium and multiplies
(c) whole virus enter the bacterium and multiplies
(d) None of the above

EXAM POINTS

- The term '*Contagium vivum fluidum*' was coined by MW Beijerinck. **[CBSE AIPMT 2015]**
- Viruses are obligate parasites. **[NEET 2019]**
- Cell theory does not apply for viruses, as they are considered as both living and non-living organisms. **[BHU 2012]**
- Viruses have free RNA without protein coat. **[NEET 2020]**
- A virus differs from a bacterium as it contains DNA or RNA as genetic material with no ribosomes. **[J&K CET 2011]**
- Virus penetration is done in host. **[EAMCET 2015]**
- Through DNA or RNA, viruses infect their host organisms. **[NEET 2019]**
- WM Stanley showed that viruses could be crystalised. **[CBSE AIPMT 2015]**
- A viruses consists of DNA or RNA surrounded by a protective coat of a protein called capsid. **[CBSE AIPMT 2014]**
- Most plant viruses are RNA viruses. **[BCECE 2012]**
- Tobacco mosaic virus is a single-stranded RNA virus. **[BCECE 2014]**
- Tobacco mosaic virus shows coiled RNA and capsomeres. **[CBSE AIPMT 2015]**
- Most animal viruses are DNA viruses. **[BCECE 2012]**
- Mosaic disease in tobacco and AIDS in human being are caused by viruses. **[CBSE AIPMT 2015]**
- Bacterial viruses usually have double-stranded DNA. **[AMU 2015]**
- Bacteriophage has a double-stranded DNA molecule. **[BCECE 2012]**
- HIV is a retroviridae. **[EAMCET 2015]**
- Double-stranded RNA is found in reovirus. **[Odisha JEE 2011]**
- Lysogeny is process in prophage. **[EAMCET 2015]**
- Recombination is the process of transfer of genetic material from the bacterium to another viruses. **[AMU 2015]**
- The sequence of events in the viral replication process is Adsorption $\rightarrow$ Penetration $\rightarrow$ Eclipse $\rightarrow$ Maturation $\rightarrow$ Assembly Lysis. **[AFMC 2014]**
- The viroids were discovered by T Dioner. **[CBSE AIPMT 2015]**
- Viroid RNA is of low molecular weight. They are smaller than viruses. They cause infections. **[NEET 2016]**
- Pilus is donor. **[EAMCET 2015]**
- Prion causes Craetzfeldt Jacob disease. **[EAMCET 2014]**
- Prions are abnormally folded proteins which cause mad cow disease in cattle. **[NEET 2019]**

Chapter Exercises

(A) Topical Exercise

Characteristics of Virus

1 The non-living characteristic of viruses is
(a) ability to multiply only inside the host
(b) ability to cause diseases in the host
(c) ability to undergo mutation
(d) ability to be crystallise

2 The latest view for the origin of viruses is
(a) these have arisen from nucleic acid and protein found in primitive soup
(b) these arose from bacteria as a result of the loss of cell wall, ribosome, etc.
(c) these arose from some bacteria, which had developed a nucleus only
(d) these are modified plasmids, which are infact the fragments of the nucleic acids of the host

3 The coevolution theory for the origin of viruses is
(a) these have arisen simultaneous and evolution with other cells
(b) these arose from bacteria as a result of the loss of cell wall, ribosome, etc.
(c) these arose from some bacteria, which had developed a nucleus only
(d) these are modified plasmids, which are in fact the fragments of the nucleic acids of the host

4. Match the following columns.

Column I	Column II
A. Felix x' Herelle	1. Bacteriophage
B. Loeffler and Frosch	2. Foot and mouth disease
C. Robert Gallo	3. Prions
D. Stanley Prusiner	4. HIV

Codes

	A	B	C	D		A	B	C	D
(a)	1	2	3	4	(b)	4	3	2	1
(c)	2	1	3	4	(d)	1	2	4	3

5 Consider the following statements.
I. The TMV is a double-stranded RNA.
II. Heinz Fraenkel-conrat *et. al.*, demonstrated that RNA is a genetic material of TMV by using two strains.
Choose the correct option.
(a) Statement I is true, but II is false
(b) Statement I is false, but II is true
(c) Both statements I and II are true
(d) Both statements I and II are false

6 Study the following statements about virus.
I. Nucleic acid as genetic material.
II. Produce their own copies.
III. Cellular organisation.
IV. Autotrophic nutrition.
The living characteristic of virus is
(a) II and III
(b) I and II
(c) I and IV
(d) Only I

7 Virus is an infectious agent which may consist of
(a) DNA or RNA
(b) proteins
(c) lipids
(d) Both (a) and (b)

8 Which among the following is not a characteristic feature of viruses?
(a) They can pass through bacterial filters
(b) On getting favourable conditions, they can reproduce outside host cells
(c) They undergo mutations *via* making changes in their nucleic acids
(d) They can be preserved as crystals outside host cells

9 Choose the characteristic features of a virus.
(a) Can multiply itself without a host
(b) Unaffected by antimicrobial drugs
(c) They may undergo mutations
(d) Both (b) and (c)

10 Viruses are
(a) intracellular, obligate parasites
(b) intercellular, obligate parasites
(c) intracellular, facultative parasites
(d) intercellular, facultative parasites

11 Which one of the following statements about viruses is correct?
(a) They possess their own metabolic system
(b) All viruses contain both RNA and DNA
(c) Viruses are obligate parasites
(d) Nucleic acid of viruses is known as capsid

12 One of the interesting features of viruses is that they
(a) occur only inside the bacteria
(b) multiply only in the host cytoplasm
(c) are made up of proteins only
(d) behave as if they were plants

13 Most specificity of a virus is due to
(a) its nucleic acid
(b) an external envelope present in the virus
(c) its protein coat
(d) mesosome

14 The coiled or polyhedral structure that encloses the nucleic acid of a virus
(a) genetic material of virus
(b) protein cover of virus
(c) extra genetic material of bacterium
(d) house-keeping genome of bacterium

15 Which among the following is not a characteristic feature of protein coat of virus?
(a) The identical subunits of capsid are called peplomers
(b) The shape and arrangement of subunits determines the shape and symmetry of a virus
(c) It protects the underlying nucleic acid from enzyme nuclease present freely in the environment
(d) It serves as a vehicle of transmission for transfer of viral nucleic acid from one host to another

16 Enveloped virus containing glycoproteins are helpful in
(a) binding to the host surface
(b) phagocytosis
(c) protection
(d) fusion with the plasma membrane of host

17 A structure which has a hexagonal baseplate and six tail fibres is
(a) bacteria (b) bacteriophage
(c) retroviruses (d) fungi

18 A portion of a virus that is integrated into the genetic material of a host cell and by replication it can be transmitted from one cell generation to the next without causing lysis is
(a) a free virus
(b) primitive virus
(c) viral capsid
(d) viral DNA

19 Match the following columns.

	Column I		Column II
A.	M13 bacteriophage	1.	*ds*RNA
B.	Rice dwarf virus	2.	*ss*RNA
C.	Cauliflower mosaic virus	3.	*ss*DNA
D.	Polio virus	4.	*ds*DNA

Codes

	A	B	C	D
(a)	3	1	4	2
(b)	2	1	3	4
(c)	3	4	2	1
(d)	4	3	1	2

20 Choose the incorrect match from the following.
(a) Smallest plant virus – Satellite tobacco necrosis
(b) Smallest animal virus – Foot and mouth disease virus
(c) Largest animal virus – Smallpox virus
(d) Largest plant virus – Tobacco mosaic virus

21 Consider the following statements.
I. Wound tumour virus and reovirus show the double stranded RNA.
II. Tobacco mosaic virus and HIV also show double-stranded RNA.
Choose the correct option.
(a) Statement I is true, but II is false
(b) Statement I is false, but II is true
(c) Both statements I and II are true
(d) Both statements I and II are false

22 Cauliflower mosaic virus has
(a) single-stranded DNA as a genetic material
(b) double-stranded RNA as a genetic material
(c) single-stranded RNA as a genetic material
(d) double-stranded DNA as a genetic material

23 The correct cryptogram for T_4-bacteriophage is
(a) R/1, 2.5/30, S/S, V/0
(b) D/2, 160/5-7.5, X/*, V/O
(c) R/1:2/5 : E/E : S/S
(d) D/2, 130/40, X/X, B/O

Description of Some Important Viruses and Viral Reproduction

24 The diagram of a virus is given below. In which of the options *A*, *B* and *C* are correct?

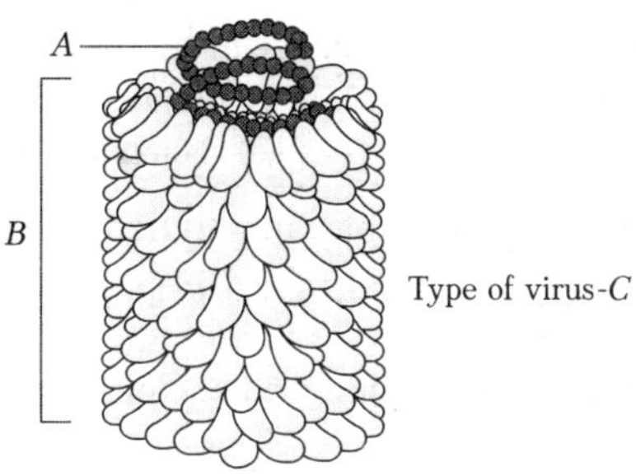

	A	*B*	*C*
(a)	RNA	Protein	HIV
(b)	RNA	Lipid	Tobacco mosaic virus
(c)	DNA	Capsid	Tobacco mosaic virus
(d)	RNA	Capsid	Tobacco mosaic virus

25 Choose the statement which is incorrect about animal viruses.
(a) These are generally polyhedral or spherical in shape
(b) In some animal viruses, the capsid is surrounded by an envelope
(c) SARS is a dermotropic virus
(d) Polio virus is a neurotropic virus

26 Which part of an animal virus is not reproduced in multiple copies?
(a) Capsid (b) Proteins
(c) Envelope (d) Ribosomes

27 Which of the following cause a severe highly contagious viral disease of livestock that has a significant economic impact

(a) *Citrus tristeza* virus (b) foot and mouth disease virus
(c) smallpox virus (d) rabies virus

28 Which of the following statements is not true for retroviruses?

(a) DNA is not present at any stage in the life cycle of retroviruses
(b) Retroviruses carry gene for RNA-dependent DNA polymerase
(c) The genetic material in mature retroviruses is RNA
(d) Retroviruses are causative agents for certain kinds of cancer in man

29 The virus responsible for a disease of the human immune system that is characterised cytologically by reduction in the numbers of CD4 bearing helper T-cells to 20 per cent or less thereby rendering him highly vulnerable to life threatening condition is an example of a/an

(a) adenovirus (b) mosaic virus
(c) T-even virus (d) retrovirus

30 During lysogenic phage, bacteriophages incorporate what into the chromosome of the host cell?

(a) Proteins (b) Nucleic acid
(c) Nucleoproteins (d) Lipids

31 A 'T-series bacteriophage' can be recognised by its

(a) tadpole shape (b) rounded shape
(c) irregular shape (d) rhomboidal shape

32 A T_4-bacteriophage has an polyhedral head with

(a) 2500 capsomeres with a 63 μm long double-stranded DNA in it
(b) 2000 capsomeres with a 53 μm long double-stranded DNA in it
(c) 1500 capsomeres with a 40 μm long double-stranded DNA in it
(d) 1000 capsomeres with a 53 μm long double-stranded DNA in it

33 Mark the correct labelling in the T_4- bacteriophage from top to the bottom.

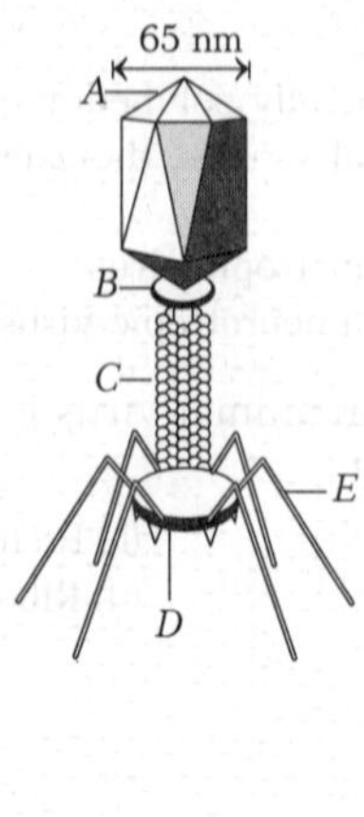

	A	*B*	*C*	*D*	*E*
(a)	Collar	Head	Base plate	End plate	Tail fibre
(b)	Head	Collar	Sheath	Base plate	Tail fibre
(c)	Head	Sheath	Tail fibre	Collar	Base plate
(d)	Head	Collar	Base plate	Sheath	Tail fibre

34 Viral genome, incorporated and integrated with bacterial genome is referred to as

(a) prophages (b) RNA
(c) DNA (d) Both (a) and (c)

35 Which of the following diseases are known to be caused by viruses?

I. Burkitt's lymphoma
II. Adult II cell leukaemia
III. Phenylketonuria

Select the correct answer from the codes given below.

(a) I and II (b) I, II and III
(c) I and III (d) II and III

36 Consider the following statements.

I. Coliphage $\phi \times 174$ has *ss*DNA.
II. Coliphage $\phi \times 174$ infects *E. coli*.

Choose the correct option.

(a) Statement I is true, but II is false
(b) Statement I is false, but II is true
(c) Both statements I and II are true
(d) Both statements I and II are false

37 Which one of the following is correctly matched?

(a) National Institute of Virology – Pune
(b) National Institute of Communicable Diseases – Lucknow
(c) Central Drug Research Institute – Kasauli
(d) National Institute of Nutrition – Mumbai

38 Match the following columns.

Column I	Column II
A. Filovirus	1. Mumps
B. Paramyxoviridae	2. Ebola
C. Deltaviridae	3. Influenza virus
D. Orthomyxoviridae	4. Hepatitis-D

Codes

	A	B	C	D		A	B	C	D
(a)	2	1	4	3	(b)	3	4	1	2
(c)	1	2	3	4	(d)	2	3	4	1

39 Latest discovered disease SARS is caused by

(a) picorna DNA virus (b) retrovirus
(c) corona RNA virus (d) corona DNA virus

40 Enveloped virus enters into host cells by

(a) injecting own nucleic acid inside host cells
(b) by contact with cell receptor and endocytosis
(c) by phagocytosis
(d) fusion with the plasma membrane of host

41 All of the following are true for virus replication except
(a) only a subset of viruses contains all of the enzymes necessary for transcribing their own genome
(b) no viruses contain the *t*RNA necessary to translate *m*RNA
(c) viruses can make ribosomes necessary to translate *m*RNA
(d) once a virus has entered a cell, it cannot begin making copies of itself immediately

42 Which of the following statements is correct?
(a) Viruses carry with them their own ribosomes for protein formation
(b) New viral ribosomes are formed after viral DNA enters the cell
(c) Viruses use the host ribosomes for their own protein formation
(d) Viruses do not need ribosomes for protein formation

43 The antibiotics have no effect on viruses because
(a) they kill all bacteria, which act as host of virus
(b) viruses are too small in size for antibiotics to act upon them
(c) viruses show no metabolism of their own
(d) viruses produce a thick covering and encyst themselves as endospores

44 Which of the methodologies is incorrect in case of viruses, for their entry into the host cell?
(a) Viruses can enter animal cells after a sucking insect has disrupted the cell wall with its mouthparts
(b) Viruses can enter bacterial cells by binding to a specific molecule on the cell membrane or cell wall
(c) Viruses can enter animal cells by binding to a specific molecule on the cell membrane
(d) Plant viruses can enter host cells after a sucking insect has disrupted the cell wall with its mouthparts

45 Which is the first step of pinocytic reproduction?
(a) Viral RNA packaging and capsid assembly
(b) Viral protein synthesis
(c) Receptor binding and membrane fusion
(d) Reverse transcription

46 Choose the statement which is incorrect about pinocytic reproduction.
(a) In pinocytic reproduction, the whole virus including the envelope passes into the host cell
(b) Retroviruses are the best examples to describe pinocytic reproduction
(c) The reverse transcriptase uses the *t*RNA as template to form a RNA-DNA hybrid
(d) The host RNA polymerase directs the synthesis of *m*RNA and new *t*RNA gene

47 A viral DNA can be made radioactive
(a) by culturing the viruses on a medium of potato, dextrose and P^{32}
(b) by culturing the viruses in a medium of P^{32}
(c) by providing P^{32} to a bacterium, which has been infected by a virus
(d) by providing P^{32} to viruses when they are about to attach the bacteria

48 Sometimes virus attacks a bacterium, but none of them dies. The phenomenon is
(a) transduction (b) penetration
(c) lysogeny (d) adsorption

49 In which of the following patterns of viral replication, viruses enter a cell, replicate and then cause the cell to burst, releasing new viruses?
(a) Lytic (b) Lysogenic
(c) Repreogenic (d) Both (a) and (b)

50 The process which cannot take place in the absence of virus is
(a) transformation (b) conjugation
(c) translocation (d) transduction

51 Identify the correct pair of events when temperate phages infect bacteria.
I. No prophages are formed.
II. Bacterial cell undergoes many divisions.
III. Bacterial cell undergoes immediate lysis.
IV. Prophages are formed.
The correct pair is
(a) I and II (b) II and III
(c) III and IV (d) II and IV

52 Which statement is correct for bacterial transduction?
(a) Transfer of some genes from one bacteria to another bacteria through virus
(b) Transfer of genes from one bacteria to another bacteria by conjugation
(c) Bacteria obtain DNA directly
(d) Bacteria obtain DNA from other external source

Subviral Agents

53 A naked RNA particle causing the symptoms like that of a virus disease is known as
(a) viroid (b) virion
(c) viral (d) None of these

54 Name the pathogen of potatoes which causes stunting of the plant and malformation and cracking of tubers.
(a) Nematode (b) Virus
(c) Bacterium (d) Viroid

55 Viroids have
(a) *ss*RNA not enclosed by protein coat
(b) *ss*DNA not enclosed by protein coat
(c) *ds*DNA enclosed by protein coat
(d) *ds*RNA enclosed by protein coat

56 Consider the following statements.
I. Viroids have free DNA.
II. Viroids cause potato spindle tuber disease.
Choose the correct option.
(a) Statement I is true, but II is false
(b) Statement I is false, but II is true
(c) Both statements I and II are true
(d) Both statements I and II are false

57 Difference between virus and viroid is
(a) absence of protein coat in viroid, but present in virus
(b) presence of low molecular weight RNA in virus, but absent in viroid
(c) Both (a) and (b)
(d) None of the above

58 The agents which are known to cause CJD are
(a) protein particles
(b) a class of bacteria
(c) a class of viruses
(d) fungi

59 A disease causing blisters in the mouth of cattle, horses and mules and on the snouts and feet of swine is caused by
(a) bacteria
(b) virus
(c) prion
(d) mycoplasma

60 Complete the analogy.
Viroids : Citrus exocortis : : Prions : ?
(a) Kuru disease
(b) AIDS
(c) Polio
(d) Mumps

(B) Cumulative Exercise

(Assorted Questions of the Chapter for Advanced Level Practice)

1 Consider the following statements about virus.
I. M Beijerinck (1898) called virus a living infectious fluid or *contagium vivum fluidum.*
II. Only ribosomes are present in viruses.
III. Viruses are obligate intracellular parasites.
IV. Viruses use single ATP as an energy source.
Choose the option containing correct statements.
(a) II and III (b) I, II and III
(c) III and IV (d) I and III

2 Study the diagram given below.

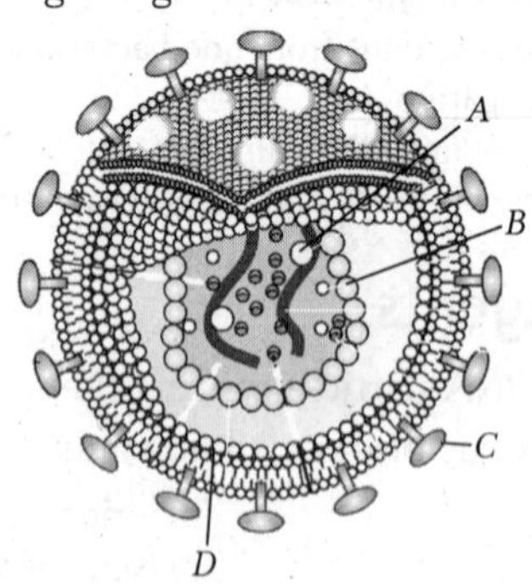

Choose the correct pair(s) from the following.
I. *A*-Reverse transcriptase
II. *B* and *C*-Capsid
III. *D*-External capsid
IV. *C*-External capsid
Select the correct option from the codes given below.
(a) I and III
(b) II and IV
(c) I, III and IV
(d) I, II, III and IV

3 Consider the following statements about retrovirus.
I. It has an enzyme located in its coat.
II. In it, the newly formed DNA integrates in host genome.
III. It has a *ds*DNA as a genetic material.
IV. Some of them are highly oncogenic.
Choose the correct answer using the codes given below.
(a) II, III and IV (b) I, III and IV
(c) I, II and IV (d) I, II and III

4 Match the following columns.

Column I	Column II
A. Double-stranded DNA viruses	1. Reovirus of humans
B. Single-stranded DNA viruses	2. TMV
C. Double-stranded RNA viruses	3. Polyoma virus
D. Single-stranded RNA viruses	4. Adeno associated viruses

Codes

	A	B	C	D
(a)	1	2	3	4
(b)	4	3	2	1
(c)	3	4	1	2
(d)	2	1	3	4

5 Select the correct difference between bacteriophage and animal virus multiplication.

	Bacteriophages	Animal viruses
I.	Biosynthesis occurs in cytoplasm.	Biosynthesis occurs in nucleus (DNA viruses) or cytoplasm (RNA viruses).
II.	Tail fibres attach to the cell wall proteins.	Attachment sites are plasma membrane proteins or glycoproteins.
III.	Uncoating is necessary.	Uncoating is not necessary.
IV.	Viral DNA enters into host cell by endocytosis.	Viral DNA injected into host cell.

Choose the correct option from the codes given below.

(a) I and II
(b) III and IV
(c) II, III and IV
(d) I, II, III and IV

6 Which of the following feature(s) are categorised under lysogeny?

I. A lysogenic cell has lesser chances of reinfection by the same phage.
II. Lysogeny favours specialised transduction.
III. Viral DNA in lysogeny never forms episome, but always integrates into the host's DNA.
IV. The lysogenic cell can never undergo lytic cycle.

Choose the correct option from the under given codes.

(a) I and II
(b) I, II and III
(c) II and IV
(d) I, II, III and IV

7 Match the following columns.

Column I	Column II
A. Plant virus	1. Potato spindle tuber disease
B. Animal virus	2. Kuru disease
C. Viroid	3. Tobacco necrosis disease
D. Prions	4. Polio

Codes

	A	B	C	D
(a)	2	1	3	4
(b)	3	4	1	2
(c)	4	3	2	1
(d)	1	2	3	4

8 Which of the following are correct to describe viruses?

I. Simple and unicellular organisms.
II. Contain DNA or RNA and enclosed by protein coat.
III. Possess own metabolic system and respond to stimuli.
IV. Maintain genetic continuity and undergo mutations.

(a) I and II
(b) II and IV
(c) II and III
(d) I and III

9 Consider the following statements.

I. Viruses are capable of growth and division independently.
II. Viruses remain unaffected by antimicrobial antibiotics.
III. HIV is the AIDS causing virus, having DNA as genetic material.
IV. Viruses have their own cellular machinery.

Choose the option containing incorrect statement(s).

(a) I, III and IV
(b) I, II and III
(c) Only IV
(d) I and II

10 Consider the following statements.

I. It is the outermost thin loose covering present in certain animal viruses and rarely in plants or bacterial viruses.
II. It is a virus without an envelope.
III. These are the identical protein subunits in viral capsid.

Choose the correct words with respect to the above statements.

(a) I-Capsid, II-Loose, III-Capsomeres
(b) I-Capsid, II-Empty, III-Peplomeres
(c) I-Envelope, II-Naked, III-Capsomeres
(d) I-Envelope, II-Spike, III-Peplomeres

11 Consider the following statements.

I. T-even phages possess lytic cycle.
II. Lysogeny is a common feature of lambda (λ) phages.
III. Latent period follows the eclipse phase.
IV. Attachment, coating and release come under eclipse phase.

Choose the correct statement(s) from the codes given below.

(a) I and II
(b) I, III and IV
(c) Only IV
(d) I, II and III

12 Match the following columns.

Column I		Column II
A. Prions	1.	Cauliflower mosaic disease
B. Viroids	2.	*Chrysanthemum* stunt
C. Plant virus	3.	Kuru
D. Animal virus	4.	Polio

Codes

	A	B	C	D		A	B	C	D
(a)	1	2	3	4	(b)	4	3	1	2
(c)	1	4	2	3	(d)	3	2	1	4

13 Consider the following statements.

I. It is an entire virus particle, consisting of an outer protein shell called capsid and an inner core of nucleic acid.

II. It is a circular single-stranded RNA dependent on virus for replication and encapsidation.

III. It is an infectious, non-protein coding highly structured small circular Ribonucleic Acids (RNAs) and induce diseases in plants.

Select the correct option referring to the statements given above.

(a) I-Prion, II-Virion, III-Virusoid
(b) I-Virion, II-Virusoid, III-Viroid
(c) I-Virion, II-Viroid, III-Virusoid
(d) I-Viroid, II-Prion, III-Virusoid

14 Read the following statements.

I. The term virus was given by Pasteur.
II. Virus means poisonous fluid or venom.
III. Viruses can pass through bacteria proof filters.
IV. W M Stanley showed that viruses could be crystallised.
V. Viruses are inert outside the host cell.

Choose the correct answer using the codes given below.

(a) I, II and III
(b) II, III, and IV
(c) I, IV and V
(d) I, II, III, IV and V

15 Find the incorrect match with respect to possession of nucleic acid material.

(a) Plant viruses – *ss*RNA usually
(b) Animal viruses – *ds*DNA/*ds*RNA/*ss*RNA
(c) Plant viruses – *ds*DNA usually
(d) Bacteriophages – *ds*DNA usually

Assertion and Reason

Direction (Q. Nos. 16-24) *In each of the following questions, a statement of Assertion (A) is given followed by the corresponding statement of Reason (R). Of the statements, mark the correct answer as*

(a) Both A and R are true and R is the correct explanation of A
(b) Both A and R are true, but R is not the correct explanation of A
(c) A is true, but R is false
(d) A is false, but R is true
(e) Both A and R are false

16 **Assertion** (A) Flu is a bacterial disease.

Reason (R) Flu viruses are animal retroviruses.

17 **Assertion** (A) Viruses can contribute to various types of cancer in human.

Reason (R) It is estimated that about 15% of all human cancer cases worldwide are due to viruses.

18 **Assertion** (A) HIV does not follow central dogma of molecular biology.

Reason (R) It has enzyme reverse transcriptase.

19 **Assertion** (A) The clinical symptoms of HIV infection generally begin after 8-10 years.

Reason (R) The latency period of HIV is 8-10 years.

20 **Assertion** (A) Viruses possess only a portion of the properties of organisms.

Reason (R) They are simpler than bacteria.

21 **Assertion** (A) Prions are proteinaceous infectious particles.

Reason (R) They possess no nucleic acid.

22 **Assertion** (A) Interferons are a type of antibodies produced by body cells infected by bacteria.

Reason (R) They stimulate inflammation at the site of injury.

23 **Assertion** (A) DNA serves as hereditary material.

Reason (R) It functions as blue-print for building and running cellular machinery.

24 **Assertion** (A) Primitive atmosphere was formed by the lightest atoms.

Reason (R) The primitive atmosphere was reducing in nature.

(C) Medical Entrances' Gallery

(Collection of Questions Asked in NEET & in Various Medical Entrance Exams)

1 Which of the following is correct about viroids?

(a) They have free RNA without protein coat **[NEET 2020]**
(b) They have DNA with protein coat
(c) They have free DNA without protein coat
(d) They have RNA with protein coat

2 Which of the following statements is incorrect?

(a) Viruses are obligate parasites **[NEET 2019]**
(b) Infective constituent in viruses is the protein coat
(c) Prions consist of abnormally folded proteins
(d) Viroids lack a protein coat

3 Mad cow disease in cattle is caused by an organism which has **[NEET 2019]**

(a) inert crystalline structure
(b) abnormally folded protein
(c) free RNA without protein coat
(d) free DNA without protein coat

4 Which of the following statements is incorrect for viroids? **[NEET 2016]**

(a) They are smaller than viruses
(b) They cause infections
(c) Their RNA is of high molecular weight
(d) They lack a protein coat

5 Select wrong statement. **[CBSE AIPMT 2015]**

(a) The viroids were discovered by DJ Ivanowsky
(b) WM Stanley showed that viruses could be crystallised
(c) The term *'contagium vivum fluidum'* was coined by MW Beijerinck
(d) Mosaic disease in tobacco and AIDS in human being are caused by viruses

6 Which of the following statement(s) is/are correct for transduction? **[WB JEE 2015]**

I. It is observed in Gram positive and Gram negative bacteria.
II. Bacteria should be in state of competence.
III. Transfer of DNA by a bacteriophage takes place.
IV. Packaging of both host and phage DNA takes place.

Choose the correct option.

(a) I, II and III (b) II, III and IV
(c) I, III and IV (d) I, II and IV

7 Match the following columns.

	Column I		Column II
A.	HIV	1.	Host
B.	Pilus	2.	Prophage
C.	Virus penetration	3.	Retroviridae
D.	Lysogeny	4.	Donor
		5.	Polyhedral virus

[EAMCET 2015]

Codes

	A	B	C	D
(a)	1	2	3	4
(b)	5	4	3	2
(c)	5	3	2	1
(d)	3	4	1	2

8 Bacterial viruses usually have **[AMU 2015]**

(a) single-stranded RNA
(b) double-stranded RNA
(c) single-stranded DNA
(d) double-stranded DNA

9 The transfer of genetic material from the bacterium to another *via* viruses is called **[AMU 2015]**

(a) transformation (b) conjugation
(c) recombination (d) transduction

10 Which of the following shows coiled RNA strand and capsomeres? **[CBSE AIPMT 2014]**

(a) Polio virus (b) Tobacco mosaic virus
(c) Measles virus (d) Retrovirus

11 Viruses have **[CBSE AIPMT 2014]**

(a) DNA enclosed in a protein coat
(b) prokaryotic nucleus
(c) single chromosome
(d) Both DNA and RNA

12 Identify the correct sequence of events in the viral replication process. **[WB JEE/AFMC 2014]**

I. Eclipse II. Maturation
III. Adsorption IV. Assembly
V. Penetration VI. Lysis

Codes

(a) I, II, III, IV, V, VI (b) II, I, III, IV, V, VI
(c) III, V, I, II, IV, VI (d) III, V, VI, IV, II, I

13 Identify the correct pairs of combination.

I. Viroid – Bovine spongiform encephalitis
II. Prion – Creutzfeldt-Jakob disease
III. Measles virus – Glycoprotein projections
IV. Rabies virus – Polyhedral symmetry

Choose the correct option from the codes given below. **[EAMCET 2014]**

(a) III and IV (b) I and III (c) I and II (d) II and III

14 Which of the following statement is false? **[BCECE 2012]**

(a) TMV has a double-stranded RNA molecule
(b) Most plant viruses are RNA viruses
(c) The bacteriophage has a double-stranded DNA molecule
(d) Most animal viruses are DNA viruses

15 Cell theory is not applicable to [BHU 2012]
(a) bacteria (b) algae
(c) fungus (d) virus

16 LPP-1 is a [Manipal 2012]
(a) bacterium (b) fungus
(c) bacteriophage (d) cyanophage

17 Find out the correct statement. [Kerala CEE 2011]
(a) In lichens, the algal component is called phycobiont and fungal component is known as mycobiont, which are heterotrophic and autotrophic, respectively
(b) Viroid contains RNA of low molecular weight and protein coat
(c) A virus contains both RNA and DNA
(d) Viruses are obligatory parasites
(e) All of the above

18 A virus differs from a bacterium as it contains [J&K CET 2011]
(a) a cell wall
(b) cytosol
(c) DNA as genetic material
(d) DNA or RNA as genetic material with no ribosomes

19 Double-stranded RNA is found in [Odisha JEE 2011]
(a) reovirus (b) TMV
(c) $\phi \times 174$ (d) None of these

20 Which one of the following viruses contains both DNA and RNA? [WB JEE 2010]
(a) Cyanophage (b) Herpesvirus
(c) *Leukovirus* (d) Polio virus

21 Bacteriophage kill [WB JEE 2010]
(a) fungi (b) parasite (c) bacteria (d) viruses

22 Virus envelope is known as [CBSE AIPMT 2010]
(a) core (b) capsid
(c) virion (d) nucleoprotein

23 Infectious proteins are present in [JCECE 2012, CBSE AIPMT 2010]
(a) gemini viruses (b) prions
(c) viroids (d) satellite viruses

24 Which one of the following does not follow the central dogma of molecular biology? [CBSE AIPMT 2010]
(a) HIV (b) Pea
(c) *Mucor* (d) *Chlamydomonas*

25 Phage genome site on bacterial chromosome resulted in the structure [Odisha JEE 2010]
(a) nucleic acid (b) heterocyst
(c) prophage (d) None of these

ANSWERS

Check Point 3.1

1. (c)	2. (d)	3. (c)	4. (d)	5. (b)	6. (b)

Check Point 3.2

1. (c)	2. (a)	3. (d)	4. (a)	5. (a)	6. (b)

(A) Topical Exercise

1. (d)	2. (d)	3. (a)	4. (d)	5. (b)	6. (b)	7. (d)	8. (b)	9. (d)	10. (a)
11. (c)	12. (b)	13. (c)	14. (b)	15. (a)	16. (a)	17. (b)	18. (d)	19. (a)	20. (d)
21. (a)	22. (d)	23. (d)	24. (d)	25. (c)	26. (d)	27. (b)	28. (a)	29. (d)	30. (b)
31. (a)	32. (b)	33. (b)	34. (a)	35. (a)	36. (c)	37. (a)	38. (a)	39. (c)	40. (d)
41. (c)	42. (c)	43. (c)	44. (d)	45. (c)	46. (a)	47. (c)	48. (c)	49. (a)	50. (d)
51. (d)	52. (a)	53. (a)	54. (d)	55. (a)	56. (b)	57. (a)	58. (a)	59. (c)	60. (a)

(B) Cumulative Exercise

1. (d)	2. (a)	3. (c)	4. (c)	5. (a)	6. (a)	7. (b)	8. (b)	9. (a)	10. (c)
11. (b)	12. (d)	13. (b)	14. (d)	15. (c)	16. (d)	17. (b)	18. (a)	19. (a)	20. (b)
21. (a)	22. (d)	23. (a)	24. (b)						

(C) Medical Entrances' Gallery

1. (a)	2. (b)	3. (b)	4. (c)	5. (a)	6. (c)	7. (d)	8. (d)	9. (d)	10. (b)
11. (a)	12. (c)	13. (d)	14. (a)	15. (d)	16. (d)	17. (d)	18. (d)	19. (a)	20. (c)
21. (c)	22. (b)	23. (b)	24. (a)	25. (c)					

Explanations

(A) Topical Exercise

1 (*d*) All the options given are living characteristic of viruses except (d) ability to be crystallise because

They can remain in crystallise stage for billion of years and still can survive.

2 (*d*) Many scientists believe that viruses are modified plasmids, which are the fragments of the nucleic acids of the host. These genome fractions can escape and get induced into new host cells.

3 (*a*) The coevolution theory for the origin of viruses was first proposed by d' Herelle who claimed that viruses are ancestral to cells. Evolution of life started with a virus like stage and the advent of modes type cells was a late event. However, it suggests that the origin of viruses is these have arisen simultaneous and evolution with other cells.

5 (*b*) Statement I is false, but II is true. Incorrect statement can be corrected as

The tobacco mosaic virus is a single-stranded RNA virus.

6 (*b*) Virus are considered as both living and non-living. Viruses have nucleic acid as genetic material. They can also produce their own copies in the host cell.

They lack cellular organisation and are considered obligate parasites.

7 (*d*) Viruses are tiny infectious agents that replicate only in the living cells of other organisms. Viruses have a very simple structure consisting of genetic material in the form of DNA or RNA within a protein capsule.

8 (*b*) Option (b) is not a characteristic feature of viruses because

Viruses reproduce by infecting their host cells and reprogramming them to become virus making 'factories'. So, even in getting favourable conditions, they cannot reproduce outside host cells.

11 (*c*) Statement in option (c) is correct for virus. Viruses are obligate intracellular parasite, it conveys the idea that viruses must carryout their reproduction by parasitising a host cell. They cannot multiply outside a living cell, they can only replicate inside of a specific host.

15 (*a*) All of the characteristics are about protein coat of virus except (a). It can be corrected as

The identical subunits of capsid are called capsomeres.

20 (*d*) Option (d) is incorrect match and can be corrected as

Citrus tristeza virus also called as CTV or *tristeza* belong to genus-*Closterovirus*. This causes damage to *Citrus* plants. It is a flexuous rod virus with dimensions of 2000 nm long and 12 nm in diameter. It is one of the largest known plant RNA viruses.

21 (*a*) Statement I is true, but II is false. False statement can be corrected as

Tobacco mosaic virus and HIV show single-stranded RNA.

25 (*c*) Statement in option (c) is incorrect. It can be corrected as

Mumps, chicken pox and rubella are all dermotropic viral disease. SARS is not dermotropic. Dermotropic virus enter by the way of skin in human.

28 (*a*) Statement in option (a) is not true and can be corrected as

DNA is involved in the life cycle of retrovirus as they replicate by first converting single-stranded RNA to double-stranded DNA, using enzyme reverse transcriptase.

Rest statements are true for retroviruses.

37 (*a*) Option (a) is correctly matched. Other incorrect matches can be corrected as

National Institute of Communicable Diseases-Kasauli

Central Drug Research Institute-Lucknow

National Institute of Nutrition-Secunderabad, Telangana

40 (*d*) Enveloped virus enters into host cell by fusion with the plasma membrane of the host.

This can be explained by the fact that some viruses have a lipid bilayer membrane around them (enveloped viruses), while the other viruses are naked.

During penetration within a host, fusion between specific proteins takes place. These proteins are present on the membranes of both the virus and the bacterial host.

46 (*a*) Statement in option (a) is incorrect. It can be corrected as

In pinocytic reproduction the whole virus enters the host cell except the envelope.

51 (*d*) Statements II and IV are correct whereas I and III are incorrect. Incorrect statements can be corrected as

Temperate phages upon being injected into the host bacteria do not immediately enter into the lytic cycle, but integrate their genome into bacterial genome as prophages.

52 (*a*) Statement in option (a) is correct for bacterial transduction. Rest of the statements are incorrect and can be corrected as

Transduction involves the picking up of DNA by bacteriophage from one bacterial cell and carrying it to another where the DNA fragment may get incorporated into the bacterial host's genome.

56 (*b*) Statement I is false, but II is true. Incorrect statement can be corrected as

Viroids have small, circular RNA not DNA.

58 (*a*) The pathogenic protein particles called prions are known to cause several diseases like CJD or Creutzfeldt-Jakob disease.

(B) Cumulative Exercise

1 (*d*) Statements I and III are correct whereas II and IV are incorrect about virus. Incorrect statements can be corrected as

Viruses do not have their own ribosome they hijack the ribosome of the human cell to make more virus. Viruses cannot generate or store energy in the form of adenosine triphosphate, but have to derive their energy and all other metabolic functions from the host cell.

3 (*c*) All given statements are correct except III. Incorrect statement can be corrected as

It has a single-stranded RNA as a genetic material.

5 (*a*) Only correct differences are I and II. Incorrect option can be corrected as

In bacteriophage, penetration of nucleic acid through cell wall occurs. There is no uncoating of nucleic acid.
In animal virus, penetration occur as whole virus is engulfed and uncoated, or virus surface fuses with cell membrane, nucleic acid is then released.

6 (*a*) Statements I and II are categorised as lysogeny. Other incorrect statements can be corrected as

The lysogenic cell can follow the lytic cycle also. Viral DNA in lytic cycle never forms episome, but does not integrate into the host's DNA.

8 (*b*) Statements II and IV are correct whereas I and III are incorrect. Incorrect statements can be corrected as

Viruses are non-cellular organisms characterised by having an inert crystalline structure outside the living cell. They do not have their own metabolic system and use host machinery to grow and multiply.

9 (*a*) Statements I, III and IV are incorrect and can be corrected as

Viruses are incapable of growth and division independently.

HIV is the AIDS causing virus, having RNA as genetic material.

Viruses do not have their own cellular machinery.

11 (*b*) Statements I, III and IV are correct. Statement II is incorrect and can be corrected as

Lytic cycle is a common reproductive cycle in lambda (λ) phages.

15 (*c*) Option (c) is incorrect. It can be corrected as

Plant viruses are usually RNA not DNA viruses.

16 (*d*) A is false, but R is true. A can be corrected as

Flu (influenza) is a viral disease. Its viruses are animal retroviruses. Wild Asian ducks, chicken and pigs are major reservoirs.

17 (*b*) Both A and R are true, but R is not the correct explanation of A.

Viruses can cause cancer in humans. They do this by mixing their DNA with the cell's DNA, thereby triggering changes that make the cell grow and multiply.

It has been suggested that the viruses contribute to about 15% of all human cancer cases worldwide.

18 (*a*) Both A and R are true and R is the correct explanation of A.

HIV (Human Immunodeficiency Virus) does not follow the central dogma of molecular biology as it has the ability to reverse transcribe the RNA to DNA *via* enzyme reverse transcriptase.

19 (*a*) Both A and R are true and R is the correct explanation of A.

Clinical symptoms begin generally 8-10 years after the initial infection with HIV. The reason why HIV remains hidden for so long seems to be that its infectious cycle continues throughout 8-10 years latent period without doing serious harm to the infected person.

20 (*b*) Both A and R are true, but R is not the correct explanation of A.

Viruses possess only a portion of the properties of organisms. They are literally parasitic chemicals, segments of DNA or RNA wrapped in a protein coat. They are simpler than bacteria and cannot reproduce on their own.

21 (*a*) Both A and R are true and R is the correct explanation A.

Prions are proteinaceous infectious particles and are known to cause Bovine Spongiform Encephalopathy (BSE), Creutzfeldt-Jakob disease, etc. They are unique as they do not possess nucleic acid of their own.

23 (*a*) Both A and R are true and R is the correct explanation A.

DNA is the genetic material in most of living organisms except the plant viruses and some bacteriophages. It is the only molecule which can replicate itself or can form its own carbon copy. The phenomenon is called molecular reproduction. All the information required for growth differentiation, running cellular machinery and reproduction is contained in DNA molecules.

These genetic information of DNA are like the blue-print. During cell division (which involve DNA replication), the daughter cell receives the same blue-print or genetic material as in the parental cell.

24 (*b*) Both A and R are true, but R is not the correct explanation of A.

The lightest atom of nitrogen, hydrogen, carbon, etc., formed the primitive atmosphere. Hydrogen atoms were most numerous and most reactive in primitive atmosphere. First hydrogen atoms combined with all oxygen atoms to form water and leaving no free oxygen.

Thus, primitive atmosphere was reducing atmosphere (without free oxygen) unlike the present oxidising atmosphere (with free oxygen). Presence of huge amount of free hydrogen in primitive atmosphere also proves its reducing nature.

(C) Medical Entrances' Gallery

1 (*a*) Option (a) is correct for viroids. Viroids are generally known as plant pathogens. They are small, single-stranded, RNA particles.

They do not have a capsid or outer envelope of protein making them simpler than virus, but like viruses can reproduce only within a host cell. They do not have a capacity to make their own protein. They produce only a single, specific RNA molecule.

2 (*b*) The statement 'infective constituent in viruses is protein coat' is incorrect. It can be corrected as

Viruses infect their host organisms through their genetic material, i.e. either DNA or RNA and not protein. They take over the biosynthetic machinery of the host cell and produce chemicals required for their own multiplication.

Rest of the statements are correct.

3 (*b*) Mad cow disease in cattle is caused by prions which are abnormally folded proteins. It is also known as Bovine Spongiform Encephalopathy (BSE) and is a progressive neurological disorder of cattle.

4 (*c*) Statement in option (c) is incorrect and can be corrected as

Viroid RNA is of low molecular weight.
Rest of the statements are correct.

5 (*a*) The statement in option (a) is incorrect. It can be corrected as
Viroids were discovered by T.O. Diener.

6 (*c*) Statements I, III and IV are correct for transduction. Statement II is incorrect and can be corrected as

Bacteria should not be in the state of competence.

9 (*d*) Transduction is the process by which a virus transfers genetic material from one bacterium to another.

10 (*b*) TMV or Tobacco Mosaic Virus, shows coiled RNA and capsomeres. In TMV, RNA is a single-stranded (*ss*), helically coiled structure containing about 2130 capsomeres which are the basic subunit of capsid. The latter is an outer covering of protein that protects the genetic material of a virus.
There are about 16 capsomeres present in each helical turn.

11 (*a*) Viruses can have either DNA or RNA as their genetic material. A complete virus particle called virion, consists of nucleic acid (DNA or RNA) surrounded by a protective coat of a protein called capsid.

12 (*c*) The correct sequence of events in the viral replication process is
Adsorption $\rightarrow$ Penetration $\rightarrow$ Eclipse $\rightarrow$ Maturation $\rightarrow$ Assembly $\rightarrow$ Lysis
Thus, option (c) is correct.

13 (*d*) II and III are correctly paired whereas I and IV are incorrectly paired. These can be corrected as

- Bovine spongiform encephalitis or mad cow disease is caused by prions and not viroids.
- Rabies virus possesses helical symmetry and not polyhedral symmetry.

14 (*a*) Statement in option (a) is false and can be corrected as

Reovirus is a RNA virus whose genome is made up of double-stranded RNA. TMV or Tobacco Mosaic Virus have single-stranded RNA (*ss*RNA) as its genome. The genome of $\phi \times 174$ is single-stranded DNA (*ss*DNA).

Rest statements are correct.

17 (*d*) Statement in option (d) is correct.

Viruses are obligate intracellular parasites. Rest statements are incorrect and can be corrected as

- In lichens, the algal component is known as phycobiont and fungal component as mycobiont, which are autotrophic and heterotrophic, respectively.
- Though viroids contain RNA of low molecular weight, protein coat is absent in them.
- No virus contains both RNA and DNA.
  Viruses that infect plants have single-stranded RNA.

18 (*d*) The genetic material in viruses could be either RNA or DNA but not both. The genome of plant viruses is usually RNA and that of animal viruses is DNA. Viruses have no cell wall, cytosol, ribosomes, etc., whereas bacteria have cell wall, ribosomes and have both DNA and RNA.

20 (*c*) *Leukovirus* contains both DNA and RNA. It is known to cause tumours in animals. Cyanophage are viruses that infect cyanobacteria. They contain double-stranded DNA. Herpes virus also contains double-stranded DNA. Polio virus contains single-stranded RNA.

22 (*b*) Virus envelope is known as capsid. The nucleic acid of virus is always located within the virion particle and is surrounded by a protein shell called capsid.

23 (*b*) Infectious proteins are present in prions, as they are known to cause a variety of diseases in animals. The genetic material (DNA or RNA) of gemini virus and satellite virus is infectious. Viroids are infectious agents that do not contain protein coat.

25 (*c*) Prophage is the genetic material of a bacteriophage that is incorporated into the genome of the host bacterium.

CHAPTER 04

Monera

Kingdom—Monera was created by **Copeland** in 1956. It contains all the prokaryotic organisms including Eubacteria, Archaebacteria, Mycoplasma, Actinomycetes, Cyanobacteria and Rickettsia. **Prokaryotes** (Gr. *Pro*–primitive, *karyos*-nucleus) are primitive nucleus containing organisms. The word primitive nucleus implies to an unorganised nucleus. It lacks associated proteins and membranes.

These organisms are microscopic and contribute a major part in microbiology, i.e. the biology of microorganisms.

In this chapter, we will discuss all the characteristics, types, physiology and economic importance of kingdom-Monera in detail.

Inside

General characteristics

Following major characteristics are shown by the members of kingdom-Monera

(i) **Genetic material** It is naked with circular DNA. It occurs inside the cytoplasm called **nucleoid.** It normally remains attached to the cell membrane.

(ii) **Cell organelles** It lacks major membrane bound cell organelles such as mitochondria, Endoplasmic Reticulum (ER), plastids, etc. But ribosomes of 70S type, with 50S and 30S subunits are present.

(iii) **Cell wall** The cell wall, if present is a rigid structure that provides protection. It is made up of polysaccharide having polypeptide cross links with peptidoglycan. Archaebacteria and Mycoplasma are exceptions to it. Cellulosic cell walls are not seen in this kingdom.

Classification

In 1977, **Carl Woese** and **George Fox** reported two distinct groups in kingdom-Monera. These groups are Archaebacteria and Eubacteria.

(i) **Archaebacteria** (Gr. *Archae*-primitive; bacteria) includes methanogens, halophiles and thermoacidophiles.

(ii) **Eubacteria** (Gr. *Eu* —true; bacteria) includes cyanobacteria, mycoplasma, actinomycetes, rickettsia and spirochaetes.

EUBACTERIA

- Bacteria were discovered by **Av Leeuwenhoek** in 1676. Thus, he is considered as the **'Father of Bacteriology'**.
- **Ferdinand Cohn** (1872) is considered as the **'Father of Modern Bacteriology'**.
- The term bacteria was introduced by **Christian Gottfried Ehrenberg** in 1838.
- The classification of bacteria was given by **DH Bergey** in Manual of Determinative Bacteriology.

Habitat

- Bacteria are cosmopolitan, i.e. found everywhere.
- They can survive almost everywhere including bottom of oceans, our mouth, acidic and alkaline environment, hot springs on glaciers, surface of plants, surface of animals, intestinal tracts of animals, etc.
- Approximately, 4000 species of bacteria have been described from all over the world out of which about 850 species are found in India.

Size

- Though, the size of bacteria greatly varies, but most of them are 0.1-1.5 μm in diameter and 2-10 μm in length. Mycoplasma are often only 0.2 μm across. Some photosynthetic cyanobacteria are up to 3 mm long.
- The smallest known bacterium is *Dialister pneumosintes* (0.15-0.3 μm long).
- A recently discovered largest known bacterium (non-photosynthetic) is *Epulopiscium fishelsoni* (0.3 mm). *Beggiatoa mirabilis* is the largest among filamentous sulphur bacteria.

Shape

Bacteria are generally enclosed in a porous, but rigid cell wall. It gives them characteristic shapes. Depending upon the shape, bacteria can be of various types as discussed below

1. Cocci (Sing. *coccus*; Gr. *kokkos* – berry)

These are **spherical** or **oval-shaped** bacteria. These do not possess flagella (atrichous).

They are subdivided into six groups

(i) **Monococci** (Sing. *Monococcus*) When coccus occur singly, e.g. *Micrococcus luteus*.

(ii) **Diplococci** (Sing. *Diplococcus*; Gr. *diplos*-double) When coccus occur in a pair of two, e.g. *Diplococcus pneumoniae*.

(iii) **Streptococci** (Sing. *Streptococcus*; Gr. *streptos*-curved) When coccus occur in the form of chains, e.g. *Streptococcus lactis*.

(iv) **Tetracocci** (Sing. *Tetracoccus*; Gr. *tetra*-four) When coccus occur in a group of four, e.g. *Tetracoccus cechii*.

(v) **Sarcinae** (L. *sarcio*-tie) When coccus occur in cubical packets of 8, 27 or more cells, e.g. *Sarcina*.

(vi) **Staphylococci** (Sing. *Staphylococcus*; Gr. *staphylos* - cluster of grapes) When coccus occur in irregular bunches of cells. It resemble a cluster of grapes, e.g. *Staphylococcus aureus*.

2. Bacilli (Sing. *bacillus*)

These are **rod-like** or **cylindrical forms** that are most commonly found in nature. These too, do not possess any flagella. They are further subdivided into various groups as follows

(i) **Monobacillus** Only one rod-like structure represents single bacillus, e.g. *Rhodospirillum sodomense*.

(ii) **Diplobacilli** These are arranged in pairs, e.g. *Bacillus subtilis*.

(iii) **Streptobacilli** These form a chain of rods, e.g. *Bacillus anthracis*.

(iv) **Palisade like** If the cells are lined side by side like matchsticks and at angles to one another, e.g. *Corynebacterium diphtheriae*.

(v) **Trichomes** The larger area of cells of the chain is in contact with each other, e.g. *Beggiatoa*.

(vi) **Hyphae** Cells are arranged as long, branched unicellular filaments, e.g. *Streptomyces*.

3. Spiral (Sing. *spirillum*)

These bacteria are **spiral-shaped**. These are usually curved with one or more twists. They are less common in nature and mostly occur singly, e.g. *Spirillum rubrum*.

Vibrio (Sing. Vibrion)

These bacteria are small and often curved-like a **comma**. They possess one flagellum, e.g. *Vibrio cholerae* and *Vibrio comma.*

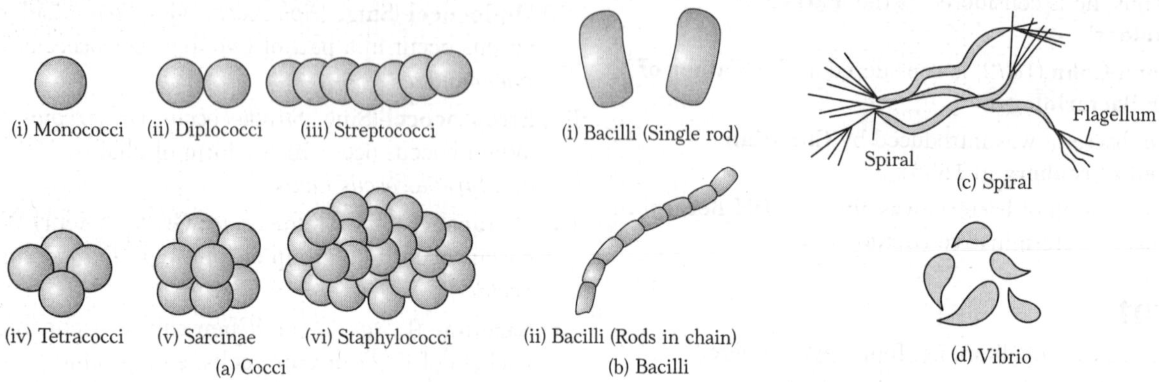

Fig 4.1 Various shapes of bacteria

Structure

The structure of a typical bacterial cell is shown in the figure given below

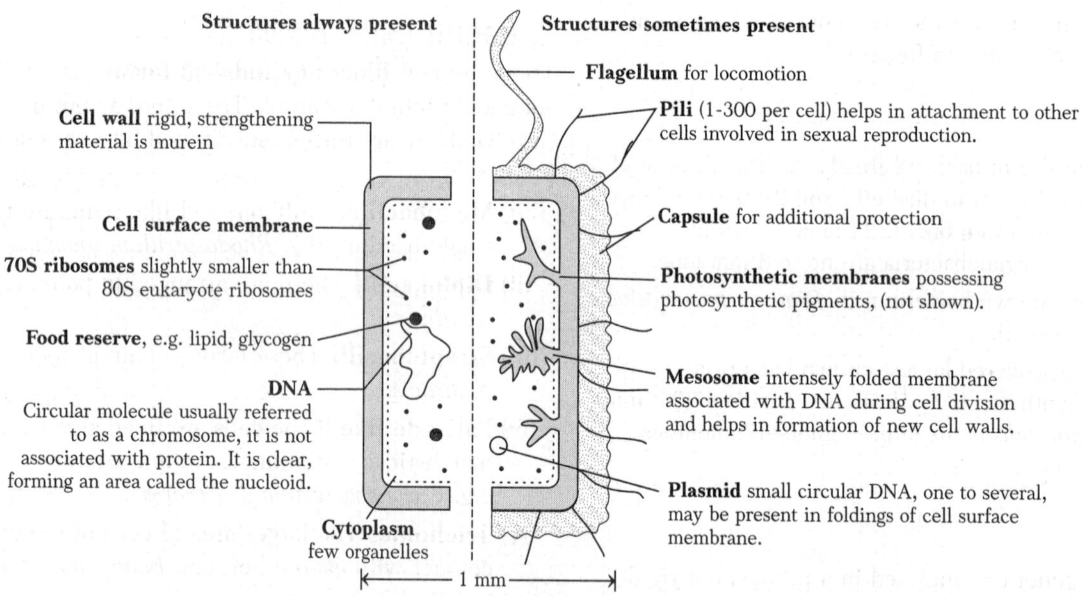

Fig 4.2 Structure of a generalised rod-shaped bacterium (a typical prokaryotic cell)

As shown in above diagram, not all the structures are always found in a bacterial cell. The detailed description of these structure is given below

For the sake of convenience in study, we will discuss these structure in three parts

1. structures external to the cell wall
2. cell wall
3. structures internal to the cell wall

Structures external to the cell wall

The structures external to the cell wall in a bacterial cell are as follows

(i) Glycocalyx

It is present external to cell wall as an outermost covering of bacteria. It protects the cell wall and also helps in adhesion. It may occur in two forms, i.e. a capsule or a slime layer.

(a) **Capsule** It is a distinct thick gelatinous layer composed of polysaccharides and D-glutamic acid. It may protect bacteria from phagocytosis and from toxic chemicals.

R and S type Bacteria

- Capsulated bacteria form smooth colonies and are known as **S-type** bacteria. These forms are highly virulent.
- Bacteria having no capsule form rough colonies and are known as **R-type** bacteria.

(b) **Slime layer** It is a diffused sheath around cell wall. It is composed of polysaccharides such as dextrin, dextran and levans. Slime layer provides protection against desiccation and loss of nutrients.

(ii) Flagella

- It is a single-stranded structure similar to microtubular fibre. It has (9+0) arrangement instead of typical (9+2) in eukaryotic cells.
- It is about 20 nm in diameter and 1-7 μm in length.
- It originates from a granular structure and protrudes out through cell wall.
- Bacterial flagella performs rotation movement, i.e. backward pushing of medium, hence it always moves forward.

Structure of flagella

A flagella comprises of three parts, i.e. basal body, hook and filament, as shown below

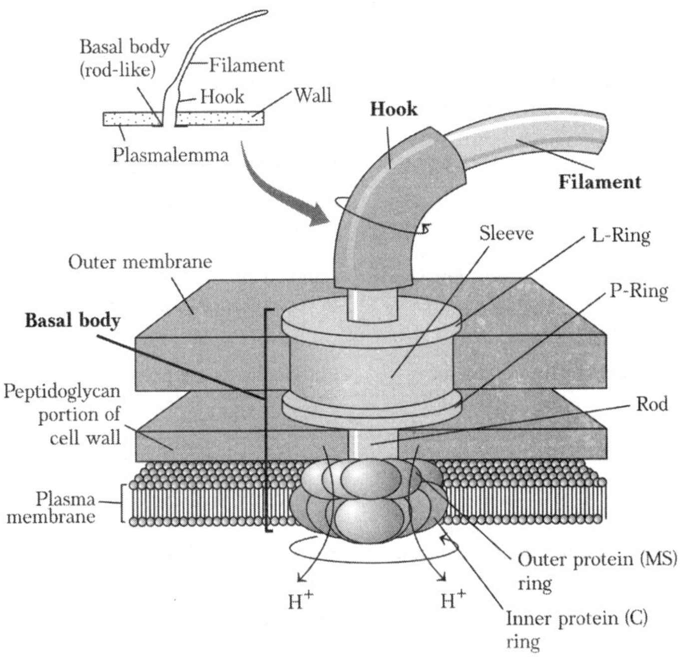

Fig 4.3 Structure of flagella

(a) **Basal body** It is the most complex portion having ring like swelling in membrane cell wall.
- There are two pairs of rings in Gram-negative bacteria (i.e. L-P ring and MS-C ring) and only a single pair of ring in Gram-positive bacteria (i.e. MS-C ring).
- The L-ring is embedded in lipid bilayer.
- The P-ring is embedded in peptidoglycan layer of cell wall.
- Likewise, MS-ring is embedded in cytoplasmic membrane and C-ring in cytoplasm.
- Out of these, most of the ring proteins function as flagellar motor, while C-proteins work as flagellar motor switch.

(b) **Hook** It is made up of different protein subunits. It is the thickest part of flagellum, that connects filament with basal body.

(c) **Filament** Long tubular structure causes turbulence in fluid medium. Made up of globular protein. In flagella these proteins are arranged in spiral manner.

Types of flagellar arrangement

Depending upon the arrangement of flagella, bacterial cell can be of following types

- **Atrichous**, i.e. flagellum is absent, e.g. *Pasteurella.*
- **Monotrichous**, i.e. with single polar flagellum, e.g. *Vibrio cholerae, Thiobacillus* and *Xanthomonas oryzae.*
- **Amphitrichous**, i.e. with single flagellum at each end of the cell, e.g. *Nitrosomonas.*
- **Lophotrichous**, i.e. with two or more flagella at one end of the cell, e.g. *Spirillum serpens, Pseudomonas.*

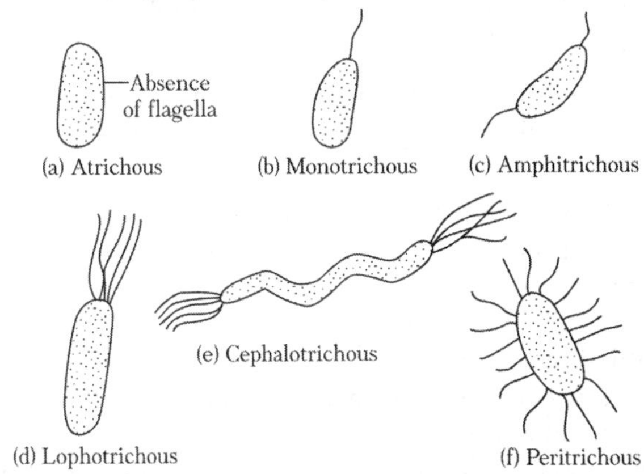

Fig 4.4 Forms of flagellation

- **Cephalotrichous**, i.e. a tuft or group of flagella present at each of the two ends or poles, e.g. *Pseudomonas fluorescence.*
- **Peritrichous**, i.e. flagella distributed over the entire cell, e.g. *Escherichia coli, Clostridium* and *Salmonella.*

(iii) Axial filaments

These are similar to flagella except that they wrap around the cell, e.g. spirochaetes move by means of axial filaments.

(iv) Pili

These are longer, fewer and thicker tubular outgrowths. These develop in response of F^+ or fertility factor in Gram negative bacteria. They are made up of a protein called **pilin**. True pili have been reported only in Gram negative bacteria.

(V) Fimbriae

These are small bristle-like fibres sprouting from cell surface in large number. These are made up of fimbrillum protein.

They are helpful in attaching bacteria to solid surface or host tissues. In the absence of fimbriae colonisation cannot occur. These are present in both Gram positive and Gram negative bacteria.

Cell wall

It is the actual covering of bacterial cell. Its structures and functions are given below

Structure

- The cell wall of bacteria is made up of peptidoglycans (amino acid + sugars). Peptidoglycan or mucopeptide or murein.
- It has long glycan strands formed of repeating units of **N-Acetyl Glucosamine** (NAG) and **N-Acetyl Muramic** (NAM) acid. They are cross-linked by small peptide chains and have 1-4 linkage.
- The bacterial cell wall contains some special amino acids, such as L-alanine, D-alanine, D-glutamic acid and either L-lysine or diaminopimelic acid.
- It also contains magnesium (Mg^{2+}) ions binded to teichoic acids. This binding protects the bacteria from thermal injuries.

Functions

- It prevents the cell from bursting, when it absorbs water as a result of osmosis (osmotic rupture).
- Tiny pores are present in the cell wall. These pores aid in movement of water, ions and molecules into the bacterial cell.

Cell wall and Gram staining

- Bacteria are classified as Gram positive (+ve) or Gram negative (–ve) based on their reaction with **Gram's stain**.
- This technique of Gram staining was introduced by **Hans Christian Gram** in 1884.
- The basis of Gram staining is the lipid content of bacteria.
- The steps involved in Gram staining are as follows
  - (i) A thin smear of bacterial culture is made. It is then air dried.
  - (ii) Smear is stained with alkaline solution of crystal violet or gentian violet. Thus, cells take up the stain and become blue. Excess stain is washed off.
  - (iii) Now, the cells are treated with 0.5% iodine solution. Iodine fixes the dye to the cells wall, i.e. it acts as mordant.
  - (iv) Cells are then washed with 95% alcohol or acetone. After washing, the bacterial cells are observed under microscope.
- The bacteria, which retain their original crystal violet stain are called **Gram positive bacteria.**
- The bacteria that lose stain and get discolourised are called **Gram negative bacteria.**
- After counter staining with safranin, former appear purple and latter appear pink.

The major differences between Gram positive and Gram negative bacteria are tabulated below

Gram positive	Gram negative
Retain Gram stain.	Does not retain Gram stain.
Rigid cell wall due to the presence of high proportion of peptidoglycan ~80%.	Less rigid and more elastic cell wall. Peptidoglycans account for only about 3-12%.
Single layered thick cell wall (100-200 Å) containing teichoic acid. Lipopolysaccharide is absent.	Triple layered thin cell wall (70-120 Å) with absence of teichoic acid. The outer layer that surrounds the cell wall is made of Lipopolysaccharide (LPS).
Basal body of flagellum has two rings of swellings.	Four rings of swellings occur in basal body of flagellum.
Lipid content is very low (2-4%) due to the absence of outer membrane.	Lipid content is high (20-30%) due to the presence of outer membrane.
Phospholipids are absent.	Phospholipids are present.
L-lysine is present.	L-lysine is absent.
Cell wall is sensitive to antibiotics such as penicillin and cephal osporin.	Cell wall is insensitive to penicillin and cephalosporin.
Pentaglycine chains are present in cell wall. [As shown in fig. 4.5]	Pentaglycine chains are absent. [As shown in fig. 4.5]
They produce exotoxins.	They produce endotoxins.
e.g. *Streptomyces, Actinomyces* and *Clostridium.*	e.g. *E. coli, Salmonella, Nostoc* and *Rhizobium.*

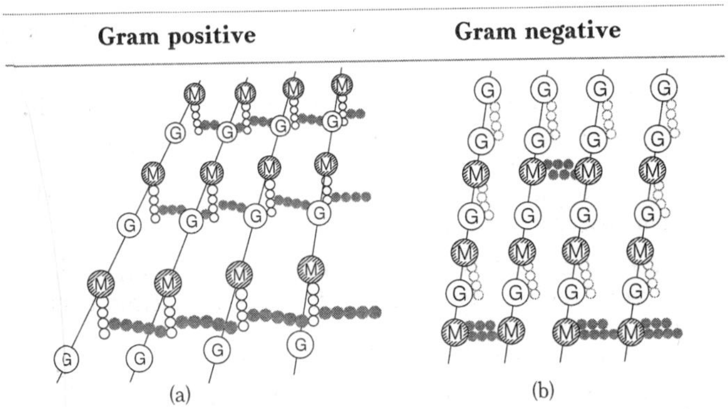

Fig 4.5 (a) Cell wall of Gram (+ve) bacteria (b) Cell wall of Gram (-ve) bacteria

Key → Ⓖ = N-acetyl glucosamine

Ⓜ = N-acetyl muramic acid,

⦂ = Tetrapeptide chains,

●●●●● = Pentaglycine bridges

Structures internal to the cell wall

The structures internal to the cell wall are as follows

(i) **Plasma membrane** (cytoplasmic membrane) The structure and functions of this cell surface membrane are similar to those in eukaryotic cells. It is also the site for processing some of the respiratory enzymes.

(ii) **Cytoplasm** It contains about 80% water along with proteins (enzymes), carbohydrates, lipids, inorganic ions, etc.

It is thick, aqueous, semitransparent and elastic. The major structures in the cytoplasm are DNA, different RNA's, ribosomes and reserve deposits called **inclusions**.

(iii) **Nucleoid or Genophore** Bacteria lack a well-defined nucleus. However, an amorphous lobular mass of fibrillar chromatin type material (DNA) is present in bacteria. It occupies about 10-20% of the cell. This chromatin material is not bounded by any membrane. It is found in cytoplasm near the centre of the cell.

(iv) **Genetic or Chromatin material** It is also called bacterial chromosome or chromatin. The bacterial DNA is a single, circular, double-stranded molecule of about 5 million base pairs. It is much less than that of a eukaryotic cell. It does not have associated proteins.

(v) **Plasmids** These were discovered by **William Hays** and **Joshua Lederberg** in 1952. A plasmid is a small, self-replicating, circular, extrachromosomal DNA.

It possesses only a few genes. It is known to confer resistance to antibiotics. Plasmids are also called as **minichromosomes** because these are capable of autonomous replication.

Types of plasmids

On the basis of their function, following types of plasmids are reported

(a) **F-plasmid** It carries fertility factor, which is responsible for conjugation.

(b) **R-plasmid** It carries antibiotic resistant genes. It aid in survival from antibiotics.

(c) **Ti-plasmid** It carries tumour inducing genes. It is large in size (200 kbp). Ti-plasmid of *Agrobacterium* is used in genetic engineering.

(d) **Col-plasmid** It carries genes that produce special proteins such as colicin to kill other bacteria.

(e) **Degradative plasmid** These have genes that help in decomposition of complex organic compounds or hydrocarbons.

(vi) **Ribosomes** Bacterial cytoplasm contains numerous 70S ribosomes having 30S + 50S subunits. These are made up of *r*RNA and protein. Several antibiotics such as streptomycin, neomycin and tetracycline work by inhibiting protein synthesis through the ribosomes. When many ribosomes read the codes of single *m*RNA, they forms polyribosomes or polysomes.

(vii) **Mesosomes** These were discovered by **Fitz James** in 1960. They are more prominent in Gram negative bacteria. These structures are considered as equivalent to mitochondria of eukaryotic cells.

Based on their position two types of mesosomes are

(a) **Central mesosome** Found near the nucleoid, help in cell division by sapta formation.

(b) **Peripheral mesosome** Found at the periphery of cell near cell membrane, help in storing respiratory enzymes, e.g. cytochrome oxidase and succinic dehydrogenase.

(viii) **Inclusions** These are reserve deposits. Bacteria contain metachromatic granules (usually glycogen or starch), lipid inclusions, sulphur granules, carboxysomes and gas vacuoles.

CHECK POINT 4.1

1. Read the following statements about bacteria and select the correct option.
   (a) Bacteria are simple in structure, but complex in behaviour
   (b) Bacteria are complex in structure, but simple in behaviour
   (c) Bacteria are simple in both structure and behaviour
   (d) Bacteria are complex in both structure and behaviour
2. Select the statement which differentiate slime layer from capsule of bacteria.
   (a) Slime layer provides virulent properties, but capsule do not
   (b) Slime layer present outside the cell wall
   (c) Slime layer is present as diffused layer, but capsule is present as distinct layer
   (d) Slime layer is present as distinct layer, but capsule is present as diffused layer
3. Peptidoglycan is a characteristic constituent of the cell wall in
   (a) Eubacteria and Archaebacteria
   (b) Bacteria and Cyanobacteria
   (c) Only Archaebacteria
   (d) All members of Monera
4. Select the correct one with respect to cell wall of bacteria.
   (a) Glycocalyx is present internal to cell wall as an innermost covering of bacteria
   (b) Cell wall contains Ca^{2+} ions binded to teichoic acids
   (c) The cell wall is made up of amino acid and sugars
   (d) Peptidoglycan or mucopeptide do not form the structural network of bacterial cell wall
5. A bacterial species possess high lipid content and phospholipid in outer membrane. It can also produce endotoxins. It is most likely to be
   (a) *Streptomyces* (b) *Clostridium*
   (c) *Salmonella* (d) All of these

Nutrition

On the basis of mode of nutrition, bacteria can be autotrophic and heterotrophic.

Less than 1% population of bacteria shows autotrophic mode of nutrition. Majority of bacteria are heterotrophic in nature.

1. Autotrophic eubacteria

- They are also known as self-feeders as they can synthesise their own food from simple inorganic substances.
- Autotrophic bacteria are further of following two types

Photosynthetic or Photolithoautotrophic bacteria

- They possess two types of photosynthetic pigments, **bacteriopurpurin** and **bacterioviridin.** Both of these pigments are different from the chlorophyll found in higher plants.
- These pigments trap solar energy and utilise it for the synthesis of complex food material. Bacterial photosynthesis is **anoxygenic**, i.e. it takes place without the involvement of O_2.
- The source of electron in bacterial photosynthesis is not water, instead they utilise compounds like hydrogen sulphide, thiosulphate, etc., as electron source.

Photosynthetic bacteria can be further classified as follows

Table 4.1 Classification of photosynthetic bacteria

Groups	Electron and hydrogen donor	Examples
Purple sulphur bacteria (with bacteriopurpurine pigment)	Inorganic sulphur compounds	*Chromatium, Thiospirrilum*
Purple non-sulphur bacteria (with bacteriopurpurine pigment)	Non-sulphur aliphatic compounds	*Rhodospirillum, Rhodopseudomonas*
Green sulphur bacteria (with bacterioviridin pigment)	H_2S	*Chlorobium*
Green non-sulphur bacteria (with bacterioviridin pigment)	Non-sulphur aliphatic organic compound	*Chloronema, Chloroflexus*

Chemosynthetic bacteria

- These bacteria do not possess photosynthetic pigments.
- They obtain energy from organic compounds made by other organisms.
- They include nitrifying bacteria (*Nitrosomonas, Nitrobacter*), iron bacteria (*Ferrobacillus, Leptothrix*) and sulphur bacteria (*Thiobacillus, Beggiatoa*), hydrogen bacteria (*Hydrogenomonas*).
- Chemosynthetic bacteria oxidise various inorganic substances, e.g. nitrates, nitrites and ammonia.
  They utilise the released energy for their ATP production.
- They play very important role in recycling of nutrients, such as nitrogen, phosphorus, iron and sulphur.

2. Heterotrophic eubacteria

- These are also known as **organotrophs**, i.e. obtain electron from organic substrates. They are most abundant in nature.
- They cannot synthesise their own food and are dependent on external organic material for their growth and energy.
- Most of these are important decomposers in nature.
- Heterotrophic bacteria can be categorised into different groups as shown below

Table 4.2 Types of heterotrophic bacteria

Groups	Characteristics	Examples
Saprophytic bacteria	Free-living bacteria, which obtain their food from organic remains, e.g. animal excreta, fallen leaves, vegetables, jams, jellies, etc.	*Pseudomonas, Bacillus mycoides*
Symbiotic bacteria	Live in mutually beneficial associations with other organisms.	*Escherichia coli, Rhizobium Azotobacter, Beijerinckia, Clostridium pasteurianum*
Parasitic bacteria	Live in contact with other organisms for obtaining nourishment. Sometimes casue disease.	*Streptococcus, Clostridium, Mycobacterium*

Respiration

Bacteria derive energy *via* following two modes of respiration

1. **Aerobic respiration** (In the presence of oxygen)
   These are of following types
   (i) **Obligate aerobes** These bacteria grow exclusively in the presence of molecular oxygen and fail to survive in its absence, e.g. *Bacillus subtilis.*
   (ii) **Facultative anaerobes** These are aerobic bacteria that can also survive when oxygen is absent, e.g. *Aerobacter, Klebsiella, Pseudomonas.*
   (iii) **Anaerotolerant aerobes** These bacteria continue to perform aerobic respiration even in the complete absence of oxygen, e.g. denitrifying bacteria.
2. **Anaerobic respiration** (In the absence of oxygen)
   These are of following types
   (i) **Obligate anaerobes** These bacteria grow and multiply in the absence of free oxygen. They fail to survive under aerobic conditions, e.g. *Clostridium botulinum.*
   (ii) **Facultative aerobes** The anaerobic bacteria, which can also survive in the presence of oxygen, e.g. *Chlorobium limicola.*
   (iii) **Aerotolerant anaerobes** They perform anaerobic respiration even in the presence of O_2, e.g. lactic acid bacteria.

Reproduction

Most of the bacteria reproduce by a simple form of asexual reproduction and very rarely by sexual reproduction.

Asexual reproduction

It takes place by the following two methods

1. Vegetative reproduction

It takes place by following method

- **Binary fission** It is the most common means of bacterial multiplication under favourable conditions. The complete process involves following steps

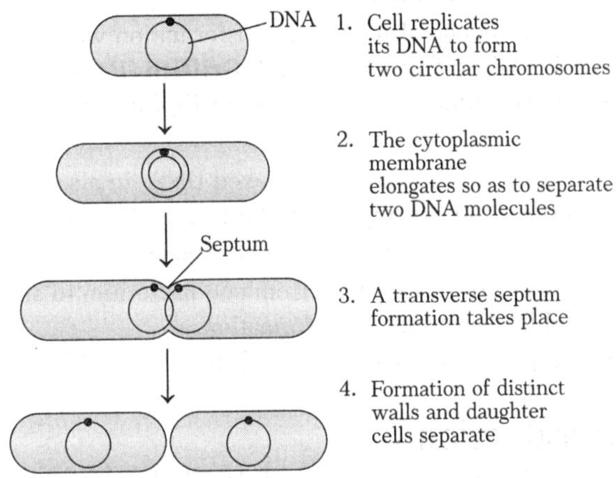

***Fig 4.6** The general process of binary fission in a rod-shaped bacteria*

This mechanism of binary fission was suggested by **Cairns** (1963) and therefore, it is also known as **Cairns' model.**

- **Budding** It is common in a group of bacteria called budding bacteria, i.e. Hyphomicrobiales.
- **Cysts** In some bacteria, the entire protoplast of the cell recedes from the cell wall. It further becomes rounded, thick-walled and resistant. This structure is called as cyst. Under suitable environmental conditions, the cyst germinates to produce a new bacterial cell, e.g. *Azotobacter.*

2. Sporulation

- It is very common and considered as the best mode of perennation.
- These spores are non-motile in nature, i.e. conidia (e.g. *Streptomyces*), oidiospores (e.g. *Actinomyces*), sporangiospores, endospores, etc.
- Out of these, **endospores** are most common. **Endospores** are highly resistant, physiologically dormant, single-celled structures. They are formed usually inside a bacterium mother cell under adverse conditions. Endospore consists of a central core, made up of nuclear material and spore cytoplasm.
- The cytoplasm includes DNA, RNA and proteins.

- The central core is surrounded by a delicate 3-4 layered membrane called spore wall. These are commonly found in *Bacillus*, *Clostridium*, etc.

Sexual reproduction

Bacteria exhibits a primitive mode of sexual reproduction. In this mode, there is no formation of gametes and cell fusion. Thus, the process is better known as genetic recombination.

These prokaryotes show following methods of genetic recombination

1. Transformation

- The first evidence of bacterial transformation was reported by the British scientist **Frederick Griffith** (1920) in a series of experiments with bacterium *Diplococcus pneumoniae.*
- It is the process by which free naked DNA (in a solution) is incorporated into a recipient bacterial cell. This in turn becomes responsible for the genetic change.
- Thus, genes gets transferred from one bacterium to another as naked DNA through transformation.
- It does not involve contact of two cells in any way.

2. Transduction

- It was first reported in *Salmonella* by **Zinder** and **Lederberg** in 1952.
- In transduction, a small piece of DNA is transferred from one cell to another by a bacterial virus, i.e. bacteriophage. Transduction is further of three types as follows
  (i) **Generalised transduction** Bacteriophage can transfer any gene of donor cell, e.g. T_4 bacteriophage.
  (ii) **Specialised or Restricted transduction** Specific segment of donor's DNA is transferred by bacteriophage, e.g. prophages.
  (iii) **Abortive transduction** Donor's DNA do not integrate with recipient's DNA and thus, the former gets lost.

3. Conjugation

- It is the only process of genetic recombination that involves cell-to-cell contact. It was reported in *E. coli* by **Tatum** and **Lederberg** in 1946.
- It occurs between a male or donor cell and female (F^-) or recipient cell.
- The former contains plasmid that possess fertility factor (F^+) and latter do not contain it.
- Fertility factor contains genes for producing sex pili and other characters needed for gene transfer.
- The gene exchange in conjugation can occur by following two methods
  (i) **Sterile male method** Plasmid with F-factor replicates and transfers its one copy to recipient *via* conjugation tube.
  Thus, it converts the donor into recipient cell. This phenomenon of reversibility of sex is called **seduction.**
  (ii) **Fertile male method** Plasmid gets integrated into bacterial chromosome. As a result, the bacterial cell is converted into **HFR** (High Frequency of Recombination) cell or super male or fertile males.
- The orginal complete genome of a bacterium before the addition of new genetic material is called **endogenate.**
- The complete structure, i.e. plasmid + bacterial nucleoid is called **episome.**
- When this episome containing HFR come in contact with a recipient cell, a part or whole of bacterial chromosome is also transferred.
- The percentage of recombination in this method is 1000 times higher as compared to F^+ or sterile male method.
- An outline of complete process is given below

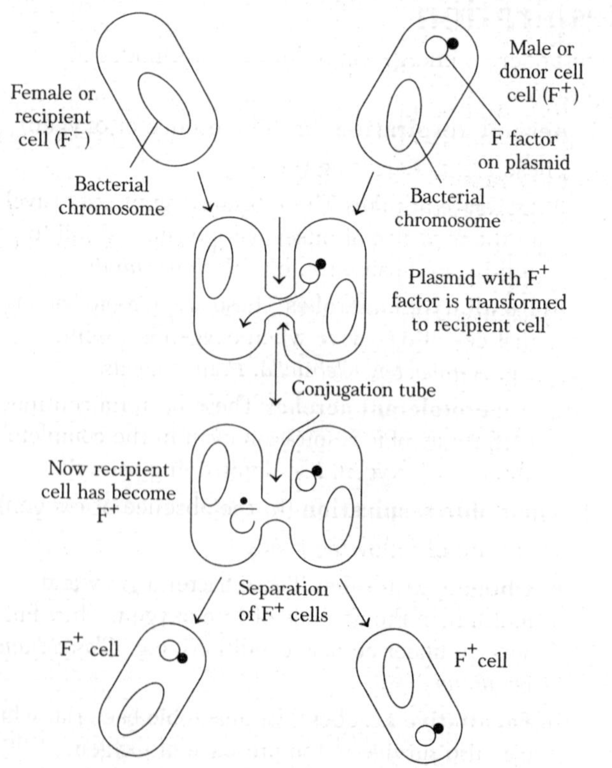

Fig 4.7 Conjugation in bacteria

Economic importance

Bacteria serve as both boon and bane to human kind. Therefore, their activites can be categorised into harmful as well as in useful categories.

Both of these categories are discussed below

Harmful bacterial activities

(i) **Water pollution** Several bacterial forms cause water pollution, by spoiling drinking water, e.g. *Vibrio cholerae, Salmonella typhi.* They convert mercury compounds released by industries into methyl mercury. The famous 'minamata disease' is caused due to methyl mercury poisoning in water resources.

(ii) **Reduction of soil fertility** It is caused due to the conversion of soil nitrates into molecular nitrogen i.e. **denitrification.** It is brought about by bacterial forms like *Bacillus subtilis, Bacillus denitrificans, Thiobacillus denitrificans,* etc.

(iii) **Food spoilage** Bacteria like *Micrococcus, Salmonella, Staphylococcus, Clostridium,* etc., spoil the food stuffs. They can cause **botulism** (*Clostridium botulinum*) and **food poisoning** (*Clostridium perfringens, E. coli, Staphylococcus aurens*) by producing endotoxins and exotoxins. Marine foods such as sea weeds, oysters, crabs and fishes are destroyed by *Spirillum rectiphysetaris.*

(iv) **Decomposers** The species of *Mycobacterium, Pseudomonas* destroy cork, rubber and cords or ropes.

(v) **Bioweapons** Pathogenic bacteria such as those causing anthrax, botulism, cholera, tuberculosis, etc., are very strong bioweapons to be used against the enemy countries.

(vi) **Plant diseases** Bacteria causes several diseases in plants. Few plant diseases are given in the table below

Table 4.3 Some common plant diseases caused by bacteria

Diseases	Causal bacterial species
Angular leaf spot of cotton	*Xanthomonas malvacearum*
Black rot of cabbage	*Xanthomonas campestris*
Blight of paddy	*Xanthomonas oryzae*
Brown rot/wilt of potato	*Pseudomonas solanacearum*
Canker of citrus	*Xanthomonas citri*
Crown gall of apple	*Agrobacterium tumefaciens*

(vii) **Animal diseases** Bacteria can also cause various diseases in human and other animals.

Few of these diseases are given in the table below

Table 4.4 Important human diseases caused by bacteria

Diseases	Pathogens	Epidemiology
Anthrax	*Bacillus anthracis*	Transmitted through contact or ingestion. Rare, except in sporadic outbreaks. May be fatal.
Chlamydia	*Chlamydia trachomatis*	Urogenital infections that can spread to eyes and respiratory tract.
Cholera	*Vibrio cholerae*	Causes severe diarrhoea that can lead to death by dehydration.
Dental caries	*Streptococcus*	Leads to the secretion of acids that destroy minerals in tooth enamel; sugar alone cannot cause caries.
Hansen's disease (leprosy)	*Mycobacterium leprae*	Chronic infection of the skin; spread through contact with infected individuals.
Peptic ulcers	*Helicobacter pylori*	Most peptic ulcers appear to be caused by this bacterium; can be treated with antibiotics.
Plague	*Yersinia pestis*	Killed 1/4 of the population of Europe in the 14th century; endemic in wild rodent population of the Western US today.
Pneumonia	*Streptococcus, Mycoplasma, Chlamydia*	Acute infection of the lungs, often fatal without treatment.
Tuberculosis	*Mycobacterium tuberculosis*	An acute bacterial infection of the lungs, lymph and meninges. Complicated by the development of new strains of the bacteria that are resistant to antibiotics.
Typhoid fever	*Salmonella typhi*	Spread through contaminated water or food (such as improperly washed fruits and vegetables). Vaccines are available for travellers.

Useful bacterial activities

(i) **Antibiotic production** Antibiotics are the drugs, which work against microbial pathogens. About 2100 antibiotics have been reported, so far from Actinomycetes alone. Of these, maximum antibiotics have been reported from *Streptomyces.*

Some antibiotics are given in the table below

Table 4.5 Some important antibiotics and their sources

Antibiotic	Sources	Active against
Gramicidin and tyrocidine	*Bacillus brevis*	Gram positive
Polymixin-B	*B. polymyxa*	Gram negative
Bacitracin, eumycin, bacillomycin, etc.	*B. subtilis*	Gram positive
Terramycin	*S. rimosus*	Gram positive
Chloromycetin	*S. venezuelae*	Gram positive and Gram negative

(ii) **Soil fertility in agriculture** Bacteria are important agents in maintaining soil fertility by nitrogen-fixation (*Clostridium, Rhizobium*) and ammonification (*Bacillus mycoides*).

(iii) **Industrial uses**

- Acetic acid is produced by the bacterium *Acetobacter aceti* from ethanol.
- *Clostridium acetobutylicum* converts acetic acid into acetone and butyric acid to butanol.
- *Streptococcus lactis* converts sugar into lactic acid and milk into curd.
- *Acetobacter* is also used for producing butylene glycol, that is used in the production of synthetic rubber.
- *Escherichia coli* produces the amino acids threonine.
- Flavouring of tobacco and tea leaves is done by *Bacillus megaterium.*
- The decomposing and fermentation capability of bacteria is utilised for removing fats, hairs and other tissues from the hides in leather industry. The clean hides are then tanned to prepare leather.
- Vitamin-B_{12} is obtained as a byproduct of *Streptomyces* antibiotic fermentation. It can also be produced commercially by direct fermentation using *Propionibacterium shermanii* or *Pseudomonas denitrificans.* Riboflavin is a byproduct of acetone-butanol fermentation.
- Retting of flex, jute and hemp is done by the bacterial forms like *Clostridium butyricum* and *C. felsineum.*
- The milk contains bacterial forms-like *Streptococcus lactis, Escherichia coli, Lactobacillus lactis, Clostridium* sp., *Pseudomonas* sp., etc. Yogurt is prepared by *Lactobacillus sanfrancisco, L. bulgaricus* and *Streptococcus thermophilus.*

(iv) **Environment protection**

- Bacteria are capable of decomposing some hazardous pesticides, e.g. 2, 4-D by *Achromobacter, Flavobacterium,* Dalapon (2, 2-dichloropropionic acid) by *Agrobacterium* and DDT partially by *Acetobacter aerogenes,* etc.
- Biological control of some insects is done by bacteria, e.g. *Bacillus papillae* kills Japanese beetle, *B. sphaericus* kills malarial mosquitoes.
- The disposal of sewage is also done by the microbes including bacteria, e.g. *E. coli, Pseudomonas.*
- *Methanobacillus* produces methane under anaerobic condition by the fermentation of organic matter. It is used as the source of power in the form of biogas.

CHECK POINT 4.2

1. Bacterial photosynthesis takes place without the involvement of O_2. Which source of electron in bacterial photosynthesis is not utilised?
(a) Hydrogen sulphide
(b) Non-sulphur aliphatic compound
(c) Thiosulphate
(d) Organic sulphur compounds

2. Choose the incorrectly matched pair.
(a) *Aerobacter* – Facultative aerobe
(b) *Clostridium* – Parasitic bacteria
(c) *Chlorobium* – Bacterioviridin pigment
(d) *Beggiatoa* – Sulphur bacteria

3. Which of the following is not present in bacteria?
(a) Meiosis
(b) Sexual reproduction
(c) Asexual reproduction
(d) Parasexual method of gene recombination

4. During conjugation process in bacteria, sex pilus is not formed by
(a) donor bacterium (b) F^+ bacterium
(c) F^- bacterium (d) Both (a) and (c)

5. Choose the odd one out with respect to economic importance of given bacteria.
(a) *Salmonella* (b) *Listeria* (c) *Brucella* (d) *Pseudomonas*

OTHER MONERANS

The group eubacteria include several other monerans like Cyanobacteria, Mycoplasma, Actinomycetes, Rickettsiae, Chlamydiae, Spirochaetes, etc. In detail, Mycoplasma and Cyanobacteria are discussed below

Mycoplasma (PPLO)

- These are the simplest and the smallest free-living prokaryotes.
- Mycoplasma were first isolated by **E Nocard** and **ER Roux** in 1898 from pleural fluid of cattles suffering from bovine pleuropneumonia. Hence, they are also called **Pleuropneumonia Like Organisms** (PPLO).
- Later, the name mycoplasma was given by **Nowak** in 1929 (about 0.1 μm).
- PPLO are Gram negative bacteria. They are larger than the largest viruses, but smaller than smallest bacteria. The smallest mycoplasma is *Mycoplasma gallisepticum.*

General characteristics

- They are self-replicating forms that are ubiquitous in environment.
- They are heterotrophic. Some are saprophytic but mostly parasitic, e.g. *M. gallisepticum.* They can survive without oxygen.
- They can change their shape (pleomorphic) as cell wall is absent in them. Hence, they are also called as the **'jokers of microbiology'**.
- Mycoplasma are sensitive to those antibiotics which affect metabolic pathways, e.g. tetracyclin, streptomycin and erythromysin. However, they are insensitive to penicillin as it acts on the cell wall.
- They can easily pass through bacterial filters.
- Mycoplasma are resistant to osmotic lysis and can withstand conditions under which a protoplast may lyse due to the presence of sterols.
- They infect animals like dog, sheep, mice and human. *Mycoplasma bovis* can cause considerable economic losses in cattle production.
- They cause pleuropneumonia, rheumatism, inflammation in genitals, etc., in humans. In plants, they cause clover draft, potato witch's broom, maize stunt, etc.

Structure

- The plasma membrane is three-layered. It is made up of lipoprotein.
- The genetic material is a single, linear, double-stranded molecule of DNA, without a nuclear envelope.

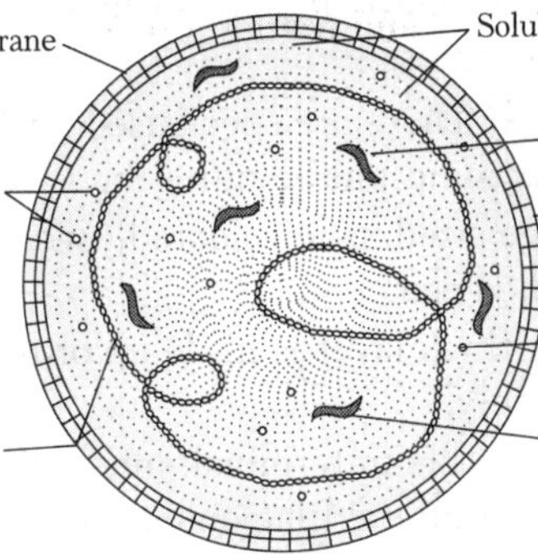

***Fig 4.8** Structure of a typical mycoplasma*

Reproduction

- They reproduce by budding, fragmentation and binary fission.
- They are generally non-motile, but motility is provided by gliding movement.

Cyanobacteria

- These are Gram negative prokaryotes.
- The members of this group are popularly known as blue-green algae, myxophyceae or cyanophyceae.
- They perform oxygenic photosynthesis.

General characteristics

- These are predominantly freshwater forms, but a few are marine as well.
- The protoplast lacks membrane bound organelles. It contains 70S ribosome.
- Photosynthetic pigments include chlorophyll-*a*, carotenoids, C-phycocyanin (blue coloured) and C-phycoerythrin (red coloured). These pigments together impart characteristic blue-green colour to the cells.
- Food is stored in the form of cyanophycean starch or α-granules, β-granule (fat droplets), volutin body (reserve phosphate), polyhedral body (RuBisCO rich).
- Flagella are completely absent, movement occurs by gliding motion.
- They reproduce asexually by fission, heterocysts, hormogonia, akinetes and fragmentation. Sexual reproduction is totally absent.

Structure

(i) **Outer layers** They possess four-layered cell wall. The cell membrane contain protein and phospholipid in 2 : 1 ratio. Sterols are not found in it.

(ii) **Trichomes** Cyanobacteria can be unicellular, colonial or filamentous. Each filament consists of a sheath of mucilage and one or more cellular strands called trichomes.

(iii) **Heterocyst** These are specialised structures that help to fix atmospheric nitrogen.

- It is a large sized, pale coloured, two layered, thick-walled cell.
- The outer thick layer is made up of pectin and cellulose only.
- It usually occurs in terminal, intercallary or lateral position, within the filamentous cyanobacteria.
- Heterocysts possess enzyme nitrogenase that perform N_2-fixation, e.g. *Nostoc, Anabaena, Aulosira,* etc.
- Some filamentous forms are without heterocyst, e.g. *Oscillatoria*, etc.

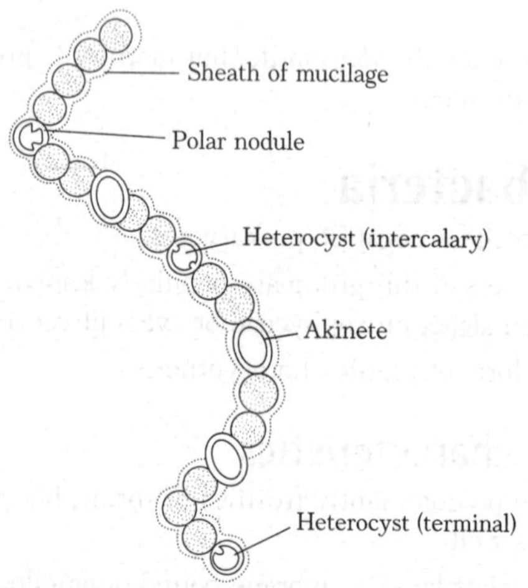

Fig 4.9 Structure of Nostoc showing heterocyst

Economic importance

- *Oscillatoria* is highly pollution resistant and grows in freshwater with high organic matter. Thus, it acts as indicator of water pollution.
- *Spirullina* is cultivated in tanks to obtain protein rich animal food, thus acts as food supplement.
- *Nostoc commune* is used as food called Yuyuho in Japan and China.
- Few species of *Anabaena* and *Aulosira* are inoculated in ponds to check the development of mosquito larvae.
- Cyanobacteria are one of the early colonisers of bare and baren areas thus, help in succession.
- They produce water blooms, imparting bad odour and colour to waterbodies.
- They grow, photosynthesise, multiply and ultimately die, resulting in addition of organic matter to the soil. Thus, they increases soil fertility.
- Red colour of red sea is due to cyanobacteria *Trichodesmium erythraeum.* It is produced due to the complementary chromatic adaptation or Gaidukov phenomenon.

Actinomycetes

- These are Gram positive bacteria.
- They form branching filaments often mistaken for fungi, e.g. *Streptomyces* and *Actinomyces.*
- They are one of the most common types of soil bacteria that act as decomposers.
- They reproduce asexually by means of conidia.
- These are used in the production of many commonly used antibiotics, including streptomycin and tetracyclin.
- They also cause dental plaque in humans.

Spirochaetes

- These are chemoheterotrophic bacteria.
- These are free inhabitants of mud and water and are spiral or helicoid in shape.
- They remain covered by flexible cell wall.
- They swim actively with flagella present at both the poles or ends. Many diseases are caused by them such as, *Treponema pallidum* causes syphilis, *Leptospira* causes infectious jaundice and *Borrelia* causes relapsing fever.

Rickettsia

- These are small, Gram negative intracellular parasites.
- They were first reported by **HT Rickets** in 1909.
- These are considered as intermediate between true bacteria and viruses.
- Their life cycle involves both mammals and arthropods (such as fleas and ticks) as their host.
- Rickettsia are responsible for many fatal human diseases including typhus (*Rickettsia prowazekii*).
- Rickettsia are known to cause Rocky mountain **spotted fever.**

Enterobacteria

- These include *Escherichia coli, Salmonella* and *Vibrio.*
- They are Gram negative, rod-shaped bacteria.
- They usually do not form spores.
- They are aerobic heterotrophs.
- They cause many important diseases, including **bubonic plague** and **cholera.**

ARCHAEBACTERIA

- It is a relatively small and primitive prokaryotic group.
- They are the oldest unicellular organisms and belong to the kingdom—Archaea.
- These are considered to be oldest living fossil.

General characteristics

- They are obligate anaerobes.
- They are found in harsh habitats such as extreme salty areas (**halophiles**), hot springs (**thermophiles**) and marshy areas (**methanogens**).
- They have lipids in their cell membranes that contain branched hydrocarbon chains, i.e. phytanyl side groups connected to glycerol by other linkages.
- Cell wall altogether lacks murmaric acid and peptidoglycan.
- Their genetic material floats freely in the cytoplasm as they do not contain nuclei.
- They do not contain membrane-bound organelles.
- Unlike eukaryotes, their translation process is sensitive to diphtheria toxin and not to chloramphenicol.

Table 4.6 Three categories of Archaebacteria

Methanogens	Halophiles	Thermoacidophiles
They are mostly Gram negative and some are Gram positive.	Most of them are Gram negative.	They all are Gram negative.
They are strict anaerobes.	They are mainly aerobes.	They are mainly aerobes.
They occur in marshy areas where they convert formic acid and CO_2 into methane with the help of hydrogen.	They occur in salt rich substrata like salt pans, Great Salt Lake, dead sea and salt marshes.	They occur in hot sulphur springs where the temperature may be 80°C or more and pH as low as 2.
Some of them live as symbionts in the rumen of the ruminant's stomach and help in fermentation of cellulose.	They can trap solar energy but do not utilise the same in the manufacturing of organic food due to the presence of a purple-pigmented (bacteriorhodopsin) membrane.	Under aerobic conditions, they oxidise sulphur to sulphuric acid and when the conditions are anaerobic. They reduce sulphur to H_2S.
e.g. *Methanothrix*, *Methanogenium* and *Methanoplanus*, *Methanococcus methanobacterium*.	e.g. *Halococcus*, *Haloferax* and *Haloarchaea*, *Halobacterium*.	e.g. *Sulfolobus*, *Thermoproteus* and *Desulphurococcus*, *Thermoplasma*.

CHECK POINT 4.3

1. Choose the odd one out with respect to cyanobacteria.
(a) They lack flagella throughout the life cycle
(b) They are biflagellated or multiflagellated prokaryotes
(c) Cell wall contain peptidoglycan and lipopolysaccharide
(d) Cell wall is typically four layered

2. *Nostoc* is known to perform
(a) Only photosynthesis
(b) Photosynthesis and nitrogen fixation simultaneously
(c) Only nitrogen fixation
(d) Either photosynthesis or nitrogen fixation at a time

3. All the given monerans are Gram negative, except
(a) *Aulosira* (b) *Salmonella*
(c) *Streptomyces* (d) *Rickettsia*

4. A few organisms are known to grow and multiply at temperatures of 100-105°C. They belong to
(a) Thermophilic subaerial fungi
(b) Marine archaebacteria
(c) Thermophilic sulphur bacteria
(d) Hot spring blue-green algae

5. Choose the correct statement with respect to primitive bacteria.
(a) Ester lipids found eukaryotic cell membrane is more stable than ether lipids found in the cell membrane of archaebacteria
(b) Cell membrane possess branched lipid in the bacteria
(c) Pseudomurein is absent in the cell wall of Archaebacteria
(d) Methanogens are obligate aerobes

EXAM POINTS

- Eubacteria are true bacteria, as they have rigid cell walls. **[NEET-I 2016]**
- Staphylococci appears like grapes under microscope. **[WB JEE 2015]**
- Glycocalyx gives the sticky characteristic to the bacterial cell. **[NEET 2017]**
- Pili or fimbriae are used by bacteria for attachment. **[NEET-I 2016]**
- All monerans have DNA and RNA, DNA is not enclosed in a nuclear membrane. **[JIPMER 2017]**
- In *Anabaena* true nucleus are absent. **[CBSE AIPMT 2015]**
- 23 *sr*RNA acts as structural RNA as well as ribozyme in bacteria. **[NEET-I 2016]**
- Muramic acid is the cell wall material present only in bacteria and blue-green algae. **[AMU 2015]**
- Mitochondria and chloroplast are bacterial endosymbiont. **[AIIMS 2017]**
- Green sulphur bacteria are anaerobic bacteria. **[NEET 2018]**
- Transduction is a process of transfer of DNA by bacteriophage virus in bacteria. **[AIIMS 2017]**
- Cyanobacteria perform oxygenic photosynthesis because they have the ability of nitrogen fixation due to the presence of heterocyst. **[NEET 2020]**
- *Nostoc* contain heterocysts which are the specialised nitrogen-fixing cells formed during nitrogen starvation by some filamentous cyanobacterium. **[JIPMER 2019]**
- Heterocysts create a microanaerobic environment to facilitate nitrogen fixation. **[AIIMS 2017]**
- Inclusion bodies of blue-green, purple and green photosynthetic bacteria are gas-vacuole which are the inclusion bodies in many aquatic prokaryotes. **[NEET 2020]**
- Archaebacteria live in extreme diverse habitats, e.g. extreme hot temperature, saline condition, variable pH, etc. **[NEET 2017]**
- Halophiles live in salty areas, thermoacidophiles live in hot springs, methanogens live in guts of ruminants. **[NEET 2020]**
- Methanogens grow anaerobically on cellulosic material and produce large amount of methane. **[NEET 2019]**
- Methanogens are involved in methanogenesis, producing methane gas in rumen of cattles. **[NEET-I 2016]**
- Mycoplasma is smallest and wall less living organism. **[NEET 2019]**
- Mycoplasmas are called joker of microbiology. **[CBSE AIPMT 2015]**

Chapter Exercises

(A) Topical Exercise

General Characteristics of Kingdom–Monera

1 Bacteria are
(a) sole members of kingdom—Monera
(b) without glycocalyx layer
(c) with either DNA or RNA as genetic material
(d) eukaryotic in cell nature

2 Consider the following statements.
I. Bacteria are unicellular, cell wall bound, spherical or rod-shaped microorganisms.
II. Bacteria is usually smaller than 5 micron in size and lack an organised nucleus.

Choose the correct option.
(a) Statement I is true, but II is false
(b) Statement I is false, but II is true
(c) Both statements I and II are true
(d) Both statements I and II are false

3 Bacteria are regarded as primitive organisms because they
(a) possess incipient nucleus
(b) are small, microscopic plants, which are not visible to the naked eyes
(c) cause serious diseases to human beings, domesticated animals and crop plants
(d) produce endospores, which are resistant to adverse conditions

4 Choose the incorrect statement with respect to kingdom-Monera.
(a) Membrane bound organelles are absent
(b) 70S type ribosomes are present
(c) Cell walls are cellulosic in nature
(d) None of the above

5 Choose the odd one out.
(a) Actinomycetes (b) Rickettsia
(c) Spirochaetes (d) Halophiles

Eubacteria

6 Choose the incorrectly matched pair.
(a) Father of Bacteriology - Leeuwenhoek
(b) Smaller bacterium - *Beggiatoa*
(c) Cocci bacteria - Lack flagella
(d) All of the above

7 Sarcina is a type of
(a) Spiral bacteria (b) Vibrio
(c) Cocci (d) Bacilli

8 Study the figures given below.

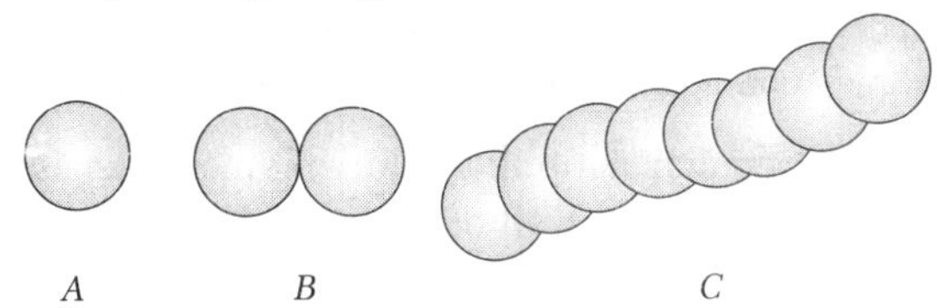

Choose the correct option.

	A	B	C
(a)	*Monococcus*	*Diplococcus*	*Streptococcus*
(b)	*Bacillus*	*Diplobacillus*	*Streptobacillus*
(c)	*Micrococcus*	*Bacillus*	*Streptobacillus*
(d)	*Streptococcus*	*Diplococcus*	*Bacillus*

9 Bacteria are grouped under four categories based on their shape. Study the given figures and identify *A*, *B*, *C* and *D*.

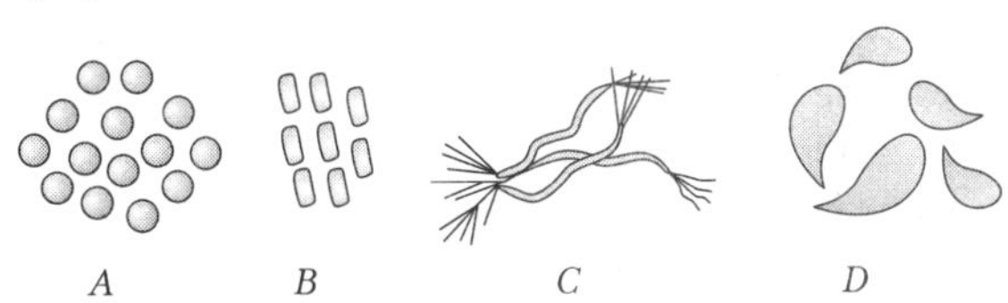

	A	*B*	*C*	*D*
(a)	Vibrio	Cocci	Bacilli	Spirilla
(b)	Cocci	Bacilli	Spirilla	Vibrio
(c)	Bacilli	Spirilla	Vibrio	Cocci
(d)	Spirilla	Vibrio	Cocci	Bacilli

10 Which of the following is an atrichous, rod-like bacteria, found arranged in pairs?
(a) *Bacillus subtilis*
(b) *Pneumococcus pneumoniae*
(c) *Vibrio cholerae*
(d) *Streptococcus nigricans*

11 The layer outside the bacterial cell wall is composed of
(a) chitinous capsule, mucopolysaccharide
(b) slimy capsule, mucopolysaccharide
(c) suberised capsule, cellulose
(d) lignified capsule, cellulose

12 The microtubular organisation in cilia or flagella of moneran is
(a) 9 + 2
(b) 9 + 3
(c) 9 + 0
(d) 9 + 1

13 Many bacteria bear minute hairy structures on their cell wall, these are called
(a) hair (b) flagella
(c) pili (d) cilia

14 Binding of magnesium ions by teichoic acids in bacterial cell wall protects bacteria from
(a) drying injuries (b) thermal injuries
(c) radiation injuries (d) physical injuries

15 Which peculiar naturally occurring monomer is present in the cell wall of bacteria?
(a) Cadipicoliniate (b) Diaminopimelic acid
(c) Glutamate (d) None of these

16 The main function of bacterial cell well is to
(a) protect the internal structures of cell
(b) provide shape to the cell
(c) prevent cell from bursting
(d) control the movement of substances into and out of the cell

17 Consider the following statements.
I. Gram staining is a technique developed by Dutch scientist Gram.
II. In Gram staining of bacteria, iodine is used as mordant.

Choose the correct option.
(a) Statement I is true, but II is false
(b) Statement I is false, but II is true
(c) Both statements I and II are true
(d) Both statements I and II are false

18 Consider the following statements.
I. Pentaglycine chains are found in the cell walls of Gram positive bacteria.
II. Gram negative bacteria appears colourless on staining with safranin.

Choose the correct option.
(a) Statement I is true, but II is false
(b) Statement I is false, but II is true
(c) Both statements I and II are true
(d) Both statements I and II are false

19 In Gram positive bacteria, the peptidoglycan layer constitutes
(a) 10% of the cell wall (b) 30% of the cell wall
(c) 70% of the cell wall (d) 90% of the cell wall

20 In Gram negative bacterial cell wall, the peptidoglycan chains are laterally linked by short chains of four amino acids which are constituted of
(a) L-Alanine, D-Glutamic acid, L-Lysine, D-Alanine
(b) L-Alanine, D-Glutamic acid, diaminopimelic acid, D-Alanine
(c) L-Alanine; L-Lysine; D-Alanine, diaminopimelic acid
(d) L-Alanine; D-Glutamic; L-Lysine, diaminopimelic acid

21 Bacterial cell contains
I. 70S ribosomes II. 80S ribosomes
III. plasmids IV. mitochondria

Codes
(a) I, II and III (b) II and IV
(c) I and IV (d) I and III

22 An irregularly shaped region within the cell of a prokaryote that contains most of the genetic material consists of
(a) one *ss*DNA (b) one *ds*DNA
(c) two *ds*DNA (d) many *ds*DNA

23 Plasmids
(a) replicate independently of the main chromosome
(b) often contain antibiotic resistance genes
(c) are transferred from one bacterium to another by conjugation
(d) All of the above

24 The fertility or F^+ factor that plays a role in conjugation is a
(a) Ti-plasmid
(b) F-plasmid
(c) col-plasmid
(d) R-plasmid

25 Folds of plasma membrane in bacterial cells where oxidative enzymes are also found are known as
(a) episomes (b) mesosomes
(c) spherosomes (d) acrosomes

26 Which of the following is not inclusion or reserve deposits in bacteria?
(a) Metachromatic granules
(b) Carboxysomes
(c) Lipid inclusions
(d) Protein granules

Nutrition and Respiration

27 Which statement is correct?
(a) All bacteria are autotrophic
(b) All bacteria are heterotrophic
(c) All bacteria are photosynthetic
(d) Mostly bacteria are heterotrophic, but some are autotrophic

28 The chemosynthetic bacteria are
I. *Nitrosomonas* II. *Nitrobacter*
III. *Rhodospirillum* IV. *Chlorobium*

Codes
(a) I, II and III (b) I and II
(c) II and IV (d) I and III

29 Bacterio purpurine pigment is found in
(a) *Thiospirillum* (b) *Chlorobium*
(c) *Beggiatoa* (d) *Ferrobacillus*

30 Chemosynthetic bacteria do not need sunlight to grow because
(a) they prepare their food without the help of light
(b) they do not like sunlight brightness
(c) due to the absence of chlorophyll, they are incapable of manufacturing their own food
(d) they use other kinds of light for manufacturing their own food

31 In an experiment, a microbiologist put equal number of each of the following organisms into a flask of sterile broth, consisting mostly of sugar and a few amino acids. He, then placed the flask in the dark. Which of the organism would be most likely to survive?
(a) Chemoheterotrophic bacteria
(b) Cyanobacteria
(c) Photoheterotrophs
(d) Photoautotrophs

32 Choose the incorrectly matched pairs.
(a) Saprophytic bacteria - *Pseudomonas*
(b) Aerotolerant anaerobe - Denitrifying bacteria
(c) Parasitic bacteria - *Mycobacterium*
(d) Peritrichous form - *E. coli*

Reproduction

33 Which of the following is not correct?
(a) Heterotrophic bacteria are most abundant in nature
(b) Chemosynthetic bacteria oxidise various inorganic substances such as nitrates, nitrites and ammonia and use the released energy for their ATP production
(c) Cyanobacteria have chlorophyll-*a* like green plants
(d) Bacteria reproduce only by fission

34 In unfavourable adverse conditions, bacteria produce resting spores called
(a) exospores (b) chlamydospores
(c) microspores (d) endospores

35 Select the incorrect pair(s).
I. Endospore-Actinomycetes
II. Cyst formation-*Azotobacter*
III. Conidia-*Clostridium tetani*
Choose the correct option.
(a) I and II (b) II and III
(c) I and III (d) All of these

36 Sexual reproduction in bacteria occurs by
I. transduction II. conjugation
III. sporulation IV. fission
Codes
(a) I, II and III (b) I and II
(c) II and IV (d) I and III

37 A process by which free naked DNA is incorporated into a recipient bacterial cell was seen in
(a) *Diplococcus pneumoniae*
(b) *Streptococcus*
(c) *E. coli*
(d) *Salmonella typhimurium*

38 When a bacteriophage, in its lytic phase, carries some of the bacterium's partially digested chromosome with it to another host cell, the process is called
(a) transformation
(b) generalised transduction
(c) restricted transduction
(d) conjugation

39 Bacteriophage helps in transferring a small piece of DNA from one cell to another. This process in bacteria was discovered by
(a) Zinder and Lederberg
(b) Wallace and Jacob
(c) Herelle and Twort
(d) Lederberg and Tatum

Economic Importance

40 Match the following columns.

Column I	Column II
A. *Nitrobacter*	1. Free-living aerobic nitrogen-fixer
B. *Clostridium*	2. Converts ammonia into nitrite
C. *Nitrosococcus*	3. Changes nitrite to nitrate
D. *Azotobacter*	4. Free-living anaerobic nitrogen- fixer

Codes

	A	B	C	D		A	B	C	D
(a)	2	4	3	1	(b)	3	4	2	1
(c)	4	2	1	3	(d)	1	2	3	4

41 The bacteria that cause tetanus can be killed only by prolonged heating at temperature considerably above boiling. Tetanus is caused by
(a) *Agrobacterium tumefaciens*
(b) *Vibrio cholerae*
(c) *E. coli*
(d) *Clostridium tetani*

42 A person had street food, he is suffering from diarrhoea, dizziness, indigestion, nausea, vomiting. It is caused by the infection of
(a) *Bacillus megaterium*
(b) *Salmonella typhi*
(c) *E. coli*
(d) *Clostridium botulinum*

43 Bacterium, which is concerned with whooping cough is
(a) *Bordetella pertussis* (b) *Bacillus*
(c) *Diplococcus* (d) *Mycobacterium tuberculum*

44 Botulism is a
(a) type of food poisoning due to saprophytic bacterium
(b) type of food poisoning due to spore forming aerobic bacterium
(c) type of food poisoning due to uncooked food
(d) All of the above

45 In the following table identify the correct matching of the crop, its disease and the corresponding pathogen.

	Crops	Diseases	Pathogens
(a)	Citrus	Canker	*Pseudomonas rubrilineans*
(b)	Potato	Late blight	*Fusarium udum*
(c)	Brinjal	Root-knot	*Meloidogyne incognita*
(d)	Pigeon pea	Seed gall	*Phytophthora infestans*

46 Which of the following is a bacterial plant disease?
(a) Tikka disease of groundnut
(b) Downy mildew of grapes
(c) Ring rot of potato
(d) Red rot of sugarcane

47 Vinegar is prepared from fermented sugar solution by the activities of
(a) *Acetobacter aceti* (b) *Bacillus aceti*
(c) *B. subtilis* (d) *Diplococcus*

48 Tetanolysin is produced by
(a) *Mycobacterium leprae* (b) *Clostridium botulinum*
(c) *Clostridium tetani* (d) None of these

49 *Streptococcus lactis* is responsible for the
(a) conversion of molasses into alcohol
(b) conversion of milk into curd
(c) tanning of leather
(d) flavouring the leaves of tea and tobacco

50 *Clostridium acetobutylicum* has been used in the synthesis of
(a) vitamin-B (b) vitamin-A
(c) vitamin-C (d) vitamin-D

51 Match the following columns.

Column I	Column II
A. Actinomycin-D	1. Inhibits *m*RNA synthesis
B. Tetracycline	2. Inhibits 50S subunits of ribosomes
C. Streptomycin	3. Inhibits binding of *t*RNA with ribosomes
D. Chloramphenicol	4. Alters 30S subunits of ribosomes

Codes

	A	B	C	D
(a)	1	2	3	4
(b)	3	4	2	1
(c)	1	3	2	4
(d)	1	3	4	2

Other Monerans and Archaebacteria

52 Which one of the following statements about mycoplasma is incorrect?
(a) They are also called PPLO
(b) They are pleomorphic
(c) They are sensitive to penicillin
(d) They cause diseases in plants

53 What is incorrect for mycoplasma?
(a) They are osmotically inactive
(b) They show absence of cell wall
(c) They are sensitive to modern antibiotics
(d) They are obligate intracellular parasites

54 Which one is the smallest organism capable of autonomous growth and reproduction?
(a) Virus (b) Viroid
(c) Mycoplasma (PPLO) (d) None of these

55 Which of the following is effective against mycoplasmas diseases?
(a) Vancomycin (b) Penicillin
(c) Chloramphenicol (d) All of these

56 Which of the species of mycoplasma causes human sterility?
(a) *M. hominis* (b) *M. fermentans*
(c) Both (a) and (b) (d) None of these

57 Cyanobacteria are
(a) mosses which attack bacteria
(b) bacteria which attack cyanophyceae
(c) autotrophic organism with phycocyanin
(d) None of the above

58 Which of the following important features are found in blue-green algae?
(a) Abundant secretion of pectin
(b) Presence of phycocyanin-*c* as dominant pigment
(c) No plastids
(d) All of the above

59 Pigment phycocyanin and phycoerythrin are found in
(a) Bacillariophyceae
(b) Archaebacteria
(c) Eubacteria
(d) Cyanobacteria

60 What is the role of heterocysts in a cyanobacterial filament?
(a) They carryout photosynthesis and nitrogen-fixation
(b) They carryout nitrogen fixation
(c) They carryout photosynthesis
(d) They oxidise inorganic substances to obtain energy

61 The similarity between bacterium and cyanobacterium is in the
(a) presence of nucleoid
(b) presence of flagella
(c) presence of 80S ribosomes
(d) None of the above

62 Which of the following is not correct?
(a) Cyanobacteria are photosynthetic autotrophs
(b) Both photosynthesis and nitrogen fixation take place in cyanobacteria
(c) Heterocysts are found in archaebacteria
(d) Mesosomes are found in eubacteria

63 Which of the following has polar nodule on both the ends?
(a) Akinetes (b) Hormogonia
(c) Heterocysts (d) None of these

64 Consider the following statements.
I. Cyanobacteria has flagellated cells.
II. Mycoplasma is a wall less microorganism.
Choose the correct option.
(a) Statement I is true, but II is false
(b) Statement I is false, but II is true
(c) Both statements I and II are true
(d) Both statements I and II are false

65 Common mode of multiplication in cyanobacteria and the material present in the cell walls of blue-green algae and bacteria is
(a) heterocyst and muramic acid
(b) exospore and cellulose
(c) hormogonia and muramic acid
(d) trichomes and cellulose

66 Which one is true about domain Archaea?
(a) They differ from both prokaryotes and eukaryotes
(b) They completely differ from prokaryotes
(c) They resemble eukarya in all aspects
(d) They have some novel features absent in other prokaryotes and eukaryotes

67 Which of the following is not a characteristic of Archaebacteria?
(a) The ability to produce methane from carbon dioxide
(b) The absence of a nuclear envelope
(c) The presence of a circular chromosome
(d) The presence of peptidoglycan

68 Which of the following bacteria cannot thrive at relatively high temperatures, between 41 and 122°C.
(a) *Sulfolobus*
(b) *Haloferax*
(c) *Thermoproteus*
(d) *Desulfurococcus*

69 Those organism which are extremophiles that thrive in high salt concentration can comfortably live in
(a) Dead sea (b) Dal lake
(c) Arabian sea (d) Godavari

70 Which of the following is a difference between bacteria and Archaea?
(a) Archaea are unicellular and bacteria are colonial
(b) The genes of Archaea have introns, whereas bacteria lack introns
(c) They have different chemicals in their cell wall
(d) Bacteria are autotrophic and Archaea are heterotrophic

(B) Cumulative Exercise

(Assorted Questions of the Chapter for Advanced Level Practice)

1 Bacteria cannot survive in a highly salted pickle because
(a) salts inhibit reproduction
(b) bacteria do not get enough light for photosynthesis
(c) they become plasmolysed and consequently killed
(d) the pickle does not contain nutrients necessary for bacteria to live

2 Consider the following statements.
I. Plasmid is a small circular piece of DNA, located in the cytoplasm of many bacteria.
II. Distilled water is usually free from bacteria.
III. Red colour of red sea is due to the presence of a cyanobacterium *Trichodesmium erythraeum.*
IV. Bacteria are eukaryotes.
Choose the option containing correct statements.
(a) I, II and III
(b) II and III
(c) I and II
(d) I and III

3 Match the following columns.

Column I	Column II
A. Actinomycetes	1. *Streptomyces*
B. Enterobacteria	2. *Salmonella*
C. Spirochaetes	3. *Treponema*
D. Chemoautotrophs	4. *Nitrosomonas*

Codes

	A	B	C	D		A	B	C	D
(a)	4	3	2	1	(b)	1	2	3	4
(c)	3	4	1	2	(d)	2	1	4	3

4 Consider the following statements.

I. Amphitrichous (single flagellum at each end of the cell) is found in *Nitrosomonas.*
II. Lophotrichous (two or more flagella at one end of the cell) is found in *Pseudomonas.*
III. Peritrichous (flagella distributed over the entire cell) is found in *E. coli.*
IV. Monotrichous (single end flagellum) is found in *Clostridium.*

Choose the option containing correct statements.

(a) I, II and III (b) I and II
(c) II and IV (d) I and III

5 According to the mode of respiration, bacteria can be aerobic or anaerobic. Each of them is further of two types, obligate and facultative.

Regarding above information match the following columns.

Column I	Column II	Column III
A. Obligate aerobes	1. Respire anaerobically under normal condition, but can respire aerobically when oxygen is available	(i) *Bacillus subtilis*
B. Facultative anaerobes	2. Respire only anaerobically	(ii) Halophiles
C. Obligate anaerobes	3. Generally respire aerobically, but switch over to anaerobic mode of respiration if oxygen becomes deficient	(iii) *Clostridium botulinum*
D. Facultative aerobes	4. Respire only aerobically	(iv) Rhodopseudomonas

Codes

	A	B	C	D
(a)	4-(i),	3-(iii),	1-(iv),	3-(ii)
(b)	4-(i),	3-(ii),	2-(iii),	1-(iv)
(c)	1-(ii),	2-(i),	3-(iv),	4-(iii)
(d)	4-(iii),	3-(iv),	2-(i),	1-(ii)

6 Consider the following statements.

I. *Staphylococcus* occurs in cluster-like a bunch of grapes.
II. Some species of *Streptococcus* infect the lower respiratory tract.
III. *Staphylococcus aureus* lives in the sinuses.
IV. Different strains of *Staphylococcus aureus* cause boils fururculosis, pneumonia, food poisoning and other diseases.

Which of the following statements are correct?

(a) I, II and IV (b) II and III
(c) II, III and IV (d) I and IV

7 Read the following statements and choose the correct statements from the options given below.

I. Archaebacteria are of three types methanogens, halophiles and thermoacidophiles.
II. About 850 species of bacteria are found in India.
III. Some photosynthetic bacteria may be up to 3 mm long.

Codes

(a) I and III (b) I and II
(c) II and III (d) I, II and III

8 Match the following columns.

Column I	Column II
A. Archaebacteria	1. One or two per cell
B. Mesosomes	2. *E. coli*
C. Pili	3. Discovered by Carl Woese
D. Bacilli bacteria	4. Obtain carbon from organic compounds
E. Chemosynthetic bacteria	5. Respiratory enzymes

Codes

	A	B	C	D	E
(a)	1	2	3	4	5
(b)	3	5	1	2	4
(c)	2	1	3	4	5
(d)	5	3	2	1	4

9 Consider the following statements about Gram staining.

I. All bacteria whether Gram positive (+) or Gram negative (–) will stain bluish-purple when stained with crystal violet.
II. Gram negative bacteria appears pink on staining with safranin.
III. Iodine in Gram staining acts as a mordant.
IV. Alcohol treatment is required after Crystal Violet Iodine (CVI) complex.

Choose the correct option.

(a) I and II (b) II and IV
(c) II, III and IV (d) I, II, III and IV

10 Consider the following statements and mark them as True (T)/False (F).

I. Transformation requires contact of two bacterial cells.
II. Cell-to-cell contact is required in conjugation.
III. Transduction is carried out by bacteriophages.
IV. High frequency recombinant is a super or fertile male.

Codes

	I	II	III	IV		I	II	III	IV
(a)	T	F	T	F	(b)	F	T	T	T
(c)	F	T	T	F	(d)	T	F	F	T

11 Consider the following statements about mycoplasma and mark as true/false.

I. They lack cell wall and pili.
II. They contain circular *ss*DNA as genetic material.
III. They lack histones.
IV. They cause maize stunt in plants.

Choose the correct option.

	I	II	II	IV		I	II	II	IV
(a)	F	T	T	T	(b)	T	T	T	F
(c)	T	F	T	T	(d)	F	F	F	T

12 Match the following columns.

Column I	Column II
A. Archaebacteria	1. Do not possess photosynthetic pigments
B. Mesosomes	2. Atrichous
C. Pili	3. More prominent in Gram negative bacteria
D. Cocci bacteria	4. One or two per cell
E. Chemosynthetic bacteria	5. Lack peptidoglycan in cell wall

Codes

	A	B	C	D	E		A	B	C	D	E
(a)	5	3	4	2	1	(b)	1	2	3	4	5
(c)	4	2	1	3	5	(d)	5	4	3	2	1

13 Select the correct combination of statements (I-IV) regarding the characteristics of certain organisms.

I. Methanogens are archaebacteria, which produce methane in marshy areas.
II. *Nostoc* is a filamentous blue-green algae, which fixes atmospheric nitrogen.
III. Chemosynthetic autotrophic bacteria synthesise cellulose from glucose.
IV. Mycoplasma lacks a cell wall and can survive without oxygen.

The correct statements are

(a) II and III
(b) I, II and III
(c) II, III and IV
(d) I, II and IV

14 Consider the following given bacteria and classify them as Gram positive (+) and Gram negative (–) bacteria.

Choose the correct option from the codes that follows.

I. *Acetobacter* II. *Escherichia*
III. *Vibrio* IV. *Salmonella*
V. *Streptococcus* VI. *Mycobacterium*
VII. *Enterococcus* VIII. *Pneumococcus*

	Gram positive	Gram negative
(a)	V, VI, VII, VIII	I, II, III, IV
(b)	I, II, III, IV	V, VI, VII, VIII
(c)	II, IV, VIII	I, III, V, VI, VII
(d)	I, VIII, III, II	IV, V, VI, VII

15 Select the correct pair(s).

(a) Methanogens—*Methanogenium, Methanothrix*
(b) Halophiles—*Halococcus, Haloarchaea*
(c) Thermoacidophiles—*Sulfolobus, Desulfurococcus*
(d) All of the above

16 Match the following columns.

Column I	Column II
A. Mycoplasma	1. Ether linked lipids
B. Transduction	2. Sex pili
C. Conjugation	3. Requires virus
D. Thermophiles	4. Cell wall-less bacteria

Codes

	A	B	C	D		A	B	C	D
(a)	4	3	2	1	(b)	1	2	3	4
(c)	2	3	4	1	(d)	3	4	1	2

17 *Thermococcus*, *Methanococcus* and *Methanobacterium* examplify the

(a) archaebacteria that contain protein homologous to eukaryotic core histones
(b) archaebacteria that lack any resemblance those found in eukaryotes, but whose DNA is negatively supercoiled
(c) bacteria whose DNA is relaxed or positively supercoiled but, which was a cytoskeleton as well as mitochondria
(d) bacteria that contain a cytoskeleton and ribosomes

18 The main difference between photosynthetic and chemosynthetic bacteria is

(a) photosynthetic bacteria are seen in green plants, while chemosynthetic bacteria are seen in chemical substances
(b) photosynthetic bacteria are parasites within green plant cells, while chemosynthetic bacteria are saprophytes on decaying food substances
(c) H_2O is used by photosynthetic bacteria, but not by chemosynthetic bacteria
(d) energy of sunlight is used in photosynthetic bacteria, whereas in chemosynthetic bacteria energy is derived by the oxidation of inorganic substances

19 Select the correct pair(s).
I. Atrichous-*Pasteurella*
II. Cephalotrichous-*Pseudomonas*
III. Monotrichous-*Vibrio*
IV. Lophotrichous-*Spirillum*
Choose the correct option.
(a) I and III (b) II and IV
(c) I, III and IV (d) I, II, III and IV

20 Certain bacteria living in soil, poor in oxygen, convert nitrates into nitrite and then to free nitrogen, such bacteria are known as
(a) nitrogen-fixing bacteria
(b) denitrifying bacteria
(c) ammonifying bacteria
(d) saprophytic bacteria

21 Select the correct pair(s).
I. Glycocalyx–Polypeptide
II. Cell wall–Pseudomurien
III. Mesosomes–Abundant in Gram positive bacteria
IV. Ribosomes–70S type
Choose the correct option.
(a) I and II (b) II and III
(c) I and IV (d) All of these

22 Among the statements given below find out which statement (s) is/are incorrect about Gram bacteria and select the appropriate code.
I. The technique of Gram staining was introduced by Hans Christian Gram in 1874.
II. The basal body of flagellum of Gram positive bacteria has four rings of swellings.
III. The lipid content in Gram positive bacteria cell wall is very low.
IV. *Escherichia coli* and *Salmonella* are Gram negative bacteria.

Codes
(a) I, III and IV (b) I and II
(c) Only III (d) III and IV

23 The statements given below describe some of the characteristic features of Gram negative bacteria. Read them carefully and mark them as True (T)/False (F).
I. Peptidoglycans in the cell wall only form 3-12%.
II. They produce exotoxins.
III. The examples of Gram negative bacteria are *Streptomyces, Staphylococcus* and *Streptococcus.*
IV. The cell wall in these bacteria is single-layered containing teichoic acid.

Codes

	I	II	III	IV
(a)	F	T	T	F
(b)	F	T	F	T
(c)	T	F	F	F
(d)	T	F	T	F

24 Read the following statements and mark them as True (T)/False (F).
I. The cell wall strengthening material in a bacterial cell is murein.
II. The food reserve present in a bacterial cell is starch and maltose.
III. The ribosomes present are of 70S type.
IV. The mesosome helps in sexual reproduction.
Choose the correct option.

	I	II	III	IV
(a)	T	F	T	F
(b)	F	T	T	F
(c)	T	T	F	T
(d)	F	T	T	F

25 Consider the following statements and find the correct statements from the options given below.
I. Amphitrichous flagella is found in *Nitrosomonas.*
II. Lophotrichous flagella is found in *Pseudomonas.*
III. Red colour of red sea is due to the presence of *Trichodesmium erythraeum.*
IV. DNA of bacteria is known as genophore.

Codes
(a) I and II (b) II and III
(c) III and IV (d) All of these

Assertion and Reason

Direction (Q. Nos. 26-32) *In the following questions, a statement of Assertion (A) is given followed by the corresponding statement of Reason (R). Of the statements, mark the correct answer as*
(a) Both A and R are true and R is the correct explanation of A
(b) Both A and R are true, but R is not the correct explanation of A
(c) A is true, but R is false
(d) A is false, but R is true
(e) Both A and R are false

26 **Assertion** (A) Gram negative bacteria do not retain the stain when washed with alcohol.
Reason (R) The outer surface of the membrane of Gram negative bacteria contains lipopolysaccharides, a part of which is integrated into the membrane lipids.

27 **Assertion** (A) Autotrophic bacteria obtain their carbon from inorganic CO_2.
Reason (R) Autotrophs that obtain their energy from sunlight, are called photoautotrophs.

28 **Assertion** (A) Cell wall of bacteria is characterised by having mucopolysaccharides.
Reason (R) Acetyl muramic acid is an example of mucopolysaccharide.

29 **Assertion** (A) Plague is a bacterial disease.

Reason (R) It is caused by *Borrelia burgdorferi*, the vector of reservoir for plague is fleas of wild rodents.

30 **Assertion** (A) Gram negative (–ve) bacteria release exotoxins and cause diseases in animals.

Reason (R) WBCs of animals get destroyed by exotoxins.

31 **Assertion** (A) Eubacteria and archaebacteria have a same outer cell wall structure.

Reason (R) They both lack peptidoglycan in their cell wall.

32 **Assertion** (A) A true nucleus is generally absent in *E. coli* and other prokaryotes.

Reason (R) An undifferentiated unorganised fibrillar nucleus without any limiting membrane is observed in prokaryotic cells.

(C) Medical Entrances' Gallery

(Collection of Questions Asked in NEET & in Various Medical Entrance Exams)

1 Inclusion bodies of blue-green, purple and green photosynthetic bacteria are **[NEET 2020]**

(a) contractile vacuole
(b) gas vacuole
(c) centrioles
(d) microtubules

2 Which of the following is incorrect about cyanobacteria? **[NEET 2020]**

(a) They are photoautotrophs
(b) They lack heterocysts
(c) They often form blooms in polluted water bodies
(d) They have chlorophyll-a similar to green plants

3 The size of Pleuro Pneumonia Like Organism (PPLO) is **[NEET 2020]**

(a) 0.02 μm (b) 1-2 μm
(c) 10-20 μm (d) 0.1 μm

4 Which statement is correct? **[NEET 2019]**

(a) Mycoplasma is smallest and wall less living organism
(b) Influenza and herpes are caused by virus having DNA and RNA, respectively
(c) *Nostoc* and *Anabaena* are important decomposer
(d) Methanogen are methane producing bacteria in wheat crop

5 Match the organisms in Column I with habitats in Column II. **[NEET 2019]**

Column I	Column II
A. Halophiles	1. Hot springs
B. Thermoacidophiles	2. Aquatic environment
C. Methanogens	3. Guts of ruminants
D. Cyanobacteria	4. Salty area

Codes

	A	B	C	D
(a)	4	1	3	2
(b)	1	2	3	4
(c)	3	4	2	1
(d)	2	4	3	1

6 Which of the following statements about methanogens is not correct? **[NEET 2019]**

(a) They can be used to produce biogas
(b) They are found in the rumen of cattle and their excreta
(c) They grow aerobically and breakdown cellulose rich food
(d) They produce methane gas

7 Identify the organism containing heterocyst. **[JIPMER 2019]**

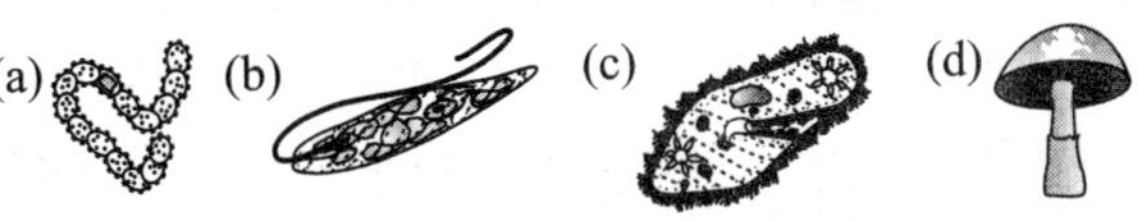

(a) (b) (c) (d)

8 Oxygen is not produced during photosynthesis by **[NEET 2019]**

(a) *Cycas* (b) *Nostoc*
(c) Green sulphur bacteria (d) *Chara*

9 Which of the following are found in extreme saline conditions? **[NEET 2018]**

(a) Archaebacteria (b) Eubacteria
(c) Cyanobacteria (d) Mycobacteria

10 Which of the following components provides sticky character to the bacterial cell? **[NEET 2018]**

(a) Cell wall (b) Nuclear membrane
(c) Plasma membrane (d) Glycocalyx

11 Read the following statements regarding bacteria. **[AIIMS 2018]**

I. Bacteria exchange their genetic matter through conjugation which involves cell to cell contact.
II. Transduction in '*Salmonella* is reported by Tatum and Lederberg in 1952.
III. Citrus canker disease is caused by bacterium *Xanthomonas citri.*
IV. Hans Christian Gram staining method is based on cell wall composition of bacteria.

Choose the correct option with true statements.

(a) I and III (b) I, III and IV
(c) I, II and III (d) II and IV

12 Heterocyst present in *Nostoc* is specialised for [AIIMS 2017]

(a) fragmentation (b) nitrogen-fixation
(c) symbiotic relation (d) food storage

13 All monerans [JIPMER 2017]

(a) contain DNA and RNA
(b) demonstrate a long circular strand of DNA, not enclosed in a nuclear membrane
(c) are bacteria
(d) All of the above

14 The primitive prokaryotes responsible for the production of biogas from the dung of ruminant animals, include the [NEET-I 2016]

(a) thermoacidophiles
(b) methanogens
(c) eubacteria
(d) halophiles

15 Which one of the following statements is incorrect? [NEET-I 2016]

(a) Golden algae are also called desmids
(b) Eubacteria are also called false bacteria
(c) Phycomycetes are also called algal fungi
(d) Cyanobacteria are also called blue-green algae

16 Select the incorrect statement. [NEET-I 2016]

(a) Bacterial cell wall is made up of peptidoglycan
(b) Pili and fimbriae are mainly involved in motility of bacterial cells
(c) Cyanobacteria lack flagellated cells
(d) Mycoplasma is a wall-less microorganism

17 Methanogens belong to [NEET-II 2016]

(a) Eubacteria (b) Archaebacteria
(c) Dinoflagellates (d) Slime moulds

18 Which of the following *r*RNAs acts as structural RNA as well as ribozyme in bacteria? [NEET 2016]

(a) 5 *sr*RNA (b) 18 *sr*RNA
(c) 23 *sr*RNA (d) 5.8 *sr*RNA

19 Of the following statements which are not relevant to Archaebacteria? [Kerala CEE 2015]

I. They live in some of the most harsh habitats.
II. They are present in the gut of several ruminant animals.
III. They are characterised by the presence of a rigid cell wall.
IV. They include mycoplasma.
V. They are also referred to as blue-green algae.

Codes

(a) I, II and III (b) I, III and V
(c) III, IV and V (d) I, III and IV
(e) II, III and V

20 The organisms which completely lack a cell wall and can live without oxygen are [Karnataka CET 2015, CBSE AIPMT 2015, AIIMS 2012, AFMC 2009]

(a) mycoplasmas (b) archaebacteria
(c) methanogens (d) thermoacidophiles

21 True nucleus is absent in [CBSE AIPMT 2015]

(a) *Anabaena* (b) *Mucor* (c) *Vaucheria* (d) *Volvox*

22 The guts of cow and buffalo possess [CBSE-AIPMT 2015]

(a) *Fucus* sp. (b) *Chlorella* sp.
(c) methanogens (d) cyanobacteria

23 Which one of the following cocci appears like grapes under microscope? [WB JEE 2015]

(a) Streptococci (b) Diplococci
(c) Staphylococci (d) Pneumococci

24 Which one of the following is a causative agent of plague? [WB JEE 2015]

(a) *Shigella flexneri* (b) *Bordetella pertussis*
(c) *Staphylococcus aureus* (d) *Yersinia pestis*

25 The cell wall material present only in bacteria and blue-green algae is [AMU 2015]

(a) pectin (b) cellulose
(c) chitin (d) muramic acid

26 Which is not properly matched? [CG PMT 2015]

(a) *Azotobacter*—Nitrogen fixation
(b) *Streptococcus thermophilus*—Yogurt preparation
(c) *Chlorobium*—Photosynthesis
(d) *Streptomyces rimosus*—Chloromycetin

27 Peptidoglycan (mucopeptide) is found in the walls of [CG PMT 2015]

(a) Only Gram +ve bacteria
(b) Both Gram +ve and –ve bacteria
(c) Gram +ve and –ve bacteria and fungi
(d) Bacteria and blue-green algae

28 Archaebacteria differ from eubacteria in [CBSE AIPMT 2014]

(a) cell membrane structure (b) mode of nutrition
(c) cell shape (d) mode of reproduction

29 The motile bacteria are able to move by [CBSE AIPMT 2014]

(a) flagella (b) cilia (c) pili (d) fimbriae

30 *Trichodesmium erythraeum*, which gives colour to Red Sea is [EAMCET 2014]

(a) green algae (b) blue-green algae
(c) red algae (d) brown algae

31 Anoxygenic photosynthesis is characteristic of [CBSE AIPMT 2014]

(a) *Spirogyra* (b) *Chlamydomonas*
(c) *Ulva* (d) *Rhodospirillum*

32 Match the following columns. **[EAMCET 2013]**

Column I	Column II
A. Syphilis	1. *Acetobacter*
B. Pathogen of cattle	2. *Agrobacterium*
C. Crown gall of apple	3. *Corynebacterium*
D. Diphtheria	4. *Mycobacterium*
	5. *Treponema*

Codes

	A	B	C	D		A	B	C	D
(a)	3	1	4	2	(b)	5	4	2	3
(c)	5	3	2	1	(d)	2	4	5	3

33 Besides paddy fields, cyanobacteria are also found inside the vegetative part of **[NEET 2013]**
(a) *Pinus* (b) *Cycas*
(c) *Equisetum* (d) *Psilotum*

34 Pigments containing membranous extensions in some cyanobacteria are **[NEET 2013]**
(a) heterocysts (b) basal bodies
(c) pneumatophores (d) chromatophores

35 Which of the following are likely to be present in deep sea water? **[NEET 2013]**
(a) Archaebacteria (b) Eubacteria
(c) Blue-green algae (d) Saprophytic fungi

36 Bacteria differ from plants in that they do not have
(a) DNA (b) RNA **[UP CPMT 2013]**
(c) cell wall (d) a well-define nucleus

37 Identify the bacterium that appears violet after Gram staining **[WB JEE 2013]**
(a) *Salmonella enterica*
(b) *Escherichia coli*
(c) *Mycobacterium tuberculosis*
(d) *Rhizobium meliloti*

38 N-acetyl muramic acid is found in **[WB JEE 2013]**
(a) cell wall component of plant
(b) cell wall component of Gram positive bacteria
(c) cell wall component of fungi
(d) viral coat material

39 Spirochaetes is/are **[DUMET 2012]**
(a) a class of insect (b) a class of viruses
(c) bacteria (d) fungi

40 Which of the following is a Gram negative bacterium? **[DUMET 2012]**
(a) *Escherichia coli* (b) *Bacillus subtilis*
(c) *Streptomyces coelicolor* (d) *Ampycolatopsis orientalis*

41 The site of photosynthesis in blue-green algae is
(a) chromatophores (b) mitochondria **[BCECE 2012]**
(c) chloroplast (d) root hair

42 Cyanobacteria is also referred to as **[CBSE AIPMT 2012]**
(a) protists (b) slime moulds
(c) green algae (d) blue-green algae

43 A peculiar odour that prevails in marshy areas and cow-sheds is an account of a gas produced by **[DUMET 2012]**
(a) Mycoplasma (b) Archaebacteria
(c) Slime moulds (d) Cyanobacteria

44 Maximum nutritional diversity is found in the kingdom **[CBSE AIPMT 2012]**
(a) Fungi (b) Animalia
(c) Monera (d) Plantae

45 How many organisms in the list given below are autotrophs?
Lactobacillus, Nostoc, Chara, Nitrosomonas, Nitrobacter, Streptomyces, Saccharomyces, Trypanosoma, Porphyra, Wolffia. **[CBSE AIPMT 2012]**
(a) Four (b) Five
(c) Six (d) Three

46 Fin rot of fish is caused by **[WB JEE 2012]**
(a) *Aeromonas* (b) *Pseudomonas*
(c) *Branchiomyces* (d) *Xenopsylla*

47 Which of the following is an important source of edible protein? **[Haryana PMT 2012]**
(a) *Spirogyra* (b) *Porphyra*
(c) *Spirulina* (d) *Cephaleuros*

48 The red colour of 'red sea' is due to which of the following blue-green algae? **[RPMT 2012]**
(a) *Chlamydomonas* (b) *Anabaena*
(c) *Microcystis* (d) *Trichodesmium*

49 Teichoic acid is present in **[UP CPMT 2011]**
(a) cell wall of Gram positive bacteria
(b) cell wall of Gram negative bacteria
(c) capsid of virus
(d) protoplasm of mycoplasma

50 According to five kingdom classification, bacteria belong to **[J&K CET 2011]**
(a) Protista (b) Monera (c) Plantae (d) Archaea

51 The main difference between Gram positive and Gram negative bacteria is in there **[WB JEE 2011, CBSE AIPMT 2001]**
(a) cell membrane (b) cell wall
(c) ribosome (d) mitochondria

ANSWERS

Check Point 4.1

1. (a)	2. (c)	3. (b)	4. (c)	5. (c)

Check Point 4.2

1. (d)	2. (a)	3. (a)	4. (c)	5. (d)

Check Point 4.3

1. (b)	2. (b)	3. (c)	4. (b)	5. (b)

(A) Topical Exercise

1. (a)	2. (c)	3. (a)	4. (c)	5. (d)	6. (b)	7. (c)	8. (a)	9. (b)	10. (a)
11. (b)	12. (c)	13. (c)	14. (b)	15. (b)	16. (c)	17. (c)	18. (a)	19. (c)	20. (b)
21. (d)	22. (b)	23. (d)	24. (b)	25. (b)	26. (d)	27. (d)	28. (b)	29. (a)	30. (a)
31. (a)	32. (b)	33. (d)	34. (d)	35. (c)	36. (b)	37. (a)	38. (b)	39. (a)	40. (b)
41. (d)	42. (b)	43. (a)	44. (a)	45. (c)	46. (c)	47. (a)	48. (c)	49. (b)	50. (a)
51. (d)	52. (c)	53. (d)	54. (c)	55. (c)	56. (c)	57. (c)	58. (d)	59. (d)	60. (b)
61. (a)	62. (c)	63. (c)	64. (b)	65. (c)	66. (d)	67. (d)	68. (c)	69. (a)	70. (c)

(B) Cumulative Exercise

1. (c)	2. (a)	3. (b)	4. (a)	5. (b)	6. (d)	7. (d)	8. (b)	9. (c)	10. (b)
11. (c)	12. (a)	13. (d)	14. (a)	15. (d)	16. (a)	17. (a)	18. (d)	19. (d)	20. (b)
21. (c)	22. (b)	23. (c)	24. (a)	25. (d)	26. (b)	27. (b)	28. (e)	29. (c)	30. (d)
31. (c)	32. (a)								

(C) Medical Entrances' Gallery

1. (b)	2. (b)	3. (d)	4. (a)	5. (a)	6. (c)	7. (a)	8. (c)	9. (a)	10. (d)
11. (b)	12. (b)	13. (d)	14. (b)	15. (b)	16. (b)	17. (b)	18. (c)	19. (c)	20. (a)
21. (a)	22. (c)	23. (c)	24. (d)	25. (d)	26. (d)	27. (b)	28. (a)	29. (a)	30. (b)
31. (d)	32. (b)	33. (b)	34. (d)	35. (a)	36. (d)	37. (c)	38. (b)	39. (c)	40. (a)
41. (a)	42. (d)	43. (b)	44. (c)	45. (c)	46. (b)	47. (c)	48. (d)	49. (a)	50. (b)
51. (b)									

Explanations

(A) Topical Exercise

1 (*a*) Bacteria constitute large domain of prokaryotic microorganisms. They were among the first life forms to appear on earth. Bacteria are the sole members of the kingdom-Monera. Organisms of kingdom-Monera are called lower level organisms because no tissue level organisation is present in them.

2 (*c*) Both statements I and II are true because

Bacteria are unicellular organisms that lack internal membrane bound organelles. They do not have a true nucleus. Instead they have the incipient nucleus called as nucleoid. These organisms are smaller than 5μ in size.

4 (*c*) Statement in option (c) is incorrect and can be corrected as Cellulosic cell walls are not found in kingdom-Monera.

5 (*d*) Halophiles is odd one out as it belongs to Archaebacteria whereas other three are Eubacteria.

13 (*c*) Pili are minute hair like appendages, that emerges from surface of Gram -ve bacteria. They have similar structure as flagella and often involved in attachment to surface and making of biofilm.

15 (*b*) The amino acid composition varies in different bacteria, the short peptides are typically composed of L-Ala, unusual D-amino acid (D and D-Alu) and the dibasic amino acid, meso-diaminopimelic which form the cross links.

18 (*a*) Statement I is true, but II is false. False statement can be corrected as

Gram negative bacteria appears pink in colour after counter staining with safranin.

24 (*b*) The fertility or F^+ factor that plays a role in conjugation is a F plasmid. It contain genes for the formation of a specialised pilus that initiates the formation of a conjugation bridge between two cells for the purpose of transferring genetic material.

33 (*d*) Statement in option (d) is incorrect and can be corrected as

Bacteria reproduces asexually as well as sexually. Asexual means include fission, cyst formation, budding, etc.

35 (*c*) Pairs I and III are incorrectly matched and can be corrected as

Actinomycetes have oidiospores and *Clostridium* possess endospores.

52 (*c*) Statement in option (c) is incorrect and can be corrected as

Mycoplasma is not sensitive to penicillin due to the absence of cell wall.

53 (*d*) Statement in option (d) is incorrect and can be corrected as

Mycoplasma are not obligate parasites. They exist as free-living forms also.

62 (*c*) Statement in option (c) is incorrect and can be corrected as

Heterocysts are found in cyanobacteria or BGA.

64 (*b*) Statement I is false, but II is true. False statement can be corrected as

Cyanobacteria do not possess flagella.

66 (*d*) Domain Archaea have single celled microorganisms. They do not have well defined nuclei and other membrane bound organelles. Archaebacteria are placed in different kingdoms from plants and animals. They have some properties which are not common to the organisms placed in bacteria and eukaryotes. However, they have some novel features absent in other prokaryotes and eukaryotes. Thus, correct option is (d).

67 (*d*) Option (d) is not a characteristics of archaebacteria because

The cell wall of archaebacteria do not contain peptidoglycans, but these are present in eubacteria.

(B) Cumulative Exercise

6 (*d*) Statements I and IV are correct. Statements II and III are incorrect and can be corrected as

Some species of *Streptococcus* infect both upper and lower respiratory tract. *Staphylococcus aureus* is a Gram positive round-shaped bacterium that is a member of the firmicutes and it is a usual member of the microbiota of the body, frequently found in the upper respiratory tract and on the skin. It is not found in the sinuses.

10 (*b*) Only statement I is false and can be corrected as

Transformation do not require cell to cell contact.

11 (*c*) Statement II is false and can be corrected as

Mycoplasma contain linear, double stranded DNA as genetic material.

13 (*d*) All the statements are correct except III. The correct form is Chemosynthetic autotrophic bacteria use carbon as a source of energy for its production of food. Carbon dioxide is an example of a carbon source used by the chemosynthetic autotrophic bacteria.

17 (*a*) Archaea has two sub-divisions, i.e. the Euryarchaeota and the Crenarchaeota. The Euryarchaeota includes Methanogens, Halophiles, Thermophiles, e.g. *Methanobacterium, Methanococcus, Thermococcus.*

These microorganisms contain protein which is homologous to eukaryotic core histones.

20 (*b*) Denitrification is the reverse process of nitrification. During this process, nitrates are reduced to nitrites and then to nitrogen gas and ammonia.

These compounds are gaseous compounds and are not readily available for microbial growth, therefore they are typically released to the atmosphere. Certain bacteria living is soil, poor in oxygen convert nitrates into nitrites and then to free nitrogen, such bacteria and termed as denitrifying bacteria.

22 (*b*) Statements I and II are incorrect and can be corrected as

The technique of Gram staining was introduced by Hans Christian Gram in 1884.

In Gram negative bacteria there are two pairs of rings, i.e. four rings are L (Lipopolysaccharide) ring, P (Peptidoglycan) ring, S (super membrane) ring and M (Membrane) ring, Gram positive bacteria have a single pair of rings.

26 (*b*) Both A and R are true, but R is not the correct explanation of A.

Gram negative bacteria do not retain the stain when washed with alcohol as they possess a very thin layer of peptidoglycan between two membranes.

27 (*b*) Both A and R are true, but R is not the correct explanation of A.

Bacteria have evolved many mechanisms to acquire the energy and nutrients they need for growth and reproduction. Many autotrophs obtain their carbon from inorganic CO_2. But autotrophs, which obtain their energy from sunlight are called photoautotrophs.

28 (*e*) Both A and R are false. These can be correct as Chemically bacterial cell wall consists of acetylglucosamine, acetyl muramic acid and a peptide chain of four or five amino acids. All these chemicals form a polymer called peptidoglycan or murein. Other chemicals, which are present on the cell wall of bacteria are teichoic acid, protein, polysaccharides, lipoproteins and lipopolysaccharides.

29 (*c*) A is true, but R is false because

Plague is a bacterial disease caused by *Yersinia pestis* (*Pasteurella pestis*). The vector or reservoir for this disease are fleas of wild rodents, rats and squirrels.

30 (*d*) A is false, Reason is true because

Exotoxins are released by Gram positive bacteria. These are proteinaceous and heat labile compound, which are excreted by bacteria outside the cell. They kill the animals WBC and cause disease in them.

31 (*c*) A is true, but R is false because

Both eubacteria and archaebacteria have cell wall covering the plasma membrane that strengthen the cell. The cell wall of eubacteria is made up of carbohydrate-protein complexes called peptidoglycan, which link together to create a strong mesh that gives the eubacterial cell wall a great strength. The cell wall of archaebacteria lacks peptidoglycan.

32 (*a*) Both A and R are true and R is the correct explanation of A. *E. coli* is a well-studied prokaryotic cell. These lack true nucleus and membrane bound organelles like mitochondria, chloroplast, ER, Golgi body, etc. There is an undifferentiated, unorganised fibrillar nucleus without any nuclear membrane in prokaryotic cells. The genetic material of prokaryotic cell is dispersed in cytoplasm and is known as nucleoid.

(C) Medical Entrances' Gallery

1 (*b*) Gas vacuoles are the inclusion bodies in many aquatic prokaryotes like blue-green, purple and green photosynthetic bacteria. These are generally small, hollow cylindrical structure which facilitates air permeability. Gas vacuoles are membrane bound inclusion bodies that contain an array of sub-structures referred to as gas vesicles. The membrane of gas vacuoles is rigid, impermeable to water and freely permeable to all gases.

2 (*b*) Cyanobacteria or blue-green algae are photosynthetic organisms which perform oxygenic photosynthesis. They have the ability of nitrogen fixation due to the presence of large pale cells called heterocyst in their filaments. Due to the presence of thick walls, heterocysts are impermeable to oxygen.

3 (*d*) The size of Pleuro Pneumonia Like Organism (PPLO) is 0.1 μm.

The sizes of other organisms are as follows

- Viruses - 0.02-0.2 μm
- Bacterial cell - 1-2 μm
- Eukaryotic cell - 10-20 μm

4 (*a*) Statement in option (a) is correct. Rest of the statements are incorrect and can be corrected as

Influenza and herpes are caused by virus having RNA and DNA, respectively

Nostoc and *Anabaena* are examples of free-living nitrogen fixers.

Methanogens are methane producing bacteria found in marshy areas and paddy fields, not in wheat crops.

6 (*c*) Statement in option (c) is incorrect and can be corrected as

Certain bacteria, which grow anaerobically on cellulosic material, produce large amount of methane along with CO_2 and H_2. These bacteria are collectively called methanogens and one such example is *Methanobacterium.*

Rest of the statements are correct.

7 (*a*) The diagram in option (a) is of *Nostoc* which contains a heterocysts. These are the specialised nitrogen-fixing cells formed during nitrogen starvation by some filamentous cyanobacterium such as *Nostoc*. Heterocyst creates microanaerobic environment around the cell that helps in the fixation of free atmospheric nitrogen.

The organisms shown in other options are as follows

(b) *Euglena* (c) *Paramecium* (d) Mushroom (*Agaricus*)

These do not contain heterocyst.

8 (*c*) Green sulphur bacteria are anaerobic bacteria. They do not evolve oxygen during photosynthesis. Such type of photosynthesis is known as anoxygenic photosynthesis. They do not use water as a source of reducing power. Instead, hydrogen and sulphide ions are obtained from hydrogen sulphide which act as the electron donors.

9 (*a*) Archaebacteria are the most primitive form of bacteria. These live in extreme diverse habitats, e.g. extreme hot temperature, saline condition (deep sea water), variable pH, etc. The type of archaebacteria found in saline conditions are called halophiles (e.g. *Halobacterium, Halococcus*).

10 (*d*) Glycocalyx is the outer most mucilage layer of the cell envelope. It gives the sticky characteristic to the bacterial cell.

11 (*b*) All statements are true except II. The correct form is as follows

Transduction is a process of transfer of DNA by bacteriophage virus in bacteria. It was first reported in the *Salmonella* by Zinder and Lederberg in 1952.

12 (*b*) Cyanobacteria like *Nostoc* can fix nitrogen through a specific structure called heterocysts. They fix atmospheric using nitrogenase enzyme which gets inactivated in the presence of oxygen. Hence, the heterocysts create a microanaerobic environment to facilitate nitrogen-fixation.

13 (*d*) All the given statements about monerans are correct.

14 (*b*) Methanogens are group of obligate anaerobic ancient and primitive bacteria. They are involved in methanogenesis and produce methane gas in rumen of cattle.

15 (*b*) All the statements are correct except (b). Its correct form is

Eubacteria are also called true bacteria. They have rigid cell walls and flagella.

16 (*b*) Statement in option (b) is incorrect and can be corrected as

Fimbriae or pili are fine hair-like appendages used by bacteria for attachment rather than mobility. These are formed of protein called pilin.

17 (*b*) Methanogens belong to the group-Archaebacteria which contain three major classes of primitive bacteria, i.e. methanogens, halophiles and thermoacidophiles.

18 (*c*) Bacterial cells use their 23 *sr*RNA as an enzyme during protein synthesis. This is the only non-proteinaceous enzyme known so far.

19 (*c*) Statements III, IV, V are not relevant to Archaebacteria because

Archaea have a semi-rigid cell wall that protects them from the environment. There are three major known groups of archaebacteria; Methanogens, halophiles and Thermophiles. *Mycoplasma* comes under Eubacteria. Blue-green algae is alternative name of Cyanobacteria.

20 (*a*) Mycoplasmas are organism that completely lacks cell wall. It can live without oxygen and its mode of nutrition is heterotrophic.

21 (*a*) *Anabaena* is prokaryotic blue-green algae having no true nucleus and membrane bound organelles. Rather a nucleoid region containing genetic material is present in the cytoplasm. The other cells like *Mucor*, *Vaucheria* and *Volvox* belong to fungi and algae.

22 (*c*) Methanogens are microorganisms that produce methane as a byproduct of their metabolism. These are both archaeans and obligate anaerobes. They are commonly found in wetlands and in guts of animals such as ruminants (cow, buffalo) where they form marsh gas.

23 (*c*) Staphylococci are species of Gram positive bacteria that cause a wide variety of infections in humans and other animals through infestation or by production of toxins. When viewed under the microscope, they appear in 'grape-like' clusters. It is the morphological characteristic of these bacteria that has given rise to their name, staphyle in Greek refers to a bunch of grapes.

24 (*d*) Plague is bacterial disease caused by *Yersinia pestis*. It is a Gram negative, rod-shaped, cocco bacillus, facultative anaerobic bacterium that can infect humans as well as animals. It gets transmitted from rodents to humans by the bite of infected fleas.

25 (*d*) Bacteria and blue-green algae have peptidoglycan cell wall. It is made of N-Acetyl Glucosamine (NAG) and N-Acetyl Muramic Acid (NAM).

26 (*d*) Option (d) shows incorrectly matched pair and can be corrected as

Streptomyces rimosus produces antibiotic called terramycin. Chloromycetin is produced by *Streptomyces venezuelae.*

27 (*b*) The bacteria are classified depending upon their reaction to Gram stain. They are Gram positive if, they take stain and Gram negative if do not take stain. Peptidoglycan forms a structural network of cell wall, known as mucopeptide in both Gram +ve and -ve bacteria. Murein or mucopeptide is 70-80% in Gram +ve and 10-20% in Gram –ve bacteria.

28 (*a*) Archaebacteria differ from eubacteria in cell membrane structure. The cell membrane of eubacteria is composed of glycerol-ester lipids whereas archaebacteria have cell membrane, which is composed of glycerol ether lipids. Unlike eubacteria, archaebacteria also lack peptidoglycan in their cell wall.

29 (*a*) Most motile bacteria move by using flagella. It is the organ of motility in bacteria. Bacterial flagella are unistranded, equivalent to a single microtubular fibre and formed of protein called flagellin.

30 (*b*) *Trichoderma erythraeum*, a member of group-Cyanobacteria, i.e. blue-green algae, which is responsible for providing red colour to red sea.

31 (*d*) In *Rhodospirillum*, electron donor is organic compound instead of water, hence no oxygen is released, i.e. anoxygenic photosynthesis occurs. In other plants water is used as electron donor and H^+ and O_2 are produced during photolysis of water.

33 (*b*) Besides paddy fields, cyanobacteria are also found in the vegetative parts of *Cycas*. Cyanobacteria within the coralloid roots of *Cycas* are chemoautotrophic and specifically adapted to live in symbiosis.

34 (*d*) Chromatophores are the pigment containing membranous extensions in some cyanobacteria. These pigments mainly include chlorophyll-*a*, in some species carotenoids, phycobilins and phycoerythrin.

35 (*a*) Archaebacteria flourish in hot springs and deep sea water. They differ from other bacteria in having a rigid cell wall with diverse structure and this feature is responsible for ensuring their survival in extreme conditions.

36 (*d*) Prokaryotic cell is found in bacteria. These cells lack nucleus and membrane bound cell organelles, which are present in plant cell (eukaryotic type).

37 (*c*) The bacterium that appears violet after Gram staining is *Mycobacterium tuberculosis*, a Gram positive bacterium, containing a thick layer of peptidoglycan in its cell wall and hence it resists decolourisation. Bacteria given in other options do not appear violet after Gram staining. These are Gram negative bacteria.

38 (*b*) The bacterial cell wall is composed of protein-lipid-polysaccharides (peptidoglycan) with two important chemical components namely NAG (N-Acetyl Glucosamine) and NAM (N-Acetyl Muramic Acid).

39 (*c*) Spirochaetes are slender and helically coiled bacteria varying in length from 3-500 μm. Some of them are saprophytes and the other are parasites. The spirochaete, *Treponema pallidum* causes syphilis disease.

40 (*a*) The Gram stain is named after the developer Christian Gram. About 75% of known bacteria are Gram negative, e.g. *Salmonella, Pseudomonas, Vibrio, Helicobacter, Haemophilus* and *Escherichia.*

43 (*b*) Archaebacteria is primitive group of bacteria. The three main groups of Archaebacteria are Methanogens, halophiles and Thermoacidophiles. Methanogens are obligate, anaerobic Archaebacteria, which oxidise CO_2 during cellular respiration to produce methane as a waste product. They are found in the musk of swamps and marshes, the rumen of cattle, sewage sludges and gut of termites, e.g. *Methanococcus jannaschii* and *Methanobacterium.*

44 (*c*) Maximum nutritional diversity is found in the kingdom-Monera. Some of them are autotrophs (e.g. photosynthetic autotrophic or chemosynthetic autotrophic) while the vast majority of them are heterotrophs (e.g. saprotrophic or parasitic). Ecologically, these are either producers or decomposers.

45 (*c*) Autotrophs are organisms which are capable of synthesising their food themselves from organic and inorganic substances. In the given list, six are autotrophs. They are *Nostoc*, *Chara*, *Nitrosomonas*, *Nitrobacter*, *Porphyra* and *Wolffia.*

46 (*b*) Fin rot disease of fish is caused by the bacterial infection of *Pseudomonas fluorescens.* Its symptoms include white fin edges fins ray, inflamed bases of fins, etc.

47 (*c*) *Spirulina* is highly rich in protein, vitamin-B complex and minerals also. Powdered *Spirulina* is being used in herbal tonics and biscuits.

48 (*d*) Due to the presence of *Trichodesmium*, a blue-green alga 'red sea' have their specific red colour.

49 (*a*) Teichoic acid is present in the cell wall of Gram positive bacteria. It is an acidic polymer consisting of carbohydrate, phosphate and an alcohol. It maintains the low pH to prevent degradation of cell wall by self produced enzymes.

51 (*b*) The main difference between Gram positive and Gram negative bacteria lies in their cell wall. The cell wall of Gram negative bacteria contains peptidoglycan (10%), lipopolysaccharides, lipoprotein and phospholipid, while cell wall of Gram positive bacteria contains peptidoglycan (60-90%), teichoic acid and lipids.

CHAPTER 05

Protista

The term 'Protista' was given by **Ernst Haeckel** in 1866. Protistans were the first eukaryotes to evolve, about 1000 million years ago. This kingdom includes eukaryotic unicellular organisms that cannot be classified as plants, animals or fungus. Thus, protistans form a connecting link between kingdom—Monera, i.e. prokaryotes and complex eukaryotic organisms, i.e. plants, fungi and animals.

Besides their relatively simple levels of organisation, protistans do not necessarily have much in common. These are the most primitive living organisms and are found wherever life exists.

Characteristic features

The main characteristics of Protista are as follows

- They are solitary or colonial unicellular eukaryotic organisms. They are mostly aquatic.
- Protists form nearly 95% of the total plankton population.
- Their single cell may be naked or covered by cell wall, pellicle, shell or cuticle. Cell wall, if present, is made up of cellulose or silica.
- Protists possess well-defined nucleus in them. These may be uninucleate, binucleate or multinucleate.
- Their cytoplasm contains a variety of other organelles like mitochondria, plastids (in photosynthetic protists), lysosomes, ribosomes, cytoskeleton, endoplasmic reticulum and Golgi bodies.
- Their cytoplasm is always in motion. This phenomenon is called **cytoplasmic streaming** or **cyclosis.**
- They have all possible types of body symmetries, along with asymmetrical forms.
- No organs or tissues are present in the body, i.e. subcellular or physiological grade of body organisation is seen.
- Locomotion can occur through flagella (e.g. *Euglena*), cilia (e.g. *Paramecium*), pseudopodia (e.g. *Amoeba*) along with contractions and mucilage extrusion.
- Flagella and cilia, when present, have 9 + 2 pattern of microtubular strands.
- Protistans have variable nutritive modes, as follows
  - Photosynthetic (holophytic)
  - Ingestive (holozoic and phagotrophic)
  - Absorptive (saprobic and parasitic).
- Food is reserved in the form of starch, glycogen, paramylon, chrysolaminarin and fat.

Inside

- Some protistans act as parasites, while some live symbiotically in the gut of other animals and few act as decomposers also.
- Asexual reproduction occurs through budding, binary fission, multiple fission, plasmotomy and spore formation.
- Plasmotomy occurs by division of multinucleate protozoans into several small multinucleate daughters without nuclear division.
- Sexual reproduction is believed to have originated in primitive protists. It involves syngamy (isogamy, anisogamy and oogamy) and conjugation.

Classification of protists

They are broadly divided into three groups, i.e. photosynthetic protists, fungi like protists and animal like protists.

Protista

- **Photosynthetic protists**
  - Phylum-Pyrrophyta (Dinoflagellates)
  - Phylum-Chrysophyta (Diatoms and yellow green algae)
  - Phylum-Euglenophyta (Euglenoid flagellates)
- **Fungus like protists or Slime moulds**
  - Myxomycota (Acellular slime moulds)
  - Acrasiomycota (Cellular slime moulds)
- **Animal like protists or Protozoan**
  - Phylum-Mastigophora (Zooflagellates)
  - Phylum-Sarcodina or Rhizopoda (*Amoeba*)
  - Phylum-Sporozoa (*Plasmodium*)
  - Phylum-Ciliophora (*Paramecium*)

PHOTOSYNTHETIC PROTISTS

These are plant like protists or phytoplanktons. Their common habitats are lakes and oceans. They are responsible for about 70% photosynthesis of the biosphere that occurs in marine habitat. Hence, they are the major producers of food in the aquatic ecosystem.

These protists can be further classified into the following phyla

Phylum–Pyrrophyta (Dinoflagellates)

The general characteristics of these organisms are as follows

- These include unicellular, golden-brown photosynthetic protists. They are mostly marine, but some are found in freshwater.
- These are commonly called as **dinoflagellates** due to the presence of two flagella in them.
- Out of the two flagella, one of the flagella is longitudinal, while the other one is transverse. The condition with two different flagella is known as **heterokontae**.
- The combined action of two flagella may cause the dinoflagellate to turn on its axis causing spinning movement. Therefore, these protists are also called **whirling whips**.
- They appear yellow, green, brown, blue or red depending on the pigments present in their cell.
- The cell wall of dinoflagellates is composed of sculptured plates made up of cellulose and pectin. It is called **theca** or **lorica**. These are often called as **armoured dinoflagellates** due to the presence of these sculptured plates.
- The theca contains two grooves, a **longitudinal sulcus** and a **transverse annulus**.
- Their mode of nutrition is mainly holophytic or photosynthetic. Pigments present in them are chl-*a*, chl-*c* and xanthophylls.
- Reserve food material is in the form of starch in freshwater forms and oil in marine forms.
- They have a non-contractile vacuole called **pusule**. This vacuole helps in osmoregulation.
- They possess eyespot or stigma (light sensitive organelle) and trichocysts (e.g. *Peridinium*) or cnidoblasts or nematocysts (e.g. *Nematidium*) as specialised structures.
- Their nucleus is large in size. It possesses condensed chromosomes even in interphase. It has been named as mesokaryon by **Dodge** in 1966. Their chromosomes are without histones.
- They also exhibit the phenomenon of **bioluminescence**, i.e. emit light (e.g. *Noctiluca, Gonyaulax,* etc). So, these dinoflagellates are also called **night light**.

Reproduction in dinoflagellates

- Asexual reproduction occurs by cell division and encystment.
- Sexual reproduction is reported in some dinoflagellates like *Ceratium*. It may be isogamous or anisogamous type.
- Life cycle is haplontic. In this type of life cycle during sexual reproduction, gametes are formed by mitosis. These gametes fuse to form a diploid zygote. After that meiosis takes place in zygote, as a result haploid cells are formed again.

Noctiluca

- It is a colourless dinoflagellate, which is an important constituent of coastal plankton of both temperate and tropical seas.
- Bioluminescence was recorded for the first time in this organism. It is thus referred to as **sea-ghost**.
- Their body is oval or rounded and is covered by a pellicle.

- Transverse flagellum is reduced to a tooth. Longitudinal flagellum is very small.
- Mesokaryon (also called **dinokaryon**) appears during gametogenesis.
- Cytoplasm is rich in food vacuoles and photogenic granules.

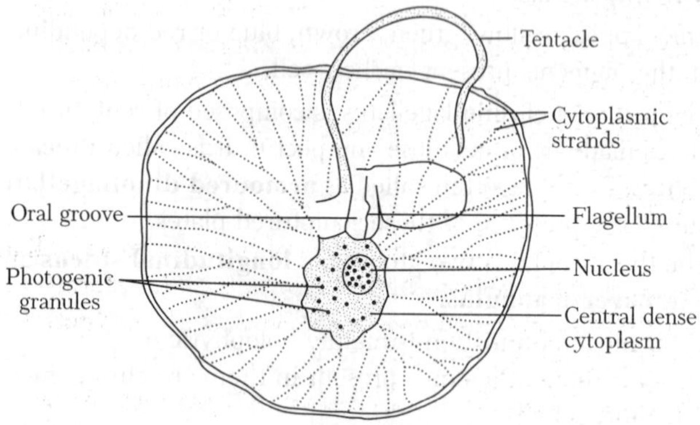

Fig 5.1 Noctiluca

Gonyaulax

- It is an armoured dinoflagellate, known for its bioluminescence.
- Its mesokaryon is large and central.
- Chromatophores are brownish and peripheral.
- It grows in such abundance over the sea surface that makes the sea appear red. It is called as **'red tide'**.
- It reproduces asexually by binary fission and cyst formation. Sexual reproduction is not reported in them.
- *Gonyaulax* produces the neurotoxin called **saxitoxin** (in water bodies) when grown in large number. These toxins are poisonous and kill a large number of fishes.

Phylum–Chrysophyta

The members of this phylum store food as oil. Some forms give unpleasant odours to the freshwater lakes where they inhabit.

It includes these two groups

1. Xanthophyta (Yellow-green algae)

- This is the group of freshwater algae.
- It contains xanthophyll, besides chlorophyll-*a*, *e* and β-carotene as photosynthetic pigments.
- Plant body is unicellular, coccoid, filamentous or siphonaceous. The siphonaceous forms are branched, aseptate or coenocytic and multinucleate.
- Each cell has a distinct nucleus and many discoid chromatophores.

Reproduction in yellow-green algae

Vegetative reproduction is by fragmentation, asexual reproduction by zoospore/aplanospore and sexual reproduction is either oogamous or isogamous type, e.g. *Botrydium*, *Vaucheria*, etc.

2. Bacillariophyta (Diatoms)

- This is the group of unicellular, golden-brown autotrophic protists.
- Their cell wall is called **frustule** or **shell**. It contains silica and is constructed in two overlapping halves or theca.
- Both of these fit together like two parts of a soap box (in pinnate diatoms) or like a pair of petridishes (in centric diatoms).
- Their upper half is called **epitheca** and lower half is called **hypotheca**.
- The siliceous shells of diatoms do not decay easily and thus pile up at the bottom of water bodies. Due to this, big heaps known as **diatomite** or **diatomaceous earth** or **kieselguhr** are formed.
- Each cell of diatoms has large central vacuole.
- Their reserve food materials are oil and leucosin.

Reproduction in diatoms

- Asexual reproduction is by binary fission. In them, each daughter cell receives epitheca from parent cell because of which the size goes on reducing in successive generation. Extrusion of protoplasm may serve to restore normal cell size. Resting spores called **statospores** may be formed by centric diatoms.
- Sexual reproduction varies from isogamy to oogamy. It results to form zygote. This zygote grow in size and forms a rejuvenescent cell called **auxospore** (diploid).
- Life cycle is diplontic. In this type of life cycle, meiosis takes place in diploid cell, due to which haploid gametes are formed. Now haploid gametes fuse to form diploid zygote and after that mitosis takes place in zygote, due to which diploid cells are formed again.

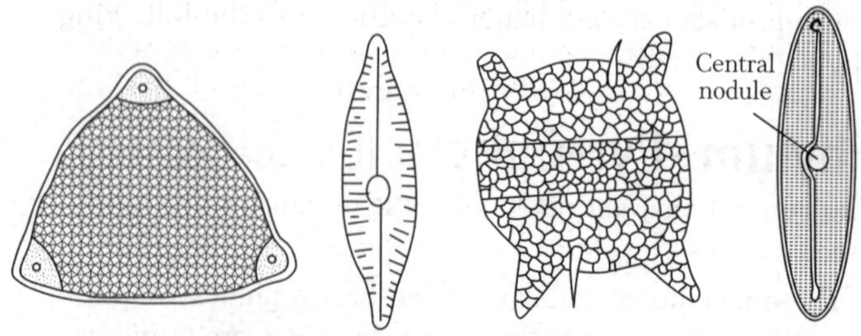

Fig 5.2 Some common diatoms

Economic importance of diatoms

- They are very good pollution indicators.
- They are an important source of food to animals in an aquatic ecosystem.
- Diatomite is porous and chemically inert, therefore used in filtration of sugar, alcohols and antibiotics. It is employed as an insulation material in refrigerators, boilers and furnaces.
- Diatomite is also used as cleaning agent in toothpastes and metal polishes, for sound proofing and for enhancing night visibility of paint.

Phylum–Euglenophyta

(Euglenoid flagellates)

The general characteristics of phylum-Euglenophyta are as follows

- These are unicellular, flagellate protists found in water or damp soil. Majority of them are freshwater organisms found in stagnant water.
- Cell wall is absent in euglenoids. A protein rich layer called **pellicle** or **periplast** is present as their outermost covering. It makes their body flexible.
- They have two flagella, usually one long and one short.
- They show euglenoid movement/metaboly (creeping movement) and flagellar movement.
- The anterior end of the organism has an invagination with three parts, namely **mouth** (cytostome), **canal** (cytopharynx) and **reservoir**.
- At the anterior region of the body, the flagellate possesses an eyespot. This eyespot contains a red pigment **astaxanthin**. It acts as a photoreceptor.
- Nutrition is either holophytic (photoautotrophic) or saprobic (e.g. *Rhabdomonas*) or holozoic (e.g. *Peranema*). Even holophytic forms can pick up organic compounds from the surrounding medium.
- The photoautotrophs or holophytic forms possess chloroplasts with or without pyrenoids (proteinaceous bodies).
- The photosynthetic pigments include chlorophyll-*a* and *b*.
- The reserve food material is stored inside the cytoplasm as **paramylon** or **paramylum granules**.
- A single large nucleus lies roughly in the middle of the cell.
- These include *Rhabdomonas, Petalomonas, Peranema* and *Euglena.*

Reproduction in euglenoid flagellates

- Asexual reproduction occurs under favourable conditions, *via* longitudinal binary fission. While under unfavourable conditions, they undergo multiple fission in an encysted condition. This resembles palmella stage of many algae.
- Sexual reproduction is not seen in them.

Euglena

- This organism shares the characters of both plants as well as animals. It has chlorophyll just like plants and performs photosynthetic nutrition. But in the absence of sunlight, it feeds like animals also.
- A thread-like flagellum arises from its **blepharoplasts** or **basal granule.** It is an organelle formed from centriole. The base of flagellum bears a swelling called **paraflagellar body** (photoreceptors).

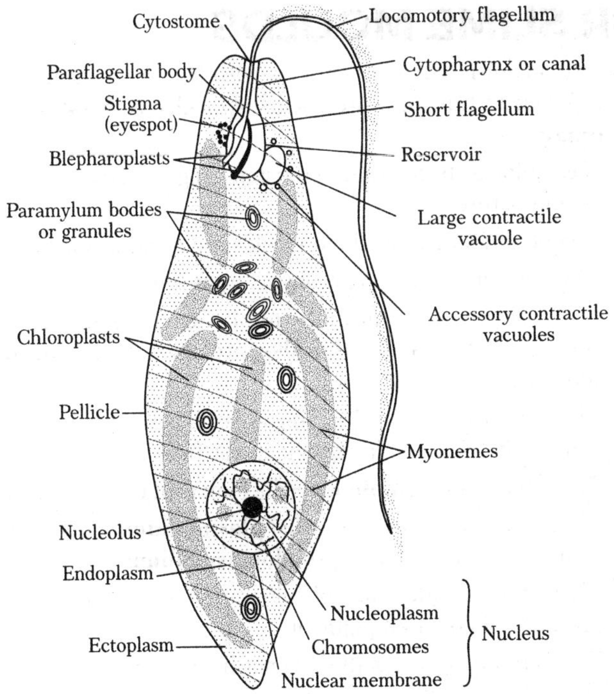

Fig 5.3 Euglena

CHECK POINT 5.1

1. The protistans are unicellular organisms with
   (a) a number of cell organelles without membrane
   (b) they only undergo mixotrophic nutrition
   (c) their cytoplasm is always in motion
   (d) they only survive in colonial forms

2. What is/are the characteristics of dinoflagellates?
   (a) Planktonic golden yellow algae with soapbox like structure
   (b) Marine red biflagellate protists
   (c) Appear orange, black, pink, purple and green in colour
   (d) Diflagellate organisms with pellicle

3. Which of the following is not present in dinoflagellates?
   (a) most of them are unicellular
   (b) They are biflagellate
   (c) Food reserve in them is fucoxanthin
   (d) Pusule in them function in food storage

4. In which of the following organisms, the cell wall is made up of two overlapping shells?
   (a) Chrysophytes
   (b) Euglenoids
   (c) Slime moulds
   (d) Dinoflagellates

5. Which of the following option for diatoms is correct?
   (a) Pecto cellulosic cell wall
   (b) Silicified cell wall
   (c) Multicellular eukaryotes
   (d) Produce saxitoxin

6. The cell wall encloses the peripheral layer of cytoplasm surrounding a large central vacuole in
   (a) *Euglena* (b) *Biddulphia*
   (c) *Amoeba* (d) *Aulosira*

FUNGUS LIKE PROTISTS OR SLIME MOULDS

- These organisms share characters with both animals and fungi. Due to this, they are commonly called as **fungus animals.**
- Modern biologists call them protistan fungi because of their protistan nature.
- The general features of slime moulds are discussed below
  - They are unicellular saprophytic organisms.
  - These organisms occur in the form of slimy mass during their vegetative phase, thus they are also called as **slime moulds.**
  - They usually grow at decaying areas. They are commonly found on lawns and moist fields.
  - They move through pseudopodia (amoeboid movement).
  - Their acellular forms are called **plasmodium**, while the cellular forms are called **pseudoplasmodium.**
  - They exhibit wide range of colouration due to the presence of coloured pigments.
  - They are not green as they do not have chlorophyll.
  - They possess numerous diploid nuclei.
  - They lack cell wall and are surrounded by plasma membrane only. However, the spores have cellulosic cell wall.
  - They produce spores by reduction division. These spores act as gametes and are known as **myxamoebae** (non-motile) or **swarm cells** (motile).
  - Actually during development, swarm cells swim freely in water and undergo repeated division. After a period, they loose their flagella and become amoeboid like, myxamoebae.
  - They show isogamous type of sexual reproduction.

These are of two main types as follows

1. Myxomycota

(Acellular slime moulds)

- About 500 species of plasmodial (acellular) slime moulds are known.
- The somatic structure of Myxomycota is a free-living, diploid, multinucleate, mobile mass of protoplasm called **plasmodium**. It is surrounded by a thick layer called **sclerotium.**
- The sporangium usually contains cytoplasmic strands called **capillitium.**
- They absorb dissolved organic substances from the substratum showing saprotrophic nutrition, e.g. *Physarum*, *Physarella*, *Tubifera* and *Dictydium.*

The various events in the life cycle of acellular slime moulds are shown in the diagram below

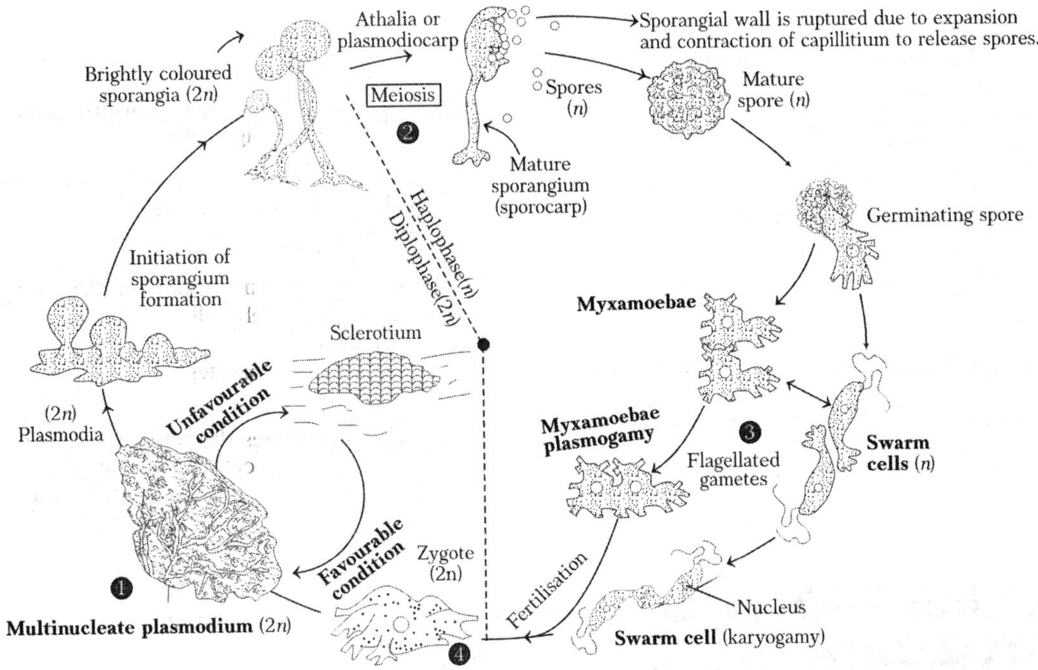

Fig 5.4 Life cycle of acellular slime moulds

2. **Acrasiomycota** (Cellular slime moulds)

- The somatic structure of Acrasiomycetes is a haploid uninucleate cell covered by the plasma membrane only, cell wall absent.
- These somatic cells are called **myxamoebae** (sing. myxamoeba).
- Myxamoebae form microcyst (cellulose layer) as a means of perennation. This helps in the dispersal of mould. It is formed during unfavourable condition, but get ruptured by forming a pore through which protoplasm oozes out.
- The cellular slime moulds are characterised by complete absence of flagella in their life cycle, e.g. *Dictyostelium* and *Polysphondylium.*
- Cellular slime moulds are cosmopolitan and feed over other microorganisms through ingestion.
- The various events taking place at different stages, during the life cycle of cellular slime moulds are shown below

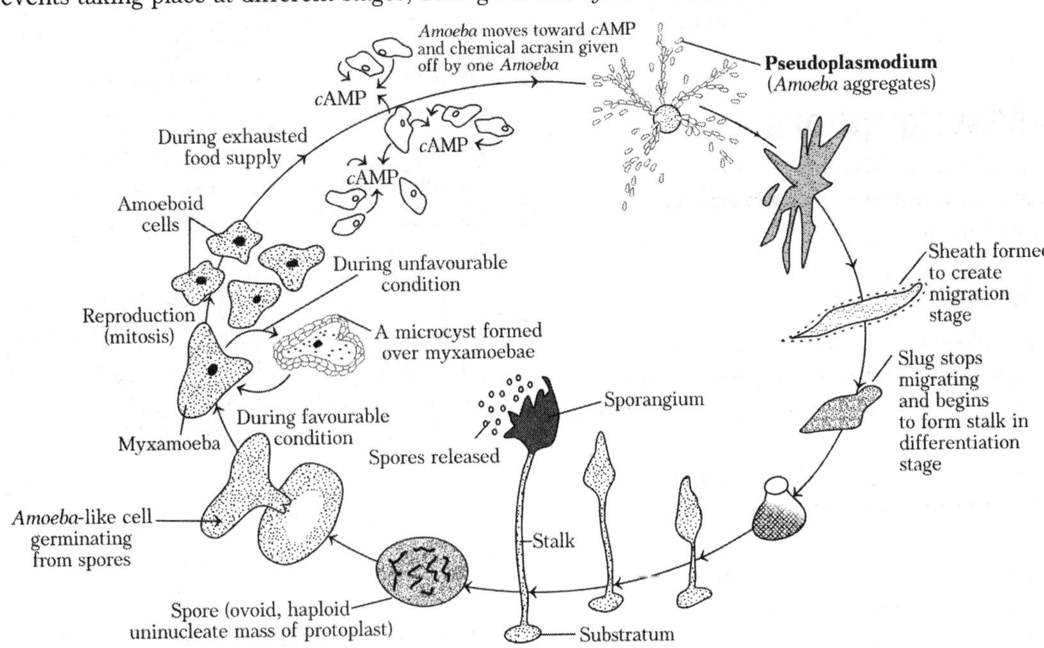

Fig 5.5 Life cycle of cellular slime moulds

CHECK POINT 5.2

1. A feature that categorises a 'true slime mould' is
   (a) occurs as an independent cell
   (b) a multinucleate plasmodial stage
   (c) secretes mucoidal exudate
   (d) None of the above
2. Slime moulds differ from fungi in having
   (a) phagocytosis mode of nutrition
   (b) amoeboid cells
   (c) flagella
   (d) Both (a) and (b)
3. Which of the following combination of characters is true for slime moulds?
   (a) Parasitic plasmodium without walls, spores dispersed by air currents
   (b) Saprophytic plasmodium with walls spores dispersed by water
   (c) Parasitic plasmodium without walls, spores dispersed by water
   (d) Saprophytic plasmodium without walls, spores dispersed by air currents
4. The multinucleate slimy mass of protoplasm which forms the body of slime mould is called as
   (a) plasmodium (b) myxamoeba
   (c) sporocytes (d) periplasmodium
5. In acellular slime moulds, sporangium has an intricate network of cytoplasmic threads called
   (a) peridium (b) pseudopodic
   (c) capillitium (d) trichocyst
6. The structure formed in the life cycle of cellular slime mould due to chemotactic movement is
   (a) macrocyst (b) plasmodium
   (c) pseudopodium (d) swarm cells

ANIMAL LIKE PROTISTS OR PROTOZOAN

The term 'Protozoa' means first animals. It was coined by **Goldfuss** in 1817. But, their presence was first reported by **Av Leeuwenhoek** in 1677.

These include all unicellular or acellular eukaryotic animals. These are called **acellular** because they have the potential to perform all vital activities, which a single-cell cannot perform.

These are considered as animals because of their heterotrophic mode of nutrition and motility.

On the basis of locomotory organelles and nuclear apparatus, protozoans are divided into following four phyla

Phylum – Mastigophora

These are animal-like flagellated protists. These are also called as **zoomastigophora** or **zooflagellates**, e.g. *Trypanosoma*, *Leishmania*, etc.

Trypanosoma gambiense

- It is found in the blood of man and causes the West and Central African **sleeping sickness** (Trypanosomiasis). It is a fatal disease. This disease was discovered by **Forde** in 1901.
- *Trypanosoma* is polymorphic animal, i.e. having four forms namely, *Leishmania*, Leptomonad, Crithidial and Metacyclic.
- Their body has eukaryotic nucleus with a flagellum. The flagellum arises from the posterior end and runs anteriorly with pellicular undulating membrane. This membrane is a parasitic adaptation to swim in the plasma of blood.
- It is **digenetic parasite**, i.e. it completes its life cycle in two hosts. The primary host is man and intermediate host or vector is tse-tse fly (*Glossina palpalis*).
- *Trypanosoma* have a complex developmental cycle as shown below

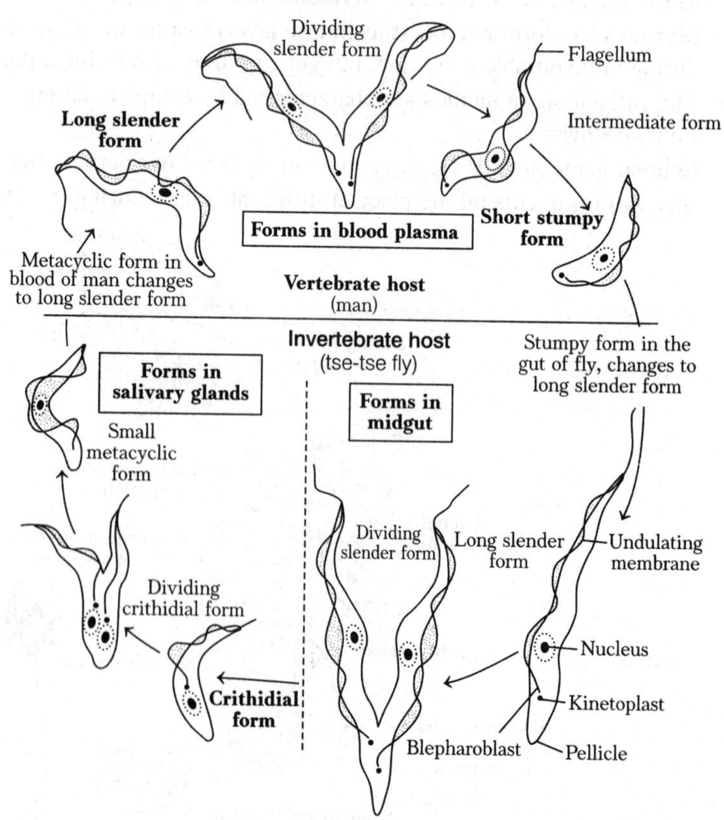

Fig 5.6 Life cycle of Trypanosoma in man and tse-tse fly

- The animal takes its nourishment by osmosis from blood plasma, because a well-defined mouth is absent.
- Its long slender form continuously absorbs glucose from the bloodstream.
- This glucose is converted into pyruvic acid along with ATP *via* glycolysis. ATP produced in glycolysis is utilised by the animal for its metabolic needs, while pyruvic acid is stored in the body for future requirements.
- Asexual reproduction occurs by longitudinal binary fission.

Different species of *Trypanosoma* infecting man and other animals

- *T. rhodensia* – Causes East African or Rhodensian sleeping sickness. Vector–*G. morsitans*.
- *T. cruzi* or *Schizotrypanum cruzi*–Causes South and Central American sleeping sickness or **Chaga's** disease, Vector-A bug (*Triatoma*).
- *T. brucei* – Causes Nagana in domestic animals of Africa. It is basically sleeping sickness of domestic animals. Vector *G. norsitansitams*.

Leishmania donovani

- It is an important pathogenic genus closely related to *Trypanosoma*.
- The genus-*Leishmania* was discovered by **Ross** in 1903.
- Its various species infect man, cattle, dog, sheep, horse, etc., and cause serious diseases collectively known as **Leishmaniasis**.
- All *Leishmania* species are intracellular parasites in leucocytes of blood or in cells of liver and spleen.
- In all types of diseases the vectors are blood sucking sandflies of the genus, *Phlebotomus*.
- *Leishmania donovani* causes a malaria, like oriental disease in man called **kala-azar** or **dumdum fever** or **black fever**.
- The parasite primarily infects the reticuloendothelial system.
- Nourishment is obtained saprozoically through body surface from host cells.
- The parasite exists in two forms **amastigote** (leishmanial) form occurring in man and **promastigote** (leptomanial) form occurring in the gut of the vector (sandfly).
- Gaseous exchange and excretion also occur by diffusion through body surface.
- Sexual reproduction is unknown and asexual reproduction always takes place by longitudinal binary fission in the same manner as in case of *Trypanosoma*.

Table 5.1 Some other important species of Mastigophora

Giardia	Parasite causing diarrhoea
Trichomonas	Parasite causing vaginitis in human females
Trichonympha	Symbiont found in gut of termite causing cellulose digestion
Proterospongia	Colonial form with collar cells, a connecting link between Protozoa and Porifera

Phylum–Sarcodina or Rhizopoda

The body of the organisms is amoeboid, i.e. without definite pellicle. Some are with internal or external skeleton. These are mostly free-living forms. Locomotion takes place by pseudopodia, e.g. *Amoeba, Entamoeba.*

Pseudopodia (*Pseudo*–false, *podia*–feet)

These are temporary structures formed by streaming flow of cytoplasm. These are of following types

- **Lopopodia** These are large and broad pseudopodia with rounded tips, e.g. *Amoeba*.
- **Filopodia** These are filamentous (with tapering base) or branched pseudopodia forming complex network, e.g. shelled *Amoeba*.
- **Reticulopodia** These are also called as **rhizopodia** or **myxopodia**. They are filamentous and interconnected pseudopodia, forming a network, e.g. in foraminiferans (shelled protozoans).
- **Axopodia** These are straight pseudopodia radiating from the body surface, e.g. *Actinosphaerium* and *Actinophyrus* (radiolarians).

Amoeba proteus

- It is the most popular, free-living protozoan. It was first discovered by **RV Rosenhoff** in 1755. Later in 1962, **HR Hirschfield** gave a comprehensive and detailed account of *Amoeba.*
- Pseudopodia (lopopodia) are seen at the anterior region, while posterior region is marked by a wrinkled region called **uroid**. It represents degenerating pseudopodia.

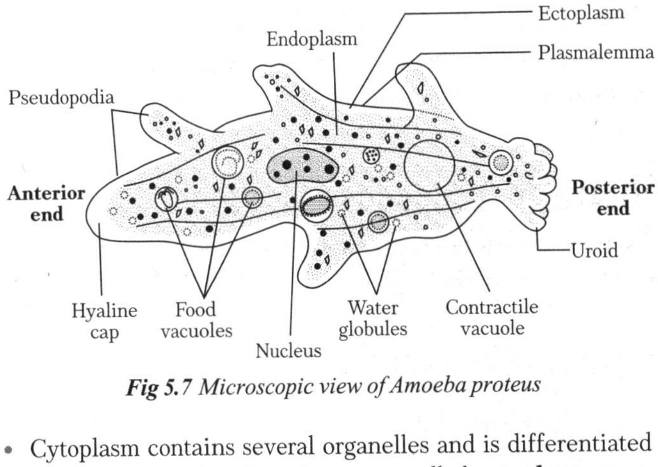

Fig 5.7 Microscopic view of Amoeba proteus

- Cytoplasm contains several organelles and is differentiated into two zones, i.e. the outer one is called **ectoplasm** or **ectosarc** and the inner one is called **endoplasm** or **endosarc.**
- The temporary contractile vacuole is present in animal, which helps in osmoregulation and excretion.
- Food is digested inside the food vacuole. Food vacuole of *Amoeba* is analogous to the alimentary canal of a vertebrate animal or gastrovascular cavity of *Hydra.*

- *Amoeba* moves by the formation of pseudopodia. These pseudopodia are formed as a result of forward flowing of cytoplasm. This mechanism was explained in **Sol-Gel theory**. This theory was first given by **Hyman** (1917) and later supported by **Pantin** (1923-26) and **Mast** (1925). It is the most widely accepted theory now-a-days.
- *Amoeba* is omnivorous and the mode of nutrition is holozoic. It feeds on bacteria, diatoms, algae, other protozoans and even metazoans, e.g. nematodes and rotifers. The ingestion of solid food is called **phagocytosis**.

  Amoeba ingests food by the following ways
  - **Import**, i.e. taking in of food on contact.
  - **Circumvallation**, i.e. to engulf an active prey by forming a cup-like structure called **food cup** around it, e.g. flagellate and ciliates.
  - **Circumfluence**, i.e. for less active or motionless organism, e.g. bacteria, etc.
  - **Invagination**, i.e. the adherence of food due to toxic and sticky substances (secretions) of ectoplasm.
  - **Pinocytosis**, i.e. the cell drinking phenomenon, which assists in the ingestion of liquid food.
- **Digestion** in *Amoeba* is intracellular, by forming food vacuole. The digestive medium is first acidic than alkaline.
- **Egestion** of undigested residue takes place through temporary rupture from any point on the body surface.
- *Amoeba* shows various responses to stimuli called as **taxes**. These are as follows
  - Thermotaxis (response to temperature)
  - Phototaxis (response to light)
  - Thigmotaxis (response to touch)
  - Chemotaxis (response to chemicals)
  - Rheotaxis (response to water current)
  - Galvanotaxis (response to electric currents)
  - Geotaxis (response to gravity)
- *Amoeba* reproduces by the asexual modes like binary fission, multiple fission, sporulation, etc.
- *Amoeba* has great power of regeneration and is also considered immortal.

Entamoeba histolytica

- It is a microscopic monopodial endoparasite found in upper part of large intestine, i.e. colon in man. It causes **amoebic dysentery** or **amoebiasis.**
- It is cosmopolitan in distribution, but its prevalence is greater in tropics and subtropics than in the temperate zone.
- The animal was first discovered by **Lambl** in 1859. Its pathogenic nature was described by **Losch** in 1875, who discovered this organism in the faeces and intestinal ulcers of dysentery patients.
- *Entamoeba* occurs in three distinct forms, the trophozoite or magna form, precystic or minuta form and cystic form.
  - The **trophozoite form** is the most active, motile and feeding form, which is pathogenic to man.
  - The **minuta forms** are small, spherical, non-motile and non-feeding forms.
  - The **cystic form** is formed when minuta form undergoes encystation and becomes rounded. It serves the purpose of transmission of the parasite from one host to another.

Life cycle

- *Entamoeba histolytica* is a monogenetic parasite, i.e. man is the only host.
- It forms tetranucleate or quadrinucleate cysts constituting the transmissive or infective stage. These pass out of the host through its faeces. These are highly resistant to desiccation and survive for about 12 days.
- These cysts if ingested by another host pass through the stomach and reach the small intestine.
- After 5-6 hours, the cyst wall ruptures and a tetranucleate encystic amoebae (metacyst) emerges out.
- The metacyst then divides by binary fission, to produce eight small, uninucleate metacystic trophozoites. These pass into large intestine and grow into mature trophozoites.

Entamoeba hartmanni

It resembles minuta form of *E. histolytica*. The trophozoites measure between 9-14 μ in diameter. The nucleus is more dense. *E. hartmanni* also causes amoebiasis, but is supposed to be less harmful.

Entamoeba coli

This *Entamoeba* is said to be non-pathogenic as it does not invade the host tissue. It lives in the lumen of the intestine and may produce digestive complications. *Entamoeba coli* infects more than 50% of human population.

Entamoeba gingivalis

This *Entamoeba* dwells in nearly 70% of human population. Its incidence is seen in patients suffering from pyorrhoea. The pathogen dwells in gum pockets of mouth and destroys tissues in gums. It was the first parasitic *Amoeba* reported in humans. It was reported by **Gros** in 1849.

CHECK POINT 5.3

1. Which of the following is not a characteristic of protozoans?
(a) Reproduction by binary fission
(b) Presence of contractile vacuole
(c) Cell membrane as an outer body covering
(d) Presence of pseudopodia

2. At which stage *Trypanosoma* is/are able to initiate an infection to its host?
(a) Long and slender form
(b) Crithidial form
(c) Metacyclic form
(d) Intermediate form

3. Consider a locomotion type in which a single pseudopod is formed on the surface of the body. This locomotion is not seen in
(a) *Entamoeba histolytica* (b) *Entamoeba coli*
(c) *Entamoeba gingivalis* (d) *Entamoeba ovale*

4. Egestion of undigested food in *Amoeba* takes place through
(a) circumvallation
(b) a temporary rupture of its surface membrane
(c) a hyaline cap formed at its advancing end
(d) pinocytosis

5. A single-celled animal that catches food and moves about by extending finger like projections of protoplasm. Which is the most widely accepted theory of locomotion in such organism?
(a) Walking movement theory
(b) Rolling movement theory
(c) Sol-gel theory
(d) Surface tension theory

6. The small, spherical, non-motile and non-feeding form of a human parasite is
(a) minuta (b) cystic
(c) sporubts (d) Magna

Phylum–Sporozoa

- These are the protozoans, which lack locomotory organs and contractile vacuole. All of these are endoparasites.
- A thick pellicle is present around their body.
- Reproduction occurs by multiple fission (asexually) or by spore formation (sexually). The multiple fission is called **schizogony** in these protozoans. e.g. *Monocystis, Gregarina, Eimeria, Plasmodium, Babesia bigmina*, etc.

Plasmodium

- It is a sporozoan genera whose species causes malaria in humans and other animals.
- Among these, *P. vivax, P. ovale, P. malariae* and *P. falciparum* cause malaria in humans.
- The parasite resides in the RBCs, reproduces in them and destroys them. Mosquitoes are the transmitting agents or vectors of these malarial parasites.

Life cycle

- It is the most common human malaria causing species in human.
- This parasite is digenetic with human as its primary host and female *Anopheles* mosquito as its secondary host.
- Its life cycle occurs in three phases as follows
  (i) **Schizogony** (Aexual phase) refers to multiple fission. It is of common occurrence in protozoan parasites under favourable conditions, while free-living protozoans show this process normally during unfavourable conditions.
  (ii) **Gametogony** (Sexual phase) means gamete formation. These gametes are not well-defined as the sperms and ovum of metazoans, but can be considered as microgamete and megagamete.
  (iii) **Sporogony** means formation of spores. As the animal belongs to Sporozoa, hence spores serve as the main media of dispersal.
- The infective stage of the parasite for man is sporozoite, which is a sickle-shaped motile individual, 6-15 μm with a distinct nucleus as shown below

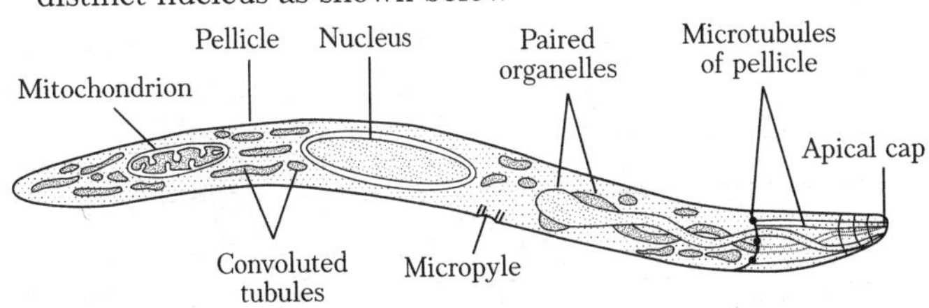

Fig 5.8 Plasmodium structure of a sporozoite as revealed by electron microscope

Asexual cycle

It is divided into following stages

(i) **Liver schizogony** Within the human blood, the sporozoite circulates and enters into liver cells. Liver schizogony has two phases pre-erythrocytic and exoerythrocytic cycles.

Both of these phases are shown in the figure given below

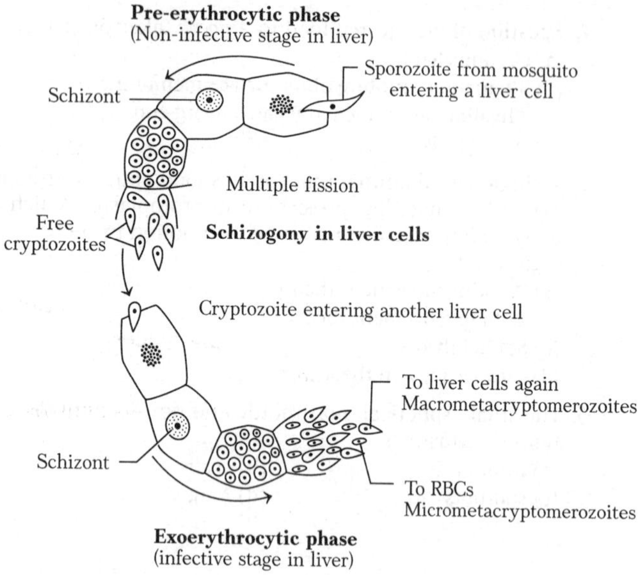

Fig 5.9 Liver schizogony in Plasmodium

The pre-erythrocytic and exoerythrocytic cycles remain immune to the antimalarial drugs and a very little damage occurs to the host during these phases.

The exoerythrocytic cycle is absent in *P. falciparum*.

(ii) **Erythrocytic schizogony** Micrometacrypto- merozoites enter the RBCs after being released in bloodstream. These then become rounded disc-like and called **trophozoites.** It is the feeding stage of parasite.

The complete journey of this trophozoite inside the RBCs is given below

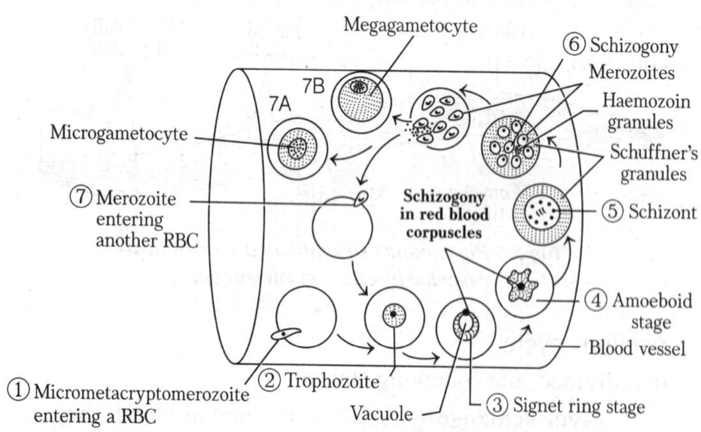

Fig 5.10 Blood schizogony of Plasmodium in asexual cycle

- The digestive enzymes of trophozoite bring about proteolysis of blood haemoglobin. The haemoglobin breaks down into its protein component and haematin.
- Protein is used as food by the trophozoite, while the unused haematin forms the toxic malarial pigment called **haemozoin.**
- This release of haemozoin in the blood causes chills and body pain along with fever.
- At the time of haemozoin release small red granules also appear in the cytoplasm of the corpuscle. These are known as **Schuffner granules.** These are characteristic for any *Plasmodium* sp.
- Merozoites are the end product of this cycle. These are liberated along with haemozoin in blood plasma. These again enter fresh RBC and repeat the erythrocytic cycle to produce large number of merozoites.

(iii) **Gametocyte formation** When the conditions become unfavourable to the parasite a few merozoites do not repeat erythrocytic cycle. Instead, they attack RBCs and get converted into gametocytes.

These gametocytes are common in the subcutaneous blood capillaries of malarial patient. The formation of gametocytes is the need of parasite for its survival.

Sexual cycle

- Sexual life cycle of *Plasmodium* in mosquito is also called **Ross cycle.** When a mosquito sucks the blood of infected person, it also ingests erythrocytes present in the form of gametocytes.
- Their journey inside the mosquito body involves two major processes namely **gametogony,** i.e. the formation of gametes and **sporogony,** i.e. the formation of spores.
- Both of these processes are bridged with fusion of gametes or fertilisation in between.
- Thus, the events inside mosquito involve the following
  (i) gametogony
  (ii) fertilisation
  (iii) sporogony
- Sporogony results to sporozoites, which reside in the salivary gland of mosquito and wait for reinfecting the humans.

This complete phase of *Plasmodium* is shown in figure given below

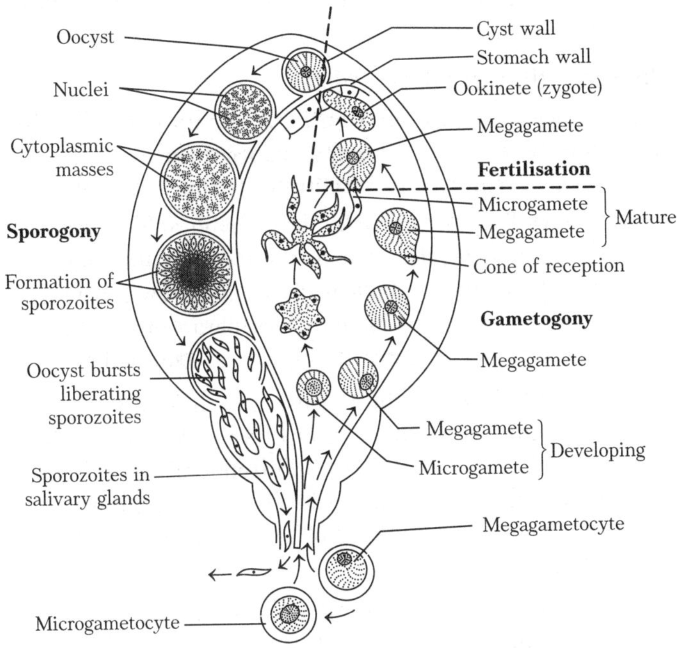

Fig 5.11 Sexual cycle of Plasmodium in mosquito

As this phase involves fusion of gametes that is why we call it sexual cycle. In mosquito, the whole cycle is completed in 2-4 weeks depending upon the temperature.

Malaria

- Term 'malaria' was coined by **McCulloch** in 1827. It is the most important parasitic disease of man.
- **Charles Laveran** (1880) first discovered this parasite in the blood of man.
- **Golgi** and **Celli** (1885) observed the malarial parasite in RBCs and also found that by injecting the blood of one malarial patient to other, the second one also gets the disease.
- **Sir Ronald Ross** (1897) was the person, who discovered that the disease was transmitted by female *Anopheles* mosquito.
- The complete life cycle of malarial parasite was discovered by **Grassi** (1898) in the stomach of *Anopheles* mosquito.

Types of malaria

Following four types of malaria are recognised on the basis of the period of recurrence of fever and infecting species

(i) **Tertian, benign tertian** or *vivax* malaria, with 48 hours fever cycle, caused by *P. vivax.*

(ii) **Quartan malaria**, with 72 hours fever cycle, caused by *P. malariae.*

(iii) **Ovale** or mild tertian malaria, with 48 hours fever cycle, caused by *P. ovale.*

(iv) **Subtertian** or malignant tertian malaria also called **Aestivo autumnal malaria, cerebral malaria** or **blackwater fever**, with 48 hours fever cycle, caused by *P. falciparum.*

Control and therapy

(i) Elimination of mosquitoes either by destruction of adult or elimination at breeding places or by destructing larvae and pupae of mosquito.

(ii) Vaccination is not possible but various synthetic drugs like, quinine, chloroquine, atabrine, daraprim, etc., can be given.

Besides, Paludrin and Resochin (chloroquine) are said to be more effective as these kill every stage of parasite. Daraprim is most promising drug, but is slow in action.

Table 5.2 Some other important Sporozoa

Monocystis	Parasite in the seminal vesicles of earthworm
Eimeria	Causes coccidiosis in rabbit and chicken
Babesia	Causes haemoglobinuric fever/tick fever in cattle

Phylum–Ciliophora

- It is the largest and the most advanced group of protozoans with about 600 species.
- Locomotory organelles are cilia or associated ciliary organelles.
- Nuclear dimorphism, i.e. more than two nuclei are present with different structure.
- This is the only subphylum, which contains animals with a definite cell mouth (cytostome) and cell anus (cytopyge).
- The contractile vacuoles are permanent.
- This phylum contains a single class—Ciliata, including all ciliophora, e.g. *Balantidium, Chromidina, Paramecium* and *Tetrahymena,* etc.

Paramecium

- *Paramecium* is a ciliate discovered by **Hill** in 1752.
- It contains about eight known species. Few common are *P. caudatum, P. aurelia* and *P. multimicronucleatum.*
- It is commonly called **'slipper animalcule'** because of its slipper-like shape. The body is elongated, blunt and rounded at the anterior end and somewhat pointed at the posterior end.

- Heterokaryotic condition is seen with two different nuclei. One is small and spherical called **micronuclei** while other one is large and oval called **macronuclei**. Out of these, the micronucleus works for reproduction while macronucleus is the controller of day to day activities.
- Contractile apparatus occupies fixed position in endoplasm. These are two in number, i.e. one at anterior end and other at posterior end.
- Peculiar rod-like or oval organelles for offence and defence are seen in the ectoplasm. These are called **trichocyst**. These were first seen and reported by **Ellis**.
- The ingestion of food takes place through a well-defined cell mouth or cytostome. The oral apparatus contains an oral groove, which leads to a tubular structure called **vestibule.**
- The animal is a ciliary filter feeder, i.e. feeds on microorganisms like bacteria, algae and yeasts. The cilia of buccal apparatus help in ingestion by mouth.
- Many non-contractile vesicles or vacuoles are found in the cytoplasm, called **gastrioles**. These contain food particles along with the digestive enzymes. These vacuoles are always seen during streaming movement (cyclosis).
- The posterior region has a cytopyge or cell anus or cytoproct for expelling undigested waste.

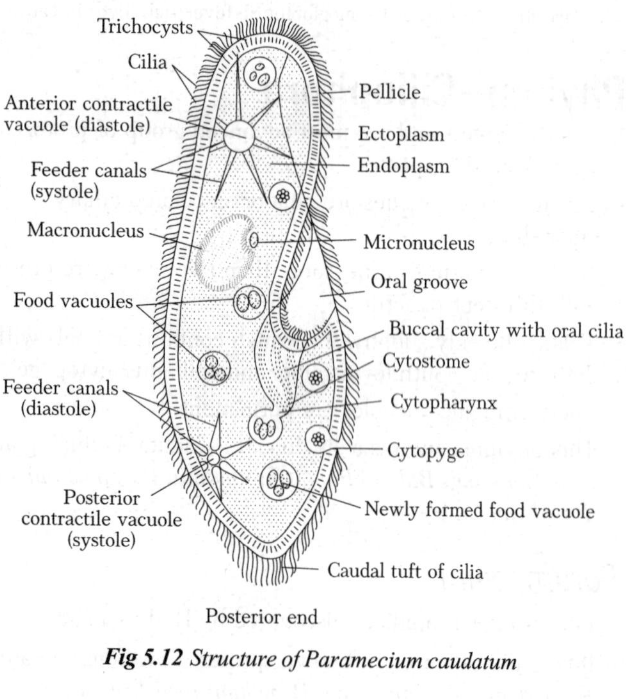

Fig 5.12 Structure of Paramecium caudatum

- Certain specialised particles are found in the endoplasm of *Paramecium*. These particles are Kappa particles (containing paramecin toxin), Pi particles, Mu and Lambda particles.

Reproduction in *Paramecium*

It reproduces asexually by transverse binary fission. It also undergoes several other **'nuclear reorganisation processes'** to maintain its vitality for fission. These are

(i) **Binary fission** It occurs under favourable conditions. In the process, the animal stops feeding, becomes spindle-shaped and starts dividing. After transverse fission, two transverse daughter paramecia namely, proter (upper) and opisthe (lower) are formed.

(ii) **Conjugation** It is the process, which occurs frequently in between binary fission. It is an initial form of cross-fertilisation and frequently referred to as sexual reproduction. It is a simple temporary union of two individuals of same species with different strains.

A cross exchanging of micronuclear parts is the characteristic feature. In *P. caudatum*, it is considered as the major process for nuclear reorganisation under favourable conditions.

(iii) Other modes of reproduction are **autogamy** (self-fertilisation), **cytogamy** (sexual reproduction without nuclear exchange), **endomixis** (total internal nuclear reorganisation) and **hemixis** (an act of purification on behalf of macronucleus).

Useful activities of protozoans

Usually protozoans are disease causing, but some of them are useful in nature and to humans also. There uses are as follows

- Many protozoans feed on purifying bacteria, thus help indirectly in the purification of water.
- They form first trophic level in food chains for the ocean world.
- Some protozoans are examples of symbionts like intestinal flagellates (*Trichonympha, Colonympha*, etc.) of termites and wood roaches. These provide help in their digestion process.
- Protozoans due to their minute size and quick reproduction are studied by geneticists for heredity variations and physiological studies.

Table 5.3 Summary of some additional protozoan parasites pathogenic to humans

Phylum	Genus	Species	Habitats	Pathogenic effects
Mastigophora	*Giardia* (intestinal flagellate)	*intestinalis*	Small intestine of man and other animals	Diarrhoea, giardiasis, loss of appetite and weight
	Trichomonas (genital flagellate)	*vaginalis*	Vagina	Vaginitis
Sporozoa	*Isospora*	*hominis*	Intestine	Diarrhoea
		belli	Intestine	Diarrhoea
	Toxoplasma	*gondii*	RBCs, parenchyma cell	Encephalomyelitis, chorioretinitis, etc.
Ciliophora	*Balantidium* (largest protozoan infecting humans)	*coli*	Large intestine	Balantidiasis (in human), cause ulcer in colon, diarrhoea and constipation

Table 5.4 Symbiotic protozoan parasites that are non-pathogenic to humans

Phylum	Genus	Species	Habitats
Sarcodina	*Entamoeba*	*gingivalis*	Mouth
		coli	Large intestine
	Endolimax	*nana*	Large intestine
	Iodamoeba	*butschlii*	Large intestine
	Dientamoeba	*fragilis*	Large intestine
Mastigophora (Zoomastigophora)	*Chilomastix*	*mesnili*	Caecum
	Trichomonas	*hominis*	Ileocaecal region
		tenax	Teeth and gum
	Enteromonas	*hominis*	Large intestine
	Embadomonas	*intestinalis*	Large intestine
	Trypanosoma	*rangeli*	Blood

CHECK POINT 5.4

1. An active feeding stage in the life cycle of certain protozoan and those of the *Giardia* group is
   (a) trophozoite (b) sporozoite
   (c) merozoite (d) metacryptozoite
2. Prepatent period in malarial infection is
   (a) interval between inoculation and fever
   (b) duration between inoculation and start of blood cycle
   (c) interval between two life cycles
   (d) duration between repetition of fever
3. A structure in the cortex of certain ciliate and flagellate protozoans consisting of a cavity and long thin threads is required for
   (a) osmoregulation (b) nutrition
   (c) reproduction (d) defence
4. In *Paramecium*, metabolic activities and growth is regulated by
   (a) macronucleus (b) micronucleus
   (c) Both (a) and (b) (d) contractile vacuole
5. In *Paramecium*, both conjugation and autogamy are sexual processes because of
   (a) involvement of two individuals
   (b) fusion of two haploid nuclei
   (c) gene recombination
   (d) rejuvenation
6. A ciliated protozoan parasite is responsible for causing a large intestinal infection.
   (a) *Giardia* (b) *Trichomonas*
   (c) *Balantidium* (d) *Toxoplasma*

EXAM POINTS

- All single-celled eukaryotic organisms like chrysophytes, euglenoids, dinoflagellates and slime moulds are included in the kingdom–Protista. **[NEET 2016, Kerala CEE 2013]**
- Protists obtain their food as photosynthesisers and heterotrophs. **[AIIMS 2013]**
- *Chlorella* and *Chlamydomonas* are placed in kingdom–Protista. They are green phytoplanktons and photosynthetic and have chloroplast for carrying out photosynthesis. **[Chhattisgarh PMT 2013]**
- Chrysophytes are commonly called dinoflagellates. They are parasitic forms which cause diseases in animals. They have a protein rich layer called pellicle. They are saprophytic protists. **[Kerala CEE 2011]**
- In dinoflagellates, organisms have two overlapping shells which fit together to form cell walls. **[CBSE AIPMT 2015]**
- Diatoms are chief producers in the oceans. They are microscopic and float passively in water. 'Diatomaceous earth' is formed by the cell walls of diatoms. **[NEET 2016]**
- The walls of diatoms cannot be easily destroyed because their wall is made up of nano-silica (SiO_2). **[NEET 2016]**
- The diatoms produce auxospores. **[EAMCET 2015]**
- Euglenoids possess a protein rich layer called pellicle. **[Kerala CEE 2014]**
- *Euglena, Amoeba* and *Noctiluca* are all free-living protozoans. **[Bihar CECE 2015]**
- Slime moulds are fungus like organisms in one phase of their life cycle and *Amoeba* like organisms in another phase of their life cycle. **[AIIMS 2012]**
- Motile zygote of *Plasmodium* occurs in gut of female *Anopheles.* **[CBSE AIPMT 2012]**
- Pseudopodium is the motile stage of protozoans helpful in feeding. **[JIPMER 2018]**
- *Trypanosoma* is polymorphic organism. **[UP CPMT 2012]**
- Excretion in *Amoeba* occurs through plasma membrane. **[UP CPMT 2012]**
- The method of dispersal in *Amoeba* is encystment. **[Manipal 2012]**
- *Triatoma infestans* is the intermediate host in the life cycle of *Trypanosoma cruzi*. **[Bihar CECE 2012]**
- Ciliates differ from all other protozoans in having two types of nuclei. **[NEET 2020]**
- Sporozoite is the infectious stage of *Plasmodium* that enters the human body. **[NEET 2020, KCET 2014]**
- Schizonts are seen in *Plasmodium* and are formed by multiple fission this process is known as schizogony. **[Gujarat CET 2017]**
- In *Paramecium*, cytoplasmic bridge is formed at the time of exchange of gamete nuclei during sexual reproduction. **[Gujarat CET 2015]**
- During the life cycle of *Plasmodium*, sporozoites multiply in liver cell, not in blood. **[KCET 2015]**
- Babesiosis is a malaria like disease caused by protozoan, *Babesia bigemina*. **[CBSE AIPMT 2015]**
- Protozoan with heliopodia as locomotory structure is *Actinophrys*. **[EAMCET 2014]**
- The type of syngamy seen in *Trichonympha* is hologamy. **[EAMCET 2014]**
- Unicellular algae, diatoms and protozoans are members of Protista. **[J&K CET 2013]**
- Macronucleus and micronucleus are the characteristics feature of *Paramecium* and *Vorticella*. **[UP CPMT 2012]**
- Schizogony occurs in liver of human. **[BHU 2012]**

Chapter Exercises

(A) Topical Exercise

Characteristics of Protista

1 Protists are
(a) unicellular and prokaryotes
(b) unicellular and eukaryotes
(c) multicellular and eukaryotes
(d) multicellular and saprotrophic

2 The major productive factors of the aquatic ecosystem are members of
(a) Pyrrophyta (b) Chrysophyta
(c) Euglenophyta (d) All of these

3 Consider the following statements.
I. The phylum–Protozoa was first observed by Anton van Leeuwenhoek.
II. Protozoans are known as grazers of microbial world.
Choose the correct option.
(a) Statement I is true, but II is false
(b) Statement I is false, but II is true
(c) Both statements I and II are true
(d) Both statements I and II are false

4 Lifestyles represented by Protista are
(a) plant like (b) animal like
(c) fungal like (d) All of these

Photosynthetic protists

5 Photosynthetic protists belong to
(a) Bacillariophyceae
(b) Bacillariophyceae and Euglenophyceae
(c) Bacillariophyceae, Euglenophyceae and Dinophyceae
(d) Zooflagellates

6 A nucleus with condensed chromosome is found in
(a) prochloron (b) dinoflagellates
(c) diatoms (d) euglenoids

7 A dinoflagellate with venom containing stinging organelles is
(a) *Nematodium* (b) *Ceratium*
(c) *Gonyaulax* (d) *Symbodinium*

8 Consider the following statements.
I. The mode of asexual reproduction in dinoflagellates is only by cell division.
II. A dinoflagellate in which mesokaryon does not occur in somatic cells is *Peridinium.*
Choose the correct option.
(a) Statement I is true, but II is false
(b) Statement I is false, but II is true
(c) Both statements I and II are true
(d) Both statements I and II are false

9 How many of the following features are common in dinoflagellates?
Cell wall, mesokaryon, flagella, pusule, silica, paraflagellar body, swarm cells, trichocysts
(a) Four (b) Five (c) Seven (d) Three

10 Which species of protists are known as the whirling whips because of the spins produced by two flagella beating in opposing grooves along their hard surfaced bodies?
(a) Golden-brown algae
(b) Chrysophytes
(c) Diatoms
(d) Dinoflagellate

11 *Noctiluca* swims with the help of
(a) transverse flagellum (b) longitudinal flagellum
(c) tentacle (d) Both (a) and (b)

12 In *Noctiluca*, mesokaryotic stage appears in
(a) somatic stage
(b) at the time of first binary fission
(c) after a few binary fission
(d) midway during gametogenesis

13 Consider the following statements.
I. *Peridinium* exhibits the phenomenon of bioluminescence
II. *Gonyaulax* exhibits the phenomenon of bioluminescence.
Choose the correct option.
(a) Statement I is true, but II is false
(b) Statement I is false, but II is true
(c) Both statements I and II are true
(d) Both statements I and II are false

14 A theca is made up of cellulose and also contains many texture plates made up of
(a) silica (b) lime
(c) cellulose (d) protein

15 In Dinophyceae, a structure in the cortex of certain ciliates and flagellates occurs in
(a) *Peridinium* (b) *Nematodinium*
(c) *Ceratium* (d) *Gonyaulax*

16 Identify the statement which is incorrect about *Gonyaulax*?
(a) It is an armoured dinoflagellate
(b) Its mesokaryon is large and subterminal
(c) It reproduces by binary fission
(d) It produces a patent neurotoxin called saxitoxin

17 Xanthophytes are
(a) yellow-green algae
(b) unicellular and coccoid
(c) having many discoid chromatophores
(d) All options are correct

18 Consider the following statements.
I. Diatoms belong to Chrysophyta.
II. Diatoms belong to Myxophyta.
Choose the correct option
(a) Statement I is true, but II is false
(b) Statement I is false, but II is true
(c) Both statements I and II are true
(d) Both statements I and II are false

19 Choose the correct statement about diatoms.
(a) They belong to class-Bacillariophyceae
(b) Have fucoxanthin pigment which gives brown colour
(c) Male gamete has one whiplash flagellum
(d) All of the above

20 Consider the following statements
I. Diatoms have leucosin as reserve material.
II. Diatoms cell size decreases with each cell division.
Choose the correct option.
(a) Statement I is true, but II is false
(b) Statement I is false, but II is true
(c) Both statements I and II are true
(d) Both statements I and II are false

21 To survive in unfavourable condition, several diatoms and few cyanobacteria form
(a) homocysts and auxospores, respectively
(b) auxospores only
(c) auxospores and homocysts, respectively
(d) homocysts only

22 Diatomaceous earth is used for insulating boilers and steam pipes because it is
(a) cheap and easily available
(b) good conductor of heat
(c) bad conductor of heat
(d) made up of $CaCO_3$

23 Identify the statement which is incorrect about euglenoids.
(a) The apical end of the organism has an invagination, which bears a mouth, canal and reservoir
(b) The eyespot present on the anterior region of the body contains the pigment fucoxanthin
(c) In the absence of light, they behave as heterotrophs
(d) The reserve food material is stored in cytoplasm as paramylon granules

24 ...... is not true for *Euglena.*
(a) The presence of chlorophyll-*a* and *b*
(b) The presence of protoplasmic capsule
(c) The presence of cellulose cell wall
(d) The presence of proteinaceous pellicle

25 Metaboly is a creeping movement of euglenoids which occurs by
(a) contraction of the body
(b) sliding of the body
(c) slithering movement of the body
(d) expansion and contraction of the body

26 Consider the following statements.
I. *Euglena* are multicellular organisms.
II. Mixotrophic nutrition is present in *Euglena.*
Choose the correct option.
(a) Statement I is true, but II is false
(b) Statement I is false, but II is true
(c) Both statements I and II are true
(d) Both statements I and II are false

27 The eyespot or stigma of *Euglena* occurs at the attachment of the membranes of
(a) contractile vacuole (b) cytopharynx
(c) reservoir (d) Both (b) and (c)

28 Consider the following statements.
I. *Euglena* can be considered as a plant due to the presence of chlorophyll.
II. *Euglena* cannot be classified on the basis of two kingdom system of classification.
Choose the correct option.
(a) Statement I is true, but II is false
(b) Statement I is false, but II is true
(c) Both statements I and II are true
(d) Both statements I and II are false

Fungus Like Protists or Slime Moulds

29 Sporangium of acellular slime moulds is characterised by
(a) absence of covering (b) presence of covering
(c) capillitium (d) Both (b) and (c)

30 Consider the following statements.
I. The plasmodium of slime moulds which lacks veins is known as protoplasmodium.
II. Slime moulds are ecologically consumers.
Choose the correct option.
(a) Statement I is true, but II is false
(b) Statement I is false, but II is true
(c) Both statements I and II are true
(d) Both statements I and II are false

31 Which group of organisms is represented by the given figure?

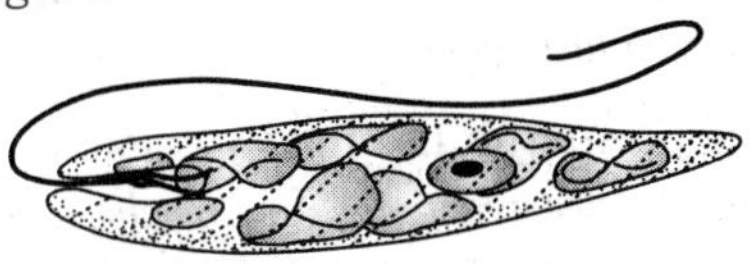

(a) Dinoflagellates (b) Protozoans
(c) Slime mould (d) Euglenoids

32 Slime moulds are included in Protista, otherwise these are very close to
(a) plants (b) fungi (c) algae (d) bacteria

33 Choose the incorrect pair about slime moulds.
(a) Pigment - Chlorophyll
(b) Male gametes - Sporocarp
(c) Amoeboid shaped cells - Swarm cells
(d) Spore dispersal - Water

34 Which of the following statement is incorrect for cellular slime moulds?
(a) Microcyst is for perennation in slime moulds
(b) Communal slime moulds are cellular slime mould
(c) Cellular slime mould feed over other microorganisms through ingestion
(d) Cellular slime mould is *Dictydium*

35 Acellular slime moulds are
(a) communal slime moulds (b) haploid slime moulds
(c) plasmodial slime moulds (d) aplano slime moulds

36 Mass of streaming protoplasm in acellular slime moulds is
(a) plasmodium (b) pseudoplasmodium
(c) sorophorm (d) scleroteum

37 A cytoplasm without cell wall and cell membrane, multinucleated and saprophytic are the characteristics of
(a) Monera (b) Protista
(c) Fungi (d) Slime moulds

38 Consider the following statements.
I. The lime knots of slime moulds are capillitia.
II. The pigment present in some slime moulds is chlorophyll.
Choose the correct option.
(a) Statement I is true, but II is false
(b) Statement I is false, but II is true
(c) Both statements I and II are true
(d) Both statements I and II are false

39 An aggregation of myxamoebae resembling a plasmodium, but without protoplasmic fusion is formed in
(a) cellular slime moulds (b) acellular slime moulds
(c) zooflagellates (d) dinoflagellates

Animal like Protists or Protozoans

40 The term 'Protozoa' was coined by
(a) Goldfuss
(b) Anton van Leeuwenhoek
(c) Ernst Haeckel
(d) None of the above

41 Protozoans
(a) are unicellular or acellular eukaryotes
(b) show heterotrophic nutrition
(c) have locomotory organelles
(d) All of the above

42 The characteristic not associated with *Trypanosoma* is
(a) longitudinal binary fission
(b) digenetic parasite
(c) flagellum
(d) free-living protozoan

43 Consider the following statements.
I. Black fever is caused by *Trypanosoma.*
II. Cause of 'Surra' disease in cattle is *T. evansi.*
Choose the correct option.
(a) Statement I is true, but II is false
(b) Statement I is false, but II is true
(c) Both statements I and II are true
(d) Both statements I and II are false

44 Sometimes, *Trypanosoma* passes into internal organs like spleen, lungs, where they lose the flagella and become oval-shaped. What is the resulting form's name?
(a) Parabasal body
(b) Latent bodies
(c) Rhizoplast
(d) *Leishmania* form

45 In the body of tse-tse fly, three morphological forms of *Trypanosoma gambiense* are observed, they are metacyclic form, long and slender form and
(a) megalospheric form (b) microspheric form
(c) crithidial form (d) primary form

46 The number of flagella are more than four in
(a) some parasitic zooflagellates
(b) some free-living zooflagellates
(c) some dinoflagellates
(d) some euglenoids

47 Choose the incorrect match.
(a) *Trypanosoma rhodensia* — East African sleeping sickness
(b) *T. cruzi* — Chaga's disease
(c) *T. brucei* — South American sleeping sickness
(d) *T. evansi* — Surra disease

48 Phylum-Sarcodina is characterised by the presence of
(a) cell wall
(b) pseudopodia for locomotion
(c) endoparasites
(d) All of the above

49 *Amoeba* is placed in phylum—Protozoa, because of
(a) the presence of cell wall
(b) phagocytic mode of nutrition
(c) acellular body
(d) the presence of contractile vacuole

50 Match the following columns.

Column I (Types of pseudopodia)	Column I (Protozoan species)
A. Lobopodia	1. Euglypha
B. Filopodia	2. *Globigerina* (Foraminiferans)
C. Reticulopodia	3. *Amoeba*
D. Axopodia	4. *Actinophyrus*

Codes

	A	B	C	D
(a)	3	1	2	4
(b)	4	3	2	1
(c)	1	2	3	4
(d)	2	4	1	3

51 In *Amoeba*, the contractile vacuole is present
(a) near the trailing end
(b) near the advancing end
(c) at the middle of the body
(d) anywhere inside the body

52 The function of contractile vacuole is increased when
(a) it is needed to remove waste products
(b) it has stopped feeding
(c) the cell membrane is permeable to water
(d) the concentration of water outside is higher than that inside

53 A full-grown *Amoeba* undergoes binary fission. The total surface area of one daughter *Amoeba* soon after division is likely to be
(a) slightly less than that of parent *Amoeba*
(b) equal to the half of that of the parent *Amoeba*
(c) slightly more than half of that of the parent *Amoeba*
(d) less than half of that of the parent *Amoeba*

54 If an *Amoeba* is kept in distilled water, its contractile vacuole will
(a) disappear
(b) become swollen and full of water
(c) work faster
(d) have no effect

55 Encysted stage is not known to occur in the life of
(a) *Entamoeba gingivalis* (b) *Entamoeba coli*
(c) *Entamoeba histolytica* (d) *Amoeba*

56 RBCs are present in the food vacuole of
(a) *Entamoeba histolytica* (b) *Amoeba proteus*
(c) *Trypanosoma gambiense* (d) *Leishmania donovani*

57 Carriers of *Entamoeba histolytica* are
(a) mosquito of genus-*Anopheles*
(b) cattle
(c) *Musca domestica* (housefly)
(d) healthy human host

58 The minuta form of *Entamoeba histolytica* differs from *Amoeba proteus* in having
(a) one pseudopodium formed of ectoplasm
(b) one pseudopodium formed of endoplasm
(c) no food vacuoles
(d) small contractile vacuole

59 What would happen, if some freshwater protozoans are placed in a medium of high osmotic potential?
(a) The contractile vacuole will work faster
(b) The contractile vacuole grows larger
(c) The contractile vacuole shows no change
(d) The contractile vacuole does not form

60 Sporozoa includes diverse organisms that have
(a) thousands of cilia on the body surface
(b) flagella on their body
(c) infectious spore-like stage in their life cycle
(d) silica shells on their surface

61 One of the characteristic features of *Telosporea* is
(a) uninuclear trophozoites
(b) multinuclear trophozoites
(c) the presence of shells
(d) the presence of an eyespot

62 Which is an incorrect combination?
(a) Haemocyanin—Prawn
(b) Haemocyanin—RBC
(c) Haemoglobin in plasma—*Pheretima*
(d) Haemozoin—*Plasmodium* cytoplasm

63 Sporozoite, infective state of *Plasmodium* parasite contains
(a) a nucleus
(b) two nucleus and a vacuole
(c) vacuoles and chloroplasts
(d) one nucleus and several vacuoles

64 What is cryptomerozoite of *Plasmodium*?
(a) A product of gametogamy
(b) A product of schizogony
(c) A product of sporogony in mosquito
(d) None of the above

65 Schuffner's dots in a patient of malaria, are found in
(a) RBCs (b) gametocyte of *Plasmodium*
(c) liver cells (d) signet ring trophozoite

66 Beside RBCs, *Plasmodium* attacks one or more types of human body cells during its normal courses of life cycle. These human body cells are
(a) liver cells (b) muscle cells
(c) columnar cells of intestine (d) nerve cells

67 Consider the following statements.
I. Signet ring stage of *Plasmodium* is found in the erythrocytic cycle.
II. Schuffner's granules are found in schizont.
Choose the correct option.
(a) Statement I is true, but II is false
(b) Statement I is false, but II is true
(c) Both statements I and II are true
(d) Both statements I and II are false

68 Sporozoite of *Plasmodium* is
(a) amoeboid and uninucleate
(b) pelliculate, uninucleate and with numerous mitochondria
(c) capable to pass across a blood capillary
(d) multinucleate

69 Malaria parasite can be best obtained from its patient
(a) one hour before rise of the temperature
(b) five hours after rise of the temperature
(c) when temperature rises with rigor
(d) anytime

70 Shivering with cold, i.e. the attack of the disease is experienced by a patient of malaria when
(a) merozoites and toxic substances are released in its blood on completion of an erythrocytic cycle
(b) state of high fever has subsided
(c) patient is very weak
(d) cold wind is blowing

71 Consider the following statements.
I. Cell anus or cytopyge is present at the posterior of the *Paramecium*.
II. Numerous food vacuoles are present in *Paramecium*.
Choose the correct option.
(a) Statement I is true, but II is false
(b) Statement I is false, but II is true
(c) Both statements I and II are true
(d) Both statements I and II are false

72 The function of the neuromotor system of *Paramecium* is
(a) coordination of ciliary beat
(b) coordination of various stimuli
(c) control of digestion
(d) coordination of respiratory movement

73 In *Paramecium*, the coordinated movement of row of cilia causes the water laiden with food to be
(a) expelled out from the cell
(b) steered into the gullet
(c) cycled into the protoplasm
(d) None of the above

74 What is incorrect about *Paramecium*?
(a) Macronucleus can be more than one
(b) Micronucleus can be more than one
(c) Macronucleus is vegetative
(d) Micronucleus is reproductive

75 During conjugation in *Paramecium*
(a) out of the 4 micronuclei formed 3 degenerate
(b) out of 6 macronuclei formed 4 degenerate
(c) zygote nucleus undergoes 8 successive divisions in each conjugant
(d) out of 16 nuclei, only 4 degenerate

(B) Cumulative Exercise

(Assorted Questions of the Chapter for Advanced Level Practice)

1 Read the following statements regarding *Plasmodium* species causing malaria.
I. *Plasmodium* performs gametogony and sporogony inside the mosquito.
II. Schizont stage occurs in RBCs and liver.
III. Cerebral malaria is drug resistant and fatal.
IV. Schuffner's dots are present in *Plasmodium vivax*.
Choose the incorrect statement(s).
(a) Only I
(b) III and IV
(c) II and IV
(d) None of the above

2 Read the following statements about *Paramecium*.
I. *Paramecium* has Kappa, Lambda, Mu and Pi particles in cytoplasm.
II. *Paramecium aurelia* shows autogamy.
III. *Paramecium* has trichocyst present on its body for offence and defence.
IV. Conjugation of *Paramecium* involves exchange of nuclei.
Select the above statements as True (T) or False (F).

	I	II	III	IV		I	II	III	IV
(a)	T	F	T	F	(b)	F	T	F	T
(c)	F	F	T	T	(d)	T	T	T	T

3 Read the following statements and fill in the blanks by choosing the correct option given below.

I. Mode of nutrition in *Trypanosoma* is ....
II. Infective stage of *Trypanosoma gambiense* is ...... form.
III. ...... is the common mode of reproduction in *Euglena* and *Trypanosoma*.
IV. *Trypanosoma* belongs to phylum ....

	I	II	III	IV
(a)	parasitic	metacyclic	Binary fission	Mastigophora
(b)	saprophytic	crithidial	Sexual reproduction	Sarcodina
(c)	phototrophic	leptomonas	Budding	Mastigophora
(d)	holozoic	metacyclic	Asexual reproduction	Sarcodina

4 Read the following statements.

I. Kala-azar caused by *Leishmania donovani* is also called as black fever.
II. *Leishmania* species divides asexually by budding.
III. *Leishmania* species obtains its nourishment saprozoically through body surface of host cells.

Choose the incorrect statement(s).

(a) Only I (b) Only II
(c) II and III (d) I and III

5 Match the following columns.

Column I	Column II
A. Protozoan	1. *Gonyaulax*
B. Bacillariophyceae	2. Diatoms
C. Euglenoid	3. *Paramecium*
D. Dinoflagellates	4. *Euglena*

Codes

	A	B	C	D		A	B	C	D
(a)	2	3	4	1	(b)	4	2	3	1
(c)	1	3	2	4	(d)	3	2	4	1

6 Which one of the following statements is correct?

(a) Tse-tse fly spreads kala-azar
(b) Sandfly spreads sleeping sickness
(c) *Trichonympha*, a symbiotic protozoan is found in the gut of termite
(d) *Pediculus humanus corporis* is an endoparasite

7 Which of the following are the characters of dinoflagellates?

I. Planktonic golden-yellow algae with soapbox-like structure.
II. Marine red biflagellated protists.
III. Appear yellow, green, brown, blue or red in colour.
IV. Biflagellated organisms with pellicle.
V. Saprophytic or parasitic unicellular forms.

Choose the correct option.

(a) I, II and III (b) II, IV and V
(c) II and III (d) II and V

8 Match Column I with Column II and select the correct option from the codes given below.

	Column I		Column II
A.	Chief producers in oceans	1.	Euglenoids
B.	Red tides	2.	Diatoms
C.	Mixotrophic nutrition	3.	Slime moulds
D.	Plasmodium	4.	Dinoflagellates

Codes

	A	B	C	D		A	B	C	D
(a)	2	4	1	3	(b)	2	4	3	1
(c)	2	3	1	4	(d)	1	4	3	2

9 Choose the option with the correct statements about *Entamoeba histolytica*.

I. It causes amoebic dysentery in man.
II. Its adult form is called trophozoite.
III. Its infection occurs chiefly in spleen and liver.
IV. It belongs to phylum—Mastigophora.

Codes

(a) I and II (b) III and IV (c) I and III (d) II and IV

10 Match the following columns.

Column I (Structures)		Column II (Features)
A. Magna form	1.	Precystic stage of *Entamoeba*
B. Minuta form	2.	Most active form of *Entamoeba histolytica*
C. Trophozoite	3.	Infective stage of *Entamoeba histolytica*
D. Tetranucleate cyst	4.	The normal trophozoites in *Entamoeba histolytica*

Codes

	A	B	C	D		A	B	C	D
(a)	1	2	3	4	(b)	4	1	3	2
(c)	4	1	2	3	(d)	3	4	1	2

11 Select the correct combination of characteristic for slime moulds.

(a) Parasitic, spore dispersal by water, plasmodium with walls
(b) Saprophytic, spore dispersal by air currents, plasmodium without walls
(c) Parasitic, spore dispersal by air currents, plasmodium without walls
(d) Saprophytic, spore dispersal by water, plasmodium with walls

12 A biologist was experimenting with a protozoan and noticed that the animal's contractile vacuole stopped working although the other parts of the organism seemed to be in good health and exhibited normal activity. Which of the following experiment was likely to have produced this result?

(a) Changing the pH of the medium from 7.00-6.50
(b) Cooling of the medium from 20-10°C
(c) Transferring the protozoan from sea water to pond water
(d) Transferring the protozoan from pond water to sea water

13 Choose the correct option for the parts labelled *A-E* of the given figure of *Trypanosoma*.

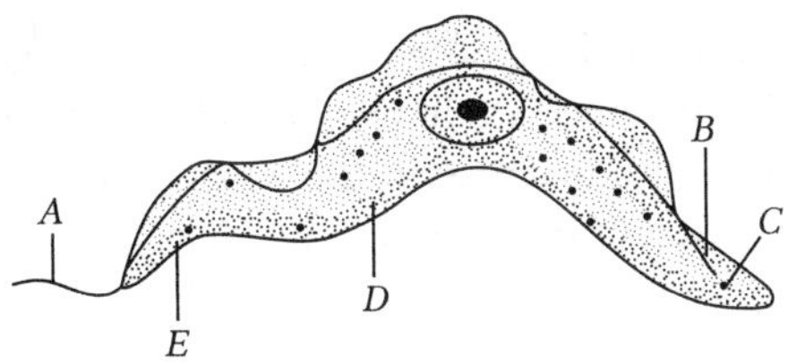

	A	*B*	*C*	*D*	*E*
(a)	Attached flagella	Nucleus	Food granule	Endoplasm	Ectoplasm
(b)	Undulating membrane	Blepharoplast	Pellicle	Ectoplasm	Kinetoplast
(c)	Free flagellum	Blepharoplast	Kinetoplast	Endoplasm	Pellicle
(d)	Free flagellum	Nucleus	Kinetoplast	Pellicle	Food granule

14 Which are true about malarial parasite?
I. Sporogony occurs in stomach cavity of mosquito.
II. After erythrocytic cycle, gametes are formed.
III. Gametogony is found in salivary glands of mosquito.
IV. The recurrence of fever is after every 48 hours.

Codes
(a) I and IV
(b) II and III
(c) II and IV
(d) III and IV

15 Given below are some statements regarding harmful protozoans. Identify the ones which is/are correct.
I. *Uroglenopsis* produces bad foul odour in water bodies.
II. *Balantidium* causes ulcer in colon.
III. *Isospora* causes vaginitis.
IV. *Trypanosoma rhodensia* causes Chaga's disease.

Codes
(a) Only II
(b) I, III and IV
(c) I and II
(d) I, II, III and IV

16 Which of the following is not a character of Protista?
(a) Protists are prokaryotic
(b) Some protists have cell walls
(c) Mode of nutrition is both autotrophic and heterotrophic
(d) Body organisation is cellular

17 Match the following columns.

Column I	Column II
A. Dinoflagellate	1. *Dictyostelium*
B. Plasmodium	2. *Noctiluca*
C. Myxomycota	3. Slime mould
D. Acrasiomycota	4. *Fuligo*

Codes

	A	B	C	D		A	B	C	D
(a)	3	2	1	4	(b)	1	4	3	2
(c)	2	3	4	1	(d)	4	3	1	2

18 Losch discovered that
(a) *Entamoeba histolytica* causes amoebic dysentery
(b) trophozoite of *Plasmodium* has brownish-yellow pigment granules, called haemozoin granules
(c) *Trypanosoma* causes sleeping sickness
(d) digestion in *Amoeba* is similar to that of higher animals

19 Match the following columns.

Column I	Column II
A. *E. histolytica*	1. Kala-azar
B. *E. gingivalis*	2. Sleeping sickness
C. *P. vivax*	3. Amoebic dysentery
D. *T. gambiense*	4. Malaria
	5. Pyorrhoea

Codes

	A	B	C	D		A	B	C	D
(a)	4	3	2	1	(b)	3	4	5	1
(c)	3	5	4	2	(d)	3	5	2	1

Assertion and Reason

Direction (Q. Nos. 20-27) *In the following questions, a statement of Assertion (A) is given followed by the corresponding statement of Reason (R). Of the statements, mark the correct answer as*

(a) Both A and R are true and R is the correct explanation of A
(b) Both A and R are true, but R is not the correct explanation of A
(c) A is true, but R is false
(d) A is false, but R is true
(e) Both A and R are false

20 **Assertion** (A) Ciliates use cilia for locomotion.
Reason (R) Ciliates lack nuclear dimorphism.

21 **Assertion** (A) *Euglena* lacks nuclear membrane.
Reason (R) It is an obligatory phototroph.

22 **Assertion** (A) *Giardia* is called 'Grand old man' of the intestine.
Reason (R) It is a protozoan parasite.

23 Assertion (A) There is no chance of transmission of malaria to a man on the bite of a male *Anopheles* mosquito.

Reason (R) It carries a non-virulent strain of *Plasmodium.*

24 Assertion (A) Macronucleus in *Paramecium* is polyploid.

Reason (R) Micronucleus in *Paramecium* is diploid.

25 Assertion (A) Contractile vacuole is present in the endoplasm of *Amoeba.*

Reason (R) They are responsible for osmoregulation.

26 Assertion (A) *Paramecium* takes food by exocytosis.

Reason (R) In *Paramecium*, ejection of undigested material occurs through endocytosis.

27 Assertion (A) Conjugation is a temporary union between two ciliates belonging to two different mating types for the exchange and reconstitution of nuclear material.

Reason (R) It occurs between two inactive individuals, which have gained their vigour and vitality due to chromosomal imbalance in their macronuclei, caused by repeated amitotic division.

(C) Medical Entrances' Gallery

(Collection of Questions Asked in NEET & in Various Medical Entrance Exams)

1. The infectious stage of *Plasmodium* that enters the human body is **[NEET 2020]**

(a) sporozoites (b) female gametocytes
(c) male gametocytes (d) trophozoites

2 Match the following Columns I and II. **[AIIMS 2019]**

	Column I		Column II
A.	Chrysophyte	1.	*Gonyaulax*
B.	Dinoflagellate	2.	*Euglena*
C.	Euglenoids	3.	Diatom
D.	Slime moulds	4.	Plasmodium

Codes

	A	B	C	D
(a)	1	3	2	4
(b)	1	4	2	3
(c)	3	2	4	1
(d)	3	1	2	4

3 Ciliates differ from all other protozoans in **[NEET 2018]**

(a) using pseudopodia for capturing prey
(b) having a contractile vacuole for removing excess water
(c) using flagella for locomotion
(d) having two types of nuclei

4 Which motile-stage of protozoans is helpful in feeding? **[JIPMER 2018]**

(a) Pseudopodium (b) Cilia
(c) Flagella (d) Tentacles

5 Refer to the following figures. **[AIIMS 2017]**

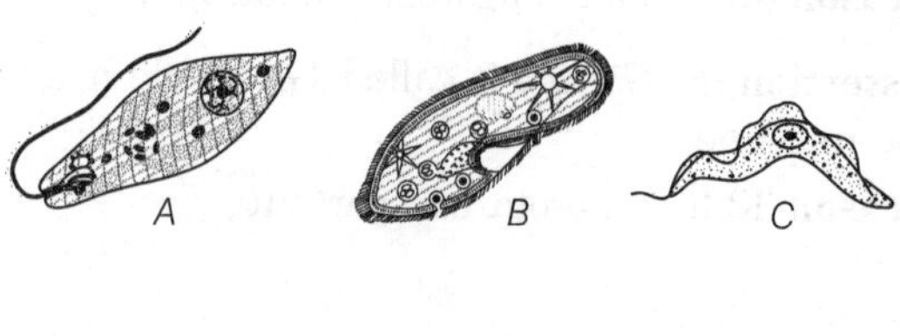

Match the following columns and choose the correct option from the codes given below.

Column I	Column II	Column III
A	1. *Paramecium*	I. Kinetoplast
B	2. *Euglena*	II. Trichocysts
C	3. *Trypanosoma*	III. Astaxanthin

Codes

	A	*B*	*C*
(a)	3-II	1-III	2-I
(b)	1-II	3-I	2-III
(c)	2-III	1-II	3-I
(d)	3-I	2-II	1-III

6 Match the following columns.

	Column I		Column II
A.	Pinocytosis	1.	*Euglena gracilis*
B.	Holozoic	2.	*Paramecium*
C.	Parasitic	3.	*Amoeba proteus*
D.	Mixotrophic	4.	*Monocystis*

Codes **[JIPMER 2017]**

	A	B	C	D
(a)	3	2	4	1
(b)	2	3	4	1
(c)	4	3	1	2
(d)	1	4	2	3

7 Which option contains the incorrectly matched group? **[Gujarat CET 2017]**

(a) Gemmules-Spongilla-Budding
(b) Pseudopodiospores-*Plasmodium*-Sporulation
(c) Zoospores-*Aspergillus*-Sporulation
(d) Conidia-*Penicillium*-Asexual reproduction

8 Chrysophytes, euglenoids, dinoflagellates and slime moulds are included in the kingdom **[NEET 2016]**
(a) Protista (b) Fungi
(c) Animalia (d) Monera

9 Select the incorrect statement. **[NEET 2016]**
(a) The walls of diatoms are easily destructible
(b) 'Diatomaceous earth' is formed by the cell walls of diatoms
(c) Diatoms are chief producers in the oceans
(d) Diatoms are microscopic and float passively in water

10 Identify the characteristic features of diatoms. **[EAMCET 2015]**
(a) Cell wall is made of cellulose, floating on water and produce auxospores
(b) Cell wall is made of chitin, fixed form and produce auxospores
(c) Cell wall is made of silica, floating on water and produce auxospores
(d) Cell wall is made of silica, symbionts and produce zoospores

11 Green phytoplanktons are kept in which kingdom of five kingdom classification system? **[Chhattisgarh PMT 2015]**
(a) Kingdom—Monera
(b) Kingdom—Protista
(c) Kingdom—Plantae
(d) Kingdom—Fungi

12 Which is not a free-living protozoan? **[BCECE 2015]**
(a) *Euglena* (b) *Amoeba*
(c) *Giardia* (d) *Noctiluca*

13 In which group of organisms, the cell walls form two overlapping shells which fit together? **[CBSE AIPMT 2015]**
(a) Chrysophytes (b) Euglenoids
(c) Dinoflagellates (d) Slime moulds

14 Match the following columns.

Column I	Column II
A. *Entamoeba histolytica*	1. Cytozoic parasite
B. *Plasmodium*	2. Coelozoic parasite
C. *Wuchereria*	3. Rhabditiform larva
D. Extraintestinal migration	4. Monogenic parasite
	5. Lymphadenitis

Codes **[EAMCET 2015]**

	A	B	C	D
(a)	4	2	1	3
(b)	4	1	5	3
(c)	5	1	2	3
(d)	4	2	5	1

15 Which structure is formed at the time of exchange of gamete nuclei in given animal during sexual reproduction?

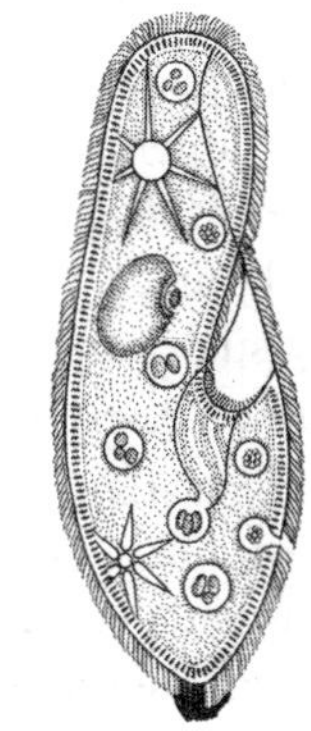

[Gujarat CET 2015]
(a) Cytoplasmic bridge (b) Cytoplasmic filaments
(c) Internal tubule (d) Plasmodesmata

16 Which one of the following statements is not correct with respect to malaria? **[Karnataka CET 2015]**
(a) RBCs rupture and release haemozoin which causes chills
(b) Sporozoites multiply in blood
(c) Female *Anopheles* mosquito is the vector
(d) Malignant malaria is caused by *Plasmodium falciparum*

17 Which one of the following diseases is caused by a protozoan? **[CBSE AIPMT 2015]**
(a) Syphilis (b) Influenza
(c) Babesiosis (d) Blastomycosis

18 Which group of organisms possesses a protein rich layer called pellicle? **[Kerala CEE 2014]**
(a) Chrysophytes (b) Euglenoids
(c) Dinoflagellates (d) Slime moulds
(e) None of the above

19 Sandflies play significant role in spreading kala-azar because they **[WB JEE 2014]**
I. suck blood only from the patients suffering from kala-azar.
II. convert amastigote into promastigote.
III. engulf amastigote at the time of blood sucking from infected persons.
IV. inject promastigote into the body of non-infected person at the time of blood sucking.

Codes
(a) I and II (b) II and III
(c) II, III and IV (d) III and IV

20 The protozoan with heliopodia as locomotory structures is **[EAMCET 2014]**
(a) *Actinophrys* (b) *Entamoeba*
(c) *Elphidium* (d) *Euglypha*

21 The type of syngamy seen in *Trichonympha* is **[EAMCET 2014]**
(a) hologamy (b) anisogamy
(c) isogamy (d) conjugation

22 The time interval of appearance of fever in the malaria patients depends on the types of malaria. The previous evidences suggest that such time intervals are (1) 36 hours, (2) 48 hours and (3) 72 hours. If any such person experiences fever at an interval of 48 hours, then the patient suffers from
(a) only benign tertian malaria **[WB JEE 2014]**
(b) quarantan malaria or mild tertian malaria
(c) malignant tertian malaria or benign tertian malaria
(d) mild tertian malaria or benign tertian malaria

23 Unicellular algae, diatoms and protozoans are members of **[J&K CET 2013]**
(a) Monera (b) Protista
(c) Fungi (d) Plantae

24 From the options below, select the group that includes protozoans, protistan algae and slime moulds. **[Odisha JEE 2013]**
(a) Fungi (b) Plantae (c) Monera (d) Protista

25 Protists obtain their food as **[AIIMS 2013]**
(a) photosynthesisers
(b) chemosynthesisers
(c) heterotrophs
(d) Both (a) and (c)

26 Diatoms do not decay easily because they **[AIIMS 2013]**
(a) have siliceous walls
(b) body is impervious to water
(c) are chitinous
(d) are abundant in saline soil

27 These organisms are fungus like in one phase of their life cycle and *Amoeba* like in another phase of their life cycle. **[AIIMS 2012]**
(a) Diatoms
(b) Slime moulds
(c) Dinoflagellates
(d) Water moulds

28 Which one single organism or the pair of organisms is correctly assigned to its or their named taxonomic group? **[CBSE AIPMT 2012]**
(a) *Paramecium* and *Plasmodium* belong to the same kingdom as that of *Penicillium*
(b) Lichen is a composite organism formed from the symbiotic association of an alga and a protozoan
(c) Yeast used in making bread and beer is a fungus
(d) *Nostoc* and *Anabaena* are examples of Protista

29 Motile zygote of *Plasmodium* occurs in **[CBSE AIPMT 2012]**
(a) gut of female *Anopheles* (b) salivary glands of *Anopheles*
(c) human RBCs (d) human liver

30 Excretion in *Amoeba* occurs through **[UP CPMT 2012]**
(a) lobopodia (b) uroid portion
(c) plasma membrane (d) contractile vacuole

31 Macro and micronucleus are the characteristic feature of **[UP CPMT 2012]**
(a) *Paramecium* and *Vorticella*
(b) *Opalina* and *Nictothirus*
(c) *Hydra* and *Balantidium*
(d) *Vorticella* and *Nictothirus*

32 What is true about *Trypanosoma*? **[UP CPMT 2012]**
(a) Polymorphic (b) Monogenetic
(c) Facultative parasite (d) Non-pathogenic

33 The methods of dispersal in *Amobea* is **[Manipal 2012]**
(a) locomotion (b) sporulation
(c) encystment (d) binary fission

34 *Triatoma infestans* is the intermediate host in the life cycle of **[BCECE 2012]**
(a) *Leishmania donovani* (b) *Trypanosoma cruzi*
(c) *Leishmania tropica* (d) *Schistosoma haematobium*

35 Schizogony occurs in **[BHU 2012]**
(a) RBC of human (b) intestine of parasite
(c) liver of human (d) liver of parasite

36 Which of the following is not correctly matched? **[Kerala CEE 2011]**
(a) *Anabaena* — Cyanobacteria
(b) *Amoeba* — Protozoa
(c) *Gonyaulax* — Dinoflagellated
(d) *Thermoacidophiles* — Archaebacteria
(e) *Albugo* — Chrysophytes

37 Which one of the following is a characteristic feature of chrysophytes? **[Kerala CEE 2011]**
(a) They are parasitic forms which cause diseases in animals
(b) They have a protein rich layer called pellicle
(c) They have indestructible wall layer deposited with silica
(d) They are commonly called dinoflagellates
(e) They are saprophytic protists

38 Where will you look for the sporozoites of the malarial parasite? **[CBSE AIPMT 2011]**
(a) Red blood corpuscles of human suffering from malaria
(b) Spleen of infected humans
(c) Salivary glands of freshly moulted female *Anopheles* mosquito
(d) Saliva of infected female *Anopheles*

ANSWERS

Check Point 5.1

1. (c)	2. (b)	3. (d)	4. (a)	5. (b)	6. (b)

Check Point 5.2

1. (b)	2. (d)	3. (d)	4. (a)	5. (c)	6. (c)

Check Point 5.3

1. (c)	2. (c)	3. (b)	4. (b)	5. (c)	6. (a)

Check Point 5.4

1. (a)	2. (b)	3. (d)	4. (a)	5. (b)	6. (c)

(A) Topical Exercise

1. (b)	2. (d)	3. (c)	4. (d)	5. (c)	6. (b)	7. (a)	8. (b)	9. (b)	10. (d)
11. (d)	12. (b)	13. (c)	14. (c)	15. (a)	16. (b)	17. (d)	18. (a)	19. (d)	20. (c)
21. (c)	22. (c)	23. (b)	24. (c)	25. (d)	26. (b)	27. (c)	28. (c)	29. (d)	30. (a)
31. (d)	32. (b)	33. (d)	34. (d)	35. (c)	36. (a)	37. (d)	38. (a)	39. (a)	40. (a)
41. (d)	42. (d)	43. (b)	44. (d)	45. (c)	46. (a)	47. (c)	48. (b)	49. (c)	50. (a)
51. (a)	52. (c)	53. (b)	54. (c)	55. (c)	56. (a)	57. (d)	58. (b)	59. (d)	60. (c)
61. (a)	62. (d)	63. (a)	64. (b)	65. (d)	66. (a)	67. (c)	68. (c)	69. (c)	70. (a)
71. (c)	72. (a)	73. (b)	74. (a)	75. (a)					

(B) Cumulative Exercise

1. (d)	2. (d)	3. (a)	4. (b)	5. (d)	6. (c)	7. (c)	8. (a)	9. (a)	10. (c)
11. (b)	12. (d)	13. (c)	14. (c)	15. (c)	16. (a)	17. (c)	18. (a)	19. (c)	20. (c)
21. (e)	22. (b)	23. (c)	24. (b)	25. (a)	26. (a)	27. (c)			

(C) Medical Entrances' Gallery

1. (a)	2. (d)	3. (d)	4. (a)	5. (c)	6. (a)	7. (b)	8. (a)	9. (a)	10. (c)
11. (b)	12. (c)	13. (c)	14. (d)	15. (a)	16. (b)	17. (c)	18. (b)	19. (c)	20. (a)
21. (a)	22. (d)	23. (b)	24. (d)	25. (d)	26. (a)	27. (b)	28. (c)	29. (a)	30. (c)
31. (a)	32. (a)	33. (c)	34. (b)	35. (c)	36. (e)	37. (c)	38. (d)		

Explanations

(A) Topical Exercise

5 (*c*) Photosynthetic protists belong to Bacillariophyceae, Euglenophyceae and Dinophyceae because they are capable of preparing their own food by the process of photosynthesis.

6 (*b*) A nucleus with condensed chromosome known as mesokaryon is found in dinoflagellates. It exhibits the characters of both prokaryotic as well as eukaryotic nucleus.

8 (*b*) Statement I is false, but II is true. Incorrect statement can be corrected as

The mode of asexual reproduction in dinoflagellates are cell division and encystment.

12 (*b*) In *Noctiluca*, mesokaryotic stage appears at the time of first binary fission because mesokaryotic stage possesses intermediate resemblance to both prokaryotic and eukaryotic cells.

14 (*c*) A theca that is made up of cellulose and also contains many texture plates are dinophycean theca and it is made up of cellulosic plates. Silica is mainly found in diatomaceous earth obtained from diatoms.

16 (*b*) Statement in option (b) is incorrect and can be corrected as

In *Gonyaulax*, the mesokaryon is large and central.

Rest statements are correct about *Gonyaulax.*

18 (*a*) Statement I is true, but II is false. Incorrect statement can be corrected as

Diatoms do not belong to Myxophyta.

23 (*b*) Statement in option (b) is incorrect and can be corrected as

The eyespot present on the anterior region of the body contains a red pigment astaxanthin.

Rest statements are correct about euglenoids.

25 (*d*) Metaboly is a creeping movement of euglenoids which occurs by expansion and contraction of the body. *Euglena* and other euglenoids are best known for their characteristic undulating, shape changing motion called metaboly. Many euglenoids can also glide and swim using their flagella. Members of this group are either unicellular or colonial.

26 (*b*) Statement I is false, but II is true. Incorrect statement can be corrected as

Euglena are unicellular organisms.

30 (*a*) Statement I is true, but II is false. Incorrect statement can be corrected as

Slime moulds are decomposers.

33 (*d*) Option (d) is incorrect pair and can be corrected as

The spores are dispersed by air currents in slime moulds.

Rest options are correct pairs about slime moulds.

34 (*d*) Statement in option (d) is incorrect and can be corrected as

Cellular slime mould is *Dictyostelium. Dictydium* is acellular slime mould.

Rest statements are correct for cellular slime moulds.

38 (*a*) Statement I is true, but II is false. Incorrect statement can be corrected as

Chlorophyll is not present in slime moulds.

42 (*d*) The characteristic not associated with *Trypanosoma* is free. living protozoan because

As *Trypanosoma* belongs to phylum-Protozoa and locomotion in it, takes place by flagellum. It is digenetic in character, i.e. it acts as an endoparasite, so cannot be free-living protozoan.

43 (*b*) Statement I is false, but II is true. Incorrect statement can be corrected as

Sleeping sickness is caused by *Trypanosoma.*

46 (*a*) The number of flagella are more than four in some parasitic zooflagellates. In case of some free-living zooflagellates, the number of flagella is two. In some dinoflagellates, there are two flagella and in case of some euglenoids, there is only one flagellum.

51 (*a*) In *Amoeba*, the contractile vacuole is present in the posterior part (tailing end) at the end which helps in dragging the organism or moving the organism. This vacuole is filled with water.

52 (*c*) The function of contractile vacuole is increased when the cell membrane is permeable to water. Contractile vacuoles keep water entry in check. They expand when water enters and contract when filled with water, carrying water along with wastes outside of the cell.

54 (*c*) If an *Amoeba* is kept in distilled water, its contractile vacuole works faster due to increased rate of endosmosis.

57 (*d*) *E. histolytica* is monogenetic, i.e. only one host is required for its complete life cycle. Pigs, dogs, rats and rabbits are supposed to be the reservoir host. It undergoes reproduction and completes its life cycle only in man. Thus, carriers of *Entamoeba histolytica* are healthy human host.

59 (*d*) If some freshwater protozoans are placed in a medium of high osmotic potential, the contractile vacuole does not form because its function is to remove the excess water outside by the process of osmoregulation.

If a medium of high osmotic potential, it is necessary to retain the water inside the body instead of expelling out. So, it would not be formed.

64 (*b*) Cryptomerozoite of *Plasmodium* is a product of schizogony. It infects host such as human body as in case of malarial parasite, initiates asexual of malarial parasite, initiates asexual life cycle thereby causing several infection in hosts.

(B) Cumulative Exercise

4 (*b*) All the statements are correct except II. Incorrect statement can be corrected as

Leishmania species divides asexually by binary fission.

6 (*c*) Statement in option (c) is correct. Other statements are incorrect and can be corrected as

- Tse-tse fly spreads sleeping sickness.
- Sandfly spread kala-azar.
- *Pediculus humanus corporis* is an ectoparasite.

7 (*c*) Statements II and III are characters of dinoflagellates. Other statements are incorrect characters of dinoflagellates and can be corrected as

- Diatoms not dinoflagellates are planktonic golden-yellow algae with soapbox like structure.
- Pellicle is present in euglenoid flagellates.
- Dinoflagellates are parasitic, symbiotic or decomposers.

9 (*a*) Statements I and II are correct. Other statements are incorrect about *Entamoeba histolytica* and can be corrected as

Entamoeba belong to phylum-Sarcodina. Its infection occurs chiefly in intestine.

14 (*c*) Statements II and IV are true. Other statements are false and can be corrected as

- Sporogony occurs in salivary gland of mosquito.
- Gametogony occurs in stomach cavity of mosquito.

15 (*c*) Statements I and II are correct regarding harmful protozoans. Statements III and IV are incorrect and can be corrected as

- *Trichomonas* causes vaginitis
- *Trypanosoma cruzi* causes Chaga's disease.

16 (*a*) Option (a) is not a character of protista and can be corrected as

Protists are eukaryotes.

Rest options are characters of protista.

20 (*c*) A is true, but R is false because

In ciliates, cilia are used for locomotion and driving food. They show nuclear dimorphism, i.e. two types of nuclei (macronucleus and micronucleus).

21 (*e*) Both A and R are false and can be corrected as

Euglena have eukaryotic cell organisation (i.e. well-organised nucleus and cell organelles). It is a facultative autotroph and makes its own food through photosynthesis.

22 (*b*) Both A and R are true, but R is not the correct explanation of A.

A protozoan, *Giardia* is commonly nicknamed as the 'Grand old man' of the intestine. It is a diplomonadida parasitic flagellate occurring in man and other animals. *Giardia intestinalis* is a parasite of small intestine of man and it stays there for a long period.

23 (*c*) A is true, but R is false because

Malaria cannot be transmitted by the bite of male *Anopheles* mosquito as it does not carry any active stage of *Plasmodium.*

24 (*b*) Both A and R are true, but R is not the correct explanation of A.

Paramecium contains two nuclei. The larger bean-shaped macronucleus is polyploid and controls metabolic activities, while micronucleus is diploid and controls reproduction.

25 (*a*) Both A and R are true and R is the correct explanation of A.

A fixed contractile vacuole is present in the endoplasm of *Amoeba.* It is responsible for osmoregulation, i.e. the maintenance of a constant water potential inside the cell.

26 (*a*) Both A and R are true and R is the correct explanation of A.

In *Paramecium,* food particles are injected into a food vacuole formed by the endoplasm (endocytosis). The vacuole follows a distinct pathway through the endoplasm, where undigested material is egested (exocytosis).

27 (*c*) A is true, but R is false because

Conjugation is a temporary union between two ciliates belonging to two different mating types for the exchange and reconstitution of nuclear material.

This happens probably due to the fact that macronucleus loses its potentialities in performing its manifold metabolic activities. Its replacement by a new macronucleus brings renewed vigour and vitality to accelerate the metabolic activities.

(C) Medical Entrances' Gallery

1 (*a*) The infectious stage of *Plasmodium* that enters the human body is sporozoites, present in salivary gland of *Anopheles* mosquito.

The sporozoites grow and multiply in the liver to become merozoites.

These merozoites invade the erythrocytes (RBCs) to form trophozoites, schizonts and gametocytes, during which the symptoms of malaria are produced.

3 (*d*) Ciliates differ from all other protozoans in having two types of nuclei.

These two nuclei are usually of different size, i.e. one is meganucleus and the other is micronucleus. The former controls metabolism whereas the latter is concerned with reproduction, e.g. *Paramecium.*

4 (*a*) Pseudopodium helps protozoans in feeding. It is a temporary cytoplasm filled projection of the eukaryotic cell membrane. It can be used for motility or for ingesting nutrients or other particulate matter.

5 (*c*) The correct matches are as follows

A-Astaxanthin is a keto carotenoid, present in *Euglena.* The red colour of the *Euglena* eyespot is due to this pigment.

B-Trichocyst is a structure in the cortex of certain ciliate, e.g. *Paramecium.*

C-Kinetoplast is a network of circular DNA inside a large mitochondrion that contains many copies of the mitochondrial genome, e.g. *Trypanosoma.*

7 (*b*) Option (b) contains the incorrectly matched group and can be corrected as

Schizont-*Plasmodium*-Schizogony (multiple fission).

Schizogony is a common mode of asexual reproduction that occurs in protozoan parasites under favourable conditions.

Rest options are the correctly matched groups.

8 (*a*) All single-celled eukaryotic organisms like chrysophytes (diatoms and desmids), euglenoids (*Euglena*), dinoflagellates and slime moulds are included in the kingdom-Protista.

9 (*a*) Statement in option (a) is incorrect and can be corrected as

Diatoms are single-celled plant-like protists that produce intricately structured cell wall made up of nano-silica (SiO_2). Thus, the walls are indestructible.

All other statements are correct.

11 (*b*) Green phytoplanktons are photosynthetic and have chloroplast for carrying out photosynthesis. Since, these are unicellular eukaryotic organisms. They are placed in kingdom—Protista, e.g. *Chlorella, Chlamydomonas,* etc.

15 (*a*) *Paramecium* reproduces sexually by lateral conjugation in which cytoplasmic bridge is formed between male and female individuals.

16 (*b*) Statement in option (b) is not correct and can be corrected as

During the life cycle of *Plasmodium*, sporozoites multiply in liver cell, not in blood.

Rest statements are correct with respect to malaria.

17 (*c*) Babesiosis is a malaria like parasitic disease caused by infection with *Babesia bigemina*, a genus of Protozoa piroplasms.

Other options are explained as

Syphilis —*Treponema pallidum* (bacterium)
Influenza —Influenza virus
Blastomycosis—*Blastomyces dermatitidis* (fungus)

19 (*c*) Kala-azar is a slow progressive indigenous disease caused by a protozoan parasite *Leishmania* and spread by the bite of certain types of vector in sandflies. If this sandfly takes a blood meal from the patient of kala-azar and injects its macrophages infected with amastigotes transmission occurs. These amastigotes in the midgut of sandfly get transformed into promastigotes and migrate to proboscis. Amastigotes again get transferred into the skin of host (non-infected person) at the time of blood sucking. Thus, statements II, III and IV are correct.

20 (*a*) The protozoan with heliopodia as locomotory structures is *Actinophrys* (Protozoa).

These axopodia consist of a central rigid rod which is coated with thin layer of ectoplasm and is rooted in the endoplasm.

21 (*a*) *Trichonympha* is a protist and the type of syngamy seen in them is hologamy. In hologamy, the gametes involved resemble somatic cells in size and form.

26 (*a*) In diatoms, the cell walls form two thin overlapping shells, which fit together as a soapbox. The walls are embedded with silica and thus, the walls of diatoms do not decay easily.

28 (*c*) Option (c) is correct. Other options are incorrect and can be corrected as

- *Paramecium* and *Plasmodium* belong to kingdom—Protista, *Penicillium* belongs to kingdom-Fungi.
- Lichen is a composite organism formed from the symbiotic association of algae and fungi.
- *Nostoc* and *Anabaena* are examples of cyanobacteria or blue-green algae.

34 (*b*) *Trypanosoma cruzi* is the digenetic parasite, i.e. its life cycle is completed in two hosts. The primary host is man, cat, dog or monkey and secondary or intermediate host of this parasite is *Triatoma infestans.*

35 (*c*) Schizogony is an asexual reproduction of *Plasmodium* (malarial parasite), in which schizont is formed from the human blood sporozoites, which enter the liver cells. The sporozoite grows in size to become a rounded schizont called cryptozoite.

They later divide to form cryptomerozoites (merozoites). The merozoites are asexual haploid forms. The host liver cells rupture to liberate the cryptomerozoites into the liver sinusoids.

36 (*e*) Option (e) is not correctly matched and can be corrected as

Albugo is a phycomycetous fungus, while chrysophytes include diatoms and golden algae (desmids).

Rest options are correctly matched pairs.

37 (*c*) Presence of indestructible wall layer deposited with silica is a characteristic feature of chrysophytes.

In chrysophytes, the cell walls are formed by two thin overlapping shells, which fit together as in a soapbox. The walls are embedded with silica and thus, the walls are indestructible.

38 (*d*) Sporozoites are the infective stage of malarial parasite. They are present in the saliva of infected female, *Anopheles* mosquito. This infective stage is capable of passing across blood capillaries and can directly goes to parenchyma cells of liver.

CHAPTER 06

Fungi

Fungi are eukaryotic and achlorophyllous (i.e. without chlorophyll) organisms. They show heterotrophic nutrition due to the absence of chlorophyll. Their plant body is thallus-like (thalloid) and non-vascular, i.e. without xylem and phloem.

The term 'Fungi' was coined by **Gaspard Bauhin**. The study of fungi is known as **Mycology**. **PA Micheli** is known as the **Father of Mycology**. **Anton de Bary** is regarded as the Father of Modern Mycology and Plant Pathology. **EJ Butler** is regarded as the Father of Indian Mycology and Plant Pathology.

GENERAL CHARACTERISTICS

These are as follows

Habitat

- Fungi are cosmopolitan in distribution, i.e. they can be found everywhere.
- These are mostly terrestrial, but a few forms are aquatic.
- These mostly grow in warm and humid places, e.g. tree barks (corticolous), dung (coprophilous), wood (epixylic) and on hairs or horns of animals (keratinophilous).

Body and cellular structure

- The body is haploid (*n*) and thalloid. The thalloid plant body is not differentiated into distinct roots, stems and leaves. In simplest form, it is also called **thallus**.
- This thallus may be **unicellular** (*Chytrids*) or **filamentous** (multicellular) in most of the fungi.
- Thallus consists of large number of fine, tubular, thread-like filaments called **hyphae**. These hyphae are interwoven loosely to form a **mycelium**.
- The hyphae may be
  (i) **Septate** (with walls) which are uninucleate (primary mycelium), e.g. Ascomycetes.
  (ii) **Aseptate** (without walls) which are multinucleated (coenocytic), e.g. Oomycetes, Zygomycetes.
  Both of these are shown below

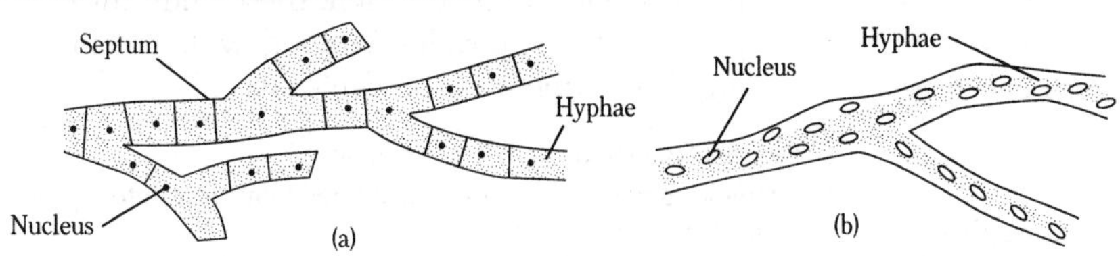

Fig. 6.1 Mycelium of a fungus with (a) Septate hyphae, (b) Aseptate hyphae

Inside

- The cell wall of hyphae are made up of chitin or fungal cellulose.
- Fungal mycelium may be of following types
  (i) **Plectenchyma** A false tissue formed by the loose aggregation of hyphae.
  It is further of two types
    (a) **Pseudoparenchyma** It is similar to parenchyma of higher plants in appearances, e.g. it is present in the fungal part of lichens and mycorrhiza.
    (b) **Prosenchyma** In this, hyphae are composed of elongated cells parallel to each other.
  (ii) **Rhizomorph** It is root like compactly interwoven hyphae with distinct growing tip. It helps in absorption and perennation, e.g. *Armillaria mellea.*
  (iii) **Sclerotium/Sclerotia** It is a compact hard dormant bodies of hyphae. These are externally thickened to protect themselves from unfavourable conditions, e.g. *Claviceps purpurea* and *Rhizoctonia solani.*

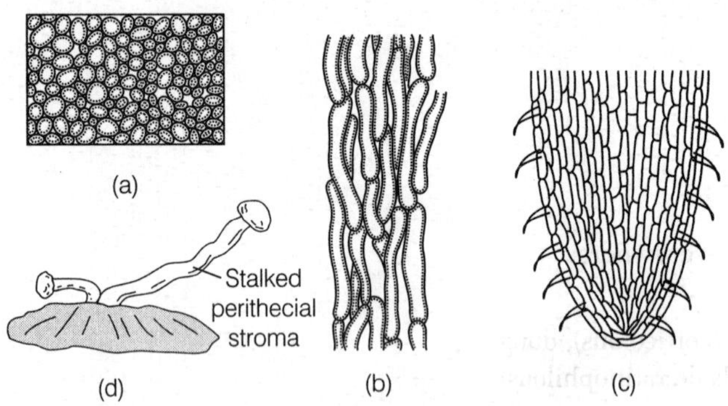

***Fig. 6.2** (a) Pseudoparenchyma, (b) Prosenchyma, (c) Rhizomorph, (d) Sclerotia*

Nutrition

- Fungi are heterotrophic. Since, they cannot ingest food material like animals. They absorb nutrients from the substratum through their cell membranes. This process is known as **osmotrophy** and related fungus are called **osmotrophs.**
- The reserve food material is stored as glycogen or oil.
- Depending upon the mode of nutrition, fungi can be **saprophytic** and **parasitic.**

Saprophytic

- These fungi grows on dead, decaying organic matter.
- They derive nutrition by secreting some special enzymes for this purpose.
- They are further of two types
  (a) **Obligate saprophytes** live strictly as saprophytes, e.g. *Mucor*, toadstools, mushrooms and most of the species of *Penicillium.*
  (b) **Facultative saprophytes** normally live as parasites but they are able to grow as saprophytes under emergent circumstances, e.g. *Taphrina deformans* (peach leaf curl), *Phytophthora infestans* (late blight of potato).

Parasites

- These fungi live inside or on the body of a host organisms e.g. plants and animals.
- They derive their nutrition by means of a **haustorium.** These haustoria are peg-like outgrowth that penetrate the host tissue.
- They are further of two types
  (a) **Obligate parasites** live strictly as parasites, e.g. *Peronospora* causing downy mildews in cucurbits.
  (b) **Facultative parasites** are normally saprophytic in their habit, but under forced conditions may become parasites, e.g. *Fusarium* and *Botrytis cinerea.*
- The parasitic fungi can also be classified on the basis of degree of destruction into following
  (a) **Destructive parasite** It produces toxic products and kills the host tissue completely.
  (b) **Balanced parasite** It derives the nutrition from the host in such a way that the host is able to supply the needs of the parasite and continues to live itself too.
- Some fungi are symbiotically associated with algae (lichens) and some with roots of higher plants (mycorrhiza).

Reproduction

Reproduction in fungi may take place by vegetative, asexual or sexual means.

Vegetative reproduction

Vegetative reproduction in fungi may occur through following methods

(i) **Fragmentation** In some fungi, mycelium breaks into fragments due to decay or injury. Each of these fragments grow into independent mycelium, e.g. in *Rhizopus, Mucor*, etc.

(ii) **Fission** Splitting of vegetative cell into two daughter cells through amitosis or less defined mitosis, e.g. in *Schizosaccharomyces* (yeast), etc.

(iii) **Budding** In few fungi, small buds arise from the parent cells. On maturation, these buds cut-off from parent cell and develop into a new organism, e.g. *Saccharomyces, Ustilago*, etc.

> The vegetative body of a fungus can be
> - **Holocarpic**, i.e. the entire fungal body transforms into reproductive body, e.g. *Synchytrium*.
> - **Eucarpic**, i.e. only a part of body transforms into reproductive structures, e.g. *Phytophthora*.

Asexual reproduction

Asexual reproduction in fungi takes place by the formation of following asexual spores

(i) **Sporangiospores** These are thin-walled, uni or multinucleated, non-motile spores. Their production is endogenous, i.e. formed inside the sporangium, e.g. *Rhizopus* and *Mucor.*

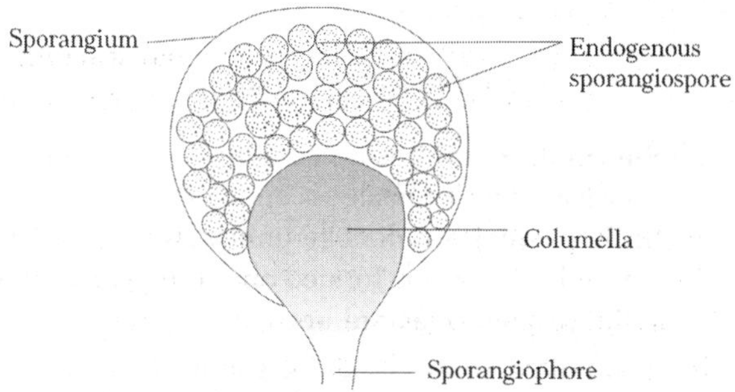

***Fig. 6.3** Sporangium containing sporangiospores*

(ii) **Zoospores** These are formed by the lower group of fungi like Phycomycetes. These are thin-walled and uninucleated structures. Their production is endogenous.

They can be uniflagellated, e.g. *Synchytrium* or biflagellated, e.g. *Saprolegnia.*

The flagella are always of unequal length known as **heterokont type**.

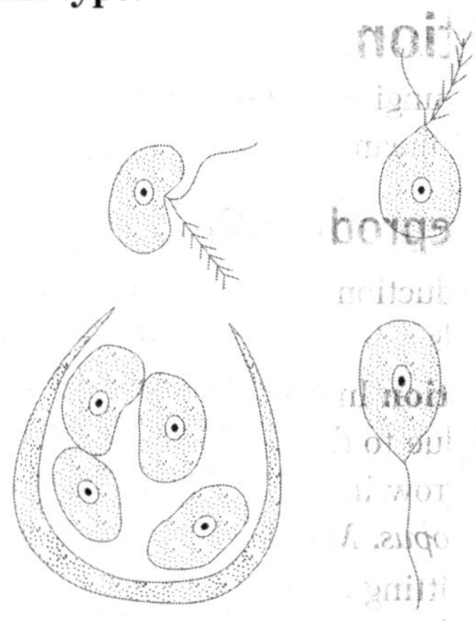

***Fig. 6.4** Zoospores*

(iii) **Chlamydospores** These are thick-walled, highly resistant spores. These are formed under unfavourable conditions.

They germinate to form mycelium under favourable condition, e.g. *Rhizopus.*

(iv) **Aplanospores** These are thin-walled, non-motile and non-flagellated structures. They are produced inside the sporangium, e.g. *Rhizopus.*

(v) **Conidia** These are non-motile and thin-walled spores. They are produced exogenously, i.e. generated freely on the tips of **conidiophores** (special aerial branches) and are arranged in chains, e.g. *Aspergillus* and *Penicillium.*

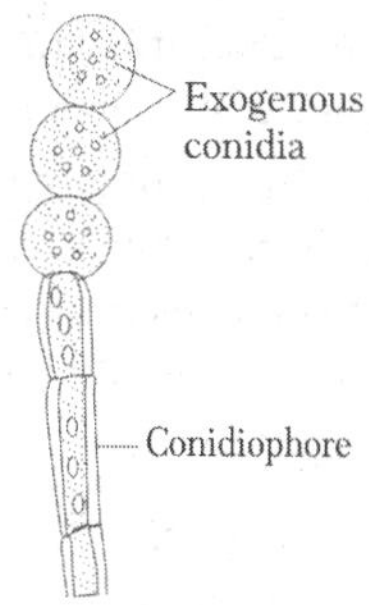

***Fig. 6.5** Conidia*

(vi) **Binucleate spores** Cells which contain two nuclei, e.g. aecidiospores, uredospores and teleutospores of Basidiomycetes.

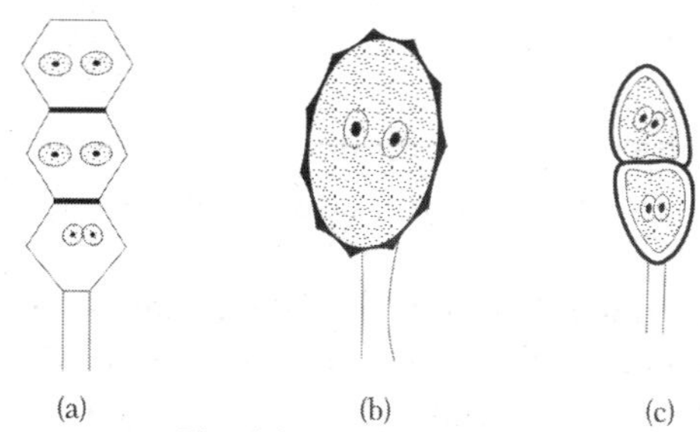

***Fig. 6.6** Binucleate spores*
(a) Aecidiospores, (b) Uredospores, (c) Teleutospores

(vii) **Oidia** These are non-motile thin-walled spores developing under sugar rich conditions. These oidia further increase by budding in yeast like fashion. This is called **torula stage**, e.g. *Mucor, Rhizopus.*

Sexual reproduction

- Sexual reproduction in fungi takes place by the fusion of two compatible nuclei. These two nuclei belong to two different parents or strains.
- The phenomenon of sexual reproduction occurs at a definite stage in the life cycle of a fungus and involves three processes, i.e. plasmogamy, karyogamy and zygotic meiosis.

(i) Plasmogamy

It is the fusion of protoplast of two parent cells, without the fusion of nuclei.

It occurs in following five ways

(a) **Planogametic copulation** Flagellated gametes or planogametes of opposite strains fuse with each other. They lead to the formation of a diploid zygote.

Fusion of planogametes can occur by either of following three ways

- **Isogamy** It is the fusion of motile gametes that are similar in appearance, e.g. *Synchytrium.*
- **Anisogamy** It is the fusion of motile gametes that differ in size and structure. The larger gamete is female and smaller one is male, e.g. *Allomyces.*
- **Oogamy** It is the fusion of larger, non-motile female gamete with smaller, motile male gamete.
  The male gamete is formed inside the antheridium. The female gamete is formed inside the oogonium, e.g. Chytridiomycetes.

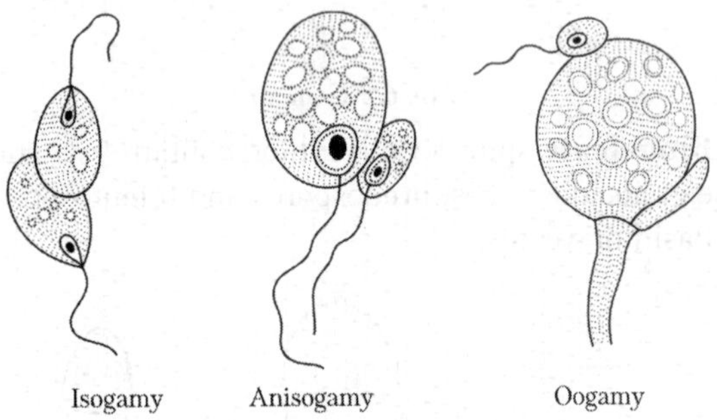

***Fig. 6.7** Planogametic copulation*

(b) **Gametangial contact** It is the fusion of two gametangia. These are the structures in which gametes are formed.

The male gametangium is called **antheridium** and the female one is called **oogonium** (in Oomycetes) or **ascogonium** (in Ascomycetes). A fertilisation tube is formed first at gametangial contact followed by karyogamy to form zygote ($2n$) or zygospore, e.g. *Albugo, Phytophthora.*

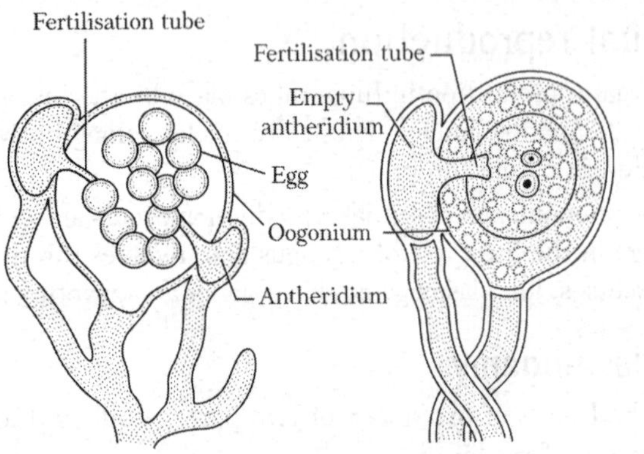

***Fig. 6.8** Gametangial contact*

(c) **Gametangial copulation** It is the direct fusion of male and female gametangia forming diploid zygospore, e.g. *Rhizopus, Mucor* and *Zygomycetes.*

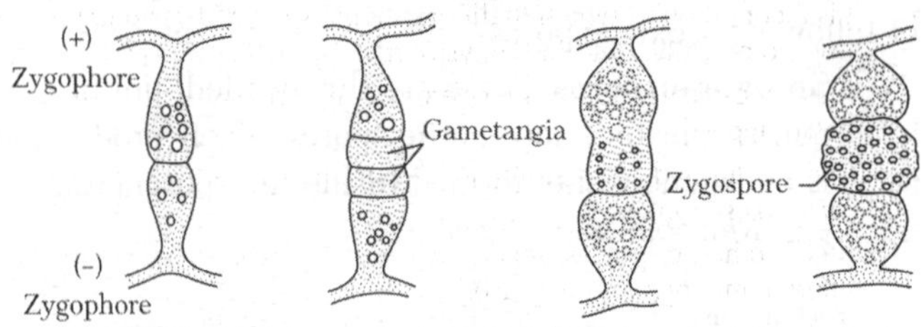

***Fig. 6.9** Gametangial copulation*

(d) **Spermatisation** It is a sexual fusion between spermatia and a female receptive hyphae. Spermatia are conidia like uninucleated, small male sex cells. These are formed on reproductive hyphae called the **spermatiophores.**

Upon maturation, spermatia get detached from spermatiophores and are carried to female receptive hyphae. The karyogamy is delayed and a dikaryotic stage ($n + n$) begins, e.g. *Neurospora,* Ascomycetes and Basidiomycetes.

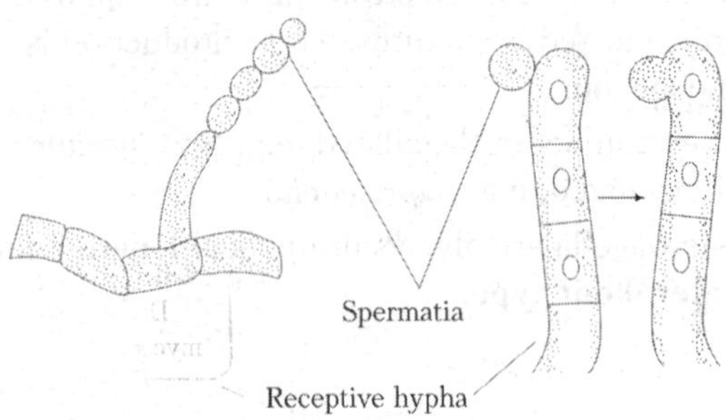

***Fig. 6.10** Spermatisation*

(e) **Somatogamy** It is the fusion of two somatic cells. Though distinct sex organs are absent. In these, two vegetative hyphae from mycelia of opposite strains (+ and –) come in contact and fuse. The nuclei of one migrate to the other. Thus, dikaryons are formed, e.g. *Ustilago, Agaricus* and some Ascomycetes.

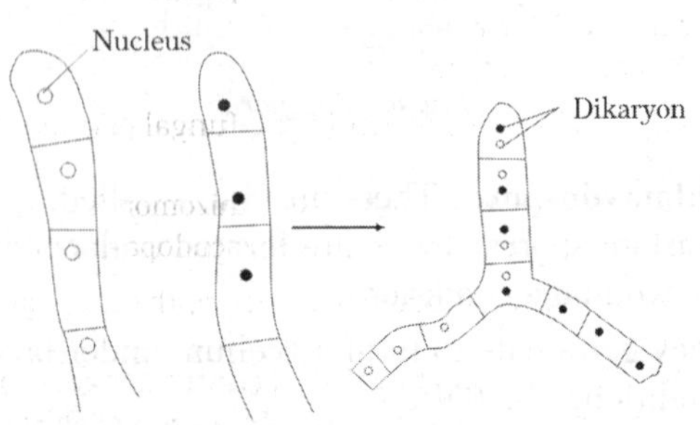

***Fig. 6.11** Somatogamy*

Heterothallism and sexual reproduction in fungi

- Heterothallism in fungi was discovered by **AF Blakeslee** (1904).
- It is a condition in which thalli of a single species are similar morphologically, but differ physiologically and genetically.
- These two distinct forms are referred as two sexual strains or mating types.
- They are termed as '+' strain and '–' strain, e.g. *Rhizopus stolonifer*.
- In heterothallic species, sexual reproduction occurs only when two different mating types are fused.
- Thus, occurrence of two different sexual structures on a single body begins from lower fungi.

(ii) Karyogamy

In this, haploid nuclei from two different parents fuse together to form diploid zygote (2*n*).

(iii) Zygotic meiosis

In this, zygote divides to form haploid spores (*n*).

These three processes divide the life cycle of a fungus into three mycelial phases. These are as follows

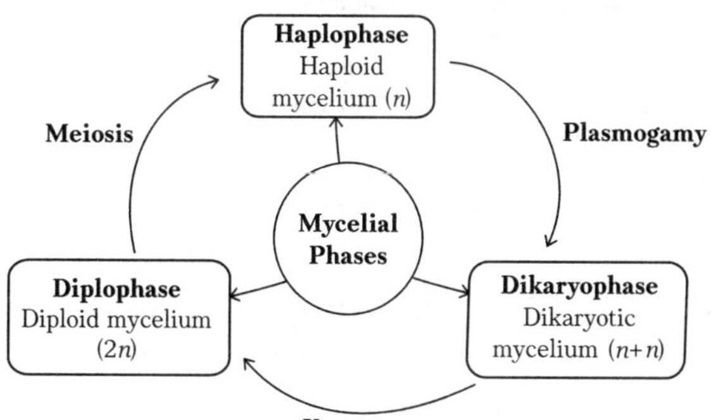

Note *In some fungi, karyogamy is delayed and as a result a dikaryon (n + n) is formed. This phase showing presence of dikaryotic stage is called **dikaryophase**.*

Sexual spores

Sexual reproduction leads to the formation of following haploid sexual spores

(a) **Ascospores** These are formed endogenously in sac like structures called **ascus**. Each ascus contains two haploid nuclei. These nuclei fuse during sexual reproduction. The zygote undergoes meiosis to form four or eight ascospores, e.g. Ascomycetes.

(b) **Basidiospores** They are formed exogenously on specialised structures called **basidium**. Inside basidium, two nuclei fuse to form zygote. This zygote undergoes meiosis later on to form four basidiospores. Basidiospores are attached on a peg-like outgrowth basidium through sterigmata, e.g. Basidiomycetes.

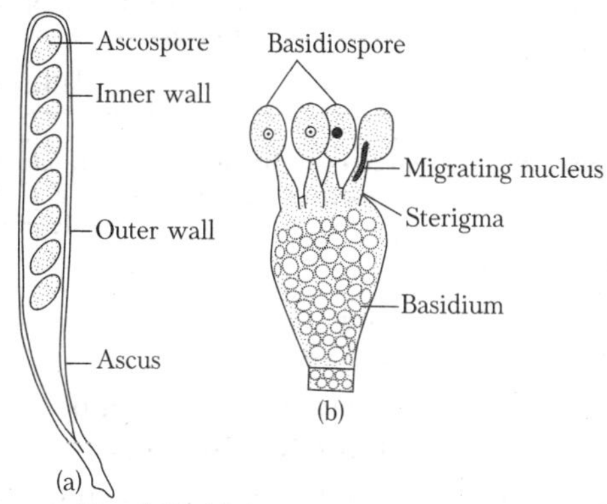

Fig. 6.12 *(a) Ascospores in ascus and (b) Basidiospores on basidium*

(c) **Zygospores** These are thick-walled spores. They are produced by the fusion of entire gametangia, e.g. Zygomycetes.

(d) **Oospores** These are formed within female structures called **oogonium**, e.g. Oomycetes. All these haploid sexual spores form haploid hyphae after germination.

CHECK POINT 6.1

1. Which of the following options describes the coenocytic condition in fungus?
(a) Uninucleate hypha without septum
(b) Multinucleate hypha without septum
(c) Multinucleate hypha
(d) Multiciliate hypha

2. Which mycelium is present in the fungal part of lichens and mycorrhiza?
(a) Sclerotia (b) Rhizomorph
(c) Prosenchyma (d) Pseudoparenchyma

3. According to their modes of nutrition, the fungi are classified into
(a) obligate and saprophytic (b) parasitic and facultative
(c) parasitic and saprophytic (d) parasitic and obligate

4. The plant makes organic molecules by photosynthesis and supplies them to the fungus and the fungus supplies water and mineral nutrients to the plant. This relation exists between
(a) fungi and vascular plants
(b) fungi and non-vascular plants
(c) fungi and roots of higher plants
(d) fungi and bryophytes

5. Which one is not related to perfect stage of fungus?
(a) Zygospores (b) Meiospores
(c) Ascospores (d) Mitospores

6. Fungal spores produced asexually at the tips of hyphae are called
(a) sporangiospores (b) endospores
(c) conidia (d) meiospores

7. A type of life cycle in which plasmogamy, karyogamy haplodisation take place, but not at specific place in life cycle of an organism is called as
(a) parasexuality (b) heterozygosity
(c) homozygosity (d) asexuality

CLASSIFICATION OF FUNGI

Many botanists have classified fungi in different ways

1. **Oswald Tippo** (1942) divide entire plant kingdom into two kingdoms, i.e. Thallophyta and Embryophyta. He classified 8, 9 and 10 phyla of Thallophyta as Fungi.
2. **Martin's classification** (1961) is one of the most accepted classification for Fungi.

   The outline of this classification is given below

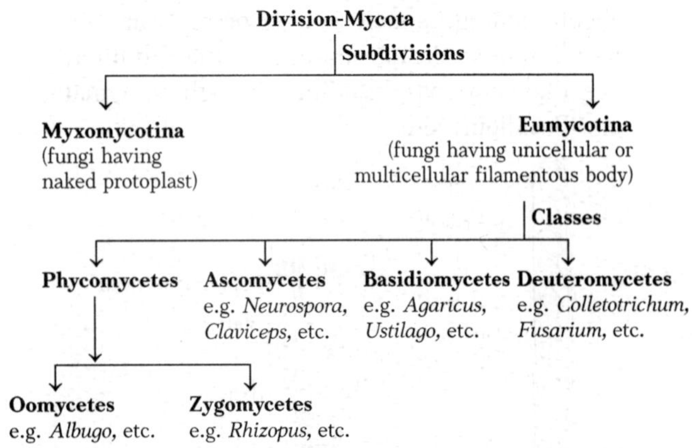

Earlier division-Mycota was placed into kingdom-Plantae. But **RH Whittaker** in his five kingdom classification system (1969) considered Mycota as a separate kingdom. He gave the name Fungi to this kingdom.

Characteristics of various groups of fungi are as follows

1. Phycomycetes (Lower fungi)

- They live either as obligate parasite or obligate saprophytes.
- Somatic phase is constituted by aseptate and coenocytic hyphae.
- Asexual reproduction occurs by the formation of zoospores or aplanospores or conidiosporangia.
- Sexual reproduction may occur by isogamy or anisogamy. Karyogamy occurs immediately after plasmogamy. Thus, dikaryophase is absent.
- These are divided into two groups

Oomycetes or Algal fungi

- Most of them are parasites.
- Unlike true fungi, cell wall of hyphae cells are made up of cellulose and not chitin.
- Asexual reproduction takes place by means of pear-shaped, biflagellate zoospores. These zoospores require water for their movement.
- Sexual reproduction is oogamous and occurs through gametangial contact resulting into the formation of oospores, e.g. *Phytophthora, Albugo candida, Pythium debaryanum*, etc.

Phytophthora infestans

- This fungus causes a serious disease called **late blight of potato**. This disease is also seen in tomatoes.
- Blight is the appearance of brownish to black dead areas on the plant tissues.
- These are first formed on the margins and tips of leaflets. Later on, the whole foliage becomes blighted. Due to this disease, tuber yield is reduced in potato.

Albugo candida

This fungus causes **white rust on crucifers**. In this disease, white blisters are formed on the foliage part. It causes serious damage and yield reduction.

White fungus disease

- **White fungus** also known as **Canidiasis** is caused by *Candida albicans*. It affects the oral cavity, gastrointestinal tract. Till now four cases of white fungus are registered in Bihar.
- **Symptoms** Appearance of whitish spot on the mucosal lining of mouth and throat.
- **Medication** No specific antibiotic known, only antifungul drugs are being used for the treatment of white fungus.

Zygomycetes or Conjugation fungi

- These are terrestrial and saprophytic fungi. Unlike Oomycetes, their cell wall contains chitin.
- There is complete absence of any motile cell in the life cycle.
- Asexual reproduction takes place by the formation of sporangiospores.
- Sexual reproduction occurs through gametangial conjugation. That is why, Zygomycetes are also called as **conjugation fungi**.
- The gametes are multinucleated and are called as **coenogametes**. These are basically the hyphae, that directly act as gametes.
- As a result of sexual reproduction, thick-walled zygospores ($2n$) are formed.
- These zygospores do not germinate directly rather they produce germ spores inside the germ sporangium *via* meiosis. Germ spores give rise to mycelia. Mycelia are coenocytic.
- Few common members are *Pilobolus crystallinus* (dung mould/squirting fungus), *Rhizopus stolonifer*, etc.

Rhizopus

- It is also known as **black mould/pin mould/bread mould**.
- It is a saprophytic fungus mostly found growing on decaying fruits, vegetables and food materials (like pickles, jams and jellies).
- Thallus of *Rhizopus* consists of white cottony mycelium. It bears tangled, aseptate, coenocytic hyphae.
- Three types of hyphae are recognised in mycelium. These are as follows
    - Stoloniferous hyphae (horizontal)
    - Rhizoidal hyphae (repeatedly branched)
    - Sporangiophores (aerial, erect)

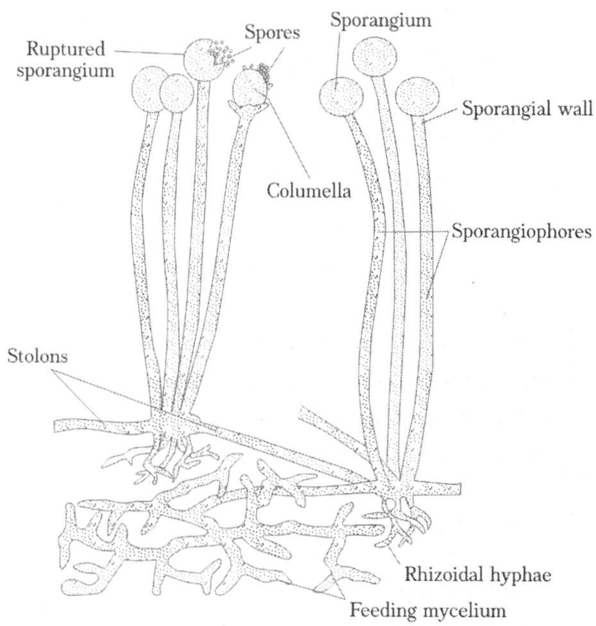

Fig 6.13 Rhizopus stolonifer showing mycelium stolons, rhizoidal hyphae and sporangiophore

- **Reproduction** in *Rhizopus* may take place by vegetative, asexual and sexual methods.
    - **Vegetative reproduction** occurs by fragmentation and chlamydospore formation.
    - **Asexual reproduction** occurs by the formation of minute, non-motile sporangiospore. During development the tip of sporangiophore swells into a vesicle (young sporangia). This vesicle accumulates food and cytoplasm to grow in size.
    Its nucleus divides mitotically. Each daughter nucleus develops to form a separate spore at maturity. Spores are wind disseminated. After dehiscence, spores germinate to form young mycelia.
    - **Sexual reproduction** occurs by gametangial copulation.
    - The zygospore formed in this method are remains dormant for 5-9 months and liberated free after this period.
    - After liberation, zygospores begin to divide *via* meiosis. Thus, initially four germ spores are formed. Two of them are of (+) while other two are of (–) spores.
    - The zygospore at this stage acts as germ sporangium. These germspores germinate through promycelium.
    - The outline of its complete life cycle is given below

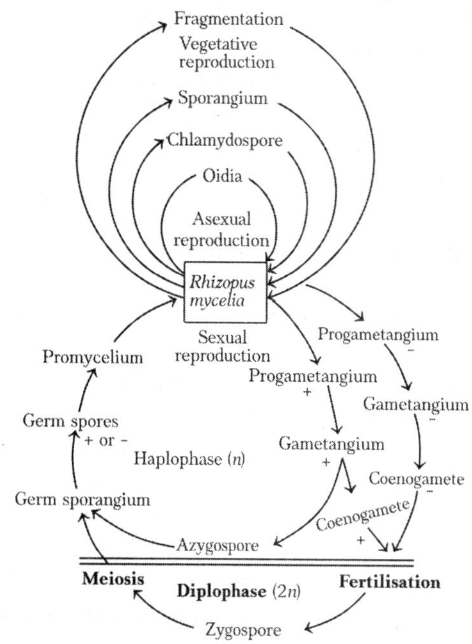

Fig. 6.14 Graphic representation of life cycle of Rhizopus

- Sometimes, the gametangia fail to fuse. In this condition, gametangium functions as a parthenospore by developing a thick-wall around itself. Thus, a zygospore like structure called the **azygospore** is produced.
- Azygospore is haploid in nature and it develops through promycelium formation.

Black fungus disease

- Black fungus is the rare but dangerous infection. It is also known as **Mucormycosis** caused by Mucormycetes.
- These moulds are present everywhere in the environment and cause infection when come in contact through nose, mouth, skin, etc.
- The connectivity of this disease with COVID-19 patients is due to following two reasons
    - Usage of steroid to counter corona infection.
    - Diabetes.
- Both these are responsible for enhancing sugar levels in patients due to which the patient becomes more prone to mould infection.

Symptoms

- Blockage of nose, nasal discharge with blood stain.
- Headache, swelling in eyes and blurring vision.
- **Medication** Amphotericin B and antifungal therapy.

2. Ascomycetes–Sac fungi

- This group includes pigmented moulds (brown, green, blue and pink). Ascomycetes are said to be our worst fungal enemies.
- These are mainly saprophytes, a few may be parasites also. Most of them are multicellular, but yeast is unicellular.
- Thallus consists of branched, septate hyphae. The cross walls in hyphae possess simple pores. Mycelium is generally monokaryotic.
- Dikaryotic mycelium is also seen, but only for a shorter duration. There is complete absence of any motile structures in the life cycle.
- Asexual reproduction is through conidia formation (exception budding in yeast).
- Sexual reproduction may occur by gametangial contact (e.g. Penicillium, Neurospora), conjugation (e.g. yeast), spermatisation (e.g. *Ascobolous)* or somatogamy (e.g. *Peziza*).
- Delayed karyogamy may results in the formation of dikaryotic cells. These cells produce ascus which is a special sporangial sac that produces ascospores.
- Ascus group together form a definite fruiting body called **ascocarps**. An ascocarp consists of a wall called as **periderm**.
- When an opening is present in ascocarp, it is called **ostiole**. It also consists of sterile hyphae called **paraphysis**.
- Depending upon their shapes, ascocarps can be of four types

Ascocarps
- → **Apothecium** (cup-shaped) e.g. *Peziza*
- → **Perithecium** (flask-shaped) e.g. *Claviceps*
- → **Cleistothecium** (globose without opening) e.g. *Penicillium*
- → **Ascostroma** (cushion-like) e.g. *Pleospora*

- Some of the common Ascomycetes are
  (i) **Morels** (sponge mushroom) These are edible ascocarps, e.g. *Morchella esculenta, M. deliciosia*
  (ii) **Truffles** Edible tuber-like subterranean ascocarp, e.g. *Tuber aestivum*

Yeast (*Saccharomyces cerevisiae*)

- It is a saprophytic fungus found on sugar rich substratum, e.g. sugarcane juice, sweet fruits and in the nectar of flowers and milk.
- It is unicellular, non-mycelial fungi that form no fruiting body. Sometimes, chains of buds are formed during their rapid growth. It looks like a mycelium and called as **pseudomycelium**.
- It is the smallest fungal cell. Its cell wall is thin, delicate and chitinous in nature. It is made up of about 6-8% proteins, 8.5-13.5% lipid, 1-2% chitin, 30-35% glucan and 30% mannan.
- Yeast cell helps in fermentation. It is done by the secretion of zymase enzyme which changes the starch or complex sugars into simpler sugars.
- In aerobic conditions, the sugar is completely oxidised by yeast. However in anaerobic condition, the yeast multiplies slowly and the sugar is converted into carbon dioxide (CO_2) and ethyl alcohol (C_2H_5OH) within the cytoplasm.
- True yeasts do not develop ascocarps.

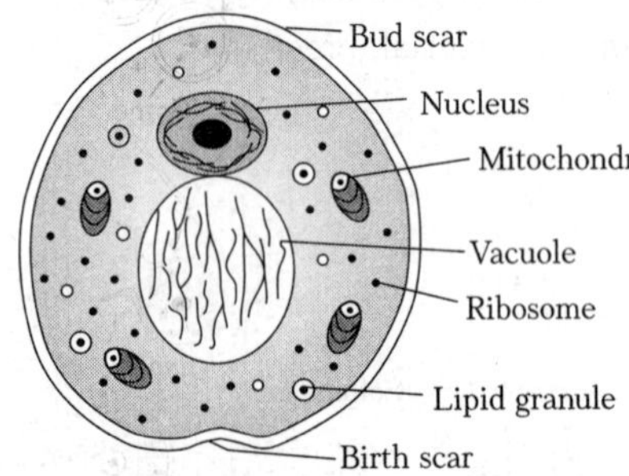

Fig. 6.15 Morphology of yeast

- **Types of yeast** These are as follows
  (i) ***Saccharomyces*** These are budding yeast.
  (ii) ***Schizosaccharomyces*** These are fission yeast.
  (iii) ***Saccharomycodes*** These are helobial yeast.
- In the life cycle of *Saccharomyces cerevisiae* (Baker's yeast or Brewer's yeast), there are two types of cells
  (i) **Dwarf yeast cells** Smaller in size, haploid, formed by germination of haploid ascospores and are rounded in shape.
  (ii) **Large yeast cells** Larger in size, oval in shape, diploid (being zygotes) and multiply by budding.
- **Reproduction** in yeast is either vegetative or *via* sexual reproduction. Asexual reproduction is not seen.
  (i) **Vegetative reproduction** takes place in dwarf yeast cells. These cells multiply by budding and form two dwarf cells of opposite strains.

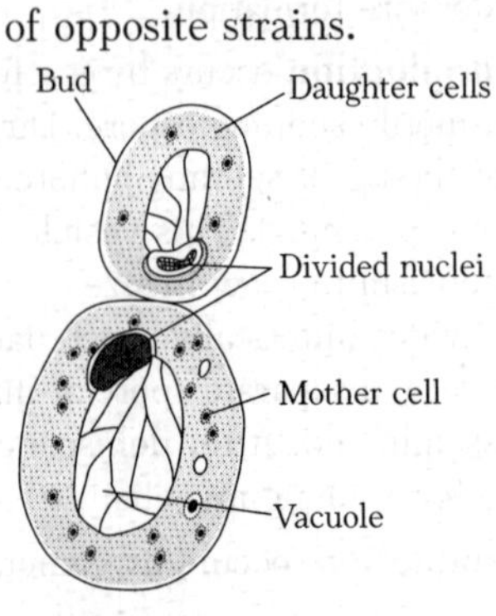

Fig. 6.16 Budding in yeast

(ii) **Sexual reproduction** is without definite sex organs. The gametangia of opposite strain undergo gametangial copulation forming zygote which directly act as ascus mother cell (2*n*).

These cells are called as **ascus** and they produce four ascospores after meiosis.

Two of them are of '+' strain, while other two are of '–' strain.

The complete life cycle of yeast is given below

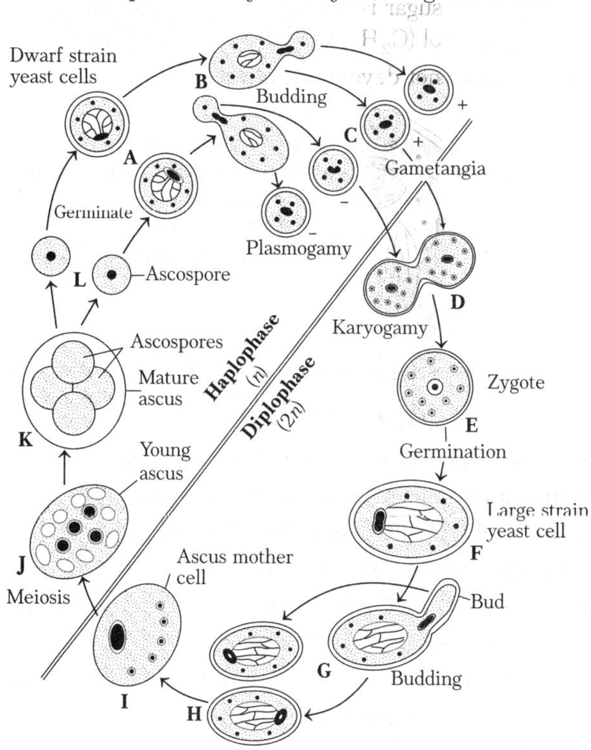

Fig. 6.17 Sexual cycle in Saccharomyces cerevisiae

- In life cycle of *Saccharomyces cerevisiae,* both haploid and diploid phases are equally represented and thus life cycle pattern is haplo-diplobiontic or haplodiplontic.
- **In fission yeast** (*Schizosaccharomyces octosporus*), life cycle pattern is haplobiontic, i.e. diploid phase is greatly reduced and is represented by zygote whereas haploid phase is more prominent. Further, number of ascospores produced per ascus is 8.
- **In helobial yeast** (*Saccharomycodes ludwigii*), life cycle pattern is diplobiontic as haploid phase is greatly reduced. It is represented only by ascospores the diploid phase is more prominent. Further, number of ascospores produced per ascus are 4.

Penicillium (A blue mould)

- It is a saprophytic, fungus commonly occurring on decaying oranges and lemons.
- It is used for producing wonder drug penicillin, which was discovered by **Alexander Fleming** in 1929 from *P. notatum.*
- Mycelium is profusely branched and septate. Cells are thin-walled and contain two to many nuclei in them.
- **Reproduction** in *Penicillium* can be vegetative, asexual or sexual.

  (i) **Vegetative reproduction** takes place by accidental breaking of vegetative mycelium into two or more fragments. Each fragment then grows individually like the mother mycelium.

  (ii) **Asexual reproduction** takes place by means of conidia.

  (iii) **Sexual reproduction** occurs in three phases, i.e. plasmogamy, delayed karyogamy (due to which dikaryophase occurs) and meiosis. The ascocarp (cleistothecium) is associated with 4-8 haploid ascospores.

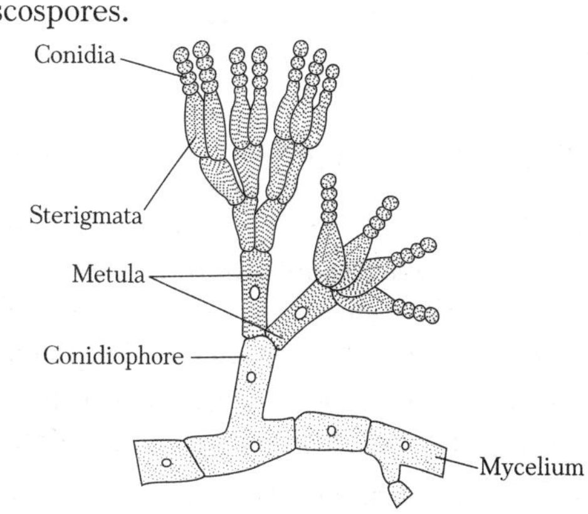

Fig. 6.18 Penicillium: Asexual reproduction branched conidiophore with chains of conidia

CHECK POINT 6.2

1. Choose the incorrect match with respect to different classes of fungi.
   (a) Oomycetes – Gametic copulation
   (b) Zygomycetes – Zygophore
   (c) Ascomycetes – Monokaryotic aseptic mycelium
   (d) Phycomycetes – Coenocytic mycelium

2. Which of the following hyphae is recognised in mycelium of *Rhizopus*?
   (a) Stoloniferous
   (b) Rhizoidal
   (c) Sporangiophores
   (d) All of the above

3. Which of the following is a good example of heterothallism?
   (a) *Spirogyra*
   (b) *Rhizopus*
   (c) *Pinus*
   (d) Castor bean

4. In yeast, the sexual reproduction is by conjugation. After fusion, the zygotic nucleus divides meiotically and forms haploid spores. The unicellular structure with haploid spores is
   (a) sporangium (b) gametangium
   (c) ascus (d) zoosporangium

5. The bread becomes soft and porous when the yeast cells are mixed in the lump of dough of wheat flour, because
   (a) yeast is soft and flour also becomes soft
   (b) yeast produces acetic acid and alcohol which gives softness to the bread
   (c) evolution of CO_2 makes the bread spongy
   (d) yeast produces benzoic acid

6. Which of the following classes of kingdom–Fungi are characterised by the presence of coenocytic, multinucleate and branched mycelium?
   (a) Basidiomycetes (b) Phycomycetes
   (c) Ascomycetes (d) Deuteromycetes

3. Basidiomycetes–Club fungi

- Members of this group are best decomposers of wood. They are the most advanced fungi.
- These are also called as club fungi due to the presence of club-shaped structure, i.e. basidium.
- Motile structures are completely absent. Mycelium is of two types
  (i) **Primary mycelium** is monokaryotic and short-lived. It multiplies by conidia-like spores called **pycniospores.**
  (ii) **Secondary mycelium** is long-lived and dikaryotic. This mycelium is produced as a result of heterokaryosis. It is a process where cytoplasm and nucleus of two genetically different cells fuses. It leads to the formation of a dikaryon or heterokaryon. Secondary mycelium is branched and septate.
- The cells of almost all members of Basidiomycetes contain specialised dividing walls between cells called **dolipore septa** as shown in the given figure.

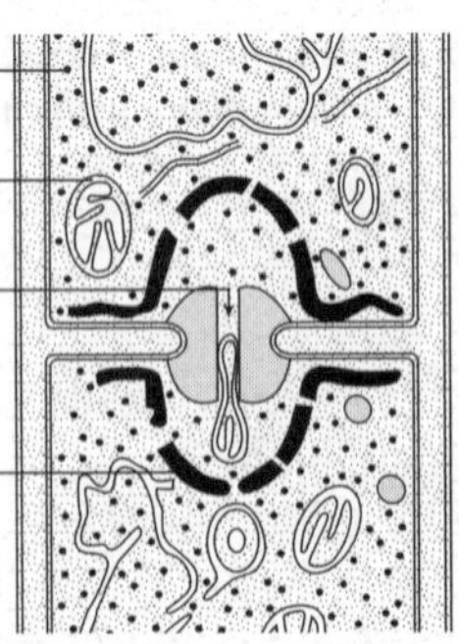

Fig. 6.19 Dolipore septum

- Sexual reproduction does not involve any sex organs. It occurs by the formation of basidiospores on club-shaped basidium exogenously.
- Fruiting bodies, i.e. basidiocarps are seen. Presence of dolipore septum and clamp connections are some distinguishing features of the class.
- This class includes the following fungi
  (i) Rusts, i.e. *Puccinia* sp.
  (ii) Mushrooms, i.e. *Agaricus* sp.
  (iii) Smuts, i.e. *Ustilago* sp.
  (iv) Toad stool, i.e. *Amanita* sp.
  (v) Puffball, i.e. *Lycoperdon* sp.
  (vi) *Armillaria*, i.e. Largest fungi
  (vii) Bracket fungi appearing on tree trunks, e.g. *Polyporus sulphureus.*

Rusts or Rust fungi

- The fungi belonging to class-Basidiomycetes and order-Uridinales are called rust fungi.
- These are obligate parasites.
- These are highly specialised pathogens.
- These can be heteroecious (require two hosts) or autoecious (require single host).
- Mycelium is septate and devoid of clamp connection.
- Rusts produce five different types of spores to complete their life cycle,
  (i) pycniospores (stage-0), (ii) aeciospores (stage-I),
  (iii) uredospores (stage-II), (iv) teliospores (stage-III)
  (v) basidiospores (stage-IV)

- Indian scientist **KC Mehta** is famous for his work on black rust of wheat. The stage-0 and stage-I are produced on the colateral host, Barberry leaves, while II, III and IV on wheat leaves.
- Three types of rusts produced on wheat are
  (i) Black (stem) rust of wheat, i.e. *Puccinia graminis tritici.*
  (ii) Brown rust of wheat, i.e. *Puccinia recondita.*
  (iii) Yellow rust of wheat, i.e. *Puccinia striiformis* or *Puccinia glumarum.*

Smuts or Smut fungi

- Smuts are common and serious diseases of wheat, barley, maize, oat and sugarcane. These are caused by different species of *Ustilago*
- Smuts are of two types
  (i) **Loose smut**, where the spores are exposed while they are attached to the host, e.g. *Ustilago tritici.*
  (ii) **Covered smut**, where the spores remain covered, e.g. *Ustilago hordei.*
- This fungus mainly attacks the stem, flowers and often the whole inflorescence.

Mushrooms

- These are found in moist humus rich soil, open fields, grasslands and piles of straws. They develop in rings called **fairy rings**.
- Their secondary mycelium later develops a tertiary mycelium through **clamp connections**. A **clamp** is a pouch-like outgrowth through which division of nuclei is facilitated.

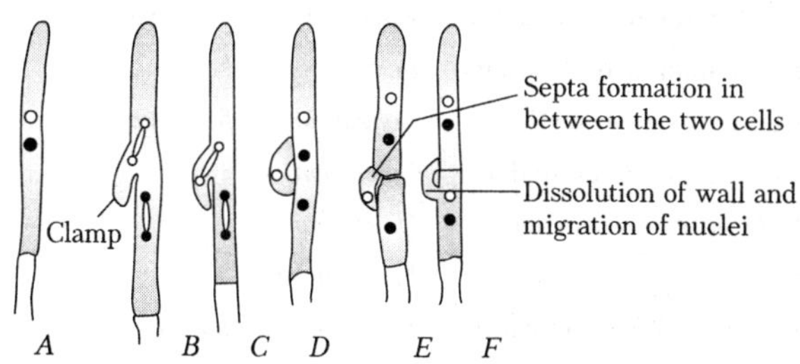

B. Formation of outgrowth, i.e. clamp
C. Division of nuclei
D. Migration of nuclei to upper cell

Fig. 6.20 *Illustration of clamp connections (A mean of conjugate division in secondary mycelium)*

- Tertiary mycelium forms the fruiting body, i.e. basidiocarp. Each basidiocarp consists of a stalk called **stipe** and as umbrella-shaped cap, i.e. **pileus.**
- Stipe is fleshy. It is swollen at the base.
- Under surface of pileus, there are 300-600 radiating rows or gills. Gills bear basidium and basidiospores.

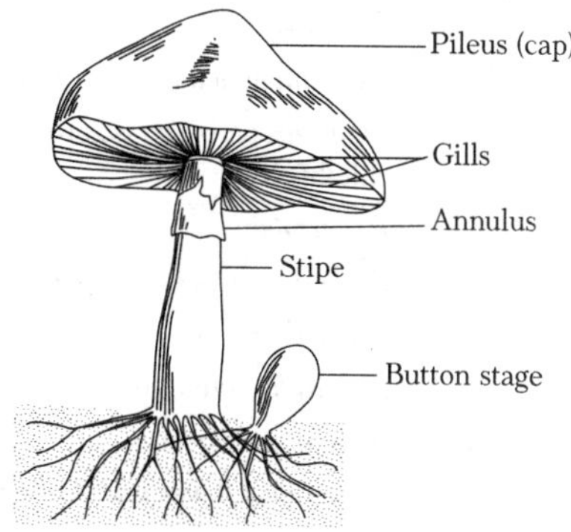

Fig. 6.21 *Basidiocarp of Agaricus*

4. Deuteromycetes (Fungi imperfecti)

- It is an artificial class of fungi.
- Some members are saprophytic or parasitic while a large number of them are decomposer of litter and help in mineral cycle.
- It includes all those fungi whose sexual stage is not known.
- It is believed that most of the members of Deuteromycetes are actually Ascomycetes in which sexual stage is either absent or is not known.
- Their vegetative mycelium is septate and profusely branched.
- These fungi reproduce by asexual means only, like conidia, e.g. *Colletotrichum, Helminthosporium, Alternaria, Cercospora, Fusarium* and *Trichoderma*, etc.

Useful fungi

- The fructifications of certain fungi are used as nutritious and delicious foods, e.g. *Agaricus campestris* (common field mushroom), *Morchella* (used in Himachal Pradesh and Jammu and Kashmir) and *Calvatia gigantea* (giant puffball).
- Mushrooms are rich source of vegetable protein (21-30% protein on dry weight basis), vitamins (B, C, D and K) and minerals (Ca, Na, K, P, Fe), hence mushroom cultivation is done in different parts of our country. *Agaricus bisporus* also known as **white button mushroom.** It is a saprophytic.
- Some yeasts are an important source of vitamin-B and some Deuteromycetes like *Endomyces, Rhodotorula, Torulopsis* are rich in proteins. A food called **sufu** is produced from *Mucor.*
- Fungi are used in the production of different **antibiotics.** These are the substances of microbial origin and have antimicrobial activities.

Table 6.1 Some important antibiotics and their sources

Antibiotics	Sources
Penicillin	*Penicillium notatum* and *P. chrysogenum*
Ergotine	*Claviceps purpurea* (sclerotia). Given after childbirth
Citrinine	*P. citrinum*
Campestrin	*Psaliotta campestris* (*Agaricus campestris*)
Glyotoxin	*Trichoderma* species
Proliferin	*Aspergillus proliferans*
Clavicin (anti-cancerous)	*Aspergillus clavatus*
Flavicin	*Aspergillus flavus* and *A. fumigatus*
Griseofulvin	*Penicillium griseofulvum*
Bacitracin	*Bacillus subtilis*
Fumagillum	*Aspergillus fumigatus*

- Fungi are responsible for the production of several useful products in industries. These are
  - (i) Yeast (*Saccharomyces*) performs fermentation, hence it is used in the production of ethyl alcohol. Ethyl alcohol is the basis of brewing industry.
  - (ii) In baking industry, yeast produces CO_2, which is used to raise **dough** while in making bread.
  - (iii) Digestive enzymes like diastase, digestin and polyzyme are produced from *Aspergillus flavus-oryzae.*
  - (iv) *Penicillium camemberti* and *P. roqueforti* are used for flavouring **cheese.**
  - (v) Organic acids, production, e.g.
    - **Citric acid** By *Aspergillus niger* and *Aspergillus wentii, Mucor,* etc.
    - **Gluconic acid** By *Aspergillus niger* and *Penicillium purpurogenum.*
    - **Gallic acid** By *Penicillium glaucum* and *Aspergillus allomyces.*
    - **Kojic acid** By *Aspergillus oryzae.*
    - **Fumaric acid** By *Rhizopus stolonifer.*
- Gibberellin was first discovered in the extract of *Gibberella fujikuroi* growing on rice. This hormone acts as plant growth promotor.
- *Neurospora sitophila* (red mould) has a short life cycle and hence used in genetical and biochemical studies. *Neurospora* is also known as *Drosophila* of plant kingdom.
- Some strains of *Aspergillus niger* can detect trace elements like Cu, Zn, Mo, Mn, Cd, Pb, etc.
- Fungi is used to increase soil fertility and as biological control, e.g. *Trichoderma* and *Rhizoctonia.*
- Predacious fungi like *Dactylella, Dactylaria, Zoophagus* and *Arthrobotrys*, etc., destroy certain nematodes, eelworms and thus act as good biological control agents.

Yellow fungus disease

- It is caused by *Aspergillus flavus*.
- According to researches, this disease is usually found in animals. It is considered more fatal than the black fungus as it can affect internal organs.

Symptoms

- Lethargy, loss of appetite, weight loss and sunken eyes.
- Slow healing of wounds, oozing of pus from wound.

Medication

- No specific antibiotic known only antifungal drugs are being used for the treatment of white fungus.

Harmful fungi

- Fungi are causal organisms of many diseases in living things. Some harmful fungal diseases in plants, humans and animals are as follows

Table 6.2 Fungi cause several diseases in plants

Diseases	Fungus responsible
Black wart disease of potato	*Synchytrium endobioticum*
White rust of crucifers	*Albugo candida* or *Cystopus candidus*
Late blight of potato	*Phytophthora infestans* (famous famine of Ireland caused death of million of people in 1845)
Early blight of potato	*Alternaria solani*
Powdery mildews	*Erysiphe* sp.
Damping off disease and seedlings of cereals and vegetables	*Pythium* spp.
Loose smut of wheat	*Ustilago tritici*
Downy mildew in cereals	*Sclerospora graminicola*
Downy mildew of cucurbits	*Peronospora cubensis*
Red rot of sugarcane	*Colletotrichum falcatum*
Tikka disease of groundnut	*Cercospora* sp.
Whip smut of sugarcane	*U. scitaminea*
Wilt of arhar	*Fusarium* sp.
Green ear disease of bajra	*Sclerospora graminicola*
Flag smut of wheat	*Urocystis agropyri*
Maize smut	*Ustilago maydis*
Loose smut of oat	*Ustilago avenae*
Covered smut of oat	*Ustilago kolleri*
Covered smut of jowar	*Sphacelotheca sorghii* (*Sorghum*)
Leaf spot of rice	*Helminthosporium oryzae* (caused severe famine of Bengal in 1943).
Red rot of apple	*Armillaria mellea*
Wood rot	*Polyporus*
Loose smut of wheat	*Ustilago tritici*
Covered smut of barley	*Ustilago hordei*
Soft rot of sweet potato	*Rhizopus stolonifer*

Table 6.3 Some important fungal diseases in humans

Diseases	Fungus
Athlete foot	*Tinea rubrum*
Ringworm	*Trichophyton*, *Microsporum* and *Epidermophyton*
Mucomycosis	*Mucor* and *Rhizopus*
Penicillosis	*Penicillium* sp.
Ear infection (automycosis)	*Aspergillus flavus* and *Aspergillus nidulans*
Thrush disease of throat	*Monilia* species
Candidiasis	*Candida albicans*

Table 6.4 Some important fungal diseases in animals

Diseases	Fungus
Aspergillosis (lung disease)	*Aspergillus niger*
Salmon disease of gills in fishes	*Saprolegnia*
Mastistis	*Aspergillus* sp.
Histoplasmosis	*Histoplasma*

- Spores of *Rhizopus*, *Mucor* and *Aspergillus* germinate on jam, jellies, pickles, bread, fruits, etc., and destroy them. Likewise, *Penicillium*, *Mucor* and *Aspergillus* destroy meat. Thus, they cause food spoilage.
- *Alternaria*, *Penicillium*, *Trichoderma*, *Mucor* and *Fusarium* destroy leather, cloth, rubber, paper, camera lenses, etc.
- *Amanita phalloides* (like *Agaricus*) is highly poisonous and fatal to man. *Boletus* is also a poisonous fungi. It causes distortion of perception. LSD or lysergic acid hallucinogen is extracted from *Claviceps* fungus.

Common fungicides and their composition

- **Bordeux mixture** ($CuSO_4$, $Ca(OH)_2H_2O$) First fungicide discovered by RMA Millardet commonly known as holy water of plant pathology.
- It has been used for decades to control diseases of fruits and nut trees, vine fruits and ornamental plants.
- **Burgundy mixture** Mixture of $CuSO_4 + NaCO_3 + H_2O$.
  Prevost (1807) First used as fungicide for the control of bunt or smut disease of wheat.
- **Chestnut mixture** Mixture of ammonia carbonate + copper sulphate. It is used against the damping off disease.

LICHENS

- These are symbiotic associations between algae and fungi.
- The algal component (autotrophic) is called **phycobiont** and fungal component (heterotrophic) is **mycobiont**.
- Lichens are found in least polluted habitats like walls, window panes, barren rocks, cooled volcanic lava, tree bark, soil and aquatic habitats.
- Lichens vary in colour like orange, brown, dark brown, grey, yellowish green, etc.
- The major part of lichen body constitutes fungus and only 5% part is algal component.
- The body is divided as upper cortex, algal zone (gonidial layer), middle medulla, lower cortex and rhizines.
- Only the algal zone contains photosynthetic partner.
- In about 98% lichens, the fungal partner belongs to Ascomycetes. In the rest, the fungal partner belongs to Basidiomycetes and Deuteromycetes.
- Lichens may reproduce vegetatively or sexually. They reproduce vegetatively by the following ways such as fragmentation, death and decay, etc.
- Lichens reproduce sexually by following ways
  (i) Formation of ascomycetous fruiting body.
  (ii) Formation of basidiocarp.
- Sexual reproduction is not considered to be a common means of reproduction to form a new lichen.

Types of lichens

Based on their morphology, lichens are of three types

(i) **Foliose lichens** are leaf-like in appearance and structure, e.g. *Parmelia.*

(ii) **Crustose lichens** are crust-like. They are tightly attached to or embedded in their substrate, e.g. *Graphic scripta.*

(iii) **Fruticose lichens** have no distinct top and bottom, and are often round in cross-section, e.g. *Usnea, Cladonia*, etc.

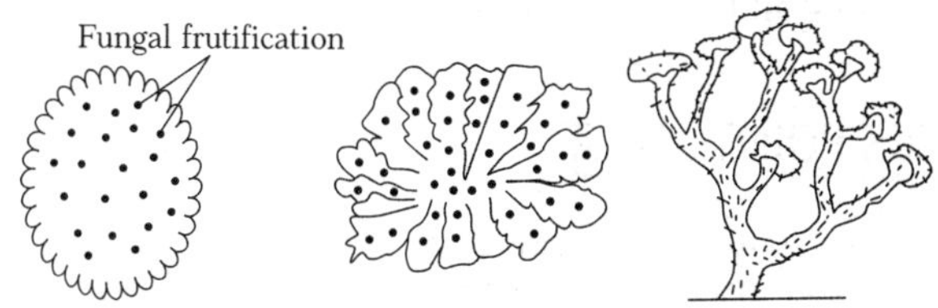

***Fig. 6.22** Forms of lichens*

Based on their habitat, lichens can be of following types

(i) **Saxicolous** Lichen growing on rock, e.g. *Dermatocarpon.*

(ii) **Corticolous** Lichen growing on bark, e.g. *Usnea.*

(iii) **Lignocolous** Lichen growing on wood, e.g. *Cyphelium.*

(iv) **Terricolous** Lichen growing on soil, e.g. *Cladonia* and *Lecidea.*

Economic importance of lichens

- Lichens are used as indicators of pollution as they are highly sensitive to SO_2.
- These are used in the preparation of dyes.
- Litmus is produced from lichen *Rocella.*
- Lichens are the pioneer of succession in xerosere.
- *Cladonia* is used as fodder.

MYCORRHIZA

- It is a symbiotic association between a fungus and roots of higher plants. In this association, both the partners, i.e. plants and fungi are benefitted.
- Fungi being heterotrophic, derive nutrition and shelter from plant roots. In turn, it helps the root in dissolution and absorption of inorganic nutrients locked in organic matter. For example, absorption of phosphorus.
- Mycorrhizae can be of following two types
  (i) **Ectomycorrhiza** In these, the fungus remains confined to intercellular spaces between cortical cells producing a net, called **Hartig's net**, e.g. *Boletus*, etc.
  (ii) **Endomycorrhiza** In these, the fungal hyphae passes the cortex and can be intercellular or intracellular, e.g. *Rhizoctonia*, *Endogone*, etc. They are also called as **Vesicular Arbuscular Mycorrhiza** (VAM) as they form vesicles inside the cortical cells. This type of mycorrhiza are mostly seen in orchid root cells.

Economic Importance of mycorrhiza

- Mycorrhiza help in absorption of minerals from the soil and provide to the root.
- These help in increased absorption of water.
- These produce growth inducing hormones. In the absence of fungus, *Pinus* and *Betula* show stunted growth despite providing all types of nutrients.
- Mycorrhiza enable the plants to grow in mineral deficient soil, saline soil and unfavourable pH and temperature.
- In case of orchids, the fungus absorbs nourishment from outside and deliver to the germinating seed as the same is devoid of stored food.
- Mycorrhiza have antimicrobial substances which protect the root from parasitic fungi and harmful bacteria by secreting antimicrobial substances.

Table 6.5 Resemblances and differences between algae and fungi

Resemblances	Differences
Presence of unicellular sex organs in both.	Chlorophyll is found in algae, while it is absent in fungi.
Presence of thalloid plant body in both.	Algae are autotrophic (photosynthetic) in nutrition, whereas fungi are heterotrophic.
Lack of tissue differentiation of any kind in both.	True parenchyma is found in algal thallus, while pseudoparenchyma is found in fungal thallus.
No embryo formation after gametic union in both.	Cell is the unit of algal thallus, whereas hyphae are the unit in fungal thallus.
	Algal cell wall is composed of cellulose, while fungal cell wall is of chitin or fungal cellulose.
	In algae, reserve food is starch whereas in fungi, reserve food is glycogen.

CHECK POINT 6.3

1. Which statement is true?
   (a) Toadstool is an edible fungus
   (b) Rust fungi are homoecious
   (c) Parathecium is fruiting body
   (d) In mushroom, gills produce basidia
2. The disease which occurs at irregular intervals and locations are known as
   (a) epiphytotic diseases
   (b) endemic diseases
   (c) sproadic diseases
   (d) epidemic diseases
3. The chemical produced by the host plant to protect themselves against fungal infection is
   (a) toxin
   (b) phytoalexin
   (c) phytotoxin
   (d) hormone
4. Which of the following fungi is found useful in the biological control of plant disease?
   (a) *Penicillium notatum*
   (b) *Phytophthora parasitica*
   (c) *Mucor mucedo*
   (d) *Trichoderma viridae*
5. The disease produced by fungus *Ustilago* are known as smuts because
   (a) they are parasitic cereals
   (b) the affected host beomes completely black
   (c) their mycelium is black
   (d) they produce sooty mass of spores
6. Which lichens are leaf-like in appearance and structure?
   (a) Saxicolous (b) Foliose
   (c) Crustose (d) Fruticose

EXAM POINTS

- Fungi are eukaryotic. They are both unicellular and multicellular. Their mode of nutrition is either saprotrophic or parasitic. **[NEET-II 2016]**
- The cell wall of fungi is composed of chitin. **[JIPMER 2018]**
- Yeast have short temporary filamentous structure called pseudomycelium. They lack hyphae. **[NEET 2019]**
- The generalise life cycle of a fungus is that it undergoes meiosis followed by mitosis and then fertilisation. **[AIIMS 2017]**
- In *Mucor,* zygospore germinates through the formation of meiospore. **[Chhattisgarh PMT 2015]**
- Asexual reproduction in fungi occurs by conidia. **[Odisha JEE 2013, 10]**
- Heterothallism is found in *Mucor*. **[Chhattisgarh PMT 2015]**
- Ascomycetes are saprophytic, decomposers, parasitic or coprophilous. These are commonly called sac fungi, e.g. *Saccharomyces*. **[Kerala CEE 2014, 12, 11]**
- Morels and truffles groups of fungi are classified under Ascomycetes. **[AMU 2015]**
- Clamp connections are observed in Basidiomycetes. **[UP CPMT 2012]**
- Edible part of mushroom is basidiocarp. **[WB JEE 2011, 12]**
- The imperfecti fungi which are decomposed of litter and help in mineral cycling belong to Deuteromycetes. **[CBSE AIPMT 2015]**
- In Deuteromycetes sexual dimorphism is absent. **[JIPMER 2012]**
- Yeast in used in the production of bread and beer. **[J&K CET 2015]**
- *Glomus* helps in the absorbtion of phosphorus from soil. **[CBSE AIPMT 2012]**
- Red rot of sugarcane is caused by *Colletotrichum falcatum.* **[OJEE 2012]**
- Rust of wheat, covered smut of bajra are some other fungal disease. **[Gujarat CET 2014]**
- Lichens are a complex form of life that is a symbiotic association of a fungus and an alga. They act as a pollution indicator as they do not grow in polluted area. **[NEET 2019]**
- Mycorrhiza is a symbiotic association betweens a fungus and roots of higher plants. In this, fungi derive nutrient from plant root and in return fix nutrient like phosphorus for plant. **[NEET 2019, UP CPMT 2012]**

Chapter Exercises

(A) Topical Exercise

General Characteristics of Fungi

1 A facultative parasite is one which
(a) always requires a living host
(b) is normally a saprophyte, but can also become a parasite
(c) is normally parasitic, but can also become saprophyte
(d) always requires dead organic matter

2 The absorptive nutrition of fungi is aided by
(a) dikaryon formation
(b) spore formation
(c) the fact that they are all parasites
(d) their large surface area to volume ratio

3 Consider the following statements.
I. Thread-like filaments that form the plant body of fungi are hyphae.
II. Plectenchyma is a tissue composed of collenchyma.

Choose the correct option.
(a) Statement I is true, but II is false
(b) Statement I is false, but II is true
(c) Both statements I and II are true
(d) Both statements I and II are false

4 Which of the following have cells that imitate parenchyma of higher plants?
(a) Sclerotia
(b) Rhizomorph
(c) Pseudoparenchyma
(d) Prosenchyma

5 The false fungal tissue formed by loose aggregation of hyphae is
(a) parenchyma (b) plectenchyma
(c) chlorenchyma (d) None of these

6 Which statement about fungi is true?
(a) Fungal hyphae constitute a body called mycelium
(b) Hyphae are composed of individual mycelia
(c) Many fungi tolerate highly hyperosmotic environments
(d) Many fungi can tolerate extreme low temperature

7 Heterothallism is found in
I. *Rhizopus* II. *Mucor*
III. *Puccinia graminis*
Choose the correct option.
(a) Only II (b) I and II
(c) I and III (d) Only I

8 The holocarpic fungus is
(a) *Pythium debaryanum* (b) *Phytophthora infestans*
(c) *Peronospora destructor* (d) *Synchytrium endobioticum*

9 The asexual spores of fungi produced at terminal ends or at intercalary places during extremely unfavourable conditions are
(a) zoospores (b) chlamydospores
(c) conidia (d) oidia

10 Consider the following statements.
I. Fungal spores produced asexually at top or side of hyphae exogenously are called as zoospores.
II. The fusion of protoplasm between two motile or non-motile gametes are called plasmogamy.

Choose the correct option.
(a) Statement I is true, but II is false
(b) Statement I is false, but II is true
(c) Both statements I and II are true
(d) Both statements I and II are false

11 Gametangial copulation is seen in
(a) *Candida* (b) *Albugo*
(c) *Rhizopus* (d) *Puccinia*

12 A conidia-like uninucleate male sex cell formed in a structure called spermatiophores is
(a) spermatia (b) pycnidium
(c) trichogyne (d) gametangium

13 Choose the incorrect pair.
(a) Isogamy - *Synchytrium* (b) Anisogamy - *Allomyces*
(c) Aplanospores - *Mucor* (d) Oogamy - *Synchytrium*

14 In laboratory preparations, *Rhizopus* does not produce zygospore because of
(a) oxygen deficiency
(b) light deficiency
(c) absence of + and - strains in the mycelium
(d) presence of + and - strains in the mycelium

15 The major difference between an ascospore and a basidiospore is
(a) ascospore is a sexual spore, while basidiospore is asexual
(b) ascospore is characteristic to Basidiomycetes, while basidiospore is to Zygomycetes
(c) ascospores are endogenously produced, while basidiospores are exogenously produced
(d) ascospore is diploid, while basidiospore is haploid

16 Match the different types of spores listed under Column I with the names of the organisms given under Column II.

Column I (Spores)	Column II (Organisms)
A. Ascospores	1. Diatoms
B. Endospores	2. *Agaricus*
C. Auxospores	3. Bacteria
D. Basidiospores	4. Yeast
	5. *Nephrolepis*

Codes

	A	B	C	D		A	B	C	D
(a)	4	3	1	2	(b)	4	1	3	2
(c)	4	2	5	1	(d)	4	5	1	2

17 Choose the best option with respect to fungi.

	Cell walls	Reserve food materials	Thallus	Habitats
(a)	Cellulose	Starch	Parenchymatous	Terrestrial
(b)	Chitin	Glycogen	Loose tissue, mycelium	Moist, damp
(c)	Glycogen	Starch	Chlorenchymatous	Aquatic
(d)	Cellulose	Fructose	Sclerenchymatous	Aquatic

Classification of Fungi

18 Members of Phycomycetes are found
(a) on decaying wood
(b) in moist and damp places
(c) as obligate parasites on plants
(d) All of the above

19 Consider the following statements.
I. Unlike true fungi cell wall of hyphae cells in Oomycetes are made up of cellulose.
II. Cup-shaped ascocarp is apothecium.
Choose the correct option.
(a) Statement I is true, but II is false
(b) Statement I is false, but II is true
(c) Both statements I and II are true
(d) Both statements I and II are false

20 Which of the following option is true about *Rhizopus*?
(a) The zygospore in *Rhizopus* develops into promycelium
(b) In *Rhizopus*, sexual reproduction involves the formation of two club-shaped structures called coenogametes
(c) Both (a) and (b)
(d) Vegetative reproduction occurs by budding

21 Fusion cell formed as a result of plasmogamy of two gametangia in *Rhizopus* is called
(a) pro-zygosporangium
(b) pregametangium
(c) progametangium
(d) prosporangium

22 Fungus producing eight spores in a sac belongs to
(a) Phycomycetes
(b) Ascomycetes
(c) Basidiomycetes
(d) Deuteromycetes

23 Consider the following statements.
I. Chains of buds attached to parent a yeast cell form pseudomycelium.
II. In yeast, vegetative reproduction takes place by budding.
Choose the correct option.
(a) Statement I is true, but II is false
(b) Statement I is false, but II is true
(c) Both statements I and II are true
(d) Both statements I and II are false

24 Yeast is used in the production of
(a) ethyl alcohol (b) fumaric acid
(c) leather (d) rubber

25 Scientific name of fission yeast is
(a) *Saccharomycodes ludwigii*
(b) *Schizosaccharomyces octosporus*
(c) *Saccharomyces octosporus*
(d) *Saccharomyces cerevisiae*

26 Which of the following is false for Ascomycetes?
(a) Ascomycetes involve variously pigmented moulds
(b) Thallus of Ascomycetes consists of aseptate, unbranched hyphae
(c) In *Ascobolous*, sexual reproduction occurs by spermatisation
(d) Process of karyogamy is usually delayed to begin a dikaryophase

27 Consider the following statements with respect to *Penicillium*.
I. *Penicillium* is commonly known as blue-green mould.
II. The conidia of *Penicillium* are uninucleate pigment.
Choose the correct option.
(a) Statement I is true, but II is false
(b) Statement I is false, but II is true
(c) Both statements I and II are true
(d) Both statements I and II are false

28 Which of the following option is true about *Penicillium*?
(a) It is commonly occurring on decaying oranges and lemons
(b) Vegetative reproduction occurs by fission
(c) Asexual reproduction takes place by the formation of sporangiophores
(d) It is affective on viruses

29 Fruiting body in *Penicillium* is known as
(a) cleistothecium (b) apothecium
(c) perithecium (d) pycnidium

30 Conidia bearing structures of *Penicillium* are
(a) metulae (b) rami
(c) sterigmata (d) asci

31 In *Penicillium*, the mycelium bears sex organs. The male organ is called antheridium, while the female one is known as
(a) archegonium (b) spermatangium
(c) ascogonium (d) sporogonium

32 Which statement is false regarding the life cycle of *Rhizopus* and *Spirogyra*?
(a) Asexual reproduction takes place in both by spores
(b) When conjugation fails azygospores and parthenospores are formed in both
(c) Zygotic meiosis occurs in both
(d) Sexual reproduction takes place by conjugation in both

33 Which pair of the following belongs to Basidiomycetes?
(a) Birds nest fungi and puffballs
(b) Puffballs and *Claviceps*
(c) *Peziza* and stick horns
(d) *Morchella* and mushrooms

34 Dikaryotisation in Basidiomycetes occurs during
(a) plasmogamy
(b) basidiospore formation
(c) clamp connection formation
(d) basidium formation

35 Out of the following features given in box, how many are seen in Basidiomycetes?

Dolipore septum, Ascospore, Clamp connection, Dikaryotisation, Ascocarp, Ascospores

Choose the correct option.
(a) Two (b) Four
(c) Six (d) Five

36 Sexual spores produced by Basidiomycetes are
(a) basidiospores (b) ascospores
(c) oospores (d) zygospores

37 The stage-O and stage-I of *Puccinia* are produced on the
(a) barberry (b) wheat
(c) maize (d) rice

38 Which of the following is true for *Agaricus*?
(a) The primary mycelia of *Agaricus* is long-lived
(b) The secondary mycelium in *Agaricus* is formed as a result of karyogamy
(c) Clamp connections aid in the formation of tertiary mycelium
(d) Basidiocarp in *Agaricus* consists of umbrella-shaped stipe and fleshy stalk pileus

39 Which one of the following statements is incorrect with respect to *Puccinia graminis*?
(a) It is a fungal parasite of wheat
(b) It produces uredospores and teleutospores on stem and foliage
(c) Teleutospores can reinfect wheat plants
(d) Uredospores can infect wheat plants

40 Which of the following option consists of all Deuteromycetes?
(a) *Fusarium, Colletotrichum, Helminthosporium*
(b) *Cercospora, Trichoderma, Penicillium*
(c) *Aspergillus*, puffballs, *Mucor*
(d) *Peronospora, Pythium*

41 Fungi imperfecti is a group of fungi which
(a) do not have a sexual phase
(b) include toad stool and puffballs
(c) have species that pray upon some nematodes
(d) All of the above

42 Match the following columns.

Column I	Column II
A. *Rhizopus*	1. Eurotiomycetes
B. *Penicillium*	2. Ustilaginomycetes
C. *Ustilago*	3. Deuteromycetes
D. *Alternaria*	4. Zygomycetes

Codes

	A	B	C	D
(a)	4	3	1	2
(b)	2	3	4	1
(c)	4	1	2	3
(d)	3	4	2	1

43 Which of the following pairs are unrelated?
(a) Ascomycetes — Ascogonium
(b) Conidiophore — Sterigmata
(c) Basidiocarp — Archegonium
(d) Ascocarp — Ascus

44 Consider the following statements.
I. Fungi lacking cross wall in the mycelium belong to Basidiomycetes.
II. Yeast like budding in oidia in *Mucor/Rhizopus* is called torula stage.
Choose the correct option.
(a) Statement I is true, but II is false
(b) Statement I is false, but II is true
(c) Both statements I and II are true
(d) Both statements I and II are false

Useful and Harmful Fungi

45 Beadle and Tatum gave 'one gene-one enzyme' hypothesis, while working on which fungus?
(a) *Saccharomyces*
(b) *Neurospora*
(c) *Penicillium*
(d) *Aspergillus*

46 Which of the following is used for flavouring cheese?
(a) *Penicillium camemberti*
(b) *Aspergillus niger*
(c) *Penicillium roqueforti*
(d) Both (a) and (c)

47 Consider the following statements.
I. Riboflavin can be commercially produced from *Ashbya gossypii.*
II. The fungus that is edible is *Rhizopus.*
Choose the correct option.
(a) Statement I is true, but II is false
(b) Statement I is false, but II is true
(c) Both statements I and II are true
(d) Both statements I and II are false

48 Consider the following statements.
I. Digestive enzymes like diastase, digestin and polyzyme are produced from *Aspergillus wentii.*
II. Bakane disease is caused by fungus, *Gibberella fujikuroi.*
Choose the correct option.
(a) Statement I is true, but II is false
(b) Statement I is false, but II is true
(c) Both statements I and II are true
(d) Both statements I and II are false

49 Which one is wrong pairing for the disease and its causal organism?
(a) Red rot of apple — *Armillaria mellea*
(b) Black rust of wheat — *Puccinia graminis*
(c) Loose smut of wheat — *Ustilago nuda*
(d) Root knot of vegetables — *Meloidogyne* sp.

50 Early blight and late blight are respectively caused by .... and ....
(a) *Alternaria, Phytophthora*
(b) *Ustilago, Puccinia*
(c) *Albugo, Phytophthora*
(d) *Erysiphe, Phytophthora*

51 Identify the incorrect match pair.
(a) Predacious fungi — *Dactylella*
(b) Damping off disease — *Pythium*
(c) Polyzyme — *Puccini*
(d) *Neurospora* — *Drosophila* of plant kingdom

52 Sclerotia extract of .... fungus produce a powerful hallucinogenic drug LSD and drug ergotine.
(a) *Claviceps purpurea*
(b) *Aspergillus flavus*
(c) *Agaricus*
(d) All of these

53 Identify the correct match pair.
(a) Puffball — *Polyporus*
(b) Rusts — *Ustilago*
(c) Mushrooms — *Lycoperdon*
(d) Toad stool — *Amanita*

54 Which of the following is not correctly matched.
(a) Root knot disease - *Meloidogyne jauanica*
(b) Smut of bajra - *Tolysporium penicillariae*
(c) Covered smut of barley - *Ustilago nuda*
(d) Late blight of potato - *Phytophthora infestans*

55 Red rot of sugarcane and white rust of radish are respectively caused by
(a) *Colletotrichum* and *Fusarium*
(b) *Pythium* and *Phytophthora*
(c) *Albugo candida* and *Puccinia graminis*
(d) *Colletotrichum* and *Albugo candida*

Lichen and Mycorrhiza

56 *Cladonia rangiferina* is a/an
(a) alga
(b) lichen
(c) fungus
(d) angiosperm

57 Which statement about lichen is true?
(a) The fungus is protected by a network of algal filaments
(b) Algal cells supply food to the fungal component
(c) The algal component is parasitic on fungus
(d) Lichens always need organic substratum for growth

58 Most of lichens are composed of
(a) red algae and Zygomycetes
(b) green algae and Phycomycetes
(c) blue-green algae and Oomycetes
(d) blue-green algae and Ascomycetes

59 Small circular depressions irregularly scattered on the underside of some lichens are called

(a) cyphellae (b) soredia
(c) corticolous (d) terricolous

60 Consider the following statements.

I. The carbohydrate used as stored food matter in lichens is lichenin.

II. *Usnea* a fructicose lichen is commonly known as reindeer moss.

Choose the correct option.

(a) Statement I is true, but II is false
(b) Statement I is false, but II is true
(c) Both statements I and II are true
(d) Both statements I and II are false

61 When two host species are required for the completion of the life cycle of a parasitic fungus, the condition is described as

(a) autoecious (b) heteroecious
(c) autotrophic (d) heterokaryotic

(B) Cumulative Exercise

(Assorted Questions of the Chapter for Advanced Level Practice)

1 Arrange the following in correct sequence with reference to sexual reproduction in *Rhizopus*.

I. Formation of germ tube
II. Formation of zygophores
III. Formation of warty wall layer of zygospore
IV. Secretion of trisporic acid

(a) IV, III, II and I (b) IV, II, III and I
(c) II, I, IV and III (d) I, III, II and IV

2 A teacher was explaining about a constant physical contact involving almost equal physiological interdependence in two different thalloid forms. He was trying to explain

(a) mycorrhizal association
(b) establishment of heterothallism
(c) operation of heterothallism
(d) advent of lichen formation

3 Read the statements given below. Which of these is incorrect?

(a) Sporangiospores borne in the sporangium of *Rhizopus* are diploid structures
(b) *Rhizopus* belongs to the class-Zygomycetes
(c) Dominant phase in the life cycle of *Chlamydomonas* is haploid
(d) Zoospores of *Chlamydomonas* are haploid

4 In the table below, some of the crop plants, their diseases and the pathogens are given. Match the three columns and identify the correct choice.

Crops	Diseases	Pathogens
A. Pigeon pea	I. Root knot	1. *Pseudomonas*
B. Brinjal	II. Ear cockle	2. *Fusarium*
C. Sugarcane	III. Wilt	3. *Anguina*
D. Wheat	IV. Red stripe	4. *Meloidogyne*
E. Citrus	V. Canker	5. *Xanthomonas*

(a) A-III-2 B-I-4 C-IV-1 D-II-3 E-V-5
(b) A-I-2 B-V-4 C-II-5 D-IV-1 E-III-3
(c) A-V-3 B-I-5 C-III-1 D-II-3 E-IV-2
(d) A-II-1 B-IV-5 C-V-2 D-III-4 E-I-3

5 Match the following columns.

	Column I		Column II
A.	Blast disease of rice	1.	Dikaryotic mycelium
B.	Citrus canker	2.	Single-celled conidiophores
C.	Grain smut of *Sorghum*	3.	Gram positive bacteria
D.	Red rot of sugarcane	4.	Septate conidiophores
		5.	Gram negative bacteria

Codes

	A	B	C	D
(a)	4	5	3	2
(b)	4	5	1	2
(c)	2	3	5	1
(d)	4	5	3	2

6 In the figure given below, sporangiophore, mycelium and stolon are indicated by

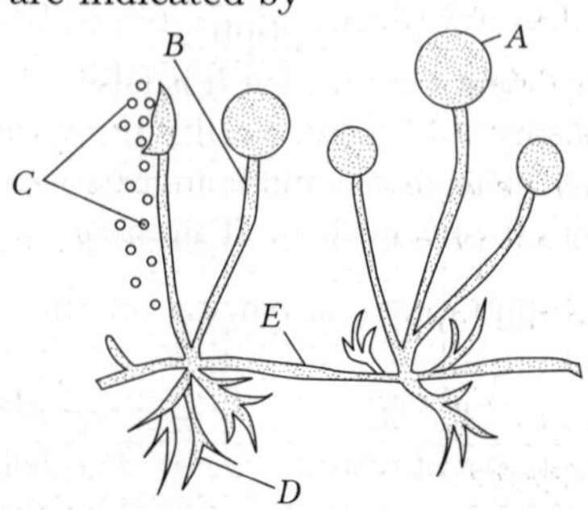

(a) *B*, *D* and *E*, respectively
(b) *A*, *B* and *C*, respectively
(c) *C*, *A* and *D*, respectively
(d) *B*, *A* and *D*, respectively

7 Which one of the following pairs is correctly matched?

(a) *Rhizobium* — Parasite in the roots of leguminous plants
(b) Mycorrhizae — Mineral uptake from soil
(c) Yeast — Production of biogas
(d) Myxomycetes — The disease ringworm

8 Select the incorrect statement.

(a) Gaspard Bauhin coined the term fungi
(b) Mycology is the study of fungi and PA Micheli is known as Father of Mycology
(c) In algal fungi, sexual reproduction occurs through gametangial contact
(d) Phycomycetes are called club fungi because of a club-shaped end of mycelium known as basidium

9 Which of the following environmental conditions are essential for optimum growth of *Mucor* on a piece of bread?

I. Temperature of about 25°C
II. Temperature of about 5°C
III. Relative humidity of about 5%
IV. Relative humidity of about 95%
V. A shady place
VI. A brightly illuminated place

Codes

(a) I, III and V (b) I, IV and V
(c) II, IV and V (d) II, III and VI

10 Yeast is not included in protozoans, but in fungi because

(a) it has no chlorophyll
(b) it shows saprophytic mode of nutrition
(c) it has eukaryotic organisation
(d) cell wall is made up of cellulose and reserves food material as starch

11 The sexual stages of pathogens of blast of rice and red rot of sugarcane are named respectively as

(a) *Magnaporthe grisea* and *Colletotrichum falcatum*
(b) *Colletotrichum falcatum* and *Pyricularia oryzae*
(c) *Glomerella tucumanensis* and *Magnaporthe grisea*
(d) *Magnaporthe grisea* and *Glomerella tucumanensis*

12 Match the following columns.

	Column I		Column II
A.	*Aspergillus niger*	1.	Ethanol
B.	*Clostridium butylicum*	2.	Statins
C.	*Saccharomyces cerevisiae*	3.	Citric acid
D.	*Trichoderma polysporum*	4.	Butyric acid
E.	*Monascus purpureus*	5.	Cyclosporin-A

Codes

	A	B	C	D	E
(a)	4	5	2	1	3
(b)	5	4	1	2	3
(c)	3	4	1	5	2
(d)	3	4	5	1	2

13 A type of life cycle in which plasmogamy, karyogamy, haploidisation takes place but not at specific place in life cycle of an organism is called as

(a) parasexuality (b) heterozygosity
(c) homozygosity (d) asexuality

14 With respect to fungal sexual cycle, choose the correct sequence of events. **[NCERT Exemplar]**

(a) Karyogamy → Plasmogamy → Meiosis
(b) Meiosis → Plasmogamy → Karyogamy
(c) Plasmogamy → Karyogamy → Meiosis
(d) Meiosis → Karyogamy → Plasmogamy

15 If food sources begin to run out, hyphae respond by making spores. Why?

(a) Spore production attracts animals, which can be used to transport the fungi to an area where more food is available
(b) Fungi change feeding mechanisms once food can no longer be absorbed and the spore production is a byproduct of this change
(c) Spore production is favoured by natural selection when hyphae are under nutritional stress
(d) Spore production allows fungi to harvest food particles that are floating in the air

16 How many of the given features belong to fungi?

Zygotic meiosis, Sexual spores, Asexual spores, Loose tissue, Thalloid body, Sporic meiosis, Cellulose in cell wall, Starch as reserved food , Chlorophyll, Absence of cell wall.

(a) Three (b) Five (c) Four (d) Six

17 Match the following columns.

	Column I	Column II
A.	A domain of plant-like, heterotrophic organisms that work primarily as decomposers	1. Lichen
B.	The slender, reproductive structures of fungi	2. Dikaryotic
C.	Indicator of SO_2	3. Hyphae
D.	Cells with two nuclei	4. Fungi

Codes

	A	B	C	D
(a)	4	3	1	2
(b)	1	2	3	4
(c)	4	3	2	1
(d)	3	2	1	4

18 Consider the following statements regarding Deuteromycetes.

I. It is an artificial class of fungi.
II. Asexual stages of these fungi are unknown.
III. Deuteromycetes are called as fungi imperfecti.
IV. Deuteromycetes are actually Ascomycetes.

Choose the correct option.

(a) Statement II is incorrect, while others are correct
(b) Statements III and IV are incorrect, while others are correct
(c) Statement I is correct and all others are incorrect
(d) All statements are correct

19 The given below figure shows a generalised life cycle of a fungus. The suitable terms for *A*, *B*, *C*, *D* and *E* are

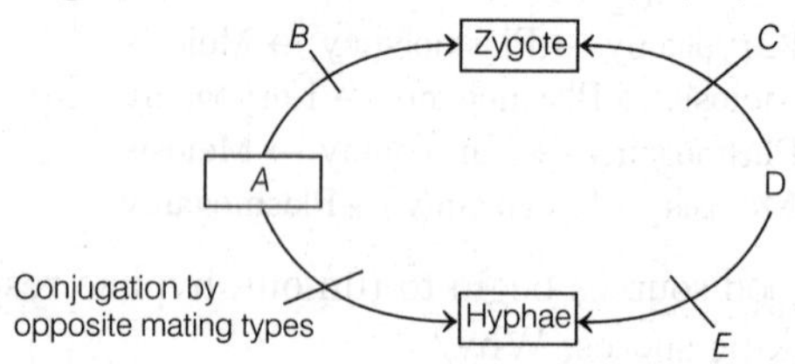

	A	*B*	*C*	*D*	*E*
(a)	Meiosis	Mitosis	Spore	Fertilisation	Dikaryotic phase
(b)	Dikaryotic phase	Fertilisation	Meiosis	Spore	Mitosis
(c)	Fertilisation	Meiois	Mitosis	Dikaryotic phase	Amitosis
(d)	Mycelium	Mitosis	Meiosis	Fertilisation	Spore

20 The sequence of spore production in *Puccinia* is

(a) Pycniosproe → Aeciospore → Uredospore → Teliospore
(b) Basidiospore → Aeciospore → Pycniospore → Teliospore
(c) Teliospore → Aeciospore → Uredospore → Basidiospore → Pycniospore
(d) Teliospore → Basidiospore → Aeciospore → Pycniospore → uredosproe

21 Which of the following statements is incorrect?

I. There is complete absence of motile structures in Ascomycetes.
II. *Armillaria* is the largest fungi.
III. Deuteromycetes sexually reproduce *via* conjugation.
IV. *Synchytrium* possesses holocarpic vegetative body.
V. Rust fungi belong to order-Ustilaginales.

Codes

(a) I, III and IV (b) III, IV and V
(c) II and III (d) III and V

Assertion and Reason

Direction (Q. Nos. 22-28) *In each of the following questions, a statement of Assertion (A) is given followed by the corresponding statement of Reason (R). Of the statements mark the correct answer as*

(a) Both A and R are true and R is the correct explanation of A
(b) Both A and R are true, but R is not the correct explanation of A
(c) A is true, but R is false
(d) A is false, but R is true
(e) Both A and R are false

22 **Assertion** (A) Budding occurs in yeast.
Reason (R) It is similar to fission process of bacteria.

23 **Assertion** (A) Heterothallism prevents inbreeding.
Reason (R) It helps the fungi to adapt to unfavourable conditions.

24 **Assertion** (A) Fragmentation is easy to grow fungi in the laboratory.
Reason (R) Fungi show fragmentation.

25 **Assertion** (A) Spermatia are also conidia.
Reason (R) They are produced on spermatophore or spermatogonia.

26 **Assertion** (A) Penicillin is obtained from *Penicillium griseofulvum.*
Reason (R) Ergot is obtained from *Penicillium notatum.*

27 **Assertion** (A) Invertase is manufactured from *Aspergillus oryzae.*
Reason (R) Amylase is produced by *Saccharomyces cerevisiae.*

28 **Assertion** (A) The fungi are widespread in distribution and they even live on or inside other plants and animals.
Reason (R) Fungi are able to grow anywhere on land, water or on other organisms, because they have a variety of pigments, including chlorophyll, carotenoids, fucoxanthin and phycoerythrin.

(C) Medical Entrances' Gallery

(Collection of questions Asked in NEET & Various Medical Entrance Exams)

1 Match Column I with Column II.

Column I	Column II
A. Saprophyte	1. Symbiotic association of fungi with plant roots
B. Parasite	2. Decomposition of dead organic materials
C. Lichens	3. Living on living plants or animals
D. Mycorrhiza	4. Symbiotic association of algae and fungi

Codes **[NEET 2019]**

	A	B	C	D		A	B	C	D
(a)	3	2	1	4	(b)	2	1	3	4
(c)	2	3	4	1	(d)	1	2	3	4

2 Which of the following statements is correct?
(a) Lichens do not grow in polluted areas **[NEET 2019]**
(b) Algal component of lichens is called mycobiont
(c) Fungal component of lichens is called phycobiont
(d) Lichens are not good pollution indicators

3 Which of the following statements is incorrect? **[NEET 2019]**
(a) *Claviceps* is a source of many alkaloids and LSD
(b) Conidia are produced exogenously and ascospores endogenously
(c) Yeasts have filamentous bodies with long thread-like hyphae
(d) Morels and truffles are edible delicacies

4 After karyogamy followed by meiosis, spores are produced exogenously in **[NEET 2018]**
(a) *Agaricus* (b) *Alternaria*
(c) *Neurospora* (d) *Saccharomyces*

5 Cell wall of fungi is composed of **[JIPMER 2018, NEET-I 2016]**
(a) chitin (b) pectin
(c) cellulose (d) mannans

6 The given below figure shows a generalised life cycle of a fungus. Identify *A*, *B* and *C* from the given option.

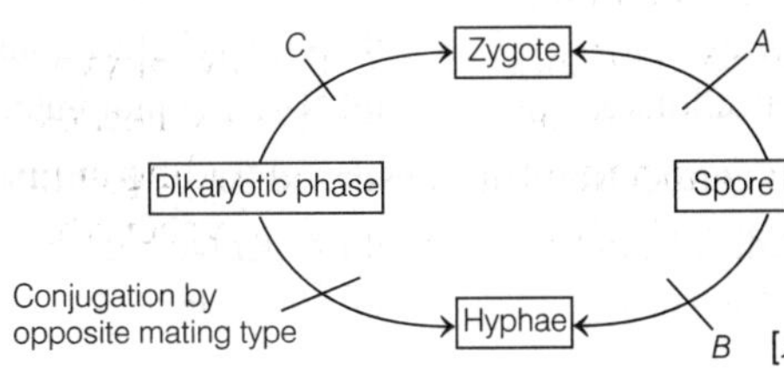

[AIIMS 2017]

	A	*B*	*C*
(a)	Meiosis	Fertilisation	Meiosis
(b)	Meiosis	Mitosis	Fertilisation
(c)	Mitosis	Meiosis	Fertilisation
(d)	Mitosis	Fertilisation	Meiosis

7 Which one of the following is wrong for fungi?
(a) They are eukaryotic **[NEET-II 2016]**
(b) All fungi possess a purely cellulosic cell wall
(c) They are heterotrophic
(d) They are both unicellular and multicellular

8 Which one of the following matches is correct? **[CBSE AIPMT 2015]**

(a)	*Phytophthora*	Aseptate mycelium	Basidiomycetes
(b)	*Alternaria*	Sexual reproduction absent	Deuteromycetes
(c)	*Mucor*	Reproduction by conjugation	Ascomycetes
(d)	*Agaricus*	Parasitic fungus	Basidiomycetes

9 Yeast is used in the production of **[J&K CET 2015, CBSE AIPMT 2015]**
(a) bread and beer
(b) cheese and butter
(c) citric acid and lactic acid
(d) lipase and pectinase

10 The imperfecti fungi which are decomposers of litter and help in mineral cycling belong to **[CBSE AIPMT 2015]**
(a) Deuteromycetes (b) Basidiomycetes
(c) Phycomycetes (d) Ascomycetes

11 Morels and truffles groups of fungi are classified under **[AMU 2015]**
(a) Phycomycetes (b) Deuteromycetes
(c) Basidiomycetes (d) Ascomycetes

12 Match the following columns and choose the correct combination from the options given below.

Column I	Column II
A. Eubacteria	1. *Trichoderma*
B. Dinoflagellates	2. *Albugo*
C. Phycomycetes	3. *Gonyaulax*
D. Deuteromycetes	4. *Anabaena*

Codes **[EAMCET 2015]**

	A	B	C	D
(a)	1	2	3	4
(b)	2	3	4	1
(c)	4	3	2	1
(d)	4	3	1	2

13 Which one of the following natural polymer is found both in insects and fungi? **[WB JEE 2015]**
(a) Pectin (b) Chitin
(c) Cellulose (d) Suberin

14 In *Mucor*, zygospore germinates through the formation of **[Chhattisgarh PMT 2015]**
(a) meiospores (b) mitospores
(c) zoospores (d) planospores

15 Heterothallism is found in **[Chhattisgarh PMT 2015]**
(a) *Mucor* (b) *Cycas* (c) *Selaginella* (d) *Volvox*

16 Which of the following does not apply to Ascomycetes?
(a) Mycelium coenocytic and aseptate **[Kerala CEE 2014]**
(b) Commonly known as sac fungi
(c) Asexual spores called conidia are produced exogenously
(d) Sexual spores called ascospores are produced endogenously
(e) They are saprophytic, decomposers, parasitic or coprophilous

17 Which one of the following is not a fungal disease? **[Gujarat CET 2014]**
(a) Rust of wheat (b) Covered smut of bajra
(c) Black rot of crucifers (d) Red rot of sugarcane

18 Asexual reproduction in fungi occurs by **[Odisha JEE 2013, 2010]**
(a) ascospores (b) basidiospores
(c) conidia (d) oospores

19 Which one single organism or the pair of organisms is correctly assigned to its or their named taxonomic group? **[CBSE AIPMT 2012]**
(a) *Paramecium* and *Plasmodium* belong to the same kingdom as that of *Penicillium*
(b) Lichen is a composite organism formed from the symbiotic association of an alga and a protozoan
(c) Yeast used in making bread and beer is a fungus
(d) *Nostoc* and *Anabaena* are examples of Protista

20 Sexual dimorphism is absent in **[JIPMER 2012]**
(a) Ascomycetes (b) Deuteromycetes
(c) Basidiomycetes (d) Phycomycetes

21 Which one of the following helps in the absorption of phosphorus from soil by plants? **[CBSE AIPMT 2012]**
(a) *Rhizobium* (b) *Frankia*
(c) *Anabaena* (d) *Glomus*

22 Which one of the following is wrongly matched pair? **[CBSE AIPMT 2012]**
(a) *Puccinia* — Smut
(b) Root — Exarch protoxylem
(c) *Cassia* — Imbricate aestivation
(d) Root pressure — Guttation

23 The association mycorrhiza is **[UP CPMT 2012]**
(a) relationship of algae and fungi
(b) relationship of fungi and higher plants
(c) relationship of algae and higher plants
(d) None of the above

24 Which of the following is not correctly matched pair? **[Kerala CEE 2012]**
(a) *Amoeba* — Protozoa
(b) *Gonyaulax* — Dinoflagellate
(c) *Thermoacidophiles* — Archaebacteria
(d) *Albugo* — Chrysophyte
(e) *Anabaena* — Cyanobacteria

25 Which of the following is a unicellular sac-fungus? **[Kerala CEE 2012, 2011]**
(a) *Claviceps* (b) *Saccharomyces*
(c) *Penicillium* (d) *Neurospora*
(e) *Aspergillus*

26 Match the following columns.

Column I	Column II
A. Morels	1. Deuteromycetes
B. Smut	2. Ascomycetes
C. Bread mould	3. Basidiomycetes
D. Imperfecti fungi	4. Phycomycetes

Codes **[Kerala CEE 2012]**

	A	B	C	D		A	B	C	D
(a)	3	4	1	2	(b)	2	3	4	1
(c)	4	1	2	3	(d)	3	4	2	1
(e)	2	1	4	3					

27 Edible part of mushroom is **[WB JEE 2012, 11]**
(a) basidiocarp (b) primary mycelium
(c) fungal hyphae (d) basidiospores

28 VAM is **[WB JEE 2012]**
(a) symbiotic bacteria
(b) saprophytic bacteria
(c) saprophytic fungi
(d) symbiotic fungi

29 Chitin is a constituent of fungus cell wall. Its hydrolysis yields **[UP CPMT 2012]**
(a) cellulose (b) N-acetyl glucosamine
(c) hemicellulose (d) N-acetyl muramic acid

30 Clamp connections are observed in **[UP CPMT 2012]**
(a) Basidiomycetes (b) Zygomycetes
(c) Ascomycetes (d) Oomycetes

ANSWERS

Check Point 6.1

1. (b)	2. (d)	3. (c)	4. (c)	5. (d)	6. (c)	7. (c)

Check Point 6.2

1. (c)	2. (d)	3. (b)	4. (c)	5. (c)	6. (b)

Check Point 6.3

1. (d)	2. (c)	3. (b)	4. (b)	5. (d)	6. (b)

(A) Topical Exercise

1. (b)	2. (d)	3. (a)	4. (c)	5. (b)	6. (a)	7. (b)	8. (d)	9. (b)	10. (b)
11. (c)	12. (a)	13. (d)	14. (c)	15. (c)	16. (a)	17. (b)	18. (d)	19. (c)	20. (c)
21. (a)	22. (b)	23. (c)	24. (a)	25. (b)	26. (b)	27. (c)	28. (a)	29. (a)	30. (c)
31. (a)	32. (b)	33. (a)	34. (d)	35. (b)	36. (a)	37. (a)	38. (c)	39. (c)	40. (a)
41. (a)	42. (c)	43. (c)	44. (b)	45. (b)	46. (d)	47. (c)	48. (b)	49. (c)	50. (a)
51. (c)	52. (a)	53. (d)	54. (c)	55. (d)	56. (b)	57. (b)	58. (d)	59. (a)	60. (a)
61. (d)									

(B) Cumulative Exercise

1. (b)	2. (d)	3. (a)	4. (a)	5. (b)	6. (a)	7. (b)	8. (d)	9. (b)	10. (b)
11. (d)	12. (c)	13. (a)	14. (c)	15. (b)	16. (c)	17. (a)	18. (d)	19. (b)	20. (a)
21. (d)	22. (c)	23. (b)	24. (b)	25. (d)	26. (e)	27. (e)	28. (c)		

(C) Medical Entrances' Gallery

1. (c)	2. (a)	3. (c)	4. (a)	5. (a)	6. (b)	7. (b)	8. (b)	9. (a)	10. (a)
11. (d)	12. (c)	13. (b)	14. (a)	15. (a)	16. (a)	17. (c)	18. (c)	19. (c)	20. (b)
21. (d)	22. (a)	23. (b)	24. (d)	25. (b)	26. (b)	27. (a)	28. (d)	29. (b)	30. (a)

Explanations

(A) Topical Exercise

3 (*a*) Statement I is true, but II is false. Incorrect statement can be corrected as

Plectenchyma are tissue composed of mycelial threads of fungi (hyphae). The two most common types of tissues are prosenchyma and pseudoparenchyma.

4 (*c*) Pseudoparenchyma is a tissue that superficially resembles plant parenchyma, but is made up of an interwoven mass of hyphae or filaments. Examples of pseudoparenchymatous structure are the fruiting bodies or certain fungi and the thalli of certain red and brown algae. Thus, we can conclude that pseudoparenchyma imitate parenchyma of higher plants.

10 (*b*) Statement I is false, but II is true. Incorrect statement can be corrected as

The fungal spores produced asexually at top or side of hyphae exogeneously are called conidiophores. e.g. *Aspergillus* and *Penicillium.*

13 (*d*) Option (d) is incorrect pair and can be corrected as

Oogamy is the sexual reproduction in which fusion of large non-motile female gamete with smaller motile female gamete. It is seen in chytridiomycetes.

Rest options are correct pairs.

26 (*b*) Statement in option (b) is false and can be corrected as

The mycelium of Ascomycetes is branched and septate.

Rest options are true for Ascomycetes.

34 (*d*) Dikaryotisation in Basidiomycetes occurs during basidium formation. The teliospores are initially dikaryotic, but become diploid *via* karyogamy. Meiosis takes place at the time of germination.

A promycelium is formed that consists of a short hypha, Smuts with both a yeast phase and infectious hyphal state are the example of dimorphic Basidiomycota.

39 (*c*) Statement in option (c) is incorrect and can be corrected as *Puccinia graminis tritici* belong to Basidomycotina and caused black rust of wheat.

Puccinia is a macrocyclic and heteroecious rust fungus. It produce uredia and felia stage on wheat plant. The spores produced on wheat are uredospores (stage II) and teleutospores (stage III). Uredospores can reinfect wheat but teleutospores cannot reinfect the wheat plant.

44 (*b*) Statement I is false, but statement II is true. Incorrect statement can be corrected as

Fungi lacking cross wall in the mycelium belong to Phycomycetes as they grow like a single cell.

45 (*b*) The one gene-one enzyme hypothesis is the idea that each gene encodes a single enzyme. Beadle and Tatum confirmed Garrod's hypothesis using genetic and biochemical studies of the bread mould *Neurospora.* Beadle and Tatum identified bread mould mutants that were enable to make specific amino acids.

46 (*d*) *Penicillium camemberti* and *Penicillium roqueforti* are used for flavouring cheese while *Aspergillus niger* are used for the porduction of organic acids. Thus, option (d) is correct.

48 (*b*) Statement I is false, but statement II is true. Incorrect statement can be corrected as

Digestive enzymes like diastase, digestin and polyzyme are produced from *Aspergillus flavus-oryzae.*

55 (*d*) Red rot of sugarcane is caused by the pathogen, *Colletotrichum falcatum* which is a fungus of class-Deuteromycetes.

White rust of radish or white rust of crucifers is caused by *Albugo candida* or *Cystopus candidus,* which is an algal fungi (Phycomycetes or Oomycetes).

58 (*d*) Lichens consist of the fungal component mycobiont, mainly Ascomycotina and the algal component phycobiont which are mostly blue-green algae or green algae. Thus, most of lichens are composed of blue-green algae and Ascomycetes.

60 (*a*) Statement I is true, but statement II is false. Incorrect statement can be corrected as

Cladonia rangiferina is commonly known as reindeer moss, it is a fruticose lichen seen in great amounts in Arctic lands.

(B) Cumulative Exercise

1 (*b*) Option (b) is correct. The correct sequence is IV, II, III, and I.

During sexual reproduction in *Rhizopus,* (+) and (–) strains of mycelia stimulate each other by releasing trisporic acid and form zygophores. Zygophores of two strains come in contact to form progametangia, then coenogametangia.

After gametangial copulation, zygospores with warty wall layers are formed which germinate in favourable conditions and form a germ tube.

2 (*d*) The teacher was trying to explain the advent of lichen formation. Lichen is a symbiotic relationship between algae (phycobiont) and fungi (mycobiont). Both the partners are in a constant physical contact and have almost equal physiological interdependence. The fungal partner takes part in reproduction and protection, while the algae synthesise food through photosynthesis.

3 (*a*) Statement in option (a) is incorrect and can be corrected as Sporangiospores of *Rhizopus* are haploid structures. The young sporangium of *Rhizopus* contains certain amount of cytoplasm and many nuclei. The sporangium is divided into the denser, peripheral sporiferous zone and the central dome-shaped zone, columella.

The protoplast of columella is continuous with that of sporangiophore. The sporiferous zone undergoes cleavage and forms haploid sporangiospores.

Rest options are correct statements.

4 (*a*) Option (a) is correctly matched of the crops with their diseases and pathogens.

Crops	Diseases	Pathogens
Pigeon pea	Wilt	*Fusarium exysporum*
Brinjal	Root knot	*Meloidogyne incognita*
Sugarcane	Red stripe	*Pseudomonas*
Wheat	Ear cockle	*Anguina*
Citrus	Canker	*Xanthomonas citri*

8 (*d*) Statement in option (d) is incorrect and can be corrected as 'Club fungi' is the common name given to the fungi of class-Basidiomycetes (not Phycomycetes) because of club-shaped end of mycelium known as basidium.
Rest of the statements are correct.

9 (*b*) Option (b) is correct.
Mucor shows the best growth on a piece of bread at a temperature of about 25°C, relative humidity of about 95% and in a moist and shady place.
Mucor is a saprophytic fungus belonging to the order—Mucorales and family- Mucoraceae and it grows on decaying dung and on some food items.

10 (*b*) Though yeast is unicellular and non-filamentous, but they are placed in Fungi instead of Protozoa because the other characters of yeast are same as fungi like they are eukaryotic, their mode of nutrition is saprophytic, have chitinous cell wall. Hence, they are placed in fungi.

11 (*d*) Blast of rice or paddy is caused by the fungus *Pyricularia oryzae* of class-Deuteromycetes. *Magnaporthe grisea* is the sexual stage of *P. oryzae.* Red rot of sugarcane is caused by the fungus *Colletotrichum falcatum* and its the sexual stage is *Glomerella tucumanensis.*

13 (*a*) Parasexuality is a type of life cycle in which plasmogamy, karyogamy and haploidisation take place, but not at a specific place in the life cycle. It was discovered in fungi, *Aspergillus nidulans* by Pontecarvo and Roper in 1952.

21 (*d*) Statements III and V are incorrect and can be corrected as

- Deuteromycetes reproduce by asexual means only like conidia.
- Rust fungi belong to class-Basidiomycetes and order—Uridinales. Smut fungi belong to order—Ustilaginales.

Rest statements are correct.

22 (*c*) A is true, but R is false because
Budding of vegetative cells is common in yeasts. It is different from fission, in which cell divides in transverse plane into two daughter cells.

23 (*b*) Both A and R are true, but R is not the correct explanation of A.
Heterothallism prevents inbreeding and promotes outbreeding. It also helps fungi to overcome unfavourable conditions.

24 (*b*) Both A and R are true, but R is not the correct explanation of A.
Fungi show fragmentation and are very easy to grow in laboratory on suitable culture media.

25 (*d*) A is false, but R is true because
Spermatia differ from conidia in being smaller and incapable to germinate and give rise to new thallus. Though, spermatia are produced on spermatophore or spermatogonia.

26 (*e*) Both A and R are false and can be corrected as
Penicillin is obtained from *Penicillium notatum* and *Penicillium chrysogenum,* whereas ergot is obtained from *Claviceps purpurea.*

27 (*e*) Both A and R are false and can be corrected as
Amylase is produced by *Aspergillus oryzae,* while invertase from *Saccharomyces cerevisiae.*

28 (*c*) A is true, but R is false. R can be corrected as
Fungi can be found everywhere as they just require, shady moist places to flourish.

(C) Medical Entrances' Gallery

2 (*a*) Statement in option (a) is correct. Rest of the statements are incorrect and can be corrected as

- Algal component of lichens is called phycobiont.
- Fungal component of lichens is called mycobiont.
- Lichens are good pollution indicators.

3 (*c*) The statement in option (c) is incorrect and can be corrected as
Yeast is a unicellular sac fungus which lacks filamentous structures or hyphae. However, they may form short temporary filamentous structure called pseudomycelium.
Rest of the statements are correct.

4 (*a*) *Agaricus* belongs to class—Basidiomycetes in which meiospores are produced exogenously after karyogamy and meiosis. Other options are explained as
Alternaria belongs to the class-Deuteromycetes and they lack sexual reproduction.
Neurospora and *Saccharomyces* belong to the class-Ascomycetes and they produce ascospores endogenously.

5 (*a*) The cell wall of fungi is composed of chitin. It is the second most abundant carbohydrate and is a homopolymer of N-Acetyl Glucosamine (NAG) joined with β 1-4 linkages. NAG is a modification of glucose molecule.

7 (*b*) Option (b) is wrong for fungi and can be corrected as
In fungi, cell wall contains chitin or cellulose along with other polysaccharides, proteins and lipids. Only in some fungi, e.g. *Phytophthora* or other Oomycetes, a purely cellulosic cell wall is present. Hence, only option (b) is wrong.
Rest of the options are correct for fungi.

8 (*b*) Option (b) is correct match. Other options are incorrect matches and can be corrected as

- *Phytophthora* belongs to Phycomycetes (algal fungi). They contain either unicellular thallus or non-septate coenocytic mycelium.
- *Mucor* belongs to Phycomycetes. Their mycelium is coenocytic (multinucleate) and profusely branched. They reproduce vegetatively *via* conjugation.
- *Agaricus* belongs to Basidiomycetes. They contain well-developed filaments, branched and septate mycelium. They are saprophytic, but not parasitic.

9 (*a*) Yeast (*Saccharomyces cerevisae*) is used to produce bread and beer as it is involved in the fermentation of wheat flour and extract of barley.

10 (*a*) The fungi imperfecti which are decomposers of litter and help in mineral cycling belongs to Deuteromycetes. They are fungi which do not fit into the commonly established taxonomic classification of fungi. They include all those fungi in which the perfect stage (sexual stage) is not reported.

11 (*d*) Morels and truffles are edible fungi and are considered delicacies. They are classified under class-Ascomycetes of kingdom-Fungi.

13 (*b*) The natural polymer present in both insects and fungi is chitin. It is a long polymer of N-acetyl glucosamine, a derivative of glucose.
It is characteristic component of the cell wall of fungi, the exoskeletons of arthropods like crustaceans and insects, radulae of molluscs and internal shells of cephalopods.

14 (*a*) In *Mucor*, the sexual reproduction occurs by the union of compatible complementary gametangia which develop into zygosporangium. Zygosporangium contains single zygospore where meiosis occurs and meiospores are formed. Meiospores form haploid mycelium after germination.

15 (*a*) Heterothallism refers to the presence of two types of mycelia in a fungus. *Mucor* is a heterothallic fungi as it produces two types of mycelia which are physiologically different.

16 (*a*) Statement in option (a) does not apply to Ascomycetes (sac-fungi) because
Mycelium is branched and septate in Ascomycetes (rather than coenocytic and aseptate). It may consist of distinct hyphae or the same may aggregate to produce parenchyma and pseudoparenchyma.
Rest of the statements apply to Ascomycetes.

17 (*c*) Black rot of crucifers is not a fungal disease as it is caused by bacterium *Xanthomonas campestris.* Remaining three are fungal diseases. Rust of wheat is caused by *Puccinia graminis,* covered smut of bajra (barley) is caused by *Ustilago hordei* and red rot of sugarcane is caused by *Colletotrichum falcatum.*

19 (*c*) Statement in option (c) is correct whereas other statements are incorrect. Incorrect statements can be corrected as
- *Paramecium* and *Plasmodium* belong to kingdom-Protista, *Penicillium* belongs to kingdom-Fungi.
- Lichen is a composite organism formed by the symbiotic association of algae and fungi.
- *Nostoc* and *Anabaena* are examples of algae.

20 (*b*) In Deuteromycetes, sexual phase of reproduction is absent and it reproduces only by asexual means. Therefore, Deuteromycetes are also called as fungi imperfecti.

21 (*d*) *Glomus* is a genus of Vesicular Arbuscular Mycorrhiza (VAM) fungi. It helps in nutrient uptake and mainly the absorption of phosphorus from the soil.

22 (*a*) Option (a) is wrongly matched pair and can be corrected as *Puccinia* is commonly called rust fungus and smut fungus is *Ustilago.* Both rust and smut fungi belong to the class-Basidiomycetes.
Rest options are correctly matched pairs.

23 (*b*) Mycorrhiza is the symbiotic association between fungus and roots of higher plants. The mycorrhizal roots are usually covered with a (wooly) fungal outgrowth. Fungal growth does not cause any harm to the plant, instead help in absorption of water, phosphorus and nitrogen.

24 (*d*) Option (d) is not correctly matched pair and can be corrected as
Albugo is a phycomycetous fungus, whereas chrysophyte includes diatoms and golden algae (desmids).
Rest of the options are correctly matched pairs.

25 (*b*) *Saccharomyces* is a unicellular ascomycetous fungus. The members of class-Ascomycetes are called sac-fungi because they produce spores in a distinctive sporangium called as an ascus.

27 (*a*) The edible part of mushrooms is the fruiting body called basidiocarp. It is a multicellular structure on which spores are formed.

28 (*d*) VAM or Vesicular Arbuscular Mycorrhiza is a symbiotic fungi. It is formed by the symbiotic association between certain phycomycetous fungi and angiosperm roots. The fungus colonises the root cortex forming a mycelial network and characteristic vesicles (bladder-like structure) and arbuscles (branched finger-like hyphae).

30 (*a*) Clamp connections are formed in dikaryotic mycelium in the class-Basidiomycetes. This is a kind of conjugation division found in this class.

CHAPTER 07

Plantae

Plantae is the kingdom of eukaryotic and multicellular organisms that grow indefinitely in their lifetime. The study of plants is called **phytology**.

The main features of plants are as follows

1. Plants are autotrophic due to the presence of chlorophyll. Thus, they are also referred to as **primary producers**. Some plants also show heterotrophism.
2. Plant cells have cellulosic cell wall, big sap vacuole and plastids with several photosynthetic pigments.
3. Reproduction takes place by vegetative, asexual and sexual means.
4. A multicellular embryo is formed during development from the zygote in plants. In lower plants, embryo formation is absent, e.g. algae.
5. Plants show alternation of generation, i.e. gametophytic and sporophytic generations alternate with each other.

In this chapter, we will discuss the detailed classification of plant kingdom, including its characteristics and examples.

Systems of plant classification

- The branch of science, dealing with classification of plants is called **plant taxonomy**.
- Plants are classified into various divisions and subdivisions based on their similarities and dissimilarities.
- A number of scientists including **Eichler, Endlicher, Engler, Tippo, Bessey, Smith, Alexopoulos, Bentham** and **Hooker**, etc., have given various plant classification systems.
- Among the various classification systems, a few have been discussed below

(i) **Linnaeus' system** (1758) He classified plants into two subkingdoms, i.e.

- **Cryptogamae** It includes plants without any visible sex organs, i.e. algae, fungi, liverworts, mosses and ferns.
- **Phanerogamae** It includes plants that bear visible sex organs inside their flowers, i.e. gymnosperms and angiosperms.

(ii) **AP de Candolle's system** (1819) He classified plants on the basis of the presence or absence of vascular system into two subkingdoms, i.e. Vascularae and Avascularae.

(iii) **Endlicher's system** (1836) This system divided plants into Protophyta (algae and lichens), Histerophyta (fungi) and Cormophyta (all plants with roots, stems and leaves), etc.

Inside

(iv) **Engler's system** (1886) In this, plants are divided into two categories
- **Thallophyta** with thalloid plant body and no embryo formation.
- **Embryophyta** with a well-differentiated plant body and embryo formation.

(v) **Oswald Tippo's system** He divided Embryophyta, into Bryophyta and Tracheophyta. This distinction was based on the presence or absence of vascular tissue in embryophytes.
- Tracheophyta was divided into Pteridophyta and Spermatophyta based on the presence or absence of seeds.
- Spermatophyta was further divided into Gymnosperms and Angiosperms, based on the nature of seeds.

(vi) **Bentham and Hooker's classification** They divided plant kingdom into two subkingdoms, i.e.
- **Cryptogamia** includes non-seeded plants.
- **Phanerogamia** includes seeded plants. It was further divided into three classes, i.e. Dicotyledonae, Gymnospermae and Monocotyledonae.

CLASSIFICATION OF PLANT KINGDOM

A generalised classification of kingdom-Plantae is as follows

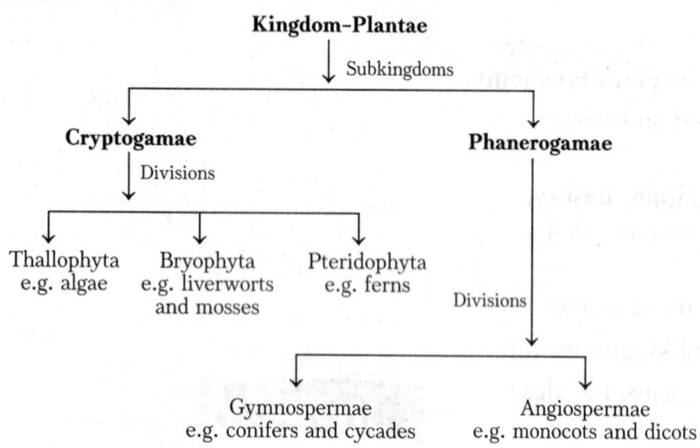

Subkingdom–Cryptogamae

These are lower plants with hidden sex organs. Seeds and flowers are absent. Three divisions under this subkingdom are as follows

(i) **Division–Thallophyta** Aquatic plants having gametophytic (n) thalloid plant body, e.g. algae.

(ii) **Division–Bryophyta** Terrestrial plants with gametophytic thalloid or leafy plant body without conductive tissue, e.g. *Marchantia.*

(iii) **Division–Pteridophyta** These plants possess a sporophytic ($2n$), well-developed plant body with conductive tissue, e.g. *Selaginella.*

Subkingdom – Phanerogamae

These are higher plants, which bears seed and have prominent sex organs. It includes two divisions as follows

(i) **Division–Gymnospermae** Plants with naked seeds or seed plants without flowers, e.g. *Cycas.*

(ii) **Division–Angiospermae** Plants in which seeds are formed inside fruits and sporophylls are organised into flowers, e.g. mango.

THALLOPHYTA : ALGAE

- This is a group of thalloid plants in which body is not differentiated into true roots, stems and leaves.
- The term 'Algae' was first used by **Linnaeus** in 1755.
- The study of algae is called as **Phycology**. **MOP Iyengar** is known as the **Father of Indian Phycology**.

General characteristics

The general characteristics of algae are as follows

1. **Habitat** Most algae occur in aquatic habitats. These can be
   - Freshwater forms, e.g. *Spirogyra* and *Hydrodictyon.*
   - Marine forms, e.g. *Macrocystis, Nereocystis, Vaucheria* and *Butrydium* are terrestrial forms.
2. **Plant body** It is thalloid. It can be of various types based on their organisation, i.e.
   - Unicellular motile, e.g. *Chlamydomonas.*
   - Unicellular non-motile, e.g. *Chlorella.*
   - Colonial motile, e.g. *Volvox.*
   - Colonial non-motile, e.g. *Hydrodictyon.*
   - Filamentous unbranched, e.g. *Spirogyra.*
   - Filamentous branched, e.g. *Cladophora.*
   - Heterotrichous, e.g. *Coleochaete.*
3. **Cell structure** Algal cells exhibit eukaryotic cell structure with cellulosic cell wall.
   - Though, cell wall of red algae possesses pectin and calcium salts.
   - Similarly in brown algae, alginic acid and sulphated polysaccharides are found.
4. **Nutrition** Mostly algae are autotrophic, but few are heterotrophic as well.

5. The reserve food material varies from one class to another, i.e.
   - **Green algae** Starch
   - **Brown algae** Laminarin and mannitols
   - **Red algae** Floridean starch
6. **Pigments** Algal thalli are variously coloured due to the presence of pigments.
   - These are organic molecules that absorbs light. Following three types of pigments are found in algae
   - **Chlorophylls** Chl-*a* and Chl-*b* are responsible for green colour of green algae.
   - **Carotenoids** These are responsible for brown-yellow colour of brown algae.
   - **Phycobillins** These are abundant in red algae and cyanobacteria.
7. **Locomotion** It is absent in most alga. Though it can be seen by means of flagella in some lower forms. The flagella can be of two types, i.e. acronematic (whiplash) and pleuronematic (tinsel type).
8. **Reproduction** It may occur vegetatively, asexually or sexually.

   (i) **Vegetative reproduction** It takes place by various methods as follows
   - **Fragmentation**, e.g. *Ulothrix* and *Spirogyra.*
   - **Budding**, e.g. *Protosiphon.*
   - **Formation of akinetes**, i.e. vegetative cells with thick wall and huge amount of reserve food material, e.g. *Cladophora.*

   (ii) **Asexual reproduction** It is accompanied by the formation of spores. These spores are produced mitotically. These include
   - **Zoospore**, i.e. motile, flagellated spores, e.g. *Ectocarpus.*
   - **Aplanospores**, i.e. non-motile spores, e.g. *Ulothrix.*
   - **Hypnospores** (red coloured), i.e. non-motile, thick-walled spores, e.g. *Vaucheria.*
   - **Autospores,** i.e. minute, non-motile spores, e.g. *Chlorella.*
   - **Palmella spores**, i.e. green, non-motile spores, e.g. *Chlamydomonas.*

   (iii) **Sexual reproduction** It is of three types
   - **Isogamy** It involves union of morphologically undistinguishable gametes, e.g. *Spirogyra.*
   - **Anisogamy** It involves union of two morphologically different gametes. One of them is small, while other is large, e.g. *Chlamydomonas.*
   - **Oogamy** It involves two gametes. One of them is large and non-motile (female), while other is smaller and motile (male), e.g. *Fucus.*

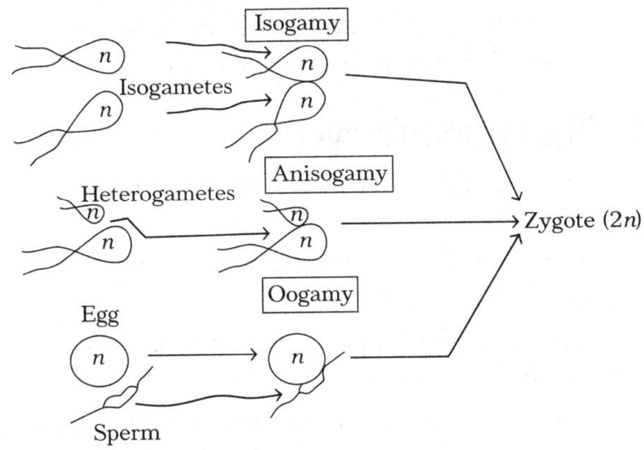

***Fig 7.1** Different types of sexual reproduction*

Life cycle in algae

- The sexual life cycle in algae alternates between a diploid or sporophytic ($2n$) phase and haploid or gametophytic (n) phase.
- The sporophyte produces gametes by meiosis, which represent the haploid or gametophytic generation in algae.
- The haploid gametes further fuse to make a zygote ($2n$) from which another diploid phase is started.
- Unlike other group of plants, the diploid and haploid phases of algae are free-living.
- Three distinct versions of life cycle occur among algae. These are as follows

1. Haplontic life cycle

- In this life cycle, sporophytic generation or diploid phase is represented only by the one-celled zygote.

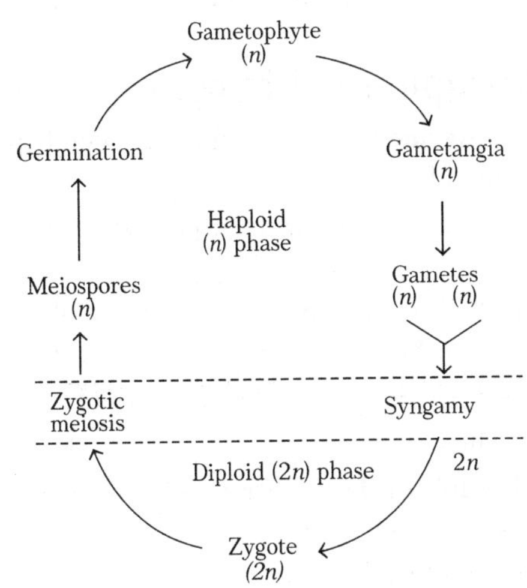

***Fig 7.2** Graphical representation of a haplontic life cycle*

- The haploid phase is formed when the zygote undergoes meiosis (zygotic meiosis).
- Thus, the dominant phase of life cycle is haploid or gametophytic, e.g. most of the green algae, i.e. *Spirogyra*, *Chlamydomonas*, *Volvox*, etc.

2. Diplontic life cycle

- In this life cycle, diploid sporophytic phase is dominant.
- It is photosynthetic and independent of the gametophytic phase.
- Some cells of diploid plant undergoes meiosis, as a result, haploid gametes are formed.
- These gametes represent the gametophytic phase, e.g. some green algae and *Fucus* (brown algae).
- Such life cycle is also seen in most gymnosperms and angiosperms.

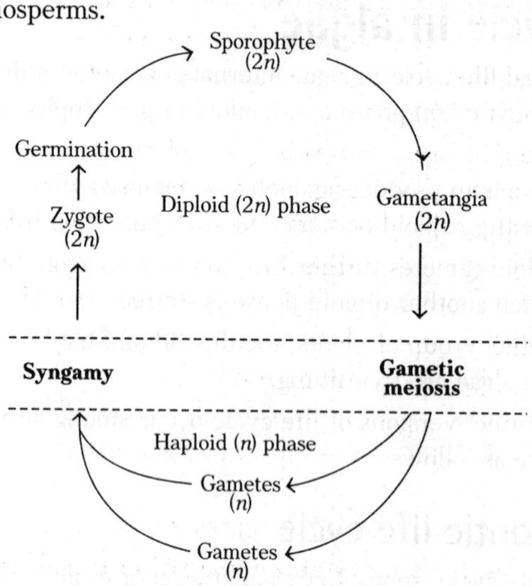

Fig 7.3 Graphical representation of a diplontic life cycle

3. Haplo-diplontic life cycle

- This is an **intermediate condition** between haplontic and diplontic life cycles.
- Both the phases are multicellular photosynthetic and free-living.
- In this, haploid phase alternates with the diploid phase, e.g. *Laminaria*, *Kelps*, *Ulva*, *Cladophora* and *Ectocarpus*.
- Haplo-diplontic life cycle can be isomorphic or heteromorphic.
- **Isomorphic life cycle** It has morphologically similar sporophytes and gametophytes.
- **Heteromorphic** It has morphologically different sporophyte and gametophyte life cycle.

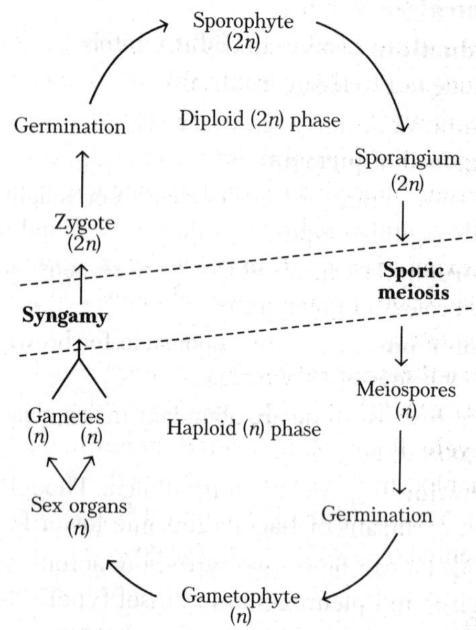

Fig 7.4 Graphical representation of a haplo-diplontic life cycle

Classification of algae

- Algae has been divided by **Whittaker** into three major subgroups based on their pigment composition and reserved food material.
- These groups are Chlorophyceae, Phaeophyceae and Rhodophyceae.

Chlorophyceae (Green algae)

The members of this group are called green algae due to the abundance of green coloured chlorophyll pigment over carotenoids.

1. **Habitat** These are mostly freshwater and only few are marine.
2. **Thallus** The plant body may be unicellular, colonial or filamentous.
3. **Cell organisation** Cell wall contain cellulose (inner layer) and pectose (outer layer).
    - The chloroplast may be discoid, plate-like, reticulate, cup-shaped, spiral or ribbon-shaped. The chloroplasts contain photosynthetic pigments.
    - Most of the members have one or more storage bodies called **pyrenoids** located in the chloroplast. Pyrenoids contain protein besides starch.
    - Some algae may store food in the form of oil droplets also.

4. **Flagellation** Flagella if present is isokont type, i.e. two flagella of equal length, which shows whiplash movement.
5. **Reproduction** The members of Chlorophyceae reproduce vegetatively, asexually and sexually by various methods.
   (i) **Vegetative reproduction** occurs through cell division (unicellular forms), fragmentation, stolons, tubers, storage cells, etc.
   (ii) **Asexual reproduction** occurs by zoospores, aplanospores, hypnospores, akinetes and daughters colonies.
   (iii) **Sexual reproduction** may be isogamous, anisogamous or oogamous.
6. **Life cycle** It can be haplontic, diplontic and diplohaplontic. In haplontic life cycle, there is a single somatic phase, which is haploid. Diploid stage is represented by a single cell or zygote, e.g. in *Spirogyra.*

Spirogyra

The characteristic features of *Spirogyra* are as follows

1. **Habitat** It is a freshwater, free-living green alga.
   It is usually found to be floating in standing water supported by bubbles of oxygen, hence called **pond scum.**
   It forms bright green silky masses and is slimy to touch. Thus, it is also known as **pond silk** or **water silk** or **mermaid's tresses.**
2. **Cell structure** Its thallus is multicellular, unbranched and filamentous. The cell contains cell wall surrounding the protoplast.

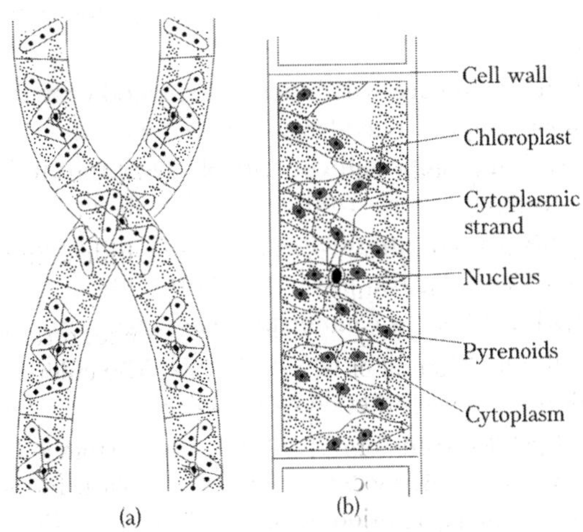

***Fig 7.5** Spirogyra : (a) Two filaments, (b) Detailed structure of a cell of Spirogyra*

The cell wall consists of two layers, i.e. inner and outer layers. The inner layer is made up of cellulose, while the outer is of pectose. Pectose is water soluble, due to this, it forms mucilage around algae.

3. **Growth** The growth of *Spirogyra* is intercalary, in contrast to apical growth of higher plants.
4. **Chloroplast** In *Spirogyra*, it is ribbon-shaped and coiled. Coiling is mostly left-handed. Many pyrenoids are also found to be present in chlorolplast. Usually 1-16 chloroplasts are found in each cell of *Spirogyra.*
5. **Reproduction** *Spirogyra* reproduces by means of vegetative, asexual and sexual methods.
   (i) **Vegetative reproduction** takes place mainly by fragmentation.
   (ii) **Asexual reproduction** takes place by asexual reproductive bodies, i.e. spores such as hypnospores, aplanospores and akinetes (e.g. *S. farlowii*).
   (iii) **Sexual reproduction** takes place through conjugation. It involves the fusion of two morphologically identical, but physiologically dissimilar gametes.
   The conjugation is of two types, i.e. lateral conjugation and scalariform conjugation.
   (a) **Lateral conjugation** It takes place between the two adjacent cells of the same filament.
   - It is primitive and uncommon type of conjugation.
   - It can be of following two types
     - **Indirect lateral conjugation** In this, fusion between the two protoplasts takes place by means of a conjugation tube. As result of conjugation, zygospore is formed. This takes place in bisexual and monoecious species, e.g. *Spirogyra affinis.*
     - **Direct lateral conjugation** In this, a perforation develops in the transverse septum. Then, the male gamete moves into the other cell having female gamete by means of this perforation. Afterwards, male and female gametes fuse to form a zygospore, e.g. *Spirogyra mirabilis, Spirogyra gratiana* and *Spirogyra jogensis.*

   (b) **Scalariform conjugation or ladder type conjugation** This takes place in both monoecious and dioecious species.

In this, the two filaments of *Spirogyra*, come close to each other and fuse directly, e.g. *S. tuwensis*.

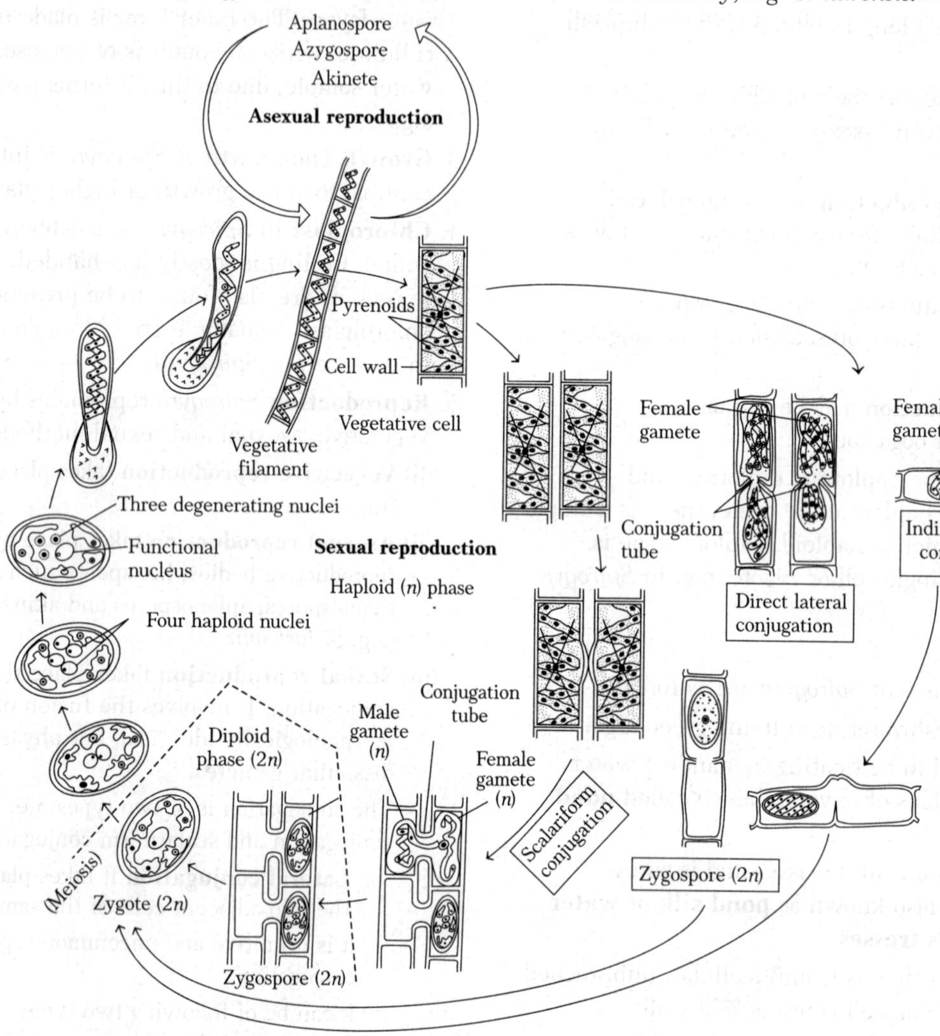

Fig 7.6 Life cycle of Spirogyra

Chlamydomonas

Salient features of *Chlamydomonas* are as follows

1. **Habitat** It is found in both freshwater and marine habitats, generally rich in ammonia salts.

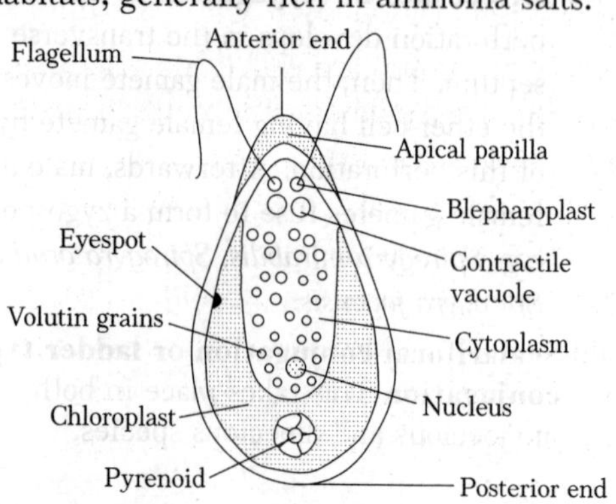

Fig. 7.7 Chlamydomonas

2. **Cell structure** It is unicellular, motile, pyriform or pear-shaped. The alga is covered by a sheath of mucilage. Cell wall consists of glycoprotein. Flagella is isokont (equal sized) and whiplash type. There is presence of neuromotor apparatus, i.e. basal granules, paradesmos, rhizoplast and centrosome.
3. **Chloroplast** In *Chlamydomonas* single cup-shaped chloroplast with single pyrenoid is present. It also has one eyespot or stigma as photoreceptor organ.
4. **Reproduction** *Chlamydomonas* reproduce either by asexual or by sexual method.
   (i) **Asexual reproduction** occurs by the formation of zoospore, aplanospore, hypnospores and palmella stage.
   - **By zoospores** Under favourable condition.

- **By aplanospores** (thin walled spores) or **hypnospores** (thick walled spores) These spores are produced under drought conditions and are without flagella.
- **By palmella stage** Under unfavourable conditions like shortage of water, excess salts, etc. The protoplast of parent cell divides to make many daughter proplasts, but they do not form zoospores. The parent cell gets covered by mucilaginous sheath with daughter protoplast known as **palmellospores.** This stage is called palmella stage.
- When favourable condition arise these palmellospores develop flagella and gets released.

(ii) **Sexual reproduction** It occurs by isogamy, anisogamy or oogamy. It also reproduce by hologamy, i.e. fusion of young cells.

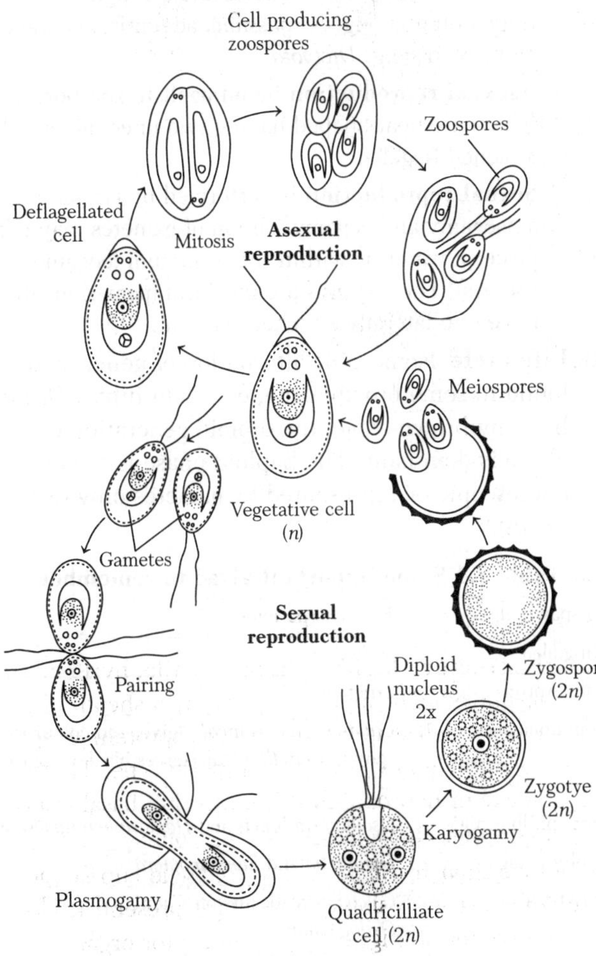

Fig. 7.8 Chlamydomonas : Diagrammatic life cycle

Ulothrix

The characteristic features of *Ulothrix* are as follows

1. **Habitat and cell structure** It is an unbranched, multicellular filamentous green alga. It is found in cold and slow-flowing freshwater. Its each filament consists of row of cells placed end to end. Its lower most cell is called **holdfast**, which helps to attach the plant body to the substratum. It possesses girdle-shaped chloroplast unlike *Spirogyra.*

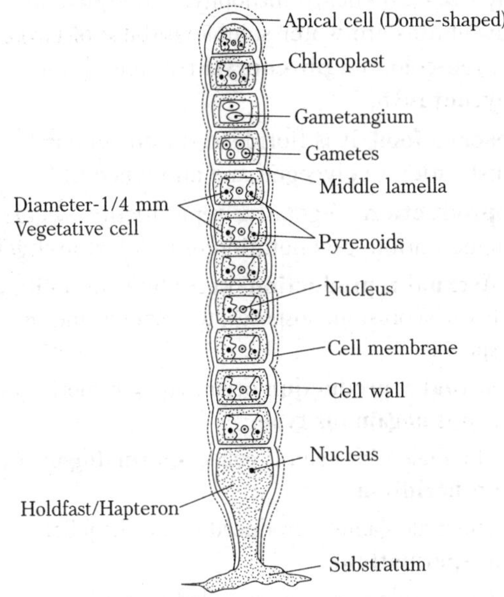

Fig. 7.9 Ulothrix

2. **Reproduction** Asexual and sexual reproductions take place differently in this algae than the rest of Chlorophyceae.
   (i) **Asexual reproduction**, there is the formation of zoospores, aplanospores, hypnospores, akinetes and palmella stage. Further zoospores are of two types, i.e.
   - **Macrozoospores** quadriflagellate, uninucleate and pear-shaped.
   - **Microzoospores** uninucleate, ovoid, bear 2–4 flagella and are produced in large number.

   (ii) **Sexual reproduction**, *Ulothrix* shows isogamy and forms zygospore (2*n*). This zygospore forms four zoospores (motile) or aplanospores (non-motile) by meiosis. Each spore gives rise to a new plant.

Rhodophyceae (Red algae)

The members of this group are called red algae, due to the predominance of the red pigment, *r*-phycoerythrin in them.

1. **Habitat** These are mostly marine with abundance in warmer areas. They also exhibit a high degree of parasitism (e.g. *Harveyelia*) and epiphytism.
2. **Thallus** The plant body is mostly multicellular. These multicellular forms can be filamentous, parenchymatous, feathery or ribbon-like.
3. **Cell organisation** The cell wall is composed of cellulose, pectin and hydrocolloids like agar, carrageenan and funori.

The flagellated motile stages are completely absent in their life cycle.

4. **Photosynthetic pigments** These include chlorophyll-*a*, carotenes, xanthophylls and phycobilins. Phycobilins are water soluble and are of two types, i.e. red-coloured **phycoerythrin** and blue-coloured **phycocyanin.**
5. **Reserve food** It is floridean starch similar in constitution to glycogen and amylopectin.
6. **Reproduction** Vegetative reproduction occurs by fragmentation, regeneration of holdfast and gemmae.
   - **Asexual reproduction** occurs by non-motile spores (carpospores, monospores, tetraspores and neutral spores).
   - **Sexual reproduction** occurs by non-motile gametes and is **oogamous type**.
   - The male sex organ is called **spermatogonium** or **antheridium.**
   - The male gamete produced is non-flagellated, called as **spermatium.**
   - The female sex organ is called **carpogonium.**
   - After fertilisation, a new structure called **carposporophyte** is produced. It remains attached to the parent alga.
7. **Life cycle** There is alternation of haploid and diploid multicellular generations in many red algae.

Phaeophyceae (Brown algae)

The members of this group are called brown algae, due to the presence of a brown pigment, fucoxanthin in addition to chlorophyll-*a* and *c* in the chromatophores.

1. **Habitat** About 99.7% members of brown algae are marine and a few grow in freshwater, e.g. *Lithoderma, Pseudobodanella*, etc.
   They are mainly found in cold water except *Sargassum,* which is found in warm water and form sargasso sea.
2. **Thallus** Plant body is immobile, multicellular and highly differentiated.
   - Unicellular forms are completely absent.
   - They range from simple microscopic heterotrichous filament (*Ectocarpus*) to the largest alga (*Macrocystis pyrifera*) about 100 meters.
   - The largest forms are known as **kelps** or **rockweeds.**
   - The body of brown algae is differentiated into **holdfast** (for attachment to substratum), **stipe** (stalk) and **frond** (leaf-like photosynthetic organ).
   - The large forms (sea weeds) often possess air bladders for providing buoyancy.
3. **Cell organisation** Cell wall is composed of cellulose, pectose and phycocolloids. The cellulosic wall of vegetative cells is usually covered on the outside by a gelatinous coating of algin.
4. **Photosynthetic pigments** These include chlorophyll-*a, c* and carotenoids. They range in colour from olive green to various shades of brown depending upon the amount of the xanthophyll and fucoxanthin present in them.
5. **Food reserve** It remains in the form of complex carbohydrates such as laminarin or mannitol.
6. **Flagellation** These contain heterokont flagellation with one smooth (whiplash) and one tinsel flagella.
7. **Reproduction**
   - **Vegetative reproduction** occurs through fragmentation (e.g. *Sargassum*), adventitious branches and stolons (e.g. *Dictyota*).
   - **Asexual reproduction** by biflagellate zoospores, which are pear-shaped having two unequal laterally attached flagella.
   - **Sexual reproduction** is performed by isogamy, anisogamy and oogamy. Union of gametes may take place in water or within the oogonium (oogamous species). The gametes are pyriform (pear-shaped) and have two laterally attached flagella.
8. **Life cycle** Isomorphic alternation of generation is found in some brown algae, e.g. *Ectocarpus, Dictyota.* In many brown algae, the diploid generation or phase is dominant. The haploid phase is either microscopic or represented by gametes only (e.g. *Fucus*).

Table 7.1 Some important algae to remember

Frog spawn alga	*Batrachospermum*
Rolling alga	*Volvox*
Devil's Apron	*Laminaria*
Cryophilous alga	*Chlamydomonas nivalis* (gives red colour to snow), *Scotiella nivalis* (gives black colour to soil).
Thermophilous alga	*Oscillatoria brevis* and *Haplosiphon lignosum*
Halophytic alga	*Dunaliella* and *Stephanoptera*
Epiphytic alga	*Coleochaete* (occurs on *Chara* and *Oedogonium*)
Symbiotic alga	*Nostoc* and *Anabaena*
Parasitic alga	*Cephaleuros* (causes red rust of tea), *Polysiphonia* is (semiparasitic on *Ascophyllum*).
Epizoic alga	*Cladophora* (found on shells of molluscs).
Endozoic alga	*Zoochlorella* (found inside the body of *Hydra*).
Smallest alga	*Micromonas*
Largest alga	*Macrocystis* (brown alga) (60 m)

Various characteristics of all three groups of algae are given in the table

Table 7.2 General comparison of three classes of algae

Characters	Chlorophyceae	Phaeophyceae	Rhodophyceae
Common name	Green algae	Brown algae	Red algae
Major pigments	Chlorophyll-*a, b,* carotenoids	Chlorophyll-*a, c,* fucoxanthin	Chlorophyll-*a, d,* *r*-phycoerythrin, *r*-phycocyanin
Habitat	Fresh, saline water	Fresh (rare), generally saline	Mostly saline
Body structure	Unicellular, colonial or filamentous	Branched, filamentous or profusely branched	Multicellular
Cell wall composition	Cellulose	Cellulose and algin	Cellulose, pectin, phycocolloids, polysulphate esters and carrageenin
Number of flagella	2-8 and equal	2 and unequal	Absent
Position of flagella	Apical	Lateral	Absent
Stored food	Starch	Starch, mannitol and laminarin	Floridean starch and fluoridoside (a galactoside of glycerol)
Thylakoid	Stacked in groups of 2-20	Occur in group of three	Unstacked
Asexual reproduction by vegetative part	Fragmentation	Fragmentation	Fragmentation
Asexual reproduction by spores	Flagellated zoospore	Biflagellated zoospore, motile or non-motile	Non-motile spores
Sexual reproduction	Isogamous, anisogamous and oogamous	Isogamous, anisogamous and oogamous	Oogamous
Examples	*Chlamydomonas, Volvox, Ulothrix, Spirogyra, Chara,* etc.	*Ectocarpus, Dictyota, Laminaria, Sargassum, Fucus,* etc.	*Polysiphonia, Porphyra, Gracilaria, Gelidium,* etc.

Economic importance of algae

- At least a half of the total CO_2 fixation on earth is carried out by algae through photosynthesis.
- **Food sources** Algae is an important source of food for fishes, aquatic amphibians, mammals and other animals.
  - *Porphyra* is rich in protein (25-30%), vitamins and mineral salts.
  - *Gracilaria* and *Gelidium* are used as food in coastal regions of India.
  - In Tamil Nadu, *Gracilaria edulis* is used for making porridge or kanji.
  - Carrageenan obtained from red algae and algin from brown algae are used as food source.
- *Chlorella* and *Spirulina* are used by astronauts as food. These algae multiply rapidly and synthesise a rich harvest of food. They utilise CO_2 (carbon dioxide) and liberate oxygen as a byproduct. This O_2 is used by the astronauts.
- **Fooder** The important genera used as fodder are *Sargassum, Laminaria, Fucus, Ascophyllum, Rhodymenia,* etc.
- **Medicinal uses** Iodine (obtained from brown algae such as *Laminaria, Macrocystis, Fucus,* etc.) is used to prepare various anti-goitre medicines.
  - Chlorellin (an antibiotic) is obtained from *Chlorella.* Carrageenan also acts as a blood coagulant.
  - An effective vermifuge is obtained from the extracts of *Digenea, Codium, Alsidium* and *Durvillea.* Many algae are used in the treatment of lung, kidney and bladder ailments.
  - Algal colloid called **phycocolloid** is obtained from the cell walls of sea weeds. It has haemostatic properties and has been used in emergency transfusion in the treatment of shock.
- **Nitrogen fixation** Some blue-green algae, such as *Anabaena, Nostoc* can fix atmospheric nitrogen into nitrates and nitrites. Thus, they increase the fertility of soil.
- Agar-agar is obtained from certain red algae, such as *Gelidium, Gracilaria, Gigartina.* It is used in laxatives, certain medicines, cosmetics, culture media, leather industry, textile industry, baked foods, meat industry and as an emulsifier in dairy products.
- **Industrial uses** Alginic acid or alginate is found in sea weeds, such as *Laminaria, Macrocystis.* Alginates are used as thickners in food industry, in cosmetic, in textile industry as printing paste, in plastic and artificial fibre production, in rubber industry and in latex production.

Harmful effects of algae

- The luxuriant growth of algae in water reservoirs makes the water unfit for drinking. It imparts bad taste, odour and colouration to the water.
- Water blooms are formed by the growth of *Anabaena, Nostoc, Microcystis,* etc., in pond water. This may cause fish mortality by causing depletion of oxygen.
- The green alga, *Cephaleuros virescens* is parasitic on tea leaves. It causes a disease called **red rust of tea.**
  It also causes leaf spot disease in papaya and guava. Similarly, *Cephaleuros coffea* is parasitic on coffee plants.

CHECK POINT 7.1

1. Select the correct match of algal class with its characteristic reserve food.
   (a) Chlorophyceae – Starch
   (b) Phaeophyceae – Mannitol, laminarian
   (c) Rhodophyceae – Floridean starch
   (d) All of the above
2. A group of autotrophic plants in which sex organs are non-jacketed and zygote secrete thick wall are called
   (a) phycophytes
   (b) lichens
   (c) bryophytes
   (d) thallophytes
3. Which sexual reproduction method involves union of undistingushible gametes?
   (a) Oogamy (b) Isogamy
   (c) Anisogamy (d) None of these
4. Which type of conjugation takes place between the two adjacent cells of the same filament?
   (a) Scalariform conjugation (b) Lateral conjugation
   (c) Direct conjugation (d) Ladder conjugation
5. Which pigment is responsible for colour of algae?
   (a) Carotenoids (b) Fucoxanthin
   (c) Chlorophyll-*a*, *c* (d) *r*-phycoerythrin

BRYOPHYTA

(Gr. *Bryo*–moss; *phyta*–plant)

- The term 'Bryophyta' was given by **Robert Braun.**
- These are non-vascular plants that do not possess flowers or seeds.
- The study of bryophytes is called as **Bryology. Hedwig** is known as **Father of Bryology.**
- Bryophytes have evolved as the simplest and foremost forms among land inhabiting (terrestrial) plants.
- They occupy an intermediate position between algae and Pteridophyta.

General characteristics

Some characteristic features of bryophytes are as follows

1. **Habitat** Most of the bryophytes are terrestrial and are found in moist and shady places. Some forms are aquatic (e.g. *Riccia fluitans, Fontinalis*), saprophytic (e.g. *Cryptothallus mirabilis*) or epiphytic (e.g. *Frullanaia*).
2. **Alternation of generation** Bryophytes show heteromorphic alternation of generations, i.e. two morphologically different forms (or phases) of plant alternates with each other.
   - The two morphologically distinct phases of bryophytes are gametophytic (n) phase and sporophytic ($2n$) phase.
   - The free-living **gametophyte** is dominant, green and independent.
   - The gametophytic plant body can be either thalloid (liverworts) or leafy (mosses). The plant remains fixed to substratum *via* unicellular or multicellular root like **rhizoids.**
   - **Sporophyte** is dependent upon the gametophyte for nutrition.
3. **Vegetative reproduction** It takes place by fragmentation, formation of adventitious branches, tubers, bulbils and gemmae.

   Gemmae are multicellular, asexual, reproductive bodies. They are formed inside gemma cups.
4. **Sex organs** These are stalked and multicellular. These are surrounded by a single-layered jacket of sterile cells.
   - Male sex organ is globular and called as **antheridium.**
   - Each antheridium produces several biflagellated (motile) male gametes, i.e. sperms or **antherozoids.**
   - Female sex organ is **archegonium.** It is flask-shaped and made up of three parts, i.e. slot, venter and neck.
   - The neck is tubular and venter is swollen.
   - The single-layered wall of neck has 5-6 rows of cells.
   - Neck encloses a few sterile neck canal cells.
   - The venter wall is 1-2 layered and it encloses a sterile venter canal cell and a fertile egg cell.
5. **Sexual reproduction** It is oogamous type.
   - Fertilisation is internal and produces diploid zygote.
   - The zygote is retained inside the venter surrounded by calyptra (a protective covering).
   - Zygote (oospore) never develops into perennating/resting spore called **zygospore** as there is no resting period.

- The zygote divides repeatedly and an embryo is formed.
- The embryo develops into a sporophyte that consists of foot, seta and a capsule.

6. **Spore formation and dispersal** In the capsule, spores are formed by meiosis.
   - Elaters are tubuler cells which help in spores dispersal in some bryophytes such as liverworts.
   - The spores after liberation, germinates into a young gametophytic plant.
   - In a moss, germination of spores indirectly forms the leafy gametophyte, i.e. a protonema is formed, which develops into adult gametophyte whereas in liverworts, spores directly germinate into young thalli.
7. Bryophytes are often called as **amphibians of plant kingdom** because they require water for fertilisation. They show adaptations to both land and water conditions.
8. **Adaptations to land conditions** It include development of compact plant structures, rhizoids for absorption, jacketed reproductive structures, retention of zygote in archegonium and embryo formation.
9. **Adaptations to water conditions** It include the absence of true roots, vascular system and chemotactic movement of sperms.

Note
- *Zoopsis is the **smallest bryophyte**.*
- *Dawsonia is the **largest bryophyte**.*
- *Western Himalayas are called as **Golden mine of liverworts.***
- ***Polytrichum commune** was employed in removing kidney stones.*
- *Funaria has the largest archegonium in plant kindgom.*

Life cycle

- The life cycle of bryophytes is heteromorphic or heterologous haplo-diplontic type.
- In bryophytes, the sporophyte and gametophyte are morphologically distinct from each other.
- The haploid spores formed by meiosis begin the gametophytic generation. These spores germinate to form a gametophyte, which is more diverse in bryophytes than in any other group of plants.
- Gametophyte produces gametes that fuse and form a diploid ($2n$) sporophytic phase. The latter forms haploid spores by meiosis.

An outline of bryophytic life cycle is given below

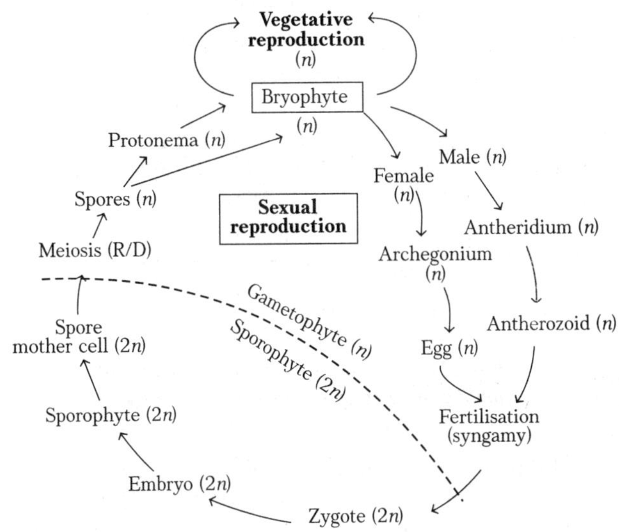

Fig 7.10 Sexual life cycle of bryophytes

Classification of bryophytes

According to the most accepted classification system, bryophytes are classified into three main classes as follows

Bryophyta

Classes

Hepaticopsida	**Anthocerotopsida**	**Bryopsida**
(Hepaticae) (Liverworts), *Riccia, Marchantia* and *Pellia*	(Anthocerotae) (Hornworts) *Anthoceros*	(Musci) (Mosses) *Funaria, Sphagnum, Polytrichum* and *Tokakia*

Some important bryophytes

Funaria (Moss)

Some characteristic features of *Funaria* are as follows

1. **Habitat** *Funaria* grows in dense patches in moist conditions, such as damp soil, on the wall, on moist rocks and on lands burnt by fire.
2. **Plant body** The gametophyte plant body is leafy, green and upright. The leaves are small, sessile, ovate with acute apex. They have a distinct midrib.

The moss gametophyte is the most highly developed of all the bryophytes.

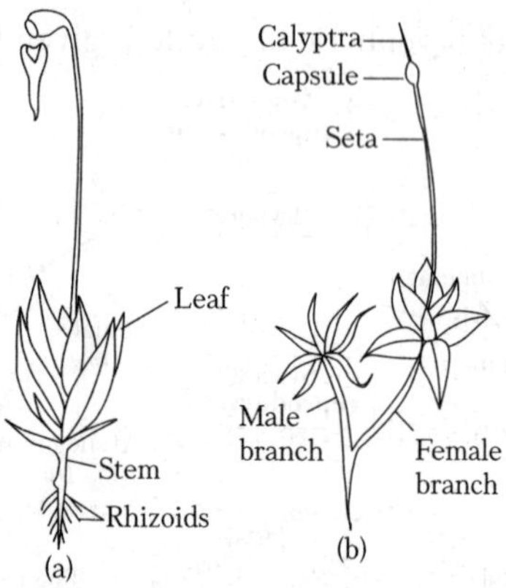

Fig 7.11 Funaria : (a) Leafy gametophyte of Funaria showing sporophyte, (b) Gametophyte depicting male and female branch

3. The plants are attached to substratum by rhizoids. These are multicellular and branched. Scales are absent in *Funaria.*
4. Internally, the cells are arranged in three distinct regions, i.e. epidermis, cortex and central cylinder.
5. **Vegetative reproduction** In *Funaria*, vegetative reproduction takes place by fragmentation, formation of secondary protonema, tubers (on rhizoids) and gemmae.
6. **Sex organs** These are borne in clusters at the tip of male and female shoots.
7. **Antheridium** The antheridium is club-shaped and orange-coloured. Within jacket layer, a mass of sperm producing cells is present. The liberated sperms are minute, biflagellate and spirally-coiled structures.

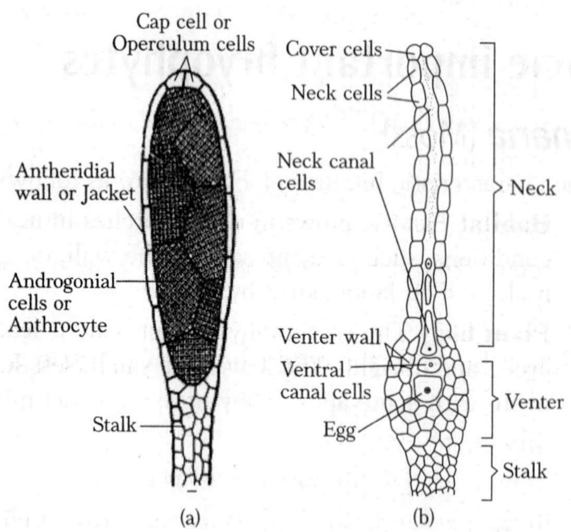

Fig 7.12 Funaria : (a) Mature antheridium ready for dehiscence, (b) A nearly mature archegonium is section showing the reproductive structure

8. **Archegonium** The archegonium is flask-shaped.
   - It consists of only two parts, i.e. a neck and a venter.
   - The neck is long and twisted.
   - It possesses numerous (8-10) neck canal cells.
   - The venter wall is two-layered thick.
   - The venter cavity contains an egg and a ventral canal cell above it.
   - The protective sheaths (perigynium and perichaetium) are absent.
9. **Sporophyte** It consists of a small conical foot, a long and slender seta and a pear-shaped capsule.
   - It is dependent upon parent gametophyte for the supply of water and mineral salts.
   - Thus, it is semiparasitic (parasitic). Here, peristome teeth are responsible for spore dispersal. After the spore, dispersal sporophyte dies.
10. **Capsule** The capsule of *Funaria* consists of three portions, i.e. basal apophysis, central theca and terminal operculum.
    - **Apophysis** is the green photosynthetic region.
    - **Theca** is the middle fertile zone with outer epidermis, two layered hypodermis and middle sterile columella, spore sac is present outer to columella and encloses spores.
    - **Operculum** is the upper region of capsule that covers peristome.
      Peristome has 32 peristomial teeth in 2 rows of 16 each. The teeth of outer rows are called **exostomes** while that of inner rows are called **endostome.** These teeth are hygroscopic and help in dehiscence of spores after the opening of operculum.

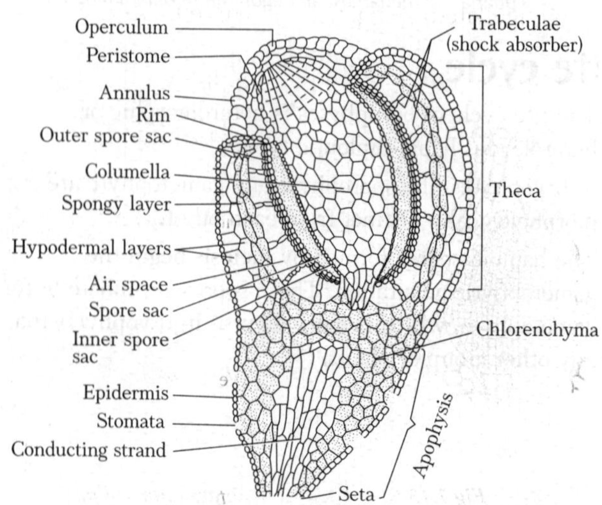

Fig 7.13 LS capsule structural details of Funaria sp.

11. **Protonema** It is the juvenile filamentous stage of moss gametophyte.
    - It results from the germination of spores.
    - Protonema is a vegetative and transitory stage.
    - Protonema has non-green rhizoidal branches as well as green prostate branches.
    - Prostate branches develop buds which grow to form moss plants.

Riccia

Some characteristic features of *Riccia* are as follows

1. **Habitat** It is commonly found in moist terrestrial habitats. *R. fluitans* is an exclusively aquatic species.
2. **Plant body** The main plant body is gametophytic (n).
3. **Thallus** It is dorsiventral and dichotomously branched.
    - Both types of rhizoids, i.e. smooth and tuberculate are found in *Riccia*. These are present in the midrib on the ventral surface.
    - Thallus is divided into photosynthetic and storage regions. Plant may be monoecious (e.g. *R. billardieri*, *R. gangetica*) or dioecious (e.g. *R. discolor*).

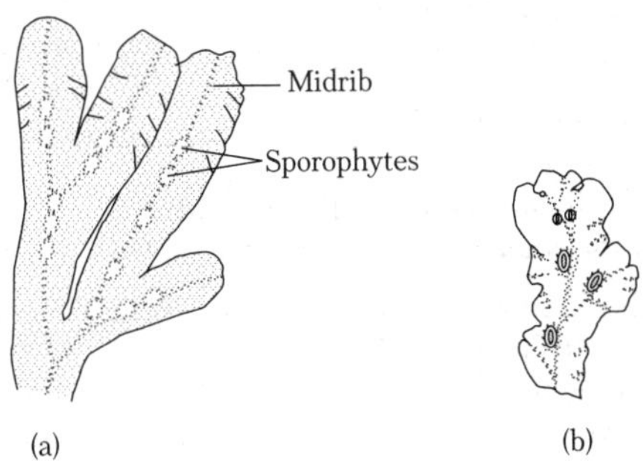

Fig 7.14 Riccia : TS of thallus (a) Female, (b) Male

4. **Antheridium** It is oval-shaped. It produces biflagellated, elongated and curved antherozoids.

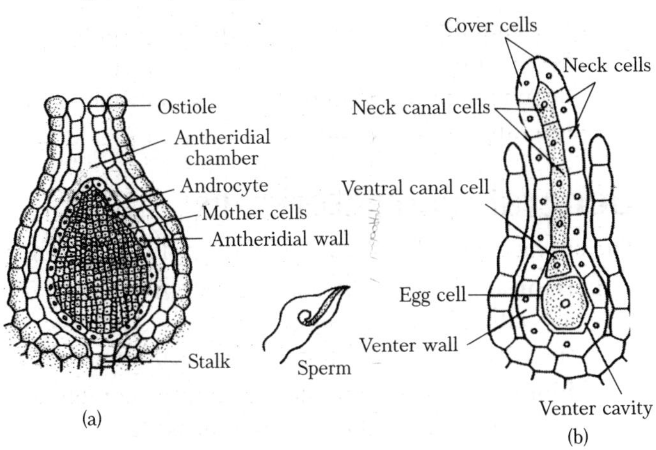

Fig 7.15 Riccia sp. (a) A mature antheridium, (b) A mature archegonium

5. **Archegonium** It is flask-shaped having 4-6 neck cells and a ventral canal cell.
6. **Fertilisation** Antherozoids enter into the archegonia, fuse with the egg cell and form zygote ($2n$). Embryo forms sporophyte, which consists of a globular capsule.
7. Foot and seta are absent in *Riccia*. Sporogenous cells divide mitotically and form tetrahedral spores.
   These spores get dispersed and germinate to form next generation of gametophyte thalli (n).

Table.7.3 Some bryophytes with their common name

Bryopsida	Common names
Polytrichum	Hair cap moss
Sphagnum	Peat/Bog moss
Adrea	Granite moss
Grimmia	Black moss
Tortula	Twisted moss
Ptilium	Feather moss
Rhodobryum	Rose moss
Merceya	Copper moss
Climacium	Tree moss
Dicranum	Wind blown moss

Economic importance of bryophytes

- *Sphagnum* is an economically very important alga. It is the chief constituent of peat, which is used as fuel.
  It is also used in the preparation of ethyl alcohol, peat tar, ammonia, paraffin, etc. Due to its capacity to retain water, it is used to cover the plant roots during transportation.
- Some mosses grow in pools and settle to the bottom when they die. They help in soil formation.
- Some bryophytes also work as soil binders when they grow in aggregations.
- Some bryophytes are used as fodder for the herbivorous animals.

Resemblances between bryophytes and algae

- Thalloid plant body.
- Absence of roots.
- Absence of complex vascular tissues.
- Need of water during fertilisation due to swimming habit of gametes.

Table 7.4 Differences between algae and bryophytes

Algae	Bryophytes
Usually found in aquatic habitat.	They grow in moist and shady places.
Plant body can be unicellular or multicellular, filamentous or thalloid.	Plant body can be thalloid or leafy.
Well-developed tissue differentiation is only seen in higher forms.	Tissue differentiation is well-developed.
Sexual reproduction is isogamous, anisogamous or oogamous type.	Sexual reproduction is only oogamous type.
Sex organs are not covered by sterile jacket.	Sex organs are covered by a sterile jacket.
Female sex organ is oogonium.	Female sex organ is archegonium.
Zygote is usually liberated after sexual reproduction.	Zygote is not liberated, instead it is retained in the sporophyte.
Embryo formation is absent.	Embryo formation is seen.
Sporophyte is independent of gametophyte.	Sporophyte is dependent upon gametophyte.
Sporophyte is not differentiated.	Sporophyte is differentiated into foot, seta and capsule.
Isomorphic alternation of generation.	Heteromorphic alternation of generation.

CHECK POINT 7.2

1. Which one of the following statements is incorrect about bryophytes?
   (a) Fertilisation takes place in the presence of water
   (b) Gametophytic phase is dominant in life cycle
   (c) Sporophyte is physiologically dependent on gametophyte
   (d) Zygote undergoes meiosis to produce sporophyte
2. Calyptra develops from
   (a) venter wall of archegonium
   (b) outgrowth of gametophyte
   (c) neck wall of archegonium
   (d) paraphysis of the archegonial branch
3. Which of the following in moss capsule is haploid/gametophytic tissue?
   (a) Annulus and peristome
   (b) Calyptra and spore
   (c) Columella and theca
   (d) Operculum, foot and seta
4. Moss peat is used as a packaging material for sending flowers and live plants to distant places because
   (a) it is easily available
   (b) it is hygroscopic
   (c) it reduces transpiration
   (d) it serves as a disinfectant
5. Which type of the rhizoids are present in *Riccia*?
   (a) Unicellular smooth
   (b) Multicellular smooth
   (c) Unicellular smooth and tuberculated
   (d) Multicellular smooth and tuberculated
6. Which of the given species has the water absorbing capacity and also used as a fuel?
   (a) *Marchantia*
   (b) *Riccia*
   (c) *Sphagnum*
   (d) *Funaria*

PTERIDOPHYTA

(Gr. *Pteron*–feather; *phyton*–plant)

- The term 'Pteridophyte' was given by **E Haeckel.**
- These are seedless, vascular plants commonly called **horse tail, ferns** and **lycopods.**
- Pteridophytes are known as **reptiles of plant kingdom.**
- These plants dominated the landscape during the Carboniferous period (from 286 million to 360 million years ago).
- Though, pteridophytes have vascularised roots, stem and leaves, but they could not achieve desired success in land conditions due to following reasons
  (i) Requirement of moist, wet place for spore germination.
  (ii) Susceptibility of prothallus (gametophytes) to desiccation.
  (iii) Requirement of a film of water by male gamete for reaching to female gamete.

General characteristics

The characteristic features of Pteridophyta are as follows

1. **Habitat** These plants flourish well in a variety of habitats. These are mostly found in wet, moist, shady places. Some grow in xeric conditions also.
2. **Plant body** The main plant body is a sporophyte, which represents the dominant phase. It is differentiated into true roots, stem and leaves or fronds.

3. **Sporophyte** The sporophyte of pteridophytes can be of two types
   (i) **Megaphyllous type** That bears larger leaves, e.g. Ferns (*Pteris*).
   (ii) **Microphyllous type** that bears smaller leaves, e.g. lycopods (*Selaginella*).
4. **Vascular tissue** consists of xylem (without true vessels) and phloem (without companion cells).
5. **Secondary growth** is absent in pteridophytes.
6. **Vegetative propagation** It takes place by means of bulbils, adventitious buds and fragmentation.
   - The most unique method of vegetative reproduction is seen in walking fern, *Adiantum caudatum*.
   - It produces a new plant when its leaf tips come in contact with the soil.
7. **Asexual reproduction** It is occurs by the means of spores that are produced in sporangia.
   - These sporangia are found to be present on leaves.
   - The leaves bearing sporangia are called **sporophylls**.
8. **Sporangium** It has a single-layered jacket. It has 12 spore mother cells ($2n$).
   The marginal cells of jacket of sporangium are differentiated into annulus and stomium.
   Sporangium development is of two types
   - **Leptosporangiate** Sporangium develops from single superficial cell, e.g. *Pteris, Adiantum*.
   - **Eusporangiate** Sporangium develops from group of cells, e.g. *Selaginella, Equisetum*.
9. **Gametophyte** Haploid spore germinates to form a prothallus (gametophyte), which is monoecious, i.e. has both antheridia and archegonia.
   Gametophyte is independent and autotrophic.
10. **Sexual reproduction** It occurs through the development of male and female reproductive structures.
11. **Antheridium** Male sex organ, i.e. antheridium is a hemispherical or globose structure with 3-celled jacket. It contains 32-48 sperm mother cells, which produce multiciliated sperm.
    These sperms swim in water to reach the female sex organs.
12. **Archegonium** It is flask-shaped and embedded in prothallus. The neck of archegonia opens and mucilage comes out to which the sperms are attracted. The mucilage contains malic acid.
13. **Fertilisation and development** The male gamete fuses with an egg and diploid zygote is formed.
    - Fertilisation is facilitated by water.
    - Zygote undergoes divisions repeatedly and forms embryo, which further gives rise to the young sporophyte.
    - In pteridophytes, embryogeny is of two types
      (i) **Exoscopic** Axis of embryo is directed towards archegonial neck, e.g. *Equisetum*.
      (ii) **Endoscopic** Axis of embryo is directed inward towards venter, e.g. *Selaginella*.
    - **Development of embryo** occurs in following ways in pteridophytes
      - **Holoblastic** Entire zygote is involved in embryonic development, e.g. ferns.
      - **Meroblastic** Part of zygote is involved in embryonic development, e.g. *Selaginella*.
14. **Life cycle** Ferns exhibit alternation of dominant sporophyte generation with an inconspicuous gametophyte generation (heteromorphic). Ferns exhibit phenomenon of **apospory** and **apogamy**.
    - **Apospory**, sporophyte directly develops into gametophyte ($2n$) without the spore production.
    - **Apogamy**, gametophyte directly develops into sporophyte (n) without the fusion of gametes.

Note
- ***Club moss***–*Lycopodium*
- ***Smallest pteridophyte***–*Azolla*
- ***Largest pteridophyte***–*Cyathea*

Stelar system

Stele is the central core of axis, which includes the vascular system, i.e. xylem and phloem. It is of primitive type. The stelar theory was proposed by **Van Tiegham** and **Douliot**.

Steles may be of the following types in pteridophytes and other plants

(i) **Protostele** It is the simplest and most primitive type of stele. In this, vascular cylinder consists of solid core of xylem surrounded by phloem, pericycle and endodermis. There is no pith tissue found in this stelar system.

Protostele may be of the following types

(a) **Haplostele** Here, the solid central core of xylem is surrounded by a uniform layer of phloem and pericycle, e.g. *Rhynia, Selaginella kraussiana*, etc.

(b) **Actinostele** Here, the xylem core is star-shaped with radiating arms and phloem is present as separate patches, e.g. *Lycopodium serratium, Psilotum, Asteroxylon*, etc.

(c) **Plectostele** Xylem is broken into parallel plates that are immersed in the phloem, e.g. *Lycopodium volubile* and *Lycopodium clavatum*, etc.

(d) **Mixed protostele** Here, xylem is embedded as discrete units in the mass of phloem, e.g. *Lycopodium cernuum*, etc.

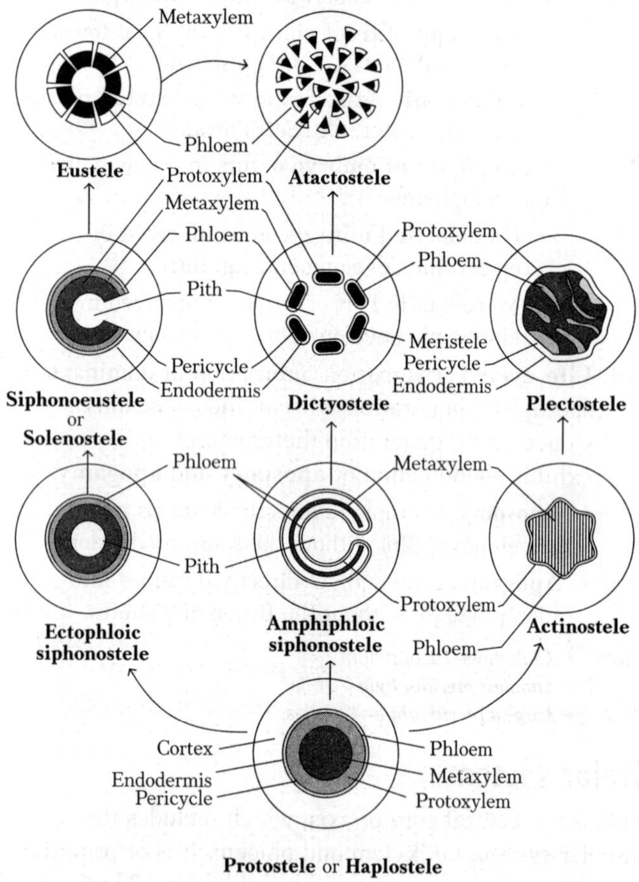

Fig 7.16 Various types of steles

(ii) **Siphonostele** Protostele with pith in the centre is known as siphonostele.The leaf gaps are absent. According to the positions of the phloem and xylem, siphonostele is further classified into the following two types

(a) **Ectophloic siphonostele** In this case, phloem is present only external to the xylem, e.g. *Osmunda, Equisetum, Botrychium* etc.

(b) **Amphiphloic siphonostele** In this type, phloem surrounds the xylem internally as well as externally, e.g. *Adiantum, Marselia,* etc.

(iii) **Solenostele** It is similar to siphonostele, but has scattered leaf gaps. Thus, the siphonostele with leaf gaps is called solenostele. It may be ectophloic or amphiphloic depending upon the type of siphonostele from which it is derived, e.g. *Adiantum pedatum.*

(iv) **Dictyostele** Here, a number of leaf gaps breaks up vascular bundles into smaller units called **meristeles**, e.g. *Dryopteris, Pteris* and *Pellaea rotundifolia,* etc.

(v) **Eustele** It consists of a number of parallel vascular bundles separated by medullary rays or perforations, e.g. gymnosperms and dicots stem.

(vi) **Atactostele** Vascular strands are scattered in vascular tissue, e.g. monocot stem.

Heterospory

- It is the production of two different kinds of spores. It is a characteristic feature of pteridophytes.
- These spores may have different sizes and different development patterns.
- Among these spores, smaller ones are called as **microspores** (male spores) and larger ones as **megaspores** (female spores). These spores are produced within microsporangia and megasporangia, respectively.
- Heterospory is the most important evolutionary development in the vascular plants. It ultimately leads to the seed development.
- *Selaginella, Isoetes, Marselia, Salvinia* and *Azolla* are the heterosporous genera of pteridophytes.

Seed habit

- The ability of a plant to form seeds is called seed habit.
- This is considered to be originated in pteridosperms (seeded ferns) during the Devonian-Carboniferous periods.
- *Selaginella* is a pteridophyte that shows some advances towards the seed habit without forming a seed.
- The conditions that are pre-requisite for the formation of the seeds are as follows
  (i) Production of two types of spores (heterospory).
  (ii) Early development of spores with sporangium.
  (iii) Production of two kinds of gametophytes.
  (iv) Retention of megaspore within megasporangium.
  (v) Development of receptive areas like pollen chamber, stigma, etc., for catching microspores.
  (vi) Development and retention of embryo inside the megasporangium leading to the formation of seed.

Classification of pteridophytes

The most widely accepted and convenient system of classification of vascular cryptogams is based on **Smith** in 1955; **Bold** in 1957-1967; **Benson** in 1957 and **Zimmermann** in 1957.

It is as follows

1. **Subdivision-Psilophyta**
   Most primitive, protostele, homosporous, living genera—*Psilotum* and *Tmesipteris.*

2. **Subdivision–Lycophyta** (Club moss)
   These possess a differentiated plant body, protostele sometimes siphonostele, homosporous (*Selaginella*), or heterosporous.
3. **Subdivision– Sphenophyta** (Horse tail)
   All are fossils except *Equisetum.* They have differentiated plant body and are homosporous.
4. **Subdivision– Filicophyta** (Pterophyta)
   Most widely distributed vascular cryptogams. They have both homosporous (*Pteris*) and heterosporous (*Marselia*) species.

Some important pteridophytes

Dryopteris (Fern)

Some characteristic features of *Dryopteris* are as follows

1. **Habitat** It usually grows in moist, wet, shady and cool places. It is found abundantly in warm tropical regions.
2. **Plant body** Sporophyte ($2n$) represents the main plant body.
   - It is differentiated into root, stem and leaves.
   - Stem is an underground hairy, rhizome, which expands and bears large bipinnate leaves.
   - Fern leaves (fronds) appear as tightly coiled structure during early stages, i.e. circinate vernation. Adventitious roots arise from the rhizome.
3. **Vegetative reproduction** It takes place by the formation of adventitious buds and fragmentation of rhizome.
4. **Asexual reproduction** It occurs by means of spores. Spores are produced in sporangia.
   - Some fertile leaves, i.e. sporophylls bear group of sori. Each sori contains groups of sporangia.
   - Sori remains covered by a kidney-shaped layer called **indusium.** Each sporangium is stalked and produces small, thick-walled spores.
   - Spore wall is made up of two layers, i.e. outer thick exine and inner thin intine.
   - Spores are liberated after the dehiscence of sporangia.
   - The spores germinate to form monoecious prothallus, which bears both reproductive structure, i.e. both antheridium and archegonium.
5. **Antheridium** It is a globular and sessile, i.e stalkless structure. It is jacketed. Jacket is made up of three cells (two ring cells and single cover cell). It contain about 32 sperm mother cells. Antherozoids are spirally coiled, large and multiflagellated.
6. **Archegonium** It is flask-shaped. It is differentiated into a neck and a venter. Neck protrudes out and consists of single binucleate neck canal cell whereas venter consists of an egg cell and a venter canal cell.

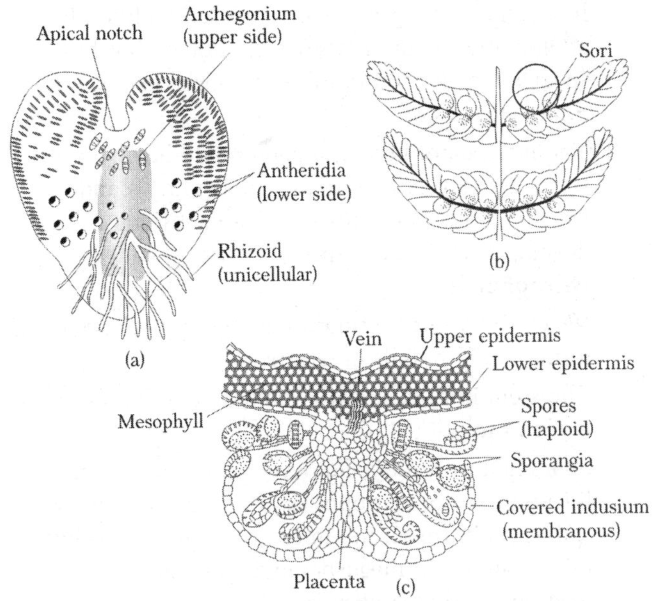

Fig 7.17 Dryopteris : (a) Prothallus, (b) Macrophyllous leaves with sori and (c) TS of sorus showing stalked sporangia

7. **Fertilisation** Water is required for fertilisation.
   - Sperms move towards archegonium by means of chemotactic movement in the water.
   - This movement is facilitated by chemicals secreted by the disintegration of neck and ventral canal cell.
   - Sperms gets attached to opened archegonia by malic acid present in their mucilage.
   - Egg and antherozoid fuse to form zygote ($2n$). It further divides and form embryo.
   - Embryo develops to form the sporophyte.

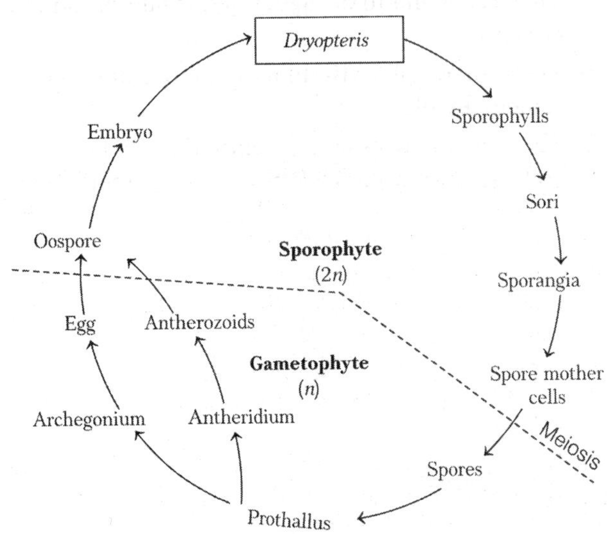

Fig 7.18 Life cycle of Dryopteris (fern)

Selaginella

The characteristic features of *Selaginella* or spike moss are as follows

1. It is a small evergreen vascular pteridophyte. It is abundantly found in tropical rainforest, but few species are also found in xeric conditions, e.g. *S. bryopteris*, etc.
2. The main plant body is a sporophyte having roots, stem and leaves. Leaves are small, simple and lanceolate in shape. Roots are adventitious and develop at the tips of special structure called **rhizophores**.
3. Vegetative reproduction occurs by fragmentation and formation of bulbils or tubers.
   The stem in *Selaginella* is distelic. Rarely, the stem is monostelic.
4. Sporangia in *Selaginella* develops in the axil of fertile leaves or sporophylls. There are two types of sporangia in these, i.e. microsporangia producing microspores and megasporangia producing megaspores (heterospory).

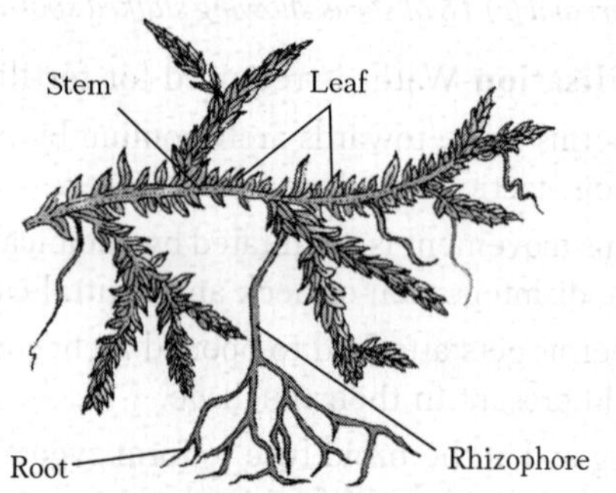

Fig. 7.19 Selaginella

5. They are found to be aggregated inside the strobilus or cone.
6. These spores give rise to male and female gametophytes.
7. The gametophytes contain antheridia and archegonia, respectively. Sex cell of antheridia and archegonia fertilise to form a diploid oospore. This grows further to form a diploid sporophyte of *Selaginella*.

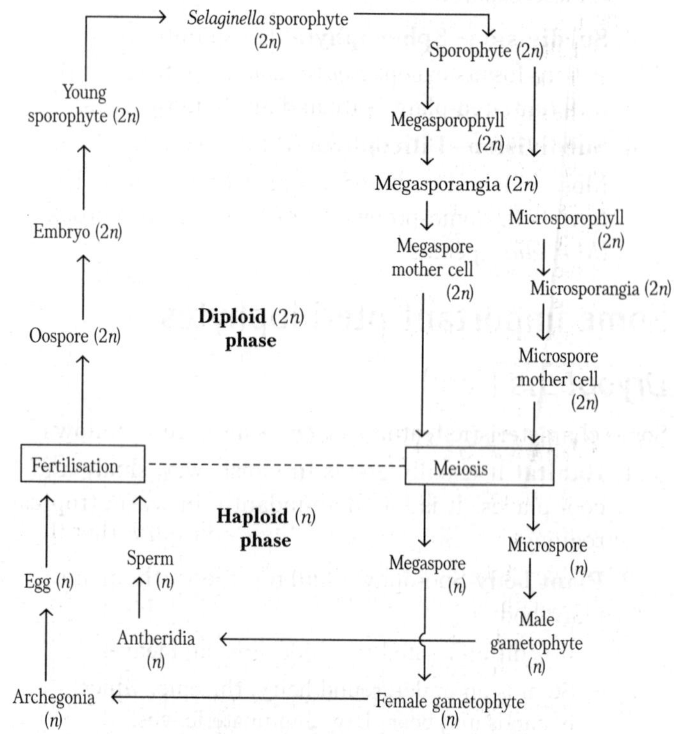

Fig 7.20 Life cycle of Selaginella

Economic importance of pteridophytes

- Pteridophytes are used as a source of food, i.e. *Marsilea*, a water fern is used to extract starch, which is cooked and consumed by tribal people.
- Species of *Equisetum* are utilised in scouring and polishing of metals.
- Some plants, such as *Azolla* are efficient biofertilisers.
- *Dryopteris* is used to yield an antihelminthic drug.
- Ferns, such as *Adiantum, Pteris, Pteridium* are ornamentally valuable plants.

CHECK POINT 7.3

1. In pteridophytes, spore germinates to give rise to
   (a) thalloid gametophyte called prothallus
   (b) thalloid sporophyte called prothallus
   (c) thalloid sporocarp
   (d) thalloid, photosynthetic sporophyte
2. The heterosporous pteridophytes are
   (a) *Lycopodium* and *Pteris* (b) *Selaginella* and *Psilotum*
   (c) *Selaginella* and *Salvinia* (d) *Dryopteris* and *Adiantum*
3. Choose the incorrect matched pair.
   (a) Haplostele – *Psilotum*
   (b) Plectostele – *Lycopodium clavatum*
   (c) Eustele – Dicot stem
   (d) Siphonostele – *Adiantum*
4. 'X' is widely distributed vascular crytogam having both homosporous and heterosporous species. 'X' is
   (a) Sphenophyta
   (b) Lycophyta
   (c) Psilophyta
   (d) Filicophyta
5. Mosses and ferns are found in moist and shady places because both
   (a) require the presence of water for fertilisation
   (b) do not need sunlight for photosynthesis
   (c) depend for their nutrition on microorganisms, which can survive only at low temperature
   (d) cannot compete with sun loving plants

GYMNOSPERMS

(*Gymnos*—naked; *sperma*—seeds)

- These are naked seeded plants. These evolved earlier than the flowering plants. They all are perennial and woody.
- Among gymnosperms, conifers (class-Coniferophyta) with 500 species still dominate on large areas of the earth. Conifers are found in hills and mountains.
- Other gymnosperms, such as Ginkgos (class-Ginkgophyta) and Cycads (class-Cycadophyta) have declined in their abundance.
- The Ginkgos were probably the first of the surviving seed plants to evolve. They became widespread during the Jurassic period that began 208 million years ago.
- Today, Ginkgos are represented by a single species *Ginkgo biloba* (maiden hair tree). *Ginkgo biloba* is also known as living fossil. *Ginkgo* trees are either male or female. The female trees bear cherry-sized, foul smelling, fleshy seeds.

General characteristics

The characteristics features of gymnosperms are as follows

1. **Habitat** Gymnosperms are worldwide in distribution. These plants abundantly grow in temperate region.
2. **Plant body** The main plant body is sporophytic ($2n$). It is differentiated into true roots, stems and leaves.
3. **Root system** It is well-developed with a tap root system.
   - Roots of some gymnosperms are symbiotically associated with algae.
   - These roots are called **coralloid roots** like in *Cycas*.
   - Roots of some gymnosperms have symbiotic relations with fungi also. These are known as **mycorrhizae**, e.g. *Pinus*.
4. **Stem** In gymnosperms it is usually erect, profusely branched and woody. Leaf scars are usually present on stem. Stems possess collateral, endarch and open vascular bundles.
5. **Leaves** These are dimorphic, i.e. of two types. These are as follows
   - **Foliage leaves** are green, simple, needle-shaped or pinnately compound.
   - **Scaly leaves** are minute and deciduous. The leaves have a thick cuticle layer and many sunken stomata.
6. **Conducting tissues** In gymnosperms, xylem contains only tracheids and xylem parenchyma. Vessels are absent in xylem. Phloem is also devoid of companion cells, but sieve cells are present.
7. **Wood** It is homoxylous and is of following types
   - **Manoxylic** Soft wood, vascular tissue possess medullary rays and less commercial importance, e.g. *Cycas*.
   - **Pycnoxylic** Compact wood, sith or without medullary rays and commercially important, e.g. *Pinus*.
   - **Monoxylic** Single persistent cambium rings and bundles, e.g. *Pinus*.
   - **Polyxylic** With several persistent cambium rings and bundles, e.g. *Cycas*.
8. **Spore formation** Gymnosperms are heterosporous.
   - They produce two types of spores, i.e. microspores and megaspores.
   - Microspores and megaspores are borne on microsporangia and megasporangia that are present on microsporophyll and megasporophyll, respectively.

- Microspore possesses highly reduced gametophyte. It is called as **pollen grain** (n).
- Pollen grain possesses male gametes.
- Male gametes are usually non-motile (male gamete of *Cycas* and *Ginkgo* are motile).

9. **Ovules or Megasporangia** These are borne naked on the megasporophylls.
    - They consist of a body called **nucellus** protected by a single envelope or integument.
    - Ovary is absent and thus fruit formation is also absent.
    - The ovule bears female gametophyte that possesses archegonium. Each archegonium contains a single egg and a venter canal cell.
10. **Pollination** It causes the pollen grains to be carried away with air currents to reach the opening of the ovules (micropylar end).
11. **Fertilisation** It is siphonogamous, i.e. it occurs through the formation of a pollen tube.
    - The male and female gamete fuse with each other. This leads to the formation of a zygote, which develops into haploid embryo.
    - There is a tendency of polyembryony in gymnosperms. It is the formation of more than one embryo inside a single seed.

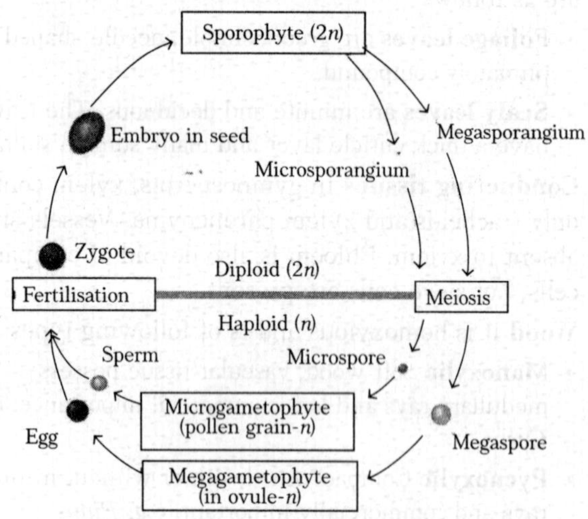

Fig 7.21 Life cycle of gymnosperm

Note
- *Gnetum and Ephedra show evolutionay tendencies towards angiosperms. Among these,* ***Ephedra*** *shows largest archegonium, largest pollen chamber and double fertilisation.*
- *The* ***smallest*** *gymnosperms is* ***Zamia pigmaea.***
- *The* ***tallest*** *gymnosperms is* ***Sequoia wellingtonia*** *(Sequoia dendron).*

Classification of gymnosperms

Chamberlain (1934) divided gymnosperms into two classes, i.e. Cycadophyta and Coniferophyta.

1. **Class-Cycadophyta** It includes both living and extinct forms, e.g. *Cycadeoidea, Lyginopteris* (extinct), *Cycas* and *Zamia.*
2. **Class-Coniferophyta** It also includes both extinct and living species, e.g. *Cordaites, Ginkgo biloba, Pinus, Gnetum* and *Ephedra.*

Some important gymnosperms

Cycas

The general characters of *Cycas* are as follows

1. **Habitat** *Cycas* is a medium-sized palm-like evergreen plant. It is mainly found in Australia and Asia. In India, *Cycas* trees are found in Odisha, Bengal, Assom, Chennai, Karnataka and Andaman Islands.
2. **Main plant body** is sporophytic ($2n$). A fully grown *Cycas* plant may attain a height of about 2-5 m. Plant body is divided into true roots, stem and leaves.

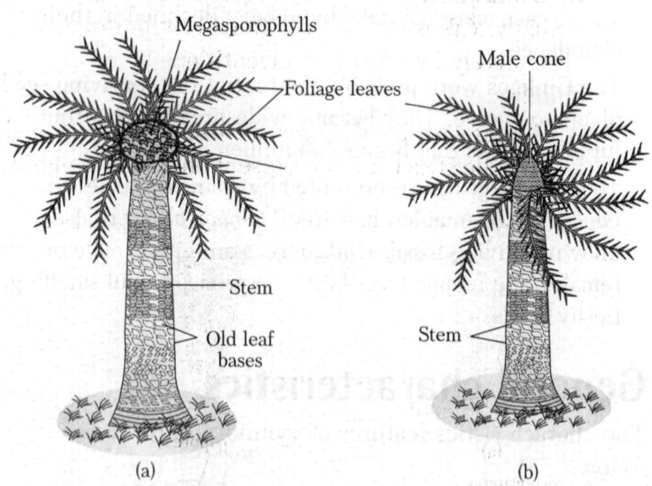

Fig 7.22 Cycas circinalis : (a) Female plant, (b) Male plant

3. **Roots** arise from the lower part of stem.
    - The two types of roots found in *Cycas* are tap roots and coralloid roots.
    - **Coralloid roots** or corallorhiza are dichotomously branched, apogeotropic and bluish-green in colour.

Coralloid roots are inhabited by a blue-green alga, i.e. *Anabaena cycadacearum.*

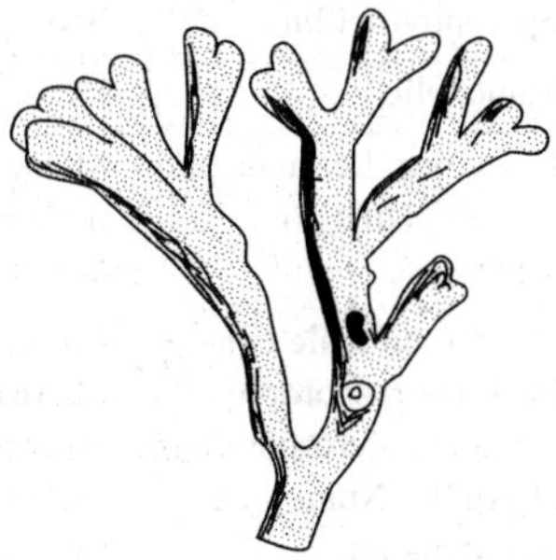

Fig 7.23 *Portion of coralloid root of Cycas*

4. **Stem** is thick, cylindrical and columnar. It is aerial and unbranched.
   - The older parts of the stem remain covered with an armour of persistent woody leaf bases. Stem bears adventitious buds at its base. These buds are called as **bulbils.**
   - Vascular bundles in stem are conjoint, collateral and open. Secondary growth occurs in *Cycas.*
5. The wood of *Cycas* is polyxylic, i.e. possess more than one xylem rings.
6. *Cycas* shows dimorphism in leaves. There are two types of leaves, i.e. foliage leaves and scale leaves.
7. **Reproduction** takes place vegetatively as well as sexually. *Cycas revoluta* is extensively propagated vegetatively by means of adventitious buds or bulbils.
8. All living species of *Cycas* are **dioecious,** i.e. male and female structures are present on separate plants.

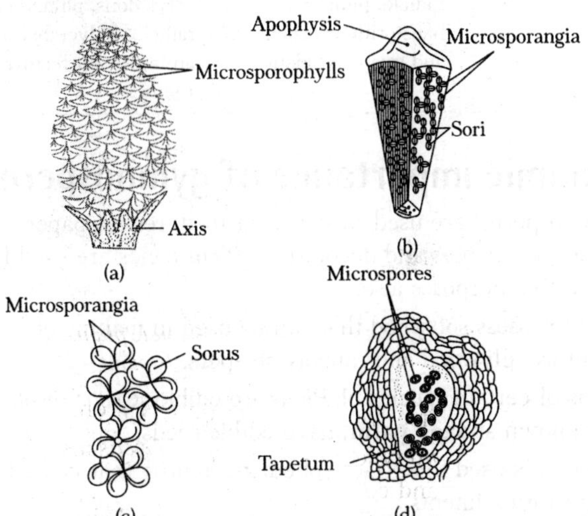

Fig 7.24 *Male cone : (a) Entire male cone, (b) One microsporophyll, (c) Sori microsporangia, (d) LS of single microsporangium*

9. Male plant bears male strobilus. It is formed by the aggregation of microsporophylls. Microsporophyll of *Cycas* is a flat, wedge-shaped and triangular structure. The upper tip of microsporophyll is pointed and sterile. It is called as **apophysis**. On each microsporophyll, numerous microsporangia are borne.
10. Inside the microsporangia, microspores are formed. Microspore (pollen grain) is the first cell of the male gametophytic generation. Inside the microspore, male gametes are formed from generative cell of microspore.

    Male gametes are large bicilliate and reniform in shape. The microspores shed at 3-celled (prothalial cell, tube cell and generative cell) stage.
11. The megasporophylls are loosely arranged on a female strobilus. 2-4 pairs of ovule are found to be present on megasporophyll. The ovules of *Cycas* are large, naked and orthotropous (erect). They are unitegmic, i.e. enclosed in a thick coat, the integument. The latter consists of outer and inner fleshy layers with middle stony layer.

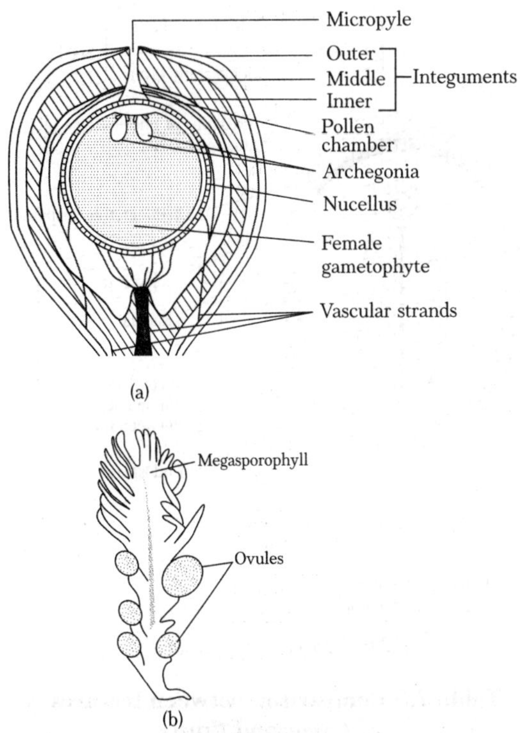

Fig 7.25 *Cycas : (a) LS of ovule, (b) Megasporophyll of Cycas revoluta*

12. The megaspore is the first cell of female gametophyte. It undergoes repeated divisions to form

a multicellular female gametophyte. Due to abundant food reserves, it serves as an endosperm. Endosperm is haploid and formed before fertilisation. The archegonia are embedded in the female gametophyte of ovule.

13. *Cycas* has the largest male cone, largest ovule, largest egg and the largest sperm.
14. The spermatozoids after their release reaches the archegonium where it fuses with the egg cell to form the zygote.
15. The diploid zygote develops *in situ* into a multicellular proembryo which further differentiates into an embryo.
16. Embryo is made up of radicle, plumule, haustorium, suspensor and two cotyledons.
17. After fertilisation, the ovular structures transform to form following structures as follows
Nucellus → Perisperm, Female gametophyte → Endosperm, Zygote → Embryo, Integuments → Seed coat, Ovule → Seed

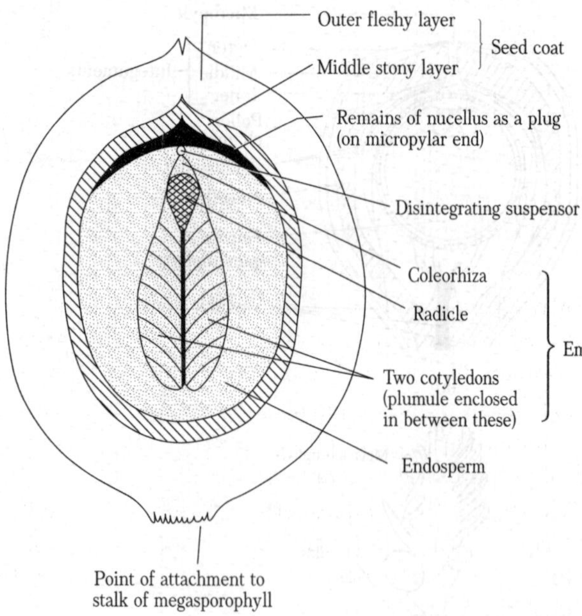

Fig 7.26 LS of Cycas seed

Table 7.5 Comparison between features of *Cycas* and *Pinus*

Features	*Cycas*	*Pinus*
Roots	Coralloid roots besides normal root system (tap root system).	Tap root system with mycorrhiza.
Leaf	Young leaflets with circinate vernation.	Foliage (needle-like) leaves and scale leaves occur.
Whole plant body	Sporophytic (2*n*)	Sporophytic (2*n*)
Wood	Manoxylic	Pycnoxylic
Sexual diamorphism	Dioecious, i.e. male and female plants are separate.	Monoecious, i.e. both male and female cones are on same plant.
Microsporophyll and microsporangia	Present on male cone. Each microsporophyll contains microsporangia abaxially. Numerous sporangia on microsporophyll.	Present on male cone. Microsporophylls are arranged spirally and each microsporophyll has two microsporangia abaxially.
Microspore or male gametophyte	Shed at three cell stages, i.e. prothallial, generative and tube cell.	Shed at four cell stages, i.e. 2 vegetative prothallial cells, a generative and one tube cell. Wings are found on microspore.
Pollination	By wind (anemophily)	By wind (anemophily)
Antherozoid formation	Generative cell forms two motile antherozoids.	Generative cell forms body cell, which produces non-motile antherozoids.
Megasporophyll	Arranged loosely and no true cone is formed.	Elongated cylindrical structure with elongated central axis.
Type of ovule	Large, naked and orthotropous (erect), one integument is found (unitegmic).	Ovules are anatropous (inverted).
Embryo	Zygote forms embryo *in situ*. It is made up of radicle, plumule, haustorium, suspensor and two cotyledons.	Zygote forms embryo *in situ* made up of 3-18 cotyledons, plumule and radicle. Polyembryony is an important feature of *Pinus*.

Economic importance of gymnosperms

- Gymnosperms are used as source of food, resins, paper, timber, medicines and decoration. *Pinus* cones are used for decorative purposes also.
- These possess softwood that can be used in making of furniture, plywood and railway sleepers.
- Seeds of certain species of *Pinus* are edible. *Cycas revoluta* also known as **sago palm**, have edible seeds.
- *Ephedra* is used to extract ephedrine, a drug that is used in respiratory ailments.
- Woods of *Picea*, *Pinus*, *Larix* and *Abies* are used in paper manufacturing.
- Canada balsam is a turpentine extracted from *Abies balsamea*. It is used as mounting medium in permanent slide preparations.

CHECK POINT 7.4

1. Which of the following is not the feature of gymnosperms?
   (a) Parallel venation (b) Perennial plants
   (c) Distinct branches (d) Xylem with vessels
2. Conifers differ from grasses in the
   (a) production of seeds from ovules
   (b) lack of xylem tracheids
   (c) absence of pollen tubes
   (d) formation of endosperm before fertilisation
3. The sclerenchyma of the hypodermis in the *Pinus* needle helps in
   (a) increasing the absorptive surface of the cell
   (b) checking transpiration
   (c) mechanical support
   (d) photosynthesis
4. In *Cycas* stem, open vascular bundle is characterised by
   (a) phloem being sandwitched between xylem
   (b) cambium present in between xylem and phloem
   (c) xylem being sandwitched between phloem
   (d) xylem and phloem occurring on different radii
5. Though *Cycas* has two cotyledons, this is not included in dicot because
   (a) of naked ovule (b) they have megaspore
   (c) appears as palm tree (d) has compound leaves
6. Select the mismatched pair.
   (a) *Cycas* – Living fossil
   (b) *Pinus* – Resin production
   (c) *Abies* – Paper manufacturing
   (d) *Picea* – Sago palm

ANGIOSPERMS

(*Angio*—covered; *sperma*—seed)

- These are plants in having seeds, fruit and flowers.
- The presence of flower is most important feature of angiosperms. It contains the sex organs of the plant.
- These are most highly evolved plants on earth.
- They appeared about 130 million years ago in Jurassic (dicot plant) and Cretaceous (monocot plants) period of Mesozoic era; long before the evolution of man.
- They flourished well in Cenozoic era (called **Age of angiosperms**).

General characteristics

1. **Habitat** Angiosperms are found in every possible habitat on earth like oceans, plains, deserts, hot spring, at altitude of 6000 feet, extreme cold, etc.

   Zostera and *Thalassia* are marine angiosperms, *Tillandsia* is epiphytic, *Neottia* is saprophytic, etc.
2. **Plant body** Plants may be herbs, shrubs or trees based on their habit.
   - They may be annuals, biennials or perennials, based on the duration of their life.
   - Plant body is sporophytic (2*n*) and differentiated into root, stem and leaves.
   - Roots and stem develop from radicle and plumule, respectively. Leaves arise from nodes.
3. **Vascular bundles** These consist of prominent xylem consisting of both tracheids and vessels.
   - The phloem consists of both sieve tubes and companion cells.
   - Almost all dicots and some monocots show secondary growth.
4. **Sex organs** The angiosperms lack both archegonia and antheridia.
   - The archegonium is replaced by gynoecium (pistil) and antheridium by pollen grain. Instead of cone, they bear flowers.
   - The ovules remain enclosed in the ovary. The wall of pollen grain is differentiated into **intine** and **exine**.
   - Intine is further differentiated into a foot layer, columellar layer and tectum.
   - Two male gametes are usually found in pollen grains.
5. **Pollination** It is achieved through many agencies, i.e. wind, water, insects, animals, etc. Pollen grains germinate on stigma of female reproductive body.
6. **Fertilisation** The two male gametes fuse with two distinct cells, i.e. egg cell and central cell. Hence, double fertilisation occurs.

   Food storing tissue endosperm is formed as a result of triple fusion [central cell (2*n*) + male gamete (*n*)]. This triploid (3*n*) endosperm provides nourishment to the developing embryo.
7. **Developments** After fertilisation, ovary becomes the fruit, ovule transforms into the seed and zygote becomes the embryo.

   Fruit formation is peculiar to angiosperms only.

Classification of angiosperms

- Different classification schemes were given to classify angiosperms. Out of which the most famous is the natural system of classification by **Bentham** and **Hooker**.

- This system of classification is based on external features showing (similarities and dissimilarities) between various plant groups.
- They classified phanerogams into three classes, namely Dicotyledonae, Gymnospermae and Monocotyledonae.
- Out of these, the Dicotyledonae and Monocotyledonae combine to form Angiospermae.

Class–Dicotyledonae

1. It is divided into three subclasses, 14 series, 165 natural orders (families), 2610 genera and 31874 species, e.g. *Hibiscus, Gossypium, Citrus, Helianthus, Solanum, Petunia, Salvinia*, etc.
2. Dicotyledonous leaves show reticulate venation (exceptions *Calophyllum, Eryngium*).
3. The vascular bundles in stem are open and arranged in a ring.
4. Seeds of these plants have mostly two cotyledons.
5. Flowers are pentamerous or tetramerous.
6. Pollen grains have three germ pores.
7. Secondary growth is prominent.

Class–Monocotyledonae

1. Class-Monocotyledonae is grouped into 7 series, 34 natural orders (families), 1495 genera and 18576 species, e.g. *Triticum, Oryza, Allium, Musa, Vanda, Lamna*, etc.
2. Monocots mostly depict parallel venation (exceptions *Smilax, Colocasia*).
3. The vascular bundles are mostly scattered in stem.
4. These plants have seeds with one cotyledon.
5. Flowers are trimerous. Pollen grains have only one germinal furrow. Except a few, secondary growth is usually absent.

Resemblances between gymnosperms and angiosperms

- Root system is well-developed and may be diarch, triarch, tetrarch or polyarch.
- The xylem is exarch (xylem develops from center outwards) in roots.
- The stem is eustelic. A true stele, in which vascular bundle is found in a broken ring.
- Secondary growth is present in stems and secondary wood formed can be monoxylic (wood formed by one ring due to persistant) or polyxylic (wood formed by more than one ring).
- Formation of ovules surrounded by integument and seeds.
- Presence of heterospory.
- Presence of endosporic and reduced gametophytes.
- Formation of pollen tube.
- Male gametes are non-motile in the majority of gymnosperms and all angiosperms.
- Embryogeny is endosporic and germination of spores is precocious.

Table 7.6 Differences between gymnosperms and angiosperms

Gymnosperms	Angiosperms
These are mostly woody trees.	These can be herbs, shrubs or trees.
The reproductive structures are cones, which are unisexual.	The reproductive structures are flowers, which may be both unisexual or bisexual.
The ovules are exposed, which develops into naked seeds.	The ovules are enclosed in the ovary, which after ripening forms fruit.
The microspores and megaspores are produced by male and female cones, respectively.	The microspores are produced in anther, while megaspores are produced in ovules of the ovary in flowers.
The pollen grains fall and germinate directly on the micropyle of the ovules.	The pollen grains fall on the stigma, germinate and the pollen tube carries the male gamete to the ovary.
Fertilisation is simple.	Fertilisation is double.
Endosperm is haploid.	Endosperm is triploid.
In gymnosperms, xylem vessels and companion cells are absent.	In angiosperms, xylem vessels and companion cells are present.

CHECK POINT 7.5

1. Choose the odd one out.
   (a) *Vanda* (b) *Lemna*
   (c) *Petunia* (d) *Allium*
2. Angiosperms and gymnosperms resemble in having
   (a) vessels in wood (b) mode of fertilisation
   (c) siphonogamy (d) nature of endosperm
3. The stele found in monocot is
   (a) haplostele (b) atactoslele
   (c) dictyoslete (d) actinostele
4. Which of the following is a saprophytic angiosperm?
   (a) *Cuscuta*
   (b) *Neottia*
   (c) *Agaricus*
   (d) Mango
5. All of these are the characters of angiosperms except they
   (a) are most advanced
   (b) bear true fruit
   (c) bear covered seeds
   (d) are cryptogams

EXAM POINTS

- *Chlorella* and *Spirulina* is a pair of unicellular algae. **[NEET 2020, 18]**
- *Dryopteris* have rhizome, *Cycas* have coralloid roots and *Volvox* is a colonial form algae. **[JIPMER 2017]**
- Zygotic meiosis is a characteristic of *Chlamydomonas*. **[NEET 2017]**
- Palmella stage is present in *Chlamydomonas*. **[JIPMER 2018]**
- Agar-agar is obtained from *Gelidium* and *Gracilaria*. **[CBSE AIPMT 2015]**
- Biflagellate zoospores are present in brown algae. **[NEET 2018]**
- *Chlamydomonas* exhibits both isogamy and anisogamy and *Fucus* shows oogamy. **[NEET 2013]**
- Isogamous condition with non-flagellated gametes is found in *Spirogyra*. **[NEET 2013]**
- The life cycle of algae such as *Spirogyra* is haplontic. **[MHT CET 2014]**
- In *Ulothrix* chloroplast is girdle-shaped. **[JIPMER 2018]**
- Male gametes are flagellated in *Ectocarpus*. **[CBSE AIPMT 2015]**
- *Ectocarpus* show haplo-diplontic life cycle. **[UK PMT 2014]**
- An alga which can be employed as food for human being is *Chlorella*. **[CBSE AIPMT 2014]**
- *Chlorella* and *Spirulina* are used as space food. **[CBSE AIPMT 2015]**
- *Laminaria* and *Sargassum* are used as food. **[NEET 2016]**
- Algae increase the level of dissolved oxygen in the immediate environment. **[NEET 2016]**
- The presence of pyrenoid is characteristic feature of class–Chlorophyceae. **[UK PMT 2014]**
- Algin is extracted from brown algae and carrageenan is extracted from red algae. **[NEET 2016]**
- *Chara* has upper oogonium and lower round antheridium. It has globule and nucule is present on the same plant. The globule is male reproductive structure. **[CBSE AIPMT 2014]**
- *Cephaleror virescens* is a parasitic alga. **[JIPMER 2018]**
- Mannitol is stored food in Phaeophyceae. In red algae, storage food is floridean starch. **[CBSE AIPMT 2015]**
- Mosses and lichens are the first organisms to colonise a bare rock. **[CBSE AIPMT 2015]**
- Gemma cups are present in *Marchantia*. **[NEET 2018]**
- *Sphagnum* is responsible for peat formation. **[CBSE AIPMT 2014]**
- In *Funaria, Marchantia* and *Pteris* the gametophyte is independent and free-living. **[CBSE AIPMT 2015]**
- *Marchantia* is considered as a heterothallic plant because it is dioecious. **[Karnataka CET 2014]**
- Mosses and liverworts are members of bryophytes. **[Karnataka CET 2014]**
- Pseudoelators are found in sporophyte of *Riccia*. **[JIPMER 2017]**
- From evolutionary point of view, retention of the female gametophyte with developing young embryo on the parent sporophyte for some time is first observed in pteridophytes. **[NEET 2019]**
- The plant part which consists of two generations, one within the other is pollen grains inside the anther and embryo sac inside the ovule. **[NEET 2020]**
- Botanical snakes are refered to pteridophytes. **[CBSE AIPMT 2014]**
- Pteridophyta is the first group of vascular plants. **[MHT CET 2014]**
- Heterospory and retention of female gametophyte are responsible for origin of seed habit in *Selaginella*. **[AIIMS 2019]**
- *Psilotum* is a living fossil. **[AIIMS 2019]**
- *Salvinia* is heterosporous and *Equisetum* is homosporous. **[NEET 2017]**
- Phloem in gymnosperms lacks both sieve tube and companion cells. **[NEET 2019]**
- *Pinus* is monoecious and *Cycas* is dioecious. **[NEET 2017]**
- Conifers are adapted to tolerate extreme environmental conditions because of thick cuticle. **[NEET 2016]**
- Strobili or cones are found in *Equisetum*. **[NEET 2020]**
- Sequoia is one of the tallest trees. **[NEET 2016]**
- In gymnosperms, male and female gametophytes are present within sporangia located on sporophyte. **[CBSE AIPMT 2015]**
- The leaves of gymnosperms are well-adapted to extremes of climate. **[NEET 2016]**
- Zooidogamy is found in *Cycas*. **[JIPMER 2019]**
- Male and female gametophytes do not have an independent free-living existence in angiosperms. **[NEET 2020]**

Chapter Exercises

(A) Topical Exercise

Plants and Their Classification

1 Which of the following is a characteristic of members of kingdom-Plantae?
(a) Presence of chlorophyll
(b) Presence of cellulosic cell wall
(c) Alternation of generation
(d) All of the above

2 In Bentham and Hooker's system of classification, main emphasis has been given to
(a) embryological characters
(b) reproductive characters
(c) vegetative characters
(d) Both (a) and (b)

3 Chief merit of Bentham and Hooker's classification is that
(a) it is a system mostly based on evolutionary concepts
(b) it is a natural systems of classification of all groups of plants
(c) the description of the taxa are based on actual observation of the specimen
(d) it also considers the phylogenetic aspects

4 Consider the following statements.
I. Natural system of classification was based upon anatomy only.
II. Phylogenetic system of classification is based upon evolutionary relationship of organisms.

Choose the correct option.
(a) Statement I is true, but II is false
(b) Statement I is false, but II is true
(c) Both statements I and II are true
(d) Both statements I and II are false

5 Identify the scientists who worked extensively on chlorophyllous and non-chlorophyllous thallophytes, respectively.
I. Iyengar
II. Swaminathan
III. Mehta
IV. Maheshwari

Choose the correct option.
(a) I and IV (b) I and III
(c) II and III (d) III and IV

Algae

6 Consider the following statements about Thallophyta.
I. Sex organs are unjacketed and one celled gametangia.
II. The vascular system is well-developed.

Choose the correct option.
(a) Statement I is true, but II is false
(b) Statement I is false, but II is true
(c) Both statements I and II are true
(d) Both statements I and II are false

7 Algae are characterised by the
(a) lack of vascular tissues (b) lack of seeds
(c) lack of embryo (d) All of these

8 Life cycle of *Fucus* is
(a) haplontic (b) haplo-diplontic
(c) diplontic (d) None of these

9 If you are asked to classify the various algae into distinct groups, which of the following characters you should choose?
(a) Types of pigments present in the cell
(b) Nature of stored food materials in the cell
(c) Structural organisation of thallus
(d) Chemical composition of the cell wall

10 In algae, aplanospores are produced during
(a) favourable conditions
(b) unfavourable conditions
(c) mild conditions
(d) Both (a) and (b)

11 Which of the following is not correctly matched?
(a) *Laminaria* – Flattened leaf-like thallus
(b) *Chlorella* – Unicellular non-flagellated
(c) *Spirogyra* – Filamentous structure
(d) *Volvox* – Colonial form, non-flagellated

12 Consider the following statements.
I. Fusion of two gametes which are dissimilar in size is termed as anisogamy.
II. Holdfast, stipe and frond constitute the plant body in case of Phaeophyceae.

Choose the correct option.
(a) Statement I is true, but II is false
(b) Statement I is false, but II is true
(c) Both statements I and II are true
(d) Both statements I and II are false

13 Which of the following option is correct about Rhodophyceae?
(a) The stored food material is starch
(b) The major pigments are chlorophyll-a and d
(c) Laminarin is the stored food and major pigments are chlorophyll-a and b
(d) Floridean starch is the stored food and the major pigments are chlorphyll-a, d and phycoerythrin

14 Which one of the following is incorrect matched?
(a) *Spirogyra* – Motile gametes
(b) *Sargassum* – Chlorophyll-c
(c) Basidiomycetes – Puffballs
(d) *Nostoc* – Water blooms

15 Match the following columns.

	Column I (Types of chloroplast)		Column II (Algae)
A.	Cup-shaped	1.	*Ulothrix*
B.	Girdle-shaped	2.	*Oedogonium*
C.	Stellate	3.	*Chlamydomonas*
D.	Reticulate	4.	*Zygnema*

Codes

	A	B	C	D
(a)	2	4	3	1
(b)	3	1	4	2
(c)	3	4	2	1
(d)	4	3	1	2

16 Two adjacent filaments of *Spirogyra affinis*, each having 10 cells participating in reproduction. How many new *Spirogyra* plants are produced during sexual reproduction?
(a) 5 (b) 10 (c) 20 (d) 40

17 Zygote of *Spirogyra* produces four haploid nuclei in which
(a) one is functional (b) two are functional
(c) three are functional (d) all are functional

18 The cell wall in *Spirogyra* is
(a) 3-layered with outermost mucilaginous layer
(b) 3-layered with innermost mucilaginous layer
(c) 2-layered with outer mucilaginous layer
(d) 2-layered with inner mucilaginous layer

19 Consider the following statements.
I. In *Spirogyra*, vegetative reproduction occurs by fragmentation.
II. Conjugation in *Spirogyra* involves physiologically dissimilar gametes.

Choose the correct option.
(a) Statement I is true, but II is false
(b) Statement I is false, but II is true
(c) Both statements I and II are true
(d) Both statements I and II are false

20 Incorrect character of brown algae is
(a) chlorophyll-a and b present
(b) they remain attached
(c) chlorophyll-a and c present
(d) presence of fucoxanthin

21 Carrageenan, a jelly-like substance is obtained from
(a) green algae
(b) brown algae
(c) red algae
(d) None of the above

22 From which of the following algae, agar-agar is commercially extracted?
(a) *Gracilaria* (b) *Fucus*
(c) *Sargassum* (d) *Turbinaria*

Bryophyta

23 Which one of the following statements is not true for Bryophyta?
(a) They lack tracheids and sieve tubes
(b) They are photosynthetic
(c) Their zygote undergoes meiosis and then produces the sporophyte
(d) Their spores germinate producing gametophytes

24 Bryophytes are different from algae in
(a) having multicellular jacketed sex organs
(b) having no sporophyte
(c) having zygotic meiosis
(d) having oogamous reproduction

25 Bryophytes are not tall plants due to the
(a) absence of meristem
(b) absence of vascular tissues
(c) presence of root system
(d) All of the above

26 In bryophytes, organs are referred to as leaf-like and stem-like and not the true leaf and stem because
(a) they lack vascular tissues
(b) they are non-green
(c) they do not function as leaf and stem
(d) All of the above

27 Bryophytes are called amphibians of plant kingdom because
(a) their reproductive phase requires water
(b) their sex organs are multicellular and jacketed
(c) they have tracheids
(d) All of the above

28 How are gametes produced in bryophytes?
(a) By mitosis of gametophyte cells
(b) By meiosis of gametophyte cells
(c) By meiosis of sporophyte cells
(d) By mitosis of spores

29 In bryophytes, gametophyte is formed by

(a) sporophyte (b) zygote
(c) spores (d) vascular tissue

30 Which of the following is true for bryophytes?

(a) It is the only plant group that shows an alternation of generation
(b) Bryophytes exhibit extensive vascular tissue
(c) The sporophyte (multicellular diploid) is the dominant stage
(d) The gametophyte (multicellular haploid) is the dominant stage

31 Consider the following statements.

I. Gametophyte is the predominant phase in the life cycle of the bryophyte.
II. The first cell of sporophytic generation in bryophytes is spore mother cell.

Choose the correct option.

(a) Statement I is true, but II is false
(b) Statement I is false, but II is true
(c) Both statements I and II are true
(d) Both statements I and II are false

32 Match the following columns.

Column I	Column II
A. Liverworts	1. *Funaria*
B. Hornworts	2. *Anthoceros*
C. Mosses	3. *Riccia*
D. Gnetales	4. *Ephedra*

Codes

	A	B	C	D		A	B	C	D
(a)	4	1	2	3	(b)	3	1	2	4
(c)	3	2	1	4	(d)	2	1	3	4

33 Consider the following statements regarding bryophytes.

I. The first plant group to survive on the land was probably bryophyte.
II. The gemmae in *Funaria* are formed on the axis of rhizoids or leaves or protonema.
III. *Funaria* is monoecious.

Which of the following are correct statements?

(a) I and II (b) I and III
(c) Only III (d) I, II and III

34 Consider the following statements.

I. Spores of *Funaria* represent gametophytic generation.
II. In bryophytes, simplest sporophyte occurs in *Riccia*.

Choose the correct option.

(a) Statement I is true, but II is false
(b) Statement I is false, but II is true
(c) Both statements I and II are true
(d) Both statements I and II are false

35 The sporophyte in *Funaria* is

(a) the dominant and main phase of the plant body
(b) differentiated into distinct foot, seta and capsule
(c) parasite on the gametophyte
(d) Both (b) and (c)

36 Consider the following statements.

I. Foot of moss sporophyte is a well-organised structure.
II. In the capsule of moss, shock absorbers are peristome teeth.

Choose the correct option.

(a) Statement I is true, but II is false
(b) Statement I is false, but II is true
(c) Both statements I and II are true
(d) Both statements I and II are false

37 Choose the incorrect pair.

(a) Hepaticopsida – *Marchantia*
(b) Sporophyte – *Columella*
(c) Bryopsida – *Anthoceros*
(d) *Riccia* – Biflagellated antherozoids

38 Protonema is **[NCERT Exemplar]**

(a) haploid and is found in mosses
(b) diploid and is found in liverworts
(c) diploid and is found in mosses
(d) haploid and is found in liverworts

39 In a moss, the sporophyte

(a) is partially parasitic on the gametophyte
(b) produces gametes that give rise to the gametophyte
(c) arises from a spore produced from the gametophyte
(d) manufactures food for itself, as well as for the gametophyte

40 Match the following columns.

Column I	Column II
A. Protonema	1. Numerous neck canal cells in the capsule
B. *Columella*	2. Bryophyte of economic importance
C. *Sphagnum*	3. Haploid structrue of *Funaria*
D. *Funaria*	4. Middle sterile region in moss capsule

Codes

	A	B	C	D		A	B	C	D
(a)	1	2	3	4	(b)	3	4	2	1
(c)	1	2	4	3	(d)	3	2	1	4

41 Sexual reproduction in liverworts takes place by

(a) the formation of sex organs always on the same thallus
(b) the formation of male and female sex organs always on different thallus
(c) sexual reproduction is absent
(d) male and female sex organs may be present on same or different thallus

42 An archegonium of *Riccia* has
(a) 4-6 neck canal cells, 1 venter canal cell and one oosphere
(b) 13 neck canal cells, 2 venter canal cells and one oosphere
(c) 8 neck canal cells, one venter canal cell and two oospheres
(d) 6 neck canal cells, 2 venter canal cells and one oosphere

43 Read the statements given below with respect to *Riccia* and choose the incorrect statement.
(a) Most of the species are terrestrial except *R. fluitans* that is aquatic
(b) Scales of *Riccia* are unicellular
(c) Neck of archegonium is made up of 4-6 neck canal cells
(d) Fertilisation requires the presence of water

44 Consider the following statements.
I. During development of embryo in archegonium of Bryophyta, its posterior part form an embryo cover, called as calyptra.
II. Calyptra protects the young capsule and acts as transpiration shield around the immature capsule.

Choose the correct option.
(a) Statement I is true, but II is false
(b) Statement I is false, but II is true
(c) Both statements I and II are true
(d) Both statements I and II are false

45 Which of the following is the correct statement?
(a) No diploid structure is found in sporogonium of *Riccia*
(b) Sporogonium of *Riccia* is globular and undifferentiated
(c) Sporogonium and spore mother cells of *Riccia* have chloroplast
(d) None of the above

46 Match the following columns.

Column I	Column II
A. Coleorhiza	1. Development of sporophyte directly from gametophyte without intervention of gametes.
B. Apogamy	2. Development of gametophyte directly from sporophyte without the involvement of reduction division.
C. Indusium	3. An unbranched columnar stem with a crown of leaves.
D. Caudex	4. Protective covering of radicle.
	5. Protective structure of a sorus.

Codes

	A	B	C	D
(a)	5	2	4	1
(b)	4	1	5	3
(c)	3	5	2	4
(d)	2	3	1	5

47 *Sphagnum* is used as a packing material for transporting living plants because of its
(a) acidic nature, it does not undergo decay
(b) creeping capacity
(c) water absorbing capacity
(d) Both (a) and (c)

Pteridophyta

48 Which statement is incorrect for pteridophytes?
(a) Xylem tissue lacks true vessels
(b) Gametophyte of pteridophytes is autotrophic
(c) Sperms of pteridophytes are non-motile
(d) Pteridophytes exhibit heteromorphic alternation of generation

49 Which of the following is correct regarding pteridophytes?
(a) *Selaginella* and *Salvinia* are heterosporous
(b) Heterosporous pteridophytes have mega (large) and micro (small) spores
(c) The development of zygote within female gametophyte is the precursor to the seed habit
(d) All of the above

50 Consider the following statements.
I. Pteriodophytes are avascular and seedless.
II. *Adiantum* is called walking fern because of its walking spores.

Choose the correct option.
(a) Statement I is true, but II is false
(b) Statement I is false, but II is true
(c) Both statements I and II are true
(d) Both statements I and II are false

51 Gametophyte of pteridophytes is
(a) short-lived, free-living and sexual organ bearing
(b) heart-shaped, dependent on sporophyte and sex-organ bearing
(c) fibre-like and dependent on sporophyte
(d) All of the above

52 Sex organs in pteridophytes are formed on the
(a) multicellular sporophyte
(b) specialised cells originating from sporophyte
(c) photosynthetic, free-living gametophyte
(d) parasitic gametophyte

53 The gametophytes of pteridophytes grow in damp, moist and shady places because
(a) they are limited and restricted to a narrow geographical region
(b) they need water for fertilisation of gametes
(c) water is required for gamete formation
(d) egg cell swims in water to reach to the antheridia

54 Prothallus in the life cycle of pteridophyte represents
(a) the gametophytic phase
(b) the sporophytic phase
(c) the neutral phase
(d) no phase

55 A prothallus is **[NCERT Exemplar]**
(a) a structure in pteridophytes formed before the thallus develops
(b) a sporophytic free-living structure formed in pteridophytes
(c) a gametophyte free-living structure formed in pteriodophytes
(d) a primitive structure formed after fertilisation in pteridophytes

56 Prothallus of fern has
(a) antheridia and archegonia on lower surface
(b) antheridia and archegonia on upper surface
(c) antheridia on upper surface and archegonia on lower surface
(d) antheridia on lower surface and archegonia on upper surface

57 Possible advantage of antheridia occurring on the under surface for fern prothallus is
(a) protection from wind
(b) protection from direct rays
(c) easy diffusion of nutrients from prothallus
(d) accumulation of capillary water

58 Which of the following taxa shows zooidogamous oogamy?
(a) *Funaria* (b) *Pteris*
(c) *Cycas* (d) All of these

59 What is true for siphonostele?
(a) It has pith
(b) *Adiantum* and *Marsilea* have siphonostele
(c) Leaf gaps are absent
(d) All of the above

60 The given figure represents a type of stelar system.

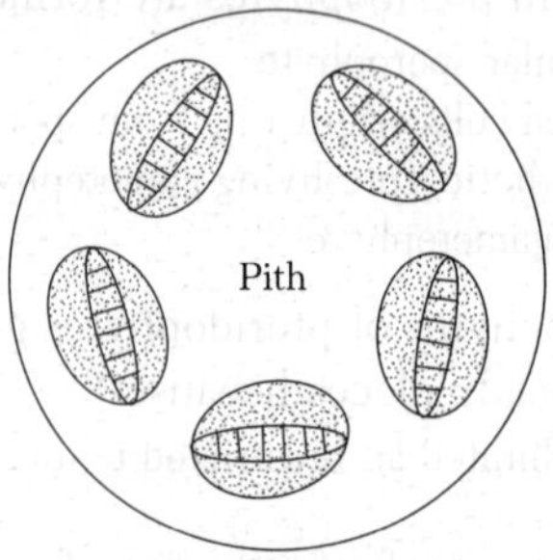

Choose the plants having such an organisation.
(a) *Dryopteris, Pteris* and *Pellaea*
(b) *Adiantum*
(c) *Marsilea* and *Platyzoma*
(d) Gymnosperms and dicots

61 Ectophloic siphonostele is found in
(a) *Adiantum* and Cucurbitaceae
(b) *Osmunda* and *Equisetum*
(c) *Marsilea* and *Botrychium*
(d) *Dicksonia* and maiden hair fern

62 A fern differs from a moss in having
(a) swimming archegonia
(b) swimming antherozoids
(c) independent gametophytes
(d) independent sporophytes

63 Antherozoids of *Dryopteris* are
(a) multiciliated and coiled
(b) multiciliated and sickle-shaped
(c) biciliated and coiled
(d) multi flagellate and spirally coiled

64 Which of the following is not correct with reference to *Selaginella*?
(a) It is commonly distributed on hills and plains
(b) Some species are xerophytic
(c) Vascular cylinder is protostelic
(d) It is heterosporous

65 The 13-celled male gametophyte in *Selaginella* is
(a) 12 cells of antheridium + 1 prothallial cell
(b) 10 cells of antheridium + 3 prothallial cells
(c) 8 cells of antheridium + 2 prothallial cells
(d) None of the above

Gymnosperms

66 Gymnosperms are characterised by
I. the presence of naked seeds.
II. haploid endosperm.
III. herbs and annual life cycle.
IV. recently evolved plants.
Choose the correct option from the codes given below.
(a) Only I (b) I and III
(c) I and II (d) Only III

67 How many of these features are correct for gymnosperms?

Heterosporous, Homosporous, Bisporangiate, Zooidogamy, Entomophilly, Siphonogamy, Roots with Mycorrhizae, Ramenta, Circinate vernation

Choose the correct option.
(a) Seven (b) Five (c) Six (d) Three

68 Phanerogamae are
(a) seed plants with evident sex organs
(b) non-seed plants
(c) non-seed plants with hidden sex organs
(d) None of the above

69 Consider the following statements.

I. Xylem in gymnosperms is devoid of vessels.

II. The phloem in gymnosperms lacks companion cells.

Choose the correct option.

(a) Statement I is true, but II is false
(b) Statement I is false, but II is true
(c) Both statements I and II are true
(d) Both statements I and II are false

70 Most of the gymnosperms have

(a) antheridia
(b) archegonia but no antheridia
(c) Only corpophore
(d) Neither antheridia nor archegonia

71 Flagellated male gametes are present in all the three of which one of the following sets?

(a) *Anthoceros, Funaria* and *Spirogyra*
(b) *Zygnema, Saprolegnia* and *Hydrilla*
(c) *Fucus, Marsilea* and *Calotropis*
(d) *Riccia, Dryopteris* and *Cycas*

72 In gymnosperms, the pollination is

(a) anemophilous-micropylar
(b) anemophilous-stigmatic
(c) entomophilous-micropylar
(d) entomophilous-stigmatic

73 In gymnosperms, the pollen chamber represents

(a) a cell in the pollen grain in which the sperms are formed
(b) a cavity in the ovule in which pollen grains are stored after pollination
(c) an opening in the mega gametophyte through which the pollen tube approaches the egg
(d) the microsporangium in which pollen grains develop

74 Male gametes of *Cycas* are

(a) small, biciliate and circular
(b) small, uniciliate and circular
(c) large, biciliate and reniform
(d) large, multiciliated and top-shaped

75 Which one of the following is lacking in *Cycas* ?

(a) a well-organised flower
(b) circinate vernation
(c) scale and foliage-like leaves
(d) roots with blue-green algae

76 Conifers differ from grasses in the

(a) production of seeds from ovules
(b) lack of xylem tracheids
(c) absence of pollen tubes
(d) formation of endosperm before fertilisation

77 Which one of the following is a primitive gymnosperm that produces motile sperms, bears ovulate and microsporangiate cones on different plants and has foul-smelling, fleshy seeds?

(a) *Pinus* (b) *Cycas*
(c) *Ginkgo* (d) *Gnetum*

78 Which is the source of turpentine oil?

(a) Gymnospermic wood
(b) Angiospermic wood
(c) Gymnospermic seed
(d) Angiospermic seed

79 Consider the following statements.

I. In gymnosperms, the vascular system consist of xylem without vessels and phloem without companion cells out.

II. Vascular bundle of *Pinus* are conjoint, collateral and open.

Choose the correct option.

(a) Statement I is true, but II is false
(b) Statement I is false, but II is true
(c) Both statements I and II are true
(d) Both statements I and II are false

Angiosperms

80 All of these are the characters of angiosperms except they

(a) are most advanced
(b) bear true fruit
(c) bear covered seeds
(d) are cryptogams

81 In angiosperms, the

(a) gametophyte is prominent and the sporophyte is dependent upon the gametophyte
(b) sporophyte is prominent with the sporophyte and gametophyte living independently
(c) sporophyte is prominent and the gametophyte is highly reduced
(d) gametophyte is prominent and the sporophyte stage has disappeared

82 Consider the following statements.

I. Monocots exhibits prominent secondary growth.

II. The xylem of angiosperms consists of tracheids only.

Choose the correct option.

(a) Statement I is true, but II is false
(b) Statement I is false, but II is true
(c) Both statements I and II are true
(d) Both statements I and II are false

(B) Cumulative Exercise

(Assorted Questions of the Chapter for Advanced Level Practice)

1 Which of the following is not a difference between *Cycas* and *Pinus*?

(a) In *Cycas*, entomophilly is seen while in *Pinus*, anemophilly is seen
(b) *Cycas* possesses algal associations, while *Pinus* is symbiotically associated with some fungi
(c) Wood in *Cycas* is manoxylic while in *Pinus*, it is pycnoxylic
(d) Microspore of *Cycas* shed at 3-cell stage while that of *Pinus* at 4-cell stage

2 Which one of the following is the major difference between mosses and ferns?

(a) Ferns lack alternation of generation, while mosses show the same
(b) Mosses are facultative aerobes, while ferns are obligate aerobes
(c) Vascular bundles of ferns show xylem vessels, while those of mosses lack it
(d) Mosses require water for fertilisation, while ferns do not require water

3 The diagrams given below represent the structures of *Cycas*. Identify *A-D* from the codes given.

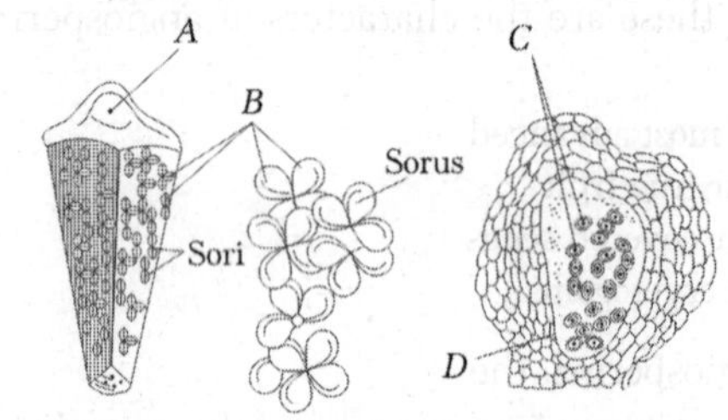

	A	*B*	*C*	*D*
(a)	Microspores	Apophysis	Microsporangia	Pectic wall
(b)	Apophysis	Microsporangia	Microspores	Tapetum
(c)	Microsporangia	Strobili	Microspores	Cellulose wall
(d)	Microsporophyll	Annulus	Megaspores	Exine

4 The given figure represents transverse section of sorus of a fern. Identify the parts labelled as *A-D* by choosing correct alternative from the codes.

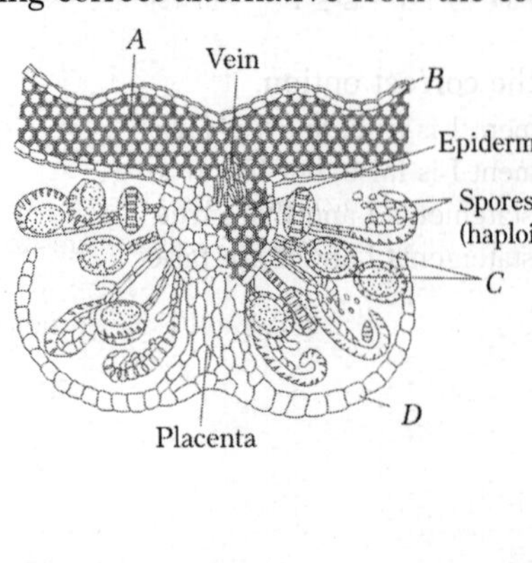

	A	*B*	*C*	*D*
(a)	Parenchymatous layer	Cuticle	Megasporangia	Calyptra
(b)	Mesophyll	Upper epidermis	Microsporangia	Endothecium
(c)	Chlorenchymatous cells	Upper epidermis	Sporangia	Perithecium
(d)	Mesophyll	Upper epidermis	Sporangia (stalked)	Indusium

5 Consider the following statements regarding gymnosperms.

I. In gymnosperms, the male and female gametophytes have an independent existence.
II. The multicellular female gametophyte is retained within the megasporangium.
III. The gymnosperms are heterosporous.

Choose the correct option.

(a) I and III are true, but II is false
(b) II and III are false, but I is true
(c) I and III are false, but II is true
(d) II and III are true, but I is false

6 Match the following columns.

Column I	Column II
A. Largest ovules	1. *Cannabis sativa*
B. *Cuscuta*	2. *Papaver sominiferum*
C. *Drosera*	3. *Cycas*
D. Morphine	4. Carnivorous plant
E. Opium	5. Parasitic plant

Codes

	A	B	C	D	E		A	B	C	D	E
(a)	3	5	4	1	2	(b)	1	3	2	4	5
(c)	3	2	1	5	4	(d)	4	2	1	3	5

7 Go through the figure given below and identify the parts labelled as *A*, *B*, *C* and *D* by choosing the correct alternative.

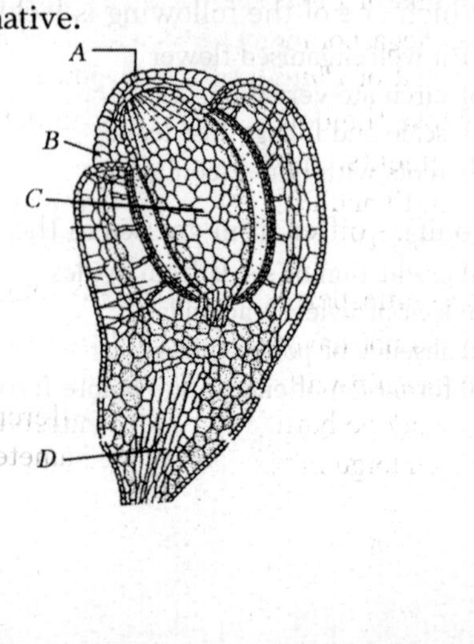

	A	*B*	*C*	*D*
(a)	Operculum	Annulus	Columella	Conducting strand
(b)	Conducting tissue	Operculum	Annulus	Columella
(c)	Chlorenchyma	Operculum	Columella	Conducting tissue
(d)	Columella	Apophysis	Vaginula	Operculum

8 Match the following columns.

	Column I		Column II
A.	Cycadales	1.	Vessels present
B.	Cycadofilicales	2.	*Zamia*
C.	Ginkgoales	3.	Extinct genera
D.	Gnetales	4.	Single living species

Codes

	A	B	C	D
(a)	2	3	4	1
(b)	3	2	1	4
(c)	4	2	3	1
(d)	1	2	3	4

9 Read the following statements regarding male and female structures of *Cycas*.

I. Microsporophyll of *Cycas* is flat and wedge-shaped.
II. Microspores of *Cycas* shed at 4 cell (2 vegetative cells and 2 prothallial cells) stage.
III. Female plants of *Cycas revoluta* are raised by means of bulbils.
IV. The megasporangia of *Cycas* are large naked and orthotropous.

Choose the option containing incorrect statements.

(a) II and III (b) Only IV
(c) Only II (d) Only III

10 Following statements represent comparative account of *Pinus* and *Cycas*.

I. The whole plant body of both the plants is sporophytic (2*n*).
II. Pollination in both the plants is carried out by the agency of insects.
III. Microspores of *Cycas* shed at 4-cell stage whereas that of *Pinus* shed at 3-cell stage.
IV. Megasporangium of *Cycas* is orthotropous whereas that of *Pinus* is circinotropous.

Choose the option having combination of all correct statements.

(a) II, III and IV (b) II and III
(c) Only IV (d) Only I

11 In gymnosperms, male and female gametophytes

I. do not have independent free-living existence.
II. remain within the sporangia.
III. may be borned on same or different tree.
IV. undergo mitosis to produce gametes.

Choose the correct option from the codes given below.

(a) I and II (b) II and III
(c) III and IV (d) I, II, III and IV

12 Which of the following statements is/are true?

I. Trimerous condition of floral whorl is characteristic of dicotyledons.
II. *Adiantum* is also called walking fern.
III. In gymnosperms, the vascular system consists of xylem without vessels and phloem without companion cells.
IV. *Riccia* and *Marchantia* are liverworts.

Choose the correct option from the codes given below.

(a) II and IV (b) II, III and IV
(c) I and III (d) Only IV

13 In gymnosperms, the pollen chamber represents

(a) a cell in the pollen grain, in which the sperms are formed
(b) a cavity in the ovule, in which pollen grains are stored after pollination
(c) an opening in the megagametophyte through, which the pollen tube approaches an egg
(d) the microsporangium, in which pollen grains develop

14 Study the figure given below and identify them by choosing correct option.

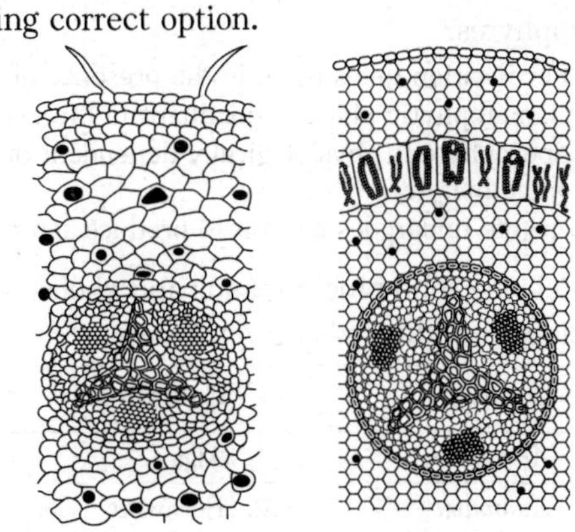

	A	B
(a)	Stem of *Dryopteris*	Normal root of an angiosperm
(b)	Normal root of *Cycas*	Coralloid root of *Cycas*
(c)	Adventitious root of *Cycas*	Dicot stem infected by algae
(d)	Coralloid root of *Cycas*	Adventitious root of *Pinus*

15 In *Cycas*,

(a) microsporangia and ovule are present in same sporophyll
(b) micro and megasporophylls are present in same cone
(c) microspores and megaspores are present on separate strobili
(d) male cone and megasporophylls are borne on the same plant

16 The given figure represents the plant body of *Funaria*. Identify *A-D*.

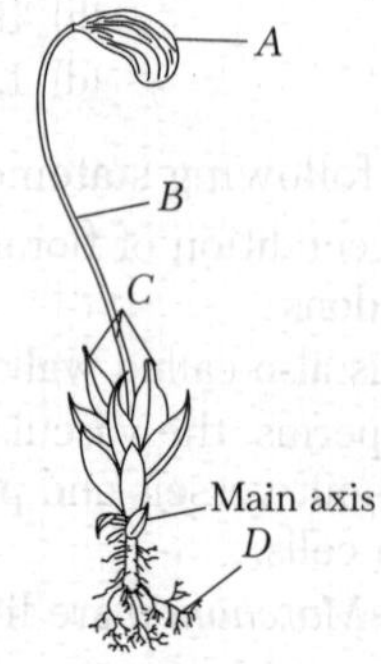

Codes

	A	B	C	D
(a)	Capsule	Seta	Leaves	Rhizoids
(b)	Gametangium	Neck	Ligule	Roots
(c)	Capsule	Stem	Leaf bases	Rhizods
(d)	Indusium	Seta	Leaf bud	Corralloid roots

17 In which of the following features, *Cycas* resembles with angiosperms?
(a) Presence of vessels
(b) Circinate vernation
(c) Dichotomously branched leaves
(d) Pollen tube is the carrier of male gametes

18 Which of the following statements is incorrect about bryophytes?
(a) Fertilisation takes place in the presence of water
(b) Gametophytic phase is dominant in life cycle
(c) Sporophyte is physiologically dependent on gametophyte
(d) Zygote undergoes meiosis to produce sporophyte

19 Match the following columns and choose the correct option from the codes given below.

Column I	Column II
A *Anthoceros*	1. Alga
B *Adiantum*	2. Hornwort
C *Sargassum*	3. Gametophyte
D Prothallus	4. Inferae
E Asterales	5. Walking fern

Code

	A	B	C	D	E
(a)	2	5	1	3	4
(b)	5	4	3	2	1
(c)	5	1	2	4	3
(d)	3	2	1	5	4

20 Study the figure given below depicting a prothallus of fern and identify structures labelled as *A-D*.

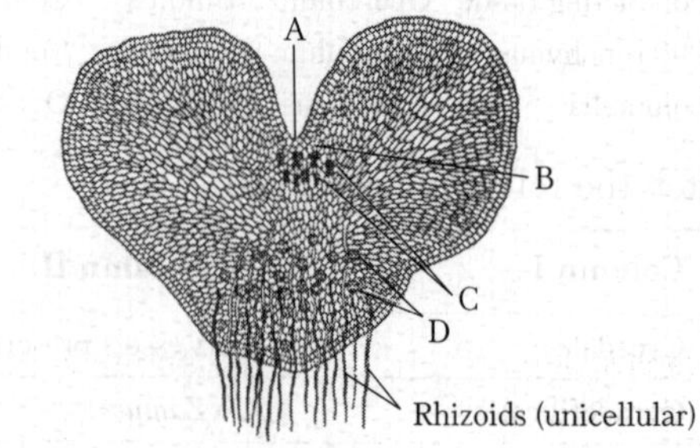

	A	B	C	D
(a)	Apical notch	Cushion	Archegonia	Antheridia
(b)	Cushion	Apical notch	Antheridia	Archegonia
(c)	Basal body	Apical notch	Antheridia	Sporophyte
(d)	Cushion	Basal body	Capsule	Gemmule

21 Select one of the following pairs of important features distinguishing *Gnetum* from *Cycas* and *Pinus* and showing affinities with angiosperms
(a) the absence of resin duct and leaf venation
(b) the presence of vessel elements and the absence of archegonia
(c) perianth and two integuments
(d) embryo development and apical meristem

22 The characteristic features of bryophyte are
I. dominant gametophytic plant body.
II. dominant sporophytic plant body.
III. requirement of water for fertilisation.
IV. lack of sterile jacket around the sex organs.
Choose the correct option from the codes given below.
(a) Only I (b) I, II and III
(c) I and III (d) II and IV

23 The type of stele shown below in the diagram is

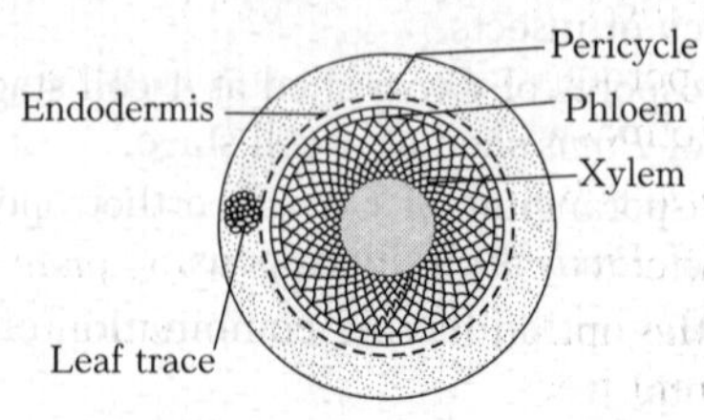

(a) Haplostele (b) Actinostele
(c) Dictyostele (d) Plectostele

24 The type of stele shown in the given diagram is

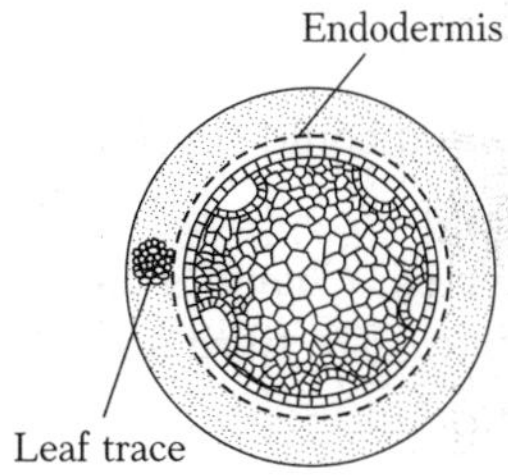

(a) Haplostele (b) Actinostele
(c) Plectostele (d) Siphonostele

25 Mosses are ecologically important because

I. first organism to colonise bare rocks.
II. decompose rocks making suitable for growth of higher plant.
III. form dense mats on soil.
IV. reduce impact of falling rain and soil erosion.

Choose the correct option from the codes given below.

(a) Only I (b) Only II
(c) III and II (d) I, II, III and IV

26 Which of the following are correct about pteridophytes?

I. The sporophyte bears sporangia formed on sporophylls.
II. They are frequently grown as ornamentals.
III. They are the first terrestrial plants to forms seeds.
IV. Vascular tissues are present in pteriodophytes.
V. They do not form embryo.

Choose the correct option from the codes given below.

(a) IV and V
(b) I, II and IV
(c) III, IV and V
(d) I and II

27 Consider the following statements.

I. The sporophyte in liverworts is more elaborate than that in mosses.
II. *Funaria* is heterosporous.
III. The life cycle in all seeds bearing plants is diplontic.
IV. In *Pinus*, male and female cones are borne on different trees.

Choose the option containing incorrect statement(s).

(a) I and III (b) I and IV
(c) II and III (d) Only I

28 Read the following statements (I-V) and answer the question which follows them.

I. In liverworts, mosses and ferns gametophytes are free-living.
II. Gymnosperms and some ferns are heterosporous
III. Sexual reproduction in *Fucus*, *Volvox* and *Albugo* is oogamous.
IV. The sporophyte in liverworts is more elaborate than that in mosses.
V. Both *Pinus* and *Marchantia* are dioecious.

How many above statements are correct?

(a) One (b) Two
(c) Three (d) Four

29 *Selaginella* a pteridophyte shows some advances towards the seed habit. Consider the following statements.

I. Develoment and retention of embryo inside megasporangium.
II. Homospory
III. Formation of several megaspores within a megasporangium.
IV. Enhancement in the size of male gametophyte.

Choose the incorrect statements with regard to pre-requisites for seed habit.

(a) I and III (b) I and IV
(c) II, III and IV (d) None of these

30 Find out the mismatched pair.

(a) Agar	—	Polymer of glucose and sulphur containing carbohydrates
(b) Chitin	—	Polymer of glucosamine
(c) Peptidoglycan	—	Polysaccharide linked to peptides
(d) Lipopolysaccharides	—	A complex of lipid and polysaccharide
(e) Glycogen	—	Polymer of glucose

31 The correct statements about bryophytes are

I. the sperms are biflagellate.
II. the female sex organ is archegonium.
III. the bryophytes are heterosporous.
IV. the body is divided into true roots, stems and leaves.

Choose the correct option from the codes given below.

(a) I and II (b) I and III
(c) II and III (d) I, II and III

Assertion and Reason

Direction (Q. Nos. 32-40) *In each of the following questions, a statement of Assertion (A) is given followed by the corresponding statement of Reason (R). Of the statements, mark the correct answer as*

(a) Both A and R are true and R is the correct explanation of A
(b) Both A and R are true, but R is not the correct explanation of A
(c) A is true, but R is false
(d) A is false, but R is true
(e) Both A and R are false

32 **Assertion** (A) Bryophytes are the amphibians of plant kingdom.

Reason (R) They are found in swamps and the areas, where water and land meet.

33 **Assertion** (A) The term 'Bryophyta' was used first time by Braun (1864).

Reason (R) In bryophytes, sex organs are non-stalked and unicellular.

34 **Assertion** (A) Bryophytes are also known as vascular cryptogams.

Reason (R) Pteridophytes show heterospory.

35 **Assertion** (A) Mosses have evolved from algae.

Reason (R) Protonema of mosses is similar to some green algae.

36 **Assertion** (A) Endosperm is haploid in gymnosperm.

Reason (R) In *Cycas*, stem is usually branched.

37 **Assertion** (A) The seeds of gymnosperms are naked.

Reason (R) Seed consists of three generations, one within the other.

38 **Assertion** (A) *Chlorella* could serve as a potential source of food and energy.

Reason (R) When dried, *Chlorella* has 15% protein, 45% fat, 10% carbohydrate, 20% fibre and 10% minerals and vitamins.

39 **Assertion** (A) In *Spirogyra,* some cells in one of the two filaments become empty after conjugation.

Reason (R) The aplanogametes from the cells of one filament pass through the conjugation tubes into the cells of the other filament.

40 **Assertion** (A) Both apospory and apogamy are observed in *Pteris.*

Reason (R) They are evident by the production of haploid sporophyte by apospory and diploid gametophyte by apogamy.

(C) Medical Entrances' Gallery

(Collection of questions Asked in NEET & Various Medical Entrance Exams)

1 Which of the following statements is incorrect about gymnosperms? **[NEET 2020]**

(a) They are heterosporous
(b) Male and female gametophytes are free-living
(c) Most of them have narrow leaves with thick cuticle
(d) Their seeds are not covered

2 The plant part which consists of two generations, one within the other. **[NEET 2020]**

I. Pollen grains inside the anther.
II. Germinated pollen grain with two male gametes.
III. Seed inside the fruit.
IV. Embryo sac inside the ovule.

Choose the correct option.

(a) I, II and III (b) III and IV
(c) I and IV (d) Only I

3 Which of the following pairs is of unicellular algae? **[NEET 2020]**

(a) *Gelidium* and *Gracilaria*
(b) *Anabaena* and *Volvox*
(c) *Chlorella* and *Spirulina*
(d) *Laminaria* and *Sargassum*

4 Strobili or cones are found in **[NEET 2020]**

(a) *Pteris* (b) *Marchantia*
(c) *Equisetum* (d) *Salvinia*

5 Male and female gametophytes do not have an independent free-living existence in **[NEET 2020]**

(a) pteridophytes (b) algae
(c) angiosperms (d) bryophytes

6 Which one is the oldest in following? **[JIPMER 2019]**

(a) Gymnosperms (b) Angiosperms
(c) Bryophytes (d) None of these

7 From evolutionary point of view, retention of the female gametophyte with developing young embryo on the parent sporophyte for some time, is first observed in **[NEET 2019]**

(a) mosses (b) pteridophytes
(c) gymnosperms (d) liverworts

8 **Assertion** (A) *Selaginella* and *Salvinia* are homosporous.

Reason (R) In pteridophytes, *Lycopodium* is the precusor of seed habit. **[AIIMS 2019, Kerala CEE 2014]**

(a) Both A and R are true and R is the correct explanation of A
(b) Both A and R are true, but R is not the correct explanation of A
(c) A is true, but R is false
(d) Both A and R are false

9 **Assertion** (A) Heterospory and retention of female gametophyte are responsible for origin of seed habit in *Selaginella.* **[AIIMS 2019]**

Reaseon (R) *Psilotum* is a living fossil.

(a) Both A and R are true and R is the correct explantion of A
(b) Both A and R are true, but R is not the correct explanation of A
(c) A is true, but R is false
(d) Both A and R are false

10 Which of the following are homosporous? **[AIIMS 2019]**

(a) *Salvinia, Equisetum*
(b) *Salvinia, Lycopodium*
(c) *Selaginella, Salvinia*
(d) *Lycopodium, Equisetum*

11 Phloem in gymnosperms lacks **[NEET 2019]**

(a) sieve tubes only
(b) companion cells only
(c) Both sieve tubes and companion cells
(d) albuminous cells and sieve cells

12 Zooidogamy is found in **[JIPMER 2019]**

(a) *Cedrus* (b) *Pinus*
(c) *Cycas* (d) Both (b) and (c)

13 Which is the most common type of embryo sac in angiosperms? **[NEET 2019]**

(a) Tetrasporic with one mitotic stage of divisions
(b) Monosporic with three sequential mitotic divisions
(c) Monosporic with two sequential mitotic divisions
(d) Bisporic with two sequential mitotic divisions

14 Which one is incorrectly matched? **[NEET 2018]**

(a) Gemma cups - *Marchantia*
(b) Biflagellate zoospores - Brown algae
(c) Uniflagellate gametes - *Polysiphonia*
(d) Unicellular organism - *Chlorella*

15 Shape of chloroplast of *Ulothrix* is **[JIPMER 2018]**

(a) star-shaped (b) band-shaped
(c) girdle-shaped (d) spinal-shaped

16 Which one is a parasitic alga? **[JIPMER 2018]**

(a) *Oedogonium* (b) *Cephaleuros*
(c) *Spirogyra* (d) *Cladophora*

17 Palmella stage is present in **[JIPMER 2018]**

(a) *Aspergillus* (b) *Cystopus*
(c) *Chlamydomonas* (d) None of these

18 Heterotrichous thallus is shown by which organism? **[JIPMER 2018]**

(a) *Chlamydomonas* (b) *Ectocarpus*
(c) *Spirogyra* (d) *Volvox*

19 Which one of the following statements is correct? **[NEET 2018]**

(a) Horsetails are gymnosperms
(b) *Selaginella* is heterosporous, while *Salvinia* is homosporous
(c) Ovules are not enclosed by ovary wall in gymnosperms
(d) Stems are usually unbranched in both *Cycas* and *Cedrus*

20 An example of colonial alga is **[NEET 2017]**

(a) *Chlorella* (b) *Volvox*
(c) *Ulothrix* (d) *Spirogyra*

21 In the given diagrams, some of the algae have been labelled as *A*, *B*, *C*, *D* and *E*. Choose the correct option to identify these algae. **[AIIMS 2017]**

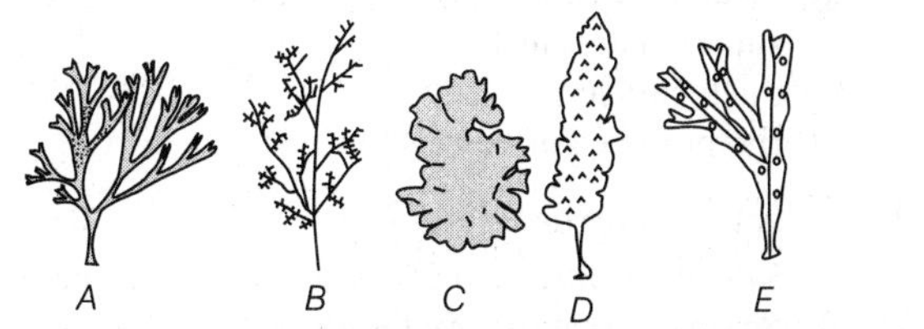

	A	B	C	D	E
(a)	*Dictyota*	*Polysiphonia*	*Porphyra*	*Fucus*	*Laminaria*
(b)	*Porphyra*	*Dictyota*	*Laminaria*	*Fucus*	*Polysiphonia*
(c)	*Dictyota*	*Polysiphonia*	*Porphyra*	*Laminaria*	*Fucus*
(d)	*Fucus*	*Porphyra*	*Dictyota*	*Polysipnonia*	*Laminaria*

22 Identify the incorrect combination. **[JIPMER 2017]**

(a) *Dryopteris* — Rhizome
(b) *Cycas* — Coralloid roots
(c) *Volvox* — Colonial form
(d) *Marchantia* — Pseudoelaters

23 Select the mismatch. **[NEET 2017]**

(a) *Pinus* — Dioecious
(b) *Cycas* — Dioecious
(c) *Salvinia* — Heterosporous
(d) *Equisetum* — Homosporous

24 Zygotic meiosis is characteristic of **[NEET 2017]**

(a) *Marchantia*
(b) *Fucus*
(c) *Funaria*
(d) *Chlamydomonas*

25 Life cycle of *Ectocarpus* and *Fucus* respectively are **[NEET 2017]**

(a) haplontic, diplontic
(b) diplontic, haplodiplontic
(c) haplodiplontic, diplontic
(d) haplodiplontic, haplontic

26 Which one of the following statements is incorrect? **[NEET-II 2016]**

(a) Algae increase the level of dissolved oxygen in the immediate environment
(b) Algin is obtained from red algae and carrageenan from brown algae
(c) Agar-agar is obtained from *Gelidium* and *Gracilaria*
(d) *Laminaria* and *Sargassum* are used as food

27 In bryophytes and pteridophytes, transport of male gametes requires **[NEET 2016]**

(a) insects (b) birds
(c) water (d) wind

28 Conifers are adapted to tolerate extreme environmental conditions because of [NEET-II 2016]
(a) broad hardy leaves
(b) superficial stomata
(c) thick cuticle
(d) the presence of vessels

29 Select the correct statement. [NEET-I 2016]
(a) *Salvinia*, *Ginkgo* and *Pinus*, all are gymnosperms
(b) *Sequoia* is one of the tallest trees
(c) The leaves of gymnosperms are not well-adapted to extremes of climate
(d) Gymnosperms are both homosporous and heterosporous

30 Read the following five statements (I toV) and select the option with all correct statements.
I. Mosses and lichens are the first organisms to colonise a bare rock.
II. *Selaginella* is a homosporous pteridophyte.
III. Coralloid roots in *Cycas* have VAM.
IV. Main plant body in bryophytes is gametophytic, whereas in pteridophytes, it is sporophytic.
V. In gymnosperms, male and female gametophytes are present within sporangia located on sporophyte. [CBSE AIPMT 2015]

Codes
(a) I, III and IV (b) II, III and IV
(c) I, IV and V (d) II, III and V

31 In which of the following, gametophyte is not independent free-living? [CBSE AIPMT 2015]
(a) *Funaria* (b) *Marchantia*
(c) *Pteris* (d) *Pinus*

32 Which one of the following statements is incorrect? [CBSE AIPMT 2015]
(a) Algin and carrageenan are products of algae
(b) Agar-agar is obtained from *Gelidium* and *Gracilaria*
(c) *Chlorella* and *Spirulina* are used as space food
(d) Mannitol is stored food in Rhodophyceae

33 Male gametes are flagellated in [CBSE AIPMT 2015]
(a) *Polysiphonia* (b) *Anabaena*
(c) *Ectocarpus* (d) *Spirogyra*

34 'Botanical snakes' are [CBSE AIPMT 2014]
(a) algae (b) fungi
(c) bryophytes (d) pteridophytes

35 Which one of the following is the first group of vascular plants? [MHT CET 2014]
(a) Thallophyta (b) Bryophyta
(c) Pteridophyta (d) Spermatophyta

36 Male gametophyte with least number of cells is present in [CBSE AIPMT 2014]
(a) *Pteris* (b) *Funaria*
(c) *Lilium* (d) *Pinus*

37 Which of the following is responsible for peat formation? [CBSE AIPMT 2014]
(a) *Marchantia* (b) *Riccia* (c) *Funaria* (d) *Sphagnum*

38 Choose the correct statement. [Kerala CEE 2014]
(a) Bryophytes can live in soil, but are dependent on water for sexual reproduction
(b) The sex organs in bryophytes are unicellular
(c) In bryophytes, the main plant body is a gametophyte which is differentiated into true root, stem and leaves
(d) Common example of liverwort is *Polytrichum*
(e) None of the above

39 *Marchantia* is considered as a heterothallic plant because it is [Karnataka CET 2014]
(a) heterogametic (b) bisexual
(c) monoecious (d) dioecious

40 Mosses and liverworts are members of [Karnataka CET 2014]
(a) gametophytes (b) chlorophytes
(c) bryophytes (d) pteridophytes

41 Choose the incorrect statement. [Kerala CEE 2014]
(a) Gymnosperms lack vessels in their xylem
(b) The cell wall of collenchyma is made up of cellulose, hemicellulose and pectin
(c) The first formed primary xylem elements are called protoxylem
(d) The cell wall of parenchyma is made up of pectin
(e) Gymnosperms have albuminous cells and sieve cells in their phloem

42 *Ectocarpus* shows [UK PMT 2014]
(a) haplontic life cycle
(b) diplontic life cycle
(c) haplodiplontic life cycle
(d) diplontic-haplontic life cycle

43 The life cycle of algae such as *Spirogyra* is [MHT CET 2014]
(a) haplontic (b) diplontic
(c) haplo-diplontic (d) diplo-haplontic

44 An alga which can be employed as food for human being is [CBSE AIPMT 2014]
(a) *Ulothrix* (b) *Chlorella*
(c) *Spirogyra* (d) *Polysiphonia*

45 Which one of the following shows isogamy with non-flagellated gametes? [CBSE AIPMT 2014]
(a) *Sargassum* (b) *Ectocarpus*
(c) *Ulothrix* (d) *Spirogyra*

46 Which one of the following is incorrect about *Chara*? [CBSE AIPMT 2014]
(a) Upper oogonium and lower round antheridium
(b) Globule and nucule present on the same plant
(c) Upper antheridium and lower oogonium
(d) Globule is male reproductive structure

47 Which of the following groups of algae belongs to class-Rhodophyceae? **[Kerala CEE 2014]**

(a) *Laminaria, Fucus, Porphyra, Volvox*
(b) *Gelidium, Porphyra, Dictyota, Fucus*
(c) *Gracilaria, Gelidium, Porphyra, Polysiphonia*
(d) *Volvox, Spirogyra, Ulothrix, Sargassum*
(e) *Sargassum, Laminaria, Fucus, Dictyota*

48 Match the following columns and choose the correct option from the codes given below.

	Column I (Alga types)		Column II (Examples)
A.	Green algae	1.	*Dictyota*
B.	Brown algae	2.	*Porphyra*
C.	Red algae	3.	*Spirogyra*

Codes **[Kerala CEE 2014]**

	A	B	C		A	B	C
(a)	3	2	1	(b)	3	1	2
(c)	2	3	1	(d)	1	2	3
(e)	1	3	2				

49 The presence of pyrenoid is characteristic feature of class **[UK PMT 2014]**

(a) Phaeophyceae (b) Chlorophyceae
(c) Rhodophyceae (d) Poaceae

50 Food is stored in the form of mannitol in the class of algae **[UK PMT 2014]**

(a) Rhodophyceae
(b) Phaeophyceae
(c) Chlorophyceae
(d) Poaceae

51 Select the incorrect statement. **[NEET 2013]**

(a) Isogametes are similar in structure, function and behaviour
(b) Anisogametes differ either in structure, function and behaviour
(c) In oogamous, female gamete is smaller and motile, while male gamete is larger and non-motile
(d) *Chlamydomonas* exhibits both isogamy and anisogamy and *Fucus* shows oogamy

52 Isogamous condition with non-flagellated gametes is found in **[NEET 2013]**

(a) *Chlamydomonas* (b) *Spirogyra*
(c) *Volvox* (d) *Fucus*

53 Monoecious plant of *Chara* shows occurrence of **[NEET 2013]**

(a) antheridiophore and archegoniphore on the same plant
(b) stamen and carpel on the same plant
(c) upper antheridium and lower oogonium on the same plant
(d) upper oogonium and lower antheridium on the same plant

54 Alginic acid is found in the cell wall of **[UP CPMT 2013]**

(a) *Gigartina* (b) *Laminaria*
(c) *Gelidium* (d) *Scytonema*

55 Algae, which form motile colony, is **[Manipal 2013]**

(a) *Volvox* (b) *Nostoc*
(c) *Spirogyra* (d) *Chlamydomonas*

56 Non-motile, greatly thickened, asexual spore in *Chlamydomonas* is **[EAMCET 2013]**

(a) carpospores (b) aplanospores
(c) akinetes (d) hypnospores

57 Which one of the following is an example of chlorophyllous thallophyte? **[Kanataka CET 2013]**

(a) *Volvariella* (b) *Spirogyra*
(c) *Nephrolepis* (d) *Gnetum*

58 Elaters are absent in **[AIIMS 2013]**

(a) *Funaria* (b) *Marchantia*
(c) *Pellia* (d) *Porella*

59 Vegetative reproduction in *Funaria* takes place by **[UP CPMT 2013]**

(a) primary protonema (b) gemmule
(c) secondary protonema (d) All of these

60 *Pinus* belongs to the class **[Karnataka CET 2013]**

(a) Gnetopsida (b) Cycadopsida
(c) Coniferopsida (d) Sphenopsida

61 Which one of the following is common to multicellular fungi, filamentous algae and protonema of mosses? **[CBSE AIPMT 2012]**

(a) Diplontic life cycle
(b) Members of kingdom-Plantae
(c) Mode of nutrition
(d) Multiplication by fragmentation

62 In bryophytes, the sporophytic phase is represented by **[UP CPMT 2012]**

(a) spores (b) antheridium
(c) spore mother cell (d) egg

63 Gymnosperms are also called soft wood spermatophytes because they lack **[CBSE AIPMT 2012]**

(a) cambium (b) phloem fibres
(c) thick-walled tracheids (d) xylem fibres

64 *Cycas* and *Adiantum* resemble each other in having **[CBSE AIPMT 2012]**

(a) seeds (b) motile sperms
(c) cambium (d) vessels

65 In *Pinus*, male cone bears a large number of **[UP CPMT 2012]**

(a) ligules (b) anthers
(c) microsporophylls (d) megasporophylls

66 Which one of the following is a correct statement? **[CBSE AIPMT 2012]**
(a) Pteridophytic gametophyte has a protonemal and leafy stage
(b) In gymnosperms, female gametophyte is free-living
(c) Antheridiophores and archegoniophores are present in pteridophytes
(d) Origin of seed habit can be traced in pteridophytes

67 The plants commonly called vascular cryptogams are **[UP CPMT 2012]**
(a) bryophytes (b) pteridophytes
(c) algae (d) angiosperms

68 The plants having vascular tissue, but lacking seeds are placed under **[Manipal 2012]**
(a) algae (b) bryophytes
(c) pteridophytes (d) gymnosperms

69 Pteridophytes are called vascular cryptogams, because they are non-seeded plants containing **[Karnataka CET 2012]**
(a) only phloem (b) only xylem
(c) Both (a) and (b) (d) None of these

70 The cell wall of algae is chemically composed of **[BHU 2012]**
(a) hemicellulose, pectins and proteins
(b) cellulose, galactans and mannans
(c) pectin, cellulose and proteins
(d) chitin

71 Which one of the following is algal parasite? **[UP CPMT 2012]**
(a) *Volvox* (b) *Ulothrix*
(c) *Porphyra* (d) *Cephaleuros*

72 Bryophytes resemble algae in one of the following aspect **[BHU 2012]**
(a) filamentous body, presence of vascular tissue and autotrophic nutrition
(b) differentiation of plant body into root, stem and leaves
(c) thallus-like plant body, presence of root and autotrophic nutrition
(d) thallus-like plant body, lack of vascular tissue and autotrophic nutrition

73 Which of the following group of plants are generally called amphibians of plant kingdom? **[BHU 2012]**
(a) Algae (b) Gymnosperms
(c) Bryophytes (d) Pteridophytes

74 Spore of *Funaria* on germination gives rise to **[Manipal 2012]**
(a) protonema (b) embryo
(c) antheridia (d) archegonia

75 Which of the following is true about bryophytes? **[BCECE 2012]**
(a) They are thalloid (b) They contain chloroplast
(c) They possess archegonia (d) All of these

76 Pyrenoids are made up of **[UP CPMT 2011]**
(a) core of starch surrounded by sheath of protein
(b) core of protein surrounded by fatty sheath
(c) proteinaceous centre and starchy sheath
(d) core of nucleic acid surrounded by protein sheath

77 Iodine is found in algae **[Odisha JEE 2011]**
(a) *Ulva* (b) *Ulothrix*
(c) *Chlorella* (d) *Laminaria*

78 Floridean starch is reserve food in **[Odisha JEE 2011]**
(a) Rhodophyceae (b) Phaeophyceae
(c) Chlorophyceae (d) Xanthophyceae

79 Select the correctly matched ones. **[Kerala CEE 2011]**

I. Phaeophyceae	—	Mannitol
II. Rhodophyceae	—	*Dictyota*
III. Chlorophyceae	—	Non-motile gametes
IV. Rhodophyceae	—	*r*-phycoerythrin

Choose the correct option form the codes given below.
(a) II, III and IV (b) I and III
(c) III and IV (d) I and IV
(e) I, II and III

80 Archegoniophore is present in **[CBSE AIPMT 2011]**
(a) *Chara* (b) *Adiantum*
(c) *Funaria* (d) *Marchantia*

81 Moss spore germinates to form **[Odisha JEE 2011]**
(a) sporophyte (b) protonema
(c) seta (d) capsule

82 A prokaryotic autotrophic nitrogen-fixing symbiont is found in **[CBSE AIPMT 2011]**
(a) *Cycas* (b) *Cicer* (c) *Pisum* (d) *Alnus*

83 The type of pollination in *Cycas* is **[UP CPMT 2011]**
(a) entomophily (b) hydrophily
(c) anemophily (d) malacophily

84 Which of the following is correct for *Cycas* reproduction? **[UP CPMT 2011]**
(a) Zooidogamy is followed by siphonogamy
(b) Siphonogamy is followed by zooidogamy
(c) Siphonogamy
(d) Zooidogamy

85 Both heterospory and circinate ptyxis occur in **[UP CPMT 2011]**
(a) *Dryopteris* (b) *Pinus* (c) *Cycas* (d) *Funaria*

86 Gametophyte is the dominant phase in the life cycle of **[Karnataka CET 2011]**
(a) *Hibiscus* (b) *Nephrolepis*
(c) *Cycas* (d) *Riccia*

87 Gametophyte is dominant stage in the life cycle of **[UP CPMT 2011]**
(a) bryophytes (b) pteridophytes
(c) angiosperms (d) gymnosperms

88 Seed habit is linked to **[Odisha JEE 2011]**
(a) homospory (b) heterospory
(c) parthenogenesis (d) parthenocarpy

89 Which of the following pteridophytes belong to class- Pteropsida? **[Kerala CEE 2011]**
(a) *Equisetum* and *Psilotum*
(b) *Lycopodium* and *Adiantum*
(c) *Selaginella* and *Pteris*
(d) *Pteris* and *Adiantum*
(e) *Dryopteris* and *Psilotum*

ANSWERS

Check Point 7.1

1. (d)	2. (d)	3. (b)	4. (b)	5. (d)

Check Point 7.2

1. (d)	2. (a)	3. (b)	4. (c)	5. (d)	6. (c)

Check Point 7.3

1. (a)	2. (c)	3. (a)	4. (d)	5. (a)

Check Point 7.4

1. (d)	2. (d)	3. (c)	4. (b)	5. (a)	6. (d)

Check Point 7.5

1. (a)	2. (c)	3. (b)	4. (b)	5. (d)

(A) Topical Exercise

1. (d)	2. (d)	3. (b)	4. (b)	5. (b)	6. (a)	7. (d)	8. (c)	9. (a)	10. (b)
11. (d)	12. (c)	13. (d)	14. (a)	15. (b)	16. (b)	17. (a)	18. (c)	19. (c)	20. (a)
21. (c)	22. (a)	23. (c)	24. (a)	25. (b)	26. (a)	27. (a)	28. (c)	29. (c)	30. (d)
31. (a)	32. (c)	33. (d)	34. (c)	35. (d)	36. (a)	37. (c)	38. (a)	39. (a)	40. (b)
41. (d)	42. (a)	43. (b)	44. (c)	45. (b)	46. (b)	47. (c)	48. (c)	49. (d)	50. (a)
51. (a)	52. (c)	53. (b)	54. (a)	55. (c)	56. (a)	57. (d)	58. (d)	59. (d)	60. (a)
61. (b)	62. (d)	63. (d)	64. (a)	65. (a)	66. (c)	67. (c)	68. (a)	69. (c)	70. (b)
71. (d)	72. (a)	73. (b)	74. (c)	75. (a)	76. (d)	77. (c)	78. (a)	79. (c)	80. (d)
81. (c)	82. (d)								

(B) Cumulative Exercise

1. (a)	2. (c)	3. (b)	4. (d)	5. (d)	6. (a)	7. (a)	8. (a)	9. (c)	10. (d)
11. (d)	12. (b)	13. (b)	14. (b)	15. (c)	16. (a)	17. (d)	18. (d)	19. (a)	20. (a)
21. (b)	22. (c)	23. (a)	24. (c)	25. (d)	26. (b)	27. (b)	28. (c)	29. (c)	30. (a)
31. (a)	32. (a)	33. (c)	34. (d)	35. (a)	36. (c)	37. (b)	38. (c)	39. (a)	40. (c)

(C) Medical Entrances' Gallery

1. (b)	2. (c)	3. (c)	4. (c)	5. (c)	6. (c)	7. (b)	8. (d)	9. (b)	10. (d)
11. (c)	12. (c)	13. (b)	14. (c)	15. (c)	16. (b)	17. (c)	18. (b)	19. (c)	20. (b)
21. (c)	22. (d)	23. (a)	24. (d)	25. (c)	26. (b)	27. (c)	28. (c)	29. (b)	30. (c)
31. (d)	32. (d)	33. (c)	34. (d)	35. (c)	36. (c)	37. (d)	38. (a)	39. (d)	40. (c)
41. (d)	42. (c)	43. (a)	44. (b)	45. (d)	46. (c)	47. (c)	48. (b)	49. (b)	50. (b)
51. (c)	52. (b)	53. (d)	54. (b)	55. (a)	56. (d)	57. (b)	58. (a)	59. (d)	60. (c)
61. (d)	62. (c)	63. (d)	64. (b)	65. (c)	66. (d)	67. (b)	68. (c)	69. (c)	70. (b)
71. (d)	72. (d)	73. (c)	74. (a)	75. (d)	76. (c)	77. (d)	78. (a)	79. (d)	80. (d)
81. (b)	82. (a)	83. (c)	84. (b)	85. (c)	86. (d)	87. (a)	88. (b)	89. (d)	

Explanations

(A) Topical Exercise

2 (*d*) In Bentham and Hooker's system of classification, main emphasis has been given to both embryological and reproductive characters. They proposed the natural system of classification on the basis of floral characters, which is based on embryological and reproductive characters.

Thus, option (d) is correct.

4 (*b*) Statement I is false, but II is true. Incorrect statement can be corrected as

Natural system of classification was based upon both morphology and anatomical relationships and affinities.

6 (*a*) Statement I is true, but II is false. Incorrect statement can be corrected as

Thallophyta are most simple and primitive plants. They do not possess a vascular system.

9 (*a*) Types of pigments present in the cell of algae is the most important character for their classification. On the basis of these pigments, algae are classified as red, brown and green algae.

11 (*d*) The pair in option (d) is not correctly matched and can be corrected as

Volvox is a freshwater green alga. It occurs in colonies or coenobium (in definite number or group), surrounded by a pellicle (gelatinous glycoprotein) layer. Each pyriform-shaped cell has two long similar and smooth flagella, eye spot, cup-shaped chloroplast with pyrenoids and contractile vacuoles.

Rest pairs are correctly matched.

14 (*a*) Option (a) is incorrectly matched and can be corrected as

In *Spirogyra*, gametes are non-motile. They do not possess flagella and known as aplanogametes.

Rest pairs are correctly matched.

16 (*b*) In *Spirogyra affinis*, sexual reproduction occurs through conjugation (indirect lateral) in which adjacent cells of same filament conjugate the protoplast of one cell (male gamete), migrates to the other (female gamete). These protoplasts then fuse to form zygospore which on germination forms, a single new filament. Thus, from two adjacent filaments with 10 cells participating in reproduction, 10 new filaments will be formed.

17 (*a*) The united protoplasmic mass of two gametes is called zygospore (zygote). Prior to germination, the diploid zygospore nucleus undergoes meiosis and forms four nuclei. Three of these disintegrate and only one is functional. It undergoes transverse division to give rise to a single filament.

18 (*c*) The cell wall of *Spirogyra* is 2-layered with outer mucilaginous layer. It consists of an inner layer of cellulose and an outer layer of pectin, which is responsible for the slippery texture of the *Spirogyra*. The outermost lining of pectose turn into pectin and gets dissolved in water to form mucilage. It surrounds the filament and forms mucilaginous sheath.

20 (*a*) Incorrect option is (a) and can be corrected as

Chlorophyll-*b* is absent in brown algae. It possess as the pigment fucoxanthin ($C_{40}H_{54}O_6$) in their chromatophores in addition to chlorophyll-*a*, chlorophyll-*c*, carotene and xanthophylls.

23 (*c*) Statement in option (c) is not correct and can be corrected as

Meiosis produce haploid sporophyte, whereas the body of a bryophyte is a gametophyte. Therefore, a bryophyte undergoes mitosis to produce a diploid sporophyte.

28 (*c*) Gametes are produced in bryophytes by meiosis of sporophytic cells. It is produced from cell division of spores, which are further produced by meiosis in sporophytes. The fusion of male and female gametes produce a diploid zygote, which develops by repeated cell division into a multicellular sporophyte.

31 (*a*) Statement I is true, but II is false. Incorrect statement can be corrected as

The antherozoid penetrates the egg in the archegonium and fertilisation is effected with the result of fertilisation the diploid zygote is formed. The zygote begins to develop into multicellular embryo just after the act of fertilisation, without going under any rest period. Ultimately, the sporogonium develops. The spore mother cell is produced within the sporangium. They are diploid and represent the last stage of the sporophytic generation.

36 (*a*) Statement I is true, but II is false. Incorrect statement can be corrected as

In a capsule of moss, shock absorbers are trabeculae. Trabeculae are equivalent to endodermis and found outside the spore sac.

43 (*b*) Statement in option (b) is incorrect and can be corrected as

Riccia belongs to the category of Bryophyta. *Riccia fluitans* are exclusively aquatic plants. Scales of *Riccia* are multicellular.

Other statements are correct.

47 (*c*) *Sphagnum* plant has capacity to retain ample amount of water for longer duration. This is why, they are used as packing materials, so that the living materials continue to get water during the shipment.

48 (*c*) Statement in option (c) is incorrect and can be corrected as

Sperms of pteridophytes are motile, they swim in water to reach the female sex organ.

50 (*a*) Statement I true, but II is false. Incorrect statement can be corrected as

Adiantum which is a fern belonging to pteriodophytes is named as 'walking fern' because it propagates vegetatively by its leaf tips. In *Adiantum* adventitious buds develop at leaf tips. When such leaf tips touch the soil, they form new plants. The process helps in spreading the fern over a larger area.

59 (*d*) All given options are true for siphonostele.

Siphonostele is an advanced type of stele than protostele. The siphonostele is derived from protostele by the formation of pith in the centre. The centrally placed xytem core in a protostele is replaced by parenchymatous pith. Leaf gaps are absent, *Adiantum* and *Marsilea* have siphonostele.

Thus, option (d) is correct.

61 (*b*) In the ectophloic siphonostele, the xylem surrounds pith and this xylem is surrounded by phloem, pericycle and endoderm, respectively, e.g. *Osmunda* and *Equisetum.*

62 (*d*) A fern (pteridophyte) differs from a moss (bryophyte) in the having independent sporophyte while in moss, the sporophyte is simpler than the gametophyte and remains attached to the parent gametophyte throughout its life. This sporophyte is dependent upon gametophyte partially or wholly for its nutrition.

65 (*a*) The 13-celled microspore of male gametophyte in *Selaginella* is shed from microsporangium, which is having 1 prothallial cell + 8 jacket cells + 4 androgonial cells, i.e. 8 + 4 = 12 antheridial cells.

67 (*c*) Among given options, six features are found in gymnosperms. These include heterosporous, bisporangiate, siphonogamy, roots with mycorrhizae, ramenta and circinate vernation.

71 (*d*) Flagellated male gametes are present in *Riccia*, *Dryopteris* and *Cycas.*

73 (*b*) In gymnosperms, the pollen chamber represents a cavity in the ovule in which pollen grains are stored after pollination. In these, below micropylar beak some of the cells of nucellus of ovule disintegrates to form pollen chamber.

74 (*c*) Male gametes of *Cycas* are large, top-like, spirally twisted with cilia. In the male strobiles (cones), there are microsporophylls that contain microsporangia within the microsporangium spore mother cells undergo meiosis and generate haploid microspores.

76 (*d*) The conifers (gymnosperm) differ from the grasses (angiosperm) in the formation of endosperm (haploid) before fertilisation.

77 (*c*) *Cycas* and *Ginkgo* both are primitive gymnosperms with motile gametes. They both bear cones microsporangia and megasporangia on different plants. *Ginkgo* possesses foul smelling seeds not *Cycas.*

78 (*a*) *Pinus* (gymnosperm) woody stem is used as a source of turpentine and resin. Turpentine is used as solvent in paints, polishes and wax.

80 (*d*) Angiosperms are not cryptogams, instead they are phanerogams which produce flower and seeds. Whereas, cryptogams include Thallophyta , Bryophyta and Pteridophyta.

82 (*d*) Both statements are false. These can be corrected as

Almost all monocots do not possess secondary growth, except a few exceptions. The xylem of angiosperms contain both tracheids and vessels.

(B) Cumulative Exercise

5 (*d*) Statements II and III are true, but I is false. Incorrect statement can be corrected as

In gymnosperms, the male and female gametophytes do not have an independent existence.

9 (*c*) Statement II is incorrect and can be corrected as

Microspores of *Cycas* shed at 3-celled stage. These include prothalial cell, tube cell and generative cell.

12 (*b*) All the statements are correct except I. Incorrect statement can be corrected as

Trimerous condition of floral whorl is a characteristic of monocotyledons. They have three floral parts in each whorl.

18 (*d*) Statement in option (d) is incorrect and can be corrected as

In Bryophyta, the union of two gametes during fertilisation produces a diploid zygote which divides mitotically to form a new sporophyte. Thus, in the higher plants, the sporophyte is the dominant phase in the life cycle, whereas in the more primitive non-vascular plants (bryophytes), the gametophyte remains dominant.

Rest statements are correct.

21 (*b*) Option (a) is important feature distinguishing *Gnetum* from *Cycas* and *Pinus.*

The presence of vessels in the xylem and the absence of archegonia are angiospermic characters. These are also found in *Gnetum,* but differentiates it from *Cycas* and *Pinus.*

27 (*b*) Statements I and IV are incorrect and can be corrected as

The sporophyte is more developed in mosses rather than liverwort. The simplest sporophyte is found in *Riccia*, a liverwort where it is represented by capsule only. The most complex sporophyte is found in *Funaria*, a moss. However, the most advance sporophyte is found in *Anthoceros* (a hornwort) due to the presence of intercalary meristem. *Pinus* (a gymnosperm) is a monoecious plant in which male and female cones are borne on different branches of the same plant.

Rest statements are correct.

30 (*a*) The pair in option (a) is mismatched and can be corrected as

Agar-agar is polymer of D-galactose 3-6 anhydro L-galactose, having sulphate esterification after every tenth galactose unit.

Rest pairs are correctly matched.

31 (*a*) Statements I and II are correct whereas III and IV are incorrect because

Bryophytes are a class of mosses and liverworts. They lack true roots, stems and leaves. The plant body of bryophytes is haploid and they produce gametophytes. The males sex organ is antheridium which produced biflagellate antherozoids. The female sex organ is archegonium which produces an egg. The gamete fuse to produce a zygote that produces sporophytes. Heterospores are not present in bryophytes. They are present in some pteridophytes and gymnosperms. Bryophytes are homosporus.

32 (*a*) Both A and R are true and R is the correct explanation of A.

Bryophytes are the dwellers of transitional habitat between the aquatic (e.g. algae) and terrestrial (e.g. spermatophytes) habitats. It is represented by the swampy areas, where water and land meet.

It is also called as amphibious zone and mosses, liverworts and hornworts collectively called as bryophytes usually grow here. They cannot complete their life cycle without external water, thus they are called as the amphibians of plant kingdom.

33 (*c*) A is true, but R is false because

The word Bryophyta was coined and used for the first time by Braun (1864). In bryophytes, sex organs are stalked and multicellular.

34 (*d*) A is false, but R is true because

Cryptogams include all the plants that reproduce by means of spores and do not produce seeds, e.g. algae, bryophytes and pteridophytes. The pteridophytes have a well-developed conducting system, therefore they are called as vascular cryptogams. On the other hand, bryophytes lack vascular system. Heterospory is observed in pteridophytes.

35 (*a*) Both A and R are true and R is the correct explanation of A.

According to some botanists, mosses originated from algae because protonema of mosses is similar to certain algae.

36 (*c*) A is true, but R is false because

In gymnosperms, endosperm is haploid and develops before fertilisation, while in angiosperms, endosperm is triploid and develops after fertilisation. The stem of *Cycas* is unbranched and in *Zamia*, stem is tuberous.

37 (*b*) Both A and R are true, but R is not the correct explanation of A.

Gymnosperms are commonly known as naked seeded plants, because their ovules (which later become seeds) are not covered by the ovary wall. Seed is a complex structure because it contains cells from three generations *viz.*,

(i) Testa, tegmen and perisperm represent parent sporophyte.

(ii) Endosperm represents female gametophyte.

(iii) Embryo the next sporophytic generation.

38 (*c*) A is true, but R is false because

Chlorella could serve as a potential source of food and energy because when it is dried it has about 45% protein, 20% fat, 20% carbohydrate, 5% fiber and 10% mineral and vitamins which is much higher than the other highly efficient crops nutrient.

39 (*a*) Both A and R are true and R is the correct explanation of A.

Sexual reproduction in *Spirogyra* is known as conjugation. In this process, one cell called gametangium functions as female and receives male gamete of adjacent cell, called aplanogamete, which is simply a non-flagellated mass of protoplasm formed in the aplanogametangium. This transfer of aplanogamete to gametangium leaves behind an empty cell.

40 (*c*) A is true, but R is false because

Both apospory and apogamy are observed in *Pteris*. The development of diploid gametophytes from the vegetative parts of the diploid sporophyte without the intervention of spores is known as apospory. Apogamy is the development of sporophyte directly from the gametophyte without the intervention of sex organs and gametes.

(C) Medical Entrances' Gallery

1 (*b*) Statement in option (b) is incorrect and can be corrected as

In gymnosperms, the male and female gametophytes do not have an independent free-living existence. The male gametophyte remains within the sporangia, retained on the sporophytes and is dependent on sporophyte.

Rest options are correct statements about gymnosperms.

2 (*c*) The plant part which consists of two generations, one within the other are pollen grains inside the anther and embryo sac inside the ovule.

This can be explained as follows

In an ovule or megasporangia (2*n*), there occurs an embryo sac or female gametophyte which consists of an egg cell (*n*). This egg cell or female gamete is the part of next generation which form zygote after syngamy. Likewise an anther (2*n*) is a male reproductive structure of a plant. As the anther mature and dehydrate, it forms pollen grains *via* microsporogenesis. These pollen grains represent the male gametophyte comprising of male gametes (*n*). So in the way, there occur two generations one with the other.

Thus, option (c) is correct,

3 (*c*) *Chlorella* and *Spirulina* are unicellular algae as they are rich in proteins and hence, used as food supplements by space travellers. In the other hand, *Gelidium, Gracilaria, Laminaria* and *Sargassum* are multicellular. *Volvox* is colonial.

4 (*c*) Strobili or cones are found in *Equisetum. Equisetum* species have rhizomes that grow deep below the ground surface. Some species have cone-like structures called strobili that produce and house the plants spores for reproduction.

5 (*c*) Male and female gemetophytes do not have an independent free-living existence in gymnosperms and angiosperms. In them, they remain within the sporangia retained on the sporophytes. The pollen grain is released from the microsporangium and are carried in air currents and come in contact with the opening of the ovules borne on megasporophylls. The pollen tube carrying the male gametes grows towards archegonia in the ovules and discharge their contents near the mouth of the archegonia.

6 (*c*) In terms of evolution, bryophytes are the oldest (most primitive) plant type. These are the simplest and first among land inhabiting plants occupying an intermediates position between algae and Pteridophyta. They are non-vascular terrestrial autotrophic, non-flowering and seedless plants.

7 (*b*) From the evolutionary point of view, pteridophytes were the first to show the retention of female gametophyte with young embryo on the parent sporophyte for some time. They also represent the origin of seed habit during the course of evolution of seed plants.

8 (*d*) Both A and R are false and can be corrected as

Most of the pteridophytes are homoporous meaning they produce similar spores, *Selaginella* and *Salvinia* are exception. They produce two kinds of spores (microspores and megaspores). *Selaginella* plant shows the traits of the seed habit, so it is marked as precursor of seed habit evolution in plants.

9 (*b*) Both A and R are true, but R is not the correct explanation of A.

Heterospory and retention of female gametophyte are responsible for origin of seed habit in *Selaginella* (Pteridophyta). Heterospory is a phenomenon of producing two types of spores (microspores and megaspores) by the same plant. *Psilotum* is a pteridophyte plant with primitive pteridophyte characters, so it is known as living fossil.

10 (*d*) *Lycopodium, Equisetum* are homosporous.These plants produce only one kind of spores.

Salvinia and *Selaginella* are examples of heterosporous plants that produce two kinds of spores, macrospore and microspore.

11 (*c*) Phloem in gymonosperms lacks both sieve tube and companion cells.

In gymnosperms, the phloem has albuminous cells and sieve cells only. On the other hand, the phloem of angiosperms is composed of four elements namely, sieve tube, companion cells, phloem parenchyma and phloem fibres.

12 (*c*) Zooidogamy is found in *Cycas.* It is a type of plant reproduction in which male gametes, called antherozoids, reach the female gametes, called archegonium, by swimming in a path of water. It is also found in algae, bryophytes, pteridophytes and some gymnosperms.

13 (*b*) The most common type of female gametophyte (embryo sac) in angiosperms is the monosporic embryo sac in which the embryo sac develops form a single functional megaspore (n), while the other three megasproes degenerate. The functional megaspore undregoes three sequential mitotic divisions and gives rise to 8-nucleate and 7-celled mature embryo sac.

14 (*c*) Option (c) is incorrectly matched and can be corrected as

Polysiphonia is a red alga. In its sexual reproduction, the male sex organ, spermatogonium produces non-flagellate male gametes, spermatium.

15 (*c*) The chloroplast of *Ulothrix* is girdle-shaped containing one or more pyrenoids. The thallus of *Ulothrix* is filamentous, long, unbranched and multicellular. It is a freshwater alga growing on substratum like rocks, sand, etc.

16 (*b*) *Cephaleuros* is a parasitic thalloid green alga. Its common name is red rust. This alga is parasitic on some important economic plants of the tropics and subtropics, e.g. tea, guava, mango, coffee, etc. It damages the area of algal growth on leaves or kills new shoots or disfigures fruits.

17 (*c*) Palmella stage is present in *Chlamydomonas.* In adverse conditions, the parent cell wall gelatinises and forms a mucilaginous sheath around the daughter protoplasts. All cells of palmella stage are devoid of flagella. They escape from the gelatinous matrix on the arrival of favourable conditions.

18 (*b*) Heterotrichous thallus is shown by *Ectocarpus.* It is a filamentous marine brown alga having both upright and prostrate regions. This type of growth is called heterotrichous.

19 (*c*) The statement in option (c) is correct. In gymnosperms, ovules are not enclosed by ovary wall. Seeds do not occur inside a fruit. They are naked.

Rest statements are incorrect and can be corrected as

- Horsetail is the common name of *Equisetum*, which is a pteridophyte.
- *Selaginella* and *Salvinia* are heterosporous and possess two types of spores, i.e. microspores and megaspores.
- *Cycas* has an unbranched columnar stem, while *Cedrus* possesses branched stem.

20 (*b*) *Volvox* is a freshwater, green, hollow ball-like colonial alga. Its colony has a fixed number of cells (500 to 60,000) and is called coenobium.

21 (*c*) Option (c) is the correct identification of the diagrams.

A-Dictyota, B-Polysiphonia, C-Porphyra, D-Laminaria and *E-Fucus.*

Based on their pigments, flagellation and storage product algae are divided into three main classes, i.e. Chlorophyceae, Phaeophyceae and Rhodophyceae. *Laminaria, Fucus* and *Dictyota* belong to Phaeophyceae (brown alage). *Porphyra* and *Polysiphonia* belong to Rhodophyceae (red algae).

22 (*d*) Option (d) is the incorrect combination and can be corrected as

Marchantia is a bryophyte belonging to the class–Hepaticopsida, the class of liverworts. They have elaters, which are hygroscopic structures that develop inside the capsule along with spores. Elaters help in dehiscence of capsule and liberation of spores.

Rest options are correct combinations.

23 (*a*) Among the given options, option (a) is mismatched and can be corrected as

Pinus is a monoecious plant as it bears male and female cones on the same plant.

The male pollen cones of *Pinus* occur in clusters subterminally on lower long branches. The female or seed cones develop in groups of 2-6 on upper long branches of the tree.

Rest options are matched.

24 (*d*) Zygotic meiosis is observed in the haplontic life cycle of many algae including *Chlamydomonas.* In such a life cycle, all cells are haploid except the zygote.

This is because meiosis occurs in the zygote itself resulting into four haploid cells that give rise to haploid plants.

All algae except *Fucus, Ectocarpus, Dictyota* and a few others, show haplontic life cycle or zygotic meiosis. Bryophytes and pteridophytes show haplo-diplontic life cycle and undergo sporic meiosis.

25 (*c*) *Ectocarpus* and *Fucus* show haplo-diplontic and diplontic life cycle, respectively. In *Ectocarpus*, sporic meiosis occurs and haploid biflagellate meiozoospores are formed. They germinate to produce gametophytic thalli.

The gametophytes liberate gametes, which fuse to form diploid zygote, which gives rise to a diploid plant.

In *Fucus*, diploid sporophytic cell produces haploid gametes. They fuse during fertilisation to give rise to a diploid individual.

26 (*b*) The statement in option (b) is incorrect and can be corected as

Algin extracted from brown algae, e.g. *Laminaria*, etc., is a hydrocolloid which is used in shaving creams, jellies, flameproof plastic, etc.

Carrageenan is extracted from red algae like *Chondrus* and is used as emulsifier and clearing agent.

Rest of the statements are correct.

27 (*c*) In several primitive simple plants like algae, bryophytes and pteridophytes, water is the medium through which male gametes are transferred to the female reproductive organ or gamete to bring about fertilisation.

28 (*c*) Conifers are gymnosperms. Their leaves show xerophytic adaptations. The leaves are needle-like with thick-walled, single-layered epidermal cells covered with thick cuticle. This enables them to tolerate extreme climatic conditions.

29 (*b*) The statement in option (b) is correct.

Sequoia is one of the tallest tree species, known as red wood tree having height up to 100 m.

Rest statements are incorrect and can be corrected as

- *Salvinia* is a pteridophyte, but *Ginkgo* and *Pinus* are gymnosperms.
- Gymnosperms are well-adapted to extremes of climate and are heterosporous.

30 (*c*) All given statements are correct except II and III.

Incorrect statements can be corrected as

- *Selaginella* species are creeping lycopods and produce heterosporous (megaspores and microspores).
- Coralloid roots are developed in *Cycas*. They are produced as cluster at base of stem. They are dichotomously branched and greenish-brown in colour. They contain algal zone in the cortex. This algal zone contains *Nostoc* and *Anabaena* which grow in symbiotic association with coralloid root.

31 (*d*) Option (d) is correct because *Pinus* does not have independent and free-living gametophyte, but is dependent on sporophyte.

Other options are explained as

In bryophytes (*Funaria* and *Marchantia*), fertilisation is dependent on water for the transfer of sperm to the egg. The gametophytes are fertilised outside the body.

Therefore, they have independent free-living gametophyte. In this respect, pteridophytes (*Pteris*) are similar to bryophytes, i.e. they still have free-living gametophytes.

32 (*d*) The statement in option (d) is incorrect and can be corrected as

Mannitol is stored food in Phaeophyceae (not in Rhodophyceae). The brown algae store their food in compounds called laminarin (starch) and transport it throughout their bodies in the form of compound called mannitol whereas in red algae, floridean starch is the storage food.

Rest of the statements are correct.

33 (*c*) Male gametes are flagellated in *Ectocarpus* belonging to the class—Phaeophyceae. The flagella of male gamete play an important role in establishing the initial sexual contact with the female gamete. However, the female gamete too, is flagellated but has a different structure.

Other options are explained as

In *Polysiphonia* (Rhodophyceae), flagellated gametes are not observed, in *Anabaena*, sexual reproduction through gametes is absent while in *Spirogyra*, the male gametes are non-flagellated and show amoeboid movement.

34 (*d*) Botanical snakes are pteridophytes. Like reptiles (first true land animals that evolved after amphibians), they are considered as the first true land plants that evolved from bryophytes. Hence, pteridophytes are sometimes called 'Botanical snakes' or 'Snakes of plant kingdom'.

35 (*c*) Pteridophytes are the most primitive vascular plants on earth and are known as vascular cryptogams.

They were the first vascular plants to grow and began their life period from leafless or rootless individuals. They played an important part in establishment of the early land flora as they emerged shortly after the evolution of land plants and are much larger than Bryophyta (which are non-vascular plants).

36 (*c*) *Lilium* (angiosperm) psosesses the male gametophyte with least number of cells. The number of cells in male gametophyte shows the pattern of reduction from bryophytes to angiosperms. In angiosperms, it is reduced to about 2-3 celled and called as pollen grains.

The number of cells in male gametophyte decreases in the following order

Funaria > *Pteris* > *Pinus* > *Lilium*

37 (*d*) *Sphagnum* is responsible for peat formation. The older parts of moss slowly get compressed and fossilised over thousands of years to produce peat (a dark spongy mass). It is used as a fuel and packaging material for trans-shipment of living material because of its great water holding capacity.

38 (*a*) The statement in option (a) is correct.

Bryophytes can live in soil, but are dependent on water for sexual reproduction. Thus, are called amphibians of plant kingdom.

Rest statements are incorrect and can be corrected as

- The sex organs in bryophytes are multicellular and jacketed.
- In bryophytes, true roots are absent, instead rhizoids occur.
- Common examples of liverwort and moss are *Marchantia* and *Polytrichum*, respectively.

39 (*d*) *Marchanita* is considered as a heterothallic plant because it is dioecious in nature whose male plant bearing antheridiophore and female plant bearing archegoniophore are present on different plants.

40 (*c*) Mosses and liverworts are members of Bryophyta. The members of class–Hepaticopsida are known as liverworts. The members of class–Bryopsida are commonly known as mosses.

41 (*d*) The statement in option (d) is incorrect and can be corrected as

The cell wall of parenchyma is made up of cellulose only. Parenchyma is responsible for photosynthesis, secretion and storage.

Rest statements are correct.

42 (*c*) *Ectocarpus* shows haplodiplontic life cycle which is represented below

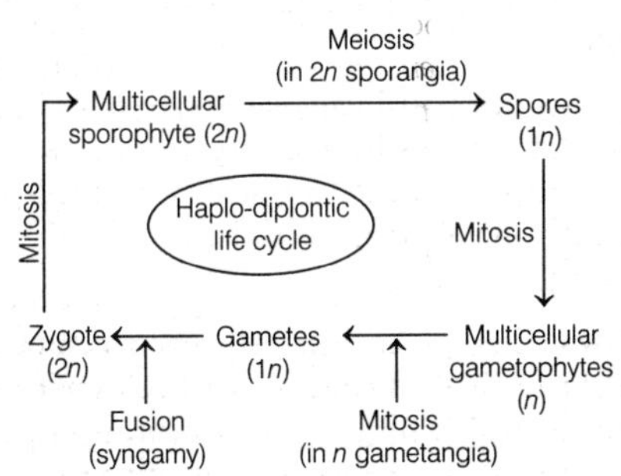

43 (*a*) *Spirogyra* shows haplontic life cycle. It is the type of plant life cycle, in which the main plant body is haploid. It is mainly characterised by a single predominant haploid vegetative phase, with meiosis taking place on germination of the zygote. This type of life cycle is also seen in *Volvox*, *Chlamydomonas*, etc.

44 (*b*) Green alga, *Chlorella* is an attractive potential food source now-a-days because it is rich in protein and other essential nutrients. When dried, it contains about 50% proteins, 20% fats, 20% carbohydrates, 5% fibres and 10% minerals along with vitamins. It is mainly used as a food supplement for astronauts.

45 (*d*) Isogamy with non-flagellated gametes is seen in *Spirogyra*.

It can reproduce both by sexual and asexual (vegetative) means.

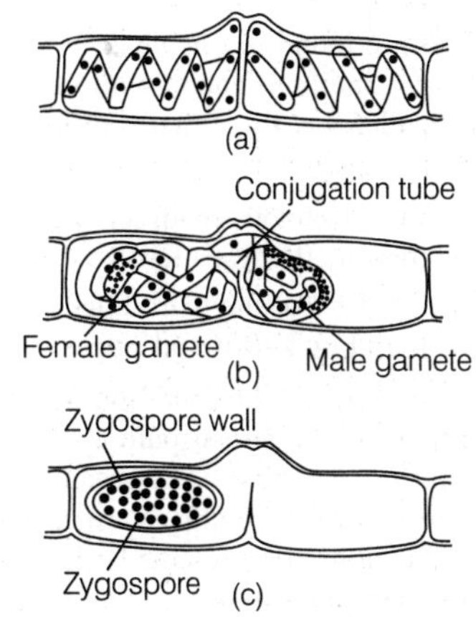

Spirogyra sp. (a-c) stages in lateral conjugation of indirect types

They reproduce sexually by conjugation in which two non-flagellated morphologically similar, but physiologically different gametes (isogamous) fuse together.

46 (*c*) Statement in option (c) is incorrect about *Chara* and can be corrected as

Antheridium and oogonium are the male and the female reproductive structures, respectively. They have sterile jackets on their surface. In *Chara*, the structure like globule (male sex organ) is rounded and present on lower side, while the nucule (female sex organ) is present on upper side of sterile vegetative (leaf-like) structure.

Rest statements are correct about *Chara.*

47 (*c*) Out of the following groups of algae given in the options, the group that belongs to Rhodophyceae is *Gracilaria*, *Gelidium*, *Porphyra*, *Polysiphonia.*

All these algae are economically important in various fields. *Gracilaria* and *Gelidium* are agar yielding algae and are called agarophytes. *Porphyra* is used as food and *Polysiphonia* has medicinal properties.

49 (*b*) The presence of pyrenoid is the characteristic feature of class-Chlorophyceae. In Chlorophyceae, one or more proteinaceous bodies (pyrenoids) are present in chloroplasts, which manufacture and store starch.

50 (*b*) Mannitol is the white crystalline solid, stored food product of Class–Phaeophyceae (brown algae). On the other hand, Poaceae is the family of angiospermic plants especially cereals. Chlorophyceae and Rhodophyceae store their food as starch and floridean starch, respectively.

51 (*c*) Statement in option (c) is incorrect and can be corrected as

Oogamy is the fusion of non-motile, larger female gamete with motile, smaller male gamete. The gametes differ both morphologically as well as physiologically. It occurs in *Chlamydomonas*, *Fucus*, *Chara*, *Volvox*, etc.

Rest of the statements are correct.

- Isogametes are found in algae like *Ulothrix*, *Chlamydomonas*, *Spirogyra*, etc., which are similar in structure, function and behaviour.
- Anisogametes are found in *Chlamydomonas* in which one gamete is larger and non-motile, while the other one is motile and smaller, i.e. the gametes differ in size and structure.

52 (*b*) *Spirogyra* shows isogamous condition with non-flagellated (non-motile) gametes.

Chlamydomonas shows isogamy with flagella. Oogamy along with flagella is found in *Volvox* and *Fucus*.

53 (*d*) Monoecious or homothallic condition in *Chara* (green algae) is used to denote upper oogonium and lower antheridium on the same plant. The organisms, which possess both the reproductive organs are called bisexual, while unisexual condition represents both male or female organs in different organisms.

54 (*b*) Alginic acid or alginate is found in the middle lamella and primary cell walls of sea weeds such as *Laminaria*, *Macrocystis*, *Ascophyllum*, etc.

55 (*a*) *Volvox* forms motile colony (coenobium). A typical *Volvox* colony is a hollow-sphere of mucilage having 500 or more biflagellate algal cells.

The combined beating of their flagella causes the colony to glide slowly through the water. Large colonies can reach a diameter of a millimeter or so.

56 (*d*) Hypnospores are a mode of asexual reproduction in *Chlamydomonas*. During unfavourable conditions, the protoplasts develop a thick wall to form the hypnospores (i.e. non-motile, greatly thickened, asexual spore). Aplanospores are also produced during unfavourable conditions, but are thin-walled as compared to hypnospores.

57 (*b*) *Spirogyra* is an unbranched filamentous green thallophyte alga, having green pigments, i.e. chlorophyll-*a* and *b* and yellow pigments, i.e. carotene and xanthophyll.

58 (*a*) Elaters are absent in *Funaria*. They are the characteristic feature of Class-Hepaticopsida, which includes *Marchantia*, *Pellia*, *Porella*, etc.

59 (*d*) Vegetative reproduction in *Funaria* takes place by fragmentation of primary protonema, secondary protonema, gemmae and bulbils.

Thus, option (d) is correct.

60 (*c*) *Pinus* belongs to class–Coniferopsida. A mature *Pinus* plant is large tree growing up to 30-70 m in height and differentiated into root, stem and leaves. Its branches are arranged in acropetal order thus, giving pyramid or conical-shaped appearance to the tree.

61 (*d*) Multicellular fungi, filamentous algae and protonema of mosses, all show multiplication or asexual reproduction by means of fragmentation.

62 (*c*) In bryophytes, the sporophytic phase is represented by the spore mother cell. There are two generations in life cycle of bryophytes. The main plant body is gametophytic. The gametophytic phase starts with the formation of spores, which generate bryophytic thallus.

The antheridium (male sex organ) produces antherozoids and archegonium (female sex organ) produces egg, which undergoes fertilisation to form zygote. The sporophytic phase starts with zygote and ends with reduction division in spore mother cells.

63 (*d*) Gymnosperms lack xylem fibres. Large amount of parenchymatous cells are present with secondary xylem tracheids. So, these are also known as soft wood spermatophytes.

64 (*b*) Multiciliated motile sperms are found in both *Cycas* (gymnosperm) and *Adiantum* (pteridophyte or walking fern).

65 (*c*) In *Pinus*, each male cone consists of an elongated axis, bearing a number of spirally arranged microsporophylls. On the underside of it, two microsporangia develop and get filled with microspores (pollen grains).

66 (*d*) The statement in option (d) is correct.

Origin of seed habit can be traced in pteridophytes, which show heterospory and retention of female gametophyte with young embryo on the parent sporophyte for nutrition.

Rest statements are incorrect and can be corrected as

- Bryophytic gametophyte has a protonema stage and a leafy stage.
- In pteridophytes, the female gametophyte is free-living.
- Antheridiophores and archegoniophores are present in bryophytes.

67 (*b*) The plants commonly called vascular cryptogams are pteridophytes. These are called so, because these are the seedless plants having vascular tissues, i.e. xylem and phloem.

68 (*c*) Pteridophyta is a group of non-seed forming, vascular cryptogams, i.e. these contain vascular tissues, i.e. xylem and phloem for the conduction of minerals and water, but do not produce seeds.

69 (*c*) Pteridophytes are called vascular cryptogams, because these are the non-seeded plants having vascular tissues, i.e. xylem and phloem. Here, xylem consists of tracheids and phloem is made up of sieve cells and albuminous cells.

Thus, option (c) is correct.

70 (*b*) The cell wall of algae is chemically composed of cellulose, galactans and mannans. In general, algal cell walls are made up of two components, i.e. fibrillar component, which forms the skeleton of the wall and the amorphous component, which forms a matrix within which the fibrillar component is embedded.

The fibrillar components inlcude cellulose, mannan and xylan, while the amorphous components alginic acid, fucoidan and galactans.

71 (*d*) *Cephaleuros* is a genus of parasitic thalloid alga, commonly known as red rust. The algae is parasitic on some important economic plants of the tropics and subtropics such as tea, coffee, mango and guava. Members of this genera may also grow with a fungus to form a lichen that does not damage the plants.

72 (*d*) Bryophytes resemble algae in the following characters

- Thalloid plant body
- Absence of tap roots
- Absence of complex vascular tissue
- Autotrophic mode of nutrition
- Starch as reserve food material

73 (*c*) Bryophytes are called amphibians of plant kingdom. They have various characters for land adaptation (compact plant structure, rhizoids, sterile jacket, large number of spores, wide dispersal of spores) and water adaptation (no special supportive structure, depend on water for reproduction).

74 (*a*) Spore (meiospore) of *Funaria* germinates and produces protonema (juvenile phase) which is a slender, green and branching system of filaments.

75 (*d*) All given options are true about bryophytes.

They are a group of embryo producing plants, which do not bear fruits, seeds and any vascular tissue. Body is thalloid and green (due to the presence of chloroplast). Male sexual organ is antheridium and female sexual organ is archegonium.

Thus, option (d) is correct.

76 (*c*) Pyrenoids are proteinaceous bodies surrounded by starchy sheath, present in chloroplasts. These are considered to be associated with storage of starch. In members of the class Chlorophyceae, pyrenoids are present.

77 (*d*) Iodine is found in *Laminaria* (brown algae) which is about 7% of the total world production. Iodine is obtained from kelps (brown algae) in Japan.

78 (*a*) Option (a) is correct.

Floridean starch — Rhodophyceae

Other options are explained as

Laminarian and mannitol — Phaeophyceae

Starch — Chlorophyceae

Xanthophyceae — Chrysolaminarin

79 (*d*) Option (d) is correctly matched.

In the members of Phaeophyceae or brown algae, food is stored in the form of D-mannitol. The members of Rhodophyceae or red algae contain the red pigment, *r*-phycoerythrin in their body.

Other pairs are incorrect and can be corrected as

Dictyota belongs to class—Phaeophyceae. Chlorophyceae have both motile and non-motile gametes.

80 (*d*) In *Marchantia* (a bryophyte), the archegonia (female sex organs) are borne on special branches called archegoniophore or female receptacles.

Each archegoniophore has rows of archegonia protected by perichaetium.

81 (*b*) The spores of moss germinate to form protonema. The cells of protonema contain chloroplasts. A protonema is a thread-like chain of cells that forms the earliest stage of the life cycle of mosses and liverworts.

82 (*a*) The coralloid root of *Cycas* is symbiotically associated with nitrogen-fixing blue-green algae, *Anabaena cycadae* and *Nostoc punctiforme*. These blue-green algae (cyanobacteria) are prokaryotic, photosynthetic and autotrophic.

83 (*c*) *Cycas* shows anemophily. In *Cycas*, the 3-celled microspores are shed in the air after the dehiscence of the sporangium. They are very light in weight and are carried by air currents (anemophily).

84 (*b*) Fertilisation in *Cycas* is siphonogamous followed by zooidogamy. During fertilisation, the pollen tube discharges spermatozoids into the archegonial chamber (siphonogamy). One of these reaches the archegonium by swimming in the fluid present in the archegonial chamber (zooidogamy).

85 (*c*) Circinate ptyxis and heterospory are found in *Cycas*.

Dryopteris has circinate vernation of leaves, but it is homosporous.

86 (*d*) *Riccia* is one of the bryophytes and the gametophytic phase is the dominant phase in their life cycle. The bryophytes represent two morphologically distinct generations, i.e. gametophytic and sporophytic. The gametophytic phase is dominant, while the sporophytic phase is dependent upon as well as being permanently attached to the gametophyte.

87 (*a*) In bryophytes, the gametophytic phase is more conspicuous, independent and dominant, while the sporophyte depends on gametophyte.

88 (*b*) Seed habit is linked to heterospory, i.e. some of the pteridophytes produce smaller spores (microspores) and larger ones called (megaspores), which germinate and give rise to male and female gametophytes, respectively.

89 (*d*) Pteridophytes are classified into four classes, i.e. Pteropsida (*Pteris, Adiantum, Dryopteris*), Psilopsida (*Psilotum*), Lycopsida (*Selaginella, Lycopodium*) and Sphenopsida (*Equisetum*).

CHAPTER 08

Animalia-I (Lower Invertebrates)

Animals are classified according to a broader range of characteristics. These include primarily their anatomy, pattern of development and genetic makeup. These features not only provide more reliable guidance to place an animal in the living world, but also helps to show how different species are linked through evolution. In recent years, animal systematics had started arranging animals into groups that reflect their evolutionary relationships. Of these 3 main groups are

(i) **Monophyletic group** It includes a single ancestral species and all of its descendants.

(ii) **Polyphyletic group** It has members that can be traced to many ancestral groups.

(iii) **Paraphyletic group** It includes some, but not all members of a lineage. This type of group results when the knowledge about the group is insufficient.

So, basis of animal classification are more defined and more precise today as compared to past.

BASIS OF CLASSIFICATION

The fundamental features used as basis of the animal classification are discussed below

Levels of body organisation

Body organisation is the first basic character used in classifying animals. Animal bodies range from acellular or protoplasmic levels up to organ system levels. Different levels of body organisation found in animals (in increasing order of complexity) are as follows

Level of body organisation

- **Acellular or protoplasmic level** Body consists of mass of protoplasm. All activities are performed by different cell organelles and are confined within the limit of plasma membrane, e.g. Protozoa.
- **Cellular level** Body consists of cells only. Animal body at this level can be seen either as a well organised colony or a simple aggregation.
  - **Cellular colony** e.g. protists and some algae.
  - **Cellular aggregate** e.g. sponges, where cells are not organised into tissue.
- **Tissue level** Group of similar cells form tissue which serve for specific function, e.g. Coelenterata.
- **Tissue-organ level** Some tissues join and function as a unit of an organ, like alimentary canal e.g. Platyhelminthes.
- **Organ-system level** Group of organs work together to form an organ system, such as digestive, circulatory, respiratory system, e.g. Nemathelminthes to Chordata.

Inside

Patterns of complexities of various organ systems

Organ system level of body organisation started to appear in animals like Platyhelminthes and kept on complexing from lower to higher phyla.

These complexities are used as the basis of classification in animals as follows

1. Digestive system

(i) **Incomplete digestive system** When the digestive system has only a single opening to the outside of the body that serves as both mouth and anus. It is called incomplete digestive system, e.g. coelenterates, ctenophores and platyhelminths.

(ii) **Complete digestive system** When the digestive system has two openings, i.e. mouth and anus, it is called complete digestive system.

Hence, the entry of food and exit of waste takes place from separate opening, e.g. Aschelminthes to Chordata.

2. Circulatory system

(i) **Open circulatory system** When the blood is pumped out of the heart and the cells and tissues are directly bathed in it. It is called open circulatory system, e.g. arthropods. molluscs, echinoderms, hemichordates and some lower chordates like tunicates.

(ii) **Closed circulatory system** When the blood is circulated through a series of vessels of varying diameters, i.e. arteries, veins and blood capillaries. It is called closed circulatory system, e.g. annelids, cephalopods, molluscs, vertebrates, etc.

Body symmetry

It refers to the distribution of body parts in animals in reference with the body axis. On the basis of symmetry, animals can be of following two types

1. Asymmetrical

When any plane passing through the centre does not divide the body into equal halves. An animal with this condition is called asymmetrical, e.g. sponges.

***Fig. 8.1** Asymmetry*

2. Symmetrical

- When the parts of an animal are arranged in such a way that its body can be cut into two similar halves by one or more planes. An animals with this condition is called symmetrical.
- These animals may have any one of the following symmetry types

(i) **Spherical symmetry** Found very rarely in animals. It occurs in organisms, which can be cut into two identical halves through any plane. Their body is more or less of spherical shape, e.g. *Volvox* and *Heliozoa.*

(ii) **Radial symmetry** In this type of symmetry, the individual can be divided into two equal halves by any plane passing through the centre. This symmetry is found in sponges (*Sycon*), coelenterates (*Hydra*, jellyfish; tetramerism) and echinoderms (starfish; pentamerism).

(iii) **Biradial symmetry** In this symmetry the animal body can be cut into two similar halves by one or two vertical planes only.

- Animals showing radial and biradial symmetry have **oral** and **aboral** sides.
- Oral side bears mouth while aboral side lie opposite to it, e.g. sea walnuts (phylum–Ctenophora) and sea anemones (Anthozoa).

(iv) **Bilateral symmetry** Here, a lengthwise vertical (sagittal) plane divides the animals into equal and opposite halves.

- These two halves can be either seen as **anterior-posterior** or **dorsal-ventral.**
- Usually the body end that contains mouth is called **anterior**, while its opposite end is called **posterior.**

- Likewise, the back or upper surface is termed as **dorsal** and the under surface is termed as **ventral**. This surface usually faced toward the ground.
- Bilateral symmetry is developed due to **cephalisation**, i.e. differentiation of definite head into the anterior end.
- Bilateral symmetry is found in almost all animals from invertebrates, e.g. Nemathelminthes to Chordata.

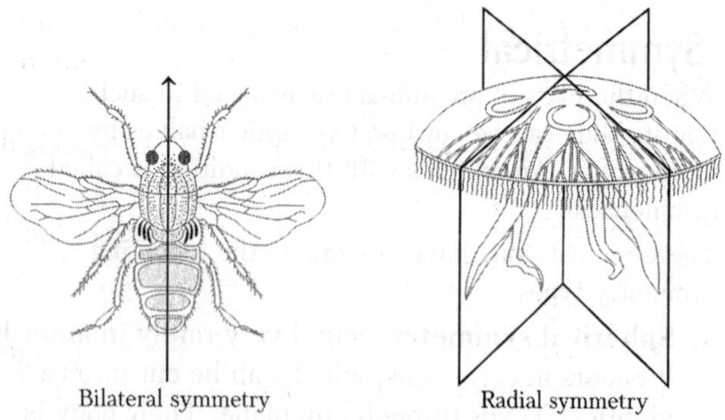

Fig. 8.2 Types of symmetry in animals

Note *Bilateral symmetry is considered as the second major evolutionary advancement in the animal body plan. It allows an animal to have a more stream-lined shape to concentrate the power of its muscles and appendages for producing motion in one direction.*

Germ layers

These are primary cell layers that differentiate in animal embryos at the gastrulation stage. These are responsible for the development of different tissues and organ systems in the adult.

On the basis of germ layers, animals are of two types

1. Diploblastic animals

- These animals develop from a two-layered embryo, i.e. the outer layer or ectoderm and inner layer or endoderm.
- A non-cellular mesenchyme or mesoglea layer is present in between them, e.g. sponges and coelenterates.

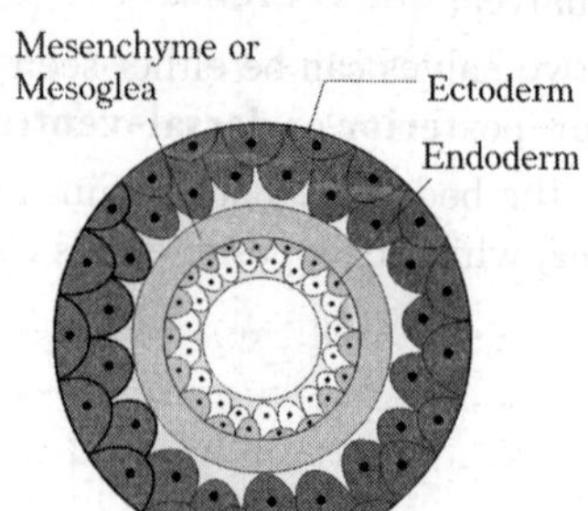

Fig. 8.3 Diploblastic germ layer

2. Triploblastic animals

- These animals develop from a three-layered embryo. In addition to endoderm and ectoderm, they have an extra middle layer, i.e. the mesoderm.
- These animals can develop complex body organs and organ systems. All other animals after Coelenterata are triploblastic.
- Thus in kingdom-Animalia, only two phyla are diploblastic and rest other phyla are triploblastic.

Body cavity or Coelom

- The presence or absence of a body cavity or coelom is an important basis of classification in animals.
- Coelom refers to a large fluid-filled space (or cavity) lined by mesoderm. It lies between outer body wall and inner visceral organs.

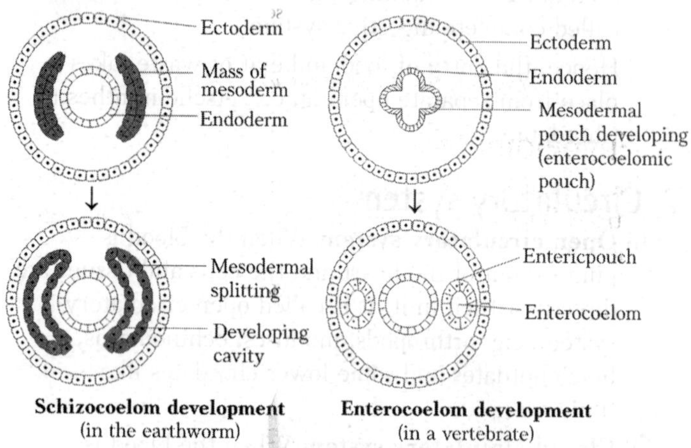

Fig. 8.4 Development of true coelom

- On the basis of coelom, animals are of following types

1. Acoelomates

- Animals without coelom.
- Region between the ectodermal epidermis and the endodermal digestive tract is completely filled with mesoderm in the form of a spongy mass of space filling cells called **parenchyma**, e.g. Platyhelminthes.

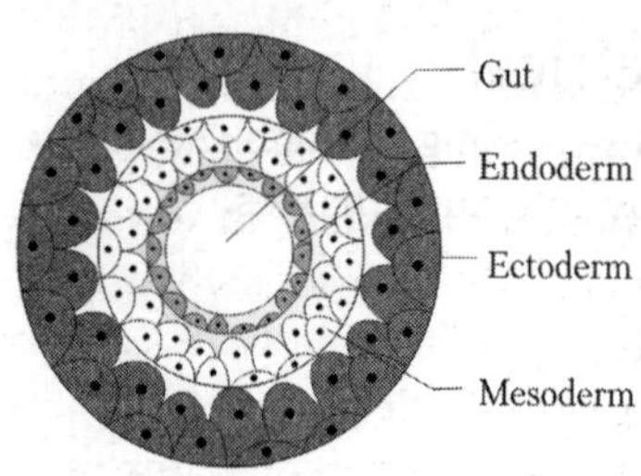

Fig. 8.5 Triploblastic acoelomate

2. Pseudocoelomate

- These animal have the presence of false coelom or perivisceral cavity. This body cavity is not lined by mesoderm.
- Instead, the mesoderm is present as scattered pouches in between the ectoderm and endoderm.
- Developmentally, pseudocoelom is the persistent blastocoel of blastula stage, e.g. Rotifera, Aschelminthes and Nematoda.

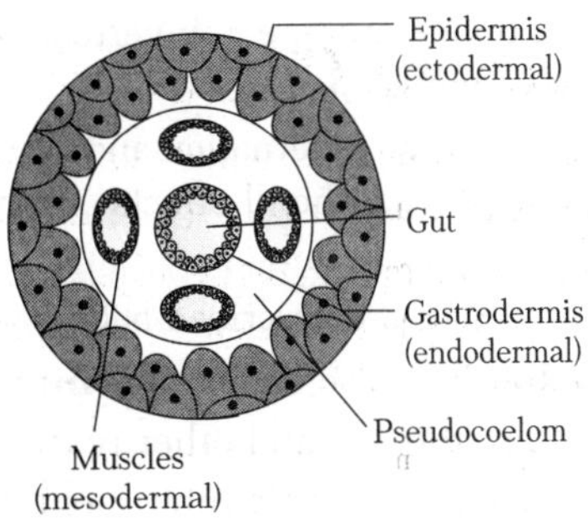

***Fig. 8.6** Triploblastic pseudocoelomate*

3. Eucoelomate

- These animals have a true body cavity. This cavity is entirely lined with mesoderm, e.g. higher invertebrates to Chordata.

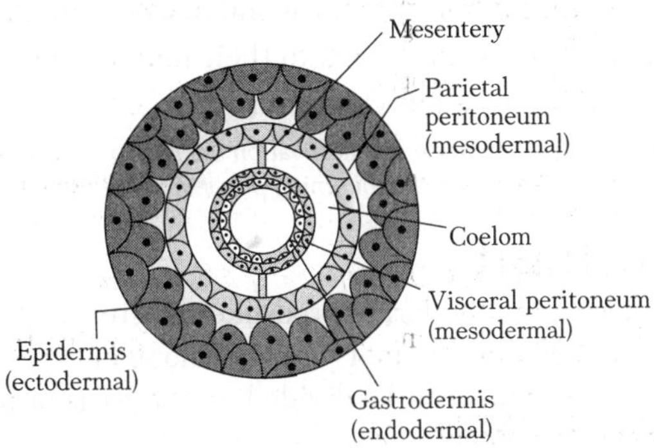

***Fig. 8.7** Triploblastic eucoelomate*

- A true coelom can be formed in two ways
  (i) **By the splitting of mesodermal layer** This type of coelom is called **schizocoelic** (*schizo*-split off).

  The group of animals which bear this type of coelom are called **schizocoelomates**, e.g animals of phyla-Annelida, Arthropoda and Mollusca.

  (ii) **By the fusion of gastric or enteric pouches** These pouches are formed from archenteron or primitive gut. This kind of coelom is called **enterocoelic** (*Entero*-gastro). The group of animals with this type coelom are called **enterocoelous** or **enterocoelomates**, e.g. animals of phyla-Echinodermata and Chordata.

  The mesodermal coelomic layers in both the cases are called **peritoneum** (in vertebrates) or **mesothelium** (in invertebrates).

Note *In some animals, the actual or true coelom is reduced, but spaces between viscera grow and form a large cavity filled with blood called **haemocoel.** It is found in arthropods and non-cephalopod molluscs.*

Body plan

Animals have three types of body plans as given below

1. **Cellular** or **cell aggregate body plan** Body consists of a cluster of cells. Division of labour is limited. Tissue or organs are absent, e.g. sponges.
2. **Blind sac body plan** Cells are properly organised into tissues. There is division of labour and each tissue performs a specific task. Single opening of the body for ingestion and egestion, e.g. coelenterates, flatworms and ctenophores.
3. **Tube within a tube plan** Organisation with organ system level. Proper digestive canal with two openings, one for ingestion and other for egestion, e.g. Aschelminthes onwards.

Development of mouth

All higher invertebrates and vertebrates are categorised into following two types on the basis of developmental patterns of mouth.

(i) **Protostomes** (*Proto* - Primitive or first; *stoma* - mouth) In these animals, mouth develops first from the developmental opening of embryo called **blastopore** and anus develops later.
- These animals show **spiral determinate cleavage** with schizocoelous coelom formation.
- The group of protostomes is called protostomia, e.g. Nematoda, Annelida, Arthropoda and Mollusca.

(ii) **Deuterostomes** In these animals, anus develops first from the blastopore region and mouth develops later.
- The animals show **radial** and **indeterminate cleavage** and **enterocoelous coelom** formation.
- The group of deuterostomes is called **Deuterostomia**, e.g. echinoderms and chordates.

Segmentation

- It is the serial repetition of similar parts along the length of an animal. These parts are called **somites** or **segments** and the process of their formation is called **segmentation.**
- On the basis of segmentation, the body of animals could be unsegmented, pseudosegmented or metamerically segmented.

- These can be described as follows

1. Unsegmented body

The unsegment or without segment body is found in most of the Platyhelminthes and roundworms.

2. Pseudosegmented body

- It is seen in tapeworms. In them, the body is made up of specialised segments called **proglottids.**
- Each proglottid is special in possessing its own reproductive machinery. Thus, each of these is considered as a separate individual and the animal body as the colony (accumulation) of such individuals.
- The technical name given to this colony is strobila and process of formation of new segments is called **strobilisation.**
- In tapeworms, the anterior body region (Neck) is responsible for strobilisation. It simply means that the oldest segment lies at the posterior most end of the body.

3. Metameric segmentation

- It is seen in Annelida onwards higher invertebrates and chordates.
- This is the true segmentation. Earthworm is an ideal example to analyse this phenomenon.
- In metameric animals, each segment is completely divided externally as well internally. It has the equal participation of all the systems of the body.
- Thus, we can say that, all the segments work in a well coordinated way and are more or less similar in structure.

Tagmatization

In Arthropoda, Mollusca and Chordata, metameric segmentation is a bit modified. This condition is called **tagmatization.** It is defined as **regionalisation of metameric segments**, i.e. metameric segments show their appearance regionwise. In humans and other vertebrates body muscles, vertebrae, mesentries, etc., are the examples of such, tagmatization.

Body temperature

In animals, body temperature varies according to the surrounding environment for normal vital activities. On the basis of this, animals are classified as follows

1. Warm-blooded animals

- These animals are also called **endotherms** (Gr. *Endon*-within or internal) or **homoeothermic** animals.
- They have constant body temperature, which is independent of their external environment.
- Their body temperature is derived from the heat produced by the oxidative metabolism of their food.
- The presence of insulating hair, feathers and blubbers (fat layer) prevent the internally generated heat, from free diffusion into the external environment, e.g. few fishes, mammals and birds.

2. Cold-blooded animals

- These animals are also called as **ectotherms** (Gr. *Ektos*-outside or external) or **poikilothermic** (Gr. *Poikilo*-varied or irregular) animals.
- Their body temperature varies with the change in environmental temperature, e.g. invertebrates, fish, amphibians and reptiles.
- These animals undergo hibernation and aestivation to prevent their body from the ill effects of temperature variance.
- These both processes are described below
  (i) **Hibernation** It is also called as **winter sleep.** In this, animals store fat and other reserve food and become inactive during winter months. When the weather changes to warmer conditions, these animals continue their normal lifestyle, e.g. amphibians (frogs) and lizards.
  (ii) **Aestivation** It is also called as **summer sleep.** In this, some animals hide themselves in shady areas or mud during hot summer months, e.g. amphibians. They too continue back to their normal lifestyle when the weather conditions change to their favour.

Note *In both hibernation and aestivation the metabolic activities of animal reaches to the minimum possible value during the sleep.*

Notochord

It is a mesodermally derived rod-like structure, formed on the dorsal side during embryonic development in chordates. Animals can be divided on the basis of presence or absence of notochord as

1. **Chordates** Animals with notochord, e.g. urochordates, cephalochordates and vertebrates.
2. **Non-chordates** Animals without notochord, e.g. Porifera to Hemichordata.

Table 8.1 Differences between notochord and nerve cord

Notochord	Nerve cord
It is a supporting structure which is a part of the internal skeleton of animal.	It is a part of nervous system.
It is found only in chordates.	It is found in chordates as well as non-chordates.
It is formed on the dorsal side in chordates.	It is found in ventral side in non-chordates and in dorsal side in chordates.

Outline classification of kingdom-Animalia

Animal kingdom is broadly classified as follows

Animal kingdom

Sub-kingdoms

- **Protozoa**
  Unicellular, heterotrophic, eukaryotes.
  Now aggregated as a separate kingdom, i.e. Protista
- **Metazoa**
  Multicellular eukaryotes
  - Infra-kingdoms or Branches
    - **Mesozoa** (extinct) Worm-like small, bilaterally symmetrical animals, parasitic on cephalopods and other invertebrates, e.g. phylum-Mesozoa.
    - **Parazoa** Animals with loosely organised cells, no organs, diploblastic and no digestive cavity, e.g. phylum-Porifera.
    - **Enterozoa** True tissues or organs are found, a digestive cavity present. Entero word in itself signifies digestive cavity.
      - **Divisions**
        - **Radiata**
          Radial or biradial symmetry, no organs, diploblastic, e.g. phylum-Coelenterata
        - **Bilateria**
          Bilateral symmetry, organ systems present, triploblastic, digestive tract complete.
          - **Subdivisions**
            - **Protostomia**
              Blastopore forms mouth; cleavage is spiral and determinate.
              - Sections
                - **Acoelomata**
                  No coelom, mesoderm as parenchyma, e.g. phylum-Platyhelminthes.
                - **Pseudocoelomata**
                  Pseudocoelom present, e.g. phylum-Nematoda.
                - **Eucoelomata** (Schizocoela) True coelom which is schizocoelous, e.g. phylum-Annelida, Arthropoda and Mollusca.
            - *** Deuterostomia**
              Blastopore do not form mouth, mesoderm develops from archenteron.

* **Subdivision-Deuterostomia** contains one section, i.e. **Eucoelomata** in which coelom is enterocoelous thus also called **Enterocoela.** This section has following **three phyla**

- **Echinodermata**
  Radially symmetrical adults, but larva with bilateral symmetry, exclusively marine, body with true endoskeleton, e.g. starfish.
- **Hemichordata**
  Marine animals, sister group of echinoderms, connecting link between non-chordates and chordates, e.g. acorn worms.
- **Chordata**
  Animals with following 3 characters any where in their life.
  Presences of dorsal tubular notochord.
  Presence of dorsal nerve cord. Presence of pharyngeal gill slits.

CHECK POINT 8.1

1. Which of the following feature is common to all animals without any exception?
(a) Locomotion
(b) Herbivorous mode of nutrition
(c) Specialised nervous system
(d) Multicellularity

2. Tissue level of body organisation present in
(a) coelenterates (b) protists
(c) poriferans (d) All of these

3. In which symmetry a lengthwise vertical plane divides the animals into equal and opposite halves?
(a) Radial (b) Bilateral
(c) Spherical (d) Biradial

4. In which group body cavity is not lined by mesoderm?
(a) Acoelomates
(b) Pseudocoelomates
(c) Enterocoelomates
(d) Schizocoelomates

5. Besides Annelida and Arthropoda metamerism is found in
(a) Cestoda
(b) Acanthocephala
(c) Chordata
(d) Mollusca

6. Hibernation and aestivation is the exclusive feature of
(a) endotherms (b) poikilotherms
(c) ectotherms (d) Both (b) and (c)

The animal kingdom consists of over one million described species of animals. Of these, about 5 per cent animals possess a backbone or vertebral column and thus called **'vertebrates'**. The remaining animals which lack vertebral column are referred to as **'invertebrates'**.

These invertebrates can be further categorised into lower and higher invertebrates. The lower invertebrates include phylum-Porifera, Coelenterata, Ctenophora, Platyhelminthes and Aschelminthes.

PHYLUM-PORIFERA

- It consists of most primitive multicellular animals called **sponges**. They are also known as **pore bearers**.
- About 5000 living species of sponges are known today. The noting **'Evolutionary dead end'** is assigned to sponges because their ancestral forms do not give rise to any other major group.
- Poriferans are the only animal group without nerve cells. Although a rudimentary means of communication from cell to cell may exist.

Parazoology

The body organisation of sponges shows certain features, which are not seen in other metazoan group. This leads to placing of the sponges in an isolated group, called **Parazoa** by **Sollas** and **Delage**. That is why the study of poriferans is called **Parazoology**.

- **John Ellis** first recognised sponges as animals in 1765. But, the name 'Porifera' was given by **Robert E Grant** to this group in 1836.

General characteristics

1. Sponges are aquatic animals. Most of the sponges are marine and only few are freshwater (e.g. *Spongilla*).
2. They are either solitary (e.g. *Sycon*) or colonial (e.g. *Leucosolenia*) attached forms.
3. The body forms of sponges vary considerably. They may be vase-like, cylindrical, cup-shaped, spherical or disc-like with a porous body.
4. They are either asymmetrical (mostly) or radially symmetrical.
5. The sponges have cellular level of body organisation. The cells in their body do not form tissues. However, some division of labour is visible among the cells.
6. The sponges are the **first multicellular diploblastic** organism, i.e. their body is derived only from two embryonic primordial layers, *viz.*, ectoderm and endoderm.
7. The body wall of sponge consists of following layers
   (i) **Pinacoderm** (Outer layer) is also called dermal epidermis. It consists of flattened **pinacocytes** and oval porocyte.
   (ii) **Choanoderm** (Inner layer) is also called gastral epidermis. It consists of specialised flagellated cells called **choanocytes** or **collar cells.** These are called so because they contain a **central flagellum** surrounded by a **collar** of **microvilli** which are connected by a thin membrane.
   A typical collar cell is shown below

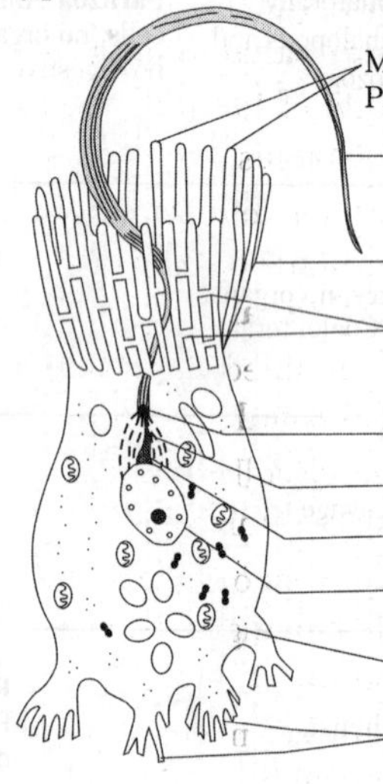

***Fig. 8.8** A choanocyte*

Mesenchyme or Mesohyl layer

It is a non-cellular layer present between pinacoderm and choanoderm. It is secreted by choanoderm and is not considered as a true body layer. It has fine dispersed spongin fibres and numerous spicules. It also contains *Amoeba*-like cells called **amoebocytes**. These amoebocytes have the capability to get modified further into following major types

- **Archaeocytes** These are undifferentiated **totipotent** cells. These may differentiate into all other types of cells. Thus, play an important and crucial role in regeneration.
- **Myocytes** These are highly contractile spindle-shaped cells. They form a circular ring around osculum and help in closing and opening of the osculum.
- **Chromocytes** These contain pigmented granules and excretory substances.
- **Thesocytes** These are mainly storage cells with food reserves.
- **Scleroblast** These are skeleton manufacturing cells. They secrete spicules or spongin *via* calciblasts, silicioblasts or spongioblasts. Their names are given depending upon the skeletal element these are producing.
- **Gland cells** These secrete an adhesive substance to provide the means of attachment for the animal.
- **Trophocytes** These provide food to the developing cells and are also called as **nurse cells**.
- **Collencytes** These secrete collagen fibres and form connective tissue.
- **Phagocytes** These collect food from choanocytes through their pseudopodia. These also engulf excreta and damaged tissue.
- **Germ cells** (sex cells) These form sperms and ova during the breeding season.

8. In sponges throughout the body a transverse cavity is present. It is called **spongocoel** or **paragastric cavity**.
    - The spongocoel communicates with the outside by openings or pores formed due to the folding of body walls.
    - These pores are of two types, i.e. inhalent pores called **ostia** (sing. ostium) and exhalent pores called **oscula** (sing. osculum).
    - Both of these are connected through a network of channels. This complete system along with both the openings is called **water canal system**.
9. Due to the attached mode of life, this **canal system** is the characteristic of sponges.
    - It helps in their respiration, circulation, excretion, nutrition and reproduction, etc. The channels involved in it are called **canals**.
    - Some of these canals are lined by choanocytes. These are called **radial canal**.
    - By constant beating of flagella of choanocyte cells, a water current is produced, which has a fixed direction.
    - The entering water delivers oxygen and food to the inner body cells while moving and collects excretory material, sperms, larvae, etc.
    - This water along with other material passes out of the body through osculum.
    - Thus, all exchange between sponge body and external medium is maintained by means of this current.

Depending upon the complexity of canal network, the canal system can be categorised as follows

(i) ***Ascon* type or asconoid canal system** It is the simplest type of canal system, which is found in *Leucosolenia* and few other sponges. It is the only canal system type in which the spongocoel is directly lined by choanocytes.

The course of water current in ascon type of canal system is shown below

Ingressing water $\xrightarrow[\text{ostia}]{\text{Through}}$ Spongocoel $\xrightarrow[\text{osculum}]{\text{Through}}$ To outside

Here, the ostia are represented by cells, called **porocytes**, which have intracellular canal inside them. These cells extend radially into the mesenchyme and open directly into the spongocoel.

(ii) ***Sycon* type or syconoid canal system** It is found in *Sycon* and some other sponges.

- It develops from *Ascon* type, primarily by horizontal folding of asconoid body wall.
- Here, the spongocoel is lined by pinacoderm.
- The choanocytes are restricted to small chambers, called as **radial chambers**.
- Incurrent canals are additional chambers seen along with radial chambers. These open to outside *via* **dermal pore** or **dermal ostia**.

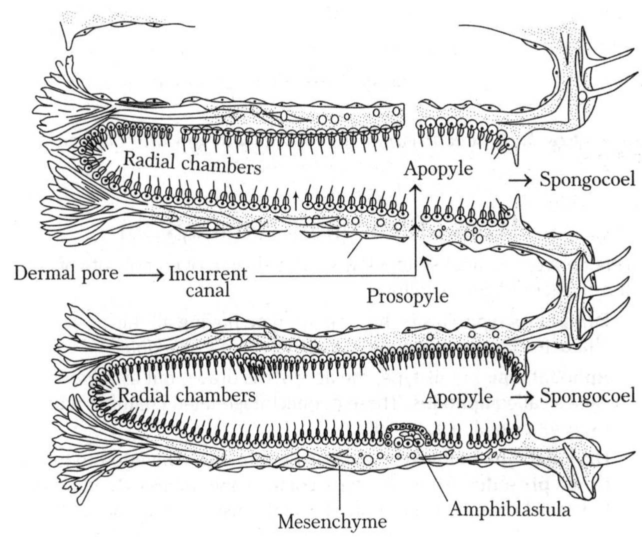

(a)

Ingressing water $\xrightarrow{\text{Dermal ostia}}$ Incurrent canals $\xrightarrow{\text{Prosopyles}}$ Radial canals $\xrightarrow{\text{Apopyles}}$ Spongocoel $\xrightarrow{\text{Osculum}}$ To outside

(b)

***Fig. 8.9** Sycon canal system : (a) A portion of body wall showing one incurrent canal and two radial canals, (b) Course of water-current*

- The openings of incurrent canals into radial chamber are called **prosopyle**. The openings of radial chamber into spongocoel are called **apopyles**.

(iii) ***Leucon* type or leuconoid canal system** It is considered as the most complex canal system.

- It is found in *Spongilla* and some other sponges.
- This system develops from syconoid type by further folding of walls of radial canals.
- Here, the radial symmetry of the body is lost and the canal system becomes very irregular. In the place of spongocoel, excurrent canals are visible.
- These canals are connected to osculum directly, e.g. *Plakina, Geodia, Spongilla* and *Oscarella.*
- In *Spongilla*, the leuconoid condition is derived from a larval stage called the **rhagon**. In this larva, radial chambers are present on side walls only.

- In other members, development of *Leucon* type canal system occurs *via Ascon* and *Sycon* types.
- Generalised course of water current in leuconoid canal system is given below

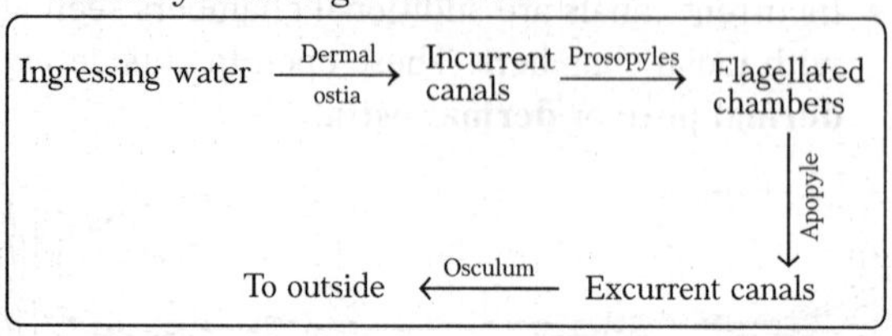

***Fig. 8.10** Leucon canal system : Course of water current*

Types of leuconoid canal system

Leucon type of canal system has varying degree of complexity of canal. It is further divided as follows

- **Euryplours type** This is the simplest type having all the characteristics of leuconoid canal system, e.g. *Plakina*.
- **Aphodal type** In this type, the apopyle is drawn out as narrow canals called **aphodus**. These connect flagellated chamber with excurrent canal, e.g. *Geodia*.
- **Diplodal type** In this type, prosopyles are also converted to tubes called **prosodus**. Thus, this type contains two tubes extending from the radial chambers, i.e. Aphodus and prosodus e.g. *Spongilla*.

10. Almost all sponges possess an endoskeleton. The endoskeleton provides support to the body of sponge. It consists of either **spicules** or **spongin fibres** or both.
    Further, the spicules may be **calcareous** (made up of calcium carbonate) or **siliceous** (made up of silica).
    These may be **monaxon, triaxon, tetraxon** and **polyaxon** also depending upon the number of radiating arms they possess.
11. Sponges are the **filter-feeders**, i.e. the microvilli of choanocytes filter the microscopic plants and animal particles from the water entering their body.
    These food particles then enter into their food vacuoles and gets digested thereafter within these vacuoles. Thus, digestion is **intracellular**, i.e. the food is digested inside the cells.
12. In **respiration**, gaseous exchange occurs by diffusion through the body surface.
13. **Excretion** also occurs by diffusion through the body surface. Their excretory product is mainly **ammonia**, thus sponges are **ammonotelic animals**.
14. **Reproduction** occurs by both asexual and sexual modes. Asexual mode occurs mainly by budding or fragmentation or fission under favourable conditions.
    Asexual **gemmules** or internal buds are formed under unfavourable conditions. Each gemmule has a mass of archaeocytes inside them.
    All sponges are **hermaphrodite**, but cross-fertilisation is the rule.

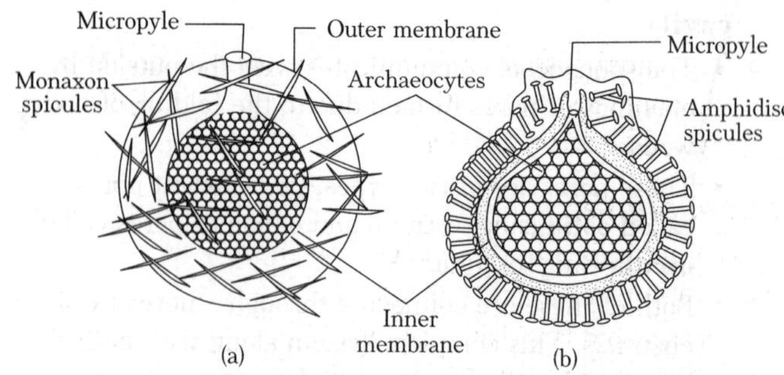

***Fig. 8.11** Gemmules : (a) Complete gemmule of Spongilla (b) Gemmule of Ephydatia in section*

15. **Protandrous** condition is present. Fertilisation is defined as *in situ.*
    Sperm leaves the body of one sponge and carried by water current to fertilise attached exposed ovum inside another sponge, i.e. **internal fertilisation.**
    Cleavage of egg is **equal** and **holoblastic.**
16. Sponges have enormous power of regeneration. This is due to the presence of archeocytes in them.
17. **Development** is **indirect**, i.e. it is mediated by ciliated and freely swimming larval forms called **parenchymula** (*Leucosolenia* and *Clathrina*), **amphiblastula** (*Sycon*), etc.

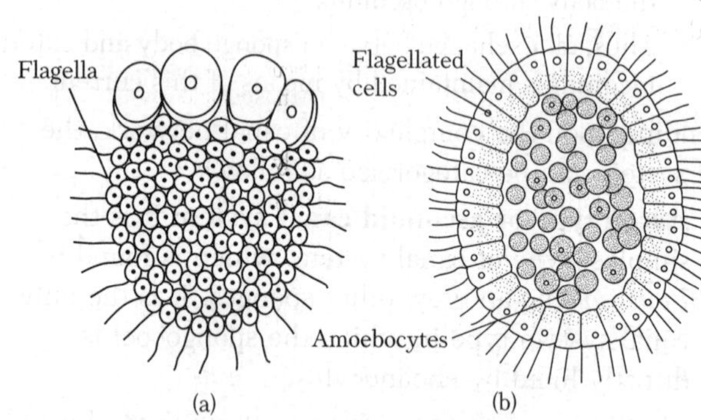

***Fig. 8.12** (a) Amphiblastula larva of Sycon, (b) Parenchymula larva of Leucosolenia*

Note
- *In syconoid sponges, the larva produced is called **stomoblastula** which after growing changes into amphiblastula.*
- *Olynthus is a stage during sponge development that looks like a little sponge.*

Classification

On the basis of endoskeleton phylum-Porifera is classified into following three classes

Class – Calcarea (Calcispongiae)

1. These are dull-coloured and exclusively marine animals.
2. Body is radially symmetrical, cylindrical or vase-like.

3. The skeleton consists of calcareous spicules.
4. There spicules are monaxon, triaxon and tetraxon type.
5. The canal system is Ascon or Sycon type and the choanocytes are relatively large, e.g. *Leucosolenia*, *Sycon*, *Grantia*, *Clathrina*, etc.

Leucosolenia

- It is a marine, colonial sponge, attached to the substratum by adhesive discs.
- It has asconoid type canal system.
- *Leucosolenia* reproduces asexually by external budding.
- Equal and holoblastic cleavage in zygote results to **stereogastrula** or **parenchymula larva**, which swims freely and gets attached to substratum. It finally develops into an adult sessile sponge.

Sycon

- It is commonly called as **urn sponge** or **crown sponge**. *Sycon* is a marine sponge, which possesses a vase-like shape and radially symmetrical body.
- Its canal system is of *Sycon* type.
- The larval stage of this animal is called **amphiblastula** with one-half of the larva have flagella.
- It is a freely swimming stage until the larva settles down and undergoes gastrulation.

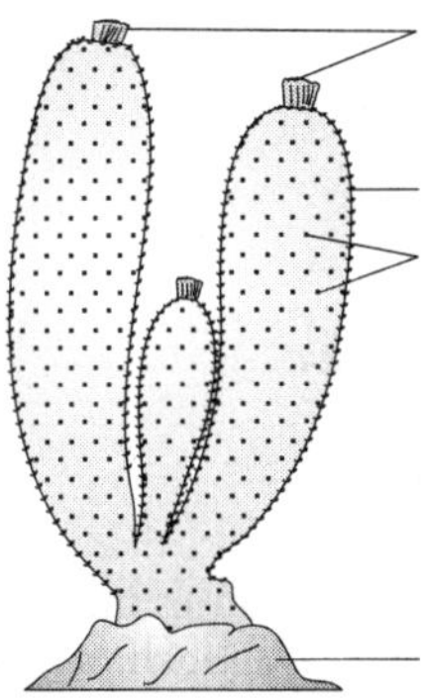

Fig. 8.13 Sycon

Class–Hexactinellida (Hyalospongiae)

1. These are also called **glass rope sponges**.
2. Body shape is cup, urn or vase-like.
3. Spicules are siliceous and six-rayed or triaxon.
4. They possess *Leucon* type canal system with small choanocytes present in finger-shaped chambers called as **radial chambers**.
5. Exclusively marine forms and many of them are seen in deep sea, e.g. *Euplectella*, *Hyalonema* (the glass rope sponge).

Euplectella

- It is popularly called as '**venus flower basket**'.
- The body is cylindrical and the osculum contains a sieve plate formed of fused spicules. The surface possesses sieved areas or parietal gaps.
- The base or root tuft has siliceous threads to anchor the sponge.
- *Euplectella* shows commensalism with certain shrimps. The dried cleaned sponge is highly decorative.

Fig. 8.14 Euplectella : Dried skeleton

The sponge with its locked shrimps is given as a wedding gift in Japan, symbolising 'Remaining together till death'.

Class–Demospongiae

1. Body is asymmetrical, often massive and brightly coloured.
2. Skeleton of siliceous spicules or spongin or both or may be absent in some.
3. Spicules are mostly monaxon or tetraxon, but never triaxon.
4. They possess *Leucon* or *Rhagon* canal system and the choanocytes are small.
5. They occur in marine as well as in freshwater, e.g. *Euspongia*, *Spongilla*, *Cliona* (boring sponge), *Chalina* (dead man's fingers), etc.

Spongilla

- It is a **freshwater sponge** which grows on submerged plants and twigs.
- It appears green because of the presence of symbiotic green alga in it.
- Canal system in *Spongilla* is essentially of rhagon type with choanocytes, which are restricted to small radial chambers on the side walls.
- Spicules are of monaxon type, with larger ones called **megasceleres**. These are held by spongin fibres and provide support to the colony.

- The smaller spicules are called **microsceleres**. These are found embedded in the cells of sponges.

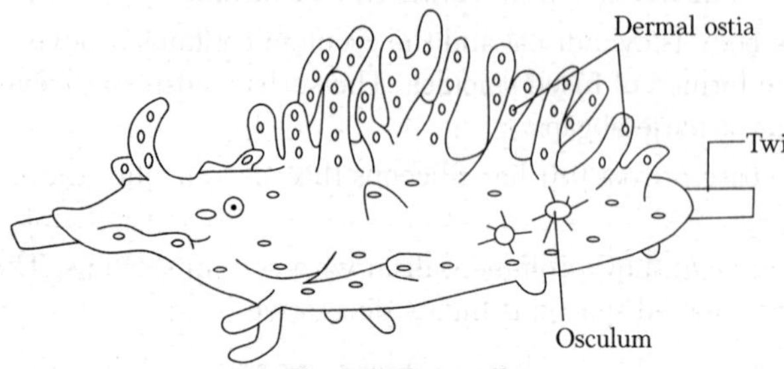

Fig. 8.15 Spongilla : A colony growing on a submerged twig

Euspongia

- It is a marine demosponge and also called as **bath sponge**.
- Its surface is raised into small projections called the **connules**.

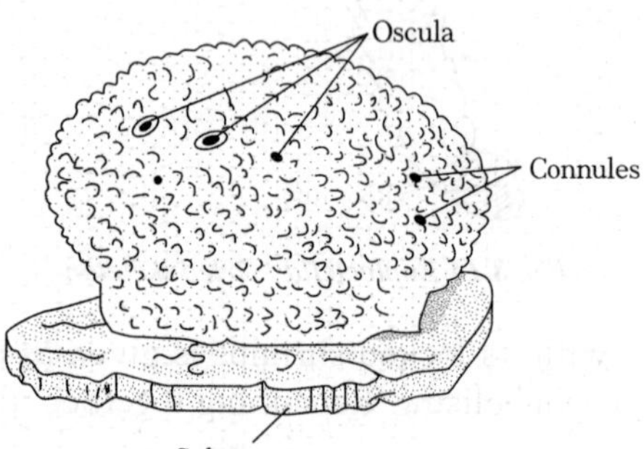

Fig. 8.16 Euspongia

- These connules bear small pores in between them called **dermal ostia**.
- It has spongin skeleton with high water holding capacity thus used for bathing, washing, cleaning, etc.

Note ***Class–Sclerospongiae*** *The 4th class given by* ***Storer****. It has been recently abandoned and its members are assigned to Calcarea and Demospongiae. Rest of the classification is exactly similar to the one given above by Storer.*

Economic importance of sponges

Sponges can either be beneficial or harmful. These are as following

Beneficial sponges

- Skeleton of some silicious sponges like *Euplectella* with its imprisoned shrimps is given as a popular marriage gift in Japan.
- Some marine sponges such as *Euspongia* and *Hippospongia*, are used for bathing, washing, mopping and ceremic work, etc., after removal of the living parts.
- Sponge fishing is done at a large scale in mediterranean Florida and West Indies for commercial purposes.

Harmful sponges

Sponge like *Cliona* destroys several molluscan shells by boring them. It is therefore harmful for pearl oyster industry. This way several edible molluscs such as oysters and clams are destroyed.

CHECK POINT 8.2

1. Which of the following features is universally present in all sponges?
(a) Marine habitat (b) Presence spicules
(c) Presence of spongin fibres (d) Presence of spongocoel

2. In sponges, canal system develops due to the
(a) development of a gestrovascular system later
(b) folding of inner walls
(c) presence of porous walls
(d) reproduction purpose

3. Structure of sponges corresponding to mouth of other animals is
(a) incurrent canal
(b) ostium
(c) osculum
(d) excurrent canal

4. Choose the correct pair.
(a) Parenchymula – *Leucosolenia*
(b) Amphiblastula – *Sycon*
(c) Stomoblastula – Clathrina
(d) All of the above

5. What is left when bath sponges dries up?
(a) Spicules (b) Holdfast
(c) Tentacles (d) Spongin fibres

6. What will happen if a sponge is cut into maximum possible pieces?
(a) These will die
(b) These will differentiate
(c) Every piece will form a sponge
(d) Some pieces will develop in organs

PHYLUM-CNIDARIA or COELENTERATA

The name Cnidaria is derived from cnidoblast or cnidocytes, which are characteristic stinging capsule or nematocyst containing cells of this phyla.

Leuckart coined the term 'coelenterate' for these animals. The study of coelenterates is called **Cnidology**.

General characteristics

1. Aquatic forms, mostly marine, but few of them are freshwater also, e.g. *Hydra*.
2. **Cellular-tissue grade of body** organisation is present.
3. Most of the coelenterates are colonial, e.g. *Physalia*, but some are solitary also, e.g. *Hydra*.
4. Cnidarians are without cephalisation, i.e. distinct head and segmentation in their body.
5. The body is diploblastic with an external epidermis and an inner gastrodermis. A non-cellular mesoglea is present between these two layers. **Mesoglea** consists of proteinaceous matrix devoid of cellular elements.
6. Various types of cells belonging to epidermis and gastrodermis can pass through this layer. It serves for the attachment of the cellular layers, thus serving as the supporting lamella or elastic framework of body.
7. The unique cells of phylum are **cnidoblast** or **stinging cells** or **nematoblasts**, which contain a capsular nematocyst inside them. **Nematocyst** contains a poisonous fluid or **hypnotoxin** inside it.

   The cnidoblasts are found only in epidermis mainly on tentacles. They lack basal-disc. Their structure is shown below

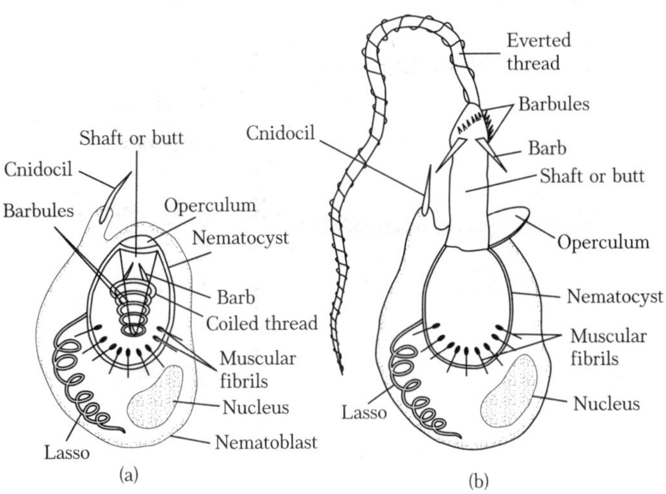

Fig. 8.17 Cnidoblast cells

   These cells are related to food capture, defence and locomotion. Their action depends upon certain enzymes, which brings about increase in osmotic pressure inside the capsule of cnidoblast.

8. The characteristic totipotent cells of this phylum are **interstitial cells**.
9. Just like sponges, coelenterates also have very high power of regeneration. This very high regenerative power of these animals is because of interstitial cells only.
10. The body cavity of coelenterates is called **gastrovascular cavity**. This cavity opens to outside *via* a single opening. Thus, this opening performs the function of mouth and anus both.Thus, they have **blind sac body plan**.
11. Most of the body forms are soft and some are with transparent appearance also. Some of the forms show bioluminescence as well.
12. Cnidarians are mostly carnivorous. The mode of digestion is partly extracellular and partly intracellular. The food enters into the gastrovascular cavity and partially digested there. It is then taken up by epithelial cells for final intracellular digestion.
13. No specialised respiratory, circulatory and excretory organs are present. Gaseous exchange takes place through general body surface. Circulation of various materials takes place only at the cellular level, i.e. from cell to cell.
14. No nervous system is seen, but non-polar neurons are scattered irregularly in the body. In free-swimming forms, statocysts or tentaculocysts are present as the organs of balance.
15. Coelenterates show the phenomenon of **polymorphism**, i.e. occurrence of more than one type of morphological forms within the life cycle of same individual.
16. The two basic forms of coelenterates are **polyp** and **medusa**. Some other animlas, e.g. *Physalia, Vellila* have more morphological forms also like **dactylozooids, tentaculozooids, skeletozooids,** etc., in their colony.
17. The **polyp** is a sessile or attached form. It mainly provides nourishment to the colony. It reproduces through asexual means only especially *via* budding.
18. The **medusa** forms are free-swimming, umbrella-shaped forms. They reproduce by sexual means and can be monoecious (bisexual) or dioecious (unisexual).
19. In unisexual forms, although sexes are separate, but sexual dimorphism is not visible.

20. Another important phenomenon shown by coelenterates is **metagenesis**, i.e. alternation of generation. Here, one diploid generation like polyp alternates with the other diploid generation like medusa. Hence, it is quite different from the alternation of generation found in plants.

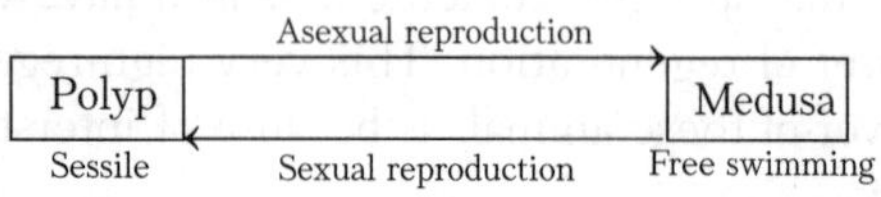

21. Reproduction commonly occurs with asexual or external budding in polyp forms during favourable conditions.
22. For sexual reproduction the gonads are simple, but without sex ducts.
23. Fertilisation is *in situ* and cleavage is **holoblastic**.
24. Development can be direct or indirect. The common larval forms in the case of indirect development are **planula** and **ephyra**.

Note
- *Sensory and nerve cells appeared first time in phylum–Cnidaria.*
- *During unfavourable conditions, some members of phylum–Cnidaria reproduce by branching and reduction bodies also.*

Classification

Coelenterata can be classified on the basis of polyp and medusa shape. It is divided into three classes, i.e. Hydrozoa, Scyphozoa and Anthozoa.

Class–Hydrozoa

(Gk. *Hydros*-water; *zois*-animal)

1. These are the animals with dominant polyp stage and reduced or absent medusa stage.
2. Gastrovascular cavity is without partitions and nematocyst. However, nematocysts are present over tentacles.
3. Mouth is seen at the tip of a conical elevation called **hypostome**.
4. Polymorphism of highest degree is seen in this class. Gametes are epidermal in origin.
5. Some solitary, some colonial and most living in salt water, e.g. *Hydra, Obelia, Physalia* and *Ephelota.*

Hydra vulgaris (Freshwater polyp)

- **Trembley** a Swiss biologists discovered *Hydra.* It is commonly called as **freshwater solitary polyp**.
- It is negatively chemotropic, i.e. remains away from chemicals, hence avoids chlorinated water. It is found attached to solid objects in water such as leaves, stones, weeds, etc.

Body wall and its cells

- The body wall of *Hydra* consists of following cells
  (i) **Epidermis** The various epidermal cells are
  - **Epitheliomuscular cells** Each of these cell has
    - **Outer epithelial** part extending to the body surface.
    - **Basal muscular** part containing contractile fibril myoneme. Due to muscular part *Hydra* can increase or decrease its body dimensions.
  - **Glandulomuscular cells** These are present in the pedal disc region. These cells secrete sticky material for attachment of animal to substratum.
  - **Interstitial cells** They are **totipoent cells** which give rise to all the different cells of the body.
  - **Sensory cells** They are the most numerous and help receive and transmit impulses.
  - **Nerve cells** These cells occur for the first time in coelenterates. There main function is to conduct nerve impulses.
  - **Germ cells** They proliferate to form gonads.
  - **Cnidoblast** Stinging cells of *Hydra* which act as organs for offence and defence.

  (ii) **Gastrodermis** The various gastrodermal cells are
  - **Endotheliomuscle or Nutritive muscle cells** They help in contraction of the body and nutrition.
  - **Endotheliogland cells** They secrete digestive enzymes in *Hydra.*
  - It also contains interstitial cells, sensory cells and nerve cells, but no cnidoblasts.

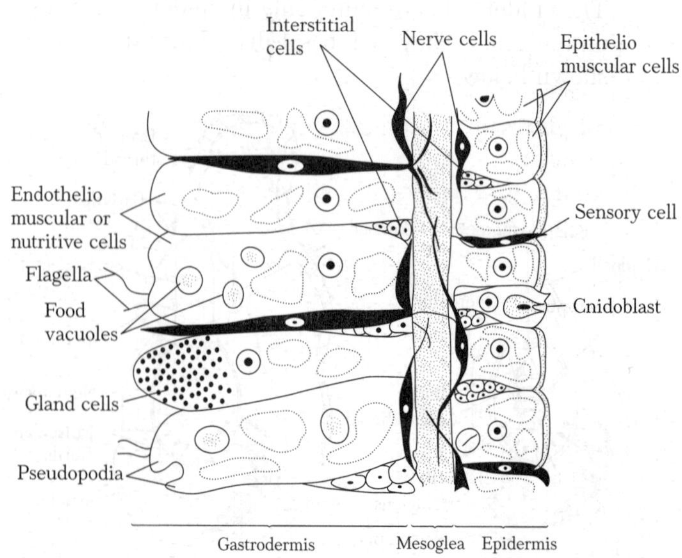

Fig. 8.18 Hydra : A portion of body wall in longitudinal section (magnified)

External morphology

- The body of *Hydra* has mouth over hypostome. It is the most sensitive region of *Hydra's* body.
- The basal part of the body has a pedal or basal disc.
- The complete animal is shown below

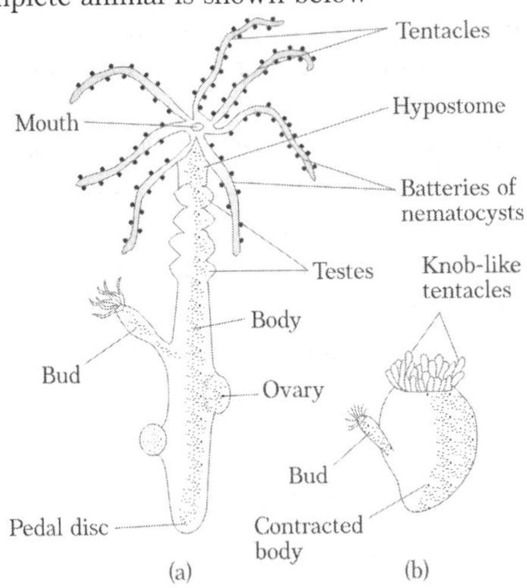

***Fig. 8.19** External morphology; (a) Expanded body with bud and gonads, (b) Contracted body bearing bud*

- Mouth is surrounded by 6-10 tentacles, which help in locomotion. Hence, these are analogous to pseudopodia of *Amoeba*.
- The tentacles can be stretched to several millimetres for obtaining food.

Cnidoblasts

Stinging cells or **Cnidoblasts** are especially abundant on tentacles. Where these are arranged in clusters or batteries. These contain oval nematocyst and are of following four types

- **Penetrant** or stenotele (largest cnidoblast).
- **Large glutinant** or holotrichous isorhizous or streptoline glutinant.
- **Small glutinant** or atrichous isorhizous or steroline glutinant.
- **Volvent** or desmoneme (smallest cnidoblast).

Table 8.2 Differences between four types of nematocysts

Cause of difference	Penetrants	Large glutinant	Small glutinant	Volvents
Size of filament	Pointed, long	Round, long	Round, small	Round, long
Butt	With spines	Absent	Absent	Absent
Cnidocil	Present	Present	Present	Present
Material in capsule	Hypnotoxin, poisonous	Mucus substance and hypnotoxin	Mucus substance	–
Functions	Help in offence and defence	Help in defence and offence	Help in locomotion	Round up around prey

Movement

- The contraction and expansion of *Hydra's* body called **contraction burst**.
- It is initiated by a pacemaker located near the base of its hypostome.
- The other movement types shown by the *Hydra* are looping, somersaulting, gliding, walking, climbing, floating, surfacing and swimming.
- All these movements are attributed to the circular and longitudinal muscle coats of its body wall.

Food, Feeding and Digestion

- *Hydra* is exclusively carnivorous and feeds on all sorts of small animals. It likes water fleas, e.g. *Daphnia, Cyclops*, etc., as its food.
- *Hydra* feeds only on those substances, which possess glutathione as their body component. *Hydra* can digest proteins, fats and some carbohydrates, but not starch.
- The **digestion** in *Hydra* takes place in two phases, i.e. first extracellular digestion (in gastrovascular cavity) and then intracellular digestion (in endodermal cells).
- The **undigested** remains are left in gastrovascular cavity and from there these remains are flushed out periodically by sporadic strong constrictions of the body, through the mouth.

Other activities

- The exchange of gases and elimination of nitrogenous wastes (ammonia) occur through the general body surface.

Nervous system

- *Hydra* has a very primitive type of nervous system. Brain is absent in *Hydra*. It includes bipolar and multipolar cells, which form **nerve net** or **nerve plexus**.
- The axons and dendrites are not differentiated in nerve cells. *Hydra* is one of the first animal among invertebrates to show beginning of nervous system.

Reproduction

- It reproduces amitotically or asexually as well as sexually or mitotically. **Asexual reproduction** mainly occurs through external budding during summer season and when plenty of food is available.
- In natural habitat, the animal **reproduces sexually** in the autumn season.
- Several species of *Hydra* are dioecious or unisexual, (e.g. *Pelmatohydra oligactis*), while some of the species are monoecious or bisexual (e.g. *Hydra vulgaris*). There is no free larval stage seen in the development of *Hydra*.
- *Hydra* has enormous power of regeneration but unlike sponges, it does not use its power of regeneration as a means of reproduction.
- One characteristic feature of regenerating piece in *Hydra* is that it retains its polarity.

Obelia (Sea fur)

- It is commonly called as **sea fur.**
- It is typically a sedentary, marine and colonial hydroid having cosmopolitan distribution (abundant in both Atlantic and Pacific coastal water).
- In *Obelia*, life cycle shows **metagenesis**, that includes a fixed polypoid phase represented by the hydroid colony and a pelagic medusoid phase.

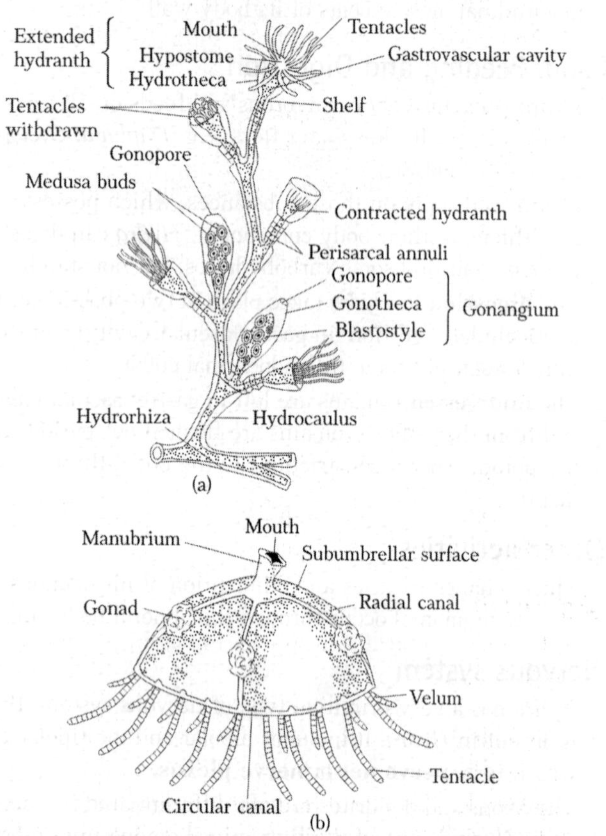

Fig. 8.20 Obelia : (a) A portion of colony, (b) Swimming with everted

Table 8.3 Differences between polyp and medusa forms

	Polyp	Medusa
Shape	Cylindrical in shape	Umbrella-like
Nature	Mostly sessile, but some are motile	Free swimming
Habit	May be solitary or colonial	Always solitary
Polym-orphic forms	• Gastrozooids (Hydranth) : For nutrition	• Phyllozooids : For protection
	• Dactylozooids : For protection	• Nectophore : For swimming
	• Gonazooids : For reproduction	• Goinophore : For reproduction
		• Phenumatophores : For swimming

Note *The hydranth of Obelia bears 24 tentacles, while medusae bear 16 tentacles in addition to tentaculocytes.*

Physalia (Portugese man of war)

- This animal is considered as one of the best example of **polymorphism** in Coelenterata.
- The animal body is a well-defined colony. It has a large (up to 30 cm), semi-transparent and gas-filled, bladder-like bright blue float or pneumatophore, which always keeps the colony at water surface.
- The up and down movements of *Physalia* body in water occur due to the secretion and absorption of gases like N_2 and CO.
- These gases are secreted by gas glands located in inner lining of the float.
- The complete colony looks like as shown below

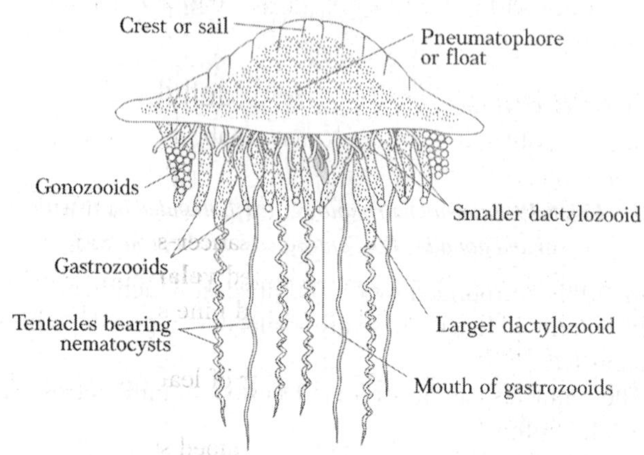

Fig. 8.21 Physalia : A colony

- Several zooids in small group hang down from float into water. Each group of zooids is called **cormidium.**

The cormidium includes following four types of zooids

Zooid types	Features	Functions
Gastrozooids	Have mouth and tentacles for feeding	To capture and digest the food
Small dactylozooids	Without mouth, but carry tentacles	In feeding and defence
Large dactylozooids	Without mouth, but with a very long fishing tentacle with large number of cnidocytes	Most important in feeding and defence
Gonozooids or **blastostyle** or **gonodendron**	Stalk-like structure without mouth and tentacles	Carries several small medusoid zooids (gonophores), bearing male or female gonads.

Class–Scyphozoa or Scyphomedusae

(Gk. *Skyphos*–cup; *zois*–animals)

1. It is the class of **jellyfishes** or true medusae.
2. The polyp stage is very much reduced. It is being represented as **scyphistoma** stage.
3. Symmetry is radial and mesoglea is thick and usually cellular with more amoebocytes.

4. Mouth is present at the tip of tube-like structure called **manubrium**, which is present as a protruding structure from the subumbrellar surface of the body.
5. In scyphozoans, a row of filament on the surface of gastric cavity is present. These are known as **phacella**. These functions in killing or paralysing the live prey (in stomach).
6. Notches in bell margin are with sense organs called **rhopalia**. These have endodermal statolith. The tentacles also contain tentaculocysts in their base.
7. The medusae is sexual, dioecious with endodermal gonads in gastric cavity.
8. Sexual dimorphism is not seen.
9. Development is indirect with the involvement of a free-swimming **ephyra** larvae, e.g. *Aurelia* (jellyfish) and *Chiropsalmus* (sea wasp).

Aurelia aurita (Moonjelly or Jellyfish)

- It is one of the larger jellyfishes inhabiting in the warmer and temperate seas.
- It is carnivorous, unisexual, bioluminescent and free-living organisms, with a soft, gelatinous saucer-shaped body.
- The outer edge of the body is called **velarium**. Its margin bears numerous short tentacles and nine sense organs called as **rhopalia**.
- Each rhopalia is enclosed by a pair of leaf-like structure called **lappets**.
- Gonads are visible as horse shoe-shaped structures placed at subumbrellar surface. The larval stage is 8 arm containing ephyra larva.

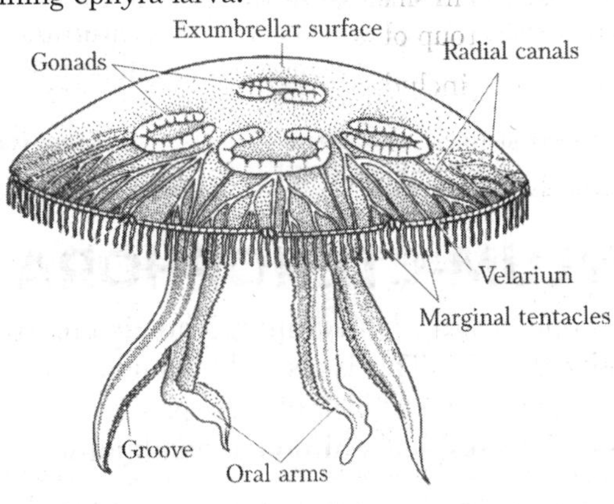

Fig. 8.22 Aurelia aurita (adult male medusa)

- The polyp stage in life cycle of *Aurelia* is represented by a reduced stage **scyphistoma.**

Class – Anthozoa or Actinozoa

(Gk. *anthos*-flower; *zois*-animal)

1. This class includes corals, sea anemones, etc.
2. All of these are marine, solitary or colonial animals with only attached forms or polyps. The medusoid forms are altogether absent.
3. Mouth present as a slit at the surface. It leads to a **tubular stomadaeum.**
4. Gastrovascular cavity is divided by vascular septa or mesenteries (i.e. membrane folds) bearing cnidoblasts on inner margin.
5. Mesoglea is seen as connective tissue and contains cells and fibres.
6. No alternation of generation occurs in Anthozoa.
7. The gonads are endodermal and shed gametes into digestive tract, where they escape *via* mouth.

Some anthozoan with their common names are given below

Examples	Commonly called
Tubipora	Organ pipe coral
Fungia	Mushroom coral (solitary coral)
Gorgonia	Sea fan
Astraea	Star coral (colonial coral)
Meandrina	Brain coral
Alcyonium	Dead man's finger
Heliopora	Blue coral
Corallium nobile	Red coral (most precious coral, vern-moonga)
Pennatula	Sea pen or sea feather
Renilla	Sea pansies
Adamsia	Sea anemone
Madrepora	Staghorn coral

Adamsia (Sea anemone)

- It is small sedentary and solitary cylindrical coelenterate of shallow coastal water.
- It has a slit-like mouth and several whorls of tentacles.
- Externally, the appearance of anemone is that of a radially symmetrical animal, but looks like biradial symmetry.
- Anemones feed on invertebrates and fishes.
- *Adamsia* is found attached to the empty shell of gastropods (molluscs) occupied by hermit crab. Thus, it shows **commensalism.**

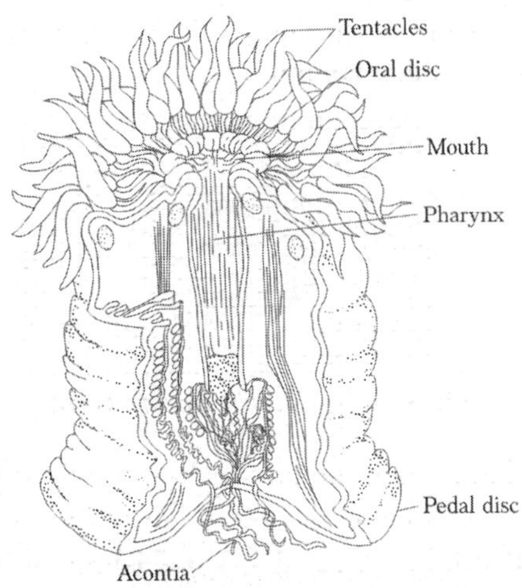

Fig. 8.23 Diagrammatic representation of Adamsia (sea anemone)

Corals

- These are marine mostly colonial, polyploid group of cnidarians. These belong to class-Anthozoa.
- Skeleton of solitary coral is known as **corallite**, secreted by epidermis.
- Coral colonies grow continuously in size by budding and often form extensive masses known as **coral reefs**.

Coral reefs

- These are ridges or mounds of limestone. The upper surface of which is formed cheifly of $CaCO_3$. It is secreted by polyps.
- The principle builders of coral reef are stony coral, but they cannot build a reef alone.
- Many other important organisms, which help in reef growth are ***Zooxanthellae***, encrusting crystalline algae like *Porolithon, Lithamnion*, green algae like *Halimeda* and coral rubbles, i.e. fragments of broken coral.

Types of coral reefs

There are three main categories of coral reefs

(i) **Fringing reefs** These are the simplest and most common reef type. They grow as a narrow band along the shore of volcanic island or continents. These are most relevant type of reef found in Hawaiian Island.

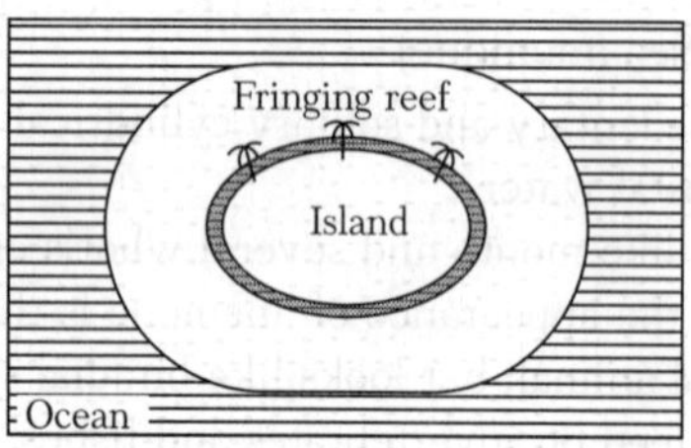

Fig. 8.24 Typical structure of fringing reef

(ii) **Barrier reefs** They are similar to fringing reefs as both grow along the coast, but are considerably separated from the shore (100 km or more) by a relatively deep **lagoon**.

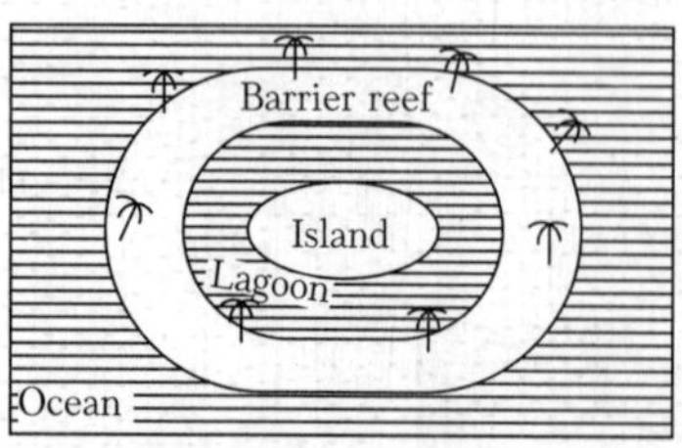

Fig. 8.25 Typical structure of a barrier reef

The largest and the famous barrier reef is great barrier reef along the North-East coast of Australia.

(iii) **Atoll** It is also termed as **coral island** or **lagoon island**. It is a ring-like or horse shoe-shaped reef that encircles a lagoon (not an island) either completely or partially, e.g. Lakshadweep and Maldive Islands.

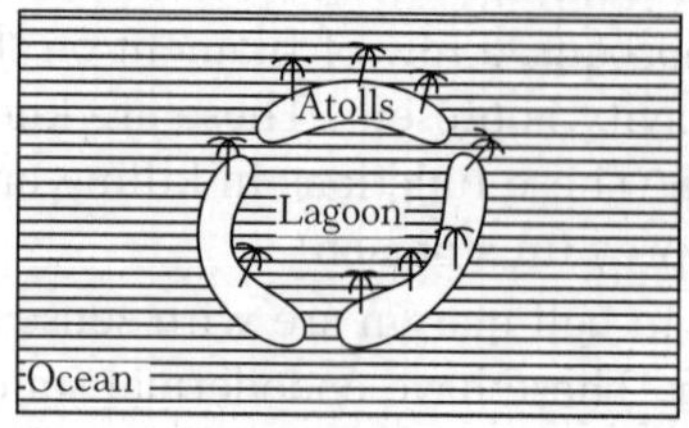

Fig. 8.26 Typical structure of atoll

To sum up, the major differences between all the three classes of phylum-Coelenterata, refer the table given below

Table 8.4 Differences between class-Hydrozoa, Scyphozoa and Anthozoa

Hydrozoa	Scyphozoa	Anthozoa (Actinozoa)
Freshwater as well as marine.	Exclusively marine.	Exclusively marine.
Polyp form, dominant medusa reduced.	Medusa dominant polyp reduced.	Only polyp form, medusa absent altogether.
Mesoglea is acellular.	Mesoglea is thick gelatinous and contains cells.	Mesoglea contains cells and fibres.
Mouth on hypostome.	Mouth on manubrium.	Mouth within the level of body, but extends inside as stomodaeum.
Gastrovascular cavity undivided.	Gastrovascular cavity undivided.	Gastrovascular cavity has compartments, i.e. mesenteries.
Cnidoblasts are present only in epidermis.	Cnidoblasts are present in epidermis and gastrodermis.	Cnidoblasts are present in epidermis and gastrodermis.
Gonads are ectodermal.	Gonads are endodermal.	Gonads are endoermal.

PHYLUM-CTENOPHORA

The ctenophores as a distinct group were first recognised by **Escsholtz** in 1892. This is a small phylum of exclusively marine animals. The animals of Ctenophora are also called **sea goose berries, sea walnuts** or **comb jellies**.

General characteristics

1. The members are generally marine, solitary, free-swimming or pelagic.
2. These animals have flat, conical, round or oval body shapes.
3. Their symmetry is **biradial type** along with an oral aboral axis, involving a mixing of radial and bilateral symmetries.
4. They are acoelomate and diploblastic animals with tissue grade level of body organisation.

5. A gelatinous **mesogloea** is present between epidermal and gastrodermal tissue layers.
6. Ctenophores are transparent and show **bioluminescence**.
7. Nematocysts are absent. Instead, specialised adhesive and sensory cells called **colloblasts** or **lasso cells** are present on tentacles, which help in food capture.
8. All the members of this phylum contain **eight vertical rows of ciliary bands** called **comb plates** and two stout tentacles. The name comb jelly is given just because of these comb plates.
9. Digestive system is with mouth, stomodaeum, complex gastrovascular canals and two aboral anal pores.
10. Skeletal, circulatory, respiratory and excretory organs are completely absent.
11. Nervous system is similar to that of coelenterates with statocyst (in aboral region) as the organ of balance.
12. Asexual reproduction and alternation of generation are absent. But regeneration and paedogenesis, i.e. reproduction by young forms are common.
13. All hermaphrodite (monoecious) forms, gonads develop from endoderm, with orientation side by side on digestive canal. Fertilisation is external.
14. Development is indirect through a characteristic **cydippid larva**.

Classification

The phylum is also known as Acnidaria and is divided into following two classes

Class–Tentaculata

- It comprises of animals with simple, rounded or oval or ribbon-like body.
- They contain two long aboral tentacles.
- Examples of Tentaculata include *Ctenoplana*, *Velamen*, (venus girdle), *Pleurobrachia*, *Cestum*, etc.

Pleurobrachia

- These are commonly called as **sea walnuts**.
- These are common marine pelagic forms, found all over the world.
- Their body is transparent and pear or walnut-shaped.

Class–Nuda

The members of Nuda do not contain tentacles and have a highly branched gastrovascular cavity. Example of Nuda is *Beroe*.

Beroe

- It has a thimble-shaped body, which is somewhat compressed along the sagittal axis.
- It is commonly called as **swimming eye of cat**.

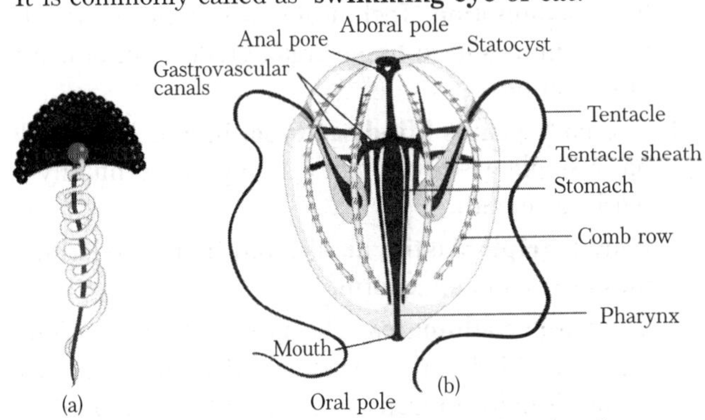

Fig. 8.27 *Ctenophora : (a) Colloblast (b) Pleurobrachia* sp.

CHECK POINT 8.3

1. The mesoglea of cnidarian is
   (a) cellular and contain mesenchyme cells
   (b) cellular and give rise to germ cells
   (c) proteinaceous but devoid of cellular elements
   (d) proteinaceous with cellular elements
2. Hypostome is
   (a) conical mouth in hydrozoans
   (b) conical elevation in scyphozoan
   (c) conical elevation with mouth in hydrozoans
   (d) Both (a) and (c)
3. Out of the given options, choose the character present only in the coelenterates.
   (a) Hermaphroditism (b) Flame cells
   (c) Polymorphism (d) Nematocysts
4. Which pair of cells are present in the epidermis of *Hydra*, but not in its endodermis?
   (a) Gland cells and germ cells
   (b) Stinging cells and gland cells
   (c) Stinging cells and germ cells
   (d) Stinging cells and interstitial cells
5. If a *Hydra* is cut into two parts,
   (a) mouth and basal disc develop at any end
   (b) mouth and basal disc develop according to mouth and basal disc of parent *Hydra*
   (c) *Hydra* will be formed without mouth and basal disc
   (d) None of the above
6. Zoochlorellae and zooxanthallae present in *Hydra* are
   (a) symbionts in nutritive cells
   (b) symbionts in the gut
   (c) symbionts in chidoblasts
   (d) organisms that provide hypnotoxin
7. Why does Ctenophora is a minor phylum?
   (a) It includes small-sized animals
   (b) It includes only few genera
   (c) It does not include animals of economic importance
   (d) It was included earlier in Cnidaria

PHYLUM–PLATYHELMINTHES

- The term Platyhelminthes (Gr. *Platys*—flat; *helminths*—worms) was first proposed by **Gegenbauer** in 1859.
- This animal phylum contains over 13,000 species.
- Some of these are free-living, occupying marine, freshwater and even damp terrestrial habitat. However, most of the species of this phylum are endoparasites, i.e. they live inside the body of other animals called **hosts**.

General characteristics

1. The body of these animals is dorsoventrally flattened and often elongated. Hence, they are commonly called as **flatworms**.
2. The animals are **bilaterally symmetrical** with definite polarity, i.e. they have well-defined anterior and posterior (tail) ends. Head is still not well defined.
3. It is the first phylum with preliminary organ level of body organisation. Technically, it is better to say that these animals have tissue organ grade level of body organisation.
4. The body plan is **blind sac type**, in which the body has a single gastrovascular opening that primarily works as mouth.
5. Body is **triploblastic**, i.e. derived from ectoderm, mesoderm and endoderm.
6. They are **acoelomates**. The space between various organs is filled with special mesodermal parenchyma tissue. It helps in transportation of food materials.
7. Body is unsegmented except in the class-Cestoda.
8. Beginning of **cephalisation**, i.e. differentiation in head region is visible in the form of structures like scolex (*Taenia*), oral cone (*Fasciola*) and auricle in (*Dugesia*).
9. Digestive tract is incomplete, i.e. having a single inlet opening, i.e. mouth, but any opening comparable with anus is absent.
10. Respiration is aerobic in free-living forms as well as anaerobic in parasitic forms. Specialised organs of gaseous exchange are absent and gaseous exchange takes place through the general body surface *via* diffusion.
11. Parasitic flatworms have special adaptations in their body to obtain nutrition from host. These include presence of **hooks** as adhesive structure, **suckers** as adhesive and ingestive structure and **tegument** as protective layer.
12. These animals lack well-defined circulatory and skeletal system.
13. Excretion and osmoregulation are done by a system of specialised cells called **protonephridia** or **flame cells** or **solenocytes with ducts.**
14. A **flame cell** is a hollow cell having irregular shape which send out the pseudopodial processes into the surrounding tissue. Its broad end encloses a long cilia or flagella.
These cilia beat continuously to give the cell a flame-like appearance. Thus, the name flame cell is given to them.

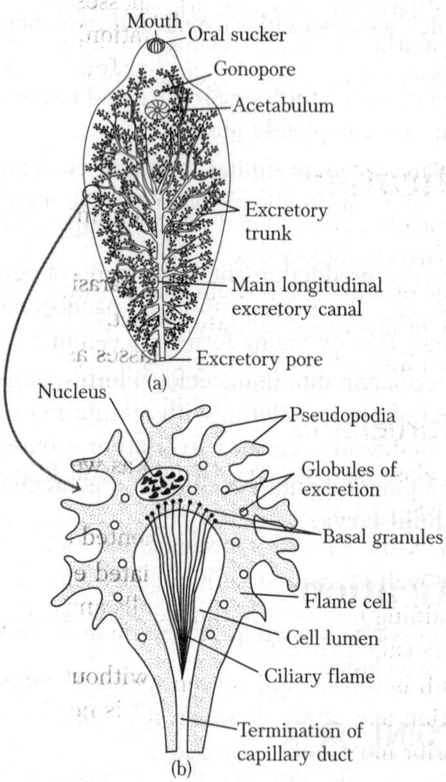

***Fig. 8.28** Fasciola : (a) Protonephridial system and (b) A flame cell*

15. A well-defined nervous system is seen for the first time in flatworms. It is **ladder-like** in appearance. Structurally, it consists of a brain, a pair of anterior ganglia and two main longitudinal nerve cords. These cords are connected at intervals by the transverse commissures.
16. Sensory organs are well-developed in free-living forms, but in parasitic forms these are comparatively less developed.
The sensory organs of free-living forms include eyes. These are not image forming, but can detect the changes in the intensity of light.
Other sensory organs of free-living forms include auricles, papillae, etc.

17. They mainly reproduce sexually by producing sperms and ova. However, some forms reproduce asexually also like planaria by transverse binary fission.
18. The flatworms are hermaphrodite or monoecious except *Schistosoma*. Fertilisation is internal in flatworms.
19. The life cycle of parasitic forms is remarkable. It involves the processes like **polyembryony**, i.e. more than one larval forms within the life cycle of one individual and **paedogenesis**, i.e. larval forms with the power of multiplication.
20. Some flatworms like planaria possess a well-marked power of regeneration, i.e. an ability to develop a complete organism from a part of the body.

Classification

The classification of phylum-Platyhelminthes is based on following two characteristics

- The mode of life, i.e. free-living or parasitic.
- Presence or absence of digestive tract.

It is divided into following three classes as

Class–Turbellaria

1. This class includes free-living as well as parasitic species.
2. Body of the animal is unsegmented and leaf-like.
3. Body wall is covered with ciliated epidermis containing mucus secreting cells and rod-shaped bodies called **rhabdites**.
4. Mouth is ventral and is seen without suckers. Its position may vary. However, it is not be seen at the anterior most end of the body.
5. Alimentary canal is seen in its incomplete form with a **protrusible pharynx**.
6. Intestine is also present, but it ends blindly.
7. Tango, chemo and photoreceptors are common as most forms show free-living habit.
8. Development is direct usually with asexual mode like binary fission in some, e.g. *Microstomum*, *Dugesia* (planaria), *Bipalium* (only terrestrial planarian), *Euplanaria*, *Mesostoma* and *Convoluta*.

Dugesia (Planaria)

- It is commonly found in freshwater ponds, lakes, streams and shallow rivers.
- These are gregarious in habit, i.e. live in groups.
- Anterior most portion of the body has a pair of eye spots and also bears a pair of lateral projections called **auricles**. Some scientists may consider it as head, but it does not bear mouth.

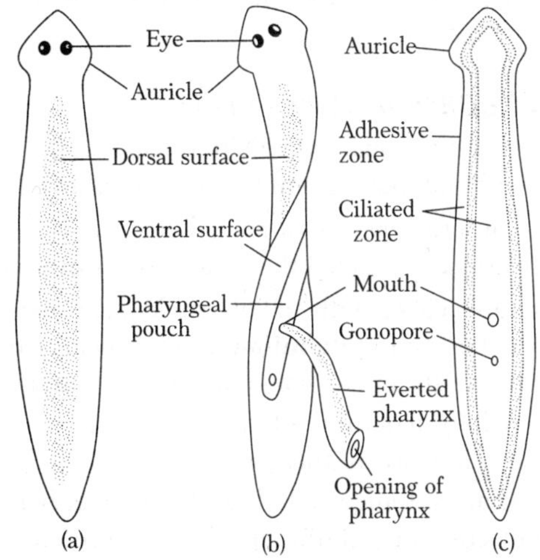

***Fig. 8.29** Dugesia (Planaria) External features (a) Dorsal view, (b) Folded showing protrusible pharynx and (c) Ventral view*

- Their pharynx is a tubular structure that can be everted or protrusible beyond the mouth.
- A common genital aperture or gonopores is present only in sexually mature individuals just below the mouth on the ventral side.
- **Neoblast cells** are the characteristic feature of planarians. These are pleuripotent cells which helps in regeneration of lost body parts. Pieces of planarian can also be grafted on to another with the help of these.

Class–Trematoda

1. Members of this class are ecto or endoparasitic flatworms called as **flukes**.
2. Body is unsegmented. Cellular epidermis is absent. Its portective function is taken up by thick cuticle. called **tegument**. The living nature of tegument is proven due to the presence of cellular organelles like mitochondria in it.
3. Suckers and sometimes hooks are present for attachment to the host.
4. Digestive tract is incomplete, mouth anterior, but not at anteriormost region. **Pharynx suctorial** and intestine is with two branches.
5. Three pairs of sensory nerve cords are present in ectoparasitic and free-living larval forms.
6. Mostly monoecious, but some unisexual forms are also found, e.g. *Schistosoma*.

7. Development is direct in ectoparasitic forms and indirect in endoparasitic forms.
8. All of them have alternation of hosts in their life cycle, e.g. *Fasciola, Polystomum, Opisthorchis, Schistosoma,* etc.

Fasciola hepatica (Liver fluke)

- It is a digenetic endoparasite. Body is soft, oval, flat, leaf-like with dull pink-ground colour and brown border.
- The broad anterior portion of animal bears a distinct triangular projection called the **head lobe** or **oral cone.**
- This structure bears mouth and two suckers, i.e. **oral sucker** (anterior sucker) and **acetabulum** (ventral sucker). Both the suckers are the structures containing radial muscles.
- With the help of mouth, the fluke feeds on bile, blood, lymph and cell debris of the host.
- There is no anus. Indigestible food matter is either probably ejected through the mouth or diffused into the excretory system.
- Respiration is anaerobic type, i.e. without oxygen.
- In between the acetabulum and the oral sucker, there is a midventral gonopore or genital opening.

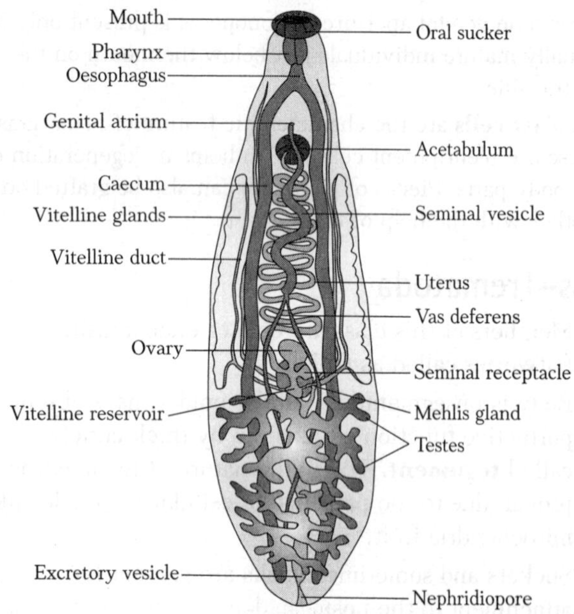

Fig. 8.30 Structure of liver fluke

- At the posterior most end of the body, a single excretory pore is present.
- The animal is hermaphrodite or monoecious.
- The male reproductive system consists of two testes, one pair of vas deferens, seminal vesicle, an ejaculatory duct and a penis.
- The female reproductive system consists of an ovary with its oviduct, vitelline glands along with their ducts and yolk reservoir, shell glands Mehlis gland, ootype, ovovitelline duct, uterus and the female genital pore.
- The male as well as female genital ducts open into a common chamber called **genital atrium.**
- During breeding season, a temporary opening called **opening** of **Laurer's canal** is also seen on the dorsal surface.
- Fertilisation is cross type. Fertilisation is internal and takes place at the distal end of oviduct.

Life cycle of liver fluke

Life cycle of *Fasciola* is complex as it is completed in two host. **Primary host** in which adult fluke lives is sheep, while the **intermediate** or **secondary host** in which numerous stages are passed is a snail.

The life cycle of liver fluke passes through the following developmental stages

(i) **Miracidium** An embryo is ejected out with the faeces of primary host. Under suitable conditions, this embryo differentiates to form **miracidium larva.** It is a free-swimming larva. It does not feed.

(ii) **Sporocyst** When miracidium comes in contact with suitable intermediate host (snail), it enters into snail's tissues and starts absorbing nutrition from it. Now, it is converted into sporocyst larva. Each sporocyst larva produces 5-8 rediae.

(iii) **Rediae** It is another larval stage. It comes out of sporocyst and feeds upon snail's tissue. Its further courses of action is season dependent.

(a) **During winter**, every rediae produces 14-20 cercaria, i.e. next larval stage.

(b) **During summer**, it replicates giving rise to the same form, i.e. daughter rediae.

(c) Both of these activities help in increasing the number of larvae.

(iv) **Cercaria** It is again a free-swimming larval stage. It escapes from the secondary host into surrounding water. After 2-3 days, it attaches itself on grass or aquatic plant and is now called **metacercaria.**

(v) **Metacercaria** This is the encysted infective stage of the fluke. When primary host feeds on grasses it enters the body and establishes itself in the liver, where it becomes mature.

- The metacercaria is infect a juvenile fluke also called **marita.**
- The tough thick and white cyst provides protection against the short periods of desiccation.
- The metacercaria's cyst survives the actions of gastric juices of host because its cyst wall is insoluble in them.
- It phenomenon of appearance multiple larval forms of *F. hepatica* is referred to as **polyembryony**. It is an adaptation for parasitic mode of life.

Some facts about liver fluke

- The liver fluke causes **liver rot** (acute fascioliasis) in sheep. The symptoms appear about one month after the infection.
- ***Fasciola gigantica*** or ***F. indica*** is the common liver fluke of cattle.
- ***Fasciliopsis buski*** seen as intestinal parasite in man and pig.
- ***Clonorchis sinensis*** or ***Opisthochis sinensis*** is the Chinese liver fluke found in human beings and seen particularly in the countries like Japan, Vietnam, Korea, South-East China, etc.

Schistosoma

- It is commonly called as **blood fluke** of man and mammals.
- It was discovered by **Theodor Bilharz** in 1851. Unique feature of this animal is its **dioecious** in nature, i.e. male and female are separate and show sexual dimorphism.
- Secondly, the female lives permanently in a ventral groove running along the complete length of male. This groove is called **gynaecophoric canal** of males.

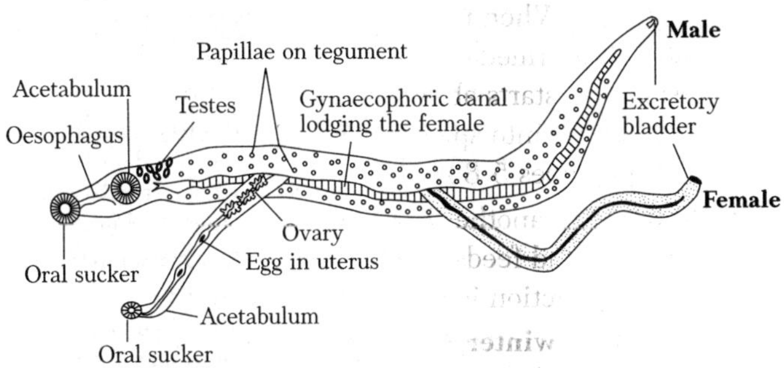

***Fig. 8.31** Male and female Schistosoma*

- It is digenetic and involves two hosts, i.e. man and intermediate host (snail).
- Different larval stages of blood flukes are **miracidium, sporocysts** and fork tailed **cercaria**. The infective stage for humans is cercaria and that for snail is miracidium.
- Blood fluke causes schistosomiasis or Bilharzia. It shows symptoms like allergy, itching, rashes, aches, fever and eosinophilia, etc.

Class–Cestoda

1. It is the class of endoparasitic segmented tapeworms.
2. Body cuticularised, ribbon-like and with false segmentation.
3. Anterior end of the body is in the form of well-defined **scolex**, containing suckers or hooks or both for attachment to the host.
4. Scolex is not considered as the true head because it does not contain mouth over it. However, its presence is considered as the beginning of cephalisation in animal world.
5. Digestive tract and sense organs are absent.
6. Each mature segment or proglottids have a complete set of male and female reproductive organs in it.
7. Life cycle is complicated involving one or more intermediate hosts.

   Embryos formed are with hooks, e.g. *Taenia, Echinococcus, Moneizia, Hymenolepis*, etc.

Taenia solium

- It is commonly called as **pork tapeworm** due to its ribbon-like flat appearance. It is a digenetic parasite.
- Two unique adaptations are seen in tapeworms regarding their parasitic mode of life, i.e.
  - It lacks a mouth and digestive tract in all their life stages. Thus, absorption of food occurs through their general body surface.
  - The body of all the tapeworms is segmented. These segments are called **proglottids.** These segments show structural and physiological differences within a single body. This type of segmentation is called **pseudometamerism** or **strobilisation.**

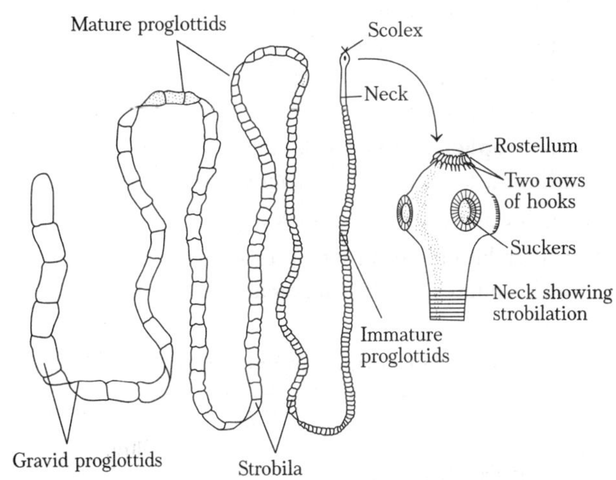

***Fig. 8.32** Morphology of Taenia solium*

- Humans are considered as the primary host of this parasite, while the animals like pig (or dog and sheep) are considered as the secondary or intermediate hosts.
- The body is flattened, elongated and consists of a linear series of segments.
- The body of animal is divided into three regions **scolex, neck** and **strobila.**
- The **scolex** is a knob-like, roughly quad balloon-like structure. It bears oval cup-shaped suckers one at each corner. Its conical top is called **rostellum**, which bears chitinous hooks in two rows.
- The suckers and hooks are organs of attachment and do not play any role in food ingestion.

- The **neck** is a short, narrow and unsegmented region behind scolex. It is the **area of proliferation** or **area of segmentation** from where the budding or segmentation of strobila begins.
- The **strobila** forms the main bulk of the body consisting of 800-1000 segments. A strobila contains three kinds of proglottids, i.e. **immature, mature** and **gravid proglottids**.
- The absence of any reproductive system is the unique feature of the immature proglottids.
- All tapeworms are hermaphrodite. Mature proglottids contain both male and female reproductive system, while gravid proglottids contain only uterus filled with eggs.
- Gravid segments usually detach one by one from the body and passout with the faeces of the host. This phenomenon is known as **apolysis**.
- This helps to limit the body size of animal and transfer the embryos outside, with the faeces of the host for their further development.
- Protandrous condition is visible. Fertilisation within the same proglottid or two different proglottids of same individuals is considered as cross-fertilisation.
- **Onchosphere**, **hexacanth** and **cysticerus** are the different larval stages of *Taenia solium*.
- Onchosphere are the 6 hook eggs.
- The embryonic form which emeges out from onchospheres is called **hexacanth** or six hooked embryo.
- Hexacanth after boring through intestinal mucosa reaches to striated muscles. Here it develops to next larval stage called **cysticerus** or bladder worm. It is the infective stage of human.

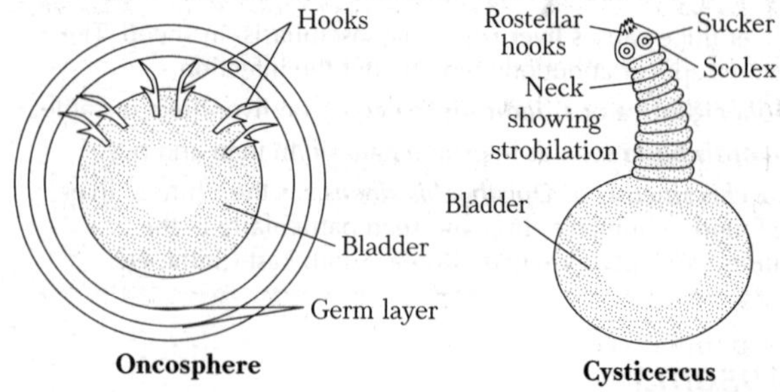

***Fig. 8.33** The life cycle of Taenia solium depicting different larval stages*

On the basis of primary hosts, the tapeworms can be of Pork. Beef, goat or sheep. A brief comparative account of these is given below

Table 8.5 Comparative study of *T. solium*, *T. saginata* and *Moniezia expansa*

T. solium (Pork tapeworm)	***T. saginata*** (Beef tapeworm)	***Moniezia expansa*** (Goat or sheep tapeworm)
Spreads when humans eat improperly cooked pork.	Spreads when humans eat improperly cooked beef.	Infects only sheep and goat.
Primary hosts are humans, while secondary hosts are pigs.	Primary hosts are humans, while secondary hosts are cows or buffaloes.	Primary hosts are sheep and goats, while secondary hosts are orbatid mites.
Ribbon-like 2-4 mm long.	Comparatively longer.	Reaches up to 6-10 m in length.
Rostellum present with 6 suckers.	Rostellum absent, 4 suckers are present	Rostellum and suckers altogether absent.
Proglottids 850 in number.	Approximately 1000-1500 proglottids.	Enormous proglottids.

Other tapeworms are

- *Echinococcus granulossus* or **Hydatid** or **Dog worm.**
- **Hymenolepis nana** or **Dwarf tapeworm** It is the smallest tapeworm infecting human.

CHECK POINT 8.4

1. Platyhelminthes members are generally parasites. Which of the following class is exempted?
   (a) Trematoda (b) Cestoda
   (c) Turbellaria (d) None of these
2. Which one of the following is an actively swimming, free-living and non-feeding stage of liver fluke?
   (a) Sporocyst (b) Rediae
   (c) Cercaria (d) Metacercaria
3. Unlike other trematodes, *Schistosoma haematobium* is dioecious and
   (a) male lives in the pelvic veins
   (b) female lives in the pelvic veins
   (c) female is larger than male and encloses male in its gynaecophoral canal
   (d) the larger male retains the smaller female in its gynaecophoral canal
4. What is common amongst tapeworm, liver fluke and planarian?
   (a) They all are segmented (b) They all are found in gut
   (c) They all have flattened bodies (d) They all have a coelom
5. The metacercaria's cyst of liver fluke survives the actions of gastric juices of host because
   (a) it produces multiple larval forms
   (b) cyst wall is insoluble in them
   (c) it secretes an additional tough coat around it
   (d) it undergoes segmentation
6. Choose the false statement about multiplication in flatworms.
   (a) Parasitic species show polyembryony
   (b) Larval forms may show paedogenesis
   (c) Some forms like planaria reproduce asexually by longitudinal binary fission
   (d) Some flatworms have the capacity to regenerate

PHYLUM-ASCHELMINTHES

Members of Aschelminthes (Gk. *Askos*-cavity; *helminth*-worm) are called **pseudocoelomates.**

Earlier this phylum was called **Nemathelminthes** (Gk. *Nema*-thread; *helminth*-worm) because of their thread like appearence.

These worms are still popularly known as **roundworms** or **threadworms** or **trueworms.**

General characteristics

1. Animals of this phylum may be free-living or parasitic. Free-living forms are few and found in sea water, freshwater or soil, e.g. *Rhabditis* lives freely in soil rich in organic matter.
2. Most of these animals are parasitic, which live inside humans, animals and plants as endoparasites.
3. The animal body is triploblastic with bilateral symmetry.
4. They have **organ-system level of body organisation**. The organs associate to form a distinct system concerned with a specific physiological function such as digestion, excretion and reproduction, etc.
5. Their body is cylindrical, often tapered towards ends and cuticularised. It is non-degenerative to digestive juices of the host.

   Chemically, the cuticle is composed of scleroprotein. It is a protective covering and may bear spines, scales or other forms of ornamentation that protects the animal. Its shedding is called **ecdysis** or **moulting**.
6. No segmentation and cilia are present over the body. In some of the cases, the cuticle is striated transversely, thus giving the appearance of segmentation. This segmented appearance is sometimes called **pseudosegmentation.**
7. Epidermis is **syncytial**, i.e. cells contain many nuclei. A muscular layer is present below the epidermis. It is mainly composed of **longitudinal muscles** only.
8. The roundworms possess **pseudocoelom**. This pseudocoelom is often fluid-filled or may contain a gelatinous substance with mesenchymal cells.
9. Digestive system is complete having two separate openings. The alimentary canal is straight with mouth, specialised suctorial pharynx, non-muscular intestine and anus.
10. Respiration is usually aerobic in free-living forms and anaerobic in parasitic forms. No specialised organs are present for gaseous exchange. In general, it occurs through general body surface in aerobic forms.
11. The excretory organs are formed by the fusion of intracellular canals.

    The cells, which form excretory system in *Ascaris* is formed of **renette cells.**

    In some cases, glandular excretory organs **protonephridia** are seen, while in others combination of both, i.e. canals and glandular excretory organs are seen. Flame cells are altogether absent.
12. Nervous system is with either **cerebral ganglia** or **circumcentric ring** and nerve cords. The number of nerve cords are usually more than two.
13. Following sense organs are also seen in roundworms
    - (i) **Papillae** are raised structures, which occur on the lips. These are present on both sides of mouth in male and female.

      Their presence is also visible in front and behind the cloacal apperture. All papillae are tactile (touch sensitive) in function.
    - (ii) **Amphids** or pits are present on the lips and are chemoreceptors.
    - (iii) **Phasmids** are unicellular gland-like structure located on the lateral sides of the posterior end and are glandulosensory in nature.
14. No asexual reproduction or regeneration is seen.
15. Sexes are separate, i.e. dioecious, sexual dimorphism is usually seen as male is usually smaller than female.
16. Fertilisation is internal. Eggs are microscopic with chitinous shell.
17. Development is direct or indirect. When indirect have many juvenile stages or larval forms, e.g. **filariform larva** is present in *Ancylostoma* (hookworm), **microfilaria larva** is found in *Wuchereria* (filarial worm) and **rhabditiform larva** is present in *Ascaris* and *Enterobius* (pinworm).
18. Many species of Aschelminthes show **eutely**, i.e. a condition in which the number of cells or number of nuclei remains constant both for entire animal as well as for each organ, e.g. the number of body (somatic) cells in all adults of *Caenorhabditis elegans* is 959 and the number of cells in the pharynx of every worm is constant at 80.

Classification

The two important classes of phylum-Aschelminthes (Nematoda) are as follows

Class–Aphasmida (Adenophorea)

1. Phasmids are absent.
2. Amphids are present which are spiral, cord-like or disc-like and rarely pore-like.
3. Excretory system is absent, caudal glands are present.
4. Mostly free-living, e.g. *Trichinella* (triching worm), *Trichuris* (whipworm), etc.

Trichuris trichiura

- It is commonly called as **whipworm**.
- This parasite resides in caecum and colon part of alimentary canal in man.
- The anterior two-third part of body is narrow and has a **thread-like** or **whip-like appearance** and hence, it is referred to as whipworm.
- Its whip-like portion contains a long oesophagus. This helps in the anchoring of worm to the wall of host's intestine.
- Infection usually causes gastrointestinal problems, anaemia and even severe diarrhoea. The parasite is more common in children of tropical region.

Class–Phasmida (Secernentea)

1. Phasmids are present.
2. Amphids are present, which are pore-like.
3. Excretory system is present.
4. Mostly parasitic, e.g. *Anguina*, *Ancylostoma*, *Ascaris*, *Enterobius*, *Dracunculus*, *Wuchereria*, *Loa loa*, etc.

Ascaris lumbricoides

- It is commonly called as roundworm.
- It is an endoparasite of small intestine of man and some other vertebrates like pigs, cattle, sheep, squirrels, etc.
- The digestive system is complete. Mouth is guarded by three **denticulate lips** or **labia**, one dorsal and two ventro-lateral, thus referred to as **triradiate**.
- Amphids are present in associated with middorsal lip.
- Pharynx is seen as a thick-walled, muscular and suctorial structure.
- Intestine extends almost along the entire body length. Its free inner margins contain microvilli and form brush-bordered epithelia.
- The animal feeds upon blood and on partially digested food of host.
- Rectum is lined internally by cuticle. In male, it opens into cloaca to which an ejaculatory duct also opens. In female, it opens separately through the anus which is guarded by an anterior and a posterior lip.
- **Respiratory system is absent** and respiration is mostly anaerobic.
- In the presence of free oxygen, aerobic respiration may occur which is probably indicated by the presence of a small amount of cytochrome in them.
- Excretory system is H-shaped. It is formed by a single excretory cell called **renette cell.**
- This system contains two longitudinal excretory canals connected anteriorly by a transverse canalicular network.
- Sexes are separate and **sexual dimorphism** is well-defined as shown in table below

Male	Female
Body smaller	Body larger
Recurved tail	Straight tail
Presence of single cloacal aperture	Separate genital aperture and anus present
Pre-anal and post-anal papilla present	Absent
Penial setae extends through cloacal aperture	Absent

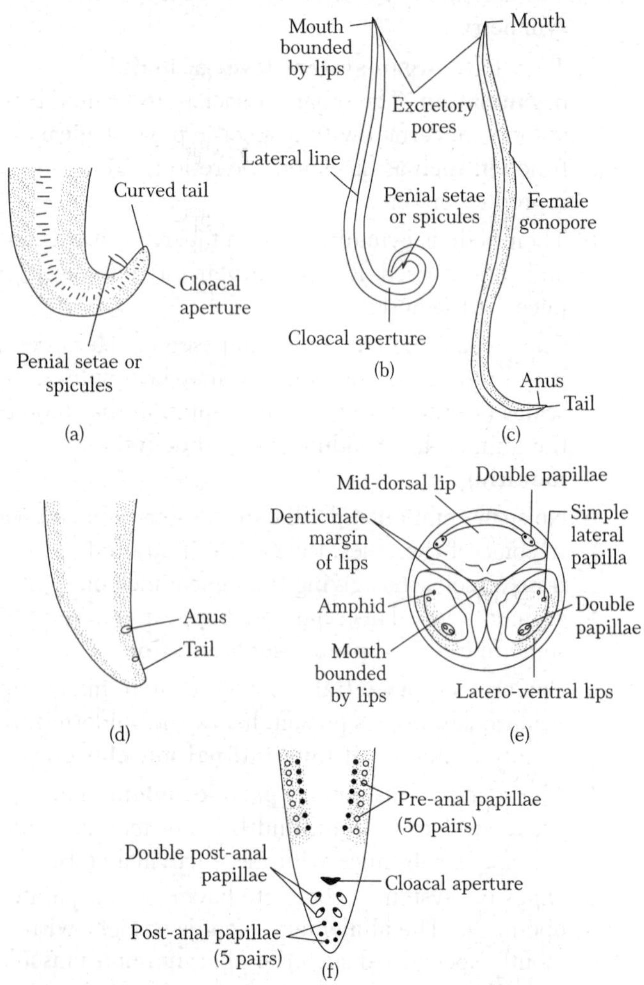

***Fig. 8.34** Ascaris lumbricoides : (a) Anterior end in male, (b) Adult male, (c) Adult female, (d) Anterior end in female, (e) Face view of mouth and lips and (f) Posterior end of male showing papillae*

- Female *Ascaris* is **didelphic**, i.e. having two uteri, two oviducts and two ovaries.
- Male is **monorchic** with single testis.
- The formation of gametes in *Ascaris* is **telogonic**, i.e. the gametes or the gametogonia are budded out only from the proximal or the apical tip of the gonads.
- Fertilisation is preceded by copulation and occurs within the small intestine of host.

Life cycle

- These parasites are monogenetic and have a direct life cycle with no intermediate host.
- The adult attains sexual maturity within 8-10 weeks.
- The average lifespan of animal in the host is about nine months to one year.
- After this, it dies and disintegrates.
- The disease caused by the animal is called **ascariasis.** It is caused *via* contaminated food and water.
- It includes symptoms such as intestinal pain, nausea, diarrhoea and an allergic reaction to toxins produced by the worms.
- For treatment, drugs such as oil of *Chenopodium*, hexylresorcinol, tetrachloroethylene, piperazine hydrate or citrate are used.

Larval moults in life cycle

- **First moult** within the intestine of first human host.
- **Second** and **third moult** within the lungs of second human host.
- **Fourth moult** within the intestine of second human host

Note *Sometimes, larva instead of moving to the gut may migrate to other vital organs like brain, spinal cord, etc. In such organs, they soon die and become encysted in calcified cysts. This is called* **aberrant migration**. *The study of worms causing parasitic infestation in human is called* **helminthology**.

Ancylostoma duodenale

- It is commonly called as hookworm and is the most dangerous roundworms.
- The name hookworm is given, because of the fact that the male have a hook-shaped body.
- These worms live as endoparasites of human small intestine and possess a monogenetic life cycle.
- They differ from roundworms like *Ascaris* in the absence of lips and in possessing a copulatory bursa in male.
- The mode of infection to the host is specific. The larva of animal enters into the host body after boring through the skin. Generally, they penetrate the soft skin of sides of feet and hands.
- The adult worm sucks blood, lymph, bits of mucous membrane and tissue fluid from the lining of intestinal wall.
- This results into mental and physical deficiencies in humans (children).
- The disease can be treated by the administration of drugs like carbon tetrachloride, thymol and oil of *Chenopodium*, etc.

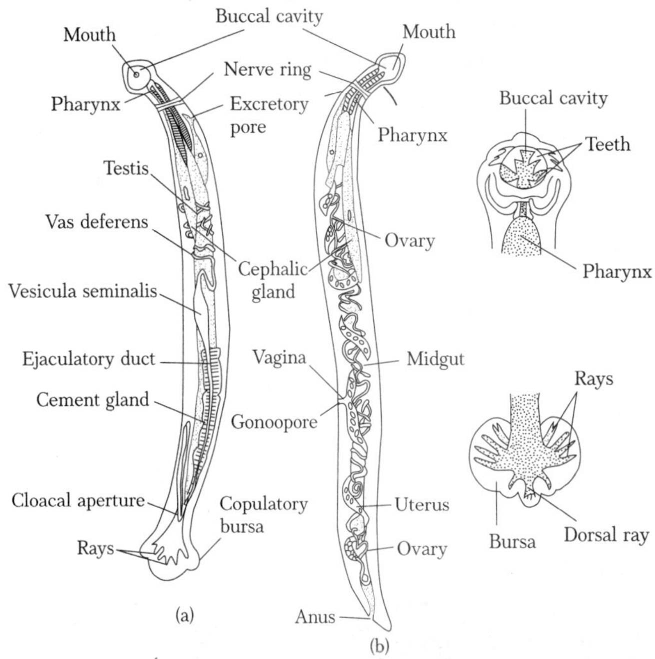

Fig. 8.35 *Ancylostoma duodenale : (a) Adult male, (b) Adult female*

Wuchereria bancrofti

- It is the causative organism of **elephantiasis** or **filariasis.** Thus, also called as **filarial worm.** This is a disease characterised by the immense swelling of lower limbs and genitalia due to the blockage of lymph transport.
- The worm (*Wuchereria*) lives in the lymphatic system and muscles of humans.

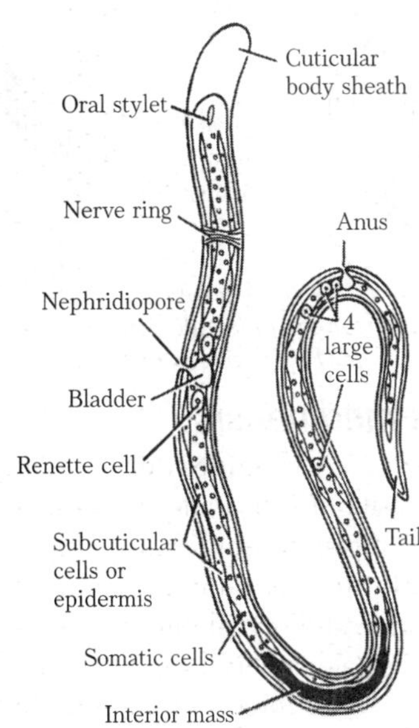

Fig. 8.36 *Wuchereria bancrofti : Juvenile or microfilaria*

- The animal exhibits a digenetic life cycle, i.e. man and *Culex* or *Aedes* mosquito are the two hosts.
- It shows sexual dimorphism where male worm is smaller and female worm is larger. The worms are also **ovoviviparous**, i.e. eggs are produced and hatched inside the parent body.
- Larvae are called as **microfilaria** and excretory system of larva is represented by a **V-spot**. No proper treatment is known for **elephantiasis**, but preventive measures can be used such as eradication of larvae.

Loa loa

- It is a filarial roundworm of Central and Western Africa. Its common name is **African eyeworm**.
- It is commonly found in the subcutaneous connective tissue of human host.
- The parasite is digenetic. The mango flies (*Chrysops dimidia and C. silecea*) act as the intermediate host or vector.
- Larvae are called as microfilariae. During their migration in human host, they pass across the eyeball and may live in conjunctiva or in the eyelids. Migration of microfilariae is accompanied by intense itching and swelling.
- Thus *Loa loa* causes loiasis a disease characterised by subcutaneous swelling called **calabar swelling** mostly around the eyes.

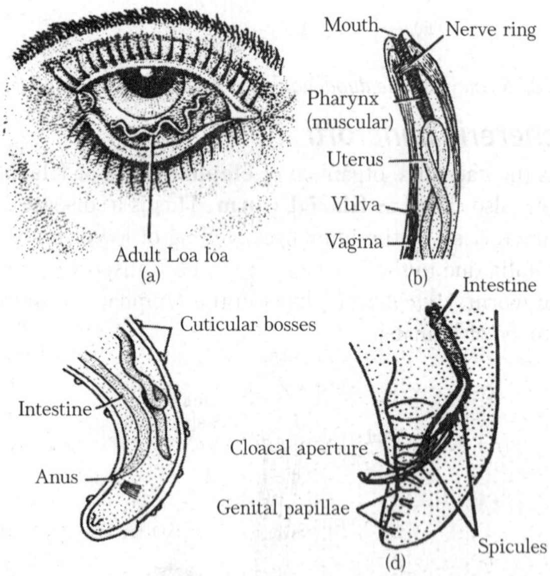

Fig. 8.37 Loa loa : (a) Adult across the bulbar conjunctiva of human eye, (b) Anterior end of female, (c) Posterior end of female, (d) Posterior end of male

Dracunculus medinensis

- It is commonly called as **guineaworm** or **fiery serpent**. It occurs as a parasite of subcutaneous tissue of man.
- The animal is a digenetic endoparasite and its secondary host is *Cyclops* (water flea), a microscopic arthropod.
- Sexes are separate and sexual dimorphism is well-defined. Female is very long, while male is short.
- Infection occurs when infected intermediate host is taken in by the man along with the larva.
- Formation of blisters and itching on arms, shoulders and legs are characteristic symptoms.

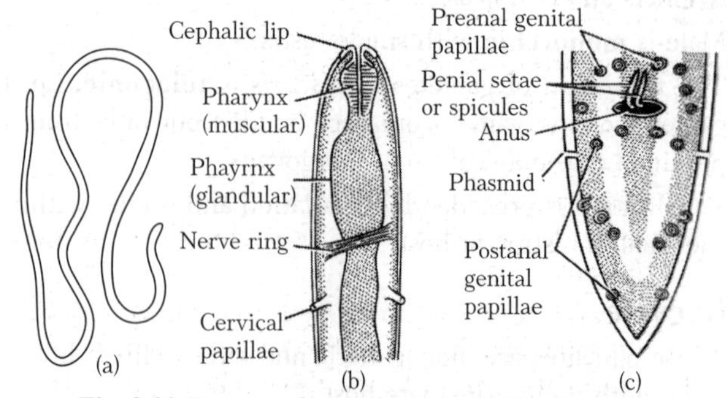

Fig. 8.38 Dracunculus medinensis : (a) Entire worm, (b) Anterior end of male, (c) Posterior end of male in ventral view

Enterobius vermicularis

- This endoparasite is also called **pinworm** or **seatworm**. It perhaps the most common nematode parasite of humans throughout the world.
- It causes enterobiasis in humans especially in children. It inhabits caecum, colon, appendix and rectum.
- Male and female are well-differentiated. The posterior end of male is curved, while the posterior end of female is straight and pointed.
- The life cycle of animal is simple and without any intermediate host, i.e. monogenetic.
- The mode of transmission is very interesting. At night, the female reaches near anus and deposit eggs. This causes severe itching in the rectal and perineal regions and the host tempted to scratch the affected region. The eggs are thus associated with the fingers and nails of host. These finally reach the gut of host with food, etc., to cause reinfection.
- The symptoms of infection include loss of appetite, insomnia, hysteria, restlessness and inflammation of mucous membrane of the infected region. Good sanitary habits are the best weapon against the infection.

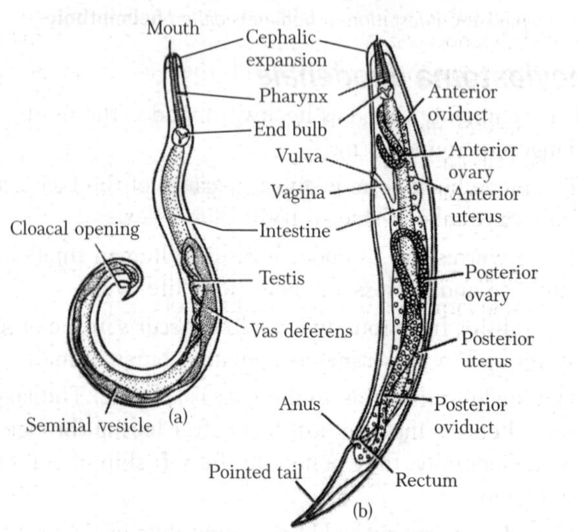

Fig. 8.39 Enterobius vermicularis : (a) Adult male, (b) Adult female

CHECK POINT 8.5

1. The roundworm can be called specialised and non-degenerative with reference to parasitism because
   (a) it has no respiratory organs
   (b) it is dioecious
   (c) it has a straight, uncoiled alimentery canal
   (d) the cuticle over its body wall is resistant to digestive juices of the host

2. The body cavity of Aschelminthes is pseudocoel beause
   (a) it contains large cells termed pseudocoelocytes
   (b) it is bound externally by muscle layer and internally by intestines
   (c) it has very little parenchyma
   (d) it is filled with pseudocoelomic fluid

3. Phasmids are
   (a) multicellular glands located on lateral sides of the posterior end of the worms
   (b) unicellular glands located on medial sides of the anterior
   (c) unicellular glands located on lateral sides of the posterior end of the worms
   (d) multicellular glands located on medial side of the anterior end of worms

4. Aerobic respiration may occur in *Ascaris lumbricoides* which is supported by the presence of
   (a) cytochrome (b) haemoglobin
   (c) haemocyanin (d) vanadium

5. Differentiating trait of *Ascaris* is
   (a) sexual dimorphism and rhabditiform larva
   (b) unisexual and digenetic parasite
   (c) pseudocoelom and metameric segmentation
   (d) hermaphrodite and pseudocoelom

6. Choose the correct statement.
   (a) Guinea worm is a monogenetic parasite
   (b) Pinworm is a monogenetic parasite
   (c) Calabar swellings are associated with fiery serpent
   (d) Microfilariae are the larval form of hook worm

EXAM POINTS

- Cellular organisation of body is present in Porifera and Protista. **[MHT CET 2014]**
- Coelom is found between mesoderm and endoderm. **[BHU 2012]**
- A bilaterally symmetrical deuterostome is classified in the phylum–Chordata. **[COMEDK 2010]**
- *Sycon* belongs to a group of animals, which are best described as multicellular without any tissue organisation. **[BCECE 2012]**
- Body having meshwork of cells, internal cavities lined with food filtering flagellated cells and indirect development are the characteristics of phylum–Porifera. **[CBSE AIPMT 2015]**
- In case of poriferans, the spongocoel is lined with flagellated cells called choanocytes. **[NEET 2017]**
- Animals belonging to the phylum–Porifera have nematocyst. **[WB JEE 2015]**
- Choanocytes are present in Porifera. **[AFMC 2013]**
- Collar cells are useful for feeding in sponges. **[Manipal 2012]**
- In most simple type of canal system of Porifera, water flows through Ostia $\rightarrow$ Spongocoel $\rightarrow$ Osculum $\rightarrow$ Exterior. **[WB JEE 2012]**
- Porifera animal phyla possesses spicules. **[WB JEE 2012]**
- Gemmule formation is a common mode of asexual reproduction in sponges. **[AIIMS 2017]**
- The non-cellular layer present between pinacoderm and choanoderm in body wall of poriferans is known as mesenchyme. **[OJEE 2012]**
- The skeleton of sponges is secreted by scleroblasts. **[OJEE 2012]**
- *Euplectella* is referred to as 'venus flower basket'. **[Kerala CEE 2014]**
- Symmetry in Cnidaria is radial. **[JCECE 2012]**
- Cnidaria represents both marine and freshwater species. **[CBSE AIPMT 2014]**
- Metagenesis refers to alternation of generation between asexual and sexual phases of an organism. **[CBSE AIPMT 2015]**
- Jellyfish belongs to class–Scyphozoa. **[UP CPMT 2012]**
- Bilaterally symmetrical and acoelomate animals are exemplified by Platyhelminthes. **[NEET 2020]**
- Flatworms excrete through protonephridia. **[Manipal 2012]**
- The cercarial stage of liver fluke is produced by asexual multiplication. **[AMU 2015]**
- Platytelminthes have hooks and suckers and are endoparasite on other animals. **[NEET 2019]**
- Roundworms (Aschelminthes) are pseudocoelomates. **[BCECE 2012]**
- Third stage larva of *Wuchereria bancrofti* carried by *Culex* mosquito is called microfilariae . **[WB JEE 2015]**
- Infection of *Ascaris* usually occurs by drinking water containing eggs of *Ascaris*. **[NEET 2013]**

Chapter Exercises

(A) Topical Exercise

Basis of Classification

1 A paraphyletic group includes
(a) a single ancestral species and all of its descendents
(b) some, but not all members of a lineage
(c) members that can be traced to many ancestral groups
(d) different species which are linked through evolution

2 Subcellular physiological division of labour occurs in
(a) Protozoa (b) Porifera
(c) Cnidaria (d) All of these

3 Which level of organisation is found in majority of animals?
(a) Cellular grade (b) Cell-tissue grade
(c) Tissue-organ grade (d) Organ-system grade

4 The grade of organisation in sponges is
(a) cellular grade (b) cell-tissue grade
(c) protoplasmic grade (d) organ system grade

5 Tissue grade of organisation originated from
(a) Protozoa (b) Porifera
(c) Coelenterata (d) Platyhelminthes

6 Which of the following phyla possess multicellular organ grade level of organisation?
I. Platyhelminthes II. Porifera
III. Nematoda IV. Protozoa
Codes
(a) I, II and III (b) I and II
(c) II and IV (d) I and III

7 Match the following columns.

Column I (Phylum)	Column II (Levels of organisation)
A. Porifera	1. Tissue
B. Protozoa	2. Protoplasmic
C. Cnidaria	3. Organ system
D. Chordata	4. Cellular

Codes

	A	B	C	D		A	B	C	D
(a)	4	2	1	3	(b)	2	3	4	1
(c)	4	3	2	1	(d)	3	2	4	1

8 Choose the odd one out with repect to complexity of organ system.
(a) Platyhelminthes (b) Aschelminthes
(c) Ctenophora (d) Coelenterata

9 Which of the following is rare in animals?
(a) Radial symmetry (b) Spherical symmetry
(c) Bilateral symmetry (d) All of these

10 Organisms attached to substratum generally possess
(a) radial symmetry
(b) asymmetrical body
(c) one single opening of digestive canal
(d) cilia on the surface to create water current

11 Radial symmetry occurs in
(a) Porifera and Coelenterata
(b) Arthropoda and Mollusca
(c) Coelenterata and Echinodermata
(d) Coelenterata and Platyhelminthes

12 The body of an individual can be divided into equal halves by any plane passing through the center from top to bottom. This type of symmetry is found in
I. Cnidaria II. Echinodermata
III. Porifera IV. Mammals
Codes
(a) I, II and III (b) I and II
(c) II and IV (d) I and III

13 Which of the following may exhibit biradial symmetry?
(a) Ctenophora (b) Man
(c) Radiolarian (d) Sponge

14 A bilaterally symmetrical animal can be divided into mirror images by
(a) any cut from its anterior of its posterior end
(b) any cut from its dorsal to its ventral surface
(c) only a cut through the midline of its body from its anterior to its posterior end
(d) only a cut through the midline of its body from its dorsal to its ventral surface

15 Match the following columns.

Column I (Symmetry)	Column II (Phylum)
A. Asymmetry	1. Arthropoda
B. Radial symmetry	2. Ctenophora
C. Biradial symmetry	3. Cnidaria
D. Bilateral symmetry	4. Porifera

Codes

	A	B	C	D		A	B	C	D
(a)	4	2	3	1	(b)	2	3	4	1
(c)	3	2	4	1	(d)	4	3	2	1

16 During which stage in embryonic development, primary germ layers appear for the first time?

(a) Zygote (b) Morula
(c) Blastula (d) Gastrula

17 Find the incorrect match with respect to body plan.

(a) Tube-within-tube body plan	-	Nemathelminthes, Annelida, Arthropoda, Mollusca, Echinodermata, Chordata
(b) Cell-aggregate type body plan	-	Coelenterata
(c) Blind-sac type body plan	-	Platyhelminthes and Coelenterata
(d) Cell-aggregate type body plan	-	Porifera

18 Consider the following statements.

I. Cnidarians, flatworms and roundworms have blind sac body plan.
II. Cephalisation is strongly associated with bilateral symmetry.

Choose the correct option.

(a) Statement I is true, but II is false
(b) Statement I is false, but II is true
(c) Both statements I and II are true
(d) Both statements I and II are false

19 Body cavity is the cavity present between body wall and gut wall. In some animals, the body cavity is not lined by mesoderm. Such animals are called **[NCERT Exemplar]**

(a) acoelomate (b) pseudocoelomate
(c) coelomate (d) haemocoelomate

20 The coelom not lined by mesoderm and directly connected to the archenteron is found in

(a) platyhelminths (b) nematodes
(c) earthworm (d) echinoderms

21 Choose the incorrect statement for coelom.

(a) A large cavity with blood called haemocoel is found in Platyhelminthes
(b) A coelom (body cavity) derived from blastocoel is known as pseudocoelom
(c) Enterocoelom appeared first in the course of evolution in Echinodermata
(d) Schizocoelomates and enterocoelomates are true coelomates

22 The cross section of animal body is given below.

Which of the following group will satisfy the above corss section?

(a) Platyhelminthes (b) Ctenophora
(c) Aschelminthes (d) Annelida

23 Refer to the diagram showing different coeloms and identify the phylums (A, B, C) in which these are likely to be found.

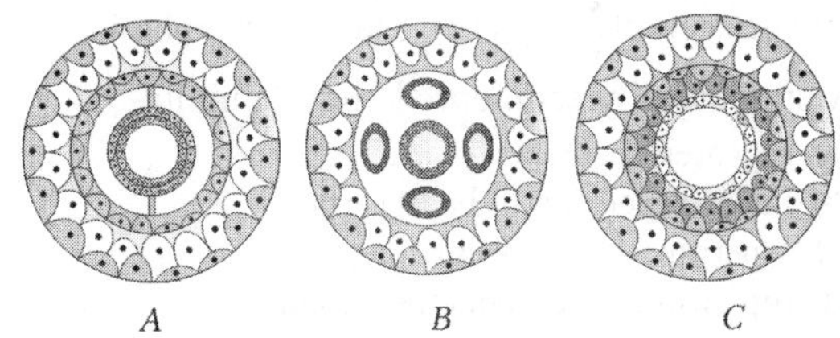

(a) *A*-Platyhelminthes, *B*-Aschelminthes, *C*-Annelida
(b) *A*-Platyhelminthes, *B*-Annelida, *C*-Aschelminthes
(c) *A*-Annelida, *B*-Platyhelminthes, *C*- Aschelminthes
(d) *A*-Annelida, *B*-Aschelminthes, *C*-Platyhelminthes

24 Which one of the following statements is incorrect? **[NCERT Exemplar]**

(a) Mesoglea is present in between ectoderm and endoderm in *Obelia*
(b) *Asterias* exhibits radial symmetry
(c) *Fasciola* is a pseudocoelomate animal
(d) *Taenia* is a triploblastic animal

25 Consider the following statements.

I. Protostomes are schizocoelus.
II. Platyhelminthes, annelids, arthropods and molluscans are protostomes.

Choose the correct option.

(a) Statement I is true, but II is false
(b) Statement I is false, but II is true
(c) Both statements I and II are true
(d) Both statements I and II are false

26 Protostomes and deuterostomes differ in

(a) type of cleavage
(b) origin of mouth and anus
(c) mode of coelom formation
(d) All of the above

27 Which of the following characters is not found in deuterostomes?

(a) Embryo divides by radial cleavage
(b) Anus develops from the blastopore
(c) Schizocoelous coelom
(d) Includes echinoderms and chordates

28 Deuterostome condition and indeterminate radial cleavage are the characteristics of

(a) chordates, arthropods and annelids
(b) arthropods and echinoderms
(c) chordates and echinoderms
(d) chordates and arthropods

29 Enterocoelous coelom is found in
(a) Deuterostomia (b) Astomia
(c) Protostomia (d) Blastostomia

30 In some animal groups, the body is found divided into compartments with at least some organs. This characteristic feature is called **[NCERT Exemplar]**
(a) segmentation (b) metamerism
(c) metagenesis (d) metamorphosis

31 Metamerically segmented body exhibits
(a) repeated segments
(b) linear repetition of body parts
(c) repetition of nephridia
(d) repetition of reproductive organs

32 Metameric segmentation is the characteristic of
(a) Mollusca and Chordata
(b) Annelida and Arthropoda
(c) Echinodermata and Annelida
(d) Platyhelminthes and Arthropoda

33 Heteronomous segmentation, i.e. the division of labour resulting in dissimilarity of the metameres of the different regions of the body, is found in the phyla
(a) Cnidaria and Ctenophora
(b) Ciliophora and Sporozoa
(c) Protista and Porifera
(d) Annelida and Arthropoda

34 Which one is the group of bilaterally symmetrical and triploblastic animals?
(a) Sponges (b) Coelenterates
(c) Ctenophores (d) Aschelminthes

35 Temperature changes in the environment affect most of the animals which are
(a) poikilothermic (b) homothermic
(c) aquatic (d) desert living

36 Poikilothermy is a phenomenon observed in
(a) vertebrates like human beings and reptiles
(b) invertebrates like earthworms and roundworms
(c) invertebrates, fish, amphibians and reptiles
(d) frogs and lizards

37 Consider the following statements.
I. Cold blooded animals falls under the category of endotherms.
II. Warm blooded animals falls under the category of ectotherms.

Choose the correct option.
(a) Statement I is true, but II is false
(b) Statement I is false, but II is true
(c) Both statements I and II are true
(d) Both statements I and II are false

38 Choose the correctly matches pair.
I. Summer sleep - Aestivation
II. Tagmatisation - Arthropoda
III. Strobilisation - Tapeworms

Codes
(a) I only (b) I and II only
(c) I, II and III (d) I and III only

39 Notochord is a
(a) endodermally derived rod-like structure, formed on the ventral side
(b) ectodermally derived rod-like structure formed on the dorsal side
(c) mesodermally derived rod-like structure formed on the dorsal side during embryonic development
(d) mesodermally derived rod-like structure formed on the ventral side during embryonic development

40 Animals with notochords are
(a) hemichordates (b) urochordates
(c) cephalochordates (d) Both (b) and (c)

41 Choose the false option.
(a) *Amoeba* - Asymmetrical
(b) Coelentrates - Diploblastic, radial symmetry
(c) Chordata - *Petromyzon, Equus*
(d) Annelids - Pseudocoelomate

42 Which one of the following phyla are included in Enterozoa?
(a) Annelida, Mollusca, Porifera
(b) Porifera, Arthropoda, Mollusca
(c) Mollusca, Arthropoda, Hemichordata
(d) Echinodermata, Hemichordata, Porifera

43 Consider the following statements.
I. Animals which have well-marked digestive cavity are put under Parazoa.
II. An acoelomate animal with bilateral symmetry is liver fluke.

Choose the correct option.
(a) Statement I is true, but II is false
(b) Statement I is false, but II is true
(c) Both statements I and II are true
(d) Both statements I and II are false

Phylum-Porifera

44 Consider the following statements.
I. The scientist who gave the name 'Porifera' to the sponges was Robert E Grant.
II. The animal nature of Porifera was described by Ellis.

Choose the correct option.
(a) Statement I is true, but II is false
(b) Statement I is false, but II is true
(c) Both statements I and II are true
(d) Both statements I and II are false

45 Sponges form a sideline of evolution, hence called
(a) Mesozoa (b) Porifera
(c) Parazoa (d) Eumetazoa

46 Sponges are primitive multicellular animals and have
(a) cellular level of organisation
(b) tissue level of organisation
(c) organ level of organisation
(d) organ system level of organisation

47 The middle layer of the body wall of Porifera is
(a) mesenchyme (b) mesoderm
(c) mesoglea (d) mesentery

48 A sponge can be distinguished from other animals in having
(a) all possible body symmetries
(b) subcellular body organisation
(c) asexual mode of reproduction
(d) numerous pores in their body

49 Sponges are poriferans, because their bodies have
(a) canal system (b) several pores
(c) spicules in skeleton (d) All of these

50 Which of the following features is universally present in all sponges?
(a) Marine habitat
(b) Presence of spicules
(c) Presence of spongocoel
(d) Presence of spongin fibres

51 Which among the following is not a characteristic feature of sponges?
(a) They are bilaterally symmetrical
(b) They are the first multicellular diploblastic organisms
(c) The body wall of sponges consists of pinacoderm and choanoderm
(d) They show the presence of trophocytes which provide food to the developing cells

52 Which one of the following is the most distinctive character of sponges?
(a) They are acellular
(b) They possess special cells called choanocytes
(c) They reproduce asexually
(d) They are all marine

53 Which of the following is a false statement?
(a) All sponges are hermaphrodites
(b) Choanocytes are reproductive in function
(c) Sponges are multicellular organisms
(d) Porocytes allow incurrent of water

54 Consider the following statements.
I. Myocytes are cells that surround the osculum in poriferans.
II. Amoebocytes are gametic cells in sponges.

Choose the correct option.
(a) Statement I is true, but II is false
(b) Statement I is false, but II is true
(c) Both statements I and II are true
(d) Both statements I and II are false

55 Match the following columns.

	Column I		Column II
A.	Archaeocytes	1.	Storage cells
B.	Chromocytes	2.	Amoebocytes with branched pseudopodia
C.	Thesocytes	3.	Totipotent
D.	Phagocytes	4.	Pigmented amoebocytes

Codes

	A	B	C	D
(a)	2	3	1	4
(b)	3	4	1	2
(c)	1	2	3	4
(d)	4	3	2	1

56 Consider the following statements.
I. The inter-communicating cavities in the body of a sponge constitute canal system.
II. Canal system in sponges helps in respiration, excretion and circulation.

Choose the correct option.
(a) Statement I is true, but II is false
(b) Statement I is false, but II is true
(c) Both statements I and II are true
(d) Both statements I and II are false

57 Porocytes are special cells for the passage of
(a) excretory products within body of flatworms
(b) sweat upon surface of mammalian epidermis
(c) incoming water current in the body of sponges
(d) outgoing water current on the top of sponges

58 In sponges, prosopyle is
(a) contractile element at the osculum
(b) small canal which connects incurrent canal with radial canal
(c) cell which forms lining of spongocoel
(d) cell at the opening of incurrent canal

59 The simplest type of canal system in the phylum-Porifera is
(a) *Sycon* type (b) *Ascon* type
(c) radial type (d) *Leucon type*

60 Evolutionary sequence of canal system in sponges is
(a) *Sycon* → *Ascon* → Leucon
(b) *Leucon* → *Sycon* → *Ascon*
(c) *Ascon* → *Sycon* → *Leucon*
(d) *Sycon* → Leucon → *Ascon*

61 A *Leucon* type canal system is the characteristic of
(a) *Sycon* (b) *Leucosolenia*
(c) *Spongilla* (d) None of these

62 Depending on the degree of complexity, the *Leucon* type of canal system is further divided into
(a) diplodal type and aphodal type
(b) euryplours type and diplodal type
(c) aphodal and euryplours type
(d) euryplours type, aphodal and diplodal type

63 Rhagon type canal system is present in
(a) Echinodermata
(b) All sponges
(c) adult sponge
(d) larval stage of some sponges

64 Choose the incorrectly matched pair.
(a) Excretion in sponges - Ammonotelism
(b) Aphodal canal system - *Geodia*
(c) Filter feeding - Microvilli of choanocytes
(d) *Plakina* - *Ascon* type canal system

65 The sponge body is supported by a skeleton made up of
(a) calcareous spicules (b) siliceous spicules
(c) spongin fibres (d) All of these

66 After drying, a bath sponge contains
(a) hold fast (b) tentacles
(c) spicules (d) spongin fibres

67 Consider the following statements.
I. Digestion in sponges takes place in paragastric cavity.
II. In sponges, first food is ingested by the choanocytes and digestion is completed in amoebocytes.
Choose the correct option.
(a) Statement I is true, but II is false
(b) Statement I is false, but II is true
(c) Both statements I and II are true
(d) Both statements I and II are false

68 Which of the following is not true for reproduction in sponges?
(a) Asexual reproduction by gemmules only
(b) Trapping of sperms by choanocytes
(c) Gametes production by archaeocytes
(d) Internal fertilisation

69 A ciliated and freely swimming larval form called parenchymula is seen in
(a) *Sycon* and *Clathrina*
(b) *Leucosolenia* and *Sycon*
(c) *Leucosolenia* and *Clathrina*
(d) *Leucosolenia*, *Clathrina* and *Sycon*

70 In the members of phylum-Porifera,
(a) sexes are not separated
(b) fertilisation is internal and development is indirect
(c) larval stage is found, which is morphologically distinct from the adult
(d) All of the above

71 Point out the statement which is not a salient feature of the class-Calcarea.
(a) The skeleton consists of calcarous spicules
(b) Body is radially symmetrical, cylindrical or vase-like
(c) They are bright coloured and freshwater or marine organisms
(d) The choanocytes are large

72 The zygote of *Leucosolenia* develops into a hollow blastula called
(a) stereogastrula (b) coeloblastula
(c) amphiblastula (d) holoblastula

73 The internal skeleton of *Leucosolenia* is formed of calcareous
(a) monaxon (single-rayed) spicules
(b) monaxon and tri-radiate (three-rayed) spicules
(c) tri-radiate and quadri-radiate (four-rayed) spicules
(d) monaxon, tri-radiate and quadri-radiate spicules

74 The principle cell types present in the body wall of *Leucosolenia* are the
(a) pinacocytes, porocytes, choanocytes, amoebocytes
(b) pinacocytes, choanocytes, amoebocytes, nephrocytes
(c) choanocytes, nerve cells, amoebocytes, nephrocytes
(d) choanocytes, porocytes, nephrocytes, amoebocytes

75 Digestion in *Leucosolenia* and other sponges is
(a) first intracellular, then extracellular
(b) Only intracellular
(c) Only extracellular
(d) first extracellular, then intracellular

76 Consider the following statements.
I. The inhalent aperture in *Leucosolenia* is prosopyle.
II. The chief excretory product of *Leucosolenia* is urea.
Choose the correct option.
(a) Statement I is true, but II is false
(b) Statement I is false, but II is true
(c) Both statements I and II are true
(d) Both statements I and II are false

77 The crown sponge measures
(a) 5-6 cm in height and 1-3 cm in diameter
(b) 1-3 cm in height and 5-6 mm in diameter
(c) 1-3 mm in height and 5-6 cm in diameter
(d) 2-4 mm in height and 8-10 mm in diameter

78 *Sycon* is also called as
(a) urn sponge (b) bath sponge
(c) crown sponge (d) Both (a) and (b)

79 Choose the statement which is not a characteristic feature of class-Hexactinellida.
(a) The spicules are monoaxon or 3-4 branched
(b) Choanocytes are present in radial chambers
(c) A lot of forms of this class are seen in deep sea
(d) Spicules are siliceous

80 Find out the mismatched pair.
(a) *Euspongia*—Connules
(b) *Spongilla*—Symbiotic green alga
(c) *Euplectella*— Spherical body
(d) *Leucosolenia*—Internal fertilisation

81 The sponge showing commensalism with shrimps is
(a) *Adamsia* (b) *Eupagurus*
(c) *Euplectella* (d) *Euspongia*

82 Identify the incorrect matched pair with respect to glass rope sponges.
(a) Canal system - *Leucon* type
(b) Habitat - Exclusively marine
(c) Spicules - Calcareous
(d) Class - Hexactinellida

83 The type of canal system exhibited by *Spongilla* is
(a) *Ascon* (b) *Sycon*
(c) rhagon (d) *Leucon*

84 The marine sponges which are used for bathing, washing, mopping and ceramic work is
(a) *Euplectella* (b) *Euspongia*
(c) *Chalina* (d) *Cliona*

85 Which of the following examples given below belongs to largest class of sponge. Identify the correct option.
I. *Spongia* II. *Euspongia*
III. *Cliona* IV. *Euplectella*
(a) I, II and III (b) II, III and IV
(c) Only IV (d) III and IV

86 Match the following columns.

	Column I		Column II
A.	Venus flower basket	1.	*Spongilla*
B.	Glass rope sponge	2.	*Euspongia*
C.	Freshwater sponge	3.	*Cliona*
D.	Boring sponge	4.	*Hyalonema*
E.	Bath sponge	5.	*Euplectella*

Codes

	A	B	C	D	E		A	B	C	D	E
(a)	5	4	1	3	2	(b)	1	2	4	5	3
(c)	4	1	3	2	5	(d)	3	5	2	4	1

87 Choose the parts of the world where sponge fishing is practised on a large scale.
(a) South America and Atlantic sea
(b) Central America and Mediterranean sea
(c) Mediterranean Florida and West Indies
(d) South Africa and Mediterranean sea

88 A sponge harmful to oyster industry is
(a) *Cliona* (b) *Euspongia*
(c) *Hyalonema* (d) *Spongilla*

Phylum-Cnidaria

89 Cnidarians exhibit
(a) tissue level of organisation
(b) diploblastic body
(c) central gastrovascular cavity with a single opening
(d) All of the above

90 Mesoglea of Cnidaria is
(a) cellular and gives rise to germ cells
(b) cellular and contains mesenchyme cells
(c) proteinaceous matrix devoid of cellular elements
(d) acellular and cannot be crossed by interstitial cells

91 Cnidoblasts are used for
(a) anchorage (b) locomotion and defence
(c) capture of prey (d) Both (b) and (c)

92 Which is correct about nematocyst in *Hydra*?
(a) It is reused
(b) Ejection is conditioned reflex
(c) Ejection occurs in response to contact and it pierces the prey
(d) Prevents coming in contact with other *Hydra*

93 Consider the following statements with respect to cnidarians.
I. They are herbivorous and mode of digestion is intracellular.
II. Non-polar neurons are scattered irregularly in body.

Choose the correct option.
(a) Statement I is true, but II is false
(b) Statement I is false, but II is true
(c) Both statements I and II are true
(d) Both statements I and II are false

94 Cnidarians exhibit two basic body forms called polyp and medusa. The polyp represents
(a) asexually reproducing, sessile, cylindrical form
(b) sexually reproducing, free-swimming form
(c) sexually reproducing, cylindrical form
(d) asexually reproducing, umbrella-shaped form

95 Those cnidarians which exist in both polyp and medusa forms, exhibit
(a) metagenesis (b) alternation of generation
(c) polymorphism (d) All of the above

96 Medusa of coelenterates is
(a) sexually reproducing (b) free-swimming
(c) umbrella-shaped (d) All of these

97 In metagenesis,
(a) polyps produce medusae asexually
(b) medusae produce polyps sexually
(c) one diploid generation alternates with other diploid generation
(d) All of the above

98 The two larval forms seen in phylum-Cnidaria are
(a) planula and trochophore
(b) ephyra and trochophore
(c) planula and ephyra
(d) trochophore and cysticercus

99 Coelenterata can be classified into
(a) Hydrozoa, Actinozoa, Polychaeta
(b) Scyphozoa, *Hirudinaria*, Hydrozoa
(c) Scyphozoa, Actinozoa, Trematoda
(d) Hydrozoa, Actinozoa, Scyphozoa

100 Select the correct statement with respect to Hydrozoa.
(a) Unlike other coelenterates, hydrozoans do not possess mesoglea
(b) They inhabit freshwater habitat
(c) They possess mouth at the tip of hypostome
(d) All of the above

101 Which is a freshwater cnidarian?
(a) *Hydra* (b) *Spongilla*
(c) *Dugesia* (d) None of these

102 Gland cells for secreting adhesive material in *Hydra* are present in
(a) pedal disc (b) stomach region
(c) growth region (d) hypostome

103 The number of tentacles in *Hydra* is
(a) 6-10 (b) 10-16
(c) 16-20 (d) 20-26

104 The smallest cnidoblast in the body of *Hydra* is
(a) penetrant (b) volvent
(c) Astrichous isorhiza (d) holotrichous isorhiza

105 Consider the following statements.
I. *Hydra* can move by gliding, climbing, etc.
II. All movements of *Hydra* are attributed to circular and longitudinal muscles of its body wall.
Choose the correct option.
(a) Statement I is true, but II is false
(b) Statement I is false, but II is true
(c) Both statements I and II are true
(d) Both statements I and II are false

106 The digestion of food within the gastrovascular cavity of *Hydra* is called
(a) intercellular digestion (b) intracellular digestion
(c) extracellular digestion (d) extracorporeal digestion

107 Excretion of nitrogenous wastes in *Hydra* takes place through the
(a) flame cells
(b) nephridia
(c) nematocysts
(d) diffusion across the general body surface

108 Regeneration in *Hydra* is
(a) epimorphosis by archaeocytes
(b) epimorphosis by interstitial
(c) epimorphosis by glandular cells
(d) morphallaxis by interstitial cells

109 *Hydra* will regenerate from a fragment, if it contains
(a) tentacles
(b) epidermis and gastrodermis
(c) tentacles, epidermis and gastrodermis
(d) epidermis, hypodermis and gastrodermis

110 Match the following columns.

	Column I (Cells)		Column II (Functions)
A.	Epitheliomuscular cells	1.	Totipotent
B.	Glandular cells	2.	Unique cells of phylum-Coelenterata
C.	Interstitial cells	3.	Main cells of epidermis with the ability of contraction
D.	Cnidoblasts	4.	Secretion
E.	Sensory cells	5.	Conduction of stimulus

Codes

	A	B	C	D	E
(a)	3	4	1	2	5
(b)	1	2	3	4	5
(c)	5	4	3	2	1
(d)	4	3	2	1	5

111 Given below are types of cells present in some animals. Each one is specialised to perform a single specific function except
(a) choanocytes (b) interstitial cells
(c) gastrodermal cells (d) nematocytes

112 In *Hydra*, egestion of undigested food and excretion of nitrogenous wastes occur through
(a) mouth and tentacles (b) mouth and body wall
(c) mouth and mouth (d) body wall and body wall

113 What structure in *Hydra* functionally corresponds to the pseudopodia of *Amoeba*?
(a) Tentacles (b) Mouth
(c) Hypostome (d) Nematocysts

114 A chemical substance called hypnotoxin is found in
(a) *Entamoeba histolytica* (b) ectoderm of *Olynthus*
(c) nematocyst of *Hydra* (d) venom of snake

115 Cnidoblasts present on the tentacles contain
(a) nematocysts (b) spines
(c) stings (d) None of these

116 In *Hydra*, both flagella and pseudopodia are found in
(a) gland cells (b) epitheliomuscular cells
(c) interstitial cells (d) nutritive-muscular cells

117 Select the incorrect statement.
(a) There are about 6-10 tentacles in *Hydra*
(b) Stenotele is nematocyst containing structure on tentacles
(c) *Hydra* feeds on *Daphnia* and *Cyclops*
(d) *Hydra* use regeneration power as mean of reproduction

118 The proximal end of *Hydra* has no adhesive disc, which secretes a sticky substance for
(a) protection
(b) defence from enemies
(c) sexual attraction
(d) attachment to the substratum

119 Spermatogonia in *Hydra* develop from
(a) interstitial cells (b) gastrodermis
(c) musculoendothelial cells (d) gland cells

120 Zoochlorellae and zooxanthallae present in *Hydra* are
(a) symbionts in nutritive cells
(b) symbionts in the gut
(c) symbionts in cnidoblasts
(d) organisms that provide hypnotoxin

121 *Hydra/Obelia* is
(a) diploblastic, radially symmetrical and acoelomate
(b) diploblastic, bilaterally symmetrical and acoelomate
(c) triploblastic, radially symmetrical and coelomate
(d) triploblastic, bilaterally symmetrical and coelomate

122 Most appropriate term to describe the life cycle of *Obelia* is
(a) metagenesis
(b) neoteny
(c) metamorphosis
(d) alternation of generation

123 Gonads of *Obelia* occur
(a) on blastocyst
(b) in hydrula stage
(c) in radial canals of medusa
(d) in bases of tentacles of medusa

124 In *Physalia*, nematoblasts paralyse the prey by
(a) haemotoxin (b) neurotoxin
(c) myotoxin (d) Both (b) and (c)

125 The larva of *Physalia* is called
(a) a planula (b) an ephyra
(c) a cydippid larva (d) a scyphistoma

126 In *Physalia*, zooids occur in groups called
(a) conulae (b) palpons
(c) cormidia (d) dactyloids

127 The zooid having mouth and tentacles for feeding is
(a) dactylozooids (b) gonozooids
(c) blastostyle (d) gastrozooids

128 The function of a dactylozooid in polymorphic colony is
(a) offence
(b) reproduction
(c) locomotion
(d) feeding and defence

129 Consider the following statements.
I. Gastrozooid and dactylozooid are modified form of polyps.
II. *Physalia* possess gas glands in the innerlining of pneumatophore.

Choose the correct option.
(a) Statement I is true, but II is false
(b) Statement I is false, but II is true
(c) Both statements I and II are true
(d) Both statements I and II are false

130 The animals belonging to this class consist the following characteristics—Medusae prominent in the life cycle, polyp small, gametes gastrodermal in origin, medusae cuboidal in shape, with tentacles long from each corner of the bell. Identify the class
(a) Anthozoa (b) Scyphozoa
(c) Hydrozoa (d) None of these

131 An example of Scyphozoa is
(a) catfish (b) jellyfish
(c) silverfish (d) cuttlefish

132 The *Aurelia aurita* have nine sense organs called as
(a) lappets (b) rhopalia
(c) velarium (d) None of these

133 Medusa of *Obelia* differs from *Aurelia* in having
(a) gonads (b) a tentacle
(c) a rhopalium (d) tentaculocysts

134 Ephyra is
(a) a poriferan (b) cnidarian
(c) young *Hydra* (d) young *Aurelia*

135 Jellyfishes, sea anemones and their relatives have .... symmetry and their cells form ....
(a) radial, tissues (b) bilateral, tissues
(c) radial, mesoderm (d) bilateral, mesoderm

136 In the life history of *Aurelia*, the correct chronological sequence of larval stages is

(a)	Ephyra	Scyphistoma	Planula
(b)	Scyphistoma	Ephyra	Planula
(c)	Planula	Scyphistoma	Ephyra
(d)	Scyphistoma	Planula	Ephyra

137 Tentaculocyst in a jellyfish helps in
(a) balancing (b) photoreception
(c) chemoreception (d) All of these

138 The polyp stage in the life cycle of *Aurelia* is represented by a reduced stage called
(a) *Siphonoa* (b) *Acontia*
(c) Scyphistoma (d) *Euplectella*

139 Commensalism is found in which coelenterate?
(a) *Adamsia* (b) *Heliopora* (c) *Tubipora* (d) *Astraea*

140 Choose the correct match.
(a) *Aurelia* - Jellyfish
(b) *Physalia* - Portugese man of war
(c) *Pennatula* - See pen
(d) All of the above

141 Consider the following statements.
I. *Pennatula* and *Gorgonia* are anthozoans.
II. Gastrovascular cavity is divided into compartments in Anthozoa.

Choose the correct option.
(a) Statement I is true, but II is false
(b) Statement I is false, but II is true
(c) Both statements I and II are true
(d) Both statements I and II are false

142 A coral island with a central shallow lake is known as
(a) atoll (b) corallite
(c) coral reef (d) diatomaceous sheath

143 Choose the correct option about corals.
(a) Corals have a skeleton composed of calcium carbonate
(b) Coral formation is connected to *Adamsia*
(c) Corals are mostly found in the warm parts of Arabian sea
(d) Fringing reef is usually located away from the shore

144 Which of the following statements is true?
(a) Coral reefs occur in tropical seas
(b) Cnidarians have 'tissue grade' organisation
(c) Anthozoa have no medusa in the life cycle
(d) All of the above

Phylum-Ctenophora

145 Which of the following is intermediate between diploblastic and triploblastic body structure?
(a) Poriferans (b) Coelenterates
(c) Ctenophores (d) Platyhelminthes

146 Which of the following statement is incorrect about Ctenophora?
(a) Statocysts are present in aboral region
(b) They have bilateral symmetry
(c) Tentacles contain cnidoblasts
(d) They are acoelomates

147 What is present in mesoglea of ctenophores, which makes it a loose layer of cells?
(a) Amoebocytes
(b) Elastic cells
(c) Nerve cells and muscle cells
(d) All of the above

148 Comb plates of ctenophores possess
(a) cilia for locomotion (b) cilia for filter feeding
(c) teeth for crushing (d) teeth for filtering

149 Consider the following statements.
I. Ctenophores show bioluminescence.
II. Lasso cells on tentacles of ctenophores release toxins.

Choose the correct option.
(a) Statement I is true, but II is false
(b) Statement I is false, but II is true
(c) Both statements I and II are true
(d) Both statements I and II are false

150 The larval form present in ctenophorans is
(a) planula (b) cydippid
(c) amphiblastula (d) parenchymula

151 Biradial symmetry and lack of cnidoblasts are the characteristics of
(a) *Hydra* and starfish (b) *Ctenoplana* and *Beroe*
(c) *Aurelia* and *Paramecium* (d) Starfish and sea anemone

152 The generic name of 'venus girdle' is
(a) *Velamen* (b) *Coeloplana*
(c) *Ctenoplana* (d) *Hormiphora*

153 Which of the following pairs is not correctly matched?
(a) *Beroe*-Nuda
(b) Jellyfish-Scyphozoa
(c) Sea anemone-Anthozoa
(d) *Hydra*-Tentaculata

154 Among the statements given below find out which one is not true about *Pleurobrachia*?
(a) They have a transparent body
(b) They have short, hollow tentacles
(c) Their body is pear or walnut-shaped
(d) They are common marine pelagic forms

155 *Nuda* and *Beroe* are members of class-Nuda of phylum-Ctenophora. The statements given below give some of their important features. Pick out the one which is incorrect.
(a) They have a highly branched gastrovascular cavity
(b) *Nuda* possess three highly branched tentacles.
(c) *Beroe* has a thimble-shaped body
(d) *Beroe* is commonly called swimming eye of cat

Phylum-Platyhelminthes

156 The characteristic feature of the members of phylum-Platyhelminthes is
(a) all are free-living
(b) all are found in damp terrestrial habitats
(c) most of the species are endoparasites
(d) Both (a) and (b)

157 Choose the correct set of characteristics about Platyhelminthes.
(a) Bilateral symmetrical, triploblastic, blind sac body plan
(b) Biradial symmetry, diploblastic, acoelomates
(c) Bilateral symmetry, diploblastic, pseudocoelomate
(d) Radial symmetry, tube within tube plan, eucoelomate

158 Parasitic adaptation of flatworms is
(a) anaerobic respiration
(b) underdeveloped sense organ
(c) resistant covering, cuticle or integument
(d) All of the above

159 Select the incorrect match pair with respect to flatworms.
(a) Hooks - Help in food ingestion
(b) Tegument - Protective layer
(c) Solenocytes - Excretory function
(d) None of the above

160 The nervous system in Platyhelminthes consists of
(a) a brain, a pair of dorsal ganglia and four longitudinal nerve cords
(b) a brain, a pair of anterior ganglia and two longitudinal nerve cords
(c) a brain, two pairs of anterior ganglia and two longitudinal nerve cords
(d) a brain, two pairs of dorsal ganglia and four longitudinal nerve cords

161 The sensory organs in free-living forms include
(a) eyes, which are image forming
(b) only eyes, which can detect the changes in the intensity of light
(c) eyes, auricles, papillae
(d) compound eyes and antennae

162 Phylum-Platyhelminthes does not include
(a) Turbellaria (b) Chilopoda
(c) Trematoda (d) Cestoda

163 Select the statement which is incorrect about the class-Turbellaria.
(a) Body wall has rod-shaped bodies called rhabdites
(b) Incomplete alimentary canal
(c) Animals are free-living or parasitic
(d) They reproduce asexually by budding

164 Select the statement which is incorrect about *Dugesia*.
(a) It is a common inhabitant of freshwater ponds, lakes, streams and shallow rivers
(b) Locomotion is altogether absent in it
(c) A common genital aperture is present in sexually mature individuals
(d) A gonopore is present just below the mouth on the ventral side

165 Choose the incorrectly matched pair.
(a) Liver fluke - Class-Cestoda
(b) Neoblast cells - Regeneration
(c) Laurer's canal - Breeding season
(d) None of the above

166 Metacercaria of liver fluke is
(a) free-swimming
(b) encysted stage
(c) non-encysted and non-motile stage
(d) free-floating stage

167 During its life cycle, *Fasciola hepatica* (liver fluke) infects its intermediate host and primary host at the following larval stages, respectively.
(a) Rediae and miracidium
(b) Cercaria and rediae
(c) Metacercaria and cercaria
(d) Miracidium and metacercaria

168 *Fasciola hepatica* is a digenetic parasite. Sheep and snail are two hosts. Snail is
(a) vector host (b) reservoir host
(c) parasite host (d) intermediate host

169 Which one of the following is not a nematode parasite?
(a) *Trichinella* (b) *Ascaris*
(c) *Dracunculus* (d) *Schistosoma*

170 The disease Bilharzia is an infectious disease leading to skin rash, bronchial cough, diarrhoea, etc. The causal organism of the disease is
(a) *Fasciola* (b) *Schistosoma*
(c) *Echinococcus* (d) *Taenia*

171 Choose the correct statement(s) related to *Schistosoma*.
I. Male is longer than female.
II. Male has gynaecophoric canal for holding female.
III. Male is broader than female.

Codes
(a) Only I (b) I and II (c) II and III (d) Only III

172 Consider the following statements.
I. The larva of *Schistosoma* are cercaria, sporocyst and miracidium.
II. *Fasciola hepatica* larva are rediae, sporocyst and cercaria.

Choose the coırect option.
(a) Statement I is true, but II is false
(b) Statement I is false, but II is true
(c) Both statements I and II are true
(d) Both statements I and II are false

173 Among Platyhelminthes, body segmentation is found in
(a) Turbellaria (b) Cestoda
(c) Nematoda (d) Trematoda

174 Gut is found in all, but absent in one of the following taxonomic groups of Platyhelminthes.
(a) Digenea (b) Cestoda
(c) Turbellaria (d) Aspidogastrea

175 Match the following columns.

Column I	Column II
A. *Fasciola hepatica*	1. Sheep liver fluke
B. *Fasciola gigantica*	2. Cattle liver fluke
C. *Fasciolopsis buski*	3. Intestinal fluke
D. *Schistosoma mansoni*	4. Human blood fluke

Codes

	A	B	C	D		A	B	C	D
(a)	1	4	3	2	(b)	1	2	3	4
(c)	4	1	2	3	(d)	3	2	1	4

176 Absence of the alimentary canal is the characteristic feature of
(a) *Taenia* and *Schistosoma* (b) *Ascaris* and *Fasciola*
(c) *Taenia* and *Echinococcus* (d) *Trichuris* and *Fasciola*

177 Among the statements given below, select the one which is incorrect about tapeworm.
(a) The body is divided into a number of proglottids
(b) Each proglottid is able to perform all the vital functions of the body
(c) Each repeated unit is called somite
(d) Body grows by the addition of new segments from the anterior end

178 A proglottid is called gravid proglottid, when it has
(a) Both male and female reproductive units well-developed
(b) Only female reproductive unit well-developed
(c) Only male reproductive unit well-developed
(d) Branched uterus filled with fertilised eggs

179 Tapeworm does not possess digestive system as it
(a) does not require solid food
(b) obtains food through general surface
(c) does not require food
(d) lives in intestine

180 Tapeworm infesting intestine of animals obtain their nutrition
(a) by ingesting food particles through their suckers
(b) by scraping the intestinal walls with their hooks
(c) by preparing food in their own bodies
(d) by absorption of liquid food through their general body surface

181 Suckers in tapeworm are meant for
(a) attachment to intestinal wall
(b) ingestion of food
(c) attachment during locomotion
(d) All of the above

182 In *Taenia solium*, the proximal portion of the oviduct that leads into the vagina is called
(a) receptaculum seminis
(b) uterus
(c) fertilisation canal
(d) vitellarium

183 Apolysis is a process in *Taenia* for
(a) regular detachment of proglottid from strobila
(b) embryonic development
(c) ingestion
(d) infection to man

184 *Taenia solium* is characterised by the
(a) absence of digestive tract
(b) presence of hooks of adhesion
(c) externally divided body
(d) All of the above

185 *Taenia saginata* differs from *Taenia solium* as it lacks
(a) rostellum
(b) suckers
(c) scolex
(d) None of the above

186 Identify the animals and their intermediate host in the given figures.

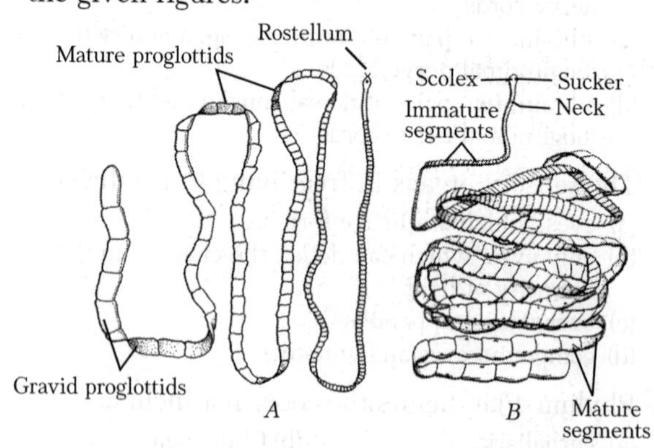

(a) A–*Taenia solium*–Sheep; B–*Taenia saginata*–Cattle
(b) A–*Taenia saginata*–Pig; B–*Taenia solium*–Buffalo
(c) A–*Taenia solium*–Cattle; B–*Taenia saginata*–Dog
(d) A–*Taenia solium*–Pig; B–*Taenia saginata*–Sheep

187 In *Taenia solium*, the scolex top has a conical projection called
(a) rostellum which bears chitinous hooks in two rows
(b) acetabulum which has radial muscles
(c) gonopore with a wide canal
(d) strobilus with proglottids

188 Tapeworms are hermaphrodites and show self-fertilisation between
(a) same proglottids
(b) two proglottids of different species
(c) two proglottids of same individual
(d) Both (a) and (c)

189 What is correct about *Taenia*?
(a) Male organs occur in posterior proglottids
(b) Male organs occur in anterior proglottids
(c) Female organs occur in anterior proglottids
(d) Mature proglottids contain both male and female organs

190 In hydatid worm, the scolex has
(a) two suckers, a rostellum and a double row of hooks
(b) four suckers, a rostellum and a double row of hooks
(c) six suckers, a rostellum and four rows of hooks
(d) two suckers, a rostellum and four rows of hooks

191 The primary hosts of *Moniezia expansa* are
(a) sheep and orbated mites
(b) goats and orbated mites
(c) orbated mites
(d) sheep and goats

192 Select the incorrect statement.
(a) *Hymenolepis nana* is the largest human tapeworm
(b) The scolex of *Echinococcus* has four suckers
(c) *Moniezia* infect only sheep and goat
(d) All tapeworms are hermaphrodite

Phylum-Aschelminthes

193 Which of the following is not a feature of phylum-Aschelminthes?
(a) They are triploblastic and pseudocoelomate
(b) The roundworms bear a straight alimentary tract with a terminal mouth and anus
(c) The body is segemented
(d) A circulatory system and respiratory organs are absent

194 Identify the true statement about members of phylum-Aschelminthes.
(a) Free-living forms are found in sea water, freshwater and soil
(b) They show an organ system level of organisation
(c) They have tube within a tube body plan
(d) All of the above

195 The pseudocoelom of roundworms is/may
(a) fluid-filled
(b) packed with chitin
(c) contain a gelatinous substance with mesenchyme cells
(d) Both (a) and (c)

196 Select the statement which is incorrect about digestive system in Aschelminthes.
(a) The digestive system is complete
(b) The alimentary canal is straight
(c) A specialised pharynx is present
(d) Intestine is muscular

197 Which one of the following characteristics is not correct for Nematoda?
(a) Elongated cylindrical body
(b) Fixed number of cells in the body
(c) Hermaphroditism
(d) Pseudocoelomate

198 The two longitudinal excretory canals in roundworm are connected
(a) posteriorly by a transverse canalicular network
(b) anteriorly by a transverse canalicular network
(c) posteriorly by a longitudinal canalicular network
(d) anteriorly by a medial transverse section

199 Amphids in Aschelminthes are
(a) Chemoreceptors
(b) Glandulosensory in nature
(c) Tactile in function
(d) Organs of locomotion

200 Eutey is a condition in which
(a) the number of cells or number of nuclei remains constant for entire animal
(b) the number of cells or number of nuclei remains constant for entire animal as well as for each organ
(c) number of nuclei remains constant for entire animal
(d) number of cells remains constant for entire animal

201 Match the following columns.

Column I	Column II
A. Eyeworms	1. *Trichuris*
B. Whipworm	2. *Dracunculus medinensis*
C. Guineaworm	3. *Enterobius*
D. Seatworm	4. *Loa loa*

Codes

	A	B	C	D		A	B	C	D
(a)	4	1	2	3	(b)	1	2	3	4
(c)	3	4	1	2	(d)	2	3	4	1

202 The body of *Ascaris* is
(a) cuticularised and ciliated
(b) cylindrical and ciliated
(c) cylindrical, cuticularised and non-ciliated
(d) round, cuticularised and non-ciliated

203 *Ascaris* is characterised by
(a) the presence of true coelom and metamerism (metamerisation)
(b) the presence of neither true coelom nor metamerism
(c) the absence of true coelom, but presence of metamerism
(d) the presence of true coelom, but absence of metamerism

204 Identify the statement which is not true about the reproduction in *Ascaris.*
(a) Female *Ascaris* is didelphic
(b) Male *Ascaris* is monorchic with a single testis
(c) The formation of gametes in *Ascaris* is telogenic
(d) Fertilisation occurs in the stomach of host

205 Indicate the correct statement pertaining to *Ascaris.*
(a) Snail is the secondary host of *Ascaris*
(b) A larval stage of *Ascaris* is swallowed by cattle which in turn infects human beings
(c) The eggs of *Ascaris* containing second stage juvenile larva when swallowed by man become infective
(d) Once inside the intestine, *Ascaris* larva directly develops into an adult without entering other organs of the host

206 Consider the following statements with respect to life cycle of *Ascaris.*
I. Average lifespan of animal in host is about 10 days.
II. Animal undergo 3-4 moultings in host body.

Choose the correct option.
(a) Statement I is true, but II is false
(b) Statement I is false, but II is true
(c) Both statements I and II are true
(d) Both statements I and II are false

207 Identify the figure of the endoparasite given below and name the disease, which is caused by this parasite.

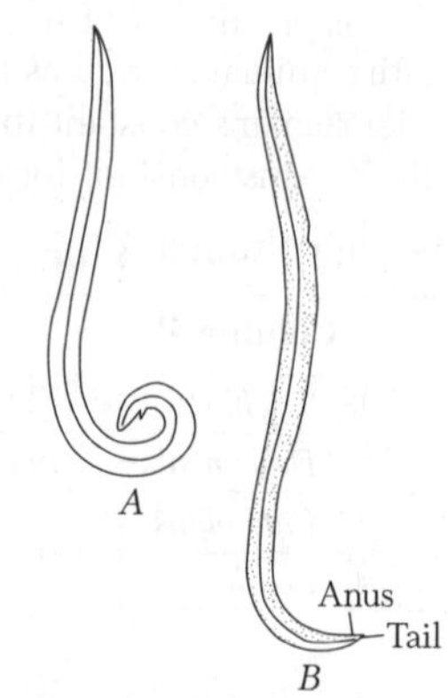

(a) *Ascaris* : *A*-Male, *B*-Female; Ascariasis
(b) *Ascaris* : *A*-Female, *B*-Male; Ascariasis
(c) *Wuchereria* : *A*-Male, *B*-Female; Elephantiasis
(d) *Wuchereria* : *A*-Female, *B*-Male; Elephantiasis

208 The adult male of *Ascaris* can be identified externally by the presence of
(a) pineal setae
(b) fifty pairs of pre-anal papillae
(c) five pairs of post-anal papillae
(d) All of the above

209 The sequence of layers of body wall in *Ascaris* is
(a) Cuticle—Epidermis—Longitudinal muscle layer
(b) Cuticle—Epidermis—Circular muscle layer
(c) Cuticle—Epidermis—Longitudinal muscle layer—Circular muscle layer
(d) Cuticle—Epidermis—Circular muscle layer—Longitudinal muscle layer

210 Choose the incorrect statement.
(a) Number of lips of *Ascaris* are three
(b) The mouth of *Ascaris* is guarded by labrum
(c) Amphids for olfactoreception are present in *Ascaris* on ventrolateral lips
(d) The female *Ascaris* is distinguished from the male by the absence of pre and post-anal papilla

211 Which of the following is not a characteristic feature of *Ascaris*?
(a) Endoparasite (b) Monogenic
(c) Free-living (d) Dioecious

212 The route through which developmental stages of *Ascaris* serially pass in man
(a) Outside → Trachea → Lung → Heart → Liver → Intestine → Outside
(b) Outside → Intestine → Liver → Heart → Lung → Intestine → Outside
(c) Outside → Oesophagus → Stomach → Heart → Lung → Liver → Intestine → Outside
(d) Outside → Mouth → Intestine → Liver → Lung → Heart → Trachea → Outside

213 Choose the incorrrect option.
(a) *Ascaris* has segmented body
(b) *Ascaris* have phasmids and amphids
(c) The cells which form excretory system in *Ascaris* are called renette cells
(d) Pineal setae in male *Ascaris* are found in cloaca

214 Which is not true for *Ascaris* infection?
(a) More common in children
(b) Does not produce tonsilitis
(c) Number can be 500-5000
(d) Infection is cured even without medication

215 The reproductive organs of *Ascaris* comprise
(a) one pair of testes in male and one pair of ovaries in female
(b) a single testis in male and one pair of ovaries in female
(c) one pair of testes in male and a single ovary in female
(d) a single testis in male and a single ovary in female

216 The larva found in *Ancylostoma* is
(a) microfilaria
(b) bipinnaria
(c) filariform
(d) rhabditiform

217 The adult *Wuchereria bancrofti* lives in or attacks
(a) human subdermal spaces
(b) muscles of *Culex*
(c) salivary glands of *Culex*
(d) human lymph glands

218 *Wuchereria bancrofti* is carried by

(a) tse-tse fly
(b) male *Anopheles*
(c) *Culex* mosquito
(d) female *Anopheles*

219 Mark the odd one out with respect to *Loa loa*.

(a) Found in subcutaneous tissue of host
(b) Monogenetic parasite
(c) Larval form is microfilarie
(d) Cause calabar swelling around the eyes.

220 *Dracunculus medinensis* is transmitted to man by

(a) mosquito (b) fly
(c) crustacean (d) infected cyclops

221 In *Dracunculus medinensis*,

(a) female is short, while male is long
(b) secondary host is an arthropod
(c) posterior end of female is straight and pointed
(d) infection is transmitted through eggs

222 Pinworm is

(a) *Ancyclostoma* (b) *Wuchereria bancrofti*
(c) *Dracunculus medinensis* (d) *Enterobius vermicularis*

223 *Enterobius vermicularis* causes

(a) loss of appetite
(b) insomnia and hysteria
(c) restlessness and inflammation of mucous membrane of infected region
(d) All of the above

(B) Cumulative Exercise

(Assorted Questions of the Chapter for Advanced Level Practice)

1 In understanding different types of symmetry, the term used as principal axis means

(a) a flat area that runs through any axis
(b) an imaginary straight line joining the two opposite points at the ends
(c) an imaginary straight line joining the mid-point at one end and the mid-point at the opposite end
(d) an animal having its body parts arranged in such a manner to exhibit symmetry

2 Structures present in a mature proglottid are

I. a pair of longitudinal nerve cords (lateral nerve cords).
II. Mehlis gland.
III. cirrus sac.
IV. one ovary and two oviducts.

Codes

(a) I, II and III are correct
(b) I and II are correct
(c) II and IV are correct
(d) I and III are correct

3 Identify and tick the option which is the incorrect feature of medusae.

I. The medusae have marginal tentacles.
II. The mouth is ventrally located on the surface of a small umbrella-like structure.
III. They are modified zooids meant for sexual reproduction.
IV. The medusae of *Obelia* bear 16 tentacles in addition to tentaculocyst.
V. Manubrium and velum are the parts of medusa.
VI. The hydranth of *Obelia* bears 24 tentacles.

Codes

(a) I and III (b) Only II (c) III and VI (d) V and VI

4 Identify among the statements given below which one is incorrect about phylum-Ctenophora.

I. They are acoelomate and diploblastic animals with tissue grade level of body organisation.
II. They exhibit bilateral symmetry.
III. They show the presence of a gelatinous mesoglea between epidermal and gastrodermal tissue layers.
IV. They have transparent ctenophores which show bioluminescence.

Codes

(a) I, II and III (b) Only II
(c) II and IV (d) I and III

5 Which of the following statements are true for guineaworm?

I. Digenetic parasite whose intermediate host is *Cyclops*.
II. It causes elephantiasis.
III. It is also called fiery serpent.
IV. Digenetic parasite whose intermediate host is female *Culex* mosquito.

Codes

(a) I and IV (b) II and III
(c) I and III (d) III and IV

6 Select the statements which is correct about *Ancylostoma duodenale*.

I. These worms live in the intestinal wall of humans.
II. They follow a digenetic life cycle.
III. The larva of the animal enters into the host body after boring through the skin.
IV. No proper treatment for the disease caused by its infection is known.

Codes

(a) I, II and III (b) II, III and IV
(c) I, II and IV (d) I and III

7 In the figure given below, a discharged cnidoblast is shown. Identify which labelled part will be

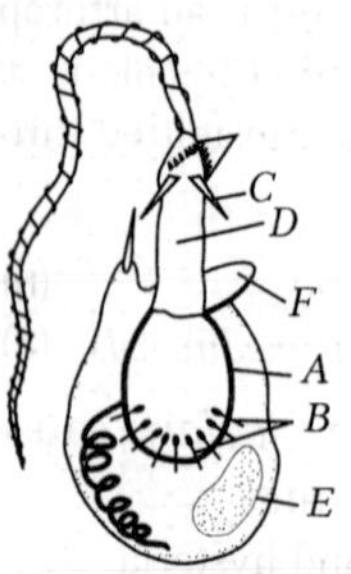

I. Barb II. Shaft
III. Nematocyst IV. Muscular fibrils
V. Nucleus VI. Operculum

Choose the correct option.

	I	II	III	IV	V	VI
(a)	C	B	A	E	F	D
(b)	C	D	A	B	E	F
(c)	B	E	A	F	D	C
(d)	C	F	D	A	E	B

8 The diagram given below shows the typical features of a *Physalia* colony. Label *A-F* and choose the correct option.

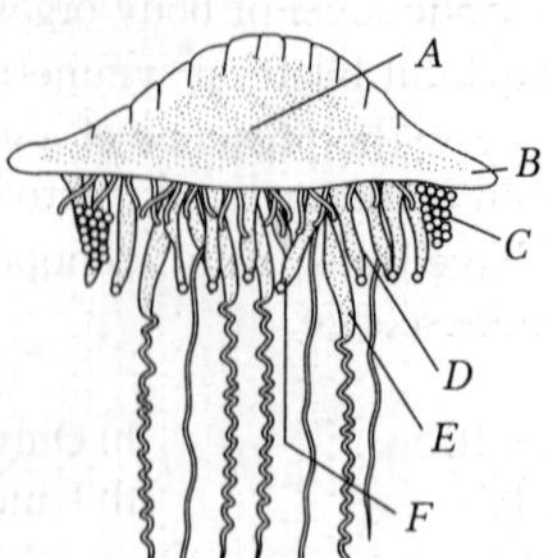

	A	*B*	*C*	*D*	*E*	*F*
(a)	Gonozooids	Gastrozooids	Cormidium	Pore	Mouth of gastrozooids	Float
(b)	Pore	Float	Gonozooids	Mouth of gastrozooid	Cormidium	Crest
(c)	Float	Pore	Gonozooids	Gastrozooids	Larger dactylozooid	Mouth of gastrozooids
(d)	Crest	Pore	Smaller dactylozooid	Larger dactylozooid	Cormidium	Float

9 Choose the statement which is incorrect about homeothermic animals.

(a) The body temperature in them is independent of the external environment
(b) These animals undergo hibernation and aestivation
(c) Their body temperature is derived from heat produced by the organism's own oxidative metabolism
(d) The presence of body parts like feathers and blubbers prevent the diffusion of internally generated heat to the external environment

10 Among the examples given below, select the set which includes animals developing from a two-layered embryo (diploblastic).

(a) Sea anemone, sea urchin, feather star
(b) *Hydra*, starfish, basket star
(c) Jellyfish, sea anemone and *Hydra*
(d) Sea urchin, cake urchin and jellyfish

11 Select the statement which is false regarding bilateral symmetry.

(a) Bilateral symmetry consists of the second major evolutionary advance in the animal body plan
(b) Bilateral symmetry enables an animal to have a more stream-lined shape
(c) It allows an animal to concentrate the power of its muscles and appendages for producing motion in one direction
(d) It is very common in *Cestum* and *Velamen*

12 Observe the given diagram of *Pleurobrachia* sp. and match *P, Q, R, S, T* (Column II) with statements (Column I).

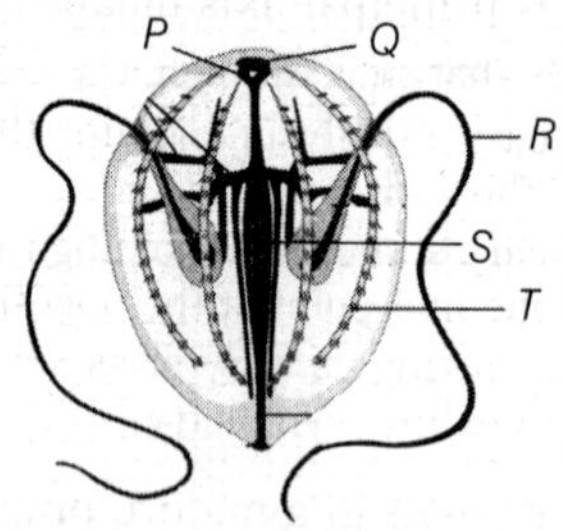

Column I	Column II
A. Organ of balance	*S*
B. Help in food capture	*Q*
C. Anal pore	*P*
D. Eight vertical rows of ciliary bands	*T*
E. Stomach	*R*

Codes

	A	B	C	D	E
(a)	*S*	*Q*	*P*	*T*	*R*
(b)	*Q*	*T*	*P*	*R*	*S*
(c)	*T*	*Q*	*S*	*P*	*R*
(d)	*Q*	*R*	*P*	*T*	*S*

13 Choose the incorrect pair from the following.

(a) *Olynthus*–Solitary hypothetical ancestral stage of sponge
(b) *Euplectella*–Symbolises 'Till death we do us apart'
(c) *Trochophore*–Free-swimming larva of sponge
(d) *Leucosolenia*–Sexual reproduction by syngamy

14 Sponges are multicellular organisms yet they do not have respiratory organs. This is due to the reason that

I. the mode of respiration is anaerobic.
II. the presence of highly porous body through which water flows.
III. limited requirement of oxygen, neglecting need of separate system for respiration.
IV. the respiratory organs become non-functional in marine water.

Choose the correct option.

(a) I and III (b) Only II (c) Only IV (d) III and IV

15 Which one of the following statements about all the four of *Spongilla*, leech, dolphin and penguin is correct?

(a) Penguin is homeothermic, while the remaining three are poikilothermic
(b) Leech is a freshwater form, while all others are marine
(c) *Spongilla* has special collar cells called choanocytes, which are not found in the remaining three
(d) All are bilaterally symmetrical animals

16 Match the following columns.

Column I (Phyla)	Column II (Body cavities)
A. Platyhelminthes	1. Enterocoel
B. Nematoda	2. No body cavity
C. Annelida	3. Pseudocoel
D. Echinodermata	4. Schizocoel

Codes

	A	B	C	D
(a)	4	2	3	1
(b)	2	3	4	1
(c)	4	3	2	1
(d)	3	2	4	1

17 The cross section of the body of an invertebrate is given below. Identify the animal, which has this body plan.

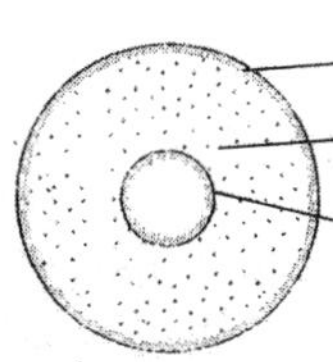

(a) Cockroach (b) Roundworm
(c) Planaria (d) Earthworm

18 Read the following four statements (I-IV) for Ctenophora.

I. Diploblastic with tissue level of organisation.
II. The body bears eight external row of ciliated comb plates.
III. Digestion is both extracellular and intracellular.
IV. Reproduction takes place only by sexual means.

How many of the above statements are correct?

(a) Four (b) Three
(c) Two (d) One

19 Match the following columns.

	Column I	Column II
A.	10 hooked larva	1. *Taenia*
B.	Acetabulum	2. Cycliophora
C.	Bladderworm	3. Sucker in Cestoda
D.	Hydatid worm	4. *E. granulossus*

Codes

	A	B	C	D
(a)	2	3	1	4
(b)	1	2	3	4
(c)	4	3	2	1
(d)	3	1	4	2

20 Match the following columns.

Column I	Column II
A. Flame cells	1. Sponges
B. Collar cells	2. *Hydra*
C. Stinging cells	3. Planaria
	4. *Ascaris*

Codes

	A	B	C
(a)	3	1	2
(b)	3	1	4
(c)	3	4	1
(d)	3	2	4

21 Consider the following four statements (I-IV) and select the option which includes all the correct ones only.

I. All members of Animalia are multicellular.
II. In sponges, the cells are arranged as loose cell aggregates.
III. Platyhelminthes exhibit cellular level of organisation.
IV. All multicellular animals do not exhibit the same pattern of organisation of cells.

(a) Statements II, III and IV (b) Statements I and II
(c) Statements III and IV (d) Statements I, II and IV

22 Identify the symmetry shown below as well as the related explanation and select the right option for the two together.

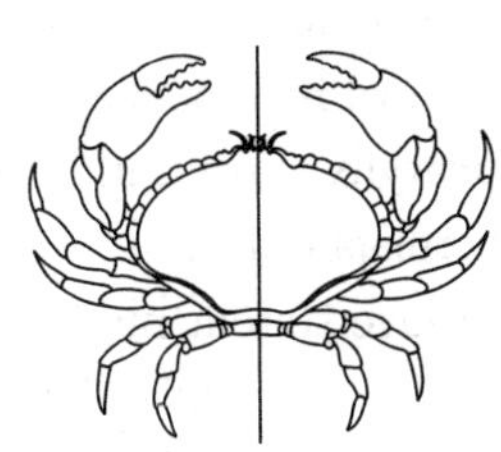

Symmetry	Explanation
(a) Asymmetrical	Any plane that passes through the centre does not divide them equal halves
(b) Radial	The body can be divided into identical left and right halves in only one plane
(c) Bilateral	Any plane pass through the central axis of the body divides the organism into two identical halves
(d) Bilateral	The body can be divided into identical left and right halves in only one plane

23 Match the following columns.

Column I	Column II
A. Porifera	1. *Velamen*
B. Cnidaria	2. *Leucosolenia*
C. Ctenophora	3. *Physalia*
D. Platyhelminthes	4. *Fasciola*

Codes

	A	B	C	D
(a)	1	3	4	2
(b)	2	3	1	4
(c)	1	2	3	4
(d)	4	3	2	1

24 Given below are the three figures of sponges. Which of the following sponges are found in freshwater?

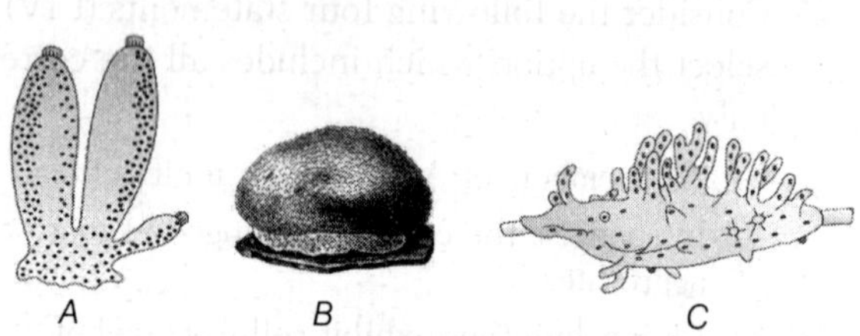

Select the correct option.

(a) Only *A* (b) Only *C*
(c) *A* and *B* (d) *A* and *C*

25 Read the following statements with respect to Aschelminthes.

I. They are diploblastic and pseudocoelomate animals.
II. Their body is circular in cross-section so, they are called roundworms.
III. Alimentary canal is incomplete.
IV. Muscular pharynx is present.
V. They are hermaphrodites.

Which of the above statements are correct?

(a) II and IV (b) I, II and V
(c) III, IV and V (d) II, III and IV

26 Read the following statements with respect to Aschelminthes/Nemathelminthes.

I. Development may be direct or indirect.
II. Fertilisation is internal.
III. Females are often shorter than males.
IV. *Wuchereria* causes elephanitiasis or filariasis.
V. *Ascaris* possesses cuticle which is resistant to the digestive enzymes of host.

Which of the above statements are correct?

(a) II, III, IV and V (b) I, II, IV and V
(c) I, II and III (d) III, IV and V

27 Match the following columns.

Column I	Column II
A. Calcarea	1. Six-rayed siliceous spicules
B. Demospongiae	2. Spongin fibres aragonite and siliceous spicules
C. Sclerospongiae	3. Spongin fibres and siliceous spicules
D. Hexactinellida	4. Aragonite spicules

Codes

	A	B	C	D
(a)	1	2	3	4
(b)	2	4	3	1
(c)	3	2	1	4
(d)	4	3	2	1

28 Read the following statements.

I. True coelom is a body cavity which arises as a cavity in the embryonic mesoderm.
II. Digestive cavity is found in acoelomates, pseudocoelomates as well as coelomates.
III. The body cavity of arthropods and non-cephalopod molluscs is called haemocoel.
IV. Aschelminthes possess true body cavity.

Which of the above statements are correct?

(a) I and III
(b) I, II and III
(c) III and IV
(d) I and IV

29 Read the following statements about the phylum-Porifera.

I. Commoly known as sponges.
II. Generally found in freshwater and are mostly symmetric animals.
III. Possess water transport canal system.
IV. The body is supported by endoskeleton made up of spicules or spongin fibres.
V. Fertilisation is external.

Which of the above statements are correct?

(a) III and V (b) I, II and III
(c) I, III and IV (d) II and V

30 How many of the examples mentioned below possess closed (C) and open (O) circulatory system.

Cephalopod molluscs, Non-cephalopod molluscs, Chordates, Annelids, Tunicates, Hemichordates and Arthropods

(a) C = 2, O = 5 (b) C = 4, O = 3
(c) C = 5, O = 2 (d) C = 3, O = 4

31 The type of body symmetry shown below can be found in

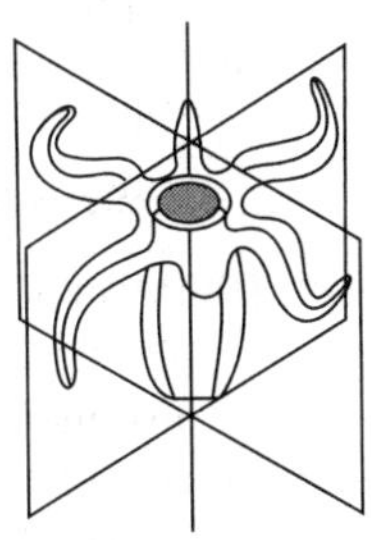

(a) All sponges
(b) platyhelminths, aschelminths and annelids
(c) coelenterates, ctenophores, adult echinoderms and some sponges
(d) arthropods, molluscs, hemichordates and chordates

32 Read the following statements.

I. In open type circulatory system, the blood is pumped out of the heart and the cells and tissues are directly bathed in it.
II. Radially symmetrical animals are mostly cylindrical and central axis is longitudinal in their body.
III. Cephalisation is considered an evolutionary trend which eventually produces a head region with sensory organs.
IV. The great majority (~99%) of animals are radially symmetric.
V. The phylum-Echinodermata is unique among animals in having bilateral symmetry in the larval stage, but radial symmetry at the adult stage.

Which of the above statements are correct?

(a) III, IV and V (b) I, II, III and V
(c) I, II and III (d) III and V

33 Identify the layers/germ layers -*A*, *B*, *C* and *D* in diploblastic (I) and triploblastic (II) organisation in animals.

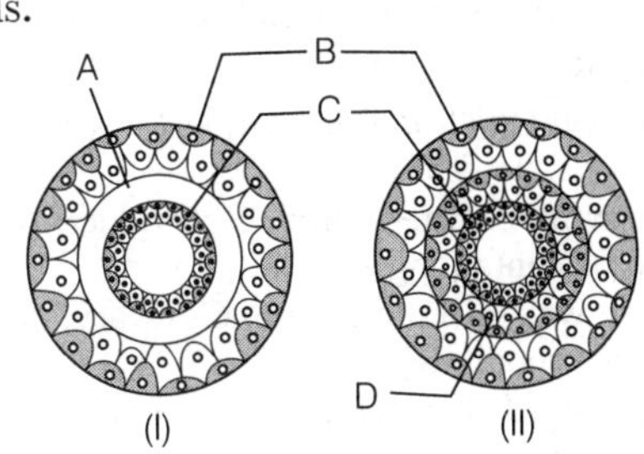

	A	*B*	*C*	*D*
(a)	Mesoderm	Ectoderm	Endoderm	Mesoglea
(b)	Mesoglea	Ectoderm	Endoderm	Mesoeerm
(c)	Mesoglea	Mesoderm	Ectoderm	Endoderm
(d)	Mesoderm	Mesoglea	Ectoderm	Endoderm

34 Identify the organisms *A*, *B* and *C*.

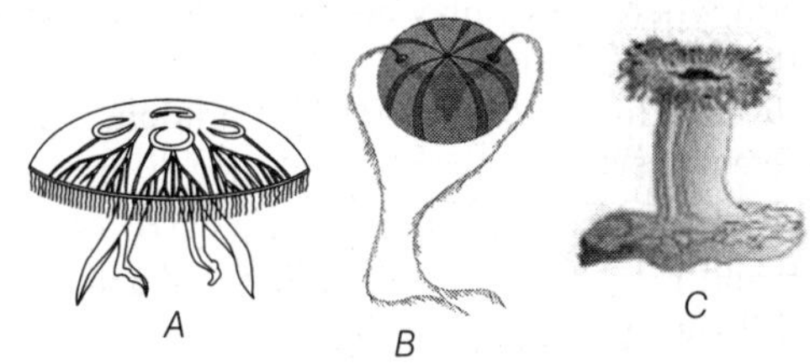

	A	B	C
(a)	*Pleurobrachia*	*Adamsia*	*Aurelia*
(b)	*Aurelia*	*Adamsia*	*Pleurobrachia*
(c)	*Pleurobrachia*	*Aurelia*	*Adamsia*
(d)	*Aurelia*	*Pleurobrachia*	*Adamsia*

35 Read the following statements with respect to cnidarians.

I. Cnidarians exhibit tissue level of organisation and are triploblastic.
II. Digestion is extracellular and intracellular.
III. Corals secrete calcium bicarbonate to form a soft skeleton commonly.
IV. Corals may harbour some photosynthetic dinoflagellates for taking nutrition.
V. They possess a central gastrovascular cavity with a single opening mouth on hypostome.

Which of the above statements are correct?

(a) II and III
(b) I, II and III
(c) II, IV and V
(d) III and IV

36 Match the generic name listed under Column I with the common name given under Column II. Choose the answer which gives the correct combination of the alphabets of the two columns.

	Column I (Genus)		Column II (Common name)
A.	*Millepora*	1.	Sea wasp
B.	*Pennatula*	2.	Blue coral
C.	*Heliopora*	3.	Sea pen
D.	*Chiropsalmus*	4.	Stinging coral

Codes

	A	B	C	D		A	B	C	D
(a)	4	2	3	1	(b)	2	3	4	1
(c)	4	3	2	1	(d)	3	2	4	1

37 Read the following statements.

I. Some cnidarians exists only in polyp forms like *Adamsia* and *Hydra.*
II. Nervous system of Ctenophora is similar to that of coelenterates.
III. Beginning of cephalisation is seen in flatworms.
IV. Some cnidarians exist as medusa as the dominant form where polyp may be absent or reduced like *Aurelia* or jellyfish.
V. *Obelia* exhibit alternation of generation, i.e. metagenesis.

Which of the above statements are correct?

(a) I, II and V (b) I, II, III, IV and V
(c) II, IIII and IV (d) I, III, IV and V

38 Consider the following pairs.

I. Parazoa - No digestive cavity
II. Mesozoa - Parasitic on cephalopods
III. Enterozoa - True tissues and organs

Choose the correctly matched pair.

(a) I and II (b) II and III
(c) I, II and III (d) None of these

39 From the given set of statements, identify the correct ones.

I. Rediae, cercaria, miracidium and bipinnaria are the larval stages of flatworms.
II. Mehli's gland in tapeworm is associated with reproduction.
III. Cestodes are distinguished from other flatworms by the absence of digestive system.
IV. Pseudocoelom is primary coelom derived from blastocoel of the embryo and occurs in *Taenia.*

Codes

(a) I and II (b) II and III
(c) I, III and IV (d) II, III and IV

40 Read the following statements.

I. In Porifera, excretion mainly occurs by diffusion.
II. Sponges are ammonotelic animals.
III. Sponges are filter feeders or suspension feeders.

Which of the statements given above are correct?

(a) I and II (b) I and III (c) II and III (d) I, II and III

41 From the given set of statements, identify the statements as true or false.

I. Opening of osculum is regulated partly by myocytes.
II. Scleroblasts of sponges occur in the mesohyl layer.
III. Sponges do not possess muscular tissue.
IV. Pinacocytes are involved in capturing of food in sponges.

Codes

(a) I and II are true, III and IV are false
(b) II and IV are true, I and III are false
(c) I, IV, III are true, II is false
(d) I, II, III are true, IV is false

42 Read the following statements.

I. Notochord is ectodermally derived rod-like structure.
II. Notochord is formed on the dorsal side during embryonic development.
III. Notochord is present in most of the phylums.
IV. The animal from Porifera to Echinodermata are without notochord.
V. In some chordates, notochord is replaced by the vertebral column and these chordates are called vertebrates.

Which of the above statements are correct?

(a) III, IV and V (b) I, II and III
(c) I, III and IV (d) II, IV and V

43 Scypha belongs to a group of animals which are best described as

I. multicellular without any tissue organisation.
II. mostly marine and have collared cells.
III. universal presence of spongocoel.
IV. circulatory system constituted by intercommunicating canals in the body.

Codes

(a) I and II
(b) I, II and IV
(c) III and IV
(d) I, II, III and IV

44 Which of the following pairs is correctly matched?

I. Amoebocytes —Transport food to non-feeding cells
II. Collar cells — Movement of water and filtering food
III. Spicules — Skeletal supporting element

Codes

(a) I only (b) I and II
(c) I, II and III (d) II and III

45 The figure given below shows the female reproductive system in *Ascaris.*

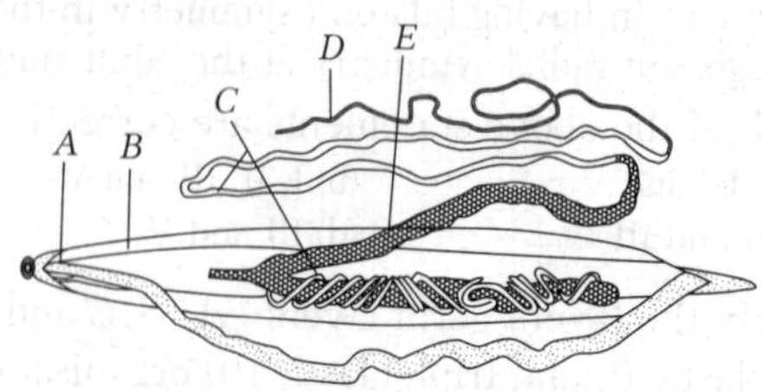

Identify the parts indicated (*A-E*) and select the correct code.

(a) *A*-Cut body wall, *B*-Pharynx, *C*-Lips, *D*-Oviduct, *E*-Ovary
(b) *A*-Pharynx, *B*-Lateral line, *C*- Oviduct, *D*-Ovary, *E*-Uterus containing eggs
(c) *A*-Body wall, *B*-Lateral line, *C*- Ovary, *D*-Oviduct, *E*-Uterus containing eggs
(d) *A*-Pharynx, *B*-Lateral line, *C*-Oviduct, *D*-Ovary, *E*-Intestine

46 The figure given below illustrates the male reproductive system of *Ascaris*. Identify the parts indicated *A-E* and mark the correct code.

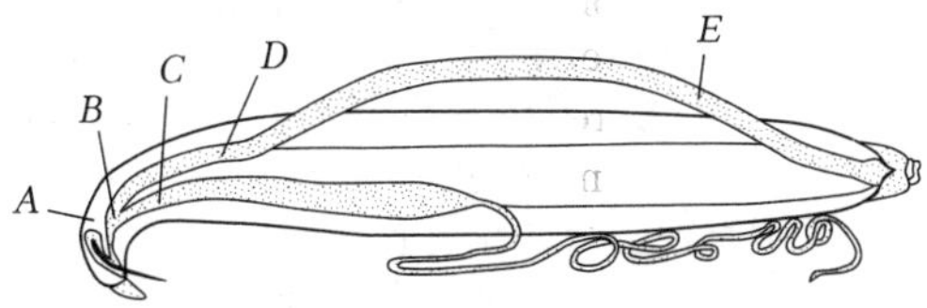

(a) *A*-Penial spicule, *B*-Cloaca, *C*-Seminal vesicle *D*-Rectum, *E*-Cut body wall
(b) *A*-Penial sac, *B*-Penial spicule, *C*-Rectum, *D*-Ejaculatory duct, *E*-Intestine
(c) *A*-Penial sac, *B*-Cloaca, *C*-Ejaculatory duct, *D*-Rectum, *E*-Intestine
(d) *A*-Penial spicule, *B*-Penial sac, *C*-Cloaca, *D*-Ejaculatory duct, E-Rectum

47 Pick out the statement which is incorrect about the filarial worm (*Wuchereria bancrofti*) or the disease caused by it.
(a) The adults live in human lymph vessel and lymph glands
(b) Its common intermediate host is female *Culex* mosquito
(c) It can be treated by the drug albendazole, diethylcarbamazine, etc.
(d) The disease caused by this worm shows non-specific symptoms such as eosinophilia, nausea, asthma, fainting, etc.

48 The common name for *Enterobius vermicularis* is pin worm. Identify the statement among the ones given below which is incorrect about this worm.
(a) This worm inhabits human caecum, colon, appendix and rectum
(b) Its larvae appear in cutaneous blood in midnight
(c) Pinworm causes itching of anal parts
(d) It is monogenetic

49 Which one of the following statements about the characteristic of given sponges is not related to it?
(a) *Euspongia* found in warm shallow water of the sea, used as sound absorbers in cinema halls
(b) *Leucosolenia* is the smallest and simplest colonial sponge, Asconoid and *Olynthus* type of organisation
(c) *Spongilla* a freshwater sponge, often green due to symbiotic algae
(d) *Euplectella* show commensalism with shrimps of genus *Spongicola*. The siliceous spicules when present are never six-rayed

50 Which of the following statements is incorrect about the class-Hydrozoa of phylum-Cnidaria?
(a) They lack a stomodaeum and the gastrovascular cavity is without partitions and nematocysts
(b) The mesoglea is cellular and seen as a thick transient layer
(c) The highest degree of polymorphism in phylum-Cnidaria is shown by the organisms of this class
(d) Gametes in the organisms of this class are epidermal in origin

51 Which of the following is not correctly matched?
(a) *Loa loa*-Subdermal — Connective tissues
(b) *Dracunculus medinensis* — Monogenetic
(c) *Enterobius vermicularis* — Oxyurasis
(d) *Ancylostoma* — Feed upon blood

52 *Ascaris* is monogenetic, i.e. no intermediate host is involved for development. The egg of *Ascaris* shows various characteristic features.
Among the statements given below, identify the one which is false about the egg of *Ascaris lumbricoides*.
(a) The egg develops into rhabditiform larva
(b) Embryonated egg is an infective stage
(c) The egg has got three protective coverings
(d) The inner membrane of the egg is made of chitin

53 Select the statement which is incorrect about liver fluke (*Fasciola hepatica*).
(a) It has a flat, leaf-like body with dull pink-ground colour and brown border
(b) The oral cone present contains mouth and two suckers
(c) In between the acetabulum and the oral sucker, a midventral gonopore or genital opening is present
(d) During breeding season, a temporary opening called opening of kauler canal is seen on the ventral surface

54 Identify the given canal system by the course of water current in sponges, represented as *A*, *B* and *C*. Choose the correct code.

I. Ingressing water $\xrightarrow[\text{ostia}]{\text{Through}}$ Spongocoel $\xrightarrow[\text{osculum}]{\text{Through}}$ To outside.

II. Ingressing water $\xrightarrow[\text{ostia}]{\text{Dermal}}$ Incurrent canals $\xrightarrow[\text{Prosopyles}]{}$ Flagellated chambers $\longrightarrow$ Apopyle $\longrightarrow$ Excurrent canals $\xrightarrow{\text{Osculum}}$ To outside.

III. Ingressing water $\xrightarrow[\text{ostia}]{\text{Dermal}}$ Incurrent canals $\xrightarrow{\text{Prosopyles}}$ Radial canals $\xrightarrow{\text{Apopyles}}$ Spongocoel $\xrightarrow{\text{Osculum}}$ To outside.

Codes

	A	B	C
(a)	*Sycon*	*Leucon*	*Ascon*
(b)	*Leucon*	*Ascon*	*Sycon*
(c)	*Ascon*	*Leucon*	*Sycon*
(d)	*Sycon*	*Ascon*	*Leucon*

55 Study the following statements about *Adamsia* and select your answer from the codes given below.
I. It is a parasite.
II. It is an example of commensalism.
III. It is called sea anemone.
IV. It is called hermit crab.

Codes
(a) II and III (b) I and II
(c) III and IV (d) I and IV

56 The following are the features associated with Cnidaria.

I. Radial symmetry.
II. The presence of gastrovascular cavity.
III. Animals are in either of the two forms, polyp and medusa or both.
IV. Alternation of generations in their history.

Which of the above are true of *Metridium*?

(a) I, II and IV (b) I and II
(c) II and III (d) I, II, III an IV

57 Select the total number of correct statements from the following about coelenterates.

I. Most of the coelenterates are solitary.
II. Coelenterates have four types of nematoblasts.
III. The interstitial cells present are undifferentiated mesenchyme cells with a capability to convert in any type of epithelial cells.
IV. Non-polar neurons are scattered irregularly in the body.
V. They show medusoid forms which are always dioecious.
VI. They exhibit alternation of generation.

(a) 4 (b) 3
(c) 1 (d) 5

58 Given below are the pair of phyla with their body cavities.

Phyla		Body cavities
I. Platyhelminthes	–	Pseudocoel
II. Annelida	–	Schizocoel
III. Nematoda	–	No body cavity
IV. Echinodermata	–	Enterocoel

Choose the option with the correct codes.

(a) I, II and III (b) I and II
(c) II and IV (d) I and III

59 Find out the incorrect match.

(a) *Wuchereria bancrofti* – Ovoviviparous
(b) *Loa loa* – *Cyclops* is the secondary host
(c) *Dracunculus medinensis* – Formation of blisters on arms
(d) *Enterobius vermicularis* – Monogenetic

60 Which of the following requires an invertebrate intermediate host?

I. *Dugesia*
II. *Schistosoma*
III. *Echinococcus*
IV. *Ancylostoma*
V. *Wuchereria*

Codes

(a) III and IV (b) I and IV
(c) III and V (d) II and V

61 Observe the diagram given below and select the correct statements.

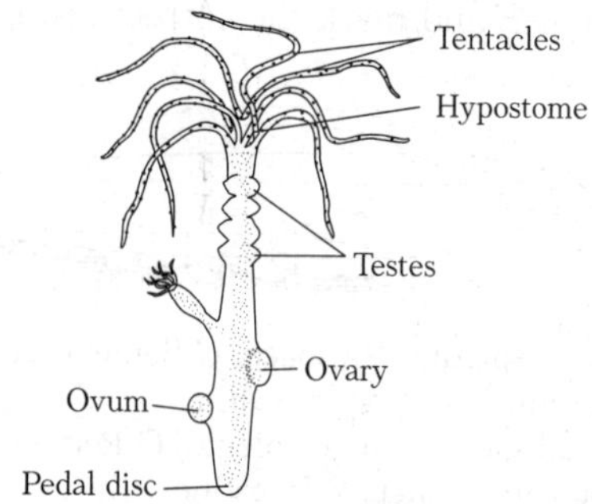

I. The nematocysts are present in epidermis as well as basal disc of *Hydra*.
II. *Hydra* shows the presence of 8-12 tentacles.
III. Medusae of *Hydra* reproduce both by sexual and asexual method.
IV. Hypostome is the most sensitive region in the body of *Hydra*.
V. The pedal or basal disc helps in attachment as it shows the presence of a glandular zone which secretes adhesives.

(a) I, IV and V (b) I, III and V
(c) II, IV and V (d) II, III and IV

62 Choose the correct statements regarding *Taenia solium*.

I. Scolex is present at the posterior most end of the body.
II. Alimentary canal is absent in all stages of its life cycle.
III. Aerobic respiration is found and gaseous exchange takes place through general body surface.
IV. The excretory system consists of long canals, capillaries and the protonephridia or flame cells.

Codes

(a) II and IV (b) I, II and IV (c) I and IV (d) I and II

63 Given below are few characteristic features of phylum – Platyhelminthes.

I. The presence of ectodermal, endodermal and mesodermal layer.
II. The presence of hydroskeleton.
III. Bilateral symmetry.
IV. Complete digestive system.

Codes

(a) II and IV (b) I, II and III (c) Only III (d) Only I

64 Read the following statements about *Ascaris*.

I. It spreads through contaminated food and water.
II. Its symptoms include nausea, vomiting, swelling in the gums and enlargement of the lymph glands.
III. It can be treated by drugs such as oil of *Chenopodium*.
IV. It exhibits aberrant migration.

Choose the option with correct statements.

(a) I, II and III (b) I and II (c) II and IV (d) I, III and IV

Assertion and Reason

Direction (Q. Nos. 65-77) *In each of the following questions, a statement of Assertion (A) is given followed by corresponding statement of Reason (R). Of the statements, mark the correct answer as*

(a) Both A and R are true and R is the correct explanation of A
(b) Both A and R are true, but R is not the correct explanation of A
(c) A is true, but R is false
(d) A is false, but R is true
(e) Both A and R are false

65 Assertion (A) Metamerism is the characteristic of annelids.
Reason (R) It is one type of body segmentation.

66 Assertion (A) Endotherms are cold-blooded animals.
Reason (R) Ectotherms can tolerate broad range of environmental temperature.

67 Assertion (A) *Hydra* is a diploblastic animal.
Reason (R) *Hydra* is derived from two germ layers.

68 Assertion (A) Archaeocytes help in regeneration.
Reason (R) In sponges, archaeocytes are totipotent.

69 Assertion (A) *Spongilla* and *Leucosolenia* are structurally similar sponges.
Reason (R) In *Spongilla* and *Leucosolenia*, body is radially symmetrical.

70 Assertion (A) Sponges are less specialised ones.
Reason (R) They show least power of regeneration.

71 Assertion (A) Sponges do not show any animal nature.
Reason (R) They are sessile with no apparent way of capturing food or eliminating water.

72 Assertion (A) In *Hydra*, both pseudopodia and flagella occur in sensory cells.
Reason (R) Pseudopodia and flagella are used for sexual reproduction in *Hydra*.

73 Assertion (A) The coelenteron of *Hydra* is called gastrovascular cavity.
Reason (R) *Hydra* is a monoblastic animal.

74 Assertion (A) Medusa of *Obelia* is carnivorous.
Reason (R) Digestion is both extracellular and intracellular.

75 Assertion (A) Coelenterates show alternation of generation.
Reason (R) In coelenterates, asexual generation is followed by sexual generation.

76 Assertion (A) Hydatid vesicle is the characteristic of *Echinococcus* for absorption.
Reason (R) Hydatid cyst can survive for several years within the human body.

77 Assertion (A) *Taenia solium* has a characteristic of uterus.
Reason (R) The uterus is present in gravid proglottid.

(C) Medical Entrances' Gallery

(Collection of questions Asked in NEET & Various Medical Entrance Exams)

1 Match the following (columns) group of organisms with their respective distinctive characteristics and select the correct option from the codes given below.

Column I (Organisms)	Column II (Characteristics)
A. Platyhelminthes	1. Cylindrical body with no segmentation
B. Echinoderms	2. Warm-blooded animals with direct development
C. Hemichordates	3. Bilateral symmetry with incomplete digestive system
D. Aves	4. Radial symmetry with indirect development

Codes [NEET 2020]

	A	B	C	D		A	B	C	D
(a)	3	4	1	2	(b)	2	3	4	1
(c)	4	1	2	3	(d)	1	2	3	4

2. Bilaterally symmetrical and acoelomate animals are exemplified by **[NEET 2020]**
(a) Platyhelminthes
(b) Aschelminthes
(c) Annelida
(d) Ctenophora

3. Animals of which phylum have hooks and suckers and are endoparasite on other animals? **[NEET 2019]**
(a) Platytelminthes
(b) Annelida
(c) Aschelminthes
(d) Arthropoda

4. What is the symmetry of medusa? **[JIPMER 2018]**
(a) Bilateral
(b) Radial
(c) Asymmetrical
(d) Biradial

5. **Assertion** (A) *Taenia solium* and *Dugesia* belong to Platyhelminthes.

Reason (R) Platyhelminthes are coelomates. **[AIIMS 2018]**

(a) Both A and R are true and R is the correct explanation of A
(b) Both A and R are true, but R is not the correct explanation of A
(c) A is true, but R is false
(d) Both A and R are false

6. In case of poriferans, the spongocoel is lined with flagellated cells called **[NEET 2017]**

(a) ostia (b) oscula
(c) choanocytes (d) mesenchymal cells

7. Gemmule formation is a common mode of asexual reproduction in **[AIIMS 2017]**

(a) *Paramecium* (b) *Hydra*
(c) sponges (d) yeast

8 Body having meshwork of cells, internal cavities lined with food filtering flagellated cells and indirect development are the characteristics of phylum **[CBSE AIPMT 2015]**

(a) Coelenterata (b) Porifera
(c) Mollusca (d) Protozoa

9 Metagenesis refers to **[CBSE AIPMT 2015]**

(a) the presence of different morphic forms
(b) alternation of generation between asexual and sexual phases of an organism
(c) occurrence of a drastic change in form during post-embryonic development
(d) the presence of a segmented body and parthenogenetic mode of reproduction

10. Third stage larva of *Wuchereria bancrofti* carried by *Culex* mosquito is called **[WB JEE 2015]**

(a) cysticercus (b) merozoite
(c) microfilariae (d) trophozoite

11 Which one of the following statements is incorrect?

(a) Insects have one pair of antenna **[WB JEE 2015]**
(b) Millipedes possess two pairs of appendages in each segment of the body
(c) Prawns have two pairs of antenna
(d) Animals belonging to the phylum-Porifera have nematocyst

12 Which one is also called hookworm? **[BCECE 2015]**

(a) *Ancylostoma* (b) *Enterobius*
(c) *Rhabditis* (d) *Ascaris*

13 The cercarial stage of liver fluke is produced by **[AMU 2015]**

(a) sexual reproduction
(b) asexual multiplication
(c) binary fission
(d) parthenogenesis

14 Which of the following is a mismatch pair? **[AMU 2015]**

(a) Scales—Reptilia (b) Comb plates—Mollusca
(c) Choanocytes—Porifera (d) Parapodia—Annelida

15 Which of the following exhibits metagenesis? **[AMU 2015]**

(a) *Hydra* (b) *Adamsia*
(c) *Aurelia* (d) *Obelia*

16 Select the taxon mentioned that represents both marine and freshwater species. **[CBSE AIPMT 2014]**

(a) Echinodermata (b) Ctenophora
(c) Cephalochordata (d) Cnidaria

17 Cellular organisation of body is present in **[MHT CET 2014]**

(a) Annelida (b) Platyhelminthes
(c) Porifera (d) Urochordata

18 The kind of coelom represented in the diagram is characteristic of **[KCET 2014]**

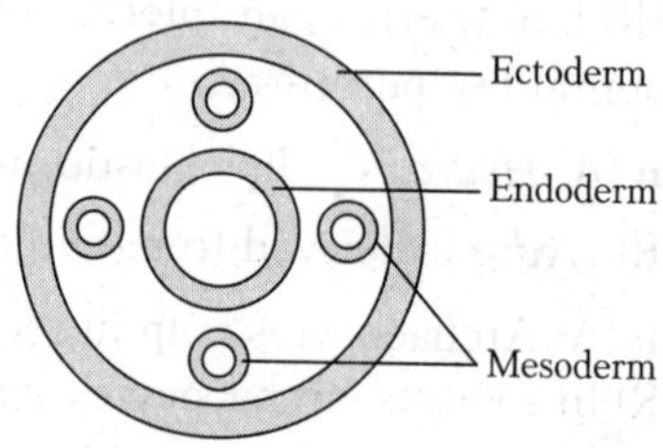

(a) earthworm (b) cockroach
(c) roundworm (d) tapeworm

19 Match the following columns.

Column I	Column II
A. Incomplete digestive system	1. Sponges
B. Cellular level of organisation	2. Coelenterates
C. Radial symmetry	3. Annelids
D. Pseudocoelomate	4. Platyhelminthes
E. Metamerism	5. Aschelminthes

Codes **[Kerala CEE 2014]**

	A	B	C	D	E
(a)	3	4	1	2	5
(b)	4	5	2	3	1
(c)	4	1	2	5	3
(d)	1	2	3	4	5
(e)	2	3	4	1	5

20 Which of the following statement(s) regarding coelenterates is/are wrong? **[Kerala CEE 2014]**

I. Cnidocytes are present on the tentacles and on the body.
II. Diploblastic with cellular level of organisation.
III. Polyp forms are free-swimming.
IV. Exhibits metagenesis.
V. Polyps produce medusae sexually and medusae form polyps asexually.

Codes

(a) II and IV (b) III and V
(c) I, II and III (d) Only III
(e) II, III and V

21 Which of these is referred to as 'venus flower basket'? [Kerala CEE 2014]
(a) *Spongilla* (b) *Sycon*
(c) *Euplectella* (d) *Cliona*
(e) *Proterion*

22 In the cycle of *Ascaris lumbricoides*, rhabditiform larva undergoes 2nd and 3rd moultings in [EAMCET 2014]
(a) liver (b) heart
(c) alveoli of lungs (d) small intestine

23 Infection of *Ascaris* usually occurs by [NEET 2013]
(a) eating imperfectly cooked pork
(b) tse-tse fly
(c) mosquito bite
(d) drinking water containing eggs of *Ascaris*

24 Choanocytes are present in [AFMC 2013]
(a) Coelenterata (b) Porifera
(c) Echinodermata (d) Mollusca

25 Match the columns with reference to *Taenia*. [EAMCET 2013]

Column I	Column II
A. Mehli's glands	1. Tegument formation
B. Vitelline gland	2. Osmoregulation and excretion
C. Mesenchymal cells	3. Lubricate passage of capsules into uterus
D. Flame cells	4. Secretion of embryophore
	5. Capsule formation around zygote

Codes

	A	B	C	D
(a)	3	5	1	2
(b)	3	5	2	4
(c)	5	1	2	4
(d)	4	3	1	2

26 Match the following columns.

Column I	Column II
A. Organ level	1. *Pheretima*
B. Cellular aggregate level	2. *Fasciola*
C. Tissue level	3. *Spongilla*
D. Organ system level	4. *Obelia*

Codes [NCERT Exemplar, KCET 2013]

	A	B	C	D
(a)	4	3	1	2
(b)	4	2	3	1
(c)	2	4	3	1
(d)	2	3	4	1

27 Which one of the following statements about certain given animals is correct? [BCECE 2012]
(a) Roundworms (Aschelminthes) are pseudocoelomates
(b) Molluscs are acoelomates
(c) Insects are pseudocoelomates
(d) Flatworms (Platyhelminthes) are coelomates

28 Biradial symmetry and lack of cnidoblasts are the characteristics of [BCECE 2012]
(a) Starfish and sea anemone (b) *Ctenoplana* and *Beroe*
(c) *Aurelia* and *Paramecium* (d) *Hydra* and starfish

29 *Sycon* belongs to a group of animals, which are best described as [BCECE 2012]
(a) multicellular with a gastrovascular system
(b) multicellular having tissue organisation, but no body cavity
(c) unicellular or acellular
(d) multicellular without any tissue organisation

30 The true statement regarding the coral is [CBSE AIPMT 2012]
(a) they form branched colonies
(b) they are solitary or colonial polyploid
(c) they grow as massive bodies
(d) All of the above

31 Jellyfish belongs to class [UP CPMT 2012]
(a) Hydrozoa (b) Scyphozoa
(c) Anthozoa (d) None of these

32 Nematoblasts are found in [UP CPMT 2012]
(a) ectoderm (b) endoderm
(c) Both (a) and (b) (d) mesoderm

33 Match the following columns. [Kerala CEE 2012]

Column I	Column II
A. *Physalia*	1. Sea anemone
B. *Meandrina*	2. Brain coral
C. *Gorgonia*	3. Sea fan
D. *Adamsia*	4. Portuguese man of war

Codes

	A	B	C	D
(a)	3	2	1	4
(b)	4	3	2	1
(c)	4	2	3	1
(d)	2	3	1	4
(e)	1	2	3	4

34 Flatworms excrete through [Manipal 2012]
(a) kidney (b) nephridia
(c) protonephridia (d) Malpighian tubules

35 Which of the following cells are useful for feeding in sponges? [Manipal 2012]
(a) Thesocytes (b) Collar cells
(c) Pinacocytes (d) Sclerocytes

36 In most simple type of canal system of Porifera, water flows through which one of the following ways? [WB JEE 2012]
(a) Ostia $\rightarrow$ Spongocoel $\rightarrow$ Osculum $\rightarrow$ Exterior
(b) Spongocoel $\rightarrow$ Ostia $\rightarrow$ Osculum $\rightarrow$ Exterior
(c) Osculum $\rightarrow$ Spongocoel $\rightarrow$ Ostia $\rightarrow$ Exterior
(d) Osculum $\rightarrow$ Ostia $\rightarrow$ Spongocoel $\rightarrow$ Exterior

37 Which one of the following animal phyla possesses spicules? **[WB JEE 2012]**
(a) Annelida (b) Mollusca
(c) Porifera (d) Platyhelminthes

38 The non-cellular layer present between pinacoderm and choanoderm in body wall of poriferans is known as **[OJEE 2012]**
(a) mesoderm (b) middle lamellae
(c) mesenchyme (d) None of these

39 The skeleton of sponges is secreted by **[OJEE 2012]**
(a) chromocytes (b) myocytes
(c) scleroblasts (d) phagocytes

40 Symmetry in Cnidaria is **[JCECE 2012]**
(a) radial (b) bilateral (c) pentamerous (d) spherical

41 The animals that rely on the heat from environment than metabolism to raise their body temperature are in strict sense, called **[AMU 2012]**
(a) ectothermic (b) poikilothermic
(c) homeothermic (d) endothermic

42 Which of the following has biradial symmetry? **[BHU 2012]**
(a) *Paramecium* (b) Jellyfish
(c) Cockroach (d) Sea anemone

43 Coelom is found between **[BHU 2012]**
(a) ectoderm and endoderm
(b) mesoderm and ectoderm
(c) body wall and ectoderm
(d) mesoderm and endoderm

44 Which one of the following is not a characteristic feature of sponges? **[Kerala CEE 2011]**
(a) Cellular level of organisation
(b) Presence of ostia
(c) Intracellular digestion
(d) Body supported by chitin
(e) Indirect development

45 The anterior V-spot in microfilaria of *Wuchereria* represents **[WB JEE 2011]**
(a) nerve ring (b) cervical papilla
(c) excretory system (d) reproductive system

46 Which of the following animals have a single opening to the outside that serves both as mouth as well as anus? **[CBSE AIPMT 2010]**
(a) *Octopus* (b) *Asterias* (c) *Ascidia* (d) *Fasciola*

47 Which one of the following animals possesses high regeneration capacity? **[Kerala CEE 2010]**
(a) Planaria (b) *Taenia* (c) *Salpa* (d) *Periplaneta*
(e) *Ascidia*

48 The pseudocoelomate among these is **[Kerala CEE 2010]**
(a) Porifera (b) Annelida
(c) Aschelminthes (d) Mollusca
(e) Platyhelminthes

49 A bilaterally symmetrical Deuterostome is classified in the phylum **[COMEDK 2010]**
(a) Annelida (b) Chordata
(c) Arthropoda (d) Echinodermata

ANSWERS

Check Point 8.1

1. (d) 2. (a) 3. (b) 4. (b) 5. (c) 6. (d)

Check Point 8.2

1. (d) 2. (b) 3. (b) 4. (c) 5. (d) 6. (c)

Check Point 8.3

1. (c) 2. (c) 3. (d) 4. (b) 5. (b) 6. (d) 7. (b)

Check Point 8.4

1. (d) 2. (c) 3. (d) 4. (c) 5. (b) 6. (c)

Check Point 8.5

1. (d) 2. (b) 3. (c) 4. (a) 5. (a) 6. (b)

(A) Topical Exercise

1. (b)	2. (a)	3. (d)	4. (a)	5. (c)	6. (d)	7. (a)	8. (b)	9. (b)	10. (a)
11. (c)	12. (a)	13. (a)	14. (c)	15. (d)	16. (d)	17. (b)	18. (b)	19. (b)	20. (b)
21. (a)	22. (a)	23. (d)	24. (c)	25. (c)	26. (d)	27. (c)	28. (c)	29. (a)	30. (b)
31. (b)	32. (b)	33. (d)	34. (d)	35. (a)	36. (c)	37. (d)	38. (c)	39. (c)	40. (d)
41. (d)	42. (c)	43. (b)	44. (c)	45. (c)	46. (a)	47. (a)	48. (d)	49. (d)	50. (c)
51. (a)	52. (b)	53. (b)	54. (c)	55. (b)	56. (c)	57. (c)	58. (b)	59. (b)	60. (c)
61. (c)	62. (d)	63. (d)	64. (d)	65. (d)	66. (d)	67. (b)	68. (a)	69. (c)	70. (d)
71. (c)	72. (a)	73. (d)	74. (a)	75. (b)	76. (d)	77. (b)	78. (d)	79. (a)	80. (c)
81. (c)	82. (c)	83. (c)	84. (b)	85. (a)	86. (a)	87. (c)	88. (a)	89. (d)	90. (c)
91. (d)	92. (c)	93. (b)	94. (a)	95. (d)	96. (d)	97. (d)	98. (c)	99. (d)	100. (d)
101. (a)	102. (a)	103. (a)	104. (b)	105. (c)	106. (c)	107. (d)	108. (d)	109. (b)	110. (a)
111. (b)	112. (b)	113. (a)	114. (c)	115. (a)	116. (d)	117. (d)	118. (c)	119. (a)	120. (a)
121. (a)	122. (a)	123. (c)	124. (b)	125. (a)	126. (c)	127. (d)	128. (d)	129. (c)	130. (b)
131. (b)	132. (b)	133. (c)	134. (d)	135. (a)	136. (c)	137. (a)	138. (c)	139. (a)	140. (d)
141. (c)	142. (a)	143. (a)	144. (d)	145. (c)	146. (c)	147. (d)	148. (a)	149. (a)	150. (b)
151. (b)	152. (a)	153. (d)	154. (b)	155. (b)	156. (c)	157. (a)	158. (d)	159. (a)	160. (b)
161. (c)	162. (b)	163. (d)	164. (b)	165. (a)	166. (b)	167. (d)	168. (d)	169. (d)	170. (b)
171. (c)	172. (c)	173. (b)	174. (b)	175. (b)	176. (c)	177. (c)	178. (d)	179. (d)	180. (d)
181. (a)	182. (a)	183. (a)	184. (d)	185. (a)	186. (d)	187. (a)	188. (d)	189. (d)	190. (b)
191. (d)	192. (a)	193. (c)	194. (d)	195. (d)	196. (d)	197. (c)	198. (b)	199. (a)	200. (b)
201. (a)	202. (c)	203. (b)	204. (d)	205. (c)	206. (b)	207. (a)	208. (d)	209. (a)	210. (b)
211. (d)	212. (b)	213. (a)	214. (d)	215. (b)	216. (c)	217. (d)	218. (c)	219. (b)	220. (d)
221. (b)	222. (d)	223. (d)							

(B) Cumulative Exercise

1. (b)	2. (c)	3. (b)	4. (b)	5. (c)	6. (d)	7. (b)	8. (c)	9. (b)	10. (c)
11. (d)	12. (d)	13. (a)	14. (b)	15. (c)	16. (b)	17. (c)	18. (a)	19. (a)	20. (a)
21. (d)	22. (d)	23. (b)	24. (b)	25. (a)	26. (b)	27. (d)	28. (b)	29. (c)	30. (d)
31. (c)	32. (b)	33. (b)	34. (d)	35. (c)	36. (c)	37. (b)	38. (c)	39. (b)	40. (d)
41. (d)	42. (d)	43. (d)	44. (c)	45. (b)	46. (c)	47. (d)	48. (b)	49. (d)	50. (b)
51. (b)	52. (d)	53. (d)	54. (c)	55. (a)	56. (b)	57. (a)	58. (c)	59. (b)	60. (d)
61. (c)	62. (a)	63. (b)	64. (a)	65. (b)	66. (e)	67. (a)	68. (a)	69. (e)	70. (c)
71. (d)	72. (e)	73. (c)	74. (b)	75. (a)	76. (b)	77. (b)			

(C) Medical Entrances' Gallery

1. (a)	2. (a)	3. (a)	4. (b)	5. (c)	6. (c)	7. (c)	8. (b)	9. (b)	10. (c)
11. (d)	12. (a)	13. (b)	14. (b)	15. (d)	16. (d)	17. (c)	18. (c)	19. (c)	20. (e)
21. (c)	22. (c)	23. (d)	24. (b)	25. (a)	26. (d)	27. (a)	28. (b)	29. (d)	30. (d)
31. (b)	32. (a)	33. (c)	34. (a)	35. (b)	36. (a)	37. (c)	38. (c)	39. (c)	40. (a)
41. (a)	42. (d)	43. (d)	44. (d)	45. (c)	46. (d)	47. (a)	48. (c)	49. (b)	

Explanations

(A) Topical Exercise

6 (*d*) Platyhelminthes and nematodes possess tissue organ and organ system, respectively. Thus, option d (i.e. I and III) is correct.

8 (*b*) Aschelminthes is odd one out as it possess complete digestive system whereas other three possess incomplete digestive system.

12 (*a*) Radial symmetry is found in cnidarian, echinoderms and porifers whereas mammals belonging to Chordata possess bilateral symmetry.

17 (*b*) Option (b) represents incorrectly matched pair. It can be corrected as Coelentrates possess tissue level of body organisation.

18 (*b*) Statement I is false, but II is true. Incorrect statement can be corrected as

Cnidarians and flatworms have blind sac body plan whereas roundworms have tube within tube plan.

21 (*a*) Statement in option (a) is incorrect and can be corrected as Platyhelminthes are acoelomates, i.e. without coelom. Haemocoel is found in arthropods.

24 (*c*) Statement in option (c) is incorrect. It can be corrected as *Fasciola* or liver fluke belongs to phylum-Platyhelminthes which are acoelomate animals.

37 (*d*) Both statements I and II are false and can be corrected as Cold blooded animals falls in the category of ectotherms and warm blooded animals falls in the category of endotherms.

41 (*d*) Option (d) is incorrect match and can be corrected as Annelids are coelomates.

Rest options are correct.

43 (*b*) Statement I is false, but II is true. Incorrect statement can be corrected as

Infra kingdom-Parazoa contain animals that do not possess digestive cavity, e.g. Porifera.

53 (*b*) Statement in option (b) is false and can be corrected as Choancytes are not reproductive cells. These are flagellated cells that maintain the flow of water in poriferans.

64 (*d*) Option (d) represents incorrectly matched pair because *Plakina* possess leuconoid canal system and more specifically, eurypylous type.

67 (*b*) Statement I is false, but II is true. Incorrect statement can be corrected as

Digestion in sponges takes place in food vacuoles.

68 (*a*) Statement in option (a) is not true because Asexual reproduction in sponges occur through budding, fragmentation and gemmule formation.

71 (*c*) Statement in option (c) is incorrect about class-Calcarea. It can be corrected as

The members of this class are dull coloured and exclusively marine animals.

76 (*d*) Both statements I and II are false. These can be corrected as *Leucosolenia* possess asconoid type canal system, therefore the prosopyle is altogether absent in it. Thus, the inhalent aperture is ostia only. Like other poriferons, *Leucosolenia* is ammonotelic and excretory product is ammonia.

79 (*a*) Statement given is option (a) is incorrect and can be corrected as

The spicules in members of calss-Hexactinellida are siliceous and triaxon.

80 (*c*) Option (c) represents mismatched pair. It can be corrected as The shape of *Euplectella* is basket-like, curved and tubular.

82 (*c*) Option (c) represent incorrectly matched pair because Spicules in glass rope sponges or *Hyalonema* are siliceous and these are six-rayed or triaxon.

92 (*c*) Statement in option (c) is correct while other three statements are incorrect about nematocyst. These can be corrected as

Once ejected, it cannot be reused and its ejection is a type of automatic response, not a conditioned reflex. It is used to paralyse the prey or predator.

93 (*b*) Statement I is false, but II is true. Incorrect statement can be corrected as

Cnidarians are mostly carnivorous. The mode of digestion is partially extracellular and partially intracellular.

117 (*d*) Statement in option (d) is incorrect. It can be corrected as *Hydra* has great power of regeneration, but it does not use it as means of reproduction.

146 (*c*) Statement in option (c) is incorrect. It can be corrected as Cnidoblasts are not found in Ctenophora. Instead, tentacles contain specialised sensory cells called colloblasts.

149 (*a*) Statement I is true, but II is false. Incorrect statement can be corrected as

Lasso cells secrete adhesive material that help in capturing prey.

153 (*d*) Option (d) is not correctly matched and it can be corrected as Tentaculata is a class of ctenophorans, whereas *Hydra* is a cnidarian.

154 (*b*) Statement in option (b) is incorrect. It can be corrected as *Pleurobranchia* belong to class-Tentaculata which possess two long aboral tentacles.

155 (*b*) Statement in option (b) is incorrect about class-Nuda and can be corrected as

The member of class-Nuda do not contain tentacles, but have highly branched gastrovascular cavity.

159 (*a*) Option (a) represents incorrectly matched pair. It can be corrected as Hooks helps in the adhesion of animal of with host body.

163 (*d*) Statement in option (d) is incorrect. It can be corrected as Members of class-Tubellaria reproduce asexually by binary fission.

164 (*b*) Statement in option (b) is incorrect. It can be corrected as In *Dugesia*, locomotion occurs by gliding movements (through cilia) and crawling (muscular contractions).

165 (*a*) Option (a) is incorrectly matched pair because
Liver fluke belongs to class-Trematoda of phylum-Platyhelminthes.

171 (*c*) Statements II and III are correct whereas I is incorrect.
Incorrect statement can be corrected as
Schistosoma are not hermaphrodite but dioecious. Males are shorter than females. Males have longitudinal cleft or gynaecophoric canal, in which the longer slender female lies folded.

177 (*c*) Statement in option (c) is incorrect. It can be corrected as
Each repeated unit of body in tapeworm is called proglottid and it shows pseudometamerism.

192 (*a*) Statement in option (a) is incorrect. It can be corrected as
Hymenolepis nana is the smallest human tapeworm and therefore, it is commonly called dwarf tapeworm.

196 (*d*) Statement in option (d) is incorrect. It can be corrected as
The members of phylum-Aschelminthes possess complete digestive system. The intestine is non-muscular in these animals.

204 (*d*) Statement in option (d) is incorrect and can be corrected as
The fertilisation in *Ascaris* occurs within the small intestine of host, i.e. human.

205 (*c*) Only statement in option (c) is correct, while other three are incorrect. These can be corrected as
- *Ascaris* is monogenetic, i.e. have a single host *viz.*, human.
- Humans are infected by ingesting contaminated food and water.
- Within host body, *Ascaris* larva develop and invade different organs like intestine, liver and lungs.

206 (*b*) Statement I is true, but II is false. Incorrect statement can be corrected as
Average lifespan of *Ascaris* in host human body is nine months to one year.

210 (*b*) Statement in option (b) is incorrect. It can be corrected as
The mouth of *Ascaris* is guarded by three denticulate lips or labia.

219 (*b*) Option (b) is not correct for *Loa loa.* It can be corrected as
It is a digenetic parasite involving human or primary host and mango flies as intermediate host or vector.

(B) Cumulative Exercise

3 (*b*) Statement II is incorrect. It can be corrected as
Medusa has umbrella shaped body with marginal tentacles and mouth is centrally located in a projection called manubrium.

4 (*b*) Statement II is incorrect. It can be corrected as
Ctenophores do not have bilateral or radial symmetry. They rather have biradial symmetry.

5 (*c*) Statements I and III are correct whereas II and IV are incorrect. These can be corrected as
Guineaworm do not cause elephantiasis, it is caused by filarial worm. It is digenetic parasite with secondary host, *Cyclops.*

6 (*d*) Statements I and III are correct, while statements II and IV are incorrect. These can be corrected as
Ancylostoma possess a monogenetic life cycle. It causes mental and physical deficiencies that can be treated by drugs like tetrachloride, thymol, etc.

9 (*b*) Statement in optiopn (b) is incorrect. It can be corrected as
Homeothermic animals can regulate internal body temperature and thus, they do not undergo hibernation and aestivation.

11 (*d*) Statement in option (d) is incorrect. It can be corrected as *Cestum* and *Velamen* belongs to phylum-Ctenophora which possess biradial symmetry.

13 (*a*) Option (a) is incorrect pair becasue
Olynthus is a short lived stage in the development of some calcareous Porifera and it looks like little sponge.

21 (*d*) Statements I, II and IV are correct while statement III is incorrect. It can be corrected as
Platyhelminthes exhibit tissue-organ level of organisation.

25 (*a*) Statements II and IV are correct while I, III and V are incorrect. These can be corrected as
Aschelminthes are triploblastic, alimentary canal is complete and sexes are separate, i.e. dioecious.

26 (*b*) Statements I, II, IV and V are correct, while III is incorrect and can be corrected as Females are longer while males are shorter in Aschelminthes.

29 (*c*) Statements I, III and IV are correct, while II and V are incorrect. These can be corrected as
Poriferans are mostly found in marine habitat and these are asymmatric animals. Fertilisation occur internally in poriferans.

32 (*b*) Statements I, II, III and V are correct, while IV is incorrect. It can be corrected as
The majority of animals are bilaterally symmetrical, ranging from phylum-Nemathelminthes to Chordata.

35 (*c*) Statements II, IV and V are correct, while I and III are incorrect. These can be corrected as
Cnidarians are diploblastic animals. Skeleton of corals is rigid and calcareous.

41 (*d*) Statements I, II, III are true, while IV is false. It can be corrected as Pinacocytes helps in movement, cell adhesion, signaling, etc. They have no role in capturing prey.

48 (*b*) Pinworm is not found in the blood of host. It remain near the anal areas and cause oxyurasis, which result in intense itching in anal parts.

51 (*b*) Option (b) represents incorrectly matched pair. It can be corrected as *Dracunculus medinensis* is a digenetic worm with human being primary and *Cyclops* as secondary host.

58 (*c*) Pairs II and IV are correct while I and II are incorrect. These can be corrected as Platyhelminthes are acoelomate whereas Nematoda possess pseudocoelom.

65 (*b*) Both A and R are true, but R is not the correct explanation of A.
The body of annelids is divided into segments called metameres, externally by ring-like grooves (annuli) and internally by vertical partitions called septa. This phenomenon is called metameric segmentation.

66 (*e*) Both A and R are false and can be corrected as Ectotherms are cold-blooded animals. Endothermy allows animals to maintain a high body temperature even when the environment is very cold, so endotherms can tolerate a broad range of environmental temperatures.

67 (*a*) Both A and R are true and R is the correct explanation of A.

Hydra is a diploblastic animal, i.e. it is derived from two germ layers, the ectoderm and endoderm. These germ layers form two distinct cellular layers, i.e. outer epidermis and inner gastrodermis, respectively.

68 (*a*) Both A and R are true and R is the correct explanation of A.

Archaeocytes are cells of sponge which play an important role in reproduction and regeneration. They may be converted into other types of cells as these are undifferentiated totipotent cells.

69 (*e*) Both A and R are false and can be corrected as *Spongilla* and *Leucosolenia*, both belongs to phylum-Porifera. *Spongilla* belongs to Haplosclerina order of class-Demospongia and *Leucosolenia* belongs to Asconosa order of class-Calcarea. *Spongilla* is a freshwater sponge and asymmetrical, whereas *Leucosolenia* is a white or dull grey colour, cylindrical individual having radial symmetry.

70 (*c*) A is true, but R is false because.

Sponges are less specialised, because they have no tissue system, organ system, etc. All animals particularly the less specialised ones can replace their lost or injured part. The process is known as regeneration. Sponges have a high power of regeneration.

71 (*d*) A is false, but R is true because.

Robert Grant was the first to recognise and prove the true animal nature of sponges.

The animal nature of sponge was well established on the following facts

- Their mode of nutrition is truely holozoic.
- They are devoid of cell walls (characteristic feature of plants).
- Life cycle includes free-swimming, ciliated larval stages resembling those of other animals.
- Sponges are sessile and digestion is very simple without any apparent way of capturing food or eliminating wastes.

72 (*e*) Both A and R are false because

In *Hydra*, both pseudopodia and flagella occur in nutritive cell to capture and engulf food material.

73 (*c*) A is true, but R is false because

Hydra is diploblastic animal. Its body wall is composed of two layers, i.e.epidermis and gastrodermis separated by a non-cellular layer (mesoglea).

74 (*b*) Both A and R are true, but R is not the correct explanation of A.

Medusoid form of *Obelia* is carnivorous. It feeds on minute worms, nematodes, insects, etc. It shows both extracellular as well as intracellular digestion.

75 (*a*) Both A and R are true and R is the correct explanation of A.

Alternation of generation may be defined as a phenomenon where asexual phase and sexual phase regularly alternates with each other. This type of alternation of generation is also called metagenesis. In coelenterates, an asexual polyploid generation appears to alternate regularly with a sexual medusoid generation.

76 (*b*) Both A and R are true, but R is not the correct explanation of A.

The cysticercus larva of *Echinococcus* develops into a watery bladder or hydatid vesicle or hydatid cyst having a two-layered cyst wall. This vesicle exerts suppressive effect on T-lymphocytes and thus animal can survive for several years within the human body.

77 (*b*) Both A and R are true, but R is not the correct explanation of A.

Uterus is the characteristic feature of reproductive system of tapeworms. It is meant to accommodate thousands of fertilised eggs and in the process of accomplishing this, it forms 7 to 13 lateral branches on each side. The branched uterus is the only genital structure persisting in a gravid proglottid.

(C) Medical Entrances' Gallery

3 (*a*) Animals of phylum-Platyhelminthes have hooks and suckers and are endoparasites on other animals. These are triploblastic, bilaterally symmetrical, acoelomate animals with blind sac body plan. They live as endoparasites in intestine of other animals. For this purpose, they bear hooks and suckers near their mouth region which help them to cling on the intestinal walls of the host to obtain nutrition.

4 (*b*) The medusa of *Obelia* shows radial symmetry. It is a free-swimming, umbrella-like sexual generation. It is formed by blastostyle. It exhibit sexual reproduction and produce new hydroid colonies.

5 (*c*) A is true, but R is false because

Taenia solium belongs to the class-Cestoda and *Dugesia* (Planaria) belongs to class-Turbellaria of phylum-Platyhelminthes. These are acoelomates, i.e. they do not possess true body cavity.

6 (*c*) In case of poriferans, the spongocoel is lined with flagellated cells called choanocytes. The body wall of a common sponge consists of three layers, i.e. pinacoderm, choanoderm and mesenchyme. Choanoderm is inner cellular layer which consists of highly specialised flagellated cells called choanocytes. The beating of flagella creates water current.

7 (*c*) Gemmule is a type of spore, formed during unfavourable conditions. It is helpful in asexual mode of reproduction and produces new individuals in sponges. It is formed within the parent body by internal budding.

8 (*b*) In Porifera (sponges), bodies are asymmetrical. Body lacks tissue or organs, but forms a meshwork of cells surrounding channels that open to the outside through pores. These pores extend into internal cavities lined with food filtering flagellated cells (choanocytes).

10 (*c*) The third stage larva of *Wuchereria bancrofti* carried by *Culex* mosquito is microfilariae. These are discharged in lymph vessels of man and soon reaches the blood stream.

11 (*d*) Statement in option (d) is incorrect and can be corrected as follows

Porifera do not have nematocyst. These are the stinging capsules produced by the cnidocytes or cnidoblasts present on the tentacles and the body of the animals beloning to the Phylum—Coelenterata (Cnidaria).

14 (*b*) Option (b) is a mismatch pair and can be corrected as Comb plates are the characteristic feature of phylum—Ctenophora. These animals have cilia in each row, arranged to form a stack of combs called comb plates.

15 (*d*) Metagenesis is the phenomenon, in which one generation of certain plants and animals reproduces asexually followed by sexually reproducing generations. It is found in the life history of *Obelia.*

16 (*d*) Cnidarian members are found both in freshwater and marine environments, while members of Ctenophora, Cephalochordata and Echinodermata are found exclusively in marine environment.

18 (*c*) The coelom given in the diagram is found in roundworm (*Ascaris*), which belongs to the phylum—Aschelminthes. The body cavity in the diagram is not lined by mesoderm, instead the mesoderm is present in the form of scattered pouches in between the ectoderm and endoderm.

Such a body cavity is called pseudocoelom and the animals possessing these cavity are called pseudocoelomates.

20 (*e*) The statements, i.e. II, III and V are incorrect, which can be corrected as

Coelenterates are diploblastic with tissue grade organisation and medusae forms are free-swimming. Polyps produce medusae through vegetative budding and medusae form polyp after sexual reproduction.

21 (*c*) *Euplectella* is referred to as venus flower basket. Its skeleton is a very costly marriage gift in Japan.

22 (*c*) In the life cycle of *Ascaris lumbricoides,* rhabditiform larva is the infective stage of man. It reaches the alimentary canal through contaminated food and water. In the small intestine, the shell gets dissolved to release 2nd stage larva.

After that, larva reaches the liver through the hepatic portal vein and then to heart through post vena cava vein. It goes to lungs through the pulmonary arteries. In the alveoli of lungs, it undergoes the 2nd moulting and produces the 3rd moulting in the alveoli only and then enters the small intestine again. In the small intestine, it undergoes 4th and 5th moulting to become a young one, attaining sexual maturity within 8-10 weeks.

28 (*b*) *Ctenoplana* and *Beroe* lack cnidoblasts and have biradial symmetry. These belong to phylum-Ctenophora.

39 (*c*) Scleroblasts are skeleton manufacturing cells and these may be calciblasts, silicioblasts or spongioblasts, depending upon the skeletal elements produced by them.

44 (*d*) Option (d) is not a characteristic feature of sponges. The sponges possess an endoskeleton in the form of calcareous spicules, siliceous spicules and spongin fibres, But sponges do not possess chitin.

46 (*d*) *Fasciola hepatica* (sheep liver fluke) belongs to phylum-Platyhelminthes.

These worms have incomplete alimentary canal and there is a single opening for both digestion and egestion.

This is also called as blind sac body plan.

47 (*a*) Planaria (*Dugesia*) has a remarkable power of regeneration. If an individual is cut transversely into two parts, the anterior fragment will regenerate a new tail and a posterior piece will develop a new head.

48 (*c*) Aschelminthes are triploblastic, bilaterally symmetrical, pseudocoelomate (false coelom derived from embryonic blastocoel and is not lined by mesoderm), unsegmented organisms.

CHAPTER 09

Animalia-II (Higher Invertebrates)

Higher invertebrates phyla consist of animals from Annelida, Arthropoda, Mollusca and Echinodermata phylums. These animals are generally larger in size and complex in body organisation. In this chapter, we will focus on the characteristic features and classification of these animal groups.

Inside

1 Phylum-Annelida
2 Phylum-Arthropoda
3 Phylum-Mollusca
4 Phylum-Echinodermata

PHYLUM-ANNELIDA

Lamarck (1809) coined the term 'Annelida' (L. *Annellus*-little ring) for these ringed animals. He also established Annelida as the first phylum having **eucoelomate, schizocoelous** condition. This phylum contains over 9000 species.

The annelid worms were thought to have evolved from a coelomate **worm-like** ancestor which developed **metameric segmentation**. These segments are called **somites** or **metameres**. Mostly, the external and internal parts of annelids are repeated with each segment and this condition is also called as **serial homology**.

General characteristics

The characteristics of this phylum are as follows

1. Annelids are aquatic (marine and freshwater) or terrestrial animals. For example, *Nereis* is marine and earthworm is terrestrial animal. Most of the annelids are free-living and some of them are parasitic also.
2. These are bilaterally symmetrical, triploblastic animals.
3. Body of annelids is elongated and metamerically segmented.
   - Their body is covered by thin moist **albuminoid cuticle** over columnar epithelium. This epithelium also contains unicellular gland cells and sensory cells.
   - The body appears to have a **tube within a tube plan.**
4. They possess true coelom or schizocoelom. In leeches, coelom is reduced and restricted to excretory and reproductive organs only. Instead a new system of channels called **haemocoelomic system** is present in leeches. Coelom contains coelomic fluid that acts as hydrostatic skeleton in initial segment of the body.

5. The first segment of their body is called **peristomium**. It contains mouth covered with an outgrowth from the top. This outgrowth is called **prostomium**.
6. Digestive tract is complete and contains distinct mouth and anus. Digestion is entirely extracellular. The gut has both circular and longitudinal muscles.
7. Respiration is **aerobic** in free-living forms. Gaseous exchange mostly occurs through moist body surface and by gills in some tube dwellers. Gaseous exchange can also occur through **parapodia**.
8. A proper **blood vascular system** develop for the first time in the annelids. It is of closed type. It has longitudinal blood vessels (coelomic sinuses) with lateral branches in each segment.
   Blood plasma contains dissolved **haemoglobin** or **erythrocruorin** as the respiratory pigment. RBCs are altogether absent.
9. Different types of corpuscles floating in coelomic plasma of annelids and their functions are given below

Names	Shapes	Functions
Granulocytes or Phagocytes (most numerous)	Largest and subspherical	Serve to store nutrients
Leucocytes	Smaller, flattened and circular	Uncertain
Amoebocytes	Small star-shaped	Phagocytes so destroy harmful microorganisms and debris.
Mucocytes	Elongated	Uncertain
Chloragogen	–	Centre of synthesis of glycogen and fat deamination and urea formation.

10. Their excretory system consists of metamerically distributed coiled tubes called **nephridia**. These are ectodermal in origin. Nephridia remove nitrogenous waste (mainly urea and ammonia) from the coelom and blood stream of the body. These organs also help in **osmoregulation**.
11. Locomotory organs are either minute rod-like **chitinous setae** or **suckers**. The setae are found in all the animals except leeches. In some annelids like *Nereis*, setae are aggregated over unjoined fleshy appendages called the **parapodia**.
12. Nervous system of annelids consists of a pair of **cerebral ganglia** or **brain** with many commissures and connectives.
    Commissures connect similar nervous parts, while connectives connect dissimilar nervous parts.
13. Sensory cells and organs for touch, taste and light perception are well-developed.
14. Sexes are united and development is direct in amphibious forms like earthworm and leeches.
    - Sexes are separate and development is indirect with **trochophore larval stage** in aquatic burrowing and free-swimming forms like *Nereis*.
    - Annelids are oviparous and the eggs are laid in ootheca/cocoons/egg cases.
15. Fertilisation is external, either in water (in aquatic forms) or in cocoon (in terrestrial forms and leeches).
    - The cocoon is formed with the help of specialised highly glandular, thickened regions of some body segments. These specialised regions are collectively called as **clitellum**. It is completely absent in aquatic forms, temporary in leeches and is permanent in terrestrial forms like earthworms.

Classification

In this phylum, classification is primarily based on
- Position and arrangements of setae.
- Presence or absence of setae.
- Presence or absence of sense organ.
- On the basis of above characteristics, this phylum is divided into following four classes

Class–Archiannelida (Gr. *Arch*–first)

- It is the group of **primitive polychaete worms**.
- The characteristic features of this class are as follows
  1. It includes primitive and small-sized annelids.
  2. These are exclusively marine.
  3. Body is elongated and worm-like.
  4. External segmentation is less observable, but internal segmentation is perfect and with septa.
  5. Setae and parapodia are absent. Clitellum is also absent.
  6. Unisexual or hermaphrodite and gonads are numerous.
  7. **Trochophore larva** is present, e.g. *Polygordius*, *Dinophilus*, etc.

Note • *According to new classification, Archiannelida is considered as an order of class–Polychaeta.*

• *The name Archiannelida was given by* ***Berthold Hatschek Ray Lankester*** *named this class as Haplodrili.*

Class–Polychaeta

(Gr. *Polys*-many; *chaeta*-setae)

- These are also called **Bristle worms**. This class contains about 8000 species.
- Polychaetes are considered as the most common marine organisms.
- The characteristic features of this class are as follows
  1. It is the **biggest class** of phylum—Annelida. It includes marine and some freshwater forms.
  2. Body is segmented externally and internally, but segments may be similar (e.g. *Nereis*) or dissimilar (e.g. *Chaetopterus*).
  3. Setae are numerous on lateral parapodia.
  4. A well-defined head region with tentacles, palps and eyes are present.
  5. Clitellum is absent in all forms.
  6. Pharynx is protrusible or irreversible.
  7. Sexes are usually separate, but sexual dimorphism may not be found.
  8. Gonads develop temporarily during breeding season.
  9. Development occurs through a **trochophore larva**.
  10. Asexual budding is also seen in some species, e.g. *Nereis*, *Eunice viridis* (palolo worm), *Aphrodita* (sea mouse), *Syllis*, *Eurythoe* (fireworm), *Chaetopterus* (exhibit great power of regeneration), etc.

Nereis or *Neanthes*

- It is commonly called **clamworm** or **sandworm**. It is a marine, nocturnal and carnivorous animal. The animal lives in burrows on tidal areas of sea shores.
- The body of animal is divisible into head, trunk and pygidium (terminal part).
- The trunk region have about 80-200 segments. Each segment bears a pair of parapodia, except first and last segments.
- The **pygidium** is the last or anal segment. It bears anus along with one pair of anal cirri (sensory) and several minute sensory papillae.
- Coelom communicates with exterior through nephridia and coelom ducts.
- The sexual form of *Nereis* is called ***Heteronereis***. The name itself suggest the presence of two different regions in the body, i.e. anterior **atoke** or non-sexual region and posterior **epitoke** or sexual region. The phenomenon of formation of *Heteronereis* is called **epitoky**.
- *Heteronereis* leaves burrow during breeding season and becomes free swimmer. This behaviour of *Nereis* is called **swarming**, which generally occurs at night.
- Female produces a substance, called **fertilium**. This attracts the male and stimulates the shedding of sperm. Fertilisation occurs in sea water.
- The species of animal includes *N. virens, N. pelagica, N. cultrifera, N. diversicolor*, etc.

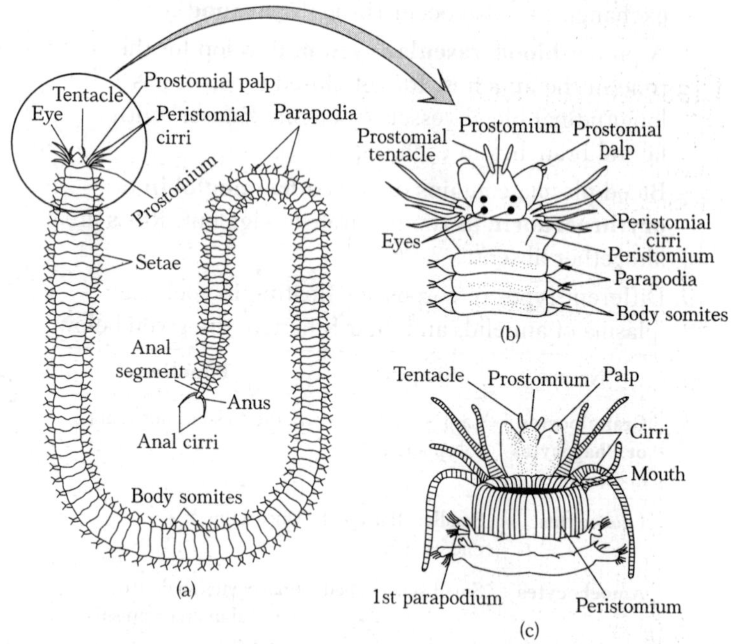

Fig. 9.1 Nereis : (a) External morphology, (b) Head in dorsal view and (c) Head in ventral view

Class–Oligochaeta

(Gr. *Oligos*-few; *chaeta*-setae)

The characteristic features of this class are as follows

1. Representatives of this class are commonly found in freshwater and some in moist soil.
2. Both external and internal segmentations are present.
3. Clitellum is present and permanent. It helps in secreting cocoon for egg development.
4. Setae are few which are embedded within the skin.
5. Parapodia are absent.
6. Hermaphrodite (monoecious) animals, fertilisation is external.
7. No larval stage is found. Development is direct. e.g. Earthworm.

8. Common names of some oligochaetes and their clitellar regions are given below

Scientific names	Common names	Clitellar regions
Pheretima posthuma	Common Indian earthworm	14th to 16th
Megascolex	South Indian earthworm	14th to 17th
Lumbricus terrestris	North American earthworm	33rd to 37th
Tubifex tubifex	Sludge worm or Blood worm	11th and 12th

9. Hermaphrodite, clitellum develops temporarily in 9th, 10th and 11th segments only during the breeding season.
10. Fertilisation is internal. Eggs are usually present in cocoons. No larval stage is seen, e.g. *Hirudo, Hirudinaria, Pontobdella* (ectoparasite on elasmobranch), etc.

Class– Hirudinea (L. *Hirudo*–leech)

This class includes approximately 500 species of leeches. Biologists consider this class as the one having most specialised annelids.

The characteristic features of this class are as follows

1. Animals are generally ectoparasitic, sanguivorous, i.e. blood sucking or carnivorous.
2. Body is pigmented and provided with an anterior and posterior sucker.
3. Tentacles, parapodia or setae are altogether absent.
4. The number of body segments is fixed, i.e. 33.
5. Each segment is subdivided externally into annuli.
6. Coelom is restricted only to the excretory and reproductive organs of the body due to its filling by **botryoidal tissue**. It is a type of connective tissue.
7. A system of channels called **haemocoelomic system,** which is present for the purpose of circulation.
8. Locomotory organs are suckers. It occurs by looping and swimming.

Hirudinaria granulosa

- It is commonly called common **Indian cattle leech.**
- The mouth of leeches is triradiate, which is guarded by a structure called **velum.** Pharynx is suctorial.
- The salivary secretion of animal contains an anticoagulant called **hirudin** or **anticoagulin.** It prevents the clotting of blood during ingestion.
- The blood gets haemolysed during its storage in crop, i.e. the blood corpuscles show breakdown.
- The animal contains anterior and posterior suckers for adhesion and locomotion, respectively.
- Each segment in the body of animal from 6th-22nd bears a pair of nephridiopore.
- It is a **hermaphrodite,** but cross-fertilisation is a rule.
- Male organs include eleven pairs of testes sacs in the segments 12th-20th and accessory organs in 10th segment. Female organs includes a pair of ovisacs, oviducts and vagina, all located in 11th segment.
- Male and female genital pores are present between 2nd and 3rd annuli of 10th and 11th segments, respectively.
- Fertilisation is internal and occurs in vagina. It is followed by cocoon formation by the temporary clitellum in 9th-11th segment. Each cocoon contains 1-24 developing embryos. Development is direct.

CHECK POINT 9.1

1. The phylum name 'Annelida' signifies that the organisms of this phylum
(a) possess parapodia (b) possess metameric segments
(c) lack metameric segments (d) lack nephridia

2. ………… in the earthworm perform the function similar to flame cells in *Taenia*.
(a) Parapodia (b) Nephridia
(c) Metameres (d) Ganglia

3. *Heteronereis* stage is
(a) transformation of sexual individual into asexual
(b) sexually matured with two regions, i.e. atoke and epitoke
(c) sexually immatured with two regions, i.e. atoke and epitoke
(d) All of the above

4. Trochophore larva is associated with the classes …… of Annelida.
(a) Polychaeta and Oligochaeta
(b) Polychaeta and Hirudinea
(c) Polychaeta and Archiannelida
(d) Hirudinea and Archiannelida

5. Which of the following lacks parapodia or setae ?
(a) *Pheretima* (b) *Hirudinaria*
(c) *Nereis* (d) *Lumbricus*

6. In which of the following classes of Annelida, one pair of ovaries and several pairs of testes are found?
(a) Archiannelida (b) Hirudinea
(c) Oligochaeta (d) Polychaeta

PHYLUM-ARTHROPODA

- These are invertebrate animals equipped with **jointed appendages** (Gk. *Arthros*-joint; *podos*-foot or appendage). In 1845, **von Siebold** with **H Stannius** proposed the term 'Arthropoda' and established them as a separate phylum.
- It is the **largest phylum** of kingdom—Animalia with about 900000 species. Over two-third of identified species on earth are arthropods.

General characteristics

These are as follows

1. These animals inhabit a variety of habitats as these are cosmopolitan in distribution, e.g. oceans (e.g. *Limulus*), freshwater (e.g. *Daphnia*), land (e.g. cockroach) and air (e.g. housefly).
2. They are **triploblastic, schizocoelous coelomate** and **bilaterally symmetrical** animals. But, true coelom is reduced and restricted to excretory and reproductive organs only. It is largely replaced by a blood-filled cavity called **haemocoel.**
3. These animals are metamerically segmented with regionalisation of segments called **tagmatisation.** The body regions in tagmatisation are called **tagmata** (sing. tagma).
4. Their body is covered with a hard exoskeleton called **cuticle.** It is made up of chitin (a polymer of N-acetylglucosamine). The chitinous plates present over the body are called **sclerites.**
5. Sclerites shed at regular intervals during growth and development by the process of **ecdysis** or **moulting.**
6. Body is divisible into **head, thorax** and **abdomen.** These structures are variously distinct or fused. Head and thorax are often fused to form **cephalothorax.**
7. The lateral outgrowths of body segments are called **appendages.** In the case of arthropods, these appendages are jointed, i.e. formed by the joining of many small pieces or sclerites. These sclerites are formed with the help of arthrodial membranes.
8. Under ideal conditions, one pair of appendages are present per segment. These are modified for various functions like feeding, sensory reception, locomotion and reproduction, etc.
9. Muscles are striated and usually capable of rapid contraction.
10. Digestive tract is **complete** and has glands. In most of the cases, it is equipped with specialised structures like **crop** or **gizzard** or **protrusible pharynx**, etc.
    - Specialised mouth parts are present to help in the process of ingestion. The digestion of food is extracellular.
11. Circulatory system is **open type** with sinuses or blood spaces. In most of the cases, a segmented dorsal blood vessel is present with contractile activity and pumping action. It is comparable with heart of other animals.
    - The circulating blood is called **haemolymph.** It may or may not contain respiratory pigment.
    - The usual respiratory pigment found in arthropods is copper containing **erythrocruorin.**
12. Gaseous exchange occurs through general body surface or specialised respiratory structures. The latter include **gills** (in prawn), **trachea** (in insects), **book lungs** (in scorpion) and **book gills** (in horse-shoe crab).
13. Excretion occurs by **coxal** or **green glands** (e.g. arachnids) or **Malpighian tubules** (e.g. insects). Nephridia is altogether absent. In crustaceans, the **antennary glands** also aid in excretion. These are considered as the modified green glands.
14. Nervous system is well-defined with central, peripheral and autonomic nervous system. The CNS consists of dorsal nerve ring and a double ventral nerve cord.
15. The sensory organs are well-developed in the free-living forms. These include antennae and sensory hairs as tactile and chemoreceptive organs, statocyst as balancing organs along with simple, compound eyes and sound receptors in insects.
16. An endocrine system in arthropods includes **Corpora cardiaca** secretes prothoraciotropic hormone, **ecdyson** or moulting hormone.
    - **Corpora allata** secretes juvenile hormone.
17. Fertilisation is generally **internal.** Animals are mostly **dioecious**, i.e. with separate male and female individuals. Sexual dimorphism is also well-established.
    - These animals are either oviparous (lay eggs) or **ovoviviparous,** i.e. egg hatch within the female body (e.g. scorpions).
18. Development is direct or indirect. It usually possesses one to several larval stages. Sometimes, parthenogenesis, i.e. development of adult from an unfertilised egg is also seen, e.g. male honeybees (drones).

19. Arthropods usually show a great diversity in their larval forms. Types of larva in different arthropods is given below

Organisms	**Larvae**
Ants, bees, beetles	Grub
Housefly	Maggot (its pupa is coarctate)
Hermit crab	Glaucothoe
Lepas	Cypris
Butterfly	Caterpillar
Mosquito	Wriggler (pupa is tumbler)
Peneaus (marine prawn)	Nauplius, Zoea, Mysis
Sacculina	Cypris
Squilla	Alima
Crab	Zoea, Megalopa
Cockroach	Nymph
Dragonfly and Hayfly	Naiad
Palinurus	Phyllosoma (glass crab)
Silkworm moth	Caterpillar (its pupa is chrysalis)

Classification

The phylum-Arthropoda is divided into following four subphyla.

Subphylum–Trilobitomorpha (Extinct)

1. These were mostly marine and bottom dwelling animals.
2. Over 4000 were species, e.g. *Triarthrus, Dalmanites.*
3. These had evolved during Cambrian to Permian period.
4. Body was divided by two lengthwise furrows into three lobes as shown below

Fig. 9.2 *Trilobitomorpha fossil showing its three lobes*

5. Head was distinct. Abdomen had 2nd-29th segments and a fused caudal plate was present.
6. Development was indirect with larval stages.

Subphylum–Chelicerata

1. Body is divisible into two parts, i.e. anterior cephalothorax (prosoma) and a posterior abdomen (opisthosoma).
2. Cephalothorax is unsegmented and covered with a hard shell covering called **carapace.**
3. **Antennae** and **true jaws** are absent.
4. Six pairs of prosomatic appendages are present.
5. First pair of appendages is feeding **chelicerae** with **claws** followed by second pair of **pedipalps** and **4 pairs of walking legs.**
6. Sexes are separate, female is oviparous, development is direct or through a larval stage.

Class–Merostomata

1. It includes aquatic animals like extinct giant aquatic *Eurypterus* and *Limulus* (horse-shoe crab).
2. Cephalothorax possesses compound lateral eyes. It is broadly joined to abdomen.
3. 5-6 pairs of appendages with **book gills** for gaseous exchange are present.
4. Excretion occurs by **coxal gland**. Malpighian tubules are absent.

Limulus

- It is a large-sized marine animal.
- It is living member of very ancient (200 million years), primitive chelicerates and hence called a **'living fossil'**.
- Its prosoma (cephalothorax) is covered by a large horseshoe-shaped carapace, thus it is also called **horse-shoe crab**.
- It bears a median simple eyes (ocelli) and two lateral compound eyes.

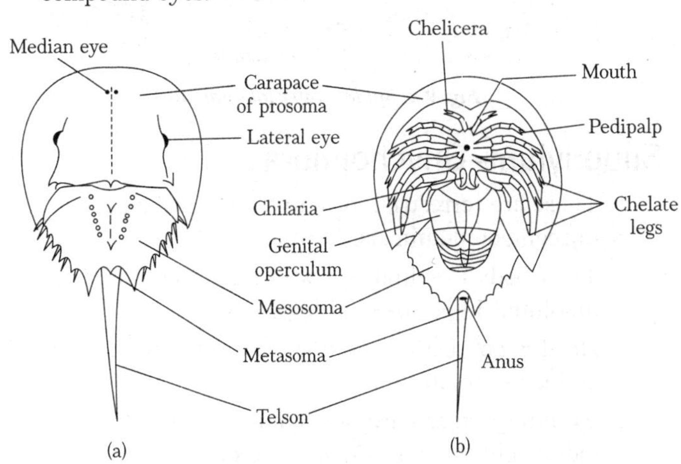

Fig. 9.3 *Limulus : (a) Dorsal view, (b) Ventral view*

Class–Arachnida

1. These are terrestrial or aquatic animals with simple eyes.
2. Cephalothorax (prosoma) with **two chelicerae, two pedipalps** and **4 pairs of walking legs.**
3. Exoskeleton often possesses sensory hairs or scales.
4. Respiratory organs are **book lungs** or **tracheae** or both.
5. Excretion occurs by coxal gland or Malpighian tubules or both.
6. Sexes are separate, i.e. dioecious. Mostly oviparous. Development is direct.
7. Many species contain poison glands and poison fangs, e.g. *Palamnaeus* (scorpion), spiders, etc.

Spider

- It is a common arachnid which is found weaving web at various places.
- Body is divided into **prosoma** and **opisthosoma.**
- The opisthosoma bears **spinnerets** or spinning organs and underlying **'silk glands'**. These glands produce silken-threads for the construction of spider web. Chelicerae are with poison glands.
- Excretion occurs by **Malpighian tubules.**
- **Pedipalps** in males function as copulatory organs.
- Different species of spider are *Araneus* (garden spider), *Latrodectus* (black-widow spider), *Achaearanea* (house spider), etc.

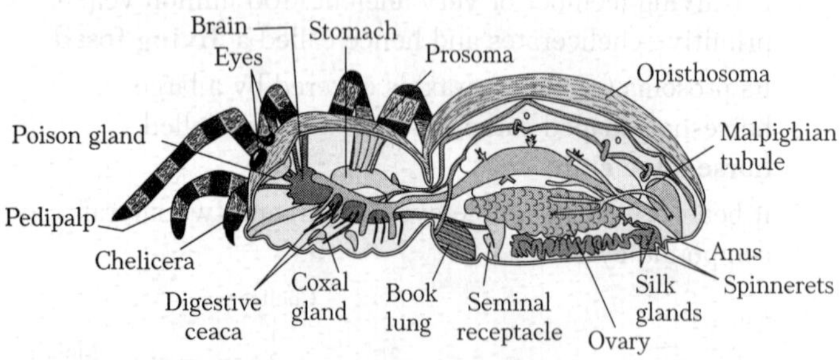

Fig. 9.4 Spider: Internal anatomy

Subphylum–Onychophora

1. These are terrestrial, primitive, worm-like, unsegmented animals.
2. Their body is cylindrical with numerous stumpy, unjointed 14-43 pairs of legs.
3. Head is not distinct. A single pair of antennae, eyes and jaws are present.
4. Excretory organs are segmentally arranged metanephridia, e.g. *Peripatus*, etc.

Peripatus (Walking worm)

This animal is a **connecting link between arthropods** and **annelids**. These bear nephridia and unjointed legs (just like parapodia of Polychaeta) as annelidan characters. The presence of antennae, haemocoel, trachaeal respiration are their arthropodan characters.

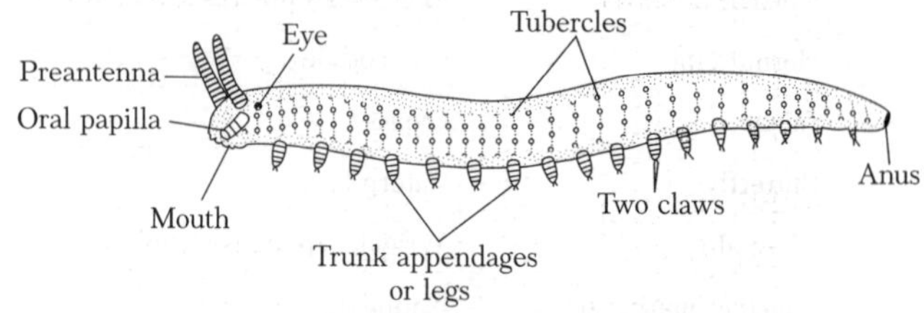

Fig. 9.5 Peripatus : External view

Subphylum–Mandibulata (Antennata)

1. Their body consists of either two parts (i.e. head and trunk) or three parts (i.e. head, thorax and abdomen).
2. 1 or 2 pairs of **antennae**, 1 pair of **jaws** (mandibles) in place of **chelicerae**, 1 or more pairs of **maxillae** and 3 or more pairs of **walking legs** are present.
3. Respiration occurs by **gills** or **tracheae.**
4. Excretion occurs by **Malpighian tubules** or **antenal glands.**
5. Sexes are usually separate.
6. Oviparous or ovoviviparous. Usually with larval stages, i.e. exhibit indirect development.

Class–Crustacea

1. They include mainly aquatic forms, e.g. prawns, *Cyclops.*
2. The body is divisible into cephalothorax (i.e. fused head and thorax) along with abdomen.
3. Body usually has a dorsal shield or carapace. It ends into a telson or posteriormost segment of the body containing the anus.
4. Head is made up of 5 fused segments. It has 2 pairs of **antennae**, 1 pair of **jaws** and 2 pairs of **maxillae.**
5. Exoskeleton is chitinous, hard and calcareous.
6. Respiration occurs by **gills** or body surface.
7. Excretion occurs by **antennal glands.**
8. Sexes are usually separate. Sexual dimorphism is also seen.

9. Development is indirect with larval stages.
10. Tiny crustaceans like *Daphnia* and *Cyclops* act as zooplanktons in aquatic food chain.
11. Some important crustaceans and their common names are given below

Crustaceans	Common names
Daphnia	Water flea
Palaemon	Prawn
Cypris	Shrimp
Palinurus	Lobster
Eupagurus	Hermit crab
Cancer	True crab

Astacus (Cray fish)

- It is dark green in colour and resembles prawn.
- It is found in Europe.
- Cephalothorax is larger than abdomen.
- Antennules are small. Three pairs of legs are chelate, the middle one being massive.

Class–Chilopoda

1. These are terrestrial, nocturnal animals with long and dorsoventrally flattened segmented body.
2. This body is divisible into head and trunk.
3. Each trunk segment bears a pair of legs.
4. First pair of legs are modified into poisonous claws.
5. Respiration occurs by tracheae.
6. Excretion occurs by Malpighian tubules.
7. Unisexual, females oviparous or viviparous, development is direct, e.g. *Scolopendra*, etc.

Scolopendra

- It is commonly called **centipede** or **hundred legger**.
- It is cosmopolitan in occurrence and lives in damp places.
- Head bears two long antennae, two group of ocelli and mouth appendages.
- It is not harmful to man. But, some species inflict a painful bite that cause fever and headache.

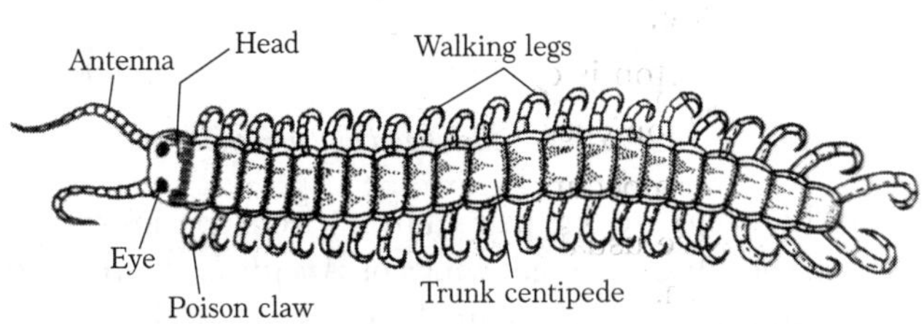

Fig. 9.6 Scolopendra

Class–Diplopoda

1. These are terrestrial animals with long and usually cylindrical body.
2. Their body is divisible into head, thorax and trunk.
3. They possess single pair of antennae and ocelli.
4. Thorax bears four segments out of which first is legless, 2nd-4th segments have one pair of legs, each.
5. Abdomen bears 9-100 segments, each with 2 pairs of legs.
6. Respiration occurs by tracheae.
7. Excretion occurs by Malpighian tubules.
8. Development is direct, e.g. *Julus* (millipedes).

Class–Pauropoda

1. Minute terrestrial animals. Present in dark damp places under logs, stones or leaves and in soil.
2. Soft worm-like body is divisible into head and trunk.
3. Head bears one pair of antennae, but without eyes.
4. Body is cylindrical with 9-10 pairs of legs, e.g. *Pauropus huxleyi.*

Class–Symphylla

1. Body is cylindrical, divided into head and trunk.
2. Head with antennae, jaws and two pairs of maxillae, but without eyes.
3. Trunk with 15-12 somites, bears 12 pairs of legs, e.g. *Scutigerella* (garden centipede).

Class–Insecta (Hexapoda) [Largest Class of Arthropods]

1. Body is divisible into head, thorax and abdomen.
2. There is a single pair of antennae and a pair of compound eyes.
3. Mouth parts in insects are of following types
   - **Bitting and chewing type** Grasshopper, cockroach, termites, caterpillars.
   - **Piercing-sucking type** Mosquitoes, bugs, fleas.
   - **Chewing-lapping type** Bee.
   - **Sponging type** Housefly, fruitfly.
   - **Siphoning type** Butterflies, moths.
4. Thorax consists of three segments typically with
   - 3 pairs of jointed legs (one on each segment)
   - 2 pairs of wings, which may be absent in few species (e.g. *Diptera*).
5. Heart is tubular and divided into chambers.
6. Respiration usually occurs by branched tracheal tubes.

7. Excretion occurs by Malpighian tubules. Excretory product is **uric acid.**
8. Sexes are separate, sex opening is usually single at the end of abdomen.
9. Usually oviparous. Fertilisation is internal and development is indirect that involves metamorphosis.
10. Insects communicate with each other by releasing special hormone like substances out of their body. These substances are called as **pheromones.**
11. Some important insects and their respective orders are given below

Insect groups	Orders
Beetles	Coleoptera
Flies, mosquitoes	Diptera
Butterflies and moths	Lepidoptera
Bees, wasps, ants	Hymenoptera
Bed bugs (*Cimex*)	Hemiptera
Grasshoppers or locust, crickets, cockroach, *Mantis*	Orthoptera
Silverfish (*Lepisma*)	Thysanura

12. **Dipterans** are the only insects which bear a single pair of wings. Actually, their hindwings are modified into balancing organs called **halters.**

Mosquitoes

- These belong to the class–Insecta and order–Diptera, which also includes flies. Mosquitoes differ from flies in following ways
  - In having short but thin body
  - Long and many segmented antennae
  - Long slender legs
  - In having piercing and sucking type of mouth parts.
- The common genera of mosquitoes are *Culex* and *Anopheles.*
- The general characteristic features of mosquitoes are as follows

(i) These are nocturnal creatures. Females are sanguivore, but the males suck the nectar of flowers only. Thus, females only act as secondary hosts for some diseases.

(ii) Body is divided into head, thorax and abdomen.

(iii) The head bears

- **Compound eyes** as paired, kidney-shaped structures.
- **Antenna** is present below the eyes. It bears several hairs arranged in several whorls. **Plumose** in male and **pilose** in female.
- **Johnston's organ** is present in pedicel of antenna to perceive sound vibrations.
- The **mouth parts** are of piercing and sucking type.
- The labrum, epipharynx and hypopharynx together form a food channel. This is helpful in sucking up of blood by pharyngeal suction force.
- The saliva contains an anticoagulant (heparin) which enters into the wound of host through hypopharynx.
- The mouthparts of male and female *Anopheles* are shown below

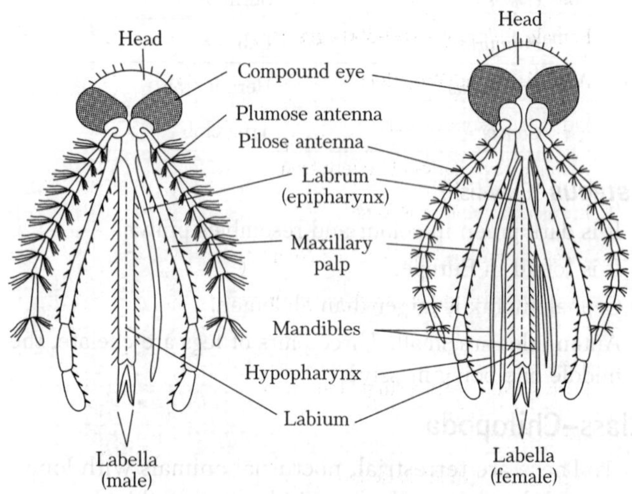

Fig. 9.7 *Anopheles head and mouthparts*

- Likewise the mouth parts of male and female *Culex* mosquito are shown below

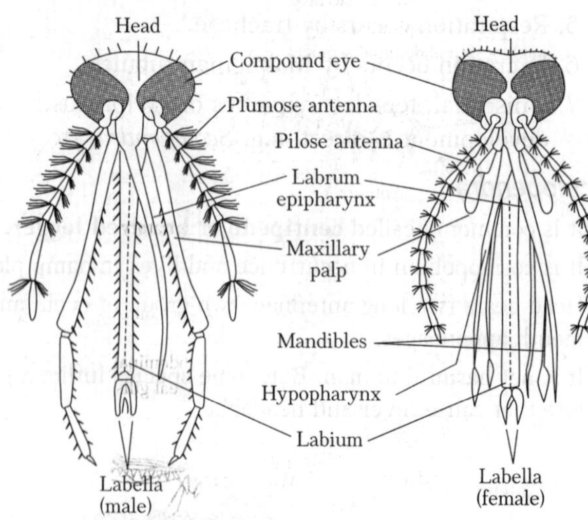

Fig. 9.8 *Culex head and mouthparts*

(iv) The wings are one pair and vary from species to species. For example, wings of *Anopheles* **bear dark spots,** while *Aedes* have **black-white stripped body.**

(v) All mosquitoes are dioecious and their reproductive organs resemble as those of typical cockroach.

(vi) Differences between the life cycle of *Culex* and *Anopheles* are given below

Culex	***Anopheles***
Egg	
Eggs are laid vertically on the surface of water in clusters forming **rafts.**	Eggs are laid horizontally and singly on the surface of water.
They are cigar-shaped.	They are boat-shaped.
The eggs do not have lateral air floats.	Each egg has two lateral air floats.
Eggs are laid in dirty water near human habitation.	Eggs are laid in clear water.
Female *Culex* lays 150-300 eggs at a time.	Female *Anopheles* lays 40-100 eggs at a time.
Aquatic larva (Wriggler)	
Larva is bottom feeder.	Larva is surface feeder.
At rest, larva hangs obliquely from the surface of the water.	At rest, the larva lives parallel to the surface of water.
Respiratory siphon is long.	Respiratory siphon is absent.
Pupa (Tumbler; Non-feeding; comma-shaped)	
Respiratory trumpets are long and narrow.	Respiratory trumpets are short and broad.
Greyish in colour.	Greenish in colour.
Each paddle bears a single long spine.	Each paddle bears two unequal spines.
Adult (Imago)	
Wings are without dark spots.	Wings have dark spots.
At rest, the body lies more or less parallel to the surface.	At rest, body lies at an angle of 45° to the surface.

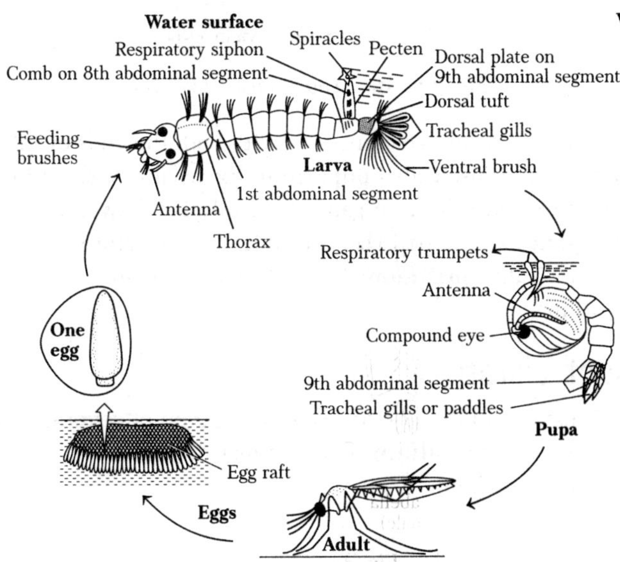

Fig. 9.9 Life cycle of Culex

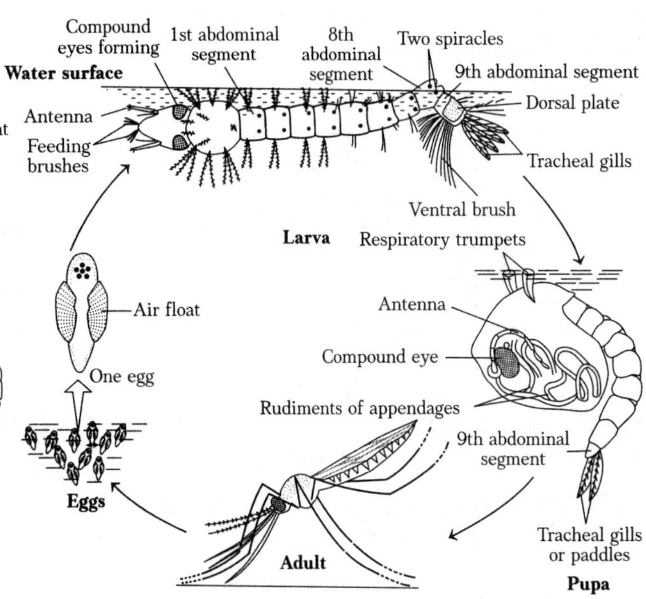

Fig. 9.10 Life cycle of Anopheles

Housefly

- This is also a genera of order—Diptera. Common housefly (*Musca domestica*) is worldwide in distribution. It is present anywhere and especially in filthy or dirty places.
- Its three species, i.e. *domestica*, *nebula* and *vicinia* are well-represented in India.
- It is a dangerous house pest as it serves as the carrier of many diseases like cholera, typhoid, diarrhoea, dysentery, etc.
- The insect is diurnal and very active feeder.

Mouthparts and feeding

- Mouthparts are of **sponging type**. These are adapted only to suck liquid and not to puncture the tissues. These mainly consist of a long fleshy proboscis formed by the labium.
- Labellum consists of a pair of large, fleshy oral lobes traversed by a fine network of grooves or channels called **pseudotrachea**.
- The oral lobes have teeth-like structures for scratching of solid food and sensory organs to examine its taste and smell.
- Mandibles are altogether absent.
- If food is solid-like sugar then the housefly first liquefy it with the help of its saliva and then suck it *via* pseudostrachae mouth and suctorial pharynx.

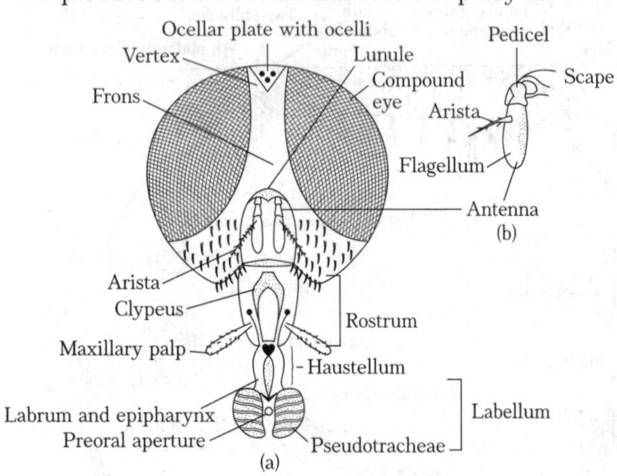

Fig. 9.11 (a) Head and mouthparts of housefly (Frontal view), (b) Single antenna with arista

Note *Arista is a three jointed process of large distal flagellum of antenna.*

Life cycle

- Sexes are separate and female is slightly bigger than the male.
- The breeding season of housefly is spring and summer, i.e. June to October.
- Female lays fertilised eggs on decaying organic matter in clusters of 500-600 eggs, each in a batch of 100-125 eggs in one season.
- The metamorphosis is complete and shows all the four stages of development, e.g. egg, larva, pupa and adult stage. While growing up, the larva moults twice, thus including three stages or instars.
- The complete life cycle of housefly is given below

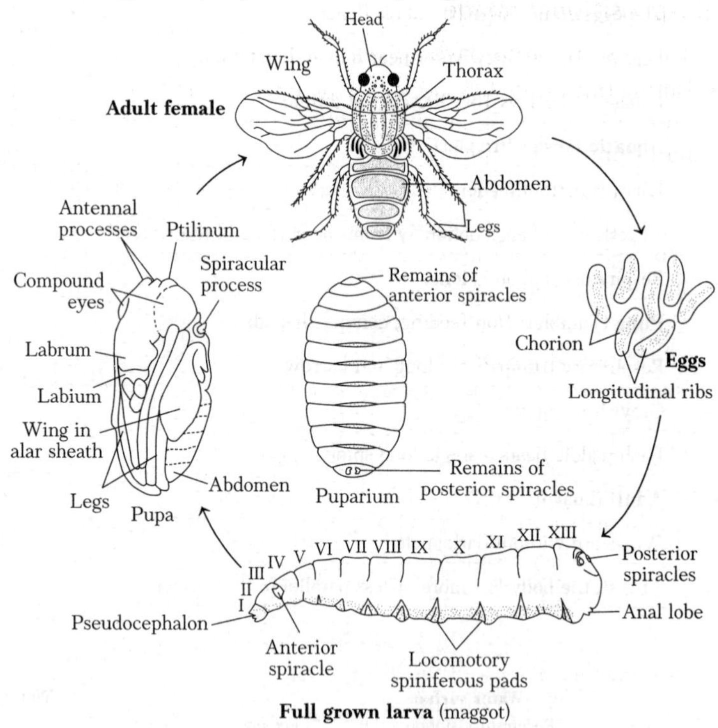

Fig. 9.12 Musca domestica : Life cycle

- The pupa is as an ovoid creature with both the ends evenly rounded in a case called **puparium**. Externally, it is motionless but internally, it forms organs *via* histolysis followed by **histogenesis**. The pupa has a bladder-like structure at its tip which is called **ptilinum**. This helps the puparium in breathing.

Insect vision

Insects have compound eyes as it is made up of several smaller units called **ommatidium**. Each ommatidium consists of following components

- One lens.
- Two corneagen cells.
- One crystalline cone secreted by cone cells.
- Four cone cells surrounding the cone.
- Seven retinular cells.
- A long refractile structure called **rhabdome**. It is formed by microvilli of retinular cells.

- Two pigment sheaths. First one is **iris pigment sheath** around cone cells. Second one is **retinular pigment sheath** around retinular cells. These sheaths are responsible for the separation of ommatidia and absorption of light.
- The process of image formation involves two types of images, i.e. appositional or mosaic image and superpositional image.

(i) Appositional image

- This is seen during daytime.
- During this image formation, the pigment sheaths are present in extended condition and a separate image is formed by every ommatidium.
- It is due to the fact that these pigment sheaths are light absorbing in nature. Thus, only parallel or nearly parallel light rays take part in image formation. Rest others are absorbed by these sheaths.
- The vision associated with this type of image formation is called **mosaic vision** because here the image of a single object is seen in small pieces jointed just like a mosaic pattern.
- This kind of vision is much more clear and detailed.

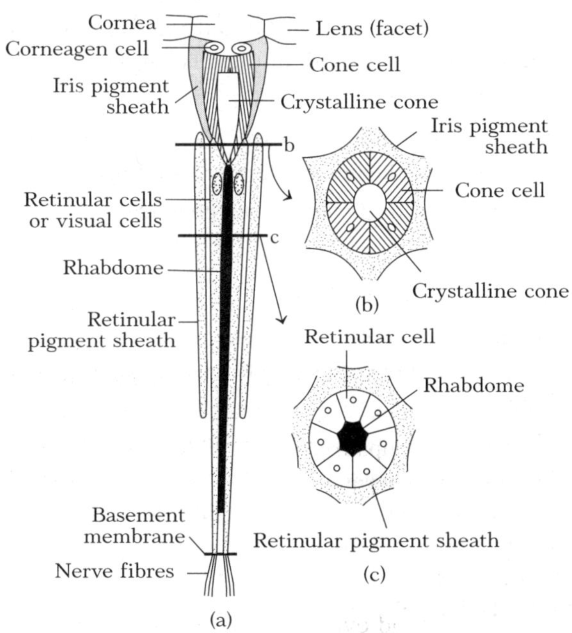

Fig. 9.13 *Vision : (a) VS of ommatidium, (b) TS through region of cone cells, (c) TS through region of retinular cells*

(ii) Superpositional image

- This image is seen in dark, when the pigment sheath of ommatidia gets contracted or shrink to their bases.
- In this condition, light rays can cross over one ommatidium to other. Thus, many images are formed, which overlap with each other.
- It results in blurred superposition image. Their refraction increases total light content of the eye as well. Thus, it enables the eyes to see in dark.
- In cockroach and some other related insects, pigment sheaths are permanently non-contractile. Hence, inspite of being nocturnal, these insects possess **mosaic vision** only.

Metamorphosis in insects

Transformation of an immature individual into a sexually mature reproducing adult is called **metamorphosis.** In insects, various degree of metamorphosis is seen depending upon the structural details of immature stages.

Types of metamorphosis

Following types of metamorphosis occurs in arthropods

(i) **Ametabolous** (no metamorphosis) Juveniles resemble adult except in size, sex organs or armature, e.g. silverfish (*Lepisma*).

(ii) **Paurometabolous** (gradual metamorphosis) Immatures (usually called **nymphs**) resemble adults (imago), but lack wings, e.g. grasshopper and cockroach.

(iii) **Hemimetabolous** (simple or incomplete metamorphosis) Immatures usually called **naiads** or **nymphs** may or may not resemble adults, e.g. dragonflies, stoneflies, bedbug, etc.

(iv) **Holometabolous** (complete metamorphoris) In this metamorphosis, four distinctive stages are seen in the life cycle, i.e. egg larva, pupa and adult, e.g. butterfly, mosquito and housefly.

(v) **Hypermetabolous** (hypermetamorphosis) In this, many larval stages are present. Each larval stage or instar differs from the other in habits, food and mode of living, e.g. blister beetle.

Economic importance of insects

The economic importance of insects in different spheres is given below

Useful insects

Insects giving commercial products

- The most common commercial product provider is **honeybee** with common species like *Apis indica, A. dorsata, A. laboriosa, A. florea*, etc. The products of honeybees include honey, royal jelly, bee pollen, wax, etc. Along with this, the bee poison is also useful to some extent in treating arthritis.
- The **lac** or **scale insect**, e.g. *Tachardia lacca, Kerria lacca* secrete lac, a secretion of integumental glands.

- Another **scale insect** *Dactylopius coccus* or *Coccus cacti* is famous for its cochineal dye. This dye is used in dyeing of wool, silk and cotton. It is also used in the formation of alta or mahavar.
- The silk is produced by various species of **silkworm**. These species include
  - *Bombyx mori* race *Nistari* for mulberry silk,
  - *Samia synthia ricini* (*Philosomia ricini*) for eri silk
  - *Antheraea myllita* for tropical tassar silk
  - *Antheraea assama* (*A. assamensis*) for muga silk
- Elytra (dried) of *Lytta vesicatoria* (Spanish fly) and *Mylabris cinchorii* (blister beetle) is used for making cantharidin oil. This oil is considered best for nourishing human hairs.
- Red ants give formic acid.

Useful predaceous insects

- The most common insect predators are dragonflies, praying mantis, mud wasps, robber flies, hoverflies, ladybird beetles, water scorpions and few pentatomid bugs, etc.
- Some of the common examples, of insects which are used in pest management are as follows
  - *Carabidae beetles* and *Stagomantis* (praying mantis) being voracious, feed on flies, grasshoppers and caterpillars which are injurious to crops.
  - *Cheilomenes* (ladybird beetle) feeds on aphids that infect cotton plants.
  - *Epicanta* (blister beetle) preys upon many kinds of lepidopteron larvae, which destroy cereals and cotton.

Beneficial parasitic insects

- *Trachina* larvae are parasites of injurious armyworms that are harmful to cereals.
- *Apanteles* (hymenoptera fly) lays eggs inside the body of armyworms and bollworms.
- *Leptomastix* is natural enemy of citrus mealy bug, i.e. *Planococcus citri*.

Other beneficial activities

- The insect scavengers feed on decomposing plants or animals or on dung. These normally include termites, silverfish (*Lepisma*), housefly, blowfly, maggots, dung beetles, carrion beetles, fleas, cockroaches and many larvae.
- Insects actually have a high nutritional value. They are rich in protein. Thus, these are eaten as a staple or as a delicacy food.
- Many crickets and cicadas in Japan are kept in small cages (also called **singing insects**), because of their unique singing ability.
- The wings of butterflies and other large insects are used for decoration purposes.

Harmful insects

The harmful activities of insects are summarised below.

Plant pests

These mostly belong to order-Lepidoptera, Coleoptera, Diptera and Hemiptera.

Pests of trees and crops

- *Euproctis chrysorrhoea* (brown tail moth), *Lymantria dispar* (Gypsy moth) are serious pests of shade and foliage trees. Their larvae also destroy forest trees.
- *Mayetiola* (Hessian fly) is a small-sized midge and its larva damages wheat plants.
- *Chilo* (Indian), *Diatraea* (American) are **'sugarcane stem borer'** while *Pyrilla perprusilla* is **sugarcane leafhopper**.
- *Nephotettix* is an Indian rice leafhopper and *Leptocorisa vericornis* is an oriental pest of rice and millet.
- *Anthonomus* (cotton boll weevil), *Dysdercus* (Indian cotton bug), *Aphis* (a serious cotton pest in India) can stain and destroy cotton bolls.

Pests of vegetables

- *Agrotis* (larvae), *Gnorimoschema* (larvae) are called **potato cut worms** (India). Both of them belongs to order-Lepiodoptera.
- *Anasa* (squash bug) is destructive to cucurbita plant.
- *Earias* (spotted bollworm) destroys ladyfingers.
- *Aulacophora* and *Limonius* (red pumpkin beetle) feed on pumpkins.

Pests of fruits

- *Drosicha mangifera* (mealy bug) causes destruction of mangoes, plums, papaya, jackfruit, citrus fruits and pears in India.
- *Psylla* (apple bug) lays eggs on apple trees thus, damage blossom and shoots.
- *Anthonomus* (beetle) destroys apple blossom and prevents the formation of fruit.

Stored grain pests

- *Tenebrio* and *Tribolium* (ghun), both have similar habits and are commonly found in stores and granaries. They destroy stored wheat and grains.
- *Calandra* (weevil) bores through the grains of *Oryza sativa* or rice and other stored grains.
- *Sitophillus oryzae* is one of the most common pest of the stored grains in Indian houses.

Disease spreading insects

Mosquitoes (Female)

- *Anopheles* is the vector of human **malaria**.
- *Culex* is the vector of **malaria** in birds.
- *Culex fatigans* and *Culex pipiens* spread *Wuchereria bancrofti*, the organism responsible for **filaria/elephantiasis**.
- *Aedes aegypti* is the vector of virus of **yellow fever, chikungunya** and **dengue fever** or **backbone fever**.

Flies

- *Glossina palpalis* and *G. tachinoides* are the vectors of *T. gambiense* causing **gambian sleeping sickness**.
- *G. marsitans* is the vector of *Trypanosoma rhodesience* causing **rhodesian sleeping sickness**.
- *Phlebotomus papetasia* is the vector of **sandfly fever** or **kala-azar** caused by *Leishmania donovani. P. major* in China and *P. argentipes* in India are also the vector of *Leishmania donovani.*
- *P. papatasia* and *P. sergenti* are the vectors of *Leishmania tropica* causing oriental sore in India and E. Asia.
- *Tabanus* (gadfly) and *Stomoxys* (stablefly) are the vectors of *Trypanosoma evansi* causing **Surra** (a wide spread tropical disease of horses and cattle).
- *Stomoxys* is the vector (intermediate host) of larvae of *Hymenolepis* (tapeworm of poultry) and larvae of *Habronema* (nematode of horse).

Fleas

- *Xenopsylla cheopis* and *X. astia* are rat fleas and are the vectors of *Bacillus* causing **bubonic plague to man**. *Cheopis* also transmits germs of **epidemic typhus** from rat to man.
- *Xenopsylla* is also an intermediate host of larvae of two tapeworms, i.e. *Dipylindium caninum* of dogs and cats and *Hymenolepis* of rats and man.

Bugs

- *Cimex* (bedbug) may carry and transmit germs of **relapsing fever** and **plague**.
- *Triatoma* (assassin bug), *Strongylus* and *Rhodinus* are the vectors of **Chagas disease** in South America.
- *Pediculus* human lice spread epidemic and endemic typhus (germ *Rickettsia*) and cause **French fever**.

CHECK POINT 9.2

1. Which of the following is not an exclusive trait of Arthropoda?
(a) Presence of wings
(b) Jointed appendages
(c) Compound eyes
(d) Chitinous exoskeleton

2. What is common character between earthworm and cockroach?
(a) Nephridia
(b) Solid ventral nerve cord
(c) Closed circulatory system
(d) Malpighian tubule

3. Which of the following is largest class of Arthropoda?
(a) Hexapoda (b) Pauropoda
(c) Symphylla (d) Diplopoda

4. Which one of the following genus of insects prefers to breed in clean water and their larvae lie parallel to the surface of water?
(a) *Anopheles* (b) *Culex*
(c) *Phlebotomus* (d) *Aedes*

5. *Anopheles* larva differs from that of *Culex* in having
(a) a developed respiratory siphon
(b) two respiratory siphons
(c) no respiratory siphon
(d) None of the above

6. Housefly differs from mosquito in having
(a) compound eyes
(b) only 3 pairs of jointed legs
(c) all legs resting upon surface while sitting
(d) head and thorax fused to form a cephalothorax

PHYLUM-MOLLUSCA

- It is the **second largest phylum** of animals after Arthropoda (L. *Molluscs*-soft bodied). It includes more than 60,000 species. The term 'Mollusca' was proposed by **Johnson** in 1650. The scientific study of Mollusca is called **Malacology** and the study of the shells of Mollusca is called **Conchology**.
- Molluscs represent rich fossil records because of their hard shell. Ordovician period is also called **Age of Giant Molluscs.**

General characteristics

The characteristics of phylum—Mollusca are as follows

1. Animals are mostly aquatic, marine as well as freshwater. Some forms are terrestrial also.
2. They are **triploblastic, schizocoelomate** animals with **organ system level of body organisation.**
3. Body is **bilateral symmetrical.** But in some molluscs like *Pila*, the bilateral symmetry is lost in adult because of the phenomenon of **torsion** (twisting) during growth.
4. The body is unsegmented and distinguished into three regions
   (i) **Head** is the anterior part of the body. It bears mouth, eyes and tentacles. Eyes are generally borne on stalks called **ommatophores.**
   (ii) **Visceral hump** contains the digestive tract and some visceral organs.
   (iii) **Foot** is a large muscular part of the body which is variously modified for crawling, burrowing, swimming or food capture.
5. **Mantle** or **Pallium** is soft and spongy layer of skin that covers the visceral hump or pallial complex. It is a characteristic feature of molluscs.
6. In most molluscs, mantle secretes a shell of calcium carbonate that protects and supports their soft tissues. This shell may be external, internal or absent. The space between the mantle and the visceral mass is called the **mantle cavity.**
7. True coelom is reduced to cavities called **haemocoel**. The true coelom is seen in nephridia, gonads and pericardium only.
8. **Digestive tract** is complete. It has ciliary tracts for sorting small food particles. In most of the cases, the mouth bears transverse rows of minute chitinous teeth called **radula**. This radula helps in food rasping.
   - Anus opens into the mantle cavity. The digestion is extracellular.
9. **Circulatory system** is of **open type** (except in the Cephalopoda and *Sepia*). The heart is dorsal, with one or two atria and one ventricle. Blood has copper containing blue respiratory pigment called **haemocyanin.**
10. **Respiration** is of **aerobic** type only. Gaseous exchange is performed by **ctenidia**, the feather-like gills present in the mantle cavity of aquatic molluscs.
    In terrestrial forms, gaseous exchange occurs by pulmonary sac or diffusion lungs or body surface.
11. **Excretion** occurs by paired sac-like kidneys or metanephridia. These kidneys are called **organ of Bojanus** in *Unio*. This animal also contains **Keber's organ** or **pericardial gland** for excretion. Gills also help in excretion up to some extent.
    - The nitrogenous waste in aquatic forms is ammonia while in terrestrial forms, it is urea or uric acid.
    - Guanine is the excretory material of *Sepia.*

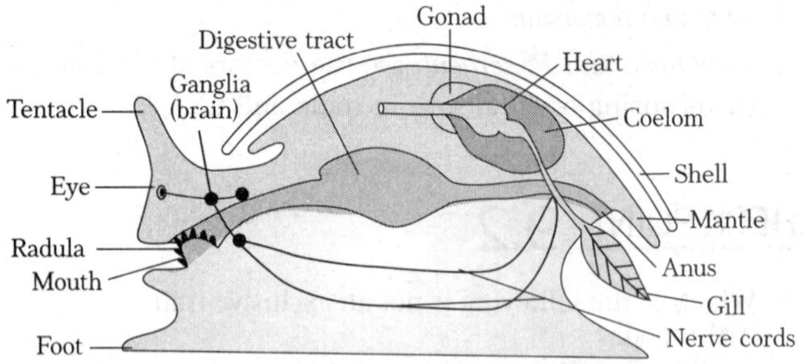

Fig. 9.14 A generalised mollusc

12. **The nervous system** consists of **paired ganglia,** i.e. cerebral, pleural, pedal and visceral. These are joined by the nerve connectives form dissimilar ganglion and commissures form similar ganglion.
13. Molluscs also possess specialised organs for sight, touch, smell or taste. Eyespots or complex eyes are present for sight. The eyes in cephalopods are comparable with the eyes of vertebrates.
    The statocysts are present for equilibrium. **Osphradium** helps in testing chemical and physical nature of water.
14. **Sexual reproduction** is present. Asexual reproduction is completly absent. Molluscs are dioecious animals, i.e. sexes are usually separate. But some are monoecious also with protandric condition. Fertilisation can be external or internal.
15. They are oviparous. The development is either direct or indirect by metamorphosis.When the development is indirect, it includes a characteristic larva, i.e. **veliger** (*Pila*), **trochophore** or **glochidium** (*Unio*). **Glochidium larva** is also known as **point of an arrow**.

Classification

Phylum—Mollusca is divided into following six classes based on their shell, mantle, foot, nerve, etc.

Class–Monoplacophora

(Gk. *Monos*–one; *plax*–plate; *pherein*–bearing)

1. All are marine forms.
2. Body is flat, bilaterally symmetrical and oval-shaped.
3. They possess shell, tentacles, mantle, foot, etc., like other molluscs.
4. The animals bear nephridia and internal segmentation like annelids.
5. Sexes are separate, e.g. ***Neopilina*–living fossil** and a connecting link between Annelida and Mollusca as it shows body segmentation.

Class–Polyplacophora *or* Amphineura

(Gk. *Amphi*–two; *neuron*–nerve)

1. All are marine forms.
2. Body is bilaterally symmetrical and elongated.
3. The ventral surface of the body has flat foot.
4. The head one to lacks eyes and tentacles.
5. Shell has one to eight calcareous dorsal plates or pieces.
6. **Radula** is present.
7. Nervous system is non-ganglionated.
8. Larva is **trochophore**, e.g. *Chiton* (coat of mail shell or sea mouse) and *Cryptochiton* (largest living *Chiton*).

Chiton

- It is nocturnal and herbivorous animal.
- It has a very long and wide foot which helps in clinging to coastal rocks.
- It has boat-shaped shell composed of **overlapping eight symmetrical plates**.
- Sexes are separate.

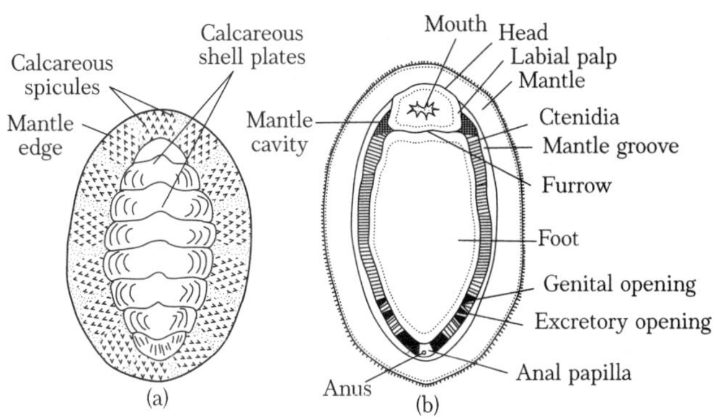

Fig. 9.15 *Chiton : (a) Dorsal view, (b) Ventral view*

Class–Aplacophora (No shells)

1. Worm like forms.
2. No shell or foot. Some forms have vestigial foot.
3. Integument is thick with minute spicules.
4. Solenogasters, e.g. *Chaetoderma.*

Class–Scaphopoda

(Gk. *Scapha*–boat; *podos*–foot)

1. All are marine.
2. Commonly called as **tooth shells** or **tusk shells.**
3. No head, no true tentacles, no eyes.
4. **Foot is conical** or boat-shaped and used for digging.
5. Shell is tubular and opens at both ends. It is the most distinctive character of this class.
6. Respiration occurs by mantle, ctenidia or gills are absent.
7. Gonad is single and kidney is paired.
8. Development is indirect through **trochophore larva,** e.g. *Dentalium* (the tusk shell), *Pulsellum.*

Dentalium

- It is found burrowing in sand.
- It has filamentous **captacula** instead of true tentacles. These are used for feeding.

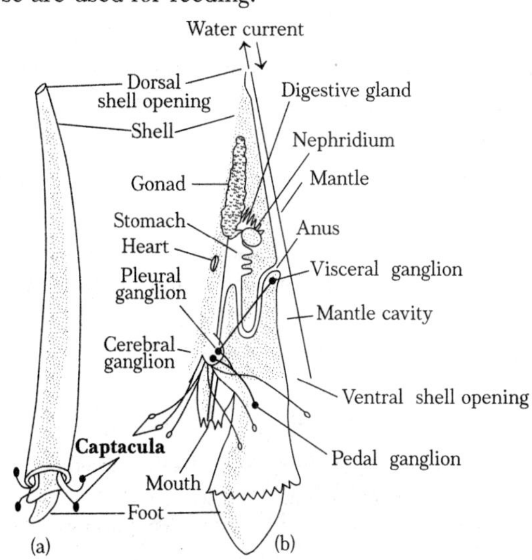

Fig. 9.16 *Dentalium : (a) External features, (b) Internal (anatomical) features*

Class–Gastropoda

(Gk. *Gaster*–belly; *podos*–foot)

1. It is the largest class of Mollusca, comprising snail, slugs and limpets.
2. They are marine, freshwater or terrestrial. These all three forms found in this class only.

3. The early embryo is symmetrical with anterior mouth and posterior anus. But during development, the body twists by 180 degree called **torsion**. This brings anus near the mouth and the adult becomes asymmetrical.
4. The shell is made up of one piece enclosing the visceral mass. The shell is spiral, uncoiled, reduced or absent in some classes.
5. Head distinct with tentacles and eyes.
6. The foot is large, flat, muscular and massive.
7. The buccal cavity has an **odontophore** as a cartilaginous projection with a **radula** bearing chitinous teeth.
8. **Ctenidia** are two or one in some cases. These are replaced by secondary gills or diffusion lungs or both. Ctenidia when present, are placed anterior to heart.
9. Sexes are separate. Fertilisation is internal.
10. **Trocophore** and **veliger** larval stages are seen during development.
    For example, *Pila* (apple snail), *Aplysia* (sea hare), *Doris* (sea lemon), *Patella* (limpet), *Limax* (grey slug, slowest invertebrate, terrestrial), *Achatina* (land snail), *Helix* (garden snail), etc.

Pila

- It is a **herbivorous** animal.
- It inhabits pond, lakes and streams.
- Asymmetrical body is covered by a spiral shell. The coils of the shell around the axis are called **whorls**.
- The digestive system is complete.
- It has **osphradium** (for testing nature of water) and **radula** (rasping organ).
- It leads an amphibious life. Gaseous exchange occurs by means of gills or ctenidia (in water) and by lungs (when on land).
- *Pila* bears 8-shaped nervous system.

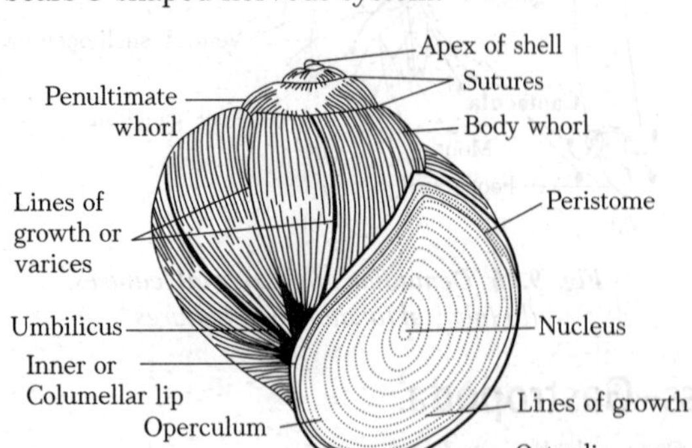

***Fig. 9.17** Pila (apple snail) : The shell with operculum (ventral view)*

Class–Bivalvia or Pelecypoda or Lamellibranchia

(Gk. *Pelekus*- hatchet; *podos*-foot)

1. It is the second largest molluscan class comprising of claws, mussels, oysters, scallops and shipworms. They are basically aquatic.
2. Body is bilaterally, symmetrical and laterally compressed.
3. Mantle is present with flat, right and left lobes.
4. No head, tentacles, eyes, jaws and radula.
5. Animals of this class have flattened foot which is hatchet-shaped.
6. These are mostly filter feeding forms.
7. Gonads open into the mantle cavity. They are usually dioecious.
8. Development involves **veliger** or **glochidium** larval stage. For example, *Unio* (freshwater mussel), *Pecten* (scallop), *Mytilus* (sea mussel), *Teredo* (shipworm), *Pinctada* (pearl oyster) and *Lamellidens* (mussel).
9. Commercially, this class is very important for pearl formation.

Unio

- It is found in rivers, lakes and ponds.
- The animal is **omnivorous**.
- Its shell consists of two halves called the **valves**. Both the valves are joined *via* a hinge ligament. An elevation in each valve is called **umbo**.
- Clear lines are visible over the valves. These are called **lines of growth**. These lines indicate the age of the individual.
- Its glochidium larva is parasitic on fish.

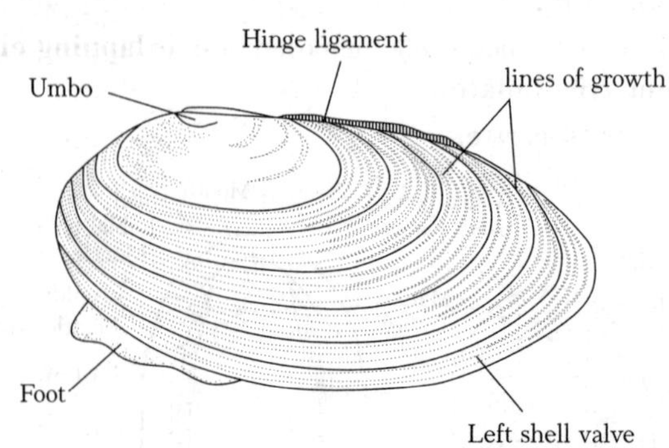

***Fig. 9.18** Unio : External features (Lateral view)*

Pearl formation

Pearl (mainly $CaCO_3$) formation was introduced in Japan by **Mikimoto**. It is naturally formed by oyster when some external particles get entangled in its skin which causes irritation to animal.

The oyster produces a secretion from its mantle around these sand particles. Technically, this secretion is called **nacre**. The amount of its deposition is in direct proportion to the degree of irritation. The nacre ultimately changes sand particle into the pearl. Best quality pearls are obtained from *Pinctada vulgaris*, the Indian pearl oyster. It is a sedentary mollusc which is now reared artificially.

The diagrammatic view of pearl formation is shown below

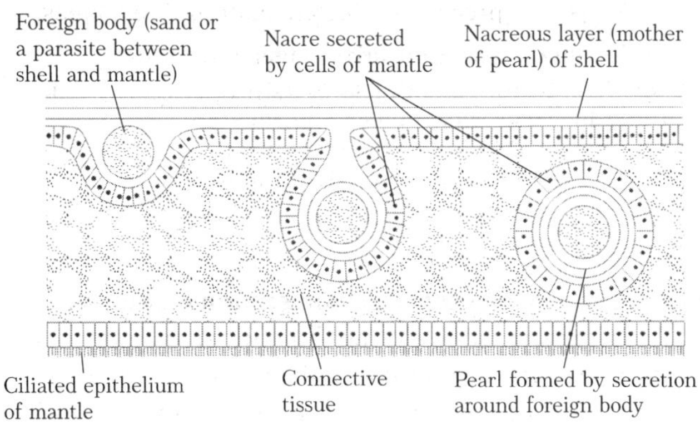

Fig. 9.19 Pearl formation

Class–Cephalopoda or Siphonopoda

(Gk. *Kephale*-head; *podos*-foot)

1. These are most evolved molluscs and exclusively marine.
2. This class comprises **squids** (largest cephalopods), *Octopus*, cuttlefish and nautiloids (a living fossil).
3. Body is bilaterally symmetrical.
4. Shell may be external (*Nautilus*), internal (*Sepia*) or absent (*Octopus*).
5. Head is large, eyes are conspicuous and resemble to vertebrates. Head is surrounded by a ring of tentacles. Mouth possesses horny jaws and **radula**.
6. The foot is located on the head. It is modified in the form of oral arms. Part of the foot forms a funnel-shaped **siphon**.
7. Locomotion occurs by expelling water as jet through siphon, i.e. jet propulsion.
8. Ink glands are present in some forms for offence and defence. The animal releases a cloud of ink fluid through its siphon to escape from enemies.
9. Chromatophores help cephalopods in colour change for concealment.
10. Animals are dioecious. Male has one arm specialised for sperm transfer into the female. This arm is called **hectocotylised arm**.
11. Development is direct. No larval form seen.

For example, *Sepia* (cuttle fish), *Loligo* (squid or sea arrow), *Octopus* (devil fish), *Nautilus*, *Spirula* (spiral shell), etc.

Note ***Nautilus*** *is the only cephalopod with external shell, so in it ink glands are absent.*

Octopus (Devil fish)

- It is found at the bottom of the sea.
- It feeds on prawns, crabs and small fishes. The prey is killed with its poisonous fluid.
- The shell is absent.
- It has eight oral arms.
- Suckers occur in two rows.
- Ink glands are present.
- It can change its colour.
- *Octopus* has good learning power and thus can be trained.

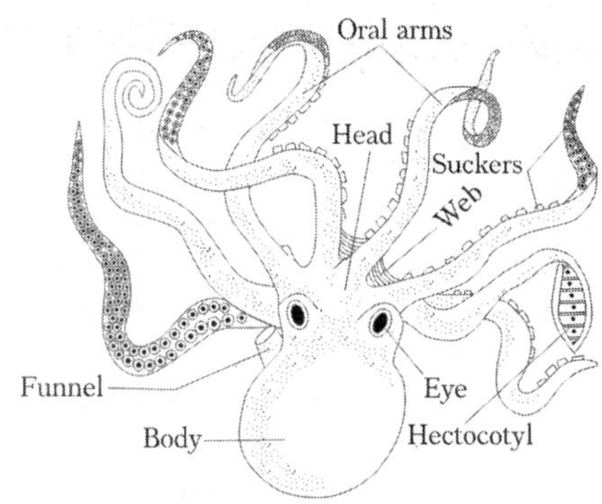

Fig. 9.20 Octopus

Squids

- These are found in warm seas.
- Body is distinguished into head, neck and trunk.
- Mouth is surrounded by ten oral arms, two of which are tentacles.
- They bear internal shell. Siphons and ink glands are present.
- They bear two lateral fins that help in swimming. These are also called **parapodium**.
- **Largest living invertebrate is giant squid** (*Architeuthis*) which is up to 18 m long. The giant squid has the **largest eye** in the animal kingdom.
- *Loligo* is an edible squid.

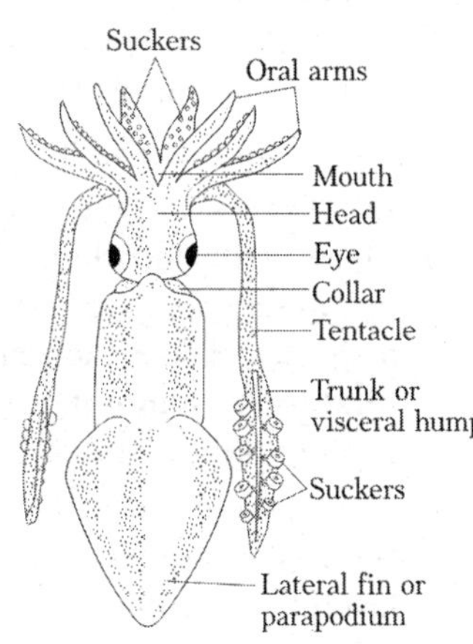

Fig. 9.21 Loligo

CHECK POINT 9.3

1. Phylum—Mollusca can be distinguished from other invertebrates by the presence of
(a) bilateral symmetry and exoskeleton
(b) shell and non-segmented body
(c) mantle and non-segmented body
(d) mantle and gills

2. In mollusc, excretion occurs by
(a) metanephridia
(b) organ of Bojanus
(c) Keber's organ
(d) All of the above

3. The mollusc which is considered to be a living fossil and also shows characters of annelids like nephridia and internal segmentation is
(a) *Pinctada vulgaris* (b) *Nautilus*
(c) *Neopilina galatheae* (d) None of these

4. The coelom of those animals which are commonly known as 'Tooth shells' originates
(a) by enterocoelic process
(b) by schizocoelic process and is filled with coelomic fluid
(c) from the blastocoel
(d) by splitting of embryonic mesoderm and is filled with blood

5. The molluscan which is dioecious with a single gonad and trochophore larva in its life history is
(a) Octopus (b) *Sepia*
(c) *Dentalium* (d) *Loligo*

6. The swift locomotion in *Sepia* and *Octopus* is affected by
(a) a fast moving neighbouring animal
(b) undulation of their own lateral fins
(c) lashing movement of their cephalic arms
(d) water in jets expelled through siphon

PHYLUM-ECHINODERMATA

The term 'Echinodermata' (Gk. *Echinos*–spines; *derma*–skin) was coined by **Jacob Klein** in 1738. **Leucport** in 1847 established this phylum. It contains about 6000 living species and 20,000 extinct species. This phylum appeared in the early Cambrian period. Echinoderms are closely related to the chordates because they possess true coelom called **enterocoelom.**

General characteristics

The characteristics of phylum–Echinodermata are as follows

1. All the animals of this phylum are **exclusively marine** and **non-parasitic.**
2. They generally live at sea bottom. Some are free-swimming and a few are attached forms (sessile).
3. The body is **polymorphic**, i.e. variously shaped and unsegmented. The body lacks distinct head.
4. Body wall is triploblastic and they possess organ system level of body organisation.
5. The **adult** echinoderm have **pentamerous radial symmetry**, i.e. body parts are arranged in five or multiples of five. However, their larvae are **bilaterally symmetrical**. Thus, we can say the bilateral symmetry of larva is lost or modified to radial pentamerous symmetry of adult.
6. Body is covered by delicate, ciliated and single layered epidermis.
7. Calcareous plates or ossicles form their **mesodermal endoskeleton.** These mesodermal ossicles are considered as the precursors of cartilage and bones of chordates.
8. Presence of a **water vascular system** is the characteristic feature of phylum–Echinodermata. This complete system is of coelomic origin.

Water vascular system

- It is a system of canals which are filled with water. It is also called **ambulacral system**, because its canals are present in the ambulacral groove within the body of animal.
- Ambulacral groove is a narrow groove which extends radially from mouth and runs all along the ventral surface of the corresponding arms.
- The water vascular system is composed of a porous diaphragm, called **madreporite**. It works as a water inlet of the system. The madreporite also works as pressure equalisation plate as the complete system is a hydraulic system. Madreporite is connected to a ring canal through a stone canal. From ring canal, five radial canals arise, each of which runs through the length of the arm. These radial canals bear tubefeet over their lateral branches called **lateral canals**.
- Each tubefeet is made up of **ampulla**, **podium** and a **sucker**. Tubefeet help in locomotion, capture and transport of food and respiration.
- This system is also associated with **Tiedemann's bodies** or racemose glands of unknown function.
- There are also present pear-shaped contractile bladders called as **polian vesicles**. These help in regulating water in the water vascular system.

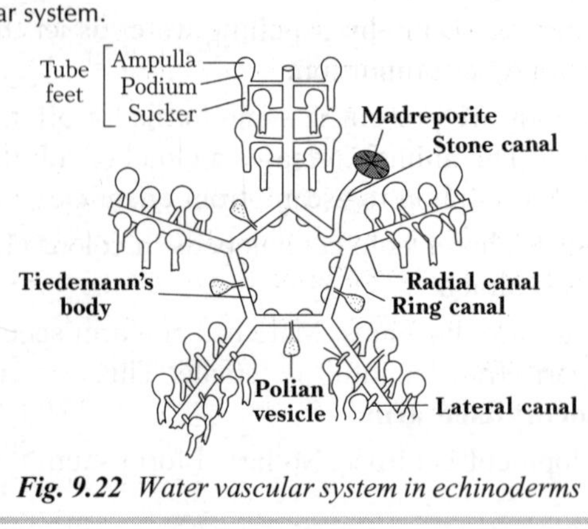

Fig. 9.22 Water vascular system in echinoderms

9. Body bears spines and jaw-like structures called **pedicellariae**. The spines are protective in function. The pedicellariae help in protection as well as clearing the body surface from debris and small organisms.
10. They possess **enterocoelom**, i.e. their body cavity is developed from digestive tract. It is lined with ciliated peritoneum.

    The surface with mouth is called **oral**, while that with anus is called **aboral** surface. However, Class—Ophiuroidea have **incomplete** digestive system.
11. A complex haemal and perihaemal system for circulation is present in the place of blood vascular system. The haemal system is enclosed inside the coelomic perihaemal channels. This complete system is thus of **open type**. Blood often lacks a respiratory pigment. There is no heart.
12. Respiration is **aerobic.** Gaseous exchange takes place by **dermal branchae** (skin gills) or **papulae** in starfish, by **peristomial gills** in sea urchins, by **genital bursae** in brittle stars and by **cloacal respiratory tree** in *Holothuria.* The movement of gases takes place through tubefeet.
13. Specialised excretory organs are absent. Nitrogenous waste is ammonia (ammonotelic). Removal of nitrogenous waste takes place through gills.
14. Nervous system consists of a nerve ring and radial nerve. No brain-like structure is seen. Sense organs are poorly developed.
15. Sexes are separate, but sexual dimorphism is absent. Testes and ovaries are morphologically similar and occupy similar positions in the body. Fertilisation takes place **externally** in sea water.
16. Phenomenon of **autotomy** i.e. ability to break a part of body and **regeneration** occur in these animals. A combination of both of these processes serves as a means for asexual reproduction.
17. Development is indirect with **free-swimming bilaterally symmetrical** larvae. These larvae are ciliated and lost their symmetry during metamorphosis.
    - The larval forms are **dipleurula, brachiolaria, bipinnaria, pluteus, ophiopluteus, echinopluteus, auricularia** and **doliolaria**. Their presence is visible across the different classes of this phylum.
    - **Dipleurula** larva is the common ancestral larva form of hemichordates, echinoderms and chordates.

Classification

Phylum-Echinodermata is divided into following two subphylum and further five classes

Subphylum–Eleutherozoa

(Gk. *Eleutheros*-free; *zoos*-animal).

This subphylum contains following four classes

Class–Asteroidea

(Gk. *Aster*-star; *eidos*-form)

1. Body is flat and star-shaped or pentagonal.
2. Endoskeleton is flexible due to separate ossicles.
3. Spines are short. **Pedicellariae** are present.
4. 2-4 rows of tubefeet with orally placed suckers are present in ambulacral grooves.
5. Aboral surface bears **anus** and **madreporite**.
6. Mostly free-living and predaceous. Larval forms are dipleurula, bipinnaria and brachiolaria, e.g. *Asterias* (starfish), *Pentaceros* (starfish), etc.

Asterias

- It is a benthonic i.e. bottom dwelling, nocturnal and carnivorous animal.
- Equal radial parts of starfish are called **actinomeres**.
- The **stomach is protrusible** in nature.
- Starfish feeds on molluscs and destroys pearl oyster.
- It shows **autotomy** and **regeneration**.
- Dried and powdered starfishes form a calcium rich fertiliser.

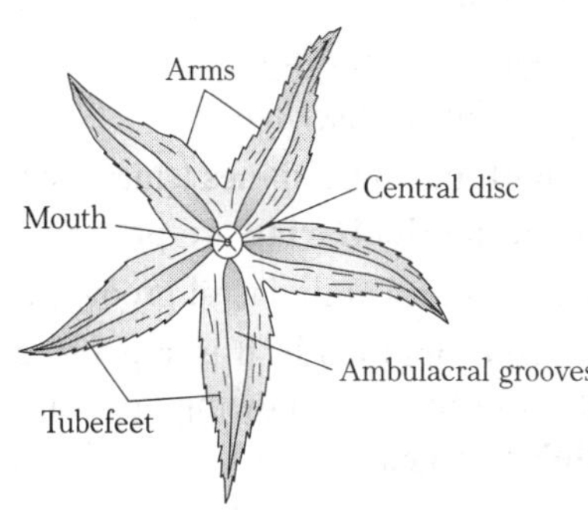

***Fig. 9.23** Asterias : Oral view*

Class–Ophiuroidea

(Gk. *Ophis*-snake; *oura*-tail; *eidos*-form)

1. Body is flat and star-like.
2. Arms are sharply marked off from central disc.
3. Terminal ends of arms resemble snake tails.
4. Spines are short. **Pedicellariae** are **absent**.

5. **Tubefeet** are present in two rows, but **without suckers**.
6. Oral surface has mouth and madreporite.
7. Anus and ambulacral grooves are absent.
8. Free-living and active larval form is called **ophiopluteus**.

   For example, *Ophiothrix* (the brittle star), *Gorgonocephalus* (the basket star), *Ophiura* (britte star), etc.

Ophiothrix

- It has small rounded central disc and five slender jointed arms.
- These arms are covered on all sides by the plates or shields with spines.

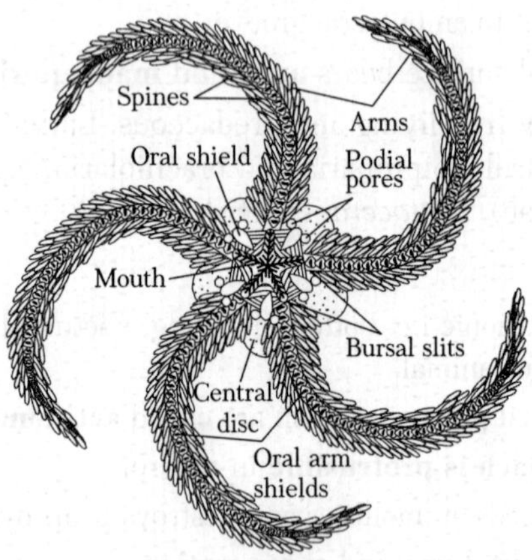

Fig. 9.24 Ophiothrix

Class–Echinoidea

(Gk. *Echinos*-hedgehog; *eidos*-form)

1. Body is globular or disc-like.
2. Skeleton with movable spines (small or large) and **three jawed pedicellariae** are present.
3. Tubefeet, arms and ambulacral grooves are absent.
4. Oral surface bears mouth. It is supported with masticatory biting and chewing apparatus called **Aristotle's lantern**.
5. Aboral surface has anus and madreporite.
6. Gonads are usually **five** or **less**.
7. Larval forms are **pluteus** and **echinopluteus**.

   For example, *Echinus* (sea urchin), *Clypeaster* (cake urchin), *Echinarachnius* (sand dollar), *Echinocardium* (heart urchin), etc.

Echinus

- It moves with the help of tubefeet and spines.
- It is an omnivorous animal.
- The mouth has a group of five specialised calcareous plates, known as an **Aristotle's lantern** which acts as a jaw.
- *Echinus esculentus* is edible.

Class–Holothuroidea

(Gk. *Holothurion*-sea cucumber; *eidos*-form)

1. Body is elongated and cylindrical.
2. Arms, pedicellariae and spines are absent.
3. Skeleton is made up of microscopic plates.
4. Madreporite is internal. Tubefeet are present with suckers. But, ambulacral grooves are absent.
5. Oral end has mouth surrounded by retractile tentacles.
6. Aboral end has anus.
7. Rrespiratory trees for gaseous exchange are present.
8. Larval form is **auricularia**, e.g. *Holothuria* (the sea cucumber), *Cucumaria* (the sea cucumber), etc.

Cucumaria

- This animal is very common in shallow tropical water.
- It has fifteen sets of feeding arms that fan out into bushy tentacles.
- Respiration occurs through system of tubules called **respiratory tree.**
- It feeds upon detritus and planktons.

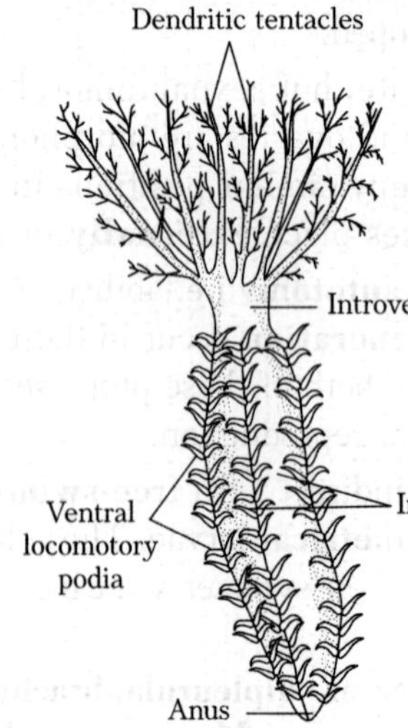

Fig. 9.25 Cucumaria

Subphylum–Pelmatozoa

(Gk. *Pelmatos*-stalk; *zoos-animal)*
This subphylum contains only one class

Class–Crinoidea (Gk. *Crinon*–lily; *eidos*–form)

1. Body enclosing a central disc, calyx is attached to the substratum by stalk.
2. Arms are branched with **pinnules**.
3. Tubefeet is without suckers.
4. Madreporite, spines and pedicellariae are absent.
5. Oral surface has mouth and anus both.
6. Ciliated ambulacral grooves are present on the aboral surface.
7. The larval form is called **doliolaria**.
8. Sedentary echinoderms are grouped under this class. For example, *Antedon* (feather star or sea-lily), *Ptilocrinus* (sea lily with cirri), *Bathycrinus* (sea lily without cirri), *Neometra*, etc.

Antedon

- Its aboral surface bears cylindrical appendages called **cirri.** These are meant for gripping the substratum.
- It feeds on microscopic plankton and debris.
- It has a good power of autotomy and regeneration.

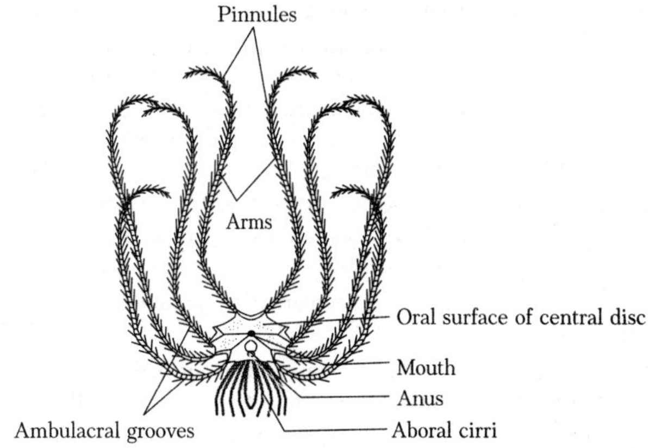

Fig. 9.26 *Antedon : Ventral view*

CHECK POINT 9.4

1. Echinoderms are heartless, brainless, headless, yet from evolutionary point of view, they have been placed on the top of the invertebrate phyla because of
(a) power of reproduction
(b) power of regeneration
(c) presence of enterocoel
(d) exclusively marine habitat

2. Identify the animal that acquires an adult secondarily radial symmetry when its bilaterally symmetrical larva undergoes metamorphosis.
(a) *Pila* (b) *Gorgonia*
(c) *Gorgonocephalus* (d) *Polygordius*

3. The formation of mesoderm as evaginated sacs from the archenteron is a characteristic of
(a) *Pila*
(b) *Pentaceros*
(c) *Physalia*
(d) *Palamnaeus*

4. Which of the following is not the larval form of echinoderms?
(a) Doliolaria
(b) Auricularia
(c) Bipinnaria
(d) Glochidia

5. Choose the correct statement.
(a) Aristotle's lantern helps in respiration
(b) Respiratory tree is present in rectum
(c) Dipleura larva is seen in molluscans, echinoderms and chordates
(d) Anus is absent in brittle star

6. *Antedon* belongs to the class–Crinoidea because
(a) it has calyx for attachment and arms bear pinnules
(b) the mouth is situated orally and the anus is found on the aboral surface
(c) the arms are not distinctly separated from the central disc
(d) the tubefeet are present in ambulacral groove

EXAM POINTS

- Annelida are triploblastic, segmented and bilaterally symmetrical animals. They are true coelomates. **[NEET 2020]**
- The body of annelids appears to have tube within a tube plan. **[Bihar BCECE 2012]**
- In class-Polychaeta, the setae are numerous. **[NEET 2013, AFMC 2013]**
- Trochophore is the larval form of *Nereis*. **[AIIMS 2017]**
- *Hirudinaria* (freshwater leech), *Haemadipsa* (tiger leech), *Pontobdella* (marine leech), *Arenicola*. (lung worm) are some members of phylum–Annelida. **[EAMCET 2015]**
- Arthropods have metameric segmentation, jointed appendages and chitinous exoskeleton. **[NEET 2016]**
- The chitinous exoskeleton of arthropods is formed by the polymerisation of N-acetyl glucosamine. **[CBSE AIPMT 2015]**
- Arthropods have open circulatory system. In them, the fluid is composed of blood mixed with tissue fluid.
  It oozes through spaces or cavity that surround the organ.
  This mixture of fluid is referred to as haemolymph. **[AIIMS 2013]**
- King crab is an example of living fossil. **[WB JEE 2014]**
- Juvenile hormone is secreted from corpora allata. **[WB JEE 2014]**
- Larva of hermit crab is glaucothoe and larva of housefly is maggot. **[AIIMS 2017, JIPMER 2014]**
- Classification of housefly is Phylum–Arthropoda, class –Insecta, Order – Diptera, Family – Muscidae. **{CBSE AIPMT 2016]**
- Scorpion has book lungs for respiration. **[NEET 2020]**
- Tse-tse fly, housefly, rat flea, sand fly are some members of phylum–Arthropoda which spread disease in humans, plants and animals. **[Chhattisgarh PMT 2015]**
- Torsion of visceral mass is seen in animals belonging to Gastropoda, e.g. *Pila*. **[JIPMER 2017]**
- Radula is a part of phylum–Mollusca which helps in food rasping, e.g. *Pila*. **[JIPMER 2019, NEET 2020]**
- Glochidium is the larval stage of *Unio*. **[AIIMS 2017]**
- *Chiton* belongs to class–Polyplacophora. **[AMU 2015]**
- *Octopus* is a molluscan without shell. **[EAMCET 2015]**
- Veliger larva occurs in phylum–Mollusca. **[UPCPMT 2013, 09]**
- Most advanced invertebrates are Cephalopoda. **[JIPMER 2019]**
- Echinoderms are marine with organ system level of organisation. They are triploblastic and coelomates. **[Kerala CEE 2015]**
- Echinoderms are dioecious, sexes are separate. They lay eggs, fertilisation is external and indirect development is observed. **[Kerala CEE 2015]**
- Some classes of Echinodermata have spine-like structure called pedicellariae which protect as well as clean the body surface, e.g. members of classes–Asteroidea and Echinoidea. **[EAMCET 2014]**
- The larva of phylum–Echinodermata shows bilateral symmetry while the adult shows radial symmetry, e.g. *Asterias*. **[NEET 2020, Karnataka CET 2012]**
- Sea urchin has mouth on lower side and anus on the upper side. **[Gujarat 2015]**
- *Ophiura* is an echinoderm. **[NEET 2019]**
- Tubefeets are the locomotory organ of phylum–Echinodermata. **[Odisha JEE 2010]**
- Water vascular system is found in phylum–Echinodermata, e.g. sea cucumber. **[WB JEE 2010]**
- Members of phylum–Echinodermata do not have excretory system. **[Kerala CEE 2011**

Chapter Exercises

(A) Topical Exercise

Phylum-Annelida

1 Phylum-Annelida includes invertebrates which have
(a) metameric segmentation (b) triploblastic body
(c) coelomate organisation (d) All of these

2 Metamerism in Annelida,
(a) increases the efficiency of body movement
(b) is also called as serial homology
(c) plays an important role in respiration
(d) Both (a) and (b)

3 Which one of the following statements is true regarding phylum-Annelida?
(a) They are radially symmetrical
(b) The coelom is well-developed
(c) Digestive tract is incomplete
(d) Respiration is anaerobic in free-living forms

4 Which one of the following structure is not present in phylum-Annelida?
(a) Nephridia
(b) Stomach
(c) Coelomic sinuses
(d) Cerebral ganglia

5 The respiratory pigment present in Annelida is/are
(a) haemoglobin (b) haemocyanin
(c) erythrocruorin (d) Both (a) and (c)

6 The eggs of annelids are laid in
(a) ootheca (b) cocoons
(c) egg cases (d) All of these

7 Identify the statement which is false regarding the characteristic features of phylum—Annelida.
(a) The locomotory organs are minute rod-like chitinous setae or suckers
(b) A closed circulatory system of longitudinal blood vessels is present with lateral branches in each segment
(c) Nephridia are present as excretory organs, which are endodermal in origin
(d) Fertilisation is external

8 The fleshy appendages of annelids which bear setae are called ...... and these are found in class- ....
(a) legs, Oligochaeta
(b) parapodia, Polychaeta
(c) ciri, Hirudinea
(d) None of the above

9 Consider the following statements.
I. *Nereis* are dioecious.
II. In *Nereis*, coelom duct performs the function of both reproductive and excretory organs.

Choose the correct option.
(a) Statement I is true, but II is false
(b) Statement I is false, but II is true
(c) Both statements I and II are true
(d) Both statements I and II are false

10 Which one of the following characterises *Heteronereis* stage of *Nereis*?
(a) Very active asexual stage
(b) Quite inactive sexual stage
(c) Sexual stage with epitoke which is very active
(d) Asexual stage with epitoke which is very active

11 In *Nereis*, swarming occurs at night for
(a) copulation
(b) shedding of sperms and ova
(c) attaining sexual maturity
(d) None of the above

12 Consider the following statements for *Nereis*.
I. Parapodia are meant for exchange of gases.
II. Peristomium is composed of six embryonic segments.

Choose the correct option.
(a) Statement I is true, but II is false
(b) Statement I is false, but II is true
(c) Both statements I and II are true
(d) Both statements I and II are false

13 In leech, number of annular receptor in each annulus is
(a) 18 on the dorsal surface only
(b) 18 on the dorsal and ventral surface, respectively
(c) 18 on the ventral surface only
(d) None of the above

14 A temporary clitellum is formed in *Hirudinaria* on the segment
(a) 7-9 th (b) 9-11th (c) 11-13th (d) None of these

15 Consider the following statements.
I. *Hirudinaria* bears a pair of nephridiopore in each segment of the body from 6th to 22nd segment.
II. The male and female genital pores are present between 2nd and 3rd annuli of 10th and 11th segments.

Choose the correct option.
(a) Statement I is true, but II is false
(b) Statement I is false, But II is true
(c) Both statements I and II are true
(d) Both statements I and II are false

16 Suckers of leech are located at
(a) anterior and posterior ends of the body
(b) anterior end of the body
(c) posterior end of the body
(d) all over the body

17 Which one assists in locomotion?
(a) Trichocysts in *Paramecium*
(b) Penae spicules in *Ascaris lumbricoides*
(c) Clitellum in *Pheretima*
(d) Posterior sucker in *Hirudinaria*

18 Development is direct in the class
(a) Oligochaeta and Polychaeta
(b) Polychaeta and Archiannelida
(c) Archiannelida and Hirudinea
(d) Hirudinea and Oligochaeta

19 Find out the statement which is false about classification of annelids.
(a) It is based on position and arrangement of setae, their presence and absence and presence of sense organs
(b) They are divided into four classes
(c) The class–Polychaeta is divided into three subclasses–Essantia, Sedentaria and Hirudinea
(d) Class–Archiannelida consists of primitive and small-sized annelids

20 Match the following columns.

Column I (Annelid common names)	Column II (Scientific names)
A. Sandworm	1. *Eurythoe*
B. Fireworm	2. *Tubifex*
C. Earthworm	3. *Nereis*
D. Cattle leech	4. *Hirudinaria*
E. Sludge worm	5. *Pheretima*

Codes

	A	B	C	D	E
(a)	3	1	5	4	2
(b)	1	4	3	2	5
(c)	5	3	1	4	2
(d)	4	2	5	3	1

Phylum-Arthropoda

21 Which of the following is not a characteristic of arthropods?
(a) Organ system level of organisation
(b) Bilaterally symmetrical, triploblastic animals
(c) Segmented, coelomate animals
(d) Body is always covered by calcareous shell

22 Chitinous plates making the exoskeleton in arthropods are called
(a) tergites (b) sternness
(c) pleurites (d) sclerites

23 Which of the following statement(s) is/are correct about arthropods?
(a) They are mostly dioecious and oviparous
(b) Fertilisation is internal
(c) Development may be direct or indirect
(d) All of the above

24 Green glands present in some arthropods help in
(a) respiration (b) excretion
(c) digestion (d) reproduction

25 Trilobites evolved during
(a) Cambrian (b) Precambrian
(c) Ordovician (d) Silurian

26 A characteristic feature of subphylum-Chelicerata is
(a) first appendage is feeding chelicerae
(b) presence of covering called carapace
(c) presence of four pairs of walking legs
(d) All of the above

27 Consider the following statements.
I. *Limulus* belongs to class-Merostomata.
II. In *Limulus* (king crab), the respiratory organs are book gills.
Choose the correct option.
(a) Statement I is true, but II is false
(b) Statement I is false, but II is true
(c) Both statements I and II are true
(d) Both statements I and II are false

28 Arachnids possess
(a) 3 pairs of legs (b) 4 pairs of legs
(c) 5 pairs of legs (d) 6 pairs of legs

29 The coxal glands in arachnids serve as
(a) digestive glands (b) excretory glands
(c) reproductive glands (d) poison glands

30 *Peripatus* has
(a) Both arthropod and roundworm characters
(b) parapodia
(c) jointed legs
(d) haemocyanin

31 Consider the following statements for *Peripatus.*
I. Connecting link between arthropods and annelids.
II. A single pair of antennae, eyes and jaws are present.
Choose the correct option.
(a) Statement I is true, but II is false
(b) Statement I is false, but II is true
(c) Both statements I and II are true
(d) Both statements I and II are false

32 The characters like hard exoskeleton, cephalothorax, gills, for respiration are present in which of the following?
(a) Insecta (b) Myriapoda
(c) Crustacea (d) Polychaeta

33 The balancing organ equal to our ear in aquatic arthropods as in prawn is
(a) otolith (b) lateral line organ
(c) otocyst (d) statocyst

34 Classes of mandibulate are
I. Crustacea II. Merostomata
III. Insecta IV. Arachnida
Choose the correct option.
(a) I and III (b) II and III
(c) I, II and III (d) III and IV

35 Shrimps, crayfish and lobsters belong to the class
(a) Insecta (b) Crustacea
(c) Arachnida (d) Chilopoda

36 Which one of the following is known as water flea?
(a) *Daphnia* (b) *Aedes* (c) *Laccifer* (d) *Mysis*

37 Consider the following statements.
I. *Scolopendra* are centipede, they are cosmopolitan.
II. *Paurapus* are animals having one pair of antennae and no eyes.
Choose the correct option.
(a) Statement I is true, but II is false
(b) Statement I is false, but II is true
(c) Both statements I and II are true
(d) Both statements I and II are false

38 Match the following columns.

	Column I (Scientific names)		Column II (Common names)
A.	*Palaemon*	1.	Hermit crab
B.	*Eupagurus*	2.	True crab
C.	*Palinurus*	3.	Prawn
D.	*Cancer*	4.	Lobster

Codes

	A	B	C	D		A	B	C	D
(a)	1	4	2	3	(b)	3	2	4	1
(c)	3	1	4	2	(d)	1	3	2	4

39 Body of class-Insecta members is divided into
(a) head, thorax
(b) head, abdomen, tail
(c) head, thorax and abdomen
(d) antennae, head, thorax

40 In insects, excretory product is
(a) ammonia (b) urea
(c) uric acid (d) All of these

41 Johnston's organ is associated with
(a) stigma of housefly (b) head of housefly
(c) antenna of mosquito (d) intestine of cockroach

42 Mouthparts of mosquito are
(a) chewing type (b) sucking type
(c) sponging type (d) piercing and sucking type

43 Thorax of larval mosquito is
(a) unsegmented (b) two-segmented
(c) three-segmented (d) multi-segmented

44 Male mosquitoes can be recognised from female mosquitoes by
(a) longer legs
(b) larger heads
(c) their larger bodies
(d) more numerous bristles on antennae

45 Which mouthparts occur in female *Anopheles*, but not in male?
(a) Proboscis (b) Mandibles (c) Maxillae (d) Antennae

46 Larva of *Anopheles* is a
(a) surface feeder with long and conical respiratory siphon
(b) surface feeder with undeveloped respiratory siphon
(c) bottom feeder with undeveloped respiratory siphon
(d) bottom feeder with long and conical respiratory siphon

47 Which of the following is the respiratory organ in wriggler?
(a) Trachea (b) Spiracles
(c) Integument (d) Respiratory siphon

48 Eggs of *Culex* are found in the form of a
(a) raft (b) loose clusters
(c) Both (a) and (b) (d) None of these

49 Respiratory siphon is found in
(a) larva of housefly (b) larva of *Culex*
(c) larva of *Anopheles* (d) nymph of cockroach

50 Male and female adult *Culex* can be identified by their
(a) body size
(b) length of wings
(c) length of proboscis
(d) length of maxillary palps

51 Adult female *Anopheles* differs from adult female *Culex* in having
(a) wings spotted, maxillary palps long, body inclined at 45° to surface at rest
(b) wings plain, maxillary palps short, body parallel to surface at rest
(c) wings plain, maxillary palp long, body inclined at 45° to surface at rest
(d) wings spotted, maxillary palps short, body parallel to surface at rest

52 *Culex* and *Anopheles* show similarity in that
(a) females of both transmit diseases
(b) larvae float in water
(c) respiratory trumpets are present in pupa stage
(d) All of the above

53 The common housefly belongs to order
(a) Lepidoptera (b) Hemiptera
(c) Diptera (d) Homoptera

54 The flagellum on the antennae of housefly is
(a) pilose (b) filiform
(c) plumose (d) arista

55 How many simple ocelli a housefly has?
(a) One (b) Two (c) Three (d) Many

56 Consider the following statements.
I. Mouthparts in housefly are sponging type.
II. Haustellum is the middle part of proboscis of housefly.
Choose the correct option.
(a) Statement I is true, but II is false
(b) Statement I is false, but II is true
(c) Both statements I and II are true
(d) Both statements I and II are false

57 In housefly, pseudotracheae are found in
(a) rostrum (b) labellum
(c) basiproboscis (d) haustellum

58 In housefly, ... are present.
(a) 1 pseudotrachea and 1 spiracle
(b) 2 pseudotracheae and 1 spiracle
(c) 2 pseudotracheae and 2 spiracles
(d) None of the above

59 Consider the following statements.
I. The hatching of eggs of housefly occurs in 12-24 hours.
II. The outermost covering of the egg of housefly is chorion.
Choose the correct option.
(a) Statement I is true, but II is false
(b) Statement I is false, but II is true
(c) Both statements I and II are true
(d) Both statements I and II are false

60 Common housefly (*Musca nebulo*) lays eggs on
(a) stagnant water (b) hanging ropes
(c) cow/horse dung (d) open meats and sweets

61 Housefly feeds on sugar by
(a) sucking sugar crystals through pseudotracheae
(b) dissolving sugar crystals by saliva and sucking the solution
(c) crushing sugar crystals with mandibles before sucking it
(d) crushing sugar crystals with its legs and then transferring it to pseudotracheae

62 A pair of claspers with a penis in between is present in .... segment of housefly.
(a) 10th (b) 9th (c) 5th (d) 16th

63 The common house-cricket, which produces the familiar sound with its stridulating organs is
(a) *Gryllus* (Acheta) (b) *Locusta*
(c) *Lepisma* (d) praying mantis

64 The second instar larva of housefly has
(a) one pair of abdominal and two pairs of prothoracic spiracles
(b) two pairs of abdominal and one pair of prothoracic spiracles
(c) one pair of abdominal and one pair of prothoracic spiracles
(d) one pair of abdominal spiracles

65 Which of the following is not used in the locomotion of housefly larva?
(a) Ptilinum
(b) Anal lobes
(c) Mandibular sclerites
(d) Spiniferous pads

66 The order of metamorphosis in housefly is
(a) egg → pupa → larva → adult
(b) egg → larva → pupa → adult
(c) egg → nymph → pupa → adult
(d) egg → larva → nymph → adult

67 Consider the following statements.
I. In insects, the cuticle lined tracheae are the tubes which carry oxygen to the cells and tissues.
II. The insect blood has no role in transportation of oxygen and there is no link between respiratory and circulatory systems.
Choose the correct option.
(a) Statement I is true, but II is false
(b) Statement I is false, but II is true
(c) Both statements I and II are true
(d) Both statements I and II are false

68 The cells present in the eye of an insect are
(a) cone cells (b) rod cells
(c) retinular cells (d) Both (a) and (c)

69 Equivalent to eye lens of human eye, each ommatidium of insect has
(a) retinal cells (b) rhabdome
(c) cone cells (d) All of these

70 The two pigment sheaths of eye in insects are responsible for
(a) separation of ommatidia and absorption of light
(b) processing of light entering in the eye
(c) daytime vision
(d) night time vision

71 Consider the following statements.

I. Rhabdome in the eye of insect is formed of retinular cell.

II. The image formed in insect eye is mosaic and superpositional image.

Choose the correct option.

(a) Statement I is true, but II is false
(b) Statement I is false, but II is true
(c) Both statements I and II are true
(d) Both statements I and II are false

72 Mark the incorrect match.

	Organisms		Larvae
(a)	Housefly	–	Maggot
(b)	Mosquito	–	Wriggler
(c)	Ant	–	*Cypris*
(d)	Cockroach	–	Nymph

73 Maximum number of economically important species belong to the class

(a) Insecta (b) Crustacea (c) Chilopoda (d) Diplopoda

74 Common characteristics of cockroach, housefly and mosquito are

(a) compound and simple eyes
(b) one pair of wings and halteres
(c) three pairs of legs and one pair of antennae
(d) two pairs of legs and two compound eyes

75 What is common between mosquito and cockroach?

(a) Both feed on liquid food
(b) Both have pupal stages
(c) Both have simple eyes
(d) Both are nocturnal

76 Which one of the following is not a lepidopteran insect pest?

(a) Jute semilooper (b) Gram cut worm
(c) Sugarcane borer (d) Sugarcane white fly

77 An arthropod, which acts as a gregarious pest is

(a) *Apis* (b) *Anopheles*
(c) *Locusta* (d) *Laccifer*

78 Scientific name of oak silkworm is

(a) *Bombyx mori* (b) *Attacus ricini*
(c) *Antheraea pernyi* (d) *Antheraea assamensis*

79 The scientific name of the brown plant hopper is

(a) *Nilaparvata lugens*
(b) *Carausius morosus*
(c) *Helicoverpa armigera*
(d) *Calliphora crythrocephala*

80 Choose the incorrect match.

(a) *Pyrausta larvae* – corn borer
(b) *Nephotettix* – rice leafhopper
(c) *Leptocorisa* – pest of rice and millet
(d) *Apis indica* – sugarcane leafhopper

81 Squash bug is a destructive pest of

(a) ladyfinger (b) cucurbitous plants
(c) apple (d) All of these

82 Termite

I. is a harmful colonial insect.
II. shows caste system.
III. has a pair of respiratory trumpets.
IV. are blood suckers of vertebrates.

Choose the correct option.

(a) I and II (b) II and III
(c) III and IV (d) II and IV

83 Which of the following acts as a vector for disease?

(a) *Anopheles* (b) *Culex*
(c) *Aedes* (d) All of these

84 *Phlebotomus* is responsible for spreading

(a) dumdum fever (b) oriental sores
(c) kala-azar (d) All of these

85 The common predator insect is

(a) praying mantis (b) ladybird
(c) dragonflies (d) All of these

86 Which one of these is a set of useful insects?

(a) Sandfly, butterfly, honeybee
(b) Honeybee, silkmoth, tse-tse fly
(c) Silkmoth, honeybee, locust
(d) Honeybee, lac insect, cochineal insect

87 Cantharidine oil, which prevents hair loss is extracted from

(a) honeybee (b) blister beetles
(c) lac insects (d) maxillae

88 Which of the following is not an economically important insect?

(a) *Apis* (b) *Bombyx* (c) *Locusta* (d) *Laccifer*

Phylum-Mollusca

89 The scientific study of Mollusca is called

(a) Conchology (b) Molluscology
(c) Seal Biology (d) Malacology

90 The term 'Mollusca' was proposed by

(a) Johnson in 1680 (b) Johnson in 1650
(c) Hutchinson in 1750 (d) Copeland in 1720

91 Mostly molluscs are

(a) bilaterally symmetrical (b) radial symmetrical
(c) asymmetrical (d) None of these

92 In molluscs, the body is unsegmented with a distinct

(a) head, thorax and abdomen
(b) head, thorax and muscular foot
(c) head, muscular foot and abdomen
(d) head, muscular foot and visceral hump

93 Consider the following statements.
I. Body of molluscs is covered by calcareous shell.
II. The tentacles of Mollusca secretes its shell.
Choose the correct option.
(a) Statement I is true, but II is false
(b) Statement I is false, but II is true
(c) Both statements I and II are true
(d) Both statements I and II are false

94 Many molluscs have a water-testing organ present in the mantle cavity, this is called
(a) statocyst (b) osphradium
(c) tentaculocyst (d) amphid

95 In Mollusca, eye is present over a stalk called
(a) osphradium (b) ostracum
(c) ommatophore (d) operculum

96 In the members of phylum-Mollusca, the anterior head region has
(a) sensory tentacles (b) sharp hook
(c) sticky disc (d) radula

97 Haemocyanin, the blue colouring pigment of molluscan blood contains
(a) iron (b) magnesium (c) copper (d) manganese

98 In the mantle cavity, feather-like gills are present, which have
(a) masticatory function (b) respiratory function
(c) excretory function (d) Both (b) and (c)

99 What are the respiratory organs of molluscs?
(a) Gills or Ctenidia (b) Mantle
(c) Pulmonary sac (d) All of these

100 The mouth of the molluscs bears transverse rows of minute chitinous teeth called
(a) matula (b) chitinula (c) radula (d) macula

101 The characteristic larva formed in case of indirect development in molluscs is called
(a) glochidium (b) trochophore
(c) veliger (d) All of these

102 Under which of the following groups is *Neopilina* included?
(a) Amphineura (b) Monoplacophora
(c) Polyplacophora (d) Scaphopoda

103 Connecting link between Annelida and Mollusca is
(a) *Neopilina* (b) *Peripatus*
(c) *Periplaneta* (d) *Limulus*

104 The most remarkable characteristic of *Neopilina* is
(a) presence of shell (b) presence of gill
(c) metameric segmentation (d) nephridia

105 Animals having ...... are termed as Polyplacophora.
(a) two shells (b) one shell
(c) many shells (d) eight shell pieces

106 Elephant's tusk shell belongs to the class
(a) Monoplacophora (b) Scaphopoda
(c) Gastropoda (d) Pelecypoda

107 In Scaphopoda, the foot is
(a) shoe-shaped (b) hatchet-shaped
(c) dome-shaped (d) boat-shaped

108 Class ...... includes snail, slug and limpets.
(a) Cephalopoda (b) Scaphopoda
(c) Gastropoda (d) Lamellibranchiata

109 The most important characteristic feature of gastropod molluscs is
(a) ventral muscular and massive foot
(b) vertebrate-like eyes
(c) head well-formed
(d) radula absent

110 Torsion is a characteristic in *Pila* belonging to class
(a) Pelecypoda (b) Scaphopoda
(c) Gastropoda (d) Amphineura

111 Consider the following statements with respect to *Pila*.
I. They have amphibious type of respiration.
II. They have 8-shaped nervous system.
Choose the correct option.
(a) Statement I is true, but II is false
(b) Statement I is false, but II is true
(c) Both statements I and II are true
(d) Both statements I and II are false

112 Hatchet-shaped foot, extending between the mantle lobes are present in
(a) Pelecypoda (b) Scaphopoda
(c) Gastropoda (d) Amphineura

113 Which of the following is a distinctive character of class-Bivalvia of Mollusca?
(a) Rudimentary head (b) Absence of gills
(c) Tentacles around the mouth (d) Presence of coiled shell

114 The animal causing holes in boats and ship, thus referred as threat to ship industry is
(a) *Pinctada* (b) *Neopilina* (c) *Doris* (d) *Teredo*

115 *Teredo* is
I. commonly known as shipworm.
II. destructive to wood in seawater.
III. a living fossil.
IV. commonly called as sea mouse.
Choose the correct option.
(a) I and II (b) I, II and III (c) II and IV (d) I and III

116 Glochidium and veliger both are the larval forms of
(a) *Pila* (b) *Nereis*
(c) *Asterias* (d) Bivalves

117 Pericardial gland of *Unio* is called
(a) Johnston's organ (b) organ of Bojanus
(c) Keber's organ (d) None of these

118 Which of the following do not have myogenic heart?
(a) Man (b) *Unio*
(c) Frog (d) All of these

119 Consider the following statements with respect to pearl.
I. The main component of pearl is $CaCO_3$.
II. The most precious Asian pearl is produced by *Pinctada.*

Choose the correct option.
(a) Statement I is true, but II is false
(b) Statement I is false, but II is true
(c) Both statements I and II are true
(d) Both statements I and II are false

120 Pearls are produced in an oyster around
(a) tears of sea mermaids falling into the oyster
(b) first drop of rain falling into the oyster in a particular month
(c) the eggs of oyster which fail to leave its body
(d) some external particles that get embedded in the skin of oyster

121 Sedentary filter feeding molluscs are
I. oyster II. pecten
III. mussels IV. *Chaetoderma*

Codes
(a) I, II and III (b) I and II
(c) II and IV (d) I and III

122 What is the function of statocyst of snail?
(a) Tasting (b) Locomotory
(c) Equilibrium (d) Reproductive

123 Which of the following molluscs were formerly used as currency (money) since ancient time?
(a) *Pila* (b) *Turbinella*
(c) *Octopus* (d) *Cypraea*

124 Which of the following cephalopods possesses an external shell?
(a) *Octopus* (b) *Sepia* (c) *Nautilus* (d) *Loligo*

125 Jet propulsion is the mode of locomotion in
(a) dogfish (b) starfish
(c) jellyfish (d) cuttlefish

126 Ammonites are the fossil shell remains of which class of Mollusca?
(a) Cephalopoda (b) Gastropoda
(c) Scaphopoda (d) Pelecypoda

127 Ink glands are found in
(a) sponge (b) starfish
(c) *Sepia* (d) jellyfish

128 Foot is present in the form of oral arms in
(a) Pelecypoda (b) Scaphopoda
(c) Cephalopoda (d) Gastropoda

129 The most evolved molluscs are
(a) cephalopods (b) gastropods
(c) scaphopods (d) None of these

130 Consider the following statements.
I. The eyes in cephalopods are comparable with the eyes of invertebrates.
II. In class-Scaphopoda, they have single gonad, whereas kidneys are paired.

Choose the correct option.
(a) Statement I is true, but II is false
(b) Statement I is false, but I is true
(c) Both statements I and II are true
(d) Both statements I and II are false

131 Which of the following is not correct about phylum—Mollusca?
(a) Mollusca is the second largest animal phylum
(b) In Mollusca, the body surface is distinctly marked out into segments
(c) Members of phylum—Mollusca are usually dioecious and oviparous
(d) In Mollusca, the development is indirect

132 The largest and heaviest living invertebrate is *A* belonging to the class *B*. Identify *A* and *B*.
(a) *A*-*Sepia* *B*-Cephalopoda
(b) *A*-*Architeuthis* *B*-Cephalopoda
(c) *A*-*Tridacna* *B*-Pelecypoda (bivalvia)
(d) *A*-*Lamellidens* *B*-Pelecypoda

133 The incorrect match amongst the following is
(a) Aplacophora – *Chaetoderma*
(b) Polyplacophora – *Chiton*
(c) Cephalopoda – *Dentalium*
(d) Gastropoda – *Achatina*

134 Match the following columns.

Column I (Mollusc scientific names)	Column II (Common names)
A. *Achatina*	1. Land snail
B. *Dentalium*	2. Squid
C. *Loligo*	3. Tusk shell
D. *Pila*	4. Apple snail

Codes

	A	B	C	D
(a)	1	2	3	4
(b)	3	2	4	1
(c)	2	4	3	1
(d)	1	3	2	4

135 The incorrect match among the following is
(a) Glochidium – Point of an arrow
(b) *Chiton* – Coat of mail shell
(c) *Cryptochiton* – Spiral shell
(d) *Dentalium* – Curved tubular shell

136 Which one of the following is unique to Mollusca?
(a) Nacre and radula
(b) Ommatophores
(c) Mantle and unsegmented soft body
(d) All of the above

Phylum–Echinodermata

137 The endoskeleton of echinoderms is unique with respect to other invertebrates by being ...... in origin.
(a) ectodermal
(b) mesodermal
(c) endodermal
(d) mesoendodermal

138 Consider the following statements.
I. Spiny skin, pedicellariae and water vascular system are characteristics of Echinodermata.
II. Trochophora larvae is not a characteristic of Echinodermata.
Choose the correct option.
(a) Statement I is true, but II is false
(b) Statement I is false, but II is true
(c) Both statements I and II are true
(d) Both statements I and II are false

139 The echinoderms exhibit ...... development.
(a) incomplete (b) direct
(c) indirect (d) parthenogenetic

140 The echinoderms have an enterocoelom made from the digestive tube which is lined by a
(a) flagellated perineum
(b) ciliated peritoneum
(c) ciliated mesoderm
(d) flagellated pericardium

141 Nervous system of echinoderms comprises
(a) radial nerve cord
(b) presence of brain
(c) lateral nerve cord
(d) chain of ganglia

142 In which class of Echinodermata, bipinnaria larva is found?
(a) Crinoidea (b) Pentameria
(c) Asteroidea (d) Holothuroidea

143 Blastopore of the embryo of starfish forms
(a) anus of larva (b) mouth of adult
(c) madreporite (d) polian vesicle

144 Consider the following statements with respect to *Asterias.*
I. Equal redial parts of starfish are actinomere.
II. They are ammonotelic.
III. Autonomy is followed by regeneration.
(a) I and II are correct
(b) I and III are correct
(c) Only II is correct
(d) I, II and III are correct

145 Which echinoderm among the following yields a calcium rich fertiliser in its dried and powdered form?
(a) Sea urchin (b) *Cucumaria*
(c) Starfish (d) *Holothuria*

146 Which of the following is not a part of ambulacral system of a starfish?
(a) Ring canal
(b) Madreporite
(c) Radial canal
(d) Excurrent canal

147 Pedicellariae of sea star is a pincer-like structure which
I. is used for biting and rasping.
II. helps in keep the body surface clean.
III. helps in cloacal respiration.
IV. acts as organ of offence.
Choose the correct option.
(a) I and II (b) I, II and III
(c) II and IV (d) None of these

148 Consider the following statements for brittle star.
I. Oral surface has mouth and madreporite.
II. Tubefeets are present with suckers orally placed in ambulacral grooves.
Choose the correct option.
(a) Statement I is true, but II is false
(b) Statement I is false, but II is true
(c) Both statements I and II are true
(d) Both statements I and II are false

149 *Echinus* (sea urchin)
I. is omnivorous.
II. has a masticatory apparatus called Aristotle's lantern.
III. has tubefeet without suckers.
Choose the correct option.
(a) I, II and III (b) II and III
(c) I and II (d) Only II

150 In Ophiuroidea, branched arms are seen in
(a) *Gorgonocephalus*
(b) *Clypeaster*
(c) *Salmacis*
(d) *Gorgonia*

151 Pick out the statement which is correct about *Echinus*.

(a) It is a herbivorous animal
(b) It moves with the help of tubefeet only
(c) Mouth is supported with masticatory (biting and chewing) apparatus
(d) The larval form is echinocardius

152 Select the statement which is false about the member of the class-Echinoidea.

(a) Tubefeet, arms and ambulacral grooves are absent
(b) One of the larval form is echinopluteus
(c) Sand dollar, heart urchin, sea lily without ciri are some members of class-Echinoidea
(d) Both is globular or disc-like

153 Which of the following feeds upon detritus and plankton?

(a) Sea urchin (b) Feather star
(c) Sea cucumber (d) Brittle star

154 Find which of the following is incorrectly paired.

(a) Ophiuroidea — Ophiopluteus
(b) Crinoidea — Bipinnaria
(c) Echinoidea — Pluteus
(d) Holothuroidea — Auricularia

155 ... has no arms, no spines but has an oral and aboral axis.

(a) Echinoidea (b) Holothuroidea
(c) Crinoidea (d) Asteroidea

156 Sedentary echinoderms are grouped under class-

(a) Crinoidea (b) Ophiuroidea
(c) Echinoidea (d) Holothuroidea

157 Match the following columns.

	Column I (Echinoderm common names)	Column II (Scientific names)
A.	Starfish	1. *Gorgonocephalus*
B.	Cake urchin	2. *Asterias*
C.	Feather star	3. *Clypeaster*
D.	Basket star	4. *Antedon*

Codes

	A	B	C	D
(a)	3	4	1	2
(b)	3	1	4	2
(c)	2	3	4	1
(d)	2	1	4	3

158 Choose the statement which is incorrect about reproduction in echinoderms.

(a) Fertilisation takes place externally in sea water
(b) Echinoderms may undergo asexual reproduction by self division
(c) The testes and ovaries are morphologically dissimilar
(d) Sexual dimorphism is absent in echinoderms

159 Which of the following is not true about echinoderms?

(a) An excretory system is absent
(b) Sexes are separate and reproduction is sexual
(c) Development is indirect with free-swimming larva
(d) Mouth is present on the upper (dorsal) side and anus on the lower (ventral) side

160 The porous plate present in the aboral side of starfish is

(a) umbelicus (b) genital pore
(c) madreporite (d) mouth

(B) Cumulative Exercise

(Assorted Questions of the Chapter for Advanced Level Practice)

1 Among the statements given below about *Nereis*, select the incorrect statements about the organism and mark the correct code.

I. The phenomenon of sexual maturation of *Nereis* is called heterotoky.
II. The pygidium of *Nereis* is called anal segment.
III. The body of the organism has two distinct regions.
IV. Fertilisation is external and occurs in sand.
V. Mesenteron of *Nereis* has oesophagus, stomach and intestine.

Codes

(a) I and III (b) II and V
(c) II, III and IV (d) I, II, III and IV

2 Among the statements given below about leech (*Hirudinaria granulosa*) find out the one, which are correct about the organism and select the appropriate code.

I. It is sanguivorous in habit.
II. It is a hermaphrodite.
III. It is narrowest near the anterior end and broadest near the posterior end.
IV. Its eggs are deposited in cocoons.

Codes

(a) I and IV (b) I and III
(c) Only I (d) I, II, III and IV

3 Identify the correct statements, which are the characteristic features of leech and select the appropriate code.

I. The body of *Hirudinaria granulosa* is soft, vermiform, elongated and dorsoventrally flat.
II. The segments 3rd-26th are further divided into rings or annuli.
III. The mouth of leeches is guarded by a structure called velum.
IV. The anterior and posterior suckers are meant for adhesion and locomotion.

Codes
(a) III and IV (b) I and IV
(c) II and IV (d) I, II, III and IV

4 In exoskeleton of arthropods,

I. chitinous plates present over the body are called sclerites.
II. sclerites are joined with the help of soft membranous structures called arthrodial membrane.
III. the dorsal sclerites are called tergum, ventral sclerites as sternum and lateral sclerites are called pleura.
IV. epidermis is present above the cuticle, called hypodermis.

Choose the correct option.
(a) Only IV (b) II and III
(c) I, II and IV (d) I, II and III

5 Which one of the following is a matching pair of an animal and a certain phenomenon it exhibits?

(a) *Chameleon* – Mimicry
(b) *Taenia* – Polymorphism
(c) *Pheretima* – Sexual dimorphism
(d) *Musca* – Complete metamorphosis

6 Which of the following have biting and chewing type of mouth parts?

I. *Cimex*
II. Larvae of silkmoth
III. Larvae of butterfly
IV. Grasshopper
V. Tse-tse fly

Codes
(a) II, IV and V (b) II, V and IV
(c) I, II and IV (d) III, IV and V

7 Match the following columns.

	Column I (Animals)	Column II (Characteristics)
A.	*Ophiothrix*	1. Has pair of respiratory tree in the coelom which opens into the rectum
B.	*Asterias*	2. Benthonic, carnivorous animal
C.	*Antedon*	3. It is also called feather star
D.	*Cucumaria*	4. It has ophiopluteus larval stage in its life cycle

Codes

	A	B	C	D		A	B	C	D
(a)	1	3	2	4	(b)	4	2	3	1
(c)	3	4	1	2	(d)	2	1	3	4

8 Which one of the following phyla is correctly matched with its two general characteristics?

(a) Chordata – Notochord at some stages and separate anal and urinary openings to the outside
(b) Mollusca – Normally oviparous and development through a trochophore or veliger larva
(c) Arthropoda – Body divided into head, thorax and abdomen; and respiration by tracheae
(d) Echinodermata – Pentamerous radial symmetry and mostly internal fertilisation

9 Which of the following traits is not a characteristic of echinoderms?

I. They have a complete digestive system with two separate openings.
II. They possess a complex haemal and perihaemal system for circulation.
III. Removal of nitrogenous waste takes place through cloacal aperture.
IV. Sense organs are poorly developed.

Codes
(a) I and II (b) II and III
(c) Only III (d) I and IV

10 Which among the following are the characteristic features of *Cucumaria*?

I. It is an orange coloured animal which lives in shallow tropical water.
II. It has twenty sets of feeding arms that far out into thorny tentacles.
III. Respiration occurs through system of tubule called respiratory tree.
IV. Its larval form is auricularia.

Codes
(a) I, II and III (b) I, III and IV
(c) II, III and IV (d) I and IV

11 Match the following columns.

Column I	Column II
A. Ommatidia	1. Reduced hindwings
B. Trochanter	2. For vision
C. Halteres	3. Forming exoskeleton
D. Sclerites	4. Present between coxa and femur

Codes

	A	B	C	D		A	B	C	D
(a)	4	3	1	2	(b)	1	2	3	4
(c)	2	4	1	3	(d)	3	1	2	4

12 Match the following columns.

Column I	Column II
A. *Pecten*	1. Used in India in an indoor game
B. *Chiton*	2. Presence of eight articulated, shell plates borne on the mantle dorsally
C. *Dentalium*	3. Commonly known as scallop
D. *Cypraea moneta*	4. Presence of filamentous tentacles called captacula

Codes

	A	B	C	D		A	B	C	D
(a)	3	2	4	1	(b)	1	2	3	4
(c)	3	1	2	4	(d)	2	4	1	3

13 Read the following statements regarding the class—Crinoidea of phylum—Echinodermata.

I. Ciliated ambulacral grooves are present on the aboral surface.
II. *Antedon* belonging to this class has a good power of regeneration.
III. The oral surface bears mouth and cloaca.
IV. The tubefeet lack suckers.

Choose the correct ones.

(a) I and II (b) III and IV (c) I, II and IV (d) II and III

14 The statements given below mention certain characteristic features of *Nereis*. Select the correct statements and mark the correct code.

I. The body of *Nereis* is divisible into head, trunk and pygidium.
II. Anus is mid-dorsal in position present on the 26th segment.
III. Prostomium bears a pair of short, cylindrical, sensory tentacles.
IV. It is an omnivorous annelid.

Codes

(a) I and II (b) II, III and IV
(c) II and IV (d) I and III

15 Which of the following is incorrect for excretory and respiratory structures of animal mentioned?

Animals	Excretory structures	Respiratory structures
(a) Insects	Malpighian	Trachea
(b) *Asterias*	Tubefeet	Papulae
(c) *Nereis*	Nephridia	Parapodia
(d) *Pila*	Metanephridia	Ctenidia

16 Read the statement about characteristic features of class—Monoplacophora.

I. Body is flat, bilaterally symmetrical and oval-shaped.
II. The animals bear nephridia and internal segmentation like annelids.
III. This is the only class among molluscs which includes monoecious species.
IV. The body is enclosed in a mantle which secretes a dome-shaped outer shell.

Choose the correct option.

(a) I, II and IV (b) I and II
(c) II and IV (d) I and III

17 Which of the following is properly matched?

Scientific names	Common names	Larval stages
(a) *Cucumaria*	Sea cucumber	Auricularia
(b) *Holothuria*	Sea urchin	Doliolaria
(c) *Ophiothrix*	The basket star	Dipleurula
(d) *Antedon*	Feather star	Bipinnaria

18 Which of the following is not correct matching of phylum and its three examples?

(a) Annelida – *Tubifex, Nereis, Hirudo*
(b) Arthropoda – *Limulus, Peripatus, Enterobius*
(c) Mollusca – *Chiton, Dentalium, Loligo*
(d) Echinodermata – *Echinus, Ophiothrix, Cucumaria*

19 Match the following columns.

Column I (Common names)	Column II (Scientific names)
A. Sea lemon	1. *Doris*
B. Sea mussel	2. *Mytilus*
C. Garden snail	3. *Helix*
D. Grey slug	4. *Limax*
E. Sea hare	5. *Aplysia*

Codes

	A	B	C	D	E
(a)	1	2	3	4	5
(b)	4	3	5	1	2
(c)	4	2	5	1	3
(d)	3	4	5	2	1

20 In the external apertures of body of *Hirudinaria granulosa*,

I. mouth is triradiate aperture, opens into preoral chamber of anterior sucker.
II. posterior sucker is formed by the complete fusion of last seven body segments.
III. anus is present in 31st segment, which is mid-dorsal in position.
IV. female genital pores are mid-ventral in position between 3rd and 4th annuli of 11th segment.

Codes

(a) I, II and III (b) I and II
(c) II and IV (d) III and IV

21 Which of the following statements are correct?

I. The respiratory trumpets in *Culex* are long and narrow, while the respiratory trumpets of *Anopheles* are short and broad.
II. The pupa of *Culex* is without any palmate hairs on the abdomen, while the pupa of *Anopheles* is with palmate hairs on the abdomen.
III. In the case of *Culex*, each paddle bears a long bristle, while in the case of *Anopheles* pupa, each paddle bears short bristle.
IV. The pupa of *Culex* is smaller and green, while that of *Anopheles* is larger and colourless.

Codes

(a) I, II and III (b) I and III
(c) II and IV (d) I and IV

22 Match the following columns.

Column I	Column II
A. Pollen basket	1. Butterfly
B. Pseudotracheae	2. *Laccifer lacca*
C. Shellac	3. Mosquito
D. Dutton's membrane	4. *Musca*
E. Well-developed galen	5. Worker bee

Codes

	A	B	C	D	E
(a)	3	2	4	5	1
(b)	1	5	3	4	2
(c)	4	5	3	2	1
(d)	5	4	2	3	1

23 Which of the following is true for the organism shown below.

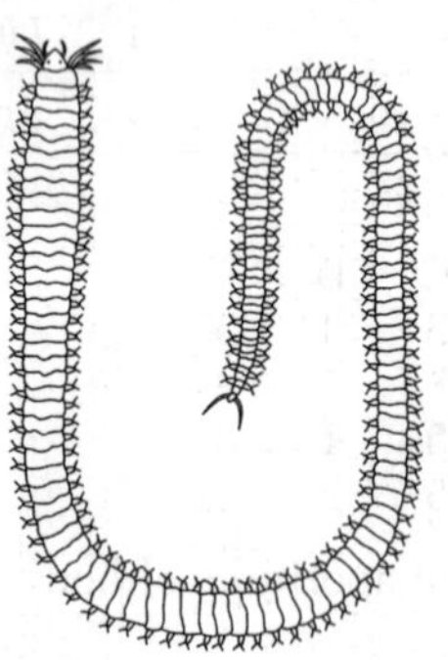

(a) Exhibits organ system level of body organisation
(b) Possess bilateral symmetry
(c) Is triploblastic, metamerically segmented and coelomate animal
(d) All of the above

24 Read the following statements.

I. Excretory organs of arthropods are Malpighian tubules and green glands.
II. Blood of arthropods is yellow.
III. In insects, respiratory gaseous exchange occurs through tracheae.
IV. *Bombyx* is commonly called silkworm.
V. King crab or horse-shoe crab is *Limulus.*

Which of the above statements are correct?

(a) I, III and IV (b) I, II and III
(c) I, III, IV and V (d) II, IV and V

25 Read the following statements with respect to molluscs.

I. Eyes of cephalopods resembles vertebrate eye.
II. Head is the exterior part of the body.
III. Possess circulatory system.
IV. Blood has haemocyanin is respiratory pigments.
V. Respiration is of anaerobic type.

Which of the above statements are correct?

(a) II and IV (b) I, III and IV
(c) II, III and V (d) IV and V

26 Identify *A, B, C* and *D* along with their phyla.

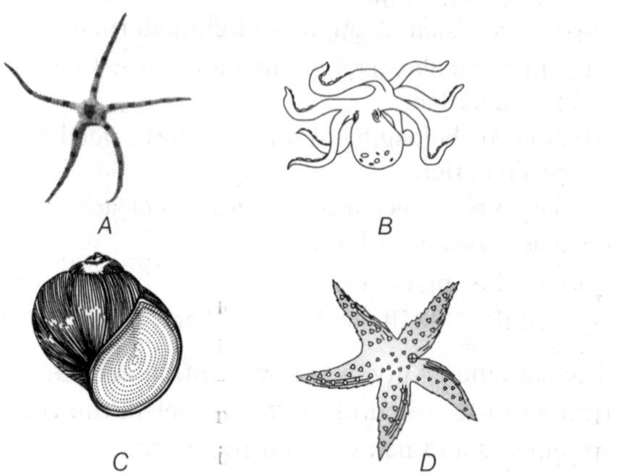

	A	B	C	D
(a)	*Ophiura,* Mollusca	*Octopus,* Mollusca	*Pila,* Echinodermata	*Asterias,* Echinodermata
(b)	*Ophiura,* Echinodermata	*Octopus,* Mollusca	*Pila,* Mollusca	*Asterias,* Echinodermata
(c)	*Ophiura,* Echinodermata	*Octopus,* Echinodermata	*Pila,* Mollusca	*Asterias,* Mollusca
(d)	*Ophiura,* Mollusca	*Octopus,* Echinodermata	*Pila,* Echinodermata	*Asterias,* Mollusca

27 Read the following statements with respect to molluscs.

I. A soft and spongy layer of skin forming a thin delicate covering over the visceral hump is called mantle or pallium.
II. *Octopus* possesses the calcareous shell secreted by mantle.
III. Molluscs can have gills or lungs.
IV. Ganglia is absent.
V. Direct development occurs in *Octopus, Sepia* and *Loligo.*

Which of the above statements are incorrect?

(a) II and IV (b) I, III and V
(c) I and III (d) III and IV

28 Echinodermata

I. is the largest animal phylum to lack any freshwater or terrestrial representatives.
II. constitutes the only major group of Deuterostome.
III. are diploblastic animals.
IV. have direct development without any larval stage.

Choose the correct option.

(a) I, II and III
(b) I and II
(c) II and IV
(d) I and III

29 Class—Hirudinea of Annelida

I. lacks parapodia and setae.
II. includes leeches which lead ectoparasitic life on vertebrate animals.
III. possesses clitellum as temporary organ.
IV. show external fertilisation.

Codes

(a) I, II and III
(b) II and IV
(c) III and IV
(d) I, III and IV

30 Which of the following are correctly matched?

I. Trophallaxis — Insect colony
II. *Limulus* — Class—Merostomata
III. Statocyst — Prawn
IV. Insect compound eye — Mosaic image in dim light

Codes

(a) I and II (b) I, II and III
(c) III and IV (d) II and IV

31 In Mollusca,

I. digestive tract is a complete and complex with ciliary tracts for sorting small particles.
II. respiration takes place by one to many uniquely structured ctenidia.
III. circulatory system is closed type and typically includes a dorsal heart.
IV. excretion takes place by flame cells on green glands.

Choose the correct option.

(a) Statements I and III are correct
(b) Statements I and II are correct
(c) Statements I, II and IV are correct
(d) Only statement IV is correct

32 Pick out the statement which is incorrect about Pelecypoda.

(a) *Pinctada vulgaris* is a sedentary mollusc which is reared artificially for its pearl oyster
(b) Keber's organ is a pericardial gland in *Unio* which helps in excretion
(c) *Unio* bears a myogenic heart
(d) The shell of *Unio* bears two halves called valves

33 Select the statement which is correct about the echinoderms.

I. Their body is polymorphic.
II. Their body bears jaw-like structures which are called oral arms.
III. All are marine with cellular level of organisation.
IV. They are triploblastic and coelomate animals.
V. This phylum appeared in early Cambrian period.

Codes

(a) I and II (b) I and IV
(c) III and IV (d) II and V

34 Which one of the following statements is true about an organism and its classification?

(a) Blue-green alga is a kind of fungus
(b) Sea horse is closely related to dolphin
(c) Maiden hair tree is a kind of angiosperm
(d) Sea lily is a kind of echinoderm

35 Consider the following statements.

I. In cockroaches and prawns, excretion of waste material occurs through Malpighian tubules.
II. Earthworms are hermaphrodites and yet cross-fertilisation takes place among them.

Choose the correct option.

(a) Statement I is true, but II is false
(b) Statement I is false, but II is true
(c) Both statements I and II are true
(d) Both statements I and II are false

36 Consider the following statements.

I. Clitellum is absent in *Nereis, Aphrodite* and *Chaetopterus.*
II. The members of class-Hoilothuroidea lack pedicellariae, ambulacral grooves and spines.
III. Holothuroidea has no arms, no spines but has an oral and aboral axis.
IV. Leech is with velum, suctorial mouth and haemocoelomic system.

Choose the option containing correct statements.

(a) I, II and III (b) I, II and IV
(c) II and III (d) I, II, III and IV

Assertion and Reason

Direction (Q. Nos. 37-52) *In each of the following questions, a statement of Assertion (A) is given followed by corresponding statement of Reason (R). Of the statements, mark the correct answer as*

(a) Both A and R are true and R is the correct explanation of A
(b) Both A and R are true, but R is not the correct explanation of A
(c) A is true, but R is false
(d) A is false, but R is true
(e) Both A and R are false

37 **Assertion** (A) Metamerism is the characteristic of Annelida.

Reason (R) It is a type of body segmentation.

38 **Assertion** (A) Blood is red in Annelida.

Reason (R) RBCs are absent in them.

39 **Assertion** (A) Chloragogen cells behave like vertebrate liver.

Reason (R) They are present in Annelida.

40 **Assertion** (A) Spermathecae help in sperm transfer.

Reason (R) They are the main part of reproductive system of Annelida.

41 **Assertion** (A) Epitoky is seen in *Nereis*.

Reason (R) Individual showing epitoky is called *Heteronereis*.

42 **Assertion** (A) Haemocoel in Arthropoda is not lined by the mesodermal epithelium.

Reason (R) Arthropods possess only true coelom.

43 **Assertion** (A) Prawn have walking legs.

Reason (R) It can swim with the help of walking legs.

44 **Assertion** (A) *Pila* has a dual mode of respiration.

Reason (R) It leads an amphibious life.

45 **Assertion** (A) Keber's organ in *Unio* is situated in front of the pericardium.

Reason (R) It has an important role in circulation.

46 **Assertion** (A) Point of an arrow larva is seen in some molluscs.

Reason (R) It is the characteristic of freshwater mussel.

47 **Assertion** (A) *Pila* is ammonotelic.

Reason (R) It excretes uric acid.

48 **Assertion** (A) Tubefeet are characteristic organs of Echinodermata.

Reason (R) They have only role in respiration.

49 **Assertion** (A) Water vascular system is the characteristic of echinoderms.

Reason (R) Main function of water vascular system is locomotion.

50 **Assertion** (A) Ossicles around the mouth of sea urchin from the Aristotle's lantern.

Reason (R) Teeth of Aristotle's lantern are analogous to human teeth, as both are masticatory in function.

51 **Assertion** (A) In many gastropods, the anus and the mantle cavity are placed anteriorly above the head.

Reason (R) During embryonic development in many gastropods, one side of the visceral mass grows faster than the other side. This uneven growth rotates the visceral organs up to 180° in many gastropods.

52 **Assertion** (A) Arthropoda is the most successful of all known animal groups.

Reason (R) It exhibits the greatest adaptive radiations and have adapted to diverse habitats.

(C) Medical Entrances' Gallery

(Collection of Questions Asked in NEET & in Various Medical Entrance Exams)

1 Match the following columns and select the correct option.

Column I	Column II
A. Gregarious polyphagous pest	1. *Asterias*
B. Adult with radial symmetry and larva with bilateral symmetry	2. Scorpion
C. Book lungs	3. *Ctenoplana*
D. Bioluminescence	4. *Locusta*

Codes [NEET 2020]

	A	B	C	D
(a)	4	1	2	3
(b)	3	2	1	4
(c)	2	1	3	4
(d)	1	3	2	4

2 Which of the following options does correctly represent the characteristic features of phylum-Annelida? [NEET 2020]

(a) Triploblastic, unsegmented body and bilaterally symmetrical
(b) Triploblastic, segmented body and bilaterally symmetrical
(c) Triploblastic, flattened body and acoelomate condition
(d) Diploblastic, mostly marine and radially symmetrical

3 Which of the following animals are true coelomates with bilateral symmetry? [NEET 2020]

(a) Adult echinoderms
(b) Aschelminthes
(c) Platyhelminthes
(d) Annelids

4 Match the following organisms (in Column I) with their respective characteristics (in Column II).

Column I		Column II
A. *Pila*	1.	Flame cells
B. *Bombyx*	2.	Comb plates
C. *Pleurobrachia*	3.	Radula
D. *Taenia*	4.	Malpighian tubules

Codes **[NEET 2020]**

	A	B	C	D
(a)	3	4	2	1
(b)	2	4	3	1
(c)	3	2	4	1
(d)	3	2	1	4

5 Choose the correct option for larva of housefly. **[JIPMER 2019]**

(a) Grub (b) *Cypris*
(c) Maggot (d) Wriggler

6 Most advanced invertebrates are **[JIPMER 2019]**

(a) Arthropoda (b) Annelida
(c) Mollusca (d) Cephalopoda

7 Radula is a part of which animal? **[JIPMER 2019, WB JEE 2013, 10]**

(a) Mollusca (b) Porifera
(c) Coelenterata (d) Annelida

8 Match the following genera (in Column I) with their respective phylum (in Column II).

Column I	Column II
A. *Ophiura*	1. Mollusca
B. *Physalia*	2. Platyhelminthes
C. *Pinctada*	3. Echinodermata
D. *Planaria*	4. Coelenterata

Select the correct option from the following codes given below. **[NEET 2019]**

	A	B	C	D
(a)	4	1	3	2
(b)	3	4	1	2
(c)	1	3	4	2
(d)	3	4	2	1

9 Which one of the following animals does not undergo metamorphosis? **[NEET 2018]**

(a) Moth (b) Tunicate
(c) Earthworm (d) Starfish

10 Torsion of visceral mass is seen in animals belonging to class **[JIPMER 2017]**

(a) Cephalopoda (b) Scaphopoda
(c) Amphineura (d) Gastropoda

11 Match the larval stages (in Column I) with their corresponding animals (in Column II) and select the correct option from the codes given below.

Column I		Column II
A. Planula	1.	*Holothuria* (sea cucumber)
B. Trochophore	2.	Hermit crab
C. Glochidium	3.	*Obelia*
D. Glaucothoe	4.	*Nereis*
	5.	*Unio*

Codes **[AIIMS 2017]**

	A	B	C	D
(a)	1	3	5	4
(b)	3	4	5	2
(c)	1	2	3	4
(d)	3	1	2	5

12 Which one of the following features is not present in the phylum—Arthropoda? **[NEET I 2016, J&K CET 2012, DUMET 2011]**

(a) Metameric segmentation
(b) Parapodia
(c) Jointed appendages
(d) Chitinous exoskeleton

13 Match Column I with Column II for housefly classification and select the correct option using the codes given below.

Column I	Column II
A. Family	1. Diptera
B. Order	2. Arthropoda
C. Class	3. Muscidae
D. Phylum	4. Insecta

Codes **[NEET-II 2016]**

	A	B	C	D
(a)	3	1	4	2
(b)	3	2	4	1
(c)	4	3	2	1
(c)	4	2	1	3

14 Choose the correct set of disease transmitting arthropods. **[Chhattisgarh PMT 2015]**

(a) tse-tse fly, housefly, rat flea, sandfly
(b) *Anopheles*, louse, housefly, termite
(c) cockroach, louse, termite, housefly
(d) rat flea, butterfly, housefly, tse-tse fly

15 Which is gynandromorph type of animal? **[Gujarat CET 2015]**

(a) *Drosophila*
(b) Beetles
(c) Silkworms
(d) All of the above

16 The chitinous exoskeleton of arthropods is formed by the polymerisation of **[CBSE AIPMT 2015]**
(a) keratin sulphate and chondroitin sulphate
(b) D-glucosamine
(c) N-acetyl glucosamine
(d) lipoglycans

17 The first body segment of earthworm is **[J&K CET 2015]**
(a) peristome (b) peristomium
(c) protostomium (d) protostome

18 Match the following columns.

Column I	Column II
A. *Arenicoloa*	1. Terrestrial leech
B. *Hirudinaria*	2. Marine leech
C. *Pontobdella*	3. Lungworm
D. *Haemadipsa*	4. Freshwater leech

[EAMCET 2015]

Codes

	A	B	C	D
(a)	1	2	3	4
(b)	3	4	2	1
(c)	2	3	4	1
(d)	1	3	4	2

19 Which of the following matching pairs is incorrect? **[WB JEE 2015]**
(a) Mollusca — Pseudocoel
(b) Cnidaria — Nematocyst
(c) Annelida — Chloragogen cells
(d) Echinodermata — Water vascular system

20 Shell is spiral in **[Chhattisgarh PMT 2015]**
(a) Pelecypoda
(b) Gastropoda
(c) Cephalopoda
(d) Scaphopoda

21 Which one belongs to class—Polyplacophora? **[AMU 2015]**
(a) *Chiton* (b) *Patella*
(c) *Pila* (d) *Murex*

22 The molluscan without a shell is **[EAMCET 2015]**
(a) *Sepia* (b) *Octopus*
(c) *Loligo* (d) *Nautilus*

23 Match and choose the correct pair. **[EAMCET 2015]**
I. Cephalopoda — Closed circulatory system— *Pinctada*
II. Echinoidea — Aristotle's lantern — Sea urchin
III. Tentaculata — Cnidoblasts —*Pleurobrachia*
IV. Hexactinellida—Siliceous spicules—Venus flower basket

Codes
(a) II and IV (b) I and IV
(c) III and IV (d) I and III

24 Read the following statements regarding echinoderms and choose the correct option.
I. All are marine with organ system level of organisation.
II. Adults are bilaterally symmetrical.
III. They are dioecious.
IV. Fertilisation is internal and indirect development is observed.
V. Triploblastic and acoelomate animals.

Codes **[Kerala CEE 2015]**
(a) I and III (b) Only V (c) I, III and V (d) I and V
(e) I and II

25 This animal has mouth on lower side and anus on the upper side. **[Gujarat CET 2015]**
(a) *Limulus* (b) *Buthus*
(c) *Daphnia* (d) Sea urchin

26 With reference to phylum—Echinodermata, identify the classes which have pedicellariae. **[EAMCET 2014]**
(a) Crinoidea and Holothuroidea
(b) Holothuroidea and Echinoidea
(c) Asteroidea and Echinoidea
(d) Ophiuroidea and Holothuroidea

27 Which one is an example of living fossil? **[WB JEE 2014]**
(a) Coral (b) Ascidia
(c) *Octopus* (d) King crab

28 Juvenile hormone in insect is released from **[WB JEE 2014]**
(a) protocerebrum (b) corpora cardiaca
(c) corpora allata (d) thoracic gland

29 Pentaradial symmetry is found in **[Odisha JEE 2013]**
(a) Echinodermata (b) Annelida
(c) Porifera (d) Arthropoda

30 In Polychaeta, the setae are **[NEET, AFMC 2013]**
(a) numerous
(b) singly arranged in bundles
(c) radially arranged in bundles
(d) fused

31 Veliger larva occurs in phylum **[UP CPMT 2013, 2009]**
(a) Mollusca (b) Echinodermata
(c) Arthropoda (d) Cnidaria

32 **Assertion** (A) Open circulatory system is found in most arthropods.

Reason (R) Arthropods contain haemolymph, which directly bathes internal tissues and organs. **[AIIMS 2013]**
(a) Both A and R are true and R is the correct explanation of A
(b) Both A and R are true, but R is not the correct explanation of A
(c) A is true, but R is false
(d) Both A and R are false

33 One of the representative of phylum-Arthropoda is **[NEET 2013, Rajasthan PMT 2011]**
(a) cuttle fish (b) silverfish
(c) puffer fish (d) flying fish

34 Which group of animals belongs to the same phylum? **[NEET 2013, Odisha JEE 2010]**
(a) Earthworm, pinworm, tapeworm
(b) Prawn, scorpion, *Locusta*
(c) Sponge, sea anemone, starfish
(d) Malarial parasite, *Amoeba*, mosquito

35 What is the similarity between housefly, mosquito and cockroach? **[Manipal 2012]**
(a) All have 13-chambered heart
(b) All have 4 pairs of legs
(c) All have closed circulatory system
(d) All belong to class–Insecta

36 The stages between larval moults in an insect are called **[J&K CET 2012]**
(a) pupae (b) grubs
(c) caterpillars (d) instars

37 Which one of the following exhibits concentric 'tube within tube' plan? **[Bihar BCECE 2012]**
(a) Cnidaria (b) Annelida
(c) Platyhelminthes (d) Nematoda

38 *Pheretima* and its close relatives derive nourishment from **[CBSE AIPMT 2012]**
(a) sugarcane roots
(b) decaying fallen leaves and soil
(c) soil insects organic matter
(d) small pieces of fresh fallen leaves of maize

39 Glochidium larva is absent in **[UP CPMT 2012]**
(a) *Unio* (b) *Pinctada*
(c) *Sepia* (d) None of these

40 The adult animal in this phylum is radially symmetrical, but its larva exhibits bilateral symmetry. **[Karnataka CET 2012]**
(a) Protozoa (b) Echinodermata
(c) Coelenterata (d) Arthropoda

41 The members of which of the following phyla are exclusively marine? **[Odisha JEE 2012, 2006]**
(a) Arthropoda
(b) Echinodermata
(c) Annelida
(d) Mollusca

42 Hemimetabolus insect, which is nocturnal and sanguivorous is **[EAMCET 2012]**
(a) *Cimex* (b) *Musca*
(c) *Anopheles* (d) *Apis*

43 Study the lists and choose the correct ones.

	A	*B*	*C*
I.	Sea urchin	Aristotle's lantern	Echinoidea
II.	Sea cucumber	Respiratory tree	Holothuroidea
III.	Sea biscuit	Anus is absent	Asteroidea
IV.	Sea star	Pedicelleriae with two jaws	Ophiuroidea

[EAMCET 2012]

Codes
(a) I and II (b) II and III (c) I and III (d) II and IV

44 Which of the following is a crustacean? **[Gujarat CET 2011]**
(a) Prawn (b) Snail
(c) Sea anemone (d) *Hydra*

45 Open circulatory system occurs in **[Odisha JEE 2011]**
(a) earthworm (b) snail
(c) cockroach (d) Both (b) and (c)

46 Which one is not an insect? **[Rajasthan PMT 2011]**
(a) Spider (b) *Locusta* (c) Mosquito (d) Ant

47 A female *Anopheles* mosquito can be recognised as its **[WB JEE 2011]**
(a) proboscis and palpi are long and more or less of equal length
(b) proboscis long and palpi short
(c) proboscis short and palpi long
(d) Both proboscis and palpi short

48 Which of the following have the highest number of species in nature? **[CBSE AIPMT 2011]**
(a) Insects (b) Birds
(c) Angiosperms (d) Fungi

49 The arthropods exoskeleton is composed of **[DUMET 2011]**
(a) several kinds of polysaccharides
(b) layers of proteins and a polysaccharide called chitin
(c) several kinds of proteins
(d) single complex protein called arthropodin

50 Which one of the following does not have an excretory system? **[Kerala CEE 2011]**
(a) *Myxine* (b) *Carcharodon*
(c) *Balanoglossus* (d) *Asterias*
(e) *Catla*

51 Tubefeet are the locomotory organs of **[Odisha JEE 2010]**
(a) Platyhelminthes (b) Echinodermata
(c) Mollusca (d) Arthropoda

52 Water vascular system is found in **[WB JEE 2010]**
(a) sea anemone (b) sea pen
(c) sea cucumber (d) sea horse

ANSWERS

Check Point 9.1

1. (b)	2. (b)	3. (b)	4. (c)	5. (b)	6. (b)

Check Point 9.2

1. (a)	2. (b)	3. (a)	4. (a)	5. (c)	6. (c)

Check Point 9.3

1. (c)	2. (d)	3. (c)	4. (d)	5. (c)	6. (d)

Check Point 9.4

1. (c)	2. (c)	3. (b)	4. (d)	5. (d)	6. (a)

(A) Topical Exercise

1. (d)	2. (d)	3. (b)	4. (b)	5. (d)	6. (d)	7. (c)	8. (b)	9. (c)	10. (c)
11. (b)	12. (c)	13. (b)	14. (b)	15. (c)	16. (a)	17. (d)	18. (d)	19. (c)	20. (a)
21. (d)	22. (d)	23. (d)	24. (b)	25. (a)	26. (d)	27. (c)	28. (b)	29. (b)	30. (b)
31. (c)	32. (c)	33. (d)	34. (a)	35. (b)	36. (a)	37. (c)	38. (c)	39. (c)	40. (c)
41. (c)	42. (d)	43. (c)	44. (d)	45. (b)	46. (b)	47. (d)	48. (a)	49. (b)	50. (d)
51. (a)	52. (d)	53. (c)	54. (d)	55. (c)	56. (c)	57. (b)	58. (c)	59. (c)	60. (c)
61. (b)	62. (b)	63. (a)	64. (c)	65. (a)	66. (b)	67. (c)	68. (d)	69. (b)	70. (a)
71. (c)	72. (c)	73. (a)	74. (c)	75. (d)	76. (d)	77. (c)	78. (c)	79. (a)	80. (d)
81. (b)	82. (a)	83. (d)	84. (d)	85. (d)	86. (d)	87. (b)	88. (c)	89. (d)	90. (b)
91. (a)	92. (d)	93. (a)	94. (b)	95. (c)	96. (a)	97. (c)	98. (d)	99. (d)	100. (c)
101. (d)	102. (b)	103. (a)	104. (c)	105. (d)	106. (b)	107. (d)	108. (c)	109. (a)	110. (c)
111. (c)	112. (a)	113. (a)	114. (d)	115. (a)	116. (d)	117. (c)	118. (b)	119. (c)	120. (d)
121. (a)	122. (c)	123. (d)	124. (c)	125. (d)	126. (a)	127. (c)	128. (c)	129. (a)	130. (b)
131. (b)	132. (b)	133. (c)	134. (d)	135. (c)	136. (d)	137. (b)	138 (c)	139. (c)	140. (b)
141. (a)	142. (c)	143. (a)	144. (d)	145. (c)	146. (d)	147. (c)	148 (a)	149. (a)	150. (a)
151. (b)	152. (c)	153. (c)	154. (b)	155. (b)	156. (a)	157. (c)	158 (c)	159. (d)	160. (c)

(B) Cumulative Exercise

1. (a)	2. (d)	3. (d)	4. (d)	5. (d)	6. (a)	7. (b)	8. (b)	9. (c)	10. (b)
11. (c)	12. (a)	13. (c)	14. (d)	15. (b)	16. (a)	17. (a)	18. (b)	19. (a)	20. (b)
21. (b)	22. (d)	23. (d)	24. (c)	25. (b)	26. (b)	27. (a)	28. (b)	29. (a)	30. (b)
31. (b)	32. (c)	33. (b)	34. (d)	35. (b)	36. (d)	37. (b)	38. (b)	39. (b)	40. (b)
41. (b)	42. (c)	43. (b)	44. (a)	45. (c)	46. (b)	47. (b)	48. (c)	49. (b)	50. (b)
51. (a)	52. (a)								

(C) Medical Entrances' Gallary

1. (a)	2. (b)	3. (d)	4. (a)	5. (c)	6. (d)	7. (a)	8. (b)	9. (c)	10. (d)
11. (b)	12. (b)	13. (a)	14. (a)	15. (d)	16. (c)	17. (b)	18. (b)	19. (a)	20. (b)
21. (a)	22. (b)	23. (a)	24. (a)	25. (d)	26. (c)	27. (d)	28. (c)	29. (a)	30. (a)
31. (a)	32. (a)	33. (b)	34. (b)	35. (d)	36. (d)	37. (b)	38. (b)	39. (c)	40. (b)
41. (b)	42. (a)	43. (a)	44. (a)	45. (d)	46. (a)	47. (a)	48. (a)	49. (b)	50. (d)
51. (b)	52. (c)								

Explanations

(A) Topical Exercise

7 (*c*) Statement in option (c) is incorrect because

Nephridia are present as excretory organs, but these are ectodermal in origin.

19 (*c*) Statement in option (c) is incorrect and can be corrected as

The class-Polychaeta is divided into two subclasses-Errantiata (free moving forms) and Sedentaria (sedentary or tube dwelling forms).

34 (*a*) Crustacea and Insecta are the classes of subphylum-Mandibulata while Merostomata and Arachnida are class of subphylum-Chelicerata.

41 (*c*) Johnston's organ is associated with antenna of mosquito, these are collection of sensory cells which perceives sound vibrations.

75 (*d*) Both mosquito and cockroach are nocturnal. Other characters are not common because

- Mosquito feed on liquid food therefore, it has sucking type mouth parts while cockroach have biting and chewing type mouth parts.
- The pupal stage is found in mosquito only while cockroach have nymph.
- Members of phylum-Arthropoda have compound eyes.

80 (*d*) Option (d) is incorrect match and can be corrected as *Apis indica* is the Indian honeybee. Rest other options are correct match.

82 (*a*) Statements I and II are correct, while III and IV are incorrect and can be corrected as

Respiratory trumpets are the tubular structure present in pupa stage of mosquitoes. These are not blood sucker of vertebrates like *Pediculus* (head louse). Termites are wood shredders, they feed on wood.

93 (*a*) Statement I is true, but II is false. Incorrect statement can be corrected as

The mantle wall of Mollusca secretes shell of calcium carbonate that protects and supports their soft tissues.

121 (*a*) Oyster, pecten and mussels are the sedentary filter feeder molluscs. They feed by straining suspended matter and food particles from water.

130 (*b*) Statement I is false, but II is true. Statement I can be corrected as

The eyes in cephalopods are comparable with the eyes of vertebrates as they consist of an iris, a circular lens, vitreous cavity and pigment cell like vertebrates.

131 (*b*) Statement in option (b) is incorrect and can be corrected as

In Mollusca the body is divided into head, muscular foot and visceral hump.

133 (*c*) Matched pair in option (c) is incorrect and can be corrected as

Dentalium belong to class-Scaphopoda.

135 (*c*) Option (c) is incorrect match and it can be corrected as

Spiral shell (*Spirula*) belong to class-Cephalopoda.

147 (*c*) Pedicellariae of a sea star is a pincer-like structure which help in protection, hence act as offensive organ. It also helps in keeping the body surface clean.

148 (*a*) Statement I is true, but II in incorrect and it can be corrected as

Brittle star have tube feet present in two rows, but without suckers. Ambulacral grooves are also absent in them.

152 (*c*) Sand dollar, heart urchin are the example of class-Echinoidea, but sea lily is an example of subphylum- Pelmatozoa and class-Crinoidea.

154 (*b*) Option (b) is incorrectly paired. It can be corrected as

The larval form of class-Crinoidea is called doliolaria.

(B) Cumulative Exercise

1 (*a*) Statements I and III are incorrect and can be corrected as

The phenomenon of sexual maturation of *Nereis* in called *Heteronereis.* The body of *Nereis* is divided into three regions head, trunk and pygidium.

4 (*d*) Statements I, II and III are correct only statement IV is incorrect and can be corrected as

Arthropods do not have hypodermis on their epidermis.

6 (*a*) Larvae of silkmoth, grasshopper and tse-tse fly have biting and chewing type of mouth part. Larvae of butterfly have sucking type which help to draw nectar from the flowers. *Cimex* have pointed beak or rostrum which is used for piercing and sucking blood.

9 (*c*) Statement III is incorrect and can be corrected as

In echinoderms, removal of nitrogenous waste takes place through their body surface.

Rest statements are correct.

13 (*c*) Statements I, II, IV are correct whereas III is incorrect. Incorrect statement can be corrected as

Cloaca is not present in Crinoidea class. The oral surface instead bears mouth and anus.

20 (*b*) Statements I and II are correct while statements III and IV are incorrect and can be corrected as

Male and female genital pores are present between 2nd and 3rd annuli of 10th and 11th segments.

21 (*b*) Statements I and III are correct while statements II and IV are incorrect and can be corrected as

Pupa of *Culex* has palmate hairs on the abdomen and are large and greyish while the pupa of *Anopheles* are greenish in colour.

25 (*b*) Statements I, III and IV are correct as

Eyes of cephalopods resembles to the eye of vertebrate as they have iris, lens, etc. They posses circulatory system, the blood has copper containing blue respiratory pigment called haemocyanin. Other statements are incorrect and can be corrected as

Head of Mollusca is the anterior part of body which bears mouth, eye and tentacles.

Respiration is of aerobic type in mollusca.

27 (*a*) Statements II and IV are incorrect and can be corrected as

The calcareous shell is absent in *Octopus.*

The nervous system consist of paired ganglia.

28 (*b*) Statements I and II are correct while III and IV are incorrect and can be corrected as

Echinodermata are triploblastic animals and have indirect development with larval stage. The larva of echinoderms show bilateral symmetry while the adult shows radial symmetry.

30 (*b*) I, II and III are correctly matched while IV is incorrectly matched and can be corrected as

Insects have compound eyes, which form mosaic vision in bright light.

31 (*b*) Statements I and II are correct while statements III and IV are incorrect and can be corrected as

The circulatory system in Mollusca is open type except in the cephalopoda and *Sepia*.

Excretion occur by paired sac-like kidney called organ of Bojanus (in *Unio*). They also contain Keber's organ or pericardial gland for excretion.

35 (*b*) Statement I is false, but II is true. False statement can be corrected as

Malpighian tubules are excretory structures in most insects, but in crustaceans such as prawn, green glands are the excretory structure.

37 (*b*) Both A and R are true, but R is not the correct explanation of A.

The body of annelids is divided into segments called metameres, externally by ring-like grooves and internally by vertical partitions called septa. This phenomenon is called metameric segmentation. Phylum– Annelida represents the first group of metazoan animals developing a true coelom with metameric segmentation.

38 (*b*) Both A and R are true, but R is not the correct explanation for A.

In Annelida, blood is red due to the presence of haemoglobin or erythrocruorin dissolved in plasma. RBCs are absent in them. Blood corpuscles are colourless.

39 (*b*) Both A and R are true, but R is not the correct explanation of A.

Chloragogen cells are present in the coelomic fluid of earthworm. These cells are the centres of synthesis of glycogen and fat, deamination of excess amino acids and formation of urea. Because of the connection with storage and synthesis of glycogen and fat, deamination and urea formation, chloragogen cells behave like 'vertebrate liver'.

40 (*b*) Both A and R are true, but R is not the correct explanation of A.

In Annelida, four pairs of flask-shaped sacs, each with a diverticulum for storage of sperms and large ampulla for their nourishment is present. Spermathecae occur in 6th-9th segments. They receive sperms during copulation.

As cross-fertilisation occurs in earthworm, the sperms of one worm get transferred to spermathecae of the other.

41 (*b*) Both A and R are true, but R is not the correct explanation of A.

In *Nereis*, most of the posterior segments are filled with gametes, constituting the sexual region of worm or epitoke. Few anterior segments do not take part in gamete formation, these are the asexual region or atoke. Sexually mature worm, with these two regions is known as *Heteronereis* and the phenomenon involving transformation of non-sexual individual into sexual individual is referred to as epitoky.

42 (*c*) A is true, but R is false because

Arthropods have a much reduced coelomic cavity. The space between the gut and body wall is mostly occupied by muscles and organs with blood containing interspaces. These spaces together form the haemocoel, which is not a true coelom as it is not lined by the mesodermal epithelium. However, greatly reduced true coelom exists in the form of a number of separate spaces, enclosing excretory and genital organs.

43 (*b*) Both A and R are true, but R is not the correct explanation of A.

The prawn crawls at the bottom of the river or pond by means of its walking legs. It can swim forward in a leisurely manner by beating its swimmerets or the abdominal appendages.

44 (*a*) Both A and R are true and R is the correct explanation of A.

Pila is adapted to lead an amphibious life, i.e. it lives both in land and water. The animal possesses a gill or ctenidium for aquatic respiration. This respiration takes place when the animal is at the bottom or attached to the plants and weeds in water. Aerial respiration is observed when the animal is on land.

45 (*c*) A is true, but R is false because

The Keber's organ or the pericardial gland is a large reddish-brown, glandular mass situated in front of the pericardium.

It helps in excretion, discharging waste products into the pericardial cavity to be finally eliminated through the kidneys.

46 (*b*) Both A and R are true, but R is not the correct explanation of A.

In molluscs, some freshwater mussels have the characteristic glochidium larva, which is also known as point of an arrow. It is so named because of its minute size (0.1-0.4 mm). The glochidium is a special microscopic larval stage of larger freshwater mussels.

47 (*b*) Both A and R are true, but R is not the correct explanation of A.

During aquatic phase *Pila* excretes ammonia, but during terrestrial phase it excrete uric acid. So, it is both ammonotelic as well as uricotelic.

48 (*c*) A is true, but R is false because

Each ambulacral groove of echinoderms contains two double rows of short, tubular retractile projection, called as podia or tube feet, that end in suckers. Tube feet are characteristic organs of echinoderms serving functions like locomotion, capturing of food, respiration, etc.

49 (*b*) Both A and R are true, but R is not the correct explanation of A.

Water vascular system or ambulacral system is a unique system of echinoderms, which helps mainly in locomotion. It is in fact, a modified part of coelom consisting of a system of canals containing sea water and amoeboid corpuscles. Tube feet of this system also help in anchoring the body to substratum, in capturing and ingesting the food and respiration.

50 (*b*) Both A and R are true, but R is not the correct explanation of A.

Aristotle's lantern is a group of ossicles around the mouth that has been highly modified for feeding mechanism in sea urchin. It is masticatory in function like human teeth.

51 (*a*) Both A and R are true and R is the correct explanation of A.

During development in most gastropods, because of torsion, the animal's anus and mantle cavity, both positioned above the head.

Since, one side of visceral mass grows more quickly than other side. Uneven growth rotates visceral mass which ends up 180° rotation relative to head.

52 (*a*) Both A and R are true and R is the correct explanation of A. The success of arthropods is largely due to their unique cuticle, segmentation and jointed appendages. Cuticle being tough, protect the arthropods against predators and forms jointed exoskeleton for muscle attachment. Being light, it does not add to body weight. Being impervious to water, it checks to dry up the body. Group of segments and jointed appendages have become specialised for a great variety of functions.

(C) Medical Entrances' Gallery

2 (*b*) Option (b) correctly represents the characteristic features of phylum-Annelida.

Animals belonging to phylum—Annelida are triploblastic, bilaterally symmetrical and metamerically segmented. They exhibit organ system level of body organisation with presence of coelom. They may be aquatic (marine and freshwater) or terrestrial, free-living and sometimes parasitic.

3 (*d*) Annelids are true coelomates with bilateral symmetry. They exhibit organ system level of body organisation with true coelom. They are triploblastic, metamerically segmented and coelomate animals, e.g. earthworm.

5 (*c*) Larva of housefly is called maggot. Larvae of certain beetles and other insects are called grub. Cypris is a larval form of class-Crustacea. Wriggler is a larval form of mosquitoes.

6 (*d*) Cephalopod molluscs such as *Octopus*, cuttle fish and squid are believed to have the most advanced cognitive behaviours of all invertebrates, rivaling and abilities of many vertebrates. Cephalopod class of molluscs is considered as the most intelligent invertebrates.

7 (*a*) Radula is a part of Mollusca group of animals. It is a rasping organ of molluscs situated in a sac on the underside of buccal cavity. It is used for tearing plant material by rubbing it against the hardened surface of the mouth.

9 (*c*) All the given animals except earthworm undergo metamorphosis. Earthworm exhibits direct development where no larval stage is involved. Metamorphosis is usually seen in animals exhibiting indirect development involving a larval stage which later transforms into an adult. Larval form of moth is caterpillar and that of tunicates is tadpole. In starfish, bipinnaria larva occurs.

10 (*d*) Animals of class—Gastropoda of phylum—Mollusca undergo twisting or torsion of the visceral mass during development, leading to the conversion of symmetrical embryo into an asymmetrical adult.

19 (*a*) Option (a) is incorrect matching pair and can be corrected as Molluscs are coelomate animals. The body cavity is a true coelom which is often reduced to haemocoel and surround the heart, nephridia and gonads.

Rest options are correctly matching pairs.

20 (*b*) In Gastropoda, the shell is univalve and spirally coiled due to torsion.

26 (*c*) Pedicellariae are small wrench or claw-shaped structures commonly found in echinoderms, particularly in class—Asteroidea, e.g. starfishes and Echinoidea, e.g. sea urchins.

28 (*c*) Juvenile hormone in insects is secreted by corpora allata. This hormone regulates morphogenesis and promote metamorphosis (development of larvae into adult through pupal stage).

29 (*a*) Pentaradial symmetry is found in Echinodermata, whereas phylum—Annelida and Arthropoda have bilateral symmetry and Porifera has radial symmetry.

30 (*a*) Setae in Polychaeta are numerous and are borne upon lateral prominences of the body wall known as parapodia.

32 (*a*) Both A and R are true and R is the correct explanation of A.

Most arthropods, certain molluscs and tunicates have open circulatory system. In them, a fluid composed of blood mixed with tissue fluid, bathes the internal tissues and organs directly. It oozes through spaces or cavities that surround the organs, this mixture of fluid is usually referred to as haemolymph.

33 (*b*) Silverfish is representative of phylum-Arthropoda

Cuttle fish	—	Mollusca
Puffer fish	—	Chordata (phylum), class—Pisces
Flying fish	—	Pisces

37 (*b*) In annelids, alimentary canal is straight with anterior mouth and posterior anus. Due to spacious, fluid-filled body cavity between body wall and alimentary canal, the body appears like a tube within a tube in section.

38 (*b*) Food of earthworm (*Pheretima* sp.) consists of organic matter, fallen decaying leaves, algae, etc., present in the soil. Food is swallowed along with soil by sucking action.

39 (*c*) Glochidium larva is the characteristic feature of Mollusca. It is found in the members of the class—Pelecypoda (*Unio* and *Pinctada*), but absent in Cephalopoda (*Sepia*).

This is because in *Sepia*, the development is direct thus, no metamorphosis takes place and the larval stage is absent.

44 (*a*) Prawn (*Palaemon*) belongs to class—Crustacea of phylum—Arthropoda. *Hydra* and sea anemone are coelenterates, snail belongs to class—Gastropoda of phylum—Mollusca.

48 (*a*) More than 70% of all the species recorded are animals. Among animals, insects are the most species rich taxonomic group, making more than 70% of the total. It means out of every 10 animals on this planet 7 are insects.

49 (*b*) Arthropods are the most successful group of animals. Their success is due to unique chitinous cuticle. Exoskeleton is light weight, tough and composed of structural polysaccharide chitin. It is strengthened with proteins and calcium carbonate occurring on the outside. It usually occurs in the forms of plate called sclerites.

50 (*d*) In the members of phylum—Echinodermata like *Asterias* (starfish), *Echinus* (sea urchin), *Antedon* (sea lily), *Cucumaria* (sea cucumber) and *Ophiothrix* (brittle star), excretory system is absent.

51 (*b*) Tube feet are the locomotory organs of echinoderms. These are sac-like protrusions of the body wall, used for locomotion, feeding and respiration.

52 (*c*) Sea cucumber (*Cucumaria*) is an echinoderm. These have water vascular system (ambulacral system) as a characteristic feature.

CHAPTER 10

Animalia-III (Hemichordates and Chordates)

Hemichordates are burrowing worms. They show affinities with chordates as well as with echinoderms. All these three are deuterostome phyla possessing enterocoel. In this chapter, we will explore about phylum-Hemichordata and phylum-Chordata in detail.

PHYLUM-HEMICHORDATA

(Gk. *Hemi*-half; *chorde*-cord)

It was earlier considered as a subphylum under phylum-Chordata. But now it is considered as a separate well defined phylum under non-chordates.

This is because they lack one major characteristics of chordates, i.e. notochord. However, pharyngeal gill slits and embryonic invagination of dorsal nerve cord are present in them. Hence, they were given the name Hemichordata or half chordates.

General characteristics

1. Hemichordata includes marine worm-like animals called **acorn worms**.
2. They are solitary or colonial, but mostly ubicolous, i.e. found in tubes.
3. Body is soft fragile, worm-like and unsegmented.
4. They are bilaterally symmetrical animals with triploblastic body wall.

Inside

Their body is divisible into three regions, i.e.

(i) **Proboscis** It is also called **protosome**. It forms the conical anterior most part of the body.

(ii) **Collar** It is the muscular middle region. It bears ventral mouth below the proboscis stalk.

(iii) **Trunk** It forms the lower most part and is further divisible into following three regions

(a) The anterior branchiogenital region.

(b) The middle hepatic region.

(c) The posterior abdominal region.

5. Coelom is enterocoelous, usually divided into **protocoel**, **mesocoel** and **metacoel**, corresponding to three body regions.
6. Digestive tube is complete and straight. Sometimes it may be U-shaped also. Proboscis contains a hollow outgrowth from the gut.
7. Respiration takes place through branchial portion of pharynx bearing gill slits.
8. Circulatory system is simple or open type including a dorsal heart. Blood is with few cells and without respiratory pigment.
9. Excretion is by a single **proboscis gland** or glomerulus connected to the blood vessels.
10. Nervous system is primitive consisting mainly of a subepidermal nerve plexus. The sensory cells of epidermis act as sense organ.
11. Reproduction is mainly sexual. Sexes are usually separate, but sexual dimorphism not visible. Gonads are one to several pairs.
12. Fertilisation is external (in sea water). Development is mostly indirect with a free-swimming **tornaria larva**. This larva resembles the auricularia larva of echinoderms, e.g. *Balanoglossus, Cephalodiscus* and *Saccoglosssus* larva.

Note *In hemichordates, a notochord-like structure is also found. It is known as* ***stomochord****. This is a hollow outgrowth arising from the buccal cavity called* ***buccal diverticulum*** *in proboscis.*

Balanoglossus (Acorn or Tongue worm)

- It is a sluggish marine animal that moves with the help of cilia present all over its body.
- It feeds on organic matter ingested along with sand.
- **Power of regeneration** is well marked in them.
- Their body is uniformly ciliated without exoskeleton or external appendages.

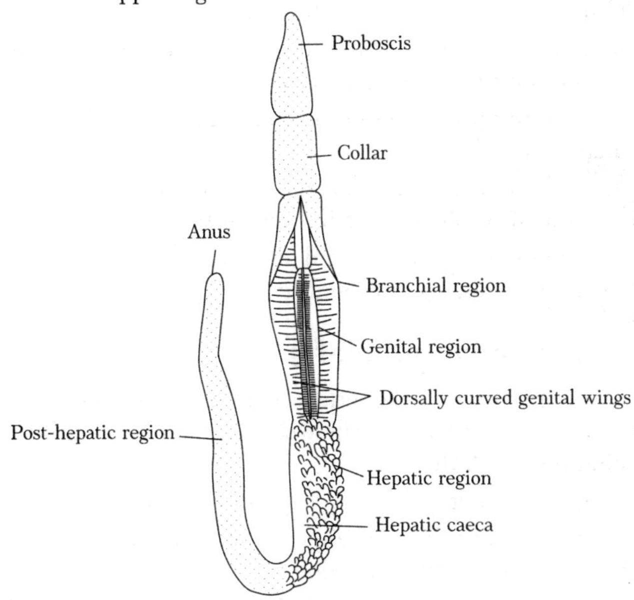

***Fig 10.1** Balanoglossus : External features*

PHYLUM-CHORDATA

It is the largest phylum of subdivision-Deuterostomia. It was created by **Balfour** in 1880. This phylum evolved during Ordovician period. It comprises of a vast variety of living and extinct animals. Most intelligent and highly evolved animals like birds and apes also belong to this phylum.

Fundamental characteristics

Three fundamental characteristics of Chordata are given below

1. Notochord

- It is an **elastic**, solid, ensheathed rod-like structure. It is made up of large vacuolated cells of embryonic mesoderm. It lies below the **nerve cord** in the mid-dorsal line and above the alimentary canal.
- It may persist throughout life in Cephalochordata, Cyclostomata and some fishes. But in most of the other chordates it is replaced partially or completely by a backbone or vertebral column.

2. Dorsal tubular nerve cord

- It is a hollow fluid-filled, non-ganglionated canal that runs from one end of the body to the other.
- It lies above the notochord and outside the coelom. It serves for the integration and coordination of the body activities.
- The anterior part of nerve cord became specialised to form brain.
- The posterior part of nerve cord becomes the spinal cord and is protected within the vertebral column.
- This dorsal tubular nerve cord degenerates before maturity in some chordates, e.g. urochordates.

3. Pharyngeal gill clefts

- These are present as the series of paired narrow openings on the lateral sides of pharynx. These clefts are also called **gill slits** or **pharyngeal clefts**.
- They appear during the development of all chordates.
- In many aquatic forms, these are lined with vascular lamellae, which form gills for respiration.
- In terrestrial chordates, these are absent, but traces of gill clefts are present during early development. Most of these traces disappear before adult life.
- Such gill clefts, which do not bear gills are called **visceral clefts**.

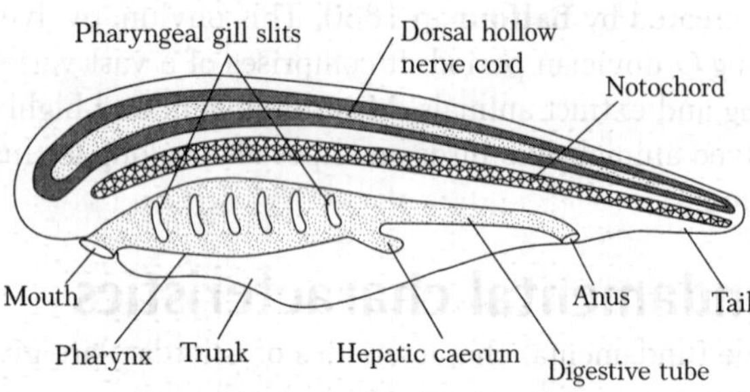

***Fig 10.2** Generalised diagrammatic representation of a chordate*

- In lower chordates, the visceral clefts are used for filtering food particles from the water-current. But in higher chordates, these are modified and used primarily to form their jaw suspensions.
- Besides above 3 characteristics some scientists also consider **post anal tail** as the fundamental characteristic of chordates. It is the simple extension of muscles, nerve cord and notochord.
- It acts as organ of balance and locomotion. However, presence of notochord in it prevent us to count this characteristic as a separate one.

Other characteristics

1. These animals are free-living, aquatic, terrestrial or aerial in nature.
2. Their body is bilaterally symmetrical with triploblastic body wall. There is presence of true coelom of enterocoelous type and metameric segmentation.
3. Endoskeleton is made up of cartilage or bones.
4. Alimentary canal and digestive system is complete. Digestive glands are present.
5. Respiration is of aerobic type, i.e. oxygen is required for the oxidation of nutrients. Gaseous exchange occurs through specialised organs which may be gills in aquatic forms or lungs in terrestrial and aerial forms. In some forms slimy skin also works for gaseous exchange, e.g. amphibians.
6. Blood vascular system is of closed type. It consists of a heart as pumping organ along with blood vessels. Two major blood vessels are dorsal and ventral blood vessels, which usually run along the complete length of the body.
7. Heart is always ventral in position. In higher chordates the heart lies in a pericardial cavity, which is a modified form of the coelom.
8. The flow of blood in the dorsal vessel is from anterior to the posterior end. On the other hand in the ventral vessel, it is from the posterior to the anterior end.
9. Presence of portal system is a characteristic feature. A portal system is a network of veins, which opens into any organ other than heart. The most common type of portal system found in vertebrates is **hepatic portal system**. Through this system blood from alimentary canal reaches to the liver of vertebrates.
10. Most chordates possess red blood corpuscles or erythrocytes containing haemoglobin as a respiratory pigment.
11. Excretion occurs through kidneys. These can be of proto/meso/metanephric types.
12. Sexual reproduction is predominant. Sexes usually separate. Sexual dimorphism is also of usual occurrence.

Classification

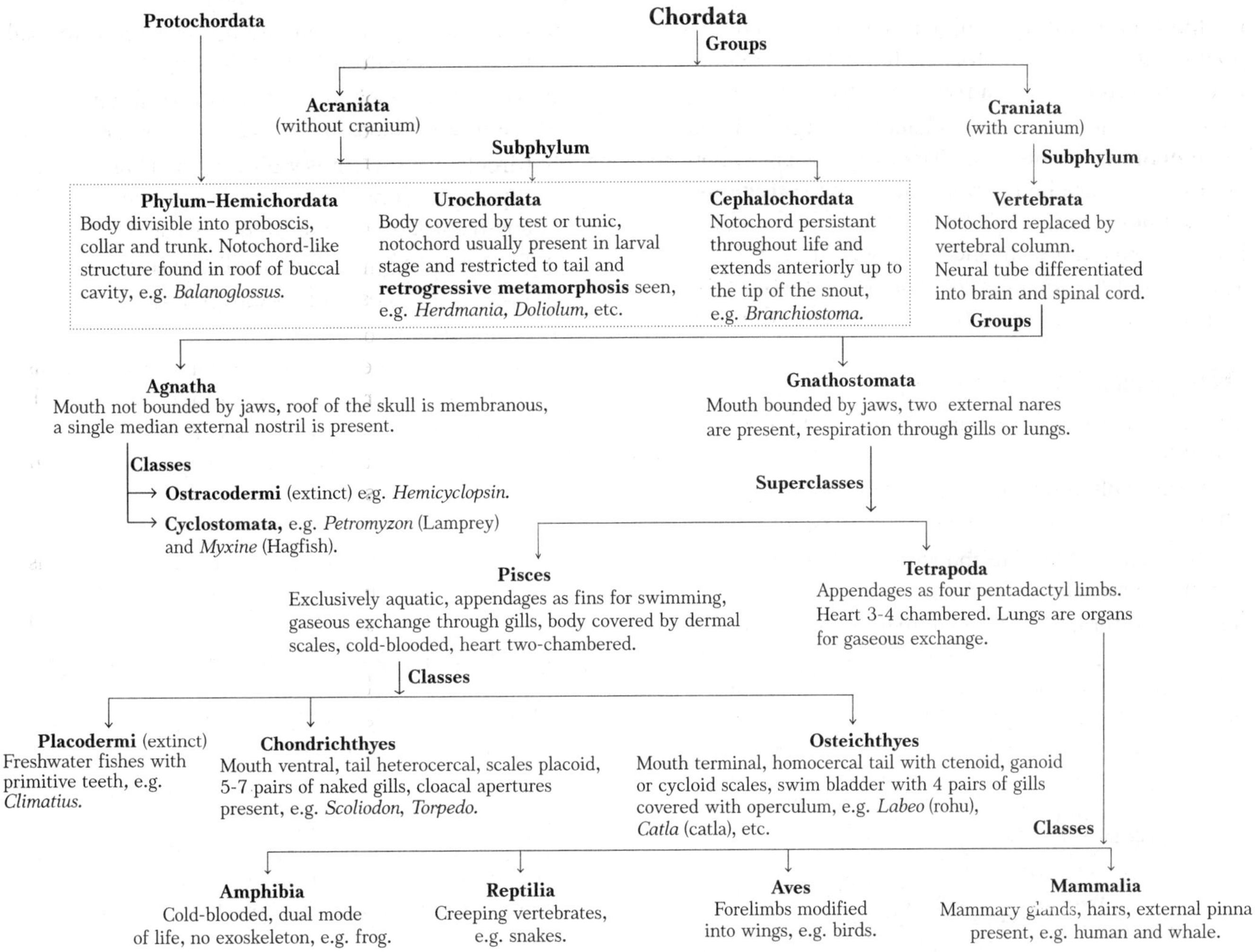

* Protochordates or primitive chordates is a primitive (presently invalid) group of animals. It consists of one phylum and two subphylum of Chordata.

SUBPHYLUM–UROCHORDATA

(Tunicates) (Gk. *Uro*-tail; *chorde*-cord)

These are **marine filter feeders**. According to their fossil records, they were originated in early Cambrian period.

General characteristics

1. Urochordates are exclusively marine, mostly sessile, filter-feeding animals.
2. Most of them are sedantary, i.e. fixed to substratum.
3. Some are simple (solitary), aggregated in groups or composite, i.e. colonial.
4. Body is sac-like and unsegmented. They lack appendages and are without tail.
5. The adult body is enclosed within a **leathery test** or **tunic**. It is composed of cellulose-like organic substance called **tunicin**. The name **tunicata** refers to this covering.
6. Notochord present only in larval tail and disappears in the adult. Hence, they are named as urochordates.
7. Alimentary canal is complete. Pharynx or branchial sac is perforated by numerous apertures called **stigmata** formed by the larval gill slits.
8. Respiration is aerobic. Gaseous exchange occur through gill slits.
9. Blood vascular system is open with blood vessels. Heart is simple, tubular and ventral. The blood of *Herdmania* is green due to the presence of **vanadium** in blood.
10. Excretion is by neural gland, pyloric gland and nephrocytes.
11. Dorsal tubular nerve cord is present only in larval stage. It is reduced to a single dorsal nerve ganglion in adult.

12. Asexual reproduction is also present and occurs by budding.
13. Most forms are hermaphrodite or monoecious. Protogynous condition is visible in them and external cross-fertilisation is a rule.
14. Development is indirect including a free-swimming **tadpole larva** with tail. The ascidians are famous for the fact that, their larval forms show **retrogressive metamorphosis.** It is the formation of less advanced adult from more advanced larval form, e.g. *Herdmania* (sea squirt), *Salpa, Doliolum, Pyrosoma* (colonial bioluminiscent form), *Botryllus, Ascidia, Ciona,* etc.

Herdmania

- It is a sedentary filter feeding tunicate found in marine habitats.
- The animal often seen in gregarious form.
- The whole body is covered with a transluscent tunic or test.
- Animal is hermaphrodite and cross-fertilisation is seen, which is external.
- Larval stage or tadpole show retrogressive metamorphosis.

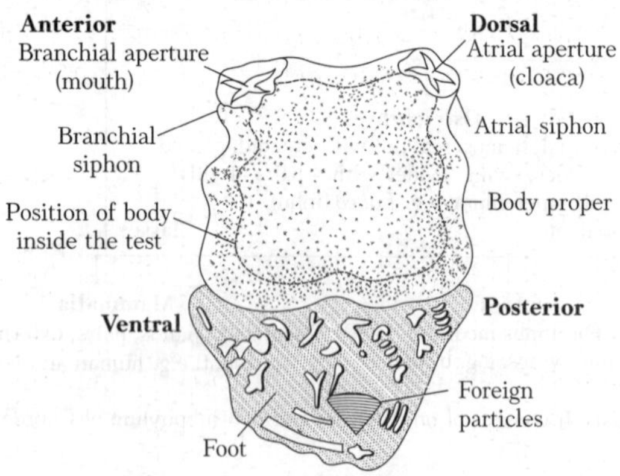

Fig 10.3 Herdmania : External features

SUBPHYLUM-CEPHALOCHORDATA

(Gk. *Cephalo*-head; *chorde*-cord)

These are the only chordates which have all three basic chordate characteristics throughout their life. Their name in itself signifies the presence of notochord in head region.

General characteristics

1. These are marine animals found in shallow water.
2. They have lance-like pointed body from both ends, so commonly called **sea lancelets.**
3. Body is fish-like, transparent and lacks a head. It is only divisible into trunk and a tail.
4. Paired appendages absent, but fins are present.
5. Notochord is rod-like, extending from head to tail. It persists throughout the life.
6. Digestive tract is complete. Pharyngeal gill slits are more numerous and better developed.
7. No specialised respiratory organs are present. Gaseous exchange is through general body surface.
8. Circulatory system is well-developed and closed type. It is without heart and their blood is without any respiratory pigment.
9. Excretion is by protonephridia with solenocytes.
10. Sexes are separate (dioecious). Asexual reproduction is absent.
11. Gonads are numerous and metamerically repeated in the body. Gonoducts are absent.
12. Fertilisation is external in sea water.
13. Development is indirect including a free-swimming fish like larva.
14. They show **progressive metamorphosis**, i.e. change from less developed larva to more developed adult, e.g. *Branchiostoma* and *Asymmetron.*

Branchiostoma (Amphioxus – Lancelet)

- This is one of the few living genera of lancelets.
- These are small eel or snake like animals.
- Their scientific name means 'gill mouth' referring to their anatomy.
- Unlike other vertebrates they do not have a true head, but have a mouth surrounded by gill slits with slightly enlarged anterior end of the dorsal nerve cord.

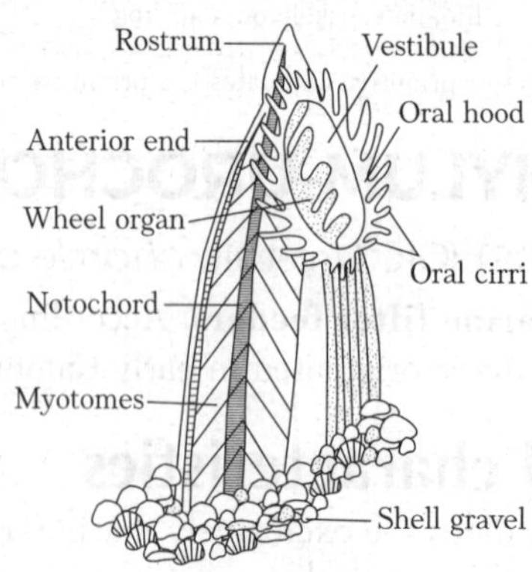

Fig 10.4 Branchiostoma in natural habitat, partly buried in shell gravel with anterior end protruding

Non-chordates *Vs* Invertebrates

It is very clear from the above text that words Non-chordata and Invertebrata have different meanings.

Uro and cephalochordates are excluded from non-chordates, but included in group invertebrates. Thus, we can say that, all vertebrates are chordates, but all chordates are not vertebrates.

CHECK POINT 10.1

1. The feature(s) which the hemichordates share with chordates are
   (a) presence of pharyngeal gill slits
   (b) a bilaterally symmetrical body
   (c) dorsal tubular nerve cord
   (d) Both (a) and (c)
2. In higher chordates, pharyngeal gill slits are
   (a) absent
   (b) present in the embryo
   (c) replaced by abdominal gill slits
   (d) replaced by thoracic gill slits
3. Non-chordates have a central nervous system, which is ventral, solid and double, while in chordates, it is
   (a) ventral, hollow and double
   (b) ventral, hollow and single
   (c) dorsal, hollow and single
   (d) dorsal, solid and single
4. Protochordates are considered a connecting link between
   (a) invertebrates and vertebrates
   (b) vertebrates and chordates
   (c) invertebrates and chordates
   (d) hemichordates and cephalochordates
5. Cephalochordates are termed as sea lancelets because
   (a) they have a lance-like pointed body at one end
   (b) their body bears lance-like tail
   (c) they have a lance-like fins
   (d) they have a lance-like pointed body from both the ends
6. How cephalochordates are different from urochordates?
   (a) Cephalochordates possess notochord throughout their life, but urochordates have it only in larval stage
   (b) Cephalochordates are exclusively marine, but urochordates are also found in freshwater
   (c) Cephalochordates possess dorsal nervous system, but urochordates have ventral nervous system
   (d) Cephalochordates have ventral heart, but urochordates have dorsal heart

SUBPHYLUM-VERTEBRATA

(L. *vertebrus*–backbone)

Vertebrates are **advanced chordates**. Their name is derived from the serially arranged 'vertebrae' which comprise of a major part of their endoskeleton, i.e. the **vertebral column**.

General characteristics

1. Brain of vertebrates is enclosed within a cranium.
2. Notochord is seen only in the embryonic stage and afterwards it is incorporated within the vertebral column or the backbone.
3. High degree of cephalisation, i.e. formation of head is seen.
4. Skin is covered by protective exoskeleton, comprising of scales, feathers, hairs, etc.
5. Coelom is large and well-developed. It is of enterocoelous type. It is partitioned in accordance with visceral internal organs, e.g. pericardial or pleural cavities.
6. Respiration is aerobic. Gaseous exchange by gills, skin, buccopharyngeal cavity or lungs.
7. Circulation is of closed type. Circulatory system is with a pumping organ, i.e. heart and blood vessels. Heart may be 2, 3 or 4-chambered.
8. Blood is usually with a respiratory segment.
9. Organs of excretion and osmoregulation are called kidneys. Depending upon the structural complexity these may be of pro, meso and metanephric types.
10. Nervous system consists of
    (i) **Central nervous system**, i.e. brain and spinal cord.
    (ii) **Peripheral nervous system**, i.e. cranial and spinal nerves.
    (iii) **Autonomic nervous system**, i.e. sympathetic and parasympathetic nervous systems.
11. Sense organs include eyes, ears, tongue, nasal chambers and skin. In some vertebrates, lateral line system is also present.
12. Chemical coordination is always through endocrine glands.
13. Sexes are separate (unisexual) except in hagfish, which is bisexual. Gonads are paired, with gonoducts.

On the basis of presence or absence of jaws this subphylum has following two groups

Group I–Agnatha (Jawless vertebrates)

1. They are the most primitive of craniates. Their mouth does not possess jaws hence named Agnatha. They lack exoskeleton and paired appendages.
2. They bear a single median nostril.

3. Notochord persists throughout their life. Vertebral column is small and not well-developed.
4. They are cold-blooded or poikilotherms. This group includes an extinct class-Ostracodermi and a living class-Cyclostomata.

Class–Ostracodermi (Extinct)

(Gk. *ostrakon*–shell; *derma*–skin)

- They are the earliest known vertebrates which appeared in Ordovician period.
- They had well-developed dermal scales. This led to their names 'Ostracoderms' or bony skin.
- They are also called **'armoured fishes'**. All are extinct, e.g. *Cephalaspis, Hemicyclaspis, Pteraspis,* etc.

Class–Cyclostomata (Living)

(Gk. *Cyklos*-circular; *stoma*-mouth)

- This is a group of chordates that includes **living jawless fishes**, i.e. lampreys and hagfishes.
- The name Cyclostomata signifies round mouth. Being jawless they do not have the ability to close their mouth.
- A continuous water-current is always maintained through this open mouth.

General characteristics

1. They are marine fishes, but some migrate to freshwater for spawning. This condition is known as **anadromous migration**.
2. Body is elongated, eel-like with median fins having cartilaginous fin rays. Paired appendages are present.
3. Notochord persists throughout the life. Imperfect neural arches present over notochord. These represent rudimentary vertebrae.
4. Skin is soft, smooth and contains unicellular mucous glands. Scales are absent.
5. Mouth is ventral, suctorial, circular and the tongue is with denticles.
6. Digestive system lacks stomach and pancreas. Intestine is with a fold, called **typhlosole** to increase absorptive surface area.

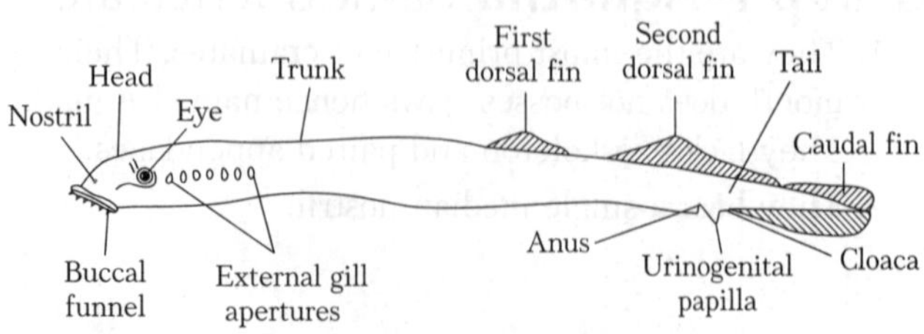

Fig 10.5 *Petromyzon*

7. They may be parasitic, i.e. sucks out blood of host species (e.g. lamprey) or scavenging, i.e. feeds mostly on dead fishes (e.g. hagfish).
8. 5-16 pairs of gills are present in lateral sac-like pouches of pharynx. Gill-slits are 1-16 pairs.
9. Air bladder is absent. These are mostly poikilotherms.
10. Heart is two-chambered with one auricle and one ventricle.
11. A pair of mesonephric kidneys with ducts and urinogenital papilla are present.
12. Dorsal nerve cord with well differentiated brain, as its front portion. About 8-10 pairs of cranial nerves also present. Its lower portion is developed as spinal cord.
13. Sensory organs are well-developed.
    (i) The olfactory sac is single and median.
    (ii) Ear consists only of membranous labyrinth, i.e. internal ear.
    (iii) A well-developed lateral line system is also present for coordinating activites during swimming.
14. There tail is **diphycercal**, i.e. a tail fin in which the vertebral columns extend to the tip, with symmetrical upper and lower parts.
15. Sexes are separate or united. Gonads are single, large and without gonoducts.
16. Fertilisation is external.
17. Development is indirect with a prolonged larval stage, i.e. **ammocoete**, e.g. lampreys (*Petromyzon*), *Lampetra, Entospheus, Ichthyomyzon, Myxine* (hagfish) and *Eptatretus* (*Bdellostoma*).

Table 10.1 Differences between lamprey and hagfish

Characters	Lamprey (*Petromyzon*)	Hagfish (*Myxine*)
Habitat	Marine as well as freshwater.	Exclusively marine, burrowing in sand.
Parasitism	Both parasitic and non-parasitic species.	Primarily scavenger, but can attack injured living fishes also.
Feeding	Sucks out blood of host fishes.	Feeds mostly upon the flesh of dead injured fishes.
Breeding	Marine primarily, migrates to freshwater river and stream for spawning.	Spawn on sea floor.
Skin	Less slimy.	More slimy, hence called 'slime eel'.
Gills	Both gill slits and gill pouches are 7 pairs each. These open in a separate respiratory pharynx.	Gill pouches 6-15 pairs and gill slits 1-15 pairs.

Characters	Lamprey (*Petromyzon*)	Hagfish (*Myxine*)
Paired eyes	Functional	Degenerate
Salivary glands	Present and secrete an anticoagulant also	Absent
Sexes	Separate (unisexual).	United (hermaphrodite-bisexual).
Development	Indirect with a larval stage called ammocoete. Metamorphosis is present.	Direct, without larva, no metamorphosis.

Group II– Gnathostomata

(Jawed vertebrates)

1. It includes advanced vertebrates.
2. Mouth has jaws hence the name **Gnathostomata** is given to them.
3. The notochord is usually replaced by vertebral column.
4. They possess paired appendages, exoskeleton and paired nares.
5. Olfactory sac is double. Internal ear has three semicircular ducts.
6. They appeared in Silurian period and flourished in Carboniferous period.
7. The Gnathostomata is divided into two superclasses-Pisces and Tetrapoda.

Table 10.2 Differences between Agnatha and Gnathostomata

Features	Agnatha	Gnathostomata
Mouth	Circular without jaws	With jaws
Paired appendages	Lacking	Present
Notochord	Persistent	Replaced by vertebral column
Vertebral column	Under developed	Fully developed
Cranial nerves	8-10 pairs	10-12 pairs
Internal ear	1 or 2 semicircular canals	3 semicircular canals
Gonads	Single	Paired
Gonoduct	Absent	Present
Pancreas and spleen	Not organised	Present as distinct organ
Nostril and Olfactory sac	Single	Paired

Aquatic gnathostomes are kept is superclass-**Pisces** while, terrestrial and aerial gnathostomes are kept in superclass-**Tetrapoda**.

SUPERCLASS-PISCES

The superclass-Pisces includes **true fishes.** It is the first true jawed vertebrate group.

It is further categorised into three classes-Placodermi, Chondrichthyes and Osteichthyes.

Class–Placodermi (Extinct)

(Gr. *Plakos*-plates; *derma*-skin)

1. It includes the earliest fossil fishes which were mainly found in freshwater.
2. There body was with an external protective armour of bony scales or plates.
3. Primitive jaws with teeth were present.
4. Fins were mostly formed of large spines, e.g. *Climatius*, *Coccosteus* and *Palaeospondylus*.

Class–Chondrichthyes

(Gr. *Chondros*-cartilage; *ichthys*-fish)

This is the class of **cartilaginous fishes** and contains two major groups, i.e. **Elasmobranchii** (sharks, rays, skater and sawfish) and **Holocephali** (chimaeras or ghostfishes).

General characteristics

1. Cartilaginous fishes are mostly marine and predaceous in nature.
2. Their endoskeleton is made up of cartilage and exoskeleton is of **placoid scales.**
3. These are poikilothermous.
4. Except in *Chimaera*, gills are not covered by an operculum or gill cover.
5. Skin is tough, containing minute placoid scales.
6. Body is laterally compressed or dorsoventrally flattened. It consists of a head, trunk and tail.
7. Mouth is ventrally placed over the head. Jaws present. Teeth are modified placoid scales.
8. Stomach is J-shaped. A **spiral valve** or **scroll valve** is usually present in the intestine to increase absorptive surface area. The digestive tract leads into the cloaca, i.e. a common urinogenital opening.
9. External nares are present on the ventral side of the head and internal nares are absent.
10. Paired fins are broad and caudal fin is mostly **heterocercal**, i.e. asymmetrical.
11. Lung or Air bladder is absent.
12. Heart is two-chambered with one auricle and one ventricle. It is associated with a very well-developed

sinus venosus and conus arteriosus. As only venous blood flows through **venous heart.**

13. Kidneys are mesonephric and urea is the chief nitrogenous waste.
14. Brain is with large olfactory lobes and a large cerebellum.
15. External ears are absent. But internal ears are present, which have three semicircular canals.
16. Cranial nerves are 10 pairs. **Lateral line system** is well-developed which detects waves and currents in water.
17. Sexes are separate and sexual dimorphism is seen.
18. Gonads paired. Gonoducts opens into cloaca. In males, cloacal opening is seen with **claspers**, as copulatory organs.
19. Development is direct without any larval stage.
20. Fertilisation is internal. They are oviparous or ovoviviparous, i.e. produce eggs, which hatch inside mother's uterus, e.g. sharks, *Scoliodon, Torpedo, Chimaera, Tetrahedron* (most poisonous fish), *Rhinobates* (guitarfish), *Pristis* (sawfish), etc.

Scoliodon (Dogfish)

- It is also known as **dogfish** due to its great sense of smell.
- It is a carnivorous fish.
- Lateral line system is well-developed. It is a group of sense organs that are made from epidermal cells. It helps in responding to the displacements caused by motion, electric impulse, etc.
- **Ampulla** of **Lorenzini** are present as thermoreceptors in *Scoliodon.*
- These are viviparous and their egg case is called **Mermaid's purse.**
- *Scoliodon* is edible and is rich in vitamin-A and D.
- *Scoliodon dumerili* is seen in the Bay of Bengal.

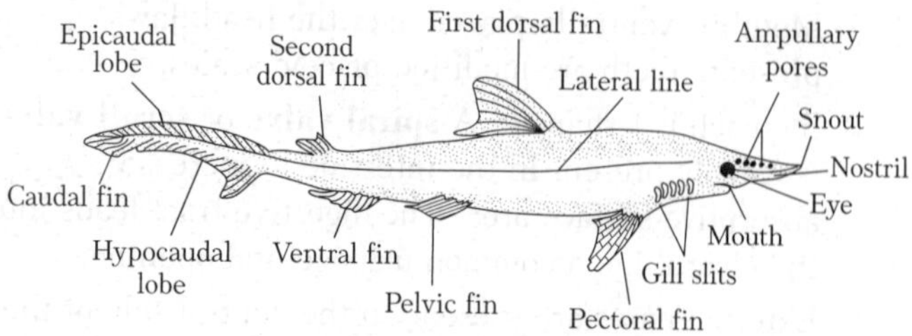

Fig 10.6 Scoliodon

Torpedo (Electric ray)

- It is a bottom-dwelling carnivorous fish.
- It is famous for discharging electricity from its electric organs, that is why also called **electric ray**.
- These electric organs are infact modified muscles. The fish uses this ability to stun its preys and enemies.
- It has smooth skin (without scales).
- One pair spiracles are present behind the eyes and are used for taking water in for gaseous exchange.

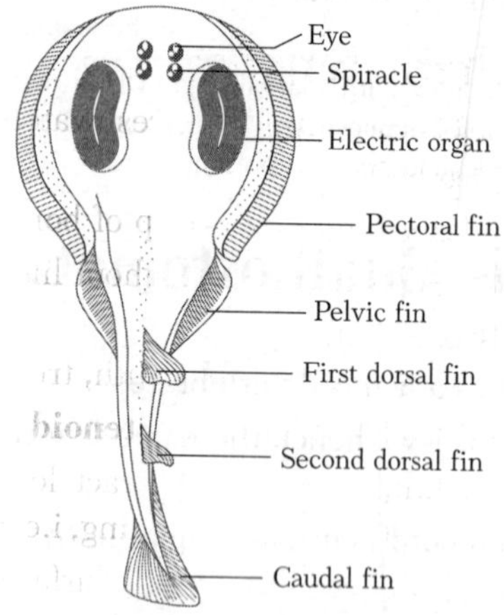

Fig 10.7 Torpedo (Dorsal view)

Trygon (Sting ray)

- It is also called **sting ray** and resembles the electric ray in many aspects. It is devoid of electric organs.
- It is carnivorous. The body is somewhat kite-shaped.
- Its whip-like tail bears a spine or sting as an organ of defence. This sting is a modified dorsal fin.

Chimaera (King of herrings)

- It is commonly called **rabbit fish, rat fish, ghostfish** or **king of herring**. It is found in deep sea.
- It is considered as the connecting link between cartilaginous and bony fishes.
- Just like cartilaginous fishes it has claspers and cartilagenous endoskeleton. On the other hand, like bony fish, it possesses operculum on each side over gill slits.
- The tail is long and filamentous looking like whip.
- Male also has a frontal or cephalic clasper on the dorsal surface of head, with unknown function.
- Cloaca is absent, i.e. it has distinct anus and urinogenital apperture.

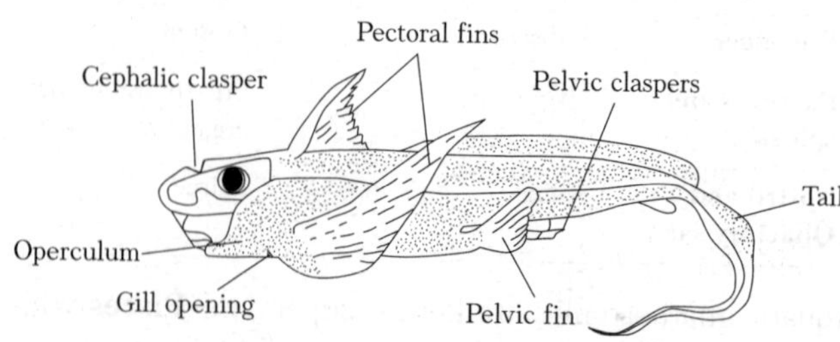

Fig 10.8 Chimaera

Class–Osteichthyes

(Gr. *Osteon*—bone; *ichthyes*—fish)

It is a diverse taxonomic group of bony fishes. It is considered as the largest class of vertebrates in existance today.

General characteristics

1. It includes both marine and freshwater fishes with streamlined body.
2. Their endoskeleton is made up of bones.
3. Notochord is persistent throughout life as vertebral column.
4. Body has three regions *viz* mouth, trunk and tail.
5. Skin is covered with **cycloid**, **ctenoid** or **ganoid** scales.
6. Mouth is terminal, digestive tract leads into an anus. The common urinogenital opening, i.e. cloaca is absent.
7. External nares lie on the dorsal surface of snout. However, in lung fishes internal nares are also present.
8. Caudal fine usually homocercal or symmetrical.
9. They have four pairs of gill, which are covered by an operculum on each side.
10. **A swim bladder** or **air bladder** is present, which regulates buoyancy. In some cases it is modified to help in gaseous exchanges, i.e. capable to work equivalent to lungs of tetrapods.
11. Heart is two-chambered and has sinus venosus and conus arteriosus. Lung fishes have **three-chambered** heart with two auricles and one ventricle. They also have well-developed **renal portal** system.
12. Kidneys are **mesonephric** and **ammonia** is the **chief nitrogenous waste** product.
    Freshwater bony fishes maintain their water balance by excreting hypotonic urine. They survive in freshwater by getting rid off water and gaining salts.
13. Brain bears relatively small olfactory lobes and cerebellum. 10 pairs of cranial nerves are present.
14. Body possess a well-marked **lateral line system**.
15. Claspers are absent. Fertilisation is generally external. Most forms are oviparous, e.g. *Gambusia*.
16. Development is direct. Some bony fishes show parental care as well, e.g. *Labeo rohita*, *Anabas*, *Mystus*, lung fishes, seafish, etc.

Labeo (Rohu)

- *Labeo rohita* and *Labeo calbasu* (edible carps) are found in clear, sluggish rivers and streams.
- The adult fish show chiefly herbivorous habit, but the young ones are carnivorous in nature.
- Head is scaleless with a subterminal mouth surrounded by very thick and fringed lips.

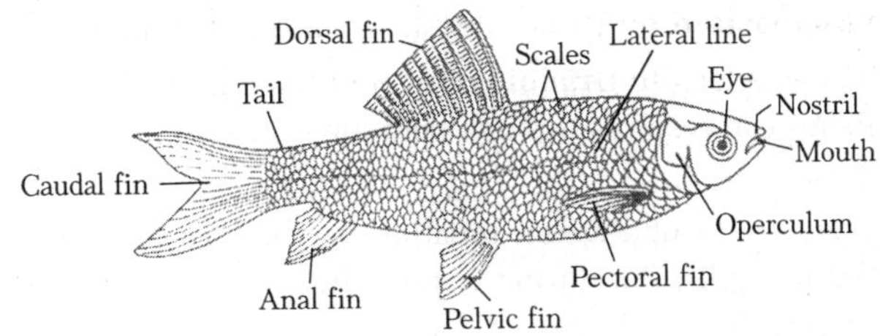

Fig 10.9 Labeo

Hippocampus (Seahorse)

- It inhabits the tropical and temperate seas including Indian ocean.
- Head is large at right angle to the body and produced into a tubular snout, thus resembling the head of horse.
- Mouth is toothless and suctorial.

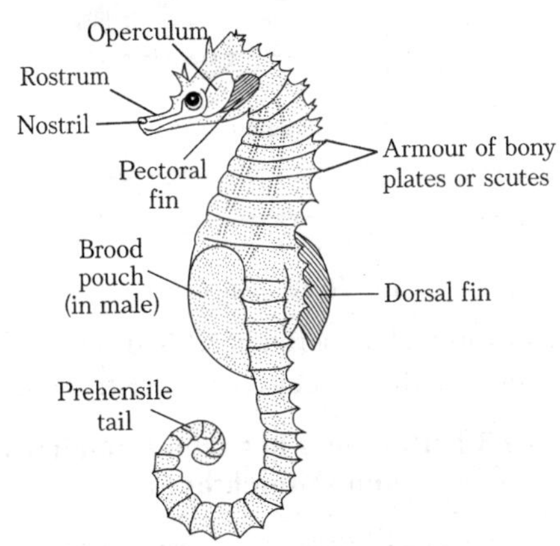

Fig 10.10 Seahorse (male) : Hippocampus

- It possesses a long, prehensile tail.
- Sexual dimorphism is seen. On the belly of male is a **brood pouch** is present for incubating the eggs. On the belly of female only a small anal fin is present.

Anabas (Climbing perch)

- It is a common freshwater fish, which can also live outside water for sometime. Outside water, it respires through the accessory respiratory organs lying in front of the gills.

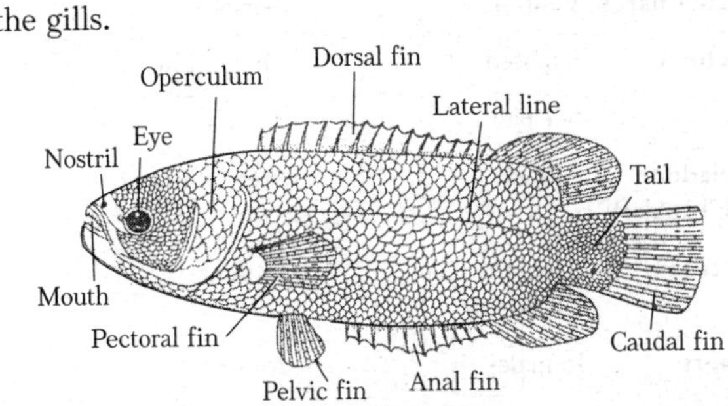

Fig 10.11 Anabas

Lung fishes

- Those fishes in which **swim bladder is modified** or **vascularised** for gaseous exchange are called lung fishes.
- These belong to **Dipnoi** group of Osteichthyes. These fishes can respires through both lungs or swim bladder and gills.
- They are facultative air breathers as they normally breathe through gills. But during water deficient condition, they respire through lungs.
- Therefore, they are also called **uncle of amphibians**.
- There are three genera of living lung fishes, i.e. *Neoceratodus*, *Lepidosiren* and *Protopterus*.

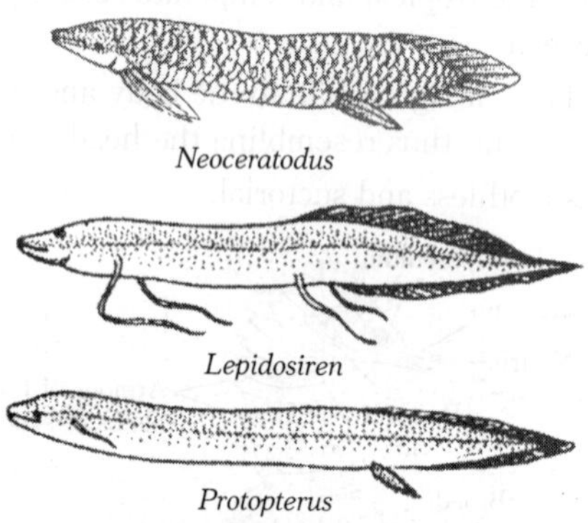

***Fig 10.12** Lung fishes*

- Presence of three-chambered heart and well-defined renal portal system are their other characteristic features.

Table 10.3 Differences between Chondrichthyes and Osteichthyes

Characters	Chondrichthyes (Cartilaginous fishes)	Osteichthyes (Bony fishes)
Habitat	Mostly marine	Both marine and freshwater
Endoskeleton	Wholly cartilaginous	Mostly bony
Exoskeleton (scales)	Placoid	Cycloid, ctenoid and ganoid
Caudal fin	Heterocercal (asymmetrical)	Homocercal (symmetrical)
Mouth	Ventral on head	Usually terminal or subterminal on head
External nares	Ventral	Dorsal
Gill clefts	Exposed	Covered by operculum
Gills	5-7 pairs	4 pairs
Air bladder (swim bladder)	Absent	Present in some
Operculum	Absent (except *Chimaera*)	Present
Claspers	In males	Absent
N_2 waste	Urea	Ammonia
Cloaca	Present	Absent
Spiral valves	Present in intestine	Usually absent in intestine
Development	Direct without metamorphosis	Indirect with metamorphosis

Latimeria (Living fossil)

- It resembles lung fishes in having internal nares, lobed fins and lung or air sac.
- All its fins except the two dorsal ones are lobed.
- *Latimeria* is a living fossil as it has remained unchanged for several million years.
- Lobe finned lung fishes are believed to be fore runners of tetrapods.

Table 10.4 Some other important bony fishes

Scientific name	Common Name
Echeneis	Sucker fish
Gambusia	Mosquito fish
Exocoetus	Flying fish
Tetradon	Globe fish
Diodon	Porcupine fish
Cyanoglossus	Flat fish
Syngnathus	Pipe fish
Fistularia	Flute fish
Carassius	Gold fish
Anguilla	Freshwater eel
Oncorhynchus	Pacific salmon
Lophius	Angler fish
Betta	Fighting fish
Pterophyllum	Angel fish
Clarias batrachus	Walking cat fish

Unique features of Pisces

- The fins and their generalised functions in fishes are given below
  (i) **Paired pectoral** and **pelvic fins** Balance and breakes counteract rolling and yawing.
  (ii) **Dorsal fin** and **anal fins** Help fish to not role over onto their sides.
  (iii) **Caudal fins** Propels the body and provide lift.

- Some fishes like blue fish tuna and sword fish are able to maintain their body temperature especially in cold waters.
- Some fishes show parental care as well. For example,
  - **Nest building** is seen in *Amia* (bow fish).
  - In lung fishes *Protopterus* and *Lepidosiren* nests are built by males.
  - Males of catfish, *Arius* provide shelter to young ones in their mouth.
  - **Brood pouch** is visible in the males of *Hippocampus*.
- Stone fish is most poisonous fish.
- **Catadromous** fishes live in freshwater and go down to sea for breeding, e.g. *Anguilla* (European eel).
- Pomfret is most common edible fish of India.
- Weberian ossicles are specialised skeletal structures of cartilaginous fishes which connect swim bladder with internal ear.
- The commercial name of integument of *Scoliodon* is **shegreen**.
- **Isinglass** is a gelatinous product of certain fishes like carps, salmons, catfishes, etc. It is obtained from their air bladder and used for making cement and jellies. It as also used for the clarification of wines and beers.

Table 10.5 False fishes in animal kingdom

Common name	Genus	Phylum
Jellyfish	*Aurelia*	Coelenterata
Silverfish	*Lepisma*	Arthropoda
Crayfish	*Astacus*	Arthropoda
Cuttlefish	*Sepia*	Mollusca
Devilfish	*Octopus*	Mollusca
Starfish	*Asterias*	Echinodermata
Hagfish	*Myxine*	Chordata
Whale	*Balaenoptara*	Aquatic mammals

CHECK POINT 10.2

1. Which of the following statements is/are true?
   (a) All chordates are vertebrates
   (b) All vertebrates are chordates
   (c) All chordates are not vertebrates
   (d) Both (b) and (c)
2. The cyclostomes are
   (a) marine and non-migratory
   (b) freshwater form and non-migratory
   (c) marine and migrate to freshwater for spawning
   (d) freshwater form and migrate to sea for spawning
3. Gnathostomata differ from Agnatha in having
   (a) no gonads
   (b) four semicircular ducts in internal ear
   (c) pancreas and spleen as distinct organs
   (d) 6-8 pairs of cranial nerves
4. The organisms of the class–Chondrichthyes swims constantly
   (a) to avoid sinking (b) as they lack air-bladder
   (c) as they possess air-bladder (d) Both (a) and (b)
5. Freshwater bony fishes maintain water balance by
   (a) excreting hypotonic urine
   (b) excreting salt across their gills
   (c) drinking small amount of water
   (d) excreting waste in the form of uric acid
6. The character common to both cartilaginous and bony fishes, is the presence of
   (a) two-chambered heart
   (b) terminal mouth and placoid scales on skin
   (c) air bladder
   (d) external fertilisation

SUPERCLASS–TETRAPODA

It includes land vertebrates with two pairs of pentadactyl limbs, cornified skin and lungs. It is further categorised into four classes-Amphibia, Reptilia, Aves and Mammalia.

Class – Amphibia

(Gr. *Amphi*-dual; *bios*-life)

These are the first animals to attempt transition from aquatic to terrestrial life. But, they and their eggs have no protective coat to check water loss through evaporation.

Therefore, they inhabit damp places and breed in water. This dual mode of life is expressed in their name also, i.e. amphibians.

They were evolved in the Devonian era.

> **Anamniotes and amniotes**
>
> Along with fishes these animals are categorised as anamniotes, i.e. animal without amnion. It is an embryonic covering membrane seen in rest other tetrapods, i.e. reptiles, birds and mammals. These tetrapods are thus called **amniotes**.

General characteristics

1. From evolutionary point of view they are the first cold-blooded poikilothermic or ectothermic vertebrates which came to the land.
2. They are amphibious in nature, i.e. can live on land as well as in freshwater. However, amphibians are

generally not found in sea water, except *Rana cancrivora*.

3. They are the most primitive and earliest tetrapods with two pair of limbs. Each limb is pentadactyl in nature and used for locomotion.
4. Body is usually divisible into head and trunk, while tail may be present in some. The skin is smooth, moist, rich in multicellular mucous and poison glands.
5. Body with well-defined endoskeleton, having skull, vertebrae and limb bones. Exoskeleton (scales) are absent. But if scales are present they are hidden beneath the skin, e.g. caecilians.
6. Skull is **dicondylic**, i.e. with two occipital condyles and usually platybasic, i.e. a type of axial skull with well-defined projected orbits lying in same plane with the brain.
7. The centrum of vertebrae is procoelous in general. However, acoelous and amphicoelous conditions are also seen.
8. Alimentary canal complete. Mouth terminal, usually with non-masticating teeth. Alimentary canal and urinogenital tracts open into a common chamber called **cloaca** which opens to the exterior.
9. Aerobic animals, perform gaseous exchange by gills during larval stage. In adults it occurs by skin, lungs and mouth cavity or even by gills in some cases. Lungs are absent in some salamanders.
10. Circulatory system closed. Heart is three-chambered having two auricles and one ventricle. Mixed blood flows through the heart.
11. RBCs are nucleated, biconvex and oval. *Proteus* and *Amphiuma* have largest RBCs of animal kingdom.
12. Kidney is mesonephric with nephrostome. Larva is ammonotelic, but adults are ureotelic.
13. Ear consists of internal and middle ear. A tympanum or outer membrane covers the middle ear. The middle ear has a single ear ossicle called **columella auris**.
14. Brain is well-developed with ten pairs of cranial nerves. Lateral line system is present, but in non-functional form, i.e. vestigial.
15. Sexes are separate. Sexual dimorphism seen. Male with copulatory organs. Most females are oviparous and return to water for breeding.
16. Development is indirect. During development, a **tadpole larva** is formed which undergoes metamorphosis.
17. **Neoteny**, i.e. animal with larval and adult characters and paedogenesis, i.e. reproduction by larval forms are of common occurrence.
18. Metamorphosis in amphibians is controlled by a hormone called **thyroxine**.

Classification

Class-Amphibia is divided into two subclasses

Amphibia

Subclasses

Stegocephalia (extinct)

- Had pentadactyl limbs.
- Skin with scales and bony plates.
- Crocodile and salamander like amphibians.
- Enlarged teeth, with greatly folded dentine in both jaws, e.g. *Eryops* and *Brachiosaurus*.

Lissamphibia

- Modern living amphibians.
- No bony dermal skeleton.
- Teeth small and simple.

Orders

Apoda/Gymnophiona/Caecilians

- Also called **limbless amphibians**.
- Long, worm-like body, burrowing, dermal scales present in skin.
- Tail short or absent, cloaca terminal.
- Males have protrusible copulatory organ.
- Larva has three pairs of external persistent gill, e.g. *Ichthyophis* and *Uraeotyphlus*.

Anura/Salientia

- Also called **tailless amphibians.**
- Commonly includes frogs and toads.
- Forelimbs are shorter than hindlimbs.
- Skin is loosely fitting, scaleless with teeth either present only on upper jaw or absent.
- Vertebral column is very small with 5-9 procoelous vertebrae and a slender urostyle. Adults without gills.
- Fertilisation is always external.
- Full metamorphosis without neotenic forms, e.g. *Rana, Bufo, Hyla* and *Rhacophorus.*

Urodela/Caudata

- Also called **tailed amphibians**.
- Commonly called **newts** and **salamanders**.
- Lizard like, two pairs of weak and equal sized limb.
- Skin devoid of scales and tympanum. Possess largest RBC. Gills permanent or lost in adult. *Necturus, Proteus, Siren* and axolotl larva have external gills.
- Fertilisation is internal. Larva is aquatic, adult-like with teeth, e.g. *Necturus, Salamandra* and *Ambystoma.*

Ichthyophis

- It is a limbless amphibian or caecilian.
- It feeds on underground worms and insects. It lacks tongue.
- Their eyes are small, degenerated and hide below skin. That is why they are also known as **blind worms**.

- Female lays eggs in wet burrows and guard them, i.e. parental care is seen.

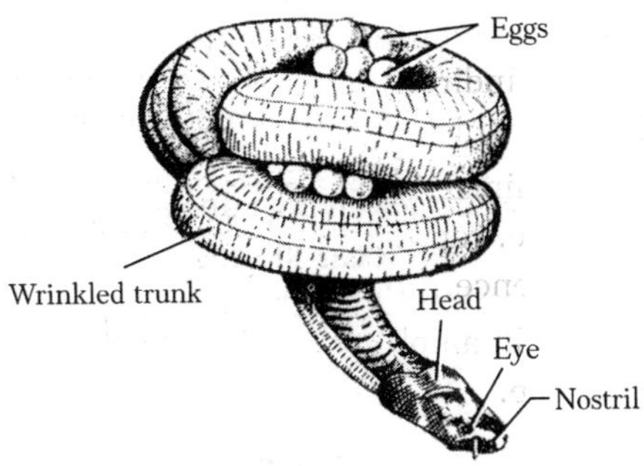

Fig. 10.13 Ichthyophis

Salamandra

- It is semiterrestrial lizard like tailed amphibian.
- Male lays sperms in a capsule called **spermatophore,** which is picked up by female with her cloacal lips to fertilise her eggs internally. These eggs are retained and develop inside the body of female, who gives birth to larvae.
- These larvae complete their further development in water.

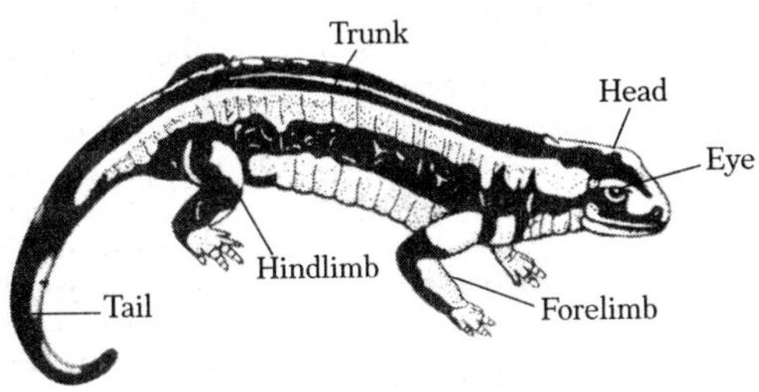

Fig 10.14 Salamandra

Table 10.6 Other important examples of amphibians

Scientific name	Common name
Gymnophiona	
Uraeotyphlus	Blindworm
Urodela	
Ambystoma	Tiger salamander
Amphiuma	Congo eel
Necturus	Mud puppy or newt
Proteus	Cave salamander
Siren	Mud eel
Triton	Newt
Salamandra	European spotted or fire salamander
Anura	
Alytes	Midwife toad
Bufo melanostictus (Poisonous amphibian)	Indian toad
Pipa	Surinam toad
Hyla	Tree frog
Rhacophorus	Flying frog
Bombinator	Fire bellied toad
Xenopus laevis	African clawed toad
Ascaphus	Bell toad
Astylosternus	Hairy frog
Nototrema (Gastrotheca)	Marsupial frog

Parental Care in Amphibia

- The female Surinam toad (*Pipa*) carries the tadpoles in special pits on its back till tadpoles become little toads.
- The male *Alytes* carries eggs around its thighs till the tadpoles are ready to hatch.
- *Rhinoderma darwinii* (Darwin's frog) carries the eggs and tadpoles in its vocal sacs.

Unique features of Amphibia

- Urodela forms a transitional group between fishes and amphibians.
- Animals in amphibia themselves show a transition from water to land as
  - *Necturus* like animals are completely aquatic.
  - *Amphiuma, Rana* like animals are transitional forms.
  - *Amphiuma* adult is a complete lung breathing form which prefers to live on land.
- *Rana tigrina* shows formation of middle ear first time in animal kingdom.
- Seymouria is a connecting link between amphibians and reptiles.
- Tadpole skin secretes diastase enzyme. It does not feeds at the time of metamorphosis.
- The most poisonous frog is **golden dart poison frog** of South America. Poision is secreted from their skin and can cause instant death. One adult frog contains enough poison to kill 200 persons.

Class – Reptilia

(L. *repre* or *reptum*—to crawl or creep)

These are the first fully adapted terrestrial chordates. Four factors have made reptiles true land animals and independent of moist environment. These are

- Horny scales which check loss of water.
- Internal fertilisation as gametes cannot survive on land.
- Shell around the egg to check desiccation.
- Fluid-filled amnion around the embryo to provide aquatic environment for development on land.

The study of reptiles is called **Herpetology**.

General characteristics

1. They are cold-blooded or poikilothermic animals.
2. Reptiles are the creeping and burrowing vertebrates with cornified skin.
3. Their skin is rough and dry, i.e. usually without glands. An exoskeleton of horny epidermal scales or scutes containing a waterproof protein keratin, is always present. It is periodically sloughed off in pieces or as a whole. The process is called **moulting**. The toes end into claws, which are thick and are known as **corneoscutes.**
4. They have a completely ossified endoskeleton.
5. Skull is **monocondylic** with single median occipital condyle.
6. Alimentary canal is complete and leads into the **cloaca**. It is three-chambered structure to adapted to conserve water. The mouth is terminal and usually armed with conical teeth in both the jaws. These teeth are **homodont** (similar), **acrodont** (root less) and **polyphyodont** (having many sets of teeth in succession).
7. Respiration aerobic and gaseous exchange occurs by lungs. The ventilating movements of lungs are due to the intercostal muscles of the ribs. Cloacal respiration is seen in turtles.
8. Heart is incompletely four-chambered with two auricles and partly divided ventricles. Members of *Crocodilia* have complete four-chambered heart. RBCs are nucleated, oval and biconvex.
9. Kidneys are **metanephric**. A thin-walled urinary bladder is present in lizard and helps in water retention. It is altogether absent in snakes and crocodiles. The crocodiles are **ammonotelic**, turtles and alligators are **ureotelic**, lizards and snakes are **uricotelic**.
10. They possess 12 pairs of cranial nerves. **Jacobson's organ** is present in the roof of buccal cavity. It is concerned with smell and is very well-developed in snakes and lizards.
11. Males possess hemipen is as their copulatory organs. These are double in snakes and lizards; and single in crocodiles and turtles. Hemipenis is absent in *Sphenodon.*
12. Most forms are oviparous. Some are ovoviviparous or even viviparous. Eggs are macrolecithal with calcareous shells, i.e. **cleidoic** eggs.
13. Fertilisation is internal, metamorphosis is absent. Development is direct.
14. Embryonic membranes namely **amnion, chorion, yolk sac** and **allantois** appear during development.

Classification

Class-Reptilia is divided into following subclasses

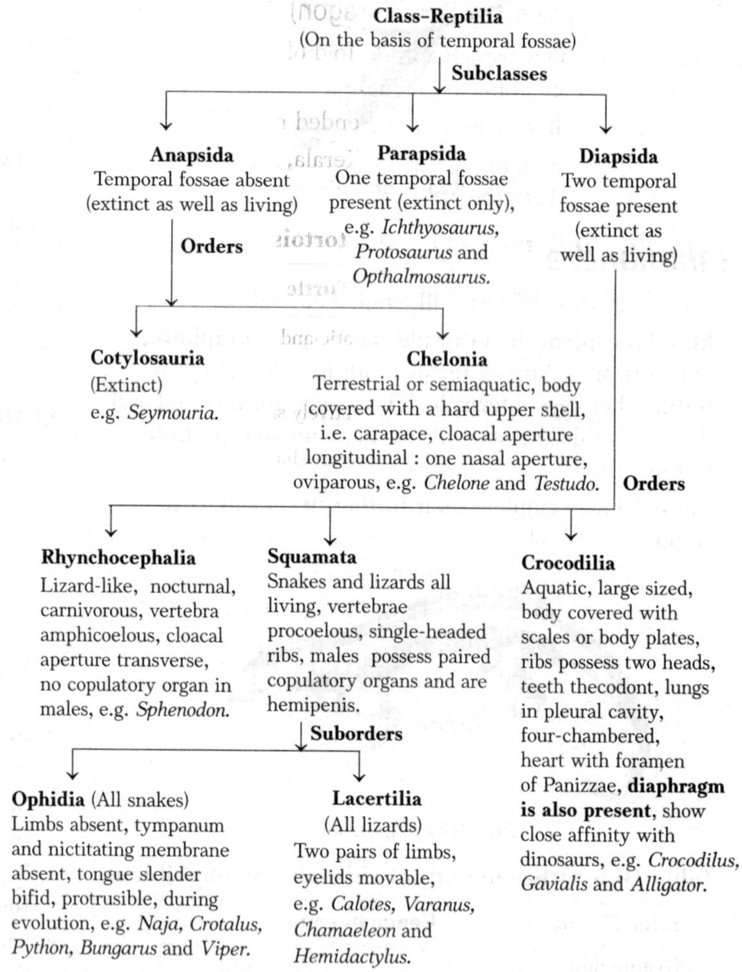

Table 10.7 Common names of some important reptiles

Scientific name	Common name
Testudo	Land tortoise
Chelone mydas	Turtle (green turtle)
Sphenodon punctatum	Tuatara–a living fossil
Hemidactylus	Common wall lizard
Calotes versicolor	Garden lizard or girgit
Varanus (V. komodoensis) (komado dragon) (largest living lizard)	Monitor lizard (Goh)
Phrynosoma	Horned toad lizard
Ophiophagus hannah	King cobra
Enhydrina schistosa (marine snake)	Beaked sea snake
Crocodilus	Crocodile
Gavialis gangeticus	Gharial or gavial

Heloderma (Gila moster)

- It is the only poisonous lizard. It is found in South-West United States.
- Its scales resemble beads, hence it is also called **beaded lizard**.

Draco (Flying lizard or Flying dragon)

- It has membranous parachute like fold of skin called **patagium**, which helps in gliding.
- Patagium is well supported by extended ribs.
- This lizard is commonly found in Kerala, Assom, Myanmar and Malaysia.

Table 10.8 Differences between tortoise and turtle

Tortoise (*Testudo*)	**Turtle** (*Chelone*)
A semi-terrestrial and freshwater form.	An aquatic and marine form.
Usually, large in size.	Comparatively smaller in size.
It has ovoid carapace (leathery shell).	It has heart-shaped carapace.
Forelimbs and hindlimbs are not modified into paddles.	Forelimbs and hindlimbs are modified into paddles.
Head and neck of these can be completely withdrawn into the shell.	Head of these is partially withdrawn into the shell.
Tortoise feeds almost entirely on vegetation.	Turtle are flesh eaters.

Snakes

- The study of snakes is called as **Serpentology**.
- Sternum, pectoral girdle, tympanum and urinary bladder are absent in snakes.
- Eyelids are mostly absent in them but, if present, they are immovable. Nictitating membrane is also absent.
- The tongue is long, slender, forked, protrusible and sensitive to both odours and vibrations.
- Some of them produce hissing sound through the forceful expulsion of air from nares (nostrils).
- Both upper and lower jaw bear pleurodont teeth.

Some unique features of few snakes are given below

- Pit vipers (species of rattle snakes and copper head viper) have highly sensitive heat detectors or thermoreceptor known as **pit organs** or loreal pit. These are used to sense and strike at prey.
- Snakes do not have external ear.
- Most of the snakes are oviparous except *Vipera russelli* and *Hydrophis* which are viviparous.
- Vestigial hindlimbs occur in *Python.*
- The tail of sea snake is laterally compressed.
- Hedonic glands present in male snakes secrete a sticky substance, which hardens like spine and helps to hold the mate firmly.

Identification of poisonous and non-poisonous snakes

Tail, ventral side scales, head and jaw scales help in the identification of poisonous and non-poisonous snakes, as shown below

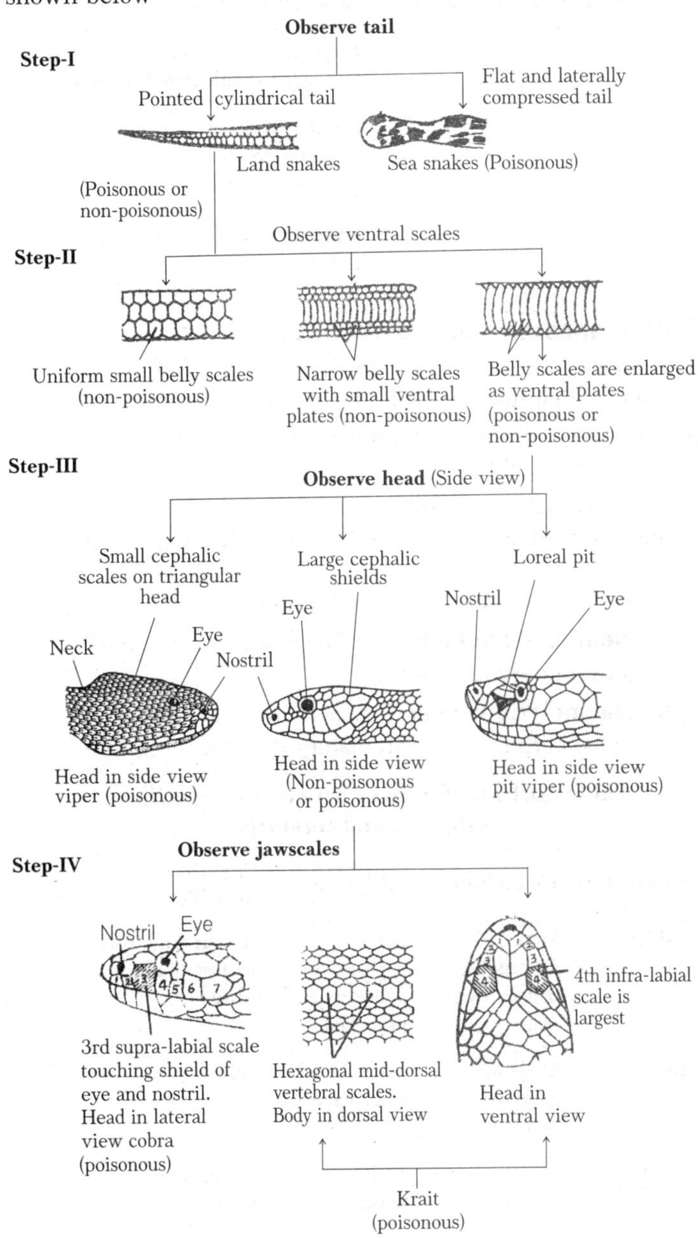

Table 10.9 Examples of poisonous and non-poisonous snakes

Poisonous snakes	Non-poisonous snakes
Naja naja (cobra)	Typhlops (blind snakes)
Viper russell (Russell's viper)	*Python molurus* (*Python*) (Ajgar)
Bungarus caeruleus (common krait)	*Ptyas mascasus* (rat snake)
Ancistrodon himalayansis	*Dryophis* (whip snake)
Hydrophis (sea snakes)	*Dendrophis* (tree snake)
Crotalus horridus (rattle snake)	*Natrix* (water snake)
	Uropeltus (similar to blind snakes, also called **rough tailed snakes**)
	Lycodon (wolf snake, mimics the poisonous krait, but is non-poisonous)
	Tropidontus (common pond or grass snake)

Poison apparatus and snake venom

- All poisonous snakes have a pair of poison glands which are modified superior labial or parotid salivary glands.
- The poison injecting fangs (teeth) are maxillary teeth located on maxilla.
- Snake venom is a mixture of enzymes and specific toxin's and is a good digestive juice. It is fatal when mixed with blood.
  There are two categories of snake venom
  (i) **Neurotoxic venom** (of cobra, krait and sea snake) It acts on nervous system.
  (ii) **Haemolytic venom** (in vipers) It breaks down red blood corpuscles of circulatory system.

Table 10.10 Differences between *Crocodilus*, *Alligator* and *Gavialis*

Characters	*Crocodilus*	*Alligator*	*Gavialis* (unique crocodile)
Habits	More aggressive, dangerous to man.	Less aggressive, attack when provoked.	Can eat only fish because of narrow throat 6 m.
Snout	Moderately long and pointed.	Short, broad and rounded.	Very long and slender.
Teeth	18-19/15, unequal.	17-22, unequal.	27-29, almost equal.
Fourth mandibular tooth	Fits into a maxillary groove and visible externally.	Fits into a pit and not visible externally.	No such structure seen.

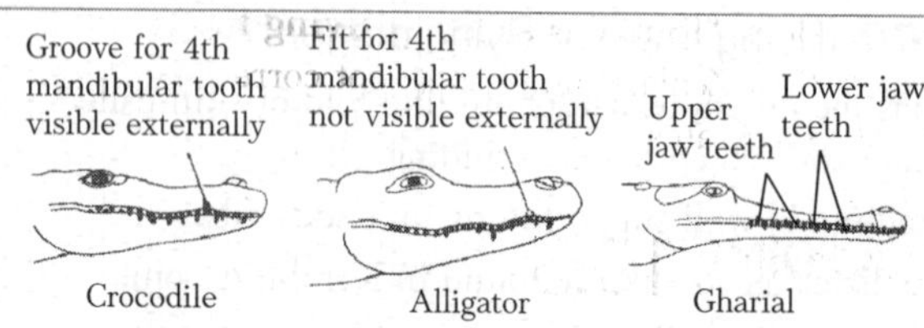

Characters	*Crocodilus*	*Alligator*	*Gavialis* (unique crocodile)
Hump	Such structure does not develop.	Such structure does not develop.	During breeding, a hollow hump (ghara-like) develops at the end of snout of the male. Hence, it is commonly called **gharial**.
Distribution	Central America, Africa, Asia (including India), Malaya, Indonesia, North Australia.	North America, China.	India, Myanmar (Burma), Pakistan and Nepal.

Note • *Saurology is the study of lizards.*
- *Benadryl is used as an antidote to countract the effect of hemotoxins.*
- *Snake bites are treated with antivenum serum containing antibodies.*
- *Hafkin Institute of Bombay is the main centre where antivenom is produced.*

Unique features of Reptiles

- Important points about dinosaurs
  - Longest dinosaur – *Diplodocus*
  - Smallest dinosaur – *Compsognathus*
  - Haviest dinosaur – *Brontosaurus*
  - Horned dinosaur – *Stegosaurus*
  - Biggest flesh eater dinosaur – *Tyrannosaurus*
- Important points about living reptiles
  - Smallest lizard – Dwarf gecko (2 species)
  - Largest snake – *Python*
  - Smallest snake – Thread snake
  - Largest living carnivorous reptile – *Crocodilus*
  - Largest Indian viper – Russel's viper
  - Most poisonous snake of the world – Peninsular tiger snake

- Besides snakes urinary bladder is also not present in alligators and crocodiles.
- The king cobra of India is the only snake in the world which builds a nest.
- *Phrynosoma* or Horned toad have hygroscopic skin. It takes water in the form of dew. It also has the ability to exude red blood like stream from its eyes whenever terrified.
- *Sphenodon* or 'Tautra' is called as **living fossil**. It is found only in New Zealand and has a spot corresponding to third eye of vertebrates.
- Thermomorpha is a group of reptiles, which are considered as the ancestors of mammals.
- *Chameleon* show syndactyly or fusion of digits as an adaptation of arboreal life.
- The tail of most lizards like home and glass lizard (*Ophisaurus attenuatus*) is easily broken off when threatened or seized by predator. This ability is known as **autotomy**. This process is used to confuse enemy.
- *Seymouria* is a connecting link between amphibians and reptiles.

CHECK POINT 10.3

1. Which is the smallest taxonomic group having cranium, vertebral column, ventral heart, pulmonary respiration and two pairs of legs?
(a) Chordata (b) Gnathostomata
(c) Vertebrata (d) Tetrapoda

2. Only one order of amphibians have exoskeleton (dermal scales). These are called
(a) Urodela (b) limbless amphibian
(c) tailless amphibian (d) tailed amphibian

3. Axolotl larva of *Ambystoma* normally fails to metamorphosis due to the
(a) lack of Ca^{2+} and Mg^{2+} ions in water
(b) absence of phosphorus in water
(c) lack of iodine in water or diet
(d) high concentration of iodine in body

4. The greatest evolutionary change that enabled the land vertebrates to be completely free from water, was the development of
(a) four appendages
(b) lungs
(c) cleidoic eggs
(d) four-chambered heart

5. Which animal ejects a stream of blood on being irritated?
(a) Horned toad, *Phrynosoma* (b) Glass snake, *Ophisaurus*
(c) Indian monster, *Varanus* (d) Garden lizard, *Calotes*

6. Some reptiles show autotomy which means
(a) voluntary breaking of tail to confuse enemy
(b) signal for charging
(c) signal for courtship
(d) state of starvation prior to death

Class–Aves

(L. *Avis*-bird; Gr. *ornis*-bird)

This class includes living birds. The study of birds is known as **Ornithology**.

These were originated in Jurassic period and became modernised in Cretaceous period. Their characters seem to be advanced reptilian characters, that is why these are also called as **glorified reptiles** by **Huxley**.

General characteristics

1. These are air breathing, warm-blooded or homeothermic, oviparous, bipedal, feather-covered vertebrates.
2. The body is boat-shaped and is divisible into head, neck, trunk and tail. On the head a **toothless beak** or **bill** is also visible. Infact, it develops as a modification of teeth.
3. The forelimbs are modified into wings and have powerful flight muscles, which bear three clawless digits. Three main categories of flight muscles are
   (i) **Pectoral muscles** These are pectoralis major (most powerful flight muscle) and pectoralis minor.
   (ii) **Accessory muscles** These are coracobrachialis longus and coracobrachialis brevis.
   (iii) **Tensor muscles** These are tensor longus, tensor brevis and tensor accessorius.
4. The hindlimbs are adapted for walking, perching and swimming. These bear four claws.
5. They have epidermal exoskeleton in the form of feathers, scales, claws and beaks. Epidermal scales are present only on the hindlimbs. It is a reptilian characteristic.
6. The skin is dry and without glands, except the cutaneous oil glands or **preen glands** or **uropygial glands.** These are located at the base of the tail. These glands are absent in parrot and ostrich.
7. The skull is **monocondylic**. It is developed from the reptilian diapsid skull.
8. They have **pneumatic**, or air filled bones. These bones contain air cavities to reduce their apperent weight while flying.
9. Most of the bones of the body are fused to provide additional support during flight. The vertebrae are heterocoelous, i.e. vertebrae with saddle-shaped

articulatory centrum. Last thoracic, all lumbar, all sacral and a few caudal vertebrae fuse to form a **synsacrum.** This synsacrum fuses with pelvic girdle and supports hindlimbs. Last caudal vertebrae fuses to form pygostyle. It supports tail feathers.

10. The sternum is large and it bears a keel for the attachment to flight muscles.
11. Clavicle and interclavicle bones fuse to form V-shaped furcula also known as **wish bone.** It is absent in flightless birds.
12. Eyes are powerful and have nictitating membranes, sclerotic ossicles and pecten. The pecten helps in the nutrition of eyeball and is found in all birds except kiwi.
13. Two special structures crop and gizzard are present in alimentary canal. The crop stores and softens the food. The gizzard is muscular to crush the softened food. Some birds keep stones in their gizzard to effectively crush seeds and grains. Gall bladder is absent. Cloaca is 3 chambered and adapted for water conservation as seen in reptiles.
14. The lungs are provided with air sacs. The birds exhibit **double-respiration** that takes place in lungs. Their air sacs are spongy and inelastic. These work to supplement the gaseous exchange. They also extend into the bones.
15. A syrinx or sound box is present at the caudal end of the trachea along with larynx.
16. The heart is four-chambered with two auricle and two ventricle and without sinus venosus. This keeps the oxygenated and deoxygenated blood separate. Only right systemic arch is present. RBCs are oval, biconvex and nucleated.
17. Birds are uricotelic with metanephric kidneys. The urinary bladder is absent. *Rhea americana* (The American *Rhea*) is the only bird to have a urinary bladder.

    Urine is semisolid in the form of uric acid crystals. Excreta of marine birds contains guanine, hence they are also called as **guano.**
18. Birds have twelve pairs of cranial nerves.
19. They exhibit sexual dimorphism. The females have a single left ovary and lay telolecithal (large yolk), cleidoic and amniotic eggs.
20. Fertilisation is internal, but lacks copulatory organs except Ratitae, geese and ducks.

    Development of all the four embryonic membranes is the characteristic feature of embryonic development.
21. *Dromaius* (Emu) is the second largest living bird, after Ostrich *(Struthio)*. Largest egg also belongs to ostrich. Parental care is well-marked feature of members of class-Aves.

Classification

The members of class-Aves have been classified as follows

Class-Aves

Subclasses ↓

Archaeornithes (Extinct)
Homodont (same type) teeth in both the jaws, long tapering tail weak, amphicoelous vertebrae, keeled sternum, non-pneumatic bones, hand with clawed fingers, wings primitive with little power of flight, e.g. *Archaeopteryx* (ancient or lizard bird) and *Archaeornithes.*

Neornithes
Includes extinct as well as living birds. Teeth are absent except in some fossil birds, wings are well-developed and adapted for flight, tail short and reduced, fingers of the wings are without claw and with heterocoelous vertebrae.

Superorders

Odontognathae (Extinct)
Jaw bear teeth for catching fish. Keel in sternum absent, e.g. *Hesperornis* and *Ichthyornis.*

Palaeognathae
flightless running birds
- Wings vestigial or rudimentary, feathers without any interlocking mechanism.
- Oil glands are absent, except in *Tinamus* and kiwi.
- Sternal keel is vestigial, flat or raft-like. Pygostyle is absent or reduced.
- Syrinx is absent.
- Male has a penis, e.g. *Struthio camelus* (African ostrich), *Rhea americana* (American ostrich), *Dromaius* (Emu), *Casuarius* (cassowary) and *Apteryx* (kiwi).

Impennae
- Modern aquatic flightless birds with paddle-like wings or flippers.
- Feet are webbed. The skeleton is solid air sacs are absent.
- Teeth are absent.
- Sternum without keel.
- The integument is with a fatty insulating layer, e.g. *Aptenodytes* (emperor penguin) and *Eudyptes* (rock hopper penguin).

Neognathae (Carinatae)
- Modern flying birds, with well-developed wings and feathers having interlocking mechanism.
- Sternum with well-developed keel. Males have no copulatory organ.
- Some important orders of flying birds are
  Anseriformes, e.g. swans, geese and ducks.
  Falconiformes, e.g. vultures, eagles, hawks and falkons.
  Columbiformes, e.g. pigeons.
  Psittaciformes, e.g. parrots.
  Cuculiformes, e.g. cuckoo.
  Coraciiformes, e.g. kingfishers.
  Passeriformes, e.g. crow and thrashers.

Flight adaptation in birds

In birds almost every system is modified to support flight as described below

- The **boat-shaped body** helps to propel through the air easily.
- The **feathers** constitute very smooth covering over the body to reduce the friction of air. Due to non-conducting nature of these, body temperature is maintained. Feathers of tail (rectrices) form fan-like structure and steer the body during flight.
- **Wings** (remiges) act as main organ of flight with association of feathers. They are responsible for supporting the bird during the flight. Remiges are attached by ligament or directly to the bone.
- Birds have vary characteristic phenomenon of **perching**, i.e. holding of twig locks automatically after sitting over hindlimbs.
- Extremely powerful muscles are developed to enable the wings to work during flight.

  The arrangement of these flight muscles is given in the figure below

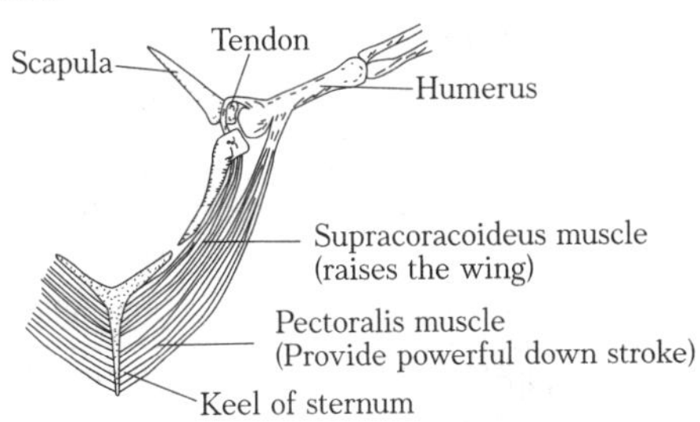

***Fig 10.15** Major muscles of flight in birds*

- The **bones** are light, hollow and provide more space for the muscle attachment. Due to **pneumatic** nature, weight of the body is reduced, which is helpful in flight. The bones are without bone marrow.
- Most of the bones of the body are fused. This helps in bearing additional pressure during flight.
- Sternum supports the viscera and provide the surface for the attachment of flight muscles. It bears a **prominent keel** where the flight muscles are attached.
- The **furcula** (wishbone) serves as a brace during the flight stroke. It is derived from the pectoral girdle.
- Lungs are provided with air sacs. These air sacs serve dual purpose as
  (i) They supplement the additional oxygen demand created during flight.
  (ii) Their presence in bones makes the body lighter.
- Optic lobes and cerebrum are well-developed, as visual sense and intelligence are much more required in birds. Cerebellum is also well-formed for balancing.
- The power of accommodation of eyes is well-developed due to the presence of a comb-like structure pecten, which projects from the blind spot into the vitreous chamber. The sclerotic of eye is bony here. It helps in protection of eye.
- Another weight reducing adaptation in birds is the absence of a urinary bladder and a urethra. Water is reabsorbed in urinary tubules and coprodaeum (a chamber of cloaca). So, the outcome of the nitrogenous waste is the insoluble uric acid, which is thrown at once along with faecal matter.
- Reproductive system has single left ovary and oviduct to reduce the weight of the body.

Feathers and their types

Body in birds is covered by feathers made up of keratin protein. An arrangement of feathers on the body of birds is called **pterylosis**.

An outline of these feathers are as follows

(i) **Contour feathers** These are small feathers that cover the body, wings and tail. Each contour feather has a central axis and a vane.

(ii) **Flight feathers or Quills** These are useful in flights and can be of following types

(a) **Remiges** These are large wing feathers and further categorised to

- **Primaries** which are attached to the bones of the hand.
- **Secondaries** which are attached to the bones of the forearm.
- **Tertiaries** which are attached to the humerus of upper arm bone.

(b) **Retrices** These are large tail feathers.

(c) **Coverts** These are found at the edge of remiges and rectrices.

(iii) **Filoplumes** These are hair-like feathers scattered over body surface and lie between the contour feathers. These act as sensory organs, registering pressure and vibration.

(iv) **Bristles** Modified filoplumes found in certain birds near nostrils and eyes. These are used as a touch sensor or funnel that make the bird reflexively snap up food.

(v) **Down feathers** Found only in the newly hatched birds. These form their first feathery covering, which provides insulation.

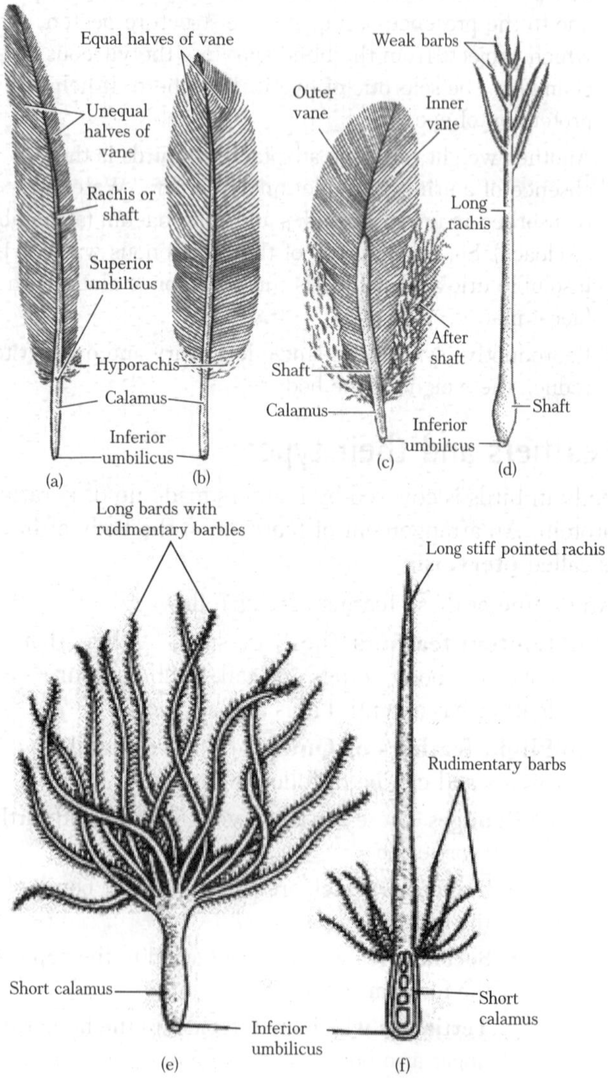

Fig. 10.16 Feathers : (a) Remiges, (b) Retrices, (c) Contour, (d) Filoplume, (e) Down, (f) Rictal bristle

Unique features of Aves (Birds)

- Largest bird-*Struthio camelus*-8 feet height, 300 lbs weight.
- Smallest bird-*Mellisuga helenae*-5.5 cm long.
- Largest sea bird-*Diomedea epomorphora* (royal albatross); wing stretch 200-375 cm.
- Deepest diver bird-*Aptenodytes torsteri* (emperor penguin)
- Fastest swimmer bird-*Pygoscelis papua* (Penguin)
- Fastest runner bird-*Struthio* -60 km/h speed.
- Second largest bird is emu (*Dromaeus*) as is the most primitive living bird.
- Though polygamy is common in birds, emu is strictly monogamous.
- Largest egg of bird-*Struthio* - 1.5-L capacity.
- Smallest egg of bird : *Mellisuga* (hummingbird).
- The hummingbird is the only bird which can fly backward as well as forward.
- Recently extinct bird is *Dodo* (passenger birds) of mauritius.
- The incubation period of a hen's egg at 102°F (38.8°C) is 21 days.
- Wings of penguins are paddle like in structure. These act as flippers during swimming and are covered by scale like feathers.
- The **bursa of Fabricii** is a lymphoid organ of birds. It produces lymphocytes and also called **cloacal thymus.**
- Kiwi is the only bird known to use the sense of smell for finding food. It lays longest eggs in proportion to its own size.
- True penis is present in ostrich, duck, swan and goose.

Table 10.11 List of flightless birds

Common name	Genus	Distribution
African ostrich	*Struthio*	Africa and Arabia
South American ostrich	*Rhea*	South America
Cassowary	*Casuarius*	Australia and New Guinea
Emu	*Dromaius*	Australia
Tinamou	*Tinamus*	South America
Kiwi	*Apteryx*	New Zealand

Class–Mammalia (L. *mamma*–breast)

These are considered as most advanced and most intelligent land vertebrates found in the world.

The study of mammals is known as **Mammalogy.**

Coenozoic era is considered as the Golden Age of Mammals.

General characteristics

1. These are terrestrial, warm-blooded or endothermal or homeothermal animals.
2. The body is variously-shaped and generally divisible into head, neck, trunk and tail.
3. The body cavity is divided into anterior thoracic cavity and posterior abdominal cavity by a **muscular diaphragm.**
4. Their skin is covered by horny epidermal hair coat called **pelage** or **fur coat**. It acts as an insulator and helps to maintain the high body temperature.

5. The skin has two kinds of glands, **sudoriferous** or **sweat** glands and **sebaceous** glands.
6. The sweat glands produces sweat. Main function of sweat glands is temperature regulation. In females some sweat glands are modified to mammary glands.

   These help in nourishing young ones. It is a unique identifiable feature of mammals.
7. Sebaceous glands produce an oily substance sebum which makes the hair water resistant.
8. Movable eyelids are present. Ear has three parts-external, middle and internal. External ear has a large, fleshy **pinna** for collection of sound waves.
9. Two pairs of limbs, pentadactyle, each with 5 or fewer digits are present. These are adapted variously to facilitate different locomotory movements such as running, walking, climbing, burrowing, swimming or flying.

   However, in whales and dolphins, limbs are modified to flippers or paddles. The tips of limbs are provided with claws or nails or hooves.
10. Mammals show three types of locomotion, i.e.
    (i) **Digitigrade** The condition in which an animal stands or walks on its digits or toes, e.g. dog, cat, etc. These animals generally move more quickly and quietly than other animals.
    (ii) **Unguligrade** The condition in which mammals use the tips of their toes to sustain their whole body weight while moving or swimming. These are also called **hoofed mammals**, e.g. cow, buffalo, etc.
    (iii) **Plantigrade** means walking on the sole with the heel touching the ground, e.g. human, bear, etc.
11. Endoskeleton is bony. Skull is **dicondylic**; jaw suspension craniostylic and bones of suspensoria moved to middle ear as ear ossicles.
12. Cervical vertebrae are seven in number except in sloths and sea cow.
13. **Alimentary canal is complete** with mouth at its proximal end and anus at its distal end. The cloaca is absent in most cases.
14. Buccal cavity has true salivary glands. Teeth in both jaw can be of
    - **Thecodont** teeth develop in sockets.
    - **Heterodont** teeth are four types, incisors, canines, premolars and molars.
    - **Diphyodont** teeth grow in two sets; temporary or milk teeth and permanent teeth type.
15. Respiration aerobic, gaseous exchange occurs inside spongy and elastic lungs. These lungs are proper ventilating lungs.
16. **Heart is four-chambered** with two auricles and two ventricles. It is myogenic in nature. Double circuit circulation with oxygenated blood flow in left and deoxygenated blood flows in right half.
17. Sinus venosus and conus arteriosus are absent. Only left systemic arch is present. Renal portal system is absent.
18. RBCs are round, biconcave and non-nucleated except camel, llama, where RBCs are oval, flat and nucleated.
19. **Kidneys are metanephric** and bean-shaped. Urinary bladder is present. Chief excretory product is urea (ureotelic). The urinary and genital apertures are separate except in monotremes.
20. Brain is highly developed and has large cerebrum and cerebellum. **Corpus callosum** connects the two halves of the cerebrum. Four optic lobes (corpora quadrigemina) are present. There are 12 pairs of cranial nerve.
21. Sexes are separate, the testes lie in scrotal sacs in most of the cases, but in elephants and prototherians, they are abdominal in position.
22. Fertilisation is internal. They are viviparous except in egg laying monotremes. Development is direct. Mammalian eggs are small and microlecithal, i.e. with little yolk.
23. All mammals except monotremes have three or four extraembryonic membranes, i.e. amnion, chorion and allantois. The yolk sac may or may not be present. Most of the times it is underdeveloped.

Classification

The members of class-Mammalia have been classified as follows

Class–Mammalia

↓ **Subclasses**

Prototheria
(Most primitive mammals)
- Oviparous
- No pinna
- No nipples
- No marsupial pouch
- Digestive and urinogenital tracts open into cloaca, which opens to outside through cloacal aperture.
- Corpus callosum is feebly developed or absent.
- Testes abdominal, no scrotum.
- Females without uterus and vagina.
- No placenta

↓ **Order**

Monotremata
(Connecting link between reptiles and mammals)
e.g. only 3 examples *Ornithorhynchus* (duck billed platypus), *Tachyglossus* or *Echidna* (spiny anteater).

Theria

↓ **Infraclasses**

Metatheria
(Marsupial mammals)
- Viviparous
- Pinna present
- Nipples abdominal
- Marsupial or abdominal pouch often present.
- Anus and urinogenital aperture open into a shallow cloaca surrounded by a common sphincter.
- Corpus callosum is feebly developed or absent.
- Placenta is less developed.

↓ **Order**

Marsupialia
Pouched mammals
Macropus (kangaroo), *Phascolarctos* (kola bear) and *Didelphys* (opossum).

Eutheria
(Placental mammals)
- Viviparous
- Pinna usually present.
- Nipples abdominal or thoracic.
- Marsupial pouch absent.
- Digestive and urinogenital tracts open out by separate apertures.
- Corpus callosum is well-developed connecting two cerebral hemispheres of brain.
- Testes are extra-abdominal in scrotum and lie penis.
- Placenta is well-developed.
- e.g. humans, elephant, tiger and whales, etc.

The important orders and examples of Eutheria are given below

Infraclass–Eutheria

Eutheria has many orders out of which 13 important ones are given below

1. **Order–Insectivora** The most primitive nocturnal eutherians with sharp teeth, prolonged snout and plantigrade feet with claws, e.g. *Sorex* (shrew), *Talpa micura* (mole) and *Paraechinus* (hedgehog).
2. **Order–Chiroptera** It includes **true flying nocturnal mammals**, i.e. bats. The pinna is large and the sense of hearing is very acute. Bats possess **echolocation** apparatus, which helps to locate objects with the help of sound waves during flight, e.g. *Rhinolopus* (horse-shoe bat), *Megaderma* (Indian vampire) and *Pteropus* (flying fox).
3. **Order–Edentata** They are solitary, nocturnal and arboreal mammals. Teeth are few or absent, tongue is long and protrusible, leprosy bacteria is cultured in armadillo, e.g. *Dasypus* (armadillo), *Bradypus tridactylus* (sloth), *Myrmecophaga* (giant anteater).
4. **Order–Pholidota** It includes terrestrial and burrowing forms. Their upper surface is covered with horny scales. The snout is long and teeth are absent. Their tongue is also long, protrusible and sticky. It is used to capture insects, e.g. *Manis* (pangolin or scaly anteater).
5. **Order–Rodentia** It includes gnawing mammals with with sharp chisel like incisors and diastema, i.e. gap between teeth.
   Most of them are small and terrestrial, e.g. *Rattus rattus* (rat), *Funambulus* (squirrel), *Mus booduga* (Indian field mouse), *Hystrix indica* (porcupine), *Cavia* (Guinea pig), etc.
6. **Order–Lagomorpha** Their pinnae are large, upper jaw contains two pairs of incisor, while lower jaw has one pair of incisor, e.g. *Oryctolagus* (rabbit) and *Lepus* (common hare).
7. **Order–Cetacea** It includes **true aquatic mammals**, i.e. whales and dolphins the body is stream-lined and uniformly smooth from head to trunk.
   There are no external projections like limbs, pinnae or hair. They have many floating ribs, i.e. ribs not attached to the sternum.
   The forelimbs are modified to form paddles. The hindlimbs are completely absent. These teeth are homodont (same shape) and called **baleen**.
   A thick subcutaneous layer of fat is called **blubber**, e.g. *Delphinus* (dolphin), *Platanista* (Ganges dolphin),

Physeter (sperm whales), *Orcinus* (killer whale), *Phocaena* (porpoise) and *Balaenoptera* (blue whales).

- **Blue whale** (*Balaenoptera indica*) A gregarious marine mammal of very large size.
  - It feeds on small fishes, molluscs, crustaceans, etc., including planktons.
  - Whale swims with the help of tail and flippers.
  - Gaseous exchange occur through lungs only. It sends out a thick cloud of air mixed with water drops to several metres in height. The phenomenon called **echolocation**.
  - It is also used by them for prey hunting and communication.
  - Its upper jaw bears a whale bone, consisting of two rows of 600-800 plates called **baleens**. It is hunted for its blubber.

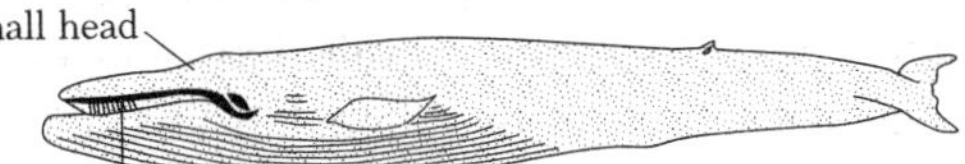

Fig 10.17 Blue whale

- **Dolphin** (*Platanista gangetica*) Highly intelligent aquatic mammal, which can imitate human laughter. It feeds on arthropods, molluscs and fishes.
  - The body is 2-3 metres long with beak like rostrum having numerous teeth.
  - Forelimbs are modified into flippers. Hindlimbs are absent.
  - Gangetic dolphin (*Platanista gangetica*) is the national aquatic animal of India. It is found in Ganga and Brahmaputra rivers. It is also found in Pakistan, China and Latin America. Echolocation is also seen in these fishes.

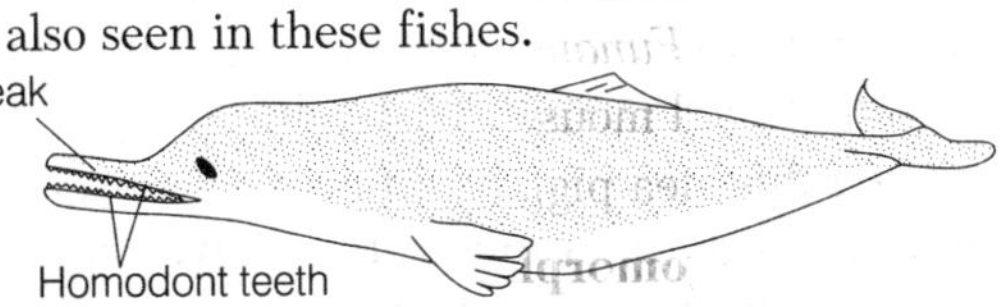

Fig 10.18 Gangetic dolphin

8. **Order-Carnivora** These are carnivorous with large canines powerful jaws and neck muscles.

   These are digitigrade or plantigrade in foot-posture. Mammae are abdominal, e.g. *Panthera leo* (lion), *Panthera tigris* (tiger), *Hyaena hyaena* (the striped hyaena), *Odobenus* (walrus), *Canis familiaris* (dog) and *Felis domestica* (cat).

   - **Tiger** (*Panthera tigris*) They are found in India, China and Indonesia and prefer the places with humid evergreen forests.

     Tiger is the National Animal of India. It has a fawn colour with dark vertical stripes.

     Tiger is polygamous and carnivorous. It is a fast runner and a good swimmer.

9. **Order-Perissodactyla** It includes odd-toed hoofed and herbivorous mammals.

   Their canines are reduced, e.g. *Equus equus* (horse), *Equus asinus* (donkey), *Rhinoceros indicus* (the great one horned rhinoceros).

10. **Order-Artiodactyla** It includes the even toed unguligrade and herbivores mammals.

    Stomach may or may not be ruminating, e.g. *Sus cristatus* (Indian wild boar), *Sus scrofa* (Western wild boar), *Camelus dromedarius* (single-humped camel), *Llama pacos* (llama), *Axis axis* (cheetal or spotted deer), *Cervus unicolor* (sambar deer), *Bos bubalis* (wild buffalo), etc.

11. **Order-Sirenia** It includes sluggish aquatic herbivorous mammals popularly known as **'sea cow'** browsing on sea-weeds and aquatic plants.

    Their forelimbs are modified into flippers to assist in locomotion, e.g. *Halicore* (dugong), *Trichechus manatus* (manatee) and *Rhytina* (sea cow).

12. **Order-Proboscidea** It includes elephants that are largest land animals having a long prehensile, powerful and very mobile proboscis.

    It is a muscular expansion of the nose and bears the nostrils at its tip.

    Upper incisors enlarge to form tusks. Locomotion is digitigrade.

    There is no gall bladder, the mammae are thoracic and the testes remain in the abdomen.

    These have only two living representative species, e.g. *Loxodonta africana* (African elephant) and *Elephas maximus* (the Indian elephant).

13. **Order-Primata** These are primarily arboreal mammals, but some have secondarily returned to land. Hand and feet have five digits with flat nails.

    The hairs cover the entire body except the palms, soles and parts of the face.

    They have binocular or 3D vision. The cheek teeth are commonly crowned with blunt cusps (bunodont).

    Testes are enclosed in scrotum outside abdomen.

    The olfactory region in the brain is reduced, but the cerebrum is large and convoluted.

    They show parental care and their post-natal development is slow, e.g. *Loris malabaricus* (loris), tarsius, lemur, baboon, gibbons, *Pan* (chimpanzee), *Macaca* (monkey) and *Homo sapiens sapiens* (human).

    - **Rhesus Monkey** (*Macaca mulatta*)

      It is a common monkey that is arboreal as well as terrestrial, omnivorous and gregarious.

      Body is yellowish-brown.

Head is rounded. Face is red. Cheek pouches are present.

Digits bear nails. Heads and feet are of grasping type. Tail is present, but is not prehensile.

- **Baboon** (Langur, *Semnopithecus/Presbytis entellus*)
It is terrestrial, arboreal, gregarious and herbivorous primate.
It can leap a distance of 6-9 metres. Face, pinnae, hands and feet are black.
Body is greyish. Cheek pouches are absent.
Head is round. Eyebrows possess long stiff hair.
Tail is very long and slightly prehensile.

Important members of Proto and Metatheria

Duck-billed platypus (*Ornithorhynchus anatinus*)

- It is a semi-aquatic prototherian found in Tasmania and Australia.
- It lives in burrows on the banks of river and streams.
- The animal has a broad bill or beak, having teeth only in the young stage, along with flat tail, webbed and clawed digits.
- Females possess mammary glands, but teats or nipples are absent.
- Milk collects in two abdominal grooves from where the young ones obtain the same through lapping. Male has a poisonous spur on hindlimbs.
- Platypus feeds on freshwater animals. It swims with the help of webbed feet and flattened tail.
- It is a connecting link between reptiles and mammals.
- It is an insectivorous prototherian of Tasmania, Australia and New Guinea.

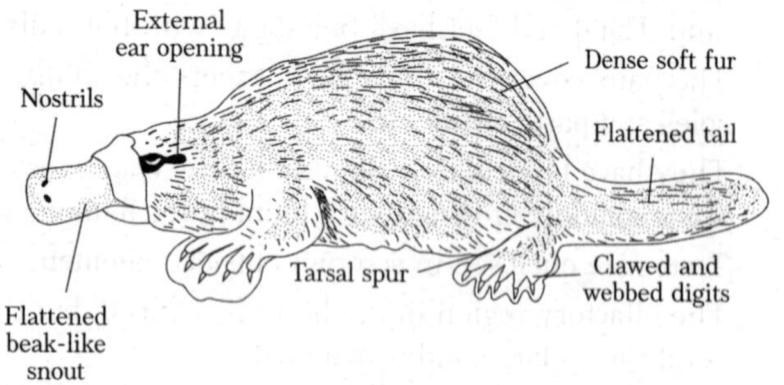

***Fig 10.19** Duck-billed platypus*

Spiny anteater
(*Tachyglossus aculeatus/Echidna aculeata*)

- Jaws are prolonged into beak-like muzzle. Teeth are absent. Tongue is sticky.
- Hair are present. Second claw of each hindlimb is longer and curved. It is used for cleaning spines and fur hence, called **toilet claw**.
- Male has a grooved poison spur on hindlimbs. Milk glands are functional in both sexes.

Kangaroo (*Macropus major*)

- It is the National animal of Australia. It has a small head with pinnae, a long muscular tail, shorter forelimbs with five digits and longer hindlimbs with four digits (one free, three fused).
- Kangaroo can walk on the forelimbs, but gallops on hindlimbs. Its locomotion is called **saltatorial locomotion**.
- Young one is born singly in blind, naked and helpless state. It is kept in the ventral pouch called Marsupial pouch of the female till it grows to become independent.

Unique features of some mammals

- Three-toed sloth (*Bradypus*) have 9 cervical vertebrae.
- *Manatee* have 6 cervical vertebrae.
- In humans sweat glands are absent at lip borders, clitoris, glans penis, nail beds, undersurface of prepuce.
- Spiny anteater, scaly anteater, moles, cetaceans, sirenians and some rodents do not possess sweat gland.
- Fourth premolar of upper jaw and first molar of lower jaw of carnivores are called **carnassial teeth**. These are special shearing teeth for cracking bones.
- The horns of rhinoceros, scales of scaly anteater and spine of porcupine are derivatives of hairs.
- Just like retina of owls, the retina of shrews, hedgehog, rodents and bats also contains rods only. Thus, their eyes are suitable for night vision only.
- The retina of squirrels have cones only. It is similar to fowls among birds. Hence, both of these can see in day time only.
- In some insectivores, chiropteres and rodents scrotum is temporary, i.e. formed during breeding season only. Their testes later moves to their abdominal cavity.
- Edentates, proboscideans (elephants), cetaceans (whales) always have their testes inside the abdominal cavity.

CHECK POINT 10.4

1. Which one is well developed in Aves?
   (a) Right systemic arch
   (b) Left systemic arch
   (c) Both (a) and (b)
   (d) A complex of three aorta

2. Birds have bipedal locomotion as it
   (a) reduces body weight
   (b) increases rate of locomotion
   (c) provides more support to the body
   (d) spare forelimbs for flight

3. Which one of the following in birds, indicates their reptilian ancestry?
   (a) Two special chambers crop and gizzard in their digestive tract
   (b) Eggs with a calcareous shell
   (c) Scales on their hindlimbs
   (d) Four-chambered heart

4. Which of the following is an exclusive character of class—Mammalia?
   (a) Homoiothermy
   (b) Internal fertilisation
   (c) Presence of a 4-chambered heart
   (d) Presence of a muscular diaphragm

5. Bats can fly at night in the dark because of
   (a) strong vision
   (b) strong olfaction
   (c) capability of producing high frequency sound
   (d) capability of echolocation of obstacles

6. The difference between Rhesus monkey and baboon is
   (a) head is rounded and head is with snout
   (b) one is arboreal and other is terrestrial
   (c) tail is long and tail is short
   (d) cheek pouches are present in one and cheek pouches are absent in the other

EXAM POINTS

- In vertebrates, notochord is present during the embryonic period only. **[NEET 2020]**
- Central nervous system is dorsal and hollow in phylum–Chordata. **[NEET 2020]**
- All vertebrates are chordates, but all chordates are not vertebrates because notochord in replaced by vertebral column in adult of some chordates. **[NEET 2020]**
- The chordate features shared by non-chordata are bilateral symmetry, triploblastic condition, metamerism. **[NEET 2020]**
- Retrogressive metamorphosis occurs in Urochordata. **[WB JEE 2013]**
- Cyclostomes do not posses jaws and paired fins. **[CBSE AIPMT 2015, NEET 2016]**
- Cyclostomes are ectoparasites. **[Kerala CEE 2015]**
- *Petromyzon* is a jawless fish which lays egg in freshwater and whose ammocoete larvae after metamorphosis return to ocean. **[EAMCET 2015, CBSE AIMPT 2015]**
- Fishes belonging to Chondrichthyes do not have air bladder, but it is present in bony fishes. They have heterocercal caudal fin. **[NEET 2020, EAMCET 2016]**
- The Chondrichthyes have lamelli form gills not filamentous gills. **[EAMCET 2015]**
- Jaws of shark contain acrodont teeth. **[CBSE AIPMT 2011]**
- 6-15 pairs of gill slits are present in cyclostomes. **[NEET 2020]**
- Air bladder is present in Osteichthyes. **[NEET 2020]**
- *Scoliodon* and *Ichthyophis* are amphicoelous vertebrat. **[EAMCET 2010]**
- *Exocoetus* is a flying fish. **[BCECE 2015]**
- All bony fishes have four pairs of gills and an operculum. **[Uttrakhand PMT 2015]**
- Poison sting is present in *Trygon*. **[NEET 2020]**
- Lung fish *Protopterus* is a connecting link between fish and Amphibia. **[Chhattisgarh PMT 2015]**
- *Nasikabatrachus* is an amphibian discovered from Western ghats. **[Chhattisgarh PMT 2015]**
- Member of *Crocodilus* have four-chambered heart and are poikilothermal. **[HPMT 2012]**
- The body of reptiles is covered with dry and cornified skin, scales over body are epidermal, they do not have external ears. **[NEET 2013, KCET 2013]**
- The fangs of viper when pulled out can come again with same length. **[Manipal 2012]**
- Aves have crop and gizzard in its digestive system. **[NEET 2018]**
- Aves are oviparous and warm blooded. **[NEET-I 2016]**
- *Syrinx* is a sound producing organ found in birds. **[WB JEE 2012]**
- Crow do not have urinary bladder. **[AIIMS 2017]**
- Mammals have originated from reptiles. **[WB JEE 2012]**
- *Echidna* is the mammal which is oviparous, do not give birth to young ones. Young ones suck the milk from mammary gland. **[MHT CET 2011]**
- Mammals have seven cervical vertebrae and myogenic heart. **[J&K CET 2011, RPMT 2011]**
- Whales, dolphins and seals are aquatic mammal. **[NEET 2017]**
- River-dolphin is the national aquatic animal of India. **[NEET-I 2016]**
- Metatheria is a group of animals having marsupium. **[RPMT 2011]**

Chapter Exercises

(A) Topical Exercise

Phylum–Hemichordata

1 Hemichordata was earlier considered as a subphylum under phylum-Chordata. But now, it is placed as a separate phylum under
(a) Urochordata (b) Cephalochordata
(c) Protochordata (d) Non-chordata

2 Which of the following statements is not correct about Hemichordata?
(a) It consists of a worm-like marine animals with organ system level of organisation
(b) They are bilaterally symmetrical, triploblastic and coelomate animals
(c) In hemichordates, body is cylindrical and is composed of an anterior proboscis, a collar and a long trunk
(d) In hemichordates, respiration takes place through general body surface

3 Consider the following statements.
I. The blood vascular system in Hemichordata is of open type.
II. The larval form in Hemichordata is called tornaria.
Choose the correct option.
(a) Statement I is true, but II is false
(b) statement I is false, but II is true
(c) Both statements I and II are true
(d) Both statements I and II are false

4 Regarding the hemichordates, which of the following is true?
(a) Sexes are separate
(b) Fertilisation is external
(c) Development is indirect
(d) All of the above

5 In which of the following sets, all examples represent hemichordates?
(a) *Branchiostoma, Balanoglossus, Ascidia*
(b) *Balanoglossus, Saccoglossus*, Amphioxus
(c) *Salpa, Doliolum, Branchiostoma*
(d) *Balanoglossus, Saccoglossus*, Acorn worm

6 The free-swimming larva of hemichordates resembles the
(a) bipinnaria larva of echinoderms
(b) miracidium larva of platyhelminths
(c) auricularia larva of echinoderms
(d) pluteus larva of echinoderms

7 The body of *Balanoglossus* is divisible into
(a) proboscis, tunic and trunk
(b) collar, trunk and tunic
(c) proboscis, collar and trunk
(d) proboscis, stomochord and trunk

Phylum–Chordata

8 Notochord persists throughout the life in
(a) Gnathostomata and Agnatha
(b) Cephalochordata and Cyclostomata
(c) Urochordata and Cephalochordata
(d) Urochordata and Cyclostomata

9 Gill clefts which do not bear gills in chordates are called
(a) pharyngeal-clefts (b) ephemeral-clefts
(c) visceral-clefts (d) gill slits

10 The post-anal tail present in most chordates is characterised by the absence of
(a) coelom and viscera (b) nerves
(c) blood vessels (d) muscles

11 Which of the following is the fundamental characteristic present in chordates?
(a) Ventral spinal cord
(b) Dorsal heart
(c) Pharyngeal gill slits
(d) Blood flow in forward direction in ventral blood vessels

12 Which of the following is a characteristic of chordates?
(a) Closed circulatory system
(b) Post-anal tail
(c) Both (a) and (b)
(d) Dorsal heart

13 Connecting link between Echinodermata and Chordata is
(a) *Peripatus* (b) *Archaeopteryx*
(c) *Balanoglossus* (d) None of these

Urochordata and Cephalochordata

14 Subphylums—Urochordata and Cephalochordata are often referred to as
(a) Protochordata
(b) Hemichordata
(c) Chordata
(d) Non-chordata

15 Any one of the following set of characters is definitely found at some stage in the life cycle of a chordate.
(a) Mammary glands, hair, gill cleft
(b) Notochord, vertebral column, gill cleft
(c) Notochord, gill cleft, dorsal tubular nervous system
(d) Dorsal tubular nervous system, vertebral column, scales

16 In Urochordata, the notochord is
(a) present only in larval tail
(b) present only in adult
(c) persistent throughout their life
(d) None of the above

17 Choose the animals that belong to subphylum-Urochordata of class-Protochordata from the options given below.
(a) *Balanoglossus, Salpa, Doliolum*
(b) Amphioxus, *Balanoglossus, Salpa*
(c) *Herdmania, Doliolum, Salpa*
(d) *Herdmania, Balanoglossus*, Amphioxus

18 Retrogressive metamorphosis is present in
(a) *Herdmania* (b) Amphioxus
(c) *Balanoglossus* (d) All of these

19 Which of the following is not a salient feature of tunicates?
(a) They are mostly sessile, filter feeding invertebrates
(b) The adult body is enclosed within a test composed of cellulose-like organic substance
(c) The branchial sac is perforated by numerous apertures called stigmata
(d) The blood of *Herdmania* is blue because of the presence of copper in blood

20 Cephalochordates have a notochord, which
(a) extends from head to tail region
(b) is persistent throughout their life
(c) Both (a) and (b)
(d) None of the above

21 In Cephalochordates, excretion occurs by
(a) proboscis gland
(b) pyloric gland
(c) protonephridia with solenocytes
(d) glomerulus

22 Select the statement which is incorrect about reproduction in cephalochordates.
(a) Sexes are separate
(b) They do not show asexual reproduction
(c) They show the presence of numerous gonads which are metamerically repeated in the body
(d) Fertilisation is internal

23 Identify the three defining characteristics of chordates on the diagram of amphioxus.

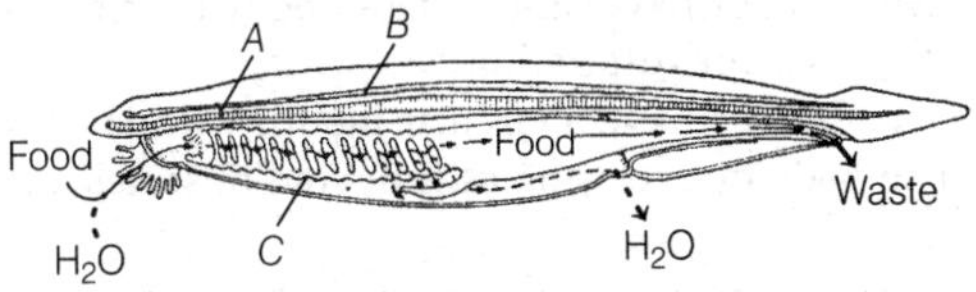

(a) *A*–Lateral line, *B*–Notochord, *C*–Pharynx with gill slits
(b) *A*–Notochord, *B*–Dorsal hollow nerve cord, *C*–Pharynx with gill slits
(c) *A*–Lateral line, *B*–Nerve cord, *C*–Gill slits
(d) *A*–Nerve cord, *B*–Notochord, *C*–Pharynx with gill slits

Subphylum–Vertebrata

24 Which among the following is not a feature of vertebrates?
(a) The notochord observed in embryonic stage is replaced by a vertebral column in adult stage
(b) They show a high degree of cephalisation
(c) Coelom is of schizocoelus type
(d) Chemical coordination is always through endocrine glands

25 In adult vertebrates, the notochord is
(a) degenerated or reduced
(b) replaced by cartilaginous or bony vertebral column
(c) disappeared
(d) persistent

26 Besides the basic chordate characters, vertebrates also have
(a) a ventral muscular heart with two, three or four chambers
(b) kidneys for excretion and osmoregulation
(c) paired appendages, which may be fins or limbs
(d) All of the above

27 What is false about the group-Agnatha?
(a) They lack exoskeleton and paired appendages
(b) They bear a single median nostril
(c) They are warm-blooded animals
(d) It includes the extinct class-Ostracodermi and a living class-Cyclostomata

28 Among the given sets, select the one which are included in extinct class–Ostracodermi.
(a) *Lampetra, Eptatretus, Pteraspis*
(b) *Entosphorus, Ichthyomyzon, Myxine*
(c) *Cephalaspis, Pteraspis, Hemicyclaspis*
(d) *Hemicyclapsis, Eptatretus, Entospherus*

29 Cyclostomes are marine, but migrate to freshwater
(a) for spawning (b) after spawning
(c) after metamorphosis (d) None of these

30 Which one of the following statements is not correct about cyclostomes?
(a) Cyclostomes have an elongated body bearing 6-15 pairs of gill slit for respiration
(b) Cyclostomes have a circular mouth with jaws and teeth
(c) The body of cyclostomes is devoid of scales and paired fins
(d) In cyclostomes, circulation is of closed type

31 Which one of the following pairs comprises 'jawless fishes?'
(a) Mackerels and rohu (b) Lampreys and hagfishes
(c) *Pristis* and hagfishes (d) Lampreys and eels

32 *Petromyzon* is not a true fish because of
(a) absence of median fins
(b) presence of gills
(c) absence of paired fins and true jaws
(d) absence of operculum

33 Gnathostomata are characterised by the presence of
(a) exoskeleton (b) jaws with teeth
(c) paired appendages (d) All of these

34 Select the statement which is incorrect about jawed vertebrates.
(a) They have paired appendages, exoskeleton and paired nares
(b) The internal ear has three semicircular ducts
(c) A single olfactory sac is present
(d) The gonads are paired and have gonoducts

35 Find out the statement which is true about the group-Gnathostomata.
(a) It comprises of four classes
(b) They use hindlimbs for locomotion
(c) This group has organisms having an internal ear with two semicircular canals
(d) The animals of this group have paired appendages and paired nares

36 The greatest problem for the primitive vertebrates reaching freshwater was
(a) absorption of too much water through their skin
(b) loss of water through their skin
(c) availability of less food
(d) encountering new predators

Pisces

37 Consider the following statements.
I. Internal nares are absent in both Condrichthyes and Osteichthyes.
II. Cartilagious fishes have heterocercal tail.
Choose the correct option.
(a) Statement I is true, but II is false
(b) Statement I is false, but II is true
(c) Both statements I and II are true
(d) Both statements I and II are false

38 The body of members of class-Chondrichthyes is
(a) antero-posteriorly flattened
(b) laterally compressed
(c) consist of head, trunk and tail
(d) Both (b) and (c)

39 Consider the following statements with respect to class-Chondrichthyes.
I. Gill slits are separate and without operculum.
II. They have the capacity to regulate the body temperature.
Choose the correct option.
(a) Statement I is true, but II is false
(b) Statement I is false, but II is true
(c) Both statements I and II are true
(d) Both statements I and II are false

40 In class-Chondrichthyes, the mouth is located ventrally and the teeth are modified placoid scales, which are
(a) forwardly directed
(b) backwardly directed
(c) sparsely arranged
(d) None of the above

41 The members of Chondrichthyes have to swim constantly to avoid sinking due to
(a) non-functional anterior fins
(b) high pressure of water
(c) absence of air bladder
(d) heavy body weight

42 Consider the following statements with respect to cartilagious fishes.
I. Teeth like placoid scales are found on their skin.
II. Most fishes have colour vision.
Choose correct option.
(a) Statement I is true, but II is false
(b) Statement I is false, but II is true
(c) Both statements I and II are true
(d) Both statements I and II are false

43 Operculum with spiracle is absent in
(a) *Hippocampus* (b) *Syngnathus*
(c) *Scoliodon* (d) *Gastrosteus*

44 *Scoliodon* is commonly called dogfish due to one of its following characteristics.
(a) Gait
(b) Mouth
(c) Carnivorous
(d) Power of smell

45 Ampullae of Lorenzini of *Scoliodon* is a
(a) mechanoreceptor
(b) thermoreceptor
(c) locomotory organ
(d) copulatory organ

46 Select the statement which is incorrect about *Scoliodon.*

I. It is omnivorous.
II. It is rich in vitamin-A and D.
III. It lacks spiracles.

Choose the correct option.

(a) I and II (b) II and III (c) III only (d) I only

47 In *Torpedo* or electric ray, the electric organs are modified form of

(a) nerve nets
(b) cartilage complexes
(c) bronchial muscles
(d) respiratory centres

48 The spiracles in *Torpedo* are situated

(a) it front of the eyes (b) behind the fins
(c) over the operculum (d) behind the eyes

49 The fish famous for parental care is

(a) *Hippocampus* (b) *Gambusia*
(c) Labeo (d) *Scoliodon*

50 Class-Osteichthyes includes both marine and freshwater fishes with

(a) cartilaginous endoskeleton
(b) claspers
(c) Both (a) and (b)
(d) bony endoskeleton

51 As compared to cartilaginous fishes, bony fishes have dorsal fin

(a) single instead of two (b) two instead of one
(c) single in both (d) two in both

52 The skin of bony fishes is covered with

(a) cycloid and pelvic scales
(b) cycloid, ctenoid or ganoid scales
(c) ctenoid, pelvic and ganoid scales
(d) pectoral, ganoid and pelvic scales

53 Select the statement which is false about *Labeo rohita.*

(a) It is found in clear, sluggish rivers and streams
(b) Head is covered with scales and bears a centrally placed mouth
(c) The mouth has very thick and fringed lips
(d) The young ones of *Labeo* are carnivorous in nature

54 Identify the statement which is incorrect about *Hippocampus.*

(a) It inhabits tropical and temperate seas
(b) Its head resembles the head of a horse
(c) It possesses a long, prehensile tail
(d) On the belly of male, a small anal fin and brood pouch is present

55 *Gambusia* is

(a) parasitic fish
(b) pest of fishes
(c) fish predator of mosquito larvae
(d) a mosquito spreading yellow fever

56 Lung fishes are

(a) marine inhabitants
(b) marine and freshwater inhabitants
(c) freshwater and terrestrial inhabitants
(d) freshwater inhabitants

57 The three genera of living lung fishes are

(a) *Scoliodon, Torpedo* and *Neoceratodus*
(b) *Lepidoserin, Torpedo* and *Anabas*
(c) *Neoceratodus, Protopterus* and *Lepidosiren*
(d) *Clarias, Anabas* and *Neoceratodus*

58 *Latimeria*

(a) is a living fossil
(b) has two lobed dorsal fins
(c) possesses two pairs of external nares
(d) All of the above

Class-Amphibia

59 Which one of the following characters is not related with the amphibians?

(a) Two pairs of limbs
(b) Moist skin without scales
(c) Poikilothermy or cold-bloodedness
(d) Incomplete four-chambered heart

60 Which of the following is/are correct about amphibians?

(a) Body is divisible into head and trunk, while tail may be present in some
(b) Eyes have eyelids and tympanum represents the ear
(c) Oviparous animals with external fertilisation and indirect development
(d) All of the above

61 In amphibians, the alimentary canal, urinary and reproductive tracts open into a common chamber called

(a) urinary bladder (b) anus
(c) cloaca (d) rectum

62 Consider the following statements.

I. The largest RBCs have been seen in amphibians.
II. Frog RBCs are nucleated.

Choose the correct option.

(a) Statement I is true, but II is false
(b) Statement I is false, but II is true
(c) Both statements I and II are true
(d) Both statements I and II are false

63 See the following figures and choose the correct option with their respective orders.

(a) *A*-Caecilians *B*-Caudata *C*-Anura
(b) *A*-Caudata *B*-Anura *C*-Caecilians
(c) *A*-Anura *B*-Caecilians *C*-Caudata
(d) *A*-Anura *B*-Caudata *C*-Caecilians

64 Pick out the set of amphibians which belong to order-Anura.
(a) *Ambystoma, Proteus, Triton, Alytes*
(b) *Alytes, Bombinator, Xenopus, Nototrema*
(c) *Salamandra, Amphiuma, Ambystoma, Cryptobranchus*
(d) *Triton, Ichthyophis, Uraeocotyphlus, Hyla*

65 Which of the following subclass of amphibians are extinct?
(a) Lissamphibia (b) Stegocephalia
(c) Caecilians (d) Anura

66 Select the correct order of classification of *Rana tigrina* up to genus.
(a) Chordata, Craniata, Amphibia, Gnathostomata, *Rana*
(b) Chordata, Craniata, Gnathostomata, Amphibia, *Rana*
(c) Chordata, Amphibia, Gnathostomata, Craniata, *tigrina*
(d) Chordata, Craniata, Amphibia, Gnathostomata, *tigrina*

67 The glands present in the skin of frog are
(a) mucous and poisonous
(b) sweat and mammary
(c) sweat and sebaceous
(d) mucous and sweat

68 Which of the following vertebrates show the formation of middle ear (Eustachian recess) for the first time?
(a) *Rana* (b) *Echis*
(c) *Exocoetus* (d) *Hippocampus*

69 Gymnophiona are
(a) vermiform without limbs
(b) have dermal scales
(c) *Ichthyophis* and *Hyla*
(d) Both (a) and (b)

70 In *Alytes obstetricans*,
(a) the female collects the fertilised eggs and hides the same in safe place
(b) the female after collecting the eggs and hides itself in a safe place
(c) the male collects eggs around its neck and hides itself
(d) the male collects eggs around its legs and hides itself

71 In the amphibian order-Urodela
(a) limbs are small infront and long behind
(b) limbs are of equal size
(c) hindlimbs are absent in adults
(d) limbs absent altogether

72 Salamanders and newts belongs to the order .... and they are ....
(a) Urodela, viviparous
(b) Anura, oviparous
(c) Apoda, sterile
(d) Stegocephalia, parthenogenic

73 Which of the following characters is not shown by a salamander?
(a) It is a semiterrestrial lizard like animal
(b) It is oviparous
(c) Its male discharge sperms in a capsule called spermatophore
(d) It is a tail-less amphibian

74 Consider the following statements.
I. Salamander is cold-blooded and tailed amphibian.
II. *Ichthyophis* is a limbless amphibian and feeds on underground worms and insects.

Choose the correct option.
(a) Statement I is true, but II are false
(b) Statement I is false, but II is true
(c) Both statements I and II are true
(d) Both statements I and II are false

75 From given set of statements choose the option, which is not a characteristic feature of class-Amphibia.
(a) Amphibia originated during Devonian period from Stegocephalia like animals
(b) Ovoviviparity is seen in *Typhlonectes,* which is a caecilian
(c) Stratum germinatium is present in the integument of frog, but not in mammals
(d) Axolotl shows neoteny

Class-Reptilia

76 Which of the following is not a characteristic of reptiles?
(a) They are poikilothermic animals
(b) The ventilatory movement is due to intercostal muscles of the ribs
(c) They possess ten pairs of cranial nerve
(d) The kidneys are metanephric

77 Which of the following statement(s) is/are true about reptiles?
(a) They have creeping or crawling mode of locomotion
(b) They are mostly terrestrial and body is covered by dry cornified skin, epidermal scales or scutes
(c) Heart is three or partially four-chambered
(d) All of the above

78 Which of the following characters does not belong to class-Reptilia?
(a) They lays macrolecithal eggs
(b) They are warm-blooded animals
(c) They are creeping and burrowing type
(d) Monocondylic skull

79 Choose the incorrect statement about reptiles.
(a) Jacobson's organ in reptiles is related to touch
(b) *Sphenodon* is a living fossil
(c) Rhynchocephalia lacks penis
(d) *Sphenodon* has a median eye

80 Which one of the following has exoskeleton of scales and paired copulatory organ or penis?
(a) Sharks (b) Lizards
(c) Urodela (d) Urochordata

81 A house lizard can climb a smooth wall due to the presence of
(a) suckers and mouth
(b) claws on digits
(c) adhesive lamellar pads on digits
(d) sticky ventral side of the animal

82 Syndactly, prehensile tail and long protrusible tongue are the unique features of
(a) Rhesus monkey (b) *Archaeopteryx*
(c) horse fish (d) *Chamaeleon*

83 In which reptile does the patagium is present, which helps in gliding?
(a) *Draco* (b) *Chelone* (c) Snakes (d) *Alligator*

84 Which of the following is not a feature of turtle?
(a) It is an aquatic; mostly marine animal
(b) It is large in size than tortoise
(c) It is viviparous
(d) Forelimbs and hindlimbs are modified into paddles

85 The missing link between amphibians and reptiles is/are
(a) *Archaeopteryx* (b) *Ichthyostega*
(c) *Seymouria* (d) All of these

86 Snake produces hissing sound through
(a) vocal cord
(b) forceful expulsion of air from lungs
(c) forceful expulsion of air from nares
(d) forceful inhalation

87 Choose the correct statement about snakes.
(a) Loreal pit of a pit viper is a chemoreceptor
(b) *Lycodon* is a snake having large head shields and fangs
(c) Poison fangs in snakes are present on their maxillary bone
(d) Cobra is viviparous

88 Consider the following statements.
I. Vestigial hindlimbs occur in *Python.*
II. Post anal tail is present in cobra.
Choose the correct option.
(a) Statement I is true, but II are false
(b) Statement I is false, but II is true
(c) Both statements I and II are true
(d) Both statements I and II are false

89 The sea snakes have
(a) cylindrical tail
(b) dry honey scale at tail end
(c) laterally compressed tail
(d) dorsoventrally flattened tail

90 Consider the following statements.
I. Venom of cobra affects the circulatory system of its prey.
II. Venom of viper affects the nervous system of its prey.
Choose the correct option.
(a) Statement I is true, but II is false
(b) Statement I is false, but II is true
(c) Both statements I and II are true
(d) Both statements I and II are false

91 Choose the correct pair.
(a) Common wall lizard - *Hemidactylus*
(b) Indian krait - *Viper russelli*
(c) Study of lizards - Herpentology
(d) Study of snakes - Nidology

Class-Aves

92 The hindlimbs of birds generally have scales and are variously modified for
(a) walking
(b) swimming
(c) perching
(d) All of the above

93 Which of the following is not a characteristic of Aves?
(a) Presence of feathers, wings and beak
(b) Fully ossified endoskeleton with pneumatic bones
(c) Completely four-chambered heart
(d) Lack the capacity to regulate their body temperature

94 Select the correct statement about class-Aves.
(a) They are warm-blooded (homoiothermous) animals and are able to maintain a constant body temperature
(b) Respiration occurs through lungs and air sacs connected to lungs
(c) They are oviparous with separate sexes, internal fertilisation and direct development
(d) All of the above

95 Consider the following statements.
I. Air sacs of birds reduce body weight.
II. Bird vertebraes are heterocoelous.
Choose the correct option.
(a) Statement I is true, but II is false
(b) Statement I is false, but II is true
(c) Both statements I and II are true
(d) Both statements I and II are false

96 The character of birds, without exception is
(a) omnivorous
(b) flying wings
(c) beak without teeth
(d) lay eggs with calcareous shell

97 Both male and female pigeons secrete milk through
(a) mammary glands (b) crop glands
(c) salivary glands (d) gizzard glands

98 Consider the following statements.
I. Bone marrow is absent in birds.
II. The digestive tract of birds has additional chambers, which are crop and gizzard.
Choose the correct option.
(a) Statement I is true, but II is false
(b) Statement I is false, but II is true
(c) Both statements I and II are true
(d) Both statements I and II are false

99 What is common among dodo, penguin and *Rhea*?
(a) These are flightless birds
(b) These are running birds
(c) These are migratory birds
(d) These have four toes in their feet

100 Remiges are
(a) flight feathers of wings
(b) flight feathers of tail
(c) small feathers similar to quills
(d) small feathers that cover the body

101 Sternum of birds is prolonged to form boat-shaped structure called
(a) keel (b) retrices (c) remiges (d) pygostyle

102 A flying bird lands by
(a) folding the wings back and dropping on the ground or other substratum
(b) folding the wings above and back of the body and coming down
(c) lowering and fanning out the retrices
(d) pronation and forward movements of wings

103 The feathers at the edges of remiges and retrices are called
(a) down feathers (b) coverts
(c) barbules (d) filoplumes

104 Identify the bone derived from the pectoral girdle in bird.
(a) Wish bone (furcula) (b) Xiphisternum
(c) Suprascapula (d) Coracoid

105 In flying birds, which of the following are useful in flight besides pneumatic bones?
(a) Crop and gizzard (b) Syrinx
(c) Quill feathers (d) Preen glands

106 The largest and the most powerful adductor muscle in flying bird is the
(a) coracobrachialis longus (b) pectoralis major
(c) pectoralis minor (d) tensor longus

107 Choose the incorrect pair from the following.
(a) Emu-Flightless bird and confined to Australia
(b) Kiwi-National bird of New Zealand
(c) Ostrich-Wings modified into paddles (flippers)
(d) Dodo-Recently extinct bird

108 Choose the incorrect pair with respect to birds.
(a) Skull – Dicondylic
(b) Study of birds – Ornithology
(c) Eggs of birds – Telolecithal
(d) Rudimentary wings – Albatross

109 Birds differ from bats in the absence of
(a) 4-chambered heart (b) homoeothermy
(c) diaphragm (d) tracheae

Class-Mammalia

110 Which one of the following orders of mammals possesses homodont dentition?
(a) Monotremata (b) Cetacea
(c) Marsupialia (d) Chiroptera

111 Identify the mammals, in which both male and female have functional mammary glands and feed the young one.
(a) Rabbit (b) *Tachyglossus*
(c) *Macropus* (d) Dolphins

112 Which of the following feature(s) is/are unique to mammals?
(a) Milk producing mammary glands
(b) Presence of hair on skin
(c) Presence of pinnae or external ears
(d) All of the above

113 Select the incorrect statement about class-Mammalia.
(a) They are warm-blooded or homoeothermous
(b) Sexes are separate and fertilisation is internal
(c) They are viviparous with exceptions
(d) The development is indirect

114 Select the correct set of animals of class—Mammalia.
(a) Lion, bat, whale, ostrich
(b) Lion, *Hippopotamus*, penguin, bat
(c) Whale, bat, kangaroo, *Hippopotamus*
(d) *Hippopotamus*, penguin, whale, kangaroo

115 Find out which statement does not characterise a mammal.
(a) Mammalian brain differs from amphibian brain in having corpus callosum
(b) Haversian system is characteristic of mammalian bones
(c) Rhesus monkey has the smallest cranial capacity
(d) Long neck of camel is due to increase in number of cervical vertebrae

116 All mammals without any exception are characterised by
(a) viviparity and biconcave red blood cells
(b) diphyodont teeth and 12 pairs of cranial nerves
(c) a muscular diaphragm and milk producing glands
(d) extra abdominal testes and a four-chambered heart

117 Consider the following statements.
I. Eutherian mammals are characterised by true placentation.
II. True placentation is present in *Perameles.*

Choose the correct option.
(a) Statement I is true, but II is false
(b) Statement I is false, but II is true
(c) Both statements I and II are true
(d) Both statements I and II are false

118 In which of the following organisms, testes descend into scrotum in breeding season, but in non-breeding season goes up?
(a) Frog (b) Kangaroo
(c) Shrew (d) Bat

119 Which one of the following characters is not typical of the class-Mammalia?
(a) Seven cervical vertebrae
(b) Thecodont dentition
(c) Ten pairs of cranial nerves
(d) Alveolar lungs

120 Two chief features of mammals which distinguish them from other vertebrates are
(a) hair and viviparity
(b) viviparity and mammary glands
(c) viviparity and internal fertilisation
(d) mammary glands and internal fertilisation

121 Point out the mammalian characters.
(a) Vertebral column, four-chambered heart, diaphragm
(b) Hairy skin, viviparity, mammary glands
(c) Fins, gills, viviparity
(d) Kidneys, lungs, four-chambered heart

122 In which one of the following sets of animals, do all the four give birth to young ones?
(a) Lion, bat, whale, ostrich
(b) Platypus, penguin, bat, *Hippopotamus*
(c) Shrew, bat, cat, kiwi
(d) Kangaroo, hedgehog, dolphin, *Loris*

123 Which of the following is not correct about Metatheria?
(a) Testes are extra-abdominal
(b) Placenta is well-developed
(c) Scrotum in front of penis
(d) Corpus callosum is small or absent

124 Which one of the following statements is true in regard to a certain mammal and its feature?
(a) Bats have feathers
(b) Platypus is oviparous
(c) Elephant is ovoviviparous
(d) Camel has biconcave RBCs

125 See the following figure and choose the correct option with their respective orders.

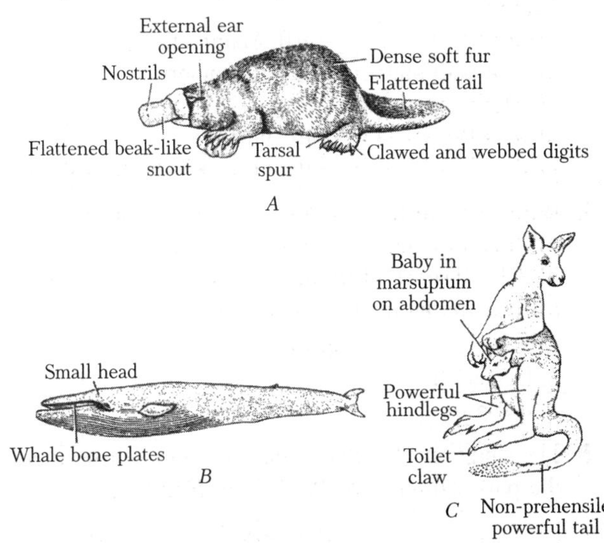

	A	B	C
(a)	Marsupialia	Monotremata	Cetacea
(b)	Monotremata	Cetacea	Marsupialia
(c)	Monotremata	Marsupialia	Cetacea
(d)	Marsupialia	Cetacea	Monotremata

126 Which one is correctly matched?

	Animals	Characteristics	Orders
(a)	Elephant	Digitigrade locomotion	Sirenia
(b)	Rabbit	Shows coprophagy	Lagomorpha
(c)	Blue whale	Presence of pinnae	Cetacea
(d)	Giant anteater	Teeth without enamel	Chiroptera

127 Choose the incorrect pair.
(a) Hairless mammals - Cetacea
(b) Sea cows - Sirenia
(c) Bats - Chiroptera
(d) Whales and dolphins - Rodentia

128 Which one of the following sets of animals share a four-chambered heart?
(a) Amphibian, reptiles, birds
(b) Crocodiles, birds, mammals
(c) Crocodiles, lizards, turtles
(d) Lizards, mammals, birds

129 Which one of the following sets of animals belong to a single taxonomic group?
(a) Cuttlefish, jellyfish, silverfish, dogfish, starfish
(b) Bat, pigeon, butterfly
(c) Monkey, chimpanzee, man
(d) Silkworm, tapeworm, earthworm

130 Identify the aquatic mammal(s) from the following.
I. *Balaenoptera* II. *Equus caballus*
III. *Delphinus* IV. *Pteropus*
V. *Felis catus*
Choose the correct option.
(a) I and III (b) II and IV (c) Only V (d) IV and V

131 Animals which have well-developed echolocation system like that of bats are
(a) whales and dolphins (b) beavers
(c) primates (d) hedgehogs

(B) Cumulative Exercise

(Assorted Questions of the Chapter for Advanced Level Practice)

1 Identify the correct sequence of classification of the following.
I. Eutheria II. Mammalia
III. Tubulidentata IV. Lagomorpha
V. Hyracoidea
(a) II → IV → I → V → III (b) II → I → IV → III → V
(c) II → I → IV → V → III (d) I → V → III → II → IV

2 Which of the following is/are correct for axolotl larva?
I. It shows neoteny and paedogenesis.
II. The absence of thyroxin affects metamorphosis.
III. It is the larva of Hemichordata.
Choose the correct option.
(a) I, II and III (b) I and II
(c) II and III (d) Only III

3 The diagram of *Labeo rohita* is given below. Identify the parts labelled *A, B, C, D, E, F* and *G*.

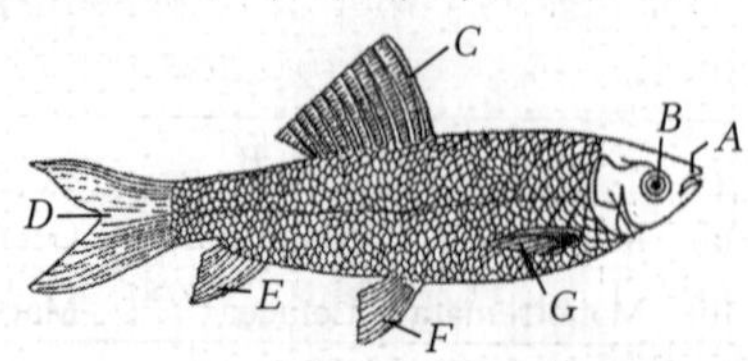

	A	B	C	D	E	F	G
(a)	Stimulus	Receptor	Sensory nerve	Motor nerve	Effector	Pectoral fin	Pelvic fin
(b)	Nostril	Eye	Anal fin	Caudal fin	Dorsal fin	Pectoral fin	Pelvic fin
(c)	Nostril	Eye	Dorsal fin	Caudal fin	Anal fin	Pelvic fin	Pectoral fin
(d)	Nostril	Eye	Dorsal fin	Caudal fin	Pectoral fin	Anal fin	Pelvic fin

4 Match the following columns.

Column I (Birds)	Column II (Types of beak)	Column III (Types of feet)
A. Duck	(i) Tearing and piercing	1. Raptorial
B. Sparrow	(ii) Water straining	2. Swimming
C. Owl	(iii) Seed eating	3. Perching
	(iv) Spatulate	4. Wading

Codes

	A	B	C
(a)	iii, 3	iv, 2	ii, 4
(b)	ii, 4	iii, 3	iv, 1
(c)	i, 4	iv, 1	iii, 3
(d)	ii, 2	iii, 3	i, 1

5 What is true about mammals?
I. Skin is covered by epidermal hair.
II. Evolved during Triassic period of Coenozoic era.
III. Kidneys are metanephric.
IV. Fertilisation is external.
Choose the correct option.
(a) I, II and III (b) I and II
(c) II and IV (d) I and III

6 The order–Primata have
I. testes enclosed in scrotum.
II. reduced cerebrum.
III. animals like lemurs, monkeys and apes.
IV. absence of haemochorial placenta.
Choose the correct option.
(a) I, II and III (b) I and II
(c) II and IV (d) I and III

7 Identify the false statement from the following.

I. The presence of uropygial gland is associated with pigeon.

II. Migration of birds is studied under 'Phenology'.

III. The presence of bone marrow and preen gland are fundamental characteristics of class-Aves.

Codes

(a) Only I (b) II and III
(c) Only III (d) All of these

8 Which is/are not correct about flightless birds?

I. These birds show discontinuous distribution.

II. They have well-developed legs, head is small, rudimentary eyes and wings.

III. Oil glands, keel, pygostyle and syrinx are present.

IV. The male has a large penis.

Codes

(a) I and II (b) II and III
(c) Only III (d) Only IV

9 Identify the statements as True (T) or False (F) and choose the correct option.

I. Chief excretory product in mammals is ammonia.

II. Cervical vertebrae are seven in number in mammals except is slots.

III. True flying nocturnal mammals like bats belong to Chiroptera.

IV. Mammalian eggs are small and microlecithal.

Codes

	I	II	III	IV
(a)	T	T	F	F
(b)	F	F	F	T
(c)	T	F	T	F
(d)	F	T	T	T

10 Match the following amphibians with their correct scientific names.

Column I	Column II
A. Flying frog	1. *Xenopus laevis*
B. African clawed toad	2. *Alytes*
C. Congo eel	3. *Rhacophorus*
D. Caecilian	4. *Amphinima*
E. Midwife toad	5. *Ichthyophis*

Codes

	A	B	C	D	E
(a)	2	5	3	1	4
(b)	3	1	4	5	2
(c)	4	3	2	1	5
(d)	2	4	5	3	1

11 Match the Column I with Column II and select your correct answer from the following codes.

Column I	Column II
A. *Arius*	1. Four eyed fish
B. *Hippocampus*	2. Mouth breeder
C. *Gambusia*	3. Sucker fish
D. *Echeneis*	4. Larvivorous fish
E. *Anableps*	5. Male broods eggs

Codes

	A	B	C	D	E
(a)	1	2	3	4	5
(b)	2	5	4	3	1
(c)	2	5	3	4	1
(d)	5	4	3	2	1

12 Identify the correct statement with respect to structures marked as X and Y in the diagram given below.

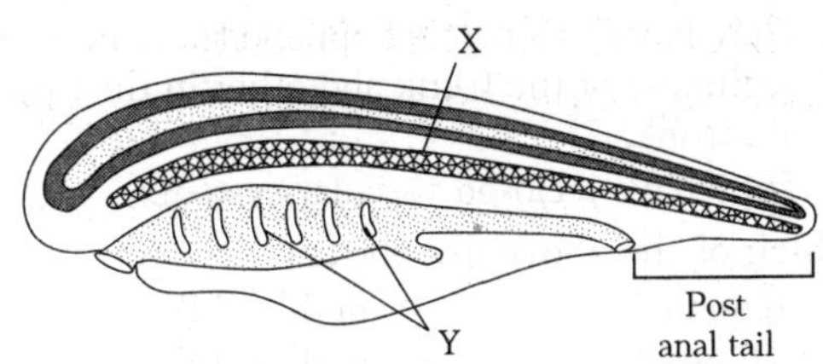

Chordate characters

(a) Both X and Y are present throughout the life in Urochordates

(b) Both X and Y takes part in circulation of water for respiration

(c) In vertebrates, X is replaced by a cartilaginous or bony vertebral column in adults and Y occurs only in embryonic stages in higher chordates

(d) X may be present in few non-chordates, but Y is completely absent in them

13 The characteristics of Vertebrata are

I. high degree of cephalisation is seen in the group.

II. coelom is well-developed and of enterocoelous type.

III. cranial nerves either 8, 10 or 12 pairs.

IV. notochord is seen throughout the life of all animals.

Codes

(a) I, II and III (b) I and II (c) II and IV (d) I and III

14 Which of the following are correct regarding Osteichthyes?

I. They include both marine and freshwater fishes with bony endoskeleton.

II. They are also called bony fishes.

III. They are also called cartilaginous fishes.

IV. They include sharks, rays and rat fishes.

Codes

(a) I and II (b) II, III and IV
(c) II and IV (d) I and III

15 Which of the following characters belong to Cyclostomata?

I. Their tail is diphycercal.
II. Metanephric kidney with ducts to urinogenital papilla are present.
III. They are divided into Pisces and Tetrapoda.
IV Have typhlosole to increase absorptive surface area.

Codes

(a) I, II and III (b) I and IV
(c) II and III (d) I and III

16 The statements given below highlight some of the chordates features.

I. They have a hollow nervous system which is present ventrally above the notochord.
II. They possess four pharyngeal gill slits present in the form of perforations on the lateral sides of pharynx.
III. Epidermis is made up of stratified epithelium.
IV. They have two pairs of appendages, one pair at the beginning of the trunk and other in the middle of the trunk.
V. They have a closed vascular system.

Which of the above are correct?

(a) I, II and IV (b) III and IV
(c) II, III and V (d) III and V

17 Match the following columns.

	Column I (Common names)	Column II (Scientific names)
A.	Cat fish	1. *Clarias batrachus*
B.	Milk fish	2. *Chanos chanos*
C.	Indian prawn	3. *Mugil cephalus*
D.	Grey mullet	4. *Penaeus indicus*

Codes

	A	B	C	D		A	B	C	D
(a)	1	3	2	4	(b)	2	4	3	1
(c)	1	2	4	3	(d)	2	4	3	1

18 *Balanoglossus* is a saprozoic hemichordate which lives in shallow coastal waters inside U-shaped burrows. Given below are certain features of this tongue worm.

I. It has a cylindrical body which is ciliated for most of the part.
II. Body is divisible into a mouth, collar and trunk.
III. The trunk is differentiated into branchiogenital, hepatic and abdominal regions.
IV. They possess cells called vanadocytes.
V. They obtain food from swallowed mud.

Which of the above are correct?

(a) I, II and IV (b) II, IV and V
(c) I, III and IV (d) I, III and V

19 Which of the following statements is/are not true?

I. In Urochordata, notochord is present only in larval tail.
II. In Cephalochordata, notochord extends from head to tail region.
III. *Branchiostoma* belongs to Hemichordata.
IV. Only one class of living members class–Cyclostomata represents the superclass–Agnatha.

Choose the correct option.

(a) Only III (b) I and IV
(c) III and IV (d) I, II and IV

20 Which of the following statements are correct regarding class–Reptilia of phylum–Chordata?

I. Males possess copulatory organs (hemipenis). They are double in snakes and lizards; and single in crocodiles and turtles.
II. Fertilisation is internal, metamorphosis is absent.
III. Cranial nerves 10 pairs.
IV. Kidney mesonephric, urea is chief nitrogenous waste.

Choose the correct option.

(a) I, II and III (b) I and II
(c) II and IV (d) III and IV

21 Amphibians have

I. three-chambered heart with two auricles and one ventricle.
II. two-chambered heart with one auricle and one ventricle.
III. kidney is mesonephric with nephrostome.
IV. females are viviparous.

Choose the correct option.

(a) I, II and III (b) I and III
(c) II and IV (d) I and IV

22 *Sphenodon*

I. belongs to Rhynchocephalia order of Anapsida class.
II. have pineal or third eye as their characteristic feature.
III. male have double copulatory organs (hemipenis).
IV. is a living fossil.

Codes

(a) I, II and III (b) I and II
(c) II and IV (d) III and IV

23 Match the following columns.

Column I	Column II
A. Nucleated RBCs	1. Aves
B. Sweat glands	2. Mammary glands
C. Carnivora	3. Dogs, cats, tigers
D. Cetacea	4. Aquatic mammals

Codes

	A	B	C	D
(a)	1	2	3	4
(b)	4	3	2	1
(c)	3	2	1	4
(d)	1	2	4	3

24 The statements given below describe certain features of some common amphibians. Read them carefully and identify which statements are True (T) and False (F).

I. The male *Pipa* carries eggs around its thighs till the tadpoles are ready to hatch.
II. *Alytes* carries the eggs and tadpoles in its vocal sac.
III. *Salamandra atra* is a ovoviviparous animal.
IV. The eggs of *Ichthyophis* are guarded by the male.

Choose the correct option.

	I	I	III	IV
(a)	F	T	T	F
(b)	F	T	F	T
(c)	F	F	T	T
(d)	F	F	T	F

25 Among the following statements, identify the ones which have made reptiles true land animals.

I. The presence of horny scales, which check loss of water.
II. The eggs of the reptiles have shells around them to check desiccation.
III. They lay their eggs in wet burrows/holes.
IV. The secretion or presence of a fluid-filled amnion around the embryo in reptiles.
V. They have a completely ossified endoskeleton.

Which of the statements are true? Mark the correct code.

(a) III and V
(b) I, III and IV
(c) II, III and IV
(d) I, II and IV

26 Which of the following statements are correct about the class-Amphibia?

I. They are the most primitive and earliest tetrapods with two pairs of limbs.
II. Their skin is rich in multicellular mucous and poison glands.
III. The brain is well-developed with 12 pairs of cranial nerves.
IV. Gaseous exchange occurs always by skin or lungs.
V. RBCs are nucleated and biconvex.

Codes

(a) II and III
(b) III and IV
(c) I, II and IV
(d) I, II and V

27 The class-Reptilia is further subdivided into three subclasses. Among the statements provided below, select the statements which are True (T) and False (F) about the classification of the class-Reptilia.

I. The order-Chelonia of subclass-Anapsida consists of organisms with movable pterygoids and absence of tympanum and a nictitating membrane.
II. The order-Squamata of subclass-Diapsida has organisms with single-headed ribs and the males have paired copulatory organs.
III. The organisms of subclass-Parapsida have become extinct.
IV. *Hemidactylus* belongs to suborder-Ophidia.

Choose the correct option.

	I	II	III	IV		I	II	III	IV
(a)	F	F	T	T	(b)	F	T	T	F
(c)	F	F	F	T	(d)	T	T	F	F

28 Match the following columns.

Column I	Column II
A. Only poisonous lizard	1. *Calotes versicolor*
B. Flying lizard	2. *Draco*
C. Largest non-poisonous snake	3. *Heloderma*
D. Garden lizard	4. *Python*

Codes

	A	B	C	D		A	B	C	D
(a)	1	2	3	4	(b)	4	3	2	1
(c)	3	2	4	1	(d)	2	1	4	3

29 Consider the following statements about snakes?

I. *Python molurus* is a non-poisonous snake.
II. *Bungarus* have head shield and elongated hexagonal vertebrats.
III. King cobra is also called snake eating snake.

Which of the statements are correct? Mark the correct code.

(a) I is correct, but II and III are incorrect
(b) II and III are correct, but I is incorrect
(c) I and III is correct, but II is incorrect
(d) All above are correct

30 The characteristic features of class-Aves are

I. they are warm-blooded and oviparous.
II. they have pneumatic bones.
III. syrinx is not present.
IV. fertilisation is external.
V. heart is four-chambered.
VI. mesonephric kidneys are present.
VII. bipedal movement is present.

Choose the correct option.

(a) I, II and VI
(b) I, II, V and VII
(c) II, IV, V and VI
(d) III, VI and VII

Assertion and Reason

Direction (Q. Nos. 31-42) *In each of the following questions, a statement of Assertion (A) is given followed by the corresponding statement of Reason (R). Of the statements, make the correct answer as*

(a) Both A and R are true and R is the correct explanation of A
(b) Both A and R are true, but R is not the correct explanation of A
(c) A is true, but R is false
(d) A is false, but R is true
(e) Both A and R are false

31 **Assertion** (A) Lateral line system is found in fishes and aquatic larval amphibians.
Reason (R) Lateral line system has receptor of sensory cells derived from ectoderm.

32 **Assertion** (A) The copulatory organs of sharks and reptiles is called claspers.
Reason (R) Uterus is also called womb.

33 **Assertion** (A) Cyclostomes belong to Agnatha.
Reason (R) True jaws are absent in cyclostomes.

34 **Assertion** (A) Urochordates are known as tunicates.
Reason (R) Their body is covered with a protective tunic or sac.

35 **Assertion** (A) *Heloderma* is called beaded lizard.
Reason (R) Its scales resemble beads.

36 **Assertion** (A) Kraits are poisonous snakes.
Reason (R) Belly scales are small and continuous with dorsal in poisonous snakes.

37 **Assertion** (A) All reptiles have three-chambered heart.
Reason (R) Reptilian heart is more advanced than avian heart.

38 **Assertion** (A) In mammals, hair covers the entire body.
Reason (R) Hairs are not present on palms, soles and some parts of face in mammals.

39 **Assertion** (A) Birds have one ovary.
Reason (R) This reduces the body weight for flight.

40 **Assertion** (A) The duck-billed platypus and the spiny anteater, both are egg-laying animals yet they are grouped under mammals.
Reason (R) Both of them have seven cervical vertebrae and 12 pairs of cranial nerves.

41 **Assertion** (A) Koel (cuckoo) build nests for retiring and taking rest during night time and day time for nocturnal birds.
Reason (R) Koel lays egg in the nest of tailor bird.

42 **Assertion** (A) Bats and whales are classified as mammals.
Reason (R) Bats and whales have four-chambered heart.

(C) Medical Entrances' Gallery

(Collection of Questions Asked in NEET & in Various Medical Entrance Exams)

1 Which of the following statements are true for the phylum–Chordata?

I. In Urochordata notochord extends from head to tail and it is present throughout their life.
II. In Vertebrata, notochord is present during the embryonic period only.
III. Central nervous system is dorsal and hollow.
IV. Chordata is divided into 3 phyla, Hemichordata, Tunicata and Cephalochordata. **[NEET 2020]**

(a) III and I
(b) I and II
(c) II and III
(d) IV and III

2 Match the following columns and select the correct option.

	Column I		Column II
A.	6-15 pairs of gill slits	1.	*Trygon*
B.	Heterocercal caudal fins	2.	Cyclostomes
C.	Air bladder	3.	Chondrichthyes
D.	Poison sting	4.	Osteichthyes

Codes **[NEET 2020]**

	A	B	C	D		A	B	C	D
(a)	3	4	1	2	(b)	4	2	3	1
(c)	1	4	3	2	(d)	2	3	4	1

3 Match the following columns and select the correct option from the codes given below.

Column I		Column II	
A.	*Aptenodytes*	1.	Flying fox
B.	*Pteropus*	2.	Angel fish
C.	*Pterophyllum*	3.	Lamprey
D.	*Petromyzon*	4.	Penguin

Codes **[NEET 2020]**

	A	B	C	D		A	B	C	D
(a)	3	4	2	1	(b)	3	4	1	2
(c)	4	1	2	3	(d)	2	1	4	3

4 All vertebrates are chordates, but all chordates are not vertebrates, why? **[NEET 2020]**
(a) Notochord is replaced by vertebral column in adult of some chordates
(b) Ventral hollow nerve cord remains throughout life in some chordates
(c) All chordates possess vertebral column
(d) All chordates possess notochord throughout their life

5 Which one of these animals is not a homeotherm? **[NEET 2018]**
(a) *Camelus* (b) *Chelone*
(c) *Macropus* (d) *Psittacula*

6 Identify the vertebrate group of animals characterised by crop and gizzard in its digestive system. **[NEET 2018]**
(a) Aves (b) Reptilia
(c) Amphibia (d) Osteichthyes

7 Which among these are the correct combination of aquatic mammals? **[NEET 2017]**
(a) Seal, dolphins, sharks (b) Dolphins, seals, *Trygon*
(c) Whales, dolphins, seals (d) *Trygon*, whales, seals

8 Among which of the animals urinary bladder is absent? **[AIIMS 2017]**
(a) Frog (b) Crow (c) Snake (d) Camel

9 Choose the correct statement. **[NEET-II 2016]**
(a) All mammals are viviparous
(b) All cyclostomes do not possess jaws and paired fins
(c) All reptiles have a three-chambered heart
(d) All Pisces have gills covered by an operculum

10 Which one of the following characteristics is not shared by birds and mammals? **[NEET-I 2016]**
(a) Breathing using lungs (b) Viviparity
(c) Warm-blooded nature (d) Ossified endoskeleton

11 Which is the National Aquatic Animal of India? **[NEET-I 2016]**
(a) River dolphin (b) Blue whale
(c) Seahorse (d) Gangetic shark

12 **Assertion** (A) Fishes belonging to Chondrichthyes have heterocercal caudal fin.
Reason (R) These fishes have an air bladder that acts as a hydrostatic organ. **[EAMCET 2016]**
(a) Both A and R are true and R is the correct explanation of A
(b) Both A and R are true, but R is not the correct explanation of A
(c) A is true, but R is false
(d) A is false, but R is true

13 Identify the pair having amphicoelous vertebrae. **[EAMCET 2016]**
(a) *Scoliodon* and *Ichthyophis*
(b) *Crocodylus* and *Hemidactylus*
(c) *Ornithorhynchus* and *Macropus*
(d) *Ophiophagus* and *Tropidonotus*

14 Study the following statements about Chordata. **[EAMCET 2015]**
I. *Columella auris* present in the middle ear of amphibians is modified hyomandibula of fishes.
II. Reptiles lay cleidoic eggs.
III. *Sphenodon* is a connecting link between amphibians and reptiles.
IV. *Panthera tigris* is an extinct animal of India.
Identify the correct statements.
(a) I and IV (b) II and IV (c) I and II (d) II and III

15 This class of animals are all ectoparasites on some fishes. **[Kerala 2015]**
(a) Amphibia (b) Osteichthyes
(c) Reptilia (d) Cyclostomata
(e) Chondrichthyes

16 Which one is flying fish? **[BCECE 2015]**
(a) *Scoliodon* (b) *Exocoetus*
(c) *Pristis* (d) *Trygon*

17 A jawless fish which lays eggs in freshwater and whose ammocoete larvae after metamorphosis return to ocean is **[CBSE AIPMT 2015, EAMCET 2015]**
(a) *Entatretus* (b) *Myxine*
(c) *Neomyxine* (d) *Petromyzon*

18 The character not related to Chondrichthyes is **[EAMCET 2015]**
(a) claspers (b) placoid scales
(c) filamentous gills (d) heterocercal caudal fin

19 Match the following columns.

Column I		Column II	
A.	Ostracodermi	1.	Lobe finned fishes
B.	Coelacanth	2.	*Exocoetus*
C.	Cyclostomes	3.	Extinct jawless fishes
D.	Osteichthyes	4.	Hagfish

[EAMCET 2015]

Codes

	A	B	C	D
(a)	3	1	4	2
(b)	2	3	1	4
(c)	1	3	2	4
(d)	3	2	4	1

20 Common character of all vertebrates without exception is **[J&K CET 2015]**
(a) body divided into head, trunk and tail
(b) two pairs of limb
(c) exoskeleton
(d) the presence of skull

21 Which of the following categories of animals is correctly described without any exception? **[Uttarakhand PMT 2015]**
(a) All reptiles possess scales, have three-chambered heart and are cold-blooded
(b) All bony fishes have four pairs of gill and an operculum on each side
(c) All sponges are marine and have collared cells
(d) All mammals are viviparous and possess muscular diaphragm for breathing

22 Connecting link between fish and Amphibia is **[Chhattisgarh PMT 2015]**
(a) lung fish *Protopterus*
(b) *Latimeria*
(c) *Seymouria* and *Sphencodon*
(d) All of the above

23 Anapsid skull is found in **[Chhattisgarh PMT 2015]**
(a) Crocodiles (b) Chelonia
(c) Squamata (d) Rhynchocephalia

24 The number of digits in the limb used for walking in reptiles is **[BCECE 2015]**
(a) four (b) two
(c) five (d) six

25 *Nasikabatrachus* is a ...... discovered from... . **[Chhattisgarh PMT 2015]**
(a) reptile, Western Ghats
(b) amphibian, Western Ghats
(c) fish, Eastern Ghats
(d) cavernicole, Eastern Ghats

26 The red jungle fowl is **[AMU 2015]**
(a) *Gallus gallus* (b) *Ravo cristatus*
(c) *Choriotis nigriceps* (d) *Raja hansa*

27 Pick the odd one out. **[AMU 2015]**
(a) *Chelone, Calotes, Naja*
(b) *Pavo, Psittacula, Ornithorhynchus*
(c) *Canis, Felis, Rattus*
(d) *Bufo, Rana, Hyla*

28 The snow leopard is **[AMU 2015]**
(a) *Panthera pardus* (b) *Panthera uncia*
(c) *Panthera tigris* (d) *Hylobates hoolock*

29 Which of the following represents the correct combination without any exception? **[CBSE AIPMT 2015]**

	Characteristics	Classes
(a)	Mammary gland; hair on body; pinnae; two pairs of limbs.	Mammalia
(b)	Mouth ventral; gills without operculum; skin with placoid scales; persistent notochord.	Chondrichthyes
(c)	Sucking and circular mouth; jaws absent integument without scales; paired appendages	Cyclostomata
(d)	Body covered with feathers; skin moist; and glandular lungs with air sacs; forelimbs for wings.	Aves

30 Which of the following animals is not viviparous? **[CBSE AIPMT 2015]**
(a) Flying fox (bat) (b) Elephant
(c) Platypus (d) Whale

31 Which one of the following is not a characteristic feature of mammals? **[WB JEE 2015]**
(a) Diphyodont tooth
(b) Ten pairs of cranial nerves
(c) Seven cervical vertebrae
(d) Left aortic arch in the circulatory system

32 *Antelope cervicapra* is (One or more options are correct) **[WB JEE 2015]**
(a) a mammal
(b) commonly known as black buck
(c) an animal under data deficient category of wildlife
(d) a threatened Indian wildlife

33 *Proteus anguinus* is an example for **[EAMCET 2014]**
(a) circannual rhythms (b) effect of light pigmentation
(c) phototaxis (d) photokinesis

34 Identify and select the correct match in the columns I, II and III. **[MHT CET 2014]**

	Column I	Column II	Column III
(a)	Earthworm	Annelida	Superclass
(b)	Frog	*Rana*	Species
(c)	Lancelet	Vertebrata	Division
(d)	Walrus	Mammalia	Class

35 Oviparous mammal is **[MHT CET 2014]**
(a) *Equus* (b) *Macropus*
(c) *Ornithorhynchus* (d) *Pteropus*

36 Which of the following animal has enucleated erythrocytes? **[MHT CET 2014]**
(a) Earthworm (b) *Sepia*
(c) Frog (d) Rat

37 Intra-abdominal testes are found in **[EAMCET 2014]**
(a) *Panthera* and *Equus*
(b) *Macaca* and *Macropus*
(c) *Balaenoptera* and *Delphinus*
(d) *Canis* and *Felis*

38 Retrogressive metamorphosis occurs in **[WB JEE 2013]**
(a) Hemichordata
(b) Cephalochordata
(c) Urochordata
(d) Vertebrata

39 The chordate features shared by the non-chordates are **[AIIMS 2013]**
(a) bilateral symmetry
(b) triploblastic condition and bilateral symmetry
(c) metamerism
(d) All of the above

40 **Assertion** (A) The earliest fossils of Amphibia are from Devonian and called labyrinthodontia.
Reason (R) Dentine of earliest amphibian fossils was not of folded nature. **[AFMC 2013]**
(a) Both A and R are true and R is the correct explanation of A
(b) Both A and R are true, but R is not the correct explanation of A
(c) A is true, but R is false
(d) Both A and R are false

41 The characteristics of class-Reptilia are **[NEET, KCET 2013]**
(a) body covered with moist skin which is devoid of scales, the ear is represented by a tympanun, alimentary canal, uninary and reproductive tracts open into a common cloaca
(b) freshwater animals with bony endoskeleton, air bladder to regulate buoyancy
(c) marine animals with cartilaginous endoskeleton, body covered with placoid scales
(d) body covered with dry and cornified skin, scales over the body are epidermal, they do not have external ears

42 Choose a pair of statements true for Gymnophiona.
I. Skin contains minute dermal scales. **[EAMCET 2012]**
II. Teeth are absent on both jaws.
III. Vertebrae are numerous and amphicoelous.
IV. Fertilisation is external.
Choose the correct option.
(a) II and III (b) II and IV
(c) I and II (d) I and III

43 The fangs of viper when pulled out, can **[Manipal 2012]**
(a) not come again
(b) come again with half the length
(c) come again with same length
(d) come again with double length

44 Four-chambered heart is characteristic of which poikilotherm? **[HPMT 2012]**
(a) *Psittacula* (b) *Hemidactylus*
(c) *Pteropus* (d) *Crocodilus*

45 Mammals have originated from which of the following? **[WB JEE 2012]**
(a) Pisces (b) Amphibia (c) Reptilia (d) Aves

46 Jaws of shark contain **[AMU 2012]**
(a) thecodont teeth (b) acrodont teeth
(c) pleurodont teeth (d) None of these

47 Which of the following belongs to Hemichordata? **[CBSE AIPMT 2011]**
(a) *Balanoglossus* (b) Lampreys
(c) *Pikaia* (d) *Petromyzon*

48 Which sound producing organ is found in bird?
(a) Pharynx (b) Larynx **[WB JEE 2011]**
(c) Syrinx (d) Trachea

49 Among the following which one lay eggs and does not produce young ones directly? **[MHT CET 2011]**
(a) *Echidna* (b) Kangaroo (c) Polcapine (d) Whale

50 All mammals without any exception are characterised by **[MHT CET 2011]**
(a) viviparity and biconcave red blood cell
(b) extra-abdominal testis and four-chambered heart
(c) heterodont teeth and 12 pairs of cranial nerves
(d) a muscular diaphragm and milk producing glands

51 Animals of class-Mammalia have **[J&K CET 2011]**
(a) seven cervical vertebrae
(b) seven cranial nerve
(c) single ventricular chamber
(d) fourteen cervical vertebrae

52 Mammal's heart is **[RPMT 2011]**
(a) myogenic (b) neurogenic
(c) voluntary (d) sympathetic

53 A group of animals having marsupium is **[RPMT 2011]**
(a) Monotremata (b) Eutheria
(c) Metatheria (d) Pantotheria

ANSWERS

Check Point 10.1

1. (d)	2. (b)	3. (c)	4. (a)	5. (d)	6. (a)

Check Point 10.2

1. (d)	2. (c)	3. (c)	4. (d)	5. (a)	6. (a)

Check Point 10.3

1. (d)	2. (a)	3. (c)	4. (c)	5. (a)	6. (a)

Check Point 10.4

1. (a)	2. (d)	3. (c)	4. (d)	5. (d)	6. (d)

(A) Topical Exercise

1. (d)	2. (d)	3. (c)	4. (d)	5. (d)	6. (c)	7. (c)	8. (b)	9. (c)	10. (a)
11. (c)	12. (c)	13. (c)	14. (a)	15. (c)	16. (a)	17. (c)	18. (a)	19. (d)	20. (c)
21. (c)	22. (d)	23. (b)	24. (c)	25. (b)	26. (d)	27. (c)	28. (c)	29. (a)	30. (b)
31. (b)	32. (c)	33. (d)	34. (c)	35. (d)	36. (a)	37. (c)	38. (d)	39. (a)	40. (b)
41. (c)	42. (a)	43. (c)	44. (d)	45. (b)	46. (d)	47. (c)	48. (d)	49. (a)	50. (d)
51. (a)	52. (b)	53. (b)	54. (d)	55. (c)	56. (d)	57. (c)	58. (a)	59. (d)	60. (d)
61. (c)	62. (c)	63. (d)	64. (b)	65. (b)	66. (b)	67. (a)	68. (a)	69. (d)	70. (d)
71. (b)	72. (a)	73. (d)	74. (c)	75. (c)	76. (c)	77. (d)	78. (b)	79. (a)	80. (b)
81. (c)	82. (d)	83. (a)	84. (b)	85. (c)	86. (c)	87. (c)	88. (c)	89. (c)	90. (d)
91. (a)	92. (d)	93. (d)	94. (d)	95. (c)	96. (d)	97. (b)	98. (c)	99. (a)	100. (a)
101. (a)	102. (d)	103. (b)	104. (a)	105. (c)	106. (b)	107. (c)	108. (d)	109. (c)	110. (b)
111. (d)	112. (d)	113. (d)	114. (c)	115. (d)	116. (c)	117. (c)	118. (d)	119. (c)	120. (d)
121. (a)	122. (d)	123. (b)	124. (b)	125. (b)	126. (b)	127. (d)	128. (b)	129. (c)	130. (a)
131. (a)									

(B) Cumulative Exercise

1. (b)	2. (b)	3. (c)	4. (d)	5. (a)	6. (d)	7. (c)	8. (c)	9. (d)	10. (b)
11. (b)	12. (c)	13. (a)	14. (a)	15. (b)	16. (d)	17. (c)	18. (d)	19. (a)	20. (b)
21. (b)	22. (c)	23. (a)	24. (d)	25. (d)	26. (d)	27. (b)	28. (c)	29. (d)	30. (b)
31. (b)	32. (b)	33. (a)	34. (a)	35. (a)	36. (c)	37. (e)	38. (b)	39. (a)	40. (a)
41. (c)	42. (b)								

(C) Medical Entrances' Gallery

1. (c)	2. (d)	3. (c)	4. (a)	5. (b)	6. (a)	7. (c)	8. (b)	9. (b)	10. (b)
11. (a)	12. (c)	13. (a)	14. (b)	15. (d)	16. (b)	17. (d)	18. (c)	19. (a)	20. (d)
21. (b)	22. (a)	23. (b)	24. (c)	25. (b)	26. (a)	27. (b)	28. (d)	29. (c)	30. (c)
31. (b)	32. (a, b, d)	33. (b)	34. (d)	35. (c)	36. (d)	37. (c)	38. (c)	39. (d)	40. (c)
41. (d)	42. (d)	43. (c)	44. (d)	45. (c)	46. (b)	47. (a)	48. (c)	49. (a)	50. (d)
51. (a)	52. (a)	53. (c)							

Explanations

(A) Topical Exercise

2 (*d*) Statement in option (d) is incorrect and can be corrected as
In Hemichordata, respiration takes place through branchial portion of pharynx bearing gill slits.

19 (*d*) Statement in option (d) is incorrect and can be corrected as
The blood of *Herdmania* is green due to the presence of vanadium is blood.

22 (*d*) Statement in option (d) is incorrect and can be corrected as
Fertilisation in Cephalochordata is external and it occur in sea water.

24 (*c*) Statement in option (c) is incorrect and can be corrected as
The coelom in vertebrates is well-developed and it is of enterocoelous type.

27 (*c*) Statement in option (c) is false about Agnatha because
They are cold-blooded or poikilotherms.
Rest statements are true.

30 (*b*) Statement in option (b) is incorrect and can be corrected as
Cyclostomes are jawless fishes, hence they have ventral, circular and suctorial mouth.

34 (*c*) Statement in option (c) is incorrect and can be corrected as
Gnathostomata have double olfactory sac.
Rest statements are correct.

35 (*d*) Statement in option (d) is correct. Other options are incorrect and can be corrected as
Gnathostomata are divided into two superclasses-Pisces and Tetrapoda. They do not have hindlimbs. They have internal ears and has three semicircular ducts.

39 (*a*) Statement I is true, but II is false. Incorrect statement can be corrected as
Chondrichthyes (cartilaginous fishes) have gill slits without operculum (gill cover), they lack the capacity to regulate body temperature.

42 (*a*) Statement I is true, but II is false. Incorrect statement can be corrected as
Cartilaginous fishes have placoid scales, mediating vision, however, they do not have colour vision.

46 (*d*) Statements II and III are correct, whereas I is incorrect and can be corrected as
Scoliodon is a carnivorous fish.

53 (*b*) Statement in option (b) is incorrect and can be corrected as
The head of *Labeo rohita* is scaleless with a subterminal mouth with very thick and fringed lips.
Rest statements are correct.

54 (*d*) Statement in option (d) is incorrect and can be corrected as
On the belly of male *Hippocampus*, a brood pouch is present for incubating eggs while on the belly of female a small anal fin is present.

59 (*d*) Statement in option (d) is incorrect and can be corrected as
Amphibians have closed circulatory system. They have three chambered heart with two auricles and one ventricle.

75 (*c*) Statement (c) is incorrect and can be corrected as
Stratum germinatium is absent in frog, but present in mammals.

76 (*c*) Statements (c) is incorrect because
Reptiles posesses twelve pairs of cranial nerves.

79 (*a*) Jocobson's organ in reptiles is concerned with smell. It is present on the roof of buccal cavity and is well developed in snakes and lizard. Hence, option (a) is incorrect.
Rest options are correct.

87 (*c*) Statement in option (c) is correct while other options are incorrect and can be corrected as

- Loreal pit of a pit viper is a thermoreceptor.
- *Lycodon* is a Indian wolf snake which is non-poisonous.
- Cobra is oviparous.

90 (*d*) Both statements I and II are false and can be corrected as
Venom of cobra is neurotoxic and it affects the nervous system of its prey. The venom of viper is haemolytic, which affects the circulatory system of its prey.

91 (*a*) The correct pair is in option (a). Other options are incorrect and can be corrected as
Indian krait - *Bungarus caeruleus*
Study of lizard - Saurology
Study of snake - Serpentology

107 (*c*) The matched pair in option (c) is incorrect and can be corrected as
The wings of penguins are modified into paddles, which act as flippers during swimming.

108 (*d*) Option (d) is incorrect pair. It can be corrected as
Ostrich is a flight less bird as they have rudimentary wings. Albatross is a seabird with large wings.
Rest matched pairs are correct.

115 (*d*) In mammals, cervical vertebrae are seven in number except in sloth and sea cow. Hence, option (d) is incorrect.

119 (*c*) Characteristic given in option (c) is incorrect and can be corrected as
Their are 12 pairs of cranial nerves in the members of class-Mammalia.

124 (*b*) Statement in option (b) is correct while others are incorrect and can be corrected as

- Bats do not have feathers.
- Camels have highly ovoid RBCs.
- Elephants are viviparous.

127 (*d*) Matched pair in option (d) is incorrect and can be corrected as
Whales and dolphins belong to order-Cetacea.

(B) Cumulative Exercise

2 (*b*) Statements I and II are correct while III is incorrect.

The axolotl also known as Mexican salamander or a Mexican walking fish is a neotenic salamander, closely related to the tiger salamender.

It exhibit neoteny, i.e. it attain sexual maturity without undergoing metamorphosis.

This occur due to the absence of thyroxine.

It is not a larva of Hemichordata, it belong to amphibians.

6 (*d*) Statements I and III is correct while II and IV are incorrect for Primata it can be corrected as

Cerebrum is well developed and enlarged. Haemochorial placenta is present in mammals as the chorionic villi are directly connected with maternal blood which help in exchange of gases and nutrients.

7 (*c*) Statement III is false it can be corrected as

Preen gland is present in aves, but the bones of birds are hollow, which help them to fly easily hence, their bones lack bone marrow.

8 (*c*) Statement III is incorrect with respect to flightless birds, it can be corrected as

Flightless birds like penguins and ostrich have modified wings or no wings. They do not have keel, oil glands, pygostyle and syrinx.

Rest statements are correct.

9 (*d*) Statement I is false and can be corrected as. In mammals, chief excretory product is urea.

Rest statements are true.

12 (*c*) X is notochord, which is replaced by a cartilaginous or bony vertebral column in adult of vertebrates.

Y is pharyngeal gill clefts, which only occur in embryonic stages of higher Chordata and disappear before adult life.

13 (*a*) Statements I, II and III are correct, but IV is incorrect and can be corrected as

Notochord persists throughout life in the lower vertebrates but in the higher vertebrates, it is replaced by vertebral column.

15 (*b*) Statements I and IV are correct, while II and III are incorrect and can be corrected as

Cyclostomata are jawless fishes, they have mesonephric kidney with ducts to urinogenital papilla.

Gnathostomata are divided into two super class, i.e. Pisces and Tetrapoda.

16 (*d*) Statements III and V are correct, while I, II, IV are incorrect and can be corrected as

- They have a hollow nervous system which is present dorsally above the notochord.
- Pharyngeal gill slits can differ in different chordates.
- Animals of super class-Tetrapoda have four pairs of appendages, but Pisces do not have four pairs of appendages.

18 (*d*) Statements I, III and V are correct, while statements II and IV are incorrect and can be corrected as

Body of *Balanoglossus* is divided into proboscis, collar and trunk.

Vanadocytes are found in the blood of tunicates, e.g. *Herdmania* and not in *Balanoglossus.*

19 (*a*) Statement III is incorrect while other statements are correct. Incorrect statement can be corrected as

Branchiostoma belong to subphylum-Cephalochordata.

20 (*b*) Statements I and II are correct, while III and IV are incorrect and can be corrected as

Reptiles have 12 pairs of cranial nerves.

They have metanephric kidney.

24 (*d*) Statements I, II and IV are false and can be corrected as

- The female suriname toad (*Pipa*) carries the tadpoles in special pits on its back till the tadpoles becomes the little toads.
- The male *Alytes* carries eggs around its thighs till the tadpoles are ready to hatch.
- The eggs of *Ichthyophis* are guarded by the female.

26 (*d*) Statements I, II and V are correct about Amphibia while statements III and IV are incorrect and can be corrected as

- The brain is well-developed with 10 pairs of cranial nerves.
- Gaseous exchange occur by gills during larval stage and in adults, it occur by skin, lungs and mouth cavity or even by gills in some cases.

27 (*b*) Statements I and IV are false and can be corrected as

- The order-Chelonia of subclass-Anapsida consist of organism with movable pterygoids and presence of tympanum and nictitating membrane.
- *Hemidactylus* belongs to sub order-Lacertilia.

30 (*b*) Statements I, II, V and VII are correct regarding Aves, but statements III, IV and VI are incorrect and can be corrected as

- Syrinx is present at the caudal end of the trachea along with larynx.
- Fertilisation is internal.
- Metanephric kidneys are present.

31 (*b*) Both A and R are true, but R is not the correct explanation of A.

Lateral line system of fishes and aquatic larval amphibians have receptors that are the group of sensory cells derived from ectoderm.

32 (*b*) Both A and R are true, but R is not the correct explanation of A.

Clasper is the copulatory organ of a male shark. Uterus is the organ in female mammals, in which the embryo develops.

33 (*a*) Both A and R are true and R is the correct explanation of A.

Cyclostomes are jawless animals, hence belong to Agnatha.

34 (*a*) Both A and R are true and R is the correct explanation of A.

Urochordates are known as tunicates because there body is covered by a protective covering called tunic or test. Gut is composed largely of tunicine, similar to cellulose.

35 (*a*) Both A and R are true and R is the correct explanation of A.
Heloderma is the only poisonous lizard in the world. It is commonly called beaded lizard, because its scales resemble as beads.

36 (*c*) A is true, but R is false because
Kraits, cobras and true vipers are poisonous snakes, while *Python, Typhlops, Uropeltis* and *Lycodon* are non-poisonous snakes. Belly scales are small in non-poisonous snakes.

37 (*e*) Both A and R are false because
Crocodile, alligator and *Gavialis* have four-chambered heart with two auricles and two ventricles. Avian and mammalian hearts are most advanced. They possess completely four-chambered heart. There is separation of oxygenated and deoxygenated blood and there is a complete double blood circulation.

38 (*b*) Both A and R are true, but R is not the correct explanation of A.
Skin is covered by epidermal hairs. These act as insulators. Hairs are not present in like palm, soles and some parts of the face, etc.

39 (*a*) Both A and R are true and R is the correct explanation of A.
Birds have many adaptations for flight. They have pneumatic bones and only one ovary, which reduces the body weight.

40 (*a*) Both A and R are true and R is the correct explanation of A.
Spiny anteater and duck-billed platypus belong to subclass-Prototheria of class-Mammalia. These are connecting link between reptiles and mammals. Prototherians are egg laying mammals. Seven cervical vertebrae and 12 pairs of cranial nerves is characteristic feature of mammals.

41 (*c*) A is true, but R is false because
Koel is the most beautiful bird among all the animals. They show courtship, nest building, parental care, migration and territorial behaviour. Koel (*Eudynamis*) does not make any nest, but lays eggs in the crow nest. In this way, koel is a nest parasite.

42 (*b*) Both A and R are true, but R is not the correct explanation of A.
Bats resembles birds in having wings for flying. But, they share more features to mammals like presence of teeth, hair, urea excretion, mammary glands, mammalian middle ear, placenta and giving birth to young ones. Therefore, bats are grouped under mammals. Similarly, whale resembles to fishes as they have flipper to swim in water, but they show parental care and give birth to their young ones.

(C) Medical Entrances' Gallery

1 (*c*) Statements II and III are correct. Statements I and IV are incorrect and can be corrected as

- In Urochordata, notochord is present only in larval tail, while in Cephalochordata it extend from head to tail region and is persistent throughout their life.
- Phylum—Chordata is divided into three subphyla, i.e. Urochordata or Tunicata, Cephalochordata and Vertebrata.

4 (*a*) The members of sub-phylum-Vertebrata possess notochord during the embryonic period. The notochord is replaced by a cartilaginous or bony vertebral column in the adult. Thus, all vertebrates are chordates, but all chordates are not vertebrates.

5 (*b*) Among the given animals, *Chelone* is not a homeotherm. It is green sea turtle belonging to class—Reptilia which is ectotherms or cold blooded and its internal body temperature varies according to the ambient environment.

6 (*a*) Crop and gizzard are found in the digestive tract of birds (Aves). Crop helps in storage and softening of food particles, whereas gizzard (muscular stomach) helps in its crushing and churning.

7 (*c*) Option (c) with whales, dolphins and seal are all aquatic mammals. *Trygon* and sharks are fishes which belong to class-Chondrichthyes of superclass—Pisces.

8 (*b*) Crow is a member of class-Aves. In this class, all animals are birds. The absence of urinary bladder is an adaptation in order to make their body light for flight purpose. Hence, the urine is released in semi-solid form.

9 (*b*) Cyclostomata is a class belonging to group-I-Agnatha of subphylum-Vertebrata. It is paraphyletic superclass of jaweless fishes. They lack paired fins too.

10 (*b*) Mammals are viviparous while birds are oviparous. Viviparous means giving birth to offspring that develops within the mother's body oviparous means producing eggs that hatch outside the body of mother.

12 (*c*) A is true, but R is false and R can be corrected as
Chondrichthyes or cartilaginous fishes do not have air bladder, but it is present in bony fishes. In cartilaginous fishes, heterocercal caudal fin is present. Air bladder helps in floating and maintaining balance.

13 (*a*) *Scoliodon* and *Ichthyophis* have amphicoelous vertebrae. This type of vertebrae is concave at both ends and present in fishes, apoda and cropoda members.

15 (*d*) The animals of class-Cyclostomata are all ectoparasites on some fishes. The animal possess jawless suctorial mouth surrounded by suckers. By means of suckers they attach to the larger animals such as fishes and the tongue punctures the body of prey and sucks the blood.

16 (*b*) *Exocoetus evolans* is the flying fish. It is called so because of its ability to leap into air by its tail and cover long distances in air by gliding the large pectoral fins.

17 (*d*) *Petromyzon* (lamprey) is jawless fish belonging to class-Cyclostomata which lays eggs in freshwater. For breeding, these fishes migrate from marine to freshwater like rivers and streams. After breeding they undergo indirect development with a prolonged larval stage, the ammocete larvae. After metamorphosis they return to ocean.

18 (*c*) The Chondrichthyes have lamelli form gills not filamentous gills. Chondrichthyes or cartilaginous fishes are jawed fishes with paired fins, paired nares, scales, a heart with its chambers in series and skeletons made up of cartilage rather than bone.

20 (*d*) All vertebrates without any exception have skull which protects the brain well-developed internal skeleton of cartilage and bone, separated into axial skeleton (skull, vertebra, ribs and sternum) and appendicular skeleton (girdles and appendages).

22 (*a*) Lung fish, *Protopterus* (Dipnoi species) is a living fossil. It is a connecting link between the fishes and amphibians. It possesses characteristic features of both the classes.

23 (*b*) Anapsid skull is found in Chelonia (turtles and tortoises). An anapsid is an amniotic whose skull does not have openings near the temples. The anapsid skull has a solid bony roof found in primitive reptilians.

24 (*c*) In reptiles, two pairs of limbs are present and are pentadactyle. Hence, 5 digits in the limbs are used for walking in reptiles.

25 (*b*) *Nasikabatrachus sahyadrensis* is a frog species (an amphibian) belonging to family-Sooglossidae. It is found in the Western Ghats in India.

27 (*b*) *Pavo, Psittacula, Ornithorhynchus* are odd ones as *Pavo* and *Psittacula* belong to class–Aves, whereas *Ornithorhynchus* belongs to class—Mammalia of phylum—Chordata.

29 (*c*) Cyclostomata bears sucking and circular mouth with no jaws, integument without scales and gills (5-16) are present as paired appendages in lateral sac-like pouches of pharynx. Persistent notochord is a feature of Cyclostomata and not Chondrichthyes. Mammals lack pinnae and in Aves, the skin is dry and skin glands are absent except the oil glands.

32 (*a, b, d*) *Antelope cervicapra* (the black buck) is an ungulate species of antelope, native to the Indian subcontinent that has been threatened by IUCN, since 2003. It is a mammal belonging to order-Artiodactyla.

33 (*b*) *Proteus anguinus* is a cave salamander (a rare amphibian). It is an example for effect of light pigmentation. They have generally depigmented skin, but its skin retains the ability to produce melanin. When exposed to light, it will gradually turn dark. Its skin lacks pigment, giving its body a white appearance. It also has a pink head due to blood capillaries near the skin and its translucency shows contours of the internal organs.

34 (*d*) Among the given options, the correct one is Walrus. It is a large flippered marine mammal which belongs to class-Mammalia.

Correct matches for the three options are

Earthworm	–	Annelida	–	Phylum
Frog	–	*Rana*	–	Genus
Lancelet	–	Vertebrata	–	Subphylum

37 (*c*) Marine mammals cretacean like *Balaenoptera* (whale) and *Delphinus* (dolphin) have intra-abdominal testes. They are surrounded by tissues that can generate heat. Terrestrial mammals holds their testes in scrotum outside the body for maintaining ideal temperature for sperm activity.

45 (*c*) Mammals evolved from the reptiles during Triassic period of Mesozoic era.

47 (*a*) Hemichordata includes marine worm-like animals called acorn worms. *Balanoglossus* is a sluggish marine animal, which is also called as acorn worm or tongue worm.

48 (*c*) Human larynx contains vocal cords, the sound producing elastic fibres called voice box. The sound producing organ in birds is syrinx.

49 (*a*) *Echidna* belongs to Prototheria group of class-Mammalia. It is oviparous and only female incubates the eggs. Young ones suck the milk from mammary glands.

50 (*d*) Any animal of the Mammalia a large class of warm-blooded vertebrates having milk producing mammary gland in the female, a thoracic diaphragm and a four-chambered heart.

53 (*c*) Metatherians are pouched mammals. The complete development of embryo takes place in abdominal pouch or marsupium.

CHAPTER 11

Morphology of Flowering Plants

Plant morphology (*Morphe*–form; *logos*–study) is a branch of botany, which deals with the study of external forms and features of different plant organs such as roots, stems, leaves, flowers, seeds, fruits, etc. Morphological characteristics of plants play a significant role in their identification. These characteristics also aid in classifying plants in their respective groups and establishing a phylogenetic relationships, etc.

Types of flowering plants

Many different types of flowering plants are found on the earth. They vary in their forms, lifespan, habitat and mode of nutrition. Based on these criteria, the plants are classified as given below

1. On the basis of shape and size

(i) **Herbs** These are non-woody plants which do not have persistent aerial parts. These have short, delicate and green stem, e.g. wheat, gram, spinach, radish, mustard, tomato, potato, etc.

(ii) **Shrubs** These are woody plants. The branching of stem occurs near the ground level in them. The stem does not develop into clear trunk, e.g. rose, China rose, *Duranta*, etc.

(iii) **Trees** These are woody plants with well-developed trunk. The branching of stem occurs at same height. These plants can be conical in shape, palm-like or deliquescent (branches spread in different directions), e.g. mango, neem, etc.

2. On the basis of lifespan

(i) **Annuals** These plants complete their life cycle within a year or a season, e.g. wheat, rice, etc.

(ii) **Biennials** These plants complete their life cycle in two growing seasons. In the first season they grow vegetatively and in second season they complete their reproductive phase, e.g. carrot, radish, turnip, cabbage, etc.

(iii) **Perennials** These plants grow for many years and produce flowers and fruits every year, e.g. mango, jamun, papaya, etc.

Inside

3. On the basis of mode of nutrition

(i) **Autotrophic plants** These are green plants which synthesise their own food from inorganic molecules *via* photosynthesis. Angiosperms are mostly autotrophic.

(ii) **Heterotrophic plants** These are plants which obtain whole or a part of nourishment from outside. These are further categorised as follows

(a) **Insectivorous plants** These plants grow in nitrogen deficient soil, so as to make up their nitrogen deficiency. They have adapted to catch, kill and digest insects, e.g. *Nepenthes, Utricularia, Drosera,* etc.

(b) **Parasitic plants** These derive their nutrition from other (host) plants by special roots called **haustoria**. Some plants are total parasites (*Cuscuta*) whereas some are partial parasites (*Santalum*) on host plant.

Parts of a flowering plant

- Flowering plants consist of two major organ systems, i.e. the root system and the shoot system.
- The shoot system is usually located above the ground. It comprises of leaves, buds, flowers and fruits.
- The root system is located below the ground. It has branches and buds only.
- Roots, stems and leaves are called as **vegetative parts** of the plants. Flowers, fruits and seeds are **reproductive parts** of a plant.

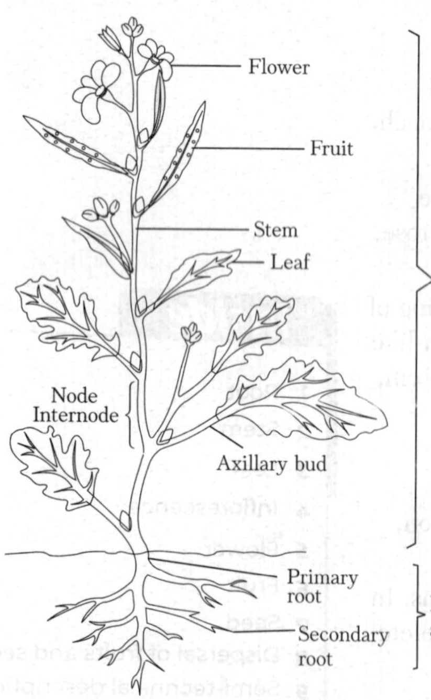

Fig 11.1 Parts of a flowering plant

ROOT

It is the underground non-green part of a plant. It develops from the radicle of germinating seed. Its primary function is absorption of water and minerals. It also helps in anchoring plants to the soil.

Main characteristic features of roots are as follows

- Roots are descending, cylindrical and non-photosynthetic.
- They lack nodes, internodes, leaf buds and flower buds. They possess many unicellular hairs.
- A root cap covers the growing point of a functional root.
- Roots show branching endogenously, i.e. the lateral branches of roots develop from the internal part of the main root.
- Roots are mostly positively geotropic, i.e. move towards the ground/soil. They are also negatively phototropic, i.e. move away from light and positively hydrotropic, i.e. move towards water.

Regions of root

It consists of the following four zones

(i) **Root cap zone** It is a smooth cap-shaped, multicellular protective layer. It is made up of parenchymatous cells. It is present at the apex of the root. Root cap secretes mucilage that softens the passage of growing root through the compact layers of soil.

(ii) **Meristematic zone** It is also called as the **zone of cell division**. It is small (about 1mm) and present just next to the root cap. The cells of this zone are thin-walled with no intercellular spaces. These cells actively perform cell division. Therefore they are responsible for the growth of roots.

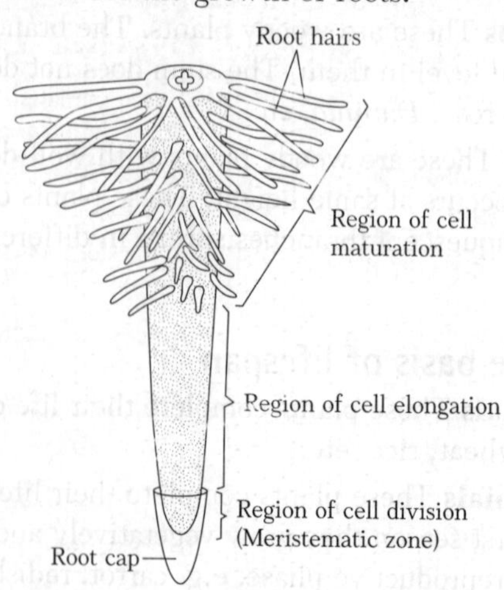

Fig. 11.2 Regions or Zones of roots

(iii) **Zone of cell elongation** It is situated just behind the meristematic zone. Cell become elongated and develop a large central vacuole filled with cell sap. Cells of this region loose their power of division. These cells absorb water and minerals from the soil.

(iv) **Zone of cell maturation** This region consists of mature differentiated cells. Few cells of outer layer get extended to form unicellular **root hairs**. These root hairs increase surface area of roots for absorption of mineral salts and water.

Types of root system

There are mainly two types of root system present in plants, i.e. tap root system and adventitious root system.

1. Tap root system

- Tap root or primary root develops from radicle of germinating seeds.
- It forms lateral branches or secondary roots. These branches may further divide to form tertiary roots.
- The primary roots, secondary roots and tertiary roots altogether form the tap root system.
- Tap roots are common in dicots, e.g. *Mangifera indica.*

2. Adventitious root system

- The roots produced from any other part of the plant except the radicle are called **adventitious roots** (L. *adventitious*-extraordinary).
- These parts may be hypocotyl, base of the stem, nodes of stem, etc.
- Adventitious root system is usually found in monocots, e.g. grasses, cereals, etc.

Table 11.1 Differences between tap root and adventitious root

Tap root	Adventitious root
These arise from radicle of embryo.	These originate from any other part of the plant except radicle.
Always remain underground except pneumatophores.	May be underground, subaerial or aerial.
Consists of a single primary (main) root.	Consists of roots forming a cluster.
Primary (main) root is persistent.	Primary root though develops, but is short-lived.
It may be **surface feeder** or **deep feeder**. Deep feeders are more common.	If underground, it is usually a surface feeder.

Note • *Pandanus has multiple root caps.*
- *Eichhornia, Pistia, Lemna have no root caps instead they have root pockets (i.e. non-functional root caps).*
- *Many hydrophytic plants, e.g. Utricularia, Wolffia, Ceratophyllum do not develop any roots as they need little absorption of water and mineral salts, this need is fulfilled by absorption through general body surface.*

Modification of tap root

In addition to their normal function, tap roots perform several other modified functions also.

These modified forms of tap roots are as follows

1. Fleshy tap roots

These roots modify to store food materials. The primary root becomes swollen due to the presence of stored food.

Depending upon the shape, these fleshy or storage, tap roots can be classified into following types

(i) **Conical** These are conical in shape, i.e. widest at top, tapering smoothly towards the bottom, e.g. carrot (*Daucus carota*).

(ii) **Fusiform** The root is broader in the middle and narrower at both the ends giving a spindle-shaped appearance, e.g. radish (*Raphanus sativus*).

(iii) **Napiform** Such a root swells extraordinarily in the middle and becomes abruptly narrower towards the apical region, e.g. turnip (*Brassica rapa*) and chukandar (*Beta vulgaris*).

(iv) **Tuberous** These are thick and fleshy tubers of indefinite and irregular shape, e.g. roots of *Mirabilis jalapa* and *Cassava.*

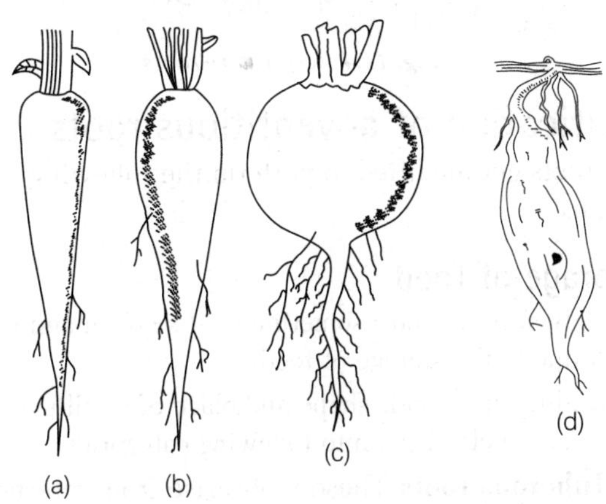

Fig. 11.3 Tap root modifications : (a) Conical (carrot), (b) Fusiform (radish) (c) Napiform (turnip), (d) Tuberous (Mirabilis jalapa)

2. Nodulated tap root

- The plants of Leguminosae family are characterised by the presence of nodules on roots. These roots are called as **nodulated** or **tuberculated roots.**
- Nodules on these roots bear N_2-fixing bacteria, i.e. *Rhizobium.* These roots help in N_2-fixation, e.g. pea, gram, arhar, etc.

3. Respiratory roots (Pneumatophores)

- These roots are found in plants growing in swampy/marshy areas. The underground roots of such plants start growing up and become aerial (negatively geotropic).
- These roots possess minute pores on tips. These help in respiration and are called **pneumathodes.**
- These roots bring about exchange of gases so, these are also called as **pneumatophores** or **respiratory roots,** e.g. *Avicennia, Rhizophora,* etc.
- The pneumatophores of *Rhizophora* are exceptional as the prop roots are non-hygroscopic.

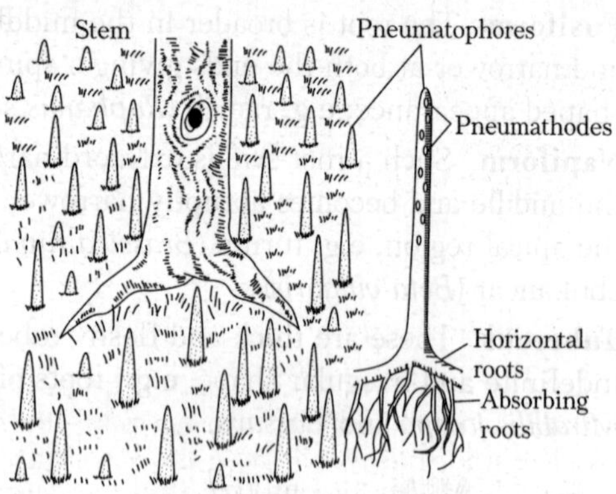

Fig. 11.4 Respiratory roots

Modification of adventitious roots

These roots get modified to perform the following functions

1. Storage of food

- The adventitious roots sometimes get swollen and become thick due to the storage of food.
- Depending upon their shape and place of swellings, these roots can be classified into following categories
  (i) **Tuberous roots** These roots arise from internodes of stem. They grow and become fleshy and tuberous as a result of food storage. They do not have a definite shape, e.g. sweet potato (*Ipomoea batata*).
  (ii) **Fasciculated roots** These roots are produced from base or lower node of stem and become thick and fleshy for storage of food, e.g. *Dahlia, Asparagus.*
  (iii) **Beaded** or **Moniliform roots** These roots are swollen at regular intervals giving a bead-like appearance of necklace, e.g. *Momordica, Portulaca,* some grasses, etc.
  (iv) **Annulated roots** Formed by a number of swollen discs placed one above the other. It looks like a series of ring-like outgrowths or swellings, e.g. *Cephaelis* or *Psychotria* (ipecac), etc.
  (v) **Palmate roots** These roots are branched and appear like palm of human hand, e.g. *Orchis, Aconitum,* etc.
  (vi) **Nodulose roots** These roots are swollen at their tips. *Curcuma amada* (mango ginger), *Curcuma domestica* (turmeric) and *Maranta* (arrow-root), etc.

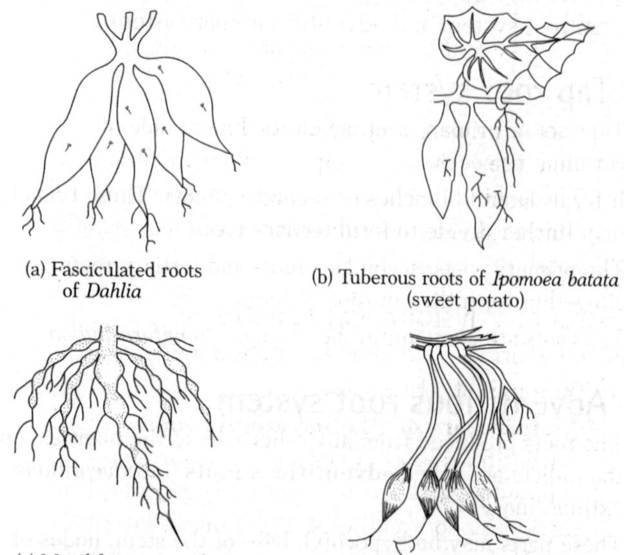

(a) Fasciculated roots of *Dahlia*
(b) Tuberous roots of *Ipomoea batata* (sweet potato)
(c) Moniliform roots of grass
(d) Nodulose roots of *Curcuma amada*

Fig. 11.5 Modification of adventitious root (storage)

2. Mechanical support

The adventitious roots also modify into following types to provide mechanical support to plants

(i) **Buttress roots** These roots radiate from the base of the tree (like finger projections) to provide mechanical support, e.g. *Ficus, Bombax, Terminalia, Salmalia,* etc.

(ii) **Stilt roots or Brace roots** These roots develop from the basal nodes of the stem near the soil. These provide mechanical support by fixing the plant firmly in soil, e.g. *Saccharum officinarum* (sugarcane), *Zea mays* (maize), *Pandanus tectorius,* etc.

(iii) **Prop or Pillar roots** These roots develop from the large horizontal aerial branches. These grow vertically downward, penetrate the soil and become thick. Hence, they give a pillar-like appearance, e.g. banyan.

(iv) **Contractile roots** These roots become thick, fleshy and contractile. These remain associated with underground stems, e.g. *Crocus.*

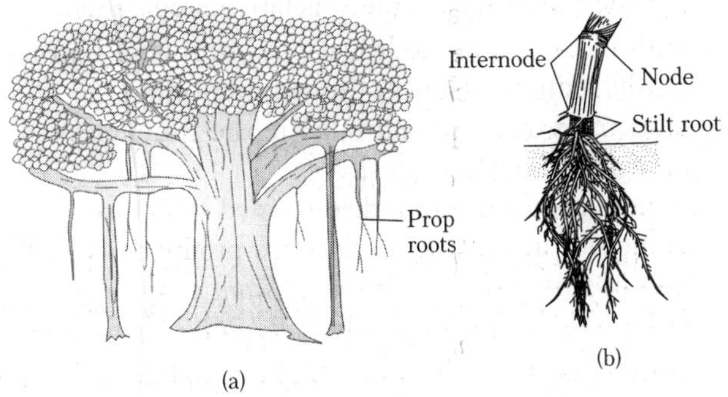

***Fig 11.6** (a) Prop roots of banyan tree, (b) Stilt roots of sugarcane*

(v) **Climbing or Clinging roots** These roots arise from the nodes and help the plant in climbing. These stick firmly to a support by secreting a cementing viscous substance, e.g. *Ficus pumila*, betel, etc.

3. Vital or Physiological functions

The adventitious roots modify into following types to perform some vital functions

(i) **Assimilatory or Photosynthetic roots** Adventitious roots of certain plants develop chlorophyll and become green. These roots can carry out photosynthesis, e.g. *Trapa, Tinospora* and *Podostemon.*

(ii) **Parasitic or Haustorial roots** These roots are developed by the parasitic plants. These roots penetrate host tissue for absorbing nourishment, e.g. *Cuscuta* is a parasitic plant on plants like *Duranta, Acacia*, etc.

(iii) **Epiphytic roots** The roots of the epiphytes hang in the air. These absorb hygroscopic water present in the atmosphere by special spongy tissue called **velamen.** These cells of velamen have pores for the absorption of atmospheric moisture. e.g *Vanda, Dendrobium*, Orchids.

(iv) **Reproductive roots** In some plants, roots help in vegetative reproduction, e.g. parwal. In case of sweet potato the adventitious buds act as reproductive roots and give rise to new plants.

(v) **Floating roots** These are adventitious roots present in some floating plants. These arise from nodes. Floating roots store air, become inflated and spongy. Thus, they make a plant possess buoyancy by functioning as floats, e.g. *Jussiaea.*

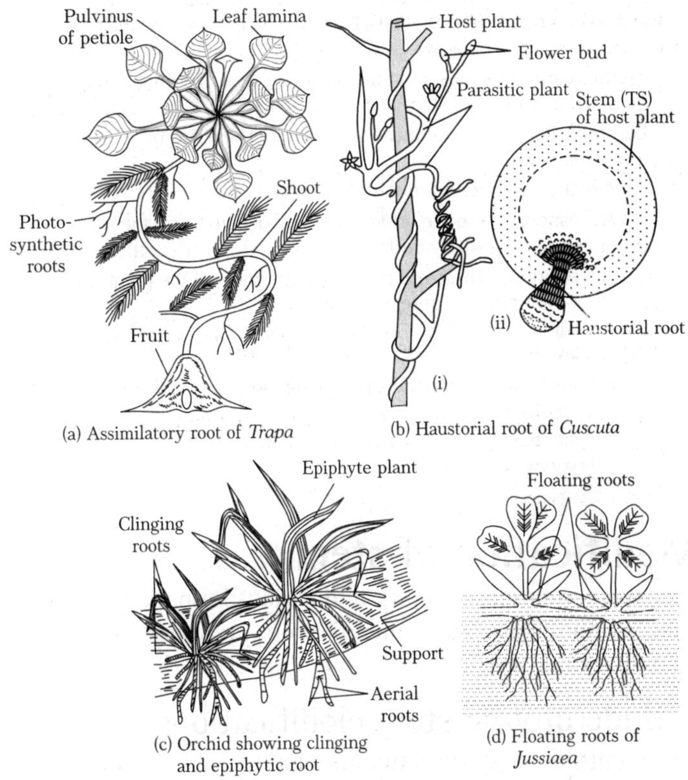

***Fig 11.7** Showing different types of root according to their vital functions*

Note ***Plants that are all roots*** *In Podostemon a hydrophyte, the main plant body is a root, which carries assimilation as well as reproduction. Small flowering shoots are formed from the roots from time to time.*

Rootless plants *In Wolffia, a floating plant (smallest angiosperm) and Utricularia, a submerged hydrophyte, there are no roots.*

STEM

- Stem develops from the plumule of embryo. It is the ascending part of main plant axis. It bears branches, leaves, flowers and fruits, etc.
- Stem generally grow above the ground (negatively geotropic) and hence is the aerial part of the plant.
- It conducts water and mineral salts from roots to leaves and fruits.
- A stem is usually differentiated into nodes and internodes. Nodes are regions where leaves are born, whereas internode is the part of stem between two nodes.
- The leaf bearing part of the stem or its branch is called **shoot.**

> **Bud**
>
> It is defined as a young, immature and underdeveloped embryonic shoots. It normally occurs in the axil of a leaf or at the tip of a stem. Thus, buds are generally of two types based on their position. These are as follows
>
> 1. **Terminal bud** It is present at the tip of the stem. Terminal bud is involved in growth of stem and its branches.
> 2. **Lateral buds** These are lateral in position. These are of following types
>   - (i) **Axillary** It is present in the axial of leaf.
>   - (ii) **Accessory** These are additional buds occurring on the side or above the axillary buds.
>   - (iii) **Adventitious** These are formed on the parts of the plant other than nodes. These are as follows
>   - **Cauline buds** They arise directly from stem, e.g. *Artocarpus* (Jack fruit).
>   - **Radicle buds** These develop on the roots, e.g. *Dalbergia, Ipomoea*, etc.
>   - **Foliar buds** These buds develop on the leaves, e.g. *Bryophyllum, Begonia*, etc.

Modification of stem

Like roots, stems also modify into various forms to perform different functions. These modifications are classified into following categories

1. Underground stem modifications

Certain plants have stems that develop under the soil. These are called as underground stems. They store food material and become thick and fleshy.

This type of underground stem is divided into following categories

(i) **Rhizome** These underground stem grows horizontally. It is fleshy, dorsiventral and perennial. Rhizomes possess nodes and internodes. These nodes bear scale leaves in the axil of which axillary buds are born, e.g. *Dryopteris, Musa paradisiaca* (banana), *Nelumbo* (lotus), *Zingiber officinale* (ginger), *Canna indica* and *Curcuma domestica* (turmeric).

(ii) **Tuber** In plants like potato, the lower branches of stem become positively geotropic and get buried in soil.

- The food (reserve) gets accumulated in the tip of such underground branches. Later they swell up and become oval/spherical in shape.
- There are no adventitious roots on the tubers.
- Tubers possess a number of spirally arranged depressions called **eyes** or **buds**. Each eye represents a node and bear a small scale leaf having a bud in its axil.
- The food reserve is starch in potato (*Solanum tuberosum*) and inulin in Jerusalem artichoke (*Helianthus tuberosus*).
- Tubers are used for the vegetative propagation.

(iii) **Bulb** In this stem modification, internodes become highly reduced. Hence, the leaves become condensed (rosette) and take up the function of storage and become fleshy.
Stem is reduced into a disc-shaped structure. Numerous adventitious roots are produced from the base of this disc, e.g. onion (*Allium cepa*), *Tulipa*, garlic (*Allium sativum*), lily (*Lilium bulbiferaum).*

(iv) **Corm** It is short, thick and highly condensed form of rhizome. The corm bears buds at nodes and adventitious roots at the base, e.g. *Amorphophallus* (elephant foot, vern. zamikand, suran), *Crocus* (vern. kesar), *Alocasia.*

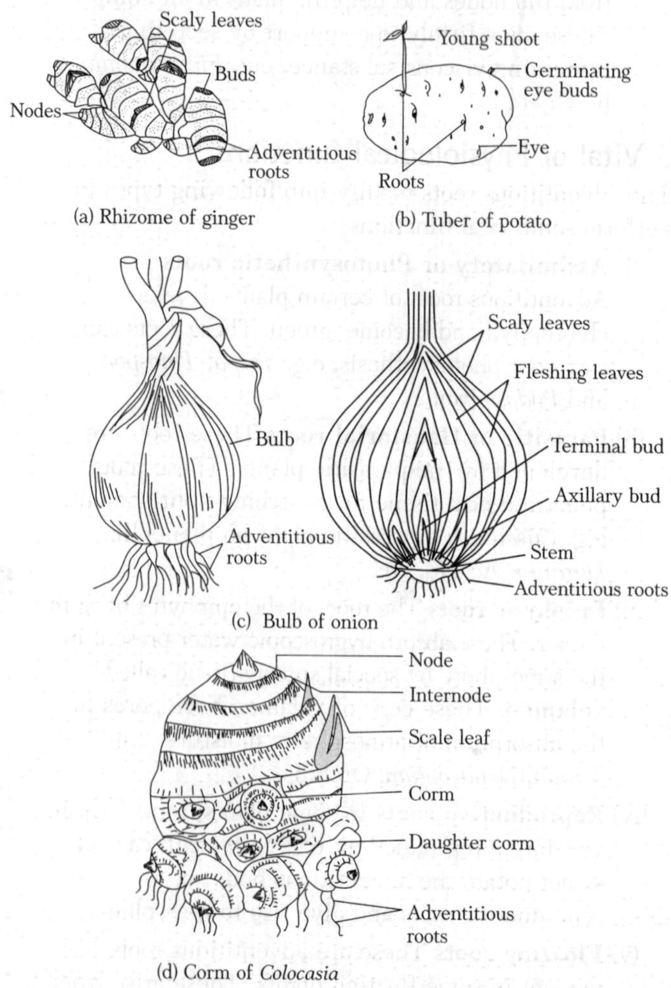

Fig 11.8 Underground stem modification

2. Subaerial stem modification

In these modifications, a part of stem remains underground and the remaining part becomes aerial.

The dormant bud present on the stem become active and develop lateral branches for vegetative reproduction.

Subaerial stem is modified into the following types

(i) **Runners** These stems have weak, long and thin internodes. These creep over the surface of soil, e.g. doob grass or lawn grass (*Cynodon dactylon*), brahmi booti (*Centella asiatica*), wood sorrel (*Oxalis*).

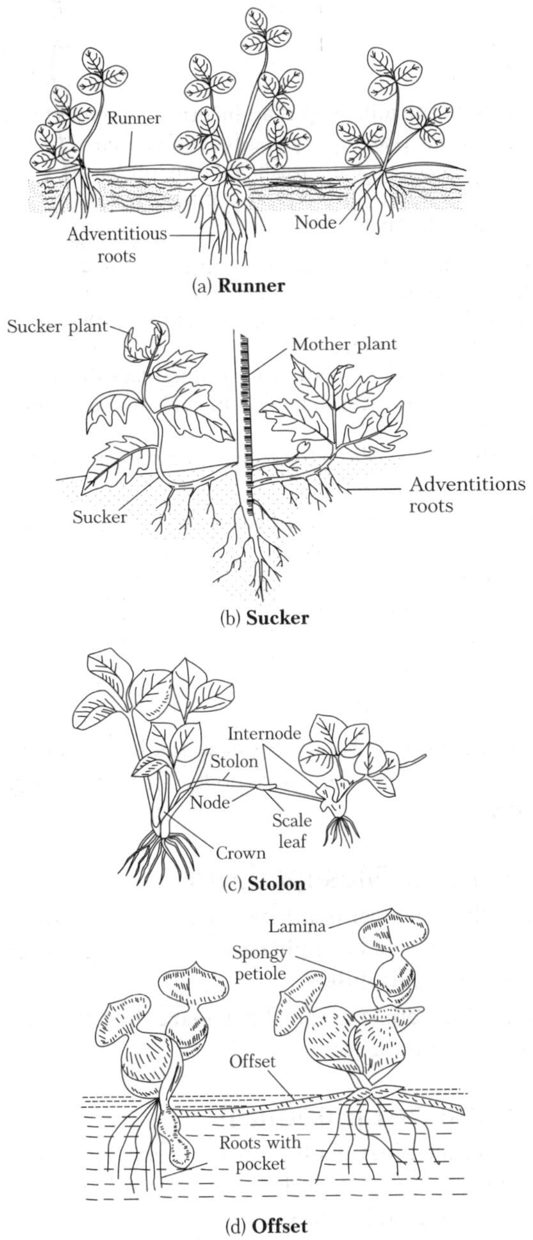

Fig. 11.9 Subaerial stem modification

(ii) **Suckers** Some axillary branches that arise from the node of underground stem grow horizontally for sometime. These branches then become obliquely upward to serve as sucker, e.g. *Chrysanthemum*, banana.

(iii) **Stolons** These are long, lateral branch that arise from base of the stem. They grow aerially for some time and then arch downward to touch the ground, and give rise to daughter plant, e.g. wild strawberry (*Fragaria*), jasmine (*Jasminum*), peppermint (*Mentha piperita*).

(iv) **Offsets** These are lateral branches that bear short internodes. Each node bears a rosette of leaves upward (aerial) and cluster of roots downwards, e.g. water lettuce (*Pistia*), water hyacinth (*Eichhornia crassipes*).

Note *Stolons and runners are quite similar to each other. The only difference between them is that stolons grow below the surface of soil, while runners grow above the surface of soil.*

3. Aerial stem modification

In this, the real stem gets so much modified that it is difficult to identify it, as a stem.

Aerial stem modifications are of following types

(i) **Stem tendrils** When stem is weak, the axillary or terminal bud gets modified into tendril. These tendrils perform the function of climbing, e.g. *Vitis* (modified apical bud), *Passiflora* (modified axillary bud), etc.

Note *In cucurbits, the lower part of tendril is modified stem, while the upper part of tendril is modified leaf.*

(ii) **Stem thorns** These are axillary buds which got modified into stiff and sharp structures called thorns. They are formed in the axil of leaf or apex of stem. These perform the function of either climbing or defence, e.g. *Duranta*, climbing rose (with curved prickles), *Asparagus* (with leaf spines), *Bougainvillea* (with extra axillary thorns), *Calamus* (with leaf sheath spines).

Note *Thorns are modified stem while spikes are modified leaves.*

(iii) **Phylloclades** These are modified flattened, thick, succulent, green and photosynthetic stem. They help the plants to grow in dry and xerophytic habitats. The phylloclades can be flattened, e.g. *Opuntia, Muehlenbeckia, Coccoloba, Epiphyllum*, e.g. *Casuarina, Euphorbia royleana, E. tirucalli.*

(iv) **Cladodes** (Cladophylls) These are axillary bud which developed into green one internode long stems. They are photosynthetic, hence perform the function of leaves. The true leaves of such plants are usually reduced to scales or spines, e.g. *Asparagus* and *Ruscus.*

(v) **Bulbil** It is a modified stem developed either from a vegetative bud or a flower bud. It gets swollen because of accumulation of food. It gives rise to a new plant after being detached from parent plant, e.g. *Dioscorea bulbifera.*

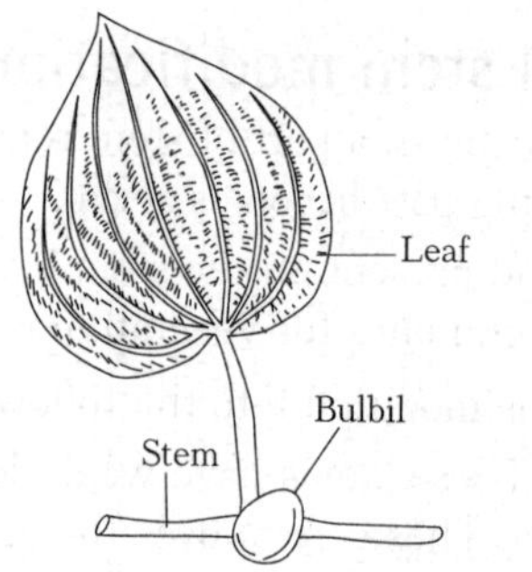

Fig. 11.10 Bulbil of Dioscorea

CHECK POINT 11.1

1. A fibrous root system is better adapted than tap root system for
   (a) storage of food
   (b) anchorage of plant to soil
   (c) absorption of water and minerals
   (d) transport of water and organic food
2. Stilt roots, which grow obliquely from basal nodes of culm stem and acting as brace are found in
   (a) *Saccharum*
   (b) Orchid
   (c) *Tinospora*
   (d) *Pothos*
3. Epiphytes like *Vanda* develop special layer of absorptive tissue velamen consisting of 4 or 5 layers of long polygonal cells. Velamen is formed by
   (a) absorbing roots
   (b) stem
   (c) clinging roots
   (d) hanging roots
4. Which of the following is an incorrect statement?
   (a) Rhizome is an underground modified stem
   (b) Runners have long, weak and thin internodes
   (c) Each eye in a tuber represent internode
   (d) All of the above
5. Stolons and runners are quite similar to each other. The only difference between them is
   (a) stolons grow below the surface of soil, while runners grow above the surface of soil
   (b) stolon possess nodes, while runners lack them
   (c) stolon show tendril modification also, while runner lack them
   (d) stolon grow above the ground and runner grow below ground
6. Thorn is a stem structure because it
   (a) develops from trunk
   (b) develop from axillary bud
   (c) grows from external surface
   (d) is pointed

LEAF

It is an aerial, lateral, green and flattened structure. It is borne on the stem. It develops at nodes laterally and bears a bud in its axil. Leaves perform photosynthesis.

Parts of leaf

A typical foliage leaf consists of three distinct parts, leaf base, petiole and lamina.

These are described below

1. Leaf base or Hypopodium

- It is the part of leaf, which attaches leaf to the stem/branch.
- It protects the young axillary bud.
- In the leaves of family–Fabaceae, basal part becomes swollen and forms special structure called **pulvinus.**
- It is also found in plants like *Mimosa.*
- In the members of family-Poaceae the leaf base becomes broader and covers the entire node. This structure is called **sheath.**

2. Petiole or Mesopodium

- The stalk of the leaf that joins the lamina with the stem or its branch, is called **petiole.** A small petiole in compound leaf is called **petiolule.**
- Leaf having petiole is called **petiolate.** It is termed **sessile,** if the petiole is absent.

3. Lamina or Epipodium

- It is also called **leaf blade.** It is the terminal, thin, flattened part of a leaf. It is green and specialised to manufacture food (photosynthesis).
- Leaf lamina is supported by the veins and veinlets. The prominent vein present in the middle of lamina is called **midrib.**

- The veins provide rigidity and strength to the leaf blades. These also act as a channel for transport of water, minerals and food.
- The pattern of veins in leaves varies from plant to plant.

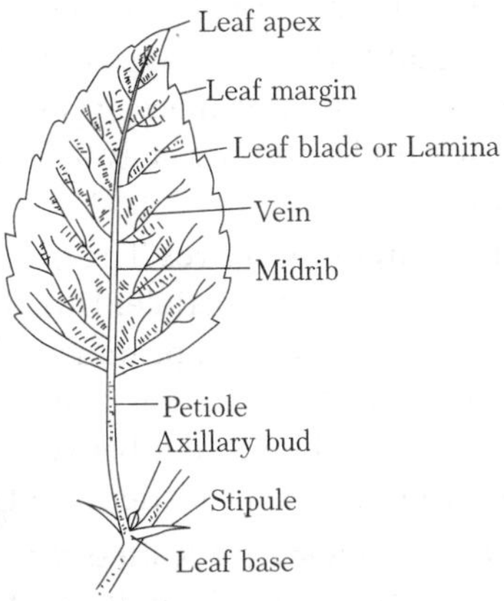

Fig. 11.11 A typical foliage leaf

Stipules

On either side of leaf base lateral appendages are present. These are known as **stipules**. Leaf with stipules is known as **stipulate leaf**, e.g. Fabaceae. On the other hand, a leaf without stipule is called **ex-stipulate leaf**, e.g. Solanaceae, Liliaceae.

Stipules are of various types

- **Free-lateral** They are present on both the sides of leaf base, e.g. *Hibiscus rosa-sinensis* (China rose).
- **Interpetiolar** The adjacent stipules of two opposite leaves are fused to form interpetiolar stipules. In this way only two stipules of two leaves are found in place of four, e.g. *Ixora, Anthocephalus*.
- **Intrapetiolar** In this type, both stipules of a single leaf fuse to form a single stipule, e.g. *Gardenia*.
- **Foliaceous** These type of stipules are leaf-like, e.g. pea *(Lathyrus aphaca)*.
- **Scaly** These stipules are dry, small and paper-like, e.g. *Desmodium*.
- **Spiny** These stipules are modified into spines, e.g. *Ziziphus, Acacia*.
- **Ochreate** When both stipules of a leaf combine together and form a tube-like structure, then it is called as **ochreate**, e.g. *Polygonum, Rumex*.
- **Adnate** In this type, both stipules are attached to the petiole, e.g. rose, lupin, etc.
- **Tendrillar** These stipules are modified into tendril-like structure, e.g. *Smilax* for climbing.

Venation

- It refers to the arrangement of veins and veinlets on the leaf lamina.
- The following types of venation are found on leaves of different plants

1. Reticulate venation

- In this venation, main vein divides into few veinlets.
- These veinlets are irregularly distributed and form a network or reticulum on the lamina.
- It is the characteristic feature of dicotyledonous plants.
- Reticulate venation is of two types
  (i) **Pinnate or Unicostate reticulate venation** The lamina consists of a single principle vein or midrib in its centre. This midrib ultimately branches into veinlets forming a fine network, e.g. *Mangifera indica* (mango), *Ficus religiosa* (peepal), *Syzygium* (*Eugenia*, vern. jamun), *Hibiscus rosa-sinensis* (shoe-flower).
  (ii) **Palmate or Multicostate reticulate venation** The lamina consists of a number of midribs arising from the top of the petiole. All these midribs divide into veinlets giving an extensively reticulated pattern, e.g. *Ziziphus jujuba* (ber), *Ricinus communis* (castor), *Cucurbita, Vitis vinifera* (grapevine), *Carica papaya* (papaya), *Luffa*, etc.

2. Parallel venation

- It is the arrangement of veins in parallel fashion.
- This venation is the characteristic of monocotyledons with few exceptions, e.g. *Smilax, Dioscorea, Alocasia, Colocasia*, etc.
- It is also found in few dicots, such as *Calophyllum, Corymbium* and *Eryngium*.
- Parallel venation is of two types
  (i) **Pinnate or Unicostate parallel venation** The lamina possesses a single midrib in its centre running from base to the apex, e.g. *Musa paradisiaca* (banana), turmeric, *Zingiber officinale* (ginger), etc.
  (ii) **Palmate or Multicostate parallel venation** The lamina consists of a number of midribs arising from its base. It can be of following two types
    (a) **Convergent** The midribs run parallel to each other and converge (unite) only towards the apex, e.g. *Bambusa* (bamboo), grasses, *Eichhornia, Oryza sativa* (rice).
    (b) **Divergent** The midribs diverge towards the margins of the lamina just near the petiole, e.g. *Borassus flabellifer* (fan palm).

Note ***Furcate venation*** *Leaflets of most ferns, show venation in which veins divide dichotomously and branches never meet to form a network. In higher plants, Circeaster (an angiospermic plant) shows furcate venation.*

Types of leaves

On the basis of incision of lamina, the leaves are divided into two types

1. Simple leaf

- In these leaves, lamina is not divided, i.e. it is entire and it bears an axillary bud at the axil of petiole, e.g. mango and guava leaves. Sometimes simple leaves can be incised also.

- These incised simple leaves are of two types
  (i) **Pinnate simple leaf** Incisions do not reach up to the midrib, e.g. *Raphanus sativus* (radish).
  (ii) **Palmate simple leaf** Incisions do not reach up to the petiole, e.g. *Ricinus communis.*

2. Compound leaf

- A leaf is said to be compound, when its lamina is completely incised up to the level of midrib or a petiole.
- The incisions result into formation of a number of distinct segments called **leaflets.**
- The compound leaves are of two types
  (i) **Pinnately compound leaf** In this type, the leaflets are borne laterally on an elongated axis. These axis may be the main midrib or any other lateral vein of a simple leaf. Four different kinds of pinnately compound leaves are recognised.
  These are as follows
  - **Unipinnate leaf** The lamina divides only once and the leaflets are directly attached on rachis or midrib. These leaves can be **paripinnate** (even number of leaflets), e.g. *Tamarindus* or **imparipinnate** (odd number of leaflets), e.g. *Rosa indica* (rose), *Clitoria.*
  - **Bipinnate leaf** Rachis divides twice and secondary leaflets are borne on secondary axis or veins, e.g. *Mimosa pudica* (touch-me-not), *Caesalpinia* (vern. gold mohur), *Acacia nilotica* (vern. kikar).

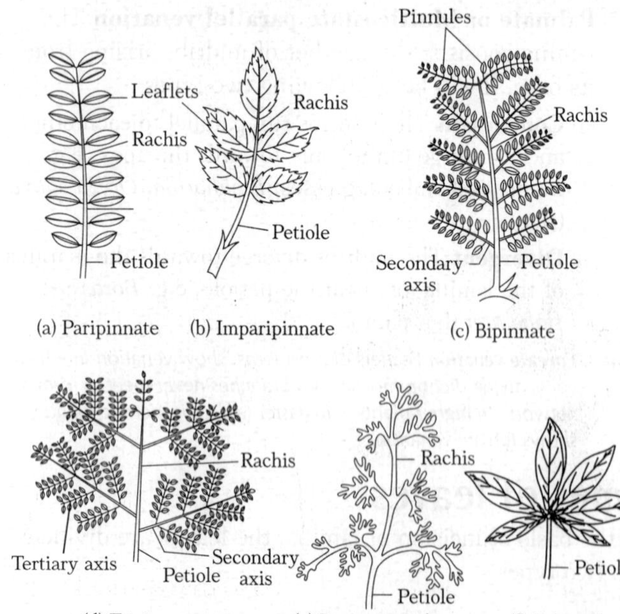

Fig. 11.12 Compound leaf

  - **Tripinnate leaf** In these leaves, the secondary axis (rachillae or rachules) too divides giving rise to tertiary axis. On this axis leaflets are borne, e.g. *Melia azadirachta* (vern. dharek, drek), *Moringa* (vern. sahinjana, soonjana), *Oroxylon* (vern. arlu).
  - **Decompound leaf** In this type of leaf, the rachis divides more than three times, e.g. *Coriandrum sativum* (coriander, vern. dhaniya), *Daucus carota* (carrot, vern. gajar).

  (ii) **Palmately compound leaves** The incisions of leaves are directed towards the petiole. Due to this type of incisions, all leaflets give appearance of fingers of human palm. There is no rachis found in such leaves. A joint may be present between the tip of the petiole and from this joint leaflets are borne.
  On the basis of number of leaflets present at the tip of petiole, five types of palmately compound leaves are recognised.
  - **Unifoliate** Only one leaflet separated from the petiole by a constriction, e.g. *Citrus* (lemon).
  - **Bifoliate** Two leaflets are attached side by side at the tip of petiole, e.g. *Balanites, Hardwickia, Prinsepia.*
  - **Trifoliate** (Ternate) Three leaflets are joined at the petiole, e.g. *Aegle marmelos* (wood apple, vern. bel), *Butea* (vern. dhak), *Oxalis, Trigonella* (vern. methi).
  - **Quadrifoliate** (Quadrinate) Four leaflets are attached at the tip of petiole, e.g. *Paris quadrifolia.*
  - **Multifoliate** (Digitate) Five or more leaflets are present at the tip of petiole, e.g. *Bombax* (red silk cotton, vern. simbal), *Cleome*, lupin.

Bracts

- These are specialised modification of leaves present at the base of flower axis. Different forms of bracts are as follows
- **Foliaceous**, e.g. *Adathoda, Grossandro.*
- **Spath**, (spathe), e.g. *Colocasia, Cocos, Musa, Typhorium.*
- **Petaloid**, e.g. *Bougainvillea, Euphorbia.*
- **Involucre**, e.g. Coriander.
- **Epicalyx**, e.g. *Hibiscus rosa-sinensis.*
- **Scaly**, e.g. *Tridex helianthus.*
- **Glumen**, e.g. Sterile or fertile (Lemme, e.g. *Oryza sativa*).

Phyllotaxy

It refers to the arrangement of leaves on the stem or its branches. Types of phyllotaxy are as follows

(i) **Alternate or Spiral** In this, a single leaf is present at each node. These leaves are arranged alternatively in same plane, e.g. *Hibiscus rosa-sinensis.*

(ii) **Opposite type** When two leaves are borne on each node opposite to each other, it is called opposite arrangement. It is of following two types

(a) **Opposite decussate** When the opposite leaves of two successive nodes lie at right angles to each other, e.g. *Calotropis* (vern. aak), *Ocimum* (vern. tulsi).

(b) **Opposite superimposed** When the opposite leaves of two successive nodes lie in the same plane, i.e. they are superimposed on each other, e.g. *Syzygium* (*Eugenia*; vern. jamun).

Note *Quisqualis indica (Rangoon creeper) is an exception, because it shows both opposite superimposed and opposite decussate phyllotaxy.*

(iii) **Whorled or Verticillate type** In this type three (e.g. *Nerium*, vern. kaner) or more than three (e.g. *Alstonia, Galium*) leaves are borne on a single node in a whorled manner.

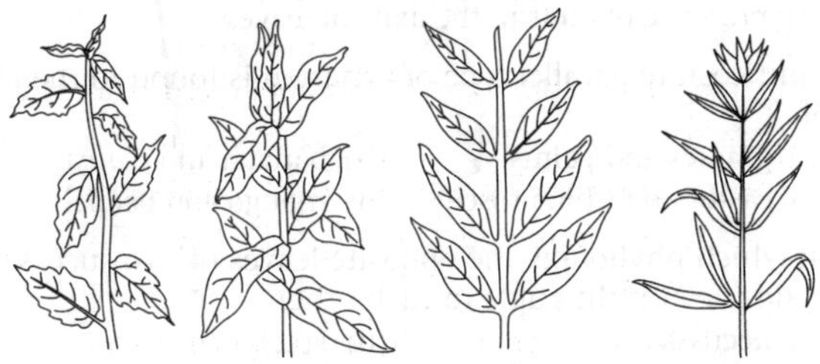

(a) (b) (c) (d)

***Fig. 11.13** Types of phyllotaxy : (a) Alternate, (b) Opposite decussate, (c) Opposite superimposed, (d) Whorled or verticillate*

Modification of leaf

Leaves are occasionally modified into specialised structures to perform various functions other than photosynthesis.
These modifications are as follows

1. Whole leaf tendrils

- Tendrils are coiled structures that help the plants to climb up. These are long, slender and thread-like.
- Different parts of a leaf can get transformed into tendrils such as upper leaflet (*Pisum sativum, Lathyrus odoratus*), petiole (*Tropaeolum majus*), stipules (*Smilax*), leaf-tip (*Gloriosa superba*), rachis and petiolules (*Clematis*), etc.
- Tendrils are sensitive to touch. They coil around a support to which they come in contact to help plant in climbing.

2. Leaf pitchers

- In this, leaf lamina (e.g. *Nepenthes*) and whole leaf (e.g. *Sarracenia*) get modified into pitcher-like structure to trap insects. In *Dischidia*, pitcher harvest rainwater.
- Some of the leaf segments of aquatic insectivore plants, e.g. *Utricularia* are modified into bladder for trapping insects.

3. Leaf bladders

- They occur in the aquatic carnivorous plants of *Utricularia* (Bladderwort). Some of the leaf segments are modified to form small bladders (1-3 mm in diameter). A bladder has sensitive hair, branched trigger bristles, a trap valve, internal and external glands for trapping and digesting small animals (e.g. water fleas).

4. Phyllodes

The petiole of leaves gets modified into photosynthetic leaf-like structure called phyllode, e.g. *Acacia.*

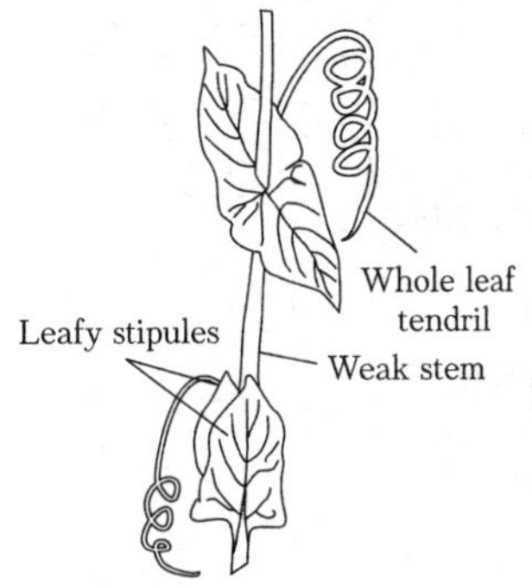

(a) Whole leaf tendril *of L. aphaca*

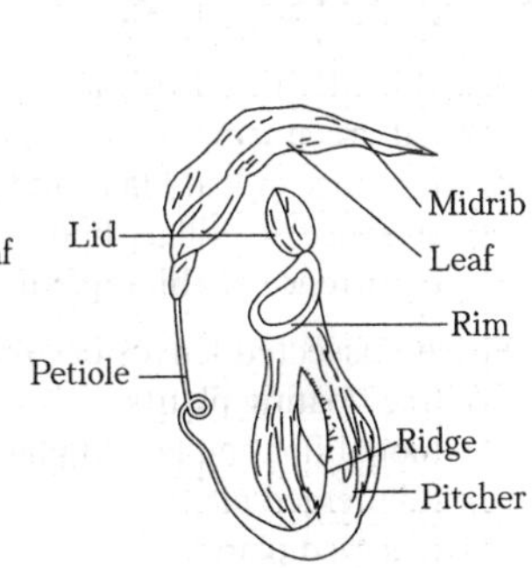

(b) Leaf pitcher of *Nepenthes*

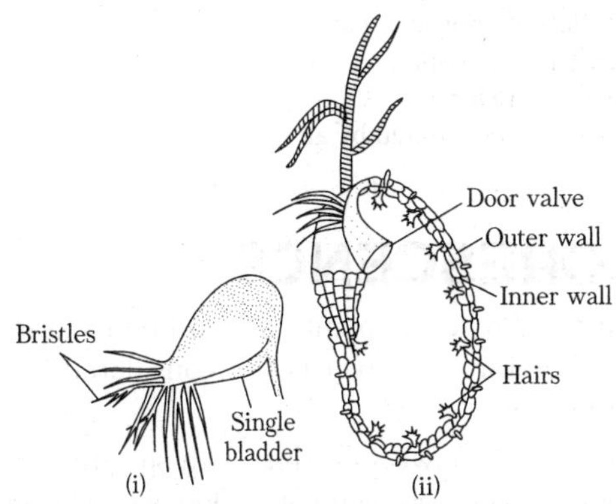

(c) (i) Single leaf bladder of *Utricularia*, (ii) LS of bladder

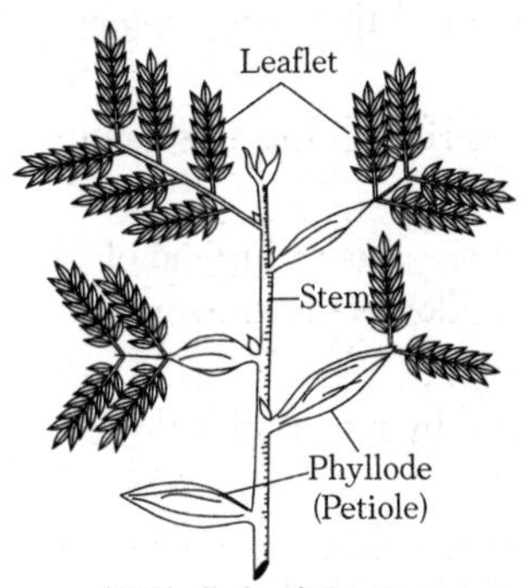

(d) Phyllode of *Acacia*

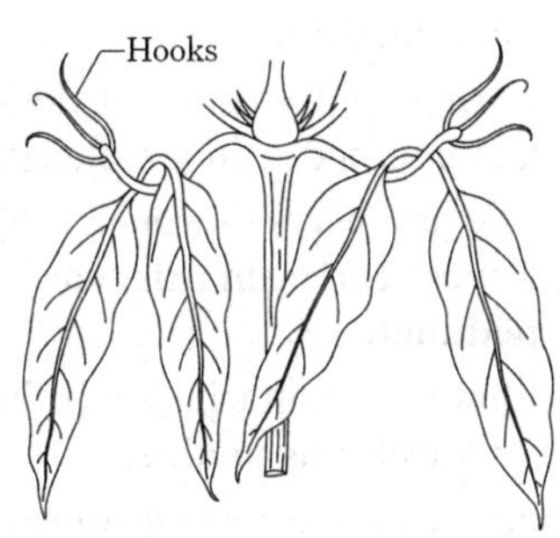

(e) Leaf hooks of water fleas

***Fig 11.14** Modification of leaf*

5. Leaf hooks

These are claw-like structures. They are formed from three terminal leaflets to help in climbing, e.g. *Bignonia.*

6. Scaly leaf

In this, leaves become small, dry and membranous. They are non-photosynthetic and help in protection, e.g. *Ruscus*. Some scaly leaves also store food, e.g. onion and garlic.

7. Storage or Succulent leaf

Some plants of xerophytic habitats and members of the family-Crassulaceae have thickened succulent leaves with water storage tissue. This helps the plant to resist desiccation (drying up), e.g. *Crocus*, *Aloe*, *Agave*, etc.

Heterophylly

- It is the presence of more than one type of leaf on the same plant. *Generally four types of heterophylly are recognised*
  (i) Developmental (ii) Habitual
  (iii) Adaptive (iv) Environmental
- Example of heterophylly is *Limnophila heterophylla*.

Note • *Phyllode and phylloclade are two different structures.*
• *Phylloclade is a modified stem serving as a photosynthetic and storage organ, while phyllode is a modified petiole performing photosynthesis.*

CHECK POINT 11.2

1. A leaf is identified from
(a) flat green lamina
(b) presence of leaf blade and petiole
(c) presence of axillary bud
(d) occurrence of chlorophyll

2. Finely dissected leaves occurs in
(a) free floating plants
(b) rooted floating leaved plants
(c) submerged plants
(d) emerged plants

3. Imparipinnate leaf is the one where
(a) leaflets are borne in pairs
(b) leaflets are small
(c) leaflets are large
(d) rachis is terminated by an odd leaflet

4. A unipinnate componud leaf can be differentiated from a branch having simple leaves by the
(a) presence of terminal bud in compound leaf
(b) absence of veins in the leaflets
(c) presence of buds in the axile of leaflets
(d) presence of buds in the axile of leaves

5. Multicostate parallel type of venation is found in the leaves of
(a) grasses and palms (b) banana and *Canna*
(c) castor and China rose (d) mango and peepal

6. In which phyllotaxy the opposite leaves of two successive nodes lie at right angle to each other
(a) decussate (b) spiral
(c) superimposed (d) whorled

INFLORESCENCE

- The shoot of flowering plants is modified to form the flower. Therefore, a flower is also considered as a modified shoot.
- Each plant bears flowers on shoot in a specific arrangement. The arrangement of flowers on shoot is known as **inflorescence.**
- Shoot bears flowers, when its apical meristem changes to floral meristem.
- During the formation of floral meristem, the axis condenses and internodes do not elongate.
- The axis bears flowers at successive nodes instead of leaves. The main axis that bears flowers is known as **peduncle.**
- Flowers are attached to peduncle by means of a flower stalk called as **pedicel.**

Note *Sometimes single flowers are present on the branches. If this single flower is found on the terminal part of a branch it is called* **solitary terminal flower.** *If it is present in the axil of a leaf it is called* **solitary axillary flower.**

- Inflorescence can be of following two types depending upon whether the shoot apex continuously growing or convert into a flower.

Racemose inflorescence

- In this type, the floral axis grows continuously producing flowers in acropetal manner (younger towards the apex, older towards the base). It is also called as **indefinite inflorescence.**
- It is divided into following sub-types

1. Simple racemose

These have unbranched peduncle. It is further categorised into following types

(i) **Raceme** In this type, the peduncle grows indefinitely. It bears bisexual and pedicellate flowers. These flowers usually are bracteate, e.g. mustard, radish, larkspur, *Lupinus*, etc.

(ii) **Spike** It bears bisexual and sessile (without pedicel) flowers, e.g. bottle brush, *Achyranthes* and *Adhatoda.*

(iii) **Catkin** The peduncle is weak and the whole inflorescence drops down. Flowers are bracteate and sessile, e.g. mulberry, oak, *Birch* and *Populus.*

(iv) **Spadix** The peduncle is fleshy and contains female (basal), male (upper) and neutral (middle) flowers. These flowers lack pedicel.

Terminal part of the peduncle does not have any flower and is called **appendix.**

Axis is surrounded by an enlarged bract which is brightly coloured and protective in function, e.g. banana, *Colocasia, Alocasia, Acalypha.*

(v) **Corymb** Flowers are borne on slightly condensed peduncle. The older flowers have longer pedicels than the newer ones. In this way, all the flowers reach the same level, e.g. candytuft, *Capradla.*

(vi) **Umbel** Tip of the peduncle is drastically reduced. It bears numerous bracteate and pedicellate flowers. These flowers appear to be arising from the same point, e.g. *Centella*, pronus, etc.

(vii) **Capitulum** In this, the peduncle is modified into disc-like and is called as **receptacle**.

It bears sessile bracteolate centripetally arranged florets of two types, i.e. **ray florets** (female, zygomorphic) and **disc florets** (bisexual, actinomorphic), e.g. sunflower, marigold, etc.

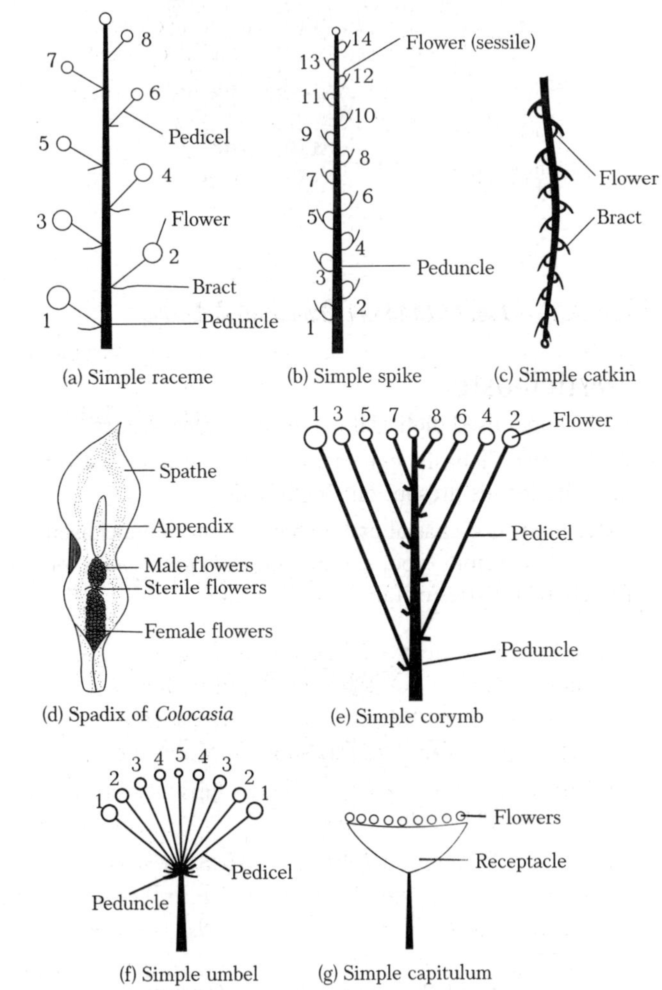

***Fig. 11.15** Simple racemose inflorescence*

2. Compound racemose

When the peduncle of simple racemose is branched, it is known as compound racemose.

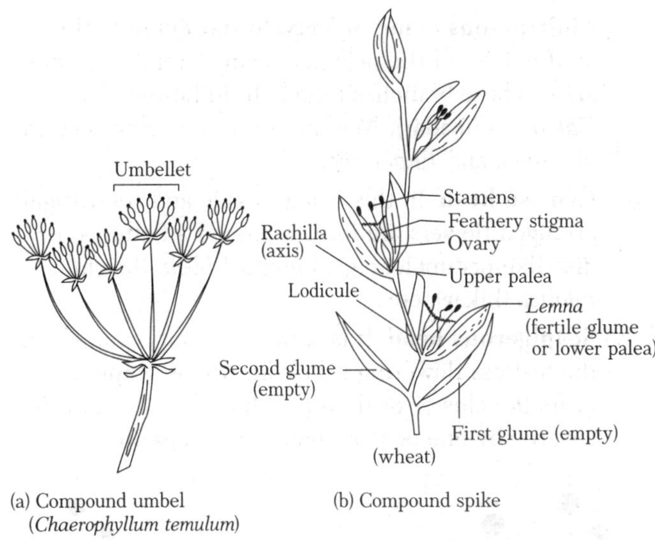

***Fig. 11.16** Compound racemose inflorescence*

It is further categorised into following types

- **Panicle** or **Compound raceme**, e.g. gold mohur (*Delonix*), amaltas, *Asparagus.*
- **Corymbose raceme**, e.g. cauliflower, *Pyrus malus.*
- **Compound umbel**, e.g. carrot, fennel, cumin, coriander.
- **Compound spadix**, e.g. coconut, datepalm, banana.
- **Compound spike**, e.g. wheat, oat.

Cymose inflorescence

In cymose inflorescence, peduncle terminates into a flower that develops in basipetal succession, i.e. younger flowers occupy basal position. It can be of following types

(i) **Uniparous or Monochasial cyme** Peduncle ends up in a flower producing new single floral lateral axis at the base of older flower. This new lateral axis also ends in a flower. This branching is continued.

It is two types

(a) **Helicoid cyme** When all lateral branches develop on the same side of peduncle, it is called as helicoid cyme, e.g. *Solanum nigrum, Saraca, Begonia* and *Drosera.*

(b) **Scorpioid cyme** In this type, one lateral branch develops on one side and the other branch develops opposite to first one. In this way, these two branches lie alternate to each other, e.g. *Ranunculus* and *Heliotropium.*

(ii) **Dichasial cyme or Biparous cyme** In this, at the basal part of the peduncle, two lateral branches arise. These new branches also end in a flower, e.g. *Bougainvillea, Dianthus* and *Mirabilis.*

(iii) **Multiparous cyme or Polychasial cyme** In this, at the base of the peduncle many lateral branches arise. These branches terminate in flower, e.g. *Calotropis* (madar), *Nerium, Asclepias,* some species of *Croton* and *Euphorbia.*

(iv) **Cymose head** In this, many sessile and centrifugally arranged flowers are formed around a flattened or disc-like peduncle, e.g. *Albizzia lebbek, Acacia nilotica* (kikar), etc.

(v) **Scapigerous head** It is seen in *Allium cepa,* where the leafless flowering axis is known as **scape**. It bears the cluster of flowers, which form head. It is covered by one or more bract called **spathe**.

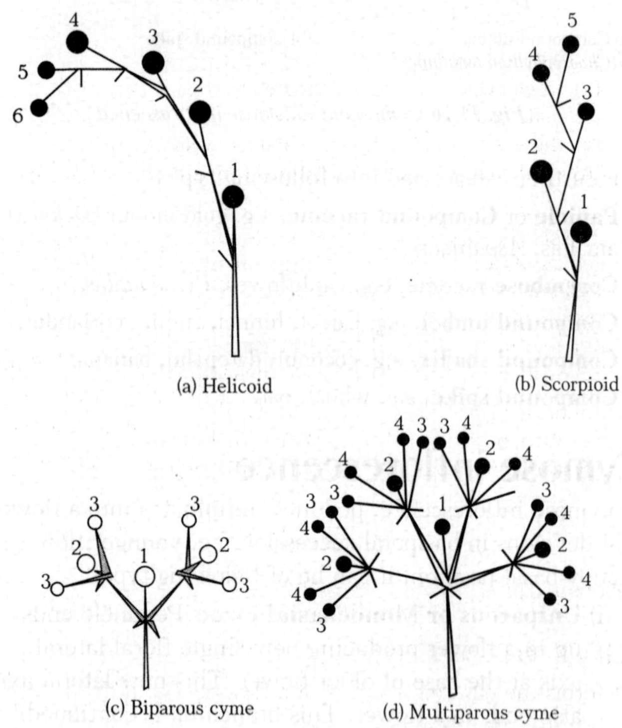

Fig. 11.17 *Different kind of cymose inflorescence*

Table 11.2 Differences between racemose and cymose inflorescence

Racemose	Cymose
Indefinite growth of flower bearing shoot.	The apex terminate into flower, so growth of flower bearing shoot is definite.
Flowers borne in acropetal manner (older flower at base).	Flowers arranged in basipetal manner (older flower at apical position).

Some special inflorescence

These are given below

1. Cyathium

- It is the characteristic inflorescence of family—Euphorbiaceae. It looks like a single flower.
- In this, a cup-shaped structure is formed by the fusion of **involucre of bracts** (i.e. whorl of bracts). This cup encloses a large, achlamydeous (sepals and petals are absent), pedicellate female flower.
- This flower is tricarpellary and syncarpous.
- It is surrounded by numerous, achlamydeous, pedicellate, centrifugally arranged and scorpioid male flower.

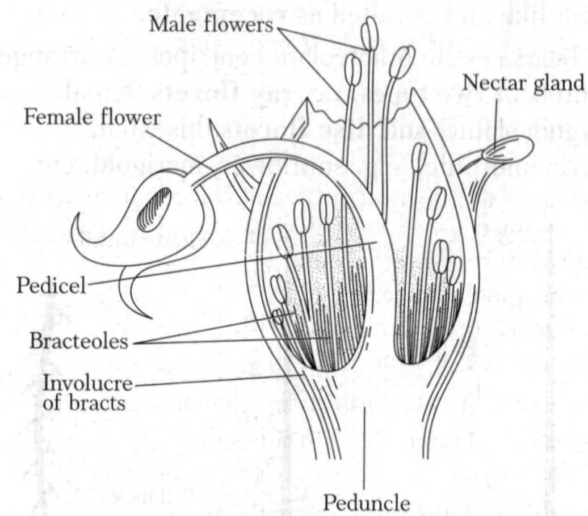

Fig. 11.18 *LS of cyathium of Euphorbia*

2. Verticillaster

- It is a condensed dichasial cyme-like inflorescence.
- In this, two dichasial cymes are borne in the axil of two opposite leaves present on single node.
- Later on, this dichasial cyme changes into monochasial cyme of scorpioid type, e.g. *Salvia, Ocimum, Coleus,* etc., is the characteristics of family-Lamiaceae.

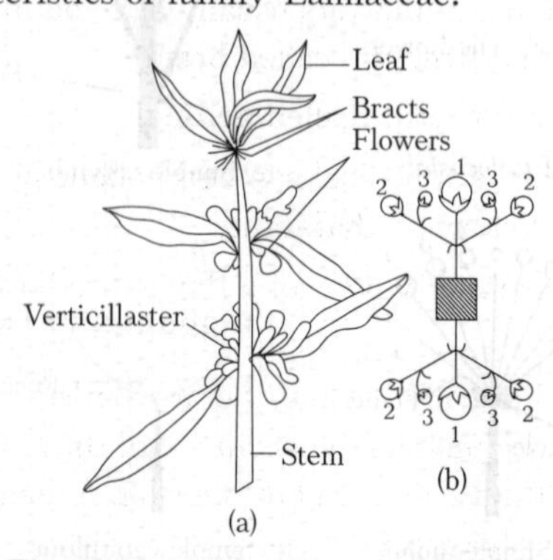

Fig. 11.19 *(a) Verticillaster of Ocimum, (b) Diagrammatic view of verticillaster*

3. Hypanthodium

- In this, the peduncle becomes a cup-shaped receptacle. It contains a hollow central cavity and a hairy opening called ostiole at their anterior side.

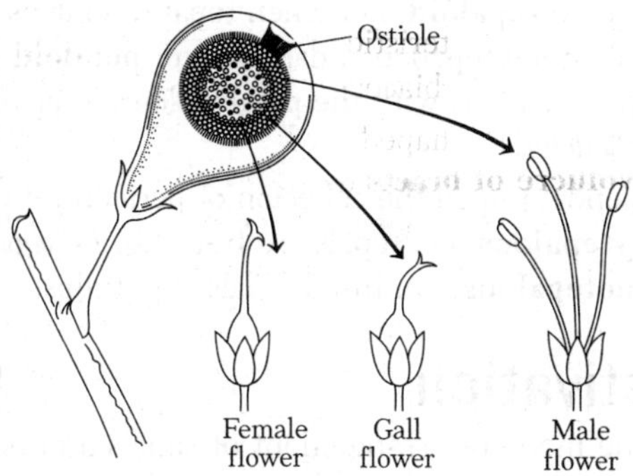

Fig. 11.20 Hypanthodium of Ficus

- The hollow cavity bears many male flowers at the ostiole side and female flowers at the basal side.
- The sterile female flowers with short style are called gall flowers, e.g. *Ficus religiosa* (peepal), banyan, etc.

4. Coenanthium

In this, the receptacle is saucer-shaped and contains curved margins. It bears flowers (florets) in a similar arrangement as hypanthodium, e.g. *Dorstenia*.

Note *Certain inflorescence are misinterpreted as flowers, e.g. cauliflower (compound corymb), Euphorbia (cyathium) and sunflowers (capitulum) are not flowers, but whole inflorescence.*

CHECK POINT 11.3

1. Most accurate function in the following statement about inflorescence is
(a) dispersal of seeds
(b) formation of more fruit
(c) formation of pollen grain
(d) dispersal of pollens

2. Inflorescence in *Musa paradisiaca* (banana) is a
(a) raceme (b) catkin
(c) spadix (d) verticillaster

3. In cymose inflorescence, the main axis
(a) has unlimited growth
(b) bears a solitary flower
(c) has unlimited growth, but lateral branches end in flowers
(d) terminates in a flower

4. In cyathium inflorescence,
(a) single male flower is surrounded by female flowers
(b) male and female flowers are borne in different plants
(c) there is one male and one female flowers
(d) single female flower surrounded by many peripheral male flowers

5. The special types of inflorescence found in *Ficus* where the female flower are at bottom and male flower near ostiole and enclosed within a cup-shaped fleshy thalamus with ostiole is called
(a) cyathium (b) verticillaster
(c) spadix (d) hypanthodium

6. Inflorescence is edible in *Brassica oleracea*
(a) varn. botrytis (b) varn. capitata
(c) varn. gougyloids (d) varn. genifera

FLOWER

- A flower represents the reproductive part of an angiospermic plant. Flowers usually develop in the axil of a small leaf-like structure called **bract**.
- The stalk of a flower, is called **pedicel**.
- The upper part of a pedicel can be swollen, spherical or conical. It is known as **thalamus**.
- The various parts or whorls of a flower are borne on thalamus.
- Each flower normally has four floral whorls, i.e. calyx, corolla, androecium and gynoecium. All these whorls are arranged on the swollen end of the stalk or pedicel called **thalamus** or **receptacle**.

Various parts of flowers are as follows

Calyx or Sepals

- It forms the outermost whorl of a flower. The units of calyx are called **sepals**.
- Each sepal is a green and leaf-like structure in most of the flowers.
- Sepals can be free from each other, i.e. polysepalous or fused with each other, i.e. gamosepalous.
- The various types of sepals are as follows
  (i) **Caducous sepals** fall down at the time of opening of flower, e.g. poppy.
  (ii) **Deciduous sepals** fall down at the time of flower maturation, e.g. mustard.

(iii) **Persistent sepals** do not fall even after the formation of fruit.
 (a) **Marcescent sepals** become dry in fruit, e.g. *Piper*, guava.
 (b) **Accrescent sepals** become associated with the fruit, e.g. tomato.

Corolla or Petals

- It occurs next to the calyx in a flower. It is non-essential colourful whorl of the flower. It helps in attracting the pollinating insects.
- The units of corolla are called **petals.**
- All petals of a flower can be fused, i.e. gamopetalous, e.g. *Ipomoea.*
- The petals can also be free from each other, i.e. polypetalous, e.g. mustard.
- Polypetalous corolla can be of following types depending upon their shape
 (i) **Cruciform** It consists of 4 petals arranged crosswise, each petal is clawed e.g. mustard.
 (ii) **Caryophyllaceous** It consists of 5 petals having long claws and the limb of petals are placed at right angles to the claw, e.g. *Dianthus.*
 (iii) **Papillionaceous** It consists of five asymmetrical petals, arranged like a butterfly shape, e.g. pea, gram.
 (iv) **Rosaceous** It consists of 5 petals. Petals are with short claws or none. The limbs spread regularly outwards, e.g. rose.
 (v) **Campanulate** or **bell-shaped** In this case the shape of the corolla is like a bell, e.g. *Physalis.*
- Similarly, gamopetalous corolla can also be of following types depending upon their shapes
 (i) **Tubular**, e.g. disc florets of sunflower.
 (ii) **Infundibulum**, e.g. *Datura.*
 (iii) **Rotate** or **wheel-shaped**, e.g. brinjal.
 (iv) **Star-shaped**, e.g. *Ixora.*
 (v) **Bilabiate** (two lipped), e.g. *Ocimum, Adhatoda.*
 (vi) **Ligulate** (Strap-shaped), e.g. ray florets of sunflower.

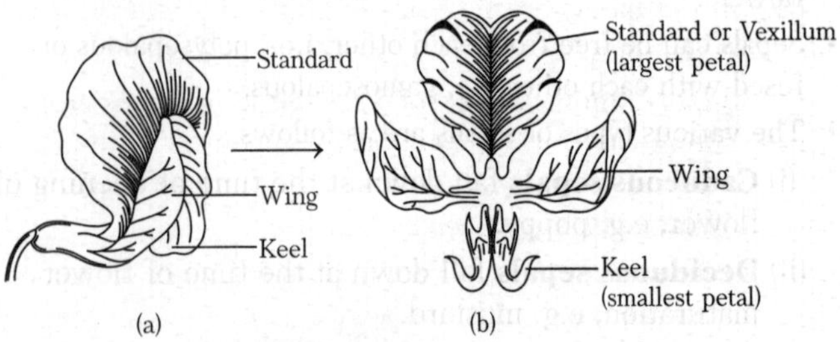

Fig. 11.21 (a) Papillionaceous corolla, (b) Different parts separated

Perianth

- When sepals and petals are not differentiated, the structure present is known as perianth.
- The unit of perianth is called **tepal.**
- It can be **sepaloid**, i.e. when tepal resembles with the sepal (green tepal), e.g. datepalm or **petaloid**, i.e. when tepal resembles with the petal (coloured tepal), e.g. *Asphodelus.*
- Depending upon the cohesion of perianth, it can be **polytepalous**, i.e. tepals are free, e.g. *Polygonum* or **gamotepalous**, i.e. fused tepals, e.g. tulips.

Aestivation

It is the mode of arrangement of sepals and petals on the thalamus.

It is of following types

(i) **Valvate** Petals are close to each other, but do not overlap, e.g. mustard (*Brassica*).
(ii) **Twisted** Petals overlap each other regularly. Margin of one petal overlaps the next, e.g. China rose (*Hibiscus rosa-sinensis*).
(iii) **Imbricate** There are five petals, arranged in such a way that one petal is completely external and another petal is completely internal. The remaining three petals are partially external and partially internal, e.g. *Cassia, Callistemon* and *Caesalpinia.*
(iv) **Quincuncial** There are five petals, in which two are completely external and other three are partially external and partially internal, e.g. *Cucurbita* (Cucurbitaceae).
(v) **Vexillary** It is the characteristic feature of family–Papillionaceae, posterior petal overlaps the two smaller lateral petals and the later in turn overlap the two anterior petal, e.g. *Artobotrys* and *Polyalthea.*

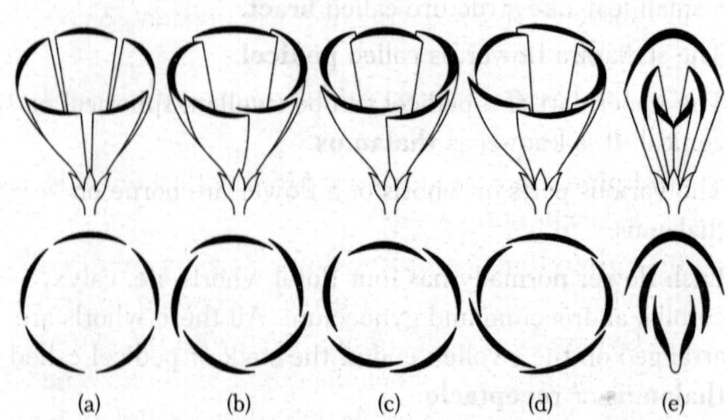

Fig. 11.22 Different types of aestivation of calyx and corolla : (a) Valvate, (b) Twisted, (c) Imbricate, (d) Quincuncial, (e) Vexillary

Stamens or Androecium

- It forms the third essential whorl of the flower. It is male part of the flower.
- The basic unit of androecium is a **stamen.**
- The number and arrangement of stamen is specific and constant for each plant species.
- Each stamen (microsporophyll) has two parts, a long filament and a terminally situated bag-shaped anther.
- Most of the anther of angiospermic plants are **dithecous,** i.e. have two lobes, e.g. pea, but some have only one lobe, i.e. **monothecous**, e.g. China rose (Malvaceae).
- The two lobes of a dithecous anther, adhere to each other by means of connective. It is a band of sterile, parenchymatous cells, present between the two lobes.

Types of stamens

1. On the basis of attachment of anthers to the filament, it can be of following types
   (i) **Basifixed** Filament adheres to the basal part of anther, e.g. *Cassia* and mustard.
   (ii) **Adnate** Filament runs through the anther, e.g. *Michelia* and *Ranunculus.*

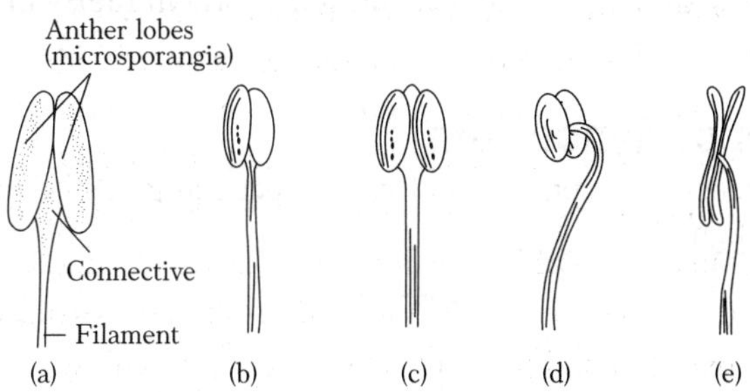

Fig. 11.23 *Androecium : (a) Parts of stamen, (b) Basifixed, (c) Adnate, (d) Dorsifixed, (e) Versatile*

   (iii) **Dorsifixed** Filament attaches to the dorsal side of an anther and it is immobile, e.g. *Citrus* and *Passiflora.*
   (iv) **Versatile** A dorsifixed type attachment, but here anther is free to swing over the filament, e.g. grasses and *Eucalyptus.*
2. On the basis of adhesion of stamens with other floral whorls, it is of following types
   (i) **Epipetalous** Stamens are fused with the petals, e.g. *Solanum,* sunflower.
   (ii) **Epiphyllous or Epitepalous** Stamens are fused with the tepals, e.g. *Asphodelus, Asparagus,* etc.
   (iii) **Gynandrous** Stamens are fused with the carpel throughout their whole length or by their anthers only, e.g. *Calotropis* (Asclepiadaceae family).
3. On the basis of fusion of stamens or its parts, it can be of following types
   (i) **Polyandrous** Stamens are free, e.g. mustard, *Papaver,* lily, radish, etc.
   (ii) **Adelphous** Filaments of stamens become fused, but anthers remain free. It is of three types
   - **Monadelphous** All filaments become fused and form a single group, characteristics feature of Malvaceae (China rose), *Achyranthes, Abutilon.*
   - **Diadelphous** Two separate bundles of united filaments are formed, e.g. Papillionaceae, *Pisum* and *Tephrosia.*
   - **Polyadelphous** More than two separate bundles of filaments are formed, e.g. castor (*Ricinus*), lemon (*Citrus*) and silk cotton (*Bombax*).

   (iii) **Synandrous** In this, stamens are united throughout their whole length to form a complete single structure, e.g. members of family—Cucurbitaceae (*Cucurbita maxima*).

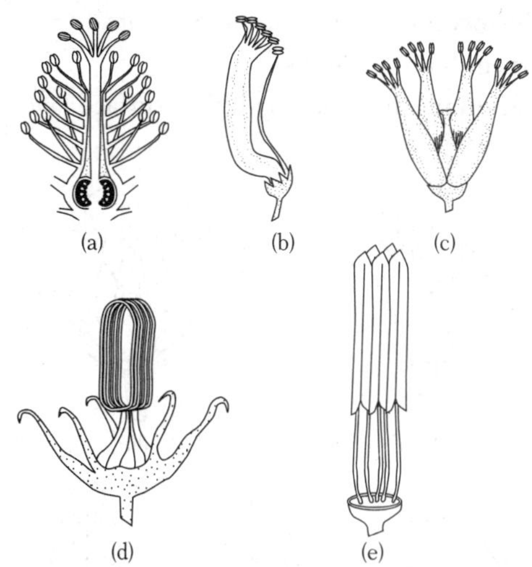

Fig. 11.24 *Types of stamens : (a) Monadelphous, (b) Diadelphous, (c) Polyadelphous, (d) Synandrous, (e) Syngenesious*

   (iv) **Syngenesious or Synantherous** All anthers are united, but their filaments remain free. Characteristic feature of family—Asteraceae, e.g. sunflower, marigold.
4. On the basis of length of filaments of stamens, it can be of following types
   (i) **Didynamous** Here, total number of stamens are four. Out of these four, two stamens are with shorter filaments, while rest two contain longer filaments, e.g. *Ocimum, Salvia.*
   (ii) **Tetradynamous** Here, the total number of stamens are six. The four stamens of inner side have longer

filament than the two stamens of outer side, e.g. members of family—Brassicaceae.

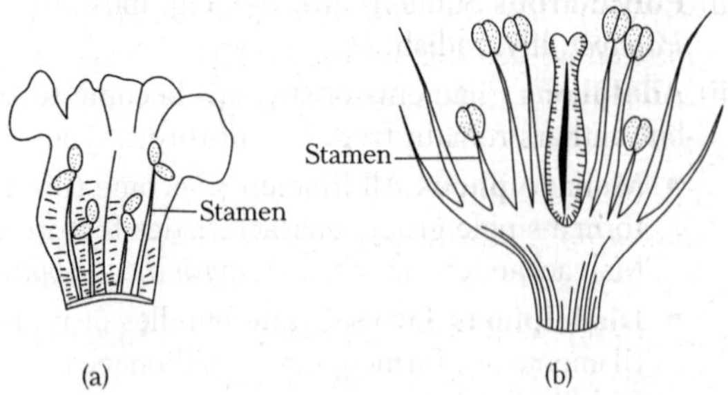

Fig 11.25 Types of stamens : (a) Didynamous, (b) Tetradynamous

5. On the basis of dehiscence of anthers, it can be of following types
   (i) **Extrorse** anther opens towards periphery, e.g. poppy.
   (ii) **Introrse** opening towards the centre, e.g. *Citrus.*
   (iii) **Transverse** breadthwise opening, e.g. *Althaea.*
   (iv) **Porous** through pores, e.g. *Solanum.*
   (v) **Irregular**, e.g. *Najus.*
   (vi) **Valvular** lobes open by lifting surface layers, e.g. berberry.

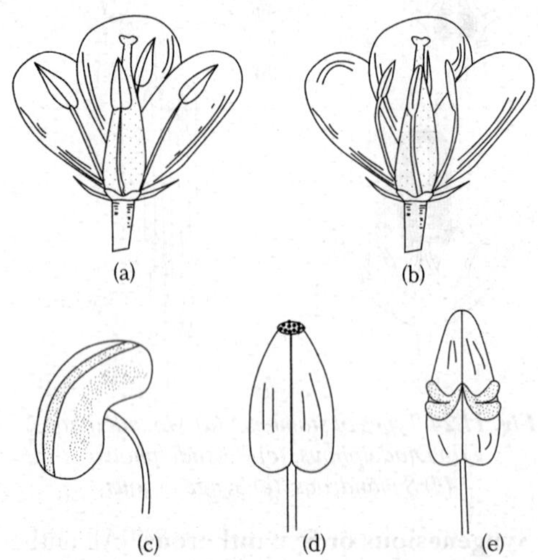

***Fig. 11.26** Dehiscence of anthers : (a) Extrorse, (b) Introrse, (c) Transverse, (d) Porous, (e) Valvular*

Gynoecium (Pistil)

- It is the innermost essential whorl of a bisexual flower that forms female reproductive parts.
- Gynoecium is made up of one or more units called **carpels.**
- Each carpel consists of a basal swollen part, i.e. ovary that contains ovules, a cylindrical long stalk style and a pollen receptive structure stigma.

Types of gynoecium

- Depending upon the number of carpels, gynoecium can be of following types
  (i) **Monocarpellary gynoecium** consists of only one carpel, e.g. pea (Papilionaceae).
  (ii) **Bicarpellary gynoecium** has two carpels, e.g. *Helianthus.*
  (iii) **Tricarpellary gynoecium** has three carpels, e.g. *Stellaria.*
  (iv) **Tetracarpellary gynoecium** has four carpels, e.g. Rhamnaceae.
  (v) **Pentacarpellary gynoecium** has five carpels, e.g. *Melia.*
  (vi) **Multicarpellary gynoecium** has many carpels, e.g. *Papaver.*
- Depending upon cohesiòn of carpels, i.e. interaction between members of gynoecium can be of two types
  (i) **Apocarpous** The number of carpels are two to many and they are free from each other, e.g. *Ranunculus* (butter cup), *Aconitum, Clematis* (Magnoliaceae).
  (ii) **Syncarpous** The number of carpels are two to many, but they are fused, e.g. mustard (Cruciferae), China rose (Malvaceae).

Parts of gynoecium

- Gynoecium is comprised of various parts as given below in the figure

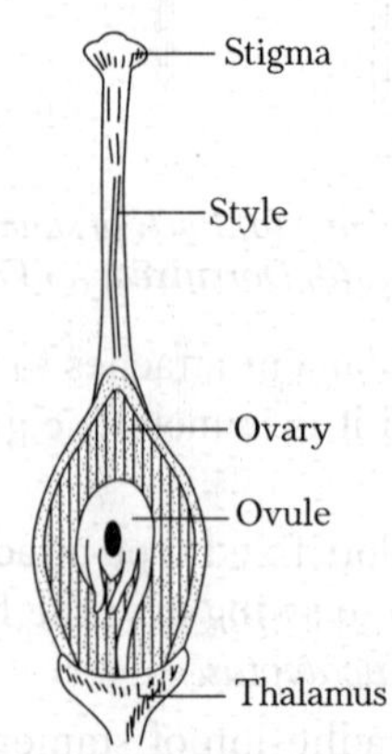

***Fig. 11.27** Gynoecium (single carpel)*

1. Stigma

It is the terminally situated, modified part of style. It is adapted for receiving pollen grains during pollination. Stigma contains various exudates such as sugar, proteins, lipid, water, etc. These help in the germination of pollen grains.

Stigma is of following types

- **Capitate** Like a head or cup-shaped, e.g. lemon and *Cleome.*
- **Plumose** Like a feather, e.g. grasses.
- **Bifid** With two longitudinal forks, e.g. *Sonchus* and *Ixora.*
- **Discoid** Like a disc, e.g. China rose and *Melia.*
- **Lamellate** With many fan-like blades, e.g. *Mazus.*
- **Dumb-bell shaped** Like a dumb-bell, e.g. *Ipomoea fistulosa.*
- **Sticky** Stigma with sticky liquid, e.g. *Cleome viscosa.*
- **Linear** In the form of small lines.
- **Knob-like** Having knob-shape, e.g. *Achyranthes* and *Cryptostegia.*
- **Radiate hood-like** A hood-like structure with radiate septae, e.g. poppy.

2. Style

It is a cylindrical more or less elongated hollow or solid part of carpel. It arises from top of the ovary. Style is of following types depending upon its origin

- **Terminal style** It occurs upright and arises on the tip of ovary, e.g. China rose.
- **Lateral style** It arises from one lateral side of an ovary, e.g. mango (*Mangifera indica*), strawberry.
- **Gynobasic style** It arises directly from the centre of ovary or from thalamus, e.g. *Ocimum, Salvia.*

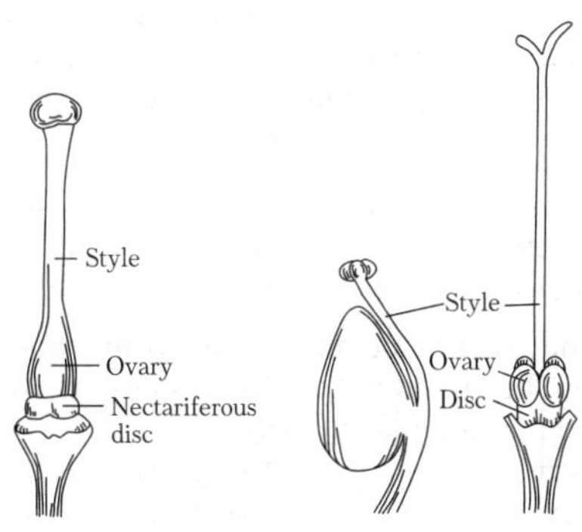

***Fig. 11.28** Various types of style : (a) Terminal, (b) Lateral, (c) Gynobasic*

3. Ovary

It is the basal, swollen part of the gynoecium in which ovules are borne. It can be of following types

(i) Depending upon the number of locules, i.e. chambers in which ovules are found, ovary can be of following types

- **unilocular** (one locule), e.g. *Stellaria*
- **bilocular** (2 locules), e.g. *Solanum*
- **trilocular** (3 locules), e.g. *Asphodelus*
- **multilocular** (many locules), e.g. *Citrus.*

(ii) Depending upon the position of ovary on thalamus, ovary can be of following types

(a) **Superior** The ovary occupies the highest (superior) and central position on the thalamus. The other floral whorls like androecium, petals and sepals arise successively below ovary.

Flowers containing superior ovary are called hypogynous flowers, e.g. *Citrus,* buttercup, *Stellaria,* etc.

(b) **Semi-superior or Semi-inferior** In this situation, margins of thalamus grow upward and a disc-shaped or cup-shaped thalamus is formed. Here, the position of all the whorls are more or less at same plane, e.g. rose, pea, etc.

(c) **Inferior** Here, the position of ovary is inferior (lowermost), while the other whorls grow above the ovary, e.g. coriander, carrot, apple, sunflower, *Cucurbita,* guava, etc.

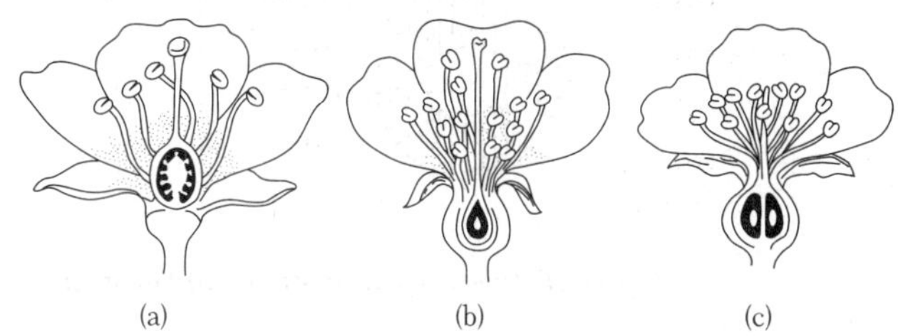

***Fig. 11.29** Position of ovary on thalamus : (a) Superior (Hypogyny) (b) Semi-inferior (Perigyny) (c) Inferior (Epigyny)*

Placentation

The parenchymatous tissue by means of which ovules remain attached to the ovary is called **placenta**. The mode of arrangement of ovules inside the ovary is called **placentation.**

It is of following types

(i) **Marginal** Here, monocarpellary, one-chambered (unilocular) ovary bears a single, longitudinal ovule. The placenta, arises from the margins of the unilocular ovary, e.g. pea (*Pisum sativum*), *Lathyrus odoratus.*

(ii) **Axile** It occurs in multicarpellary and syncarpous ovary. There is inward growth of margins of carpels which form a multilocular condition, that contains an axis in the centre.
Placenta arise from this central axis, that bears ovules, e.g. *Citrus* (Rutaceae), *Solanum nigrum* (Solanaceae), China rose (Malvaceae), Liliaceae, etc.

(iii) **Parietal** This situation arises in a syncarpous and unilocular ovary. It contains placenta towards the inner side from the fused margins of two carpels, e.g. *Argemone maxicana* (Papaveraceae), radish, *Funaria indica,* mustard, *Cucurbita, Iberis amara,* etc.

(iv) **Free-central** This ovary contains only one chamber (unilocular) and the placenta bearing ovules arise from the central axis, e.g. *Stellaria, Dianthus, Silene,* etc.

(v) **Basal** The ovary is unilocular. It bears the placenta at the base and contains a single large ovule, e.g. sunflower (Asteraceae), wheat, maize (Poaceae), *Ranunculus* (Ranunculaceae).

(vi) **Superficial** The ovary is multicarpellary, multilocular and syncarpous. The ovule bearing placenta arises all over the inner partition walls of the ovary, e.g. *Nymphaea* (water lily).

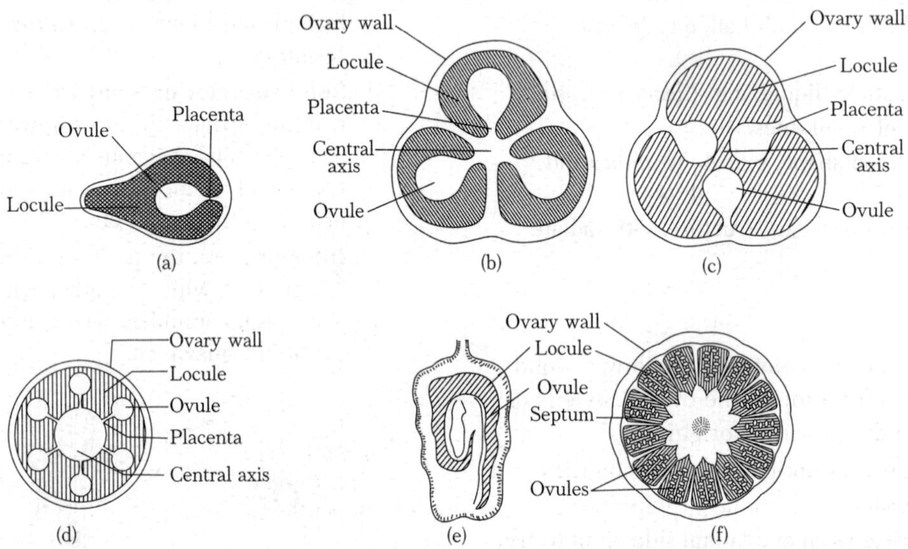

Fig 11.30 Types of placentation : (a) Marginal, (b) Axile, (c) Parietal, (d) Free-central, (e) Basal, (f) Superficial

CHECK POINT 11.4

1. Flower is complete when it has
   (a) calyx, corolla, androecium and gynoecium
   (b) calyx and corolla
   (c) androecium and gynoecium
   (d) corolla, androecium and gynoecium

2. In monadelphous condition, stamens have
   (a) filaments of all united in one group, but anthers are free
   (b) filaments united in groups, but all anthers are free
   (c) anthers are fused, but filaments are free
   (d) Both anthers and filaments are fused

3. The expression 'gynoecium is apocarpous' implies that the
   (a) gynoecium comprises only one pistil which is fused with stamen
   (b) gynoecium comprises more than one carpel, all of which are free
   (c) gynoecium comprises only one carpel which is free
   (d) gynoecium comprises more than one carpel which are fused

4. On the basis of position of the ovary mustard plants are
   (a) hypogynous (b) perigynous
   (c) epigynous (d) zygomorphic

5. A flower having polypetalous, cruciform corolla, valvate aestivation, basifixed polyandrous stamens and superior ovary with parietal placentation is
   (a) mustard (b) guava
   (c) *Stellaria* (d) wheat

6. Superficial placentation is seen in
   (a) *Nymphaea* (b) *Helianthus*
   (c) *Nerium* (d) Wheat

FRUIT

- After the process of fertilisation, ovary transforms into fruit. It is also called as a **ripened ovary**. A fruit mainly consists of two parts, i.e. a pericarp and seeds.
- Pericarp is the fruit wall. It is differentiated into epicarp, mesocarp and endocarp. Seeds are hard structures that develop from ovules.
- The fruits that develop directly from ovary are known as **true fruits**. On the other hand, fruits that develop from any other floral part except ovary are called as **false fruits**, e.g. apple and pear develop from thalamus.
- Depending upon the origin of flowers, fruits can be of three types.

These are as follows

Table 11.3 Examples of types of true fruits and their floral origin

Types of fruit	Floral origin	Examples
Simple	Single ovary of one flower	Legumes (e.g. pea), cypsela (e.g. *Sonchus*), capsule (e.g. cotton), silicula (e.g. *Iberis*), nut (e.g. litchi), caryopsis (e.g. wheat), cremocarp (e.g. fennel), etc.
Aggregate	Many ovaries of one flower	Etaerio of follicles, e.g. *Michelia* and *Catharanthus*. Etaerio of berries, e.g. *Polyalthea, Anona and Squamosa.* Etaerio of achenes, e.g. *Naravelia* and *Clematis.* Etaerio of drupes, e.g. *Rubus idaeur.*
Multiple or Composite	Many ovaries of many flowers	Sorosis, e.g. *Artocarpus intigrifolia, Ananas sativus* and *Morus indica,* Syconus, e.g. *Ficus carica.*

Simple fruits

Simple fruits, develop from the monocarpellary or multicarpellary syncarpous ovary. In these, only one fruit is formed from one gynoecium.

These are of two types as given below

1. Dry fruits

These are classified into three categories

Capsular (Multiseeded dehiscent fruits)

In them, the pericarp split open (ruptures) after ripening and seeds become exposed. These fruits are of following sub-types

(i) **Legume or Pod** These develop from unilocular superior and monocarpellary ovary, e.g. members of family—Leguminosae (*Pisum,* beans such as *Phaseolus, Glycine* and *Abrus*).

(ii) **Follicle** These are similar to legumes in characters, but dehiscence by only one suture, usually the ventral, e.g. *Delphinium* (larkspur), *Calotropis* (madar), *Michelia* (champa) and *Vinca* (periwrinkle).

(iii) **Siliqua** These fruits develop from bicarpellary, bilocular and superior ovary with parietal placentation. These have false septum called **replum.** Each locules has many seeds. Siliqua is a characteristic of family—Brassicaceae (Cruciferae), e.g. *Brassica* (mustard).

(iv) **Silicula** It resembles siliqua. It is short, broad (wide) and flat and possess few seeds, e.g. *Capsella* (Shepherd's purse) and *Iberis* (candytuft).

(v) **Capsule** They develop from a multicarpellary and superior ovary with axile placentation, e.g. poppy, cotton and lady's finger.

(vi) **Pyxis** It is a kind of capsule which undergo transverse dehiscence, so the apical half of the ovary wall is thrown off as a lid exposing the seeds, e.g. *Celosia cristata* (cock's comb), etc.

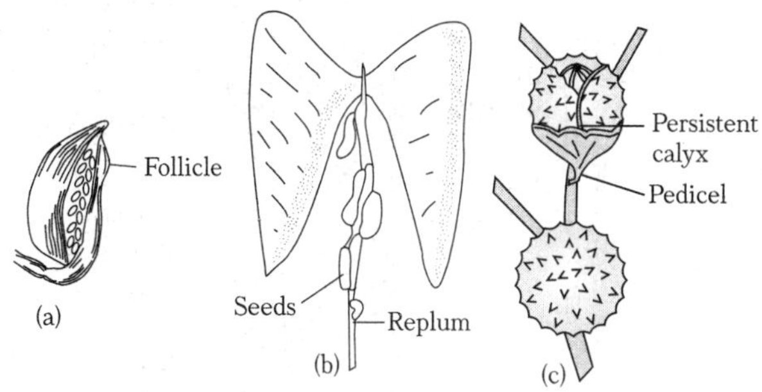

Fig. 11.31 Types of capsular fruit : (a) Follicle of Calotropis, (b) Silicula of Capsella, (c) Capsule of Datura

Achenial fruits

These are simple, dry, single-seeded, indehiscent fruits. They develop from monocarpellary superior ovary. These fruits are of following types

(i) **Achene** These fruits develop from the unilocular, superior ovary, e.g. *Mirabilis jalapa* (4'O clock), *Ranunculus, Clematis* and *Naravelia.*

(ii) **Caryopsis** (grain) They develop from unilocular, single-ovuled, superior ovary. These fruits commonly occur in family—Gramineae, e.g. *Triticum aestivum* (wheat), *Oryza sativa* (rice), *Zea mays* (maize), etc.

(iii) **Cypsela** This fruit develops from a unilocular, inferior bicarpellary ovary. In these fruits a hairy calyx is present, which is called **pappus**, e.g. *Helianthus* (sunflower), *Taraxacum* (dandelion), *Sonchus* and *Cosmos.*

(iv) **Nut** These fruits have hard and stony pericarp. They develop from bi or multicarpellary ovary, e.g. *Anacardium* (cashew nut), *Litchi chinensis* (*Nephelium litchi*; Litchi), chestnut, *Trapa* and *Juglans regia* (walnut).

(v) **Simple samara** Its pericarp becomes membranous and produces wing-like structures that help in dispersal, e.g. *Ulmus* (elm) and *Holoptelia indica* (Indian elm, vern. chilbil).

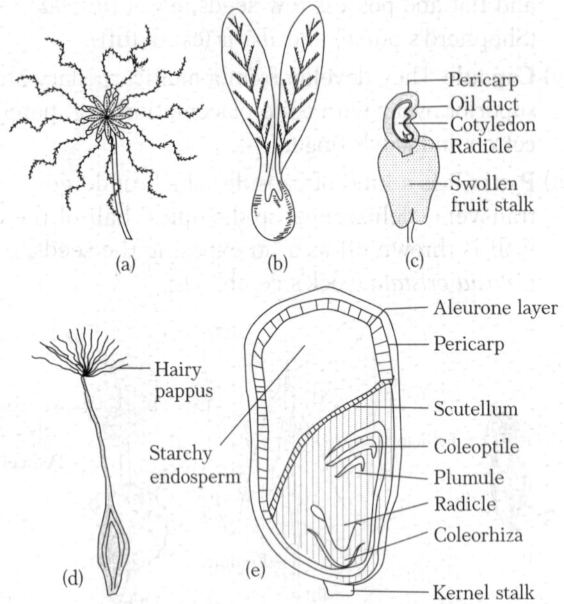

***Fig. 11.32** Types of achenial fruits : (a) Achene of Narvelia with persistent feathery styles, (b) Samara of Dipterocarpus (c) Nut of Anacardium, (d) Cypsela of Tarcaxacum, (e) Caryopsis of maize*

Schizocarpic fruits

These are also called as **splitting fruits.** These develop from uni, bi or multi carpellary ovary. After ripening they break open into segments, i.e. mericarp, each containing one or two seeds.

(i) **Lomentum** These are bisutural fruits. They are constricted or divided into one seeded compartments or mericarps. Initially they look like legumes, but can be distinguished after ripening, e.g. *Raphanus sativus* (radish), *Arachis hypogea* (groundnut), *Mimosa pudica* (touch-me-not, vern. chhui mui), *Acacia* (kikar) and *Tamarindus* (imli).

(ii) **Cremocarp** These fruits develop from an inferior and bicarpellary ovary possessing persistent stylopodium. For example, members of family-Apiaceae, e.g. *Foeniculum vulgare* (fennel), *Coriandrum sativum* (coriander) and *Daucus carota* (carrot).

(iii) **Regma** These develop from multicarpellary, superior ovary. Each fruit breaks up into single-seeded dehiscent parts called cocci. These cocci remain attached onto carpophore (the central axis for sometime), e.g. *Ricinus communis* (castor) and *Geranium.*

(iv) **Carcerulus** This fruit is formed from a superior ovary of bicarpellary to multicarpellary ovary. It breaks up into single-seeded indehiscent mericarps at maturity, e.g. *Ocimum, Abutilon* (flowering maple), *Althea rosea* (holly hock) and garden nesturtium.

(v) **Compound samara** These fruits arise from multicarpellary, superior ovary fruit breaks up into winged mericarp, e.g. *Acer* and hiptage (double samara) or *Dodonaea* (triple samara).

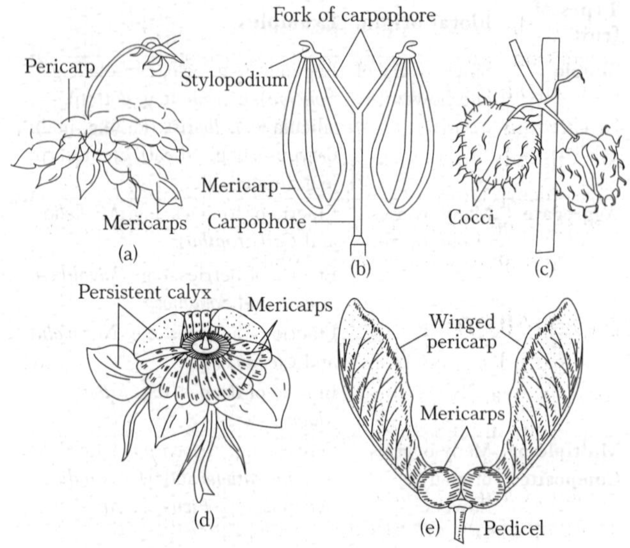

***Fig. 11.33** Types of schizocarpic fruit : (a) Lomentum of Mimosa pudica, (b) Cremocarp of coriander, (c) Regma of castor, (d) Carcerulus of Althea, (e) Double samara of Acer*

2. Succulent or Fleshy fruits

These are simple fruits. Pericarp and associated structures become fleshy. Fruits are always indehiscent.

Succulent or fleshy fruits are of following types

Drupe or Stone fruits

- It is a fleshy, single-seeded fruit.
- The seeds of drupe are enclosed in a stony endocarp.
- It develops from mono or polycarpellary, superior ovary. In this only one or more (rarely) ovules develop into seeds.
- The mesocarp is fleshy (e.g. mango) or fibrous (e.g coconut).
- It is edible in most of the drupes. The epicarp forms the outer skin, which may be edible, e.g. *Prunus persica*-peach; *Prunus domestica*; *Ziziphus jujuba.* Eaicarp it may not be edible, as in *Mangifera indica* and *Cocos nucifera.*

Berries

- It is a pulpy, few to multi-seeded fruit derived from the multicarpellary ovary.
- Its endocarp (innermost layer) can be pulpy (e.g. *Vitis vinifera*-grape; *Solanum melongena*—brinjal; *Musa paradisiaca*-banana; *Lycopersicon esculentum*-tomato; *Psidium guajava*-guava), or papery (e.g. *Phoenix dactylifera*-datepalm).

Pepo

- It develops from tricarpellary, inferior ovary. The epicarp (rind) is not separable from mesocarp and many seeds are present within the placenta.
- Usually the endocarp, mesocarp as well as placenta (exception being *Cucumis melo,* musk melon, vern. kharbuja) are edible.
- In *Citrullus lanatus* (*Citrulus vulgaris*, watermelon, vern. tarbuj), only endocarp and placentae are juicy.
- The epicarp (rind) of the fruit may or may not be edible, e.g. *Cucumis sativus* (cucumber, vern. khira), *Cucurbita moschata* (pumpkin) vern. sitaphal.

Hesperidium

- This fruit develops from a multicarpellary, multilocular, superior ovary with axile placentation.
- Numerous unicellular juicy edible hairs (placental hairs) are present on inner side of the endocarp, e.g. *Citrus lemon* (lemon), *Citrus reticulata* (vern. santara).
- Many oil glands are also present in epicarp.

Pome

- It is a false fruit. Here, the thalamus becomes swollen, fleshy and edible.
- It develops from the fleshy thalamus, multicarpellary, inferior ovary.
- Pericarp is not differentiated into epicarp, mesocarp and endocarp, e.g. *Pyrus malus* (apple), pear.

Balausta

- This is a special type of false or pseudocarpic berry. It develops from a multilocular, inferior ovary.
- Plate-like infoldings are developed by mesocarp.
- The papery endocarp covers the individual group of seeds. The seeds possess bright-red juicy testa that forms edible part of the fruit.
- The fruit has persistent calyx, e.g. *Punica granatum* (pomegranate, vern. Anar).

Amphisarca

- This is also a special type of berry. It develops from multicarpellary, multilocular, (multi-seeded) and superior ovary.
- The epicarp is hard and woody.

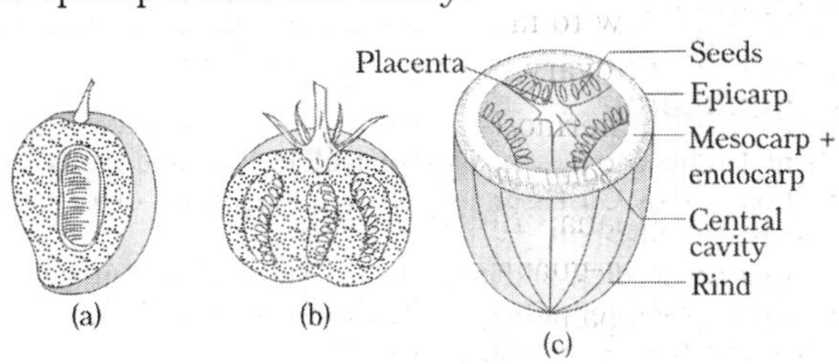

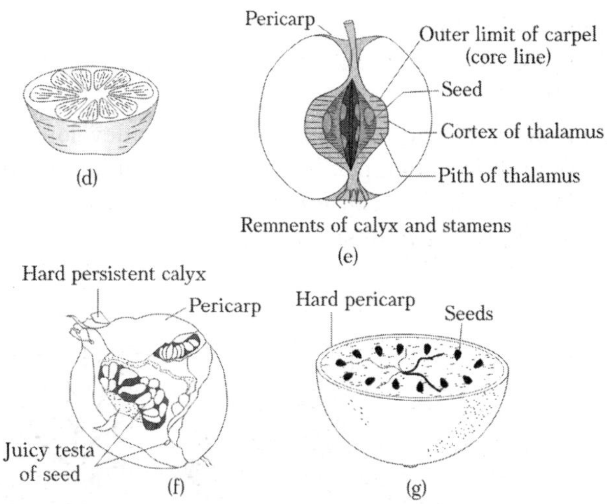

Fig. 11.34 *Types of succulent fruits: (a) Drupe of mango, (b) Berry of tomato, (c) Pepo of Cucurbita, (d) Hesperidium of orange, (e) Pome of apple, (f) Balausta of pomegranate, (g) Amphisarca of wood apple*

- Mesocarp is fibrous, while the endocarp and placenta are pulpy. Mesocarp, endocarp and swollen placenta of this fruit are edible, e.g. *Aegle marmelos* (wood apple).

Table 11.4 Differences between dry and succulent (fleshy) **fruits**

Dry fruits	**Succulent** (Fleshy) **fruits**
These are dry.	These are fleshy.
Pericarp not differentiated.	Pericarp (fleshy) differentiated into epicarp, mesocarp and endocarp.
These are single-seeded or multiseeded.	These are few to multiseeded.
Fruits may be dehiscent or indehiscent.	Fruits are always indehiscent.
Pericarp gets separated from the seed.	Pericarp and seeds are separated after decay of the fruit.

Aggregate fruits

- It is a group of fruitlets. It develops from the multicarpellary, apocarpous (free) ovaries of a single flower. Such type of aggregate fruits developed from a single flower are known as **etaerio.**
- Important aggregate fruits or etaerio of fruitlets are discussed below
  (i) **Etaerio of achenes** Each fruitlet is an achene, e.g. *Fragaria* sp. (strawberry), *Ranunculus* (buttercup), *Clematis* (virgin's flower), *Naravelia* (traveller's joy), *Nelumbo nucifera, Nelumbium speciosum* (lotus) and *Rosa indica* (rose).
  (ii) **Etaerio of follicles** Each fruitlet is a follicle arranged on the enlarged thalamus in a bunch. The number of follicles in an etaerio varies greatly from two (e.g. *Calotropis* and *Catharanthus*) to many, (e.g. *Michelia*).

(iii) **Etaerio of berries**, e.g. *Annona squamosa* (custard apple, vern. sharifa), *Artabotrys* (kantli champa), *Polyalthia.*

(iv) **Etaerio of drupes** Many small drupes developed from different carpels are arranged in groups on the fleshy thalamus, e.g. *Rubus fructicosus* (Black-berry) and *Rubus ideus* (raspberry).

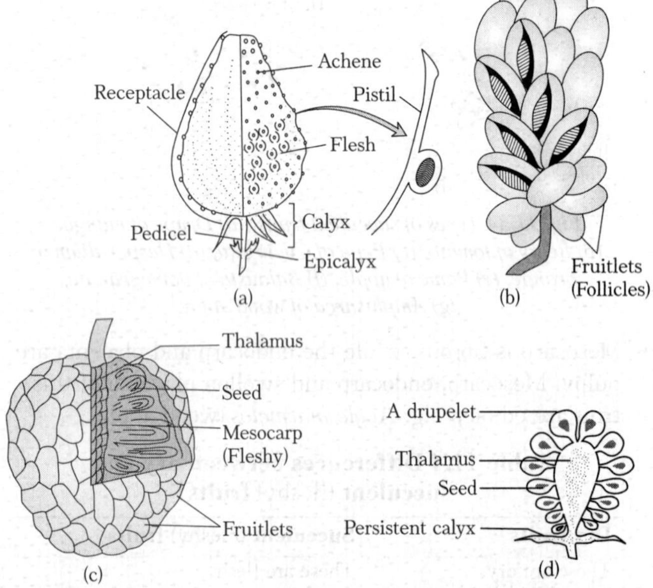

***Fig. 11.35** Types of aggregate fruits : (a) Etaerio of achenes, (b) Etaerio of follicles, (c) Etaerio of berries, (d) Etaerio of drupes*

Composite or Multiple fruits

These develop from complete inflorescence. These are mainly of two types

(i) **Sorosis** It is developed from a spike (pineapple), a female catkin (mulberry) or spadix (*Artocarpus*).

(ii) **Syconus** It is a composite fruit developed from hypanthodium inflorescence, e.g. *Ficus carica* (fig vern. anjir), *Ficus religiosa* (peepal) and *Ficus benghalensis* (banyan).

The flask-shaped fleshy receptacle encloses female flowers that give rise to achene like fruitlets. This fruit possesses a small pore protected by the scaly leaves.

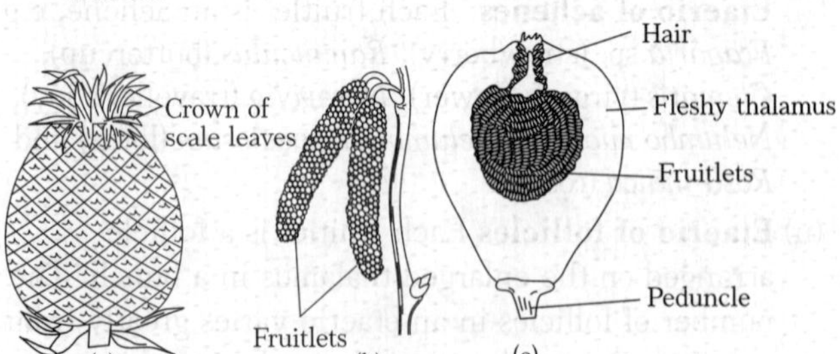

***Fig. 11.36** Types of composite fruits : (a) Sorosis of pineapple, (b) Mulberry, (c) Syconus of fig.*

Note *In **Ananas sativus** (pineapple), the fruit develops from an intercalary spike of sterile flowers possessing persistent bracts that appear as small projections on the fruit. Peduncle (that enlarges and becomes fleshy), bracts, perianth and pericarp are edible.*

Table 11.5 Different types of fruits and their edible parts

Edible parts	Types of fruits	Fruits
Endosperm and embryo	Caryopsis	Rice, wheat, maize
Fleshy aril	Nut	*Litchi chinensis* (litchi)
Cotyledons and fleshy pedicel	Nut	*Anacardium occidentale* (cashew nut)
Pericarp and aril	Nut	*Trapa bispinosa*
Seeds	Berry Legume Lomentum Drupe	*Areca catechu* (beetle nut). pea *Arachis hypogaea* (groundnut) almond
Epicarp and mesocarp	Drupe	Plum (*Prunus domestica*), peach (*P. persica*)
Pericarp and placenta	Berry	Grape (*Vitis vinifera*), brinjal, tomato, guava (*Psidium guajava*)
Mesocarp and endocarp	Pepo Amphisarca Berry	Cucumber (*Cucumis melo*), muskmelon, bottle gourd, watermelon. Elephant apple (*Feronia elephantum*), wood apple (*Aegle marmelos*), banana.
Mesocarp	Drupe Berry Etaerio of berries	Mango, *Pistacia vera* (Pistacia nut) Papaya, date. Custard apple (*Annona squamosa*).
Pericarp	Berry	*Eugenia jambolana* (jamun)
Entire fruit	Capsule Pepo	*Abelmoschus esculentus* (lady finger) kakri.
Endocarpic juicy hairs	Hesperidium	Lemon (*Citrus medica*). Sweet orange (*C. sinensis*)
Fleshy thalamus	Pome Etaerio of achenes Etaerio of drupes	Pear, apple. Strawberry Raspberry
Juicy testa	Balausta	Pomegranate
Perianth	Sorosis	Mulberry
Fleshy receptacle	Syconus	Peepal
Peduncle and seeds	Syconus	Fig
Peduncle, bracts, perianth and seeds	Sorosis	Pineapple
Bracts, perianth and seeds	Sorosis	Jackfruit
Thalamus and seeds	Etaerio of achenes	Lotus
Endosperm, testa cotyledon and embryo	Drupe	Coconut

Parthenocarpy

The term 'Parthenocarpy' was used by **Noll** in 1902 and it was first time introduced by **Gustafson** in 1939. It is the process of formation of fruits without fertilisation.

These fruits are seedless and in most cases get aborted. Parthenocarpy is induced in plants for raising seedless fruits. It can be induced using hormones, such as auxins and gibberellins.

CHECK POINT 11.5

1. In caryopsis type of fruit
   (a) seed is absent
   (b) three layers of pericarp are distinct
   (c) seed coat and pericarp are fused
   (d) autochory occurs
2. Aggregate fruit develops from
   (a) multicarpellary, apocarpous ovary
   (b) multicarpellary ovary
   (c) multicarpellary, syncarpous ovary
   (d) monocarpellary ovary
3. The fruit in *Datura* is
   (a) loculicidal capsule (b) septifragal capsule
   (c) septicidal capsule (d) porous capsule
4. Pineapples (*Ananas*) fruit develops from
   (a) a cluster of compactly borne flowers on a common axis
   (b) a multilocular monocarpellary flower
   (c) a unilocular polycarpellary flower
   (d) a multipistillate syncarpous flower
5. Fruits developed from polycarpellary apocarpous ovary of a flower are categorised as
   (a) simple fruits
   (b) aggregate fruits
   (c) multiple fruits
   (d) achenial fruits
6. The fruit of regma breaks up into single seeded dehiscent parts called cocci. These cocci remain attached onto
   (a) winged appendages (b) pappus
   (c) stylopodium (d) carpophore
7. A fruit that develop from a spike or spadix inflorescence is called
   (a) caryopsis (b) etaerio of spike or spadix
   (c) cornus (d) sorosis

SEED

- A seed can be defined as a mature fertilised ovule enclosing a small embryonic miniature plant by a covering called **seed coat**. The formation of seed completes the process of reproduction in seed plants.
- Seed reserves food material for its future development. A seed contains an embryo, an endosperm and a seed coat.

Embryo

- The embryo of seed consists of an axis called **tigellum** to which one or two embryonic leaves or cotyledons are attached.
- The part of tigellum below the point of attachment of cotyledons is called **hypocotyl** (hypo—below, cotyl—cotyledon).
- On the other hand, part of tigellum above the point of attachment is called **epicotyl** (epi-above, cotyl—cotyledon).
- The lower portion of hypocotyl and upper portion of epicotyl are called **radicle** and **plumule**, respectively. The radicle (embryonic root) gives rise to the root system, while the plumule (embryonic stem) develops the shoot system.

Endosperm

- It is the nutritive tissue of seed.
- It provides nutrition to the developing embryo.
- It is highly rich in proteins, starch and other nutrients.
- It is usually gets consumed in the process of embryo development.

Seed coat

- It is the protective covering of the seed made up of either one or two layers. The outer layer, is called **testa**, while the inner layer is called **tegmen**.
- There is a minute opening called **micropyle** present at one end of seed coat. Through this opening water enters the seed.
- A scar called **hilum** is also present on the seed coat. It represents the point of attachment of seed with the funiculus (stalk).
- Sometimes, an outgrowth occurs along the hilum which is called **straphiole**.
- Some seeds also possess **chalaza** (the place of origin of seed coats) and **raphe** (the part of funiculus fused with seed wall).

Types of seeds

1. Depending upon the number of cotyledons present in the embryo, seeds can be classified into different types
   (i) **Dicotyledonous seeds** The seeds that possess two cotyledons, e.g. *Pisum* (pea), *Cicer* (gram), *Brassica* (mustard), *Lycopersicon* (tomato) and *Mangifera* (mango).
   (ii) **Monocotyledonous seeds** The seeds that have only one cotyledon, e.g. *Oryza* (rice), *Triticum* (wheat), *Zea* (maize), *Orchis* (orchids), etc.

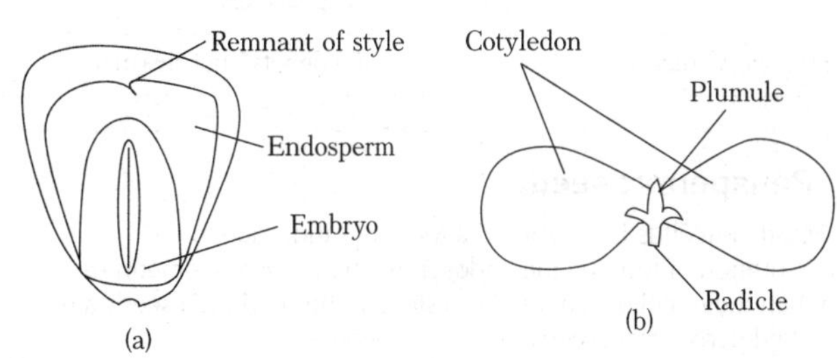

Fig. 11.37 Seeds : (a) Monocotyledons, (b) Dicotyledons

2. Depending upon the presence or absence of endosperm, seeds are categorised as
   (i) **Endospermic** (Albuminous) **seeds** The seeds that possess endosperm and store their food in it, e.g. *Ricinus communis, Zea mays,* etc.
   (ii) **Non-endospermic** (Exalbuminous) **seeds** The seeds that do not possess endosperm.
   - Their endosperm is completely consumed in embryo nourishment.
   - These store their food material in cotyledons, e.g. *Pisum sativum, Mangifera indica* and orchids.

Note *• Smallest and lightest seed : Orchids (0.004 mg)*
- *Largest and heaviest seed : Lodoicea maldivica (6 kg)*
- *Dicot seed without cotyledons : Cuscuta*
- *Dicot endospermic seeds : Castor*
- *Monocot non-endospermic seeds : Orchids*

Table 11.6 Differences between bean seed and maize grain

Bean seed (Dicot)	**Maize grain** (Monocot)
Formed within a multiseeded fruit called legume or pod.	It is a single-seeded fruit called caryopsis.
Pericarp (fruit wall) is free from testa.	Pericarp (fruit wall) inseparably fused with testa.
Seed coat consists of two-layers, i.e. testa and tegmen.	Single-layered seed coat bearing testa only.
Non-endospermic	Endospermic
Two cotyledons	One cotyledon
Food stored in cotyledons is in the form of protein grains, fats and starch grains.	Food stored as protein grains or aleurone grains in outer cell-layers of endosperm called aleurone, while as starch grains in rest of the endosperm.
Chalaza, hilum and micropyle are clearly visible on seed surface.	Chalaza, hilum and micropyle absent on seed surface.
Raphe clearly distinct.	Ridge-like raphe absent.
Coleoptile and coleorhiza absent.	The plumule and radicle are well protected by coleoptile and coleorhiza sheaths, respectively.
Hypocotyl distinct.	Hypocotyl not distinct.

Perispermic seed

Mostly nucellus is consumed after fertilisation due to the absorption of food by the endosperm and embryo. Sometimes when the nucellus remains persistent in the seed such seeds are called perispermic seeds, e.g. black pepper.

DISPERSAL OF FRUITS AND SEEDS

- Dissemination of fruits and seeds from their parent plants to various distances is called dispersal.
- Some plants have built in devices for dispersing their seeds to considerable distances (autochory).
- Some plants depend on certain foreign agents, such as animals (zoochory), water (hydrochory) and air (anemochory). Man also works as an agent to disperse fruits and seeds from one place to other.

Autochory

- It is a mechanism in which fruits are dispersed by plants only.
- The following are some examples of mechanism of autochory
   (i) **Curling mechanism** It is seen in *Impatient passiflora.* A touch or mild jerking of capsule breaks the carpels. The five carpels curl inward. As a result, seeds are thrown to a distance up to 2.5 m.
   (ii) **Piston mechanism** This mechanism is mainly seen in *Ecballium elaterium.* Here, when the fruit matures, some cells surrounding the seeds produce mucilage. Due to this, high turgor pressure is generated inside the fruit. Thus, seeds are thrown away.
   (iii) **Bursting mechanism** It is seen in some legumes, such as *Abrus*, *Pisum* and beans. The valves of pod twist spirally and coil with a force. Due to this seeds are thrown out of the fruit. The pods of *Entada gigas* open with a very loud noise.
   (iv) **Jaculator mechanism** In this, the dry and capsular fruits burst suddenly when they come in contact with water. The seeds possess, jaculators (curved hooks).
   These are hygroscopic in nature, e.g. *Andrographis* , *Ruellia tuberosa* and *Barleria.*

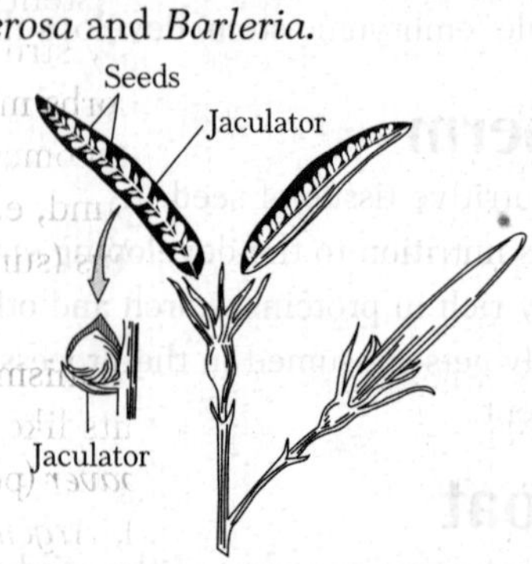

***Fig. 11.38** Jaculator mechanism*

- Some morphological structure of seeds and fruits that help in autochory are given below
  - (i) **Tiny and light seeds** Some plants such as orchids and grasses produce small, dry, dust-like seeds. These seeds are produced in a very high number.
  - (ii) **Winged fruits and seeds** Winged appendages (expansions of pericarp) are a common adaptation for wind dispersal of fruits and seeds, e.g. *Acer* (mapple), *Hiptage, Ulmus* (elm), *Holoptelia indica* (Indian elm, vern. chilbil), *Dodonaea, Dioscorea;* persistent calyx, e.g. *Shorea robusta* (sal).

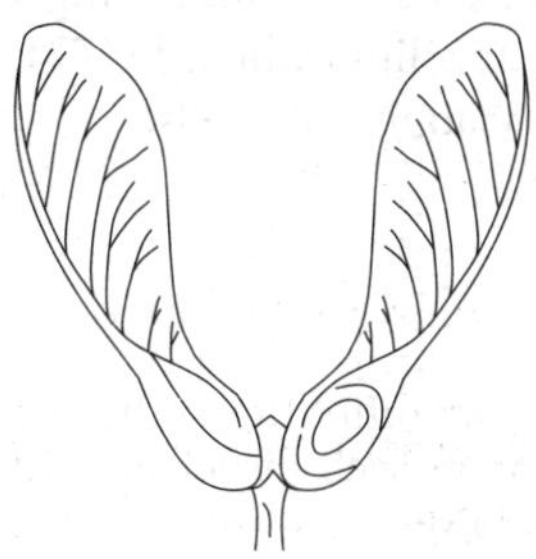

Fig 11.39 *Winged seed of Acer*

  - (iii) **Hair** These hair help the seed to float in air, e.g. *Launaea, Gossypium* (cotton), *Calotropis* (milk-weeds), *Adenium, Alstonia,* etc.
  - (iv) **Persistent styles** Some fruits (e.g. *Clematis, Naravelia, Anemone*) possess persistent hairy (feathery) styles. It help in flying of seed in air during dispersal by wind.

Anemochory

- It takes place by wind and it is the main seed-dispersing agency in angiosperms.
- Their fruits and seeds possess some special morphological structures, e.g. wings, hairs, pappus, etc., that help in their dispersal.
- Some mechanism of anemochory are given below
  - (i) **Parachute mechanism** The cypsela fruits of family-Asteraceae possess persistent calyx (sepals). The calyx is modified into hairy structures called **pappus**. The hairy pappus absorbs moisture and becomes heavy. Thus, the fruit comes down just like a parachute landing on the ground, e.g. *Taraxacum* (dandelion), *Sonchus, Helianthus* (sunflower), *Tagetes* (marigold), etc.
  - (ii) **Censer mechanism** This mechanism commonly takes place in multiseeded fruits like legumes, siliquas and capsules, e.g. *Papaver* (poppy), *Aristolochia* (Indian brithwort), *Argemone* (prickly poppy), *Datura* (thornapple), *Antirrhinum* (snapdragon or dog flower), *Brassica* (mustard), etc.

    In this mechanism, seeds come out through the pores as the fruits are shaken by air.

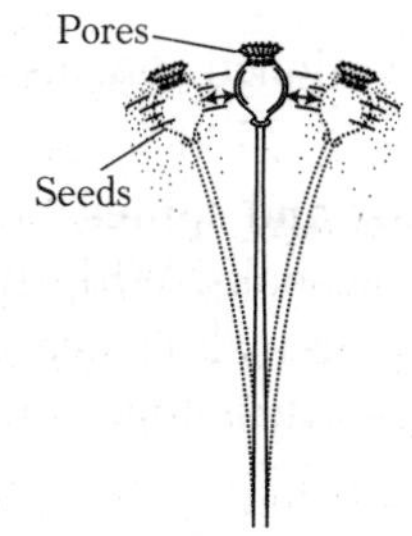

Fig 11.40 *Censer mechanism*

Hydrochory

It is dispersal method by water. It occurs in fruits and seeds, which grow in water or so close to it that their fruits and seeds can easily fall in it.

Some ways and methodology of hydrochory are given below

- Coconut (*Cocos nucifera*) dispersal is a commonly sited example. It has a smooth and waterproof epicarp and a fibrous and buoyant mesocarp. It is filled with air spaces. Thus, it becomes buoyant.
- The small seeds of *Polygonum* are carried over the surface of water through surface tension.
- The fruits of *Alisma, Nymphaea,* possess aril having air spaces with the help of which, they can float on water.
- In *Nelumbo* (lotus), the spongy thalamus with air-chambers helps in fruit dispersal. The thalamus floats in water streams. The fruits settle down at the bottom of the pond and their seeds germinate.
- Fruits and seeds of aquatic plants like *Heritiera, Mucuna,* etc., possess an impermeable coating and air-cavities. These help the seeds to float.

Zoochory

It is a dispersal mechanism by animals. It can be of following types

- (i) **Myrmecochory** It is done by ants, e.g. *Trilium* and *Anemone* have oily covering, that makes them rich in lipids, amino acid, etc., that attract ants.
- (ii) **Ornithochory** It is done by birds, e.g. mulberry, nutmeg, etc.
- (iii) **Dispersal by humans** (Anthropochory) Man is considered as the most potential mean to disperse seeds and fruits. Potato (*Solanum tuberosum*) was brought from **New World** by the Portuguese. Maize (*Zea mays*), groundnut (*Arachis hypogea*), tea (*Thia*), cabbage, cauliflower, rubber and cinchona are plants which have been dispersed by man.

Many weeds, e.g. *Parthenium* (congress grass), *Lantana, Argemone maxicana* got dispersed by man along with useful plants.

Zoochory by above mentioned animals is done with the help of following means

(a) **Hooks, hard hair and spines** Some fruits have hard hair and spines or hooks. With the help of these structure grazing animals can carry these fruits away, e.g. *Xanthium, Martynia* (tiger's nail), *Bidens* (begger tricks), etc.

(b) **Sticky glands** These are present on the surface of some fruits and seeds. These structure help to stick to the body of animals and thus disperse, e.g. *Boerhavia, Cleome, Viscosa, Datura stramonium, Plumbago,* etc.

(c) **Edible fruits** Some fruits such as tomato, papaya, apple, mulberry, etc., are edible. Animals eat these fruits and thus, dispersion of their seeds takes place.

Viability of seeds

- The ability of a seed to germinate when provided with moist, wet, i.e. favourable conditions is called **seed viability**. It reduces in dry or high temperature conditions.
- It may also get lost when the embryo is damaged or the enzymes required for germination get denatured. Triphenyl Tetrazolium Chloride (TTC) is used for checking the viability of seeds.

CHECK POINT 11.6

1. An example of a seed with endosperm, perisperm and caruncle is
(a) cotton (b) coffee
(c) lily (d) castor

2. The scutellum observed in a grain of wheat or maize is comparable to which part of the seed in other monocotyledons
(a) plumule (b) cotyledon
(c) endosperm (d) aleurone layer

3. Curling mechanism of autochory is found in
(a) *Impatient passiflora* (b) *Barleria*
(c) *Ruellia* (d) *Shorea*

4. Pods of *Entada gigas* differ from other leguminous pod at the time of autochorous seed-dispersal as they
(a) open by two valves
(b) twist spirally
(c) open with a loud noise
(d) throw out their seeds

5. Seeds of opium poppy (*Papaver*) and prickly poppy (*Argemone*) disperse by
(a) explosive mechanism
(b) parachute mechanism
(c) censer mechanism
(d) jaculator mechanism

SEMI–TECHNICAL DESCRIPTION OF FLOWERING PLANTS

- Various features of a flowering plant can be described using semi-technical language. For each character a sign or symbol is assigned and thus a floral formula is created.
- Floral diagrams are also constructed for giving extra informations like placentation, aestivation, position of mother axis, etc.

Flower

The flower of angiospermic plants show great variations in its shape, size and structure. It is important to know certain important terms related to flower for its complete, good understanding.

These terms are given below

1. **Actinomorphic flower** It is a radially symmetrical flower, i.e. can be divided into two equal halves in any plane, e.g. mustard (Cruciferae), onion (Liliaceae) and brinjal (Solanaceae).
2. **Zygomorphic flower** It is the term used for a bilaterally symmetrical flower. It can be divided into two equal halves only in a single vertical plane, e.g. pea, sem.
3. **Asymmetric flower** Flowers that cannot be divided into two similar halves in any plane, e.g. orchids, *Canna.*
4. **Hermaphrodite or Bisexual flower** It contains both male and female reproductive organs, e.g. China rose, mustard and pea.
5. **Unisexual or Dioecious flower** These flowers has only one essential whorl. It can be androecium (staminate) or gynoecium (pistillate or pistalloid), e.g. *Norus alba*, papaya and *Luffa.*
6. **Complete flower** A flower that contains all the floral whorls, i.e. calyx, corolla, androecium and gynoecium, e.g. *Hibiscus,* mustard, pea.
7. **Incomplete flower** These flowers lack one or more floral whorls, e.g. *Luffa.*

8. **Achylamydeous flower** These flowers lack both the accessory floral whorls, i.e. calyx and corolla, e.g. Piperaceae members.
9. **Monochlamydeous flowers** Flower that contains only one accessory whorl, i.e. either a calyx or corolla. It is called as **monochlamydeous flower**. Sometimes these instead have a perianth, e.g. *Polygonum,* onion.
10. **Dichlamydeous flowers** In these flowers, both the accessory whorls are present, e.g. in most of the flowers.
11. **Isomerous flowers** In these flowers, all floral whorls are found in a particular number or its multiple. These can be **dimerous**, (e.g. poppy), **trimerous** (e.g. onion) or **tetramerous** (e.g. mustard).
12. **Heteromerous flowers** When floral whorls of a flower are found in different numbers, e.g. *Achyranthus.*
13. **Hypogynous flowers** In these flowers, gynoecium occupies the highest position while other floral whorls are present below it, e.g. *Brassica,* China rose, *Papaver, Citrus.*
14. **Perigynous flowers** In these flowers, all floral whorls lie at same level of height, e.g. rose, peach and *Prunus.*
15. **Epigynous flowers** Here, the position of gynoecium is inferior and other whorls arise above the ovary, e.g. sunflower, *Cucurbita* and coriander.
16. **Bracteolate** A flower with bracteole (i.e. green, sepal-like structure arising from pedicel) is called bracteolated.
17. **Ebracteolate** A flower without a bracteole is called as ebracteolated.
18. **Epicalyx** It is a term used for bracteole-like structure found externally to the sepals, e.g. China rose and cotton (Malvaceae family).

Symbols used in floral formula

I.	**Bracts (Br)**	**V.**	**Symmetry of the flower**
Br.	Bracteate	⊕	Actinomorphic
Ebr.	Ebracteate	% or +	Zygomorphic
II.	**Sex**	**VI.**	**Caylx (K)**
♂	Staminate (male)	K_4	4 sepals, polysepalous (Sepals free)
♀	Pistillate (female)	$K_{(4)}$	4 sepals, gamosepalous (Sepals fused)
⚥	Hermaphrodite	K_{2+2}	4 sepals in 2 whorls of 2 each
III.	**Corolla (C)**	**VII.**	**Perianth (P)**
C_4	4 petals, polypetalous	P_6	6 tepals, polytepalous
$C_{(4)}$	4 petals, gamopetalous	P_{3+3}	6 tepals, in 2 whorls of 3 each
C_{2+2}	4 petals in 2 whorl of 2 each	$P_{(3)+(3)}$	6 tepals, gamotepalous
IV.	**Androecium (A)**	**VIII.**	**Gynoecium (G)**
A_6	6 stamens, polyandrous	G_0	Gynoecium absent
A_{2+4}	6 stamens in 2 whorls of 2 and 4 each	G_2	2 carpels, apocarpous
A_0	Stamens absent	$G_{(2)}$	2 carpels, syncarpous
A_∞	Stamens indefinite	$\overline{G}$	Inferior ovary
A_{10}	Stamens 10, monadelphous	$\underline{G}$	Superior ovary
$A_{1+(9)}$	Stamens 10, diadelphous, 1 free, 9 united	G-	Semi-inferior or half-inferior ovary
$\overset{\frown}{C\quad A}$	Epipetalous		
$\overset{\frown}{P\quad A}$	Epiphyllous		

Description of some important angiospermic families

Characteristics	Fabaceae	Solanaceae	Liliaceae	Brassicaceae	Malvaceae
General description	The family is also termed as **pea family**. It is distributed all over the world.	Commonly known as **potato family.** It is distributed in tropics and sub-tropics.	Commonly called as **lily family**. It is a representative of monocots.	It is known as **mustard family** or Brassicaceae. Mainly distributed in tropics or subtropics.	Also known as **mallow family**. Found in tropic and subtropics.
Habit	Tree, shrub, herb.	Herb, shrub and rarely small trees.	Perennial herbs, shrubs or trees.	Annual or perennial herbs.	All herb, shrub and trees.
Root structure	Tap root with root nodules.	Tap root system.	Adventitious root often modified as fasciculated or nodulated roots.	Tap root, fusiform and napiform roots may found. Oil glands are present.	Profusely branched tap root.
Stem structure	Erect or climber, herbaceous.	Herbaceous, rarely woody, hairy, solid, underground tubers (potato).	Stem may be underground partially, with bulb, corm and rhizome.	Herbaceous stem with pungent watery fluid. Have stellate hairs.	Erect, branched with star-shaped trichomes.
Leaves	Alternate, pinnately compound or simple, leaves, reticulate venation, pulvinate, stipulate.	Alternate, simple ex-stipulate, unicostate, reticulate venation.	Mostly basal, alternate and linear parallel venation.	Simple rarely compound alternate, sessile exstipulate reticulate venation.	Simple, alternate palmately lobed, stipulate reticulate venation.
Inflorescence	Racemose, (panicle in *Dalbergia*).	Solitary axillary or terminal, cymose, monochasial scorpioid or helicoid.	Solitary/cymose often umbellate clusters.	Corymose raceme or corymb.	Solitary terminal or axillary cymose or raceme.
Flower	Bisexual, zygomorphic, bracteate or ebracteate, pentamerous.	Bisexual, complete actinomorphic, hypogynous bracteate or ebracteate.	Bisexual, complete actinomorphic, bracteate, trimerous, hypogynous.	Bisexual, actinomorphic (may be zygomorphic), hypogynous, ebracteate.	Bisexual, bracteate, actinomorphic, complete pentamerous, hypogynous.
Calyx	Five, gamosepalous imbricate/valvate.	Five united persistent, sepals with valvate aestivation.	Six tepals arranged in two whorls (3 + 3). Free or rarely united valvate aestivation.	Four polysepalous in two whorls.	Calyx like whorl called epicalyx is formed by bracteoles. It is protective in nature.
Corolla	Five, polypetalous, papilionaceous (posterior standard, two lateral wings, keel enclosing pistil and stamen), vexillary aestivation.	Five united petals with valvate aestivation.		Four, polypetalous cruciform and reduced.	Five, petal free, but basally adnate, twisted aestivation.
Androecium	Ten, diadelphous, anther dithecous, introrse.	Stamens five, polyandrous epipetalous and bithecous anthers.	Stamens Six (3 + 3) epiphyllous.	Six stamens arranged in two whorls, polyandrous tetradynamous dithecous.	Numerous stamens monadelphous epipetalous, monothecous extrorse anthers reniform.
Gynoecium	Ovary superior monocarpellary, unilocular with many ovules, single style, marginal placentation.	Bicarpellary, bilocular syncarpous, superior, ovary with axil placentation.	Tricarpellary, syncarpous trilocular, superior ovary with axile placentation.	Bicarpellary syncarpous superior ovary with parietal placentation.	A compund pistil of one to many carpel, syncarpous, axile placentation.
Fruit	Legume	Berry or capsule	Capsule, rarely berry	Fruit siliqua or silicula	Capsule or schizocarp.

Characteristics	Fabaceae	Solanaceae	Liliaceae	Brassicaceae	Malvaceae
Seed	One to many and non-endospermic.	Many are endospermous.	Endospermous	Small non-endospermic	Seed with curved embryo and scanty endosperm.
Floral formula	Br % ⚥ $K_{(5)}C_{1+2+(2)}$ $A_{(9)+1}\underline{G}_1$	⊕ ⚥ $K_{(5)}C_{(5)}A_5\underline{G}_{(2)}$	Br⊕ ⚥ $\overset{\frown}{P_{3+3}A_{3+3}}\underline{G}_{(3)}$	Ebr. ⊕ ⚥ $K_{2+2}C_4\ A_{2+4}\underline{G}_{(2)}$	Br ⊕ ⚥ $Epi_{3-9}K_{5\text{ or }(5)}$ $\overset{\frown}{C_5\ A_{(\infty)}}\underline{G}_{(2-\infty)}$
Floral diagram	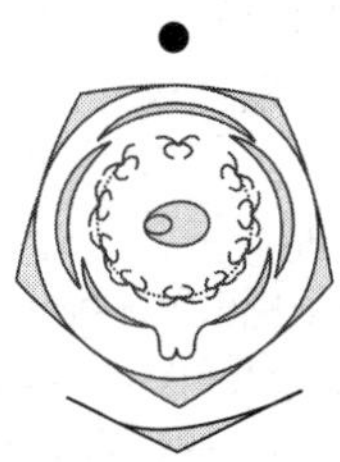 Fabaceae	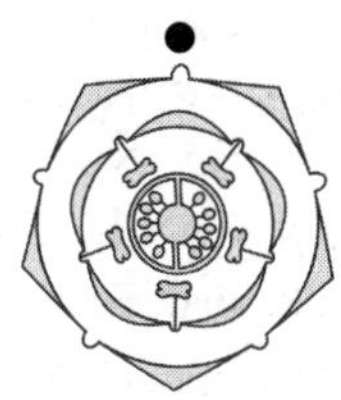 Solanaceae	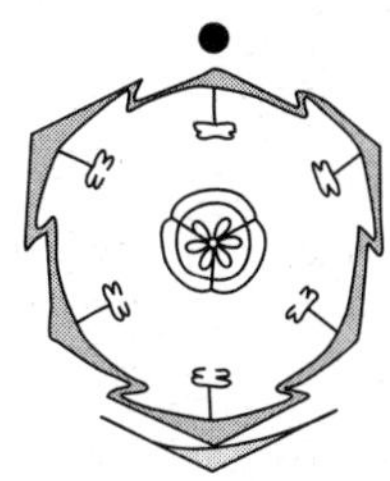 Liliaceae	 Cruciferae or Brassicaceae	 Malvaceae
Economic Importance	(i) Members of this family are mainly used as **food**, i.e. pulses, e.g. gram, arhar, sem, moong, soybean, etc.	Members of this family are sources of food, e.g. potato, tomato, brinjal.	Members of this family are used as vegetables, e.g. young shoots and root tubers of *Asparagus*, bulbs of onion and garlic.	The members of this family are sources of food, e.g. *Brassica compestris* (mustard), *B. oleracea* (cauliflower), *Brassica rapa* (turnip) and *Raphanus sativus* (radish).	These plants are used as food, e.g. *Abelmoschus esculentus* (ladyfinger).
	(ii) Fodder, e.g. *Trifolium*, *Sesbania*, etc.	Spices, e.g. *Capsicum annum* and *Capsicum frutescence.*	Perfumes, e.g. *Convallaria majalis* and *Hyacinthus orientalis.*	Oils, e.g. *B. campestris*, *B. juncea* (rye), *B. nigra* (black mustard).	Fibres, e.g. *Gossypium* sp. (cotton). Edible oil, e.g. cotton.
	(iii) Edible oil, e.g. soybean, groundnut, etc. (iv) For fibres, e.g. sunhemp.	Medicines, e.g. *Withania somnifera*, *Atropa belladonna* (Atropine-painkiller), fumigatory, e.g. *Nicotiana.*	Fibres, e.g. leaves of *Phormium tenax* (New Zealand hemp), *Agave*, *Yucca* and *Sansevieria.*	Medicines, e.g. *Iberis amara* (gout and rheumatism), *Lobularia* (gonorrhoea), *Lepidium sativum* (asthma and piles).	Ornamentals, e.g. *Althaea rosea* (gulekhera), *Hibiscus rosa-sinensis* (China rose), etc., roots and bark of *Urena lobata* are used as a cure for hydrophobia.
	(v) Ornamentation, e.g. lupin, sweet pea.	Ornamentals, e.g. jasmine, *Night jasmine*, *Petunia hybrida*, etc.	Resins, e.g. *Xanthorrhoea* and *Dracaena.*	Ornamentals, e.g. *Iberis amara* (candytuft), *Cherianthus cheiri* (wall flower), etc.	
	(vi) For the extraction of medicines, e.g. mulaithi.		Ornamentals, e.g. *Gloriosa*, lily and *Smilax.* Medicines, e.g. *Aloe* is applied on boils and burns (*Aloe*), colchicine is used to induce polyploidy (*Colchicum autumnale*), etc.		

CHECK POINT 11.7

1. The two families dominate in having maximum useful plants
(a) Fabaceae and Poaceae
(b) Liliaceae and Solanaceae
(c) Malvaceae and Brassicaceae
(d) Liliaceae and Poaceae

2. Mark the correct statement for Poaceae.
(a) The carpal has two styles
(b) Spikelets are always in pairs
(c) Awn is an appendage of the palea
(d) Palea is the bracteole

3. $\oplus \text{⚥}\ K_{2+2}\ C_{X4}\ A_{2+4}\ \underline{G}_{(2)}$ is the floral formula of
(a) *Allium cepa* (b) *Solanum nigrum*
(c) *Helianthus annuus* (d) *Brassica niger*

4. Which of the following set of families depict the given floral formulas?

$$\oplus \text{⚥}\ K_{(5)}\ C_{(5)}\ A_5\ \underline{G}_2 \text{ and } \oplus \text{⚥}\ K_{2+2}\ C_4\ A_{2+4}\ \underline{G}_{(2)}$$

(a) Fabaceae and Solanaceae (b) Malvaceae and Cruciferae
(c) Solanaceae and Malvaceae (d) Solanaceae and Cruciferae

5. Which among the following alternatives represent the floral formula of family–Malvaceae?
(a) Br. $\% \text{⚥}\ K_{(5)}\ C_{(1+2+2)}\ A_{9+1}\ \underline{G}_1$
(b) Br. $\% \text{⚥}\ K_{(5)}\ C_{(4)}\ A_{(1)}\ \underline{G}_{(2)}$
(c) Br. $\oplus \text{⚥}\ P_{(3+3)}\ A_{(3+3)}\ \underline{G}_3$
(d) Br. $\oplus \text{⚥}\ \text{Epi.}_{3-9}\ K_{5 \text{ or } (5)}\ \overparen{C_{(5)}\ A}_{(\infty)}\ \underline{G}_{(2-\infty)}$

6. Which statement is wrong for Asteraceae?
(a) 5- lobed stamens (b) Syngenesious stamens
(c) Basal ovule (d) Ligulate ray florets

EXAM POINTS

- Roots originating from the base of the stem are fibrous roots. **[NEET 2020]**
- Roots play insignificant role in water absorption in *Pistia*. **[CBSE AIPMT 2015]**
- Roots are absent in *Wolffia*. **[UP CPMT 2011]**
- Sweet potato is a modified adventitious roots. **[NEET 2018]**
- Pneumatophores occur in halophytes. **[NEET 2018]**
- Ginger multiplies vegetatively by rhizome. **[UP CPMT 2013]**
- Offsets are produced by mitotic divisions. **[NEET 2018]**
- Stem modified into flat green organs performing the functions of leaves are known as phyloclades. **[NEET 2016]**
- In *Bougainvillea*, thorns are the modification of stem. **[NEET 2016]**
- Pitcher of *Nepenthes* is not a stem modification. **[NEET 2016]**
- Whorled phyllotaxy was shown by *Alstonia*. **[NEET 2019]**
- Leaf tendrils are found in peas. **[AIIMS 2018]**
- Pattern of arrangement of leaves on the stem is called phyllotaxy. **[MHT CET 2014]**
- Cauliflower shows corymbose inflorescence. **[Odisha JEE 2011]**
- Verticillaster inflorescence is found in *Ocimum*. **JIPMER 2019]**
- Keel is the characteristic feature of flower of *Indigofera*. **[CBSE AIPMT 2015]**
- The term polyadelphous related to androecium. **[NEET 2016]**
- China rose shows monadelphous condition of stamens. **[CBSE AIPMT 2015]**
- Bicarpellary ovary with obliquely placed septum is seen in *Solanum*. **[NEET 2019]**
- Tricarpellary, syncarpous gynoecium is found in flowers of Liliaceae. **[NEET 2016]**
- Multilocular ovary is found in case of China rose. **[EAMCET 2015]**
- Ray florets have inferior ovary. **[NEET 2020]**
- In mustard plant, gynoecium occupies the highest position and other parts are situated below it. **[CBSE AIPMT 2015]**
- Perigynous flowers are found in rose. **[CBSE AIPMT 2015]**
- Placentation in which ovules develop on the inner wall of the ovary or in peripheral part is parietal. **[NEET 2019]**
- Tomato and lemon show axile placentation. **[CBSE AIPMT 2012]**
- Apple is a false fruit. **[Manipal 2012]**
- Caryopsis is a form of a simple and dry fruit which is unicarpellate, it is observed in Poaceae family. **[JIPMER 2018]**
- Cereals belong to Poaceae family. **[Odisha JEE 2013]**
- Coconut fruit is a drupe. **[NEET 2017]**
- The morphological nature of the edible part of coconut is endosperm. **[NEET 2016]**
- Persistent nucellus in the seed is known as perisperm. **[NEET 2019]**
- Cotyledon of maize grain is called scutellum. **[NEET 2016]**
- A flower which can be durided into two equal halves by only one plane is zygomorphic. **[WB JEE 2011]**
- Radial symmetry is found in the flowers of *Brassica*. **[NEET 2016]**
- *Datura* belongs to Solanaceae family and have epipetalous stamens. **[EAMCET 2015]**
- Compositae is the most advanced of all the families known. **[BHU 2012]**
- Pea, bean, *Cassia*, gulmohar represents the zygomorphic symmetry. **[AIIMS 2019]**
- Floral formula of family–Fabaceae is
  $\% \text{⚥}\ K_{(5)}\ \overparen{C_{1+2+(2)}\ A}_{(9)+1} + \underline{G}_1$ **[AMU 2015]**
- The floral formula of Liliaceae is Br. $\oplus \text{⚥}\ \overparen{P_{(3+3)}\ A}_{3+3}\ \underline{G}_{(3)}$ **[NEET 2020]**

Chapter Exercises

(A) Topical Exercise

The Root

1 The zone of cell division is

(a) root cap zone
(b) zone of elongation
(c) zone of maturation
(d) meristematic zone

2 Identify the correct order from the base of root to its apex.

I. Mineral absorption zone.
II. Soil penetration zone.
III. Cell number increasement zone.
IV. Area of root apex protection.

(a) II, I, IV and III (b) I, II, III and IV
(c) IV, III, II and I (d) III, IV, I and II

3 Which of the following is incorrect match pair?

(a) Root cap — Thimble like structure
(b) Meristematic zone — Slow cell division
(c) Elongation zone — Lengthwise growth
(d) Maturation zone — Root hairs

4 The origin of lateral root arising from primary root is

(a) endogenous
(b) exogenous
(c) superior
(d) from region of elongation

5 Tall, erect plants with large shoot masses generally have a

(a) tap root system (b) adventitious root
(c) fibrous root system (d) moniliform root

6 In majority of dicotyledonous plants, the direct elongation of the ...*A*... leads to the formation of ...*B*... . Select the correct option to fill the blanks.

(a) *A*-Radicle, *B*-Root hairs
(b) *A*-Plumule, *B*-Primary root
(c) *A*-Radicle, *B*-Primary root
(d) *A*-Plumule, *B*-Adventitious root

7 Which one is not related to roots in monocotyledon plants?

(a) The primary root is short lived
(b) Root originates from the radicle part of embryo
(c) Presence of fibrous root system
(d) Roots originate from the base of the stem

8 Match the following columns.

Column I (Types of fleshy tap root)	Column II (Examples)
A. Conical	1. *Brassica rapa*
B. Fusiform	2. *Daucus carota*
C. Napiform	3. *Raphanus sativus*
D. Tuberous	4. *Mirabilis jalapa*

Codes

	A	B	C	D		A	B	C	D
(a)	2	3	1	4	(b)	3	2	1	4
(c)	2	1	3	4	(d)	2	3	4	1

9 In the given figure, *A*, *B* and *C* show different types of roots. Identify *A*, *B* and *C*.

A *B* *C*

(a) *A*-Tap root, *B*-Adventitious root, *C*-Fibrous root
(b) *A*-Tap root, *B*-Fibrous root, *C*-Adventitious root
(c) *A*-Adventitious root, *B*-Fibrous root, *C*-Tap root
(d) *A*-Adventitious root, *B*-Tap root, *C*-Fibrous root

10 Consider the following statements.

I. In *Curcuma*, fasciculated roots are found.
II. Moniliform roots are found in *Momordica*.

Choose the correct option.

(a) Statement I is true, but II is false
(b) Statement I is false, but II is true
(c) Both statements I and II are true
(d) Both statements I and II are false

11 Identify the correctly matched pair.

(a) Prop root – Banyan
(b) Conical root – *Ficus*
(c) Sucking root – *Zea mays*
(d) Climbing root – *Crocus*

12 Which of the following matches correctly?

(a) *Curcuma* – Moniliform root
(b) *Portulaca* – Nodulose root
(c) *Asparagus* – Fasciculated root
(d) *Ipomoea* – Annulated root

13 Consider the following statements.

I. Respiratory roots help in the exchange of gases with the help of pneumathodes or lenticels.

II. Respiratory roots do not possess any definite organ for the exchange of gases.

Choose the correct option.

(a) Statement I is true, but II is false
(b) Statement I is false, but II is true
(c) Both statements I and II are true
(d) Both statements I and II are false

14 Find out the incorrect match pair.

(a) Tuberous tap root – *Mirabilis jalapa*
(b) Respiratory root – *Heritiera*
(c) Epiphytic roots – *Jussiaea*
(d) Clinging roots – *Pothos*

15 Which of the following statements is correct?

(a) In *Vanda*, the velamen in roots helps in exchange of gases
(b) Radish is an example of napiform root
(c) Tuberous roots are found in *Ficus*
(d) Fibrous roots are found in maize

16 Which of the following pairs does not match?

(a) Epiphytic roots – *Vanda*
(b) Velamen in roots – Absorption of moisture from air
(c) Climbing roots – *Ficus*
(d) Sucking roots – Carrot

17 Match the following columns.

	Column I		Column II
A.	Prop root	1.	*Ficus benghalensis*
B.	Stilt root	2.	*Zea mays*
C.	Respiratory root	3.	*Cicer*
D.	Nodulated root	4.	*Rhizophora*

Codes

	A	B	C	D		A	B	C	D
(a)	1	2	3	4	(b)	4	3	2	1
(c)	2	1	3	4	(d)	1	2	4	3

18 Choose the incorrect statement.

(a) Epiphytic roots of *Vanda* have well-developed root cap and root hair
(b) *Tinospora* and *Trapa* possess assimilatory roots
(c) *Cuscuta* is a parasitic plant
(d) *Saccharum* and *Zea mays* possess stilt roots

19 Which of the following pairs does not match?

(a) Prop root – Banyan tree
(b) Stilt root – *Zea mays*
(c) Buttress root – *Bombax*
(d) Storage root – *Piper*

20 Which one of the following sets represents the example of stilt roots?

(a) *Bryophyllum, Begonia, Loranthus*
(b) *Dahlia, Asparagus, Pandanus*
(c) *Cuscuta, Loranthus, Bryophyllum*
(d) *Zea mays, Saccharum, Pandanus*

21 Which one of the following pairs is not correct?

(a) Carrot and turnip – Modified tap root
(b) Sweet potato – Prop root
(c) Maize and sugarcane – Stilt root
(d) Rhizophora – Pneumatophores

Stem

22 The structure of plants develops from plumule of the embryo of a germinating seed is called

(a) root (b) stem
(c) flower (d) bud

23 A stem is an ascending, aerial, negatively geotropic part of a plant. Identify which of the following statements is correct regarding stem and its modifications?

(a) Stolons grow above the surface of soil
(b) Offsets are several internode long and bear a tap root
(c) Bulbils are organs of vegetative reproduction
(d) Phylloclade of *Opuntia* is cylindrical while that of *Casuarina* is flattened

24 Match the Column I with Column II and select the correct option.

	Column I (Types of buds)		Column II (Features)
A.	Vegetative buds	1.	Buds develop in axils of leaves
B.	Floral buds	2.	Buds produce leafy shoots
C.	Axillary buds	3.	Reproductive buds that produce flowers
D.	Accessory buds	4.	Additional buds borne at leaf bases

Codes

	A	B	C	D
(a)	2	3	1	4
(b)	3	2	1	4
(c)	4	3	2	1
(d)	1	2	4	3

25 Which of the following group of plants represent underground fleshy, dorsiventral stem possessing nodes and internodes?

(a) *Dryopteris, Zingiber, Nelumbo*
(b) *Solanum, Helianthus tuberosus*
(c) *Allium, Tulipa, Lilium*
(d) *Crocus, Alocasia*

26 The stem modification in which internodes become highly reduced, the leaves become condensed and flesh, to take up the function of storage are known as

(a) corm (b) rhizome
(c) bulb (d) suckers

27 Identify the correctly matched pair.

(a) Offset – Mint
(b) Stolon – *Pistia*
(c) Runner – Jasmine
(d) Sucker – *Chrysanthemum*

28 Select the option that is correct for the description of subaerial modification.

I. A slender lateral branch arises from the base of the main axis and after growing aerially for some time, arches downwards to touch the ground.
II. A lateral branch originates from the basal and underground portion of the main stem, grows horizontally beneath the soil and comes out obliquely upwards, giving rise to leafy shoot.

(a) I-Sucker, II-Runner (b) I-Runner, II-Stolon
(c) I-Stolon, II-Sucker (d) I-Stolon, II-Offset

29 In lily, the axillary bud is modified into

(a) thorns (b) phylloclades
(c) bulbils (d) tendril

30 Identify the modified structure in the given figure.

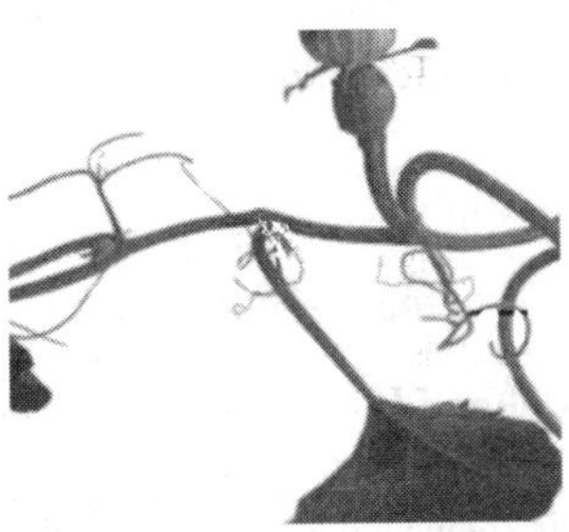

(a) Leaf tendril (b) Axillary stem tendril
(c) Leaflet tendril (d) Subaerial stem

31 The stem tendrils are modified extra-axillary branches in

I. *Cucurbita* II. *Opuntia*
III. *Luffa* IV. *Nerium*

Choose the correct option.

(a) Only I (b) I and II
(c) III and IV (d) I, II and III

32 Which of the following statements is incorrect?

(a) Underground stems of turmeric, zamikand and *Colocasia* modifies to store food
(b) Stem tendrils develop from the internodes of stem
(c) Thorns in *Citrus* and *Bougainvillea* are modified axillary buds
(d) Stem tendrils help cucumber and watermelon plants to climb

33 Identify the modification of stem given below.

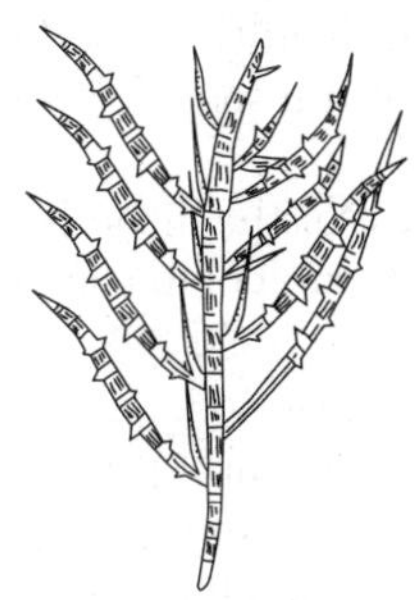

(a) Phylloclade of *Muehlenbeckia*
(b) Phylloclade of *Opuntia*
(c) Cladode of *Asparagus racemosis*
(d) Phylloclade of *Ruscus*

34 Which of the following is correct according to the information provided below in the columns.

	Cladode	Phylloclade
(a)	Limited growth	Unlimited growth
(b)	Both main stem and branches are modified to function as leaves.	Only branches are modified to function as leaves.
(c)	It possesses several nodes and internodes.	It is usually one internode long.
(d)	Common examples are *Euphorbia* and *Opuntia*.	Common examples are *Asparagus* and *Ruscus*.

35 Mark the mismatched pair with respect to modification of stem.

(a) Rhizome - Turmeric (b) Corm - Banana
(c) Bulb - Onion (d) Phylloclade - *Euphorbia*

Leaf

36 Fill in the blanks and choose the correct option.
Leaves develop at the ...*A*..., originate from ...*B*..., arranged in ...*C*... order.

	A	*B*	*C*
(a)	nodes	axillary bud	basipetal
(b)	axillary bud	nodes	acropetal
(c)	stem	nodes	basipetal
(d)	nodes	shoot apical meristem	acropetal

37 Which of the following structure of leaf is terminal, thin and specialised to manufacture food?

(a) Hypopodium (b) Epipodium
(c) Hesopodium (d) Endopodium

38 Leaf base is swollen to form *A* in some *B*.

	A	*B*
(a)	Pulvinus	Leguminous plants
(b)	Pulvinus	Crucifers
(c)	Phyllode	Leguminous plants
(d)	Phylloclade	Plants of arid regions

39 Consider the following statements.

I. Veins provide rigidity to the leaf blade and act as the channel of transport for water, minerals and food materials.

II. Leaves of most of the dicotyledonous plants bear veins that run parallel to each other within a lamina.

Choose the correct option.

(a) Statement I is true, but II is false
(b) Statement I is false, but II is true
(c) Both statements I and II are true
(d) Both statements I and II are false

40 Identify the incorrectly matched pair relating to the part of leaf modified into spine with the example of the plant.

(a) Margin of lamina — *Argemone*
(b) Midrib of lamina — *Phoenix*
(c) Stipules — *Acacia*
(d) Margin and apex of lamina — *Aloe*

41 Fill in the blanks and choose the correct option.

I. ...*A*... provides rigidity to leaf blade.

II. The leaf is attached to the stem by the ...*B*... and may bear two lateral leaf-like structures called ...*C*... .

	A	*B*	*C*
(a)	Petiole	Leaf base	Bracts
(b)	Veins	Leaf base	Stipules
(c)	Veins	Petiole	Axillary bud
(d)	Petiole	Hypopodium	Stipules

42 Which one of the following statements is incorrect?

(a) Arrangement of veins in leaf lamina is called venation
(b) In reticulate venation, veinlets form a network
(c) In parallel venation, veins run parallel to each other within a lamina
(d) Leaves of *Mangifera indica* plant possess parallel venation

43 Pinnately compound leaves possess a number of leaflets attached laterally on the

(a) margin (b) veins
(c) rachis (d) petiole

44 Find the incorrect statement.

(a) In pinnately compound leaf, rachis is present
(b) Rachis, actually represents the midrib of the leaf a pinnately compound leaf
(c) In palmately compound leaf, rachis is present
(d) Silk cotton has got palmately compound leaf

45 Select the incorrectly matched pair.

(a) Palmately compound leaf-*Alstonia*
(b) Pinnately compound leaf-Neem
(c) Whorl phyllotaxy-*Nerium*
(d) Fleshy leaves-Garlic

46 Identify the incorrectly matched pair of leaf.

(a) *Rosa indica* — Unipinnate leaf
(b) *Daucus* — Decompound leaf
(c) *Acacia* — Tripinnate
(d) *Moringa* — Tripinnate

47 Identify the correctly matched pair.

(a) Multifoliate — Silk cotton tree
(b) Bifoliate — *Cleome*
(c) Quadrifoliate — *Marsilea*
(d) Trifoliate — *Citrus*

48 A unipinnate compound leaf can be distinguished from a branch having simple leaves by

(a) the presence of terminal buds in compound leaf
(b) the absence of veins in the leaflets
(c) the presence of buds in the axils as leaflets
(d) the presence of buds in the axils of leaves

49 Among the following plants, how many show bipinnately compound leaves?

Mimosa pudica, *Caesalpinia*, Kikar, *Moringa*, *Daucus*, *Citrus*

(a) 5 (b) 3 (c) 2 (d) 4

50 Which of the following is correctly matched pair?

(a) Largest leaves — *Victoria regia*
(b) Free-lateral stipules — Rice
(c) Stipules modified into tendril — *Pisum sativm*
(d) Ochreate stipule — Rose leaf

51 Match the following columns.

	Column I (Leaf modifications)		Column II (Examples)
A.	Pinnately compound leaf	1.	Silk cotton
B.	Stem thorns	2.	Rose
C.	Palmately compound leaves	3.	*Duranta*
D.	Prickles	4.	*Bombax*

Codes

	A	B	C	D
(a)	4	1	3	2
(b)	2	3	1	4
(c)	3	2	4	1
(d)	2	1	4	3

52 Which of the following correctly represents the parts of leaf modified into tendrils with the example of plant?

(a) Stipule — *Smilax*
(b) Whole leaf — *Clematis*
(c) Leaf tip — *Tropaeolum*
(d) Upper leaflet — *Naravelia*

53 Match the following columns with respect to phyllotaxy.

	Column I		Column II
A.	Alternate	1.	*Psidium*
B.	Opposite superimposed	2.	*Hibiscus*
C.	Opposite decussate	3.	*Calotropis*
D.	Whorled	4.	*Alstonia*

Codes

	A	B	C	D		A	B	C	D
(a)	1	4	3	2	(b)	2	4	1	3
(c)	3	4	1	2	(d)	2	1	3	4

Inflorescence and Flower

54 In racemose inflorescence
(a) the flowers are borne laterally in an acropetal succession
(b) the main axis is of limited growth
(c) the flowers are borne in a basipetal order
(d) the main axis terminates in a flower

55 Consider the following statements.
I. In corymb inflorescence, all flowers reach the same level though they arise from different points.
II. The composite fruit develops from the hypanthodium inflorescence and occurs in *Ficus carica.*

Choose the correct option.
(a) Statement I is true, but II is false
(b) Statement I is false, but II is true
(c) Both statements I and II are true
(d) Both statements I and II are false

56 Consider the following statements.
I. In cymose inflorescence, the main axis terminate into flower.
II. In cymose inflorescence, young flowers are present towards the base and older at the apex.

Choose the correct option.
(a) Statement I is true, but II is false
(b) Statement I is false, but II is true
(c) Both statements I and II are true
(d) Both statements I and II are false

57 If in an inflorescence, two lateral branches develop on either side of the terminal flower, which again branch, is known as
(a) monochasial cyme (b) scorpioid cyme
(c) helicoid cyme (d) dichasial cyme

58 The compound umbel type of inflorescence is found in
I. *Coriandrum sativum* II. *Delonix*
III. *Helianthus annuus* IV. *Daucus carota*

Choose the correct option.
(a) I and IV (b) Only I
(c) Only III (d) I, II and III

59 Match the following columns.

	Column I (Inflorescence types)		Column II (Examples)
A.	Racemose	1.	*Euphorbia*
B.	Cyathium	2.	Jasmine
C.	Cymose	3.	Peepal
D.	Hypanthodium	4.	*Calotropis*

Codes

	A	B	C	D		A	B	C	D
(a)	2	1	4	3	(b)	1	3	4	2
(c)	2	4	1	3	(d)	4	3	2	1

60 The inflorescence of poplar and willow are
(a) amentum only
(b) catkin and spikes, respectively
(c) spadix and catkin, respectively
(d) spike and catkin, respectively

61 In cyathium inflorescence, the ratio between male and female flowers is
(a) one : many
(b) many : one
(c) one : one
(d) many : many

62 Verticillaster is a type of ...*A*... which is found in ...*B*... .
(a) *A*—Fruit, *B*—*Ocimum*
(b) *A*—Inflorescence, *B*—*Ocimum*
(c) *A*—Fruit, *B*—*Ficus*
(d) *A*—Flower, *B*—*Salvia*

63 A flower is a modified shoot because
(a) certain flowers have well-developed nodes and internodes
(b) it arises in the axil of bracts
(c) the growing point of thalamus may give rise to shoots
(d) All of the above

64 A typical flower has four different kinds of whorls arranged on the swollen end of the pedicel called
(a) thalamus (b) pedicel
(c) Bract (d) peduncle

65 Find the incorrect statement with respect to flower.
(a) A flower is modified shoot wherein the shoot apical meristem changes into floral meristem
(b) Flower is always solitary when shoot bud transforms into flower
(c) The arrangement of flowers on the floral axis is called inflorescence
(d) The apex produces different floral appendages horizontally and internodes elongates

66 Match the following columns.

Column I (Parts of flower)	Column II (Description)
A. Calyx	1. Female reproductive part
B. Corolla	2. Male reproductive part
C. Androecium	3. Outermost whorl of flower
D. Gynoecium	4. Composed of petals

Codes

	A	B	C	D		A	B	C	D
(a)	4	1	3	2	(b)	3	2	1	4
(c)	3	4	2	1	(d)	2	1	3	4

67 Identify the given diagram and choose the incorrect option.

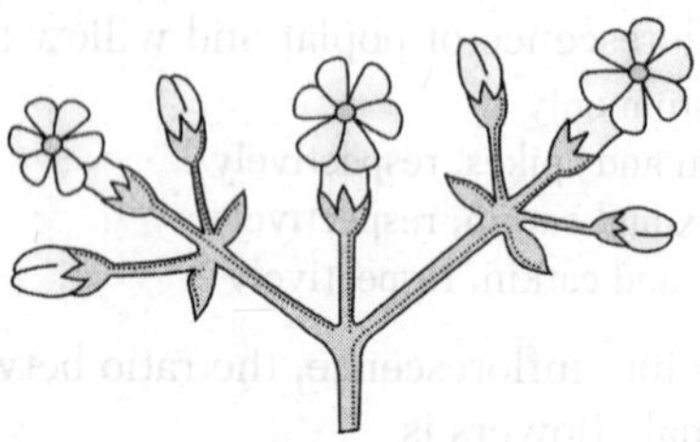

(a) The main axis terminates into a flower
(b) Flowers are basipetally arranged
(c) Growth of the peduncle is determinate
(d) The main axis continues to grow

68 Consider the following statements.

I. Petals are usually brightly coloured and can be united or free in a flower.
II. Petals protect the inner reproductive structure of flower.

Choose the correct option.
(a) Statement I is true, but II is false
(b) Statement I is false, but II is true
(c) Both statements I and II are true
(d) Both statements I and II are false

69 Find the incorrectly matched pair.
(a) Caducous sepals - Poppy
(b) Marcescent sepals - *Piper*
(c) Cruciform petals - Mustard
(d) Campanulate petals - *Dianthus*

70 Find the incorrect match with respect to aestivation.
(a) Pea and beans —Vexillary
(b) *Calotropis* — Twisted
(c) *Cassia* — Imbricate
(d) Lady's finger — Twisted

71 Valvate type of aestivation of calyx is found in
I. *Solanum* II. *Iberis amara*
III. mustard IV. *Dianthus*
Choose the correct option.
(a) I and III (b) II and III (c) III and IV (d) Only IV

72 The aestivation in petals of China rose and gulmohar plants can be shown, respectively by

A B C D

(a) *B* and *A* (b) *B* and *C* (c) *C* and *D* (d) *D* and *C*

73 The given aestivation is ......*A*...... and found in ......*B*...... .

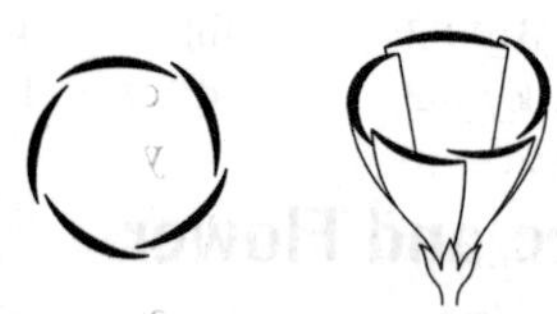

	A	*B*
(a)	Valvate	*Calotropis*
(b)	Twisted	Lady's finger
(c)	Imbricate	China rose
(d)	Vexillary	Pea

74 Match the following columns with respect to placentation.

Column I	Column II
A. Axile	1. *Dianthus, Primrose*
B. Parietal	2. Sunflower, marigold
C. Free-central	3. China rose, tomato, lemon
D. Basal	4. Mustard, *Argemone*

Codes

	A	B	C	D		A	B	C	D
(a)	3	4	2	1	(b)	4	3	1	2
(c)	3	4	1	2	(d)	2	3	4	1

Fruits and Seeds

75 Which of the following statements is incorrect regarding capsular dry fruits?
(a) Siliqua and silicula resemble each other, but siliqua is shorter, wider and flat
(b) Pod develops from monocarpellary ovary, while capsule develops from polycarpellary ovary
(c) Legumes are characteristic fruits of family —Leguminosae
(d) Follicle and legumes both show dehiscence only by one suture (ventral)

76 Two dry fruits *X* and *Y* are observed X developed from monocarpellary and *Y* developed from bicarpellary ovary. *X* is the characteristic fruit of Leguminosae family whereas *Y* is the characteristic fruit of Brassicaceae family *X* and *Y* are
(a) legume, capsule (b) legume, siliqua
(c) capsule, follicle (d) capsule, siliqua

77 Achenial fruits are simple, dry, single-seeded, indehiscent fruits, which do not dehisce after ripening and seeds remain inside the pericarp. They may be

I. caryopsis II. siliqua
III. nut IV. capsule

Choose the correct option.

(a) I and II (b) II and IV (c) Only II (d) I and III

78 Fruits of which of the following sets develop from unilocular, single-ovuled, superior ovary of polycarpellary gynoecium?

(a) Achene and cypsela (b) Achene and caryopsis
(c) Caryopsis and cypsela (d) Cypsela and nut

79 Consider the following statements.

I. The fruits of *Annona squamosa* (custard apple) develop from multicarpellary, apocarpous.
II. An achenial fruit, in which the seed is attached to pericarp only at one point and rest of the pericarp remains free from the seed occurs in *Taraxacum.*

Choose the correct option.

(a) Statement I is true, but II is false
(b) Statement I is false, but II is true
(c) Both statements I and II are true
(d) Both statements I and II are false

80 The fruit develops from an inferior and bicarpellary ovary possessing persistent stylopodium is known as

(a) lomentum (b) regma
(c) cremocarp (d) drupe

81 Lomentum fruits are considered to be modified legumes, but few lomentum fruits internally resemble siliqua. Such a fruit is found in

(a) *Raphanus sativus* (b) *Mimosa pudica*
(c) *Arachis hypogea* (d) *Tamarindus indica*

82 The character of a nut fruit of *Anacardium occidentale* is

(a) husky exocarp, woody pericarp and single seed possessing edible cotyledons
(b) husky exocarp, woody pericarp and two seeds each possessing single large edible cotyledon
(c) hard, brittle and somewhat spiny pericarp enclosing a non-edible seed
(d) hard pericarp enclosing edible seed with two large cotyledons

83 How many plants have drupe fruits?

Citrus, coffee, mango, apricot, apple, coconut, mustard, pea, date, olive, almond, peach, cherry, pistachio.

(a) 12 (b) 10
(c) 16 (d) 7

84 The figures of some fruits are given below, write the type of fruit and name for each fruit.

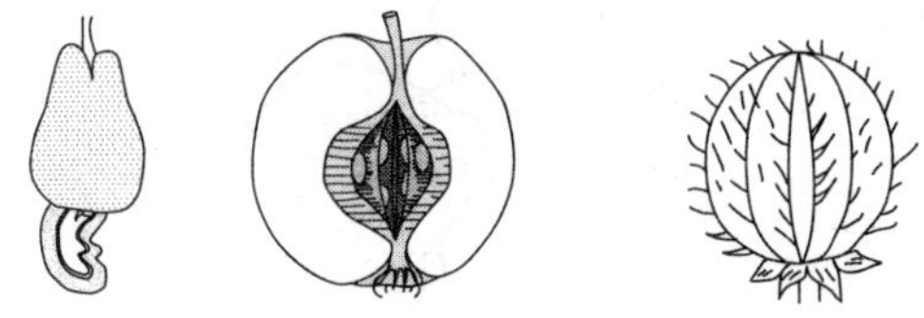

(a) *A*-Sorosis (*Morus*), *B*-Regma (apple), *C*-Siliqua (*Ficus*)
(b) *A*-Nut (*Anacardium*), *B*-Pome (apple), *C*-Regma (*Ricinus*)
(c) *A*-Syconus (*Ficus*), *B*-Pome (apple), *C*-Silicula (*Iberis*)
(d) *A*-Samara (Elm), *B*-Pome (apple), *C*-Syconus (*Ficus*)

85 In which of the following fruits, the edible parts are endosperm and embryo?

I. *Oryza sativa*
II. *Triticum aestivum*
III. *Litchi chinensis*
IV. *Lycopersicon esculentum*

Choose the correct option.

(a) I and II (b) I and III
(c) II and III (d) Only IV

86 Find out the correct sequence of labelling.

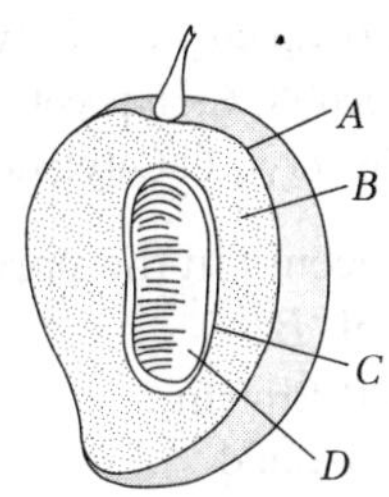

(a) *A*-Epicarp, *B*-Seed, *C*-Mesocarp, *D*-Endocarp
(b) *A*-Epicarp, *B*-Mesocarp, *C*-Endocarp, *D*-Seed
(c) *A*-Epicarp, *B*-Seed, *C*-Endocarp, *D*-Mesocarp
(d) *A*-Seed, *B*-Endocarp, *C*-Mesocarp, *D*-Epicarp

87 Edible parts in apple (*A*), mango (*B*) and tomato (*C*) are

(a) *A*-Pericarp, *B*-Entire fruit, *C*-Thalamus
(b) *A*-Thalamus, *B*-Mesocarp, *C*-Pericarp and placenta
(c) *A*-Thalamus, *B*-Mesocarp and endocarp, *C*-Mesocarp
(d) *A*-Entire fruit, *B*-Mesocarp, *C*-Pericarp

88 The plants in which only the pericarp and placenta of fruits are edible?

I. *Vitis vinifera*
II. *Lycopersicon esculentum*
III. *Juglans regia*
IV. *Mangifera indica*

Choose the correct option.

(a) I and II (b) II and III
(c) Only III (d) Only IV

89 Identify *A*, *B* and *C* in the given figure.

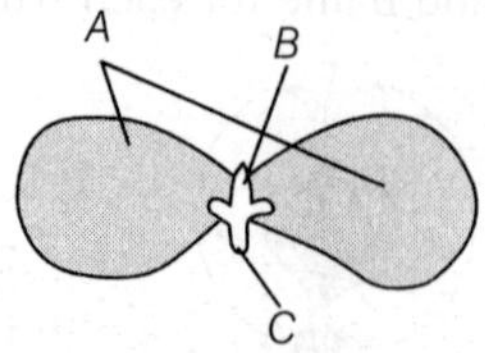

(a) *A*-Cotyledon, *B*-Plumule, *C*-Radicle
(b) *A*-Cotyledon, *B*-Radicle, *C*-Plumule
(c) *A*-Radicle, *B*-Plumule, *C*-Cotyledon
(d) *A*-Plumule, *B*-Radicle, *C*-Cotyledon

90 Identify the structures *A*-*C* indicated in the diagram of a seedling given below.

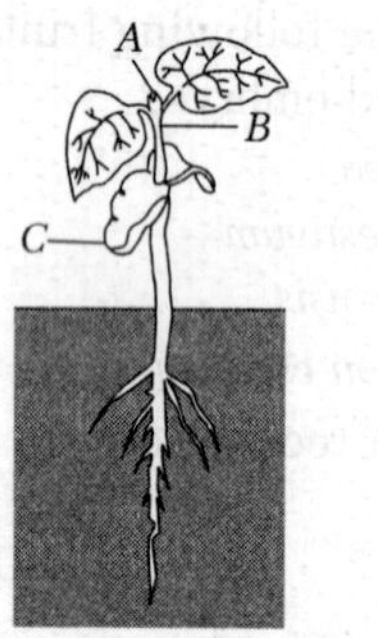

(a) *A*-Epicotyl, *B*-Hypocotyl, *C*-Cotyledon
(b) *A*-Hypocotyl, *B*-Epicotyl, *C*-Cotyledon
(c) *A*-Shoot, *B*-Epicotyl, *C*-Hypocotyl
(d) *A*-Shoot, *B*-Epicotyl, *C*-Cotyledon

91 Aleurone layer occurs in the peripheral region of '*A*' and is made up of '*B*'.
(a) *A*-Coleoptile, *B*-Protein
(b) *A*-Scutellum, *B*-Lipid
(c) *A*-Endosperm, *B*-Protein
(d) *A*-Endosperm, *B*-Lipid

92 The mature seeds of plants such as gram and peas possess no endosperm, because **[NCERT Exemplar]**
(a) these plants are not angiosperms
(b) there is no double fertilisation in them
(c) endosperm is not formed in them
(d) endosperm gets used up by the developing embryo during seed development

93 The examples of dicotyledonous seeds are
I. *Cicer*
II. *Brassica*
III. *Lycopersicon*
IV. *Triticum*

Choose the correct option.
(a) Only II (b) II and III
(c) Only III (d) I, II and III

94 Generally monocotyledonous seeds are ...*A*..., but some as in ...*B*... are ...*C*....
(a) *A*-non-endospermic, *B*-orchids, *C*-endospermic
(b) *A*-endospermic, *B*-orchids, *C*-non-endospermic
(c) *A*-non-endospermic, *B*-castor, *C*-endospermic
(d) *A*-endospermic, *B*-castor, *C*-non-endospermic

95 Name the mechanism of seed dispersal shown in the given figure.

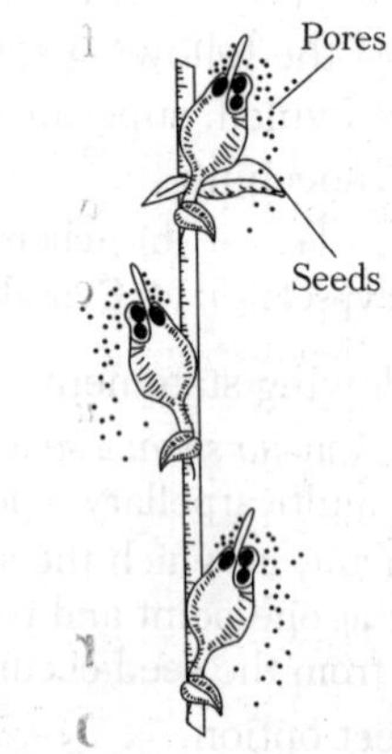

(a) Censer mechanism of seed dispersal in *Papaver somniferum*
(b) Censer mechanism of seed dispersal in *Antirrhinum*
(c) Parachute mechanism of seed dispersal in *Taraxacum*
(d) Parachute mechanism of seed dispersal in *Naravelia*

96 Flowers of which of the following plant sets are zygomorphic?
I. *Adhatoda* and pea
II. Larkspur and *Ocimum*
III. Mustard and onion
IV. Brinjal and *Canna*

Choose the correct option.
(a) Only II (b) I and II
(c) III and IV (d) I, II, III and IV

97 Select the incorrect match.
(a) Zygomorphic flower – Gulmohar
(b) Asymmetric flower – *Canna*
(c) Actinomorphic flower – Mustard
(d) Pentamerous flower – 4 calyx

98 Which of the following represents hypogynous condition of a flower?

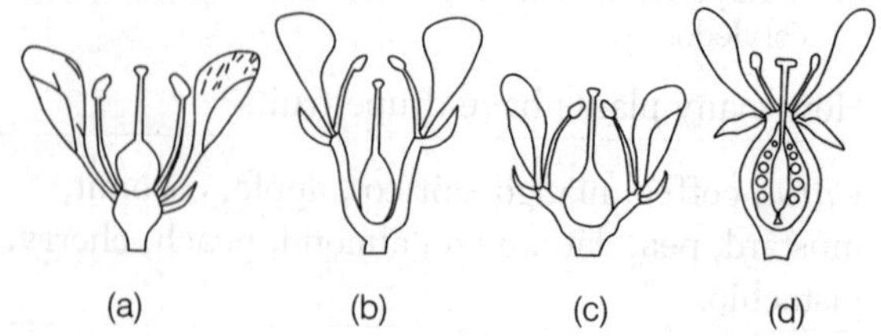

(a) (b) (c) (d)

99 All statements are correct with reference to the given figure below, except

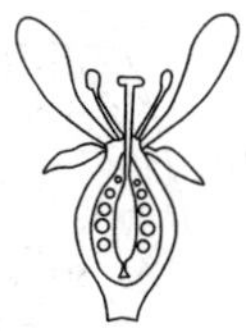

(a) the margin of thalamus grows upwards enclosing the ovary completely and getting fused with it
(b) ovary is said to be inferior
(c) the flower is said to be perigynous
(d) it is seen in the flowers of guava and cucumber

100 Which of the following is not correctly matched?
(a) Actinomorphic flower–brinjal
(b) Bilateral symmetric flower–*Datura*
(c) Trimerous flower–Lily
(d) Asymmetric flower–Orchids

Semi-technical Description of a Typical Flowering Plants and Important Families

101 Consider the following statements.
I. In the floral formula, a superior ovary is represented by a line drawn below G.
II. Adhesion is indicated by enclosing the figure within a bracket.

Choose the correct option.
(a) Statement I is true, but II is false
(b) Statement I is false, but II is true
(c) Both statements I and II are true
(d) Both statements I and II are false

102 Which of the following floral features is not represented by symbols in the floral formula of a plant family?
(a) Symmetry of flower
(b) Aestivation of calyx and corolla
(c) Cohesion and adhesion between floral whorls
(d) Relative position of ovary with respect to other parts

103 Match the following columns.

Column I (Symbols of floral formula)		Column II (Description)	
A.	$\oplus$	1.	Zygomorphic
B.	%	2.	Actinomorphic
C.	K	3.	Calyx
D.	C	4.	Corolla

Codes

	A	B	C	D
(a)	4	3	2	1
(b)	2	1	3	4
(c)	3	4	2	1
(d)	1	2	3	4

104 Find the incorrect match with respect to the floral formula concerned families.
(a) Fabaceae $\% \text{⚥} K_{(5)} C_{1+2+(2)} A_{(9)}$
(b) Liliaceae $Br \oplus \text{⚥} P_{(3+3)} A_{3+3} \underline{G_{(3)}}$
(c) Solanaceae $\oplus \text{⚥} K_{(5)} \overset{\frown}{C_{(5)} A_5} \underline{G_{(2)}}$
(d) Brassicaceae $\% \text{⚥} K_{(2+2)} C_4 A_{2+4} \overline{G}_{(2)}$

105 Observe the floral formula given below.

$\% \text{⚥} K_{(5)} C_{1+2+(2)} A_{(9)+1} \underline{G_1}$

Identify the plants which possess the above floral formula.
(a) *Lupinus, Pisum*
(b) *Solanum, Tabacum*
(c) *Lilium, Aloe*
(d) *Brassica, Solanum*

106 How many of the following plants belong to the family whose formula is given below?

$\% \text{⚥} K_{(5)} C_{1+2+(2)} A_{(9)+1} \underline{G_1}$

Soybean, groundnut, *Indigofera*, chilli, *Solanum*, *Trifolium*, *Sesbania*, tulip, sunhemp, *Colchicum*, *Asparagus*, *Petunia*

(a) Six (b) Five
(c) Seven (d) Four

107 Match the following columns.

Column I (Members of Fabaceae)		Column II (Economic importance)	
A.	*Dalbergia sisso*	1.	Fibres
B.	Moong	2.	Fodder
C.	*Sesbania*	3.	Timber
D.	Sunhemp	4.	Pulses

Codes

	A	B	C	D
(a)	3	2	1	4
(b)	3	2	4	1
(c)	1	2	3	4
(d)	3	4	2	1

108 Which of the following is not a floral character of Liliaceae family?

(a) Tepals are six in number (3 + 3)
(b) Stamens are six in number (3 + 3)
(c) Ovary is six in number
(d) Syncarpous superior ovary is present

109 I. Plants belonging to the family...*A*... possesses a persistent calyx.
II. The family ...*B*... mainly contains monocotyledonous plants.

Choose the correct option to fill in the blanks *A* and *B*.

(a) *A*-Fabaceae, *B*-Poaceae
(b) *A*-Solanaceae, *B*-Fabaceae
(c) *A*-Solanaceae, *B*-Liliaceae
(d) *A*-Asteraceae, *B*-Solanaceae

110 Select the incorrectly matched pair.

(a) *Petunia*- $\widehat{C_{(5)}\,A_5}$
(b) *Brassica*- K_{2+2}
(c) *Sesbania*- $A_{(9)+1}$
(d) *Gloriosa*- $G_{(2)}$

111 What is true for the floral diagram given below?

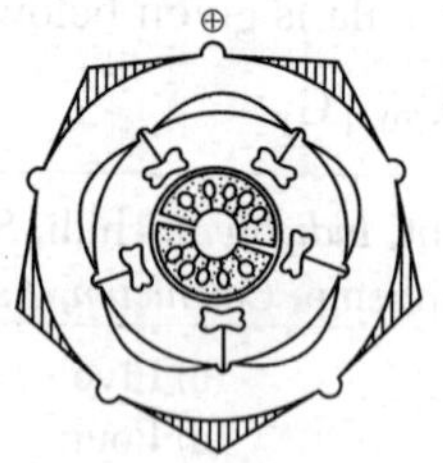

(a) It is found in the *Aloe* plant
(b) It is found in the family where fruit is commonly berry or capsule
(c) It is found in the Fabaceae family
(d) It is present in the family with flowers having cruciform corolla

112 Select the characteristic feature of plants that have the given floral diagram.

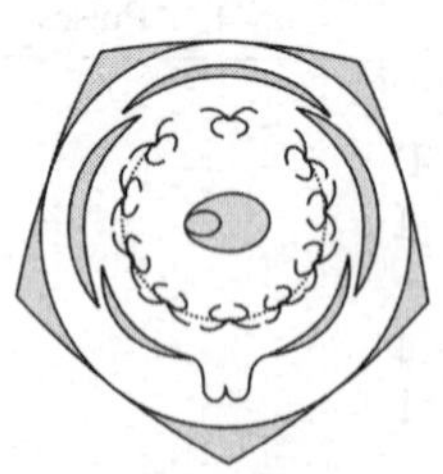

(a) Bisexual actinomorphic flower
(b) Monocarpellary unilocular ovary with many ovules
(c) Epigynous flower with marginal placentation
(d) Monadelphous condition of androecium

113 Floral diagram of the family to which tulip and *Gloriosa* belong is

(a)  (b) 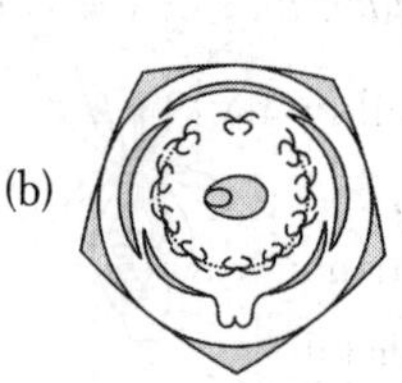

(b) 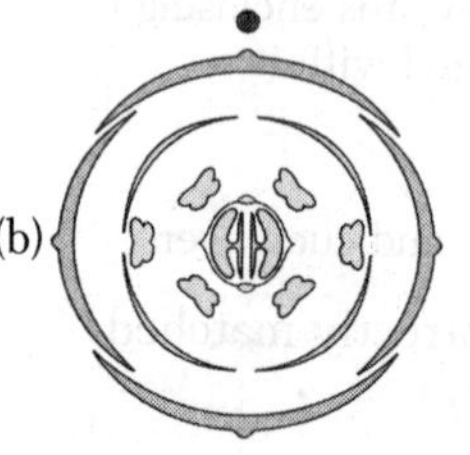 (d) 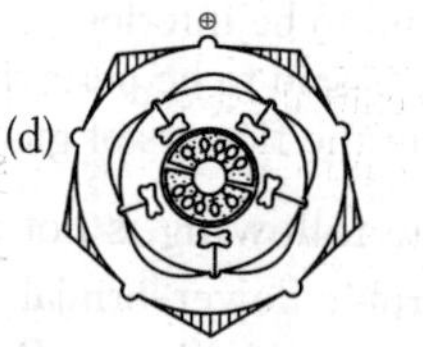

114 Study carefully the given floral diagram and select the option, which correctly represents the related floral formula (FF).

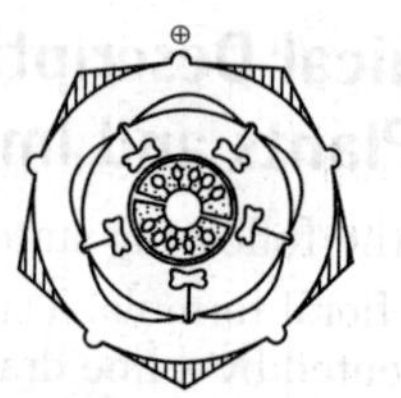

(a) % ⚥ $K_{(5)}C_{1+2+(2)}A_5\bar{G}_{(2)}$
(b) $\oplus$ ⚥ $K_{(5)}\widehat{C_{(5)}\,A_5}\underline{G}_{(2)}$
(c) $\oplus$ ⚥ $P_{5+5}\,A_{(5)}\underline{G}_{(2)}$
(d) $\oplus$ ⚥ $K_{(5)}\widehat{C_{(5)}\,A_5}\underline{G}_{(2)}$

115 Study carefully the given floral diagram and select the option, which correctly represents the related floral formula (FF).

(a) $\oplus$ ⚥ $\widehat{P_{(3+3)}}A_{3+3}\underline{G}_{(3)}$ (b) $\oplus$ ⚥ $P_6A_6\underline{G}_{(3)}$
(c) $\oplus$ ⚥ $P_{5+5}\,A_{(5)}\underline{G}_{(2)}$ (d) $\oplus$ ⚥ $K_{(5)}\widehat{C_{(5)}\,A_5}\underline{G}_{(2)}$

116 Choose the correct option.

(a) Members of Solanaceae family are used as spices (chilli), medicine (Belladonna, ashwagandha)
(b) Members of Liliaceae include *Aloe, Asparagus*
(c) Members of Fabaceae like *Lupin*, sweet pea are used for ornamental purposes
(d) All of the above

117 Consider the following statements.

I. Colchicine is obtained from the members of family-Liliaceae.
II. Colchicine is used in chromosomal doubling.

Choose the correct option.

(a) Statement I is true, but II is false
(b) Statement I is false, but II is true
(c) Both statements I and II are true
(d) Both statements I and II are false

(B) Cumulative Exercise

(Assorted Questions of the Chapter for Advanced Level Practice)

1 Match the following columns.

Column I (Modification in leaves)	Column II (Examples)
A. Leaf tip tendrils	1. Pitcher plant and Venus flytrap
B. Bud is absent in the axil of leaflets	2. *Cacti*
C. Spines (modified leaves)	3. Compound leaf
D. Leaves modified to catch insects	4. *Gloriosa*
E. Fleshy leaves with stored food	5. Garlic and onion

Codes

	A	B	C	D	E
(a)	1	2	3	4	5
(b)	5	4	3	2	1
(c)	4	3	2	1	5
(d)	4	2	3	1	5

2 Read the following statements with respect to economic importance of various angiospermic families.

I. Family–Liliaceae is a source of New Zealand hemp.
II. A member of family–Solanaceae is a great source of painkillers.
III. Fumigatory product nicotin is extracted from a member of family–Solanaceae.
IV. Candytuft and wall flower belongs to family–Brassicaceae and are widely used as ornamental plants.

Choose the correct statements from the option given below.

(a) III and IV (b) Only III
(c) I and III (d) None of these

3 Identify the fruits given in figure.

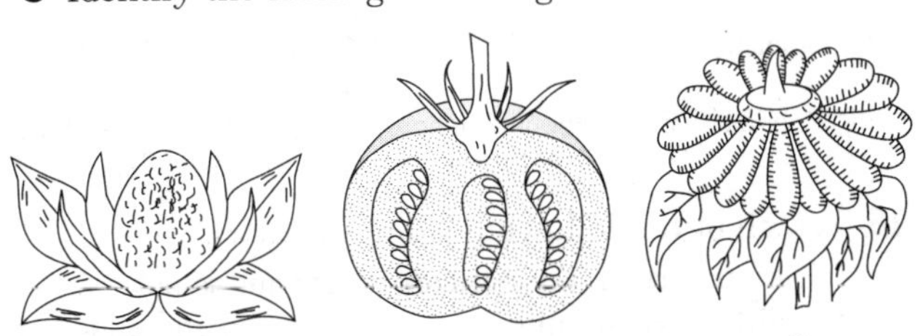

A B C

(a) *A*-Achene, *B*-Berry, *C*-Carcerulus
(b) *A*-Capsule, *B*-Carcerulus, *C*-Achene
(c) *A*-Berry, *B*-Achene, *C*-Carcerulus
(d) *A*-Carcerulus, *B*-Berry, *C*-Achene

4 Consider the following statements.

I. Fruits such as *Clematis* have hard hair and spines for hydrochory.
II. Pappus of *Taraxacum* helps in wind dispersal.
III. Jaculator mechanism of self-dispersal is shown by *Ruellia*.
IV. Myrmechory is dispersal of fruits or seed by ants.

Choose the incorrect statement(s).

(a) Only I (b) I and II
(c) Only IV (d) I, II, III and IV

5 Consider the given statements regarding roots and choose the option having most correct statement(s).

I. Epiphytic roots are hygroscopic.
II. Climbing roots of *Ficus pumila* secrete a gummy substance to stick to the support.
III. Roots are positively geotropic, negatively phototropic and negatively hydrotropic.
IV. Cells of zone of elongation in a root loose their power to divide.

Choose the correct option.

(a) I, II and III (b) Only II
(c) I, II and IV (d) Only III

6 Read the following statements about stem and identify them as true (T) and false (F).

I. Ascending part of plant axis.
II. Develops from radicle of the embryo of a germinating seed.
III. Generally green when older, but brown when young.
IV. It may preform function of vegetative propagation.

	I	II	III	IV
(a)	T	F	F	T
(b)	T	T	F	T
(c)	T	T	F	F
(d)	T	F	F	F

7 Which of the stipule type does not match with the example of plant?

(a) Foliaceous – Sweet pea
(b) Free-lateral – *Hibiscus*
(c) Interpetiolar – *Morinda*
(d) Adnate – *Zizyphus*

8 Identify the types of root given in figure from the given options.

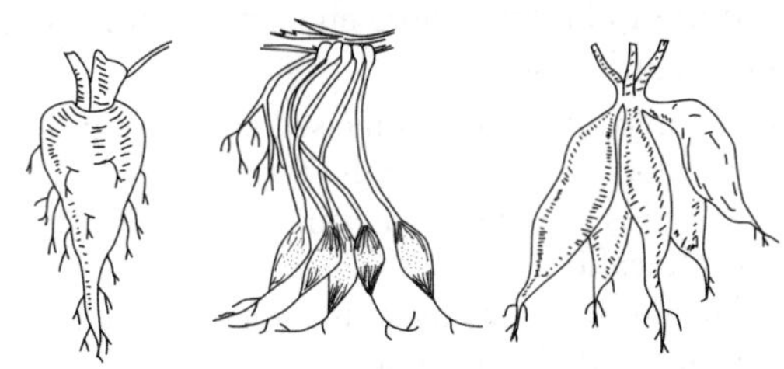

A B C

(a) A – Fusiform, radish; B – Napiform, turmeric; C – Tuberous, sweet potato
(b) A – Conical, turnip; B – Nodulated, sweet potato; C – Tuberous, *Curcuma amada*
(c) A – Conical, carrot; B – Nodulose, turmeric C – Fasciculated, *Dahlia*
(d) A – Napiform, carrot; B – Nodulated, tamarind; C – Tuberous, turmeric

9 Read the statement(s) given below and identify the incorrect statement(s).

I. In cymose inflorescence main axis terminates into a flower.
II. In racemose inflorescence, main axis does not terminate into a flower.
III. In racemose inflorescence, flowers are borne basipetally.
IV. In cymose inflorescence, flowers are borne acropetally.

Choose the correct option.

(a) Only IV (b) III and IV
(c) Only III (d) None of these

10 Fill in the blanks for A, B, and C and select the correct option.

I. Epiphyllous stamens are present in flowers of ...A...
II. Pappus is ...B... .
III. Ovary in plum is ...C... .

(a) A-Brinjal, B-Modified calyx, C-Inferior
(b) A-Lily, B-Modified calyx, C-Half inferior
(c) A-Lily, B-Modified perianth, C-Half inferior
(d) A-Brinjal, B-Modified corolla, C-Half superior

11 Identify the fixation of filament in the following diagrams of stamens.

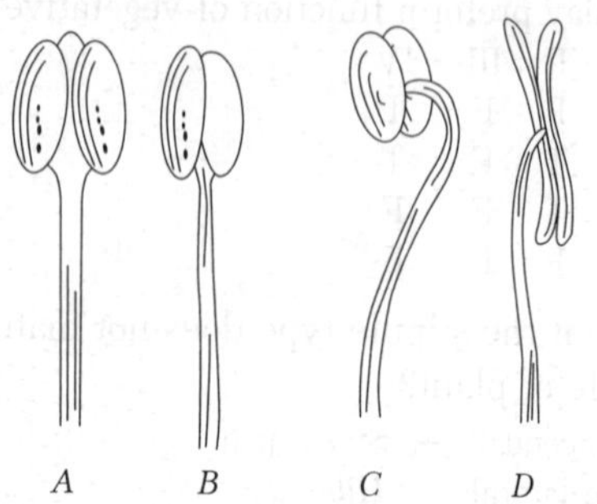

(a) A-Adnate, B-Basifixed, C-Dorsifixed, D-Versatile
(b) A-Basifixed, B-Dorsifixed, C-Adnate, D-Versatile
(c) A-Adnate, B-Dorsifixed, C-Versatile, D-Basifixed
(d) A-Basifixed, B-Versatile, C-Dorsifixed, D-Adnate

12 Read the following statements and identify them as true (T) and false (F).

I. A compound leaf has incised lamina and the incision reaches up to the midrib.
II. A bud is present in the axile of a leaf as well as each leaflet.
III. Simple leaves always arise singly at each node and bear undivided lamina.
IV. The leaves are large and short lived in Australian acacia.

	I	II	III	IV
(a)	T	F	F	F
(b)	T	F	T	T
(c)	T	F	F	T
(d)	T	T	F	F

13 Match the following columns.

Column I (Stem modification)	Column II (Found in)
A. Underground stem	1. *Euphorbia*
B. Stem tendril	2. *Opuntia*
C. Stem thorn	3. Potato
D. Flattened stem	4. *Citrus*
E. Fleshy cylindrical stem	5. Cucumber

Codes

	A	B	C	D	E
(a)	1	2	3	5	4
(b)	2	3	4	5	1
(c)	3	4	5	1	2
(d)	3	5	4	2	1

14 The given figure represents, a specific kind of roots modification found in mangroves.

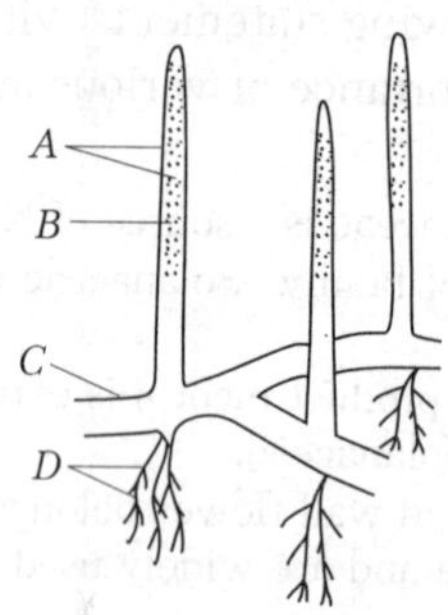

Identify A-D by selecting correct alternative.

	A	B	C	D
(a)	Pneumatophores	Lenticels	Absorbing roots	Pneumathodes
(b)	Pores	Vertical roots	Photosynthetic root	Clinging root
(c)	Stomata	Stem	Assimilatory root	Tap root
(d)	Pneumathodes	Pneumatophores	Horizontal root	Absorbing root

15 Which of the following is/are correct?

I. In dry fruits, the pericarp is dry and is not differentiated into epicarp, mesocarp and endocarp.
II. In succulent fruits, the pericarp is fleshy and is differentiated into epicarp, mesocarp and endocarp.
III. Capsular fruits are those simple, dry, multi-seeded dehiscent fruits, in which the pericarp split ruptures after ripening and seeds are exposed.
IV. Achenial fruits are multi-seeded dehiscent, in which pericarp ruptures and seeds remain inside.

Codes

(a) Only IV (b) I, II and III
(c) I, II, III and IV (d) None of these

16 Which of the following statements is correct?
(a) *Ocimum* shows inflorescence which is initially a dichasial cyme, but soon changes into monochasial helicoid type
(b) *Ligustrum* shows a mixed umbel in which flowers are borne acropetally
(c) In a hypanthodium, male flowers are borne at the basal side, while female flowers at the ostiole side
(d) Coenanthium is a saucer-shaped hypanthodium

17 Study the given figure that depicts a stem modification and identify the missing links *A*, *B*, *C* and *D* by selecting the correct option.

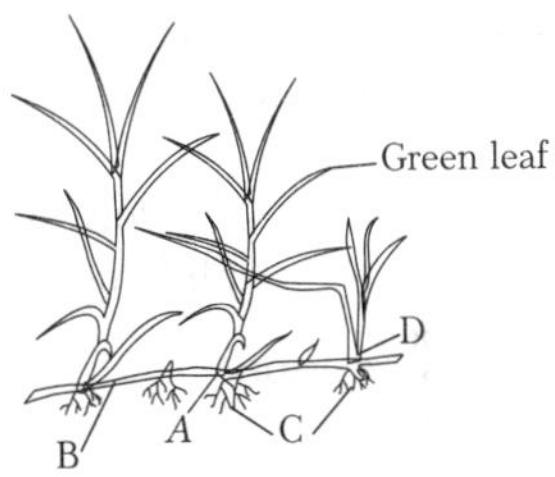

	A	*B*	*C*	*D*
(a)	Internode	Node	Tap root	Green leaf
(b)	Node	Internode	Fibrous adventitious roots	Scale leaf
(c)	Spines	Runner	Tap root	Green leaf
(d)	Node	Stolon	Adventitious roots	Internode

18 The main functions of stem are that
I. it bears and supports foliage, flowers and fruits.
II. it stores food in its aerenchyma cells.
III. it adds new cells, tissues and organs every year to keep continuing the functioning of the plant for long period of time.
IV. leaf-like, flattened, modified stems are called the thallus.

Choose the correct option.
(a) I and III (b) I and II
(c) II and IV (d) Only IV

19 Spadix is a type of inflorescence. Read the following statements regarding spadix.
I. Spadix is a spike with fleshy axis.
II. *Salix* and *Morus* possess spadix.
III. Spadix has both male and female flowers.
IV. Spadix possess bright coloured bracts.

Codes
(a) II and IV
(b) II and III
(c) Only I
(d) I, III and IV

20 Match the following columns.

Column I		Column II	
A.	Thorn	1.	Modified leaf
B.	Prickle	2.	Fleshy aquatic vegetative bud
C.	Spine	3.	Modified axillary bud
D.	Turion	4.	Superficial outgrowth

Codes

	A	B	C	D
(a)	1	4	3	2
(b)	2	3	1	4
(c)	4	3	1	2
(d)	3	4	1	2

21 Consider the following statements.
I. The veins provide rigidity to the leaf lamina.
II. Leaf bears a bud in its axil.
III. In *Polygonum*, leaf base may become swollen and is called pulvinus.
IV. Leaf takes its origin from root apical meristem.

Choose the correct set of statements.
(a) Only I
(b) II, III and IV
(c) I and II
(d) I, II, III and IV

22 Select the correct set of option with respect to position of floral parts on thalamus.

	Flowers	**Ovary**	**Examples**
(a)	Hypogynous	Half inferior	Guava
(b)	Epigynous	Superior	Cucumber
(c)	Perigynous	Inferior	Ray florets of sunflower
(d)	Hypogynous	Superior	China rose

23 Study the figures given below and choose the option that represents correct example of each phyllotaxy.

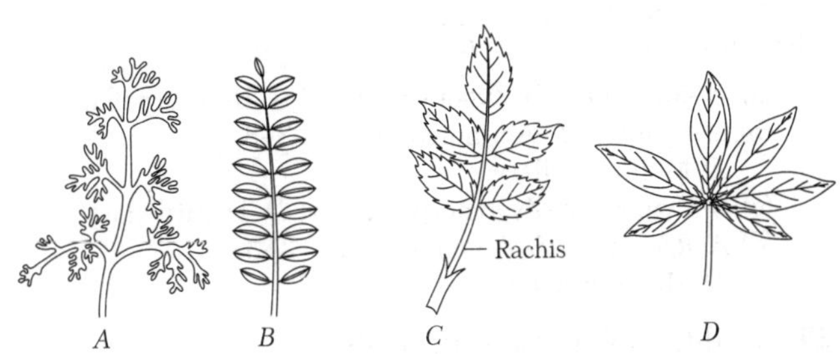

	A	B	C	D
(a)	*Daucus carota*	*Tamarindus*	*Rosa indica*	*Bombax*
(b)	*Acacia*	*Coriandrum sativum*	*Citrus*	*Moringa*
(c)	*Albizzia*	*Rosa indica*	*Murraya*	*Mimosa pudica*
(d)	*Prinsepia*	*Balanitis*	*Cassia*	Kikar

24 The following statements given below are in context of simple fruits. Read the statements.

I. Pericarp is fleshy in dry fruits.
II. Aggregate fruits formed from whole inflorescence.
III. Succulent fruits are either single-seeded or multiseeded.
IV. Simple fruits develop from apocarpous ovary.

Choose the correct option.

(a) I, II, III and IV (b) Only IV
(c) II and III (d) None of these

25 The following figure shows the LS of inflorescence of *Helianthus annuus* flower. Find out the correct sequence of labelling.

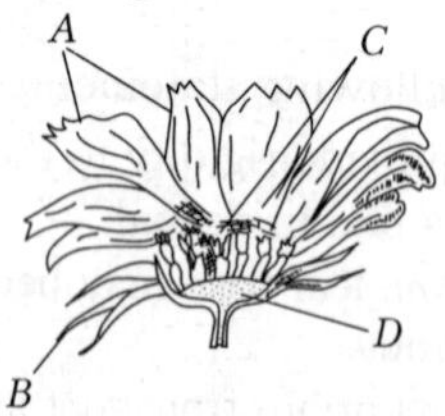

(a) *A* – Disc florets, *B* – Receptacle, *C* – Ray florets, *D* – Involucre
(b) *A* – Ray florets, *B* – Involucre, *C* – Disc florets, *D* – Receptacle
(c) *A* – Disc florets, *B* – Involucre, *C* – Receptacle, *D* – Ray florets
(d) *A* – Ray florets, *B* – Disc florets, *C* – Receptacle, *D* – Involucre

26 Identify the missing characters (*A*, *B*, *C* and *D*) and choose the correct option.

Family	Inflorescence	Flower	Stamens	Gynoecium
Liliaceae	Solitary, cymose or racemose	Actinomorphic	*C*	Tricarpellary
Solanaceae	Solitary, axillary or cymose	Actinomorphic	5	Bicarpellary
Fabaceae	*A*	*B*	10	*D*

(a) *A*-Racemose, *B*-Zygomorphic, *C*-5, *D*-Bicarpellary
(b) *A*-Cymose, *B*-Actinomorphic, *C*-3+3, *D*-Multicarpellary
(c) *A*-Cymose, *B*-Zygomorphic, *C*-5, *D*-Tricarpellary
(d) *A*-Racemose, *B*-Zygomorphic, *C*-3+3, *D*-Monocarpellary

27 Read the following statements.

I. Inflorescence is spikelet type.
II. Versatile anthers present.
III. Basal placentation.
IV. Feathery stigma.
V. Perianth represented by two lodicules.

The above statements are true for which family?

(a) Asteraceae (Compositae)
(b) Brassicaceae (Cruciferae)
(c) Fabaceae (Leguminosae)
(d) Poaceae (Graminae)

28 'A flower represents, the reproductive unit of a plant'.
Read the following statements regarding flowers and select the option with correct statement(s).

I. A zygomorphic flower is radially symmetrical.
II. China rose, cotton, *Cucurbita*, *Morus alba* are examples of unisexual flowers.
III. Epicalyx is a structure found in flowers lacking calyx.
IV. Poppy is a dimerous flower.

Codes

(a) Only III (b) Only IV
(c) Only I (d) I and II

29 Identify the correct option.

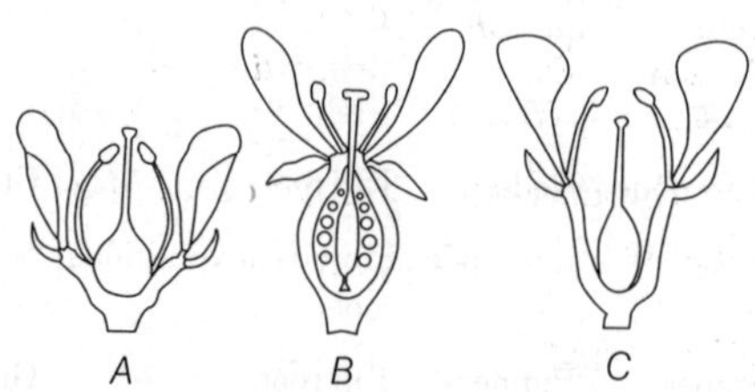

(a) *A*-This type of flower is hypogynous
B-This type of flower is seen in mustard
C-This type of flower is with inferior ovary
(b) *A*-This type of flower is seen in plum, rose and peach
B-This is epigynous flower with inferior ovary
C-In this type of flower, ovary is half-inferior
(c) *A*-This type of flower is seen in plum, rose and peach
B-This is epigynous flower with superior ovary
C-In this type of flower gynoecium occupies the highest position
(d) *A*-This type of flower is perigynous
B-This type of flower is epigynous
C-This type of flower possess inferior ovary

30 Read the following statements.

I. A sterile stamen is called staminode.
II. Epipetalous stamen is found in lily and epiphyllous stamen in brinjal.
III. China rose possess polyadelphous stamen.
IV. *Salvia* and mustard possess stamens filaments of variable length.
V. Pea possess monadelphous as well diadelphous stamen.

How many of the above statements are correct?

(a) Three (b) Two (c) Four (d) Five

31 Study the figures given below and choose the alternative that represent correct examples of each type.

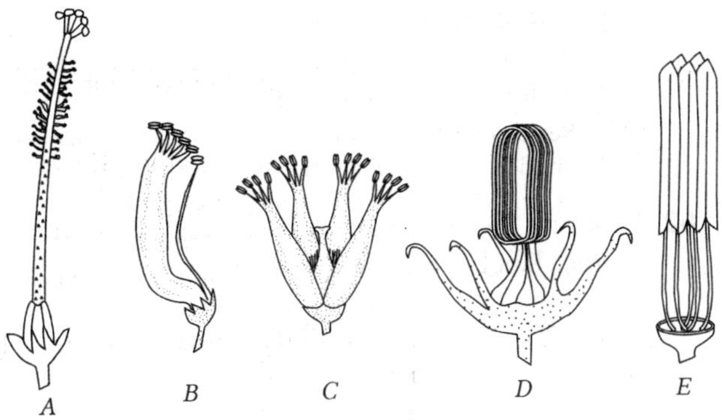

	A	B	C	D	E
(a)	Bean	*Luffa*	*Colocasia*	Maize	Lily flowers
(b)	China rose	*Pisum sativum*	*Ricinus*	*Cucurbita*	*Helianthus*
(c)	Lemon	*Ricinus*	*Pisum sativum*	*Lathyrus*	*Ocimum* Bean
(d)	Gram	Mustard	Pea	Gram	Bean

32 Find out the correct sequence of labelling of diagram given below.

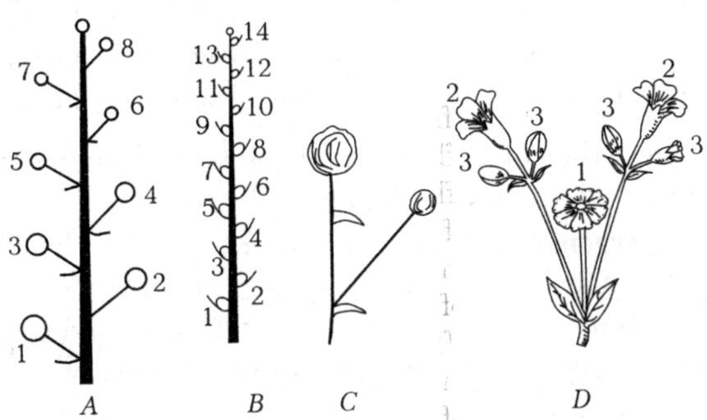

(a) *A*-Spike; *B*-Raceme; *C*-Dichasial cyme; *D*-Monochasial cyme

(b) *A*-Raceme; *B*-Spike; *C*-Monochasial cyme; *D*-Dichasial cyme

(c) *A*-Dichasial cyme; *B*-Monochasial cyme; *C*-Raceme; *D*-Spike

(d) *A*-Spike; *B*-Dichasial cyme; *C*-Monochasial cyme; *D*-Raceme

33 Identify the correct statement.

(a) Largest leaves are found in *Victoria regia*

(b) Free-lateral stipules are found in rice

(c) In *Pisum sativum*, the stipules are modified into tendril

(d) Rose leaf has ochreate type of stipule

34 Study the figures given below and identify the type of inflorescene shown in *A*, *B*, *C* and *D* by selecting correct option.

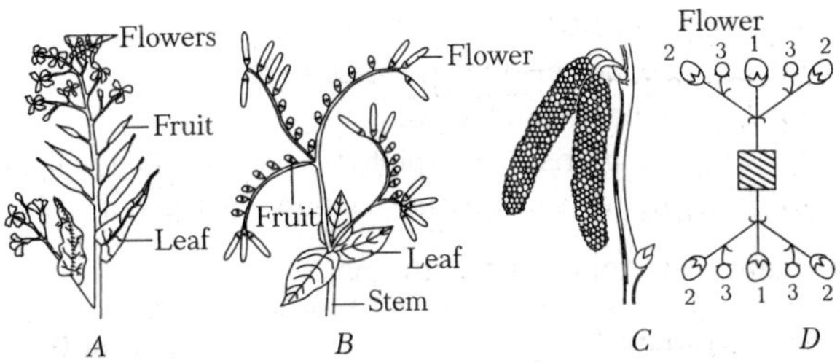

	A	B	C	D
(a)	Compound spike	Compound corymb	Spike	Hypanthodium
(b)	Catkin	Umbel	Spadix	Coenanthium
(c)	Raceme	Helicoid cyme	Catkin	Verticillaster
(d)	Corymb	Scorpioid cyme	Raceme	Capitulum

35 Identify the following types of fruits.

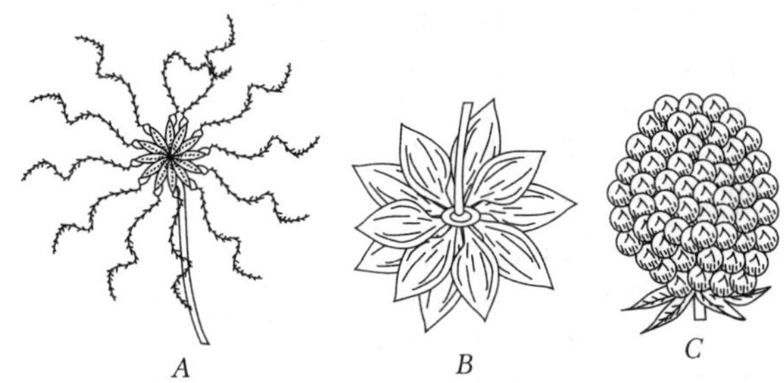

(a) *A*-Etaerio of follicle, *B*-Etaerio of achenes, *C*-Etaerio of berries

(b) *A*-Etaerio of berries, *B*-Etaerio of follicle, *C*-Etaerio of drupes

(c) *A*-Achene, *B*-Etaerio of berries, *C*-Etaerio of drupes

(d) *A*-Etaerio of follicle, *B*-Etaerio of berries, *C*-Etaerio of achenes

36 Choose the incorrect matched pair from given pairs.

(a) *Anthocephalus* – Interpetioler stipules

(b) *Lathyrus* – Foliaceous stipules

(c) *Polygonum* – Ochreate stipules

(d) *Acacia* – Tendrillar stipules

37 Consider the following statements and mark true (T) or false (F).

I. The fruits consist of pericarp and seeds.

II. Most of the fruits are eucarpic.

III. Fruits formed without fertilisation are called parthenocarpic.

IV. Parthenocarpic fruits are useless for the fruit industry.

Codes

	I	II	III	IV
(a)	T	F	F	T
(b)	T	T	T	F
(c)	T	T	F	F
(d)	F	F	T	T

38 Identify the incorrect statement.

(a) Lateral buds of stem give rise to flower
(b) Parwal has reproductive roots
(c) The spongy tissue of epiphytic root is called velamen
(d) Floating roots of *Cascuta* contain chlorophyll

39 Match the following columns.

	Column I		Column II
A.	Coleorhiza	1.	Grapes
B.	Food storing tissue	2.	Mango
C.	Parthenocarpic fruit	3.	Maize
D.	Single-seeded fruit developing from monocarpellary superior ovary	4.	Radicle
E.	Membranous seed coat	5.	Endosperm

Codes

	A	B	C	D	E
(a)	3	1	4	2	5
(b)	4	5	1	2	3
(c)	5	1	3	4	2
(d)	1	3	2	5	4

40 Read the following statements.

I. Small supporting roots growing obliquely from the basal node of main stem.
II. The root arises from horizontal aerial branches of the tree and provides support to the spreading branches of tree.
III. Negatively geotropic roots present in mangrove plants.

The plant root described above can be represented by

	I	II	III
(a)	Stilt root	Prop root	Assimilatory roots
(b)	Stilt root	Prop root	Pneumatophores
(c)	Prop root	Stilt root	Respiratory roots
(d)	Stilt root	Fibrous root	Respiratory roots

41 The given figure shows the regions of root tip with labelling *A*, *B* and *C*. Identify the correct labelling.

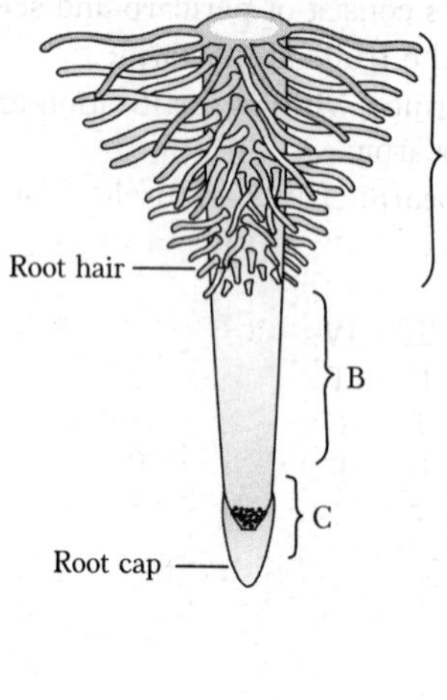

	A	*B*	*C*
(a)	Region of maturation	Region of elongation	Region of meristematic activity
(b)	Region of differentiation	Region of cell division	Region of meristematic activity
(c)	Region of elongation	Region of differentiation	Region of cell division
(d)	Region of cell division	Region of elongation	Region of maturation

42 Examine the figure given below and select the correct labels for *A*-*D*.

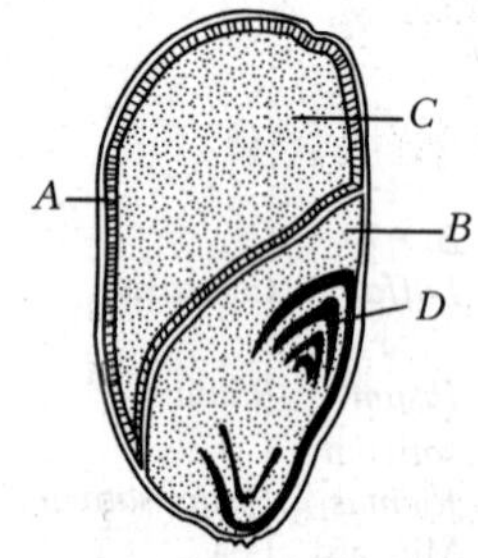

	A	*B*	*C*	*D*
(a)	Endosperm	Aleurone layer	Scutellum	Coleorhiza
(b)	Aleurone layer	Scutellum	Endosperm	Coleoptile
(c)	Scutellum	Seed coat	Plumule	Coleoptile
(d)	Embryo	Aleurone layer	Endosperm	Perisperm

43 Select the correct option for the floral formula of the plant which has the following features.

I. Zygomorphic, bisexual flowers.
II. Petals five with vexillary aestivation.
III. Ten stamens arranged in diadelphous condition.
IV. Gynoecium monocarpellary superior ovary.

(a) ⊕ ⚥ $K_{(5)}C_{1+2+(2)}A_{(9)+1}\underline{G_1}$

(b) % ⚥ $K_{(5)}C_{1+2+(2)}A_{(9)+1}\underline{G_1}$

(c) % ⚥ $K_{(5)}C_{1+2+(2)}A_{(5)+(5)}\underline{G_1}$

(d) ⊕ ⚥ $K_{(5)}C_{1+2+(2)}A_{(5)+(5)}\underline{G_1}$

44 Consider the placentation from the given diagrams and choose the correct set of placentation with example.

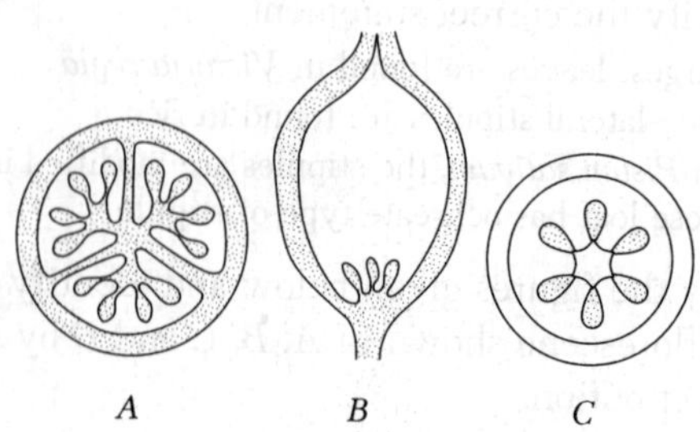

(a) A-Marginal, pea; B-Parietal, *Argemone;* C-Basal, wheat
(b) A-Axile, *Citrus* ; B-Basal, wheat; C-Free-central, *Silene*
(c) A-Basal, sunflower; B-Superficial, *Nymphaea;* C-Free central, *Stellaria,*
(d) A-Parietal, radish; B-Basal, maize; C-Superficial, *Nymphaea*

45 Glumes are
(a) calyx like, green leafy bracteoles present in whorls or whorls below the calyx occurring in most members
(b) small dry scaly bracts found in spikelet of the members of family-Poaceae
(c) green-coloured and in one or more whorls around and below the entire inflorescence
(d) brightly coloured appearing like the petals of a flower

Assertion and Reason

Direction (Q. Nos. 46-54) *In each of the following questions, a statement of Assertion (A) is given followed by corresponding statement of Reason (R). Of the statements, mark the correct answer as*

(a) Both A and R are true and R is the correct explanation of A
(b) Both A and R are true, but R is not the correct explanation of A
(c) A is true, but R is false
(d) A is false, but R is true
(e) Both A and R are false

46 **Assertion** (A) Caryopsis fruits differ from typical achenes with respect to the fusion of pericarp with the seed coat (testa).
Reason (R) Caryopsis fruits commonly occur in the members of family—Poaceae.

47 **Assertion** (A) In synconus fruit, the achenes formed are fewer then the total number of flower in the inflorescence.
Reason (R) Upper and middle flower cannot develop into fruits.

48 **Assertion** (A) The banyan trees are able to extend over large areas.
Reason (R) The trees have prop roots.

49 **Assertion** (A) Ture fruits develop from ovary.
Reason (R) In true fruits, seeds also develop from the ovary.

50 **Assertion** (A) It is possible to identify nodes on a stem.
Reason (R) Leaves arise from the nodes of a stem.

51 **Assertion** (A) Ginger has a prostrate growing rhizome.
Reason (R) The growth of ginger rhizome is not affected by gravity.

52 **Assertion** (A) The number of follicles in the etaerio is many in *Calotropis.*
Reason (R) Each fruitlet of an aggregate fruit is called a etaerio.

53 **Assertion** (A) Kidney-shaped fruit of cashewnut develops from multi-carpellary ovary.
Reason (R) The edible seed with two large cotyledons is enclosed by a hard pericarp in the fruit of cashewnut.

54 **Assertion** (A) Cremocarp fruits possess persistent stylopodium or stylopod.
Reason (R) Fruits of *Ricinus* and *Ocimum* are the examples of cremocarp.

(C) Medical Entrances' Gallery

(Collection of Questions Asked in NEET & in Various Medical Entrance Exams)

1 Correct position of floral parts over thalamus in mustard plant is **[NEET 2020]**
(a) gynoecium occupies the highest position, while the other parts are situated below it
(b) margin of the thalamus grows upward enclosing the ovary completely and other parts arise below the ovary
(c) gynoecium is present in the centre and other parts cover it partially
(d) gynoecium is situated in the centre and other parts of the flower are located at the rim of the thalamus, at the same level

2 Which of the following is the correct floral formula of Liliaceae? **[NEET 2020]**

(a) % ⚥ $C_{1+2+(2)}A_{(g)+1}G_1$
(b) ⊕ ⚥ $K_{(5)}\overset{\frown}{C_{(5)}\ A_1}\underline{G}_{(1)}$
(c) Br⊕ ⚥ $\overset{\frown}{P_{(3+3)}A}_{3+3}\underline{G}_{(3)}$
(d) ⊕ ⚥ $K_{(5)}\overset{\frown}{C_{(5)}\ A_5}\underline{G}_{(2)}$

3 Vegetative propagule in *Agave* is termed as **[NEET 2020]**

(a) rhizome
(b) bulbil
(c) offset
(d) eye

4 Identify the correct features of mango and coconut fruits.

I. Both fruits are drupe.
II. Endocarp is edible in both.
III. Mesocarp in coconut is fibrous and in mango, it is fleshy.
IV. In both, fruit develops from monocarpellary ovary.

Select the correct option from the codes given below. **[NEET 2020]**

(a) I, III and IV (b) I, II and III
(c) I and IV (d) I and II

5 Ray florets have **[NEET 2020]**

(a) superior ovary (b) hypogynous ovary
(c) half inferior ovary (d) inferior ovary

6 The roots that originate from the base of the stem are **[NEET 2020]**

(a) primary roots
(b) prop roots
(c) lateral roots
(d) fibrous roots

7 Match the following columns. **[AIIMS 2019]**

Column I	Column II
A Tap root	1. Sweet potato
B Adventitious root	2. Turnip
C Stem	3. Wheat
D Fibrous root	4. Potato

Codes

	A	B	C	D
(a)	1	2	3	4
(b)	2	3	1	4
(c)	2	1	4	3
(d)	2	3	4	1

8 Which of the following shows whorled phyllotaxy? **[NEET 2019]**

(a) Mustard (b) China rose
(c) *Alstonia* (d) *Calotropis*

9 Verticillaster inflorescence is found in **[JIPMER 2019]**

(a) Mustard (b) *Ocimum*
(c) China rose (d) Sunflower

10 Which of the following represents the zygomorphic symmetry? **[AIIMS 2019]**

(a) *Canna*, mustard, chilli, *Datura*
(b) Mustard, *Canna*, pea, *Datura*
(c) Pea, bean, *Cassia*, gulmohar
(d) Pea, bean, *Canna*, chilli

11 Which of the following is correct? **[AIIMS 2019]**

(a) Perigynous – Plum, peach, rose
(b) Epigynous – Guava and cucumber
(c) Perigynous – Mustard, rose
(d) Both (a) and (b)

12 Bicarpellary ovary with obliquely placed septum is seen in **[NEET 2019]**

(a) *Brassica* (b) *Aloe*
(c) *Solanum* (d) *Sesbania*

13 Match the placental types (Column I) with their examples (Column II). **[NEET 2019]**

Column I	Column II
A Basal	1. Mustard
B Axile	2. China rose
C Parietal	3. *Dianthus*
D Free central	4. Sunflower

Codes

	A	B	C	D
(a)	2	3	4	1
(b)	1	2	3	4
(c)	4	2	1	3
(d)	3	4	1	2

14 Placentation in which ovules develop on the inner wall of the ovary or in peripheral part, is **[NEET 2019]**

(a) axile (b) parietal
(c) free central (d) basal

15 Which of the following is incorrect? **[AIIMS 2019]**

(a) Mango and coconut are drupe fruit
(b) According to Euro Norm IV, sulphur content for petrol is 50 ppm
(c) CO_2 and other poisonous gases cause pollution
(d) Apple is a true fruit

16 King of spices is **[AIIMS 2019]**

(a) *Brassica nigra* (b) *Piper nigrum*
(c) *Piper longum* (d) *Curcuma longa*

17 Persistent nucellus in the seed is known as **[NEET 2019]**

(a) perisperm (b) hilum
(c) tegmen (d) chalaza

18 Pneumatophores occur in **[NEET 2018]**

(a) carnivorous plants
(b) free-floating hydrophytes
(c) halophytes
(d) submerged hydrophytes

19 Sweet potato is a modified **[NEET 2018]**

(a) tap root (b) adventitious root
(c) stem (d) rhizome

20 Offsets are produced by **[NEET 2018]**

(a) parthenocarpy (b) mitotic divisions
(c) meiotic divisions (d) parthenogenesis

21 Leaf tendrils are found in **[AIIMS 2018]**

(a) grapevine (b) peas
(c) cucumber (d) All of these

22 Arrange in correct order according to the given figures. **[AIIMS 2018, Kerala CEE 2011]**

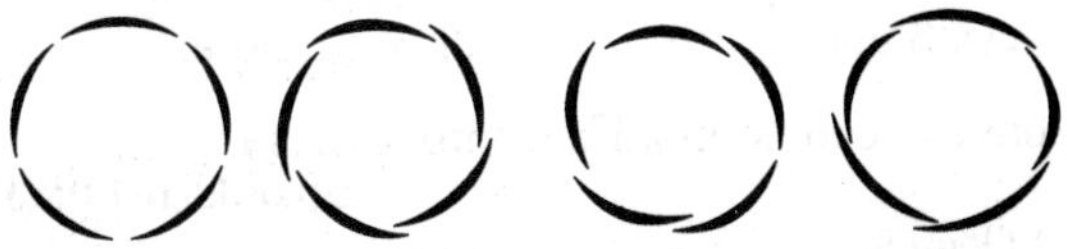

A *B* *C* *D*

(a) *A*-Imbricate, *B*-Quincuncial, *C*-Valvate, *D*-Twisted
(b) *A*-Vexillary, *B*-Valvate, *C*-Twisted, *D*-Imbricate
(c) *A*-Quincuncial, *B*-Twisted, *C*-Vexillary, *D*-Imbricate
(d) *A*-Valvate, *B*-Twisted, *C*-Imbricate, *D*-Quincuncial

23 Caryopsis is present in **[JIPMER 2018]**
(a) wheat (b) groundnut (c) coconut (d) mango

24 In *Bougainvillea*, thorns are the modifications of **[NEET 2017]**
(a) stipules (b) adventitious root
(c) stem (d) leaf

25 Study the following statements.
I. Pulvinus leaf base is present in some leguminous plants.
II. In *Eichhornia*, the petioles expand, become green and synthesise food.
III. Opposite phyllotaxy is seen in guava.
Select the option with correct statements. **[JIPMER 2017]**
(a) I and II (b) I and III
(c) II and III (d) I, II and III

26 Which one of the following options is not correctly matched? **[AIIMS 2017]**
(a) Cymose — *Acacia*
(b) Hypanthodium — Banyan
(c) Cyathium — *Euphorbia*
(d) Verticillaster — *Calotropis*

27 Coconut fruit is a **[NEET 2017]**
(a) drupe (b) berry (c) nut (d) capsule

28 The morphological nature of the edible part of coconut is **[NEET 2017]**
(a) perisperm (b) cotyledon
(c) endosperm (d) pericarp

29 Cotyledons and testa are edible parts of **[JIPMER 2017]**
(a) groundnut and pomegranate
(b) walnut and tamarind
(c) French bean and coconut
(d) cashewnut and litchi

30 How many plants among *Indigofera, Sesbania, Salvia, Allium, Aloe*, mustard, groundnut, radish, gram and turnip have stamens of different lengths in their flowers? **[NEET-II 2016]**
(a) Three (b) Four (c) Five (d) Six

31 Radial symmetry is found in the flowers of **[NEET-II 2016]**
(a) *Brassica* (b) *Trifolium*
(c) *Pisum* (d) *Cassia*

32 Free-central placentation is found in **[NEET-II 2016]**
(a) *Dianthus* (b) *Argemone*
(c) *Brassica* (d) *Citrus*

33 The term 'Polyadelphous is related to **[NEET-II 2016]**
(a) gynoecium (b) androecium
(c) corolla (d) calyx

34 Tricarpellary, syncarpous gynoecium is found in flowers of **[NEET-I 2016]**
(a) Solanaceae (b) Fabaceae
(c) Poaceae (d) Liliaceae

35 The coconut water from tender coconut represents **[NEET-I 2016]**
(a) fleshy mesocarp (b) free nuclear proembryo
(c) free nuclear endosperm (d) endocarp

36 Which of the following is not a stem modification? **[NEET-I 2016]**
(a) Thorns of *Citrus*
(b) Tendrils of cucumber
(c) Flattened structures of *Opuntia*
(d) Pitcher of *Nepenthes*

37 Stems modified into flat green organs performing the functions of leaves are known as **[NEET-I 2016]**
(a) phyllodes (b) phylloclades
(c) scales (d) cladodes

38 The standard petal of a papilionaceous corolla is also called **[NEET-I 2016]**
(a) pappus (b) vexillum
(c) corona (d) carina

39 Identify the plant with multilocular ovary. **[EAMCET 2015]**
(a) Cucumber (b) Cashew
(c) China rose (d) Chilli

40 Identify the correct combination. **[EAMCET 2015]**
(a) Carrot - Tuberous root - Head inflorescence
(b) Onion - Bulb - Corymb
(c) *Colocasia* - Rhizome - Spadix
(d) *Bougainvillea* - Thorns - Cymule

41 The chief edible part in (*A*) groundnut, (*B*) jackfruit, (*C*) apple and (*D*) mango fruits are **[EAMCET 2015]**

	A	*B*	*C*	*D*
(a)	Endosperm	Inflorescence	Mesocarp	Endocarp
(b)	Cotyledons	Succulent perianth	Fleshy thalamus	Mesocarp
(c)	Cotyledons	Endosperm	Mesocarp	Fleshy thalamus
(d)	Cotyledons	Fleshy thalamus	Pericarp	Mesocarp

42 Identify the correct combination. **[EAMCET 2015]**

(a) Neem–The absence of buds in the axils of leaflets–Pinnately compound leaf
(b) Sunflower–Opposite phyllotaxy–Inferior ovary
(c) Carrot–Flowers brought to same height due to varied lengths of pedicels–Involucre of bracts
(d) *Pistia*–Discoid stem–Lateral branch with many internodes

43 **Assertion** (A) Subaerial modifications participate in vegetative reproduction.

Reaction (R) They store food and water. **[EAMCET 2015]**

(a) Both A and R are true and R is the correct explanation of A
(b) Both A and R are true, but R is not the correct explanation of A
(c) A is true, but R is false
(d) A is false, but R is true

44 Identify the correct combination. **[EAMCET 2015]**

(a) *Datura*–Solanaceae–Epipetalous stamens
(b) *Allium*–Solanaceae–Tricarpellary gynoecium
(c) *Petunia*–Solanaceae–Didynamous stamens
(d) *Crotalaria*–Fabaceae–Diadelphous stamens

45 An example of axile placentation is **[CBSE AIPMT 2015, 09]**

(a) marigold (b) *Dianthus* (c) lemon (d) argemone

46 Monadelphous condition of stamens is found in **[CBSE AIPMT 2015]**

(a) pea (b) China rose
(c) *Citrus* (d) None of these

47 Perigynous flowers are found in **[CBSE AIPMT 2015]**

(a) guava (b) cucumber
(c) China rose (d) rose

48 Keel is the characteristic feature of flower of **[CBSE AIPMT 2015]**

(a) tulip (b) *Indigofera* (c) *Aloe* (d) tomato

49 Among China rose, mustard, brinjal, potato, guava, cucumber, onion and tulip, how many plants have superior ovary? **[CBSE AIPMT 2015]**

(a) Five (b) Six (c) Three (d) Four

50 Flowers are unisexual in **[CBSE AIPMT 2015]**

(a) pea (b) cucumber
(c) China rose (d) onion

51 Roots play insignificant role in absorption of water in **[CBSE AIPMT 2015]**

(a) sunflower (b) *Pistia*
(c) pea (d) wheat

52 Edible part in the fruit of litchi is **[Uttarakhand PMT 2015]**

(a) mesocarp (b) aril
(c) fleshy thalamus (d) cotyledons

53 The inflorescence in coconut plant is **[Uttarakhand PMT 2015]**

(a) compound spadix (b) hypanthodium
(c) cyathium (d) verticillaster

54 Leaves become modified into spines in **[Uttarakhand PMT 2015]**

(a) *Opuntia* (b) pea
(c) onion (d) silk cotton

55 Match the following columns.

Column I	Column II
A. Pneumatophores	1. Axillary buds
B. Tendrils in pea	2. Roots
C. Thorns in *Citrus*	3. Leaves

Codes **[Kerala CEE 2015]**

	A	B	C		A	B	C
(a)	2	1	3	(b)	3	2	1
(c)	3	1	2	(d)	2	3	1
(e)	1	2	3				

56 In one plant, underground stems are modified to store food and in another plant, the stem tendrils develop from axillary buds to help plant climb. They are **[Kerala CEE 2015]**

(a) ginger, cucumber (b) carrot, jasmine
(c) sweet potato, *Bougainvillea* (d) *Opuntia*, *Eichhornia*
(e) sweet potato, mint

57 Consider the following statements.

I. In leguminous plants, leaf base becomes swollen, called pulvinus.
II. The fleshy leaves of onion and garlic store food.
III. The buds in Australian *Acacia* become green and synthesise food.
IV. In *Alstonia*, leaves show alternate phyllotaxy.

Choose the option containing correct statements. **[Kerala CEE 2015]**

(a) II and IV (b) I and III
(c) I and II (d) I and IV
(e) II and III

58 Leaf tendrils are found in **[AMU 2015]**

(a) peas (b) cucumber
(c) grapevine (d) All of these

59 Floral formula of family–Fabaceae is **[AMU 2015]**

(a) $\oplus\ K_{(5)}\ C_{1+2+(2)}\ A_{(9)+1}\ \underline{G}_1$
(b) $\%\ ⚥\ K_{(5)}\ C_{1+2+(2)}\ A_{(9)+1}\ \underline{G}_1$
(c) $\%\ ⚥\ K_{(5)}\ C_{1+(2)+2}\ A_{(9)+1}\ \underline{G}_1$
(d) $\%\ ⚥\ K_5\ C_{1+2+(2)}\ A_{(9)+1}\ \underline{G}_1$

60 Multicostate divergent reticulate venation is seen in ...... leaf. **[MHT CET 2014]**

(a) *Ziziphus* (b) bamboo
(c) castor (d) mango

61 The pattern of arrangement of leaves on the stem is known as **[MHT CET 2014]**
(a) heterophylly (b) phyllode
(c) phyllotaxy (d) phylloclade

62 The type of placentation seen in argemone and primrose are respectively, **[Kerala CEE 2014]**
(a) axile and free-central
(b) parietal and free-central
(c) parietal and basal
(d) marginal and free-central
(e) basal and parietal

63 In China rose, the flowers are **[NEET 2013]**
(a) actinomorphic, hypogynous with twisted aestivation
(b) actinomorphic, epigynous with valvate aestivation
(c) zygomorphic, hypogynous with valvate aestivation
(d) zygomorphic, epigynous with twisted aestivation

64 Among bitter gourd, mustard, brinjal, pumpkin, China rose, lupin, cucumber, sunhemp, gram, guava, bean, chilli, plum, *Petunia*, tomato, rose, *Withania*, potato, onion, *Aloe* and tulip, how many plants have hypogynous flower? **[NEET 2013]**
(a) Six (b) Ten
(c) Fifteen (d) Eighteen

65 Perianth is represented by **[UP CPMT 2013]**
(a) glums (b) lemma
(c) lodicules (d) palea

66 Ginger multiplies vegetatively by **[UP CPMT 2013]**
(a) bud (b) tuber
(c) stem (d) rhizome

67 Cyathium inflorescence is found in **[Odisha JEE 2013]**
(a) *Morus* (b) *Dorstenia*
(c) *Ficus* (d) *Euphorbia*

68 Ruminate endosperm is found in the seeds of family **[UP CPMT 2013]**
(a) Compositae
(b) Cruciferae
(c) Euphorbiaceae
(d) Annonaceae

69 Which of the following is a wheat fruit? **[Manipal 2013]**
(a) Achene (b) Cypsela
(c) Caryopsis (d) Endosperm

70 Parts of two plants were observed. Structure-*A* develops aerially and produces roots when comes in contact with the soil structure-*B* develops from underground part of the stem, grows obliquely becomes aerial and produces roots on its lower surface. Identify, respectively the structure of *A* and *B*. **[Manipal 2013]**
(a) Sucker, stolon (b) Stolon, runner
(c) Stolon, sucker (d) Runner, stolon

71 Match the types of the fruits listed in Column I with the examples listed in Column II. Choose the answer which gives the correct combination of the two columns.

Column I	Column II
A. Capsule	1. Paddy
B. Berry	2. Mango
C. Drupe	3. Sunflower
D. Cypsela	4. Tomato
	5. Lady's finger

Codes **[Karnataka CET 2013]**

	A	B	C	D
(a)	5	4	2	3
(b)	5	3	1	2
(c)	4	5	2	3
(d)	1	2	3	5

72 Inflorescence of Liliaceae is **[AIIMS 2013]**
(a) actinomorphic (b) trimerous
(c) pentamerous (d) imperfect

73 A fruit developed from a condensed inflorescence is **[AIIMS 2013]**
(a) simple fruit
(b) aggregate fruit
(c) composite fruit
(d) etaerio

74 Cereals are mostly belong to family **[Odisha JEE 2013]**
(a) Cruciferae (b) Brassicaceae
(c) Poaceae (d) Asteraceae

75 In *Nepenthes* (pitcher plant), pitcher is the modification of **[Odisha JEE 2013]**
(a) petiole (b) base
(c) lamina (d) All of these

76 Cymose inflorescence is present in **[CBSE AIPMT 2012]**
(a) *Solanum* (b) *Sesbania*
(c) *Trifolium* (d) *Brassica*

77 The gynoecium consists of many free pistils in flowers of **[CBSE AIPMT 2012]**
(a) *Aloe* (b) Tomato (c) *Papaver* (d) *Michelia*

78 Phyllode is present in **[CBSE AIPMT 2012]**
(a) *Asparagus* (b) *Euphorbia*
(c) Australian *Acacia* (d) *Opuntia*

79 Placentation in tomato and lemon is **[CBSE AIPMT 2012]**
(a) parietal (b) free-central
(c) marginal (d) axile

80 Which one of the following is correctly matched? **[CBSE AIPMT 2012]**
(a) Onion – Bulb
(b) Ginger – Sucker
(c) *Chlamydomonas* – Conidia
(d) Yeast – Zoospores

81 How many plants in the list given below have composite fruits that develop from an inflorescence? Walnut, poppy, radish, fig, pineapple, apple, tomato, mulberry **[CBSE AIPMT 2012]**
(a) Four (b) Five
(c) Two (d) Three

82 The coconut water and the edible part of coconut are equivalent to **[CBSE AIPMT 2012]**
(a) endosperm (b) endocarp
(c) mesocarp (d) embryo

83 Vexillary aestivation is characteristic of the family **[CBSE AIPMT 2012]**
(a) Fabaceae (b) Asteraceae
(c) Solanaceae (d) Brassicaceae

84 Which one of the following organisms is correctly matched with its three characteristics? **[CBSE AIPMT 2012]**
(a) Pea—C_3-pathway, endospermic seed, vexillary aestivation
(b) Tomato—Twisted aestivation, axile placentation, berry
(c) Onion—Bulb, imbricate aestivation, axile placentation
(d) Maize—C_3-pathway, closed vascular bundles, scutellum

85 How many plants in the list given below have marginal placentation?
Mustard, gram, tulip, *Asparagus*, arhar, sunhemp, chilli, colchicine, onion, moong, pea, tobacco, lupin **[CBSE AIPMT 2012]**
(a) Four (b) Five (c) Six (d) Three

86 Which type of placentation is seen in sweet pea? **[UP CPMT 2012; AFMC 2012]**
(a) Basal (b) Axile
(c) Free-central (d) Marginal

87 The mesocarp in coconut is **[Manipal 2012]**
(a) fleshy (b) stony (c) fibrous (d) milky

88 Which of the following is false fruit? **[Manipal 2012]**
(a) Mango (b) Apple
(c) Banana (d) Jackfruit

89 In which of the following fruits, the edible part is the aril? **[Manipal 2012]**
(a) Mango (b) Pomegranate
(c) Orange (d) Litchi

90 In which of the following plants, leaf apex changes into tendril? **[Manipal 2012]**
(a) *Gloriosa* (b) *Smilax*
(c) *Pisum sativum* (d) Australian *Acacia*

91 Usually the whorl in a flower that attracts insects and protects the essential part is **[Karnataka CET 2012]**
(a) corolla (b) calyx
(c) androecium (d) gynoecium

92 Given below is the floral diagram of a flower. Which of the following descriptions of the flower matches the floral diagram?

[Karnataka CET 2012]
(a) Homochlamydeous, gamopetalous, tetramerous and unisexual
(b) Heterochlamydeous, gamopetalous, pentamerous and bisexual
(c) Heterochlamydeous, gamopetalous, tetramerous and bisexual
(d) Homochlamydeous, polypetalous, pentamerous and bisexual

93 Identify the incorrect statements from the following.
I. Cymose inflorescence found in *Hibiscus* sp.
II. Hypanthodium is found in *Ficus benghalensis.*
III. Synandrous stamen is found in *Calotropis.*
IV. Hesperidium type of fruit is mango. **[WB JEE 2012]**
Codes
(a) III and IV (b) I and II
(c) II and III (d) I and IV

94 A drupe develops in **[AFMC 2012]**
(a) wheat (b) pea
(c) tomato (d) mango

95 The plant having monadelphous stamens and axile placentation is **[AFMC 2012]**
(a) lemon (b) pea (c) tomato (d) China rose

96 The cotyledons of monocots (grasses) is generally called **[BHU 2012]**
(a) scutellum (b) radicle
(c) plumule (d) endosperm

97 The inflorescence in *Ocimum* is **[JCECE 2012]**
(a) cyathium (b) verticillaster
(c) hypanthodium (d) raceme

98 Drupes are called stony fruits because they have hard **[JCECE 2012]**
(a) epicarp and mesocarp (b) mesocarp
(c) epicarp (d) endocarp

99 A student observed 39 inflorescence in *Bougainvillea* and 42 inflorescences in *Poinsettia.* Find out the number of flowers in *Bougainvillea* and number of female flowers in *Poinsettia*, respectively, **[BHU 2012]**
(a) 34 and 126
(b) 68 and indefinite
(c) 204 and 164
(d) 117 and 42

100 The inflorescence of family-Poaceae is [AFMC 2012]
(a) spikelet (b) spike
(c) raceme (d) capitulum

101 The most advanced family is [BHU 2012]
(a) Cruciferae
(b) Cucurbitaceae
(c) Compositae
(d) Euphorbiaceae

102 Keel is a characteristic of the flowers of [BCECE 2012]
(a) gulmohur (b) *Cassia* (c) *Calotropis* (d) bean

103 Simple, cluster of radial leaves, stipulate and parallel venation leaves and cyme or umbel inflorescence are the characteristics of [BCECE 2012]
(a) Poaceae (b) Liliaceae
(c) Asteraceae (d) Fabaceae

104 Which one of the following statements is correct? [CBSE AIPMT 2011]
(a) Seeds of orchids have oil-rich endosperm
(b) Placentation in primrose is basal
(c) Flower of tulip is a modified shoot
(d) In tomato, fruit is a capsule

105 Flowers are zygomorphic in [CBSE AIPMT 2011]
(a) goldmohur (b) tomato
(c) *Datura* (d) mustard

106 The ovary is half-inferior in flowers of [CBSE AIPMT 2011]
(a) cucumber (b) cotton
(c) guava (d) peach

107 Pome fruit is found in [UP CPMT 2011]
(a) mango (b) apple
(c) litchi (d) peach

108 Roots are absent in [UP CPMT 2011]
(a) *Wolffia* (b) *Podostemon*
(c) *Pistia* (d) *Lemna*

109 Match the following columns.

Column I (Placentation types)		Column II (Represented in)	
A.	Basal	1.	*Dianthus*
B.	Free-central	2.	Pea
C.	Parietal	3.	Lemon
D.	Axile	4.	Marigold
E.	Marginal	5.	*Argemone*

Codes [Kerala CEE 2011]

	A	B	C	D	E
(a)	1	2	3	4	5
(b)	2	3	4	5	1
(c)	4	1	5	3	2
(d)	4	3	5	1	2
(e)	5	4	3	2	1

110 Which of the following plants has the floral characters like zygomorphic flower, vexillary aestivation, diadelphous androecium and marginal placentation? [Kerala CEE 2011]
(a) *Pisum* (b) Belladonna
(c) Brinjal (d) *Asparagus*
(e) *Aloe*

111 Study the following statements. [Kerala CEE 2011]
I. Buds are present in the axil of leaflets of the compound leaf.
II. Pulvinus leaf-base is present in some leguminous plants.
III. In *Alstonia*, the petioles expand, become green and synthesise food.
IV. Opposite phyllotaxy is seen in guava.

Choose the correct option from the codes given below.
(a) II and IV are correct, but I and III are incorrect
(b) I and III are correct, but II and IV are incorrect
(c) I and IV are correct, but II and III are incorrect
(d) II, III and IV are correct, but I is incorrect
(e) I and II are correct, but III and IV are incorrect

112 Which of the following plants have long slender and coiled stem tendrils developed from axillary buds? [Kerala CEE 2011]
(a) Grapevine and pumpkins
(b) Australian *Acacia* and watermelon
(c) *Bougainvillea* and cucumber
(d) Strawberry and grapevine
(e) *Alstonia* and pumpkins

113 Match the following columns.

Column I (Stem modification)	Column II (Found in)	
A. Underground stem	1.	*Euphorbia*
B. Stem tendril	2.	*Opuntia*
C. Stem thorns	3.	Potato
D. Flattened stem	4.	*Citrus*
E. Fleshy cylindrical stem	5.	Cucumber

Codes [Kerala CEE 2011]

	A	B	C	D	E
(a)	1	2	3	5	4
(b)	2	3	4	5	1
(c)	3	4	5	1	2
(d)	3	5	4	2	1
(e)	5	3	4	1	2

114 In cauliflower, the inflorescence is [Odisha JEE 2011]
(a) corymbose (b) cymose
(c) raceme (d) capitulum

115 Aggregate fruit formed from [Odisha JEE 2011]
(a) multicarpellary apocarpous ovary
(b) multicarpellary syncarpous ovary
(c) monocarpellary apocarpous ovary
(d) monocarpellary syncarpous ovary

116 Thalamus of hypogynous ovary is **[Odisha JEE 2011]**
(a) concave (b) convex
(c) biconcave (d) biconvex

117 Tetradynamous condition is found in **[Odisha JEE 2011]**
(a) Asteraceae
(b) Malvaceae
(c) Papilionacae
(d) Brassicaceae

118 Some features of plant leaves are **[Odisha JEE 2011]**
(a) hair on the lower surface and waxy cuticle
(b) hair on the upper surface and no cuticle
(c) epidermis without stomata
(d) the presence of endodermis and Casparian strips

119 Multicostate parallel venation of leaf is found in **[Odisha JEE 2011]**
(a) Grass, palm (b) *Dalbergia*
(c) *Argemone* (d) *Mangifera*

120 120° phyllotaxy is found in **[Odisha JEE 2011]**
(a) distichous condition
(b) tristichous condition
(c) monostichous condition
(d) None of the above

121 Subaerial stem modification with long internode is **[Odisha JEE 2011]**
(a) tuber (b) phyllode
(c) phylloclade (d) runner

122 Match the types of the fruits in Column I with the examples listed in Column II. Choose the answer which gives the correct combination of the two columns.

Column I	Column II
A. Sorosis	1. Orange
B. Pome	2. Cucumber
C. Pepo	3. Grape
D. Berry	4. Pear
	5. Jackfruit

Codes **[Karnataka CET 2011]**

	A	B	C	D
(a)	5	4	2	3
(b)	5	3	1	2
(c)	4	5	2	3
(d)	1	2	3	5

123 A flower which can be divided into two equal halves by only one plane is **[WB JEE 2011]**
(a) zygomorphic (c) actinomorphic
(e) regular (d) perfect

124 Caryopsis is found in **[J&K CET 2011]**
(a) sunflower (b) maize
(c) pea (d) *Datura*

125 In unilocular ovary with a single ovule, the placentation is **[CBSE AIPMT 2010]**
(a) marginal (b) basal
(c) free-central (d) axile

ANSWERS

Check Point 11.1

1. (b)	2. (a)	3. (d)	4. (b)	5. (a)	6. (b)

Check Point 11.2

1. (b)	2. (c)	3. (d)	4. (d)	5. (a)	6. (a)

Check Point 11.3

1. (b)	2. (a)	3. (b)	4. (a)	5. (a)	6. (d)

Check Point 11.4

1. (a)	2. (a)	3. (b)	4. (a)	5. (a)	6. (d)

Check Point 11.5

1. (c)	2. (a)	3. (c)	4. (a)	5. (b)	6. (c)	7. (d)

Check Point 11.6

1. (d)	2. (b)	3. (a)	4. (c)	5. (c)

Check Point 11.7

1. (a)	2. (d)	3. (a)	4. (d)	5. (d)	6. (a)

(A) Topical Exercise

1. (d)	2. (c)	3. (b)	4. (a)	5. (a)	6. (c)	7. (b)	8. (a)	9. (b)	10. (b)
11. (a)	12. (c)	13. (c)	14. (c)	15. (d)	16. (d)	17. (d)	18. (a)	19. (d)	20. (d)
21. (b)	22. (b)	23. (c)	24. (a)	25. (a)	26. (c)	27. (d)	28. (c)	29. (d)	30. (b)
31. (a)	32. (b)	33. (a)	34. (a)	35. (b)	36. (d)	37. (b)	38. (a)	39. (a)	40. (b)
41. (b)	42. (d)	43. (c)	44. (c)	45. (a)	46. (c)	47. (c)	48. (d)	49. (b)	50. (a)
51. (b)	52. (a)	53. (d)	54. (a)	55. (c)	56. (c)	57. (d)	58. (a)	59. (a)	60. (a)
61. (b)	62. (b)	63. (d)	64. (a)	65. (d)	66. (c)	67. (d)	68. (c)	69. (d)	70. (b)
71. (a)	72. (b)	73. (b)	74. (c)	75. (a)	76. (b)	77. (d)	78. (b)	79. (c)	80. (c)
81. (a)	82. (d)	83. (b)	84. (b)	85. (a)	86. (b)	87. (b)	88. (a)	89. (a)	90. (d)
91. (c)	92. (d)	93. (d)	94. (a)	95. (b)	96. (b)	97. (d)	98. (a)	99. (c)	100. (b)
101. (a)	102. (b)	103. (b)	104. (d)	105. (a)	106. (a)	107. (d)	108. (c)	109. (b)	110. (d)
111. (b)	112. (b)	113. (a)	114. (d)	115. (a)	116. (d)	117. (c)			

(B) Cumulative Exercise

1. (c)	2. (a)	3. (a)	4. (a)	5. (c)	6. (a)	7. (d)	8. (c)	9. (b)	10. (b)
11. (a)	12. (a)	13. (d)	14. (d)	15. (b)	16. (d)	17. (b)	18. (a)	19. (d)	20. (d)
21. (c)	22. (d)	23. (a)	24. (c)	25. (b)	26. (d)	27. (d)	28. (b)	29. (b)	30. (b)
31. (b)	32. (b)	33. (a)	34. (c)	35. (c)	36. (d)	37. (b)	38. (d)	39. (b)	40. (b)
41. (a)	42. (b)	43. (b)	44. (b)	45. (b)	46. (b)	47. (a)	48. (a)	49. (c)	50. (a)
51. (a)	52. (a)	53. (b)	54. (d)						

(C) Medical Entrances' Gallery

1. (a)	2. (c)	3. (b)	4. (a)	5. (d)	6. (d)	7. (c)	8. (c)	9. (b)	10. (c)
11. (d)	12. (c)	13. (c)	14. (b)	15. (d)	16. (b)	17. (a)	18. (c)	19. (b)	20. (b)
21. (b)	22. (d)	23. (a)	24. (c)	25. (b)	26. (d)	27. (a)	28. (c)	29. (a)	30. (b)
31. (a)	32. (a)	33. (b)	34. (d)	35. (c)	36. (d)	37. (b)	38. (b)	39. (c)	40. (b)
41. (b)	42. (a)	43. (c)	44. (a)	45. (c)	46. (b)	47. (d)	48. (b)	49. (b)	50. (b)
51. (b)	52. (b)	53. (a)	54. (a)	55. (d)	56. (a)	57. (c)	58. (d)	59. (b)	60. (c)
61. (c)	62. (a)	63. (a)	64. (c)	65. (c)	66. (d)	67. (d)	68. (d)	69. (c)	70. (c)
71. (a)	72. (b)	73. (c)	74. (c)	75. (c)	76. (a)	77. (d)	78. (c)	79. (d)	80. (a)
81. (d)	82. (a)	83. (a)	84. (b)	85. (c)	86. (d)	87. (c)	88. (b)	89. (d)	90. (a)
91. (a)	92. (b)	93. (a)	94. (d)	95. (d)	96. (a)	97. (b)	98. (d)	99. (d)	100. (a)
101. (c)	102. (d)	103. (b)	104. (c)	105. (a)	106. (d)	107. (b)	108. (a)	109. (c)	110. (a)
111. (a)	112. (a)	113. (d)	114. (a)	115. (a)	116. (b)	117. (d)	118. (a)	119. (a)	120. (b)
121. (d)	122. (a)	123. (a)	124. (b)	125. (b)					

Explanations

(A) Topical Exercise

7 (*b*) Statement in option (b) is not related to roots in monocotyledon and can be corrected as
Monocots have adventitious roots which do not originate from radicle of the embryo.
Rest options related to monocotyledon plants.

10 (*b*) Statement I is false, but II is true. Incorrect statement can be corrected as
In *Curcuma,* nodulose roots are found.

11 (*a*) Option (a) shows correctly matched pair while other options are incorrectly matched pairs.
These options can be corrected as
Buttress root - *Ficus*
Brace roots - *Zea mays*
Contractile roots - *Crocus*

12 (*c*) Option (c) is correctly matched pair. Other options are incorrectly matched pairs and can be corrected as
Curcuma - Nodulose roots
Portulaca - Moniliform root
Ipomoea - Tuberous root

14 (*c*) Option (c) is incorrectly paired and can be corrected as
Floating root - *Jussiaea*
Rest all options are correctly paired.

16 (*d*) Option (d) is incorrectly paired and can be corrected as
Tap root system - Carrot
Rest all options are correctly paired.

19 (*d*) Opion (d) shows incorrect match and can be corrected as
Piper plant is associated with climbing roots. These roots help in giving support to raise the plant. The climbing roots arise from the stem, so they are adventitious roots.
Rest all options show correct match.

21 (*b*) Opion (b) is incorrectly paired and can be corrected as
Sweet potato - Tuberous root
Rest all options are correctly paired.

27 (*d*) Option (d) is correctly matched pair. Other options are incorrectly matched pairs and can be corrected as
In mint, the vegetative propagation occurs through suckers. In *Pistia,* vegetative propagation occurs by offset. Jasmine is a stolon and not a runner.

32 (*b*) Statement in option (b) is incorrect and can be corrected as
Stem tendrils develop from axillary buds.
Rest statements are correct.

35 (*b*) Option (b) shows mismatched pair and can be corrected as
Sucker - Banana
Rest all options are correctly matched pairs.

39 (*a*) Statement I is true, but II is false. Statement II can be corrected as
Leaves of most of the dicotyledonous plants bear networks of veinlets in their lamina and constitute reticulate venation.

40 (*b*) Option (b) shows incorrectly matched pair and can be corrected as
Phoenix leaf tip get modified into a spine.
Rest all options show correctly matched pairs.

42 (*d*) Statement in option (d) is incorrect and can be corrected as
Leaves of *Mangifera indica* plants possess reticulate venation.
Rest statements are correct.

45 (*a*) Option (a) is incorrectly matched pair and can be corrected as
Palmately compound leaf is found in silk cotton. *Alstonia* possess whorled leaf arrangement of pinnate leaves.
Rest options are correctly matched pairs.

46 (*c*) Option (c) is incorrectly matched pair and can be corrected as
Acacia - Bipinnate leaf
Rest all options are correctly matched pairs.

47 (*c*) Option (c) is correctly matched pair whereas other options are incorrectly matched pairs.
Incorrectly matched pairs can be corrected as
- Multifoliate - *Cleome*
- Bifoliate - *Hardwickia*
- Trifoliate - Wood apple

49 (*b*) Three plants, i.e. *Mimosa, Caesalpinia* and kikar show bipinnately compound leaves.

50 (*a*) Option (a) is correctly matched whereas other options are incorrect. Incorrectly matched pair can be corrected as
Rose leaf has an adnate stipule. Free lateral stipules are found in China rose. In *Pisum sativum,* tendrils are modified upper leaflets.

52 (*a*) Option (a) is correctly matched whereas other options are incorrectly matched pairs. These can be corrected as
Tendrils are spirally coiled thread-like or hair-like structures and sensitive to solid support. Different parts modified into tendrils, i.e.
- Whole leaf in *Lathyrus.*
- Leaf tip in *Gloriosa.*
- Upper leaflet in sweet pea.

65 (*d*) Statement (d) is incorrect and can be corrected as
The apex produces different floral appendages laterally at successive nodes.

69 (*d*) Option (d) is incorrectly matched pair and can be corrected as
Campanulate petals-*Physalis*
Rest options are correctly matched pairs.

70 (*b*) Option (b) is incorrectly matched pair and can be corrected as
Calotropis - Valvate
Rest all options are correctly matched pairs.

71 (*a*) I and III are correct, while II and IV are incorrect. These can be corrected as
Iberis amara - Imbricate aestivation
Dianthus - Imbricate

75 (*a*) Statement (a) is incorrect and can be corrected as

Silicula resembles siliqua. Silicula is short, broad and flat.

76 (*b*) When two dry fruits *X* and *Y* are observed, *X* developed from monocarpellary and *Y* developed from bicarpellary ovary. *X* is characteristic fruit of Leguminosae family whereas *Y* is the characteristic fruit of Brassicaceae family. *X* and *Y* are legume and siliqua, respectively.

83 (*b*) 10 out of 14 are drupe fruits. These are coffee, mango, apricot, coconut , date, olive, almond, peach, cherry and pistachio.

97 (*d*) Option (d) is incorrectly matched pair and can be corrected as

A flower may be trimerous, tetramerous or pentamerous when the floral appendages are in multiple of 3, 4 or 5, respectively.

Rest all options are correctly matched pairs.

99 (*c*) Option (c) is incorrect because the given diagram represents epigynous condition.

100 (*b*) Option (b) is not correctly matched pair and can be corrected as

In bilateral symmetry, the flower can be divided into two similar halves only in one plane. *Datura* possesses actinomorphic symmetry.

Rest all options are correctly matched.

101 (*a*) Statement I is true, but II is false, Statement II can be corrected as

Fusion is indicated by enclosing the figure within bracket and adhesion by a line drawn above the symbols of the floral parts.

104 (*d*) Option (d) is incorrectly matched pair and can be corrected as

Brassicaceae Ebr. ⊕ ⚥ $K_{2+2}C_4A_{2+4}\underline{G}_{(2)}$

Rest all options are correctly matched pairs.

106 (*a*) The given floral formula is of Fabaceae. Among given plants soybean, groundnut, *Indigofera*, *Sesbania*, *Trifolium* and sunhemp belongs to this family.

110 (*d*) Option (d) is incorrectly matched pair and can be corrected as

Gloriosa – $G_{(3)}$

Rest all options are correctly matched pair.

(B) Cumulative Exercise

2 (*a*) Statements III and IV are correct, but I and II are incorrect. These can be corrected as

Hemp is a termed used for *Cannabis sativa* and it is not a source of family-Liliaceae.

Solanaceae is known as 'Potato family' and mainly used for obtaining food products. Hence, it is not a source of painkillers.

4 (*a*) Only statement I is incorrect and can be corrected as

The fruits of *Clematis* are dispersed by wind. The air cavities are required for floating of fruits and seeds.

Rest all statements are correct.

5 (*c*) Statements I, II and IV are correct. While III is incorrect and can be corrected as

The roots are positively hydrotropic and geotropic while negatively phototropic.

6 (*a*) Statements I and IV are true whereas statements II and III are false. These can be corrected as

Stem develops from the plumule of the embryo of the seed. It is generally green when young and later often become woody and dark brown.

7 (*d*) Option (d) is incorrect pair and can be corrected as

Spiny – *Zizyphus*

Rest all options are correct pairs.

9 (*b*) Statements I and II are correct while III and IV are incorrect. These can be corrected as

- In racemose inflorescence, flowers are borne in an acropetal order.
- The basipetal order is seen in cymose inflorescence.

12 (*a*) Only statement I is true. Statements II, III and IV are false and can be corrected as

- A bud is present in the axile of leaf, not in the axile of leaflets.
- A simple leaf may have incisions, but incisions do not touch the midrib.
- The leaves are small and short lived in Australian *Acacia.*

15 (*b*) Statements I, II and III are correct, while statement IV is incorrect. It can be corrected as

Achenial fruits are unilocular and single seeded, indehiscent, pericarp does not rupture and seeds remain enclosed within it.

16 (*d*) Statement (d) is correct. Rest statements are incorrect and can be corrected as

- *Ocimum* shows verticillaster inflorescence which is initialy a dichasial cyme, but soon changes into monochasial scorpioid type.
- *Lingustrum* shows panicle in which flowers are borne acropetally.
- In hypanthodium, male flowers are borne at the ostiole, while female flowers at the basal side.

18 (*a*) Statements I and III are correct. Statements I and IV are incorrect and can be corrected as

- Some parenchyma cells in stem have colourless plastids that store starch.
- Leaf like, flattened modified stems are called phylloclade.

19 (*d*) Statements I, III and IV are correct. Statement II is incorrect and can be corrected as

Morus and *Salix* possess catkin inflorescence.

21 (*c*) Statements I and II are correct.

- Veins provide support to the leaf and also transport food, water, minerals to the leaf and rest of the palnt. It provides rigidity and strength to the leaf blade.
- Axillary bud is embryonic and is located in the axil of leaf.

Statements III and IV are incorrect and can be corrected as

- Pulvinus is found in the bean family (Fabaceae).
- *Polygonum* belongs to family-Polygonaceae.
- Leaf takes its origin from shoot apical meristem.

24 (*c*) Statements II and III are correct. Statements I and IV are incorrect and can be corrected as

- In dry fruits, the pericarp becomes dry at maturity.
- Simple fruits develop from a single carpel or fused carpels (syncarpous) of a ovary.

28 (*b*) Only statement IV is correct. Statements I, II and III are incorrect and can be corrected as

- A flower having only one plane of symmetry is known as bilateral symmetrical. These are also known as zygomorphic flowers.
- China rose is a bisexual flower and *Cucurbita* is a monoecious flower. The flower of *Morus alba* is unisexual or bisexual.
- An epicalyx is an additional whorl around the calyx of a flower.

30 (*b*) Statements I and IV are correct while II, III and V are incorrect. These can be corrected as

- Epipetalous stamen is found in brinjal whereas epiphyllous stamen is found in lily.
- China roses possess monadelphous stamen.
- Pea has a diadelphous androecium, e.g. all the stamens unite to form two bundles.

33 (*a*) Statement in option (a) is correct. Rest statements are incorrect and can be corrected as

- Free lateral stipules are found in China rose.
- The upper leaflets of *Pisum sativum* are modified into tendril.
- *Polygonum* has ochreate type of stipule. Rose has adnate type of stipule.

36 (*d*) Option (d) is incorrect pair and can be corrected as

Tendrillar stipules are found is *Smilax*.

Rest all options are correct.

37 (*b*) Statements I, II and III are true. Statement IV is false and can be corrected as

Parthenocarpic fruits are seedless with great economic importance.

38 (*d*) Statement in option (d) is incorrect and can be corrected as

The *Cuscuta* contains no chlorophyll. These have haustoria, a root like organs which penetrate the tissue of a host.

Rest statements are correct.

46 (*b*) Both A and R are true, but R is not the correct explanation of A.
Caryopsis and achenes develop from monocarpellary unilocular, one-seeded and superior ovaries. The only difference in them is that, in caryopsis, pericarp and seed coat are fused and in achenes, pericarp is free from seed except at one point. Caryopsis is common in family–Poaceae.

47 (*a*) Both A and R are true and R is the correct explanation of A. Syconus fruit develops from hypanthodium inflorescence. Hence, the receptacle forms a hollow cavity with three types of flowers, i.e. upper (anterior) male flower, lower (basal) female flower and sterile female flower called gall flowers between male and female flowers. Male and gall flowers cannot develop into fruits. So, in syconus type of fruit, the achenes formed are fewer than the total number of flowers in inflorescence.

48 (*a*) Both A and R are true and R is the correct explanation of A. Prop roots or Pillar roots are thick, pillar-like adventitious roots, which support heavy horizontal branches of banyan tree. Great banyan tree growing in Indian Botanical Garden, Kolkata has 1775 prop roots. This tree is over 200 years old and crown of the tree has circumference of 404 m.

49 (*c*) A is true, but R is false because

True fruits develop from ovaries and their seeds from ovules. The pericarp of fruits is formed of ovary wall.

50 (*a*) Both A and R are true and R is the correct explanation of A.
Stems have nodes and internodes. Leaves and branches usually develop from the node, when the stem or the branches end in a vegetative bud. It continues to grow, but if it produces floral bud, the growth stops. Nodes on a stem may be identified as the place from where the leaves arise.

51 (*a*) Both A and R are true and R is the correct explanation of A.
Ginger is an example of rhizome, i.e. prostrate stem creeping horizontally and thus, the growth of ginger rhizome is not affected by gravity.

52 (*a*) Both A and R are true and R is the correct explanation of A.
Calotropis is an aggregate fruit. In such fruit, a multicarpellary ovary becomes fruit and each ovary turns into fruit. *Calotropis* is an etaerio or bunch of follicles.

53 (*b*) Both A and R are true, but R is not the correct explanation of A.
Cashewnut is a nut. It is formed from unilocular, one-seeded bi or multicarpellary ovary. Pericarp is hard and stony.

54 (*d*) A is false, but R is true because
Cremocarp fruits do possess persistent stylopodium, but they are found in family–Umbelliferae.

(C) Medical Entrances' Gallery

1 (*a*) Option (a) is correct. It can be explained as
A flower has four different whorls arranged successively on the swollen end of the stalk or pedicel called thalamus or receptacle. In mustard, flower is hypogynous, i.e. the gynoecium occupies the highest position, while the other parts are situated below it. The ovary in such flowers is said to be superior.

2 (*c*) The floral formula of Liliaceae is

$$\text{Br} \oplus ⚥ \; \widehat{P_{(3+3)} \; A_{(3+3)}} \; \underline{G}_{(3)}$$

It represents bracteate, actinomorphic, bisexual flowers having six tepals and six epitepalous stamens. Ovary is superior with tricarpellary and syncarpous.

3 (*b*) Vegetative propagule in *Agave* is bulbil. It develops as a small bud on the axial leaf and after falling on the soil, it develops into new plants.

4 (*a*) Features I, III and IV are correct. Feature II is incorrect because

Both mango and coconut fruits are drupe.

They develop from monocarpellary ovary but in both fruits, edible part is different as shown below

		Mango	Coconut
1.	Epicarp	Leathery	Membranous
2.	Mesocarp	Edible	Fibrous
3.	Endocarp	Hard and stony	Hard and stony
4.	Endocarp	Not edible	Edible

5 (*d*) Ray florets have inferior ovary as epigynous flowers have an inferior ovary with the flattened thalamus as seen in the ray florets of sunflowers. In epigynous flowers, all the floral parts are present above the gynoecium. Ovary is covered completely by thalamus, e.g. guava, apple, sunflower.

6 (*d*) The roots that originate from the base of the stem are fibrous roots. In fibrous root system, the primary roots stop growing and becomes rudimentary, while new roots for anchorage and absorption develop from the base of the stem. This type of root formation is observed in monocot like wheat plant.

7 (*c*) A–2, B–1, C–4, D–3

8 (*c*) *Alstonia* shows whorled phyllotaxy. In whorled phyllotaxy, more than two leaves arise at a node and form a whorl.

Mustard and China rose show alternate phyllotaxy. *Calotropis* shows opposite phyllotaxy.

9 (*b*) Verticillaster inflorescence is found in *Ocimum*. It is the characteristic feature of tulsi (*Ocimum*) belonging to family—Lamiaceae or Labiatae. Here, flowers are arranged in two opposite cymose groups on each node.

In mustard, raceme and in sunflower, capitulum or head type of racemose inflorescence is found. China rose has solitary axillary flower.

10 (*c*) Pea, bean, *Cassia* and gulmohar plants have zygomorphic flowers. The flowers of these plants are said to be zygomorphic as these can be divided into two equal halves along a single vertical plane.

11 (*d*) Both options (a) and (b) are correct.

Plum, peach and rose plants have perigynous flowers. In perigyny, only the base of the ovary gets enclosed by the thalamus (thalamus hollows out to form hypanthium) and the upper part remains free, carrying the sepals, petals and stamens with it. This forms a half-inferior ovary.

Guava and cucumber have epigynous flowers. In epigyny, the entire ovary gets enclosed by the hypanthium and both the ovary and hypanthium fuse together. The sepals, petals and stamens borne by the hypanthium get located at a position above the ovary. This forms a completely inferior ovary.

Thus, option (d) is correct.

12 (*c*) *Solanum* (family-Solanaceae) has bicarpellary ovary with obliquely placed septum. *Brassica* (family-Brassicaceae) has bicarpellary ovary with false septum. In *Sesbania* (subfamily Fabaceae), ovary is monocarpellary, *Aloe* (family-Liliaceae) shows tricarpellary ovary.

13 (*c*) A–4, B–2, C–1, D–3

14 (*b*) In parietal placentation, the ovules develop on the inner wall of the ovary or in peripheral part. In this type, the ovary is compound or syncarpous type which is either unilocular or falsely two or more locular, e.g. mustard, *Argemone*, *Fumaris*, etc. Other options are explained as

In axile placentation, ovules are placed along the central axis of the ovary.

In basal placentation, ovules are placed at the base of the ovary.

In free central placentation, placenta develops at the centre of ovary as prolongation of floral axis. On this axis, the ovules are attached.

15 (*d*) The statement in option (d) is incorrect and can be corrected as

Apple is a false fruit. Any fruit which develops from any other part of flower except ovary is called false fruit.

In apple, the thalamus of flower becomes juicy and is edible.

Other statements are correct.

16 (*b*) *Piper nigrum* (black pepper) is regarded as king of spices. It is a flowering vine in the family—Piperaceae. It is cultivated for its fruit which is used as a spice for seasoning in the food industry.

Other options are explained as

Brassica nigra is annual black mustard cultivated as oil seed crop.

Piper longum is used as medicinal plant as it contains piperine alkaloid.

Curcuma longa is commonly called turmeric and it has medicinal properties. It is also used in cosmetic and dyeing industry.

17 (*a*) Persistent nucellus in the seed is known as perisperm. The albuminous seeds usually retain a part of endosperm as it is not completely used up during embryo development.

But in some seeds, remnants of nucellus are also persistent, e.g. black pepper and beet.

18 (*c*) Pneumatophores are breathing or respiratory roots which are found in halophytes like mangroves. Halophytes grow in saline swamps, therefore respiratory roots come out of water to pick up oxygen for respiration and excess CO_2 is also given out. It occurs through small pores, called lenticels.

Carnivorous plants, free-floating hydrophytes and submerged hydrophytes do not possess pneumatophores.

19 (*b*) Sweet potato is a modified adventitious root, which is meant for the storage of food. It does not assume a definite shape and occurs singly.

20 (*b*) Offsets are produced by mitotic divisions. They are one internode long runners that occur in some aquatic plants. Offsets help in vegetative propagation and give rise to new plants, on contact with soil, e.g. *Eichhornia*.

Other options are explained as

Meiotic divisions occur in only germ cells. Parthenocarpy is the development of seedless fruits without fertilisation.

Parthenogenesis can be defined as the development of an egg into a complete individual without fertilisation.

21 (*b*) Leaf tendrils are found in pea. Grapevine and cucumber have stem tendrils. Tendrils are coiled thread-like, long, slender leaf modification. They help the plants in climbing up by coiling around a solid support.

22 (*d*) Valvate (*A*) petals are close to each other and do not overlap. In twisted (*B*), petals overlap each other regularly.

In imbricate (*C*), there are five petals, arranged in such a way that one petal is completely external and one is completely internal. In quincuncial (*D*), two out of five petals are completely external, one is completely internal and remaining two are partially external and partially internal.

23 (*a*) Caryopsis is present in wheat. It is a type of simple dry fruit. It is also called as grain. In such fruits, the ovary wall and seed coat fuse, thus making it difficult to separate the two.

24 (*c*) In *Bougainvillea*, thorns are the modifications of stem. They are stiff, sharp structures, which have lost their growing point and become hard. They reduce transpiration. They also protect plants from browsing animals.

25 (*b*) Statements I and III are correct. Statement II is incorrect and can be corrected as

Eichhornia is an aquatic plant, in which the petiole becomes spongy and helps in floating.

26 (*d*) Option (d) is not correctly matched and can be corrected as

The inflorescence of *Calotropis* is an example of multiparous cymose.

Verticillaster is a modified, condensed and dichasial cyme inflorescence observed in *Salvia, Ocimum, Coleus,* etc.

Rest options are correctly matched.

27 (*a*) Coconut fruit is a drupe. It has a membranous epicarp, fibrous mesocarp and a stony endocarp, whereas example of berry is tomato, nut includes litchi and capsular fruits are found in cotton.

28 (*c*) The edible part of coconut is endosperm. Coconut water is free nuclear endosperm and white kernel is the cellular endosperm.

29 (*a*) Cotyledons and testa are edible parts of groundnut and pomegranate, respectively. The edible part of walnut is cotyledon, tamarind-mesocarp, French beans-seeds on ripening, coconut-endosperm, testa, cotyledons and embryo, cashewnut-cotyledons and fleshy pedicels and of litchi is fleshy aril.

30 (*b*) Only *Salvia*, mustard, radish and turnip have stamens of different length in their flowers.

31 (*a*) The *Brassica* flower can be cut into two equal halves from any plane. So, it shows the radial symmetry. These flowers are referred to as actinomorphic flowers. Other flowers given in the option (*Trifolium, Pisum* and *Cassia*) can be cut into two equal halves only at one plane, so they are called zygomorphic flowers. These three flowers belong to family–Fabaceae.

35 (*c*) Coconut milk represents free nuclear endosperm where the division of primary endosperm nucleus is not followed by formation of cell walls (cytokinesis) thus all nucleus remain free in liquid form. It is rich in plant hormone cytokinin.

38 (*b*) Vexillum is a term used for standard petal of a papilionaceous corolla.

40 (*b*) Onion has bulb which is a short, underground modified stem surrounded by fleshy modified leaves that store food for shoot formation. It has corymb inflorescence.

43 (*c*) A is true, but R is false.

The subaerial stems are modified for vegetative reproduction as the stem is partly above ground and partly underground. The nodes near the ground develop short aerial branches and adventitious roots. The underground stem modification are meant for food storage not the subaerial roots.

44 (*a*) Option (a) is correct combination.

Datura belongs to Solanaceae family with epipetalous stamens. Other options are not correct combinations and can be corrected as

Allium is of Amaryllidaceae family with epipetalous stamens. *Petunia* belongs to Solanaceae family with epipetalous stamens. *Crotolaria* belongs to Fabaceae family with monadelphous stamens.

48 (*b*) Keel represents petal in vexillary aestivation, which is a characteristic of family–Fabaceae, i.e. *Indigofera*.

Papilionaceae (e.g. *Indigofera*, pea) is subfamily–Fabaceae, which has bilateral symmetry. It follows vexillary aestivation. A single upper posterior petal is known as standard. Two lateral compressed petals called wings and a pair of anterior small fused petal forms keel. Keel has concave sides which make them fused at bases.

49 (*b*) Option (b) is correct.

China rose, mustard, brinjal, potato, onion and tulip are the plants with superior ovary.

50 (*b*) Cucumber is a member of the family–Cucurbitaceae. In the members of this family, the flowers are unisexual (monoecious), with male and female flowers on different plants (dioecious) or on the same plant (monoecious).

54 (*a*) Leaves are modified into spines in *Opuntia* (xerophytes) so as to reduce the surface area and minimise rate of water loss by evaporation and transpiration. Besides, they also provide protection to the plant from grazing animals.

57 (*c*) Statements I and II are correct, whereas III and IV are incorrect. Incorrect statements can be corrected as

- In Australian *Acacia*, the leaves are small and short-lived and petioles expand, become green and synthesise food.
- In *Alstonia*, leaves show whorled phyllotaxy.

58 (*d*) Leaf tendrils are specialised stem, leaf or petiole with a thread-like shape that is used by climbing plants for support, attachment and cellular invasion by parasitic plants, generally by twining around suitable hosts, e.g. peas, cucumber and grapevine.

64 (*c*) The plants have hypogynous flower mustard, brinjal, China rose, lupin, sunhemp, gram, chilli, *Petunia*, tomato, *Withania*, potato, onion, *Aloe* and tulip.

Thus, among given flowers, fifteen plants have hypogynous flower.

65 (*c*) In family–Gramineae (Poaceae), the perianth is represented by membranous scales called lodicules. The lodicules are situated above and opposite the superior palae.

66 (*d*) Ginger (*Zingiber officinale*) is a straggling sympodial rhizome, which is a perennial, fleshy, dorsiventral, horizontal, usually branched, underground stem growing beneath the surface of soil. It possesses nodes and internodes, scaly leaves, axillary buds and roots at their nodes.

68 (*d*) Mature endosperm with any degree of irregularity and unevenness in its surface is called ruminant endosperm. It is known to occur in about 32 families of angiosperms. In family–Annonaceae, the ruminate endosperm is found.

70 (*c*) Stolons (*A*) are special kind of runner, which initially grow upwards like ordinary branches and then arch down to develop new daughter plants on coming in contact with the soil. Sucker (*B*) is a subaerial branch that arises from the main stem. Initially, it grows horizontally below the soil surface and later grow obliquely upwards.

75 (*c*) *Nepenthes* (pitcher plant) is an insectivorous plant of tropical region. Pitcher is a modified lamina. These are alternate and modified with a foliaceous leaf base. Upper part of petiole is elongated and tendrillar, whereas leaf blade (lamina) is modified into pitcher, which collects small amount of water containing digestive enzyme. Insects that slip into water are not allowed to come out by the hair near the rim.

77 (*d*) Apocarpous condition arises when the number of carpels is two or more and they are free from each other, e.g. *Clematis, Michelia* (Magnoliaceae), *Aconitum, Ranunculus* (buttercup), etc.

80 (*a*) Option (a) is correctly matched.

Onion is a simple tunicated layered bulb, while ginger is a straggling rhizome having uniparous cyme branching with sympodial axis.

81 (*d*) Fig, pineapple and mulberry are composite fruits.

Plants	Botanical name	Fruits	Inflorescences
Fig	*Ficus carica*	Syconus	Hypanthodium
Pineapple	*Annanas sativus*	Sorosis	Spike
Mulberry	*Morus* sp.	Sorosis	Catkin

85 (*c*) Gram, arhar, sunhemp, moong, pea and lupin have marginal placentation.

Thus, option (c) is correct.

87 (*c*) In *Cocos nucifera* (coconut), the epicarp is membranous, mesocarp is fibrous (hence, called fibrous drupe and yields coir) and the endocarp is stony.

88 (*b*) False fruits are developed when some other floral parts participate in the formation of fruits, e.g. thalamus in apple form the fruit. Thus, it is a false fruit.

89 (*d*) Aril is the edible part in the fruit of litchi. The aril is an accessory seed covering formed from an outgrowth at base of the ovule.

Other options are explained as

- Fleshy mesocarp is the edible part of mango (*Mangifera indica*)
- Juicy testa is the edible part of pomegranate (*Punica granatum*).
- Endocarpic juicy hair are edible part of orange (*Citrus* sp).

90 (*a*) In wild pea (*Lathyrus aphaca*), whole leaf is transformed into tendril. In *Pisum sativum*, terminal leaflet; in *Gloriosa*, leaf apex is transformed into tendril.

92 (*b*) The given figure shows that the flower is heterochlamydeous (i.e. having both the perianth segments, sepals and petals in distinct whorls).

Gamopetalous (i.e. all the petals of the flower are fused), pentamerous and bisexual (having both the sex organs, male and female).

93 (*a*) Statements III and IV are incorrect and can be corrected as

In *Calotropis*, the stamens are gynandrous, i.e. fused with the gynoecium. The mango is juicy stone fruit (drupe).

In *Hibiscus*, flower is solitary axillary which is a type of cymose. Hypanthodium inflorescence is found in *Ficus benghalensis*.

Rest statements are correct.

99 (*d*) In *Poinsettia*, cyathium inflorescence is found. In this, a single central female flower remains surrounded by numerous male flowers. So, in 42 inflorescence, the number of female flowers is 42. While in *Bougainvillea*, inflorescence is dichasial cyme, where medialy situated peduncle itself finishes in a flower and bears two lateral flower branches at the base of its origin. Hence, in 39 inflorescence the number of flower is 117. One terminal and two lateral flowers, i.e $3 \times 39 = 117$.

Thus, option (d) is correct.

101 (*c*) Some taxonomists believed that Compositae is the most advanced family. It is characterised by the capitulum or head inflorescence.

103 (*b*) In family–Liliaceae, the leaves are simple, in the form of a cluster of radial leaves, cauline and ramal, exstipulate (but stipulate in *Smilax*), have parallel venation (but reticulate in *Smilax*), sessile or petiolate with the sheathing leaf base. The inflorescence may be racemose or sometime solitary (e.g. tulip, *Gloriosa*) or umbellate condensed cymes (umbel, cyme), e.g. onion.

104 (*c*) Statement in option (c) is correct.

Flower of tulip is highly condensed and modified shoot meant for sexual reproduction (Dr. Goethe; 1790). During the course of evolution, the nodes of the axis of shoot came in contact so, that internodes got reduced and leaves got modified and specialised to form floral leaves.

Rest statements are incorrect and can be corrected as

- Orchid seeds do not have endosperm.
- Placentation in primrose is free central.
- In tomato, fruit is berry.

105 (*a*) When a flower can be divided into two similar halves only in one particular vertical plane, it is called zygomorphic, e.g. bean, pea, goldmohur, *Cassia*, etc.

106 (*d*) If gynoecium is situated in the centre and other parts of the flowers are located on the rim of the thalamus almost at the same level, it is called perigynous. The ovary here is said to be half inferior, e.g. plum, rose, peach, etc.

110 (*a*) *Pisum* belongs to family–Fabaceae. In this family, flower is bisexual and zygomorphic, corolla is polypetalous, papilionaceous and with vexillary aestivation; androecium is diadelphous with dithecous anther and gynoecium has monocarpellary, unilocular and superior ovary with marginal placentation having many ovules.

111 (*a*) Statements II and IV are correct, but I and III are incorrect.

In some legumes, the leaf base may become swollen, which is called the pulvinus.

In opposite phyllotaxy, a pair of leaves arises at each node and lie opposite to each other as in *Calotropis* (akon/madar) and guava (*Psidium*) plants.

112 (*a*) Long, slender and spirally coiled stem tendrils developing from axillary buds and helping plants to climb up are found in gourds (cucumber, pumpkins, watermelon) and grapevines.

120 (*b*) The arrangement of leaves on a stem or branch is called phyllotaxy. The number of vertical rows in which leaves are arranged is called as orthostichies. 120° phyllotaxy is found in tristichous condition, i.e. three leaves in one spiral.

121 (*d*) In subaerial modification, the stems are delicated, thin, weak and unable to stand erect. Runners grow prostrate in all directions above the soil level. It has a creeping stem with long internodes. On the lower side, nodes bear adventitious roots.

CHAPTER 12

Anatomy of Flowering Plants

Plant anatomy is the study of internal structure of various organs of plants. More precisely, it is the study of the structure and arrangement of tissues and their organisations in the plant body.
It is the study of organisation and structure of tissues. **N Grew** and **M Malpighi** are regarded as **'Founders of Plant Anatomy'**.

PLANT TISSUES

Tissue is a group of similar or dissimilar cells of common origin and performs a common function. The study of tissues is called **Histology**. Plant tissues can be categorised into following three groups

1. Meristematic tissues 2. Permanent tissues 3. Secretory tissues

Meristematic tissues

The term 'Meristem' (Gk. *Meristos* - divisible) was introduced by **Nageli** (1858). A meristem is a group of cells with the ability to divide continuously. They are active tissues that form new cells in the plant body continuously.

The characteristic features of meristematic cells are

- These cells are undifferentiated, small and can be variously shaped, e.g. isodiametric, round, oval, polygonal or rectangular.
- Absence of intercellular spaces.
- Cell walls are thin, elastic and made up of cellulose.
- They possess dense cytoplasm and a large nucleus.
- They do not store reserve food material as are always in the active state of metabolism.
- The cells of meristematic tissues have small endoplasmic reticulum, simple mitochondria, immature plastids and very small or no vacuoles.
- These cells are totipotent and divide mitotically.
- The meristematic cells have high surface area per unit volume and nucleocytoplasmic ratio.
- Colourless protoplastids are present in the cells.

Inside

Classification of meristematic tissues

The meristematic tissues can be classified as follows

1. On the basis of origin and development

(i) **Promeristem or Primordial meristem** A promeristem is the primary stage of meristematic cells.

- It originates from the embryonic cells. These cells are thin-walled and isodiametric.
- They have dense cytoplasm and large nuclei.
- These are present in a small region of shoot and root apex.
- Promeristem differentiates into primary meristem.

(ii) **Primary meristem** Cells of primary meristem are always in active state of division.

- These are found below promeristem at root and shoot apices.
- They are also present in **intercalary parts** and **intrafascicular cambium**.
- They give rise to primary permanent tissues.
- Due to primary meristem, plants increase lengthwise and widthwise.

(iii) **Secondary meristem** These meristems develop from the primary permanent tissues by dedifferentiation.

- They are always lateral in position. Such meristem develops during later stages of plant growth.
- They give rise to secondary permanent tissues, e.g. interfascicular cambium, cork cambium, etc.

2. On the basis of position in plant body

(i) **Apical meristem** It is found at the apices of shoots and roots.

- This tissue is responsible for the apical growth in length of roots and shoots. This kind of growth is called as **primary growth**.
- All primary tissues of the plant body originate from apical meristem.
- Technically, these include both promeristem and primary meristem.

(ii) **Intercalary meristem** These meristems lie between the permanent tissues.

- They may be present at the base of the internodes, e.g. in stem of grasses and wheat.
- They are also found at the base of leaves, e.g. *Pinus* or at the base of a node, e.g. mint *(Mentha arvensis)*.
- These are specially known to also increase the length of the plant and its organs.
- They originate from the apical meristems during development.

(iii) **Lateral meristem** This meristem is present along the lateral sides of stems and roots.

- These are especially present in the mature regions of roots and shoots.
- Cells of lateral meristems divide tangentially and give rise to secondary permanent tissues on the inner and outer side. Thus, they increase the girth of the plants.
- These are responsible for secondary growth in plants. For example, intrafascicular cambium, interfascicular cambium and cork cambium.

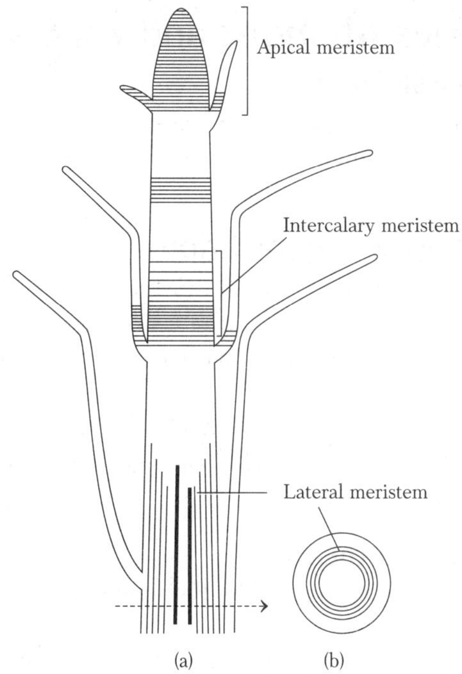

Fig 12.1 Meristems on the basis of position : (a) Longitudinal view, (b) Cross-section

3. On the basis of functions

(i) **Protoderm** It is the outermost layer of meristematic tissues. It develops into epidermal tissue system. For examples, epidermis, stomata and hairs.

(ii) **Procambium** It is the innermost layer of meristematic cells. It is composed of narrow and elongated cells. These cells develop into primary vascular tissue, i.e. phloem, xylem and cambium.

(iii) **Ground meristem** These cells are large and thin-walled. These produce hypodermis, cortex, endodermis, pericycle, pith and medullary rays. Ground tissues include all tissues except epidermis and vascular strands.

4. On the basis of plane of cell division

(i) **Mass meristem** The cells of this meristem divide in all planes, so that mass of cells is formed, e.g. formed cortex and pith.

(ii) **Plate meristem** The cells of this meristem divide in two planes. Thus, increase the plate-like area, e.g. epidermis and lamina formation.

(iii) **Rib/file meristem** Here, the cells divide in one plane, hence row or columns of cells are formed, e.g. formation of lateral roots.

Theories of shoot and root apex organisation

Different theories have been suggested to understand the structure of root and shoot apices. These are studied as follows

Shoot apex organisation

- Shoot apex is a dome-shaped structure.
- It is present above the youngest leaf primordium on the shoot. It also occurs in the inactive state in the axil of leaves as lateral buds.
- The shoot apex remains covered by young leaves.

Following theories have been put forward to explain the shoot apex organisation

1. Apical cell theory

- This theory was proposed by **Hofmeister** and **Nageli** (1858).
- It states that a single apical cell is the structural and functional unit of apical meristem.
- This apical cell governs the complete process of the primary growth.
- This theory is applicable to lower plants only, e.g. algae, bryophytes and pteridophytes.

2. Histogen theory

This theory was proposed by **Hanstein** (1868). According to this theory, shoot apex has following three zones

(i) **Dermatogen** It is the outermost layer or **histogen.** It forms the epidermis and epidermal tissue system.

(ii) **Periblem** It is the middle layer that forms the cortex and the endodermis.

(iii) **Plerome** It is the innermost layer. It forms the pith, vascular bundles, pericycle and the medullary rays.

Histogen is the generalised term used for these layers separately.

3. Tunica-Corpus theory

This theory was proposed by **Schmidt** (1924). It is based on the plane of division of cells.

According to this theory, the shoot apex consists of following two layers

(i) **Tunica** It is a single outer layer that forms the epidermis. Cells of this layer are smaller than that of other layers. Tunica divides by anticlinal divisions only.

(ii) **Corpus** It is the central core of the shoot apex. Corpus has larger cells than tunica. It can divide in all planes (i.e. anticlinal or periclinal) to form cortex and stele of the shoot.

Root apex organisation

- The root apex or root apical meristem is a small cup-shaped region at the tip of a root.
- In this region, all cells are capable of repeated division and from here all primary root tissues are derived.
- It is subterminal and is covered by the root cap.
- Following theories have been proposed to given to understand the organisation of root apex

1. Histogen theory

- According to this theory, the root apex consists of four histogens, e.g. dermatogen, periblem, plerome and calyptrogen.
- In case of root apex, dermatogen forms epiblema and calyptrogen gives rise to root cap in monocots.
- The function of other layers is same as seen in shoot apex.

2. Korper-Kappe theory

- This theory was given by **Schuepp** (1917).
- It states that a root system is formed by differentiation of two zones, i.e. Korper and Kappe. The Korper zone forms the body and Kappe forms the cap.
- This theory is comparable to tunica-corpus theory of the shoot apex.

3. Quiescent centre theory

- This theory was suggested by **Clowes** (1961) in maize.
- According to this theory, root apex consists of a quiescent centre. It is an inverted cup-like structure located between calyptrogen and dermatogen.
- It has low DNA, RNA and proteins. Cell divisions are very few in quiescent centre. The cells of quiescent centre divide only when the root apex gets injured.

CHECK POINT 12.1

1. Meristematic cells have
   (a) thick cell wall and large intercellular spaces
   (b) thick cell wall and no intercellular spaces
   (c) thin cell wall and large intercellular spaces
   (d) thin cell wall and no intercellular spaces
2. Select the incorrect statements with respect to meristem.
   (a) It is a group of small, isodiametric immature cells
   (b) It is the area of cell division in plants
   (c) Cell possesses dense cytoplasm and large prominent nucleus
   (d) Cell shows very low metabolic activities
3. The tissue that occurs in grasses and regenerates parts removed by grazing herbivores is
   (a) axillary meristem (b) apical meristem
   (c) intercalary meristem (d) lateral meristem
4. Apical cell theory is applicable to
   (a) algae, bryophytes and gymnosperms
   (b) angiosperms only
   (c) pteridophytes and angiosperms
   (d) algae, bryophytes and pteridophytes
5. Histogen theory is more applicable for
   (a) root apex (b) shoot apex
   (c) meristematic tissue (d) None of these
6. According to Korper-Kappe theory, Kappe zone is responsible for the formation of
   (a) root elongation zone
   (b) root cap
   (c) root hair
   (d) root maturation zone

Permanent tissues

- These tissues are formed as a result of division and differentiation in meristematic tissues.
- Permanent tissues lose their ability to divide.
- They become structurally and functionally specialised to perform specific functions in the plant body.

Permanent tissues are of the following types

I. Simple permanent tissue
II. Complex permanent tissue

I. Simple permanent tissues

These tissues have one type of cells only. These cells are structurally and functionally similar. They coordinate with each other to perform similar function.

These tissues are further classified as

1. Parenchyma

(Gk. *Para* - beside; *enchyma* - tissue)

- The term 'Parenchyma' was given by **N Grew** (1682).
- It is the most abundant and common tissue in plants.

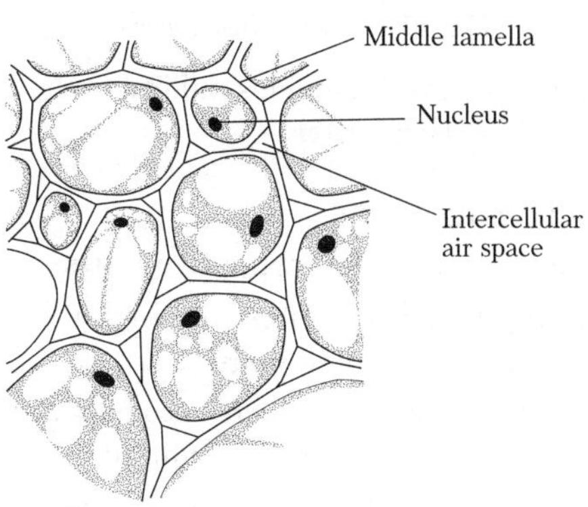

Fig. 12.2 Structure of parenchyma cells

- Parenchyma cells are thin-walled. They can be oval, round, polygonal or isodiametric.
- Parenchyma cells do not loose their protoplasm during differentiation.
- They possess a distinct nucleus.
- These cells usually have large intercellular spaces in between them.
- The parenchyma tissue is generally present in roots, stems, leaves, flowers, fruits and seeds.

Types of parenchyma

Parenchyma cells are modified into following types to perform some specialised functions

(i) **Chlorenchyma** It is the chloroplast containing parenchymatous tissue. It is specialised to perform photosynthesis. It consists of palisade and spongy parenchyma.

(ii) **Aerenchyma** Parenchyma enclosing the air cavity is known as aerenchyma.

These spaces allow gaseous exchange in aquatic plants (e.g. *Hydrilla*) and also provide buoyancy to the plants. Aerenchyma is also found in petioles of banana and *Canna.*

(iii) **Prosenchyma** These are elongated parenchyma cells.

They provide strength to plants. Prosenchyma are found in the pericycle of roots.

(iv) **Storage parenchyma** It is made up of large-sized vacuolated cells which store water, mucilage and food, e.g. *Aloe*, *Opuntia*, etc.

(v) **Idioblasts** These are non-living parenchyma cells. They store tannins, resins, oils and crystals, etc.

Functions of parenchyma

The functions performed by parenchyma are

- **Storage of food** It is the primary function of parenchyma. This tissue stores food materials like carbohydrates, proteins and fats, etc.
- **Secretion** Parenchyma cells also secrete substances like enzymes, resins, nectar and oil, etc.
- **Transport** Parenchyma cells transport nutrients and other substances inside the plant.
- **Totipotency** Parenchyma cells have the ability to act as precursor and transform to other types of cells.
- **Photosynthesis** Chloroplast in chlorenchyma helps in photosynthesis.
- **Photosynthesis** Chloroplast in chlorenchyma helps in photosynthesis.
- **Gaseous exchange and buoyancy** These functions are provided by aerenchyma cells.

2. Collenchyma

(Gk. *Kolla* - glue; *enchyma*-tissue)

- The term collenchyma was given by **Schleiden** (1839).
- It is an elastic, living and mechanical tissue. Cells of collenchyma are oval, spherical or polygonal.
- These cells do not have intercellular spaces. The cells are closely packed.
- Cell walls of collenchyma possess thickenings at the corners which may be of cellulose, hemicellulose or pectin.

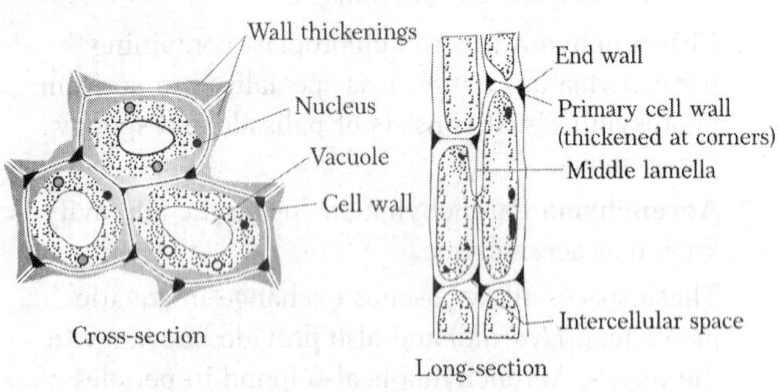

Fig. 12.3 Collenchyma tissue

- Collenchyma forms the hypodermis in the plant body.
- It is not found in monocot plants and dicot roots.

Types of collenchyma

On the basis of wall thickenings, collenchyma is of following three types

(i) **Lamellar** Thickenings are symmetrically around the walls, e.g. stems of *Raphanus*, sunflower and *Sambucus*.

(ii) **Lacunar** Thickenings are present around the intercellular spaces, e.g. aerial roots of *Monstera* and stems of *Cucurbita*.

(iii) **Angular** Thickenings are at the corners or angles of cells and side walls remain thin, e.g. *Ficus*, *Vitis*.

Functions of collenchyma

- It gives mechanical strength to young parts of plants.
- It provides elastisity and flexibility to plants. They are peripherally located in stems of plants. Therefore resist bending and pulling of plants by wind. It also helps some plants like *Cucurbita* in bending.
- It prevents tearing of leaves.
- It performs photosynthesis as contains chlorophyll.
- It can store food materials.
- In herbaceous plants, these are important supporting tissues.

3. Sclerenchyma

(Gk. *Skeleros*-hard, *enchyma*-tissue)

- The term 'Sclerenchyma' was given by **Mettenius** (1805).
- The cells in this tissues are elongated with narrow ends.
- They are dead and without protoplasts. Their cell walls become thickened due to deposition of lignin.
- At some places, unthickened areas called **pits** are present in cell wall.

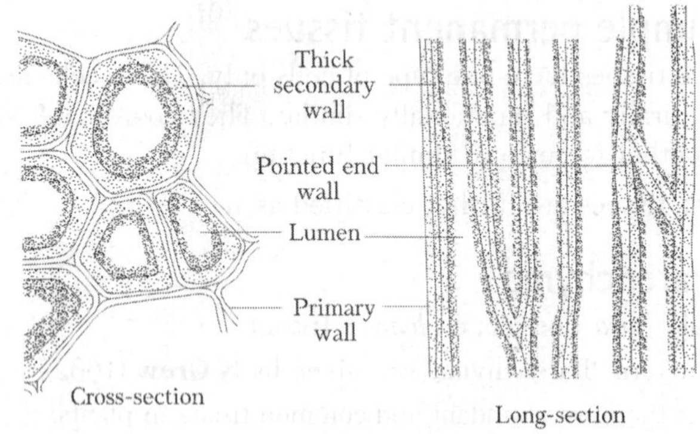

Fig. 12.4 Sclerenchyma tissue

- Lignification of cell wall provides rigidity to the cells and also protects them from water.

Types of sclerenchyma

Sclerenchyma cells are of following two types

(i) **Sclerenchyma fibres** These are highly elongated and narrow with pointed ends.

- The cell walls of these fibers become much lignified, hence their lumen is greatly reduced.
- Some plants with long fibres are *Linum*, *Chorchorus*, *Cannabis*, *Cocos* and *Crotalaria*.

- Sclerenchyma fibres occur in parts of plants where mechanical strength is required. For example, leaves, petioles, cortex, pericycle, phloem, xylem and around vascular bundles of monocot stems.
- On the basis of their presence in tissues, the sclerenchyma fibres can be cortical, pericyclic or phloem fibres.
- Due to its presence, outside the xylem, these are also called **extraxylary fibres**.

(ii) **Sclereids or Stone cells** These are spherical, cylindrical or oval-cells with thick wall.

- Due to much lignification lumen is almost reduced.
- Sclereids occur single as well as in groups.

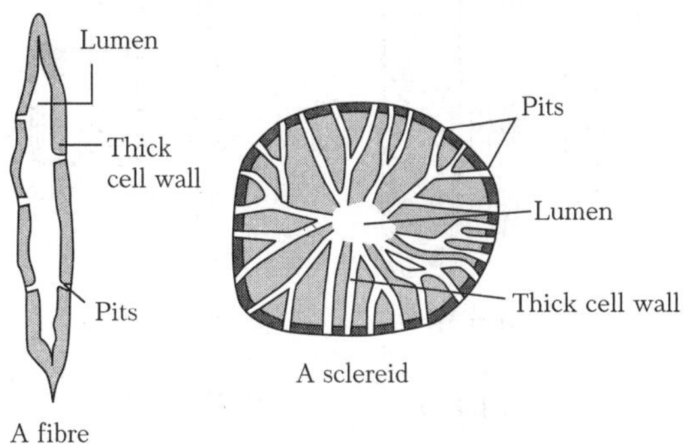

Fig. 12.5 Sclerenchyma

- They are mostly found in hard parts of the plants, e.g. endocarp of coconut. However, they may occur in soft parts of the plants, e.g. pulp of the fruits.
- Sclereids cell are of following types
  (a) **Brachysclereids** (stone cells) They cause grittiness in fruits, e.g. pear.
  (b) **Macrosclereids** (rod cells) These are present in seed coat, e.g. leguminous seeds.
  (c) **Asterosclereids** (star cells) Most commonly found in stems and leaves of dicot plants, e.g. tea, petioles of lotus.
  (d) **Osteosclereids** (prop cells) These are present in leaves and seed coat of monocot plants and the sub-epidermis of seed coats in legumes.
  (e) **Trichosclereids** (internal hairs) These are long hair-like branched sclereids. Usually these are found in hydrophytes and in the aerial roots of *Monstera.*

Functions of sclerenchyma

- Main function of sclerenchyma is to provide mechanical strength to plants.
- Sclereids provide grittiness in the pulp of fruits like guava and pear, etc. They also provide stiffness to seed coats and nuts.

Table 12.1 Differences between parenchyma, collenchyma and sclerenchyma

Parenchyma	Collenchyma	Sclerenchyma
Cells are living.	Cells are living.	Cells are dead.
No wall thickening.	Uneven wall thickenings only at corners of cells and consists of pectin.	Wall thickening uniform and consists of lignin.
Found in both, the outer and inner parts of plant organs.	Restricted only to the outer parts of plant organs.	Found in both the outer and inner parts, restricted to the areas, which have stopped elongation.
Storage and secretory in function	Provide mechanical strength as well as elasticity.	Provide only mechanical strength.
Ability to dedifferentiate and produce secondary meristem.	Dedifferentiate ion is almost absent.	No dedifferentiation at all.

II. Complex permanent tissues

This tissue is made up of different types of cells. However, they perform a common function.

They constitute the conducting tissues of plants, i.e. xylem and phloem.

1. Xylem

(Nageli; 1858) (Gk. *Xylos*-wood)

- Xylem is the main water conducting tissue of plant. It is also known as **hadrome**. It conducts water and solutes from roots upto the top of the plants.
- Xylem also provides mechanical strength to the plant.
- Xylem consists of four types of elements, i.e. tracheids, vessels, xylem or wood fibres and xylem or wood parenchyma.

(i) **Tracheids** These are most primitive conducting elements in xylem.

- These cells are elongated and with pointed ends. These cells have thickened lignified wall.
- The wall thickenings may be annular, scariform, spiral reticulate or pitted.
- At maturity tracheids are dead and possess bordered pits.
- In pteridophytes and gymnosperms, wood consists of tracheids only (no vessels).
- In angiosperms, tracheids are associated with vessels.
- The key function of tracheids is conduction of water in plants.

(ii) **Vessels** These are elongated and placed end to end in the plants.

- The partition walls between two vessels may be perforated or become disappear. Therefore, these cells appear as an elongated tube.
- The walls of vessels are lignified and with various types of thickenings as in tracheids. These cells do not possess nuclei and are also dead at maturity.
- The main function of vessels is conduction of water in plants.
- Vessels are absent in pteridophytes and gymnosperm therefore in these groups of plants, non-porous wood is found. However, primitive vessels are found in some gymnosperms, e.g. *Gnetum* and *Ephedra*.
- Vessels are characteristic features of angiospermic plants. Hence, the wood of flowering plants are known as **porous wood**.
- Some vesselless angiospermic families are Winteraceae, Tetracentraceae and Trochodendraceae.
- The distribution and size of vessels determine the types of porous wood in angiospermic plants
  - **Diffuse porous** It is primitive type of wood. The vessels of this wood are of same size and are distributed uniformly throughout the plant, e.g. *Betula*, *Pyrus*.
  - **Ring porous** It is advanced type of wood. In early wood, large vessels are formed when the requirement of water is more in plants. Later on small vessels are forward in late wood, e.g. *Morus quercus*.

Table 12.2 Differences between tracheids and vessels

Tracheids	Vessels
These are elongated cells with tapering ends.	It is a long cylindrical tube-like structure. End walls of vessels are generally oblique and are called as **perforation plates**.
It is formed from a single cell.	It is formed from a row of cells placed one above the other.
Its cells have comparatively less lignified walls, hence lumen is wide lumen. Tracheids have varied lignin thickenings.	Its cells have more lignified and thick walls with wide lumen.
Bordered pits are present in cross-walls.	Bordered pits are absent.
These are found in all groups of vascular plants.	Vessels are characteristics of angiosperms only. Gymnosperms and pteridophytes lack vessels. Some primitive angiosperms lack vessels, e.g. Winteraceae, Trochodendraceae and Tetracentraceae.

(iii) **Xylem fibres** (wood fibres) These are dead elongated cells with tapering ends.

- They have thick lignified walls, hence lumen is highly reduced.
- They provide mechanical support to plant.
- Xylem fibres are classified into libriform fibres and fibre tracheids.

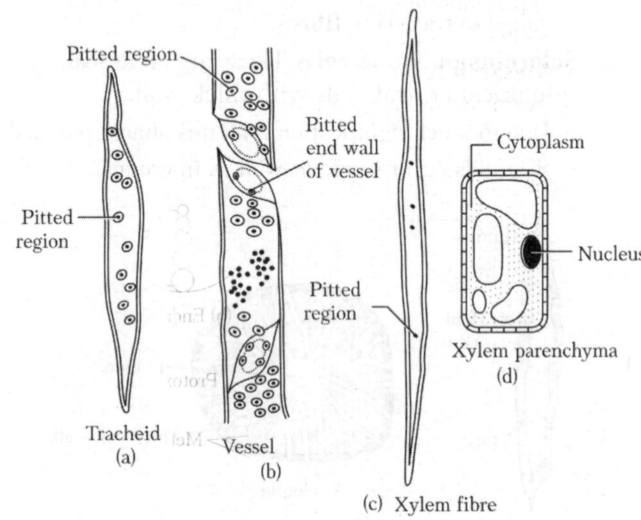

Fig. 12.6 Components of xylem

- The libriform fibres are characterised by the presence of very thick secondary walls and simple pits, while the fibre tracheids, have comparatively thin walls and bordered pits.

(iv) **Xylem parenchyma** (wood parenchyma) These are living cells.

- These make up the ground mass of xylem tissue. Their cell walls are made up of cellulose.
- They store food material and tannins.
- They help in radial conduction of water.
- Parenchyma cells that help in radial conduction are called **ray parenchyma**.

Types of xylem

On the basis of origin, xylem tissue is of following two types

(i) **Primary xylem** It arises from procambium during the primary growth of plant body. It can further be of two types

- **Protoxylem** This is the first formed primary xylem. It has vessels with a narrow diameter.
- **Metaxylem** It is formed after protoxylem. It has vessels with a broader diameter.

The arrangement of primary xylem can be of following three types

- **Exarch** Protoxylem lies towards the periphery and metaxylem lies towards the centre (pith). This arrangement is common in roots.
- **Endarch** This arrangement is exactly opposite to exarch one. Here, protoxylem is present towards the centre and metaxylem towards the periphery, e.g. stems.
- **Mesarch** A midway condition in which protoxylem lies in between the metaxylem.

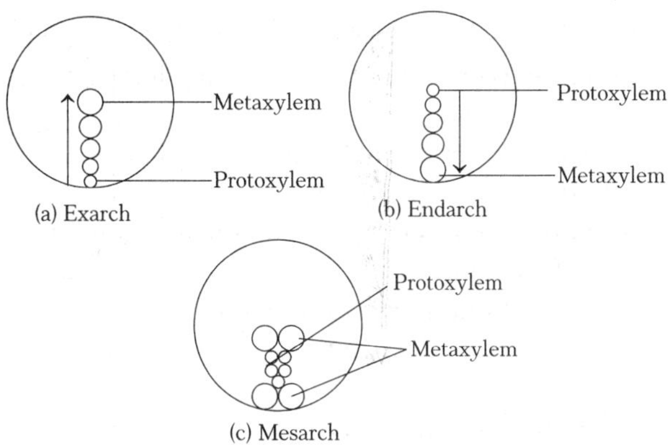

Fig. 12.7 *Position of protoxylem and metaxylem*

(ii) **Secondary xylem** It is formed from cambium during secondary growth of plants.

2. Phloem

(Nageli, 1858) (Gk. *Phloios* - inner bark)

- Phloem transports organic food produced in the leaves to stem and root in downward direction, i.e. from source to sink. The place of source is the area of food production. It is required to the sink, which is area of food consumption.
- It is also called **bast**.
- The conducting part of phloem is known as **leptome** (term given by **Haberlandt**).
- Phloem is made up of following four kinds of cells (elements).

(i) **Sieve tube elements** These are elongated and tube-like structures formed by many cells.

- These cells are collectively called **sieve tube elements.**
- These are arranged one above the other in longitudinal rows.
- Cell walls of sieve tube element are thin-walled and cellulosic.
- They lack nucleus and possess a large vacuole with peripheral cytoplasm.
- The functions of sieve tubes are controlled by companion cells, which possess nucleus.
- End walls of sieve tube elements are perforated. These are called **sieve plates** and possess many sieve pores. These **sieve pores** are involved in the movement of food.
- Sieve tubes are found in angiosperms only. These are absent in pteridophytes and gymnosperms. In these plant groups, sieve cells are present to perform almost same function.

(ii) **Companion cells** These are elongated parenchymatous thin-walled living cells.

- They possess cytoplasmic content with a conspicuous nucleus. These are found to be attached to the lateral sides of sieve tubes.
- The common wall between sieve tubes and companion cells have pits. Through these pits cytoplasmic connections are maintained between these two elements.
- These cells help in maintaining the pressure gradient in the sieve tubes.
- Companion cells are living and found in angiosperms only. These are not seen in pteridophytes and gymnosperms. In gymnosperms, albuminous cells are present.

(iii) **Phloem fibres** (Bast fibres) These are sclerenchymatous, elongated unbranched only dead cells among phloem tissue.

- These fibres possess thick lignified walls.
- These are absent or fewer, but more is number in secondary phloem.
- They differ from the xylem fibres in having simple pits. Bast fibres of jute (*Corchorus*), flax (*Linum*) and hemp (*Cannabis*) are commercially used.

(iv) **Phloem parenchyma** These are the living cells of phloem tissue.

- They are found in both primary and secondary phloem, but not seen in monocots.

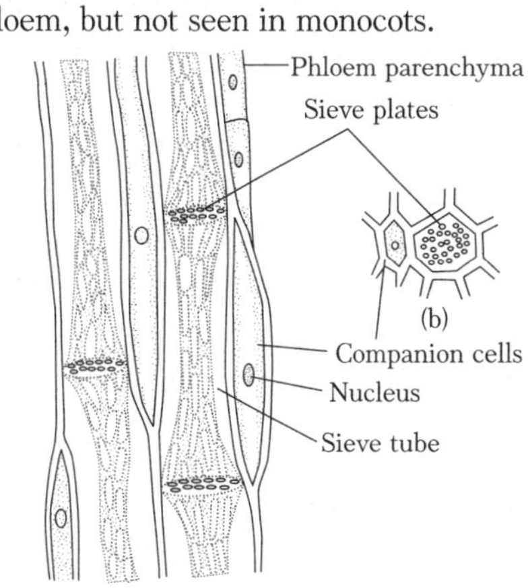

Fig. 12.8 *Phloem : (a) Longitudinal section, (b) Transverse section*

- The cells of phloem parenchyma are elongated and cylindrical. Their cell walls are with pits. Their main function is storage of food. They are also involved in lateral transport of soluble food materials and water.
- They also accumulate tannins, resins, crystals and mucilage, etc.

Types of phloem

On the basis of origin, phloem is of following two types

(i) **Primary phloem** These are formed during primary growth of plant body, hence are referred to as early formed phloem. These are of following two types

- **Protophloem** is the first formed primary phloem. It consists of narrow sieve tubes.
- **Metaphloem** is formed later during the development. It has bigger sieve tubes.

(ii) **Secondary phloem** It is formed due to secondary growth of plant body. Cambium is involved in the formation of secondary phloem.

Secondary phloem consists of two distinct systems

- **Vertical or Axial systems**
  - **Bast fibre** It provides support.
  - **Sieve elements** It includes sieve tube and companion cells. Its main function is conduction of food in plants.
  - **Bast parenchyma** It stores food in plants.
- **Horizonal** or **Ray system**
  - **Ray parenchyma** It also stores food in plants.

On the basis of position, phloem is of following three types

(i) **External phloem** It is of normal type and is present outside the xylem, e.g. mostly angiosperms and gymnosperms.

(ii) **Internal or Intraxylary phloem** It originates from procambium. It is the primary phloem which occurs on the innerside of primary xylem in bilateral bundles, e.g. membres of Apocynaceae, Asclepiadaceae, Convolvulaceae, Solanaceae, etc.

(iii) **Included or Interxylary phloem** It originates from cambium and is a secondary phloem which occurs in groups within the secondary xylem, e.g. *Leptadenia, Salvadora, Chenopodium, Boerhaavia, Amaranthus,* etc.

Note **Transfer cell** *is a new cell type recently reported, in the phloem. These cells are involved in short distance transport of food.*

Secretory tissues

- These are group of cells that 'secrete' different substances in plants. The substances secreted by these plant tissues are economically very important. These include rubber, resins, oils and gums, etc.

Secretory tissues are classified as follows

1. Laticiferous tissues

- It is mainly composed of thin-walled and elongated cells.
- These cells are multinucleated and contain colourless or milky fluid called **latex**.
- They are scattered throughout the ground tissue of the plant and contain stored organic matter in the form of starch, rubber, tannins, alkaloids, mucilage and enzymes, etc.
- Laticiferous tissue is further of following two types

(i) **Latex cells** (non-atriculated laticifers) are highly branched cells.

They do not form networks by fusing, e.g. *Vinca* and *Calotropis.*

(ii) **Latex vessels** (atriculated laticifers) are formed due to fusion of cells.

These form networks in mature parts of plant, e.g. *Opium, Carica papaya, Argemone.*

2. Glandular tissues

These consist of glands that secrete other substances, such as nectaries and digestive enzymes in insectivorous plants. Glandular tissues are of following two types

(i) **External glands** These occur as an outgrowth on the epidermis. These are of following types

- **Hydathodes** (water secreting glands) in tomato, etc.
- **Glandular hairs** (with a stalk and head) in tobacco, etc.
- **Stinging hairs** (secrete poisonous substance) in *Urtica dioica.*
- **Nectar secreting glands** (secretes sugary substances) in *Jatropha*, etc.
- **Digestive glands** (produce digestive enzymes) in insectivorous plants, e.g. *Utricularia*, etc.

(ii) **Internal glands** These are spherical or tubular and are found inside the plant body. It can be of following types

- **Oil glands** in rind (outer skin) of fruits like *Citrus* and orange. These develop lysigenously, i.e. by the breaking down of adjoining cells. Thus, they are also called as **lysigenous cavities.**
- **Mucilaginous glands** These store mucilage in leaves, e.g. Vine, Betel.
- **Resin or Tannin or Gum glands** These are formed schizogenously, i.e. by splitting in common walls of adjoining cells. These store resin in *Pinus* and gum in *Acacia.*

CHECK POINT 12.2

1. Which of the following is not a characteristic of permanent tissue?
   (a) They are speacialised to perform specific functions
   (b) They cannot undergo cell divisions
   (c) These all are dead cells
   (d) All of the above
2. Collenchyma differs from sclerenchyma in
   (a) retaining protoplasm at maturity
   (b) having thick walls
   (c) having wide lumen
   (d) being meristematic
3. In the following pairs, where will your find deposition of lignin in both the elements?
   (a) Tracheid and collenchyma
   (b) Sclerenchyma and sieve tube
   (c) Sclerenchyma and trachea
   (d) Parenchyma and endodermis
4. Tracheid, vessel and sclereids are similar in that they all
   (a) lack secondary wall
   (b) translocate water and minerals
   (c) function when dead
   (d) are primary permanent tissue
5. Tracheid differs from vessels in having
   (a) thick wall
   (b) bordered pits
   (c) discontinuous intercalary wall
   (d) spiral thickening
6. Which of the following represents a schizogenous cavity?
   (a) Resin cavity in *Pinus*
   (b) Oil gland in *Citrus*
   (c) Water secreting glands in tomato
   (d) Nectar secreting glands in *Jatropha*

PLANT TISSUE SYSTEM

Various tissues in the plant may perform similar functions irrespective of their position in the plant body. The group of tissues performing similar functions is known as tissue system.

Sachs (1875), classified the plant tissue system into three types, namely epidermal tissue system, ground (fundamental) tissue system and vascular (fascicular) tissue system.

Epidermal tissue system

- Epidermal tissue system or dermal tissue system is derived from embryonic protoderm.
- It forms the outer covering of a plant. Therefore, it is in direct contact with the external environment.
- This tissue system mainly provides mechanical support to the plants. It also performs functions like, absorption, secretion, gaseous exchange and protection, etc.

Different components of this tissue system are discussed below

Epidermis (Gk. *Epi*–upon; *derma*–skin)

- It is the outermost covering of a plant body. Usually it is a single layer of cells. However, it may be multilayered as in leaves of *Nerium* and *Ficus*.
- Epidermal cells are elongated living cells. These are compactly arranged parenchyma cells.
- The outer wall is thick and is lined by cuticle (deposition of waxy material).
- In leaf epidermis, specialised bulliform cells (motor cells) are often found. These help in stomatal movements and also aid in rolling of leaves to reduce transpiration, e.g. *Ammophila*.
- Some epidermal cells contain calcium carbonate crystals. These are called as **cystoliths** (e.g. *Ficus*).
- Epidermal cells having cystoliths are called **lithocytes**.
- Epidermis of roots is called as **epiblema** or **piliferous** layer.
- Root epidermis does not have stomata and cuticle.

Stomata

- They are minute openings in the epidermis of green aerial parts of plant.
- Each stomata is surrounded by a pair of kidney-shaped guard cells. These cells in turn are surrounded by 4 or 8 subsidiary cells.
- The guard cells are living and contain chloroplast.
- Inner walls of guard cells are thicker than the outer walls.
- Guard cells regulate the opening and closing of stomatal pores.

- Guard cells are dumb-bell-shaped in some monocots, e.g. doobgrass, maize.

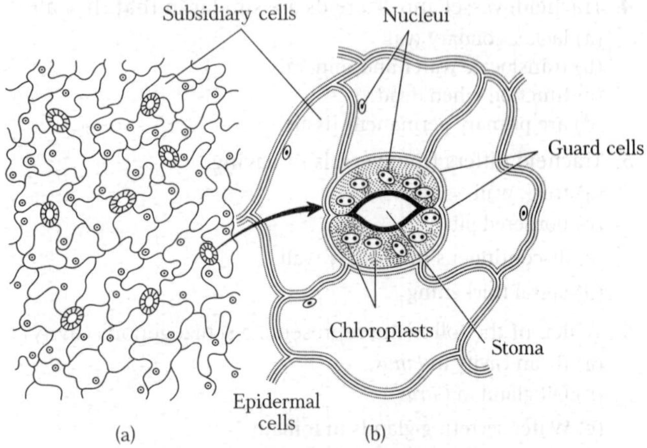

Fig. 12.9 Stomata : (a) Stomata on the lower surface of a dicot leaf, (b) Detailed structure of one stoma

Epidermal outgrowths

These are of following types

(i) **Trichomes** These are unicellular or multicellular outgrowths which vary in shape, structure and function.

These may be simple or branched and can be of different types such as hairs (unicellular or multicellular) or scales (multicellular), etc.

They prevent water loss due to transpiration. These can also be of secretory nature in some plants.

(ii) **Root hairs** These are tubular and unbranched outgrowths. These develop from trichoblast cells of epiblema.

Usually root hairs are unicellular. Their presence forms specialised root hair zones in roots. These increase the absorptive surface area of roots.

(iii) **Emergences** These are the multicellular outgrowths like prickles (e.g. spines of *Rosa indica*). They are sharp and stiff.

They lack any vascular supply. Prickles protect plants from excessive transpiration and grazing animals. These also help plants in climbing.

Ground tissue system

- It constitutes the main bulk of the plant body. Therefore, it is the largest tissue system. It is also called as the **fundamental tissue system.**
- It includes all the tissues except epidermis and the vascular bundles.
- It is partly derived from the periblem and partly from the plerome.
- Ground tissue of leaves is not differentiated. It consists of mesophyll cells only.
- The ground tissue system is differentiated into the following three main zones

1. Cortex

- It is the zone lying between the epidermis and pericycle.
- This zone varies in thickness and is made up of primary tissues.
- In roots of monocot and dicots, the cortex consists of parenchymatous cells.
- In stems, cortex is differentiated into the three subzones, i.e. hypodermis, general cortex and endodermis.

(i) **Hypodermis** It is also called **outer cortex**. This zone is present below the epidermis.

- It is the outermost zone of cortex. It consists of collenchyma cells in dicots, others as in monocot stems, sclerenchyma cells form hypodermis.
- This tissue protects the internal tissues. The hypodermis of dicot stem may contain chloroplast.
- Therefore, it can perform photosynthesis also. It provides mechanical strength and rigidity.

(ii) **General cortex** In stems, next to hypodermis general cortex is present. On the other hand in roots, it is present inner to epidermis.

- General cortex consists of several layers. Cells of this tissue are of parenchyma cells.
- Different types of crystals can also be found in the cells of general cortex. The cortex serves as a protective zone.
- It also helps in storage of food, absorption and translocation of water.

(iii) **Endodermis** It is also called as **starch sheath** due to the presence of starch. It is the innermost layer of cortex. It separates cortex from the stele.

- Endodermis consists of a single layer of parenchymatous cells. These cells are barrel-shaped and without any intercellular spaces.

Note ***Casparian strips*** *These are thickenings present in the radial walls of endodermal cells of roots. These thickenings are strips of lignin, suberin or cutin. Presence of Casparian strips makes endodermis impervious to water.*

2. Pericycle

- It is present in between endodermis and stele.
- Cells of pericycle can be parenchymatous or sclerenchymatous.
- In dicot stems, it is often multilayered, whereas not distinct in monocot stems.

- It is usually single-layered in both (moncot and dicot) roots. However, in the root of *Smilax* (a monocot), it is multilayered.
- It forms an outer boundary of vascular bundle and pith.
- Pericyle is absent in roots of parasitic plants and hydrophytes.
- In angiosperms, it gives rise to lateral roots. It is involved in the formation of vascular cambium in dicot roots.

3. Medulla or Pith

- It is the central core of the dicot stem and root. Pith is not distinct in monocot stems.
- It is made up of parenchymatous cells.
- Cells of pith are sclerenchymatous in family—Asteraceae.
- Stem becomes hollow when pith is torn (e.g. *Cucurbita*).
- The medulla or pith stores excretory substances, such as tannins, phenols, calcium oxalate, etc., along with reserve food material.
- The pith cells between the vascular bundles in dicot stems become radially elongated. These are known as medullary or pith rays these help in lateral translocation in plants.

Vascular tissue system

- It is derived from the procambium.
- It consists of varying number of strands or bundles called the **vascular bundles**.
- These vascular bundles are found scattered in ground tissue, monocot stem. They are properly arranged in dicot stems and root.
- Each vascular bundle is primarily made up of xylem and phloem. It also contains cambium in dicot stems.
- The cambium is a thin strip of meristematic tissue. Due to its presence inside the vascular bundle, cambium is also called as **intrafascicular cambium**. It is involved in secondary growth of dicot plants.
- Monocots do not possess cambium.

Depending upon the arrangement of xylem and phloem, vascular bundles can be classified as

1. Radial

These are mainly found in roots. In these, xylem and phloem are arranged alternately on different radii.

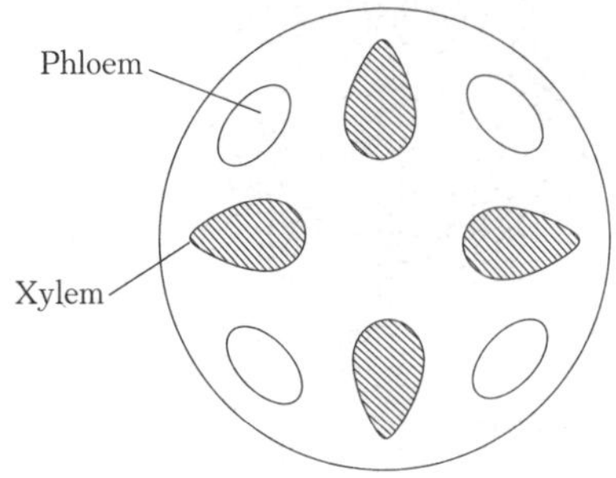

Fig. 12.10 Radial vascular bundle

2. Conjoint

These are commonly found in stems and leaves. In these, xylem and phloem are situated at the same radius.

These vascular bundles are further of two types

(i) **Collateral** It is the condition in which xylem lies towards the inner side and phloem lies towards the outer side.

A collateral vascular bundle can either be open or closed. Open collateral bundles have cambium in between xylem and phloem. While closed bundles lack cambium.

(ii) **Bicollateral** In this vascular bundle, the xylem is sandwiched between the two layers of phloem. In these bundles, two strips of cambium are found, e.g. *Cucurbita*, *Lagenaria*, etc.

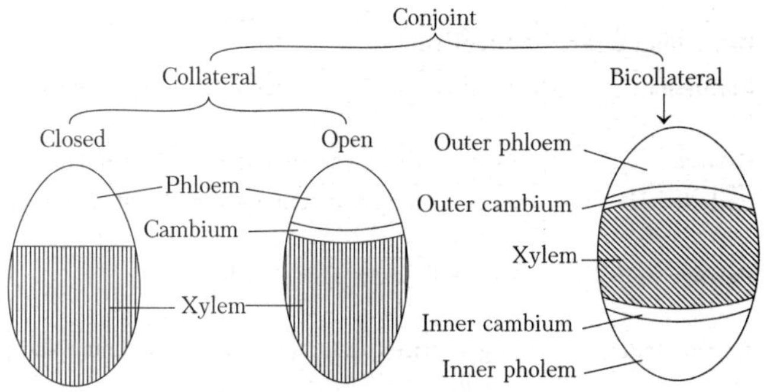

Fig. 12.11 Types of conjoint vascular bundles

3. Concentric

In concentric vascular bundles, either phloem surrounds the xylem or xylem surrounds the phloem completely. These can be further categorised as

(i) **Amphicribral** (hadrocentric) It is when xylem is present in centre. It is surrounded by a ring of phloem, e.g. ferns, *Prunus* (a dicot), etc.

(ii) **Amphivasal** (leptocentric) It is when phloem is present at the centre and is surrounded by xylem, e.g. *Dracaena, Yucca*, etc.

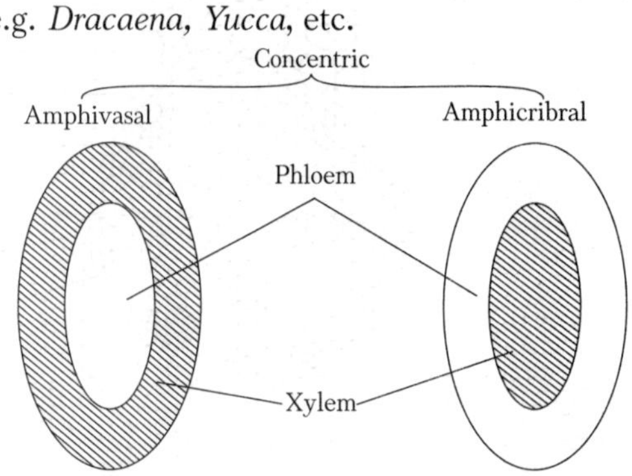

Fig. 12.12 Types of concentric vascular bundles

ANATOMY OF DICOT AND MONOCOT PLANTS

Root (Internal structure)

Three distinct zones are found in both monocot and dicot roots

(i) **Epidermis** (Epiblema/Rhizodermis) It consists of single (uniseriate) and thin-layered cells. Cutin is absent. The epidermal layer in roots is called **epiblema** or **rhizodermis** or **piliferous layer**.

(ii) **Cortex** It consists of thin-layered parenchymatous cells. In monocots, cortical cells become suberised and form a protective tissue called **exodermis.** The cortical cells may also store food material, e.g. carrot.

The innermost layer of cortex is endodermis, which consists of closely packed living cells. These cells possess band-like thickenings of lignin or suberin on its radial and transverse walls. These bands are called **Casparian strips.**

Some cells of endodermis which lie opposite to protoxylem remain thin-walled. These are called **passage cells**, which help in radial distribution of water.

(iii) **Vascular bundles** It is radial and exarch in roots. In monocot roots, central part of vascular bundle consists of parenchymatous pith.

Some basic differences between dicot and monocot roots can be summarised as given below

Table 12.3 Differences between dicot and monocot roots

Basic characters	Dicot roots	Monocot roots
Epiblema	Short-lived, single- layered (uniseriate) and with unicellular root hairs.	Generally persistent, single-layered (uniseriate) and with unicellular root hairs.
Cortex	Comparatively narrow and not differentiated into exodermis.	Very wide and differentiated into exodermis.
Endodermis	Less thickened. It contains prominent Casparian strips.	Casparian strips are visible only in young roots as the endodermal cells later become highly thickened.
Passage cells	Generally lacks passage cells.	Generally occur in endodermis opposite to the xylem bundle region.
Protoxylem	It gives rise to lateral roots, cork cambium and part of vascular cambium.	It gives rise to lateral roots only.
Vascular bundles	Radial, xylem exarch and vary from diarch to hexarch (2-6 in number).	Radial, xylem exarch and polyarch condition is seen (8 or more in number).
Secondary growth	Takes place with cambium formation.	Cambium and secondary growth are absent.
Pith (medulla)	Small or absent.	Large and well-developed.
	e.g. sunflower (*Helianthus*), pea (*Pisum*) and gram (*Cicer*).	e.g. maize (*Zea mays*), Orchid (*Vanda*) and *Sorghum*, etc.

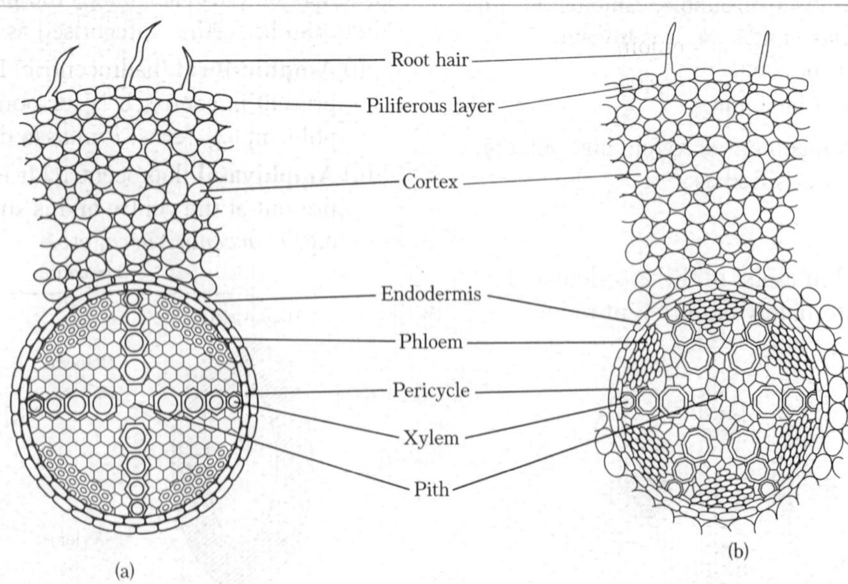

Fig. 12.13 TS of (a) dicot root, (b) monocot root

Stem (Internal structure)

The internal structures of dicot and monocot stems are as follows

Dicot stem

The dicot stem may be circular (e.g. *Helianthus*, (sunflower) or angular (e.g. *Cucurbita*).

The various primary tissues of a dicot stem are arranged in concentric fashion.

The transverse section of young sunflower (*Helianthus annuus*) stem reveals the following structures

(i) **Epidermis** It is made up of compactly arranged, rectangular and barrel-shaped single-layered parenchymatous cells. Cuticle is present. Many unbranched, multicellular hairs (trichomes) are also present.

(ii) **Hypodermis** It is below the epidermis and consists of 3-4 layers of collenchymatous cells.

(iii) **General cortex** It is made up of parenchymatous cells with intercellular spaces. The general cortex of sunflower contains many oil ducts. Each oil duct is surrounded by a glandular parenchymatous layer.

(iv) **Endodermis or Starch sheath** It is the single layer of innermost layer of the general cortex. It consists of single-layer of barrel-shaped cells, without intercellular spaces. Casparian strips are absent.

(v) **Pericycle** It is multilayered and has alternating bands of parenchyma and sclerenchyma cells. These bands lie in between the endodermis and vascular bundles.

(vi) **Vascular bundles** These are conjoint, collateral and are of open type. These are arranged in a ring and in a definite number. Xylem shows endarch condition.

(vii) **Medullary rays** These are also seen as few layers of parenchymatous cells, helping in the radial conduction of food and water. They also transport gases from pith to cortex and *vice-versa*.

(viii) **Pith** It is present in the central portion of stem. It is made up of polygonal, oval or rounded parenchymatous cells with large intercellular spaces.

Pith in some dicot stems often contains additional vascular bundles. These are called **medullary vascular bundles**. These are common in members of genus–*Mirabilis* and *Amaranthus*, etc.

Monocot stem

In monocot stems, vascular cambium is absent. Therefore, they do not show secondary growth. The monocot stems are solid (as in *Zea mays and Asparagus*) or can also be fistular, i.e. with central cavity (e.g. grass, wheat). The internal structure of maize is discussed below

(i) **Epidermis** It is the outermost layer of compactly arranged, barrel-shaped cells. It is uniseriate and is covered with a thick cuticle. Epidermal hairs are absent. Stomata are present at some places in the epidermis.

(ii) **Hypodermis** It lies below the epidermis. It is constituted by the sclerenchymatous cells. It is usually multiseriate.

(iii) **Ground tissue** It lies below the hypodermis. There is no differentiation into cortex, endodermis, pericycle and pith. The ground tissue is made up of rounded parenchymatous cells.

(iv) **Vascular bundles** These are scattered throughout the ground tissue. They are smaller in size. These are more in number towards the periphery than towards the centre.

Each vascular bundle is surrounded by a sheath of sclerenchymatous tisssue. It is called as **bundle sheath.** The vascular bundles are conjoint and collateral. They are closed as cambium is absent. Xylem shows endarch condition.

Table 12.4 Differences between dicot and monocot stems

Basic characters	Dicot stems	Monocot stems
Guard cells	Kidney-shaped.	Dumb-bell-shaped.
Hypodermis	Collenchymatous.	Sclerenchymatous.
Ground tissue	Differentiated into cortex, endodermis, pericycle and pith.	Uniform, no differentiation.
Vascular bundles	Arranged in a ring. They are conjoint and open.	Scattered in ground tissue. These are conjoint and closed.
Medullary rays	Present in between vascular bundle.	Absent.
Secondary growth	Takes place, cambium present.	Normally no secondary growth occurs, cambium absent. Few exceptions are *Dracaena, Aloe*.
Xylem vessels	Polygonal in outline. They are usually arranged in rows.	Oval or rounded in outline. Protoxylem and metaxylem are arranged in Y-shaped manner.

Basic characters	Dicot stems	Monocot stems
Fate of vascular tissue	The older vascular tissues cease functioning. After sometimes, it is replaced by newer ones.	The primary vascular tissue remains functional throughout the life of the plant.
Girth size	The stem shows more increase in diameter with age due to secondary growth.	There is little increase in diameter with age.

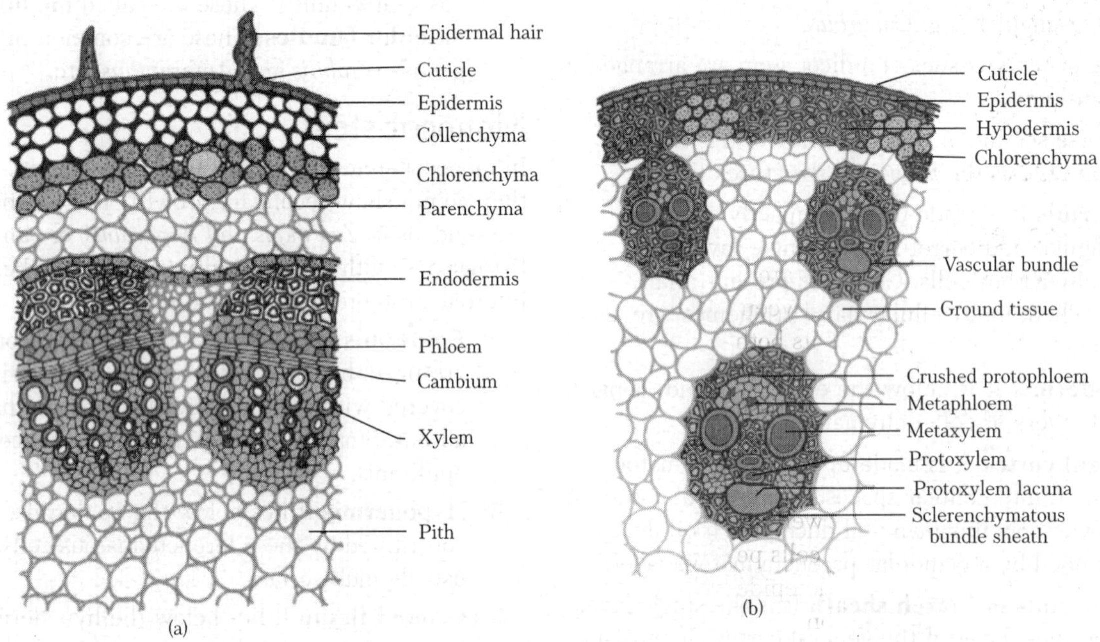

Fig. 12.14 TS of stem (a) Dicot, (b) Monocot

Leaf (Internal structure)

- Depending on the internal structure, the angiospermic leaves are of two types, i.e. **dorsiventral** (bifacial) in dicot plants or **isobilateral** (equifacial) in monocot plants.
- Leaf has epidermis on both the surfaces.
- In leaf, xylem always faces towards the upper epidermis, while phloem is towards the lower epidermis.

Dicot leaves

Most of the dicotyledonous leaves are dorsiventral or bifacial. They have distinct upper (dorsal/adaxial) and lower (ventral/abaxial) surfaces.

The dorsal surface is more strongly illuminated than ventral. This unequal illumination results in a difference in the internal structure.

A typical dorsiventral leaf shows following regions

(i) **Epidermis** It is present on both upper and lower surfaces.

(a) **Upper epidermis** It is the outermost layer. It is made up of a single layer of parenchymatous cells. It is mostly cuticularised. Upper epidermis consists of stomata and chloroplast. In some cases, stomata are absent too.

(b) **Lower epidermis** It possesses same features as upper epidermis. However, stomata are more in lower epidermis.

(ii) **Mesophyll** It is the ground tissue of leaf present between the upper and lower epidermis. It is differentiated into palisade and spongy parenchyma.

(a) **Palisade parenchyma** The cells of this tissue are elongated and tightly packed. They form an angle of 90° with the upper epidermis. These cells possess chloroplasts. Thus, they actively take part in photosynthesis.

(b) **Spongy parenchyma** This tissue is present below the palisade parenchyma. These cells are spherical or oval in shape. They are irregularly arranged with large air spaces in between. These cells also possess fewer chloroplasts than palisade parenchyma.

(iii) **Vascular bundles** They are found in the veins and midrib of leaves. They are conjoint, collateral and closed. Xylem is present towards upper epidermis. Whereas, phloem is towards lower epidermis. Each vascular bundle is surrounded by a layer of parenchymatous bundle sheath.

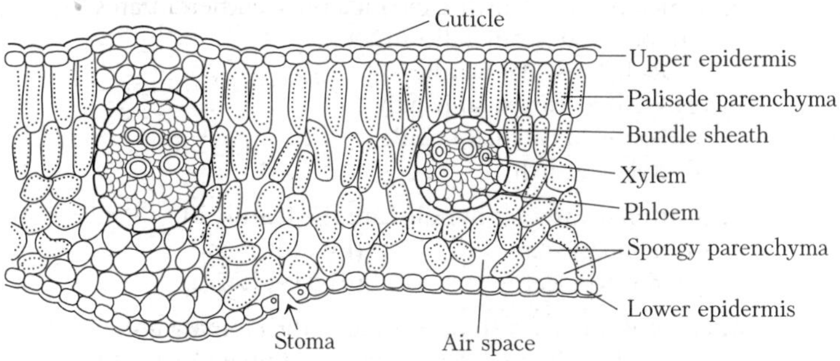

Fig. 12.15 Detailed structure of a part of TS of a dicot leaf

Monocot leaves

Most of the monocotyledonous leaves are isobilateral or equifacial. There is no distinction between the dorsal (upper) and ventral (lower) surface. Its both surfaces receive equal amount of sunlight. Therefore, there is no differentiation of mesophyll into palisade and spongy tissues.

The internal structure of isobilateral leaf, show following features

(i) **Epidermis** Both upper and lower epidermis are made up of single layer of cells. These cells possess stomata. Bulliform cells are found in the epidermis. These cells reduce the rate of transpiration.

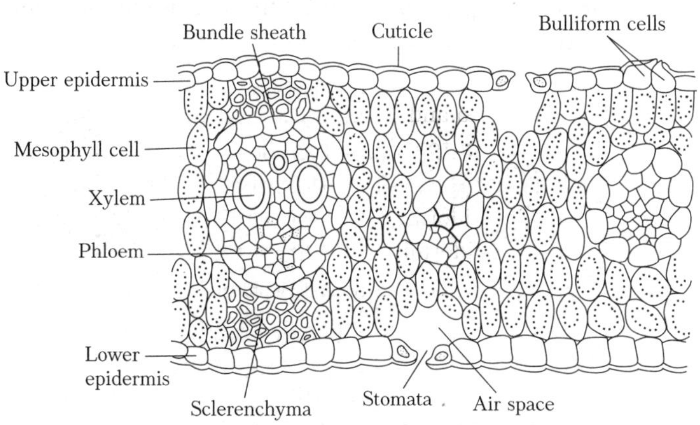

Fig. 12.16 Detailed structure of a part of TS of monocot leaf

(ii) **Mesophyll** It is not differentiated into palisade and spongy parenchyma. Cells of mesophyll are spherical and irregularly arranged.

(iii) **Vascular bundles** These may vary in size. Each vascular bundle is surrounded by bundle sheath. Cells of bundle sheath contain starch. Bundles are conjoint and closed.

Note *In xeric leaves, undifferentiated mesophyll occurs in concentric layers around the vascular bundles. They also possess large centrifugal chloroplast in bundle sheath cell. This specific arrangemant is called* ***Kranz anatomy****.*

Table 12.5 Differences between dorsiventral (dicot) **and isobilateral** (monocot) **leaves**

Basic characters	Dorsiventral leaves	Isobilateral leaves
Stomata	Usually absent or less abundant in upper epidermis. Stomata are numerous on lower epidermis.	Equally distributed in upper and lower epidermis.
Guard cells	Kidney-shaped	Dumb-bell-shaped
Bulliform cells (motor cells)	Absent	Present
Mesophyll	Differentiated into palisade and spongy parenchyma.	Undifferentiated
Vascular bundles	Arranged irregularly	Arranged in a row
Bundle sheath cells	Colourless	Chlorophyllous
Bundle sheath extensions	Parenchymatous	Sclerenchymatous
Protoxylem	Almost indistinguishable, e.g. *Nerium, Helianthus.*	Distinguishable, e.g. *Zea mays,* grasses.

CHECK POINT 12.3

1. The epidermal tissue system is characterised by all of these except
(a) parenchymatous cells without intercellular space
(b) unicellular root hair
(c) thick cuticle on epiblema
(d) trichomes in stem epidermis

2. The ground tissue system is made up of
(a) epidermis and vascular tissue
(b) simple and complex permanent tissues
(c) simple tissues such as parenchyma, collenchyma and sclerenchyma
(d) only thin-walled simple permanent tissue

3. Amphivasal vascular bundles are found in
(a) *Triticum* and *Zea* (b) *Dracaena* and *Yucca*
(c) *Cucurbita* and *Helianthus* (d) None of these

4. Which of the following characteristics differentiates monocot root from a dicot root?
(a) Radial vascular bundle
(b) Well-developed large pith
(c) Scattered vascular bundle
(d) Xylem exarch

5. Scattered vascular bundles are the feature of
(a) monocot stem (b) dicot stem
(c) monocot leaf (d) dicot leaf

6. In an isobilateral (monocot) leaf, the stomata are
(a) usually absent or less abundant in upper epidermis
(b) usually absent or less abundant in lower epidermis
(c) almost equally distributed in upper and lower epidermis
(d) usually more on upper epidermis

SECONDARY GROWTH

- The secondary growth is the increase in the girth (diameter) and thickness of a plant. It occurs as a result of activities of primary and secondary lateral meristems.
- **Intrafascicular cambium, interfascicular cambium** and **cork cambium** are the lateral meristems that participate in the secondary growth.
- These meristems become active after the formation of primary tissue.
- They divide to form secondary tissues of plant, namely secondary xylem, secondary phloem, secondary medullary rays, cork and secondary cortex.
- This phenomenon is seen in perennial gymnosperms and dicots only.

Secondary growth in dicot roots

The secondary growth in dicot roots takes place both in stelar region and extrastelar region. In stelar region, it occurs by vascular cambium.

On the other hand in extrastelar region, it occurs by cork cambium.

The whole process can be discussed as under

1. Formation of vascular cambium

- The vascular cambium develops from the pericycle cells.
- In this process, the pericycle cells present below each phloem bundle becomes meristematic, i.e. these regain their ability to divide.
- They give rise to separate strips of cambia.
- Later on the pericycle cells present outside the protoxylem also become meristematic. As a result of this fusion, a complete cambial ring is formed.
- The entire, circular ring of cambium produces secondary xylem towards the inner side and secondary phloem towards the outer side.
- The primary xylem bundles remain in their original position.
- At this stage, the dicot roots possess exarch protoxylem strands.
- The primary phloem, endodermis and cortex normally are crushed due to the enormous increase in the volume of the secondary tissue.

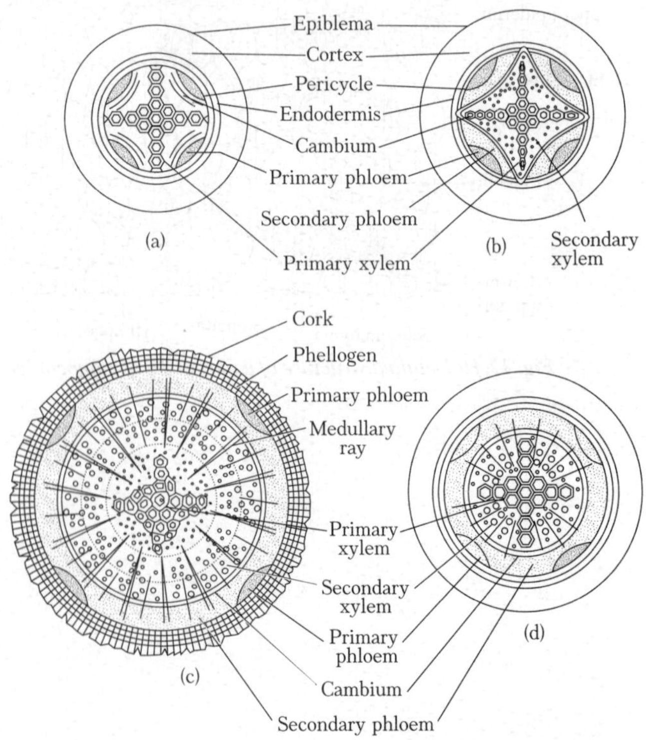

Fig. 12.17 Various stages of secondary growth in a dicot root

2. Formation of cork cambium

- In extrastelar region, the outer layers of pericycle also become meristematic. They give rise to thin-walled, rectangular cells, called **cork cambium** or **phellogen**.
- It produces cork or phellem towards the outer side and secondary cortex or phelloderm towards the inner side.
- The endodermis, general cortex and epiblema disorganise. Thus, the phellem becomes the outermost protective layer.
- Lenticels may also develop in the roots during extrastelar secondary growth.

Secondary growth in dicot stems

The secondary growth in a typical dicot stem also takes place in stelar and extrastelar regions.

1. Secondary growth in stelar region

The secondary growth in stelar region occurs in following steps

Formation of cambial ring

- Vascular cambium in dicots is arranged in a ring.
- Vascular cambium consists of two types of cells. These are called (a) **fusiform initials** and (b) **ray initials.**
- **Fusiform initials** are large with pointed end. During secondary growth, they form secondary xylem and phloem.
- **Ray initials** are small and isodiametric cells. They develop into medullary rays.
- Cambial cells are present between primary xylem and primary phloem in the intrafascicular cambium.
- The cells of medullary rays, adjacent to intrafascicular cambium become meristematic and form the interfascicular cambium. Together these form a vascular cambial ring.

Formation of secondary xylem and phloem

- The cambial ring divides periclinally to form new cells towards both outer and inner side.
- Cells towards pith mature into secondary xylem.
- On the other hand, cells towards periphery mature into secondary phloem. Thus, secondary xylem is present inner to cambial ring and outer to primary xylem.
- Likewise, secondary phloem is formed towards outside the cambial ring and inner to primary phloem.
- Secondary xylem is produced more than the secondary phloem.
- It mainly occurs due to two reasons, i.e.
  (i) Cambium is more active on inner side than on the outer side.
  (ii) Primary and secondary phloem are gradually crushed due to accumulation of secondary tissues. Hence, they contribute to more visibility of the secondary xylem.

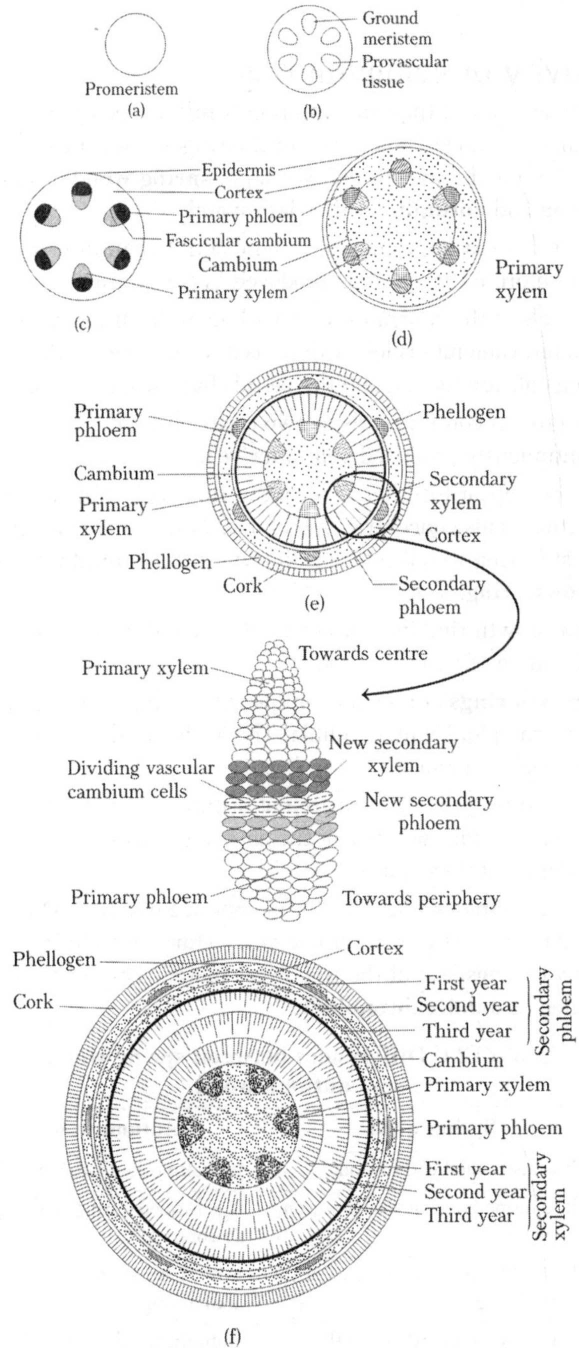

Fig. 12.18 Various stages of secondary growth in a dicot stem (a-f)

Formation of secondary medullary ray

- At some places, cambium forms a narrow band of parenchyma cells.
- These bands are continuous from secondary xylem to secondary phloem. They are called as **secondary medullary rays** or **vascular rays**. They are arranged radially.

Activity of cambium ring

- The activity of the cambium ring is influenced by the climatic changes. These seasonal changes result in formation of two types of wood, i.e. **spring wood** or **early wood** and **autumn** wood or **late wood.**
- In spring season, cambium is more active than in winter. Therefore more xylem is produced in this season.
- Vessels of this xylem have wider lumens in spring wood. In winter, cambium is less active. As a result, fewer xylary elements are formed and the vessels have narrow lumens.
- As growth continues, these two types of wood are continuously produced year after year.
- These appear together as a concentric ring in transverse section. This concentric ring of dark (autumn wood) and light (spring wood) colour is known as the **annual ring** or **growth ring**.
- One growth ring includes one circle of spring wood and one circle of autumn wood.
- **Growth rings** are distincts in plants of temperate region. For example, Nanital, Shimla. This is due to the clear seasonal variation at these places.
- In contrast, growth rings are not distinct in plants of tropical regions because of the absence of seasonal variation at these places.
- **Annual ring** can be used to find out age of a tree. One ring represents growth of one year. Thus, by counting growth rings, age of the tree is calculated. This study is called as **dendrochronology**.

Table 12.6 Differences between spring wood and autumn wood

Spring wood	Autumn wood
It is also called early wood.	It is also called late wood.
It is formed during spring season.	It is formed during winter season.
It constitutes the major part of the annual ring.	It is a narrow strip in the annual ring.
Spring wood is present in the beginning of an annual ring.	Autumn wood is present at the end of an annual ring.
Forms plenty of xylem vessels with wider lumen.	The lumen of xylem vessels is narrow.
Xylem fibres are fewer in number.	Abundant xylem fibres are produced.
Wood is light in colour.	Wood is dark in colour.

2. Secondary growth in extrastelar region

- Secondary growth in stelar region causes plant to increase in girth. Due to this, epidermis and cortex ruptures. Therefore, to provide protection, new tissues are formed in plant.
- The new protective layers of cells are formed by activity of **cork cambium** or **phellogen.** It develops in cortical region during secondary growth.
- Phellogen has two layers of meristematic cells. The outer layer of cells differentiates into cork or phellem. The inner cells differentiate into secondary cortex.

Cork or Phellem

- These are closely packed, thick-walled cells. They are more or less rectangular in shape.
- As they mature, suberin is deposited on to their cell walls. It makes cork impervious to water.
- Cork cells are airtight and used as stopper in bottles as cork. Bottle cork is obtained from ***Quercus suber*** (oak).

Phelloderm or Secondary cortex

- Cells of secondary cortex are parenchymatous.
- The phellem, phellogen and phelloderm are together called as **periderm**.
- Periderm is secondary protective tissue

Bark

- The bark includes all the tissues lying outside the vascular cambium.
- It also involves phellem, phellogen and phelloderm along with, crushed primary cortex, secondary and primary phloem.
- All these tissues may be included in the **outer bark** whereas pericycle, primary and secondary phloem in the **inner bark**.
- The outer dead region of the bark is called **Rhytidome**.
- Bark is protective in nature and protects phloem from external injuries.
- When bark is removed, phloem of stem gets damaged. Due to this, root dies of starvation (girdling experiment).

- Bark may be of two types
  - (a) **Scaly bark** It develops in strips, e.g. *Eucalyptus*, *Psidium*.
  - (b) **Ring bark** It develops in the form of rings, e.g. *Betula* (Bhojpatra).

Lenticels

- These are aerating pores and present on the surface of the bark as scars.
- They represent a small region of periderm. In this region, the phellogen activity is more than anywhere else.
- Lenticels serve for exchange of gases with the atmosphere and is also responsible for loss of water vapour (lenticular transpiration). They are also called as **breathing pores**.

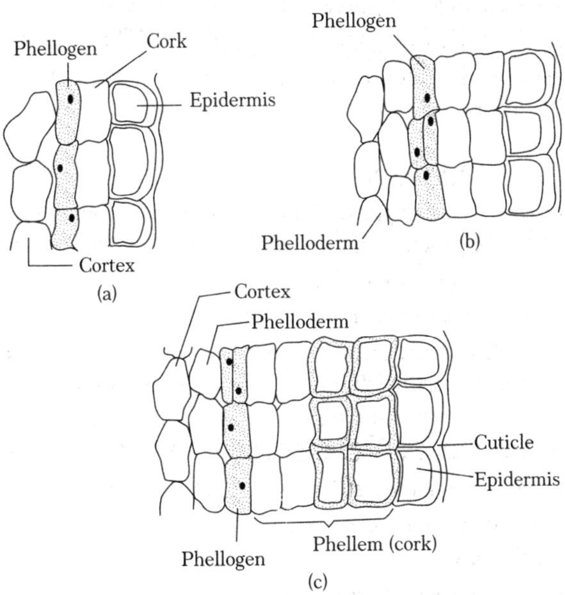

Fig. 12.19 Formation of (a) Phellogen, (b) Phelloderm and (c) Phellem

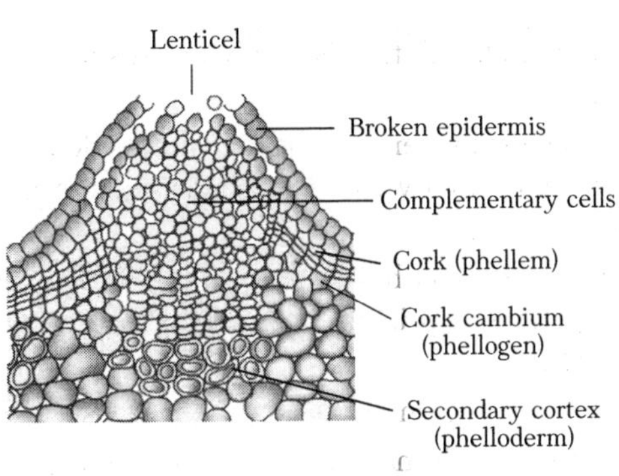

Fig. 12.20 Structure of lenticels

Table 12.7 Differences between secondary growth in dicot stem and dicot root

Secondary growth in dicot stem	Secondary growth in dicot root
The vascular cambial ring develops as a circular ring since beginning.	The vascular cambial ring is wavy in the beginning and becomes circular later on.
The vascular cambial ring is completed by both the primary (intrafascicular) strip and secondary (interfascicular) strip.	The vascular cambial ring is secondary in origin. It is formed of conjunctive parenchyma and pericycle that becomes meristematic.
The vascular rays are narrow since beginning.	The vascular rays are wider in the beginning.
Annual rings or growth rings occur commonly.	Annual rings or growth rings are absent.
Outward layer of the cortex gives birth to cork cambium (phellogen).	Generally, the pericycle gives birth to cork cambium (phellogen).

Secondary growth in monocot stems

Secondary growth is absent in monocot stems as cambium is not found in monocot plants. However, as an exception some monocots such as *Dracaena*. *Yucca*, *Agave* and *Aloe* show secondary growth.

In these plants, vascular cambium is formed from outer region of ground tissue.

In plants like palm, *Musa* and *Tulipa*, primary thickening of meristem contributes to secondary growth.

Heartwood and Sapwood

In old trees, wood is differentiated into heartwood and sapwood.

Heartwood

- It is the centrally located darker and harder wood. It is also known as **duramen**.
- It is physiologically inactive (almost dead).
- Heartwood does not conduct water. It is resistant to microorganisms and insects. A well-developed heartwood is found in *Morus alba* whereas it is absent in *Populus* and *Salix*.
- The dark colour of wood is due to the deposition of tannins, resins, gums and essential oils, etc., in cell wall and cavities.

Sapwood

- It is also known as **alburnum.** It is peripherally located and conducts the water and dissolved minerals from roots to leaves.

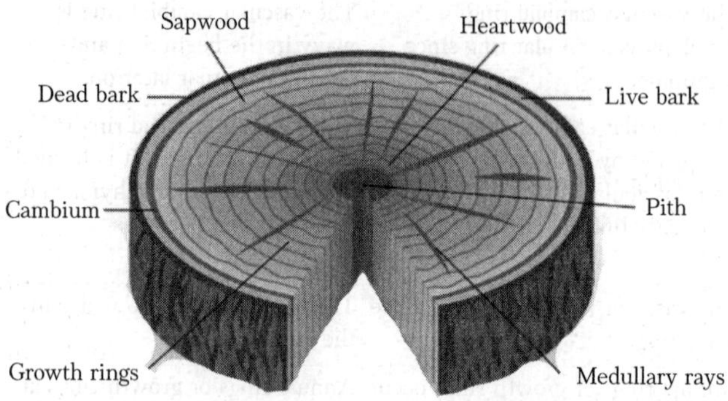

Fig. 12.21 Heartwood and sapwood

- In due course of time, new outer rings of secondary xylem are added to wood. In this way, sapwood keeps on changing into heartwood.
- During the conversion of sapwood into heartwood, **tyloses** are developed in heartwood. These are balloon-like and develop from xylem parenchyma.
- Tyloses block the passage of xylem vessels, hence also known as **tracheal plug.**

Softwood and Hardwood

- Based on the presence or absence of tracheids and vessels, wood can also be of following two types
- **Softwood** which consists of tracheids (90-95%) with few xylem fibres. This wood is also known as **non-porous wood,** e.g. gymnospermic wood.
- **Hardwood** which consists of both tracheids and vessels. Due to the presence of vessels, it is also known as **porous wood**. This wood consists of 5% tracheids with xylem fibres.

CHECK POINT 12.4

1. Secondary growth in dicot roots is not characterised by
   (a) wavy ring of vascular cambium
   (b) dedifferentiation of cells just below the phloem bundles
   (c) vascular cambium is half in primary and half in secondary origin
   (d) purely secondary origin of cork cambium
2. Amount of secondary xylem is more than secondary phloem because
   (a) cambium is more active on the outer side
   (b) cambium is more active on inner side
   (c) cambium has no role
   (d) cambium is active equally so on both sides, but xylem is required
3. An annual ring refers to
   (a) spring ring and autumn ring of 2 years
   (b) early wood and late wood of 1 year
   (c) spring wood of 1 year
   (d) autumn wood of 1 year
4. During secondary growth in dicot stem at certain regions, cork cambium cuts off parenchymatous cell on the outer side instead of ....
   (a) cork cells
   (b) phelloderm
   (c) phellogen
   (d) secondary phloem
5. The inner darker and harder portion of secondary xylem that cannot conduct water in an elder dicot stem, is called
   (a) alburnum
   (b) bast
   (c) wood
   (d) duramen
6. Sapwood is synonymous with
   (a) peripheral part of secondary xylem
   (b) bark
   (c) central part of secondary xylem
   (d) periderm

EXAM POINTS

- The girth of the stem increases due to activity of lateral meristems. **[COMED-K 2015]**
- Interfascicular cambium is a secondary meristematic tissue. **[WB JEE 2013]**
- Sclerenchyma tissues provides mechanical strength to plants. **[BHU 2012]**
- Regeneration of damaged growing grass following grazing is largely due to intercalary meristem. **[NEET 2019]**
- The balloon-shaped structures called tyloses are extensions of xylem parenchyma cells into vessels. **[NEET II 2016]**
- Medullary rays are tissues made up of xylem parenchyma. **[WB JEE 2015]**
- The cell wall of parenchyma is made up of pectin. **[[J&K CEE 2014]]**
- Stomata in grass leaf are dumb-bell-shaped. **[NEET 2018]**
- Specialised epidermal cells surrounding the guard cells are called subsidiary cells. **[NEET-I 2016]**
- Casparian strips occur in endodermis. **[NEET 2018]**
- Cortex is the region found between epidermis and stele. **[NEET-II 2016]**
- In dicot root, the vascular cambium originates from tissue located below the phloem bundles and a portion of pericycle tissue above protoxylem. **[NEET 2019]**
- Transport proteins of endodermal cells are control points where a plant adjusts the quantity and types of solutes that reach the xylem. Root endodermis is able to actively transport ions in one direction only because of the layer of suberin. **[COMED-K 2015]**
- Monocotyledonous stem has large number of scattered vascular bundles surrounded by bundle sheath, large conspicuous parenchymatous ground tissue, vascular bundles are conjoint and closed and phloem parenchyma is absent. **[NEET 2020]**
- In dicotyledonous stem, the sequence of tissues from inside to outside is pith, protoxylem, metaxylem, cambium, phloem, pericycle, endodermis, parenchyma, collenchyma and epidermis. **[J&K CEE 2014]**
- Secondary xylem and phloem in dicot stem are produced by vascular cambium. **[NEET 2018]**
- Interfascicular cambium develops from the cells of medullary rays. **[NEET 2013]**
- For a critical study of secondary growth in plants, teak and pine are suitable. **[CBSE AIPMT 2012]**
- Phellogen and phellem, respectively denote cork cambium and cork **[NEET 2017]**
- Phellem is made up of dead cells. **[NEET 2017]**
- The vascular cambium normally gives rise to secondary xylem. **[NEET 2017]**
- Lenticels are involved in gaseous exchange. **[NEET 2013]**
- Sapwood is outermost secondary xylem and is lighter in colour. **[NEET 2020]**
- No vessels are found in the wood of pine. **[AMU 2015]**
- When one wood is lighter in colour with a lower density, the other wood is darker with a higher density. They are spring wood and autumn wood. **[Kerala CEE 2014]**
- Duramen is present in inner region of secondary wood. **[UP CPMT 2012]**
- Vessels are multicellular and with wide lumen. **[BCECE 2012]**

Chapter Exercises

(A) Topical Exercise

Plant Tissues

1 The type of tissues found in plants is/are
(a) meristematic tissues
(b) permanent tissues
(c) secretory tissues
(d) All of the above

2 Meristematic tissues are
(a) premature having ability of division
(b) mature do not have ability of division
(c) premature not having ability of division
(d) complex differentiation in xylem, phloem and cambium

3 Cells of meristematic tissue possesses
(a) small endoplasmic reticulum
(b) immature plastids
(c) simple mitochondria
(d) All of the above

4 Consider the following statements with respect to vegetative shoot apical meristem.
I. Produces buds, leaves and stem tissues.
II. Protected by young leaves.
Choose the correct option.
(a) Statement I is true, but II is false
(b) Statement I is false, but II is true
(c) Both statements I and II are true
(d) Both statements I and II are false

5 Apical meristem and intercalary meristem are primary meristem as
(a) these are formed early in the life of a plant
(b) these are responsible for the formation of primary plant body
(c) these help in forming the early wood in a plant body
(d) Both (a) and (b)

6 The length of different internodes in a culm of sugarcane is variable because of
(a) shoot apical meristem
(b) position of axillary buds
(c) size of leaf lamina at the node below each internode
(d) intercalary meristem

7 The Tunica-Corpus theory was proposed by
(a) Hofmeister (b) Nageli
(c) Strasburger (d) Schmidt

8 According to Tunica-Corpus theory,
I. the cells of tunica give rise to epidermis.
II. the cells of corpus form cortex and stele of shoot.
Choose the correct option.
(a) Statement I is true, but II is false
(b) Statement I is false, but II is true
(c) Both statements I and II are true
(d) Both statements I and II are false

9 The chief function of chlorenchymatous tissue is to
(a) conduct water and minerals
(b) synthesise food
(c) store food material
(d) provide mechanical support

10 Parenchyma cells are found in
(a) petioles of *Hydrilla*, in the pericycle of roots
(b) stems of *Raphanus* in aerial roots of Monstera
(c) *Cucurbita* stem, in pulp of guava
(d) in the stem of *Cannabis*, in the pericycle of roots

11 Aerenchyma provides
(a) flexibility to plants
(b) mechanical strength of plants
(c) buoyancy to hydrophytic plants
(d) promoting nature of photosynthesis

12 Aerenchyma is present in which of the following plants?
(a) *Potamogeton* (b) *Vallisneria*
(c) Both (a) and (b) (d) *Bryophyllum*

13 A plant tissue, when stained, showed the presence of hemicellulose and pectin in cell wall of its cells. The tissue represents **[NCERT Exemplar]**
(a) collenchyma (b) sclerenchyma
(c) xylem (d) meristem

14 Collenchyma cells are absent in the stems and leaves of many monocotyledonous plants. It is so because
(a) they do not need any mechanical support
(b) dedifferentiate into secondary meristem in the early stage
(c) their wall thickening get dissolved, hence not easy to be identified
(d) sclerenchyma appears at an early stage of development

15 Sclereids found in the seed coat of pulses are
(a) macrosclereids (b) brachysclereids
(c) osterosclereids (d) asterosclereids

16 Consider the following statements.

I. The specialised non-green, large-sized parenchyma cells containing tannins, oils, crystals, etc., are called idioblasts.

II. Collenchyma are oval, spherical or polygonal cells, thickening at corners due to deposition of cellulose, hemicellulose and pectin.

Choose the correct option.
(a) Statement I is true, but II is false
(b) Statement I is false, but II is true
(c) Both statements I and II are true
(d) Both statements I and II are false

17 Select the correct difference between the given types of simple permanent tissue.

	Parenchyma	Sclerenchyma
I.	Possesses thin cellulosic cell wall	Possesses thick lignified cell wall
II.	Cells are usually living and isodiametric	Cells are usually dead and without protoplast
III.	Provides mechanical strength to the plant	Performs the function of storage and secretion

(a) I and II (b) Only I (c) Only II (d) I, II, and III

18 Collenchyma differs from sclerenchyma in
(a) being dead
(b) possessing protoplasm
(c) the ability to not dedifferentiate
(d) All of the above

19 Match the following columns.

	Column I		Column II
A.	Meristem	1.	Photosynthesis, storage
B.	Parenchyma	2.	Mechanical support
C.	Collenchyma	3.	Actively dividing cells
D.	Sclerenchyma	4.	Sclereids

Codes

	A	B	C	D
(a)	1	3	4	2
(b)	3	1	2	4
(c)	2	4	3	1
(d)	1	4	3	2

20 Choose the incorrect statement.
(a) Permanent tissues having all cells similar in structure and function are called simple tissue
(b) Complex tissue has same types of cells performing different functions altogether
(c) Secondary meristem is lateral meristem
(d) The cells of permanent tissues do not generally divide further

21 Identify the various types of xylem shown in the following figures.

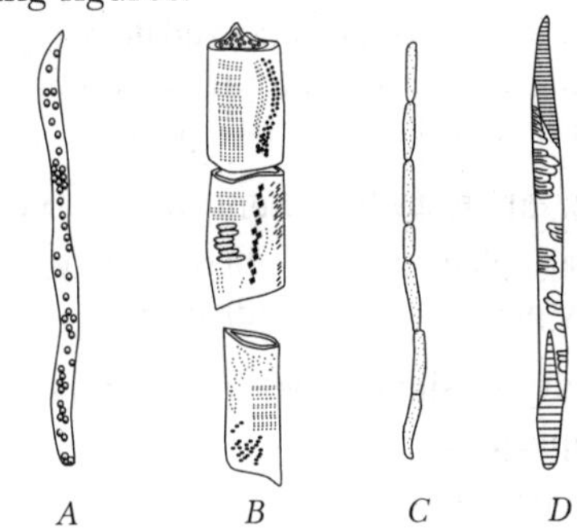

(a) *A*–Xylem tracheids, *B*–Xylem vessels, *C*–Xylem parenchyma, *D*–Wood fibres
(b) *A*–Xylem vessels, *B*–Xylem parenchyma, *C*–Xylem tracheids, *D*–Wood fibres
(c) *A*–Xylem parenchyma, *B*–Xylem tracheids, *C*–Xylem vessels, *D*–Wood fibres
(d) *A*–Xylem vessels, *B*–Wood fibres, *C*–Xylem tracheids, *D*–Xylem parenchyma

22 Sieve tubes differ from the sieve cells because they
(a) have sieve pores mainly at the end walls
(b) are shorter
(c) lack nuclei
(d) are dead

23 A common structural feature of vessel element and sieve tube element is
(a) thick secondary walls
(b) pores on end walls
(c) presence of P-protein
(d) nucleate condition

24 The secondary walls of tracheids and vessel element are hardened with
(a) cellulose (b) lignin
(c) pectin (d) Both (a) and (b)

25 A vessel-less piece of stem possessing prominent sieve tubes would belong to **[NCERT Exemplar]**
(a) *Pinus* (b) *Eucalyptus*
(c) Grass (d) *Trochodendron*

26 Read the following statements.

I. Vessel members are interconnected through perforations in their end walls.

II. Presence of vessels is a characteristic feature of gymnosperms.

Choose the correct option.
(a) Statement I is true, but II is false
(b) Statement I is false, but II is true
(c) Both statements I and II are true
(d) Both statements I and II are false

27 In sieve elements, the possible function of P-proteins is
(a) autolytic enzymes
(b) prevent leakage of phloem exudate
(c) providing energy for active translocation
(d) deposition of callose on sieve plates

28 Fibres are likely to be absent in **[NCERT Exemplar]**
(a) secondary phloem (b) secondary xylem
(c) primary phloem (d) leaves

29 The activity of sieve tubes is remotely controlled by the protoplasm of
(a) phloem parenchyma
(b) companion cells
(c) phloem fibres
(d) Both phloem parenchyma and phloem fibres

30 Consider the following statements.
I. Phloem fibres are generally absent in the primary phloem but are found in the secondary phloem.
II. Phloem fibres are much elongated, branched and have needle-like apices.
Choose the correct option.
(a) Statement I is true, but II is false
(b) Statement I is false, but II is true
(c) Both statements I and II are true
(d) Both statements I and II are false

31 A transverse section of stem is first stained with safranin and then with fast green following the usual schedule of double staining for the preparation of a permanent slide. What would be the colour of the stained xylem and phloem? **[NCERT Exemplar]**
(a) Red and green (b) Green and red
(c) Orange and yellow (d) Purple and orange

Plant Tissue System

32 Identify the tissue system from among the following. **[NCERT Exemplar]**
(a) Parenchyma (b) Xylem
(c) Epidermis (d) Phloem

33 In which of the following pairs of parts of a flowering plant is epidermis absent? **[NCERT Exemplar]**
(a) Root tip and shoot tip (b) Shoot bud and floral bud
(c) Ovule and seed (d) Petiole and pedicel

34 Select the incorrectly matched pair.
(a) Dumb-bell-shaped guard cells—Grasses
(b) Trichomes—Reduce water loss
(c) Piliferous layer—Root epidermis
(d) Starch sheath—Endodermis of dicot root

35 Radial vascular bundle is found in ....
(a) stem (b) leaf
(c) Both (a) and (b) (d) root

36 Pericycle in dicot root
(a) is collenchymatous
(b) give rise to lateral roots
(c) is multilayered
(d) Both (a) and (b)

37 (i) Xylem and phloem are present in alternating separate radii. Such vascular bundles are called ...*A*... .
(ii) Xylem and phloem are present on same radius. Such vascular bundles are called ...*B*... .
Select the correct option.

	A	*B*
(a)	Radial	Conjoint
(b)	Radial	Concentric
(c)	Conjoint	Radial
(d)	Conjoint	Concentric

38 Consider the following statements.
I. In conjoint vascular bundle, the phloem is located on the outer side of xylem.
II. In conjoint vascular bundle, cambium is never found.
Choose the correct option.
(a) Statement I is true, but II is false
(b) Statement I is false, but II is true
(c) Both statements I and II are true
(d) Both statements I and II are false

39 Identify the vascular bundles given in the following figures.

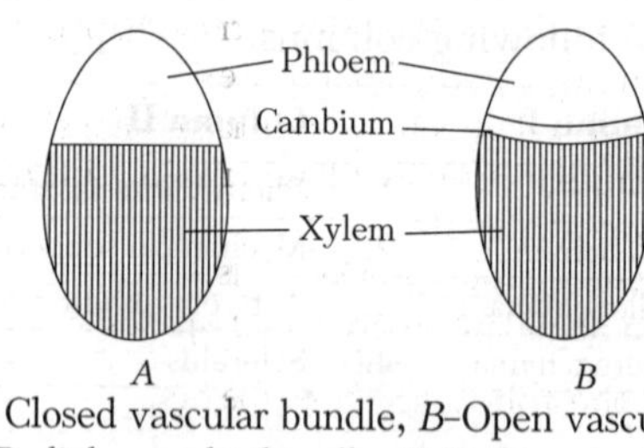

(a) *A*- Closed vascular bundle, *B*-Open vascular bundle
(b) *A*-Radial vascular bundle, *B*-Open vascular bundle
(c) *A*-Amphivasal vascular bundle, *B*-Amphicribral vascular bundle
(d) *A*-Amphicribral vascular bundle, *B*-Amphivasal vascular bundle

40 The type of vascular bundle in a Cucurbita is
(a) collateral (b) amphicribral
(c) bicollateral (d) amphivasal

41 A leptocentric or amphivasal vascular bundle has
(a) phloem flanked by xylem on interior and exterior sides only
(b) xylem flanked by phloem on interior and exterior sides only
(c) xylem surrounded by phloem
(d) phloem surrounded completely by xylem

42 Match the following columns.

Column I	Column II
A. Collateral and open	1. *Cucurbita* stem
B. Radial	2. Fern
C. Bicollateral	3. Maize root
D. Concentric	4. Sunflower

Codes

	A	B	C	D		A	B	C	D
(a)	5	4	3	1	(b)	4	3	1	2
(c)	4	1	3	2	(d)	4	3	2	1

Anatomy of Monocotyledonous and Dicotyledonous Plants

Roots

43 Epiblema of roots is equivalent to **[NCERT Exemplar]**
(a) pericycle (b) endodermis
(c) epidermis (d) stele

44 Cambium in dicot roots
(a) is formed during secondary growth and not present from the beginning during primary growth
(b) it is a group of slowly dividing cells
(c) initiates its formation from pericycle but not from conjunctive tissue
(d) never show differentiation

45 The innermost layer of cortex in a dicot root
(a) comprises a single layer of barrel-shaped cells without any intercellular spaces
(b) are composed of cells which have depositions of impermeable cutin and lignin mainly in their radial and tangential walls
(c) comprises a group of dead cells
(d) contain abundant intercellular spaces

46 Consider the following statements.
I. Pericycle of both dicot and monocot roots produces vascular cambium during secondary growth.
II. Xylem is produced in centripetal manner in roots of fruit bearing plants.

Choose the correct option.
(a) Statement I is true, but II is false
(b) Statement I is false, but II is true
(c) Both statements I and II are true
(d) Both statements I and II are false

47 The condition where the first formed xylem element faces the periphary is called ...*A*... and it is seen in ...*B*... .

	A	*B*		*A*	*B*
(a)	exarch	root	(b)	endarch	root
(c)	exarch	stem	(d)	mexarch	root

48 In monocot roots, which type of vascular bundles are found?
(a) Collateral, conjoint and closed
(b) Radial with exarch xylem
(c) Bicollateral, conjoint and closed
(d) Radial with endarch xylem

Stem

49 Anatomically fairly old dicotyledonous root is distinguished from the dicotyledonous stem by the
(a) absence of secondary xylem
(b) absence of secondary phloem
(c) presence of cortex
(d) position of protoxylem

50 In a dicotyledonous stem, the sequence of tissues from the outside to the inside is
(a) Phellem-Pericycle-Endodermis-Phloem
(b) Phellem-Phloem-Endodermis-Pericycle
(c) Phellem-Endodermis-Pericycle-Phloem
(d) Pericycle-Phellem-Endodermis-Phloem

51 Differentiation of ground tissue into endodermis, cortex, pericycle and pith is observed in
(a) *Helianthus* (b) *Zea mays*
(c) *Asparagus* (d) *Triticum aestivum*

52 Hypodermis observed in dicot stem is
(a) Sclerenchymatous (b) Collenchymatous
(c) Both (a) and (b) (d) None of these

53 Choose the correct statement regarding dicot stem.
(a) Dumb-bell-shaped stomata, medullary rays absent, little increase in diameter
(b) Secondary growth takes place, ground tissue differentiated, conjoint and open vascular bundle
(c) Guard cell dumb-bell-shaped, no secondary growth, ground tissue undifferentiated
(d) Medullary rays absent, xylem vessels are polygonal

54 What is the correct about monocot stem?
(a) Hypodermis is sclerenchymatous, vascular bundles are closed, phloem parenchyma is absent
(b) Hypodermis is sclerenchymatous, vascular bundles are open, phloem parenchyma is absent
(c) Hypodermis is collenchymatous, vascular bundles are closed, phloem parenchyma is present
(d) Hypodermis is sclerenchymatous, vascular bundles are closed, phloem parenchyma is present

55 Vascular cylinder in dicot stems occurs in the form of distinctly separated vascular bundles. But, in some dicot stems, e.g. *Linum*, *Lonicera*, it appears to be a continuous cylinder because
(a) interfascicular region is not developed at all
(b) the interfascicular region is there, but its cells are not differentiable from the cells of vascular bundles
(c) interfascicular region is extremely narrow
(d) continuous vascular cylinder differentiates since the early stage of development

Leaves

56 From the given set of statements characterising the internal structure of dicot leaf, identify the incorrect one.

(a) Type of leaf—Dorsiventral
(b) Mesophyll—Made up of palisade parenchyma and spongy parenchyma
(c) Bundle sheath—Made up of parenchyma cells
(d) Abaxial epidermis—Bears less stomata

57 Consider the following statements.

I. The adaxially placed palisade parenchyma is made up of elongated cells.
II. In a dorsiventral leaf, location of palisade tissue and phloem, respectively are adaxial and abaxial.

Choose the correct option.

(a) Statement I is true, but II is false
(b) Statement I is false, but II is true
(c) Both statements I and II are true
(d) Both statements I and II are false

58 In the vascular bundle of dorsiventral leaf, xylem lies towards ...*A*... and phloem lies towards ...*B*.... .

	A	*B*
(a)	adaxial epidermis	palisade tissue
(b)	adaxial epidermis	abaxial epidermis
(c)	abaxial epidermis	palisade tissue
(d)	spongy parenchyma	abaxial epidermis

59 Which of the following is true for the leaf anatomy shown below?

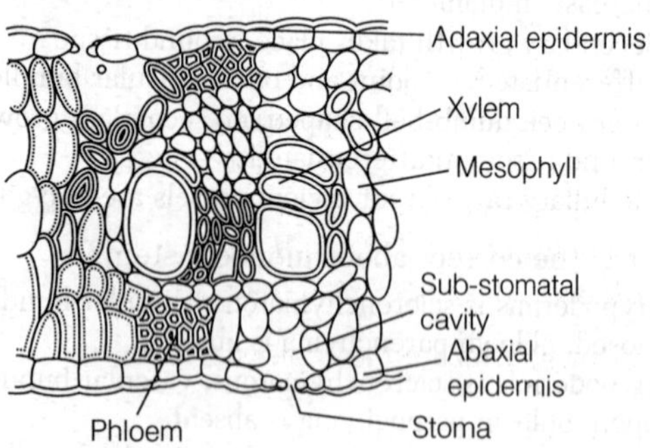

(a) It is with respect to a dorsiventral leaf
(b) It is with respect to a dicot leaf
(c) It shows the leaf anatomy of plants like *Zea mays*
(d) It shows the leaf anatomy of plants like *Pisum sativum*

60 Which one of the following sets represents the xerophytic characters of a monocot (isobilateral) leaf?

(a) Thick cuticle, sclerenchymatous patches, bulliform cells
(b) Thin cuticle, sclerenchymatous patches, bulliform cells
(c) Thick cuticle, parenchymatous patches, bulliform cells
(d) Thick cuticle, sclerenchymatous patches, myosin cells

Secondary Growth

61 Consider the following statements for secondary growth.

I. It helps in increase in girth or diameter of plant organs.
II. There are two types of lateral tissues involved in it called as vascular cambuim and cork cambium.

Choose the correct option.

(a) Statement I is true, but II is false
(b) Statement I is false, but II is true
(c) Both statements I and II are true
(d) Both statements I and II are false

62 Which of the following statements is incorrect?

(a) The tissue involved in secondary growth are the two lateral meristems, i.e. vascular cambium and cork cambium
(b) In dicot roots, cambium is completely secondary in origin
(c) In dicot stem, cambium is partially secondary and partially primary
(d) All lateral meristems are present at the time of primary growth of a plant body

63 What is the fate of primary xylem in a dicot root showing extensive secondary growth? **[NCERT Exemplar]**

(a) It is retained in the centre of the axis
(b) It gets crushed
(c) May or may not get crushed
(d) It gets surrounded by primary phloem

64 Increment borer is an instrument used to count

(a) the number of xylem vessels of early wood
(b) the number of annual rings
(c) the number of cambium rings
(d) the number of tyloses in each vessel

65 A portion of trunk of a dicot tree is shown. At which part, maximum number of growth rings can be observed?

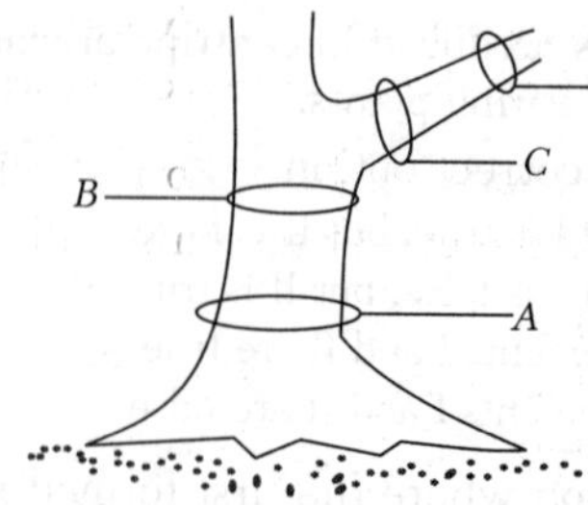

(a) *A* (b) *B*
(c) *C* (d) *D*

66 Applying dendrochronology, calculate the approximate age of a tree with respect to secondary growth in the given transverse section of a log of wood?

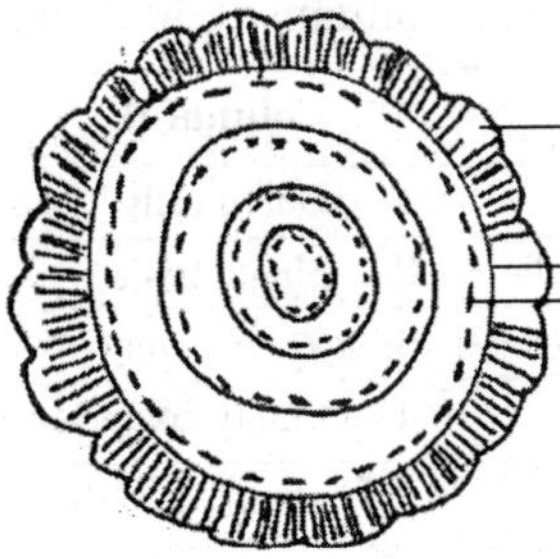

(a) 4 years old (b) 8 years old
(c) 2 years old (d) 7 years old

67 Autumn wood is ...*A*... in colour and has ...*B*... density.

	A	*B*
(a)	Darker	Higher
(b)	Lighter	Lower
(c)	Darker	Lower
(d)	Lighter	Higher

68 Cork cambium gives rise to
(a) phellogen and secondary cortex
(b) phellogen, phelloderm and secondary cortex
(c) cork and phellogen
(d) cork and secondary cortex

69 Removal of cork for commercial use is to be done with care. Otherwise the tree can die. This is because
(a) the xylem layer transporting water and minerals can be damaged
(b) the primay rays giving strength can be damaged
(c) the pith with storage cells can be damaged
(d) the phloem used in transporting the sugar can be damaged

70 Mark the incorrect statements with respect to phellogen.
(a) Couples of layers thick
(b) Cuts off cells on both sides
(c) Narrow, thin-walled and nearly rectangular cells
(d) Develops from differentiation of cortical cells

71 Bark formed early in the season is called ...*A*... bark and towards the end of season is called ...*B*... bark.

	A	*B*
(a)	ring	scaly
(b)	scaly	ring
(c)	soft	hard
(d)	hard	soft

72 Identify the mismatched pair.
(a) Phelloderm-Secondary cortex
(b) Phellem-Cork
(c) Spring wood-Lighter in colour
(d) Intrafascicular cambium-Secondary in origin

73 Choose the incorrect statement.
(a) Cork develops by secondary growth in extrastelar region
(b) Phelloderm is also known as secondary cortex
(c) Bark includes all the tissues lying outside the vascular cambium
(d) Lenticels are present on leaves

74 Consider the following statements.
I. Wood without vessel is called heteroxylous and wood with vessels is called homoxylous.
II. Heartwood is absent in *Populus* and *Salix.*
Choose the correct option.
(a) Statement I is true, but II is false
(b) Statement I is false, but II is true
(c) Both statements I and II are true
(d) Both statements I and II are false

75 In the following, how the sapwood is converted into heartwood?
(a) By degeneration of protoplast of living cells
(b) Tylosis formation
(c) By deposition of resins, oil, gums
(d) All of the above

76 Consider the following statements.
I. The bottle cork is prepared from cork of *Quercus suber.*
II. *Cedrus deodara* is the most durable wood.
Choose the correct option.
(a) Statement I is true, but II is false
(b) Statement I is false, but II is true
(c) Both statements I and II are true
(d) Both statements I and II are false

77 Centrally located and physiologically inactive wood is
(a) bark
(b) sapwood
(c) heartwood
(d) early wood

78 Heartwood differs from sapwood in
(a) the presence of rays and fibres
(b) the absence of vessels and parenchyma
(c) having dead and non-conducting elements
(d) being susceptible to pests and pathogens

(B) Cumulative Exercise

(Assorted Questions of the Chapter for Advanced Level Practice)

1 The number of stomata and epidermal cells in 1 mm^2 leaf area of lower epidermis of the leaves of *X, Y* and *Z* plants are given below. Arrange the plants in decreasing order of their stomatal index.

Plants	Numbers of stomata	Numbers of epidermal cell
X	30	150
Y	60	240
Z	90	400

(a) *X, Y* and *Z*
(b) *Y, Z* and *X*
(c) *Z, Y* and *X*
(d) *Y, X* and *Z*

2 Find out the correct sequence of labelling.

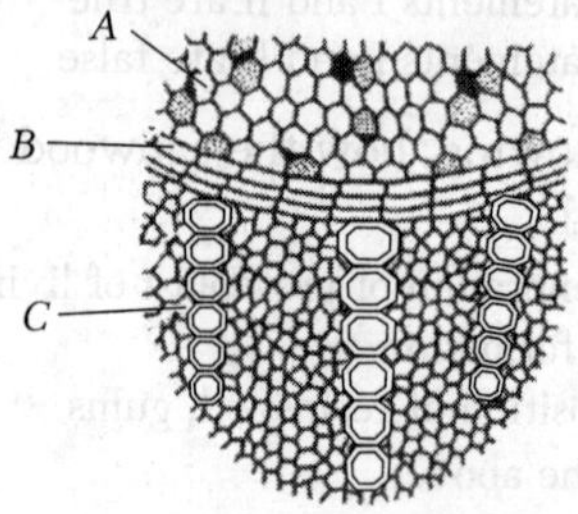

(a) *A*-Parenchyma, *B*-Cambium, *C*-Phloem
(b) *A*-Phloem, *B*-Cambium, *C*-Parenchyma
(c) *A*-Phloem, *B*-Cambium, *C*-Xylem
(d) *A*-Parenchyma, *B*-Cambium, *C*-Xylem

3 In the diagram of lenticel, identify the parts indicated as *A, B, C and D*.

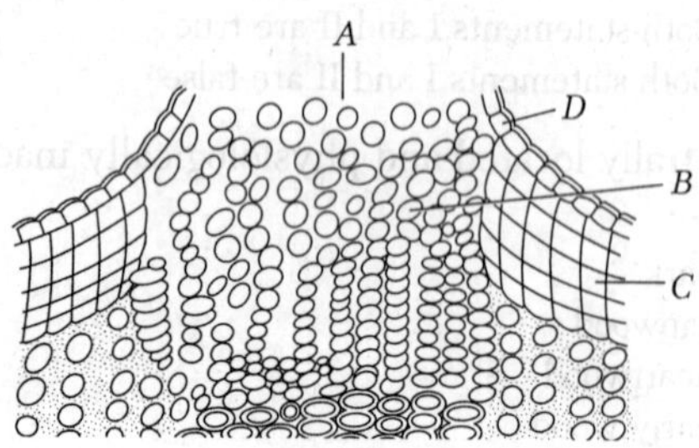

(a) *A*-Lenticel, *B*-Complementary cell, *C*-Phellogen, *D*-Epidermis
(b) *A*-Phellem, *B*-Complementary cells, *C*-Phelloderm, *D*-Periderm
(c) *A*-Complementary cell, *B*-Lenticel, *C*-Phelloderm, *D*-Periderm
(d) *A*-Complementary cells, *B*-Lenticel, *C*-Periderm, *D*-Phelloderm

4 Match the following columns.

	Column I		Column II
A.	Cuticle	1.	Guard cells
B.	Bulliform cells	2.	Single layer
C.	Stomata	3.	Waxy layer
D.	Epidermis	4.	Empty colourless cell

Codes

	A	B	C	D		A	B	C	D
(a)	3	4	1	2	(b)	1	2	3	4
(c)	3	2	4	1	(d)	3	2	1	4

5 Consider the following statements and state true (T) or false (F).

I. Aerenchyma is a type of parenchyma found in aquatic plants.
II. Sclereids are found in tea leaves.
III. The collenchyma occurs in layers below the epidermis in the monocotyledon stem.

	I	II	III		I	II	III
(a)	T	T	T	(b)	T	T	F
(c)	F	T	T	(d)	T	F	F

6 Consider the following statements and select the option that correctly fills in blanks.

I. ...*A*... are elongated or tube-like cells with thick lignified walls and tapering end.
II. ...*B*... have highly thickened wall and obliterated lumen.
III. The radial conduction of water takes place by the ...*C*... .
IV. In roots, primary xylem is in ...*D*... position.

	A	*B*	*C*	*D*
(a)	Tracheids	Xylem fibres	ray parenchyma	exarch
(b)	Tracheids	Xylem fibres	vessels	endarch
(c)	Vessels	Tracheids	ray parenchyma	exarch
(d)	Vessels	Xylem fibres	tracheids	endarch

7 Read the following statements.

I. Epidermal cells have lower number of chloroplasts or lack them completely, except the guard cells.
II. Epidermis is mostly single-layered except in leaves of *Ficus* and *Peperomia.*
III. Epidermis lacks intercellular spaces.
IV. Root epidermis lacks cuticle.
V. In plants with secondary growth, the epidermis of roots and stems is usually replaced by a periderm through the action of cork cambium or phellogen.

How many of the above statements are correct?

(a) One (b) Two (c) Four (d) Five

8 How many of the following plants given below possess open (X) and closed vascular bundles (Y), respectively?

Brassica, Tulipa, Capsicum, Gloriosa, Nicotiana, Aloe, Cicer, Asparagus, Petunia, Colchicum, Sesbania, Zea.

(a) X = 6, Y = 6 (b) X = 4, Y = 8
(c) X = 5, Y = 7 (d) X = 9, Y = 3

9 In a tree, 30 rings of autumn wood and 30 rings of spring wood were observed in 2002. What will be the total number of growth rings (annual rings) in 2018?

(a) 38 (b) 31 (c) 15 (d) 46

10 Match the following columns.

	Column I	Column II
A.	Extrafoliar nectaries	1. *Achras*
B.	Schizogenous cavities	2. *Tropaeolum*
C.	Laticiferous ducts	3. *Passiflora*
D.	Hydathodes	4. *Eucalyptus*
		5. *Pinus*

Codes

	A	B	C	D		A	B	C	D
(a)	3	1	2	4	(b)	3	5	1	2
(c)	2	1	3	4	(d)	5	2	1	3

11 Which is true for the structures marked as *A* and *B*.

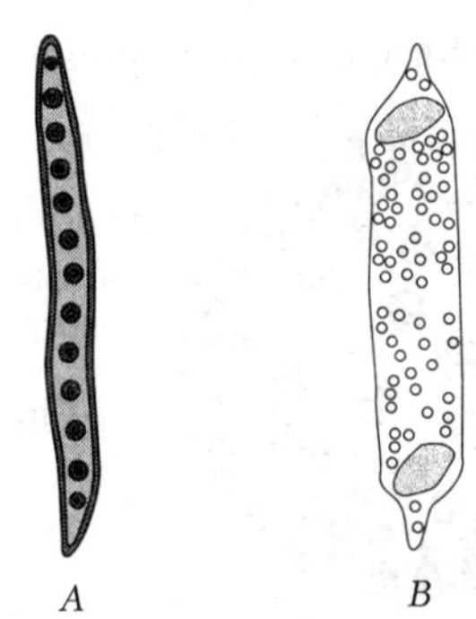

(a) *A* is evolutionarily more advanced than *B*
(b) *A* possesses pits but *B* lacks pit formation
(c) *A* is devoid of protoplasm whereas *B* possesses it
(d) *A* transports water and minerals and *B* also transports water and minerals

12 Read the following statements.

I. The first formed primary xylem elements called protoxylem and the later formed primary xylem is called metaxylem.
II. Primary xylem is endarch in stems.
III. Primary xylem is exarch in roots.
IV. Phloem transports food materials, usually from leaves to other plant parts.
V. Vessel members are interconnected through perforations in their common walls.
VI. Tracheids are elongated or tube-like with thick, lignified walls and tapering ends.

How many of the above statements are correct?

(a) Four (b) Six
(c) Two (d) Three

13 What is the reason for the formation of cork cambium or phellogen?

(a) It is beacuse cortical cells start showing redifferentiation to form cork cambium
(b) It is because vascular cambium starts to degenerate
(c) It is because vascular cambium stops its meristematic activity
(d) As the stem continues to increase in girth due to the activity of vascular cambium, the outer cortical and epidermal layer gets broken and need to be replaced

14 Read the following statements.

I. Annual rings can be distinctly observed in plants growing in temperate regions.
II. In temperate regions, the climatic conditions are not uniform throughout the year.

Choose the correct option.

(a) Statement I is correct, but II is incorrect
(b) Both the statements I and II are correct and statement II is the correct explanation of statement I
(c) Both statements I and II are correct but statement II is not the correct explanation of statement I
(d) Both statements I and II are incorrect

15 Which of the following is true for bark?

(a) It represents tissues exterior to vascular cambium
(b) It includes periderm, secondary phloem and secondary xylem
(c) Bark formed early in season is called hard bark whereas those formed towards end of the season is called first bark
(d) It includes secondary xylem and secondary medullary rays

16 Identify *A*, *B* and *C* from the given figure of phloem.

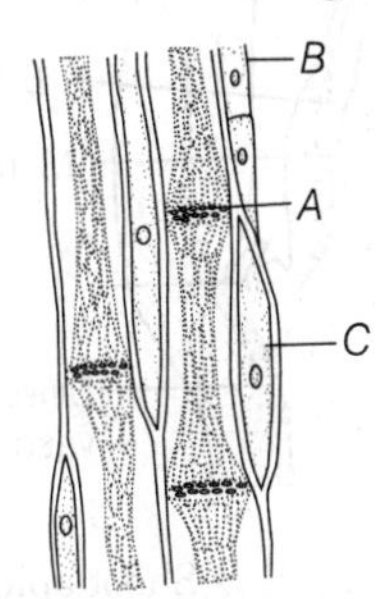

(a) *A*-Sieve tube, *B*-Phloem parenchyma, *C*-Companion cells
(b) *A*-Companion cells *B*-Phloem parenchyma, *C*-Sieve tube
(c) *A*-Phloem parenchyma, *B*-Companion cells, *C*-Sieve tube
(d) *A*-Companion cells, *B*-Sieve tube, *C*-Phloem parenchyma

17 Match the following columns.

	Column I		Column II
A.	Tracheid	1.	Individual members are interconnected through perforations in their common walls.
B.	Trachea	2.	Elongated tube-like cells with thick, lignified walls and tapering ends.
C.	Xylem fibres	3.	Cells possess highly thickened walls with obliterated central lumen.
D.	Xylem parenchyma	4.	Cells are living with thin cellulosic cell walls.

Codes

	A	B	C	D		A	B	C	D
(a)	2	1	3	4	(b)	1	2	3	4
(c)	4	3	1	2	(d)	3	4	2	1

18 Read the following statements.

I. The sieve tube elements and companion cells are connected by pit field present between their common longitudinal walls.

II. Phloem parenchyma is made up of elongated non-tapering cylindrical cells which have scanty cytoplasm.

III. Phloem fibres are made up of parenchymatous cells.

IV. Companion cells help in maintaining the pressure gradient in the sieve tube.

V. The phloem parenchyma stores food material and other substances like resins, latex and mucilage.

How many of the above statements are correct?

(a) Three (b) Two (c) One (d) Four

19 Identify the type of apical meristem and structures marked as *A*, *B* and *C* in the diagram given below.

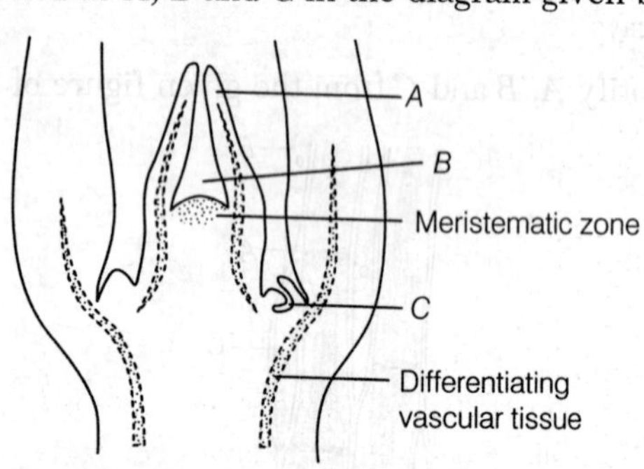

(a) *A*-Leaf primordium, *B*-Shoot apical meristem, *C*-Axillary bud

(b) *A*-Leaf primordium, *B*-Shoot apical meristem, *C*-Apical bud

(c) *A*-Root hair primordium, *B*-Root apical meristem, *C*-Axillary bud

(d) *A*-Root hair prmordium, *B*-Root apical meristem, *C*-Terminal bud

20 Which one is correct for the tissue depicted in the diagram given below?

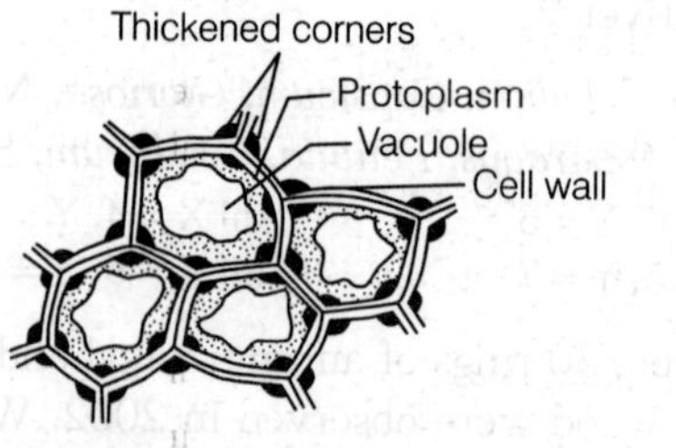

(a) Cells lack deposition of pectin

(b) Cells are never photosynthetic

(c) Intercellular spaces are absent

(d) These are absent in hypodermis of dicot stems

21 Match the following columns.

Column I	Column II
A. Bulliform cells	1. Stomata
B. Guard cells	2. Aerating pore
C. Lenticel	3. Accessory cells
D. Subsidiary cell	4. Isobilateral leaf

Codes

	A	B	C	D
(a)	4	1	2	3
(b)	1	4	2	3
(c)	4	2	3	1
(d)	1	2	3	4

22 What is true for the diagram given below?

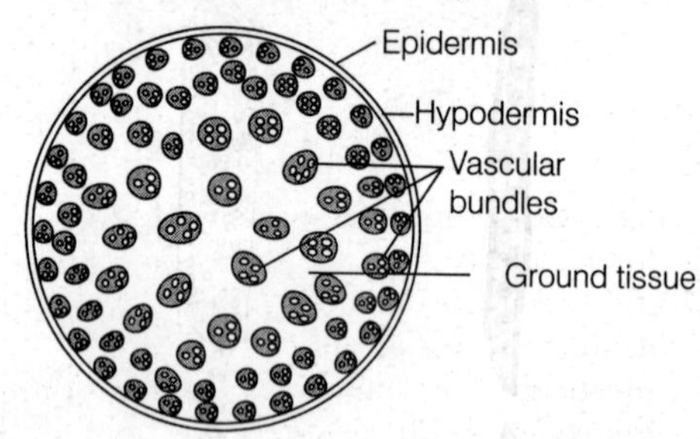

(a) Vascular bundles are present in the form of a ring

(b) Pith is distinct

(c) Hypodermis is collenchymatous

(d) Epidermis lacks trichomes

23 What is true for the structures marked as *X* and *Y*?

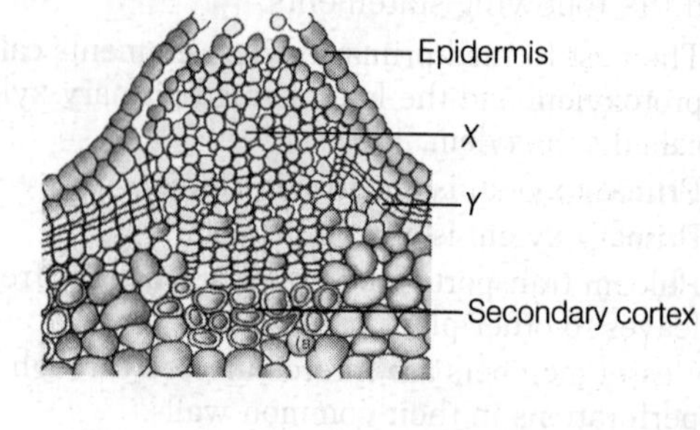

(a) X is the result of redifferentiation whereas Y is the result of dedifferentiation
(b) X is the result of differentiation whereas Y is the result of redifferentiation
(c) Both X and Y are the result dedifferentiation
(d) X is the result of dedifferentiation whereas Y is the result of redifferentiation

24 Choose the incorrectly matched pairs.

I.	Spring wood	–	Abundant xylem fibres
II.	Rhytidome	–	Living tissue inside the bark
II.	Heartwood	–	Duramen
IV.	Sapwood	–	Alburnum

Codes

(a) I and II (b) III and IV
(c) I and IV (d) II and IV

25 Which one of the following statements pertaining to plant structure is correct?
(a) Cork lacks stomata, but lenticels carryout transpiration
(b) Xylem helps in transfer of food from cortex to phloem
(c) Sieve tube elements possess cytoplasm and nuclei
(d) The shoot apical meristem has a quiescent centre

26 In the diagram of TS of stele of dicot root, the different parts have been indicated by the alphabets. Choose the correct combination.

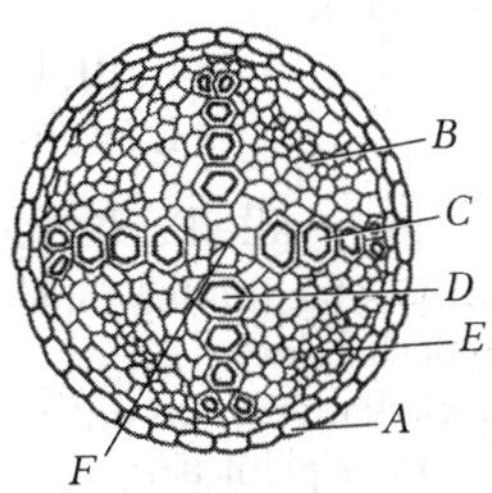

(a) A-Pericycle, B-Conjunctive tissue, C-Metaxylem, D-Protoxylem, E-Phloem, F-Pith
(b) A-Endodermis, B-Conjunctive tissue, C-Protoxylem, D-Metaxylem, E-Phloem, F-Pith
(c) A-Endodermis, B-Conjunctive tissue, C-Metaxylem, D-Protoxylem E-Phloem, F-Pith
(d) A-Endodermis, B-Pith, C-Protoxylem, D-Metaxylem, E-Phloem, F-Conjunctive tissue

27 Which of the following statement(s) is/are true?
I. Uneven thickening of cell wall is a characteristic of sclerenchyma.
II. Periblem forms the cortex of the stem and the root.
III. Tracheids are the chief water transporting elements in gymnosperms.
IV. Companion cell is devoid of nucleus at maturity.
V. The commercial cork is obtained from *Quercus suber.*

Choose the correct option from the codes given below.
(a) I and IV (b) II and V
(c) II, III and V (d) I, II and III

28 Which of the following statement(s) is/are correct?
I. Ground meristem is the outermost meristematic layer of young growing region.
II. Except intrafascicular cambium cells, primary meristem cells give rise to tissues of primary growth.
III. Primary meristems add growth in thickness.
IV. In mass meristem, cell division takes place in all planes resulting in the formation of massive plant body or organ.

Choose the correct option from the codes given below.
(a) Only I (b) II and IV
(c) I, II and IV (d) I, II, III and IV

29 Which of the following statement(s) is/are correct?
I. In amphicribral vascular bundle, xylem lying in centre is surrounded by a ring of phloem.
II. In amphivasal vascular bundle, phloem lying in centre is surrounded by xylem.
III. In amphivasal vascular bundle, xylem lying in centre is surrounded by phloem.
IV. In amphicribral vascular bundle, phloem lying in centre is surrounded by ring of xylem.

Choose the correct option from the codes given below.

Codes

(a) I and III (b) I and II
(c) I, II and III (d) None of these

30 Match the following columns.

Column I	Column II
A. Trichomes	1. Gaseous exchange
B. Tracheids	2. Mechanical strength
C. Bast fibres	3. Epidermal outgrowth
D. Guard cells	4. Water conduction

Codes

	A	B	C	D
(a)	3	1	4	2
(b)	3	4	2	1
(c)	1	4	2	3
(d)	1	3	4	2

31 Which one of the following statements is true?
(a) The collenchyma occurs in layers below the epidermis in monocotyledonous plants
(b) Sclerenchyma cells are usually dead and without protoplasts
(c) Xylem parenchyma cells are living and thin-walled and their cell walls are made up of lignin
(d) The companion cells are specialised sclerenchymatous cells
(e) Phloem fibres are generally present in the primary phloem

32 Match the following columns.

Column I	Column II
A. Endodermis	1. Companion cells
B. Stomata	2. Lenticels
C. Sieve tube	3. Palisade cells
D. Periderm	4. Passage cells
E. Mesophyll	5. Accessory cells

Codes

	A	B	C	D	E
(a)	4	5	2	1	3
(b)	5	3	1	2	4
(c)	4	5	1	2	3
(d)	2	5	3	4	1

33 Consider the following statements and choose the correct option.

I. The thread-like cytoplasmic strands, running from one cell to other is known as plasmodesmata.
II. Xylem and phloem constitute the vascular bundle of the stem.
III. The first formed xylem elements are described as metaxylem.
IV. Radial vascular bundles are mainly found in the leaves.

(a) I is true, but II, III and IV are false
(b) II is true, but I, III and IV are false
(c) IV is true, but I, II and III are false
(d) I and II are true, but III and IV are false

34 Match the following columns.

	Column I		Column II
A.	Bean-shaped guard cell	1.	Dicot stem
B.	Dumb-bell-shaped guard cell	2.	Monocot leaf
C.	Trichome	3.	Dicot leaf
D.	Exarch xylem	4.	Dicot and monocot roots

Codes

	A	B	C	D
(a)	2	3	1	4
(b)	3	2	1	4
(c)	4	3	1	2
(d)	3	4	2	1

35 How many shoot apical meristems are likely to be present in a twig of a plant possessing, 4 branches and 26 leaves?

(a) 26 (b) 1
(c) 5 (d) 30

36 Choose the incorrect statement.

(a) Lenticels are aerating pores present in the bark of plants and are associated with gaseous exchange
(b) Wood is the common name used for secondary xylem
(c) In dicot stem, pericycle is usually multilayered
(d) Sieve tube lacks peripheral cytoplasm, a large vacuole and possesses nucleus

37 Match the following columns.

	Column I	Column II
A.	Tyloses	1. Coenocytic
B.	Periderm	2. Adaxial epidermis
C.	Motor cells	3. Complementary cells
D.	Laticifers	4. Heartwood
		5. Conjunctive tissue

Codes

	A	B	C	D		A	B	C	D
(a)	3	2	1	5	(b)	2	5	1	3
(c)	4	3	2	1	(d)	4	1	3	5

38 Arrange the following in the order of their location from periphery to centre in the entire dicotyledonous plant body.

I. Fusiform cells II. Trichoblasts
III. Collocytes IV. Tyloses

The correct sequence is

(a) IV, I, II, III (b) II, III, I, IV
(c) III, II, I, IV (d) I, IV, III, II

39 Passage cells are thin-walled cells found in

(a) endodermis of roots facilitating rapid transport of water from cortex to pericycle
(b) phloem elements that serve as entry points for substances for transport to other plant parts
(c) testa of seeds to enable emergence of growing embryonic axis during seeds germination
(d) central region of style through which the pollen tube grows towards the ovary

40 Read the following statements and mark them as true (T) or false (F).

I. Intrafascicular cambium develops in between secondary xylem and secondary phloem during the secondary growth of plant organ.
II. Interfascicular cambium develops from cell of medullary rays by means of differentiation.
III. The vascular cambium is generally more active on the inner side.
IV. The vascular cambium ring cuts-off secondary xylem away from the centre.

	I	II	III	IV		I	II	III	IV
(a)	T	T	T	F	(b)	F	T	T	F
(c)	F	T	T	T	(d)	T	F	F	T

41 Which of the following is incorrect?

(a) The root hairs are unicellular

(b) The first formed primary phloem consists of narrow sieve tubes and is called protophloem whereas the later formed phloem has bigger sieve tube and is called metaphloem

(c) In stems, the protoxylem lie towards the pith and metaxylem lie towards the periphery whereas in roots, it is *vice-versa*

(d) Trichomes in shoot systems are usualy unicellular

42 These types of vascular bundles (*A*, *B* and *C*) are present in

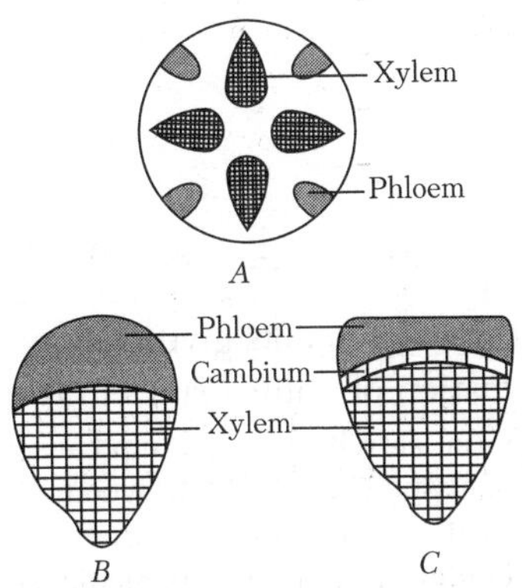

	A	*B*	*C*
(a)	Monocot leaf	Dicot stem	Monocot stem
(b)	Root	Dicot stem	Monocot stem
(c)	Root	Monocot stem	Dicot stem
(d)	Monocot root	Monocot leaf	Monocot stem

43 Select the incorrect option with respect to given anatomy of plant part.

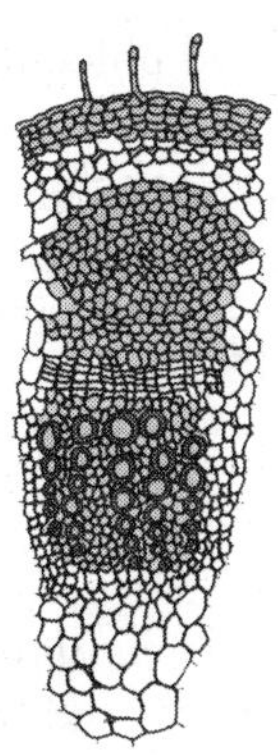

(a) Presence of well-developed pith in centre

(b) Ring-like arrangement of conjoint, collateral and closed vascular bundles

(c) Centrifugal xylem with endarch condition

(d) Cortex is well-differentiated into hypodermis, general cortex and endodermis

44 Find the incorrectly matched pair.

(a) Casparian strip - Root endodermis

(b) Well-developed pith - Dicot stem

(c) Closed vascular bundle - Cambium absent

(d) Isobilateral leaf - Hypostomatic

45 Consider the following statements and select the option that correctly fills in the blank.

I. In spring season, cambium is ...*A*... active and produces ...*B*... wood.

II. In autumn season, cambium is ...*C*... active and produces ...*D*... wood.

	A	*B*	*C*	*D*
(a)	More	Early	Less	Late
(b)	More	Late	Less	Early
(c)	Less	Early	More	Late
(d)	Less	Late	More	Early

46 Examine the figure given below and select the correct levels for *A*, *B* and *C*.

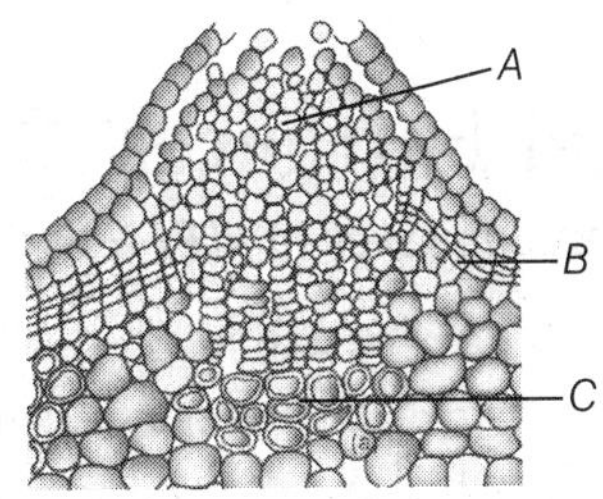

	A	*B*	*C*
(a)	Complementary cells	Phellogen	Phellem
(b)	Phellem	Cork cambium	Periderm
(c)	Complementary cells	Cork cambium	Secondary cortex
(d)	Phellem	Phellogen	Phelloderm

Assertion and Reason

Direction (Q. Nos. 47–54) *In the following questions, a statement of Assertion (A) is given followed by the corresponding statement of Reason (R). Of the statements, mark the correct answer as*

(a) Both A and R are true and R is the correct explanation of A

(b) Both A and R are true, but R is not the correct explanation of A

(c) A is true, but R is false

(d) A is false, but R is true

(e) Both A and R are false

47 **Assertion** (A) All tissues lying inside the vascular cambium are called as bark.

Reason (R) Bark is made up of phellogen, phellem and phelloderm lying outside the secondary phloem.

48 **Assertion** (A) Cambium is a lateral meristem and causes growth in width.

Reason (R) It is made up of fusiform and ray initials in stem.

49 **Assertion** (A) Thick cuticle is mostly present in disease resistant plants.

Reason (R) Disease causing agents cannot grow on cuticle and cannot invade the cuticle.

50 **Assertion** (A) Permanent tissue is composed of mature cells.
Reason (R) Meristematic tissue is a group of actively dividing cells.

51 **Assertion** (A) Intercalary meristem increases length of plant like apical meristems.
Reason (R) It originates from apical meristem.

52 **Assertion** (A) The lenticel is meant for gaseous exchange.
Reason (R) Lenticel checks excessive evaporation of water.

53 **Assertion** (A) In angiosperms, the conduction of water is more efficient because their xylem has vessels.
Reason (R) Conduction of water by vessel elements is an active process in which energy is supplied by xylem parenchyma rich in mitochondria.

54 **Assertion** (A) All the endodermal cells of the root do not contain Casparian thickenings on their radial and transverse walls.
Reason (R) Passage cells are found in the root endodermis.

(C) Medical Entrances' Gallery

(Collection of questions Asked in NEET & Various Medical Entrance Exams)

1 Identify the incorrect statement. **[NEET 2020]**
(a) Sapwood is involved in the conduction of water and minerals from root to leaf
(b) Sapwood is the innermost secondary xylem and is lighter in colour
(c) Due to deposition of tannins, resins. olis, etc., heartwood is dark in colour
(d) Heartwood does not conduct water but gives mechanical support

2 The transverse section of a plant shows following anatomical features.
I. Large number of scattered vascular bundles surrounded by bundle sheath.
II. Large conspicuous parenchymatous ground tissue.
III. Vascular bundles are conjoint and closed.
IV. Phloem parenchyma absent.
Identify the category of plant and its part. **[NEET 2020]**
(a) Monocotyledonous root
(b) Dicotyledonous stem
(c) Dicotyledonous root
(d) Monocotyledonous stem

3 Regeneration of damaged growing grass following grazing is largely due to **[NEET 2019]**
(a) lateral meristem
(b) apical meristem
(c) intercalary meristem
(d) secondary meristem

4 In dicot root, the vascular cambium originates from **[NEET 2019]**
(a) tissue located below the phloem bundles and a portion of pericycle tissue above protoxylem
(b) cortical region
(c) parenchyma between endodermis and pericycle
(d) intrafascicular and interfascicular tissue in a ring

5 Which of the statements given below is not true about formation of annual rings in trees? **[NEET 2019]**
(a) Differential activity of cambium causes light and dark bands of tissue early and late wood, respectively
(b) Activity of cambium depends upon variation in climate
(c) Annual rings are not prominent in trees of temperate region
(d) Annual ring is a combination of spring wood and autumn wood produced in a year

6 Plants having little or no secondary growth are **[NEET 2018]**
(a) conifers (b) deciduous angiosperms
(c) grasses (d) cycads

7 Secondary xylem and phloem in dicot stem are produced by **[NEET 2018]**
(a) phellogen (b) vascular cambium
(c) apical meristems (d) axillary meristems

8 Casparian strips occur in **[NEET 2018]**
(a) cortex (b) pericycle
(c) epidermis (d) endodermis

9 Stomata in grass leaf are **[NEET 2018]**
(a) rectangular (b) kidney-shaped
(c) dumb-bell-shaped (d) barrel-shaped

10 Identify the permanent tissues shown in the following figures.

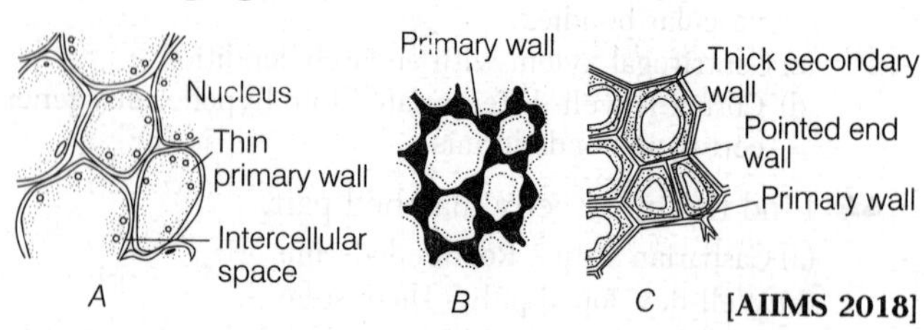

[AIIMS 2018]

(a) *A*-Collenchyma, *B*-Parenchyma, *C*-Sclerenchyma
(b) *A*- Sclerenchyma, *B*-Collenchyma, *C*-Parenchyma
(c) *A*-Collenchyma, *B*-Sclerenchyma, *C*-Parenchyma
(d) *A*-Parenchyma, *B*-Collenchyma, *C*-Sclerenchyma

11 Phellogen and phellem respectively denote **[JIPMER 2017]**
(a) cork and cork cambium
(b) cork cambium and cork
(c) secondary cortex and cork
(d) cork and secondary cortex

12 Identify the incorrect statement in context of heartwood. **[NEET 2017]**
(a) Organic compounds are deposited in it
(b) It is highly durable
(c) It conducts water and minerals efficiently
(d) It comprises dead elements with highly lignified walls

13 Which of the following is made up of dead cells? **[NEET 2017]**
(a) Xylem parenchyma (b) Collenchyma
(c) Phellem (d) Phloem

14 The vascular cambium normally gives rise to **[NEET 2017]**
(a) phelloderm (b) primary phloem
(c) secondary xylem (d) periderm

15 The balloon-shaped structures called tyloses **[NEET-II 2016]**
(a) originate in the lumen of vessels
(b) characterise the sapwood
(c) are extensions of xylem parenchyma cells into vessels
(d) are linked to the ascent of sap through xylem vessels

16 Cortex is the region found between **[NEET-II 2016]**
(a) epidermis and stele
(b) pericycle and endodermis
(c) endodermis and pith
(d) endodermis and vascular bundle

17 Specialised epidermal cells surrounding the guard cells are called **[NEET-I 2016]**
(a) subsidiary cells (b) bulliform cells
(c) lenticels (d) complementary cells

18 The girth of the stem increases due to activity of **[COMED-K 2015]**
(a) apical meristems
(b) intercalary meristems
(c) lateral meristems
(d) parenchyma cells

19 Transport proteins of endodermal cells are control points where a plant adjusts the quantity and types of solutes that reach the xylem. Root endodermis is able to actively transport ions in one direction only because of the layer of **[COMED-K 2015]**
(a) actin (b) lignin
(c) suberin (d) cellulose

20 Medullary rays are tissues made up of **[WB JEE 2015]**
(a) phloem parenchyma (b) xylem parenchyma
(c) sieve tubes (d) sclerenchyma

21 No vessels are found in the wood of **[AMU 2015]**
(a) pine (b) *Eucalyptus*
(c) teak (d) sheesham

22 Read I-IV and tell the correct order of components from outer side to inner side in a woody dicot stem.
I. Secondary cortex II. Wood
III. Secondary phloem IV. Phellem
[CBSE AIPMT 2015]

Codes
(a) III, IV, II and I (b) I, II, IV and III
(c) IV, I, III and II (d) IV, III, I and II

23 Select the characters which are not applicable to the anatomy of dicot roots.
I. Conjunctive tissue present.
II. Presence of protein compounds in Casparian strips.
III. Polyarch xylem bundles.
IV. Presence of pericycle. **[Kerala CEE 2015]**

Codes
(a) I and II (b) II and IV
(c) III and IV (d) I and IV
(e) II and III

24 You are given a fairly old piece of dicot stem and a dicot root. Which of the following anatomical structures will you use to distinguish between the two? **[CBSE AIPMT 2014]**
(a) Secondary xylem (b) Secondary phloem
(c) Protoxylem (d) Cortical cells

25 When one wood is lighter in colour with a lower density, the other wood is darker with a higher density. They are **[Kerala CEE 2014]**
(a) spring wood and autumn wood
(b) heartwood and late wood
(c) spring wood and early wood
(d) sapwood and spring wood
(e) autumn wood and spring wood

26 Which of these characters does/do not apply to the vascular bundles of monocot stems?
I. Conjoint
II. Endarch protoxylem
III. Open
IV. Phloem parenchyma is absent

Select the correct answer using the codes given below. **[Kerala CEE 2014]**
(a) I and II (b) II and III
(c) III and IV (d) Only III
(e) I and IV

27 Which of the following parts of dicot root is made up of cells with suberin deposition in tangential as well as radial walls? **[Kerala CEE 2014]**
(a) Epidermis (b) Endodermis
(c) Cortex (d) Pericycle
(e) Xylem

28 Which of the following characters are not applicable to the anatomy of dicot stem? **[Kerala CEE 2014]**
I. Collenchymatous hypodermis
II. Polyarch xylem
III. The presence of Casparian strips on the endodermis
IV. Open vascular bundle
V. The presence of medullary rays

Select the correct answer using the codes given below.
(a) I, IV and V (b) II and III
(c) II and V (d) I, II and III
(e) III, IV and V

29 Choose the incorrect statement. **[Kerala CEE 2014]**
(a) Gymnosperms lack vessels in their xylem
(b) The cell wall of collenchyma is made up of cellulose, hemicellulose and pectin
(c) The first formed primary xylem elements are called protoxylem
(d) The cell wall of parenchyma is made up of pectin
(e) Gymnosperms have albuminous cells and sieve cell in their phloem

30 In dicotyledonous stem, which of the following is the sequence of tissues from inside to outside? **[J&K CET 2014]**
(a) Pith, phloem, cambium, protoxylem, metaxylem, pericycle, parenchyma, collenchyma, endodermis and epidermis
(b) Pith, cambium, phloem, protoxylem, metaxylem, pericycle, endodermis, parenchyma, collenchyma and epidermis
(c) Pith, phloem, protoxylem, metaxylem, cambium, pericycle, endodermis, parenchyma, collenchyma and epidermis
(d) Pith, protoxylem, metaxylem, cambium, phloem pericycle, endodermis, parenchyma, collenchyma and epidermis

31 Which one of the following characters is not found in transverse section of monocot stem? **[EAMCET 2014]**
(a) Sclerenchyma bundle sheath
(b) Lysigenous cavity
(c) Sclerenchymatous hypodermis
(d) Starch sheath

32 Identify the correct pair of statements.
I. Functions of sieve tubes are controlled by the nucleus of companion cells.
II. Albuminous cells are present in angiosperms.
III. In dicot root, the vascular cambium is completely secondary in origin.
IV. Cylindrical meristems contribute to the formation of primary plant body. **[EAMCET 2014]**

Choose the correct option from the codes given below.
(a) I and III (b) III and IV (c) I and II (d) II and III

33 Interfascicular cambium develops from the cells of
(a) xylem parenchyma (b) endodermis **[NEET 2013]**
(c) pericycle (d) medullary rays

34 Lenticels are involved in **[NEET 2013]**
(a) gaseous exchange (b) food transport
(c) photosynthesis (d) transpiration

35 Interfascicular cambium is a **[WB JEE 2013]**
(a) primary meristematic tissue
(b) primordial meristem
(c) type of protoderm
(d) secondary meristematic tissue

36 Hydrophytes are characterised by the
(a) presence of sclerenchyma **[Odisha JEE 2013]**
(b) presence of aerenchyma
(c) absence of aerenchyma
(d) presence of root nodules

37 Which one of the following pairs is an example for lateral meristem? **[Karnataka CET 2013; JCECE 2012]**
(a) Phellogen and phelloderm
(b) Phellogen and fascicular cambium
(c) Procambium and phelloderm
(d) Interfascicular cambium and phellem

38 Which one of the following is dead, but work efficiently? **[UP CPMT 2012]**
(a) Sieve tube (b) Companion cells
(c) Vessels (d) Both (b) and (c)

39 Duramen is present in **[UP CPMT 2012]**
(a) inner region of secondary wood
(b) part of sapwood
(c) outer region of secondary wood
(d) region of pericycle

40 Which of the following is true? **[BCECE 2012]**
(a) Vessels are unicellular and with narrow lumen
(b) Vessels are multicellular and with wide lumen
(c) Tracheids are unicellular and with wide lumen
(d) Tracheids are multicellular and with narrow lumen

41 Two cross-sections of stem and root appear simple, when viewed by naked eye. But under microscope, they can be differentiated by **[BCECE 2012]**
(a) exarch condition of root and stem
(b) endarch condition of stem and root
(c) endarch condition of root and exarch condition of stem
(d) endarch condition of stem and exarch condition of root

42 For a critical study of secondary growth in plants, which one of the following pairs is suitable? **[CBSE AIPMT 2012]**
(a) Sugarcane and sunflower
(b) Teak and pine
(c) Deodar and fern
(d) Wheat and maiden hair fern

43 Which of the following plant tissues provides mechanical strength to plants? **[BHU 2012]**
(a) Parenchyma (b) Collenchyma
(c) Sclerenchyma (d) Aerenchyma

44 The function of a vessel is **[Manipal 2012]**
(a) conduction of food
(b) conduction of water and minerals
(c) conduction of hormones
(d) All of the above

45 Radial conduction of water takes place by **[AMU 2012]**
(a) vessels
(b) vessels and tracheids
(c) phloem
(d) ray parenchymatous cells

46 Which of the following plants shows multiple epidermis? **[Manipal 2011]**
(a) *Croton* (b) *Allium*
(c) *Nerium* (d) *Cucurbita*

47 Casparian strips are present in the ...... of the root. **[Karnataka CET 2013; Odisha JEE 2011]**
(a) epiblema (b) cortex
(c) pericycle (d) endodermis

48 Ground tissue includes **[CBSE AIPMT 2011]**
(a) all tissues except epidermis and vascular bundles
(b) epidermis and cortex
(c) all tissues internal to endodermis
(d) all tissues external to endodermis

49 Lateral roots develop from primordia originated by the division of **[UP CPMT 2011]**
(a) pericycle cells lying opposite to protoxylem points
(b) pericycle cells lying between two protoxylem points
(c) endodermal cells lying between two protoxylem points
(d) endodermal cells lying opposite to protoxylem points

50 A monocot stem with secondary growth is **[UP CPMT 2011]**
(a) *Lilium* (b) *Cocos*
(c) *Yucca* (d) *Asparagus*

51 The age of tree by counting annual rings is called **[UP CPMT 2011]**
(a) dendrochronology (b) ageing
(c) chronology (d) countrology

52 In grasses, certain adaxial epidermal cells along the veins modify themselves into large empty, colourless cells called **[Kerala CEE 2011]**
(a) bulliform cells (b) companion cells
(c) guard cells (d) subsidiary cells
(e) albuminous cells

53 Read the following statements.
I. Many organs of aquatic plants float in water.
II. Large air gaps are present in the collenchyma tissues of lotus leaf.

Select the correct option. **[Karnataka CET 2011]**
(a) Statement I is true, but II is false
(b) Statement I is false, but II is true
(c) Both statements I and II are true
(d) Both statements I and II are false

54 Vascular bundles in monocotyledons are considered closed because **[DUMET 2011]**
(a) xylem is surrounded all around by phloem
(b) there are no vessels with perforations
(c) a bundle sheath surrounds each bundle
(d) there is no secondary growth

55 In an annual ring, the light coloured part is known as **[DUMET 2011]**
(a) early wood (b) late wood
(c) heartwood (d) sapwood

56 Jute is mainly composed of **[Odisha JEE 2011]**
(a) xylem (b) secondary bast fibre
(c) phloem (d) cortex

57 The stele is composed of **[Odisha JEE 2011]**
(a) vascular bundle (b) pith and vascular bundle
(c) cortex and endodermis (d) pith and cortex

58 The beneficial use of epidermal layer is **[Odisha JEE 2011]**
(a) bast fibre (b) mesocarp
(c) cotton fibre (d) jute

59 Vascular bundle having phloem at the centre encircled by xylem is known as **[J&K CET 2011]**
(a) bicollateral (b) conjoint collateral
(c) amphivasal (d) amphicribral

60 The chief water conducting elements of xylem in gymnosperms are **[CBSE AIPMT 2010]**
(a) vessels (b) fibres
(c) transfusion tissue (d) tracheids

61 Which one of the following is not a lateral meristem? **[CBSE AIPMT 2010]**
(a) Intrafascicular cambium (b) Interfascicular cambium
(c) Phellogen (d) Intercalary meristem

ANSWERS

Check Point 12.1

1. (d)	2. (d)	3. (c)	4. (d)	5. (a)	6. (b)

Check Point 12.2

1. (c)	2. (a)	3. (c)	4. (c)	5. (b)	6. (a)

Check Point 12.3

1. (c)	2. (c)	3. (b)	4. (b)	5. (a)	6. (c)

Check Point 12.4

1. (c)	2. (b)	3. (c)	4. (a)	5. (d)	6. (a)

(A) Topical Exercise

1. (d)	2. (a)	3. (d)	4. (c)	5. (d)	6. (d)	7. (d)	8. (c)	9. (b)	10. (a)
11. (c)	12. (c)	13. (a)	14. (d)	15. (a)	16. (c)	17. (a)	18. (b)	19. (b)	20. (b)
21. (a)	22. (a)	23. (b)	24. (b)	25. (d)	26. (a)	27. (b)	28. (c)	29. (b)	30. (a)
31. (a)	32. (c)	33. (a)	34. (d)	35. (d)	36. (b)	37. (a)	38. (a)	39. (a)	40. (c)
41. (d)	42. (b)	43. (c)	44. (a)	45. (a)	46. (b)	47. (a)	48. (b)	49. (d)	50. (c)
51. (a)	52. (b)	53. (b)	54. (a)	55. (d)	56. (d)	57. (c)	58. (b)	59. (c)	60. (a)
61. (c)	62. (d)	63. (a)	64. (b)	65. (b)	66. (a)	67. (a)	68. (d)	69. (d)	70. (d)
71. (c)	72. (d)	73. (d)	74. (b)	75. (d)	76. (c)	77. (c)	78. (c)		

(B) Cumulative Exercise

1. (b)	2. (c)	3. (a)	4. (a)	5. (b)	6. (a)	7. (d)	8. (a)	9. (d)	10. (b)
11. (d)	12. (b)	13. (d)	14. (b)	15. (a)	16. (a)	17. (a)	18. (a)	19. (a)	20. (c)
21. (a)	22. (d)	23. (a)	24. (a)	25. (a)	26. (b)	27. (c)	28. (b)	29. (b)	30. (b)
31. (b)	32. (c)	33. (d)	34. (b)	35. (c)	36. (d)	37. (c)	38. (b)	39. (a)	40. (b)
41. (d)	42. (c)	43. (b)	44. (d)	45. (a)	46. (c)	47. (e)	48. (b)	49. (a)	50. (b)
51. (a)	52. (b)	53. (a)	54. (a)						

(C) Medical Entrances' Gallery

1. (b)	2. (d)	3. (c)	4. (a)	5. (c)	6. (c)	7. (b)	8. (d)	9. (c)	10. (d)
11. (b)	12. (c)	13. (c)	14. (c)	15. (c)	16. (a)	17. (a)	18. (c)	19. (c)	20. (b)
21. (a)	22. (c)	23. (e)	24. (c)	25. (a)	26. (d)	27. (b)	28. (b)	29. (d)	30. (d)
31. (d)	32. (a)	33. (d)	34. (a)	35. (d)	36. (b)	37. (b)	38. (c)	39. (a)	40. (b)
41. (d)	42. (b)	43. (c)	44. (b)	45. (d)	46. (c)	47. (d)	48. (a)	49. (b)	50. (c)
51. (a)	52. (a)	53. (a)	54. (d)	55. (a)	56. (b)	57. (b)	58. (c)	59. (c)	60. (d)
61. (d)									

Explanations

(A) Topical Exercise

20 (*b*) Statement in option (b) is incorrect and can be corrected as
The complex tissue consists of more than one type of cells which with together function as a unit. It constitutes the conducting tissue of plant, i.e. xylem and phloem.

26 (*a*) Statement I is true, but II is false. Incorrect statement can be corrected as
Presence of vessels is a characteristic feature of angiosperms.

30 (*a*) Statement I is true, but II is false. Incorrect statement can be corrected as
Phloem fibres are much elongated, unbranched and have pointed, needle-like apices.

34 (*d*) Option (d) is incorrectly matched pair and can be corrected as
Starch sheath-Endodermis of dicot stem.
Rest options are correctly matched pairs.

38 (*a*) Statement I is true, but II is false. Incorrect statement can be corrected as
In conjoint vascular bundle, cambium may be present and conjoint vascular bundles are further, either open or closed on the bases of presence or absence of cambium, respectively.

46 (*b*) Statement I is false, but II is true. Incorrect statement can be corrected as
In dicot root, pericycle produces lateral roots and vascular cambium during secondary growth, but in monocot root, pericycle produces only lateral roots during primary growth.

56 (*d*) Option (d) is incorrect pair and can be corrected as
The abaxial epidermis generally bears more stomata than the adaxial epidermis.
Rest options are correct pairs.

62 (*d*) All the statements are correct except (d) and it can be corrected as
All lateral meristems appear later at the time of secondary growth of a plant body.

70 (*d*) All of the given statements are correct except (d) and it can be corrected as
Phellogen develops from dedifferentiation of cortical cells.

72 (*d*) Option (d) is mismatched pair and can be corrected as
Intrafascicular cambium is primary in origin and developed from procambium during primary growth.
Rest options are correctly matched pairs.

74 (*b*) Statement I is false, but II is true. Incorrect statement can be corrected as
Wood with vessels are called heteroxylous wood and without vessels are called homoxylous wood.

(B) Cumulative Exercise

1 (*b*) Stomatal index is the percentage in which the number of stomata forms to the total number of epidermal cells, both per unit area. Stomatal index can be calculated using the equation,

$$SI\% = \left(\frac{S}{(E+S)}\right)\times 100$$

Where, S = Number of stomata per unit area
E = Number of epidermal cells in the same unit area
From the equation, it is given that
Plant X : S = 30, E = 150

$$SI\% = \left(\frac{30}{30+150}\right)\times 100 = \left(\frac{30}{180}\right)\times 100 = 0.1666\times 100 = 16.66$$

Plant Y : S = 60, E = 240

$$SI\% = \left(\frac{60}{60+240}\right)\times 100 = \left(\frac{60}{300}\right)\times 100 = 0.2\times 100 = 20$$

Plant Z : S = 90, E = 400

$$SI\% = \left(\frac{90}{90+400}\right)\times 100 = \left(\frac{90}{490}\right)\times 100 = 0.1836\times 100 = 18.36$$

Therefore, decreasing order of their stomatal index is 20, 18.36 and 16.66, i.e. Y, Z and X
Herce, the correct option is (b).

5 (*b*) Statements I and II are true whereas III is false. Incorrect statement can be corrected as
Collenchyma occurs in layers below the epidermis in most of the dicotyledonous plants.

8 (*a*) Open vascular bundles are found in dicotyledonous stems, e.g. *Brassica*, *Capsicum*, *Nicotiana*, *Cicer*, *Petunia* and *Sesbania*.
Closed vascular bundles are found in monocotyledonous stems, e.g. *Tulipa*, *Gloriosa*, *Aloe*, *Asparagus*, *Colchicum* and *Zea*.
So, the correct option is (a) X = 6, Y = 6

9 (*d*) Every year, one spring ring and one autumn ring make one annular ring.
Number of rings in 2002 = 30
Number of ring in 2018 would be = 30 + 16 = 46

18 (*a*) Statements I, IV and V are correct whereas II and III are incorrect. These can be corrected as
- Phloem parenchyma is made up of elongated tapering cylindrical cells which have dense cytoplasm.
- Phloem fibres are made up of sclerenchymatous cells.

24 (*a*) I and II are incorrectly matched pairs and can be corrected as
- Xylem fibres are fewer in number in spring wood.
- Rhytidome is the outer dead region of the bark.

Rest others are correctly matched pairs.

25 (*a*) All of the given statements are incorrect except (a). Incorrect statements can be corrected as

- Xylem conducts water and minerals from root to the shoot.
- Sieve tube elements possess cytoplasm, but lack nuclei
- Root apical meristem has a quiescent centre.

27 (*c*) Statements II, III and V are true, whereas I and IV are incorrect.

- Tracheids are most primitive type of conducting elements in xylem. The xylem of gymnosperms consists of tracheids only.
- Companion cells are thin-walled elongated cells in phloem. They are living and contain dense protoplasm with large elongated nucleus.
- According to histogen theory, periblem is the middle dermatogen which give rise to cortex of root and stem.

Incorrect statements can be corrected as

- Uneven thickening of cell wall is a characteristic of collenchyma.
- Sieve tube is devoid of nucleus at maturity.

28 (*b*) Statements II and IV are correct, whereas I and III are incorrect. Incorrect statements can be corrected as

- The outermost layer of cells is the protoderm which forms a single sheet of meristematic cells that give rise to the epidermis. The ground meristem forms the pith and cortex tissues of the stem and the mesophyll tissue of the leaf.
- Secondary growth is characterised by an increase in thickness or girth of the plant. It is caused by cell division in the lateral meristem.

31 (*b*) All of the given statements are false except (b). The false statements can be corrected as

- Collenchyma cell forms the hypodermis and is not found in monocotyledonous plants.
- Xylem parenchyma cells are living and their cell walls are made up of cellulose.
- Companion cells are specialised parenchymatous cells.
- Phloem fibres are present in secondary phloem.

33 (*d*) Statements I and II are true, whereas III and IV are false. Incorrect statements can be corrected as

- As a young vascular plant grows, one or more strands of primary xylem form in its stems and roots. The first xylem to develop is called protoxylem.
- Radial vascular bundles are typically found in the roots of monocots and dicots. Typical vascular bundles are conjoint vascular bundles.

36 (*d*) Statement in option (d) is incorrect and can be corrected as

Sieve tube lacks nucleus and possesess a large vacuole with peripheral cytoplasm.

40 (*b*) Statements II and III are true whereas I and IV are false. Incorrect statements can be corrected as

- Intrafascicular cambium develops in between primary xylem and primary phloem from procambium during the primary growth of the plant organ.
- Vascular cambium ring cuts off secondary xylem towards the inner side or centre.

41 (*d*) Statement in option (d) is incorrect and can be corrected as

The stem epidermal hairs are called trichomes. These are usually multicellular.

44 (*d*) Option (d) is incorrectly matched pair and can be corrected as

Isobilateral leaf - Amphistomatic (stomata present on both surfaces of epidermis).

Rest options are correctly matched pairs.

47 (*e*) Both A and R are false and can be corrected as

Bark consists of all the tissues outside the vascular cambium, i.e. phellem, phellogen and phelloderm constitute periderm. Phloem is present inside whereas phellem, phellogen and phelloderm are present outside the phloem, phellogen is called cork cambium and phelloderm is called as secondary cortex.

48 (*b*) Both A and R are true, but R is not the correct explanation of A.

Cambium is a lateral meristem. Its activity causes increase in width by forming layers of cells between xylem and phloem and thus, causing secondary growth. It is composed of fusiform and ray initial.

49 (*a*) Both A and R are true and R is the correct explanation of A.

Disease resistant plants possess thick cuticle. Due to this, infectious organisms cannot grow or invade cuticle.

50 (*b*) Both A and R are true, but R is not the correct explanation of A.

A meristematic tissue is a group of cells that are in a continuous state of division or retain their power of division. Permanent tissues are composed of mature cells that after undergoing complete growth, have assumed a definite shape, size and function and have temporarily or permanently lost the power of division.

51 (*a*) Both A and R are true and R is the correct explanation of A.

Intercalary meristems are intercalated in between the permanent tissues. The activities of this meristems, also add to the length of the plant or its organs. They originate from the apical meristems, when their portions get detached due to the growth of the organs.

For example, in the grasses, when the internodes complete their elongation, some cells at the base retain their meristematic activity and function as intercalary meristems. They lie just above the node.

52 (*b*) Both A and R are true, but R is not the correct explanation of A.

The lenticel is filled up by loosely arranged thin-walled rounded and suberised or unsuberised cells called complementary cells. They enclose intercellular spaces for the gaseous exchange.

The complementary cells are formed from loosely arranged phellogen cells and division of substomatal parenchyma cells. The suberised nature of complementary cells checks excessive evaporation of water.

53 (*a*) Both A and R are true and R is the correct explanation of A.

In angiospermic plants, the conduction of water and minerals is done by xylem vessels. These vessels conduct the water in upward direction which are coming from roots. Xylem parenchyma helps in storage of food, tracheids help in conduction of water but in gymnosperms, xylem fibres help in providing mechanical strength.

54 (*a*) Both A and R are true and R is the correct explanation of A.

Casparian strips are found only on those cells that have suberin depositions. Passage cells are non-suberised cells of endodermis.

(C) Medical Entrances' Gallery

1. (*b*) Statement in option (b) is incorrect and can be corrected as

Sapwood is outermost secondary xylem and is lighter in colour. Their cell walls are not lignified and there is no deposition of organic compounds. Sapwood is involved in the conduction of water and minerals from root to leaf.

Rest statements are correct.

2. (*d*) The transverse section of monocotyledonous stem shows following anatomical features

The monocot stem have conjoint and closed vascular bundles, scattered in the ground tissue containing the parenchyma. Each vascular bundle is surrounded by sclerenchymatous bundle sheath cells. Phloem parenchyma and medullary rays are absent in monocot stems.

3 (*c*) Regeneration of damaged growing grass following grazing is largely due to intercalary meristem. It is the meristem which occurs between mature tissues. It is found in grasses and regenerates parts damaged by the grazing herbivores.

4 (*a*) In dicot root, the vascular cambium originates from tissues located below the phloem bundles and a portion of pericycle tissue above protoxylem.

Vascular cambium is the meristematic layer that is responsible for cutting off vascular tissues (xylem and phloem). In young stem, it is present in patches as a single layer between the xylem and phloem.

5 (*c*) The statement annual rings are not prominent in trees of temperate region is incorrect. Correct information about the statement is as follows

Annual rings are formed due to the seasonal activity of cambium. In the plants of temperate region, cambium is highly active in spring and less active in autumn season.

6 (*c*) Secondary growth occurs due to the presence of vascular cambium. Grasses are monocot and lack vascular cambium. Therefore, they do not show secondary growth.

7 (*b*) Secondary vascular tissues, i.e. secondary xylem and pholem in stems are formed by the vascular cambium. It is produced by two types of meristems, i.e. fascicular or intrafascicular and interfascicular cambium.

Intrafascicular cambium is a primary meristem which occurs as strips in vascular bundles. It divides to form secondary phloem on outer side and secondary xylem on the inner side. Interfascicular cambium arises secondarily from the cells of medullary rays.

8 (*d*) Casparian strips are found in endodermis of roots. It is a band of thickening strips, which run along the radial and tangential walls of endodermal cells. It is made up of suberin and lignin. Casparian strips prevent plasmolysis of endodermal cells.

9 (*c*) Stomata in grass leaf are dumb-bell-shaped. Epidermis of all green aerial parts of plants contains minute openings called stomata which are surrounded by guard cells.

Kidney-shaped or bean-shaped guard cells are present in dicots only, while in monocots like grasses, these cells are dumb-bell-shaped.

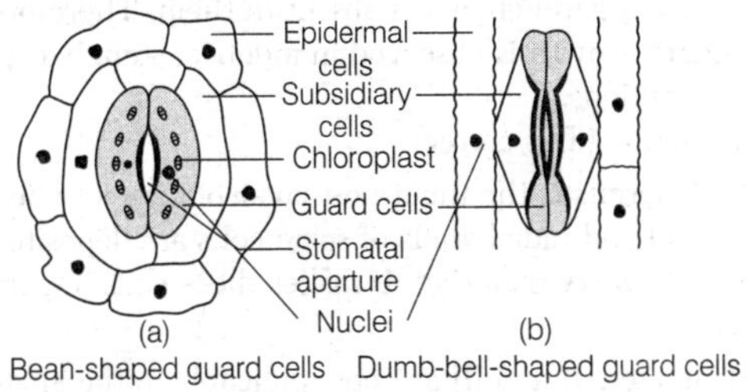

Stomatal apparatus

10 (*d*) In the given figures, *A* represents Parenchyma, *B* is Collenchyma and *C* is sclerenchyma.

The walls of the parenchymatous cells are thin and have intercellular spaces. The cell walls of collenchyma possess thickenings of the corners. Sclerenchyma cells have thick and lignified walls with very narrow cavity.

11 (*b*) In the dicot stem, the cortical cells get differentiated to give rise to another meristematic tissue, which is called cork cambium or phellogen. On the other side, it forms cork or phellem and in the inner region, it forms secondary cortical cells called phelloderm.

12 (*c*) The incorrect statement in the context of heartwood is option (c). The correct information regarding the statement is as follows

Heartwood cannot conduct water and minerals because its tracheids and vessels are blocked by the ingrowth of adjacent parenchyma cells called tyloses. Other statements hold true for heartwood.

13 (*c*) Phellem or cork is a tissue formed on the outer side of cork cambium. It is composed of dead cells. The cell wall becomes impermeable due to suberisation.

14 (*c*) Vascular cambium is located between the xylem and phloem in the stems and roots of vascular plants. It produces secondary xylem towards the pith and secondary phloem towards the bark.

15. (*c*) The tyloses are the structures found in the woody tissues of dicot stems. These are the extension of xylem parenchyma cells into the vessel elements.

16. (*a*) Cortex tissue is found in between the epidermis and stele. It is multilayered and is made up of parenchymatous cells with large intercellular spaces.

21. (*a*) Wood of gymnosperms does not contain vessels. Thus, pine is vessel-less. Vessels are characteristic feature of angiosperms like teak (*Tectona grandis*), mango (*Mangifera indica*) and coconut (*Cocos nucifera*).

24. (*c*) Protoxylem is the major anatomical structural difference which distinguish dicot stem from dicot root. In dicot stem, the protoxylem is present towards centre (pith) and metaxylem is present towards the periphery of the organ. This type of xylem is known as endarch. In dicot root, the protoxylem is present towards the periphery. This condition is called exarch.

25. (*a*) Spring wood contains vessels with wider lumen and are formed in spring season. It is light in colour and of lower density. Autumn wood contains vessels with narrow lumen and formed in autumn season. It is darker with a higher density.

26. (*d*) Vascular bundles of monocot stem are conjoint, closed and phloem parenchyma is absent in them. Therefore, no secondary growth is observed in monocot stem except for few exceptions.
Thus, option (d) is correct.

27. (*b*) Endodermis is the innermost layer of cortex, where tangential and radial walls of some cells are deposited with suberin (a waxy material). It makes these cells impervious to water.

28. (*b*) Characters II and III are not applicable to the anatomy of dicot stem. The presence of polyarch xylem and Casparian strip is the feature of root. So, it is not found in stem. Rest other characters are applicable to the anatomy of dicot stem. Thus, option (b) is correct.

29. (*d*) Out of the statements given in the options, all are correct except option (d) that the cell wall of parenchyma is not made up of pectin because its cell wall is made up of cellulose and hemicellulose.

30. (*d*) In dicot stem, central part is pith. Protoxylem is towards the centre, then come the metaxylem, cambium, phloem, pericycle, endodermis, parenchyma, collenchyma and then lastly epidermis.
Thus, option (d) is correct.

32. (*a*) Statements I and III are correct pairs, whereas II and IV are incorrect.
Sieve tubes are components of phloem. It is devoid of nucleus at maturity and its metabolic activities are controlled by nucleus of companion cells. Albuminous cells are not produced by angiosperms. They are present in gymnosperms only. In dicot roots, the vascular cambium formed during secondary growth. Primary meristem contributes to the formation of primary plant body.

33. (*d*) In dicot stems, the cells of cambium present between primary xylem and primary phloem is the intrafascicular cambium. The cells of medullary rays adjoining these intrafascicular cambium become meristematic and form the interfascicular cambium.

36. (*b*) Aerenchyma is a parenchyma, which contains wide air spaces that help to make the plant light and buoyant. It is a charecteristic of hydrophytes.

40. (*b*) Option (b) is true.
Vessels are composed of a row of cells (multicellular) placed one above the other. They form a pipe-like structure (wide lumen) due to the absence of the septa walls between the cells.
Rest options are incorrect.

41. (*d*) Stems are charecterised by endarch condition of xylem, while roots have exarch condition.

45. (*d*) Ray parenchymatous cells are responsible for lateral movement of water and food materials.

48. (*a*) All tissues except epidermis and vascular bundles constitute the ground tissue of fundamental tissue. It consists of simple tissues such as parenchyma, collenchyma and sclerenchyma. Ground tissue includes cortex, pericycle, medullary rays. In leaves, the ground tissue consists of mesophyll.

49. (*b*) Root is the part of plant body which grows and down into earth. The primary roots develop from radicle and give secondary and tertiary roots. Lateral roots develop endogenously, i.e. from pericycle between two protoxylems.

50. (*c*) Secondary growth is the growth in girth of stem and roots. Anomalous or abnormal secondary growth is found in some monocot stems such as *Yucca, Dracaena, Aloe, Agave*, etc.

51. (*a*) Dendrochronology is the determination of age of a tree by counting annual rings (these rings are formed by the activity of cambium in dicot root and stem).

52. (*a*) In grasses, certain adaxial epidermal cells along the veins modify themselves into large empty, colourless cells called bulliform cells. The bulliform cells in the leaves absorb water and they become turgid. Thus, the leaf surface is exposed. When they are flaccid due to water stress, they make the leaves curl inwards to minimise water loss.

53. (*a*) Statement I is true, but II is false. Incorrect statement can be corrected as
Many organs of aquatic plants float in water due to special parenchyma tissue known as aerenchyma. It consists of a network of parenchyma cells which enclose very large air cavities. Lotus is an aquatic plant having aerenchyma. Collenchyma is a simple permanent tissue with no intercellular gaps.

54. (*d*) The vascular bundles containing cambium are said to be open, but if no cambium develops, they are referred as closed. In most monocotyledons, closed vascular bundles are found, i.e. there is no secondary growth.

55. (*a*) Spring wood plus autumn wood of a year constitute annual ring. The spring wood (also called early wood) is light in colour and constitutes major part of annual ring. The autumn wood (also called late wood) is darker in colour.

56. (*b*) Jute fibres occur in long wedge-shaped bundles outside the xylem (i.e. phloem fibres). These fibres are called as secondary bast fibres.

57. (*b*) Stele is composed of pith and vascular bundle. On the basis of orientation of xylem and phloem, the vascular bundles of stele are classified as collateral, bicollateral and concentric. Pith is formed of large parenchymatous cells.

58. (*c*) Cotton is chiefly composed of epidermal hairs of seeds. Chemically, it is cellulose.

59. (*c*) When phloem is surrounded by xylem on all sides, such concentric vascular bundle is called amphivasal or leptocentric. Such vascular bundle is found in *Dracaena, Yucca, Aloe,* etc.

61. (*d*) Intercalary meristems are present in apical meristem only (not a lateral meristem). These are separated from the apex during the growth of axis and formation of permanent tissues. Thus, option (d) is correct.

CHAPTER 13

Animal Tissues

The life on earth has evolved from single-celled organisms. However, single-celled organisms cannot perform different activities simultaneously. In order to attain more efficiency and to perform such activities multicellular organisms were evolved. They contain group of cells that perform different functions. The group of cells along with intercellular substances which are specialised to perform a specific function is called **tissue**.

The term 'Tissue' was given by **Bichat** in 1801. He is also known as **Father of Histology**. The study of tissues is known as **Histology**. This term was introduced by **Mayer** in 1819.

In adult vertebrates, there are four kinds of tissues, i.e. epithelial, connective, muscular and nervous. The origin and basic functions of these tissues are given in table below

Types of tissues	Origins	Functions
Epithelial tissue	Ectoderm, endoderm, mesoderm	Protection, secretion and absorption
Connective tissue	Mesoderm	Support, binding, storage, protection, circulation of material
Muscular tissue	Mesoderm	Contraction of muscles, movement and locomotion
Nervous tissue	Ectoderm	Conduction of nerve impulse and control of various body functions

In this chapter, we will be focussing on various aspects of these tissues.

EPITHELIAL TISSUE

The term 'Epithelium' was given by Dutch anatomist **Ruysch** (1638-1731). Epithelial tissues are present as the lining of

- external and internal exposed parts of the body.
- open and closed cavities of the body.
- all blood and lymph vessels of the body.
- all glands and their ducts.

Generalised structure

- The cell epithelial tissues are present very close to each other, but these are separated by extracellular material.
- The cells of these tissues are placed in a **non-cellular** supporting layer called **basement membrane**. This basement membrane separates the epithelial tissue from underlying connective tissue.
- Basement membrane is made up of two layers, namely **basal** and **fibrous** or **reticular lamina**.
  (i) **Basal lamina** It forms upper thin layer of basement membrane. It is made up of mucopolysaccharides and glycoproteins secreted by the epithelial cells only. This layer is selectively permeable in nature.
  (ii) **Fibrous** or **Reticular lamina** It forms lower thin layer of basement membrane. It is made up of reticulin and collagen fibres, which are a part of underlying connective tissue.
- As stated earlier, epithelial tissue is derived from all three embryonic germ layers.
- Epithelium derived from the **ectoderm** is generally squamous epithelium. It is mainly found on the layer of skin and in the lining of all hollow organs.
- Epithelium, which is derived from the **mesoderm** is called **coelomic epithelium** or **mesothelium.** It is found as the peritoneum that underlines the body wall (parietal peritoneum) and the one that covers the viscera (visceral peritoneum).
- The mesothelium of the inner surface of the blood vessels and heart, is called **endothelium.**
- Epithelium derives from the **endoderm** include linning of digestive tract, except at the open end. It produces and releases different secretory products such as saliva, digestive enzyme, etc.

Junctions between epithelial cell

The cells of epithelial tissue are tightly packed together through following specialised intracellular cell junctions.

Tight junctions or Zonula occludens

- **Location** These junctions are found mainly in apical regions of cell.
- **Features** In these junctions, plasma membranes of adjacent cells become tightly packed together. They even get fused at series of points. These points contain **sealing strands.**
- **Function** These junctions check the flow of materials between the cells and prevent them from leaking across the tissues. These are essential for normal embryonic development.

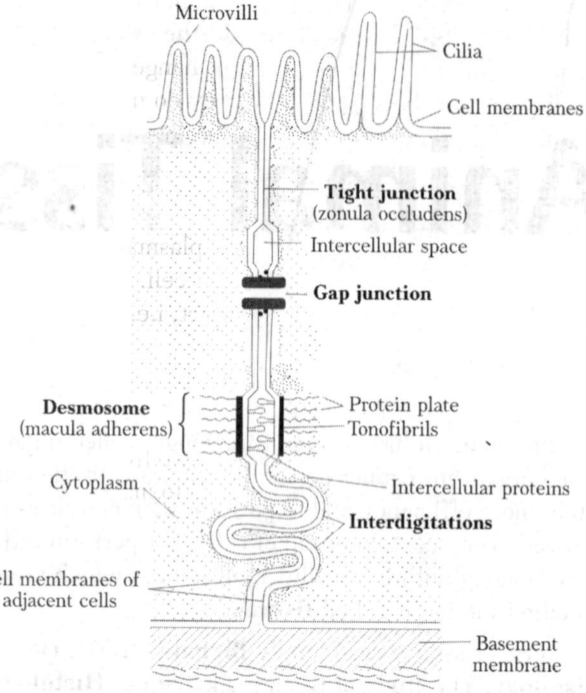

Fig 13.1 A diagram showing different junctions and specialisation of covering epithelia

Anchoring junctions

- These are widely distributed in animals, but absent in plants.
- They are mostly present in tissues that are subjected to severe mechanical stress, such as heart muscle and skin epithelium.
- These are further of following types
  (i) **Zonula adherens or Adherens junction** These occur below tight junctions and intercellular spaces. In these junctions, microfilaments made of actin extend into the cytoplasm forming a dense plaque. These plaques are present on the cytoplasmic side of each plasma membrane. They perform the anchoring functions.
  (ii) **Desmosomes or Macula adherens** These are thicker, stronger, disc-like protein plates present on the cytoplasmic side of the plasma membrane of adjacent cells. They give mechanical support to attached cells. These are usually seen in those areas where more cohesion is required.
  (iii) **Hemidesmosomes** These connect the basal surface of epithelial cells to the underlying basal lamina.

Gap junctions

These junctions are made up of protein called **connexon**. These junctions connect the cytoplasm of adjacent cells for rapid chemical exchange of ions and small molecules between them. They do not allow passage of larger molecules like protein.

Interdigitations

These are finger-like projections of plasma membrane that enter into cytoplasm of adjacent cell. These help to increase the surface area of contact, i.e. adhesion between adjacent cells.

Intercellular bridges

- These are minute projections that arise from adjacent cell membranes. They help the cells to make contact with one another.
- Before exploring further about epithelium, it is better to learn about certain specialised structures associated with epithelial cells.

Specialised structures related to epithelial cells

The free surface of epithelial plasma membrane gets modified to form following four types of functional structures

(i) **Microvilli** These are tiny finger-like projections covered by thick coat of glycoprotein called **glycocalyx**. They increase the absorptive surface of epithelium, e.g. wall of intestine, proximal convoluted tubule, etc.

(ii) **Stereocilia** These are elongated, non-motile protoplasmic extensions that are broad at the base and pointed at the top. These are found in the epithelial cells of epididymis and vas deferens. They are also present in the hair cell of the internal ear.

(iii) **Cilia and flagella** They are motile, hair-like, thin projections arising from the basal granule. Cilia present in epithelium of respiratory tract, oviduct, etc. These beat back and forth to propel fluid or particles in one direction. Flagella aid in locomotion like in spermatozoa.

(iv) **Endocytic, pinocytic and exocytic vesicles** They are associated with cell eating, cell drinking and cell excretion, respectively.

Classification of epithelia

On the basis of their function, epithelial tissue is classified into covering and glandular epithelia.

Covering epithelium

- The covering epithelial tissue is nearest to the environment. Hence, it plays an important role in the life activities of the animal. It is mainly concerned with protection, respiration, secretion, assimilation and elimination of waste products.
- Based upon 'number and arrangement of cell layers' along with 'shape of the cells', the covering epithelium is categorised as **simple** or **unilaminar** (single-layered), **stratified** or **compound** or **multilaminar** (multilayered) and **specialised epithelium.**

Table 13.1 Structures, locations and functions of simple and stratified epithelia

Structures	Locations	Functions
1. **Simple epithelium**		
(i) **Simple squamous** It is composed of single layer of flattened cells. • The edges of these cells fit closely together just like the tiles in a floor. • That is why the name **pavement epithelium** is also assigned to it. • These cells are usually polygonal and help in easy exchange of materials, (i.e. diffusion) across them. It is also called **tessellated epithelium.**	• It lines coelomic cavities, terminal bronchioles, air sacs, alveoli of lungs. • Bowman's capsules, descending limb of Henle's loop, rete testes. • Membranous labyrinth (internal ear) and some small gland ducts.	• Protection • Excretion • Gas exchange • Secretion of coelomic fluid

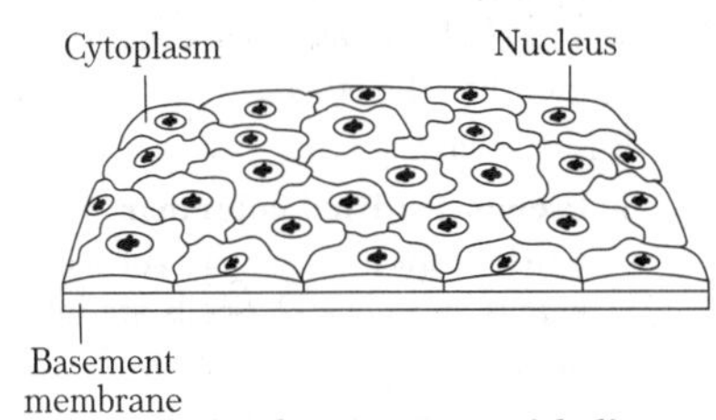

Simple squamous epithelium

Structures	Locations	Functions	
(ii) **Simple cuboidal** It consists of cuboidal cells with equal height and width, lying on the basement membrane. • Their nuclei are round and situated within the centre of cells. • The free surface of these cells bear microvilli. • The epithelium with microvilli is known as **brush bordered epithelium.**	• It lines the salivary and pancreatic ducts. • Vesicles of thyroid gland, ovaries, seminiferous tubules of testes, convoluted parts of uriniferous tubules. • Ciliary body, iris and choroid of eye, thin bronchioles • Some portions of ducts of sweat glands. • When cuboidal epithelium present as lining of ovaries and seminiferous tubules is called **germinal epithelium** because it produces gametes.	• Protection • Secretion • Absorption • Excretion • Gamete formation	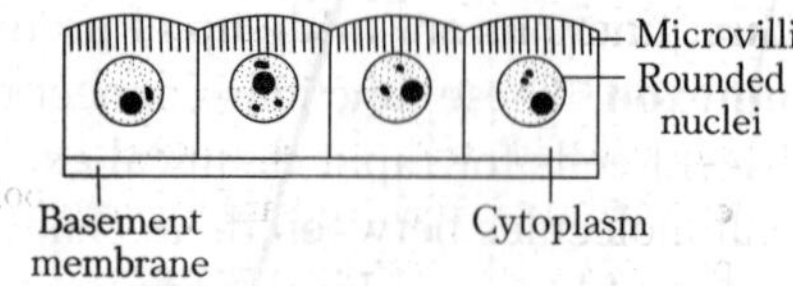 **Simple cuboidal epithelium**
(iii) **Simple columnar** It consists of tall and slender (column-like) cells lying on a basement membrane side by side in the form of columns. • Their nuclei are elongated and lie near to its base. At some places, columnar epithelium is seen with Goblet cells like in intestine. • At some places, these cells may bear microvilli and called **brush border columnar epithelium.** • These cells secrete mucinogen, which mixes with water to form slimy mucus.	• It lines gut epithelium from stomach to anus. • Bile duct, linings of gall bladder, uterine tubes. • Tympanic cavity, auditory tube or Eustachian tube. • Small bronchi of lungs.	• Absorption • Secretion of mucus, enzymes and other substances • Protection	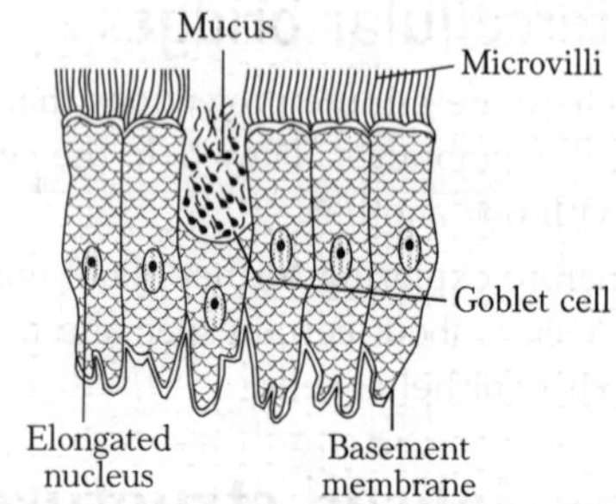 **Simple columnar epithelium**
(iv) **Simple ciliated** It is the epithelial layer having numerous delicate hair-like outgrowths called **cilia**. This epithelium is of two types (a) **Ciliated cuboidal epithelium** with ciliated cubical cells. (b) **Ciliated columnar epithelium** with ciliated column-like cells.	(a) **Ciliated cuboidal epithelium** occurs in many parts of ependyma of nervous system, etc. (b) **Ciliated columnar epithelium** lines the respiratory tract, Fallopian tube, etc.	• It provides protection. • Helps in the movement of mucus, urine, egg cells, cerebrospinal fluid in particular direction.	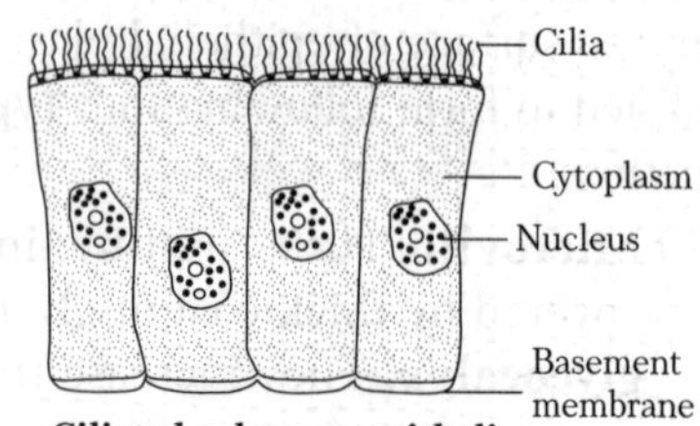 **Ciliated columnar epithelium**
(v) **Pseudostratified epithelium** It is a modified form of simple epithelium. • This epithelium is made up of a single layer of columnar cells, but it appears two-layered, because some cells are shorter than the others. • It is of following types (a) **Pseudostratified columnar epithelium** It lacks cilia but secrete mucus hence, Goblet cells are found in between the epithelium. (b) **Pseudostratified ciliated epithelium** It possesses cilia at free surface	(a) **Pseudostratified columnar epithelium** lines the large ducts of glands. • Parotid salivary glands • The urethra of human male • In the olfactory mucosa (b) **Pseudostratified ciliated columnar** lines the trachea and large bronchi.	• Protection • Movement of secretion from glands like urine and semen from urethra; mucus with dust and bacteria from trachea towards the larynx.	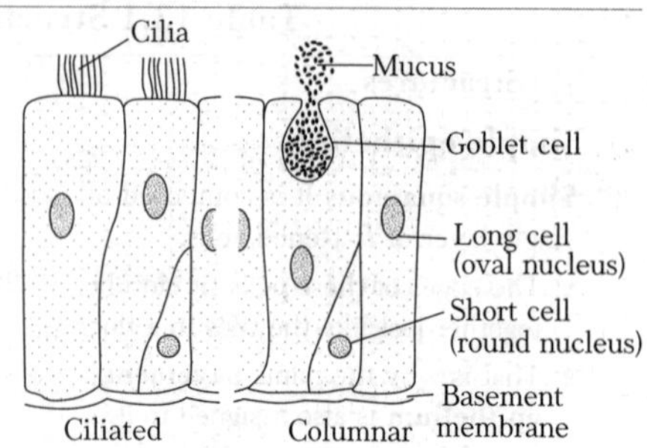 **Pseudostratified epithelium**

<table>
<tr><th>Structures</th><th>Locations</th><th>Functions</th><th></th></tr>
<tr><td colspan="4">2. Stratified (compound) epithelium</td></tr>
<tr><td>(i) Stratified squamous epithelium It is further of two types
(a) Keratinised stratified squamous epithelium It contains hard waterproof protein called keratin.
• Uppermost layer is non-nucleated and dead hence, called horny layer.
• Lowermost layer is living.</td><td>• Present in the epidermis of skin, hair and nails in land vertebrates.</td><td>• These cells are helpful in protection against abrasion and infection.
• Check the water loss.</td><td>

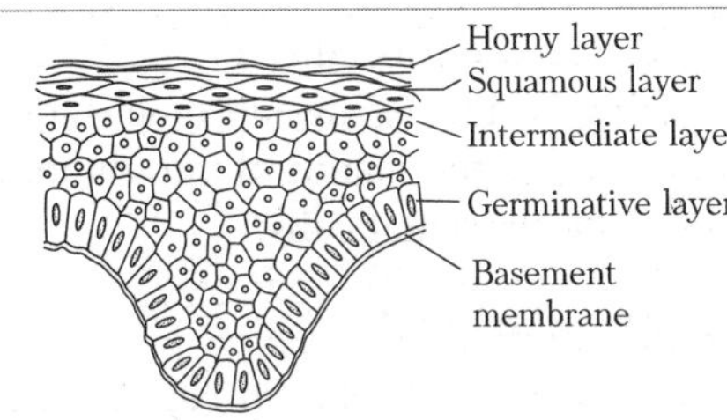

Keratinised stratified epithelium

</td></tr>
<tr><td>(b) Non-keratinised stratified squamous epithelium
• It does not have keratin.
• Cells are nucleated and living.</td><td>• Present in the oral cavity, tongue, pharynx (lower part), oesophagus, lining of vagina and cervix.
• Lower parts of nasal cavities, conjunctiva, inner surface of eyelids and cornea of eye.</td><td>• It is unable to check water loss.
• Provides moderate protection against abrasion.</td><td>

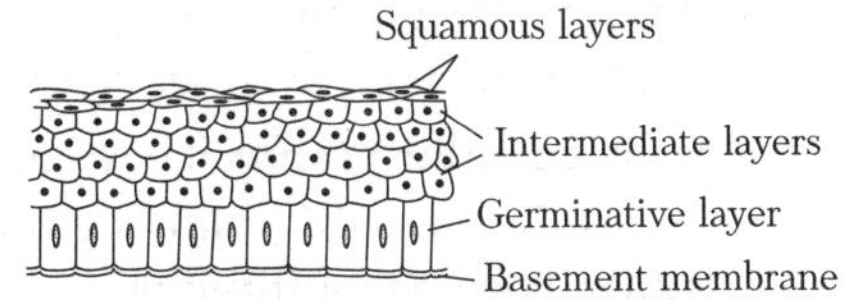

Non-keratinised stratified epithelium

</td></tr>
<tr><td>(ii) Stratified columnar epithelium Uppermost and basal layers consist of columnar cells. It can be ciliated too.</td><td>• Present in the covering of epiglottis.
• Lines mammary gland ducts and parts of urethra.
• The ciliated ones of this epithelium is present as lining of larynx and soft palate.</td><td>• It forms epidermis of fishes and many urodeles.
• Helps in protection and secretion.</td><td>

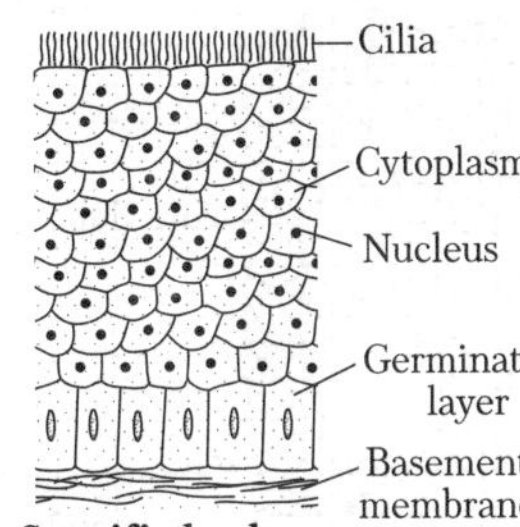

Stratified columnar ciliated epithelium

</td></tr>
<tr><td>(iii) Stratified cuboidal epithelium Uppermost layer is of cuboidal and basal layer is of columnar cells.</td><td>It is found in the lining of
• Ducts of sweat glands
• Female urethra
• Some parts of anal canal
• Pancreatic and salivary gland</td><td>It helps against mechanical and chemical stress.</td><td>

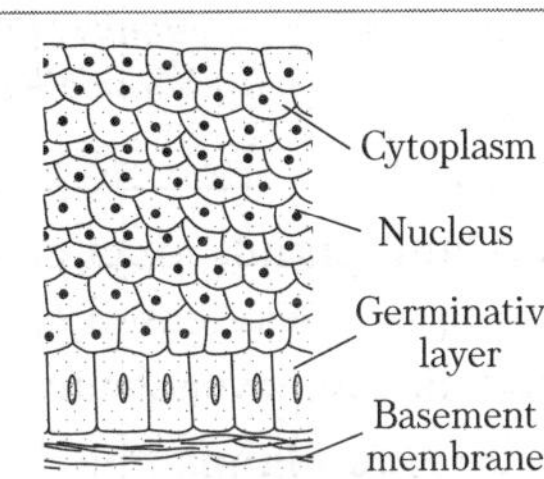

Stratified cuboidal epithelium

</td></tr>
</table>

Specialised epithelium

These are of following types

(i) **Transitional epithelium** (Urothelium) The transitional epithelium appears stratified with five or more layers of pear-shaped cells during relaxed condition.

- These cells contain many interdigitations in between them.
- At the time of stretch, this epithelium appears as trilaminar.
- It allows the expansion of the organs, it lines. Thus, it has the capability of modifying its shape without being damaged, e.g. in the wall of urinary bladder, ureters and upper part of urethra as well as in the wall and ducts of gall bladder.

(ii) **Neurosensory epithelium** The cells of this epithelium are modified as sensory cells.

- These sensory cells bear sensory hairs at their free ends, which are the main receivers of stimuli.
- In their basal portions, the sensory nerves are connected to fine nerve endings of sensory nerves through synapse.
- It is present as sensory lining of retina of eye, tongue, ear and Schneiderian membrane of nose.

(iii) **Pigmented epithelium** The cells of this epithelium contain pigment granules inside them. For example, epithelial cells of basal layer of retina and posterior part of iris.

(iv) **Myoepithelium** These cells are capable of contraction, mainly found around the mammary and sweat glands.

Glandular epithelium

These cells are specialised to perform the function of secreting hormones, sweat, mucus, enzymes and other products. All the glands of vertebrate body arises as the folding of epithelial cells.

Types of glands

Based upon the number of cells in secretory units, the glands can be of following types

(i) **Unicellular glands** They contain only one cell as the secretory unit, e.g. Goblet cell or Chalice cell.

(ii) **Multicellular glands** They contain more than one cell as the secretory unit, e.g. most of the glands of human body are of this type.

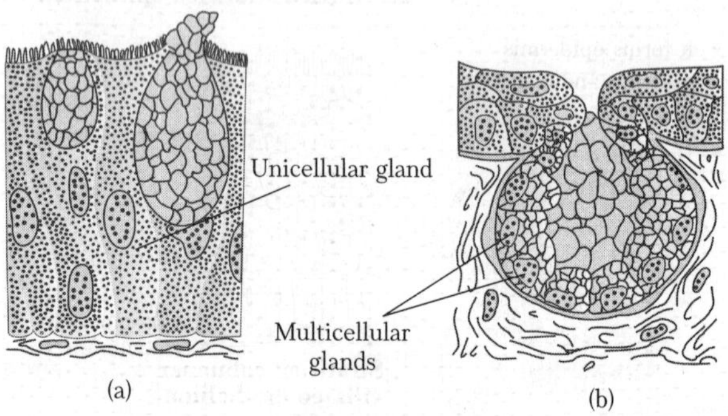

Fig. 13.2 Types of glands : (a) Unicellular gland and (b) Multicellular gland

Glands present in vertebrate body are of three basic types

(i) **Exocrine glands** (duct containing glands) Their secretions are released at specific sites with the help of ducts, e.g. sweat, salivary gland, etc.

(ii) **Endocrine glands** (ductless glands) They pour their secretions directly into the blood. These specifically secrete hormones only, e.g. thyroid, hypothalamus gland.

(iii) **Heterocrine gland** Glands having both exocrine and endocrine secretion, e.g. pancreas, etc.

The exocrine glands are further classified as follows

(i) **On the basis of nature of secretion**

(a) **Mucous glands** Their secretion is in the form of a viscous material, called **mucous fluid**, e.g. Goblet cells of intestine. Their cells are also termed as **mucocytes.**

(b) **Serous glands** Their secretion is in the form of a clear watery fluid, which may or may not contain enzymes, e.g. sweat or sudoriferous glands. These cells are also termed as **serocytes.**

(c) **Mixed glands** Some glands have both mucocytes and serocytes and hence produce both kinds of secretions, e.g. gastric glands and pancreatic glands.

(ii) **On the basis of types of secretory units present, within a single gland**

(a) **Simple** In these glands all the secretory units are of same type and secretion passes through a single unbranched duct. For example,

- **Simple straight tubular** In crypts of Lieberkuhn in intestine.
- **Simple coiled tubular** In sweat glands as terminal part is tubular.
- **Simple branched tubular** Gastric glands in stomach.
- **Simple alveolar or Sacular** Mucous glands in the skin of frog, poison glands in toad.
- **Simple branched alveolar or saccular** Sebaceous glands.

(b) **Compound** In these glands more than one type of secretory units are present and the secretion passes through many branched ducts.

For example,

- **Compound tubular gland** Mammary glands of protherians.
- **Compound saccular** Submaxillary and sublingual salivary glands.
- **Compound tubular sacccular** Mammary glands, pancreas, parotid, salivary gland, Cowper's gland, Bartholin's gland.

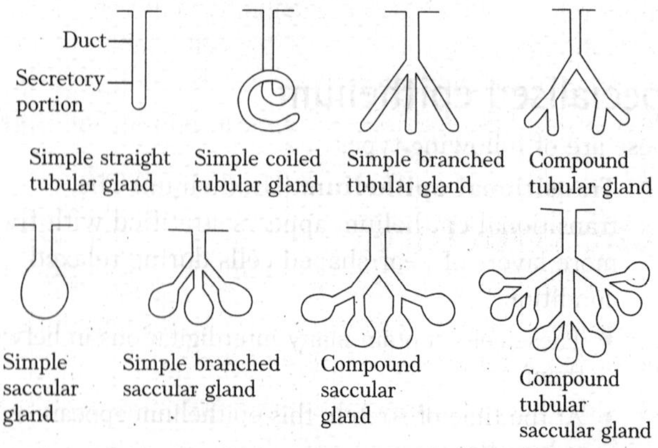

Fig. 13.3 Types of glands on the basis of structure

(iii) **On the basis of secretory nature of glands**

(a) **Holocrine** The complete secretory unit disintegrates and comes out along with the secretion. For example, sebaceous gland of mammals.

(b) **Merocrine** or **Epicrine** The secretion is discharged by diffusion. This simply means that the secretory units remain intact at the time of secretion.

For example, Goblet cells, most sweat glands, salivary and intestinal glands.

(c) **Apocrine** The secretory product accumulates in the apical part of secretory unit. Later, apical part breaks off from the cell and comes out along with the secretion. For example, mammary gland and some sweat glands.

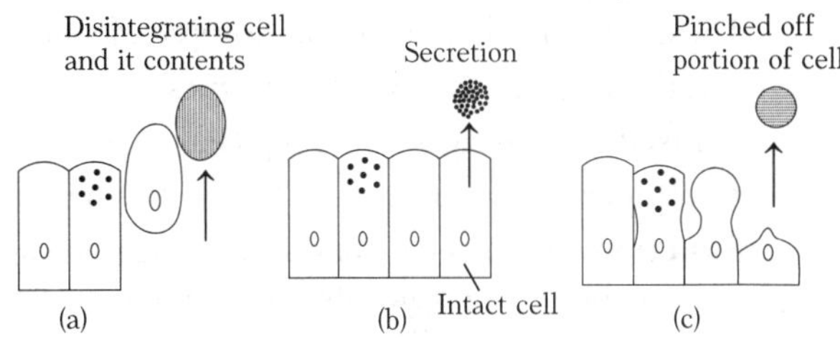

Fig. 13.4 Types of glands : (a) Holocrine gland, (b) Merocrine gland, (c) Apocrine glnad

CHECK POINT 13.1

1. Epithelial tissue always has an exposed outer surface and an inner surface anchored to connective tissue by a thin, non-cellular layer called the
   (a) non-stratified layer (b) stratified layer
   (c) basement membrane (d) fibroblast

2. The inability to absorb digested nutrients may be due to damage of which type of epithelium?
   (a) Ciliated columnar
   (b) Simple columnar
   (c) Simple squamous
   (d) Simple cuboidal

3. Trapped dust particles are pushed out of respiratory tract by
   (a) squamous epithelium
   (b) glandular epithelium
   (c) ciliated epithelium
   (d) compound epithelium

4. The protein deposited in the dead superficial cells that make the skin epithelum impervious to water is
   (a) mucus (b) elastin
   (c) keratin (d) collagen

5. Which of the following is an example of gland, where the entire cell filled with its secretory products disintegrates to discharge its products?
   (a) Sebaceous gland
   (b) Mammary gland
   (c) Sweat gland
   (d) Salivary gland

6. The glands in which the secretory products accumulate in the apical part of cells that breaks off from the cell to discharge them as secretions called
   (a) merocrine (b) apocrine
   (c) holocrine (d) exocrine

CONNECTIVE TISSUE

- It is the most abundant and widely distributed tissue in the body of animals. It consists of living cells embedded in abundant non-living intercellular matrix.
- It connects different tissues or organs and provides support to various structures of animal body.

Basic structure

The basic structure of connective tissue involves the suspension of following three types of fibres and some cells within a gelatinous amorphous matrix.

Following types of fibres are present in connectivc tissue

(i) **White collagenous fibres** These are long unbranched, soft and elastic fibres of tropocollagen protein. They have great tensile strength and are most numerous of all fibres. On boiling in water, collagen protein is converted into gelatin. These are found in **tendons**.

(ii) **Reticulin fibres** These are long, unbranched, soft, flexible, but inelastic fibre of reticulin protein. These are found more in newly born babies, embryos and at healing or regenerating wounds. Later on, these are usually replaced by collagen fibres. These are mainly found in spleen, lymph nodes, etc.

(iii) **Yellow elastin fibres** These are less numerous branched, flexible elastic fibre made up of elastin protein. These are found in **ligaments**.

Following types of cells are present in connective tissue

(i) **Fibroblasts and fibrocytes** These are numerous and largest cells of connective tissue. Vitamin-C is essential for the proper functioning of these cells.

Function These have the ability to synthesise protein fibres, i.e. collagen, reticulin, elastin and mucopolysaccharides.

They also form various proteins of the amorphous ground substance. Their aggregations are seen in healing wounds, repairing tissues and in tissues where growth is taking place.

(ii) **Macrophages or Histiocytes or Clasmatocytes** These are amoeboid cells bearing pseudopodia. These can be of following two types

- Fixed macrophages with filopodia.
- Wandering macrophages with simple lobopodia.

Function These cells work as scavengers. They have the capability to phagocytes the foreign particles, dead and injured cells.

(iii) **Mast cells or Histaminocytes or Mastocytes** These are numerous, amoeboid, small-sized cells that are similar to basophils. These are present along the blood vessels, lymphatic vessels and nerves.

Function These cells secrete heparin (proteoglycan), histamine and serotonin (both proteins).

(iv) **Lymphocytes** These are the smallest cells of connective tissue. These are less in number in connective tissue. They have amoeboid shape, large nucleus and less cytoplasm.

Function These cells synthesise antibodies.

(v) **Plasma cells or Plasmocytes** These are small rounded cells. Their lifespan is small (2-3 days). **Function** These are considered as the most potential antibody forming cells of the body.

(vi) **Fat or Adipose cells or Adipocytes or Lipocytes** These cells contain fat globules or fat droplets inside them. On the basis of number of lipid droplets inside them, these can

- monolocular, i.e. with single fat droplet
- multilocular, i.e. with many fat droplets

Note
- ***Hertvig*** *called connective tissue as* ***'mesenchyme'*** *because they originated from embryonic mesoderm.*
- *In certain organs of our body, specialised macrophages are present which are known by their specialised names, e.g. dust cells of lungs, Kupffer cells of liver, microglia of brain and spinal cord, mesangial cells of kidneys, etc.*
- *The elastin protein is probably the most resistant of all body protein to any chemical change.*
- *In mummies of Egypt, the blood vessels are preserved due to well- preserved elastin fibres in them.*

Classification

The connective tissue can be categorised as follows

Connective tissue
- Connective tissue proper
  - Loose
  - Dense fibrous
  - Specialised
- Skeletal connective tissue
- Fluid or Vascular connective tissue

Connective tissue proper

On the basis of cells and fibres, it is of three types

1. Loose connective tissue
2. Dense fibrous connective tissue
3. Specialised connective tissue

Table 13.2 Structures, locations and functions of different types of connective tissue proper

Structures	Locations	Functions	
1. **Loose connective tissue** It consists of cells and fibres loosely arranged in a semi-fluid ground substance (matrix). It is further categorised as follows			
(i) **Areolar connective tissue** It consists of matrix of modified polysaccharides (mucopolysaccharides) and proteins (glycoproteins). • Several kinds of irregular cells are also scattered in the matrix. • These are fibroblasts, macrophages, mast cells and plasma cells. • Apart from these cells, the matrix contains white collagen and yellow elastic fibres.	• Beneath the skin, in between and around muscles. • Nerves and blood vessels. • In the submucosa of gastrointestinal tract. • Within and in between lobes of compound glands.	• Provides strength, elasticity, support and packing tissue between organs. • Facilitates migration of cells at the site of inflammation. • Helps to repair tissues after injury.	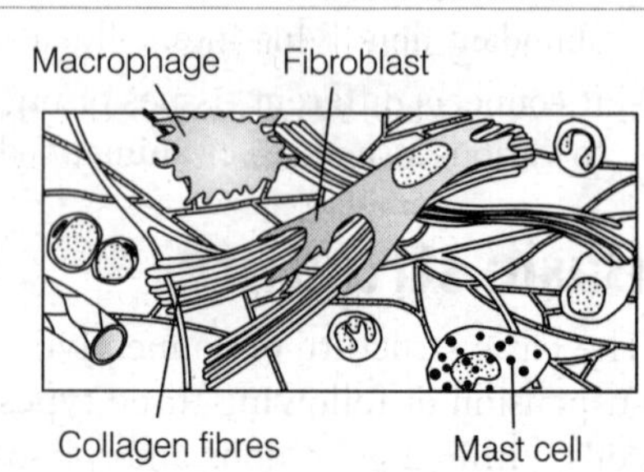

Structures	Locations	Functions	
(ii) **Adipose or Fatty connective tissue** It is basically an aggregation of fat cells or adipose cells called **adipocytes**. • Each fat cell is rounded or oval in shape. • The protein fibres are few in number and form a loose network for supporting the fat-laden cells. • Following two types of fats are seen in relation with adipose tissue.	• Beneath the skin. • Around heart, kidneys. • It also forms panniculus adiposus (subcutaneous fat in mammals). • Blubber of whales, elephant, camel hump, yellow bone marrow, etc.	• Fat storage, insulation, shock absorption.	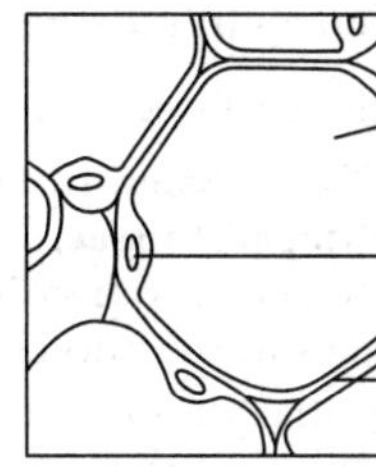
(a) **White fat** It is composed of monolocular adipocytes. In these cells, single large fat globule, peripheral cytoplasm and peripheral nucleus is present. These have less number of mitochondria. **The penniculus adiposus** (human fat layer) is also responsible for the difference in the body countours of male and females.	• It is present beneath skin and around internal organs.	• Stores energy	
(b) **Brown fat** It is composed of multilocular adipocytes. In these, more fat globules are present in each cell. These have more number of mitochondria. The colour of brown fat is due to high concentration of iron containing cytochrome pigment.	• It is present in newborn babies and hibernating animals.	• Resistance to shivering in infants and generate heat.	
2. **Dense fibrous connective tissue** It consists of fibres and fibroblast cells, which are compactly packed in the matrix. It is the principal component of tendons and ligaments. It is of following two types			
(i) **Dense irregular connective tissue** It has fibroblasts and numerous collagen fibres which are oriented in different directions.	• Skin • Perimysium, perineurium and periosteum (around bones)	• Provide strength • Makes skin resistant to tearing by streching forces.	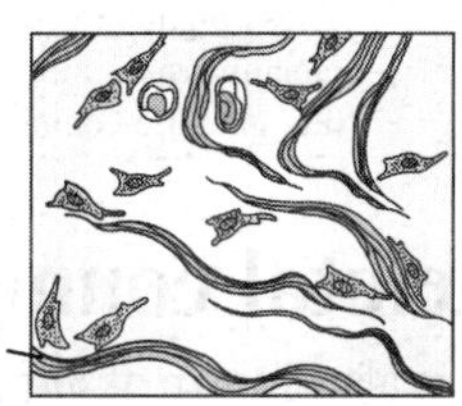
(ii) **Dense regular connective tissue** Collagen fibres are present in rows between the parallel bundles of fibres. It is further of two types			
(a) **White fibrous connective tissue** In this tissue, collagen fibres are more in quantity. Yellow fibres and reticular fibres are completely absent. Among cells, fibroblast and mast cells are more in number.	• Dermis of skin, • Perichondrium of the cartilage • Periosteum of the bone • Muscles and nerves • Fibrous capsules of penis, testes • Blood vessels, glands and other organs.	• Forms tendon (non-elastic, but flexible tissue), which joins muscle to bones.	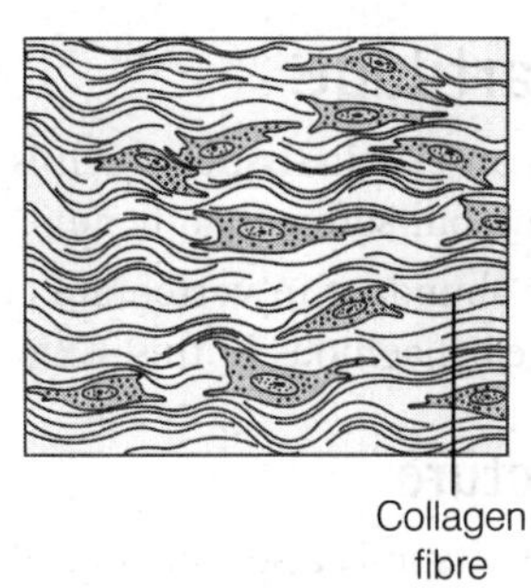
(b) **Yellow fibrous connective tissue** In this tissue, yellow elastic fibres are more in quantity, but collagen fibres are also present. Reticular fibres are absent.	• Vocal cords • Capsules of spleen and lungs • Elastic membranes of arteries and trachea, etc.	• Forms ligament (elastic and flexible tissue), which joins bones to bones.	

Structures	Locations	Functions	
3. Specialised connective tissue			
These are of following types			
(i) **Reticular fibrous connective tissue** It is also called **lymphoid tissue.** Its cells are primitive with stellate appearance due to the presence of reticulin network. • Matrix of this tissue is **lymph.** Lymphocyte cells are also more in number. • In various organs, this tissue is responsible for the formation of supportive network. • These cells are called **reticuloendothelial cells** and the system associated with them is called **reticuloendothelial system.**	• Liver • Spleen • Lymph nodes • Thymus and tonsils	• Provides strength and support. The reticular cells are phagocytic and form defence mechanism of the body.	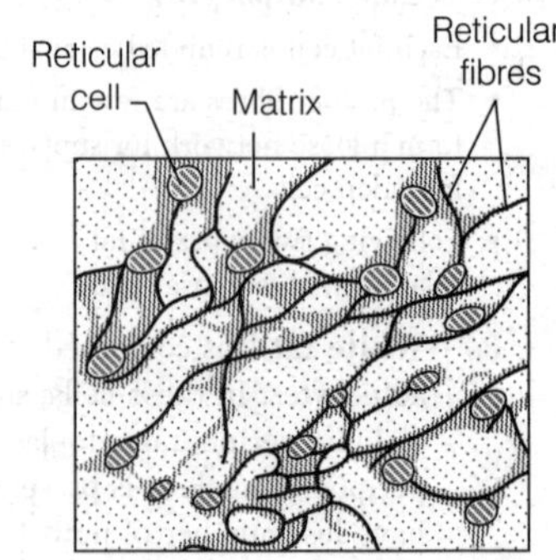
(ii) **Mucoid connective tissue** It is also called embryonic tissue, because it is mainly found during embryonic life. • Matrix is in abundance. • Few collagen fibres and fibroblast cell may be present. • Matrix is composed of jelly-like material called **Wharton's jelly.**	• Umbilical cord • Cock's comb • Vitreous body of eyeballs	• It forms the umbilical cord to connect placenta with foetus and vitreous humour.	
(iii) **Pigmented connective tissue** It is a modification of areolar connective tissue. • It contains more number of pigmented cells called **chromatophores,** which provide colouration. • These pigment cells are **melanophore** (i.e. melanin containing), **guanophore** (i.e. guanine containing), etc.	• Choroid • Ciliary body, iris of the eye • Dermis of the human skin	• Gives colour to the structures.	

Skeletal connective tissues

- It is also known as **supporting tissue** as it provides support to the body, protects various organs and helps in locomotion.
- It constitutes the endoskeleton of the vertebrates.
- It is mainly of two types, namely **cartilage** and **bone.**

1. Cartilage

- It is an elastic tissue which is harder than dense connective tissue but softer than bone.
- It is abundant in vertebrate embryos as most of the bones are cartilaginous in this stage.

Structure

- This tissue is also called **gristle.** It is a solid, semi-rigid, tough, but a flexible connective tissue.
- The core protein of cartilage is called **aggrecan.**
- The outer fibrous sheath of cartilage is called **perichondrium.** This contains an outer fibrous layer and an inner layer of **fibroblasts.**
- The fibroblasts are converted into **chondroblasts** (immature cells of cartilage), which in turn converted into the **chondrocytes** (mature cells of cartilage). These cells secrete matrix.
- Its matrix is called **chondrin.** It consists of water, proteoglycans, some lipids, collagen and non-collagenous protein. It is solid and pliable in nature.
- Due to continuous secretion of matrix, they seem to be surrounded by a vacuole-like empty space within the matrix. This space is called **lacuna.** Within the matrix, lacunae may contain more than one chondrocyte.

Formation

- The increase in the thickness of cartilage due to the increase in number of chondrocytes from fibroblasts is called the **appositional growth** or **perichondrial growth** (exogenous).
- Chondrocytes have the power of division. So, the growth of cartilage is due to the multiplication of chondrocytes during the period of growth is called the **interstitial growth** (endogenous).
- Chondrocytes receive nutrition through diffusion either from blood vessels in perichondrium or in cartilage canals.

Function

Cartilage provides support and flexibility to the body parts and resists compression. It smoothens the surface joints and thus, prevents the wear and tear of bones due to friction.

Types

The cartilage is of four types

(i) Hyaline cartilage

- **Location** It is found in the skeleton of foetus, as thin cushions or pads upon joints of long bones (called as **articular cartilage**), sternum, larynx, sternal parts of ribs, rings of trachea, bronchi, nasal cartilage and hyoid.
- **Structure** It is the most prevalent cartilage. Its matrix is glassy and purely white or bluish. They make approximately 43% of dry weight.
- **Function** Inside the bone, hyaline cartilage serves as a centre of ossification or bone growth. It forms skeleton of chondrocytes along with embryonic skeleton in amphibious and land vertebrates.

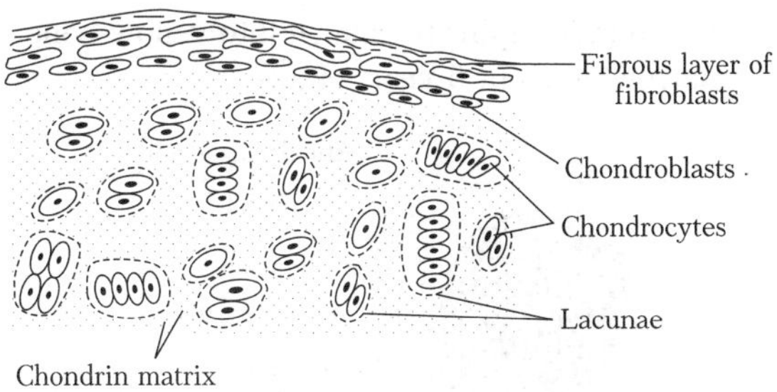

Fig 13.5 Hyaline cartilage

(ii) Calcified cartilage

- **Location** It is found in suprascapula of pectoral girdle, pubis of pelvic girdle of frog and vertebrae of cartilaginous fishes.
- **Structure** This tissue is formed by the calcification of hyaline cartilage due to the deposition of calcium salts in the matrix.
- **Function** The calcification reduces elasticity of cartilage and makes it more rigid.

(iii) White fibrous cartilage

- **Location** It is found in intervertebral discs and pubic symphysis of pelvic girdle in man.
- **Structure** It contains thick bundles of collagen fibres between rows of chondrocytes in lacunae. This tissue has great tensile strength, flexibility and rigidity. White fibrous cartilage is the strongest cartilage in vertebrate body. It lacks perichondrium.
- **Function** Fibrocartilage acts as shock absorber in intervertebral discs. It also helps in parturition (childbirth).

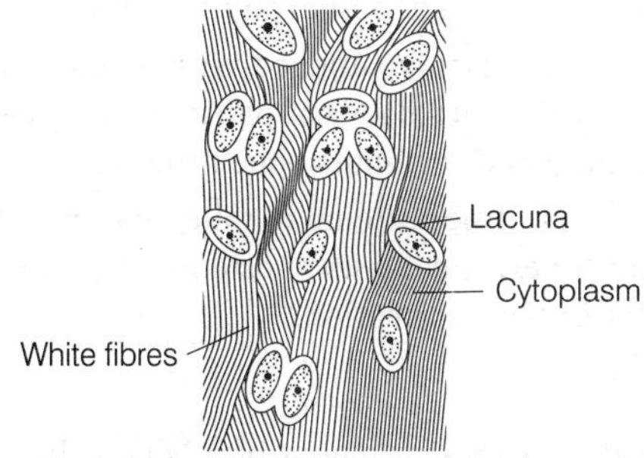

Fig 13.6 White fibrous cartilage

(iv) Yellow elastic cartilage

- **Location** It forms ear pinna, pharyngeotympanic tube or Eustachian tube, epiglottis, tip of nose along with some laryngeal and bronchiolar cartilages in man.
- **Structure** It contains a dense network of elastic fibres between scattered chondrocytes. Perichondrium is present in this cartilage.
- **Function** The elastic fibres make these structure considerably elastic and flexible.

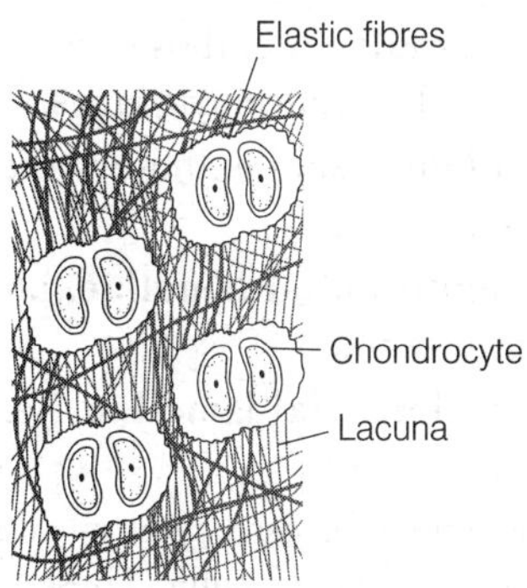

Fig 13.7 Yellow elastic cartilage

2. Bone

It is also called **osseous tissue**. It is the hardest tissue of vertebrate body. This tissue forms the major frame work of the vertebrate body especially in tetrapods. It is also considered as the homeostatic reservoir of Ca^{2+}, Mg^{2+}, phosphorus, etc.

Structure of bone

Bone consists of five parts, namely periosteum, matrix, bone cells, endosteum and bone marrow.

(i) Periosteum

- It is a thick and tough sheath that forms an envelope around the bone.

- Periosteum comprises of two distinct layers. The outermost layer consists of **fibres** and **fibroblasts**. The fibres of periosteum are also called **Sharpey's fibres**.
- The outer layer is richly supplied with blood vessels and lymphatics. This layer does not participate in the formation of bone thus, this layer limits bone growth.
- The inner layer of periosteum is called **cambium layer** or **osteogenic layer.** It consists of osteoblasts, thus participates in bone formation.

(ii) Matrix

- The ground substance or matrix of a bone is hard and non-pliable. It is majorly made up of a protein called **ossein**. It consists of 62% of inorganic matter and about 38% of organic matter.
- The organic matter of a bone contains collagen fibres with the ground substance of sulphated mucopolysaccharides, i.e. glycosaminoglycans or glycoproteins.
- The salts present in the matrix include calcium phosphate (maximum), calcium carbonate, magnesium phosphate and fluorides, within the bone the matrix.
- Matrix appears in form of thin layers called **lamellae**, which are of following four types on basis of their distribution
  (a) **Haversian lamellae**, i.e. those occur around Haversian canal as concentric rings.
  (b) **Interstitial lamellae**, i.e. those present between two Haversian canals.
  (c) **Outer circumferential lamellae,** i.e. those occur inner to periosteum.
  (d) **Inner circumferential lamellae**, i.e. those occur outer to endosteum.
- The matrix has numerous osteocytes surrounded by a fluid-filled space, the bone lacuna. These lacunae lead into fine radiating channels called the **canaliculi**.

(iii) Bone cells

These are of following types

(a) **Osteoblasts** These are bone forming cells. They synthesise and secrete unmineralised organic portion of bone matrix osteoid. Formation of bones through osteoblasts is called **osteogenesis**.

(b) **Osteoprogenitor cell** These cells develop into osteoblast cells through mitotic division.

(c) **Osteocytes** These are the mature and non-dividing osteoblast cells. They remain permanently in G_0-phase of cell cycle. These cells lie in lacunae and are surrounded by matrix.

(d) **Osteoclasts** These are the bone destroying cells. These cells are derived from osteoblasts and osteocytes. They contain acid phosphatase enzyme, which helps in cell destruction.

(iv) Endosteum

- It is present outside the marrow cavity. Endosteum is basically a membrane that line the marrow cavity.
- Its composition is similar to periosteal layer, but the white fibrous tissue is present here.

(v) Bone marrow

- The long bones of body contain two parts namely, **heads** or **epiphysis** and **shaft** or **diaphysis.**
- The marrow or medullary cavity is present in the shaft. It is filled with bone marrow.
- Bone marrow is a semisolid, fatty, neurovascular tissue also called **myelogenous** or **myeloid tissue.**
- It takes part in the production of blood cells hence, known also as **'haemopoietic tissue'**. It is composed of adipose tissue, areolar tissue and blood.
- It is of following two types

  (a) **Red bone marrow** It is red in colour and associated with the production of blood.
  - The red bone marrow is seen throughout the endoskeleton during foetal life and at birth.
  - It gradually begins to change into yellow bone marrow at about 5 years of age.

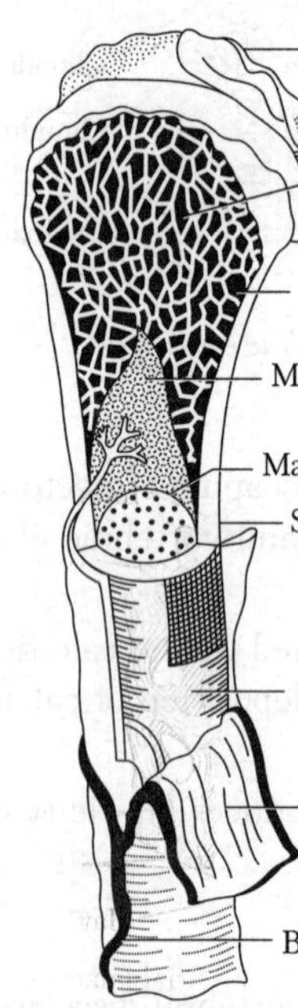

Fig. 13.8 The cancellous and compact bone regions

  - At the age of about 20-25 years, the red bone marrow persists only in the vertebrae, sternum, ribs, clavicles, scapula, pelvis, cranial bones and in the proximal portions of long bones like femur and humerus.
  - After this age, bone marrow starts degenerating especially in the bones of cranial region where it is called **gelatinous matrix.**

(b) **Yellow bone marrow** It is yellow in colour and has much fatty tissues. It is present in shaft of long bones. It produces blood cells in emergency, i.e. during excessive blood loss.

Mammalian bone

- In mammalian long bones, the bony region of epiphysis is composed of spongy or cancellous or trabecular bone, while that of diphysis is compact or periosteal bone.
- The **spongy bone** consists of a network of thin and irregularly arranged longitudinal and transverse bony bars called **trabeculae**.
- Trabeculae is covered by endosteum. The depression like spaces in between the trabeculae are called **cancelli**.
- These spaces are filled with red bone marrow. This structure is then covered up by a periosteal layer.
- Spongy bones are comparatively stronger and lighter than compact bones.

The Haversian system

- The compact bone region of mammalian bone contains a system of canals called **Haversian** and **Volkmann's canals**.

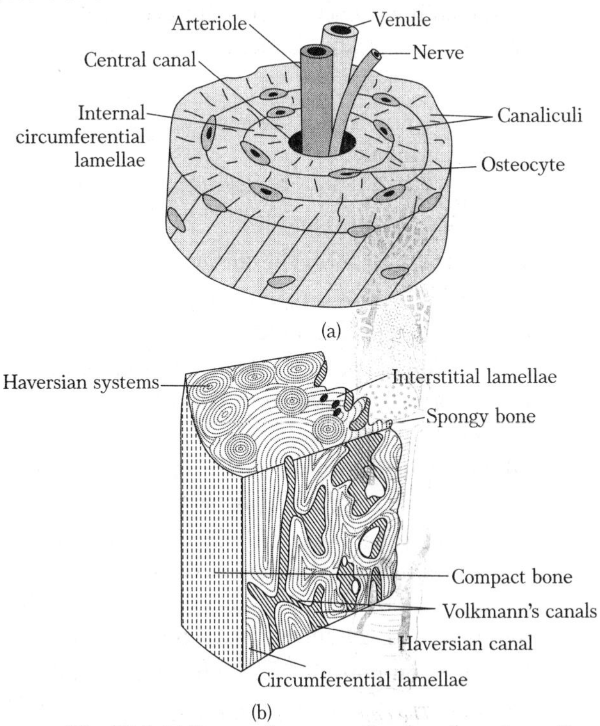

Fig. 13.9 (a) Zones of a mammalian long bone in section, (b) An osteon

- The Haversian canals are vertical canals present parallel to the length of bones. These are connected to each other by Volkmann's canals.
- About 4-20 concentric rings of lamellae surrounds one Haversian canal. These canals contain an arteriole, a venule and a nerve fibre.
- The requirement of such a system in mammalian bones is to supply sufficient blood to osteocytes.
- This complete system of lamella along with one Haversian canal is called **one osteon.**
- In a bone, these osteons are separated by specialised bright thin lines called **cementing lines of Ebner.**

Types of bones

Bones are of following types

(i) **Cartilaginous** or **Replacing bone or Endochondrial bones** These bones are formed by the ossification of cartilage, e.g, vertebrae, humerus, femur, fibula and girdles.

(ii) **Membranous or Investing bones or Dermal bones** These bones are formed by the ossification of connective tissue strip or mesentery, e.g. skull bones, phalanges, clavicles.

(iii) **Sesamoid bones** These bones are formed by the ossification of tendons. Patella or Knee cap is the best example of such a bone.

(iv) **Visceral bones** These bones are formed in the soft internal or visceral organs, e.g.

- **os penis** in the penis of most bats, insectivores, rodents like rats, carnivores like dog, whales.
- **os cordis** in the heart of some ruminants like deer.
- **os clitoris** in the clitoris of many carnivores.
- **os palpebrae** in the eyelids of crocodiles.

Note
- ***When a bone is burnt,*** *its organic phase burns away as CO_2 and H_2O (steam), but the inorganic phase is completely left behind as ash. It is called* ***dried bone.***
- ***If a bone is kept in KOH,*** *it remains unaffected only the surrounding muscles and connective tissues get dissolved in KOH, the bone become clean and bright hence called* ***bright bone.***
- ***If a bone is kept in dilute acid*** *for a few days, the inorganic salts get dissolved leaving behind the soft, flexible organic matter hence, called* ***decalcified bone.***

A table depicting the differences between bone and cartilage is given below

Table 13.3 Differences between bone and cartilage

Bone	Cartilage
Bone is surrounded by a tough sheath called periosteum.	Cartilage is surrounded by a firm sheath called perichondrium.
Matrix is composed of tough inflexible material called ossein.	Matrix is composed of a frim, but flexible material called chondrin.
Matrix is impregnated with salts, chiefly calcium phosphate and carbonate.	Matrix is impregnated with calcium salts only in calcified cartilage.
Matrix occurs in concentric lamellae.	Matrix occurs in a homogeneous mass.
Matrix contains fibres but these are indistinguishable.	Matrix may contain fibres, which may or may not be distinguishable.
Osteocytes are found lacunae singly.	Chondrocytes are found singly or in groups of two or four.
Osteocytes are irregular and give off branching processes.	Chondrocytes are oval and devoid of processes.
Lacunae send out canaliculi for the processes of bone cells, which extend into these minute canals.	Lacunae lack canaliculi.
There are outer and inner layers of special bone forming cells, the osteoblasts which produce new osteocytes, which secrete new lamellae of matrix.	Cartilage grows by division of chondroblasts.
Growth in bone is bidirectional.	Growth in cartilage is unidirectional.

Fluid or Vascular connective tissue

Blood and lymph are two fluid connective tissues, which flow through our cardiovascular system.

Blood

- It is a motile connective tissues consisting mainly of fluid matrix or ground substance along with free cells.
- The blood is slightly heavier than water with specific gravity 1.04-1.07. The viscosity of blood is 4.2-5.2 times more than water.
- It is slightly alkaline in nature with pH range 7.3-7.5.
- In human beings, it makes up 7-8 % of total body weight. Percentage of blood in females is slightly lower.
- It is the chief transport system of our body and has an important role in the defence against disease.

Two major components in blood are as follows

1. Blood plasma

- It is a yellow transparent colloidal mixture that forms 55-60% of blood and constitutes about 5% of body weight.
- It contains 90-92% of water with inorganic and organic constituents.
- **Inorganic salts** make 0.9% of plasma. Main inorganic salts include Cl^- and bicarbonates of Na, K, Mg and Ca. These salts remain as ions or electrolytes and help in maintaining the alkalinity of plasma.
- Blood plasma contains following **organic constituents**

(i) **Proteins** These are about 7-8% as colloidal particles. These include

- **Albumins** are the smallest proteins of blood. These are synthesised in the liver and are responsible for the maintenance of osmotic pressure in plasma along with osmoregulation in cells and tissue fluids. These constitute about 4-7% of total plasma.
- **Globulins** constitute 1.3-2.5% of plasma protein count. These are mainly formed in plasma cells of lymphoid organs. These are categorised as
  (a) α_1-have inhibitory effects on trypsin and many other proteases.
  (b) α_2- helpful in the transport of various proteins and hormones.
  (c) β-helpful in the transport of insoluble lipids, soluble vitamins and iron.
  (d) γ - works as antibodies or immunoglobulins.
- **Prothrombin** is essential for blood clotting. It is one of the largest plasma proteins.
- **Fibrinogen** is one of the largest plasma proteins. A clot is formed only when the fibrinogen of plasma is converted into fibrin.
- **Digested nutrients** are also present in plasma. Blood is responsible for the supply of these nutrients.
- **Excretory substances** include ammonia along with certain other nitrogenous wastes in plasma. These are chiefly liberated by body cells into the blood and eliminated by excretory organs.
- **Defence compounds** include immunoglobulins, lysozymes and properdin (a protein). They help to destroy bacteria, viruses and other foreign substances.

(ii) **Anticoagulant** It is a conjugated polysaccharide in human body. It is named as **heparin**. It prevents coagulation of blood inside the blood vessels. Some other anticoagulants are **hirudin, lamphredin** and **coumarin**, etc. Among these, coumarin is of plant origin and is also known as **dicumarol**. It simply acts as vitamin-K.

2. Blood cells

- The blood cells are the formed elements of blood. These constitute about 40-45 % of blood within the body.
- These have the specific gravity of about 1.09, i.e. these are slightly heavier than plasma.
- Blood is mainly composed of following three types of cellular components.

(i) RBCs or Erythrocytes

- RBCs was first described by **Leeuwenhoek** in 1674.
- In mammals, the RBCs are roughly circular and non-nucleated except in camels and llama where they are oval and nucleated. The absence of nuclei imparts a biconcave, disc-like shape to mammalian RBCs.
- The shape (concave surface) of RBC increases the cell's surface area and facilitates diffusion of O_2 and CO_2 into and out of the cell.
- Likewise, the lack of nuclei and organelles contribute to increased haemoglobin content and gas carrying capacity.
- In humans, the RBCs are 6.5-8 μ in diameter and 1-2 μ thick. Their count is about 52 lac per cubic mm (in males) and about 47 lac per cubic mm (in females).
- The total surface area of RBCs is approximately 1500-2000 times more than the surface area of body of a person.
- Human erythrocytes of variable shape and size are called **poikilocytes**. The composition of RBCs is 64% water, 20% haemoglobin, 7% lipids and 3% other substances.

Haemoglobin

- It is a reddish respiratory pigment present in RBC. It is a chromoprotein with four iron or haem parts and the globin or protein part.
- It is the oxygen carrier of vertebrate body.
- The normal haemoglobin count is about 15 g/100 mL of blood in males and 13 g/100 mL for females.

- The average lifespan of a human RBCs is 120 days. In transfused blood (oxalated), it is 60 days and in foetal RBCs (with nucleus), it is 180 days.
- The RBCs after completion of its lifespan is destroyed by the graveyard of RBCs, i.e. **spleen**. Here, these RBCs are ingested by free macrophages.
- Within these cells, the haemoglobin is dissociated to haem and globin. The iron of haem is released, which is further utilised in the process of blood formation.
- The unused haem is converted into a yellow or brown pigment called **bile pigment** (in particular bilirubin).
- The blood haemoglobin count is done by **haemoglobinometer**. The RBC count is done by **haemocytometer.**

Erythropoiesis

- The process of RBC formation is called erythropoiesis or **haemopoiesis** or **haematopoiesis**.
- Erythropoietic organ in foetus is liver, lymph nodes and spleen, whereas after birth, it is bone marrow.
- Iron, vitamin-B_{12} and folate are essential for the production of RBCs. Erythropoiesis gets completed in 72 hours.

Erythrocyte Sedimentation Rate (ESR)

- If blood containing anticoagulant (oxalate) is allowed to stand in a vertical tube, the erythrocytes settle to the bottom half of the tube.
- The rate at which this occurs is called ESR. This happens, because the density of erythrocyte is greater than that of plasma.
- ESR is used to diagnose various diseases such as TB.
- Its value is higher in infections and decreases in anaemia, sleep, menstruation and pregnancy.
- Sometimes in abnormal conditions, the RBCs loose their content by diffusion and thus, their emptied covers remain intact. These empty covers are called **shadows** or **ghosts of RBCs**.

Note
- *RBCs of amphibians are the largest among vertebrates.*
- *Among amphibians, Proteus and Amphiuma, have largest RBCs, i.e. these two vertebrates have largest RBCs in the animal kingdom.*
- *Mammalian RBCs are the smallest among vertebrates. Musk deer, Tragulus javanicus have the smallest RBCs among mammals (1.5 μm).*
- *In mammals, largest RBCs are found in elephant (9.4 μm).*
- *Aggregation of RBCs is called* **Rauleaux formation**. *It occurs due to surface tension of RBCs. The rauleaux of RBCs looks like piles or stack of coin.*
- *In cases of anaemia, the RBCs count or haemoglobin decreases, while RBCs count increases under condition called* ***polycythemia.***

(ii) WBCs or Leucocytes

- White Blood Cells (WBCs) are colourless, nucleated and motile cells of blood.
- These are much larger than the RBCs, but much smaller in number. Their normal count is 5000-9000 cells/mm^3 of blood with an average of about 8000 cells/mm^3 of blood.

- White blood cells are divided into two categories as shown in the table given below

Table 13.4 Types of WBCs

	Agranulocytes		Granulocytes		
Characters	**Lymphocytes** (20-25%)	**Monocytes** (6-8%)	**Eosinophils** (Acidophils) (2-3%)	**Basophils** (0-5-1%)	**Neutrophils** (60-65%)
Staining of cytoplasm	Basophilic	Basophilic	Eosinophilic	Basophilic	Neutrophilic
Nucleus	Rounded	Bean-shaped/kidney shaped	Bilobed	S-shaped Three-lobed	Multilobed
Site of formation	Lymph nodes, spleen, thymus, tonsils, Peyer's patches, bone marrow	Bone marrow	Bone marrow	Bone marrow	Bone marrow
Lifespan	Few days or even years	• 10-20 hrs in the blood and in tissue, months or even years	• 4-8 hrs in blood and 4-5 days in tissue	• 4-8 hrs in blood and 4-5 days in tissue	• 4-8 hrs in blood and 4-5 days in tissue
Functions	• Antibody fomation • Number increases in viral infection	• Phagocytic • Number increases in tuberculosis	• Important role in immunity antiallergic reactions and parasitic infection • Number increases in allergies	• Secretion of heparin histamine and serotonin • Number incrcases in chickenpox	• Phagocyte • Number incrcases in bacterial infections

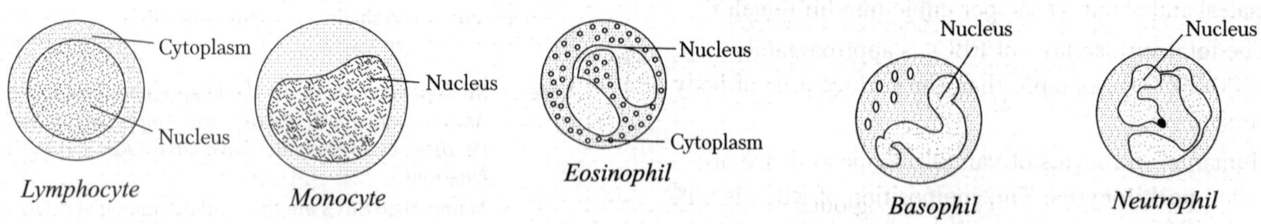

(iii) Blood platelets or Thrombocytes

- The name blood platelets was given by **Bizzozero**. Platelets occur only in mammals.
- Their count varies from 2-5 lac/mm^3 of blood.
- These are small cells with the size of about 2-4 μ.
- They are highly contractile, round, oval or biconvex with granular cytoplasm, but without nucleus.
- The formation of these cells occurs from certain large cells of bone marrow called **megakaryocytes** as buds or pinched vesicles.
- In all other vertebrates, spindle cells are present in the place of **thrombocytes**. These are spindle-shaped cells with spherical or oval nucleus.
- Their cytoplasm is granular and their function is similar to blood platelets.

- The production of blood in various organs of body and during different age period in life cycle of human beings are as follows
  - During few early weeks of embryonic life—yolk sac.
  - During later stages of life—liver (main organ) along with spleen and lymph nodes.
  - During later stage of gestation and after birth—red bone marrow of all bones until the child is 5 years old.
  - Up to the age of 20-25 years—bone marrow of long bones (except proximal portions of humerus and tibia).
  - After this blood is produced in membranous bones like vertebrae, sternum, ribs, etc.

Note
- *In humans, the relation of number of RBCs : WBCs = 600 : 1.*
- *Rise in WBCs count is called **leucocytosis**, while decrease in WBCs count is called **leukopenia**.*
- *Decrease in the number of circulating thrombocytes is called **thrombocytopenia**.*

CHECK POINT 13.2

1. Intervertebral disc consists of a shock absorber connective tissue is known as
   (a) hyaline cartilage (b) elastic cartilage
   (c) fibrocartilage (d) reticulocartilage
2. Thousands of years, old mummies are still in their condition as they were before due to the non-destruction of
   (a) yellow elatin fibres (b) white elastin fibres
   (c) collagen fibres (d) veins
3. The kind of tissue that forms the supportive structure in our pinna (external ears) is also found in
   (a) vertebrae (b) nails
   (c) ear ossicles (d) tip of the nose
4. Dense irregular connective tissue has fibroblasts and many fibres (mostly collagen) that are oriented differently and is present in
   (a) skin (b) tendon (c) cartilage (d) ligament
5. In all connective tissues except ...., the cells secrete fibres of structural proteins called collagen or elastin.
   (a) blood (b) bone
   (c) cartilage (d) areolar tissue
6. A drop of each of the following is placed separately on four sides. Which of them will not coagulate?
   (a) Blood plasma
   (b) Blood serum
   (c) Sample from the thoracic duct of lymphatic system
   (d) Whole blood from pulmonary vein
7. Which of the following substances if introduced in the blood stream, would cause coagulation at the site of its introduction?
   (a) Dicumarol
   (b) Hirudin
   (c) Heparin
   (d) Thromboplastin

MUSCULAR TISSUE

- These are derived from the **mesodermal layer** of embryonic germ cells.
- These make up 40% of body weight in mammals.
- The muscle cells are called **myocytes** or **sarcocytes**.
- These are developed from the cells are called **myoblasts**. These are spindle-shaped muscle precursor cells.
- They are richly supplied with blood vessels, which provide nutrition and take away wastes.
- This tissue is responsible for carrying out movement of the body and its various parts.
- Muscles are made up of highly specialised thin and elongated cells called **muscle fibre**.
- The membrane of a single muscle fibre is known as **sarcolemma**.
- Each muscle fibre contains proteinaceous fibrils known as **myofibrils**, present in their cytoplasm or **sarcoplasm**. The muscle fibres are capable of contraction and relaxation.
- On the basis of their structure, location and function, the muscles are classified into following three types

1. Striated or Skeletal muscles

- The striated or skeletal or voluntary muscle tissue forms the major muscle component of vertebrate body.
- These muscles are mainly responsible for the movements of skeleton of the body.
- Their muscle fibres are multinucleated, cylindrical structures.
- These consist of bundles of striated muscle fibres (fascicles or fasciculi), which are attached to the bones and tendons by connective tissue.
- Each fasciculi is covered by three coverings of connective tissue.
- These coverings are **epimysium** (outermost), **perimysium** (middle) and **endomysium** (innermost) as shown below

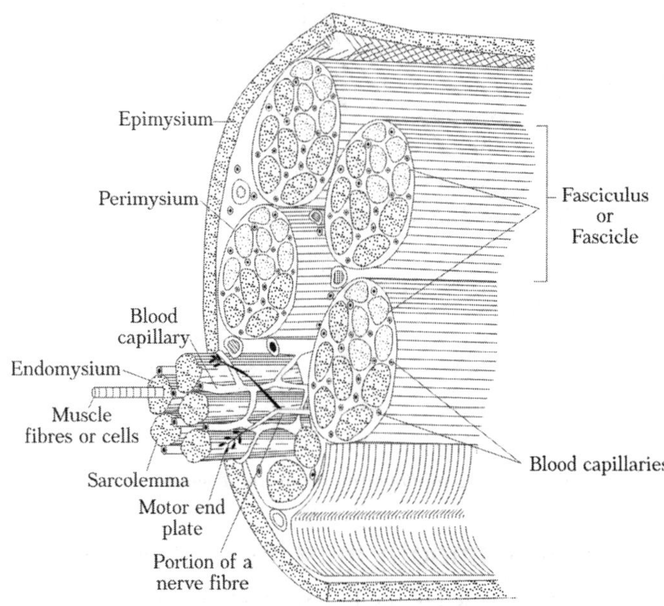

Fig 13.10 Sectional view of skeletal muscle

- Detailed study of this tissue is given in chapter-28; Movement and Locomotion.

2. Smooth muscles

- Smooth or involuntary or non-striated muscle is made up of spindle-shaped cells. These cells are elongated cells, averaging 40-80 μ in length and 6-7 μ in width.
- Each cell contains a single nucleus surrounded by sarcoplasm.

- In the sarcoplasm, the myofibrils are arranged longitudinally.
- The smooth muscles are mesenchymal in origin and have the capability to increase in size as well as in bulk when the physiological and pathological need arises.
- These are present in alimentary canal, urinary bladder, gall bladder, spleen, trachea, eyes, skin, uterus, vagina, penis, genital ducts and blood vessels.
- Smooth muscles are also known as visceral muscle as they occur in visceral organs.
- These can be further categorised in following two types

Single-unit smooth muscles

- They are composed of muscle fibres arranged in bundles of sheets and are bound with gap junctions.
- These muscles are **myogenic**, i.e. self-excitatory in nature.
- In these type of muscle, the nerve fibre terminates at the surface of the muscle.
- These are present in urinary bladder, stomach and intestine.

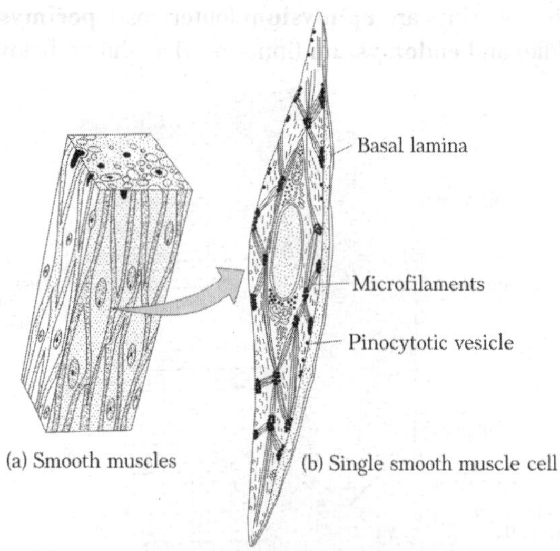

Fig 13.11 Smooth muscles with single muscle cell

Multi-unit smooth muscles

- They are composed of the fibres, which remain separated from each other. Each fibre is innervated by a separate nerve ending.
- The separation of muscle fibre occurs through a membrane like layer of **glycoprotein**. Hence, each fibre works independently of the other.
- These muscles are principally **neurogenic**, i.e. show contraction on nervous stimulation.
- For example, ciliary muscles, muscles of iris in eye, muscles of nictitating membrane of lower vertebrates, muscles of walls of larger blood vessels, etc.

3. Cardiac muscles

- These are found exclusively in heart where they form **myocardium.**
- Cardiac muscles are striated, involuntary and myogenic, i.e. the impulse of contractions are generated within the muscles. Each muscle fibre has many mitochondria, myofibrils with sarcolemma and many nuclei.
- It has abundant Ca^{2+} ions and glycogen granules.
- Each cardiac muscle fibre is formed of several cells connected to each other by special *zig-zag* junctions called **intercalated discs**. These discs are meant for the transportation of Ca^{2+} due to which its rhythmicity in contraction is maintained.
- Cardiac muscle tissues contract and relax throughout life. They never get fatigued. But, they have no regenerative capacity. That is why during **myocardial infarction** (heart attack), cardiac muscle tissue dies and is replaced by fibroblast rich scar tissue.

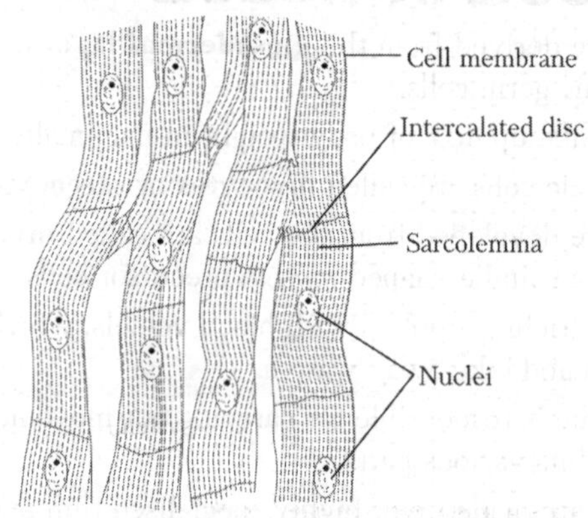

Fig. 13.12 Cardiac muscle fibres

A table showing the major differences in these three muscles is given below

Characteristics	Skeletal muscles	Cardiac muscles	Smooth muscles
Cell appearance and features	Long cylindrical fibre with many peripherally located nuclei, striated unbranched	Branched cylindrical usually with one centrally located nucleus, striated, intercalated discs join neighbouring fibres	Spindle-shaped fibre with one centrally positioned nucleus, no striations
Location	Attached primarily to bones	Heart	Walls of hollow viscera, blood vessels, iris and ciliary body of eye, errector pili of hair follicles.

Characteristics	Skeletal muscles	Cardiac muscles	Smooth muscles
Fibre diameter	Very large (10-100 μm)	Large (14 μm)	Small (3-8 μm)
Connective tissue components	Epimysium, perimysium and endomysium	Endomysium	Endomysium
Fibre length	100 μm to 30 cm	50-100 μm	30-200 μm
Organisation of contractile proteins into sarcomeres	Yes	Yes	No
Sarcoplasmic reticulum	Yes	Yes	Scanty
Transverse tubules	Yes, aligned with each A-I band junction	Yes, aligned with each Z-disc	Absent
Gap junctions between fibres	No	Yes	Yes, in single-unit smooth muscle, not in multiunit smooth muscles
Autorhythmicity	No	Yes	Yes (in visceral smooth muscle only)
Speed of contraction	Fast	Moderate	Slow
Nervous control	Voluntary (somatic nervous system)	Involuntary (autonomic nervous system)	Invulountary (autonomic nervous system)

NERVOUS TISSUE

- The nervous or neural tissue is the specialised tissue with the property of excitability and conductivity. This tissue originates from the embryonic ectoderm. Brain, spinal cord and nerves, all are composed of this tissue.
- The main cells of this tissue are called **neurons**. Within the tissue, these are present in association with neuroglial, ependymal and neurosecretory cells.

Neurons

- The word neuron was coined by the German scientists **Heinrich Wilhelm Gottfried von Waldeyer Hartz** in 1981.
- These are considered as the longest cells of the body.
- They form the structural and functional unit of nervous system.
- Each neuron arises in the embryo, from a single neuroblast cell.

Generalised structures

- A neuron has main **cell body**, which is also called **cyton**, **perikaryon** or **soma**. The cytoplasmic processes arising from it are called **neurites**.
- These neurites can be divided functionally and morphologically into two groups, i.e. **dendrites** and **axons**.
- Bundles of axons and dendrites are called **nerves**. They are
  (a) sensory (if they contain dendrites only)
  (b) motor (if they consist axons only)
  (c) mixed (if they consist of both axon and dendrites) types.
- Neurons are without centrosome. Once fully formed, these never divide and remain in the G_0-phase of interphase throughout life (unless they are cancerous). However, these grow to a larger size with the growth of body.

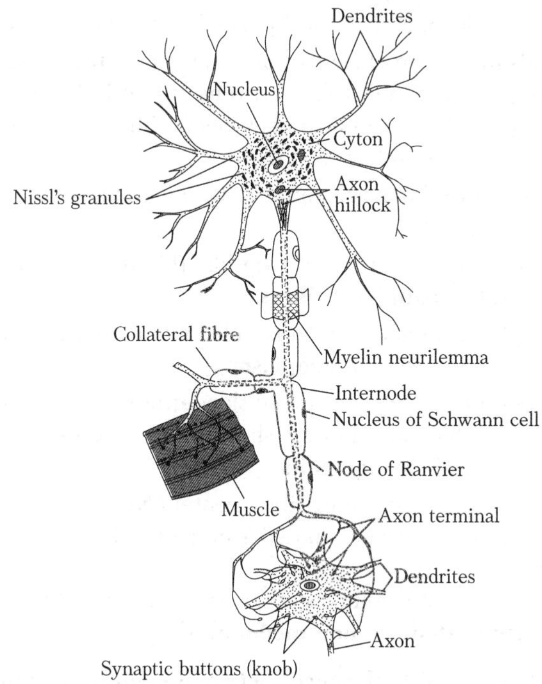

Fig. 13.13 Structure of a neuron

Molecular structure

1. Cyton or Cell body

- Cyton of most neurons occurs in the grey matter of central nervous system (particularly in brain).
- Like a typical cell, it contains mitochondria, Golgi bodies, RER, abundant cytoplasm (called **neuroplasm**) and a relatively large nucleus with a distinct **nucleolus**. This nucleolus is rich in proteins.
- The neuroplasm contains neurofibrils, neurotubules, fat globules, pigment and granules, namely **Nissl's granules** or **Nissl's bodies** or **trigoid granules**.

- These are large irregular masses of ribosomes and RER. They synthesise proteins for the cell. Thus, these are believed to be equivalent to ribosomes.
- The neurofibrils form an intricate network both in cyton and axon and are used in transmission of impulses.
- Cyton is concerned with metabolic maintenance and growth of a neuron.

2. Dendrites

- These may be one to several, usually shorter, tapering and branched processes.
- Nissl's bodies, neurofibrils and mitochondria are present in dendrites.
- They conduct nerve impulse towards the cell body and are called **afferent processes** or **receiving processes.**

3. Axon or Axis cylinder

- It is a single long cylindrical projection of uniform thickness.
- It ends in a group of slender branches called **telodendria** having knobbed ends called **buttons** or **end bulbs** or **terminals**. These terminal ends meet with dendrites of another neuron to form a **synapse**.
- The **synaptic knobs** contain mitochondria and secretory vesicles.
- Certain axons also give out twig-like thin branches at right angles called **collateral fibres.**
- The cell membrane of axon is called **axolemma** and its cytoplasm is known as **axoplasm.**
- Nerve fibres are elongated and slender processes of the neurons, which may have ensheathing of **myelin**. It is produced by **Schwann cells.**
- Myelin sheath containing fibres contain **Nodes of Ranvier** over them. These are exposed axon parts where myelin sheath is absent.
  On the basis of myelin sheath, nerve fibres are of two types, i.e. **medullated** and **non-medullated.**

Table 13.5 Differences between medullated and non-medullated nerve fibres

Medullated nerve fibres	Non-medullated nerve fibres
Medullary or myelin sheath is present.	Medullary or myelin sheath is absent.
These appear white.	These appear grey.
Nodes of Ranvier and internodes are present.	Nodes and internodes are absent.
These occur in white matter of brain and spinal cord, in cranial and spinal nerves.	These occur in grey matter of brain and spinal cord and in autonomous nervous system.
Impulse conduction is faster in comparison to non-medullated fibres.	Impulse conduction is slower in comparison to medullated fibres.

Types of neurons

Neurons can be classified into various types on the basis of structural nature and functions.

1. On the basis of structural nature

(i) **Apolar neurons**, i.e. neurons without polarity. Here, the fibres of neuron are not differentiated into afferent and efferent types. All the fibres are of same nature and can carry information towards or away from the cell body, e.g. neurons of *Hydra.*

(ii) **Unipolar neurons**, i.e. neurons with unidirectional flow of information. These have only one process, i.e. an axon. Most sensory neurons are unipolar. These are common in invertebrate and vertebrate embryos.

(iii) **Bipolar neurons**, i.e. neurons with unidirectional flow of information, but with one dendron and one axon at opposite poles. These occur in the retina of eyes, olfactory epithelium, cochlear ganglia, etc.

(iv) **Multipolar neurons**, i.e. neurons with unidirectional flow of information, but with one axon and many dendrites. They occur in the nervous system of adult vertebrates.

(v) **Pesudounipolar neurons** These are bipolar neurons that appear unipolar. It has a single process which divides to form dendrite and axon. It is found in the dorsal root ganglion of spinal cord.

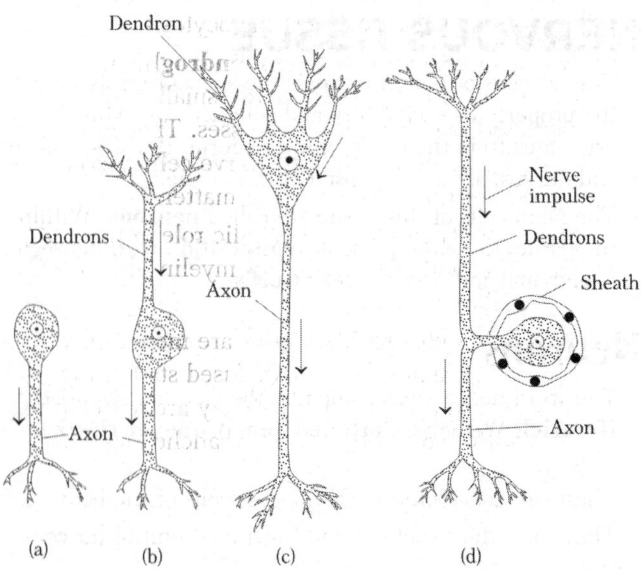

Fig. 13.14 Different types of neuron : (a) Unipolar, (b) Bipolar, (c) Multipolar, (d) Pseudounipolar

2. **On the basis of functions**
   (i) **Sensory or Afferent neurons** connect sensory or receptor cells to the CNS and conduct sensory impulses. These are present at cutaneous sense organs or **exteroceptors**. Similar endings scattered in skeletal muscles, bone joints, ligaments and tendons serve as **interoceptors**. Cell bodies of sensory neurons occur in clusters called **ganglia** next to the spinal cord.
   (ii) **Motor or Efferent neurons** connect the CNS to effectors (muscles and glands) and conduct motor impulses.
   (iii) **Internuncial or Adjuster or Interneurons** occur only in the CNS. They serve to connect two or more neurons for the distant transmission of impulses. These have the capability to adjust themselves in both efferent and afferent system.

Neuroglial cells (Glial cells)

Neuroglial cells or supporting cells make 90% of brain cells. They do not carry nerve impulse.

In vertebrates, neuroglia are of following three types

(i) **Astrocytes or Macroglia** are ectodermal in origin. These are most abundant of all supporting cells in the CNS. These are of two types, namely **fibrous astrocytes** of white matter and the **protoplasmic astrocytes** of grey matter.

Function They provide repair mechanism and replace lost tissue, provide physical support to neurons and also perform phagocytosis.

(ii) **Oligodendrocytes or Oligodendroglia** are ectodermal in origin. These are smaller than the astrocytes and have few processes. They are found in close association with large nerve cells and between bundles of fibres in the white matter.

Function They play a metabolic role in the formation and preservation of myelin sheath in the nerve fibre of CNS.

(iii) **Microgliocytes or Microglia** are mesodermal in origin. These are found in diffused state through both grey and white matter. They are normally small cells with two or more finely branched feathery processes.

Function These cells are specialised macrophages, i.e. the scavengers of the nervous system.

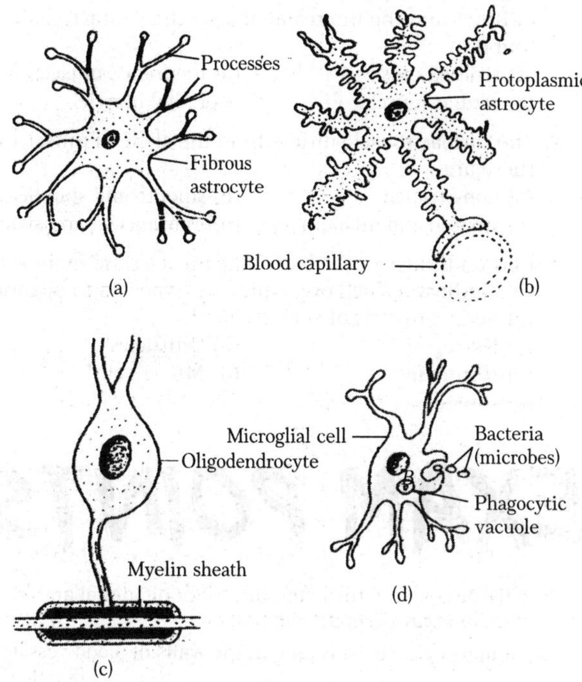

Fig. 13.15 Different kinds of neuroglial cells (a) Fibrous astrocyte, (b) Protoplasmic astrocyte, (c) Oligodendrocyte, (d) Microglial cell

Ependymal cells

- These are the epithelial cells that line the CSF (Cerebro Spinal Fluid) filled ventricles in the brain and the central canal of the spinal cord. The cells form ciliated simple cuboidal epithelium.
- Within the brain ventricles, modified ependymal cells and capillaries together form a system called the **choroid plexus**, which produces the CSF.

Neurosecretory cells

- They produce secretions called as **neurohormones** that travel along the axon and are typically released into the blood stream at neurohaemal products.
- Neurosecretory cells are present in most multicellular animals. They are usually distinguished from other neurons by there unusual large-sized nucleus, axon endings and the cell itself.

CHECK POINT 13.3

1. The oblique cross connections to form a contractile network of fibres and the intercalated discs are characteristically found in
   (a) striated muscle (b) unstriated muscle
   (c) cardiac muscle (d) radial muscle
2. The cardiac muscle differs from the skeletal muscle because these are
   (a) non-striated (b) smooth and spindle-shaped
   (c) striated and involuntary (d) unbranched and syncytial
3. Certain tissues in the body, like muscles are more active than others. Which organelles are expected to be more numerous in cells of such tissues?
   (a) Nuclei (b) Centrioles
   (c) Lysosomes (d) Mitochondria
4. Which one of the following pairs of structures distinguishes a nerve cell from other types of cells?
   (a) Perikaryon and dendrites
   (b) Vacuoles and fibres
   (c) Flagellum and medullary sheath
   (d) Nucleus and mitochondria
5. The neurons lack ability to divide as they
   (a) have myelin (b) lack centrioles
   (c) have nucleus (d) lack mitochondria
6. A dense layer of connective tissue that surrounds the entire nerve, made up of number of fasciculi, is called
   (a) epineurium (b) endoneurium
   (c) perineurium (d) epimysium

EXAM POINTS

- The tissue covering the external surface of animal and internal surface of organs is epithelial tissue. **[J&K CET 2015]**
- Squamous epithelium is seen in the walls of blood vessels. **[Kerala CEE 2011]**
- The cavities of alveoli of lungs are lined by squamous epithelium. **[BCECE 2012]**
- Simple cuboidal epithelium perfoms functions like secretion and absorption of substances. **[NEET 2020]**
- Tubular part of nephrons are lined by cuboidal epithelium. **[CBSE AIPMT 2014]**
- The ciliated epithelial cells are required to move particles or mucus in a specific direction. In humans, these cells are mainly present in branchioles and Fallopian tubes. **[AFMC 2013, NEET 2019]**
- Cuboidal epithelium with brush border or microvilli is found in proximal convoluted tubule or nephron. **[NEET 2020]**
- Tight junctions establish a barrier to prevent leakage of fluid across epithelium cells. **[NEET 2019]**
- Adhering junctions cement neighbouring cells together to form sheet. **[Odisha JEE 2013]**
- Gap junctions are cytoplasmic channels to facilitate communication between adjacent cells. **[NEET 2019]**
- The inner lining of ducts of sweat glands and pancreatic duct is formed of stratified cuboidal epithelium. **[EAMCET 2014]**
- Vagina, oesophagus, urethra contain stratified squamous epithelium tissue. **[Manipal 2013]**
- Columnar, ciliated epithelia are found in the lining of the nasal passage, bronchioles and Fallopian tubes. **[JCECE 2012]**
- Fibrinogen helps in blood clotting, globulin helps in defence mechanism and albumin helps in osmotic balance. **[NEET 2018]**
- Dense regular connective tissue is present in ligament and tendon. **[JIPMER 2017, EAMCET 2014]**
- Adult human RBCs are enucleate as all their internal space is available for oxygen transport. **[NEET 2017]**
- The lining of intestine and kidneys in humans has brush border. **[Manipal 2013]**
- Hyaline cartilage is bluish-white translucent and glass-like cartilage. It forms embryonic endoskeleton in cyclostomes and adult cartilage fishes. **[EAMCET 2016]**
- The outer covering of cartilage is called perichondrium. **[WB JEE 2013]**
- Bone marrow is absent in birds. **[AIIMS 2013]**
- Fibroblasts, macrophages and mast cells are seen in connective tissue. **[BCECE 2012]**
- Agranulocytes are lymphocytes and monocytes. **[BCECE 2012]**
- Eosinophils release histamine destructive enzymes. **[NEET 2020]**
- Basophils release granules containing histamine. **[NEET 2020]**
- Neutrophils has phagocytosis activity. **[NEET 2020]**
- Lymphocytes produce immune response by producing B-lymphocytes and T-lymphocytes. **[NEET 2020]**
- Histamine, serotonin and heparin are secreted by basophils. **[Kerala CEE 2011]**
- RBCs do not contain cell organelles. **[WB JEE 2011]**
- Haemoglobin is an oxygen carrier in human blood. **[AIIMS 2012]**
- Wharton's Jelly present in umbilical cord is an example of mucous connective tissue. **[EAMCET 2015]**
- Smooth muscles are spindle-shaped uninucleated cells. Their contractions are under the control of autonomous nervous system. They exhibit prolonged contractions. **[EAMCET 2015]**
- Smooth muscles are involuntary, fusiform and non-striated . **[NEET 2016]**
- Smooth muscles are present in intestinal walls. **[NEET 2016]**
- Skeleton muscle fibre is a syncytium which means it is multinucleated. **[Kerala CEE 2015]**
- Cardiac muscle is found in myocardium. **[Odisha JEE 2013]**
- Synaptic junctions transmit information through chemical to another cell. **[NEET 2019]**
- Fibroblasts are found in areolar tissue. **[NEET 2020]**
- Osteocytes are bone cells. **[NEET 2020]**
- Chondrocytes are cartilage cells. **[NEET 2020]**
- Neurons are nerve cells. **[NEET 2020]**

Chapter Exercises

(A) Topical Exercise

Epithelial Tissue

1 Who is known as the father of histology?
(a) Ruysch (b) Mayer
(c) Bichat (d) Watson

2 Which layer forms upper thin layer of basement membrane?
(a) Reticular lamina (b) Basal lamina
(c) Fibrous lamina (d) Both (a) and (b)

3 Identify the substance, which keeps the epithelial cells together.
(a) Hyaluronic acid (b) Collagen
(c) Mucin (d) Glycoprotein

4 The cellular junctions that meant for chemical exchange between adjacent cells are
(a) intercellular bridges (b) gap junctions
(c) anchoring junctions (d) tight junctions

5 Simple squamous epithelium is not found in
(a) blood vessels (b) ciliary body of eyes
(c) lymph vessels (d) alveoli (air sacs) of the lungs

6 Identify *A*, *B* and *C* in the following figures of simple epithelium tissue.

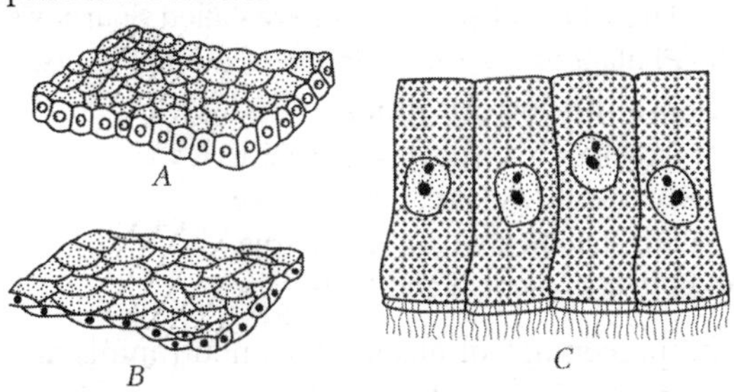

(a) *A*-Ciliated columnar, *B*-Squamous, *C*-Cuboidal
(b) *A*-Cuboidal, *B*-Squamous, *C*-Ciliated columnar
(c) *A*-Squamous, *B*-Ciliated columnar, *C*-Cuboidal
(d) *A*-Ciliated columnar, *B*-Cuboidal, *C*-Squamous

7 Of the following choices, the epithelium with shortest diffusion distance is
(a) simple cuboidal epithelium
(b) simple columnar epithelium
(c) pseudostratified ciliated columnar epithelium
(d) simple squamous epithelium

8 Select the correctly matched pair.
(a) Cuboidal epithelium – Blood vessels
(b) Columnar epithelium – Uterine endometrium
(c) Simple epithelium – Nasal cavity
(d) Stratified epithelium – Thyroid follicles

9 The specialised function shared by the cells that line the lungs and those that line the lumen of gut is that both types of cells.
(a) Receive their oxygen directly from inhaled air and ingested foods
(b) Provide abundant exchange surface
(c) Have a lowered basal metabolic rate due to cooperative exchange between cells
(d) Have exceptionally high numbers of cellular organelles in the cytoplasm

10 Find the incorrect match pair.
(a) Holocrine – Sebaceous glands
(b) Unicellular gland – Goblet cell
(c) Ciliated columnar – Iris of eye
(d) Germinal epithelium – Seminiferous tubules

11 Match the following columns.

Column I	Column II
A. Pseudostratified ciliated columnar epithelium	1. Trachea and large bronchi
B. Stratified squamous epithelium	2. Lining the oral cavity
C. Transitional epithelium	3. Wall of urinary bladder
D. Simple ciliated columnar epithelium	4. Lining the respiratory tract

Codes

	A	B	C	D		A	B	C	D
(a)	1	2	3	4	(b)	3	1	2	4
(c)	4	3	1	2	(d)	2	4	1	3

12 Which of the following statements is incorrect?
(a) Glandular epithelium lines the inner surface of urinary bladder
(b) Transitional epithelium lines the inner surface of urinary bladder
(c) Endocrine glands are duct containing glands
(d) Ciliated columnar epithelium is present in the lining of larynx

13 Goblet cells secrete
(a) mucus (b) dilute HCl
(c) polypeptide hormones (d) enzymes

14 An example of mixed gland is
(a) gastric gland (b) sweat gland
(c) hypothalamus (d) None of these

15 Which of the following is an epidermal structure of mammals?
(a) Sebaceous gland (b) Receptors
(c) Hair (d) Sweat gland

16 Which of the following is a compound alveolar gland?
(a) Gastric gland in stomach
(b) Parotid salivary gland
(c) Mammary gland of prototherians
(d) Sebaceous glands

17 An example of holocrine gland is a
(a) sweat gland (b) salivary gland
(c) pancreatic gland (d) sebaceous gland

18 All the given glands are merocrine except
(a) goblet cells (b) salivary glands
(c) mammary glands (d) most sweat glands

Connective Tissue

19 Which of the following is not a specialised type of cell found in connective tissues?
(a) Lymphocyte (b) Macrophage
(c) Goblet cell (d) Mast cell

20 Areolar tissue
(a) is made of threads crossing each other and leaving a space between them
(b) is polymorphic
(c) is undifferentiated mesenchyma
(d) is found in hyaline cartilage

21 Which statement is correct regarding areolar tissue?
I. It consist of mucopolysaccharides.
II. They continue to grow throughout the life.
Choose the correct options.
(a) Statement I is true, but II is false
(b) Statement I is false, but II is true
(c) Both statements I and II are true
(d) Both statements I and II are false

22 Consider the following statements.
I. Adipose tissue stores fat droplets.
II. Adipocytes are also known as fat cells and lipocytes.
Choose the correct option.
(a) Statement I is true, but II is false
(b) Statement I is false, but II is true
(c) Both statements I and II are true
(d) Both statements I and II are false

23 Elastin is found in
(a) yellow fibres of areolar tissue
(b) adipose tissue
(c) cartilage
(d) spongy bone

24 White fibrous tissues are present at the joints of
(a) cranial bones
(b) humerus and glenoid cavity
(c) femur and acetabulum
(d) carpals and metacarpals

25 Wharton's jelly
(a) is found in mucoid connective tissue
(b) contains collagen fibres
(c) is found in vitreous chamber of eyes
(d) All of the above

26 Select the correctly matched pair.
(a) Chondroblast – Form chondrocyte in cartilage
(b) Elastic cartilage – Pubic symphysis
(c) Fibrous cartilage – Pinna of ear
(d) Hyaline cartilage – Intervertebral disc

27 Calcified cartilage is found in
(a) head of femur (b) pectoral girdle
(c) ribs and sternum (d) Both (a) and (b)

28 Bones of old persons are brittle and fracture more readily occur because they contain more
(a) lacunae
(b) organic matter
(c) inorganic matter
(d) bone marrow

29 Which of the statement is incorrect?
(a) Cartilage contains chondrin, which makes the matrix
(b) Matrix of bone is formed by ossein
(c) Haversian canal system is the characteristic of mammalian bone
(d) Volkmann's canal connects the lacuna present in the cartilage

30 Consider the following statements.
I. The fibres of periosteum are called Sharpey's fibres.
II. Phalanges are the example of dermal bones.
Choose the correct option.
(a) Statement I is true, but II is false
(b) Statement I is false, but II is true
(c) Both statements I and II are true
(d) Both statements I and II are false

31 The percentage of eosinophils, neutrophils and monocytes in our blood are .... respectively.
(a) 2.8%, 65% and 5%
(b) 2%, 5% and 26%
(c) 5%, 10% and 40%
(d) 30%, 18% and 5%

32 In man, the number of RBC per cubic mm of blood is
(a) 1-3 million
(b) 5-5.4 million
(c) less than one million, but more than 50 thousand
(d) less than 50 thousand

33 What happens to the unused haem of haemoglobin after blood formation?
(a) It leads to rauleaux formation
(b) It leads to formation of RBC
(c) It converts into bile pigment
(d) None of the above

34 Read the following statements.

I. Formed elements constitute 45% the blood and rest 55% is constituted by plasma.

II. The formed elements mainly present in the human blood are erythrocytes, leucocytes and platelets.

Choose the correct option.

(a) Statement I is true, but II is false
(b) Statement I is false, but II is true
(c) Both statements I and II are true
(d) Both statements I and II are false

35 The blood haemoglobin count is done by

(a) haemometer (b) spirometer
(c) haemoglobinometer (d) haemoalbuminometer

36 Agranulocytes normally develop in

(a) bone marrow (b) liver
(c) lymph glands and spleen (d) None of these

37 Granulocytes are produced in

(a) liver (b) bone marrow (c) spleen (d) None of these

38 Blood cells are formed in bone marrow. This process is known as

(a) haemolysis (b) erythroblastosis
(c) haemopoiesis (d) None of these

39 Find the correct descending order of percentage proportion of leucocytes in human blood.

(a) Neutrophils → Basophils → Lymphocytes → Acidophils (Eosinophils) → Monocytes
(b) Monocytes → Neutrophils → Lymphocytes → Acidophils → Basophils
(c) Neutrophils → Lymphocytes → Monocytes → Acidophils → Basophils
(d) Lymphocytes → Acidophils → Basophils → Neutrophils → Monocytes

40 Which blood corpuscle is largest in diameter?

(a) Erythrocyte (b) Monocyte
(c) Neutrophil (d) Lymphocyte

41 Select the option having all the correct characteristics.

	Structures	Percentage of WBCs	Functions
(a)	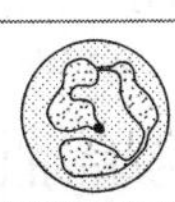	0.3 – 0.5	Phagocytic
(b)	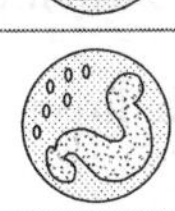	0.5 – 1.0	Secrete histamine and serotonin
(c)	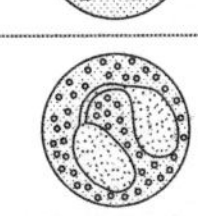	30 – 40	Defence against parasites
(d)	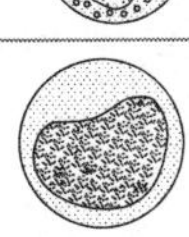	30 – 40	Allergic reactions

Muscular Tissue

42 The muscle cells are called as

(a) myofibrils
(b) chondrocytes
(c) myocytes
(d) osteocytes

43 All types of muscle tissue have

(a) Intercalated discs that allow cells to communicate
(b) Interaction between actin and myosin
(c) Striated banding patterns seen under microscope
(d) A respones that can be consciously controlled

44 What is sarcomere?

(a) Part between two Z-lines
(b) Part between two A-lines
(c) Part between two I-bands
(d) Part between two H-lines

45 Which of the muscles are attached to bones?

(a) Smooth muscles (b) Skeletal muscles
(c) Cardiac muscles (d) All of these

46 Striated muscle fibre has

(a) one nucleus
(b) two nuclei
(c) many nuclei
(d) no nucleus

47 Which of the following statements is incorrect?

(a) Smooth muscles are also called visceral muscles
(b) Intercalated discs and cross connections are the characteristics of cardiac muscles and skeletal muscles
(c) Sarcoplasm is the cytoplasm of myocyte
(d) Myosin is the contractile protein of muscle

48 Each fasciculi of skeletal muscle has three coverings from outside to inside as

(a) Endomysium → Epimysium → Perimysium
(b) Perimysium → Epimysium → Endomysium
(c) Epimysium → Perimysium → Endomysium
(d) Epimysium → Endomysium → Perimysium

49 Characteristics of smooth muscle fibres are

(a) spindle-shaped, unbranched, unstriated, uninucleate and involuntary
(b) spindle-shaped, unbranched, unstriped, multinucleate and involuntary
(c) cylindrical, unbranched, unstriped, multinucleate and involuntary
(d) cylindrical, unbranched, striated, multinucleate and voluntary

50 All the given structures contain single unit smooth muscles except

(a) urinary bladder (b) intestine
(c) stomach (d) iris

51 Consider the following statements.

I. Differentiated smooth muscle cells retain the ability to undergo cell division.

II. Smooth muscle is multinucleate.

Choose the correct option.

(a) Statement I is true, but II is false
(b) Statement I is false, but II is true
(c) Both statements I and II are true
(d) Both statements I and II are false

52 Which one of the following is correct pairing of a body part and kind of muscle tissue that moves it?

(a) Heart wall – Involuntary unstriated muscle
(b) Biceps of upper arm – Smooth muscle fibres
(c) Abdominal wall – Smooth muscle
(d) Iris – Involuntary smooth muscle

53 Which of the following statements about muscle is true?

(a) Smooth muscle is multinucleate
(b) Differentiated smooth muscle cells retain the ability to undergo cell division
(c) Cardiac muscle is voluntary
(d) None of the above

54 Achilles tendon is associated with

(a) gluteus muscle (b) hamstring muscle
(c) quadriceps muscle (d) gastrocnemius muscle

55 The cardiac muscle differs from the skeletal muscle because these are

(a) non-striated (b) smooth and voluntary
(c) voluntary in nature (d) striated and involuntary

56 Identify the type of muscle in the given figure.

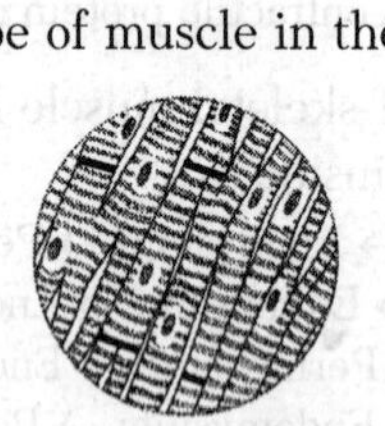

(a) Smooth muscle (b) Unstriated muscle
(c) Cardiac muscle (d) None of these

57 Cardiac muscles are

I. striated II. voluntary
III. involuntary IV. non-striated

(a) I, II and III (b) I and II
(c) II and IV (d) I and III

Nervous Tissue

58 Nervous tissues are originated from which of the following germ layer?

(a) Ectoderm (b) Endoderm
(c) Mesoderm (d) Both (a) and (b)

59 Nerve fibres differ from the muscle fibres in having

(a) dendrites
(b) sarcolemma
(c) myofibrils
(d) striations

60 Which of the following is correct with respect to neuron?

(a) It acts as biological transducers
(b) It may be sensory or motor
(c) It cannot regenerate or divide
(d) All of the above

61 Which of the following parts of a neuron is covered by fatty sheath?

(a) Axon (b) Cyton
(c) Dendrite (d) Node of Ranvier

62 Consider the following statements.

I. Schwann cells are present where nerve is covered with myelin sheath.

II. Bipolar neuron occur in the retina of eye.

Choose the correct option.

(a) Statement I is true, but II is false
(b) Statement I is false, but II is true
(c) Both statements I and II are true
(d) Both statements I and II are false

63 Which of the following is correct for neuroglia?

(a) They are pesent in the brain, but absent in spinal cord
(b) They are essentially ectodermal
(c) They make up more than one-half the volume of neural tissue in our body
(d) They play important role in neurotransmission

64 Which statement is not true about nerve fibres in mammals?

(a) All nerve fibres are non-medullated
(b) Afferent (sensory) nerves transmit impulses to brain and spinal cord
(c) Efferent (motor) nerve carry signals from brain and spinal cord to effector organs
(d) Mixed nerves carry signals in both directions

(B) Cumulative Exercise

(Assorted Questions of the Chapter for Advanced Level Practice)

1 Identify the animal tissues in the given figure.

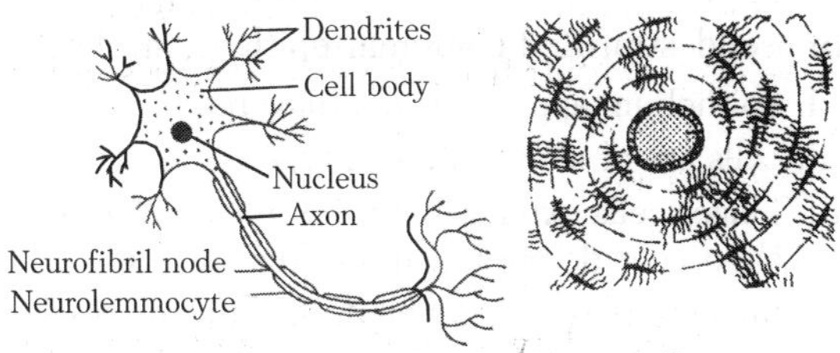

A *B*

(a) *A*-Nervous tissue, *B*-Muscular tissue
(b) *A*-Nervous tissue, *B*-Epithelial tissue
(c) *A*-Nervous tissue, *B*-Connective tissue
(d) *A*-Nervous tissue, *B*-Cardiac tissue

2 Consider the following statements.

I. Erythropoiesis in foetus occurs in liver.
II. Intervertebral discs are composed of fibrous cartilage.
III. Desmosomes serve as the anchoring junctions between the cells.
IV. Cartilage is a non-vascular connective tissues having matrix of chondrin.

Which of them are correct?

(a) I and II
(b) I, II, III and IV
(c) II and III
(d) III and IV

3 Which word combination applies to stratified squamous epithelium?

(a) Mesoderm – Calcification
(b) Ectoderm – Keratinisation
(c) Mesoderm – Ossification
(d) Endoderm – Cornification

4 Match the following columns.

Column I	Column II
A. Neutrophil	1. Single large nucleus
B. Eosinophil	2. Usually trilobed nucleus
C. Basophil	3. Kidney-shaped nucleus
D. Lymphocyte	4. Multilobed nucleus
E. Monocyte	5. Bilobed nucleus

Codes

	A	B	C	D	E
(a)	4	1	3	5	2
(b)	2	5	1	4	3
(c)	4	5	2	1	3
(d)	2	4	5	3	1

5 Identify the animal tissues in given figures.

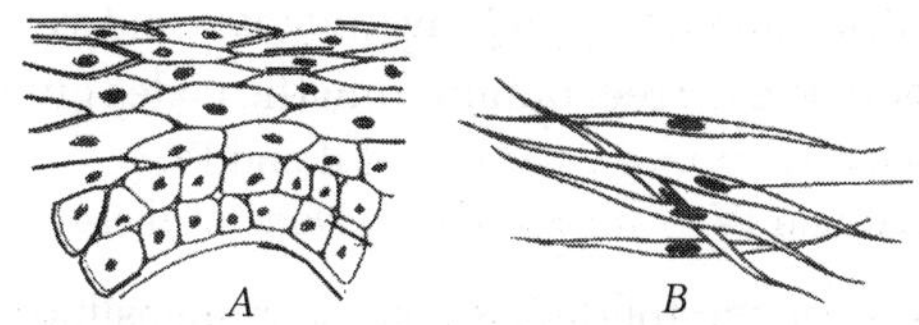

A *B*

(a) *A*-Connective tissue, *B*-Muscular tissue
(b) *A*-Epithelial tissue, *B*-Muscular tissue
(c) *A*-Muscular tissue, *B*-Connective tissue
(d) *A*-Epithelial tissue, *B*-Connective tissue

6 Choose the odd pair out in the following options.

(a) Areolar connective tissue – Collagen
(b) Epithelium – Keratin
(c) Neuron – Melanin
(d) Muscle fibre – Actin

7 Study the diagram given below and identify the cells labelled as *A*, *B* and *C* choose the correct option.

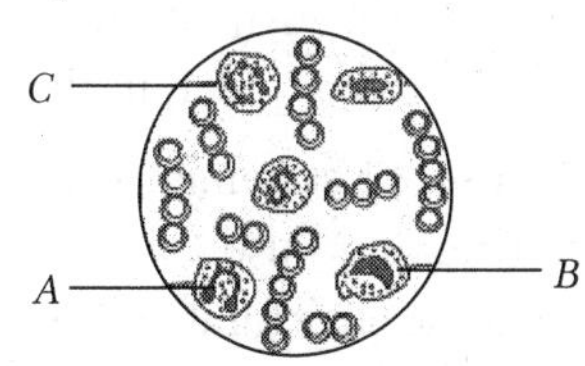

(a) *A*-Eosinophil *B*-Erythrocyte, *C*-Neutrophil
(b) *A*-Eosinophil, *B*-Lymphocyte, *C*-Neutrophil
(c) *A*-Erythrocyte, *B*-Basophil, *C*-Neutrophil
(d) *A*-Eosinophil, *B*-Monocyte, *C*-Neutrophil

8 Identify the pair that does not match.

(a) Sarcomere – Ultrastructural unit of striated muscle
(b) Copper containing respiratory pigment of Arachnida – Haemocyanin
(c) Osteoclast – Bone destroying cells
(d) Volkmann's canal – Which drain blood from bones

9 Read the statements carefully and identify the incorrect ones.

I. The skeletal tissue consists of organic matter called as chondroblast.
II. If a femur is kept in dilute HCl for three days, it becomes brittle.
III. Haversian system is found in pigeon.
IV. Hyaline cartilage is present at the end of long bones.

Choose the correct option from the codes given below.

(a) I and II (b) II and III
(c) Only III (d) III and IV

10 Which of the following is/are incorrect?

I. The palm muscle fibres are called fasciculi.
II. Cardiac muscles in the embryo are originated from mesenchyma cell.
III. Neurons are without centrosome.
IV. RBCs are seen in all invertebrates and vertebrates.

Choose the correct option from the codes given below.

(a) I and II (b) II and IV
(c) I, II and III (d) None of these

11 Which of the following layer is responsible for the difference in the body contours of male and female?

(a) Plasma membrane (b) The penniculus adiposus
(c) Basement membrane (d) Canaliculi

12 Which of the following is/are correct?

I. Tendon connects skeletal muscle with bones.
II. Ligaments connect bones with bones.
III. An infection would most likely increase phagocytic activity in reticular tissue.
IV. In thyroid follicle, transitional epithelial tissue is present.

Codes

(a) I and II (b) II, III and IV
(c) Only IV (d) I, II, III and IV

13 From the given set of statements, identify the statements as true and false.

I. Sarcomere is the structural and functional unit of striated muscle fibres.
II. Muscles present in thigh are striated and voluntary.
III. EDTA is the anticoagulant used in blood cell counting.
IV. Multi-unit smooth muscles are principally neurogenic.

Codes

(a) Statement I is true, but II, III and IV are false
(b) Statements II and III are true, but I and IV are false
(c) Statements I and IV are true, but II and III are false
(d) All statements are true

14 Identify *A*, *B* and *C* in the table given below.

Tendons	Ligaments
It is composed of white fibrous tissue	It is composed of yellow elastic tissue
A	It is strong and elastic
The fibroblast lies in almost continuous rows	The fibroblasts are scattered
B	*C*

(a) *A* - It is tough and inelastic
B - It connects skeletal muscle to bones
C - It connects a bone to another bone
(b) *A* - It is weak and inelastic
B - It connects skeletal muscles to viscera
C - It connects a bone to another bone
(c) *A* - It is strong and hightly elastic
B - It connects skeletal muscles to bone
C - It connects a bone to another bone
(d) *A* - It is tough and inelastic
B - It contains scanty collagen fibres
C - It contains abundant collagen fibres

15 In pseudostratified columnar epithelium, the

I. epithelium appears as multilayered.
II. cells are of more than one type.
III. longer cells extend up to the free surface, while the short cells do not reach the free surface.
IV. membrane is multilayered, but appears single layered.

Choose the option with the correct statements.

(a) I, II and III (b) I and III
(c) II and IV (d) I, II, III and IV

16 Four healthy people in their twenties got severe injuries resulting in damage and death of a few cells in body. Which of the cells are least likely to be replaced by new cells?

(a) Osteocytes (b) Malpighian layer of the skin
(c) Liver cells (d) Neurons

17 Read the following statements and choose the correct option.

I. Blood cells secrete fibres of structural proteins, collagen or elastin.
II. Neuroglial cells protect and support the nephrons.
III. Osteocytes are present in spaces called bone matrix osteoid.
IV. Striated muscle fibres are bundles together in parallel fashion.
V. Biceps are involuntary and striated.

Codes

(a) III and IV (b) I, II and IV
(c) II and IV (d) I and III

18 Choose the correct statement regarding diagram *A* and *B* given below.

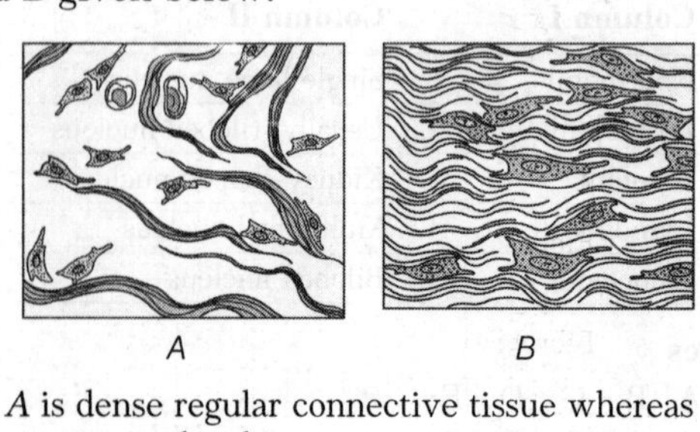

A B

(a) *A* is dense regular connective tissue whereas *B* is tissue present in the skin
(b) *A* and *B* both are dense irregular connective tissues
(c) *A* is persent in skin whereas tendons and ligaments are made of *B*
(d) *A* and *B* both are dense regular connective tissue

19 Consider the following statements.

I. Goblet cells are the unicellular glands.
II. The products of exocrine glands are hormones and are secreted directly into the fluid bathing the gland.
III. Some of the squamous cells get specialised for secretion and are called glandular epithelium.
IV. Compound epithelium is made of more than one layer of cells and thus has a limited role in secretion and absorption.

Which of the above statements are correct?

(a) I and II
(b) III and IV
(c) II and III
(d) I and IV

20 Consider the following diagram and identify the character corresponding to it.

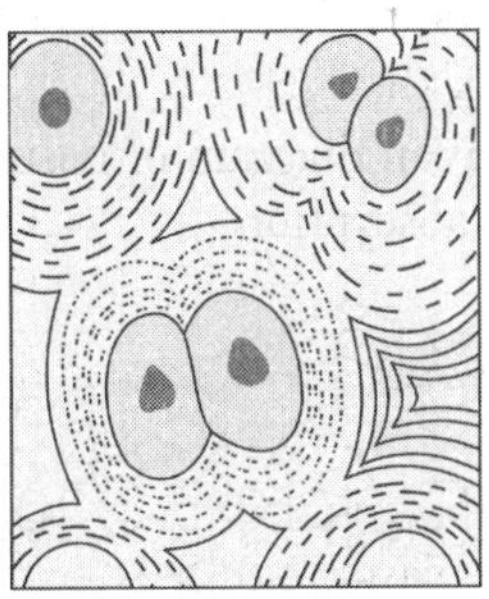

(a) It has hard and non-pliable ground substance
(b) Its intercellular material is solid and pliable and resists compression
(c) It is an example of loose connective tissue
(d) The cells of this tissue are specialised to store fats

21 Identify the labelling *A*-*E* in the diagram of areolar tissue given below.

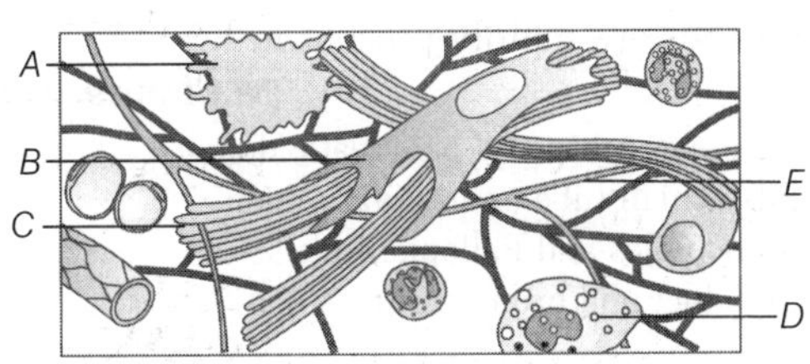

	A	*B*	*C*	*D*	*E*
(a)	Mast cell	Fibroblast	Macrophage	Collagen fibres	Elastic fibres
(b)	Macrophage	Fibroblast	Collagen fibres	Mast cells	Elastic fibres
(c)	Fibroblast	Mast cell	Collagen fibres	Macrophage	Elastic fibres
(d)	Fibroblast	Macrophage	Elastic fibres	Mast cells	Collagen fibres

22 Read the following statements and identify the incorrect one with respect to the diagram given below.

(a) It provides structural framework to the body
(b) It supports and protects softer tissues and organs
(b) It is rich in calcium salts and collagen fibres
(d) It has soft and pliable ground substance

23 Match the following columns.

Column I	Column II
A. Adipose tissue	1. Nose
B. Stratified epithelium	2. Blood
C. Hyaline cartilage	3. Skin
D. Fluid connective tissue	4. Fat storage

Codes

	A	B	C	D		A	B	C	D
(a)	1	2	3	4	(b)	4	3	1	2
(c)	3	1	4	2	(d)	2	1	4	3

Assertion and Reason

Direction (Q. Nos. 24-31) *In each of the following questions, a statement of Assertion (A) is given followed by corresponding statement of Reason (R). Of the statement, mark the correct answer as*

(a) Both A and R are true and R is the correct explanation of A
(b) Both A and R are true, but R is not the correct explanation of A
(c) A is true, but R is false
(d) A is false, but R is true
(e) Both A and R are false

24 **Assertion** (A) Connective tissue contains a large amount of non-living intercellular or extracellular matrix.

Reason (R) Intercellular substance is usually made up of protein fibres.

25 **Assertion** (A) Pleura and pericardium are serous membranes.

Reason (R) These membranes prevent friction between surface of different organs.

26 **Assertion** (A) Epithelium of mucous membrane have mucous glands.

Reason (R) Mucus provides protection to various internal organs from drying out.

27 **Assertion** (A) White collagen fibres and yellow elastin fibres are two types of areolar fibres.

Reason (R) Collagen fibres are found at the site of injury for repair.

28 **Assertion** (A) WBCs accumulate at the site of wounds by diapedesis.

Reason (R) The squeezing of leucocytes from endothelium is called diapedesis.

29 **Assertion** (A) Histamine is related with allergic and inflammatory reactions.

Reason (R) Histamine is a vasodilator.

30 **Assertion** (A) Thick layers of muscles are present in the wall of alimentary canal.

Reason (R) These muscles help in the mixing of food materials with the enzymes coming from different glands in the alimentary canal.

31 **Assertion** (A) In cardiac muscles, impulses are transmitted rapidly.

Reason (R) Gap junctions allow quick passage of ions from one cell to the other.

(C) Medical Entrances' Gallery

(Collection of Questions Asked in NEET & in Various Medical Entrance Exams)

1 Select the incorrectly matched pair from following. **[NEET 2020]**

(a) Chondrocytes-Smooth muscle cells
(b) Neurons-Nerve cells
(c) Fibroblast-Areolar tissue
(d) Osteocytes-Bone cells

2 Cuboidal epithelium with brush border of microvilli is found in **[NEET 2020]**

(a) ducts of salivary glands
(b) proximal convoluted tubule of nephron
(c) Eustachian tube
(d) lining of intestine

3 Goblet cells of alimentary canal are modified from **[NEET 2020]**

(a) columnar epithelial cells (b) chondrocytes
(c) compound epithelial cells (d) squamous epithelial cells

4 Match the following columns and select the correct option from the codes given below.

Column I	Column II
A. Eosinophils	1. Immune response
B. Basophils	2. Phagocytosis
C. Neutrophils	3. Release of histaminase destructive enzyme
D. Lymphocytes	4. Release granules containing histamine

[NEET 2020]

Codes

	A	B	C	D
(a)	4	1	2	3
(b)	1	2	4	3
(c)	2	1	3	4
(d)	3	4	2	1

5 Identify the given diagram of tissue performing secretion and absorption. **[AIIMS 2019]**

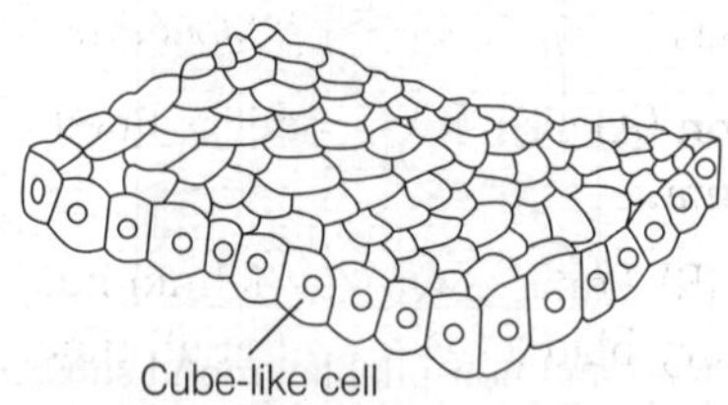

(a) Simple cuboidal epithelium
(b) Simple columnar epithelium
(c) Stratified cuboidal epithelium
(d) Stratified columnar epithelium

6 The ciliated epithelial cells are required to move particles or mucus in a specific direction. In humans, these cells are mainly present in **[NEET 2019, AFMC 2013]**

(a) Fallopian tubes and pancreatic duct
(b) Eustachian tube and salivary duct
(c) Bronchioles and Fallopian tubes
(d) Bile duct and bronchioles

7 Match the following cell structures in Column I with their characteristic features in Column II.

Column I	Column II
A. Tight junctions	1. Cement neighbouring cells together to form sheet
B. Adhering junctions	2. Transmit information through chemical to another cells
C. Gap junctions	3. Establish a barrier to prevent leakage of fluid across epithelial cells
D. Synaptic junctions	4. Cytoplasmic channels to facilitate communication between adjacent cells

[NEET 2019]

Codes

	A	B	C	D		A	B	C	D
(a)	2	4	1	3	(b)	4	2	1	3
(c)	3	1	4	2	(d)	4	3	1	2

8 Match the items given in Column I with those in Column II and select the correct option from the codes given below.

Column I	Column II
A. Fibrinogen	1. Osmotic balance
B. Globulin	2. Blood clotting
C. Albumin	3. Defence mechanism

[NEET 2018]

Codes

	A	B	C		A	B	C
(a)	1	3	2	(b)	1	2	3
(c)	3	2	1	(d)	2	3	1

9 Rouleaux formation is related to which of the cell/tissue? **[JIPMER 2018]**

(a) RBCs (b) WBCs
(c) Platelets (d) Monocytes

10 **Assertion** (A) Urinary bladder is lined by transitional epithelium.

Reason (R) Transitional epithelium keeps the size of the urinary bladder constant at all time. **[AIIMS 2017]**

(a) Both A and R are true and R is the correct explanation of A
(b) Both A and R are true, but R is not the correct explanation of A
(c) A is true, but R is false
(d) Both A and R are false

11 Dense regular connective tissue is present in **[JIPMER 2017, EAMCET 2014]**

(a) ligament and tendon
(b) joint capsule and Wharton's jelly
(c) periosteum and endosteum
(d) pericardium and heart valves

12 Adult human RBCs are enucleate. Which of the following statements is/are most appropriate explanation for this feature?

I. They do not need to reproduce.
II. They are somatic cells.
III. They do not metabolise.
IV. All their internal spaces are available for oxygen transport. **[NEET 2017]**

Choose the correct option from the codes give below.

(a) Only IV (b) Only I
(c) I, III and IV (d) II and III

13 Choose the correct statements with reference to hyaline cartilage. **[EAMCET 2016]**

I. It is bluish-white translucent and glass-like cartilage.
II. Matrix shows collagen fibres and elastic fibres.
III. It forms embryonic endoskeleton in cyclostomes and adult cartilage fishes.
IV. It is the strongest cartilage without perichondrium.

Choose the correct option from the codes given below.

(a) I, II and IV
(b) II and III
(c) II, III and IV
(d) I and III

14 Smooth muscles are **[NEET-II 2016]**

(a) involuntary, fusiform, non-striated
(b) voluntary, multinucleate, cylindrical
(c) involuntary, cylindrical, striated
(d) voluntary, spindle-shaped, uninucleate

15 Which type of tissue correctly match with its location? **[NEET-I 2016]**

	Tissues	Locations
(a)	Areolar tissue	– Tendons
(b)	Transitional epithelium	– Tip of nose
(c)	Cuboidal epithelium	– Lining of stomach
(d)	Smooth muscle	– Wall of intestine

16 **Assertion** (A) In cardiac muscles, impulses are transmitted rapidly.

Reason (R) Gap junctions allow quick passage of ions from one cell to the other. **[EAMCET 2015]**

(a) Both A and R are true and R is the correct explanation of A
(b) Both A and R are true, but R is not the correct explanation of A
(c) A is true, but R is false
(d) A is false, but R is true

17 The tissue covering the external surface of animal and internal surface of visceral organs is **[J&K CET 2015]**

(a) adipose tissue (b) connective tissue
(c) cartilaginous tissue (d) epithelial tissue

18 All the following are examples of connective tissue except **[UK PMT 2015]**

(a) tendons (b) ligaments
(c) muscle (d) adipose tissue

19 Wharton's jelly present in umbilical cord is an example of **[EAMCET 2015]**

(a) adipose tissue (b) mucous connective tissue
(c) areolar tissue (d) elastic connective tissue

20 Skeletal muscle fibre is a 'syncytium' which means it is **[Kerala CEE 2015]**
(a) made up of many fibres
(b) made up of many proteins
(c) long and slender
(d) swollen in the middle with tapered ends
(e) multinucleated

21 Choose the wrong statement with reference to smooth muscles. **[EAMCET 2015]**
(a) Myofibrils show alternate dark and light bands
(b) They are spindle-shaped uninucleate cells
(c) Their contractions are under the control of autonomous nervous system
(d) They exhibit prolonged contractions

22 Which one is a specialised connective tissue among these? **[Kerala CEE 2015]**
(a) Adipose tissue (b) Bone
(c) Areolar tissue (d) Fibroblasts
(e) Muscles

23 Choose the correctly matched pair. **[CBSE AIPMT 2014]**
(a) Moist surface of buccal cavity — Glandular epithelium
(b) Tubular parts of nephrons — Cuboidal epithelium
(c) Inner surface of bronchioles — Squamous epithelium
(d) Inner lining of salivary ducts — Ciliated epithelium

24 Choose the correctly matched pair. **[CBSE AIPMT 2014]**
(a) Adipose tissue — Dense connective tissue
(b) Areolar tissue — Loose connective tissue
(c) Cartilage — Loose connective tissue
(d) Tendon — Specialised connective tissue

25 The inner lining of ducts of sweat glands and pancreatic duct is formed of epithelium
(a) pseudostratified **[EAMCET 2014]**
(b) transitional
(c) stratified cuboidal
(d) stratified non-keratinised squamous

26 Read the following statements and choose the correct option.
I. Gap junctions cement adjacent cells together.
II. Areolar tissue contains fibroblasts, macrophages and mast cells.
III. Tight junctions facilitate the cells to communicate with each other.
IV. Adhering junctions help to stop substances from leaking across tissues.
V. Cells of connective tissue except blood secrete fibres of structural proteins called elastin.

Choose the correct option from the codes given below. **[Kerala CEE 2014]**
(a) I, II and III (b) I, III and IV
(c) III and V (d) I, II and V
(e) II, IV and V

27 The lining of intestine and kidneys in human is **[Manipal 2013]**
(a) keratinised (b) brush border
(c) ciliated (d) None of these

28 Vagina, oesophagus, urethra contain which type of tissue? **[Manipal 2013]**
(a) Stratified squamous epithelium
(b) Simple squamous epithelium
(c) Ciliated epithelium
(d) Columnar epithelium

29 The outer covering of cartilage is called **[WB JEE 2013]**
(a) peritoneum (b) periosteum
(c) endosteum (d) perichondrium

30 Bone marrow is absent in **[AIIMS 2013]**
(a) reptilians (b) amphibians
(c) fishes (d) birds

31 Cardiac muscle is found in **[Odisha JEE 2013]**
(a) pericardium
(b) endocardium
(c) myocardium
(d) peritoneum

32 Debove's membrane is a layer of **[Manipal 2013]**
(a) muscular tissue cells
(b) epithelial tissue cells
(c) connective tissue cells
(d) None of the above

33 The type of tissue lining the nasal passage, bronchioles and Fallopian tubes is **[JCECE 2012]**
(a) columnar ciliated epithelium
(b) cuboidal epithelium
(c) neurosensory epithelium
(d) germinal epithelium

34 The cell membranes of adjacent cells are fused at this cell junction **[AMU 2012]**
(a) macula adherens
(b) zonula adherens
(c) zonula occludens
(d) nexus

35 Keratinised dead layer of skin is made up of
(a) stratified squamous epithelium **[BCECE 2012]**
(b) simple cuboidal epithelium
(c) simple columnar epithelium
(d) stratified columnar epithelium

36 The cavities of alveoli of lungs are lined by
(a) cuboidal epithelium **[BCECE 2012]**
(b) columnar epithelium
(c) stratified cuboidal epithelium
(d) squamous epithelium

37 Fibroblasts, macrophages and mast cells are seen in **[BCECE 2012]**
(a) epithelial tissue (b) connective tissue
(c) skeletal muscle tissue (d) smooth muscle tissue

38 Agranulocytes are **[BCECE 2012]**
(a) lymphocytes and monocytes
(b) eosinophils and basophils
(c) lymphocytes and eosinophils
(d) basophils and monocytes

39 Haemoglobin is **[AIIMS 2012]**
(a) an oxygen carrier in human blood
(b) a protein used as food supplement
(c) an oxygen scavenger in root nodules
(d) a plant protein with high lysine content

40 The type of epithelium seen in the walls of blood vessels is **[Kerala CEE 2011]**
(a) squamous epithelium (b) columnar epithelium
(c) ciliated epithelium (d) cuboidal epithelium
(e) compound epithelium

41 Which of the following cells are round and biconcave in shape? **[Kerala CEE 2011]**
(a) White blood cells (b) Red blood cells
(c) Columnar epithelial cells (d) Nerve cells
(e) Mesophyll cells

42 Histamine, serotonin and heparin are secreted by **[Kerala CEE 2011]**
(a) lymphocytes (b) monocytes
(c) neutrophils (d) eosinophils
(e) basophils

43 Which one of the following human cells do not contain mitochondria? **[WB JEE 2011]**
(a) Nerve cell (b) Red blood cells
(c) Liver cell (d) White blood cells

44 Which of the following are phagocytic in nature? **[Gujarat CET 2011]**
(a) Neutrophil, monocyte and basophil
(b) Neutrophil, monocyte and macrophage
(c) Neutrophil, basophil and macrophage
(d) Acidophil, basophil and lymphocyte

ANSWERS

Check Point 13.1

1. (c)	2. (b)	3. (c)	4. (c)	5. (a)	6. (b)

Check Point 13.2

1. (c)	2. (a)	3. (d)	4. (a)	5. (a)	6. (a)	7. (d)

Check Point 13.3

1. (c)	2. (c)	3. (d)	4. (a)	5. (b)	6. (a)

(A) Topical Exercise

1. (c)	2. (b)	3. (d)	4. (b)	5. (b)	6. (b)	7. (d)	8. (c)	9. (b)	10. (c)
11. (a)	12. (c)	13. (a)	14. (a)	15. (a)	16. (b)	17. (d)	18. (c)	19. (c)	20. (a)
21. (c)	22. (c)	23. (a)	24. (a)	25. (d)	26. (a)	27. (b)	28. (c)	29. (d)	30. (c)
31. (a)	32. (b)	33. (c)	34. (c)	35. (c)	36. (c)	37. (b)	38. (c)	39. (c)	40. (b)
41. (b)	42. (c)	43. (b)	44. (a)	45. (b)	46. (c)	47. (b)	48. (c)	49. (a)	50. (d)
51. (a)	52. (d)	53. (b)	54. (d)	55. (d)	56. (c)	57. (d)	58. (a)	59. (a)	60. (d)
61. (a)	62. (c)	63. (c)	64. (a)						

(B) Cumulative Exercise

1. (c)	2. (b)	3. (b)	4. (c)	5. (b)	6. (c)	7. (d)	8. (d)	9. (b)	10. (b)
11. (b)	12. (a)	13. (d)	14. (a)	15. (a)	16. (d)	17. (b)	18. (c)	19. (d)	20. (b)
21. (b)	22. (d)	23. (b)	24. (b)	25. (a)	26. (a)	27. (b)	28. (b)	29. (b)	30. (a)
31. (a)									

(C) Medical Entrances' Gallery

1. (a)	2. (b)	3. (a)	4. (d)	5. (a)	6. (c)	7. (c)	8. (d)	9. (a)	10. (c)
11. (a)	12. (a)	13. (d)	14. (a)	15. (d)	16. (a)	17. (d)	18. (c)	19. (b)	20. (e)
21. (a)	22. (b)	23. (b)	24. (b)	25. (c)	26. (d)	27. (b)	28. (a)	29. (d)	30. (d)
31. (c)	32. (c)	33. (a)	34. (c)	35. (a)	36. (d)	37. (b)	38. (a)	39. (a)	40. (a)
41. (b)	42. (e)	43. (b)	44. (b)						

Explanations

(A) Topical Exercise

10 (*c*) Option (c) represents incorrectly matched pair. It can be corrected as

Ciliated columnar epithelium lines respiratory tract. Iris of eye is lined by simple cuboidal epithelium.

12 (*c*) Statement in option (c) is incorrect. It can be corrected as Endocrine glands are duct less glands. They pour their secretions directly into the blood.

29 (*d*) Statement in option (d) is incorrect. It can be corrected as Volkmann's canal are the small blood vessels that connect the Haversian canal in the mammalian bones.

47 (*b*) Statement in option (b) is incorrect. It can be corrected as Intercalated discs are special junctions in cardiac muscle fibres only. These are not found in skeletal muscles.

51 (*a*) Statement I is true, but II is false. Incorrect statement can be corrected as

Smooth muscles are uninucleate, i.e. they contain single nucleus only.

64 (*a*) Statement in option (a) is not true. It can be corrected as

Not all nerve fibres are non-medullated. Medullated nerve fibres that are lined by myelin sheath are found in CNS.

(B) Cumulative Exercise

6 (*c*) Neurons do not contain melanin instead some neurons possess myelin.

8 (*d*) The option (d) represents incorrect pair. It can be corrected as

Volkmann's canal connect one Haversian canal of bone to another.

9 (*b*) Statements II and III are incorrect and can be corrected as

If a femur is kept in dilute HCl for 3 days it becomes soft and elastic. Haversian system or osteon is a characteristic feature of mammalian bone. It is a basic unit of compact bone. It is absent in birds, but present in panther.

Rest statements are correct.

10 (*b*) Statements II and IV are incorrect and can be corrected as

Cardiac muscle in the embryo are originated from epicardial cells. Not all invertebrates have RBCs.

17 (*b*) Statements III and V are incorrect and can be corrected as

- Osteocytes are present in space called lacunae.
- Biceps are voluntary and striated muscles.

22 (*d*) Given diagram represents TS of bone. Bone have hard and non-pliable ground substance as it contain abundant of calcium salts and collagen fibres.

24 (*b*) Both A and R are true, but R is not the correct explanation of A.

Connective tissue has extracellular (intercellular) and intracellular matrix. Intracellular matrix plays an active defence role, while intercellular or extracellular serves a number of mechanical functions of support and protection against mechanical stress. This is made up of protein fibres.

25 (*a*) Both A and R are true and R is the correct explanation of A.

Pleura and pericardium are serous membranes, which cover the lungs and the heart, respectively. Epithelial layer of a serous membrane secretes a clear fluid similar to intercellular fluid. This fluid provides lubrication, which prevents friction between surfaces of different organs.

26 (*a*) Both A and R are true and R is the correct explanation of A.

Mucous membrane lines the digestive, respiratory, urinary and reproductive system. The epithelium of a mucous membrane has mucous glands, which secrete mucus to moisten and lubricate the membrane. Thus, it provides protection to the internal organs from drying out.

27 (*b*) Both A and R are true, but R is not the correct explanation of A.

Three types of cells are present in areolar tissue (connective tissue), namely fibroblasts, mast cells, macrophages. Areolar tissue consists of two types of fibres, i.e. yellow elastin fibres and white collagen fibres.

White collagen fibres are found at the site of injury during tissue repair and wound healing process. These fibres are composed of collagen protein.

28 (*b*) Both A and R are true, but R is not the correct explanation A.

In wounds, germs are removed by WBCs accumulation at wound site by diapedesis. It is the squeezing of leucocytes out from endothelium of capillaries to fight against external agent.

29 (*b*) Both A and R are true, but R is not the correct explanation of A.

Histamines are involved in allergic and inflammatory reactions. They also dilate capillaries. Hence, they are called as vasodilators.

30 (*a*) Both A and R are true, but R is not the correct explanation of A.

Thick layer of muscles are present in the wall of alimentary canal, these help in the mixtue of food material with the enzymes coming from different glands.

31 (*a*) Both A and R are true and R is the correct explanation of A.

Gap junctions allow various molecules and ions to pass freely between cells. These junctions are expressed in virtually all tissues and cells, but most notably in cell types that are involved in direct electrical communications, such as neurons and cardiac muscles.

(C) Medical Entrances' Gallery

1 (*a*) Option (a) is the incorrectly matched pair and can be corrected as

Chondrocytes are cartilage cells. The intercellular material of cartilage is solid, pliable and resists compression. Cells of this tissue are called chondrocytes, which are enclosed in small cavities within the matrix secreted by them.

Rest options are correctly matched pairs.

2 (*b*) Cuboidal epithelium with brush border of microvilli is found in PCT (Proximal Convoluted Tubule) of nephron.

The epithelium fills the lumen and the microvilli increases the surface area by 30-40 folds. A brush border is a name for the microvilli covered surface of simple cuboidal epithelium and simple columnar epithelium cells found in certain locations of the body.

3 (*a*) Goblet cells of alimentary canal are modified from columnar epithelial cells, which secrete mucus.

These cells are found in the lining of organs like intestine and respiratory tract.

These secrete mucin (glycoprotein) and maintains the layer of mucus. As each cell secretes mucin for mucus production, it is called as a unicellular mucous gland.

5 (*a*) The given diagram represents the simple cuboidal epithelium tissue, as it has a free surface and is composed of a single layer of cube-like cells resting on the basement membrane. These are compactly packed with little intercellular spaces.

6 (*c*) In humans, ciliated epithelial cells are present in the bronchioles and Fallopian tube. In bronchioles, these cells help in the movement of mucus and in Fallopian tubes, these are required to move the egg towards the uterus.

On the other hand, salivary and pancreatic ducts are lined by simple cuboidal epithelium. Bile duct is lined by simple columnar epithelium and Eustachian tube is lined by columnar epithelium having ciliated cells.

9 (*a*) Rouleaux formation is related to RBCs. In resting and slow flowing blood, the RBCs aggregate to form rouleaux, i.e. they pileup on top of each other. Fibrinogen favours rouleaux formation.

10 (*c*) A is true, but R is false. R can be corrected as

The transitional epithelium appears stratified with five or more layers of pear-shaped cells during relaxed condition.

It lines the inner surface of urinary bladder and permits distension. In the urinary bladder, it can be stretched considerably without being damaged.

11 (*a*) Dense regular connective tissue is present in ligament and tendon. In these, the collagen fibres are present in rows between many parallel bundles of fibres. Tendons attach skeletal muscles to bones and ligaments attach one bone to another.

12 (*a*) Only statement IV is the correct explanation for adult human RBCs being enucleate.

The absence of nucleus in RBCs is an adaptation that allows it to contain more haemoglobin and carry more oxygen by providing empty space. This adaptation also aids in effective diffusion of oxygen.

13 (*d*) Statements I and III are correct whereas II and IV are incorrect. Incorrect statements can be corrected as

Hyaline cartilage is translucent and glass-like and it forms strongest cartilage with perichondrium. Hyaline cartilage consists of predominantly collagenous fibres. It forms the embryonic endoskeleton in cyclostomes and cartilaginous fishes.

14 (*a*) Smooth muscles are located in the inner walls of hollow visceral organs of the body-like alimentary canal, reproductive tract, etc. They do not exhibit any striations and are smooth in appearance. Their activities are under autonomic and hormonal control and thus are described as involuntary muscles (non-striated or fusiform).

15 (*d*) Option (d) represents correct match with its location. Rest options are incorrectly matched with their locations and can be corrected as

Columnar epithelium is present in the lining of stomach.
Tendon is dense connective tissue.
Tip of nose consists of elastic cartilage.

16 (*a*) Both A and R are true and R is the correct explanation of A.

Gap junctions allow various molecules and ions to pass freely between cells. These junctions are expressed in virtually all tissues and cells, but most notably in cell types that are involved in direct electrical communications, such as neurons and cardiac muscles.

17 (*d*) The tissue covering the external surface of animal and internal surface of visceral organs is epithelial tissue.

21 (*a*) Smooth muscles do not show alternate dark and light bands. Such striations are found in skeletal muscles.

26 (*d*) Statements I, II and V are correct whereas III and IV are incorrect. Incorrect statements can be corrected as

Gap junctions allow cellular communication *via* passage of electrical and chemical signals between adjacent cells. Tight junctions help to stop substances from leaking across tissues.

27 (*b*) Lining of intestine and kidney in human is formed by columnar epithelium, which has cells with microvilli on their surface and forms brush border. Brush border surface increases the absorptive area of the surface.

28 (*a*) Stratified squamous epithelium consists of two to many layers of cells adopted to withstand mechanical abortion. This type of epithelium lines the oral cavity, oesophagus and the vagina of mammals.

29 (*d*) Perichondrium is a dense layer of fibrous connective tissue that covers the surface of cartilage.

31 (*c*) The wall of heart has three layers, i.e. epicardium, myocardium and endocardium. Myocardium consists of cardiac muscles. These muscles contract quickly and do not get fatigued. Pericardium is the membrane surrounding the heart.

32 (*c*) Debove's membrane is a layer of connective tissue present between the epithelium and basement tissue of respiratory and intestinal epithelium.

34 (*c*) At tight junction or zonula occludens, plasmalemmae of two cells fuse to form impermeable or occluding junctions, e.g. epithelial cells or capillaries and brain cells.

35 (*a*) Stratified squamous epithelium is seen in the adult human body. It may be keratinised or non-keratinised. In keratinised stratified squamous epithelium, the outer few layers contain a hard waterproof in their cytoplasm.

36 (*d*) There are 300 million of alveoli (also called acini) in two lungs. The alveoli have very thin wall consisting of squamous epithelium.

37 (*b*) Loose connective tissue contains fibroblasts (cells that produce secrete fibres), macrophages (phagocytic in nature) and mast cells (which secrete heparin, serotonin and histamine).

38 (*a*) Agranulocytes formed in spleen and lymph nodes are non-granular, white blood cells that contain non-lobulated nuclei. These form about 35% of total leucocytes (3.5×10^9 / L). These are of two types, i.e. monocytes and lymphocytes.

39 (*a*) Haemoglobin is a Fe-S protein, which carries O_2 in human blood. It is a tetrameric protein having 2 α and 2 β-chain. Fe^{2+} present in the haem structure binds to the CO_2.

43 (*b*) Mature RBCs in mammals do not have any cell organelles, not even a nucleus or mitochondria.

44 (*b*) Neutrophil, monocyte and macrophages are types of white blood cells. The granular white blood cells neutrophils, eosinophils and agranular leucocytes including monocytes and tissue macrophages are phagocytic in nature. Basophils are non-phagocytic and involved in allergic reactions.

CHAPTER 14

Structural Organisation in Some Animals

Animal kingdom comprises of a large variety of complex multicellular organisms. Their basic functions are carried out by different groups of cells in a well-organised manner.

In this chapter, we will study about morphology and anatomy of earthworm, cockroach and frog in order to understand the evolutionary trend of body organisation in animals.

EARTHWORM (*Pheretima posthuma*)

There are several species of earthworms known today. Approximately 13 species of earthworm are found in Indian soil. *Pheretima posthuma* is the common Indian earthworm. Other genera include *Drawida* and *Megascolex* occurring in Southern India and *Eutyphaeus* which is found in the Gangetic plain of North India. The common genus found in Europe and North America is *Lumbricus*.

Pheretima posthuma is a fossorial animal, i.e. it lives in burrows, made by boring and swallowing the upper layer of soil.

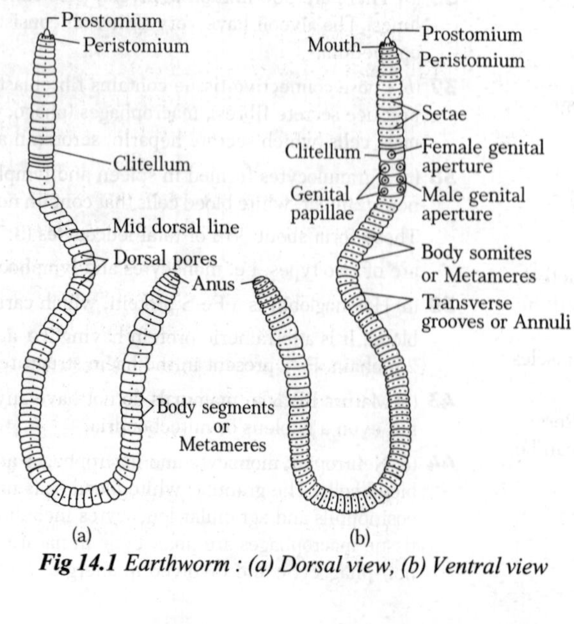

Fig 14.1 Earthworm : (a) Dorsal view, (b) Ventral view

They are found in wet soil, rich organic matter, i.e. humus. It is nocturnal in habit and it comes out of burrow during night only. *Pheretima* is a cold-blooded animal.

The natural lifespan varies from three and a half year to 10 years.

Morphology

- The animal has a glistening dark brown body, due to pigment porphyrin. It protects the earthworm from UV radiation.
- Body of earthworm is divided into 100-120 segments or **metameres** or **somites**.
- The first segment is buccal segment or **peristomium** which bears a slit-like mouth.
- It is dorsally covered with a fleshy process called **prostomium**. It helps in digging burrows and is also sensory in nature.

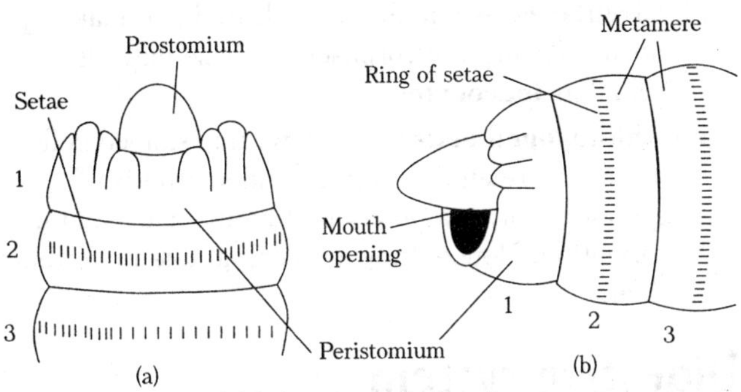

***Fig 14.2** Anterior end of earthworm; (a) Dorsal view, (b) Lateral view*

- Earthworm lacks distinct head and sense organs like eyes, cirri and tentacles.
- In a mature worm, a thick band of glandular tissue called **clitellum** is found around 14-16 segments.
- Due to its presence, the body of earthworm is divided into **preclitellar**, **clitellar** and **postclitellar** regions.
- Clitellar region has glands whose secretions help in the formation of cocoon or egg capsule. It is a protective structure in which eggs are fertilised.
- The dorsal surface of body is marked by a dark median mid-dorsal line. It represents dorsal blood vessel. The ventral surface possesses the genital pores.

  The body segments of earthworm and their associated structures are given in the table below

Table 14.1 Number of segments along with associated structures in earthworm's body

Number of segments	Structures
1st (Peristomium)	Mouth
6th, 7th, 8th, 9th	Spermathecae
14th to 16th	Clitellum
4th	Pharynx
5th to 7th	Oesophagus
8th	Gizzard
15th to last	Intestine
15th to last	Septal nephridia
3rd onwards	Integumentary nephridia
10th and 11th	Testes, testes sacs
13th	Ovary
17th and 19th	Accessory glands
16th to 21st	Prostate gland
11th and 12th	Seminal vesicles
26th to 95th	Typhlosole
7th and 9th	Lateral hearts
12th and 13th	Lateral oesophagal hearts
4th to 6th	Pharyngeal nephridia or branched nephridia
10th and 11th	Anterior loops
4th to 6th	Blood glands
14th	One female genital opening
18th	A pair of male genital opening
17th and 19th	Genital papillae
5/6, 6/7, 7/8, 8/9	Spermathecal openings

Body wall

The body wall of *Pheretima* has four layers from outside to inside as follows

(i) **Cuticle layer** is the outermost thin elastic and non-cellular layer. It is made up of collagenous protein and a polysaccharide with small amount of gelatin. It is a protective layer secreted by underlying epidermis.

(ii) **Epidermis layer** is single layer of columnar cells resting on basement membrane. It contains supporting cells, gland cells (mucous and albumin), basal cells and sensory cells.

(iii) **Muscle layer** has an outer thin, circular muscle fibre layer and an inner thick longitudinal muscle fibre layer. All muscle fibres are of smooth type.
- The circular muscle layer forms a continuous sheet. Numerous granules of porphyrin pigment are found scattered in circular muscle layer.
- The longitudinal layer is discontinuous. It is broken into several longitudinal strips/bands, separated by thin connective tissue partitions.

(iv) **Parietal peritoneum** or **Parietal layer** or **Coelomic epithelium** lies below the longitudinal muscle layer. It secretes coelomic fluid. Along with all these layers, the locomotory organs, i.e. setae are also present. These are S-shaped, chitinous structures, present in specialised pouches called **setal sacs**.

Functions of body wall

- The body wall protects internal organs and prevents excessive evaporation.
- It also aids in sensory functions and excretion through the nephridiopores present in it.

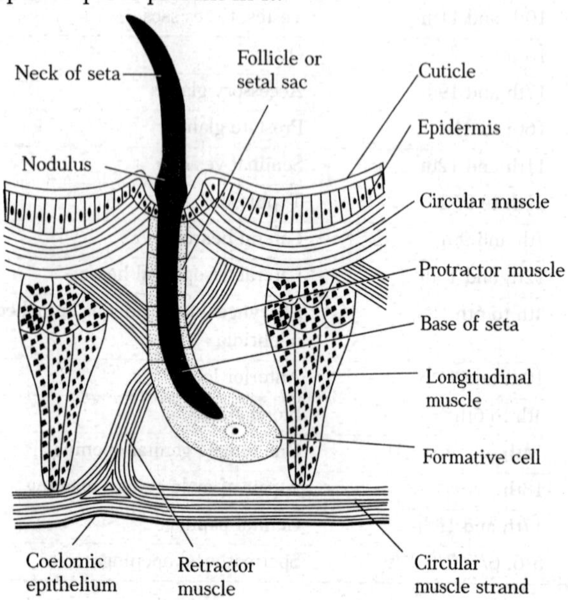

Fig 14.3 Structure of body wall

Coelom

- The body cavity of earthworm is schizocoel type, i.e. it is formed by the splitting of mesodermal band.
- It lies between the body wall and alimentary canal.
- Peritoneum lines both the surface. Outer parietal is in contact with body wall and inner visceral is in contact with the alimentary canal.
- The coeloms of first four segments are continuous or not segmented.
- It is divided from fourth and fifth segments onward with the help of intersegmental septa.
- **Coelomic fluid** The coelomic fluid is alkaline and colourless. It serves as the hydraulic or hydrostatic skeleton in initial segments. Thus, it helps in locomotion. Coelomic fluid is released through the minute dorsal pores present in each intersegmental groove. These pores are not found in first eleven segments and last annulus.
- Coelomic fluid has fluid matrix of watery plasma. It contains salts, proteins and numerous nucleated corpuscles.
- It also contains following four types of coelomic corpuscles. These are as follows
  - (i) **Phagocytes** or **Eleocytes** are the largest and most numerous cells. These are amoeboid in nature and engulf harmful germs.
  - (ii) **Mucocytes** are elongated cells with fan-like processes. The narrow end of cells contains nucleus. Their function is unknown.
  - (iii) **Circular cells** which are nucleated and make up about 10% of total corpuscles. These may also be called as **leucocytes.**
  - (iv) **Chloragogen cells** are yellow, star-shaped cells. These are related to storage and synthesis of glycogen, fat, deamination of amino acids and urea formation. Hence, they are comparable to vertebrate liver.

Digestive system

- The alimentary canal of *Pheretima* is completed and well-differentiated.
- It comprises of buccal chamber, pharynx, oesophagus, gizzard, stomach and intestine.
- **Gizzard** is the main grinding organ of alimentary canal. Its wall has hard and thick circular muscle layer.
- **Stomach** contains calciferous glands to neutralise the humic acid of soil.
- The longest and widest part of alimentary canal is intestine.

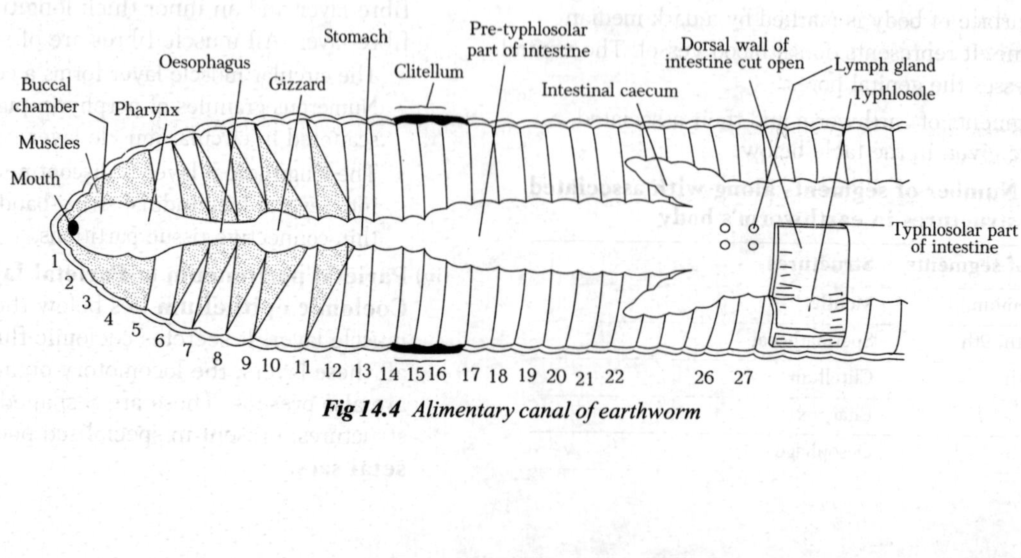

Fig 14.4 Alimentary canal of earthworm

- In most of the part of intestine, a large villi-like internal fold is visible, called **typhlosole**. Due to typhlosole, intestine is divided into three regions
  (i) **Pre-typhlosolar regions,** i.e. the region before typhlosole (15th-26th segment).
  (ii) **Typhlosolar region,** i.e. region in which typhlosole is present (27th-54th segment approximately).
  (iii) **Post-typhlosolar region,** i.e. region after typhlosole (last 23rd-25th segments).

Lymph glands

These are white coloured bodies on either side of dorsal vessel from 26th segment. These glands might help to produce phagocytes of coelomic fluid.

Process of digestion

- Earthworm feeds on dead organic matter, i.e. humus.
- The suctorial action of buccal chamber and oesophagus are helpful in the ingestion of food. Digestion is extracellular.
- Surrounding the pharynx, pharyngeal glands are present. These are the masses of chromophil cells that produce mucin for the lubrication of food. They also produce a proteolytic enzyme that helps in protein digestion.
- Undigested food and soil are eliminated through the anus to outside in the form of worm castings, i.e. faecal pellets.

Respiratory system

Earthworm do not possess any special respiratory organ.

The animal is aerobic and gaseous exchange takes place through the general body surface (cutaneous gaseous exchange). The skin is kept moist by the moisture of the soil, secretion of mucus by epidermal gland cells and the coelomic fluid oozing out through the dorsal pores.

Note *In sunlight, skin dries up and earthworm dies because of asphyxia, i.e. severe deficient supply of oxygen.*

Circulatory system

- Earthworm has closed type of circulatory system consisting of blood, blood vessels, heart and plexus.

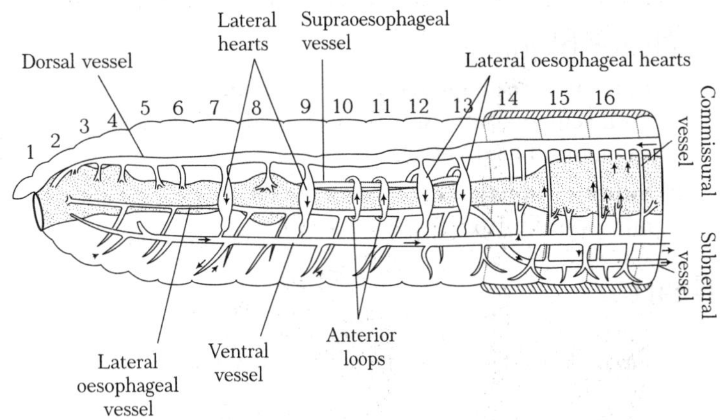

Fig 14.5 *Closed circulatory system of earthworm*

- In earthworm, blood flows in well-organised vessels, which run all along the length of the body. Major blood vessels of earthworm are as follows

1. Dorsal blood vessel

- It is the largest blood vessel of the body. It is present mid dorsally and can be recognised as a dark median line.
- It is comparable to the heart of vertebrates because its highly thick muscular wall has rhythmic contractile activity. In fact, this blood vessel is considered as dorsal tubular heart.
- Internally, this blood vessel contains a pair of valves in each segment in front of septum.
- Dorsal blood vessel is a collecting blood vessel behind 13th segment, while in initial 13 segments it works as the distributing vessel.
- It is the only blood vessel in which blood flows from last segment towards the first segment.

2. Ventral blood vessel

- It is the second largest vessel of earthworm's body. It is present mid ventrally behind the alimentary canal, but above the nerve cord.
- Its wall is thin and non-contractile. Internally the valves are absent in it.

- In this blood vessel, blood flows from anterior to the posterior end.
- This blood vessel is mainly distributing although in the initial 13 segments, it receives blood also.

3. Other blood vessels

(i) **Lateral oesophageals** are two in number up to the 13th segment, afterwards these fuse to form a single subneural blood vessel.

(ii) **Supraoesophageal** is a single blood vessel, present mid-dorsally from 9th-13th segments.

Heart

- Along with the above blood vessels, the circulatory system of earthworm contains hearts and anterior loops.
- There are four pairs of heart present in 7th, 9th, 12th and 13th segments (one pair each) as shown in figure 14.5. In real sense, these are blood vessels only which transversely connect larger blood vessels. They do not have any pulsatile activity.
- Hearts of 7th and 9th segments are called **lateral hearts,** while those of 12th and 13th segments are called **lateral oesophageal hearts.**
- Anterior loops are present in 10th and 11th segments for the same purpose.
- Behind 13th segment, ventrotegumentries, commissurals, along with dorsal and ventrointestinals collect or distribute the blood.
- The complete course of blood circulation in earthworm is given below

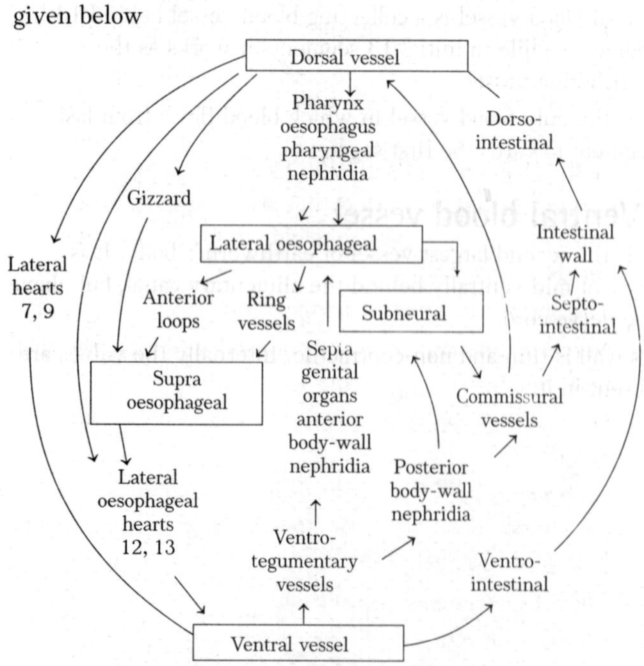

Fig 14.6 Pheretima; Course of circulation of blood

- **Earthworm's blood** is red due to dissolved haemoglobin in plasma.
- It has leucocytes only as corpuscle.
- The pharyngeal mass of 4th, 5th and 6th segments has small red coloured follicular bodies. These bodies are called **blood glands** and are involved in the manufacture of blood.

Excretory system

- Earthworm is ureotelic animal. Their excretory products contain about 40% urea, 20% ammonia and 40% amino acids.
- Excretory system is made up of many segmentally arranged microscopic coiled tubes called **nephridia**. In *Pheretima*, three types of nephridia are seen. They are

1. Pharyngeal nephridia

- These are seen as paired tufts on either side of pharynx and oesophagus in the 4th, 5th and 6th segments.
- These open into alimentary canal, i.e. buccal cavity and pharynx, so these are enteronephric.
- These are without nephrostome.

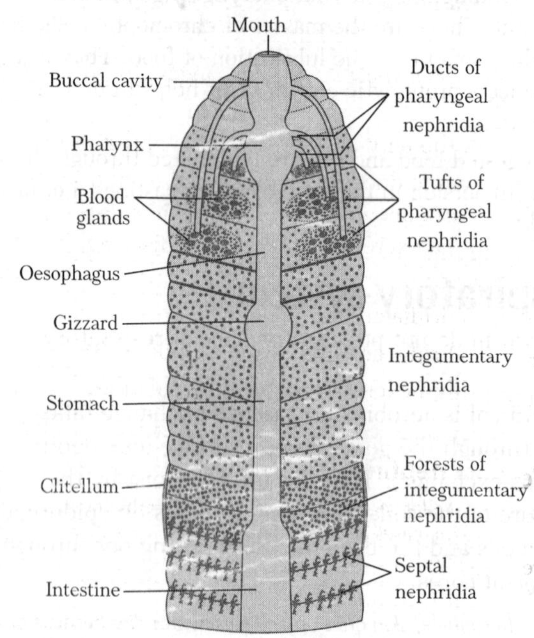

Fig. 14.7 Pheretima. Location of three types of nephridia

2. Integumentary nephridia

- These are V-shaped nephridia seen from 3rd segment onwards. They are usually 200-250 in numbers lying along the entire inner or parietal surface of animal's body wall.
- These are exonephric type, i.e. open to outside through the nephridiopores (minute openings) present on the body surface.

3. Septal nephridia

- These are the largest nephridia in earthworm. These are arranged along the sides of the intersegmental septum from 15th segment onwards.

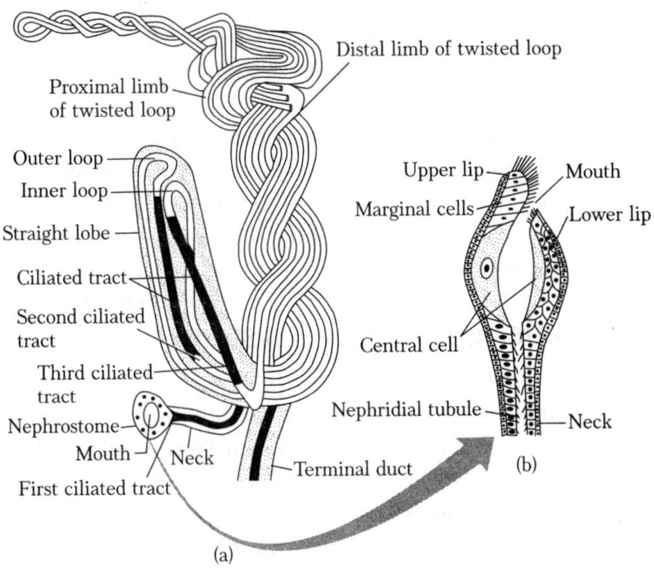

Fig 14.8 (a) A septal nephridium of Pheretima, (b) LS of Nephrostome

- The number of these nephridia ranges from 80-100 in each segment.
- These are the only nephridia with nephrostome.
- Their terminal duct opens into septal excretory canal. These inturn, open into supraintestinal excretory canals. Later, all the excretory products are discharged into intestine.
- Septal nephridia are enteronephric.
- Septal nephridia are abundantly supplied with blood vessels.
- These help in excretion and osmoregulation.

> **Forest of nephridia**
> Within the clitellar segments (14th, 15th and 16th), the total number of nephridia increases about 10 times. Hence, we call these segments forests of nephridia.

Nervous system

All the activities of the earthworm are under the control of nervous system. Nervous system is well-developed and basically comprises of the following three systems

1. Central Nervous System (CNS)

- It consists of a paired ventral nerve cord. It lies ventral to the alimentary canal and is enclosed in a common sheath. It enlarges in each segment to form a segmental ganglion.
- In the anterior region, two subpharyngeal, two cerebral or suprapharyngeal ganglion and two circumpharyngeal connectives form a nerve ring in third and fourth segments, as shown in the following figure

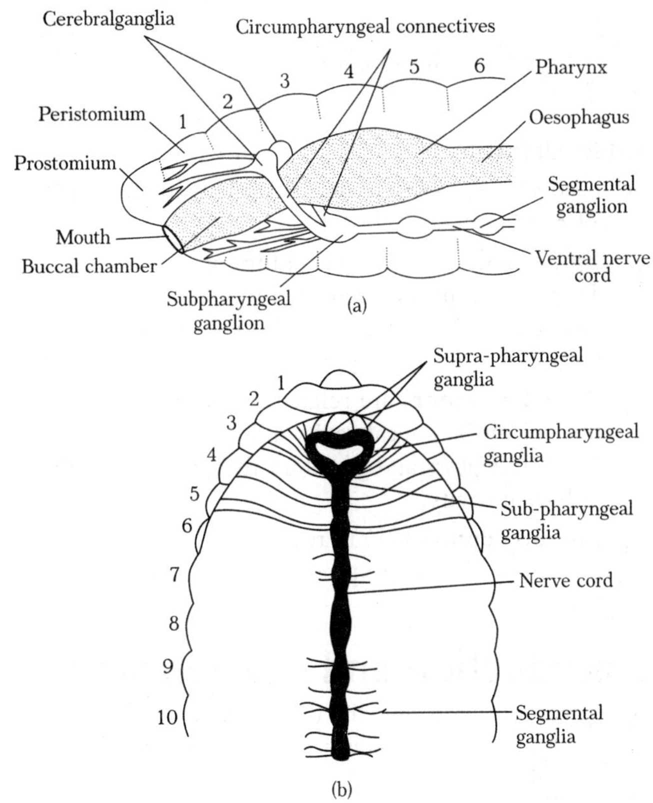

Fig. 14.9 Pheretima nerve ring and ventral nerve cord in the anterior end of the body : (a) Lateral view, (b) Dorsal view

- The suprapharyngeal ganglion is also called **brain.**
- Presence of nerve ring is a characteristic feature of nervous system of earthworm.

2. Peripheral Nervous System (PNS)

- It is formed by the nerves that arise from the cerebral ganglia, subpharyngeal ganglia and other segmental ganglia.
- The peripheral nerves are mixed in nature.
- Nerves of circumpharyngeal connectives supply the first two segments, i.e. segments 1 and 2.
- Subpharyngeal ganglia supplies to 2nd, 3rd and 4th segments.
- Each segmental ganglion seeds out three pairs of segmental peripheral nerve. One pair from anterior part and two pairs from posterior part.
- These nerves supply various organs in each segment.
- All the segmental nerves are mixed in nature.

3. Sympathetic Nervous System (SNS)

- It consists of nerve plexus within the wall of alimentary canal and some other organs.
- These plexus are connected with the pre-pharyngeal connectives by fine nervules.
- They help in coordinating the functions of associated structures.

Sense organs

The sensory receptors of animal include following three types of receptors, i.e.

(i) **Epidermal** or **Tactile receptors** or **Thigmoreceptors** are related to touch. These are distributed on the lateral and ventral surface of the body.

(ii) **Buccal receptors** are related to digestion. These are present in the buccal cavity along with olfactoreceptors and gustatory receptors in the lining of buccal cavity.

(iii) **Photoreceptors** help to respond to light through phaosome. These are present on dorsal surface of the body.

Reproduction and development

- Earthworm is hermaphrodite (bisexual), i.e. male and female reproductive organs are present in the same individual.
- Self-fertilisation does not occur as the male and female reproductive organs (testes and ovaries) do not mature at the same time. The testes mature earlier than the ovaries and such a condition is known as **protandry**. Therefore, cross-fertilisation occurs in earthworm.

Male and female reproductive system in earthworm are as follows

1. Male reproductive system

It consists of the following structures

(i) **Testes sacs** are two pairs present in the 10th and 11th segments. Each sac encloses a testis and a spermiducal funnel.

(ii) **Testes** are two pairs present in 10th and 11th segments. These are the major sites of gametogenesis.

(iii) **Seminal vesicles** are two pairs present in 11th and 12th segments. The process of development to maturation of spermatozoa takes place inside them.

(iv) **Spermiducal funnel** are present in 10th and 11th segments. These are gateways of sperm reception and lead into fine tubes, i.e. vasa deferentia.

(v) **Vasa deferentia** are tubes that runs posteriorly from the 12th to the 18th segments.

(vi) **Prostate glands** are large solid, flat and irregular glands that lie one in each side of the gut from 16th to 21st segments. They produce a secretion which serves as a medium for transfer of sperms.

(vii) **Accessory glands** are paired whitish glandular masses, present in 17th to 19th segments.

- Their secretion helps the two worms to unite at the time of copulation.
- In earthworm, male genital pores are paired structures. These are situated in 18th segment.
- Vas deferens along with spermatic and prostatic duct forms a common male duct, which opens to outside through male genital pore.

2. Female reproductive system

It consists of the following structures

(i) **Ovaries** are a pair of white minute masses found attached to the septum present between 12th and 13th segments. These are the major site of gametogenesis in females.

(ii) **Oviducts and oviducal funnel** are comparable with spermiducal funnels and vas deferens of males.

These collect eggs and convey them outside into the cocoon through a female genital aperture.

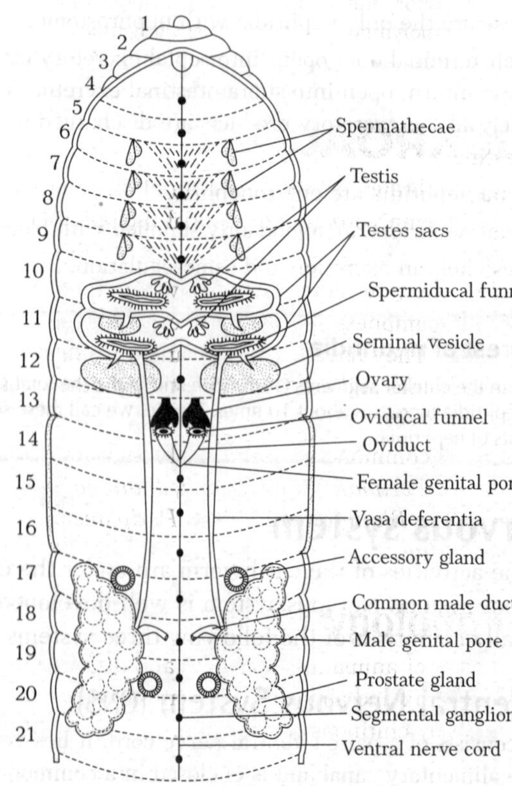

Fig 14.10 Reproductive organs of Pheretima

(iii) **Spermathecae** are present in 6th, 7th, 8th and 9th segments. They store the sperms received from another earthworm during copulation.

- **Copulation** occurs during rainy season and lasts for about one hour. After the mutual exchange of sperms, the two worms separate and lay their eggs in cocoon.
- **Fertilisation** is external occurring in cocoon.
- **Cleavage** is holoblastic and unequal.
- **Development** is direct, i.e. without any larval stage. Thus, metamorphosis do not occur in it.
- A cocoon may contain many fertilised eggs, but only one embryo develops from it at the expense of other eggs.
- Newly hatched young worm do not receives parental care and resembles the adult except for size and the absence of clitellum.

Interaction with mankind

Earthworms are of great economic importance in agriculture, thus are also called 'friends of farmers'.

They help in **vermicomposting** (a process to increase soil fertility) loosening up of soil, etc. They are also used as baits in fishing and in medicine for bladderstone.

CHECK POINT 14.1

1. Blood of earthworm is
   (a) Red in colour, due to dissolved haemoglobin in corpuscle
   (b) Red in colour, due to dissolved haemoglobin in plasma
   (c) Blue in colour, due to dissolved haemocyanin in plasma
   (d) Blue in colour, due to dissolved haemocyanin in corpuscles
2. A median mid-dorsal line along the longitudinal axis of the body of earthworm is represented by
   (a) ventral blood vessel
   (b) dorsal blood vessel
   (c) subneural vessel
   (d) typhlosole
3. If a live earthworm is pricked with a needle in its outer surface without damaging its guts, the fluid that comes out is
   (a) excretory fluid (b) coelomic fluid
   (c) haemolymph (d) slimy mucus
4. If the dorsal pores of the earthworm are blocked, then which of the following functions are affected first?
   (a) Gaseous exchange (b) Blood circulation
   (c) Copulation (d) Locomotion
5. Choose the incorrectly matched pair.
   (a) Pharyngeal nephridia – Exonephric
   (b) Forest of nephridia – 14th-16th segments
   (c) Subpharyngeal ganglia – Supplies 2nd-4th segments
   (d) None of the above
6. In a copulating pair of earthworms, which of the two processes take place?
   (a) Reciprocal fertilisation and internal fertilisation
   (b) Cross-fertilisation and reciprocal fertilisation
   (c) Internal fertilisation and cross-fertilisation
   (d) External fertilisation and internal fertilisation

COCKROACH

- These are brown coloured, nocturnal (i.e. more active at night), omnivorous and cursorial (i.e. fast runners) insects.
- Cockroaches are found in places, where there is warmth, dampness and plenty of organic food available. They have become common residents of human homes and are serious pests and vectors of many diseases.
- The four common species of cockroach found in India are *Periplaneta americana*, *Blattella germanica*, *Blatta orientalis* and *Periplaneta australasiae*.

Morphology

- The body of animal is segmented and divisible into head, thorax and abdomen. The entire body is covered by chitinous exoskeleton.
- Each segment possesses hardened plates called **sclerites.** Among these, the dorsal ones are called **tergites**, ventral ones as **sternites** and lateral ones as **pleurites.**
- These are joined together by a thin, soft, flexible articular membrane called **arthrodial membrane.**

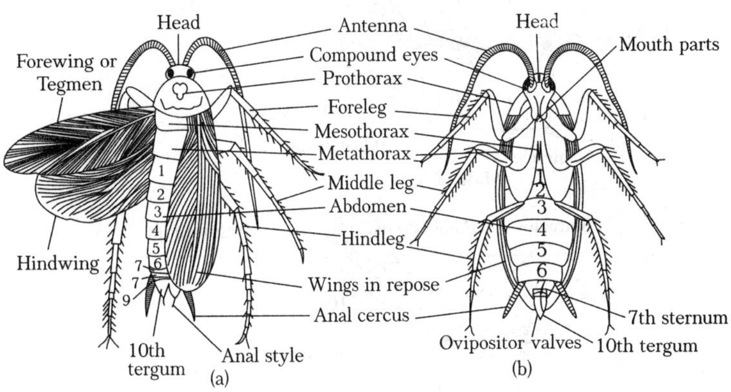

Fig 14.11 External features of Periplaneta: (a) Male in dorsal view, (b) Female in ventral view

Head

- Head is hypognathous (facing downward) and is covered by endoskeleton called **tentorium**.
- It is triangular in shape and formed by the fusion of **six** embryonic segments.
- It is attached to the flexible neck and can move in all directions. It bears a pair of thread-like antennae, a pair of compound eyes and a pair of fenestra or occilar spot or simple eyes.
- Anterior part of head bears bitting and chewing type of mouth parts.

Thorax

- It has three distinct segments namely prothorax, mesothorax and metathorax. These are covered by thicker and larger tergites called **nota**. The notum of prothorax, i.e. pronotum is very large and it covers the neck also. Exoskeleton of each segment consists of terga, sterna and pleura.
- A pair of wings arises from mesothorax, which are thick and leathery. These are called **elytra** or **tegmina**.
- A pair of membranous wings arises from metathorax. These are used for flying.

Abdomen

- The posterior region of the body is called abdomen. It is broader than thorax and dorsoventrally flattened.
- The arthrodial membrane between 5th and 6th abdominal terga possesses stink gland. They produce secretions that give characteristic stinky smell to cockroach.
- Both male and female have 10 segments in their abdomen out of 11 present during embryonic stage.
- Posterior portion of animal is helpful in identifying males and females.

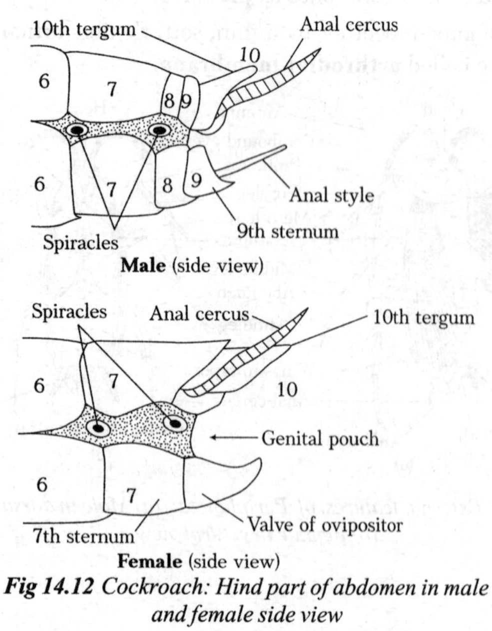

Fig 14.12 Cockroach: Hind part of abdomen in male and female side view

- Ventrally, the abdomen has nine sternites in male and seven in females.
- In **females**, 7th sternum is boat-shaped and forms a brood/genital pouch by combining with the 8th and 9th segments called as **gynatrium**. The posterior part of this pouch is called **oothecal chamber**. The anterior portion of this pouch has **gonopore**.
- In **male**, genital aperture is present in between 9th and 10th sterna. It is surrounded by some sclerites known as **gonapophyses** or **phallomeres.**
- Ninth sternite in males bears a pair of short thread-like anal styles, which are absent in females.
- Both sexes bear jointed filamentous anal cerci on the 10th segment.

Body wall and exoskeleton

The body wall of cockroach is composed of cuticle, hypodermis and basement membrane. These are described as follows

(i) **Cuticle** it is a thick, brown, melanin containing non-living, hard and chitinous covering. It is present on the entire body of cockroach including the appendages and foregut, hindgut, tracheae and genital ducts. Hardening of cuticle is called **scleralisation.**

(ii) **Hypodermis** is present below the cuticle and is single layered. It contains following components
- **Glandular cells** or **Dermal glands**, which secrete certain digestive enzymes and new cuticle at the time of moulting (ecdysis).
- **Trichogen cells**, which secrete setae, spines and bristles, etc.
- **Tormogen cells**, which invaginate to form a shallow socket.

(iii) **Basement membrane** is present below hypodermis.

All these structures are shown in the figure given below

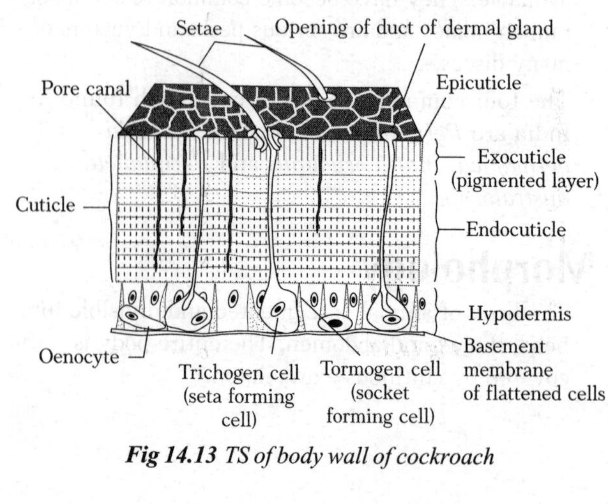

Fig 14.13 TS of body wall of cockroach

Appendages

These are outgrowths of body segments. These are classified as cephalic and thoracic appendages.

Cephalic appendages

These include antennae and the mouth parts.

(i) **Antennae** These are thread-like mobile, tactile, olfactory and thermal receptors. These are considered as the appendages of second segment of the head. Each of these is formed of several small segments called **podomeres**, i.e.

- **Scape** the first largest basal podomere.
- **Pedicel** the second narrow and elongated podomere.
- **Flagellum** rest of the part of antenna. It is narrow elongated and many-jointed structure.

Cockroach locates its food with the help of antennae.

(ii) **Mouth parts** of cockroach are of bitting and chewing type. These include

- **Labrum** or **Upper lip** It has epipharynx on its sides. It acts as chemoreceptors.
- **Labium** or **Lower lip** It bears tactile and gustatory sensory setae.
- **Mandibles** It bears pointed teeth-like structures called **denticles.**
- **First maxilla** It helps to hood food particles between mandibles. It also bears olfactory receptors.
- **Hypopharynx** It bears sensory setae on its free ends. It also possesses the opening of salivary duct at its base.

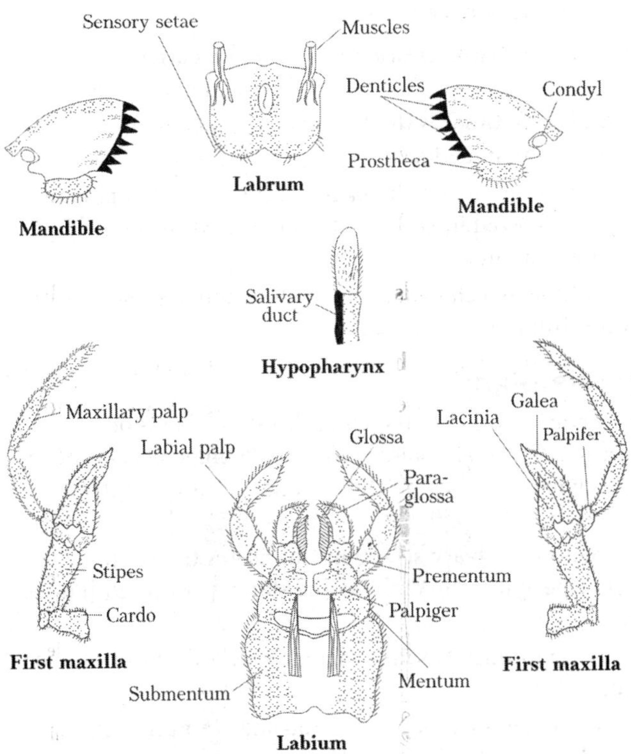

Fig 14.14 Mouth parts of cockroach

Note *If any of the mandibles or maxillae are removed, the cockroach will die because of hunger in due course of time.*

Thoracic appendages

The thorax is made up of three segments, i.e. prothorax, mesothorax and metathorax. Their appendages include three pairs of walking legs and two pairs of thoracic wings.

Walking legs

- In all the three pairs of walking legs, the following five segments are seen
  - Coxa
  - Trochanter (a triangular short rod-like)
  - Femur (long and spiny)
  - Tibia (longest segment)
  - Tarsus
- Tarsus contains five segments called **tarsomeres** or **podomeres.** The last tarsomere is called **pretarsus.** It forms the claws and bears an adhesive pad called **arolium** or **pulvillus.**
- A similar, but smaller adhesive pad is also present at each joint of tarsus. These are called **plantula.**

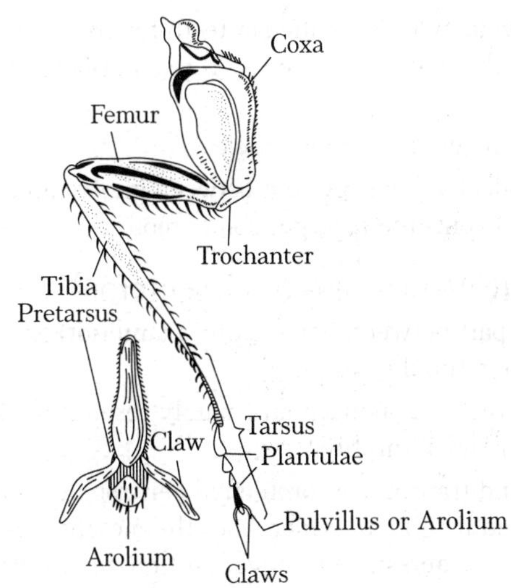

Fig 14.15 A leg of cockroach and its pretarsus

Wings

The wings of cockroach originate from mesothorax and metathorax.

(i) **Forewings** (mesothoracic wings) are thick, hard and leathery. These are also called as **wing cover** or **elytra** or **tegmina**. These are used for protection.

(ii) **Hindwings** (metathoracic wings) are delicate, membranous and are helpful in flying. They are folded like a fan and covered by elytra while at rest.

Digestive system

The digestive system of cockroach consists of **alimentary canal** and **digestive glands**.

Alimentary canal

It is a 6-7 cm long tube and extends from the mouth to anus. It is subdivided into foregut, midgut and hindgut.

1. Foregut (Stomodeum)

- It has a buccal chamber having mouth on one side of it and pharynx (a short tubular structure) on the other.
- The oesophagus enters into prothorax and gradually dilates into a pear-shaped sac called **crop**. It is the largest part of foregut. It serves for food storage.
- **Gizzard** or **proventriculus** is a small, cone-shaped, muscular, thick-walled chamber which makes the end of foregut.
  It is provided with six cuticular teeth for crushing and grinding the food. Filter feeding is also attributed to this part.
- Entire foregut is lined by cuticle.
- A stomodeal valve is present between gizzard and midgut which prevents the regurgitation of food.

2. Midgut (Ventriculus or Mesenteron)

- It is the part between hepatic caeca (anterior) and Malpighian tubule (posterior).
- It forms the true stomach and mainly responsible for digestion and absorption.
- A thin and transparent, protective peritrophic membrane is secreted by gizzard which lines the mesenteron. It prevents the abrasion of midgut portion due to the friction of food particles. It is permeable to digested food and enzymes.
- Midgut is lined with glandular endoderm.
- **Hepatic caecae** are short, blind, glandular tubes formed at the anterior end of midgut. These secrete digestive enzymes.

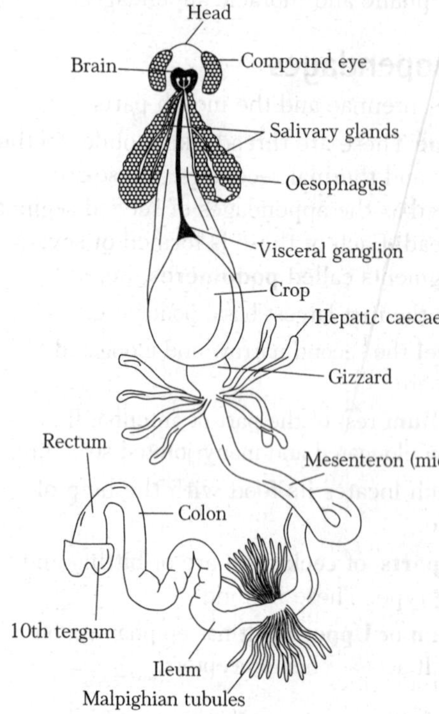

Fig 14.16 Alimentary canal of cockroach

3. Hindgut (Proctodeum)

- It is broader than midgut.
- It is divided into ileum, colon and rectum. The ileum is the first part followed by colon (the longest part) and lastly rectum is present.
- Rectum has six rectal papillae which help in the conservation of water from the faecal matter.

Digestive glands

- The salivary glands are the chief digestive glands of cockroach. These are paired structures lying on each side of the oesophagus and crop. These glands release digestive enzymes.
- The mucus of salivary secretions lubricates the food, while the enzyme zymase hydrolyses the starch present within the food.
- The other enzymes of saliva include amylase, chitinase and cellulose.
- Most of the digestion occurs in crop and absorption of digested food begins in mesenteron (midgut).

Respiratory system

- The blood does not transport gases in cockroach. It serves only as a medium for exchange of gases.
- A complicated system of **air tubes** or **tracheae** is present for gaseous exchange in cockroach.
- The tracheae receive air through 10 pairs of **spiracles** or **stigmata**. These are located along the lateral sides of the body of insect.
- Each trachea divides repeatedly to form finer tubes called **tracheoles** or **tracheal capillaries**. There are six longitudinal tracheal tubes in cockroach. These are interconnected by transverse commissures. Each tracheal tube is lined by a thin strip of cuticle called **taenidia.** These supply air to all parts of the body. This cuticular lining prevents the collapse of tracheal walls during gaseous exchange.
- The spiracles in cockroach are divided into two pairs of **thoracic spiracles** (larger) and the rest are eight pairs of **abdominal spiracles**.
- The opening of spiracles is guarded by annular sclerite, the peritreme. It has a simple valve that acts as a **sphincter**. Each spiracle internally leads into a short tracheal chamber or atrium, from which the main tracheal trunk arises.

Mechanism of respiration

- The tracheoles end in tracheal capillaries from where gases diffuse into and out of the target cells *via* diffusion.
- The first and third pair of spiracles always remain open, while rest of the spiracles open only during inspiration.
- At rest, tracheoles remain filled with fluid. The oxygen diffuses slowly into the body cells and tissues.
- When metabolic activities are high, osmotic pressure of cells increases. Thus, the fluid from tracheoles is withdrawn into surrounding tissues.
- This exposes more surface walls of tracheoles. Due to this, more oxygen is supplied to surrounding tissues (inspiration).
- The CO_2 thus, produced by body cells is moved out of the body through tracheae and spiracles.
- The process of inspiration and expiration is associated with the contraction and relaxation of abdominal muscles called **tergo-sternal muscle**.
- In the case of cockroach, the process of inspiration is considered as the passive process.

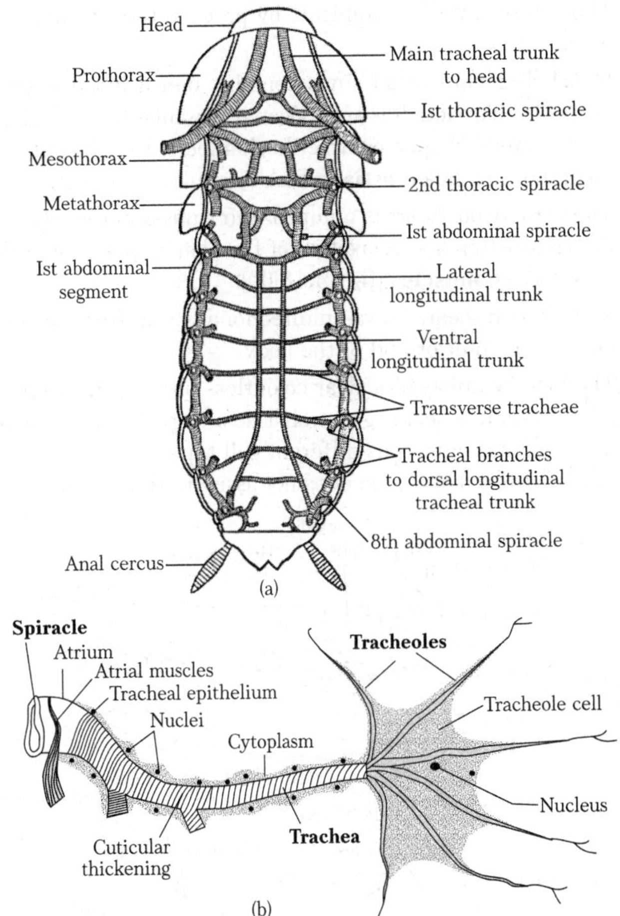

Fig 14.17 In cockroach : (a) Tracheal system (ventral view) (b) Spiracle, traheae and tracheoles

Circulatory system

- It is of open and lacunar type with blood flowing in the open spaces. The blood flows freely in the body cavity called **haemocoel.**
- **Blood sinuses** The haemocoel of cockroach is divided in sinuses or blood spaces with the help of dorsal and ventral diaphragms. These sinuses are pericardial sinus (dorsal), perivisceral sinus (middle), perineural or sternal sinus (ventral) and head sinus as shown in [fig 14.18 (a)].
- The partition between pericardial and perivisceral sinuses is called **dorsal diaphragm.**

- The partition between perivisceral and perineural sinuses is called **ventral diaphragm.**
- The sinuses intercommunicate by pores in their respective diaphragms.
- **Heart** The pericardial sinus contains dorsal blood vessel or heart of cockroach. It is an elongated muscular tube consisting of 13 funnel-shaped contractile chambers. Its inhalent openings are called **ostia**, which are guarded by valves.
- The blood from heart is pumped into sinuses anteriorly by the contraction and relaxation of fan-shaped paired muscles called **alary muscles** [fig 14.18 (b)].
- The blood in heart flows unidirectionally, i.e. from posterior end to the anterior end of the body.
- The blood consists of a clear colourless plasma, rich in a characteristic insect sugar, trehalose along with amino acids, uric acid and numerous different cell types called **haemocytes.** This blood does not assist in respiration.

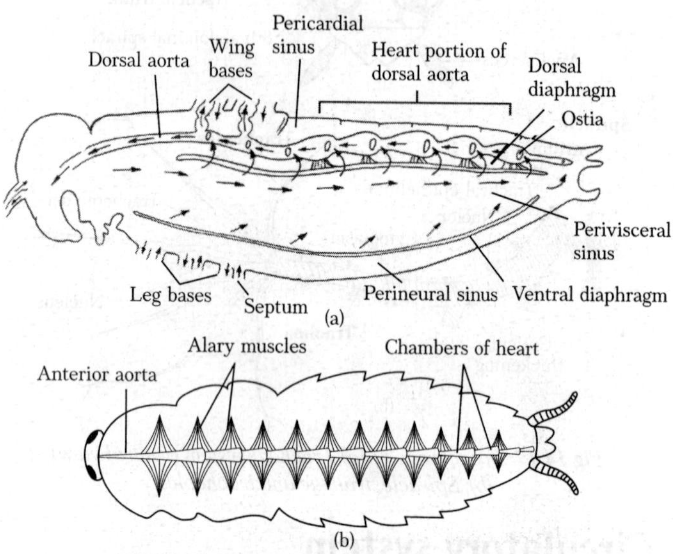

Fig 14.18 Cockroach : (a) Circulatory system, (b) Heart and Alary muscles

Excretory system

- Cockroach is uricotelic, i.e. excretes uric acid. Its excretion occurs through Malpighian tubule, fat body, nephrocytes and uricose gland or mushroom gland (in male only).
- Fat bodies contain urate cells, oenocytes, trophocytes and mycetocytes.
- Uricose glands are found at the periphery of mushroom gland in male cockroach. They store uric acid.
- Nephrocytes are found in the lateral wall of heart. They help to store and excrete nitrogenous wastes.
- The main organs of excretion are Malpighian tubules. These are present at the junction of midgut and hindgut. They lie freely in the haemolymph. Nitrogenous wastes absorbed by them are converted into uric acid and excreted through hindgut afterwards.

Nervous system

The nervous system of animal is well-developed and differentiated into following three types

1. Central Nervous System (CNS)

- It contains a brain or supraoesophageal ganglion, a suboesophageal or subpharyngeal ganglion and a Ventral Nerve Cord (VNC).
- The brain is a bilobed mass which is formed of three pairs of fused ganglion, i.e. **protocerebrum, deuterocerebrum** and **tritocerebrum.**
- Both of the large ganglions are connected by a circumoesophageal or circumpharyngeal connective.

2. Peripheral Nervous System (PNS)

- It consists of those nerves which are given off from the ganglia of CNS, so as to innervate all the parts of the body.
- The nerves from cephalic ganglion innervate eyes, antenna and other cephalic appendages.
- The nerves from thoracic ganglion innervate the thoracic muscles, especially of wings and legs.
- First five abdominal ganglions innervate the dorsal and ventral musculature of the body wall along with spiracles.
- The last abdominal ganglion, on the other hand, supplies to the nerves and musculature of last three segments along with the reproductive organs, copulatory appendages and anal cerci.
- VNC is doublet and bears nine ganglion out of which three are thoracic and six are abdominal as shown below

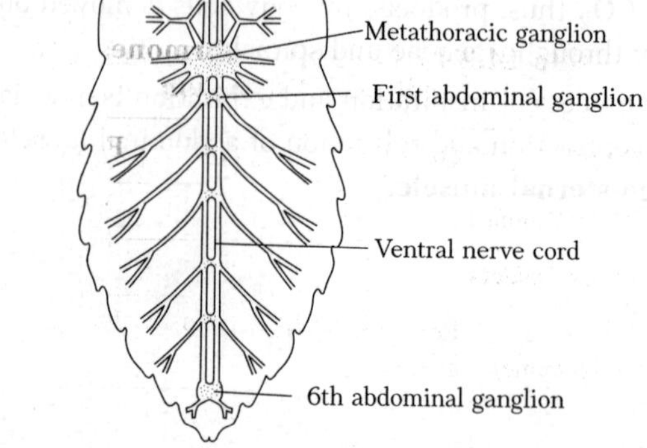

Fig 14.19 Central and peripheral nervous system of cockroach

3. Autonomic Nervous System (ANS)

- This system is of sympathetic (somatogastric) type. It consists of frontal, hypocerebral, ingluvial (visceral) and proventricular ganglion.
- One retrocerebral complex is present above the hypocerebral ganglion. It is formed of two paired neuroendocrine masses, i.e. corpora cardiaca and corpora allata.
- All of these help in synchronisation and control of endocrine system as well.

Sense organs

Sense organs in cockroach are of following types

(i) **Photoreceptors** In cockroach, the unit of photoreception is a compound eye. Each compound eye consists of about 2000-2500 units called **ommatidium.**

(ii) **Thigmoreceptors** Tactile hairs on the antennae along with bristles on the body and appendages are sensitive to touch. They mainly help in the search of food.

(iii) **Chemoreceptors** The receptors to perceive taste (gustatory receptors) are mainly confined to the tips of maxillary and labial palps, labium and hypopharynx.

(iv) **Auditory receptors** (Proprioreceptors) The major sensory activities of antennae and anal cerci for receiving sound vibrations are attributed to specialised receptors called **sensillae.**

(v) **Thermoreceptors** These are located on the 1st, 2nd and 3rd segments of the tarsus of legs.

Endocrine glands

In insects, well-defined endocrine system is also found for coordination. The endocrine system of cockroach contains **corpora allata, corpora cardiaca** and **prothoracic gland.** An outline summary of their hormones is given in the table below

Table 14.2 List of insect hormones, their origin and function

Hormones	Origins	Targets	Functional roles
Non-neural hormones			
(a) **Immature insects**			
Ecdysone (moulting hormone)	Prothoracic gland	Epidermis	Initiates moulting
Juvenile Hormone (JH) (neotinin)	Corpora allata	Epidermis	Regulates metamorphosis at moulting
(b) **Adult insects**			
Ecdysone	Ovarian tissue	Fat body	Initiates and regulates the production of vitellogenin (protein of egg yolk)
Juvenile hormone	Corpora allata	Fat body	Induces fat body to produce vitellogenin
Juvenile hormone	Corpora allata	Accessory reproductive gland	Production of glandular secretion
Juvenile hormone	Corpora allata	Follicle cells	Activates potency and uptake of vitellogenin by follicle cells

Reproductive system

Cockroach is dioecious, i.e. sexes are separate and sexual dimorphism is also seen.

1. Male reproductive system

The male reproductive organs are located in the abdomen. They include

(i) **Testes** as a pair of male gonads, situated in between 4th-6th abdominal segment.

(ii) **Vas deferens** are paired structures, each one arises from each testis. They run posteriorly to open into the ejaculatory duct.

(iii) **Mushroom gland** as a large gland present in between the 6th to 7th abdominal segments. Their secretion nourishes the sperms.

(iv) **Ejaculatory ducts** are paired ducts. These extend posteriorly and open into the genital pouch through a male genital pore.

(v) **Seminal vesicles** are present as numerous sacs and meant for storage of sperms. They are present towards the ventral surface of anterior ejaculatory duct.

(vi) **Phallic or Conglobate gland** is a large elongated sac-like structure that lies beneath the ejaculatory duct.

(vii) **The external genitalia** or **Gonapophysis** are highly branched, three, asymmetrical chitinous structures. These are also known as **phallomeres.**

- These help in reproduction by forming covering of **spermatophore** with the help of secreted alkaline fluid.
- Spermatophore is formed when all sperms of seminal vesicles are glued together into a large bundle. It is pear-shaped, three-layered tough structure.

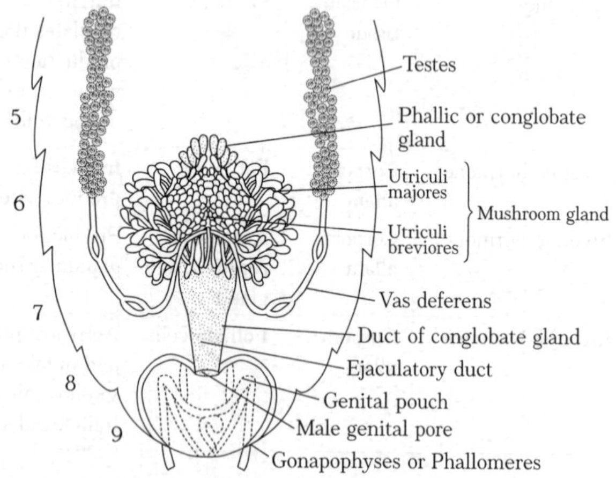

Fig 14.20 P. americana : Male reproductive organs in dorsal view

2. Female reproductive system

The female reproductive organs are also confined to abdominal region and consist of following structures

(i) **Ovaries** as paired structure situated from 2nd to 6th segments on either side of hindgut within the fat bodies. Each ovary consists of eight ovarioles or ovarian tubules, which looks-like beaded blind tubules.

(ii) **Oviducts and vagina** Posteriorly, all the ovarioles of an ovary unite to form a single oviduct and both the oviducts combine to form a single tube called **vagina** or **common oviduct**. The vagina afterwards opens into a large genital pouch through the vulva or female genital pore on the sternum of 8th segment.

(iii) **Spermatheca or Seminal receptacle** It opens by a median aperture in the dorsal wall of genital pouch on 9th sternum. It contains a left sac-like and right filamentous caecum. The sac-like caecum is the site where spermatozoa are stored after the process of copulation.

(iv) **Collaterial gland** These are a pair of branched tubular glands present just behind the ovaries. Out of the pair, the left one secretes milky protein while the right one secretes watery dihydroxyphenol. These secretions combine to form the brown scleroprotein of oothecal egg case.

(v) **Genital pouch** It is located at the high end of abdomen. It is divided into the upper, small genital chamber or genital atrium and the lower, large oothecal chamber.

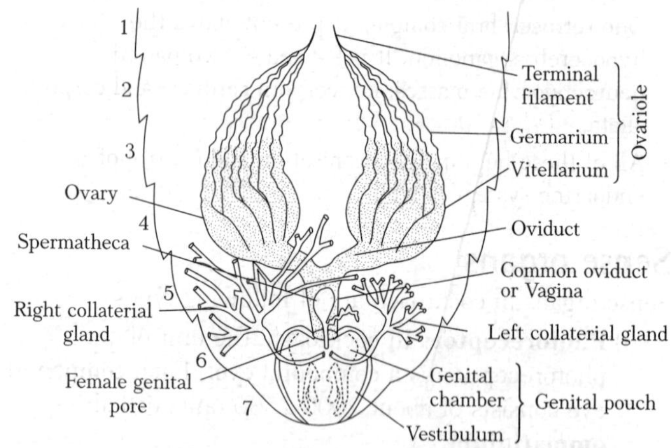

Fig 14.21 P. americana : Female reproductive organs in dorsal view

Note *According to **Roth** and **Willis** (1952), the mature female cockroach secretes a pheromone as the sex attractant. This sex attractant is detected by the antennal chemoreceptors of male.*

Life cycle

- The breeding season of animal is from March to September. Copulation normally takes place at night.
- From spermatophore within 24 hours, the spermatozoa enter the spermatheca. They remain stored there until required, while the empty spermatophore drops down.
- Fertilisation occurs in the genital chamber of genital pouch, where one egg from each ovariole is arranged.
- On an average, female cockroach produces 10 oothecae and each ootheca contains about 14-16 fertilised eggs.
- The fertilised eggs are arranged in two rows of 7-8 eggs each and become surrounded by the secretions of collaterial glands. These secretions harden to form the **ootheca egg** case.
- The eggs of cockroach are centrolecithal, i.e. the yolk is concentrated in the middle of egg.
- The cleavage is holoblastic, spiral and determinate with the hatching stage is called **nymph**.
- Freshly hatched nymphs are delicate, transparent and almost colourless creatures with black eyes. These contain almost all the adult characteristics except being sexually immature and without wings.
- The growth of nymph is associated with the successive moultings.
- **Metamorphosis** in cockroach is incomplete or paurometabolous type, i.e. occurs through a series of nymphal stages.

CHECK POINT 14.2

1. Cockroach has a stomodeal valve between
   (a) ileum and colon
   (b) crop and gizzard
   (c) mesenteron and ileum
   (d) gizzard and mesenteron
2. If head of a cockroach is cut off, it will still live for as long as one week because
   (a) head holds a bit of nervous system
   (b) the rest of nervous system is situated along the ventral (belly-side) part of its body
   (c) Both (a) and (b)
   (d) there is no control of brain over the body activities
3. *Periplaneta americana* has thermoreceptors on
   (a) 1st, 2nd and 3rd segments of tarsus of legs
   (b) 3rd, 4th and 5th segments of tarsus of legs
   (c) 15th segment of anal cerci
   (d) pedicel of antennae
4. Stink gland is found in
   (a) 4th and 5th terga of cockroach
   (b) 5th and 6th terga of cockroach
   (c) 5th and 6th sterna of cockroach
   (d) 4th and 5th sterna of cockroach
5. If corpora allata is removed from an instar, then
   (a) it dies immediately
   (b) it gets arrested in nymphal stage
   (c) it develops adult characteristics
   (d) it remains unaffected
6. Which of the following set includes the structures which are same in number in cockroach body?
   (a) Tergites and sternites
   (b) Abdominal spiracles and ovarioles
   (c) Malpighian tubule and gastric caecae
   (d) Salivary glands and tergites

FROG (*Rana tigrina*)

- Frog belongs to class–Amphibia of vertebrates. It is a cold-blooded or poikilothermic animal.
- It lives in freshwater tanks, pools and shallow margins of rivers. It can also live in moist places such as shores and banks of the different freshwater bodies.
- Frog leaps on land by means of its elongated. It swims in water by using its webbed feet as oars.
- Frogs undergo **aestivation** (summer sleep) and **hibernation** (winter sleep).
- They are carnivorous and poikilothermic.
- They show camouflage, i.e. develop protective colonisation to hide in surroundings.
- Frogs show sexual dimorphism. This difference is visible during breeding season when the male develops a nuptial or amplexusory pad in the first digit of forelimbs for grasping the female during mating.
- Males also have vocal sacs which help them to produce mating calls for copulation (amplexus).

Morphology

- The body of frog consists of a head and a trunk without neck and tail.
- **Head** is triangular in shape with a slightly pointed snout. Its tip bears a pair of external nostrils which open into buccopharyngeal cavity through internal nostrils.
- A large mouth is situated at the anterior end of the head. It is bounded by well defined jaws.
- A pair of prominent eyes are found behind the external nostrils, one on each side of the head. Each eye is covered by an immovable upper eyelid and slightly movable lower eyelid.
- From the lower eyelid, a thin transparent fold called **nictitating membrane** arises. It keeps the eyeballs moist when frog is on land and also offers protection in water.
- Eyebrows and eyelashes are absent in frog.
- Behind and below each eye, a patch of skin called **eardrum** or **tympanum** is present. It receives sound waves.
- The broad flattened part behind the head is trunk. The trunk bears a pair of short forelimbs and a pair of long hindlimbs on its ventro-lateral sides.
- The dorsal surface of body has vertebral column. The total number of bones in frog is 153.

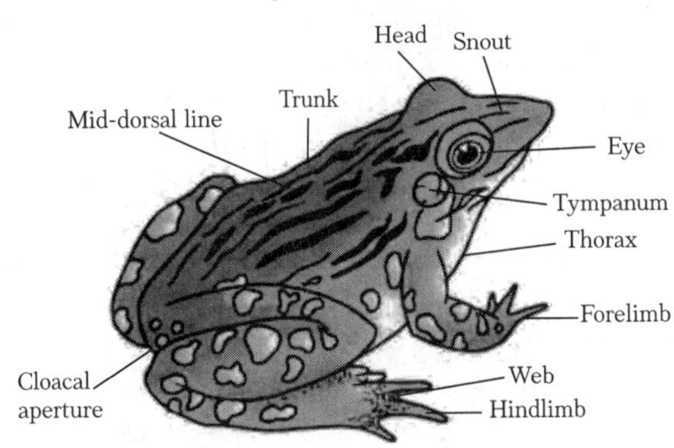

***Fig. 14.22** External features of Rana tigrina*

Body wall or Skin

- Frogs have smooth, moist and scaleless skin. The colour of skin is greenish-yellow with dark spots on the dorsal surface. On ventral surface, the skin shows pale yellow colour.
- Skin consists of two types of glands, i.e. mucous and poison glands besides chromatophores. These glands are the derivatives of epidermis, but they lie in the dermis.
- Mucous glands secrete a colourless watery fluid that keeps the skin moist, glistening and sticky.
- Poison glands secrete a mild poison that protects the animal from the enemies. Poison glands are present in toad, but absent in frog.
- Skin forms the chief respiratory organ as its moist surface brings about an exchange of respiratory gases.

Digestive system

- The alimentary canal of frog is short, as they are carnivorous.
- The alimentary canal of tadpole larva is long as it is herbivorous.
- It consists of buccal cavity, pharynx, oesophagus, stomach, intestine, rectum, which opens into cloaca through cloacal aperture.
- Mouth is present as a terminal wide opening. It opens into buccopharyngeal cavity.
- The buccopharyngeal cavity is bounded by upper and lower jaws. The upper jaw bears a row of small uniform and hook-like pointed maxillary teeth. The lower jaw lacks teeth.

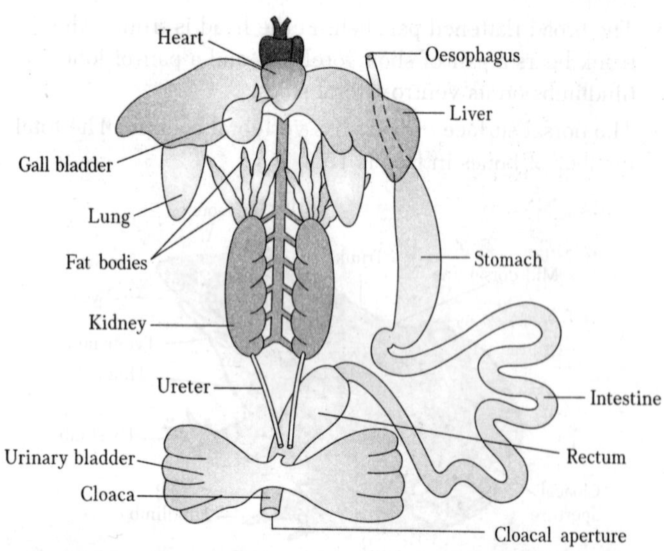

***Fig 14.23** Diagrammatic representation of internal organs of frog showing complete digestive system*

- A sticky bilobed tongue can be thrown out and retracted suddenly to capture insects as food.
- The stomach secretes gastric juice containing HCl and proteolytic enzymes.
- Intestine is divisible into an anterior duodenum and a posterior ileum. Ileum is the longest part of alimentary canal. Rectum is also known as large intestine.
- A bilobed sac called urinary bladder is produced from the ventral wall of cloaca.
- The main digestive glands are liver and pancreas which secrete several digestive enzymes.

Respiratory system

- Frog is amphibious in nature. It can thus, utilise free oxygen of air as well as dissolved oxygen in water.
- The respiration in adult frog occurs *via* following three modes

  (i) **Cutaneous**, i.e. through skin. This is considered as the most common respiratory mode.

  (ii) **Buccopharyngeal**, i.e. through buccopharyngeal cavity. It simply involves oscillatory movements (alternate lowering and raising of buccal cavity floor) of buccal cavity. Respiration occurring through buccopharyngeal and cutaneous mode is approximately 65%.

  (iii) **Pulmonary**, i.e. through lungs. It accounts for about 35% of total gaseous exchange especially when frog is outside the water.

  - The lungs are a pair of elongated, pink coloured sac-like hollow structures present in the upper part of the trunk (thorax).
  - Lungs of frog are positive pressure type, hence breathing involves gulping movement. Diaphragm is not present in frogs.
  - Movements of buccopharyngeal cavity during pulmonary respiration are carried out by a set of paired sternohyal and petrohyal muscles.
  - The sound producing organ of frog is laryngotracheal chamber. One cricoid, two arytenoids and two pre-arytenoid cartilages are associated with it.
  - The exchange of gases occurs by diffusion in all three modes of respiration.

Circulatory system

Circulatory system in frog is of closed type. It comprises of heart, blood vessels and blood. Incomplete double circulation of blood occurs in it.

Heart

- It is a muscular structure situated in the upper part of body cavity. It has **three chambers**, i.e. two atria (sing. atrium) and one ventricle.
- The heart is covered by a membrane called **pericardium**.
- There are two accessory chambers in the frog's heart,
  (i) a tubular **truncus arteriosus** placed on the ventral side of the right auricle.
  (ii) a triangular, **sinus venosus** present on the dorsal side of the heart.
- Sinus venosus receives blood through the major veins called **vena cava**. It opens into the right auricle.
- The left auricle receives oxygenated blood from the lungs by a common pulmonary vein. The auricles send their blood into ventricle where the blood gets mixed up.
- The ventricle opens into a sac-like conus arteriosus on the ventral side of the heart.
- Special venous connections between liver and intestine as well as the kidney and lower parts of the body are present in frogs. The former is called **hepatic portal system** and the latter is called **renal portal system**.
- The RBCs of frog are nucleated and biconvex. The other cells of blood are WBCs and spindle cells.

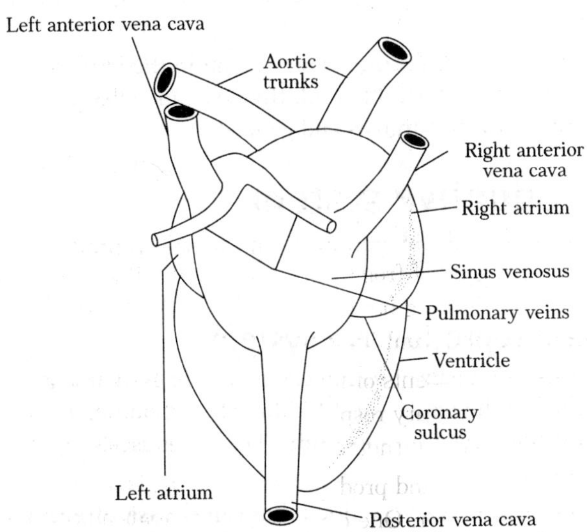

Fig. 14.24 Structure of heart (Frog)

Excretory system

- The excretory system in frog is well-developed and helps in elimination of nitrogenous waste. It consists of a pair of kidneys, a pair of ureters, the urinary bladder and cloaca.
- The kidneys are compact, dark red and bean-like structures, situated little posterior in the body cavity on both sides of vertebral column.
- Each kidney is composed of several structural and functional units called **uriniferous tubules** or **nephrons**.
- From posterior outer margin of each kidney leaves a urinary duct also called as **ureter**. It is a fine tube which then opens into the dorsal wall of cloaca.
- The ureters are also called **urinogenital ducts** in male frog since they pass out sperms as well as urine. That is why in male frog, the excretory and reproductive systems are together referred to as **urinogenital system**.
- In females, the ureters and oviduct open separately into the cloaca. The thin-walled urinary bladder is present ventral to the rectum which also opens into the cloaca.
- The frog excretes urea and thus, is a **ureotelic** animal. The excretory wastes are carried by blood into the kidney where these are separated and excreted.

Nervous system

The nervous system of frog is well-developed and is divided into three parts

1. Central Nervous System (CNS)

It consists of brain and spinal cord. The brain is enclosed in a bony structure called **cranium** or **brain box**. It is further divided into forebrain, midbrain and hindbrain.

(i) **Forebrain** consists of olfactory lobes, paired cerebral hemispheres and unpaired diencephalon.

(ii) **Midbrain** comprises of paired optic lobes and crura cerebri.

Crus cerebrum or crura cerebri are thick bands of nerve fibres which carry the stimuli between the cerebral hemisphere and medulla oblongata.

(iii) **Hindbrain** comprises of cerebellum and medulla oblongata. The spinal cord extends posteriorly from medulla oblongata through foramen magnum and lies protected within neural canal of its vertebral column. It is mainly concerned with the reflex actions.

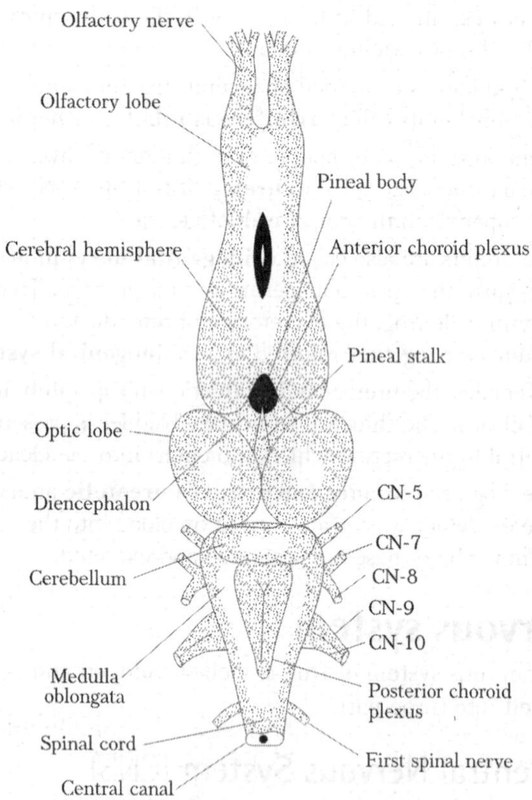

Fig 14.25 Nervous system of frog

2. Peripheral Nervous System (PNS)

It is made up 10 pairs of cranial nerves (i.e. those originating from brain) and 10 pairs of spinal nerves (i.e. those arising from either side of spinal cord).

3. Autonomous Nervous System (ANS)

It is a system of nerve fibre and ganglia. This system controls the involuntary activities of visceral organs of the body like heart, lungs and intestine.

Two type of nerves are present in it *viz*,

(i) **Sympathetic nerve** In general, it increases the rate of activity of autonomous organs or visceral organs.

(ii) **Parasympathetic nerve** In general, it decreases the rate of activity of autonomous organs.

Sense organs

The sense organs in frog are as follows

(i) **Organ of touch** (tangoreceptors), i.e. sensory papillae.

(ii) **Organ of taste** (gustatoreceptors), i.e. taste buds.

(iii) **Organ of smell** (olfactoreceptors), i.e. nasal epithelium, Jacobson's (vomeronasal) organ.

(iv) **Organ of vision**, i.e. eyes.

(v) **Organ of hearing**, i.e. tympanum with internal ears.

Out of these, eyes and ears are well-organised structures and the rest are cellular aggregations around the nerve endings.

(i) **Eyes** There are two large simple eyes seen in frog. Each eyeball is roughly spherical and lodged inside an orbit on the dorsolateral side of head protected by eyelids.

(ii) **Ears** A pair of ears are attached to skull posterio-laterally in all vertebrates. They are statoacoustic organ, i.e. helpful in both hearing and equilibrium of body. An ear of frog mainly consists of two parts middle and internal ear. External ear is represented by **tympanum**.

Endocrine system

- The chemical coordination of various organs of the body is carried out by hormones, which are secreted by the endocrine glands.
- The prominent endocrine glands found in the frogs are pituitary, thyroid, parathyroid, thymus, pineal body, pancreatic islets, adrenals and gonads.

Reproductive system

Frogs have well-developed male and female reproductive organs.

1. Male reproductive system

- It consists of a pair of yellowish, ovoid testes. These are attached to the upper part of kidney by a double fold of peritoneum called **mesorchium**. The testes produce spermatozoa.
- From testes, 10-12 very fine tubes called **vasa efferentia** arise. These enter the kidneys on either side and open into **Bidder's canal.** It is present longitudinally in the median region of kidney.
- The sperms produced by testes pass into Bidder's canal *via* vasa efferentia and then carried finally to the urinogenital duct. It is a common duct for passage of urine and sperm out of the body.
- The urinogenital duct comes out of the kidneys and opens into the cloaca.

- The **cloaca** (also known as cloacal chamber) is a small, median chamber that is used to pass faecal matter, urine and sperms outside the frog's body.

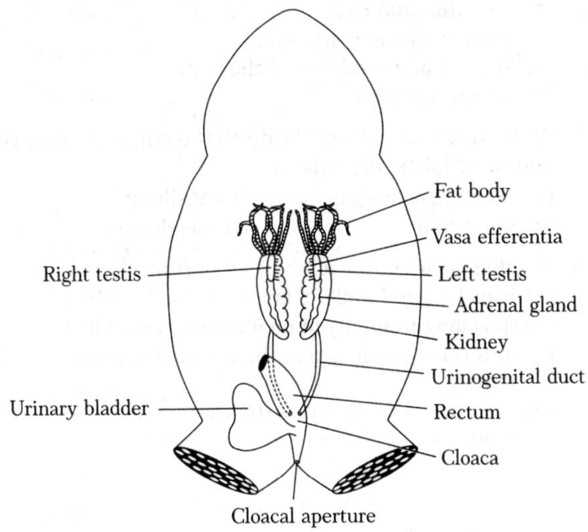

Fig 14.26 *Urinogenital system of male frog (ventral view)*

2. Female reproductive system

- It consists of a pair of ovaries, situated near the kidneys. Ovaries have no functional connection with kidneys.

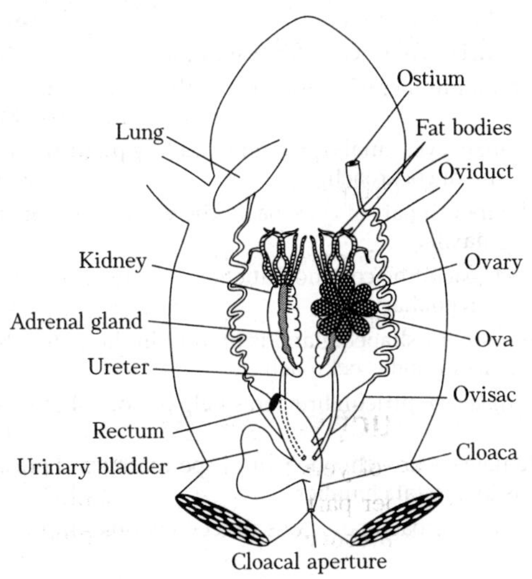

Fig 14.27 *Female reproductive system of frog (ventral view)*

- The ovaries are attached to the dorsal abdominal wall, supported by a peritoneal fold called **mesovarium.**
- On either side of the body cavity lies a pair of long, coiled tubes called **oviducts.**
- The posterior ends of oviduts are called **ovisacs.** These are broad and open dorsally into cloaca.
- A mature female frog can lay 2500-3000 ova at a time. Eggs are temporarily stored in ovisacs.
- The mass of sperms liberated by male frog is called **milt.** The mass of eggs liberated by female frog is called **spawn.**

Fertilisation and development

- With the onset of rainy season, frogs emerge out of aestivation and immediately start breeding. Males gather in shallow waters and start croaking to attract females for mating.
- The male mounts upon the back of female and grasps her firmly with the help of copulatory pads. This process is called **amplexus.** Then female releases hundreds of egg in water and male discharges sperms over these eggs in order to fertilise them.
- Thus in frog, fertilisation is external and takes place in water.
- Zygote undergoes cleavage and develops into morula and blastula.
- The cleavage is holoblastic and unequal.
- The blastula changes into gastrula by epiboly, modified invagination and involution.
- The gastrula has three germ layers, i.e. endoderm, ectoderm and mesoderm. These germ layers further differentiate and give rise to various organs.
- The life history of frogs includes three stages, egg, larva and adult.
- Within two weeks, fertilised eggs or zygote develop into free swimming, aquatic larvae called **tadpoles.**
- The tadpole undergoes metamorphosis to become adult. Metamorphosis is controlled by thyroxine hormone of thyroid gland.

CHECK POINT 14.3

1. A frog must swallow air to expand the lungs because frog
   (a) has no diaphragm
   (b) has no vagus nerve
   (c) is relatively primitive vertebrate
   (d) normally breathe through its skin

2. Dissection of frog is done from ventral side because
   (a) ventral skin is soft
   (b) abdominal vein is present ventrally
   (c) vertebral column is present on dorsal side
   (d) All of the above

3. Choose the incorrect pair.
   (a) Tadpole – Short alimentary canal
   (b) Cutaneous respiration – Most common in frog
   (c) Truncus arteriosus – Ventral side of right auricle
   (d) All of the above

4. Frog has a well-developed renal portal system between
   (a) forelimbs and liver
   (b) alimentary canal and liver
   (c) kidneys and hind part of the body
   (d) kidney and liver

5. 'X' is structure in frog's brain that comprises crura cerebri and optic lobes. Identify 'X'.
   (a) Medulla oblongata (b) Midbrain
   (c) Cerebrum (d) Cerebellum

6. A bidder's canal in each kidney of frog
   (a) runs longitudinally in lateral region of kidney
   (b) runs longitudinally in medial region of kidney
   (c) runs transversely across the width of anterior part of kidney
   (d) runs transversely across the width of posterior part of kidney

EXAM POINTS

- The structure in earthworm which serves as a wedge to force open cracks in the soil is prostomium. **[Kerala CEE 2015]**
- Role of calciferous glands in the stomach of earthworm is to neutralise the humic acid in the humus. **[Kerala CEE 2017]**
- In the earthworm, 6th to 9th segments have spermatheca, 10th and 11th segments consist of testes, intersegmental septum of 12th to 13th segments. **[Kerala CEE 2014]**
- Circulatory system of earthworm consists of lateral heart commissural vessel, subneural vessel and lateral oesophageal vessel. **[Kerala CEE 2015]**
- The nerve cord in earthworm originates from suprapharyngeal ganglia and has a fused pair of ganglia in each segment from 5th to the last. **[AMU 2012]**
- 14th segment of earthworm consists of female genital pore. **[Kerala CEE 2014]**
- The correct sequence of the part of the foregut of cockroach is Mouth $\rightarrow$ Pharynx $\rightarrow$ Oesophagus $\rightarrow$ Crop $\rightarrow$ Gizzard. **[NEET 2020]**
- If the head of cockroach is removed, it may live for few days because the head holds a small portion of nervous system, while the rest is situated along the ventral part of its body. **[NEET 2020]**
- In cockroach, proventriculus helps in grinding the food particles, hepatic caecae secrete gastric juice, 10 pairs of spiracles are present and anal cerci is present at 10th body segment. **[NEET 2020]**
- The exoskeleton of *Periplaneta americana* is composed of N-acetylglucosamine. They have schizocoelom as body cavity. **[NEET-I 2016]**
- Cockroach exhibits mosaic vision with less resolution. **[NEET 2019]**
- The terga, sterna and pleura of cockroach body are jointed by arthrodial membrane. **[CBSE AIPMT 2015]**
- Cockroach have trachea for respiration. **[UK PMT 2015]**
- The total number of podomeres in the leg of cockroach are five. **[MHT CET 2014]**
- Antennae, eyes, anal cerci and maxillary palps are the sensory structure of cockroach. **[Kerala CEE 2014]**
- In the mouth part of cockroach, the galea and lacinia form parts of the maxilla. **[AMU 2012]**
- In cockroach, the common duct of salivary reservior opens at the base of hypopharynx. **[MHT CET 2014]**
- A mushroom-shaped gland is present in the 6-7th abdominal segments of male cockroach. **[NEET 2019]**
- In *Periplaneta*, utriculi breviores help in nourishing the sperms. **[EAMCET 2014]**
- In female cockroach, one pair of spermatheca is present and opens in genital chamber. **[AIIMS 2016]**
- Female cockroach possesses sixteen ovarioles in the ovaries. **[NEET 2019]**
- After the last moult of a cockroach nymph, both forewings and hindwings develop. **[NEET 2013]**
- Frog have lymphatic system. **[Kerala CEE 2015]**
- Sinus venosus receives blood through major veins called vena cava. The ventrical opens into a sac-like conus arteriosus. The erythrocytes (RBCs) are nucleated. **[CBSE AIPMT 2011, Kerala CEE 2014]**
- The tympanic cavity in frog communicates with pharynx ventrally through Eustachian tube. **[AMU 2017]**
- The correct route for the passage of sperms in male frogs is Testes $\rightarrow$ Vasa efferentia $\rightarrow$ Kidney $\rightarrow$ Bidder's canal $\rightarrow$ Urinogenital duct $\rightarrow$ Cloaca **[NEET 2017]**

Chapter Exercises

(A) Topical Exercise

Earthworm

1 Earthworm not found in India is
(a) *Megascolex* (b) *Lumbricus*
(c) *Eutyphaeus* (d) *Drawida*

2 The characteristic brown coloured body of earthworm is due to
(a) haemoglobin (b) porphyrin
(c) haemocyanin (d) melanin

3 The function of porphyrin which imparts colour to earthworm is to
(a) help in respiration
(b) help in reproduction
(c) protect against harmful light rays
(d) protect against harmful germs

4 Clitellum of *Pheretima posthuma* is thick girdle that is
(a) non-glandular around 14th-16th segments
(b) glandular around 14th-16th segments
(c) glandular around 16th-18th segments
(d) non-glandular around 16th-18th segments

5 The body wall of the earthworm is covered by which of the following layers (externally-internally)?
(a) Epidermis, cuticle, coelomic epithelium, longitudinal muscle, circular muscles
(b) Cuticle, epidermis, circular muscles, coelomic epithelium, longitudinal muscles
(c) Non-cellular cuticle, epidermis, circular muscles, longitudinal muscles, coelomic epithelium
(d) Coelomic epithelium, epidermis, cuticle, circular muscles, longitudinal muscles

6 Which of the following is absent in the coelomic fluid of earthworm?
(a) Proteins (b) Salts (c) Haemoglobin (d) Corpuscles

7 Chloragogen cells are involved in
(a) digestion
(b) carbohydrate metabolism
(c) respiration
(d) fat storage

8 Which one will excrete silicates consumed by earthworm along with food?
(a) Intestinal cells
(b) Basal cells
(c) Chloragogen cells
(d) Flame cells

9 Chitogenous glands are found in
(a) clitellar region forming wall of cocoon
(b) prostomium
(c) suckers
(d) epidermis

10 In earthworm, calciferous glands are found in
(a) intestine (b) stomach
(c) oesophagus (d) gizzard

11 Which one of the following is correctly matched regarding earthworm?
(a) Buccal cavity – 1st-5th segments
(b) Stomach – 11th-12th segments
(c) Typhlosole – 26th-95th segments
(d) Testes – 10th-14th segments

12 Intestine of earthworm lies between
(a) 22nd-26th segments (b) 20th-24th segments
(c) 15th to last segment (d) 33th-36th segments

13 The characteristic feature of earthworm intestine is
(a) gizzard (b) calciferous glands
(c) typhlosole (d) prostomium

14 Salivary gland in earthworm is found in
(a) dorsal wall of buccal cavity
(b) ventral wall of buccal cavity
(c) pharyngeal wall
(d) None of the above

15 Which one is correct about respiration in earthworm?
(a) O_2 from atmosphere diffuses into blood and combines with haemoglobin of blood plasma
(b) Blood does not have any important role in O_2 transport
(c) Respires anaerobically
(d) O_2 diffused from air combines with haemoglobin of RBCs

16 Identify the correct option with respect to the labelling given below.

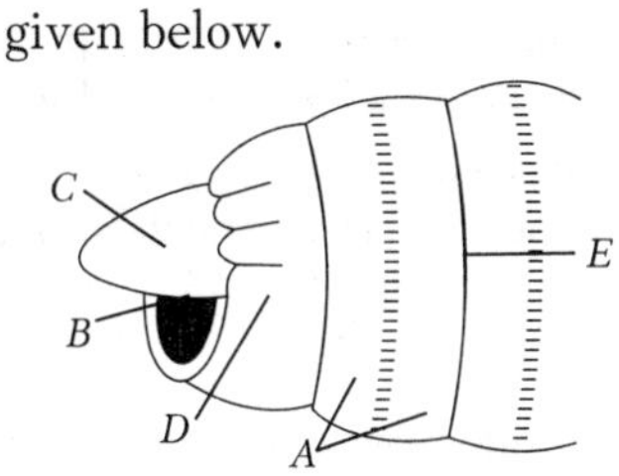

(a) *A*-Metamere, *B*-Mouth opening, *C*-Prostomium
(b) *A*-Ring of setae, *B*-Prostomium, *D*-Peristomium
(c) *B*-Mouth opening, *C*-Peristomium, *D*-Prostomium
(d) *A*-Ring of setae, *D*-Pharynx, *E*-Prostomium

17 The tracheal tubes in earthworm are lined by
(a) stigmata (b) epithelial tissue
(c) chitin (d) taenidia

18 The smallest longitudinal blood vessel in earthworm is
(a) dorsal vessel (b) ventral vessel
(c) supraoesophageal vessel (d) subneural vessel

19 Hearts in earthworm occur in the segment
(a) 6th, 7th and 9th, 10th
(b) 7th, 9th and 12th, 13th
(c) 9th, 10th and 14th, 15th
(d) None of the above

20 In earthworm, blood from seminal vesicle is collected by
(a) ventral vessel
(b) lateral oesophageal vessel
(c) dorsal vessel
(d) supraoesophageal vessel

21 Nephridia of earthworm
(a) regulate the volume and composition of the body fluids
(b) help in excretion and osmoregulation
(c) Both (a) and (b)
(d) are absent in the pharyngeal region

22 Consider the following statements.
I. Nephridia of *Pheretima* are micrometanephridia.
II. Excretory material contains 40% urea, 20% ammonia and 40% amino acid.

Choose the correct option.
(a) Statement I is true, but II is false
(b) Statement I is false, but II is true
(c) Both statements I and II are true
(d) Both statements I and II are false

23 Earthworms are
(a) uricotelic under conditions of water scarcity
(b) ammonotelic, when plenty of water is available
(c) ureotelic, when plenty of water is available
(d) uricotelic, when plenty of water is available

24 Select the incorrect match with respect to types of nephridia and their location in earthworms.
(a) Integumentary nephridia - Segment 3 to last
(b) Pharyngeal nephridia - Segments 4th, 5th and 6th
(c) Septal nephridia - Segment 15 to last
(d) None of the above

25 Which of the following structure is not a part of nerve ring in earthworm?
(a) Subpharyngeal ganglion
(b) Suprapharyngeal ganglion
(c) Circumpharyngeal connectives
(d) Segmental ganglion

26 Choose the incorrect statement regarding reproductive system of earthworm.
(a) Four pairs of spermatheca are present
(b) Two pairs of accessory glands, one pair each is present in the 19th and 20th segments
(c) Two pairs of testes are present
(d) One pair of ovaries are attached at the intersegmental septum of 12th and 13th segments

27 In earthworm, ovary is situated in segment
(a) 13th (b) 9th (c) 10th (d) 26th

28 Where does fertilisation takes place in earthworm?
(a) Vasa deferentia (b) Spermathecae
(c) Cocoon (d) Clitellum

29 Cleavage in *Pheretima* zygote is
(a) meroblastic (b) holoblastic
(c) holoblastic unequal (d) holoblastic equal

30 Consider the following statements.
I. Mature sperm, egg cells and nutritive fluid are deposited in cocoons produced by the gland cells of clitellum.
II. Development in earthworm occurs *via* larva formation.

Choose the correct option.
(a) Statement I is true, but II is false
(b) Statement I is false, but II is true
(c) Both statements I and II are true
(d) Both statements I and II are false

31 Choose the incorrect statement.
(a) The function of typhlosole is to increase the effective area of absorption in the intestine
(b) In earthworm, blood cells are phagocytic in nature
(c) Earthworm possesses open circulatory system consisting of blood vessels, capillaries and heart
(d) The epidermis of earthworm is made up of columnar epithelial cells which contain secretory gland cell

32 The secretions of which gland nourishes the sperms in earthworm?
(a) Phallic gland (b) Gonopophysis
(c) Mushroom gland (d) Seminal vesicle

33 What acts as medium for sperm transfer in male earthworm?
(a) Testes sacs (b) Prostate glands
(c) Seminal vesicles (d) None of these

34 Which one correctly describes the location of body parts in *Pheretima?*
(a) Two pairs of testes in 12th-13th segments
(b) Two pairs of accessory glands in 17th-19th segments
(c) Four pairs of spermathecae in 4th-7th segments
(d) One pair of ovaries attached over septum 14th-15th segments

Cockroach

35 Two common Indian cockroaches are
(a) *Periplaneta americana* and *Blatta indica*
(b) *Periplaneta orientalis* and *Blatta americana*
(c) *Periplaneta americana* and *Blatta orientalis*
(d) *Periplaneta indica* and *Blatta orientalis*

36 Consider the following statements.
I. Cockroach are nocturnal, omnivorous and cursorial.
II. They are found in warmth, damp place.
Choose the correct option.
(a) Statement I is true, but II is false
(b) Statement I is false, but II is true
(c) Both statements I and II are true
(d) Both statements I and II are false

37 The sclerites of cockroach body are joined together by
(a) mesentery (b) pleuron
(c) arthrodial membrane (d) chitin

38 Consider the following statements.
I. In cockroach, each eye consists of 4000 octagonal ommatidia.
II. Compound eyes of arthropods form apposition image in bright light.
Choose the correct option.
(a) Statement I is true, but II is false
(b) Statement I is false, but II is true
(c) Both statements I and II are true
(d) Both statements I and II are false

39 Tongue of cockroach is
(a) labrum (b) mandibles (c) labium (d) hypopharynx

40 Identify *A*, *B*, *C* and *D* in a given figure related to mouth parts of the cockroach.

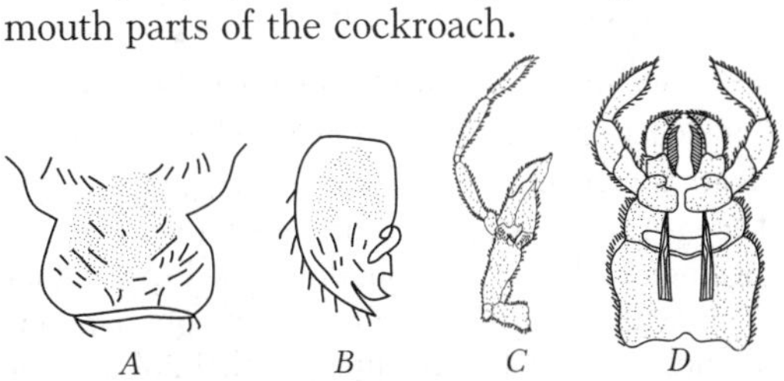

(a) *A*-Mandible, *B*-Labium, *C*-Labrum, *D*-Maxilla
(b) *A*-Labrum, *B*-Mandible, *C*-Labium, *D*-Maxilla,
(c) *A*-Labrum, *B*-Mandible, *C*-Maxilla, *D*-Labium
(d) *A*-Maxilla, *B*-Labium, *C*-Labrum, *D*-Mandible

41 The pincer-like structure present on the stripes of maxilla of cockroach are
(a) cardo (b) lacinia
(c) galea (d) paraglossa

42 The basal podomere of cockroach is known as
(a) frons (b) pedicle
(c) scape (d) None of these

43 Hypopharynx of cockroach
(a) is a tongue-like structure
(b) lies within the cavity enclosed by the mouth parts
(c) Both (a) and (b)
(d) is represented by the mandibles

44 In the following diagram of a leg of cockroach, parts have been indicated by alphabets. Choose the answer in which these alphabets have been correctly matched with the parts which they indicate.

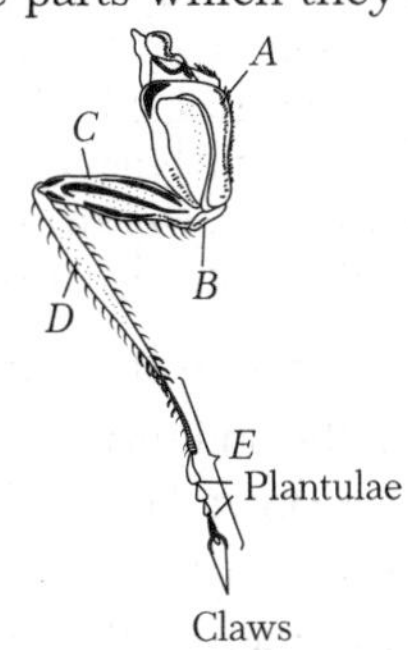

(a) *A*-Coxa, *B*-Femur, *C*-Trochanter, *D*-Tarsus, *E*-Tibia
(b) *A*-Coxa, *B*-Tibia, *C*-Tarsus, *D*-Femur, *E*-Trochanter
(c) *A*-Coxa, *B*-Trochanter, *C*-Femur, *D*-Tibia, *E*-Tarsus
(d) *A*-Coxa, *B*-Tarsus, *C*-Femur, *D*-Tibia, *E*-Trochanter

45 Wings of cockroach arise one pair each from
(a) prothorax and mesothorax
(b) prothorax and metathorax
(c) mesothorax and metathorax
(d) metathorax and first abdominal segment

46 Forewings of cockroach are articulated to
(a) first abdominal segment (b) prothorax
(b) metathorax (d) mesothorax

47 Elytra is also known as
(a) endoskeletal element (b) forewing
(c) stink gland (d) genital organ

48 Consider the following statements.
I. Crop is the largest part of the foregut of cockroach.
II. Hepatic caecae of cockroach produce digestive enzyme.
Choose the correct option.
(a) Statement I is true, but II is false
(b) Statement I is false, but II is true
(c) Both statements I and II are true
(d) Both statements I and II are false

49 Fat bodies of cockroach are analogous to vertebrate
(a) spleen (b) adipose tissue
(c) kidney (d) liver

50 Haemolymph of cockroach contains
(a) erythrocytes and plasma
(b) respiratory pigments only
(c) haemocytes, leucocytes and plasma
(d) haemocytes and plasma

51 Cockroach has no RBCs or haemoglobin because
(a) cockroach does not respire
(b) cockroach is invertebrate
(c) its blood does not transport O_2 and CO_2 and tissues have direct exchange of gases from air
(d) cockroach respires anaerobically

52 Which of the following is not incorrect?
(a) The alimentary canal of cockroach is divided into two regions, i.e. foregut and midgut
(b) Pharynx is absent in cockroaches
(c) Malpighian tubules help in the removal of excretory products from haemolymph
(d) The hindgut is narrower than midgut and is differentiated into ileum, colon and rectum

53 Consider the following statements.
I. Sensillae is the auditory receptors of cockroaches.
II. Anal cerci occur in both male and female.
Choose the correct option.
(a) Statement I is true, but II is false
(b) Statement I is false, but II is true
(c) Both statements I and II are true
(d) Both statements I and II are false

54 Diagram represents posterior region of male cockroach. Choose the correct combination of labelling.

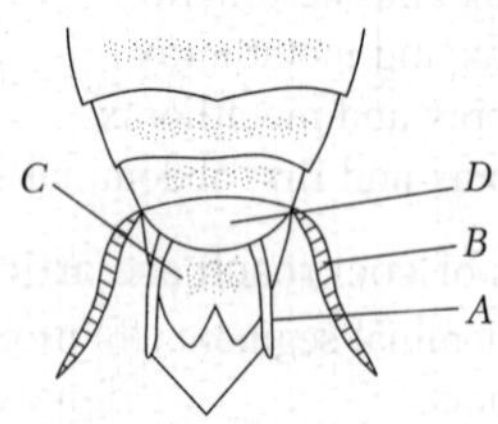

Male

(a) *A*-9th sternum, *B*-Anal style, *C*-10th tergum, *D*-Anal circi
(b) *A*-Anal style, *B*-Anal cerci, *C*-10th tergum, *D*-9th sternum
(c) *A*-9th sternum, *B*-Anal cerci, *C*-10th tergum, *D*-Anal style
(d) *A*-9th sternum, *B*-Anal style, *C*-10th sternum, *D*-Anal circi

55 In cockroach, the brain hormone secreted by the intercerebral glands is responsible for the activation of
(a) corpora allata
(b) cerebral ganglia
(c) prothoracic gland
(d) salivary glands

56 The bristles on the body of the cockroach arise from
(a) basement membrane
(b) epidermis
(c) trichogen cell
(d) Malpighian cell

57 Phallomeres or Gonapophyses are chitinous structures which
(a) are found only in male cockroaches
(b) are found only in female cockroaches
(c) form the external genitalia in male and female cockroaches
(d) form the egg shell

58 Which of the following is correct for the nervous system of cockroach?
(a) Six ganglia lie in thorax and three in the abdomen
(b) Three pairs of ganglia lie in the thorax and six pairs in the abdomen
(c) It is restricted to the head region
(d) In the head region, the brain is represented by supra-oesophageal ganglion that supplies nerves to eyes and antenna

59 Eggs of cockroach are fertilised in
(a) Fallopian tube (b) ootheca
(c) genital pouch of female (d) cocoon

60 Choose the odd one out with respect to male cockroach.
(a) Mushroom gland (b) Spermatheca
(c) Phallic gland (d) Ejaculatory duct

61 The development of cockroach is
(a) hemimetabolous (b) ametabolous
(c) paurometabolous (d) holometabolous

62 Choose the correct option given below.
(a) The fertilised eggs are encased in capsules called oothecae
(b) The development from egg to adult is indirect presence of larval stage
(c) On the lateral side of cockroach possesses 12 pairs of spiracles
(d) Crop helps in grinding the food

63 Choose the incorrect statement for cockroach.
(a) Conglobate glands occur in males
(b) Heart is myogenic
(c) The sperms are stored in the seminal vesicles and glued together in the form of bundles called spermatophores
(d) Each ovary is formed of a group of eight ovarian tubules or ovarioles containing a chain of developing ova

64 Select the incorrect match for cockroach.
(a) Crop - Food grinding
(b) Metathoracic wings - Flight
(c) Ommatidia - Vision
(d) Spiracles - Respiration

65 Choose the incorrect match pair.

Structures	Segments
(a) Gizzard	8-9
(b) Testes	10 and 11
(c) Bifurcated nerve cord	12 and 13
(d) Spermatheca	6-9

Frog

66 The most common species of frog found in India is
(a) *Rana cyanophlictis* (b) *Rana pipiens*
(c) *Rana sylvatica* (d) *Rana tigrina*

67 Which of the following is present in the skin of frog?
(a) Mucous gland (b) Poison gland
(c) Chromatophores (d) All of these

68 In the buccopharyngeal cavity of frog,
(a) the lower jaw possesses teeth
(b) the upper jaw lacks teeth
(c) Both (a) and (b)
(d) oscillatory movement is seen

69 Chromatophores in frog's skin are controlled by
(a) hormones
(b) environment
(c) nervous activity
(d) nervous and hormonal activities

70 The body temperature of frog does not remain constant as it varies with the temperature of the surrounding, so frog will be called
(a) homeotherm (b) poikilotherm
(c) mesotherm (d) psychotherm

71 Metamorphosis in frog is controlled by
(a) oxytocin (b) parathyroid
(c) thyroxine (d) testosterone

72 One of the main functions of frog's skin is
(a) diffusion of respiratory gases
(b) absorption of ultraviolet rays to produce vitamin-D
(c) storage of excess food in the form of subcutaneous fat
(d) excretion of nitrogenous waste in the form of uric acid

73 Consider the following statements.
I. Male frogs can croak louder than females because of vocal sacs.
II. Croaking of frog serves all functions except musical tone.

Choose the correct option.
(a) Statement I is true, but II is false
(b) Statement I is false, but II is true
(c) Both statements I and II are true
(d) Both statements I and II are false

74 Read the following statements carefully, which of them is correct for the eye of frog?
I. Eyes of frogs possess nictitating membrane that protects them while in water.
II. Frogs have bulged compound eyes.
III. Eyes of frog are bulged which provide them 180° vision.

Codes
(a) I and II (b) II and III
(c) Only I (d) Only III

75 Location of eyes in frog is
(a) behind the external nostrils on each side of head
(b) in front of external nostrils on each side of head
(c) behind the external nostrils on one side of head
(d) in front of external nostrils on one side of head

76 Oxygen carrier or the respiratory pigment in blood of frog and other vertebrates is
(a) myoglobin (b) cytochrome
(c) haemoglobin (d) haemocyanin

77 Choose the correct statement for the digestive system of frog.
(a) Oesophagus is very long
(b) Food is captured by the bilobed tongue
(c) Both (a) and (b)
(d) Intestine is divisible into two parts only, i.e. duodenum and rectum

78 The most important respiratory organ of frog is
(a) buccopharyngeal cavity
(b) lungs
(c) gills
(d) skin

79 In frog, digestion of fats occurs mostly in
(a) rectum (b) stomach
(c) duodenum (d) large intestine

80 Read the following statements for frog and answer accordingly.
I. Duodenum receives bile and pancreatic juice through hepatopancreatic duct.
II. Salivary glands are absent in frog, so digestion starts in stomach.

Choose the correct option.
(a) Both the statements I and II are correct and II is the correct explanation of I
(b) Both the statements I and II are correct, but II is not the correct explanation of I
(c) Statement I is correct, but II is incorrect
(d) Both statements I and II are incorrect

81 In frog, glottis is controlled by the muscles of
(a) sternum (b) vocal cords
(c) pectoral girdle (d) arytenoid cartilages

82 Which of the following statements is incorrect about circulatory system of frog?
(a) It is of closed type
(b) There is incomplete double circulation present
(c) It comprises of heart, blood vessels and blood
(d) None of the above

83 The heart of frog has
(a) two atria and one ventricle
(b) one atrium and one ventricle
(c) one atrium and two ventricles
(d) two atria and two ventricles

84 Consider the following statements.

I. In the heart of frog, the ventricles open into conus arteriosus.

II. Chordae tendineae in heart of frog is found in the wall of ventricle.

Choose the correct option.

(a) Statement I is true, but II is false
(b) Statement I is false, but II is true
(c) Both statements I and II are true
(d) Both statements I and II are false

85 Fenestra ovalis in frog is the

(a) air-filled cavity of middle ear
(b) communication between pharynx and tympanic cavity
(c) external opening of tympanic cavity covered by tympanic membrane
(d) opening of auditory capsule which separates middle ear from internal ear

86 Which of the following statements is incorrect regarding frog?

(a) Renal portal vein in frog carries blood from lower part of the kidney
(b) Arterial system of frog begins with conus arteriosus or truncus
(c) External ears are absent and only tympanum can be seen
(d) Four pairs of lymph heart are found in frog

87 In frog, each kidney composed of several functional units called

(a) Uriniferous tubules
(b) Bidder's canal
(c) Malpighian tubules
(d) Mesorchium

88 Ureters act as urinogenital ducts in

(a) Both male and female frog (b) male frog
(c) human males (d) human females

89 Mesorchium in frog refers to

(a) fold of peritoneum between a kidney and testis
(b) internal tissue of testes
(c) capsules of testes
(d) None of the above

90 Consider the following statements.

I. Kidney of tadpole is pronephros.

II. Excretory product of frog is urea.

Choose the correct option.

(a) Statement I is true, but II is false
(b) Statement I is false, but II is true
(c) Both statements I and II are true
(d) Both statements I and II are false

91 How many pairs of cranial nerves are present in frog?

(a) 10 (b) 9
(c) 8 (d) 12

92 Which of the following is correct for frogs?

(a) Has ureter to carry both urine and sperm
(b) Bidder's canal is present inside the kidneys which carry sperms
(c) Both (a) and (b)
(d) These possess monocondylic skull

93 Match the following columns.

Column I	Column II
A. Forebrain	1. A pair of optic lobes
B. Midbrain	2. Cerebellum and medulla oblongata
C. Hindbrain	3. Olfactory lobe, paired cerebral hemispheres, unpaired diencephalon

Codes

	A	B	C		A	B	C
(a)	1	3	2	(b)	2	1	3
(c)	3	1	2	(d)	1	2	3

94 The nervous system of frog controlling the involuntary activities of heart and lungs is

(a) CNS (b) ANS
(c) PNS (d) All of these

95 Choose the correct statement.

(a) The stomach of frog contains calciferous glands
(b) Frogs exhibit sexual dimorphism
(c) On the either side of eye a membranous tympanum is present which recieves the sound signals
(d) Both (b) and (c)

96 Observe the following figure indicating the male reproductive system of frog. Identify *A*, *B*, *C* and *D*.

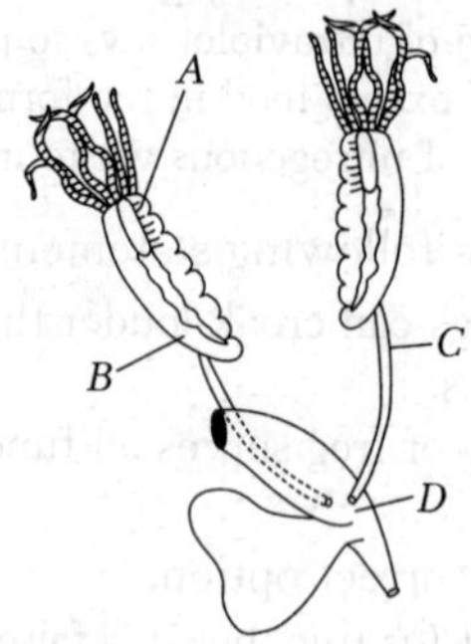

	A	*B*	*C*	*D*
(a)	Vasa efferentia	Thyroid gland	Urinogenital duct	Cloaca
(b)	Vasa efferentia	Adrenal gland	Urinogenital duct	Cloaca
(c)	Bidder's canal	Adrenal gland	Urinogenital duct	Cloaca
(d)	Bidder's canal	Adrenal gland	Urinogenital duct	Rectum

97 Testes in frogs

(a) are attached to the lower part of kidney by mesorchium
(b) are compact mass of highly coiled seminiferous tubules
(c) Both (a) and (b)
(d) are triangular-shaped

98 In frogs, jelly around the eggs is deposited in
(a) oviduct
(b) ovary
(c) the external medium having water
(d) Either of the above three

99 The development of frog is
(a) direct (b) indirect
(c) parthenogenetic (d) None of these

100 What is the sexual embrance of the frog called?
(a) Copulation
(b) Coitus
(c) Conjugation
(d) Amplexus

101 A female frog can lay how many ova at a time?
(a) 1000-1200 (b) 500-600
(c) 2500-3000 (d) 4000-6000

(B) Cumulative Exercise

(Assorted Questions of the Chapter for Advanced Level Practice)

1 Choose the correct statement regarding frog from the following.
(a) Frogs do not have a lymphatic system
(b) Frogs are ammonotelic animals
(c) Hindlimbs of frog ends in five digits and forelimbs ends in four digits
(d) Female frog contains sound producing vocal sacs which are absent in male frog

2 The given figure is related to diagrammatic representation of internal organs of frog. Identify *A* to *E*.

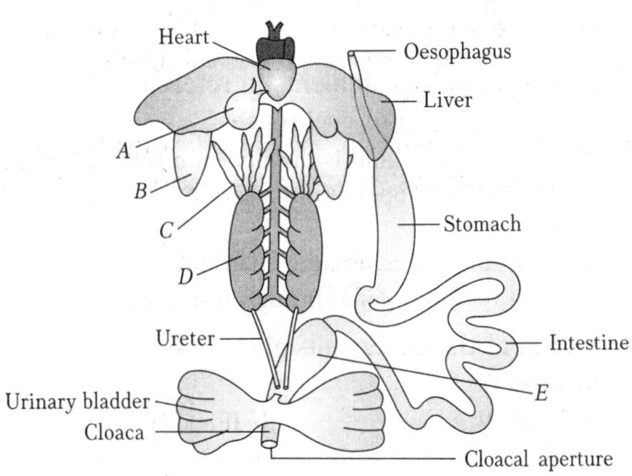

	A	*B*	*C*	*D*	*E*
(a)	Gall bladder	Lung	Fat bodies	Kidney	Rectum
(b)	Gall bladder	Lung	Testes	Kidney	Rectum
(c)	Gall bladder	Lung	Fat bodies	Testes	Rectum
(d)	Gall bladder	Lung	Ovary	Testes	Rectum

3 Which of the following correctly describes the location of some body parts in the earthworm (*Pheretima*)?
(a) Two pairs of accessory glands in 16th-18th segments
(b) Four pairs of spermathecae in 4th-7th segments
(c) One pair of ovaries attached at intersegmental septum of 14th and 15th segments
(d) Two pairs of testes in 10th and 11th segments

4 Match the following columns.

Column I	Column II
A. Hermaphrodite	1. Produces blood cells and haemoglobin
B. Direct development	2. Testis and ovary in the same animal
C. Chemoreceptor	3. Larval form absent
D. Blood gland in earthworm	4. Sense of chemical substances

Codes **[NCERT Exemplar]**

	A	B	C	D		A	B	C	D
(a)	2	3	4	1	(b)	3	2	4	1
(c)	1	3	2	4	(d)	2	4	3	1

5 Consider the following statements related to *Pheretima* and select the correct option stating, which ones are True (T) or False (F)?
I. It exhibits closed type of blood vascular system.
II. It lacks specialised breathing device.
III. Typhosole increases the effective area of absorption in intestine.
IV. There are two pair of testes present in 10th and 11th segments.

Codes

	I	II	III	IV		I	II	IIII	IV
(a)	T	F	T	F	(b)	F	F	T	T
(c)	T	T	T	F	(d)	T	T	T	T

6 Which of the following match are correct in earthworm?
I. Pharynx – 1st segment
II. Stomach – 10th to 14th segments
III. Gizzard – 26th segment
IV. Mouth – 1st segment

Choose the correct option.
(a) I and II (b) II and III
(c) II and IV (d) I and IV

7 Which of the following statements are not correct regarding earthworm?

I. Brown or clay coloured skin of the earthworm is due to the presence of pigment haemocyanin.
II. Chloragogen cells are yellow cells conserved with storage of reserve food, deamination of proteins, formation of urea and also excretory.
III. Setae and contraction of muscles help in digestion.
IV. Dorsal blood vessel is distributed in segments 1st-13th.

Codes

(a) I, II and III (b) I and II
(c) II and IV (d) I and III

8 Consider the following statements in accordance to the excretory system of the earthworm.

I. Nepridia are segmentally arranged coiled tubule.
II. Nephridia regulate the volume and composition of the body fluids.
III. There are three types of nephridia found in the earthworm.
IV. Pharyngeal nephridia are present as three paired tufts in the 4th, 5th and 6th segments.

Which of the above statement(s) is/are correct?

(a) Only I (b) I and IV
(c) I, II and III (d) I, II, III and IV

9 Match the following columns.

Column I		Column II
A. Phallomere	1.	Chain of developing ova
B. Gonopore	2.	Bundles of sperm
C. Spermatophore	3.	Opening of the ejaculatory duct
D. Ovarioles	4.	The external genitalia

Codes **[NCERT Exemplar]**

	A	B	C	D		A	B	C	D
(a)	3	4	2	1	(b)	4	3	2	1
(c)	4	2	3	1	(d)	2	4	3	1

10 Following is the diagrammatic representation of reproductive system of cockroach. Choose the correct option with respect to labelling of the given structure.

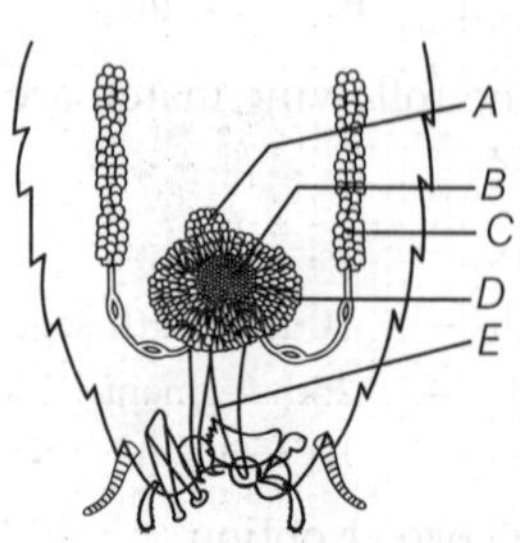

(a) *A*-Testis, *B*-Phallic gland, *D*-Vas deferens
(b) *A*-Phallic gland, *B*-Mushroom gland, *E*-Vas deferens
(c) *A*-Phallic gland, *B*-Small tubules, *E*-Ejaculatory duct
(d) *A*-Testis, *B*-Vas deferens, *D*-Ejaculatory duct

11 Find out the correct identification and descriptions of the given structure.

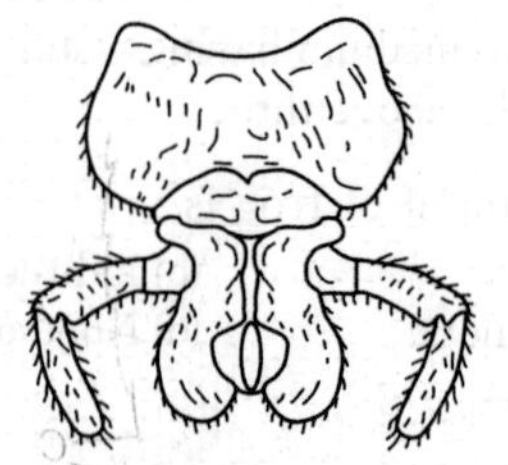

(a) Mandible–Mouth part having grinding and incising region
(b) Maxilla–Present beneath the mandible on either side of head capsule
(c) Labrum–Upper lip with sensory setae
(d) Labium– Lower lip with labial palps

12 Common salivary duct of salivary gland of cockroach opens near the base of which of the following structure?

(a)  (b) 

(c)  (d) 

13 Read the given statements in reference to the digestive system of cockroach.

I. Alimentary canal is divided into three regions.
II. Oesophagus opens into a sac-like structure called crop.
III. The hindgut is broader than midgut.
IV. The rectum opens through the anus.

Which of the statements given above is/are incorrect?

(a) I and IV (b) II and III
(c) III and IV (d) None of these

14 Consider the given statements about *Periplaneta* and select the correct option stating which ones are True (T) or False (F)?

I. Blood vascular system is of open type.
II. Malpighian tubules help in the removal of excretory products from the haemolymph.
III. They bear no eyes.
IV. Female bears mushroom glands and male bears collaterial glands.

Codes

	I	II	III	IV
(a)	T	F	T	F
(b)	T	T	F	F
(c)	F	F	T	T
(d)	F	T	T	T

15 Following is the diagram of reproductive system of female cockroach with parts labelled *A, B, C* and *D*. Which of the following is incorrect identification with its features/functions?

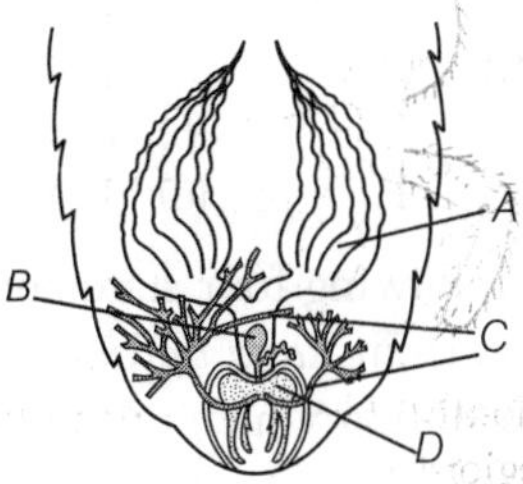

(a) *A*-Pair of ovaries lying laterally in 2nd-6th abdominal segments
(b) *B*-Pair of spermatheca in 6th segment
(c) *C*-Phallic gland which secretes ootheca
(d) *D*-Genital chamber which is a part of genital pouch in female

16 Special feature of *Pheretima is*

(a) S-shaped setae are defensive weapons against enemy
(b) it has long, dorsal tubular heart
(c) fertilisation of eggs occurs inside the body
(d) typhlosole increases effective absorptive area for digested food

17 Which one of the following statements is true for cockroach? **[NCERT Exemplar]**

(a) The number of ovarioles in each ovary are ten
(b) The larval stage is called caterpillar
(c) Anal styles are absent in females
(d) They are ureotelic

18 Choose the correct combination of labelling.

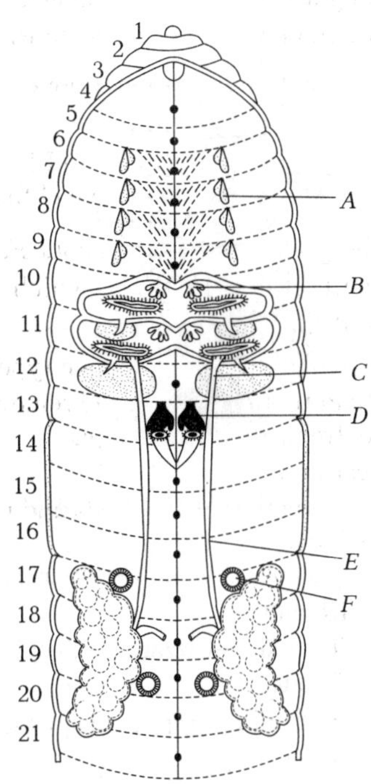

(a) *A*–Testis, *B*–Spermatheca, *C*–Seminal vesicle, *D*–Ovary, *E*–Vas deferens *F*–Accessory gland
(b) *A*–Spermatheca, *B*–Testis, *C*–Ovary, *D*–Seminal vesicle, *E*–Vas deferens, F–Accessory gland
(c) *A*–Spermatheca, *B*–Testis, *C*–Accessory gland, *D*–Ovary, *E*–Vas deferens, *F*–Seminal vesicle
(d) *A*–Spermatheca, *B*–Testis, *C*–Seminal vesicle, *D*–Ovary, *E*–Vas deferens, *F*–Accessory gland

19 Read the given statements about blood vascular system of cockroach.

I. It is of closed type.
II. It contains no blood vessels except aorta.
III. Heart is 6 chambered.
IV. The haemolymph is composed of colourless plasma and haemocytes.

Choose the option with containing incorrect statements.

(a) Only I (b) I, II and III (c) I and III (d) Only IV

20 Respiratory system of cockroach

I. consists of tracheae, tracheoles and spiracles.
II. are coordinated and regulated by nerve centres in thoracic ganglia which are stimulated by low O_2 and higher CO_2 concentration in tissue fluids.
III. contains mainly four types of cells trophocytes, mycetocytes, oenocytes and urate cells.
IV. is open or lacunar type.

Choose the option with containing correct statements.

(a) I, II and III (b) I and II (c) II and IV (d) I and III

21 Which one of the following is not a characteristic feature of frog?

(a) The skin is moist and slimy
(b) Each of the forelimbs and hindlimbs end in five digits
(c) Hepatic portal and renal portal systems are present
(d) Skin, buccal cavity and lungs are the respiratory organs

22 The given figure is related to female reproductive system of frog. Identify *A* to *D*.

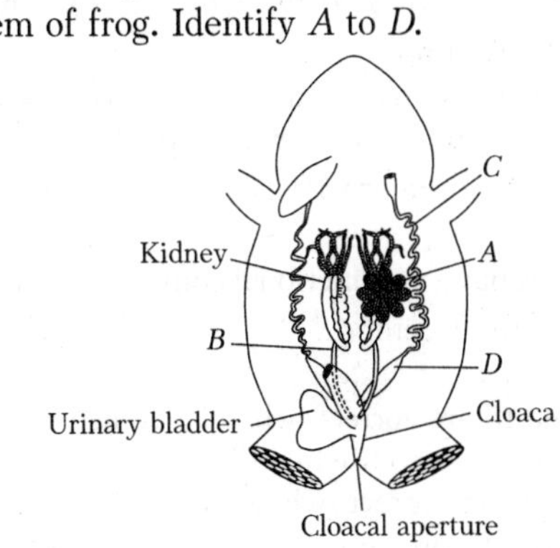

	A	*B*	*C*	*D*
(a)	Ovary	Urinogenital duct	Bidder's canal	Oviduct
(b)	Ovary	Urinogenital duct	Ovisac	Oviduct
(c)	Ovary	Urinogenital duct	Bidder's canal	Ovisac
(d)	Ovary	Ureter	Oviduct	Ovisac

23 Match the following columns.

Column I (Parts of reproductive system of earthworm)	Column II (Respective segments)
A. Testes	1. 10th-11th
B. Seminal vesicles	2. 11th-12th
C. Accessory glands	3. 17th-19th
D. Spermathecae	4. 6th-9th

Codes

	A	B	C	D		A	B	C	D
(a)	1	2	3	4	(b)	4	3	2	1
(c)	3	1	4	2	(d)	2	4	1	3

24 Which of the following statements are correct with reference to the frog?

I. Eyes are bulged and covered by nictitating membrane.
II. Membranous tympanum receives the sound signals.
III. The frog never drinks water.
IV. A pair of nostrils is present above the mouth.

Codes

(a) I and II (b) III and IV
(c) I and IV (d) I, II, III and IV

25 Which of the following statement is correct about the respiration in frog?

I. In frog, cutaneous and pulmonary respirations are found.
II. A pair of elongated pink hollow lungs are found in thorax.
III. During aestivation and hibernation, gaseous exchange takes place through skin.

Choose the correct option.

(a) I, II and III (b) I and III
(c) II and III (d) I and II

26 Read the statements with regard to frog and choose the correct option.

I. The medulla oblongata passes out through foramen of Monro and continues into spinal cord.
II. Vasa efferentia are 10-12 in number that arise from testes.
III. Ovaries have no functional connection with kidneys.
IV. Frogs are uricotelic.

Codes

(a) I, II and III are correct, while IV is incorrect
(b) I and II are correct, while III and IV are incorrect
(c) II and III are correct, while I and IV are incorrect
(d) II, III and IV are correct, while I is incorrect

27 Consider the following statements related to *Rana tigrina* and select the correct option stating which are True (T) and which are False (F).

I. Hindlimbs are larger and muscular than forelimbs.
II. The alimentary canal of frog is short.
III. They respire on the land through skin only.
IV. They contain two-chambered heart.

Codes

	I	II	III	IV		I	II	III	IV
(a)	T	F	T	F	(b)	F	F	T	T
(c)	F	T	T	F	(d)	T	T	F	F

28 Consider the following four statements (I-IV) related to the common from *Rana tigrina* and select the correct option stating which ones are True (T) and which ones are False (F).

I. On dry land, it would die due to lack of O_2 if its mouth is forcibly kept closed for a few days.
II. It has four-chambered heart.
III. On dry land, it turns uricotelic from ureotelic.
IV. Its life history is carried out in pond water.

Codes

	I	II	III	IV
(a)	F	F	T	T
(b)	F	T	T	F
(c)	T	F	F	T
(d)	T	T	F	F

29 Given below the figure of alimentary canal of cockroach. Identify *A* to *E* and choose the correct combination of *A* to *E*.

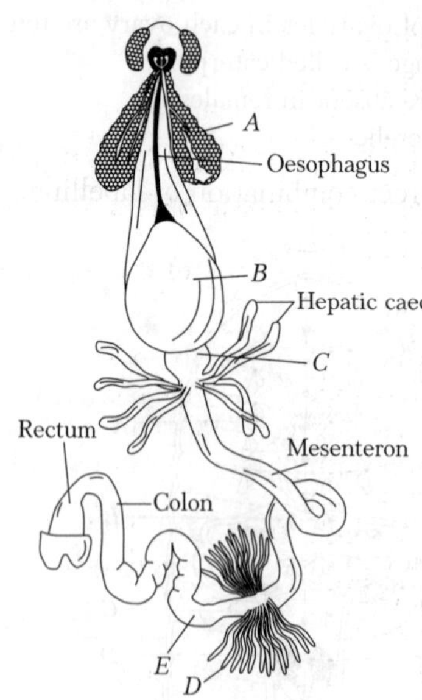

(a) *A*-Salivary gland, *B*-Gizzard, *C*-Crop, *D*-Villi, *E*-Caecum
(b) *A*-Salivary gland, *B*-Crop, *C*-Gizzard, *D*-Malpighian tubules, *E*-Ileum
(c) *A*-Salivary gland, *B*-Gizzard, *C*-Malpighian tubules, *D*-Cilia, *E*-Ileum
(d) *A*-Salivary gland, *B*-Crop, *C*-Malpighian tubules, *D*-Gizzard, *E*-Ileum

30 Match the following columns with respect to cockroach.

	Column I		Column II
A.	Anal cerci	1.	4th and 6th segments
B.	Tegmina	2.	10th segment
C.	Testes	3.	Forewings
D.	Ommatidia	4.	Sclerites
E.	Exoskeleton	5.	Visual unit

Codes

	A	B	C	D	E		A	B	C	D	E
(a)	2	3	1	5	4	(b)	4	3	2	5	1
(c)	3	4	5	2	1	(d)	5	4	3	2	1

31 Match the following columns with respect to frog.

	Column I	Column II
A.	Brain box	1. Unpaired diencephalon
B.	Midbrain	2. Cerebellum
C.	Hindbrain	3. Optic lobes
D.	Forebrain	4. Cranium
E.	Peripheral nervous system	5. Cranial nerves

Codes

	A	B	C	D	E		A	B	C	D	E
(a)	1	2	3	4	5	(b)	3	2	1	4	5
(c)	4	3	2	1	5	(d)	5	4	3	2	1

Assertion and Reason

Direction (Q. Nos. 32-35) *In each of the following questions, a statement of Assertion (A) is given followed by the corresponding statement of Reason (R). Of the statements, mark the correct answer as*

(a) Both A and R are true and R is the correct explanation of A
(b) Both A and R are true, but R is not the correct explanation of A
(c) A is true, but R is false
(d) A is false, but R is true
(e) Both A and R are false

32 **Assertion** (A) Chloragogen cells behave like vertebrate liver.

Reason (R) They are present in earthworm.

33 **Assertion** (A) In earthworm, setae are formed by a horny nitrogenous organic substances.

Reason (R) Setae are present in peristomium, pygidium and clitellum only.

34 **Assertion** (A) Body cavity of earthworm has coelomic fluid within it.

Reason (R) Coelom has no such significance.

35 **Assertion** (A) Earthworms possess both male and female reproductive organs.

Reason (R) Self-fertilisation is absent in earthworm.

(C) Medical Entrances' Gallery

(Collection of questions Asked in NEET & Various Medical Entrance Exams)

1 In cockroach, identify the parts of the foregut in correct sequence. **[NEET 2020]**

(a) Mouth → Oesophagus → Pharynx → Crop → Gizzard
(b) Mouth → Crop → Pharynx → Oesophagus → Gizzard
(c) Mouth → Gizzard → Crop → Pharynx → Oesophagus
(d) Mouth → Pharynx → Oesophagus → Crop → Gizzard

2 Match the following columns with reference to cockroach and select the correct option from the codes given below.

	Column I		Column II
A.	Grinding of the food particles	1.	Hepatic caeca
B.	Secrete gastric juice	2.	10th segment
C.	10 pairs	3.	Proventriculus
D.	Anal cerci	4.	Spiracles
		5.	Alary muscles

[NEET 2020]

Codes

	A	B	C	D		A	B	C	D
(a)	3	1	4	2	(b	4	3	5	2
(c)	1	4	3	2	(d)	2	3	1	4

3 If the head of cockroach is removed, it may live for few days because **[NEET 2020]**

(a) the cockroach does not have nervous system
(b) the head holds a small proportion of a nervous system, while the rest is situated along the ventral part of its body
(c) the head holds a 1/3rd of a nervous system, while the rest is situated along the dorsal part of its body
(d) the supraoesophageal ganglia of the cockroach are situated in ventral part of abdomen

4 Characteristic of female cockroach **[AIIMS 2019]**

(a) presence of anal style
(b) each ovary is made up of six ovarioles
(c) one pair spermatheca present and opens in genital chamber
(d) genital pouch is made up of 9th, 10th tergum and 9th sternum

5 Select the correct sequence of organs in the alimentary canal of cockroach starting from mouth **[NEET 2019]**

(a) Pharynx → Oesophagus → Gizzard → Crop → Ileum → Colon → Rectum
(b) Pharynx → Oesophagus → Gizzard → Ileum → Crop → Colon → Rectum
(c) Pharynx → Oesophagus → Ileum → Crop → Gizzard → Colon → Rectum
(d) Pharynx → Oesophagus → Crop → Gizzard → Ileum → Colon → Rectum

6 Which of the following statement is incorrect? **[NEET 2019]**

(a) Cockroaches exhibit mosaic vision with less sensitivity and more resolution
(b) A mushroom-shaped gland is present in the 6-7th abdominal segments of male cockroach
(c) A pair of spermatheca is present in the 6th segment of female cockroach
(d) Female cockroach possesses sixteen ovarioles in the ovaries

7 Select the correct route for the passage of sperms in male frog. **[NEET 2017]**

(a) Testes → Bidder's canal → Kidney → Vasa efferentia → Urinogenital duct → Cloaca
(b) Testes → Vasa efferentia → Kidney → Seminal vesicle → Urinogenital duct → Cloaca
(c) Testes → Vasa efferentia → Bidder's canal → Ureter → Cloaca
(d) Testes → Vasa efferentia → Kidney → Bidder's canal → Urinogenital duct → Cloaca

8 Frog's heart when taken out of the body continues to beat for some time.

Select the best option from the following statements.

I. Frog is a poikilotherm.
II. Frog does not have any coronary circulation.
III. Heart is 'myogenic' in nature.
IV. Heart is autoexcitable. **[NEET 2017]**

(a) Only III
(b) Only IV
(c) I and II
(d) III and IV

9 Role of calciferous glands in the stomach of earthworm is to **[Kerala CEE 2017]**

(a) neutralise the humic acid in the humus
(b) produce HCl
(c) emulsify fats
(d) increase the absorptive area
(e) secrete slime

10 Match the following columns regarding the morphology of cockroach and choose the correct option.

	Column I		Column II
A.	Tergites	1.	Forewings
B.	Tegmina	2.	Upper lip
C.	Hypopharynx	3.	Dorsal sclerite
D.	Labium	4.	Tongue
E.	Labrum	5.	Lower lip

Codes **[Kerala CEE 2017]**

	A	B	C	D	E
(a)	4	5	3	2	1
(b)	2	3	5	4	1
(c)	5	1	3	4	2
(d)	3	1	4	5	2
(e)	3	2	5	1	4

11 In male cockroaches, sperms are stored in which part of the reproductive system? **[Kerala CEE 2016]**

(a) Seminal vesicles (b) Mushroom glands
(c) Testes (d) Vas deferens

12 Which of the following features is not present in *Periplaneta americana*? **[NEET-I 2016]**

(a) Exoskeleton composed of N-acetylglucosamine
(b) Metamerically segmented body
(c) Schizocoelom as body cavity
(d) Indeterminate and radial cleavage during embryonic development

13 Which one of the following statements is incorrect pertaining to cockroach? **[EAMCET 2015]**

(a) Ostia have valves which allow the blood to pass only into the heart
(b) Thoracic and abdominal spiracles are valvular
(c) The wall of the trachea is made with lamina propria endothelium and taenidia
(d) The taenidia keep the trachea always open and prevent it from collapsing

14 Which among the following are the fat secreting cells present in the haemocoel of cockroach? **[EAMCET 2015]**

(a) Trophocytes (b) Urate cells
(c) Mycetocytes (d) Oenocytes

15 The terga, sterna and pleura of cockroach body are joined by **[CBSE AIPMT 2015]**

(a) cementing glue (b) muscular tissue
(c) arthrodial membrane (d) cartilage

16 Type of respiration in cockroach is **[UK PMT 2015]**

(a) cutaneous (b) tracheal
(c) pulmonary (d) brahcheal

17 In the circulatory system of *Pheretima*, *A*, *B*, *C* and *D* represent.

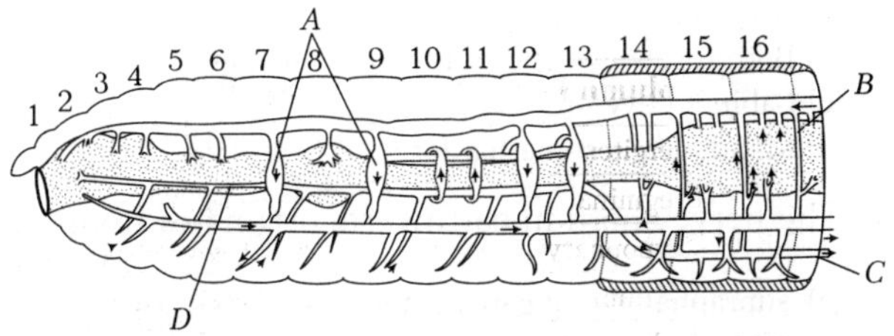

[Kerala CEE 2015]

(a) *A*-Lateral hearts, *B*-Subneural vessel, *C*-Commissural vessel, *D*-Lateral oesophageal vessel
(b) *A*-Lateral hearts, *B*-Lateral oesophageal vessel, *C*-Subneural vessel, *D*-Commissural vessel
(c) *A*-Lateral hearts, *B*-Commissural vessel, *C*-Subneural vessel, *D*-Lateral oesophageal vessel
(d) *A*-Commissural vessel, *B*-Lateral hearts, *C*-Lateral oesophageal vessel, *D*-Subneural vessel
(e) *A*-Commissural vessels, *B*-Lateral vessel, *C*-Subneural vessel, *D*-Lateral oesophageal vessel

18 The structure in earthworm which serves as a wedge to force open cracks in the soil is [Kerala CEE 2015]

(a) peristomium (b) setae
(c) clitellum (d) typhlosole
(e) prostomium

19 Which among these is not involved in excretion in cockroaches? [Kerala CEE 2015]

(a) Malpighian tubules (b) Nephrocytes
(c) Uricose glands (d) Maxillary palps
(e) Fat body

20 Frogs [Kerala CEE 2015]

(a) are uricotelic
(b) have olfactory lobes in midbrain
(c) do not have renal portal system
(d) have lymphatic system
(e) have gall bladder which secretes bile

21 In cockroach, the common duct of salivary reservior opens at the base of [MHT CET 2014]

(a) pharynx (b) maxilla
(c) mandible (d) hypopharynx

22 The total number of podomeres in the leg of cockroach are [MHT CET 2014]

(a) 5 (b) 6
(c) 7 (d) 8

23 Which is not a sensory structure in cockroach? [Kerala CEE 2014]

(a) Antennae (b) Eyes
(c) Anal cerci (d) Maxillary palps
(e) Proventriculus

24 Choose the incorrectly matched pair regarding the position of reproductive structure in earthworm.

(a) Testes - 10th-11th segments [Kerala CEE 2014]
(b) Spermathecae - 6th-9th segments
(c) Male genital pore - 9th segment
(d) Female genital pore - 14th segment
(e) Ovaries - Intersegmental septum of 12th-13th segments

25 In cockroach, the arthrodial membrane [Kerala CEE 2014]

(a) forms the hindwings (b) covers the compound eyes
(c) forms the hypopharynx (d) forms the tegmina
(e) joins the sclerites

26 Choose the incorrect statement regarding the circulatory system of frog. [Kerala CEE 2014]

(a) Sinus venosus receives blood through major veins called vena cava
(b) The ventricle opens into a sac-like conus arteriosus
(c) The erythrocytes are nucleated
(d) Special venous connection between liver and intestine called renal portal system is present
(e) Lymphatic system consists of lymph, lymph channels and lymph nodes

27 In *Periplaneta*, which one helps in nourishing the sperms? [EAMCET 2014]

(a) Ejaculatory duct (b) Vas deferens
(c) Utriculi majores (d) Utriculi breviores

28 In *Periplaneta*, ductus ejaculatorius of male reproductive system lies in [EAMCET 2014]

(a) 7th segment (b) 8th segment
(c) 5th segment (d) 6th segment

29 Consider the following statements.

I. Secretion of colleterial gland forms the egg case in cockroach.
II. The development of cockroach is hemimetabolous.

Choose the correct option. [Karnataka CET 2014]

(a) Both the statements I and II are correct and II is the reason for I
(b) Both the statements I and II are correct, but II is not the reason for I
(c) Statement I is correct, but statement II is incorrect
(d) Statement II is correct, but statement I is incorrect

30 Flow of haemolymph in cockroach is [AMU 2013]

(a) Heart → Ostia → Perivisceral sinus → Pericardial sinus → Head → Heart
(b) Heart → Pericardial sinus → Head → Perivisceral sinus → Ostia → Heart
(c) Heart → Head → Perivisceral sinus → Pericardial sinus → Ostia → Heart
(d) Heart → Head → Perivisceral sinus → Ostia → Pericardial sinus → Heart

31 Which part of the cockroach leg is attached to thorax ventrally?
(a) Trochanter (b) Claw (c) Femur (d) Coxa

32 In the male reproductive system of cockroach, *A*, *B*, *C*, *D* and *E* represent.

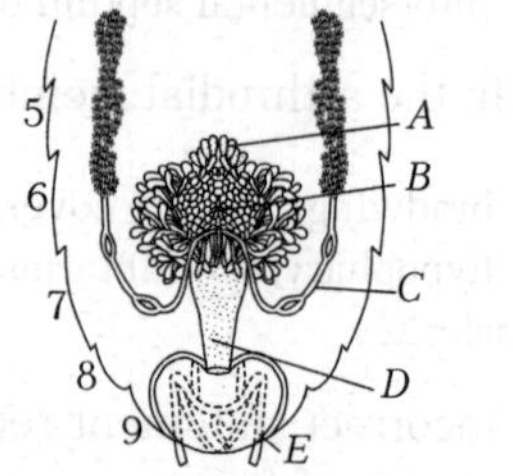

[Kerala CEE 2013]

(a) *A*-Phallic gland, *B*-Small tubules, *C*-Vas deferens, *D*-Ejaculatory duct. *E*-Right phallomere
(b) *A*-Phallic gland, *B*-Vas deferens, *C*-Small tubules, *D*-Ejaculatory duct, *E*-Right phallomere
(c) A Phallic gland, *B*-Ejaculatory duct, *C*-Vas deferens, *D*-Right phallomere, *E*-Small tubules
(d) *A*-Small tubules, *B*-Phallic gland, *C*-Vas deferens, *D*-Ejaculatory duct, *E*-Testes
(e) A-Caudal style, *B*-Phallic gland, *C*- Small tubules, *D*-Vas deferens, *E*-Ejaculatory duct

33 What external changes are visible after the last moult of a cockroach nymph? **[NEET 2013]**
(a) Anal cerci develop
(b) Both forewings and hindwings develop
(c) Labium develops
(d) Mandibles become harder

34 The nerve chord in earthworm originates from **[AMU 2012]**
(a) suprapharyngeal ganglia and has a fused pair of ganglia in each segment from 3rd to the last
(b) suprapharyngeal ganglia and has a fused pair of ganglia in each segment from 4th to the last
(c) suprapharyngeal ganglia and has a fused pair of ganglia in each segment from 5th to the last
(d) suprapharyngeal ganglia and has a fused pair of ganglia in each segment from 6th to the last

35 In the mouth parts of cockroach, the galea and lacinia form parts of the **[AMU 2012]**
(a) mandibles (b) maxilla (c) labium (d) labrum

36 The tympanic cavity in frog communicates with pharynx ventrally through **[AMU 2011]**
(a) Bidder's canal (b) horizontal canal
(c) Eustachian tube (d) semicircular canal

ANSWERS

Check Point 14.1

1. (b)	2. (b)	3. (b)	4. (d)	5. (a)	6. (b)

Check Point 14.2

1. (d)	2. (c)	3. (a)	4. (b)	5. (c)	6. (b)

Check Point 14.3

1. (a)	2. (c)	3. (a)	4. (c)	5. (b)	6. (b)

(A) Topical Exercise

1. (b)	2. (b)	3. (c)	4. (b)	5. (c)	6. (c)	7. (d)	8. (c)	9. (a)	10. (b)
11. (c)	12. (c)	13. (c)	14. (c)	15. (a)	16. (a)	17. (d)	18. (c)	19. (b)	20. (b)
21. (c)	22. (c)	23. (b)	24. (d)	25. (d)	26. (b)	27. (a)	28. (c)	29. (c)	30. (a)
31. (c)	32. (c)	33. (b)	34. (b)	35. (c)	36. (c)	37. (c)	38. (b)	39. (d)	40. (c)
41. (b)	42. (c)	43. (c)	44. (c)	45. (c)	46. (d)	47. (b)	48. (c)	49. (d)	50. (d)
51. (c)	52. (c)	53. (c)	54. (b)	55. (c)	56. (c)	57. (a)	58. (d)	59. (c)	60. (b)
61. (c)	62. (a)	63. (b)	64. (a)	65. (c)	66. (d)	67. (d)	68. (d)	69. (a)	70. (b)
71. (c)	72. (a)	73. (c)	74. (c)	75. (a)	76. (c)	77. (b)	78. (d)	79. (c)	80. (b)
81. (d)	82. (d)	83. (a)	84. (c)	85. (d)	86. (a)	87. (a)	88. (b)	89. (a)	90. (c)
91. (a)	92. (b)	93. (c)	94. (b)	95. (d)	96. (b)	97. (b)	98. (a)	99. (b)	100. (d)
101. (c)									

(B) Cumulative Exercise

1. (c)	2. (a)	3. (d)	4. (a)	5. (d)	6. (c)	7. (d)	8. (d)	9. (b)	10. (c)
11. (d)	12. (d)	13. (d)	14. (b)	15. (c)	16. (d)	17. (c)	18. (d)	19. (c)	20. (b)
21. (b)	22. (d)	23. (a)	24. (d)	25. (a)	26. (c)	27. (d)	28. (c)	29. (b)	30. (a)
31. (c)	32. (b)	33. (c)	34. (c)	35. (b)					

(C) Medical Entrances' Gallery

1. (d)	2. (a)	3. (b)	4. (c)	5. (d)	6. (a)	7. (d)	8. (d)	9. (a)	10. (d)
11. (a)	12. (d)	13. (b)	14. (c)	15. (c)	16. (b)	17. (c)	18. (e)	19. (d)	20. (d)
21. (d)	22. (a)	23. (e)	24. (c)	25. (e)	26. (d)	27. (d)	28. (b)	29. (b)	30. (c)
31. (c)	32. (a)	33. (b)	34. (c)	35. (b)	36. (c)				

Explanations

(A) Topical Exercise

30 (*a*) Statement I is true, but II is false. Incorrect statement can be corrected as

Development in earthworm is direct, i.e. without any larval stage.

31 (*c*) Option (c) represents an incorrect statement. It can be corrected as

Earthworm has closed type of circulatory system.

34 (*b*) Earthworm has two pairs of accessory glands are present in 17th to 19th segments. Rest three options are incorrect and can be corrected as

- Two pairs of testis present in the 10th and 11th segments.
- Four pairs of spermathecae are present in 6th, 7th, 8th and 9th segments.
- A pair of ovaries attached to the septum present between 12th and 13th segments.

38 (*b*) Statement I is false, but II is true. Statement I can be corrected as

The eyes of cockroach consist of 2000 octagonal ommatida.

60 (*b*) Option (b) is the odd one out because

Spermatheca are the part of female reproductive structure in case of cockroach.

62 (*a*) Option (a) is correct. Rest options are incorrect and can be corrected as

- Development from egg to adult is direct presence of larval stage. It involves egg, larva, pupa and adult.
- Spiracles present in cockroach are 10 in pairs.
- Crop serves for food storage.

63 (*b*) Statement in option (b) is incorrect and can be corrected as

The heart of cockroach in neurogenic.

64 (*a*) The given option (a) is incorrect and can be corrected as Crop serves as food storage.

Rest options are correct.

65 (*c*) Option (c) is incorrect and can be corrected as

Bifurcated nerve cord is present in the 7th segment.

(B) Cumulative Exercise

1 (*c*) Statement in option (c) is correct while other options are incorrect and can be corrected as

- Frogs do have a lymphatic system, they are ureotelic as they mainly live on land while the tadpoles are ammonotelic as they live in water and excrete ammonia as nitrogenous waste.
- Male frog contains sound producing vocal cord.

3 (*d*) Statement is option (d) is correct while other statements are incorrect and can be corrected as

- Two pairs of accessory glands are present each in 17th and 18th segments.
- Four pairs of spermatheae are located in 6th - 9th segments (one pair in each segment).
- One pair of ovaries is attached at the intersegmental septum of the 12th and 13th segment.

6 (*c*) Statements II and IV are correctly matched while others are incorrectly matched. They can be corrected as

- Pharynx - 4th segment
- Gizzard - 8th segment

7 (*d*) Statements I and III are incorrect and can be corrected as

- The brown or clay coloured skin of the earthworm is due to the presence of pigment porphyrin.
- Setae and contraction of muscles help them in locomotion.

14 (*b*) Statements I and II are true, while III and IV are false and can be corrected as

Periplaneta have a pair of compound eyes and a pair of simple eyes.

The male bears mushroom gland which is whitish in colour and situated at the junction of vasa deferentia with ejaculatory duct while colleterial gland is found in female cockroaches, which produce secretion that cover two parallel rows of eggs during oviposition and in oviparous species.

15 (*c*) Option (c) is incorrect and can be corected as C-colleterial gland which produces secretion that covers two parallel rows of eggs during oviposition.

17 (*c*) Statement in option (c) is correct, while other statements are incorrect and can be corrected as

- The number of each ovarioles in each ovary are eight.
- Larval stage is absent in cockroach instead nymphal stage is present and the nymph look like adult. They are uricotelic.

19 (*c*) Statements II and IV are correct, while statements I and III are incorrect. Incorrect statements can be corrected as

- The blood circulatory system is open type in cockroach.
- They have a three-chambered tubular heart. Oxygenated blood enters each chamber-through a pair of a slit-like opening called ostia.

20 (*b*) Statements I and II are correct, while III and IV are incorrect and can be corrected as

- The circular system of cockroach is open type and called as lacunar criculatory system.
- Respiratory system is a network of trachea hence known as the tracheal system.
- Trophocytes, mycetocytes, oenocytes and urate cells are the fat bodies of cockroach.

26 (*c*) Statements II and III are correct, while statements I and IV are incorrect and can be corrected as

- Medulla oblongata continues into spinal cord through wide aperture at the posterior end of skull called foramen magnum.
- Frog secretes urea and hence are ureotelic.

27 (*d*) Statements I and II are correct, while statements III and IV are incorrect and can be corrected as

- On land the buccal cavity, skin and lungs act as respiratory organ.
- They have three-chambered heart.

32 (*b*) Both A and R are true, but R is not the correct explanation of A.

Chloragogen cells are present in the coelomic fluid of earthworm. These cells are small, star-shaped yellow cells concerned with storage of reserve food, deamination of proteins, formation of urea and also excretion (analogous to the liver) of vertebrates.

33 (*c*) A is true, but R is false. R can be corrected as

In the middle of each segment of earthworm, there is a ring of tiny curved bristles called setae or chaetae. This is formed of horny nitrogenous organic substance known as chitin. Setae are 80th-120th in each segment except in peristomium, pygidium and the clitellum. These help in locomotion in association with musculature of body wall.

34 (*c*) A is true, but R is false. R can be corrected as

Coelom of earthworm is filled with an alkaline, colourless or milky coelomic fluid containing water, salts, some proteins and corpuscles. Coelomic fluid works as a hydrostatic skeleton and aids in locomotion.

35 (*b*) Both A and R are true, but R is not the correct explanation of A.

Earthworms are hermaphrodite/bisexual annelids, i.e. possessing both male and female reproductive organs in a single organism during their lifespan. Having this bisexual character, self-fertilisation does not occur in earthworm. This is mainly due to possessing protandrous condition. In this condition, testes mature earlier than ovaries which prevent self-fertilisation and lead to cross-fertilisation between the two worms.

(C) Medical Entrances' Gallery

1 (*d*) In cockroach, the parts of the foregut in correct sequence are as follows

Mouth → Pharynx → Oesophagus → Crop → Gizzard

The alimentary canal in cockroach has three regions, i.e. foregut, midgut and hindgut. The foregut comprises of the mouth, which opens into a short tubular pharynx, leading to a narrow tubular passage called crop used for storing food. The crop is followed by gizzard or proventriculus which helps in grinding the food particles.

2 (*a*) A–3, B–1, C–4, D–2

3 (*b*) Option (b) is correct because the nervous system of cockroach consists of a series of fused segmentally arranged ganglia joined by paired longitudinal connectives on the ventral side. Three ganglia, i.e. in the thorax and six in the abdomen. In this way, cockroach nervous system is spread throughout the body. The head holds a bit of a nervous system, while the rest is situated along the ventral (belly side part of its body. Therefore if the head region of a cockroach is removed it may live for few days.

4 (*c*) Option (c) is correct as

Female cockroach is characterised by the presence of one pair spermatheca, which opens in genital chamber. Other statements are incorrect and can be corrected as

- Anal styles are characteristic feature of male cockroach.
- Each ovary in female is made up of '8' ovarioles or ovarian tubules.
- Genital pouch is a large boat-shaped structure formed by 8th and 9th sternum.

5 (*d*) The correct sequence of organs in the alimentary canal of cockroach starting from mouth is

Pharynx → Oesophagus → Crop → Gizzard → Ileum → Colon → Rectum.

Both crop and gizzard are the structures of foregut. The former serves as the food reservoir, while the latter helps to masticate the food due to the presence of six chitinous teeth in it.

6 (*a*) Statement in option (a) is incorrect and can be corrected as

Cockroach vision is very sensitive, but provides less resolution. Such vision is called mosaic vision. It has compound eye and each eye contains about 2000 ommatidia.

Rest of the statements are correct.

7 (*d*) In male frogs, germinal epithelium of seminiferous tubules produce sperms, which are transferred to kidney *via* vasa efferentia. From the kidney, the sperms enter into Bidder's canal from where the sperms are carried to the transverse collecting tubules, longitudinal collecting tubule and then to urinogenital duct. The latter carries the sperms to seminal vesicle where, they are stored temporarily. From here, the sperms are carried to cloaca and then these are discharged into water.

Thus, the correct route of sperms passage in male frog is

Testes $\longrightarrow$ Vasa efferentia $\longrightarrow$ Kidney $\longrightarrow$ Bidder's canal $\longrightarrow$ Urinogenital duct $\longrightarrow$ Cloaca

8 (*d*) Statement in option (d) is correct.

When frog's heart taken out of the body, it continues to beat for some time, till the time ATP is available, because
Frog's heart is myogenic and autoexcitable.

Statements I and II are not correct because

Frogs have a closed circulatory system and though frog is a poikilotherm (cannot regulate its body temperature) it does not explain the statement.

9 (*a*) The function of the calciferous glands is primarily excretory in nature, but these glands are capable of neutralising any humic acid present in the humus ingested by the earthworm. These glands secrete calcium carbonate to maintain the pH of the ingested food.

10 (*d*) A–3, B–1, C–4, D–5, E–2

11 (*a*) In male cockroaches, sperms are stored in numerous small sacs called seminal vesicles. These are located on the ventral surface of the anterior part of the ejaculatory duct.

12 (*d*) Statement in option (d) is not present in *Periplaneta americana* and can be corrected as

In *Periplaneta americana*, cleavage is superficial. In all insects, cleavage during embryonic development is superficial.

Rest of the statements are correct.

13 (*b*) Statement in option (b) is incorrect pertaining to cockroach and can be corrected as
Respiratory system of insects (cockroach) is made of tracheae, which are cuticularised (taenidia) and non-collapsible (character common with mammals). Tracheoles are not cuticularised. They carry air (oxygen) to all cells of body. Cockroach has two pairs of thoracic and eight pairs of abdominal spiracles. These thoracic and abdominal spiracles do not have valves.
Rest statement are correct.

15 (*c*) Terga (dorsal sclerites), sterna (ventral sclerites) and pleura (lateral sclerites) are joined together by soft, colourless and flexible membrane called arthrodial membrane also known as intersegmental membrane. It is also seen at the base of sensory hairs, wings, etc.

24 (*c*) Option (c) is incorrectly matched pair and can be corrected as
Male genital pore is present on the 18th segment in an earthworm.

26 (*d*) Option (d) is incorrect and can be corrected as
Special venous connection between liver and intestine is hepatic portal system.

29 (*b*) Both statements I and II are correct, but statement II is not the reason for statement I.
Paired colleterial gland, present behind ovaries in female cockroaches, open into genital pouch separately secreting a hard substance that encases the egg.
Since, the development of cockroach lacks pupal stage, it can be called hemimetabolous.

CHAPTER 15

Cell: The Unit of Life

The cell is the fundamental unit of structure and function in all living organisms. This widely accepted definition of cell was given by **Lowey** and **Sikewitz**. All living organisms, whether plants or animals are made up of cells.

Anton van Leeuwenhoek was the first to describe a living cell. **Robert Hooke** (1665) saw the honeycomb-like structures in dead cork sections and called them cell. The invention of different microscopes and other tools and techniques proved very useful in studying various aspects of cell.

Some organisms are composed of a single cell. They are called **unicellular organisms**. The organisms which are composed of many cells are called **multicellular organisms**.

The study of cells is called **cytology**. It deals with the study of cells in terms of structure, function and its biochemistry. **Robert Hooke** is considered as the **Father of Cytology**.

Inside

CELL THEORY

This theory was formulated by two German scientists **MJ Schleiden** (Botanist) and **Theodor Schwann** (Physiologist) in the year **1838-39**.

Their original theory contained following two statements

1. All living beings are made up of cells and their products.
2. Cells are the structural and functional units of life.

However, the origin of cells was not explained in this theory. A major expansion in this theory was done by **Virchow** (in 1855). He gave the statement ***'Omnis cellula-e-cellula'***, i.e. all cells arise from pre-existing cells. This statement completed the cell theory. Thus, the cell theory contains three postulates

- The first two are related with structure and function of life.
- The third one is related with the origin of life.

The statement *'Omnis cellula-e-cellula'* is also known as **cell lineage theory**. This concept was the actual idea of **Nageli** (1846). It was experimentally proved by **Louis Pasteur** in **1862**.

Exceptions to cell theory The major exception to cell theory are viruses, as they are not made up of cells. Some bacteria, RBCs, sieve tubes and some coenocytic organisms are also considered as exception to cell theory.

CELL : SHAPE AND SIZE

Cells differ from each other in size, shape and their activities.

Shape

The shapes of different cells are mostly fixed. But, some cells constantly changes their shapes. Such cells are called **pleomorphic cells**, e.g. WBCs, *Amoeba*, etc.

Cells that have fixed shape are of following types

- **Flattened**, i.e. scale-like cells, e.g. squamous epithelium, endothelium and upper layers of epidermis, etc.
- **Cubical**, i.e. cube-shaped cells, e.g. most of the secretory cells.
- **Columnar** These cells are rectangular with more length and less width, e.g. cells of body parts in which absorption or ciliary movement takes place.
- **Spherical** cells are sphere-shaped, e.g. most of the plant cells of this type and eggs of many animals.
- **Spindle**, i.e. spindle-shaped, e.g. smooth muscle cell fibres.
- **Subglobose or Polyhedral,** i.e. cell with many corners, e.g. plant cells.
- **Elongated**, i.e. rectangular cells, e.g. parenchyma in plants.
- **Branched**, i.e. the cells with branched margins, e.g. pigment cells of skin, etc.

Size

The size of different cells varies to greater extent as follows

- *Mycoplasma* is the smallest cell which is about 0.3 μm in length.
- A bacterial cell could be 3-5 μm in size. The egg of an ostrich is the largest isolated single cell of about 6 inches.
- The longest cell in plant cell is sclerenchymatous fibre and in animal is neuron.

The factors that affect the size of a cell are

- The nucleoplasmic index, i.e. the ratio between the volume of nucleus and the cytoplasm.
- The ratio of cell surface to the cell volume.
- Metabolism.

TYPES OF CELLS

Usually all living cells are classified into two groups as follows

Prokaryotic cell

(Gr. *Pro*—primitive or pre or before; *karyon*—nucleus)

Prokaryotic cells contain a primitive type of nucleus which is not bounded by a membrane. It is also called as **nucleoid** or **genophore**.

Bacteria, methanogens, mycoplasma and blue-green algae, etc., are the examples of prokaryotes.

Eukaryotic cell

(Gr. *Eu*—true; *karyon*—nucleus)

Eukaryotic cells contain a true double layered membrane bound nucleus. Protists, algae, fungi, plants and animals have eukaryotic cells in their body.

Both the types of cells possess cytoplasm as semi-fluid matrix.

Mesokaryotic cells

Sometimes, a third type of cells, i.e. mesokaryotic cells are also found in some organisms. These cells have well-defined nucleus, but their DNA lacks histone proteins, e.g. some algae and protozoans.

STRUCTURAL ORGANISATION OF AN EUKARYOTIC CELL

The component of a typical eukaryotic cell can be broadly categorised to following three categories

1. **Outer covering**, i.e. cell wall and cell membrane.
2. **Protoplasm**, i.e. the fluid content of cell.
3. **Cell organelles**, i.e. structures that float within the protoplasm like, mitochondria, chloroplast, ribosome, etc.

Cell wall

- It is the outermost, rigid and non-living covering of the cell. It was discovered by **Robert Hooke** in **1665** in the cork cells.
- It is protective and supportive in nature.
- Cell wall is found in plant cells, fungi, bacteria, cyanobacteria and some protists.
- Cell wall is absent in animal cells, mycoplasmas, gametes and spores.

Structure

Cell wall is made up of pectin or hemicellulose.

It is a multilayered structure with following layers

(i) Middle lamella

- It is the outermost thin, sticky, amorphous layer of cell wall.
- It is a cementing layer between primary cell walls of adjacent cells. Its major constituents are **Ca** and **Mg pectates**.
- In ripening fruits, the pectate compounds solubilise to a jelly like materials thus, making the fruits soft.

(ii) Primary cell wall

- It is thin (1-3 μm) and elastic layer which is capable of growth.
- It is found in cell walls of younger plant cells and disintegrates as cell matures.
- Primary cell wall is present inner to the middle lamella.
- It is composed of approximately 2-5% cellulose and 50% hemicellulose. Glycoproteins are also present in primary cell wall.
- Meristematic and parenchymatous cells of plants have only primary cell wall.

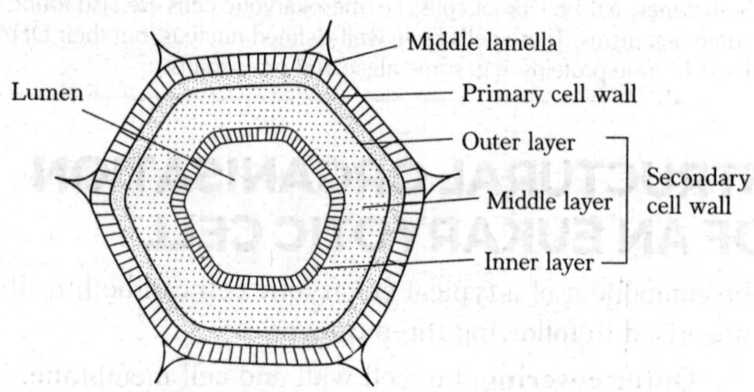

Fig. 15.1 Layers of the cell wall

(iii) Secondary cell wall

- This layer is present inner to primary cell wall and just outside the plasma membrane.
- It is 3-10 μm thick and further differentiated into outer, middle and inner layers as shown in **fig. 15.1**.
- The main constituents of this layer are cellulose (50-95%) and hemicellulose (5-25%). The other components include xylan, lignin, etc.
- Due to more cellulose component and deposition of suberin, lignin, etc., this layer is non-elastic and rigid.
- It is found in collenchyma, sclerenchyma and xylem vessels.

(iv) Tertiary cell wall

- It may present occasionally on the inner side of secondary cell wall.
- This cell wall is purely cellulosic in nature.
- Sometimes, xylan is also present in this layer, e.g. tracheids of gymnosperms.

Structures associated with cell wall

Various structures associated with cell wall are as follows

(i) Pits

- In the cell wall, there are some areas that lack secondary wall. These are unthickened regions and are called pits.
- Pits of two neighbouring cells are complementary to each other. Together these constitute a pit pair.
- A pit membrane is made up of primary wall and middle lamella.
- These pits can be of following types
  (a) **Simple pits** In these pit, pore is uniform in diameter.
  (b) **Bordered pits** In these pit, pore is more or less flask-shaped.

(ii) Plasmodesmata

- These are the cytoplasmic strands found in the cell wall of plants.
- They are the only living component of the dead cell walls.
- Plasmodesmata are made up of a fine tubules called **desmotubule**. These are considered as the extensions of SER (Smooth Endoplasmic Reticulum).
- Plasmodesmata connects the cytoplasm of one cell to its adjacent cell.
- The term plasmodesmata was coined by **Tangle** and **Strassburger.**

Formation and growth of cell wall

The formation and growth of cell wall occur either by

(i) **Intussusception** It is the deposition of cell wall material in the form of fine granules. These granules gets embedded within the existing components of cell walls. Primary cell wall is formed by this method only.

(ii) **Apposition** In this method, the new cell wall material gets deposited in the form of thin plates one over the other. Secondary cell wall is formed by this method.

Functions

- It maintains the shape of plant cell and protects them from mechanical injury.
- It acts as a protective barrier against pathogens.
- It provides the mechanical support to plant due to which they stand erect.
- It forms the protoplasmic connections. This continuation is important for the transport of substances between the cells.
- It prevents the osmotic bursting of the cell.
- It helps in cell to cell adhesion and prevents the entry of undesirable macromolecules.

Cell membrane or Plasma membrane

- Cell (or plasma) membrane is a thin, delicate and elastic living boundary around the cell. However, when this boundary was observed under electron microscope, a clear differentiation of two layers was noticed.
- The term 'cell membrane' was given by **Nageli** and **Cramer** in 1855. This term was used when cell was observed under simple microscope.
- Out of these, one layer was the real membrane (i.e. **plasma membrane** or **plasmalemma** by **Plower** in 1931). It was surrounded by a layer of cell cement.
- Thus, it was concluded that the term cell membrane is misleading. The appropriate term which can be used for this membrane is plasma membrane.

Chemical composition

The chemical composition of the plasma membrane is as follows

Components	Composition
Protein	58-59%
Lipid	40%
Carbohydrate	1-2%

The ratio of lipid and protein may vary depending upon the cell types.

The detailed structure of the plasma membrane when studied under electron microscope revealed the following compounds

(i) Lipids

- These form the continuous structural framework of the cell membrane.
- These are amphipathic molecules. They have a hydrophilic head and two hydrophobic fatty acid chains as tail.
- Phospholipids constitute the major category of membrane lipids. Some of these lipids are rich in unsaturated fatty acids. It provides fluidity to the membrane.
- Phosphatidylserine, phosphatidylcholine, *p*-ethanol amine (cephalin) are common phospholipids of plasma membrane.
- Sterols, glycolipids and sphingolipids are also found in these membranes.

(ii) Proteins

- The proteins of plasma membrane are called **membrane proteins**.
- These are of following two types
  (a) **Extrinsic** or **Peripheral membrane protein** These proteins are present mainly on the outer surface of the membrane. Thus, they are also called as **extrinsic proteins.**
    - They are easy to extract.
    - These constitute about 30% of total protein of the membrane.
    - They are responsible for cell interaction and recognition, e.g. spectrin.

  (b) **Intrinsic or Integral membrane protein** These proteins are partially or completely embedded in the plasma membrane. Thus, they are also called as **transmembrane** or **tunnel proteins** or **intrinsic proteins.**
    - They are difficult to extract.
    - These constitute about 70% of total protein of the cell.
    - They are involved in the transfer of solute or polar molecules across the lipid bilayer as they act as channel proteins, e.g. glycophorin and porins.

(iii) Carbohydrates

Oligosaccharides (sialic acid) are the main carbohydrates present in the cell membrane. These are the key components of cell recognition.

Models of plasma membrane

Following models have been proposed to explain the arrangement of constituents of plasma membrane.

I. Sandwich or Lamellar model

Danielli-Davson in **1935**, first suggested the lipid protein structure of the cell membrane. They proposed that the plasma membrane is 75Å thick and is made up of three layers. In this arrangement, a bilipid layer is sandwiched between two globular α type protein layers, each having a thickness of 20 Å. Such arrangement of lipid (L) and protein (P) appear as **PLLP**.

The complete arrangement in outline view is shown below

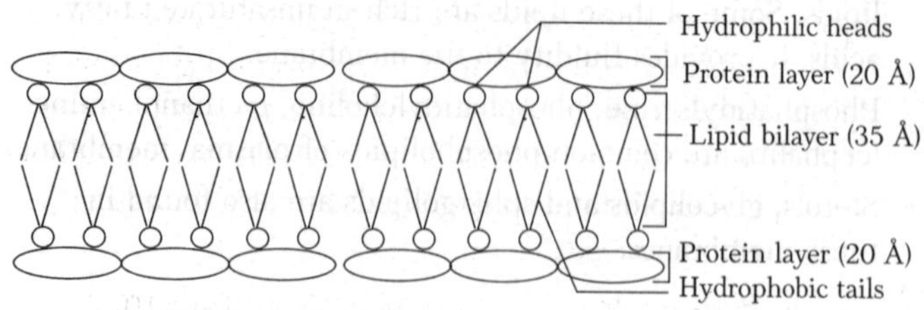

Fig. 15.2 Sandwich or Lamellar model

II. Unit membrane model

This model was proposed by **Robertson** in **1959.**

According to this model, plasma membrane has a trilaminar (three layered) structure as shown below

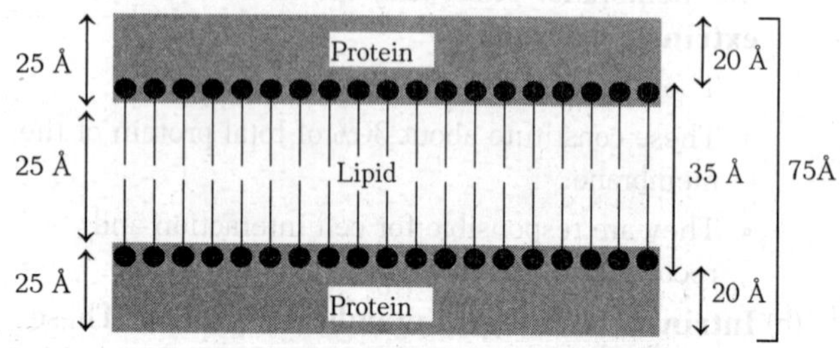

Fig. 15.3 The unit membrane model

Notice the arrangement of lipid and β type protein layers. It is a bit different from the one proposed in Danielli-Davson model. In the unit membrane model, the lipid heads are embedded in the protein layers.

III. Fluid mosaic model or Quasifluid model

- This model was proposed by **Singer** and **Nicolson** in **1972.** It is the most accepted model of plasma membrane structure.
- According to this model, the membrane consists of a continuous bilayer of phospholipid molecules. Globular proteins and sterols are embedded in lipid layer.
- The **intrinsic membrane proteins** cross the lipid layer whereas the **extrinsic proteins** are found attached on the outer surface of the lipid layer.

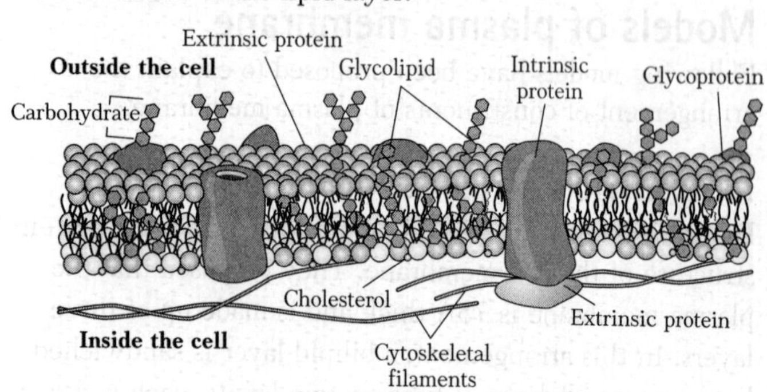

Fig. 15.4 Fluid mosaic model of plasma membrane

- This arrangement resembles as the protein icebergs are floating in a sea of phospholipids. Thus, it gives quasi (false) fluid-like appearance to the membrane.

The main features of a membrane, according to fluid mosaic model, can be summarised as follows

- It is about 7nm thick following phospholipid layer.
- The hydrophilic heads of phospholipids face outward into the aqueous environment.
- The hydrophobic tails of phospholipids face inward towards each other.
- Proteins found in the membrane have regions containing hydrophobic amino acids which interact with fatty acid tails. Rest of the protein molecule is hydrophilic, which faces into or outward, towards the aqueous environments.
- Some short branched carbohydrate (oligosaccharides) chains are attached with protein and lipid molecules.
- Cholesterol are also seen in membranes. Cholesterol disturbs the close packing of phospholipid layer and keeps them more fluid. Its bonding with phospholipids is essential to provide stability.
- In prokaryotes, cholesterol is absent and its function is performed by hopanoides pentacyclic sterols.
- The two sides of membrane differ in structure and function.

Fluidity of membrane

It is very clear from the fluid mosaic model that both the constituents of membrane show fluid properties, i.e. motion.

(i) Motion in lipid molecules

Lipid molecules show following four possible motions

(i) **Flexing**, i.e. contraction and relaxation of tail.

(ii) **Lateral diffusion**, i.e. lateral movement within the same layer.

(iii) **Flip-flop**, i.e. transfer of lipid molecules from one side of the bilayer to the other.

(iv) **Rotation**, i.e. movement of lipid molecules about their own axis.

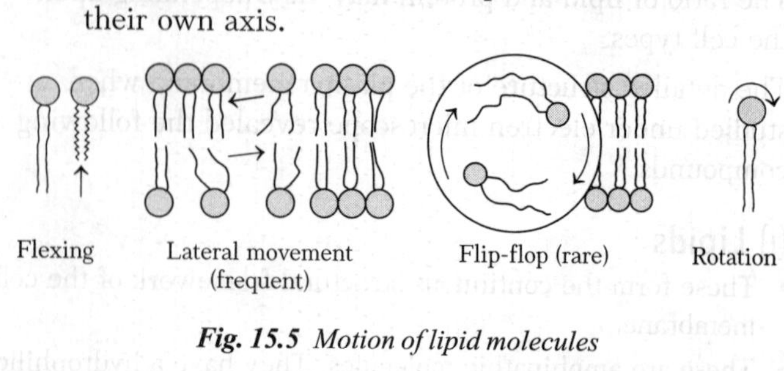

Fig. 15.5 Motion of lipid molecules

(ii) Motion in protein molecules

- The intrinsic proteins are tightly bound to lipid bilayer. Therefore, they do not dissociate from the membrane easily. Their separation requires treatments like the use of detergents.
- These proteins however, perform motion in association with their functions like transport, etc.
- Extrinsic proteins on the other hand, can easily dissociate from lipid bilayers and hence, show easy and rapid motion.

Functions

The cell membrane performs following important functions

(i) **Transport** The transport of metabolites across the plasma membrane may take place either of the following ways

(a) **Passive transport** It is the transport of molecules along the concentration gradient, i.e. from a high concentration region to a low concentration region. This transport does not require any energy.

Passive transport may occur *via*

- **Simple diffusion** It is the movement of non-polar substances from region of higher concentration to their region of lower concentration.
- **Facilitated diffusion** This is a special kind of diffusion. Movement of polar molecules across plasma membrane is facilitated by some carrier proteins. The process is carried out with the formation of a metabolites carrier complex, e.g. entry of glucose in erythrocytes.

(b) **Active transport** Molecules move against the concentration gradient, i.e. from the region of their low concentration to the region of their high concentration. This type of transport requires energy in the form of ATP. Active transport is of following two types

- **Primary active transport**, e.g. transport of Na^+/K^+ with the help of ATPase complex.
- **Secondary active transport**, e.g. glucose transport in *E. coli.*

(ii) **Transportation at cellular level** Membranes of cell organelles carry out transportation of various substances by the process of **blebbing** (formation of vesicles or blebs). This process involves the repeated fusion and detachment of vesicles. The fusion and detachment helps in the maintenance of membrane flow.

(iii) **Bulk transport** Transport of materials from outside the cell to inside and *vice versa* is carried out by bulk transport system. This may occur through following ways

(a) **Endocytosis** It is the process through which materials are transported into the cell. The term endocytosis includes two processes, namely

- **Phagocytosis**, i.e. cell eating, means intake of solid particles by the cell.
- **Pinocytosis**, i.e. cell drinking, means intake of liquid particles by the cell.

(b) **Exocytosis** It is a process through which materials are transported out of the cells.

(iv) **Cell recognition and adhesion** The oligosaccharides of glycolipids and glycoproteins like sialic acid are the sites for cell recognition. These lie at the surface of cell.

For example, macrophages recognise the surface oligosaccharides of worn out RBCs. Hence they, destroy them with their phagocytic ability.

(v) **Hormone reception** The cell membrane contains receptors that recognise specific hormones and convey the information to the interior of the cell. This in turn, causes changes in the metabolism of the cell.

(vi) **Secretion** Plasma membrane plays an important role in secretion as well. Polypeptide chains are synthesised by ribosomes and transported to cell surface from where these are secreted out *via* plasma membrane.

CHECK POINT 15.1

1. Most prokaryotic cells particularly bacterial cells have
(a) a chemically simple cell envelope
(b) a chemically complex cell envelope
(c) no cell envelope
(d) cell envelope in the form of cell wall and cell membrane only

2. An animal cell differs from plant cell in possessing
(a) vacuoles (b) centrosomes
(c) distinct cell wall (d) cytoskeleton

3. If the cell wall of a plant cell is removed, which of the following ability will be lost by the cell?
(a) Shrinking of protoplasm
(b) Exosmosis by cell
(c) Prevention of over expansion of cell
(d) Prevention of flaccidity

4. The feature which differentiate cells of human cheek with the cells of onion peel is
(a) the type of outer boundary
(b) cytoplasm which is the main arena of cellular activities
(c) absence of centriole
(d) presence of dense membrane bound structure

5. Mesokaryotic cells are different from eukaryotic cell in
(a) lacking well-defined nucleus
(b) lacking cell membrane
(c) lacking histone proteins
(d) having DNA

6. The fruit softens and ripens due to
(a) conversion of starch from sugar
(b) jelly formation of an acid pH
(c) incorporation of pectate in the middle lamella
(d) dissolution of middle lamella

Protoplasm

(*Proto*-first; *plasm*-fluid)

- The complex viscous and colourless fluid part of the cell is called protoplasm. It is present in all living cells and performs vital functions.
- **Huxley** defined the protoplasm as physical basis of life. The term 'protoplasm' was given by **Purkinje**.
- Protoplasma can be differentiated into
  (i) **Cytoplasm** It is present between nucleus and the plasma membrane of the cell.
  (ii) **Nucleoplasma** The protoplasma of nucleus is called nucleoplasm. It is denser than cytoplasm.
- Cytoplasm occupies the major part of the cell. Cell organelles such as mitochondria, endoplasmic reticulum, Golgi apparatus, ribosomes, etc., are found in the cytoplasm.
- Some non-living substances that lack biological properties and are produced by the protoplasm itself constitute the **deutoplasm**, e.g. yolk bodies, lipid droplets, secretory granules, pigments, etc.

Note • ***Protoplast*** *It is the cell without cell wall.*

Cell organelles

- An organelle is a specialised subunit, usually within a cell that has a specific function.
- The name organelle comes from the idea that these structures are parts of cells, as organs are to the body.
- Organelles are present within cell as
  (i) separately enclosed within their own lipid bilayers (also called membrane-bound organelles), e.g. mitochondria, Golgi apparatus, lysosomes, etc.
  (ii) spatially distinct functional units without a surrounding lipid bilayer (non-membrane bound organelles), e.g. ribosomes, centrioles, nucleolus, etc.
- Although most organelles are functional units within cells, some functional units that extend outside of cells are also termed as organelles, e.g. cilia, the flagellum, etc.

The detailed description of cell organelles is given below

Nucleus

- Nucleus is the largest component of a cell. It is a double membrane bound structure.
- It controls all cellular metabolism and encloses the genetic information of the cell in the form of DNA.
- **Robert Brown** (1831) studied orchid root cells and described nucleus. **Hammerling** studied *Acetabularia* (largest unicellular green alga) and proved the importance of nucleus in its heredity and growth.
- A true nucleus is absent in bacteria, blue-green algae (prokaryotes) a few dinoflagellates and mature RBCs (among eukaryotes). These possess an incipient or false nuclei.
- Some cells can have more than one nucleus. These cells can be
  (i) **Binucleate** cells with two nucleus per cell, e.g. *Paramecium*.
  (ii) **Multinucleate** cells with many nucleus, e.g. *Opalina*.

- The multinucleate condition can be further of two types
  - (i) **Coenocyte** multinucleate condition is found in plants. It occurs due to repeated karyokinesis which is not followed by cytokinesis.
  - (ii) **Syncytium** multinucleate condition is found in animals. Here also, cytokinesis does not follow the repeated karyokinesis.

Shape and position

- The shape of the nucleus may be spherical, disc-shaped, ovoid, bilobed, multilobed, kidney-shaped (pyriform), C-shaped or string of beads (moniliform), etc.
- Position of the nucleus varies from cell to cell, but it is mostly found in metabolically active regions. For example, nucleus occupies central position in embryonic and glandular cells while it is peripheral in position in fat cells.

Structure

Under electron microscope following components are seen in a nucleus.

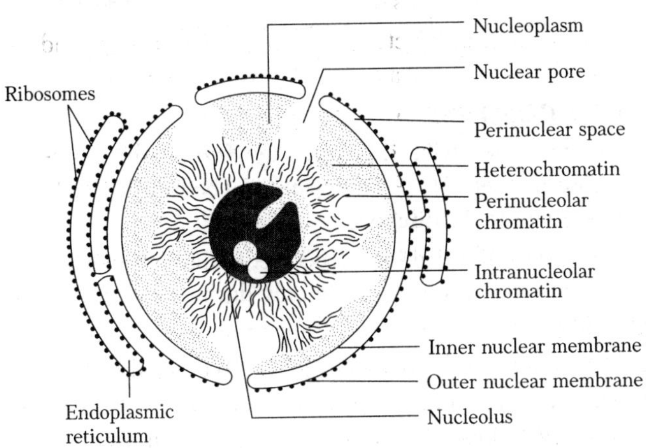

Fig. 15.6 Structure of nucleus

Nuclear membrane

- It is also called as **nuclear envelope**, nucleolemma or **karyotheca**. It is composed of two lipoprotein membranes, i.e. the outer and the inner membrane.
- These are separated by means of a perinuclear space (100-300° thick).
- The outer membrane of nucleus is continuous with ER (Endoplasmic Reticulum) which may also bear ribosomes.
- Nuclear membrane contains octagonal nuclear pores. These pores are formed by the fusion of outer and inner membranes.
- **Callan** and **Tamlin** (1950) were the first to observe nuclear pore in the nuclear membrane.
- These pores control the inward and outward movement of different materials such as *m*RNA and nucleoproteins, to and from the nucleus.
- The nuclear pores are enclosed by circular structures called as **annuli**. The pore and annuli together constitute the **nuclear pore complex** or **pore basket**.
- The inner surface of nuclear membrane is called **nuclear lamina** or **nuclear matrix**. It contains a network of intermediated filaments.
- Nuclear lamina serves as the attachment point of telomere during cell division.

Nucleoplasm

- Nuclear sap or nucleoplasm is the fluid part of a nucleus. It mainly contains DNA and associated proteins along with water, ions, enzymes, soluble molecules and nucleotides.
- It is considered as the main site for the synthesis of DNA. The DNA and associated proteins are visible either as chromatin or as chromosomes.

Nucleolus

- The nucleolus was first discovered by **Fontana**. The term nucleolus was given by **Bowman**.
- It is a spherical structure within the nucleus.
- It is the largest part of nucleus and occupies approximately 35% the nucleus. It is a non-membranous structure.
- Its constituents are RNA, DNA and proteins.
- The nucleoli are formed from specific sites on chromosomes. These sites are called **Nucleolar Organiser Regions** (NORs). A nucleolus can be the best visualised at interphase.
- Nucleolus may be plasmosome (true nucleolus) or karyosome (false nucleolus).
- According to recent studies nucleolus consists of following parts
  - (i) **Pars amorpha**, i.e. the homogenous matrix.
  - (ii) **Pars chromosome**, i.e. nucleolar chromatin that synthesises *r*RNA.
  - (iii) **Pars fibrosa or Nucleonema**, i.e. nucleolar fibrils.

(iv) **Pars granulosa**, i.e. granules which represent ribosome subunits.

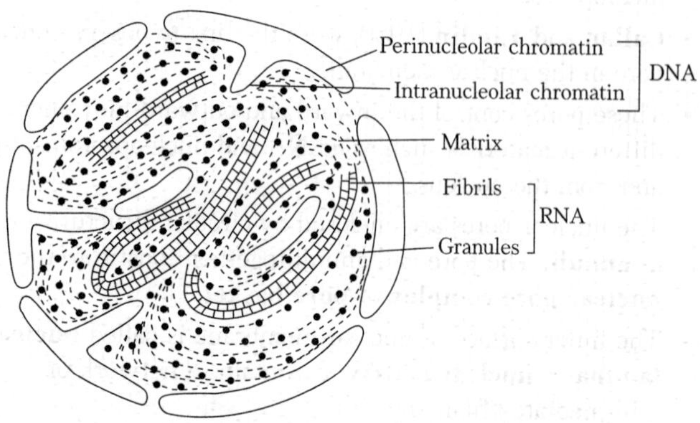

Fig. 15.7 A typical vertebrate nucleolus (Diagrammatic)

- Nucleolus is the most important site of ribosomal RNA synthesis. It is thus, called **ribosome factory of cell.** It comprises about 80% of total RNA content of the cell.
- Therefore, nucleolus is mostly found in cells that are involved in protein synthesis.
- Usually one nucleolus is found in haploid cells, whereas in diploid cells the number may vary from ore to four.
- Maximum number of nucleoli (1600) is found in the oocyte of *Xenopus*.
- It is absent in yeast, bacteria, some alga, embryonic cells, sperm cells and erythrocytes.

Chromatin

- The chromatin (meaning coloured fibres) are fine fibres composed of DNA, some proteins and transcribing RNA.
- The proteins associated with chromatin are mainly histone proteins. These are positively charged basic proteins mainly composed of lysine and arginine. However, in certain types of sperms, histones are replaced with protamines.
- During cell division, the chromatin condenses and becomes thick to form specialised structures called **chromosomes.**
- It was observed that when chromosomes are stained with basic dyes like acetocarmine or fuelgen stain, two types of regions can be observed *viz* heterochromatin region and euchromatin region.

Table 15.1 Differences between heterochromatin and euchromatin

Heterochromatin	Euchromatin
It is compactly coiled regions of the chromatin.	It is loosely coiled regions of the chromatin.
It stains deeply.	It stains less deeply.
It contains more DNA and replicates at the end of S-phase of mitotic cycle.	It contains less DNA and replicates during the early stages of S-phase.
It is less stable and gets affected by temperature, sex, age (of parents), proximity to the centromere, etc.	It is more stable.
It is genetically inert and does not transcribe. Thus, it is not involved in protein synthesis.	It is genetically active and is usually undergo active transcription. Hence it actively participates in protein synthesis.
It has low cross over frequency.	It has high cross over frequency.

Heterochromatin is further of two types

(i) **Facultative** It is condensed and inactive part of chromosome. It takes part in the formation of Barr body, Y-spot and drumstick.

(ii) **Constitutive** It is rich in repetitive DNA and found in the some region of both chromosomes of homologous pair.

Note
- *The differential coiling and staining of chromosomal parts is called **heteropycnosis.** It is usually carried out to detect heterochromatic and euchromatic regions in a chromosome.*
- *The area of chromosome near centromere, telomere and NOR region is rich in constitutive heterochromatin.*

Chromosome

- Chromosomes were discovered by **Hofmeister** in the pollen mother cell of *Tradescantia.* The term chromosome was given by **Waldeyer.**
- It is a rod or thread like structure, found in the eukaryotic nucleus and becomes visible only during the cell division.
- The length and thickness of a normal chromosome varies from 1-30 μ and 0.2-2 μ, respectively.
- The shape of chromosomes varies according to the stages of cell division.
- Chromosome shape is observed during anaphase. However, its size is measured during metaphase due to maximum coiling.
- Somatic cells always possess two sets of chromosomes. Hence, they are diploid ($2n$). On the other hand, gametic cells usually possess one set of chromosomes, i.e. haploid (n).
- The size and number of chromosomes vary from species to species and genera to genera. Usually, within a species, these remains constant.
- Among lower plants, *Ophioglossum reticulatum* has maximum number (i.e. 1260) and *Mucor* has minimum number (2) of chromosomes.
- Among higher plants, *Foa litorosa* has maximum number (266) of chromosomes and *Haplopappus gracilis* has minimum number (i.e. 2) of chromosomes.

- Among animals, maximum number (1600) of chromosomes are found in radiolarian protozoan, *Aulacantha* and *Ascaris megalocephala* has minimum number (i.e. 2) of chromosomes.

Structure of chromosome

The thickest and highly condensed chromosome is visible at metaphase stage of cell division. Following structures are visible in a chromosome during metaphase.

(i) **Pellicle** It is the outermost covering or sheath of chromosome. It is thin and proteinaceous in nature.

(ii) **Matrix** It is the colourless and non-genetic ground substance of chromosome. It has different types of enzymes, minerals, water, proteins, lipids, etc.

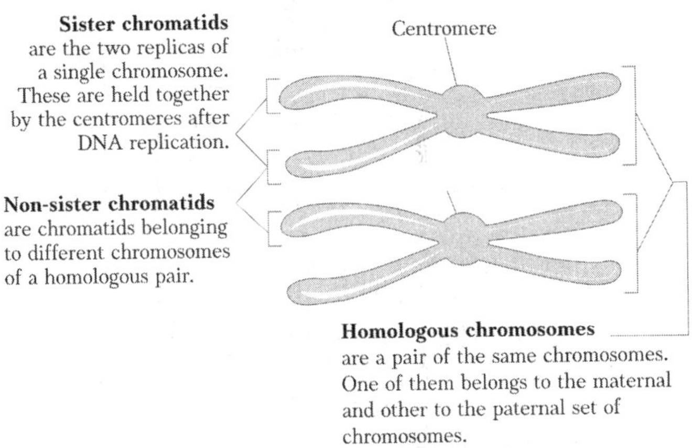

Fig. 15.8 The differences between homologous chromosomes, sister chromatids and non-sister chromatids

(iii) **Chromonema** (Sing. chromonemata) It is a highly coiled thread of chromatin. It is extended throughout the length of a chromosome. Sometimes, bead-like structures are also visible on chromonema fibres. These structures are called **chromomeres.**

(iv) **Chromatid** These are two similar thread-like structures attached to each other by a centromere.

(v) **Centromere** It is the narrow, non-chromatid part of chromosome where two chromatids are attached to each other. It is also called as **primary constriction.**

- Centromere is a non-stainable region as it is composed of constitutive heterochromatin. The DNA present at the centromere is called as **alphoid DNA.**
- On the either side of centromere, proteinaceous discs are present. These discs are called as **kinetochores.** During cell division, spindle fibres attach to kinetochores only. Centromere is functional only during cell division.

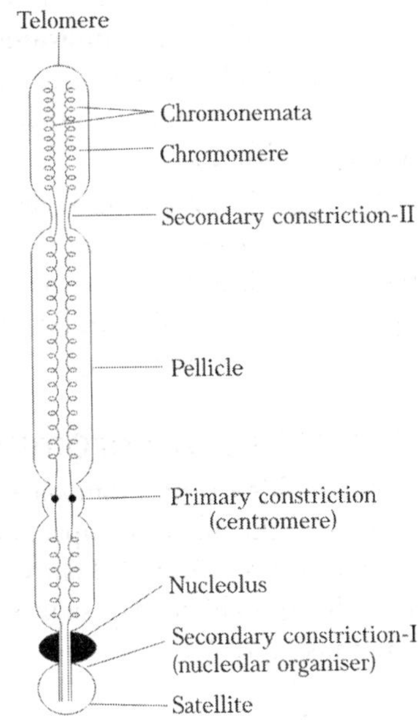

Fig. 15.9 Structure of chromosome

(vi) **Secondary constriction** These are usually present near the chromosomal ends and lack kinetochore. Hence, spindle fibres never attach to secondary constrictions during cell division. A chromosome may contain following secondary constrictions.

- **Secondary constriction-I** or **Nucleolar organiser** Normally, each diploid set of chromosome contains two chromosomes (i.e. one homologous pair) with secondary constriction-I.
- The number of secondary constriction can be more than one in a chromosome. Secondary constriction is also called **Nucleolar Organiser Rregion** (NOR) as it is necessary for nucleolus formation. In humans, these are found in chromosome number 13, 14, 15, 21, 22 and Y.
- Secondary constriction-II These secondary constrictions are different from the nucleolar organisers. Their location is constant for a particular chromosome. That is why, these are used for the identification of chromosomes.
- In humans, secondary constriction-II are found on the long arms of chromosome numbers 1, 10, 13, 16 and Y.

(vii) **Satellite or Trabant** It is a short, sphere-like structure present beyond the secondary constriction-I. The chromosomes that possess satellites are called as **SAT chromosomes**. The word SAT refers to Sine Acid Thymonucleinico. It is the region where thymine containing nucleic acid, i.e. DNA is lacking or absent.

(viii) **Telomere** The tips of chromosomes are called telomeres. These are rich in Guanine base. Telomeres prevent the ends of chromosomes from sticking together. It also facilitates attachment of chromosomes with nuclear envelope.

Molecular structure of chromosome

In eukaryotes, long size DNA is accommodated in a very small nucleus through packaging or compaction or simply coiling. This coiling of DNA within the chromosome is best explained by 'Nucleosome model' suggested by **Kornberg** and **Thomas** in 1974.

Nucleosome model

- This model is also called **'super solenoid model'**. According to this model, chromosomal threads have beaded appearance at their grass root level. These beads are produced by wrapping of DNA over specialised low molecular weight proteins called **histones**.

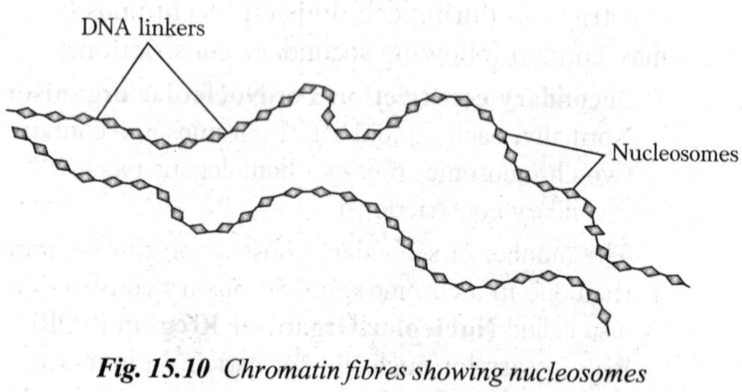

Fig. 15.10 Chromatin fibres showing nucleosomes

- There are five type of histone proteins namely, **H1, H2A, H2B, H3 and H4. H2A**, H2B, H3, H4 are found in pairs and together these eight molecules form a histone octamer or histone core. The negatively charged DNA is wrapped around this histone core. This complete structure (i.e. core + DNA) is called **nucleosome.**
- The length of wrapped DNA is found to be constant in all individuals, i.e. 140 base pairs. This takes approximately $1\frac{3}{4}$ turn around the histone core.
- The two adjacent nucleosomes are connected with the help of linker DNA. It bears H1 histone protein. That is also called as **linker histone** or **plugging protein**.
- When chromatin is in partly extended state, the diameter of beads is reduced. These beads with reduced diameter are called **'nu bodies'**.

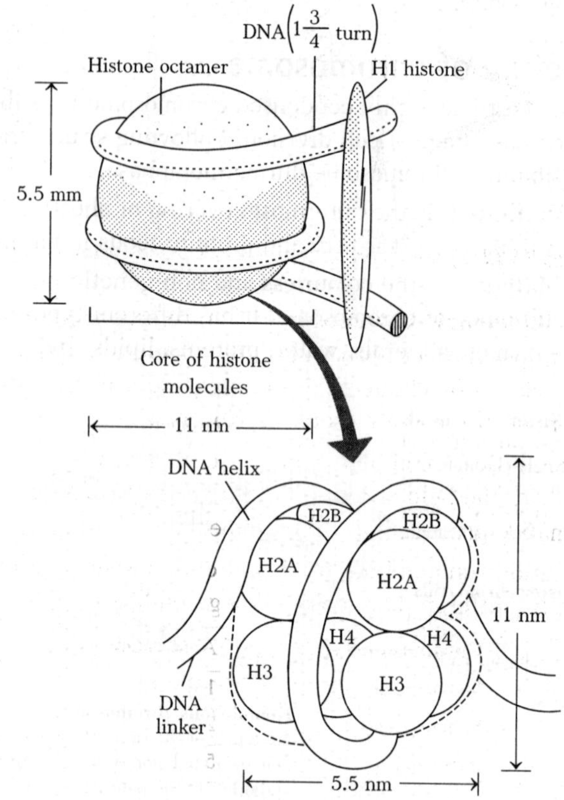

Fig. 15.11 Structure of nucleosome

Classification of chromosome

Chromosomes in a cell can be classified on the basis of number of centromeres and position of centromere as follows

I. On the basis of number of centromere

(i) **Acentric** without centromere.
(ii) **Monocentric** with a single centromere.
(iii) **Dicentric** with two centromeres.
(iv) **Polycentric** with many centromeres diffused along the entire length.

II. On the basis of position of centromere

(i) **Metacentric chromosomes** The centromere is median in position. It appears V-shaped during anaphasic movement.

(ii) **Submetacentric chromosomes** The position of centromere is submedian. Thus, one arm of chromosome is smaller and the other one is larger. It appears 'L-shaped' during anaphasic movement.

(iii) **Acrocentric chromosomes** Centromere is subterminal in position. Thus, one arm is very small and other is very long.

Fig. 15.12 *Types of chromosomes based on the position of the centromere*

(iv) **Telocentric chromosomes** Centromere is terminal, in position, i.e. situated in the telomeric region. Telocentric chromosomes are very rare. In humans, none of the chromosome is telocentric.

The identification of above written chromosomal types is done with the help of arm ratio, i.e. the ratio of lengths of chromatids present in a chromosome.

This ratio is unique for a particular chromosome and its range for different chromosomes is given below

Chromosome type	Arm ratio
Metacentric	1.07 -1.95
Submetacentric	2.05-3.75
Acrocentric	5.00-11.94

Sex chromosome and autosomes

There are two type of chromosomes present in unisexual individules. These are

(i) **Sex-chromosomes** or **Allosomes**, i.e. chromosomes which are responsible for sex-determination, e.g. X and Y-chromosomes. Terms like idiosomes, allosomes and heterosomes are often used for these sex-chromosomes.

(ii) **Autosomes**, i.e. somatic chromosomes, apart from sex-chromosomes. These contain genes, that determine the somatic characters.

Special types of chromosomes : Giant chromosomes

- Some tissues of certain organisms contain abnormal chromosomes. These chromosomes differ significantly from the normal chromosomes in structure and function.
- Such chromosomes are too large under normal conditions. Therefore, they are called **giant chromosomes.**
- Following two types of giant chromosomes are described in living organisms.

I. Polytene or Salivary gland chromosome

Polytene chromosomes were first reported by **Balbiani** (1881) from the salivary glands of *Chironomus* larvae (insect). Later on, these were studied by **Heitz** and **Bauer** in *Drosophila.*

Polytene chromosomes have also been reported in the gut epithelium, Malpighian tubules and fat bodies of some animals of order-Diptera.

In plants, these are reported in the antipodal, endosperm and suspensor cells.

Polytene chromosomes show following salient features

- The fibres present in polytene chromosomes are comparable with interphasic chromatin. Here, these are called **unit chromatids**. A polytene chromosome contains 1000-4000 such chromatids.
- These are very long and thick. Their size may increase up to 200 times than a normal metaphasic chromosome.
- Structurally, these contain several dark stained (heterochromatic) regions called **chromonemata** or **bands.** These are separated with light stained (euchromatic) regions called **interbands.**
- The chromosomal threads radiate from a deeply stained amorphous structure called **chromocentre**. It is formed by the fusion of centromeric regions of chromosomes which is involved in polytene chromosome formation.

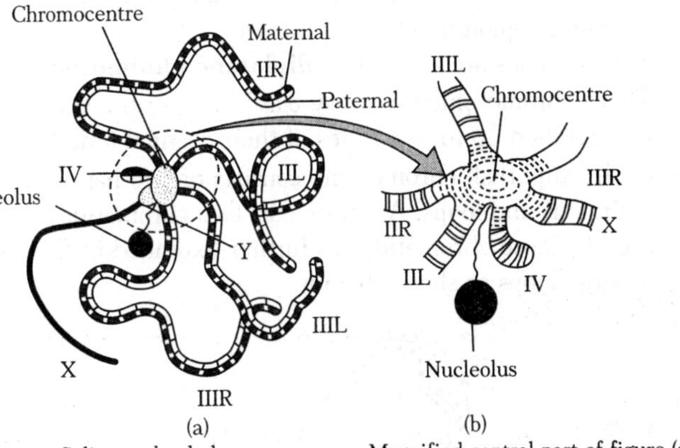

Fig. 15.13 *Salivary gland chromosomes of male Drosophila*

- Polytene chromosomes are formed by polytenisation or endoreduplication. It is the process in which DNA replicates repeatedly, but separation of nuclear material does not take place. The most common form of endoreduplication is called **endomitosis.**
- The bands of polytene chromosomes sometimes enlarge to form **Puffs** or **Balbiani rings**. These are considered as active genes and represent the site of RNA synthesis.

II. Lampbrush chromosome

These were first seen by **Flemming** (1882) and later on by **Ruckert** (1892) in the oocytes of frog and *Triturus*.

In general, these are found in the oocytes of many invertebrates and all vertebrats except some mammals.

They are so named because, under light microscope, they resemble the brush used for cleaning chimneys.

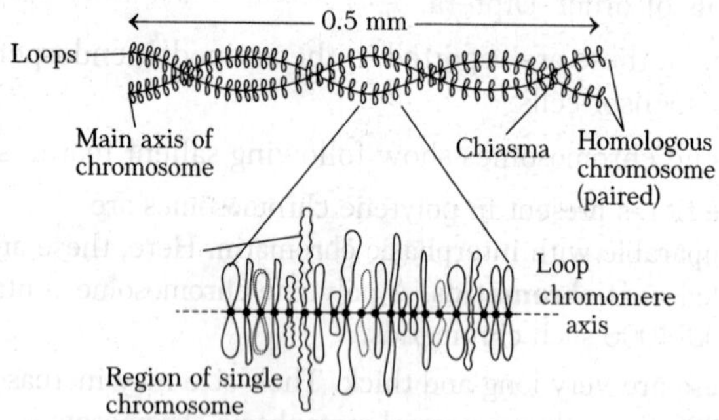

Fig. 15.14 Structure of lampbrush chromosome

Salient features of lampbrush chromosomes are as follows

- These chromosomes are approximately three times larger than the polytene chromosomes.
- Their length and width can be up to 5000 μ and 20 μ, respectively.
- Lampbrush chromosome is a synapsed (bivalent) chromosome. It is found and permanently remains in the extended diplotene phase of meiosis-I. Thus, these chromosomes are also called **diplotene chromosomes.**
- These chromosomes are very elastic in nature. These can be stretched up to two times of their original length.
- Each lampbrush chromosome consists of two parts, i.e. a main axis and loops. At main axis, two chromatids are found. These chromatids are further extended in the form of large loops as shown below

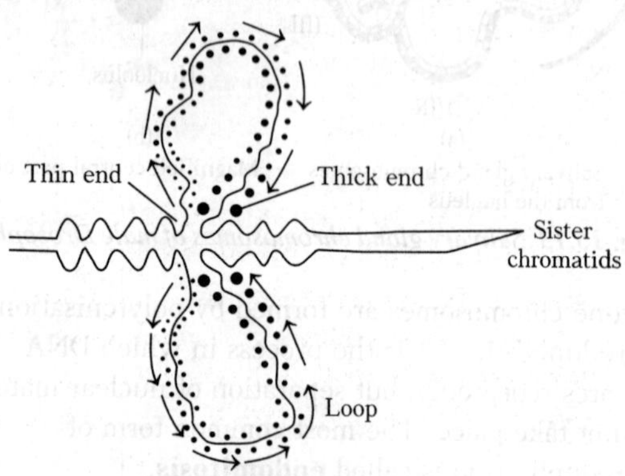

Fig. 15.15 A loop highly magnified showing thick and thin ends

- The loop is believed to represent one long operon consisting of a series of identical structural genes.
- Electron microscopic studies of **Miller** and **Beatty** (1969) on lampbrush chromosome suggested the presence of dense granules on DNA axial fibres.
  - These granules are probably large molecules of enzyme RNA polymerase which is responsible for RNA synthesis. That is why, a high rate of RNA (*m*RNA) synthesis is possible in these loops.
  - If the loop is treated with DNase, the loop axis dissolves. It suggests that the loop is composed of DNA.
  - Similarly when the loop is treated with RNAase, the matrix gets dissolved. It indicates that the matrix is composed of RNA and protein.

B-chromosome

B-chromosome is much smaller than normal chromosome. It is found in maize. It is heterochromatin and genetically inert. Thus, the rate of replication is slow in B-chromosome. It gets destroyed later on. B-chromosome affect the viability of seed and fertility of chromosome.

Karyotype

The characteristic pattern of chromosomes present in a somatic cell of any organism is known as karyotype. It is of following two types

(i) **Asymmetric karyotype** It is the karyotype in which there is a large size difference between the largest and the smallest chromosome. Such karyotype consists of fewer metacentric chromosomes. It indicates advanced features of organisms.

(ii) **Symmetric karyotype** This karyotype shows a small size difference between largest and smallest chromosome. It has many metacentric chromosomes. It indicates primitive type of karyotype.

Study of karyotype : Banding technique

Chromosome banding was discovered by **Caspersson** *et al.* In the study of karyotypes, chromosomes are stained with specific fluorescent dyes. These dyes have different binding affinities for different chromosomal regions.

Different banding techniques are as follows

- **Q-banding** It stains the AT rich region of chromosome.
- **C-banding** It stains the constitutive heterochromatin.
- **R-banding** It stains the sulphur-deficient region.
- **G-banding** It stains the AT rich region of chromosome. Giemsa stain is used in G-banding.

Idiogram

The diagrammatic or pictorial representation of chromosomes in decreasing order of their size is called an idiogram. The idiogram of human karyotype is given below

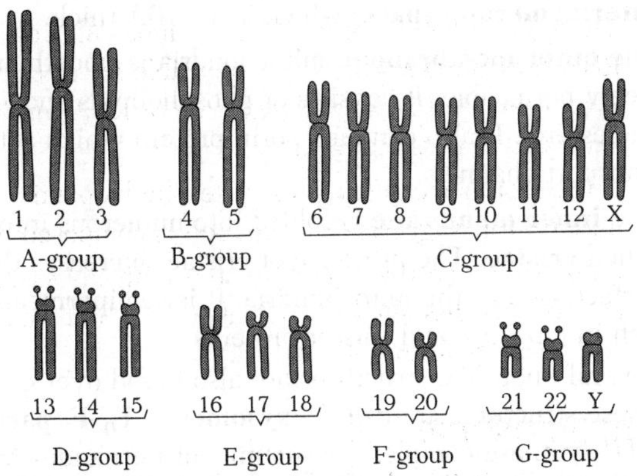

Fig. 15.16 Idiogram of human karyotype (male and female)

Table 15.2 Characteristics of the chromosomes in human karyotype

Groups	Size	Shapes	Number in sets	Number in a cells
A	Largest	Metacentric, Submetacentric	1-3	6
B	Larger	Submetacentric	4-5	4
C	Medium	Submetacentric	6-12 and X	15 male, 16 female
D	Medium	Acrocentric	13-15	6
E	Small	Metacentric or Submetacentric	16-18	6
F	Small	Metacentric	19-20	4
G	Smallest	Acrocentric	21-21 and Y	5 male, 4 female
				Total = 46

- **X-chromosome** is placed in group-C due to its large size and submedian centromere.
- **Y-chromosome** is placed in group-G due to its short size and acrocentric centromere.

Uses of karyotype and idiogram

- Karyotype is used to detect the primitive or advanced features of an organism.
- Karyotype and its idiogram is helpful in the detection of chromosomal abberation and polyploidy.
- Karyotype of different species can be studied and their idiograms can be compared to establish evolutionary relationships.

Functions of nucleus

- Nucleus stores genetic information in the form of genes.
- It act as the main controller of all the cell activities. DNA replication, DNA transcription take place in the nucleus.
- It possesses all the information required for growth, metabolism, reproduction and development of an organism.
- Nucleolus provides a site for the synthesis of ribosomes and *r*RNAs.

CHECK POINT 15.2

1. In nucleus, nuclear pores are formed by the
(a) absence of nuclear membrane
(b) breakage of nuclear membrane
(c) fusion of nuclear membranes
(d) movement of cytoplasm through nuclear membrane

2. The minimum chromosome number, 2 is found in
(a) *Ascaris megalocephala*
(b) *Haplopappus gracilis*
(c) *Ophioglossum reticulatum*
(d) None of the above

3. The outermost covering or sheath of chromosome is known as
(a) pellicle (b) membrane
(c) matrix (d) None of these

4. According to nucleosome model, chromatin consists of
(a) non-histone proteins and DNA
(b) histone proteins and DNA
(c) RNA and histone protein
(d) DNA and acidic proteins

5. Lampbrush chromosome is
(a) very elastic in nature (b) a synaped chromosome
(c) upto 5000 μ in length (d) All of these

6. Karyotyping helps in the detection of
(a) evolutionary relationships
(b) chromosomal abberation
(c) primitive features of an organism
(d) All of the above

Mitochondria

- Mitochondria (Gr. *Mitos*—thread; *chondrion*—granule) are called **powerhouse of the cell.** These organelles are involved in cellular respiration, i.e. oxidation of cellular nutrients and energy generation of the cell.
- **Altman** discovered mitochondria. **Benda** (in 1897), gave the name mitochondria to these organelles.
- These are unique organelles as they are formed by binary fission and contain their own DNA.
- Mitochondria are mostly found in actively metabolising cells and tissues, e.g. wings of insect, cardiac muscles, etc.

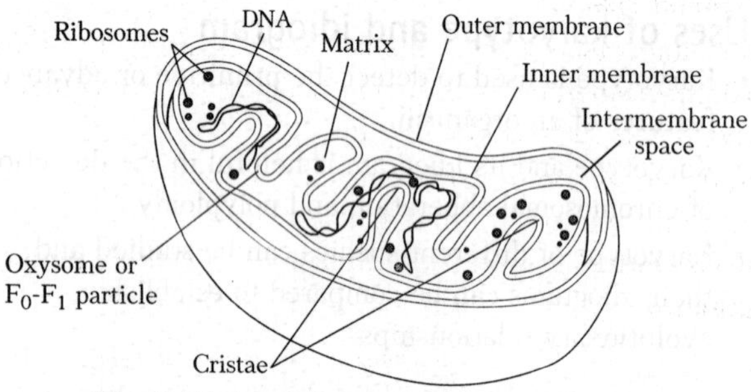

Fig. 15.17 Internal structure of a mitochondrion

Note *Mitochondria of a cell are collectively known as **chondriome**. Mitochondria is also known as **chondrioplast, plastosome, plasmosomes** and **plastochondrione**.*

- **Chemical composition** Mitochondria consist of approximately 70% proteins (dry weight) and 25-30% lipids. Among lipids 90% are phospholipids. The rest 10% includes cholesterol vitamin-E, etc.
- **Shape** The shape of mitochondria is highly variable. It ranges from short rod-shaped to elongated filamentous form. These organelles have the ability to change their shapes according to the need of cell. This ability to change shapes is known as **pleomorphism.**
- **Size** The smallest mitochondria is found in yeast and the largest one is found in the oocyte of *Rana pipens.*
- **Number** Different organisms have different number of mitochondria. Some of them are given in the table below

Structures organisms	Number of mitochondria
Chlorella, Trypanosoma and *Microasterias*	1
Human sperm cell	25
Human kidney cell	300-400
Liver cell	500-1000
Giant *Amoeba, Chaos chaos*	50,000
The egg of sea urchin	1,40,000-1,50,000
Flight muscle cells of insect	500,000

Structure

- Mitochondria are bound by two membranes, namely the outer and the inner membrane.
- These membranes are separated by a space called **intermembrane space**, which is 40-70Å thick.
- The **outer membrane** of mitochondria is smooth and freely permeable. It consists of phospholipids and cholesterol. It also contains porin protein which acts as transport channels.
- The **inner membrane** is folded into numerous infoldings called **cristae.** The presence of cristae increases the surface area of the mitochondria. It is semipermeable and rich in enzymes and carrier proteins.
- Several knob-like structures are also found over cristae. These structures are called **oxysomes** or **F_0-F_1 particles** or ATP synthase complex (or subunits of Fernandez Moran or elementary particles). They are found to be involved in the formation of ATP.
- The space inside the inner membrane is filled with **matrix.** The matrix contains dense granules along with ribosomes (70S) and mitochondrial DNA. Thus, mitochondria are a semiautonomous organelle.
- Out of the total enzymes present in a cell, mitochondria alone contribute to about 70%. The enzymes of Krebs' cycle are present in the matrix of mitochondria. Enzymes required for ETS and oxidative phosphorylation are found in cristae.

Note *Succinic dehydrogenase is found in inner mitochondrial membrane.*

Mitochondria are also called as **'Cell within cell'** or an intracellular prokaryotic parasite or bacterial endosymbiont because mitochondria are similar to bacterial cell in following ways

- Presence of porin proteins in outer membrane.
- Binary fission or division.
- Presence of 70S type of ribosomes.
- Double stranded, circular and naked DNA.

Mitochondrial DNA (*m*DNA)

- Mitochondria has its own DNA. It contributes 1% of total cellular DNA.
- It is a closed, circular molecule in most higher animals, but it can be linear in several eukaryotic cells and plants.
- It is located in mitochondrial matrix or may be attached to the membranes. The *m*DNA is equivalent to prokaryotic nucleoid.
- It contains several genes that produce enzymes used in oxidative metabolism.
- Mitochondria is called a **semiautonomous** organelles as it has its own protein synthesising machinery.

Functions

- Formation of ATP through oxidative phosphorylation. In this process, ATP is produced through Krebs' cycle and electron transport system.
- Intermediate products of cell respiration are used in the formation of steroids, cytochromes, chlorophyll and amino acids such as glutamic acid, aspartic acid, nucleic acids, etc.
- Mitochondria also regulate the calcium ion concentration inside the cell. It is a site for thermogenesis, i.e. the production of heat in body.
- Yolk nucleus with a mitochondrial cloud and Golgi bodies control vitellogenesis in eggs.
- During spermiogenesis, required energy is supplied by mitochondria.

Plastids

- The term 'Plastid' is derived from Greek word *plastikas*, i.e. formed or moulded.
- These were discovered by **E. Haeckel.**
- Plastids are small bodies found freely in most plant cells. These are absent in fungi, certain bacteria and animals.
- These are double membrane bound structures, mainly used for trapping radiations and storage purposes.

Types of plastids

Schimper discovered different types of plastids and categorised them into following types

(i) **Chloroplasts** These are green coloured plastids containing chlorophyll as pigment. These change into chromoplasts during fruit ripening.

(ii) **Chromoplasts** These plastids are variously coloured, e.g. yellow, orange, reddish, orange, etc.

The chromoplasts are further of following types

- **Phaeoplasts**, i.e. brown plastids, which contain fucoxanthin as brown pigment.
- **Rhodoplasts**, i.e. red plastids, which contain phycoerythrin as red pigment.

(iii) **Leucoplasts** These are colourless plastids used for storage purposes.

These can be further categorised as

- **Elaioplasts**, i.e. these store oils.
- **Amyloplasts**, i.e. these store starch.
- **Aleuroplasts**, i.e. these store proteins.

All these plastids have more or less similar structure. It is due to the fact that different types of plastids are formed from one another by means of binary fission.

Chloroplast

- Among plastids, the most common one are chloroplasts. The term 'Chloroplast' was coined by Schimper (in 1833).
- These may be spherical, ovoid or disc-shaped. These are involved in photosynthesis due to the presence of chlorophyll in them.
- A chloroplast is bounded by two membranes, i.e. an outer and an inner membrane. Each of these membrane is about 90-100Å thick with an intermembrane space in between them.
- The space enclosed by inner membrane is called **stroma** or **stromal space.**
- It contains matrix with soluble enzymes, proteins, lipids, 70S ribosomes, circular double-helical DNA molecule, RNA and metals such as Mn, Mg, Fe, etc.
- In the stroma, closed, flattened and disc-shaped, membranous sacs are found. These are called as **thylakoids.** Group of these thylakoids is called **grana** or **granum.**

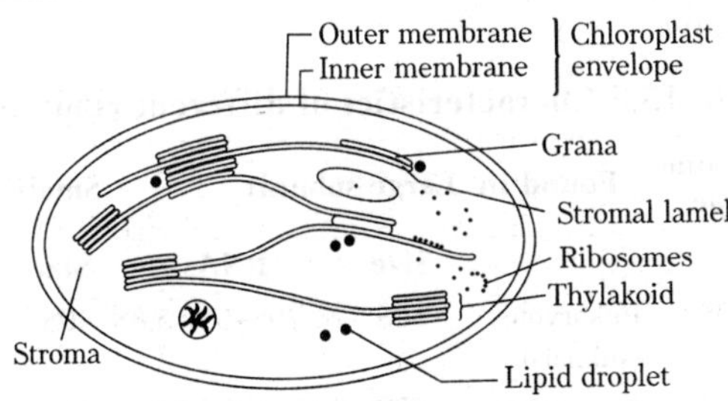

***Fig. 15.18** Chloroplast*

- The number of grana varies from 1-100 in different chloroplasts of different plants.
- The grana are interconnected with the help of network of tubules called **intergrana** or **stromal lamella.**
- On the surface of grana, photosynthetic units, i.e. quantosomes are found which contain chlorophyll.
- Chloroplast contains its own DNA and protein synthesis machinery. Hence, it is also semi-autonomous like mitochondria. Therefore, various functions of chloroplast are under the control of nuclear as well as chloroplast DNA.

Functions of chloroplast

- Chloroplasts are prime sites for photosynthesis, i.e. formation of organic compounds from inorganic raw materials.
- These are involved in the uptake of CO_2 and release of O_2, thus maintaining the gaseous balance in the atmosphere.
- They contain chlorophyll that provides the characteristic green colour to the leaves.

Ribosomes

- These were discovered by **Palade** in animal cells in the year 1955. Ribosomes are dense granules of 150-200Å diameter. Ribosomes are positively charged RNA-protein complexes. They do not possess any membrane.
- These are found in all cells that synthesise proteins. Ribosomes can be found either in free state in cytoplasm or attached to a membrane.
- Generally these are found attached to ER or nuclear membrane through specialised proteins called **ribophorins**. These are also found in mitochondria and chloroplast (70S).

Structure

- Ribosomes are composed of two subunits, i.e. a large dome-shaped subunit and a small cap-shaped subunit. The assemblage of these two units depends on a specific concentration of Mg^{2+}. Any variation in its concentration causes their disassembly. Thus, Mg^{2+} ions are essential for their stability.

The summary of characteristics of various ribosomal types is given below

Table 15.3 Characteristics of different ribosome types

Ribosome type	Found in	Large subunit		Small subunit	
		Size	RNAs	Size	RNAs
80S	Eukaryotes (animals)	60S	28-29S+5S+ 5.8S	40S	18S
80S	Eukaryotes (plants)	60S	25S + 5S + 5.8S	40S	16-18S
70S	Prokaryotes	50S	23S + 5S	30S	16S
55S	Mitochondria and plastids	40S	16-17S + 5S	30S	12-13S

- The size of ribosomes is measured in terms of **Svedberg (S) unit**. It is the speed at which ribosomes sediment during ultracentrifugation of a cell.
- Ribosomes may occur singly as **monosomes** or in groups complexed with *m*RNA as polyribosomes or polysomes, e.g. in eukaryotes.
- A bacterial cell contains an average of about 10000 ribosomes while eukaryotic cell contains 1 million to 10 million ribosomes.

Association and dissociation of ribosomes

- Both subunits of ribosomes are present separately in cytoplasm.
- The association and dissociation of ribosomes depends on the concentration of Mg^{2+}, ions in the cell.
- At high concentration of Mg^{2+}, both subunits associate to form a complete ribosome called **dimer.**
- At low concentration of Mg^{2+}, both the subunits dissociate. They reassociate again during protein synthesis.

Functions

- The **ribosomes are the protein factories** of the cell.
- The membrane bound ribosomes are involved in the synthesis of secretory proteins, integral membrane proteins and lysosomal proteins.
- The free-ribosomes synthesise soluble cytoplasmic proteins, peripheral membrane proteins, mitochondrial proteins, chloroplast proteins, peroxisomal and nuclear proteins, etc.

CHECK POINT 15.3

1. The ability of mitochondria to change its shape is called
   (a) totipotency (b) conformation
   (c) pleomorphism (d) polymorphism
2. What is common character between chloroplasts, chromoplasts and leucoplasts?
   (a) Presence of pigments
   (b) Possession of thylakoids and grana
   (c) Storage of starch, proteins and lipids
   (d) Ability to multiply by a fission-like process
3. If, we separate the cell organelles of a living cell, which part should be alive?
   (a) Endoplasmic reticulum (b) Chloroplast
   (c) Cell wall (d) Ribosomes
4. Polyribosomes are the aggregate of
   (a) 80S ribosomes on the surface of ER membrane
   (b) 70S types of ribosomes with 80S types of ribosomes
   (c) 80S ribosomes with a single *m*RNA
   (d) Ribosomal subunits with each other
5. The proteins that function outside the cytoplasm are synthesised by
   (a) Golgi apparatus
   (b) Ribosomes on RER
   (c) 70S ribosomes in mitochondria and plastids
   (d) Nucleolus
6. Ribosome is made up of
   (a) *r*RNA + Protein (b) *r*RNA + *m*RNA + Protein
   (c) *m*RNA + Protein (d) *r*RNA + *t*RNA + Protein

Endoplasmic Reticulum (ER)

- These were discovered by **Porter**, **Claude** and **Fullam** in **1945**. The term 'Endoplasmic reticulum' was first used by **Porter** and **Kallmann** in 1952.
- It is an irregular network of double membrane bound hollow tubules. These extend from the nucleus on one end to the plasma membrane on the other end by crossing the cytoplasm.
- ER are considered as one of the most important components of cytoskeleton.
- ER is absent in mature erythrocytes and prokaryotes.

Structure

The ER membrane is thinner (50 Å) than that of plasma membrane (80-100 Å thick). It consists of 30-50% lipids and 70% phospholipid. They are found in three forms

(i) **Cisternae** These are unbranched broad, flat membrane bound spaces arranged parallel to each other to form lamellae. These are about 40-50 μm in diameter. These are usually associated with ribosomes.

(ii) **Tubules** These are branched tube-like structures which are 50-100 Å thick. They do not have ribosomes on their surface.

(iii) **Vesicles** These are usually round or ovoid isolated sacs or blebs having diameter of about 25-500 μm. They also do not possess ribosomes.

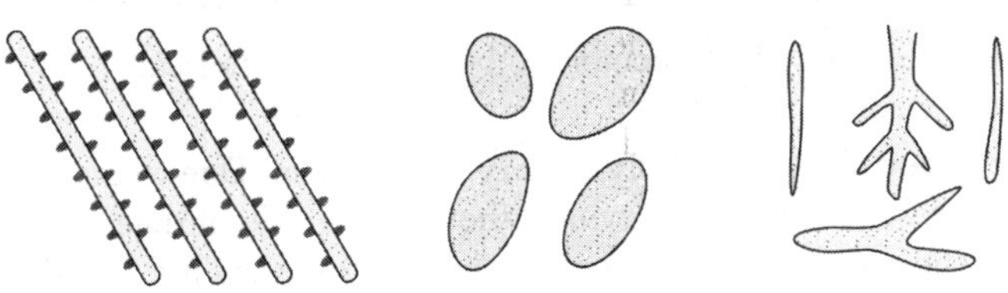

Fig. 15.19 Components of endoplasmic reticulum

Types of ER

(i) **Smooth Endoplasmic Reticulum** (SER) It originates from RER. It is composed of only tubule and vesicle. It does not possess ribosomes on its surface and thus, have smooth surface. It is mainly involved in the synthesises of lipids and steroids.

(ii) **Rough Endoplasmic Reticulum** (RER) It originates from nuclear membrane.

- It possesses ribosomes attached to its surface. RER composed of only cisternae and tubule. It forms two third of total ER of cell.
- Only large subunit (60S) of ribosome is attached with RER with the help of two types of glycoproteins namely, **Ribophorin-I** and **Ribophorin-II**.
- Due to high contents of ribosomes, RER is basophilic in nature.
- It is mainly found in cells that synthesise proteins actively.

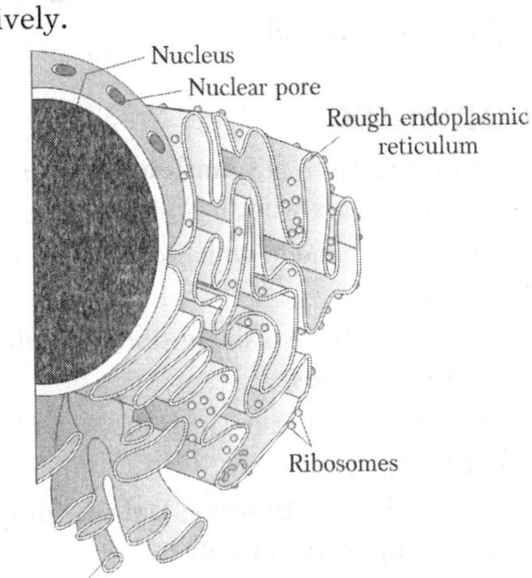

Fig. 15.20 Structure of endoplasmic reticulum

Functions

- It provides additional mechanical support to the cell by dividing the fluid content of the cell into compartments. That is why, it is considered as one of the elements of cytoskeleton.
- Proteins synthesised by ribosomes enter the RER cisternae through channels present in the membrane. In RER, processing of these proteins takes place.
- RER helps in synthesising lysosomal enzymes and SER.
- ER is found to be greatly involved in glycogen and glucose metabolism. SER plays a major role in glycogenesis.
- The major site of cholesterol synthesis is SER. These are also associated with the storage of cholesterol.
- SER helps in detoxification of drugs in liver. It also protects the cell from the toxic effect of chemicals by using cytochrome P456.
- In testes, ovary and adrenal cortex, SER helps in the synthesis of steroid hormones.
- ER acts as the member of intracellular transport system for various substances, which may pass across the ER membranes by diffusion and active transport.
- A modified form of SER is sarcoplasmic reticulum of striated muscles. These are storehouses of Ca^{2+} ions, which are necessary for muscle contraction.
- SER helps in the formation of organelles like spherosome and glyoxysome.

- Common functions of RER and SER are as follows
  - ER gives mechanical support to cell as it forms a network inside the cell.
  - Transport of material within the cell.
  - Helps in the formation of nuclear membrane.
  - Both RER and SER increase the surface area for enzymatic action the cell.

Golgi complex

- These were first reported by **Camileo Golgi** in **1897**. He got a Nobel Prize for this work in 1906.
- It is a membranous organelle which consists of flattened sac-like structures.
- Plant and fungal cells have a single stack of Golgi apparatus. It is called as **dictyosomes.**
- The Golgi complex is absent in sieve tubes of plants, RBCs, sperm cells of bryophytes, etc.
- These are also known as **idiosome**, **Dalton complex**, **lipochondria** or **apparato reticulare**.

Structure

The electron microscopic studies revealed the presence of three membranous components in Golgi complex.

(i) **Cisternae** forms the central flattened part of complex. It consists of double membranous flattened sac-like structures. These structures are stacked on one another. Each cisternae has a convex *cis* face and a concave *trans* face.

(ii) **Tubules** arise from the periphery of the cisternae. These are small and flat structures.

(iii) **Vesicles** are large round structures. These are found on the edges of *cis* and *trans* faces.
It is constantly being formed, changed, broken down and reformed. It may be formed from plasmalemma, nuclear envelope, annulate lamellae and ER.

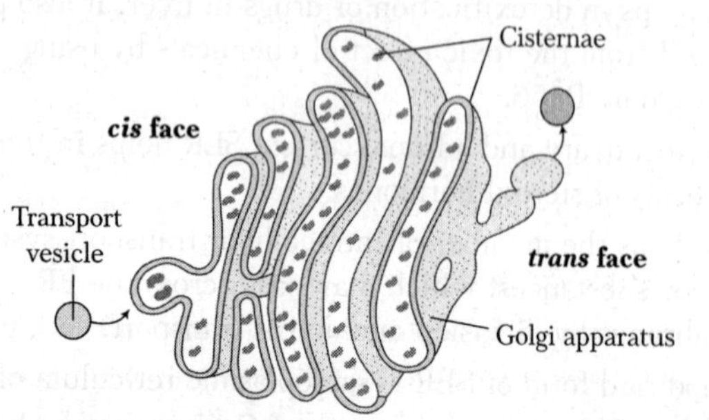

Fig. 15.21 Golgi apparatus showing a stack of cisternae

Note *Golgian vacuole is formed by the budding of terminal end of maturing face of Golgi apparatus. Golgi body is present in the cytoplasmic area and is free from other cell organelles. Hence, this area is called as zone of cell ex-clusion.*

Functions

- Golgi complex is a secretory organelle. It is involved in storage, packaging and secretion of different materials.
- Proteins in the form of zymogen granules originate from Golgi complex and fuse with plasmalemma. This helps in the renewal of constituents of membranes as well.
- It supplies glycosylated molecules to plasma membrane.
- Some polysaccharides are formed in Golgi complex. These are then deposited on the new cell wall of growing cells.
- It helps to the formation of acrosome in sperms.
- Proteins and lipids are formed in RER and SER, respectively. These combine with carbohydrates to form glycoproteins (glycosylation) and glycolipids (glycosidation), respectively. This reaction takes place in Golgi complex.
- Primary lysosomes are formed by Golgi complex.
- Precursors of neurosecretory products are found between the lamella of Golgi complex.
- Root caps are rich in Golgi body as it secrete extracellular secretion, mucilage.

Lysosomes

- **C de Duve** (1949) first reported these rounded bodies in liver cells. He initially called them as **pericanalicular dense bodies**, but later on (in 1955), Novikoff gave the term lysosomes (*Lyso*-digestive; *soma*-body) to them.
- These are absent in prokaryotic cells, RBCs and plant cells.
- These are single membrane bound small spherical organelles.
- They contain hydrolytic enzymes in them. Due to the presence of such enzymes, they are capable of digesting the own substances of cell. That is why, they are called as the **suicidal bags** or **garbage disposal system** of a cell.

Structure

The lysosomes are highly variable in their shape and morphology, i.e. show polymorphism even within a single cell during its various stages. A polymorphic form of lysosome may include following types

(i) **Primary lysosomes** These are formed from ER and Golgi complex. They store and secrete inactive hydrolytic enzymes.

(ii) **Secondary lysosomes** They are digestive vacuoles, formed by the fusion of phagosome (food containing vesicles) and primary lysosomes (enzyme containing vesicles). These are also called **heterophagosomes**.

(iii) **Tertiary lysosomes or Residual bodies** These are lysosomes containing indigestible materials.

(iv) **Autophagic vacuoles** These are formed by the union of several primary lysosomes around degenerated cells or organelles. They digest such cells parts.

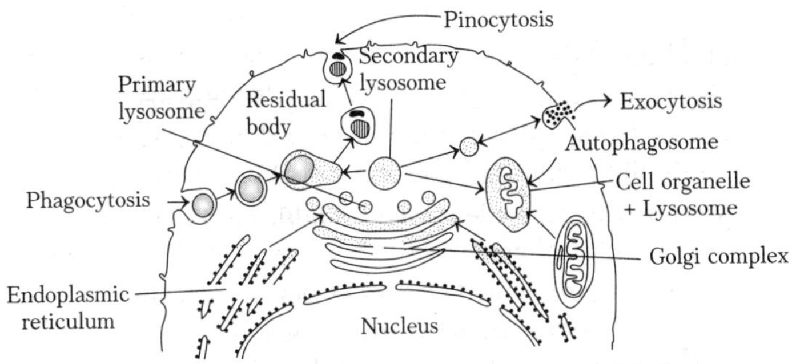

***Fig. 15.22** Figure representing the dynamic aspects of the lysosome system. Observe the relationship between the processes of phagocytosis, pinocytosis, exocytosis and autophagy*

Functions

- Lysosomal enzymes when released, take part in hydrolysis of extracellular materials.
- Intracellular digestion occurs through **autophagy** (self-eating) and **heterophagy** (eating of other substances). Both autophagy and heterophagy may occur simultaneously in a single vesicle called **ambilysosomes**.
- Malfunctioning of lysosome results in enzyme spills, which can cause following effects
  - Intracellular release of hydrolytic enzymes. These lead to damage of various cell structures. It may result in disorganisation of cellular content. The main damages include chromosomal breaks due to enzyme nuclease. It results into mutations.
  - Spilling of lysosomal contents may cause arthritis (joint pain), e.g. cathepsin-D present in lysosomes causes the breakdown of cartilage. It results into arthritis.
- Under some conditions, lysosomes may store indigestible materials. These materials can be silica, asbestos, etc. This may cause diseases such as silicosis, asbestosis, etc.
- Acrosome is a giant lysosome that releases hydrolytic enzymes during fertilisation. These enzymes help in easy penetration of sperm into the egg.

Note • *Because of close relationship between ER, Golgi complex and lysosomes, Novikott et al. (1961-64) denoted **endomembrane system** as GERL system, i.e. Golgi complex, endoplasmic reticulum, lysosome system.*
• *In old age, number of lysosomes increases in body.*

Vacuoles

- About 90% of total volume of a plant cell is occupied by a single membrane-bound vacuole.
- They store macromolecules including ions, sugars, amino acids, proteins and carbohydrates.
- The membrane that surrounds the vacuole is called **tonoplast**.
- The vacuole contains cell sap in it. This cell sap has high osmotic pressure which generates turgor pressure in plant cell.

Types

The types of vacuoles found in plant cells are

(i) **Sap vacuoles** It contains dissolved salts and helps to maintain osmotic and turgor pressure.

(ii) **Contractile vacuole** It helps in osmoregulation and excretion.

(iii) **Food vacuole** These are formed by engulfing the food particles. Thus, they contain hydrolases along with food.

(iv) **Gas vacuoles** These are found in green bacterial cells.

Functions

- Storage of various substances including waste products.
- Important contribution to the osmotic properties of the cell.
- Sometimes they function as lysosomes.

Microbodies

These are small, single membrane bound organelles. These are associated with oxidative reactions other than respiration.

Microbodies can be of following types

1. Peroxisomes

- They were discovered by **de Duve** in the leaf of spinach in 1967.
- These are round to oval bodies and primarily occur in the mesophyll cells of of C_3-plants.
- They contain enzymes like catalase (largest enzyme), urease, oxidase, luciferase, peroxidase (smallest enzyme), etc. These are found to be involved in oxidation of amino acid and uric acid.
- Peroxidation, i.e. hydrolysis of peroxide (H_2O_2) also takes place in peroxisomes. It is a method of detoxification in cells. They are also responsible for the generation of light in fireflies with the help of luciferase enzyme.
- Peroxisomes are also involved in photorespiration.

2. Glyoxysomes

- These were first reported by **Beevers** in 1969, in the endosperm of germinating seeds of castor.
- They help in β-oxidation of fatty acids.
- These are associated with the triglyceride metabolism through glyoxylate cycle. Two enzymes of this cycle, i.e. isocitrate lyase and malate synthase are found in glyoxysome.
- These are also found in yeast and *Neurospora* cells.

3. Spherosomes

- These microbodies are also called oleosomes and these were discovered by **Hanstein.**
- These are small spherical vesicles and are formed from smooth ER.
- These contain enzymes such as esterases, nucleases, hydrolases, phosphatases, proteases, etc.
- They are capable of synthesising and storing lipids.

4. Lomasomes

- They were discovered by **McAlear** in 1961.
- The lomasomes are small vesicular bodies. They are present between the cell wall and plasma membrane in the plant cells.
- These bodies are associated with cell wall formation in plant cells.

Note *Some scientists include lysosomes within microbodies because of their small size. But, these can be differentiated on the basis of mode of formation. Lysosomes are formed by blebbing while all microbodies are formed by aggregation of proteins and lipid molecules in a cell.*

Cell inclusions

Besides the cell organelles, the cytoplasm of a cell contains various substances known as inclusions. These include

(i) **Starch grains** These are spherical, oval, elliptical or polyhedral bodies. They are found in plant cells only, particularly in their storage organs like seeds, fruits, rhizomes and tubers.

(ii) **Glycogen granules** These are small, spherical or large rosette-shaped particles. They occur near SER in liver and muscle cells. These are also found in blue-green algae, slime moulds, fungi and bacteria.

(iii) **Fat droplets** These are found in adipocytes (fat storing cells) of animals, the endosperm of castor and coconut along with cotyledons of groundnut and mustard seeds.

(iv) **Aleurone grains** These contain proteins for storage. They are present in all cereal grains, e.g. wheat, maize and barley below seed coat.

(v) **Crystals** There are crystals of various chemical compounds found in plant cells. Inorganic intracellular crystals, like calcium salts of carbonate, oxalate or phosphates are common. Calcium oxalate forms a bundle of needle-shaped crystals. These are known as **raphides**. They can be found in leaves, roots, fruits, etc.

Specialised structures of cell

Some specialised structures of cells are as follows

Centrosome

- The term 'centrosome' was introduced by **Boveri** in **1888.** These are specialised structures of animal cells, lower plants and primitive flagellates only.
- A centrosome is made of two cylindrical, rod-shaped structures called **centrioles.**
- Centrioles are oriented at right angles to each other. Together they are called as **diplosome.**
- A centriole can be differentiated into two regions, i.e. central and peripheral. The peripheral region contains the fibrils of tubulin protein. Each fibril contains three subfibres. Hence, it is called **triplet fibril.**
- Adjacent triplet fibrils are connected to each other thus, forming a complete cylinder.
- The central part of the centriole consists of one or two longitudinally arranged strands. This part of centriole is called **hub**. It is connected to the peripherel fibres by radial spokes.
- During cell division, centrioles develop a number of radiating threads or rays. The whole structure is called **aster.**

Functions

- Centriole takes part in the formation of cilia and flagella.
- Centrosome helps in the formation of spindle fibres during cell division.
- Some scientists suggest their role in microtubule generation also.

Cytoskeleton

- All eukaryotic cells possess a *criss-cross* network of interconnected filaments and tubules called cytoskeleton.
- It extends from the nucleus to the plasma membrane and supports the shape of the cell. It also helps in anchoring cell organelles to fixed locations.
- The cytoskeleton of a cell is a dynamic system that keeps on changing always.
- It consists of following different types of elements

  (i) **Microtubules** These are hollow and unbranched cylinders. They are about 25 nm in size.

  Microtubules are made up of 13 parallel protofilaments of tubulin proteins. These are the constituents of spindle fibres, centrioles, cilia and flagella.

(ii) **Microfilaments** These are solid, unbranched, rod-like fibres. They are about 7 nm in size. Microfilaments consist of actin and some amount of myosin protein.

(iii) **Intermediate filaments** They are filaments of intermediate size, i.e. 10 nm. These are composed of acidic proteins.

Cilia and flagella

- Cilia and flagella are thread-like outgrowths extending from the cell surface. Both of these possess similar basic structure and function.
- They are distinguishable on the basis of their size and number.
- Flagella are long and less in number as compared to cilia which are smaller in size and more in number.
- Cilia and flagella both consists of three parts namely, i.e. basal body, a hook and a filament.
- The filament is a cylindrical structure with a core called **axoneme**. It is made up of several microtubules. The arrangement of microtubules is called **9+2 arrangement**. It is composed of 9 peripheral doublet fibres and 2 central singlet fibres.
- The central microtubules are connected to each other through bridges and enclosed in a central sheath. The central sheath in turn, is connected to peripheral tubules by means of radial spokes.
- These help in locomotion, capturing food, circulation of food in gastrovascular cavity of coelenterates, etc.
- Energy required for the movement of cilia and flagella is provided by dynein protein. This protein has ATPase enzyme.

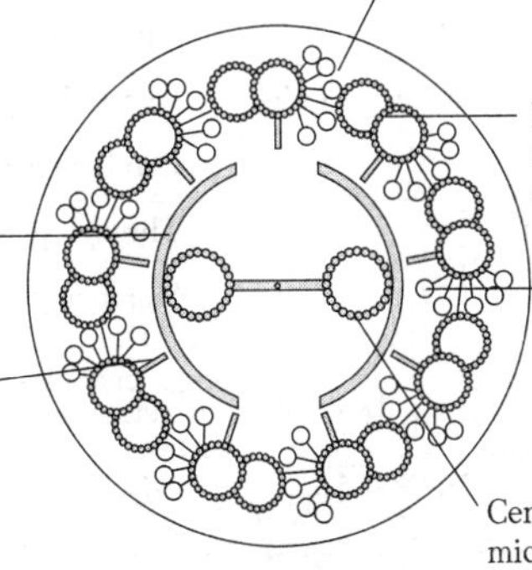

Fig. 15.23 Internal structure of cilia/flagella

Note *Cilia may beat in metachronous (beat one after one) or synchronous (beat simultaneously rhythm).*

Some comparative study

The comparison between prokaryotic and eukaryotic cell is as follows

Table 15.4 Differences between prokaryotic and eukaryotic cells

Prokaryotic cell	Eukaryotic cell
The cell size is usualy small (0.1 - 5.0 μm) in diameter.	The cell size is comparatively larger (10-100 μm) in diameter.
An organised nucleus is absent. Instead a nucleoid is found. No differentiation in nucleoid is found.	An organised nucleus is found. It is differentiated into nuclear envelope, chromatin, one or more nucleoli and nucleoplasm.
DNA is naked or without an association with histones. DNA is usually circular.	Nuclear DNA is associated with histone proteins. Nuclear DNA is linear while extranuclear DNA is circular.
DNA lies freely in the cytoplasm. It is not associated with any organelle.	Most of the cell DNA lies in the nucleus. A small quantity is also found in the plastids and mitochondria.
Additional small circular DNA segments or plasmids may occur.	Plasmids are usually absent.
ER, mitochondria, Golgi apparatus, lysosomes and centrioles (centrosome, central apparatus) are absent.	ER, mitochondria, Golgi apparatus and lysosomes or their equivalents are present in all the eukaryotic cells. Centrioles are usually present in animal cells.
Cell wall contains muramic acid.	Cell wall, if present, does not contain muramic acid.
Cell membrane may be infolded to form a complex structure called **mesosome**.	Mesosome like structures are absent.
Ribosomes are of 70S type.	Ribosomes are of 80S type. 70S ribosomes, however occur in plastids and mitochondria.
Transcription occurs in the cytoplasm.	Transcription occurs in the nucleus.
The flagella are single-stranded and without differentiation of axoneme and sheath.	The flagella, if present are 11-stranded (9+2 arrangement). They show differentiation of axoneme and sheath.

A eukaryotic cell can be either plant cell or animal cell. Both these cells can be differentiated as follows

Cell organelles	Plant cell	Animal cell
Cell wall	Present	Absent
Nucleus	Lies near periphery due to the presence of a large vacuole	Lies near the centre
Chloroplast	Present	Absent
Plastid	Present	Absent
Vacuole	Single and large	Multiple and small
Centrioles	Only present in lower plant forms	Present in all animal cells
Cilia	Very rare	Present
Spindle (formed during nuclear division)	Anastral	Amphiastral
Lysosome	Rarely present	Present
Cellular connection	Through plasmodesmata	Through number of junctions
Reserve food	Starch and fat	Glycogen and fat

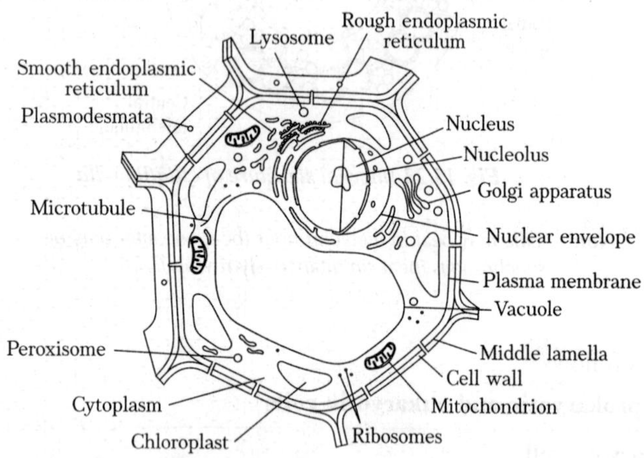

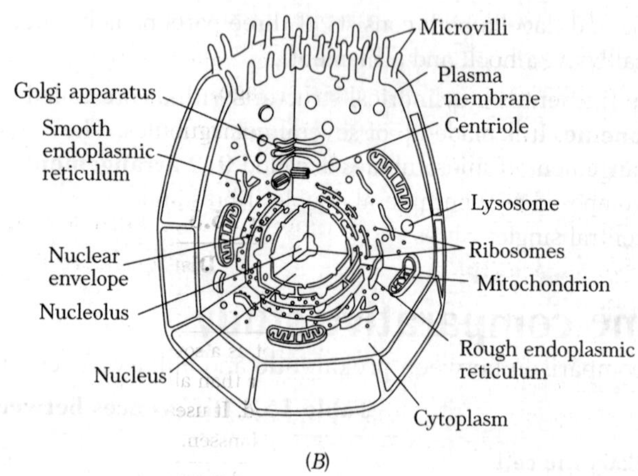

Fig. 15.24 Structure of (A) Plant cell, (B) Animal cell

CHECK POINT 15.4

1. The cytoplasm of animal cell and plant cell is transversed by a network of tiny tubular membrane system that divides intracellular space into two compartments (luminal and extra luminal). The above statement is attributed to
   (a) Endoplasmic reticulum
   (b) Golgi body
   (c) Vacuole
   (d) Mitochondria
2. The main organelle involved in modification and routing of newly synthesised proteins to their destinations is
   (a) mitochondria (b) endoplasmic reticulum
   (c) lysosome (d) chloroplast
3. The organelle responsible for the supply of glycosylated molecules to plasma membrane is
   (a) Golgi complex (b) mitochondria
   (c) ribosomes (d) peroxisomes
4. Which of the following structure of a plant cell has the greatest volume?
   (a) Vacuole (b) Lysosome
   (c) Glyoxysome (d) Ribosome
5. If the cells are broken up and sedimented, the new structure formed in one of the fraction is
   (a) ribosome
   (b) lysosome
   (c) microsome
   (d) centrosome
6. Centriole, cilium and flagellum resemble one another in that
   (a) All possess 9 + 0 organisation
   (b) All possess 9 + 2 organisation
   (c) All of them are made up of microtubules
   (d) All can duplicate themselves

TOOLS AND TECHNIQUES

In the course of history of biology, various new tools and techniques have been developed to explore more and ease out the studies. These include instruments and processes like microscopy, chromatography and electrophoresis, etc. In this section, we will explore all the important tools and techniques used in cell biology.

Microscope

(Gk. *Micron*-small; *scope*-aim)

- It is an instrument used to view objects and organisms which are too small to be seen by naked eyes. It was discovered by **Antony van Leeuwenhoek**. He is considered as the **Father of Microscopy**.
- Fundamentally, there are two types of microscopes namely, the light microscope and electron microscope.
- The **light microscope** uses a series of glass lenses to focus visible light on object to form an image.
- The electron microscope uses electromagnetic lenses to focus a beam of electrons for visualisation.

Magnifying power

The magnifying power of a microscope can be defined as the ratio of magnified image of a microscope to that formed in the retina of an unaided eye.

Thus,

$$\text{Magnifying power} = \frac{\text{Size of image formed with instrument}}{\text{Size of image with unaided normal eye}}$$

Resolving power

- The resolving power is the ability to differentiate two close points as distinct and separated. The resolving power of a **compound microscope** is 3000Å.
- It depends upon the numerical aperture. If the numerical aperture is higher, the resolution is better.
- Oil immersion is a technique used to increase the resolution of a microscope. It inturn increases numerical aperture value.

Different types of microscopes used for various purposes are discussed in the given table

Table 15.5 A summary of various types of microscopes

Microscope type	Distinguishing features	Principal uses
1. **Light**		
(a) Bright-field or compound microscope	Uses visible light as a source of illumination; cannot resolve structures smaller than about 0.2 μm; specimen appears against a bright background. It uses combination of two lens systems. It was discovered by Z Janssen.	To observe various stained specimens and to count microbes. Does not resolve very small specimens, such as viruses.
(b) Dark-field	Uses a special condenser with an opaque disc that blocks light from entering the objective lens directly; light reflected by specimen enters the objective lens and the specimen appears bright against a black background.	To examine living microorganisms that are invisible in bright-field microscopy or do not stain easily or are distorted by staining.
(c) Phase-contrast	Uses a special condensor containing an annular (ring-shaped) diaphragm. The diaphragm allows direct light to pass through the condenser, focussing light on the specimen and a diffraction plate in the objective lens. Direct and reflected or diffracted light rays are brought together to produce the image. No staining required. Fritz-Zernike invented the first phase-contrast micrograph.	To facilitate detailed examination of the internal structures of living specimens.
(d) Differential Interference Contract (DIC)	Like phase-contrast, uses differences in refractive indexes to produce image. Uses two beams of light separated by prisms; the specimen appears coloured as a result of the prism effect. No staining required.	To provide three-dimensional images.
(e) Fluorescence	Uses an ultraviolet or near ultraviolet source of illumination that causes fluorescent microbes in a specimen to emit light. Fluorescent dyes such as acridine orange is used.	In fluorescent-antibody techniques, (immunofluorescence). Rapidly detect and identify microbes in tissues or clinical specimens.
(f) Confocal	Uses laser light to illuminate one plane of a specimen at a time.	To obtain two and three-dimensional images of cells for biomedical applications.

Microscope type	Distinguishing features	Principal uses
2. Electron		
(a) Transmission	Electron microscope was developed by Knoll and Ruska. Uses a beam of electrons instead of light; electrons pass through the specimen; because of the shorter wavelength of electrons, structures smaller than 0.2 μm can be resolved. The image produced is two-dimensional. The entire apparatus is enclosed in a vacuum chamber.	To examine viruses or the internal ultrastructure in thin sections of cells (usually magnified 10000–100000 X).
(b) Scanning	Uses a beam of electrons instead of light; electrons are reflected from the specimen; because of the shorter wavelength of electrons, structures smaller than 0.2 μm can be resolved. The image produced appears three-dimensional. It was discovered by Marvin Minsky.	To study the surface features of cells and viruses (usually magnified 1000-10000 X).
3. Scanned-probe		
(a) Scanning tunneling	Uses a thin metal probe that scans a specimen and produces an image revealing the bumps and depressions of the atoms on the surface of the specimen. Resolving power is much greater than that of an electron microscope. No special preparation required.	Provides very detailed views of molecules inside cells.
(b) Atomic force	Uses a metal and diamond probe gently forced down along the surface of the specimen. Produces a three-dimensional image. No special preparation required.	Provides images of biological molecules, nearly atomic detail and molecular processes.

Transmission Electron Microscope

- A transmission electron microscope uses magnetic lens carrying current.
- The condensing lens forms a parallel beam of electrons that strikes the object.
- The objective lens forms an intermediate image that serves as the object for the final real image formed by the projection lens.

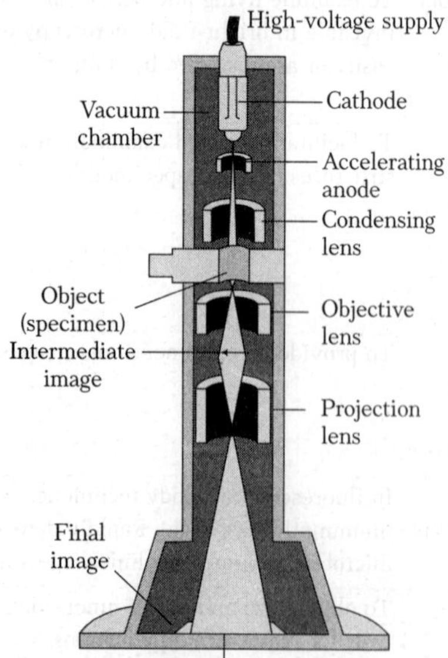

Fig. 15.25 Transmission electron microscope

Scanning electron microscope

- In a scanning electron microscope, primary electrons sweep across the specimen and knock electrons from its surface.
- The secondary electrons are picked up by a collector, amplified and transmitted onto a viewing screen or photographic plate.

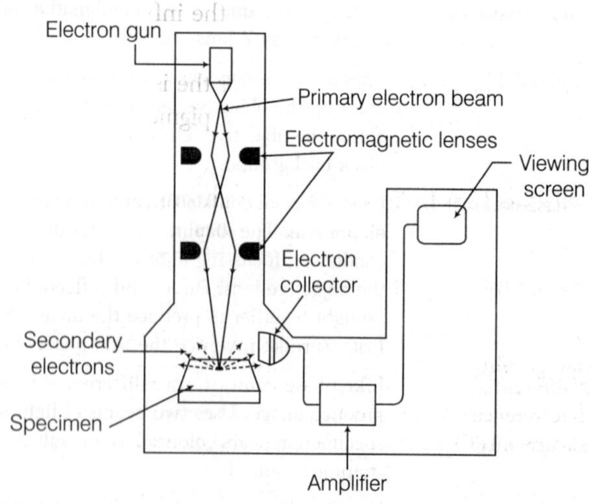

Fig. 15.26 Scanning electron microscope

Microtome

It was invented by **W His**. It is a machine with a sharp metal blade that operates like a meat slicer. It is used for obtaining uniform and extremely thin sections for light and electron microscopy.

Cell fractionation

It is the process of rupturing cells and isolating its constitutents in isotonic medium to study, their structural and functional properties. It involves the following steps

(i) **Homogenisation** is rupturing of cells to release various constituents and keeping them in natural state (by means of 0.25M sucrose solution kept at 0-4°C). Cell can be broken by forcing them at high pressure through any chemical or physical means of sound waves.

(ii) **Centrifugation** is the separation of different components of a homogenate by subjecting it to succesively higher centrifugal force. Particles, which differ in density, shape or size can be separated and become sediment at different rates in the centrifugal field. The lighter constituents settle down at lower speed, while the heavier ones settle at higher speeds.

Chromatography

- Chromatography (*Chroma*—colour; *graphy*—writing) was invented by a Russian scientist **Michael Tswett** (1906).
- It is an important technique of cell fractionation and is employed for the separation of constituents of a mixture.
- It based upon the principle of partition coefficient.
- It involves differential movement of each component through a stationary medium under the influence of a moving solvent.
- The technique is particularly used for the isolation and purification of materials such as plant pigments to amino acids.

Table 15.6 Different types of chromatography and their inventors

Chromatography	Inventor	Uses
Column chromatography	Tswett (1906)	For separating mixture of tissue lipids
Partition chromatography	Martin and Synge (1942)	For separating the solutes between two liquid phases.
Gas chromatography		For separating substances that can be vapourised.
Affinity chromatography	Pedro Cuatrecasas and Meir Wilch (1987)	For the separation of immunoglobulins, cellular enzymes and *m*RNA
Ion exchange chromatography		Purification of insulin.
Gel filtration chromatography		For determining the molecular weight of proteins.

X-ray crystallography

- It was discovered by **Max Perutz** and **John Kendrew** in **1950**.
- By this technique, arrangement of atoms in a molecule can be determined. In this technique, an X-ray beam is allowed to strike on a portion of crystal, which causes the light beam to spread in many directions producing a diffraction pattern or discrete spots on photographic film, as shown below in **fig 15.27**

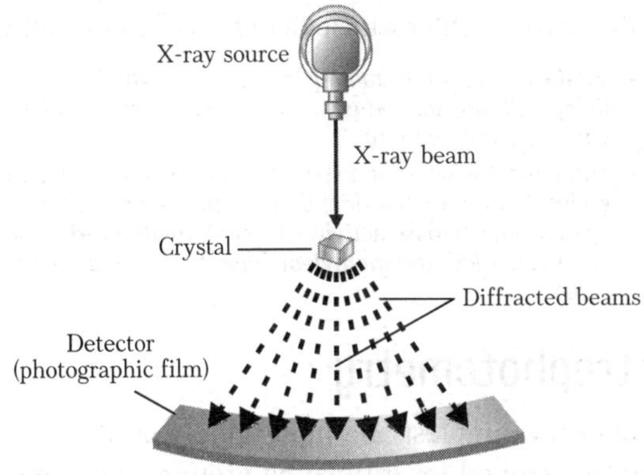

***Fig. 15.27** X-ray crystallography provides diffraction data from which the three-dimensional structure can be determined*

The distance between different spots gives information about the relative positions of various components of the specimen.

Cytological staining

- The word staining simply means colouring of specimen with a dye to get a clear view of certain structures. These dyes are called **stains.**
- **Simple stain** is an aqueous or alcoholic solution of a single basic dye. The primary purpose of a simple stain is to highlight the entire specimen, so that basic structures are visible.
- Simple stains commonly used in laboratory are methylene blue, carbol fuchsin, crystal violet and safranin.
- **Differential stains** react differently with different kinds of specimen of same category, e.g. bacteria through Gram stain, acid fast stain, etc.
- **Special stains** are used to colour the isolated specific parts of specimen. For example, endospore, flagella and capsule of microorgansisms.

Stains used for chromosomes

- Acetocarmine (pink)
- Leucobasic Fuchsin (Warm Schiff's reagent) (pink)
- Crystal violet (violet)
- Carbol fuchsin (magenta)
- Gomori's haematoxylin (violet)
- Toluidine blue stain

Stains used for other cellular components

- Lactophenol, cotton blue or methylene blue or aniline blue for fungal hyphae (blue)
- Eosin for cytoplasm (pink)
- Phloroglucinol and HCl for lignin (red)
- Ruthenium red for pectin and proteins
- Sudan III and IV for suberin, cutin and oil (red)
- Sudan black for fatty substances
- Schultz reagent (chlorozinciodine) for cellulosic cell wall

Note
- *Vital staining is the staining technique in which structures of living cells are stained either in vivo (within the body) or in vitro (in a preparation).*
- *Three most widely used stains for vital staining are **Janus green** B (for mitochondria, used for the determination of cytochrome oxidase activity in a cell), **neutral red** (to stain plant vacuoles) and **methylene blue** (for staining the dividing cells).*

Spectrophotometry

It is one of the simplest and most widely used quantitative mehod for estimating protein or nucleic acid contents present in a given solution.

The basis of this technique is the absorption of light of specific wavelength by a given solution. The instrument used in this techinque is called **spectrophotometer.**

Electrophoresis

It is the technique of separating molecules from their mixture under the influence of an applied electric field. Dissolved molecules migrate at a speed determined by their charge/mass ratio under the influence of an electric field.

Autoradiography

It is used in studying metabolic pathways, functions of cellular organelles, transport pathways, DNA replication, transcription and translation. It is carried out with the help of radioactive isotopes called **tracers**. These are incorporated into the cell and with the help of radiations emitted by tracers their path is traced. These tracers can be located by Geiger counter.

Flow cytometry

In 1968, first fluorescence based flow cytometry device was developed by **Wolfgang Gohde**. This technique counts and identifies cells and chromosomes. In this technique, the cells are suspended in fluid and then passed by an electronic detection apparatus. The flow cytometry technique was originally called **pulse cytophotometry**.

Besides counting and identification, flow cytometery has several uses. It is used to detect abnormality in genes like aneuploidy, duplication and deletion, etc.

CHECK POINT 15.5

1. The use of oil immersion can
   (a) increase the value of refractive index
   (b) decrease the value of refractive index
   (c) increase the value of numerical aperture
   (d) decrease the value of numerical aperture
2. The microscope that use laser light to illuminate the specinen is
   (a) confocal
   (b) scanning tunneling
   (c) phase contrast
   (d) dark field
3. DNA synthesis can be specifically estimated by measuring the incorporation of radiolabelled
   (a) adenine (b) uracil
   (c) thymidine (d) deoxyribose sugar
4. Which of the following development was most instrumental in enabling us to determine the function of mitochondria?
   (a) Technique of autoradiography
   (b) Technique of culturing bacteria
   (c) The phase contrast microscope
   (d) The bright field microscope
5. One of the important techniques used to study the synthesis of molecules and to trace the metabolic pathways or events in cells is
   (a) calorimetry (b) autoradiography
   (c) zonal centrifugation (d) microspectrophotometry
6. The technique that counts and identifies cells and chromosomes is
   (a) electrophoresis (b) flow cytometry
   (c) Both (b) and (c) (d) X-ray crystallography

EXAM POINTS

- The concept of *'Omnis cellula-e-cellula'* regarding cell division was first proposed by Rudolf Virchow. **[NEET 2019]**
- Sterol provides stability in cell membrane. **[JIPMER 2018]**
- Mycoplasma is the smallest living cell without a definite cell wall. **[NEET 2017]**
- Non-membranous nucleoplasmic structures in nucleus are the site for active synthesis of *r*RNA. **[NEET 2019]**
- The biosynthesis of ribosomal RNA occurs in nucleolus. **[NEET 2020]**
- The shorter and longer arms of a sub-metacentric chromosomes, are referred to as *p*-arm and *q*-arm, respectively. **[NEET 2019]**
- Double-stranded circular naked DNA is present in an organism having 70S ribosomes only. **[NEET 2019]**
- Mitochondrial matrix contains single circular DNA molecule and ribosomes. **[NEET 2019]**
- Outer membrane of mitochondria is permeable to monomers of carbohydrates, fats and proteins. **[NEET 2019]**
- Inner membrane of mitochondria is convoluted with infoldings. **[NEET 2019]**
- Mitochondria are responsible for extracting energy from carbohydrates to form ATP. **[NEET 2017]**
- Chloroplast, mitochondria and nucleus have DNA. **[NEET 2019]**
- Chloroplast is present in algae, plants and releases oxygen. **[JIPMER 2018]**
- Ribosomes help in synthesis of protein. **[NEET 2019]**
- Smooth ER helps in lipid synthesis. **[NEET 2020]**
- Rough ER helps in protein synthesis. **[NEET 2020]**
- 'RER–Cell membrane–Secretory vesicles–Golgi body'. The above sequence is of modification and transport of secretory vesicels. **[JIPMER 2019]**
- Cleavage of signal peptide, protein glycosylation and protein folding occurs in rough ER. **[NEET 2018]**
- Golgi complex helps in glycosylation. **[NEET 2020]**
- Golgi bodies is the site of formation of glycoproteins and glycolipids in eukaryotic cells. **[NEET 2020]**
- Golgi complex is present in the highest number in secretory cells. **[NEET 2019]**
- Golgi complex participates in the formation of secretory vesicles. **[NEET 2018]**
- Lysozymes are membrane bound structures, are active under acidic pH and have numerous hydrolytic enzymes. **[NEET 2019]**
- Lysosomes helps in digesting biomolecules. **[NEET 2019]**
- Lysosomes contain hydrolytic enzymes. **[NEET 2016]**
- Vacuoles help in traping waste and excretory products. **[NEET 2019]**
- Glyoxysomes are found in plant cell only. **[JIPMER 2019]**
- Microtubules are the constituents of spindle fibres, centrioles and cilia. **[NEET 2016]**
- Centriole helps in spindle formation. **[NEET 2020]**

Chapter Exercises

(A) Topical Exercise

Cell Theory, Cell Shape and Size, Cell Wall and Cell Membrane

1 Cell theory as understood today describes that
(a) All living organisms are composed of cells and products of cells
(b) All cells arise from pre-existing cells
(c) Plants and animals possess exactly same type of cellular components
(d) Both (a) and (b)

2 In view of the current status of our knowledge about the cell structure, which of the following statements about cell theory is correct?
(a) The cell theory needed modification due to discovery of subcellular structures such as chloroplasts and mitochondria
(b) The cell theory does not hold good, since all living organisms are not cellular in their organisation, e.g. virus
(c) The cell theory in its modified form, means that all living objects are made of cells capable of reproducing
(d) The cell theory means that all living objects of cell, whether or not capable of reproducing

3 Which of the following statements was not explained in the cell theory given jointly by Schleiden and Schwann?
(a) All living organisms are composed of cells and their products
(b) Formation of new cells
(c) Cell is the structural and functional unit of living organisms
(d) None of the above

4 Cell theory was formulated by
(a) Virchow
(b) Nageli
(c) Schleiden and Schwann
(d) Louis Pasteur

5 The transport of polar molecules across the cell membrane requires
(a) an ion of positive charge
(b) hydrophilic carrier ion
(c) hydrophobic carrier ion
(d) a carrier protein

6 How many of the given features are similar for both eukaryotic and prokaryotic cells?

Cell membrane, constituents of chromatin, mitochondria, cell wall with muramic acid, 70S ribosomes, porous nuclear membrane

(a) One (b) Two (c) Five (d) Three

7 Which one of these is not a eukaryote? **[NCERT Exemplar]**
(a) *Euglena* (b) *Anabaena* (c) *Spirogyra* (d) *Agaricus*

8 Different cells have different sizes. Arrange the following in an ascending order of their size.
I. Mycoplasma II. Ostrich eggs
III. Human RBC IV. Bacteria

Codes
(a) I, IV, III and II (b) I, II, III and IV
(c) II, I, IIII and IV (d) III, II, I and IV

9 Identify the mismatched pair with respect to different shapes of the cells.

	Cells	Shapes
(a)	Tracheids	Elongated
(b)	White blood cells	Round and biconcave
(c)	Nerve cells	Branched and long
(d)	Mesophyll cell	Round and oval

10 Fungal cell wall is made up of
(a) cellulose (b) chitin
(c) hemicellulose (d) glycoproteins

11 Cell wall
(a) provides mechanical support
(b) acts as barrier to pathogens
(c) maintain the shape of plant
(d) All of the above

12 Consider the following statements.
I. The cell wall of bacteria provides a strong structural support to prevent them from bursting and collapsing.
II. Bacteria can be classified as Gram positive and Gram negative on the basis of difference in cell wall.

Choose the correct option.
(a) Statement I is true, but II is false
(b) Statement I is false, but II is true
(c) Both statements I and II are true
(d) Both statements I and II are false

13 The primary cell wall is composed of
(a) 8-10% cellulose and 60% hemicellulose
(b) 2-5% cellulose and 50% hemicellulose
(c) 10-15% starch and 80% galactose
(d) 7-18% glucose and 50% hemicellulose

14 Which of the following features is common to prokaryotes and many eukaryotes? **[NCERT Exemplar]**
(a) Chromosomes present
(b) Cell wall present
(c) Nuclear membrane present
(d) Subcellular organelles present

15 Lomasomes are border organelles found in between cell wall and cell membrane. These are involved in
(a) cell elongation (b) cell elaboration
(c) cell wall formation (d) All of these

16 Which one of the following statements is not true for plasma membrane? **[NCERT Exemplar]**
(a) It is present in both plant and animal cells
(b) Lipid is present as a bilayer in it
(c) Proteins are present integrated as well as loosely associated with the lipid layer
(d) Carbohydrate is never found in it

17 Plasma membrane has
(a) phospholipid, extrinsic and intrinsic proteins
(b) no proteins, glycolipids and phospholipids
(c) histone proteins and lipoproteins
(d) None of the above

18 In a cell membrane, peripheral proteins
(a) lie on the surface of membrane
(b) lie in the matrix of membrane
(c) buried in the membrane
(d) None of the above

19 In a cell membrane, integral proteins are
(a) partially or totally buried in the membrane
(b) partially attached to the surface of the membrane
(c) permanently attached to the outer surface of the membrane
(d) temporarily attached to the surface of the membrane

20 A phospholipid molecule is amphipathic and bilayered structure. The head of phospholipid molecule is
(a) at an angle of 40°
(b) towards the outer side
(c) between the surfaces
(d) embedded in protein molecules

21 Who proposed the fluid mosaic model of plasma membrane? **[NCERT Exemplar]**
(a) Camillo Golgi (b) Schleiden and Schwann
(c) Singer and Nicolson (d) Robert Brown

22 In fluid mosaic model of plasma membrane
(a) upper layer is non-polar and hydrophilic
(b) polar layer is hydrophobic
(c) phospholipids form the middle part of bimolecular layer
(d) proteins form a middle layer

23 The factors, which influence membrane fluidity are
I. cholesterol
II. carbohydrates
III. percentage of unsaturated fatty acids
IV. receptors
Codes
(a) II and IV (b) III and IV (c) Only I (d) I and III

24 According to which model, a cell membrane consists of a phospholipid bilayer with a single continuous layer of protein molecules on either surface?
(a) Sandwich model (b) Lamellar theory
(c) Danielli and Davson model (d) All of these

25 The site for cell recognition in a plasma membrane is
(a) protein part of glycoproteins
(b) oligosaccharide of glycoproteins
(c) Both carbohydrate and protein components of glycoprotein
(d) lipid portion of glycoproteins

26 The important functions of the plasma membrane include
(a) transport of the molecules (b) cell recognition
(c) secretion (d) All of these

Protoplasm and Nucleus

27 Choose the incorrect matched pair.
(a) Cytoplasm - Present between nucleus and plasma membrane
(b) Nucleoplasm - Protoplasm of nucleus
(c) Deutoplasm - Non-living substance of protoplasm
(d) Huxley - Gave the term protoplasm

28 Identify *A*, *B*, *C* and *D* in the given figure.

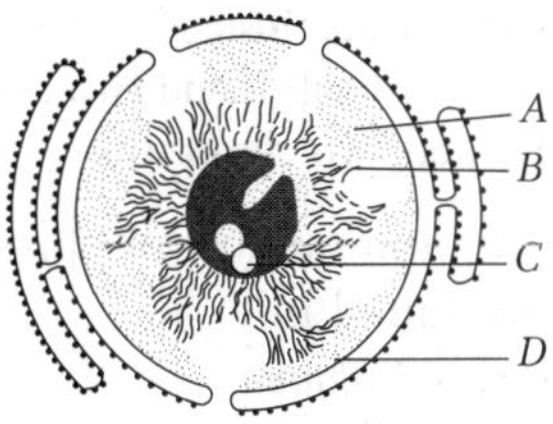

(a) *A*-Nucleoplasm, *B*-Chromatin fibre, *C*-Intranucleolar chromatin, *D*-Nuclear membrane
(b) *A*-Nucleolus, *B*-Nucleoplasm, *C*-Nuclear membrane, *D*-Nuclear pore
(c) *A*-Nuclear pore, *B*-Nuclear membrane, *C*-Nucleoplasm, *D*-Nucleolus
(d) *A*- Nuclear membrane, *B*-Nucleoplasm, *C*-Nuclear pore, *D*-Nuclear membrane

29 Nuclear membrane is continuous with
(a) rough endoplasmic reticulum
(b) smooth endoplasmic reticulum
(c) cell membrane
(d) Golgi bodies

30 Nucleolus is formed from specific sites on chromosomes known as ....
(a) nucleolar reorganising region
(b) nucleus forming region
(c) nucleolar organiser region
(d) None of the above

31 Which of the following statements regarding karyotheca is incorrect?
(a) Perinuclear space is 200-400 Å thick
(b) Outer and inner membrane of nucleus possess ribosomes
(c) Inner membrane has a dense layer called nuclear lamina
(d) Nuclear pore complex is consist of a pore and two rings called annuli

32 The term 'heterochromatin' was given by
(a) Bowman (b) Waldeyer
(c) Emil Heitz (d) Balbiani

33 Which of the following is not a difference between euchromatin and heterochromatin?
(a) Heterochromatin stains deeply while euchromatin lightly
(b) Heterochromatin is more condensed than euchromatin
(c) Heterochromatin is transcriptionally inactive while euchromatin is active
(d) Heterochromatin lies close to the nucleolus while euchromatin to nuclear lamina

34 Choose the correct one for the chromosome given below

(a) It is with one chromatid and one molecule of *ds*DNA
(b) It is with two chromatids and two molecules of *ds*DNA
(c) It is with two chromatids and two arms
(d) It is one arm and two chromatids

35 In plants, maximum number of chromosomes ($2n = 1262$) are found in
(a) *Ophioglossum reticulatum*
(b) *Haplopappus gracillis*
(c) *Saccharum officinarum*
(d) None of the above

36 An angiospermic plant with minimum number of chromosomes is
(a) *Haplopappus gracillis*
(b) *Ophioglossum reticulatum*
(c) *Asparagus ramosus*
(d) *Pisum sativum*

37 Identify the correct match between types of chromosomes in Column I and their descriptions in Column II.

Column I (Chromosomes)	Column II (Position of centromere)
A. Metacentric	1. At the tip
B. Submetacentric	2. Almost near the tip
C. Acrocentric	3. At the middle
D. Telocentric	4. Slightly away from the middle or L-shaped

Codes

	A	B	C	D		A	B	C	D
(a)	1	3	2	4	(b)	4	3	2	1
(c)	1	2	3	4	(d)	3	4	2	1

38 Consider the following statements
I. Centromere does not take up the stain.
II. Centromere is that part of chromosome where chromatids are attached.

Choose the correct option.
(a) Statement I is true, but II is false
(b) Statement I is false, but II is true
(c) Both statements I and II are true
(d) Both statements I and II are false

39 Identify the figure indicated by label *X* an choose the correct option.

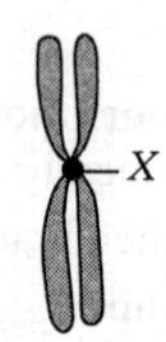

(a) Kinetochore–Site of spindle fibre attachment
(b) Centromere –Structure which provides the shape of chromosome
(c) Primary constriction–A disc-shaped structure on chromosome
(d) Secondary constriction–Non-staining part of chromosome at constant location

40 In human, secondary constrictions II is found on chromosome
(a) 1, 2, 7 and X (b) 1, 10, 13, 16 and Y
(c) 12, 14, 17, 20 and X (d) 5, 6, 8 and Y

41 Dicentric chromosomes have two centromeres because they possess additional centromere named as
(a) neocentromere (b) lethal centromere
(c) diffused centromere (d) None of these

42 Which of the following stains is not used for staining chromosomes? **[NCERT Exemplar]**
(a) Basic fuchsin (b) Safranin
(c) Methylene blue (d) Carmine

43 Polytene chromosomes are also called
(a) lampbrush chromosome
(b) B-chromosome
(c) salivary gland chromosome
(d) sex chromosome

44 Which of the following statement(s) is incorrect regarding the band region of polytene chromosome?
(a) Feulgen negative area
(b) Absorbs ultraviolet light at 2600 Å
(c) Chromonemata is tightly packed
(d) Stains intensely with basic stains

45 Karyotyping helps in detection of
(a) evolutionary relationships
(b) chromosomal aberration
(c) primitive features of an organism
(d) All of the above

46 Which of the following are used to define the karyotype of a species?
(a) The number of chromosomes
(b) The chromosome length
(c) The positions of the centromeres
(d) All of the above

47 Karyotype is
(a) division of nucleus
(b) chromosomes complement specific for each species
(c) all organisms possessing same type of chromosome
(d) None of the above

48 In a cell that is not dividing, the chromosomes are visible as
(a) chromomere (b) microtubules
(c) chromatin (d) chromosomes

Mitochondria and Plastids

49 Consider the following statements.
I. Mitochondria and plastids are semiautonomous cell organelles.
II. Mitochondria and plastids have their own-replication, transcription and translation machinery.
Choose the correct option.
(a) Statement I is true, but II is false
(b) Statement I is false, but II is true
(c) Both statements I and II are true
(d) Both statements I and II are false

50 The 'Powerhouse' of cell
(a) is bound by a single membrane
(b) possess cristae which are extension of its outer membrane
(c) are sites of formation of energy currency of the cell
(d) is found in eukaryotic and prokaryotic cells

51 The highest number of mitochondria are present in
(a) parenchyma cells
(b) sieve tubes
(c) meristematic tissues
(d) None of the above

52 The stain used to visualise mitochondria is **[NCERT Exemplar]**
(a) fast green (b) safranin
(c) acetocarmine (d) janus green

53 Which of the following correctly explains mitochondrial function and results?
(a) Oxidative phosphorylation, dephosphorylation, metabolic water production
(b) Dephosphorylation, metabolic water production, CO_2 production
(c) Oxidative phosphorylation, metabolic water production, CO_2 production
(d) Oxidative phosphorylation, dephosphorylation, CO_2 production

54 Select the incorrect statement from the following.
(a) Both chloroplasts and mitochondria contain an inner and an outer membrane
(b) Both chloroplasts and mitochondria have an internal compartment, the thylakoid space bounded by the thylakoid membrane
(c) Both chloroplasts and mitochondria contain DNA
(d) The chloroplasts are generally much larger than mitochondria

55 Plastid differs from mitochondria on the basis of one of the following features **[NCERT Exemplar]**
(a) presence of two layers of membrane
(b) presence of ribosome
(c) presence of chlorophyll
(d) presence of DNA

56 Among the used options, lowest number of mitochondria is found in
(a) human sperm
(b) human kidney cell
(c) liver cell
(d) egg of sea urchin

57 All plastids have essentially the same structure because
(a) they have to perform the same function
(b) one type of plastids can differentiate into another type of plastids depending upon the requirements
(c) all plastids have to store starch, lipids and proteins
(d) they are localised in the aerial parts of the plants

58 Plastids in a plant cell
(a) were discovered by Haeckel
(b) are found freely
(c) are double membrane bound structures
(d) All of the above

59 Different types of plastids (chloroplast, chromoplast and leucoplast) are quite related as
(a) these all store lipids and starch
(b) these are restricted to the green parts of a plant
(c) these all perform almost same function
(d) these are interconvertible depending upon the requirement of an organ

60 The pigment molecules of a chloroplast are located within
(a) its thylakoid membranes
(b) the space between the inner and outer membranes
(c) the inner membrane
(d) intra thylakoid spaces

61 Find out the correct sequence of the labelling of the given figure.

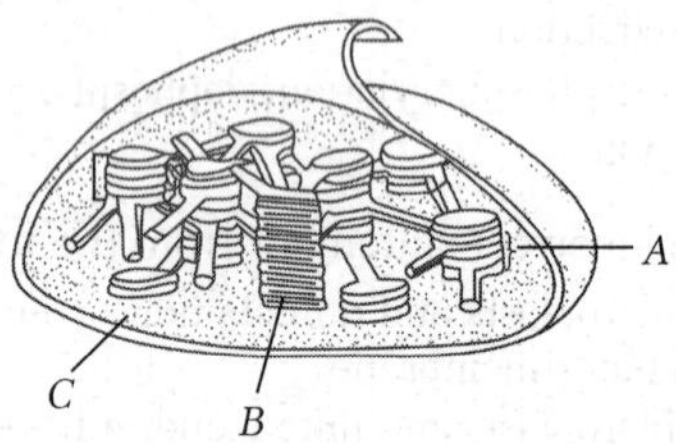

(a) *A*-Outer membrane, *B*-Inner membrane, *C*-Thylakoid
(b) *A*-Thylakoid, *B*-Granum, *C*-Thylakoid space
(c) *A*-Granum, *B*-Thylakoid, *C*-Stroma
(d) *A*-Granum, *B*-Thylakoid, *C*-Thylakoid space

62 Choose the wrongly matched pair.
(a) Amyloplasts – Starch
(b) Elaioplasts – Oils and fats
(c) Proteinoplasts – Proteins
(d) Chromoplasts – Colourless

63 Which of the following is not true for a eukaryotic cell? **[NCERT Exemplar]**
(a) It has 80S type of ribosome present in the mitochondria
(b) It has 80S type of ribosome present in the cytoplasm
(c) Mitochondria contain circular DNA
(d) Membrane bound organelles are present

Ribosomes

64 Ribosomes are so called because of high
(a) ribose content (b) RNA content
(c) Both (a) and (b) (d) None of these

65 Sedimentation coefficient or Svedberg unit is equal to
(a) 1×10^{-12}S (b) 1×10^{-7}S
(c) 1×10^{-13}S (d) 1×10^{-20}S

66 Select one, which is not true for ribosome.
(a) Made up of two subunits **[NCERT Exemplar]**
(b) Form polysome
(c) May attach to *m*RNA
(d) Have no role in protein synthesis

67 If the ribosomes of a cell are destroyed, then
(a) fats will not stored
(b) proteins will not be formed
(c) carbon assimilation will not occur
(d) respiration will not take place

68 The proteins associated with ribosomes are
(a) negatively charged
(b) positively charged
(c) amphoteric
(d) None of the above

69 The two subunits of ribosome remain united at a critical ion level of
(a) copper (b) manganese
(c) magnesium (d) calcium

70 Polysome is formed by
(a) several ribosomes attached to a single *m*RNA
(b) many ribosomes attached to a strand of endoplasmic reticulum
(c) a ribosome with several subunits
(d) ribosomes attached to each other in a linear arrangement

71 In an animal cell, protein synthesis takes place
(a) only on the ribosomes present in the cytosol
(b) only on the ribosomes attached to nuclear envelope and endoplasmic reticulum
(c) on ribosomes present in the nucleolus as well as in cytoplasm
(d) on ribosomes present in the cytosol as well as in the mitochondria

72 Identify the following figure and select the incorrect statement.

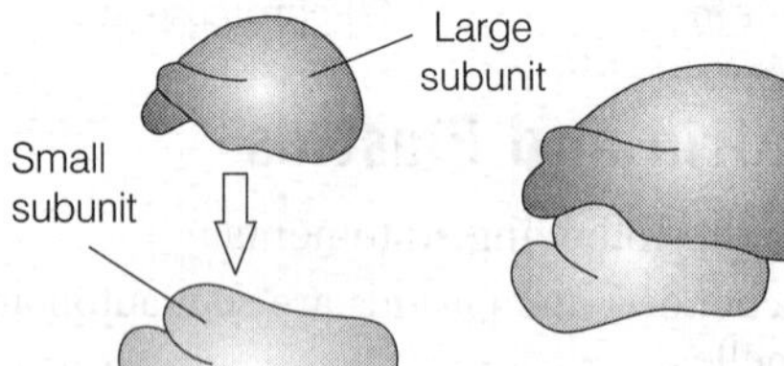

(a) First observed by George Palade under the electron microscope as dense particle
(b) They are composed of *r*RNA and proteins
(c) Single membrane bound cell organelles
(d) They are also known as protein factories

Endoplasmic Reticulum, Golgi Complex and Lysosomes

73 What is true about endoplasmic reticulum?
(a) ER membrane is 80-100Å thick
(b) ER consists of 70% phospholipids
(c) ER vesicles are free of ribosomes
(d) All of the above

74 The site for steroidogenesis is
(a) rough endoplasmic reticulum
(b) ribosomes
(c) smooth endoplasmic reticulum
(d) polyribosomes

75 Which of the following statements is true for a secretory cell? **[NCERT Exemplar]**
(a) Golgi apparatus is absent
(b) Rough Endoplasmic Reticulum (RER) is easily observed in the cell
(c) Only Smooth Endoplasmic Reticulum (SER) is present
(d) Secretory granules are formed in nucleus

76 A structure found traversing the cells and discovered by Porter, Claude and Fullam in 1945 is the
(a) Golgi apparatus (b) mitochondria
(c) endoplasmic reticulum (d) lysosomes

77 The RER in the cell, synthesised a protein which would be later used in building the plasma membrane. But, it is observed that the protein in the membrane is slightly different from the protein made in the another cell organelle. Identify that organelle in the given diagram.

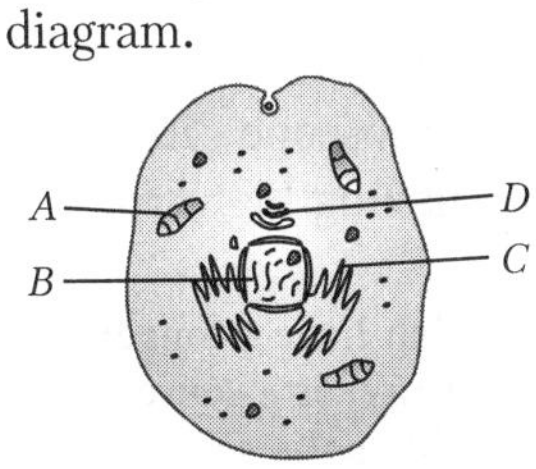

(a) *D* (b) *A* (c) *B* (d) *C*

78 How many of the given features are true for both endoplasmic reticulum and Golgi complex?

Presence of membrane, formation or modification of proteins, endomembrane system, DNA, attached ribosomes, tubules and vesicles

(a) Four (b) Six
(c) Five (d) Three

79 Consider the following statements.
I. Golgi apparatus is absent in bacteria and blue-green algae.
II. Golgi body is responsible for the origin of lysosome.
Choose the correct option.
(a) Statement I is true, but II is false
(b) Statement I is false, but II is true
(c) Both statements I and II are true
(d) Both statements I and II are false

80 Lysosomes contain a rich repertoire of
(a) hydrolases (b) isomerases
(c) transferases (d) oxido-reductases

81 Which of the following organelles are called as 'garbage disposal systems'?
(a) Peroxisomes
(b) Golgi complex
(c) Phagosomes
(d) Lysosomes

82 Which type of lysosomes contribute to the ageing process?
(a) Primary lysosomes
(b) Secondary lysosomes
(c) Autophagic vacuoles
(d) Residual bodies

83 Some scientists include lysosomes in microbodies, but these can be differentiated due to
(a) their mode of formation (b) their small size
(c) presence of enzymes (d) All of these

84 Lysosomes are single membrane bound organelles. Which of the following statements is correct?
(a) Contain 40-50 types of hydrolytic enzymes
(b) Content has a pH of 8.5
(c) Membrane is highly prone to self-digestion
(d) All of the above

Vacuoles, Microbody Cell Inclusions and Specialised Cell Structures

85 Vacuole in a plant cell
(a) is membrane bound and contains storage proteins and lipids
(b) is membrane bound and contains water and excretory substances
(c) lacks membrane and contains air
(d) lacks membrane and contains water and excretory substances

86 In a prokaryotic cell, the inclusion bodies
(a) are not bounded by any membrane system
(b) lie freely in the cytoplasm
(c) Both (a) and (b)
(d) None of the above

87 Which of the following are cell inclusions?
I. Aleurone grains II. Fat droplets
III. Crystals IV. Vacuoles
Codes
(a) I, II and III (b) I and III
(c) III and IV (d) All of these

88 The organelle which consists of enzymes such as catalase, urease, oxidase, luciferase, etc., and involved in a waste process is ....
(a) peroxisomes (b) lysosomes
(c) lomasomes (d) None of these

89 Which of the following is not a function of cytoskeleton in a cell? **[NCERT Exemplar]**
(a) Intracellular transport
(b) Maintenance of cell shape and structure
(c) Support of the organelle
(d) Cell motility

90 Cytoskeleton is made up of
(a) calcium phosphate granules
(b) callose deposits
(c) cellulosic microfibrils
(d) proteinaceous filaments

91 The cytoskeleton in a cell is involved in many functions such as
(a) mechanical support
(b) motility
(c) maintenance of the shape
(d) All of the above

92 Number of protofilaments in a microtubule of mammalian is
(a) 13 (b) 12 (c) 5 (d) 10

93 The filaments that specialise in moving organelles are
(a) microfilaments (b) intermediate filaments
(c) actin (d) microtubules

94 The cross-section of a centriole show

	Peripheral microtubule (triplet)	Central microtubule	Hub	Radial spoke
(a)	27	0	2	9
(b)	9	0	1	9
(c)	9	2	1	9
(d)	27	2	1	18

95 Section of eukaryotic cilia or flagella shows

	Peripheral microtubule (doublets)	Central microtubule	Radial spoke	Central sheath
(a)	9	2	9	1
(b)	9	2	9	2
(c)	18	2	9	1
(d)	18	1	7	1

Tools and Techniques

96 Consider the following statements.
I. The resolution of microscope can be increased by increasing numerical aperture.
II. The resolving power of electron microscope is 1 million times greater than that of a light microscope.

Choose the correct option.
(a) Statement I is true, but II is false
(b) Statement I is false, but II is true
(c) Both statements I and II are true
(d) Both statements I and II are false

97 The limit of resolution of an instrument is
(a) the ability to distinguish two distant objects as close
(b) the ability to distinguish two close objects as distant
(c) ability to distinguish two or more objects as close
(d) None of the above

98 Electron microscope is used for
(a) counting microbes in a sample
(b) whole mount study
(c) separating constituents of mixture
(d) viewing ultrastructure of the cell

99 The compound microscope is based upon the principle that
(a) the image formed by one lens acts as object for second lens
(b) a single lens system magnifies all diameters of the object being viewed
(c) Both (a) and (b)
(d) None of the above

100 The electron microscope differs from the ordinary microscope in having high resolving power due to the use of
(a) electrons, which have a longer wavelength than ordinary light ray, as a source of illumination
(b) ultraviolet light as the source of illumination
(c) electrons, which have a shorter wavelength than ordinary light ray, as a source of illumination
(d) gamma rays as the source of illumination

101 In interference microscope, the source of light is
(a) beam of electrons (b) daylight
(c) ultraviolet rays (d) None of these

102 Dark field microscopy that is used to observe bacteria makes use of
(a) no light
(b) visible light
(c) invisible infrared
(d) invisible UV-light

103 Depth of an object can be studied through
(a) dark field microscope
(b) compound microscope
(c) phase contrast microscope
(d) scanning tunneling microscope

104 Phase contrast microscopy is
(a) related to retardation and thickness of the object
(b) based on light scattering and uses a dark field condenser
(c) used for the study of living cells
(d) the best method for studying non-living ultrastructure

105 Why cannot the living cells be observed under electron microscopy?
(a) Tremendous voltage kills the cells
(b) Cells get killed during the preparation of material
(c) Dehydration of cells is must to avoid water vapours, which disturb the vacuum
(d) All of the above

106 Confocal scanning microscope was invented by
(a) Lehmann (b) Max Haitinger
(c) Marvin Minsky (d) Knoll and Ruska

107 The resolving power of scanning electron microscope is
(a) 0.001 nm (b) 0.01 nm
(c) 20 nm (d) 0.0001 nm

108 Match the following columns.

	Column I		Column II
A.	W His	1.	First confocal microscope
B.	Marvin Minsky	2.	Phase contrast microscope
C.	Zernike	3.	Microtome
D.	Zsigmondy	4.	Ultramicroscope

Codes

	A	B	C	D
(a)	3	1	2	4
(b)	2	1	3	4
(c)	4	3	2	1
(d)	3	2	1	4

109 A fluorescence microscope is used to
(a) obtain structure of DNA
(b) increase resolving power
(c) identify microbes in tissues
(d) study ultrastructural details

110 An instrument used for obtaining extremely thin sections for microscopy is
(a) diamond knife (b) microtome
(c) meat slicer (d) tissue slicer

111 Cell fractionation involves
(a) centrifugation (b) homogenisation
(c) Both (a) and (b) (d) microtomy

112 Ion exchange chromatography is used for
(a) purification of insulin
(b) for separating mixture of tissue lipids
(c) for separation of *m*RNA
(d) high separation of *m*RNA

113 The chromatography used to separate substances that can be vaporised is ....
(a) gas chromatography
(b) paper chromatography
(c) gel filtration chromatography
(d) thin layer chromatography

114 For separating amino acids, which of the following types of chromatography is used?
(a) Column chromatography
(b) Ion exchange chromatography
(c) Paper chromatography
(d) Gel filtration chromatography

115 Which of the following techniques is used to determine the molecular weight of proteins?
(a) Affinity chromatography
(b) Thin layer chromatography
(c) Ion exchange chromatography
(d) Gel filtration chromatography

116 Watson and Crick used .... technique for determining the double-helical structure of DNA.
(a) two-dimensional PAGE (b) equilibrium sedimentation
(c) X-ray crystallography (d) spectrophotometry

117 In recent years, considerable attention is being given to DNA protein interactions because the expression of genes is regulated through banding of specific proteins on the regulatory DNA sequences. This pattern of protein banding on DNA can be studied by
(a) light microscope (b) ultracentrifugation
(c) electron microscope (d) X-ray crystallography

118 The process of rupturing cells and isolating its constituents is
(a) chromatography
(b) SEM
(c) cell fractionation method
(d) phase contrast microscope

119 The arrangement of atoms in a molecule can be studied using
(a) centrifugation (b) histochemistry
(c) spectrophotometer (d) X-ray crystallography

120 Protein crystallography using X-ray was discovered by
(a) Max Perutz (b) Knoll and Ruska
(c) John Kendrew (d) Both (a) and (c)

121 Consider the following statements.
I. Fungal hyphae can be specifically stained by methylene blue.
II. Fatty substances are stained using sudan black.
Choose the correct option.
(a) Statement I is true, but II is false
(b) Statement I is false, but II is true
(c) Both statements I and II are true
(d) Both statements I and II are false

122 Protein substances are generally stained with
(a) iodine (b) light green
(c) cotton blue (d) ruthenium red

123 Lignin (red) is stained by which dye?
(a) Sudan black (b) Chlorozinciodine
(c) Phloroglucinol (d) Eosin

124 The instrument used in spectrophotometry is
(a) Spectrophotometer (b) Microtome
(c) Tracer (d) None of these

125 Which is the suitable technique to obtain correct information regarding the location and synthesis of new DNA?
(a) Electron microscopy
(b) X-ray crystallography
(c) Immunoelectrophoresis
(d) Use of radioactive precursors of DNA and autoradiography

126 Autoradiography technique is used to
(a) find out the absorption maximum of chlorophyll pigments
(b) study photorespiration in certain plants
(c) establish that the oxygen evolved during photosynthesis is by photolysis of water molecules
(d) trace the path of carbon in photosynthesis

127 Phase contrast microscope was invented by
(a) Knoll and Ruska (b) W His
(c) Fritz-Zernike (d) Z Janssen

128 Which of the following techniques, other than microscopy, is used for the study of cell?
(a) Maceration (b) Plasmolysis
(c) Chromatography (d) Autoradiography

129 If a biochemical analysis of mitochondria is to be done, the best procedure would be to
(a) plasmolyse the cells, filter it and take the debris
(b) select cells having large mitochondria
(c) grind the cell and filter out the mixture
(d) subject the cells to fractionation (centrifuge) and obtain mitochondria

130 Cell organelle can be separated by
(a) X-ray diffraction (b) autoradiography
(c) thin-section microtomy (d) differential centrifugation

131 Name the technique other than microscopy that can be used for the study of a cell.
(a) Obliteration (b) Plasmolysis
(c) Flow cytometry (d) None of these

(B) Cumulative Exercise

(Assorted Questions of the Chapter for Advanced Level Practice)

1 Consider the following statements and choose the correct one.

I. Plant cells have centrioles, which are absent in almost all animal cells.
II. Ribosomes are the site of protein synthesis.
III. The middle lamella is a layer mainly of calcium carbonate, which holds the different neighbouring cells together.
IV. In animal cells, steroidal hormones are synthesised by smooth endoplasmic reticulum.

Codes
(a) I and III (b) II and IV
(c) Only IV (d) I, II, III and IV

2 Match the following columns.

Column I	Column II
A. Nucleolus	1. Lipid storage
B. Spherosomes	2. Detoxification
C. Peroxisomes	3. Transport of macromolecules
D. Plasmodesmata	4. RNA synthesis

Codes

	A	B	C	D		A	B	C	D
(a)	4	1	3	2	(b)	1	2	4	3
(c)	4	1	2	3	(d)	1	2	3	4

3 The given figure shows a fluid mosaic model of plasma membrane. Identify correctly the four labels *A*, *B*, *C* and *D*.

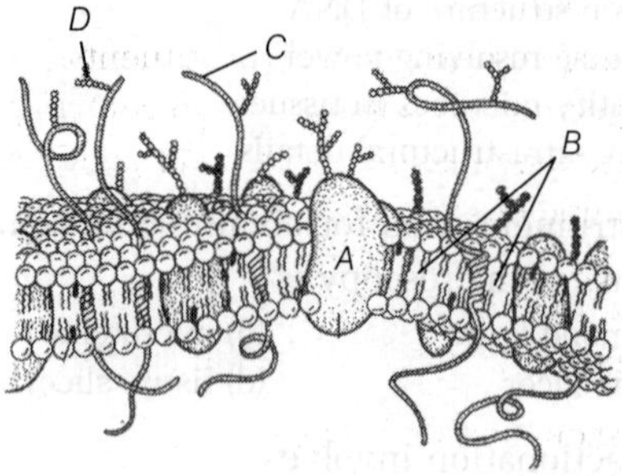

	A	B	C	D
(a)	Integral protein	Phospholipid bilayer	Peripheral protein	Sugar
(b)	Transmembrane protein	Phospholipid bilayer	Cholesterol	Oligosaccharide
(c)	Integral protein	Peripheral protein	Glycocalyx	Cholesterol
(d)	Transmembrane protein	Phospholipid bilayer	Oligosacchar-ide	Glycolipids

4 Keeping in view, 'the fluid mosaic model' for the structure of cell membrane, which one of the following statements is correct with respect to the movement of lipids and proteins from one lipid monolayer to the other (described as flip-flop movement)?
(a) Both lipids and proteins can flip-flop
(b) While lipids can rarely flip-flop, proteins cannot
(c) While proteins can flip-flop, lipids cannot
(d) Neither lipids, nor proteins can flip-flop

5 See the following figure and identify it.

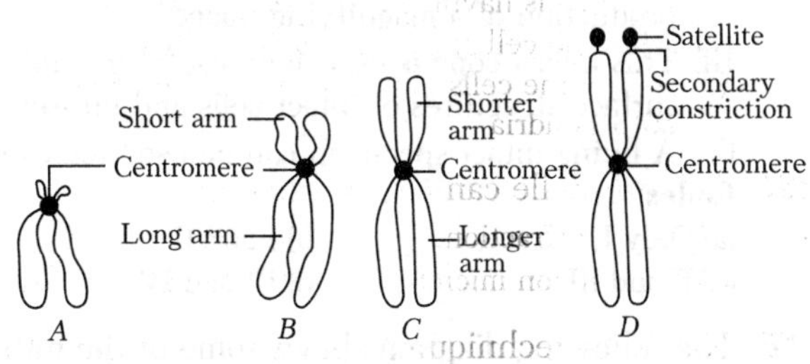

	A	B	C	D
(a)	Metacentric chromosome	Submetacentric chromosome	Acrocentric chromosome	Telocentric chromosome
(b)	Submetacentric chromosome	Metacentric chromosome	Telocentric chromosome	Acrocentric chromosome
(c)	Acrocentric chromosome	Telocentric chromosome	Metacentric chromosome	Submetacentric chromosome
(d)	Telocentric chromosome	Acrocentric chromosome	Submetacentric chromosome	Metacentric chromosome

6 Cell fractionation involves
I. homogenisation of cell constituents.
II. subjecting the homogenate to centrifugal force.
III. detection of slight variation of amino acids.
IV. staining of structures of living cells.

Codes
(a) Only II (b) I and II
(c) III and IV (d) II and IV

7 Fluorescence microscope
I. used to detect specific proteins or other molecules in cells and tissues.
II. uses UV light with higher wavelengths (3500-4000Å).
III. is useful mainly for viewing highly ordered objects such as crystals.
IV. depends on changes in the speed of light for its functioning.

Codes
(a) I and III (b) I and II
(c) Only II (d) I and IV

8 Correlate the given features of animal cells (I-V) with their respective parts (*A-N*).

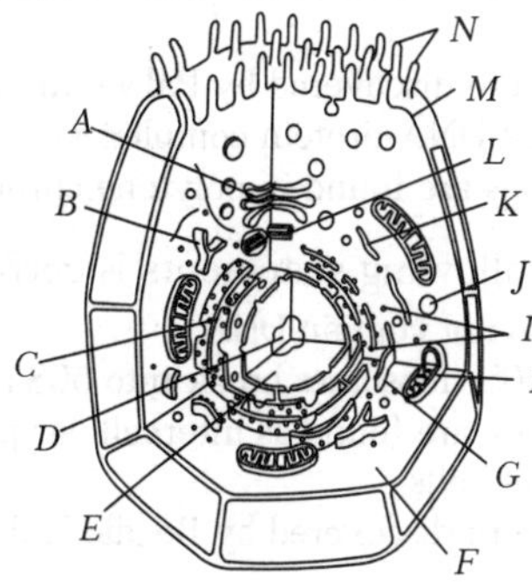

I. Major site for synthesis of lipid.
II. Powerhouse of the cell.
III. Storehouse of digestive enzyme.
IV. Site of glycolysis.
V. Site for active ribosomal RNA synthesis.

The correct option is

	I	II	III	IV	V
(a)	G	H	E	F	D
(b)	G	H	J	F	D
(c)	B	H	J	F	D
(d)	A	H	J	F	D

9 Match the following columns.

	Column I		Column II
A.	Dictyosomes	1.	Osmoregulation
B.	Mitochondria	2.	Photosynthesis
C.	Vacuoles	3.	Transport
D.	Grana	4.	Secretion
		5.	Respiration

Codes

	A	B	C	D
(a)	4	5	1	2
(b)	3	4	2	1
(c)	4	5	3	2
(d)	4	3	1	2

10 Match the following columns.

Column I		Column II
A. Karyolymph	1.	Nucleolus
B. Ribonucleoprotein	2.	Nucleus
C. Spindle fibre	3.	DNA
D. Genes	4.	Centrioles
E. Rough endoplasmic reticulum	5.	Protein synthesis

Codes

	A	B	C	D	E
(a)	1	3	2	4	5
(b)	4	2	3	5	1
(c)	2	1	4	3	5
(d)	1	2	3	4	5

11 Which of the following statements is correct?
(a) The genetic material of prokaryotes is not enclosed in a cell
(b) Ribosomes were discovered by Palade in plant cells
(c) Ribosomes are DNA-protein complexes
(d) Balbiani's rings are found in polytene chromosomes

12 Which of the following statements is correct?
(a) Ribosomes do not contain DNA
(b) Eukaryotic 80S ribosomes break into 50S and 30S
(c) Plasmodesmata are found as intercellular junction between plant cells
(d) Ribosomes were discovered by Beadle and Tatum

13 Read the given statements and identify the incorrect statement.
I. Nucleus controls cytoplasm functioning by sending out RNAs.
II. Nucleus was discovered by Robert Brown in orchid root cells.
III. Nuclear membrane often possesses attached ribosomes.
Codes
(a) Only I (b) II and III
(c) I, II and III (d) None of these

14 Read the following statements and choose the option with all correct statements.
I. Size of a cell depends upon nucleoplasmic index.
II. Occurrence of different types of tissues, organs and organ system results in division of labour.
III. '*Omnis cellula-e-cellula*' was experimentally demonstrated by Nageli.
IV. Viruses are an exception to cell theory.
Codes
(a) I, II, III and IV (b) II and III
(c) I, II and IV (d) None of these

15 Consider the following statements and choose the incorrect ones.
I. Yolk bodies, lipid droplets, pigments, etc., constitute the deutoplasm.
II. Leeuwenhoek experimentally proved the role of nucleus in control of heredity.
III. Cells that possess single nuclei are also called as coenocyte.
IV. Nucleus occupies basal position in adipose tissues.
Codes
(a) Only III (b) I and II
(c) III and IV (d) None of these

16 Read the following statements regarding microbodies and choose the incorrect statements.
I. Peroxisomes were discovered in brain cells of animals.
II. Spherosomes have affinity for fat stains.
III. Lomasomes take part in cell membrane formation.
IV. Glyoxysomes are involved in glycolysis.
Codes
(a) I and II (b) I, III and IV
(c) Only III (d) None of these

17 Which of the following statements are correct for the transmission electron microscope?
I. Electron beam passes through the specimen and an electromagnetic objective lens which magnifies the image.
II. A single electron beam is used throughout for the production of a magnifying image.
III. This microscope is especially useful in studying the surface structures of intact cells and viruses.
IV. A living intact specimen can be used for viewing.
Codes
(a) Only I (b) I and II
(c) II and III (d) III and IV

18 The following diagram shows some of the missing structures in a plant cell (*A-E*). Identify the structures.

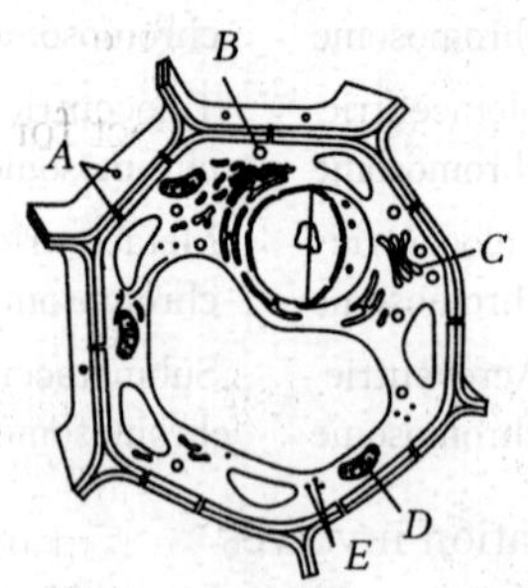

(a) *A*-Plasmodesmata, *B*-Rough endoplasmic reticulum, *C*-Golgi apparatus, *D*-Mitochondrion, *E*-Ribosomes
(b) *A*-Desmosome, *B*-Rough endoplasmic reticulum, *C*-Golgi apparatus, *D*-Mitochondrion, *E*-Ribosomes
(c) *A*-Plasmodesmata, *B*-Lysosome, *C*-Golgi apparatus, *D*-Mitochondrion, *E*-Ribosomes
(d) *A*-Tight junction *B*-Rough endoplasmic reticulum, *C*-Golgi apparatus, *D*-Mitochondrion, *E*-Ribosomes

19 Choose the correct statement.
(a) Ribophorins are proteins that attach ER membrane with ribosomes
(b) Y-chromosome is placed in group-C due to large size
(c) H3 is plugging protein
(d) SAT are regions where DNA content is highest

20 Identify the incorrect statement.
(a) Glycoprotein and glycolipids of plasma membrane facilitate cellular recognition and adhesion
(b) The semipermeable membrane surrounding the vacuole is called tonoplast
(c) Gametes of plants are without cell wall
(d) Ingestion of solid particles is called pinocytosis

21 Which of the following are correct for inner mitochondrial membrane?

I. The electron transfer and ATP synthesising system are present in inner mitochondrial membrane.
II. It bears small granules called oxysomes.
III. It contains protein called porins.
IV. It contains less than 10% of the total protein of mitochondria.

Codes

(a) I and IV (b) I and II
(c) I, II and III (d) II and III

22 Read the following statements carefully with respect to lysosomes.

I. Lysosomes are membrane less vehicular structure.
II. They are formed by the process of packaging in Golgi apparatus.
III. They are rich in hydrolytic enzyme.
IV. The enzymes found in them are active at basic pH.
V. They are also called triggers for cell division.
VI. They contain minerals, sap, vitamins and sucrose.

How many of the above statement are correct?

(a) Two (b) Three
(c) Four (d) Five

23 Which of the following is correct for the diagram given below?

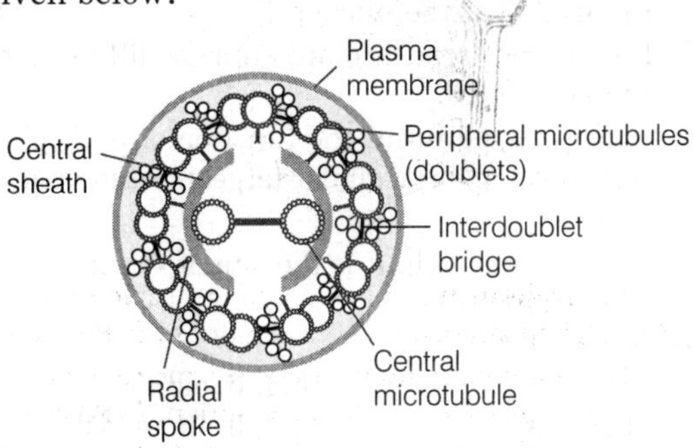

(a) It represents prokaryotic flagella
(b) It possess 9 + 0 microtubular organisation
(c) Both (a) and (b)
(d) It represents hair-like outgrowths of cell membrane in eukaryotic ciliates and flagellates

24 Match the following columns.

Column I	Column II
A. Sap-vacuole	1. Buoyancy
B. Contractile vacuole	2. Anthocyanin
C. Food-vacuole	3. Excretion
D. Gas-vacuole	4. Engulfing of food

Codes

	A	B	C	D		A	B	C	D
(a)	1	4	3	2	(b)	2	3	4	1
(c)	2	3	1	4	(d)	3	1	4	2

25 Correctly arrange the sequence of Gram staining in bacteria.

I. Treated with alcohol.
II. Excess stain washed with water.
III. Stain with crystal violet dye.
IV. 0.5% iodine added.
V. Observe under microscopic field for violet/colourless bacteria.

Codes

(a) II $\rightarrow$ III $\rightarrow$ IV $\rightarrow$ I $\rightarrow$ V
(b) III $\rightarrow$ I $\rightarrow$ II $\rightarrow$ V $\rightarrow$ IV
(c) I $\rightarrow$ II $\rightarrow$ III $\rightarrow$ IV $\rightarrow$ V
(d) III $\rightarrow$ IV $\rightarrow$ I $\rightarrow$ II $\rightarrow$ V

26 Read the following statements.

I. Peroxisomes are involved in photorespiration of the non-green cell of plants and also perform β-oxidation of fatty acids in animal cells.
II. Mitochondria possess enzymes for the β-oxidation of fatty acids.
III. The fluid-dynamic mosaic model of plasma membrane explains both structural and functional aspects.
IV. The movement of cilia and flagella is due to the presence of dynein proteins.
V. Lysosomes originate from Golgi.

How many of the above statements are correct?

(a) Two (b) Three (c) Four (d) Five

27 Read the following statements.

I. Cilia work like oars.
II. Cilia are relatively longer than flagella.
III. Prokaryotic and eukaryotic flagella are structurally similar.
IV. Cilia and flagella are covered with plasma membrane in eukaryotes.
V. The core of cilia/flagella is called axoneme which has nine pairs of doublets of radially arranged peripheral microtubules and a pair of centrally located microtubules.

How many of the above statements are correct?

(a) Three (b) Two (c) Four (d) Five

28 Match the following columns.

Column I	Column II
A. ER	1. Lipid/steroid synthesis
B. Golgi	2. cis/trans face
C. Lysosome	3. Tonoplast
D. Vacuole	4. Hydrolases

Codes

	A	B	C	D		A	B	C	D
(a)	1	2	3	4	(b)	4	3	2	1
(c)	2	1	4	3	(d)	1	2	4	3

29 Error/s is/are made by the student during pictorial representation of chromosome.

Study the diagram and select which of the following holds true regarding error.

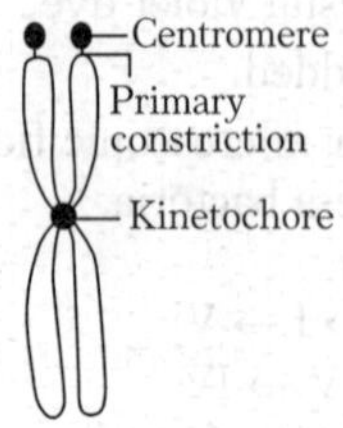

I. Centromere shown in figure should be labelled as satellite.

II. Primary constriction labelled in figrue is actually secondary constriction.

III. Kinetochore labelled in figure must be replaced by centromere.

Codes

(a) Only I (b) Only III (c) I and II (d) I, II and III

30 Choose the correct statement.

(a) Peripheral fibril of centriole is doublet
(b) Axoneme is not covered by plasma membrane
(c) Kinetochores are found on secondary constriction also
(d) Fungi possess unicisternal Golgi

31 Select the false statement.

(a) Ribosomes and centriole are non-membrane bound organelles of a cell
(b) Ribosomes are enveloped by a double membrane
(c) Plasmodesmata are the cytoplasmic connections between plant cells
(d) Cell membrane is semipermeable or selectively permeable

32 The given diagram shows a chromosome.

Which of the following table refers correctly to the chromosome?

	No. of centromere	No. of kinetochore	No. of arms
(a)	1	2	2
(b)	2	2	4
(c)	1	2	4
(d)	2	1	4

33 Choose the incorrect match.

(a) Active transport – Na^+/K^+ ATPase
(b) Flip-flop – Movement of protein
(c) Heteropycnosis – Differential staining
(d) Alphoid DNA – Centromere

34 The correct match is

(a) Central nucleus – Fat cells
(b) Peripheral nucleus – Embryonic cell
(c) Nucleonema – Nucleolus
(d) Hammerling – Orchid cells

35 Read the following statements.

I. It is the extension of plasma membrane into the cytoplasm.

II. It helps in cell wall formation, DNA replication and secretion processes.

The above features are attributed to bacterial

(a) genophore (b) mesosome
(c) pili (d) fimbriae

36 Identify the figure and choose the correct option which has correct set of statements regarding the same.

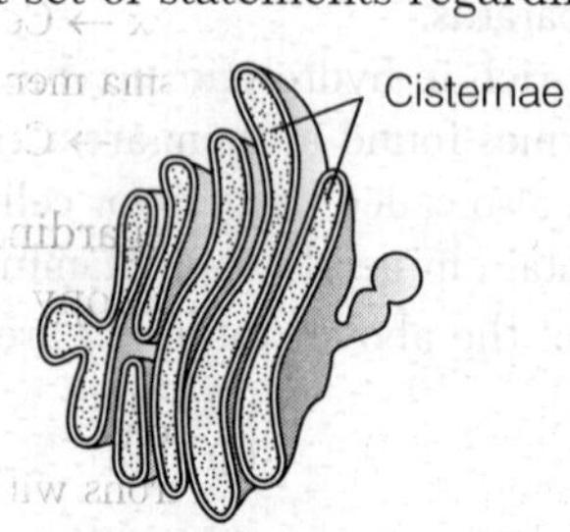

I. Flat disc-shaped cisternae are concentrically arranged near the nucleus.

II. The *cis* and *trans* face are entirely different, but interconnected.

III. Principally performs the function of packaging materials for intracellular targets or secreted outsied the cell.

IV. In the animal cell, it is the site of steroid hormone synthesis.

Codes

(a) Only I (b) I, II and IV
(c) I, II and III (d) I, II, III and IV

37 The figure given below shows the structure of a chloroplast with its four parts labelled as *A*, *B*, *C* and *D*. Select the part incorrectly matched with its function.

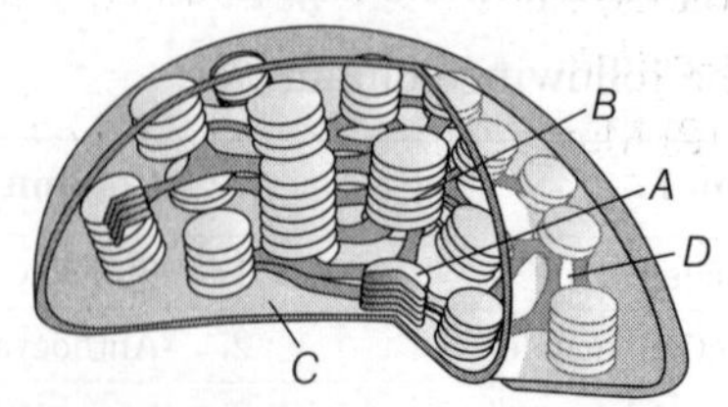

(a) *A*–Thylakoid–Contains chlorophyll pigment in its membrane
(b) *B*–Granum–Site of light independent phase
(c) *C*–Stroma–Contains enzymes required for synthesis of carbohydrates and proteins
(d) *D*–Inner membrane–Relatively less permeable than outer membrane

38 Match the following columns.

	Column I		Column II
A.	Mycoplasma	1.	Longest cell
B.	Human RBC	2.	7 μm in diameter
C.	Nerve cell	3.	10-20 μm
D.	A typical eukaryotic cell	4.	0.3 μm in length
		5.	Largest isolated single cell

Codes

	A	B	C	D		A	B	C	D
(a)	4	2	1	3	(b)	4	3	5	2
(c)	2	3	5	4	(d)	2	4	5	1

39 The bacterial cell envelope consisting of a tightly bound three layered structure from outside to inside is

(a) Cell wall → Glycocalyx → Plasma membrane
(b) Plasma membrane → Glcocalyx → Cell wall
(c) Glcocalyx → Cell wall → Plasma membrane
(d) Glcocalyx → Plasma membrane → Cell wall

40 Choose the incorrect answer regarding why should the material for electron microscopy be thin, dry and be kept in vacuum.

(a) To get straight beam of electrons
(b) To avoid the collision of electrons with the atoms of oxygen and nitrogen
(c) To avoid multiple scattering
(d) To overcome the problem of isotropy and anisotropy

Assertion and Reason

Direction (Q. Nos. 41-50) *In each of the following questions, a statement of Assertion (A) is given followed by the corresponding statement of Reason (R). Of the statements, mark the correct answer as*

(a) Both A and R are true and R is the correct explanation of A
(b) Both A and R are true, but R is not the correct explanation of A
(c) A is true, but R is false
(d) A is false, but R is true
(e) Both A and R are false

41 **Assertion** (A) Rudolf Virchow stated *'Omnis cellula-e-cellula'.*

Reason (R) Max Schultze proposed the 'protoplasm theory'.

42 **Assertion** (A) A cell membrane shows fluid behaviour.

Reason (R) A membrane is a mosaic or composite of diverse lipids and proteins.

43 **Assertion** (A) Fluid mosaic model was proposed by Singer and Nicolson.

Reason (R) The 'mosaic' is the intricate composite of proteins and lipids of the membrane.

44 **Assertion** (A) Transmembrane proteins penetrate the lipid bilayer.

Reason (R) Extrinsic proteins are loosely bound to lipid layers.

45 **Assertion** (A) Chloroplast is a cell organelle.

Reason (R) An organelle is a distinct part of a cell, which has a particular structure and function.

46 **Assertion** (A) Germinating seeds show more mitochondria than dormant seeds.

Reason (R) Mitochondria are endosymbiotic in origin.

47 **Assertion** (A) The bands of polytene chromosomes become enlarged at certain regions to form puffs or Balbiani rings.

Reason (R) These enlarged bands are active genes and represent the site of RNA synthesis.

48 **Assertion** (A) Lampbrush chromosomes are very elastic in nature.

Reason (R) Lampbrush chromosomes are found in oocytes of invertebrates only.

49 **Assertion** (A) Ribosomes are considered as the most primitive cell organelles.

Reason (R) Ribosomes occur in all eukaryotic organisms.

50 **Assertion** (A) Eukaryotic cells have the ability to adopt a variety of shapes and carry out directed movements.

Reason (R) There are three principal types of protein filaments actin filament, microtubules and intermediate filaments, which constitute the cytoskeleton.

(C) Medical Entrances' Gallery

(Collection of Questions Asked in NEET & in Various Medical Entrance Exams)

1 The biosynthesis of ribosomal RNA occurs in **[NEET 2020]**

(a) ribosomes (b) Golgi apparatus
(c) microbodies (d) nucleolus

2 Match the following columns and select the correct option from the codes given below.

	Column I		Column II
A.	Smooth endoplasmic reticulum	1.	Protein synthesis
B.	Rough endoplasmic reticulum	2.	Lipid synthesis
C.	Golgi complex	3.	Glycosylation
D.	Centriole	4.	Spindle formation

Codes **[NEET 2020]**

	A	B	C	D
(a)	2	1	3	4
(b)	3	1	2	4
(c)	4	2	1	3
(d)	1	2	3	4

3 Which is the important site of formation of glycoproteins and glycolipids in eukaryotic cells? **[NEET 2020]**

(a) Peroxisomes (b) Golgi bodies
(c) Polysomes (d) Endoplasmic reticulum

4 The concept of '*Omnis cellula-e-cellula*' regarding cell division was first proposed by **[NEET 2019]**

(a) Theodor Schwann (b) Schleiden
(c) Aristotle (d) Rudolf Virchow

5 Which of the following statements regarding mitochondria is incorrect? **[NEET 2019]**

(a) Enzymes of electron transport are embedded in outer membrane
(b) Inner membrane is convoluted with infoldings
(c) Mitochondrial matrix contains single circular DNA molecule and ribosomes
(d) Outer membrane is permeable to monomers of carbohydrates, fats and proteins

6 Which of the following statements is not correct? **[NEET 2019]**

(a) The hydrolytic enzymes of lysosomes are active under acidic pH
(b) Lysosomes are membrane bound structures
(c) Lysosomes are formed by the process of packaging in the endoplasmic reticulum
(d) Lysosomes have numerous hydrolytic enzymes

7 The shorter and longer arms of a submetacentric chromosome are referred to as **[NEET 2019]**

(a) *p*-arm and *q*-arm, respectively
(b) *q*-arm and *p*-arm, respectively
(c) *m*-arm and *n*-arm, respectively
(d) *s*-arm and *l*-arm, respectively

8 Which of the following pairs of organelles does not contain DNA? **[NEET 2019]**

(a) Chloroplast and vacuoles
(b) Lysosomes and vacuoles
(c) Nuclear envelope and mitochondria
(d) Mitochondria and lysosomes

9 Which of the following cell organelles is present in the highest number in secretory cells? **[NEET 2019]**

(a) Mitochondria (b) Golgi complex
(c) Endoplasmic reticulum (d) Lysosomes

10 Non-membranous nucleoplasmic structures in nucleus are the site for active synthesis of **[NEET 2019]**

(a) protein (b) *m*RNA
(c) *r*RNA (d) *t*RNA

11 Which of the following nucleic acids is present in an organism having 70S ribosomes only? **[NEET 2019]**

(a) Single-stranded DNA with protein coat
(b) Double-stranded circular naked DNA
(c) Double-stranded DNA enclosed in nuclear membrane
(d) Double-stranded circular DNA with histone proteins

12 Match the Column I with Column II. **[NEET 2019]**

	Column I		Column II
A.	Golgi apparatus	1.	Synthesis of protein
B.	Lysosomes	2.	Trap waste and excretory products
C.	Vacuoles	3.	Formation of glycoproteins and glycolipids
D.	Ribosomes	4.	Digesting biomolecules

Codes

	A	B	C	D		A	B	C	D
(a)	3	4	2	1	(b)	4	3	1	2
(c)	3	2	4	1	(d)	1	2	4	3

13 Which one is only found in plant? **[JIPMER 2019]**

(a) Ribosome
(b) Mitochondria
(c) Glyoxysomes
(d) Lysosome

14 Which of the following is the correct sequence of modification and transport of secretory vesicles? **[JIPMER 2019]**

(a) RER → Cell membrane → Secretory vesicles → Golgi body
(b) RER → Secretory vesicles → Cell membrane
(c) RER → Cell membrane → Secretory vesicles → Golgi body
(d) SER → Golgi body → Cell membrane → Secretory vesicles

15 The Golgi complex participates in **[NEET 2018]**

(a) respiration in bacteria
(b) formation of secretory vesicles
(c) fatty acid breakdown
(d) activation of amino acid

16 Which of the following is true for nucleolus?

(a) It takes part in spindle formation **[NEET 2018]**
(b) It is a membrane-bound structure
(c) Larger nucleoli are present in dividing cells
(d) It is a site for active ribosomal RNA synthesis

17 Select the incorrect match. **[NEET 2018]**

(a) Submetacentric chromosomes - L-shaped chromosomes
(b) Allosomes - Sex chromosomes
(c) Lampbrush chromosome - Diplotene bivalents
(d) Polytene chromosomes - Oocytes of amphibians

18 Which one of the following events does not occur in rough endoplasmic reticulum? **[NEET 2018]**

(a) Cleavage of signal peptide
(b) Protein glycosylation
(c) Protein folding
(d) Phospholipid synthesis

19 Which is incorrect with reference to chloroplast?

(a) Presence in algae and plants **[JIPMER 2018]**
(b) Releases O_2
(c) Occurs only in cells with aerobic respiration
(d) None of the above

20 What is the role of sterol in cell membrane?

(a) Stability **[JIPMER 2018]**
(b) Communication with other cells
(c) Secretion
(d) Transport

21 Which of the following cell organelles is responsible for extracting energy from carbohydrates to form ATP? **[NEET 2017]**

(a) Lysosome (b) Ribosome
(c) Chloroplast (d) Mitochondrion

22 **Statement I** Microtubules are formed only in animal cells.

Statement II Microtubules are made up of a protein called myosin. **[JIPMER 2017]**

Choose the correct option.

(a) Statement I is correct and statement II is incorrect
(b) Statement II is correct and statement I is incorrect
(c) Both statements are correct
(d) Both statements are incorrect

23 Which among the following are the smallest living cells, known without a definite cell wall, pathogenic to plants as well as animals and can survive without oxygen ? **[NEET 2017]**

(a) *Bacillus* (b) *Pseudomonas*
(c) *Mycoplasma* (d) *Nostoc*

24 A cell organelle containing hydrolytic enzymes is

(a) lysosomes (b) microsome **[NEET-II 2016]**
(c) ribosome (d) mesosomes

25 Select the mismatch. **[NEET-II 2016]**

(a) Gas vacuoles—Green bacteria cell
(b) Large central vacuoles—Animal cells
(c) Protists—Eukaryotes
(d) Methanogens—Prokaryotes

26 Microtubules are the constituents of

(a) spindle fibres, centrioles and cilia **[NEET-I 2016]**
(b) centrioles, spindle fibres and chromatin
(c) centrosome, nucleosome and centrioles
(d) cilia, flagella and peroxisomes

27 Mitochondria and chloroplast are **[NEET-I 2016]**

I. semiautonomous organelles.
II. formed by division of pre-existing organelles and they contain DNA, but lack protein synthesising machinery.

Which one of the following options is correct?

(a) Statement I is true, but II is false
(b) Statement I is false, but II is true
(c) Both I and II are false
(d) Both I and II are true

28 These types of vacuoles contain hydrolases.

(a) Sap (b) Contractile **[BCECE 2015]**
(c) Food (d) Air

29 In iodiogram, chromosomes of an organism are arranged according to their **[BCECE 2015]**

(a) increasing size (b) decreasing size
(c) position of centromere (d) number of chromosomes

30 Pectic polysaccharides are present in **[BCECE 2015]**

(a) cytoskeleton
(b) plasma membrane
(c) primary cell wall
(d) tertiary cell wall

31 Protoplast lacks **[Chhattisgarh PMT 2015]**

(a) cytoplasm (b) nucleus
(c) mitochondria (d) cell wall

32 Mitochondria performs all functions except [Chhattisgarh PMT 2015]
(a) nucleic acid synthesis
(b) steroid synthesis
(c) ATP synthesis
(d) polysaccharide degradation

33 The function of peroxisome is [Chhattisgarh PMT 2015]
(a) to convert H_2O_2 into H_2O and O_2
(b) utilisation of O_2 gas
(c) to break toxic molecules of a cell
(d) All of the above

34 GERL system is formed of [Chhattisgarh PMT 2015]
(a) Golgi body, endoplasmic reticulum, ribosome and lysosome
(b) Golgi body, endoplasmic reticulum and lysosome
(c) Golgi body, endoplasmic reticulum and ribosome
(d) Golgi body, ribosome and lysosome

35 Smallest unit in the plant cell wall is [AMU 2015]
(a) Micelle (b) Microfibril
(c) Fibril (d) None of these

36 Golgi complex is composed of [UK PMT 2015]
(a) Cisternae, tubules and vacuoles
(b) Cisternae, tubules and vesicles
(c) Cisternae, vesicles and vacuoles
(d) None of the above

37 70S ribosomes are found in [UK PMT 2015]
(a) eukaryotic cell (b) prokaryotic cell
(c) mitochondria (d) Both (b) and (c)

38 Match the following columns. [Kerala CEE 2015]

Column I	Column II
A. Rough ER	1. Synthesis of glycoproteins
B. Smooth ER	2. Aerobic respiration
C. Mitochondria	3. Synthesis of lipids
D. Golgi apparatus	4. Protein synthesis

Codes

	A	B	C	D
(a)	1	2	3	4
(b)	2	3	4	1
(c)	3	4	2	1
(d)	4	3	2	1
(e)	4	3	1	2

39 One type of chromosomes has middle centromere whereas the other has a terminal centromere. They are [Kerala CEE 2015]
(a) metacentric and acrocentric
(b) metacentric and telocentric
(c) submetacentric and telocentric
(d) telocentric and acrocentric
(e) acrocentric and metacentric

40 Match the following columns. [CBSE AIPMT 2015]

Column I	Column II
A. Thylakoids	1. Disc-shaped sacs in Golgi apparatus
B. Cristae	2. Condensed structure of DNA
C. Cisternae	3. Flat membranous sacs in stroma
D. Chromatin	4. Infoldings in mitochondria

Codes

	A	B	C	D		A	B	C	D
(a)	4	3	1	2	(b)	3	4	1	2
(c)	3	1	4	2	(d)	3	4	2	1

41 Cellular organelles with membranes are [CBSE AIPMT 2015]
(a) nuclei, ribosomes and mitochondria
(b) chromosomes, ribosomes and endoplasmic reticulum
(c) endoplasmic reticulum, ribosomes and nuclei
(d) lysosomes, Golgi apparatus and mitochondria

42 Ribosomes are essential for protein synthesis, but they are present in mitochondria and plastids, the sites of respiration and photosynthesis. What is the role of ribosomes in these organelles? [EAMCET 2015]
(a) Ribosomes transport ATP formed in respiration and photosynthesis to cytoplasm through ER
(b) Subunits of some required proteins are synthesised in these organelles
(c) Ribosomes transport RNA and DNA to cytoplasm
(d) All of the above

43 Which one of the following is correct for transmembrane proteins in lipid bilayer? [WB JEE 2015]
(a) They are absent in animal cells
(b) They act as channel proteins
(c) They are absent in plant cells
(d) They are only externally located

44 What kind of microscopy uses acridine orange? [BCECE 2015]
(a) Phase contrast
(b) Fluorescence
(c) Transmission microscope
(d) Scanning electron microscope

45 In order to sediment free ribosomes, centrifugation required is [BCECE 2015]
(a) 300000 g (b) 50000 g (c) 12000 g (d) 800 g

46 The following is generally used for creating density gradient during centrifugation. [BCECE 2015]
(a) NaCl (b) KCl (c) CsCl (d) $MgCl_2$

47 The technique separates proteins according to their molecular weight. [BCECE 2015]
(a) PAGE
(b) Affinity chromatography
(c) Ion exchange chromatography
(d) Gel filtration

48 At what speed, the mitochondria can be separated out by differential centrifugation? **[AMU 2015]**
(a) 200 × *g* (b) 500 × *g* (c) 800 × *g* (d) 8000 × *g*

49 Which of the following is always absent in prokaryotic cells? **[WB JEE 2014]**
(a) Ribosome (b) Mitochondria
(c) DNA (d) Cell wall

50 The primary cell wall is mainly made up of **[WB JEE 2014]**
(a) lignin (b) pectin
(c) cellulose (d) protein

51 Match the following columns. **[CBSE AIPMT 2014]**

Column I		Column II
A. Centriole	1.	Infoldings in mitochondria
B. Chlorophyll	2.	Thylakoids
C. Cristae	3.	Nucleic acids
D. Ribozymes	4.	Basal body of cilia or flagella

Codes

	A	B	C	D		A	B	C	D
(a)	4	2	1	3	(b)	1	2	4	3
(c)	1	3	2	4	(d)	4	3	1	2

52 Which structure performs the function of mitochondria in bacteria? **[CBSE AIPMT 2014]**
(a) Nucleoid (b) Ribosomes
(c) Cell wall (d) Mesosomes

53 Which of these organelles does not contain ribosomes?
I. Rough endoplasmic reticulum
II. Chloroplast
III. Golgi apparatus
IV. Mitochondria **[Kerala CEE 2014]**
Select the correct answer using the codes given below.
(a) I and II (b) I and IV
(c) Only IV (d) Only III
(e) II, III and IV

54 A major site for synthesis of lipid is **[CBSE AIPMT 2014]**
(a) RER (b) SER
(c) symplast (d) nucleoplast

55 The solid linear cytoskeletal elements having a diameter of 6 nm and made up of a single type of monomer are known as **[CBSE AIPMT 2014]**
(a) microtubules (b) microfilaments
(c) intermediate filaments (d) lamins

56 The osmotic expansion of a cell kept in water is chiefly regulated by **[CBSE AIPMT 2014]**
(a) mitochondria (b) vacuoles
(c) plastids (d) ribosomes

57 The Golgi complex plays a major role **[NEET 2013]**
(a) in trapping the light and transforming it into chemical energy
(b) in digesting proteins and carbohydrates
(c) as energy transferring organelles
(d) in post-translational modification of proteins and glycosidation of lipids

58 Who gave the term 'plasmalemma'? **[NEET 2013]**

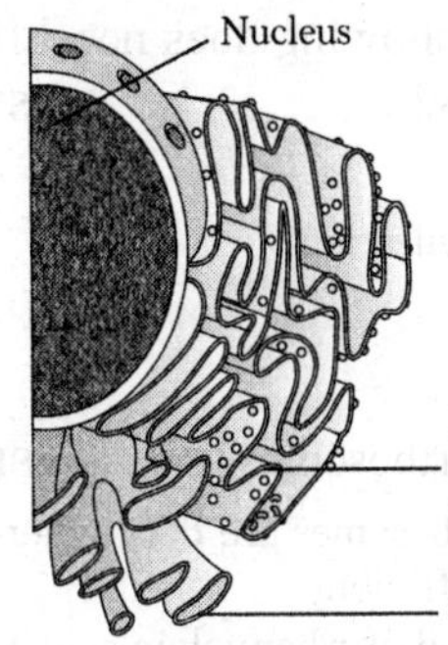

(a) C Nageli (b) Janet Plowe
(c) C Kramer (d) Nicolson

59 The difference between rough endoplasmic reticulum and smooth endoplasmic reticulum is that rough endoplasmic reticulum **[NEET 2013]**
(a) does not contain ribosomes
(b) contains ribosomes
(c) does not transport proteins
(d) transports proteins

60 Ultrastructure of cell is observed by **[MP PMT 2013]**
(a) simple microscope (b) light microscope
(c) phase contrast microscope (d) electron microscope

61 Robert Hooke used the term 'cell' in the year **[Manipal 2013]**
(a) 1650 (b) 1665
(c) 1865 (d) 1960

62 Protein synthesis takes place in **[Manipal 2013]**
(a) ribosomes (b) chloroplasts
(c) mitochondria (d) Golgi bodies

63 The cell theory is not applicable on **[Manipal 2013; WB JEE 2010]**
(a) algae (b) fungi
(c) viruses (d) lichens

64 Mitochondria will be found in abundance in cells of tissues having **[Manipal 2013]**
(a) minimum activity (b) average activity
(c) maximum activity (d) None of these

65 The types of ribosomes found in prokaryotic cells are **[KCET 2013]**
(a) 100S (b) 80S
(c) 60S (d) 70S

66 Which of the cell organelle lacks membrane? **[WB JEE 2013]**
(a) Mesosome (b) Mitochondria
(c) Ribosome (d) Liposome

67 Ribosomal RNA is actively synthesised in **[CBSE AIPMT 2012]**
(a) lysosomes (b) nucleolus
(c) nucleoplasm (d) ribosomes

68 Which one of the following does not differ in *E. coli* and *Chlamydomonas*? **[CBSE AIPMT 2012]**
(a) Ribosomes
(b) Chromosomal organisation
(c) Cell wall
(d) Cell membrane

69 What is true about ribosomes? **[CBSE AIPMT 2012]**
(a) The prokaryotic ribosomes are 80S, where S stands for sedimentation coefficient
(b) These are composed of ribonucleic acid and proteins
(c) These are found only in eukaryotic cells
(d) These are self-splicing introns of some RNAs

70 Select the correct statement from the following regarding cell membrane. **[CBSE AIPMT 2012]**
(a) Na^+ and Ka^+ ions move across cell membrane by passive transport
(b) Proteins make up 60-70% of the cell membrane
(c) Lipids are arranged in a bilayer with polar heads towards the inner part
(d) Fluid mosaic model of cell membrane was proposed by Singer and Nicolson

71 Which one possesses giant chromosomes? **[WB JEE 2012]**
(a) *Drosophila* (b) *Xenopsylla*
(c) *Branchiomyces* (d) Mouse

72 Centromere is a part of **[AFMC 2012]**
(a) ribosomes (b) chromosome
(c) mitochondria (d) endoplasmic reticulum

73 Telomere repetitive DNA sequence control the function of eukaryotic chromosome, because they **[AFMC 2012]**
(a) act as replicans
(b) are RNA transcription factor
(c) help chromosome pairing
(d) prevent chromosome loss

74 B-chromosome is also known as **[Manipal 2012]**
(a) somatic chromosome
(b) megachromosome
(c) limited chromosome
(d) supernumerary chromosome

75 Polytene chromosome was discovered in **[BHU; MP PMT 2012]**
(a) *Chironomus* (b) *Drosophila*
(c) *Musca* (d) *Culex*

76 An interconnecting membranous network of cell composed vesicles, flattened sacs and tubules is **[KCET 2012]**
(a) lysosome (b) nucleus
(c) mitochondrion (d) endoplasmic reticulum

77 Dictyosome is absent in **[AIIMS 2012]**
(a) cyanobacteria (b) *Mycoplasma*
(c) bacteria (d) All of these

78 Extranuclear chromosome occurs in **[UP CPMT 2012]**
(a) peroxisome and ribosome
(b) chloroplast and mitochondria
(c) mitochondria and ribosome
(d) chloroplast and lysosome

79 Plant cells normally lack **[AMU 2012]**
(a) ribosomes (b) Golgi bodies
(c) centrioles (d) cell membrane

80 Detailed structure of the membrane was studied after the advent of electron microscope during **[AMU 2012]**
(a) 1930s (b) 1950s
(c) 1970s (d) 1990s

81 The nucleus is separated from surrounding cytoplasm by a nuclear membrane, which is **[Manipal 2012]**
(a) single layered with pores
(b) single layered without pores
(c) double layered with pores
(d) double layered without pores

82 Which one is stained using carmine? **[MP PMT 2012]**
(a) Bacteria (b) Chromosomes
(c) Diatoms (d) Viruses

83 Which one of the following organisms is not an example of eukaryotic cells? **[CBSE AIPMT 2011]**
(a) *Escherichia coli* (b) *Euglena viridis*
(c) *Amoeba proteus* (d) *Paramecium caudatum*

84 In eubacteria, a cellular component that resembles eukaryotic cells is **[CBSE AIPMT 2011]**
(a) nucleus (b) ribosomes
(c) cell wall (d) plasma membrane

85 Comparing small and large cells, which statement is correct? **[DUMET 2011]**
(a) Small cells have a small surface area per volume ratio
(b) Exchange rate of nutrients is fast with large cells
(c) Small cells have a large surface area per volume ratio
(d) Exchange rate of nutrients is slow with small cells

86 Middle lamella is present **[UP CPMT 2011]**
(a) inside the secondary wall
(b) inside the primary wall
(c) outside the primary wall
(d) in between secondary and tertiary walls

87 Plasma membrane is made up of [OJEE 2011]
(a) lipid, protein and water
(b) lipid, protein and manganese
(c) lipid and carbohydrate
(d) lipid, protein and carbohydrates

88 Middle lamella is composed of [OJEE 2011; CBSE AIPMT 2009]
(a) carbohydrate (b) calcium pectate
(c) protein (d) peptidoglycan

89 What is a genophore? [WB JEE 2011]
(a) DNA in prokaryotes
(b) DNA and RNA in prokaryotes
(c) DNA and protein in prokaryotes
(d) RNA in prokaryotes

90 Tracer isotopes can be located by [VITEEE 2011]
(a) Muller counter (b) Geiger counter
(c) Both (a) and (b) (d) Morgan counter

ANSWERS

Check Point 15.1

1. (b)	2. (b)	3. (c)	4. (a)	5. (c)	6. (d)

Check Point 15.2

1. (c)	2. (a)	3. (a)	4. (b)	5. (d)	6. (d)

Check Point 15.3

1. (c)	2. (c)	3. (b)	4. (c)	5. (b)	6. (a)

Check Point 15.4

1. (a)	2. (b)	3. (a)	4. (b)	5. (c)	6. (c)

Check Point 15.5

1. (c)	2. (a)	3. (c)	4. (c)	5. (b)	6. (b)

(A) Topical Exercise

1. (d)	2. (b)	3. (b)	4. (c)	5. (d)	6. (a)	7. (b)	8. (a)	9. (b)	10. (b)
11. (d)	12. (c)	13. (b)	14. (b)	15. (d)	16. (d)	17. (a)	18. (a)	19. (a)	20. (b)
21. (c)	22. (c)	23. (d)	24. (d)	25. (b)	26. (d)	27. (d)	28. (a)	29. (a)	30. (c)
31. (b)	32. (c)	33. (d)	34. (a)	35. (a)	36. (a)	37. (d)	38. (c)	39. (a)	40. (b)
41. (a)	42. (b)	43. (c)	44. (a)	45. (d)	46. (d)	47. (b)	48. (c)	49. (c)	50. (c)
51. (c)	52. (d)	53. (c)	54. (b)	55. (c)	56. (a)	57. (b)	58. (d)	59. (d)	60. (a)
61. (c)	62. (d)	63. (a)	64. (b)	65. (c)	66. (d)	67. (b)	68. (b)	69. (c)	70. (a)
71. (d)	72. (c)	73. (d)	74. (c)	75. (b)	76. (c)	77. (a)	78. (a)	79. (c)	80. (a)
81. (d)	82. (c)	83. (a)	84. (a)	85. (b)	86. (c)	87. (a)	88. (a)	89. (a)	90. (d)
91. (d)	92. (a)	93. (d)	94. (b)	95. (a)	96. (c)	97. (b)	98. (d)	99. (a)	100. (c)
101. (b)	102. (b)	103. (d)	104. (c)	105. (d)	106. (c)	107. (c)	108. (a)	109. (c)	110. (b)
111. (c)	112. (a)	113. (a)	114. (c)	115. (d)	116. (c)	117. (d)	118. (c)	119. (d)	120. (d)
121. (c)	122. (d)	123. (c)	124. (a)	125. (d)	126. (d)	127. (c)	128. (d)	129. (d)	130. (d)
131. (c)									

(B) Cumulative Exercise

1. (b)	2. (c)	3. (a)	4. (b)	5. (d)	6. (b)	7. (b)	8. (c)	9. (a)	10. (c)
11. (d)	12. (a)	13. (d)	14. (c)	15. (a)	16. (b)	17. (b)	18. (c)	19. (a)	20. (d)
21. (b)	22. (b)	23. (d)	24. (b)	25. (d)	26. (c)	27. (a)	28. (d)	29. (d)	30. (d)
31. (b)	32. (c)	33. (b)	34. (c)	35. (b)	36. (c)	37. (b)	38. (a)	39. (c)	40. (d)
41. (b)	42. (a)	43. (b)	44. (b)	45. (a)	46. (b)	47. (a)	48. (c)	49. (c)	50. (a)

(C) Medical Entrances' Gallery

1. (d)	2. (a)	3. (b)	4. (d)	5. (a)	6. (c)	7. (a)	8. (b)	9. (b)	10. (c)
11. (b)	12. (a)	13. (c)	14. (c)	15. (b)	16. (d)	17. (d)	18. (d)	19. (c)	20. (a)
21. (d)	22. (d)	23. (c)	24. (a)	25. (b)	26. (a)	27. (a)	28. (c)	29. (b)	30. (c)
31. (d)	32. (d)	33. (d)	34. (b)	35. (a)	36. (b)	37. (d)	38. (d)	39. (b)	40. (b)
41. (d)	42. (b)	43. (b)	44. (b)	45. (a)	46. (c)	47. (a)	48. (d)	49. (b)	50. (c)
51. (d)	52. (d)	53. (d)	54. (b)	55. (b)	56. (b)	57. (d)	58. (b)	59. (b)	60. (d)
61. (b)	62. (a)	63. (c)	64. (c)	65. (d)	66. (c)	67. (b)	68. (d)	69. (b)	70. (d)
71. (a)	72. (b)	73. (d)	74. (d)	75. (a)	76. (d)	77. (d)	78. (b)	79. (c)	80. (b)
81. (c)	82. (b)	83. (a)	84. (d)	85. (c)	86. (c)	87. (d)	88. (b)	89. (a)	90. (c)

Explanations

(A) Topical Exercise

6 (*a*) Among given options, only cell membrane is found to be similar in prokaryotic and eukaryotic cells.

16 (*d*) Statement (d) is incorrect. Plasma membrane contains 1-2% of carbohydrates in it. These are mainly oligosaccharides.

27 (*d*) Option (d) represents incorrectly matched pair. The term protoplasm was given be Purkinje. Huxley defined the protoplasm as the physical basis of life.

42 (*b*) Safranin is used as a counterstain in Gram staining and endospore staining. Thus, it is not used to stain chromosomes.

44 (*a*) Statement in option (a) is incorrect. Polytene chromosomes possess dark bands and interbands. The band is Feulgen positive, i.e. it gets stained by feulgen stain whereas the interband is stain negative.

54 (*b*) Statement in option (b) is incorrect. Thylakoids are found within the chloroplast only and not in mitochondria instead, mitochondria contain cristae on the inner side and it is the site of aerobic respirations.

62 (*d*) Option (d) represents incorrectly matched pair. Chromoplasts are the coloured plastids, e.g. rhodoplast and phaeoplasts.

78 (*a*) Among given options both ER and Golgi complex possess and common features, *viz* presence of membrane protein modificaiton, endomembrane system, presence of tubules and vesicles.

(B) Cumulative Exercise

6 (*b*) Statements I and II are correct.

Cell fractionation is a process that involves the rupturing of cells, separation and suspension of cells using an isotopic medium. It involves three steps, i.e. extraction, homogenisation and centrifugation.

7 (*b*) Statements I and II are correct, while III and IV are incorrect and can be corrected as follows

A fluorescence microscope uses UV light with higher wavelength. It is used to detect specific proteins or other molecules in cells and tissues. It also helps to study the properties of organic and inorganic substance.

14 (*c*) Statements I, II and IV are correct, only statement III is incorrect and can be corrected as follows

Omnis cellula-e-cellula was demonstrated by Virchow.

15 (*a*) Statements I, II and IV are correct. Statement III is incorrect because

Coenocytes are multinucleate cells.

16 (*b*) Statement II is correct, while statements I, III and IV are incorrect and can be corrected as follows

Peroxisomes were discovered in liver cells.

Lomasomes are not involved in membrane formation.

Glyoxysomes function to oxidise fatty acids with the help of β-oxidation.

17 (*b*) Statements I and II are correct, while statements III and IV are incorrect and can be corrected as

Transmission electron microscope is useful for studying ultra thin structure of living cells.

20 (*d*) Statement in option (d) is incorrect and can be corrected as

Ingestion of solid particles is called phagocytosis.

21 (*b*) Statements I and II are correct, while statements III and IV are incorrect and can be corrected as

Porins are present in the cell membrane, outer membrane of Gram negative bacteria, some Gram positive mycobacteria, outer mitochondrial membrane and outer chloroplast membrane.

It contains more than 10% of the matter filled pores of mitochondria.

22 (*b*) Statements II, III, IV are correct where I, IV, VI are incorrect. Lysosomes are bounded by single membrane, contain various enzymes to degrade various cellular components and these enzymes are active at acidic pH of 5.0 only.

26 (*c*) Statements II, III, IV and V are correct. Statement I is incorrect and can be corrected as

Photorespiration involves green cells having chloroplast, peroxisomes and mitochondria.

27 (*a*) Statements I, IV and V are correct, but statements II and III are incorrect and can be corrected as follows

Flagella are generally few in numbers (<5) and relatively long. Cilia are typically short and are present in many copies (>100) in a cell.

33 (*b*) Option (b) is incorrect pair and can be corrected as follows

The flip-flop movement refers to the movement of molecules from one side of the membrane to other side transversely. According to the fluid mosaic model, phospholipid in lipid bilayer are able to flip-flop. Proteins cannot flip-flop because they have large polar region which is unable to cross the energy barrier provided by the hydrophobic core of the membrane.

34 (*c*) Pair in option (c) is correct, but other pairs are incorrect. The incorrect matches can be corrected as

Central nucleus - DNA, RNA, protein

Peripheral nucleus - Skeletal muscle cell

Hammerling - Acetabularia

36 (*c*) Statements I, II and III are correct, while IV is incorrect and can be corrected as follows

Given diagram represents Golgi complex. In animal cells, SER is the site of lipid like steroidal hormones.

41 (*b*) Both A and R are true, but R is not the correct explanation of A.

Rudolf Virchow has done many important discoveries. He is best known for his contribution to the cell theory. He gave the statement that *Omnis cellula-e-cellula.*

According to him, all living cells arise from pre-existing living cells. He first explained how new cells are formed from existing cells he modified the hypothesis of Schleiden and Schwann to give cell theory a final shape. According to Schleiden and Schwann, the cells budded out from the nucleus.

Max Schultze proposed cell theory in 1861.

Huxley (1868) defined protoplasm as the physical basis of life. O Hertwig (1892) proposed protoplasmic theory which states that all living matter of which animals and plants are formed in protoplasm.

42 (*a*) Both A and R are true and R is the correct explanation of A.

The fluid mosaic model was put forward in 1970 by SJ Singer and GL Nicolson. According to this model, plasma membrane is composed of phospholipids, extrinsic proteins (peripheral proteins) and intrinsic proteins (integral proteins). The plasma membrane is highly fluidic due to the movement of phospholipids and proteins present in it.

43 (*b*) Both A and R are true, but R is not the correct explanation of A.

In 1972, as per SJ Singer and GL Nicolson model, membrane is assumed to be 'fluid' because of the movements and packing variation of its lipid components. The lipids themselves give the membrane its basic structure and its relative impermeability to water soluble molecules. The mosaic is the intricate composite of proteins and lipids of the membrane.

44 (*b*) Both A and R are true, but R is not the correct explanation of A.

In a plasma membrane, two types of proteins are found. Transmembrane or intrinsic proteins are embedded in plasma membrane whereas extrinsic proteins are found on the outer surface of the membrane.

45 (*a*) Both A and R are true and R is the correct explanation of A.

An organelle is a distinct part of a cell, which has a particular structure and function. Chloroplast, mitochondria, ER, Golgi body, etc., are the main cell organelles of plant cell.

46 (*b*) Both A and R are true, but R is not the correct explanation of A.

Mitochondria are the site of aerobic respiration in eukaryotic cells. Cells of germinating seeds show more number of mitochondria as compared to those of dormant seeds because these seeds require more energy.

47 (*a*) Both A and R are true and R is the correct explanation of A.

The puffs represent region where tightly coiled chromosomal fibres open out to form many loops. Puffs are active genes and represent the sites of RNA synthesis.

48 (*c*) A is true, but R is false because

Lampbrush chromosomes are with uncoiled DNA forming looping structure. Thus, are very elastic in nature and can be stretched up to two times of their original length. Lampbrush chromosomes have been also reported in some mammals-like humans.

49 (*c*) A is true, but R is false because

Ribosomes are considered as the most primitive cell organelles and they occur in both prokaryotic and eukaryotic cells.

50 (*a*) Both A and R are true and R is the correct explanation of A.

Eukaryotic cells have the ability to adopt a variety of shapes and carry out directed movements because they have a network of protein fibres that support the shape of the cell and anchor organelles to fixed location. This network is called cytoskeleton which is made up of actin filaments, microtubules and intermediate filaments.

(C) Medical Entrances' Gallery

1 (*d*) The biosynthesis of ribosomal RNA occurs in nucleolus of nucleus. It helps the nucleus of the cell to control cell metabolism and other activities. The other two types of RNA, i.e. *m*RNA and *t*RNA are also synthesised here.

2 (*a*) A-2, B-1, C-3, D-4

3 (*b*) Golgi bodies are site of formation of glycoproteins and glycolipids in eukaryotic cells.

4 (*d*) Rudolf Virchow proposed the concept of *Omnis cellula-e-cellula*, i.e. all cells are derived from the pre-existing cells. Schleiden and Theodor Schwann initially put forward the cell theory in 1839, but this theory did not explain how new cells were formed.

5 (*a*) The statement 'enzymes of electron transport are embedded in outer membrane' is incorrect. The correct form of statement is

Enzymes of electron transport are embedded in the inner membrane of mitochondria. An electron transport chain is a series of coenzymes and cytochromes that take part in the passage of electrons from a chemical to its ultimate acceptor. Rest statements are correct.

6 (*c*) The statement 'lysosomes are formed by the process of packaging in the endoplasmic reticulum' is incorrect. The correct form of the statement is

'Lysosomes are actually formed by the budding off from the *trans* face of Golgi bodies. These membrane bound structures contain hydrolytic enzymes whose precursors are synthesised by rough endoplasmic reticulum. Rest statements are correct.

7 (*a*) The shorter and longer arms of submetacentric chromosome are designated as *p* and *q* arm, respectively. Here, '*p*' signifies petite or short. In a submetacentric chromosome, centromere is located slightly away from the centre due to which the two arms appear unequal in length.

8 (*b*) Lysosomes and vacuoles do not contain DNA. Lysosomes are single membrane bound small vesicles which contain hydrolytic enzymes. Vacuoles are a large membranous sac found in the cytoplasm. These contain substances that are not essentially useful for the cell like water, sap, excretory products and other materials. Chloroplast and mitochondria are semi-autonomous organelles because they contain their own DNA and are believed to be prokaryotic symbionts.

9 (*b*) Golgi complex (Golgi apparatus) is a cell organelle present in highest number in secretory cells. These are the site of modification, packaging and secretion of secretory proteins and glycoproteins outside the cell.

10 (*c*) Non-membranous nucleoplasmic structure in the nucleus of the cells are the site for active synthesis of *r*RNA. These structure are called nucleolus. Larger and more numerous nucleoli are present in the cell actively carrying out protein synthesis.

11 (*b*) Double-stranded circular naked DNA type of nucleic acid is present in an organism having 70S ribosomes. These are present in prokaryotic organisms or cells. All prokaryotic cells have a single double-stranded (double helix), circular DNA molecule for their genetic material. The DNA is attached to the inner cell membrane where the DNA replicating machinery is located. The DNA is naked it does not have proteins associated with it as eukaryotic DNA does.

12 (*a*) A–3, B–4, C–2, D–1

13 (*c*) Out of the given options, glyoxysomes are exclusively found in plant cells. They were discovered by Beevers (1919) in endosperm of germinating seeds. They are involved in fat metabolism through glyoxylate cycle.

14 (*c*) The correct sequence of modification and transport of secretory vesicles is follows

RER → Golgi body → Secretory vesicles → Cell membrane

Protein destined for secretion are made on ribosomes bound to the RER. The proteins move through the endomembrane system and are dispatched from the *trans* face of the Golgi apparatus in transport vesicles that move through the cytoplasm and then fuse with the plasma membrane releasing the protein to the outside of the cell.

15 (*b*) Golgi complex participates in the formation of secretory vesicles. It is a cytoplasmic structure found in eukaryotic cells. It is made up of four parts, i.e. cisternae, tubules, vesicles and vacuoles. The forming face or cisternae receives vesicles from endoplasmic reticulum. Their contents pass through various cisternae with the help of coated vesicles and intercisternal connectives. They ultimately reach the maturing face where they are budded off as coated secretory or Golgian vesicles or vacuoles.

16 (*d*) Statement (d) is true for nucleolus. It is the site for active ribosomal RNA synthesis. Other statements are not true for nucleolus. The correct information about the statements is as follows

Nucleolus does not take part in spindle formation. Nucleolus is not a membrane bound structure. Large nucleoli are present in cells that actively carry out protein synthesis.

17 (*d*) Polytene chromosomes are giant chromosomes that are quite common in the salivary glands of insects therefore, they are popularly called as salivary chromosomes.

The oocyte of amphibians contains lampbrush chromosomes.

18 (*d*) Phospholipid synthesis does not occur in RER. It occurs inside Smooth Endoplasmic Reticulum (SER). A signal peptide is a short peptide present at the N-terminus of the newly synthesised proteins. It targets them to the ER and is then cleaved off.

19 (*c*) Statement (c) is incorrect about chloroplasts. The correct information about the statement is as follows

They are not associated with any type of respiration, i.e. aerobic and anaerobic and can be found in any type of cell. Rest other statements are correct with reference to chloroplasts.

20 (*a*) The role of sterol in cell membrane is to provide stability. Steroles are steroids with 8-10 carbon long aliphatic side chain at carbon 17 and at least one alcoholic hydroxyl. Cholesterol ($C_{27}H_{45}OH$) is a common sterol found in many animal cell membranes.

21 (*d*) Out of given cell organelles, mitochondrion is responsible for extracting energy from carbohydrates to form ATP. It oxidises carbohydrate to produce ATP molecules in the process of aerobic respiration. Chloroplast is responsible for trapping light energy and converting it into chemical energy during photosynthesis. Lysosomes and ribosomes are not related with extraction of energy in any way.

22 (*d*) Both statements are incorrect. They are corrected as follows

Microtubules are found in both animal and plant cells. They are formed of protein tubulin and not myosin.

23 (*c*) *Mycoplasma* is the smallest living cells. It does not have definite cell wall. It is an anaerobic organism and it causes diseases in plants as well as in animals.

24 (*a*) Lysosomes are membrane bound cell organelles. These contain many hydrolytic enzymes which work at high pH. They bring about the intracellular digestion of cell debris and worn and torned cell organelles. These loose their existence while doing so, that is why they are also called as suicidal bags.

25 (*b*) Animal cells do not have large central vacuole. Instead, these have 2-3 small vacuoles. The presence of large central vacuoles is the characteristics feature of plant cells.

26 (*a*) Microtubules are structures present in cilia, flagella, centrioles and spindle fibres. They are also the part of fibres found in cytoskeleton.

27 (*a*) Statement I is correct, but II is incorrect and can be corrected as

Mitochondria are considered as semiautonomous cell organelle due to the following features

(i) Mitochondria have their own DNA which can replicate independently.

(ii) The mitochondrial DNA produces its own *m*RNA, *t*RNA and *r*RNA. The organelle possess their own ribosomes, called mitoribosomes.

31 (*d*) Protoplast do not contain cell wall. Cell membrane, cytoplasm, nucleus constitutes protoplast.

33 (*d*) Peroxisome performs the function of detoxification. It also degrades toxic products into less harmful substances, e.g. conversion of H_2O_2 into H_2O and O_2.

35 (*a*) Micelle is considered to be the smallest unit of cell wall. About 100 cellulose molecules together constitute to make micelle. About 20 micelle form microfibril and around 25 microfibrils associate and form one fibril.

42 (*b*) Ribosomes are found inside the chloroplast and mitochondria, besides nucleus. Mainly, these organelles are semiautonomous, i.e. some of the proteins that are utilised for organelle functioning are synthesised in these organelles as well as in nucleus, e.g. large subunit of RuBisCO is synthesised in chloroplast while small in nucleus.

44 (*b*) Acridine orange is a fluorescent dye that is why it is used in fluorescent microscopy.

49 (*b*) Out of the given options, prokaryotic cell does not have mitochondria to generate energy. A typical prokaryotic cell (i.e. bacterial cell) has cell wall just below the glycocalyx composed of peptidoglycan. They also have ribosomes and DNA just like eukaryotes. They generate few energy or ATP *via* glycolysis (when oxygen is available) and lactic acid fermentation (when oxygen is unavailable).

52 (*d*) Mesosomes perform the function of mitochondria in bacteria (prokaryotic cells). They are characteristic specialisation of cell membrane of bacteria that develop due to ingrowth of plasma membrane. It consists of vesicles, tubules and lamellae.

54 (*b*) Synthesis of lipids takes place in SER while that of protein in RER.

55 (*b*) Microfilaments are the solid linear cytoskeletal elements having a diameter of 6 nm and made up of single type of monomer. These active filaments are the thinnest filaments of the cytoskeleton and are found in the cytoplasm of the eukaryotic cells.

56 (*b*) The osmotic expansion of a cell kept in water is chiefly regulated by vacuole. It is the single and large organelle which constitutes about 20% of plant cells and is small and multiple in animal cells. Vacuole stores water and macromolecules including ions, sugar, amino acids, proteins and carbohydrates. The membrane that surrounds the vacuole is called tonoplast. The vacuole contains cell sap in it. The cell sap has high osmotic pressure which generates turgor pressure in plant cells.

57 (*d*) Golgi complex plays a major role in post-translational modification of proteins and glycosidation of lipids. Chloroplasts contain chlorophyll, which traps light energy and transforms it into chemical energy. Lysosomes are involved in digesting proteins, fats and carbohydrates. Mitochondria are energy transferring organelles.

59 (*b*) Endoplasmic reticulum is a network of interconnected cisternae, tubules and vesicles present in cytoplasm. Depending on the presence or absence of ribosomes, it is of two types

(i) Rough ER has ribosomes attached to its surface by ribophorin.

(ii) Smooth ER does not have ribosomes.

62 (*a*) Protein synthesis is also known as translation. It takes place in ribosomes. Ribosomes are non-membrane bound cell organelles, made up of protein and RNA.

64 (*c*) The cells which are metabolically active contain mitochondria in abundance.

65 (*d*) Ribosomes are electron microscopic ribonucleoprotein particles found to be attached either on RER or freely in cytoplasm. The ribosomes found in prokaryotes are of 70S type while in eukaryotes are of 80S type.

66 (*c*) Ribosomes were first observed by George Palade (1953) under the electron microscope as dense particles.

67 (*b*) In eukaryotes, the site of synthesis of most of the ribosomal RNA (*r*RNA) is nucleolus. The nucleolar organiser contains many copies of ribosomal DNA (repetitive DNA).

68 (*d*) Cell membrane of prokaryotes is structurally similar to that of eukaryotes, So, it does not differ in *E. coli* and *Chlamydomonas.*

69 (*b*) Ribosomes are large, non-membranous, RNA-protein complexes, which are necessary for protein synthesis.

70 (*d*) In 1972, Singer and Nicolson proposed fluid mosaic model for internal structure of plasma membrane. This is most widely accepted model for plasma membrane. According to this model, a membrane consists of a continuous bilayer of phospholipid molecules in which globular proteins are embedded. The lipids are arranged in a bilayer with polar heads toward outside and non-polar tails towards inside.

74 (*d*) B-chromosomes are also known as supernumerary or accessory chromosomes. These are mainly or entirely heterochromatic chromosomes.

76 (*d*) Endoplasmic reticulum is a network of tubules, vesicles and flattened sacs that are interconnected. They may serve specialised functions in the cells including protein synthesis.

77 (*d*) Dictyosomes are Golgi bodies that are absent in prokaryotic microorganisms like cyanobacteria, mycoplasma and bacteria, but present in plants.

79 (*c*) Centrioles are minute submicroscopic, cylindrical structures occurring in most animal cells, but are absent in higher plant. These are required to form basal bodies, cilia, flagella and astral spindle poles.

80 (*b*) The detailed structures of the membrane was studied only after the advent of the electron microscope in the 1950s.

81 (*c*) The nuclear membrane of the nucleus which separates nucleus and cytoplasm is double membranous and porous.

83 (*a*) The bacterium *E. coli* is a prokaryote.

84 (*d*) The plasma membrane of eubacteria resembles to that of eukaryotic cell. It is made of phospholipid, protein and some amount of polysaccharides. However, it lacks sterol, which is a characteristic of eukaryotic cell membrane.

85 (*c*) Small cells have a large surface area per volume ratio as compared to large cells.

86 (*c*) Middle lamella is a thin binding layer between the cell wall of adjacent plant cells. It is chemically formed of pectates of calcium and magnesium. It is present outside the primary wall.

87 (*d*) Every living cell is externally covered by a thin transparent, electron microscopic, elastic semipermeable membrane called cell membrane. It is composed of lipids (mostly phospholipid), proteins (peripheral and integral) and carbohydrates (glycoprotein and glycolipids).

88 (*b*) Middle lamella is a thin binding layer between the cell wall of adjacent plant cell. It is chemically formed of calcium and magnesium pectate.

89 (*a*) The characteristic feature of prokaryotic or bacterial nucleus is absence of nuclear membrane, nucleolus and nuclear sap and such a nucleus is called nucleoid or genophore. It contains DNA and RNA.

90 (*c*) Geiger counter are used to detect radiations. It is also known as Muller counter

CHAPTER 16

Cell Cycle and Cell Division

Growth and reproduction are the two important characteristics of all living forms. Life of all multicellular organisms begins with a single cell, i.e. zygote. It is then developed into a complete individual by continuous growth and division in it.

We all know that new cells arise by the division of pre-existing cells (**Rudolf Virchow** in 1858). Hence, we can also say that the process of cell division works as the foundation stone for the continuity of life.

CELL CYCLE

A cell can be seen either in a dividing phase or in a non-dividing phase. In **dividing phase**, it is said to be in the process of mitosis or meiosis, while when in **non-dividing phase**, it is said to be in interphase. The sequential repetition of dividing and non-dividing phases in an orderly manner is called **cell cycle**.

The concept of cell cycle was first described by **Howard** and **Pelc** in 1953. It involves those sequence of events by which a cell duplicates its genome, synthesises other cell constituents, undergoes growth and finally divides into two daughter cells.

All the above mentioned events (i.e. dividing and non-dividing) occur in a coordinated manner and are under complete genetic control.

Generation Time

- Period of cell cycle varies from organism to organism and also from cell to cell, e.g. a yeast cell divides in about every 90 minutes and a typical human cell divides every 24 hours.
- In humans, during an average duration of 24 hours, dividing phase lasts for about an hour, i.e. 5% of the cell cycle and interphase lasts for almost 23 hours, i.e. more than 95% of the duration of the cell cycle.
- This period between two cell divisions in a cell is known as **generation time**.

Inside

The broader outline view of a cell cycle is given below

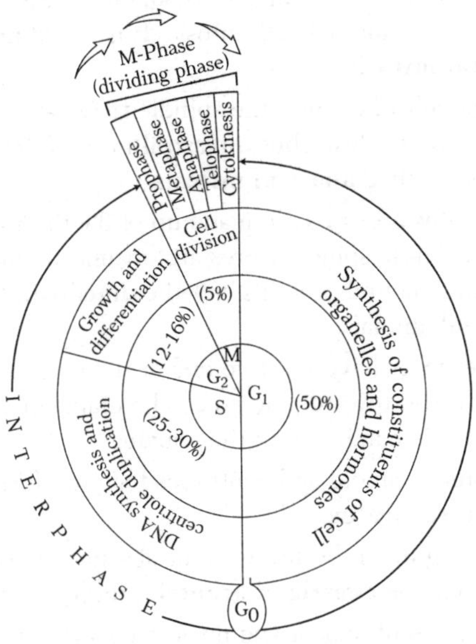

Fig 16.1 The eukaryotic cell cycle

Phases of cell cycle

There are broadly two phases of a cell cycle which are discussed in detail, as follows

1. Interphase

- It is also known as **intermitosis** or **phase-I** or **preparatory phase.**
- It looks like the long resting phase or non-dividing phase of the cell cycle. However, it is the period of intense cytogenic activities.
- During interphase, chromosomal material is seen in the form of very loosely coiled threads called **chromatin.**
- The interphase is further subdivided as follows

(i) **G_1-phase** It is also called the **first gap phase** or **first growth phase** or **post-mitotic gap phase.**

During this phase, the cell remains metabolically active and acquires nutrients from its environment to perform its specialised functions.

These functions involve synthesis of hormones and the cell organelle constituents mainly.

In this phase, following events are visible

- The cell grows in size.
- Synthesis of RNA, ribosomes, proteins, lipids, etc., takes place. No change is visible in the DNA content of the cell, i.e. DNA is not synthesised during this substage.

After G_1-phase, a cell can take one of the following two pathways, i.e.

- it may continue cell cycle and enter **S-phase**
- it may enter **G_0-phase.**

The deciding factor for its selection is storage of energy rich compounds and **mitogens**, i.e. mitosis inducing factors.

G_0-phase

- It is an extension of G_1-phase and is also known as **quiescent stage.**
- The term 'G_0-phase' was introduced by **Lajtha** in 1963. A cell attains this phase due to **the absence of cyclin proteins** which control the cell cycle.
- During G_0-phase, the cell neither divides nor synthesises organelles, but the metabolic activities of the cell continues. Liver, muscle and nerve cells of human body are in G_0-phase.
- Cells present in the G_0-phase usually function as reserve cells. These can rejoin the cell cycle any time on receiving a proper stimulus through cyclin proteins, e.g. liver cells can resume G_1-phase in response to certain chemicals. However, muscles and nerve cells of human body remain in G_0-phase permanently.

(ii) **S-phase** It is also called the **synthesis phase.** In this phase, following events are visible

- The cell synthesises a replica of its genome by the process of DNA replication. Thus, DNA content of the cell becomes doubled in this phase but no change in chromosome number occurs. Therefore in this phase, ploidy level of cell remains same.
- The synthesis of histone proteins and kinetochores is also seen.
- This phase is also called **invisible phase** as the replicated chromosomes are not visible.

(iii) **G_2-phase** It is also called as the **pre-mitotic gap phase** or **post-synthetic phase** or **second growth phase** or **second gap phase.**

It is the gap between DNA synthesis and next division.

In this phase, following events are visible

- The preparations are made for genomic separation.
- Synthesis of molecules other than DNA, which are required for cell division.
- Replacing of damage DNA.
- Duplication of mitochondria and other organelles.
- Replicated chromosomes condense.
- Microtubules begin to assemble to form spindles.

Note ***Rabl orientation*** *In some eukaryotes during interphase, a specific arrangement of chromosome is seen. In this arrangement, the centromeres cluster near one pole of the nucleus forming a rosette whereas chromosomal arms are arranged more or less parallely extending towards the opposite pole. This arrangement is known as Rabl orientation. It helps in cell division via increased coordination between chromosomes and centrioles.*

2. Dividing phase

Interphase is followed by dividing phase. This phase is also known as **mitotic phase** or **M-phase**. It is the phase of actual cell division.

Cell division can be of following two types. On the basis of cell type in which division is occurring.

1. **Mitosis** that occurs in somatic cells.
2. **Meiosis** that occurs in germ line cells.

Both the types of cell divisions involve two stages, i.e.

(a) **Karyokinesis** (*Karyos*—nucleus, *kinema*—breakage)

(b) **Cytokinesis** (cytoplasmic breakage)

The cytokinesis in both the types of division is more or less same hence, we will discuss it after discussing the karyokinesis.

MITOSIS

- Mitosis was first observed by **Strasburger** (1876) in plant cells.
- The term 'Mitosis' was given by **W Flemming** in 1882.
- It is also called the **equational** or **equal** division, i.e. the division in which the number of chromosomes in the daughter cell nuclei are same as that of the parental cell nuclei.
- It is also known as **somatic cell division**, because it occurs in the somatic or body cells of living organisms.
- In general, this division is related to the growth of all the organs and their systems in animals.
- On the other hand in plants, it is related to growth through meristematic tissues.

Karyokinesis

- The term 'Karyokinesis' was given by **Schneider** (1887).
- It is the process of division of nucleus.

It consists of following four substages

1. Prophase

This is the longest phase of karyokinesis. This phase is marked by following events

- Chromatin fibres and chromosomes become shorten and thicken by coiling to form compact chromosomes.
- Each chromosome consists of two chromatids held together by a centromere.
- In early prophase, the chromosomes appear like a ball of wool, this stage is known as **spireme** stage.
- In animal cells, the centrioles move to opposite poles of the cell and short microtubules may be seen radiating from them.
- These radiating microtubules are called **asters**, present towards the margins. The central counter parts of asters are called as **spindles**.
- Hence, the centrioles are responsible for the formation of spindle fibres in animal cell. Also, due to the formation of astral rays in animal cell mitosis. It is called **amphiastral** or **astral mitosis**.
- In plant cells, however the spindle fibres are formed without centrioles. Thus, no astral rays are formed. This mitosis is called **anastral mitosis**.
- Chemically, spindles are made up of 3% RNA and 97% tubulin. These spindle fibres are organised with the help of calcium bound proteins called **calmodulin** both in plant and animals.
- Spindle fibres have positive end at the equator and negative end towards the pole. The diameter of spindle apparatus is maximum at the equator.
- The asters, spindle and centrioles together form the **mitotic apparatus**.
- Depending upon the presence or absence of centrioles, mitosis can be **centric** or **acentric**, respectively.
- The nucleoli disappear and nuclear envelope is no longer visible.
- Golgi bodies, endoplasmic reticulum, etc., are not visible in this phase.

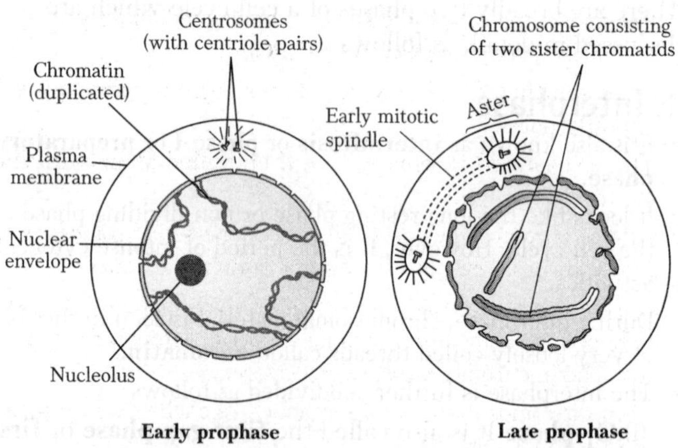

Fig. 16.2 Prophase

A transitional phase between prophase and metaphase is called **prometaphase**.

2. Metaphase

It is the best stage to observe the shape, size and number of chromosomes. This phase is marked by following events

- Complete disintegration of **nuclear envelope**.
- Spindle fibres get attached to kinetochore of chromosomes and bring the chromosomes on the equator of the spindle. This phenomenon is called **congression**.
- The centromeres of all the chromosomes lie on the equator which forms an apparent plate called **metaphasic plate** or **equatorial plate**.

- In plant cell, chromosomes are irregularly arranged on the equatorial plane.
- On the other hand, in animal cells, smaller chromosomes are usually central and larger ones are peripheral in position.

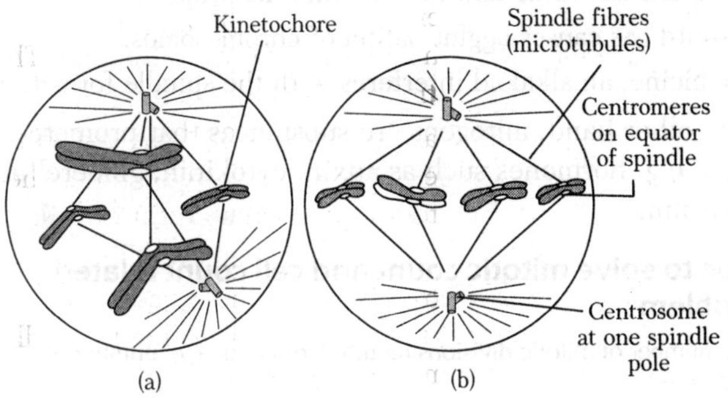

Fig 16.3 (a) Prometaphase, (b) Metaphase

3. Anaphase

It is the **shortest phase of M-phase**. This phase is marked by following events

- Shortening of spindle fibres occurs. This is due to slow removal of proteinaceous monomers in the polar region.
- The centromeres of chromosomes also start splitting due to the shortening of spindle fibres. This is mediated by Anaphase Promoting Complex (APC).
- After splitting, daughter centromeres begin to move towards opposite poles.
- These separated chromatids are then pulled along with the centromeres.
- All the chromosomes of a cell except the sex chromosomes, move towards the poles, i.e. away from the equator.
- During their migration chromosomes may appear V, L, J or I-shaped in metacentric, submetacentric, acrocentric and telocentric chromosomes, respectively.
- These chromosomes remain attached to each other by means of interzonal fibres.

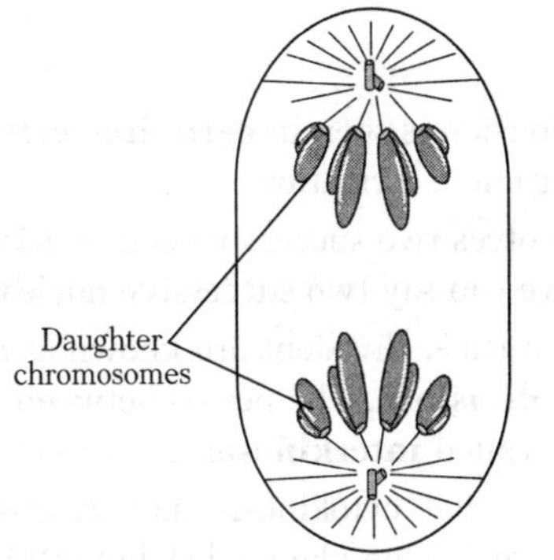

Fig. 16.4 Anaphase

4. Telophase

It is the last phase of karyokinesis and is normally followed by cytokinesis, to give two daughter cells. This phase is marked by following events

- The chromatids reach the poles of the cell, uncoil and lengthen to form chromatin again.
- The spindle fibres disintegrate and the centrioles (in animal cells) replicate.
- A nuclear envelope reappears around the chromosomes at each pole and the nucleoli, Golgi bodies, endoplasmic reticulum reappear. Thus, this phase is considered as **reverse of prophase.**
- The reappearance of nucleoli is associated with the nucleolar organisers.

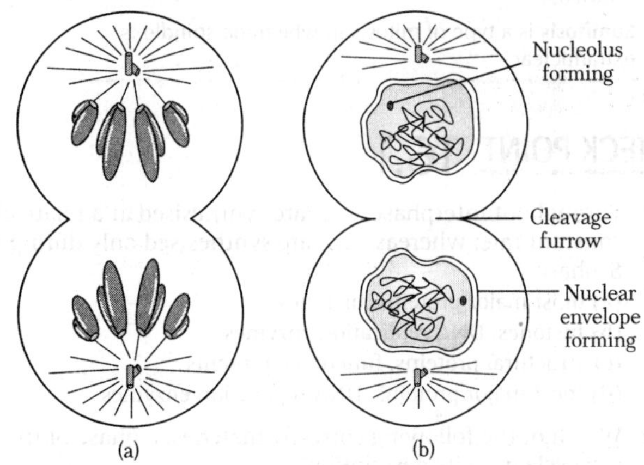

Fig. 16.5 Telophase : (a) Early telophase, (b) Late telophase

Significance of mitosis

- It results in the production of diploid daughter cells with identical genetic components.
- Growth in multicellular animals occurs owing to mitosis only.
- It contributes in cell repair and asexual reproduction as well.
- The continuous growth of plants is due to mitosis in their meristematic tissues. That is why study of meristematic tissues is very useful for the study of mitosis in the laboratory.
- In humans, uncontrolled mitotic division leads to cancer.
- An active and efficient cell has a high **nucleocytoplasmic ratio.** Increase in size lowers this ratio. A high nucleocytoplasmic ratio is brought back through division of cells especially mitosis.

Some special types of mitosis

- **Amitosis** is a comparatively rare cell division which occurs without the formation of spindles and appearance of chromosomes. It is also called **direct division**. This phenomenon was discovered by **Remak** (1841) in the RBCs of chick embryo. It is seen in prokaryotic cells.
- **Dinomitosis** is seen all dinoflagellates. In this, there is persistence of nuclear envelope during mitosis. Thus, the phenomenon of mitosis in the presence of nuclear envelope is called dinomitosis.
- **Endomitosis** is a type of mitosis in which there is chromosome duplication without nuclear division. It is seen in the salivary glands of *Drosophila* and other flies.
- **Premitosis** is a type of mitosis in which spindle apparatus is formed inside the nuclear membrane, before the disappearance of nuclear membrane. It is seen in certain protozoan.
- **Eumitosis** is a type of mitosis in which the spindle is extranuclear.

Mitotic poisons and mitogens

Mitotic poisons are chemicals or substances that interfere with the normal mitotic division. For example,

- Ribonuclease, azide and cyanide inhibits prophase.
- Mustard gas causes agglutination of chromosomes.
- Colchicine, an alkaloid interferes with the spindle formation.

On the other hand, mitogens are substances that promote mitosis, e.g. hormones such as auxin, cytokinin, gibberellins and insulin.

Tips to solve mitotic count and cell count related problem

Total number of mitotic divisions required to produce '*n*' number of cell s= $n - 1$

We can also calculate total number of mitotic generations (x) to produce '*n*' number of cells by the formula $2^x = n$, Where x = number of mitotic generations.

CHECK POINT 16.1

1. Throughout interphase, ... are synthesised at a relatively constant rate, whereas ...... are synthesised only during the S-phase.
(a) most major proteins, histones
(b) histones, DNA replication enzymes
(c) structural proteins, functional proteins
(d) most major proteins, DNA replication enzymes

2. Which of the following correctly matches, a phase of the cell cycle with its description?
(a) S-immediately precedes cell division
(b) M-duplication of DNA
(c) G_1-synthesis of RNA
(d) G_2-cell division

3. Rabl orientation of chromosomes during cell cycle occurs to
(a) help in breakage of chromosomes into chromatids
(b) increase coordination between chromosomes and centrioles
(c) help in interaction of non-sister chromatids
(d) acts as a stimulus for condensation of chromosomes

4. In plant cells during metaphase, chromosomes are
(a) shortened due to slow removal of proteinaceous monomers
(b) pulled along behind the centromeres towards the poles
(c) irregularly arranged on the equatorial plane
(d) arranged peripherally on the equatorial plane

5. Which of the following is a function of mitotic cell division in animals?
(a) Sexual reproduction
(b) Growth of gonads
(c) Production of gametes
(d) None of the above

6. During dinomitosis,
(a) only nuclear envelope disintegrates during nuclear division
(b) nuclear envelope disappears during spindle organisation
(c) nuclear envelope persists during nuclear division
(d) spindle formation does not occur

MEIOSIS

- This division was first discovered and described by **Oscar Hertwig** (1876) in sea urchin eggs. It was described again at the level of chromosomes by **Van Benedon** (1883). Term 'Meiosis' was given by **Farmer** and **Moore** in 1905. It refers to the **reductional division** also.
- Such a name is given, as in this, the number of chromosomes becomes half in the daughter nuclei as compared to the parental nuclei.
- This cell division is seen in **germ line cells** during the process of gamete formation.
- Meiosis involves two successive cell divisions. More precisely we can say two successive nuclear divisions.
- These two nuclear divisions are known as **meiotic-I** and **meiotic-II divisions**. The period between these two divisions is called **interkinesis**.
- During this period, cytokinesis may occur (as seen in all animal cells and some plant cells), but DNA does not replicate.

Meiosis-I

This division is also called **heterotypic** or **true reductional division**. Karyokinesis in meiosis-I includes following substages

1. Prophase-I

- It is the longest phase of meiosis and requires 90% of the total time required for entire meiosis.

It is further divided into following five substages

Leptotene or Leptonema

(Gk. *Leptos*-slender; *tania*-band; *nema*-thread)

This stage is marked by following events

- Chromosomes shorten and become visible as single beaded structures.

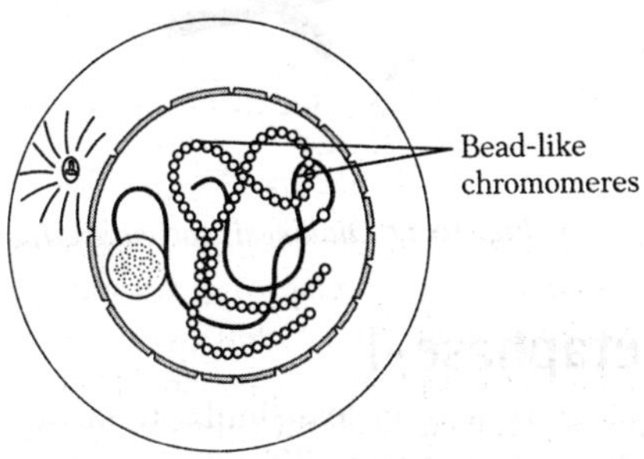

Fig. 16.6 *Leptotene (prophase-I)*

- These chromosomes may be irregularly arranged or polarised towards the centriole forming a 'bouquet'.
- During leptotene, the cytoplasm has many polyribosomes, but endoplasmic vesicles are few.

Zygotene or Zygonema

(Gk. *Zygon*-yoke; *tania*-band)

This stage is marked by following events

- Homologous chromosomes pair up. This pairing is called **synapsis** and resulted into a **complex bivalent** structure.
- Each pair of bivalent is the association of four chromatids and two centromeres.
- One chromosome of the pair comes from the male parent and another from the female parent.
- The members of the pair are homologous.
- This means, they have same length, their centromeres are in the same positions and they usually have the same number of genes arranged in the same order.
- Chromosomal synapsis is supported by a structure called **synaptonemal complex**. It stabilises the homologous chromosomes of bivalents.
- The bivalents shorten and thicken, partly by coiling. Each chromosome and its centromeres can now be seen clearly.

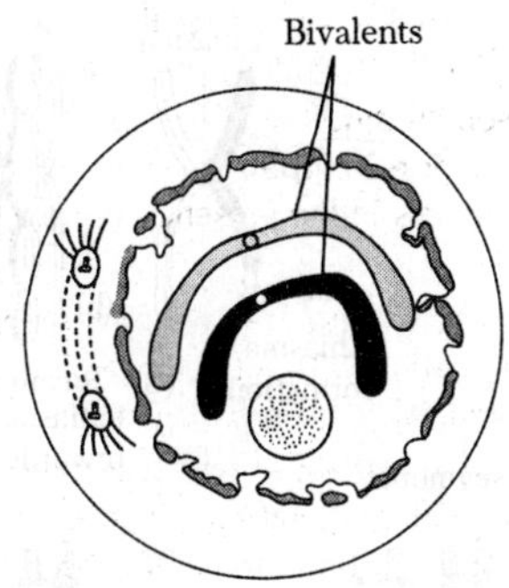

Fig. 16.7 *Zygotene (prophase-I)*

Pachytene or Pachynema

(Gk. *Pachy*-thick; *tania*-band, *nema*-thread)

It is the longest stage of prophase-I and is said to begin when synapsis is completed.

This stage is marked by following events

- It is characterised by the appearance of **recombination nodule** which is the dense area present over the bivalents.
- Bivalent chromosomes appear as tetrad.
- Non-sister chromatids of homologous chromosomes exchange segments between themselves. This exchange of segments is called **crossing over.**
- Crossing over is an enzyme mediated process involving enzyme **recombinase**, which is found to be present in recombination nodule.
- By the end of pachytene, the recombination between the homologous chromosome is completed and two chromatids are linked at the site of crossing over.

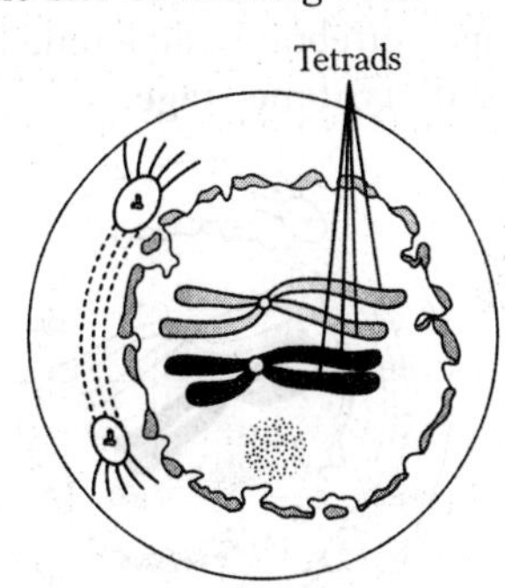

Fig.16.8 *Crossing over in pachytene (prophase-I)*

Crossing over

It is the most important process of living beings. From evolution point of view, it is considered as one of the fundamental processes involved in the fomation of a new species. It begins with synapsis and ends with terminalisation. The phenomenon in stepwise manner is given below

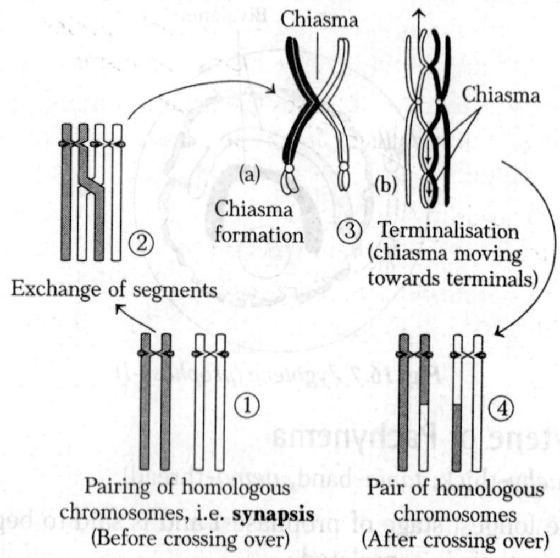

Fig 16.9 Crossing over

Diplotene or Diplotenema

This stage is marked by following events

- The beginning of this stage is marked by **chiasma formation.** This X-shaped structure was first described by **Johnson in amphibians**.
- The chiasma is formed as the result of dissolution of **synaptonemal complex**. Due to this dissolution, similar chromosomal parts begin to repel each other except in the regions of crossing over.
- Thus, the resulting structure acquires 'X' shape.
- Chiasma formation is necessary for the separation of homologous chromosomes which have undergone crossing over.
- In oocytes of some vertebrates, diplotene lasts for months or years. It is called **dictyotene stage.**

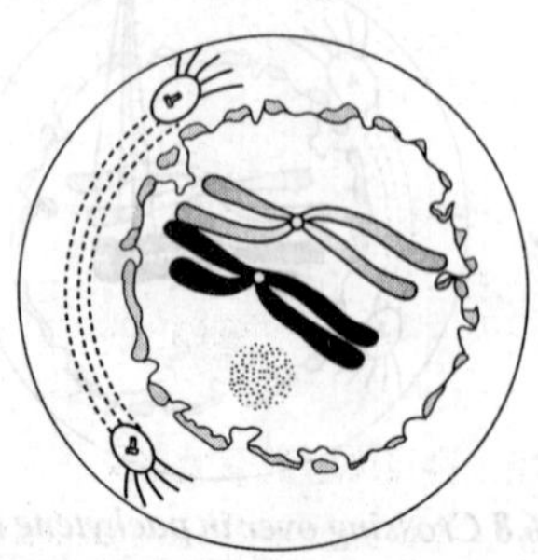

Fig. 16.10 Diplotene (prophase-I)

Achiasmatic meiosis

Meiosis without chiasmata, i.e. achiasmatic meiosis has been reported in some organisms, e.g. female of higher *Diptera* (including *Drosophila*), scorpion fly, some grasshoppers and scorpion species of Copepoda and Lepidoptera.

Diakinesis

This stage is marked by following events

- The terminalisation of chiasmata, i.e. the chiasmata start shifting towards the terminals of the chromosomes.
- It occurs due to the continuation of repulsion between the similar portions causing the chiasmata to displace along the length of chromosome.
- The nucleolus disappears and the nuclear membrane breaks completely.
- Along with all these the formation of spindle begins.

Fig. 16.11 Diakinesis (prophase-II)

2. Metaphase-I

- This stage is more or less similar to metaphase stage of mitosis. Here, the only difference is appearance of paired chromosomes at metaphasic plate.
- The attachment of spindle fibres with the kinetochore of chromosomal pair is also unique as shown below

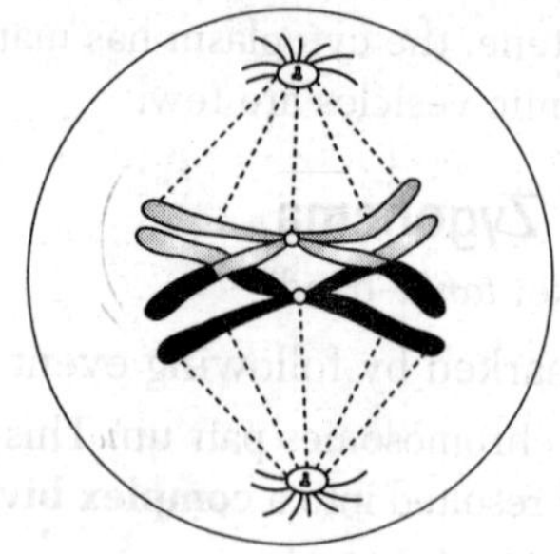

Fig. 16.12 Metaphase-I

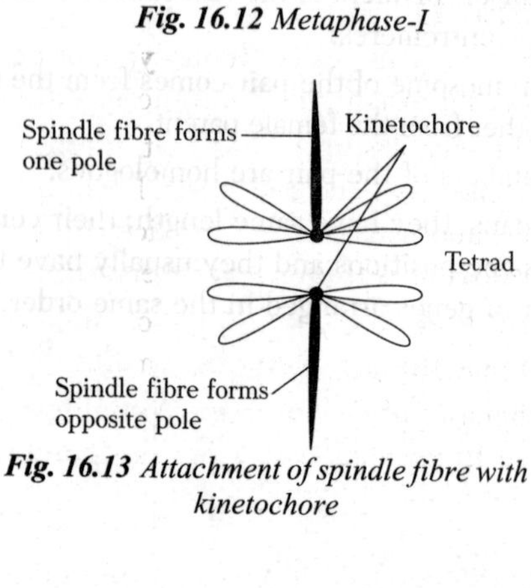

Fig. 16.13 Attachment of spindle fibre with kinetochore

3. Anaphase-I

- It is the stage in which chromosome number becomes half.
- Spindle fibres pull homologous chromosomes towards the opposite poles of the spindle.
- This process of separation is termed as **disjunction**.

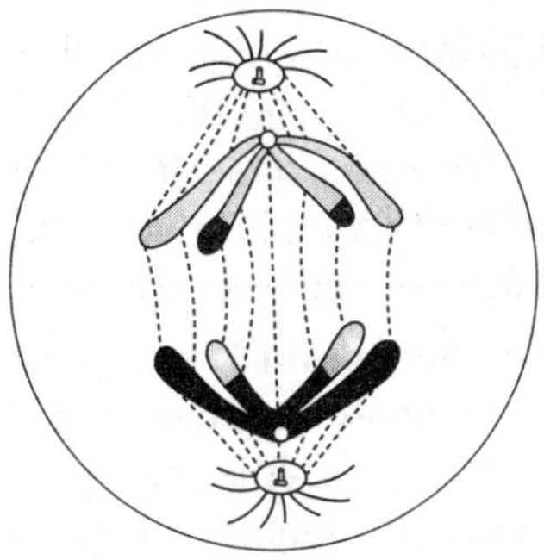

Fig. 16.14 Anaphase-I

4. Telophase-I

- It is almost similar to the telophase of mitosis. The only difference lies in the presence of bivalent chromosomes which usually remain coiled now onwards.
- However, in animals and some plants, the chromatids usually uncoil and a nuclear envelope reforms at each pole.

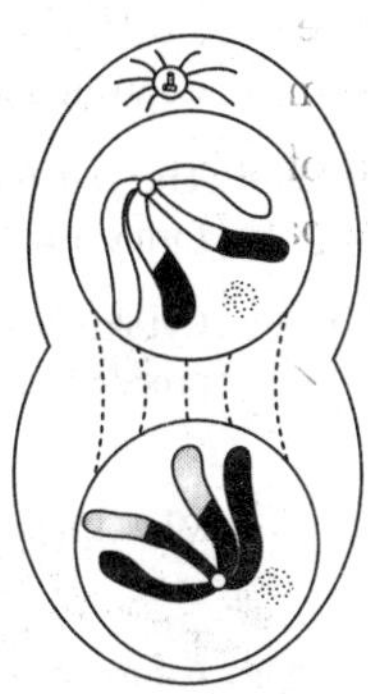

Fig. 16.15 Telophase-I

Meiosis-II

- This division is also called **homotypic** or **equational division** as it exactly resembles normal mitosis.
- The difference lies in the fact that interphase is not seen between meiosis-I and meiosis-II. Even, if it is seen, it is not associated with the DNA replication. Here, it is called **interkinesis** or **intrameiotic interphase**.
- Meiosis-II leads to the separation of chromatids of a chromosome through their centromeres.
- The subphases of meiosis-II are accordingly named as prophase-II, metaphase-II, anaphase-II and telophase-II.
- Since, meiosis-II resembles mitosis exactly so, there is the formation of four haploid daughter cells at the end of meiosis.

Significance of meiosis

- By reducing the chromosome number to half, it maintains the specific chromosome number of each species across generations in sexually reproducing organisms.
- By crossing over, it increases the genetic variability in the population of organisms from one generation to the next generation.
- Meiosis leads to the formation of haploid sex cells or gametes essential for reproduction.
- It leads to independent assortment of maternal and paternal chromosomes.

Cytokinesis

It is a process of the division of cytoplasm. The term 'Cytokinesis' was coined by **Whitman** in 1887. After karyokinesis, cell divides itself into two daughter cells by cytokinesis and cell division gets completed.

Process of cytokinesis is different in animal and plant cells.

Cytokinesis in animal cells

Following events are marked in cytokinesis of animal cells

- A furrow moves from peripheral plasma membrane to the centre of the cell. It ultimately divides the cell into two parts.
- This kind of cytoplasmic division is called **centripetal division** as shown below

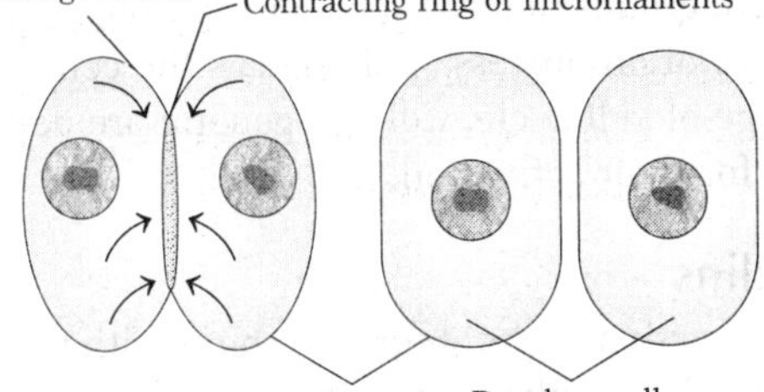

Fig 16.16 Centripetal cytokinesis in animals

Cytokinesis in plant cells

Following events are marked in cytokinesis of plant cells

- Separation starts by the formation of **cell plate** or **phragmoplast** at the centre of the cell.
- This later on expands towards the peripheral plasma membrane.

This kind of cytoplasmic division is called **centrifugal division** as shown below

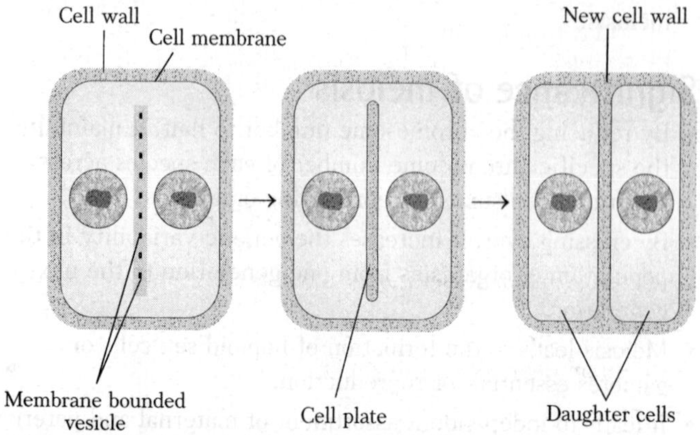

Fig 16.17 Centrifugal cytokinesis in plants

Tips to solve meiosis count related problems

- Number of meiosis requried for
- Formation of n number of pollen grains = $n/4$
- Formation of n number of eggs = n
- Formation of n number of seeds = $n + \frac{n}{4}$
- Formation of n number of seeds in Cypraceae is $n + n = 2n$

Control of cell cycle

The cell cycle is controlled by certain proteins at certain points in a cell cycle. These points are called **check points** and the proteins are called **Cyclin-dependent protein kinases** (Cdks) and **Cyclins.**

(i) Cyclin dependent protein kinases (Cdks)

A Cdk is infact an enzyme that adds negatively charged phosphate groups to other molecules through phosphorylation process. This signals the cell to enter the next stage of cell cycle. Cdks in general are dependent on Cyclins for their activation.

(ii) Cyclins

These are activating proteins that bind with Cdks to form a Cyclin-Cdk complex. They exit from a particular phase of cell cycle takes place when Cyclin gets degraded thereby deactivating Cdks.

The checkpoints

Cdks and Cyclins involved in cell cycle regulation occur at various checkpoints. These are

(i) **G_1-checkpoint** It is present at G_1/S boundary and regulated by Cdk_4/cyclin-D, Cdk_6/cyclin-D, etc. It is controlled by cell size, growth factors and cell environment, etc.

(ii) **G_2-checkpoint** It is present at G_2/M boundary and regulated by Cdk_2/cyclin B also, known as **MPF**, i.e. M-phase Promoting Factor. It is controlled by completion of DNA replication, DNA damage/mutation, cell size, etc.

(iii) **Metaphase checkpoint** It is present at metaphase/anaphase boundary and is regulated by Cyclin-B degradation. It is controlled by spindle fibre or microtubule attachment to chromosomes.

Table 16.1 Differences between cell division in plants and in animals

Cell division in plants	Cell division in animals
It is controlled by cytokinin.	It is controlled by mitogens like lymphokines and growth factors.
Plant cells do not change shape before the division.	Animal cells become rounded before cell division.
Spindle is anastral, i.e. no aster ray formation.	Spindle is amphiaster, i.e. one aster at each pole.
Centrioles are not found.	Centrioles are present.
The equatorial region of spindle forms phragmoplast.	The equatorial region of the spindle forms mid body.
Cytokinesis occurs by cell plate formation that grows centrifugally.	Cytokinesis occurs by cleavage which proceeds centripetally.
Microfilaments do not have much role.	Microfilament causes cleavage.
A solid middle lamella develops in between the two daughter cells for permanent adhesion.	A furrow is formed between two daughter cells.
The spindle usually persists during cytokinesis.	The spindle degenerates after anaphase.
New cell membrane is derived from vesicles of Golgi apparatus.	New cell membrane is derived from ER.

CHECK POINT 16.2

1. During leptotene, the cytoplasm has
   (a) few polyribosomes
   (b) few endoplasmic vesicles and many polyribosomes
   (c) complex bivalent structure
   (d) many endoplasmic vesicles and few polyribosomes

2. A bivalent in meiosis-I consists of
   (a) two chromatids and one centromere
   (b) two chromatids and two centromeres
   (c) four chromatids and two centromeres
   (d) four chromatids and four centromeres

3. In crossing over, exchange of chromosomes occurs between
   (a) non-sister chromatids of homologous chromosomes
   (b) sister chromatids of homologous chromosomes
   (c) non-sister chromatids of non-homologous chromosomes
   (d) sister chromatids of non-homologous chromosomes

4. Chiasma is formed by
   (a) rearrangement of synaptonemal complex
   (b) dissolution of synaptonemal complex
   (c) repulsion of chromosomal parts
   (d) separation of homologous chromosomes

5. During anaphase-I,
   (a) homologous chromosomes separate and migrate towards opposite poles
   (b) sister chromatids separate and migrate towards opposite poles
   (c) nuclei reform
   (d) chromosome forms straight line

6. Cytokinesis is the
   (a) exchange of homologous regions of non-sister chromatids
   (b) formation of tetrads
   (c) division of one cell into two
   (d) independent assortment of chromosomes

EXAM POINTS

- The correct sequence of phases of cell cycle is $G_1 \rightarrow S \rightarrow G_2 \rightarrow M$ **[NEET 2020]**
- In G_1-phase of interphase, cell is metabolically active, grows but does not replicate its DNA. **[NEET 2020]**
- In G_1-phase, cell grows and organelles duplicate. **[NEET 2020]**
- Cells in G_0-phase exit the cell cycle. **[NEET 2019]**
- Some dividing cells exit the cell cycle and enter vegetative inactive stage. This is called quiescent stage (G_0). This process occurs at the end of M-phase. **[NEET 2020]**
- In S-phase, DNA replication and chromosome duplication occurs. **[NEET 2020, 16]**
- In S-phase, of cell cycle, amount of DNA doubles in each cell. **[CBSE AIPMT 2014, AIIMS 2013]**
- In G_2-phase, cytoplasmic growth occurs. **[NEET 2020]**
- During interphase of eukaryotic cell division, increase in ATP synthesis, RNA synthesis and DNA synthesis occurs. **[WB JEE 2015]**
- Daughter and parent cells have same chromosome numbers in mitosis. **[WB JEE 2015]**
- In metaphase in M-phase, alignment of chromosomes on the equatorial plate occurs. **[NEET 2020]**
- Spindle fibres attach on to kinetochore of the chromosome. **[NEET 2016]**
- **Anaphase Promoting Complex** (APC) is a protein degradation machinery necessary for proper mitosis of animal cells. If APC is defective in a human cells, chromosome will not segregate. **[NEET 2017]**
- Codensation → Nuclear membrane disassembly → Arrangement at equator → Centromere division → Segregation → Telophase is the correct sequence of events occur during mitosis. **[NEET 2017]**
- Chromosome duplication without nuclear division refers to **endomitosis**. **[NEET 2017]**
- Prophase-I is the longest phase of meiosis. **[WB JEE 2011]**
- During zygotene phase, the homologous chromosomes pair of come together and form synapsis. **[NEET 2020]**
- Synapsis does not occur in somatic cells. **[NEET 2016]**
- The complex formed by a pair of synapsed homologous chromosomes is called bivalent. **[NEET 2013]**
- During meiosis-I at zygotene stage, synapsis takes place . After meiosis-I, the resultant daughter cells have twice the amount of DNA in comparison to haploid gamete. **[NEET 2020]**
- Crossing over takes place between non-sister chromatids of homologous chromosomes at pachytene stage of prophase-I **[NEET 2019]**
- Dissolution of the synaptonemal complex occurs during diplotene. **[NEET 2020]**
- The stage during which separation of the paired homologous chromosome begins is diplotene. **[NEET 2018]**
- Chiasmata is the point of contact between the two non-sister chromatids of homologous chromosomes, chiasmata becomes visible during diplotene stage. **[NEET 2020]**
- Terminalisation of chiasmata gets completed during diakinesis phase, where chromosomes get freely distributed in the cytoplasm. **[NEET 2020]**
- The correct sequences of events during meiosis is Synapsis → Crossing over → Terminalisation of chiasmata → Disappearance of nucleolus. **[CBSE AIPMT 2015]**
- After meiosis-I, the resultant daughter cells have twice the amount of DNA in comparison to haploid gamete. **[NEET 2019]**
- When cell has stalled DNA replication fork, G_2/M check point should be predominately activated. **[NEET 2016]**

Chapter Exercises

(A) Topical Exercise

Cell Cycle

1 The concept of cell cycle was first given by
(a) Rudolf Virchow (b) Strasburger
(c) W Fleming (d) Howard and Pelc

2 Synthesis of RNA takes place in which phase of the cell cycle?
(a) S-phase (b) M-phase
(c) G_1 and G_2-phase (d) None of these

3 Which of the following correctly matches, a phase of the cell cycle with its description?
(a) G_2 – Cell division
(b) M – Duplication of DNA
(c) G_1 – Synthesis of RNA
(d) S – Immediately precedes cell division

4 Cell would normally proceed to mitosis without interruption
(a) once it has entered the S-phase
(b) once it has entered the G_2-phase
(c) at any time during cell division activity
(d) None of the above

5 Cells which are not dividing are likely to be at **[NCERT Exemplar]**
(a) G_1-phase (b) G_2-phase
(c) G_0-phase (d) S-phase

6 At what stage of the cell cycle are histone proteins synthesised in a eukaryotic cell? **[NCERT Exemplar]**
(a) During S-phase
(b) During telophase
(c) During entire prophase
(d) During G_2-stage of prophase

7 Which one of the following statements is correct for cells in G_0-stage of cell cycle?
(a) Cells in G_0-stage proliferate excessively
(b) Cells are metabolically inactive
(c) Cells are metabolically active, but no longer proliferate in normal condition
(d) Both (a) and (b)

8 Select the incorrect statements regarding S-phase of interphase.
(a) Occurs between G_1 and G_2
(b) DNA replication begins in the nucleus
(c) Centrioles duplicate in the cytoplasm
(d) As DNA is doubled, number of chromosomes also doubles

9 If the number of chromosomes in G_1-phase of a cell is 18, then the number of chromosomes after S-phase of that cell would be
(a) 9 (b) 27
(c) 36 (d) 18

10 Which one of the following statements is true?
(a) The cell lies dormant during interphase
(b) The cell increases in size during G_0-phase
(c) The cell increases in size during G_1 and G_2-phases
(d) The key event of S-phase is the synthesis of proteins required for mitosis

11 It is difficult to observe individual chromosomes during interphase because
(a) the DNA has not been replicated yet
(b) they have uncoiled to form long, thin strands
(c) homologous chromosomes do not pair up until division starts
(d) the spindle must move them to the metaphase plate before they become visible

12 The second gap phase of interphase present between S and M-phase includes all, except
(a) duplication of mitochondria, chloroplast, Golgi bodies
(b) centriole duplication
(c) synthesis of tubulin protein
(d) synthesis of RNA

13 Liver cells are highly specialised, but when the liver cell is damaged or parts of it are removed surgically, tissue grows. Liver cells fall into which category of cells?
(a) Cells permanently locked in G_0
(b) Cells subjected to continual removal
(c) Cells that can be induced to enter the G_1-phase
(d) Cannot be determined from the information given

14 Increased enzymatic activity of some protein kinases important for the regulation of the cell cycle is due to
(a) kinase synthesis by ribosomes
(b) conversion of inactive cyclins to activate kinase by means of phosphorylation
(c) activation of inactive kinases by binding to cyclins
(d) destruction of inactive kinase by cytosolic protease

15 The most dramatic period of the cell cycle is
(a) Interphase (b) M-phase
(c) S-phase (d) G_0-phase

16 The phase of actual division is
(a) G_0 (b) G_1
(c) G_2 (d) M

Mitosis

17 Longest phase of karyokinesis is
(a) metaphase
(b) prophase
(c) anaphase
(d) telophase

18 In mitosis, the number of chromosomes in the parent and progeny cells is same, so this is also called as
(a) equational division
(b) reductional division
(c) binary division
(d) None of the above

19 The role of mitosis is not merely to divide cell into two daughter cells, but to ensure genetic continuity from one cell generation to another generation by
(a) the formation of cells with new chromosomes
(b) formation of two haploid daughter cells
(c) formation of two cells with identical DNA
(d) halving the chromosome number between the new cells

20 True statement for mitosis is
(a) the cell formed by it does not restore nucleoplasmic ratio
(b) two cells formed as a result of this division are identical in all respects
(c) cell formed by it have half the number of chromosomes than that of parent cell
(d) cells formed as a result of mitosis have different genetic characters

21 A cell that is entering the M-phase of the cell cycle is
(a) always haploid, always with duplicated chromosomes
(b) either haploid or diploid, always with duplicated chromosomes
(c) always diploid, either with duplicated or unduplicated chromosomes
(d) either haploid or diploid, either with duplicated or unduplicated chromosomes

22 Which of the following processes involve(s) mitosis?
I. Propagation of carrot plants by underground roots.
II. Development of the zygote into an embryo.
III. Replacement of worn-out muscle tissue.

Choose the option with containing correct statement(s).
(a) Only II (b) Only III
(c) I, II and III (d) Only I

23 During metaphase,
(a) centrioles move towards the metaphase plate
(b) chromosomes move towards the opposite poles
(c) spindle fibres attach to kinetochores of chromosomes
(d) All of the above can be seen

24 During anaphase, the chromosomes of the future daughter nuclei, begin their migration towards the two opposite poles. These are called
(a) daughter chromatids (b) chromatids
(c) daughter chromosomes (d) chromosomes

25 Consider the following statements at anaphase stage of mitosis.
I. The centromeres split into two and the spindle fibres pull the daughter centromeres to opposite poles.
II. The separated chromatids are pulled along behind the centromeres.

Choose the correct option.
(a) Statement I is true, but II is false
(b) Statement I is false, but II is true
(c) Both statements I and II are true
(d) Both statements I and II are false

26 The different shapes of chromosomes, i.e. I, J, L, V can be clearly studied during
(a) the alignment of chromosomes at the equator
(b) the poleward movement of chromosomes
(c) the phase when Golgi complex and ER disappear
(d) the phase when synthesis of amino acids of histone protein occurs

27 During mitosis, metaphase differs from anaphase in having
(a) same number of chromosomes and half number of chromatids
(b) half number of chromosomes and half number of chromatids
(c) half number of chromosomes and same number of chromatids
(d) same number of chromosomes and same number of chromatids

28 At which stage of mitosis, do these events occur?

	Spiralisation and condensation of DNA	Nuclear envelope breakdown	Centromeres split
(a)	Interphase	Interphase	Metaphase
(b)	Interphase	Prophase	Metaphase
(c)	Prophase	Metaphase	Anaphase
(d)	Prophase	Prophase	Anaphase

29 Throughout, which phase of the mitotic cycle would DNA be present in the least amount in a nucleus?
(a) Anaphase (b) Interphase (c) Metaphase (d) Telophase

30 Which of the following is true for telophase?
(a) Reformation of nuclear envelope, Golgi, nucleolus and ER occur
(b) Chromosomes cluster at opposite spindle poles
(c) Chromosomes lose their identity as discrete units
(d) More than one option is correct

31 Which of the events listed below is not observed during mitosis?
(a) Chromatin condensation
(b) Movement of centrioles to opposite poles
(c) Appearance of chromosomes with two chromosomes joined together at the centromere
(d) Crossing over

32 Which of the following serves as a mitotic spindle poison?
(a) Ca^{2+} (b) Mg^{2+}
(c) Tubulin (d) Colchicine

33 Interzonal type of spindle fibres extends
(a) between daughter chromosomes
(b) from pole to chromosomes
(c) from pole to kinetochore
(d) from pole to pole

34 Choose the correct option for the diagram given below.

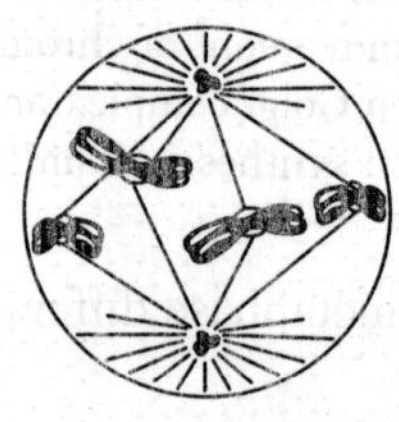

	Numbers of chromosomes	Numbers of chromatids	Numbers of centromeres	Numbers of kinetochores
(a)	4	8	8	8
(b)	4	4	4	8
(c)	4	16	8	4
(d)	4	8	4	8

35 Identify correctly the different phase of mitosis (*A*, *B* and *C*).

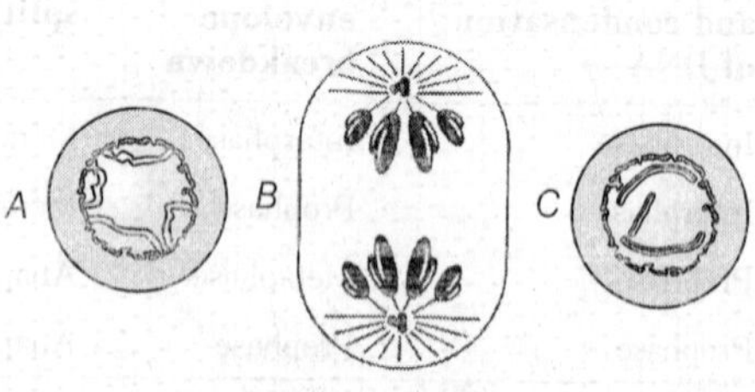

(a) *A*-Late prophase, *B*-Anaphase, *C*-Early prophase
(b) *A*-Late prophase, *B*-Anaphase, *C*-Early prophase
(c) *A*-Late prophase, *B*-Metaphase, *C*-Early prophase
(d) *A*-Early prophase, *B*-Anaphase, *C*-Late prophase

36 A diploid cell contains four chromosomes ($2n = 4$). If this cell divides by mitosis and there is no mutation, how many genotypes exist among the daughter cells?
(a) 1 (b) 2
(c) 4 (d) 8

37 The liquid endosperm of *Cocos nucifera* is the product of
(a) mitosis without involving cytokinesis
(b) mitosis involving cytokinesis
(c) amitosis without involving cytokinesis
(d) amitosis involving cytokinesis

38 Furrowing or Cleavage do not occur during cell division in a plant cell since
(a) spindle fibres are absent
(b) cell division is not required continuously
(c) it is surrounded by a relatively inextensible structure called cell wall
(d) All are correct

Meiosis

39 Prophase of meiosis includes
I. leptotene II. pachytene
III. cytokinesis IV. karyokinesis
Choose the option with correct phase(s).
(a) Only I (b) I and II
(c) I, II and IV (d) Only II

40 In which stage of meiotic prophase-I, chromosomes become thread-like and visible?
(a) Diakinesis (b) Leptotene (c) Zygotene (d) Pachytene

41 Crossing over occurs between
(a) non-sister chromatids of non-homologous chromosomes
(b) non-sister chromatids of homologous chromosomes
(c) sister chromatids of homologous chromosomes
(d) sister chromatids of non-homologous chromosomes

42 At which stage, the homologous chromosomes separate due to repulsion, but are held by chiasmata
(a) zygotene (b) pachytene (c) diplotene (d) diakinesis

43 Which stage marks the beginning chiasma formation?
(a) Pachytene (b) Diplotene
(c) Diakinesis (d) Leptotene

44 During anaphase-I of meiosis, [NCERT Exemplar]
(a) homologous chromosomes separate
(b) non-homologous autosomes separate
(c) sister chromatids separate
(d) non-sister chromatids separate

45 At which stage of meiosis is the genetic constitution of gametes is finally decided?
(a) Metaphase-I (b) Anaphase-II
(c) Metaphase-II (d) Anaphase-I

46 During anaphase-I,
(a) disjunction of homologous chromosomes occur
(b) each chromosome possesses one chromatid
(c) separated chromosomes move towards equator
(d) more than one option is correct

47 Which of the following is an event which does not occur during anaphase-I?

(a) Spindle fibres pull homologous chromosomes towards the opposite poles of the spindle
(b) The homologous chromosomes break their connection and separate out into two haploid sets or dyads
(c) One such set is present at each end of the spindle
(d) The process of separation of chromosomes which takes place is called disengagement

48 Disjunction refers to

(a) the separation of homologous chromosomes at pachytene
(b) the process of separation of homologous chromosomes
(c) incompatibility in fungi and other thallophytes
(d) modification of gene action by a non-allelic gene

49 Which type of chromosomes segregate during prophase-I of meiosis?

(a) Homologous chromosomes
(b) Non-homologous chromosomes
(c) 'A' and 'B' chromosomes
(d) Centric and acentric chromosomes

50 Identify the false statement.

(a) DNA synthesis does not occur during the interkinesis of meiotic cycle
(b) The chronological sequence of meiosis is leptotene > zygotene > pachytene > diplotene > diakinesis
(c) Synapse of homologous chromosomes occurs during diplotene
(d) Crossing over occurs during pachytene

51 The daughter cells formed as a result of meiosis differ not only from parent cell in having half the number of chromosomes but also differ among themselves qualitatively in genetic constitution as a result of

(a) crossing over only
(b) segregation and crossing over
(c) crossing over, independent assortment and segregation
(d) metaphasic arrangement of chromosomes

52 The given event occurs in

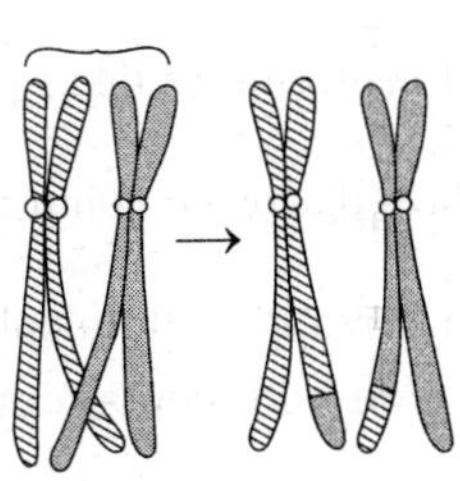

(a) prophase-I
(b) prophase-II
(c) prophase of mitosis
(d) prophase and metaphase of mitosis

53 Choose the diagram which correctly depicts anaphase-I.

(a)
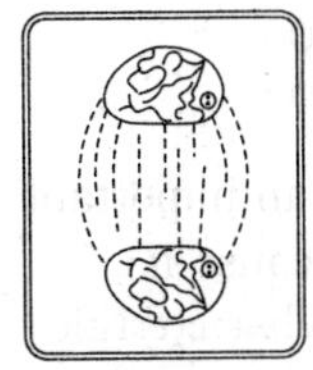

(b)

(c)
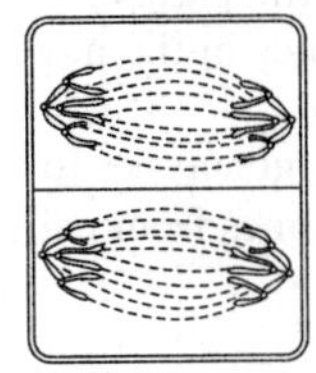

(d)
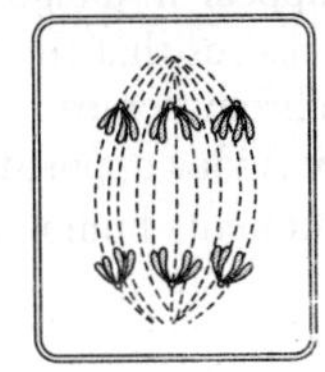

54 Identify the incorrect statement about meiosis. **[NCERT Exemplar]**

(a) Pairing of homologous chromosomes
(b) Four haploid cells are formed
(c) At the end of meiosis, the number of chromosomes are reduced to half
(d) Two cycles of DNA replication occur

55 In meiosis,

I. every chromosome behaves independently.
II. homologous chromosomes show pairing.
III. found in somatic cells.
IV. crossing over occurs.

Choose the option with containing correct statement(s).

(a) Only IV (b) I and III (c) II and IV (d) I and II

56 Arrange the different stages of meiosis (I, II, III and IV in sequence.

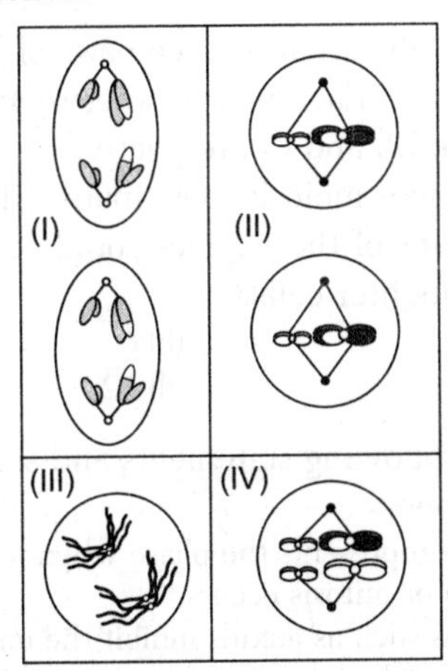

(a) I → IV → II → III
(b) III → IV → II → I
(c) IV → III → II → I
(d) II → I → III → IV

57 How many microspore mother cells are required to produce 1000 pollen grains in an anther?

(a) 250 (b) 1000 (c) 10 (d) 200

58 How many times meiotic divisions will take place to produce 512 pollen grains?
(a) 128 (b) 100
(c) 256 (d) 50

59 The formation of chiasmata is an important feature of meiotic divisions because it
(a) ensures that the same genetic characteristics appear in daughter cells as in the parents
(b) ensures that the number of genes in the new chromosomes remains constant
(c) provides opportunities for new genotypes to arise
(d) prevents homologous chromosomes from pairing

60 If one ignores the effect of crossing over, then the possible number of different haploid cells which could arise by the meiotic division of a diploid cell containing 10 chromosomes (i.e. $n = 5$) is
(a) 2 (b) 4 (c) 8 (d) 32

61 Vinblastin is a standard chemotherapeutic drug used to treat cancer. Because it interferes with the assembly of microtubules, its effectiveness must be related to
(a) suppression of cyclin production
(b) inhibition of DNA synthesis
(c) disruption of mitotic spindle formation
(d) inhibition of regulatory protein involved in phosphorylation

(B) Cumulative Exercise

(Assorted Questions of the Chapter for Advanced Level Practice)

1 The graph below shows the changes in DNA content in a single cell during the mitotic cycle.

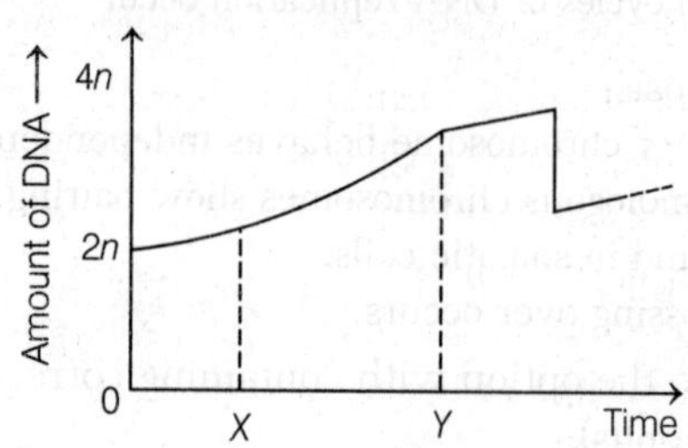

Which stage occurs between time X and time Y.
(a) interphase (b) prophase
(c) metaphase (d) anaphase

2 A diploid cell contains three pairs of homologous chromosomes. Each pair is heterozygous for a pair of alleles, Aa, Bb and Cc, respectively. If this diploid cell undergoes meiosis, how many different combinations of these genes could be produced in the haploid daughter cells?
(a) 3 (b) 6
(c) 8 (d) 12

3 Read the following statements and select the incorrect one.
(a) M-phase represents the phase when the actual cell division or mitosis occurs
(b) Mitogens such as auxins inhibit the mitosis division
(c) Mitosis can be best studied in lab using root tip mater l
(d) Spindle usually presents as phragmoplast during cytokinesis in plants

4 Match the following columns.

Column I	Column II
A. Interkinesis	1. S-phase
B. Histone synthesis	2. Diakinesis
C. Colchicine	3. Chromosome fibre
D. Terminalisation	4. Disruption of mitotic spindle

Codes

	A	B	C	D		A	B	C	D
(a)	4	1	2	3	(b)	2	1	4	3
(c)	3	1	4	2	(d)	3	2	1	4

5 If a mitocyte has 10 pg of DNA and 16 chromosomes in G_1 then which of the following is correct?
(a) S-phase will have 5 pg of DNA
(b) The cell at anaphase stage will have 20 pg of DNA and a total of 16 chromosomes
(c) G_2 phase will have 8 chromosomes in a cell
(d) The cell at metaphase will have 10 pg of DNA and a total of 16 chromosomes

6 Pick up the correct statement.
I. Synapsis of homologous chromosomes occurs during prophase-I.
II. Division of centromeres takes place during anaphase-I.
III. Spindle fibres disappear completely in telophase of mitosis.
IV. Nucleoli may reappear in telophase-I.

Choose the option with containing correct statement(s).
(a) Only I (b) Only III
(c) I and II (d) I, III and IV

7 Which of the following statement(s) is/are correct about meiosis?

I. Meiosis involves pairing of homologous chromosomes and recombination between them.
II. Two diploid cells are formed at the end of meiosis-II.
III. Meiosis involves two sequential cycles of nuclear and cell division called meiosis-I and meiosis-II, but only a single cycle of DNA replication.
IV. Meiosis-II is initiated after the parental chromosome replication, which produces identical sister chromatids at the S-phase.

(a) I and III (b) Only II
(c) II and III (d) I, II, III and IV

8 Select the statement, which is not correct.
(a) Spindle fibres arise from centriole
(b) Centromere facilitates attachment of chromosomes to spindle fibres
(c) Movement of chromosomes during cell division is due to cytoplasmic streaming
(d) Minimum coiling of chromosomes is seen during interphase

9 Identify the correct statement.
(a) Mitosis occurs in plants when there is a change from sporophyte to gametophyte
(b) Meiosis produces genetically identical daughter cells
(c) Mitosis ensures genetic continuity from generation to generation
(d) After meiosis-II, the daughter cells are diploid

10 Identify the pair which is a mismatch for plant cell division.

(a) Spindle – Amphiaster
(b) Centrioles – Absent
(c) Cytokinesis – By cell plate formation which grows centrifugally
(d) New cell membrane – Endoplasmic reticulum

11 At which stages of the mitotic cell cycle do these events occur?

	DNA replication	Breakdown of nuclear membrane	Division of centromere
(a)	Interphase	Metaphase	Metaphase
(b)	Interphase	Prophase	Anaphase
(c)	Interphase	Interphase	Anaphase
(d)	Prophase	Prophase	Anaphase

12 Human cells examined during G_2-phase of the cell cycle contained 100 units of DNA. What would be the quantity of DNA in one of the human cells produced by telophase-II of meiosis?
(a) 25 units (b) 50 units
(c) 100 units (d) Between 25 and 50 units

13 If an organism has 15 pairs of homologous chromosomes, how many chromosomes will each daughter cell have after mitosis and meiosis?
(a) 10, 15 (b) 45, 15 (c) 15, 30 (d) 30, 15

14 The graph represents the changes in the quantity of DNA present in one nucleus at different stages in the life cycle.

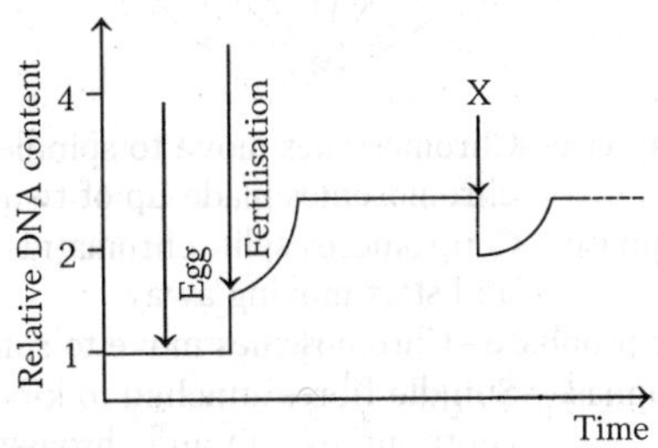

Which stage takes place at X?
(a) Interphase (b) Metaphase
(c) Prophase (d) Telophase

15 Identify the pair that does not match correctly.
(a) Gametes formed by meiosis – Meiocytes
(b) Separation of chromatids and centromere – Meiosis-II
(c) DNA duplication – Mitosis and meiosis-I
(d) Best stage to observe chromosome morphology and number – Interphase

16 Which letter indicates the end of meiosis-I as depicted in the different phases of meiosis?

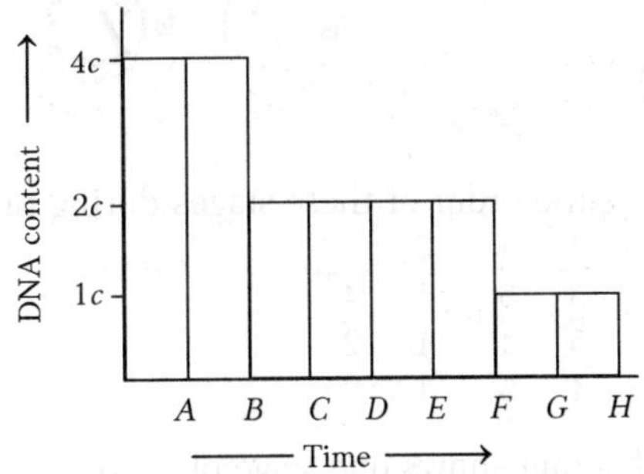

(a) D (b) G
(c) C (d) H

17 Beads on string of A are seen in B which condense to form chromosomes in C-stage of cell division. Identify the A, B and C from the given options.

	A	B	C
(a)	Chromonema	Chromatid	Metaphase
(b)	Chromatin	Chromatid	Metaphase
(c)	Chromonema	Chromosome	Anaphase
(d)	Chromonema	Chromatid	Anaphase

18 Identify the stage of mitosis with its characteristics.

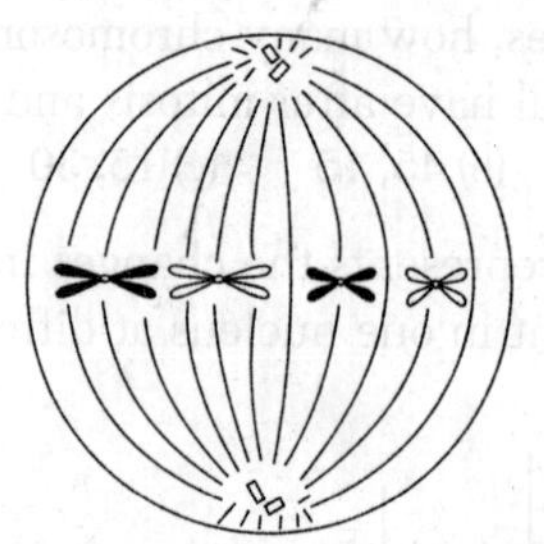

(a) Metaphase–Chromosomes move to spindle equator, chromosomes made up of two chromatids
(b) Anaphase– Centromeres split, chromatids separate and start moving away
(c) Late prophase–Chromosomes move to spindle equator
(d) Metaphase–Spindle fibres attached to kinetochores, centromeres split and chromatids separate

19 Read the following statements.

I. Interphase is absent in meiotic cell division.
II. Repeated meiosis can occur in a single cell.
III. Prophase-II is longer than prophase-I.
IV. Microsporocytes cannot undergo meiosis.
V. During mitotic metaphase, a centromere is connected with both spindle poles.

How many of the above statements are correct?

(a) One (b) Three
(c) Two (d) Five

20 The diagrams show stages of mitosis.

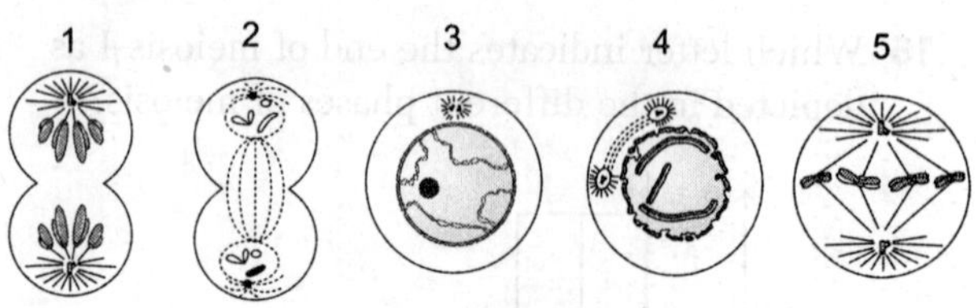

What is the order of these stages during mitosis?

(a) 1 2 4 3 5
(b) 2 3 5 1 4
(c) 3 5 4 1 2
(d) 3 4 5 1 2

21 The diagram shows one stage of mitotic cell division.

From the position of the cell organelles and the chromosomes, what stage of mitosis has been reached?

(a) Early anaphase (b) Early telophase
(c) Mid metaphase (d) Late prophase

22 Read the following statements.

I. Prophase is marked by the initiation of condensation of chromatin material.
II. The chromosomal material becomes untangled during the process of chromatin condensation.
III. In the S and G_2-phases, the new DNA molecules formed are not distinct but interwound.
IV. Nuclear envelope remains intact throughout the prophase.
V. At the end of prophase, when viewed under microscope, a cell shows distinct Golgi, ER and nucleolus.

How many of the above statements are correct?

(a) Three (b) Four
(c) Two (d) One

23 See the diagram carefully and sequentially arrange the steps of amitosis given below.

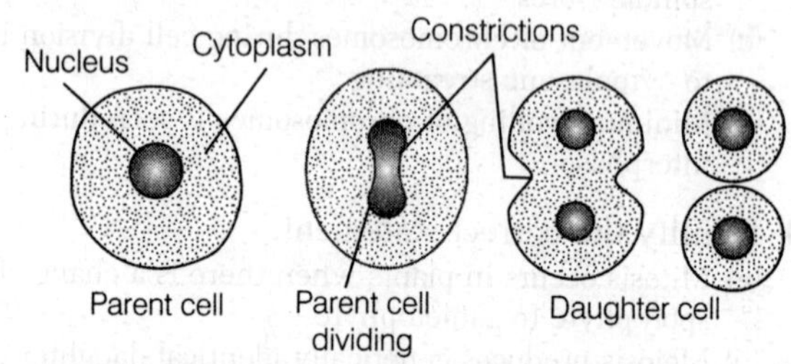

I. The constriction appears in the cytoplasm.
II. The nucleus of cell elongates and develops a constriction round its middle.
III. The constriction in nucleus gradually deepens and finally cuts the nucleus into two daughter nuclei.
IV. The cytoplasmic constriction divides the parent cell into two daughter cells, each with a nucleus.

Option containing correct sequence of events is

(a) I → III → II → IV (b) I → II → III → IV
(c) II → I → III → IV (d) II → III → I → IV

24 The events shown below occur during different phases of mitosis.

1. Spiralisation of DNA
2. Hydration of DNA
3. Centromeres split
4. Centromeres attach to spindle fibres
5. DNA replicates

Which one of the following correctly identifies each of the phases described?

	Interphase	Prophase	Metaphase	Anaphase	Telophase
(a)	1	2	3	4	5
(b)	1	5	4	2	3
(c)	2	4	1	3	5
(d)	5	1	4	3	2

25 The graph below illustrates the behaviour of the chromosomes during one mitotic cell division.

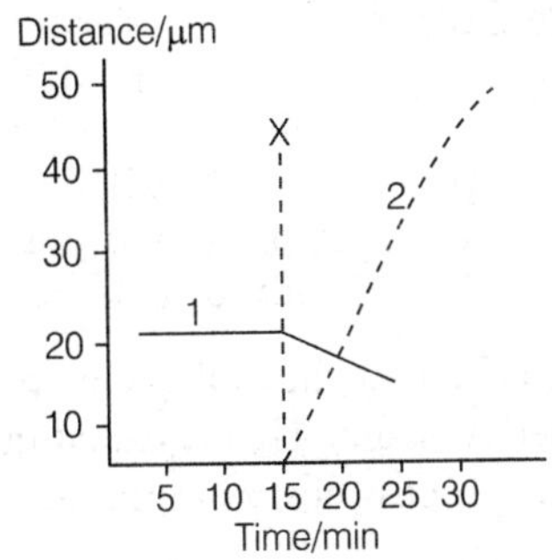

Which one of the following correctly indicates the changes illustrated by curves 1 and 2 and the stage of mitosis which is commencing at X

Stages of mitosis commencing at X	Distance between centromeres of chromosomes and poles of spindle	Distance between centromeres of sister chromatids
(a) Anaphase	1	2
(b) Anaphase	2	1
(c) Metaphase	1	2
(d) Metaphase	2	1

26 The diagram below is of a cell forming gametophyte undergoing division.

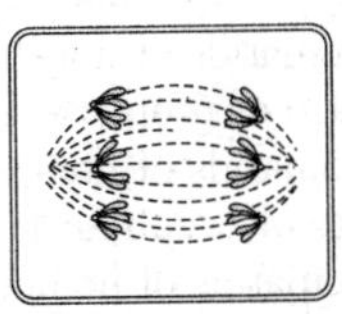

What is the type of division and the stage?
(a) Meiosis, prophase-I, $2n = 8$
(b) Meiosis, metaphase-II, $n = 8$
(c) Mitosis, prophase, $n = 8$
(d) Mitosis, early anaphase, $2n = 8$

27 The diagrams below show chromosomes in a cell undergoing mitosis and in a cell undergoing meiosis.

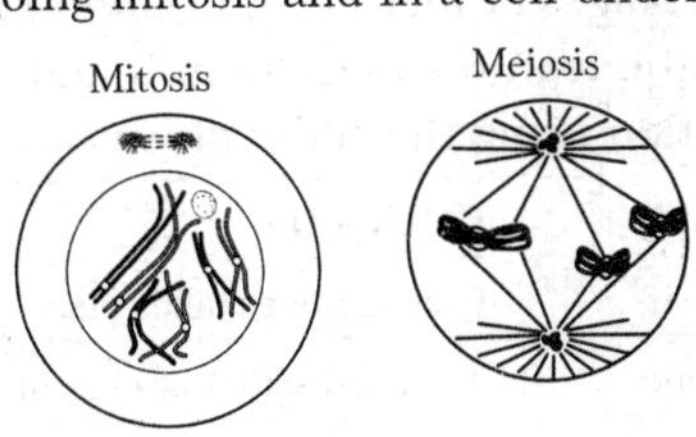

Which of the following names the stages of division correctly?

	Mitosis	Meiosis
(a)	Prophase	Prophase-I
(b)	Prophase	Metaphase-I
(c)	Metaphase	Anaphase-I
(d)	Metaphase	Metaphase-II

Assertion and Reason

Direction (Q. Nos. 28-36) *In each of the following questions, a statement of Assertion (A) is given followed by the corresponding statement of Reason (R). Of the statements, mark the correct answer as*

(a) Both A and R are true and R is the correct explanation of A
(b) Both A and R are true, but R is not the correct explanation of A
(c) A is true, but R is false
(d) A is false, but R is true
(e) Both A and R are false

28 **Assertion** (A) A cell after telophase-II does not enter another interphase.
Reason (R) Gametes or Spores are formed after telophase-II.

29 **Assertion** (A) Diakinesis is the final stage of prophase-I.
Reason (R) Terminalisation of chiasmata occurs in diakinesis.

30 **Assertion** (A) Some cells undergo G_0-phase due to inactivation of cell cycle.
Reason (R) Cells at this stage remain metabolically active, but no longer proliferate.

31 **Assertion** (A) Meiotic division occurs in reproductive cells.
Reason (R) Synapsis occurs during zygotene of meiosis.

32 **Assertion** (A) In mitosis, two identical cells are produced from a single cell and karyokinesis is followed by cytokinesis.
Reason (R) Cytokinesis is of two types.

33 **Assertion** (A) Division of protoplast is called cytokinesis.
Reason (R) This cytokinesis occurs due to cell plate formation or by cleavage.

34 **Assertion** (A) In mitosis, prophase has no substages.
Reason (R) In meiosis, prophase has six substages.

35 **Assertion** (A) Histone proteins are synthesised during S-phase.
Reason (R) They are found associated with DNA to form bead-like nucleosomes.

36 **Assertion** (A) Cell growth results in disturbing the ratio between the nucleus and the cytoplasm.
Reason (R) Mitosis helps the cell to restore the nucleocytoplasmic ratio.

(C) Medical Entrances' Gallery

(Collection of questions asked in NEET & Various Medical Entrance Exams)

1. Attachment of spindle fibres to kinetochores of chromosomes becomes evident in **[NEET 2020]**
(a) anaphase (b) telophase
(c) prophase (d) metaphase

2. In a mitotic cycle, the correct sequence of phases is **[NEET 2020, 19]**
(a) S, G_1, G_2, M (b) G_1, S, G_2, M
(c) M, G_1, G_2, S (d) G_1, G_2, S, M

3. During meiosis-I, in which stage does synapsis take place? **[NEET 2020]**
(a) Pachytene (b) Zygotene (c) Diplotene (d) Leptotene

4. Match the following (columns) events that occur in their respective phases of cell cycle and select the correct option from the codes given below.

Column I	Column II
A. G_1-phase	1. Cell grows and organelle duplication
B. S-phase	2. DNA replication and chromosome duplication
C. G_2-phase	3. Cytoplasmic growth
D. Metaphase in M-phase	4. Alignment of chromosomes

Codes **[NEET 2020]**

	A	B	C	D		A	B	C	D
(a)	2	3	4	1	(b)	3	4	1	2
(c)	4	1	2	3	(d)	1	2	3	4

5 Identify the correct statement with regard to G_1-phase (Gap 1) of interphase. **[NEET 2020]**
(a) Reorganisation of all cell components, takes place
(b) Cell is metabolically active grows, but does not replicate its DNA
(c) Nuclear division takes place
(d) DNA synthesis or replication takes place

6 Dissolution of the synaptonemal complex occurs during **[NEET 2020]**
(a) zygotene (b) diplotene (c) leptotene (d) pachytene

7 Match the following columns with respect to meiosis. Select the correct option from the codes given below.

Column I	Column II
A. Zygotene	1. Terminalisation
B. Pachytene	2. Chiasmata
C. Diplotene	3. Crossing over
D. Diakinesis	4. Synapsis

[NEET 2020]

Codes

	A	B	C	D		A	B	C	D
(a)	4	3	2	1	(b)	1	2	4	3
(c)	2	4	3	1	(d)	3	4	1	2

8 Some dividing cells exit the cell cycle and enter vegetative inactive stage. This is called quiescent stage (G_0). This process occurs at the end of **[NEET 2020]**
(a) G_1-phase (b) S-phase (c) G_2-phase (d) M-phase

9 Cells in G_0-phase **[NEET 2019]**
(a) enter the cell cycle (b) suspend the cell cycle
(c) terminate the cell cycle (d) exit the cell cycle

10 After meiosis-I, the resultant daughter cells have **[NEET 2019]**
(a) same amount of DNA as in the parent cell in S-phase
(b) twice the amount of DNA in comparison to haploid gamete
(c) same amount of DNA in comparison to haploid gamete
(d) four times the amount of DNA in comparison to haploid gamete

11 Crossing over takes place between which chromatids and in which stage of the cell cycle? **[NEET 2019]**
(a) Non-sister chromatids of non-homologous chromosomes at zygotene stage of prophase-I
(b) Non-sister chromatids of homologous chromosomes at pachytene stage of prophase-I
(c) Non-sister chromatids of homologous chromosomes at zygotene stage of prophase-I
(d) Non-sister chromatids of non-homologous chromosomes at pachytene stage of prophase-I

12 The stage during which separation of the paired homologous chromosomes begins is **[NEET 2018]**
(a) diakinesis (b) diplotene (c) pachytene (d) zygotene

13 Match the stages of meiosis in Column I to their characteristic features in Column II and select the correct option using the codes given below.

Column I	Column II
A. Diakinesis	1. Crossing over takes place
B. Pachytene	2. Terminalisation of chiasmata
C. Zygotene	3. Chromosomes align at equatorial plate
D. Metaphase-I	4. Pairing of homologous chromosomes

[AIIMS 2018]

Codes

	A	B	C	D		A	B	C	D
(a)	1	2	3	4	(b)	2	4	1	3
(c)	4	3	1	2	(d)	2	1	4	3

14 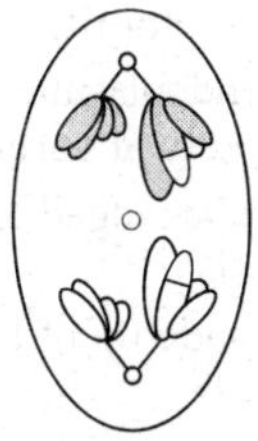

Above diagram represents **[JIPMER 2018]**

(a) anaphase-I (b) metaphase-I
(c) telophase-I (d) prophase-I

15 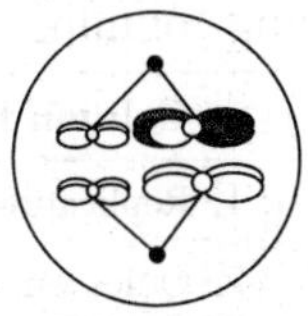

Above diagram represents **[JIPMER 2018]**

(a) metaphase-I (b) anaphase-I
(c) metaphase-II (d) anaphase-II

16 Anaphase Promoting Complex (APC) is a protein degradation machinery necessary for proper mitosis of animals cells. If APC is defective in a human cells, which of the following is expected to occur?

(a) Chromosomes will not condense **[NEET 2017]**
(b) Chromosomes will be fragmented
(c) Chromosomes will not segregate
(d) Recombination of chromosome arms will occur

17 Which of the following options gives the correct sequences of events during mitosis? **[NEET 2017]**

(a) Condensation → Nuclear membrane disassembly → Crossing over → Segregation → Telophase
(b) Condensation → Nuclear membrane disassembly → Arrangement at equator → Centromere division → Segregation → Telophase
(c) Condensation → Crossing over → Nuclear membrane disassembly → Segregation → Telophase
(d) Condensation → Arrangement at equator → Centromere division → Segregation → Telophase

18 Chromosome duplication without nuclear division refers to **[NEET 2017]**

(a) meiosis (b) mitosis
(c) androgenesis (d) endomitosis

19 During cell growth, DNA synthesis takes place in **[NEET-II 2016]**

(a) S-phase (b) G_1-phase (c) G_2-phase (d) M-phase

20 When cell has stalled DNA replication fork, which checkpoint should be predominately activated? **[NEET-II 2016]**

(a) G_1/S (b) G_2/M
(c) M (d) Both G_2/M and M

21 In meiosis, crossing over is initiated at **[NEET-I 2016]**

(a) leptotene (b) zygotene
(c) diplotene (d) pachytene

22 A cell at telophase stage is observed by a student in a plant brought from the field. He tells his teacher that this cell is not like other cells at telophase stage. There is no formation of cell plate and thus the cell is containing more number of chromosomes as compared to other dividing cells. This would result in **[NEET-I 2016]**

(a) polyploidy (b) somaclonal variation
(c) polyteny (d) aneuploidy

23 Which one of the following is not a characteristic feature during mitosis in somatic cells? **[NEET-I 2016]**

(a) Disappearance of nucleolus (b) Chromosome movement
(c) Synapsis (d) Spindle fibres

24 Spindle fibres attach on to **[NEET-I 2016]**

(a) kinetochore of the chromosome
(b) centromere of the chromosome
(c) kinetosome of the chromosome
(d) telomere of the chromosome

25 Identify the correct combination regarding anaphase, anaphase-I and anaphase-II. **[EAMCET 2015]**

(a) Anaphase-Centromere splits
Anaphase-I-Centromere splits
Anaphase-II-Centromere splits
(b) Anaphase-Chromatids move to opposite poles
Anaphase-I-Homologous chromosomes separate,
Anaphase-II-Centromere splits
(c) Anaphase-Chromosomes cluster at opposite poles,
Anaphase-I-Homologous chromosomes separate
Anaphase-II-Centromere splits
(d) Anaphase-Chromosomes move to opposite poles
Anaphase-I-Homologous chromosomes separate,
Anaphase-II-Centromere splits

26 Match the following columns.

Column I	Column II
A. Synapsis aligns homologous chromosomes	1. Anaphase-II
B. Synthesis of RNA and protein	2. Zygotene
C. Action of enzyme recombinase	3. G_2-phase
D. Centromeres do not separate, but chromatids move towards opposite poles	4. Anaphase-I
	5. Pachytene

[CBSE AIPMT 2015]

Codes

	A	B	C	D		A	B	C	D
(a)	2	1	3	4	(b)	2	3	5	4
(c)	1	2	5	4	(d)	2	3	4	5

27 A somatic cell that has just completed the S-phase of its cell cycle, as compared to gamete of the same species has **[CBSE AIPMT 2015]**
(a) twice the number of chromosomes and twice the amount of DNA
(b) same number of chromosomes, but twice the amount of DNA
(c) twice the number of chromosomes and four times the amount of DNA
(d) four times the number of chromosomes and twice the amount of DNA

28 During meiosis-I, the number of chromosomes is **[J&K CET 2015]**
(a) halved (b) tripled
(c) doubled (d) quadrupled

29 Arrange the following events of meiosis in correct sequences. **[CBSE AIPMT 2015]**
I. Crossing over
II. Synapsis
III. Terminalisation of chiasmata
IV. Disappearance of nucleolus

Codes
(a) II, I, IV, III (b) II, I, III, IV
(c) I, II, III, IV (d) II, III, IV, I

30 Which of the following does not occur in the interphase of eukaryotic cell division? **[WB JEE 2015]**
(a) Increase of ATP synthesis
(b) Increase of RNA synthesis
(c) Increase of DNA synthesis
(d) Reduction in cell size

31 Match of the item in Column I with those in Column II and the choose the correct answer.

Column I	Column II
A. Mitosis	1. Occurs in diploid cells only
B. Meiosis	2. Occurs in both haploid and diploid cells
	3. Daughter and parent cells have same chromosome numbers
	4. Synapsis of homologous chromosomes

Codes **[WB JEE 2015]**

	A	B		A	B
(a)	1	2	(b)	2	3
(c)	3	4	(d)	4	1

32 Which one of the following is the significance of mitosis? **[Kerala CEE 2015]**
(a) Restricted to haploid cells
(b) Cell repair
(c) Increase in genetic variability
(d) Leads to evolution of new genotypes
(e) None of the above

33 Find the correct statement. **[Kerala CEE 2015]**
(a) During mitosis, endoplasmic reticulum and nucleolus disappear completely at early prophase
(b) Chromosomes are arranged along the equator during prophase of mitosis
(c) Chromosome is made up of two sister chromatids at anaphase of mitosis
(d) A cell plate is laid down during interphase
(e) Small disc-shaped structures at the surface of centromeres that appear during metaphase are kinetochores

34 Match the following columns.

Column-I	Column-II
A. G_1-phase	1. Replication of DNA
B. S-phase	2. Quiescent stage
C. G_2-phase	3. Condensation of chromatin
D. G_0-phase	4. Protein synthesis
	5. Interval between mitosis and initiation of DNA replication

Codes **[EAMCET 2014]**

	A	B	C	D		A	B	C	D
(a)	3	5	1	2	(b)	5	4	1	3
(c)	5	1	4	2	(d)	5	2	3	4

35 What are spindle fibres that connect the centromere to respective poles called? **[MHT CET 2014]**
(a) Astral rays (b) Interphase fibres
(c) Chromosomal fibres (d) Interchromosomal fibres

36 Match the following columns.

Column I	Column II
A. Leptotene	1. Terminalisation of chiasma
B. Zygotene	2. Crossing over and recombination
C. Pachytene	3. Synapsis
D. Diakinesis	4. Visibility of chromosomes

Codes **[Kerala CEE 2014]**

	A	B	C	D		A	B	C	D
(a)	1	2	3	4	(b)	1	3	2	4
(c)	4	3	2	1	(d)	4	1	2	3
(e)	4	2	3	1					

37 In S-phase of cell cycle, **[CBSE AIPMT 2014, AIIMS 2013]**
(a) amount of DNA remains same in each cell
(b) chromosome number is increased
(c) amount of DNA is reduced to half in each cell
(d) amount of DNA doubles in each cell

38 The enzyme recombinase is required in which stage of meiosis? **[CBSE AIPMT 2014, 12]**
(a) Zygotene (b) Diplotene
(c) Diakinesis (d) Pachytene

39 During which phase(s) of cell cycle, amount of DNA in a cell remains at 4 C level if the initial amount is denoted as 2 C. **[CBSE AIPMT 2014]**

(a) G_1 and S (b) Only G_2
(c) G_2 and M (d) G_0 and G_1

40 **Assertion** (A) Meiosis-II is similar to mitosis.
Reasons (R) Meiosis-I cannot occur in haploid cells. **[AIIMS 2013]**

(a) Both A and R are true and R is the correct explanation of A
(b) Both A and R are true, but R is not the correct explanation of A
(c) A is true, but R is false
(d) Both A and R are false

41 Which of the following events take place during anaphase stage of mitosis?

I. Spindle fibres attach to kinetochores of chromosomes.
II. Centromeres split and chromatids separate.
III. Chromatids move to opposite poles.
IV. Nucleolus, Golgi complex and ER reform.

Codes **[Kerala CEE 2013]**

(a) I and II (b) II and III (c) III and IV (d) I and IV
(e) Only II

42 Identify the correct stage in cell division with its characteristics. **[NEET 2013]**

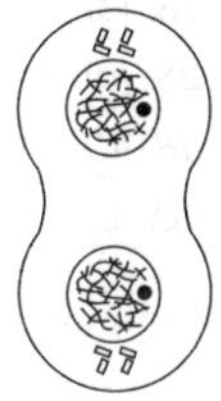

(a) Telophase — Endoplasmic reticulum and nucleolus not reformed yet
(b) Telophase — Nuclear envelop reforms, Golgi complex reforms
(c) Late anaphase — Chromosomes move away from equatorial plate, Golgi complex not present
(d) Cytokinesis — Cell plate formed, mitochondria distributed between two daughter cells

43 The complex formed by a pair of synapsed homologous chromosomes is called **[NEET 2013]**

(a) kinetochore (b) bivalent
(c) axoneme (d) equatorial plate

44 In which stage of mitosis, chromosomes arrange on equatorial plate? **[UP CPMT 2012]**

(a) Prophase (b) Metaphase
(c) Anaphase (d) Telophase

45 The substance which causes doubling of the chromosome is called **[BHU 2012]**

(a) carcinogen (b) teratogen
(c) clasteogen (d) colchicine

46 Synaptonemal complex is formed during **[BHU 2012, Odisha JEE 2011]**

(a) pachytene (b) leptotene
(c) zygotene (d) diakinesis

47 Select the correct option with respect to mitosis. **[CBSE AIPMT 2011]**

(a) Chromatids start moving towards opposite poles in telophase
(b) Golgi complex and endoplasmic reticulum are still visible at the end of prophase
(c) Chromosomes move to the spindle equator and get aligned along equatorial plate in metaphase
(d) Chromatids separate, but remains in the centre of the cell in anaphase

48 The phragmoplast is organised at the **[UP CPMT 2011]**

(a) beginning of anaphase (b) end of anaphase
(c) beginning of telophase (d) end of telophase

49 Select the matched ones.

I. S-phase	—	DNA replication
II. Zygotene	—	Synapsis
III. Diplotene	—	Crossing over
IV. Meiosis	—	Both haploid and diploid cells
V. G_2-phase	—	Quiescent stage

Codes **[Kerala CEE 2011]**

(a) I and II (b) III and IV
(c) III and V (d) I, III and V
(e) I and IV

50 The stage between two meiotic divisions is called **[Kerala CEE 2011]**

(a) interphase (b) cytokinesis
(c) interkinesis (d) karyokinesis
(e) diakinesis

51 How many chromosomes will the cell have at G_1, after S and after M-phase respectively, if it has 14 chromosomes at interphase? **[DUMET 2011]**

(a) 14, 14, 7 (b) 14, 14, 14
(c) 7, 7, 7 (d) 7, 14, 14

52 Chromatid formation takes place in **[Odisha JEE 2011]**

(a) S-phase (b) Metaphase
(c) G_1-phase (d) G_2-phase

53 Longest phase of meiosis is **[WB JEE 2011]**

(a) prophase-I (b) prophase-II
(c) anaphase-I (d) metaphase-II

ANSWERS

Check Point 16.1

1. (a)	2. (c)	3. (b)	4. (c)	5. (b)	6. (c)

Check Point 16.2

1. (b)	2. (c)	3. (a)	4. (b)	5. (a)	6. (c)

(A) Topical Exercise

1. (d)	2. (c)	3. (c)	4. (b)	5. (c)	6. (a)	7. (c)	8. (d)	9. (d)	10. (c)
11. (b)	12. (b)	13. (c)	14. (c)	15. (b)	16. (d)	17. (b)	18. (a)	19. (c)	20. (b)
21. (b)	22. (b)	23. (c)	24. (c)	25. (c)	26. (b)	27. (a)	28. (c)	29. (d)	30. (d)
31. (d)	32. (d)	33. (a)	34. (d)	35. (d)	36. (a)	37. (a)	38. (c)	39. (b)	40. (b)
41. (b)	42. (c)	43. (b)	44. (a)	45. (d)	46. (a)	47. (d)	48. (b)	49. (a)	50. (c)
51. (c)	52. (a)	53. (d)	54. (d)	55. (c)	56. (b)	57. (a)	58. (a)	59. (c)	60. (d)
61. (c)									

(B) Cumulative Exercise

1. (a)	2. (c)	3. (b)	4. (c)	5. (b)	6. (d)	7. (a)	8. (c)	9. (c)	10. (a)
11. (b)	12. (a)	13. (d)	14. (d)	15. (d)	16. (a)	17. (a)	18. (a)	19. (a)	20. (d)
21. (c)	22. (a)	23. (d)	24. (d)	25. (a)	26. (d)	27. (b)	28. (a)	29. (a)	30. (b)
31. (b)	32. (b)	33. (b)	34. (c)	35. (b)	36. (b)				

(C) Medical Entrances' Gallery

1. (d)	2. (b)	3. (b)	4. (d)	5. (b)	6. (b)	7. (a)	8. (a)	9. (d)	10. (b)
11. (b)	12. (b)	13. (d)	14. (a)	15. (a)	16. (c)	17. (b)	18. (d)	19. (a)	20. (b)
21. (d)	22. (a)	23. (c)	24. (a)	25. (b)	26. (b)	27. (c)	28. (a)	29. (b)	30. (d)
31. (c)	32. (b)	33. (e)	34. (c)	35. (c)	36. (c)	37. (d)	38. (d)	39. (b)	40. (b)
41. (b)	42. (b)	43. (b)	44. (b)	45. (d)	46. (c)	47. (c)	48. (b)	49. (a)	50. (c)
51. (b)	52. (a)	53. (a)							

Explanations

(A) Topical Exercise

8 (*d*) Statement in option (d) is incorrect and can be corrected as S-phase marks the period of DNA synthesis. During this time, the amount of DNA is doubled, but there is no increase in the chromosome number.

10 (*c*) Statement in option (c) is true. Other statements are incorrect and can be corrected as

- The cell lies dormant during G_0-phase.
- The cell size increases in G_1 and G_2-phase.
- The key event of S-phase is DNA replication.

20 (*b*) Statement in option (b) is true. Rest statements are false for mitosis. These can be corrected as

- Mitosis restores the nucleoplasmic ratio.
- Cells formed by mitosis have equal number of chromosomes to the parent cell.
- Mitosis usually results in the production of diploid daughter cells with identical genetic complement.

31 (*d*) In somatic cell which undergo mitosis, the genotype always remain unchanged and the daughter cells are all identical because there is no crossing over process or independent assortment.

50 (*c*) Statement in option (c) is false. It can be corrected as Synapse of homologous chromosomes occurs during zygotene.

54 (*d*) Statement in option (d) is incorrect and can be corrected as There is no DNA replication occurring during meiosis.

55 (*c*) Statements I and III are not correct, which can be corrected as

- In mitosis, homologous chromosomes behave independently.
- Meiosis occurs in germ cell. Somatic cells divide through mitosis.

57 (*a*) Microspore mother ($2n$) undergoes meiosis to produce pollen grains (n).

Single cell undergoing meiosis produces four haploid daughter cells.

Thus, 250 MMC will give 1000 pollen grains.

58 (*a*) Number of meiosis required for the formation of 'n' number of pollen grains is $\frac{n}{4}$. So, the required number of meiosis is 512/4 = 128 divisions.

59 (*c*) In the formation of gametes during the first meiotic division, the homologous chromosomes pair up. This process is called synapsis. The paired (maternal and paternal) homologous chromosomes cross over at the chiasmata and homologous portions of the chromosomes are exchanged, resulting in recombination to form new genotypes.

(B) Cumulative Exercise

2 (*c*) The eight ($2^n = 2^3$) different combinations of the alleles are ABC, AbC, ABc, aBC, aBc, abC, Abc and abc.

3 (*b*) Statement in option (b) is incorrect. It can be corrected as Mitogen such as colchicine inhibits the mitotic division.

5 (*b*) If a mitocyte has 10 pg of DNA and 16 chromosomes in G_1 then the amount of DNA gets double but there is no increase in chromosome number. Hence, the amount of DNA gets 20 pg and the number of chromosomes remain same at anaphase stage.

6 (*d*) Only statement II is incorrect. It can be corrected as Division of centromeres occurs during anaphase-II.

7 (*a*) Statements I and III are correct. Statements II and IV are incorrect and can be corrected as

- Four haploid cells are formed at the end of meiosis-II.
- G_2-phase occurs after S-phase, not meiosis.

8 (*c*) Statement in option (c) is incorrect. It can be corrected as Movement of chromosomes during cell division depends on spindle fibres and kinetochore.

9 (*c*) Statement in option (c) is correct. Other statements are incorrect and can be corrected as

- Meiosis occurs in plants when there is a change from sporophyte to gametophyte.
- Mitosis produces genetically identical daughter cells.
- After meiosis-II, the daughter cells are haploid.

12 (*a*) At S-phase, the amount of DNA per cell doubles. If the initial unit of DNA is 'X', it increases to '2X' at S-phase so, the initial amount of DNA in cell is 50 units and after telophase-II, it will be 25 units.

13 (*d*) If a cell has 15 pairs of homologous chromosomes (n=15), it has 30 chromosomes (2n=30). At the end of mitosis, the two daughter cells will be exact copies of the original cell. Each daughter cell have 30 chromosomes.

At the end of meiosis, each cell would have half of the original number of chromosomes, that is 15 chromosomes.

Thus, option (d) is correct.

15 (*d*) All the given matched pairs are correct except (d), which can be corrected as

Best stage to observe chromosome morphology and number-Metaphase

19 (*a*) Only one statement (V) is correct. Other statements are incorrect and can be corrected as

- Interphase is present in meiotic cell division.
- Repeated meiosis cannot occur in a single cell.
- Prophase-I is longer than prophase-II.
- Microsporophytes undergo meiosis in anthers of plant.

20 (*d*) Option (d) shows the stages in mitosis from interphase (3), condensation of chromatin (4), alignment of chromosomes in metaphase (5), separation of sister chromatids in anaphase (1) and telophase (2).

21 (*c*) The diagram shows the alignment of chromosomes on the spindle, ready for division into the sister chromatids. This is the mid metaphase stage of mitosis.

22 (*a*) Statements I, II and III are correct. Statements IV and V are incorrect and can be corrected as

- Nuclear envelope breaks down during prophase.
- At the end of prophase, when viewed under microscope a cell does not show distinct ER, Golgi, nucleolus and nuclear membrane.

25 (*a*) During anaphase, the sister chromatids of each chromosome separate at the centromere each approching opposite ends of the spindle pore. Hence, the distance between the centromere and the pole decreases, whilst that between the centromere of the sister chromatid increases.

26 (*d*) The diagram shows the separation of the sister chromatids at the early anaphase of mitosis. The resulting daughter cells are diploid will have 8 chromosomes.

27 (*b*) The centrioles in the first diagram have not separated to opposite poles yet and the sister chromatids have not yet separated. The stage is thus prophase. In the second diagram, the homologous chromosomes have undergone pairing, but have not separated into the opposite poles. The centrioles are already on opposite poles. The stage is thus metaphase-I.

28 (*a*) Both A and R are true and R is the correct explanation of A.

Telophase-II results in the formation of two groups of chromosome enveloped by a nuclear membrane. Since, gametes or spores are formed after telophase-II phase, the cell involved does not enter another interphase.

29 (*a*) Both A and R are true and R is the correct explanation of A.

Prophase-I ends with terminalisation of chiasmata during diakinesis. It represents the transition of dividing cells from prophase to metaphase.

Thus, diakinesis is the final stage of prophase-I as terminalisation of chiasmata occurs in this stage.

30 (*b*) Both A and R are true, but R is not the correct explanation of A.

Some cells undergo G_0-phase due to inactivation of cell cycle. G_0-phase is the permanent or temporary exit from the G_1-phase. Cells in this phase remain metabolically active but do not proliferate unless called out to do so depending on the requirement of the organism.

31 (*b*) Both A and R are true, but R is not the correct explanation of A.
Meiotic division is a reductional division. It occurs in reproductive cells and maintains the fixed number of chromosomes in sexually reproducing organisms.
Prophase-I of meiotic division is a long phase, which is divided into five sub-phases, i.e. leptotene, zygotene, pachytene, diplotene and diakinesis. Synapsis or pairing of chromosomes occurs during the zygotene stage.

32 (*b*) Both A and R are true, but R is not the correct explanation of A.
Mitosis is the process by which a cell nucleus divides (karyokinesis) to produce two daughter nuclei containing identical sets of chromosomes to the parent cell.
It is usually followed immediately by the division of cytoplasm (cytokinesis) to form two daughter cells.
This process is known as cell division. Cytokinesis is of two types, i.e. by cell-furrow method (e.g. animals) and cell-plate method (e.g. plants).

33 (*b*) Both A and R are true, but R is not the correct explanation of A.
Cytokinesis is the division of protoplast of a cell into two daughter cells after karyokinesis (nuclear division). Normally, cytokinesis starts towards the middle anaphase and is completed simultaneously with the telophase.
In plants, cytokinesis occurs by cell plate method in which cell plate grows centrifugally. In animals, it takes place by cleavage. Cleavage progresses centripetally.

34 (*c*) A is true, but R is false because
Prophase-I stage of meiosis is more complicated and prolonged as compared to the prophase stage of mitosis. Prophase-I is divided into five substages, i.e. leptotene, zygotene, pachytene, diplotene and diakinesis. Each stage is characterised by different events.

35 (*b*) Both A and R are true, but R is not the correct explanation of A.
Nucleosome is a structural, bead-like unit of a eukaryotic chromosome composed of a short length of DNA wrapped around a core of histone proteins. It is the fundamental subunit of chromatin. Unlike most proteins, which are made continuously throughout interphase, the histones are synthesised mainly in the S-phase (i.e. the DNA synthesis phase).

36 (*b*) Both A and R are true, but R is not the correct explanation of A.
The growth of multicellular organisms is due to mitosis cells functions are controlled by nucleus. The size of nucleus does not change, but cytoplasm increases during cell growth. Increase in size disturbs the nucleocytoplasmic ratio. This ratio is restored to efficient level through mitosis.

(C) Medical Entrances' Gallery

1 (*d*) During the metaphase stage of cell cycle, spindle fibres originating from the centrosomes attach to the kinetochore of chromosomes. Kinetochore is a disc-shaped structure at the surface of centromere through which the sister chromatids are held together. During metaphase, the chromosomes arrange themselves at the equator on metaphasic plate. Due to this arrangement, the attachment of spindle fibres to kinetochore is clearly visible.

2 (*b*) In a mitotic cycle, the correct sequence of phases is G_1, S, G2, M. The first three phases, i.e. G_1, S, G_2 occurring during interphase, whereas the M-phase is the period of actual cell division. The major event occurring in each phase is tabulated below

Phases	Activities
G_1-phase	Cell becomes metabolically active, enzymes and proteins required for replication are synthesised.
S-phase	Synthesis or replication of DNA occurs so that amount of DNA per cell gets doubled.
G_2-phase	Proteins required for mitosis are synthesised while the growth of cell continues.
M-phase	Cell divides to form daughter cells.

3 (*b*) During zygotene stage of meiosis-I, chromosomes start pairing together and this process of association is called synapsis. Such paired chromosomes are called homologous chromosomes.

5 (*b*) The statement in option (b) is correct with regard to G_1-phase of interphase because during G_1-phase, the cell is metabolically active and continuously grows, but does not replicate its DNA.
Other statements are incorrect and can be corrected as
DNA synthesis takes place in S-phase. Nuclear division occurs during karyokinesis.
Reorganisation of all cell components takes place in M-phase.

6 (*b*) Dissolution of the synaptonemal complex occurs during diplotene stage of prophase-I of meiosis-I. Prophase of meiosis-I is long and complex. It is comprised of leptotene, zygotene, pachytene, diplotene and diakinesis. During diplotene, at most places synaptonemal complex dissolves.

8 (*a*) Some dividing cells exit the cell cycle and enter vegetative inactive stage, called quiescent stage (G_0). This process occurs at the end of G_1-phase.
Cells enter G_0 for varying amounts of time and some cells enter the G_0-phase and stay there forever. This is because once they reach maturity, like nerve and heart cells, they do not divide again, so they stay in the G_0-phase.

9 (*d*) G_0-phase is the stage in which the cells exit the cell cycle. It is the quiescent phase in which the cells do not divide. It is the permanent state for some cells, e.g. neurons.

10 (*b*) After meiosis-I, the resultant daughter cells have twice the amount of DNA in comparison to haploid gamete. Meiosis-I causes segregation of homologous pairs of chromosomes. However, each chromosome is double-stranded, having two sister chromatids due to DNA replication before meiosis began.

11 (*b*) Crossing over takes place between non-sister chromatids of homologous chromosomes and occurs at the points called recombination nodules in pachytene stage of prophase-I of cell cycle.

12 (*b*) The separation of the paired homologous chromosomes begins in the diplotene stage. In this phase, the dissolution of synaptonemal complex begins.

The recombined homologous chromosomes of the bivalents separate from each other, except at the sites of crossovers and form chiasmata (X-shaped structures).

14 (*a*) The given diagram is showing anaphase-I of meiosis. It begins when the homologous chromosomes of each bivalent separate and start moving towards opposite poles of the cell. Unlike the anaphase of mitosis, the sister chromatids remain attached at their centromeres and move together towards the poles.

15 (*a*) The given diagram represents metaphase-I of meiosis-I. This stage is more or less similar to metaphase stage of mitosis. The only difference is the appearance of paired chromosomes at two parallel metaphasic plates.

16 (*c*) If Anaphase Promoting Complex (APC) is defective in a human cell, the chromosomes will not segregate during anaphase of mitosis.

APC triggers the transition from metaphase to anaphase by tagging specific proteins for degradation.

17 (*b*) Option (b) shows the correct sequence of events. The process of mitosis starts with the condensation of chromosomal material, which takes place at an early prophase stage. During late prophase, nuclear membrane disintegrates and then the chromosomes get arranged at equator in the metaphase stage. After that splitting of centromere and segregation of chromosomes occur in the anaphase stage. In telophase stage, chromosomes cluster at opposite spindle poles of the cell.

18 (*d*) Endomitosis is the replication of chromosomes in the absence of cell or nuclear division resulting in numerous copies within each cell. It occurs in the salivary glands of *Drosophila* and other flies.

20 (*b*) The G_2/M DNA damage checkpoint is an important cell cycle checkpoint in eukaryotic organisms. This ckeckpoint ensures that cells do not initiate mitosis before they have a chance to repair damaged DNA after replication.

22 (*a*) Polyploid cells have a chromosome number that is more than double the haploid number, e.g. *Triticum aestivum* (wheat) is a hexaploid (6*n*).

23 (*c*) Option (c) is not a characteristic feature during mitosis in somatic cells.

Synapsis is pairing of homologous chromosomes. It occurs during zygotene stage of meiosis-I. The homologous chromosomes come closer leading to the formation of crossover in the next stage called pachytene. These are not observed during mitosis.

25 (*b*) Anaphase is stage of mitosis, in which two groups of chromosomes move at opposite poles.

Anaphase-I is substage of meiosis-I in which the homologous chromosomes break connections and separate out. The process is named as disjunction.

Anaphase-II is stage of meiosis-II in which two chromatids of a chromosome separate completely (i.e. centromere splits) and are called daughter or new chromosomes.

27 (*c*) In general, a gamete contains '*n*' number of chromosomes (haploid) and 'X' amount of DNA. In comparison to the gamete, a (diploid) somatic cell usually have $2n$ number of chromosomes and 2X amount of DNA.

But, after the cell enters and completes the S-phase of its cell cycle, the DNA content of the cell becomes double (due to its duplication). However, the number of chromosomes remains the same.

Hence, the somatic cell that has undergone and completed its S-phase of cell cycle will have twice the number of chromosomes and four times the DNA content. It can be explained as

	Chromosome number	DNA content
Gamete	n	X
Somatic cell (diploid)	$2n$	2X
Somatic cell (after S-phase)	$2n$	4X

29 (*b*) The correct sequence of events of meiosis is Synapsis (in zygotene) $\longrightarrow$ Crossing over (in pachytene) $\longrightarrow$ Terminalisation of chiasmata (in diplotene) $\longrightarrow$ Disappearance of nucleolus (in diakinesis)

33 (*e*) Statement in option (e) is correct whereas rest statements are incorrect. They can be corrected as

- During mitosis, ER and nucleolus begin to disappear at late prophase.
- Chromosomes are arranged along the equator during metaphase of mitosis.
- Chromosome is made up of two sister chromatids at anaphase-I of meiosis.
- A cell plate is laid down during cytokinesis.

40 (*b*) Both A and R are true, but R is not the correct explanation of A.

Though meiosis-II is similar to mitosis, meiosis-II is not mitosis because it always occurs in haploid cell, it is not preceded by DNA replication, the two chromatids of a chromosome are often dissimilar, the daughter cells formed after meiosis-II are neither similar to each other nor similar to the parent.

47 (*c*) Option (c) is correct with respect to mitosis.

In metaphase of mitosis, spindle fibres attach to kinetochore of chromosomes. Chromosomes are moved to spindle equator and get alligned along metaphasic plate through spindle fibres of both pole.

51 (*b*) Replication of DNA takes place at S-phase of cell cycle. The number of chromosomes reduced only in meiosis. So, the number remains 14, 14 and 14 in G_1, after S and after M-phase.

CHAPTER 17

Biomolecules and Enzymes

About 99% of the substances of all living things are made up of four elements, namely **hydrogen**, **oxygen**, **carbon** and **nitrogen**. All these elements combine in various proportions and various ways to form **biomolecules**. These are defined as the organic molecules present in the living organisms.

These include large molecules (**biomacromolecules**) like proteins, nucleic acids, polysaccharides, lipids and small molecules (**biomicromolecules**) such as amino acids, fatty acids, etc. Biomolecules act as building blocks of life and perform important physiological processes. In this chapter, we will focus on these biomolecules.

Inside

COMPOSITION OF THE LIVING TISSUE

In order to assess the composition of living tissues or cellular pool, various experiments can be carried out. One such method includes the grinding of living tissue in **trichloroacetic acid** (CCl_3COOH). The resultant slurry is strained and it consists of following two fractions or pools, i.e. acid soluble pool (filtrate) and acid insoluble pool (retentate).

1. Acid soluble pool

It roughly represents the cytoplasmic composition. It mainly consists of **biomicromolecules**. These are the small molecules weighing 18-800 Da (Dalton). These include **amino acids, small sugars, nucleotides**, inorganic compounds like sulphate, phosphate, etc.

2. Acid insoluble pool

It represents the cytoplasmic and organelle composition. From this pool, **biomacromolecules** are obtained. These are the large-sized biomolecule weighing about 800-10,000 Da. These include four types of organic compounds, i.e.

proteins, nucleic acids, polysaccharides and **lipids.** Although, the molecular weight of lipids is less than 800 Da, they do not occur in acid soluble pool because these are non-polar compounds.

The average composition of cell includes following components

Table 17.1 Average composition of cells

Components	% of the total cellular mass
Water	70-90
Proteins	10-15
Carbohydrates	3
Lipids	2
Nucleic acids	5-7
Ions	1

Metabolites

These are the organic compounds that are formed in living beings as a result of their metabolic activity.

The essential organic compounds are also called metabolites.

They are of following two types

(i) **Primary metabolites** Biochemicals formed as intermediates and products of normal vital metabolic pathways, e.g. amino acid, nucleotides, sugars, fats, etc.

(ii) **Secondary metabolites** These are specialised products formed by alteration of normal or primary metabolic pathways, having specific functions, e.g. pigments (carotenoids), alkaloids (morphine), terpenoids (monoterpenes), toxins (ricin), drugs (vinblastin), etc.

CARBOHYDRATES

- These are among the most widely distributed compounds both in plant as well as in animal kingdom. These are also called **saccharides** because their basic components are sugars.
- Carbohydrates are the compound of carbon, hydrogen and oxygen. These have hydrogen and oxygen in the same ratio as that of water, i.e. 2 : 1.
- Generally, these are defined as **polyhydroxy aldehydes** or **ketone compounds** or the compounds which on hydrolysis give polyhydroxy aldehydes or ketones.
- They can be represented by molecular formula $[CH_2O]_n$ or $C_nH_{2n}O_n$.

Classification of carbohydrates

Carbohydrates can be classified based on the following two criteria

I. On the basis of their reducing properties, carbohydrates can be of following two types

Carbohydrates

Reducing sugar	**Non-reducing sugar**
• These are the sugars with free aldehyde (—CHO) or keto groups (>C=O).	• These sugars do not have free aldehyde or keto groups.
• They can reduce cupric (Cu^{2+}) to cuprous (Cu^+) ion.	• They cannot reduce Cu^{2+} ions to Cu^+ ions.
• They reduce Fehling's solution and Benedict reagent, e.g. all monosaccharides, maltose, lactose, etc.	• They do not reduce the Fehling's solution and Benedict reagent, e.g. sucrose.

II. On the basis of hydrolysis, products of carbohydrates can be monosaccharides, oligosaccharides and polysaccharides.

Monosaccharides (Gr. *Mono*-single)

- These are simple carbohydrates that cannot be hydrolysed further into smaller units.
- They consist of a single polyhydroxy aldehyde or ketone unit. These are mostly made up of 3-7 carbon atoms.
- They have two important chemical properties
  (i) They are reducing sugars as they can reduce sugars having a free aldehyde or ketone group, i.e. Cu^{2+} to Cu^+.
  (ii) The aldehyde or ketone group of monosaccharide can react and bind with an alcoholic group of another compound to join the two compounds together. This bond is called **glycosidic bond.**
- Monosaccharides are soluble in water, sparingly soluble in alcohol and insoluble in ether. They may exist in different isomeric forms.

Classification of monosaccharides

Monosaccharides are further subdivided, based on different aspects.

- Based on the functional group attached to them, they are of two types
  (i) **Aldoses** When the functional group in monosaccharide is aldehyde $\left[-\overset{O}{\overset{\|}{C}}-H\right]$, they are known as aldoses, e.g. glyceraldehyde, glucose, ribose, erythrose, etc.

Among these, glucose is called **universal sugar** and is also known as **grape sugar** or **corn sugar** or **dextrose**.

(ii) **Ketoses** When the functional group in the monosaccharide is keto group (— C = O), they are referred as **ketones**, e.g. ribulose, fructose, etc. Among these, fructose (6C) is most common. It is also called as **fruit sugar** or **levulose**. It is naturally occurring sweetest sugar.

Table 17.2 Classification of monosaccharides

Types of sugar based on number of carbon atoms	Aldoses	Ketoses
Trioses (3)	Glyceraldehydes	Dihydroxyacetone
Tetroses (4)	Erythrose	Erythrulose
Pentoses (5)	Ribose, xylose	Ribulose, Xylulose
Hexoses (6)	Glucose, galactose	Fructose
Heptoses (7)	Glucoheptose	Sedoheptulose

Pentoses and hexoses of monosaccharide can occur in two forms, e.g. open chain and ring form. This ring form can be of two types

(a) **Pyranose ring**, which has hexagonal shape with 5 carbon and 1 oxygen atom.

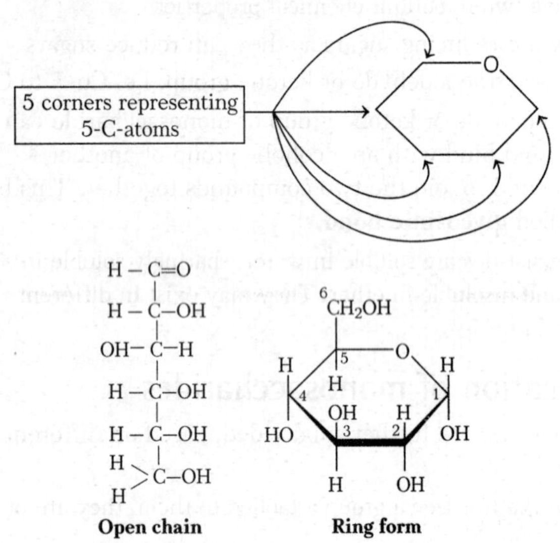

Fig 17.1 Open and ring form of glucose

(b) **Furanose ring**, which has pentagonal shape with 4 carbon and 1 oxygen atom

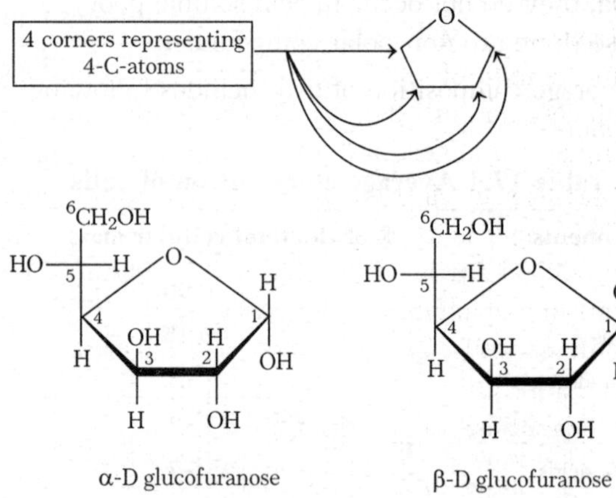

Fig 17.2 Furanose ring

Sedoheptulose is an intermediate of respiration and photosynthetic pathways.

Note *Monosaccharides are modified in various ways to form substances like amino sugar, sugar alcohol, sugar acid, etc.*

Oligosaccharides

(Gr. *Oligo*-few)

- These are formed by the condensation of 2-9 monosaccharide units.
- In oligosaccharides, these units are held together by **glycosidic bonds.**
- On hydrolysis, they produce two or more molecules of same or different monosaccharides.
- Oligosaccharides are crystalline, water soluble and sweet in taste.

Classification of oligosaccharides

On the basis of number of monosaccharide units given by them on hydrolysis, these can be of following types

- **Disaccharide** It is formed by the condensation reaction between two monosaccharides, having glycosidic bond between them, e.g. sucrose, maltose, lactose, trehalose, etc.
- **Trisaccharide** It is formed by three linked monosaccharide molecules, e.g. raffinose.
- **Tetrasaccharide** It is formed by four monomeric monosaccharide units, e.g. stachyose.

Some oligosaccharides

Examples of oligosaccharides are as follows

(i) **Lactose or Milk sugar** It is a disaccharide present in the milk of mammals. On hydrolysis, it gives one glucose and one β-D galactose unit. The glycosidic linkage of lactose is β-1, 4 type.

(ii) **Maltose or Malt sugar** It is also a disaccharide. This sugar is found in germinating seeds. It is also produced during the digestion of starch by an enzyme α-amylase. On hydrolysis, it gives 2-α-D glucose units. The glycosidic linkage of maltose is α-1, 4 type.

Note
- ***Cellobiose*** *is a disaccharide formed during the hydrolysis of cellulose. It is identical to maltose except the fact that here linkage is β-1, 4-glycosidic linkage.*
- ***Isomaltose*** *also resembles maltose except that here linkage is α-1, 6-glycosidic linkage.*

(iii) **Sucrose or Cane sugar** This disaccharide is non-reducing and widely distributed among higher plants. On commercial scale, it is obtained from sugarcane and sugarbeet. On hydrolysis, it produces one α-D glucose and one β-D fructose unit. The glycosidic linkage in sucrose is α-1, 2 type.

(iv) **Trehalose** It is a disaccharide, which is found in insects and fungi. In trehalose, two glucose units are linked with α-1, 1- glycosidic linkage.

(v) **Raffinose** It is a trisaccharide sugar found in plants It is made up of three molecules, i.e. D-glucose, D-fructose and D-galactose.

Polysaccharides or Glycans

(Gr. *Poly*-many)

- These are polymers or chains of monosaccharides (usually more than 9) bounded in linear or branched chain pattern.
- In a polysaccharide, an individual monosaccharide is linked by glycosidic bonds to another monosaccharide.
- In this structure, the right end of the polysaccharide is reducing, while the left end is non-reducing.

Classification of polysaccharides

Depending upon the chain composition, they can be classified as follows

1. Homoglycans or Homopolysaccharides

They are the polysaccharides, which are formed by the polymerisation of only one type of monosaccharide unit, e.g. starch, glycogen, cellulose, callose, etc.

Starch

- It is the major reserve food in plants.
- It is a polymer of D-glucopyranose units linked by α-1, 4-glycosidic linkages.
- It consists of a mixture of amylose (unbranched polymer) and amylopectin (branched polymer) in 1:4 ratio.
- **Amylose** is linear and consists of about 200-500 glucose units. On the other hand, **amylopectin** is branched and consists of over (2000-20,000) glucose units.

Note *The partial hydrolysis of starch results in the formation of a substance called* ***dextrin****.*

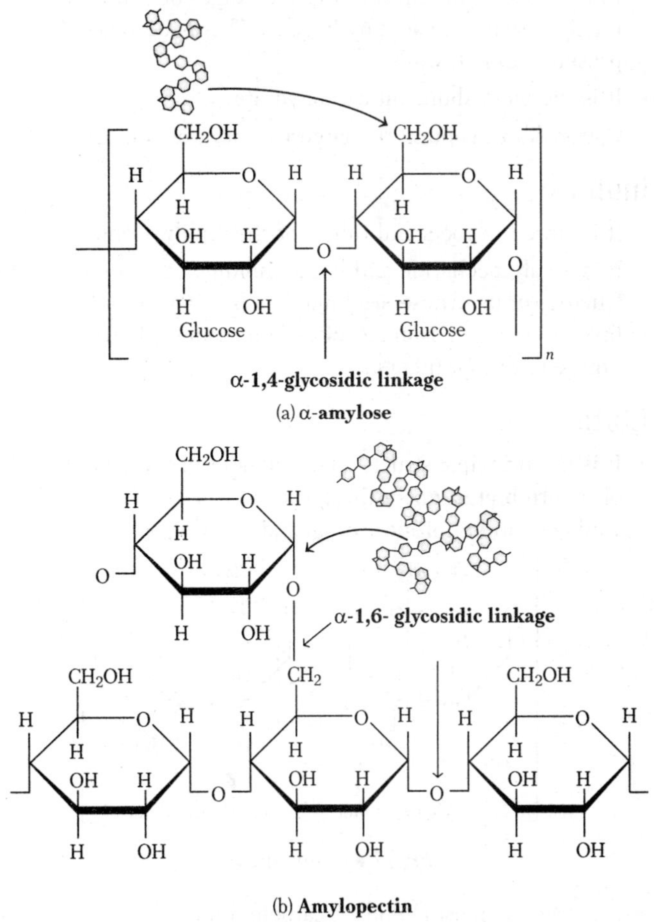

Fig 17.3 Constituents of starch

Glycogen

- It is the reserve food in animals that is why it is also called as **animal starch**. It is found mainly in liver and muscles.
- Chemically, it is analogous to starch, but it is more extensively branched than amylopectin.

- The linkage and monosaccharide units of glycogen are similar to those of starch. About 5000-15000 glucose units make up a glycogen molecule.
- It is a non-reducing sugar that gives red-violet colour with iodine.

Cellulose (Hexose polysaccharide)

- It is the most important structural component of the cell wall of plants.
- It is a linear polymer of β-D glucose units connected through β-1, 4-glycosidic linkages.
- In contrast to starch, the β-1, 4-linkages of cellulose are highly resistant to acid hydrolysis. This is because of its presence in cell walls.
- It is the most abundant carbohydrate.
- Cotton fibres contain the largest amount of cellulose (90%).

Inulin

- It is stored in roots and tubers of the family-Compositae.
- It is a polysaccharide and has a unbranched chain of 30-35 fructose units. These are linked by β -2-1 glycosidic linkage between 1 and 2 of carbon atom of D-fructose unit, e.g. *Dahlia* tubers.

Chitin

- It is the principal structural component of the exoskeleton of invertebrates, e.g. arthropods and is also a major cell wall constituent of most fungi and many algae.

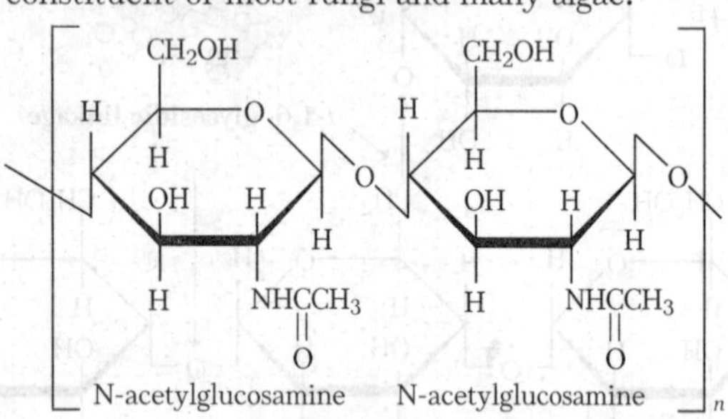

Fig 17.4 Chitin structure

- It is second most abundant carbohydrate.
- It is a homopolymer of N-acetylglucosamine (NAG) joined with β-1, 4-linkages.
- NAG is a modification of glucose molecule.

2. Heteroglycans or Heteropolysaccharides

They are the complex polysaccharides, which are formed by the polymerisation of two or more types of monosaccharide unit, e.g. agar, peptidoglycan, pectin, etc.

Heparin

It is a heteropolysaccharide. It is variably sulphated glucosaminoglycan consisting of alternating α-1, 4-linked residues of L-iduronate-2- sulphate and N-sulpho-D-glucosamine-6-sulphate.

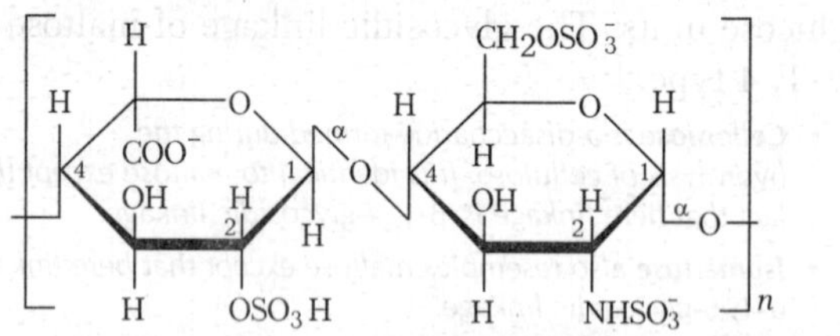

Fig 17.5 Heparin structure

Mucopolysaccharides

These are the gelatinous substances in plants and animals that possess acidic or amino sugars, uronic acid, etc.

Some important mucopolysaccharides are discussed below

- **Agar Agar** This glactan contains D-galactose, 3-6-anhydro L-galactose. It has sulphate esterification after every tenth galactose unit. It is found in the cell wall of red algae, e.g. *Gelidium.*
- **Hyaluronic acid** It is composed of glucuronic acid and acetylglucosamine. It is found in the extracellular fluid in aminal tissues, vitreous humour, CSF, synovial fluid, etc.
- **Keratin sulphate** It is composed of galactose, acetylglucosamine and sulphuric acid. It provides flexibility and strength in cornea, skin, etc.
- **Gums** These are composed of sugars like L-arabinose, D-galactose, etc. These viscous substances help to seal wounds in higher plants, e.g. gum arabic.
- **Mucoproteins** These are formed by the conjugation of proteins with mucosaccharides to form mucus. They do not contain uronic acid.
- **Murein** It is formed by linking of heteropolysaccharide to short chains of peptides. The former are formed of two alternate amine sugar molecules, *viz.* NAG and NAM.

CHECK POINT 17.1

1. The sum total composition of acid soluble and acid insoluble fraction represents the entire composition of
   (a) dead cells
   (b) gene pool
   (c) cellular pool
   (d) gene library
2. As starch is related to plant body, which of the following polysaccharides is related to animal body?
   (a) Cellulose (b) Chitin (c) Glycogen (d) Insulin
3. Plant cell walls are made up of
   (a) homopolymer of fructose
   (b) heteropolymer of glycogen
   (c) homopolymer of glucose
   (d) homopolymer of glycogen
4. Which of the following is a structural polysaccharide?
   (a) Glycogen (b) Starch
   (c) Inulin (d) Cellulose
5. Which of them represents the homopolysaccharides made up of glucose monomers?
   (a) Sucrose, lactose, maltose
   (b) Chitin, glycogen, starch
   (c) Starch, inulin, peptidoglycan
   (d) Starch, glycogen, cellulose
6. The number of ends in a glycogen molecule would be
   (a) equal to the number of branches plus one
   (b) equal to the number of branch points
   (c) one
   (d) two, one on the left side and another on the right side

AMINO ACIDS

These are small molecules made up of carbon, hydrogen, nitrogen, oxygen and in some cases sulphur also. They serve as monomers of proteins.

Structure

Amino acids are also referred as **substituted methane.** They are considered to be the first molecule formed in the atmosphere of the earth.

Their structural features are given below

- Each amino acid has a free carboxyl group, a free amino group and '*R*' as the distinctive side chain (variable).

$$H_2N-\overset{\overset{\displaystyle H}{|}}{\underset{\underset{\displaystyle R}{|}}{C}}-COOH$$

- All these components are attached to a central carbon atom called **α-carbon atom.**
- Based on the nature of *R*-group, a variety of amino acids are classified. The *R* in the amino acid could be a hydrogen or aliphatic, aromatic or heterocyclic group.
- The amino group imparts the basic character to amino acid.
- On the other hand, carboxylic group gives acidic character to amino acid.

Classification of amino acids

Out of approximately 200 amino acids known so far, only about 20 participate and stabilise proteins or polypeptide molecules.

Amino acids can be classified as follows

On the basis of their synthesis in the body

On this basis, amino acids are classified into following three categories

(i) **Essential amino acids** cannot be synthesised by any organism in the body and these are obtained from dietary sources.

These are 8 in case of animals and 7 in case of humans.

(ii) **Non-essential amino acids** can be synthesised by an organism and thus, these are not required as dietary component. These are 10 in numbers.

(iii) **Semi-essential amino acids** are required essentially by an organism during particular phase of body, i.e. growth and lactation period (in pregnant mothers). These are 2 in numbers.

Essential, non-essential and semi-essential amino acids are given in the table below

Table 17.3 Essential, semi-essential and non-essential amino acids

Name of the amino acids	Characteristic features of side chain, *R*	Three letter symbols	One letter codes
Glycine	H	Gly	G
Alanine	$—CH_3$	Ala	A
Valine*	$(H_3C)_2CH–$	Val	V
Leucine*	$(H_3C)_2CH—CH_2—$	Leu	L

Name of the amino acids	Characteristic features of side chain, *R*	Three letter symbols	One letter codes
Isoleucine*	$(H_3C)_2CH-CH_2-$	Ile	I
Arginine#	$N{=}C(NH_2)-NH-(CH_2)_3-$	Arg	R
Lysine*	$H_2N-(CH_2)_4-$	Lys	K
Glutamic acid	$HOOC-CH_2-CH_2-$	Glu	E
Aspartic acid	$HOOC-CH_2-$	Asp	D
Glutamine	$H_2N-C(=O)-CH_2-CH_2-$	Gln	Q
Asparagine	$H_2N-C(=O)-CH_2-$	Asn	N
Threonine*	$H_3C-CHOH-$	Thr	T
Serine	$HO-CH_2-$	Ser	S
Cysteine	$HS-CH_2-$	Cys	C
Methionine*	$H_3C-S-CH_2-CH_2-$	Met	M
Phenylalanine*	$C_6H_5-CH_2-$	Phe	F
Tyrosine	$(p)HO-C_6H_4-CH_2-$	Tyr	Y
Tryptophan*	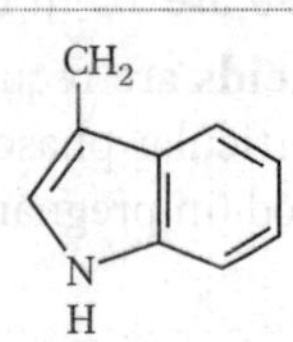	Trp	W
Histidine#	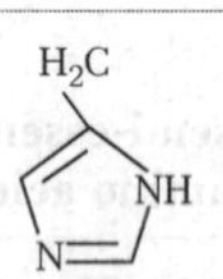	His	H
Proline (Imino acid)	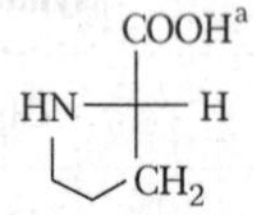	Pro	P

* *essential amino acid, # semi-essential amino acid*

On the basis of chemical nature

On this basis, amino acids are of following types

(i) **Neutral** They have one amino group and one carboxyl group (monoamino monocarboxylic) with non-cyclic hydrocarbon chain.
e.g. glycine (simple amino acid), alanine, valine, leucine and isoleucine.

(ii) **Acidic** They have an additional carboxylic group (monoamino dicarboxylic).
e.g. aspartic acid, glutamic acid, asparagine and glutamine.

(iii) **Basic** They have additional amino group without forming amides (diamino monocarboxylic).
e.g. arginine and lysine.

On the basis of side chain

On this basis, amino acids are of following types

(i) **Sulphur containing** They possess sulphur.
e.g. cysteine and methionine.

(ii) **Alcoholic** They possess alcohol or hydroxyl group.
e.g. threonine, tyrosine and serine.

(iii) **Aromatic** They possess a cyclic structure (or group) inside chain.
e.g. phenylalanine, tryptophan and tyrosine.

(iv) **Heterocyclic** They possess a cyclic ring (or group) bearing nitrogen.
e.g. histidine, proline and tryptophan.

Note • *Most amino acids are laevorotatory, while glycine is optically inactive.*

• *Non-protein amino acids are ornithine, citrulline and diamino pimelic acid.*

Zwitterion and isoelectric point

Amino acids are crystalline solids that are soluble in water, but insoluble in organic solvent. They can act as both acids and bases as they contain one amino and one carboxylic group. Hence, they are amphoteric in nature. Thus, they behave differently in solutions of different pH.

In a **neutral solution**, the amino acid molecule exists as a dipolar ion (Zwitterion), i.e. a molecule containing positive and negative ionic group. The charge on this ion changes with the pH.

The pH at which an amino acid is electrically neutral or exist as **Zwitterion** is called **isoelectric point**.

$$\underset{\text{Cation (low pH)}}{H_3^+N-CH(R)-COOH} \rightleftharpoons \underset{\text{Zwitterion (at isoelectric pH)}}{H_3^+N-CH(R)-COO^-} \rightleftharpoons \underset{\text{Anion (high pH)}}{H_2N-CH(R)-COO^-}$$

In **acid solutions** (low pH), the amino group of amino acid picks up H^+ ions and becomes positively charged.

In **alkaline solution** (high pH), the amino acid donates H^+ ions to the medium and becomes negatively charged.

PROTEINS

The word protein comes from the Greek word *'Proteios'* means first or foremost. The term 'protein' was first coined by **Berzelius** (1837) and **Mulder** (1838). Proteins are involved in structural support, storage, transport, signalling, movement, etc., in living organisms. These are the linear, unbranched polymers of amino acids.

Classification of proteins on the basis of structure

Proteins are classified on the basis of increasing complexity in their structure. These can be of following types

1. Simple proteins

These proteins are made up of amino acid only.

On hydrolysis, they yield only amino acids. Some of their important examples include

(i) **Albumins** These are water-soluble proteins. They can be coagulated by heat and precipitated by saturated salt solutions like ammonium sulphate, etc. These usually have low glycine content, e.g. lactalbumin, serum albumin and egg albumin, etc.

(ii) **Globulins** These are soluble in dilute salt solutions of strong acids and bases, but are insoluble in pure water and moderately soluble in concentrated salt solutions. Globulins can be coagulated by heat. These generally contain glycine as one of their constituents, e.g. serum globulin, lactoglobulin, tissue globulin, vegetable globulin and blood fibrinogen, etc.

(iii) **Glutelins** These are soluble in dilute acids and alkalis, but insoluble in neutral salts. These can be easily coagulated by heat. They are comparatively rich in arginine, proline and glutamic acid, e.g. glutenin of wheat, oryzenin of rice.

(iv) **Prolamins** These are soluble in 70-90% alcohol, but insoluble in absolute alcohol, water and other neutral solvents. These contain large amounts of proline, but less amount of lysine. They are found exclusively in plants, e.g. zein of maize, gliadin of wheat and hordein of barley.

(v) **Albuminoids** (Scleroproteins) These are insoluble in water and most other solvent. These are proteins of supportive tissue. They are found exclusively in mammals, e.g. keratin of hair, hoof and fibroin of silk.

2. Conjugated proteins

These proteins are formed by the binding of simple protein with a non-proteinaceous group, i.e. **prosthetic group**. It simply means on hydrolysis, they yield some other substances along with amino acids.

Some of their important examples include

(i) **Nucleoproteins**, i.e. proteins + nucleic acids, e.g. histones, protamines, etc.

(ii) **Chromoproteins**, i.e. proteins + chromophoric prosthetic group or coloured prosthetic group, e.g. haemoglobin (Fe-Haem), haemocyanin (Cu-Haem), cytochrome (Fe-Haem), flavoprotein (FMN, FAD), etc.

(iii) **Glycoproteins**, i.e. protein + carbohydrates, e.g. mucin, immunoglobulins, Major Histocompatibility Complex (MHC), etc.

(iv) **Phosphoproteins**, i.e. proteins + phosphoric acid, e.g. casein of milk, vitelline of egg yolk.

(v) **Lipoproteins**, i.e. water soluble proteins, in which prosthetic group is lipids, e.g. serum lipoproteins, chylomicrons, etc.

(vi) **Metalloproteins**, i.e. proteins + metal (generally Mg, Fe and Cu), e.g. chlorophyll, transferrin, etc.

(vii) **Mucoprotein**, i.e. mucoid carbohydrates, e.g. mucin in saliva.

3. Derived proteins

When proteins are hydrolysed by acids, alkalis or enzymes, the degradation products obtained from them are called as **derived proteins**.

These derived proteins are classified on the basis of their progressive cleavage as follows

Proteins —Heat→ Denatured proteins

Proteins ↓ Cleavage

Primary proteoses → Secondary proteoses → Peptone

Peptone ↓ Polypeptides

Polypeptides → Simple peptides → Amino acids

Classification of proteins on the basis of function

Proteins can also be classified on the basis of their functions, which are as follows

(i) **Enzymatic**, e.g. pepsin, trypsin, RuBisCO (most abundant protein in the biosphere). It is also involved in carbon fixation, etc.

(ii) **Regulatory**, e.g. insulin, growth hormone, oxytocin, etc.

(iii) **Transporting**, e.g. haemoglobin, serum, albumin, etc.
(iv) **Storage**, e.g. casein, ferritin, glutelins, globulins, etc.
(v) **Contractile** and **motile**, e.g. actin, myosin, tubulin, etc.
(vi) **Structural**, e.g. collagen, α-keratin, elastin, chondrin, ossein, etc.
(vii) **Protective**, e.g. immunoglobulin (antibody), thrombin, fibrinogen, etc.
(viii) **Exotic**, e.g. monellin, glue proteins, etc.

Structural levels in protein

- The determination of biological function of a protein lies in its conformation or 3D structure.
- In proteins, amino acids are connected to each other *via* peptide bond.
- **A peptide bond** (C–N) is formed between the nitrogen atom of amino group of one amino acid and the carbon atom of carboxylic group of another amino acids as shown below

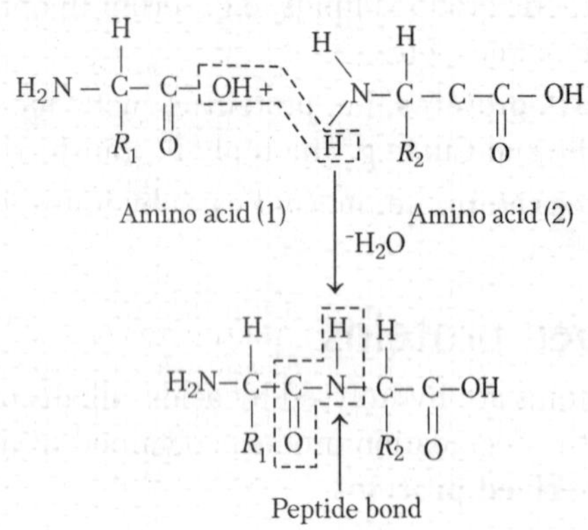

Fig 17.6 Formation of peptide bond

Proteins have following four structural levels

Primary structure

- The primary structure of a protein is its linear chain amino acids. These amino acids are connected *via* peptide bonds. It provides the positional information of amino acids in a protein.
- The first (extreme left) amino acid of the chain is known as **N-terminal** amino acid with its unbonded amino group.
- Likewise, the last one (extreme right) as **C-terminal** amino acid with its unbonded COOH group.
- Primary structure determines the specificity of protein but does not make a protein functional.

Note *In 1953, **Frederick Sanger** for the first time, determined the amino acid sequence of a protein hormone insulin.*

Secondary structure

- The longer polypeptide chains are usually folded or coiled to produce specific structures referred as secondary structures.
- These structures are produced and maintained by hydrogen bonding.
- Three major types of secondary structure **α-helix, β-pleated sheet** and **tropocollagen helix** have been identified.
- The α and β terms designate the first and second type of secondary structure discovered in proteins.
- These are discussed below

(i) **α-helix** In this, the chain is coiled spirally, generally in the right-handed manner. At places, the helix is less regular forming random coils. This structure is formed through H-bonding in single amino acid chain only.

This kind of secondary structure is found in several proteins, such as keratin (entirely helical), myosin, tropomyosin, fibrin, epidermis, etc.

(ii) **β-pleated sheets** When two or more polypeptide chains are held together by intermolecular H-bonds, the structure is called as **β-pleated sheet**. They can be classified as

β-pleated sheets

- **β-parallel sheets**
  Adjacent strand of polypeptide runs in the same direction, e.g. β-keratin.
  These are less stable.
- **β-antiparallel sheets**
  Adjacent strand of polypeptide runs in the opposite direction, e.g. fibroin of silk.
  These are more stable.

(iii) **Collagen or Tropocollagen helix** It is a special secondary structure observed in the collagen having three strands or polypeptide chains coiled around one another. The coil is strengthened by the establishment of H-bonding mainly between hydroxyproline and proline residues.

Tertiary structure

- The polypeptide chain may undergo further coiling and folding to produce the 3-dimensional structure of proteins.
- In this structure, coiling involves only a single polypeptide chain or α-helix.
- The folding brings the active amino acids together to form a distinctive cavity or cleft in which the substrate is bounded.
- The tertiary structure determines the overall shape of a protein molecule.

- Proteins are called **fibrous proteins**, when they form thin, long threads and globular proteins, if they are compact.
- These structures are stabilised by several types of bonds namely **hydrogen bond, ionic bond, Van der Waals interaction, covalent bond** and **hydrophobic bond**.

Quaternary structure

- Protein is said to be in quaternary structure, if it consists of two or more polypeptide chains united by the forces other than covalent bond (i.e. neither peptide nor disulphide).
- The forces which stabilise these aggregates are hydrogen, hydrophobic and electrostatic bonds.
- If the protein consists of identical units, it is said to have a **homogeneous quaternary structure**, e.g. the isozymes H_4 and M_4 of lactate and the enzyme phosphorylase.
- If the units are dissimilar, the protein is said to have a **heterogeneous quaternary structure**, e.g. haemoglobin, which consists of α-chains and two β-chains.

The structural levels of proteins are shown in the figures below

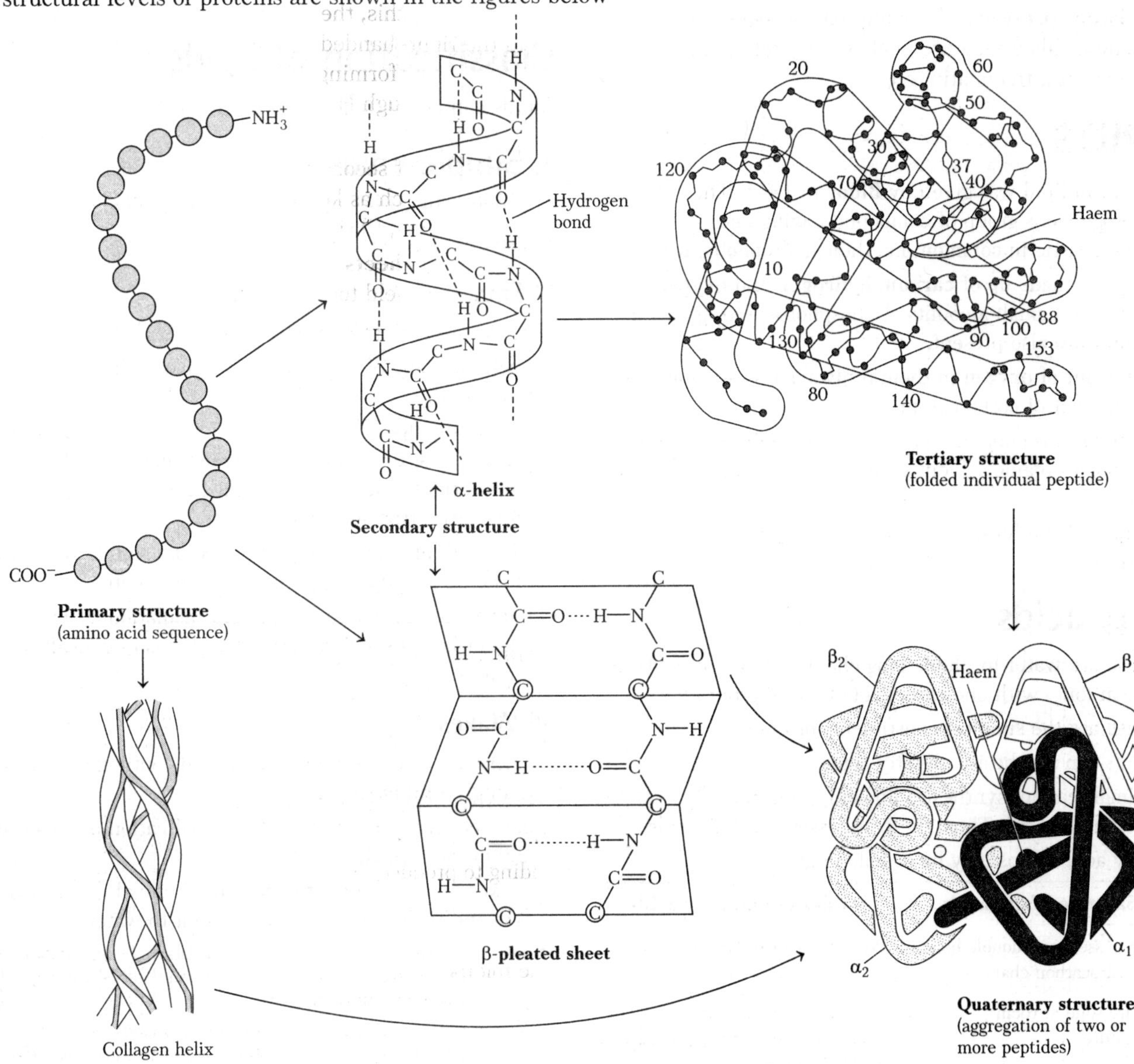

Fig 17.7 Structural level of proteins

Functions of protein

(i) **Structural proteins** They help in the formation of cell organelles, cell membrane, supporting structures like ligaments, cartilage, tendon, etc.

(ii) **Storage proteins** These include casein in milk, gluten in seeds, albumin in egg, etc.

(iii) **Transport proteins** These include haemoglobin that transport oxygen, α-globulin that transport thyroxine and bilirubin, β-globulin that transport vitamins-A, D, K, etc.

(iv) **Toxins** These are produced as defensive or offensive compounds, e.g. ricin of castor, snake venom, etc.

(v) **Sweet proteins** These are natural sweetest chemicals which are made up of proteins, e.g. brazzein, thaumatin, etc.

LIPIDS

- The term lipid was given by **Bloor** in 1943. These are heterogeneous group of compound having one characteristic in common, i.e. their hydrophobic nature.
- They are made up of **carbon, hydrogen** and **oxygen**. In some cases, small amounts of phosphorus, nitrogen and sulphur are also present.
- The count of oxygen in them is always less as compared to the number of carbon atom.
- These are insoluble in water, but are readily soluble in non-polar organic solvent such as chloroform, benzene and ether.
- In most of the forms, fatty acids are major constituents of lipids.

Fatty acids

- These are water insoluble long chain hydrocarbons (4-36 carbon long) with one carboxyl (—COOH) group.
- These are the simplest constituents of lipids.
- Plants can synthesise all fatty acids.
- Animals cannot synthesise linolenic, linoleic and arachidonic acid. Thus, these are essential fatty acids.
- Fatty acid chains may be of following two types

Saturated fatty acids	Unsaturated fatty acids
Do not possess any double bond in their hydrocarbon chain.	Possess one or more double bonds in their hydrocarbon chain.
They are solid at room temperature.	They are liquid at room temperature due to the presence of double bond in them.
These have high melting point.	They have low melting points.
They show no effect on hydrogenation.	On hydrogenation, unsaturated fats are converted into saturated.
They increase blood cholesterol.	They lower the blood cholesterol.
Their general formula is $C_n H_{2n}O_{2n}$.	Their general formula is $C_n H_{2n-2x}O_2$ [x = number of double bond].
e.g. Lauric acid, palmitic acid (16 C), stearic acid (18 C), parachidic acid (20 C), animal fats, etc.	e.g. Oleic acid (18 C), linoleic acid (18 C), α-linolenic acid and arachidonic acid with 1, 2, 3, 4 double bonds, respectively, plant fats.

Classification of lipids

Lipids can be classified into following types

1. Simple lipids

These are the esters of fatty acids with alcohol.

These are of two types

(i) Fats

- These are the esters of fatty acid with glycerol.
- Each molecule of glycerol can react wth three molecules of fatty acids.
- Depending upon the number of fatty acids attached to glycerol, the esters are called mono-di, or pure fats are triglycerides that contain similar fatty acids, e.g. tripalmitin, tristearin, triolein, etc.
- In mixed fats, three fatty acids are not similar or only two of the three fatty acids are similar, e.g. butter.
- Fats act as reserve food material, insulation coats in mammals and shock absorbing cushions in eyeballs and kidneys.

(ii) Waxes

- These are esters of higher fatty acids with alcohol other than glycerol in them.
- e.g. bee wax is formed by the esterification of palmitic acid and myricyl alcohol.
- Waxes protect, lubricate and provide flexibility to hair and skin and it also makes body resistant to water.

Note
- *Edible oils are converted to hard fats by hydrogenation. It results in the saturation of unsaturated fatty acids, e.g. vanaspati, margarine, etc.*
- *Oils containing polyunsaturated fatty acids are called polyunsaturates (PUFA). They lower the blood cholesterol level e..g. sunflower oil.*

2. Compound or Conjugated lipids

These are esters of fatty acids with alcohol, but contain some other substances also. They are of following types

(i) Phospholipids

- These are made up of a molecule of glycerol or other alcohol.
- In these, two fatty acids are attached to glycerol molecule. The third hydroxyl group of glycerol is esterified to phosphoric acid (instead of fatty acid).
- This lipid essentially consists of
  - a phosphate group joined to one of its outer—OH group.
  - two fatty acid molecules linked to each other with two—OH groups.
  - a nitrogen containing base such as choline molecule, bound to the phosphate group.
- Some common phospholipids are, e.g. lecithin and cephalin.
- Lecithin are found in cell membrane which takes part in cell permeability, osmotic tension, etc.

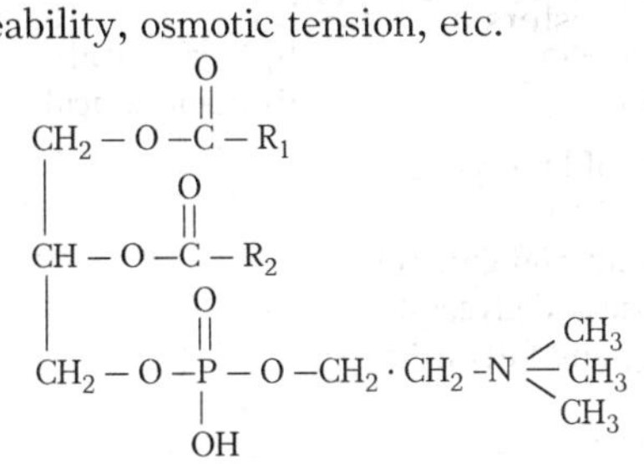

Fig 17.7 Lecithin (Phosphatidyl choline)

- Cephalin is found in the brain and acts as an insulating material for nerve and also participates in blood coagulation.

(ii) Glycolipids

- These contain fatty acids, alcohol sphingosine and a sugar (galactose).
- The sugar replaces one fatty acid molecule.
- These lipids are present in myelin sheath of nerve fibres and on the outer surface of cell membrane of chloroplast.
- Common glycolipids are cerebrosides and gangliosides.
  (a) **Cerebrosides** are composed of amino alcohol sphingosine, a fatty acid and a sugar. Glucocerebrosides (contain D-glucose) are found in cell membranes while galactocerebrosides (contain D-galactose) are found in nerve membranes.
  (b) **Gangliosides** are composed of sugar residues glucose, galactose, salic acid and N-acetyl glucosamine, sphingosine, fatty acid.
  They act as receptors for viral particles, occur in grey matter of brain, help in ion transport through membranes, etc.

(iii) Lipoproteins

- These contain lipids (mainly phospholipids) and proteins in their molecules.
- Most membranes are composed of lipoprotein.
- Lipids are transported in blood plasma and lymph as lipoprotein.
- Lipoproteins are present in milk and egg yolk.

(iv) Chromolipids

These contain pigments such as carotenoids, e.g. carotene, vitamin-A.

3. Derived lipids

These are obtained by hydrolysis of simple and compound lipids. They include fatty acids, alcohols, monoglycerides, diglycerides, prostaglandins, steroids, terpenes and cerotenoids.

Few of them are explained below

(i) Sterols

- These are the compounds that do not contain fatty acids, but they have fat like properties.
- Each type of sterol is composed of four fused carbon rings, i.e. cyclopentanoperhydrophenanthrene nucleus.
- Different sterols differ from each other in the number and position of double bonds between C-atom and in the side chain linked to the ring.
- Cholesterol is the most common sterol in animal tissues. It is found free or as cholesterol ester with a fatty acid. It is the precursor of many steroid hormones, e.g. cortisol, estradiol, etc.

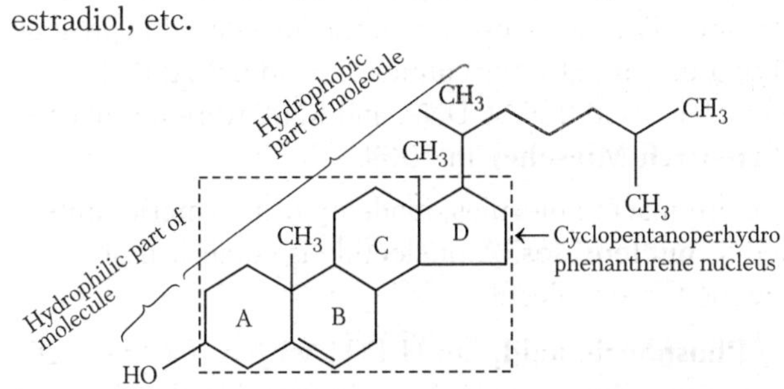

Fig 17.8 Structure of cholesterol

- Another steroid compound, diosgenin is produced by the yam plant (*Dioscorea*). It is used in the manufacturing of antifertility pills.

(ii) Prostaglandins

- These are biologically active lipids, first discovered in human seminal fluid.
- All mammalian cells except RBC produce prostaglandins.

- These are involved in binding of hormones to the membranes.
- They regulate acid production in stomach, stimulate smooth muscle contraction, used to induce labor, reduce the effect of asthma, etc.
- They are unsaturated fatty acids and are derived from arachidonic acid (C_{20} fatty acid).

(iii) Terpenes

- These are lipid like carbohydrates and are formed of isoprene units $(C_5H_8)_n$.
- These are essential oil and mostly present in oils of camphor, mint, lemon and *Eucalyptus*.
- Phytol a terpenoid is present in vitamins–A, E and K.

CHECK POINT 17.2

1. Amino acids are organic compounds and are called α-amino acids. Why?
(a) Amino acids are organic compounds containing an amino group and acidic group as substituents or two different carbons
(b) Amino acids are organic compounds containing an amino group and an acidic group as substituents on the same carbon
(c) Amino acids are inorganic compounds containing an amino group and acidic group as substituents on two different carbons
(d) Amino acids are inorganic compounds containing an amino group and acidic group as substituents on the same carbon

2. Which of the following amino acids is not optically active?
(a) Glycine
(b) Valine
(c) Leucine
(d) Isoleucine

3. The aleurone layer in monocotyledonous endospermic seed, is specially rich in
(a) lipids (b) starch (c) amino acid (d) proteins

4. Which of the following is not a conjugated protein?
(a) Peptone (b) Phosphoprotein
(c) Lipoprotein (d) Chromoprotein

5. Given below is the chemical formula of

$$CH_3(CH_2)_{14}-\overset{\overset{\displaystyle O}{\|}}{C}-OH$$

(a) Palmitic acid (b) Stearic acid
(c) Glycerol (d) Fumaric acid

6. Hydrolysis of lipid yields
(a) fats
(b) fatty acids and glycerol
(c) mannose and glycerol
(d) maltose and fatty acids

NUCLEIC ACIDS

They constitute the genetic material of all living organisms. The two most important nucleic acids present in living cell are Deoxyribonucleic Acid (DNA) and Ribonucleic Acid (RNA). DNA and RNA were discovered by **Friedrich Miescher** in 1969.

These are macromolecules, made up of polymeric units called as **nucleotides.** A nucleotide is composed of following three molecules

(i) **Phosphoric acid,** i.e. H_3PO_4 with 3 reactive –OH groups. Out of these, 2 are involved in forming sugar phosphate backbone with the help of phosphodiester linkages (— C— O—P— O— C—).

(ii) **Pentose sugar,** i.e. the carbohydrate (monosaccharide) with 5 C atoms. This may be deoxyribose in DNA or ribose in RNA.

Both of these sugars differ as shown below

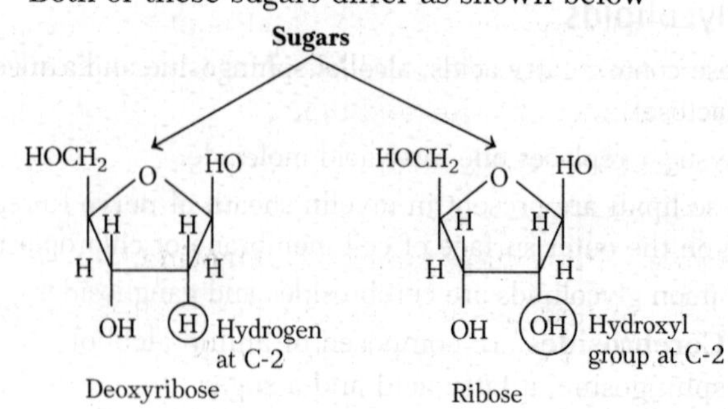

(iii) **Nitrogenous bases** are the heterocyclic organic bases that contain nitrogen in them. There are five different bases, which are categorised into two types, i.e. **purines** and **pyrimidines.**

(a) **Purines** are double-ring bases. The two purines are **adenine** and **guanine.**

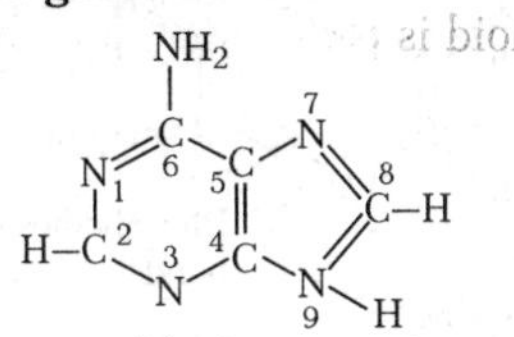

Adenine

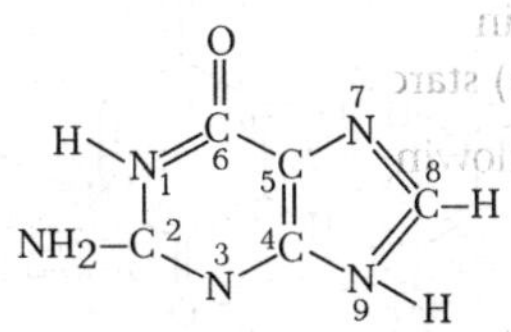

Guanine

(b) **Pyrimidines** are single ring bases, which include **cytosine, thymine** and **uracil.**

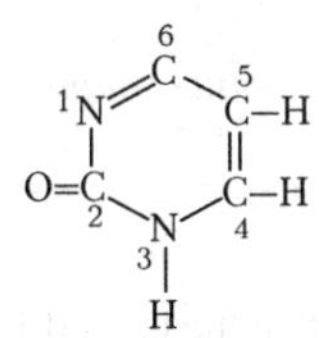

Cytosine

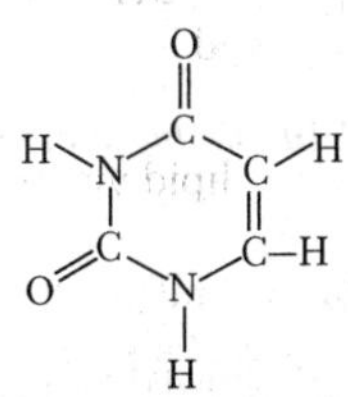

Uracil (only in RNA)

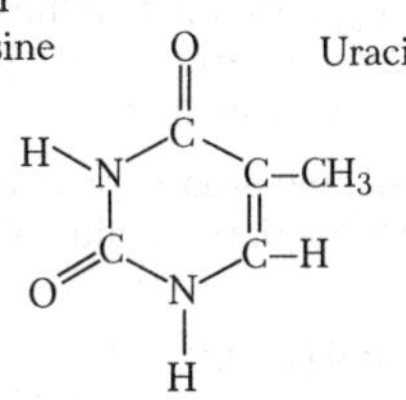

Thymine (only in DNA)

Of these nitrogenous bases, DNA consists of adenine, guanine, cytosine and thymine, whereas RNA contains adenine, guanine, cytosine and uracil.

Among the above written contituents, the sugar molecule and nitrogenous base together form the **nucleoside.**

Nucleoside + Phosphate group = Nucleotide.

Thus, these nucleosides are considered as smallest possible frame work that form units of nucleic acids.

DNA

It is the main genetic material of living beings. In prokaryotes, it is present without any association with proteins. However in eukaryotes, it is present with proteins like histones and protamines.

Structure

The most widely accepted structural model of DNA is the double-stranded double helical model of **Watson, Crick** and **Wilkins** (1953).

The four common nucleotides present in DNA are

- Deoxyadenylic acid or deoxyadenosine-3′-monophosphate and deoxyadenosine-5′-monophosphate.
- Deoxyguanylic acid or deoxyguanosine-5′-monophosphate.
- Deoxycytidylic acid or deoxycytidine-5′- monophosphate.
- Deoxythymidylic acid or deoxythymidine-5′-monophosphate.

According to the model proposed by Watson and Crick

- The DNA molecule consists of two helically twisted antiparallel strands connected together *via* base pairs.
- The twisting or coiling of strands results into the formation of deep (major) and shallow (minor) grooves as shown in the figure below

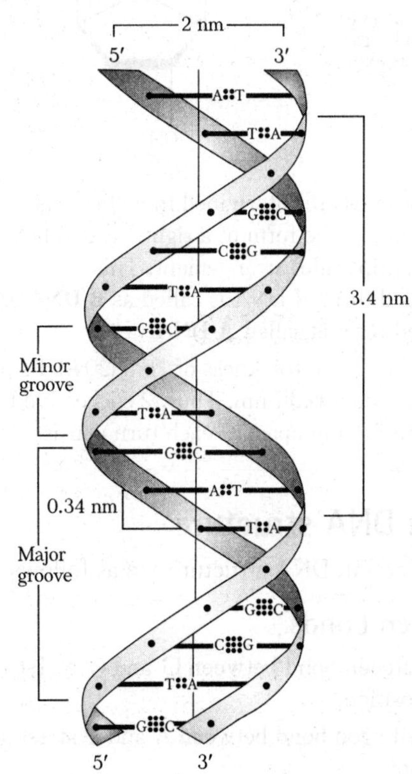

***Fig 17.9** According to Watson and Crick, double helical model of DNA*

- Each strand of DNA consists of alternating molecules of deoxyribose sugar and phosphate groups.
The linkage between the sugar and phosphate molecule is

called **phosphodiester linkage**. In this linkage, the phosphate group is attached to 3rd C-atom of one sugar molecule and 5th C-atom of other molecule as shown below

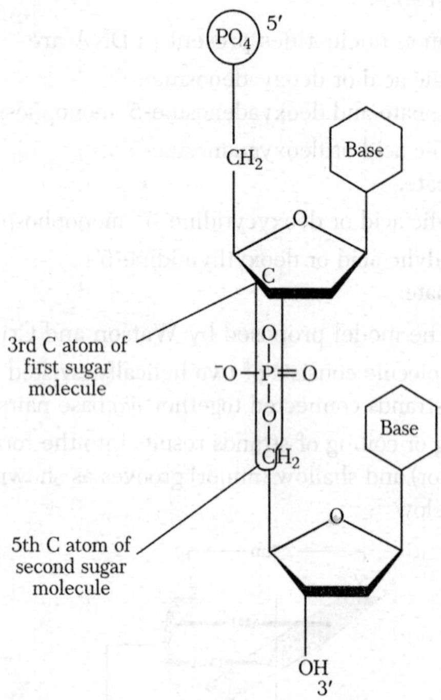

- The two strands are interwined in a clockwise direction, i.e. in the form of a right-handed helix and have antiparallel arrangement. This right-handed form of DNA is called as **B-DNA** and left-handed form is called **A-DNA**.
- DNA has a uniform thickness of 2nm (20Å). Its pitch or 1 turn is about 0.34 nm. Thus, 2 successive bases in DNA are 3.4 nm apart as each turn consists of 10 nucleotides.

Bonds in DNA structure

The bonds seen in DNA structure are as follows

1. Hydrogen bonds

- Triple hydrogen bond between G and C at 1st, 2nd and 6th position.
- Double hydrogen bond between A and T at 1st and 2nd position.

2. True covalent bonds

- Between sugar - Purine at 1 : 9 position.
- Between sugar - Pyrimidine 1: 3 position.
- Sugar-Phosphate - Sugar, i.e. phosphodiester linkage → 3-5 linkage.

These bonds are shown in the figure below

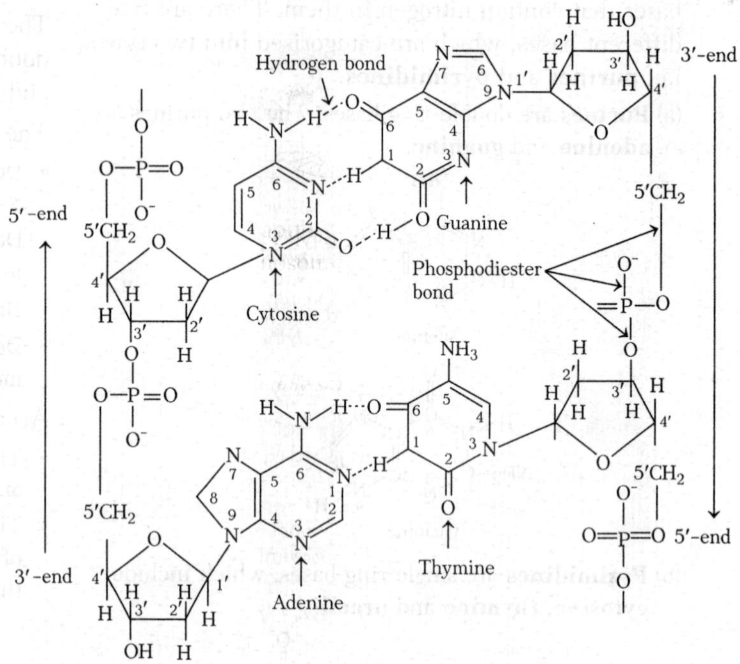

Fig. 17.10 Double standard polynucleotide chains

> **Chargaff's rule**
>
> **Erwin Chargaff** in 1950 showed through his rule that the composition of DNA molecules varies depending upon the source of DNA, but the ratio of purines and pyrimidines almost remains same in it.
>
> The Chargaff's rule states that in DNA, the proportion of adenine is always equal to that of thymine. On the other hand, proportion of guanine is always equal to that of cytosine, i.e.
>
> $$A = T \text{ and } G = C.$$
>
> Thus, there is always an equal proportion of purines and pyrimidines in DNA,
>
> i.e. $A + G = T + C$ or $\frac{A+G}{T+C} = 1$

RNA

The other nucleic acid present in the cell is RNA, i.e. Ribose Nucleic Acid. It is found predominantly in cytoplasm.

Structure

- It is mostly present in single-stranded form though some viruses like retrovirus and wound tumour virus have double-stranded RNA.
- The single RNA strand is folded upon itself either entirely or in certain regions. All normal RNA chains begin with adenine or guanine.
- RNA does not follow Chargaff's rule due to single-stranded nature and lacks of complementarity.

Types of RNA

The RNA can be of following three types

(i) Messenger RNA or *m*RNA or template RNA

(ii) Ribosomal RNA or *r*RNA

(iii) Soluble RNA or Transfer RNA (*s* or *t*RNA)

- The transfer and messenger RNAs are synthesised on chromosomal DNA. On the other hand, ribosomal RNA is synthesised on ribosomal DNA of nucleolus.
- The three types of RNA are synthesised during different stages of embryonic development. The *m*RNA is synthesised during **cleavage**, *t*RNA is synthesised at the end of cleavage, while *r*RNA is synthesised during gastrulation.

Messenger RNA or *m*RNA

The name messenger RNA was proposed by **Jacob** and **Monod** in 1961. It makes 3-5% of total cellular RNA. The sedimentation coefficient of *m*RNA is 8 S. It is always single-stranded and without any base pairing, because base pairing in *m*RNA strand can destroy its biological activity. The structural components of *m*RNA include

- A cap is present at the 5′ end. It controls the rate of protein synthesis. Without the cap, *m*RNA molecules bind very poorly to ribosomes. The cap is rich in methyl residues.
- Cap is followed by **non-coding region** of 10 to 100 nucleotides. This region is rich in A and U residues. It does not have the ability to translate protein.
- The non-coding region terminates at **initiation codon** AUG (sometimes GUG or UUG also) in both prokaryotes and eukaryotes.
- With the initiation codon, the coding region begins which contains on an average about 2500 nucleotides. The coding region terminates at **termination codon**, i.e. either of UAA, UAG, UGA.
- Other features of *m*RNA are as follows
  - The length of *m*RNA decreases with age
    *m*RNA can be of two types
    (a) **Monocistronic,** i.e. single molecule formed by one gene as in eukaryotes.
    (b) **Polycistronic,** i.e. single molecule formed by many genes as in prokaryotes.
  - The *m*RNA of prokaryotes is considered to be metabolically unstable. It does not stay for long in the body of prokaryote and gets destroyed readily even during the continuation of metabolic process itself.
  - On the other hand, the *m*RNA of eukaryotes is considered as metabolically stable, as it normally waits for the completion of process concerned.

Note
- *The transcribed mRNA in eukaryotes is found to be 20% greater than its original length. This greater RNA is called **hnRNA** (heterogeneous nuclear RNA).*
- *According to **Weinberg** (1973), all those RNAs which have sedimentation coefficient more than 8 S are called **hnRNA**.*
- *The mRNA formed in nucleus, comes out with proteins into cytoplasm and normally swim along with proteinaceous spherical balls. These RNA + protein spherical balls are called **informosomes**.*

Ribosomal RNA or *r*RNA

- The *r*RNA makes 80% or more of total cellular RNA. It is the largest RNA. This is the basic constituent of ribosomes and develops from the nucleolar organiser region of chromosomes in eukaryotes.
- In prokaryotes, it is developed from *r*DNA. This RNA is seen as a compact rod or coil or in the form of extended regions. The structure actually depends upon various factors like ionic strength, pH, temperature, etc.
- The *r*RNA is found to be stable for at least two generations.

Transfer RNA (*t*RNA) or Soluble RNA (*s*RNA)

- It makes about 10-20% of total cellular RNA with sedimentation coefficient of 3.8 S. It is the smallest RNA.
- The structure of *t*RNA is given by **Robert Holley**, contains five arms each with a loop and a stem as shown below

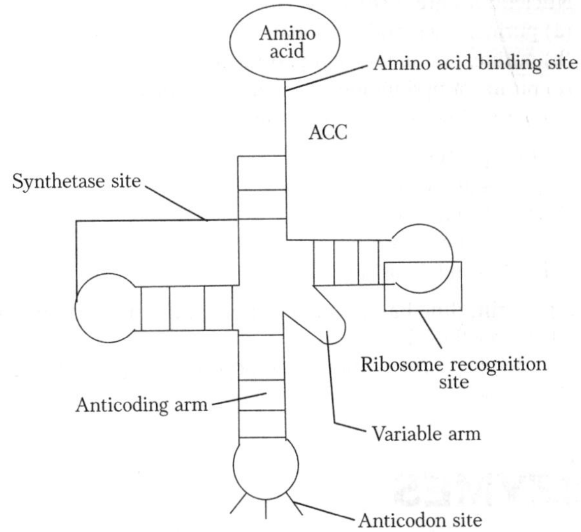

Fig 17.11 The binding sites of RNA

- A *t*RNA molecule has following four recognition sites namely,
  (i) Amino acid binding site (CCA)
  (ii) Ribosome recognition site (GT ψ CR)
  (iii) Anticodon site
  (iv) Synthetase site

- This RNA contains 73-93 nucleotides within the structure.
- The *t*RNAs, which carry amino acids for initiation codon are called **initiator *t*RNA.**
- The function of *t*RNA is to align the required amino acid according to nucleotide sequence of *m*RNA.
- During this alignment, the ribosome is also present, *t*RNA is also the carrier of the activating enzyme of amino acid.

ATP (Adenosine Triphosphate)

- ATP was discovered by **Karl Lohmann** in 1929 from muscles. **Lipman** (1941) discovered its energy carrying nature. He is called as the **Father of ATP cycle.**
- ATP is also known as **universal energy carrier** as well as **energy currency of the cell.** It is used by the cells in carrying out almost all activities.
- ATP is a high energy molecule, because it releases a large amount of usable energy when it loses its terminal phosphate group. As a result of this ATP gets transformed into ADP.
- ATP consists of a purine, adenine, ribose sugar and a row of three phosphate groups out of which the last two are attached by high energy bonds (~P ~P).
- Its general appearance is shown below

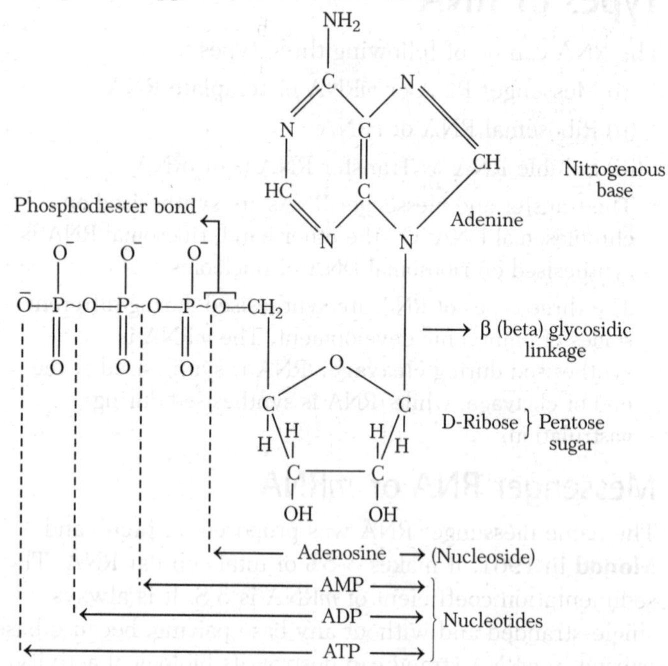

Fig 17.12 Molecular structure of Adenosine Triphosphate (ATP)

CHECK POINT 17.3

1. Nucleotides are formed by
   (a) purine, sugar and phosphate
   (b) purine, pyrimidine and phosphate
   (c) purine or pyrimidine, sugar and phosphate
   (d) pyrimidine, sugar and phosphate
2. Purines posses nitrogen at
   (a) 1, 2, 4 and 6 position
   (b) 1, 3, 5 and 7 position
   (c) 1, 3, 7 and 9 position
   (d) 1, 2, 6 and 8 position
3. The pyrimidine base, which confers additional stability to DNA over RNA is
   (a) adenine (b) guanine (c) cytosine (d) thymine
4. Which of the following is incorrect?
   (a) DNA and RNA consist of nucleotides only
   (b) Adenylic acid is a nucleoside
   (c) Uridine is a nucleoside
   (d) N-bases (A, G, C, T, U) have heterocyclic rings
5. Select the correct pair of substituted purines.
   (a) Cytosine and thymine (b) Adenine and guanine
   (c) Uracil and cytosine (d) Guanine and uracil
6. Which of the following is incorrect?
   (a) Phosphate is covalently bonded to nitrogenous base
   (b) Sugar is covalently bonded to nitrogenous base
   (c) Nitrogenous bases are hydrogen bonded to each other
   (d) Sugar and nitrogenous bases form peptide bonds

ENZYMES

- These are catalytic biological molecules.
- The term 'enzyme' was coined by **Friedrich Wilhelm Kuhne** in 1878. While working on fermentation. At that time, only the living cells of yeast were thought to be essential for fermentation of sugar.
- In 1987, a German chemist, **Eduard Buchner** isolated the enzyme **zymase** (from the yeast cell) and proved that the zymase extract has power of fermenting sugar. He was awarded with the Nobel Prize for this work.
- Later on **JB Sumner** (1926) prepared a pure crystalline form of **urease** enzyme from **Jack Beans** and suggested that enzymes are mostly proteins.
- Enzymes are proteinaceous substance, which are capable of altering the rate of chemical reaction of the biological system without undergoing any alternation in themself.

Nomenclature of enzyme

In 1961, the commission on enzymes setup by the IUB (International Union of Biochemistry) formed certain rules of enzyme nomenclature and classification.

According to IUB system of classification, the major rules are as follows

- Reactions and enzyme catalysing them are divided into 6 major classes each with 4-13 subclasses.
- The enzyme name has two parts, first name is of substrate, while the second ending in *-ase* indicates the type of reaction.
- The enzyme has a systematic code number (enzyme code/enzyme commission). The first digit denotes the class, the second subclass, the third sub-subclass and the fourth digit depicts the particular enzyme name.

Table 17.4 Major classes of enzymes according to this system are given in this table

Enzymes	Activities	Examples	Examplified reactions
Oxido-reductase	Transfer H_2 atom or ion from one molecule to another.		$S_{reduced} + S'_{oxidised} \xrightarrow{\text{Oxido-reductase}} S_{oxidised} + S'_{reduced}$
(i) Oxidases	Transfer H_2 molecule from a molecule to oxygen.	Cytochrome oxidase (involve in ETS).	$2AH_2 + O_2 \xrightarrow{\text{Oxidase}} 2A + 2H_2O$
(ii) Dehydrogenases	Transfer H_2 molecule from a molecule to a Co-enzyme such as NAD^+.	Succinic dehydrogenase	$AH_2 \xrightarrow{\text{Dehydrogenase}} A + 2(H)$
(iii) Reductase	Addition of H_2 (electron) and removal of O_2.	Nitrate reductase	$A + 2(H) \xrightarrow{\text{Reductase}} AH_2$
Transferases (group transfer reactions)	Transfers a specific functional group (e.g. amino, methyl, acyl, phosphate) from one kind of molecule to another.	Transphosphorylases, transaminases, kinases, transpeptidases, etc.	$S—G + S' \xrightarrow{\text{Transferase}} S + S'—G$
Hydrolases (hydrolytic reactions)	Addition of water to a bond such as ester, ether, peptide, glycosidic, C—C, P—N, C-halide, etc.	Digestive enzyme like lipases, amylases, esterases, phosphatases, carbohydrase, protease belong to this group.	$S + H_2O \xrightarrow{\text{Hydrolases}} X + Y$
Lyases (desmolase)	Breaks specific covalent bond and removes a group without hydrolysis, oxidation and reduction reactions.	Fumerases, carboxylases, aminases, decarboxylase, lysozyme.	$R\overset{\overset{Y}{\vert}}{C}—\overset{\overset{Y}{\vert}}{C}R \xrightarrow{\text{Lyase}} X—Y + C{=}C$
Isomerases (isomerisation reactions)	Rearrangement of molecular structure to form isomers.	Epimerase, racemase, phosphohexose isomerase (phosphoglucomutase).	$X \xrightarrow{\text{Isomerase}} Y$
Ligases or Synthetases (condensation reaction)	Join molecule together by the formation of bonds. It utilises the energy supplied from the breakdown of ATP.	DNA ligases (repair breaks in DNA molecule), amino acyl synthetase (activate *t*RNA by attaching amino acid at 3′ end.	$X + Y + ATP \xrightarrow{\text{Ligase}} XY + ADP + iP$

With the exception of small group of catalytic RNA molecules (ribozyme), all enzymes are proteinaceous in nature. They can be made of many or single polypeptide chains. The amino acids sequence in the chain impart specificity to the enzyme.

Chemical nature of enzymes

Based on its structure, enzymes can be of following two types

1. Simple enzymes

These enzymes are composed of only simple polypeptide chains, thus they have only protein part in it, e.g. urease, amylase, pepsin, etc.

2. Conjugate enzymes

These enzymes are complex in structure and these are made up of two parts

(i) **Apoenzyme** This is the protein part of the conjugate enzyme. It is made up of large molecule of protein and is heat unstable. This does not take part in the transfer of groups.

(ii) **Co-factors** The non-protein part of the enzyme is called cofactor. These are the substances, which are required by the enzyme for their efficient activity. These are stable at high temperature.

These are of three types

(a) **Inorganic ions** Certain ions of elements (like chloride for amylase) or metal ions mould either enzyme or substrate into a shape that allows an enzyme substrate complex to be formed. This increases the rate of reaction.

Table 17.5 Few enzymes that require inorganic elements as cofactors

Enzymes	Inorganic elements
Aconitase, catalase, peroxidase, cytochrome oxidase	Iron
Dehydrogenase, carbonic anhydrase, thiolase, glycine reductase	Zinc
Tyrosinase, cytochrome oxidase	Copper
Hexokinase, phosphotransferase, DNA polymerase, pyruvate kinase	Magnesium
Peptidase, decarboxylase, ribonucleotide reductase	Manganese
Nitrate reductase, dinitrogenase	Molybdenum
Urease	Nickel
Enolase	Boron
Lipase, succinic dehydrogenase	Calcium
Pyruvate kinase	Potassium
Salivary amylase	Chlorine
ATPase	Sodium

(b) **Prosthetic groups** These are organic molecules attached to enzymes, which assist catalytic activity of the enzyme, e.g. FAD (Flavin Adenine Dinucleotide), haem, biotin, etc. They remain attached to enzyme.

(c) **Coenzymes** These are organic molecules, which act as co-factor, but these do not remain attached to enzyme between reactions. These are important part of enzyme, e.g. NAD (Nicotinamide Adenine Dinucleotide).

Proenzymes or Zymogen

- It is the inactive precursor of a proteolytic enzyme. Zymogen is cleaved to form the active enzyme.
- Many enzymes are initially produced in the zymogen or proenzyme state.
- These can be activated at particular pH, in the presence of substrate or some special conditions, e.g. the protein digesting enzyme pepsinogen (inactive) gets activated in the presence of HCl and turns to pepsin.

Properties of enzymes

- Enzymes cannot initiate a reaction, rather they are involved in increasing the rate of chemical reaction.
- They are large-sized biomolecules of colloidal nature. They possess very low rate of diffusion.
- They have high molecular weight. Their large size provides large surface area and more active sites for functioning. The highest molecular weight is of pyruvate dehydrogenase.
- They are highly specific, i.e. an enzyme generally catalyses a specific reaction.
- They are thermolabile (heat sensitive). 25°-35°C is considered the best or optimum temperature for their functioning. Their activity stops at 0°C and above 80°C.
- The reactions controlled by enzyme are mostly reversible.
- Change in pH may influence the working of an enzyme. An optimum pH of 6-7.05 is required for an enzyme to show its maximum activity.
- Efficiency of an enzyme is expressed in terms of its turn over number. It is the number of moles of substance transformed into products per mole of enzyme per second. Carbonic anhydrase has highest turn over number whereas has the slowest one.
- An enzyme remains unchanged at the end of any reaction.

Site of enzyme action

The enzymes can work inside or outside the cell. Thus, with regard to the site where enzymes can act, there are two types of enzymes

(i) **Intracellular enzymes or Endoenzymes** These are those enzymes, which remain and function inside the cells. These enzymes function best at neutral pH. They include enzymes of glycolysis.

(ii) **Extracellular enzymes or Exoenzymes** These are those enzymes, which function outside the cell. They include the digestive enzymes, e.g. salivary amylase, gastric pepsin, lysozyme (in tears and nasal secretion).

Mechanism of enzyme action

Each enzyme (E) has a substrate (S) binding site. This site is called **active site.** Binding of the substrate to this site produces a highly reactive short-lived complex called **enzyme-substrate complex.** This complex is soon transformed into an enzyme-product complex. This complex breaks up into chemical products (P) and enzyme unchanged as shown below

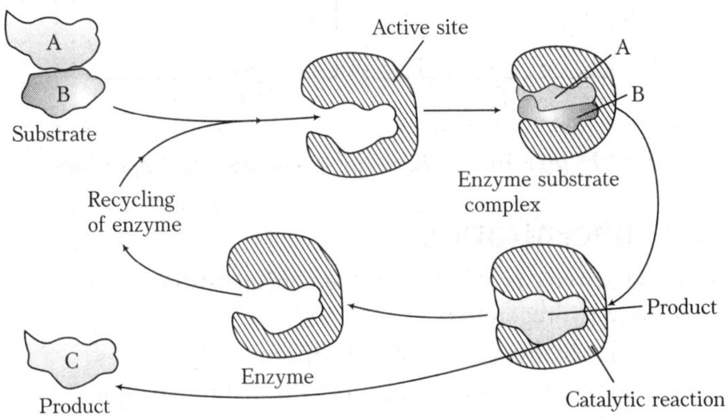

Fig 17.13 Mechanism of enzyme reaction through enzyme substrate complex formation

Active site

- An enzyme possesses distinct cavity or cleft, in which the substrate is bound. In this cavity, active centre of enzyme is present.
- In this active centre, all the reactive amino acids are grouped together to combine with the substrate.
- The polypeptide chain of these amino acids remain separated from each other. These are brought together by folding or bending to form tertiary structure.
- An enzyme may have one to several active sites. These sites are very specific for its enzyme, e.g. glycine-histidine-alanine are the active sites for aldolase, while aspartic acid-cysteine-alanines are the active sites for pyruvic oxidase.

There are two views by which enzymes are supposed to form enzyme-substrate complex.

1. Lock and key hypothesis

- This hypothesis was given by **Emil Fischer** in 1894. He proposed that the geometry of active sites of enzyme and shape of substrate molecule are complementary to each other.
- During reaction when an enzyme and substrate come together, they act more like a lock and key and just as a lock can be opened by its specific key, a substrate molecule can be acted upon by a particular enzyme.
- Thus, this theory also explains the specificity of enzyme action as well.
- The complete process in outline view is shown below

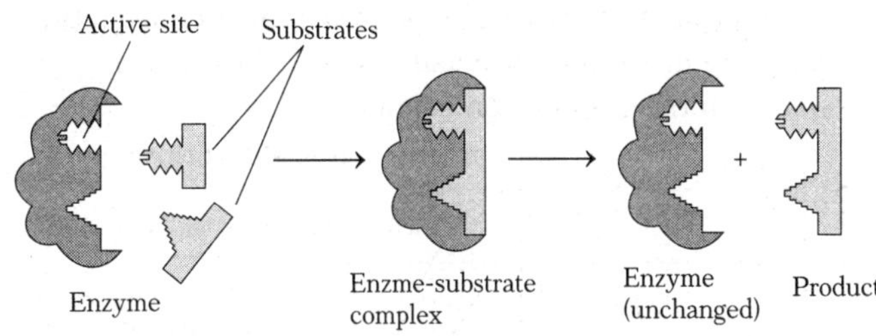

Fig 17.14 Lock and key model of enzyme action

2. Induced fit hypothesis

- This hypothesis was proposed by **Koshland** in 1960.
- According to this hypothesis, the active site of enzyme does not initially exist in a shape complementary to the substrate.
- Rather, it is induced to assume the complementary shape as the substrate become bound to the enzyme. Thus, the active site of enzyme is not rigid, but flexible.
- According to this hypothesis, the active site of the enzyme contains two groups
  - (i) **Buttressing group**, i.e. those present to support the substrate.
  - (ii) **Catalytic group**, i.e. those meant for catalysing the reaction.
- When an enzyme binds to the buttressing group, a conformation change occurs in the active site. Due to this change, the catalytic groups are brought opposite to the substrate. As a result of this, catalysis takes place.
- The complete process according to this hypothesis is shown below

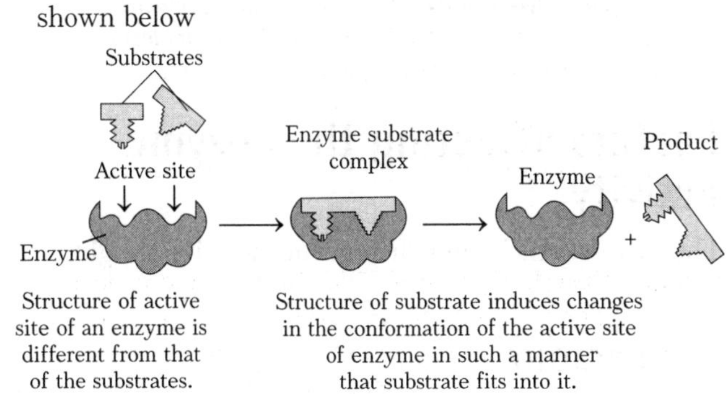

Fig 17.15 Induced fit model of enzyme action

Energy profile during enzyme action

- The transition state and the product formation from the substrate in the presence and absence of enzyme is shown in the graph given below

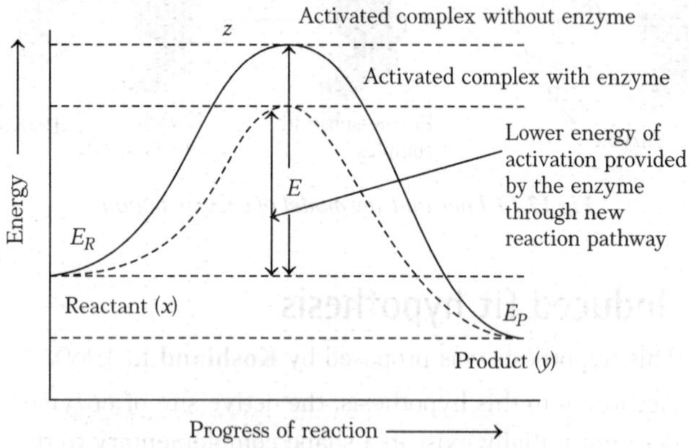

Fig 17.16 Energy profile for uncatalysed enzyme reaction

- In the above figure, the chemical compound *x* requires E as energy of activation to convert the reactant in a **highly activated state**, i.e. transition state *z*.
- The reaction then becomes spontaneous, converting z into *y*, i.e. product and energy E is liberated.
- Further, the amount of energy is released according to the difference in enthalpy (ΔH) of *x* and *y*.
- Thus, the enzyme increases the rate of conversion of *x* into *y* by providing an alternate pathway of lower activation energy.

Note
- ***Energy of activation*** *is defined as the amount of energy required by the substrate molecules to form the product. It can be calculated by subtracting energy of substrate molecules from the threshold energy of the reaction.*
- ***Turn Over Number*** *(TON) is the number of reactant molecules converted into product by one molecule of enzyme in unit time.*

Factors affecting the enzyme activity

The factors which alter the enzyme activity are given below

Temperature

- Enzymes show high activity at optimum temperature, i.e. 25-40°C.
- At high temperature, enzyme becomes denatured while at low temperature, it becomes inactive.

pH

- At optimum pH range the enzyme activity is maximum.
- Below and beyond this pH, denaturation of enzymes takes place.

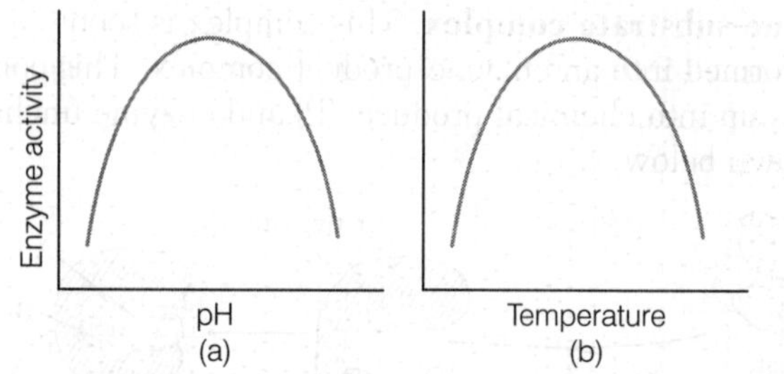

Fig. 17.17 *(a) Effect of change in : (a) pH and (b) temperature*

Ionic concentration

- Enzyme activity is affected by the concentration of ions especially hydrogen ions.
- This is so, because each molecule of enzyme possesses numerous groups, which yields H^+ (in slightly alkaline solution) and OH^- ions (in slightly acidic medium).
- Thus, enzymes are capable of ionising either as an acid or as a base depending upon the acidity of external solution (due to their amphoteric nature).

Enzyme and substrate concentration

- Enzyme concentration affects the rate of reaction in combination with substrate concentration.
- Normally, it is seen if the substrate concentration is increased, the rate of reaction also increases.

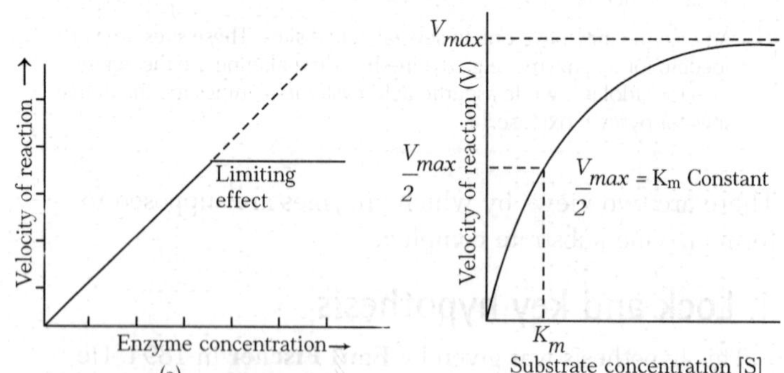

Fig. 17.18 *(a) Effect of enzyme concentration on the rate of biochemical reaction, (b) Effect of concentration of substrate on enzyme*

- However after sometime, the enzyme becomes saturated with substrate molecules. At this time further increase in reaction velocity occurs only if the enzyme concentration is increased.
- This effect of substrate concentration in combination with enzyme concentration on the rate of a reaction can be understood by **Michaelis-Menten equation.**

Michaelis-Menten equation

- A basic theory of enzyme action was proposed by **Michaelis** and **Menten** in 1913. This theory is inspired by the saturation of enzyme molecules, because of which the reaction rate becomes constant.
- This equation can describe how a reaction relatively varies with substrate concentration. They introduced a constant K_m (Michaelis constant).
- Increased enzyme levels never affect the K_m value as shown below

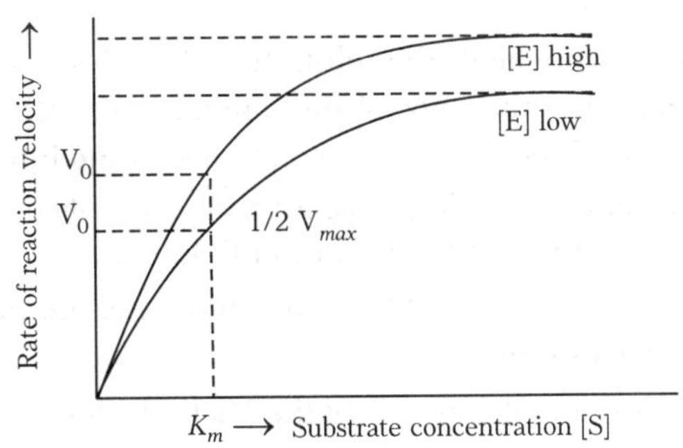

***Fig. 17.19** Reaction velocity 'V' and substrate concentration (S) for a typical enzyme catalysed reaction*

- K_m indicates the affinity of enzyme for its substrate. It is a mathematical derivative or constant, which indicates the substrate concentration at which the chemical reaction catalysed by an enzyme attains half of its maximum velocity (V_{max}).
- K_m value differs from substrate to substrate, because different enzymes differ in their affinity towards different substrates.
- A high K_m depicts low affinity, while a low K_m shows strong affinity.

Note *The concept of 1/2 V_{max} introduced as the effect of substrate concentration becomes constant after enzyme saturation. However, it is fully visible at the half way mark, i.e. 1/2 V_{max}.*

Product concentration

The reaction of an enzyme can also be controlled by its product (i.e. feedback inhibition).

Accumulation of the product usually slows down the rate of chemical reaction.

Inhibitor and enzyme inhibition

The activity of an enzyme is very sensitive to the presence of specific chemicals, which binds to an enzyme. When binding of a chemical shuts off the enzymatic activity, it is called as **inhibition**. The substance causing inhibition is called **inhibitor**. Enzyme inhibitions are of two types, i.e. competitive and non-competitive inhibitions.

Table 17.6 Competitive and non-competitive inhibitions

Competitive inhibition	Non-competitive inhibition
In this kind of inhibition, the inhibitor competes with the substrate for binding to the active site of enzyme. This type of inhibitor shares its close structural similarity with the substrate. Thus, the substrate cannot bind and enzyme action declines.	In this type, there is no competition between the molecule for the active site of enzyme. Infact, inhibition is caused by the alteration of conformational structure of enzyme by a chemical, which binds to the site other than active site (or enzyme).
The mechanism can be explained as 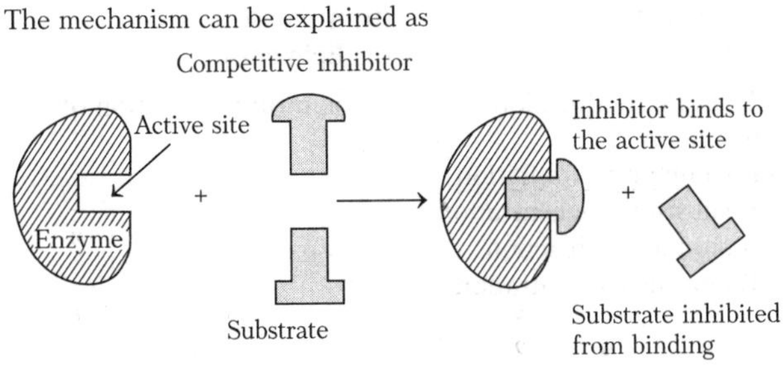	The mechanism can be explained as 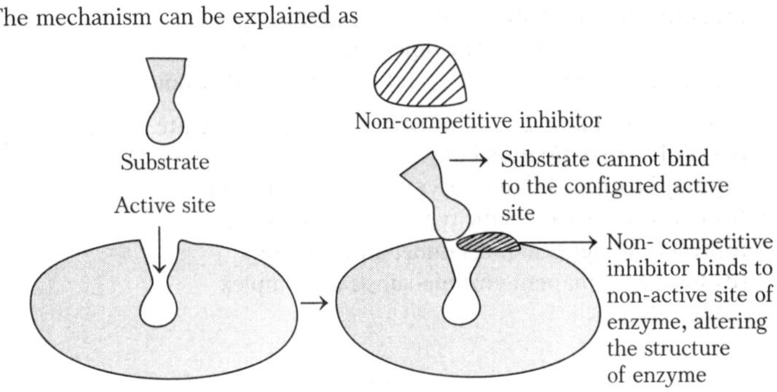
The EI (Enzyme Inhibitor complex) is formed and ES is inhibited during the reaction process.	Both EI and ES complex are formed.
The effect of competitive inhibitor on the activity of enzyme can be reversed by increasing the concentration of substrate.	The effect of inhibitor cannot be reversed by increasing the concentration of substrate, thus the inhibition is irreversible.

Competitive inhibition	Non-competitive inhibition
In this kind of inhibition, the V_{max} remains constant and K_m increases.	In this kind of inhibition, the V_{max} is lowered and K_m is changed.
e.g. Succinic acid dehydrogenase oxidises succinic acid to fumaric acid. If considerable concentration of malonic acid is added the activity of succinic dehydrogenases decreases rapidly. Thus, malonic acid acts as a competitive inhibitor due to its structural resemblance to succinic acid.	e.g. Cyanide inhibits the mitochondrial enzyme cytochrome oxidase, which is essential for cellular respiration.

Allosteric enzyme and allosteric inhibition

- In **allosteric inhibition**, two functionally different binding sites are present.
- Binding of modulator causes a conformational change in enzyme, due to which structure of active site is modified as shown below

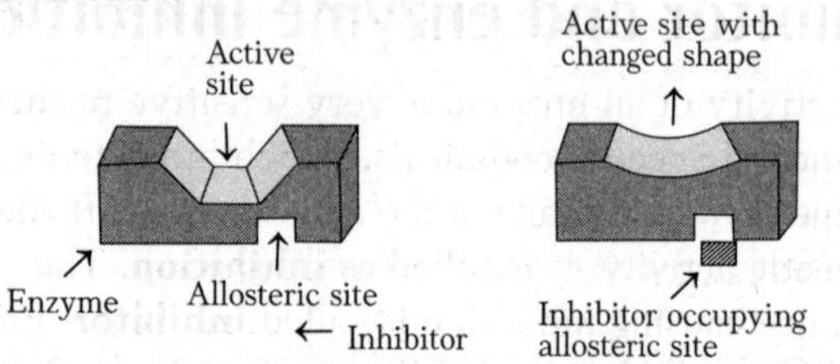

Fig 17.20 Non-competitive inhibitor changes the active site of enzyme after binding at allosteric site

- One of the site binds the substrate and catalyses the reaction. The other type of sites are known as **allosteric sites**. They bind with a molecule called as **modulator**.
- A modulator molecule can be a positive modulator (activator) or a negative modulator (inhibitor).
- Such enzymes with additional allosteric sites are called **allosteric enzyme.**
- These enzymes do not follow the Michaelis-Menten kinetics, e.g. hexokinase is an allosteric inhibitor of glucose → glucose-6- phosphate.
- Other allosteric enzymes are phosphofructokinase, hexokinase, aspartate, transcarbamylase, glycogen, phosphorylase, pyruvate kinase, etc.

CHECK POINT 17.4

1. Which of the following is true for competitive enzyme inhibition?
(a) Decrease in V_{max} and K_m
(b) Unchanged V_{max} and decrease in K_m
(c) Unchanged V_{max} and increase in K_m
(d) Increase in V_{max} and K_m

2. Enzymes that catalyse inter-conversion of optical, geometrical or positional isomers are
(a) ligases (b) lyases (c) hydrolases (d) isomerases

3. The 'lock and key' model of enzyme action illustrate that a particular enzyme molecule
(a) may be distroyed and resynthesised several times
(b) intracts with a specific type of substrate molecule
(c) reacts at identical rates under all conditions
(d) form a permanent enzyme-substrate complex

4. The effectiveness of an enzyme is affected least by
(a) temperature
(b) concentration of the substrate
(c) original activation energy of the system
(d) concentration of the enzyme

5. Choose the odd one out.
(a) Fumerase (b) Aminase
(c) Lysozyme (d) Epimerase

6. Allosteric modulation is due to the inhibition action of enzyme by
(a) competitive inhibition
(b) substrate concentration
(c) products of reaction
(d) enzyme concentration

EXAM POINTS

- Secondary metabolites are produced by plants for their defence action, e.g. nicotine, strychnine and caffeine. **[NEET 2020]**
- Carbohydrates are more suitable for the production of energy in the body than proteins and fats. **[AIIMS 2016]**
- Carbonyl and hydroxyl are two functional groups and characteristic of sugars. **[NEET 2018]**
- Sucrose is a non-reducing carbohydrate or disaccharide. It accumulates in the cytoplasm. **[NEET 2016]**
- Glucose and galactose are epimers. **[JIPMER 2019]**
- Glycogen contains 1, 4 α-glycosidic bonds. It is a polysaccharide and have right end as reducing end and its left end is called the non-reducing end. **[AIIMS 2017]**
- Cellulose is a polysaccharide. **[NEET 2016]**
- Cellulose, the most important constituent of plant cell wall is made up of unbranched chain of glucose molecules linked by β, 1-4 glycosidic bond. **[AIIMS 2014]**
- Lysine is the basic amino acids. **[NEET 2020]**
- Glycine is the simplest amino acid. **[NEET 2020, 16]**
- Proteins are needed in diet because all amino acids are not available in body, during fasting body utilised proteins and proteins act as building blocks of our body. **[JIPMER 2018]**
- Proteins are heteropolymers containing strings of amino acids. **[Kerala CEE 2015]**
- Ramachandran plot is used to confirm the structure of proteins. **[NEET 2019]**
- GN Ramachandran discovered the triple helical structure of collagen. **[Kerala CEE 2014]**
- Arginine is non-essential amino acid. **[JIPMER 2018]**
- Collagen is the most abundant protein in the animals. **[NEET 2020]**
- Hydrogen bonds, electrostatic interaction and hydrophobic interaction are involved in stabilising the three-dimensional folding of most proteins. **[NEET 2016]**
- A typical fat molecule is made up of one glycerol and three fatty acid molecules. **[NEET 2016]**
- Cholesterol is the most important animal steroid which is insoluble in water and chemically unreactive. It is important because it is structural component of cells. **[AIIMS 2017]**
- A phosphoglyceride is always made up of a saturated or unsaturated fatty acid esterified to a phosphate group, which is also attached to a glycerol molecule. **[NEET 2013]**
- Tyrosine possesses aromatic ring in its structure. **[NEET 2020]**
- Inulin has glycosidic bond whereas insulin has peptide bond in their structure. **[NEET 2020]**
- Concanavalin-A is a lectin. **[NEET 2019]**
- Vitamin-D, bile salts and steroids are all derivatives of cholesterol. **[JIPMER 2018]**
- Omega-3 fatty acid is present in flax seed oil. **[JIPMER 2018]**
- Nucleic acids, proteins and polysaccharides are all polymeric compounds. **[NEET 2017]**
- Uracil is a pyrimidine. **[NEET 2016]**
- The moiety present at the 5' end of ribose sugar in a polynucleotide is phosphate. **[Kerala CEE 2015]**
- In the ribose of RNA, unlike DNA, every nucleotide residue has an additional OH group in the 2′ position. **[Kerala CEE 2015]**
- Sulphur is an integral part of cysteine. **[NEET 2020]**
- Lecithin contains phosphorus atom in its structure. **[NEET 2020]**
- Phospholipid is the main constituent of lecithin. **[NEET 2019]**
- In the polynucleotide chain of DNA, a nitrogenous base is linked to the –OH of 1′C pentose sugar. **[NEET 2020]**
- Adenine pairs with thymine through two H-bonds. **[NEET 2020]**
- Adenylic acid, cytidylic acid and guanylic acid are all nucleotides. **[AIIMS 2019]**
- Prosthetic groups differ from co-enzymes in that they tightly bound to apoenzyme. **[NEET 2019]**
- Holoenzyme is composed of apoenzyme and coenzyme. **[NEET 2017]**
- A complete catalytic active enzyme with its bound prosthetic group is called holoenzyme. **[NEET 2019]**
- The protein component of a holoenzyme is known as apoenzyme. **[WB JEE 2014]**
- Coenzyme or metal ion that is tightly bound to enzyme protein is called prosthetic group. **[NEET 2019]**
- The enzyme which catalyses the formation of glutamine from its substrate, belongs to ligases category of enzymes. **[NEET 2016]**
- Nomenclature of enzymes is done on the basis of the type of reaction they catalyse. **[COMEDK 2015]**
- Ribozyme is a non-proteinaceous enzyme. **[NEET 2016]**
- A comepititive inhibitor declines the enzyme action. **[KCET 2016]**
- A competitive inhibitor reacts reversibly with the enzyme to form and enzyme-inhibitor. **[CBSE AIPMT 2015]**
- In competitive inhibition, the inhibitor molecule is not chemically changed by the enzyme. **[CBSE AIPMT 2015]**
- The competitive inhibitor does not affect the rate of breakdown of the enzyme substrate complex. **[CBSE AIPMT 2015]**
- The presence of the competitive inhibitor increases the K_m of the enzyme for the substrate. **[CBSE AIPMT 2015]**
- The catalytic efficiency of two different enzymes can be compared by the K_m values. **[J&K CET 2015]**
- Transition state structure of the substrate formed during an enzymatic reaction is transient and unstable. **[NEET 2013]**
- Enzyme lower the activation energy of a reaction. Enzyme speed up the attainment of equilibrium of the reaction. **[AMU 2014]**

Chapter Exercises

(A) Topical Exercise

Composition of the Living Tissue

1 Many elements are found in living organisms either free or in the form of compounds. One of the following is not found in living organisms. **[NCERT Exemplar]**

(a) Silicon (b) Magnesium (c) Iron (d) Sodium

2 The biomolecules are

I. carbohydrates II. proteins
III. lipids IV. mitochondria

Choose the correct option from the codes given below.

(a) I, II and III (b) II and III
(c) Only II (d) I, II, III and IV

3 The cellular micromolecules are

(a) amino acids, water, minerals, nucleotides and sugar
(b) glycogen, amino acids, minerals and nucleotides
(c) water, minerals, nucleic acids, amino acids and nucleotides
(d) sugars, water, minerals, proteins and nucleotides

4 The correct order of chemical composition of living tissues/cells in terms of percentage of the total cellular mass is

(a) Nucleic acid > Proteins > H_2O > Carbohydrate > Ions > Lipid
(b) H_2O > Proteins > Nucleic acid > Carbohydrate > Lipid > Ions
(c) H_2O > Proteins > Carbohydrate > Nucleic acid > Lipid > Ions
(d) Lipid > Ions > Carbohydrate > H_2O > Proteins > Nucleic acid

5 Match the following columns.

	Column I (Components)		Column II (% of the total cellular mass)
A.	Water	1.	10-15
B.	Proteins	2.	70-90
C.	Carbohydrates	3.	5-7
D.	Lipids	4.	3
E.	Nucleic acids	5.	2

Codes

	A	B	C	D	E		A	B	C	D	E
(a)	1	3	2	4	5	(b)	2	1	4	5	3
(c)	1	2	3	4	5	(d)	5	3	2	1	4

6 Consider the following statements.

I. An acid soluble compound formed by phosphorylation of nucleoside is called nucleotide.
II. When we homogenise any tissue in an acid, the acid soluble pool represents cell membrane.

Choose the correct option.

(a) Statement I is true, but II is false
(b) Statement I is false, but II is true
(c) Both statements I and II are true
(d) Both statements I and II are false

7 Which of the following metabolite is not a primary metabolite?

(a) Protein (b) Amino acid
(c) Carbohydrates (d) Rubber

8 Find out the incorrect matched pair.

(a) Primary metabolite – Ribose
(b) Secondary metabolite – Anthocyanin
(c) Protein – Insulin
(d) Cellulose – Heteropolymer

Carbohydrates

9 The components of cane sugar are

(a) glucose and fructose (b) fructose and galactose
(c) fructose and ribose (d) glucose and glucose

10 Reducing sugar

(a) can reduce Cu^{2+} to Cu^{+}
(b) have a free keto group
(c) have a free aldehyde group
(d) All of the above

11 Non-reducing sugars have

(a) free –CHO group and bound –CO group
(b) free –CO group and bound –CHO group
(c) Both –CO and –CHO free groups
(d) Neither free –CO nor free –CHO group

12 Which one of the following is the ring structure formed by pentose?

(a) Pyranose ring (b) Furanose ring
(c) Erythrose ring (d) Both (a) and (b)

13 Paper made from plant pulp and cotton fibre are respectively,

(a) cellulose and sucrose
(b) sucrose and cellulose
(c) sucrose and sucrose
(d) cellulose and cellulose

14 Consider the following statements.

I. Fructose is the sweetest natural sugar.
II. Maltose is a disaccharide.

Choose the correct option.

(a) Statement I is true, but II is false
(b) Statement I is false, but II is true
(c) Both statements I and II are true
(d) Both statements I and II are false

15 Which of the following sugars shows the same number of carbon as present in glucose? **[NCERT Exemplar]**

(a) Fructose (b) Erythrose
(c) Ribulose (d) Ribose

16 Which of the following statements concerning polymers is incorrect?

(a) Polymer synthesised from monomers during condensation
(b) Polymers are synthesised from monomers by addition of water
(c) Polymers consist of at least two types of monomers
(d) Both (b) and (c)

17 A homopolymer has only one type of building block called monomer repeated '*n*' number of times. A heteropolymer has more than one type of monomer. Proteins are heteropolymers usually made up of **[NCERT Exemplar]**

(a) 20 types of monomers (b) 40 types of monomers
(c) 30 types of monomers (d) Only one type of monomer

18 Hyaluronic acid is a heteropolysaccharide of

(a) D-glucuronic acid and D-N-acetylglucosamine
(b) D-glucuronic acid and N-acetylmuramic acid
(c) N-acetylglucosamine and N-acetylmuramic acid
(d) Glucose and fructose

19 A mucopolysaccharide is

(a) slime, phycocolloid and pectin
(b) mucin, callose and heparin
(c) hemicellulose, pectin and mucin
(d) keratin, chondroitin sulphate and heparin

20 Consider the following statements.

I. Raffinose is a trisaccharide.
II. Cellulose is formed by union of repeated residues of glucose.

Choose the correct option.

(a) Statement I is true, but II is false
(b) Statement I is false, but II is true
(c) Both statements I and II are true
(d) Both statements I and II are false

21 Even though starch and cellulose are made up of the same repeating units of glucose, they are very different in their properties. The main difference between starch and cellulose is that

(a) cellulose has all its glucose repeats oriented in same direction
(b) starch has alternate glucose repeats oriented at 180° to each other
(c) starch has all its glucose repeats oriented perpendicular to each other
(d) cellulose has alternate glucose repeats oriented at 180° to each other

22 Identify the given structure and name the compound.

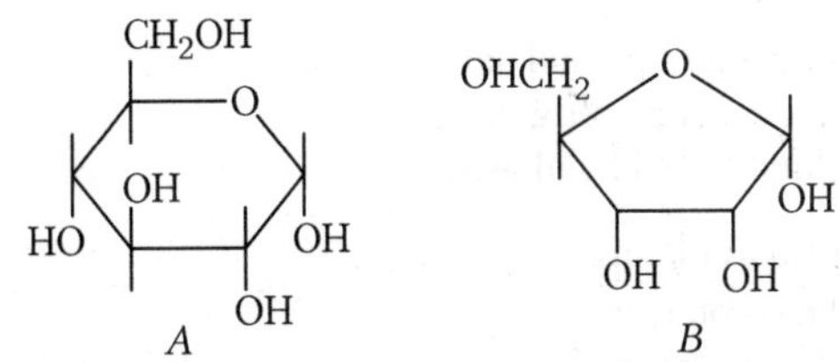

A *B*

(a) Ribose, Glucose (b) Deoxyribose, Ribose
(c) Glucose, Ribose (d) Ribose, Deoxyribose

Amino Acids, Proteins and Lipids

23 Amino acids can be classified by the

(a) number of peptide bonds
(b) number of disulphide bridges they can form
(c) number of carbon carbon double bonds
(d) characteristics of their side chains

24

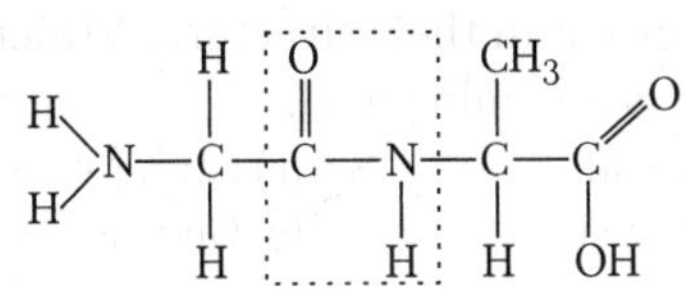

What is the structure enclosed by the above dotted box?

(a) An amino acid (b) Peptide bond
(c) Glycosidic bond (d) Zwitterion

25 A bond formed between carboxyl group of one amino acid and amino group of adjacent amino acid, is called

(a) peptide bond (b) hydrogen bond
(c) covalent bond (d) All of these

26

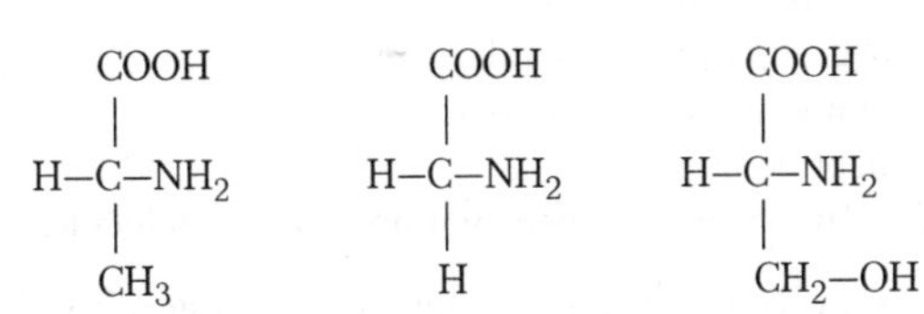

A *B* *C*

The correct names of above amino acids are

	A	*B*	*C*
(a)	Glycine	Serine	Alanine
(b)	Alanine	Glycine	Serine
(c)	Serine	Glycine	Alanine
(d)	Serine	Alanine	Glycine

27 Sulphur containing amino acids are

(a) tyrosine and tryptophan
(b) tyrosine and serine
(c) arginine or lysine
(d) cysteine and methionine

28 Among the following, acidic amino acids are
(a) glycine and alanine
(b) valine and phenylalanine
(c) glycine and methionine
(d) glutamic acid and aspartic acid

29 Amino acids are
(a) laevorotatory
(b) dextrorotatory
(c) laevorotatory except glycine, which is non-rotatory
(d) laevorotatory except glycine, which is a dextrorotatory

30 An amino acid under certain conditions have both positive and negative charges simultaneously in the same molecule. Such a form of amino acid is called **[NCERT Exemplar]**
(a) acidic form (b) basic form
(c) aromatic form (d) zwitterionic form

31 Amino acids have both an amino group and a carboxyl group in their structure. Which amongst the following is an amino acid? **[NCERT Exemplar]**
(a) Formic acid (b) Glycerol
(c) Glycolic acid (d) Glycine

32 Consider the following statements.
I. Amino acids are amphoteric in their function.
II. All amino acids are necessary for our body.
Choose the correct option.
(a) Statement I is true, but II is false
(b) Statement I is false, but II is true
(c) Both statements I and II are true
(d) Both statements I and II are false

33 The enormous diversity of protein molecules is due to the diversity of (mostly)
(a) *R*-group of amino acid
(b) amino group of amino acid
(c) peptide bonds
(d) amino acid sequences within protein molecules

34 In α-helix secondary structure, hydrogen bonds lie between amide group of one amino acid and carbonyl group of
(a) 2nd amino acid (b) 3rd amino acid
(c) 4th amino acid (d) 5th amino acid

35 The primary structure of a protein determines
(a) the sequence of amino acids
(b) the number of amino acids
(c) Both (a) and (b)
(d) None of the above

36 The shape of folded protein is often determined by
(a) its tertiary structure
(b) the sequence of its amino acids
(c) the number of peptide bonds
(d) the Chargaff's rule

37 Which of the following statements is incorrect?
(a) Quaternary structure refers to the spatial relations between individual polypeptide chains in a multichained protein
(b) The tertiary structure is absolutely necessary for many biological activities of protein
(c) Biologist describes the protein structures at three levels only
(d) Protein structure is correlated with protein function

38 A peptide chain assumes secondary structure through the formation of
(a) interchain ionic bond
(b) interchain hydrogen bond
(c) peptide chain with peptide bonds only is a secondary structure itself
(d) None of the above

39 If the subunits of protein are joined by bonds other than covalent bond then the structure is called
(a) primary structure (b) secondary structure
(c) tertiary structure (d) quaternary structure

40 During protein denaturation, which one of the following is disrupted?
(a) Peptide bond sequence of amino acids
(b) Secondary and tertiary structure
(c) Primary structure
(d) None of the above

41 The function not normally subserved by proteins is
(a) hydrolysis for energy provision
(b) structural integrity of the cell
(c) regulation of metabolism
(d) defence mechanism

42 Relation between amino acid and protein is similar to one that found in between
(a) glucose and fructose
(b) thymine and uracil
(c) nucleosides and nucleic acid
(d) nucleotides and nucleic acid

43 Proteins perform many physiological functions. For example, some functions as enzymes. One of the following represents an additional function that some proteins discharge **[NCERT Exemplar]**
(a) antibiotics
(b) pigment conferring colour to skin
(c) pigments making colour of flowers
(d) hormones

44 Which of the following can bring about the denaturation of proteins?
(a) Exposure to salts of heavy metal ions
(b) Exposure to acid and bases
(c) Exposure to temperature below –5°C
(d) All of the above

45 Give the names of the structures of proteins as shown in the figures given below.

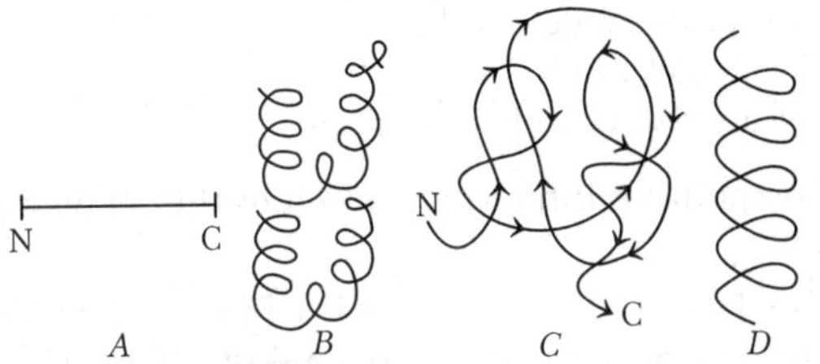

(a) $A = 1°$ structure, $B = 2°$ structure, $C = 3°$ structure, $D = 4°$ structure
(b) $A = 4°$ structure, $B = 2°$ structure, $C = 3°$ structure, $D = 1°$ structure
(c) $A = 1°$ structure, $B = 4°$ structure, $C = 3°$ structure, $D = 2°$ structure
(d) $A = 4°$ structure, $B = 3°$ structure, $C = 2°$ structure, $D = 1°$ structure

46 Fatty acids contain
(a) greater proportion of oxygen than in carbohydrates
(b) no oxygen
(c) equal oxygen in comparison to carbohydrates
(d) less oxygen than in carbohydrates

47 Which of the following is an incorrect statement?
(a) Fatty acids may be unsaturated (with one or more carbon-carbon double bonds) or a saturated (without double bonds)
(b) Fatty acids may be esterified with glycerol forming monoglyceride, diglyceride and then triglyceride
(c) Sometimes especially neural tissues have lipid in a very simple structures
(d) Fats and oils are triglycerides

48 The melting point of unsaturated fatty acids
(a) increases with increase in double bonds
(b) decreases with increase in double bonds
(c) rises in some and falls in others
(d) there is no relationship between unsaturation and melting point

49 You have isolated an unidentified liquid from a sample of beans. You add the liquid to a beaker of water and shake vigorously. After a few minutes, the water and the other liquid separate into two layers. To which class of biological molecules does the unknown liquid most likely belong?
(a) Carbohydrates
(b) Lipids
(c) Proteins
(d) Enzymes

50 Polysaccharides, polypeptides and polynucleotides have in common that they all
(a) contain amino acids
(b) are formed in condensation reactions
(c) contain nitrogen
(d) come out in acid soluble pool

51 Which of the following is made up of molecule of glycerol or other alcohol?
(a) Phospholipids (b) Lipoproteins
(c) Glycolipids (d) Chromolipids

Nucleic Acids

52 In nucleoside, nitrogen base is attached to pentose sugar at
(a) carbon-5′ of pentose sugar
(b) carbon-1′ of pentose sugar
(c) carbon-2′ of pentose sugar
(d) None of the above

53 Nucleotide constituents of RNA are
(a) adenine, guanine, cytosine, uracil
(b) adenine, guanine cytosine, thiamine
(c) thymine, cytosine, xanthine, uracil
(d) cytosine, adenine, uric acid, guanine

54 Consider the following statements.
I. The DNA molecule in eukaryotes is associated with histones and protamines.
II. The number of hydrogen bonds between adenine and thymine in DNA molecule are three.

Choose the correct option.
(a) Statement I is true, but II is false
(b) Statement I is false, but II is true
(c) Both statements I and II are true
(d) Both statements I and II are false

55 Which one is false about the diagram?

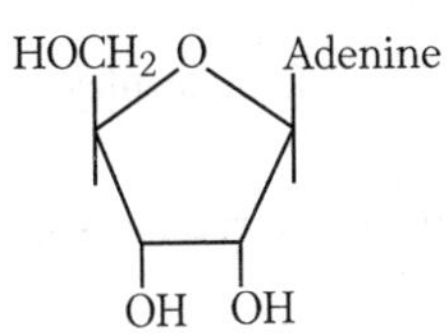

(a) It is an adenylic acid
(b) It is an adenosine
(c) It will form adenylic acid if it binds with phosphate
(d) It is a nucleoside

56 Adenosine Diphosphate (ADP) contains
(a) one high energy bond
(b) two high energy bonds
(c) three high energy bonds
(d) four high energy bonds

57 Catabolic and anabolic pathways are often coupled in cell because
(a) the intermediates of a catabolic pathway are used in the anabolic pathway
(b) Both the pathways use the same energy
(c) the free energy released from one pathway is used to drive other
(d) their enzymes are controlled by their same activators and inhibitors

58 Cell continues to function only when a metabolic equilibrium is not in reached. How do cells avoid reaching metabolic equilibrium?
(a) Providing constant supply of enzymes
(b) Cellular metabolism utilises only those reactions that are irreversible
(c) Use feedback inhibition to turn off pathways
(d) The products of one reaction become the reactant of another reaction and are unable to accumulate

Enzymes

59 What will happen to an enzyme when apoenzyme is separated from co-factor?
(a) Activity will be increased
(b) Activity will be lost
(c) Activity will be decreased
(d) There will be no change in the activity

60 Which of the following are co-enzymes?
(a) NAD, NADP, FAD, FMN (b) Vitamin, Fe, Cu
(c) $NADPH_2$, Ca, Co (d) NAD, K, Co-A

61 The function of an enzyme is to
(a) cause chemical reactions, which would not occur otherwise
(b) change the rates of chemical reactions
(c) control the equilibrium points of reactions
(d) change the direction of reactions

62 An enzyme increases the rate of a reaction by
(a) supplying the energy required to start the reaction
(b) increasing the rate of random collisions of molecules
(c) removing the product of the reaction so allowing it to continue
(d) bringing the reacting molecules into precise orientation with each other

63 Catalytic activity of enzymes is influenced by
(a) temperature
(b) pH and enzyme positions
(c) concentration, substrate and cofactors
(d) All of the above

64 At which temperature, the enzyme activity would be maximum?
(a) 20-40° C (b) 40-45° C
(c) 40-60° C (d) 60° C and above

65 Which of the following reactions is not enzyme mediated in biological system?
(a) Dissolving CO_2 in water
(b) Unwinding the two strands of DNA
(c) Hydrolysis of sucrose
(d) Formation of peptide bond

66 Most intracellular enzymes function best
(a) at neutral pH
(b) in acidic conditions
(c) in basic conditions
(d) either neutral or acidic conditions

67 The enzymatic reaction for which thiamine pyrophosphate functions as a co-factor is
(a) peptide bond formation
(b) phosphate group transfer
(c) fixation of carbon dioxide
(d) decarboxylation of a keto acids

68 In competitive inhibition, which of the following is true?
(a) $E + I \Leftrightarrow EI$ (b) $E + I \Leftrightarrow EI + S \Leftrightarrow EIS$
(c) $S + I \Leftrightarrow SI$ (d) $ES + I \Leftrightarrow ESI$

69 Which statement is true for all enzymes?
(a) They are denatured at temperatures above 60°C
(b) They catalyse the breakdown of large molecules into smaller ones
(c) They have active sites, which can bind to only one kind of substrate molecule
(d) They reduce the amount of energy required to start a reaction

70 Which of these inactivates an enzyme by changing enzyme shape?
(a) Allosteric inhibitor (b) Competitive inhibitor
(c) Irreversible inhibitor (d) Multienzyme complex

71 Decline in the activity of the enzyme hexokinase by glucose-6-phosphate is caused by
(a) non-competitive (b) competitive inhibitions
(c) allosteric modulator (d) denaturation of enzymes

72 Which of the following statement is incorrect?
(a) Enzymes hasten the completion of a reaction
(b) The two terms substrate and product signify the starting and ending materials of a reaction
(c) Enzymes are affected by the reactions they catalyse
(d) Enzymes exhibit specificity for the reactions they catalyse

73 The 'lock and key' model of enzyme action illustrates that a particular enzyme molecule
(a) may be destroyed and resynthesised several times
(b) interacts with a specific type of substrate molecule
(c) reacts at identical rates under all conditions
(d) forms a permanent enzyme-substrate complex

74 Which model of an enzyme action is seen in the given figure?

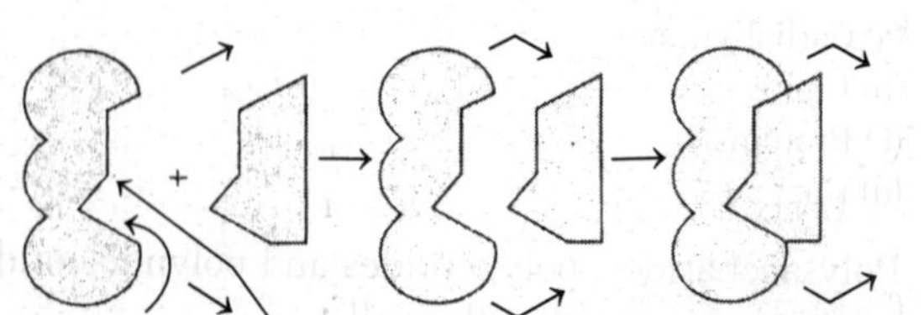

(a) Lock and key model (b) Induced fit model
(c) Enzyme-substrate model (d) None of these

75 An example of feedback inhibition is
(a) cyanide action on cytochrome
(b) sulpha drug on folic acid synthesiser bacteria
(c) allosteric inhibition of hexokinase by glucose- 6-phosphate
(d) reaction between succinic dehydrogenase and succinate

76 With reference to enzymes, turnover number means
(a) the number of substrate molecules that a molecule of an enzyme converts into products per hour
(b) the number of substrate molecules that a molecule of an enzyme converts into products per second
(c) the number of substrate molecules that a molecule of an enzyme convert into products per minute
(d) the number of substrate molecules that a molecule of an enzyme converts into products per day

77 A competitive enzyme inhibitor
(a) increases K_m without affecting V_{max}
(b) decreases K_m without affecting V_{max}
(c) increases V_{max} without affecting K_m
(d) decreases both V_{max} and K_m

78 The enzymes
I. are protein with catalytic activities.
II. are colloidal in nature.
III. working within the cell are called simple enzymes.
IV. increase the activation energy of the reactions they catalyse.

Choose the correct option from the codes given below.
(a) Only I (b) Only III (c) I and II (d) I and III

79 Match the following columns.

	Column I (Enzymes)		Column II (Characteristics)
A.	Dehydrogenases	1.	Interconversion of optical, geometrical, positional isomers
B.	Ligases	2.	Group transfer
C.	Isomerases	3.	Oxidoreduction between two substrates
D.	Hydrolases	4.	Linking together of two bonds
E.	Transferases	5.	Hydrolysis of bonds

Codes

	A	B	C	D	E
(a)	5	4	1	2	3
(b)	4	3	5	2	1
(c)	5	4	2	3	1
(d)	3	4	1	5	2

80 Which statement regarding co-enzyme is incorrect?
(a) Every co-enzyme is a co-factor and every co-factor is a co-enzyme
(b) Every coenzyme is a co-factor, but every co-factor is not a co-enzyme
(c) Most of the co-enzymes are nucleotides and are composed of vitamins
(d) Co-enzymes are the active constituents of enzymes

81 Match the following columns.

Column I (Categories)	Column II (Features)
A. Prosthetic group	1. Coordination bond with apoenzyme
B. Co-enzyme	2. Tightly bound to apoenzyme
C. Metal ions co-factor	3. Transient binding with apoenzyme

Codes

	A	B	C
(a)	2	3	1
(b)	3	2	1
(c)	2	1	3
(d)	1	2	3

82 Sulpha drugs are used for the control of bacterial pathogens, because they cause
(a) competitive inhibition of folic acid synthesis
(b) allosteric inhibition of folic acid synthesis
(c) feedback inhibition of folic acid synthesis
(d) irreversible inhibition of folic acid synthesis

83 Coenzymes
I. are needed for the function of particular enzymes.
II. are inorganic molecules.
III. are organic molecules
IV. FAD and FMN contain niacin, while NAD and NADP contain riboflavin.

Choose the correct option from the codes given below.
(a) Only I (b) II and IV
(c) Only IV (d) I and III

84 K_m value (Michaelis-Menten constant) refers to
(a) maximum reaction velocity
(b) near maximum reaction velocity
(c) one half of the maximum reaction velocity
(d) threshold value

85 Which one value is required for better enzymatic action?
(a) High K_i (b) Low K_i
(c) Low K_m (d) High K_m

86 On binding the substrate at one site, other sites on an enzyme become more reactive. This is called
(a) allosteric inhibition (b) specificity
(c) cooperativity (d) activation

87 Feedback inhibition quite often involves
(a) competitive inhibition (b) irreversible inhibition
(c) allosteric inhibition (d) All of these

88 Which of the following acts as precursor to form the active enzyme?
(a) Aspartate (b) Kinase
(c) Glycogen (d) Zymogen

(B) Cumulative Exercise

(Assorted Questions of the Chapter for Advanced Level Practice)

1 Which of the following statements are correct?

I. Monosaccharides are the simplest carbohydrates.
II. Oligosaccharides on hydrolysis, yield three to nine monosaccharide units.
III. Polysaccharides on hydrolysis, give 10-15 monosaccharide units.
IV. Polysaccharides on hydrolysis, yield many oligosaccharide units.

Choose the correct option from the codes given below.

(a) Only I (b) Only III (c) I and II (d) I and III

2 Match the following columns.

Column I	Column II
A. Cellulose	1. Insulin
B. Peptide	2. Alkaline phosphatase
C. Steroid	3. Cotton fibres
D. Phospholipid	4. Diosgenin
E. Enzyme	5. Lecithin

Codes

	A	B	C	D	E		A	B	C	D	E
(a)	3	4	5	1	2	(b)	2	1	4	5	3
(c)	3	1	4	5	2	(d)	1	4	3	5	2

3 It is said that elemental composition of living organisms and that of inanimate objects (like earth's crust) are similar in the sense that all the major elements are present in both. Then what would be the difference between these two groups? Choose a correct answer from among the following. **[NCERT Exemplar]**

(a) Living organisms have more gold in them than inanimate objects
(b) Living organisms have more water in their body than inanimate objects
(c) Living organisms have more carbon, oxygen and hydrogen per unit mass than inanimate objects
(d) Living organisms have more calcium in them than inanimate objects

4 The equation shows how the enzyme glutamine synthetase removes the ammonia produced during plant metabolism.

$$\text{Ammonia} + \text{Glutamate} \xrightarrow[\text{synthetase}]{\text{Glutamine}} \text{Glutamine}$$

Some herbicides contain an active agent, which resembles glutamate. What is the likely mode of action of this agent?

(a) It acts as an end product inhibitor
(b) It acts as a competitive inhibitor
(c) It decreases levels of ammonia
(d) It increases levels of glutamate

5 Match the following columns.

Column I	Column II
A. Triglycerides	1. Galactose
B. Lactose	2. Glycerol
C. RNA	3. Palmitic acid
D. β-pleated	4. Uracil
E. Beeswax	5. Secondary structure

Codes

	A	B	C	D	E		A	B	C	D	E
(a)	4	1	5	2	3	(b)	5	1	4	2	3
(c)	3	1	4	5	2	(d)	2	1	4	5	3

6 Which of the following statements is/are not true?

I. Glycerol is a 3 carbon alcohol with 3—OH groups that serve as binding sites.
II. Waxes are esters formed between a long chain alcohol and saturated fatty acids.
III. The term protein was coined by Gerardus Johannes Mulder.
IV. Agar is an indispensable polysaccharide and it is a complex polymer of glucose and sulphur containing carbohydrates.

Choose the correct option from the codes given below.

(a) I and III (b) I and IV
(c) I, II and IV (d) Only IV

7 Select the incorrect statement.

(a) The building blocks of lipids are amino acids
(b) Majority of enzymes contain a non-protein part called the prosthetic group
(c) Glycogen is a homopolymer made up of glucose unit
(d) ATP is a nucleotide

8 Find out the mismatched pair.

(a) Chitin—Polymer of glucosamine
(b) Peptidoglycan—Polysaccharide linked to peptides
(c) Lipopolysaccharides—A complex of lipid and polysaccharide
(d) Glycogen—Polymer of fructose

9 Arrange the steps of catalytic action of an enzyme in order and choose the correct option.

I. The enzyme releases the products of the reaction and the enzyme is free for another substrate.
II. The active site of enzyme is in close proximity of the substrate and breaks of chemical bonds of the substrate.
III. The binding of substrate induces the enzyme to alter its shape fitting more tightly around the substrate.
IV. The substrate binds to the active site of the enzyme fitting into the active site.

Codes

(a) IV, III, II, I (b) III, II, I, IV
(c) IV, II, I, III (d) II, I, IV, III

10 In which one of the following sets of three items each belong to the category mentioned against them?

(a) Lysine, glycine and thiamine — Amino acids
(b) Myosin, oxytocin and gastric — Hormones
(c) Rennin, helicase and hyaluronidase — Enzymes
(d) Optic nerve, oculomotor and vagus — Sensory nerves

11 Refer to the given figure of various levels of protein structure.

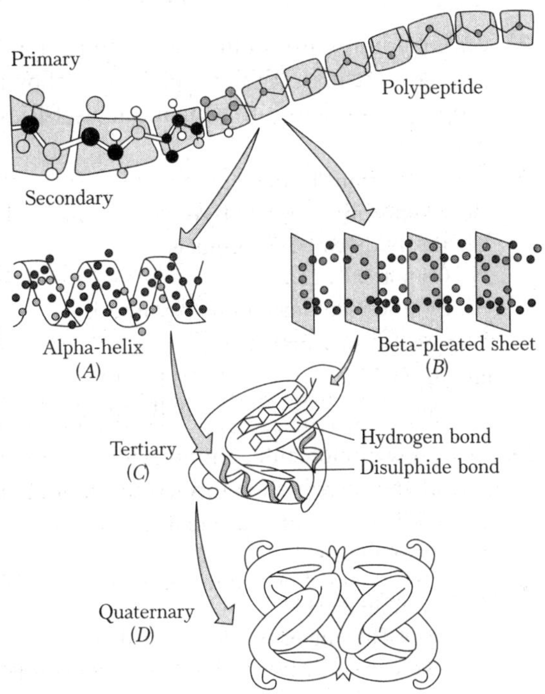

Formation of structures *A*, *B*, *C* and *D* is through

	A	*B*	*C*	*D*
(a)	H-bonding in single amino acid chain only	H-bonding in between two or more polypeptide chains	Further coiling in α-helix only	Both α-helix and β-sheet joined and coiled together
(b)	H-bonding in between two or more polypeptide chains	H-bonding in single amino acid chain only	Further coiling in α-helix only	Both α-helix and β- sheet joined and coiled together
(c)	H-bonding in between two or more polypeptide chains	H-bonding in single amino acid chain only	Both α-helix and β-sheet joined and coiled together	Further coiling in α-helix only
(d)	H-bonding in single amino acid chain only	H-bonding in between two or more polypeptide chains	Further folding of two or more secondary structures	Further folding of a number of tertiary structures

12 Based on the number of amino and carboxyl groups, amino acids are classified. Choose the correct option.

	Acidic amino acids	Basic amino acids	Neutral amino acids
(a)	Glutamic acid	Lysine	Valine
(b)	Lysine	Valine	Glutamic acid
(c)	Glutamic acid	Valine	Lysine
(d)	Lysine	Glutamic acid	Valine

13 Match the following columns.

Column I	Column II
A. Collagen	1. Binding with some chemical like hormones
B. Trypsin	2. Hormones
C. Insulin	3. Enzymes
D. Antibody	4. Intercellular ground substance
E. Receptor	5. Fight infections agents

Codes

	A	B	C	D	E
(a)	4	3	2	5	1
(b)	1	2	3	4	5
(c)	5	4	3	2	1
(d)	3	4	1	5	2

14 Which of the following statements are correct?

I. Acetic acid can form cholesterol.
II. Anabolic pathway is endergonic, while catabolic pathway is exergonic.
III. All biomolecules have a turnover, i.e. they are constantly being changed into some other biomolecules and also made from other biomolecules.
IV. Flow of metabolites through metabolic pathway has a definite rate and direction. It is called dynamic state of body constituents.

Choose the correct option from the codes given below.

(a) I and II (b) III and IV
(c) I, III and IV (d) I, II, III and IV

15 Choose the correct statements.

I. Living steady state has a self-regulatory mechanism called homeostasis.
II. Energy flow and energy transformation of living system follow law of thermodynamics.
III. Metabolism is release or gain of energy.
IV. ATP powers the cellular work by complying exergonic reaction to endergonic reactions.

Choose the correct option from the codes given below.

(a) I and IV
(b) II and III
(c) II and IV
(d) I, II, III and IV

16 Of the two chemical reactions showing the graph of reaction A is

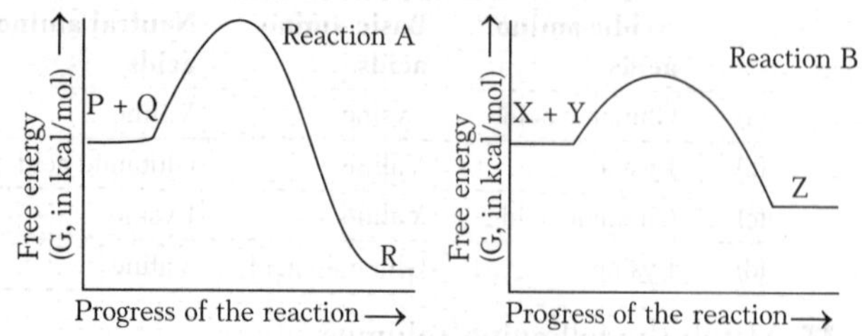

(a) Faster and more endergonic than 2
(b) Faster and more exergonic than 2
(c) Slower and more endergonic than 2
(d) Slower and more exergonic than 2

17 Which of the following statements are correct for activation energy of a reaction?

I. It is the difference in the average energy content of 'S' from that of its transition state.
II. It helps to change reactants into unstable transition state before they can be converted into products.
III. It is the minimum energy required from outside to overcome the energy barrier of reactant or to start a reaction.

Choose the correct option from the codes given below.

(a) Only I (b) Only II
(c) II and III (d) I, II and III

18 Dihydroxyacetone phosphate and glyceraldehyde-3-phosphate are reversibly interconvertible

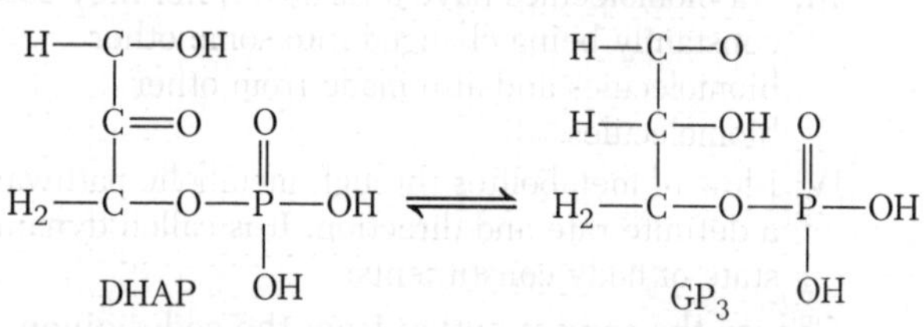

The enzyme that catalyses the above reaction is

(a) Aldolase (b) Isomerase
(c) Phosphoglycerokinase (d) Phosphoglyceromutase

19 Select the false statements for an enzyme promoting a chemical reaction by

I. lowering the energy of activation.
II. causing the release of heat, which acts as a primer.
III. increasing molecular motion.
IV. changing the free energy difference between substrate and product.

Choose the correct option from the codes given below.

(a) I and IV (b) II and III
(c) II, III and IV (d) III and IV

20 Match the Column I with Column II.

Column I (Categories)	Column II (Secondary metabolites)
A. Pigments	1. Concanavalin-A
B. Terpenoids	2. Monoterpenes, diterpenes
C. Alkaloids	3. Morphine, codeine
D. Lectins	4. Carotenoids, anthocyanin

Codes

	A	B	C	D		A	B	C	D
(a)	4	2	3	1	(b)	4	3	2	1
(c)	1	4	3	2	(d)	1	3	2	4

21 Carbohydrates are commonly found as starch in plants storage organs. Which of the following given properties of starch (I-V) make it useful as a storage material?

I. Easily translocated
II. Chemically non-reactive
III. Easily digested by animals
IV. Osmotically inactive
V. Synthesised during photosynthesis

The useful properties are

(a) I, III and V (b) I and V
(c) II and III (d) II and IV

22 In a protein molecule, amino acids are linked by a peptide bond, which is formed by the reaction of

(a) –COOH group of one amino acid with – NH_2 group of next amino acid
(b) –NH_2 group of one amino acid with – COOH group of next amino acid
(c) –COOH group of two amino acids
(d) –NH_2 group of two amino acids

23 Which one of the following is incorrectly matched?

(a) Fungi – Chitin
(b) Phospholipid – Plasma membrane
(c) Enzyme – Lipopolysaccharide
(d) ATP – Nucleotide derivative

Assertion and Reason

Direction (Q. Nos. 24-30) *In each of the following questions, a statement of Assertion (A) is given followed by the corresponding statement of Reason (R). Of the statements, mark the correct answer as*

(a) Both A and R are true and R is the correct explanation of A
(b) Both A and R are true, but R is not the correct explanation of A
(c) A is true, but R is false
(d) A is false, but R is true
(e) Both A and R are false

24 **Assertion** (A) Most of the enzymes are proteins, which catalyse biochemical reactions.
Reason (R) The enzyme itself remains unchanged in the reaction, its presence allows the reaction to takes place.

25 **Assertion** (A) Enzymes lower the activation energy of the reaction.
Reason (R) Higher activation energy helps the molecules to react with greater rate.

26 **Assertion** (A) Lactose is a non-reducing sugar.
Reason (R) Sucrose shows α-1, 2-glycosidic linkage.

27 **Assertion** (A) Protein forming amino acids are α-amino acids.
Reason (R) In all proteins forming amino acids, $-NH_2$ group is attached to second C-atom after $-COOH$ group.

28 **Assertion** (A) Sucrose is a non-reducing sugar.
Reason (R) It does not have free $-CHO$ group.

29 **Assertion** (A) Arachidic acid is an unsaturated fatty acid.
Reason (R) It has one or more double bonds between carbon atoms and unsaturated fatty acids.

30 **Assertion** (A) A co-enzyme or metal ion that is very tightly bound to enzyme protein is called prosthetic group.
Reason (R) A complete catalytically active enzyme together with its bound prosthetic group is called apoenzyme.

(C) Medical Entrances' Gallery

(Collection of questions Asked in NEET & Various Medical Entrance Exams)

1. Identify the statement which is incorrect? **[NEET 2020]**
(a) Sulphur is an integral part of cysteine
(b) Glycine is an example of lipids
(c) Lecithin contains phosphorus atom in its structure
(d) Tyrosine possesses aromatic ring in its structure

2. Identify the basic amino acid from the following. **[NEET 2020]**
(a) Glutamic acid (b) Lysine
(c) Valine (d) Tyrosine

3. Match the following columns.

	Column I		Column II
A.	Inhibitor of catalytic activity	1.	Ricin
B.	Possess peptide bonds	2.	Malonate
C.	Cell wall material in fungi	3.	Chitin
D.	Secondary metabolite	4.	Collagen

Choose the correct option from the codes given below. **[NEET 2020]**

	A	B	C	D		A	B	C	D
(a)	3	1	4	2	(b)	3	4	1	2
(c)	2	3	1	4	(d)	2	4	3	1

4. Which of the following statements is correct? **[NEET 2020]**
(a) Adenine pairs with thymine through one H-bond
(b) Adenine pairs with thymine through three H-bonds
(c) Adenine does not pair with thymine
(d) Adenine pairs with thymine through two H-bonds

5. Secondary metabolites such as nicotine, strychnine and caffeine are produced by plants for their **[NEET 2020]**
(a) growth response (b) defence action
(c) effect on reproduction (d) nutritive value

6. Identify the substance having glycosidic bond and peptide bond, respectively in their structure. **[NEET 2020]**
(a) Glycerol, trypsin (b) Cellulose, lecithin
(c) Inulin, insulin (d) Chitin, cholesterol

7. Which one of the following is the most abundant protein in the animals? **[NEET 2020]**
(a) Collagen (b) Lectin
(c) Insulin (d) Haemoglobin

8. In the polynucleotide chain of DNA, a nitrogenous base is linked to the $-OH$ of **[NEET 2020]**
(a) 2′C pentose sugar (b) 3′ C pentose sugar
(c) 5′ C pentose sugar (d) 1′ C pentose sugar

9 Glucose on reacting with Benedict's solution may give the following precipitates except **[AIIMS 2019]**
(a) violet precipitate (b) orange-red precipitate
(c) brick-red precipitate (d) green-yellow precipitate

10 Which of the following linkage is found in sucrose? **[AIIMS 2019]**
(a) 1,2-glycosidic linkage (b) 1,4-glycosidic linkage
(c) 1,3-glycosidic linkage (d) 1,1-glycosidic linkage

11 Among the following, choose the correct epimers. **[JIPMER 2019]**
(a) Glucose galactose (b) Glucose, fructose
(c) Glucose, ribose (d) Gluconic acid, glucose

12 Which of the following organic compounds is the main constituent of lecithin? **[NEET 2019]**
(a) Arachidonic acid (b) Phospholipid
(c) Cholesterol (d) Phosphoprotein

13 Which of the following are all nucleotides? [AIIMS 2019]

(a) Adenosine, Cytidylic acid, Cytosine
(b) Adenylic acid, Cytidylic acid, Guanylic acid
(c) Cytidine, Adenine, Adenylic acid
(d) Uracil, Thymidine, Thymidylic acid

14 Consider the following statements.

I. Co-enzyme or metal ion that is tightly bound to enzyme protein is called prosthetic group.
II. A complete catalytic active enzyme with its bound prosthetic group is called apoenzyme.

Choose the correct option. [NEET 2019]

(a) Statement I is true, but II is false
(b) Statement II is false, but I is true
(c) Both statements I and II are true
(d) Both statements I and II are false

15 Prosthetic groups differ from co-enzymes in that [NEET 2019]

(a) they require metal ions for their activity
(b) they (prosthetic groups) are tightly bound to apoenzymes
(c) their association with apoenzymes is transient
(d) they can serve as cofactors in a number of enzyme catalysed reactions

16 Citrulline structure is [JIMPER 2019]

(a) 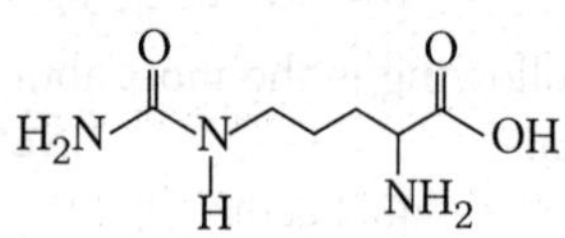

(b) 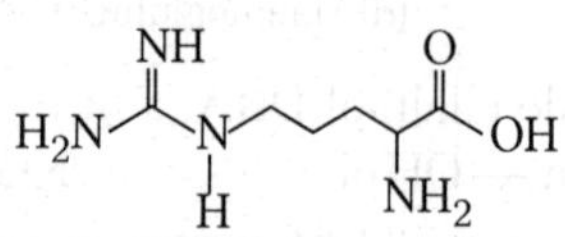

(c) 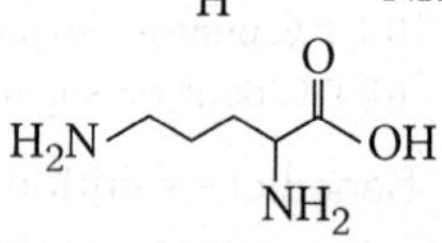

(d) 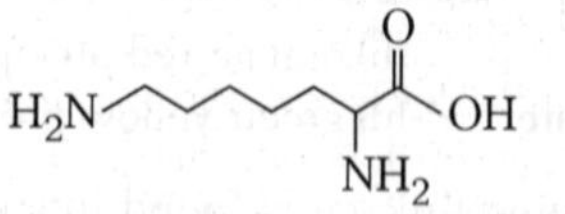

17 Concanavalin-A is [NEET 2019]

(a) an essential oil
(b) a lectin
(c) a pigment
(d) an alkaloid

18 'Ramachandran plot' is used to confirm the structure of [NEET 2019]

(a) RNA
(b) proteins
(c) triacylglycerides
(d) DNA

19 Which two functional groups are characteristic of sugars? [NEET 2018]

(a) Carbonyl and phosphate
(b) Carbonyl and methyl
(c) Hydroxyl and methyl
(d) Carbonyl and hydroxyl

20 Which one of the following is not a derivative of cholesterol? [JIPMER 2018]

(a) Vitamin-B
(b) Vitamin-D
(c) Bile salts
(d) Steroid

21 Omega-3 fatty acid is present in [JIPMER 2018]

(a) sunflower oil
(b) flax seed oil
(c) groundnut oil
(d) butter

22 Non-essential amino acid is [JIPMER 2018]

(a) valine (b) arginine (c) histidine (d) lysine

23 Proteins are needed in diet because [JIPMER 2018]

(a) all amino acids are not available in body
(b) during fasting, body utilises proteins
(c) proteins act as building blocks of our body
(d) All of the above

24 Which of the following are not polymeric? [NEET 2017]

(a) Nucleic acids
(b) Proteins
(c) Polysaccharides
(d) Lipids

25 **Assertion** (A) The structure given is the most important animal steroid which is insoluble in water and chemically unreactive.

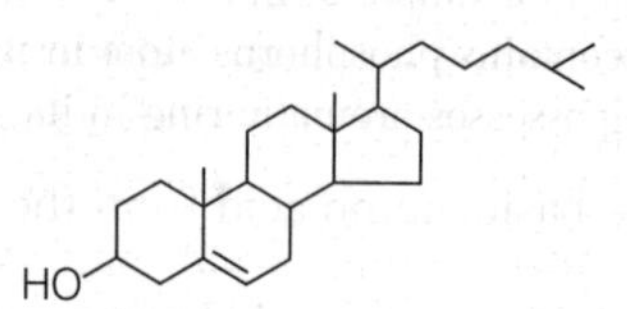

Reason (R) It is important because it is a structural component of cells. [AIIMS 2017]

(a) Both A and R are true and R is the correct explanation of A
(b) Both A and R are true, but R is not the correct explanation of A
(c) A is true, but R is false
(d) Both A and R are false

26 **Assertion** (A) The structure given below contains 1,-4 α-glycosidic bonds.

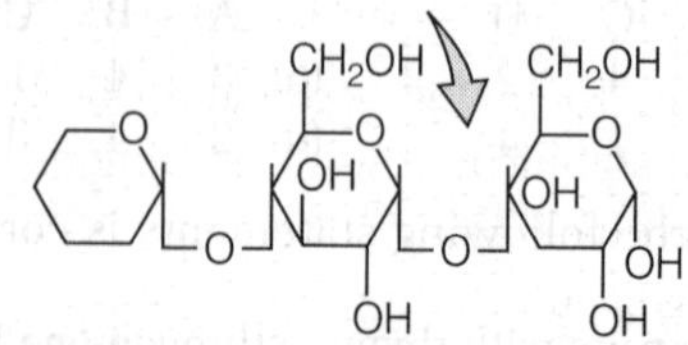

Reason (R) This is a polysaccharide and have right end as reducing end and its left end is called the non-reducing end. [AIIMS 2017]

(a) Both A and R are true and R is the correct explanation of A
(b) Both A and R are true, but R is not the correct explanation of A
(c) A is true, but R is false
(d) Both A and R are false

27 Macromolecule chitin is [JIPMER 2017]
(a) nitrogen containing polysaccharide
(b) phosphorus containing polysaccharide
(c) sulphur containing polysaccharide
(d) simple polysaccharide

28 Which one of the following statements is correct with reference to enzymes? [NEET 2017]
(a) Apoenzyme = Holoenzyme + Co-enzyme
(b) Holoenzyme = Apoenzyme + Co-enzyme
(c) Co-enzyme = Apoenzyme + Holoenzyme
(d) Holoenzyme = Co-enzyme + Co-factor

29 **Assertion** (A) Carbohydrates are more suitable for the production of energy in the body than proteins and fats.
Reason (R) They can be stored in the tissues as glycogen and can be used for the production of energy, whenever necessary. [AIIMS 2016]
(a) Both A and R are true and R is the correct explanation of A
(b) Both A and R are true, but R is not the correct explanation of A
(c) A is true, but R is false
(d) Both A and R are false

30 Which of the following is/are cellulosic?
I. Paper II. Cotton fibre
III. Chitin IV. Glycogen
Choose the correct option from the codes given below. [Kerala CEE 2016]
(a) Only I and II
(b) Only I and III
(c) Only I, III and IV
(d) Only III and IV
(e) Only III

31 A typical fat molecule (triglyceride) is made up of [NEET 2016]
(a) one glycerol and three fatty acid molecules
(b) one glycerol and one fatty acid molecule
(c) three glycerol molecules and three fatty acid molecules
(d) three glycerol molecules and one fatty acid molecule

32 Which of the following is the least likely to be involved in stabilising the three-dimensional folding of most proteins? [NEET 2016]
(a) Hydrogen bonds
(b) Electrostatic interaction
(c) Hydrophobic interaction
(d) Ester bonds

33 Which one of the following statements is wrong? [NEET 2016]
(a) Cellulose is a polysaccharide
(b) Uracil is a pyrimidine
(c) Glycine is a sulphur containing amino acid
(d) Sucrose is a disaccharide

34 Which of the following statements about amino acids is false? [Kerala CEE 2016]
(a) Based on the nature of the carboxyl group, there are many amino acids
(b) Amino acids are substituted methane
(c) Amino acids have an amino group and acidic group as substituents on the α-carbon
(d) There are four substituent groups occupying the four valency positions
(e) Tryptophan is an aromatic amino acid

35 The enzyme, which catalyses the formation of glutamine from its substrate, belongs to this category of enzymes. [NEET-II 2016]
(a) Ligases (b) Lyases
(c) Hydrolases (d) Transferases

36 Identify the given structure [AIIMS 2016]

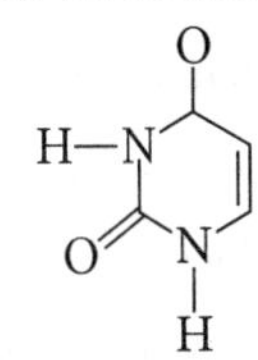

(a) Adenylic acid (b) Uracil
(c) Cholesterol (d) Adenosine

37 A non-proteinaceous enzyme is [NEET 2016]
(a) lysozyme (b) ribozyme
(c) ligase (d) deoxyribonuclease

38 Which one of the following describes the given graph correctly? [NEET 2016]

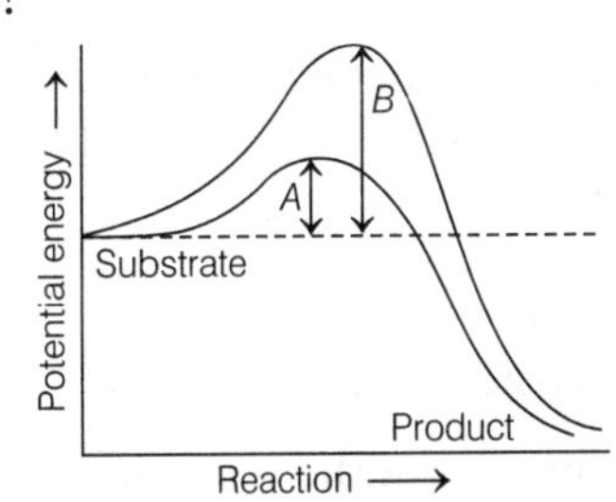

(a) Endothermic reaction with energy A in the presence of enzyme and B in the absence of enzyme
(b) Exothermic reaction with energy A in the presence of enzyme and B in the absence of enzyme
(c) Endothermic reaction with energy A in the absence of enzyme and B in the presence of enzyme
(d) Exothermic reaction with energy A in the absence of enzyme and B in the presence of enzyme

39 What is the role of competitive inhibitor during enzyme action? [Karnataka CET 2016]
(a) It alters the active site of the enzyme and prevents the binding of substrate
(b) It enhances enzyme action
(c) It inhibits breaking of chemical bonds of the substrate
(d) It declines the enzyme action

40 Which one of the following statements is incorrect? [CBSE AIPMT 2015]
(a) A competitive inhibitor reacts reversibly with the enzyme to form and enzyme-inhibitor
(b) In competitive inhibition, the inhibitor molecule is not chemically changed by the enzyme
(c) The competitive inhibitor does not affect the rate of breakdown of the enzyme-substrate complex
(d) The presence of the competitive inhibitor decreases the K_m of the enzyme for the substrate

41 Nomenclature of enzymes is done on the basis of [COMEDK 2015]
(a) the mechanism of enzyme action
(b) the type of reactions they catalyse
(c) the end products formed
(d) the co-factors they are associated with

42 The catalytic efficiency of two different enzymes can be compared by the [J&K CET 2015]
(a) molecular size of the enzymes
(b) pH optimum values
(c) $_m$ values
(d) formation of the product

43 Which of the following statements about the structure of proteins is true? [Kerala CEE 2015]
(a) The helices of proteins are always left-handed
(b) Adult human haemoglobin consists of two subunits
(c) The left end of a protein is called the C-terminal and the right end is called the N-terminal
(d) Proteins are heteropolymers containing strings of amino acids
(e) None of the above

44 Which of the following biomolecules does have a phosphodiester bond? [CBSE AIPMT 2015]
(a) Amino acids in a polypeptide
(b) Nucleic acids in a nucleotide
(c) Fatty acids in a diglyceride
(d) Monosaccharides in a polysaccharide

45 The moiety present at the 5′ end of ribose sugar in a polynucleotide is [Kerala CEE 2015]
(a) OH (b) CH_2
(c) phosphate (d) adenine
(e) 5′ methyluracil

46 In the ribose of RNA, unlike DNA, every nucleotide residue has an additional [Kerala CEE 2015]
(a) COOH group in the 2′ position
(b) OH group in the 5′ position
(c) OH group in the 2′ position
(d) phosphate group in the 2′ position
(e) None of the above

47 Which one of the following is a non-reducing carbohydrate? [CBSE-AIPMT 2014]
(a) Maltose (b) Sucrose
(c) Lactose (d) Ribose-5-phosphate

48 Cellulose, the most important constituent of plant cell wall is made up of [AIIMS 2014]
(a) branched chain of glucose molecules linked by α, 1-6 glycosidic bond at the site of branching
(b) unbranched chain of glucose molecules linked by α, 1-4 glycosidic bond
(c) branched chain of glucose molecules linked by β, 1-4 glycosidic bond in straight chain and α, 1-6 glycosidic bond at the site of branching
(d) unbranched chain of glucose molecules linked by β, 1-4 glycosidic bond

49 Which one of the following statements is incorrect for sucrose? [WB JEE 2014]
(a) It is a disaccharide
(b) It is a non-reducing sugar
(c) It accumulates in the cytoplasm
(d) It is comprised of maltose and fructose

50 The two structural formulae shown below are

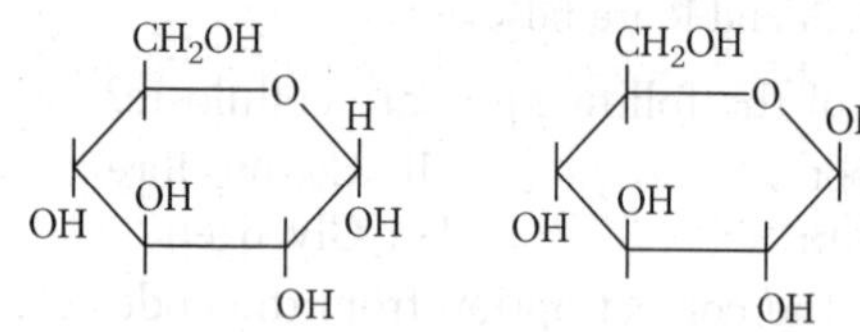

[AMU 2014]
(a) isomers (b) epimers
(c) anomers (d) All of these

51 Which of the following scientists discovered the triple helical structure of collagen? [Kerala CEE 2014]
(a) GN Ramachandran (b) Antonie van Leeuwenhoek
(c) Mathias Schleiden (d) Theodor Schwann
(e) Rudolf Virchow

52 Match the following columns and choose the correct option from the codes given below.

Column I (Chemical compounds)		Column II (Examples)
A. Nitrogen base	1.	RNA
B. Nucleoside	2.	Thymidylic acid
C. Nucleotide	3.	Cytidine
D. Nucleic acid	4.	Uracil

Codes [Kerala CEE 2014]

	A	B	C	D		A	B	C	D
(a)	1	2	3	4	(b)	1	3	2	4
(c)	4	3	2	1	(d)	4	1	2	3
(e)	4	2	3	1					

53 Select the option which is not correct with respect to enzyme action. **[CBSE AIPMT 2014]**

(a) Substrate binds with enzyme at its active site
(b) Addition of lot of succinate does not reverse the inhibition of succinic dehydrogenase by malonate
(c) A non-competitive inhibitor binds the enzyme at a site distinct from that which binds the substrate
(d) Malonate is a competitive inhibitor of succinic dehydrogenase

54 The protein component of a holoenzyme is known as

(a) co-enzyme (b) co-factor
(c) prosthetic group (d) apoenzyme

55 K_m is **[WB JEE 2014]**

(a) product (b) enzyme (c) constant (d) unit

56 Which of the following statements regarding enzymes is not true? **[AMU 2014]**

(a) Enzymes lower the activation energy of a reaction
(b) Enzymes speed up the attainment of equilibrium of the reaction
(c) Enzyme changes the value of equilibrium constant of the reaction
(d) All of the above

57 Which of the following statements regarding fats is true? **[Kerala CEE 2014]**

(a) Arachidonic acid has 20 carbons excluding the carboxyl cosbon
(b) Glycerol is trihydroxypropane
(c) Palmitic acid has 18 carbons including the carboxyl carbon
(d) Oils have higher melting points than fats
(e) Lipids are generally water soluble

58 Which of the following statements is not correct? **[AMU 2013]**

(a) Starch is a polymer of α-glucose
(b) Starch is made up of amylose and amylopectin
(c) Amylose is linear structure consisting of several glucose residues joined by 1, 4- glycosidic linkages
(d) Amylopectin is a straight chain with several glucose residues joined only by 1, 4-glycosidic linkages

59 The figure shows a hypothetical tetrapeptide portion of a protein with parts labelled *A-D*. Which one of the following options is correct?

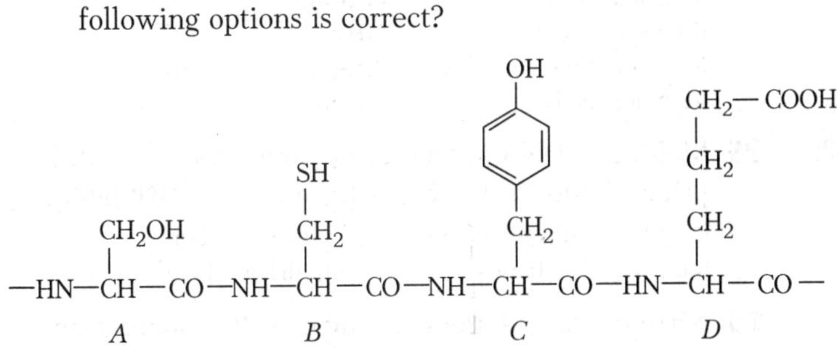

[NEET 2013]

(a) *D* is not an amino acid
(b) *C* is an aromatic amino acid, tryptophan
(c) *A* is the C-terminal amino acid and *D* is N-terminal amino acid
(d) *A* is a sulphur containing amino acid methionine

60 A phosphoglyceride is always made up of **[NEET 2013]**

(a) only a saturated fatty acid esterified to a glycerol molecule to which a phosphate group is also attached
(b) only an unsaturated fatty acid esterified to a glycerol molecule to which a phosphate group is also attached
(c) a glycerol molecule to which a phosphate group is also attached
(d) a saturated or unsaturated fatty acid esterified to a phosphate group, which is also attached to a glycerol molecule

61 Lipids, which can be found in oil based salad dressings and ice cream, during digestion are splitted into **[AIIMS, AFMC 2013]**

(a) fatty acids and glycerol
(b) glycerol and amino acids
(c) glucose and fatty acids
(d) glucose and amino acids

62 Transition state structure of the substrate formed during an enzymatic reaction **[NEET 2013]**

(a) transient but stable
(b) permanent but unstable
(c) transient and unstable
(d) permanent and stable

63 Holoenzyme is **[Odisha JEE 2013]**

(a) non-protein and apoenzyme
(b) protein and apoenzyme
(c) enzyme protein and co-enzyme
(d) enzyme non-protein and co-enzyme

64 Which of the following statements about enzymes is incorrect? **[NEET 2013]**

(a) Enzymes are denatured at high temperatures
(b) Enzymes are mostly proteins but some are lipids also
(c) Enzymes are highly specific
(d) Enzymes require optimum pH and temperature for maximum activity

65 Which of the following is false with respective to prosthetic groups? **[AMU 2013]**

(a) Proteins
(b) Non-proteins
(c) Tightly bound to enzymes
(d) Metal compounds

66 Which one out of *A-D* given below correctly represents the structural formula of the basic amino acid? **[CBSE AIPMT 2012]**

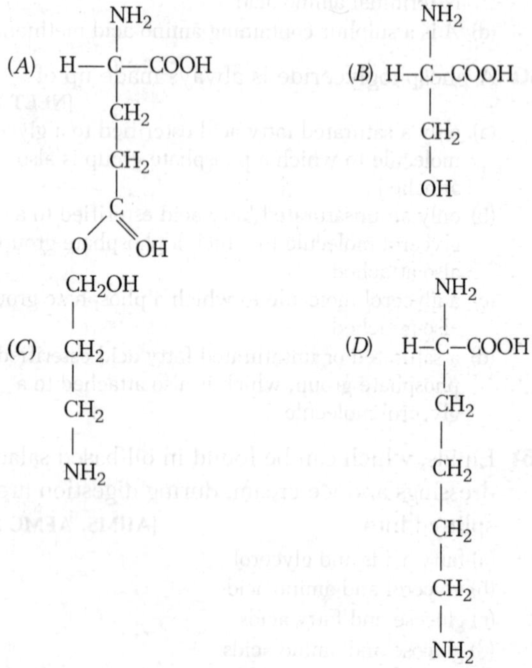

(a) *C* (b) *D* (c) *A* (d) *B*

67 Which one of the following is not true for enzymes? **[WB JEE 2012]**

(a) They act on a specific substrate
(b) They are made up of fat and sugar
(c) They act at a specific temperature
(d) They act at a specific pH

68 Inhibition of acetylcholine by DIFP (Di-Isopropyl Fluoro Phosphate) is an example of **[AMU 2012]**

(a) competitive inhibition
(b) non-competitive inhibition
(c) non-competitive irreversible inhibition
(d) allosteric inhibition

69 Thermolabile protein part of enzyme is called **[Odisha JEE 2012]**

(a) proenzyme (b) holoenzyme
(c) apoenzyme (d) isoenzyme

70 Simple storage protein that coagulates upon heating but remains soluble in dilute salt solution is correctly exemplified by **[WB JEE 2012]**

(a) globulin (b) albumin
(c) histone (d) collagen

71 .... is a globular protein of ~6 kDa consisting of 51 amino acids, arranged in a 2 polypeptide chains held together by disulphide bridge. **[AMU 2012]**

(a) Insulin (b) Keratin
(c) Glucagon (d) Fibrinogen

72 Which one of the following biomolecules is correctly characterised? **[CBSE AIPMT 2012]**

(a) Lecithin – A phosphorylated glyceride found in cell membrane
(b) Palmitic acid – An unsaturated fatty acid with 18 carbon atoms
(c) Adenylic acid – Adenosine with a glucose phosphate molecule
(d) Alanine-Amino acid contains an amino group and an acidic group anywhere in the molecule

73 Match the following columns and choose the correct option from the codes given below.

Column I (Organic compounds)	Column II (Examples)
A. Fatty acid	1. Glutamic acid
B. Phospholipid	2. Tryptophan
C. Aromatic amino acid	3. Lecithin
D. Acidic amino acid	4. Palmitic acid

[Kerala CEE 2012]

Codes

	A	B	C	D
(a)	1	2	3	4
(b)	4	3	2	1
(c)	2	4	3	1
(d)	4	3	1	2
(e)	3	2	1	4

74 Given below is the diagrammatic representation of one of the categories of small molecular weight organic compounds in the living tissues. Identify the category shown and the one blank component X in it. **[CBSE AIPMT 2012]**

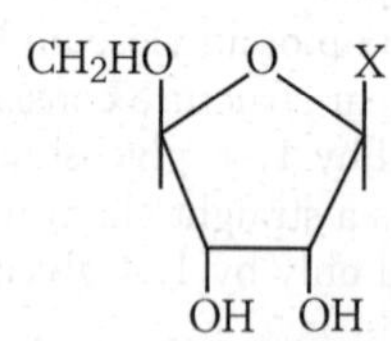

Categories		Components
(a) Cholesterol	–	Guanine
(b) Amino acid	–	NH_2
(c) Nucleotide	–	Adenine
(d) Nucleoside	–	Uracil

75 Melting of DNA at an elevated temperature (70°C) is primarily due to the breakdown of **[WB JEE 2012]**

(a) phosphodiester bonds (b) glycosidic bonds
(c) disulphide bonds (d) hydrogen bonds

76 Nitrogen base + Pentose sugar + Phosphate group is **[AMU 2012]**

(a) nucleoside (b) nucleotide
(c) nucleic acid (d) pyrimidine base

77 The pyrimidine base, which confers additional stability to DNA over RNA, is **[Kerala CEE 2011]**
(a) adenine (b) guanine (c) cytosine (d) thymine
(e) uracil

78 If a highly purified sample of rat liver DNA is dissolved in pure water, the pH of the resulting solution will be **[AMU 2011]**
(a) neutral (b) acidic (c) basic (d) highly basic

79 The curve given below shows enzymatic activity with relation to three conditions (pH, temperature and substrate concentration).
What do the two axes (X and Y) represent?

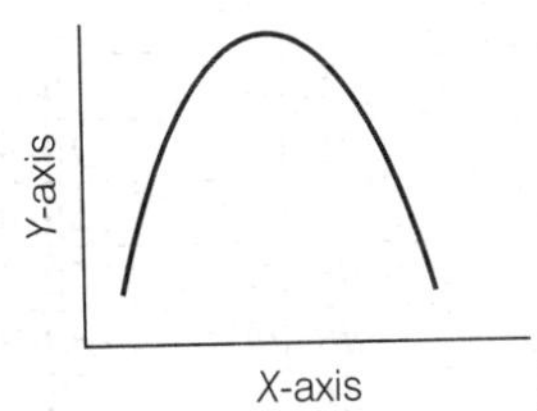

[CBSE-AIPMT 2011]

X-axis		*Y*-axis
(a) Temperature	–	Enzyme activity
(b) Substrate concentration	–	Enzymatic activity
(c) Enzymatic activity	–	Temperature
(d) Enzymatic activity	–	pH

80 An organic substance bound to an enzyme and essential for its activity is called **[UP CPMT 2011]**
(a) co-enzyme (b) holoenzyme
(c) apoenzyme (d) isoenzyme

81 Choose the type of enzyme involved in the following reaction

$$S—G + S' \rightarrow S + S'—G$$ **[Kerala CEE 2011]**

(a) Dehydrogenase (b) Transferase
(c) Hydrolase (d) Lyase
(e) Isomerase

82 Which of the following enzymes has/have haem as a prosthetic group?
I. Catalase II. Carboxypeptidase
III. Succinic dehydrogenase IV. Peroxidase
Choose the correct option from the codes given below. **[Kerala CEE 2011]**
(a) Only I (b) II and III
(c) III and IV (d) I and IV
(e) None of these

83 The most abundant molecule in cell is **[DUMET 2011]**
(a) water (b) carbohydrate
(c) lipid (d) protein

84 Inulin is a polymer of **[WB JEE 2011]**
(a) glucose (b) galactose
(c) fructose (d) arabinose

85 Mannitol is **[WB JEE 2011]**
(a) amino acid (b) amino alcohol
(c) sugar alcohol (d) sugar acid

86 Maltose gives rise to two molecules of **[DUMET 2011]**
(a) fructose (b) lactose (c) glucose (d) sucrose

87 Which one of the following is polysaccharide? **[WB JEE 2011]**
(a) Glycogen (b) Sucrose (c) Lactose (d) Maltose

88 Example of a typical homopolysaccharide is **[WB JEE 2011]**
(a) lignin (b) suberin (c) inulin (d) starch

89 Pick out the incorrect statement. **[Kerala CEE 2011]**
(a) Amino acids are substituted methane
(b) Glycerol is a trihydroxypropane
(c) Lysine is a neutral amino acid
(d) Lecithin is a phospholipid
(e) Adenosine is a nucleoside

90 Select the incorrect statement. **[Kerala CEE 2011]**
(a) Proteins are heteropolymers made up of amino acids
(b) Ribozymes are nucleic acids with catalytic power
(c) Nucleic acids serve as genetic material
(d) Proteins, nucleic acids and polysaccharides are the only three types of macromolecules found in the living system
(e) Collagen is the most abundant protein in the whole of the biosphere and RuBisCO is the most abundant proteins in animal world

91 Which one is imino acid? **[WB JEE 2011]**
(a) Pepsin (b) Proline (c) Cysteine (d) Renin

92 Which one of the following structural formula of two organic compounds is correctly identified along with its related function?

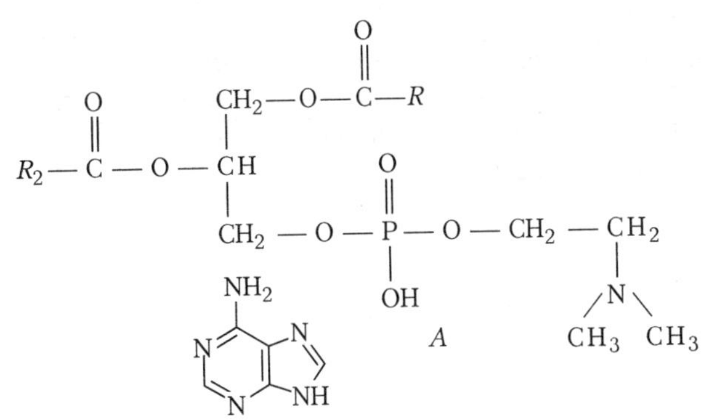

A

B

[CBSE AIPMT 2011]

(a) A-Triglyceride major	–	Source of energy
(b) B-Uracil	–	A component of DNA
(c) A-Lecithin	–	A component of cell membrane
(d) B-Adenine	–	A nucleotide that makes up nucleic acids

ANSWERS

Check Point 17.1

1. (c) 2. (c) 3. (c) 4. (d) 5. (d) 6. (a)

Check Point 17.2

1. (b) 2. (a) 3. (d) 4. (a) 5. (a) 6. (b)

Check Point 17.3

1. (c) 2. (c) 3. (d) 4. (b) 5. (b) 6. (d)

Check Point 17.4

1. (c) 2. (d) 3. (b) 4. (c) 5. (d) 6. (c)

(A) Topical Exercise

1. (a) 2. (a) 3. (a) 4. (b) 5. (b) 6. (a) 7. (d) 8. (d) 9. (a) 10. (d)
11. (d) 12. (d) 13. (d) 14. (c) 15. (a) 16. (d) 17. (a) 18. (a) 19. (d) 20. (c)
21. (d) 22. (c) 23. (d) 24. (b) 25. (a) 26. (b) 27. (d) 28. (d) 29. (c) 30. (d)
31. (d) 32. (c) 33. (a) 34. (c) 35. (d) 36. (b) 37. (c) 38. (b) 39. (d) 40. (d)
41. (a) 42. (d) 43. (d) 44. (d) 45. (c) 46. (d) 47. (c) 48. (b) 49. (b) 50. (b)
51. (a) 52. (b) 53. (a) 54. (a) 55. (a) 56. (a) 57. (c) 58. (d) 59. (b) 60. (a)
61. (b) 62. (d) 63. (d) 64. (a) 65. (a) 66. (a) 67. (d) 68. (a) 69. (d) 70. (a)
71. (c) 72. (c) 73. (b) 74. (a) 75. (c) 76. (b) 77. (a) 78. (c) 79. (d) 80. (a)
81. (a) 82. (a) 83. (d) 84. (c) 85. (c) 86. (b) 87. (d) 88. (d)

(B) Cumulative Exercise

1. (c) 2. (c) 3. (c) 4. (b) 5. (d) 6. (d) 7. (a) 8. (d) 9. (a) 10. (c)
11. (d) 12. (a) 13. (a) 14. (d) 15. (d) 16. (d) 17. (d) 18. (b) 19. (c) 20. (a)
21. (d) 22. (a) 23. (c) 24. (b) 25. (c) 26. (b) 27. (c) 28. (a) 29. (e) 30. (c)

(C) Medical Entrances' Gallery

1. (b) 2. (b) 3. (d) 4. (d) 5. (b) 6. (c) 7. (a) 8. (d) 9. (a) 10. (a)
11. (a) 12. (b) 13. (b) 14. (a) 15. (b) 16. (a) 17. (b) 18. (b) 19. (d) 20. (a)
21. (b) 22. (b) 23. (d) 24. (d) 25. (a) 26. (b) 27. (a) 28. (b) 29. (b) 30. (a)
31. (a) 32. (d) 33. (c) 34. (a) 35. (a) 36. (b) 37. (b) 38. (b) 39. (d) 40. (d)
41. (b) 42. (c) 43. (d) 44. (b) 45. (c) 46. (c) 47. (b) 48. (d) 49. (d) 50. (c)
51. (a) 52. (c) 53. (b) 54. (d) 55. (c) 56. (c) 57. (b) 58. (d) 59. (a) 60. (d)
61. (a) 62. (c) 63. (c) 64. (b) 65. (a) 66. (b) 67. (b) 68. (c) 69. (c) 70. (b)
71. (a) 72. (a) 73. (b) 74. (d) 75. (d) 76. (b) 77. (d) 78. (b) 79. (a) 80. (a)
81. (b) 82. (d) 83. (a) 84. (c) 85. (c) 86. (c) 87. (a) 88. (d) 89. (c) 90. (e)
91. (b) 92. (c)

Explanations

(A) Topical Exercise

6 (*a*) Statement I is true, but II is false. Incorrect statement can be corrected as

On homogenising any tissue in an acid, the acid soluble pool represents cytoplasm.

8 (*d*) Option (d) is incorrect matched pair and can be corrected as

Cellulose is a homopolymer of glucose which is formed by beta 1-4 linkage of D-glucose units.

Rest options are correct matched pairs.

16 (*d*) Both statements in option (b) and (c) are incorrect.

These can be corrected as

- Polymers are synthesised from monomers by removal of water molecule.
- Polymers consist of more than two monomers.

Thus, option (d) is correct.

32 (*c*) Both statements are correct.

Amino acid can be considered as an amphoteric molecule due to their tendency to either accept or donate a proton (H^+). Out of 20 amino acids, 9 are essential to diet because they cannot be synthesised by the cells. Non-essential amino acids are those which are required by the body but not in the diet because body cells can synthesise them. Thus, all amino acids are necessary for our body.

37 (*c*) Statement in option (c) is incorrect and can be corrected as

Biologist describes the protein structures at four levels.

45 (*c*) *A* represents the primary (1°) structure of proteins. *B* represents the quaternary (4°) structure of proteins. *C* represents the tertiary (3°) structure of proteins. *D* represents the secondary (2°) structure of proteins.

47 (*c*) Statement in option (c) is incorrect and can be corrected as

Lipids in neural tissues are complex in structure.

54 (*a*) Statement I is true, but II is false. Incorrect statement can be corrected as

The number of hydrogen bonds between adenine and thymine in DNA molecules are two.

62 (*d*) An enzyme increases the rate of reaction by bringing the reacting molecules into precise orientation with each other. It lowers the activation energy for a reaction, the faster the rate. Thus, enzymes speed up reaction by lowering activation energy.

72 (*c*) Statement in option (c) is incorrect and can be corrected as enzymes are not affected by the reactions they catalyse.

Rest statements are correct.

73 (*b*) Emil Fischer (1894) proposed the 'lock and key' theory to explain the mechanism of enzyme action. According to this hypothesis, the geometry of active sites of enzyme and shape of substrate molecule are complementary to each other which facilitates their binding. This model explains the specificity of enzyme action, i.e. interacts with a specific type of substrate molecule.

75 (*c*) In allosteric modulation or feedback inhibition, the product of an enzyme catalysed reaction (chain of reactions) accumulates and acts as an inhibitor of the reaction, e.g. allosteric inhibition of hexokinase by glucose-6-phosphate.

77 (*a*) A competitive enzyme inhibitor closely resembles the real substrate. This inhibitor competes with substrate and binds to the active site of the enzyme but does not undergo any catalysis. This inhibition can be reversed by increasing the substrate concentration. In competitive inhibition, the K_m value increases whereas V_{max} remains unchanged.

80 (*a*) Statement in option (a) is incorrect regarding co-enzyme and can be corrected as

Every co-enzyme is a co-factor but every cofactor is not a co-enzyme.'

Rest other statements are correct regarding co-enzyme.

85 (*c*) K_m (Michaelis Menten constant) indicates that substrate concentration attain one half of its maximum reaction velocity when enzymes catalyse the chemical reaction.

(B) Cumulative Exercise

1 (*c*) Statements I and II are correct whereas III and IV are incorrect. Incorrect statements can be corrected as Polysaccharides are very large polymers composed of tens to thousands of monosaccharides joined together by glycosidic linkages. After complete hydrolysis, they yield only one type of monosaccharides.

6 (*d*) Statement IV is not true and can be corrected as

Agar is an unbranched polysaccharide obtained from the cell walls of some species of red algae. It is a mixture of agarose and agaropectin.

Rest statements are true.

7 (*a*) Statement in option (a) is incorrect and can be corrected as

The building blocks of proteins are amino acids.

Rest statements are correct.

8 (*d*) Option (d) is mismatched pair and can be corrected as

Glycogen is a polymer of glucose.

Rest options are matched pairs.

11 (*d*) Thus, option (d) is correct.

- Alpha helix (*A*) is formed by the formation of H-bond between the NH group of one amino acid and the CO group of another amino acid located three or four residues earlier along the protein sequence.
- β-pleated sheet (*B*) is formed by the formation of at least two to three hydrogen bonds between two or more protein strands laterally.
- Tertiary protein structure (*C*) is formed by the bending and folding of two or more protein secondary structures.
- Quaternary protein structure (*D*) consists of an assembly of two or more polypeptides or subunits. It is formed by further folding in two or more tertiary protein structures.

21 (*d*) Starch serves as the storage food in plants. It is chemically non-reactive and osmotically inactive polysaccharide composed of two subunits, namely amylose and amylopectin (both of which consist of a large number of glucose units). These two properties make starch useful as a storage material in plants.

Thus, option (d) is correct.

23 (*c*) Option (c) is incorrectly matched and can be corrected as

Enzymes are not lipopolysaccharide. They are mostly proteins.

Rest options are correctly matched.

24 (*b*) Both A and R are true, but R is not the correct explanation of A.

Enzymes are either proteins or RNA (e.g. ribosome) that catalyses a specific chemical reaction. It does not affect the equilibrium of the catalysed reaction.

It just enhances the rate of a reaction by providing a reaction path with a lower activation energy. Enzyme itself remains unchanged after reaction.

25 (*c*) A is true, but R is false.

Enzymes lower the activation energy of the reaction. Lower activation energy increases the rate of reaction.

26 (*b*) Both A and R are true, but R is not the correct explanation of A.

Lactose is a non-reducing sugar, i.e. the lactose has unlinked potential aldehyde group at their first α-carbon atom. Sucrose is a non-reducing sugar that consists of D-glucose and 1, D-fructose units joined by α-1, 2-glycosidic linkage.

27 (*c*) A is true, but R is false.

Protein forming amino acids are α-amino acids because the $-NH_2$ group attached to the α-carbon atom, i.e. adjacent to $-COOH$ group not on the second one.

28 (*a*) Both A and R are true and R is the correct explanation of A.

Sucrose is a non-reducing sugar because free $-CHO$ group is not present in sucrose.

29 (*e*) Both A and R are false and can be corrected as

Arachidic acid is a saturated fatty acid because they have single bond and are solid at room temperature. It can be formed by the hydrogenation of arachidonic acid.

30 (*c*) A is true, but R is false. R can be corrected as

An apoenzyme is the protein part of the conjugate enzyme. It is made up of large molecule of protein and is heat unstable. This does not take part in the transfer of groups.

(C) Medical Entrances' Gallery

1 (*b*) Statement in option (b) is incorrect and can be corrected as

Glycine is an example of amino acid. It is a neutral amino acid that contains only one amino group and one carboxylic group with non-cyclic hydrocarbon chain.

Rest of the statements are correct.

2 (*b*) Lysine is a basic amino acid because its side chain contains nitrogen and resembles ammonia, which is a base. On the other hand, valine is a neutral amino acid, glutamic acid is an acidic amino acid, while tyrosine is an aromatic amino acid.

3 (*d*) A-2, B-4, C-3, D-1

4 (*d*) The statement in option (d) is correct because

Adenine pairs with thymine through two H-bonds, i.e. A$=$T and the cytosine pair with guanine by three hydrogen bonds, i.e. G$\equiv$C. Between the G–C base pairs, there are 3 hydrogen bonds which make this bond pair stronger than the A–T base pair.

5 (*b*) A wide variety of chemical substances (i.e. secondary metabolites) that we extract from plants on a commercial scale (nicotine, caffeine, quinine, opium, etc) are produced by them (plants) as defence against grazers and browsers.

6 (*c*) Inulin is a fructan (polysaccharide of fructose). Adjacent fructose units are linked through glycosidic bond. Insulin is a protein composed of 51 amino acids. Adjacent amino acids are attached through peptide bond.

7 (*a*) Collagen is the most abundant protein in animal world and RuBisCO is the most abundant protein in the whole of the biosphere because it is present in every plant that undergoes photosynthesis and molecular synthesis through the Calvin cycle.

8 (*d*) In a DNA polynucleotide chain, a nitrogenous base is linked to the hydroxy ($-OH$) of 1′C pentose sugar. It is represented in the structure given below

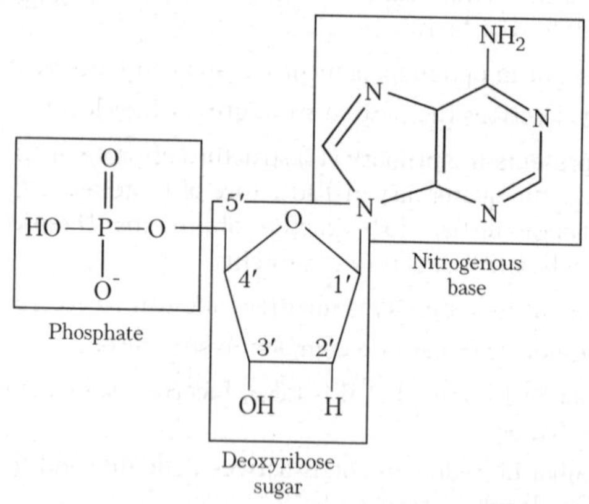

9 (*a*) Glucose on reacting with Benedict's solution does not give violet precipitate. Benedict reagent is used to detect the presence of sugar (reducing sugar) in a sample. When it is used to test monosaccharides such as glucose, the solution progress in the colours of orange, yellow, green, red and then brick red or brown precipitate within 7-10 minutes. A colour change would signify the presence of reducing sugar.

10 (*a*) 1, 2-glycosidic linkage is found in sucrose, a disaccharide, where one glucose molecule and one fructose molecule are linked together by this bond.

11 (*a*) Epimers are optical isomers that differ in the configuration of a single carbon atom. In given question, glucose and galactose sugars are examples of epimers because they differ only in the configuration at C-4.

12 (*b*) Phospholipids are main constituents of lecithin. These molecules are composed of choline and inositol. It is found in all living cells as a major component of cell membrane.

13 (*b*) Adenylic acid, cytidylic acid and guanylic acid are nucleotides. Nucleotide is composed of three components, i.e. nitrogen base (be it purines like adenine and guanine or pyrimidines like uracil, cytosine, thymine), ribose sugar and a phosphate group. These are monomer unit of nucleic acid, i.e. RNA and DNA. Adenosine, cytidine, thymidine are nucleosides.

14 (*a*) Statement I is true, but II is false. Incorrect statement can be corrected as

A complete catalytic active enzyme with its bound prosthetic group is called holoenzyme. An apoenzyme is an inactive enzyme, which gets activated by the binding of an organic or inorganic co-factor.

15 (*b*) Prosthetic groups are organic compounds and are distinguished from other co-factors in that they are tightly bound to the apoenzyme. For example, in peroxidase and catalase, which catalyse the breakdown of hydrogen peroxide to water and oxygen, haem is the prosthetic group and it is a part of the active site of the enzyme.

16 (*a*) The correct structure of citrulline is given in option (a).

Citrulline is an amino acid. It is made from ornithine and carbamoyl phosphate is one of the control reactions in urea cycle.

17 (*b*) Concanavalin-A is a lectin or a carbohydrate binding protein. It is a T-cell mitogen that can activate the immune system, recruit lymphocytes and elicit cytokine production. It can also induce programmed cell death *via* mitochondria-mediated apoptosis.

18 (*b*) 'Ramachandran plot' is used to confirm the structure of proteins. It is a plot of the angles-phi (ϕ) and psi (ψ) of amino acids found in a peptide chain. This plot was developed by GN Ramachandran, an Indian Scientist in 1963.

19 (*d*) Sugars are chemically carbohydrates. They are polyhydroxy aldoses, ketoses and their condensation products. Aldoses bear a terminal aldehyde or —CHO group, while ketoses have an internal ketone or —CO group. Thus, they possess two functional groups, i.e. carbonyl and hydroxyl.

20 (*a*) Among the given options, vitamin-B is not a derivative of cholesterol (a type of lipid molecule). Cholesterol is an essential structural component of all animal cell membranes. It also serves as a precursor for the biosynthesis of steroid hormones, bile salts and vitamin-D.

21 (*b*) Omega-3 fatty acids are present in flax seed oil. These are Polyunsaturated Fatty Acids (PUFAs, i.e. fatty acids with more than one double bond) and are important for normal metabolism. Mammals are unable to synthesis omega-3 fatty acids thus they intake it through their diet. These are recommended to individuals with high BP, cardiovascular diseases, etc., as they help in lowering the blood cholesterol.

22 (*b*) Among the given options, arginine is a non-essential amino acid. Such amino acids can be made within the body and are not essential to be incorporated in the human diet. Valine, histidine and lysine are essential amino acids. These are not synthesised within the body and as such have to be consumed through diet.

23 (*d*) Proteins are the polymers of amino acids. These can be divided into two categories, i.e. essential and non-essential amino acids. Essential amino acids cannot be synthesised in the body and therefore must be included in the diet. Proteins are called as building blocks of our body because they are used in structural components of tissues as channels, transporters, regulatory molecules and enzymes. Proteins can also be utilised as energy sources when there is not enough carbohydrate and fat in the body, e.g. during fasting.

Thus, option (d) is correct.

24 (*d*) Among the given options except lipids, all are polymers which are formed by the polymerisation of monomers. The basic units of lipids are fatty acids and glycerol molecules that do not form repetitive chains. Instead they form triglycerides from three fatty acids and one glycerol molecule.

25 (*a*) Both A and R are true and R is the correct explanation of A.

Steroids are the derivatives of a four-membered ring compound called phenanthrene. The given structure is that of cholesterol. It is one of the most important animal steroid is insoluble in water and chemically unreactive. Cholesterol is useful since it is a structural component of cells. It is synthesised from acetyl Co-A or acetate (C_2) in the liver.

26 (*b*) Both A and R are true, but R is not the correct explanation of A.

Glycogen is a polysaccharide having two types of linkage, α-1,4 linkages in the straight part and α-1,6 linkages in the area of branching.

Its right end is known as reducing end because it contains free anomeric carbon and left one is called the non-reducing end.

27 (*a*) Macromolecule chitin is a complex structural heteropolysaccharide. It has nitrogen containing glucose derivatives such as N-Acetyl Glucosamine (NAG) joined with β -1,4 linkages. It is the principal structural component of fungal cell wall and exoskeleton of arthropods.

28 (*b*) Holoenzyme is a conjugate complete catalytically active enzyme together with its co-enzyme.

- Apoenzyme is the protein part of catabolically active enzyme.
- Co-enzymes are additional organic or metallo-organic molecules which are required for the activity of other enzymes.

So, holoenzyme is a apoenzyme together with co-enzyme. Hence, option (b) is correct.

29 (*b*) Both A and R are true, but R is not the correct explanation of A.

Carbohydrates are more suitable for energy production within the body as these contain more oxygen as compared to proteins and fats. Thus, requiring less molecular oxygen for their oxidation and yielding higher amount of energy than both proteins and fats.

Also, carbohydrates can be stored in the form of glycogen in animal tissues to be used for energy production when required.

30 (*a*) Cellulose, a homopolysaccharide, is a polymer formed by joining the monomeric units of β-glucose by β-1, 4-glycosidic linkage. It is present in woody portion of the plant and in cotton. Thus, paper and cotton fibre contain cellulose.

Chitin, formed by polymerisation of N-acetylglucosamine, is found in the exoskeleton of several insects, crabs and prawns.

Glycogen serves as the stored food in animals (in liver and muscles) and is the polymer of α-glucose.

So, option (a) is correct.

31 (*a*) A typical fat molecule is a triglyceride, which is formed by the esterification of one glycerol and three fatty acid molecules. The three fatty acids can be of same type or different depending on the type of the fat molecules.

32 (*d*) Ester bonds are least likely to be involved in stabilising the 3-D folding of most proteins.

The tertiary structure (3D) structure of a protein is stabilised by several types of bonds, i.e. hydrogen bonds, ionic bonds, van der Waal's interactions, covalent bonds and hydrophobic bonds.

Ester bond is formed between sugar and phosphate in a nucleotide and is thus not involved in ensuring the stability of a polypeptide chain.

33 (*c*) Statement in option (c) is wrong and can be corrected as Glycine is the simplest amino acid, which contains carbon, hydrogen, oxygen and nitrogen atoms. Sulphur containing amino acids are cysteine and methionine.

Rest of the statements are correct.

34 (*a*) Statement in option (a) is false about amino acids and can be corrected as

Based on structure and reaction, amino acids have been classified as neutral, acidic, basic, sulphur containing, alcoholic, aromatic and heterocyclic amino acids.

Rest of the statements are true.

35 (*a*) Glutamine synthetase is a ligase that plays essential role in metabolism of N_2 by catalysing the condensation of glutamate and ammonia to form glutamine.

36 (*b*) The given structure is that of uracil. It is a nitrogenous base, i.e. one of the pyrimidines (apart from thymine and cytosine). It is present in RNA (instead of thymine) and is a six membered aromatic ring possessing nitrogen at one and three positions.

37 (*b*) Ribozyme is a type of ribosomal RNA (25S *r*RNA) which acts as a catalyst in splicing of RNA during protein synthesis. It is the only non-proteinaceous enzyme known so far. Rest all the enzymes are proteinaceous.

38 (*b*) The graph shows the exothermic reaction *A* in the presence of enzyme which lowers down the activation energy substantially.

The *B* graph shows the same reaction occurring in the absence of enzyme when activation energy is quite high.

Activation energy is the minimum amount of energy required by reactants to reach the transition state in a chemical reaction.

39 (*d*) In competitive inhibition, a substance which is similar to the substrate of the reaction binds to the active site and occupies it altering the shape of enzyme due to which the substrate cannot bind to it. This leads a decline in the activity of the enzyme catalysing the reaction.

40 (*d*) All statements are correct except (d). Incorrect statement can be corrected as

Competitive inhibitor resembles closely with the substrate and competes for the active binding site of the enzyme.

The presence of competitive inhibitor decreases the affinity of the substrate to binding site of the enzyme which results in increase of Michaelis-Menten constant (K_m).

41 (*b*) The system of enzyme nomenclature is based on the substrate involved and the type of chemical reaction being catalysed. Enzymes are named by adding a suffix, ase on the basis of the substrate involved in the reaction like lipase, amylase, sucrase, etc. Enzymes have 6 categories on the type of reaction being catalysed like transferases, hydrolases, lyases, ligases, etc.

42 (*c*) 'K_m' refers to the Michaelis-Menten constant the substrate concentration of which the chemical reaction being catalysed by the enzyme attains half of its maximum velocity.

$$K_m = \frac{1}{2} V_{max}$$

Thus, K_m indicates the affinity of an enzyme for its substrate. A high K_m value means low enzyme affinity, whereas a low K_m value means high enzyme affinity.

43 (*d*) Statement in option (d) is true about protein structures. Rest other statements are incorrect and can be corrected as

- The sequence of amino acids in a proteins represents the primary structure. The α-helices in a polypeptide chain are right-handed.
- The adult haemoglobin consists of four subunits.
- The left end of the primary structure (amino acid sequence) of a peptide chain is referred to as the N-terminal whereas the right end is called the C-terminal.

44 (*b*) DNA and RNA strands consist of a backbone made up of alternating sugar and phosphate. This phosphate group joins the two sugars through a phosphodiester bond.

Thus, nucleic acids in a nucleotide of RNA contain phosphodiester bonds.

45 (*c*) The ribose sugar is present in RNA. The polynucleotide chain contains a free phosphate moiety at its 5′-end whereas at its 3′-end it contains a free hydroxyl (OH) group.

46 (*c*) RNA contains ribose sugar whereas DNA contains deoxyribose sugar. Ribose is a normal sugar containing an attached oxygen atom to each carbon atom. But in deoxyribose sugar, one oxygen atom is absent.

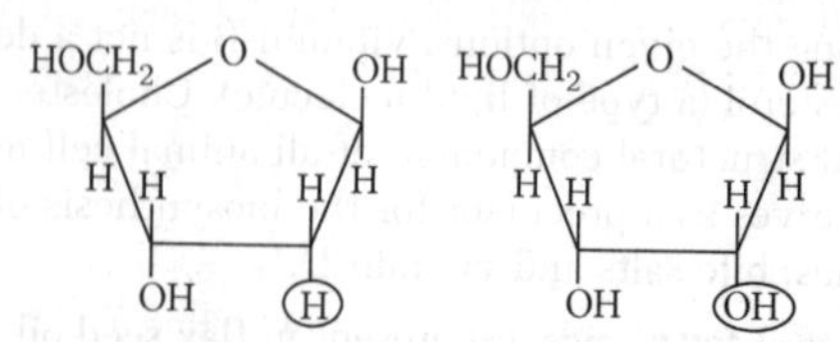

Deoxyribose   Ribose

An additional 2′-OH group present in ribose sugar of every nucleotide of RNA is a reactive group. It makes RNA highly reactive, labile and easily degradable.

47 (*b*) Sucrose is a non-reducing sugar because it does not contain any free anomeric carbon atom. It is a disaccharide of one glucose and one fructose unit.

48 (*d*) Cellulose $(C_6H_{10}O_5)n$ is a linear polymer, long unbranched chain of β-D glucose residues linked by β, 1-4 glycosidic bonds. It is the most abundant organic polymer. It forms an important structural component of plant cell walls.

49 (*d*) Statement in option (d) is incorrect and can be corrected as

Sucrose is a disaccharide sugar comprised of glucose and fructose units. It is a non-reducing sugar because it does not have a free aldehyde or ketone group.

Rest of the statements are correct.

50 (*c*) The given structural formulae are that of anomers. An anomer refers to the stereoisomer of a cyclic saccharide, differing only at the hemiacetal or hemiketal carbon in the configuration.

The given structural formulae are anomers of D-glucopyranose, the first figure is α-D-glucopyranose and the second figure is β-D-glucopyranose.

51 (*a*) GN Ramachandran is credited with the discovery of the triple helical structure of collagen. Where amino acids are wound together to form triple helices of elongated fibrils. This protein is mostly found in fibrous tissues like tendons, ligaments, etc.

52 (*c*) A–4, B–3, C–2, D–1

53 (*b*) Option (b) is not correct with respect to enzyme action and can be corrected as

Inhibition of succinic dehydrogenase by malonate is an example of competitive inhibition, which occurs when the enzyme and the inhibitor both have more or less similar structure and are present in higher concentrations. Thus, both the substrate and the inhibitor compete for the active site of the enzyme.

Rest options are correct with respect to enzyme action.

54 (*d*) Apoenzyme is the protein portion, which combines with a co-factor, i.e. either co-enzyme, metal ion activator or prosthetic group and is collectively called an active holoenzyme. A holoenzyme is a complete and catalytically active enzyme in nature.

55 (*c*) Michaelis-Menten kinetics is one of the best known and simplest models of enzyme kinetics in which K_m acts as a Michaelis-Menten constant in the equation,

$$v = \frac{V_{max}[S]}{K_m + [S]}$$

Where V_{max} represents the maximum rate achieved by the system at maximum substrate concentration and [S] indicates substrate concentration at which rate of reaction is half of the maximum velocity (V_{max}).

56 (*c*) Statement in option (c) is not true about enzymes and can be corrected as

Enzymes do not change the value of equilibrium constant. Rather they bring about equilibrium quickly by lowering the activation energy of the reaction. This subsequently leads to the attainment of equilibrium quickly.

Rest of the statements are correct.

57 (*b*) Statement in option (b) contains the true statement regarding fats. Rest other statements are not fail and can be corrected as

- Arachidonic acid contains 20 carbon atoms including the carboxyl carbon atom.
- Palmitic acid has a total of 16 carbon atoms including the carboxyl carbon atom.
- Oils have a lower melting point as compared to fats as they do not freeze at low temperature.
- Lipids are generally insoluble in water.

58 (*d*) Statement in option (d) is not correct and can be corrected as

Amylopectin is a branched chain containing more than 1000 glucose units joined by α-1, 4-glycosidic linkage in the linear chains and by α-1, 6-glycosidic linkage at linked branch points.

Rest of the statements are correct.

59 (*a*) The structure represented in *D* is not an amino acid. Hence, option (a) is correct.

The information about remaining structures is as follows

Structure *A* is Serine

Structure *B* is Cysteine

Structure *C* is Tyrosine

60 (*d*) A fat is formed of two kinds of small molecules, i.e. glycerol and fatty acids. In a single fat molecule, three fatty acid molecules are joined to glycerol by an ester linkage.

Phosphoglycerides are esters of two fatty acids namely phosphoric acid and a trifunctional alcohol-glycerol. The fatty acids are attached to the glycerol at the 1st and 2nd position through ester bonds.

61 (*a*) Digestion of all the lipids takes place in the small intestine. The fats found in foods like salad dressing, ice creams, etc., during digestion gets split into fatty acids and glycerol which serve as the digestive products of lipids.

62 (*c*) The substrate binds to the enzyme at its active site forming an enzyme-substrate complex. This complex formation is a transient and unstable phenomenon. Very soon, the final product is released from the active site. It is a fact that, all other intermediate structural states formed are unstable.

63 (*c*) Holoenzyme is an entire conjugated enzyme consisting of a protein component (an apoenzyme) and a non-protein component (a coenzyme or an activator).

64 (*b*) Statement in option (b) is incorrect about enzyme and can be corrected as

Enzymes are those organic molecules which catalyse or regulate biochemical reactions without themselves being consumed the reaction. Most of the enzymes are proteins and some are RNA derived ribozymes.

Rest of the statements are correct.

65 (*a*) Option (a) is false with respective to prosthetic group and can be corrected as

Conjugate enzymes (holoenzymes) are made up of one protein and one non-protein part. The protein part is called apoenzyme and the non-protein part can be a prosthetic group or co-enzymes.

66 (*b*) Structure *D* represents a basic amino acid, lysine, whereas structure *A* represents glutamic acid (an acidic amino acid) and structure *B* represents an alcoholic amino acid, serine. Structure *C* is not an amino acid as there is no carboxyl group in it.

67 (*b*) Statement in option (b) is not true for enzymes and can be corrected as

Enzymes or biocatalysts are exclusively proteinaceous substances. They are not made up of fat and sugar. Enzymes are capable of catalysing chemical reactions of biological systems without undergoing any change in themselves.

Rest of the statements are correct about enzymes.

68 (*c*) Inhibition of acetylcholine by DIFP is an example of non-competitive irreversible inhibition. In non-competitive inhibition, the inhibitor binds to an enzyme at a place other than substrate binding site. It results in the destruction of enzyme activity. The substrate may combine with such an enzyme but product formation is inhibited, e.g. reaction of DIFP or Di-Isopropyl Fluoro Phosphate (nerve gas) with hydroxyl group of serine of acetylcholine-esterase.

70 (*b*) Albumin is a water soluble protein which gets coagulated by heat and precipitated by saturated salt solutions.

Other options are explained as

Globulins are insoluble in pure water and moderately concentrated salt solutions. These are coagulated by heat, while histones are not coagulated by heat. Collagen is a stable protein which gets denatured upon heating.

72 (*a*) In the given options, lecithin is correctly characterised. It is a major component in the lipid bilayers of cell membranes and it consists of glycerol, two fatty acids and a phosphate groups. Rest other options are not correctly characterised and can be corrected as

- Alanine is a neutral amino acid with non-cyclic hydrocarbon chain, one amino group and one carboxyl group.
- Palmitic acid is a saturated fatty acid.
- Adenylic acid contains pentose sugar not glucose.

74 (*d*) Out of the given options, the correct category is nucleoside, which is made up of a ribose sugar and nitrogenous base only. Here, *X* represents uracil, which forms nucleoside with only ribose sugar.

78 (*b*) When a DNA sample, purified or extracted from rat liver, is dissolved in pure water, the pH of the resulting solution becomes acidic as DNA contains phosphate groups that are highly acidic. These impart acidic nature to the DNA.

79 (*a*) In the given graph, *X*-axis represents temperature, while *Y*-axis represents enzyme activity. All enzymes function best at an optimum temperature, above and below this temperature, the enzyme activity declines.

81 (*b*) Transferase catalyses the transfer of a specific group (other than hydrogen) between a pair of substrates, e.g.

$$S{-}G + S' \xrightarrow{\text{Transferase}} S + S'{-}G$$

82 (*d*) Haem (iron containing) prosthetic group is found in cytochromes, haemoglobin, myoglobin, catalase and peroxidase. A prosthetic group or co-enzymes is the non-protein part of a conjugate enzyme (the protein part being apoenzyme).

87 (*a*) Polysaccharides are the polymers of monosaccharides. Out of the given options, glycogen is polymer of α-glucose. It is stored in liver and muscles of animal cells as the main reserve food material.

89 (*c*) Option (c) depicts the incorrect statement and can be corrected as

Lysine is a basic amino acid having an amino group ($+NH_3$) at the beta carbon of the amino acid. The positively charged group on the terminal carbon atoms makes it a basic amino acid (and not a neutral amino acid).

Rest of the statements are correct

90 (*e*) Statement in option (e) is incorrect and can be corrected as

Collagen is the most abundant protein in animal world. Ribulose Bisphosphate Carboxylase Oxygenase (RuBisCO) is the most abundant protein in the whole of the biosphere and it is found in plants.

Rest of the statements are correct.

91 (*b*) An imino acid is a compound that has an imino $>C{=}NH$ group and a carboxylic acid [$—C(=O)—OH$] functional group. An imino group is formed when nitrogen is double bonded with a carbon atom and single bounded to either a hydrogen atom or sp^3 hybridised carbon atom. Proline is a heterocyclic amino acid which contains an imino (—NH) group.

92 (*c*) The structural formula *'A'* represents lecithin. It is a phospholipid found in all living cells that serve as a major component of cell membrane. It regulates the entry exit of molecules across the cell membrane. Without lecithin, the cell membrane would become less permeable.

Thus, option (c) is correct.

CHAPTER 18

Transport in Plants

In plants, transport may begin from underground part (roots) or from aerial part (leaves). The substances that would need to be transported in plants are water, mineral nutrients, organic nutrients, enzymes and growth regulators. The specialised tissues involved in the transportation of water and organic solutes are xylem and pholem, respectively.

The transportation of substances in plants is a complex multistep process that involves certain mechanisms and physiological processes such as transpiration, guttation, etc.

In this chapter, we will study how efficiently a plant transports all its essential substances along with the physiological processes related to them.

MEANS OF TRANSPORT

The process of transport in a plant takes place at the following three levels. These are

(i) The uptake and release of substances within a cell, i.e. **cellular level transport**.

(ii) Transport of substances from one cell to another, i.e. **short distance transport**.

(iii) Transport of water and suger *via* xylem and phloem, i.e. **long distance transport**.

The various means of transport used to carryout transportation in these levels are discussed below

Active transport

It is the process of transport that requires involvement of energy (ATP).

Major features of this transport are as follows

- It can allow movement of substances, molecules or ions across a membrane from a region of their low concentration to a region of high concentration, i.e. **uphill transport**.
- In active transport, the substances are transported across a membrane with the help of certain membrane carrier proteins, e.g. **Na^+–H^+ transporter protein, GLUT** (Glucose Transporter).
- These carrier proteins are highly specific. They are also sensitive to inhibitors that react with protein side chains.

Inside

Passive transport

In this type of transport, a molecule is transported along its concentration gradient without the involvement of ATP. The flow of water in and out of the plant mainly occurs by passive transport.

Processes related to passive transport

Passive transport of water and solutes in plants may take place *via* diffusion, osmosis, plasmolysis, etc.

Diffusion

It is the net movement of particles of gases, liquids and solids from a region of their higher concentration to a region of their lower concentration. This movement continues till an equilibrium is established.

Other major features of diffusion are

- It is a slow process and its rate is usually higher in gases whereas, it is relatively lower in liquid and solids.
- When more than one type of particles are present in a solution, the diffusion of particles of one substance is independent of that of another substance.

The main forces that contribute to diffusion of particles are

- Kinetic Energy (KE) present in the particles as, **Rate of diffusion $\propto$ KE**
- Differences in areas of concentrations of the particles of a particular substance.
- Difference in the size of the particles of the substance. Smaller the size, higher is the diffusion rate. So, we can say that the rate of diffusion is inversely proportional to the size of the particle.

Role of diffusion in plants

The following roles are played by diffusion in plants

- Diffusion keeps the cell wall moist.
- It helps in gaseous exchange during photosynthesis, respiration, transpiration and translocation of food materials.
- It is involved in short-distance transport of gases, liquids and some food materials.

Diffusion pressure (DP)

The pressure exerted by particles of a substance, when they move from their higher concentration to lower concentration is called as diffusion pressure. It describes the ability of a molecule to freely diffuse along a concentration gradient.

Characteristics of diffusion pressure are as follows

- Diffusion pressure is always proportional to the kinetic movement and number of diffusing particles.
- Diffusion pressure of pure water is found to be maximum.
- Its value at 20°C is 1336.40 atm.
- DP decreases as more solute is added in a solution.

Facilitated diffusion

- It is the movement of particles from a region of their higher concentration to a region of their lower concentration with the help of a membrane transport protein.
- It also occurs without the involvement of energy.
- Transport proteins involved in facilitated diffusion are of following two types
  (i) **Carrier proteins** They bind to the substance and carry it to the other side.
  (ii) **Channel proteins** They allow diffusion of solutes of appropriate size only. Some examples of channel proteins are as follows
    (a) **Porins** These are **β-barrel proteins** that form a pore-like structure in mitochondrial, chloroplast and bacterial membranes. Only specific molecules are transported through this pore.
    (b) **Aquaporins** These are membrane proteins with 8 units that unite to form a water channel. It allows transport of water.
- Based on the direction of movement, facilitated diffusion is of three types
  (i) **Uniport system** It allows diffusion of one molecule only.
  (ii) **Symport system** It allows diffusion of two types of molecules in a single direction only.
  (iii) **Antiport system** It allows diffusion of two types of molecules in opposite directions.
- All of these types are shown diagrammatically in the following figures

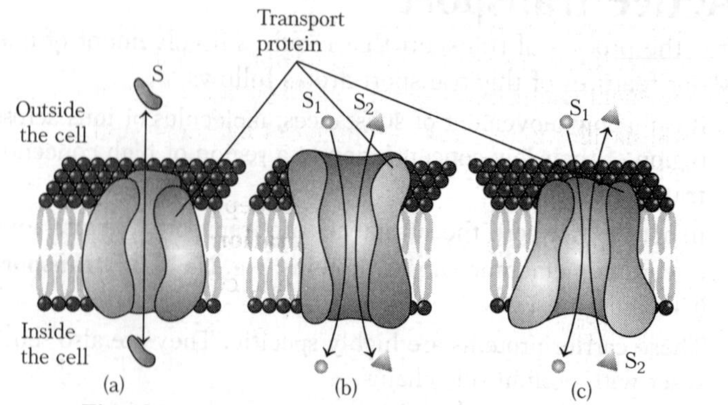

***Fig. 18.1** Types of facilitated diffusion : (a) Uniport, (b) Symport and (c) Antiport*

Comparison between different transport mechanisms

Properties	Simple diffusions	Facilitated transports	Active transports
Requirement of special membrane proteins	✗	✓	✓
Highly selective	✗	✓	✓
Transport saturates	✗	✓	✓
Uphill transport	✗	✗	✓
Requirement of ATP	✗	✗	✓
Respond to protein inhibitors	✗	✓	✓

Permeability in membranes

The term 'Permeability' refers to the degree to which a membrane permits movement of gases, liquids and dissolved substances across it. On the basis of permeability, membranes may be of following types

- **Permeable membranes** These allow diffusion of both solvent and solute molecules, e.g. cellulosic cell wall, lignified cell walls.
- **Impermeable membranes** These prohibit the diffusion of both solvents and solute particles, e.g. cutinised or suberised cell walls.
- **Semipermeable membranes** These allow diffusion of solvent molecules, but do not allow the passage of solute molecules. Such membranes form perfect partition between two osmometers, e.g. membranes of parchment paper and copper ferrocyanide, etc.
- **Differentially permeable membranes** These allow only certain specific substances to pass through them. They are also called as differentially permeable or selectively permeable membrane, e.g. all biological membranes.

Osmosis

- The phenomenon of osmosis was discovered by **Nollet** in 1748. It is a kind of diffusion that plays a role in the movement of water within the plants.
- It is defined as the diffusion of solvent molecules from a less concentrated solution to a more concentrated solution through semipermeable membrane. This movement continues until an equillibrium is attained.
- Osmosis allows free passage of solvent molecules and restricts the passage of dissolved solutes.
- The net direction and rate of osmosis depends on both the **pressure gradient** and **concentration gradient**.

Thus, in the process of osmosis, osmotic concentration plays an important role.

This can be shown by following simple experimental setup.

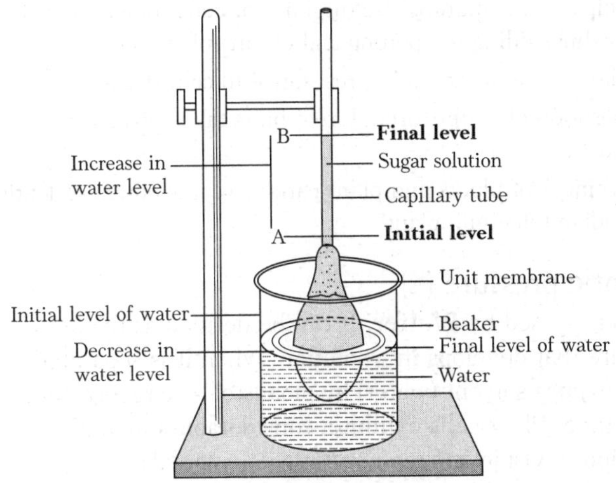

In the capillary tube, sugar solution is present, it is separated from water by a semipermeable membrane. Level of incoming water rises in the capillary tube due to osmosis.

Fig. 18.2 *Demonstration of osmosis*

Osmotic concentrations and solutions

- The amount of dissolved solutes per unit volume of a solution is called **osmotic concentration** of that solution.
- On the basis of osmotic concentrations, solutions can be of three types
  - (i) **Hypertonic solution** Having osmotic concentration higher than that of another solution or cell sap in vicinity.
  - (ii) **Hypotonic solution** Having osmotic concentration lower than that of another solution or cell sap in vicinity.
  - (iii) **Isotonic solution** Having osmotic concentration equal to another solution or cell sap in vicinity.

Types of osmosis

Depending upon the movement of water in or out of a living tissue or cell. Osmosis is of following two types

(i) **Exosmosis** It is defined as the process in which water comes out of the cell. It is usually visible when a cell is placed in a hypertonic solution. As a result of this process, cell becomes flaccid. This is called as exosmosis.

(ii) **Endosmosis** It is defined as the process in which water goes into the cell. It is usually visible when a cell is placed in a hypotonic solution and becomes turgid. This is known as endosmosis.

Note ***Reverse osmosis*** *is the reverse movement of water across semipermeable membrane, from more concentrated to less concentrated solution. It occurs due to the application of external pressure on the side of more concentrated solution.*

Significance of osmosis in plants

- It helps in maintaining the turgidity of cells such as guard cells thus, aiding in opening and closing of stomata.
- It plays a role in providing resistance to cold stress.
- Absorption of water through root hairs takes place by osmosis.
- Movement of plants and plant parts, e.g. movement of leaflet of Indian telegraph plant.

Osmotic pressure (π)

It was proposed by **Pfeffer**. It can be defined as the actual pressure that develops in a solution, when it is separated from its pure solvent (water) by means of a semipermeable membrane. Plant cells exhibit a considerable range of variation in Osmotic Pressure (OP).

Features of osmotic pressure are as follows

- The osmotic pressure of plant cell usually varies from 4-5 atmosphere.
- Osmotic pressure is directly proportional to the concentration of solution and to the temperature.
- It also depends upon the ionisation and hydration of solute particles.
- **Osmometer** is a device which measures osmotic pressure.

Plasmolysis

- When a plant cell is placed in a hypertonic solution, exosmosis takes place. The water moves out of the cells and thus, the cell wall is no longer under tension. Due to this loss of water, the protoplast shrinks and move away from the cell wall. This process of shrinkage of protoplast away from the cell wall is called as plasmolysis.
- It may occur in following three stages
  (i) **Limiting plasmolysis** It is the first stage of plasmolysis where loss of water starts. Here, the protoplast is in contact with cell wall, but does not press it (TP=0). It is characterised by the reduction in size of vacuole and cell.

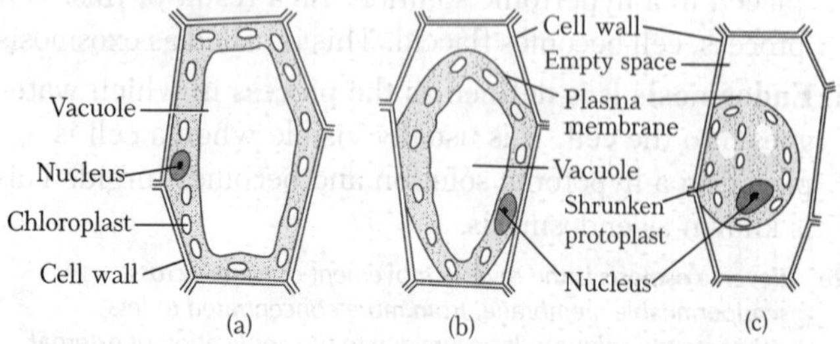

Fig. 18.3 *Plasmolysis : (a) Normal cell, (b) First stage of plasmolysis and (c) Full plasmolysed cell*

  (ii) **Incipient plasmolysis** The exosmosis is continued and size of protoplast further decreases. Therefore, contracting protoplast withdraws from cell wall. This stage is called incipient plasmolysis (TP= negative).
  (iii) **Evident plasmolysis** It is the final stage of plasmolysis. The contraction of protoplast is further continued and protoplast becomes rounded. A cell cannot remain alive in this stage (TP = more negative).

Deplasmolysis

- When a plasmolysed cell or tissue is again placed in water or hypotonic solution, endosmosis takes place.
- Due to this, water enters into the cell sap and the cell becomes turgid.
- Thus, the protoplast again get back to its normal shape and this process is called deplasmolysis.

Imbibition

- The process of adsorption of water by the solid particles of a substance without forming a solution is known as imbibition.
- The water adsorbed by solid substances results in enormous increase in their size.
- The solid substances that adsorb water are called as **imbibants**, e.g. seeds, dry wood, etc. These are mostly hydrophilic substances.
- The liquid which is imbibed is known as **imbibate.**
- As a result of imbibition
  - A pressure known as **imbibition pressure** is developed. The magnitude of this pressure is tremendously high.
  - Considerable amount of heat is generated. This heat is called as **heat of wetting**.
  - Rate of imbibition increases with the increase in temperature.
  - An imbibant with rough surface imbibes more water than the imbibant with smooth surface.
- Some examples of imbibition are as follows
  - Cellulosic cell wall of root hairs absorbs water as a result of imbibition.
  - Germination of seeds involves imbibition.
  - Wooden doors swell in rainy season, due to imbibition.

Note • *In contrast to diffusion, imbibition is a reversible process.*
• *Among plant imbibants, phycocolloids are best imbibants. Lignin and suberin do not show imbibition.*

CHECK POINT 18.1

1. Diffusion pressure decreases as
   (a) solvent (water) is added in the solution
   (b) solute is added in the solution
   (c) atmospheric pressure is decreased
   (d) concentration of solution is decreased

2. Osmosis is defined as the process in which
   (a) water diffuses from lower concentration to higher concentration
   (b) solutes diffuse from lower concentration to higher concentration
   (c) active transport of ions takes place
   (d) passive transport of ions takes place

3. The process of osmosis shall cease when
   (a) solution becomes isotonic
   (b) water concentration becomes equal
   (c) there is no light
   (d) the level of water falls

4. When cell is placed in 0.25 M concentrated sugar solution, there is no change in it. So, the external solution is called
   (a) hypertonic
   (b) isotonic
   (c) hypotonic
   (d) None of the above

5. Cell turgidity is caused by
   (a) endosmosis (b) exosmosis
   (c) plasmolysis (d) diffusion

6. Fresh grapes shall shrink when they are placed in
   (a) hot water
   (b) cold water
   (c) starch water
   (d) concentrated salt solution

PLANT WATER RELATIONS

- Water is the most abundant compound of all living cells. About 85-95% of the weight of a growing herbaceous plant is water. It is a principle determinant of vegetational distribution worldwide.
- The processes that tend to relate plants and water are imbibition, osmosis, plasmolysis, etc.

To understand plant water relation, an understanding of following standard processes is necessary, i.e. solute potential, pressure potential, water potential, etc.

Osmotic or Solute potential (ψ_s)

- Osmotic potential is the decrease in chemical potential of pure water due to the presence of solute particles in it.
- It is the potential of water molecules to move from a less concentrated solution to a more concentrated solution across a semipermeable membrane.

Differences between osmotic pressure (π) and osmotic potential (ψ_s)

- The magnitude of osmotic pressure and osmotic potential are numerically equal. The only difference is that osmotic pressure has positive sign and osmotic potential always carries a negative sign.
  Thus, osmotic pressure (π) = – Osmotic potential ($-\psi_s$)
- The value of osmotic potential or solute potential is calculated by the following formula

$$\boxed{\psi_s = C \times R \times T}$$

where,

C = Concentration of solute particles in mol/L
R = Gas constant with a value of 0.083
T =Temperature in absolute degrees

- The osmotic potential is less negative at 0°C. Due to this, water diffuses from a colder to a warmer liquid. Therefore, the warmer solution has a more negative osmotic potential than the colder one.

Turgor pressure (ψ_p)

When a plant cell or tissue is placed in hypotonic solution, water enters into the cell sap by endosmosis. As a result of entry of the water into the cell sap, a pressure is developed. This pressure presses the protoplast against the cell wall and the cell becomes turgid. This pressure exerted by the protoplast against the cell wall is called as **turgor pressure** or **hydrostatic pressure** or **pressure potential** (ψ_p).

Relation between wall pressure and turgor pressure

- As a consequence of Turgor Pressure (TP), Wall Pressure (WP) is created. It is the force exerted by the cell wall over the protoplast against the turgor pressure.
- Normally, wall pressure is equal and opposite to turgor pressure except when cell becomes flaccid. At a given time, turgor pressure equals the wall pressure.

$$\boxed{TP = WP}$$

Matric potential (ψ_m)

It is an expression of adsorption of water by colloidal particles or surfaces in plant cells. Value of matric potential is negligible.

Gravitational potential (ψ_g)

It represents the effect of gravity on the free energy of water of a cell. The value of gravitational potential is negligible.

Chemical potential

The chemical potential is a quantitative expression of the free energy associated with per mole of any chemical in a system.

Major features related to chemical potential are

- In thermodynamics, free energy represents the potential of performing a work.
- Chemical potential is a relative quantity. It can be expressed as the difference between the potential of a substance in a given state and the potential of the same substance in standard state.
- The larger the chemical potential of a substance, the greater will be its tendency to undergo chemical reaction.
- The unit of chemical potential is energy per mole of substance (J $mole^{-1}$).

Water potential (ψ_w)

(Chemical potential of water)

- Water potential is defined as the difference between the chemical potential of water molecules in the solution and that of pure water at same temperature and pressure.
- The water potential (ψ_w) of pure water at atmospheric pressure is 0 Mpa which is the maximum value.
- Addition of more and more solutes in pure water reduces its water potential to a negative value. It has a negative value and is represented by ψ or ψ_w.
  The unit of water potential is **bars** or **pascal** (1 Mpa = 10 bars).
- Quantitatively, water potential is the sum total of solute potential, the pressure potential and the matric potential.

$$\psi_w = \psi_s + \psi_p + \psi_m$$

where, ψ_s = Solute potential

ψ_p = Pressure potential

- The term ψ_m is often ignored, because it is negligible for heights under five meters.
- Water potential is related to movement of water. Water moves from a region of high water potential to a region of low water potential.
- It can be explained by different conditions as shown in given experiment.

In a U-shaped apparatus, a selectively permeable membrane separates pure water from a 0.1 Molar (M) solution. This solution contains a particular solute that cannot pass freely across the membrane. Water will move by osmosis into the solution, increasing its volume. The values of ψ_p and ψ_s are given for initial conditions, before any net movement of water.

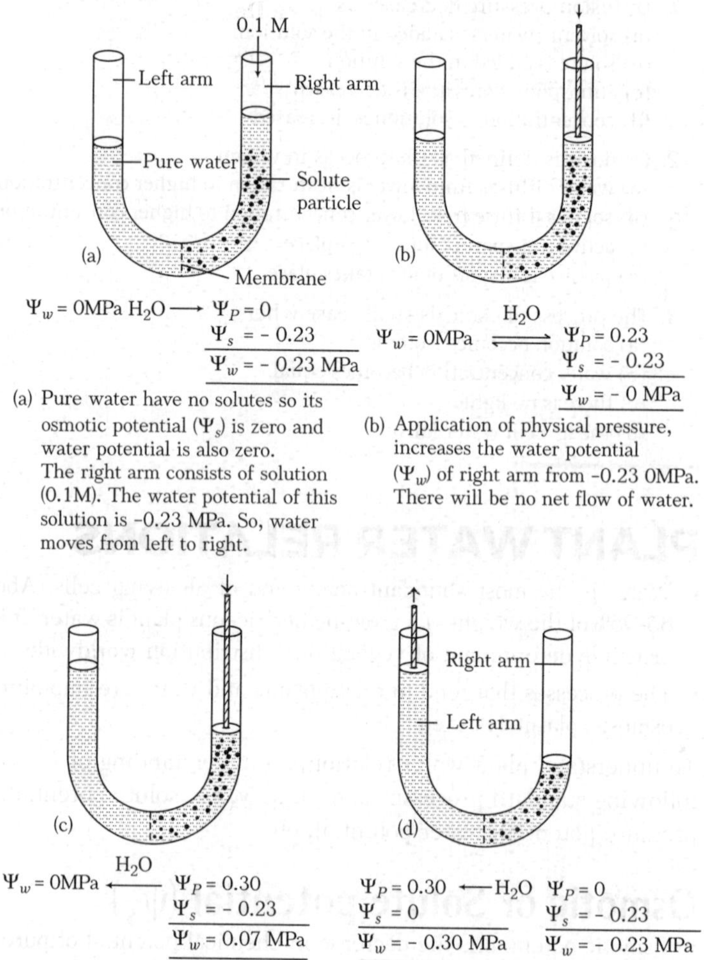

(a) Pure water have no solutes so its osmotic potential (Ψ_s) is zero and water potential is also zero. The right arm consists of solution (0.1M). The water potential of this solution is –0.23 MPa. So, water moves from left to right.

(b) Application of physical pressure, increases the water potential (Ψ_w) of right arm from –0.23 0MPa. There will be no net flow of water.

(c) Application of more pressure to increase the water potential of right arm from 0–0.07 MPa. Water moves from right to left arm.

(d) Using plunger to pull upward the pure water instead of pushing it downward on the solution. A negative pressure of –0.30 MPa and the left arm would be sufficient to draw water from right arm.

Fig. 18.4 *Water potential and water movement : A mechanical model*

Conclusion After this experiment, we conclude that water potential is a tool, which enables us to know the water status in plant cells and tissues. The lower the water potential in a plant cell or tissue, the greater is its ability to absorb water and *vice-versa.*

Thus, the water potential is used to measure water deficit and water stress in plant cells and tissues.

Diffusion Pressure Deficit (DPD)

- The term 'DPD' was coined by **Meyer** (1938). DPD has a positive value and it is directly related to the concentration of solution.

DPD ∝ Concentration of solution

- The difference between the diffusion pressure of a solution and that of its pure solvent at a particular temperature and pressure is called **Diffusion Pressure Deficit** (DPD).
- It determines the direction of net movement of water. Water always moves from low DPD to high DPD.
- DPD is also known as **Suction Pressure** (SP), because it is a measure of ability of cells to absorb water.

Relation between DPD, OP, WP and TP

DPD is related with osmotic pressure, turgor pressure and wall pressure, in the following ways

$$\text{DPD/SP} = \text{OP} - \text{WP} \qquad \text{...(i)}$$

but $\text{WP} = \text{TP}$

So, $$\text{DPD} = \text{OP} - \text{TP} \qquad \text{...(ii)}$$

In fully turgid cells

$\because$ $\text{OP} = \text{TP}$

$\therefore$ $\text{DPD} = 0$ (zero) $\qquad [\because \text{(ii)}]$

In fully plasmolysed (flaccid) cells,

$\because$ $\text{TP} = 0$

$\therefore$ $\text{SP/DPD} = \text{OP}$ $\qquad [\because \text{(i)}]$

Thus, the DPD of plant cells is dependent on both OP and TP. Hence, water is going to move from a turgid cell to flaccid cell.

A table depicting the differences between OP, TP and WP and DPD is given below

Table 18.1 Differences between OP, TP, WP and DPD

OP = ψ_s	TP = ψ_p	WP = ψ_w	DPD
It is the potential pressure that develops in a solution when it is separated from pure solvent by a semipermeable membrane.	It is the hydrostatic pressure which develops in a system due to the entry of a solvent in it through osmosis.	It is the force exerted by the wall against expansion in osmotic system.	It is the reduction in diffusion pressure or free energy of a solvent in an osmotic system so over pure solvent.
Develop in both confined and unconfined systems.	Develops only in a confined system.	Develops in a confined system.	Develops in both confined and unconfined systems.
Dependent upon the member of solute particles in the solution.	Dependent upon the amount of water that enters a system.	Dependent upon TP and the nature of wall	It is dependent upon OP (or ψ_s) of the system, Op (or ψ_s) of outside solution and the resistance that the system puts against entry of more solvent into it.
It changes slightly in an osmotic system.	It is variable and changes from zero to equal the OP.	It is commonly equal and opposite to TP.	It varies from zero to equal value of OP.
It helps in retaining water inside cells.	It provides turgidity to the cells, tissues and softer organs.	It prevents bursting of cells and limits expansion.	It controls the entry of water into cells.
It causes the movement of water across a semipermeable membrane.	It is essential for growth of cells.	Reduction of wall pressure is required for the growth of cells.	It determines whether the cell will lose or gain water from an external medium.

ABSORPTION OF WATER IN PLANTS

- The process of water absorption in plants involves the interaction of plant roots, soil and soil water.
- Most of the absorption of water from soil occurs through root hairs.
- The root hairs are present at the root apex. These are thin-walled slender extensions of root epidermis. They provide large surface area for absorption.
- The walls of root hairs are permeable and consist of pectic and cellulosic substances.
- Root hairs contain vacuoles filled with cell sap, salts, sugars and organic acid. The osmotic pressure of root hair cell is very high.
- This facililates absorption of water by cells *via* osmosis.

Pathways of water movement

Water moves from the epidermis to the endodermis of the roots *via* following two pathways

1. Apoplast pathway

- In this pathway, water passes from root hairs to xylem by osmosis through cell wall, without crossing any membranes.
- The apoplast involves a continuous system of cell walls and intercellular air spaces in plant tissues, i.e. the non-living part of cell.
- This kind of water movement occurs through mass flow along the cell wall.

2. Symplast pathway

- In this pathway, water enters the outermost root cell through the plasma membrane and then travels from one cell to another through plasmodesmata by crossing membranes.
- It may be of two types
  (i) **Non-vacuolar symplast pathway** In this pathway, water moves from one cell to another passing through the plasmodesmata of cells. Here, water does not enter the vacuole.
  (ii) **Vacuolar symplast pathway** In this pathway, apart from the plasma membranes, water also crosses the **tonoplast**, i.e. membrane of vacuole.
  Both apoplast and symplast pathways are shown in the following figure

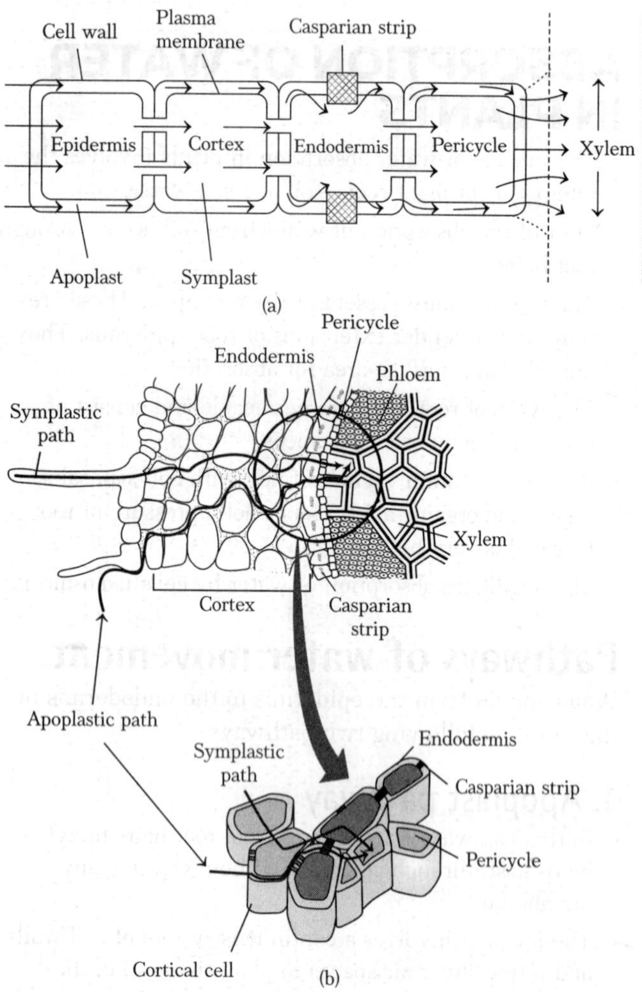

***Fig. 18.5** (a) Pathways of water movement inside the root, (b) Apoplast and symplastic pathways of water and ions transport across the root*

Casparian strip and apoplast pathway

- At the endodermis, water movement through the apoplast pathway is interrupted by the Casparian strip.
- It is a band of radial cell walls in the endodermis.
- It is impregnated with the wax-like hydrophobic substance called **suberin**, which acts as a barrier to water and solute movement.
- Casparian strip breaks the continuity of the apoplast pathway. Thus, it forces water and solutes to cross the endodermis by passing through the plasma membrane.
- Hence, despite the importance of the apoplast pathway in the root cortex and the stele, water movement across the endodermis largely occurs through the symplast.

Mechanism of water absorption

Water absorption by plants is of two types

1. Active absorption

In this type of absorption of water, root cells play active role. It is the force present in the root, which helps in water absorption. ATP is required in this type of absorption. It is further of the following two types

Osmotic absorption

- The osmotic pressure of cell sap of root hair is higher than that of soil water. Due to this, DPD or suction pressure in root hair becomes higher than soil water. Therefore, osmotic diffusion occurs and water enters into the root hair cell through plasma membrane.
- This causes the OP, suction pressure or DPD of root hair to become lower and TP to increase. Now, the cortical cells adjacent to root hairs have higher OP, suction pressure or DPD as compared to root hair cells.
  Therefore, water enters into the cortical cells by osmotic diffusion.
- In this way, the water is transported from cell to cell *via* osmotic diffusion and gradually reaches the innermost cortical cells and endodermis. In the last step, water is drawn into xylem from turgid pericycle cells.

Non-osmotic absorption

Sometimes, absorption of water takes place even when OP of soil water is higher than that of cell sap. This type of absorption is against the osmotic gradient. Hence, this type of absorption requires energy for the movement of H_2O.

2. Passive absorption

- In actively transpiring plants, mesophyll cells loose water due to transpiration. As a result of this, TP of mesophyll cells decreases, while DPD increases.

- Thus, they take up water from their neighbouring cells. As a result of this, TP of neighbouring cells decreases, while their DPD increases. It leads to the movement of water into them from their neighbouring cells.
- The continuation of this process leads to the development of a tension or suction pressure gradient.
- This tension is transmitted to the xylem of roots through the xylem of stem and thus, water rises upward into the transpiring surface. This is known as **passive absorption,** because the water is pulled through the roots and root cells remain passive during this process.

Factors affecting the rate of water absorption

- Absorption of water occurs if the amount of water present in soil is less than that of the root hair.
- Low temperature inhibits water absorption. The absorption is maximum generally between 20-30° C of soil temperature, i.e.
  Temperature α Rate of absorption
- The concentration of soil solution inhibits absorption of water.
- Increased concentration of CO_2 and deficiency of O_2 in the soil retards absorption of water.

ASCENT OF SAP

(TRANSLOCATION OF WATER)

- Sap is water with dissolved inorganic ions.
- The upward movement of this sap from roots to the aerial parts of a plant is known as **ascent of sap.**
- Various experiments demonstrate that the actual pathway of ascent of sap is through xylem tissue in terrestrial plants.
- A number of theories have been put forward to explain the mechanism of ascent of sap. These theories are as follows

Vital force theory

- This theory is also called as **pulsation theory.** This theory was supported by **Godlewski** and **Sir JC Bose.**
- According to Godlewski (1884), ascent of sap takes place due to pumping activity of xylem parenchyma, which are living components of xylem.
- According to Bose (1923), upward translocation of water takes place due to pulsatory activity of living cortical cells present just outside the endodermis. These cells absorb water from outside and pump it to xylem.
- This theory was hypothetical and indicatively discarded by the experiments of **Strasburger** which were described well before 1900. He demonstrated that the ascent of sap continues in the stems, in which xylem parenchyma living cells have been killed by the uptake of poisons. Thus, vital force theory does not hold true.

Root pressure theory

- This theory was put forward by **Priestley** in 1916.
- Root pressure is a positive hydrostatic pressure developed in xylem vessels due to the metabolic activities of roots. This pressure gets retarded under conditions of starvation, low temperature, drought and reduced availability of oxygen.
- According to this theory, root pressure is developed when the rate of water absorption is more and that of transpiration is less. This pushes the water upward in the xylem of roots.
- Root pressure theory was not accepted because of the following reasons
  - root pressure has a low magnitude (about 2 atm)
  - it is absent in gymnosperms.
  - The ascent of sap continues in the absence of root also.

Note *Ascent of sap is demonstrated by using girdling or ringing experiment.* ***Malpighi*** *introduced this experiment for the first time in 1675. In this experiment, role of xylem in ascent of sap was demonstrated by cutting or girdling a portion of bark, but not the xylem. The path of ascent of sap can be visualised using a dye like* ***eosin****.*

Cohesion-tension and transpiration pull theory

- This theory was given by **Dixon** and **Jolly** in 1894.
- Till now, it is the most accepted theory for the mechanism of ascent of sap.
- This theory is also known as **Dixon's cohesion theory** or **cohesion-tension theory**.
- This theory is based upon the following physical forces

1. Adhesion

It is defined as the force due to which water molecules remain associated to the xylem surface (polar).

2. Cohesion

The water molecules in the xylem vessels remain joined to each other as a result of cohesion, i.e. form a attraction between water molecules. It provides surface tension that accounts for high capillarity.

Role of adhesion and cohesion in transpiration

Both these properties give high tensile strength (i.e. an ability to resist a pulling force) and high capillarity (i.e. ability to rise in thin capillaries). Thus, due to adhesion and cohesion, a continuous water column is maintained in the xylem vessels. This continuous water column is then pulled upward due to a force known as **transpiration pull.**

3. Transpiration pull

- This is a force developed due to transpiration. This pull causes mesophyll cells to loose water and their DPD increases. This leads to absorption of water from adjoining cells.
- This continues transmission of water from one cell to another creating a force or pull, which is responsible for pulling the continuous water column.
- Transpiration can pull water up to a height of 130 m.

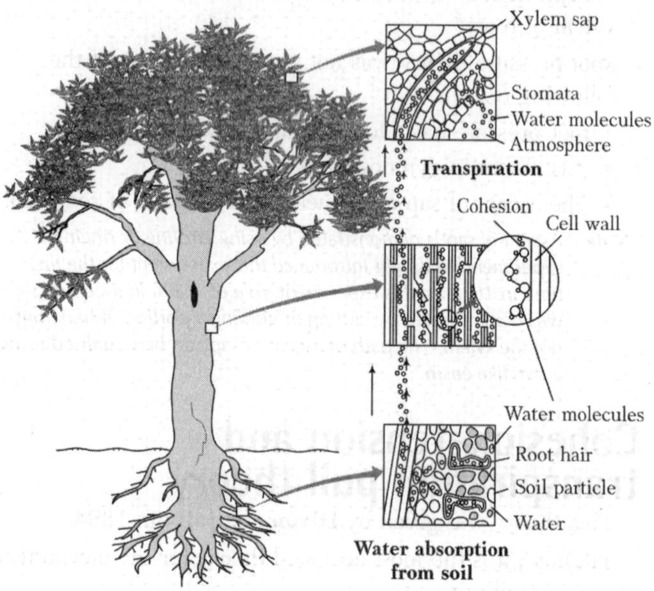

Fig 18.6 Transpiration pull

Objections to cohesion-tension theory

This is the most acceptable theory, yet there are some objections against it.

The following objections are raised against this theory

- The gases dissolved in sap shall form air bubbles under tension and high temperature. Air bubbles would break the continuity of water column. This phenomenon is known as **cavitation.**
- Overlapping cuts do not stop ascent of sap though they break the continuity of water column.

Physical force theories

Various physical forces may be involved in the ascent of sap. These are as follows

1. Capillary theory (Boehm; 1863)

- According to this theory, the tracheids and vessels act as narrow capillary tubes.
- Water moves upward in these narrow tubes due to the force of surface tension. This phenomenon is known as **capillarity** or **capillary action.**
- This upward movement of water continues untill the gravitational force balance the capillarity force in capillary tube of 1/100 mm. It can raise water up to a height of 3 m.
- There are many objections against this theory which are as follows
  - For capillarity, a free surface is required.
  - The magnitude of capillary force is low.
  - In gymnosperms, usually the vessels are absent and other xylem elements do not form continuous channels.

2. Imbibition theory (Sachs; 1878)

According to this theory, the ascent of sap could take place by imbibition through the walls of xylem. This theory was discarded, due to one major fact that it takes place through lumen of xylem elements not through walls.

3. Atmospheric theory

When water is lost by transpiration, the pressure at the transpiring surface falls below that of atmosphere. At this time, water rises upward to fill up the gap in fall of atmospheric pressure from the soil.

This theory is not acceptable as

- In case, it is working.
- It cannot act on water present in xylem in roots.
- It will not be able to raise water beyond 34 feet.

Factors affecting ascent of sap

The factors such as high temperature, low atmospheric humidity, high atmospheric pressure and wind velocity permit more rapid transpiration. Thus, these conditions promote upward movement of H_2O. Deficiency of H_2O can decrease the rate of ascent of sap.

CHECK POINT 18.2

1. The movement of water from one cell of the cortex to the adjacent one in roots is due to
   (a) accumulation of inorganic salt in the cells
   (b) accumulation of organic compound in the cell
   (c) chemical potential gradient
   (d) water potential gradient

2. DPD is equal to
   (a) OP – WP (b) OP + WP
   (c) TP (d) OP

3. Most of the water flow in the root takes place in apoplast because
   (a) cortical cells are loosely arranged
   (b) cortical cells are living cells
   (c) cortical cells are thin-walled cells
   (d) All of the above

4. Which of the following helps in the absorption of water and mineral salts?
   (a) Mycorrhiza (b) *Anabaena*
   (c) *Nostoc* (d) None of these

5. Physical force theory explains
   (a) non-living cells are not essential for ascent of sap
   (b) living cells are not essential for ascent of sap
   (c) ascent of sap may occur in both living and non-living cells
   (d) Both (b) and (c)

6. The drawback of atmospheric theory of ascent of sap is
   (a) it is not applicable to small vascular plants
   (b) atmospheric pressure does not affect pull of water
   (c) it does not account for water column during night
   (d) it cannot raise water beyond 34 feets

TRANSPIRATION

The biological process by which water is lost in the form of water vapours from the aerial portions of living plants is known as **transpiration**. Though, transpiration reduces water levels in the soil, it is necessary for water and mineral absorption and ascent of sap. Therefore, transpiration is also known as a **'necessary evil'** for plants.

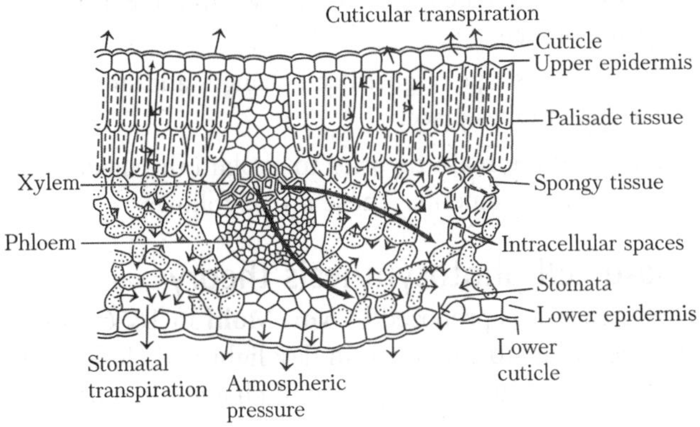

***Fig 18.7** Transpiration*

Types of transpiration

Depending upon the plant surface, transpiration is of following three types

Cuticular transpiration

- It occurs through the cuticle or epidermal cells of the leaves and other exposed parts of the plant. It is also known as **peristomatal transpiration**.
- Although cuticle is impermeable to water, still some water may be lost through it, as it is permeable to water vapours.
- Cuticle accounts for only 3-10% of the total transpiration.

Lenticular transpiration

- It is the type of transpiration which occurs through lenticel.
- Lenticular transpiration is found only in woody branches of the trees, where lenticels are present.

Stomatal transpiration

- The transpiration that takes place through stomatal opening are known as stomatal transpiration.
- About 85-90% of transpiration occurs through stomata.
- Stomata are structures found in the epidermis of leaves. These are specialised to allow gaseous exchange between plant and atmosphere.
- In monocots, e.g. grasses, they are equally distributed on both sides of leaf.
- In aquatic plants with floating leaves, they are present mainly on the upper surface.

Structure of stomata

- Stomata are the small pores found in the epidermis of leaves. Stomatal apparatus consists of a pore called **stoma** that is enclosed by kidney-shaped guard cells on the outer side.
- Stomata open and close according to the turgidity of guard cells. These guard cells contain chloroplasts that contain chlorophyll. It helps in capturing light energy.

- Around these guard cells, 2-4 subsidiary cells are also found. These subsidiary cells are present between guard cells and protect epidermal cells when the guard cells expand during stomatal opening.

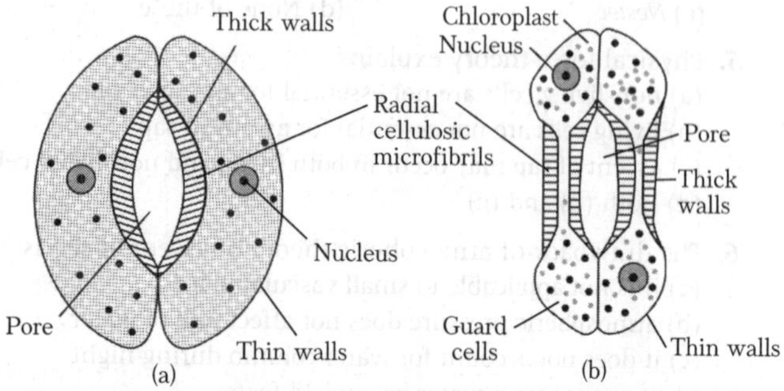

Fig. 18.8 *Open stomata : (a) Dicotyledonous (kidney-shaped), (b) Monocotyledonous (dumb-bell-shaped)*

- Number of stomata on a leaf is about 1000-60,000 per sq. cm.
- The size of stomatal pore (fully open) measures about 3-12 μm in diameter and 10-40 μm in length.
- Guard cells are small in size and are rapidly influenced by turgor changes. In the walls of guard cells, cellulosic radial thickenings are present.

Note
- *Unlike other epidermal cells, the guard cells contain small chloroplast. These have both photosystems-I and II, but they cannot assimilate food, because of the absence of Ribulose bisphosphate carboxylase and NADP linked triose phosphate dehydrogenase.*
- *In* ***Cyperus****, both kidney-shaped and dumb-bell-shaped guard cells are found.*

Types of stomata

Stomata can be of following types

1. **On the basis of position and the presence of accessory cells**
   - **Anomocytic** Accessory cells are absent, e.g. Ranunculaceae, Malvaceae, etc.
   - **Diacytic** Two accessory cells perpendicular to the pore are found, e.g. Acanthaceae, Caryophyllaceae, etc.
   - **Paracytic** Two accessory cells parallel to the stoma are found, e.g. Rubiaceous, etc.
   - **Anisocytic** Three accessory cells are found, e.g. Cruciferae, *Solanum*, etc.
2. **On the basis of distribution and position**
   - **Hypostomatic** (Apple and mulberry type) Stomata are found on the lower side, e.g. apple, nuts, etc.
   - **Anisostomatic** (Potato type) Stomata are found more on the lower side and fewer on the upper surface, e.g. potato, tomato, etc.
   - **Isostomatic** (Oat type) Number of stomata are equal on both the surfaces, e.g. wheat, maize, etc.
   - **Epistomatic** (Water lily type) Stomata are found on upper surface only, e.g. water lily, etc.
   - **Astomatic** (Potamogeton type) Stomata are absent.
3. **On the basis of opening and closing**
   - **Lucerne** (Alfalfa type) Remain open all day, e.g. pea, beans, etc.
   - **Potato type** Remain open throughout the day and night, but may close for few hours in evening, e.g. *Allium*, tulip, etc.
   - **Barley type** Open for few hours in day, e.g. wheat, maize, etc.

Stomatal movement

The opening and closing of stomatal pore is called **stomatal movement**. This movement is shown in response to entry or exit of water from the guard cells.

The process of movement is very precise and can be understood with the help of the following steps

- The guard cells differ from other epidermal cells in containing chloroplast and radial thickenings of cellulosic microfibrils in their walls that face stomatal pore.
- In response to changes in turgor pressure, these cellulosic microfibrils radiate outwards or inwards around their circumference.
- Due to this, shape of guard cell changes and stomatal movement takes place.

There are three main theories to explain the mechanism of stomatal movements. These are as follows

Guard cell photosynthesis theory

This theory was proposed by **von Mohl** in 1856. He observed that stomata open in day light and close at night. According to him, the sequence of changes in stomata are as follows

During day

Light ⟶ Photosynthesis in guard cells ⟶ Formation of sugar ⟶ Increase in OP of guard cells
↓
Stomata open ⟵ TP of guard cell increases ⟵ Water comes inside the guard cells

During night, there is no photosynthesis and hence, these changes are reversed.

Starch–sugar interconversion theory

- This theory was given by **Sayre** in 1923 and was modified by **Steward** in 1964.
- According to this theory, **during day** (light), CO_2 concentration in guard cells decreases, leading to an increase in pH.
- High pH favours hydrolysis of starch into glucose-1-phosphate, it further changes into glucose.
- Due to this, osmotic potential becomes lower in guard cells and water from the surrounding epidermal and mesophyll cells enters into the guard cells by osmotic diffusion. Guard cells become turgid and stomata open.
- **During night** (dark), reverse process occurs and glucose-1-phosphate is converted into the starch in guard cells. This increases the osmotic potential and the guard cells release water. They become flaccid and stomata are closed.

During light

$$\text{Starch} + n\text{H}_3\text{PO}_4 \xrightarrow{\text{Phosphorylase}} n\,\text{Glucose-1-phosphate}$$

$$\text{Glucose-1-phosphate} \underset{}{\overset{\text{Mutase}}{\rightleftharpoons}} \text{Glucose-6-phosphate}$$

$$\text{Glucose-6-phosphate} + \text{Water} \xrightarrow{\text{Phosphatase}} \text{Glucose} + \text{H}_3\text{PO}_4$$

During dark

$$\text{Glucose} + \text{ATP} \xrightarrow{\text{Hexokinase}} \text{Glucose-6-phosphate} + \text{ADP}$$

$$\text{Glucose-6-phosphate} \overset{\text{Mutase}}{\rightleftharpoons} \text{Glucose-1-phosphate}$$

$$\text{Glucose-1-phosphate} \xrightarrow{\text{Phosphatase}} \text{Starch} + \text{H}_3\text{PO}_4$$

Starch-sugar interconversion theory is not universally applicable. It may operate under certain circumstances only.

Objections There are many objections to this theory because

- Glucose is not found in guard cells at the time of stomatal opening.
- Starch $\rightleftharpoons$ Sugar changes are chemically slow, while opening and closing of stomata are rapid.
- Change in pH cannot be explained on the basis of CO_2 concentration.

Malate or K^+ ion pump theory

- This theory was given by **Levitt** in 1974. This is also known as the **modern theory of stomatal movement.**
- **During daytime** in the presence of light, pH of the guard cell can rise due to active H^+ uptake by guard cell in chloroplasts or mitochondria.
- A rise in pH during daytime causes hydrolysis of starch due to which organic acid, i.e. phosphoenol pyruvate is formed.
- Phosphoenol pyruvate combines with available CO_2 to produce malic acid.
- Malic acid dissociates to produce H^+ and malate.
- The protons (H^+) are pumped out from the guard cells into the subsidiary cells and in the exchange, K^+ ions are pumped into guard cells, from the adjacent subsidiary cells.
- This exchange of H^+ and K^+ ions is followed by the entry of Cl^- ions into the guard cells with the help of *c*AMP, ATP and cytokinins.
- K^+ ions immediately combine with malate and increase the osmotic concentration of the guard cells.
- As a result, guard cells absorb water from the nearby epidermal (subsidiary) cells, swell up and lead to opening of stomata.

Reverse situation prevails **during dark** when stomata are closed. There is no accumulation of K^+ ions in guard cells in dark.

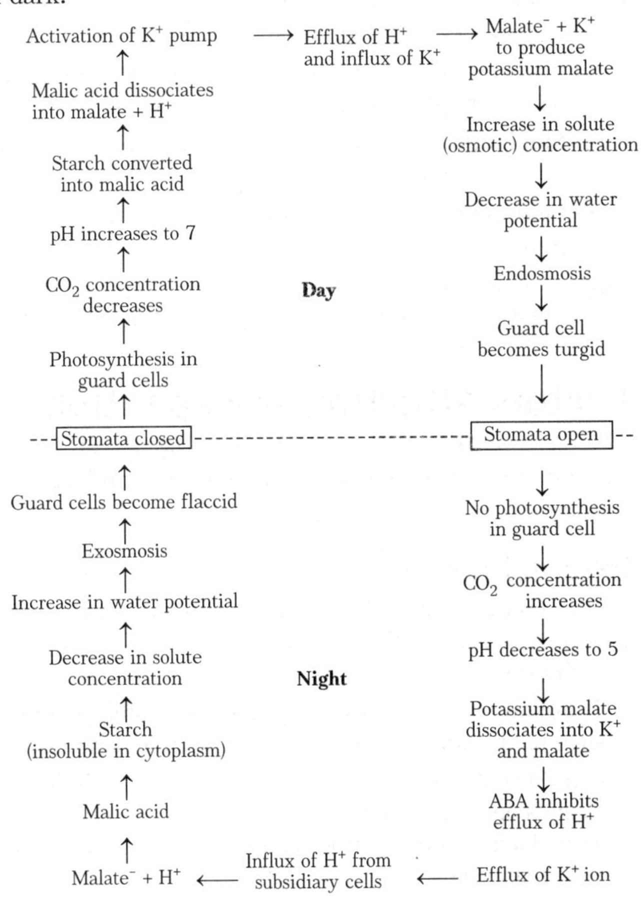

***Fig. 18.9** Active* K^+ *theory*

Factors affecting stomatal movement

(i) **Light** Stomata usually open in light and close in darkness. Therefore during day, the rate of transpiration increases, while in the night, it decreases. Blue light keeps stomata open during the day by promoting the movement of K^+ ions into guard cells.

(ii) **Temperature** Usually, an increase in temperature results in stomatal opening, because in high temperature, water evaporates rapidly.

(iii) **CO_2 concentration** Reduced CO_2 concentration favours opening of stomata, while increase in CO_2 concentration promotes closing of stomata.

(iv) **Water deficit** In leaves, it causes stomatal closure due to rise in DPD of epidermal cells.

(v) **Growth hormone** Cytokinins are essential for the opening of stomata, while abscisic acid takes part in closing of stomata.

(vi) **Atmospheric humidity** In humid environment, stomata remain open for longer period, while in dry environment, they remain closed for longer period.

(vii) **pH** High pH is required for the opening of stomata, while low pH induces closure of stomata.

(viii) **Mechanical shocks** Mechanical disturbances cause closing of stomata.

(ix) **Leaf area** It is directly proportional to the rate of transpiration.

Note *Antitranspirants retard transpiration in plants, e.g. colourless plastic, resins, silicon, oils, low viscosity waxes, phenyl mercuric acetate (fungicide).*

Factors affecting transpiration

There are various factors, which affect the rate of transpiration. These are divided as external and internal factors.

External factors

- When relative humidity is high, the rate of transpiration decreases, because the atmosphere is already saturated with moisture.
- Transpiration increases with the increase in temperature. High temperature opens the stomata even in darkness.
- Light increases the rate of transpiration because in light, stomata open and due to this, temperature increases. In dark, due to closure of stomata, stomatal transpiration is almost stopped.
- Transpiration is less in still wind. It is because water vapour accumulates around the transpiring organs and reduce the DPD of the air. Rate of transpiration increases with increase in wind velocity. It removes moisture from the transpiring parts of the plants.
- A decrease in amount of available soil water reduces absorption thus, causing a reduction in the rate of transpiration also.
- Low atmospheric pressure (as in mountain areas) enhances the rate of transpiration.
- **Antitranspirants** help to reduce rate of transpiration. These include abscisic acid, phenyl mercuric acetate, stomatal inhibitors, etc.

Internal factors

- The more the leaf surface area, the more is the rate of transpiration. Due to this xerophytic plants usually reduce the leaf surface by transforming into spines, e.g. cactus.
- An increase in the root/shoot ratio causes an increase in the rate of transpiration.
- Water content of the leaves is directly proportional to the rate of transpiration.
- More the number of stomata more is the ratio of transpiration.

Measurement of transpiration

(i) **Weighing method** This method consists of weighing the potted plants before and after transpiration.

(ii) **Cobalt-chloride method** It is based on the fact that the dry cobalt-chloride is blue in colour, while the moist cobalt-chloride is pink in colour. This method is used to measure the rate of transpiration.

(iii) **Potometer method** This method is based on the assumption that the rate of water uptake is approximately equal to the rate of transpiration. The apparatus used for this purpose is called **potometer** or **transpirometer.**

Three common types of potometer are

(a) Darwin's potometer (b) Armer's potometer
(c) Ganong's potometer

Limitations

(i) **Loss of turgidity** is quite common during noon due to transpiration being higher than the rate of water absorption.

(ii) **Reduced growth** Transpiration reduces the ability of water inside the plant. Water deficit decreases growth.

Advantages of transpiration

- It plays an important role in the upward movement of water, i.e. ascent of sap.
- It helps in absorption and translocation of mineral salts.
- Rapid evaporation of water from the aerial parts of the plant through transpiration brings down their temperature. Thus, it prevents them from excessive heating. This is also known as **cooling effect**.
- Transpiration maintains the shape and structure of plant by keeping the cells turgid.

Guttation

- It is the loss of water in the liquid state from uninjured parts of plants, usually from tips and margins of leaves.
- It occurs during night or early morning when there is high atmospheric humidity, high root pressure and transpiration is also less.
- In guttation, water exudes from the grooves of leaf cells called **hydathodes**.
- A hydathode is an opening or pore in the leaf epidermis, present usually at the margins of leaves. Around this pore, several thin-walled parenchyma cells are grouped.
- Guttation depends on root pressure. The development of root pressure in a plant leads to creation of a positive hydrostatic pressure in xylem sap throughout the plant.
- The water conducting xylem elements of a vascular bundle terminate in a hydathode. Due to this, xylem sap is forced to flow through the hydathodes.

CHECK POINT 18.3

1. The chief role of transpiration in plant is to cause
(a) loss of surplus water
(b) cooling of the plant
(c) rapid ascent of sap
(d) rapid rise of minerals

2. Stomatal transpiration accounts for
(a) 85-90% of total transpiration
(b) 3-10% of total transpiration
(c) 100% of total transpiration
(d) 1-2% of total transpiration

3. In which of the following plants would metabolism be hindered, if the leaves are coated with wax on their upper surface?
(a) *Hydrilla* (b) Lotus
(c) *Pistia* (d) *Vallisneria*

4. Stomata open during daytime because the guard cells
(a) are thin-walled
(b) are bean-shaped
(c) have to help in gaseous exchange
(d) photosynthesise and produce sugar or organic acids

5. The transpiration in plants will be lowest
(a) when there is high humidity in the atmosphere
(b) high wind velocity
(c) there is excess of water in cell
(d) environmental conditions are very dry

6. Guttation usually occurs when the plant is put
(a) in more saturated atmosphere
(b) in more humid soil
(c) in dry condition
(d) in desert

TRANSPORT OF FOOD

The transport of food from the production centre, i.e leaves to the various consumption centres through the phloem tissue is called **translocation**. The phloem sap is mainly consists of water and sucrose.

Substances transported in phloem

- About 90% of the total solutes carried in the phloem is carbohydrate in the form of sucrose (transport form of carbohydrate).
- Sucrose is a relatively inactive and highly soluble sugar. Thus, it can easily be transported in phloem.
- Sucrose can later be converted back to the more active monosaccharides, glucose and fructose easily.
- Phloem also carries some mineral elements in various forms, such as nitrogen and sulphur in the form of amino acids, phosphorus in the form of inorganic phosphate ions and sugar phosphates and potassium ions.
- Small amounts of vitamins, growth substances such as auxins and gibberellins and other components may also be present in phloem sap.

Routes of translocation

Solutes are translocated in various directions within a plant *via* different routes of translocation. These routes are described below

(i) **Downward translocation of organic solutes** It is the most common pathway of translocation. The food synthesised in leaves is translocated in the downward direction to stem and root system. This food is utilised for maintenance, metabolism, growth and storage.

(ii) **Upward translocation of organic solutes and mineral salts** Under many conditions, such as during germination of seed, tubers, etc., an upward translocation of organic solutes also takes place in plants.

- The translocation in this direction of root (storage tissue) to leaves or other apical region is termed as **upward translocation.**
- For example, during the early stages in their expansion, the leaves do not photosynthesise at a sufficient rate.
- In this condition, food moves in an upward direction from the storage tissues of the stems to the leaves.
- Similarly in the stems of woody species, when the buds resume growth in the spring, the tissues of the new shoots constructed are largely out of foods.

(iii) **Lateral translocation of solutes** Lateral translocation of phloem sap in tangential direction readily occurs in woody stems. It is a slow process that mainly occurs *via* vascular rays.

Mechanism of translocation

Several theories have been put forward to explain the mechanism of solute movement in the phloem. Some of the important ones are given below

1. Cytoplasmic streaming theory

- In 1923, **Curtis** stated that the food materials are transported through the living cells of the phloem by the streaming movement of cytoplasm, i.e. cyclosis.
- The cytoplasm carries the organic solutes within it to atleast some distance upward or downward.
- **Mason** and **Phyllis** later rejected the theory of cytoplasmic streaming because the rate of cytoplasmic streaming is too slow to account for phloem transport.

2. Diffusion theory

- **Mason** and **Maskell** (in 1928), reported that the food is translocated from its higher concentration to lower concentration by diffusion.
- Since, diffusion is a slow process, **Mason** and **Phyllis** modified this theory as **activated diffusion theory.**
- According to this theory, organic solutes are first energy activated and then transported.

3. Transcellular streaming theory

- **Thaine** (in 1962 to 1969) believed that the sieve tubes contain proteinaceous tubular strands, which translocate the solutes.
- These strands demonstrate a kind of peristaltic movement, which helps in the translocation of organic substances.
- The transcellular strands have however, not been observed in sieve tubes.

4. Mass or Pressure flow theory

- **Ernst Munch** (in 1930), firstly proposed this hypothesis of solute transport in phloem tissue which is also called as the **pressure flow hypothesis** or the **Munch hypothesis.**

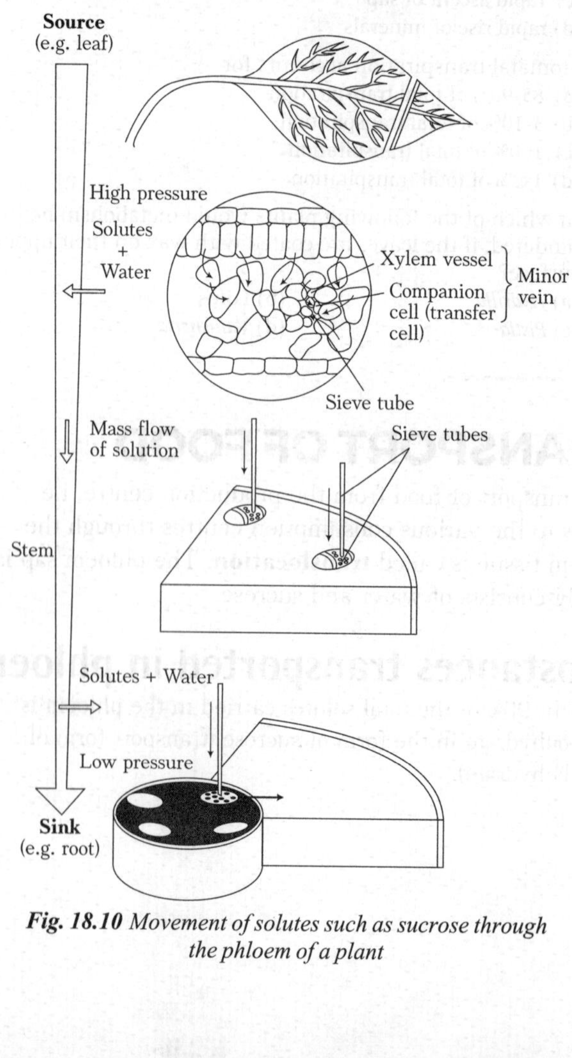

Fig. 18.10 Movement of solutes such as sucrose through the phloem of a plant

- This is the most widely accepted theory for translocation of solutes.
- According to this theory, organic solutes are translocated through the sieve tubes from the leaves which is the production site or source to the consumption site or sink such as root, fruits and tuber.
- The site of source and sink may be reversed depending on the season or plant needs.
- The direction of movement of solutes in the phloem can be upward or downward, i.e. it is **bidirectional** unlike xylem in which movement is always **unidirectional**.
- The main steps of phloem sap transport are as follows
  (i) **Source** are any region where organic solutes are produced.
  (ii) **Phloem loading** Due to active photosynthesis, the mesophyll cells contain organic solutes in them. Thus, these cells have a very high osmotic concentration and turgor pressure. Due to this, sugars dissolved in water, move from mesophyll cells into the companion cells. Hence, the OP and TP of companion cell increase and the solutes are transferred to sieve tubes *via* active transport. This is known as **loading of phloem** with solutes.
  (iii) **Transport of solutes** Pressure inside sieve tubes is greatest at the source and lowest at the sink. Sieve tubes acquires water from the xylem and their TP increases.
  This increase causes transportation of solutes to areas with low TP. In this way, solutes are transported throughout the plant.
  (iv) **Sinks** These are the regions where solutes are being used, e.g. roots, fruits, storage organs and regions of growth.
  (v) **Phloem unloading** At the sink, the osmotic concentration and TP is always low. Thus, there is a continuous pressure gradient between the source and the sink. Thus at the sink, phloem is unloaded *via* active transport.

Note *Apparatus used for various measurements.*
- *Tensiometer—Soil water potential.*
- *Atmometer—Transpiration.*
- *Psychrometer—Atmospheric pressure*

The objections to the mass flow hypothesis are as follows
- Phloem transport is not influenced by water deficit. Mass flow is not a purely physical process as described by **Munch**.
- Slime content and other fibrils of the sieve tube reduce the speed of flow of solutes even under high pressure.
- Bidirectional transport of solution to the same sieve tube needs explanation.

CHECK POINT 18.4

1. Hydathodes are also called
(a) guard cells
(b) water stomata
(c) sunken stomata
(d) subsidiary cell

2. Why sugars are transported in the form of sucrose in phloem?
(a) It is inactive and highly soluble
(b) It is active
(c) It yields high ATP
(d) It is lighter in weight

3. In downward translocation, food is transport
(a) from below to upward
(b) from roots to stem
(c) from leaves to root
(d) phloem never conduct food

4. Munch hypothesis is based on
(a) translocation of food due to TP gradient and imbibition force
(b) translocation of food due to TP gradient
(c) translocation of food due to imbibition force
(d) None of the above

5. Which of the following is the main point in distance of Munch theory?
(a) Translocation is a pure physiological process
(b) Translocation is a pure physical process
(c) Explanation of sugar transport is not given
(d) None of the above

6. Which apparatus is used to measurement of soil water potential?
(a) Psychrometer (b) Atmometer
(c) Tensiometer (d) Goniometer

EXAM POINTS

- The main difference between active and passive transport across cell membrane is passive transport requires a concentration geradient across a biological membrane whereas active transport requires energy to move solutes. **[NEET 2019]**
- Water flow into the cell when a plant cell is placed in a hypotonic solution. **[NEET 2019]**
- If the external solution balances the osmotic pressure of cytoplasm, it is said to be isotonic. **[Kerala CEE 2015]**
- When a cell is placed in a hypertonic solution, the protoplasm shrinks and leaves the cell wall due to exosmosis and cell becomes flaccid. **[KCET 2011]**
- The process by which water is absorbed by solid-like colloid causing them to increase in volume is called imbibition. **[Kerala CEE 2011]**
- Polypeptides have more imbibing capacity than polysaccharides. **[EAMCET 2015]**
- The water potential of pure water is zero. Solute potential is always negative and pressure potential is usually positive. **[NEET 2017]**
- The modified equation for water potential is $\psi_w = \psi_S + \psi_P$. **[AIIMS 2017]**
- In completely plasmolysed cell, pressure potential does not contribute to water potential. **[EAMCET 2015]**
- Movement of water through cell wall is apoplast pathway. **[AIIMS 2017]**
- Apoplast is located outside the plasma membrane. **[WB JEE 2015]**
- A mutant plant is unable to produce materials or precursors that form Casparian strips. This plant would be unable to control amount of water and solute it absorbs. **[AIIMS 2018]**
- Phloem sap is mainly made up of water and sucrose. **[AMU 2012]**
- Cohesion theory is the most accepted theory for the movement of water through plants. **[DUMET 2011]**
- The water potential of vacuolar sap is less than water potential of root hair during the water absorption by root hair. **[Gujarat CET 2011]**
- *Pinus* seed cannot germinate and establish without fungal association. This is because, it has obligate association with mycorrhizae. **[NEET 2019]**
- The process responsible for facilitating loss of water in liquid form from tip of grass blades at night and early morning is root pressure. **[NEET 2020]**
- Transpiration and root pressure cause water to rise in plants by pulling or pushing it. **[CBSE AIPMT 2015, KCET 2011]**
- The rate of transpiration in plants is dependent upon light, temperature, atmospheric humidity and wind. **[Manipal 2012]**
- Potassium influx and efflux is most recent explanation for stomatal movements. **[CBSE AIPMT 2015]**
- Radial orientation of cellulose microfibrils in the cell wall of guard cells facilitates opening of stomatal aperture. **[NEET 2017]**
- Specialised epidermal cells surrounding the guard cells are called subsidiary cells. **[NEET 2016]**
- Stomatal movement is not affected by O_2 concentration. **[NEET 2017]**
- Stomata open due to accumulation of K^+ ions. **[AIIMS 2013]**
- Active K^+ transport theory gives the latest explanation for the closure of stomata. **[Manipal 2013]**
- Malate, potassium and chloride ions get accumulated in the vacuoles of guard cells during stomatal opening. **[DUMET 2011]**
- Guard cells help in transpiration. **[AFMC 2012, CBSE AIPMT 2009]**
- Stomatal opening is affected by CO_2 concentration, temperature and light. **[DUMET 2011]**
- Stomata open at night and close during the daytime in succulents. **[Chhattisgarh PMT 2012]**
- Direction of translocation of organic food or solutes is upward, downward and radial. **[UP CPMT 2011]**
- Unloading of sucrose at sink does require utilisation of ATP. **[NEET 2020]**
- Movement of sugar in phloem is bidirectional. **[NEET 2019]**
- Guttation is a process of loss of water in liquid form containing dissolved minerals. **[WB JEE 2012]**

Chapter Exercises

(A) Topical Exercise

Means of Transport

1 The type of vascular tissue involved in unidirectional and multidirectional flow are
(a) phloem and xylem (b) xylem and phloem
(c) Only xylem (d) Only phloem

2 The porins are proteins that form large pores in the outer membranes of the
(a) plastids (b) mitochondria
(c) some bacteria (d) All of these

3 The main mechanism of determining the direction of short distance transport within a potato tuber is
(a) determined by the structure and function of the tonoplast of tuber cells
(b) diffusion due to concentration differences and bulk flow due to pressure differences
(c) not affected by temperature and pressure
(d) pressure flow through the phloem

4 Diffusion is responsible for
(a) exchange of gases during photosynthesis
(b) stomatal transpiration
(c) Both (a) and (b)
(d) turgidity to plant cells

5 Which of the following method of transport across a membrane does not involve a change in shape of transport protein?
(a) Facilitated diffusion (b) Simple diffusion
(c) Active transport (d) Na^+–K^+ pump

6 Facilitated diffusion
(a) is an active process
(b) is not stereospecific
(c) commonly allow transport of hydrophytic molecules
(d) allows hydrophobic molecules

7 Following statements are related with the diffusion of coloured molecules across a membrane. Select the correct statement which shows the fastest rate of diffusion.
(a) An internal concentration of 15% and external concentration of 10%
(b) An internal concentration of 25% and external concentration of 50%
(c) An internal concentration of 50% and external concentration of 25%
(d) Both (b) and (c) show fastest rate of diffusion

8 In the process of osmosis, water
(a) diffuses from less concentrated solution to more concentrated solution
(b) diffuses higher to lower concentrations
(c) does not more
(d) None of the above

9 Osmosis controls
(a) transportation
(b) opening and closing of stomata
(c) photosynthesis
(d) Both (a) and (b)

10 Which of the following does not apply to reverse osmosis? **[NCERT Exemplar]**
(a) It is used for water purification
(b) In this technique, pressure greater than osmotic pressure is applied to the system
(c) It is a passive process
(d) It is an active process

11 A man supplied excess fertiliser and watered the grass well. After sometime, the leaves turned brown because
(a) osmosis occurred in root and the plant dies
(b) fertilisers were drained in lower layer of soil
(c) it decreased photosynthesis
(d) due to water logging of soil

12 A thin slice of sugarbeet, when placed in a concentrated solution of sodium-chloride would
(a) loose water from its cells
(b) become turgid
(c) absorb water from salt solution
(d) Neither absorb water nor loose water

13 Match the following columns.

Column I (Surrounding medium)	Column II (Cells)
A. Hypotonic	1. No net flow of water
B. Hypertonic	2. Water moves into the cell
C. Isotonic	3. Water moves out of the cell

Codes

	A	B	C
(a)	2	3	1
(b)	3	2	1
(c)	1	2	3
(d)	2	1	3

14 Seeds when soaked in water imbibe it because
(a) osmotic pressure inside the seeds is low
(b) seed coat contains lots of salts
(c) of the processes of adsorption
(d) there are many vacuoles in endosperm

15 During osmosis, water moves through a semipermeable membrane.

	From	To
(a)	Low water potential	High water potential
(b)	High solute concentration	Low solute concentration
(c)	High osmotic potential	Low osmotic potential
(d)	A hypotonic solution	A hypertonic solution

16 When beet root cylinders are washed and then placed in cold water, anthocyanin does not come out. This indicates that most likely the plasma membrane is
(a) permeable to anthocyanin
(b) impermeable to anthocyanin
(c) differentially permeable to anthocyanin
(d) dead structure

17 If salt is present in higher concentration in a cell than
(a) water will pass from inside the cell to outside by diffusion
(b) water will enter the cell by osmosis
(c) salt will escape from the cell through the semipermeable membrane
(d) there will be no movement of substances between the cell and its environment

18 Match the following columns.

	Column I		Column II
A.	Hypotonic	1.	Water
B.	Hypertonic	2.	Sucrose
C.	Solute	3.	Lower tonicity
D.	Solvent	4.	Higher tonicity

Codes

	A	B	C	D		A	B	C	D
(a)	1	2	3	4	(b)	4	2	1	3
(c)	3	4	2	1	(d)	3	1	2	4

19 In an osmoscope of fresh and peeled potato, concentrated sugar solution is filled in the cup. It is then placed in a petridish containing water with a few drops of safranin. It shows
(a) endosmosis, inner solution becomes pink
(b) endosmosis, inner solution does not become pink
(c) exosmosis, the cup becomes limp
(d) no osmotic phenomenon is observed

20 Recognise the figure and identify cells *A*, *B*, and *C*.

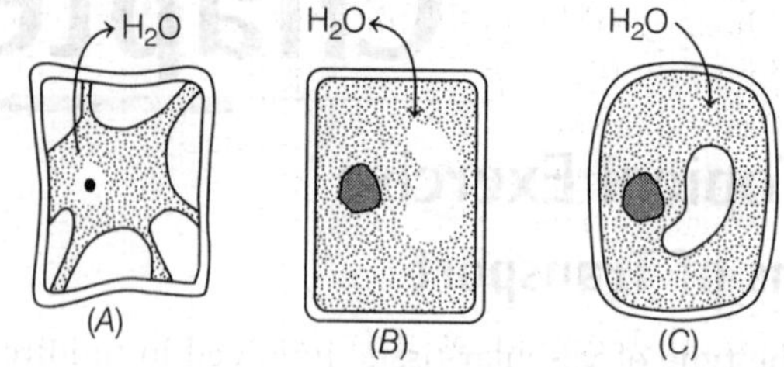

	A	*B*	*C*
(a)	Plasmolysed	Flaccid	Turgid
(b)	Plasmolysed	Turgid	Flaccid
(c)	Flaccid	Plasmolysed	Turgid
(d)	Turgid	Flaccid	Plasmolysed

21 In plasmolysis, water moves out first from the ...*A*... and then from the ....*B*.... .

	A	*B*
(a)	Vacuole	Cytoplasm
(b)	Cytoplasm	Vacuole
(c)	Cytoplasm	Cell wall
(d)	Vacuole	Cell membrane

22 Bacteria cannot survive in a highly salted pickle because
(a) salt inhibits reproduction
(b) bacteria do not get enough light for photosynthesis
(c) they become plasmolysed and consequently killed
(d) the pickle does not contain nutrients necessary for bacteria to live

23 Plasmolysis occurs when a cell is placed in a solution that has
(a) more solutes
(b) solute equivalent to cell
(c) less solutes
(d) None of the above

24 When previously moistened mustard seeds are kept in a glass bottle that is screw-capped a sudden explosion occurs after sometimes. This occurs due to
(a) osmosis
(b) imbibition
(c) diffusion
(d) plasmolysis

25 Which of the following is an example of imbibition? **[NCERT Exemplar]**
(a) Uptake of water by root hair
(b) Exchange of gases in stomata
(c) Swelling of seed when put in soil
(d) Opening of stomata

26 Which pressure is produced by swelling of wood that had been used by prehistoric human to split rocks and boulders?
(a) Osmotic pressure
(b) Turgor pressure
(c) Imbibition pressure
(d) Water potential

27 Choose the correct option in accordance to the statements given below.
I. Wall pressure is exerted to prevent any increase in protoplasm size.
II. Osmosis is the movement of substances, which takes place along a diffusion gradient.
Choose the correct option.
(a) Statement I is true, but II is false
(b) Statement I is false, but II is true
(c) Both statements I and II are true
(d) Both statements I and II are false

Plant-Water Relations

28 Mark the correct relationship.
(a) $\psi_w = \psi_p - (\psi_\pi + \psi_m)$
(b) $\psi_w = \psi_p + \psi_s + \psi_m$
(c) $\psi_w = \psi_p + \psi_\pi - \psi_m$
(d) None of the above

29 If a plant cell is placed in de-ionised water the water potential of that cell becomes
(a) more positive
(b) remain neutral
(c) more negative
(d) less negative

30 Which of the following statement is correct with respect to solute potential?
I. The effect of dissolve solute on water potential.
II. The amount by which water potential is reduced.
Codes
(a) Only I (b) Only II
(c) I and II (d) None of these

31 Water potential measures the tendency of water to
(a) evaporate
(b) condense
(c) more from one place to another
(d) adhere

32 Addition of a solute to pure water causes
(a) negative water potential
(b) negative turgor pressure
(c) positive water potential
(d) no change in water potential

33 Match the following columns.

	Column I		Column II
A.	Water potential of 10% salt solution	1.	Positive
B.	Pressure potential in a normal cell	2.	Negative
C.	Pressure potential in a plasmolysed cell	3.	Zero
D.	Metric potential on the surface of the wood		

Codes

	A	B	C	D		A	B	C	D
(a)	1	2	3	1	(b)	3	2	1	2
(c)	2	3	2	1	(d)	2	3	1	2

34 Choose the correct statement for the type of diffusion which is highly stereospecific and requires a special membrane protein.
(a) Substances are transported with the expenditure of ATP
(b) It causes net transport of molecules from low to high concentration
(c) Transport rate reaches maximum when all the protein transporters are being used
(d) It is insensitive to inbhibitors which react with protein side chains

35 Solution **A** has $\psi_s = -30$ bars and $\psi_p = 5$ bars.
Solution **B** has $\psi_s = -10$ bars and $\psi_p = 0$ bars. The two are separated by a semipermeable membrane. Flow of water will be
(a) **A** → **B** (b) **B** → **A**
(c) Equal in both directions (d) No flow of water

36 Find the direction of movement of water in the adjacent plant cells shown below.

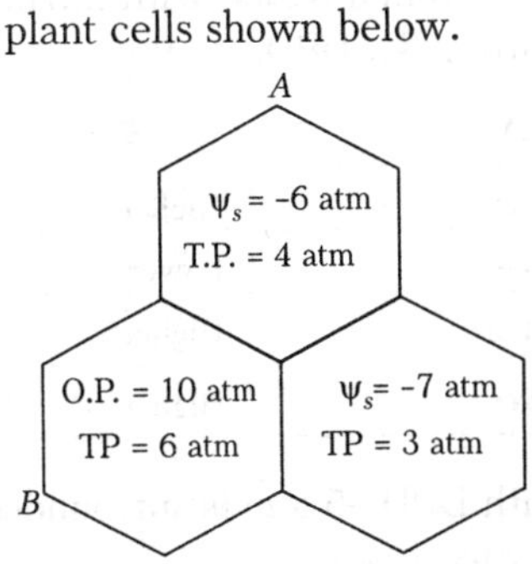

(a) 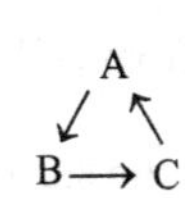

(b) 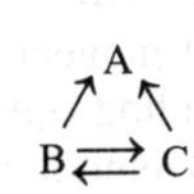

(c) 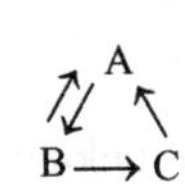

(d) 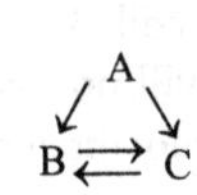

37 The cell A has an osmotic potential of −20 bars and a pressure potential of +6 bars. What will be its water potential?
(a) −14 bars (b) +14 bars
(c) −20 bars (d) +20 bars

38 If turgor pressure becomes equal to the wall pressure then
(a) water leaves the cell
(b) no exchange of water takes place
(c) water enters the cell
(d) solute goes from the cell into water

39 If the cell wall is elastic instead of being rigid and if the cell is put in a medium of sugar solution of higher concentration than in the cell then
(a) the cytoplasm will be pulled away from the wall as water moves out
(b) cell shape and size will not change
(c) the cell wall will shrink along with the cytoplasm
(d) the cell wall will rupture as the cytoplasm shrinks

40 When a cell is kept in 0.5 M solution of sucrose its volume does not alter. If the same cell is placed in 0.5 M solution of sodium chloride, the volume of the cell
(a) increases
(b) decreases
(c) cell will be plasmolysed
(d) will not show any charge

41 What is correct for a cell showing evident plasmolysis condition?
(a) DPD > OP (b) OP > DPD
(c) DPD = TP (d) DPD = OP

42 Water moves from a region with ....*A*... DPD towards a region with ...*B*... DPD.

	A	B
(a)	Lower	Higher
(b)	Higher	Lower
(c)	Higher	Higher
(d)	Lower	Higher

43 If cell A with DPD −5 atm is surrounded by many cells with DPD−4 atm.
(a) The net movement of water will be from cell A to the surrounding cells
(b) Net movement of water will be from the surrounding cells to cell A
(c) Water will not move at all
(d) Water movement will depend on other unknown factors

44 Observe the graph given below with respect to water intake by a plant cell and correctly identify the point marked X?

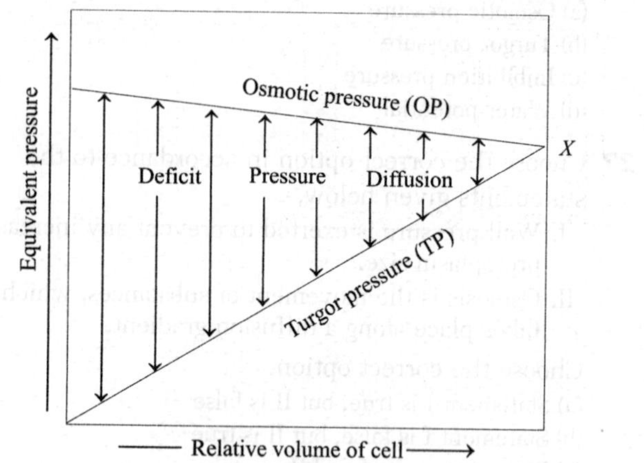

(a) X represents a flaccid state
(b) X represents a state when, DPD = zero
(c) X represents a fully turgid state
(d) Both (b) and (c)

Absorption of Water in Plants

45 Antiport system allows diffusion of
(a) one molecule is single direction
(b) two molecules in opposite direction
(c) one molecule in opposite direction
(d) It allows diffusion of two molecules in single direction

46 Pressure that pushes water and sugar from sugar source to sugar sink is referred to as
(a) translocation (b) bulk flow
(c) transpiration (d) root pressure

47 Bulk flow or mass flow
(a) can be achieved through root pressure
(b) can be achieved through transpiration pull
(c) can be achieved either through a positive hydrostatic
(d) more than one option is correct

48 Which of the following is wrong?
(a) The movement of a molecule across a typical plant cell of about 50 μm takes approximately 2.5 seconds
(b) Long distance transport of substances within a plant cannot be by diffusion alone
(c) Active transport in plants mainly helps in long distance transport
(d) All of the above

49 In apoplast pathways, water moves exclusively through the
(a) plasmodesmata (b) cell walls
(c) intercellular spaces (d) Both (b) and (c)

50 Recognise the figure and find out the correct matching.

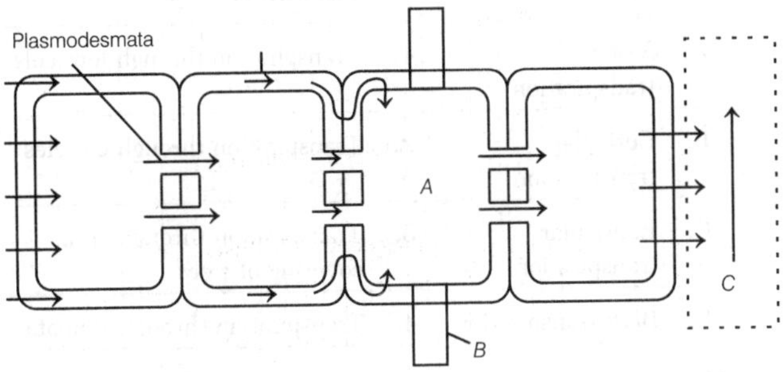

	A	B	C
(a)	Endodermis	Casparian strip	Xylem
(b)	Endodermis	Starch sheath	Xylem
(c)	Pericycle	Casparian strip	Phloem
(d)	Endodermis	Fibrous band	Sieve tube

51 In which of the following pathway, water crosses at least two membranes for each cell in its path?
(a) Apoplast (b) Symplast
(c) Transmembrane (d) Both (a) and (c)

52 Consider the following statements.
I. Primary function of Casparian strips is to force water and mineral through the membrane of endodermal cell.
II. Prevent entry of water into stele directly through the apoplast.
Choose the correct option.
(a) Statement I is true, but II is false
(b) Statement I is false, but II is true
(c) Both statements I and II are true
(d) Both statements I and II are false

Ascent of Sap (Translocation of water)

53 Ascent of sap by transpiration pull breaks due to
(a) the presence of air bubble
(b) overlapping cuts made in a tree
(c) low temperature
(d) All of the above

54 Which of the following statement is correct concerning the flow of sap in xylem of trees?
(a) In the morning, sap begins to flow first in twigs and later in trunk
(b) Flow is driven by high concentration of sugars in the vessel elements
(c) Flow from the roots to the twigs would be accelerated if the leaves are removed
(d) Rapid flow of water puts the xylem under a pressure much greater than atmospheric pressure

55 The term 'Tensile strength' represents that there is
(a) a strong cohesion force between water molecules
(b) a strong adhesion between water molecules and walls of xylem vessels
(c) absence of vacuoles in the vessels
(d) Both (a) and (b)

56 Boucherie (1940), rejected the vital force theory on the basis that the
(a) living cells are capable of growth
(b) living cells are capable for ascent of sap
(c) dead cells are capable for ascent of sap
(d) dead cells are incapable for ascent of sap

57 According to the transpiration cohesion theory, the upward pull of water is transmitted to other water molecules by cohesion, which is caused by
(a) hydrogen bonds (b) hydrophilic cell walls
(c) turgor pressure (d) osmosis

58 The basic features of cohesion-tension theory are
(a) continuous water column
(b) cohesive and adhesive properties
(c) transpiration pull
(d) All of the above

59 If cohesion-tension transpiration theory is correct then a break in water column in a xylem vessel should
(a) have no effect at all
(b) increase the rate of photosynthesis
(c) increase the water contents of leaves
(d) cause the mesophyll cells to become flaccid and result in wilting of leaves

60 What will happen when cut end of the shoot is dipped in eosin dye solution?
(a) Ascent of sap does not occur
(b) Ascent of sap does not occur, but leaves remain fresh
(c) Leaves wilt
(d) Ascent of sap occurs and tracheary elements get stained

61 According to atmospheric pressure theory, the ascent of sap is due to
(a) capillarity (b) atmospheric pressure
(c) Both (a) and (b) (d) root pressure

62 Match the following columns.

Column I		Column II	
A.	Surface tension	1.	Ability to rise in thin tube
B.	Adhesion	2.	Elastic tendency of liquid that makes them acquire the least surface area
C.	Cohesion	3.	Attraction of water molecules to tracheary elements
D.	Capillarity	4.	Mutual attraction between water molecules

Codes

	A	B	C	D
(a)	2	3	4	1
(b)	4	3	2	1
(c)	1	2	3	4
(d)	1	4	3	2

Transpiration and Mineral Uptake

63 In transpiration,
(a) upward movement of water by roots will stop
(b) turgidity of the cell will be lost
(c) all metabolic processes will stop
(d) upward movement of water by roots will continue and all the metabolic processes will occur as usual

64 Which of the following is not the significance of transpiration?
(a) Creates transpiration pull for absorption and transport of water
(b) Supplies water for photosynthesis
(c) Transport water and minerals from soil to leaf only
(d) Helps in maintaining the shape and structure of plants

65 Maximum transpiration occurs in
(a) mesophytic plants (b) xerophytic plants
(c) hydrophytic plants (d) algae

66 Rate of transpiration is
(a) increased with increase in temperature
(b) stopped in dark, due to closure of stomata
(c) increased, when relative humidity is high
(d) Both (a) and (b)

67 A tree girdled up to xylem may survive for sometime, but ultimately dies because
(a) water does not move upward
(b) sugar does not move upward
(c) sugar does not move downward
(d) water does not move downward

68 Absorption of water by roots is increased when
(a) transpiration rate is less
(b) salt absorption is increased
(c) transpiration rate increases
(d) photosynthetic rate increases

69 In both transpiration and evaporation, water is lost in the form of vapour, yet they differ because
(a) transpiration is a physical process and evaporation is a physiological process
(b) transpiration is a physiological process and evaporation is a physical process
(c) Both transpiration and evaporation are similar, but rate of water loss is different
(d) more than one option is correct

70 Match the following columns.

Column I	Column II
A. Stomatal transpiration	1. Transpiration through lenticels
B. Cuticular transpiration	2. Transpiration through cuticles
C. Lenticular transpiration	3. Transpiration through cork covering of trees
D. Bark transpiration	4. Transpiration through stomata

Codes

	A	B	C	D
(a)	3	2	1	4
(b)	4	1	2	3
(c)	1	3	2	4
(d)	4	2	1	3

71 The adaptations that reduce the transpiration are
(a) cutin
(b) sunken stomata
(c) increased surface area
(d) Both (a) and (b)

72 The common feature shared by guard cells and mesophyll cells is
(a) the presence of chloroplasts
(b) dumb-bell-shaped structure
(c) differentially thick cell wall
(d) uniformly thick cell wall

73 In the mechanism of opening of stomata, the important factor is
(a) shape of the guard cells
(b) chlorophyll contents of the cell
(c) hormone contents of the cell
(d) protein contents of the cell

74 Which one of the following will not directly affect transpiration? **[NCERT Exemplar]**
(a) Temperature
(b) Light
(c) Wind speed
(d) Chlorophyll content of leaves

75 Which of the following conditions help in opening of stomata?
(a) Darkness, pH-2, high CO_2 concentration
(b) pH-7, low CO_2 concentration and light
(c) pH-5, low CO_2 concentration and dark
(d) High CO_2 concentration, low K^+ ion concentration in guard cells

76 How many of these events occur during stomatal opening in guard cells?

Increased pH, active H^+ upake, H^+ and malate associate, K^+ enters guard cells, Cl^- exits guard cells, cellulosic fibrils become shorter

(a) Two (b) Three (c) Four (d) Five

77 According to Steward, ATP is used in stomatal mechanism during

(a) opening (b) closing
(c) Both (a) and (b) (d) None of these

78 The process of guttation takes place **[NCERT Exemplar]**

(a) when the root pressure is high and the rate of transpiration is low
(b) when the root pressure is low and the rate of transpiration is high
(c) when the root pressure is equal the rate of transpiration
(d) when the root pressure as well as rate of transpiration are high

79 A basic format of potassium pump theory is given below

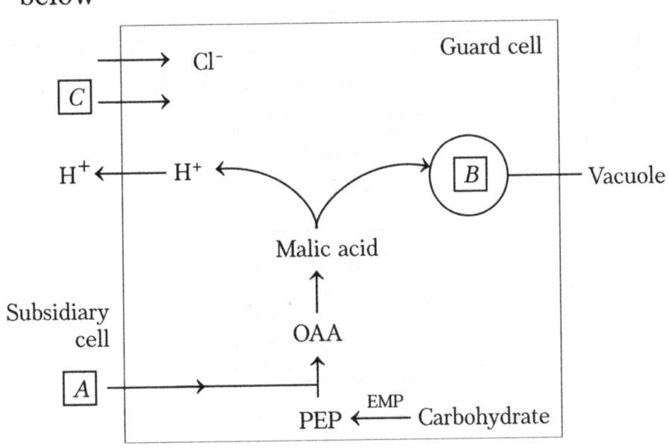

Identify *A*, *B* and *C* in the above diagram.

(a) $A - CO_2, B -$ Malate, $C - K^+$
(b) $A -$ Malate, $B - K^+$, $C - CO_2$
(c) $A - Cl^-$, $B -$ Malate, $C - CO_2$
(d) $A - K^+$, $B -$ Malate, $C - CO_2$

80 Minerals are mainly absorbed by roots by active processes as

(a) Mineral concentration in the soil is usually lower than the concentration of minerals in the root
(b) Minerals are present in the soil as charged particles
(c) ATP is not required
(d) Both (a) and (b)

81 Mineral ions are frequently remobilised

(a) particularly from older leaves to younger leaves
(b) out of which the most readily mobilised ones are Ca and S
(c) before the leaf fall in deciduous plants, minerals are removed to other plant parts
(d) more than one option is correct

82 Choose the correct option mycorrhiza is a symbiotic association of fungus with root system which helps in **[NCERT Exemplar]**

I. absorption of water II. mineral nutrition
III. translocation IV. gaseous exchange

Codes

(a) Only I (b) Only II
(c) I and II (d) II and III

83 The translocation of organic solutes through phloem is bidirectional because

(a) root acts as source and leaf acts as sink
(b) source and sink are irreversed in any season
(c) translocation is ATP regulated process
(d) source-sink relationship is variable depending upon season or needs of plant

84 Organic substances such as sugars are translocated in the phloem. It can be demonstrated by

(a) ringing the stem (b) root pressure
(c) grafting (d) defoliation

85 During phloem unloading,

(a) loss of solute produces a high water potential in the phloem
(b) water never returns to xylem
(c) gain of solute occurs in phloem
(d) Both (a) and (b)

(B) Cumulative Exercise

(Assorted Questions of the Chapter for Advanced Level Practice)

1 What is the correct labelling of the following figure?

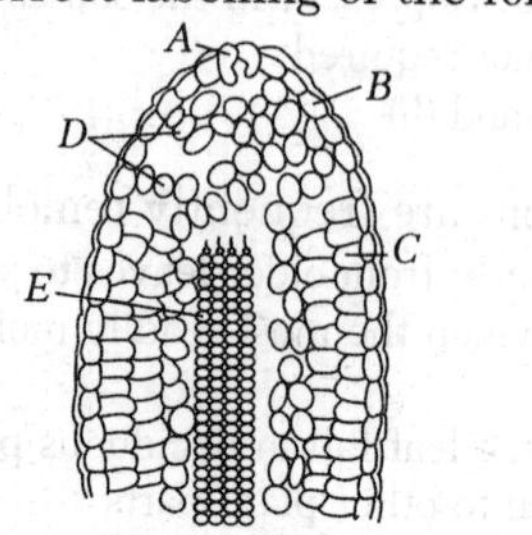

(a) *A*-Guard cell, *B*-Epithelium, *C*-Mesophyll, *D*-Epidermis, *E*-Vasculature
(b) *A*-Guard cell, *B*-Epidermis, *C*-Mesophyll, *D*-Epithelium, *E*-Vasculature
(c) *A*-Water pore, *B*-Epidermis, *C*-Mesophyll, *D*-Epithelium, *E*-Vasculature
(d) *A*-Ostiole, *B*-Epidermis, *C*-Mesophyll, *D*-Epithelium, *E*-Vasculature

2 Based on the figure given below which of the following statements is not correct?

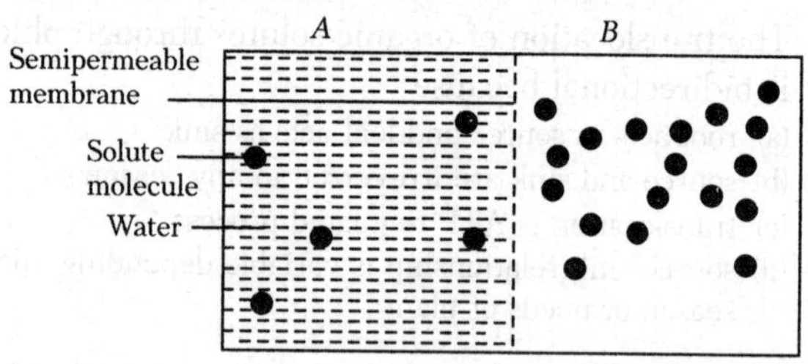

[NCERT Exemplar]

(a) Movement of solvent molecules will take place from chamber *A* to *B*
(b) Movement of solute will take place from *A* to *B*
(c) Presence of a semipermeable is a pre-requisite for this process to occur
(d) The direction and rate of osmosis depend on both the pressure gradient and concentration gradient

3 Match the following columns.

	Column I		Column II
A.	Manometer	1.	Transpiration
B.	Potometer	2.	Osmosis
C.	Osmometer	3.	Root pressure
D.	Micrometry	4.	Stomatal opening
E.	Porometer	5.	Leaf size

Codes

	A	B	C	D	E
(a)	4	1	5	2	3
(b)	2	3	4	1	5
(c)	1	5	2	3	4
(d)	3	1	2	5	4

4 Choose the correct option for the following statements.

I. Movement of substances through diffusion is active as well as passive.
II. Water potential (Ψ_w) of a solution equals the solute potential at the atmospheric pressure.
III. Inward movement of water makes the cell flaccid and *vice-versa.*

(a) I, II are true, but III is false
(b) III, II are true, but I is false
(c) II is true, but I and III are false
(d) All statements are false

5 Mark the match pair.

(a) Diffusion	Free movement of ions and gases
(b) Osmosis	Semipermeable membrane
(c) Imbibition	Hydrophilic substance
(d) Plasmolysis	Shrinkage of protoplasm

6 In the given flowchart, the flow of water is shown from soil to xylem. Mention the tissue in *A* and *B*.

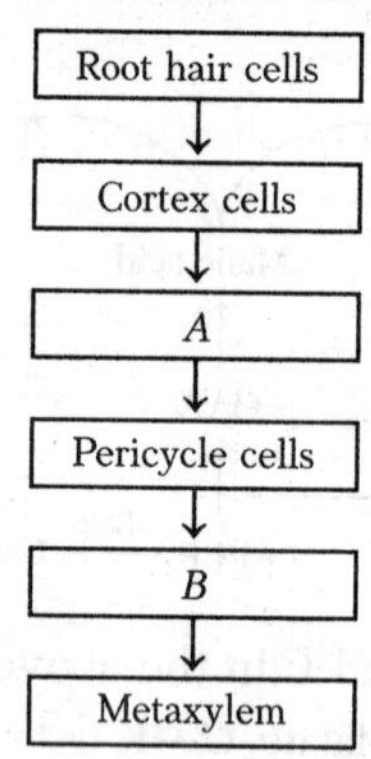

(a) *A*-Hypodermis, *B*-Protoxylem
(b) *A*-Pith, *B*-Plasmodesmata
(c) *A*-Endodermis, *B*-Protoxylem
(d) *A*-Endodermis, *B*-Protoxylem

7 The OP and TP of two pairs of cells *A*-*B* and *X*-*Y* are as under

	Cell *A*	Cell *B*
I.	OP = –10 atm TP = 4 atm	OP = –10 atm TP = 6 atm

	Cell *X*	Cell *Y*
II.	OP = –10 atm TP = 4 atm	OP = –8 atm TP = 4 atm

The net movement of water shall be from

(a) *A* to *B* and *X* to *Y*
(b) *A* to *B* and *Y* to *X*
(c) *B* to *A* and *X* to *Y*
(d) *B* to *A* and *Y* to *X*

8 As glucose is prepared at the source by photosynthesis, it is then moved into the ...*A*... and then into ...*B*... by ....*C*.... .

	A	*B*	*C*
(a)	Sieve tube cells	Companion cells	Passive transport
(b)	Companion cells	Sieve tube cells	Active transport
(c)	Companion cells	Sieve tube cells	Passive transport
(d)	Sieve tube cells	Companion cells	Facilitated diffusion

9 In which of the following options (a to d), the components labelled as *A*, *B*, *C* and *D* in given diagram are identified correctly.

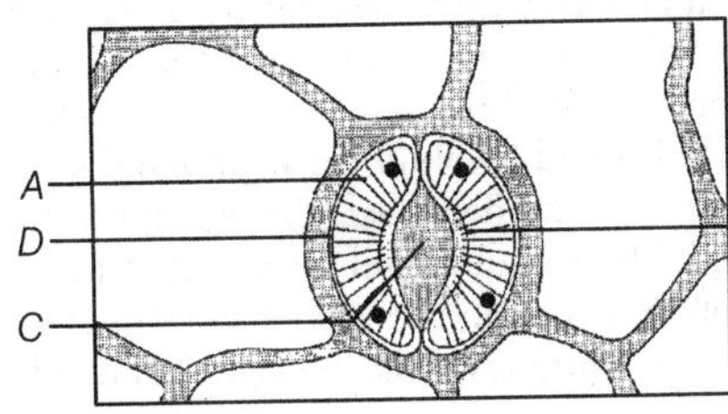

	A	*B*	*C*	*D*
(a)	Microfilament	Guard cell	Stomatal aperture	Epidermal cell
(b)	Microfilament	Guard cell	Stomatal aperture	Subsidiary cell
(c)	Microfibrils	Guard cell	Stomatal aperture	Subsidiary cell
(d)	Microtubules	Epidermal cell	Stomatal aperture	Guard cell

10 Analyse the table provided below and correctly identify *A*, *B*, *C* and *D*.

Properties	Simple diffusions	Facilitated diffusions	Active transports
Requires special membrane protein	*A*	Yes	Yes
Uphill transport	No	*B*	Yes
Highly selective	No	*C*	Yes
Transport saturates	No	Yes	*D*

	A	*B*	*C*	*D*
(a)	Yes	Yes	No	No
(b)	Yes	No	Yes	No
(c)	No	No	Yes	No
(d)	No	Yes	No	Yes

11 Given below is the pathway of water movement in the root. Identify *A*, *B*, *C* and *D* in the given diagram.

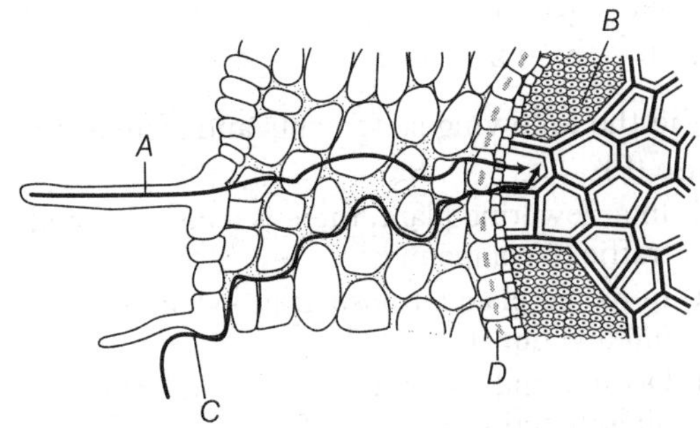

	A	*B*	*C*	*D*
(a)	Symplastic pathway	Xylem	Apoplastic pathway	Casparian strip
(b)	Symplastic pathway	Xylem	Apoplastic pathway	Starch sheath
(c)	Apoplastic pathway	Phloem	Symplastic pathway	Starch sheath
(d)	Apoplastic pathway	Xylem	Symplastic pathway	Endodermis

12 Suppose the solute concentration in three plant cells taken is 0.7(M). These are placed in three separate small test tubes T_1, T_2 and T_3 having solution with concentrations of 0.2(M), 3(M) and 0.7(M), respectively. Accordingly, which of the following is true?

(a) Cell placed in T_1 will undergo plasmolysis , that placed in T_2 wil have no change of shape and the one placed in T_3 will swell
(b) Cell placed in T_1 will swell that place in T_2 will undergo plasmolysis and the one placed in T_3 will have no change of shape
(c) Cell placed in T_1 will show no change in shape that placed in T_2 will undergo plasmolysis and the one placed in T_3 will swell
(d) Cell placed in T_1 will swell that placed in T_2 will have no change of shape and the one placed in T_3 will undergo plasmolysis

13 Match the following columns.

Column I	Column II
A. Semipermeable membrane	1. Dixon and Jolly
B. Transpiration pull	2. Holard
C. Pulsation theory	3. JC Bose
D. Whole water content in soil	4. Osmosis

Codes

	A	B	C	D
(a)	2	1	4	3
(b)	4	1	3	2
(c)	3	4	1	2
(d)	1	5	2	4

14 Read the following statements and state as true (T) or false (F).

I. In a flowering plant, there is a complex and random traffic of compounds.

II. In rooted plants, transport in xylem is essentially unidirectional.

III. Organic and mineral nutrients undergo multidirectional transport.

IV. When any plant parts undergo senescence, nutrients may be withdrawn from such regions and moved to the growing parts.

	I	II	III	IV
(a)	T	T	F	T
(b)	F	T	T	F
(c)	F	T	T	T
(d)	T	F	F	F

15 Potato slices are immersed in a series of solutions of different osmotic concentrations, no change in its volume or weight is observed with slices in a 0.4 M solution. The osmotic concentration of the vacuolar sap is therefore,

(a) 0.4 M

(b) greater than 0.4 M

(c) less than 0.4 M

(d) not related at all of the outside solution

16 If the cell *A* with OP = 10 bars and TP = 4 bars is connected to cells *B*, *C* and *D* having OP and TP respectively 4 and 4, 10 and 5 and 7 and 3 bars, the flow of water will be

(a) *C* to *A*, *B* and *D*
(b) *B* to *A*, *C* and *D*
(c) *A* to *D*, *B* and *C*
(d) *A* to *B*, *C* and *D*

17 Match the following columns.

Column I (Cells)	Column II (Locations)
A. Bulliform cells	1. Stomata
B. Guard cells	2. Aerating pore
C. Lenticel	3. Accessory cells
D. Subsidiary cells	4. Isobilateral leaf

Codes

	A	B	C	D		A	B	C	D
(a)	1	2	3	4	(b)	1	4	2	3
(c)	4	2	3	1	(d)	4	1	2	3

18 Observe a U-tube experiment shown below and identify the correct option.

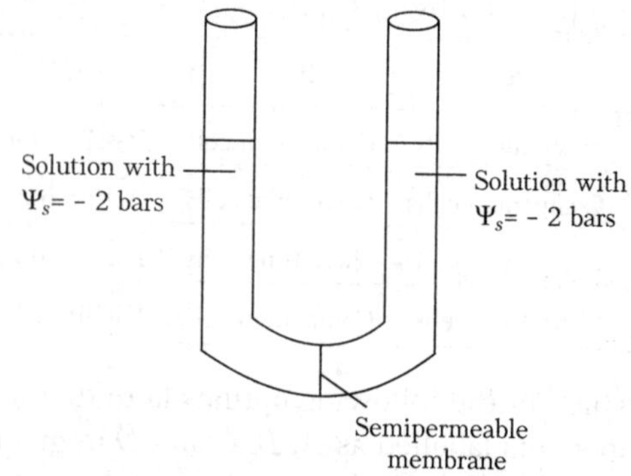

(a) Addition of solute increases the ψ_s on right hand side and rise of liquid level on the right hand side

(b) Addition of solute increases the ψ_s on right hand side and fall of liquid level on the right hand side

(c) Addition of solute decreases the ψ_s on the right hand side and fall of liquid level on the right hand side

(d) Addition of solute on the right hand side decreases ψ_s and rise of liquid level on the right hand side

19 Read the following statements.

I. The symplastic movement of water and minerals involve protoplasts and plasmodesmata.

II. Symplastic movement is slower than apoplastic movement of water.

III. Apoplastic movement of water is aided by cytoplasmic streaming.

IV. Most of the water flow in roots occurs *via* the symplast as cortical cells are loosely packed.

V. Apoplastic pathway offers less resistance than symplastic movement of water.

Which of the above statements are correct?

(a) II, III and IV
(b) I, II and V
(c) I, IV and V
(d) II, III and V

20 Aquaporins are transport channels, which

I. are found in both plants and animals.

II. allow for faster water movement between the cells than osmosis.

III. allow water molecules to pass during transpiration.

IV. do not open during drought stress.

Codes

(a) III and IV (b) Only II (c) Only III (d) I and II

21 The experiment, which demonstrates that xylem is the main water conducting tissue is

I. stain test.

II. bottle neck experiment.

III. ringing test.

IV. swan bottle neck experiment.

Codes

(a) I and III
(b) I and IV
(c) II, III and IV
(d) I, II, III and IV

22 Match the following columns.

Column I	Column II
A. Symplast	1. System of adjacent cell walls, which is continuous throughout the plant
B. Apoplast	2. System of interconnected protoplast in the plant
C. Protoplast	3. Main difference between animal and plant cells
D. Cell wall	4. Cell minus cell wall

Codes

	A	B	C	D
(a)	2	1	4	3
(b)	3	2	1	4
(c)	4	1	2	3
(d)	1	2	3	4

23 According to imbibition theory,

I. ascent of sap could take place by imbibition through the walls of xylem.

II. water can rise through the wall due to imbibitional pressure up to a certain height only.

III. the pressure at the transpiring surface falls below that of atmosphere.

IV. rate of transpiration is roughly equal to the rate of imbibition.

Codes

(a) I and II (b) II and IV
(c) Only IV (d) Only III

24 Which of the following statement(s) is/are correct regarding transpiration pull theory?

I. Eate of ascent of sap is not affected by rate of transpiration.

II. Tension created by loss of water due to transpiration can be observed with the help of an instrument called potometer.

III. Pressure inside the tracheary element when measured with the help of instrument like pressure bomb is found to be negative.

IV. During period of high transpiration, the diameter of stem increases.

Codes

(a) Only II (b) I and II
(c) Only IV (d) III and IV

25 Match the following columns.

Column I	Column II
A. Leaves	1. Antitranspirant
B. Seed	2. Transpiration
C. Roots	3. Negative osmotic potential
D. Aspirin	4. Imbibition
E. Plasmolysed cell	5. Absorption

[NCERT Exemplar]

Codes

	A	B	C	D	E
(a)	2	4	5	1	3
(b)	3	2	4	1	5
(c)	1	2	3	4	5
(d)	5	4	3	2	1

26 If a cell with a solute potential of – 0.2 MPa and a pressure potential of 0.4 MPa is placed in a chamber filled with pure water that is pressurised with 0.5 MPa what is likely to happen?

(a) Water will flow out of the cell
(b) Water will flow into the cell
(c) Cell wall will be crushed
(d) The cell will explode

27 The concentration of solute in four cells is 0.5 M. They are placed in four separate containers I, II, III and IV filled with saline water of concentrations 0.1 M, 0.5 M, 1M and 2M, respectively. In which container cell will swell?

(a) I (b) II
(c) III (d) IV

Assertion and Reason

Direction (Q. Nos 28-37) *In each of the following questions, a statement of Assertion (A) is given followed by corresponding statements of Reason (R). Of the statements, mark the correct answer as*

(a) Both A and R are true and R is the correct explanation of A
(b) Both A and R are true, but R is not the correct explanation of A
(c) A is true, but R is false
(d) A is false, but R is true
(e) Both A and R are false

28 **Assertion** (A) Imbibition capacity is maximum in phycocolloids, proteins, starch and cellulose.
Reason (R) Lignin shows imbibition of water.

29 **Assertion** (A) The adsorption of water by the solid particles of an adsorbant without forming a solution is known as imbibition.
Reason (R) The liquid, which is imbibed is known as imbibate.

30 **Assertion** (A) Cohesive and adhesive properties of water molecules form a continuous water column in the xylem.
Reason (R) Magnitude of cohesive and adhesive force is very low.

31 **Assertion** (A) Xylem is a principal water conducting tissue.
Reason (R) It has been recognised by girdling or ringing experiments.

32 Assertion (A) Plasmolysis will be severe, if the process is in the order, limiting → incipient → evident.

Reason (R) Plasmolysis is exosmosis.

33 Assertion (A) Wax and cutin coating on plant parts reduce the transpiration.

Reason (R) These adaptations are found in xerophytes.

34 Assertion (A) Guttated liquid is found on the margins of leaves.

Reason (R) Hydathodes involved in guttation are found on the margins.

35 Assertion (A) In phloem, sugars are translocated in non-reducing form.

Reason (R) Non-reducing sugars are most reactive sugars.

36 Assertion (A) In the ringing experiment, a narrow, continuous band of tissues external to the xylem is removed.

Reason (R) Ringing experiment proves that phloem is involved in food transport.

37 Assertion (A) In phloem loading, food is transported to sink.

Reason (R) Food is transported from source to sink.

(C) Medical Entrances' Gallery

(Collection of questions asked in NEET & Various Medical Entrance Exams)

1 Select the incorrect statement. **[NEET 2020]**

(a) Transport of molecules in phloem can be bidirectional
(b) Movement of minerals in xylem is unidirectional
(c) Unloading of sucrose at sink does not involve the utilisation of ATP
(d) Elements most easily mobilised in plants from one region to another are phosphorus, sulphur, nitrogen and potassium

2 The processs responsible for facilitating loss of water in liquid form the tip of grass blades at night and in early morning is **[NEET 2020]**

(a) root pressure (b) imbibition
(c) plasmolysis (d) transpiration

3 The main difference between active and passive transport across cell membrane is **[NEET 2019]**

(a) passive transport is non-selective whereas active transport is selective
(b) passive transport requires a concentration gradient across a biological membrane whereas active transport requires energy to move solutes
(c) passive transport is confined to anionic carrier proteins whereas active transport is confined to cationic channel proteins
(d) active transport occurs more rapidly than passive transport

4 Which of the following is not a feature of active transport of solutes in plants? **[NEET 2019]**

(a) Occurs against concentration gradient
(b) Non-selective
(c) Occurs through membranes
(d) Requires ATP

5 What will be the direction of flow of water when a plant cell is placed in a hypotonic solution? **[NEET 2019]**

(a) Water will flow in both directions
(b) Water will flow out of the cell
(c) Water will flow into the cell
(d) No flow of water in any direction

6 *Pinus* seed cannot germinate and establish without fungal association. This is because **[NEET 2019]**

(a) it has obligate association with mycorrhizae
(b) it has very hard seed coat
(c) its seeds contain inhibitors that prevent germination
(d) its embryo is immature

7 Cell placed in hypertonic solution is shown by **[JIPMER 2019]**

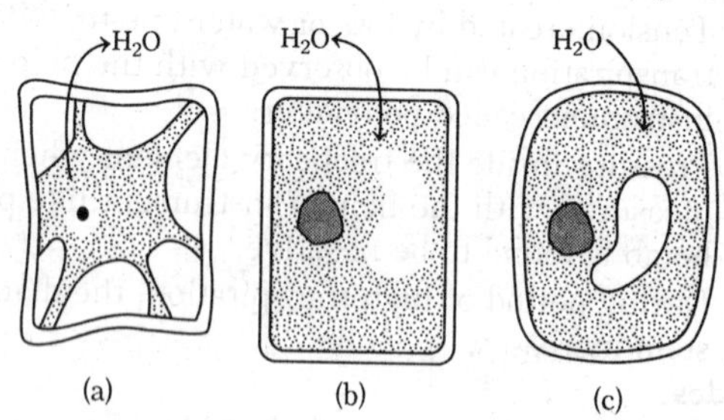

(d) Both (a) and (c)

8 Xylem translocates **[NEET 2019]**

(a) water and mineral salts only
(b) water, mineral salts and some organic nitrogen only
(c) water, mineral salts, some organic nitrogen and hormones
(d) water only

9 What is the direction of movement of sugars in phloem? **[NEET 2019]**

(a) Upward (b) Downward
(c) Bidirectional (d) Non-multidirectional

10 Cell *A* has osmotic pressure of – 20 bars and pressure potential of 5 bars whereas cell *B* has osmotic pressure of – 18 bars and pressure potential of 2 bars. The direction of flow of water will be **[AIIMS 2018]**

(a) from cell *B* to cell *A* (b) from cell *A* to *B*
(c) no flow of water (d) in both the directions

11 A mutant plant is unable to produce materials or precursors that form Casparian strips. This plant would be **[AIIMS 2017]**

(a) unable to transport water from roots to the leaves
(b) able to exert greater root pressure than the normal plant
(c) unable to transport food from leaves to roots
(d) unable to control amount of water and solute it absorbs

12 Which one of the following pathways occurs through cell wall?

Or Movement of water through cell wall is **[AIIMS 2017]**

(a) apoplast pathway (b) vascular pathway
(c) symplast pathway (d) non-vacuolar pathway

13 The water potential of pure water is **[NEET 2017]**

(a) zero
(b) less than zero
(c) more than zero, but less than one
(d) more than one

14 The modified equation for water potential is **[AIIMS 2017]**

(a) $\psi_w = \psi_s + \psi_p$ (b) $\psi_w = \psi_s - \psi_p$
(c) $\psi_w = \psi_s$ (d) $\psi_w = \psi_p - \psi_s$

15 Stomatal movement is not affected by **[NEET 2017]**

(a) O_2 concentration (b) light
(c) temperature (d) CO_2 concentration

16 Which of the following facilitates opening of stomatal aperture? **[NEET 2017]**

(a) Contraction of outer wall of guard cells
(b) Decrease in turgidity of guard cells
(c) Radial orientation of cellulose microfibrils in the cell wall of guard cells
(d) Longitudinal orientation of cellulose microfibrils in the cell wall of guard cells

17 Munch hypothesis is based on **[JIPMER 2017]**

(a) translocation of food due to Turgor Pressure (TP) gradient and imbibition force
(b) translocation of food due to Turgor Pressure (TP) gradient
(c) translocation of food due to imbibition force
(d) None of the above

18 A few drops of sap were collected by cutting across a plant stem by a suitable method. The sap was tested chemically. Which one of the following test result indicates that it is phleom sap? **[NEET-II 2016]**

(a) Acidic
(b) Alkaline
(c) Law refractive index
(d) The absence of sugar

19 Water vapour comes out from the plant leaf through the stomatal opening. Through the same stomatal opening carbon dioxide diffuses into the plant during photosynthesis. Reason out the above statements using the following options. **[NEET-I 2016]**

(a) Both processes can happen together because the diffusion coefficient of water and CO_2 is different
(b) The above processes happen only during night time
(c) One process occurs during daytime and the other at night
(d) Both processes cannot happen simultaneously

20 Which of the following is not correct in mass flow hypothesis? **[Kerala CEE 2015]**

(a) It is the accepted mechanism for translocation of sugars from source to sink
(b) As glucose is prepared at source it is converted to sucrose
(c) Sucrose is actively loaded into a sieve tube
(d) The process of loading at source produces a hypotonic condition in the phloem
(e) Water in the adjacent xylem moves into phloem by osmosis

21 Match the following columns.

Column I	Column II
A. Water potential	1. It is usually positive
B. Solute potential	2. It is zero for pure water
C. Pressure potential	3. It is always negative

Codes **[Kerala CEE 2015]**

	A	B	C
(a)	2	3	1
(b)	1	2	3
(c)	3	2	1
(d)	2	1	3

22 Which of the following statements about plasmolysis is/are true?

I. Plasmolysis occurs when water moves into cell.
II. Cells shrink in hypotonic solution.
III. If the external solution balances the osmotic pressure of cytoplasm, it is said to be isotonic.

Codes **[Kerala CEE 2015]**

(a) Only I (b) Only II
(c) Only III (d) I and II
(e) II and III

23 The correct ascending sequence with respect to their water potential is

I. $\pi = -0.8$ MPa, P = + 0.4 MPa
II. $\pi = -1.0$ MPa, P = + 0.5 MPa
III. $\pi = -0.9$ MPa, P = + 0.2 MPa
IV. $\pi = -0.3$ MPa, P = + 0.2 MPa **[EAMCET 2015]**

Codes

(a) I, II, III, IV (b) III, II, I, IV
(c) III, I, II, IV (d) III, IV, II, I

24 Find out the correct pair of statements.

I. In completely plasmolysed cell, pressure potential does not contribute to water potential.
II. If a cell is placed in hypotonic solution for longer time, the cell membrane shrinks away from its cell wall.
III. Apoplastic system comprises inter connected protoplasts.
IV. Polypeptides have more imbibing capacity than polysaccharides. **[EAMCET 2015]**

Codes

(a) I and IV (b) I amd II (c) II and IV (d) III and IV

25 Match the correct combination.

	Column I		Column II
A.	Scotoactive stomata	1.	Opening and closing of photoactive stomata
B.	Guttation	2.	Transpiration
C.	Tensile strength	3.	Water loss in liquid phase
D.	K^+ pump theory	4.	Night transpiration
		5.	Antitranspiration

Codes **[EAMCET 2015]**

	A	B	C	D
(a)	4	3	2	1
(b)	4	5	2	3
(c)	3	4	5	2
(d)	3	1	4	2

26 The number of stomatal pores per cm^2 of leaf surface area in the range of **[BCECE 2015]**

(a) 1000-60,000 (b) 10-1000
(c) 50,000-100, 000 (d) 50-100

27 A column of water within xylem vessels of tall tress does not break under its weight because of **[CBSE AIPMT 2015]**

(a) dissolved sugar in water
(b) tensile strength of water
(c) lignification of xylem vessels
(d) positive root pressure

28 Root pressure is due to **[CBSE AIPMT 2015]**

(a) active absorption/transport
(b) passive absorption/transport
(c) increased transpiration
(d) increased photosynthesis

29 Water potential gradient between adsorbent and the liquid imbibed is essential for imbibition. In addition, for any substance to imbibe any liquid, one of the following is also a pre-requisite **[COMED-K 2015]**

(a) affinity between adsorbant and the liquid
(b) molecular density of adsorbant
(c) concentration of adsorbant
(d) pressure potential of the absorbant

30 The continuity of water column in xylem is maintained due to **[Chhattisgarh 2015]**

(a) evaporation of water
(b) cohesive property of water
(c) pressure of air bubbles in water
(d) small size of xylem vessles

31 The apoplast is located **[WB JEE 2015]**

(a) outside the plasma membrane
(b) in the entire cytosol
(c) on both sides of plasma membrane
(d) in the plastidial content

32 Guttation is a process of loss of water in **[WB JEE 2015]**

(a) liquid form containing dissolved minerals
(b) liquid form without dissolved minerals
(c) vapour form with minerals
(d) vapour form without minerals

33 Swelling of wooden frames during rains is caused by **[Gujarat CET 2015, UK PMT 2014]**

(a) endosmosis (b) imbibition
(c) capillarity (d) osmosis

34 Layer of cells impervious to water because of a band of suberised matrix is called the **[UK PMT 2014]**

(a) endodermis (b) Casparian strip
(c) plasmodesmata (d) None of these

35 When a cell is fully turgid which of the following will be zero? **[UK PMT 2015, MHT CET 2014, Odisha JEE 2013, 10]**

(a) Turgor pressure/Pressure potential
(b) Wall pressure
(c) Suction pressure/DPD/Water potential
(d) Osmotic pressure (solute pressure)

36 Guttation occurs through **[MHT CET 2014]**

(a) roots (b) hydathode (c) trichome (d) stomata

37 Which of these is/are not a property of facilitated transport? **[Kerala CEE 2014]**

I. Requires special membrane proteins
II. Highly selective
III. Uphill transport
IV. Requires ATP energy

Choose the correct option.

(a) I and II (b) III and IV (c) I and III (d) II and III
(e) II and IV

38 When water moves out of the plant cell and the cell membrane of a plant shrinks away from its cell wall then this condition is known as [J&K CET 2014]
(a) plasmolysis (b) exosmosis
(c) hydrolysis (d) endosmosis

39 Choose the wrong statement. [Kerala CEE 2014]
(a) Cells swell in hypertonic solutions and shrink in hypotonic solutions
(b) Water potential is the kinetic energy of water which helps in the movement of water
(c) The absorption of water by seeds and dry wood takes by a special type of diffusion called imbibition
(d) Solute potential of ψ_s is always negative
(e) Less than 1% of the water reaching the leaves is used in photosynthesis and plant growth

40 The osmotic expansion of a cell kept in water is chiefly regulated by [CBSE AIPMT 2014]
(a) mitochondria (b) vacuoles
(c) plastids (d) ribosomes

41 Which of the following criteria does not pertain to facilitated transport? [NEET 2013]
(a) Requirement of special membrane proteins
(b) High selectivity
(c) Transport saturation
(d) Uphill transport

42 Transpiration is measured by [UP CPMT 2013]
(a) potometer (b) porometer
(c) auxanometer (d) respirometer

43 Transpiration is manifestation of [UP CPMT 2013]
(a) turgor pressure (b) wall pressure
(c) root pressure (d) None of these

44 Which of the following theory gives the latest explanation for the closure of stomata? [Manipal 2013]
(a) ABA theory
(b) Munch theory
(c) Starch-glucose theory
(d) Active K^+ transport theory

45 Translocation of photosynthetic end products in sieve tubes is [AIIMS 2013]
(a) 305 mm/h (b) 3-5 cm/h
(c) 1-15 cm/h (d) 60-100 cm/h

46 The property of semipermeability belongs to [AIIMS 2013]
(a) cell wall (b) plasma membrane
(c) mitochondria only (d) None of these

47 Cell A has osmotic potential of −18 bars and pressure potential of 8 bars, whereas, cell B has osmotic potential of −14 bars and pressure potential 2 bars. The direction of flow of water will be [AIIMS 2013, 09]
(a) from cell B to cell A (b) from cell A to cell B
(c) no flow of water (d) in both the directions

48 Stomata open due to accumulation of [Odisha JEE 2013]
(a) K^+ (b) Na^+
(c) Mg^+ (d) Ca^{2+}

49 By which mechanism, the salt resistant plants can get rid off excess Na^+ ions to the outer side, through the roots? [Manipal 2013]
(a) H^+ – ATPase uniport system
(b) Na^+ – Uniport system
(c) H^+ – Cl^- symport system
(d) Na^+ – H^+ antiport system

50 Cell A and cell B are adjacent plant cell. In cell A, $\psi_s = -20$ bars and $\psi_p = 8$ bars. In cell B, $\psi_s = -12$ bars and $\psi_p = -2$ bars. Then [Karnataka CET 2013, AFMC 2013]
(a) water moves from cell B to cell A
(b) equal amount of water is simultaneously exchanged between cell A and cell B
(c) water moves from cell A to cell B
(d) there is no movement of water between cell A and cell B

51 Stomata open at night and close during the daytime in [Chhattisgarh CET 2012]
(a) mesophytes (b) hydrophytes
(c) succulents (d) shrubs

52 Phloem sap is mainly made up of [AMU 2012]
(a) water and sucrose
(b) water and minerals
(c) oligosaccharides and hormones
(d) None of the above

53 Forces generated by transpiration can create pressure sufficient to lift water even up to height of [AMU 2012]
(a) 130 feet (b) 130 metre
(c) 230 feet (d) 230 metre

54 The rate of transpiration in plants is dependent upon [Manipal 2012]
(a) temperature and soil
(b) light and temperature
(c) wind, temperature and light
(d) light, temperature, atmospheric humidity and wind

55 Guard cells help in [AFMC 2012; CBSE AIPMT 2009]
(a) protection against grazing
(b) transpiration
(c) guttation
(d) fighting against infection

56 Direction of translocation of organic food or solutes, is **[UP CPMT 2011]**
(a) upward (b) downward
(c) radial (d) All of these

57 Stomatal opening is affected by **[DUMET 2011]**
(a) nitrogen concentration, carbon dioxide concentration and light
(b) carbon dioxide concentration, temperature and light
(c) nitrogen concentration, light and temperature
(d) carbon dioxide concentration, nitrogen concentration and temperature

58 Which of the following get accumulated in the vacuoles of guard cells during stomatal opening? **[DUMET 2011]**
(a) Water, calcium and magnesium
(b) Starch, potassium and chloride ions
(c) Malate, sodium and potassium ions
(d) Malate, potassium and chloride ions

59 Which of the following is the most accepted theory for movement of water through plants? **[DUMET 2011]**
(a) Cohesion theory (b) Capillarity
(c) Passive transport (d) Root presure

60 Which one of the following is not an antitranspirant? **[Karnataka CET 2011]**
(a) PMA (b) BAP
(c) Silicon oil (d) Low viscosity

61 Which one of the following is not related to guttation? **[Karnataka CET 2011]**
(a) Water is given out in the form of droplets
(b) Water given out is impure
(c) Water is given out during daytime
(d) Guttation is of universal occurrence

62 The force responsible for upward conduction of water against gravity comes from **[Karnataka CET 2011]**
(a) transpiration (b) photosynthesis
(c) translocation (d) respiration

63 The space between the plasma membrane and the cell wall of a plasmolysed cell surrounded by a hypertonic solution is occupied by the **[Karnataka CET 2011]**
(a) hypotonic solution (b) isotonic solution
(c) hypertonic solution (d) water

64 The process by which water is absorbed by solids like colloid causing them to increase in volume, is called **[Kerala CEE 2011]**
(a) osmosis (b) plasmolysis
(c) imbibition (d) diffusion
(e) facilitated diffusion

65 Root pressure is due to **[Odisha JEE 2011]**
(a) diffusion (b) passive transport
(c) active transport (d) osmosis

66 One factor responsible for water rise up to 100 m of tall plant, is **[Odisha JEE 2011]**
(a) root pressure (b) transpiration pull
(c) pulsation (d) diffusion

67 Which of the following does not affect water potential of water? **[Gujarat CET 2011]**
(a) Concentration of dissolved substances
(b) Atmospheric pressure
(c) Gravitation
(d) Capillarity

68 Whose water potential is less than water potential of root hair during the water absorption by root hair? **[Gujarat CET 2011]**
(a) Gravitational water (b) Soil solution
(c) Pure water (d) Vacuolar sap

69 In tall plants, because of which factor, continuous water column extends upward? **[Gujarat CET 2011]**
(a) Atmospheric pressure (b) Osmotic pressure
(c) Suction pull (d) Root pressure

ANSWERS

Check Point 18.1

1. (b)	2. (b)	3. (a)	4. (b)	5. (a)	6. (d)

Check Point 18.2

1. (d)	2. (d)	3. (a)	4. (a)	5. (b)	6. (d)

Check Point 18.3

1. (b)	2. (a)	3. (b)	4. (d)	5. (a)	6. (a)

Check Point 18.4

1. (b)	2. (a)	3. (c)	4. (b)	5. (b)	6. (c)

(A) Topical Exercise

1. (b)	2. (d)	3. (b)	4. (c)	5. (b)	6. (c)	7. (d)	8. (a)	9. (b)	10. (c)
11. (a)	12. (a)	13. (a)	14. (c)	15. (d)	16. (b)	17. (b)	18. (c)	19. (a)	20. (a)
21. (b)	22. (c)	23. (a)	24. (b)	25. (c)	26. (c)	27. (c)	28. (b)	29. (d)	30. (c)
31. (c)	32. (a)	33. (c)	34. (c)	35. (b)	36. (d)	37. (a)	38. (b)	39. (c)	40. (c)
41. (a)	42. (a)	43. (a)	44. (d)	45. (b)	46. (b)	47. (d)	48. (c)	49. (d)	50. (a)
51. (c)	52. (c)	53. (d)	54. (d)	55. (d)	56. (c)	57. (a)	58. (d)	59. (d)	60. (d)
61. (c)	62. (a)	63. (d)	64. (c)	65. (a)	66. (d)	67. (c)	68. (c)	69. (b)	70. (d)
71. (d)	72. (a)	73. (a)	74. (d)	75. (b)	76. (b)	77. (a)	78. (a)	79. (a)	80. (d)
81. (d)	82. (b)	83. (d)	84. (a)	85. (a)					

(B) Cumulative Exercise

1. (b)	2. (b)	3. (d)	4. (c)	5. (c)	6. (c)	7. (c)	8. (b)	9. (c)	10. (c)
11. (a)	12. (b)	13. (b)	14. (c)	15. (a)	16. (b)	17. (d)	18. (d)	19. (b)	20. (d)
21. (a)	22. (a)	23. (a)	24. (a)	25. (a)	26. (b)	27. (a)	28. (c)	29. (b)	30. (c)
31. (a)	32. (b)	33. (a)	34. (a)	35. (c)	36. (a)	37. (d)			

(C) Medical Entrances' Gallery

1. (c)	2. (a)	3. (b)	4. (b)	5. (c)	6. (a)	7. (a)	8. (c)	9. (c)	10. (a)
11. (d)	12. (a)	13. (a)	14. (a)	15. (a)	16. (c)	17. (b)	18. (b)	19. (a)	20. (d)
21. (a)	22. (c)	23. (b)	24. (a)	25. (a)	26. (a)	27. (b)	28. (a)	29. (a)	30. (b)
31. (a)	32. (a)	33. (b)	34. (b)	35. (c)	36. (b)	37. (b)	38. (a)	39. (a)	40. (b)
41. (d)	42. (a)	43. (a)	44. (d)	45. (d)	46. (b)	47. (b)	48. (a)	49. (d)	50. (c)
51. (c)	52. (a)	53. (b)	54. (d)	55. (b)	56. (d)	57. (b)	58. (d)	59. (a)	60. (b)
61. (c)	62. (a)	63. (c)	64. (c)	65. (d)	66. (b)	67. (d)	68. (d)	69. (c)	

Explanations

(A) Topical Exercise

4 (*c*) Option (c) is correct because

Turgidity is the pressure exerted by the osmostic flow of water. It is caused by the osmotic flow of water through a semipermeable membrane.

8 (*a*) Osmosis is the process in which water diffuses from lower concentration to higher concentration.

11 (*a*) Due to the excess fertiliser, the soil solution becomes concentrated. The concentration gradient causes osmosis and roots start to emit water. This results in root cell death due to loss of excess water.

14 (*c*) Imbibition is the process of adsorption of water by a solid substance which causes in increases in its volume. Hence, when seeds are soaked in water they adsorb water into them resulting swelling of seed.

29 (*d*) The water potential of pure water (de-ionised) is zero. When the cell placed in de-ionised water. The water moves into the cell and their water potential decreases which built pressure against. The cell wall hence pressure potential becomes more.

34 (*c*) All of the given statements are incorrect except (c). The incorrect statements can be corrected as

- In facilitated diffusion, special proteins help to move substances without the expenditure of ATP.
- It does not cause net transport of molecules from low to high concentration.
- It is sensitive to inhibitors which react with protein side chains.

35 (*b*) Solution *A*

$$\psi_s = -30 \text{ bars}$$
$$\psi_p = 5 \text{ bar}$$
$$\psi_A = \psi_s + \psi_p = -30 + 5 = -25 \text{ bar}$$

Solution *B*

$$\psi_s = -10 \text{ bar}$$
$$\psi_p = 0 \text{ bar}$$
$$\psi_B = \psi_s + \psi_p = -10 + 0 = -10 \text{ bar}$$

The water potential of solution *A* is -25 bar and solution *B* is -10 bar and the water moves from higher concentration to lower concentration hence, water will move from *B* to *A*.

36 (*d*) Water potential of cell A

$$\psi_s = -6 \text{ atm}$$
$$\text{TP } \psi_p = 4 \text{ atm}$$
$$\psi_w = \psi_s + \psi_p$$
$$= -6 + 4 = -2 \text{ atm}$$

Water potential of cell B

$$\psi_s = -\text{OP} = -10 \text{ atm}$$
$$\text{TP} = \psi_p = 6 \text{ atm}$$
$$\psi_w = -10 + 6 = -4 \text{ atm}$$

Water potential of cell C

$$\psi_s = -7\text{atm}, \text{TP} = \psi_p = 3 \text{ atm}$$
$$\psi_w = -7 + 3 = -4 \text{ atm}$$

Water always moves from higher water potential, i.e. from cell A to lower water potential, i.e. cell B and C.

40 (*c*) When the cell is placed in 0.5 M sucrose solution it does not show any change as both sucrose and cell are isotonic, but when the cell is placed in NaCl the volume of cell will get plasmolysed because NaCl solution is a hypertonic solutions and cell is hypotonic. Hence, movement of water will be from cell to NaCl solution which results in cell plasmolysis.

44 (*d*) In the given groups, X represents a state when DPD is zero and it also represents a fully turgid cell. When a cell is placed in a hypotonic solution water enters the cell. Hence, its osmotic pressure decreases and the turgor pressure increases. In a fully turgid cell, the turgor pressure becomes equal to the osmotic pressure and DPD becomes zero.

48 (*c*) Xylem elements are dead cells thus, active transport *via* membrane does not take place in them.

76 (*b*) Only three events occur during the stomatal opening, i.e. increased pH, active H^+ uptake and K^+ enters ground cells. Rest of the events support closing of stomata.

81 (*d*) Statements in options (a) and (c) are correct whereas statement in option (b) is incorrect. Incorrect statement can be corrected as

- Phosphorus, sulphur, nitrogen and potassium are readily mobilised whereas some elements that are structural components like calcium are not remobilised.

83 (*d*) Translocation of organic solutes through phloem is multidirectional because source and sink end may change depending upon the season or needs of plant.

(B) Cumulative Exercise

4 (*c*) Statement II is true, but statements I and III are false. Incorrect statements can be corrected as

- The movement of substances through diffusion is passive not active.
- Inward movement of water makes the cell turgid and *vice-versa*.

7 (*c*) **Case I**

In cell *A*,

$$\text{OP} = -10 \text{ atm}$$
$$\text{TP} = 4 \text{ atm}$$
$$\text{DPD} = \text{OP} - \text{TP}$$
$$= -10 - 4 = -14 \text{ atm}$$

In cell *B*,

$$\text{OP} = -10 \text{ atm}$$
$$\text{TP} = 6 \text{ atm}$$
$$\text{DPD} = \text{OP} - \text{TP}$$
$$= -10 - 6 = -16 \text{ atm}$$

Case II

In cell X,

$$OP = -10 \text{ atm}$$
$$TP = 4 \text{ atm}$$
$$DPD = OP - TP$$
$$= -10 - 4 = -14 \text{ atm}$$

In cell Y,

$$OP = -8 \text{ atm}$$
$$TP = 4 \text{ atm}$$
$$DPD = -8 - 4 = -12 \text{ atm}$$

Water always move from low DPD to high DPD hence in case I water will move from B to A and in case to from X to Y.

14 (*c*) All the statements are true except statement (I). Incorrect statement can be corrected as

In a flowering plants, there is a complex traffic of compounds but probably very orderly moving in different directions.

16 (*b*) DPD of cell A

$$= OP - TP = 10 - 4 = 6 \text{ bars}$$

DPD of cell B

$$= OP - TP = 4 - 4 = 0 \text{ bars}$$

DPD of cell C

$$= 10 - 5 = 5 \text{ bars}$$

DPD of cell D

$$= 7 - 3 = 4 \text{ bars}$$

As water flow from lower DPD to higher DPD

$\therefore$ Flow of water will be B to A, C and D.

19 (*b*) Statements I, II and V are correct whereas III and IV are incorrect. Incorrect statements can be corrected as

Symplastic movement of water is aided by cytoplasmic steaming.

Water flow in the roots occur *via* the apoplast because the corticlal cell are loosely packed and hence offer no resistance to water movement.

20 (*d*) Statements I and II are correct whereas III and IV are incorrect because

Aquaporines are integral membrane proteins that form pores in the membrane of biological cells, mainly facilitating transport of water between cells. The cell membrane of a variety of diffrent bacteria.

Plant cell and animal cell contain aquaporins through which water can flow more rapidly into and out of the cells than by diffusing through the phospholipid bilayer.

24 (*a*) Statement II is correct. Rest statements are incorrect and can be corrected as

- Rate of transpiration increases the ascent of sap.
- The diameter flactuates according to transpirations, during period of high transpirations, the diameter reaches its minimum.

26 (*b*) Here, the water potential of cell is $-0.2 + 0.4 = -0.2$. On the other hand, water potential of external solution is 0, but when it is pressurised with 0.5 MPa its water potential incerases.

Since, water flow from higher water potential to lower water potential, water will flow into the cell.

27 (*a*) The cell will swell if the concentration of solute is more inside the cell than the surrounding. Here, the concentration of solute inside the cell is 0.5 M and out of given solution, 0.1 M solution has a lesser concentration of solute. So, the cell which is placed in 0.1M solution will swell. This phenomenon is called osmosis.

Thus, option (a) is correct.

28 (*c*) A is true, but R is false because

Proteins, starch and cellulose show imbibition, because they are hydrophilic in nature, whereas lignin does not show imbibition, because of its hydrophobic nature.

29 (*b*) Both A and R are true, but R is not the correct explanation of A.

In imbibition, the adsorption of water by solid particles of an absorbant without forming a solution takes place. Imbibants are solid substances, which take part in absorption, while the liquid, which is absorbed (imbibed) is known as imbibate.

30 (*c*) A is true, but R is false because

Due to the cohesive and adhesive properties of water molecule from a continuous water column in xylem.

31 (*a*) Both A and R are true and R is the correct explanation of A.

Hartig performed ringing experiment to show path of water through xylem. Xylem is a basic tissue for the conduction of water in higher plants. It has vessels and tracheids, which are well-adapted for water translocation.

32 (*b*) Both A and R are true, but R is not the correct explanation of A.

Shrinkage of the protoplast of a cell from its cell wall under the influence of a hypertonic solution is called plasmolysis. Hypertonic solution causes exosmosis or withdrawl of water from the central vacuole of the cell.

The pressure on the wall is simultaneously reduced and the elastic wall contracts causing a reduction in cell size.

The first stage of plasmolysis is called limiting plasmolysis. Initially, the protoplast withdraws itself from the corners. This stage is known as incipient plasmolysis.

Due to continued exosmosis, protoplast shrinks further and withdraws from the cell wall except at one or a few points. It is known as evident plasmolysis. Cells cannot survive in such cases.

33 (*a*) Both A and R are true and R is the correct explanation of A.

Waxy coating, thick cuticle, sunken stomata, hairy surface reduce transpiration and are characters of xerophytes, which grow in places, where water is not available in much quantity.

34 (*a*) Both A and R are true and R is the correct explanation of A.

Guttation drops are restricted to tips or margins of the leaves. It takes place through the special structures called hydathodes. They are usually found on the margins and tips of the leaves.

35 (*c*) A is true, but R is false because

Sucrose is a non-reducing and most translocated sugar as phloem. Non-reducing sugars can be easily transported, because they are less reactive.

36 (*a*) Both A and R are true and R is the correct explanation of A.

In ringing experiment, a continuous band of tissues external to the xylem is removed. When such plant is placed in the light, then after sometimes, the tissue above the ring becomes swollen due to the accumulation of solutes because, due to swelling, solutes cannot transport toward the root side.

37 (*d*) A is false, but R is true because

In phloem loading, food is moved from its site of synthesis (source) to adjacent sieve elements. This loaded food is then transported to the site of consumption, i.e. sink.

(C) Medical Entrances' Gallery

1 (*c*) Statement in option (c) is incorrect and can be corrected as Unloading of sucrose at sink does require utilisation of ATP. Long distance transport of sucrose from the source to sink is carried out by phloem. It is a bidirectional movement. Both loading of food at the source and unloading of the same at sink region are active processes, i.e. require energy in the form of ATP.

2 (*a*) Root pressure is positive hydrostatic pressure. It develops in tracheary element at night and in early morning, which results loss of water in liquid form from tip of grass blades.
Other options are explained as
Imbibition is the process by which colloidal solid particles absorb water and swell without being dissolved, which results loss of water in liquid form from tip of grass blades.
Plasmolysis is the process in which cells lose water in a hypertonic solution.
Transpiration is the loss of water from a plant in the form of water vapour.

3 (*b*) The main difference between active and passive transport across the cell membrane is that passive transport requires the presence of a concentration gradient across biological membrane whereas during active transport, the movement of molecules is from low concentration to high concentration that means they move against the concentration gradient by using ATP.

4 (*b*) Option (b) is not a feature of active transport of solutes in plants. Active transport of solutes in plants is carried out by specialised membrane proteins. Like enzymes, the carrier proteins are very specific (i.e. selective) in what they carry across the membranes. Active transport uses energy (ATP) to pump molecules against a concentration gradient.

5 (*c*) The behaviour of the plant cells with regard to water movement depends on the surrounding solution. When a plant cell is placed in hypotonic solution (where the concentration of solution is lower than the cell sap) then the water will flow into the cell and the cell will swell and becomes turgid.

6 (*a*) *Pinus* has an obligate association with mycorrhizae (a symbiotic association between fungus and plant roots) due to which *Pinus* seeds are unable to germinate and establish in the absence of fungal partner.

Fungus or mycorrhizae help the *Pinus* roots to absorb water and minerals by increasing their surface area. In turn, the fungus derives food and shelter from the plant roots.

7 (*a*) The figure given in option (a) depicts that the movement of water outside of the cell. The cell is plasmolysed, i.e. the cell membrane shrinks away from the cell wall. These observations signify that the cell is placed in hypertonic solution.

The balanced inflow and outflow of water in the cell shown in option (b) indicates that it is placed in isotonic solution.

In option (c), there is net inflow of water inside the cell and the cell is turgid, thus it can be said that the cell is placed in hypotonic solution.

8 (*c*) Xylem in plants helps in the translocation of water, mineral salts, some organic nitrogen and hormones to the aerial parts of the plant. This process is known as the ascent of sap and it involves four major forces namely, root pressure, capillarity, transpirational pull and cohesion-adhesion of water molecule and cell wall.

9 (*c*) Sugars show bidirectional movements in phloem unlike unidirectional movement of water in xylem. In phloem, sugars move in both directions depending upon the source-sink relationship.

Initially leaves act as source of food from where it moves to the sink (parts requiring food). Later, the food in sink is mobilised towards the growing buds of the plant.

10 (*a*) The water potential is the chemical potential of the water, which is equivalent to diffusion pressure deficit with negative sign. Therefore, water potential (ψ_A) of cell *A* is
$\psi_A = \psi_s + \psi_p = -20 + 5 = -15$ bars
Water potential of cell *B* is
$\psi_B = \psi_s + \psi_p = -18 + 2 = -2$ bars
Since, water moves from higher water potential to lower potential, the flow of water will be from cell *B* (–2 bars) to cell *A* (–15 bars).

11 (*d*) Presence of Casparian strips in plant controls the amount of water and solute it absorbs. Thus, a plant which cannot produce Casparian strips effectively will be unable to control the amount of water and solute in plants.

12 (*a*) Out of the given pathways, only apoplast pathway occurs through cell wall. In this pathway, water passes from root hairs to xylem through cell wall only without crossing any membrane. In the symplast pathway, the water from root hairs moves into the cortex cells bounded by a continuous selectively permeable membrane through plasmodesmata.

13 (*a*) The value of water potential of pure water is highest and it is zero. Water molecules possess kinetic energy. Greater the concentration of water in a system, greater is its kinetic energy or its water potential.
If we add solute in water, the solution has fewer free water molecules and concentration of water decreases, thus reducing its water potential.

14 (*a*) The modified equation for water potential is

$$\psi_w = \psi_s + \psi_p$$

Where, ψ_w = Water potential

ψ_s = Solute potential

ψ_p = Pressure potential

15 (*a*) Stomatal movement is not affected by O_2 concentration. O_2 is essential for opening of stomata but its concentration does not affect stomatal movement. On the other hand, its movements are affected by many factors like light, temperature and CO_2 concentration. In the majority of plants, the stomata are open in light and closed in darkness. Normally high temperature above 30°C reduces stomatal opening in many species. Low CO_2 concentration usually induces opening of stomata, while high CO_2 concentration closes the same.

16 (*c*) The opening of the stomatal aperture is facilitated by the orientation of the microfibril in the cell walls of the guard cells. Cellulose microfibrils are oriented radially, rather than longitudinally making it easier for the stomata to open.

17 (*b*) Munch hypothesis proposed by Ernst Munch, is based on the movement of phloem sap along a turgor pressure gradient. The mass flow of organic solutes takes place from the site of higher osmotic pressure (source) to the site of lower osmotic pressure (sink).

18 (*b*) The phloem sap is alkaline in nature. It continuously pumps protons form its companion cells to the other neighbouring cells during transport of sugars.

19 (*a*) Diffusion of water vapour and CO_2 are independent processes. Their diffusion depends on the difference in their partial pressure in the atmosphere as well as inside the leaves.

22 (*c*) Statements III is true about plasmolysis whereas statements I and II are incorrect and can be corrected as

- Plasmolysis occurs when water moves out of the cell.
- Cells shrink in hypertonic solution.

23 (*b*) The correct ascending sequence of cells with respect to their water potential is III, II, I and IV

Because water potential (ψ_w) = π (solute potential) + P (pressure potential)

24 (*a*) Statements I and IV are correct whereas II and III are incorrect. Incorrect statements can be corrected as

- If a cell is kept in a hypotonic solution, it will absorb water and swell.
- In apoplastic system, water only passed through the cell wall and intercellular spaces.

30 (*b*) Cohesion is the property of water molecules to remain joined to each other. This property of water is responsible for the maintenance of continuous water column in xylem tissue.

34 (*b*) Casparian strip is the layer of cells, which is impervious to water as it has a band of suberised matrix. The suberin material restricts the movement of water. Plasmodesmata are transcellular connections, which help in the exchange of the protoplasm, other materials and cellular signals. Endodermis is the tissue situated just below the epidermis. It is formed by isodiametric cells.

37 (*b*) The solute can move downhill from regions of higher to lower concentration, relying on the specificity of the protein carrier to pass through the membrane. This process is called facilitated diffusion and it does not require energy. It is highly selective process as it allows the uptake of only specific and selected substances.

It requires special membrane proteins called carrier proteins for the transport of substances.

The solute can move uphill, from regions of lower to higher concentration in a process called active transport, which requires some form of chemical energy.

Thus, option (b) is correct.

38 (*a*) Plasmolysis is the phenomenon, which takes place when water from a plant cell moves out and makes the cell membrane of a plant to shrink. It occurs when a cell is placed in a hypertonic solution due to, which water is first lost from the cytoplasm and then from the vacoule.

42 (*a*) The rate of transpiration can be measured by Farmer's potometer or Ganong's potometer. These are based on the assumption that the rate of transpiration is nearly equal to the rate of absorption of water.

43 (*a*) The loss of water from the living tissue of aerial parts of plant in the form of water vapour is called transpiration. More than 95% of total loss of water takes place through stomata. Opening and closing of stomata occur due to turgor change in the guard cells.

44 (*d*) Active K^+ transport theory proposed by Levitt is the latest and most accepted theory for opening and closing of a stomata. According to this theory, pH of guard cells increase due to active H^+uptake by cell and CO_2 assimilation by guard cell.

Rise in pH of guard cell causes hydrolysis of strach to form phosphoenol pyruvate. Phosphoenol pyruvate combines with CO_2 with the help of PEPcase and forms oxalic acid, which gets changed into malic acid.

H^+ ions pass out of the guard cells, while K^+ions of epidermal cells pass into the guard cells, K^+ combines with malate. It results in absorption of water from adjacent cells, development of turgor and hence opening of stomata.

45 (*d*) Dixon (1923) reported that the sieve tube contents move at the rate of 50 cm per hour, while the simple diffusion of sugar molecules takes place at the rate of 0.2 mm per day.

47 (*b*) The water potential is the chemical potential of water, which is equivalent to DPD with negative sign. Therefore, water potential (Ψ_A) of cell A =

$$\Psi_A = \Psi_s + \Psi_P = -18 + 8 = -10$$

water potential of cell $B(\Psi_B)$ =

$$\Psi_B = \Psi_s = \Psi_P = -14 + 2 = -12$$

Since, water moves from higher water potential to lower potential the flow of water will be from cell *A* (– 10 bars) to cell *B* (–12 bars).

48 (*a*) According to active K^+ theory of Levitt, opening of stomata occurs due to influx of K^+ into the guard cells. The source of K^+ are nearby subsidiary and epidermal cells.

49 (*d*) The salt resistant plants get rid off excess Na^+ ions to the outside through the roots by Na^+-H^+ antiport system.

50 (*c*) Movement of water between cell takes place along the water potential gradient, water moves from a cell having higher water potential to a cell having lower water potential. Cell *B* has lower water potential (–12 bars - 2 bars =–14 bars), whereas cell *A* has higher water potential (–20 bars + 8 bars = –12 bars). Hence, water will move from cell *A* to cell *B*.

56 (*d*) The food materials are synthesised in leaves of green plants and are translocated to different parts of a plant. Thus, translocation may occur in upward, downward or radial direction.

57 (*b*) Carbon dioxide is an effective antitranspirant. A little rise in CO_2 concentration induces partial closure of stomata.
Light affects the rate of transpiration in two ways, firstly by controlling the stomatal opening and secondly by affecting the temperature. Increase in temperature increases the rate of transpiration.

58 (*d*) During opening of stomata uptake of K^+ takes place. The uptake of K^+ is balanced by Cl^- uptake. The malic acid dissociates into hydrogen and malate ion. The synthesis of malic acid in guard cells accompanies the influx of potassium ions.

59 (*a*) The most accepted theory of ascent of sap is transpiration pull theory or cohesion-tension theory proposed by Dixon and Jolly (1894). This theory states that water forms a continuous column from root to leaf through xylem ducts. The loss of water from mesophyll by transpiration creates transpiration pull or tension, which is transmitted downwards. The column of xylem resists breaking of water column due to force of cohesion between water molecules and adhesion between water and the wall.

60 (*b*) The rate of transpiration can be reduced by using antitranspirants. These can be of two types

- **Metabolic inhibitors** PMA, ABA, aspirin.
- **Film forming antitranspirant** Silicon, low viscosity, waxes.

BAP (Benzyl Amino Purine) is a cytokinin.

61 (*c*) In herbaceous plants, when root pressure is high and transpiration is low, plants may lose extra water in liquid drops from margins of leaves. This process is called guttation. It is very common during warm and humid nights. These water drops contain salts, amino acids, etc.

63 (*c*) When a cell is placed in hypertonic solution, the protoplasm shrinks and leaves the cell wall due to exosmosis and cell becomes flaccid. The space between plasma membrane and cell wall of plasmolysed cell is occupied by hypertonic solution.

65 (*d*) Stocking has defined root pressure as a pressure developed in the tracheary elements of xylem as a result of metabolic activities of root. It is said to be an active process that occurs due to osmosis.

67 (*d*) Water potential is the difference between free energy of water molecules in pure form and that of water in the solution. Water potential depends on solute concentration and atmospheric pressure. It is not affected by capillarity.

68 (*d*) Water always move from area of high water potential to area of low water potential, i.e. from less negative to more negative. During water absorption by root hair, the water movement is possible, if water potential of vacuolar sap is lower than root hair.

CHAPTER 19

Mineral Nutrition

Nutrition is the basic need of all living organisms whether it is plant or animal. Plants absorb inorganic nutrients from soil which act as the reservoir of mineral elements. The absorption, distribution and metabolism of various mineral elements by plants is called **mineral nutrition**. Plant ash studies are significant in the study of mineral nutrition as these determines which element is essential and in which amount it is necessary for a particular plant. In this chapter, various aspects of mineral nutrition are studied along with culture experiments, hydroponics and aeroponics.

CULTURE EXPERIMENTS

- **Julius von Sachs** (1860) and **Knop** (1865) initiated a series of culture experiments. Through these experiments, mineral requirements of plants were determined.
- **Arnon** and **Hoagland** (1938-40) later gave the description of all mineral nutrients. The culture experiments they used for their descriptions are still utilised.
- In these experiments, plants were grown in soil-free cultures.
- The nutrients were provided to them by a nutrient solution. This solution was prepared by mixing mineral nutrient in pure water and stored into sterilised jars or plastic bottles. Solution was covered with black paper to avoid contamination.
- These experiments laid the foundation of sophisticated techniques like hydroponics, aeroponics, etc.
- The outline diagram of experimental setup they used is given below

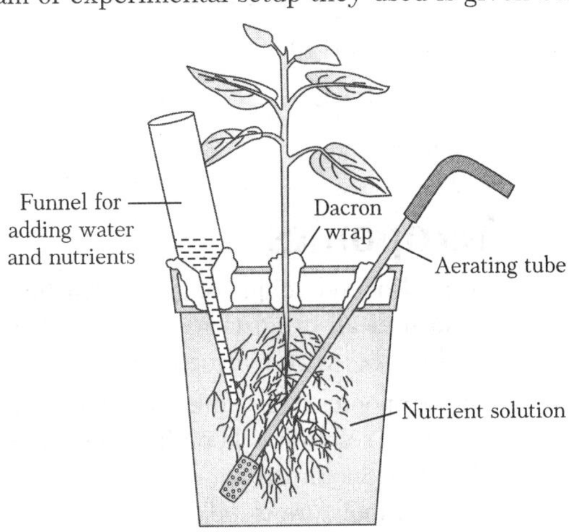

Fig. 19.1 Nutrient solution culture

Inside

Hydroponics

- This was first developed by **Gericke** is 1940.
- It is a technique of growing plants in a soil-free medium.
- In this the minerals are provided to the plant by a **nutrient solution** or **culture medium**. This culture medium is usually kept in large tanks and covered with wire mesh.
- These tanks are provided with aerating and circulating techniques.
- Plantlets are suspended in solution with the help of threads.

Types of systems

In hydroponics following two types of systems are used

(i) **Tank growth system** In this system roots of plant are immersed in nutrient solution. Air is bubbled through the solution. It circulates nutrients in the tank.

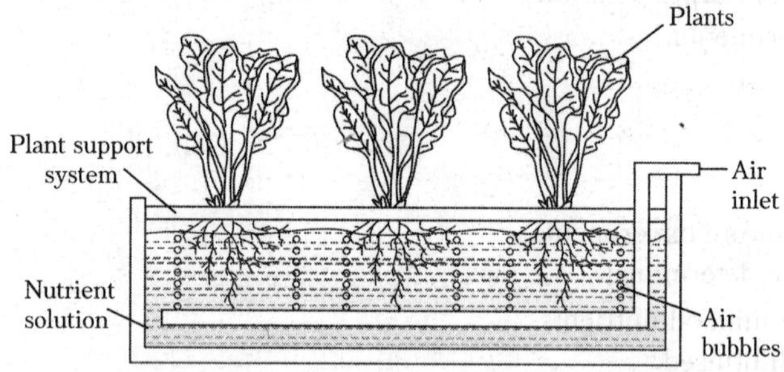

Fig. 19.2 Hydroponic tank growth system

(ii) **Film growth system** In this system plants are grown in tube. A thin film of nutrient solution is circulated in the tube with the help of a pumping system.

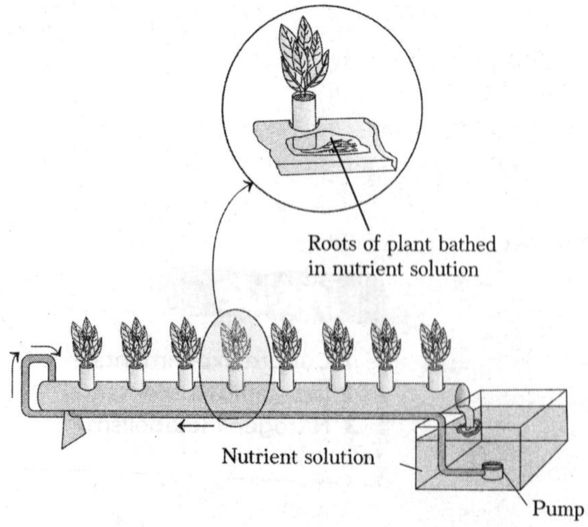

Fig. 19.3 Hydroponic film growth system

Advantages

- This technique can be used to study the role of a specific nutrient in growth and development of a plant. This is done by removing that nutrient from the solution. This removal of element will show its deficiency in plant. Thus, role of a nutrient would be predicted. Similarly, optimum concentration of a nutrient can also be predicted.
- It facilitates production of seasonal vegetables, e.g. lettuce, tomato and cucumber.
- It avoids weeding problems.
- It avoids soil-borne pathogens.
- It is useful in areas having thin, infertile and dry soils.
- It can regulate the pH at optimum for a particular crop.

Disadvantages

- The cost of these setups is very high.
- It needs skilled persons to handle it.

Culture media

Arnon and **Hoagland** in 1938-40 gave the chemical composition of nutrient media or culture media, to be used in culture experiments. This culture media is now popularly used in hydroponics. It is given in the table below

Mineral nutrients	Concentration
Potassium nitrate (KNO_3)	1.02 g/L
Calcium nitrate [$Ca(NO_3)_2$]	0.492 g/L
Ammonium phosphate $(NH_4)_3 \cdot PO_3$	0.23 g/L
Magnesium sulphate ($MgSO_4 \cdot 7H_2O$)	0.49 g/L
Boric acid (H_3BO_3)	2.86 mg/L
Manganese chloride ($MnCl_2 \cdot 4H_2O$)	1.81 mg/L
Copper sulphate ($CuSO_4 \cdot 5H_2O$)	0.08 mg/L
Strontium sulphate ($SrSO_4 \cdot 5H_2O$)	0.22 mg/L
$H_2MoO_4 \cdot H_2O$	0.09 mg/L
$FeSO_4 \cdot 7H_2O$	0.22 mg/L
Tartaric acid	0.4% mg/L

Aeroponics

- It is also a technique of growing plants in a soil-free media. It utilises air or mist environment. This system has slight differences with hydroponics.
- In aeroponics large tanks are taken. In one such tank liquid nutrient solution is kept. The roots of plants are suspended in air. A motor driven rotor is used to aid in generating mist or moisture cloud.

Thus, plants take nutrients in the form of mist.

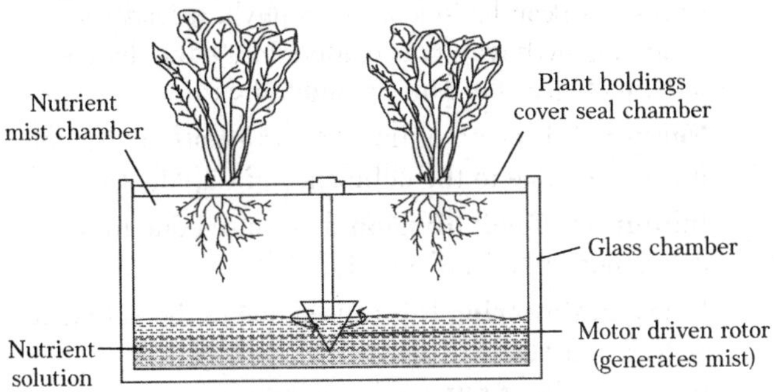

Fig. 19.4 Aeroponic growth system

INORGANIC NUTRIENTS

These are defined as the nutrients, which are taken up by living organisms in their elemental or ionic form. In general, water and mineral ions are inorganic nutrients.

Classification

- The mineral elements are grouped under two broad categories, i.e. **essential** and **non-essential elements**.
- At present 17 essential elements are known. They are C, H, O, N, P, K, Ca Mg, S, Fe, B, Mn, Cu, Zn, Mo, Cl and Ni. Apart from these Co, Si, Na and Se may be beneficial for some higher plants.
- All other elements of periodic table come under the category of non-essential elements.

Essential elements

The term 'Essential element' was proposed by **Arnon** and **Stout** (1939). The criterias for an element to be considered as essential are as follows

- The element must be essential for the normal growth and reproduction. In its absence, the plant cannot complete its life cycle or set the seeds.
- The requirement of element must be specific. It should not be replaceable by any another element.
- The element must be directly involved in metabolism of the plant.
- In the absence of these elements a disorder or malformation occur in plants. This disorder can be corrected only in the presence of that element.

These essential elements can be grouped into four broad categories, based upon their diverse functions. These categories are

- Constituent of organic biomolecules, e.g. C, H, O, N.
- Component of energy related chemical compound, e.g. Mg^{2+}, P.
- Activators and inhibitors of enzymes, e.g. Zn^{2+}, Cu^{2+}, Mg^{2+}.
- Regulators of osmotic potential of cell, e.g. K^+, Na^+ and Cl^-.

Also, based upon their quantitative requirement, essential elements are differentiated into two main categories, i.e. macroelements and microelements.

1. Macroelements

(Macronutrients or Major elements)

- Carbon, oxygen, hydrogen, nitrogen, phosphorus, potassium, calcium, sulphur and magnesium are called macronutrients.
- Out of these C, H and O are called **framework elements** being non-mineral in nature and mainly obtained from CO_2 and H_2O.
- Among these macroelements those elements which are commonly deficient in the soil are called **critical elements**, i.e. N, P and K. Whereas N, S and P are called **protoplasmic elements**.
- **Requirement** Macroelements are required by the plants in large amount. Their quantity in plants is easily detectable. It is excess of 10 mmole kg^{-1} of dry matter.
- **Function** These are usually involved in the synthesis of protoplasm. They are also responsible for the development of osmotic potential inside cell.

2. Microelements

(Trace elements or Micronutrients)

- These are eight in number, i.e. zinc, manganese, boron, copper, iron, nickel, molybdenum and chlorine.
- Some microelements become toxic to the plant when present in high concentration.
- **Requirement** Microelements are required by the plant in trace amounts. Their amount in plants can be equal or less than 10 mmole kg^{-1} of dry matter.
- **Function** They are mainly involved in functioning of enzymes. These mainly act as cofactor or metal activator of enzymes.

Note
- *The fertilisers which contain critical elements are called as* ***complete fertilisers****. They are expressed in the ratio 15:16:18 (N : P : K).*
- ***Liebig*** *gave the '****Law of minimum****'. According to this law, the yield of a crop is determined by the amount of essential element present in the minimum quantity in proportion to the demand of plant.*
- *Sodium and iodine are essential elements for salt marsh plant species, e.g.* ***Atriplex vesicaria*** *a halophyte.*

Functions and deficiency symptoms of macro and micronutrients

- As discussed essential elements are necessary for several vital functions. These functions are related to various metabolic processes occurring inside the cell.
- Whenever the supply of these essential elements becomes limited, the plant growth is retarded.
- The concentration of the essential element below which plant growth is retarded is termed as **critical concentration.**
- If any element is present below its critical concentration the plant shows deficiency symptoms related to that element.
- **Deficiency symptoms** are externally visible morphological and physiological deformities in plants. These symptoms are also called **hunger signs.**
- They differ from element to element. They disappears when the deficient mineral nutrient is provided to the plant.
- In the continuation of deficiency of essential elements, the plants may die.

Common deficiency symptoms

- **Chlorosis** It is the loss of chlorophyll. It leads to yellowing of leaves. It is caused due to the deficiency of N, K, Mg, S, Fe, Mn, Zn and Mo.
- **Necrosis** It is death of tissues, particularly in the leaf. It is caused due to the deficiency of Ca, Mg, Cu and K.
- **Inhibition of cell division** It is caused due to the deficiency of N, K, S and Mo.
- **Delay in flowering** It is caused due to the deficiency of N, S and Mo.
- **Premature leaf fall** It is caused due to the deficiency of P, Mg and Cu.
- **Stunted growth** It is caused due to the deficiency of N, K, Ca, S, Zn, B Mo and Cl.

Further, detail of essential elements, their functions and symptoms due to their deficiency are discussed below in the table.

Table 19.1 Essential elements their functions and deficiency symptoms

Elements	Absorbed as	Functions	Deficiency symptoms	Required by
Macroelements				
Nitrogen (N)	NO_3^- and sometimes as NO_2^- and NH_4^+	• Components of proteins, nucleic acids, porphyrins and coenzyme, nucleotides, vitamins, hormones. • They aid in cell division, growth and photosynthesis.	• Chlorosis appears first in older leaves • Dormancy of lateral buds • Premature leaf fall • Delaying of flowering • Interference in protein synthesis • Purple colouration of stem	Required by whole plant mainly by meristematic tissue.
Phosphorus (P)	Absorbed from soil as $H_2PO_4^-$ or PO_4^{3-}	• Components of nucleic acids, proteins, phospholipids, ATP and NADP. • Involved in energy transfer and all membrane phosphorylation reactions, thus generate metabolic energy.	• Reduced growth • Promotes premature leaf fall • Delay flowering, necrosis • Vascular tissues are poorly developed • Increased pith tissue • Pigmented leaves due to anthocyanin (purple and red spots)	Growing organ, i.e. meristems and storage organ (seed and fruit).
Sulphur (S)	SO_4^{2-}, also as SO_2 from air	• Components of vitamins (biotin, thiamin), proteins, coenzyme-A, amino acid (cysteine and methionine), ferredoxin. • It is also the component of allyl sulphide (onion, garlic) and sinigrin (mustard). • It aids in nodule formation in legumes and chlorophyll formation.	• Chlorosis in young leaves • Reduced growth • Accumulation of anthocyanin • Extensive root growth • Formation of hard and woody stem • Leaf curl • Reduction in juice content of citrus fruit **Diseases** Tea yellow	Stem, root tips (meristems) and young leaves.
Calcium (Ca)	Ca^{2+}	• Components of middle lamella of cell wall as calcium pectate. • It helps to maintain the cell permeability, stabilise the structure of chromosomes, required for the development of root and stem apex, conversion of toxic oxalic acid to calcium oxalate. • It also work for the organisation of mitotic spindle and cell wall during cell division.	• Death of root tips • Chlorosis along the margin of young leaves • Necrosis • Premature fall of flowers in legumes • Over all stunted growth **Diseases** Wither tip, blossoms end rot (tomato)	Meristematic and differentiating tissue. Excess in older leaves.

Elements	Absorbed as	Functions	Deficiency symptoms	Required by
Magnesium (Mg)	Mg^{2+}	• Components of chlorophyll and activator of enzymes involved in respiration and photosynthesis like hexokinase and phosphorylase, required for synthesis of DNA and RNA and for binding of ribosomes.	• Interveinal chlorosis first in older leaves • Purple pigmentation • Appearance of necrotic spots • Premature leaf abscission tips • Margin of leaves curls	Required in growing areas of leaves, roots, stem and seeds. Also in ageing leaves.
Potassium (K)	K^+	• It is required as cofactor for about 40 enzymes. • Essential for maintenance of turgidity of cell, anion-cation balance in cell, stomatal movement. • It also used for translocation of food essential for photosynthesis, respiration and protein synthesis.	• Inhibits protein synthesis • Causes scorching of older leaves • Rosette or bushy habit due to reduced apical bud growth • Decreased resistance to diseases • Short internodes • Loss of cambial activity	The only monovalent cation essential for plants. It is required in buds, leaves, leaf tips, root tips and meristematic tissue.
Microelements				
Iron (Fe)	Fe^{3+} (mostly) or Fe^{2+} (in acidic soil)	• Essential for chlorophyll synthesis. • Structural components of flavoproteins, cytochromes and ferredoxin. • It is an activator of catalase. Helps in energy generation. • Development of chloroplast and chlorophyll.	• Interveinal chlorosis appearing first in young leaves • Inhibition of chloroplast formation • Necrosis • Growth is reduced	Required in larger amount in comparison to any other micronutrient. Required everywhere, but more in veins.
Manganese (Mn)	Mn^{2+}	A component of oxygen evolving complex, carboxylases and dehydrogenases, reductases, i.e. enzymes of photosynthesis, respiration and nitrogen metabolism.	• Chlorotic and necrotic spots first in young leaves • Stunted growth • Flowers often sterile **Diseases** • Grey speck of oats • Speckled yellow beets • Marsh spot of peas	Leaves and seeds
Molybdenum (Mo)	MoO_2^{2+}, $HMoO_4^-$ and MoO_4^{2-}	• Involved in nitrogen metabolism including nitrogen-fixation. • It activates nitrogenase. Constituent of enzyme nitrate reductase.	• Mottled chlorosis • Marginal necrosis of leaves **Diseases** • Yellow spot of citrus • Scald of beans • Whiptail disease in cauliflower	Everywhere in plant, especially in root. It is required by plant in least quantity.
Boron (B)	BO_3^{3-} and $B_4O_7^{2-}$	Sugar metabolism and translocation, nodulation in legumes, pollen germination, cell differentiation and elongation, uptake and utilisation of Ca^{2+}, membrane functioning, synthesis of pectins, proteins and nucleic acids.	• Reduces meristematic growth • Causes premature leaf fall • Stunted growth • Small sized fruits • Disintegration of internal tissues **Diseases** • Corky core of apples • Water core of turnip • Heart rot of sugar beets • Brown heat disease of beets and turnip • Top sickness of tobacco	Leaves and seeds
Zinc (Zn)	Zn^{2+}	• Constituents of carbonic anhydrase and auxin, activates various enzymes especially carboxylases and dehydrogenases, e.g. alcohol dehydrogenase. • Involved in RNA, IAA and protein synthesis.	• Small leaf • Leaf rosettes • Decreased auxin content • White bud • Stunted growth **Diseases** • Little leaf of apple and peaches • White bud of maize • Khaira disease of rice • Sickle leaf of cocoa	Whole plant body

Elements	Absorbed as	Functions	Deficiency symptoms	Required by
Copper (Cu)	Cu^{2+}	• Component of plastocyanin, cytochrome oxidase, activator of some enzymes like ferrexidase-I. • RuBP carboxylate ceruloplasmin essential for overall metabolism in plants. • It aids in chlorophyll synthesis.	• Blackening of potato tubers. • Vegetative and reproductive growth are reduced. • CO_2 absorption is decreased. **Diseases** • Exanthema or split bark in fruit trees • Reclamation of cereals and legumes • Dieback in *Citrus*.	Whole plant body
Chlorine (Cl)	Cl^-	A component of oxygen evolving complex, solute concentration and ionic balance in cell.	• Bronze coloured leaves • Leaf wilting • Chlorosis and necrosis • Flower abscission	Whole plant body
Nickel	Ni^{2+}	It is a component of urease and hydrogenase and involved in urea metabolism.	• Necrosis of leaf tips	Leaves and roots

Note • *Insectivorous plants like Nepenthes, Utricularia usually grow in N_2 deficient soil.*
• *Parasponia is a non-leguminous plant showing symbiotic association with Rhizobium.*

Toxicity of micronutrients

- Micronutrients are required only in low amounts. When there is a slight increase in their amount, toxicity is caused.
- The mineral ion concentration at which the dry weight of a tissue is reduced by 10% is called its **toxic concentration.**
- This concentration is different for different micronutrients as well as different plants.
- For example, Mn^{2+} is toxic beyond 600 μg/g for soybean and for sunflower, it is toxic beyond 5300 μg/g.

Effect of toxicity

- Toxicity of one mineral may induce deficiency of other minerals also. For example, Mn may cause deficiency of Fe, Mg and Ca.
- The prominent symptom of Mn toxicity is brown spots surrounded by chlorotic veins.
- Due to the excess of Mn, following effects are also seen
  - Reduction in uptake of Mg and Fe.
  - Prevention of binding of Mg with enzymes.
  - Inhibition of Ca translocation to shoot apex.
- Thus, the toxicity effects of Mn cause deficiency symptom of Fe, Mg and Ca also.

Absorption of mineral elements

- Minerals are always absorbed by plants in ionic form.
- Most active areas of their absorption are elongation zone of roots and root hair.
- The movement of ions is usually called **flux**. The inward movement is **influx** and outward movement is **efflux**.
- Absorption of any mineral involves initial passive and then active phase.
- Details of these two phases of mineral absorption are given below

Passive mineral absorption (Initial phase)

- As the name suggest these process does not involve energy expenditure. The movement of ions into the roots occurs by simple diffusion.
- In this phase, ions are moved into the free or outer space of cell called **apoplast**. It is basically intercellular space along with cell walls.
- The following theories are involved to explain this phase
  (i) **Mass flow hypothesis** It was proposed by **Hylmo** (1953). According to this theory ions move across cells in a mass flow with water. This transport is influenced by transpiration. This theory was supported by **Kramer**.
  (ii) **Ion-exchange theory** It was proposed by **Jenny** and **Overstreet** (1938). It involves exchange of anions and cations between root cells and external medium.

It may occur by

(a) **Contact exchange** This includes exchanges of ions between root and soil solution. These ions oscillate in a space called **oscillation volume**. Ion exchange occurs when oscillation volume of one ion overlaps another.

(b) **Carbonic acid exchange theory** Carbonic acid dissociates into H^+ and HCO_3^- in soil solution. The cations in soil solution are absorbed on root cells in exchange for H^+ ions.

(c) **Donnan equilibrium theory** This was proposed by **Donnan** (1911). It describes the ion transport against non-diffusible or fixed ions. The membrane is impermeable to these ions. They get attached to the inner surface of plasma membrane. To balance their charge, ions with opposite charge enter through the membrane. This movement occur passively against concentration gradient without any energy expenditure. Thus, Donnan equilibrium is established.

Active mineral absorption (Metabolic phase)

The uptake of mineral ions against a concentration gradient is called active absorption. It requires expenditure of energy. In this, ions pass into the inner space of cell called **symplast**.

The following theories are involved to explain this phase

(i) **Carrier concept theory** It was given by **Van den Honert**. According to this theory, carriers are specific proteins present in membranes. These have the ability to form complexes with ions. These complexes are transported across the membrane. Ions are released from the complex on inner side. Remember, there are separate carriers for cations and anions.

(ii) **Cytochrome-pump theory** This theory was proposed by **Lundegardh** and **Burstorm**. According to this theory, mineral transportation occur through iron containing cytochromes. These are present with in the plasma membrane and carry out transport in the following manner

- They transport electrons from the inner surface to the outer surface of membrane.
- They transport anions from the outer to the inner surface.
- Anions are transported actively, i.e. with expenditive of energy. Electrons on the other hand transported passively.
- Thus, we can say in the process anion are absorbed and electrons are released.

(iii) **Protein lecithin theory** It was proposed by **Bennet** and **Clark**. According to this theory a carrier protein is associated with lecithin of membrane. It consists of a phosphate group which acts as cation binding site. A choline is also present which acts as anion binding site. Ions are released when, e.g. lecithinase hydrolyse lecithin. Thus, transport of both cation and anion is facilitated.

Factors affecting mineral absorption

The process of mineral absorption is influenced by the following factors

- **Temperature** The rate of absorption of salts and minerals is directly proportional to temperature.
- **Oxygen** Anion uptake *via* cytochrome pump increases the rate of respiration. It is now called as **salt respiration.** Deficiency of O_2 causes decrease in the rate of mineral absorption. It is probably due to unavailability of ATP.
- **pH** It affects the rate of mineral absorption by regulating the availability of ions in the medium. At normal pH monovalent ions are absorbed more rapidly whereas alkaline pH favours the absorption of bivalent and trivalent ions.
- **Light** Photosynthesis increases when there is sufficient light. As a result more food energy is available and salt uptake increases.

Soil as a reservoir of essential elements

- Majority of the inorganic nutrients that are essential for the growth and development of plants became available to the roots due to weathering and breakdown of rocks.
- This enrich the soil with dissolved ions and inorganic salts.
- Apart from supplying minerals to the roots of plants, soil plays many other roles such as supplies air to roots, holds water and act as a matrix that stabilises the plant.

Nutrients mobility in plants

Some inorganic essential nutrients have the ability to move within a given setting. Based on this, the essential nutrients/elements are divided into two categories

(i) **Mobile elements** These include elements like N, P, K, Cl, Mg which can move freely from older leaves to younger leaves and growing tips. Their deficiency affects old leaves.

(ii) **Immobile elements** These include elements like Ca, B, S, Fe, etc., unable to move freely in plants. Their deficiency symptoms first appear in young leaves.

CHECK POINT 19.1

1. Which of the following is not true about hydroponics?
   (a) Requires high investment
   (b) Technical knowledge required
   (c) Can be misused to cultivate banned crops
   (d) Plants through hydroponics cannot be cultivated everywhere
2. Which of the following option is correct about micronutrients?
   (a) Less important than macronutrients, but are present in large amount in plants
   (b) As important as macronutrients, but are present in trace amount in plant
   (c) Having a minor role in plant nutrition
   (d) Omitted from culture medium without any detrimental effect
3. The most important use of potassium is that it
   (a) provides red colour to fruit
   (b) aids photosynthesis
   (c) influences enzymatic activity, which regulate many plant process
   (d) helps in the formation of cambium
4. Name the group of nutrients that show deficiency symptoms like chlorosis and inhibit protein synthesis
   (a) N, K, Mg, S, Fe
   (b) N, K, S
   (c) Ca, Mg, Cu
   (d) N, K, Mo
5. Toxicity of Mn in plant may also cause the deficiency of
   (a) Cl and Mg
   (b) S, P and K
   (c) Ca, Fe and Mg
   (d) N, P and Mn
6. The theory which suggests that the CO_2 produces in respiration plays an important role in mineral absorption.
   (a) Contact exchange theory
   (b) Carbonic acid exchange theory
   (c) Active absorption theory
   (d) None of the above

NITROGEN METABOLISM

- Nitrogen is one of the most prevalent element in living organisms. It is a component of amino acids, proteins, hormones, chlorophyll and many vitamins.
- In the atmosphere it is present in gaseous form and is called dinitrogen, i.e. N_2.
- In soil, it is found in the form of nitrate ions (NO_3^-) and ammonium ions (NH_4^+).
- These forms are produced in the process of microbial decomposition of organic matter.
- The regular supply of nitrogen in nature is maintained through the **nitrogen cycle.**
- The outline of this cycle is given below

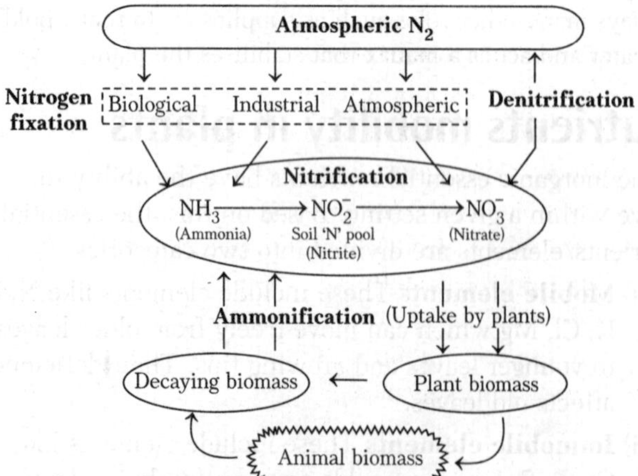

Fig. 19. 5 The nitrogen cycle showing nitrogen fixation, ammonification, nitrification, denitrification.

Nitrogen cycle mainly composed of four important processes, i.e. nitrogen-fixation, ammonification, nitrification and denitrification.

1. Nitrogen-fixation

- The atmospheric nitrogen (N_2) cannot be utilised directly by plants and animals. Thus, it has to be fixed or converted into ammonia before utilisation.
- The conversion of atmospheric nitrogen into easily absorbable froms like ammonia is called nitrogen-fixation.
- These forms include nitrate, ammonia, amino acids, etc. N_2-fixation can be abiological or biological.

Abiological nitrogen-fixation

This process does not involve any living organism. It occurs with the help of physiochemical reactions. It may occur through following ways

(i) Industrial N_2-fixation

Ammonia is industrially produced by combining N_2 and H_2. It requires a huge amount of energy. Thus, this process occurs under high pressure and temperature conditions. The reaction is as follows

$$N_2 + H_2 \rightleftharpoons 2NH_3$$

To avoid conditions of high temperature and pressure in some industries ammonia is produced from urea as shown below

$$H_2N-\overset{\overset{\displaystyle O}{\|}}{C}-NH_2 \xrightarrow{\text{Urease}} 2NH_3 + CO_2$$

Note *Enzyme urease requires Ni element to hydrolyse urea.*

(ii) Atmospheric N_2-fixation

In nature, N_2 is converted into nitrogen oxides. Energy is provided by lightning and UV-radiations. These oxides then combine with H_2O. As a result, hyponitrous or nitric acids are produced. These enter into soil *via* rain water.

$$N_2 + O_2 \xrightarrow{\text{Lightning}} \underset{\text{(Nitrogen oxide)}}{2NO} + H_2O$$

$$2NO + 2\,[O] \xrightarrow[\text{Ozonisation}]{\text{Electric discharges}} \underset{\text{(Nitrogen peroxide)}}{2NO_2}$$

$$2NO + 3\,[O] \xrightarrow{\text{Ozonisation}} N_2O_5$$

The nitrogen oxides, peroxides react with rain water due to which nitrous and nitric acid are formed.

$$\underset{\text{(Nitrogen peroxide)}}{2NO_2} + H_2O \xrightarrow[\text{rain}]{\text{Acid}} \underset{\text{(Nitrous acid)}}{HNO_2} + HNO_3$$

$$N_2O_5 + H_2O \xrightarrow[\text{rain}]{\text{Acid}} \underset{\text{(Nitric acid)}}{2HNO_3}$$

Biological nitrogen-fixation

- It is the conversion of molecular nitrogen into inorganic nitrogenous compounds by living organisms.
- It is mostly carried out by bacteria and cyanobacteria or blue-green algae.
- It is the second most important natural process after photosynthesis.
- A major portion of N_2 present in atmosphere is fixed by this process.
- Biological N_2-fixation is also called **diazotrophy**.

(i) Nitrogen-fixation by bacteria

The bacteria involved in N_2-fixation can be free-living or symbiotic.

(a) **Free-living bacteria** They fix N_2 in a free-living state. These can add 10-25 kg of N_2/ha/annum, e.g. *Azospirillum* sp.

(b) **Symbiotic bacteria** These bacteria fix N_2 in a symbiotic association with host plant.

Source of these are given below with their hosts

Table 19.2 Types of symbiotic bacteria and their hosts

Bacteria	Hosts
Rhizobium	Leguminosae family, *Sesbania* (stem nodule), *Rostrata* (root nodule), alfalfa, *Aerorhizobium.*
Frankia (filamentous bacteria)	Roots of *Casuarina* (Australian pine), *Alnus, Cycas, Caenothus, Myrica.*
Xanthomonas and *Mycobacterium*	Leaves of Rubiaceae and Myrsinaceae.
Spirullum notatum	Roots of *Digitaria, Sorghum* and maize.

Some other important points related to free-living and symbiotic nitrogen-fixing bacteria are given below

- *Rhizobium* (Gram-negative) and *Frankia* (actinomycetes) are free-living in soil. They act as symbiont only for nitrogen-fixation.
- *Azospirillum brasiliene*, is a bacterium discovered by a Brazilian microbiologist **J Dobereiner** (1978). It is the best studied bacterium associated with cereal roots.
- *Rhizobium* species are highly crop specific, e.g. in alfalfa (*R. meliloli*), in clover (*R. trifolii*), in pea (*R. leguminosarum*), in soybean (*R. phaseoli, R. japonicum*) and in lupines (*R. lupini*).
- *Parasponia* is a non-leguminous plant showing symbiotic association with *Rhizobium.*
- Symbiotic N_2-fixation requires cooperation of *nod* genes of legume and *nif* gene of bacteria.

(ii) Nitrogen-fixation by cyanobacteria

Many cyanobacteria also perform N_2-fixation. They may be also free-living or symbiotic.

(a) **Free-living cyanobacteria** They add 20-30 kg of N_2 per hectare of soil and water bodies.

- Some free living cyanobacteria are *Anabaena, Nostoc, Calothrix, Lyngbya, Aulosira, Cylindrospermum, Trichodesmium.*
- *Aulosira fertilissima* is most active N_2-fixer in rice fields.
- *Cylindrospermum* is active in maize and sugarcane fields.

(b) **Symbiotic cyanobacteria** They live in symbiotic association with host plant.

For example, *Anabaena* and *Nostoc* in host like lichens, *Anthoceros, Azolla* and *Cycas* roots.

- Symbiotic cyanobacteria fix nitrogen through their specialised structure known as **heterocyst**. These are specialised nitrogen-fixing cells formed during nitrogen starvation.
- These cells fix nitrogen from dinitrogen using enzyme **nitrogenase**. This enzyme is inactivated by oxygen, so the heterocyst must create a microanaerobic environment.
- Heterocyst is made up of three different cell wall layers that inhibit the entry of oxygen.

Symbiotic nitrogen-fixation in leguminous plants

- The members of the Leguminosae family show symbiosis with bacteria named *Rhizobium* sp.
- These form nodules in secondary and tertiary roots and fix atmospheric molecular nitrogen into absorbable nitrogen form (NO_2 and NO_3).
- The process involves following steps

Formation of root nodule

- During N_2-fixation roots of the legume plant and *Rhizobium* bacteria interact with each other. As the result of this nodules are formed.
- These root nodules act as a site for N_2-fixation. Just like cyanobacteria these also consists of enzyme **nitrogenase**. So, this enzyme is exclusive to prokaryotes N_2-fixation.
- Nitrogenous enzyme is highly sensitive for molecular oxygen. Thus, it gets inactivated when exposed to air. Hence, it requires anaerobic conditions for its functioning.
- The principal stages in the nodule formation are summarised below
  - (i) *Rhizobium* lives saprophytically in the soil.
  - (ii) Roots of legumes secrete sugars or specific amino acids. These substances attract the bactreia and induce *nod* genes in them.
  - (iii) Rhizobia in response produce *nod* factors. These factors induce the root hairs to curl.
  - (iv) The tip of root hair cell wall degenerates and a cytoplasmic infection thread is formed in root hair.
  - (v) Infection thread grows further. Bacteria multiply along this thread.

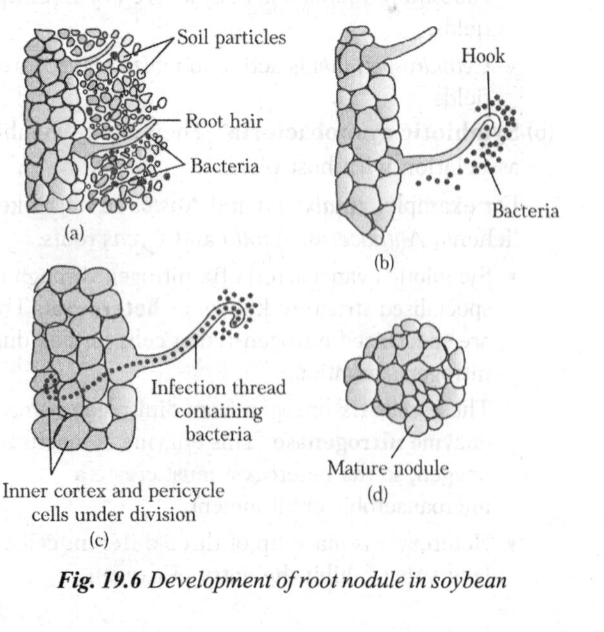

Fig. 19.6 Development of root nodule in soybean

  - (vi) It carries bacteria to cortical cells. After reaching to cortical cells this thread start repeated branching.
  - (vii) Bacteria enters into cortical cells and gets differentiated into rod-shaped bacteriods. As the result nodules also show enlargement.
  - (viii) Infected pericycle and cortical cells divide. Thus, giving a spherical shape to nodule.
  - (ix) Vascular supply between root and nodule is established. This help in transport of nitrogenous compounds from nodules to rest of the plant.

Bacteriods in a nodules

- Each nodule consists of a group of bacteriods. Each of which is surrounded by a peribacterial membrane.
- Outside this membrane **leghaemoglobin** is present. It is a pink coloured pigment.
- Its presence ensures that the enzyme nitrogenase is protected from oxygen. Thus, it works as oxygen consumer or oxygen scavanger.
- An outline diagram of a nodule cell containing bacteriod is given below

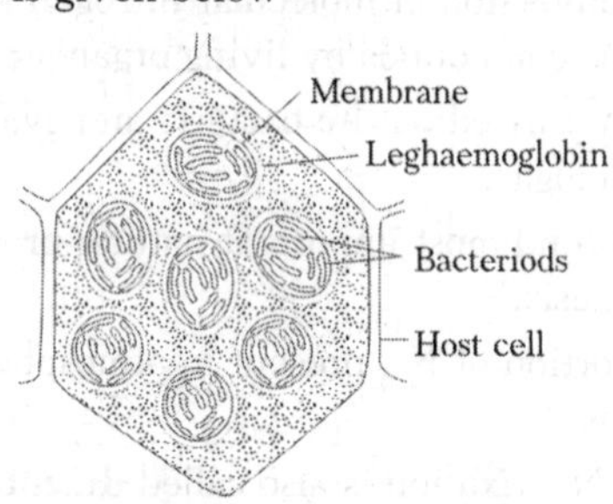

Fig 19.7 Bacteriods in a nodules

Requirement for N_2-fixation

The biological fixation of N_2 requires

- A reducing power (i.e. NADPH, $FMNH_2$)
- Source of energy (ATP)
- Dinitrogenase enzyme
- A compound for trapping ammonia after its formation from dinitrogen.

Chemical events in nitrogen-fixation

- It is believed that during the process of nitrogen-fixation, the free atmospheric nitrogen is first bound to the enzyme nitrogenase and it is not released until it completely converted to ammonia.
- Activity of nitrogenase depends on the availability of ATP. It also requires a reduced substrate that donate H_2 atom to N_2.

- Complete equation of nitrogenase working is given below

$$N_2 + 8e^- + 8H^+ + 16ATP \xrightarrow{\text{Dinitrogenase}} \underset{\text{(Ammonia)}}{2NH_3} + 2H^+ + 16ADP + 16\ iP$$

- The schematic representation of conversion of atmospheric N_2 to NH_3 by nitrogenase enzyme is given below

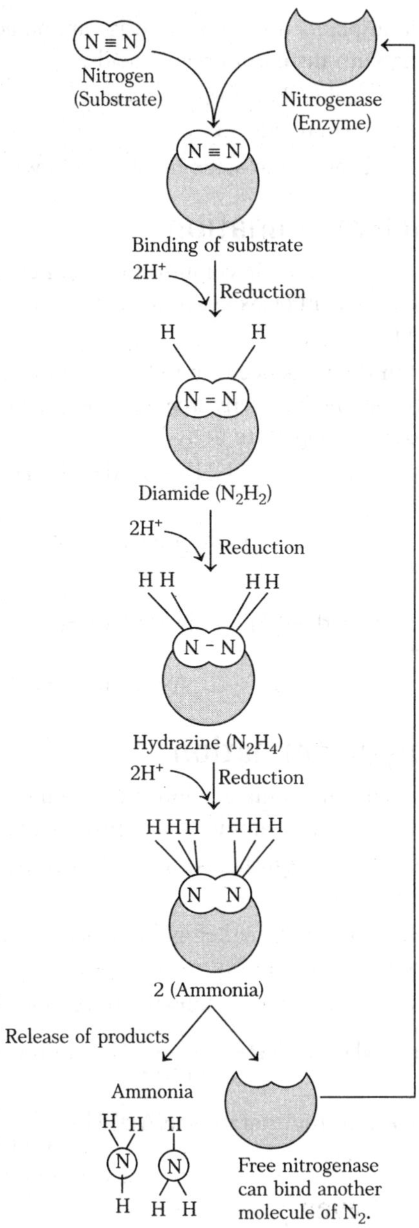

Fig. 19.8 Biological N_2-fixation by nitrogenase enzyme complex in bacteria

Symbiotic nitrogen-fixation in non-leguminous plants

- A number of non-leguminous plants are recognised to be well-endowed with nitrogen-fixing capacity. Like, *Frankia* has symbiotic association with many non-leguminous plants like *Alnus*, *Myrica*, *Casuarina*, etc.
- The cell membrane of *Frankia* contain lipid components called **hopanoids**. Membrane containing hopanoids surrounds vesicles called **diazovesicles** that contain nitrogenase.
- The process of nitrogen-fixation in non-leguminous plants is similar to leguminous plants.
- Hopanoids thicken and stabilise the walls which help to keep oxygen away from nitrogenase.

2. Ammonification

- It can be defined as 'any chemical reaction in which NH_2 groups are converted into ammonia or its ionic form ammonium ion (NH_4^+)'.
- It is produced as a result of decomposition of dead organisms.
- Proteins and nucleic acids of the dead remains of organisms are degraded by microorganisms. These include *Bacillus ramosus*, *B. vulgaris*, *Clostridium*, *Actinomyces*, etc.
- In this process proteins are first broken down into amino acids. These are then deaminated to form ammonia.
- Microbes utilise the organic acids released in this process for their metabolism.

$$\text{Protein} \xrightarrow{+H_2O} \underset{\text{(Amino acid)}}{R\text{—}NH_2} \xrightarrow{H_2O} \underset{\text{(Organic acid)}}{R\text{—}OH} + \underset{\text{(Ammonia)}}{NH_3}$$

- Ammonia changes from gaseous to ionic form in the soil. It is then utilised by the plants. It occurs when the pH of the soil is more than six.
- Very few plants, e.g. *Begonia* and *Oxalis* can store ammonium ions despite of their heavy toxicity.

3. Nitrification

- Plants usually absorb ammonium ions. However, plants are more adapted to absorb NO_3^- (nitrate) than ammonium ions (NH_4^+).
- The conversion of ammonia to nitrate is called **nitrification**. It also occurs with the help of bacteria.

- Nitrification takes place in following two steps
  (i) **Nitrite (NO_2^-) formation** In first step ammonium ions are oxidised to nitrite ion. It is done by bacteria like *Nitrosomonas* and *Nitrosococcus.*

$$2NH_3 + 3O_2 \xrightarrow{\textit{Nitrosomonas}} 2NO_2^- + 2H^+ + 2H_2O + (-84 \text{ kcal}) \text{ energy}$$

  (ii) **Nitrate formation** In the second step nitrites are oxidised to form nitrate. It is carried out by bacteria like *Nitrobacter, Nitrocystis,* etc.

$$2NO_2^- + O_2 \xrightarrow[\textit{Nitrocystis}]{\textit{Nitrobacter}} 2NO_3^- + (-17.8 \text{ kcal}) \text{ energy}$$

- *Aspergillus flavus* (a fungi) have the ability to carry both the processes.
- Most of the bacteria performing the nitrification (e.g. *Nitrosococcus, Nitrosomonas, Nitrobacter,* etc.) are chemoautotrophs.

Nitrogen assimilation

Nitrate (produced during nitrification) cannot be utilised by plants as such. They are first reduced to ammonia. After that ammonia is incorporated into organic compounds. This is called **nitrogen assimilation.**

The process occurs in following two steps

(i) Reduction of nitrate to nitrite

- It is carried out by an inducible enzyme nitrate reductase.
- This enzyme is a sulphydryl containing molybdo flavo protein.
- It occurs in the cytosol and requires a reduced coenzyme (NADH or NADPH) as electron donor for its activity.
- This reduced coenzyme make a contact with nitrate by FAD or FMN.

$$NO_3 + NADPH + H^+ \xrightarrow[\text{FAD/FMN}]{\text{Nitrate reductase}} NO_2^- + H_2O + NADP^+$$

(ii) Reduction of nitrite to ammonia

- It is carried out by the enzyme nitrite reductase.
- This enzyme is also a metallo flavo protein which contains copper and iron.
- It occurs inside the chloroplast of leaf cells. In other cells, it is found in leucoplast.
- Nitrite reductase requires reducing power. It is obtained by NADPH (in illuminated cell) and NADH (in others).
- Reduction process also requires ferredoxin of green tissues of higher plants.

$$2NO_2^- + 7NADPH + 7H^+ \xrightarrow[\text{Ferredoxin}]{\text{Nitrite reductase}} 2NH_3 + 4H_2O + 7NADP^+$$

Fate of ammonia

- Ammonia in plants used to synthesise amino acids by combining with different organic acids.
- These amino acids then form various types of nitrogenous compounds, i.e. proteins and enzymes.
- The synthesis of amino acids involves following methods

(i) Reductive amination

- In this process ammonia combines with a keto acid like α-ketoglutaric acid or oxaloacetic acid. As a result amino acids are produced.
- It occurs in the presence of a reduced coenzyme (NADH, NADPH) and enzyme dehydrogenase, e.g. glutamate dehydrogenase, aspartate dehydrogenase.

$$\alpha\text{- ketoglutaric acid} + NH_4^+ + NADPH + H^+ \xrightarrow[\text{dehydrogenase}]{\text{Glutamate}} \text{Glutamate} + H_2O + NADP$$

$$\text{Oxaloacetic acid} + NH_4^+ + NADPH + H^+ \xrightarrow[\text{dehydrogenase}]{\text{Aspartate}} \text{Aspartate} + H_2O + NADP^+$$

(ii) Catalytic amination

- In this process ammonia combines with amino acid glutamate to form an amide called **glutamine.**
- Enzyme glutamate synthetase and ATP are required in this process.
- Glutamine reacts with α-ketoglutaric acid in the presence of reduced coenzyme.
- Thus, two molecules of glutamate are produced.

$$\text{Glutamate} + NH_4^+ + ATP \xrightarrow[\text{synthetase}]{\text{Glutamine}} \text{Glutamine} + ADP + iP$$

$$\text{Glutamine} + \alpha\text{-ketoglutarate} + NADPH \xrightarrow[\text{synthetase}]{\text{Glutamate}} 2 \text{ Glutamate} + NADP$$

(iii) Transamination

- This process involves transfer of amino groups from an amino acid to keto group of a keto acid.
- The enzyme used is transaminase or amino transferase.

- Coenzyme is pyridoxal phosphate. Glutamic acid is the primary amino acid involved in the transfer of amino group.

$$\text{Glutamic acid} + \text{Oxaloacetic acid} \xrightarrow[\text{Amino transferase}]{\text{Glutamate aspartate}} \alpha\text{-ketoglutaric acid} + \text{Aspartic acid}$$

Amides : The storehouse of nitrogen

These are amino acid derivatives. In amides —OH component of carboxylic group (—COOH) is replaced by an amino group ($-NH_2$). Thus, they are double deaminated keto acids. The two most important amides are glutamine and aspargine.

Amides are the storage house of nitrogen. They contain more nitrogen than the amino acid. They are easily transported to other part of plants and transportation occurs *via* xylem vessel.

4. Denitrification

- Some microorganisms use nitrate under anaerobic conditions. They reduce nitrates into gaseous compounds of N_2 which escapes from soil.
- The bacteria causing denitrification of soil are *Pseudomonas denitrificans*, *Thiobacillus denitrificans* and *Micrococcus denitrificans*.
- Some species of *Serratia*, *Bacillus* and *Achromobacter* are also involved in the process.

$$2NO_3^- \longrightarrow 2NO_2^- \longrightarrow 2NO \longrightarrow N_2O \longrightarrow N_2$$

- This process depletes the important nutrients present in the soil. It also causes the acidification of soil.
- However, it helps in solubilisation of harmful metals present in soil.

CHECK POINT 19.2

1. Nitrogen-fixation refers to
(a) conversion of soil nitrogen into atmospheric nitrogen (N_2)
(b) conversion of atmospheric molecular nitrogen (N_2) into easily absorbable forms
(c) utilisation of nitrogen compounds in the soil by the plants
(d None of the above

2. During biological nitrogen-fixation
(a) nitrogenase requires a high input of energy about 8 ATPs for each molecule of NH_3 produced
(b) nitrogenase works without ATP utilisation
(c) NH_3 is oxidised to NO_3^-
(d) None of the above

3. Leghaemoglobin is produced in response to
(a) N_2-fixation in *Pisum sativum*
(b) N_2-fixation in *Azolla*
(c) N_2-fixation in blue-green algae
(d) N_2-fixation in *Zea mays*

4. Nitrifying bacteria are able to
(a) convert atmospheric nitrogen into soluble forms
(b) convert ammonia to nitrate
(c) ammonia to nitrogen
(d) nitrate to nitrogen

5. The reduction of nitrite to ammonia during nitrate assimilation is carried out by
(a) amino transferase and Mg
(b) nitrogenase and Zn
(c) nitrite reductase and Mn
(d) nitrite reductase and ferredoxin

6. Denitrification process deplete important nutrients from the soil. It also causes
(a) acidification of soil
(b) assimilation in the soil
(c) neutralisation of soil
(d) None of the above

EXAM POINTS

- Hydroponics is growing plants without soil and organic matter in water. **[AMU 2013]**
- Plant obtain mineral nutrients from rocks, soil and breakdown products of rocks. **[AMU 2013]**
- An essential element is that which is irreplaceable and indispensable for the growth of plants. **[BCECE 2012]**
- Minerals are absorbed by plants in ionic form. **[UP CPMT 2011]**
- Minerals known to be required in large amount for plant growth, include phosphorus, potassium, sulphur and calcium. **[CBSE AIPMT 2015]**
- Nitrogen, calcium, phosphorus, sulphur, magnesium and potassium are all macronutrients in plants. **[NEET-I 2016]**
- Boron, zinc, manganese, nickel are all micronutrients in plants. **[JIPMER 2019, AMU 2012]**
- Phosphorus is present in the structure of RNA, DNA and ATP. **[Gujarat CET 2013]**
- Sulphur is present in two amino acids-cysteine and methionine. **[Kerala CEE 2013]**
- Calcium is essential for the growth of root tips. **[NEET 2016]**
- Magnesium helps in maintaining the structure of ribosomes and ring structure of chlorophyll. **[NEET 2020, KCET 2015]**
- Manganese is involved in the splitting of water to liberate oxygen during photosynthesis. **[NEET 2020]**
- Potassium is responsible for maintaining turgor in cells. **[NEET 2018]**
- The most abundant intracellular cation is K^+. **[NEET 2013]**
- In ferric forms iron is absorbed by plants. **[NEET 2018]**
- Iron is essential for the formation of chlorophyll. **[NEET 2020]**
- Iron function as an important constituent of proteins involved in ETS, as an activator of catalase, as an important constituent of cytochrome and as an essential element for chlorophyll synthesis. **[JIPMER 2017]**
- The enzyme nitrogenase is a Mo-Fe protein. **[AMU 2015, WB JEE 2015]**
- Molybdenum is related to nitrogen metabolism. **[AMU 2015, BCECE 2015, Kerala CEE 2014]**
- Boron have a role in pollen grain germination. **[NEET 2020]**
- Boron is helpful in absorption and utilisation of calcium. **[Gujarat CET 2013]**
- Zinc is needed for synthesis of auxin. **[NEET 2020]**
- Zinc is the cofactor of alcohol dehydrogenase enzyme. **[Kerala CEE 2014]**
- 'Khaira disease' of rice is due to zinc deficiency. **[UP CPMT 2011]**
- Deficiency symptoms of nitrogen and potassium are visible first in senescent leaves. **[CBSE AIPMT 2014]**
- Nickel is a micronutrient which helps in metabolism of urea to uric acid and its deficiency causes leaf tip necrosis. **[AFMC 2013]**
- Lower level of N, K, S and Mo causes inhibition of cell division. **[Kerala CEE 2013]**
- Chlorine helps in photosynthesis and ionic balance. **[Gujarat CET 2013]**
- Mineral salt absorption is an active process and metabolic energy is used during active absorption. **[Chhattisgarh PMT 2015]**
- Pyruvate dehydrogenase activity during aerobic respiration requires magnesium. **[NEET 2020]**
- The first stable product of fixation of atmospheric nitrogen in leguminous plant is ammonia. **[JIPMER 2017, NEET 2013]**
- Microbe that fix nitrogen in the roots of non-leguminous plant *Alnus* is *Frankia*. **[Kerala CEE 2013]**
- The products of reaction catalysed by nitrogenase in root nodules of leguminous plants is/are ammonia and hydrogen. **[NEET 2020]**
- *Nitrosomonas* helps in conversion of NH_3 to nitrite. **[AIIMS 2018]**
- *Nostoc* is a free-living nitrogen-fixing cyanobacteria. **[AIIMS 2018]**
- *Azotobacter* is a free-living aerobic nitrogen-fixing bacteria. **[AIIMS 2018]**
- *Clostridium* helps in ammonification. **[WB JEE 2015, AIIMS 2018]**
- During biological nitrogen-fixation, inactivation of nitrogenase by oxygen poisoning is prevented by leghaemoglobin. **[CBSE AIPMT 2015]**
- *Rhodospirillum* is a free-living nitrogen-fixing bacteria. **[NEET 2017]**
- *Beijerinckia* is a free-living nitrogen-fixing aerobic bacterium. **[Kerala CEE 2014]**
- Besides paddy fields, cyanobacteria are also found inside vegetative parts of *Cycas*. **[NEET 2013]**
- Lectins helps in recognition of compatible *Rhizobium* by host in formation of root nodules in legumes. **[EAMCET 2013]**
- In nitrogen cycle, *Pseudomonas* produced elemental nitrogen. **[J&K CET 2013]**
- *Glomus* forms symbiotic association with plants and helps them in their nutrition. **[CBSE AIPMT 2012]**
- In *Glycine max*, the product of biological nitrogen fixation is transported from the root nodules to other parts as nitrate. **[NEET 2020]**
- *Thiobacillus* helps in denitrification. **[AIIMS 2018]**
- Anaerobic ammonium oxidation and denitrification are the two distinct micorbial processes which release dinitrogen (N_2) from fixed nitrogen. **[KPC 2013]**
- *Pseudomonas* causes denitrification. **[J&K CET 2015]**
- Denitrification is carried by bacteria *Nitrosomonas* and *Thiobacilus*. **[Kerala CEE 2013]**

Chapter Exercises

(A) Topical Exercise

Cultured Experiments

1 Who gave the description of all mineral nutrients in plants by using culture experiments?
(a) De Saussure (1804)
(b) Leibeg (1840)
(c) Glauber and Mayhon (1650)
(d) Arnon and Hoagland (1940)

2 Hydroponics is a system of growing plants in
(a) soilless culture or solution culture
(b) acidic soils
(c) soilless culture with alkaline pH
(d) soilless culture with acidic pH

3 I. The technique of hydroponics was first demonstrated by ...... *A* ....
II. Hydroponics is used for the commercial production of vegetables like ...... *B* ....
(a) *A*-Julius von Sachs, *B* - Lettuce
(b) *A*-Arnon, *B*-*Zea mays*
(c) *A*-Hoagland, *B*-Blue-berry
(d) *A*-M Calvin, *B*-Cotton

4 The given figure shows hydrophonic/soilless plant production. Plants are grown in a tube or through placed on a slight incline. The arrow indicate the duration of flow of nutrient solution.
Nutrient solution is sent to the elevated end of the tube from the reservoir by ...... and it flow back into a reservoir due to ....

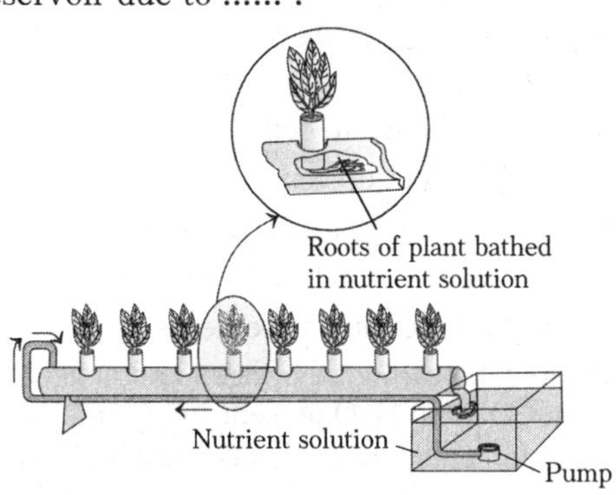

(a) Pump, gravity
(b) Gravity, pump
(c) Gravity, gravity
(d) Pump, pump

5 Consider the following statement with respect to hydroponics.
I. It determines the mineral nutrients essential for the plants.
II. Hydroponics requires purified water with non-defined minerals nutrient salts.
Choose the correct option.
(a) Statement I is true, but II is false
(b) Statement I is false, but II is true
(c) Both statements I and II are true
(d) Both statements I and II are false

6 Plants can be grown in (Tick the incorrect option)
(a) soil with essential nutrients **[NCERT Exemplar]**
(b) water with essential nutrients
(c) either water or soil with essential nutrients
(d) water or soil without essential nutrients

7 Growing plant in a nutrient solution
(a) involve use of chelating agent Aa-EDTA to keep iron unavailable to plants
(b) was first demonstrated by Arnon and Hoagland
(c) also allow growth of algae in containers
(d) requires prevention of reaction of light with roots

8 How is ash significant in the study of mineral nutrition of plants?
(a) It gives quantitative values of minerals in soil
(b) It informs, which element is essential and in which amount it is necessary for a particular plant
(c) It is of no practical significance
(d) It indicates how much irrigation is needed for a plant

9 The given diagrammatic set up is

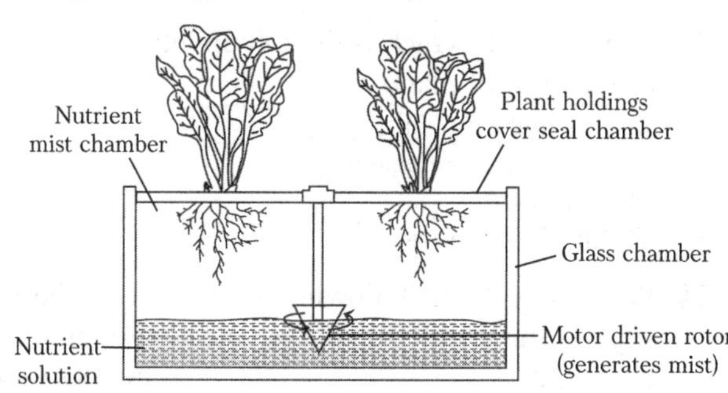

(a) Hydroponic tank growth system
(b) Hydroponic film growth system
(c) Aeroponic growth system
(d) Nutrient solution culture

10 The major portion of the dry weight of plants comprises of
(a) carbon, nitrogen and hydrogen
(b) carbon, hydrogen and oxygen
(c) nitrogen, phosphorus and potassium
(d) calcium, magnesium and sulphur

11 Consider the following statements with respect to aeroponic technique.
I. A very shallow stream of water containing dissolved nutrients is recirculated post the roots of plants in a water tight channel.
II. Root keep suspended in the air over the nutrient solution which is provided in the form of a nutrient mist.

Choose the correct option.
(a) Statement I is true, but II is false
(b) Statement I is false, but II is true
(c) Both statements I and II are true
(d) Both statements I and II are false

12 Consider the following statements.
I. In aeroponics technique, plants are grown with their roots suspended in air while being sprayed continuously with a nutrient solution.
II. Successful hydroponic culture requires a large volume of nutrient solution or frequent adjustment of the nutrient solution to prevent roots from producing radical changes in nutrient concentration and pH of the medium.

Choose the correct option.
(a) Statement I is true, but II is false
(b) Statement I is false, but II is true
(c) Both statements I and II are true
(d) Both statements I and II are false

Inorganic Nutrients

13 The element, which is directly involved in the metabolism of plants is
(a) essential element
(b) non-essential element
(c) nutrients
(d) All of the a bove

14 Which one of the following statements can best explain the term critical concentration of an essential element? **[NCERT Exemplar]**
(a) Essential element concentration below which plant growth is retarded
(b) Essential element concentration below which plant growth is enhanced
(c) Essential element concentration below which plant remains in the vegetative phase
(d) None of the above

15 Which one of the following roles is not characteristic of an essential element? **[NCERT Exemplar]**
(a) being a component of biomolecules
(b) changing the chemistry of soil
(c) being a structural component of energy related chemical
(d) activation or inhibition of enzymes

16 An element must be considered essential, when
(a) the element is necessary for supporting normal growth and reproduction of the plants
(b) the deficiency of that particular element cannot be met by supplying some other elements
(c) it is directly involved in the metabolism
(d) all of the above

17 The concentration of a macroelement per gram of dry matter in plants is at least
(a) 1000 μg/g of dry matter (b) 100 μg/g of dry matter
(c) 1500 μg/g of dry matter (d) 995 μg/g of dry matter

18 A micronutrient is one which
(a) is less important than the major essential element
(b) is present in large quantities in the soil
(c) is more important than the major essential element
(d) is needed in very small amount, but it is as important as a major essential element

19 Microelements are those essential elements, which are required by the plants in concentration
(a) less than 1 mg/g of dry matter
(b) less than 1 g/10 g of dry matter
(c) more than 1 mg/g of dry matter
(d) equal to 1 mg/g of dry matter

20 Nitrogen is required mainly by which of the following parts of the plants?
I. Meristematic tissues.
II. Differentiating tissues.
III. Apical tissues.
IV. Metabolically active cell.

Choose the correct option.
(a) Only II (b) Only I
(c) I and II (d) I and IV

21 Consider the following statements.
I. Plants take carbon, hydrogen and oxygen mainly from CO_2 and H_2O present in air and soil.
II. The elements like gold, selenium and strontium although can be absorbed by the plant, but have no role in them.

Choose the correct option.
(a) Statement I is true, but II is false
(b) Statement I is false, but II is true
(c) Both statements I and II are true
(d) Both statements I and II are false

22 About 98% of the mass of every living organism is composed of just six elements including carbon, hydrogen, nitrogen, oxygen
(a) phosphorus and sulphur
(b) sulphur and magnesium
(c) magnesium and sodium
(d) calcium and phosphorus

23 Match the following columns.

Column I	Column II
A. Mo	1. Coppery texture of leaf
B. Mg	2. Premature leaf fall
C. B	3. Interveinal chlorosis
D. P	4. Mottled chlorosis and necrosis

Codes

	A	B	C	D
(a)	4	1	2	3
(b)	1	2	3	4
(c)	4	3	1	2
(d)	4	2	1	3

24 Which of the following is a true statement?
(a) Every mineral element that is present in a cell is needed by the cell
(b) B deficiency leads to stout axis
(c) N as a nutrient element, is highly immobile in the plants
(d) It is very easy to establish the essentiality of micronutrients because they are required only in trace quantities

25 Choose the correct pair.
(a) Carbohydrate translocation - Boron
(b) Components of vitamins - Iron
(c) Synthesis of cell wall - Phosphorous
(d) Component of cytochromes - Calcium

26 Ca^{+2} is an essential element in plants. The major function it performs is
(a) selective permeability of the cell membrane
(b) maintenance of the cell turgidity
(c) energy transfer
(d) protein synthesis

27 Match the following columns.

Column I	Column II
A. Mg	1. Found in middle lamella
B. S	2. A structural component of chlorophyll
C. Ca	3. Required for enzyme activity
D. Mn	4. Found in some amino acids

Codes

	A	B	C	D		A	B	C	D
(a)	2	4	1	3	(b)	1	2	4	3
(c)	3	2	1	4	(d)	4	1	2	3

28 A gardener sows a 'X' and he wants that plant 'X' grows without any lateral branches. What does he apply to the field?
(a) Zinc to promote auxin synthesis
(b) Manganese to promote auxin synthesis
(c) Sulphur to promote auxin synthesis
(d) Iron to promote auxin synthesis

29 Match the following columns.

Column I (Elements)	Column II (Types/Functions)
A. Manganese	1. Macronutrient
B. Magnesium	2. Component of biomolecules
C. Phosphorus	3. Micronutrient
D. Nitrogen	4. In the formation of ATP

Codes

	A	B	C	D
(a)	4	3	1	2
(b)	4	2	3	1
(c)	1	3	4	2
(d)	3	1	4	2

30 Select the incorrectly matched pair.
(a) Magnesium (Mg) - Formation of mitotic spindle
(b) Iron (Fe) - Formation of chlorophyll
(c) Chlorine (Cl) - Anion-cation balance in the cell
(d) Sulphur (S) - Component of vitamin

31 Match the following columns.

Column I (Location)	Column II (Elements)
A. Found in some vitamins (thiamine, biotin, coenzyme-A)	1. Mg
B. Required for photolysis of water	2. I
C. Not important for a plants	3. S
D. Present at the the centre of porphyrin ring in chlorophyll	4. Mn

Codes

	A	B	C	D
(a)	4	1	2	3
(b)	3	4	1	2
(c)	1	2	4	3
(d)	3	4	2	1

32 The mineral involved in water splitting reaction during photosynthesis are ...... *A* ...... and ...... *B* ......

	A	*B*
(a)	Manganese	Chlorine
(b)	Copper	Chlorine
(c)	Magnesium	Chlorine
(d)	Potassium	Chlorine

33 Read the functions given below and identify the concerned nutrient.

I. Activator of catalase.
II. Important constituent of cytochrome.

Codes

(a) Mo (b) Fe
(c) Cu (d) Ca

34 Correctly match the columns A, B and C.

	A	B	C
(a)	Sulphur	Tea yellow	Cofactor of many enzyme
(b)	Calcium	Heart rot	Constituent of carbonic anhydrase
(c)	Boron	Corky core	Sugar metabolism
(d)	Chlorine	Wither tip	Death of root tip

35 Match the following columns.

Column I	Column II
A. Mg^{2+}	1. Nitrate reductase
B. Zn^{2+}	2. RuBisCO, PEPcase
C. Mo	3. Alcohol dehydrogenase

Codes

	A	B	C
(a)	2	3	1
(b)	3	2	1
(c)	1	3	2
(d)	2	1	3

36 Hunger signs in plants are

(a) symptoms due to lesser water absorption in plants
(b) symptoms due to poor photosynthesis in plants
(c) deficiency symptoms of mineral nutrients
(d) symptoms due to excess water absorption in plants

37 Deficiency symptoms of an element tend to appear first in young leaves. It indicates that the element is relatively immobile. Which one of the following elemental deficiency would show such symptoms?

(a) Sulphur
(b) Magnesium
(c) Nitrogen
(d) Potassium

38 On the basis of symptoms of chlorosis in leaves, a student inferred that this was due to deficiency of nitrogen. The inference could be correct only if we assume that yellowing of leaves appeared first in

(a) old leaves
(b) young leaves
(c) young leaves followed by mature leaves
(d) mature leaves followed by young leaves

39 Deficiency of iron causes

(a) reduced leaves and stunted growth
(b) bending of leaf tip
(c) interveinal chlorosis
(d) decrease in protein synthesis

40 The deficiency symptom(s) of nitrogen in plants is/are

(a) dormancy of lateral buds
(b) delaying of flowering
(c) premature leaf fall
(d) All of the above

41 Which of the following combination(s) is/are not correct?

I. Tea yellow — Boron
II. Whiptail of crucifers — Molybdenum
III. Yellow spot of *Citrus* — Manganese
IV. Khaira disease of rice — Zinc

Codes

(a) I and III (b) Only II
(c) I, II and III (d) Only III

42 For the elements that are actively mobilised within the plants like ...*A*..., the deficiency symptoms tend to appear first in ...*B*... tissues.

(a) *A*– N, K, S and Mg, *B*– younger
(b) *A*–Ca, *B*–older
(c) *A*–N, K and Mg, *B*–older
(d) *A*–Ca, *B*–younger

43 Necrosis or death of tissue particularly leaf tissue, is due to the deficiency of

(a) N, K and S (b) N, K, Mg and Fe
(c) Mn, Zn and Mo (d) Ca, Mg, Cu and K

44 Consider the following statements.

I. Moderate decrease or increase of micronutrients cause deficiency and toxicity symptoms, respectively.
II. Excess of manganese causes toxicity of iron, calcium and molybdenum.

Choose the correct option.

(a) Statement I is true, but II is false
(b) Statement I is false, but II is true
(c) Both statements I and II are true
(d) Both statements I and II are false

45 Which one of the following symptoms is not due to manganese toxicity in plants? **[NCERT Exemplar]**

(a) Calcium translocation in shoot apex is inhibited
(b) Deficiency in both iron and nitrogen is induced
(c) Appearance of brown spot surrounded by chlorotic veins
(d) None of the above

Absorption of Mineral Elements

46 Mineral absorption occurs
(a) if soil solution is hypotonic
(b) if soil solution is hypertonic
(c) independent of water potential
(d) independent of water absorption

47 The studies about mechanism of absorption of elements revealed that the process of absorption can be demarcated into two main phases. The first phase and second phase are

	First phase	Second phase
(a)	Active	Passive
(b)	Active	Active
(c)	Passive	Active
(d)	Passive	Passive

48 Which of the following shows the metabolic energy is required in the absorption of ions?
(a) More ions absorption in presence of oxygen
(b) Less absorption of ions in presence of oxygen
(c) More ions absorption in presence of ATP
(d) More ions absorption in presence of NAD

49 If the amount of an ion absorbed by a root hair cell at 0°C is 5 gm at 20°C is 20 gm. The amount of thin ion absorbed actively should be
(a) 25 gm (b) 20 gm
(c) 15 gm (d) 5 gm

50 Consider the following statements.
I. In the initial phase, ions are taken up into the outer space of cells, the apoplast, it is a passive process.
II. In the final phase, ions are taken slowly into the inner space, the symplast of cells and it is an active process.

Choose the correct option.
(a) Statement I is true, but II is false
(b) Statement I is false, but II is true
(c) Both statements I and II are true
(d) Both statements I and II are false

51 Match the following columns.

Column I (Mechanism)	Column II (Absorption mechanism)
A. Movement of ions into or cut of the cell	1. Diffusion
B. Outward movement is	2. Flux
C. Inward movement is	3. Efflux
D. Passive uptake	4. Influx

Codes

	A	B	C	D		A	B	C	D
(a)	1	4	3	2	(b)	4	2	1	3
(c)	2	3	4	1	(d)	3	4	1	2

52 Which of the following does not operate in the carrier concept of active absorption of minerals?
(a) Electron transport
(b) Contractile protein
(c) Protein lecithin
(d) Ascending water stream

53 In carbonic acid exchange theory, cation thus released into soil solution are adsorbed on root cells in exchange of
(a) H^+ ions (b) Na^+ ions
(c) Cl^- ions (d) K^+ ions

Nitrogen Metabolism

54 Choose the correct option.
(a) Nitrogen-fixation is the process of ammonia formation by nitrogen
(b) The regular supply of nitrogen in nature is maintained by nitrogen cycle
(c) Nitrogen is present in the atmosphere in the form of dinitrogen
(d) All of the above

55 Consider the following statements.
I. Nitrogenase is a heterodimer protein.
II. During symbiotic N_2-fixation ATP is provided by legume plant.

Choose the correct option.
(a) Statement I is true, but II is false
(b) Statement I is false, but II is true
(c) Both statements I and II are true
(d) Both statements I and II are false

56 Which two distinct microbial processes are responsible for the release of fixed nitrogen as dinitrogen (N_2) gas to the atmosphere?
(a) Anaerobic ammonium oxidation and denitrification
(b) Aerobic nitrate oxidation and nitrite reduction
(c) Decomposition of organic nitrogen and conversion of dinitrogen to ammonium compounds
(d) Enteric fermentation in cattle and nitrogen fixation by *Rhizobium* in root nodules of legumes

57 Which is not true for nitrogenase enzyme in root nodules in legumes?
(a) Synthesied by *nif* genes of *Rhizobium*
(b) Site of reduction of N_2 into NH_3
(c) It is a Mo-Fe protein
(d) Resistant to O_2 concentration

58 Which of the following diagrams correctly depicts N_2-cycle?

(a)

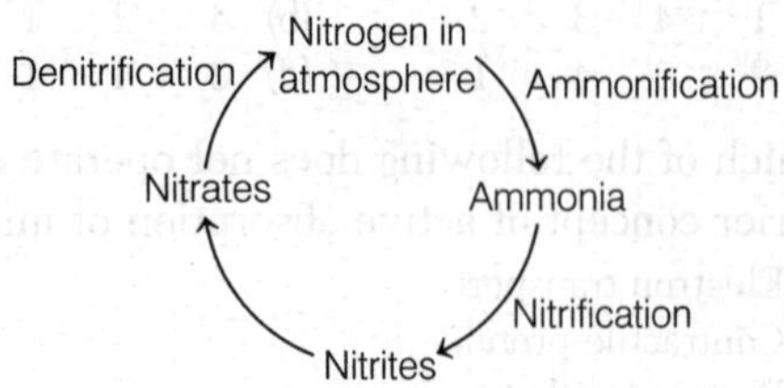

(b)

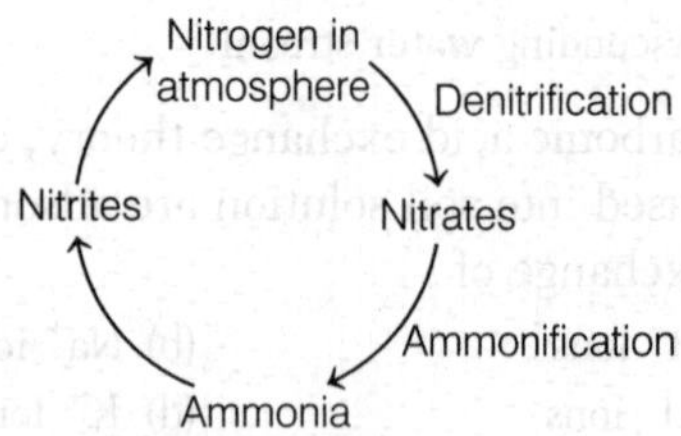

(c)

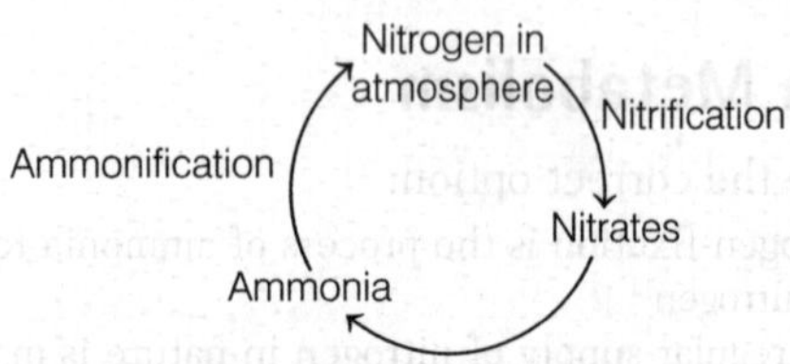

(d)

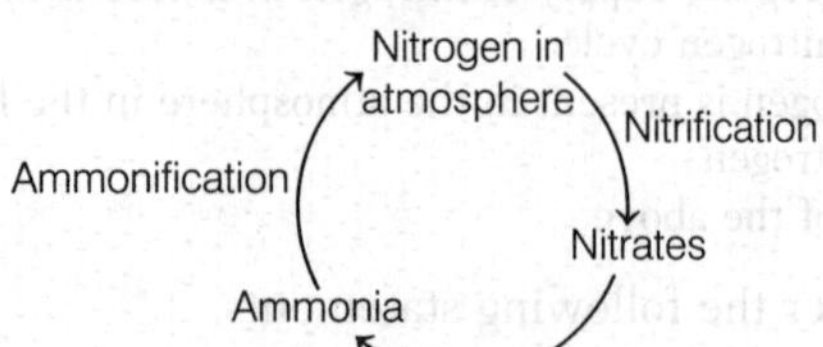

59 Conversion of $NO_3 \longrightarrow NO_2 \longrightarrow NH_3$ is called ...*A*... and is catalysed by ...*B*... .

	A	*B*
(a)	nitrate assimilation	nitrate reductase
(b)	nitrification	nitrite reductase
(c)	ammonification	glutamate dehydrogenase
(d)	denitrification	transaminase

60 Consider the following statements.

I. Legumes fix nitrogen independently of the specialised bacteria that live in their roots.

II. Root hairs curl by the action of nitrogenase and plant *nod* factors.

Choose the correct option.

(a) Statement I is true, but II is false
(b) Statement I is false, but II is true
(c) Both statements I and II are true
(d) Both statements I and II are false

61 Mark the correct sequence steps for the development of root nodules in legumes.

I. Formation of infection thread
II. Division of cortical and pericyclic cells
III. Curling of roots hairs
IV. Leghaemoglobin synthesis

Choose the correct option.

(a) I, III, II and IV
(b) II, IV, I and III
(c) III, II, I and IV
(d) III, I, II and IV

62 Which of the following association fix nitrogen through non-nodulation?

(a) Roots of maize associated with *Spirillum notatum*
(b) Roots of *Digitaria* in association with *Spirillum notatum*
(c) Coralloid roots of *Cycas* is associated with *Anabaena*
(d) All of the above

63 Leghaemoglobin is useful for legumes because it

(a) irreversibly absorbs O_2 which inhibit N_2-fixation
(b) reversibly absorbs O_2 which inhibit N_2-fixation
(c) protects plant tissues from damage by symbiotic bacteria
(d) absorbs O_2 at low oxygen tension and provides it to the tissue surrounding N_2-fixing bacteria

64 If by radiation all nitrogenase enzyme are inactivated, then there will be no

(a) fixation of nitrogen in legumes
(b) fixation of atmospheric nitrogen
(c) conversion of nitrate to nitrite in legumes
(d) conversion of ammonium to nitrate in soil

65 Consider the following statements.

I. The coordinated activities of the legume and *Rhizobium* bacteria depend on chemical interaction between the symbiotic partners.

II. Leguminous roots secrete chemical attracts that *Rhizobium* bacteria living nearby.

Choose the correct option.

(a) Statement I is true, but II is false
(b) Statement I is false, but II is true
(c) Both statements I and II are true
(d) Both statements I and II are false

66 Nitrification is done in two steps. In first step NH_3 is converted into nitrite and in second step nitrite is converted into nitrate by

(a) *Nitrosomonas* and *Nitrococcus*, respectively
(b) *Nitrosomonas* and *Nitrobacter*, respectively
(c) *Nitrococcus* and *Nitrobacter*, respectively
(d) Both (b) and (c)

67 Find out the incorrect statement.

(a) *Azotobactor* and *Beijernickia* are aerobic, free-living nitrogen-fixing microbes
(b) *Rhodospirillum* and *Bacillus polymyxa* are anaerobic, free-living bacteria
(c) *Rhizobium* and *Frankia* are symbiotic N_2 fixers in legumes
(d) Cyanobacteria (*Nostoc* and *Anabaena*) are free-living N_2 fixers

68 Match the following columns.

Column I	Column II
A. *Rhizobium leguminosarum*	1. Cereal roots
B. *Rhizobium japonicum*	2. *Azolla* (water fern)
C. *Azospirillum brasiliense*	3. Pea
D. *Anabaena* sp.	4. Soyabean

Codes

	A	B	C	D		A	B	C	D
(a)	1	3	2	4	(b)	3	4	1	2
(c)	4	3	1	2	(d)	2	1	3	4

69 Which of the following is correct with respect to root nodules of *Alnus* ?

(a) Harbours the photosynthetic *Frankia*
(b) Lacks leghaemoglobin, but posses another form of plant haemoglobin
(c) *Frankia* associated with it cannot synthesise nitrogenase enzyme
(d) All of the above

70 Denitrification is carried by bacteria

(a) *Rhizobium* and *Frankia*
(b) *Azotobacter* and *Beijernickia*
(c) *Pseudomonas* and *Thiobacillus*
(d) *Rhodospirillum* and *Bacillus*

71 What does the given reaction show?

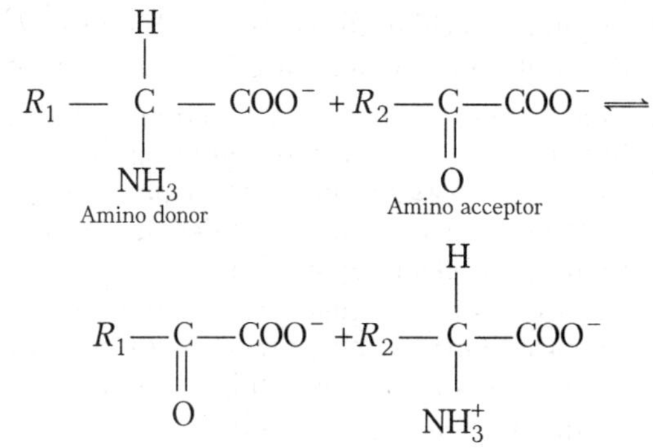

Choose the correct option.

(a) Oxidative deamination (b) Reductive amination
(c) Transamination (d) Deamination

72 During the formation of amides from amino acids

(a) COOH part of acid is replaced by NH_2
(b) NH_2 part of acid is replaced by COOH
(c) OH part of acid group is replaced by NH_2
(d) NH_2 part of acid is replaced by OH

73 When two moleules of nitrogen are converted into ammonia during nitrogen-fixation, then how many ATP are utilised in this process?

(a) 8 ATP (b) 16 ATP
(c) 24 ATP (d) 32 ATP

(B) Cumulative Exercise

(Assorted Questions of the Chapter for Advanced Level Practice)

1 Given below is the diagrammatic set-up of nutrient solution culture. Identify the aparatus marked *X* and its use

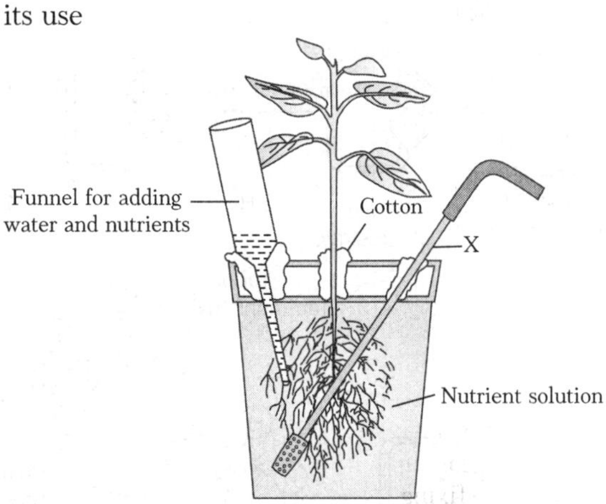

(a) *X* - Iron rod, Use : for mixing mineral salts
(b) *X* - Aerating tube, Use : aeration, to increase the dissolved oxygen in nutrient solution for optimum growth
(c) *X* - Aerating tube, Use : to add frest nutrient medium for optimum growth
(d) *X* - Copper rod, Use : to provide Cu^{2+} in the nutrient medium

2 Read the following statemetns.

I. Transfer of amino group from amino acid to keto group of a keto acid – Transamination
II. Hydroxyl part of amino acid is replaced by another amino group – Amide formation
III. Hydroxyl part of keto acid is replaced by amino group – Amide formation
IV. Transfer of amino group from keto acid to amino acid – Transamination
V. Ammonia reacts with α-ketoglutaric acid in the presence of NADPH + H^+ and glutamate dehydrogenase to form glutamic acid.

Which of the above statements are correct?

(a) III IV and V (b) I, II and V
(c) I, III and V (d) III, IV and V

3 Read the following statements.

I. In carbonic acid exchange theory, H_2CO_3 dissociates into H^+ and HCO_3^- ion in soil solution.

II. In hydroponic technique, the nutrient solution flow in a thin film over the roots ensuring that the upper part of the roots get sufficient supply of oxygen.

III. N, P and K usually do not get deficient in soil due to their low plant requirement.

IV. Nitrogen cycle is regular circulation of nitrogen amongst living organisms with its reservoir pool in lithosphere and cycling pool in atmosphere.

Select the incorrect statements.

(a) I and II (b) II and III (c) III and IV (d) I, II and III

4 Choose the incorrect statements.

I. Plant obtains Mo in the form of MoO_2^{2+} and it is the component of several enzymes including nitrogenase and nitrate reductase.

II. The concentration of the essential elements above which plant growth is retarded is termed as critical concentration.

III. Deficiency symptoms are structural changes in the plant due to certain element deficiency.

IV. Deficiency symptoms disappear if the specific nutrient is provided to the plant, also if the deprivation continues death of plant occurs.

(a) I and II (b) III and IV

(c) II and III (d) I and IV

5 Study the figure shown below and select the option which gives correct words for all the blanks.

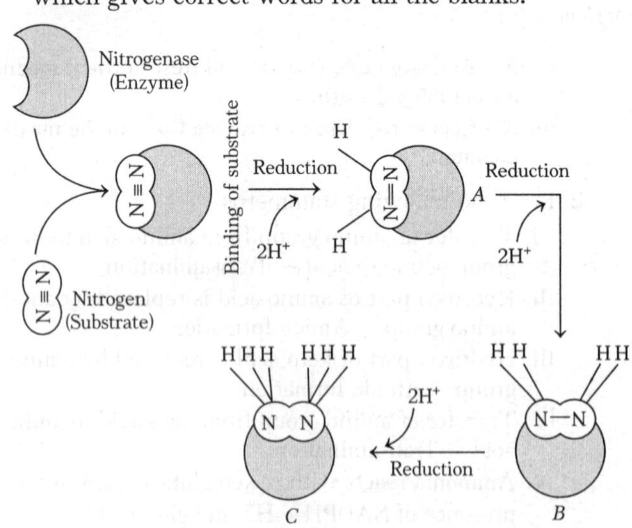

	A	B	C
(a)	Diamide	Hydrazine	Ammonia
(b)	Hydrazine	Ammonia	Diamide
(c)	Ammonia	Diamide	Hydrazine
(d)	Diamide	Ammonia	Hydrazine

6 Read the following statements.

I. A macronutrient is said to be toxic when present below a critical concentration.

II. The process of conversion of

$2NO_3^- \longrightarrow 2NO_2^- \longrightarrow 2NO \longrightarrow N_2O \longrightarrow N_2$

is called denitrification.

III. Process of nitrification and nitrate assimilation are oxidation and reduction, respectively.

IV. Tryptophan synthesis, carboxylase activity and little leaf of plants are all associated with zinc.

Choose the incorrect statement.

(a) Only I (b) I, II and III

(c) III and IV (d) I and IV

7 Match the following mineral elements with their deficiency symptoms and choose the correct option.

Column I		Column II
A. Calcium	1.	Chlorotic veins
B. Potassium	2.	Delayed germination of seeds
C. Zinc	3.	Necrosis of young leaves
D. Iron	4.	Scorched leaf tips
E. Phosphorus	5.	Malformed leaves

Codes

	A	B	C	D	E		A	B	C	D	E
(a)	3	1	5	2	4	(b)	1	4	5	3	2
(c)	3	4	5	1	2	(d)	2	3	4	1	5

8 Pick the correct set of statements for the given diagram of N_2-fixation and choose the correct option.

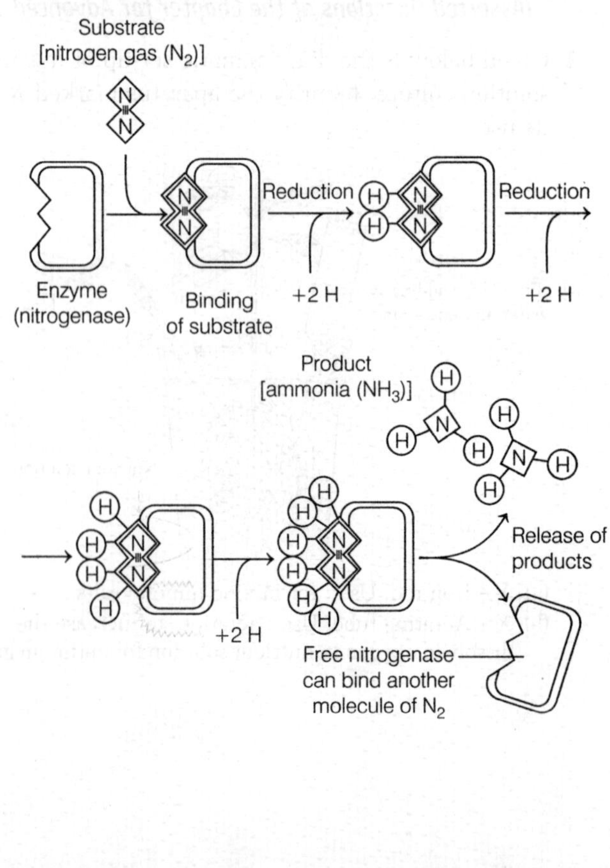

I. Nitrogenase catalyses the reaction.
II. The formation of ammonia is a reductive process.
III. One molecule of nitrogen produces two molecules of ammonia.
IV. Formation of ammonia is an oxidative reductive process.

Select the option depicting the correct statements.
(a) I, II and III (b) I, II and IV
(c) I, II and IV (d) I, III and IV

9 Match the following columns.

Column I	Column II
A. Photolysis of water	1. Zinc
B. Diazotrophy	2. Copper
C. Cytochrome-*c* oxidase	3. Manganese
D. Biosynthesis of IAA	4. Molybdenum
	5. Boron

Codes

	A	B	C	D		A	B	C	D
(a)	3	2	1	5	(b)	3	4	2	1
(c)	5	2	3	4	(d)	4	1	3	2

10 Choose the incorrect statement.
(a) Plants cannot use atmospheric nitrogen directly, though available abundantly in air
(b) Nitrogen-fixation require a strong reducing agent
(c) 8 ATPs are required for fixation of one nitrogen molecule during nitrogen-fixation
(d) The enzyme nitrogenase is Mo-Fe protein, while ferredoxin is Fe-S protein

11 Study the cycle shown below and select the option which gives correct words for all the four blanks *A*, *B*, *C* and *D*.

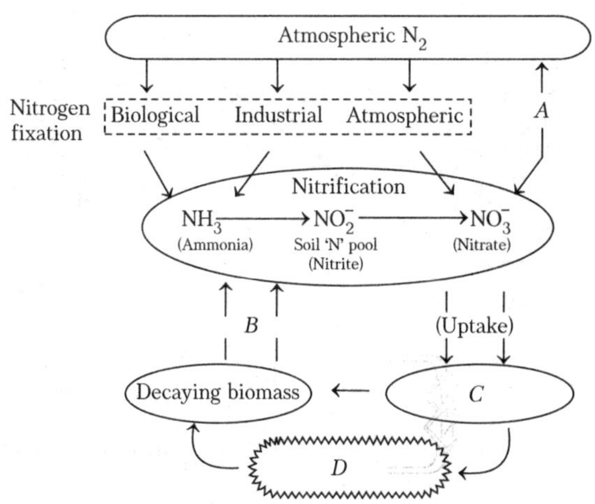

	A	*B*	*C*	*D*
(a)	Denitrification	Ammonification	Plants	Animals
(b)	Nitrification	Denitrification	Animals	Plants
(c)	Denitrification	Nitrification	Plants	Animals
(d)	Nitrification	Ammonification	Animals	Plants

12 Identify the correct match for columns I, II and III.

Column I	Column II	Column II
A. Molybdenum	1. Macronutrient	(i) Middle lamella
B. Zinc	2. Micronutrient	(ii) Stomatal movement
C. Calcium	3. Trace element	(iii) Auxin
D. Potassium	4. Macronutrient	(iv) Nitrate reductase

Codes
(a) A-4-i, B-2-ii, C-1-iii, D-3-iv
(b) A-1-iv, B-4-iii, C-3-ii, D-2-i
(c) A-2-iv, B-3-iii, C-4-i, D-1-ii
(d) A-1-ii, B-2-iv, C-3-i, D-4-iii

13 Select the correct statement(s) regarding the solution culture technique.
(a) Successful hydroponic culture requires a large volume of nutrient solution or frequent adjustment of the nutrient solution to prevent roots from producing radical changes in nutrient concentrations and the pH of the medium.
(b) In nutrient film growth system, plant root lie on the surface of a trough and nutrient solution flow in a thin layer along the trough over the roots.
(c) In aeroponics technique, plants are grown with their roots suspended in air while being sprayed continuosly with a nutrient solution.
(d) All of the above

14 Read the following steps during root nodule formation and arrange these sequentially.
I. Nodule formed
II. Infection thread is produced
III. *Rhizobium* released into cortical cells
IV. Root hairs curl
V. *Rhizobium* colonise roots
(a) I → II → IV → V → III
(b) V → IV → III → II → I
(c) V → IV → II → III → I
(d) I → II → V → IV → III

15 Mark the statements as True (T)/False (F) by choosing the correct option from the set (I-IV) given below.

I. Magnesium is a constituent of chlorophyll and helps to maintain the ribosome structure.

II. Calcium is needed during the formation of mitotic spindle.

III. Magnesium is essential for the photolysis of water.

IV. Zinc helps in sugar translocation.

Codes

	I	II	III	IV		I	II	III	IV
(a)	T	T	F	F	(b)	F	T	F	T
(c)	T	F	T	F	(d)	F	F	T	T

16 Farmers in a particular region were concerned that premature yellowing of leaves of a pulse crop might cause decrease in the yield. Which treatment could be most beneficial to obtain maximum seed yield?

(a) Frequent irrigation of the crop
(b) Treatment of the plants with cytokinins along with a small dose of nitrogenous fertiliser
(c) Removal of all yellow leaves and spraying the remaining green leaves with 2, 4, 5-trichlorophenoxy acetic acid
(d) Application of iron and magnesium to promote synthesis of chlorophyll

17 Match the following and choose the correct combination from the options given below.

	Column I		Column II
A.	Sulphur	1.	Chlorophyll
B.	Zinc	2.	Nitrogenase
C.	Magnesium	3.	Methionine
D.	Molybdenum	4.	Auxin
		5.	Middle lamella

Codes

	A	B	C	D		A	B	C	D
(a)	1	2	3	5	(b)	3	4	1	2
(c)	3	5	2	4	(d)	2	4	1	3

18 Study the figure shown below and select the option which gives correct words for all the blanks.

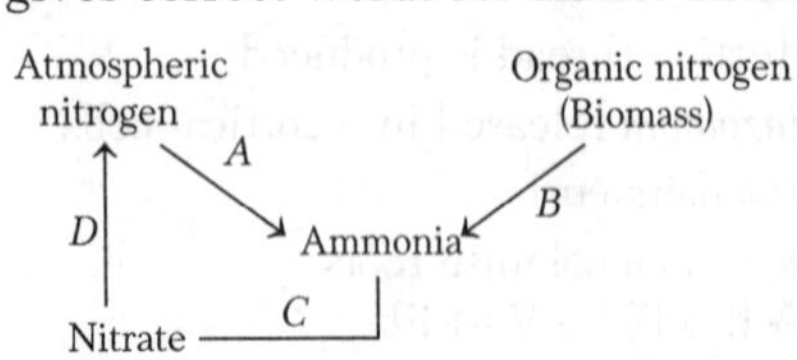

(a) *A*-Nitrification, *B*-Nitrogen-fixation, *C*-Denitrification, *D*-Ammonification
(b) *A*-Nitrogen-fixation, *B*-Nitrification, *C*-Ammonification, *D*-Denitrification
(c) *A*-Nitrogen-fixation, *B*-Ammonification, *C*-Nitrification, *D*-Denitrification
(d) *A*-Nitrogen-fixation, *B*-Nitrification, *C*-Ammonification, *D*-Denitrification

19 Study the following statements regarding leghaemoglobin and choose the option for incorrect statement(s).

I. The central portion of root nodule found in leguminous plant is white.

II. Root nodule for leguminous plant contains both enzyme nitrogenase and leghaemoglobin.

III. Leghaemoglobin traps molecular O_2 to create anaerobic conditions need for activity of nitrogenase enzyme.

(a) I is correct, II and III are incorrect
(b) I is incorrect, II is III are correct
(c) Only II is incorrect
(d) I and II are incorrect

20 Choose the incorrect statement.

(a) Nitrogenase enzyme is protected from O_2 in the nodules of legumes plants
(b) Nitrogenase enzyme is highly insensitive to the presence of molecular oxygen
(c) Leghaemoglobin is oxygen scavenger and creates anaerobic conditions in the roots of leguminous plants
(d) *Rhizobium* is an aerobic bacterium in free-living conditions

21 Identify the *A* to *D* correctly in the given diagram of root nodule development and choose the correct option accordingly.

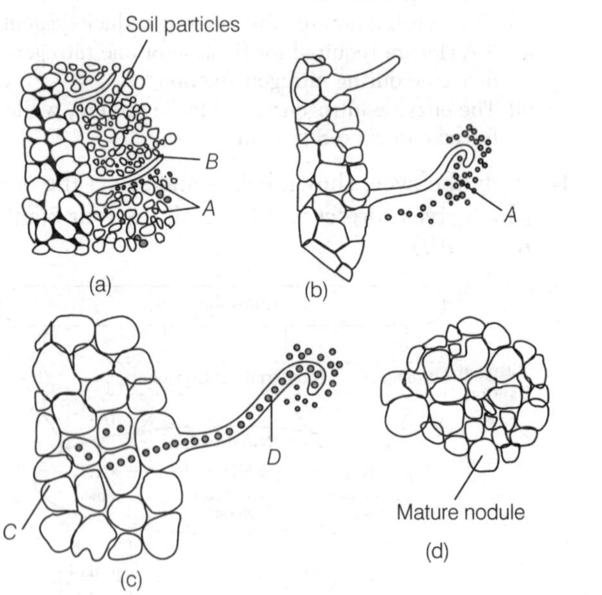

(a) *A*-*Rhizobium* bacteria, *B*-Cortex cell, *C*-Outer cortex, *D*-Infection thread
(b) *A*-*Rhizobium* bacteria, *B*-Root hair, *C*-Inner cortex, *D*-Infection thread
(c) *A*-*Rhizobium* bacteria, *B*-Endodermal cell, *C*-Inner endodermis, *D*-Infection thread
(d) *A*-*Nitrosomonas* bacteria, *B*-Root hair, *C*-Inner cortex, *D*-Infection thread

Assertion and Reason

Direction (Q. Nos. 22-34) *In each of the following questions, a statement of Assertion (A) is given followed by the corresponding statement of Reason (R). Of the statements, mark the correct answers as*

(a) Both A and R are true and R is the correct explanation of A
(b) Both A and R are true, but R is not the correct explanation of A
(c) A is true, but R is false
(d) A is false, but R is true
(e) Both A and R are false

22 Assertion (A) Leguminous plants are nitrogen fixers.
Reason (R) Leguminous plants have *Rhizobium* in their root nodules.

23 Assertion (A) Plants absorb nitrogen in the form of nitrate only.
Reason (R) Nitrogen is the most critical element.

24 Assertion (A) The leaves of cauliflower become flaccid and brown in molybdenum deficiency.
Reason (R) Cauliflower plant is affected by whiptail disease in molybdenum deficiency.

25 Assertion (A) Hydroponics is used for solution culture.
Reason (R) A balanced nutrient solution contains both essential and non-essential elements.

26 Assertion (A) Excess Mn in soil, can adversely decrease Mg, Fe and Ca concentration in soil.
Reason (R) Mn increases rate of photosynthesis thereby increasing absorption of Mg, Fe and Ca from soil.

27 Assertion (A) Both *Rhizobium* and *Frankia* are living in soil, but as symbionts can fix atmospheric N_2 in legume.
Reason (R) *Rhizobium* and *Frankia* do not occur as symbionts.

28 Assertion (A) Leghaemoglobin is universally found in legume plants and cyanobacteria.
Reason (R) Cyanobacterial nitrogenase is insensitive towards oxygen.

29 Assertion (A) Iron is a microelement.
Reason (R) Microelements are required in traces only, less than 1 mg/gm of dry matter.

30 Assertion (A) Calcium is a constituent of cell wall.
Reason (R) Calcium is required in mitotic division.

31 Assertion (A) Manganese is an activator of enzyme reductase.
Reason (R) Manganese deficient cells prefer ammonia over nitrate.

32 Assertion (A) Magnesium is important in photosynthesis and carbohydrate metabolism.
Reason (R) Mg^{2+} is involved in the synthesis of nucleic acids.

33 Assertion (A) Deficiency of sulphur causes chlorosis in plants.
Reason (R) Sulphur is a constituent of chlorophyll.

34 Assertion (A) Nitrogen-fixing bacteria of legume root nodules survive in oxygen depleted cells.
Reason (R) Leghaemoglobin completely removes oxygen from nodule cells.

(C) Medical Entrances' Gallery

(Collection of Questions Asked in NEET & in Various Medical Entrance Exams)

1 In *Glycine max*, the product of biological nitrogen fixation is transported from the root nodules to other part as **[NEET 2020]**
(a) ammonia
(b) glutamate
(c) nitrate
(d) ureides

2 The product(s) of reaction catalysed by nitrogenase in root nodules of leguminous plants is/are **[NEET 2020]**
(a) nitrate only
(b) ammonia and oxygen
(c) ammonia and hydrogen
(d) ammonia only

3 Match the following concerning essential elements and their functions in plants.

Column I	Column II
A. Iron	1. Photolysis of water
B. Zinc	2. Pollen germination
C. Boron	3. Required for chlorophyll biosynthesis
D. Manganese	4. IAA biosynthesis

Select the correct option. **[NEET 2020]**

	A	B	C	D
(a)	4	3	2	1
(b)	3	4	2	1
(c)	4	1	2	3
(d)	2	1	4	3

4 Which of the following elements helps in maintaining the structure of ribosomes? **[NEET 2020]**
(a) Magnesium (b) Zinc
(c) Copper (d) Molybdenum

5 Pyruvate dehydrogenase activity during aerobic respiration requires. **[NEET 2020]**
(a) Calcium (b) Iron
(c) Cobalt (d) Magnesium

6 Which of the following is not a micronutrient? **[JIPMER 2019]**
(a) B (b) Zn (c) Mn (d) Mg

7 Which one of the following elements is responsible for maintaining turgor in cells? **[NEET 2018]**
(a) Potassium (b) Sodium (c) Magnesium (d) Calcium

8 In which of the following forms is iron absorbed by plants?
(a) Free element (b) Ferrous **[NEET 2018]**
(c) Ferric (d) Both (b) and (c)

9 Match the organisms given in Column I to their functions given in Column II and choose the correct option. **[AIIMS 2018]**

Column I	Column II
A. *Thiobacillus*	1. Free-living nitrogen-fixing cyanobacteria
B. *Nitrosomonas*	2. Denitrification
C. *Nostoc*	3. Free-living aerobic nitrogen-fixing bacteria
D. *Azotobacter*	4. NH_3 to nitrite

Codes

	A	B	C	D		A	B	C	D
(a)	2	4	1	3	(b)	4	1	3	2
(c)	3	4	1	2	(d)	3	2	4	1

10 Some functions of nutrient element are given below.
I. Important constituent of proteins involved in ETS.
II. Activator of catalase.
III. Important constituent of cytochrome.
IV. Essential for chlorophyll synthesis.

The concerned nutrient is **[JIPMER 2017]**
(a) Cu (b) Fe (c) Ca (d) Mo

11 Select the mismatch. **[NEET 2017]**
(a) *Frankia* – *Alnus*
(b) *Rhodospirillum* – Mycorrhiza
(c) *Anabaena* – Nitrogen-fixer
(d) *Rhizobium* – Alfalfa

12 The first stable product of fixation of atmospheric nitrogen in leguminous plant is **[JIPMER 2017]**
(a) NO_2^- (b) ammonia (c) NO_3^- (d) glutamate

13 In which of the following, all three are macronutrients? **[NEET-I 2016]**
(a) Iron, copper, molybdenum
(b) Molybdenum, magnesium, manganese
(c) Nitrogen, calcium, phosphorus
(d) Boron, zinc, manganese

14 Which is essential for the growth of root tip?
(a) Zn (b) Fe **[NEET-II 2016]**
(c) Ca (d) Mn

15 Which of the following bacterium causes denitrification? **[J&K CET 2015]**
(a) *Azotobacter* (b) *Nitrobacter*
(c) *Nitrosomonas* (d) *Pseudomonas*

16 Element responsible for ring structure of chlorophyll and maintenance of ribosome structure is **[KCET 2015]**
(a) K (b) S
(c) Mg (d) Ca

17 I. **Statement** Mineral salt absorption is an active process.
II. **Explanation** Metabolic energy is not used in active absorption. **[Chhattisgarh PMT 2015]**
(a) I is correct, II is incorrect (b) I is incorrect, II is correct
(c) I and II both are correct (d) I and II both are incorrect

18 Element related to nitrogen metabolism is **[AMU 2015]**
(a) manganese (b) magnesium
(c) zinc (d) molybdenum

19 The enzyme nitrogenase is a **[AMU 2015, WB JEE 2015]**
(a) Cu-Fe protein (b) Ni-Fe protein
(c) Mo-Fe protein (d) Ni-Cu protein

20 During biological nitrogen-fixation, inactivation of nitrogenase by oxygen poisoning is prevented by **[CBSE AIPMT 2015]**
(a) leghaemoglobin (b) xanthophyll
(c) carotene (d) cytochrome

21 Match the following columns. **[Kerala CEE 2015]**

Column I (Essential elements)	Column II (Deficiency causes)
A. N, K, Mg, S, Fe, Mn, Zn and Mo	1. Inhibit cell division
B. N, K, S and Mo	2. Necrosis
C. Ca, Mg, Cu and K	3. Delay in flowering
D. N, S and Mo	4. Chlorosis

Codes

	A	B	C	D
(a)	4	2	2	1
(b)	1	2	3	4
(c)	4	1	2	3
(d)	2	3	4	1
(e)	2	4	2	1

22 Function of leghaemoglobin during biological nitrogen-fixation in root nodules of legumes is to **[J&K CET 2015, RPMT 2011, CBSE AIPMT 2011]**
(a) convert atmospheric nitrogen to ammonia
(b) convert ammonia to nitrite
(c) transport oxygen for activity of nitrogenase
(d) protect nitrogenase from oxygen

23 Match and find the correct combination. **[EAMCET 2015]**

Column I	Column II
A. Mg	1. Splitting of water during photosynthesis
B. Mo	2. Germination of pollen grain
C. Mn	3. Constituent of chlorophyll
D. B	4. Component of nitrate reductase
	5. Synthesis of auxin

Codes

	A	B	C	D
(a)	2	5	4	1
(b)	3	4	2	1
(c)	3	5	4	1
(d)	3	4	1	2

24 Conversion of organic nitrogen into ammonia, N_2 gas into ammonia, nitrates into ammonia and ammonia into nitrates are respectively known as **[EAMCET 2015]**
(a) nitrogen-fixation, nitrate reduction, nitrification, ammonification
(b) ammonification, nitrogen-fixation, nitrate reduction, nitrification
(c) nitrification, ammonification, nitrogen-fixation, nitrate reduction
(d) nitrogen-fixation, nitrate reduction, denitrification, ammonification

25 Identify the physiological function of two microelements. Of them, the former is required for synthesis and the latter for oxidation of IAA, respectively **[EAMCET 2015]**
I. constituent of chlorophyll molecule.
II. maintenance of cell turgidity.
III. splitting of water in photosynthesis.
IV. cofactor for carboxypeptidase.
V. component of methionine.

Codes
(a) II and V (b) IV and II (c) V and I (d) IV and III

26 Match the following columns and choose the correct option. **[WB JEE 2015]**

Column I	Column II
A. *Thiobacillus*	1. Nitrogen-fixation
B. *Nitrosomonas*	2. Ammonification
C. *Azotobacter*	3. Nitrification
D. *Clostridium*	4. Denitrification

Codes

	A	B	C	D
(a)	4	3	1	2
(b)	3	4	1	2
(c)	4	2	1	3
(d)	2	1	3	4

27 On which surface of cell, Donnan equilibrium occurs? **[Gujarat CET 2015]**
(a) Cell wall (b) Plasma membrane
(c) Transport (d) Nuclear membrane

28 In reductive amination, the product is **[BCECE 2015]**
(a) α-ketoglutaric acid (b) glutamine
(c) glutamate (d) alanine

29 During symbiotic nitrogen-fixation, how many ATPs are used in fixing one mole of N_2? **[BCECE 2015]**
(a) 8 (b) 16
(c) 5 (d) 10

30 Mo is part of enzyme **[BCECE 2015]**
(a) reverse transcriptase
(b) restriction endonuclease
(c) hexokinase
(d) nitrogenase

31 The process of conversion of nitrogen to ammonia by microbes is **[AMU 2014]**
(a) nitrification (b) denitrification
(c) nitrogen-fixation (d) Haber's process

32 One element is involved in opening and closing of stomata, the other helps to maintain ribosome structure. They are **[Kerala CEE 2014]**
(a) K and Ca (b) P and S
(c) K and Mg (d) Fe and Mg
(e) Ca and S

33 Match the following columns and choose the correct option. **[Kerala CEE 2014]**

Column I	Column II
A. Magnesium	1. Alcohol dehydrogenase
B. Molybdenum	2. Phosphoenol pyruvate
C. Zinc	3. Nitrogenase

Codes

	A	B	C
(a)	2	3	1
(b)	1	2	3
(c)	2	1	3
(d)	3	2	1
(e)	3	1	2

34 Free-living nitrogen-fixing aerobic bacterium is **[Kerala CEE 2014]**
(a) *Rhodospirillum* (b) *Anabaena*
(c) *Nostoc* (d) *Beijerinckia*
(e) *Rhizobium*

35 Carefully read the following reactions carried out by nitrifying bacteria. Identify the statement about these equations which is not true **[NCERT Exemplar, KCET 2014]**

$2NH_3 + 3O_2 \rightarrow 2NO_2^- + 2H^+ + 2H_2O$(I)

$2NO_2^- + O_2 \rightarrow 2NO_3^+$ (II)$^-$

(a) Step I is carried out by *Nitrosomonas* or *Nitrosococcus*
(b) Step II is carried out by *Nitrobacter*
(c) Both the steps I and II can be called nitrification
(d) Both the steps occur only in photoautotrophs

36 Deficiency symptoms of nitrogen and potassium are visible first in **[CBSE AIPMT 2014]**
(a) young leaves
(b) roots
(c) buds
(d) senescent leaves

37 Besides paddy fields, cyanobacteria are also found inside vegetative part of **[NEET 2013]**
(a) *Psilotum* (b) *Pinus*
(c) *Cycas* (d) *Equisetum*

38 The most abundant intracellular cation is **[NEET 2013]**
(a) Ca^{++} (b) H^+
(c) K^+ (d) Na^+

39 Element involved in nitrogen-fixation is **[NEET 2013]**
(a) zinc (b) boron
(c) iron (d) chlorine

40 'X' is essential micronutrient required in leaves and roots. Major function of 'X' is to perform metabolism of urea and uric acid and its deficiency causes leaf tip necrosis. It is **[AFMC 2013]**
(a) sulphur (b) magnesium
(c) zinc (d) nickel

41 Which of the following is not non-essential element for plants? **[NEET 2013]**
(a) potassium (b) sodium
(c) aluminium (d) cobalt

42 Consider the following statements. **[Kerala CEE 2013]**
I. Sulphur is present in two amino acids-cysteine and valine.
II. Lower level of N, K, S and Mo causes inhibition of cell division.
III. Microbe that fix nitrogen in the roots of non-leguminous plant *Alnus* is *Frankia*.
IV. Denitrification is carried by bacteria *Nitrosomonas* and *Nitrobacter*.

Select the correct pair of statements.
(a) I and II (b) I and III
(c) II and III (d) II and IV
(e) I and IV

43 Match the following columns and choose the correct option. **[Kerala CEE 2013]**

	Column I		Column II
A.	Free-living aerobic nitrogen-fixers	1.	*Anabaena* and *Nostoc*
B.	Anaerobic nitrogen fixers	2.	*Pseudomonas* and *Thiobacillus*
C.	Nitrogen-fixing cyanobacteria	3.	*Nitrosomonas* and *Nitrosococcus*
D.	Denitrifying bacteria	4.	*Azotobacter* and *Beijerinckia*
E.	Nitrifying bacteria	5.	*Rhodospirillum*

Codes

	A	B	C	D	E		A	B	C	D	E
(a)	4	5	1	2	3	(b)	5	4	1	3	2
(c)	4	5	2	3	1	(d)	4	3	1	2	5
(e)	4	5	3	2	1						

44 Minerals involved in carbohydrate translocation, maintaining ribosome structure and activation of nitrogenase respectively are **[Kerala CEE 2013]**
(a) Mn, B, Ca (b) Ca, Mg, Mo
(c) B, Mn, Mo (d) Cu, Mg, B
(e) B, Mg, Mo

45 Identify the role of lectins in formation of root nodules in legumes. **[EAMCET 2013]**
(a) Formation of shepherd's crook
(b) Recognition of compatible *Rhizobium* by host
(c) Formation of peribacterial membrane
(d) Formation of infection thread

46 Hydroponics is growing plants **[AMU 2013]**
(a) without soil (b) without organic matter
(c) in water (d) All of these

47 Plants obtain mineral nutrients from **[AMU 2013]**
(a) rocks (b) breakdown products of rocks
(c) soil (d) All of these

48 In nitrogen cycle, *Pseudomonas* **[J&K CET 2013]**
(a) fixes nitrogen
(b) produces elemental nitrogen
(c) oxidises ammonium nitrogen to nitrate
(d) transfer of elemental nitrogen

49 Match the Column I of minerals, Column II of functions and Column III of relative percentage and select proper option. **[Gujarat CET 2013]**

Column I	Column II	Column III
A. Phosphorus	1. For photosynthesis and ionic balance	I. 6%
B. Magnesium	2. In structure of RNA, DNA and ATP	II. 8%
C. Chlorine	3. Structural component of chlorophyll	III. 0.2%
D. Boron	4. Absorption and utilisation of calcium	IV. 0.3%

Codes

	A	B	C	D
(a)	2-II	3-II	1-I	4-IV
(b)	2-I	3-II	1-III	4-IV
(c)	3-I	2-II	4-III	1-IV
(d)	2-I	3-II	1-IV	4-III

50 Which one of the following is an incorrect statement? **[CBSE AIPMT 2012]**

(a) *Anabaena* and *Nostoc* are capable of fixing nitrogen in free-living state also
(b) Root nodule forming nitrogen fixers live as aerobes under free-living conditions
(c) Phosphorus is a constituent of cell membranes, certain nucleic acids and all proteins
(d) *Nitrosomonas* and *Nitrobacter* are chemoautotrophs

51 Which one of the following microbes forms symbiotic association with plants and helps them in their nutrition? **[CBSE AIPMT 2012]**

(a) *Azotobacter* (b) *Aspergillus*
(c) *Glomus* (d) *Trichoderma*

52 Micronutrients are **[AMU 2012]**

(a) Mn, Ni and Zn (b) O, Cu and B
(c) Mg, Mn and Mo (d) Ca, S and Fe

53 An essential element is that, which **[BCECE 2012]**

(a) improves health of the plant
(b) is irreplaceable and indispensable for the growth of plants
(c) is found in plant ash
(d) is available in the soil

54 Necrosis in crops is due to the deficiency of **[AFMC 2012]**

(a) Ca, K, S and Mo
(b) N, K, S and Mo
(c) N, S, Fe and Zn
(d) Mg, S, Mn and Ca

55 Which of the following micronutrient helps in N_2-fixation? **[BHU 2012]**

(a) Mo (b) Ca
(c) Bo (d) Na

56 Which element is required for the germination of pollen grains? **[JCECE 2012]**

(a) Boron (b) Calcium
(c) Chlorine (d) Potassium

57 Nitrifying bacteria **[CBSE AIPMT 2011]**

(a) convert free nitrogen to nitrogen compounds
(b) convert proteins into ammonia
(c) reduce nitrates to free nitrogen
(d) oxidise ammonia to nitrates

58 In root nodules of legumes, leghaemoglobin is important because it **[DUMET 2011]**

(a) transports oxygen to the root nodule
(b) acts as an oxygen scavenger
(c) provides energy to the nitrogen-fixing bacterium
(d) acts as a catalyst in transamination

59 Minerals are absorbed by plants in **[UP CPMT 2011]**

(a) colloidal form (b) ionic form
(c) precipitated form (d) None of the above

60 'Khaira disease of rice' is due to **[UP CPMT 2011]**

(a) fungus (b) bacteria
(c) Zn deficiency (d) Bo deficiency

61 Which one of the following is a micronutrient in plants? **[J&K CET 2011]**

(a) Magnesium (b) Zinc
(c) Potassium (d) Calcium

62 Which of the following minerals activate the enzymes involved in respiration? **[Kerala CEE 2011]**

(a) Nitrogen and phosphorus (b) Magnesium and manganese
(c) Potassium and calcium (d) Sulphur and iron
(e) Copper and boron

63 Select the matched ones. **[Kerala CEE 2011]**

I. *Nitrosomonas* — Nitrite to nitrate
II. *Thiobacillus* — Denitrification
III. *Nostoc* — Free-living nitrogen-fixer
IV. *Azotobacter* — Anaerobic nitrogen-fixer

(a) I and II (b) III and IV
(c) II and III (d) II and IV
(e) I, II and III

ANSWERS

Check Point 19.1

1. (d) 2. (b) 3. (c) 4. (b) 5. (c) 6. (d)

Check Point 19.2

1. (b) 2. (a) 3. (a) 4. (b) 5. (d) 6. (a)

(A) Topical Exercise

1. (d) 2. (a) 3. (a) 4. (a) 5. (b) 6. (d) 7. (d) 8. (b) 9. (c) 10. (b)
11. (c) 12. (c) 13. (a) 14. (a) 15. (b) 16. (d) 17. (a) 18. (d) 19. (a) 20. (d)
21. (c) 22. (a) 23. (c) 24. (b) 25. (a) 26. (a) 27. (a) 28. (a) 29. (d) 30. (a)
31. (d) 32. (a) 33. (b) 34. (c) 35. (a) 36. (c) 37. (a) 38. (a) 39. (c) 40. (d)
41. (a) 42. (c) 43. (d) 44. (a) 45. (b) 46. (d) 47. (c) 48. (c) 49. (c) 50. (c)
51. (c) 52. (d) 53. (a) 54. (d) 55. (c) 56. (a) 57. (d) 58. (a) 59. (a) 60. (b)
61. (d) 62. (d) 63. (d) 64. (a) 65. (c) 66. (d) 67. (c) 68. (b) 69. (b) 70. (c)
71. (c) 72. (c) 73. (d)

(B) Cumulative Exercise

1. (b) 2. (b) 3. (c) 4. (c) 5. (a) 6. (a) 7. (c) 8. (a) 9. (b) 10. (c)
11. (a) 12. (c) 13. (d) 14. (c) 15. (a) 16. (d) 17. (b) 18. (c) 19. (b) 20. (b)
21. (b) 22. (a) 23. (b) 24. (a) 25. (b) 26. (d) 27. (d) 28. (e) 29. (a) 30. (b)
31. (a) 32. (b) 33. (c) 34. (a)

(C) Medical Entrances' Gallery

1. (d) 2. (c) 3. (b) 4. (a) 5. (d) 6. (d) 7. (a) 8. (c) 9. (a) 10. (b)
11. (b) 12. (b) 13. (c) 14. (c) 15. (d) 16. (c) 17. (a) 18. (d) 19. (c) 20. (a)
21. (c) 22. (d) 23. (d) 24. (b) 25. (d) 26. (a) 27. (b) 28. (c) 29. (b) 30. (d)
31. (c) 32. (c) 33. (a) 34. (d) 35. (d) 36. (d) 37. (c) 38. (c) 39. (c) 40. (d)
41. (a) 42. (c) 43. (a) 44. (c) 45. (b) 46. (d) 47. (d) 48. (b) 49. (d) 50. (c)
51. (c) 52. (a) 53. (b) 54. (a) 55. (a) 56. (a) 57. (d) 58. (b) 59. (b) 60. (c)
61. (b) 62. (b) 63. (c)

Explanations

(A) Topical Exercise

5 (*b*) Statement I is false, but II is true. Statement I can be corrected as

Hydroponics determines the role of a specific nutrient in growth and development of a plant.

13 (*a*) The element which is directly involved in the metabolism of plants is referred as an essential element.

24 (*b*) Statement in option (b) is true. Other statements are false and can be corrected as

- Not every mineral element that is present in a cell is needed by the cell as some are needed in large quantity and some in less quantity.
- Nitrogen as nutrient element is highly mobile in the plants.
- It is very difficult to establish the essentiality of micronutrients because they are required only in trace quantities.

25 (*a*) Correct pair is option (a). Other options are incorrect pair and can be corrected as

Component of vitamins - Sulphur
Synthesis of cell wall - Calcium
Phosphorylation reactions - Phosphorus
Component of cytochrome - Iron

30 (*a*) Option (a) is incorrect and can be corrected as Magnesium (Mg) helps in synthesis of DNA and RNA not spindle fibres.

34 (*c*) Correct match is option (c). Other options are incorrect and can be corrected as

A	B	C
Sulphur	Tea yellow	Component of vitamin, protein, etc.
Calcium	Wither tip	Component of middle lamella of cell wall as calcium pectate.
Chlorine	Wilting chlorosis and necrosis	Component of oxygen evolving complex.

36 (*c*) According to the criteria of essential, there are 16 essential elements for plant growth. Hunger signs are shown by plants due to the deficiency of one or more of these mineral nutrients.

41 (*a*) I and III are incorrect. These can be corrected as

Tea yellow - Sulphur deficiency
Yellow spot of *Citrus* - Molybdeum deficiency

44 (*a*) Statement I is true, but II is false and it can be corrected as

Excess of manganese causes deficiency of Fe, Mg and Ca.

53 (*a*) Cation released during carbonic acid exchange into the soil solution from the clay particles are adsorbed on root cells in exchange of H^+ ion or an ion pairs with bicarbonate.

60 (*b*) Statement I is false, but II is true because

Legumes fix nitrogen only through the specialised bacteria that live in their roots.

67 (*c*) Statement in option (c) is incorrect and can be corrected as *Rhizobium* and *Frankia* are free living in soil, but act as symbiotic for atmospheric nitrogen-fixation.

(B) Cumulative Exercise

2 (*b*) Statements I, II and V are correct. Statements III and IV are incorrect and can be corrected as

- Hydroxyl part of amino acid not keto acid is replaced by another amino group in amide formation.
- Transfer of amino group from an amino to keto group of a keto acid is called transamination.

3 (*c*) Statements III and IV are incorrect. These can be corrected as

- N, P and K can get deficient in soil and plant will eventually show deficiency symptoms.
- Nitrogen cycle is regular circulation of nitrogen amongst living organisms with its reservoir pool in atmosphere and cycling pool in lithosphere.

4 (*c*) Statements I and IV are correct, but statements II and III are incorrect and can be corrected as

- The concentration of the essential element below which plant growth is retarded is termed as critical concentration.
- Deficiency symptoms are the morphological changes that are indicative of certain elements deficiency, e.g. chlorosis, necrosis, etc.

6 (*a*) Only statement I is incorrect because

Macronutrient is said to be toxic when present above toxic concentration.

10 (*c*) Statement in option (c) is incorrect and can be corrected as 16 ATPs are required for fixation of one nitrogen molecule during nitrogen-fixation.

16 (*d*) If a pulse crop possesses premature yellowing of leaves and decrease in yield then application of magnesium and iron to promote synthesis of chlorophyll may become most beneficial to overcome the problem and to obtain maximum seed yield.

Magnesium is an important part of the ring structure of the chlorophyll molecule and its deficiency causes chlorosis and premature leaf abscission.

In iron deficiency also, the leaves become chlorotic because iron is required for the synthesis of some of the chlorophyll-protein complexes in the chloroplast.

19 (*b*) Statements II and III are correct and statement I is incorrect. It can be corrected as

The central portion of root nodule of leguminous plant is pink due to the presence of iron (Fe) element in the protein leghaemoglobin.

20 (*b*) Statement in option (b) is incorrect. It can be corrected as Nitrogenase enzyme is highly sensitive to the presence of molecular oxygen.

22 (*a*) Both A and R are true and R is the correct explanation of A.

Leguminous plants have nodulated roots in which *Rhizobium* is present. So, these are able to fix nitrogen.

23 (*b*) Both A and R are true, but R is not the correct explanation of A.

Plants absorb nitrogen in the form of NO_3^- (nitrate) or NH_4^+ (ammonium ion). They can absorb NO_2^- (nitrite) as well, but the same does not accumulate in the soil. Only a small quantity of nitrate or ammonium is available in the lithosphere. Therefore, nitrogen is the most critical element. Other critical elements are phosphorus and potassium.

24 (*a*) Both A and R are true and R is the correct explanation of A.

Cauliflower plants very commonly show whiptail disease. The leaves first shown an interveinal motling and the leaf margins may become grey and flaccid and finally brown.

25 (*b*) Both A and R are true, but R is not the correct explanation of A.

Solution culture is being used for raising flowers and vegetables at home. This soilless production of plants is called hydroponics. A solution having all the essential elements in proper proportion is called normal or balanced nutrient solution.

26 (*d*) A is false, but R is true. A can be corrected as

Excess of Mn inside the plant creates toxicity thereby causing reduction in the uptake of Mg, Fe and Ca.

27 (*d*) A is false, but R is true. A can be corrected as

Frankia symbiotically fixes atmospheric nitrogen in non-legume plants like *Alnus* and *Casuarina*.

28 (*e*) Both A and R are false. These can be corrected as Leghaemoglobin is not found in cyanobacteria. Cyanobacteria is oxygenic photosynthetic bacteria that are capable of fixing atmospheric nitrogen. The cyanobacterial nitrogenase is responsible for the reduction of N_2 and is extremely sensitive to oxygen.

29 (*a*) Both A and R are true and R is the correct explanation of A.

Microelements are essential elements, which are required by plants in traces only, less than 1 mg/gm of dry matter, e.g. iron, zinc, copper, etc.

30 (*b*) Both A and R are true, but R is not the correct explanation of A.

Calcium is a constituent of cell wall in the form of calcium pectate. It is required in small amounts for normal mitotic division Calcium is also involved in chromatin or mitotic spindle organisation.

31 (*a*) Both A and R are true and R is the correct explanation of A.

Manganese plays an important role in nitrate reduction. Manganese acts as an activator for the enzymes nitrate reductase and hydroxylamine reductase. The preference of ammonia over nitrate as a nitrogen source by the manganese-deficient cells supports the above mentioned analysis of the role of manganese.

32 (*b*) Both A and R are true, but R is not the correct explanation of A.

Magnesium is a constituent of the chlorophyll molecule, without, which photosynthesis would not occur. Many of the enzymes involved in carbohydrate metabolism require magnesium as an activator. Magnesium is also an activator for enzymes involved in the synthesis of nucleic acids (DNA, RNA).

33 (*c*) A is true, but R is false because

Due to the deficiency of sulphur, plant shows chlorosis (i.e. yellowing due to the degradation of chlorophyll) followed by anthocyanin development. The younger leaves show chlorosis before older ones. Sulphur is not the constituent of chlorophyll. The main constituent of chlorophyll is magnesium.

34 (*a*) Both A and R are true and R is the correct explanation of A.

Enzyme nitrogenase is sensitive to O_2. Thus, root nudules have leghaemoglobin. This pigment is an oxygen scavenger.

(C) Medical Entrances' Gallery

1 (*d*) In *Glycine max*, the product of biological nitrogen fixation is transported from the root nodules to other parts as ureides.

2 (*c*) The products of reaction catalysed by nitrogenase in root nodules of leguminous plants are ammonia and hydrogen. The reaction is as follows

$$N_2 + 8e^- + 8H^{2+} + 16\,ATP \longrightarrow 2NH_3 + H_2 + 16ADP + 16\,Pi$$

Symbiotic *Rhizobium* bacteria invade the roots of leguminous plants and form root nodules in which they fix nitrogen, supplying both to the bacteria and the plants.

3 (*b*) A-3, B-4, C-2, D-1

6 (*d*) Magnesium (Mg) is a macronutrient not a micronutrient. Such nutrients are required by plants in comparatively large amount (1-10 mg per gram of dry matter) other examples are phosphorus (P), potassium (K), nitrogen (N), etc.

Macronutrients are required by plant in traces, less than 0.1 mg per gram of dry matter. Boron (B), zinc (Zn) and manganese (Mn) are micronutrients.

7 (*a*) Potassium (K^+) is responsible for maintaining turgor pressure in cell because it regulates the proton pumps involved in opening and closing of stomata.

Magnesium (Mg^{2+}) is a constituent of chlorophyll pigment which helps in photosynthesis in green plants. Calcium (Ca^{2+}) provides selective permeability to the cell membrane. Sodium (Na^+) is involved in membrane permeability.

8 (*c*) According to NCERT, plants absorb iron mostly in the form of ferric (Fe^{3+}) ions.

However, plants in acidic soil can absorb iron in ferrous (Fe^{2+}) as well as ferric (Fe^{3+}) form. It is an important constituent of proteins involved in the transfer of electrons like ferredoxin and cytochromes.

It is reversibly oxidised from Fe^{2+} to Fe^{3+} during electron transfer. It activates catalase enzyme. It is essential for the formation of chlorophyll.

10 (*b*) Iron (Fe) is the concerned nutrient associated with the given functions.

- Cu is a component or activator of plastocyanin, cytochrome oxidase, etc.
- Ca (pectate) is constituent of middle lamella, activator of enzymes connected with chromosome formation, etc.
- Mo is component of nitrogenase.

11 (*b*) Among the given options, only option (b) is mismatched. *Rhodospirillum* is a free-living nitrogen-fixing bacteria. Mycorrhiza is the symbiotic association of a fungus with the root of a higher plant.

12 (*b*) Ammonia is the first stable compound produced due to fixation of atmospheric nitrogen in leguminous plants. A nitrogen molecule is converted into two molecules of ammonia.

The fixation of atmospheric nitrogen to ammonia is given as

$$N_2 + 8e^- + 8H^+ + 16ATP \xrightarrow[\text{Nitrogenase}]{\text{MO-Fe}} 2NH_3 + H_2 + 16ADP + 16Pi$$

13 (*c*) N, Ca, P, S, Mg, K are macronutrients and Fe, Mn, MO, B, Zn, Cu, Cl, are micronutrients.

14 (*c*) Calcium is needed by the growing root tip. It is required in the formation of middle lamella of the cell wall. It is found to be present in the form of calcium pectate in the cell wall.

17 (*a*) Mineral salt absorption mainly occurs actively. Active process requires expenditure of energy. Thus, mineral salt absorption is an energy consuming process.

24 (*b*) The correct option is (b) ammonification, nitrogen-fixation, nitrate reduction, nitrification.

Ammonification It is a process of decomposition of organic nitrogen (proteins, nitrogenous excretions, etc.) of dead plants and animals into ammonia.

Nitrogen fixation It is the process of conversion of nitrogen gas to ammonia.

Nitrate reduction It is the process of conversion of nitrate into ammonia.

Nitrification It is the process of conversion of ammonia to nitrate.

27 (*b*) Donnan equilibrium occurs at the plasma membrane. It takes into account the transport of non-diffusible ions.

37 (*c*) Cyanobacterium such as *Nostoc* is found associated with the coralloid roots of *Cycas*.

40 (*d*) Nickle is a micronutrient. It helps enzyme urease to perform its activity. Thus, it is involved in metabolism of urea and uric acid.

41 (*a*) Non-essential elements are those mineral elements which are not required by the plants for their growth or they are needed in very minute quantities. They do not affect the overall growth and development of plants. Sodium, aluminium and cobalt are the non-essential elements required for metabolic activities in some plants. But potassium is an essential element. It is a macronutrient that carries out several functions in plants. Hence, it is not non-essential element. Thus, correct answer is option (a).

42 (*c*) Statements II and III are correct. Other incorrect statements can be corrected as
Sulphur is present in two amino acids cysteine and methonine.

Denitrification is carried by bacteria *Nitrosomonas* and *Thiobacillus.*

50 (*c*) Statement in option (c) is incorrect and can be corrected as

Phosphorus is a component of DNA and RNA, but not proteins.

51 (*c*) Several species of *Glomus*, including *G. aggregatum*, are cultured and sold as mycorrhizal inoculant for agricultural soils. Being endomycorrhiza, it helps the plants in the absorption of nutrients especially phosphorus from soil.

55 (*a*) Nitrogenase enzyme is required for nitrogen-fixation. It consists of two components, Fe containing unit and Mo (molybdenum) containing unit. Hence, Fe and Mo are important for nitrogenase.

57 (*d*) Nitrifying bacteria (one of the chemosynthetic bacteria) oxidise ammonia to nitrates and obtain energy for the preparation of food. This oxidation occurs in two steps. In the first step, ammonia is oxidised to nitrite by nitrite bacteria, e.g. *Nitrosomonas* and *Nitrosococcus.*

In the second step, nitrite is oxidised to nitrate by nitrate bacteria, e.g. *Nitrocystis* and *Nitrobacter.*

59 (*b*) The inorganic essential elements, which are obtained from soil are called mineral elements. Plants absorb the minerals in the form of ions. The movement of ions is called flux.

60 (*c*) Zinc deficiency causes Khaira disease in rice grown in the calcareous soil of North India. The important symptoms of this disease are discolouration of lower leaves occurring two weeks after transplanting , plants are stunted, roots are scanty, midrib at the base of growing leaves is sometimes chlorotic, etc.

62 (*b*) Magnesium activates the enzymes of respiration, photosynthesis and is involved in the synthesis of DNA and RNA. Manganese activates many enzymes involved in photosynthesis, respiration and nitrogen metabolism.

63 (*c*) The correct matches are II and III.
Nitrate present in the soil is reduced to nitrogen by the process of denitrification. Denitrification is carried out by bacteria *Pseudomonas* and *Thiobacillus.*

A number of cyanobacteria such as *Anabaena* and *Nostoc* are free-living nitrogen-fixers.
The incorrect matches I and IV can be corrected as

Nitrosomonas or *Nitrosococcus* oxidise NH_3 to nitrite. The nitrite is further oxidised to nitrate with the help of *Nitrobacter* and *Nitrocystis. Azotobacter* and *Beijernickia* are free-living nitrogen fixing aerobic microbes.

CHAPTER 20

Photosynthesis in Higher Plants

Autotrophic organisms have the ability to synthesise their organic food from inorganic raw materials. In this process, they fix various forms of energy. Plants, phytoplanktons, cyanobacteria, algae, etc., are the living organisms which are named as **photoautotrophs.** They manufacture their food inside the chlorophyll containing cells from CO_2 by utilising solar energy. This process of food production is known as **photosynthesis.** This complete process can be represented by following simple equation

$$6CO_2 + 12H_2O \xrightarrow[\text{Chlorophyll}]{hv} C_6H_{12}O_6 + 6H_2O + 6O_2\uparrow$$

Annually on an average about 75×10^{13} kg of carbon in the form of CO_2 is fixed through photosynthesis, producing about 1700 million tonnes of dry matter. About 90% of it is carried out in the oceans by phytoplanktons and algae. The studies on photosynthesis started around 300 years ago. Several landmarks in the history of photosynthesis are as follows

Table 20.1 Historical landmarks in photosynthesis

Stephen Hales (1727)	Father of plant physiology, he discovered that green plants require sunlight for their nutrition.
Joseph Priestley (1770)	Showed that plants have the ability to take up carbon dioxide from atmosphere and release oxygen.
Jan Ingenhousz (1779)	Proved release of oxygen by plants was possible in sunlight and only by the green parts of the plants.
Palletier and Caventou (1818)	Discovered and named green pigment chlorophyll in plants.
Von Mayer (1845)	Proposed that green plants transform light energy into chemical energy.

Inside

Julius von Sachs (1862)	Demonstrated that first visible product of photosynthesis is starch.
Engelmann (1888)	Plotted the action spectrum of photosynthesis. His experimental material was *Cladophora*.
Warburg (1920)	Performed flash-light experiment using *Chlorella* as experimental material.
van Niel (1931)	Demonstrated that some bacteria use H_2S in photosynthesis.
Emerson and Arnold (1932)	Performed flash-light experiment and reported the presence of two phases, i.e. dark and light in photosynthesis.
Robert Hill (1937)	He experimentally demonstrated that in photosynthesis source of O_2 is H_2O and not CO_2.
Ruben, Hassid and Kamen (1941)	Used ^{18}O and confirmed that O_2 evolved in photosynthesis comes from water.

SITE OF PHOTOSYNTHESIS

- In most generalised term, the site of photosynthesis is chlorophyll containing cells of plants.
- Chlorophyll is a photosensitive pigment which is present in specialised structures called **chloroplasts** in eukaryotic cells. So, these chloroplasts are considered as the actual site of photosynthesis whithin a plant cell.

Chloroplasts

- The most active photosynthetic tissue in higher plants is the mesophyll cells of leaves.
- These have many chloroplasts, which contain the specialised light absorbing pigment, the chlorophylls.
- A chloroplast is a double membrane bound organelle. Between its two membranes a space is present. This space is called **periplastidial space**.
- Internally chloroplast possesses a system of membranes called **thylakoids.**
- A stack of thylakoids forms a **granum** (plural grana). Adjacent grana are connected by unstacked membranes called **stroma lamellae**.
- The fluid compartment surrounding the thylakoids is called the **stroma**. The light absorbing pigments are present within the grana.
- All these structures are shown in the figure given below

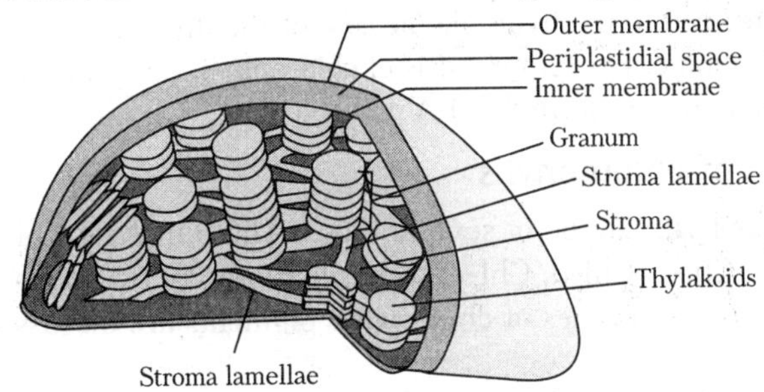

Fig 20.1 Structure of chloroplast

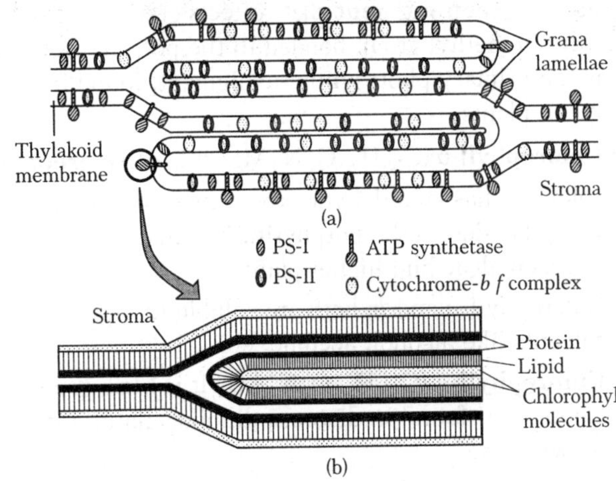

Fig 20.2 Grana membrane showing light absorbing pigments : (a) Diagrammatic overview, (b) Membrane magnified with chlorophyll distribution

- The light absorbing pigments are called **photosynthetic pigments**. Chlorophyll is one of the examples of such pigments.

Photosynthetic pigments

- These are compounds that have the ability to absorb light of specific wavelengths. Along with chloroplast, these are also found in photosynthetic bacteria.

  Various photosynthetic pigments are as follows

Chlorophylls

- These are magnesium containing green pigments.
- The molecular structure of chlorophyll consists of a porphyrin head (ring-shaped) and a phytol tail.
  - **Porphyrin head** contains four pyrrol rings with an Mg-atom in its centre. This portion of chlorophyll is hydrophilic in nature. It is the site of electron rearrangement when the chlorophyll is excited.
  - **Phytol tail** is a long hydrocarbon tail, which always remains attached to the ring structure. Phytol tail interacts with the hydrophobic regions of proteins. These proteins are present in thylakoids.

- Most of the chlorophyll molecules are found to be immersed in the hydrophobic core of the thylakoid membrane. For biosynthesis of chlorophyll, raw material required are succinyl CO-A and glycine.

Types of chlorophylls

Chlorophylls can be of several types. These include Chl-*a*, Chl-*b*, Chl-*c*, Chl-*d*, Chl-*e*, bacteriochlorophyll-*a* and *b*, etc. The main types of chlorophyll pigments are discussed below

(i) **Chlorophyll-*a*** ($C_{55}H_{72}O_5N_4Mg$) It is a **bluish-green pigment** that reflects green light. It has a methyl group bonded to the porphyrin. Chl-*a* is the primary photosynthetic pigment, found in all green plants.

(ii) **Chlorophyll-*b*** ($C_{55}H_{70}O_6N_4Mg$) It is **olive-green** in colour. It has an aldehyde group instead of a methyl group bonded to the porphyrin. It is mainly present in green algae and higher plants.
Chlorophyll-*a* and *b*, both are soluble in organic solvents like alcohol, acetone, etc.

(iii) **Chlorophyll-c** ($C_{35}H_{30}O_5N_4Mg$) It is **blue-green** in colour. It is found in diatoms, dinoflagellates and brown algae.

Carotenoids

- These are **yellow**, **brown** and **orange** pigments.
- These absorb light strongly in the blue-violet region.
- These protect plant from excessive heat and prevent photo-oxidation or photo-bleaching of chlorophyll pigments.
- Thus, they are also known as '**Shield pigments**'.
- Carotenoids are also called as **lipochromes** because of their fat-soluble nature.
- Chl-*b*, along with carotenoids are also called as **accessory pigment**. They absorb light energy and transfer it to chlorophyll-*a*.
- Carotenoids are of following two types
  (i) **Carotenes** These are hydrocarbons with molecular formula $C_{40}H_{56}$. Carotenes are **orange** in colour. They are mostly tetraterpenes. The red colour of tomato and chillies is because of carotene called **lycopene**.
  (ii) **Xanthophylls** They are yellow coloured carotenoids also called as **carotenols** or **xanthols**. Their molecular formula is $C_{40}H_{56}O_2$. Thus, they differ from carotenes in having two additional oxygen atoms, e.g. lutein and zeaxanthin. The yellow colour of autumn leaves is due to lutein. The characteristic xanthophyll of brown algae is fucoxanthin.

Phycobilins (red and blue pigments)

- These are proteinaceous pigments found in red algae and cyanobacteria.
- Phycobilins do not contain magnesium and phytol tail.
- These are soluble in hot water.
- Phycobilins are also considered as accessory pigments of photosynthesis.
- The phycobilin pigments are of two types
  (i) **Phycocyanin** found in **blue-green** algae.
  (ii) **Phycoerythrin** found in **red algae**.

Photosynthetically Active Radiation (PAR)

- A photosynthetic pigment cannot absorb all the sunlight falling on it.
- These pigments only absorb a part of this light, i.e. the visible part (390-760 nm). The remaining light is reflected back to the atmosphere.
- This visible part of light is called as **Photosynthetically Active Radiation** (PAR).

Absorption and action spectra of photosynthetic pigment

- **Absorption spectrum** is a graph, which shows the amount of absorption of different wavelengths of light by a particular photopigment.
- Following table and graph give a brief information about the light energy absorbed by different photopigments.

Pigments	Light absorbed
Chlorophyll-*a* and *b*	400-500 nm (max. at 450 nm) blue-green and 700 nm red region
Carotenoids	Maximum at 500 nm (blue-green region)
Phycocyanin	Maximum at 600 nm

- This wavelength range of absorbed light is depicted in the graph below

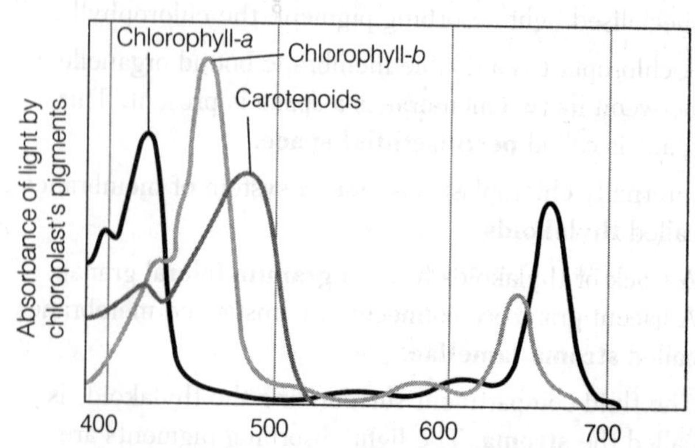

Fig. 20.3 Absorption spectrum of different pigments

- From the graph, we can see that chlorophyll-a absorbs light in blue-violet region while chlorophyll-b absorbs light in red-blue region.
- Neither a nor b absorb green light, as green light is reflected or transmitted by them. It is the reason why chlorophyll appears green.
- Carotenoids absorb light in the blue-green and violet regions and reflect longer wavelengths.

Action spectrum is a graph, which shows the degree of different wavelengths that affect the rate of photosynthesis.

- This spectrum can be given for any light driven process.
- It gives an idea that, which wavelength is responsible for the onset or shutdown of photosynthesis.

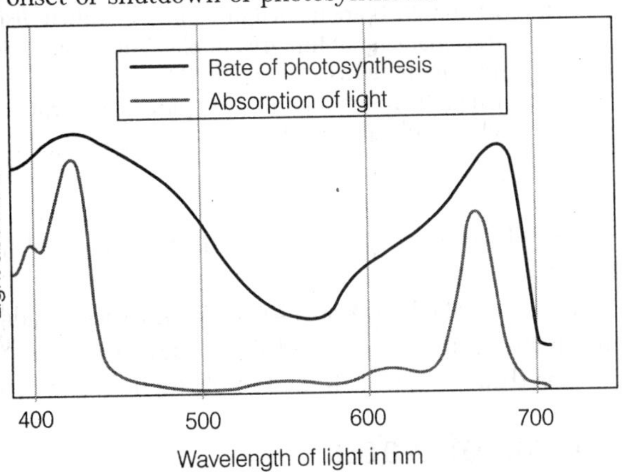

Fig. 20.4 Action spectra of photosynthesis superimposed on absorption spectrum of chlorophyll-a

Interpretation

- Action spectrum of photosynthesis corresponds closely to absorption spectra of chlorophyll-a, showing that chlorophyll-a is the chief pigment associated with photosynthesis.
- The curves in the graph depict that maximum photosynthesis occurs at that wavelength, where there is maximum absorption by chlorophyll-a, i.e. in the blue and red regions of light.

Red drop and Emerson effect

- **Emerson** (1957) reported that when a monochromatic beam of wavelength more than 680 nm is used, the rate of photosynthesis is reduced. This fall in rate above 680 nm is called **Red drop**.
- He later on found that if two monochromatic beams of wavelength 650-680 nm and 700-720 nm are used, the rate of photosynthesis increases. This is known as **Emerson's enhancement effect**.
- Emerson's effect suggests the participation of two different units in photosynthesis. One that is driven by longer wavelength and the other by shorter wavelength.

Concept of pigment system

- The smallest number of pigment molecules, required as a unit, to carryout the reactions of photosynthesis is known as a **photosynthetic unit**.
- **Park** and **Biggins** in 1964 named these units as **quantasomes.** These are present on the thylakoid membranes.
- Each quantasome consists of
  - 200-240 chlorophyll molecules
  - carotenoids
  - quinone compounds
  - sulpholipids, phospholipids, proteins, etc.
- Since, these units contain various pigments that harvest light molecules, so they are also known as **pigment systems** or **Light Harvesting Complex** (LHC) or **photosystems.**
- A photosystem consists of many **antenna molecules** and a **reaction centre**. Antenna molecules absorb light energy and transfer it to the reaction centre as shown below

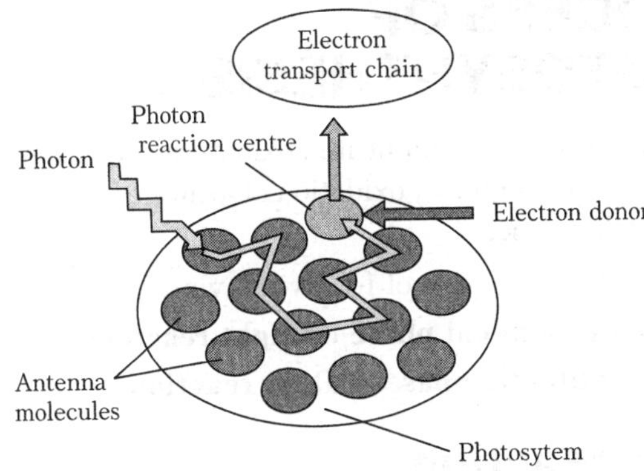

Fig 20.5 Antenna molecules and reaction centre in a photosystem

- The two types of photosystems are found in the thylakoid membrane of plants. These are known as **photosystem-I** (PS-I) and **photosystem-II** (PS–II).
- Differences between PS-I and PS-II are as follows

Photosystem-I	Photosystem-II
It is situated on the outer surface of thylakoid membrane. It is also found on stroma lamellae.	It is situated on the inner surface of thylakoid membrane.
It is involved in both cyclic and non-cyclic photophosphorylations.	It is involved in non-cyclic photophosphorylation only.
Reaction centre is called P_{700}, i.e. it absorbs 700 nm wavelength.	Reaction centre is called P_{680}, i.e. it absorbs 680 nm wavelength of light.
It is basically concerned with $NADP^+$ reduction.	It is concerned with water splitting or photolysis of water.

CHECK POINT 20.1

1. The experiment material used by Engelman to plot action spectrum of photosynthesis is
   (a) *Chlorella*
   (b) *Cladophora*
   (c) purple and green sulphur bacteria
   (d) blue-green algae
2. The main difference between chlorophyll-*a* and *b* is
   (a) chlorophyll-*a* is a linear chain compound and *b* is branched chain
   (b) Chlorophyll-*a* has no Mg^+ ion in centre of molecule
   (c) Chlorophyll-*a* has $—CH_3$ group where *b* has —CHO group
   (d) All of the above
3. Choose the correct option.
   (a) Chl-*a* is blue-green pigment with formula $C_{55}H_{70}O_6N_4$ Mg
   (b) Chl-*a* is blue-green pigment with formula $C_{55}H_{72}O_5N_4Mg$
   (c) Chl-*b* is yellow-green pigment with formula $C_{55}H_{70}O_5N_4Mg$
   (d) Xanthin is pigment with formula $C_{40}H_{60}$
4. Red colour of fruits and vegetables are due to
   (a) chlorophyll-*a* (b) phycocyanin
   (c) xanthophyll (d) lycopene
5. Emerson effect is related to
   (a) Decrease in photosynthesis in the presence of high light intensity
   (b) Decrease in photosynthesis when lights of two different wavelengths are provided together
   (c) Increase in photosynthesis in the presence of monochromatic light
   (d) Increase in photosynthesis when light of two different wavelength are used
6. Which of the following is incorrect regarding photosystems?
   (a) These are two photosystems (PS-I and PS-II)
   (b) PS-I and PS-II are named in sequence of their discovery
   (c) LHC is made up of hundred of pigments molecule which are proteins
   (d) Each photosystem has all the pigments (except one molecule of chlorophyll-*b*)

PROCESS OF PHOTOSYNTHESIS

Photosynthesis is a complicated oxidation-reduction process. In this process, oxidation of water and reduction of CO_2 take place.

Photosynthesis consists of following two phases

1. **Photochemical phase** or **Light reaction**
2. **Biosynthetic phase** or **Dark reaction**

Light reaction

- It is a light dependent reaction which includes absorption and conversion of light energy into energy rich molecules, like ATP and NADPH.
- **Site of occurrence** It takes place inside the thylakoids of grana region.

Steps

- The process of light reaction begins when photosynthetic pigments of PS absorb light energy.
- Then their antenna molecules transfer the absorbed energy to the reaction centre.
- This makes the reaction centre of PS highly energised or photoexcited and as result it emits a pair of electron.
- The electrons ejected by PS are accepted by primary electron acceptor.
- It passes them into an **Electron Transport System** (ETS).
- In the process of electron transport, reduction of NADP into NADPH + H^+ takes place.
- It also results in the formation of ATP from ADP and inorganic phosphates.
- This process is called as **photophosphorylation,** i.e. addition of phosphate in the presence of light. Thus, the flow of electrons through ETS is linked to photophosphorylation.

Photolysis of water

- The photolysis of water or **photochemical oxidation** of water was first described by **van Niel** in 1931.
- In this process, water splits into H^+, [O] and electrons. The electrons obtained from the photolysis of a water molecule are then passed on to PS-II.
- Photolysis of water also requires minerals such as Mn^{2+}, Cl^- and Ca^{2+}.
- The protons and oxygen formed in this process are released within the lumen of thylakoids. Here, O_2 is evolved as byproduct.
- Whole reaction of photolysis is summerised in the equation given below

$$2H_2O \xrightarrow[\text{Chlorophyll}]{\text{Light}} 4H^+ + 4e^- + O_2\uparrow$$

Hill Reaction

- **Robin Hill** in 1937 showed that illuminated suspension of isolated chloroplasts evolve oxygen in the presence of a suitable reducing substance.
- This substance was an electron acceptor or Hill oxidant like potassium ferrocyanide or potassium ferric oxalate.
- Other substances like methylene blue may also serve as Hill reagents.

Conclusion

Hill's discovery demonstrates that photochemical O_2 evolution and CO_2 reduction in photosynthesis are two separate processes.

> **Quantum yield and quantum requirement**
> - The number of oxygen molecules produced per quantum of light absorbed during photosynthesis is called **quantum yield**. Whereas the quantum of light required to release one molecule of O_2 is known as **quantum requirement**.
> - It is known that eight quanta are required during photosynthesis to evolve one molecule of O_2. Thus,
>
> $$\text{Quantum yield} = \frac{1}{8} = 0.125$$

Photophosphorylation

- It is the light driven or light energised synthesis of ATP molecules.
- It was discovered by **Arnon** *et al*, in 1954.
- In plants, photophosphorylation reactions are of two types
  - Cyclic photophosphorylation
  - Non-cyclic photophosphorylation

I. Cyclic-photophosphorylation

- The process of formation of ATP from ADP+iP *via* a cyclic flow of electrons is known as cyclic photophosphorylation.
- It is called, so because in this process, the donor and final acceptor of electrons is same, i.e. P_{700}.
- **Site of occurrence** This phosphorylation is marked in stromal lamellae and **unstacked thylakoids.**
- **Conditions** This process occurs in low light intensity, aerobic conditions and when less CO_2 is available.

Steps

- The energy rich electrons from the PS-I, i.e. P_{700} are first accepted by primary electron acceptor and transferred to Fd (ferredoxin).
- From Fd, the electrons are transferred to cytochrome-b_6 then, cytochrome-f and then finally to Plastocyanin (PC).
- From PC, the electrons are then returned back to the reaction centre of PS-I, i.e. P_{700}.

Outcome

- 2 molecules of ATP are synthesised.
- No reduction of NADP to NADPH + H^+ takes place.

II. Non-cyclic photophosphorylation

- When ATP formation is coupled to a non-cyclic transfer of electrons, it is called as non-cyclic photophosphorylation.
- **Site of occurrence** This phosphorylation occurs in stacked thylakoids, i.e. granum.
- **Conditions** This occurs under high light intensity, an aerobic conditions and high CO_2 availability. The process occurs with the help of many electron carriers and connects PS-I and PS-II.

Electron carriers

- Cytochrome, plastoquinone and plastocyanin are the important electron carriers of non-cyclic photophosphorylation.
- In brief, these are described below
  - (i) **Cytochromes** These are small proteins that contain a co-factor, **haem, having an Fe-atom.** Cytochromes are intrinsic membrane proteins of thylakoid membranes.
  - (ii) **Plastoquinones** These transport electrons over short distance within a membrane. Their long hydrocarbon tail is hydrophobic. Thus, they dissolve easily into the lipid component of the chloroplast membrane.
  - (iii) **Plastocyanin** This is a small protein that carries electrons with the help of **copper.**
- During this process, electrons are not cycled back to the P_{680}, i.e. PS-II.
- They are used in the reduction of NADP to $NADPH_2$. It also utilises water to release oxygen and hydrogen (photolysis).

Steps

- The electrons emitted by reaction centre P_{680} are first accepted by an electron acceptor **PQ** (plastoquinone).
- These electrons are then transferred through electron carriers like cytochrome-b_6, cytochrome-f and plastocyanin.
- From plastocyanin, electrons are finally transferred to the reaction centre of PS-I.
- This PS-I also receives photons of light directly and becomes photoexcited.
- Now, it expels electrons that enters the ETS.
- $NADP^+$ present in stroma receive electrons through ETS. It also receives protons *via* photolysis of water.
- Thus, it gets reduced into NADPH + H^+ as shown below

$$2NADP + 4e^- + 4H^+ \longrightarrow 2NADPH_2$$

Due to the transfer of electrons, energy is released during non-cyclic transfer. This energy is utilised to form ATP from ADP and iP (inorganic phosphate). This non-cyclic photophosphorylation is also known as **Z-scheme** of light reaction or photosynthesis because the shape of path in which electrons flow. This path is shown below

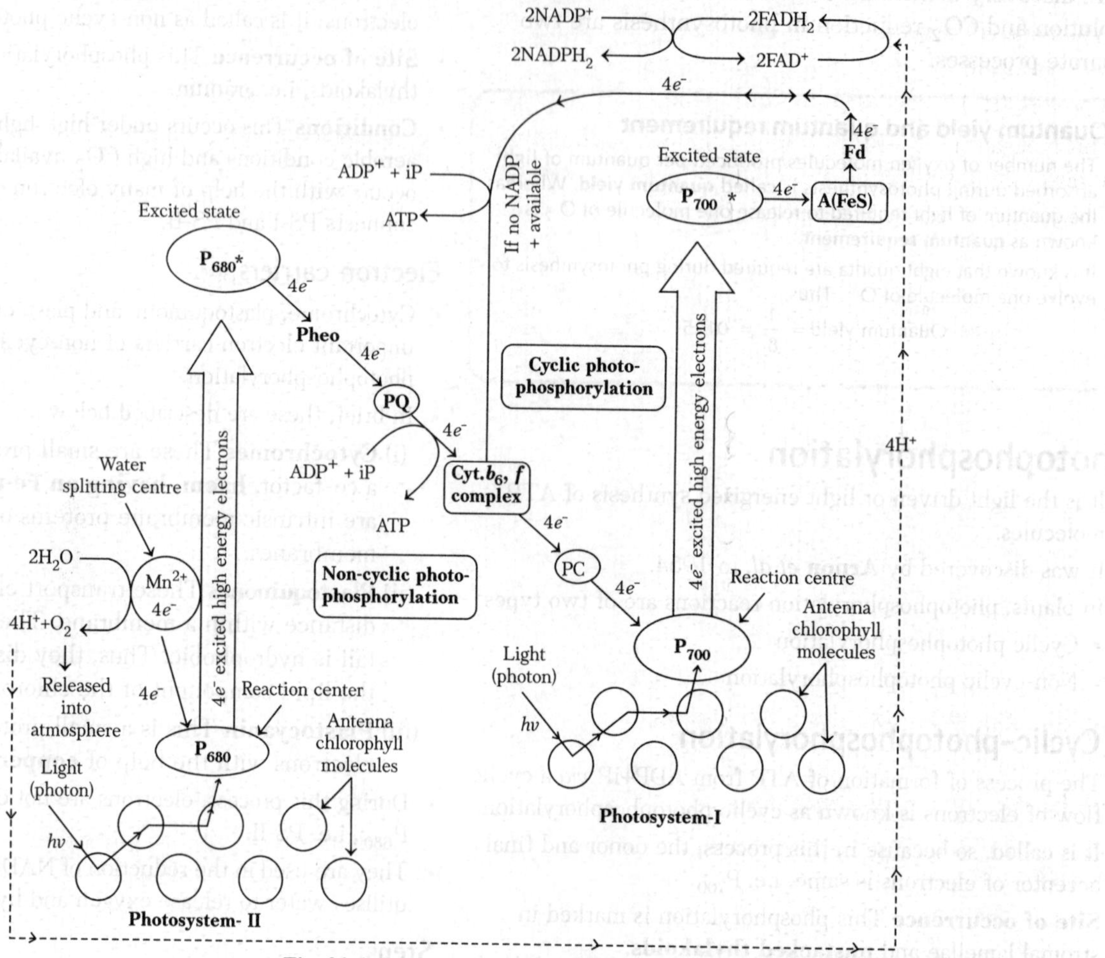

Fig. 20.6 Z-scheme of photosynthesis

Differences between cyclic and non-cyclic photophosphorylation are as follows

Cyclic photophosphorylation	Non-cyclic photophosphorylation
In this, process, only PS-I is involved.	It involves both PS-I and PS-II.
It does not require an external source of electrons.	It requires an external source of electrons as well.
In this, photolysis of H_2O does not occur.	It involves photolysis of H_2O .
End product is ATP.	End products are ATP and NADPH + H⁺.
It occurs in stroma lamellae and unstacked thylakoids.	It occurs mostly in stacked thylakoids, i.e. granum.
It is functional under low light intensity, anaerobic condition and less CO_2 availability.	It occurs in high light intensity, aerobic condition and high CO_2 availability.
It is not affected by inhibitors, e.g. diuron.	It is affected by inhibitors, e.g. diuron.

Chemiosmotic hypothesis

- This hypothesis was proposed by **Peter Mitchell** in 1961.
- According to this hypothesis, ATP synthesis during photosynthesis is linked to development of proton gradient or force across thylakoid membrane.
- Development of proton gradient occurs by following mechanisms
  (i) Splitting of water molecules takes place on the inner side of thylakoid membrane.
  Thus, protons produced during photolysis are accumulated within thylakoid lumen.
  (ii) With the transport of electrons through photosystems, protons are also transported across the membrane. It occurs during the transport of electrons *via* plastoquinone.
  Thus, the number of protons decreases in stroma and increases in lumen resulting to proton gradient.

ATP generation

This proton gradient is broken by transmembrane channel known as **CF_0-CF_1 particle** or **ATPase.** It catalyses the formation of ATP from ADP + iP as follows

- When one H^+ moves from inside to outside, a configuration change occurs in ATPase unit.
- Due to this, it starts catalysing ATP formation.
- 3 protons (H^+) are required to form a molecule of ATP through F_0-F_1 particle or ATPase.
- This mechanism of ATP formation is known as **chemiosmotic mechanism** of ATP formation. The details of chemiosmosis at molecular level will be discussed in respiration in plants.
- Flow of electrons and H^+ (protons) from PS-I and II to ATP synthase with the help of electron carriers is shown below

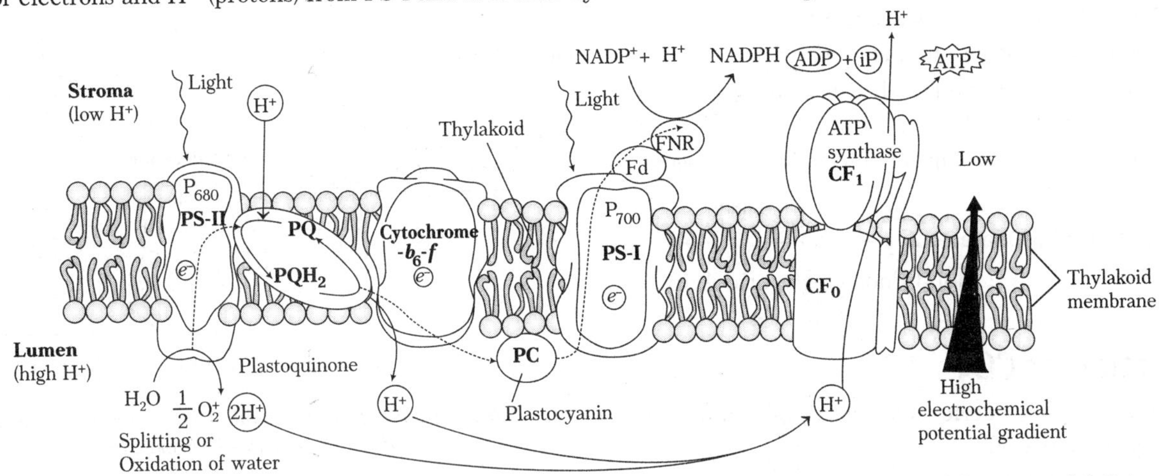

Fig. 20.7 Mechanism of chemiosmotic synthesis of ATP in chloroplast dotted line shows the flow of electrons and definite lines show pathway of H^+, (FNR = Flavoprotein ferredoxin-NADP reductase)

Products of light reaction

- The products of complete light reaction are **ATP**, **$NADPH_2$** and **O_2**.
- Among these products, cyclic photophosphorylation contributes to ATP production only. In non-cyclic photophosphorylation, ATP, $NADPH_2$ and O_2 are produced.
- All these molecules except O_2 are chemical forms of energy. These are used in the dark reactions or phase-II of photosynthesis.
- The process involves **carbon assimilation**, i.e. reduction of CO_2 to form carbohydrate. That is why these products of light reactions are also called **assimilatory power** or **reducing power**.

Inhibitor's of light reaction

There are certains compounds having inhibitory effects on photosynthesis are

(i) **DCMU** or **Diuron** (Dichlorophenyl Dimethyl Urea or Diuron) It is a herbicide that blocks the plastoquinone binding site of PS-II. Thus, inhibits electron flow from plastoquinone to cytochrome.

(ii) **Paraquat** It accepts electrons from PS-I, then reacts with O_2 to form superoxides. These can damage PS-I.

Other photosynthetic inhibitors are **atrazine, CMU** (Chlorophenyl Dimethyl Urea), **simazine**, etc.

Dark reaction

- It is light independent reaction means it may take place in the presence or absence of light.
- It involves biochemical reduction of carbon dioxide to form carbohydrates.
- This was first discovered by **Blackman** in 1905. Hence, it is also called as **Blackman's reaction.**
- The dark reaction utilises the products of light reaction, i.e. ATP and $NADPH_2$.
- Each step of this process is controlled by a specific enzyme.
- **Site of occurrence** It occurs in the stroma of the chloroplast. Dark reaction in green plants may occur as C_3-pathway, C_4-pathway or CAM pathway.

Details of these pathways are as follows

C_3-pathway or Calvin cycle

- The presence of this pathway was discovered by **Professor M Calvin**. Hence, it is also called as **Calvin cycle.**
- Dark reaction *via* this pathway takes place in majority of green plants.
- In this pathway, the first stable compound is a 3-carbon compound, **Phosphoglyceric Acid** (PGA). Hence, the name C_3-pathway is given to this process.

Calvin cycle broadly contains following three steps

1. Carboxylation of RuBP

- In this process, atmospheric CO_2 is fixed into a stable organic compound.
- It involves combining of CO_2 with a CO_2-acceptor , i.e. RuBP (Ribulose 1, 5-bisphosphate).

  Major highlights of this step are
  - The enzyme for CO_2 fixation is **RuBisCO** (Ribulose Bisphosphate Carboxylase Oxygenase).
  - It is the most abundant enzyme on earth.
  - Its name is based on the fact that its active site can bind to both CO_2 and O_2.
  - The equation of this complete process is given below

$$\text{Ribulose 1, 5-bisphosphate} + CO_2 + H_2O \xrightarrow{\text{RuBP carboxylase}} \text{2 mol 3 Phosphoglyceric Acid (3PGA)}$$

2. Reduction of CO_2

In this step, Phosphoglyceric Acid (PGA) formed in step one undergoes reduction.

Major highlights of this step are

- The result of reduction is a 3-C compound, i.e. Phosphoglyceraldehyde (PGAL).
- This reaction is catalysed by enzyme triose phosphate dehydrogenase.
- $NADPH_2$ and ATP are utilised in this reaction.
- PGAL is inter convertible to Dihydroxy Acetone Phosphate (DHAP). This reaction is catalysed by the enzyme phosphotriose isomerase.
- From PGAL further synthesis of hexose sugar Fructose 1,6-diphosphate takes place. It occurs in the presence of enzyme **aldolase**.

Fate of fructose 1, 6-diphosphate or Glycolytic reversal

- The fructose 1,6-diphosphate is formed then dephosphorylated first to fructose monophosph ite and then to fructose (6 C) in the presence of enzyme **phosphatase**.
- Some fructose monophosphate molecules may be isomerised into glucose monophosphate by the **enzyme isomerase** and then into glucose (6 C).
- These hexose sugars may be further converted to sucrose ($C_{12}H_{22}O_{11}$) or to starch $(C_6H_{10}O_5)_n$ and are stored in storage cells.
- Since, all these reactions are opposite of glycolysis so, these are collectively called as **glycolytic reversal**.

- The complete scheme of C_3-pathway is given below

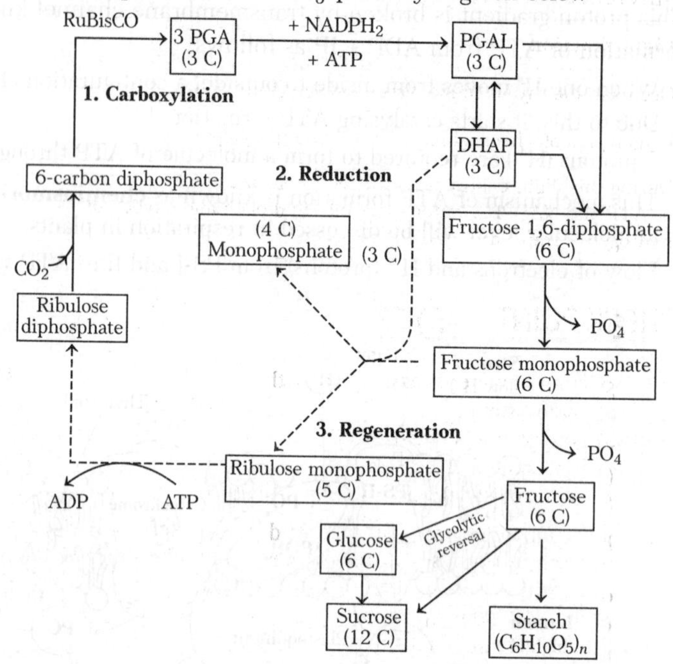

Fig. 20.8 Diagrammatic representation of Calvin cycle, regeneration of RuBP is indicated by broken lines

3. Regeneration of RuBP

- In this step, the primary CO_2 acceptor, i.e. RuBP is regenerated through another chain of reactions.
- Some molecules of triosephosphates and fructose monophosphates are used for the formation of RuBP to be used again to combine with CO_2.

Net Reaction

The net reaction of Calvin cycle can be represented by the following equation

$$6\,\text{RuBP} + 6CO_2 + 18\,\text{ATP} + 12\,\text{NADPH} \longrightarrow 6\,\text{RuBP} + C_6H_{12}O_6 + 18\,\text{ADP} + 12\,\text{NADP} + 18\,\text{iP}$$

Balance sheet of Calvin cycle

One molecule of CO_2 is fixed in one turn of C_3-cycle. Thus, 6 turns of the cycle will be required to fix 6 molecules of CO_2, i.e. to form one molecule of $C_6H_{12}O_6$ (glucose). The summary of molecules consumed and liberated in the process is given below

In	Out
6 CO_2	1 glucose
18 ATP	18 ADP
12 NADPH	12 NADP

Differences between light reaction and dark reaction are as follows

Light reaction	Dark reaction
Occurs in the thylakoid membranes of the grana.	Occurs in the stroma.
Largely a photochemical reactions that require light energy.	A series of biochemical reactions which are catalysed by specific enzymes.
During this, light energy is converted to chemical energy in the form of ATP and $NADPH_2$, water splits into hydrogen and oxygen or hydrogen is trapped in $NADPH_2$.	Carbon dioxide is converted to compounds such as carbohydrates by utilising the chemical energy of assimilating power, i.e. ATP and $NADPH_2$.

CHECK POINT 20.2

1. Splitting of water is associated with
   (a) photosystem-I
   (b) Both photosystems-I and II
   (c) photosystem-II
   (d) lumen of thylakoid
2. How many electrons, protons and photons are involved in the lysis of water to evolve one molecule of oxygen?
   (a) $4e^-$, $4H^+$ and 4 photons
   (b) $4e^-$, $4H^+$ and 8 photons
   (c) $2e^-$, $2H^+$ and 8 photons
   (d) $2e^-$, $2H^+$ and 4 photons
3. The correct sequence of flow of electrons in the light reaction is
   (a) PS-II, plastoquinone, cytochrome, PS-I, ferredoxin
   (b) PS-I, plastoquinone, cytochromes, PS-II, ferredoxin
   (c) PS-I, ferredoxin, PS-II
   (d) PS-I plastoquinone, cytochromes, PS-II, ferredoxin
4. Which condition creates a proton gradient across thylakoid membrane?
   (a) When number of proton decreases in stroma and increases in lumen
   (b) When number of proton increases in stroma and decreases in lumen
   (c) When protons are released into the stroma
   (d) When protons are released into the thylakoids
5. DCMU is a herbicide which kills plants by
   (a) inhibiting photosynthesis as it inhibits PS-I
   (b) inhibiting photosynthesis as it inhibits PS-II
   (c) inhibits photolysis of water since it inhibits PS-I
   (d) Checking respiration due to inhibitory effect on dehydrogenase
6. Which one of the following statement is incorrect?
   (a) Dark reaction is not directly light driven but is dependent on the products ($ATP+NADPH_2$) formed in light reaction
   (b) In stroma dark, reactions occur
   (c) Light reaction occurs in stroma
   (d) Light reaction occurs in grana and $ATP+NADPH_2$ are formed

C_4-pathway (Hatch and Slack cycle)

- Plants that are adapted to dry, tropical regions use C_4-pathway for CO_2 fixation.
- The first stable product of this pathway is a 4-carbon compound, i.e. **Oxaloacetic Acid** (OAA). That is why this pathway is called C_4-pathway and these plants are also called as **C_4-plants**.
- This pathway was discovered by Hatch and Slack in 1966. Therefore, it is also known as **Hatch-Slack** cycle or **pathway**.
- This pathway is observed in both monocot and dicot plants like sugarcane, maize, *Sorghum*, *Panicum pennisetum*, etc.

Anatomical peculiarities of C_4-plants

C_4-plants possess a characteristic leaf anatomy called as **Kranz anatomy**.

Specific features of this anatomy are as follows

- The mesophyll region consists of compactly arranged cells.
- It is not differentiated into palisade and spongy mesophyll as in C_3-plants.

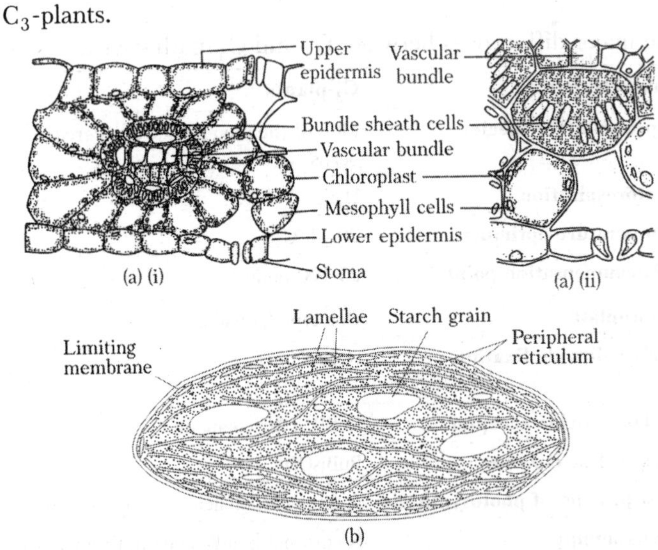

Fig. 20.9 (a) Parts of C_4-plant : (i) Leaf showing mesophyll and bundle sheath cells, (ii) Enlarged view of mesophyll bundle sheath cell and (b) Structure of a typical bundle sheath chloroplast

- The vascular bundles are surrounded by a radially enlarged bundle sheath cells.
- The mesophyll and bundle sheath cells are connected by plasmodesmata or cytoplasmic bridges.
- Dimorphic chloroplast present in leaf cells. Chloroplasts of bundle sheath cells are larger and without grana (agranal) whereas mesophyll chloroplasts are small and with grana.

Important steps in C_4-pathway

In C_4-pathway, double fixation of CO_2 occurs. It takes place as follows

(i) **First CO_2 fixation in mesophyll cells** In this step, firstly CO_2 combines with a 3 C compound Phosphoenol Pyruvic Acid (PEPA) to form a 4 C compound Oxaloacetic Acid (OAA). As OAA is a dicarboxylic acid, this is also known as the **dicarboxylic acid pathway.** This reaction is catalysed by PEP carboxylase.

OAA is then further converted to other 4 carbon compounds like **malic acid** or **aspartic acid.** These are then transported through plasmodesmata to bundle sheath cells.

(ii) **Second CO_2 fixation in the bundle sheath cells** In these cells, 4 C containing malic acid undergoes decarboxylation to form CO_2 and 3 C containing pyruvic acid. CO_2 released here combines with Ribulose-1,5 Bisphosphate (RuBP), which is present in chloroplast of bundle sheath. Thus, Calvin cycle begins here.

(iii) **Back in mesophyll cells** The bundle sheath cells are rich in the enzyme RuBisCO, but lacks PEP kinase. The pyruvic acid from bundle sheath cell is transported back to the mesophyll cells. Then, it is converted into Phosphoenol Pyruvic Acid (PEPA). The conversion requires two molecules of ATP. However, there is no net gain or loss of NADPH in C_4 cycle.

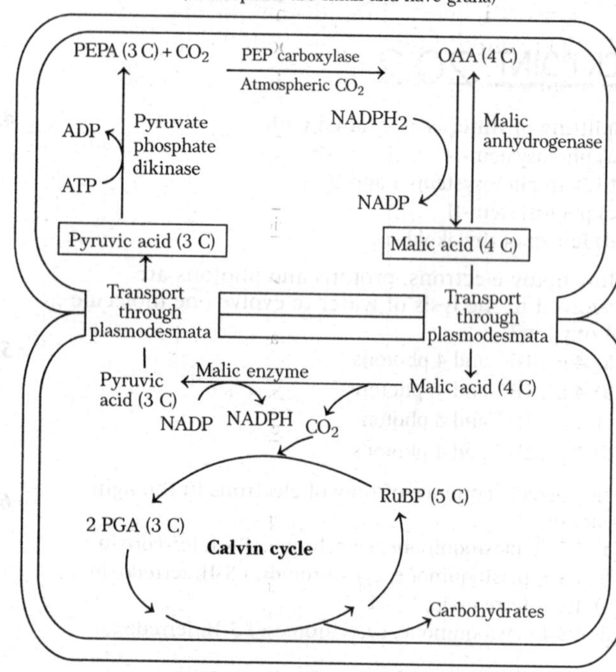

Fig. 20.10 Schematic representation of Hatch and Slack pathway

The major differences between C_3 and C_4-pathways are given below

Characteristics	C_3-plants	C_4-plants
Representative species	Most of the crop plants, e.g. cereals, tobacco and beans.	Maize and sugarcane.
Photorespiration	High	Negligible
Temperature optimum	20-25°C	30-45°C
CO_2 compensation point	25-100 ppm	0-10 ppm
Chloroplast	One type (granal only)	Two types (granal and agranal)
Carbon dioxide fixation	Occurs once	Occurs twice, first in the mesophyll cells, then in bundle sheath cells.
Carbon dioxide acceptor	RuBP, a 5C compound	PEP, a 3C compound
Carboxylase enzyme	RuBisCO	PEPCase, RuBisCO
First product of photosynthesis	A three carbon acid, e.g. phosphoglyceric acid.	A four carbon acid, e.g. oxaloacetate.
Leaf anatomy	No distinct bundle sheath, Kranz anatomy absent.	Kranz anatomy present.
Efficiency	Less efficient photosynthesis than plants. Yield is usually much lower. Require 18ATPs to form one glucose molecule.	More efficient photosynthesis than plants, but use more energy. Yield is usually much higher. Require 30 ATPs and 12 $NADPH_2$ to form one glucose molecule.

CAM (Crassulacean Acid Metabolism)

This pathway is mostly present in **succulent xerophytes** such as plants of family—Crassulaceae, Euphorbiaceae, Cactaceae, etc. Major highlights of this pathway are

- The stomata in these plants remain closed during the day and open during the night for gaseous exchange such stomata are called scotoactive **stomata.**
- In this way, CO_2 is fixed during night and sugars are formed during day.
- This is a desert adaptation that helps to check transpiration.
- This pathway is similar to C_4-pathway. However, there is no separation of initial PEP carboxylase fixation and final RuBisCO fixation of CO_2. These both the steps occur in the same cell, but at different times.
- The outline of complete pathway is shown below

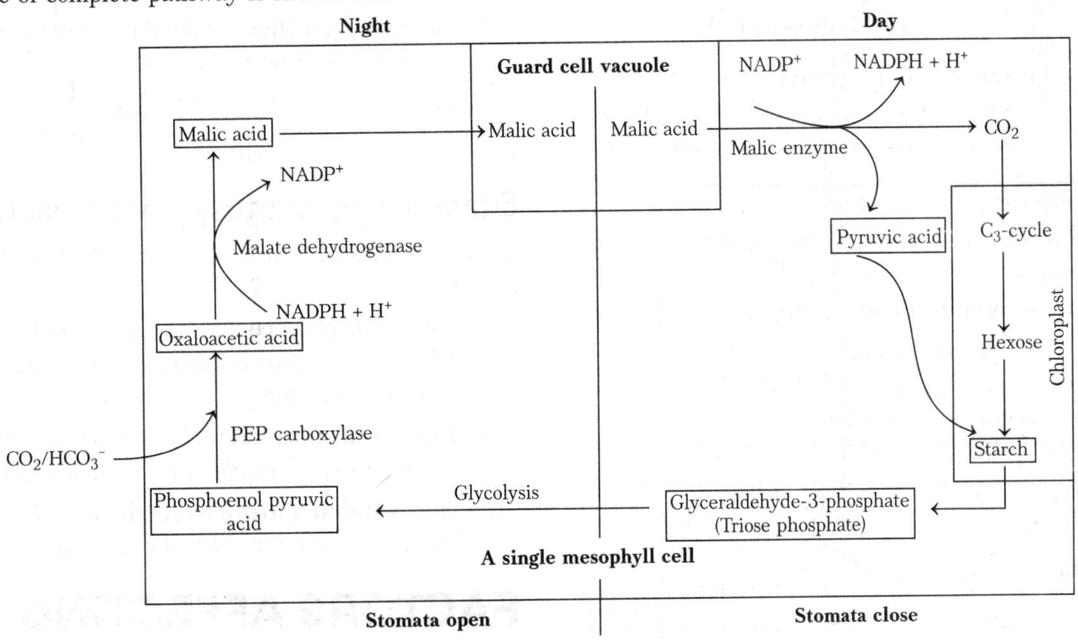

Fig. 20.11 Schematic representation of the CAM pathway

Steps during night

- CO_2 is accepted at night by Phosphoenol Pyruvic Acid (PEPA) to form Oxaloacetic Acid (OAA).
- This OAA in the presence of malic dehydrogenase is converted to malic acid.
- The malic acid is stored in vacuoles, making the pH of cell lowered during night.

Steps during daytime

- Light reactions takes place.
- This allows the stored malic acid to be used.
- This gives off CO_2 molecule, which then enter the C_3-cycle to form starch.

Photorespiration

- It was first observed by **Otto Warburg** in 1920. He showed that in the presence of high O_2, the rate of photosynthesis decreases due to increase in the oxygenase activity of RuBisCO. Later, it was also demonstrated by **Decker** and **Tijo** (1959) in tobacco.
- It is a light dependent wasteful process, which involves release of CO_2 by the photosynthetic organs of a plant. It is initiated in chloroplasts. This process does not produce ATP or NADPH.
- Photorespiration occurs usually when there is high concentration of oxygen in the atmosphere. There is neither the synthesis of sugar nor of ATP. Rather, it results in the release of CO_2 with the utilisation of ATP. It leads to a 25% loss of the fixed CO_2.

- **Site of occurrence** Photorespiration involves three organelles, i.e. **chloroplast, peroxisomes** and **mitochondria**. Loss of CO_2 takes place in mitochondria only.
- In the presence of O_2, RuBP binds to O_2. As a result of this PGA (3C) and phosphoglycolate (2C) are formed.
- In this process, the first stable product is a 2C compound thus, this process is also called **C_2-pathway**.
- The schematic representation of this process is shown below

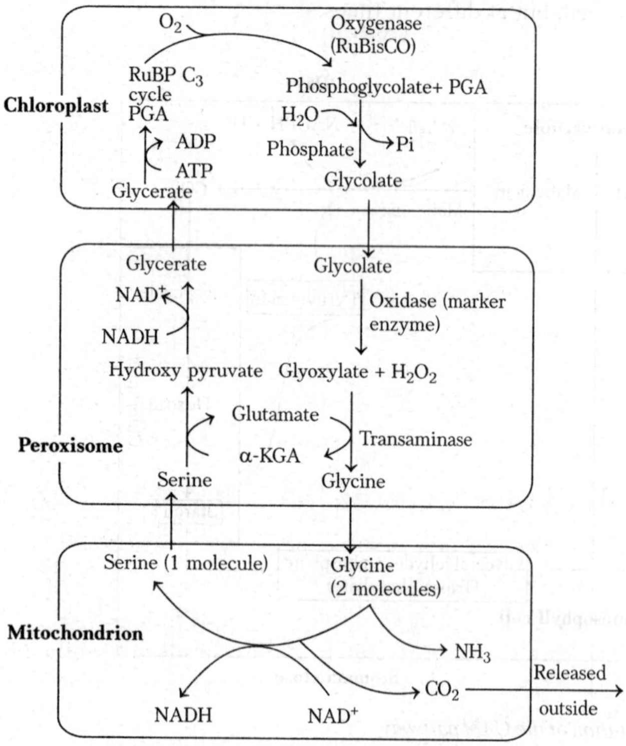

Fig. 20.12 Schematic representation of photorespiration (C_2-pathway)

- In C_4-plants, photorespiration does not take place. It is due to following two reasons
  (i) In mesophyll cell, first CO_2 fixation is carried out by PEPCase and not by RuBisCO.
  (ii) C_4-plants have a mechanism to increase CO_2 concentration at the enzyme site. When malic acid is broken into pyruvic acid, CO_2 is released. Due to this, bundle sheath cells always have high CO_2 concentration. It ensures that RuBisCO function as carboxylase and not as oxygenase.

Bacterial photosynthesis

The process of reduction of carbon dioxide to organic matter by a bacteria is called **bacterial photosynthesis**.

Major features of this phenomenon are

- In this process, hydrogen sulphide, thiosulphate, hydrogen and other inorganic reductants are used instead of H_2O.
- In photosynthetic bacteria, true chlorophyll is absent. They have **bacteriochlorophyll, bacterioviridin, bacteripheophytin** and **carotenoids** as their photosynthetic pigments.
- These pigments are found in their inter-cytoplasmic membranes.
- These bacteria utilise radiations of infrared region of the spectrum, i.e. 870 nm to 1020 nm.
- They have only one pigment system and consequently one reaction centre, i.e. P_{890}.
- Bacterial photosynthesis does not evolve O_2.
- According to **van Niel** (1931), the generalised equation for bacterial photosynthesis is as follows.

$$CO_2 + 2H_2A \xrightarrow[\text{Radiant energy (Infrared)}]{\text{Bacteriochlorophyll}} (CH_2O)_n + 2A + H_2O$$

Examples of photosynthetic bacteria

There are following three main groups of photosynthetic bacteria

(i) **Green-sulphur bacteria**, e.g. *Chlorobium* sp, *Chlorochromatium aggregatum*, etc. These lack RuBisCO enzyme.

(ii) **Purple-sulphur bacteria**, e.g. *Thiospirillum* sp, *Thiocystis* sp., *Chromatium* sp., *Thiocapsa* sp., etc.

(iii) **Non-sulphur purple bacteria**, e.g. *Rhodospirillum* sp., *Rhodobacter* sp., *Rhodocyclus* sp., etc.

FACTORS AFFECTING PHOTOSYNTHESIS

Photosynthesis is influenced by many factors. For the sake of suitability in studies these factors can be categorised as external factors and internal factors.

External factors

The external factors that affect photosynthesis are as follows

(i) **Light** Major points related to this factor are summerised as follows

- A wavelength of 400-700 nm is most effective for photosynthesis.
- Under low intensity of light, the rate of photosynthesis is low. As the intensity of light increases, rate of photosynthesis also increases.
- But very strong light may inhibit the rate of photosynthesis also. This is called as **solarisation**.
- Plants which are adapted to grow in high light intensities are called **heliophytes**.
- Plants adapted to grow in shady areas are called **sciophytes**.

(ii) **Carbon dioxide** Major points related to this factor are summerised as follows

- An increase in CO_2 concentration increases the rate of photosynthesis. However, higher concentration of this gas has inhibitory effect on photosynthesis also.
- Thus, CO_2 is a major limiting factor for photosynthesis.
- In atmosphere, concentration of CO_2 is 0.03-0.04%.
- Increase in concentration up to 0.05% can cause an increase in the rate of photosynthesis.
- The C_4-plants show saturation at 360 ppm, while C_3-plants show saturation at about 450 ppm of CO_2.

CO_2 compensation point

When the CO_2 concentration in the environment is reduced, a point comes at which plant stop absorbing CO_2 from its environment. This point is known as CO_2 **compensation point**. At this point, CO_2 fixed in photosynthesis is equal to CO_2 evolved in respiration and photorespiration.

(iii) **Temperature** Major points related to this factor are summerised as follows

- The optimum temperature range at which rate of photosynthesis is maximum comes as 25-30°C.
- Though in certain plants like *Opuntia*, photosynthesis occurs at 55°C.
- Optimum temperature for tropical plants is more than that for temperate plants.

(iv) **Water** Major points related to this factor are summerised as follows

- An increase in water content of the leaf results in increased rate of photosynthesis.
- Water maintains the turgidity of the assimilatory cells and proper hydration of protoplasm.
- During water stress, wilting of leaves occurs, due to which, surface area of leaves and their metabolic activities are reduced.

Internal factors

The internal factors that affect photosynthesis are as follows

(i) **Chlorophyll content of leaves** The chlorophyll only acts as a biocatalyst. Hence, a small quantity of chlorophyll is enough to carryout photosynthesis.

(ii) **Accumulation of byproducts** The final product in the photosynthetic reaction is sugar. Accumulation of sugars in the cells slows down the process of photosynthesis.

(iii) **Internal structure of leaf** The thickness of the cuticle and epidermis of the leaf, the size and distribution of intercellular spaces and the stomata also affect the rate of photosynthesis.

Biackman's law of limiting factor (1905)

When several factors are controlling a particular physiological process, the rate of the process is limited by pace of the slowest factor.

In other words, the rate of that process is determined by the most deficient factor. This is known as **law of limiting factor**.

CHECK POINT 20.3

1. In the Kranz anatomy of C_4 plants, the bundle sheath cells have
(a) thin walls, no intercellular spaces and several granal chloroplast
(b) thick wall, many intercellular spaces and few chloroplasts
(c) thin wall, many intercellular spaces and no chloroplast
(d) thick wall, no intercellular spaces and large number of agranal chloroplast

2. For synthesis of one glucose molecule, the number of ATP required
(a) 9 for C_3 cycle and 20 for C_4 cycle
(b) 18 for C_3 cycle and 30 for C_4 cycle
(c) 22 for C_3 cycle and 35 for C_4 cycle
(d) 24 for C_3 cycle and 36 for C_4 cycle

3. CAM photosynthesis occur in plants with
(a) thin green leaves with reticulate venation
(b) thin green leaves with parallel venation
(c) fleshy green leaves
(d) None of the above

4. Photorespiration is favoured by
(a) low light and high O_2
(b) low O_2 and high CO_2
(c) low temperature and high O_2
(d) high O_2 and low CO_2

5. Which photosynthetic bacteria lack RuBisCO enzyme?
(a) *Thiosprillum* sp.
(b) *Thiocystis* sp.
(c) *Chlorobium* sp.
(d) *Rhodobacter* sp.

EXAM POINTS

- Chloroplasts are not associated with any type of respiration, i.e. aerobic and anaerobic and can be found in any type of cell. **[JIPMER 2018]**
- The structures that are formed by stacking of organised flattened membranous sacs in the chloroplasts are grana. **[CBSE AIPMT 2015]**
- Magnesium is found in chlorophyll. **[Chhattisgarh PMT 2015]**
- Molecular formula of chlorophyll-*b* is $C_{55}H_{70}O_6N_4$ Mg. **[JIPMER 2018]**
- Water soluble pigment found in plant cell vacuole is phycobilin. **[NEET 2016]**
- Phycoerythrin is the major pigment in red algae. **[NEET 2020]**
- Visible portion of light spectrum useful in photosynthesis is referred to as PAR (Photosynthetic Active Radiation). **[MHT CET 2015]**
- Emerson's enhancement effect and red drop have been instrumental in the discovery of two photosystem operating simultaneously. **[NEET 2016]**
- CO_2 is necessary for photosynthesis. The chemical used to remove this gas most effectively from entering a control apparatus is potassium hydroxide solution. **[AIIMS 2015]**
- PS-II and PS-I are present on stroma lamella membrane. **[EAMCET 2015]**
- NADP reductase is a part of PS-I. **[EAMCET 2015]**
- Manganese and chlorine are required in photolysis of water during photosynthesis. **[CBSE AIPMT 2015]**
- The quantum yield of photosynthesis is 12%. **[Chhattisgarh PMT 2015]**
- Aerobic condition and optimum light intensity are conditions required for cyclic photophosphorylation. **[JIPMER 2018]**
- Cyclic photophosphorylation results in the formation of ATP only. **[Kerala CEE 2015]**
- When both PS-I and PS-II are functional electrons flow in an non-cyclic way. **[EAMCET 2015]**
- In light reaction, plastoquinone facilitates the transfer of electrons from PS-II to cyt-b_6f complex. **[NEET 2020]**
- The correct sequence of involvement of electron acceptor in non-cyclic photophosphorylation is Pheophytin $\rightarrow$ PQ $\rightarrow$ PC $\rightarrow$ Ferredoxin **[EAMCET 2015]**
- During non-cyclic photophosphorylation, when electrons are lost from the reaction centre at PS-II, water replaces the lost electrons. **[NEET 2020]**
- Oxygen released in photosynthesis is formed during non-cyclic photophosphorylation. **[UK PMT 2015]**
- ATPase are enzyme consisting of $F_0 - F_1$ unit. **[EAMCET 2015]**
- ATP synthesis in cell requires H^+ gradient across the membrane. **[WB JEE 2015]**
- Light reaction of photosynthesis produces NADPH, ATP and oxygen. **[NEET 2018]**
- NADH is not produced in light reaction. **[NEET 2018]**
- Dark reaction requires product of light reaction. **[JIPMER 2018]**
- Stroma in the chloroplasts of higher plants contains light independent reaction enzymes. **[CBSE AIPMT 2015]**
- In a chloroplast, the highest number of protons are found in the lumen of thylakoids. **[NEET 2016]**
- Oxygen is produced during photosynthesis by *Cycas, Nostoc* and *Chara*, but not by green sulphur bacteria. **[NEET 2018]**
- DCMU is the herbicide that blocks electron transport from PS-II to PS-I by inhibiting electron flow between plastoquinone to cytochrome. **[JIPMER 2017]**
- C_3 plants are rice, wheat and potato. **[BCECE 2015]**
- The enzyme responsible for primary carboxylation in C_3-plants is RuBP carboxylase/oxygenase. **[J&K CET 2015]**
- In C_3-plants, C_3-cycle occurs in mesophyll cell and in C_4-plants, C_3-cycle occurs in bundle sheath cell. **[AIIMS 2019]**
- Fixation of one CO_2 molecule through Calvin cycle requires 3ATP and 2NADPH$_2$. **[JIPMER 2017]**
- 36 ATP and 24 NADPH are required in the formation of two molecules of glucose. Also 12 Calvin cycles are required. **[Gujarat CET 2015]**
- 18 ATP molecules are required during the formation of one molecule of glucose. **[UK PMT 2015]**
- In C_3-plants, the first stable compound formed after CO_2 fixation is 3-phosphoglycerate. **[J&K CET 2015]**
- Kranz anatomy is found in C_4-plants. **[JIPMER 2019]**
- RuBisCO is a bifunctional enzyme. In C_4-plants, the site of RuBisCO activity is bundle sheath cells. The substrate molecule for RuBisCO activity is a 6-carbon compound. RuBisCO action requires ATP and NADPH. **[NEET 2020]**
- In Hatch and Slack pathway occurring in C_4 plants, the primary CO_2 acceptor is phosphoenol pyruvate. **[NEET 2019, 17]**
- In C_4-plants, Kranz anatomy occurs, CO_2 acceptor is PEP and PEPCase is present in mesophyll. **[JIPMER 2018]**
- C_4-plants respond to higher temperature with enhanced photosynthesis, while C_3 plants have much lower temperature optimum. **[NEET 2017]**
- C_4-pathway is advantageous over C_3-pathway in plants, because it occurs in relatively low CO_2 concentration. **[AIIMS 2018, WB JEE 2015]**
- An example of CAM plant is snake plant (*Sansevieria trifasciata*) **[AMU 2015]**
- Photorespiration is the process that makes important difference between C_3 and C_4 plants. **[NEET 2016]**
- During photorespiration compounds formed having 2C and 3C respectively in peroxisome is glycine and glycerate. **[Gujarat CET 2015]**
- Photorespiration requires oxygenase activity. **[BCECE 2015]**
- The oxygenation activity of RuBisCO enzyme in photorespiration leads to the formation of 1 molecule of 3C compound. **[NEET 2020]**
- Increasing atmospheric CO_2 concentration up to 0.05% can enhance CO_2 fixation rate. **[NEET 2017]**
- Light saturation for CO_2 fixation occurs at 10% of full sunlight. **[NEET 2017]**
- Tomato is a greenhouse crop, which can be grown in CO_2 enriched atmosphere for higher yield. **[NEET 2017]**

Chapter Exercises

(A) Topical Exercise

Photosynthesis and Its Related Experiments

1 Which of the following statement is/are correct regarding the photosynthesis?
(a) It is a process by which green plants convert light energy into chemical energy
(b) It is a catabolic, endergonic and redox process
(c) Of the total world's photosynthesis, 90% is carried out by freshwater plants
(d) All of the above

2 During photosynthesis,
(a) water gets reduced and carbon dioxide gets oxidised
(b) CO_2 gets reduced and water gets oxidised
(c) Both carbon dioxide and water get reduced
(d) Both (a) and (b)

3 Which of the following is incorrect about the experiment performed by TW Engelmann?
(a) *Cladophora* (a green alga) is used
(b) Suspension of aerobic bacteria is used
(c) Bacteria accumulated mainly in the region of orange and green light of the split spectrum
(d) Prism was used to split light into its spectral component

4 ...*A*... performed an elegant experiment with a/an ...*B*... plant to show that in bright sunlight, small bubbles were formed around the ...*C*... parts, while in dark they do not die.

	A	*B*	*C*
(a)	Ingenhousz	aquatic	green
(b)	Ingenhousz	terrestrial	green
(c)	Priestley	aquatic	green
(d)	Priestley	aquatic	non-green

5 Which is the evidence to show that oxygen release in photosynthesis comes from water?
(a) Isotopic oxygen (O^{18}) supplied as H_2O appears in the O_2 released in photosynthesis
(b) Activated chloroplast in water released O_2 if supplied potassium ferrocyanide or some other reducing agent in the absence of CO_2
(c) Photosynthetic bacteria use H_2S and CO_2 to make carbohydrates, H_2O and sulphur
(d) All of the above

6 Match the following columns.

Column I (Scientists)	Column II (Discoveries)
A. von Mayr	1. Proposed that green plants transform light energy into chemical energy
B. Engelmann	2. Plotted the action spectrum of photosynthesis
C. van Niel	3. Demonstrated that some bacteria use H_2S in photosynthesis
D. Warburg	4. Perform flash light experiment using *Chlorella*

Codes

	A	B	C	D		A	B	C	D
(a)	2	3	4	1	(b)	1	2	3	4
(c)	1	4	3	2	(d)	3	2	4	1

7 In an experiment demonstrating the evolution of oxygen in *Hydrilla*, sodium bicarbonate is added to water in the experimental setup. What would happen if all other conditions are favourable?
(a) Amount of oxygen evolved decreases as carbon dioxide in water is absorbed by sodium bicarbonate
(b) Amount of oxygen evolved increases as the availability of carbon dioxide increases
(c) Amount of oxygen evolved decreases as the availability of carbon dioxide increases
(d) Amount of oxygen evolved increases as carbon dioxide in water is absorbed by sodium bicarbonate

Site of Photosynthesis and Pigment

8 Which of the following statements is false about structure of chloroplast?
(a) It is double membrane bound organelle
(b) A stack of thylakoids in chloroplast form a lamella
(c) Adjacent grana are connected by unstacked membranes
(d) The light absorbing pigments are present in the grana

9 Select the option that correctly identifies *X*, *Y* and *Z*.

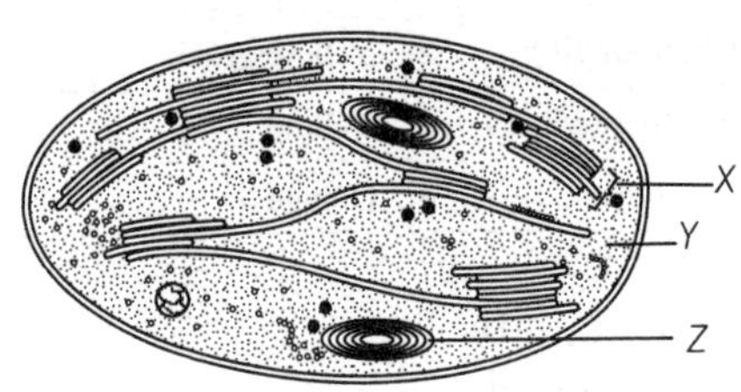

X	Y	Z
(a) Stroma	Grana	Chloroplast DNA
(b) Stroma	Grana	Starch granule
(c) Grana	Stroma	Starch granule
(d) Grana	Stroma	Chloroplast DNA

10 Select the correct match.
(a) Stroma - Light reactions
(b) Membrane system - Trapping of light energy
(c) Thylakoids - CO_2 fixation
(d) Stromal lamellae - Synthesis of ATP

11 Choose the correct option regarding chlorophyll-*a* molecule.
(a) In pure state, it is red in colour and thus it absorbs more blue wavelength of light than the red wavelength
(b) It is soluble in water as well as petroleum ether
(c) Phytol chain is the part of chlorophyll, where electron rearrangement takes place when chlorophyll get excited
(d) Molecular formula of chlorophyll-*a* is $C_{55}H_{72}O_5N_4Mg$

12 Accessory pigments
(a) play no role in photosynthesis
(b) release electron and get oxidised
(c) transfer electron to NADP
(d) allow plants to harvest visible light of under range of wavelength

13 $C_{40}H_{56}$ is the empirical formula of ... *A* ... that prevents the ...*B*... of chlorophyll molecules.

A	B
(a) Xanthophyll	Photoreduction
(b) Carotenes	Photo-oxidation
(c) Fucoxanthin	Photo-oxidation
(d) Phycobilins	Photoreduction

14 Match the following columns.

Column I (Pigments)	Column II (Colour produced on chromatograph)
A. Chlorophyll-*a*	1. Blue-green
B. Chlorophyll-*b*	2. Yellow-green
C. Xanthophyll	3. Yellow
D. Carotenes	4. Orange

Codes

	A	B	C	D
(a)	1	2	3	4
(b)	1	3	2	4
(c)	4	3	1	2
(d)	4	3	2	1

15 Carotenoids are
(a) yellow to orange pigments, which absorb light strongly in the blue-green range
(b) called shield pigments because they protect chlorophyll from photo-oxidation
(c) present in green algae and higher plants
(d) soluble in organic solvents

16 Carotene and xanthophyll pigments absorb radiant energy and
(a) carryout photosynthesis to produce carbohydrate
(b) do not take any part in photosynthesis
(c) lose the energy as heat
(d) transfer this energy to chlorophyll-*a* and *b* where it is transformed into chemical energy

17 Match the following columns.

Column I	Column II
A. Red colour of tomato	1. Chlorophyll-*a*
B. Reaction centre	2. Phycobilins
C. Phycoerythrin and phycocyanin	3. Lycopene

Codes

	A	B	C		A	B	C
(a)	3	2	1	(b)	2	1	3
(c)	1	2	3	(d)	3	1	2

18 Photosynthetic pigments such as chlorophyll-*a*, chlorophyll-*b*, xanthophyll and carotene can be separated by which of the following techniques?
(a) Paper chromatography (b) Gel electrophoresis
(c) X-rays diffusion (d) ELISA test

19 During high light intensity, the chloroplasts align themselves
(a) in vertical position along lateral walls
(b) along tangential walls
(c) in centre and get scattered
(d) perpendicular to light

20 The most abundant protein found in plant is
(a) cellulose (b) lignin
(c) pectin (d) RuBisCO

21 Consider the following statements.
I. The curve showing the amount of absorption of different wavelength of light by a photosynthetic pigment is called absorption spectrum.
II. The portion of the spectrum between 400-700 nm is also referred to as Photosynthetically Active Radiation (PAR).

Choose the correct option.
(a) Statement I is true, but II is false
(b) Statement I is false, but II is true
(c) Both statements I and II are true
(d) Both statements I and II are false

22 The absorption spectrum of chlorophyll
(a) shows that some colours of light are absorbed more than the others
(b) approximates the action spectrum of photosynthesis
(c) explain why chlorophyll is a green pigment
(d) has all the above properties

23 In below graph (a), (b) and (c) represent, respectively.

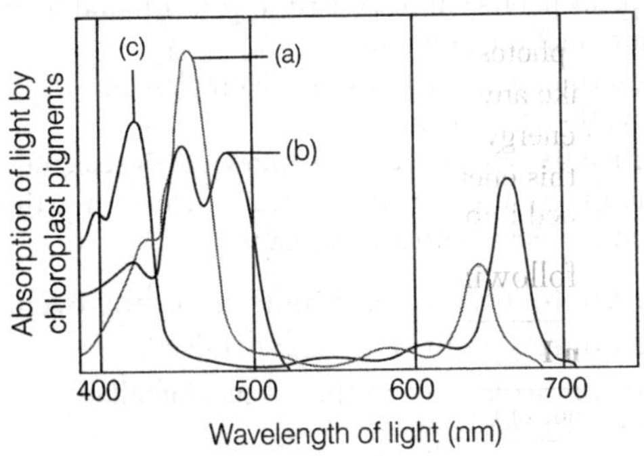

(a) Chlorophyll-*a*, *b* and *c*
(b) Chlorophyll-*b*, carotenoids and chlorophyll-*a*
(c) Chlorophyll-*a*, *b* and carotenoids
(d) Chlorophyll-*b*, *a* and carotenoids

24 Identify *A* and *B* in the following diagram.

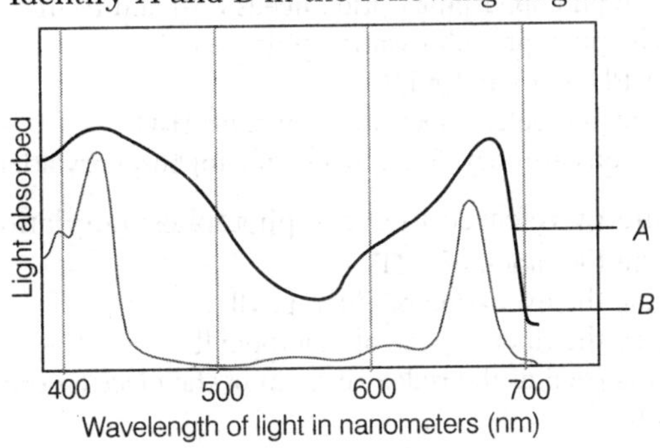

	A	*B*
(a)	Absorption spectrum of chlorophyll-*a*	Action spectrum of chlorophyll-*a*
(b)	Action spectrum of photosynthesis	Absorption spectrum of chlorophyll-*a*
(c)	Action spectrum of chlorophyll-*a*	Action spectrum of chlorophyll-*b*
(d)	Absorption spectrum of chlorophyll-*a*	Absorption spectrum of chlorophyll-*a*

25 Which option(s) correct about quantasomes?
(a) It is a group of pigment molecules required for carrying out a photochemical reaction
(b) It consists of 200-240 chlorophyll molecules, carotenoids, quinone compounds, sulpholipids, phospholipids proteins, etc
(c) It is present on the thylakoid membranes as a small unit
(d) All of the above

26 The Light Harvesting Complex (LHC) is made up of
(a) one molecule of chlorophyll-*a*
(b) very few molecule of chlorophyll-*a*
(c) hundreds of pigment molecule bound to protein
(d) chlorophyll-*a*, chlorophyll-*c*, proteins and DNA

27 Given figure depicts the Light Harvesting Complex (LHC) of photosystem-I (PS-I).

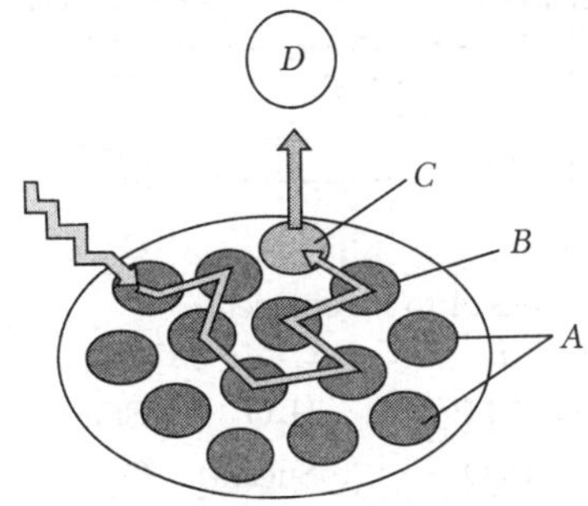

Select the correct identification for *A*, *B*, *C* and *D*.

	A	*B*	*C*	*D*
(a)	Core molecules	Antenna molecule	P_{680}	Primary e^- acceptor
(b)	Antenna molecules	Core molecule	P_{700}	Primary e^- acceptor
(c)	Antenna molecules	Core molecule	P_{700}	Plastocyanin
(d)	Core molecules	Reaction centre	P_{680}	Plastocyanin

28 PS-I occurs in
(a) appressed part of granal thylakoid
(b) appressed and non-appressed part of grana thylakoid
(c) stroma
(d) thylakoid membrane and stroma lamellae

29 Which statement about the light reactions of photosynthesis is false ?
(a) PS-I and PS-II are located in stroma of the chloroplast
(b) PS-I and PS-II are linked by e^- carriers
(c) Chlorophylls have an absorption spectrum with pronounced peaks in red and blue light
(d) Protons diffuse through protein channels which are ATP synthetase molecule

Light and Dark Reaction

30 Which one of the following is not true about the light reaction of photosynthesis?
(a) Light energy provides energy for the photolysis of water through excitation of the reaction centre of PS-II
(b) The flow of electrons from water to NADP in non-cyclic electron transport produce one ATP
(c) Reaction of the two photosystems are needed for the reduction of NADP
(d) P_{680} and P_{700} are the reaction centre of PS-I and PS-II, respectively

31 The diagram shows the movement of substances into and out of a chloroplast.

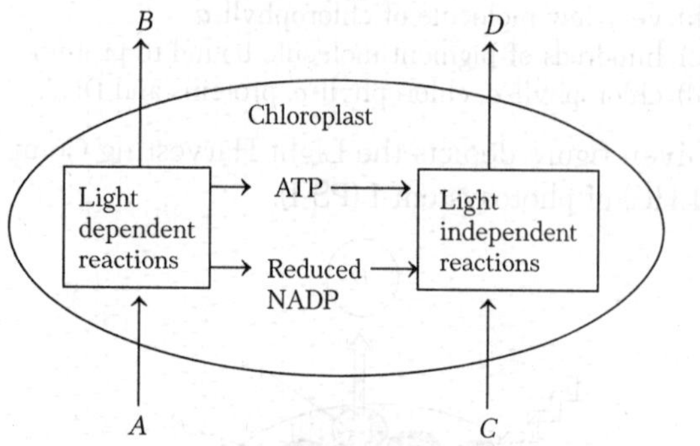

What do labels *A* to *D* represent ?

	A	*B*	*C*	*D*
(a)	CO_2	ATP	H_2O	Starch
(b)	CO_2	H_2O	Sugars	O_2
(c)	H_2O	O_2	CO_2	Sugars
(d)	Sugars	H_2O	ATP	O_2

32 Photophosphorylation is a process in which
(a) light energy is converted into chemical energy by production of ATP
(b) glutamic acid is formed
(c) 2 PGA is formed
(d) None of the above

33 Photolysis of water
(a) is associated with PS-II and is catalysed by the presence of Mn^{2+} and Cl^- ions
(b) involves a strong oxidant, which is yet unknown
(c) expels an electron with a gain of energy
(d) Both (a) and (b)

34 Photolysis of each water molecule in light reaction will yield
(a) 2 electrons and 4 protons
(b) 4 electrons and 4 protons
(c) 4 electrons and 3 protons
(d) 2 electrons and 2 protons

35 Which one of the following is a correct outline of the main events in photosynthesis?
(a) Oxygen reacts with a carbohydrate to produce water and carbon dioxide in the presence of light
(b) Light joins carbon dioxide to an acceptor compound which is then redcued by hydrogen obtained from water
(c) Light splits water and the resulting hydroxyl group combines with a compound which has incorporated carbon dioxide
(d) Carbon dioxide combines with an acceptor compound and this is reduced by hydrogen split from water by light

36 Cytochromes
(a) dissolve easily into the lipid component of chloroplast membranes
(b) are small proteins that contain a co-factor, haem, which holds an iron atom
(c) are loosely associated with chloroplast membranes
(d) Both (a) and (b)

37 The Z-scheme refers to
(a) the type of photosynthesis used in plants found in area with minimal precipitation
(b) the pattern of grana within the chloroplast of photosynthetic pigment
(c) the carbon fixation process also known as the calcium cycle
(d) the energy diagram for the transfer of electron is the light reaction of photosynthetic in plants

38 Which of the following statements regarding cyclic flow of electron during light reactions is false ?
(a) This process takes place in the stromal lamellae
(b) ATP synthesis takes place
(c) NADPH + H^+ is synthesised
(d) Take place only when light of wavelength beyond 680 nm is available for excitation

39 Which is the correct option regarding cyclic photophosphorylation?
(a) Cyclic photophosphorylation needs PS-I and PS-II
(b) Cyclic photophosphorylation produces NADPH + H^+ and ATP
(c) Cyclic photophosphorylation involves H_2O
(d) Electrons are recycled in cyclic photophosphorylation.

40 Free energy released in cyclic photophosphorylation
(a) by the formation of ATP
(b) during the excitation of chlorophyll
(c) during the fluorescence of chlorophyll
(d) during each of the redox reaction of the electron transport chain

41 The correct sequence of flow of electrons in the light reaction is **[NCERT Exemplar]**
(a) PS-II, plastoquinone, cytochromes, PS-I, ferredoxin
(b) PS-I, plastoquinone, cytochromes, PS-II, ferredoxin
(c) PS-I, ferredoxin, PS-II
(d) PS-I, plastoquinone, cytochromes, PS-II, ferredoxin

42 During the light stage of photosynthesis, the photo-activated pigment removes an electron from the hydroxylation derived from the water molecule. The fate of the free hydroxyl radical is that
(a) is broken down into oxygen and a free radical of hydrogen
(b) is used to raise the activation level of chlorophyll by donating a positive charge
(c) is used to produce adenosine triphosphate from adenosine diphosphate
(d) reduces carbon dioxide to sugar

43 In the below schematic diagram, which is plastocyanin?

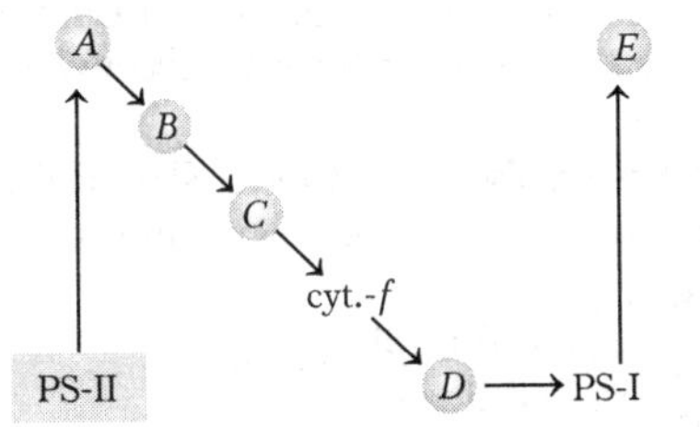

(a) *C* (b) *D* (c) *A* (d) *B*

44 The following figure is the pathway of non-cyclic photophosphorylation. In which of the following options, find the correct words for all the given blanks indicated by *A, B, C, D* and *E*.

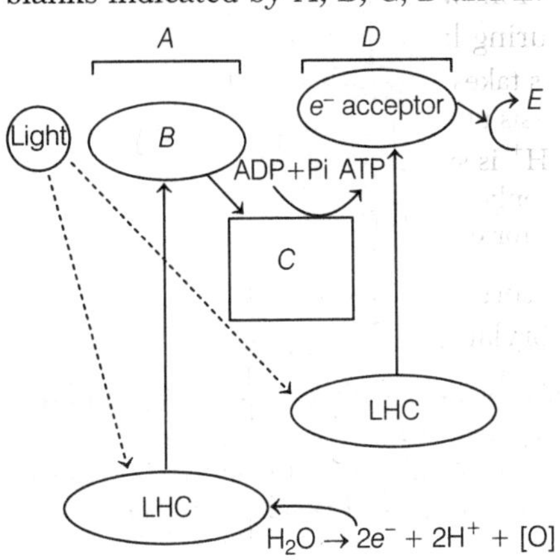

	A	B	C	D	E
(a)	PS-II	Electron acceptor	Electron transport system	PS-I	NADPH
(b)	PS-I	Electron acceptor	Electron transport system	PS-II	NAD^+
(c)	P_{680}	Proton acceptor	Proton transport system	P_{700}	NADPH
(d)	P_{700}	Proton acceptor	Proton transport system	P_{680}	NAD^+

45 Which option is correct about non-cyclic photophosphorylation?
(a) It does not require an external source of electron
(b) It occurs in stroma lamellae
(c) Its end product is ATP
(d) It is affected by diuron

46 Which of the following statement incorrect?
(a) Non-cyclic photophosphorylation has both PS-I and PS-II
(b) Non-cyclic photophosphorylation produces both ATP and NADPH + H
(c) During photolysis of water formation of ATP + NADPH + H^+ occurs
(d) Production of NADPH + H^+ is associated with PS-II, not PS-I

47 Identify *A*, *B* and *C* in the given figure of cyclic electron flow.

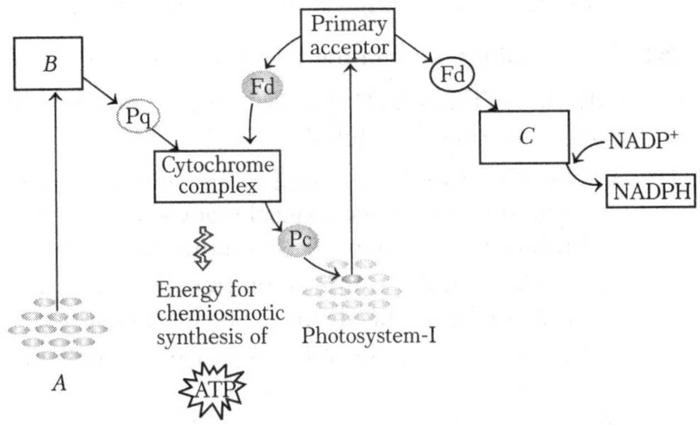

(a) *A*-Photosystem-II, *B*-Primary acceptor, *C*-$NADP^+$ reductase
(b) *A*-Electron, *B*-Primary acceptor, *C*-Oxidation
(c) *A*-Primary acceptor, *B*-Oxidation, C-Reduction
(d) *A*-PS-I, *B*-$NADP^+$, *C*-PC

48 Assume a thylakoid which is somehow punctured, so that the interior of the thylakoid is no longer separated from the stroma. This damage will have the most direct effect on which of the following processes?
(a) Splitting of water
(b) Absorption of light energy by chlorophyll
(c) flow of electrons from photosystem-II to photosystem-I
(d) Synthesis of ATP

49 Read the given statements and select the correct option.
I. In photosynthesis, during ATP synthesis, protons accumulate in the lumen of thylakoid.
II. In respiration, during ATP synthesis, protons accumulate in the intermembranal space of mitochondria.

Choose the correct option.
(a) Statement I is true, but II is false
(b) Statement I is false, but II is true
(c) Both statements I and II are true
(d) Both statements I and II are false

50 Which of the following statement is incorrect for chemiosmotic hypothesis?
(a) It explains the formation of ATP molecule
(b) It is linked with development of proton gradient
(c) It does not involve photolysis
(d) It was proposed by Peter Mitchell

51 In chemiosmotic synthesis of ATP, H^+ diffuses through ATP synthetase
(a) from the stroma into thylakoid lumen/space
(b) from thylakoid space into stroma
(c) from the cytoplasm into stroma
(d) from periplasmic space to stroma

52 The enzyme ATPase couples the synthesis of ATP to
(a) diffusion of protons (b) reduction of $NADP^+$
(c) excitation of chlorophyll (d) carbon dioxide fixation

53 The thylakoid membrane bears several F_0-F_1 particle/ATPase/ATP synthase, which of following is correct for these particles?
(a) One of its part (F_0) is embedded in the membrane and forms transmembrane channel that carries out facilitated diffusion of protons across the membrane
(b) Its other part (F_1) protrudes out from the outer surface of the thylakoid membrane facing towards stroma
(c) The catalytic sites for ATP formation are located in F_1 part
(d) All of the above

54 Dichlorophenyl Dimethyl Urea (DCMU) also called diuron, a potent herbicide stops
(a) O_2 evolution
(b) non-cyclic photophosphorylation
(c) Both (a) and (b)
(d) oxidative photophosphorylation

55 The weedkiller DCMU blocks the flow of electrons from the electron transport chains in photophosphorylation. Why does this kill the plant?
(a) Active transport of mineral ions is prevented
(b) ATP and reduced NADP are not produced
(c) Photoactivation of the chlorophyll cannot occur
(d) Photolysis of water does not occur

56 Match the columns and choose the correct option with respect to chloroplast.

Column I	Column II
A. Light reaction	1. Enzymatic process
B. Dark reaction	2. Synthesis of assimilatory power
	3. Membranous system
	4. Photolysis of water
	5. RuBisCO

Codes

	A	B		A	B
(a)	2, 3, 4	1, 5	(b)	3, 4	1, 2, 5
(c)	1, 3, 4	2, 4	(d)	2, 5	2, 3, 4

57 When RuBisCO acts as an oxygenase
(a) phosphoglycerate and phosphoglycolate are produced
(b) phosphoenol pyruvate is oxidised
(c) net carbon fixation is enhanced
(d) it must mean that the plant is deprived of CO_2

58 Read the given statements and select the correct option.
I. Carboxylation is the most crucial step of Calvin cycle where CO_2 is utilised for the carboyxlation of RuBP.
II. Carboxylation is catalysed by the enzyme RuBisCO which results in the formation of two molecules of 3PGA.
(a) Statement I is true, but II is false
(b) Statement I is false, but II is true
(c) Both statements I and II are true
(d) Both statements I and II are false

59 Complete the table given below with respect Calvin cycle.

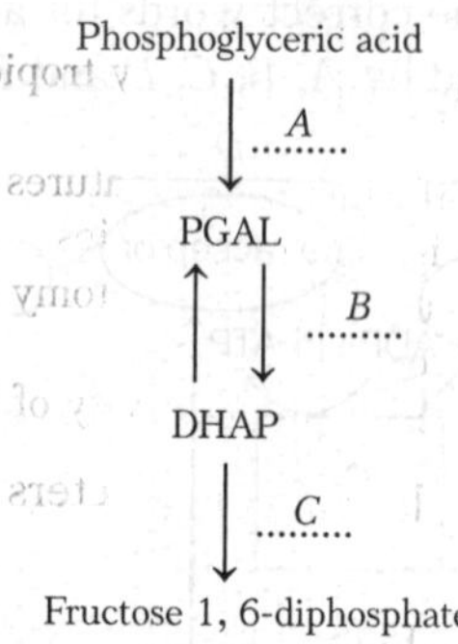

(a) Triose phosphate dehydrogenase, Phosphotriose isomerase, Aldolase
(b) Aldolase, Triose phosphate isomerase, Carboxylase
(c) Aldolase, Carboxylase, Phosphotriose isomerase substances
(d) None of the above

60 Removal of the source of carbon dioxide from photosynthesising chloroplasts results in rapid changes in the concentration of certain chemicals. Which one of the following represents the correct combination of concentration changes?

	ATP	Ribulose diphosphate	PGA
(a)	Decreases	Decreases	Increases
(b)	Decreases	Increases	No change
(c)	Increases	Increases	Decreases
(d)	Increases	No change	Increases

61 Select the correct match.
(a) Site of Calvin cycle in C_3 plants - Bundle sheath cells
(b) Primary acceptor of carbon in Calvin cycle - PEP
(c) First stable product of Calvin cycle - 3PGA
(d) Site of Calvin cycle in C_4 plants - Mesophyll cells

C_4-cycle, CAM and Photorespiration

62 Select the correct statement regarding the first stable product formed in Hatch and Slack pathway in C_4 plants.
(a) Oxaloacetate is formed by carboxylation of Phosphoenol Pyruvate (PEP) in the bundle sheath cells
(b) Oxaloacetate is formed by carboxylation of Phosphoenol Pyruvate (PEP) in the mesophyll cells
(c) Phosphoglyceric acid is formed in the mesophyll cells
(d) Phosphoglyceric acid is formed in the bundle sheath cells

63 Which of the following statements is incorrect with respect to C_4-plants?
(a) Plants that are adapted to dry tropical regions have C_4 pathway
(b) They tolerate higher temperatures and show a response to highlight intensities
(c) Their leaves show Kranz anatomy and they undergo photorespiration
(d) They have greater productivity of biomass

64 Which of the following characters is/are true for C_4 plants?
(a) They lack photorespiration
(b) Mesophyll cell consists of compactly arranged cells
(c) Mesophyll and bundle sheath cells are connected by plasmodesmata
(d) All of the above

65 PEPCase has an advantage over RuBisCO that
(a) PEPCase conserves energy, but RuBisCO does not
(b) RuBisCO combines with O_2, but PEPCase does not
(c) PEPCase combines with O_2, but RuBisCO does not
(d) PEPCase is present in both mesophyll and bundle sheath, but RuBisCO is not

66 Photosynthesis in C_4-plants is relatively less limited by atmospheric CO_2 level because
(a) 4C acids are the primary initial CO_2 fixation products
(b) the primary fixation of CO_2 is mediated *via* PEP carboxylase
(c) effective pumping of CO_2 into bundle sheath cells
(d) RuBisCO in C_4-plants has higher affinity for CO_2

67 Select the incorrect option as far as Kranz anatomy is concerned.
(a) Undifferentiated mesophyll occurs in concentric layers around vascular bundles
(b) Centrifugal chloroplasts are present in bundle sheath cells
(c) Large-sized bundle sheath cells are arranged in a wreath-like manner in one to several layers
(d) Chloroplasts of bundle sheath cells possess well-developed grana lamellae

68 Given figure represents C_4-pathway. Select the suitable option for *A*, *B* and *C*.

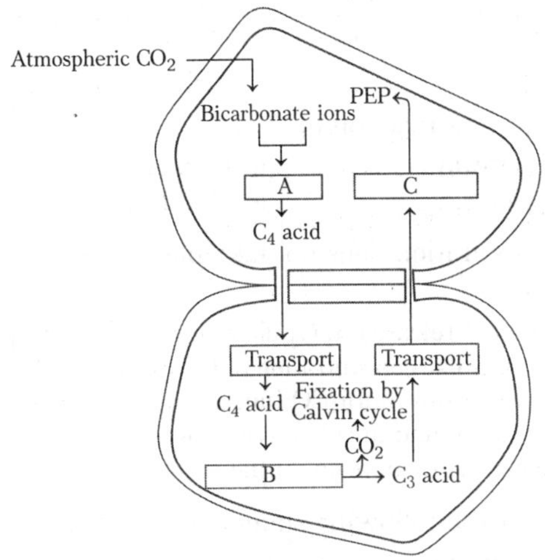

	A	*B*	*C*
(a)	Decarboxylation	Reduction	Regeneration
(b)	Fixation	Transamination	Regeneration
(c)	Carboxylation	Decarboxylation	Reduction
(d)	Fixation	Decarboxylation	Regeneration

69 Initial CO_2 fixation occurs in
(a) bundle sheath cells (b) guard cells
(c) mesophyll cells (d) palisade tissue

70 What type of stomata is present in succulent xerophytes?
(a) Photoactive stomata (b) Hypocyclic stomata
(c) Scotoactive stomata (d) None of these

71 Crassulacean acid metabolism
(a) is an alternative of the C_3 and C_4-pathway of CO_2 fixation found in plants living in dry hot climate
(b) pathway resemble C_4-pathway in that CO_2 is trapped by highly efficient PEP carboxylase
(c) is a special type of respiration show by many green plants, when they are exposed to light
(d) Both (a) and (b)

72 In CAM plants, CO_2 required for photosynthesis enters the plant body during
(a) daytime through the lenticle
(b) night through the stomata, which are kept open
(c) daytime when stomata are open
(d) night when the hydathods are open

73 Match Column I with Column II and select the correct option from the given codes.

Column I	Column II
A. C_3-plants	1. Kalanchoe, *Opuntia*
B. C_4-plants	2. Maize, sugarcane
C. CAM plants	3. *Helianthus*

Codes

	A	B	C		A	B	C
(a)	2	3	1	(b)	1	2	3
(c)	3	2	1	(d)	1	3	2

74 Refer to the given reaction.

$$RuBP + O_2 \xrightarrow[\text{Oxygenase}]{\text{RuBP}} \text{Phosphoglyceric acid} + \text{Phosphoglycolic acid}$$

It is the first reaction of

(a) C_3-pathway (b) C_4-pathway
(c) C_2-pathway (d) glycolysis

75 Photorespiration does not take place in C_4-plant because

(a) O_2 is not released in C_4 plant during photosynthesis
(b) intracellular concentration of CO_2 is high in bundle sheath ensure RuBisCO function as carboxylase
(c) they are found in cold environment
(d) stomata in them open during night

76 Consider the following statements.

I. Photorespiration interferes with the successful functioning of Calvin cycle.
II. Photorespiration oxidises ribulose-1, 5 bisphosphate which is an acceptor of CO_2 in Calvin cycle.

Choose the correct option.

(a) Statement I is true, but II is false
(b) Statement I is false, but II is true
(c) Both statements I and II are true
(d) Both statements I and II are false

Bacterial Photosynthesis and Factors Affecting Photosynthesis

77 Which of the following option(s) is true for bacterial photosynthesis?

(a) A true chlorophyll absent in photosynthetic bacteria
(b) Bacterial photosynthesis does not evolve O_2
(c) They have only one pigment system (i.e. P_{890})
(d) All of the above

78 Rate of photosynthesis is

I. greater in intense light than in diffused light.
II. minimum in red light.
III. greater in intermittent light than incontinuous light.
IV. decreased in the presence of cytokinins and gibberellins.

Choose the correct option.

(a) Only I (b) I and III (c) I, II and III (d) Only IV

79 Algae and other submerged plants bound in water during daytime and sink at night, because

(a) they come up to enjoy some time
(b) they lose weight at night
(c) they become buoyant due to accumulation of O_2 as a result of photosynthesis
(d) they become light due to food material accumulation

80 Which of the following describes conditions in a photosynthesising cell, exposed to high light intensity and low carbon dioxide concentration ?

	Concentration of CO_2 acceptor	Concentration of ATP	Concentration of GP (PGA)
(a)	high	high	high
(b)	high	high	low
(c)	high	low	low
(d)	low	high	high

81 Under water stress, the rate of photosynthesis declines because of

(a) stomatal closure leading to decrease in CO_2 supply
(b) reduced water potential that decreases leaf surface areas for photosynthesis
(c) Both (a) and (b)
(d) Turgidity of leaf

82 A point at which illuminated plant parts stop absorbing CO_2 from their environment, is known as

(a) CO_2 compensation point
(b) CO_2 saturation point
(c) CO_2 optimum point
(d) CO_2 limiting point

83 If light intensity is uniformly high, under which of the following conditions is carbon dixoide most likely to be the main limiting factor upon the rate of photosynthesis ?

	Temperature/°C	% carbon dioxide
(a)	25	4.00
(b)	25	0.04
(c)	15	0.04
(d)	5	0.01

84 A typical light response curve of photosynthesis is shown. The limiting factor/s for photosynthesis at M and N is/are

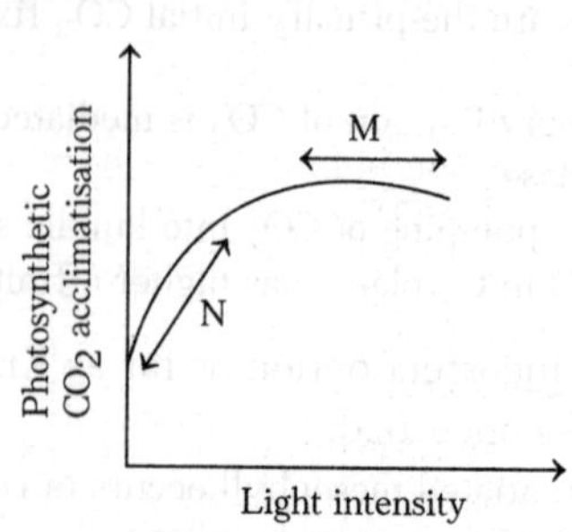

(a) temperature and CO_2 respectively
(b) CO_2 and light respectivley
(c) Only CO_2
(d) light and CO_2 respectively

(B) Cumulative Exercise

(Assorted Questions of the Chapter for Advanced Level Practice)

1 A student setsup an experiment on photosynthesis as follows. He takes soda water in a glass tumbler, adds chlorophyll extract into the contents and keeps the tumbler exposed to sunlight, hoping that he has provided all the necessary ingredients for photosynthesis to proceed (namely CO_2, water, chlorophyll and light). What do you think, will happen after, say, a few hours of exposure to light?

(a) Photosynthesis will take place and glucose will be produced and the mixture will become sweet
(b) Photosynthesis will take place and starch will be produced which will turn the mixture turbid
(c) Photosynthesis will not take place because the CO_2 dissolved in soda water escape into the atmosphere
(d) Photosynthesis will not take place because intact chloroplasts are needed for the process

2 Match Column I with Column II and select the correct option.

	Column I		Column II
A.	OEC	1.	Primary *e*-acceptor
B.	NADP reductase	2.	Photolysis of H_2O
C.	Succinyl Co-A	3.	Outer surface of thylakoid membrane
D.	Phaeophytin	4.	Chlorophyll synthesis

Codes

	A	B	C	D
(a)	2	3	4	1
(b)	2	3	1	4
(c)	3	1	2	4
(d)	1	2	3	4

3 Consider the following statements.

I. There is a clear division of labour within the chloroplast.
II. The membrane system of chloroplast is responsible for trapping the light energy and also for the synthesis of ATP and NADPH.
III. In stroma, enzymatic reactions incorporate CO_2 into the plant leading to the synthesis of sugar, which in turn forms starch.
IV. Green plants carryout photosynthesis, a physicochemical process by which they release light energy to breakdown the organic compounds.
V. Photosynthesis is responsible for the release of oxygen into the atmosphere by green plants.

How many of the above statements are incorrect?

(a) One
(b) Two
(c) three
(d) I, II, III, IV and V

4 Study the given graph showing the effect of light intensity on the rate of photosynthesis. Which of the following statements regarding this is correct?

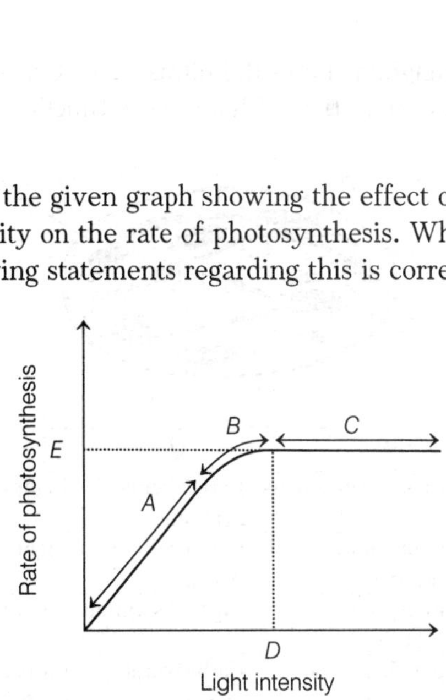

(a) Light is a limiting factor in the region *A*
(b) Region *C* represents that rate of photosynthesis is not increased further by increasing light intensity because some other factors became limiting
(c) Point *D* represents the intensity of the light at which some other factors became limiting
(d) All of the above

5 Study the following statements.

I. Red light falling in the range of wavelength 660-760 nm is the most effective for photosynthesis.
II. Green light falling in the range of wavelength 500-580 nm is the least effective for photosynthesis.
III. Chlorophyll-*a*, chlorophyll-*b*, carotenes and xanthophylls are soluble in organic solvents.
IV. Phycobilins (phycocyanin, allophycocyanin and phycoerythrin) are soluble in water.

Which of the above statements is/are incorrect?

(a) II and III (b) III and IV (c) Only I (d) None of them

6 Read the following statements and select the correct ones.

I. PS-I is involved in non-cyclic photophosphorylation only.
II. PS-II is involved in both cyclic and non-cyclic photophosphorylation.
III. Stroma lamellae membranes possess PS-I only, whereas grana lamellae membranes possess both PS-I and PS-II.
IV. Mesophyll cells possess both RuBisCO and PEPCase enzyme.

Codes

(a) I and II
(b) II and III
(c) Only III
(d) I, II, III and IV

7 The diagram shows the ultrastructure of a chloroplast as seen in section. What are the function of P, Q and R?

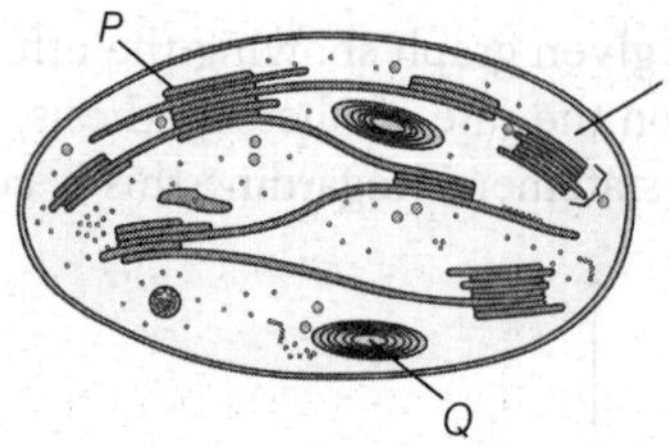

P	Q	R
(a) Carbohydrate storage	Carbohydrate synthesis	Light absorption
(b) Carbohydrate synthesis	Carbohydrate storage	Light absorption
(c) Carbohydrate synthesis	Light absorption	Carbohydrate storage
(d) Light absorption	Carbohydrate storage	Carbohydrate synthesis

8 Consider the following statements.

I. Priestly demonstrated that green plants purify the carbon dioxide air produce by breathing animals and burning candles.
II. Emerson and Arnold performed flashing light experiment of photosynthesis.
III. Ruben and Kamen confirmed that O_2 comes from water during photosynthesis.
IV. Robin Hill experimentally demonstrated that the source of O_2 is H_2O and not CO_2.

How many above statement is/are incorrect?

(a) One (b) Zero (c) Three (d) All of these

9 Study the following figure and select the incorrect statement.

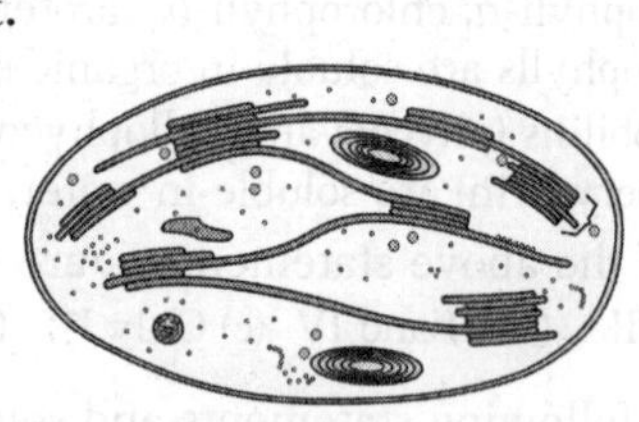

(a) It shows clear-cut division of labour
(b) It has membranous system for trapping sunlight
(c) Align themselves along the wall of mesophyll cells
(d) Has light-independent non-enzymatic reactions for CO_2 fixation

10 Match Column I with Column II and select the correct option from the codes given below.

Column I	Column II
A. C_4-plants	1. Succulents
B. Chlorophyll-b	2. Accessory photosynthetic pigment
C. PS-II	3. Photo-oxidation of H_2O
D. CAM	4. Kranz anatomy

Codes

	A	B	C	D
(a)	4	2	3	1
(b)	3	2	4	1
(c)	1	3	2	4
(d)	1	2	3	4

11 During light dependent reactions, light energy is converted to chemical potential energy through the process of chemiosmosis in the chloroplasts. Which of the following statements about this process is false ?

I. The electron carriers of phosphorylation are located in the thylakoid.
II. During phosphorylation, the chloroplast stroma becomes more acidic than the interior of thylakoid membrane.
III. Protons diffuse through the protein channels which are ATP synthetase molecules.
IV. ATP is formed from ADP + Pi on the stroma side of the thylakoid in the chloroplast.
V. During phosphorylation, water ionises to form $H^+ + 2OH^-$, yielding an e^- to PS-II.

(a) I, II and IV (b) Only II
(c) III and IV (d) Only IV

12 Consider the following statements regarding starch and sucrose synthesis during daytime and select the correct ones.

I. Triose phosphate is confined to chloroplast and is utilised for the synthesis of starch only.
II. Triose phosphate is translocated to cytosol from chloroplast.
III. Triose phosphate is utilised for the synthesis of both starch and sucrose.
IV. Triose phosphate is translocated from cytosol to chloroplast.

(a) I and III
(b) II and III
(c) II and IV
(d) III and IV

13 Match the following columns.

Column I (Saturation points)	Column II (Plant types)
A. 10% of full sunlight	1. Plant in shade or in dense forest
B. 50-70% of full sunlight	2. Most photosynthetic plants
C. 200% of full sunlight	3. Plant in dry tropical region

Codes

	A	B	C
(a)	3	2	1
(b)	3	1	2
(c)	1	3	2
(d)	1	2	3

14 Refer to the given cross section of a C_4 leaf and select the incorrect option.

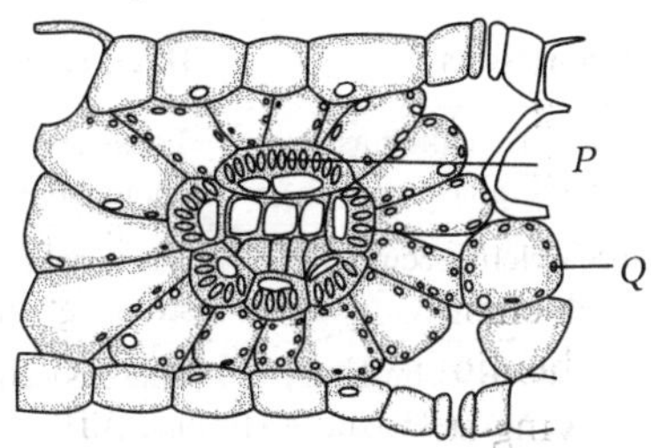

(a) *P* are the chloroplasts in which thylakoids are stacked together to form grana
(b) *P* are the chloroplasts which can perform light reaction, evolve molecular O_2 and produce assimilatory power
(c) *Q* are the chloroplasts in which thylakoids occur as stroma lamellae
(d) *Q* are the chloroplasts in which CO_2 is fixed by phosphoenol pyruvic acid to form oxaloacetic acid

15 Read the following four statements I, II, III and IV and select the correct option having both incorrect statements.

I. Antennae molecules are made up of hundreds of pigment molecules bound to proteins.
II. Antennae pigments help to make photosynthesis more efficient by absorbing different wavelength of light.
III. Only PS-II is function in cyclic photophosphorylation.
IV. In PS-I, the reaction centre is chlorophyll-*a*, which absorbs light of 680 nm.

(a) I and III (b) II and III (c) III and IV (d) II and IV

16 When two plants L and M were exposed to different light intensities and temperatures, they showed changes in their rates of photosynthesis, which have been represented in the following graph.

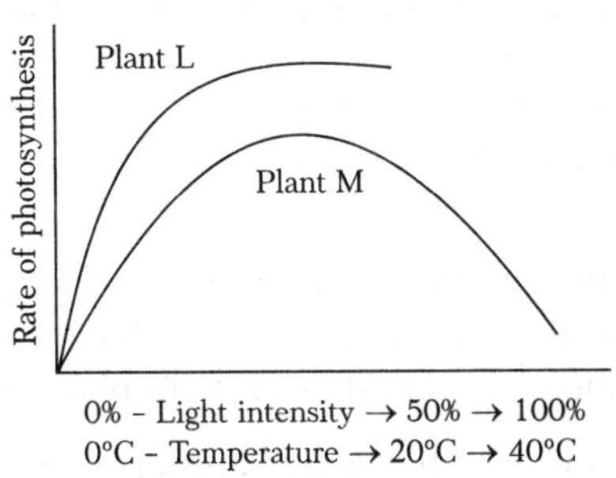

The graph indicates that
(a) plant L is a C_3-plant for which the light saturation point is 100% of full sunlight
(b) plant M is a C_4-plant for which the optimum temperatrue is around 20°C
(c) plant M is a C_3-plant which is more affected at higher temperature and higher light intensity as compared to plant L
(d) plant L is a C_4-plant and cannot function at light intensities above the saturation point

17 Read the following four statemetns (I-IV) on photosynthesis.

I. Photosynthesis is an oxidoreductive process.
II. The stroma of chloropolast is responsible for the synthesis of ATP and NADPH.
III. Membrane system of chloroplast is responsible for photochemical reaction.
IV. Photosynthesis does not take place in the ground tissue of leaves.

Select the right option having both incorrect statements.

(a) I and IV (b) II and IV (c) II and III (d) III and IV

18 Read the following statements and state them as true (T) and false (F).

I. A photosystem consists of a reaction centre, LHC, and an electron acceptor.
II. The final acceptor of electrons during non-cyclic flow of electron is $NADPH^+$.
III. The ring structure of chlorophyll consists of iron.
IV. The cytochrome system transports electron and pumps H^+ from stroma to thylakoid membrane.

Codes

	A	B	C	D		A	B	C	D
(a)	T	F	F	T	(b)	T	T	F	T
(c)	F	F	T	T	(d)	T	T	F	F

19 Match the following columns.

Column I	Column II
A. Pigment soluble in hot water	1. Chlorophyll-*a*
B. Primary pigment of chloroplasts	2. Phycobilins
C. Other light absorbing molecules	3. Thylakoid
D. Network of membranes within the chloroplasts	4. Accessory pigments

Codes

	A	B	C	D		A	B	C	D
(a)	1	2	3	4	(b)	2	1	4	3
(c)	4	2	1	3	(d)	3	1	2	4

20 The given figure represents the light harvesting complex.

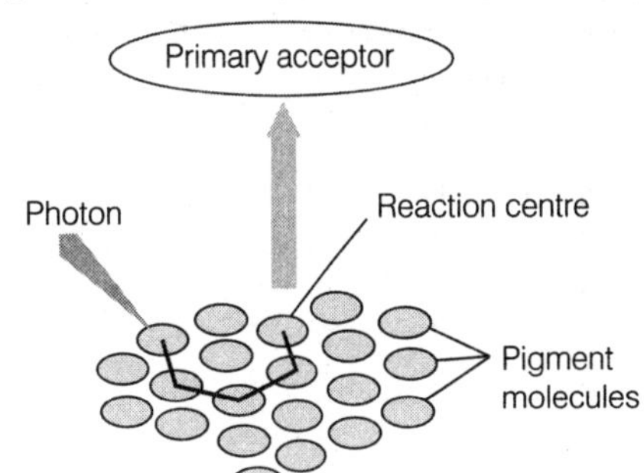

Select the incorrect statement with respect to LHC.
(a) Made up of hundreds of pigment molecules
(b) Consists of accessory pigments, including one molecule of chlorophyll-*a*
(c) Harvest different wavelengths of light
(d) Help to make photosynthesis more efficient

21 The following (I-VI) are the main steps of chemosynthetic ATP synthesis in the light reaction. Arrange them in correct order.
I. H^+ concentration gradient established.
II. H^+ diffuses through ATP synthase.
III. Carriers use energy from electrons to move H^+ across the membrane.
IV. Electrons from PS-II pass along electron transport chain.
V. Light excites electrons in PS-II
VI. Energy of H^+ flow is used by ATP synthase to make ATP.
(a) I, II, III, IV, V, VI (b) II, IV, V, III, II, VI
(c) V, IV, III, I, II, VI (d) V, VI, III, IV, II, I

22 Calvin cycle can be described under three stages. These stages are
I. Carboxylation of CO_2 into stable organic intermediate.
II. Ligation reactions leading to the formation of RuBisCO.
III. Reduction reactions leading to the formation of glucose.
IV. Regeneration of CO_2 acceptor molecule.
Choose the correct option.
(a) II, III and IV (b) I, III and IV
(c) I, II and IV (d) I, II and III

23 Match the following columns.

	Column I		Column II
A.	The 5-carbon sugar that fixes CO_2 in the Calvin cycle	1.	RuBisCO
B.	Three carbon sugar that gives Calvin cycle its nickname	2.	Glyceraldehyde 3-phosphate
C.	Activated form of 3-PGA	3.	3-phosphoglyceric acid
D.	Huge enzyme complex that brings CO_2 and a 5-carbon sugar together	4.	RuBP

Codes

	A	B	C	D
(a)	1	2	4	3
(b)	4	3	2	1
(c)	3	2	4	1
(d)	1	3	4	2

24 Following table summarises the differences between light reactions and dark reactions.

	Light reactions	Dark reactions
I.	These are also called as biosynthetic phase.	These are also called as photochemical phase.
II.	These reactions occur over thylakoids.	These reactions occur in stroma of chloroplasts.
III.	These produce assimilatory power, i.e, $NADPH_2$ and ATP.	These consume $NADPH_2$ and ATP.
IV.	These are directly dependent upon light.	These depend upon the products synthesised during light reactions.

Which of the above pairs of differences is/are incorrect?
(a) I and IV (b) III and IV
(c) Only II (d) Only I

25 Select the incorrect statement from the following.
(a) At high intensities, both C_3 and C_4-plants show increase in rate of photosynthesis by increasing CO_2 concentration
(b) C_4-plants show saturation at 360 ppm
(c) C_3-plants show saturation beyond 450 ppm
(d) Productivity of tomatoes and bell pepper cannot be increased by enriching environment by CO_2

Assertion and Reason

Direction (Q. Nos. 26-32) *In each of the following questions, a statement of Assertion (A) is given followed by corresponding statement of Reason (R). Of the statements, mark the correct answer as*
(a) Both A and R are true and R is the correct explanation of A
(b) Both A and R are true, but R is not the correct explanation of A
(c) A is true, but R is false
(d) A is false, but R is true
(e) Both A and R are false

26 **Assertion** (A) Photosynthesis is a redox process.
Reason (R) Oxidation of carbon dioxide and reduction of water takes place in photosynthesis.

27 **Assertion** (A) Chlorophyll appears green.
Reason (R) It absorbs light mainly in the region of green part of light spectrum.

28 **Assertion** (A) Red light of visible spectrum contains high energy.
Reason (R) Green light of visible spectrum contains low energy than red light.

29 **Assertion** (A) Carotenoids are accessory pigments.
Reason (R) They absorbed light energy which is transferred to reaction centre.

30 **Assertion** (A) Non-cyclic photophosphorylation occurs in the stroma of chloroplasts.
Reason (R) There is a continuous flow of electrons in this process.

31 **Assertion** (A) Carotenes and xanthophylls are soluble in ether.
Reason (R) These are accessory pigments of photosynthesis.

32 **Assertion** (A) Cyanobacteria perform oxygenic-photosynthesis.
Reason (R) Cyanobacteria possess chloroplasts.

(C) Medical Entrances' Gallery

(Collection of Questions Asked in NEET & in Various Medical Entrance Exams)

1 Phycoerythrin is the major pigment in **[NEET 2020]**
(a) red algae (b) blue-green algae
(c) green algae (d) brown algae

2 During non-cyclic photophosphorylation, when electrons are lost from the reaction centre at PS-II, what is the source which replaces these electrons?
(a) Oxygen (b) Water **[NEET 2020]**
(c) Carbon dioxide (d) Light

3 Which of the following statements is incorrect?
(a) RuBisCO is a bifunctional enzyme **[NEET 2020]**
(b) in C_4-plants, the site of RuBisCO activity is mesophyll cell
(c) The substrate molecule for RuBisCO activity is a 6-carbon compound
(d) RuBisCO action requires ATP and NADPH

4 The oxygenation activity of RuBisCO enzyme in photorespiration leads to the formation of **[NEET 2020]**
(a) 1 molecule of 3 C compound
(b) 1 molecule of 6 C compound
(c) 1 molecule of 4 C compound and 1 molecule of 2 C compound
(d) 2 molecules of 3 C compound

5 In light reaction, plastoquinone facilitates the transfer of electrons from **[NEET 2020]**
(a) Cyt-b_6f complex of PS-I (b) PS-I to $NADP^+$
(c) PS-I to ATP synthase (d) PS-II to Cyt-b_6f complex

6 Which is the site of C_3 cycle in C_3 and C_4-plants?
(a) In C_3-plants—Mesophyll cell and in C_4-plants—Bundle sheath cell **[AIIMS 2019]**
(b) In C_3 plants—Bundle sheath cell and in C_4-plants—Mesophyll cell
(c) In C_4 plants—Bundle sheath cell and in C_3-plants—Bundle sheath cell
(d) In C_3 plants—Mesophyll cell and in C_4-plants—Mesophyll cell

7 Kranz anatomy is found in **[JIPMER 2019]**
(a) C_3-plants (b) only in monocots
(c) Both C_4 and C_3-plants (d) C_4-plants

8 In Hatch and Slack pathway, the primary CO_2 acceptor is **[NEET 2019]**
(a) oxaloacetic acid (b) phosphoglyceric acid
(c) phosphoenol pyruvate (d) RuBisCO

9 Dark reaction **[JIPMER 2018]**
(a) in light
(b) in dark
(c) requires product of light reaction
(d) All of the above

10 Molecular formula of chlorophyll-*b* is **[JIPMER 2018]**
(a) $C_{55}H_{70}O_6N_4Mg$ (b) $C_{55}H_{72}O_5N_4Mg$
(c) $C_{55}H_{70}O_5N_4Mg$ (d) $C_{54}H_{70}O_6N_4Mg$

11 Which is incorrect with reference to chloroplast?
(a) Presence in algae and plants **[JIPMER 2018]**
(b) Releases O_2
(c) Occurs only in cells with aerobic respiration
(d) None of the above

12 Oxygen is not produced during photosynthesis by **[NEET 2018]**
(a) *Cycas* (b) *Nostoc*
(c) Green sulphur bacteria (d) *Chara*

13 Which of the following is not a product of light reaction of photosynthesis? **[NEET 2018]**
(a) NADPH (b) NADH
(c) ATP (d) Oxygen

14 Conditions required for cyclic photophosphorylation are
(a) aerobic condition, low light intensity **[JIPMER 2018]**
(b) aerobic condition, optimum light intensity
(c) anaerobic condition, low light intensity
(d) anaerobic condition, optimum light intensity

15 C_4-pathway is advantageous over C_3-pathway in plants, because it **[AIIMS 2018]**
(a) occurs in relatively low CO_2 concentration
(b) uses more amount of water
(c) occurs in relatively low O_2 concentration
(d) is less efficient in energy utilisation

16 Which of these is incorrect for C_4-plants? **[JIPMER 2018]**

(a) Kranz anatomy (b) CO_2 acceptor is PEP
(c) PEPCase in mesophyll (d) RuBisCO in mesophyll

17 A plant is provided with ideal conditions for photosynthesis and supplied with isotope $^{14}CO_2$. When the products of the process are analysed carefully, what would be the nature of products? **[JIPMER 2017]**

(a) Glucose and oxygen are labelled
(b) Oxygen is labelled but glucose is normal
(c) Glucose and oxygen are normal
(d) Glucose is labelled but oxygen is normal

18 Herbicide that blocks electron transport from PS-II to PS-I by inhibiting electron flow between plastoquinone → cytochrome is **[JIPMER 2017]**

(a) DCMU (b) Paraquat
(c) DCPIP (d) None of these

19 Phosphoenol Pyruvate (PEP) is the primary CO_2 acceptor in **[NEET 2017]**

(a) C -plants (b) C -plants
(c) C -plants (d) C and C -plants

20 Fixation of one CO_2 molecule through Calvin cycle requires **[JIPMER 2017]**

(a) 1 ATP and $2NADPH_2$ (b) 2 ATP and $2NADPH_2$
(c) 3ATP and $2NADPH_2$ (d) 2ATP and $1NADPH_2$

21 With reference to factors affecting the rate of photosynthesis, which of the following statements is not correct? **[NEET 2017]**

(a) Light saturation for CO_2-fixation occurs at 10% of full sunlight
(b) Increasing atmospheric CO_2 concentration up to 0.05% can enhance CO_2-fixation rate
(c) C_3-plants respond to higher temperature with enhanced photosynthesis, while C_4-plants have much lower temperature optimum
(d) Tomato is a greenhouse crop, which can be grown in CO_2 enriched atmosphere for higher yield

22 Photosynthesis in C_4-plants is relatively less limited by atmospheric CO_2 levels because **[AIIMS 2016]**

(a) there is effective pumping of CO_2 into bundle sheath cell
(b) RuBisCO in C_4-plants has higher affinity for CO_2
(c) six carbon acid are the primary initial CO_2 fixation products
(d) the primary fixation of CO_2 is mediated *via* PEP carboxylase

23 The process which makes major difference between C_3 and C_4-plants is **[NEET II 2016]**

(a) glycolysis (b) Calvin cycle
(c) photorespiration (d) respiration

24 Emerson's enhancement effect and red drop have been instrumental in the discovery of **[NEET I 2016]**

(a) two photosystems operating simultaneously
(b) photophosphorylation and cyclic electron transport
(c) oxidative phosphorylation
(d) photophosphorylation and non-cyclic electron transport

25 In a chloroplast, the highest number of protons are found in **[NEET I 2016]**

(a) lumen of thylakoids (b) intermembrane space
(c) antennae complex (d) stroma

26 CO_2 is necessary for photosynthesis. The chemical used to remove this gas most effectively from entering a control apparatus is **[AIIMS 2015]**

(a) calcium oxide (b) potassium hydroxide solution
(c) distilled water (d) sodium carbonate

27 Correct sequence of involvement of the following in non-cyclic electron transport **[EAMCET 2015]**

I. PC II. PQ III. Pheo IV. Fd

(a) II, I, III and IV (b) III, II, IV and I
(c) IV, I, II and III (d) III, II, I and IV

28 Choose the correct statement. **[EAMCET 2015]**

(a) Stroma lamella membrane lacks PS-II and PS-I
(b) When PS-I is functional electrons flow in an non-cyclic way
(c) ATPase are enzyme consists of F_0 and F_1 units
(d) NADP reductase is a part of PS-II

29 The enzyme responsible for primary carboxylation in C_3-plants is **[J&K CET 2015]**

(a) pyruvate carboxlases
(b) succinic dehydrogenase
(c) hexokinase
(d) RuBP carboxylase/oxygenase

30 In C_3-plants, the first stable compound formed after CO_2 fixation is **[J&K CET 2015]**

(a) oxaloacetic acid (b) malic acid
(c) phosphoglyceraldehyde (d) 3-phosphoglycerate

31 Which of the elements is found in chlorophyll? **[CG PMT 2015, NCERT Exemplar]**

(a) Manganese (b) Potassium
(c) Iron (d) Magnesium

32 What is the quantum yield of photosynthesis? **[CG PMT 2015]**

(a) 33% (b) 9%
(c) 12% (d) 78%

33 The structures that are formed by stacking of organised flattened membranous sacs in the chloroplasts are **[CBSE AIPMT 2015]**

(a) cristae (b) grana
(c) stroma lamellae (d) stroma

34 The oxygen evolved during photosynthesis comes from water molecules. Which one of the following pairs of elements are involved in this reaction? **[CBSE AIPMT 2015]**
(a) Manganese and chlorine
(b) Manganese and potassium
(c) Magnesium and molybdenum
(d) Magnesium and chlorine

35 In photosynthesis, the light-independent reactions take place at **[CBSE AIPMT 2015]**
(a) thylakoid lumen (b) photosystem-I
(c) photosystem-II (d) stromal matrix

36 ATP synthesis in cell requires **[WB JEE 2015]**
(a) H^+ gradient across the membrane
(b) K^+ gradient across the membrane
(c) PO_4^{3-} gradient across the membrane
(d) Ca^{2+} gradient across the membrane

37

$$4H_2O \longrightarrow 4H^+ + 4OH^-$$

In this process, which of the following play important role? **[Gujarat CET 2015]**
(a) Chlorophyll (b) Light energy
(c) Ca^{2+}, Mn^{2+}, Cl^+ (d) All of these

38 During photorespiration, which compounds are formed having 2C and 3C respectively in peroxisome? **[Gujarat CET 2015]**
(a) Phosphoglycerate and glycolate
(b) Glycine and glycerate
(c) Serine and glycine
(d) Glycolate and glycine

39 How many molecules of ATP and NADPH are required in the formation of two molecules of glucose? How many Calvin cycles are required? **[Gujarat CET 2015]**
(a) 24 ATP, 36 NADPH, 12 Calvin cycles
(b) 18 ATP, 12 NADPH, 6 Calvin cycles
(c) 36 ATP, 24 NADPH, 6 Calvin cycles
(d) 36 ATP, 24 NADPH, 12 Calvin cycles

40 An example of CAM plant is **[AMU 2015]**
(a) black night shade *(Solanum nigrum)*
(b) lemon grass *(Cymbopogon flexuosus)*
(c) sugarbeet *(Beta vulgaris)*
(d) snake plant *(Sansevieria trifasciata)*

41 Photorespiration requires this activity by an enzyme **[BCECE 2015]**
(a) hydrolase (b) oxygenase
(c) carboxylase (d) ATPase

42 This is not a C_3-plant **[BCECE 2015]**
(a) *Amaranthus* (b) rice
(c) wheat (d) potato

43 Visible portion of light spectrum useful in photosynthesis is referred to as **[MHT CET 2015]**
(a) RFLP (b) PAR
(c) VAM (d) VNTR

44 Cyclic photophosphorylation results in the formation of **[Kerala CEE 2015]**
(a) NADPH + H^+ (b) ATP and NADPH + H^+
(c) ADP (d) ATP
(e) ADP and $NADP^+$

45 The strong inhibitor substance for PS-II in photosynthesis is **[UK PMT 2015]**
(a) ethylene
(b) chloroform
(c) dichlorophenyl dimethyl urea
(d) yellow light

46 How many molecules of ATP are required during the formation of one molecule of glucose? **[UK PMT 2015]**
(a) 8 (b) 18
(c) 28 (d) 2

47 Oxygen released in photosynthesis is formed during **[UK PMT 2015]**
(a) oxidative phosphorylation
(b) cyclic photophosphorylation
(c) non-cyclic photophosphorylation
(d) carbon assimilation during dark reactions

48 Anoxygenic photosynthesis is characteristic of **[CBSE AIPMT 2014]**
(a) *Rhodospirillum* (b) *Spirogyra*
(c) *Chlamydomonas* (d) *Ulva*

49 Identify the incorrect statement(s) in relation to C_4-photosynthesis. **[WB JEE 2014]**
(a) Kranz anatomy is an essential feature for C_4-plants
(b) C_4-plants have higher water use efficiency than C_3-plants
(c) Photorespiration can be minimised when C_4-pathway is in operation
(d) Conversion of oxaloacetate to malate occurs in the bundle sheath cells

50 Find out the mismatched pair. **[Kerala CEE 2014]**

(a) C_4-plants	Kranz anatomy
(b) Primary CO_2 fixation product of C_4-plants	OAA
(c) Primary CO_2 acceptor of C_3-plants	RuBP
(d) Calvin pathway of C_4-plants occurs in	Bundle sheath
(e) C_3-plants	Maize

51 C_4-plants have bundle sheath cells which possess [J&K CET 2014]
(a) few chloroplasts with thin walls so that gaseous exchange can take place
(b) large number of chloroplasts with thick walls impervious to gaseous exchange
(c) large number of chloroplasts with thick walls and no intercellular spaces
(d) None of the above

52 Which of the following statements regarding cyclic flow of electrons during light reactions is false? [Kerala CEE 2014]
(a) This process takes place in the stromal lamellae
(b) ATP synthesis takes place
(c) NADPH + H^+ is synthesised
(d) Takes place only when light of wavelength beyond 680 nm is available for excitation
(e) PS-II is not involved in the process

53 Photosynthetic Active Radiation (PAR) has the following range of wavelengths [UK PMT 2014]
(a) 340-450 nm (b) 400-700 nm
(c) 500-600 nm (d) 450-950 nm

54 Synthesis of one glucose molecule requires .... reduced NADP molecules. [MHT CET 2014]
(a) 6 (b) 12 (c) 18 (d) 24

55 Enzymes required for phosphorylation are located in .... of chloroplast. [MHT CET 2014]
(a) peristomium (b) plastidome
(c) stroma (d) quantasome

56 C_4-photosynthetic system is present in plants which are found in [UK PMT 2014]
(a) cold region (b) hot region
(c) Both (a) and (b) (d) dry tropical region

57 Which element plays a vital role in splitting of water to liberate oxygen during photosynthesis? [UK PMT 2014]
(a) Copper (b) Boron (c) Chlorine (d) Manganese

58 The electron from photosystem-II goes to photosystem-I and finally taken up by NADP leaving an electron hole in photosystem-II. This electron hole in photosystem-II is filled up by the electron released from [AMU 2013]
(a) CO_2 (b) H_2O (c) chlorophyll (d) light

59 In C_4-plants, the first stable product of photosynthesis during the dark reaction is [NEET 2013]
(a) PGAL (b) RuBP (c) PGA (d) OAA

60 In C_4-plants, the carbon dioxide fixation occurs in [Manipal 2013]
(a) guard cells (b) spongy cells
(c) palisade cells (d) bundle sheath cells

61 CAM pathway is observed in [KCET 2013]
(a) pineapple (b) maize (c) sunflower (d) sugarcane

62 In C_4-pathway, the CO_2 fixation in mesophyll cells is carried out by the enzyme [KCET 2012; AFMC 2013]
(a) pyruvate dehydrogenase (b) pyruvate decarboxylase
(c) PEP carboxylase (d) RuBisCO

63 Which one of the following statements about the events of non-cyclic photophosphorylation is not correct? [KCET 2013]
(a) Photolysis of water takes place
(b) Oxygen is released
(c) Only one photosystem participates
(d) ATP and NADPH are produced

64 Photosynthetic bacteria have [AIIMS 2013]
(a) pigment system-I
(b) pigment system-II
(c) Both (a) and (b)
(d) some other kind of pigment, P_{890}

65 Cyclic photophosphorylation links to [OJEE 2013]
(a) PS-II (b) PS-I
(c) Both (a) and (b) (d) dark reaction

66 The correct sequence of cell organelles during photorespiration is [CBSE AIPMT 2012]
(a) Chloroplast – Golgi bodies–Mitochondria
(b) Chloroplast – Rough endoplasmic reticulum–Dictyosomes
(c) Chloroplast – Peroxisome–Mitochondria
(d) Chloroplast – Vacuole–Peroxisome

67 The essential element needed for water splitting in photosynthesis leading to O_2 evolution is [AMU 2012]
(a) Mo (b) Mn (c) Mg (d) K

68 In the overall process of photosynthesis, the number of CO_2, water, sugar and O_2 molecules utilised and produced is [AMU 2012]
(a) 12 (b) 13 (c) 19 (d) 31

69 During Calvin cycle, the total number of CO_2, ATP and NADPH molecules utilised and glucose, ADP and NADP generated is [AMU 2012]
(a) 31 (b) 38 (c) 61 (d) 67

70 Which one occurs during both cyclic and non-cyclic modes of photophosphorylation? [KCET 2012]
(a) Involvement of both PS-I and PS-II
(b) Formation of ATP
(c) Release of O_2
(d) Formation of NADPH

71 In which cell of leaf, pyruvate is converted to PEP in C_4-pathway? [AIIMS 2012]
(a) Epidermal cell (b) Mesophyll cell
(c) Bundle sheath cell (d) Guard cell

72 CAM helps the plants in [CBSE AIPMT 2011]
(a) secondary growth (b) disease resistance
(c) reproduction (d) conserving water

73 The presence of bundle sheath is a characteristic of [DUMET 2011]
(a) xerophytic plants (b) members of grass family
(c) C_4-plants (d) C_3-plants

74 In C_4-plants, the bundle sheath cells [DUMET 2011]
(a) have thin walls to facilitate gaseous exchange
(b) have large intercellular spaces
(c) are rich in PEP carboxylase
(d) have a high density of chloroplasts

75 Very strong light has a direct inhibiting effect on photosynthesis, which is known as [UP CPMT 2011]
(a) solarisation (b) etiolation
(c) chlorosis (d) defoliation

76 Plastocyanin contains [UP CPMT 2011]
(a) copper (b) iron
(c) calcium (d) potassium

77 Organelles associated with photorespiration are [OJEE 2011]
(a) chloroplast, mitochondria, peroxisome
(b) chloroplast, mitochondria, lysosome
(c) mitochondria, peroxisome, centrosome
(d) nucleus, centrosome, peroxisome

78 Light reaction of photosynthesis occurs inside [J&K CET 2011]
(a) stroma (b) grana
(c) endoplasmic reticulum (d) cytoplasm

79 Sunken stomata are usually found in [J&K CET 2011]
(a) C_3-plants (b) CAM plants
(c) insectivorous plants (d) phanerogams

80 A reduction in the quantity of oxygen evolution during photosynthesis may be observed at [J&K CET 2011]
(a) light having wavelength more than 680 nm
(b) light having wavelength less than 680 nm
(c) light having wavelength 560 nm
(d) light having wavelength less than 360 nm

81 Which one of the following statements is true with regard to the light reaction of photosynthesis? [Kerala CEE 2011]
(a) In PS-II, the reaction centre chlorophyll-*a* has an absorption peak at 700 nm and is called P_{700}
(b) In PS-I, the reaction centre chlorophyll-*a* has an absorption maximum at 680 nm and is called P_{680}
(c) The splitting of water molecule is associated with PS-I
(d) Photosystem-I and II are involved in Z-scheme
(e) Lamellae of the grana have PS-I and PS-II and stroma lamellae membranes

82 Select the incorrectly matched pair with regard to C_4-cycle. [Kerala CEE 2011]
(a) Primary CO_2 fixation — PGA product
(b) Site of initial carboxylation — Mesophyll cells
(c) Primary CO_2 acceptor — PEP
(d) C_4-plant — Maize
(e) Location of enzyme RuBisCO — Bundle sheath cells

83 In C_3-cycle for the fixation of every CO_2 molecules, the reduction and regeneration steps required are [Kerala CEE 2011]
(a) 3 ATP and 2 $NADPH_2$
(b) 2 ATP and 2 $NADPH_2$
(c) 2 ATP and 3 $NADPH_2$
(d) 3 ATP and 3 $NADPH_2$
(e) 3 ATP and 1 $NADPH_2$

84 Which of the following is formed during photorespiration? [Kerala CEE 2011]
(a) Sugar (b) Phosphoglycolate
(c) NADPH (d) ATP
(e) Oxaloacetate

85 Which of the following is true for photosynthesis? [Gujarat CET 2011]
(a) Reduction of CO_2 and water
(b) Oxidation of CO_2 and water
(c) Reduction of CO_2 and oxidation of water
(d) Oxidation of CO_2 and reduction of water

86 Which of the following is not related to photorespiration? [Gujarat CET 2011]
(a) Lysosome (b) Chloroplast
(c) Peroxisome (d) Mitochondria

87 With reference to three Calvin cycle, which of the given option is correct for the following question
I. How many gross PGAL molecules are produced?
II. Total, how many ATP molecules are required for synthesis of PGAL molecules?
III. Total, how many $NADPH_2$ molecules are required for the synthesis of obtained PGAL molecules? [Gujarat CET 2011]
(a) I-3PGL, II-3ATP, III-3 $NADPH_2$
(b) I-6 PGAL, II-6 ATP, III-6 $NADPH_2$
(c) I-18 PGAL, II-18 ATP, III-18 $NADPH_2$
(d) I-9 PGAL, II-9 ATP, III-9 $NADPH_2$

88 Energy released during movement of electrons through photosystems in photosynthesis is used to drive protons across the membrane against concentration gradient. As a result the protons accumulate in [AMU 2011]
(a) thylakoid lumen (b) stroma
(c) intrathylakoid space (d) stromal lamella

ANSWERS

Check Point 20.1

1. (b)	2. (c)	3. (b)	4. (d)	5. (d)	6. (b)

Check Point 20.2

1. (d)	2. (a)	3. (a)	4. (a)	5. (b)	6. (c)

Check Point 20.3

1. (d)	2. (b)	3. (c)	4. (d)	5. (c)

(A) Tropical Exercise

1. (a)	2. (b)	3. (c)	4. (a)	5. (d)	6. (b)	7. (b)	8. (b)	9. (c)	10. (b)
11. (d)	12. (d)	13. (b)	14. (a)	15. (b)	16. (d)	17. (d)	18. (a)	19. (a)	20. (d)
21. (c)	22. (d)	23. (b)	24. (b)	25. (d)	26. (c)	27. (b)	28. (d)	29. (a)	30. (d)
31. (c)	32. (a)	33. (d)	34. (d)	35. (d)	36. (b)	37. (d)	38. (c)	39. (d)	40. (d)
41. (a)	42. (a)	43. (b)	44. (a)	45. (d)	46. (d)	47. (a)	48. (d)	49. (c)	50. (c)
51. (a)	52. (a)	53. (d)	54. (c)	55. (b)	56. (a)	57. (a)	58. (c)	59. (a)	60. (c)
61. (c)	62. (b)	63. (c)	64. (d)	65. (b)	66. (b)	67. (d)	68. (d)	69. (c)	70. (c)
71. (d)	72. (b)	73. (c)	74. (a)	75. (b)	76. (c)	77. (d)	78. (b)	79. (c)	80. (b)
81. (c)	82. (a)	83. (a)	84. (b)						

(B) Cumulative Exercise

1. (d)	2. (a)	3. (a)	4. (d)	5. (d)	6. (c)	7. (d)	8. (b)	9. (d)	10. (a)
11. (b)	12. (b)	13. (d)	14. (d)	15. (c)	16. (c)	17. (b)	18. (b)	19. (b)	20. (b)
21. (c)	22. (b)	23. (b)	24. (d)	25. (d)	26. (c)	27. (c)	28. (e)	29. (a)	30. (d)
31. (b)	32. (c)								

(C) Medical Entrances' Gallery

1. (a)	2. (b)	3. (b)	4. (a)	5. (d)	6. (a)	7. (d)	8. (c)	9. (c)	10. (a)
11. (c)	12. (c)	13. (b)	14. (b)	15. (a)	16. (d)	17. (d)	18. (a)	19. (b)	20. (c)
21. (c)	22. (d)	23. (c)	24. (a)	25. (a)	26. (b)	27. (d)	28. (c)	29. (d)	30. (d)
31. (d)	32. (c)	33. (b)	34. (a)	35. (d)	36. (a)	37. (d)	38. (b)	39. (d)	40. (d)
41. (b)	42. (a)	43. (b)	44. (d)	45. (c)	46. (b)	47. (c)	48. (a)	49. (d)	50. (e)
51. (b)	52. (c)	53. (b)	54. (b)	55. (d)	56. (d)	57. (d)	58. (b)	59. (d)	60. (d)
61. (a)	62. (c)	63. (c)	64. (d)	65. (b)	66. (c)	67. (b)	68. (d)	69. (d)	70. (b)
71. (b)	72. (d)	73. (c)	74. (d)	75. (a)	76. (a)	77. (a)	78. (b)	79. (b)	80. (a)
81. (d)	82. (a)	83. (a)	84. (b)	85. (c)	86. (a)	87. (b)	88. (a,c)		

Explanations

(A) Topical Exercise

1 (*a*) Statement in option (a) is correct. Other statements are incorrect and can be corrected as

Photosynthesis is an anabolic, reductive and endergonic process.

Of the total world photosynthesis, 90% is performed by algae.

3 (*c*) Statement in option (c) is incorrect and can be corrected as

Bacteria accumulated mainly in the region of blue and red light of the split spectrum. Rest statements are correct about the experiment performed by TW Engelmann.

8 (*b*) Statement in option (b) is false and can be corrected as

A stack of thylakoids in chloroplast form a grana.

Rest statements are true about structure of chloroplast.

10 (*b*) Correct pair is in Option (b) is correct pair

Other options are incorrect pair and can be corrected as

Stroma – Dark reaction

Thylakoids – Synthesis of ATP

Stromal lamellae – Contains PS-I and chlorophyll

29 (*a*) Statement in option (a) is false and can be corrected as

PS-I is situated on the outer surface of thylakoid membrane and also on stroma lamellae, while PS-II is situated on the inner surface of thylakoid membrane.

Rest other statements are correct about the light reactions of photosynthesis.

30 (*d*) Statement in option (d) is not true and can be corrected as
P_{700} and P_{680} are the reaction centre of PS-I and PS II, respectively.
Rest other statements are correct.

31 (*c*) In the light dependent reaction, photolysis of water occurs to release electrons, hydrogen ions and oxygen. In the light independent reaction, CO_2 is fixed, producing carbohydrate.

38 (*c*) Statement in option (c) is incorrect and can be corrected as
Only ATP is synthesised in cyclic flow of electrons during light reaction not NADPH + H^+.

39 (*d*) Statement in option (d) is correct. Rest statements are incorrect and can be corrected as
- Cyclic photophosphorylation needs PS-I.
- Cyclic photophosphorylation produces ATP only.
- Cyclic photophosphorylation does not involve H_2O.

46 (*d*) Statement in option (d) is incorrect and can be corrected as
Photosystem-I is associated with $NADP^+$ reduction.
Rest statements are correct.

50 (*c*) Statement in option (c) is incorrect and can be corrected as
Photolysis is involved in chemiosmotic synthesis of ATP.

63 (*c*) Statement in option (c) is incorrect and can be corrected as C_4-plants show Kranz anatomy, but photorespiration in them is minimum, i.e. less than C_3-plants.

67 (*d*) Statement in option (d) is incorrect and can be corrected as Chloroplast of mesophyll cells possesses well-developed grana lamellae in Kranz anatomy.

70 (*c*) Both statements I and II are true because
Crassulacean acid metabolism is a mechanism of photosynthesis involving double fixation of CO_2 which occurs in succulents belonging to *Crassulaceae*, cacti, euphorbias and some other plants of dry habits where the stomata remain closed during the daytime and open only at night. Sunken stomata are deep seated stomata in which subsidiary cells lie above the guard cells.

78 (*b*) The correct option is (b) and can be corrected as
Rest other options are incorrect
Rate of photosynthesis is highest in red light and increases in the presence of cytokinins and gibberellins.

(B) Cumulative Exercise

3 (*a*) Only one statement IV is incorrect and can be corrected as Green plants carryout photosynthesis by which they use light energy to form organic compounds.
Rest four statements are correct.

5 (*d*) In the given question, all the statements are correct and none of them are incorrect.

6 (*c*) Only statement III is correct. Other statements are incorrect and can be corrected as
- PS-I is involved in both cyclic and non-cyclic photophosphorylation.
- PS-II is involved only in non-cyclic photophosphorylation.
- PS-II is present in the appressed (inner) part of grana thylakoids.
- PS-I is located in the non-appressed (outer) part of grana thylakoids as well as stroma thylakoids.
- RuBisCO enzyme is absent in mesophyll cells.

11 (*b*) Only statement II is incorrect and can be corrected as
During phosphorylation, the chloroplast stroma becomes less acidic than the interior of thylakoid membrane.
Rest other statements are correct.

12 (*b*) Statements II and III are correct. Other statements are incorrect and can be corrected as
Glyceraldehyde-3-phosphate (G3P) or 3-phosphoglyceraldehyde (PGAL), also known as triose phosphate, is a chemical compound that occurs as an intermediate in several central metabolic pathways of all organisms. It is an intermediate in photosynthesis. A part of triose phosphate is translocated to cytosol from chloroplast and is utilised for the synthesis of glucose, sucrose, starch, etc.

15 (*c*) Statements III and IV are incorrect and can be corrected as
- Only PS-I functions in cyclic photophosphorylation.
- In PS-II, the reaction centre is chlorophyll-*a*.

Rest statements are correct.

17 (*b*) Statements II and IV are incorrect. These can be corrected as
- The membrane system of chloroplast is responsible for the synthesis of ATP and NADPH.
- Photosynthesis takes place in the ground tissue of leaves.

18 (*b*) Only statement III is false and can be corrected as
The ring structure of chlorophyll consists of magnesium not iron.
Rest statements are true.

20 (*b*) Only statement in option (b) is incorrect and can be corrected as
The light Harvesting complex does not consist of chlorophyll-*a*. It includes all accessory pigments like chlorophyll-*b*, carotenoids and xanthophyll. The chlorophyll-*a* forms the reaction centre.
Rest statements are correct with respect to LHC.

21 (*c*) The correct order of steps in chemosynthetic ATP synthesis is as follows
- Light excites electrons in PS-II.
- Electrons from PS-II pass along electron to move H^+ across the membranes.
- Carriers use energy from electrons to move H^+ across the membrane.
- H^+ concentration gradient is established.
- H^+ diffuses through ATP synthase.
- Energy of H^+ flow is used by ATP synthase to make ATP.

25 (*d*) Statement in option (d) is incorrect and can be corrected as Growing crops in greenhouse produce higher yields. This means productivity of tomatoes and bell pepper can be increased by enriching environment by CO_2.

26 (*c*) A is true, but R is false. Reason can be corrected as
Photosynthesis is a redox process. It includes reduction of carbon dioxide to organic carbon compounds such as sugar and oxidation of water to oxygen.

27 (*c*) A is true, but R is false. Reason can be corrected as
Chlorophyll molecules appear green to our eyes because they absorb visible light mainly in the region of red and blue parts of light spectrum, so some of the light enriched in green wavelength, (i.e. about 550 nm) is reflected into our eyes.

28 (*e*) Both A and R are false and can be corrected as

Energy is opposite to the wavelength of visible light. So, red light of visible spectrum contains low energy, while blue light contains high energy. Due to this reason, absorption of blue light excites the chlorophyll molecules to a high energy state than the absorption of red light.

29 (*a*) Both A and R are true and R is the correct explanation of A.

Carotenoids are integral constituents of the thylakoid membrane and are found in all photosynthetic organisms. They absorb light energy in the mid region of visible spectrum. These are known as accessory pigments, because the light absorbed by them is transferred to chlorophyll for photosynthesis.

30 (*d*) A is false, but R is true. Assertion can be corrected as

Non-cyclic photophosphorylation occurs in granum portion of chloroplast. It is an association of photosystem-I and II. In this process, electrons continuously flow from water to PS-II to PS-I and to final electron acceptor, i.e. NADP. In this process, ATP is formed from ADP and O_2 is also evolved.

31 (*b*) Both A and R are true, but R is not the correct explanation of A.

Carotenoid is a group of carotenes and xanthophyll pigments. These are soluble in organic solvents like ether, chloroform, etc. During photosynthesis, these pigments absorb light energy and transfer it to reaction centre. Thus, the carotenoids act as an accessory pigment in photosynthesis.

32 (*c*) A is true, but R is false. Reason can be corrected as

Cyanobacteria or blue-green algae are prokaryotic organisms, but they perform oxygenic photosynthesis, i.e. they evolve oxygen during photosynthesis. They do not possess chloroplasts as found in photosynthesising eukaryotes.

(C) Medical Entrances' Gallery

1 (*a*) Phycoerythrin is the major pigment in red algae or rhodophytes. The photosynthetic pigments in this include chlorophyll-*a*, carotenoids and phycobilins. Phycoerythrin belongs to phycobilins. These pigments are soluble in water.

2 (*b*) During non-cyclic photophosphorylation, electrons expelled by the excited PS-II photocentre does not return to it. Therefore, it requires an external electron donor and that purpose is served by water.

H_2O undergoes photolysis and the electrons thus released are picked up by PS-II or P_{680} and handed over to PS-I or P_{700}.

3 (*b*) Statement in option (b) is incorrect and can be corrected as In C_4-plants, Kranz anatomy in leaf is found due to the presence of two type of cells *viz.* mesophyll cells and bundle sheath cells. The mesophyll cells are specialised to perform light reaction, evolve O_2 and produce assimilatory power. The bundle sheath cells possess RuBisCO and thus, perform RuBisCO activity at this site.

4 (*a*) The oxygenation activity of RuBisCO enzyme in photorespiration leads to the formation of 1 molecule of 3 C compound (phosphoglycerate). In C_3-plants during oxygen fixation, one molecule of PGA (3C) and one molecule of 2-phosphoglycolate (2C) are formed. The latter is then converted back to PGA in the photorespiratory cycle. Photorespiration occurs at high concentration of oxygen and temperature in the environment.

5 (*d*) In light reaction, plastoquinone facilitates the transfer of electrons from PS-II to cytochrome b_6f complex (non-cyclic photophosphorylation) process of light reaction starts with PS-II (680 nm).

6 (*a*) In C_3-plants, all the reactions of C_3 or Calvin cycle occur in mesophyll cells of leaves. In C_4-plants, Calvin cycle does not occur in the mesophyll cells. It takes place only in the bundle sheath cells.

7 (*d*) Kranz anatomy is found in C_4-plants. It refers to the presence of two types of chloroplasts, i.e. agranal in bundle sheath cells and granal in mesophyll cells, e.g. sugarcane, maize, etc. In these plants, the first stable product after the fixation of CO_2 is 4 C dicarboxylic acid called Oxaloacetic Acid (OAA).

8 (*c*) In Hatch and Slack pathway, the primary CO_2 acceptor is phosphoenol pyruvate. This occurs in C_4-plants. Phosphoenol pyruvate, a 3-carbon compound accepts CO_2 and forms oxaloacetic acid which is a 4-carbon compound.

9 (*c*) None of the option in this question is correct but option (c) seems to be close to be the correct answer. It is because in dark reaction, the products formed from light reaction are required. These are $NADPH_2$ and ATP molecules which are collectively called reducing power.

10 (*a*) The molecular formula of chlorophyll-*b* is $C_{55}H_{70}O_6N_4Mg$. It helps in photosynthesis by absorbing light energy.

It is more soluble in polar solvents than chlorophyll-*a* because of its carbonyl group. It is of yellow colour and primarily absorbs blue light.

11 (*c*) Option (c) is incorrect about chloroplasts. It can be corrected as

They are not associated with any type of respiration, i.e. aerobic and anaerobic and can be found in any type of cell. They are spherical, ovoid or disc-shaped cell organelles. They are involved in photosynthesis due to the presence of chlorophyll in them. Rest of the statements are correct.

12 (*c*) Green sulphur bacteria are anaerobic bacteria. They do not evolve oxygen during photosynthesis. Such type of photosynthesis is known as anoxygenic photosynthesis. They do not use water as a source of reducing power.

Instead, hydrogen is obtained from hydrogen sulphide.

$$\underset{\text{Hydrogen sulphide}}{H_2S} \longrightarrow 2[H] + S$$

$$6CO_2 + 12H_2 \xrightarrow[\text{Enzymes}]{\text{Light}} C_6H_{12}O_6 + 6H_2O$$

13 (*b*) During light reaction of photosynthesis, NADPH, ATP and oxygen are formed. Oxygen is liberated by the photolysis of water. The electrons released during photolysis of water are picked up by P_{680} photocentre of PS-II.

On receiving light energy, photocentre expels an electron which passes over a series of carriers. As a result assimilatory power, i.e. ATP and NADPH are produced. NADH is formed during respiration.

14 (*b*) The conditions required for cyclic photophosphorylation are aerobic environment and optimum light intensity. It is one of the types of photophosphorylation in which an electron expelled by the excited photocentre passes through a series of electron carriers and returns back to its original place. Cyclic photophosphorylation is carried out in photosystem-I only.

15 (*a*) A plant with the C_4-pathway is advantageous over a plant with a C_3-pathway as the former can occur in low CO_2 concentration.

When the weather is hot and arid, the leaves close their stomata to reduce water loss. This closure results in low CO_2 concentrations and thus, leads to both loss of RuBP and inhibition of Calvin-Benson cycle.

16 (*d*) Among the given options, option (d) is incorrect about C_4-plants. It can be corrected as

The mesophyll cells of C_4-plants do not possess RuBisCO enzyme. C_4-plants show Kranz anatomy.

Rest options are correct for C_4-plants.

In these plants, the initial fixation of CO_2 occurs in mesophyll cells.

The primary acceptor of CO_2 is Phosphoenol Pyruvate or PEP. It combines with CO_2 in the presence of PEP carboxylase or PEPCase to form oxaloacetic acid or oxaloacetate.

17 (*d*) In the process of photosynthesis,

$$6CO_2 + 12H_2O \xrightarrow[\text{Chlorophyll}]{h\nu} C_6H_{12}O_6 + 6H_2O + 6O_2 \uparrow$$

Plants manufacture glucose from CO_2 and oxygen gas is releases from H_2O. So, due to isotopic C of CO_2 only glucose is labelled, but oxygen is normal.

18 (*a*) 3, 4-Dichlorophenyl-l, l-dimethylurea (DCMU) is a selective herbicide that acts as a potent inhibitor of PS-II. Paraquat inhibits PS-I by producing super oxides. DCPIP (2, 6-Dichlorophenolindophenol) is a dye used to measure rate of photosynthesis.

19 (*b*) Phosphoenol Pyruvate (PEP) is the primary CO_2 acceptor in C_4 -plants. It is found in the mesophyll cell. It accepts the atmospheric CO_2 in C_4-plants and gets converted into oxaloacetate, a C_4 compound. It is the first stable compound of C_4-plants.

20 (*c*) 2ATP molecules are required during conversion of PGA to 1, 3- diphosphoglyceric acid and 1 ATP during conversion of glyceraldehyde phosphate to ribulose bisphosphate. 2NADP H_2 molecules are utilised for converting 1, 3 diphosphoglyceric acid to glyceraldehyde phosphate.

21 (*c*) Statement in option (c) is not correct and can be corrected as

In C_4-plants, the initial fixation of CO_2 occurs in mesophyll cells. The primary acceptor of CO_2 is Phosphoenol Pyruvate (PEP). It combines with CO_2 in the presence of enzyme PEP carboxylase to form the first stable product, i.e. Oxaloacetic Acid (OAA).

C_3-plants lack PEP carboxylase enzyme. They possess RuBisCO enzyme. At higher temperature, its affinity for CO_2 decreases and it works as oxygenase (fixation of O_2). Therefore, at higher temperature photosynthesis decreases in C_3-plants, while in C_4-plants it increases.

Rest of the statements are correct.

22 (*d*) CO_2 acceptor molecule in C_4-plants is Phosphoenol Pyruvate (PEP). C_4-plants are more efficient in picking up CO_2 even when it is found in low concentration because of high affinity of PEP. PEP carboxylase (PEPCase) is the key enzyme which catalysis the primary fixation of CO_2.

23 (*c*) Photorespiration is the process, which makes a difference between the C_3 and C_4-plants. In this process, there is a continuous loss of carbon fixed in the form of CO_2. It occurs due to the high O_2 content, high temperature conditions in which RuBP carboxylase starts working as RuBP oxygenase and normal photosynthesis does not occur.

24 (*a*) Emerson performed photosynthetic experiment on *Chlorella.* He provided monochromatic light of more than 680 nm and observed a decrease in the rate of photosynthesis. This is known as **red drop.**

Later, he provided synchronised light of 680 nm and 700 nm and observed a increase in the rate of photosynthesis. This is known as **Emerson enhancement effect.** This experiment led to discovery of two photosystems, **PS-I** (P_{700}) and **PS-II** (P_{680}) operating in photosynthesis.

25 (*a*) Proton concentration is higher in the lumen of thylakoids because of the photolysis of water, H^+ pumping and NADP reductase activity, which occurs in stroma of the chloroplast.

29 (*d*) RuBP carboxylase enzyme catalyses the primary carboxylation reaction in which CO_2 combines with RuBP (Ribulose bisphosphate) to form 2 molecules of 3 PGA (Phosphoglyceric Acid). It also shows oxygenase activity.

32 (*c*) Quantum yield of photosynthesis is derived from light intensity and rate of photosynthesis. It is 12% in grassland ecosystem.

36 (*a*) ATP synthesis takes place by the process of chemiosmosis, where ATP is generated *via* proton (H^+) gradient across the membrane. As proton moves from inside to outside across a membrane some configuration change occurs in the ATPase which start catalysing formation of ATP.

37 (*d*) Photolysis of water is an important event in the light reaction of photosynthesis. In this process water produces H^+ion to reduce CO_2 to form glucose and O_2 is released as byproduct. The pre-requisite of photolysis in the presence of chlorophyll, light energy, Ca^{2+}, Mn^{2+} and Cl^- ions.

48 (*a*) Anoxygenic photosynthesis (in which O_2 is not realised) is seen in *Rhodospirillum*, a purple non-sulphur bacterium. It helps an organism to trap light energy and store it as chemical energy. Other than this, anoxygenic photosynthesis commonly occurs in green sulphur/non-sulphur bacteria, helicobacteria, etc.

50 (*e*) The only mismatched pair in the options given is that of maize, which is a C_4 -plant. Rest all are correctly matched. As maize is a monocot plant, which uses C_4-pathway to fix CO_2, thus it is considered as C_4-plant.

52 (*c*) During cyclic photophosphorylation, synthesis of ATP takes place (i.e. not NADPH+ H^+) because the stroma lamellae does not possess enzyme NADP reductase due to which the excited electrons in the cyclic flow do not pass on to NADP instead they get back to the PS-I complex only.

53 (*b*) Photosynthetic active radiation is the range of solar radiation, which is received and absorbed by the plants for the process of photosynthesis. It has a range of 400-700 nm.

54 (*b*) In order to produce one molecule of glucose through the Calvin pathway, 12 $NADPH_2$ and 18 ATP are required. This can be easily understood by the following table given below

In	Out
6 CO_2	1 glucose
18 ATP	18 ADP
12 NADPH	12 NADP

62 (*c*) In C_4-plants, initial fixation of CO_2 occurs in mesophyll cells. The primary acceptor of CO_2 is phosphoenol pyruvate or PEP. It combines with CO_2 in the presence of PEP carboxylase to form oxaloacetic acid.

$$PEP + CO_2 + H_2O \xrightarrow{\text{PEP carboxylase}} \text{Oxaloacetic acid} + H_3PO_4$$

63 (*c*) Statement in option (c) is incorrect because

In non-cyclic phosphorylation, electrons derived from the photolysis of water are raised to higher energy levels in PS-I, PS-II and pass along an electron transport chain of carrier molecules of NADP reductase. This enzyme transfers electrons to $NADP^+$ to make NADPH, which provides reducing power for light dependent reactions of photosynthesis. So, both PS-I and PS-II participate in non-cyclic photophosphorylation.

65 (*b*) Cyclic photophosphorylation involves only PS-I and in this electrons expelled by excited photocentre are returned to it after passing over a chain of electron carriers. Two molecules of ATP are synthesised in this process.

66 (*c*) Photorespiration requires three cell organelles in sequence of chloroplast, peroxisome and mitochondria.

69 (*d*) The net reaction of carbon dioxide (Calvin cycle) is

$$6\,RuBP + 6CO_2 + 18ATP + 12NADPH \longrightarrow$$
$$6RuBP + C_6H_{12}O_6 + 18ADP + 18Pi + 12NADP^+$$

Thus, option (d) is correct.

72 (*d*) CAM plants are mostly succulent xerophytes. The stomata in these plants remain closed during the day. They help to check the transpiration. In this way, water is conserved.

73 (*c*) C_4-plants have Kranz anatomy in their leaves. In this leaf, the vascular bundle is surrounded by bundle sheath and mesophyll cells. Chloroplasts in the bundle sheath cells lack grana, while mesophyll chloroplasts are normal, e.g. sugarcane, maize, *Euphorbia, Amaranthus, Sorghum, Portulaca* and *Chenopodium.*

75 (*a*) Usually with increase in light intensity the rate of photosynthesis increased. At very high light intensity, the cells exhibit photo-oxidation by the process of solarisation and if it continues for few hours, the photosynthetic apparatus is destroyed.

76 (*a*) Copper is an activator of plastocyanin, cytochrome oxidase, RuBP carboxylase and many other enzymes. It has a major role in electron transfer, maintenance of carbohydrate, nitrogen balance and chlorophyll synthesis.

77 (*a*) Photorespiration is the light dependent utilisation of O_2 and release of CO_2 by photosynthetic organs of a plant. Photorespiration involves chloroplast, peroxisome and mitochondria to complete the process.

79 (*b*) Sunken stomata are usually found in Crassulacean Acid Metabolism (CAM) plants. Such stomata remain situated below the epidermis and open at night.

80 (*a*) Reduction in the rate of photosynthesis when beam of more than 680 nm use, it is known light as red drop.

81 (*d*) Statement in option (d) is true.

Photophosphorylation is the synthesis of ATP from ADP and inorganic phosphate in the presence of sunlight. When the two photosystems work in a series, first PS-II and then the PS-I, a process called non-cyclic photophosphorylation occurs. The two photosystems are connected through an electron tansport chain in the Z-scheme (due to the shape of path of electrons flow). Both ATP and NADH + H^+ are synthesised by this kind of electron flow.

82 (*a*) Option (a) shows incorrectly matched pair. It can be corrected as

The primary CO_2 fixation product in C_4-cycle is Oxaloacetic Acid (OAA).

Rest options are correctly matched pairs with regard to C_4 cycle.

83 (*a*) For every CO_2 molecule entering the Calvin cycle (the reduction and regeneration steps), three molecules of ATP and two molecules of NADPH + H^+ are required.

84 (*b*) In photorespiration, the RuBP instead of being converted to two molecules of PGA, binds with O_2 to form one molecule of PGA and phosphoglycolate. In the phosphorespiratory pathway, there is neither synthesis of sugars, nor of ATP.

88 (*a & c*) According to chemiosmotic hypothesis, energy released during movement, of electron through photosystem in photosynthesis is used to drive protons across the membrane against concentration gradient. As a result, protons accumulate in thylakoid lumen or intra-thylakoid space which further increases the proton gradient.

This gradient is important because the breakdown of the gradient provides enough energy to cause conformational change in the F_1-particle of the ATPase, which makes enzymes to synthesise several molecules of energy packed ATP.

CHAPTER 21

Respiration in Plants

All living cells require continuous supply of energy to perform various vital activities. This energy is released in a controlled manner for cellular use *via* the process of **respiration**. The released energy is stored in cells as ATP. Thus, **ATP acts as the energy currency** of the cell. It was discovered by **Karl Lohmann** in 1929.

CELLULAR RESPIRATION

It is simply defined as a series of metabolic processes which causes breakdown of C—C bonds of complex organic compounds within a cell. It is done with the help of oxygen and energy is released in this process. These complex organic compounds are called **respiratory substrates**.

On the basis of these substrates, respiration is classified into following two types

(i) **Floating respiration** In this type of respiration, carbohydrates are used as respiratory substrate.

(ii) **Protoplasmic respiration** In this type of respiration, proteins are used as respiratory substrate.

The generalised chemical equation representing cellular respiration is given below

$$C_6H_{12}O_6 + 6O_2 \longrightarrow 6CO_2 + 6H_2O + \text{Energy}$$

Thus, during respiration, oxygen is utilised and carbon dioxide, water and energy are released as products. This process can also occur in the absence of oxygen. Thus, in a cell the breakdown of complex organic molecules can be **aerobic** (i.e. in the presence of oxygen) or **anaerobic** (i.e. in the absence of oxygen).

AEROBIC CELLULAR RESPIRATION

This respiratory process is completed in following two phases

Glycolysis

- The word 'Glycolysis' is derived from two Greek words, i.e. *Glycos*-sugar and *lysis*-splitting.
- This process is also known as **Embden Meyerhof-Parnas** (EMP) **pathway** after the names of three German scientists who discovered it in 1930.

Inside

- **Nature** It is an oxidative process, in which 1 mole of glucose is converted into 2 moles of pyruvic acid. It is a series of ten enzyme mediated reactions, releasing energy as ATP and $NADH_2$. It is common to both aerobic and anaerobic respiration.
- **Location** Glycolysis occurs in the cytosol or cytoplasm of all living organisms.
- The reaction pathway of glycolysis is shown below

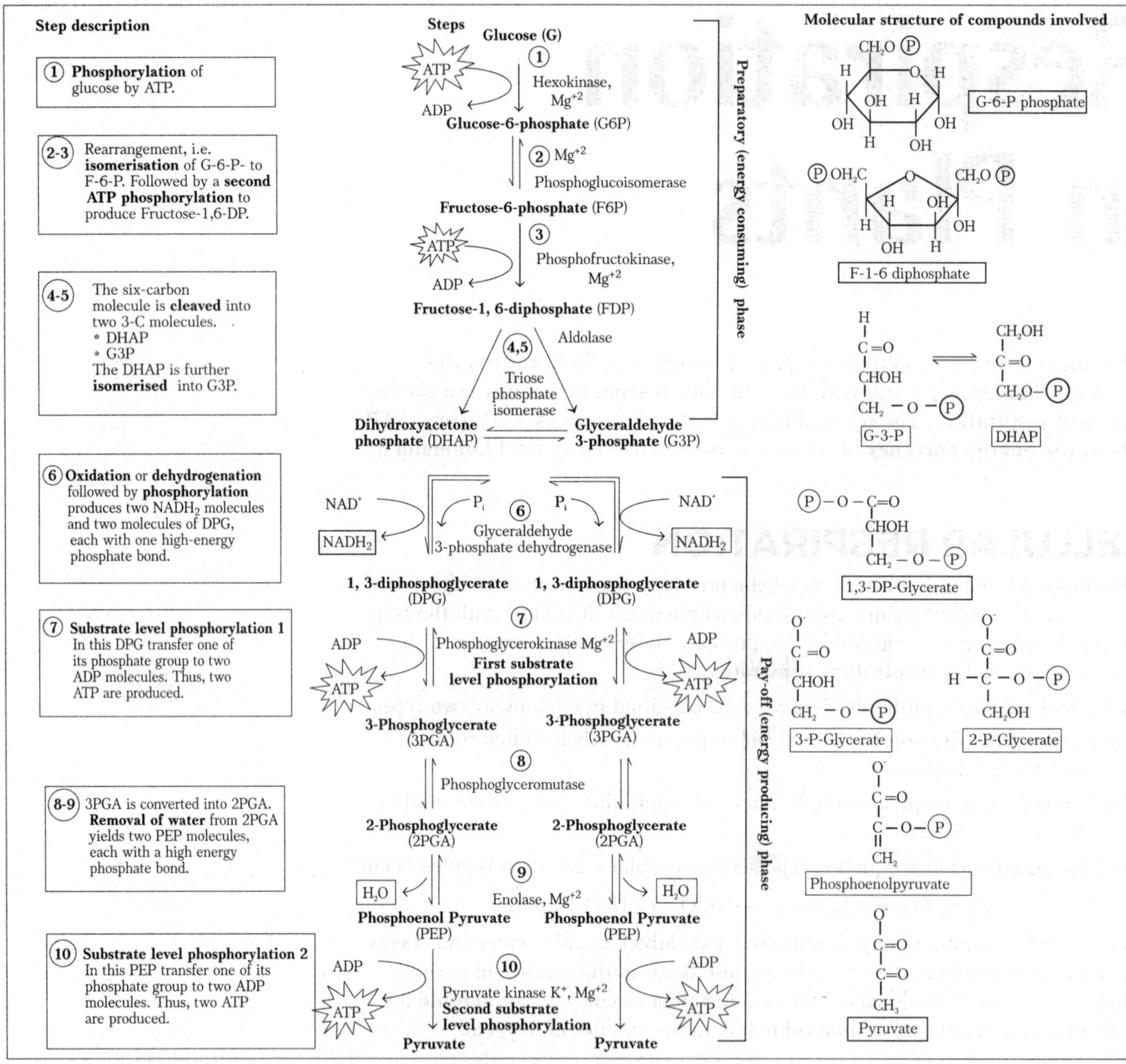

Fig. 21.1 The glycolytic pathway: The first five reactions convert a molecule of glucose into two molecules of G-3-P. The second five reactions convert G-3-P into pyruvate

Other important points

- In plants, during the third step, apart from ATP and phosphofructokinase, another enzyme, i.e. Pyrophosphate (PPi) dependent phosphofructokinase is present. It is also called as **pacemaker enzyme of EMP pathway**. It utilises pyrophosphate (PP) instead of ATP for phosphoryl transfer. In this way, 1 molecule of ATP is saved.
- G-3-P and DHAP are isomers. Likewise F-6-P and G-6-P are isomers.
- All enzymes of glycolysis require magnesium ion as a cofactor.
- Enzyme **kinases** or phosphorylases is used in such reactions where either consumption or production of ATP occurs.

- Enzyme **isomerases** carry out isomerisation of two isomeric forms.
- Enzyme **dehydrogenase** is involved in the removal of hydrogen. This is accepted by either $NADH_2$ or $FADH_2$.

Net result of glycolysis

The overall reaction of glycolysis is given below

$$\text{Glucose} + 2NAD^+ + 2ADP + 2Pi \xrightarrow{\text{Glycolysis}} 2\text{ Pyruvate} + 2NADH_2 + 2H^+ + 2ATP + 2H_2O$$

- During glycolysis, two ATP molecules are consumed in the conversion of glucose into fructose-1, 6-diphosphate. On the other hand, it produces 4 ATP and 2 $NADH_2$ molecules.
- Both the molecules of $NADH + H^+$ enter into the mitochondria and get oxidised through ETS to form 3 ATP each, i.e. a total of 6 ATP molecules.
- Thus, the total or net gain of ATP in glycolysis process is (4 ATP produced − 2 ATP consumed) + 6 ATP from $NADH_2$ = **8 ATP** molecules.

Substrate level phosphorylation

- It is a type of chemical reaction in which a high energy phosphate bond containing substrate (like DPG) transfers one of its phosphate group to ADP, thus an ATP molecule is synthesised.
- In simple words, ADP becomes ATP after gaining one of its phosphate group from the substrate.
- Glycolysis involves two such substrate level phosphorylation reactions, i.e. at step 7 and step 10 of the process.

Fate of pyruvic acid

Pyruvic acid is the end product of glycolysis. Its fate depends upon the availability of oxygen in the cell. There are three major ways in which cells handle pyruvic acid. These are lactic acid fermentation, alcoholic fermentation and aerobic respiration.

The relationship between aerobic and anaerobic respiration is represented by **Pfeffer-Kostychev** scheme shown below

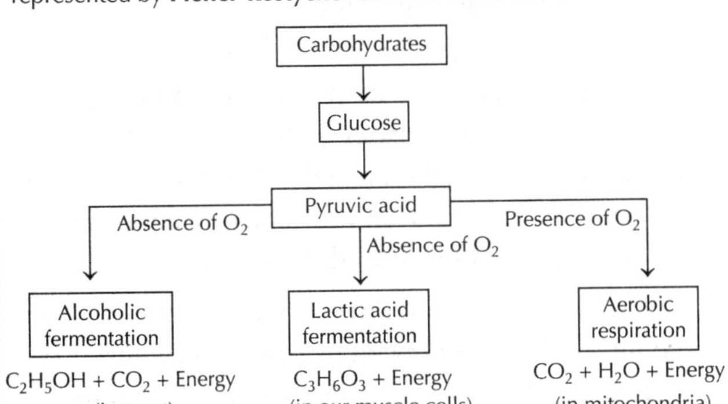

Breakdown of pyruvic acid

The breakdown of pyruvic acid formed during glycolysis is occurs in following two steps

1. Oxidative decarboxylation

The pyruvic acid formed in glycolysis moves into the mitochondria. Here, oxidative decarboxylation of pyruvic acid takes place. During this process, pyruvic acid is first decarboxylated, i.e. releases CO_2.

2. Dehydrogenation

It is then dehydrogenated by removing H-atom. Thus, 3-C pyruvic acid is converted into 2-C **acetyl coenzyme-A.** This process is catalysed by a large enzyme complex called **pyruvate dehydrogenase.** This enzyme complex consists of Mg^{2+}, thiamine pyrophosphate, NAD, coenzyme-A and lipoic acid as cofactors.

This complete process takes place in perimitochondrial or intermembrane space.

The equation for this process is given below

$$[\text{Pyruvic acid} + NAD^+ + \text{Co-A} \xrightarrow[Mg^{2+}]{\text{Pyruvate dehydrogenase complex}} \text{Acetyl Co-A} + NADH + H^+ + CO_2\uparrow] \times 2$$

Acetyl Co-A formed above reacts with oxaloacetate in Krebs' cycle. Thus, breakdown of **pyruvic acid is considered as the connecting link between glycolysis and Krebs' cycle.**

Tricarboxylic acid or Krebs' cycle

- This cycle is also called as **citric acid cycle**, because citric acid is the first product of this cycle. It was discovered by **Sir Hans Krebs'** in **1937** in the flight muscles of pigeon.
- **Nature** Krebs' cycle acts as an **amphibolic pathway,** because it contains both catabolic and anabolic processes. It is catabolic in nature as it oxidises acetyl Co-A and organic acids into CO_2 and H_2O. It is anabolic in nature as its intermediates product act as building blocks of many other molecules such as amino acids, cholesterol, nucleotide bases, etc.
- **Location** In **eukaryotic organisms**, all reactions of Krebs' cycle take place in the mitochondrial matrix.

  Thus, all enzymes of this cycle are also found in matrix except succinic or succinate dehydrogenase. This enzyme is found to be located in the inner membrane of mitochondria.
- In **prokaryotes,** Krebs' cycle occurs in cytoplasm.
- Krebs' cycle involves eight steps to oxidise two molecules of acetyl Co-A.

- The products of Krebs' cycle are CO_2, H_2O, ATP, $FADH_2$ and NADH + H^+.

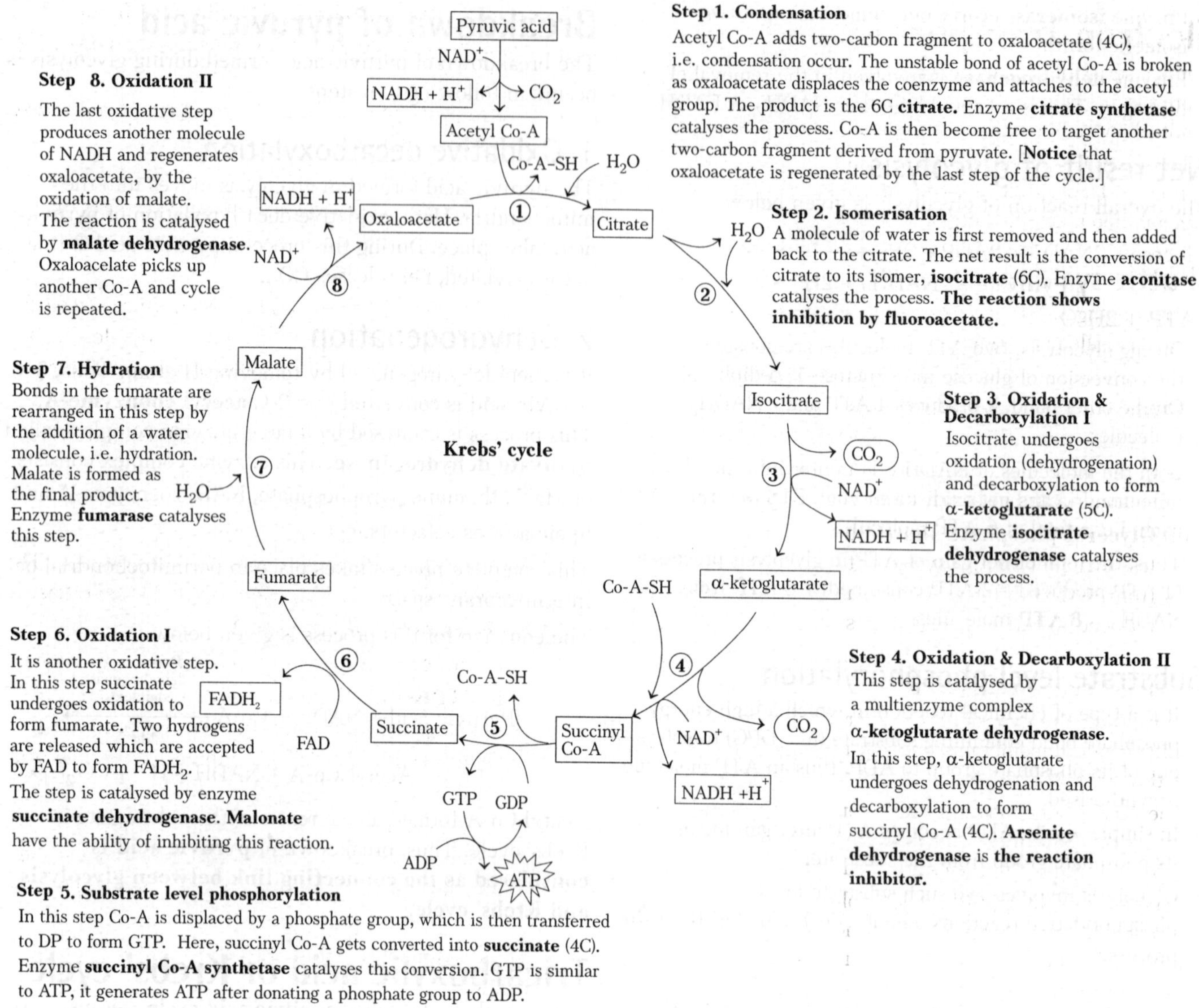

Fig. 21.2 The Krebs' cycle

Other Important points

- GTP and ATP perform similar function. Since, ATP is stable, it is used as energy currency.
- The only step in TCA cycle, where there is involvement of FAD is step 6. In this step succinate gets converted into fumarate.
- Steps involving oxidation, usually form $NADH_2$ or $FADH_2$.

Net result of Krebs' cycle

In total, 1 molecule of pyruvic acid undergoes 3 **decarboxylation** and 5 **oxidation** or **dehydrogenation** reactions. Out of which, oxidation occurs at 4 sites during Krebs' cycle.

On oxidation, in the process, 1 molecule of pyruvic acid produces 15 ATP molecules.

Thus, the 2 molecules of pyruvic acid undergo complete degradation. As the result

- 2 molecules of ATP (GTP),
- 8 molecules of $NADH_2$ and
- 2 molecules of $FADH_2$ are produced

Although the degradation of pyruvic acid is completed in above written first two steps i.e. oxidative decarboxylation and dehydrogenation. But, the 3rd step, i.e. electron transport system or chain has its own importance in the process.

This step is focussed on the generation of ATP from the molecules of hydrogen or proton acceptors, i.e. $NADH_2$ and $FADH_2$ generated in first two steps.

Electron Transport Chain (ETC)

- It is a series of various enzymes and coenzymes, which are present in the inner mitochondrial membrane. The inner mitochondrial membrane is impermeable to $NADH_2$.
- Thus, $NADH_2$ produced cannot enter into mitochondria directly. Hence, to overcome this problem, mitochondrial membrane takes the help of shuttle systems.

Shuttle system

- Functioning of ETC occur through a series of electron carriers called **shuttles.**
- These shuttle systems are of following two types
  - (i) **Malate-aspartate shuttle** It is a more efficient shuttle system. It transfers electrons from NADH to NAD. In this system, there is no loss of ATP. This shuttle occurs in heart, kidney, etc.
  - (ii) **Glycerol-phosphate shuttle** It is comparatively a less efficient shuttle system. It transfers electrons from NADH to FAD. As a result of this, FAD is reduced to $FADH_2$. It involves loss of one ATP molecule per NADH + H^+. This shuttle occurs in flight muscle cells and brain.
- Thus, there are two routes of electron transfer in mitochondrial membrane, one is through $NADH_2$ and other is through FAD.
- The ATP production outcome differ significantly in both the routes. In route or shuttle 1 where $NADH_2$ is substrate, 3ATP molecules are produced.
- On the other hand, only 2 ATP molecules are produced through shuttle 2, where FAD is the substrate. Their presence is the characteristic feature of eukaryotes.

Constituents and functioning of ETC

In total, electron transport chain consists of following nine electron carriers.

- **NAD** (Nicotinamide Adenine Dinucleotide)
- **FMN** (Flavin Mono Nucleotide)
- **FAD** (Flavin Dinucleotide)
- **Coenzyme-Q** or **Ubiquinone**
- **Cytochrome**-b, c_2, c, a and a_3

Out of these, **coenzyme-Q** and **cytochrome-c** act as mobile carriers. The above given nine electron carriers are grouped into following four large multiprotein complexes in the respiratory chain.

(i) **Complex I or NADH-Co-Q reductase** It consists of NADH dehydrogenase and prosthetic groups, i.e. one molecule of FMN (Flavin Mono Nucleotide) and eight or nine FeS (Iron sulphur centres). It passes electrons from NADH to Co-Q.

(ii) **Complex II succinate Co-Q reductase** It consists of succinate dehydrogenase and prosthetic groups FAD and Fe-S.

(iii) **Complex III reduced Co-Q-cytochrome reductase** It consists of haem as prosthetic group, cytochrome-b, cyt-c and Fe-S.

(iv) **Complex IV cytochrome oxidase** It consists of cytochrome-a, cyt-a_3, haem and copper ions as prosthetic group.

Out of these, complex I and II are alternate to each other. This simply means that either complex I, III and IV or II, III and IV function at a time. Their functioning depends upon the substrate.

The first combination, i.e. complex I, III and IV functions only, if $NADH_2$ is the substrate.

For $FADH_2$, on the other hand the second combination is effective.

Both the chains terminate into the Vth complex. This complex is called $F_0 - F_1$ complex or ATP synthase.

(v) **F_0- F_1 complex or complex V**

This complex has following two major components

(a) **F_1-ATPase** It is the head piece and is a water soluble peripheral membrane protein, containing eight subunits. These eight subunits are paired 3α, 3β, 1γ and 1ε.

(b) **F_0-component** It is water insoluble integral membrane protein. It is attached to F_1 by means of its γ subunit. It forms a channel through which proton can pass through the inner membrane.

The diagrammatic view of F_0- F_1 complex is given below

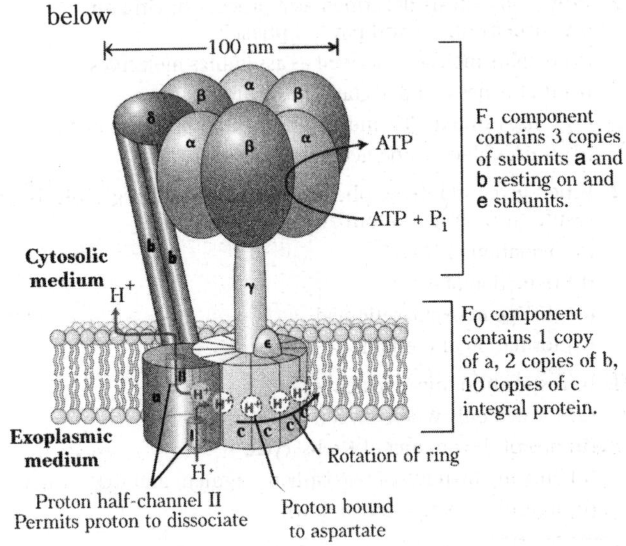

***Fig. 21.3** F_0 - F_1 complex (γ, ε and c constitutes the rotatory part and rest is static part)*

Arrangement of all these complexes within mitochondrial membrane is given below

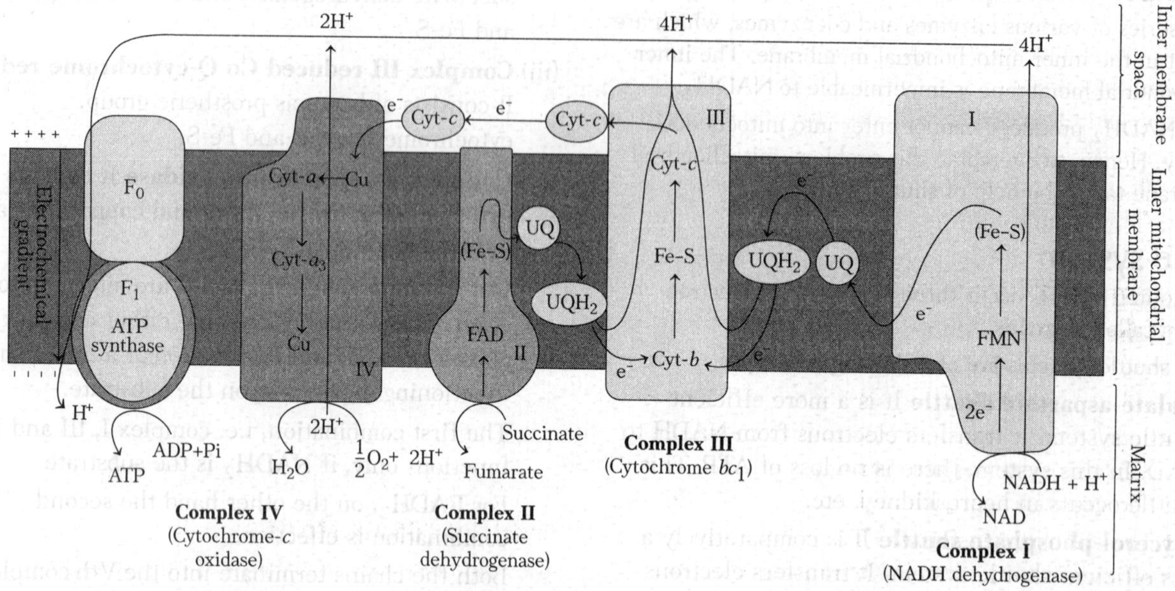

Fig. 21.4 Electron Transport System (ETS)

- **Oxygen acts as the final electron acceptor** at the end of the respiratory chain. Oxygen also receives two protons directly from the aqueous medium to form one molecule of water.
- At each electron transfer, an electron donor is oxidised, while acceptor is reduced. As the electrons move down the system, energy is released and used to form ATP.
- The acceptance of electrons at the end of respiratory chain is called **terminal oxidation**. The energy liberated during terminal oxidation is used to produce ATP.
- In mitochondria of some plants, alternative oxidase system is present, in which ETS continues even in presene of cyanides. This type of respiration is known as **cyanide resistant respiration** or **alternate electron pathway**, e.g. spinach, *Pisum*.

CHECK POINT 21.1

1. Why is glycolysis described as a process having an investment phase and pay-off phase?
(a) It splits molecules as well as assembles molecules
(b) It attaches and detaches phosphate groups
(c) It uses stored ATP and then form a net increase in ATP
(d) It uses glucose and generate pyruvate

2. When glyceraldehyde phosphate is oxidised in glycolysis, the resultant hydrogen atom and electrons cause
(a) oxidation of NAD^+
(b) reduction of NAD^+
(c) change in oxaloacetic acid
(d) formation of methane

3. In respiration, pyruvic acid is
(a) formed only when oxygen is available
(b) one of the product of Krebs' cycle
(c) broken down into a two carbon fragment and CO_2
(d) a result of protein breakdown

4. In Krebs' cycle,
(a) acetyl coenzyme-A undergoes 4 oxidations and 2 decarboxylation
(b) pyruvic acid undergoes 4 oxidation and 2 decarboxylation
(c) TCA undergoes 4 oxidation and 2 decarboxylation
(d) OAA undergoes 4 oxidation and 2 decarboxylation

5. A false statement regarding cytochrome c-oxidase complex in ETS in mitochondria is
(a) received electrons directly from ubiquinone
(b) capable of reducing O_2
(c) extended across the thickness of inner mitochondrial membrane
(d) contained Fe and Cu both

6. F_0 part of ATP synthase (Complex V) is
(a) integral membrane protein complex acts as a site of ATP synthesis
(b) peripheral membrane protein complex act as site of protein synthesis
(c) integral membrane protein complex that function as proton tunnel
(d) complex molecule made up of phospholipids

Oxidative phosphorylation

- The process of production of ATP *via* terminal oxidation of reduced coenzymes is known as oxidative phosphorylation.
- **Location** It takes place in the inner mitochondrial membrane of eukaryotes. In case of prokaryotes, it occurs in mesosomes.
- The mechanism of oxidative phosphorylation is usually explained by **chemiosmosis** or **chemiosmotic hypothesis.**

Chemiosmotic hypothesis

Peter Mitchell proposed this hypothesis in 1961. According to this hypothesis, the ATP molecule is synthesised during electron transport by the movement of hydrogen ions H^+ or protons across the mitochondrial membrane. These electrons are donated by NADH and $FADH_2$. The stepwise approach of the process is as follows

- **Step 1** The electron transfer *via* ETC pumps protons out of the mitochondrial matrix into intermitochondrial space.
- **Step 2** Thus, the pH of matrix becomes higher, while that of intermitochondrial space becomes lowered.
- **Step 3** Due to this movement of H^+ and change in pH, an electrochemical gradient is generated.
- **Step 4** This electrochemical gradient is responsible for the generation of a **Proton Motive Force** (PMF). Due to which, protons flow back to the matrix.
- **Step 5** During this flow, energy is released. This energy is used in the formation of ATP by ATP synthase.

Mechanism of ATP synthesis

- Conformational change mechanism is the most widely accepted model of ATP synthesis.
- The rotation of γ-subunit is powered by proton translocation through F_0 as shown below

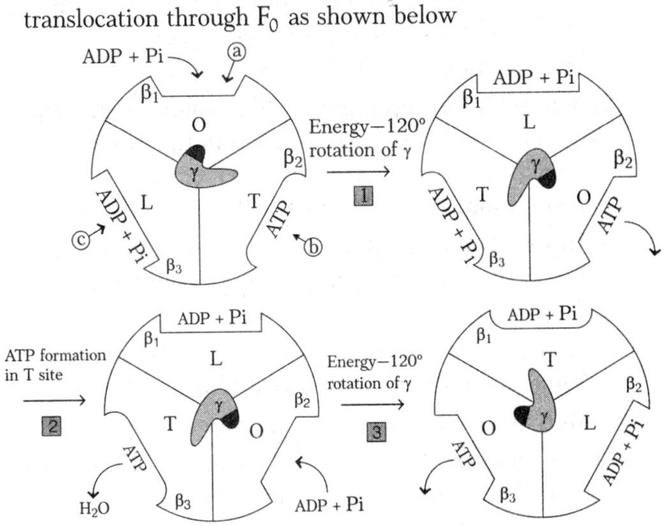

Fig.21.5 The binding-change mechanism of ATP synthesis

- According to this, ATP synthesis occurs as a result of conformation change in β-subunit of F_1.
- This change in conformation is caused by the rotation of its γ-subunit which acts as rotating shaft.
- The conformation of β-subunits alternates between following three states as visible in figure also
  - (a) **O-state** (open) ADP and Pi attach weakly.
  - (b) **T-state** (tight) Elevated tight margins binds ADP + Pi tightly to give ATP.
  - (c) **L-state** (loose) Intermediate of above two conformations, binds ADP + Pi loosely.
- These states differ in their binding to ATP, ADP and Pi. For every 360° rotation of γ-subunit, 3 ATP molecules are produced.

Total ATP yield during aerobic cellular respiration

- Net gain of ATP (*via* malate-aspartate shuttle) = 38 ATP (+ 2 GTP).
- Net gain of ATP (*via* glycerol-phosphate shuttle) = 36 ATP (+ 2 GTP).
- This net gain can be calculated as

Stages	Total no. of Net ATP/GTP produced	Total no. of NADH/$FADH_2$ produced
Glycolysis	2	2 NADH ×3 = 6 ATP
Oxidative decarboxylation	-	2 NADH ×3 = 6 ATP
Krebs' cycle	2 GTP = 2 ATP	2 $FADH_2$ ×2 = 4ATP 6 NADH ×3 = 18 ATP

Thus, in totality, $2+2+6+6+4+18=38$ ATPs are produced in the process of aerobic respiration. From these 38 molecules, 2 molecules are consumed in the transfer of NADH $+H^+$ to from cytoplasm to mitochondria. This expenditure of 2 ATP does not occur if the route of electron transfer is *via* malate-aspartate shuttle.

Efficiency of aerobic respiration

An ATP molecule yields about 34 KJ/mol of energy. Thus, during aerobic respiration 1292 kJ of energy is produced as in total 38 ATP are formed in the process. Secondly, the glucose molecule stores about 2870 KJ energy. Therefore, efficiency of aerobic respiration

$$= \frac{1292}{2870} \times 100 = 45\%$$

In terms of quantity, oxidation of 1 glucose molecule yields **686 kcal** of energy.

Production of ATP in prokaryotes

Since, in prokaryotic cells mitochondria is absent, hence NADH + H^+ oxidation takes place in mesosomes only. This results to a net gain of 38 ATP.

Pentose Phosphate Pathway (PPP) or Hexose Monophosphate Pathway (HMP)

- This pathway was discoverd by **Warburg** and **Dickens** in 1938. It is considered to be an alternate method of aerobic respiration.
- **Location** It occurs in the cytoplasm of both prokaryotes and eukaryotes.
- This pathway is also called as **shunt of glycolysis** and mostly occurs when concentration of NADP is more. This pathway is also resistant to cyanide.
- In this pathway, there are two distinct phases.
  - The first phase is **oxidative** phase in which NADPH is generated.
  - The second phase is **non-oxidative** phase. In this phase, synthesis of 5C sugars takes place.
- The outline of this process is described below

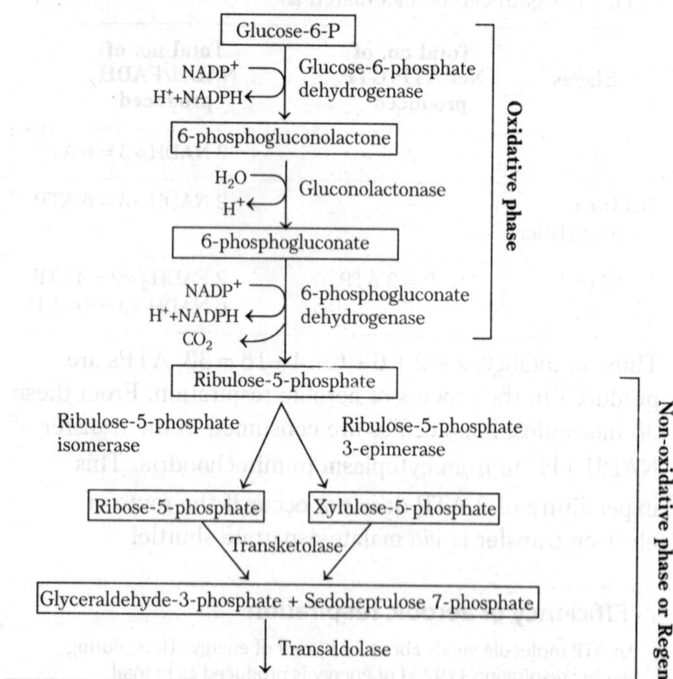

Fig 21.6 Flowchart depicting pentose phosphate pathway

Primary results of this pathway are

- Generation of 12 NADPH + H^+ for various biosynthetic reactions like fatty acid synthase. $6CO_2$ molecules are also produced.
- Production of Ribulose-5-Phosphate (R-5-P). It is used in the synthesis of nucleotides and nucleic acids.
- Production of Erythrose-4-Phosphate (E-4-P). It is used in the synthesis of aromatic amino acids.
- Net gain of ATP in PPP is 35 molecules.

Note *PPP accounts for 60% of aerobic respiration in liver, adrenal cortex, RBC and adipose tissue.*

Factors affecting aerobic cellular respiration

Respiration is considered as an **amphibolic pathway** as besides **catabolic** processes, it also works for **anabolic** processes.

The factors affecting respiration in plants are as follows

(i) **Oxygen concentration** Rate of aerobic respiration decreases, if the atmospheric oxygen concentration is below 5% for whole tissues.

(ii) **CO_2 concentration** CO_2 has a limited direct inhibitory effect on the respiration rate. The carbon dioxide concentration at which net gaseous exchange is zero, is called **carbon dioxide compensation point.**

(iii) **Light** Under suitable light, the rate of photosynthesis is optimum. Thus, it supplies respiratory substrates at a moderate rate. Therefore the rate of respiration is increased.

(iv) **Temperature** Respiration increases as the temperature increases from 0°C and 30°C. Above 30°C, the respiration rate often increases more slowly and decreases at further higher temperatures, i.e. 40-50°C and above.

(v) **Minerals** Certain minerals like magnesium, copper, etc., are required for the optimum functioning of enzymes. Deficiency of these minerals may hamper respiration.

(vi) **Injury** The respiration rate increases at the site of injury, because at injured portion, new cells are formed to heal up the wound.

Hence, they require more energy. This energy is supplied by the increased rate of respiration.

(vii) **Dehydration** slows down the rate of respiration.

(viii) The **internal factors**, which affect aerobic cellular respiration include water contents of the cell, kinds and age of tissue as well as the amount and type of substrate available.

CHECK POINT 21.2

1. All of the following is true about oxidative phosphorylation, except
   (a) in prokaryotes it occurs in mesosomes
   (b) it produces 2 ATPs for each $FADH_2$
   (c) it can occur under anaerobic condition
   (d) it involves a proton gradient
2. According to chemiosmosis, electron are donated by
   (a) NADH and $FADH_2$ (b) ATP and $FADH_2$
   (c) NADH and ATP (d) NADH and FMN
3. Which of the following is the reason for ATP synthesis?
   (a) Conformational change in β-subunit of F_1
   (b) Conformational change in γ-subunit
   (c) Conformational change in β-subunit of F_0
   (d) All of the above
4. Net gain of ATP during cellular respiration *via* glycerol phosphate shuttle is
   (a) 38 (b) 36
   (c) 37 (d) 40
5. In the pentose phosphate pathway, the major products are
   (a) Ribulose and ATP (b) Ribulose and NADPH
   (c) Ribose and NADH (d) Ribose and NAD^+
6. At temperature above 35°C
   (a) rate of respiration will decline earlier than that of photosynthesis
   (b) rate of photosynthesis will decline earlier than that of respiration
   (c) there is no fixed pattern
   (d) Both decline simultaneously

ANAEROBIC CELLULAR RESPIRATION

This type of cellular respiration has fermentation as its main process. It can occur in prokaryotes and unicellular eukaryotes.

Ratio of ATP synthesis in aerobic and anaerobic cellular respiration is 18 : 1.

Fermentation

- When O_2 is absent in respiration then after the glycolysis fermentation takes place. Thus, this process involves breakdown of pyruvic acid.
- Presence of this pathway is dependent upon the type of organism, tissue and end products.
- In particular, following three types of fermentations are most common

Alcoholic fermentation

- It is common in yeast (*Saccharomyces*).
- Cells of this organism release an enzyme in surrounding medium called **zymase.**
- This enzyme is responsible for the breakdown of substrate outside the cell.
- End products of this process are ethyl alcohol, CO_2 and energy.
- The overall equation of alcoholic fermentation as described by the **French chemist Gay-Lussac** (1815) is given below

Step 1 $\underset{\text{(Glucose)}}{C_6H_{12}O_6} \xrightarrow{\text{Glycolysis}} \underset{\text{(Pyruvate)}}{CH_3COCOOH}$

Step 2 $\underset{\text{(Pyruvate)}}{CH_3COCOOH} \xrightarrow[\text{TPP, } Mg^{2+}]{\text{Pyruvate decarboxylase}} \underset{\text{(Acetaldehyde)}}{CH_3CHO} + CO_2\uparrow$

Step 3 $CH_3CHO + NADH \xrightarrow{\text{Alcohol dehydrogenase}} C_2H_5OH + NAD^+$

Pasteur effect

It is defined as slowing down or complete cessation of alcoholic fermentation in the presence of oxygen. This effect was further confirmed by **Meyerhoff** and **Warburg.**

Lactic acid fermentation

- It is common in **Lactic Acid Bacteria** (LAB) and inside our muscles.
- LAB can ferment lactose sugar formed in milk. Enzyme involved in this fermentation is **Lactate Dehydrogenase** (LDH).
- It is produced and released by LAB only.
- Here, no water and carbon dioxide are produced in the process. The complete equation looks like

$$CH_3COCOOH + NADH_2 \xrightarrow[\text{FMN, } Zn^{2+}]{\text{LDH}} \underset{\text{(Lactic acid)}}{CH_3\,CHOHCOOH} + NAD^+$$

Acetic acid fermentation

It is carried out by bacteria *Acetobacter aceti* and requires oxygen up to some extent. Acetic acid fermentation occurs in following two steps

$$\underset{\text{(Pyruvic acid)}}{CH_3COCOOH} \xrightarrow[\textit{Acetobacter}]{[O]} \underset{\text{(Acetaldehyde)}}{CH_3CHO} \xrightarrow[\textit{Acetobacter}]{[O]} \underset{\text{(Acetic acid)}}{CH_3COOH}$$

Applications of fermentation

This process is used in various industries to produce a variety of substances. Out of these some important application are summarised below

- Antibiotics, vitamins, dairy products and alcohol etc., can be easily produced by this process for industrial and medicinal used.
- It is also used for tanning of leather, retting of fibres and in curing of tea and coffee.
- In humans, lactic acid is formed in muscles during prolonged starvation or exercies. These mucles cells respire anaerobically during this period.

Respiration as amphibolic pathway

As stated earlier, an amphibolic pathway includes both breakdown, i.e. **catabolism** and build up processes, i.e. **anabolism**. Respiration is considered as an amphibolic pathway.

Following diagram gives a clear picture of this concept.

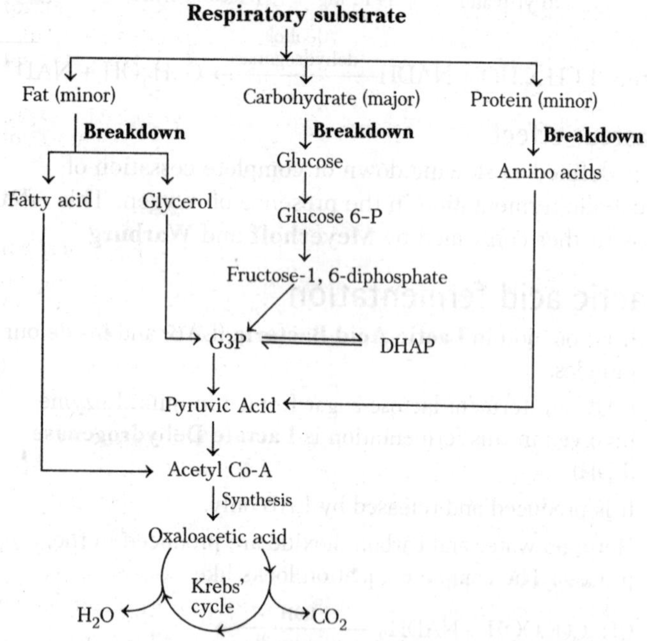

Fig 21.7 Amphibolic pathway of respiration

Inhibitors of Respiration

(i) **2,4- Dinitrophenol** (2, 4-DNP) is soluble in lipid and inhibits ATP synthesis.

(ii) **Oligomycin** inhibits oxidative phosphorylation.

(iii) ETC inhibitors

- **Rotenone and amytal** For complex 1
- **Antimycin** For complex 3
- **Cyanide and CO** For complex 4.

Respiratory Quotient (RQ) or Respiratory Ratio

It is the ratio of volume of carbon dioxide (CO_2) released to the volume of oxygen (O_2) consumed in respiration.

Its value may vary for different respiratory substrates and measured by **Ganong's respirometer.**

It is useful in knowing the type of respiration and respiratory substrates.

$$RQ = \frac{\text{Volume of } CO_2 \text{ evolved}}{\text{Volume of } O_2 \text{ absorbed}}$$

Following table contains the RQ. Values of some important substrates

RQ	Substrate
>1.0	Glucose in anaerobic respiration
1.0	Carbohydrates in aerobic respiration
0.9	Protein
0.7	Fat, such as tripalmitin
0.5	Fat associated with carbohydrate synthesis (Carbon dioxide released during this respiration is being put to other uses and therefore, not considered as released from the body)
0.3	Carbohydrate with associated organic acid synthesis
>1	Organic acids such as malic acid, citric acid
0	CAM plants (at night)

CHECK POINT 21.3

1. Fermentation takes place
   (a) under anaerobic condition in many prokaryotes and unicellular eukaryotes
   (b) under aerobic condition in many prokaryotes and unicellular eukaryotes
   (c) under anaerobic condition in all prokaryotes and unicellular eukaryotes
   (d) under aerobic condition in all prokaryotes and unicellular eukaryotes

2. Which of the following option are correct about yeast?
   (a) It respire anaerobically, due to lack of mitochondria
   (b) It respire aerobically, if it gets O_2 otherwise perform alcoholic fermentation
   (c) It respire aerobically
   (d) It can perform alcoholic fermentation

3. The enzyme LDH is used in
   (a) alcohol fermentation (b) acetic acid fermentation
   (c) Krebs' cycle (d) our muscles

4. Lactic acid fermentation does not produce
   (a) CO_2 (b) NADH
   (c) Both (a) and (b) (d) NAD^+

5. Antimycin is related to
   (a) ETC complex I inhibition (b) ETC complex IV inhibition
   (c) ETC complex III inhibition (d) ETC complex II inhibition

6. Among the following substances, which one has the highest respiratory quotient?
   (a) Glucose (b) Fat
   (c) Protein (d) CAM plants

EXAM POINTS

- $C_6H_{12}O_6 + 2NAD^+ + 2ADP + 2Pi \longrightarrow$
  $2CH_3—CO—COOH + 2NADH_2 + 2ATP + 2H^+ + 2H_2O$
  The above reaction shows the complete glycolysis. **[MHT CET 2015]**
- During glycolysis, fructose-1, 6-bisphosphate is split into dihydroxyacetone phosphate and 3-phosphoglyceraldehyde. **[J&K CET 2014]**
- Conversion of glucose-6-phosphate, the first irreversible reaction of glycolysis is catalysed by hexokinase. **[NEET 2019]**
- Oxygen content reduction make the glycolysis intensity increased due to increase in concentration of peroxides and free radicals. **[AIIMS 2013]**
- Common enzyme in glycolysis and pentose phosphate pathway is hexokinase. **[UP CPMT 2011]**
- Oxidative decarboxylation of pyruvic acid result in the formation of ATP and NADH + H^+. **[Kerala CEE 2018]**
- Conversion of pyruvate to acetyl Co-A is an example of oxidative decarboxylation. **[KCET 2013]**
- Pyruvate dehydrogenase activity during aerobic respiration require magnesium. **[NEET 2020]**
- FAD is an electron acceptor in citric acid cycle during the oxidation of succinic acid to fumaric acid. **[Kerala CEE 2015]**
- Krebs' cycle is a catabolic reaction that oxidises acetyl Co-A and organic acids into CO_2 and H_2O. It acts as an amphibolic pathway as it includes both catabolic as well as anabolic reaction. **[EAMCET 2015]**
- The number of substrate level phosphorylation in one turn of citric acid cycle is one. **[NEET 2020]**
- Biological oxidation in Krebs' cycle involves O_2. **[OJEE 2013]**
- In Krebs' cycle, GTP is formed in substrate level phosphorylation. **[UP CPMT 2011]**
- The correct sequence of electron pathway in ETS is NADH dehydrogenase $\rightarrow$ Cyt-*c* reductase $\rightarrow$ Cyt-*c* oxidase $\rightarrow O_2$. **[EAMCET 2015]**
- Respiratory electron transport system occur in the inner mitochondrial membrane. **[NEET 2019]**
- Cytochromes occur in cristae of mitochondria. **[CBSE AIPMT 2015]**
- 2ATP are produced when 1 molecule of $FADH_2$ is oxidised to FAD through electron transport system. **[WB JEE 2014]**
- A small protein attached to the outer surface of the inner membrane and which acts as a mobile carrier for transfer of electrons between complex III and IV is cytochrome-*c*. **[J&K CET 2018]**
- Oxidative phosphorylation is formation of ATP by transfer of phosphate group from a substrate to ADP. **[NEET-II 2016]**
- ATP formation occurs through photophosphorylation, oxidative phosphorylation and substrate level phosphorylation. **[AIIMS 2019]**
- Conversion of pyruvic acid into ethyl alcohol is facilitated by enzyme decarboxylase and dehydrogenase. **[COMED-K 2015]**
- During alcohol fermentation, CO_2 is releases but not in the process of lactate fermentation. **[CBSE AIPMT 2014]**
- Respiratory Quotient (RQ) of glucose is 1.0, tripalmitin is 0.7 and malic acid is 1.33. **[NEET 2019, JIPMER 2018, WB JEE 2014]**

Chapter Exercises

(A) Topical Exercise

Cellular Respiration

1 Respiration is different from the combustion in that
(a) energy from carbohydrates is released in one step
(b) energy from carbohydrates is released in different steps
(c) efficiency is very low
(d) All of the above

2 How many of the following used as respiratory substrates?
Fats, Proteins, Carbohydrates, Nucleic acids
(a) one (b) three (c) two (d) four

3 The highest rate of respiration occurs in ......*A*...... and their core resultant of respiration is released in the form of ...... *B*......
(a) *A*-germinating seeds, *B*-heat
(b) *A*-growing shoot apex, *B*-water
(c) *A*-root tip, *B*-carbon dioxide
(d) *A*-leaf bud, *B*-oxygen

4 Match the correct answer with type of respiration and respiratory substrates.

Respiration	Types of substrates
A. Floating respiration	1. Protein
B. Protoplasmic respiration	2. Fat
	3. Carbohydrate

Codes

	A	B		A	B
(a)	3	2	(b)	1	2,3
(c)	2,3	1	(d)	2	3,1

5 Consider the following statements.
I. The main purpose of cellular respiration is to convert energy stored in the chemical bonds of glucose to an energy that the cell can use.
II. It is purely a catabolic process.
Choose the correct option.
(a) Statement I is true, but II is false
(b) Statement I is false, but II is true
(c) Both statements I and II are true
(d) Both statements I and II are false

Glycolysis

6 Glycolysis causes partial oxidation of glucose (one molecule) to form 2 molecules of pyruvic acid and
(a) 3 ATP (b) 2 ATP
(c) 4 ATP (d) 1 ATP

7 The step in glycolysis, that utilises one ATP molecule is
(a) Glucose → Glucose-6-phosphate
(b) 2-phosphoglycerate → Phosphoenol pyruvate
(c) Pyruvate → Acetyl Co-A
(d) Fructose-1, 6-diphosphate → Dihydroxy acetone phosphate

8 Pick out the reaction of glycolysis in which isomerisation takes place
(a) G-3-P → DHAP
(b) Glucose→ Fructose-6-phosphate
(c) 3-phosphoglycerate → 2-phosphoglycerate
(d) Phosphoenol pyruvate → Pyruvate

9 Phosphorylation of glucose during glycolysis is catalysed by **[NCERT Exemplar]**
(a) phosphoglucomutase
(b) phosphoglucoisomerase
(c) hexokinase
(d) phosphorylase

10 First half of glycolysis is characterised as glucose activation phase or preparatory phase. It is marked by the
(a) synthesis of 2 ATPs by substrate level phosphorylation
(b) phosphorylation of glucose molecule in the presene of hexokinase and Mg^{2+}
(c) synthesis of reducing equivalent
(d) presence of one dehydrogenation step

11 How many PGAL are produced by glycolysis of three molecules of glucose? How many ATP are released by respiration of these PGAL till formation of CO_2 and H_2O?
(a) 4 PGAL - 80 ATP
(b) 6 PGAL - 160 ATP
(c) 4 PGAL - 40 ATP
(d) 6 PGAL - 120 ATP

12 ...*A*... and ...*B*... are removed by substrate at the time of glycolysis.

	A	*B*
(a)	Hydrogen	Electrons
(b)	NAD	ATP
(c)	O_2	$FADH_2$
(d)	$NADPH_2$	$NADH_2$

13 In which of the following reactions of glycolysis, a molecule of water is removed from the substrate?
(a) Fructose-6-phosphate → Fructose-1-6-diphosphate
(b) 3-phosphate glyceraldehyde→ 1, 3-diphosphoglyceric acid
(c) PEP → Pyruvic acid
(d) 2-phosphoglycerate → PEP

14 In the third step of glycolysis apart from ATP phosphofructokinase, another enzyme ......*A*...... may also used. This enzyme also known as ......*B*...... of EMP pathway.
(a) *A*–PPi dependent phosphofructokinase, *B*–Pacemaker
(b) *A*–Pyruvate kinase, *B*–Pacemaker
(c) *A*–Glyceraldehyde 3-phosphate, *B*–Pacemaker
(d) *A*–Phosphoglucoisomerase, *B*–Pacemaker

15 The flowchart given below show the steps of glycolysis. Identify the missing enzymes.

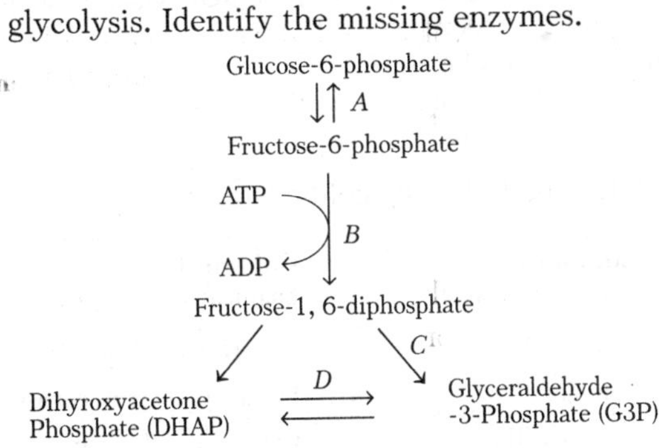

	A	*B*	*C*	*D*
(a)	Phosphogluco isomerase	Phosphofructo kinase	Aldolase	Triose phosphate isomerase
(b)	Phosphoglycero kinase	Enolase	Pyruvate kinase	Triose phosphate isomerase
(c)	Hexokinase	Phosphogluco isomerase	Phosphofruto kinase	Aldolase
(d)	Aldolase	Phosphoglycero mutase	Enolase	Pyruvate kinase

16 Pyruvic acid, the key product of glycolysis can have many metabolic fates. Under aerobic condition it forms **[NCERT Exemplar]**
(a) lactic acid (b) $CO_2 + H_2O$
(c) acetyl Co-A + CO_2 (d) ethanol + CO_2

17 All are true about glycolysis, except
(a) it is a ten-step process
(b) glucose and fructose are phosphorylated to give rise to glucose-6-phosphate by the activity of hexokinase
(c) reduces two molecules of NAD^+ for energy glucose molecule processes
(d) it produces one pyruvic acid molecule from each glucose molecule

18 Which of the following is true regarding glycolysis?
I. The step glucose → glucose-6-phosphate is energy investment phase
II. Has no connection with ETC.
Choose the correct option.
(a) Statement I is true, but II is false
(b) Statement I is false, but II is true
(c) Both statements I and II are true
(d) Both statements I and II are false

19 Match the following columns.

Column I	Column II
A. Fats made of three fatty acid chains attached to glycerol	1. Pyruvic acid
B. Glycolysis metabolite made from glycerol	2.Glucose-6-phosphate
C. First intermediate of glycolysis	3. Glyceraldehyde
D. Resultant of glycolysis	4. Triglycerides

Codes

	A	B	C	D		A	B	C	D
(a)	3	4	2	1	(b)	4	3	2	1
(c)	1	4	3	2	(d)	4	2	1	3

20 Which of the following is not true for glycolysis?
(a) End product is CO_2 and H_2O
(b) Substrate level phosphorylation
(c) Production of H_2O
(d) Expenditure of ATP

Krebs' Cycle

21 Krebs' cycle was discovered by Krebs' in pigeon muscle in 1937. Which step is called gateway step/link reaction /transition reaction in respiration?
(a) Glycolysis
(b) Formation of acetyl Co-A
(c) Citric acid formation
(d) ETS terminal oxidation

22 All enzymes of TCA cycle are located in the mitochondrial matrix except one, which is located in inner mitochondrial membranes in eukaryotes and in cytosol in prokaryotes. This enzyme is
(a) Lactate dehydrogenase
(b) Isocitrate dehydrogenase
(c) Malate dehydrogenase
(d) Succinate dehydrogenase

23 The diagram shows Krebs' cycle

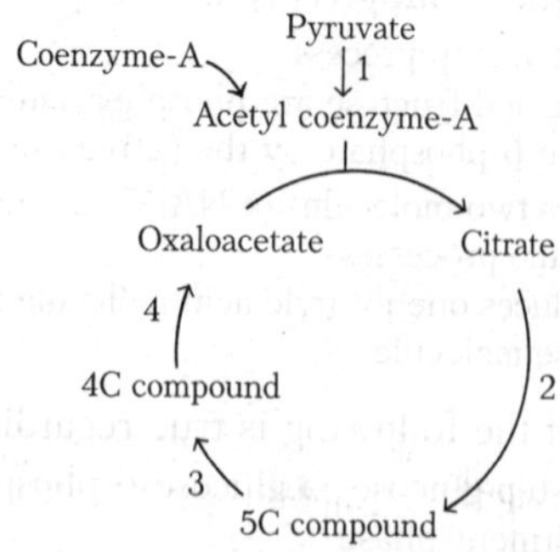

At which numbered stages does decarboxylation takes place?

(a) 1 and 2 (b) 1, 2 and 3
(c) 1, 3 and 4 (d) 1, 2, 3 and 4

24 The correct sequence of acids in TCA cycle is

(a) 1 : 3 PGA → 3 PGA → 2 PGA
(b) OAA → AA → PA
(c) Isocitric acid → Cis aconitic acid → succinic acid
(d) Succinic acid → Fumaric acid → Malic acid

25 Consider the following statements. During Krebs' cycle,

I. acetyl Co-A combines with 4 carbon oxaloacetic acid to produce 6 carbon citric acid.
II. in the presence of α-ketoglutarate dehydrogenase enzyme, α-ketoglutaric acid is converted into succinyl Co-A.

Choose the correct option.

(a) Statement I is true, but II is false
(b) Statement I is false, but II is true
(c) Both statements I and II are true
(d) Both statements I and II are false

26 The correct sequence of reaction in Krebs' cycle is

(a) isocitric acid → α-ketoglutaric acid → succinyl-Co-A
(b) isocitric acid → citric acid → α-ketoglutaric acid
(c) oxalosuccinic acid → citric acid → α-ketoglutaric acid
(d) α-ketoglutaric acid → isocitric acid → oxalosuccinic acid

27 Oxidation of succinic acid to fumaric acid in the Krebs' cycle takes place by

(a) addition of oxygen to it
(b) removal of hydrogen from it
(c) loss of electron from it
(d) None of the above

28 In which of the following steps of Krebs' cycle, substrate level phosphorylation occurs?

(a) Citrate → Isocitrate
(b) α-ketoglutarate → Succinate
(c) Succinyl Co-A → Succinate
(d) Malate → Oxaloacetate

29 Arsenite dehydrogenase can block

(a) Isocitrate → α-keto glutarate
(b) α-keto glutarate → Succinyl Co-A
(c) Malate → Oxaloacetate
(d) Pyruvate → Acetyl Co-A

30 Read the following statements with respect to Krebs' cycle and select the correct option.

I. Acetyl Co-A, a high energy compound acts as the substrate entrant for the cycle.
II. All the enzyme of Krebs' cycle are present in the mitochondrial matrix.
III. A single turn of Kerbs' cycle produces 1 $FADH_2$, 4 $NADH_2$ and 1 ATP.

(a) Only I and II are correct
(b) Only I and III are correct
(c) Only II and III are correct
(d) I, II and III are correct

Electron Transport Chain and Oxidative Phosphorylation

31 In electron transport chain, a carrier holds electron at

(a) high energy level than the previous carrier
(b) lower energy than the previous carrier
(c) same energy level as the adjacent one
(d) None of the above

32 Choose the incorrect statement.

(a) Flow of electrons in ETS is
$Fe^{+3} \longrightarrow Fe^{+2} \longrightarrow Fe^{+3}$
(b) In ETS electrons move from high negative to high positive redox potential
(c) Cyt-a_3 has Fe and Cu
(d) Cytochrome are non-proteinaceous, but ubiquinone is proteinaceous

33 Match the following columns about ETS.

Column I (ETS complexes)	Column II (Names)
A. Complex I	1. NADH dehydrogenase
B. Complex II	2. Succinate dehydrogenase
C. Complex III	3. Cytochrome-bc_1 complex
D. Complex IV	4. Cytochrome-c oxidase complex

Codes

	A	B	C	D		A	B	C	D
(a)	1	4	3	2	(b)	1	2	3	4
(c)	2	3	1	4	(d)	3	4	1	2

34 The correct sequence of electron acceptor in ATP synthesis is

(a) Cyt-a, a, b, c (b) Cyt-b, c, a, a_3
(c) Cyt-b, c, a_3, a (d) Cyt-c, b, a, a_3

35 Which can accept a hydride ion during ETS?

(a) FADH, NADH (b) FAD^+, $NADP^+$
(c) FAD^+, NADH (d) FADH, NAD^+

36 Which of the following is least electronegative?

(a) NADH (b) O_2
(c) Cyt-a_3 (d) Cyt-b

37 Identify P and Q in the given diagram of electron transport system.

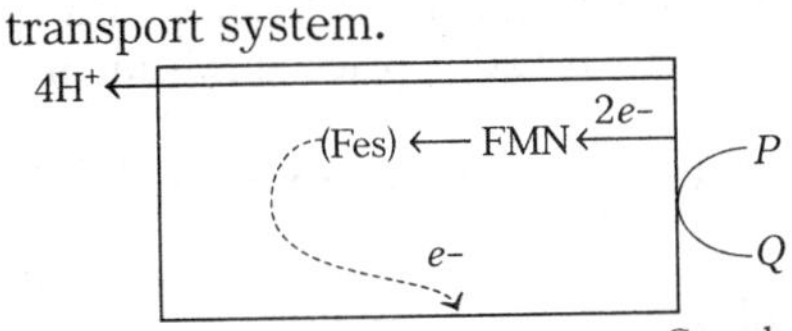

Complex I
(NADH dehydrogenase)

P	Q
(a) NADH + H^+	NAD^+
(b) NAD^+	NADH + H^+
(c) $FMNH_2$	$NADH_2$
(d) NADH + H^+	$FMNH_2$

38 Ubiquinone receives reducing equivalents *via*., $FADH_2$ (complex II) that is generated in TCA cycle by
(a) oxidation of fumaric
(b) oxidation of succinate
(c) carboxylation of citric acid
(d) dehydrogenation of citric acid

39 The ETC in cells is directly coupled with
(a) hydrolysis of hexose phosphate
(b) production of pyruvate
(c) reduction of NAD
(d) synthesis of ATP

40 Consider the following statements.
I. F_0 head piece contains the site for the synthesis of ATP from ADP + Pi.
II. F_1 part forms the channel through which protons cross the inner membrane.
Choose the correct option.
(a) Statement I is true, but II is false
(b) Statement I is false, but II is true
(c) Both statements I and II are true
(d) Both statements I and II are false

41 Oxidative phosphorylation is the formation of ATP by energy released from electrons removed during
(a) substrate reduction (b) substrate dehydrogenation
(c) substrate oxidation (d) substrate carboxylation

42 During respiration, most energy flows in the sequence
(a) Glucose → ETS → NADH → PMF → ATP
(b) Glucose → PMF → NADH → ETS → ATP
(c) Glucose → NADH → ETS → PMF → ATP
(d) Glucose → ATP → NADH → FADH → ETS

43 In oxidative phosphorylation, the last 3 steps are as follows
(a) Co-Q → Cyt-*c* is H^+ absorbing site
(b) Cyt-aa_3 → O_2, H^+ yielding site
(c) Co-Q → Cyt-*c* is H^+ yielding site and Cyt-aa_3 → O_2 is H^+ absorbing site
(d) No H^+ is absorbed or released

44 The end product of oxidative phosphorylation is **[NCERT Exemplar]**
(a) NADH (b) oxygen (c) ADP (d) ATP + H_2O

45 Choose the correct statement.
(a) Ubiquinone receives redox equivalents *via* $FADH_2$, but not NADH + H^+
(b) Oxidative phosphorylation is production of ATP by chemiosmosis
(c) Cytochrome-*c* is completely embeded in the inner mitochondrial membrane
(d) Cytochrome-*c*-oxidase possess cytochromes-a and a_3, and one copper centre

46 Which of the following is not true for oxidative phosphorylation?
(a) It uses oxygen as the initial e^- donor
(b) It involves the redox reactions of electrons transport chain
(c) It involves an ATP synthesis located in the inner mitochondrial membrane
(d) It depends on chemiosmosis

47 Chemiosmotic theory of ATP synthesis in the chloroplasts and mitochondria is based on
(a) proton gradient
(b) accumulation of K^+ ions
(c) accumulation of Na^+ ions
(d) membrane potential

48 In chemiosmotic hypothesis, for each ATP produced $2H^+$ pass through ......*A*...... from the ......*B*...... to the ......*C*...... down the electrochemical proton gradient.

	A	*B*	*C*
(a)	Intermembrane sapce	F_0	F_1
(b)	F_0	Intermembrane space	Matrix
(c)	F_0	F_1	Matrix
(d)	Matrix	F_0	F_1

Pentose Phosphate Pathway

49 Consider the following statements.
I. Pentose phosphate pathway occurs in cytoplasm of both eukaryotes and prokaryotes.
II. This pathway is resistant to cyanide.
Choose the correct option.
(a) Statement I is true, but II is false
(b) Statement I is false, but II is true
(c) Both statements I and II are true
(d) Both statements I and II are false

50 The oxidation of 3 mole of glucose by the pentose phosphate pathway may result in the production of
(a) 2 mole of pentose, 4 mole of NADH and 8 mole of CO_2
(b) 3 mole of pentose, 4 mole of NADH and 3 mole of CO_2
(c) 3 mole of pentose, 6 mole of NADPH and 3 mole of CO_2
(d) 4 mole of pentose, 3 mole of NADPH and 3 mole of CO_2

51 Pentose phosphate pathway may be referred to as
(a) Pentose phosphate cycle
(b) Warburg-Dickens Horecker shunt
(c) Hexose monophosphate cycle
(d) All of the above

52 Which one of the following statements is correct about the pentose phosphate pathway?
(a) It generates 6 mole of CO_2 for each mole of glucose consumed
(b) It generate 38 mole of ATP per mole of glucose consumed
(c) It is a reductive pathway that consume NADH
(d) Present in plants, but not in animals

53 Match the following columns.

	Column I		Column II
A.	Insulin	1.	Oxidative irreversible
B.	High concentration of NADPH	2.	Stimulate HMP shunt
C.	1st phase of HMP	3.	Non-oxidative reversible
D.	2nd phase of HMP	4.	Inhibits HMP shunt

Codes

	A	B	C	D		A	B	C	D
(a)	2	4	1	3	(b)	1	2	3	4
(c)	4	3	2	1	(d)	2	1	4	3

54 Erythrose-4-Phosphate (E-4-P) formed in pentose phosphate pathway has its applications in
(a) synthesis of nucleotides
(b) synthesis of fatty acids
(c) synthesis of carbohydrates
(d) synthesis of aromatic amino acids

Factors Affecting Aerobic Cellular Respiration

55 The graph shows the relation between light intensity and the giving and taking up of carbon dioxide by the leaves of a plant. Why is most carbon dioxide given off when the light intensity is zero units?

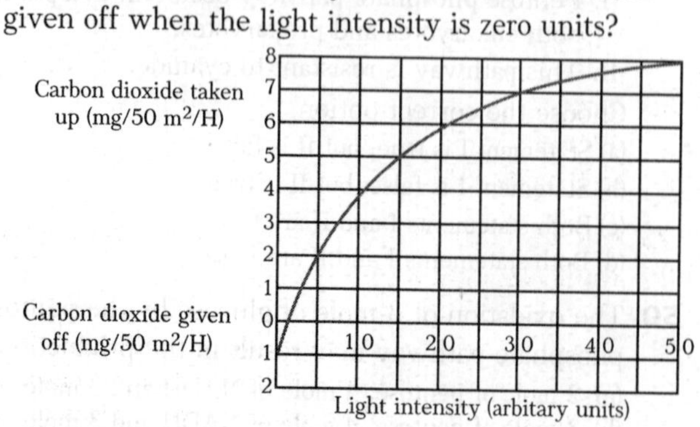

(a) Because it is just the start of the experiment
(b) Only respiration is taking place at this intensity of light
(c) Only photosynthesis is taking place at this intensity of light
(d) The rate of photosynthesis is equivalent to the rate of respiration

56 Consider the following statements.
I. Rate of respiration increases with an increase in temperature from 0°C to 30°C.
II. At very high temperature such as 50°C or more, rate of respiration decreases due to enzymatic degradation.

Choose the correct option.
(a) Statement I is true, but II is false
(b) Statement I is false, but II is true
(c) Both statements I and II are true
(d) Both statements I and II are false

57 Refrigerated fruits maintain their flavour and taste for longer period due to
(a) non-availability of oxygen
(b) the presence of excess of CO_2
(c) the presence of excess humidity
(d) slower rate of respiration

58 Dry seeds can tolerate higher temperature than the germinating seeds due to the reason
(a) dry seeds have more reserve food
(b) hydration makes the enzymes more sensitive to temperature
(c) seeds are meant for perennation
(d) None of the above

59 If the net rate of respiration is higher than that of photosynthesis, what will happen to plant?
(a) More growth will be there
(b) Plant will die due to starvation
(c) Continue to grow
(d) None of the above

Anaerobic Cellular Respiration

60 .... is a way of harvesting chemical energy without using either oxygen or ETC.
(a) Fermentation
(b) Lactic acid cycle
(c) Glycolysis in liver cells
(d) Pentose phosphate shunt in liver

61 The enzyme facilitates the conversion of pyruvic acid into ethyl alcohol is
(a) oxygenase
(b) decarboxylase
(c) decarboxylase and dehydrogenase
(d) hexokinase

62 In alcoholic fermentation,
(a) triose phosphate is the electron donor, while acetaldehyde is the electron acceptor
(b) triose phosphate is the electron donor, while pyruvic acid is the electron acceptor
(c) there is no electron donor
(d) oxygen is the electron acceptor

63 During alcoholic fermentation by yeast two molecules of glucose produce
(a) 3 molecules of ethanol + 3 molecules of CO_2
(b) 6 molecules of ethanol + 6 molecules of CO_2
(c) 2 molecules of ethanol + 2 molecules of CO_2
(d) 4 molecules of ethanol + 4 molecules of CO_2

64 Consider the following statements for the process of fermentation.
I. No ATP are produced during fermentation as it is extracellular process in some microbes.
II. Maximum concentration of alcohol in alcoholic beverage is 100%.
Choose the correct option.
(a) Statement I is true, but II is false
(b) Statement I is false, but II is true
(c) Both statements I and II are true
(d) Both statements I and II are false

65 LAB can ferment lactose sugar formed in milk. What is LAB in the statement?
(a) Lactic Acid Breakdown
(b) Lactic Acid Bacteria
(c) Lactic Adenine Breakdown
(d) None of the above

66 When a molecule of pyruvic acid is subjected to anaerobic oxidation and produced lactic acid, there is a net
(a) loss of 3 ATP molecules
(b) loss of 6 ATP molecules
(c) gain of 2 ATP molecules
(d) gain of 6 ATP molecules

67 Which of the following describes significance of fermentation?
I. Production of alcohol in brewing industry.
II. Making of dough in baking industry.
III. Curing of tea and tobacco.
IV. Production of vinegar by acetic acid bacteria.
Choose the correct option.
(a) I, II and III (b) I, II and IV
(c) II, III and IV (d) I, II, III and IV

68 Which of the following enzymes is not required during fermentation in microbes?
(a) Alcohol dehydrogenase
(b) Pyruvate dehydrogenase
(c) Lactic acid dehydrogenase
(d) Pyruvate decarboxylase

69 Alcoholic fermentation differs from lactic acid fermentation in
(a) oxidation of NADH to NAD^+
(b) net gain of 2ATP molecules
(c) involving decarboxylase activity
(d) more than one option is correct

70 What will happen if fermentation is allowed to proceed in a closed vessel?
(a) Vacuum will result
(b) No change will be there
(c) Pressure will develop because of excessive CO_2
(d) Pressure will develop because of excessive O_2

71 Which one of the following statements is false ?
(a) Less than 7% of energy in glucose is released during lactic or alcoholic fermentation
(b) Fermentation is observed in all cells
(c) O_2 is an essential requirement for aerobic respiration, but it enters the respiratory process at the end
(d) In aerobic respiration glucose is broken down into CO_2 and H_2O

Respiratory Quotient

72 RQ (Respiratory Quotient)
I. is the ratio of the volume of CO_2 taken in and volume of oxygen evolved in respiration.
II. depends upon the type of respiratory substrate used during respiration.
Choose the correct option.
(a) Statement I is true, but II is false
(b) Statement I is false, but II is true
(c) Both statements I and II are true
(d) Both statements I and II are false

73 $2C_{51}H_{98}O_6 + 145O_2 \rightarrow 102\ CO_2 + 98H_2O$ + Energy
The RQ of the above reaction is
(a) 1 (b) 0.7
(c) 1.45 (d) 1.62

74 Which of the following respiratory substrate has RQ value more than one?
(a) Tripalmitin
(b) Malic acid
(c) Oxalic acid
(d) Both (b) and (c)

75 If RQ is less than 1.0 in a respiratory metabolism, it would mean that
(a) carbohydrates are used as respiratory substrate
(b) organic acids are used as respiratory substrate
(c) the oxidation of the respiratory substrate consumed more oxygen than the amount of CO_2 released
(d) the oxidation of respiratory substrate consumed less oxygen than the amount of CO_2 is released

76 The Respiratory Quotient (RQ) of some of the compounds are 4, 1 and 0.7. These compounds are identified respectively as
(a) malic acid, palmitic acid and tripalmitin
(b) oxalic acid, carbohydrate and tripalmitin
(c) tripalmitin, malic acid and carbohydrate
(d) palmitic acid, carbohydrate and oxalic acid

77 The Respiratory Quotient (RQ) of a germinating castor seed is
(a) equal to one
(b) greater than one
(c) less than one
(d) equal to zero

78 In germinating seed, RQ falls when there is a shift from
(a) carbohydrate to fat as substrate
(b) fat to carbohydrate
(c) aerobic to anaerobic respiration
(d) protein to carbohydrate

79 Select the correct combination of the respiratory substrates and their respective RQs.

	Organic acids	Fats	Succulents
(a)	1.3	0.7	Zero
(b)	Infinity	0.7	Zero
(c)	Zero	1.3	0.7
(d)	Zero	0.7	1.3

80 The correct order of the values of respiratory quotient for glucose, fats and organic acid is
(a) Glucose > Fats > Organic acid
(b) Glucose < Organic acid < Fats
(c) Fats > Glucose > Organic acid
(d) Fats < Glucose < Organic acid

81 Which of the following compound will have RQ less than unity ?
(a) $C_{51}H_{98}O_6$
(b) $C_{57}H_{104}O_6$
(c) $C_4H_6O_5$
(d) More than one option is correct

(B) Cumulative Exercise

(Assorted Questions of the Chapter for Advanced Level Practice)

1 Which of the following pathway outlines the order of events during aerobic respiration?
(a) Glucose → triose phosphate → pyruvate → Krebs' cycle → $CO_2 + H_2O$ + ATP
(b) Glucose→ triose phosphate → pyruvate → Krebs' cycle → $CO_2 + H_2O$ + ADP + Pi
(c) Glucose→ hexose phosphate → pyruvate → Krebs' cycle → $CO_2 + H_2O$ + ADP + Pi
(d) Glucose → hexose phosphate → pyruvate → Krebs' cycle → ethanol + CO_2 + ATP

2 ATP can be formed by oxidative phosphorylation in the ETC and by glycolysis. In the complete oxidation of one molecule of glucose, approximately what percentage of ATP is formed by oxidative phosphorylation?
(a) 10%
(b) 25%
(c) 75%
(d) 90%

3 Match the following columns.

Column I		Column II
A. Double aminated compound	1.	Glycogen
B. Storage form of glucose	2.	Asparagine
C. Fats made of three fatty acid chain attached to glycerol	3.	citric acid
D. Tricarboxylic acid	4.	Triglyceride

Codes

	A	B	C	D
(a)	2	1	4	3
(b)	1	4	3	2
(c)	2	1	3	4
(d)	1	2	3	4

4 In the process of respiration in plants 180 gm of sugar + 192 gm of oxygen produce
(a) large amount of CO_2, no water and no energy
(b) 132 gm of CO_2, 54 gm of water and 337 Kcals of energy
(c) 264 gm of CO_2, 108 gm of water and 674 Kcals of energy
(d) 528 gm of CO_2, 216 gm of water and 1348 Kcal of energy

5 Respiratory balanced sheet prepared for oxidation of one molecule of glucose is based on certain assumptions. Read the following statements and state true (T) or false (F).
I. Sequential orderly functioning pathway including glycolysis, TCA cycle and ETS.
II. Only glucose is being respired, no other alternative substrate is entering in the pathway.
III. Oxidation of extra-mitochondrial NADH occurs through oxidative phosphorylation.
IV. The intermediates of TCA cycle are used in different anabolic pathway.

	I	II	III	IV
(a)	T	T	T	F
(b)	T	T	F	F
(c)	T	F	T	F
(d)	T	F	F	F

6 Match the following columns.

Column I (Stages of aerobic respiration of a glucose molecule)	Column II (ATP produced through ETS)
A. Glycolysis	1. 6
B. Formation of acetyl Co-A	2. 6
C. Krebs' cycle	3. 2

Codes

	A	B	C		A	B	C
(a)	1	2	3	(b)	2	1	3
(c)	3	2	1	(d)	3	1	2

7 The complete oxidation of one mole of glucose yields 2880 KJ of energy. The addition of one phosphate molecule to ADP requires 30.6 KJ of energy per mole. In aerobic respiration, 38 molecules of ATP are formed as a result of the breakdown of each glucose molecule. Which figure best represents the efficiency of aerobic respiration in trapping the energy released by the glucose molecule?

(a) 23% (b) 36% (c) 40% (d) 45%

8 The diagram shows a mitochondrion in a cell.

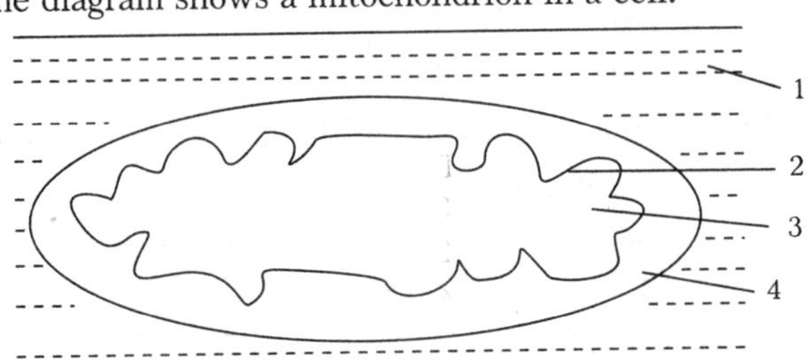

Where is ATP produced from ADP during respiration?

(a) 1 and 2 only (b) 1, 2 and 3
(c) 2 and 3 only (d) 2, 3 and 4

9 In order for glucose to be used as an energy source, it is necessary that

(a) glucose be formed from fructose
(b) glucose be degraded to CO_2
(c) 2 molecules of ATP be invested in the system
(d) None of the above

10 Consider the following statements with respect to respiration.

I. Glycolysis occurs in the cytoplasm of the cell.
II. Aerobic respiration takes place within the mitochondria.
III. Electron transport system is present in the outer mitochondrial membrane.
IV. $C_{51}H_{98}O_6$ is the chemical formula of tripalmitin a fatty acid.
V. Respiratory quotient = $\frac{\text{Volume of } O_2 \text{ evolved}}{\text{Volume of } CO_2 \text{ consumed}}$

Of the above statements.

(a) I, II and IV are correct
(b) II, III and IV are correct
(c) III, IV and V are correct
(d) I, III and V are correct

11 Match the following columns.

Column I	Column II
A. RQ	1. Chemiosmotic ATP synthesis
B. Peter Mitchell	2. Muscle fatigue
C. Cytochromes	3. Inner mitochondrial membrane
D. Lactic acid	4. Alcoholic fermentation
E. Yeast	5. Respirometer

Codes

	A	B	C	D	E		A	B	C	D	E
(a)	5	1	3	4	2	(b)	5	1	3	2	4
(c)	1	5	2	3	4	(d)	5	2	4	3	1

12 The crucial event in aerobic respiration which starts after transport of pyruvate from the cytoplasm into the mitochondria includes

(a) stepwise removal of all the H-atom from pyruvate
(b) releasing three molecules of CO_2 per molecule of pyruvate
(c) simultaneous synthesis of ATP when electrons are removed as part of hydrogen atoms to molecule oxygen
(d) All of the above

13 The table shows events that may occur in the Calvin cycle, Krebs' cycle or in glycolysis. In which of these processes do these events occur?

	ATP used	Dehydrogenation	Reduced NADP used
(a)	Calvin cycle	Glycolysis	Krebs' cycle
(b)	Glycolysis	Calvin cycle	Krebs' cycle
(c)	Glycolysis	Krebs' cycle	Calvin cycle
(d)	Krebs' cycle	Glycolysis	Calvin cycle

14 Which substances enter and leave a mitochondrion during aerobic respiration?

	Enter	Leaves
(a)	Acetyl Co-A	CO_2
(b)	ADP	Lactate
(c)	Glucose	Reduced NAD
(d)	Pyruvate	ATP

15 Which one of the following statements correctly describes relationship between the Krebs' cycle and electron transport pathway?

(a) The Krebs' cycle releases H^+ used by electron transport
(b) The electron transport pathway obtains electron from the CO_2 produced by the Krebs' cycle
(c) The Krebs' cycle and electron transport pathway, both produce ATP
(d) NADH + H^+ produced by Krebs' cycle is used to make ATP by electron transport

16 Mechanism of phosphorylation is depicted in the diagram below. Study it carefully and mark the correct option.

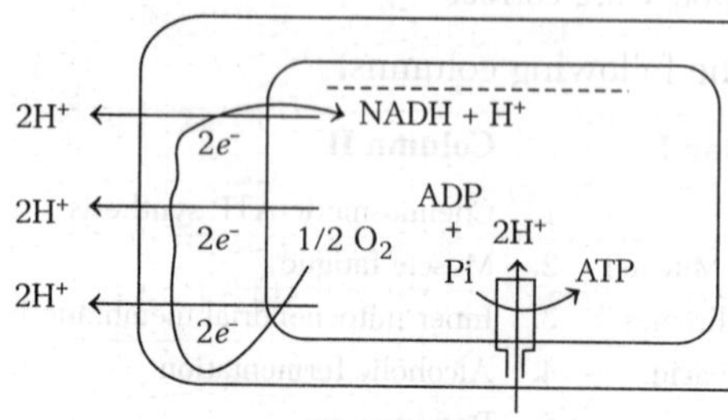

(a) The diagram depicts photosynthetic phosphorylation taking place in the chloroplast. The incident light should be shown in the diagram
(b) The diagram depicts oxidative phosphorylation taking place in mitochondria. However, the flow of electrons should be shown in reverse direction
(c) Diagram depicts the basic process of both oxidative as well as photosynthetic phosphorylation. However, the proton concentration should be high inside and low outside
(d) The diagram correctly depicts the oxidative phosphorylation occurring in all heterotrophic organisms

17 Identify the wrong step from the following.
(a) Citric acid (6C) → α-ketoglutaric acid (5C)
(b) Oxaloacetic acid (4C) → Malic acid (5C)
(c) α-ketoglutaric acid (5C) → Succinic acid (4C)
(d) Malic acid (4C) → Oxaloacetic acid (4C)

18 What happens to most of the redused NAD molecules in cell metabolism?
(a) They act as oxidising agent in glycolysis
(b) They are oxidised in mitochondrial ATP formation
(c) They are oxidised in the Calvin cycle
(d) They combine with succinic acid as part of Krebs' cycle

19 The diagram shows some of the reaction following glycolysis during aerobic respiration.

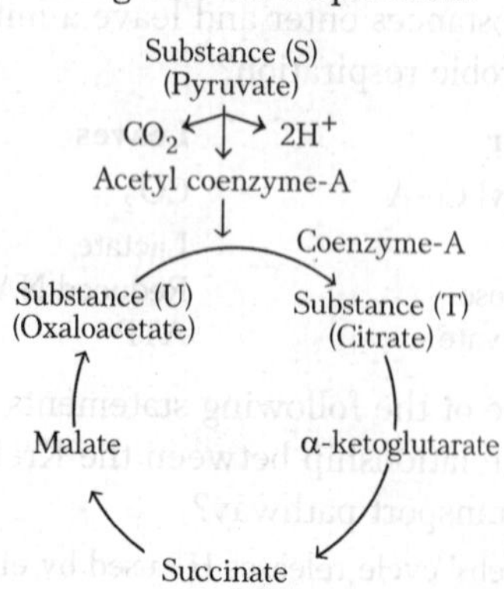

How many carbon atoms are in each of the substance S, T, and U.

	S	T	U		S	T	U
(a)	2	5	3	(b)	2	6	4
(c)	3	4	6	(d)	3	6	4

20 Select the incorrect statement with respect to the given representation.

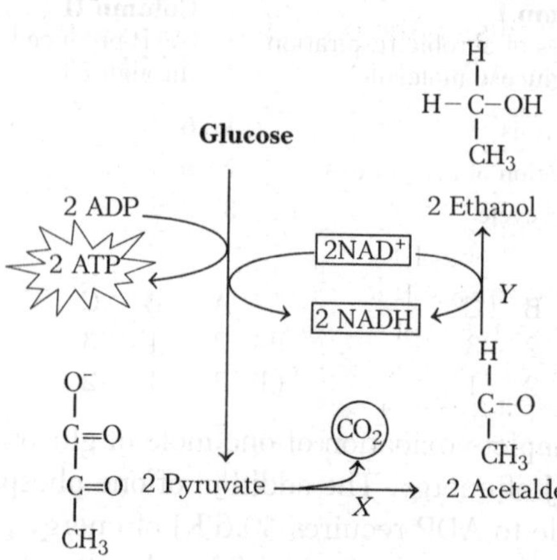

(a) X is the enzyme pyruvate dehydrogenase and Y is the enzyme ethanol decarboxylase
(b) This process is involved in brewing industry for producing beverages like beer, rum, whisky, etc.
(c) Accumulation of the end product (i.e. ethanol, during this process, in a culture of yeast stops the multiplication of yeast cells and may even lead to death of cells.
(d) None of the above

21 Match the following for the location of the steps of aerobic respiration.

Column I (Respiratory pathways)	Column II (Location)
A. Glycolysis	1. Inner mitochondrial membrane
B. TCA cycle	2. Mitochondrial matrix
C. ETS	3. Cytoplasm

Codes

	A	B	C		A	B	C
(a)	3	1	2	(b)	3	2	1
(c)	1	2	3	(d)	2	1	3

22 Which of the following is a true distinction between aerobic respiration and fermentation?
(a) Only aerobic respiration oxidises glucose
(b) In aerobic respiration NADH + H^+ is oxidised by ETC
(c) Substrate level phosphorylation is unique to fermentation, but not aerobic respiration
(d) Aerobic respiration is catabolic, but fermentation is not

23 Out of 38 molecules of ATP produced upon aerobic respiration of glucose, the breakup to ATP production in glycolysis (P), pyruvate to acetyl Co-A formation (Q) and Krebs' cycle (R) is as follows
(a) P = 1, Q = 6, R = 30
(b) P = 8, Q = 6, R = 24
(c) P = 8, Q = 10, R = 20
(d) P = 2, Q = 12, R = 24

24 The balance sheet for ATP production in glycolysis has been given below. Select the option which correctly fills up the blank for P, Q, R and S. ['X' stands for 'nil'].

Steps	ATP Utilisation	ATP Production
1. Glucose → Glucose-6-phosphate	P	X
2. Fructose-6-phosphate → Fructose-1, 6-bisphosphate	1	Q
3. 1, 3-bisphosphoglyceric acid → 3-phosphoglyceric acid	X	R
4. 2-phosphoenol pyruvic acid → Pyruvic acid	S	2

	P	Q	R	S
(a)	1	X	X	2
(b)	1	X	2	X
(c)	2	1	X	1
(d)	X	1	2	X

25 The total number of substrate level phosphorylation in cellular respiration as a result of complete oxidation of one glucose molecule is

(a) 6 (b) 3 (c) 5 (d) 4

26 Match the following and choose the correct option from those given below.

Column I	Column II
A. Molecular oxygen	1. α-ketoglutaric acid
B. Electron acceptor	2. Hydrogen acceptor
C. Pyruvate dehydrogenase	3. Cytochrome-*c*
D. Decarboxylation	4. Acetyl Co-A

Codes

	A	B	C	D
(a)	2	3	4	1
(b)	3	4	2	1
(c)	2	1	3	4
(d)	4	3	1	2

27 Match the following columns.

Column I	Column II
A. Peripheral membrane protein of ATP synthase	1. Complex-III
B. Pyruvate dehydrogenase complex	2. F_0
C. Integral membrane protein of ATP synthase	3. F_1
D. Cytochrome-*b* c_1	4. Mitochondrial matrix

Codes

	A	B	C	D
(a)	1	2	3	4
(b)	2	4	3	1
(c)	3	4	1	2
(d)	3	4	2	1

28 Match the following columns.

Column I	Column II
A. Pyruvate	1. 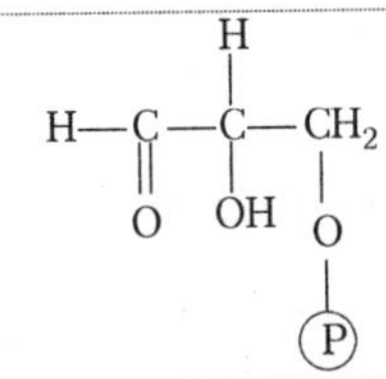
B. 3-phosphoglycerate	2. 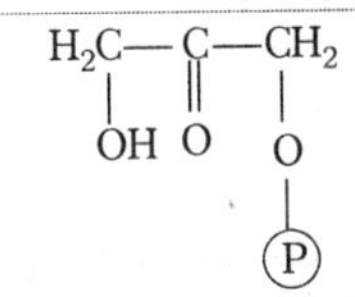
C. Dihydroxyacetone phosphate	3. 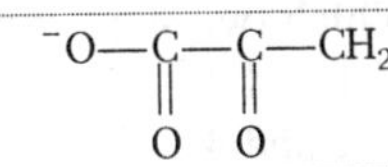
D. Glyceraldehyde-3-phosphate	4. 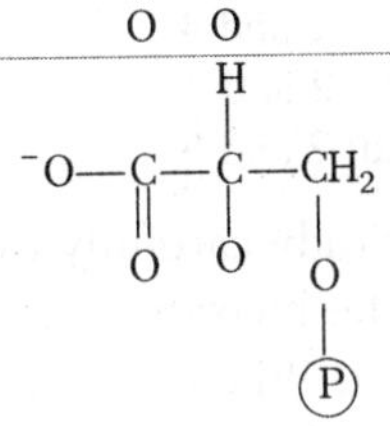

Codes

	A	B	C	D
(a)	1	4	2	3
(b)	2	1	4	3
(c)	3	4	2	1
(d)	1	2	4	3

29 Find out the correct statement.

(a) Glycolysis can be the main source of energy for plant roots in flooded soil

(b) Fermentation cannot be performed by eukaryotes

(c) For complete oxidation of glucose to CO_2 and H_2O Krebs' cycle is required, but O_2 is not a requirement

(d) Electron transport chain can occur without the presence of oxygen

30 The number of glucose molecule required to produce 38 ATP molecules under anaerobic conditions by a yeast cell is

(a) 2 (b) 4

(c) 19 (d) 25

31 In the presence of a metabolic poison that specifically and completely inhibits all functions of mitochondrial ATP synthase, what would you expect?

(a) An increase in the pH difference across the inner mitochondrial membrane

(b) A decrease in the pH difference across the inner mitochondrial membrane

(c) Increased O_2 consumption

(d) Increased ATP synthesis

32 The diagram summarises the pathway of glucose breakdown.

Hexose
↓1
Triose phosphate
↓2
Pyruvate $\xrightarrow{3}$ Lactate
↓
Acetyl Co-A
↓4
6 C compound
↓5
$H_2O + CO_2$

Which two steps result in a net increase of ATP?
(a) 1 and 3
(b) 1 and 4
(c) 2 and 4
(d) 2 and 5

33 Study carefully the following statements and select the incorrect ones.

I. When fats are used in respiration, the RQ is more than unity because fats contain more O_2 and require relatively less amount of O_2 for oxidation.
II. The most important energy carrier is ATP. This energy rich compound is mobile and can pass from one cell to another.
III. Before pyruvic acid enters Krebs' cycle, one of the two carbon atoms of pyruvic acid is reduced to carbon dioxide in the reaction called reductive carboxylation.
IV. A special electron carrier system located in the mitochondrial membrane is called shuttle system. It transfers electrons from the hydrogens of cytoplasmic NADH to the mitochondrial electron carriers across the mitochondrial membrane.
V. Zymase is a complex mixture of many enzymes which requires several coenzymes for its action. The enzyme complex-zymase catalyses series of reactions taking place during fermentation leading to the production of ethyl alcohol.

Codes
(a) I and II
(b) III and IV
(c) I, II and III
(d) III, IV and V

34 Suppose a total of 38 ATP are formed in aerobic respiration, then the number of ATP formed outside mitochondria will be
(a) 2 (b) 14
(c) 18 (d) 20

35 Refer to the following flowchart representing the cellular respiration and its fuels. Blanks *A*, *B*, *C*, *D*, *E*, *F* and *G* are respectively

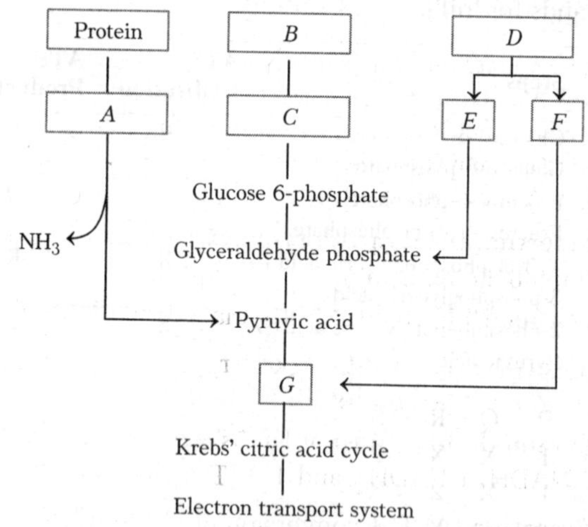

(a) amino acids, carbohydrate, glucose, fats, glycerol, fatty acid, acetyl Co-A
(b) fats, acetyl Co-A, amino acid, fatty acid, carbohydrate, glycerol, glucose
(c) fatty acid, glucose, acetyl Co-A, glycerol, fats, carbohydrate, amino acid
(d) carbohydrate, fats, glycerol, fatty acids, amino acids, glucose, acetyl Co-A

36 Read the following statements.

I. Glycerol enters the glycolytic pathway after being converted into phosphoglyceraldehyde (PGAL).
II. In fermentation, there is a net gain of only two molecules of ATP for each molecule of glucose degraded into pyurvic acid.
III. NADH + H^+ is oxidised to NAD^+ rather slowly in fermentation, however the reaction is very vigorous in case of aerobic respiration.
IV. If fatty acids are to be respired, they are first degradated into acetyl Co-A.
V. When skeletal muscle cells undergo anaerobic respiration, they become fattigued and pain occurs due to build up of lactate.

How many of the above statements are correct?
(a) Two (b) Five (c) Four (d) Three

37 There are three major ways in which different cells handle pyruvic acid produced by glycolysis. These are
(a) fermentation, TCA and ETS
(b) fermentation, aerobic respiration and TCA
(c) alcoholic fermentation, lactic acid fermentation and aerobic respiration
(d) alcoholic fermentation, lactic acid fermentation and ETS

Assertion and Reason

Direction (Q. Nos. 38-48) *In each of the following questions, a statement of Assertion (A) is given followed by corresponding statement of Reason (R). Of the statements, mark the correct answer as*

(a) Both A and R are true and R is the correct explanation of A
(b) Both A and R are true, but R is not the correct explanation of A
(c) A is true, but R is false
(d) A is false, but R is true
(e) Both A and R are false

38 **Assertion** (A) Respiratory pathway is considered as an amphibolic pathway.
Reason (R) It involves both anabolism and catabolism.

39 **Assertion** (A) Two turns of Krebs' cycle occur per glucose molecule used.
Reason (R) Each turn of Krebs' cycle produces 3 NADH, 1 $FADH_2$ and 1 ATP molecules.

40 **Assertion** (A) The conversion of glyceraldehyde-3-phosphate into 1, 3-bisphosphoglycerate is an oxidation reaction.
Reason (R) The reaction in which 1, 3-bisphosphoglycerate is formed from glyceraldehyde-3-phosphate utilises NAD^+.

41 **Assertion** (A) Carbohydrates are used first by most cells in respiration.
Reason (R) Lipids are never used in respiration.

42 **Assertion** (A) ATP formation during respiration is an exergonic reaction.
Reason (R) ATP formation requires input of energy.

43 **Assertion** (A) Aerobic respiration is bioenergetically more efficient than anaerobic glycolysis.
Reason (R) Aerobic respiration occurs in the presence of oxygen, while fermentation occurs in the absence of oxygen.

44 **Assertion** (A) ATP is the energy carrier of cells.
Reason (R) AMP is a nucleotide.

45 **Assertion** (A) To make ATP from ADP and inorganic phosphate, it requires 30.6 kJ of energy per mole.
Reason (R) ATP is found only in some cells.

46 **Assertion** (A) Glycolysis occurs in the matrix of mitochondria.
Reason (R) Krebs' cycle occurs on the cristae of mitochondria.

47 **Assertion** (A) Cytochrome oxidase enzyme contains copper.
Reason (R) Cyanide combines with the copper of cytochrome oxide and prevents oxygen to combine with it.

48 **Assertion** (A) Yeast cells perform alcoholic fermentation.
Reason (R) Fermentation is an aerobic process.

(C) Medical Entrances' Gallery

(Collection of Questions Asked in NEET & in Various Medical Entrance Exams)

1 The number of substrate level phosphorylation in one turn of citric acid cycle is **[NEET 2020]**
(a) one (b) two (c) three (d) zero

2 Pyruvate dehydrogenase activity during aerobic respiration requires **[NEET 2020]**
(a) calcium (b) iron (c) cobalt (d) magnesium

3 Conversion of glucose to glucose-6-phosphate, the first irreversible reaction of glycolysis, is catalysed by **[NEET 2019]**
(a) hexokinase (b) enolase
(c) phosphofructokinase (d) aldolase

4 In gluconeogenesis, which of the following cannot acts as substrate? **[JIPMER 2019]**
(a) Fructose (b) Palmitic acid
(c) Glycine (d) Glycerol

5 Respiratory Quotient (RQ) value of tripalmitin is **[NEET 2019, 15]**
(a) 0.7 (b) 0.07 (c) 0.09 (d) 0.9

6 ATP formation occurs through which of the following? **[AIIMS 2019]**
(a) Photophosphorylation
(b) Oxidative phosphorylation
(c) Substrate level phosphorylation
(d) All of the above

7 Where is respiratory Electron Transport System (ETS) located in plants? **[NEET 2019]**
(a) Mitochondrial matrix
(b) Outer mitochondrial membrane
(c) Inner mitochondrial membrane
(d) Intermembrane space

8 Which of the following statements regarding mitochondria is incorrect? **[NEET 2019]**
(a) Enzymes of electron transport are embedded in outer membrane
(b) Inner membrane is convoluted with infoldings
(c) Mitochondria matrix contains single circular DNA molecule and ribosomes
(d) Outer membrane is permeable to monomers of carbohydrates, fats and proteins

9 Which statement is incorrect for Krebs' cycle? **[NEET 2018, 17]**

(a) There are three points in the cycle where NAD^+ is reduced to $NADH + H^+$
(b) There is one point in the cycle where FAD^+ is reduced to $FADH_2$
(c) During conversion of succinyl Co-A to succinic acid, a molecule of GTP is synthesised
(d) The cycle starts with condensation of acetyl group (acetyl Co-A) with pyruvic acid to yield citric acid

10 Acetylation of pyruvate takes place in the **[MHT CET 2018]**

(a) perimitochondrial space
(b) mitochondrial matrix
(c) cristae
(d) F_1-particles

11 Oxidative decarboxylation of pyruvic acid results in the formation of **[Kerala CEE 2018, 14]**

I. Acetyl Co-A II. CO_2
III. ATP IV. NADH + $^+$

Select the correct answer using the codes given below.

(a) Only I (b) I and II (c) I, II and III (d) I, II and IV
(e) III and IV

12 How many ATP is/are produced when 1 molecule of $FADH_2$ is oxidised to FAD through electron transport system? **[WB JEE 2018]**

(a) 2 (b) 3 (c) 1 (d) 4

13 A small protein attached to the outer surface of the inner membrane and which acts as a mobile carrier for transfer of electrons between complexes III and IV is **[J&K CET 2018, 14]**

(a) cytochrome-*d* (b) cytochrome-*c*
(c) cytochrome-*b* (d) cytochrome-*a*

14 How many molecule(s) of pyruvic acid is/are formed in glycolysis ? **[JIPMER 2018]**

(a) 2 (b) 1 (c) 15 (d) 16

15 Which one is the correct sequence occurring in glycolysis? **[JIPMER 2018]**

(a) G-6-P $\rightarrow$ PEP $\rightarrow$ 3-PGAL $\rightarrow$ 3-PGA
(b) G-6-P $\rightarrow$ 3-PGAL $\rightarrow$ 3-PGA $\rightarrow$ PEP
(c) G-6-P $\rightarrow$ PEP $\rightarrow$ 3-PGA $\rightarrow$ 3-PGAL
(d) G-6-P $\rightarrow$ 3-PGA $\rightarrow$ 3-PGAL $\rightarrow$ PEP

16 RQ of malic acid is **[JIPMER 2018]**

(a) 1.9 (b) 1.49 (c) 1.33 (d) 1

17 What is the role of NAD^+ in cellular respiration?

(a) It is a nucleotide source of ATP synthesis **[NEET 2018]**
(b) It functions as an electron carrier
(c) It functions as an enzyme
(d) It is the final electron acceptor for anaerobic respiration

18 Which one of these statements is incorrect? **[NEET 2018]**

(a) Glycolysis operates as long as it is supplied with NAD that can pick up hydrogen atoms
(b) Glycolysis occurs in cytosol
(c) Enzymes of TCA cycle are present in mitochondrial matrix
(d) Oxidative phosphorylation takes place in outer mitochondrial membrane

19 Which one of the following biomolecules is common to respiration-mediated breakdown of fats, carbohydrates and proteins? **[NEET-II 2016, 13, KCET 2013]**

(a) Glucose -6- phosphate (b) Fructose-1, 6-bisphosphate
(c) Pyruvic acid (d) Acetyl Co-A

20 Oxidative phosphorylation is **[NEET-II 2016]**

(a) formation of ATP by transfer of phosphate group from a substrate to ADP
(b) oxidation of phosphate group in ATP
(c) addition of phosphate group to ATP
(d) formation of ATP by energy released from electrons removed during substrate oxidation

21 During anaerobic respiration, the conversion of pyruvate into acetaldehyde alongwith TPP, the cofactor is **[MHT CET 2015, 14]**

(a) Mg^{2+} (b) Mn^{2+} (c) Fe^{2+} (d) Zn^{2+}

22 What does the following equation represent?

$$C_6H_{12}O_6 + 2NAD + 2ADP + 2Pi \longrightarrow 2CH_3COCOOH + 2NADH_2 + 2ATP$$

(a) Complete glycolysis **[MHT CET 2015]**
(b) Complete aerobic respiration
(c) Complete anaerobic respiration
(d) Complete fermentation

23 How much of the energy released during aerobic respiration is approximately conserved in the form of ATP? **[MHT CET 2015]**

(a) 20% (b) 40% (c) 60% (d) 100%

24 Cytochromes are found in **[AIIMS 2015]**

(a) matrix of mitochondria (b) outer wall of mitochondria
(c) cristae of mitochondria (d) lysosomes

25 Given below are some reactions and enzymes involved. Identify the correct pairs. **[MHT CET 2015]**

A. Fructose-1, 6-diphosphate $\longrightarrow$ 3 PGAL + DHAP — 1. Enolase
B. Citrate $\longrightarrow$ *cis*-aconitase — 2. Thiokinase
C. Succinyl Co-A $\longrightarrow$ Succinate — 3. Aconitase
D. 2 PGA $\longrightarrow$ PEPA — 4. Aldolase

Codes

	A	B	C	D		A	B	C	D
(a)	4	3	2	1	(b)	1	2	3	4
(c)	2	1	4	3	(d)	3	4	1	2

26 When protein is aerobically oxidised, the RQ value will be **[Chhattisgarh PMT 2015]**

(a) one (b) zero
(c) more than one (d) less than one

27 Which of the following is the only 5-carbon compound formed during Krebs' cycle? **[Chhattisgarh PMT 2015]**

(a) Malic acid (b) Succinic acid
(c) *Cis*-aconitic acid (d) α-ketoglutaric acid

28 Which one is the first compound which is common for both glucose and fructose in glycolysis? **[BCECE 2015]**

(a) Fructose-6-P
(b) Glucose-6-P
(c) Fructose-1, 6-bisphosphate
(d) Fructose-1-P

29 Which molecule links glycolysis with fermentation as well as TCA cycle? **[BCECE 2015]**

(a) Ethanol (b) Acetaldehyde
(c) PEP (d) Pyruvic acid

30 In which one of following reactions of glycolysis, oxidation takes place? **[Kerala CEE 2015, OJEE 2013]**

(a) Glucose-6-P to fructose-6-P
(b) Fructose-6-P to fructose-1, 6-biphosphate
(c) 1, 3-bisphosphoglycerate to 3-phosphoglyceric acid
(d) 3-phosphoglyceraldehyde to 1, 3-bisphosphoglycerate
(e) 2-phosphoglycerate to phosphoenol pyruvate

31 FAD is an electron acceptor in citric acid cycle during the oxidation of **[Kerala CEE 2015]**

(a) malic acid to oxaloacetic acid
(b) succinic acid to fumaric acid
(c) citric acid to α-ketoglutaric acid
(d) α-ketoglutaric acid to succinic acid
(e) oxaloacetic acid to citric acid

32 Choose the correct combination of labelling the molecules involved in the pathway of anaerobic respiration in yeast. **[Kerala CEE 2015]**

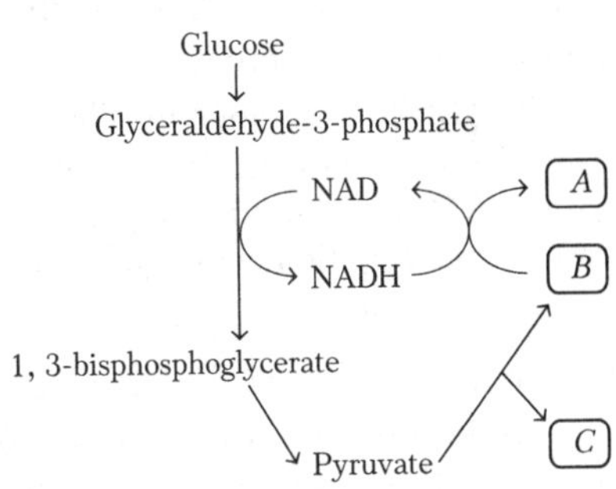

(a) *A*- CO_2, *B*- Ethanol, *C*-Acetaldehyde
(b) *A*- Ethanol, *B*-Acetaldehyde, *C*- CO_2
(c) *A*- Ethanol, *B*- CO_2, *C*-Acetaldehyde
(d) *A*- Acetaldehyde, *B*-CO_2, *C*- Ethanol
(e) *A*-CO_4, *B*-Acetaldehyde, *C*- Ethanol

33 Choose the correct statement. **[Kerala CEE 2015]**

(a) Oxygen is vital in respiration for removal of hydrogen
(b) Pyruvate is formed in the mitochondrial matrix
(c) There is complete breakdown of glucose in fermentation
(d) During the conversion of succinyl Co-A to succinic acid, a molecule of ATP is synthesised
(e) All of the above

34 Conversion of pyruvic acid into ethyl alcohol is facilitated by enzyme(s) **[COMED-K 2015]**

(a) carboxylase
(b) phosphatase
(c) dehydrogenase
(d) decarboxylase and dehydrogenase

35 During respiration .... **[Gujarat CET 2015]**

(a) 2 PGAL are evolved during glycolysis and none in Krebs' cycle
(b) 2 PGAL are evolved during glycolysis and two pyruvic acid in Krebs' cycle
(c) 2 PGAL are evolved during glycolysis and 4 pyruvic acid in Krebs' cycle
(d) PGAL is not produced during respiratory events

36 Which of the two statements together support that respiratory pathway is an 'amphibolic pathway'? **[EAMCET 2015]**

I. Fats breakdown to glycerol and fatty acids, subsequently yielding acetyl Co-A.
II. In respiration C—C bonds of complex compounds breakdown through oxidation leads to release of energy.
III. Acetyl Co-A from respiratory pathway is withdrawn for synthesis of fatty acids.
IV. Proteins are degraded by protease to amino acids and enter the respiratory pathway.

Codes

(a) I and II (b) I and IV
(c) II and IV (d) I and III

37 Choose the correct sequence of electron pathway in ETS. **[EAMCET 2015]**

(a) Cyt. oxidase $\longrightarrow$ Cyt. reductase $\longrightarrow$ Succinate dehydrogenase $\longrightarrow$ NAD dehydrogenase
(b) NADH dehydrogenase $\longrightarrow$ Succinate dehydrogenase $\longrightarrow$ Cyt-*c* reductase $\longrightarrow$ Cyt-*c* oxidase
(c) NADH dehydrogenase $\longrightarrow$ Cyt-*c* reductase $\longrightarrow$ Cyt-*c* oxidase $\longrightarrow O_2$
(d) Succinic dehydrogenase $\longrightarrow$Cyt. oxidase $\longrightarrow$ Cyt. reductase $\longrightarrow O_2$

38 In which one of the following processes CO_2 is not released? **[CBSE AIPMT 2014]**

(a) Aerobic respiration in plants
(b) Aerobic respiration in animals
(c) Alcoholic fermentation
(d) Lactate fermentation

39 How many NAD molecules get reduced in complete oxidation of one glucose molecule? **[MHT CET 2014]**

(a) Two (b) Five
(c) Ten (d) Twelve

40 Acetylation of pyruvate takes place in the **[MHT CET 2014]**

(a) perimitochondrial space
(b) mitochondrial matrix
(c) cristae
(d) F_1-particles

41 In which of the following steps of citric acid cycle, CO_2 is evolved? **[Kerala CEE 2014]**

I. Citric acid — α- ketoglutaric acid
II. Succinic acid — Malic acid
III. Malic acid — Oxaloacetic acid
IV. α- ketoglutaric acid — Succinyl Co-A

Select the correct answer using the codes given below.

(a) I and II (b) I and IV
(c) II and III (d) II and IV
(e) III and IV

42 Select the correct order of reactions in glycolysis. **[Kerala CEE 2014]**

I. Conversion of 3-phosphoglyceraldehyde to 1, 3-bisphosphoglycerate.
II. Conversion of 3-phosphoglyceric acid to 2-phosphoglycerate.
III. Conversion of-1, 3-BPGA to 3 phosphoglyceric acid.
IV. Splitting of fructose-1, 6-bisphosphate into dihydroxy acetone phosphate and 3-phosphoglyceraldehyde.

Codes

(a) IV, III, I and II (b) II, III, I and IV
(c) IV, II, I and III (d) I, IV, III and II
(e) IV, I, III and II

43 The Respiratory Quotient (RQ) of glucose is **[WB JEE 2014]**

(a) 0.5 (b) 0.7
(c) 1.0 (d) 1.5

44 How many ATP are produced when 1 molecule of $FADH_2$ is oxidised to FAD through electron transport system? **[WB JEE 2014]**

(a) Two (b) Three (c) One (d) Four

45 When respiratory quotient is less than 1.0 in a respiratory metabolism, it means that **[J&K CET 2014]**

(a) carbohydrates are used as respiratory substrate
(b) volume of carbon dioxide evolved is less than volume of oxygen consumed
(c) volume of carbon dioxide evolved is more than volume of oxygen consumed
(d) volume of carbon dioxide evolved is equal to volume of oxygen consumed

46 During glycolysis fructose-1, 6- bisphosphate is split into **[J&K CET 2014]**

(a) dihydroxyacetone phosphate and 2-phosphoglyceraldehyde
(b) dihydroxyacetone phosphate and 1-phosphoglyceraldehyde
(c) dihydroxyacetone phosphate and 2-phosphoglycerate
(d) dihydroxyacetone phosphate and 3-phosphogly ceraldehyde

47 There are three major ways in which different cells handle pyruvic acid produced by glycolysis, these are **[UK PMT 2014]**

(a) lactic acid fermentation, alcoholic fermentation, aerobic respiration
(b) oxaloacetic acid fermentation, lactic acid fermentation, aerobic fermentation
(c) alcoholic fermentation, oxaloacetic acid fermentation, citric acid fermentation
(d) citric acid fermentation, lactic acid fermentation, alcoholic fermentation

48 The three boxes in this diagram represents the three major biosynthetic pathways in aerobic respiration. Arrows represent net reactants or products.

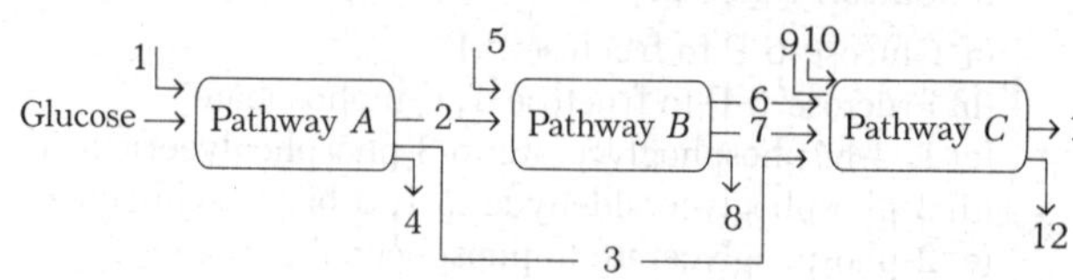

Arrows numbered 4, 8 and 12 can all be **[NEET 2013]**

(a) ATP (b) H_2O
(c) FAD^+ or $FADH_2$ (d) NADH

49 Which one of the following reactions is an example of oxidative decarboxylation? **[KCET 2013]**

(a) Conversion of succinate to fumarate
(b) Conversion of fumarate to malate
(c) Conversion of pyruvate to acetyl Co-A
(d) Conversion of citrate to isocitrate

50 Oxygen content reduction makes the glycolyse (glycogenesis) intensity increased due to **[AIIMS 2013]**

(a) increase of ADP concentration in cell
(b) increase of NAD^+ concentration in cell
(c) increase of ATP concentration in cell
(d) increase in concentration of peroxides and free radicals

51 The process by which ATP is produced in the inner membrane of a mitochondrion. The electron transport system transfers protons from the inner compartment of the outer, as the protons flow back to the inner compartment, the energy of their movements is used to add phosphate to ADP, forming ATP. **[AIIMS 2013, 2009]**

(a) Chemiosmosis (b) Phosphorylation
(c) Glycolysis (d) Fermentation

52 Biological oxidation in Krebs' cycle involves **[OJEE 2013]**
(a) O_2 (b) CO_2
(c) O_3 (d) NO_3

53 The pyruvic acid formed during glycolysis is oxidised to CO_2 and H_2O in a cycle called **[BCECE 2012]**
(a) Calvin cycle (b) nitrogen cycle
(c) Hill reaction (d) Krebs' cycle

54 FAD acts as an e^- acceptor in between **[Manipal 2012]**
(a) fumaric and malic acid
(b) succinic and fumaric acid
(c) malic and oxaloacetic acid
(d) citric and isocitric acid

55 The energy content in kcal/g of carbohydrate: protein: triglycerol, respectively is approximately in the ratio of **[AMU 2012]**
(a) 1 : 1 : 2 (b) 1 : 2 : 1
(c) 2 : 1 : 1 (d) 2 : 2 : 1

56 Anaerobic respiration is also called as **[OJEE 2011]**
(a) β-oxidation (b) fermentation
(c) oxidation (d) None of these

57 The number of ATP produced when a molecule of glucose undergoes fermentation. **[KCET 2011]**
(a) 4 (b) 36 (c) 12 (d) 38

58 In Krebs' cycle, GTP is formed in **[UP CPMT 2011]**
(a) oxidative phosphorylation
(b) substrate level phoshorylation
(c) photophosphorylation
(d) decarboxylation

59 Common enzyme in glycolysis and pentose phosphate pathway is **[UP CPMT 2011]**
(a) hexokinase (b) aconitase
(c) fumarase (d) dehydrogenase

60 Link enzyme in cellular respiration is **[UP CPMT 2011]**
(a) citrate synthetase
(b) pyruvate dehydrogenase
(c) isocitrate dehydrogenase
(d) succinyl thiokinase

61 Chemiosmosis hypothesis given by Peter Mitchell proposes the mechanism of synthesis of **[KCET 2011]**
(a) NADH (b) ATP
(c) $FADH_2$ (d) NADPH

62 Select the wrong statement. **[Kerala CEE 2011]**
(a) When tripalmitin is used as a substrate in respiration, the RQ is 0.7
(b) The intermediate compound which links glycolysis with Krebs' cycle is malic acid
(c) One glucose molecule yields a net gain of 36 ATP molecules during aerobic respiration
(d) One glucose molecule yields a net gain of 2 ATP molecules during fermentation
(e) The scheme of glycolysis was given by Embden, Meyerhof and Parnas

63 Which of these steps in Krebs' cycle indicates substrate level phosphorylation? **[Kerala CEE 2011]**
(a) Conversion of succinyl acid to α-ketoglutaric acid
(b) Conversion of succinic acid to malic acid
(c) Conversion of succinyl Co-A to succinic acid
(d) Conversion of malic acid to oxaloacetic acid
(e) Conversion of citric acid to α-ketoglutaric acid

64 In the electron transport system present in the inner mitochondrial membrane, complexes I and IV are respectively **[Kerala CEE 2011]**
(a) NADH dehydrogenase and $FADH_2$
(b) $NADH_2$ and NADH dehydrogenase
(c) NADH dehydrogenase and cytochrome-*c* oxidase complex
(d) NADH dehydrogenase and ATP synthase
(e) Cytochrome-bc_1 complex and NADH dehydrogenase

ANSWERS

Check Point 21.1

1. (c)	2. (b)	3. (c)	4. (c)	5. (b)	6. (a)

Check Point 21.2

1. (c)	2. (a)	3. (b)	4. (b)	5. (b)	6. (b)

Check Point 21.3

1. (a)	2. (d)	3. (d)	4. (c)	5. (c)	6. (a)

(A) Topical Exercise

1. (b)	2. (b)	3. (a)	4. (c)	5. (c)	6. (b)	7. (a)	8. (a)	9. (c)	10. (b)
11. (d)	12. (a)	13. (d)	14. (a)	15. (a)	16. (c)	17. (d)	18. (a)	19. (b)	20. (a)
21. (b)	22. (d)	23. (b)	24. (d)	25. (c)	26. (a)	27. (b)	28. (c)	29. (b)	30. (b)
31. (b)	32. (d)	33. (b)	34. (b)	35. (b)	36. (a)	37. (a)	38. (b)	39. (d)	40. (d)
41. (c)	42. (c)	43. (c)	44. (d)	45. (b)	46. (a)	47. (a)	48. (b)	49. (c)	50. (c)
51. (d)	52. (a)	53. (a)	54. (d)	55. (b)	56. (c)	57. (d)	58. (b)	59. (b)	60. (a)
61. (c)	62. (a)	63. (d)	64. (d)	65. (b)	66. (a)	67. (d)	68. (b)	69. (c)	70. (c)
71. (b)	72. (b)	73. (b)	74. (d)	75. (c)	76. (b)	77. (c)	78. (a)	79. (a)	80. (d)
81. (d)									

(B) Cumulative Exercise

1. (a)	2. (d)	3. (a)	4. (c)	5. (a)	6. (a)	7. (c)	8. (d)	9. (c)	10. (a)
11. (b)	12. (d)	13. (c)	14. (d)	15. (d)	16. (b)	17. (b)	18. (b)	19. (d)	20. (a)
21. (b)	22. (b)	23. (b)	24. (b)	25. (a)	26. (a)	27. (d)	28. (c)	29. (a)	30. (c)
31. (a)	32. (d)	33. (c)	34. (a)	35. (a)	36. (b)	37. (c)	38. (a)	39. (b)	40. (a)
41. (c)	42. (d)	43. (b)	44. (b)	45. (c)	46. (e)	47. (b)	48. (c)		

(C) Medical Entrances' Gallery

1. (a)	2. (d)	3. (a)	4. (a)	5. (a)	6. (d)	7. (c)	8. (a)	9. (d)	10. (b)
11. (d)	12. (a)	13. (b)	14. (a)	15. (b)	16. (c)	17. (b)	18. (d)	19. (d)	20. (a)
21. (a)	22. (a)	23. (b)	24. (c)	25. (a)	26. (d)	27. (d)	28. (a)	29. (d)	30. (d)
31. (b)	32. (b)	33. (a)	34. (d)	35. (a)	36. (d)	37. (c)	38. (d)	39. (c)	40. (a)
41. (b)	42. (e)	43. (c)	44. (a)	45. (b)	46. (d)	47. (a)	48. (a)	49. (c)	50. (a)
51. (a)	52. (a)	53. (d)	54. (b)	55. (a)	56. (b)	57. (c)	58. (b)	59. (a)	60. (b)
61. (b)	62. (b)	63. (c)	64. (c)						

Explanations

(A) Topical Exercise

6 (*b*) 4 ATP produced and 2 ATP used up. So, net gain = 2 ATP

18 (*a*) Statement I is correct, but statement II is incorrect and can be corrected as

Glycolysis is connected with ETC as the electron carriers such as NADH produced during glycolysis and the Krebs' cycle passes their electron to the electron transport chain which result in synthesis of ATP.

30 (*b*) Statement II in incorrect and can be corrected as

In eukaryotes, Krebs' cycle takes place in mitochondrial matrix hence, all enzymes are found in matrix except succinate dehydrogenase, it is found in the inner membrane of mitochondria.

32 (*d*) Statement in option (d) is incorrect. It can be corrected as

Ubiquinone or coenzyme-Q is an organic molecule dissolved in the hydrophobic region of the inner membrane of the mitochondria these are non-proteinaceous while cytochrome are proteinaceous also found in mitochondria. Rest statements are correct.

40 (*d*) Both the statements I and II are false and can be corrected as

F_1 head piece contains the site for the synthesis of ATP from ADP + Pi and F_0 part form the channel through which protons cross the inner membrane.

46 (*a*) Statement in option (a) is incorrect. It can be corrected as

During oxidative phosphorylation oxygen is used as electron acceptor not as electron donor. Rest statements are correct.

52 (*a*) Statement in option (a) is correct only other statements are incorrect and can be corrected as

Pentose phosphate pathway is present in both prokaryotes and eukaryotes. It generate 35 molecules of ATP. The first phase of PPP is oxidative in which NADPH is generated.

64 (*d*) Both statements I and II are false and can be corrected as

Fermentation can be define as the partial breakdown of glucose to produce ethyl alcohol and carbon dioxide. It releases energy in the form of ATP molecules, 2 ATP molecules are produced per glucose. The maximum concentration of alcohol in alcoholic beverages is 15-16%, only.

71 (*b*) Statement in option (b) is false and can be corrected as

In the absence of oxygen, many cells use fermentation to produce ATP by substrate level phosphorylation.

Rest statements are true.

72 (*b*) Statement I is false, but II is true. Incorrect statement can be corrected as

The ratio of the volume of CO_2 evolved to the volume of O_2 consumed in respiration is called respiratory quotient.

73 (*b*) $RQ = \dfrac{\text{Volume of } CO_2 \text{ producing}}{\text{Volume of } O_2 \text{ absorbed}}$

$$RQ = \frac{102}{145}$$

$RQ = 0.703$

The value of RQ is less than 1 hence, respiratory substrate can be protein or fat. In the given equation $C_{15}H_{98}O_6$ is tripalmitin which is a fatty acid, hence the respiratory substrate is fat.

81 (*d*) RQ of $C_{57}H_{104}O_6$ (Triolein) $= \dfrac{57\, CO_2}{80\, O_2} = 0.71$

RQ of $C_{51}H_{98}O_6$ (Tripalmitin) $= \dfrac{102\, CO_2}{145\, O_2} = 0.7$

(B) Cumulative Exercise

1 (*a*) Glucose is converted to hexose phosphate which is broken down to triose phosphate, which is subsequently converted to pyruvate. Pyruvate enters the Krebs' cycle, producing carbon dioxide, reduced NAD and ATP. Reduced NAD enters the electron transport chain, where oxygen is reduced to water and ATP is produced.

2 (*d*) Oxidative phosphorylation contributes 34ATP out of the 38 ATP produced per glucose molecule, which is 89.98%. Glycolysis and Krebs' cycle contribute 4 ATP.

4 (*c*) Reaction of respiration is

$$C_6H_{12}O_6 + 6O_2 \longrightarrow 6CO_2 + 6H_2O + 674\text{ kcal}$$

In this reaction considering molecular weight of each element, we get

$$C_6H_{12}O_6 = 6 \times 12 + 12 \times 1 + 6 \times 16 = 180\text{ gm}$$

$$6O_2 = 6 \times 32 = 192\text{ gm}$$

Now, we consider product of this reaction $6CO_2$ and $6H_2O$

$$6CO_2 = 6\,(12 + 32) = 264\text{ gm}$$

$$6H_2O = 6\,(2 + 16) = 108\text{ gm}$$

So, option (c) is correct.

5 (*a*) All statements are true except statement IV. Incorrect statement can be corrected as

None of the intermediates in the pathways of respiration are utilised to synthesise any other compounds.

7 (*c*) The total amount of energy is in forming

$$38\text{ ATP} = 38 \times 30.6\text{ kJ} = 1162.8\text{ kJ}$$

Thus, the efficiency of aerobic respiration is

$$\frac{1162.8}{2880} \times 100\% = 40\%$$

8 (*d*) 1 - Cytoplasm, 2 - Cristae, 3-matrix, 4-Intermembrane sapce

ADP is converted to ATP for the storing of energy by the addition of a high energy phosphate group. The conversion takes place in the substance between the cell membrane and the nucleus known as cytoplasma where, NAD picks up hydrogen atom in the matrix. The split produces protons, which enter the space, then flows back with the matrix *via* the stalked granules. This flow act as a driving force to combine ADP with the inorganic phosphate to form ATP.

10 (*a*) Statements I, II and IV are correct, but III and V are incorrect. Incorrect statements can be corrected as

Electron transport system is a chain of electron carriers that is present in the inner mitochondrial membrane that carry the electron from NADH and $FADH_2$ produced in the mitochondrial matrix to the final acceptor, i.e. oxygen, resulting in the formation of H_2O. It is essential to release and utilise the energy stored in NADH and $FADH_2$.

Respiratory quotient is the ratio of volume of CO_2 evolved to the volume of O_2 consumed, i.e.

$$RQ = \frac{\text{Volume of } CO_2 \text{ evolved}}{\text{Volume of } O_2 \text{ consumed}}$$

13 (*c*) At two steps, ATP is used in glycolysis, one mole used during conversion of glucose to glucose-6-phosphate and other one during formation of fructose-1, 6-bisphosphate.

Pyruvate is the end product in glycolysis, enter the mitochondrion and then is converted into acetyl Co-A, which undergoes Krebs' cycle and finally results in ATP formation.

In Krebs' cycle the oxidation or dehydrogenation reactions occur four times. In second stage of Calvin cycle, the 3-PGA molecules created through carbon fixation are converted into molecules of simple sugar. This step in called reduction.

16 (*b*) In the diagram, NADH + H^+ and 1/2 O_2 is mentioned, so this is oxidative phosphorylation taking place in mitochondria and not in chloroplast.

18 (*b*) Reduced NAD are oxidised *via* reaction in the electron transport chain, which is coupled with ATP production.

20 (*a*) Statement in option (a) is incorrect and can be corrected as X is the enzyme pyruvate decarboxylase and Y is the enzyme alcohol dehydrogenase.

Rest statements are correct with respect to the given representation.

23 (*b*) The breakup of ATP during complete aerobic respiration is given as follows

Glycolysis (P)	2 ATP directly = 2 ATP		= 8 molecules
	2 molecules of NADH	= 6 ATP	
Pyruvic acid to acetyl Co-A (Q)	2 molecules of NADH	= 6 ATP	= 6 molecules
Citric acid cycle or Krebs' cycle (R)	6 molecules of NADH	= 18 ATP	18 + 4 + 2 = 24 molecules
	2 molecules of FADH	= 4 ATP	
	2 molecules of GTP	= 2 ATP	
	Total	= 38 ATP molecules	

29 (*a*) Option (a) is correct other statements are incorrect and can be corrected as

Eukaryotes can perform fermentation, they also undergo anaerobic respiration. For example, fermentation in yeast and lactic acid in muscles of mammals.

For complete oxidation of glucose both O_2 and Krebs' cycle is required electron transfer chain cannot occur without the presence of oxygen.

33 (*c*) Statements I, II and III are incorrect and can be corrected as When fats are used as respiratory substrate RQ is usually less than one. Fats are hydrolysed during respiration to glycerol and fatty acids results in the RQ less than 1. ATP is most important energy carrier ATP mobile and transport the chemical energy to the energy consuming processes anywhere within the cell, however the ATP molecule do not pass from one cell to another. Pyruvic acid generating in the glycolysis enters into mitochondrial matrix and one of the three C-atoms of pyruvic acid undergoes oxidative decarboxylation before it enters the Krebs' cycle.

38 (*a*) Both A and R are true and R is the correct explanation of A.

Respiratory pathway is considered as an amphibolic pathway as it involves both anabolic and catabolic reactions. For example, fatty acids are broken down to acetyl Co-A before entering respiratory pathway and then many other carbonic acids an synthesised as an intermediate compounds.

39 (*b*) Both A and R are true, but R is not the correct explanation of A, it can be explained as

Two Krebs' cycles occur per glucose molecule because glycolysis results in the formation of 2 pyruvic acid molecules by glucose breakdown. Each turn of Krebs' cycle produces 3 NADH, 1 $FADH_2$ and 1 ATP molecule. On oxidation, NADH produces 3 ATP molecules and 1 $FADH_2$ produces 2 ATP molecules. Thus, in each turn of Krebs' cycle 12 ATP molecules are produced.

40 (*a*) Both A and R are true and R is the correct explanation of A.

The conversion of glyceraldehyde-3-phosphate to 1, 3-bisphosphoglycerate is an oxidation reaction.

This oxidation reaction occurs due to the use of NAD^+ which gets converted to NADH by accepting a hydrogen molecule and this reaction is catalysed by the enzyme glyceraldehyde-3-phosphate dehydrogenase.

41 (*c*) A is true, but R is false because

Carbohydrates are used first by most cells in respiration. Lipids (fats or oils) are used mainly when carbohydrate reserves have been exhausted. As proteins have other essential functions these are used only when all the carbohydrate and lipid reserves have been used up.

42 (*d*) A is false, but R is true and A can be corrected as ATP formation during respiration is an endergonic process. This is because ATP formation requries input of energy.

43 (*b*) Both A and R are true, but R is not the correct explanation of A.

The reaction of glycolysis occurs in the cytoplasm and does not require the presence of oxygen. The conversion of one molecule of glucose to two pyruvate results in a net gain of two ATP and two NADPH. The Krebs' cycle is far more efficient in the release of energy than are either glycolysis or fermentation. The reactions of the Krebs' cycle require the presence of oxygen. Total ATP produced during Krebs' cycle is 24.

Fermentation yields a total of 2 ATP molecules for each molecule of glucose degraded. So, aerobic respiration is bioenergetically more efficient than anaerobic glycolysis.

44 (*b*) Both A and R are true, but R is not the correct explanation of A.

Carbohydrates and fats are broken down gradually by a series of enzymatic reactions and release energy. Some energy is transferred to ATP. The rest of the energy is lost as heat. AMP is a nucleotide, which is made up of a sugar, nitrogenous base and phosphate.

45 (*c*) A is true, but R is false because

ATP is mobile and can carry energy to energy consuming processes anywhere in the cells. To make ATP from ADP and inorganic phosphate, it requires 30.6 kJ of energy per mole.

46 (*e*) Both A and R are false. These can be corrected as Glycolysis occurs in the cytoplasm of cells, not in the mitochondria and does not require the presence of oxygen. The Krebs' cycle takes place in the matrix of mitochondria.

47 (*b*) Both A and R are true, but R is not the correct explanation of A. The final stage of respiratory chain involves cytochrome oxidase, which contains copper. This stage can be specifically inhibited by cyanide or carbon monoxide. Cyanide combines with the copper and prevents oxygen from combining with it.

48 (*c*) A is true, but R is false because Alcoholic fermentation is an anaerobic process, in which pyruvic acid is converted into ethanol and carbon dioxide.

$$\text{Pyruvate} \xrightarrow[\text{fermentation}]{\text{Alcoholic}} \text{Ethanol} + CO_2$$

Its occurrence is in yeast. It is used for manufacture of beer, wine and other alcoholic drinks.

(C) Medical Entrances' Gallery

1. (*a*) The number of substrate level phosphorylation in one turn of citric acid cycle is 1.

During Krebs' or citric acid cycle, succinyl Co-A is acted upon by enzyme succinyl Co-A synthetase to form succinate (a 4C compound). The reaction releases sufficient energy to form ATP (in plants) or GTP (in animals) by subtrate-level phosphorylation. GTP can form ATP through a coupled reaction.

Succinyl Co-A + GDP/ADP + H_3PO_4 synthetase
Succinyl Co-A Succinate + Co-A + GTP/ATP.

2 (*d*) Pyruvate dehydrogenase enzyme is involved in the conversion of pyruvate to acetyl Co-A, after the completion of glycolysis and before the start of Krebs' cycle. This enzyme is made up of decarboxylase, lipoic acid, transacetylase and Mg^{2+} ion. The reaction occur in following way

$$\text{Pyruvate} + NAD^+ + \text{Co - A} \xrightarrow[\text{dehydrogenase, } Mg^{2+}]{\text{Pyruvate}}$$

$$\text{Acetyl Co - A} + NADH + H^+ + CO_2$$

In this reaction Mg^{2+} acts as a cofactor.

3 (*a*) Conversion of glucose to glucose-6-phosphate during glycolysis is catalysed by the enzyme hexokinase in the presence of Mg^{2+} ions (cofactor). During this step, glucose is phosphorylated to glucose-6-phosphate by ATP. It is the first step of the activation phase of glycolysis.

4 (*a*) In gluconeogenesis, fructose cannot act as substrate because it is the metabolic process by which formation of glucose from non-carbohydrate substrates such as amino acid (glycine), fatty acid (glycerol, palmitic acid), etc., takes place.

5 (*a*) The RQ value of tripalmitin is 0.7. It can be calculated as follows

$$\text{Respiratory Quotient (RQ)} = \frac{\text{Amount of } CO_2 \text{ released}}{\text{Amount of } O_2 \text{ consumed}}$$

$$\underset{\text{Tripalmitin}}{2C_{51}H_{98}O_6} + 145O_2 \rightarrow 102\,CO_2 + 98\,H_2O$$

$$RQ = \frac{CO_2}{O_2} = \frac{102}{145} = 0.7$$

It is to note that RQ of common fats is usually less than 1 under aerobic conditions.

6 (*d*) All the given mechanisms in the options lead to ATP formation. The type of method used depends on the type of organism and their energy transfer mechanisms.

Photophosphorylation is the enzymatic formation of ATP from ADP coupled to the light dependent transfer of electrons in photosynthetic cells.

Substrate level phosphorylation is the phosphorylation of ADP or some other nucleoside 5-diphosphate coupled to the dehydrogenation of an organic substrate, independent of the electron transfer chain.

Oxidative phosphorylation is the enzymatic phosphorylation of ADP to ATP coupled to electron transfer from a substrate to molecular oxygen.

7 (*c*) Respiratory Electron Transport System (ETS) in plants is located in the inner mitochondrial membrane. It serves as the site of oxidative phosphorylation through the action of ATP synthase.

8 (*a*) The statement in option (a) is incorrect. The correct form of the statement is

Enzymes of electron transport are embedded in the inner membrane of mitochondria.

Rest of the statements are correct.

9 (*d*) Statement in option (d) is incorrect because Krebs' cycle starts with the condensation of acetyl group with oxaloacetic acid and water to yield citric acid. Rest of the statements are correct.

10 (*b*) Glycolysis is a stepwise process in which one molecule of glucose breaks down into 2 molecules of pyruvic acid. This pyruvic acid then enters the Krebs' cycle and undergoes oxidative decarboxylation, i.e. removal of CO_2 and gets acetylated to form a key compound acetyl Co-A.

The complete process of the formation of acetyl Co-A takes place in mitochondrial matrix.

11 (*d*) In mitochondria, pyruvic acid (formed by the glycolytic catabolism of carbohydrates in cytosol) undergoes oxidative decarboxylation, (i.e. removal of CO_2 in aerobic conditions) forming a key compound, i.e. acetyl Co-A by the action of pyruvic acid dehydrogenase (in mitochondrial matrix) through a series of reactions. Besides acetyl Co-A, CO_2 and NADH + H^+ are also formed in this process.

$$\text{Pyruvic acid} + \text{Co-A} + NAD^+ \xrightarrow[\text{Pyruvate dehydrogenase}]{Mg^{2+}}$$

$$\text{Acetyl Co - A} + CO_2 + NADH + H^+$$

12 (*a*) 2 ATPs are poduced when one molecule of $FADH_2$ is oxidised to FAD through electron transport system.

13 (*b*) Cytochrome-*c* is the small protein attached to the outer surface of the inner membrane that acts as a mobile carrier for the transfer of electrons between complexes III and IV during Electron Transport Systems (ETS).

14 (*a*) In glycolysis, 2 molecules of pyruvic acid are formed by the breakdown of a single glucose molecule. It serves as the first step of degradation of glucose.

15 (*b*) The correct sequence of events occurring in glycolysis is as follows
Glucose-6- Phosphate → 3-Phosphoglyceraldehyde → 3-Phosphoglyceric acid → Phosphoenol pyruvate → Pyruvic acid.

16 (*c*) Respiratory Quotient (RQ) of malic acid is 1.33. It is slightly more than unity when organic acids are broken down as respiratory substrates under aerobic conditions, e.g. malic acid.

$$\underset{\text{Malic acid}}{C_4H_6O} + 3O_2 \longrightarrow 4CO_2 + 3H_2O$$

$$RQ = \frac{CO_2 \text{ released}}{O_2 \text{ consumed}} = \frac{4}{3} = 1.33$$

17 (*b*) NAD^+ functions as an electron carrier in cellular respiration. NAD is an oxidising agent which accepts electrons and then transfers them to the Electron Transport System (ETS). As a result, 3 ATP molecules are formed.

18 (*d*) Statement in option (d) is incorrect. The correct information about the statement is as follows
Oxidative phosphorylation occurs in the inner mitochondrial membrane because of its less permeability, presence of ETC proteins and ATP synthase.
The rest three statements are correct.

19 (*d*) Carbohydrates, fats and proteins, all can be used as a substrate in cellular respiration. All of them get converted to acetyl Co-A when they enter the Krebs' cycle during aerobic cellular respiration.

20 (*a*) Oxidative phosphorylation is the process of formation of ATP from ADP and inorganic phosphate (Pi) in the presence of oxygen. It occurs mainly in the Electron Transport Chain (ETC) of cellular respiration.

21 (*a*) The conversion of pyruvate is carried out by pyruvate dehydrogenase, an enzyme complex with Mg^{2+}, TPP and vitamin-B.

24 (*c*) Cytochromes are present on the cristae, i.e. inner foldings of mitochondria. Cytochromes are iron (haem) containing proteins primarily responsible for transfer of electrons in ETC from ubiquinone to complex-V for generation of ATP (energy molecules).
They are also called respiratory pigments, which are capable of performing both the oxidation and reduction reactions.

26 (*d*) Respiratory Quotient (RQ) is the ratio of CO_2 produced to O_2 consumed during aerobic oxidation of substrate (carbohydrates, fats or proteins).
The value of RQ for protein substrates is always less than one.

27 (*d*) Krebs' cycle is a part of aerobic cellular respiration. It was discovered by Hans Krebs in 1938. *Cis*-aconitic acid is a 6C compound, whereas malic and succinic acid are 4C compound. α- ketoglutaric acid is the only 5C compound formed in Krebs' cycle.

30 (*d*) Conversion of G-6-P to F-6-P is isomerisation. F-6-P to F-1, 6-bisphosphate is phosphorylation.
1, 3-bisphosphoglycerate to 3- phosphoglyceric acid leads to removal of one phosphate. Conversion of 3-phosphoglyceraldehyde to 1, 3-bisphosphoglycerate is oxidation.

33 (*a*) Statement in option (a) is correct.
Rest statements are incorrect and can be corrected as
- Pyruvate is formed by glycolytic catabolism of carbohydrates in cytosol.
- There is incomplete oxidation of glucose in fermentation.
- During the conversion of succinyl Co-A to succinic acid, substrate level phosphorylation occurs and with the use of GTP, ATP is formed.

35 (*a*) During respiration, 2-phosphoglyceraldeyde is formed in glycolysis, but no such compound is formed in TCA or Krebs' cycle.

36 (*d*) Statements I and III together support respiratory pathway. Amphibolic pathway is a collective term used for both catabolic and anabolic processes. During respiration fats are first broken down to glycerol and fatty acids yielding acetyl Co-A. This acetyl Co-A is withdrawn when required and used in fatty acid synthesis.

38 (*d*) Lactic acid or lactate fermentation is a process, by which glucose, fructose and sucrose are converted into energy and the metabolite lactate without releasing CO_2 as a byproduct.

39 (*c*) Complete oxidation of glucose involves both glycolysis and Krebs' cycle along with oxidative decarboxylation, where 2 NAD reduced in glycolysis, producing 2 pyruvic acid molecules. They undergo oxidative decarboxylation forming 2 molecules acetyl Co-A by reducing 2 NAD and these acetyl Co-A enters Krebs' cycle where they reduce 6 NAD. Thus, 10 NAD reduced in complete oxidation of glucose.

40 (*a*) Glycolysis is a stepwise process by which one molecule of glucose breaks down into 2 molecules of pyruvic acid. This pyruvic acid undergoes oxidative decarboxylation, i.e removal of CO_2 and gets acetylated to form a key compound acetyl Co-A. The complete process of the formation of acetyl Co-A takes place in perimitochondrial space.

43 (*c*) RQ for glucose or carbohydrate is 1.0 and for fat is 0.7, for protein it is 0.5 and for organic acids it is more than one.

48 (*a*) 3 boxes represent glycolysis, Krebs' cycle and oxidative phosphorylation. Arrows 4, 8 and 12 represent the ATP release.

54 (*b*) FAD converts to $FADH_2$ by accepting e^- and hydrogen ion during the formation of succinic acid to fumaric acid.

57 (*c*) In microorganisms, the term anaerobic respiration is replaced by fermentation. The pyruvic acid formed in glycolysis is transformed to ethyl alcohol and releases 12 ATP molecules.

59 (*a*) Hexokinase causes phosphorylation of glucose to glucose-6 phosphate in both glycolysis and pentose phosphate pathway. Both glycolysis and pentose phosphate pathway occur in cytoplasm.

62 (*b*) Statement in option (b) is wrong and can be corrected as The intermediate compound, which links glycolysis with Krebs' cycle is acetyl Co-A.

63 (*c*) During the conversion of succinyl Co-A to succinic acid, a molecule of GTP is synthesised. This is a substrate level phosphorylation. In a coupled reaction, GTP is converted to GDP with the simultaneous synthesis of ATP from ADP.

64 (*c*) Complex I of Electron Transport System (ETS) is NADH dehydrogenase, which oxidises NADH produced in the mitochondrial matrix during citric acid cycle. Complex IV of ETS refers to cytochrome-oxidase complex containing cytochrome-*a* and a_3, haem and copper ions.

CHAPTER 22

Plant Growth and Development

There are various types of tissues present in a plant. They occupy specific locations throughout the plant body. These tissues have different structure and function. The process of development occurs in an orderly fashion in these tissues.

They grows through various types of mitotic divisions. The cells produced after such divisions differentiate to give requisite tissue. Thus, we can say that development is a sum total of growth and differentiation. Growth in general is considered as an increase in size, volume or weight of the object. Plants growth can be visualised as an increase in stem height, girth, leaf shape, size and surface area, etc.

In this chapter, we will focus on various aspects of growth and development processes.

GROWTH

It can be defined as an irreversible increase in size of an individual cell or organ or its parts. This increase occurs as the net result of all the metabolic activities. These activities make use of that energy which is obtained during respiration in plants. In living organisms, growth usually occurs when there is synthesis of new materials inside or outside the cell.

Features

- At cellular level growth means to increase in protoplasm.
- Plants can grow throughout their life.
- Plant growth is localised due to the presence of meristematic tissue in them.
- Leaves, flowers and fruits may show determinate growth, i.e. these grow upto a size and then undergo death.
- The vegetative stem and roots show indeterminate or indefinite growth due to continuous cell division.

Inside

Types

In plants, growth is two types as follows

(i) **Primary growth** It is the formation of primary permanent tissues and organs. It is caused by the activity of apical and intercalary meristems. It principally contribute to the elongation of the plants along their axis.

(ii) **Secondary growth** It is the increase in girth. It occurs by two types of lateral meristems, i.e. vascular cambium and cork cambium.

Indicators

Growth in plants can be measured in terms

- Increase in cellular dimensions, i.e. length, diameter, surface area, etc.
- Increase in dry and fresh weight.
- Increase volume of cells.
- Increase in number of cells produced.

Phases

In plants, growth occurs in three phases depending upon the region of its occurrence.

(i) **Meristematic phase** Growth in this phase occurs in the meristematic regions, e.g. shoot apex and root apex.

The characteristics of cells of these regions are given below

- Small in size
- The cells in this phase can be easily differentiated from other cells.
- They have a **dense protoplasm** and a **large nucleus.**
- They possess a primary cell wall and are thin and cellulosic.
- They have abundant plasmodesmatal connection.
- These cells divide actively and grow in numbers.
- Their respiratory rate is higher.

(ii) **Elongation phase** Cells of this phase are found around the meristematic cells.

The characteristics of cells of these regions are given below

- These cells are **highly vacuolated.**
- Cell wall starts accumulating new material.
- These are larger in size.

(iii) **Maturation phase** Cells of this phase undergo structural and functional differentiation. Thus, these cells develop into a specialised tissue. After differentiation, no further growth occurs in them.

Growth rate

It can be defined as increase in length or weight of a plant part per unit time.

The growth rate shows an increase that may be arithmetic or geometrical.

Arithmetic growth

- In this type of growth, the rate of growth is constant. Here after mitosis, two cells are formed.
- One of these cells, differentiate into a particular type of cell. The other one undergo further division again and again.
- In this way, growth progress in arithmetic manner, i.e. 2, 4, 6, 8, etc.
- This type of growth is found in root and shoot elongation.
- On plotting the arithmetic growth against time, a linear curve is obtained as shown below

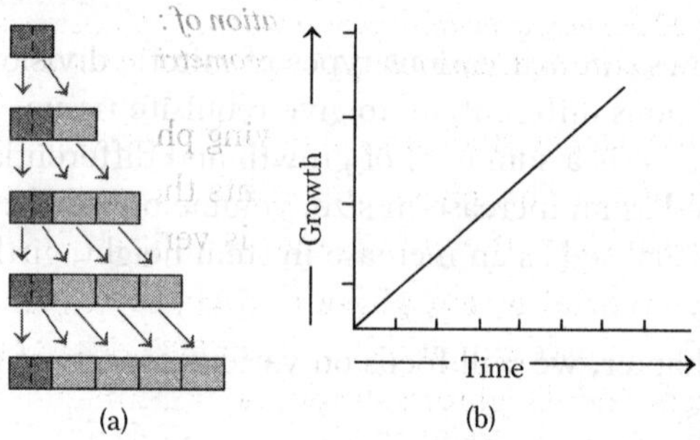

Fig. 22.1 Diagrammatic representation of: (a) Arithmetic growth, (b) Graph plotted for arithmetic growth

- Mathematically, arithmetic growth can be expressed as

$$L_t = L_0 + rt$$

Where, L_t = Length of plant after time 't'
L_0 = Length of plant at initial stage (time = 0)
t = Time taken for the growth
r = Growth rate or elongation per unit time

Geometric growth

- In this type of growth, cell divides repeatedly to form many daughter cells. Here, every cell divides again and again.
- Geometric growth rate when plotted against time, shows sigmoid or S-shaped curve.

- A sigmoid curve is a characteristic of living organism growing in a natural environment.
- It is typical for all cells, tissues and organs of a plant.

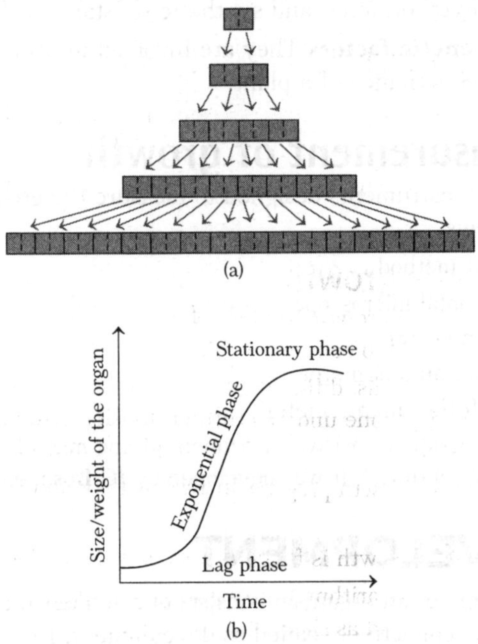

Fig 22.2 Diagrammatic representation of : (a) Geometric growth, (b) Graph plotted for geometric growth

- This curve can be studied in following phases
   (i) **Lag phase** This phase represents the beginning of growth. Here, the growth rate is very slow.
   (ii) **Exponential or log phase** In this phase, growth occurs at a very high rate, i.e. exponentially. Both parent and daughter cells divide repeatedly by utilising the available food.
   (iii) **Stationary phase** Due to rapid growth, number of cells are more in this phase. It leads to shortage of food and space. Also, there is increase in toxin accumulation. Thus, growth becomes slow and constant.

   The exponential growth can be expressed as

$$\boxed{W_1 = W_0 e^{rt}}$$

Where, W_1 = Final size (weight, height, number, etc.)
W_0 = Initial size at the beginning of the period
r = Growth rate
t = Time
e = Base of natural logarithm

Blackman in 1919 called growth rate as efficiency index of plant.

An embryo of living organism shows both types of growth, i.e. arithmetic and geometric. Initially, embryo shows geometrical growth which later passes into arithmetic phase as shown below

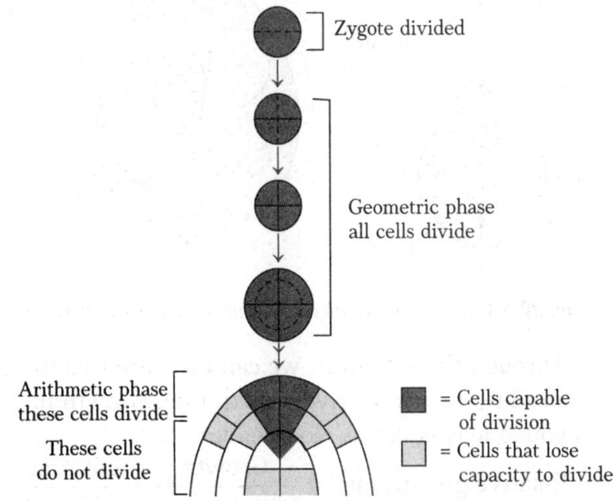

Fig. 22.3 Presence of two types of growth phases during embryo development

Quantitative comparisons of growth rate

The quantitative comparisons between the growth of living systems can be made by measuring their **absolute growth rate** and **relative growth rate**.

(i) **Absolute growth rate** It is the measurement of total growth per unit time.

A graph plotted for absolute growth is bell-shaped as shown below

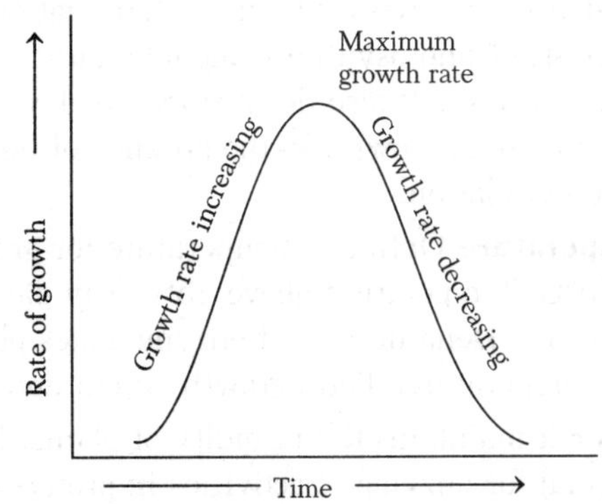

Fig. 22.4 Absolute growth rate curve

(ii) **Relative growth rate** It is the growth per unit time per unit initial parameter.

- Relative growth is generally high in young plant parts. It can be seen in following figure where the growth of one young and older leaf is compared.

- The time period taken is 7 days.

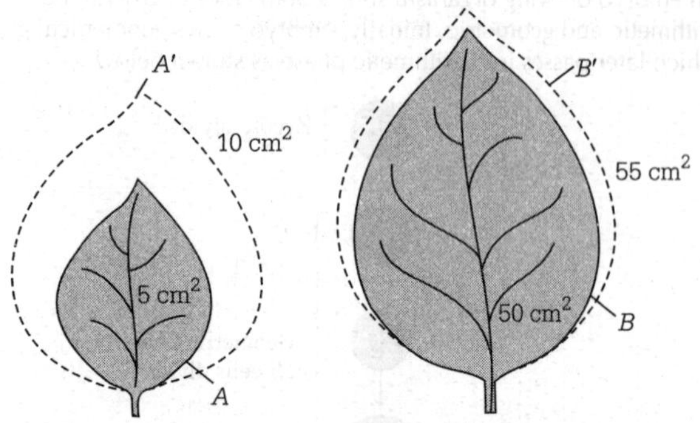

Fig. 22.5 Comparison of absolute and relative growth rates

- Through these figures, we can compare two things absolute *versus* relative growth rate. Growth in young leaves *versus* older leaves.

$$\text{Relative growth rate} = \frac{\text{Growth per unit time}}{\text{Initial parameter (e.g. size)}}$$

- In the figure 22.5, both the leaves possess same absolute growth rate, i.e. 5 cm^2 in 7 days.
- On the other hand, their relative growth varies. In leaf A, it is 100% and in leaf B, it is only 25%.

Factors influencing plant growth

Plant growth is influenced by variety of external and internal factors. These factors given below

External factors

(i) **Light** It is required for tissue differentiation, synthesis of photosynthesis pigment and photosynthesis. It also determines the direction of root and shoot growth. In the absence of light, plants exhibit etiolation.

(ii) **Temperature** Optimum temperature for growth is 20-35°C. Temperature above 45°C may damage the protoplast. Denaturation of enzyme takes place at high temperature. Thus, growth is retarded.

(iii) **Water** It maintains the turgidity of plants. It is essential for enzymatic activities in protoplasm and cell elongation.

(iv) **Oxygen** It is necessary for cellular respiration to produce metabolic energy.

(v) **Mineral nutrients** They are raw material for synthesis of protoplasm and source of energy. Nutrients like Na, K, Cu, etc., are necessary for growth and metabolism of plants.

(vi) **Gravity** It determines the direction, orientation or movements of root, stem and the branches borne over them.

Internal factors

(i) **Growth regulators** These are manufactured by living protoplasm. Growth regulators include phytohormones and synthetic substances.

(ii) **Genetic factors** They are involved in overall maintenance of a plant.

Measurement of growth

Various instruments designed to measure the growth are as follows

- Direct method
- Horizontal microscope
- Auxanometer
  (a) Arc auxanometer
  (b) Pfeffer auxanometer
- Crescograph (sensitive instrument giving magnification up to 10000 times). It was developed by **JC Bose**, etc.

DEVELOPMENT

Life cycle of an organism consists of a sequence of changes, collectively called as development. During development, structures and functions of organs or cells change. It involves a switch from one developmental phase of an organism to another.

The developmental stages in a plant are as follows

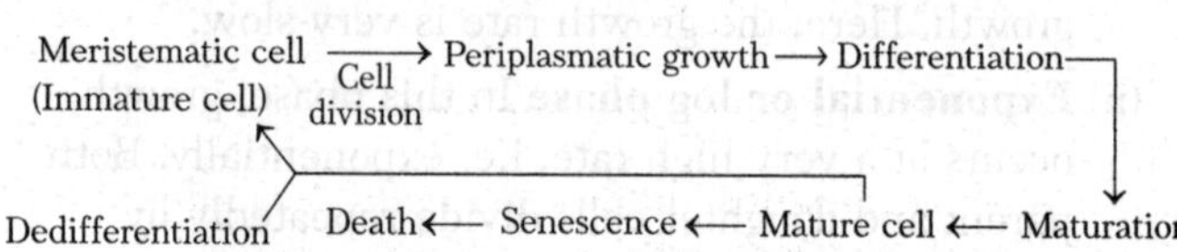

Fig. 22.6 Sequence of development of plant cell

Differentiation, dedifferentiation and redifferentiation

Differentiation

It is a permanent localised qualitative change in size, biochemistry, structure and function of cells, tissues or organs, etc.

The examples of differentiation are as follows

- Normal cells differentiate into tracheary elements. It occurs by elongating, loosening the protoplasm and development.
- Development of strong, elastic lignocellulosic secondary cell wall.
- Deposition of suberin and tannins in cells.
- Differential wall thickenings (in guard cells).
- Secretion of mucilage in root cap.

Dedifferentiation

- It occurs when the living differentiated cells (in capable of dividing) regain their capacity of division under certain conditions.
- For example, interfascicular cambium and cork cambium are formed from fully differentiated parenchyma cells.

Redifferentiation

- It is the process where dedifferentiated cells again loose their ability to divide. These become specialised (permanent) to perform for a particular function.
- For example, secondary phloem, secondary xylem, cork are formed *via* redifferentiation.
- The process of plant tissue culture occurs as a result of redifferentiation in which parenchyma cells are made to divide under controlled laboratory conditions.

Plasticity

- The development process in plant does not always take a single path. It may follow different pathways. Occurrence of different pathway is a result of varying environmental conditions.
- Thus, plants have the ability to modify itself by forming different kinds of structures to adjust with changes in its enviroment. It is called **plasticity**.
- It could be of two types, *viz.* **intrinsic plasticity** and **environmental** (extrinsic) **plasticity**.
- In both the cases, plants show heterophylly and a number of other morphological features, e.g. cotton, coriander, larkspur, etc.

Heterophylly

The occurrence of different forms of leaves on the same plant is called heterophylly. It could be due to different growth phases or varying environmental conditions. In environmental plasticity, aquatic buttercup, *Ranunculus flagellaris*, the submerged leaves are highly dissected. The emerged leaves are broad and lobed.

In larkspur, juvenile leaves are broadly lobed. These mature leaves become pinnately divided with lobes becoming linear in the region of flowers.

***Fig. 22.7** Heterophylly in; (a) larkspur, (b) buttercup (Grow in water)*

CHECK POINT 22.1

1. Which among the followng is incorrect about meristematic cells?
   (a) They are present in the root apex and shoot apex
   (b) They have a dense cytoplasm and a large nucleus
   (c) They are larger than the usual cells
   (d) They have a high respiration rate
2. The unique ability to retain the capacity of growth in plants is due to the presence of
   (a) lifetime differentiation
   (b) activator hormones
   (c) chlorophyll to utilise energy lifetime
   (d) meristems at certain locations
3. The phase of growth where cell wall starts accumulating new material is known as
   (a) phase of cell formation
   (b) phase of elongation
   (c) phase of differentiation
   (d) primary growth
4. The S-shaped growth curve and 'grand period of growth' may changed with
   (a) sudden fluctuation in light intensity
   (b) change in temperature
   (c) fluctuation in humidity
   (d) it remains unaffected
5. During differentiation of normal cells into tracheary elements, the changes which occur
   (a) elongation of cells and oxidation of food
   (b) elongation of leaves and development of an elastic lignocellulosic secondary wall
   (c) elongation of cells and toxin accumulation
   (d) elongation of cells, loosening the protoplasm and development of a strong lignocellulosic secondary wall
6. Redifferentation implies
   (a) loss of ability to divide in dedifferentiated cells
   (b) damage of protoplast
   (c) release of metabolic energy
   (d) shortage of food

PLANT HORMONES

- Growth and development in plants are regulated by special chemical substances. These substances are **plant hormones** or **phytohormones**.
- The description of this term was given by **Thimann** in 1948.
- A plant hormone is an organic compound. These hormones are present in minute quantities. They are translocated to another plant part and it causes a physiological response.
- Chemically, they are Indole compound (IAA, IBA), adenine derivative (cytokinin), derivatives of carotenoids (ABA), terpenes (GA), gases (ethylene).
- They can be placed into two groups based on their function
  (i) **Growth promoters**, e.g. Auxin, gibberellin and cytokinin.
  (ii) **Growth inhibitors**, e.g. Abscisic acid and ethylene can induce
  (a) Dormancy (b) Abscission

Auxin

- It was the first plant growth regulator to be discovered. Chemically, auxins are **indole derivatives**.
- These are produced in the growing regions, i.e. stem and root apices.
- They migrate to their sites of action. They exhibit polar transport through parenchyma.
- Auxins are synthesised from precursor **tryptophan** (amino acid).

History

- **Charles Darwin** and his son **Francis Darwin** performed experiments on oats and canary grass in 1800. They observed bending movement of coleoptile.
- It led to the discovery of auxin. **Kogl** and **Haagen Smit** (1931) isolated the first naturally occurring auxin from human urine. In plants, the first isolated auxin from coleoptile tips of oat seedlings by **FW Went**.

Examples

Names of some popular auxins are given below

Auxins

Natural auxins
- Indole-3- Acetic Acid (IAA)
- 4-chloro Indole Acetic Acid (4-chloro IAA)
- Indole-3-Butyric Acid (IBA)
- Phenyl Acetic Acid (PAA)
- Indole-3-Acetaldehyde
- Indole-3-Propionic Acid (IPA)

Synthetic auxins
- 2, 4-dichlorophenoxy acetic acid (2, 4-D)
- Naphthalene Acetic Acid (NAA)

Bioassay of auxin

A bioassay is a method to determine the activity of a substance.

For testing the activity of auxin following tests are used
- Avena curvature test
- Root growth inhibition
- Split pea test

The experiment of **FW Went** in 1926 provided a basis for understanding the mechanism of auxin action as given below

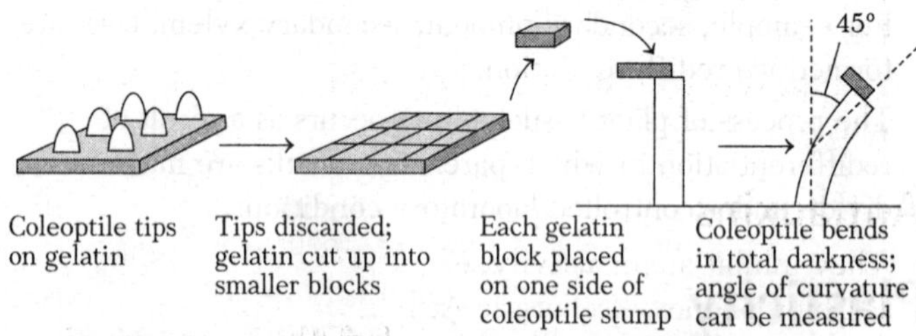

Fig 22.8 Discovery of auxins-experiments of FW Went-Avena curvature test

Went showed that the active growth promoting substance can diffuse into a gelatin block. He also devised a coleoptile bending assay for quantitative auxin analysis.

Functions

- **Cell elongation** Auxin promotes cell elongations. It is done by loosening cell wall, microfibrils, solubilisation of carbohydrate reserve and increased respiration.
- **Abscission** Auxin influences the develpoment of abscission or separation layer for delaying the falling of leaves.
- **Cell division** Auxin initiates and promotes cell division in vascular cambium.
- **Xylem differentiation** Auxin controls the differentiation of xylem.
- **Sex expression** Auxin increases the number of female flowers in plants, e.g. *Cannabis*.
- **Apical dominance** Auxin promotes apical dominance (in the presence of apical bud, the growth of lateral buds are inhibited). It is due to the auxin secreted by apical bud.
- **Parthenocarpy** Application of auxin to unpollinated pistil develops seedless fruits or parthenocarps.
- **Nodule formation** In legumes, IAA stimulates nodule formation.
- Auxin induces negative potential in cell membranes.
- Seasonal activity of cambium is promoted by auxin.

Applications

- **Rooting** NAA and IBA are used for rooting.
- **Flowering** NAA and 2,4-D induced flowering in litchi and pineapple.

- **Parthenocarpy** IAA and IBA are used to induce parthenocarpy.
- **Herbicide** 2, 4-dichlorophenoxy acetic acid is used as herbicide.
- **Prevention of lodging** Naphthalene acetic acid is used for prevention of lodging in cereals.
- **Dwarf shoots** NAA is used for increasing dwarf shoots and the number of fruits in apple.
- **Storage** Methyl ester of NAA is used for the storage of potato.
- **Vegetable crop** Chlorophenoxy propionic acid is used to improve the quality of vegetable crops by inhibiting flower formation.
- **Pre-harvest fruit drop** 2, 4-D for citrus fruit, NAA for tomato.

Antiauxins

- They inhibit auxin activity, e.g. Triidobenzoic Acid (TIBA), P-Chlorophenoxy Isobutyric Acid (PCIB), etc.

Gibberellins

- Chemically, all gibberellins are terpenes, i.e. a complex group of plant chemicals related to lipids. These are also weak acids having a gibbane skeleton.
- The precursor of gibberellin is **acetyl Co-A**.
- Gibberellins are synthesised in the apices of young leaves, embryo, buds and roots and are transported through xylem.
- Reproductive parts have a high concentration of gibberellins.

History

- Gibberellins were first discovered in Japan from a fungus *Gibberella fujikuroi.* This fungus causes **foolish seedling disease of rice** (Bakanae disease).
- **E Kurosawa** prepared an extract of this fungus. He reported the presence of an active substance in the extract.
- **Yabuta** and **Samuki** in 1935, isolated this active substance from fungal extract. They named it gibberellin. Till now 100 types of gibberellins have been reported. These are denoted as GA_1, GA_2, GA_3 and so on. $GA_3(C_{19}H_{26}O_6)$ is the first discovered gibberellin.

Bioassay of gibberellin

It is done by dwarf pea test, dwarf corn test and barley endosperm bioassay.

Functions

- **Stem and leaf growth** Gibberellins can stimulate stem elongation, e.g. in sugarcane, it elongates stem size.
- **Bolting** Gibberellins induce stem elongation in rosette plants, e.g. beet, cabbage, etc. In these plants, profuse leaf development is visible while internodes growth get retarded. Before the reproduction, the internodes start growing causing a marked increase in height. This is known as **bolting.** Bolting and flowering require a long day and cold conditions.
- **Seed germination** Gibberellins promote seed germination. They stimulate the synthesis of hydrolytic enzymes like amylases, proteases, etc.
- **Sex expression** Gibberellins are usually found to cause maleness in some plants. In this function, they are antagonistic to auxin.
  Auxins cause femaleness in plants.
- **Parthenocarpy** Gibberellins are considered more efficient than auxins to produce parthenocarpic fruits.
- **Substitute of cold treatment** Biennial plants (carrot, etc.) require a pre-cold treatment (**vernalisation**) before flowering. Gibberellins are used to induce flowering in these plants.

Applications

- **Fruit growth** Gibberellins promote growth of fruits, e.g. increase length of bunches of grapes.
  Pomalin is a combination of GA_4, GA_7 and cytokinin (6-Benzyladenine). It acts as apple enlarger.
- **Malting** Gibberellin (GA_3) increases malt production in barley grains. It acts by enhancing starch digestion.
- **Delay ripening** Gibberellins cause delay in ripening of citrus fruits.
- **Rapid maturity** Gibberellins cause quicker growth in conifers. Thus, it is used to obtain economically important seeds.
- **Breaking dormancy** Gibberellin causes mobilisation of food. Due to this, seed and bud dormancy is broken. In this function, gibberellin acts as an antagonist of ABA.
- **Elongation of dwarfs** Mutation in a particular gene causes dwarfism in plants. Dwarfism can be restored by spraying gibberellins, like in maize, pea, etc.

Antigibberellins

These are chemical substances that inhibit the activity of gibberellins, e.g. phosphon-D, B-995, malic hydrazide, etc.

Cytokinin

- It is a basic hormone and purine (adenine) derivative. It primarily acts on cell division.
- Roots are the major regions where cytokinin synthesis occurs. From there the cytokinin pass upwardly through xylem.
- Cytokinins is also synthesised in seeds and developing fruits. Coconut milk and apple fruit are rich in cytokinins.

History

- It was discovered by **Miller** in 1955.
- He was working on tobacco pith culture. He added the contents of herring fish sperms DNA to the culture medium.
- He observed that tobacco pith callus grow for longer period. He then isolated an active substance from this.
- That stimulates cell division. He named it as **kinetin**. It does not occur naturally in plants.
- The first natural cytokinin was isolated by **Lethom *et al*** from unripe maize grains (1963). It was named **zeatin** (6-aminopurine).
- There are about 18 cytokinins have been discovered, e.g. dihydro zeatin, isopentenyl adenine, benzyl adenine.
- Many cytokinins are found as constituents of *t*RNAs.

Bioassay of cytokinins

It is done by chlorophyll preservation (retention) test, Excised cell enlargement test, germination and differentiation test.

Functions

- **Promotes cell division and organ formation** Cytokinins are essential for cytokinesis (cell division). In tissue culture, cytokinins/auxin ratio determines organ formation. High auxin to cytokinin ratio favour the root development. High cytokinin to auxin ratio favour shoot development. The intermediate concentration results in the formation of undifferentiated cells.
- **Promotes lateral-bud development in dicots** Cytokinins counter act apical dominance. They promote lateral bud formation even in the presence of apical buds.
- **Development of chloroplast** Cytokinins promote chloroplast formation. They promote grana formation. They also increase the rate of chlorophyll formation.
- **Sex expression** Cytokinin causes femaleness in certain plants.
- **Transport** Cytokinins favour phloem transport.
- **Flowering** Cytokinins induce flowering in plants like in *Lemna* and *Wolffia*. It is also responsible for breaking dormancy.
- **Delay in senescence and increase nutrient-sink activities** Cytokinin can delay the ageing process of the leaves and other organs. It controls protein synthesis and mobilisation of nutrient. This is called as **Richmond-Lang effect.**
- **Salt accumulation** Cytokinin favours salt accumulation in the cells.
- **Temperature** Cytokinin increases resistance to low and high temperature.

Applications

- **Tissue culture** Cytokinins are essential for tissue culture because besides cell division they are also involved in morphogenesis.
- **Shelf life** Application of cytokinin helps keep fruits and vegetables fresh for several days.
- **Resistance** Cytokinin application help plants in developing resistance to pathogens and extremes of temperature.
- **Overcoming senescence** Cytokinins delay senescence of intact plant parts.

Ethylene

- It is the only gaseous phytohormone.
- It is produced in plants from amino acid methionine. All parts of seed plants produce ethylene.
- Maximum synthesis of ethylene occurs during the ripening of fruits and tissues undergoing senescence.
- The commercial product for providing ethylene is ethephon. It is present in liquid form.
- Ethylene gas is released from it. Hence helps in artificial ripening of fruits.

Bioassay of ethylene

It is done by **triple response test.**

Functions

- **Root initiation** Ethylene in low concentration promotes growth of roots. It can initiate root hair formation also.
- **Epinasty** Ethylene causes drooping of leaves and flowers.
- **Horizontal growth** It promotes horizontal growth of seedlings. However, it inhibits their longitudinal growth. Ethylene causes swelling in stems.
- **Transverse growth** Ethylene induces rapid growth of internodes and leaf bases.
- **Flowering** Ethylene usually inhibits flowering. But in some plants (e.g. pineapple), it induces flowering.
- **Apogeotropism** Ethylene may cause apogeotropism. It is the phenomenon in which roots lack their sensitivity to gravity.
- **Fruit ripening** Ethylene stimulates ripening of fruits such as tomatoes, lemons, oranges, etc. It also increases rate of respiration, i.e. respiratory climacteric in fruits. It causes dehiscence of dry fruits.
- **Abscission and senescence** It enhances abscission and senescence of leaves, fruits and flowers.
- **Breaking dormancy** Ethylene is responsible for breaking dormancy of different plant organs, but not of lateral buds.
- **Sprouting of storage organs** The sprouting of storage organs such as rhizomes, tubers can be enhanced by exposing them to ethylene.

Abscisic acid

- It is also known as **stress hormone** or **dormin.** It is produced when a plant is exposed to stress.
- The common stresses include lack of water, saline soils, cold temperature and frost.

History

- **Bennet Clark** and **Kefford** identified first growth inhibitor in dormant potato tuber. They named it as inhibitor-B.
- **Addicott** and **Ophuma** obtained it from cotton fruits and named it abscisin-II.
- **Waring** and **Robins** also isolated a growth inhibitor from *Betula* leaves and called it **dormin.**
- Later on, it was established that inhibitor-B, abscisin-II and dormin are same and can be called as **ABA.**
- ABA is a terpene derivative.
- In leaves, ABA is produced from chloroplast in other parts, it is produced from chromoplast, leucoplast, etc.
- It is formed from mevalonic acid or xanthophyll.

Bioassay of ABA

It is done by endosperm test and growth inhibition test.

Functions

- **Induces stomatal closure** ABA causes stomata to close during stress. It is done by inhibiting an ATP dependent proton pump.
  This is present in the plasma membrane of guard cells.
- **Bud dormancy** ABA induces axillary buds to become dormant during winters.
- **Abscission** It causes abscission in leaves, flowers and fruits.
- **Geotropism in roots** Application of ABA on root results in geotropism.
- ABA induces synthesis of carotenoids. It is applied to green-oranges. It is done to turn them yellow.
- It regulates fruit drop at the end of the season.
- ABA acts as antagonist of all growth promoters.

Thus, it causes following effects

(i) ABA delays flowering of long day plants.
(ii) ABA delays cell division and cell elongation.
(iii) ABA promotes tuberisation in potato.
(iv) ABA inhibits seed germination.
(v) ABA enhances senescence.
(vi) ABA inhibits mitosis in vascular cambium.

CHECK POINT 22.2

1. When Darwin illuminated the coleoptile of a grass from one side, it bent towards the light source. This was due to
(a) diffusion of auxin from tip to the subterminal region
(b) increased cell elongation on dark side
(c) diffusion of some chemicals from basal to tip region
(d) increased cell division on the illuminated side

2. Which of the following physiological effects is caused in plants by gibberellic acid?
(a) Shortening of genetically tall plants
(b) Elongation of genetically dwarf plants
(c) Rooting in stem cuttings
(d) Yellowing of yellow leaves

3. Which of the following is not related to cytokinin?
(a) Promotes apical dominance
(b) Promotes chloroplast maturation
(c) Cause Richmond-Lang effect
(d) Promotes nutrient mobilisation

4. Lateral stem development is controlled by the relative levels of
(a) cytokinins and auxins
(b) ABA and auxins
(c) auxins and gibberellins
(d) auxins and ethylene

5. Triple response test, inhibition of elongation of epicotyl, horizontal growth of epicotyl is shown by grown pea seedling in the presence of which hormone?
(a) Ethylene (b) Cytokinin
(c) Auxin (d) Gibberellin

6. ABA controls which of the following?
(a) Abscission
(b) Fruit climacteric
(c) Stomatal closure
(d) Cell elongation

SEED DORMANCY

Dormancy is defined as the inactive state of a seed. During dormancy, growth of the embryo remains suspended and does not resume even in the presence of favourable conditions.

If the viable seed fails to germinate due to unfavourable condition, it is called **quiescence.**

Causes of seed dormancy

- In many plants, impermeability of seed coat to water (e.g. *Chenopodium, Trigonella, Meliotus*), oxygen (e.g. *Brassica alba, Pyrus malus* (apple), *Sinapis arvensis*) and chemicals (e.g. *Xanthium*) causes dormancy.
- Seed coats can be mechanically resistant, e.g. *Capsella* and *Lepidium.*
- It occurs due to some chemical substances, such as abscisic acid, phenolic acids, short chain fatty acids, etc. These inhibit seed germination.
- Dispersed seed of wheat, barley, oat, etc., need an interval for ripening. It requires gaining the ability to germinate.
- The embryos of some seeds are immature at the time of seed germination, e.g. *Ginkgo biloba* (gymnosperm), *Eranthis hyemalis.*
- Some seeds, e.g. *Atriplex* contain high concentration of salt at the time of shedding.

Methods to breaking seed dormancy

Various methods used for breaking dormancy are

- **Scarification** It is a method of making seed coat soft and weak. It is done by using acids and alcohols. Mechanically, it is done using knives.
- **Stratification** It involves exposure of seeds to low temperature and O_2.
- **Hydrolytic** enzymes Enzymes such as pectinases, cellulases, etc., are used to degrade the tough seed coat.
- **Destruction of inhibitors** Germination inhibitors can be removed using heat, light or oxidation.
- **Microbial action** Seed coats can be broken by microbial action.
- **Growth hormones** Cytokinins, gibberellins and auxins are involved in breaking dormancy.

Biological significance

- Dormancy allows storage of seeds in viable state for longer periods.
- Dormant seeds are a means of survival. They germinate on arrival of favourable conditions.
- Dormancy aids in dispersal over long distances.

SEED GERMINATION

- It is the process by which the dormant embryo of the seed resumes active growth. It results in the formation of a seedling.
- It is an irreversible process. Here, radicle grows out first to establish root after that, plumule grows to form shoot.

Types of seed germination

There are different types of seed germination are as discussed below

Epigeal germination

- In this type of germination, cotyledons are pushed out of the soil
- The cotyledons become green, it performs phyotosynthesis in addition to food storage.
- In dicots, this happens due to the elongation of hypocotyl (stem of germinating seedling between cotyledons and radicle), e.g. bean, castor, mustard, tamarind, sunflower, etc.

Hypogeal germination

- In this type of germination, the cotyledons do not come out of the soil. It remains underground.
- The epicotyl (part of embryonic axis between plumule and cotyledons) elongates first to raise the first leaves out of the soil.
- The hypocotyl growth is restricted, e.g. all monocots such as rice, maize, wheat, etc., and dicots such as gram, pea, mango, groundnut, etc.

Viviparous germination

- It is a type of gemination found growing in salty lakes, sea coasts and deltas, i.e. in mangrove plants.
- In this process, seed germinates while attached to the parent plant.
- The radicle elongates considerably and projects out of the fruit. Its lower part becomes thick and swollen.

- The seedling breaks off the parent plant due to its increase in weight. It embedded in the muddy soil. The lateral roots develop at the basal end of the radicle.
- It is also called **aerial germination**, e.g. *Rhizophora, Ceriops, Avicennia,* etc.

PHOTOPERIODISM

- The effect of **photoperiod**, i.e. daily duration of light hours and dark period on the growth and development of plant, especially their flowering is called **photoperiodism.**
- The phenomenon of photoperiodism was first discovered by **Garner** and **Allard** in 1920. Their experimental material was '**Maryland mamoth**' a mutant variety of tobacco. They manipulated the photoperiod for these plants. Due to this change in flowering time was observed.
- Thus, they concluded that plants differ in their requirements for day length.
- Most plants flower only when they are subjected to a light phase for less or more than a critical period.
- A **critical period** is the period of light or darkness required by the plant to induce flowering.

Duration of photoperiod

Depending upon the duration of photoperiod, plants have been divided into following categories

Short-day plants (SDP)

Photoperiod of these plants is lesser than the critical photoperiod. Thus, require shorter photoperiod in order to initiate flowering, e.g. *Xanthium* (cocklebur), *Chrysanthemum, Cosmos, Dahlia,* rice, sugarcane, strawberry, tobacco, *Glycine max* (soyabean), etc.

Long-day plants (LDP)

These require a light period more than the critical length. Thus, they require longer day light period for flowering. Long night period prevents flowering in LDP. These are also called as **short-night plants**, e.g. *Hyoscyamus niger* (henbane), *Spinacia* (spinach), *Beta vulgaris* (sugarbeet), wheat, oat, raddish, lettuce, etc.

Day neutral plants (Indeterminate plants)

These plants flower in all photoperiods. Thus, the floral initiation in them is independent of photoperiodism. These can blossom throughout the year, e.g. tomato, cotton, maize, sunflower, cucumber, etc.

Long-short day plants (L-SDP)

These are short-day plants. These plants require long photoperiods for floral initiation. They require short photoperiod for blossoming, e.g. *Bryophyllum, Cestrum.*

Short-long day plants

These are long-day plants. They require short days for floral initiation. They require long day for blossoming, e.g. certain varieties of wheat (*Triticum*) and rye (*Secale*).

Perception of photoperiodic (Flowering) stimulus

The photoperiodic stimulus is perceived in leaves by phytochrome.

It was discovered by **Borthwick** and **Hendricks,** but the term 'phytochrome' was given by **Warren Butler.**

Role of phytochrome

- It is a **blue-green pigment**. It can detect light of particular wavelength.
- Phytochrome exists in two interconvertible forms. First form is Pr. It absorbs light of wavelength 660 nm (red).
- The second form is Pfr. It absorbs light of wavelength 730 nm (far-red.)
- Pr is present in cytoplasm, while Pfr is present on plasma membrane. Pfr is the physiologically active form.
- Pr is readily convertible into Pfr. It occurs when it absorbs red light.
- The conversion of these two forms is as under

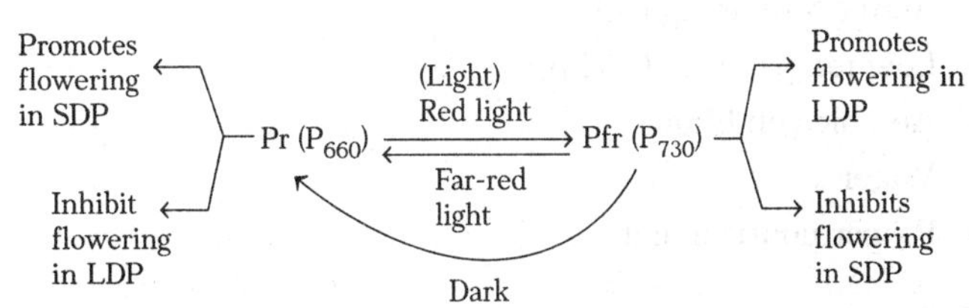

Fig. 22.9 Mechanism of two types of phytochrome

Role of florigen

- From leaves, the photoperiodic stimulus is transported to buds, where flowering occurs.
- Migration of this stimulus occurs, when plants are exposed to required photoperiod.
- This flowering stimulus is considered to be due to a hormone. It is known as **florigen.** It is a hypothetical hormone. Chemically, it is similar to gibberellins.

Importance of photoperiodism

- Photoperiodism is useful in keeping some plants in vegetative phase. By this method, high yields of potato, tubers can be obtained.
- It can be used to keep a plant in reproductive phase. Thus, more flowers and fruits can be obtained.
- Cross-breeding can be done more effectively.
- It is useful in laying out gardens. Crop pattern can also be manipulated.
- Flowering can be induced or retarded. It is done to meet the demands of market.
- Photoperiodic response of plant is under the control of genes. Manipulation of genes can enable a plant to flower in different seasons.

VERNALISATION

- Term vernalisation was coined by Russian agronomist **Lysenko**.
- It is a process of exposure of plant to low temperature, i.e.between 0-4°C. It is done to induce flowering in plants.
- Vernalisation was first studied on cereals.
- Due to vernalisation, vegetative phase of a plant is shortened. Thus, it initiates the reproductive phase in a plant.
- The sites of vernalisation are embryo, root apical meristem, shoot apical meristem and immature leaves, etc.

Conditions of vernalisation

These are as follows

- Actively dividing cells
- Low temperature (0°C-5°C)
- Aerobic condition
- Water
- Proper nourishment

Mechanism

- **G Melcher** studied vernalisation. He believed that stimulus of vernalisation is a hormone. This hypothetical hormone was named as **'vernalin'**.
- The stimulus is received by the actively dividing cells of shoot or embryo tip.
- In the presence of vernalin, physiological change is induced in the plant. It leads to flowering.
- It is believed that during vernalisation, gibberellins increase.

Devernalisation

It is the reverse process of vernalisation. During devernalisation, a vernalised plant is subjected to high temperature (30°C).

Uses

- Vernalisation shortens the vegetative period of plant. Thus, crops can be grown earlier.
- It increases yield of the plant.
- It increases resistance to cold and diseases.
- Kernal wrinkles of *Triticale* can be removed by vernalisation.
- It prepares a plant for better and early flowering.
- Plants can be grown in region where they fail to grow.

SENESCENCE

It is the consequence of ageing (i.e. the process of growing old). It is a highly ordered degenerative process and finally terminates life of an organ or an organism.

Types of senescence in plants

Senescence in plants is of following types

(i) **Whole plant senescence** As the plant matures, the whole plant dies after seed production. It occurs in monocarpic plant species which flower and fruit only once in their life cycle, e.g. wheat, rice, mustard, etc.

(ii) **Shoot senescence** It is seen in many perennial plants like *Zingiber, Musa, Chrysanthemum,* etc., in which shoot dies every year and underground stem remains alive.

(iii) **Organ senescence** Senescence of only lateral organs of plants such as leaves and fruits occur and they die prior to the death of the whole plant.
Organ senescenes may be of the following types

(a) **Simultaneous or Synchronous** It is found in the case of deciduous trees in which almost all leaves senescence and fall more of less simultaneously at a particular season of the year, e.g. banyan leaves.

(b) **Sequential** It is a progressive senescence of different parts of the plant, especially the lower and older leaves. Because the leaves have a limited lifespan, the senescence takes place in a sequential manner depending on the age of the leaf, e.g. coconut leaves.

ABSCISSION

- It is a natural separation or shedding of leaves, foliage branches, fruits, floral parts, etc., from plants. It is generally seasonal.
- A special narrow zone develops in the area of abscission called an **abscission zone**. Two distinct layers develop in abscission zone, i.e. separation layer (upper layer) and protective layer (lower suberised and lignified layer).

PLANT MOVEMENTS

They are changes in the position of plants or their parts. The movements occur due to response to a stimulus.

Stimulus may be defined as 'any change in the external environment'.

Classification

The plant movements are classified into two main types given as follows

Plant movements

- **Movements of locomotion** (Movement of whole plant body or organ from one place to another)
  - **Autonomous / Spontaneous** (seen in plants bearing cilia/flagella)
    - **Ciliary movement** (in *Chlamydomonas*, *Volvox*)
    - **Amoeboid movement** (in slime moulds)
    - **Cyclosis** (protoplasmic streaming)
      - → Rotation (in *Hydrilla*)
      - → Circulation (in *Tradescantia*)
  - **Paratonic/induced/tactic** (induced by external stimuli like light, heat, touch, etc.)
    - **Chemotactic movement** (in antherozoids of bryophytes and pteridophytes)
    - **Thermotactic movement** (in some motile algae)
    - **Phototactic movement** (in algae)
    - **Thigmotactic movement** (coiling of tendril due to contact with foreign particle)
- **Movements of curvature** (Uneven distribution of auxin causes curvature)
  - **Movement shown by vital activities of cells or organs**
    - **Spontaneous type**
      - Movement of variations (in Indian telegraph plant)
      - Movement of growth like nutation (in tendrils), Epinasty (in petal of flower during opening) and Hyponasty (in petal of flower during bud condition)
    - **Induced type**
      - **Tropic movement**
        Curvature towards direction of external stimulus like light, gravity, water, touch, temperature, etc., (directive)
        - Phototropism (in stem)
        - Geotropism (in roots)
        - Thermotropism (in shoot of some weeds)
        - Hydrotropic (in young roots)
        - Thigmotropic (tendrils)
        - Chemotropic
      - **Nastic movement**
        Curvature irrespective of direction of diffused external stimulus (non-directive)
        - Seismonasty (in *Mimosa pudica*)
        - Nyctinasty (in leaves of some legumes)
        - Thermonasty (in *Crocus* flowers)
        - Photonasty (in *Oxalis*)
  - **Movement shown by mechanical activities of dead and hygroscopic cells or tissues**

Note *Circadian rhythm in plants synchronises with the light cycle of its surrounding environemnt. It helps in the regulation of certain processes in plants, e.g. flowering, germination and leaf movement, etc.*

CHECK POINT 22.3

1. The cause of seed dormancy in *Lepidium* is
   (a) the presence of coumorin which germinate seed
   (b) immaturity of embryo at the time of seed germination
   (c) mechanical resistance of seed coat
   (d) the presence of high concentration of salt in the seed at the time of shedding

2. In lettuce, red light is most effective in breaking seed dormancy because it
   (a) stimulates water absorption
   (b) reverses inhibitory effect of far-red light
   (c) induces active transport
   (d) induces gibberellin production

3. In Short Day Plants (SDPs), flowering is induced by
   (a) long night
   (b) photoperiod less than 12 hrs
   (c) photoperiod shorter than initial value and uninterrupted long night
   (d) short photoperiod and interrupted long night

4. When a Short Day Plant (SDP) and a Long Day Plant (LDP), both kept under favourable condition for flowering are exposed to a short flash of red light in the middle of the dark period.
   (a) Both SD and LD plants will flower
   (b) Both SD and LD plants will remain vegetative
   (c) SD plant will remain vegetative and LD plant will flower
   (d) SD plant will flower and LD plant will remain vegetative

5. Vernalisation
   (a) prevents the precocious reproductive development late in the growing season and enables the plant to have sufficient time to reach maturity
   (b) is done by the application of high temperature
   (c) site is mature leaves
   (d) requires anaerobic condition

6. Slowing of winter variety of wheat in spring season after vernalisation.
   (a) Helps in shortening the vegetative period of plant and brings about early flowering
   (b) Does not enable winter rice to initiate flowering span of the flowering season
   (c) Promotes delayed reproductive development in the growing season
   (d) None of the above

EXAM POINTS

- Cells attaining their maximum size in terms of wall thickening and protoplasmic modification belong to maturation phase. **[AMU 2014]**
- The process of growth is maximum during log phase. **[NEET 2020]**
- Apical dominance is caused by auxin. **[OJEE 2013]**
- Fruit and leaf drop at early stages can be prevented by the application of auxins. **[NEET 2017]**
- The Avena curvature is used for bioassay of IAA. **[NEET 2017]**
- Removal of shoot tips is very useful technique to boost the production of tea leaves. This is because effect of auxins is removed and growth of lateral buds is enhanced. **[NEET 2019]**
- Gibberellic acid increase the yield of sugar by increasing the length of stem in sugarcane. **[AIIMS 2017]**
- Fruit and leaf drop at early stages can be prevented by the application of Gibberellic acid. **[AIIMS 2017]**
- During seed germination its stored food is mobilised by gibberellin. **[NEET 2013]**
- In order to increase the yield of sugarcane crop, gibberellins should be sprayed. **[NEET 2019]**
- Gibberellic acid is not an inhibitory substance governing seed dormancy. **[NEET 2020]**
- Cell elongation is internodal regions of the green plants takes place due to gibberellins. **[AFMC 2012]**
- Ethylene promotes senescence and abscission of plant organs. It breaks seed and bud dormancy. **[Kerala CEE 2014]**
- The term 'kinetin' was coined by Skoog and Miller. **[NEET 2020]**
- Cytokinins is used for inducing morphogenesis in plant tissue culture. **[Manipal 2012]**
- Ethylene is the only gaseous phytohormone. It induces ripening of fruits. **[EAM CET-A 2015]**
- Ethylene influence the swelling of the axis in dicot plant. **[Karala CEE 2015]**
- Auxin and ethylene can be applied to artificially induce flowering in pineapple plants throughout the year to increase yield. **[NEET 2019]**
- Auxin and cytokinin pair of hormone is used as the medium to secure shoots as well as roots. **[NEET 2016]**
- ABA acts as an antagonist to gibberellins. **[JIPMER 2017]**
- Abscisic acid stimulates the closure of stomata. **[Chhattisgarh PMT 2015, Kerala CEE 2015]**
- ABA promotes seed dormancy and bud dormancy. **[BHU 2012]**
- 2, 4-D is a dicot weedicide. **[MHT CET 2014]**
- Short day plants are affected by interruption of long day dark period by brief exposure of light. **[WB JEE 2015]**
- A plant produces flowers when exposed to alternating period of 5 hours lights and 3 hours dark in a 25-hrs cycle the plant should be LDP. **[Karnataka CET 2014]**
- Phytochrome is a chromoprotein. **[NEET-II 2016]**
- Phytochrome is a pigment sensitive for red and far-red light. **[Manipal 2012]**
- Leaves are the site of perception of photoperiod necessary for the induction of flowering in plants. **[EAMCET 2014, NEET 2019]**
- Vernalisation is low temperature treatment. **[JIPMER 2019]**
- Gibberellin can be used to replace the requirement of vernalisation. **[UP CPMT 2013]**
- Movement of tendrils in response to touch is known as thigmotropism. **[UPC PMT 2013]**

Chapter Exercises

(A) Topical Exercise

Growth and Development

1 Growth can be measured in various ways. Which of these can be used as parameters to measure growth? **[NCERT Exemplar]**
(a) increase in cell number
(b) increase in cell size
(c) increase in length and weight
(d) All of the above

2 Consider the following statements.
I. The basis to the phenomenon of growth is formation of new protoplasm.
II. Growth is maximum in the zone of cell elongation.
Choose the correct option.
(a) Statement I is true, but II is false
(b) Statement I is false, but II is true
(c) Both statements I and II are true
(d) Both statements I and II are false

3 Which of the following statements about growth rate is correct?
(a) The increased growth per unit time is termed as growth rate
(b) Rate of growth can be expressed mathematically
(c) An organism or a part of the organism can produce more cells in a variety of ways
(d) All of the above

4 The sigmoid curve shows
I. the lag phase
II. the senescence phase
III. the exponential growth phase
IV. the maturation phase
Choose the correct option from the codes given below.
(a) I, II and III (b) I and III (c) Only II (d) II and IV

5 Plant tissue culture is an example of
(a) differentiation (b) redifferentiation
(c) dedifferentiation (d) Both (a) and (c)

6 Development of plants depends upon
(a) intrinsic factors
(b) extrinsic factors only
(c) Both intrinsic and extrinsic factors
(d) None of the above

7 How many of the given extrinsic factors affect the growth of a plant?

Light, Genetic factors, Auxin, Minerals, Nutrients, Water, Oxygen, Soil, Temperature, Plasticity

(a) Four (b) Five (c) Six (d) Three

8 The developmental stages in a plant are as follows
(a) Division → Differentiation → Maturation → Senescence → Death
(b) Division → Maturation → Differentiation → Expansion → Death
(c) Division → Differentiation → Expansion → Maturation → Senescence → Death
(d) Division → Expansion → Differentiation → Maturation → Senescence → Death

9 Which of the following statement does not hold true for growth?
(a) Growth denotes a permanent increase in size of plant/organ
(b) Growth is a reversible process
(c) Plant growth is a unique feature and is confined to meristem
(d) Growth rate in plants is of two types

10 Initial rate of growth is slow in
(a) log phase
(b) lag phase
(c) stationary phase
(d) Both (a) and (b)

11 Ability of plant to change their mode of development.
(a) Adaptation (b) Growth
(c) Phototropism (d) Plasticity

12 Read the option given below.

Larkspur, Buttercup, Coriander, Sunflower

In which plant, plasticity can be observed in response to prevailing environment?
(a) Coriander and sunflower
(b) Larkspur, buttercup and coriander
(c) Buttercup and coriander
(d) Larkspur and buttercup

13 Study the following structures in the box and find out how many of them are the products of differentiation ...*A*..., dedifferentiation ...*B*... and redifferentiation ...*C*.... .

I	II	III
Cortex	Interfascicular cambium	Phelloderm
Primary medullary rays		Secondary cortex
Endodermis		Secondary phloem
		Phellem

	A	B	C		A	B	C
(a)	III	I	II	(b)	I	II	III
(c)	II	I	III	(d)	III	II	I

14 Secondary cortex $\xrightarrow[\text{Phloem}]{\text{Xylem}}$ Secondary phellem, phellogen

Which options are the odd ones associated with development from process of redifferentiation.

(a) Secondary cortex and xylem
(b) Secondary phloem and phellem
(c) Secondary cortex and secondary phloem
(d) Phellem and phellogen

15 Formation of interfascicular cambium and cork cambium from fully differentiated parenchyma cells is an example of

(a) differentiation (b) dedifferentiation
(c) redifferentiation (d) Both (a) and (b)

16 Plasticity in plant growth means that **[NCERT Exemplar]**

(a) plant roots are extensible
(b) plant development is dependent on the environment
(c) stems can extend
(d) None of the above

Plant Growth Regulators

17 Identify the correct matches associated with plant hormones with respect to their action.

(a) Growth promoters – Abscisic acid and cytokinin
Growth inhibitors – Ethylene and auxin
(b) Growth promoters – Auxin, cytokinin and gibberellin
Growth inhibitors – Abscisic acid and ethylene
(c) Growth promoters – Auxin and ethylene
Growth inhibitors – Cytokinin and abscisic acid
(d) Growth promoters – Auxin and cytokinin
Growth inhibitors – Gibberellins, abscisic acid and ethylene

18 The term 'synergistic action of hormones' refers to **[NCERT Exemplar]**

(a) when two hormones act together, but bring about opposite effects
(b) when two hormones act together and contribute to the same function
(c) when one hormone affects more than one function
(d) when many hormones bring about any one function

19 Charles Darwin and his son Francis discovered (around 1880) that a grass seedling bends towards light only if the

(a) stem is at least 10 cm long
(b) tip of the coleoptile is present and exposed to light
(c) cotyledon is present and exposed to light
(d) nights are long in comparison with the days

20 Dr. F Went noted that if coleoptile tips were removed and placed on agar for one hour, the agar would produce a bending when placed on one side of a freshly cut coleoptile stumps. Of what significance is this experiment?

(a) It is basis of quantitative determination of small amounts of growth promoting substances
(b) It supports the hypothesis that IAA is auxin
(c) It demonstrated polar movement of auxins
(d) It made possible the isolation and exact identification of auxin

21 Arrange the following in the manner of decreasing auxin activity.

(a) IAA $\rightarrow$ IBA $\rightarrow$ IPA (b) IPA $\rightarrow$ IBA $\rightarrow$ IAA
(c) IAA $\rightarrow$ IPA $\rightarrow$ IBA (d) IBA $\rightarrow$ IPA $\rightarrow$ IAA

22 6-furfurylaminopurine, 2,-4-dichlorophenoxy acetic acid and indole-3-acetic acids are the examples respectively for

(a) synthetic auxin, kinetin and natural auxin
(b) gibberellin, natural auxin and kinetin
(c) natural auxin, kinetin and synthetic auxin
(d) kinetin, synthetic auxin and natural auxin

23 Phenyl acetic acid, 2, 4-dichlorophenoxy acetic acid, ethephon, indole-3-butyric acid, indole-3-acetic acid. The naturally occurring auxins from these are given options are

(a) phenyl acetic acid, indole-3-butyric acid, indole-3-acetic acid
(b) 2-4 dichlorophenoxy acetic acid, ethephon and indole-3-butyric acid
(c) phenyl acetic acid and ethephon
(d) ethephon and indole-3-acetic acid

24 Which of the following effects of auxins on plants are the basis for commercial application?

(a) Callus formation
(b) Curvature of stem
(c) Induction of root formation in stem cuttings
(d) All of the above

25 Fruit drop is caused by

(a) more auxin in fruit than in stem
(b) less auxin in fruit than in stem
(c) equal distribution of auxin in stem and fruit
(d) absence of auxin in stem and fruit

26 Apical dominance is caused when auxin

(a) concentration is more than cytokinins
(b) concentration is less than cytokinins
(c) and cytokinin concentration are equal
(d) and cytokinin concentration are fluctuating

27 Match the following columns.

	Column I (Types of auxin)		Column II (Examples)
A.	Natural auxin	1.	IAA
B.	Synthetic auxin	2.	NAA
		3.	IBA
		4.	2,4-D

Codes

	A	B		A	B
(a)	1, 2	3, 4	(b)	3, 4	1, 2
(c)	1, 3	2, 4	(d)	2, 4	1, 3

28 In decapitated plants, axillary buds are activated because
(a) absence of auxin
(b) more food material is available to them
(c) amount of cytokinin increases in them
(d) they get more light

29 Consider the following statements.
I. Auxins promote apical dominance by suppressing the activity of lateral buds.
II. In moriculture, periodic pruning of shoot tips is done to make mulberry plants bushy.

Choose the correct option.
(a) Statement I is true, but II is false
(b) Statement I is false, but II is true
(c) Both statements I and II are true
(d) Both statements I and II are false

30 Auxins are known to induce
(a) female flowers in male plant
(b) male flowers in female plants
(c) male sterility in bisexual plants
(d) early bolting

31 Which of the following statements is incorrect?
(a) Cytokinins are formed primarily in roots
(b) Auxin and cytokinin are antagonistic in apical dominance
(c) Kinetin (a modified DNA purine) was discovered from herring sperm
(d) Zeatin is an auxin

32 Which one of the following is agent orange?
(a) Weedicide with dioxin
(b) Chemical used in luminous paint
(c) Biodegradable insecticide
(d) Colour used in fluorescent lamp

33 Auxin was isolated from tips of coleoptiles of ...*A*... by ...*B*... .

	A	*B*
(a)	Canary grass	Charles Darwn, Francis Darwin
(b)	Canary grass	FW Went
(c)	Oat seedlings	Charles Darwin, Francis Darwin
(d)	Oat seedlings	FW Went

34 How does pruning help in making the hedge dense?
(a) It induces the differentiation of new shoots from the rootstock
(b) It frees axillary buds from apical dominance
(c) The apical shoot grows faster after pruning
(d) It releases wound hormones

35 Growth hormones such as ...*A*... and ...*B*... are used in diluted form to produce parthenocarpic fruits, e.g. ...*C*... . Identify *A*, *B* and *C*.

	A	*B*	*C*
(a)	auxin	cytokinin	tomatoes
(b)	cytokinin	auxin	tomatoes
(c)	kinetin	zeatin	tomatoes
(d)	IAA	IBA	tomatoes

36 Which of the following functions are concerned with the growth hormone auxin?
I. Promotion of chloroplast development
II. Cell elongation
III. Control of stomatal closure
IV. Apical dominance

Codes
(a) Only II (b) II and IV (c) Only III (d) I and III

37 Negative geotropism in horizontal stems is caused by
(a) accumulation of auxins on the lower side
(b) accumulation of auxins on the upper side
(c) cell shrinkage on the lower side
(d) cell enlargement on the upper side

38 In 1920s, the Japanese biologist Kurosawa and his colleagues discovered the foolish-seedling disease, in which rice plants grew too tall, mainly because of
(a) a gibberellin secreting fungi
(b) the herbicide 2, 4-D
(c) degraded nucleic acids
(d) too much auxin, due to stagnant water in the fields

39 Read the given statements regarding effects of gibberellins in plants.
I. Shortening of internodes
II. Chlorophyll preservation
III. Maleness in plants
IV. Early growth for faster maturity

Codes
(a) I, II and V (b) III and IV
(c) II, III and IV (d) I, III and V

40

Stem growth, Fruit growth, Bolting, Increased mass yield, Seed germination, Overcome apical dominance

From the above given options, how many are the uses of gibberellins?
(a) Three (b) Four
(c) Six (d) Five

41 Which among the following are weak acids having gibbane ring structure?
(a) IAA (b) IBA
(c) Gibberellins (d) Auxin

42 Which of the following statements regarding gibberellins is incorrect?
(a) GA_3 was one of the first gibberellins to be discovered
(b) All GAs are acidic
(c) They increase the length of the plant axis as in grapes
(d) Sugarcane yield decreases on spraying GA

43 Gibberellin application to a plant can induce the following.
(a) Leaf fall
(b) Delay in senescence
(c) Elongation of shoot system
(d) Diversification of root system

44 The gibberellins have been commercially exploited for
(a) increasing the size of grape fruits
(b) breaking the dormancy of seeds
(c) Both (a) and (b)
(d) production of disease resistant varieties

45 Gibberellins do not cause
(a) shortening of genetically tall plants
(b) stimulation of seed germination
(c) promotion of parthenocarpy
(d) induction of α-amylase in barley

46 Gibberellins can promote seed germination because of their influence on
(a) rate of cell division
(b) production of hydrolysing enzymes
(c) synthesis of abscisic acid
(d) absorption of water through hard seed coat

47 Gibberellic acid has been successfully employed to induce flowering in
(a) short day plants under long day conditions
(b) long day plants under short day conditions
(c) the day neutral plants
(d) few plants

48 The biosynthesis of gibberellin takes place *via*
(a) citric acid cycle (b) mevalonic acid pathway
(c) glyoxylate cycle (d) Calvin cycle

49 Which of the following is not a bioassay for cytokinins?
(a) Chlorophyll preservation test
(b) Dwarf maize test
(c) Seed germination test
(d) Cell enlargement test

50 Apical bud inhibits the growth of lateral buds. This effect can be counteracted by
(a) auxin application (b) spraying auxin in root
(c) the application of 2, 4-D (d) cytokinin application

51 Which one of the following pairs is not correctly matched?
(a) Abscisic acid – Stomatal closure
(b) Gibberellic acid – Leaf fall
(c) Cytokinin – Cell division
(d) IAA – Cell wall elongation

52 The excised leaves of barley remain green for a longer period in cytokinin substituted medium due to delay in the destruction of
(a) chlorophyll (b) chlorophyll and proteins
(c) nucleic acids (d) All of these

53 Coconut milk contains **[NCERT Exemplar]**
(a) ABA (b) auxin (c) cytokinin (d) gibberellin

54 In tissue culture, differentiation of root and shoot can be controlled by
(a) modifying auxin and cytokinin ratio in the medium
(b) using tissue of right size
(c) giving temperature shocks
(d) change in light intensity

55 Identify the correct matches with respect to plant hormones and their bioassays.
(a) Avena curvature test – Cytokinin
(b) Growth inhibition test – Gibberellins
(c) Triple response test – Ethylene
(d) Barley endosperm test – Abscisic Acid

56 Apples are generally wrapped in waxed paper to **[NCERT Exemplar]**
(a) prevent sunlight from changing its colour
(b) prevent aerobic respiration by checking the entry of ethylene hormone
(c) prevent ethylene formation due to injury
(d) make the apples look attractive

57 When mature green bananas are put in a sealed plastic bag, they remain green for many days. However, if they are sealed with a ripe orange, the bananas turn yellow within a few days. The reason is orange
(a) removes CO_2, which inhibits fruit ripening
(b) removes CO_2 released from green bananas
(c) produces CO_2, which activates fruit ripening
(d) produces ethylene, which induces fruit ripening

58 Which is not the effect of ethylene?
(a) Breaks seed and bud dormancy
(b) Brings about horizontal growth of seedlings
(c) Hastens fruit ripening
(d) Helps to overcome apical dominance

59 Which of the following combination is incorrect?
(a) Cell elongation – Auxin
(b) Substitute of cold treatment – Ethylene
(c) Apogeotropism – Ethylene
(d) Lateral-bud development in dicots – Cytokinin

60 Consider the following statements.
I. Hastening the ripening of fruits.
II. Synthesis of carotenoids.
Select the option of set of hormones suitable for above functions in plants.
(a) I–Abscisic, II–Zeatin
(b) I–Zeatin, II–Ethylene
(c) I–Ethylene, II–Zeatin
(d) I–Ethylene, II–Abscisic acid

61 Consider the following statements.
I. ABA is antagonistic to GA_3.
II. ABA helps seed understand desiccation and other factors unfavourable for growth.
Choose the correct option.
(a) Statement I is true, but II is false
(b) Statement I is false, but II is true
(c) Both statements I and II are true
(d) Both statements I and II are false

62 Abscisic acid
I. induces dormancy
II. induces cell elongation
III. provides defence against salt and cold stresses
IV. inhibits abscission
Choose the correct option.
(a) Only I (b) I and II
(c) Only IV (d) I and III

63 Which one of the following hormones inhibits elongation in plants?
(a) ABA
(b) Cytokinin
(c) Auxin
(d) Ethylene

64 Increased availability of abscisic acid stimulates
(a) lipid formation
(b) protein formation in cell
(c) starch synthesis in guard cells
(d) cellulose synthesis

65 ABA acts as antagonistic to **[NCERT Exemplar]**
(a) ethylene
(b) cytokinin
(c) gibberellic acid
(d) IAA

Seed Dormancy and Germination

66 Consider the following statements.
I. During the state of seed dormancy, the growth of the embryo remains suspended.
II. Seed dormancy may be induced by short chain fatty acids.
Choose the correct option.
(a) Statement I is true, but II is false
(b) Statement I is false, but II is true
(c) Both statements I and II are true
(d) Both statements I and II are false

67 Seed dormancy can be broken by
(a) ABA and ethylene
(b) auxin and gibberellin
(c) gibberellin and cytokinin
(d) auxin and ABA

68 Consider the following statements.
I. Seed dormancy is due to the abscisic acid.
II. Rubbing of seed coat with sandpaper to break the seed dormancy is known as scarification.
Choose the correct option.
(a) Statement I is true, but II is false
(b) Statement I is false, but II is true
(c) Both statements I and II are true
(d) Both statements I and II are false

69 Seed packets give a recommended planting depth for the enclosed seeds. The most likely reason some seeds are to be covered with only 1/4 inch of soil is that these
(a) seeds do not have and etiolation response
(b) seeds of some kind require light to germinate
(c) seeds require a very high temperature to germinate
(d) do not produce hypocotyl

70 Hormone group responsible for breaking seed dormancy
I. ABA II. Cytokinin
III. Auxin IV. Gibberellin
Codes
(a) I and III (b) I, II and IV
(c) II, III and IV (d) I, II and III

71 Identify the statement which is incorrect about scarification of seed coat to break seed dormancy.
(a) It is a method of making seed coat soft and weak
(b) Scarification process entails treatment with acids and alcohols
(c) Mechanically, scarification is done using knives
(d) It also involves removal of germination inhibitors by oxidation

Photoperiodism and Vernalisation

72 Which of the following statements is correct?
(a) Short day plants require light period more than critical length
(b) Day neutral plants are independent of photoperiod
(c) Short-long day plants require long days for floral initiation
(d) Long day plants do flower only during long night periods

73 Importance of day length in flowering plants was first shown in
(a) cotton (b) *Lemna* (c) tobacco (d) rose

74 Two plants A and B flowers when given 14 hrs of light. When plant A is given more than 15.6 hours of light it does not flower and when plant B is given more than 11 hrs of light it flowers. This means
(a) B is a short day plant with a critical photoperiod of 11 hrs
(b) Both A and B are day neutral plants
(c) The critical photoperiod of plant A is 4 hrs
(d) A is a short day plant with a critical photoperiod of 15.6 hrs

75 Match the following columns.

Column I	Column II
A. SDP	1. Wheat
B. LDP	2. *Chrysanthemum*
C. DNP	3. Maize

Codes

	A	B	C		A	B	C
(a)	1	2	3	(b)	3	2	1
(c)	2	1	3	(d)	1	3	2

76 Which one of the following is correct with respect to long day plants?
(a) Flower only when amount of daylight lasts longer than critical threshold
(b) Bloom typically late in summers or autumn
(c) Examples of long day plant include rice, *Chrysanthemum*
(d) None of the above

77 Match the following columns.

Column I	Column II
A. Short day plants	1. Early long-days followed by short days
B. Long day plants	2. Dark period of 14-16 hrs
C. Day neutral plants	3. Light period of 14-16 hrs
D. Long-short day plants	4. All photoperiods

Codes

	A	B	C	D		A	B	C	D
(a)	2	3	4	1	(b)	1	3	4	2
(c)	2	4	1	3	(d)	3	2	1	4

78 Which of the following is the effect of a plant hormone, which is synthesised more in the absence of light?
(a) Inhibits the development of seedless fruits
(b) Responsible for closing of stomata
(c) Induces the dormancy of seeds
(d) Length of internodes increases

79 Which of the following is a day neutral plant?
(a) *Helianthus annuus*
(b) *Euphorbia pulcherrima*
(c) *Avena sativa*
(d) *Beta vulgaris*

80 One set of a plant was grown at 12 hours day and 12 hours night period cycles and it flowered, while in the other set, night phase was interrupted by flash of light and it did not produce flowers. Under which one of the following categories will you place this plant?
(a) Long-day
(b) Darkness neutral
(c) Day-neutral
(d) Short-day

81 Pick up the correct match pair.
(a) *Xanthium* — Long-day plant
(b) Sunflower — Short-day plant
(c) Wheat — Short-day plant
(d) Tomato — Day-neutral plant

82 Which of the following is/are short-day plants?
I. *Chrysanthemum*
II. *Dahlia*
III. *Spinacia*
IV. *Beta vulgaris*

Choose the option containing incorrect examples.
(a) I and II (b) III and IV
(c) Only I (d) Only III

83 Consider the following statements.
I. The critical photoperiod is maximum value for flowering in SDP and a minimum value for flowering in LDP.
II. A short day plant and a long day plant can never have the same value of photoperiod for flowering to occur.

Choose the correct option.
(a) Statement I is true, but II is false
(b) Statement I is false, but II is true
(c) Both statements I and II are true
(d) Both statements I and II are false

84 In short-day plants, flowering is inhibited by
(a) interruption of dark by white or red light
(b) dark interruption by far-red light
(c) dark interruption by red light followed by far-red light
(d) not possible

85 In a short-day plant, flowering is induced by
(a) long nights
(b) photoperiods less than 12 hours
(c) photoperiods shorter than initial value and uninterrupted long night
(d) short photoperiods and interrupted long nights

86 Which one of the following statements is true for the phytochromes?
(a) Phytochrome is a phytohormone
(b) Phytochrome is a photosynthetic pigment
(c) Phytochrome is a pigment that controls growth, photomorphogenesis and development of many plants
(d) Phytochrome regulates dark-dependent developmental processes

87 Consider the following statements.
I. Phytochrome is a pigment that controls growth, photomorphogenesis and development of many plants.
II. Phytochrome is a regulatory protein that controls several dark-dependent developmental processes.

Choose the correct option.
(a) Statement I is true, but II is false
(b) Statement I is false, but II is true
(c) Both statements I and II are true
(d) Both statements I and II are false

88 Phytochrome is a
(a) chemical regulating flowering
(b) chemical regulating growth from seed to adulthood
(c) hormone regulating growth from seed to adulthood
(d) regulator synthesised by plants and influencing physiological process

89 Physiologically active form of phytochrome is (A), and it is found in (B).

	A	B
(a)	Pr	Plasma membrane
(b)	Pfr	Plasma membrane
(c)	Pr	Cytosol
(d)	Pfr	Cytosol

90 In the long dark phase,
(a) Pfr form remains unchanged
(b) only 50% Pfr changes to Pr
(c) Pfr changes to Pr
(d) change from Pfr to Pr occurs after flowering

91 A biological clock measures the length of each night by the
(a) relative amount of red-absorbing and far-red absorbing phytochrome present at dawn
(b) amount of far-red absorbing phytochrome at dusk
(c) relative amount of red-absorbing and far-red absorbing phytochrome at mid day
(d) rate at which one kind of phytochrome is converted to the other

92 Identify the incorrect option for effects of the red and far-red light.

	Red light	Far-red light
(a)	Stimulates germination	Inhibits germination
(b)	Inhbits flowering in short day plants	Stimulates flowering in short day plants
(c)	Causes epicotyls hook to unbend	Maintains epicotyls hook bend
(d)	Inhibits the formation of anthocyanins	Stimulates the formation of anthocyanins

93 The seat of synthesis and seat of action of 'florigen' is
(a) root and leaf, respectively
(b) root and shoot tip, respectively
(c) leaf and shoot tip, respectively
(d) storage tissues

94 Photo-induced flowering was hypothesised to be because of
(a) vernalin (b) florigen (c) ethylene (d) cytokinin

95 The conditions necessary for vernalisation are
I. high temperature II. low temperature
III. carbon dioxide IV. oxygen and water
Choose the correct option.
(a) Only I (b) I and II
(c) II and IV (d) Only I

96 Vernalisation treatment can convert
(a) A biennial into an annual
(b) A spring variety into winter variety
(c) An annual into a perennial
(d) All of the above

97 Biennial plant $\xrightarrow{A}$ Annual plant

This conversion can be brought by A that can be
(a) light (b) high temperature
(c) GA and vernalisation (d) vernalisation only

98 Slowing of winter variety of wheat in spring season after vernalisation.
(a) Helps in shortening the vegetative period of plant and brings about early flowering
(b) Does not enable winter rice to initiate flowering span of the flowering season
(c) Promotes delayed reproductive development in the growing season
(d) None of the above

99 In many plants, the change over from vegetative to reproductive phase takes place in response to
(a) the length of the day
(b) the severity of temperature
(c) mainly, the food material available in the soil
(d) the O_2 present in the air

(B) Cumulative Exercise

(Assorted Questions of the Chapter for Advanced Level Practice)

1 Which of the following statements is incorrect?
(a) The role of PGR is only to exert intrinsic control
(b) Along with genomic control and extrinsic factor, they play an important role in plant growth and development
(c) Many of the intrinsic factors such as temperature and light control the plant growth and development *via* PGR like vernalisation, flowering, dormancy, seed germination and plant movements
(d) Both light and temperature have a role on initiating flowering

2 Pick out the correct statements.
I. Cytokinins especially help in delaying senescence.
II. Auxins are involved in regulating apical dominance.
III. Ethylene is especially useful in enhancing seed germination.
IV. Gibberellins are responsible for immature falling of leaves.

Codes
(a) I and III (b) I and IV
(c) II and III (d) I and II

3 Study the following statements.
I. Ethylene and ABA hormones accelerate the process of senescence.
II. Zeatin is physiologically active cytokinins in maize kernels.
III. Arginine is precursor for auxin biosynthesis.
IV. Auxin induces cell division.
Choose the correct option.
(a) I and II (b) II and III
(c) III and IV (d) I, II, III and IV

4 Match the following columns.

Column I	Column II
A. Girdling experiment	1. Ascent of sap
B. Cobalt-chloride paper method	2. Unequal transpiration on leaf surfaces
C. Crescograph	3. Growth
D. Bell-jar experiment	4. Transpiration

Codes

	A	B	C	D
(a)	1	2	3	4
(b)	4	3	1	2
(c)	4	1	2	3
(d)	2	1	4	3

5 Which of the following statements regarding photoperiodism is false?
(a) The response of plants to periods of light/day is termed as photoperiodism
(b) The shoot apices cannot perceive photoperiods
(c) In day neutral plants, there is no correlation between exposure to light duration and induction of flowering response
(d) The site of perception of light/dark duration is the flower

6 Spraying sugarcane crop with ...*A*... increase in the length of the stem thus increasing the yield by as much as ...*B*... tonnes per acre. ...*C*... does not occur naturally in plants search for natural substances with cytokinin like activities led to the isolation of ...*D*... from corn kernels and coconut milk.
(a) *A*-auxin, *B*-10, *C*-NAA, *D*-zeatin
(b) *A*-gibberellin, *B*-20, *C*-Zeatin, *D*-kinetin
(c) *A*-gibberellin, *B*-10, *C*-Zeatin, *D*-kinetin
(d) *A*-gbberellin, *B*-20, *C*-Kinetin, *D*-zeatin

7

Digrams *A* and *B* indicate the shape of leaves in larkspur and buttercup, respectively, choose the correct option.
(a) The juvenile and adult leaf of larkspur differ in size due to genetic and PGRs
(b) Both leaf of buttercup differ in size due to genetic and intercellular factors
(c) Both larkspur and buttercup leaf size variation is due to habitat plasticity
(d) None of the above

8 The figure given below shows growth of two leaves in one week. Select the correct option, if RGR = Relative Growth Rate and AGR = Absolute Growth Rate.

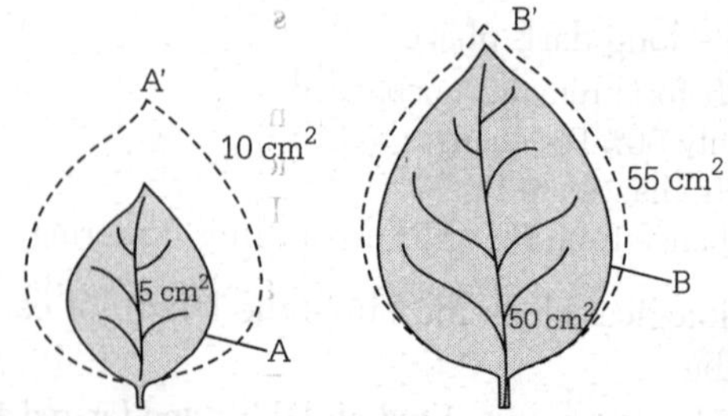

Options	RGR of A'	AGR of A	RGR of B'	AGR of B
(a)	5%	$100 cm^2$	5%	$100 cm^2$
(b)	50%	$5 cm^2$	50%	$5 cm^2$
(c)	10%	$5 cm^2$	100%	$5 cm^2$
(d)	100%	$5 cm^2$	10%	$5 cm^2$

9 Given below are the accuses of seed dormancy and methods of breaking it. Identify the incorrect match.
(a) Impermeable seed coat – Hydrolytic enzymes
(b) Immature seeds – Scarification
(c) Mechanically resistant – Microbial action
(d) Chemical substances like ABA – Growth hormones

10 The sigmoid growth and arithmetic growth can be represented graphically as

(a)
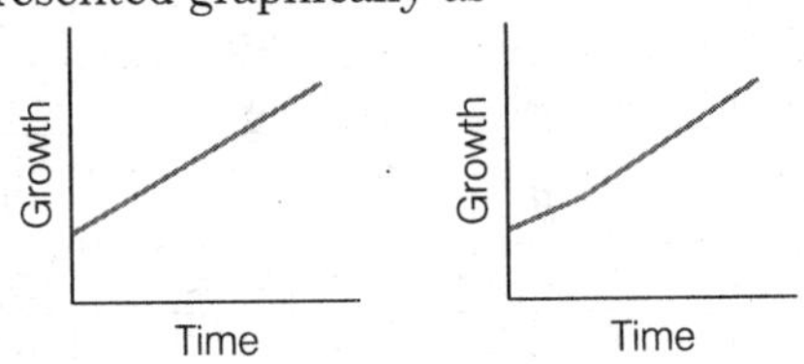

(b)
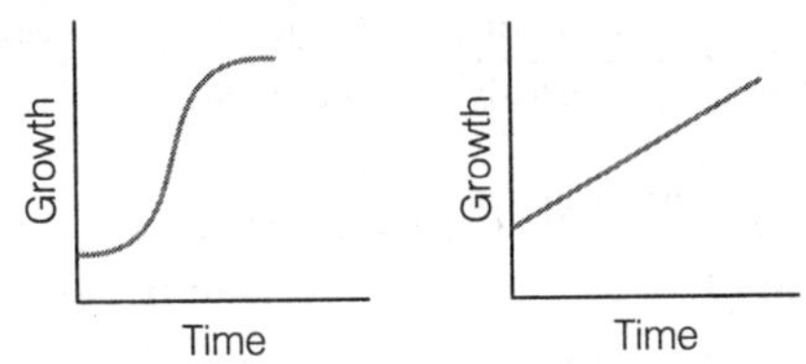

(c)
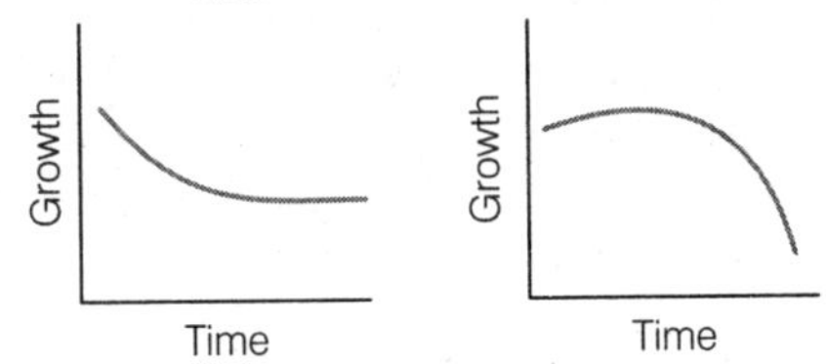

(d)
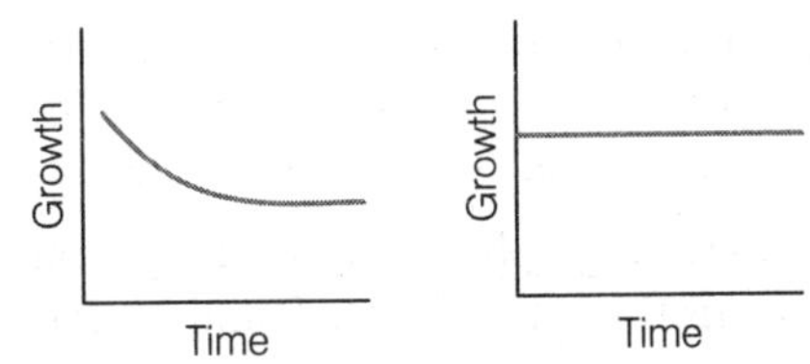

11 Read the following statements.
I. Florigens are associated with flowering in plants.
II. Phytochromes become active in green light.
III. Plant hormones are growth inhibitors.
IV. α-amylase test is a bioassay of GA.

Identify the true and false statement from above.

	True	False
(a)	I, II and III	IV
(b)	I and IV	II and III
(c)	II and III	I and IV
(d)	Only I	II, III and IV

12 Match the following columns and choose the correct combination.

Column I	Column II
A. Zeatin	1. Flowering hormone
B. Florigen	2. Synthetic auxin
C. IBA	3. Cytokinin
D. NAA	4. Natural auxin

Codes

	A	B	C	D
(a)	3	4	1	2
(b)	2	1	4	3
(c)	1	2	3	4
(d)	3	1	4	2

13 Read the following statements.
I. Positively photoblastic seeds require light for germination.
II. Non-photoblastic seeds have no influence with the presence or absence of light.
III. Negatively photoblasic seeds have inhibition of seed germination due to light exposure.
IV. Maize and beans are common examples of non-photoblastic seed.
V. Seed germination is inhibited by ABA and gibberellin.

How many above statements are correct?
(a) Four
(b) Two
(c) Five
(d) Three

14 Match the following columns.

Column I	Column II
A. IAA	1. Herring sperm DNA
B. ABA	2. Bolting
C. Ethylene	3. Stomatal closure
D. GA	4. Weed free lawns
E. Cytokinins	5. Ripening of fruit

Codes

	A	B	C	D	E
(a)	4	3	5	2	1
(b)	5	3	4	2	1
(c)	4	1	5	3	2
(d)	5	3	2	4	1

15 Observe the photoperiodic response depicted below and find out which one (*A*, *B* or *C*) is a Short Day Plant (SDP), Long Day plant (LDP) and Day Neutral Plant (DNP).

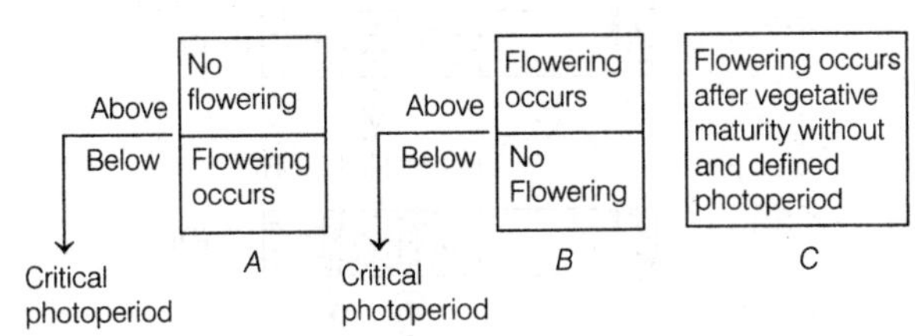

Options	*A*	*B*	*C*
(a)	SDP	DNP	LDP
(b)	SDP	LDP	SDP
(c)	DNP	LDP	SDP
(d)	SDP	LDP	DNP

16 Match the following columns.

Column I	Column II
A. Substitute of vernalisation	1. High temperature
B. Critical photoperiod	2. ABA
C. Stress hormone	3. Gibberellin
D. Devernalisation	4. SDP and LDP

Codes

	A	B	C	D
(a)	1	2	4	3
(b)	2	3	2	4
(c)	3	4	2	1
(d)	2	4	1	4

17 One hormone hastens maturity period in juvenile conifers, second hormone controls xylem differentiation, while the third hormone increases the tolerance of plants to various stresses. They are respectively

	A	*B*	*C*
(a)	Gibberellin	Auxin	Ethylene
(b)	Auxin	Gibberellin	Cytokinin
(c)	Gibberellin	Auxin	ABA
(d)	Auxin	Gibberellin	ABA

18 Given below is the representation of the effect of interruption of dark peiod in a Short Day Plant (SDP). Correctly identify the results of *A*, *B*, *C* and *D*.

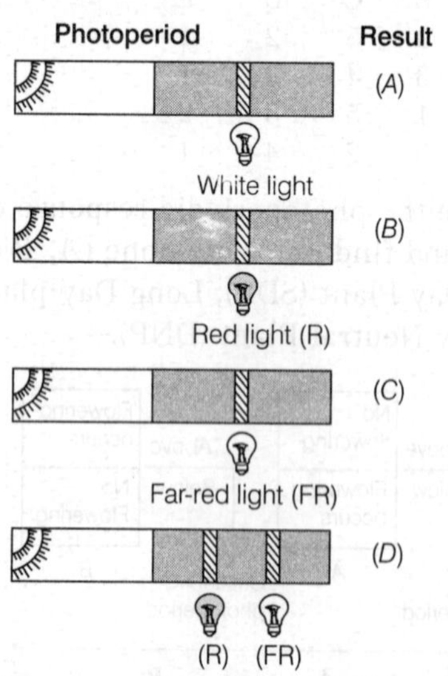

	A	*B*	*C*	*D*
(a)	Vegetative	Vegetative	Flowering	Flowering
(b)	Flowering	Flowering	Vegetative	Vegetative
(c)	Flowering	Vegetative	Flowering	Vegetative
(d)	Vegetative	Flowering	Vegetative	Flowering

19 *Xanthium* is a SDP with a critical photoperiod of 15.6 hours. Accordingly in which of the following cases (*A*, *B*, *C* and *D*) plant will remain vegetative or will flower.

Plants	Photoperiods	Results
Normal plant	12 hrs	*A*
Plant with one intact leaf	12 hrs	*B*
Deforated plant	16 hrs	*C*
Plant with one intact leaf	16 hrs	*D*

	A	*B*	*C*	*D*
(a)	Flowering	Vegetative	Flowering	Vegetative
(b)	Vegetative	Flowering	Vegetative	Flowering
(c)	Flowering	Flowering	Vegetative	Vegetative
(d)	Vegetative	Vegetative	Flowering	Flowering

20 Match the following columns.

Column I	Column II
A. Human urine	1. Cytokinin
B. *Gibberella fujikuroi*	2. Auxin
C. Herring fish DNA	3. Ethylene
D. Ripening fruits	4. Abscisic acid
E. Aged leaves of plants	5. Gibberellins

Codes

	A	B	C	D	E
(a)	2	5	1	3	4
(b)	2	3	4	5	1
(c)	1	5	2	4	3
(d)	5	4	3	2	1

21 Given below is a graph drawn on the parameters of growth *versus* time. In this, *A*, *B* and *C* respectively represent

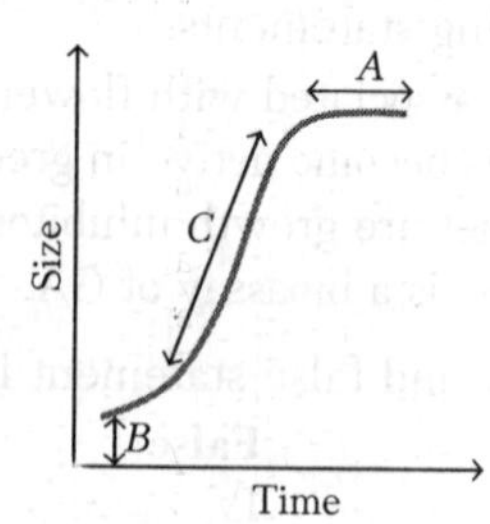

(a) Exponential phase, Log phase and Steady state phase
(b) Steady state phase, Lag phase and Log phase
(c) Slow growing phase, Lag phase and Steady state phase
(d) Lag phase, Steady state phase and Log phase

22 Read the following statements about effects of ethylene and choose the correct option.

I. More female flowers in cucumber.
II. α-amylase production in seed.
III. Acceleration of fruit ripening in tomato.
IV. Delay sprouting in potato tubers.

(a) I and II
(b) I and III
(c) II and IV
(d) III and IV

23 Match the following columns.

Column I	Column II
A. Auxin	1. Herring sperm DNA
B. Cytokinin	2. Inhibitor of growth
C. Gibberellin	3. Apical dominance
D. Ethylene	4. Epinasty
E. Abscisic acid	5. Induces amylase synthesis

Codes

	A	B	C	D	E		A	B	C	D	E
(a)	3	1	5	4	2	(b)	4	5	1	3	2
(c)	2	1	5	3	4	(d)	3	1	5	2	4

24 Match the following columns.

Column I (PGRs)	Column II (Types)
A. IAA	1. Gases
B. N_6-furfurylaminopurine	2. Terpenes
C. ABA	3. Derivatives of carotenoids
D GA_3	4. Adenine derivatives
E. C_2H_4	5. Indole compounds

Codes

	A	B	C	D	E
(a)	1	2	3	4	5
(b)	5	4	3	2	1
(c)	5	4	1	2	3
(d)	4	5	1	2	3

25 Match the following columns.

Column I	Column II
A. Differentiation	1. Acts of maturation
B. Redifferentiation	2. Acts of again loosening cell division capacity and mature to perform special function
C. Dedifferentiation	3. Acts of gaining cell division capacity after differentiation

Codes

	A	B	C		A	B	C
(a)	1	3	2	(b)	3	2	1
(c)	1	2	3	(d)	3	1	2

26 On the basis of correlation, find the correct option from columns.

Column I	Column II	Column III
I. Foolish plant	A. Volatile hormone	1. Induces dormancy
II. Induces senescence	B. GA	2. Ripens fruit
	C. Zeatin	3. Usually sterile plant

Codes

(a) I-A-2, II-C -1 (b) I-C-3, II-B -3

(c) I-B-3, II-A -2 (d) I-B-1, II-C -2

27 Abscisic acid is an antitranspirant. It prevents transpiration by

(a) forming a film impervious to water over the stomata

(b) preventing K^+ ions uptake, thus decreasing the osmotic concentration of the guard calls and keeping them flaccid

(c) inducing starch formation, thus decreasing the osmotic concentration of the guard cells and keeping them flaccid

(d) increasing respiration, thus increasing CO_2 concentration in the guard cells which help in keeping the guard cells flaccid

28 The relation of flowering to the lengths of light and dark period was investigated by two experiments. In one experiment, the plants were subjected to dark periods of various lengths, while the light periods were kept as 4 hrs. In the second experiment, plants were kept in dark periods of various lengths, while the light periods were kept at 16 hrs. The numbers of flowers formed in relation to these periods is shown in the table.

Dark period hours	Flowers formed in experimental light periods	
	4 hrs	16 hrs
8	0	0
10	0	0
12	4	6
14	5	7
16	5	8

These results indicate that flowering

(a) requires long days

(b) is initiated by a long light period

(c) is due to short light periods

(d) requires a minimum dark period

29 Which one of the plant growth regulator will we use in following causes given below?

I. Bolt a rosette plant II. Enhance senescence

III. Flowering in a plant

(a) I—Auxin, II—Cytokinin, III—Ethylene

(b) I—Gibberellins, II—Abscisic acid, III—Cytokinin

(c) I—Gibberellins, II—Abscisic acid, III—Auxin

(d) I—Auxin, II—Cytokinin, III—Gibberellins

30 Which one of the PGRs would be used by farmers if they are asked to

(A) Induce parthenocarpy in tomatoes.

(B) Hastens fruit ripening in tomatoes and apples.

(C) Induce flowering in mango.

(D) Elongation and improvement in shape of apple.

	A	B	C	D
(a)	Auxin	Ethephon	Ethylene	GA
(b)	Ethephon	GA	Auxin	Cytokinin
(c)	Auxin	Ethephon	Cytokinin	Ethylene
(d)	Cytokinin	Auxin	GA	Ethephon

Assertion and Reason

Direction (Q. Nos. 31-43) *In each of the following questions, a statement of Assertion (A) is given followed by corresponding statement of Reason (R). Of the statements, mark the correct answer as*

(a) Both A and R are true and R is the correct explanation of A
(b) Both A and R are true, but R is not the correct explanation of A
(c) A is true, but R is false
(d) A is false, but R is true
(e) Both A and R are false

31 **Assertion** (A) Gibberellin is useful in early seed production in conifers.
Reason (R) Ethephon is responsible for early ripening in tomato and apple.

32 **Assertion** (A) Photomodulation of flower is phytochrome regulated process.
Reason (R) Active form of phytochrome (Pfr) directly induces floral induction in shoot buds.

33 **Assertion** (A) Abscisic Acid (ABA) is also known as stress hormone.
Reason (R) It is produced in much higher amounts, when plants are subjected to stress.

34 **Assertion** (A) Cytokinins are antisenescent.
Reason (R) Effect of cytokinins is antagonistic to ethylene.

35 **Assertion** (A) Ethylene causes climacteric ripening of fruits.
Reason (R) Climacteric fruit shows a rise in respiration at the time of ripening.

36 **Assertion** (A) Phytochrome exists in two forms, Pr and Pfr.
Reason (R) Pr form stimulates and Pfr form inhibits flowering.

37 **Assertion** (A) Sigmoid curve of growth consists of four parts.
Reason (R) Lag phase is also called grand phase of growth.

38 **Assertion** (A) In some plants, flowering depends only an a combination of light and dark exposure.
Reason (R) The site of perception of light or dark duration are the shoot apices of plants.

39 **Assertion** (A) Vernalisation is acceleration of subsequent flowering by low temperature treatment.
Reason (R) Site of vernalisation is apical meristem.

40 **Assertion** (A) Auxin helps to prevent fruit and leaves drop at early stages.
Reason (R) It promotes the abscission of older mature leaves and fruits.

41 **Assertion** (A) The first phytohormone to be discovered was auxin.
Reason (R) Light reaction of coleoptile tip attracted the attention of Darwin in the last quarter of the nineteenth century.

42 **Assertion** (A) The shoot apical meristems are the only source of auxin synthesis.
Reason (R) Dormancy of lateral buds over apical buds is due to auxin.

43 **Assertion** (A) Hormones are also called growth adjusters.
Reason (R) They promote or inhibit plant growth.

(C) Medical Entrances' Gallery

(Collection of questions Asked in NEET & Various Medical Entrance Exams)

1 Who coined the term 'kinetin'? **[NEET 2020]**
(a) Skoog and Miller
(b) Darwin
(c) Went
(d) Kurosawa

2 Which of the following is not an inhibitory substance governing seed dormancy? **[NEET 2020]**
(a) Abscisic acid
(b) Paraascorbic acid
(c) Phenolic acid
(d) Gibberellic acid

3 The process of growth is maximum during
(a) lag phase (b) senescence **[NEET 2020]**
(c) dormancy (d) log phase

4 It takes very long time for pineapple plants to produce flowers. Which combination of hormones can be applied to artificially induce flowering in pineapple plants throughout the year to increase yield? **[NEET 2019]**
(a) Gibberellin and cytokinin
(b) Gibberellin and abscisic acid
(c) Cytokinin and abscisic acid
(d) Auxin and ethylene

5 Removal of shoot tips is very useful technique to boost the production of tea leaves. This is because **[NEET 2019]**

(a) gibberellins prevent bolting and are inactivated
(b) auxins prevent leaf drop at early stages
(c) effect of auxins is removed and growth of lateral buds is enhanced
(d) gibberellins delay senescence of leaves

6 In order to increase the yield of sugarcane crop, which of the following plant growth regulators should be sprayed? **[NEET 2019]**

(a) Ethylene (b) Auxins
(c) Gibberellins (d) Cytokinins

7 What is the site of perception of photoperiod necessary for the induction of flowering in plants? **[NEET 2019, EAMCET 2014]**

(a) Pulvinus (b) Shoot apex
(c) Leaves (d) Lateral buds

8 Vernalisation is **[JIPMER 2019]**

(a) low pH treatment
(b) low temperature treatment
(c) high temperature treatment
(d) high pH treatment

9 Which of the following is not a plant growth inhibitor? **[JIPMER 2018]**

(a) Dormin (b) IAA
(c) Ethylene (d) ABA

10 Apical dominance is caused by **[AIIMS 2018]**

(a) auxin (b) ethylene
(c) gibberellin (d) cytokinin

11 Which one of the following generally acts as an antagonist to gibberellins? **[JIPMER 2017]**

(a) Zeatin (b) Ethylene
(c) ABA (d) IAA

12 Which of the following plant growth hormones increases the yield of sugar by increasing the length of stem in sugarcane? **[AIIMS 2017]**

(a) Cytokinin (b) Ethylene
(c) Gibberellic acid (d) Auxin

13 Fruit and leaf drop at early stages can be prevented by the application of **[NEET 2017]**

(a) cytokinins (b) ethylene
(c) auxins (d) gibberellic acid

14 The Avena curvature is used for bioassay of **[NEET 2017, NEET-II 2016]**

(a) GA_3 (b) IAA
(c) Ethylene (d) ABA

15 You are given a tissue with its potential for differentiation in an artificial culture. Which of the following pairs of hormones would you add to the medium to secure shoots as well as roots?

(a) IAA and gibberellin **[NEET-II 2016]**
(b) Auxin and cytokinin
(c) Auxin and abscisic acid
(d) Gibberellin and abscisic acid

16 Phytochrome is a **[NEET-II 2016]**

(a) flavoprotein (b) glycoprotein
(c) lipoprotein (d) chromoprotein

17 Match the following columns.

Column I	Column II
A. 2, 4-D	1. Promotion of lateral shoot growth
B. GA	2. Thinning in cotton
C. Ethephon	3. Stress tolerance
D. ABA	4. Brewing industry
	5. Weed-free lawns

Codes **[EAMCET 2016]**

	A	B	C	D		A	B	C	D
(a)	5	4	1	2	(b)	3	2	1	4
(c)	2	1	5	3	(d)	5	4	2	3

18 Match the following columns.

Column I	Column II
A. Munch	1. Stomatal movement
B. Darwin	2. Hastens fruit ripening by volatiles
C. Cousins	3. Growth of coleoptile towards light
D. Levitt	4. Nature of enzyme action
	5. Bidirectional transport in pholem

Codes **[EAMCET 2015]**

	A	B	C	D
(a)	3	4	1	2
(b)	4	3	1	2
(c)	5	3	2	1
(d)	3	1	4	5

19 Identify the correct pair of combinations.

I. Zeatin—adenine derivative — overcoming apical dominance
II. C_2H_4—indole compound—effective in fruit ripening
III. ABA—carotenoid derivative — induction of parthenocarpy
IV. GA_3—terpene—bolting in cabbage **[EAMCET 2015]**

Codes

(a) II and III (b) II and IV
(c) I and IV (d) III and IV

20 **Assertion** (A) Ethylene induces ripening.
Reason (R) It is a gaseous hormone. **[EAMCET-A 2015]**

(a) Both A and R are true and R is the correct explanation of A
(b) Both A and R are true, but R is not the correct explanation of A
(c) A is true, but R is false
(d) A is false, but R is true

21 One hormone stimulates the closure of stomata and another one influences the swelling of the axis in dicot plants. They are **[Kerala CEE 2015]**

(a) gibberellins and ethylene
(b) abscisic acid and cytokinins
(c) gibberellins and cytokinins
(d) auxins and cytokinins
(e) abscisic acid and ethylene

22 Which one is correct for blooming of short day plants? **[WB JEE 2015]**

(a) The long dark period is not critical
(b) It is affected by interruption of long dark period by brief exposure to light
(c) It is not affected by interruption of long dark period by brief exposure of light
(d) It is affected if continuous light period is interrupted

23 Which of the following plant hormones reduces transpiration rate by inducing stomatal closure? **[Chhattisgarh PMT 2015]**

(a) Ethylene (b) Gibberellin
(c) ABA (d) Cytokinin

24 Which of the following is a dicot weedicide? **[MHT CET 2014]**

(a) 2, 4-D (b) NAA (c) IBA (d) IAA

25 A few normal seedlings of tomato were kept in the dark room. After a few days, they were found to become white coloured like albinos. Which of the following terms will you use to describe them? **[CBSE AIPMT 2014]**

(a) Embolised (b) Etiolated (c) Defoliated (d) Mutated

26 Which is not the effect of ethylene? **[Kerala CEE 2014]**

(a) Promotes senescence and abscission of plant organs
(b) Breaks seed and bud dormancy
(c) Brings about horizontal growth of seedlings
(d) Hastens fruit ripening
(e) Helps to overcome apical dominance

27 If a plant produces flowers when exposed to alternating periods of 5 hours light and 3 hours dark in a 24-hour cycle, the plant should be **[Karnataka CET 2014]**

(a) SDP (b) LDP (c) SLDP (d) DNP

28 Match the following columns. **[WB JEE 2014]**

Column I	Column II
A. Control of weeds	1. Gibberellin
B. Induction of germination	2. Cytokinin
C. Ripening of fruit	3. 2, 4-D
D. Delaying of senescence	4. Ethylene

Codes

	A	B	C	D
(a)	2	4	3	1
(b)	3	1	4	2
(c)	1	2	4	3
(d)	2	3	1	4

29 Match the following columns.

Columns I	Columns II
A. Early seed production in conifers	1. Indole substance
B. Seed development and maturation	2. Terpene substance
C. Lateral shoot growth	3. Volatile substance
D. Root hair formation	4. Adenine derivative
	5. Carotenoid derivative

Codes **[EAMCET 2014]**

	A	B	C	D
(a)	3	1	5	2
(b)	2	5	4	3
(c)	2	1	5	4
(d)	4	3	2	1

30 Spraying sugarcane crop with a plant hormone increases length of plants and increases yield by as much as 20 tonnes/acre. The hormone is **[AMU 2014]**

(a) gibberellin (b) auxin
(c) cytokinin (d) ABA

31 Cells attaining their maximum size in terms of wall thickening and protoplasmic modifications belong to **[AMU 2014]**

(a) meristematic phase (b) elongation phase
(c) maturation phase (d) None of these

32 During seed germination, its stored food is mobilised by **[NEET 2013]**

(a) ethylene (b) cytokinin
(c) ABA (d) gibberellin

33 Movement of tendrils in response to touch is known as **[UP CPMT 2013]**

(a) nyctinastism (b) thigmotropism
(c) seismonatism (d) haptonastism

34 The phytohormone, which helps in germination of seed is **[UP CPMT 2013]**
(a) ABA (b) auxin
(c) gibberellin (d) cytokinin

35 Compare the statements I and II.
I. Ethylene is a gas, which acts as growth regulator of plants.
II. It is the most simple plant hormone. **[Manipal 2013]**
(a) Both the statements I and II are correct and II is the correct explanation of I
(b) Both the statements I and II are correct, but II is not the correct explanation of I
(c) Statement I is correct, but II is incorrect
(d) Statement I is incorrect, but II is correct

36 Hormone replacing the requirement of vernalisation is **[UP CPMT 2013]**
(a) ethylene (b) auxin
(c) gibberellin (d) cytokinin

37 The rosette habit of cabbage can be changed by application of **[Karnataka CET 2013]**
(a) IAA (b) GA
(c) ABA (d) ethephon

38 Apical dominance is caused by **[Odisha JEE 2013; NEET 2013]**
(a) auxin (b) cytokinin
(c) ethylene (d) gibberellin

39 Which of the following hormone promotes seed dormancy and bud dormancy? **[BHU 2012]**
(a) Auxin (b) GA
(c) Cytokinin (d) ABA

40 A plant hormone used for inducing morphogenesis in plant tissue culture is **[Manipal 2012]**
(a) ethylene (b) gibberellin
(c) cytokinin (d) abscisic acid

41 The pigment sensitive for red and far-red light is **[Manipal 2012]**
(a) chlorophyll (b) phytochrome
(c) cytochrome (d) carotene

42 Fluorine ring is a feature of **[AFMC 2012]**
(a) GA (b) ethylene
(c) morphactines (d) None of these

43 Cell elongation in internodal regions of the green plants takes place due to **[AFMC 2012]**
(a) indole acetic acid (b) cytokinins
(c) gibberellins (d) ethylene

44 Day neutral plant relates to **[J&K CET 2011]**
(a) loss of activity during daytime
(b) overactive during daytime
(c) flowering in all possible photoperiods
(d) no flowering in any photoperiod

45 Auxanometer is used to measure **[J&K CET 2011]**
(a) the growth in length of a plant organ
(b) the growth in breadth of a plant organ
(c) population of the pests attacking a plant
(d) Both (a) and (b)

46 Bolting may be induced by **[J&K CET 2011]**
(a) gibberellins (b) ABA
(c) auxin (d) cytokinin

47 Which one of the following pairs is not correctly matched? **[Kerala CEE 2011]**
(a) Adenine derivative — Kinetin
(b) Carotenoid derivative — ABA
(c) Terpenes — IAA
(d) Indole compounds — IBA
(e) Gas — Ethylene

48 Photoperiodism was first characterised in **[CBSE AIPMT 2011]**
(a) tobacco (b) potato
(c) tomato (d) cotton

49 Phototropic curvature is the result of uneven distribution of **[CBSE AIPMT 2011]**
(a) gibberellin
(b) phytochrome
(c) cytokinins
(d) auxin

ANSWERS

Check Point 22.1

1. (c)	2. (d)	3. (a)	4. (d)	5. (d)	6. (a)

Check Point 22.2

1. (b)	2. (b)	3. (a)	4. (a)	5. (a)	6. (c)

Check Point 22.3

1. (c)	2. (b)	3. (c)	4. (c)	5. (a)	6. (a)

(A) Topical Exercise

1. (d)	2. (c)	3. (d)	4. (b)	5. (b)	6. (c)	7. (c)	8. (c)	9. (b)	10. (b)
11. (d)	12. (b)	13. (b)	14. (d)	15. (b)	16. (b)	17. (b)	18. (b)	19. (b)	20. (d)
21. (d)	22. (d)	23. (a)	24. (c)	25. (b)	26. (a)	27. (c)	28. (a)	29. (c)	30. (a)
31. (d)	32. (a)	33. (d)	34. (b)	35. (d)	36. (b)	37. (a)	38. (a)	39. (b)	40. (b)
41. (c)	42. (d)	43. (c)	44. (c)	45. (a)	46. (b)	47. (b)	48. (b)	49. (b)	50. (d)
51. (b)	52. (b)	53. (c)	54. (a)	55. (c)	56. (b)	57. (d)	58. (d)	59. (b)	60. (d)
61. (c)	62. (d)	63. (a)	64. (c)	65. (c)	66. (c)	67. (c)	68. (c)	69. (b)	70. (c)
71. (d)	72. (b)	73. (c)	74. (d)	75. (c)	76. (a)	77. (a)	78. (d)	79. (a)	80. (d)
81. (d)	82. (a)	83. (a)	84. (a)	85. (c)	86. (c)	87. (a)	88. (a)	89. (b)	90. (c)
91. (a)	92. (d)	93. (c)	94. (b)	95. (c)	96. (d)	97. (c)	98. (a)	99. (a)	

(B) Cumulative Exercise

1. (c)	2. (d)	3. (a)	4. (a)	5. (d)	6. (d)	7. (a)	8. (d)	9. (c)	10. (b)
11. (b)	12. (d)	13. (a)	14. (a)	15. (d)	16. (c)	17. (c)	18. (a)	19. (c)	20. (a)
21. (b)	22. (b)	23. (a)	24. (b)	25. (c)	26. (c)	27. (b)	28. (d)	29. (b)	30. (a)
31. (b)	32. (c)	33. (a)	34. (b)	35. (b)	36. (b)	37. (c)	38. (d)	39. (b)	40. (b)
41. (b)	42. (e)	43. (a)							

(C) Medical Entrances' Gallery

1. (a)	2. (d)	3. (d)	4. (d)	5. (c)	6. (c)	7. (c)	8. (b)	9. (b)	10. (a)
11. (c)	12. (c)	13. (c)	14. (b)	15. (b)	16. (d)	17. (d)	18. (c)	19. (c)	20. (b)
21. (e)	22. (b)	23. (c)	24. (a)	25. (b)	26. (e)	27. (b)	28. (b)	29. (b)	30. (a)
31. (c)	32. (d)	33. (b)	34. (c)	35. (b)	36. (c)	37. (b)	38. (a)	39. (d)	40. (c)
41. (b)	42. (c)	43. (c)	44. (c)	45. (d)	46. (a)	47. (c)	48. (a)	49. (d)	

Explanations

(A) Topical Exercise

4 (*b*) The growth curve of any organism appears to have a sigmoidal curve which includes lag phase, log phase or exponential phase, stationary phase and the death phase. The lag phase is the adaptation phase for the organism where they acclimatises themselves to the new environmental conditions provided. The growth is slow at this stage. Hence, the correct answer is option (b).

29 (*c*) Both statements I and II are true.

Auxin is produced by shoot tip and act as a growth promoter for apical buds and inhibits the growth of lateral buds. When a terminal bud is removed, the nearest axillary buds begin to grow and the plant branches rapidly. In case of mulberry plants, pruning is done to remove apical dominance and cause more branching of the main body of the plant.

31 (*d*) Statement in option (d) is incorrect and can be corrected as

Zeatin was the first natural cytokinin discovered from the corn-kernels and coconut milk.

Hence, zeatin is not a auxin.

42 (*d*) Given option (d) is incorrect and can be corrected as Spraying of gibberellic acid causes extra elongation of the stem resulting in comparatively higher yields.

55 (*c*) Option (c) is correct match.

Triple response test - Ethylene

Other options are incorrect matches and can be corrected as

Avena curvature test - Auxin

Growth inhibiton test - Abscisic acid

Barley endosperm test - Gibberellins

59 (*b*) Option (b) is incorrect and can be corrected as

Substitute of cold treatment can be gibberellins and not ethylene. They help in resistance to stresses. Other options are correct as

Ethylene is involved in apogeotropism in roots and cytokinin helps in lateral bud development in dicots. Auxins promote cell division and cell elongation.

71 (*d*) Statement is option (d) is incorrect about scarification and can be corrected as

Scarification is a process used for breaking seed dormancy by weakening or altering the seed coat in order to promote seed germination. It can be done mechanically, thermally and chemically.

72 (*b*) Statement in option (b) is correct. Other statements are incorrect and can be corrected as

- Short day plants needs less duration of light.
- Short day plants require less than 12 hours of sunlight.
- For initiation of flower, the long day plants flower when the night length falls below their critical photoperiod.

76 (*a*) Option (a) is correct in regard to long day plants. Long day plants flower only when amount of light lasts longer than the critical threshold. For example, *Henbane*, spinach. Rest all statements belong to short day plants.

83 (*a*) Statement I is true, but statement II is false. Incorrect statement can be corrected as

The flowering response in short day plants and long day plant is dependent on the duration for which these plants are exposed to light. The short day plant and long day plant can flower at the same place provided they have been given an adequate photoperiod.

92 (*d*) Option (d) is incorrect for effect of the red and far-red light. It can be corrected as

In the red light, the formation of anthocyanin is stimulated, while it is inhibited in the presence of far light.

95 (*c*) Vernalisation is defined as the dependence of plants on exposure to a low temperature to flower. Factors needed for vernalisation are low temperature, water, nutrients and aerobic respiration.

Hence, the correct option is (c).

(B) Cumulative Exercise

1 (*c*) Statement in option (c) is incorrect and can be corrected as

Temperature and light are extrinsic factors which control plant growth and development *via* PGR like vernalisation, flowering, dormancy, seed germination and plant movements.

Rest of the statements are correct.

2 (*d*) Statements I and II are correct, whereas III and IV are incorrect and can be corrected as

Gibberellins are not responsible for immature falling of leaves.

Ethylene and abscisic acid generally have a negative role and are responsible for immature falling of leaves. They play no role in seed germination.

3 (*a*) Statements I and II are correct whereas III and IV are incorrect. Incorrect statements can be corrected as

The plant hormone that induces cell division is cytokinin. *Trp* (Tryptophan) is a known precursor for auxin biosynthesis. Zeatin is a cytokinin derived from adenine was discovered in immature core kernels from the genus-*Zea*. Ethylene induces premature senescence symptoms such as leaf yellowing, abscision or desiccation/necrosis. ABA induces senescence by increasing biosynthesis of ethylene.

5 (*d*) Statement in option (e) is false and can be corrected as

Photoperiodism is the response of plants to photoperiods that is the length of day and night.

During photoperiodism, shoot apices modify themselves into flowering apices prior to flowering. However, they do not perceive photoperiods. Rather the site of perception of light and dark duration are leaves.

In day neutral plants, induction of flowering response does not depend on exposure to light duration. The site of perception of light/dark duration is the leaves.

8 (*d*) **Absolute Growth Rate** (AGR) Increase in total growth of two organs or organisms is measured and comparison of total growth per unit time is called absolute growth.

Absolute growth rate is the total growth per unit time.

Relative Growth Rate (RGR) It is growth per unit time per unit initial growth.

$$\text{RGR} = \frac{\text{Growth in given time period}}{\text{Measurement at start of time period}}$$

$$\text{AGR for leaf A} = 10\text{ cm}^2 - 5\text{ cm}^2 = 5\text{ cm}^2$$

$$\text{RGR for leaf A}' = \frac{5}{5} \times 100 = 100\%$$

$$\text{AGR for leaf B} = 55\text{ cm}^2 - 50\text{ cm}^2 = 5\text{ cm}^2$$

$$\text{RGR for leaf B}' = \frac{5}{50} \times 100 = 10\%$$

Though the absolute growth is came for both the leaves (A and B), relative rate of growth is more than leaf A because of its initial small size.

11 (*b*) Statements I and IV are true, whereas II and III are false, Incorrect statements can be corrected as

Phytochromes become active in red and far-red light. They are a class of photoreceptor in plants, bacteria and fungi. All plant hormones are not inhibitor of plant growth.

13 (*a*) Statements I, II, III and IV are correct. Only statement V is incorrect and can be corrected as

Seed germination is inhibited by ABA, but promoted by GA.

Hence, four statements are correct.

15 (*d*) Plants which flower after exposure to light above critical photoperiod are called long day plants. Plants which flower when exposed to light below critical photoperiod are called short day plants. Plants which flower after vegetative maturity are called day neutral plants.

17 (*c*) Gibberellin is a plant hormone which is responsible for plant growth and development during the juvenile period, stem elongation, germination, flowering and leaf senescence.

Auxin is a plant hormone that controls xylem differentiation and apical dominance.

Abscisic acid is a plant hormone which is also known as stress hormone which increases the tolerance power of plants during various stress condition.

20 (*a*) The correct match will be

Human urine - Auxin

Gibberella fujikuroi - Gibberellins

Herring fish DNA - Cytokinin

ripening fruit - Ethylene

Aged leaves of plant - Abscisic acid

21 (*b*) There are different phases in the stages of growth. It follows a sigmoid curve. The initial growth phase, during which cell number remains relatively constant prior to rapid growth. The phase next to it is log phase which is otherwise called as exponential phase, which has a lot of growth involved in it. The final stage is steady state phase where the growth stops.

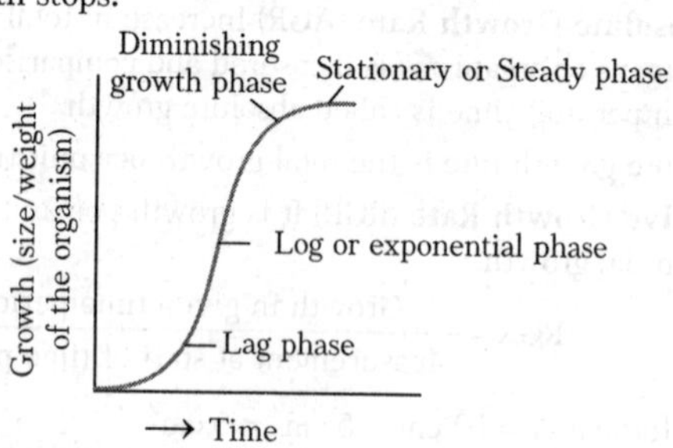

22 (*b*) Statements I and III are correct. Statements II and IV are incorrect and can be corrected as

- Amylase production at the time of seed germination is an important function of GA.
- Potato tubers can be made to sprout early by exposing them to ethylene.

27 (*b*) Abscisic Acid (ABA) also called the stress hormone, causes stomatal closure by causing efflux of K^+ ions from guard cells.

29 (*b*) Gibberellins are plant hormones which play a vital role in regulation of various developmental processes. They can stimulate stem elongation in rosette plants, causing an increase in height before reproduction. This is bolting.

Abscisic acid or the stress hormone is known to enhance senescence.

Cytokinin is a plant growth promoter which induces flowering in a plant.

30 (*a*) Induce parthenocarpy in tomatoes - Auxins (*A*)

Hastens fruit ripening in tomatoes and apples - Ethephon (*B*)

Induce flowering in mango - Ethylene (*C*)

Elongation and improvement in shape of apple - GA (*D*)

31 (*b*) Both A and R are true, but R is not the correct explanation of A.

Gibberellin is plant hormone which is useful in early seed production is conifers. Ethephon is commercial name of ethylene hormone used for early ripening of fruits like tomato and apple.

32 (*c*) A is true, but R is false.

Photomodulation of flower is a phytochrome regulated process. It is a chromoprotein existing in two interconvertible forms, i.e. Pr and Pfr.

When Pr molecules are exposed to red light, they absorb and convert to Pfr. This results in floral induction of the shoot. Alter perceiving the photoperiods, the leaves (phytochromes) produce a chemical that is passed over to the shoot apex which undergoes differentiation to produce flowers. The chemical inducing flowering is called florigen. Thus, Pfr does not directly induce flowering but through florigen.

33 (*a*) Both A and R are true and R is the correct explanation of A. Abscisic acid increases resistance of plants to cold and other types of stresses like drought and water logging etc. It is therefore known as stress hormone.

34 (*b*) Both A and R are true, but R is not the correct explanation of A.

When cytokinins are added directly to the abscission layer senescence of the zone is retarded. This delays the degradation of protein and chlorophyll of the plant parts and hence delay senescence.

They act as antisenescent, they act as antagonistic to ethylene which accelerate senescence.

35 (*b*) Both A and R are true, but R is not the correct explanation of A.

In most fruits, the rate of respiration will undergo sharp rise and then fall near the end of ripening Kidd and West termed this phenomenon as 'climacteric rise'.

The climacteric act as a trigger that sets in progress those changes rapidly transform the fruit from an unripe to a ripe condition. Finally, application of ethylene to unripe fruit will bring on a premature climacteric and accelerate ripening.

36 (*b*) Both A and R are true, but R is not the correct explanation of A.

Light energy become effective when it is absorbed by a pigment. These pigments are called phytochromes. It occurs in two forms, namely Pr and Pfr. The two forms are interconvertible. The Pr form absorbs red light of 660 nm and changes to Pfr. The Pfr form absorbs far-red light of 730 nm and changed to Pr. The Pr form stimulates whereas Pfr form inhibits flowering. Pfr form of phytochrome is the active form. The Pr form is not considered to be active.

37 (*c*) A is true, but R is false.

If total growth is plotted aginst time, an *S*-shaped or Sigmoid curve is obtained. It consists of four parts, i.e. lag phase, log phase, phase of diminishing growth and stationary phase. Growth is slow in the lag phase, rapid during log or exponential growth.

Growth stops completely during the stationary phase. Log phase is also called as grand phase of growth due to fast growth in this phase.

38 (*d*) A is false, but R is true because

Some plants require a periodic exposure to light to induce flowering. It is now also known that not only the duration of light period, but the duration of dark period is also of equal importance. Hence, it can be said that flowering in certain plants depends not only on a combination of light and dark exposures but also their relative durations. This response of plants to periods of day/night is termed photoperiodism. The shoot apices modify themselves into flowering apices prior to flowering, shoot apices cannot perceive photoperiods by themselves. The site of perception of light/dark are the leaves.

39 (*b*) Both A and R are true, but R is not the correct explanation of A.

The physiological mechanism of flowering in plants is controlled by two factors, i.e. light period and low temperature.

The cold treatment of plants to induce flowering is called as vernalisation. Term vernalisation first given by TD Lysenko (1928). As a result of vernalisation, a flowering hormone called as vernalin is formed. Site of vernalisation is apical meristem.

40 (*b*) Both A and R are true, but R is not the correct explanation of A.

Auxin delays abscission of young leaves and fruits. It effects through non-formation of abscission zone below a leaf or fruit.

Abscission zone cuts off nutrients and water supply. However, auxin promotes the abscission of mature or older leaves and fruits.

41 (*b*) Both A and R are true, but R is not the correct explanation of A.

One of the most renowned scientist Fritz performed an experiment an oat seedlings regarding the effect of the phytohormone further named as auxin hormone.

Darwin performed an experiments on coleoptiles of the canary grass and observed that the plant shows its unidirectional growth towards the source of the light and it bends towards the source from where light is coming.

42 (*e*) Both A and R are false and can be corrected as

The shoot apical meristems are not the only source of auxin synthesis.

It is produced in young developing leaves also. Auxin controls apical dominance. In this phenomenon, apical bud due to the presence of auxin inhibits the lateral bud formation.

43 (*a*) Both A and R are true and R is the correct explanation of A.

Hormones are also called growth adjusters. They are produced in one part of a plant and then migrate to another part where they promote or inhibit the growth by stimulating tissues of that location.

(C) Medical Entrances' Gallery

1 (*a*) The term 'kinetin' was coined by Skoog and Miller in 1955. Chemically, kinetin is 6-furfurylaminopurine. It was first cytokinin to be discovered from the degraded autoclaved herring sperm DNA. Kinetin does not occur naturally.

2 (*d*) Gibberellic acid is not an inhibitory substance governing seed dormancy because gibberellic acid promotes growth and elongation of cells. It affects decomposition of plants and helps plants grow if used in small amounts, but eventually plants develop tolerance to it. Action of ABA is counteracted by GA, which promotes seed germination at appropriate time. Abscisic acid, phenolic acid and para-ascorbic acid are inhibitory substances that cause seed dormancy as they occur in the seed coats and cotyledons of the embryos, e.g. apple, peach, ash, *Cucurbita*, iris, *Xanthium*.

3 (*d*) The process of growth is maximum during log phase (exponential phase) because during log phase, the growth rate of the cells gradually increases, at a maximum rate. In exponential growth, the initial growth is slow (lag phase) and thereafter it increases rapidly.

4 (*d*) Auxin and ethylene can be applied to artificially induce flowering in pineapple plants throughout the year to increase yield. Auxin induces flowering in pineapple and ethylene helps to synchronise flower and fruit growth in this plant. Though in other cases, ethylene causes fading of flowers.

5 (*c*) Removal of shoot tips is a very useful technique to boost the production of tea leaves. This is because effect of auxin is removed and growth of lateral bud is enhanced. This phenomenon is most higher plants in which growing apical buds (shoot tips) inhibit growth of lateral buds due to effects of auxin is called apical dominance.

6 (*c*) In order to increase the yield of sugarcane crop, gibberellins should be sprayed. Sugarcane stores carbohydrates as sugar in their stems. Spraying sugarcane crop with gibberellins increases the length of the stem, thus increasing the yield by as much as 20 tonnes per acre.

7 (*c*) For the induction of flowering in plants, photoperiod stimulus is perceived by the leaves of plants. As a result, floral hormones are produced in the leaves which are then translocated to the apical part and subsequently cause the initiation of floral primordial growth.

8 (*b*) Vernalisation is low temperature treatment. It is the process of shortening of vegetative phase and initiation of reproductive phase by the previous application of low temperature (0-4°C) to moistened seeds and young plants.
It is used for a number of temperate plants like winter wheat, winter rye, cabbage, henbane, *Chrysanthemum* and even certain perennials, e.g. apples.

9 (*b*) IAA is a plant growth promoter because it helps in growth and development related processes in plants. While dormin, abscisic acid and ethylene are plant growth inhibitors. IAA (Indole 3-Acetic Acid) is an auxin.

It was discovered by Koge *et al.* (1934). Auxins are synthesised in shoot apices, leaf primordia and developing seeds from amino acid, tryptophan.

10 (*a*) Apical dominance in plants is accomplished by the production of auxin hormone. Auxin inhibits the growth of lateral buds and promotes apical bud growth.

11 (*c*) Gibberellins and Abscisic Acid (ABA) are antagonistic to each other. They promote seed germination whereas ABA promotes seed dormancy.

12 (*c*) Gibberellic acid is also known as weak acidic plant growth regulator. It increases the yield of sugar by increasing the length of stem in sugarcane.

It is the only plant growth hormone that elongates the internodes to increase the height of the plant by causing an increase in the length of axis.

13 (*c*) Auxins delay abscission of leaves and fruits at early stages. Whenever leaf or fruit fall occurs, the organ concerned stops producing auxin. However, it promotes abscission of older, mature leaves and fruits.

14 (*b*) Bioassay is a quantitative and qualitative test used to determine the nature and function of a biochemical by using living material, e.g. Avena curvature test is used as bioassay usually for auxins (Indole Acetic Acid), which can measure auxin up to 300 μg/L.

15 (*b*) When a tissue with a potential of differentiation is grown in a medium containing auxin and cytokinin in a specific ratio, it starts differentiating into roots and shoots.

16 (*d*) Phytochrome is a chromoprotein, which exists in two forms, i.e. Pr and Pfr. These forms are interconvertible. When plants get red light this protein gets converted into Pfr from Pr and *vice-versa.* It controls the photoperiodism in the plants.

32 (*d*) Gibberellins hydrolyse the stored food by promoting the synthesis of hydrolytic amylases, lipases, ribonucleases and proteases during seed germination.

33 (*b*) Thigmotropism are the movements, which occur due to contact with a foreign body. Tendrils show thigmotropic movements.

36 (*c*) Application of gibberellin to plant can be a substitute for vernalisation. Thus, gibberellin can be used for replacing the requirement of vernalisation.

38 (*a*) Auxin hormone is responsible for apical dominance process as it inhibits the growth of lateral buds.

40 (*c*) Auxin in combination with cytokinin is used in tissue culture for good growth of callus.

41 (*b*) Phytochrome is sensitive to red and far-red light.

44 (*c*) Day neutral plants can flower in all possible photoperiods. Such plants can blossom throughout the year, e.g. cucumber, cotton, sunflower, tomato, some varieties of pea, etc.

47 (*c*) The pair which is not correctly matched and can be corrected as

Terpenes are usually precursors of steroid.

48 (*a*) The phenomenon of photoperiodism was first discovered by Garner and Allard (1920, 22) in tobacco plant. They observed that Maryland Mammoth variety of tobacco could be made to flower only by reducing the light hours with artificial darkening. It could be made to remain vegetative in winter by providing extra light.

CHAPTER 23

Digestion and Absorption

Food and water are the basic requirements of all living beings. Food provides energy and organic materials for growth and repair of tissues. The major components of the food are biomacromolecules, i.e. carbohydrates, proteins and fats. Vitamins and minerals are also required in small quantities. The water plays an important role in various metabolic processes.

The biomacromolecules in food are not utilised by our body in their original form. For utilisation purposes, these complex food components are converted into simpler and absorbable forms by the process of **digestion.** It is defined as 'The catabolic process of conversion of complex bimacromolecules into simpler and absorbable forms *via* various mechanical and biochemical methods.'

This chapter is devoted to various processes and organs involved in digestion of nutrients.

DIGESTIVE SYSTEM OF HUMANS

The human digestive system comprises of **alimentary canal** and **digestive glands**. An outline overview of these organs is given below

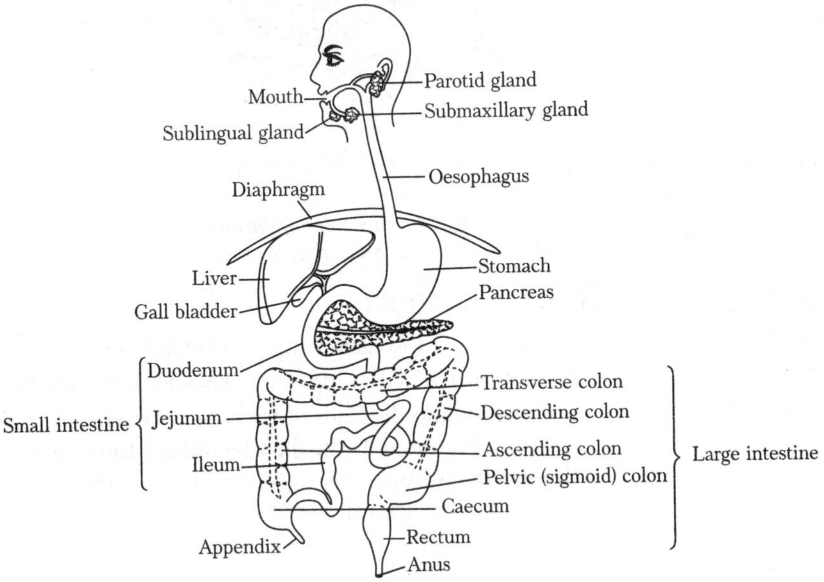

Fig. 23.1 Human digestive system

Inside

ALIMENTARY CANAL

It is a tube-like structure extended from mouth to anus. Its lumen is called **enteric cavity**. On the basis of embryonic origin, the structure of alimentary canal contains following three regions

(i) **Foregut** (stomodaeum) is ectodermal in origin. It includes mouth and buccal cavity.

(ii) **Midgut** (mesenteron) is endodermal in origin. It contains pharynx, oesophagus alongwith main digesting parts, i.e. stomach and intestine up to the colon region.

(iii) **Hindgut** (proctodaeum) is ectodermal in origin. It contains colon, rectum and anus.

These various parts of alimentary canal are discussed in detail below

Mouth

- It is a transverse slit-like apperture on the ventral side of the body. It is guarded by two movable lips, i.e. upper and lower lip.
- Upper lip has a middle ventral groove or **philtrum.** Lips are attached to gums by thin transparent fold called **labial frenulum.**
- A specific muscle called **orbicularis oris** helps in lip movement.
- The mouth opens into the buccal cavity or oral cavity.

Buccal cavity

- The buccal cavity is divided into **vestibule** and **oral cavity.**
- Vestibule is the space between lips and teeth.
- Oral cavity contains the palate, tongue and teeth.

Palate

- The palate is a horizontal structure that separates the buccal cavity and the nasal cavity. It represents the roof of oral cavity.
- The anterior and posterior parts of palate are called hard and soft palate, respectively.
- **Hard palate** is well-supported by maxilla, premaxilla and palatine bones. It contains specialised grooves and ridges over it. These ridges are called **palatine rugae** and are well-developed in carnivores as they help in gripping the prey firmly.
- **Soft palate** contains a curved fold of hanging muscle called as **velum palati** or **uvula.**
- This hanging portion is covered with four folds of mucous membrane. Out of these, two passes downwards at each side to form the membranous arches.
- The anterior arches are called **palatoglossal arches** and the posterior arches are called **palatopharyngeal arches.**
- Uvula closes the internal nostrils during swallowing of food or deglutition.

The structure of the human oral cavity is as follows

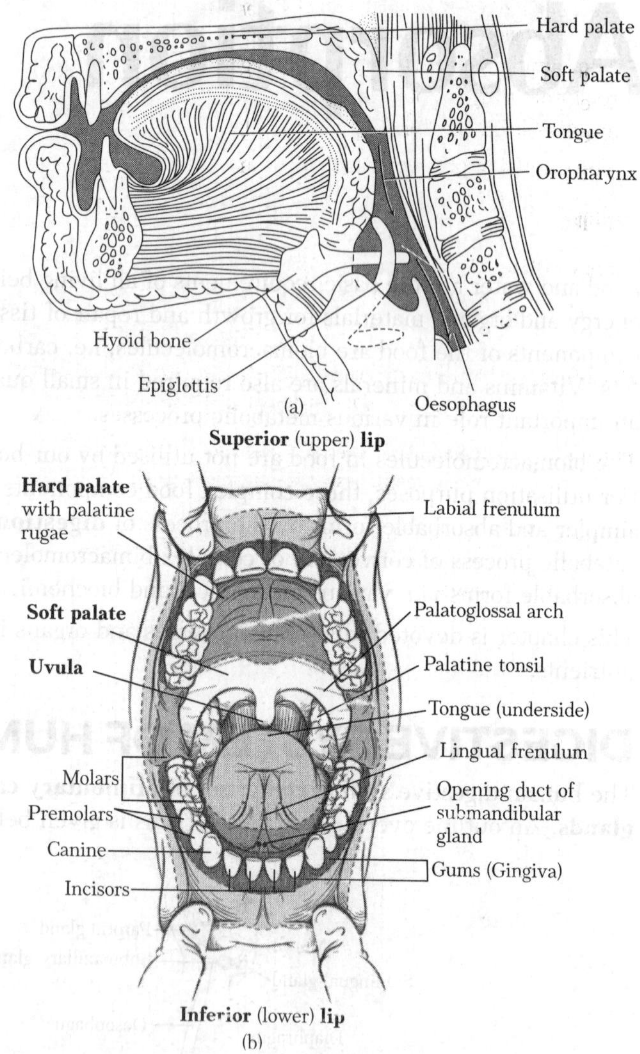

Fig. 23.2 Structure of the human oral cavity : (a) Side view, (b) Anterior view

Tongue (Linguae)

- It is a highly muscular structure containing voluntary muscles. It is lined by stratified squamous epithelium.
- It is attached to the floor of buccal cavity with the help of a connective tissue fold called **frenulum linguae** or **lingual frenulum.** This structure prevents the tongue from falling backwards.

- The anterior part of tongue is free.
- Its posterior part is connected to hyoid plate or hyoid bone.
- The dorsal surface of tongue is divided into two unequal halves by a V-shaped sulcus, called as **sulcus terminalis.**
- Anterior 2/3rd part of tongue is called **oral** or **papillary part** while posterior 1/3rd part is called **pharyngeal part.**

Note *The apex of sulcus terminalis projects backward and is marked by small median pit called foramen cecum. It is an embryological remnant and represents the upper end of thyroglossal duct.*

- The papillary part has specialised gustatory or taste receptors (lingual papillae) present on it. These receptors are visible in the form of taste buds.
- The papillary part contains following four major papillae.

Papillae	Characteristics
Circumvallate	Circular-shaped, 8-12 in number, largest papillae, found at the posterior part of tongue, extending from one side to another.
Fungiform	Mushroom-shaped, numerous, found at the anterior margins of tongue, contain about 200 taste buds.
Foliate	Leaf-shaped present along the lateral margins near molar teeth, absent in humans.
Filiform	Conical-shaped, smallest and most numerous, distributed throughout tongue, **do not contain taste buds.**

- These papillae along with other structures of tongue are shown in the diagram below

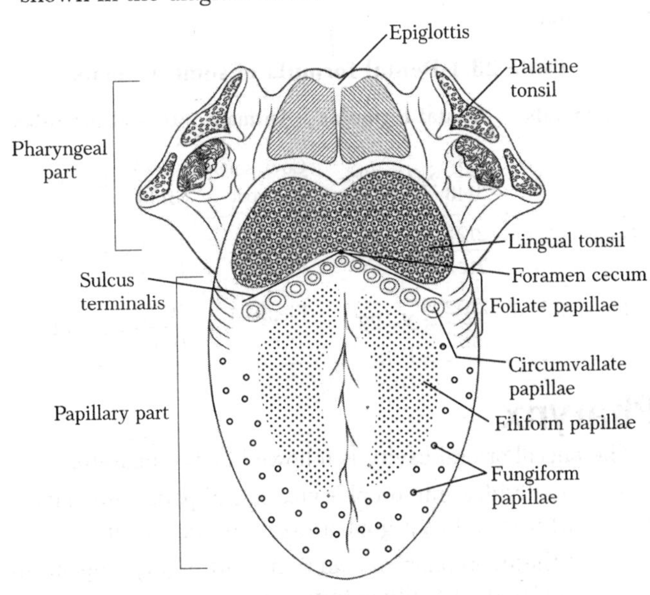

Fig. 23.3 Human tongue showing upper surface

- **Functions of tongue** The main function of tongue is to help in tasting through different taste zones and swallowing of food. It also helps in food mastication, speech articulation and cleaning of buccal cavity. Various taste zones on human tongue are shown below

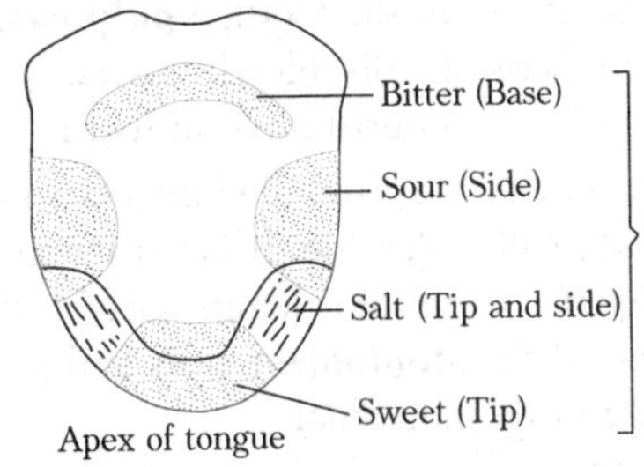

Fig. 23.4 Human tongue showing four taste zones

Note
- *There are not taste buds for tasting chillies. The burning sensation occurs due to the stimulation of pain receptors.*
- *Tongue also contains sepecialised mucus secreting* ***lingual*** *or* ***Weber's gland****.*

Dentition

Teeth are unique masticatory organs in vertebrates. These are derived from embryonic ecto and endoderm.

Teeth structure

The teeth have almost same structure in all the vertebrates. It is differentiated into three regions

- **Crown** The part of tooth projecting above the **gum line** or gingiva is called crown.
- **Neck** It is the slightly constricted and overlapping part where the crown merges with the root.
- **Root** It remains embedded inside the gums. The root is present inside the bony socket of jaw bone or alveolus.

The structure of tooth in its sectional view is given below

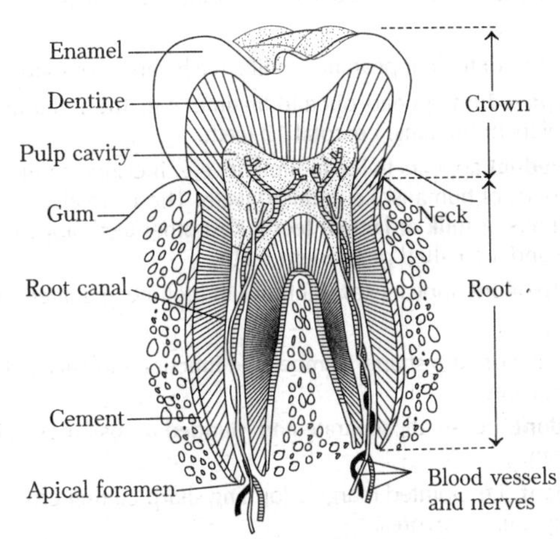

Fig. 23.5 A section of a tooth

A tooth is formed of following structures

- **Enamel** In the crown region, the outermost layer is called enamel. It contains 95% calcium and thus, it is the hardest substance of human body. It is formed by enamel forming cells called **ameloblasts.** These enamel forming cells are ectodermal in origin. In the neck and root region, the outermost layer is called **cement layer.**

- **Pulp cavity** In the centre of the tooth, a **pulp cavity** is present. It contains mass of cells, blood vessels, lymph vessels and nerves for the nourishment of teeth.
- **Dentine** The pulp cavity is surrounded by a layer of hard ivory-like substance called dentine. It is harder than bone because it has higher content of calcium salts (~ 70% of dry wt). It is secreted by **odontoblast cells** and grows throughout the life of an individual.
- **Apical foramen** Blood vessels and nerves innervate the tooth through a small foramen at the apex of each root, called as **apical foramen.**
- The **fixation of teeth** within the socket occurs with the help of thick bundles of collagen fibres covering the cementing layer. It is called as **periodontal membrane** or **periodontal ligament**. Collagen fibres of this membrane are called **Sharpey's fibres.**

Types of dentition

In lower vertebrates, teeth are usually **homodont**, i.e. similar in general appearance, whereas in tetrapods and mammals, teeth are **heterodont**, i.e. differ in general appearance. Here, each dental set contains

- **Incisors** (front) for cutting or clipping the food.
- **Canines** (next to incisors) for puncturing or holding or tearing the food.
- **Premolars** and **molars** (lateral) for crushing or grinding the food.

On the basis of attachment in the jaws, the teeth are of following types

- **Thecodont** Tooth embedded in the jaw bone socket, i.e. man, crocodile.
- **Acrodont** Tooth is fixed in the jaw bone without sockets, e.g. amphibians, fishes.
- **Pleurodont** Tooth is fused with the inner side of jaw ridge, e.g. lizards.

On the basis of their appearance in life, teeth can be of following types

- **Monophyodont** appears once in lifetime, e.g. 3rd molar and all premolars of humans, i.e. total 12 teeth.
- **Diphyodont** appears twice in lifetime, e.g. incisors, canines and 1st molars of humans, i.e. total 20 teeth. These are also called **deciduous** or **milk** or **lacteal teeth**. Rest other teeth appear once in life and are called **permanent teeth**.
- **Polyphyodont** appears several times in lifetime, e.g. teeth of shark.

On the basis of cusp shape, number and function molars can be of following types

- **Bunodont**, i.e. small, separate and rounded cusps for grinding, e.g. man.
- **Secodont**, i.e. pointed margins forming sharp cutting crowns for tearing, e.g. carnivores.
- **Lophodont**, i.e. multicuspid condition with margins irregularly drawn as ridges, e.g. horses.
- **Selenodont**, i.e. crescent-shaped cusps, e.g. sheep and cattle.

On the basis of length of crown and root, molars can be of following types

- **Hypsodont**, i.e. high crown with short roots, e.g. horses.
- **Brachydont**, i.e. short crown with deep roots, e.g. humans.

The number of roots of different teeth in humans are as follows

Number of roots	Types of teeth
One root	Incisors, canines, lower premolars
Two roots	Upper premolars, lower molars
Three roots	Upper molars

Note
- *It is often hard to differentiate between premolars and molars visually. The collective term used for both of these* ***molariform*** *or* ***cheek teeth****.*
- *The last molar in humans are called* ***wisdom teeth*** *and they are vestigial.*
- *The permanent teeth which appear first in humans are first molars while the third molars appear at the last. Also, the teeth of lower jaw appear before those of upper jaw.*

Dental formula

The number and the type of teeth in mammals are represented by an equation called dental formula. It is constant for every species. Since, two halves of each jaw are identical, the teeth of only one side are represented in this formula, i.e.

$$\frac{I\,C\,Pm\,M}{I\,C\,Pm\,M}$$

Thus, the total number of teeth in jaw = Number of teeth in dental formula × 2.

Here, I = Incisors, C = Canines, Pm = Premolars, M = Molars.

Table 23.1 Dental formula of some animals

Animals	Dental formulas	Animals	Dental formulas
Dog	$\frac{3142}{3143} \times 2 = 42$	Cow, sheep and goat	$\frac{0033}{3133} \times 2 = 32$
Human (Adult)	$\frac{2123}{2123} \times 2 = 32$	Rat	$\frac{1003}{1003} \times 2 = 16$
Human (Child)	$\frac{2102}{2102} \times 2 = 20$	Elephant	$\frac{1003}{0003} \times 2 = 14$

Pharynx

- The buccal or oral cavity is followed by the pharynx.
- It is sub-divided into oropharynx or oral part, nasopharynx or nasal part and laryngopharynx or laryngeal part.
- Out of these, oropharynx and some part of laryngopharynx are associated with alimentary canal.
- Nasopharynx and rest part of laryngopharynx is associated with the respiratory system.
- The lymphoid tissue patches of pharynx and oral cavity are called **tonsils.** Within the pharynx, these are arranged in

the form of a ring called as **Waldeyer's ring** as shown below

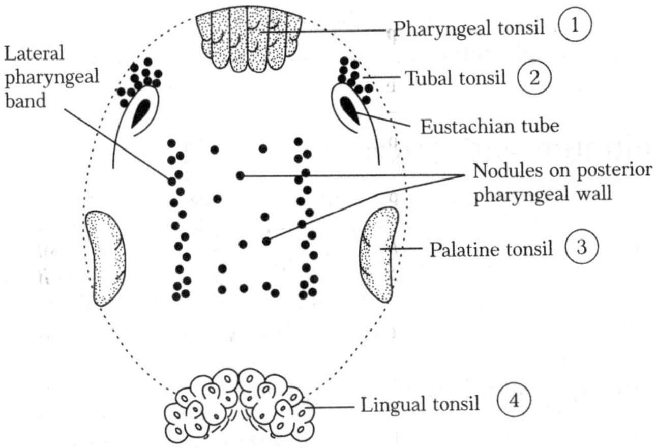

Fig. 23.6 *Waldeyer's ring (see positioning of 1, 2, 3 and 4)*

- Details of different tonsils of ring are as follows
  - (i) **Nasopharyngeal tonsils** or **Pharyngeal tonsils,** which are present in the posterior wall of the nasopharynx. These tonsils may become enlarged in children. The condition is known as **adenoids.** The resulting swelling may cause an obstruction in normal breathing.
  - (ii) **Tubal tonsils,** which are present around Eustachian tube. This tube connects the middle ear with the nasopharynx.
  - (iii) **Palatine or Faucial tonsils** are located in the lateral walls of oropharynx. These tonsils often get infected and lead to sore throat. This condition is known as **tonsilitis.** Surgical removal of these tonsils is called **tonsillectomy.**
  - (iv) **Lingual tonsils** are present behind the sulcus terminalis of tongue.

CHECK POINT 23.1

1. Choose the incorrect pair.
(a) Stomodaeum – Stomach
(b) Mesenteron – Intestine up to colon
(c) Proctodaeum – Rectum and anus
(d) Foregut – Mouth

2. During swallowing, the structure that prevents foods and liquid from entering nasopharynx is
(a) Palatine rugae
(b) Soft palate
(c) Uvula
(d) Palatoglossal arches

3. Human tongue contains four papillae having given features except
(a) give the tongue its rough texture
(b) increase the surface area for the taste buds
(c) aid in the mechanical handling of food
(d) all papillae have taste buds on them

4. Which of the following is correct for human teeth?
(a) Thecodont, but not heterodont
(b) Thecodont, but not diphyodont
(c) Thecodont and diphyodont, but not heterodont
(d) Thecodont and heterodont

5. A two years baby is passes through a dental checkup. The dentist observed that the boy had twenty teeth. Which teeth were absent?
(a) Canines (b) Premolars
(c) Molars (d) Incisors

6. Identify the region of the pharynx that extends from the soft palate to the epiglottis.
(a) Oropharynx (b) Hypopharynx
(c) Nasopharynx (d) Lingual tonsils

7. The lymphatic masses that are embedded in the lateral walls of the oropharynx are
(a) parotid gland (b) submental gland
(c) palatine tonsils (d) tubal tonsils

Oesophagus

- The oesophagus or food pipe in human beings is about 23-27 cm long. It lies behind the trachea and the heart.
- It pierces the diaphragm to enter the abdomen through the site called **oesophageal hiatus.**
- It is lined by non-keratinised stratified squamous epithelium.
- Due to the presence of submucosal mucous glands, the submucosa of oesophagus is thickest than other parts of alimentary canal.
- The upper opening of oesophagus is called **gullet.**
- The upper and lower ends of oesophagus are enclosed by following sphincter muscles
  - The upper **cricopharyngeal sphincter** prevents air from passing into the oesophagus during breathing.
  - The lower **gastro-oesophageal sphincter** or **cardiac sphincter** prevents the reflux of acidic contents of gastric juice into the oesophagus.
- **Functions** No enzymes are secreted in the oesophagus. It just conveys food to stomach through **peristalsis,** which is controlled by medulla.

Note *In new born babies, cardiac sphincter is very less developed, therefore, regurgitation of gastric content is very common.*

Stomach

- The oesophagus opens into a J-shaped dilated sac called stomach or **gaster** or **ventriculus.** It has two orifices or openings, i.e.
  - (i) cardiac orifice, joined at the lower end of oesophagus.
  - (ii) pyloric orifice that opens into the duodenum.
- The stomach is divided into four regions namely cardiac, fundus, body or corpus and pyloric part as shown in the diagram below

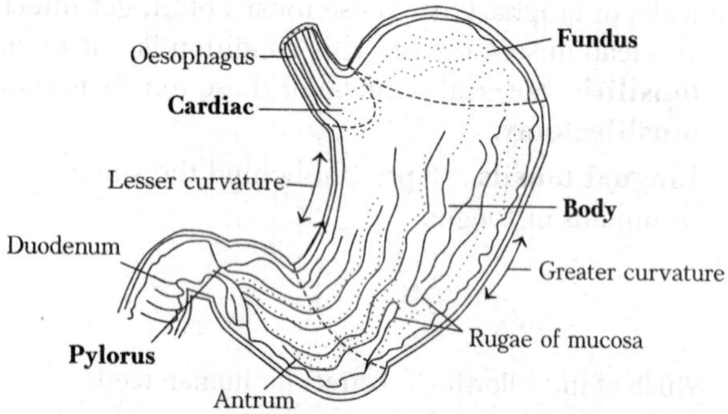

Fig 23.7 Longitudinal section of the human stomach

- The area of stomach nearby cardiac orifice is called **cardiac region.** The angle of positioning of oesophagus and stomach form a notch called **cardiaca ventriculi.**
- The convex part of stomach is called **fundus** or **fornix ventriculi**. It is situated above the level of cardiac orifice and is continuous with the **body** or **corpus ventriculi**, the largest part of stomach. This is without any apparent boundaries.
- The distal **pyloric region** is the last angular **J-shaped region** of stomach. It is divided into two parts, i.e. pyloric antrum and pyloric canal. Pyloric part of stomach opens into small intestine.
- The inner lining of stomach is raised into numerous longitudinal folds called **rugae.**

Small intestine

- It is the longest part of the digestive tract. It is about 5-7 m long and is divided into duodenum, jejunum along with distal ileum.
- The duodenum is **retroperitoneal,** i.e. present beyond the abdominal cavity or coelom, while other parts are **intraperitoneal**, i.e. present inside the abdominal cavity.

Duodenum

- The duodenum is a C-shaped, shortest and the widest structure.
- It is connected to both the liver and pancreas.
- It is 4.0-4.5 cm in diameter and is the main digesting part of human alimentary canal.
- **Pyloric sphincter** regulates the opening of stomach into duodenum.

Jejunum and ileum

- The proximal 1/3rd long coiled middle portion of small intestine is considered as **jejunum.** The rest 2/3rd, highly coiled part is considered as **ileum.**
- Ileum is the longest part of small intestine. Its distal part is considered as terminal ileum or ileum terminalis.

Villi and associated structures

- The lining of small intestine has a series of circular folds called **plicae circulares** or **valve of Kerkring** or **valvulae conniventes.**
- These circular folds contain villi over them. These villi in turn, contain microvilli or brush-bordered cells.
- For better understanding of these structures look at the figure given below

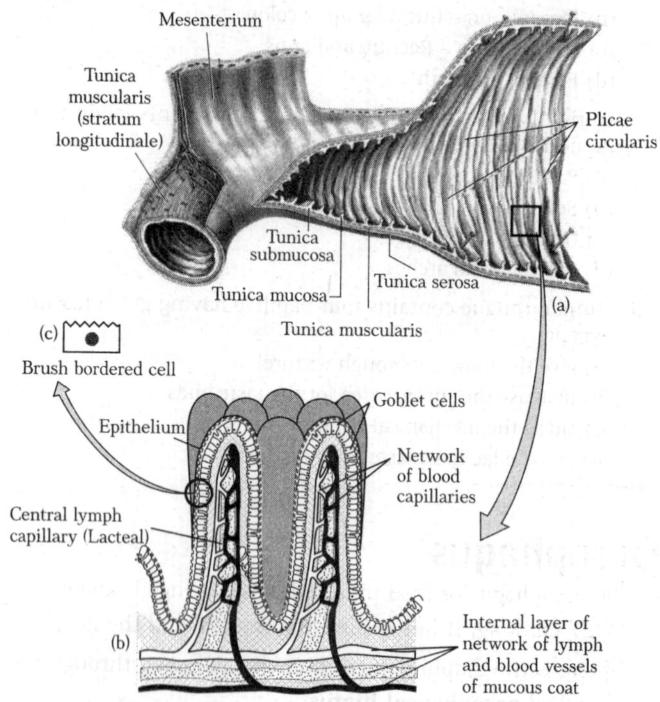

Fig 23.8 Small intestine : (a) Valvulae conniventes, (b) Villi, (c) Brush bordered cell (represented schematically)

- The circular folds and villi decrease in jejunum, but increase in ileum. Thus, indicating that the chief function of jejunum is secretion, while that of ileum is absorption.

Large intestine

It is about 100-150 cm in length and 4-6 cm in diameter. The word small and large in the name of intestine are associated with diameter of these portions.

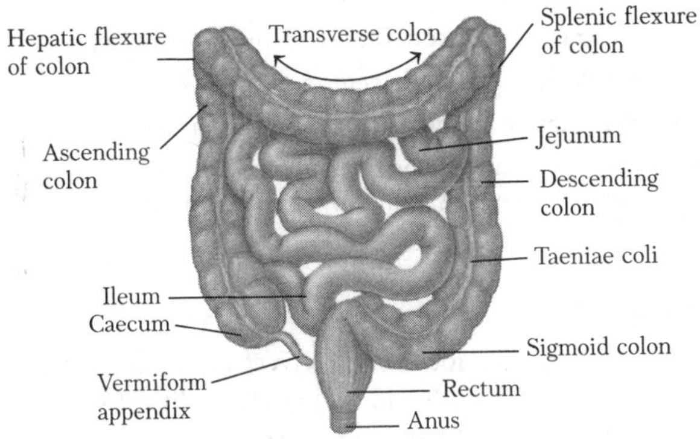

Fig. 23.9 *Parts of small intestine and large intestine*

The large intestine consists of following three parts

Caecum

- It is a blind pouch-like structure situated below the junction of ileum and colon.
- The length of caecum is about 3-8 cm with width 4-7 cm. It is considered as the widest part of large intestine after the ampulla of rectum.
- Caecum is intraperitoneal. It also hosts some symbiotic microorganisms in it.
- At the junction of ileum and caecum, the wall of ileum is thickened and became muscular to function as **ileocaecal sphincter**. This sphincter surrounds the opening of ileum into caecum called **ileocaecal orifice**.
- A mucous membranous valve called **ileocaecal valve** is also present in this assembly. It slows down the pouring of ileum content into the colon.
- Externally, caecum bears a blind tube called **vermiform appendix**. It is 3-4 mm in diameter and 2.5-15 cm in length. This lymphoid tissue containing structure is a vestigial organ in humans.

Colon

- It is placed in abdominal cavity in such a way that it borders the coils of small intestine inside.
- The colon has three longitudinal bands called **taeniae coli.**
- It also has small pouches named **haustra** and small fat-filled projections called **epiploic appendages.**
- As shown in figure above, the colon consists of following parts
  - (i) The ascending colon
  - (ii) The transverse colon
  - (iii) The descending colon

  } (All intraperitoneal)
  - (iv) The pelvic or sigmoid colon (Retroperitoneal)
- The transverse colon is considered as the longest part of large intestine.
- The ascending colon is the shortest part of colon and it lacks mesentery as well.
- The sigmoid colon is S-shaped structure situated in the pelvis region.
- The descending part of colon opens into rectum.
- The curvature or flexure present between ascending and transverse colon is called **hepatic flexure** or **right flexure** *(Flexura coli dextra).*
- Similarly, the flexure or curvature present between transverse and descending colon is called **splenic flexure** or **left flexure** *(Flexura coli sinistra).*

Rectum

It is the terminal part of large intestine as well as the digestive tract. Its length varies from 14-18 cm and is mainly composed of two parts

(i) **Pelvic part** It contains a narrow suprampullar part and a wide ampulla of rectum or ampulla recti.

(ii) **Perineal part** It is present in the perineal region and called as **anal canal** or **canalis analis**. This canal is a short canal, about 3.8 cm long in adults. Its opening to outside is called **anus.** Following two sphincter muscles control the opening of anus

- **Internal sphincter** It is made up of involuntary or smooth muscles.
- **External sphincter** It is made up of voluntary or striated muscles.

Note *Peristaltic movements are least in rectum.*

Important variations

Digestive tract adaptations in other mammals

- Whale and platypus have immovable lips.
- The upper lip of rabbit has cleft called **harelip**.
- Elephant tusks are modified upper incisors.
- Walrus tusks are modified canines.
- Rabbit and suquirrel lack canines due to which there is a space between incisors and premolars. This space is called **diastema.**
- The tongue is immovable in whale.
- In dogs, tongue is helpful in the process of **thermoregulation** due to the presence of sweat glands over it.
- Ileocaecal valve is well-developed in rabbit in the form of swollen portion called **sacculus rotundus**.
- The caecum is well-developed in herbivorous mammals like rabbit, horses, etc.

Ruminant stomach or Cellulose digesting stomach

- The stomach of a ruminant is called **compound stomach**. It has four compartments, namely rumen, reticulum, omasum and abomasum.
- Rumen (largest chamber), reticulum and omasum are involved in storage of food. These (especially, rumen and reticulum) contain cellulosic bacterial like *Ruminobacter, Streptococcus, Fibrobacter,* etc.
- Abomasum is true stomach which secretes gastric juices.
- Omasum is absent in camel and deer.

Histology of alimentary canal

The wall of alimentary canal is made up of four layers. From outer to inner side, these layers are serosa, muscularis or muscularis externa, submucosa and mucosa.

Serosa

- The serous membrane or serosa is made up of squamous epithelium and areolar connective tissue. It is continuous with mesentery and represents visceral peritoneum, i.e. membranous coelomic epithelium.
- The outer wall of oesophagus is not lined by serous membrane because oesophagus lies outside the coelom.
- Hence in this case, the outer wall is seen in the form of an irregular coat of yellow elastic dense fibrous connective tissue called **adventitia externa** or **tunica adventitia.**

Muscularis externa

- The outer muscle layer or muscularis externa is made up of two muscle layers. These are **inner circular** and **outer longitudinal smooth muscle layers.**
- Out of these, the longitudinal layer is thicker than the circular layer.
- Another (third) layer of oblique muscle fibres is also seen in stomach. It is present inside to circular muscle layer. This third layer is absent in the intestine.
- These layers are helpful in peristalsis.
- Throughout the alimentary canal, a network or plexus of neurons and nerve fibre is present in between the longitudinal and circular muscle layers.

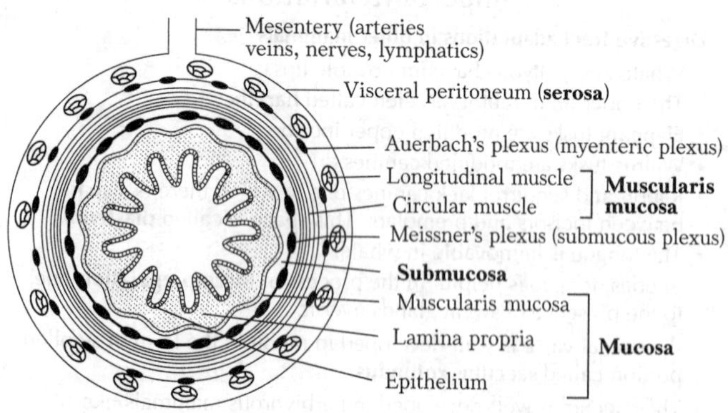

Fig. 23.10 General structure of the wall of the mammalian gut

- This plexus is made up of sympathetic and parasympathetic nerve fibres and is called **Auerbach's** or **Myenteric plexus**. This plexus controls peristalsis.
- In most of the mammals including humans, muscularis externa of oesophagus contains
  - **voluntary** or **striated fibres** in its anterior region.
  - **mixed fibres** in middle region.
  - **involuntary** or **smooth fibres** in the posterior region.

Note *In ruminants, all the fibres of oesophageal region are voluntary or striated type. This is to help the animal in the process of rumination.*

Submucosa

- The submucosa is a loose connective tissue layer. It is richly supplied with blood and lymphatic blood vessels.
- This layer contains another plexus of parasympathetic and sympathetic nerve fibres called **Meissner's plexus** or **Submucosal plexus.**
- In the intestine, this layer contains mucus secreting **Brunner's glands.**

Note *The **Auerbach's** and **Meissner's plexus** are absent in the upper one third part of oesophagus.*

Mucosa

- The mucous membrane or mucosa is the innermost layer lining the lumen of alimentary canal. It is named so because it secretes mucus.
- It is composed of three sub-layers
  (i) **Muscularis mucosa** or **Muscularis interna** consisting of outer longitudinal and inner circular muscle fibres.
  (ii) **Lamina propria** consisting of loose connective tissue, blood vessels, glands and lymphoid tissues having protective function.
  (iii) **Epithelial lining** forms rugae and gastric glands in stomach. It forms villi and intestinal glands in small intestine.
- The mucous membrane layer contains lymph nodules as well. These lymph nodules in the small intestinal region gets aggregated at some places to form well-defined **lymphoid patches** called as **Peyer's patches**. These patches are especially visible in ileum region and produce phagocytic lymphocytes.

CHECK POINT 23.2

1. Oesophageal hiatus is found on
   (a) oesophagus
   (b) stomach
   (c) plarynx
   (d) diaphragm
2. Where is the stomach situated?
   (a) Epigastric region of abdominal cavity
   (b) Right hypochondriac region of abdominal cavity
   (c) In front of pelvic cavity
   (d) Behind the pelvic cavity
3. Choose the incorrect pair.
   (a) Cardiac ventriculi – Near heart
   (b) Corpus ventriculi – Smallest part of stomach
   (c) Fornix ventriculi – Convex portion of stomach
   (d) Pyloric antrum – Distal part of stomach
4. Network of sympathetic and parasympathetic neurons and nerve fibres present between the two major layers of muscularis externa is not called
   (a) Auerbach plexus (b) Myenteric plexus
   (c) Meissiner's plexus (d) None of these
5. Which mucosal layer consists of loose connective tissue, blood vessels, glands and lymphoid tissue?
   (a) Lamina propria
   (b) Muscularis mucosa
   (c) Muscularis interna
   (d) Peyer's patches
6. Gastric glands are restricted to which of the following layer?
   (a) Adventitia
   (b) Muscle layer
   (c) Mucosa
   (d) Submucosa

DIGESTIVE GLANDS

Various types of digestive glands are found in mammals. These are salivary, gastric and intestinal glands along with pancreas and liver.

Salivary glands

These digestive glands pour their secretions into the mouth.

These glands can be categorised into the major salivary glands and accessory or minor salivary glands.

(i) The **major salivary glands** in humans include
- Two parotid glands below ears
- Two submandibular glands at the lower jaw angles
- Two sublingual glands below tongue

Position of these three glands is shown below

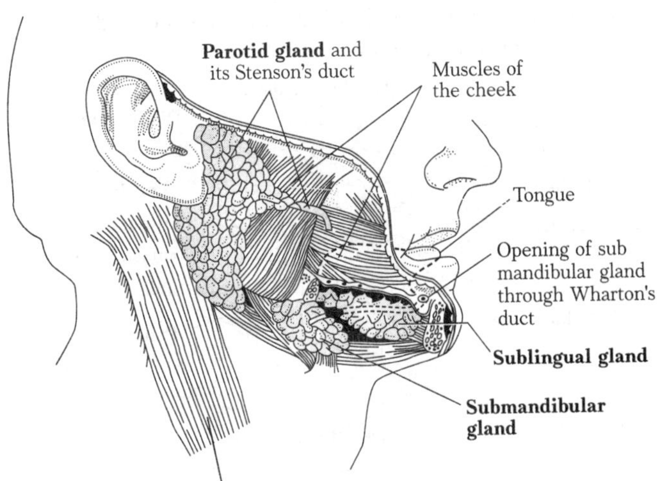

Fig 23.11 Salivary glands in humans

Table 23.2 Different salivary glands and their characteristics

	Parotid glands	Submandibular or Submaxillary glands	Sublingual glands
Location	Near ear (Cheeks)	At the junction of upper and lower jaw	Below tongue
Size	Largest	Medium sized	Smallest
Contribution in saliva	25%	70% (Maximum)	5% (Minimum)
Secretions	Salivary amylase (Ptylin)	Mucus and small amount of ptylin	Mucus and small amount of ptylin
Ducts	Stenson's ducts	Wharton's ducts (largest salivary duct)	Duct of Rivinus (shortest salivary duct)
Opening of ducts	In vestibule of upper jaw	Behind lower incisors	Floor of buccal cavity
Nerve	IXth	VIIth	VIIth

Infraorbital glands

These are not found in humans. These glands are present in animals like dog, rabbit and cat below the eyes. In them, these glands are also called **zygomatic glands**.

(ii) The **accessory** or **minor salivary glands** include
- Unicellular goblet cells
- Nuhn's glands - restricted to tongue only, mixed mucous and serous gland
- Weber's glands - restricted to tongue only, secrete mucus
- Ebner's glands.

Out of these, the first 3 does not secrete any enzyme while **Ebner's glands** secrete minute quantities of salivary lipase.

Secretions of salivary glands form **saliva**. The release of saliva is under autonomic control as follows

- The **parasympathetic** stimulation of salivary glands causes vasodilation and profuse secretion of the watery saliva.
- Their **sympathetic** stimulation inhibits the secretion of saliva.
- Along with the nervous control, an enzyme **kallikrein** is present in saliva. It acts upon plasma proteins and produces a substance called **Kallidin**. This is responsible for the vasodilation of salivary glands during secretion.

Note *According to their position within the buccal cavity, the salivary glands are also classified as*

- ***Labial*** *in the mucous membrane of lips.*
- ***Buccal*** *in the mucous membrane of cheeks.*
- ***Lingual*** *in the mucous membrane of tongue.*
- ***Palatine*** *in the mucous membrane of palates.*

Gastric glands

These are numerous microscopic, tubular glands in the mucosa of the stomach. These glands can be categorised into following types

(i) **Cardiac glands** These are simple tubular and mucus secreting glands. These are located in a narrow band around the cardiac orifice.

(ii) **Oxyntic or Main gastric glands** These HCl secreting glands occur in the fundic part and corpus or body region of stomach. These are the most important and most numerous gastric glands.

(iii) **Pyloric glands** These are present in the pyloric part of stomach. Cells present in these glands are mostly mucus secreting cells. However, some **oxyntic** and **zymogenic** G-cells are also present in these glands.

Four types of cells found in gastric glands are as follows

(a) **Mucous Neck Cells** (MNCs) or **Goblet cells** These are modified columnar epithelium cells which secrete alkaline mucus. These are located in the neck part of the glands. Secretion of these cells protects the lining of GIT from various enzymes.

(b) **Oxyntic or Parietal cells** These are large and rounded cells, located upon the surfaces of gland. These secrete 0.04% HCl and Castle's Intrinsic Factor (CIF).

Note
- *CIF is secreted in its inactive form and gets functional only in distal part of intestine where the process of absorption occurs.*
- *Non-secretion of HCl (achlorhydria) or gastrectomy (removal of stomach) can lead to iron deficiency or anaemia because HCl of gastric juice converts Fe^{3+} into Fe^{2+} and thus, help in iron absorption.*

(c) **Chief or Peptic or Zymogenic cells** They form the major part of gastric gland by secreting inactive precursors of gastric enzymes or proenzymes.

(d) **Argentaffin cells or Enteroendocrine cells** These are located in deeper parts of the gastric gland.

These cells are further differentiated into three types as follows

- **D-cells** secrete somatostatin (vasoconstrictor).
- **Enterochromaffin cells** (ECl-cells) secrete serotonin and histamine (help in acid production).
- **G-cells** secrete gastrin hormone which regulates appetite and digestion.

Intestinal glands

These glands are formed by the surface epithelium of the small intestine. These are of following two types

(i) **Crypts of Lieberkuhn** Simple tubular glands which secretes the intestinal digestive juice or **succus entericus.**

These are present between villi throughout the mucous membrane of small intestine. These crypts contain two type of cells, i.e.

(a) **Paneth cells** These are found in duodenum only. These contain zinc and acidophilic granules and secrete lysozyme.

(b) **Argentaffin cells** These cells synthesise secretin hormone and a powerful stimulant, 5-hydroxytryptamine.

(ii) **Brunner's or Duodenal glands** These are multicellular glands which are found only in the submucosa of the duodenum. These secrete a little enzyme and mucus.

Liver

- It is the largest gland of vertebrate body.
- It is situated in upper right section of abdominal cavity of the body. In humans, it is about 1.5 kg in weight.
- It is divided into two lobes, i.e. right and left. Both of these lobes are joined with the help of a **falciform ligament**.
- The diagram of liver and pancreas with associated structures is shown below

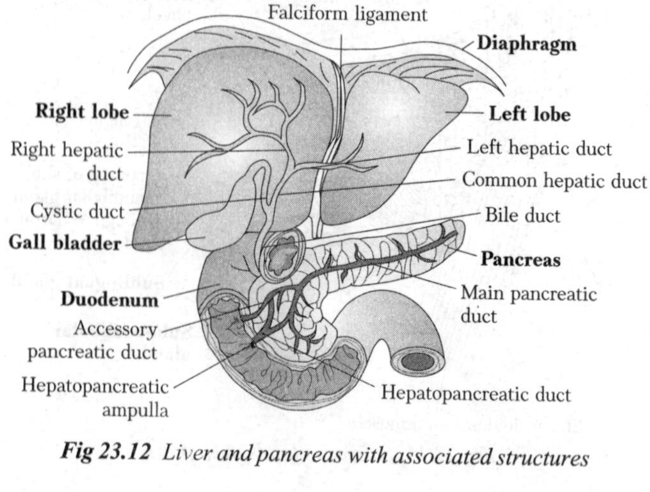

Fig 23.12 Liver and pancreas with associated structures

Basic structure

- Each liver lobe is further divided into **hepatic lobules.** These are considered as the structural and functional units of liver.
- These lobules are connected together by connective tissue called **Glisson's capsule.** It contains blood vessels along with blood sinuses and hepatic ductules.
- Each lobule consists of two radial rows of hepatic cells or hepatocytes. These rows are called **hepatic cords.**
- The hepatocytes are polygonal in shape and are separated by bile capillaries or bile canaliculi.

Ducting system

- Bile capillaries open into small **Herring's canal,** which in turn opens into interlobular bile ductules. These ductules merge with each other to form a few large hepatic ducts in each lobe.
- All hepatic ducts of a lobe combine with those of other to form a single **main hepatic duct.** This hepatic duct later joins to the **cystic duct** that arises from the gall bladder.
- The cystic duct and hepatic duct joins to form **bile duct.** It attaches with pancreatic duct to form the **hepatopancreatic duct.**
- The terminal part of hepatopancreatic duct is swollen as **hepatopancreatic ampulla** or **ampulla of vater.**
- This ampulla opens into the duodenum. This opening is guarded by the **sphincter of Oddi.**
- Another **sphincter** of Boyden surrounds the opening of the bile duct before it joins with the pancreatic duct.

Blood supply

- The blood supply of liver includes a hepatic portal vein. It joins with hepatic artery to form large hepatic sinusoids.
- A part of transverse section of mammalian liver is shown below

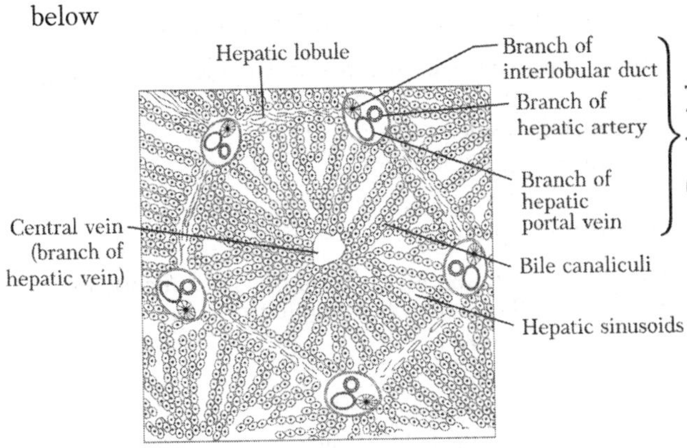

Fig 23.13 A part of transverse section of mammalian liver

- These sinusoids then drain blood into hepatic vein as shown below

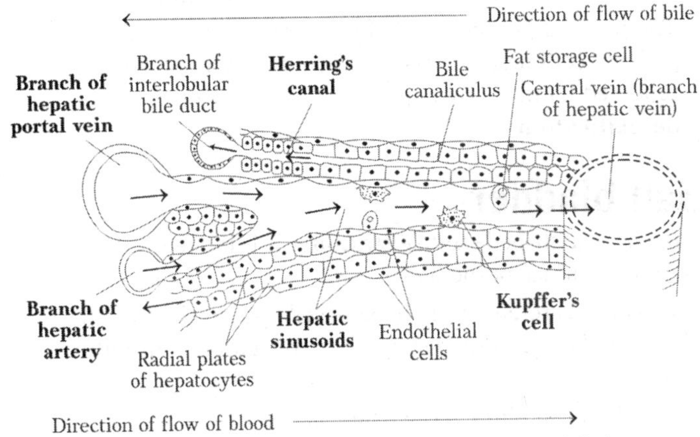

Fig. 23.14 A simplified diagram of a part of liver lobule

Functions of liver

Liver performs following functions

- Bile secretion is the main function of liver.
- Hepatic cells or hepatocytes synthesise urea (*via* deamination) for excretion.
- It helps in the elimination of certain other waste products.
  These include cholesterol, metal ions and waste products of haemoglobin in **reticulo-endothelial** tissues.
- Liver acts as a blood-filter by removing the worn-out and aged corpuscles, but it retains their iron.
- Hepatocytes secrete certain enzymes which play important roles in the metabolism of proteins, fats and carbohydrates in the body.
  They synthesise important blood proteins like **prothrombin** and **fibrinogen.** Both of these proteins are helpful in the clotting of blood.
- Hepatocytes continuously secrete a substance called **heparin** into the blood. Heparin prevents the clotting of blood within blood vessels.
- Liver synthesises vitamin-A from carotene and stores vitamins-A, D and B_{12}. It is also an important site of lymph formation.
- Liver possesses an extensive network of liver sinusoids or specialised blood vessels. Thus, besides a major haemopoitic or blood forming organ, liver acts as blood reservoir also.
- The mammalian liver has specialised phagocytes called **Kupffer cells** or **hepatic macrophages** or **stellate cells.** They engulf wornout WBCs, RBCs and bacteria.
- The excess sugars are withdrawn from portal blood by liver cells. These sugar or glucose get converted into glycogen granules for storage by insulin and this process is called **glycogenesis.**
  When normal amount of sugar in blood tends to decline between meals, liver cells convert stored glycogen into glucose and this process is called **glycogenolysis.**

- Up to some extent, liver also helps in **lipogenesis**, i.e. formation of fats from excess glucose and amino acids.
- The harmful prussic acid is formed during metabolism in all body cells and released into blood. It is absorbed and rendered harmless by hepatocytes. Thus, liver helps in **detoxification.**

Gall bladder

- It is a pear-shaped sac-like structure. It is located on the interior surface of the liver.
- The bile secreted by the hepatic cells passes through the hepatic ducts and then stored and concentrated in gall bladder.
- It is absent in birds, rat, whale and horse.

Pancreas

- It is elongated, irregularly branched, flattened, digestive gland. It is pale yellow in frog and pinkish in mammals.
- The pancreas is covered by a thin external serous capsule of visceral peritoneum and a thin internal capsule of connective tissue.
- It is a mixed or heterocrine gland, i.e. it contains both exocrine (99%) and endocrine (1%) parts. Thus, pancreas performs two major functions *viz.*, secretion of pancreatic juice (exocrine) and production of hormones (endocrine).

Exocrine part

- The exocrine part of pancreas contains numerous **acini** or **acinar cells.**
- It is a group of secretory cells surrounding a cavity. Each acini or acinar cells is lined by pyramidal-shaped cells.
- These acini cells secrete the enzymes of pancreatic juice.
- This juice is carried by the **duct of Wirsung,** i.e. the main pancreatic duct. This duct opens into the duodenum through the hepatopancreatic ampulla.
- An accessory pancreatic duct called **duct of Santorini** may sometimes lead directly into the duodenum from pancreas.

Endocrine part

- The endocrine part of pancreas is formed by the group of endocrine cells [α, β, δ and Pancreatic Polypeptide (PP) cells].
- These cells are found in between the groups of acini as **islet of Langerhans.**
- Various types of cells in islet of Langerhans, their characteristics and functions are tabulated below

Types of cells in islet of Langerhans	Characteristics	Functions
Alpha (α) cells	Constitute 15%, more numerous towards the periphery of islet.	Produce glucagon which converts glycogen into glucose in liver.
Beta (β) cells	Constitute 65% of islet of Langerhans.	Produce insulin which converts glucose into glycogen in liver and muscles.
Delta (δ) cells	Constitute 5% of islet of Langerhans.	Secrete somatostatin (SS) which inhibits the secretion of glucagon, insulin and growth hormone from anterior lobe of pituitary.
Pancreatic polypeptide (PP) cells (F-cells)	Constitute 15% of iselt of Langerhans.	Secrete pancreatic polypeptide which inhibits the release of pancreatic juice.

Diagram showing the section of pancreas is given in Chemical Control and Coordination.

CHECK POINT 23.3

1. Salivary glands are not found in
   (a) anterior part of the hard palate
   (b) posterior part of the hard palate
   (c) in mandible, posterior to the 3rd molar teeth
   (d) nasopalatine canal
2. X is zymogenic salivary gland which opens in the buccal cavity through Stenson's duct. Identify X.
   (a) Sublingual gland (b) Parotid gland
   (c) Ebner's gland (d) Submandibular gland
3. Identify the glands whose ducts opening into the floor of the mouth.
   (a) Submandibular and sublingual
   (b) Submandibular and parotid
   (c) Sublingual and parotid
   (d) Von Ebner
4. Identify the odd one out.
   (a) Oxyntic gland (b) Cardiac gland
   (c) Pyloric gland (d) Duodenal gland
5. Thin connective tissue sheath associated with mammalian liver is
   (a) Remak's capsule
   (b) Glisson's capsule
   (c) Perimysium
   (d) Auerbach's plexus
6. Alpha cells are found in .... of islets of Langerhans while beta cells are usually found in the ....
   (a) periphery, centre
   (b) centre, periphery
   (c) equally in both regions, periphery
   (d) periphery, equally in both regions

FUNCTIONING OF DIGESTIVE SYSTEM

It can be studied under three major processes *viz*, digestion, absorption and assimilation and egestion.

Digestion

- The process of digestion comprises mechanical as well as chemical processes.
- The **mechanical processes** includes mastication along with various movements of alimentary canal.
- The **chemical processes** involves the action of various digestive juices containing biological catalysts or enzymes in them.
- Four broad categories of these digestive enzymes are
  (i) **Carbohydrases** or carbohydrate digesting enzymes, e.g. salivary amylase, pancreatic amylase, maltase, etc.
  (ii) **Proteinases** or **proteolytic enzymes** are protein digesting enzymes, e.g. pepsin, trypsin, peptidase, etc.
  (iii) **Lipases** lipid digesting enzymes, e.g. lipase.
  (iv) **Nucleases** nucleic acid digesting enzymes, e.g. nuclease, nucleosidases and nucleotidase.
- Out of these, only lipases are secreted in their active forms while rest are secreted as their inactive precursors.

For the sake of convenience in study, we will study the process of digestion in different organs of alimentary canal, in the sequence of their occurrence only.

Digestion in mouth

- The process or physiology of digestion begins with mastication or chewing of food. Chewing has both voluntary and involuntary functions. It is regulated by **chewing reflex**.
- The tongue helps in tasting the food. It also mixes the food with saliva which is being continuously secreted and released into the buccal cavity by salivary glands.
- The pH of saliva is 6.3-6.8 and about 1-1.5 litre is released everyday. Saliva contains about 99% water and rest is mucus along with K^+, Na^+, Cl^- and HCO_3^- ions.
- Saliva is bacteriacidal as it contains **thiocyanate** and **lysozymes**. It also contain IgA antibody.
- It is slightly acidic (pH about 6.8) and contains **ptyalin** or **salivary amylase** enzyme. It brings about the initial digestion of starch like polysaccharide as

$$\text{Starch} \xrightarrow[\text{pH 6.8}]{\text{Salivary amylase}} \text{Maltose} + \text{Isomaltose} + \alpha\text{-Dextrins}$$

- The action of ptyalin is pH specific, i.e. it decreases with increased H^+ concentration.
- As a result of chewing and salivary action, the food in mouth is converted into **bolus**, i.e. ball of food. This bolus is then conducted to stomach through oesophagus after swallowing as shown below

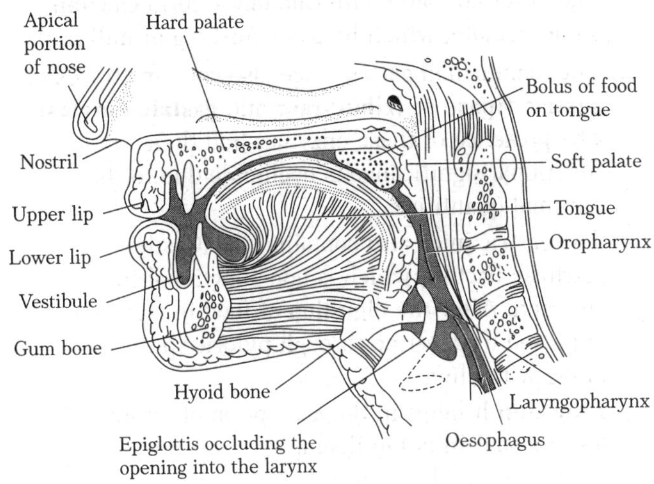

***Fig 23.15** Section of the face and neck showing the positions of structures during swallowing*

Note *Ptyalin is absent in the saliva of most of the herbivores and ruminants.*

Digestion in stomach

- The bolus reach the stomach through reflex wave of relaxation and contraction, i.e. **peristalsis** in oesophagus.
- It remains in the stomach for 3-4 hours. Afterwards, it gets gradually converted into semifluid mixture called **chyme** which is highly acidic.
- The gastric juice is secreted in considerable amount, i.e. about 2 L/day with pH range of 1.8-3.2. The pH of infant gastric juice is near about 5. It contains 97-99% water, 0.2-0.5% HCl and rest include mucus, salt and enzymes. The major digestive enzymes in gastric juice and their functions are as follows
  (i) **Pepsin** It is secreted as **pepsinogen** (inactive form) and then gets converted into its active form, i.e. pepsin by HCl.

Function It converts protein molecules (especially collagen) into proteoses, peptones and ultimately into large polypeptides. It also helps in milk coagulation in humans.

(ii) **Rennin** It is secreted only in some mammals as an inactive precursor, i.e. **prorennin.** Presence of rennin is suspicious in human beings. It is presumed to be present in initial 6-15 days in human infants.

Function This enzyme is responsible for converting casein protein of milk to paracasein. The latter combine with calcium to form calcium paracaseinate, which helps in curdling of milk.

(iii) The mammalian gastric juice also contains traces of **gastric lipase** or **tributyrase** and **gastric amylase.** The presence of these enzymes and their functional significance is negligible due to their minimal quantities.

(iv) The gastric juice also contains **CIF** (Castle's Intrinsic Factor), which is non-functional in stomach. After the digestion of food in stomach, it is passed to intestine through pyloric sphincter along with chyme.

Function It helps in the absorption of vitamin-B_{12} (cyanocobalamine) in ileum.

Digestion in small intestine

In small intestine, chyme is treated with three digestive juices, i.e. bile juice, intestinal juice and pancreatic juice.

(i) **Bile juice** It is a complex greenish yellow alkaline fluid.

- It is secreted by liver cells. It gets collected in gall bladder. From gall bladder, the bile is taken into the intestine, as and when required.
- It has a pH value of 7.6 in gall bladder and 8.0 in liver.
- **Composition** It contains mainly water (92%), fatty acids (0.3-1.2%), the bile salts (6%), bile pigment (0.3%), cholesterol (0.3-0.9%) and lecithin (0.3%). The **bile pigments** are **bilirubin** (yellow) and **bilivirdin** (green).
- Bile plays an important part in digestion. It does not contain any digestive enzyme, but being alkaline in nature (due to the presence of sodium bicarbonate), it neutralises the acidity of gastric juices of chyme. This results to an alkaline medium, which is necessary for the action of pancreatic and intestinal juice in small intestine.
- **Sodium glycocholate** and **sodium taurocholate** present in bile helps in the breakdown of large fat droplets into smaller ones, i.e. **emulsification.**
- It also prevents **putrefaction** (decay or rotting) of food by destroying harmful bacteria present in food. Beside these functions, bile increases peristalsis of intestine and activate the lipase enzyme.

(ii) **Pancreatic juice** It is highly alkaline (pH 7.3-8.6). It contains about 96% water along with 4% salts and enzymes. It is a complete digestive juice.
Its major enzymes are as follows

- **Endopeptidases**, i.e. two pepsin like enzymes namely **trypsin** and **chymotrypsin**
- **Exopeptidases**, e.g. carboxypolypeptidase
- **Steapsins** or pancreatic lipases
- **Pancreatic amylase** or amylopsin (act on starch and glycogen, but has no effect on cellulose)
- **Nucleases**, i.e. deoxyribonuclease and ribonuclease.

(iii) **Intestinal juice** or **Succus entericus** It is a transparent, yellowish and alkaline fluid. Its pH varies between 7.6-8.3. In humans about 2 L of succus entericus is secreted everyday. Its major enzymes are

- **Erepsins**, is a group of several protein digesting exopeptidases including both **aminopeptidases** and **carboxypeptidases.**
- **Carbohydrases** (Carbohydrate digesting) includes maltase, sucrase or invertase and lactase.
- **Enterokinase,** the activator enzyme.
- **Intestinal lipase** helps in fat digestion.
- **Nucleolytic enzymes** include both nucleotidases and nucleosidases.

Note
- *Pepsin and trypsin cannot hydrolyse keratine.*
- *Pepsin can hydrolyse milk proteins casein, but trypsin cannot.*
- *In predatory animals, trypsin can hydrolyse fibrinogen of blood into fibrin, leading to blood coagulation in food.*

A detailed outline of complete digestion process is given in the table below

Table 23.3 Detailed outline of the digestion process in alimentary canal of an adult

Digestive juice and Optimum pH	Source	Stimulation by	Proenzyme (Inactive)	Activator	Enzyme	Substrates	End products
Saliva (6.8)	Salivary glands	Neuronal reflex and Kallidin	...	...	Ptyalin or Salivary amylase	Starch, dextrins, glycogen	Dextrins, maltose, isomaltose
Gastric juice (1.8-3.2)	Gastric glands	Neuronal reflexes and gastrin hormone	Pepsinogen	HCl	Pepsin	Proteins	Proteoses, peptones and large polypeptides
			Prorennin (Present in infants)	HCl	Rennin	Milk protein (casein)	Calcium paracaseinate
		...	...	...	Gastric lipase, Gastric amylase	Fats, Starch	Fatty acid and glycerol (Negligible)
Bile (7.7-8.0)	Liver	Secretin and CCK hormones	...	...	...	Fats	Emulsified fats
Pancreatic juice (7.3-8.6)	Pancreas	Neuronal reflexes, Secretin and CCK hormones	...	...	Amylopsin or pancreatic amylase/ Diastase	Starch, dextrins, glycogen	Dextrins, maltose, isomaltose
					Steapsin	Emulsified fats	Monoglycerides, fatty acids, cholesterol
			Trypsinogen	Enterokinase (of intestinal juice)	Trypsin	Proteins, proteoses, fibrinogen, casein	Large peptides
					Elastase	Elastin	Large peptide
			Chymotryp-sinogen	Trypsin	Chymotrypsin	Peptones	Peptides
			Procarboxy-peptidase	Trypsin	Carboxypeptidase	Small peptides	Amino acids
				...	Deoxyribonu-clease	DNA	Nucleotides, nucleosides
			...	...	Ribonuclease	RNA	Nucleotides, nucleosides
Intestinal juice (7.6-8.3)	Intestinal glands	Neuronal reflex, enterokinin hormone	...	...	Erepsin group (exopeptidases)	Small peptides and dipeptides	Amino acids
					Aminopeptidases	Large peptides	Dipeptides
					Dipeptidases	Dipeptides	Amino acids
			...	...	Maltase	Maltose	Glucose (2 molecules)
			...	...	Sucrase	Sucrose	Glucose and fructose
			...	...	Lactase	Lactose	Glucose and galactose
					Enterokinase	Trypsinogen	Active trypsin
					Intestinal lipase	Emulsified fats	Fatty acids and glycerol
					Nucleases and nucleosidases	Nucleotides and nucleosides	Nitrogenous bases and pentose sugars
					α-dextrinase	α-Dextrin	Glucose
Symbiotic bacteria and Protozoa of caecum	...	...	...	...	...	Cellulose	Sugars

End products of digestion

The end products of digestion are monosaccharides (from complex carbohydrates), amino acids (from complex proteins), free fatty acids, monoglycerides (from lipids) cholesterol, nitrogenous bases and pentose sugars.

Movements in alimentary canal

During the process of digestion, following movements are shown by various parts of alimentary canal

- **Swallowing or deglutition** This process is carried out by the smooth and skeletal muscles in the mouth, tongue and pharynx. Process involves the pushing of food mixed with saliva in the mouth backwards, through the pharynx and finally into the oesophagus. This process is controlled by a swallowing centre in the medulla oblongata and pons Varolii of the brain.
- **Peristalsis** It is a wave of contraction and relaxation of circular muscles of the oesophagus. In gastrointestinal tract, peristaltic wave pushes the bolus or ball of food from one part to another. Very forceful and directional peristaltic movements of stomach are called **churning movements**. These are visible after the entry of food in stomach. There is least peristalsis in the rectum of human beings.
- **Segmentation** This process occurs only in the small intestine. During this, short segments of intestine contract like hands squeezing a toothpaste tube. Segmentation helps to increase the absorption of nutrients by mixing food and increasing its contact with the walls of the intestine.
- **Haustral churning** It is the characteristic movement in large intestine. While getting filled up, the haustra of large intestine remain relaxed and become distended. At certain point, the walls contract and squeeze the content into the next haustrum.

Absorption

It is the process by which the nutrients derived from the end products of digestion are passed from alimentary canal into blood and lymph. These end products are absorbed in alimentary canal in following ways

In oral cavity

No absorption of food occurs in oral cavity. Only some chemicals, medicines and alcohol are absorbed in buccal cavity.

In stomach

Water, alcohol, some salts, some drugs (such as aspirin) and moderate amounts of sugar are absorbed in the stomach.

In small intestine

Major absorption of food (90%) takes place in small intestine. Its detailed description is given below

(i) **Absorption of Monosaccharides** All the carbohydrates are absorbed as monosaccharides in the stomach and jejunum.

- Glucose and galactose are absorbed by active transport.
- Fructose is absorbed into blood capillaries by facilitated diffusion. Sodium pumps of cell membrane play supportive role in their absorption.
- Galactose is transported most rapidly out of all monosaccharides.

(ii) **Absorption of amino acids** Amino acids are also absorbed in blood stream. Most of the amino acids are absorbed by active transport while some are absorbed by facilitated transport in duodenum and jejunum.

(iii) **Absorption of lipids** The lipids are absorbed in lymph capillaries (lacteals) present within the villi in following step wise manner

- Lipids are first incorporated into small spherical and water soluble droplets called **micelles.** These micelles are formed with the help of bile salts and phospholipids in intestinal lumen.
- From micelles, fatty acids, glycerides, sterols and fat soluble vitamins are absorbed into the intestinal cells by simple diffusion.
- Here, they are resynthesised in endoplasmic reticulum and gets converted into very small fat droplets called **chylomicrons.**
- Chylomicrons are released from intestinal cells into lymph present in lacteals.
- The lacteals after absorption of lipids contain white coloured liquid inside them. This white coloured liquid is called **chyle.**

(iv) **Absorption of water** About 90% of all water absorption occurs in the small intestine. It is associated with the absorption of electrolytes. Water soluble vitamins except vitamin-B_{12} are also absorbed along with water. Vitmain-B_{12} require CIF for its absorption and it is absorbed in ileum.

In large intestine

Products of bacterial digestion (amino acids, vitamin-B complex and vitamin-K) and water are absorbed in the large intestine.

Assimilation

It is the process by which absorbed substances finally reaches the tissues. It is then utilised for various activities, e.g. glucose is utilised in the production of energy.

Egestion

- The elimination of the undigested remains of the food from the alimentary canal is called egestion. It is carried out by the defecation reflex in large intestine.

- Egestion involves coordinated action of muscles of large intestine, abdominal muscles and anal sphincter. The act is partly voluntary and partly involuntary.
- In infants, defecation occurs by reflex action in which external anal sphincter are not controlled voluntarily.
- In the colon, water is reabsorbed into the blood from chyle, thus concentrating and converting it into **faeces**.
- **Colour of faeces** is brown due to the presence of stercobilin, urobilin and stercobilinogen. These are formed due to the degradation of bilirubin.
- **Foul smell of faeces** is due to the presence of indole, scatole and H_2S in it. These are formed in colon due to the decomposition of amino acids by bacteria.

Fluid Balance in GI Tract

- A healthy person takes about 1-2 litres of dietary fluid every day and about 6-7 litres of fluid is secreted by various glands of GI (Gastro-Intestinal) tract. This simply means that our GI tract receives about 8-9 litres of fluid everyday.
- However, out of this, only 0.1-0.2 litres of fluid is excreted in faeces and rest is reabsorbed in intestinal portion of GI tract.
- The figure given below gives an overall summation of fluid balance in GI tract

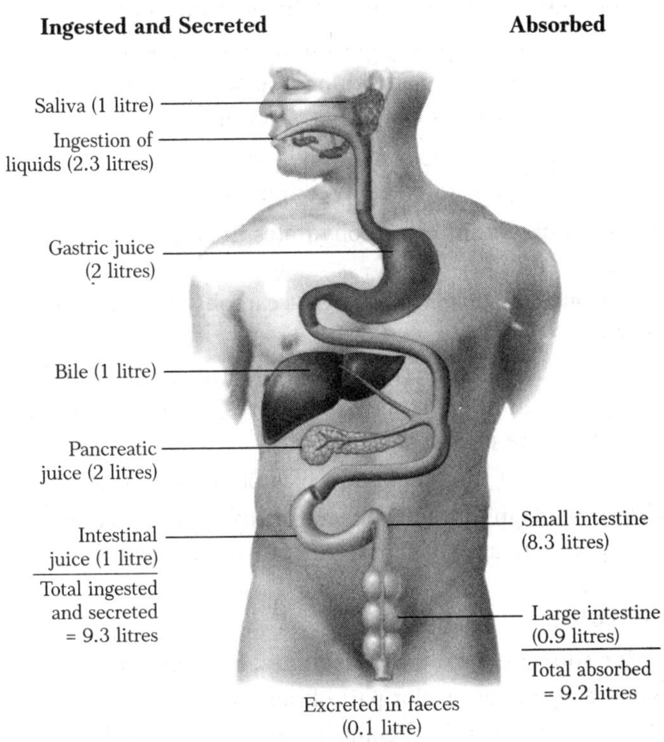

Fig. 23.16 Fluid balance in GI tract

Hormonal control of digestion

Following hormones regulate the secretions of digestive system

(i) **Gastrin** It is secreted by gastric cells, i.e G-cells. These cells are present in the pyloric region of stomach and in the duodenum. This hormone target stomach and stimulate the release of gastric juice and also increases gastric mobility.

(ii) **Enterogastrone** or **Gastric Inhibitory Peptide** (GIP) It is secreted by duodenal epithelium. It targets stomach and inhibit gastric secretions and motility.

(iii) **Secretin** It is secreted by the epithelium of duodenum especially, in response to acidity.
It was the first GI tract hormone, discovered by the scientists. It targets pancreas, liver and stomach thus, releases bicarbonates in pancreatic juice, increases bile secretion from liver and decreases gastric secretions and motility.

(iv) **Cholecystokinin Pancreozymin** (CCK-PZ) It is secreted by the epithelium of duodenum. It contract the gall bladder to release bile. It also stimulates pancreas to secrete and release digestive enzymes in pancreatic juice.

(v) **Duocrinin** It is secreted by the duodenal epithelium. It stimulates Brunner's glands of duodenum to release mucus and enzymes into the intestinal juice.

(vi) **Enterocrinin** It is secreted by the epithelium of entire small intestine. It stimulates the crypts of Lieberkuhn of small intestine to release enzymes into the intestinal juice.

(vii) **Vasoactive Intestinal Peptide** (VIP) It is secreted by small intestine. It dilates the peripheral blood vessels of stomach. It also inhibits gastric acid secretion and absorption from the lumen of small intestine.

(viii) **Villikinin** It is secreted by the epithelium of entire small intestine to accelerate the movements of villi in small intestine.

(ix) **Somatostatin** It is released by the delta cells of islets of Langerhans of pancreas. It inhibits the secretion of insulin and glucagon by pancreatic cells. It also inhibits the absorption of nutrients from the gastrointestinal tract. Small amount of somatostatin is secreted by argentaffin cells of gastric and intestinal glands as well. Here, somatostatin helps to supress the release of hormones from gastrointestinal tract.

(x) **Pancreatic Polypeptide** (PP) It is secreted by pancreatic polypeptide cells or F-cells of islets of Langerhans. It inhibits the release of pancreatic juice from the pancreas.

CHECK POINT 23.4

1. The bile juice released into the duodenum does not contain
   (a) cholesterol and phospholipids
   (b) bile salts such as taurocholate and glycocholate
   (c) bile pigments such as bilirubin and biliverdin
   (d) cholecystokinin
2. Trypsin differs from pepsin in that
   (a) trypsin digests protein in an acidic medium while pepsin does so in an alkaline medium
   (b) trypsin is secreted from the gastric glands while pepsin is secreted from the pancreas
   (c) trypsin digest protein in an alkaline medium while pepsin does so in an acidic medium
   (d) trypsin digest fat and pepsin digests protein
3. The enzymes chymotrypsin and carboxypeptidase are activated from their precursor proenzymes by the action of
   (a) pepsin (b) enterokinase
   (c) trypsin (d) enteropeptidase
4. A person consumes a starchy food. The chyme was analysed before it enters duodenum. It will show the presence of
   (a) maltose, dextrin, starch
   (b) maltose, glucose, sucrose
   (c) maltose, sucrose, starch
   (d) dextrin, maltose, sucrose
5. Gastric juice of infants contain
   (a) nuclease, pepsinogen, lipase
   (b) pepsinogen, lipase, rennin
   (c) amylase, rennin, pepsinogen
   (d) maltase, rennin, pepsinogen
6. The enzyme vital for protein digestion, but that does not directly act upon food substrate is
   (a) trypsin
   (b) pepsin
   (c) chymotrypsin
   (d) enterokinase

NUTRIENTS

These are the substances that provide nourishment essential for the maintenance of life and growth. The process by which body obtains and utilise these nutrients is called **nutrition.** Precisely, the nutrients are organic or inorganic chemicals and can be categorised into two broad categories

- **The macronutrients** are also called **proximate principles of food** because they constitute energy source for the production of heat. They include proteins, lipids and carbohydrates.
- **The micronutrients** are also called **protective principles of food** because although, these do not provide energy, yet their deficiencies result in specific abnormalities. They include vitamins and minerals.

Important terms related to Nutritional Habits of heterotrophs

- **Herbivores** Feeding on plants, e.g. rabbit, goat, cow, etc.
- **Carnivores** Feeding on other animals, e.g. tiger, lion, etc.
- **Omnivores** Feeding on all types of food, e.g. humans, cockroaches, etc.
- **Detritivores** Feeding on detritus or organic remains, e.g. earthworm, etc.
- **Scavengers** Feeding on carrions, i.e. the decaying flesh of dead animals, e.g. hawk, vulture, etc.
- **Insectivores** Eating insects, e.g. common bats, etc.
- **Cannibals** Eating other members of one's own species, e.g. many snakes, etc.
- **Frugivores** Feeding on fruits, e.g. parrot, etc.
- **Sanguivores** Taking meal of blood, e.g. leech, female mosquito, etc.
- **Coprophagous** Eats its own fecal matter, e.g. rabbit.

Macronutrients : Proximate principles of food

These organic nutrients are of following types

Carbohydrates

Major sources Bread, beans, potatoes, corns, etc.

Mostly living cells prefer glucose as the source of energy, hence glucose is also called **blood sugar.** Normal content of glucose in blood is 80-100 mg/ 100 mL.

Excess of glucose gets converted to glycogen (by glycogenesis) and stored in liver and muscles.

Daily requirement of carbohydrates in an adult is about 500 g.

Functions The major functions of carbohydrates are as follows

(i) They act as reserve food in the form of starch and glycogen in plant and animals, respectively. Both of these are polymers of glucose.

(ii) It prevent the breakdown of protein for energy.

(iii) It helps in fat metabolism as surplus carbohydrates of our body are converted into fats by lipogenesis.

Proteins

Major sources Nuts, beans, legumes, fish, milk, etc.

Proteins constitute major part of our food.

These are the structural components of tissues and present as channels, transporters, regulatory molecules, enzymes, etc.

They form about 15% of our protoplasmic portion and about 75% of dried body weight. The protein forming amino acids are categorised as essential and non-essential types.

Essential amino acids must be provided in the diet from outside and are thus, designated as essential.

Daily requirement of proteins is 70-100 g.

Non-essential amino acids are produced in our body.

Functions The major functions of proteins include

(i) Essential for growth and repair of body.

(ii) Enzymes are proteins which act as biochemical catalyst in metabolic reactions.

(iii) Antibodies are the proteins which work for the defence of the body. Thus, it forms a part of immune system.

Lipids

Major sources Vegetable oil, butter, whole milk, cheese, etc. The lipids associated with animal body includes fats, oils, waxes, sterols, etc. About 95% of the lipids in human diet are triglycerides of fatty acids. These are found attached to glycerol molecule. The fatty acids can also be essential or non-essential.

Daily requirement of lipids is 50 g.

Functions The major functions of lipids are as follow

(i) Fats mainly act as concentrated fuel. 1 g of fat provides 2.25 times more energy than the energy provided by same amount of glucose.

(ii) Fats forms the protective shock absorbing cushions around a number of vital organs.

(iii) These help in the absorption of fat soluble vitamins like A, D, E and K.

(iv) They provide insulation sheath called **myelin sheath** around nerves. It is mainly made up of phospholipids.

Micronutrients : Protective principles of food

These can be of following types

Vitamins

- These are essential compounds for the metabolic activities of our body. These are required in very small quantities. There are about 30 vitamins known.
- Vitamins do not provide energy, but control energy yielding reactions of our body. These actually work as various cofactors for the enzymes of our body.
- In general, vitamins are not synthesised in the body of animals, except some, e.g. vitamin-A and vitamin-D.
- The deficiency or excess of any of vitamin can alter various metabolic processes of the body.
- On the basis of their solubility and other physiochemical properties, these can be classified as follows

  (i) **Fat soluble vitamins**-A, D, E and K.

  (ii) **Water soluble vitamins**-B complex (having vitamin-B_1, B_2, B_3, B_6, etc), C and P, etc.

Table 23.4 An overview of various vitamins

Name of vitamins and their discovery	Sources	Functions	Diseases
Fat soluble vitamins			
Vitamin-A (Retinol or anti-infection vitamin) By McCollum and Davies in 1913	Fish liver oils, egg yolk, milk, animal liver, yellow, red and green fruits. Vegetables like carrot, orange, etc., which have β-carotene (yellow and red pigment) are helpful in vitamin-A synthesis in human body.	Synthesis of retinal pigment, promotes normal vision, growth of bones and teeth, maintenance of epithelia and prevention of keratinisation, anti-infective, β-carotene helps as an antioxidant. **Daily dietary requirement** : 2 mg	**Hypovitaminosis** Xerophthalmia, night blindness (nyctalopia) dermatosis, keratomalacia, etc. **Hypervitaminosis** Headache, discomfort, vomiting, bone pain, etc.
Vitamin-D (Calciferol) Sun-shine vitamin or poor man's vitamin. By McCollum in 1922 Two type, i.e. vitamin-D_2 (ergocalciferol) and vitamin-D_3 (cholecalciferol)	Fish liver oil, milk and its products, egg yolk, butter. In the presence of sunlight, it is synthesised in skin cells from 7-dehydrocholesterol.	Regulates metabolism of calcium and phosphorus for growth of bones and teeth. **Daily dietary requirement** : 0.01 mg	**Hypovitaminosis** Rickets in children, weak bones, bone deformities, osteomalacia in adults, dental caries (cavity in teeth), etc. **Hypervitaminosis** Weight loss, nausea, hypercalcemia, anorexia, etc.
Vitamin-E (Tocopherol or anti-sterility vitamin or beauty vitamin) By Mattill and Conklin in 1920	Vegetable oils, egg yolk, meat, green vegetables like lettuce, cereals, nuts. Found in maximally in liver.	It maintains the normal functioning of reproductive organs. Potent antioxidant, help in healing wound, etc. **Daily dietary requirement** : 15 mg	**Hypovitaminosis** Anaemia, sterility, muscular dystrophy (degeneration of muscles).

Name of vitamins and their discovery	Sources	Functions	Diseases
Vitamin-K (Anti-haemorrhagic vitamin) By H. Dam and Droisy in 1935 Three derivatives are : Vitamin-K_1 or phylloquinone, vitamin-K_2 or menaquinone and vitamin K_3 or menadione.	Green leafy vegetables, tomatoes, soybean, wheat, fish, liver, egg, etc. Vitamin-K_2 is synthesised by the intestinal bacteria.	Helps in the formation of prothrombin in liver for normal clotting of blood, also found to have role in formation of other blood clotting proteins. **Daily dietary requirement** : 0.07-0.14 mg	**Hypovitaminosis** Haemorrhage, deficiency of prothrombin in blood. **Hypervitaminosis** Anaemia with gastrointestinal discomfort.
Water soluble vitamins (with only hypovitaminosis and no hypervitaminosis as their excess is removed with urine)			
Vitamin-B_1 (Thiamine or anti-neuritic or anti beri-beri) By Eijkman in 1926	Beans, peas, green vegetables, cereals, whole wheat, yeast, milk, egg yolk, rice, etc.	Helps in increasing appetite, promotes growth of the body. **Daily dietary requirement** : 1.5 mg	Beri-beri, anorexia, i.e. loss of appetite, cardiac failure, neuritis. Wernicke's and Korsakoff's syndrome in alcoholics.
Vitamin-B_2 (Riboflavin or vitamin-G) By Warsung and Christian in 1932	Milk, liver, kidney, egg yolk, fish, yeast, germinating seeds, etc.	Required for cell growth, form pair of coenzyme (FMN, FAD). **Daily dietary requirement** : 2 mg	Cheilosis, i.e cracking corners of mouth. Dermatitis, i.e dryness of skin, glossitis, i.e inflammation of tongue.
Vitamin-B_3 (Niacin or nicotinic acid) By Conrad Elvehjem in 1937	Wheat bran, yeast, liver, vegetables, pulses, etc.	Converts food into glucose, used to produce energy, facilitate DNA repair. **Daily dietary requirement** : 20 mg	Pellagra, diarrhoea, indigestion, skin lesions, vomiting, dementia.
Vitamin-B_5 (Pantothenic acid) By Roger J. Williams in 1933	Honey, egg yolk, tomatoes, potato, yeast, etc.	Used in synthesis of coenzyme-A, helpful for healthy skin and hair. **Daily dietary requirement** : 5-10 mg	Fatigue, insomnia, premature greying of hair, burning feet syndrome, retarded growth.
Vitamin-B_6 (Pyridoxin or anti-dermatitis factor) By Gyorgyi in 1934	Meat, milk, egg yolk, fish, peas, whole grain, green leafy vegetables, liver, etc.	Involved in carbohydrate, lipid and proteins metabolism, helps in RBCs and haemoglobin synthesis, maintains nervous system and overall growth in rats and microorganisms. **Daily dietary requirement** : 2 mg	Anaemia, dermititis, paralysis, mental disorder.
Vitamin-B_7 (Biotin or vitamin-H) By Paul Gyorgyi in 1931	Yeast, vegetables, peanuts, milk, egg yolk, etc.	Serves as coenzyme of protein and fatty acid synthesises. **Daily dietary requirement** : 0.03 mg	Acne, muscular dystrophy.
Vitamin-B_{10} (Vitamin-M or folic acid or ptoroylglutamic acid) By Lucy Wills in 1935	Majorly found in spinach leaves and other sources are peanuts, wheat, yeast, soybean, liver, kidney, etc.	Required in erythrocyte maturation, DNA and RNA synthesis. **Daily dietary requirement** : 0.5 mg	Sprue, i.e. conditions involving ulceration of mouth, inflammation of bowel, etc. macrocytic anaemia, erythrocytopenia.
Vitamin-B_{12} (Cyanocobalamine or castle's intrinsic factor) By George Whipple in 1934	This cobalt containing vitamin is not found in plants (except *Spirulina*). Liver and kidney are the richest sources. Other sources are egg, milk, fish, intestinal bacteria, etc.	Required for erythrocyte maturation, DNA synthesis, myelin formation.	Pernicious anaemia, Demyelination of nerve fibres, glossitis.
Vitamin-B_{15} (Pangamic acid) By Babsky in 1982	In the seeds of almost all plants.	This vitamin is given in the treatment of cardiac and vascular disorders, prevents premature ageing, increases utilisation of oxygen in organisms.	No deficiency disease known yet.
Vitamin-C (Ascorbic acid or anti-scurvy factor) By Albert Gyorgyi in 1930	Citrus fruits like lemon, grapes, cherry, tomato, green vegetables, etc. D-sorbose is used to manufacture vitamin C.	Healing wounds and damaged bones, necessary for healthy teeth and gums. **Daily dietary requirement** : 50 mg	Scurvy or Sailor's disease, delayed wound healing, loss of appetite.
Lipoic acid (Thioctic acid or oxidative factor) By Reed and Gunsalus in 1953	Found in all common sources of vitamin-B complex.	Works as coenzyme for decarboxylation of pyruvic acid to α-ketoglutaric acid, also helps in normal growth of body.	No deficiency disease known yet.
Choline By Horowitz in 1960	Egg yolk, cereals, germ oil, kidney, liver, etc.	Works as coenzyme in lipid metabolism, helps in acetylcholine formation.	Liver cirrhosis, haemorrhage in kidney.

Name of vitamins and their discovery	Sources	Functions	Diseases
PABA or Para Amino Benzoic Acid	Green leafy vegetables, seeds, etc.	Skin pigmentation and hair growth in rats, connection with human is unknown.	Failure of lactation in rats.
Vitamin-P (Citrin or Rutin or Hesperidin) By Albert Gyorgyi in 1936	Citrus fruits, green vegetables, wild wheat leaves.	Maintenance of resistance in the walls of blood capillaries.	No deficiency disease known yet.

Other important facts regarding vitamins

- **Provitamins** are compounds which get converted into vitamins in human body, e.g. ergosterol found in food gets converted into vitamin-D in the skin, in the presence of sunlight.
- **Vitamers** or **isotelic vitamins** are different forms of same vitamins, e.g. vitamers of vitamin-A are vitamin-A_1 and vitamin-A_2.
- **Vitamin poisons** are the compounds which replace vitamins in our body, e.g. sulfa drugs replace vitamin-C.
- Vitamins-A, E, C are antineuritic vitamins.
- It was **Casimir Funk** in 1912, who extracted a crystalline substance from unpolished rice which could prevent beri-beri. This substance was chemically identified as an amine and it was vital for life. That is why, Funk used the term 'Vitamin or vital amine' for this extract. The alphabet was later on dropped as it was found that the other compounds belonging to the same class are not amines.
- **Paul Karrer** was the first person who determined the structure of a vitamin (carotene or vitamin-A). He was awarded Nobel Prize in chemistry in 1937 together with W. Norman Haworth.
- **Dorothy Crowfoot Hodgkin** won Nobel Prize in 1964 in chemistry for analysing the structure of B_{12}.
- Both plant and animals have the ability to synthesise the nicotinic acid, i.e. niacin from the amino acid tryptophan. The process of synthesis of niacin from tryptophan requires pyridoxial phosphate (an active coenzyme of pyridoxine, i.e. vitamin-B_6).
- Vit-B_{17} is recently discovered vitamin with anticancer properties. It is found in apple, peach, orange, apricot, etc.

Minerals

These come next to vitamins under the category of protective principles of foods. Minerals can be categorised as major or **macroelements** (more than 100 mg per day) and **minor or trace elements** (required in very small amounts). A brief account of minerals is given below

Macroelements

(i) **Calcium** (Ca) (Dose 1000 mg daily) is useful for bone and teeth formation. It is a major constituent for nerve functioning, muscle contraction, coagulation of blood, etc.

Sources It is found in dairy products, eggs, wheat germ, dark green vegetables and dates, etc.

Deficiency Its deficiency in blood causes **rickets**, **muscular spasm**, **haemorrhage** with more coagulation time, nervous disorders, muscular cramps, etc.

(ii) **Magnesium** (Mg) (Dose 350 mg daily) is an enzyme activator, component of bones and teeth, required for muscle relaxation, ribosomes binding and nerve functioning.

Sources It is found in dairy products (rich sources), cereals, green leafy vegetables, sea food, chocolate and wheat germ.

Deficiency Its low levels in blood may results in convulsions and hallucinations.

Note ***Water is not gained but lost from the body on drinking sea water.*** *It is so because sea water contain magnesium (Mg^{2+}). If it is drunk, Mg^{2+} ions increases solute concentration in intestinal lumen due to their very slow absorption. It prevents, water absorption from intestine and infact draws water from the blood by osmosis.*

(iii) **Potassium** (K) (Dose 1000 mg daily) is one of the principal univalent cation inside a living cell. It is one of the most important constituents for muscle and nerve activity. It also helps in cardiac functioning.

Sources It is found in most fruits especially in banana, coconut, etc.

Deficiency Its deficiency, i.e. **hypokalemia** causes nausea, muscular and nervous disorders.

(iv) **Sodium** (Na) (Dose 2400 mg daily) is most important for blood and nerve functioning. It is a principal cation of extracellular fluid (0-7%). In excitable tissues, it is responsible for the impulse transmission.

Sources It is present in table salt, most foods and wheat germ.

Deficiency Its deficiency, i.e. **hyponatremia** causes hormonal imbalance, muscular cramps, dehydration, diarrhoea and delerium means confused state of mind. As it helps in maintenance of pH of body fluids, its deficiency also causes pH disbalance.

(v) **Phosphorus** (P) (Dose 1000 mg daily) is the most important anionic element of our body. It is a part of energy carriers like AMP, ADP, ATP, CTP, GTP, etc. It also constitutes bones, teeth and biomembranes. It helps in the maintenance of body buffers, through pH maintenance.

Sources It is found in meat, fish, eggs, dairy products, whole grains and wheat germ.

Deficiency Its deficiency causes demineralisation of bones and metabolic disorders.

(vi) **Chlorine** (Cl) (Dose 3400 mg daily) is an important anion involved in HCl formation in digestive tract. It also helps in CO_2 transport from tissue to lungs and maintain acid-base balance.

Sources It is mainly found in table salt, sea food, vegetables and chlorinated water.

Deficiency Its deficiency causes pH disbalance, ulcers in stomach, etc.

(vii) **Sulphur** (S) (Dose 15 mg daily) is an important constituent of some amino acids, e.g. methionine, cysteine, etc. It is required for the synthesis of vitamin-B_1 and some hormones like insulin. It is essential for healthy hair, skin and nails.

Sources It is found in nuts, seeds, legumes, oats, walnuts, eggs, etc.

Deficiency Its deficiency interferes with the healthy growth of hair and nails.

Microelements

(i) **Fluorine** (F) (Dose 3.5 mg daily) is the constituent of our teeth.

Sources It is found in tea, fish and water.

Deficiency Its deficiency causes dental caries (a condition associated with weak teeth and their decay).

(ii) **Iodine** (I) (Dose 0.15 mg daily) is a component of hormones of thyroid gland.

Sources It is present in salt (iodised salt), sea food, onions and drinking water.

Deficiency Its deficiency is associated with malfunctioning of thyroid gland and weaken immune system.

(iii) **Iron** (Fe) (Dose 15 mg daily) is a component of respiratory pigments (like haemoglobin and myoglobin) and respiratory enzyme (like cytochromes).

Sources It is found in meat, dried nuts, cereals, beans, apple, banana, honey, dates, green leafy vegetables and pomegranate.

Deficiency Its deficiency causes anaemia.

(iv) **Other microelements** like Zn, Cr, Co, Cu, Mn, Mo and Se are present as various cofactors of enzymes involved in metabolic reactions in our cells.

The sources, functions and deficiency symptoms of these micronutrients are as follows

Micronutrients	Sources	Functions	Deficiency symptoms
Cobalt (Co)	Milk, meat	Help in RBC formation as it is a component of B_{12}	Anaemia
Copper (Cu) (2 mg daily)	Peanuts, barley, black grams	Promote iron utilisation in haemoglobin.	Anaemia
Molybdenum (Mo)	Wheat, peas, nuts, potato	Synthesis of haemoglobin and iron absorption	Disturbed iron metabolism
Zinc (Zn) (15 mg daily)	Soyabeans, barley, black grams, wheat	Act as cofactor of several enzymes, e.g. carbonic anhydrase.	Poor growth and wound healing
Manganese (Mn) (5 mg daily)	Nuts, wheat, liver, legumes	Synthesis of haemoglobin, normal reproduction and mammary gland function	Infertility, impaired fat metabolism and menstruation.
Chromium (Cr)	Meat, barley, nuts, black gram	Promote insulin action.	Diabetes
Selenium (0.35 mg daily)	Meat, cereal, sea food	Function with vitamin-E, component of several enzymes.	Premature ageing, cataract, cancer and heart disease.

WATER

It is one of the major constituents of our food. In a cell, usually 70-90% is water. As a whole, our body has 45-80% water. It acts as a medium for various metabolic reactions and their coordination in our body.

Calorific value and physiological value of nutrients

The amount of energy liberated during complete combustion of 1 g of a substance in bomb calorimeter (a closed metal chamber filled with O_2) is its **calorific** or **gross energy value.** The actual amount of energy liberated on combustion of 1g of food in our body is the **physiological value of food.**

Both the calorific and physiological values of lipids, carbohydrates and proteins are given below

Nutrients	Gross calorific value	Physiological value
Lipids	9.45 Kcal/g	9 Kcal/g
Carbohydrates	4.1 Kcal/g	4 Kcal/g
Proteins	5.65 Kcal/g	4 Kcal/g

The energy requirements of animals and the energy content of food are expressed in terms of measure of heat energy.

This is measured as Calorie (Cal) or Joule (J) which is the amount of heat energy required to raise the temperature of 1 g of water by 1°C.

CHECK POINT 23.5

1. Doctors incessantly advice our community to consume vitamins and minerals especially to protect us from COVID-19 because
   (a) they give energy to the body
   (b) they help to carry out metabolic reactions
   (c) they insulate the body organs
   (d) they withdraw heat from the body
2. Wernicke-Korsakoff syndrome in alcoholics is caused due to the deficiency of
   (a) thiamine (b) riboflavin
   (c) pyridoxine (d) folate
3. Tonics made out of the animal liver are very effective in curing haemopoietic disorder because
   (a) they contain proteins (b) they contain RBCs
   (c) they contain bile juice (d) they contain vitamin-B_{12}
4. Which of the following nutrients is needed for a healthy immune system?
   (a) Calcium (b) Iodine
   (c) Vitamin-K (d) Vitamin-C
5. Which one among the following elements/ions is essential in small quantities for the development of healthy teeth but causes molting of the teeth if consumed in higher quantities?
   (a) Iron (b) Chloride
   (c) Fluoride (d) Potassium
6. Which one among the following mineral is essential for the transmission of nerve impulses in the nerve fibres of human body?
   (a) Magnesium (b) Cobalt
   (c) Iodine (d) Sodium

NUTRITIONAL AND DIGESTIVE DISORDERS

The diet which contains the various nutrients in such proportions that can satisfy the various needs of our body is called as **balanced diet**. Every organism requires adequate supply of nutrients in their diet for proper growth and development. Any change in it could lead to either nutritional imbalance or a digestive disorder.

Nutritional disorders

Some of the nutritional disorders are given below

(i) **Hypercholesterolemia** It is caused due to intake of excess of saturated fat such as butter, ghee, etc. In this, cholesterol level in blood rises abnormally. This may cause thrombosis and heart attack.

(ii) **Fluorosis** It is caused due to the excess of fluorides which results in bone and tooth decay.

(iii) **Protein Energy Malnutrition** (PEM) It is also referred to as protein calorie malnutrition.

It is a potentially fatal disorder, leading to death in children in developing countries. It develops in children and adults whose consumption of protein and energy (measured by calories) is insufficient to fulfil the body's nutritional needs.

It is of following two types

(a) **Kwashiorkor** It is a form of severe protein energy malnutrition. It is characterised by irritability, edema (swelling), dermatosis and an enlarged liver with thinning of hair and loss of teeth. It commonly affect children between 1-3 years of age.

(b) **Marasmus** It is caused by severe energy deficiency in the body. It is characterised by stunted growth, loss of muscle mass, anaemia, etc.

It usually develops in children between age group of six months to one year.

Table 23.5 Differences between Kwashiorkor and Marasmus

Kwashiorkor	Marasmus
It is caused by the deficiency of proteins in the diet.	It is caused by the prolonged deficiency of proteins and calories in the diet.
It commonly affects babies between 1-3 years of age.	It affects infants under one year of age.
Subcutaneous fat persists.	Subcutaneous fat is used up, making ribs appearance very prominent.
Edema affects legs and face.	No edema occurs.
Change in skin and hair colour.	No change in skin and hair colour.
Body becomes under weight.	Body becomes very under weight.
Appetite is reduced.	Appetite is not affected.
Patient needs proteins to recoup.	Patient needs proteins as well as carbohydrates and fats to recover.

(iv) **Obesity** It is a luxury nutritional disorder caused by greater intake of food than the requirement of body. It results in deposition of excess fat in the body. Obesity leads to risk of diabetes, gall bladder stones, high blood pressure and cardiac problems.

Disorders of digestive system

Some of the digestive system related disorders are as follows

Indigestion

- It is also known as **dyspepsia** (e.g. upset stomach). It usually happens when people eat too much of food, which is not easily digestable.
- It may also seen due to smoking, drinking alcohol or certain stomach problems like gastritis (stomach inflammation) or an ulcer.
- Indigestion can also be accompanied by heart burn (i.e. a burning sensation deep in the chest), abdominal pain, nausea, bloating (feeling full), etc.

Constipation

- It is the infrequent or difficult evacuation of faeces.
- It is associated with slow movements of alimentary canal.
- It results normally by not taking sufficient water and roughage in daily diet.

Vomiting

- It is the ejection of stomach contents through the mouth.
- This happens due to the various conditions like
  - gastritis (gastric wall inflammation)
  - gastroentritis (intestinal inflammation)
  - pyloric stenosis (narrowing of pylorus)
  - bowel obstruction, over eating, food allergies
  - cholecystis (gall bladder inflammation)
  - pancreatitis, appendicitis, hepatitis, food poisoning, etc
- In children, it can happen also due to lactose intolerance. Excess vomiting causes dehydration and disturbed electrolyte balance in the body.

Jaundice

- It is also known as **icterus**, i.e the yellowness of the skin, mucosa membranes and eyes.
- This happens due to the abnormal functioning of liver which causes excess quantity of bilirubin in the blood.
- Concentration of bilirubin in blood plasma is normally below 1.2 mg/dL (< 25 μmol/L). A concentraion higher than 2.5 mg/dL (> 50 μmol/L) indicates jaundice.
- In fact, **it is a symptom** which indicates malfunctioning of liver.
- It is seen in conditions like hepatitis, liver cancer, gall stones (obstructive jaundice), pancreatic cancer, etc.
- **Neonatal** or **physiological jaundice** is caused due to the poor functioning of liver in initial days. It disappears as the liver matures.

Diarrhoea

- The term is commonly used for the frequent evacuation of watery stools.
- It is caused due to infection like gastroenteritis or food poisoning.
- Infection is spread through contaminated food or drinking water or from poor sanitation.

Lactose intolerance

- In this, the absorptive cells of small intestine fail to produce enough lactase that is required for lactose digestion.
- The undigested lactose in chyme causes fluid retention in faeces and bacterial fermentation of undigested lactose result in gas, bloating, abdominal cramps, etc.
- With advancing age, humans cannot digest milk due to insufficient production of lactase enzyme.
- Curd do not affect a person suffering from lactose intolerance because curd contain lactic acid.

Colitis

It is the inflammation of colon and rectum which may result in loose motions, bloody feaces and dehydration.

Appendicitis

- It is caused due to the rupture of the wall of vermiform appendix, because of which the bacteria are released in the coelomic cavity.
- Severe infection may cause death.
- It can be treated by removing appendix by surgery.

Note • ***Belching*** *Production of burping sound when stomach is over dilated so that the air rises and is expelled through the mouth.*
• ***Flatus*** *Expulsion of gases through anus accompanied by characteristic sound, when gastrointestinal tract gets filled with gases.*

Galactosemia

- It is a metabolic disorder which is caused due to the deficiency of enzyme **uridyl transferase**.
- It result in the accumulation of galactose, leading to mental retardation.
- It can be prevented by taking milk free diet.

Gall stones

- Sometimes, cholesterol, calcium carbonate and bilirubin present in the bile comes out of it and produce **gall stones** or **biliary calculi**. These can further cause pancreatitis.
- These gall stones can be removed by using ultrasonic vibrations to pulverise or crush gall stones without surgery. This is called **lithotripsy**.
- The surgical removal of gall bladder is called **cholecystectomy**.

CHECK POINT 23.6

1. 1 year 2 months old baby a with PEM, does not show
   (a) fatty liver
   (b) oedema due to water retention
   (c) changes in hair colour, lethargy and irritability
   (d) thin limbs and emaciated appearance
2. A patient is advised to include *Spirulina* in his diet only when he is suffering from
   (a) marasmus (b) thyroid
   (c) osteomalacia (d) kwashiorkor
3. Obesity in humans is most clearly linked to
   (a) Type 2 diabetes and decreased appetite
   (b) Type 2 diabetes and cardiovascular problems
   (c) Type 1 diabetes and prostate cancer
   (d) Type 1 diabetes and breast cancer
4. Choose the incorrectly matched pairs.
   (a) Diarrhoea – Infection of gastroenteritis
   (b) Colitis – Inflammation of colon
   (c) Vomiting – Less roughage in diet
   (d) None of the above
5. An obstruction in bile duct can cause a person to suffer from
   (a) cholera (b) jaundice
   (c) vomiting (d) acidity
6. Gall stones or alcohol abuse can lead to
   (a) Pancreatitis
   (b) Cystitis
   (c) Pancreatic cancer
   (d) Marasmus

EXAM POINTS

- Teeth of rabbits and humans are thecodont, diphyodont, heterodont. **[NEET 2018]**
- The primary dentition in human differs from permanent dentition in not having premolars. **[CBSE AIPMT 2015]**
- Length of oesophagus in humans is about 25 cm. **[JIPMER 2019]**
- The middle part of small intestine is jejunum. **[J&K CET 2014]**
- Sphincter of Oddi guards the opening of hepato-pancreatic duct into the duodenum. **[NEET 2016]**
- Sacculus rotandus is located between ileum and caecum in rabbit. **[JIPMER 2018]**
- Goblet cells are found in mucus secreting organs. **[AIIMS 2019]**
- Goblet cells of alimentary canal are modified columnar epithelial cells. **[NEET 2020]**
- Secretions from goblet cells protects the lining of gastro-intestinal tract from various enzymes. **[NEET 2019]**
- In the stomach, gastric acid is secreted by parietal cells. **[NEET 2016]**
- Parietal cells indirectly help in erythropoiesis. **[NEET 2018]**
- Nuclease enzyme is not present in succus entericus. **[CBSE AIPMT 2015]**
- Paneth cells secrete antibacterial lysozyme. **[NEET 2017]**
- Intrinsic factor is secreted by oxyntic cells and it helps in the absorption of vitamin-B_{12}. **[NEET 2020]**
- Salivary amylase is a digestive enzyme that begins the digestion of carbohydrates in mouth. **[OJEE 2013]**
- Salivary amylase shows maximum digestive action at slightly acidic pH, i.e. 6.8. **[MHT CET 2014]**
- Ptyalin gets inactivated by the HCl present in gastric juice. **[WB JEE 2011]**
- Gastric juice of infant contain pepsinogen, lipase and rennin. **[CBSE AIPMT 2015]**
- The proteolytic enzyme rennin is found in gastric juice. **[NEET 2020]**
- Pepsin is the main enzyme that initiates the digestion of protein in stomach. **[Gujarat CET 2014]**
- Cholecystokinin and secretin hormones stimulate the production of pancreatic juice and bicarbonate. **[NEET 2017]**
- Secretion of pancreatic juice is stimulated by secretin. **[BHU 2012]**
- Lipase, amylase, trypsinogen, procarboxy peptidase are the enzymes present in pancreatic juice. **[NEET 2017]**
- In the absence of bile salts, emulsification of fat would not occur in human body. **[UP CPMT 2011]**
- Both secretin and cholecystokinin are secreted by duodenum. **[KCET 2012]**
- Conversion of milk to curd improve its nutritional value by increasing the amount of vitamin-B_{12}. **[NEET 2018]**
- Daily requirement of vitamin-A for adult women is 700 mg. **[JIPMER 2018]**
- Calciferol deficiency in humans leads to osteomalacia. **[WB JEE 2014]**
- Vitamin-B_{12} deficiency results in pernicious anaemia. **[UK PMT 2014]**
- Watery stools containing blood and mucus is the symptom of lucerative colitis. **[Gujarat CET 2015]**
- Kwashiorkor disease is caused due to the protein deficiency and it is not accompanied by caloric deficiency. **[NEET 2019]**

Chapter Exercises

(A) Topical Exercise

Alimentary Canal

1 Among the given parts choose the one which is not originated from endoderm.
(a) stomach (b) oesophagus
(c) anus (d) small intestine

2 Which structure represents the roof of oral cavity?
(a) Uvula
(b) Labial frenulum
(c) Palate
(d) Lingual frenulum

3 All the given taste papillae are found in human tongue except
(a) circumvallate (b) foliate
(c) fungiform (d) filiform

4 Analysis of jawbones from the skeletal remains of a vertebrate animal reveal its dietary patterns owing to
(a) the position of muscle attachment sites
(b) the size of mouth opening
(c) the evidence of food molecules still present
(d) the prevalence of specific kinds of teeth

5 Which of the following teeth are lophodont?
(a) Incisor and canine
(b) Premolar and molar
(c) Canine and premolar
(d) Premolar and incisor

6 Consider the following statements.
I. Total number of canines in permanent dental set of human is six.
II. Wisdom teeth in humans are third molar and four in number.
Choose the correct option.
(a) Statement I is true, but II is false
(b) Statement I is false, but II is true
(c) Both statements I and II are true
(d) Both statements I and II are false

7 Dental formula in human beings is **[NCERT Exemplar]**
(a) $\frac{3223}{3223}$ (b) $\frac{2123}{2123}$
(c) $\frac{1232}{1232}$ (d) $\frac{2233}{2233}$

8 If the dental formula of rabbit is 2033/1023, what does it indicate?
(a) Total number of teeth in rabbit is 14
(b) Total number of incisors in rabbit is 3
(c) Diastema is present between incisors and premolars
(d) In the formula 2033 is for adult and 1023 is for young ones

9 Consider the following statements.
I. Pharynx is a junction of buccal cavity, nasal chamber, Eustachian tube and trachea.
II. Pharynx does not allow air to enter into trachea.
Choose the correct option.
(a) Statement I is true, but II is false
(b) Statement I is false, but II is true
(c) Both statements I and II are true
(d) Both statements I and II are false

10 Adenoids in children is caused due to the enlargement of
(a) tubal tonsils (b) palatine tonsils
(c) lingual tonsils (d) pharyngeal tonsils

11 Choose the incorrectly matched pairs.
I. Faucial tonsils - Lateral wall of oropharynx
II. Gullet - Lower opening of oesophagus
III. Tongue -Stratified squamous epithelium
IV. Waldeyer's ring - Arrangement of tonsils
Codes
(a) I and II (b) II only (c) III and IV (d) IV only

12 Cricopharyngeal sphincter is associated with
(a) oesophagus (b) stomach
(c) small intestine (d) pancreas

13 Select the incorrect statement.
(a) Human stomach in unilobed whereas rabbit stomach is bilobed
(b) The inner lining of stomach possesses numerous longitudinal folds called rugae
(c) The submucosa of stomach is thickest than other parts of alimentary canal
(d) Cardiac sphincter is much less developed in new born babies

14 The characteristic feature of valves of Kerckring is the
(a) presence of rugae
(b) presence of villi
(c) absence of villi
(d) presence of two sphincters

15 Label the given diagram of transverse section of mucosa of small intestine showing small finger-like projections. Choose the correct option accordingly.

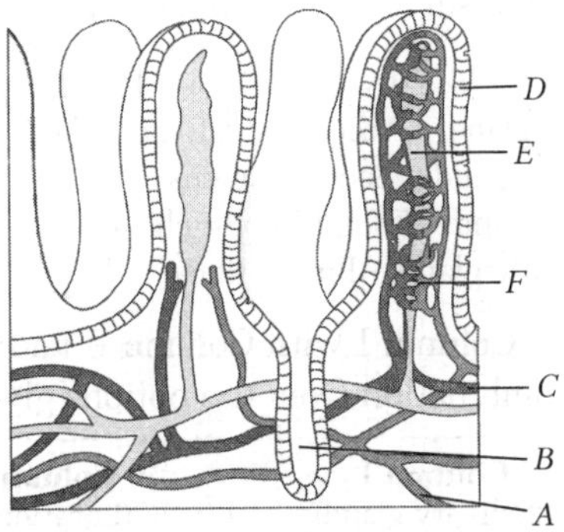

(a) *A*-Vein, *B*-Crypts, *C*-Artery, *D*-Villi, *E*-Lacteal, *F*-Capillaries
(b) *A*-Artery, *B*-Crypts, *C*-Vein, *D*-Villi, *E*-Capillaries, *F*-Lacteal
(c) *A*-Vein, *B*-Artery, *C*-Crypts, *D*-Villi, *E*-Capillaries, *F*-Lacteal
(d) *A*-Villi, *B*-Lacteal, *C*-Capillaries, *D*-Artery, *E*-Crypts, *F*-Vein

16 Which of the following is not true for intestinal villi? **[NCERT Exemplar]**
(a) They possess microvilli
(b) They increase the surface area
(c) They are supplied with capillaries and the lacteal vessels
(d) They only participate in digestion of fats

17 Microvilli in intestine serve to
(a) release digestive enzyme
(b) increase membrane area for absorption
(c) protect cells from invading microorganisms
(d) establish intercellular contact

18 Which region of alimentary canal has the largest internal surface area?
(a) duodenum
(b) stomach
(c) colon
(d) ileum

19 Besides buccal cavity, which among the following structure of human alimentary canal contain vestigial structure?
(a) Small intestine
(b) Pancreas
(c) Stomach
(d) Large intestine

20 Symbiotic microorganisms are found in
(a) vermiform appendix
(b) caecum
(c) small intestine
(d) haustra

21 Refer to the given diagram of large intestine and choose the odd one out in the following labelling.

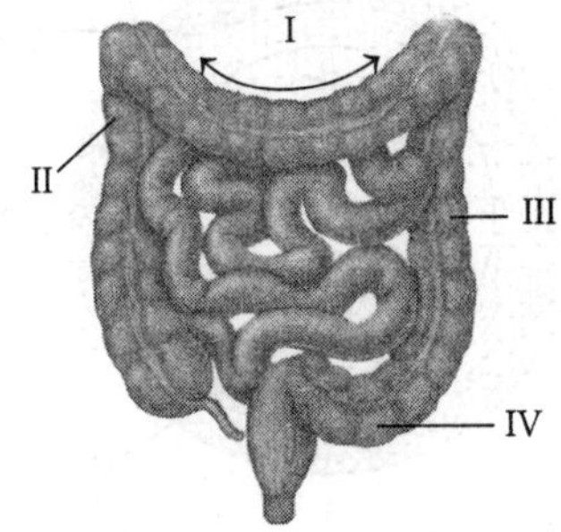

(a) I (b) II (c) III (d) IV

22 Match the two columns and select the right one among options given. **[NCERT Exemplar]**

Column I	Column II
A. Duodenum	1. A cartilaginous flap
B. Epiglottis	2. Small blind sac
C. Glottis	3. 'C'-shaped structure emerging from the stomach
D. Caecum	4. Opening of wind pipe

Codes

	A	B	C	D		A	B	C	D
(a)	1	2	3	4	(b)	4	3	2	1
(c)	3	1	4	2	(d)	2	4	1	3

23 Consider the following statements.
I. Taeniae coli are longitudinal bands in colon.
II. Ascending colon is the shortest part of colon.
Choose the correct option.
(a) Statement I is true, but II is false
(b) Statement I is false, but II is true
(c) Both statements I and II are true
(d) Both statements I and II are false

24 External sphincter of rectum is
(a) absent in humans
(b) involuntary in nature
(c) composed of striated muscles
(d) composed of smooth muscles

25 Match the two columns and select the correct among options given. **[NCERT Exemplar]**

Column I	Column II
A. Biomacromolecules of food	1. Alimentary canal and associated gland
B. Human digestive system	2. Embedded in jawbones
C. Stomach	3. Outer wall of visceral organs
D. Thecodont	4. Converted into simple substances
E. Serosa	5. J-shaped bag-like structure

Codes

	A	B	C	D	E		A	B	C	D	E
(a)	2	1	5	3	4	(b)	4	1	5	2	3
(c)	1	2	3	4	5	(d)	1	3	2	4	5

26 Recognise the figure and find out the correct matching.

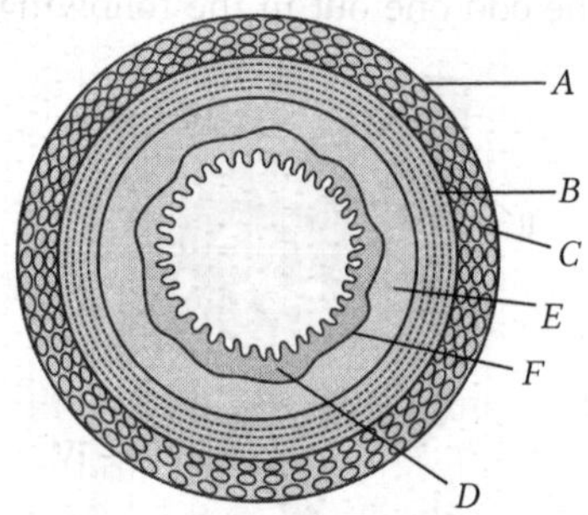

(a) A-Mucosa, B-Circular muscle, C-Longitudinal muscle, D-Submucosa, E-Muscularis, F-Serosa,
(b) A-Mucosa, B-Longitudinal muscle, C-Circular muscle, D-Submucosa, E-Muscularis, F-Serosa
(c) A-Serosa, B-Circular muscle, C-Longitudinal muscle, D-Lamina propria, E-Submucosa, F-Muscularis mucosa
(d) A-Serosa, B-Longitudinal muscle, C-Circular muscle, D-Muscularis, E-Submucosa, F-Mucosa

27 Match the following columns.

Column I (Layers of alimentary canal)	Column II (Characteristics)
A. Serosa membrane	1. Consists of Auerbach's plexus
B. Muscularis	2. Forms rugae in stomach
C. Submucosa	3. Contains Brunner's gland in intestine
D. Mucosa	4. Does not occur in oesophagus

Codes

	A	B	C	D
(a)	4	1	3	2
(b)	2	3	4	1
(c)	3	2	1	4
(d)	1	4	2	3

28 Select the incorrect statement.
(a) Meissner's plexus are found in the muscularis wall of alimentary canal
(b) Peyer's patches are found in the mucosa wall of intestine
(c) Muscularis externa contains smooth fibres throughout the oesophagus of humans
(d) Peyer's patches are abundantly found in ileum

Digestive Glands

29 A gland not associated with the alimentary canal is **[NCERT Exemplar]**
(a) pancreas (b) adrenal
(c) liver (d) salivary

30 Among the listed salivary glands, which one is not found in humans?
(a) Parotid (b) Sublingual
(c) Infraorbital (d) Submandibular

31 Which among the following enzyme is likely to be found in the Stenson's duct ?
(a) Pancreatic lipase (b) Salivary amylase
(c) Chymotrypsin (d) Enterokinase

32 Choose the incorrectly matched pair.
(a) Ebner's gland - Accessory salivary gland
(b) Weber's gland - Secrete mucus
(c) Nuhn's gland - Found in oesophagus
(d) Zygomatic gland - Found in cat

33 Match the Column I with Column II and choose the correct combination from the options given.

Column I (Salivary gland)	Column II (Location)
A. Submaxillary	1. Cheek
B. Sublinguals	2. Lower jaw
C. Parotids	3. Below the tongue

Codes

	A	B	C		A	B	C
(a)	1	2	3	(b)	3	1	2
(c)	2	3	1	(d)	2	1	3

34 Consider the following statements.
I. Minute quantity of salivary lipase is secreted by Ebner's glands.
II. Secretion of saliva is controlled automatically.

Choose the correct option.
(a) Statement I is true, but II is false
(b) Statement I is false, but II is true
(c) Both statements I and II are true
(d) Both statements I and II are false

35 Which of the following statements is incorrect with respect to gastric glands ?
(a) Oxyntic cells are also called parietal cells
(b) The secretions of goblet cells are acidic in nature
(c) G cells in pyloric region secrete gastrin
(d) Oxyntic glands are most numerous gastric glands

36 What is the function of enterochromaffin cells in human's gut ?
(a) Secretion of Castle's intrinsic factor
(b) Gastrin secretion
(c) Serotonin and histamine secretion
(d) All of the above

37 Select the incorrect statement.
(a) Crypts of Lieberkuhn are found in ileum of small intestine only
(b) Brunner's glands are exclusively found in duodenum of small intestine
(c) The pH of infant's gastric juice is more than that adults
(d) Paneth cells are found at the base of crypts of Lieberkuhn

38 Succus entericus is the name given to **[NCERT Exemplar]**
(a) functions between ileum and large intestine
(b) intestinal juice
(c) swelling in gut
(d) appendix

39 What change occur in the pH of the secretions of digestive glands in mouth, stomach and intestine?
(a) pH of secretions decreases along the mouth, stomach and intestine in linear pattern
(b) pH of secretions increases in stomach and then decreases again in intestine
(c) pH of secretions increases along the mouth, stomach and intestine in a linear pattern
(d) pH of secretions decreases in stomach and then increases again in intestine

40 Match the following columns.

Column I	Column II
A. Oxyntic cells	1. Secrete somatostatin and gastrin
B. Chief cells	2. Secrete Castle's intrinsic factor
C. Argentaffin cells	3. Secrete proenzymes

Codes

	A	B	C		A	B	C
(a)	3	1	2	(b)	2	3	1
(c)	3	2	1	(d)	1	2	3

41 Falciform ligament in human's liver joins
(a) its three lobes (b) numerous hepatic lobules
(c) its two lobes (d) Both (b) and (c)

42 The duct of gall bladder is called
(a) cystic duct (b) hepatic duct
(c) common bile duct (d) hepato-pancreatic duct

43 The cystic duct along with the hepatic duct from liver forms the
(a) hepato-pancreatic duct (b) duct of Santorini
(c) common bile duct (d) duct of Rivinus

44 The bile duct and the pancreatic duct open together into the duodenum as the
(a) hepato-pancreatic duct (b) duct of Santorini
(c) common bile duct (d) duct of Rivinus

45 Hepato-pancreatic duct opens into the duodenum and carries **[NCERT Exemplar]**
(a) bile (b) pancreatic juice
(c) Both (a) and (b) (d) saliva

46 Liver is the largest gland and is associated with various functions. Choose which is not correct from the following? **[NCERT Exemplar]**
(a) Metabolism of carbohydrate
(b) Digestion of fat
(c) Formation of bile
(d) Secretion of hormone called gastrin

47 Consider the following statements.
I. The food stored in liver is in the form of glycogen.
II. Liver store vitamin-A and B but not fats.
Choose the correct option.
(a) Statement I is true, but II is false
(b) Statement I is false, but II is true
(c) Both statements I and II are true
(d) Both statements I and II are false

48 Identify the statement which is not a function of liver.
(a) Liver acts as blood filter by removing the worn out and aged corpuscles
(b) Hepatocytes secrete heparin which prevents clotting of blood in blood vessels
(c) Liver synthesises vitamin-A from carotene and stores vitamins-B_6, B_7 and B_{10}
(d) The extensive network of liver sinusoids acts as a blood reservoir

49 The function of the gall bladder is to
(a) store bile produced in the liver
(b) produce hormones essential for maintenance of glucose levels in the blood
(c) act as a backup if the urinary tract becomes too full
(d) sense changes in blood pressure and the oxygen concentration in blood

50 After cholecystectomy,
(a) fats are not digested efficiently
(b) protein-rich diet should be avoided
(c) saturated fatty acids should be taken in diet
(d) carbohydrates should be avoided maximally

51 Which of the statements describes pancreas correctly?
(a) It is a pear-shaped flattened structure
(b) Pancreas is a microscopic, tubular gland
(c) It is an elongated, irregularly branched, flattened digestive gland
(d) Pancreas is an elevated, branched, balloon-shaped digestive gland

52 Mark the correct statement regarding pancreas.
(a) It produces three digestive enzymes and one hormone
(b) It produces three digestive enzymes and two hormones
(c) It produces two digestive enzymes and one hormone
(d) It produces two digestive enzymes and no hormone

53 Identify the statement which is incorrect about pancreas?
(a) It is a mixed gland as it secretes both enzymes and hormones
(b) its endocrine part contain numerous acinar cells
(c) It is made up of numerous acini
(d) Each acinar is lined by pyramidal cells

54 Duct of Wirsung abundantly contains
(a) salivary amylase (b) pancreatic juice
(c) intestinal juice (d) hydrochloric acid

Functioning of Digestive System

55 Which one of the following digestive enzymes is secreted in its active form?
(a) Carbohydrase (b) Proteolytic enzyme
(c) Nuclease (d) Lipase

56 What is common among pepsin, peptidase and trypsin?
(a) All are not proteins (b) Proteolytic enzymes
(c) Produced in stomach (d) Act at pH lower than 7

57 Saliva is bactericidal because it contains
(a) kallidin (b) ptyalin
(c) thiocynate and lysozyme (d) serotonin

58 The food which passes through oesophagus is called
(a) chyle (b) chylomicron
(c) chyme (d) bolus

59 Which digestive process takes place in the mouth?

	Chemical digestion	Physical digestion
(a)	✓	✓
(b)	✓	✗
(c)	✗	✓
(d)	✗	✗

60 Chyme differs from bolus in containing
(a) salivary amylase and thiocyanate
(b) gastric juice
(c) IgA antibody and HCl
(d) IgA and salivary amylase

61 Gastric juice contains **[NCERT Exemplar]**
(a) pepsin, lipase, rennin (b) trypsin, lipase, rennin
(c) trypsin, pepsin, lipase (d) trypsin, pepsin, rennin

62 Match the column I (enzymes) with column II (their respective substrate) and choose the right one among options given. **[NCERT Exemplar]**

Column I	Column II
A. Lipase	1. Carbohydrates
B. Nucleases	2. Fats
C. Carboxypeptidase	3. Nucleic acids
D. Glycosidases	4. Proteins, peptones and proteoses

Codes

	A	B	C	D
(a)	2	3	1	4
(b)	3	4	2	1
(c)	3	1	4	2
(d)	2	3	4	1

63 Mark the right statement among the following. **[NCERT Exemplar]**
(a) Trypsinogen is an inactive enzyme
(b) Trypsinogen is secreted by intestinal mucosa
(c) Enterokinase is secreted by pancreas
(d) Bile contains trypsin

64 Pepsin acts in
(a) basic medium (b) acidic medium
(c) neutral medium (d) All types of media

65 The amount of bile released is proportionate to the amount of
(a) fat in meal (b) protein in meal
(c) carbohydrate in meal (d) All of these

66 Digestive enzyme not found in human pancreatic juice is
(a) nucleases (b) nucleotidase
(c) trypsin (d) lipase

67 Which does not occur in pancreatic juice?
(a) Pepsinogen (b) Trypsinogen
(c) Chymotrypsin (d) Lipase

68 Enzyme maltase of human gut acts on food at
(a) pH less than 7, changes maltose to fructose
(b) pH more than 7, changes maltose to glucose
(c) pH less than 7, changes starch to maltose
(d) pH more than 7, changes starch to maltose

69 Just as hydrochloric acid activator of pepsinogen, so is
(a) haemoglobin activator of oxygen
(b) enterokinase activator of trypsinogen
(c) bile juice activator of fat
(d) glucagon activator of glycogen

70 Trypsin and pepsin cannot hydrolyse
(a) protein (b) keratin (c) fibrinogen (d) casein

71 Mark the incorrect statement.
(a) Lipid hydrolysis produces glycine and glycerol
(b) HCl in gastric juice inactivates ptyalin and activates pepsin
(c) pH of gastric juice is acidic than that of pancreatic juice
(d) Enzyme lactase is found in succus entericus

72 Gastric secretion and mobility is inhibited by
(a) gastrin and enterocrinin
(b) enterogastrone and vasoactive intestinal peptide
(c) villikinin and somatostatin
(d) duocrinin

73 Which one of the following hormones inhibits the release of pancreatic juice?
(a) Somatostatin (b) Secretin
(c) Enterocrinin (d) Pancreatic polypeptide

74 Given below are the enzymes found in intestinal juices and the end products yield by them.
I. Exopeptidase - Dipeptides
II. Lactase - Glucose and fructose
III. α-dextrinase - Glucose
IV. Maltase - 2 molecules of glucose
Choose the incorrectly matched pairs.
(a) I and II (b) III and IV
(c) II and III (d) I and III

75 Functional unit of food absorption is

(a) rugae (b) villi
(c) Peyer's patches (d) crypts of Lieberkuhn

76 Where do chemical digestion and absorption of food occur in the human alimentary canal ?

	Chemical digestion	Absorption
(a)	Oesophagus	Colon
(b)	Oesophagus	Stomach
(c)	Stomach	Liver
(d)	Stomach	Ileum

77 Actively absorbed food substances are

(a) amino acids (b) glucose
(c) galactose (d) All of these

78 Match the following columns.

	Column I		Column II (Absorbed by)
A.	Chloride ions	1.	Simple diffusion
B.	Fructose	2.	Facilitated transport
C.	Sodium ions	3.	Active transport
D.	Amino acids	4.	Simple diffusion or facilitated or active transport

Codes

	A	B	C	D		A	B	C	D
(a)	1	2	3	4	(b)	2	3	4	1
(c)	3	4	1	2	(d)	4	1	2	3

79 Consider the following statements.

I. Amino acids and lipids are absorbed in blood capillaries.

II. Chyle is white-coloured fluid left in lacteals after the absorption of fatty acids.

Choose the correct option.

(a) Statement I is true, but II is false
(b) Statement I is false, but II is true
(c) Both statements I and II are true
(d) Both statements I and II are false

80 Bicarbonate ions are released in exchange of chloride ions in ileum from

(a) villi
(b) crypts of Lieberkuhn
(c) brush border cell
(d) valvulae conniventes/ plicae circulares

81 Which one is absorbed in the alimentary canal without any breakdown?

(a) Protein (b) Polysaccharide
(c) Fat soluble vitamin (d) Albumin of egg

82 Consider the following statements.

I. Decomposition of amino acids in colon by bacteria produce foul smelling of faeces.

II. Faeces are brown in colour due to presence of degraded substances of bilirubin in it.

Choose the correct option.

(a) Statement I is true, but II is false
(b) Statement I is false, but II is true
(c) Both statements I and II are true
(d) Both statements I and II are false

83 In infants, external and internal anal sphincter are

(a) voluntary
(b) involuntary
(c) voluntary and involuntary, respectively
(d) involuntary and voluntary, respectively

84 Identify the statement which is incorrect about hormonal control of digestion?

(a) Enterocrinin stimulates the crypts of Lieberkuhn to release enzyme into the intestinal juice
(b) Vasoactive intestinal peptide dilates peripheral blood vessels of gut
(c) Pancreatic polypeptide inhibits release of pancreatic juice from the pancreas
(d) The pH of succus entericus varies between 6.5-7.0

85 Choose the incorrectly matched pair.

(a) Secretin - Duodenal epithelium
(b) Villikinin - Entire epithelium of large intestine
(c) Pancreatic polypeptide - F-cells of islet of Langerhans
(d) Gastrin - G-cells of stomach

86 Choose the odd one out.

(a) Somatostatin
(b) Duocrinin
(c) Secretin
(d) Vasoactive intestinal peptide

Nutrition

87 Which of the following best describes the process of nutrition?

(a) A process to obtain necessary energy and growth substances
(b) A process to obtain energy from foods
(c) A process to supply the necessary nutritive elements to body
(d) A sum total of processes which provide the necessary nutritive element for growth, maintenance and to meet the need of energy

88 Excess of carbohydrates are

(a) changed into proteins (b) changed into glycogen
(c) aminated (d) passed out in stool

89 Man cannot digest cellulose but cow and other herbivorous animals can digest food containing cellulose because

(a) they have enzyme cellulase in their stomach
(b) they masticate it well by chewing teeth
(c) they have bacteria in their alimentary canal which digest cellulose
(d) None of the above

90 Identify the incorrect statement with respect to carbohydrates.
(a) Carbohydrates are digested in the alimentary canal in the form of monosaccharides
(b) Their calorific value is less than that of proteins and lipids
(c) All polysaccharides can be digested by human beings
(d) Normal content of glucose in blood is 80-100 mg/100 mL

91 X is a macronutrient whose major sources are nuts, fish and legumes. Its daily requirement in human body is about 70-100 g. Identify X.
(a) Carbohydrate (b) Lipid
(c) Protein (d) Vitamin

92 Vitamins are
(a) inorganic substances which cannot be synthesised by animals
(b) inorganic substances which can be synthesised by animals
(c) organic substances which cannot be synthesised by animals
(d) organic substances which can be synthesised by animals

93 Vitamins are classified on the basis of
(a) physiochemical properties
(b) solubility in water and organic solvents
(c) solubility in fats and water
(d) Both (a) and (c)

94 Which one of the following is the correct matching of a vitamin, its nature and its deficiency disease?
(a) Vitamin-A—Fat soluble— Night blindness
(b) Vitamin-K—Fat soluble—Beri-beri
(c) Vitamin-A—Fat soluble—Beri-beri
(d) Vitamin-K—Water soluble—Pellagra

95 Which one of the following is a fat soluble vitamin and its related deficiency disease?
(a) Ascorbic acid – Scurvy
(b) Retinol – Xerophthalmia
(c) Cobalamin – Beri-beri
(d) Calciferol – Pellagra

96 Osteomalacia is a deficiency disease of
(a) infants due to protein energy malnutrition
(b) adults due to protein energy malnutrition
(c) infants due to vitamin-D/calcium deficiency
(d) adults due to vitamin-D/calcium/phosphorus deficiency

97 Identify the statement which is true about vitamin-A
(a) It regulates metabolism of calcium and phosphorus
(b) It plays an important role in healing of wound
(c) Animal liver is rich in vitamin-A
(d) It facilitates DNA repair

98 Select the incorrect statement.
(a) Vitamin-B_5 is recently discovered vitamin having anticancer properties
(b) Vitamin-C, E and A are antioxidant vitamins
(c) Tocopherol does not belong to vitamin-B group
(d) Vitamin-E is also known as antisterility vitamin

99 Loss of appetite, gastrointestinal disorders and muscular weakness are deficiency symptoms of
(a) vitamin-B (b) vitamin-A (c) vitamin-E (d) vitamin-C

100 The richest sources of vitamin-B_{12} are
(a) goat's liver and *Spirulina*
(b) chocolate and green gram
(c) rice and hen's egg
(d) carrot and chicken's breast

101 Consider the following statements.
I. Ergocalciferol is a type of vitamin-D.
II. Deficiency of tocopherol causes, muscular dystrophy and anaemia.

Choose the correct option.
(a) Statement I is true, but II is false
(b) Statement I is false, but II is true
(c) Both statements I and II are true
(d) Both statements I and II are false

102 Which one of the following compounds does not directly provide energy, but is still required by the body in little quantity?
(a) Antigen (b) Antibody
(c) Vitamin (d) Carbohydrates

103 Match the following columns.

Column I	Column II
A. Vitamin-B_1	1. Thiamine
B. Vitamin-B_3	2. Niacin
C. Vitamin-A	3. Retinol
D. Vitamin-B_2	4. Riboflavin

Codes

	A	B	C	D		A	B	C	D
(a)	1	2	3	4	(b)	1	3	4	2
(c)	3	4	2	1	(d)	4	3	2	1

104 Identify the statement which is incorrect about lipoic acid.
(a) Cereals and germ oil are a rich source of this vitamin
(b) It works as coenzyme for decarboxylation of pyruvic acid to α-ketoglutaric acid
(c) It helps in normal growth of body
(d) It is also termed as oxidative factor

105 Which one of the following pairs is characterised by swollen lips, thick pigmented skin of hands and legs and irritability?
(a) Iodine –Goitre (b) Protein – Kwashiorkor
(c) Thiamine – Beri-beri (d) Nicotinamide –Pellagra

106 Cheilosis is caused by deficiency of vitamin-B_2 which leads to
(a) dryness of skin
(b) inflammation of tongue
(c) cracking of corners of mouth
(d) ulceration of mouth

107 Vitamin-B_{10} plays an important role in
(a) DNA synthesis
(b) fatty acid synthesis
(c) RNA synthesis and erythrocyte maturation
(d) Both (a) and (c)

108 Which vitamin is administered for cure of cardiac and vascular disorders?
(a) Vitamin-B_{10} (b) Vitamin-B_{15}
(c) Vitamin-B_{12} (d) Vitamin-B_7

109 Vitamin-K is useful for
(a) converting prothrombin to thrombin
(b) synthesis of prothrombin
(c) calcium combination with prothrombin
(d) All of the above

110 Which of the following is an incorrect match?
(a) Vitamin-B_1—Beri-beri
(b) Niacin—Pellagra
(c) Vitamin-K—Sterility
(d) None of the above

111 Among the listed macroelements, the daily requirement in human body is maximum for
(a) Sodium (b) Calcium
(c) Potassium (d) Magnesium

112 Hyponatremia refers to the deficiency of
(a) Phosphorus (b) Calcium
(c) Vitamin (d) Sodium

113 Match the following columns.

	Column I (Minerals)		Column II (Deficiency symptoms)
A.	Selenium	1.	Diabetes
B.	Calcium	2.	Ulcers in stomach
C.	Chromium	3.	Rickets
D.	Chlorine	4.	Premature ageing

Codes

	A	B	C	D		A	B	C	D
(a)	2	1	4	3	(b)	3	2	4	1
(c)	4	3	1	2	(d)	1	4	3	2

114 Identify the statement which is false about potassium.
(a) It is the principal univalent cation inside a living cell
(b) It is found in green vegetables and dates
(c) It helps in cardiac function
(d) Its deficiency causes muscular and nervous disorders

115 Which one is the correct match?
(a) Calcium—Beri-beri (b) Na—Addison's disease
(c) Iron—Haemophilia (d) P—Bone fragility

116 Choose the incorrectly matched pairs.
(a) Cobalt - RBC formation
(b) Copper - Iron utilisation
(c) Fluorine - Prevent tooth decay
(d) None of the above

117 Deficiency of which microelement can cause infertility and irregular menstruation in females?
(a) Zinc (b) Chromium
(c) Manganese (d) Copper

Nutritional and Digestive Disorders

118 Which of the following is not a common disorder associated with digestive system? **[NCERT Exemplar]**
(a) Tetanus (b) Diarrhoea
(c) Jaundice (d) Dysentery

119 Which disease does not occur in children younger than six months?
(a) Jaundice (b) Kwashiorkor
(c) Marasmus (d) Both (b) and (c)

120 Dyspepsia is
(a) PEM (b) hypercholesterolemia
(c) indigestion (d) jaundice

121 Haustral churning is affected in
(a) jaundice (b) obesity
(c) marasmus (d) constipation

122 Intake of milk free diet can help to prevent
(a) galactosemia (b) gall stones
(c) lactose intolerance (d) Both (a) and (c)

123 Choose the incorrect statement.
(a) Vomiting may occur due to pyloric stenosis
(b) Neonatal jaundice is caused due to poor functioning of liver in initial days
(c) Belching occurs when oesophagus is over dilated
(d) Absence of enzyme uridyl transferase in the main cause of galactosemia

(B) Cumulative Exercise

(Assorted Questions of the Chapter for Advanced Level Practice)

1 Examination of blood of a person suspected of having anaemia shows large, immature, nucleated erythrocytes without haemoglobin. Supplementing his diet with which of the following is likely to alleviate these symptoms?

(a) Thiamine (b) Folic acid and cobalamin
(c) Riboflavin (d) Iron compounds

2 Which of the following is/are correct for absorption from the stomach?

I. The substances normally absorbed from the stomach are some water, glucose and considerable amount of alcohol.
II. The substances are absorbed through the walls of the stomach.
III. Drugs are not absorbed in human alimentary canal.
IV. Some amount of vitamins are absorbed.

(a) Only I (b) I and II
(c) III and IV (d) Only IV

3 Match the following nutrition/vitamin deficiencies in Column I with causes/deficiencies in Column II and choose the correct option from the codes given below.

Column I	Column II
A. Kwashiorkor	1. Iron
B. General anaemia	2. Menadione
C. Dermatitis	3. Protein
D. Marasmus	4. Pyridoxine
E. Bleeding	5. Biotin

Codes

	A	B	C	D	E
(a)	1	5	2	3	4
(b)	5	2	3	4	1
(c)	2	3	4	1	5
(d)	3	1	5	4	2

4 Identify the incorrectly matched pair.

(a) Parotid glands – Stenson's duct
(b) Submandibular glands – Wharton's duct
(c) Sublingual glands – Duct of Rivinus
(d) Infraorbital glands – Duct of Wirsung

5 Which one of the following statements is incorrect regarding digestion and absorption of food in humans?

(a) Fructose and amino acids are absorbed through intestinal mucosa with the help of carrier ions like Na^+
(b) Steapsin is the pancreatic amylase
(c) Taeniae coli are associated with small intestine
(d) Bile acids are steroids in nature

6 Go through the following statements regarding the absorption of fats and choose an option accordingly from the given below.

I. Micelles and chylomicron are concerned with the absorption of fats.
II. Chylomicrons are water soluble droplets of fat which contain triglycerides, sterols and phospholipids.
III. Micelles are water insoluble droplets of fatty acids and glycerol.
IV. Chylomicron, protein coated small vesicles are released from the intestinal cells into the blood stream by lacteals.

(a) II and IV correct, while I and III are incorrect
(b) I, III and IV are correct, while II is incorrect
(c) I, II and IV are correct, while III is incorrect
(d) IV and I are correct, while II and III are incorrect

7 Match the following columns.

Column I	Column II
A. Crypts of Lieberkuhn	1. Loop of duodenum
B. Pancreas	2. Stomach
C. Adrenal gland	3. Intestine
D. Gastric gland	4. Kidney

Codes

	A	B	C	D		A	B	C	D
(a)	3	1	2	4	(b)	3	1	4	2
(c)	2	4	3	1	(d)	1	3	4	2

8 Go through the following statements regarding starch digestion and select the correct option accordingly.

I. Digestion of starch starts from the mouth.
II. Around 30% of the starch is digested in the stomach.
III. Digestion of food requires the action of pancreatic juices.
IV. Digestion of food is completed in the longest part of the alimentary canal.

(a) All are true
(b) I, III and IV are true, while II is false
(c) II and III are false, while I and IV are true
(d) II and IV are false, while I and III are true

9 Match the following columns.

Column I (Types of cell)	Column II (Secretions)
A. Beta cells	1. Lysozyme
B. Mast cells	2. Mucus
C. Paneth cells	3. Histamine
D. Acinar cells	4. Insulin
	5. Pancreatic enzymes

Codes

	A	B	C	D		A	B	C	D
(a)	4	2	1	5	(b)	5	2	3	4
(c)	4	3	1	5	(d)	2	3	1	5

10 Consider the following statements.

I. Lower surface of the tongue has little projection which bears taste buds.

II. Sphincter of oddi guards and regulates the opening of stomach into duodenum.

Choose the correct option.

(a) Statement I is true, but II is false
(b) Statement I is false, but II is true
(c) Both statements I and II are true
(d) Both statements I and II are false

11 Read the following statements thoroughly and identify whether they are true and false. Choose the right options accordingly.

I. Bile is produced and stored in the liver and gall bladder, respectively.

II. Common bile duct is the fusion of all the right and left hepatic ducts.

III. Pancreas consists of two parts, exocrine and endocrine, which secrete insulin and glucagon hormone and pancreatic juices containing enzymes, respectively.

(a) All statements are true
(b) All statements are false
(c) Statements I and III are true, while II is false
(d) Statements I and II are true, while III is false

12 Match the following columns.

Column I	Column II
A. Retinol	1. Inhibits oxidation of unsaturated fatty acids
B. Tocopherols	2. Absorption of calcium ions
C. Calciferol	3. Essential for maintenance of epithelial tissue
D. Ascorbic acid	4. Required for amino acid metabolism

Codes

	A	B	C	D		A	B	C	D
(a)	3	2	4	1	(b)	2	3	1	4
(c)	3	1	2	4	(d)	4	1	2	3

13 Go through the following statements regarding the absorption of nutrients and choose the correct option accordingly.

I. Absorption of monosaccharides, alcohol some water and medicines like aspirin occurs in the stomach.

II. Fatty acids cannot be absorbed directly.

III. Glycerol can reach into the blood directly.

IV. Maximum absorption of water (90%) takes place in the small intestine.

(a) I, II and IV are true, while III is false
(b) I, II, III are true, while IV is false
(c) II, III and IV are false, while I is true
(d) I and II are false, while III and IV are true

14 During prolonged fasting first

(a) carbohydrates are used up and next fat is withdrawn and in the last proteins are metabolised
(b) fats are used up, next carbohydrates are withdrawn and then the proteins are withdrawn
(c) lipids are used up, proteins and finally carbohydrates
(d) enzymes are used up and next minerals are absorbed

15 Identify the correct statements and choose a correct option accordingly.

(a) If breastfeeding is replaced by less nutritive food lacking protein and calories, the infant are likely to suffer from marasmus and kwashiorkor
(b) Bile salts of bile juice activates enzyme lipase
(c) Lipase present in the pancreatic juices is the principal enzyme for digestion of fat
(d) Medulla oblongata of hindbrain controls reflex action of vomiting

16 Milk casein $\xrightarrow{Y}$ Paracasein $\rightarrow + M \rightarrow$ Calcium paracaseinate (curdling of milk),

In the above equation, letter '*Y*' and '*M*' denote.

(a) rennin and HCl, respectively
(b) Ca^{2+} and rennin, respectively
(c) rennin, HCl and Ca^{2+}, respectively
(d) rennin and Ca^{2+}, respectively

17 Which action of digestive enzyme is/are correct regarding its site of action, substrate and the end product?

Enzyme	Site of action	Substrate	End product
I. Rennin	Calf's stomach	Casein	Ca paracaseinate
II. Pepsin	Human's stomach	Proteins	Two or more molecule of amino acids
III. Nuclease	Small intestine	Nucleosides	Nucleosides and iPO
IV. Enterokinase	Small intestine	Trypsinogen	Trypsin

Codes

(a) All are correct
(b) I, II and III are correct
(c) I, II and IV are correct
(d) Only III is correct

18 The diagram shows the effect of pH on the activity of two enzymes, a protease and an amylase, in the alimentary canal.

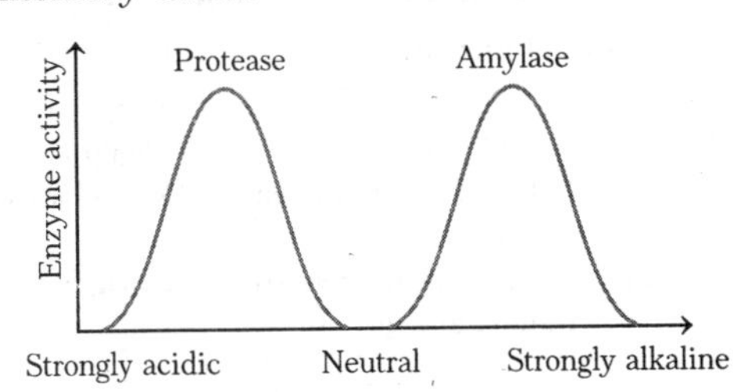

In which regions of the alimentary canal would these enzymes be most active?

	Protease	Amylase
(a)	Duodenum	Colon
(b)	Duodenum	Stomach
(c)	Stomach	Colon
(d)	Stomach	Duodenum

19 Match the following columns.

Column I	Column II
A. Kupffer's cells	1. Islets of Langerhans
B. β- cells	2. Liver sinusoids
C. Brush border cells	3. Thyroid gland
D. Paneth cells	4. Duodenum
	5. Small intestine

Codes

	A	B	C	D		A	B	C	D
(a)	3	1	4	2	(b)	4	5	1	2
(c)	2	1	5	4	(d)	3	5	4	1

20 The following is a scheme showing the fate of carbohydrates during digestion in the human alimentary canal. Identify the enzymes acting at stages indicated as *A, B, C* and *D*. Choose the correct option.

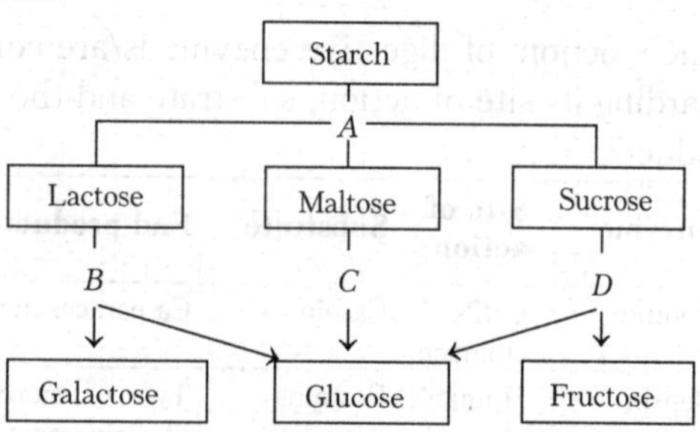

(a) *A*-Amylase, *B*- Maltase, *C*- Lactase, *D*- Invertase
(b) *A*-Amylase, *B*-Maltase, *C*- Invertase, *D*- Lactase
(c) *A*-Amylase, *B*-Invertase, *C*-Maltase, *D*-Lactase
(d) *A*- Amylase, *B*- Lactase, *C*-Maltase, *D*- Sucrase

21 Which of the following is a fat soluble vitamin and its related deficiency disease ?
(a) Retinol - Xerophthalmia
(b) Cobalamin - Beri-Beri
(c) Calciferol - Pellagra
(d) Ascorbic acid - Scurvy

22 Which of the following is correct match of vitamins, their functions and deficiency diseases?

	Vitamin	Functions	Diseases
(a)	Riboflavin	Antioxidant	Sterility
(b)	Biotin	Cell growth	Cheilosis
(c)	Tocopherol	Coenzyme of protein synthesis	Muscular dystrophy
(d)	Niacin	DNA repair, converts food into glucose	Pellagra

23 Read the following statements and identify the incorrect statements.
I. Small intestine is distinguished into three regions-a 'U' shaped duodenum, a long coiled middle portion jejunum and a shortest, uncoiled ileum.
II. Oesophagus opens into the cardiac part of stomach.
III. Arrangement of teeth in each half of upper and lower jaw is represented by dental formula.
IV. Rectum is a small sac which hosts some symbiotic microorganisms.
(a) I and IV
(b) II and IV
(c) II and III
(d) I and III

24 Match the following columns.

Column I (Structures)	Column II (Associated structures)
A. Frenulum	1. Lips
B. Rugae	2. Tongue
C. Uvula	3. Hard palate
D. Orbicularis oris muscle	4. Soft palate

Codes

	A	B	C	D
(a)	4	2	1	3
(b)	3	1	2	4
(c)	2	3	4	1
(d)	1	4	3	2

25 What will happen if secretion of parietal cells of gastric glands are blocked?
(a) Gastric juice is deficient of pepsinogen
(b) Gastric juice is deficient of chymosin
(c) Deficiency of HCl leads to non-conversion of inactive pepsinogen into active pepsin
(d) Non-release of enterokinase from duodenal mucosa and non-conversion of trypsinogen into trypsin

26 The secretion of a less than normal amount of the hormone gastrin would have which effect?
(a) Stomach pH would decrease
(b) Protein digestion in the stomach would decrease
(c) Carbohydrate digestion in the stomach would increase
(d) Secretion of alkaline mucus in the stomach lining would increase

27 Which of the following is incorrectly paired with its function?
(a) Small intestine — enzymatic digestion of lipids proteins and polysaccharides
(b) Oesophagus — transport food from the mouth to the stomach
(c) Large intestine — absorption of undigested sugars, fats and amino acids
(d) Stomach — mechanical and some chemical digestion of food

28 The diagram shows part of the human alimentary canal. Which two structures produce substances involved in the digestion of fat?

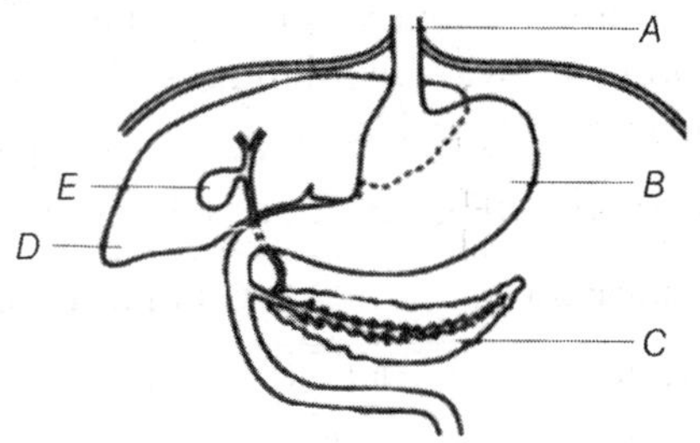

(a) A and E (b) C and D (c) B and C (d) D and E

29 Digestion of food involves breaking down of food components into smaller molecules by enzymes. These enzymes are active only at certain hydrogen ion concentrations. As a result certain food combinations can facilitate or retard the process of digestion. Of the following combinations, one that can result in very efficient digestion is

(a) meal with high proteins and acid fruits
(b) meal with high starch and high proteins
(c) meal with high starch and acid fruits
(d) meal with high fat and high proteins

30 Find out the correct sequence of substrate, enzyme and product.

(a) Proteins $\xrightarrow{\text{Pepsin}}$ Amino acid

(b) Fat $\xrightarrow{\text{Lipase}}$ Micelle

(c) Triglycerides $\xrightarrow{\text{Trypsin}}$ Monoglycerides

(d) Starch $\xrightarrow{\text{Amylase}}$ Maltose

31 Chymotrypsinogen, trypsinogen and procarboxypeptidase are

(a) inactive forms of enzyme in gastric juices
(b) active enzymes of intestinal juices
(c) inactive enzymes of pancreatic juices
(d) active enzymes of intestinal juices

32 The diagram represents a villus.

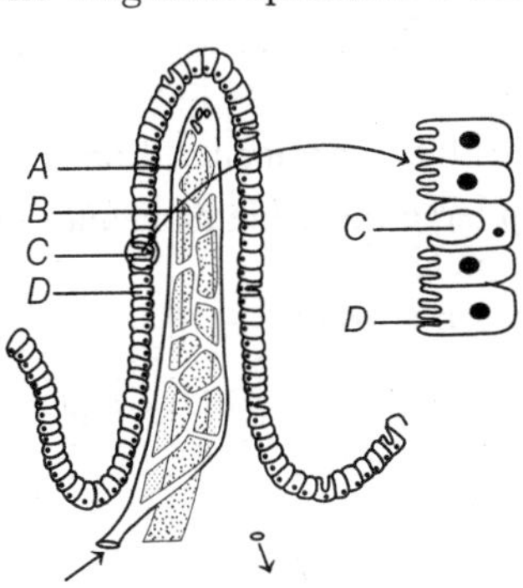

Which sequence correctly describes the functions of the labelled parts?

	A	*B*	*C*	*D*
(a)	absorbs digested fats	absorbs glucose	produces enzymes	produces mucus
(b)	absorbs digested fats	absorbs glucose	produces mucus	produces enzymes
(c)	absorbs glucose	absorbs digested fats	produces enzymes	produces mucus
(d)	absorbs glucose	absorbs digested fats	produces mucus	produces enzymes

33 Segmentation in the small intestine

(a) increases the absorption of nutrients by mixing food
(b) increases the contact of food with the walls of intestine
(c) increases the absorption of nutrients by mixing and increasing contact of food with the walls of intestine
(d) None of the above

34 The diagram shows some food moving along the alimentary canal, by peristalsis.

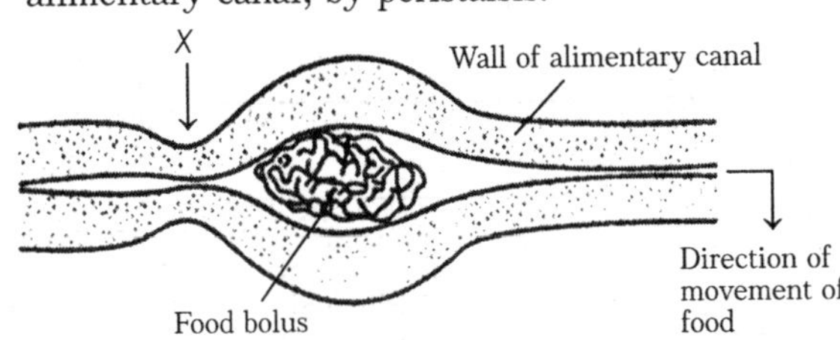

In which state are the muscles in the wall of the alimentary canal at point *X*?

	Circular muscles	**Longitudinal muscles**
(a)	Contracting	Contracting
(b)	Contracting	Relaxing
(c)	Relaxing	Contracting
(d)	Relaxing	Relaxing

35 Identify the statement which is incorrect about absorption of lipids.

(a) The lipids are absorbed in lymph capillaries present within the villi
(b) From the micelles, fatty acids and glycerol are absorbed into the intestinal cells by active transport
(c) Micelles are formed with the help of bile salts and phospholipids in intestinal lumen
(d) The lacteals after absorption of lipids contain chyle

36 Consider the following statements.

I. No absorption of food takes place in mouth and oesophagus.
II. Absorption of H_2O, alcohol, simple salts, glucose and chloride takes place in the stomach to a slight extent.

Choose the correct option.

(a) Statement I is true, but II is false
(b) Statement I is false, but II is true
(c) Both statements I and II are true
(d) Both statements I and II are false

37 Which one statement is incorrect regarding the process of digestion and absorption in humans?

(a) Small intestine is the major site for the absorption of all nutrients
(b) Around 90% of the total absorption of nutrients takes place in the distal part of the small intestine
(c) Drugs, alcohols , little water and salt are absorbed in the stomach through the mucous membrane
(d) Large intestine is the site of absorption for water and products of bacterial digestion

38 Identify the type of transport marked as *A, B, C, D* and *E.*

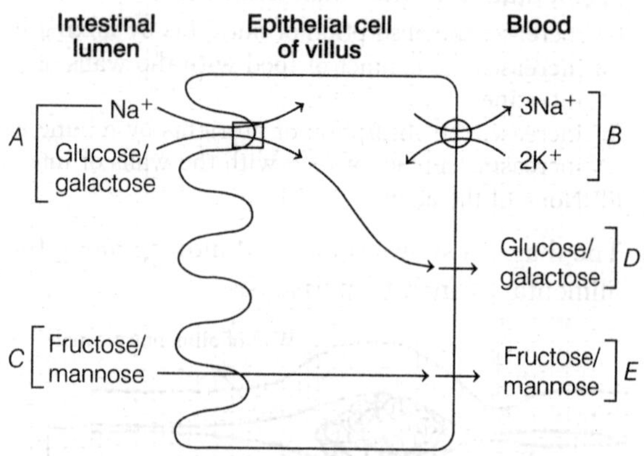

	A	*B*	*C*	*D*	*E*
(a)	Primary active transport	Secondary active transport	Facilitated diffusion	Facilitated diffusion	Facilitated diffusion
(b)	Facilitated diffusion	Secondary active transport	Primary active transport	Facilitated diffusion	Facilitated diffusion
(c)	Diffusion	Facilitated diffusion	Primary active transport	Secondary active transport	Facilitated diffusion
(d)	Secondary active transport	Primary active transport	Facilitated diffusion	Diffusion	Diffusion

39 The graph shows the pH of the saliva in the mouth after eating sugar.

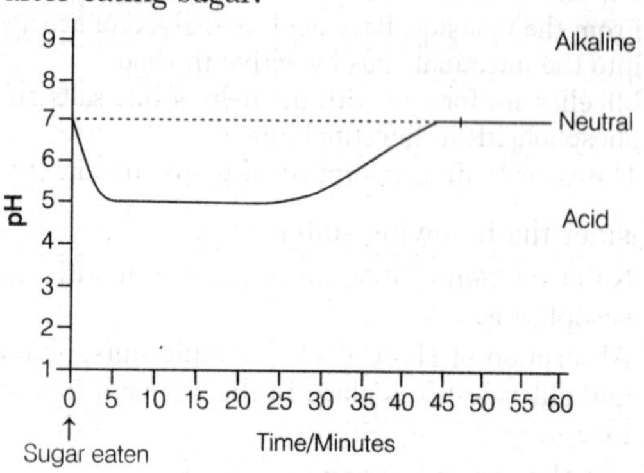

When are conditions in the mouth most likely to cause tooth decay?

(a) 0-5 minutes
(b) 5-25 minutes
(c) 25-45 minutes
(d) 45-60 minutes

40 Equal quantities of a 5% starch solution were added to equal quantities of different solutions. After 5 minutes at room temperature, the solutions were tested with iodine in potassium iodide solution and the depth of colour was recorded using a colorimeter (with an arbitrary scale of 1 to 5. 5 being the darkest blue).

The results are given in the table.

Solution	Colorimeter reading
1. Saliva and dilute HCl	4
2. Saliva and dilute $NaHCO_3$	1
3. Dilute HCl only	4
4. Saliva at 100°C	5
5. Saliva at 20°C	3

This experiment suggests that saliva contains a biologically active substance that

(a) hydrolyses starch in acid conditions
(b) is denatured by hydrochloric acid
(c) is promoted in its action by high temperatures
(d) may be activated by hydrogen carbonate ions

41 A medical scientist investigates four species of insects. He knows that one feeds on human blood and that the others feed on plants. As the insects look similar, he investigates the digestive enzymes present in their guts.

Which insect feeds only one blood?

Enzyme present in insect guts

Insects	Amylase	Lipase	Protease	Sucrase
(a)	✓	✗	✓	✓
(b)	✓	✗	✗	✓
(c)	✗	✓	✓	✗
(d)	✗	✓	✗	✗

42 Only two of the following statements accurately describe what happens in the mouth.

I. Amylase breaks down larger starch molecules into smaller maltose molecules.
II. Chewing increases the surface area of food for digestion.
III. Saliva emulsifies fats into smaller droplets.
IV. Teeth break up large insoluble molecules into smaller soluble molecules.

Which statements are correct?

(a) I and II
(b) II and III
(c) III and IV
(d) I and IV

Assertion and Reason

Direction (Q. Nos. 43-54) *In each of the following questions, a statement of Assertion (A) is given followed by the corresponding statement of Reason (R). Of the statements, mark the correct answer as*

(a) Both A and R are true and R is the correct explanation of A
(b) Both A and R are true, but R is not the correct explanation of A
(c) A is true, but R is false
(d) A is false, but R is true
(e) Both A and R are false

43 **Assertion** (A) A feeling of vomiting precedes nausea.
Reason (R) Vomiting is the ejection of stomach contents through the mouth.

44 **Assertion** (A) The stomach mucosa is not digested by its own secretions.
Reason (R) Mucin coats the mucosa of stomach.

45 **Assertion** (A) Saliva is a secretion of salivary glands.
Reason (R) Saliva is a mixture of water and electrolytes, derived from blood plasma.

46 **Assertion** (A) Oxyntic cells are minor salivary glands.
Reason (R) Oxyntic cells stain strongly with eosin.

47 **Assertion** (A) An elephant's tusk is a modified upper incisor.
Reason (R) Tusk of elephant is used in food uptake.

48 **Assertion** (A) Large intestine also shows the presence of villi like small intestine.
Reason (R) Absorption of water takes place in large intestine.

49 **Assertion** (A) Thick layer of muscles are present in the wall of alimentary canal.
Reason (R) These muscles help in the mixing of food materials with the enzymes coming from different glands in the alimentary canal.

50 **Assertion** (A) Digested material is absorbed in the intestine through the process of diffusion.
Reason (R) Diffusion is a very fast process of absorption.

51 **Assertion** (A) Jejunum is not involved in the absorption of food nutrients.
Reason (R) Villi are abundant in jejunum which decreases in ileum.

52 **Assertion** (A) Scurvy is known as vitamin-C deficiency.
Reason (R) Deficiency of ascorbic acid causes scurvy.

53 **Assertion** (A) pH in buccal cavity is 6.8, stomach is 1.8 and duodenum is 7.8.
Reason (R) pH is maintained by HCl in all the places.

54 **Assertion** (A) Absorption of simple sugar, alcohol and medicines takes place in the stomach.
Reason (R) Most of the water is absorbed in the large intestine.

(C) Medical Entrances' Gallery

(Collection of questions Asked in NEET & Various Medical Entrance Exams)

1 Intrinsic factor that helps in the absorption of vitamin-B_{12} is secreted by **[NEET 2020]**
(a) goblet cells (b) hepatic cells
(c) oxyntic cells (d) chief cells

2 The proteolytic enzyme rennin is found in **[NEET 2020]**
(a) intestinal juice (b) bile juice
(c) gastric juice (d) pancreatic juice

3 The enzyme enterokinase helps in the conversion of **[NEET 2020]**
(a) trypsinogen into trypsin (b) caseinogen into casein
(c) pepsinogen into pepsin (d) protein into polypeptides

4 Identify the correct statement with reference to human digestive system. **[NEET 2020]**
(a) Serosa is the innermost layer of the alimentary canal
(b) Ileum is a highly coiled part
(c) Vermiform appendix arises from duodenum
(d) Ileum opens into small intestine

5 Goblet cells of alimentary canal are modified from **[NEET 2020]**
(a) columnar epithelial cells (b) chondrocytes
(c) compound epithelial cells (d) squamous epithelial cells

6 Match the following structures with their respective location in organs. **[NEET 2019]**

Column I	Column II
A. Crypts of Lieberkuhn	1. Pancreas
B. Glisson's capsule	2. Duodenum
C. Islets of Langerhans	3. Small intestine
D. Brunner's glands	4. Liver

Codes

	A	B	C	D
(a)	2	4	1	3
(b)	3	4	1	2
(c)	3	2	1	4
(d)	3	1	2	4

7 Identify the cells whose secretion protects the lining of gastrointestinal tract from various enzymes. [NEET 2019]
(a) Goblet cells (b) Oxyntic cells
(c) Duodenal cells (d) Chief cells

8 Match the items given in Column I with those in Column II and choose the correct option. [NEET 2019]

Column I	Column II
A. Rennin	1. Vitamin-B_{12}
B. Enterokinase	2. Facilitated transport
C. Oxyntic cells	3. Milk proteins
D. Fructose	4. Trypsinogen

Codes

	A	B	C	D		A	B	C	D
(a)	3	4	2	1	(b)	4	3	1	2
(c)	4	3	2	1	(d)	3	4	1	2

9 Kwashiorkor disease is caused due to the [NEET 2019]
(a) simultaneous deficiency of proteins and fats
(b) simultaneous deficiency of proteins and calories
(c) deficiency of carbohydrates
(d) protein deficiency not accompanied by calorie deficiency

10 Which cell is found in mucus secreting organs? [AIIMS 2019]
(a) Goblet cells (b) Paneth cells
(c) Oxyntic cells (d) Peptic cells

11 **Assertion** (A) Cholecystokinin is released by duodenum.

Reason (R) It activates pepsinogen and bile juice. [AIIMS 2019]
(a) If both A and R are true and R is the correct explanation of A
(b) If both A and R are true, but R is not the correct explanation of A
(c) If A is true, but R is false
(d) If both A and R are false

12 **Assertion** (A) Gastrin is a hormone that is released from the gastrointestinal tract and helps in digestion.
Reason (R) It promotes secretion of HCl and trypsinogen. [AIIMS 2019]
(a) If both A and R are true and R is the correct explanation of A
(b) If both A and R are true, but R is not the correct explanation of A
(c) If A is true, but R is false
(d) If both A and R are false

13 **Assertion** (A) Deficiency of an element may lead to scurvy.
Reason (R) Daily requirement of ascorbic acid is 5 mg/day. [AIIMS 2019]
(a) If both A and R are true and R is the correct explanation of A
(b) If both A and R are true, but R is not the correct explanation of A
(c) If A is true, but R is false
(d) If both A and R are false

14 Length of oesophagus is [JIPMER 2019]
(a) 25 cm (b) 55 cm
(c) 33 cm (d) 45 cm

15 Which of the following gastric cells indirectly help in erythropoiesis? [NEET 2018]
(a) Goblet cells (b) Mucous cells
(c) Chief cells (d) Parietal cells

16 Which one of the following terms describe human dentition? [NEET 2018]
(a) Pleurodont, Monophyodont, Homodont
(b) Thecodont, Diphyodont, Heterodont
(c) Thecodont, Diphyodont, Homodont
(d) Pleurodont, Diphyodont, Heterodont

17 Diagram of large intestine is given below. Identify the parts *A*, *B*, *C*, *D*, *E* and *F*. [AIIMS 2018]

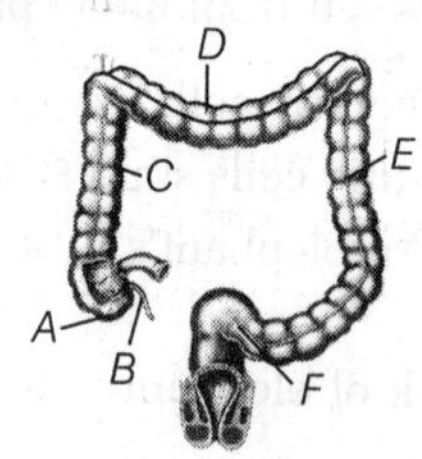

(a) *A*-Sigmoid colon, *B*-Vermiform appendix, *C*-Ascending colon, *D*-Transverse colon, *E*-Descending colon, *F*-Caecum
(b) *A*-Caecum, *B*-Vermiform appendix, *C*-Sigmoid colon, *D*-Ascending colon, *E*-Transverse colon, *F*-Descending colon
(c) *A*-Caecum, *B*-Vermiform appendix, *C*-Ascending colon, *D*-Transverse colon, *E*-Descending colon, *F*-Sigmoid colon
(d) *A*-Sigmoid colon, *B*-Vermiform appendix, *C*-Descending colon, *D*-Transverse colon, *E*-Ascending colon, *F*-Caecum

18 Daily requirement of vitamin-A for adult woman is [JIPMER 2018]
(a) 500 micrograms (b) 700 micrograms
(c) 900 micrograms (d) 300 micrograms

19 Brunner's gland is located in [JIPMER 2018]
(a) duodenum (b) jejunum
(c) ileum (d) stomach

20 What is the function of Kupffer's cells? [JIPMER 2018]
(a) Bile secretion (b) Digestion of lipid
(c) Phagocytosis (d) Digestion of protein

21 Where is sacculus rotundus located? [JIPMER 2018]

(a) Between duodenum and jejunum
(b) Between ileum and caecum
(c) Caecum and colon
(d) Colon and rectum

22 Which cells of 'crypts of Lieberkuhn' secrete antibacterial lysozyme? [NEET 2017]

(a) Argentaffin cells (b) Paneth cells
(c) Zymogen cells (d) Kupffer cells

23 A baby boy aged two years is admitted to play school and passes through a dental check up. The dentist observed that the boy had twenty teeth. Which teeth were absent? [NEET 2017]

(a) Incisors (b) Canines
(c) Premolars (d) Molars

24 Which of the following options best represents enzyme composition of pancreatic juice? [NEET 2017]

(a) Amylase, peptidase, trypsinogen, rennin
(b) Amylase, pepsin, trypsinogen, maltase
(c) Peptidase, amylase, pepsin, rennin
(d) Lipase, amylase, trypsinogen, procarboxypeptidase

25 Which hormones do stimulate the production of pancreatic juice and bicarbonate? [NEET-II 2017]

(a) Angiotensin and epinephrine
(b) Gastrin and insulin
(c) Cholecystokinin and secretin
(d) Insulin and glucagon

26 Which one of the following options is correct regarding digestion of food substrates? [AIIMS 2017]

	Substrate	Enzyme	Site of action	Product
(a)	Starch	Amylase	Stomach	Maltose
(b)	Protein	Pepsin	Duodenum	Peptones
(c)	Lipid	Lipase	Pancreas	Fat globules
(d)	Sucrose	Invertase	Duodenum	Glucose and Fructose

27 **Assertion** (A) Gall stones are caused by disturbance in the composition of the bile.

Reason (R) A change in the ratio of liver glycogen and bile salts may result in the formation of deposits. [AIIMS 2017]

(a) If both A and R are true and R is the correct explanation of A
(b) If both A and R are true, but R is not the correct explanation of A
(c) If A is true, but R is false
(d) If A is false, but R is true
(e) If both A and R are false

28 Which one of the following pairs is not correctly matched? [JIPMER 2017]

(a) Vitamin-B_{12} — Pernicious anaemia
(b) Vitamin-B_6 — Loss of appetite
(c) Vitamin-B_1 — Beri-beri
(d) Vitamin-B_2 — Pellagra

29 Which of the following guards the opening of hepatopancreatic duct into the duodenum? [NEET-I 2016, DUMET 2011]

(a) Ileocaecal valve (b) Pyloric sphincter
(c) Sphincter of Oddi (d) Semilunar valve

30 In the stomach, gastric acid is secreted by the [NEET-I 2016]

(a) parietal cells (b) peptic cells
(c) acidic cells (d) gastrin secreting cells

31 Name a peptide hormone which acts mainly on hepatocytes, adipocytes and enhances cellular glucose uptake and utilisation. [NEET-II 2016]

(a) Insulin (b) Glucagon
(c) Secretin (d) Gastrin

32 Choose the correct statement among the following. [Kerala CEE 2015]

(a) Intestinal mucosal epithelium has oxyntic cells
(b) Ptyalin converts proteins into proteoses and peptones
(c) Crypts of Lieberkuhn are seen between bases of villi in intestine
(d) Sphincter of Oddi is present at the junction of oesophagus and cardiac stomach
(e) Goblet cells secrete hydrochloric acid in stomach

33 Identify the correctly matched structure and its secretion. [Kerala CEE 2015]

(a) Brunner's glands – Salivary amylase
(b) Intestinal mucosa – Insulin
(c) Gall bladder – Bile
(d) Salivary gland – Lysozyme
(e) Goblet cells – HCl

34 Match the Column I with Column II and find the correct option from the codes given below. [EAMCET 2015]

	Column I		Column II
A.	Cardiac glands	1.	Castle's intrinsic factor
B.	Pyloric glands	2.	Gastrin
C.	Peptic cells	3.	Mucus for protein
D.	Parietal cells	4.	Pepsinogen

Codes

	A	B	C	D		A	B	C	D
(a)	3	2	4	1	(b)	3	2	1	4
(c)	3	1	2	4	(d)	1	4	2	5

35 Match the following columns and find the correct option from the codes given below **[WB JEE 2015]**

	Column I		Column II
A.	Vitamin-B_1	1.	Accumulation of fat
B.	Gastric juice	2.	Loss of fat
C.	Starvation	3.	Pepsin
D.	Obesity	4.	Beri-beri

Codes

	A	B	C	D		A	B	C	D
(a)	3	4	2	1	(b)	3	4	1	2
(c)	4	3	2	1	(d)	4	2	3	1

36 The optimum pH for pepsin is **[AMU 2015]**

(a) 11 (b) 5-6
(c) 1.6-2.4 (d) 4-7

37 Enzyme not present in pancreatic juice is **[AMU 2015]**

(a) amylase (b) chymotrypsinogen
(c) lipase (d) enterokinase

38 The proteolytic enzyme found in gastric juice of infants which helps in digestion of milk protein is **[UK PMT 2015,Wardha 2007]**

(a) rennin (b) salivary amylase
(c) pepsin (d) lactase

39 Intrinsic factor is secreted by **[UK PMT 2015]**

(a) goblet cells (b) oxyntic cells
(c) argentaffin cells (d) chief cells

40 Match the following columns and choose the correct option from the codes given below. **[UK PMT 2015]**

	Column I		Column II
A.	Cobalamin	1.	Vitamin-B_{12}
B.	Tocopherol	2.	Vitamin- E
C.	Folic acid	3.	Vitamin-B_{10}
D.	Cholecalciferol	4.	Vitamin-D_3
E.	Phylloquinone	5.	Vitamin-K

Codes

	A	B	C	D	E
(a)	1	2	3	4	5
(b)	2	3	1	5	4
(c)	5	2	3	4	1
(d)	1	3	2	4	5

41 The secretion of brush border cells of intestinal mucosa along with secretion of goblet cells constitute **[Kerala CEE 2015]**

(a) succus entericus (b) chyme
(c) gastric juice (d) chylomicrons
(e) bolus

42 Match the following columns and find the correct options. **[Gujarat CET 2015]**

	Column I	Column II	Column III
A.	Gastric juice	1. Chymotrypsinogen	I. Dipeptide to amino acids
B.	Intestinal juice	2. Ptyalin	II. Proteoses to small peptides
C.	Saliva	3. Rennin	III. Casein into paracasein
D.	Pancreatic juice	4. Pepsin	IV. Starch into maltose

Codes

	A	B	C	D
(a)	3-III	4-I	2-II	1-IV
(b)	4-IV	3-III	1-II	2-I
(c)	3-III	4-I	2-IV	1-II
(d)	2-I	1-III	3-II	4-IV

43 The enzyme that is not present in succus entericus is **[CBSE AIPMT 2015]**

(a) maltase (b) nucleases
(c) nucleosidase (d) lipase

44 Primary dentition in human differs from permanent dentition in not having one of the following teeth. **[CBSE AIPMT 2015]**

(a) Canines (b) Premolars (c) Molars (d) Incisors

45 The initial step in the digestion of milk in humans is carried out by **[CBSE AIPMT 2014]**

(a) lipase (b) trypsin (c) rennin (d) pepsin

46 Fructose is absorbed into the blood through mucosa cells of intestine by the process called **[CBSE AIPMT 2014]**

(a) active transport (b) facilitated transport
(c) simple diffusion (d) cotransport mechanism

47 The salivary amylase shows maximum digestive action at pH **[MHT CET 2014]**

(a) 3.6 (b) 6.8 (c) 7.5 (d) 8.5

48 In the following process of digestion, the enzymes at location 'X' and 'Y' are respectively,

Proteins $\xrightarrow{X}$ Protease and Peptones $\xrightarrow{Y}$ Dipeptides **[MHT CET 2014]**

Choose the correct option.

(a) Chymotrypsin and Pepsin (b) Pepsin and Trypsin
(c) Ptyalin and Pepsin (d) Trypsin and Dipeptidase

49 Choose the incorrect statement among the following. **[Kerala CEE 2014]**

(a) Trypsinogen is activated by enterokinase
(b) The optimum pH for salivary amylase activity is 8.9
(c) Rennin helps in the digestion of milk proteins
(d) Goblet cells secrete mucus
(e) Submucosal glands of the intestine are also known as Brunner's glands

50 Chylomicrons are **[Kerala CEE 2014]**
(a) small fat globules coated with protein
(b) protein molecules coated with fat
(c) small granules found in gastric juice
(d) neural signals that stimulate intestinal secretions
(e) aerobic microbes

51 Match the following columns.

Column I (Substrate)	Column II (Enzyme)	Column III (Product)
A. Lactose	1. Lipase	I. Galactose
B. Monoglycerides	2. Trypsin	II. Maltose
C. Starch	3. Lactase	III. Fatty acid
D. Peptones	4. Amylase	IV. Dipeptides

Codes **[Kerala CEE 2014]**

	A	B	C	D
(a)	1-I	3-II	2-III	4-IV
(b)	4-I	1-II	2-III	3-IV
(c)	3-I	1-III	4-II	2-IV
(d)	3-I	1-II	4-III	2-IV
(e)	2-I	4-II	3-III	1-IV

52 Which of the following does not produce any digestive enzymes? **[WB JEE 2014]**
(a) Salivary gland (b) Pancreas
(c) Liver (d) Stomach

53 The disease that occurs in mature adult human being due to deficiency of calciferol is **[WB JEE 2014]**
(a) keratomalacia (b) osteomalacia
(c) glossitis (d) pernicious anaemia

54 Absorption of vitamin-B_{12} in human requires P glycoprotein secreted from Q. The correct choices of P and Q are **[WB JEE 2014]**
(a) P-Extrinsic factor and Q-stomach
(b) P-Intrinsic factor and Q-stomach
(c) P-Intrinsic factor and Q-small intestine
(d) P-Exopolysaccharide and Q-small intestine

55 The middle part of small intestine is **[J&K CET 2014]**
(a) duodenum (b) jejunum
(c) ileum (d) pyloric region

56 Release of pancreatic juice from pancreas is stimulated by **[UK PMT 2014]**
(a) secretin (b) trypsinogen
(c) cholecystokinin (d) enterokinase

57 Pernicious anaemia results due to deficiency of **[UK PMT 2014]**
(a) vitamin-B_1 (b) vitamin-B_{12}
(c) vitamin-A (d) iron

58 Which is incorrectly matched? **[UK PMT 2014]**
(a) Rennin-liver (b) Ptyalin-mouth
(c) Pepsin-stomach (d) Trypsin-intestine

59 Emulsified fats are digested by **[EAMCET 2014]**
(a) gastric juice and pancreatic juice
(b) bile juice and intestinal juice
(c) pancreatic juice and bile juice
(d) pancreatic juice and intestinal juice

60 Which enzyme initiates the digestion of proteins? **[Gujarat CET 2014]**
(a) Trypsin (b) Pepsin
(c) Amino peptidase (d) Carboxypeptidase

61 Select the correct match of the digested products in humans given in Column I with their absorption site and mechanism in Column II. **[NEET 2013]**

Column I	Column II
(a) Glycine and glucose	Small intestine and active absorption
(b) Fructose and Na^+	Small intestine and passive absorption
(c) Glycerol and fatty acids	Duodenum and move as chylomicrons
(d) Cholesterol and maltose	Large intestine and active absorption

62 Which type of teeth are absent in rabbit? **[UP CPMT 2013, 11]**
(a) Molars (b) Premolars (c) Canines (d) Incisors

63 In rabbit, the digestion of cellulose takes place in **[UP CPMT 2013]**
(a) colon (b) ileum (c) caecum (d) rectum

64 Which part of our body secretes the hormone secretin? **[Manipal 2013]**
(a) Ileum (b) Stomach (c) Duodenum (d) Oesophagus

65 A balanced diet does not include **[KCET 2013]**
(a) carbohydrates and fats (b) nucleic acids and enzymes
(c) proteins and vitamins (d) minerals and salts

66 Column I contains names of the sphincter muscles of the alimentary canal and Column II contains their location. Match them properly and choose the correct answer. **[KCET 2013]**

Column I	Column II
A. Sphincter of aninternus	1. Opening of hepatopancreatic duct into duodenum
B. Cardiac sphincter	2. Between duodenum and posterior stomach
C. Sphincter of Oddi	3. Guarding the terminal part of alimentary canal
D. Ileocaecal sphincter	4. Between oesophagus and anterior stomach
E. Pyloric sphincter	5. Between small intestine and bowel

Codes

	A	B	C	D	E
(a)	3	2	4	1	5
(b)	2	5	1	4	3
(c)	3	4	1	5	2
(d)	4	3	1	2	5

67 Which of the following is a gastrointestinal hormone? **[WB JEE 2013]**
(a) Prolactin (b) Enterogastrone
(c) GH (d) FSH

68 The pH of stomach is 1.6, then which enzyme will digest protein? **[AIIMS 2013]**
(a) Trypsin (b) Pepsin
(c) Amylase (d) Erepsin

69 Digestion is brought about by **[AFMC 2013]**
(a) enzymes (b) hormones
(c) water (d) mucus

70 Most digestion and absorption of food take place in **[OJEE 2013]**
(a) stomach (b) small intestine
(c) caecum (d) large intestine

71 Salivary amylase, a digestive enzyme begins digestion of **[OJEE 2013]**
(a) carbohydrates (b) fats
(c) proteins (d) All of these

72 Anxiety and eating spicy food together in an otherwise normal human, may lead to **[CBSE AIPMT 2012]**
(a) indigestion (b) jaundice
(c) diarrhoea (d) vomiting

73 For activity of carboxypeptidase requires **[CBSE AIPMT 2012]**
(a) zinc (b) iron
(c) niacin (d) copper

74 Where do certain symbiotic microorganisms normally occur in human body? **[CBSE AIPMT 2012]**
(a) Caecum
(b) Oral lining and tongue surface
(c) Vermiform appendix and rectum
(d) Duodenum

75 Argentaffin cells in human beings are found in **[AMU 2012]**
(a) small intestine (b) stomach
(c) large intestine (d) liver

76 In human beings, the three pair of salivary glands and numerous buccal glands produce about **[AMU 2012]**
(a) 1.0 dm^3 of saliva per day
(b) 1.5 dm^3 of saliva per day
(c) 2.0 dm^3 of saliva per day
(d) 2.5 dm^3 of saliva per day

77. If this enzyme were to be absent in our small intestine, digestion of proteins in our body would be severely affected **[KCET 2012]**
(a) enterokinase
(b) pancreatic amylase
(c) maltase
(d) lipase

78 Secretin and cholecystokinin are the hormones secreted in **[KCET 2012; BCECE 2012]**
(a) oesophagus
(b) pyloric stomach
(c) duodenum
(d) ileum

79 This gastrointestinal hormone stimulates insulin secretion **[AMU 2012]**
(a) gastrin (b) CCK
(c) secretin (d) GIP

80 This is the common passage for bile and pancreatic juices **[AMU 2012]**
(a) ampulla of Vater
(b) ductus Choleciochus
(c) duct of Wirsung
(d) duct of Santorini

81 The mucosal layer in the stomach forms irregular folds known as **[Kerala CEE 2011]**
(a) villi (b) lumen
(c) rugae (d) crypts of Lieberkuhn
(e) lacteals

82 The back flow of faecal matter in the large intestine is prevented by the presence of **[Kerala CEE 2011]**
(a) epiglottis
(b) sphincter of Oddi
(c) ileocaecal valve
(d) gastric-oesophageal sphincter
(e) pyloric sphincter

83. In human body, the role of bile salts in digestion is to **[MHT CET 2011]**
(a) act as coenzymes during the digestion of carbohydrates
(b) emulsify fat and facilitate their absorption
(c) aid in the breakup of proteins into amino acids and their absorption
(d) stimulate the pancreas to release its enzymes

84. With reference to a normal human being, which one of the following statements is not correct?
(a) Human saliva is slightly alkaline **[MHT CET 2011]**
(b) An adult human may secrete 1-1.5 L of saliva per day
(c) Saliva is secreted by six pairs of salivary glands in human beings
(d) The salivary enzyme (ptyalin) breaks down cooked starch into maltose

85 Vitamin necessary for normal functioning of liver, clotting of blood and preventing haemorrhage is **[MHT CET 2011]**

(a) tocopherol (b) phylloquinone
(c) cyanocobalamin (d) riboflavin

86 Emulsification of fat will not occur in the absence of **[UP CPMT 2011]**

(a) lipase (b) bile pigments
(c) bile salts (d) pancreatic juice

87 The best source of vitamin-B_1 is **[UP CPMT 2011]**

(a) whole wheat bread
(b) cod liver oil
(c) egg
(d) curd

88 Characteristic of mammalian liver is **[UP CPMT 2011]**

(a) Kupffer's cells and leucocytes
(b) Leucocytes and canaliculae
(c) Glisson's capsules and Kupffer's cells
(d) Glisson's capsule and leucocytes

89 The fat soluble vitamin is **[UP CPMT 2011]**

(a) B (b) C (c) K (d) H

90 Ptyalin is inactivated by a component of gastric juice known as **[WB JEE 2011]**

(a) pepsin (b) mucus
(c) rennin (d) HCl

91 Which one is the correct option for labels *A*, *B* and *C* in the given diagram? **[Gujarat CET 2011]**

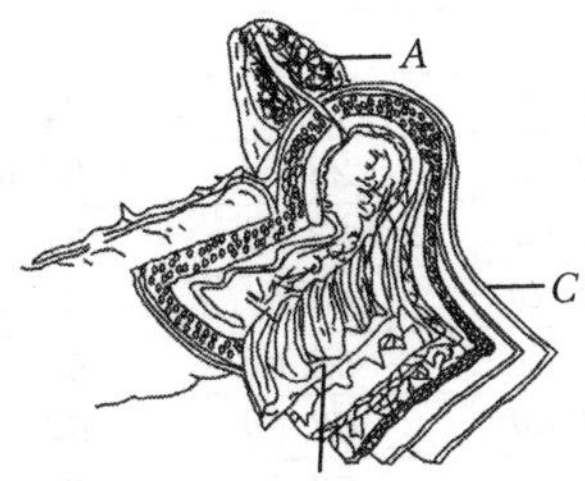

(a) *A*-Liver, *B*-Mucosa, *C*-Peritoneum
(b) *A*-Liver, *B*-Circular muscle layer, *C*-Serosa
(c) *A*-Pancreas, *B*-Mucosa, *C*-Peritoneum
(d) *A*-Pancreas, *B*-Submucosa, *C*-Serosa

92 Proportion of which of the following should be increased in diet to improve strength and growth of bones? **[Gujarat CET 2011]**

(a) Vitamin-D, Ca^{2+} and vitamin-K
(b) Vitamin-D, Ca^{2+} and iodine
(c) Vitamin-D, Ca^{2+} and vitamin-A
(d) Vitamin-A, Ca^{2+} and Mn^{2+}

93 Aggregates of lymphoid tissue present in the distal portion of the small intestine are known as **[DUMET 2011]**

(a) villi (b) Peyer's patches
(c) rugae (d) choroid plexus

ANSWERS

Check Point 23.1

1. (a)	2. (c)	3. (d)	4. (d)	5. (b)	6. (a)	7. (c)

Check Point 23.2

1. (a)	2. (a)	3. (b)	4. (c)	5. (a)	6. (c)

Check Point 23.3

1. (a)	2. (b)	3. (a)	4. (d)	5. (b)	6. (a)

Check Point 23.4

1. (d)	2. (a)	3. (c)	4. (a)	5. (b)	6. (d)

Check Point 23.5

1. (b)	2. (a)	3. (d)	4. (b)	5. (c)	6. (d)

Check Point 23.6

1. (a)	2. (d)	3. (b)	4. (c)	5. (b)	6. (a)

(A) Topical Exercise

1. (c)	2. (c)	3. (d)	4. (d)	5. (b)	6. (b)	7. (b)	8. (c)	9. (a)	10. (d)
11. (b)	12. (a)	13. (c)	14. (b)	15. (a)	16. (d)	17. (b)	18. (d)	19. (d)	20. (b)
21. (d)	22. (c)	23. (c)	24. (c)	25. (b)	26. (c)	27. (a)	28. (a)	29. (b)	30. (c)
31. (b)	32. (c)	33. (c)	34. (c)	35. (b)	36. (c)	37. (a)	38. (b)	39. (d)	40. (b)
41. (c)	42. (a)	43. (c)	44. (a)	45. (c)	46. (d)	47. (c)	48. (c)	49. (a)	50. (a)
51. (c)	52. (b)	53. (b)	54. (b)	55. (d)	56. (b)	57. (c)	58. (d)	59. (a)	60. (b)
61. (a)	62. (d)	63. (a)	64. (b)	65. (a)	66. (b)	67. (a)	68. (b)	69. (b)	70. (b)
71. (a)	72. (b)	73. (d)	74. (a)	75. (b)	76. (d)	77. (d)	78. (a)	79. (b)	80. (c)
81. (c)	82. (c)	83. (b)	84. (d)	85. (b)	86. (a)	87. (d)	88. (b)	89. (c)	90. (c)
91. (c)	92. (a)	93. (d)	94. (a)	95. (b)	96. (d)	97. (c)	98. (a)	99. (a)	100. (a)
101. (c)	102. (c)	103. (a)	104. (a)	105. (d)	106. (c)	107. (d)	108. (b)	109. (b)	110. (c)
111. (a)	112. (d)	113. (c)	114. (b)	115. (d)	116. (d)	117. (c)	118. (a)	119. (b)	120. (c)
121. (d)	122. (d)	123. (c)							

(B) Cumulative Exercise

1. (b)	2. (b)	3. (d)	4. (d)	5. (c)	6. (c)	7. (b)	8. (b)	9. (c)	10. (d)
11. (d)	12. (c)	13. (a)	14. (a)	15. (a)	16. (d)	17. (c)	18. (d)	19. (c)	20. (d)
21. (a)	22. (d)	23. (a)	24. (c)	25. (c)	26. (b)	27. (c)	28. (b)	29. (a)	30. (d)
31. (c)	32. (d)	33. (c)	34. (b)	35. (b)	36. (c)	37. (b)	38. (d)	39. (b)	40. (d)
41. (c)	42. (a)	43. (d)	44. (a)	45. (b)	46. (d)	47. (c)	48. (d)	49. (a)	50. (e)
51. (d)	52. (a)	53. (a)	54. (b)						

(C) Medical Entrances' Gallery

1. (c)	2. (c)	3. (a)	4. (b)	5. (a)	6. (b)	7. (a)	8. (d)	9. (d)	10. (a)
11. (c)	12. (c)	13. (d)	14. (a)	15. (d)	16. (b)	17. (c)	18. (b)	19. (a)	20. (c)
21. (b)	22. (b)	23. (c)	24. (d)	25. (c)	26. (d)	27. (c)	28. (d)	29. (c)	30. (a)
31. (a)	32. (c)	33. (d)	34. (a)	35. (c)	36. (c)	37. (d)	38. (a)	39. (b)	40. (a)
41. (a)	42. (c)	43. (b)	44. (b)	45. (d)	46. (b)	47. (b)	48. (b)	49. (b)	50. (a)
51. (c)	52. (c)	53. (b)	54. (b)	55. (b)	56. (a,c)	57. (b)	58. (a)	59. (d)	60. (b)
61. (a)	62. (c)	63. (c)	64. (c)	65. (b)	66. (c)	67. (b)	68. (b)	69. (a)	70. (b)
71. (a)	72. (a)	73. (a)	74. (a)	75. (b)	76. (b)	77. (a)	78. (c)	79. (d)	80. (a)
81. (c)	82. (c)	83. (b)	84. (a)	85. (b)	86. (c)	87. (a)	88. (c)	89. (c)	90. (d)
91. (d)	92. (c)	93. (b)							

Explanations

(A) Topical Exercise

6 (*b*) Statement I is false, but II is true. False statement can be corrected as
Total number of canines in permanent dental set of human is four.

9 (*a*) Statement I is true, but II is false. False statement can be corrected as
Pharynx serves as common passage for food and air. It allows air to enter into trachea and food to oesophagus.

11 (*b*) Only II is incorrectly matched pair, because gullet is the upper opening of oesophagus.

13 (*c*) Statement in option (c) is incorrect. It can be corrected as
The submucosa of oesophagus is thickest than other parts of alimentary canal due to the presence of submucosal mucous glands in this region.

21 (*d*) IV represents sigmoid colon which is retroperitoneal in position whereas other three are intraperitoneal in position.

28 (*a*) Statement in option (a) is incorrect whereas other statements are correct. Incorrect statement can be corrected as
Meissner's plexus are found in the submucosa of alimentary canal.

30 (*c*) Infraorbital is not found in humans instead it is found in animals like dog, cat below their eyes.

32 (*c*) Option (c) represents incorrectly matched pair, which can be corrected as
Nuhn's gland is accessory salivary gland that is found associated with the tongue.

35 (*b*) Statement in option (b) is incorrect because
The secretions of goblet cells are alkaline in nature. It is known as mucus.

37 (*a*) Statement in option (a) is incorrect because
Crypts of Lieberkuhn are found throughout the small intestine.

46 (*d*) Statement in option (d) is incorrect because
Gastrin hormone is secreted by gastrin cells in the pyloric region of the stomach. It stimulates gastric glands to secrete and release the gastric juice.

48 (*c*) Statement in option (c) is incorrect. Liver does not store vitamins-B_6, B_7 and B_{10}. It stores vitamins-A, D and B_{12}.

53 (*b*) Statement in option (b) is incorrect and can be corrected as
The endocrine part of pancreas contain islet of Langerhans. Acinar cells forms the exocrine part of pancreas.

71 (a) Statement in option (a) is incorrect and can be corrected as
Lipid hydrolysis produces fatty acids and glycerol. Glycine is an amino acid.

74 (*a*) I and II are incorrectly matched pairs. Exopeptidase act upon small peptides and yield amino acids. Lactase act upon lactose and yield glucose and galactose.

79 (*b*) Statement I is false, but II is true. Incorrect statement can be corrected as
Amino acids are absorbed in blood capillaries whereas lipids are absorbed in lymph capillaries or lacteals.

85 (*b*) Option (b) represents incorrectly matched pair. Villikinin is secreted by the epithelium of small intestine.

86 (*a*) Option (a) is odd one out as somatostatin is secreted by pancreas whereas other three hormones are secreted by small intestine.

90 (*c*) Statement in option (c) is incorrect because humans cannot digest all the polysaccharides, e.g. cellulose is indigestible.

94 (*a*) Matched pair in option (a) is correct. Other incorrect matches can be corrected as
Vitamin-K–Fat soluble–Haemorrhage.
Vitamin-B_1–Water soluble–Beri-beri

98 (*a*) Statement in option (a) is incorrect. It can be corrected as
Recently discovered vitamin with anticancerous properties is vitamin-B_{17}.

114 (*b*) Statement in option (b) is false because
Potassium is found in fruits like banana, coconut, etc.

123 (*c*) Belching occurs when stomach is over dilated.
Rest all the statements are correct.

(B) Cumulative Exercise

2 (*b*) I and II are correct whereas III and IV are incorrect. Small amount of drugs are absorbed in stomach whereas vitamin absorption do not occur in stomach.

4 (*d*) Option (d) represents incorrectly matched pair. Duct of wirsung is associated with pancreas. Infraorbital glands are absent in human.

5 (*c*) Statement in option (c) is incorrect and can be corrected as
Taeniae coli are smooth muscle bands in the large intestine.

6 (*c*) Statement III is incorrect because
Micelles are water soluble droplets of fatty acids and glycerol.

10 (*d*) Both statements I and II are false. These can be corrected as

- Taste bud bearing projections are found on the upper surface of tongue.
- Sphincter of oddi guards the opening of hepatopancreatic duct into duodenum.

11 (*d*) Statement III is false because
Insulin and glucagon are secreted by the endocrine part whereas pancreatic juice is secreted by exocrine part of pancreas.

13 (*a*) Statement III is false because
Glycerol is not absorbed directly. It gets converted into chylomicrons and then it is absorbed in lacteals.

23 (*a*) Statements I and IV are incorrect and can be corrected as

- Duodenum is C-shaped and ileum is the longest highly coiled part of small intestine.
- Symbiotic bacteria containing pouch is associated with caecum.

32 (*d*) A is a blood vessel lining the villus of small intestine which absorbs glucose, B is a lacteal or a branch of lymphatic vessel which absorbs digested fat (fatty acids and glycerol). C is a goblet cell of epithelial lining which secretes mucus, while D is simple epithelial cell which produces enzymes for digestion of different types of food substances.

35 (*b*) Statement in option (b) is incorrect and can be corrected as

Fatty acids and glycerol are absorbed by simple diffusion into the intestinal cells.

37 (*b*) Statement in option (b) is incorrect and can be corrected as

Around 90% of total absorption of nutrients occur in the middle portion, i.e. jejunum of small intestine. A few remaining nutrients are absorbed in distal portion, i.e. ileum.

39 (*b*) Between 5-25 minutes, pH is 5, which is due to acidic conditions that can cause tooth decay.

40 (*d*) Saliva contains amylase enzymes that function well at an optimum pH 6.8-7.4. Therefore, the presence of hydrogen carbonate ions provide the optimum pH for this enzyme.

41 (*c*) Complex carbohydrates are not found in blood so there would be no need for the enzymes, amylase and sucrase to be present in the insect's guts. Thus, option (c) is correct.

43 (*d*) A is false, but R is true. A can be corrected as

A feeling of nausea precedes vomiting. It results in ejection or expulsion of stomach content through mouth.

44 (*a*) Both A and R is true and R is the correct explanation of A. Acid secreted from stomach does not digest stomach mucosa because mucin, another gastric secretion, which is a highly viscous organic compound coats and protects the mucosa from both chemical and mechanical injuries.

45 (*b*) Both A and R are true, but R is not the correct explanation of A.

Saliva is the secretion of exocrine salivary glands that are present in mouth.

Saliva contain enzyme, water and electrolytes like Na^+, K^+, Cl^-, etc., derived from blood plasma.

46 (*d*) A is false, but R is true. A can be corrected as

Oxyntic cells are present in the epithelium of stomach and thus these are gastric glands. They stain strongly with eosin due to which these are called oxyntic cells.

47 (*c*) A is true, but R is false. It can be corrected as

Elephants tusk is modification of upper incisor which is used for defence, attack, etc. It is not used for food uptake.

48 (*d*) A is false, but R is true. A can be corrected as

Large intestine do not possess villi like small intestine because absorption of food nutrients do not occur at this site. However, water, some salts and drugs are absorbed in large intestine.

50 (*e*) Both A and R are false and these can be corrected as

Diffusion cannot account for complete absorption of any nutrient from the intestine because diffusion of any nutrient from the intestine would continue as long as the concentration difference persists.

Moreover, diffusion is a slow process of absorption and to complete the absorption of nutrients, expenditure of energy is required.

51 (*d*) A is false, but R is true. A can be corrected as

Maximum absorption of food occur in jejunum due to villi are abundant at this site.

54 (*b*) Both A and R true, but R is not the correct explanation of A.

Stomach help in the absorption of some non-polar substances such as alcohol and drugs (e.g. aspirin). Never the less much of digested food is absorbed in small intestine, but most of water is absorbed in large intestine.

(C) Medical Entrances' Gallery

1 (*c*) Parietal cells or oxyntic cells secrete HCl and intrinsic factor. These intrinsic factors are essential for absorption of vitamin-B_{12}. Goblet cells secrete mucus. Peptic or chief cells secrete the proenzyme pepsinogen. Hepatic cells secrete bile.

2 (*c*) The proteolytic enzyme rennin is found in gastric juice of infants which helps in the digestion of milk proteins, casein into paracasein.

3 (*a*) The correct option is (a), because the enzyme enterokinase helps in conversion of trypsinogen into trypsin. Trypsinogen is activated by an enzyme, enterokinase, secreted by the intestinal mucosa into active trypsin. Trypsinogen is a zymogen released from pancreas.

4 (*b*) Option (b) is correct as ileum is a highly coiled tube with reference to human digestive system. Other option can be corrected as

Serosa is the outermost layer of the alimentary canal. A narrow finger like tubular projection, the vermiform appendix arises from caecum part of large intestine. Ileum opens into the large intestine.

5 (*a*) Goblet cells of alimentary canal are modified from columnar epithelium cells, which secrete mucus. These cells are found in the lining of organs like intestine and respiratory tract.

7 (*a*) Secretions of goblet cells protect the lining of gastrointestinal tract from various enzymes. These cells secrete mucus which along with bicarbonate ions help in the lubrication and protection of the mucosal epithelium from excreation by the highly concentrated HCl. On the other hand, oxyntic or parietal cells secrete hydrochloric acid. Chief cells or peptic cells secrete proenzymes-pepsinogen and prorenin.

9 (*d*) Kwashiorkor disease is caused due to protein deficiency not accompanied by calorie deficiency in the children of age 1-5 years. Its symptoms are weak muscle, thin limbs, retarded growth of the body and brain swelling of legs due to retention of water (oedema) reddish hair, pot belly, etc.

10 (*a*) Goblet cells are found in mucous secreting organs like intestine and respiratory tract. Paneth cells are found in the lining of small intestine.

It secretes antimicrobial molecules which help to maintain the gastrointestinal barrier.

Oxyntic or parietal cells are located in gastric glands of stomach. These secrete HCl and intrinsic factor.

Peptic cells or gastric chief cells are found in the stomach where they secrete pepsinogen and chymosin.

11 (*c*) A is true, but R is false.
Cholecystokinin is a gastric hormone which is released in duodenum. The correct form of R is as follows

Cholecystokinin acts on pancreas and gall bladder and stimulates the secretion of pancreatic enzymes and bile juice.

12 (*c*) A is true and R is false.
Gastrin hormone is released from gastrointestinal tract and it helps in the digestion of food.

It does not promote secretion of HCl and trypsinogen. It stimulate gastric glands to secrete and release gastric juices and also stimulates gastric mobility.

13 (*d*) Both A and R are false. The correct form of both the sentences is as follows

The deficiency of vitamin-C may lead to scurvy. The daily requirement of ascorbic acid for healthy man in about 65-90 mg/day.

14 (*a*) The length of oesophagus is nearly 25 cm in adults. Oesophagus is a hollow muscular tube found in vertebrates through which food is transferred from pharynx to stomach. Peristaltic contractions of oesophageal muscle facilitate the passage of food through it.

15 (*d*) Parietal cells (oxyntic cells) secrete hydrochloric acid and castle intrinsic factor. HCl converts iron (in diet) from ferric to ferrous form which can be easily absorbed and used during erythropoiesis (formation of RBCs).

16 (*b*) The terms thecodont, diphyodont and heterodont describe human dentition. Humans have diphyodont condition as the teeth appear twice during their lifetime, i.e. milk or deciduous teeth and permanent teeth. Teeth in humans remain embedded in the sockets of the jaw bones (thecodont). Also, humans have four different types of teeth, incisors, canine, premolars and molars and hence are called heterodont.

18 (*b*) The recommended daily amount of vitamin-A for adult woman is 700 micrograms. Vitamin-A is a nutrient important for good vision, growth, cell division, reproduction and immunity. It is found in carrots, oranges, etc.

19 (*a*) The Brunner's glands are located in the duodenum. They secrete both enzyme (in small quantity) and mucus. The mucus protects the duodenal wall from getting digested. The Brunner's glands open into the crypts of Lieberkuhn.

20 (*c*) The main function of Kupffer's cells is phagocytosis. They are specialised macrophages located in the liver, lining the walls of the sinusoids. Kupffer's cell engulf worn out RBCs, WBCs and microorganisms.

21 (*b*) Sacculus rotundus is located between ileum and caecum. The junction of ileum and large intestine is seen in the form of an orifice called ileocolic or ostium ileocaecal valve. This portion is well-developed in rabbit in the form of a swollen portion called sacculus rotundus.

22 (*b*) The mucosa present in between the bases of villi of small intestine (crypts of Lieberkuhn) contains Paneth cells, which secrete antibacterial lysozyme.

23 (*c*) In human beings, after birth the first set of teeth that develops are deciduous teeth or temporary teeth. These are 20 in number. The dental formula of child is 2102/2102.

Thus, they have 2 incisors, 1 canine, 0 premolars and 2 molars. Therefore, the baby boy would not have premolars.

24 (*d*) Pancreas is a mixed gland which consists of an exocrine and endocrine part. Exocrine part secretes alkaline pancreatic juice. This juice contains three proenzymes, trypsinogen, chymotrypsinogen, procarboxypeptidase and some enzymes like lipase, amylase, elastase, etc.

25 (*c*) Cholecystokinin (CCK) and secretin are the peptide hormones that stimulate the production of pancreatic juice and bicarbonates within the alimentary canal.

26 (*d*) Option (d) is correct regarding digestion of food substrates. Sucrose in the intestine (duodenum) breaks down into monosaccharides, glucose and fructose in the presence on invertase enzymes. Rest of the substrates are incorrectly matched with their enzymes, site of action and product.

The correct information regarding the incorrect ones is as follows

Substrate	Enzyme	Site of action	Product
Starch	Amylase	Mouth	Maltose
Protein	Pepsin	Stomach	Peptones
Lipid	Lipase	Small intestine	Fatty acids

27 (*c*) A is true but R is false. The correct form of R is as follows Gall stones is an inflammatory condition caused by disturbances in bile composition due to change in ratio of cholesterol and bile salts.

28 (*d*) Option (d) shows the incorrect match. The correct information is as follows

Pellagra is caused by the deficiency of vitamin-B_3 (niacin), while vitamin-B_2 (riboflavin) deficiency causes dermatitis.

29 (*c*) Sphincter of Oddi guards the opening of hepatopancreatic duct into the duodenum.

Hepatopancreatic duct brings secretion of the liver as well as pancreas to the duodenum.

30 (*a*) In stomach, gastric acid (HCl) is secreted by the parietal cells of gastric gland. It makes the medium of food in stomach acidic for the activation of the proteolytic enzymes in the stomach.

31 (*a*) Insulin is a peptide hormone that is secreted by the β-cells of islet of Langerhans of pancreas. It lowers the blood glucose level by enhancing the cellular uptake and utilisation of glucose by hepatocytes and adipocytes.

32 (c) Statement in option (c) is correct whereas the other four statements are incorrect. These can be corrected as follows

- Oxyntic cells are present in the epithelium of stomach and not of intestine.
- Ptyalin acts upon carbohydrates, i.e. starch and convert it to simple forms, i.e. glucose.
- Sphincter of oddi is present at the junction of ampulla of vater and duodenum. More precisely it guards the opening of hepatopancreatic duct.
- Goblet cells secrete mucus in stomach.

33 (d) Option (d) represents correctly matched pair. Other options represents incorrect pairs and can be corrected as follows

- Brunner's gland - Mucus secretion
- Intestinal mucosa - Mucus-secreting goblet cells.
- Gall bladder - Store bile
- Goblet cell - Mucus

36 (c) The optimum pH for pepsin function is 1.6 - 2.4, i.e. it acts in highly acidic medium, maintained by HCl secreted by oxyntic cells.

37 (d) Pancreatic juice do not contain enterokinase. It is rather present in intestinal juice and it helps to convert inactive trypsinogen to active trypsin.

39 (b) Intrinsic factor for the absorption of vitamin-B_{12} is secreted by the oxyntic or parietal cells of stomach.

41 (a) the intestinal mucosal epithelium possess mucus secreting goblet cells. When this mucus combine with the secretions of brush border cells of mucosa, succus entericus or intestinal juice is formed. It contain different enzymes, e.g. lipase, nucleosidases, etc.

43 (b) Nucleases enzyme is found in pancreatic juice as DNAase and RNAase. It is not found in succus entericus.

44 (b) Primary dentition in human do not possess premolars. Therefore, the dental formula of a child is $\frac{2102}{2102}$, where zero represents premolars.

45 (d) The initial step in the digestion of milk in humans is carried by pepsin. It hydrolyses the milk protein casein into paracasein which precipitates as calcium paracasein to form semisolid curd. This is known as coagulation of milk. In infants coagulation of milk is done by rennin.

46 (b) Fructose is absorbed into the blood through mucosa cells of intestine by the process called facilitated transport.

It is the process of spontaneous passive transport of the molecules or ions across a biological membrane via specific transmembrane integral protein, but without the involvement of energy.

47 (b) Human salivary amylase is an enzyme present in the saliva and the optimum pH for its activity is 6.8. This means that it is slightly acidic.

Starch is broken down into sugars in the mouth because of the presence of salivary amylase.

48 (b) During digestion of food, proteins are digested into peptones and proteoses and finally into dipeptides. In the given equation 'X' is pepsin (enzyme of stomach which gets activated on exposure to HCl) which converts proteins into peptones and proteoses at pH 1.3.

Enzyme 'Y' is trypsin (enzyme secreted from pancreas which activates when enterokinase acts on trypsinogen), which activates trypsinogen and procarboxypeptidase into chymotrypsin carboxypeptidase, respectively and further converts proteoses and peptones to dipeptides.

49 (b) The statement in option (b) is not correct. The correct information regarding the statement is as follows

The optimum pH for activity of salivary amylase is 6.8. It brings about the initial digestion of starch like polysaccharides.

Rest of the statements are correct.

50 (a) The fatty acids, glycerides, sterols and fat soluble vitamins that are absorbed into the intestinal cells by diffusion are resynthesised in the endoplasmic reticulum and are converted into small fat molecules (droplets) coated with proteins called chylomicrons. These are release into the lymph by the intestinal cells.

51 (c) The correct matching of the end products, their simpler form and enzymes acting upon them are as follows

(i) Lactose $\xrightarrow{\text{Lactase}}$ Galactose

(ii) Monoglycerides $\xrightarrow{\text{Lipase}}$ Fatty acids

(iii) Starch $\xrightarrow{\text{Amylase}}$ Maltose

(iv) Peptones $\xrightarrow{\text{Trypsin}}$ Dipeptides

Thus, option (c) is correct.

52 (c) Liver is the largest exocrine gland of the body. It does not produce any digestive enzymes. It consists of many hepatic cells (hepatocytes) arranged in the form of cords which produce bile juice. Bile juice does not contain any digestive enzyme like the gastric juice.

53 (b) Osteomalacia occurs due to deficiency of calciferol, i.e. vitamin-D (a group of fat or lipid soluble secosteroids). It is responsible for enhancing intestinal absorption of calcium, iron, magnesium, phosphate and zinc and is obtained from diet or from sun exposure. Keratomalacia is an eye disorder that occurs due to the vitamin-A deficiency. Glossitis refers to inflammation of the tongue. Pernicious anaemia is a deficiency disorder occurring due to vitamin-B_{12} deficiency.

54 (b) Absorption of vitamin-B_{12} or cobalamin in human requires intrinsic factor (P) glycoprotein secreted from stomach (Q). Intrinsic Factor (IF), also known as Gastric Intrinsic Factor (GIF), is a glycoprotein produced by the parietal cells of the stomach. It is present in the gastric juice as well as gastric mucous membrane and is necessary for the absorption of vitamin- later in the small intestine.

55 (b) Jejunum is the highly vascular, thick-walled middle part of the small intestine. It has a diameter of about 4 cm and is about 2.5 metre long.

56 (a, c) The secretion of pancreatic juice is stimulated by both secretin and cholecystokinin (CCK). These two hormones, however, are secreted in response to different stimuli and they have different effects on the composition of pancreatic juice.

58 (a) The pair given in option (a) is incorrectly matched. The correct information regarding the incorrect match is as follows

Rennin is secreted only in some mammals by the stomach chief cells as an inactive precursor, i.e. prorennin. It is one of the major digestive enzymes of gastric juice which is responsible for converting casein protein of milk to calcium paracaseinate.

Rest of the matches are correct.

59 (*d*) Emulsified fats are digested by pancreatic and intestinal juices. First, the bile salts of the bile carry out emulsification of fat droplets.

This increases the action of lipase on fats which is present in both the pancreatic and intestinal juices.

Pancreatic lipase converts emulsified fats into monoglycerides.

The intestinal lipase then converts the remaining fats into monoglycerides, fatty acids and glycerol.

60 (*b*) Digestion of proteins in human starts in the stomach with the help of pepsin enzyme. In buccal cavity, there is no digestion of proteins because saliva contains no proteolytic (i.e. protein digesting) enzyme. When food reaches the stomach, it gets mixed with HCl and becomes acidic, which activates pepsinogen into pepsin. This activated pepsin initiates digestion of protein and converts complex proteins into proteases, peptones and larger peptides.

65 (*b*) To keep people healthy, strong, energetic and long lived, it is necessary to have a balanced diet. The balanced diet contains carbohydrate, lipids, proteins, vitamins, minerals and water. Thus, balanced diet does not include nucleic acids and enzymes.

67 (*b*) Enterogastrone hormone produced by small intestine which slows down the secretion of gastric juice.

68 (*b*) Pepsin is present in gastric juice. It acts on proteins at 1.2-1.8 pH and breaks them down into peptones.

72 (*a*) Unhealthy eating habits together with anxiety, stress or panic attacks may cause indigestion, stomach ache, stomach palpitations, nausea, etc.

73 (*a*) Carboxypeptidase is synthesised and secreted by the exocrine cells of pancreas. Zn^{2+} is required for the activity of carboxypeptidase, carbonic anhydrase and alcohol dehydrogenase.

74 (*a*) Caecum is the proximal part of large intestine. It is connected to colon (the largest part of large intestine). Colon does not contain digestive gland, but harbours a number of microorganisms, which feed on indigestible matter, some of them are symbiotic. They produce vitamin-K and vitamin-B complex.

76 (*b*) The saliva is a colourless, opalescent and sticky liquid produced by the salivary glands in quantity about 1.5 dm^3 (cubic decimetre), i.e. 1.5 L/day.

77 (*a*) Enterokinase is an activator enzyme, which converts the proenzyme trypsinogen into active trypsin, a proteolytic enzyme.

80 (*a*) The ampulla of Vater, also known as the hepatopancreatic ampulla, is formed by the union of the pancreatic duct and the common bile duct. It provides the common passage for bile and pancreatic juices.

84 (*a*) There are three pairs of salivary glands in man which secrete slightly acidic saliva of about pH 6.8.

92 (*c*) Vitamin-D affects normal growth of body and the formation of teeth and bones by depositing essential mineral in them.

Vitamin-A is essential for maintenance of epithelial cells and bones. There is correlation between vitamin-A and tooth development. Calcium is essential for the formation of bone and teeth.

CHAPTER 24

Breathing and Exchange of Gases

All activities of living organisms require energy. This energy is obtained by the breakdown of nutrients present in their food. This breakdown process is called **respiration** and usually occurs in the presence of oxygen. Thus, respiration is defined as 'the process of oxidation of nutrients to release energy'.

This energy is released in the form of ATP along with the release of water and carbon dioxide. Out of these, carbon dioxide is toxic to the body cells. So, removal of this CO_2 from the body is an essential requirement.

Thus, to carryout normal body functions, living organisms constantly require oxygen from the atmosphere. Simultaneously, they continuously remove carbon dioxide produced by their cells, out of the body. This is done by specialised organs called **respiratory organs**. This chapter is devoted to these organs and their functioning along with their disorders.

Inside

TYPES OF RESPIRATION AND RESPIRATORY ORGANS

Based on the mode of oxidation of nutrients, respiration is of the following two types

Aerobic respiration

It occurs, when the cells utilise molecular oxygen for oxidising nutrient. Its site of occurrence is mitochondria of the cells. Large amounts of ATP per glucose molecule is produced in this process. It is performed under normal circumstances by animals.

Anaerobic respiration

It occurs, when nutrients are oxidised without using molecular oxygen. It is also called **fermentation**. Its site of occurrence is the cytoplasm of the cells. It produces less ATP per glucose molecule. It is performed during oxygen deficient situations, like intense exercise session, etc. Anaerobic respiration is a common characteristic feature of yeast, bacteria, endoparasites, muscles and mature RBCs.

Note *In anaerobic respiration, when glucose acts as the substrate*

- *Yeast forms ethyl alcohol and CO_2 as the final product.*
- *Bacteria and muscles form lactic acid as the final product.*

Respiratory Organs in Animals

Respiratory organs collectively form the **respiratory system.** A well-defined respiratory system is the characteristic feature of animals only. As in plants, gaseous exchange takes place through general body surface.

Respiratory systems and modes of respiration in animals differ greatly. It majorly depends upon their mode of life. This diversification is given below in brief within the table.

Table 24.1 Respiratory organs and modes of respiration in various animals

Respiratory organs	Mode of respiration	Examples
Body surface	Gaseous exchange through general body surface	Almost all lower animals including sponges, coelenterates and free-living helminths upto phylum-Annelida
Gills	Bronchial respiration (Through diffusion from water)	Crustaceans (prawn), cartilage and bony fishes
Book gills	Book gill respiration (Through diffusion from atmospheric air)	*Limulus* (king crab) of Phylum-Arthropoda
Skin	Cutaneous respiration	Annelids (earthworm) and amphibians (frog)
Trachea (Ectodermal tubes)	Tracheal respiration	Insects (cockroach), centipedes and millipedes
Lungs	Pulmonary respiration	Most of tetrapods
Book lungs	Book lung respiration (Through diffusion from atmospheric air)	Arachnids (spiders and scorpion)

RESPIRATORY SYSTEM IN HUMANS

Diagrammatic view of human respiratory system is given below

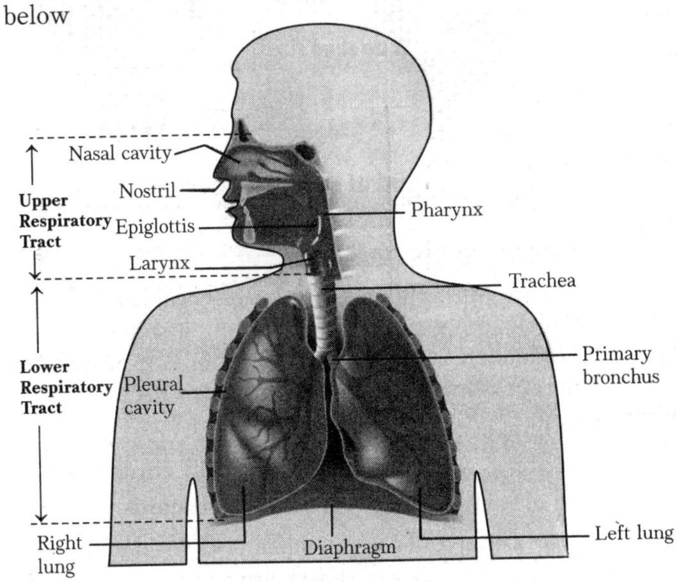

Fig. 24.1 Human respiratory system

The **special features** of human respiratory system are

- Presence of a nose.
- Elongation of nasal cavity or passage and its complete separation from the buccal passage through palate so that internal nostrils open deep into nasopharyngeal part of pharynx.
- Long windpipe due to the presence of well-defined neck.
- Spongy and solid lungs.
- Presence of a muscular diaphragm which separates thoracic cavity from abdominal cavity.

All the above structures are helpful in performing two essential functions of respiration, i.e. intake of oxygen and removal of CO_2. Hence, before exploring further, it is better to get a quick grasp of steps involved in the process of respiration. These are as follows

Step 1 **Breathing** or **Pulmonary ventilation,** i.e. the process through which atmospheric air is drawn in and CO_2 rich air is released out *via.* the following route

External nostrils ⟶ Nasal chamber ⟶ Nasopharynx ⟶ Larynx ⟶ Trachea ⟶ Primary bronchi ⟶ Secondary bronchi ⟶ Tertiary bronchi ⟶ Bronchioles network ⟶ Alveoli

Step 2 Diffusion of gases (O_2 and CO_2) across the alveolar membrane.

Step 3 Transport of these gases by blood.

Step 4 Diffusion of O_2 and CO_2 between blood and tissues.

Step 5 Utilisation of O_2 by cells and resultant release of CO_2.

Respiratory system in humans constitutes two major parts, i.e. respiratory tract and lungs.

Respiratory tract

The structures involved in the respiratory tract are as follows

External nostrils

A pair of external nostrils represents the initiation of the respiratory passage. These are the openings of nose and are present above the upper lip. These lead into the nasal passage through vestibule.

Nasal cavity

The initial portion of nasal cavity is called **vestibule.** It contains mucous lining and hairy epithelium. Its main function is to prevent the entry of large dust particles.

The portion behind the vestibule is called **nasal chamber.** The nasal chamber has

(i) **Respiratory region** It is a highly vascularised region. It involves the conditioning (i.e. temperature and moisture maintenance) of air entering through the nasal passage.

(ii) **Olfactory region** It is the upper part of nasal cavity. It is lined by olfactory epithelium, called as **Schneiderian membrane**. It performs the function of smelling.

(iii) **Nasal Turbinates** or **Concha** These are the processes of bones present to increase the surface area of this chamber. In accordance with position, their names and bones involved in their formation are given below

Concha	Position	Bone
Superior	Uppermost	Nasals
Middle	Middle	Ethmoids
Inferior	Lowermost	Maxilla

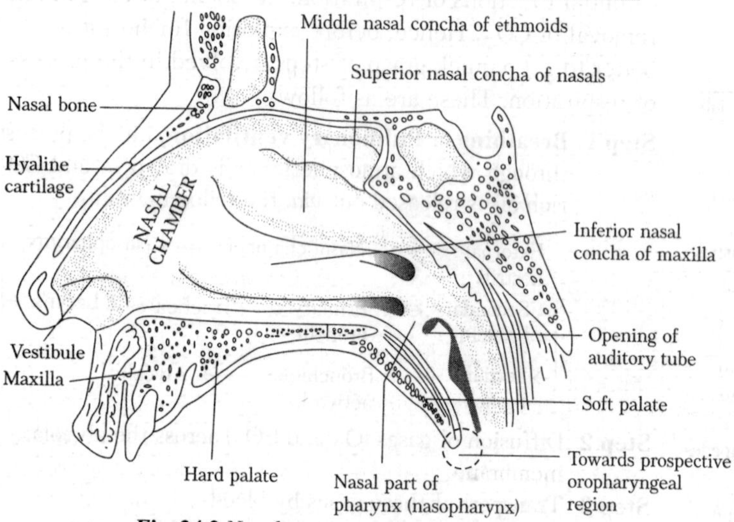

Fig. 24.2 Nasal passage and associated structures

Nasal cavity terminates at internal nostril.

Pharynx

It is the common passage of air and food. It is divided into following three regions

(i) **Nasopharynx** Internal nostril opens into this part of pharynx. It is the portion of pharynx through which only air can pass. It opens into the trachea through the glottis of laryngeal region.

(ii) **Oropharynx** It is the middle part of pharynx. It possesses the opening of oral cavity called **fauces**. It contains two pairs of tonsil namely, **palatine** and **lingual** tonsils.

(iii) **Laryngopharynx/hypopharynx** It is the lowermost part of pharynx. It leads into two tubes namely, trachea (front) and oesophagus (back).

Larynx

It is present anteriorly in the neck of mammals at the le el of 3rd to 6th cervical vertebrae. It is commonly called as the **voice box** due to its involvement in sound production. Its upper part has an opening, i.e. **glottis**.

During swallowing of food, this glottis remains covered by **epiglottis**, a leaf-shaped cartilaginous structure made up of elastic cartilage. It prevents the entry of food into the larynx.

Laryngeal cartilages

The laryngeal skeleton consists of total nine cartilages. Details of these cartilages are given in following table

Cartilage names	Paired or Single	Cartilage types	Remarks (if any)
Epiglottis	Single	Elastic	Covers glottis
Thyroid	Single	Hyaline	Most prominent (more in males), C-shaped, dorsally incomplete, forms **Adam's Apple**
Cricoid (Present below thyroid)	Single	Hyaline	Signet ring-shaped
Arytenoid	Paired	Hyaline	Forms posterior wall of larynx, roughly pyramidal in shape
Cartilage of Santorini or **Corniculate**	Paired	Elastic	Node-like, present at the tip of arytenoid cartilage
Cuneiform	Paired	Elastic fibres	Club-shaped nodules, present above and anterior to corniculate cartilage, connects epiglottis with arytenoid cartilage

Actual arrangement of these cartilages is shown in following figure

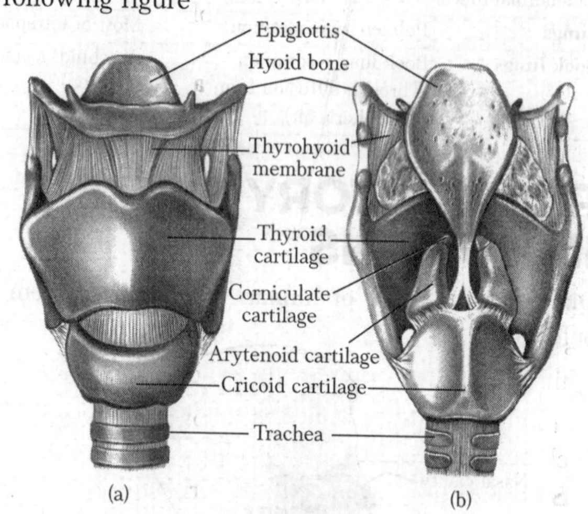

Fig. 24.3 Larynx : (a) Anterior view, (b) Posterior view

Vocal cords

The larynx contains the folds of vocal cords, which are essential for phonation or sound production. It connects the inferior part of the pharynx (hypopharynx) with the trachea.

In humans, 2 pairs of vocal folds are present,

(i) inner white coloured **true vocal cords**

(ii) outer pink coloured **false vocal cords**

True vocal cords are attached to thyroid and arytenoid cartilages. These are actually responsible for sound production. The space present between these cords is called **Rima glottidis**. This space can be further differentiated into

(i) **Intermembranous space**, i.e. the part towards thyroid cartilage.

(ii) **Intercartilaginous space**, i.e. the part in between arytenoid cartilages.

The pitch of sound is directly proportional to the tension present in vocal cords. The greater the tension, the higher the pitch.

Note *In birds, sound box is called* **syrinx**, *an additional structure present within trachea, below larynx.*

Before exploring the details of respiratory tract further, it is better to learn about the outline structure of lungs. It is so because the major portions of onwards respiratory tract are placed inside lungs.

Lungs (Pulmones)

The lungs occupy most of the part of thoracic cavity. Each human lung is pink coloured at birth but becomes dark grey and mottled in adults.

Coverings and enclosing cavity

Each lung is covered by a thin double layer of simple squamous epithelium called **pleura**. These are enclosed within a fluid-filled **pleural cavity**. The watery fluid inside the cavity is called **pleural fluid** and it is glycoproteinaceous in nature.

This fluid lubricates the lungs to prevent any friction which might produce during breathing.

Right lung (avg. weight 625 g)

This is the shorter lung. Its size is about 2.5 cm less due to raised position of diaphragm to accommodate liver.

The right human lung contains following 3 lobes which are separated by 2 fissure.

(i) **Superior lobe** Separated by horizontal fissure

(ii) **Middle lobe**
(iii) **Inferior lobe** } Separated by oblique fissure

Internally, these lobes are further divided into 10 segments.

Left lung (avg. weight 565 g)

This lung is longer but narrow than right lung. It contains a **cardiac notch** to accommodate heart.

The left human lung contains following 2 lobes which are separated by a single fissure.

(i) **Superior lobe**
(ii) **Inferior lobe** } Separated by oblique fissure

Internally, these lobes are further divided into 5 segments. Segments of both the lungs are furhter divided into lobules. Both the lungs with their associated structure are shown in the figure below

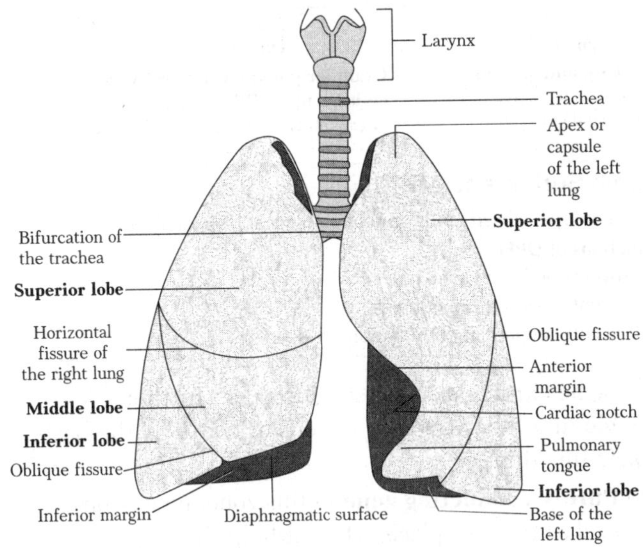

***Fig. 24.4** Human lungs and trachea*

Positive and negative pressure lungs

(i) **Positive pressure lungs** In them, the pressure inside the lungs is + ve in comparison to the atmospheric pressure at the time of inspiration, e.g. frog (hollow lungs).

(ii) **Negative pressure lungs** In them, the pressure inside the lungs is – ve as compared to atmospheric pressure at the time of inspiration, e.g. humans (solid spongy lungs).

Trachea (Windpipe)

It is an air conducting tube with non-collapsible walls due to cartilaginous C-shaped/incomplete rings of hyaline cartilage. The number of rings are 16 or 17 in human beings. Trachea is about 12 cm long.

Its first branching is visible just before its entry into lungs at the level of 5th thoracic vertebrae. These branches are called **bronchus** (Pl. bronchi).

Bronchi and bronchioles

Bronchus after entering into lungs divides repeatedly to lobar, segmental and terminal bronchus. These terminal bronchi then divide further to give the series of bronchioles namely lobular, terminal and respiratory bronchioles.

Alveoli

Respiratory bronchioles further divide to form alveolar ducts. These alveolar ducts then open into atrium containing alveolar sacs. The latter contain the primary units of gaseous exchange, i.e. **alveolus** (Pl. alveoli). Human lung contains approximately 300 million alveoli.

These are specialised air-filled sacs which are richly supplied with blood capillaries. These sacs contain two types of cells as listed below

Alveoli

Cell types

- **Type I** — **Flat squamous cells** Normal supportive epithelial cells
- **Type II** — **Granular pneumocytes or Clara cells** Secrete **DPPC** (Dipalmitoyl phosphatidylcholine), i.e. lipid surfactant

Lipid Surfactant (DPPC)

These are specialised lipids present on outer surface of alveoli.

Functions of DPPC

- Prevents oedema (swelling).
- Prevents alveoli from collapse.
- Prevents accumulation of lymph on alveoli.

The trachea and associated structures are further divisible into following two regions or zones on the basis of their functions.

1. **Purely conducting zone** In this zone, no gaseous exchange takes place. Thus, it helps in the conduction or transport of gases.
2. **Respiratory cum conducting zone** In this zone, transport associated with gaseous exchange takes place.

Both of these zones are shown in the figure below

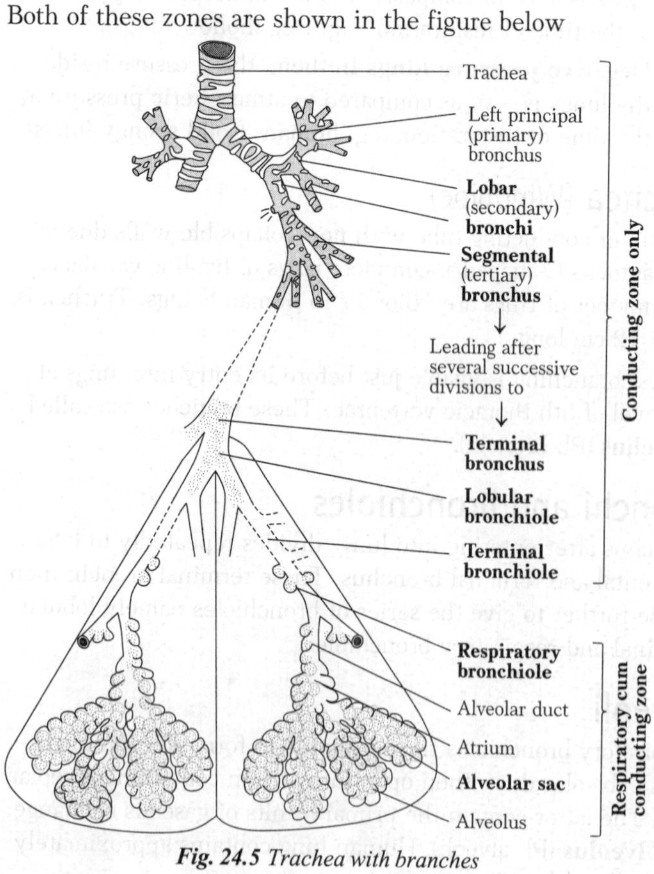

Fig. 24.5 Trachea with branches

Trachea has the diameter of 2.5 cm^2 only, while the total area of alveoli in a human lung is 70 m^2 (approximately).

Thus, such intense branching help to increase the surface area for diffusion or exchange of gases. The bronchioles of respiratory cum conducting zone, i.e. respiratory bronchioles along with alveolar ducts and alveoli do not contain cartilaginous rings.

On the other hand, all the structures of conducting zone are supported by cartilaginous rings.

Epithelium in different respiratory parts/organs

The conducting and gas exchanging properties of various respiratory organs/parts are attributed to their epithelial lining. A brief of this epithelial lining with different epithelia in different parts of respiratory tract is given below

Organs	Epithelium
Nose	
• Vestibule	Non-keratinised stratified squamous
• Respiratory region	Pseudostratified ciliated columnar
• Olfactory region	Olfactory epithelium
Pharynx	
• Nasopharynx	Pseudostratified ciliated columnar
• Oropharynx	Non-keratinised stratified squamous
• Laryngopharynx	Non-keratinised stratified squamous
Larynx	Non-keratinised stratified squamous above vocal folds; pseudostratified ciliated columnar below vocal folds.
Trachea	Pseudostratified ciliated columnar
Bronchi, Bronchioles and Alveoli	
• Primary bronchi	Pseudostratified ciliated columnar
• Secondary bronchi	Pseudostratified ciliated columnar
• Tertiary bronchi	Pseudostratified ciliated columnar
• Larger bronchioles	Ciliated simple columnar
• Smaller bronchioles	Ciliated simple columnar
• Terminal bronchioles	Non-ciliated simple columnar
• Respiratory bronchioles	Simple cuboidal to simple squamous
• Alveolar ducts	Simple squamous
• Alveoli	Simple squamous

CHECK POINT 24.1

1. A book-shaped organ which is not truly ventilating is the characteristic of
   (a) mites (b) *Opiliones* (c) sunspiders (d) scorpion
2. Choose the animal in which lungs have more pressure than atmospheric pressure.
   (a) frog (b) rat
   (c) camel (d) rabbit
3. Which of the following in given options act as a common passage between two organ systems?
   (a) Larynx (b) Gullet
   (c) Glottis (d) Pharynx
4. Name the part of the lung in which the apex of the heart lies?
   (a) Costal surface (b) Mediastinal surface
   (c) Hilum (d) Cardiac notch
5. Which one of the following has the smallest diameter?
   (a) Right terminal bronchiole (b) Left terminal bronchiole
   (c) Tertiary bronchus (d) Respiratory bronchiole
6. In which part(s) exactly the gaseous exchange takes place in mammals?
   (a) Trachea and alveolar duct (b) Trachea and bronchi
   (c) Alveolar duct and alveoli (d) Alveoli and tissues

FUNCTIONING OF RESPIRATORY SYSTEM : BREATHING

The process of respiration includes breathing, gaseous exchange and gaseous transport. Breathing or pulmonary ventilation is the process of in (inspiration) and out (expiration) movement of air from the lungs.

Muscles involved in breathing

The overall ventilation movements of the lungs in mammals are governed by **diaphragm** and **intercostal muscles.**

These muscles are attached to the borders of ribs. Other muscles, which are helpful in the breathing process include **transversus abdominis** and **rectus abdominis** as shown in the given diagram.

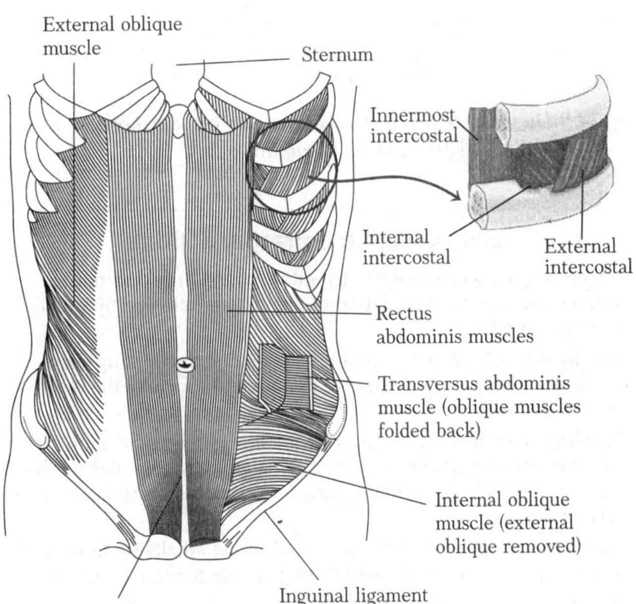

Fig. 24.6 Muscles of Pulmonary ventilation

Mechanism of breathing

Mechanism of breathing includes following two stages

I. Inspiration or inhalation

It is an active process by which fresh air enters the lungs. It is carried out mostly by the movement of diaphragm. During this process, following muscular movements are visible.

(i) **Diaphragm** is lowered by the contraction of muscle fibres and becomes flat. This causes an increase in the volume of thoracic chamber in the antero-posterior axis.

(ii) **Abdominal muscles** relax and allow compression of abdominal organs by the diaphragm.

(iii) **External intercostal muscles** contract to move ribs in upward and outward direction. Due to this movement of ribs, volume of thoracic cavity increases.

The movement of these muscle can be better understood with the help of following figure

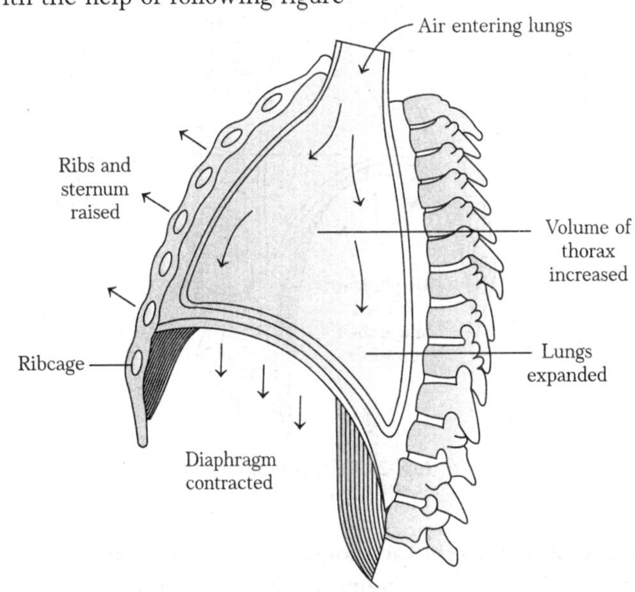

Fig. 24.7 Inspiration

Result

The overall increase in the volume of thoracic cavity causes an increase in pulmonary volume. As a result, intrapulmonary pressure gets lowered than the atmospheric pressure. Thus, negative pressure develops in the lungs and air rushes inside it. Pressure changes in lungs and thorax during inspiration are as follows

(i) –2.5 mm of Hg to –6 mm of Hg during normal breathing.

(ii) –2.5 mm of Hg to –40 to –50 mm of Hg during forceful breathing.

II. Expiration or Exhalation

It is a passive process by which air (CO_2) is expelled out from the lungs. This process involves following muscular movements.

(i) **Relaxation** of both diaphragm and external intercostal muscles.

(ii) The process begins with **contraction of rectus abdominis muscles** due to which diaphragm bends upwards and attains its normal position. It reduces the vertical length of thoracic cavity.

(iii) During the relaxation of external intercostal muscles, the ribs and ribcage move downwards and inwards, respectively. This reduces thoracic volume in antero-posterior direction.

The movement of these muscle can be better understood with the help of following figure.

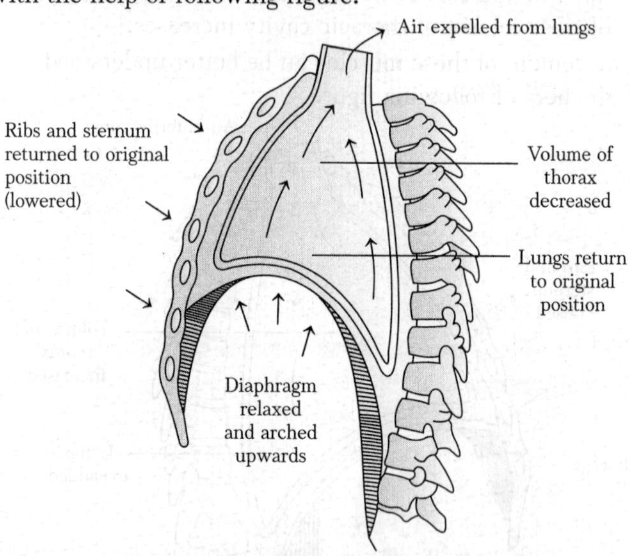

Fig. 24.8 Expiration

Result

The overall volume of the thoracic cavity decreases thereby reducing the pulmonary volume. As a result, the intrapulmonary pressure increases slightly above the atmospheric pressure. Thus, positive pressure creates in lungs and results in exhalation.

Note
- ***During inspiration**, sternum moves toward ventral and anterior direction, whereas during expiration, it attains normal position.*
- ***Forceful breathing** is the ability to increase the strength of inspiration and expiration with the help of additional muscles.*

Diagrammatic view of volume changes in thoracic cavity during breathing is given below

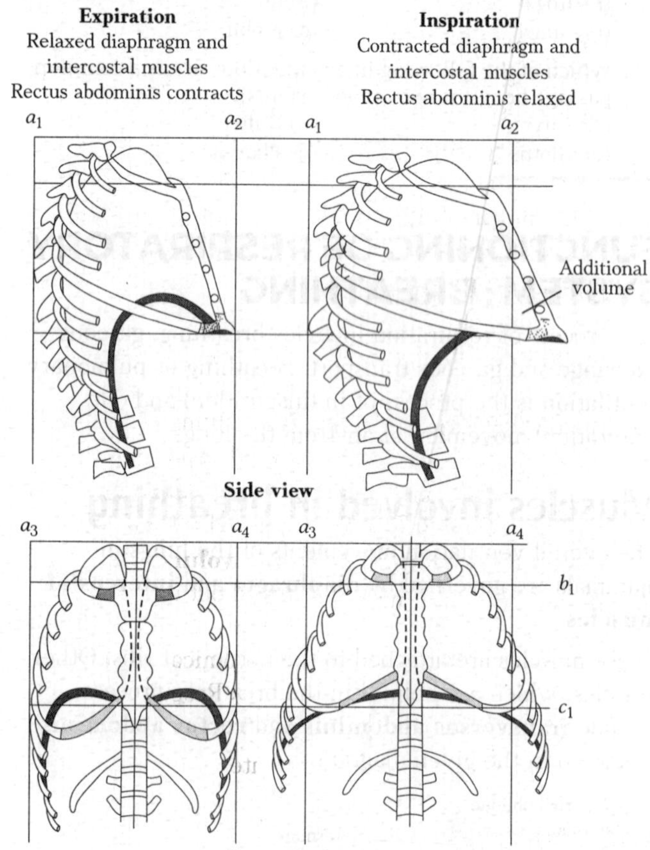

Fig. 24.9 Diagram showing changes in volume of the thoracic cavity during breathing

Specialised respiratory actions

- **Hiccups** are jerky incomplete inspiratory sound. It occurs due to sudden involuntary contraction of diaphragm followed by sudden closing of glottis.
- **Snoring** is rough rattling inspiratory noise, due to vibrating vocal cords during sleep. It occurs due to partial blockage of the upper respiratory tract.
- **Coughing** is a reflex action required for strong deep inhalation followed by a complete closure of Rima glottidis. Stimulus for this reflex is normally generated by any foreign body after entering into respiratory tract.
- **Yawning** is a reflex action for deep and strong inhalation through the widely opened mouth. It may be triggered by drowsiness or usually high CO_2 concentrations in blood.
- **Sneezing** A sudden blast of air or mucus or both expelled from mouth and nose. It is a natural way to remove any irritant from upper respiratory tract.

LUNG VOLUMES AND CAPACITIES

The functioning of lungs can be mapped with the help of following lung volumes and capacities.

Lung volumes

1. **Tidal Volume** (TV) It is the amount of air, which passes into and out of the lungs during each cycle of normal respiration. Its approximate value is 500 mL.
2. **Inspiratory Reserve Volume** (IRV) It is the extra volume of air that can be inhaled into the lungs during maximal inspiration. The average IRV is 2500-3000 mL.
3. **Expiratory Reserve Volume** (ERV) It is the maximum volume of air, which can be expelled from the lungs during forceful expiration. It is about 1000-1100 mL.
4. **Residual Volume** (RV) It is the volume of air remained in the lungs after forced expiration. It cannot be measured directly and observed only after lungs are opened. It is about 1100-1200 mL.
5. **Alveolar ventilation** or **Minute volume** It is the volume of air that moves into and out of the alveoli per minute.

Alveolar ventilation = (TV– Anatomical dead space) × Respiratory rate

$$= (500 - 150)\ \text{mL} \times 15 \text{ times per minute}$$
$$= 5.25 \text{ litre per minute}$$

Lung capacities

1. **Inspiratory Capacity** (IC) It is the amount of air that can be inspired with the maximum effort. It is about 3000-3500 mL. It is the sum total of TV and IRV.
2. **Functional Residual Capacity** (FRC) It is the amount of air remained in the air passages and alveoli at the end of quiet expiration. The functional residual volume prevents the collapse of alveoli during expiration. It is about 2500 mL. It is the sum total of ERV and RV.
3. **Vital Capacity** (VC) It is the maximum volume of air, which can be moved in and out of the lungs during forced breathing. It is about 3500-4500 mL and it is the sum total of Tidal Volume, IRV and ERV.
4. **Total Lung Capacity** (TLC) The total volume of air contained in the lungs at the end of forceful inspiration.

   It is about 5800-6000 mL. It is the sum total of RV, ERV, TV and IRV.

All these lung capacities and volumes can be measured by using a device called **spirometer**.

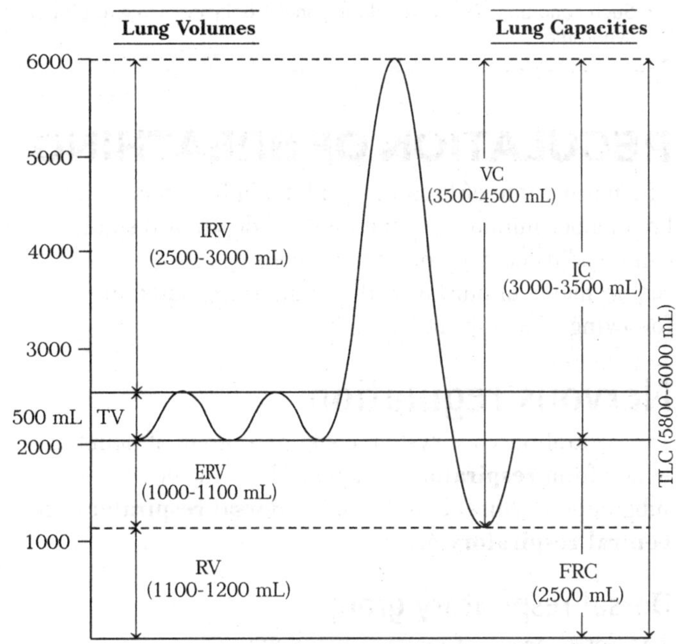

Fig. 24.10 Spirogram of lung volumes and capacities

Dead space

In lungs, it can be categorised as

(i) **Anatomical dead space** The portion of respiratory passage in which incoming and outgoing air is completely locked.

This air is not used in gaseous exchange and is about 30% of tidal volume, i.e. 150 mL.

(ii) **Physiological dead space** or **Total dead space** It is the anatomical dead space plus the volume of alveoli in which gaseous exchange is less than normal.

In a normal person, anatomical and physiological dead space are same.

Note *In the lungs of birds, dead space is absent because of which, the lungs of birds are considered more efficient than the human lungs.*

Physiological shunt

- Sometimes deoxygenated blood bypass the lungs from the right side of the heart to its left side. Subsequently, oxygenation does not occur and ventilation/perfusion ratio remains at below normal range.
  Thus, non-ventilated alveoli remain profused (usually with fluid) and function as a physiological shunt.
- The blood flowing through this shunt is called **shunted blood**.
- Such condition is observed during pneumonia and acute respiratory syndrome.

REGULATION OF BREATHING

The normal rate of breathing in human beings is 14-18 breaths per minute. But the rate and depth of respiration can be adjusted and controlled according to the requirements of our body. Breathing is regulated in following two ways in humans

Nervous regulation

The central nervous system contains a group of neurons, which form **respiratory centres**. These include two subgroups within the medulla, i.e. **dorsal respiratory** and **ventral respiratory** groups.

Dorsal respiratory group

The dorsal group of neurons functions as **inspiratory centre**. This group is connected to the muscles of diaphragm and external intercostals.

It is responsible for transmitting continuous impulses to these muscles.

Thus, due to inspiratory centre, these muscles are always in the conditions of excitation for getting **BRR**, i.e. Basic Respiratory Rhythm.

Ventral respiratory group

The ventral group of neurons are concerned with both the processes, i.e. inspiration and expiration. It is mainly helpful in faster breathing.

Respiratory subcentres

Pons also has a pontine respiratory group with two subcentres, i.e pneumotaxic centre and apneustic centre.

- The **pneumotaxic centre** is situated in the upper part of pons towards the dorsal side. The neural signals from this centre reduces the inspiration thereby, control the respiratory rate.
- The **apneustic centre** is located in the lower part of pons and it controls the depth of inspiration.

The complete arrangement of respiratory centres in human brain is shown below

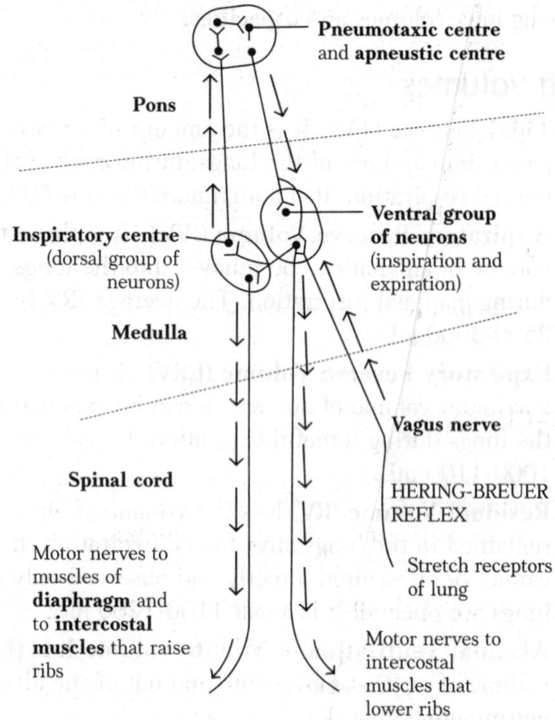

Fig. 24.11 Diagram showing nervous control of breathing

Process of regulation and Hering-Breuer Reflex

For this process, specialised receptors are present on the walls of bronchi and bronchioles. These receptors are called **stretch receptors**.

When lungs become overstretched, these stretch receptors send signals to dorsal respiratory group of neurons through cranial (vagus) nerves. In response to these signals, the inspiratory centre is inhibited. This response is called **Hering-Breuer inflation reflex.**

Chemical regulation

The receptors for chemical regulation are located in carotid bodies (largest in number), aortic bodies and in brain.

(i) **Carotid and aortic bodies** The carotid and aortic bodies are found associated with heart. These bodies are responsible for the detection of CO_2 concentration in blood and thus these affect the rate of breathing. These are considered as **peripheral chemoreceptors**.

(ii) **Receptors of brain** The chemoreceptors that mediate the hyperventilation process are situated in medulla oblongata and consequently called as **medullary chemoreceptors.**

- **Effect of O_2 and CO_2** The blood oxygen does not have significant direct effect on the respiratory centre of brain. But CO_2 levels in blood directly affect the chemoreceptors of brain.
- Thus, our body responds more quickly to blood CO_2 levels.

$$\text{Breathing rate} \propto \frac{1}{O_2 \text{ concentration}} \propto CO_2 \text{ concentration}$$

- **Effect of temperature** Along with above listed factors, temperature also affects breathing rate.

Increased temperature upto certain extent, increases metabolic activities thus, increase the breathing rate.

Terminology related to breathing and blood gaseous levels

- **Eupnea** or **Eupnoea** Normal breathing rate, i.e. 12-15 per minute.
- **Hypopnea** or **Hypopnoea** Decrease in breathing rate.
- **Hyperpnea** or **Hyperpnoea** Increase in breathing rate.
- **Dyspnea** or **Dyspnoea** Painful breathing or breathing with difficulty.
- **Tachypnea** or **Tachypnoea** Rapid shallow breathing.
- **Hypercapnia** Excess of carbon dioxide in the body.
- **Anoxia** No oxygen at tissue level.
- **Hypoxia** Lack of oxygen at tissue level. This term is preferred over anoxia.

CHECK POINT 24.2

1. Contraction of diaphragm increases the volume of thoracic chamber in the
(a) antero-posterior axis (b) dorso-ventral axis
(c) ventro-lateral axis (d) dorso-lateral axis

2. During forced expiration, actively contracting muscles include the
(a) Diaphragm muscles
(b) External intercostals
(c) Abdominal muscles
(d) Diaphragm and intestinal muscle

3. Which of the following statements about the mechanism of ventilation is incorrect?
(a) As the diaphragm relaxes, air is expelled from the respiratory system
(b) During inspiration, the lungs act as suction pump
(c) Inspiration is a passive and expiration is an active process
(d) For quiet breathing, external intercostal muscles and diaphragm play an important role

4. A person breathing normally at rest takes in and expels approximately half a litre of air during each respiratory cycle. This is called
(a) inspiratory reserve volume (b) tidal volume
(c) expiratory reserve volume (d) vital capacity

5. The respiratory centre in the brain is stimulated by
(a) more air in trachea (b) more air in lungs
(c) more air in mouth (d) more O_2 in blood

6. Chemosensitive area of respiratory centre in medulla is affected by
(a) less CO_2 and H^+ ions (b) less O_2 and H^+ ions
(c) excess CO_2 and OH^- ions (d) excess O_2 and OH^- ions

EXCHANGE OF GASES

The gaseous exchange is visible at two major sites

(i) Within the alveoli of lungs, where atmospheric air is exchanged with the blood.

(ii) At the tissue level, where gaseous exchanged between tissue fluid and blood.

The oxygen and carbon dioxide is exchanged at these sites by simple diffusion mainly based on pressure gradient.

The details and various factors involved in gaseous exchange at alveoli and tissue level are discussed below

Gaseous exchange at alveoli level

The diffusing respiratory membrane in lungs consists of alveolar surfactant, alveolar epithelium, epithelial basement membrane, a thin interstitial space, capillary basement membrane, capillary endothelial membrane as shown below

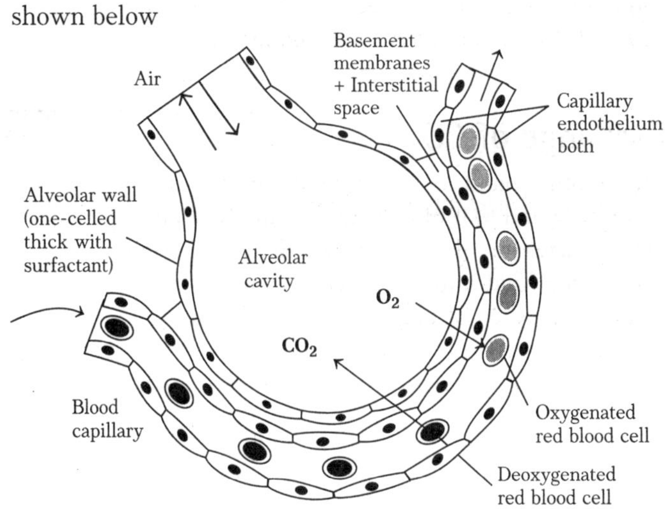

Fig. 24.12 Diagrammatic representation of a section showing an alveolus with a pulmonary capillary

The limit of exchange of gases between alveoli and pulmonary blood is called **diffusing capacity**. In lungs, this diffusing capacity is 20 times more for CO_2 than O_2.

Exchange of O_2

The partial pressure of oxygen (pO_2) in alveolar air is about 104 mm Hg and that in arterial capillaries of alveoli is 40 mm Hg.

Therefore, oxygen from the alveolar air rapidly diffuses into the blood capillaries. As the result, the pO_2 becomes 95 mm of Hg in the blood leaving lungs.

Exchange of CO_2

Similarly, pCO_2 in alveolar air is 40 mm of Hg and that in arterial capillaries of alveoli is 45 mm of Hg.

Therefore, CO_2 rapidly leaves blood capillaries and reaches alveoli. As the result pCO_2 becomes 40 mm of Hg in the blood leaving lungs.

Gaseous exchange at tissue level

The gaseous exchange between blood and body tissue fluid also occurs due to differential partial pressures as described below

Exchange of O_2

pO_2 in a respiring cell tissue fluid is 40 mm Hg, while it is 95 mm Hg in capillary blood.

Therefore, O_2 diffuses from blood into tissue fluid and from there into cells. As the result, the pO_2 becomes 40 mm of Hg in the blood leaving the tissue fluid.

Exchange of CO_2

On the other hand, pCO_2 of blood capillaries is 40 mm Hg and that of cells is 50 mm Hg. Hence, CO_2 diffuses out of the cells into tissue fluid and from tissue fluid into blood. This results in the increase of pCO_2 of blood to 45 mm of Hg.

Both of these exchanges at lung as well as at tissue level are visible in following figure

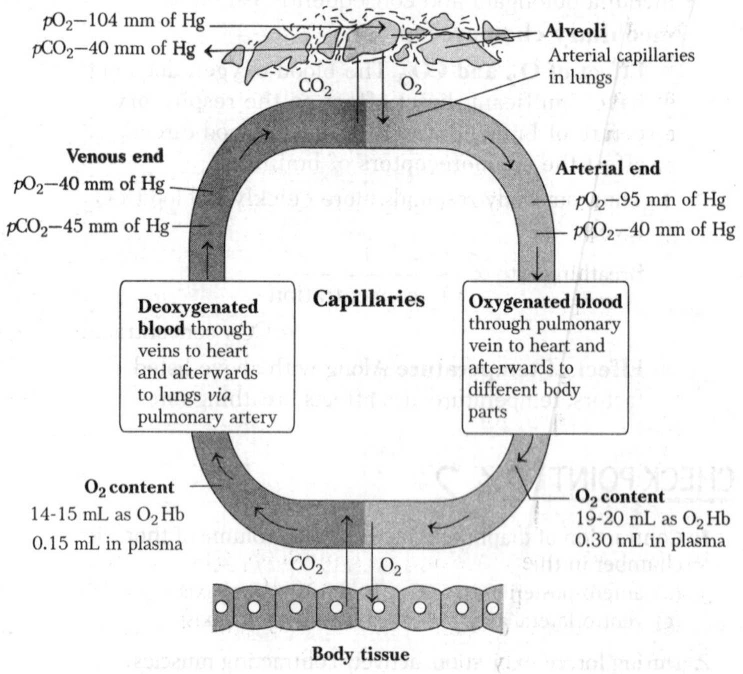

Fig. 24.13 Exchange of O_2 in the tissue and in the lungs

The table given below represents partial pressure (in mm Hg) of oxygen and carbon dioxide at different sites involved in diffusion in comparison to those in atmosphere.

Sites	pO_2 (in mm Hg)	pCO_2 (in mm Hg)
Atmospheric air	159	0.3
Alveoli	104	40
Blood (Deoxygenated)	40	45
Blood (Oxygenated)	95	40
Tissue	40	45
Inside cells	20	50

TRANSPORT OF GASES IN BLOOD

A normal male at rest requires about 250 mL of oxygen per minute. The requirement increases many folds during vigorous exercise.

All of this oxygen is transported by the blood in dissolved state. Similarly, the transportation of CO_2 also occurs *via* blood.

Transport of oxygen

Oxygen enters the venous blood within the lungs. It leaves the blood stream in the tissue capillaries and then utilised by the tissue cells.

Intake and release of O_2 from blood depend upon

- **O_2 tension**, i.e. difference in the partial pressures of O_2 between the circulating fluid and static media.
- **CO_2 tension**, i.e. difference in the partial pressure of CO_2 within the blood, i.e. circulating fluid and static media.
- H^+ ion concentration
- Electrolyte content
- Temperature
- Diphosphoglyceric acid (2, 3-DPG)

The oxygen transportation in human body occurs in the following forms

In physical solution

Under normal temperature and pressure, only 0.30 mL of oxygen is carried as physical solution in 100 mL of arterial blood. It is therefore, evident that the major part of oxygen is carried in chemically combined state.

As chemical compound

- Oxygen is transported throughout the body in combined state with haemoglobin. Thus, it is carried at the surface of red blood cells.
- **Haemoglobin** is a chromoprotein. Molecular weight of haemoglobin is 68000 daltons. It contains four atoms of iron in ferrous form.
- Each atom of iron can combine reversibly with one molecule of oxygen. Thus, four haem (iron) units of haemoglobin react with four molecules of oxygen as shown below

  (i) $Hb_4 + O_2 \rightarrow Hb_4O_2$ (ii) $Hb_4 + 2O_2 \rightarrow Hb_4O_4$

  (iii) $Hb_4 + 3O_2 \rightarrow Hb_4O_6$ (iv) $Hb_4 + 4O_2 \rightarrow Hb_4O_8$
- Thus, one haemoglobin molecule can carry 4 oxygen molecules at the maximum.
- In these reactions, the ferrous state of iron does not change due to which this process of combination is called **oxygenation** instead of oxidation.
- The importance of this carriage lies in the fact that here both the association and dissociation with haemoglobin play equivalent role in carriage,

This fact can be best understood with the help of oxyhaemoglobin dissociation curve given below

Oxyhaemoglobin-dissociation curve

The relation between partial pressure of oxygen and amount of oxyhaemoglobin formed or dissociated has been studied under various conditions. The results when plotted in the form of curve represents oxyhaemoglobin-dissociation curve.

The oxyhaemoglobin-dissociation curves under different conditions is given below

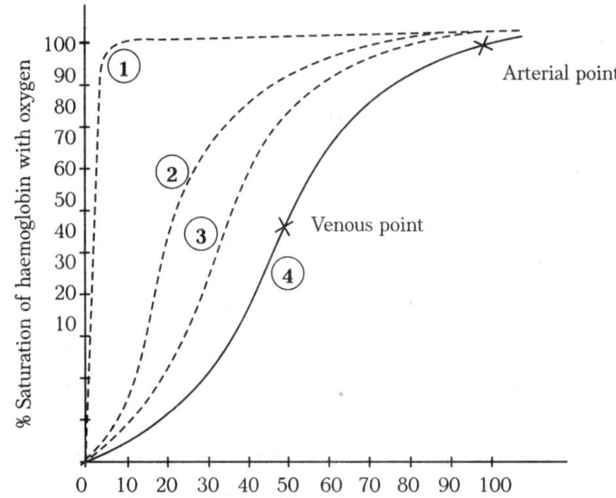

Where, (1) = At room temperature without CO_2
(2) = At body temperature without CO_2
(3) = At body temperature + 20 mm of Hg CO_2
(4) = At body temperature + 40 mm of Hg CO_2

Fig. 24.14 Oxyhaemoglobin dissociation curve at different conditions

The effects of various factors on oxyhaemoglobin formation and dissociation are as follows

(i) **Formation of oxyhaemoglobin or shift of curve to left**

- High pO_2 in blood.
- Low pCO_2 in blood.
- Less H^+ ion concentration and increase in pH means decrease in acidity.
- Low temperature and low 2, 3-DPG.

(ii) **Dissociation of oxyhaemoglobin or shift of curve to right**

- Low pO_2 in blood.
- High pCO_2 in blood.
- High H^+ ion concentration and decrease in pH means increase in acidity.
- High temperature and high 2, 3-DPG.

Conclusions

Various interpretations and conclusion that can be made in relation to the above curve are given below

(i) The curve is sigmoid or S-shaped under normal conditions, i.e. curve 4 in given figure. It is drawn with normal conditions of CO_2 and temperature present in blood capillaries.

(ii) **Under normal body conditions**, whatsoever increase occurs in partial pressure of O_2 (even upto 100 mm of Hg), the haemoglobin is never fully saturated becuase of the presence of CO_2 and temperature conditions in the body.

(iii) **2, 3-DPG** is present in red blood cells at approximately 5m mol/L concentration. It has an greater affinity to bind with deoxygenated haemoglobin molecule by producing a conformational change in it. As a result, it becomes harder for this molecule to bind with oxygen. Thus in a way, it helps in dissociation of oxyhaemoglobin.

(iv) The curve is **completely sigmoid** in the presence of **strong electrolytes**, while it is **hyperbolic** in the presence of **weak electrolytes.**

(v) The partial pressure of oxygen at which 50% saturation of haemoglobin takes place is called **P_{50} value.**

$$P_{50} \text{ value} \propto \frac{1}{\text{Affinity of blood for } O_2}$$

(vi) **Bohr effect** At tissue level, the entry of CO_2 or increased acidity or decreased pH of blood helps in the dissociation of oxyhaemoglobin *via* promoting the lesser affinity of blood for oxygen.

(vii) **Haldane effect** At the level of alveoli, the entry of O_2 into the blood, i.e. more and more formation of oxyhaemoglobin is responsible for more and more replacement of CO_2 from the venous blood.

Effect of some special factors

Following special factors affect oxyhaemoglobin dissociation curve in universal manner

(i) **Carbon monoxide** Blood haemoglobin has 250 times more affinity for CO than for O_2.

Therefore, if carbon monoxide is breathed in, it binds irreversibly with Hb to form **carboxyhaemoglobin.**

It means that Hb can no longer be available to combine with O_2. Thus, O_2 deficiency is created in cells. This O_2 deficiency, if persists for longer period, leads to immediate death.

Note *Increased CO levels in blood is known as* ***CO poisoning.*** *Its symptoms include headache, dizziness, nausea and finally death.*

(ii) **Foetal haemoglobin** The structure of foetal haemoglobin (HbF) is almost similar to adult haemoglobin (HbA) except the fact that HbF has 2α and 2γ polypeptide chains. HbA, on the other hand have 2α and 2β polypeptide chains.

Due to this difference in structures, HbF has greater affinity for oxygen as compared to HbA.

Thus, the dissociation curve for HbF is shifted towards the left hand side as compared to HbA.

(iii) **Myoglobin** Skeletal muscle contains a protein called myoglobin. It has more affinity for O_2. But since it has only one polypeptide chain, hence it is attached to one haem group only.

The dissociation curve for myoglobin is almost hyperbolic and not sigmoid.

The indicative curves for myoglobin HbF and HbA are shown below

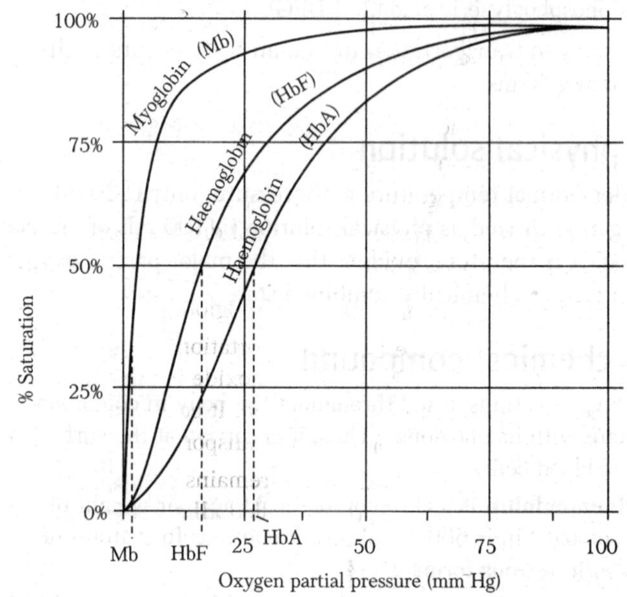

***Fig. 24.15** Factors affecting oxyhaemoglobin dissociation curve*

Transport of carbon dioxide

Carbon dioxide is produced in the tissues and enters the bloodstream within the tissue capillaries. From here, it is carried to the lungs where its liberation takes place into the alveoli.

Transportation of volume

- About 100 mL of venous blood carries 52 mL of carbon dioxide, whereas 100 mL of arterial blood carries about 48 mL of carbon dioxide.
- The average normal arteriovenous difference of carbon dioxide is therefore, about 4 mL.
- In simple words, each 100 mL of arterial blood while passing through the tissues takes up 4 mL of carbon dioxide and transport it towards the lungs.
- Similarly, each 100 mL of venous blood, while passing through the lungs, releases this 4 mL of carbon dioxide.

- Carbon dioxide is transported in human body in various forms given below

In physical solution

- Under normal conditions of temperature and pressure, only about 2.7 mL of carbon dioxide is carried in physical solution (i.e. as H_2CO_3) in 100 mL of venous blood.
- Since, this represents a very small part of total carbon dioxide carried, it is evident that the major part of carbon dioxide is transported in some form of chemical combinations.

Carriage as bicarbonates

- Carbon dioxide is converted immediately into H_2CO_3 after entering into the blood. This reaction is catalysed by enzyme **carbonic anhydrase** present at the surface of RBCs.
- This H_2CO_3 then forms bicarbonates in following four ways
  - (i) after reacting with NaPr
  - (ii) after reacting with Na_2HPO_4
  - (iii) after reacting with Na of NaCl due to 'chloride shift'
  - (iv) after reacting with KHb

Here, **Pr** is protein and **Hb** is haemoglobin.

- Bicarbonates are the main transportation compounds of CO_2. About 45.7 mL of carbon dioxide is carried in these forms per 100 mL of venous blood. It contributes to about 90% of the total carbon dioxide transport.
- A large part of this bicarbonate remains permanently in the plasma and constitutes the so called **alkali reserve** of blood.

Role of Blood Buffer

- Although, much carbon dioxide is carried in the blood, yet blood does not become acidic. Thus, during carbon dioxide transport, the blood buffers play a very important role.
- The constancy of pH during this transportation is maintained by a specialised sequence of events called chloride shift or Hamburger's phenomenon.

Chloride shift or Hamburger phenomenon

When the whole blood is saturated with carbon dioxide, following changes are seen in it

- The bicarbonate content of plasma and corpuscles increases.
- The chloride content of plasma is diminished and that of the cell is increased.
- The total base (cations) of both plasma and corpuscles remain unchanged.
- The water content and the volume of corpuscles increases.

These changes are seen due to following reactions shown by various compounds in body cells, blood plasma and at the surface of RBCs.

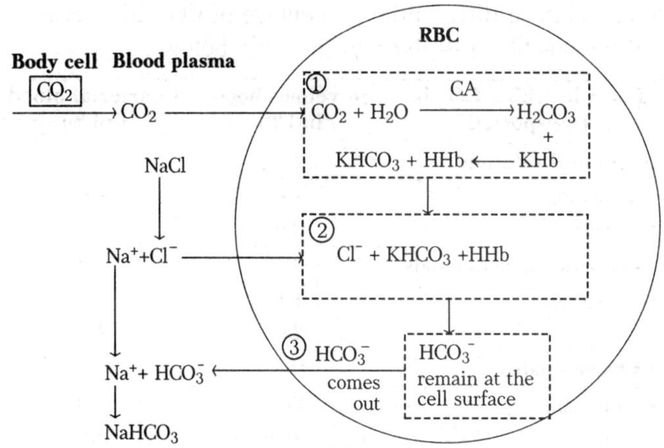

***Fig. 24.16** Schematic representation showing the steps of chloride shifts (CA = Carbonic Anhydrase)*

From these reactions, it is evident that

- (i) When carbon dioxide enters blood, Cl^- ions from plasma enter the red blood cells, while the base (Na^+) is left behind.
- (ii) The movement of Cl^- ions occurs to counter the increased HCO_3^- ion content at RBCs surface and facilitated by a **band 3** protein. This results in their migration into the plasma.
- (iii) Now, these HCO_3^- ions of plasma are balanced by Na^+ ions left from Na^+Cl^-. Carbon dioxide in this form gets transported from tissues to the alveoli.

In lungs however, the above reactions and effects become reversed, thus releasing CO_2. So, the transportation of CO_2 from tissues to lungs in bicarbonate form involves alternate shifting of Cl^- ions from plasma to RBCs surface and *vice-versa.* As it is due to this alternate movement of chloride ions, hence, this phenomenon is called **chloride shift** or **Hamburger phenomenon.**

Note *Band 3 protein is the most abundant membrane protein in human RBCs. These are found closely associated with carbonic anhydrase and help to mediate the exchange of cellular bicarbonate ions with chloride ions in the plasma.*

Carriage of carbon dioxide as carbamino compounds

Carbamino compounds are formed in the NH_2 radical of the globin part of haemoglobin and/or plasma proteins. This combines with one molecule of carbon dioxide as free gas and not as H_2CO_3. It does not require the help of carbonic anhydrase. The reaction may be represented as

$$CO_2 + Pr\cdot NH_2 \rightleftharpoons Pr\cdot NH\cdot COOH$$

About 3.7 mL of carbon dioxide is carried in this form per 100 mL of venous blood, *viz.* 2.6 mL in red blood cells and 1.1 mL in plasma.

A summary of differential percentage of CO_2 in arterial and venous blood is given in the table below

Form in which CO_2 is transported	In venous bood (mL%)	In arterial blood (mL%)
In physical solution	2.7	2.4
• In plasma	1.8	1.6
• In cells	0.9	0.8
As carbamino compounds	3.7	3.0
• In plasma	1.1	1.0
• In cell	2.6	2.0
As bicarbonates	45.7	42.9
• In plasma	35.2	33.1
• In cells	10.5	9.8
Total	52.1mL% = 52%	48.3 mL% = 48%

As we know that each 100 mL of venous blood, while passing through lungs liberates 4 mL of CO_2. Of this 4 mL, about 20-25% is contributed by the breakdown of carbamino compounds (0.7 mL), 72% from bicarbonates (2.8 mL) and 8% from CO_2 in physical solution (0.3 mL).

DISORDERS OF RESPIRATORY SYSTEM

Disorders of respiratory system are as follows

1. **Asthma** It is an allergic disorder in which expiration becomes difficult. It is associated with bronchial obstruction or spasm (contraction) of smooth muscles and the expiratory **wheeze** or **whistling** is its characteristic feature.
2. **Emphysema** It is an irreversible distension of the respiratory bronchioles, alveolar sacs and alveoli. In this condition, the wall separating adjacent alveoli breakdown and air remains filled even after expiration. It develops mainly due to tobacco (cigarette) smoking.
3. **Asphyxia** It is suffocation due to lack of oxygen and retention of carbon dioxide. It is produced by occlusion of the air ways, acute hypercapnia and hypoxia developed together. So we can say;

   Asphyxia = Hypoxia + Hypercapnia.
4. **Bronchitis** It is an inflammation of the bronchi and bronchioles. It is characterised by hypertrophy and hyperplasia of sero-mucous glands and Goblet cells lining the bronchi.

   In medical terminology, term **hypertrophy** is used when enlargement occurs due to increase in size of cell.

   Term **hyperplasia** on the other hand is used when enlargement due to increase in number of cells occurs. Its symptoms include regular coughing with thick greenish-yellow sputum that indicates the underlying infection, resulting in excessive secretion of mucus.
5. **Cyanosis** Bluish or Violet discolouration of skin and the mucous membrane. This develops due to the presence of excessive amount of deoxygenated haemoglobin.
6. **Hay fever** It is an allergic disorder of nasal lining. It develops due to hypersensitivity to pollen grains or other foreign particles. This is characterised by sneezing followed by allergy.
7. **Atelectasis** It is an inability of lungs to expand at birth. This is mainly due to the deficiency of surfactant. In normal cases, the surfactant reduces tension at alveolar surface and allows the lungs to expand at birth.
8. **Pleurisy** It is the inflammation of pleural membranes.
9. **Occupational lung diseases** or **Diseases of industrial workers** Exposure to harmful particles, chemicals, vapours or gases while at work can lead to a variety of health problems including following diseases

   (i) **Anthracosis** or **Pneumoconiosis of coal workers** It is characterised by proliferation of fibrous tissues in the lungs, which then cause bronchitis and emphysema.

   (ii) **Silicosis** It occurs due to long exposure to dust containing silicon compounds. Workers of glass industry, potters, gold and copper miners develop progressive fibrosis in their lungs.

   (iii) **Asbestosis** It is due to inhalation of asbestos-fibres, which may result in cancer of pleura.

   Other diseases related to occupation are allergies, lung irritation, **black lung disease** in coal workers, **chronic beryllium** disease, **Byssinosis**, (brown lung disease) particularly in cotton and textile workers, etc.
10. **SARS** (Severe Acute Respiratory Syndrome) It is a contagious and sometimes fatal respiratory illness. It is caused by corona virus. This virus is transmitted through air in the form of droplets that comes out from diseased person through coughs, sneezes or talks. Initial cases of SARS were reported from China in the year 2002. No known transmission of this disease is reported after 2004 till 2019. However, since 2019 spread of **COVID-19** is resulted to pandemic situation in the world.

11. **Pneumonia** It is an infectious disease of lungs. This infection is caused by certain *Streptococcus* bacteria and *Mycoplasma.*

    Gaseous exchange in lungs became impaired in the disease due to the accumulation of mucus and lymph (containing WBCs) in alveoli. Due to this, oxygen level of blood fall drastically. This disease is more prone in infants and old aged humans.

12. **Tuberculosis** (TB) A bacterial disease is caused by *Mycobacterium tuberculosis.* In this disease, elasticity of lungs decreases and their fibre content is increased. Thus, it restricts gaseous exchange in lungs.

13. **Wooping Cough or Pertussis** (Kali Khansi) Another bacterial disease caused by bacterium, *Bordetella pertussis.* Breathing becomes difficult in this disease due to the obstruction caused by the formation of a membranous covering in the laryngeal region.

14. **Sleep Apnoea Syndrome** (SAS) A disease which occurs due to bony deformities in bucco-nasal region or enlarged tonsils or both. Due to this, the person snores heavily with disturbed rhythmicity in breathing rate. This can be cured after surgery.

CHECK POINT 24.3

1. What is true for CO_2 concentration?
(a) More in alveolar air than in atmospheric air
(b) More in atmospheric air than in alveolar air
(c) More in oxygenated blood than in deoxygenated
(d) More in atmospheric air than in deoxygenated blood

2. Left shift of oxyhaemoglobin curve is noticed under
(a) normal temperature and pH
(b) low temperature and high pH
(c) low pH and high temperature
(d) low pH and low temperature

3. In Bohr's effect, the curve gets shifted to right because
(a) P_{50} CO_2 decreases and P_{50} O_2 increases
(b) P_{50} CO_2 increases and P_{50} O_2 decreases
(c) P_{50} CO_2 increases and P_{50} O_2 increases
(d) P_{50} CO_2 increases and P_{50} O_2 decreases and pH increases

4. Tobacco smoke contains carbon monoxide which
(a) reduces the oxygen carrying capacity of blood
(b) causes gastric ulcers
(c) raises blood pressure
(d) is carcinogenic

5. After fast running, man has fast heartbeat, slow pulse and shallow breathing. In such conditions, he has
(a) oxygen debt
(b) poisoning due to lactic acid
(c) no pulmonary pressure
(d) weak heart

6. Concentration of carbonic acid does not increase in blood due to the presence of
(a) Na^+ (b) K^+
(c) Ca^{2+} (d) Mg^{2+}

EXAM POINTS

- Skin is an accessory organ of respiration in frog. **[WB JEE 2013]**
- The serous membrane which covers the lungs is called pleura. **[WB JEE 2015]**
- The process of exchange of O_2 from the atmosphere with CO_2 produced by the cells is called gaseous exchange. **[J&K CET 2014]**
- During inspiration, there is contraction of diaphragm and external intercostal muscles, resulting into inhalation as volume of lungs increases. **[NEET 2020]**
- To generate pressure gradients to facilitate expiration and inspiration, human body uses the intercostal muscles and diaphragm. **[Kerala CEE 2014]**
- Inspiration occurs when intrapulmonary pressure is lower than the atmospheric pressure. **[NEET 2019]**
- Tidal volume is the volume of air inspired or expired during normal breath. Range is 500–550 mL. **[NEET 2020]**
- The volume of air that can be breathed in by maximum forced inspiration over and above the normal inspiration is called inspiratory reserve volume. **[WB JEE 2014]**
- Expiratory reserve volume is the extra amount of air that can be expired forcibly after a normal expiration. Range is 1000–1100 mL. **[NEET 2018, 2020]**
- Residual volume is the volume of air which remain in the lungs even after a forceful expiration because there is a negative intrapleural pressure pulling at the lung walls. Due to this, lungs do not collapse. **[NEET 2017, 2018, 2020]**
- Inspiratory capacity of human is equal to TV + IRV. **[UK PMT 2014]**
- Functional residual capacity prevents collapse of the alveoli during expiration. Its range is 2500 mL. It is the sum total of ERV and RV. **[Kerala CEE 2015]**
- Total lung capacity includes residual volume, expiratory reserve volume, tidal volume and inspiratory reserve volume. **[NEET 2020]**
- The volume of 'anatomical dead space' air is normally 150 mL. **[WB JEE 2012]**
- The pneumotaxic and repiratory rhythm centres are respectively present in pons and medulla oblongata. **[AMU 2015]**
- O_2 and CO_2 diffuse from areas of their higher partial pressure to their lower partial pressure. **[NEET II 2016]**
- In systemic arteries, the pO_2 is 95 mm Hg and pCO_2 is 40 mm Hg. **[Kerala CEE 2014]**
- The binding of O_2 to haemoglobin is affected by low pO_2, high acidity (pH), pCO_2 and high temperature. **[UPCPMT 2017, AMU 2015]**
- Approximately 70% of carbon dioxide absorbed by the blood is transported to the lungs as bicarbonate ions. **[CBSE AIPMT 2014]**
- Oxygen dissociation curve of haemoglobin is sigmoid (S-shaped). It gives the relationship between the saturation of haemoglobin and oxygen tension. **[JIPMER 2017]**
- Haemoglobin has having maximum affinity for CO. **[UP CPMT 2013]**
- People who have migrated from the planes to an area adjoining Rohtang pass about six months back have more RBCs and their haemoglobin has a lower binding affinity to O_2. **[CBSE AIPMT 2012]**
- Band 3 proteins help to mediate the exchange of cellular bicarbonate ions with chloride ions in the plasma. **[NEET 2019]**
- In asthma, there is inflammation of bronchioles and in emphysema, there is decreased respiratory surface. **[NEET 2018, 2019]**
- Exposure to harmful particles, chemical vapours or gases while at work can lead to disease such as anthracosis and silicosis. **[NEET 2018]**
- When the oxygen supply to the tissue is inadequate, the condition is called as hypoxia. **[WB JEE 2013]**
- Workers in grinding and stone breaking industries may suffer from lung fibrosis. **[CBSE AIPMT 2012]**
- The oxygen toxicity is related with the failure of ventilation of lungs. **[Manipal 2013]**

Chapter Exercises

(A) Topical Exercise

Types of Respiration and Various Respiratory Organs

1 Which process occurs during respiration?
(a) breakdown of nutrients
(b) release of energy
(c) oxidation of fatty acids
(d) All of the above

2 The most abundant, harmful and universal waste product of metabolism is
(a) CO_2 (b) uric acid (c) H_2O (d) None of these

3 Aerobic respiration does not
(a) take place under normal conditions
(b) take place in animals
(c) utilise molecular oxygen
(d) take place in oxygen deficient situations

4 Which is the best equation to represent aerobic respiration?
(a) Carbon dioxide + Water + Energy → Oxygen + Sugar
(b) Energy + Oxygen → Carbon dioxide + Water + Sugar
(c) Oxygen + Sugar → Carbon dioxide + Water + Energy
(d) Sugar + Carbon dioxide → Oxygen + Water + Energy

5 In anaerobic respiration,
(a) nutrients are not oxidised and molecular O_2 is used up
(b) nutrients are oxidised and molecular O_2 is used up
(c) nutrients are not oxidised and molecular O_2 is not used up
(d) nutrients are oxidised and molecular O_2 is not used up

6 What are the products of aerobic and anaerobic respiration in muscle tissue?

	Aerobic respiration	Anaerobic respiration
(a)	Carbon dioxide and water	Ethanol
(b)	Carbon dioxide and water	Lactic acid
(c)	Ethanol	Carbon dioxide and water
(d)	Lactic actic	Carbon dioxide and water

7 *Leucosolenia* respires through
(a) body surface (b) lungs
(c) gills (d) book lungs

8 Respiration through book gills occur in
(a) frog (b) *Limulus*
(c) millipede (d) crocodile

9 Choose the incorrectly matched pair of given animal and its respiratory structure.
(a) Prawn - Gills
(b) Earthworm - Skin
(c) Scorpion - Trachea
(d) Spiders - Book lungs

10 Respiration in insects is called direct because **[NCERT Exemplar]**
(a) the cell exchange O_2/CO_2 directly with the air in the tubes
(b) the tissues exchange O_2/CO_2 directly with coelomic fluid
(c) the tissues exchange O_2/CO_2 directly with the air outside through body surface
(d) tracheal tubes exchange O_2/CO_2 directly with the haemocoel which then exchange with tissues

11 Match the following columns and mark the correct option.

	Column I (Animals)		Column II (Respiratory organs)
A.	Earthworm	1.	Moist cuticle
B.	Insects	2.	Gills
C.	Fishes	3.	Lungs
D.	Birds/Reptiles	4.	Trachea

Codes **[NCERT Exemplar]**

	A	B	C	D		A	B	C	D
(a)	2	1	4	3	(b)	1	4	2	3
(c)	1	3	2	4	(d)	1	2	4	3

12 In which of the following animals, respiration occurs without the involvement of a specialised respiratory organ?
(a) Fish (b) Cockroach
(c) Tadpole (d) Earthworm

Respiratory System in Humans

13 The structure containing mucous lining and hair epithelium within the nasal passage is
(a) larynx (b) vestibule
(c) gullet (d) glottis

14 Schneiderian membrane is present at
(a) upper part of nasal passage
(b) larynx
(c) wall of trachea
(d) None of the above

15 In the nasal chamber, the processes of three bones are found to increase its surface area. These bones are
(a) nasal, hyoid, maxilla (b) stapes, maxilla, mandible
(c) nasal, ethmoids, maxilla (d) malleus, incus, stapes

16 The mucus produced by the Goblet cells of nasopharynx
I. moistens the epithelial surface.
II. traps the finest dust particles.
III. filter out largest dust particles.
IV. warm the incoming air.
Codes
(a) I and II (b) II and III (c) Only IV (d) All of these

17 Nasopharynx open into the trachea through
(a) vestibule (b) glottis
(c) internal nares (d) concha

18 The position of larynx is
(a) at the level of 3rd to 6th cervical vertebrae
(b) below the level of 2nd to 5th cervical vertebrae
(c) at the level of 6th to 8th cervical vertebrae
(d) below the level of 6th to 8th cervical vertebrae

19 The number of cartilage in larynx is
(a) 3 (b) 6 (c) 9 (d) 12

20 Which among the following set represents single laryngeal cartilage?
(a) Thyroid, cricoid, epiglottis
(b) Cricoid, cuneiform, arytenoid
(c) Epiglottis, arytenoid, thyroid
(d) Corniculate, cuneiform, cricoid

21 Choose the incorrectly matched pair.
(a) Adam's apple - C-shaped cartilage
(b) Cuneiform cartilage - Prominent in males
(c) Hyoid bone - Not the part of larynx
(d) True glottis - Responsible for phonation

22 Adam's apple represents
(a) arytenoid cartilage of larynx
(b) cricoid cartilage of larynx
(c) thyroid cartilage of larynx
(d) Both (b) and (c)

23 The diagram represents the human larynx. Choose the correct combination of labelling from the options given.

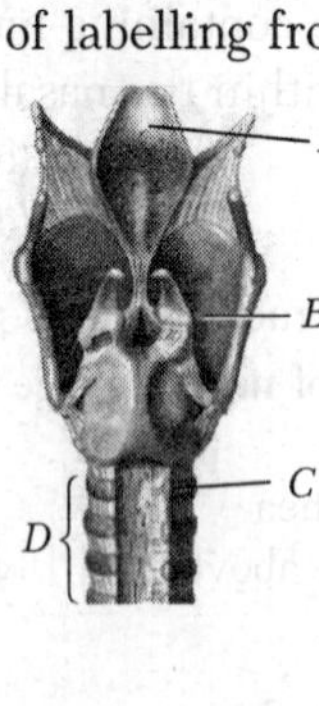

(a) A - Larynx, B -Parathyroid, C - Tracheal cartilage, D - Trachea
(b) A - Naso larynx, B - Thyroid, C - Tracheal cartilage, D - Trachea
(c) A - Trachea, B - Thyroid, C - Bronchiole, D - Tracheal cartilage
(d) A - Epiglottis, B - Thyroid, C - Tracheal cartilage, D - Trachea

24 Consider the following statements.
I. True and false vocal cords differ in appearance and attachment.
II. Rima glottidis is the space between true and false vocal cords.
Codes
(a) Statement I is true, but II is false
(b) Statement I is false, but II is true
(c) Both statements I and II are true
(d) Both statements I and II are false

25 What is the location of syrinx in birds?
(a) Within the trachea
(b) Above the trachea
(c) Adjacent to trachea
(d) None of the above

26 The covering of lung is called
(a) pericardium
(b) perichondrium
(c) pleural membrane
(d) peritoneum

27 Identify A, B, C, D and E in the diagram of human respiratory system.

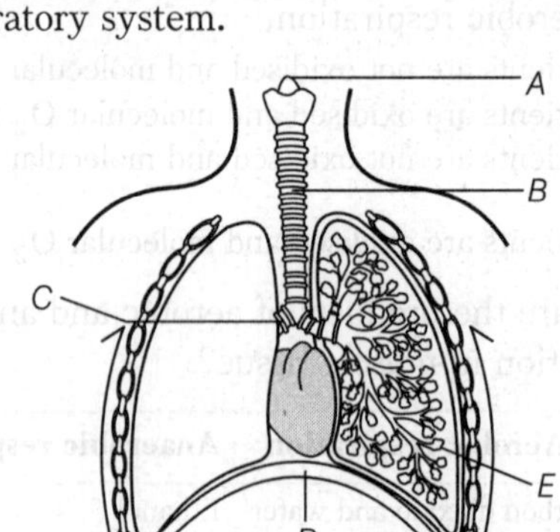

Choose the correct option.
(a) A-Epiglottis, B-Trachea, C-Glottis, D-Diaphragm, E-Bronchiole
(b) A-Glottis, B-Trachea, C-Bronchus, D-Diaphragm, E-Bronchiole
(c) A-Adam's apple, B-Trachea, C-Bronchus, D-Diaphragm, E-Bronchiole
(d) A-Epiglottis, B-Trachea, C-Bronchus, D-Diaphragm, E-Bronchiole

28 Match the following columns.

	Column I		Column II
A.	Bronchial tree	1.	Passage way for air from nasal cavity to larynx and for food from mouth cavity to oesophagus
B.	Larynx	2.	Conducts air from trachea to the alveoli to its mucous lining continues to filter air
C.	Trachea	3.	Passage way for air, prevents foreign objects from entering trachea and houses vocal cords
D.	Pharynx	4.	Flexible tube, which connects larynx with bronchial tree

Codes

	A	B	C	D		A	B	C	D
(a)	1	2	3	4	(b)	2	3	4	1
(c)	4	3	2	1	(d)	3	2	1	4

29 The left lung in humans possesses
(a) two lobes (b) three lobes
(c) four lobes (d) five lobes

30 Select the incorrectly matched pair.
I. Nasal turbinates - Concha
II. Lingual tonsils - Hypopharynx
III. Arytenoid - Hyaline cartilage
IV. Trachea - Non-keratinised stratified squamous
Codes
(a) I and III (b) II and IV
(c) I, II and III (d) III and IV

31 Positive pressure lungs are
(a) spongy and solid in nature
(b) found in humans
(c) found in frogs
(d) All of the above

32 The walls of trachea are non-collapsible due to the presence of
(a) complete rings of fibrous cartilage
(b) incomplete rings of hyaline cartilage
(c) incomplete rings of fibrous cartilage
(d) complete rings of hyaline cartilage

33 What is the approximate number of tracheal rings in humans?
(a) ~ 60 (b) ~ 120 (c) ~ 16 (d) ~ 5

34 Trachea is straight tube extending up to the mid-thoracic cavity, which divides at the level of
(a) second cervical vertebra (b) fifth cervical vertebra
(c) fifth thoracic vertebra (d) fifth lumbar vertebra

35 Common factor in the trachea of mammals and insects is
(a) ciliated inner lining (b) non-collapsible wall
(c) paired nature (d) origin from head region

36 In alveoli, surfactant is produced by
(a) type I pneumocyte (b) type II pneumocyte
(c) Kupffer's cells (d) dust cells

37 DPCC produced by pneumocytes is a type of
(a) protein (b) carbohydrate
(c) glycoprotein (d) lipid

38 A characteristic feature of purely conducting zone of trachea is
(a) it has alveoli
(b) there is maximum exchange of gases
(c) there is no gaseous exchange
(d) it has respiratory bronchioles

39 Which of the following statements is incorrect regarding respiratory system?
(a) Each terminal bronchiole gives rise to a network of bronchus
(b) The alveoli are highly vascularised
(c) The lungs are covered by a double-layered membrane
(d) The pleural fluid reduces friction on the hard surface

40 Which one of the following has the smallest diameter?
(a) Right primary bronchus (b) Left primary bronchus
(c) Trachea (d) Respiratory bronchiole

41 The terminal bronchiole is lined by
(a) non-ciliated simple columnar
(b) ciliated columnar or cuboidal epithelium
(c) stratified epithelium
(d) pseudostratified epithelium

Mechanism of Breathing

42 Which of the following does not occur during breathing? **[NCERT Exemplar]**
(a) Brings the air to body temperature
(b) Warms up the air
(c) Diffusion of gases
(d) Cleans up the air

43 Which set of muscles are involved in the normal breathing in humans? **[NCERT Exemplar]**
(a) External and internal intercostal muscles
(b) Diaphragm and abdominal muscles
(c) Diaphragm and external intercostal muscles
(d) Diaphragm and intercostal muscles

44 A person suffers punctures in his chest cavity in an accident without any damage to the lungs. Its effect could be **[NCERT Exemplar]**
(a) reduced breathing rate
(b) rapid increase in breathing rate
(c) no change in respiration
(d) cessation of breathing

45 When does the diaphragm straighten during breathing?
(a) Inspiration
(b) Expiration
(c) Both (a) and (b)
(d) Period in between (a) and (b)

46 During the process of inspiration in humans,
(a) external intercostal muscles relax
(b) external intercostal muscles contract
(c) Both external and internal intercostal muscles relax
(d) internal intercostal muscles contract

47 During inspiration, the ribs move
(a) inward and downward (b) inward and upward
(c) outward and upward (d) outward and downward

48 Mark the true statement among the following with reference to normal breathing. **[NCERT Exemplar]**
(a) Inspiration is a passive process whereas expiration is active
(b) Inspiration is an active process whereas expiration is passive
(c) Inspiration and expiration are active processes
(d) Inspiration and expiration are passive processes

49 During inspiration, air passes into lungs due to the
(a) increase in volume of thoracic cavity and fall in lung pressure
(b) fall in pressure inside the lungs
(c) increased volume of thoracic cavity
(d) muscular expansion of lungs

50 Air moves into the lungs when atmospheric pressure is
(a) less than the pressure within the lungs
(b) greater than the pressure within the lungs
(c) equal to the pressure within the lungs
(d) None of the above

51 Choose the incorrectly matched pair.
(a) Reduced volume of thoracic cavity – Inspiration
(b) Contraction of external intercostal muscles -Inspiration
(c) Relaxation of phrenic muscles – Expiration
(d) Diaphragm bend upwards – Expiration

52 Which among the following muscles contract during expiration?
(a) Phrenic muscles
(b) External intercostal muscles
(c) Both (a) and (b)
(d) None of the above

53 Expiratory muscles contract at the time of
(a) deep inspiration
(b) normal inspiration and expiration
(c) forceful expiration
(d) normal expiration

54 During forced expiration, actively contracting muscles include the
(a) diaphragm
(b) external intercostal muscles
(c) abdominal muscles
(d) diaphragm and intestinal muscles

55 The relaxation of external intercostal muscle causes
(a) relaxation of rectus abdominis muscles
(b) contraction of rectus abdominis muscles
(c) contraction of diaphragm
(d) Both (a) and (b)

Lung Volumes and Capacities

56 Tidal volume can be estimated during
(a) forceful expiration (b) forceful inspiration
(c) Both (a) and (b) (d) normal breathing

57 IRV in humans is about
(a) 2500-3000 mL (b) 3000-3500 mL
(c) 500 mL (d) 1100-1200 mL

58 Which among the following cannot be measured directly?
(a) Tidal volume
(b) Residual volume
(c) Expiratory reserve volume
(d) Inspiratory capacity

59 A person breathes in some volume of air by forced inspiration after having a forced expiration. This quantity of air taken in is **[NCERT Exemplar]**
(a) total lung capacity (b) tidal volume
(c) vital capacity (d) inspiratory capacity

60 The largest quantity of air that can be expired in a single respiration after a maximal inspiratory effort is called
(a) residual volume (b) tidal volume
(c) vital capacity (d) total lung volume

61 Calculate the amount of air that can be inspired with maximum effort if
(i) amount of air breath in and out during each normal breathing is 505 mL and
(ii) additional volume of air that can be inspired with maximum effort is 2800 mL.
(a) 3305 mL (b) 2295 mL
(c) 5.54 mL (d) ~ 141400 mL

62 Vital capacity of lung is
(a) Inspiratory Reserve Volume (IRV) + Expiratory Reserve Volume (ERV) + Tidal Volume (TV) + Residual Volume (RV)
(b) IRV + ERV + TV
(c) IRV + ERV
(d) IRV + ERV + TV– RV

63 From the following relationships between respiratory volume and capacities and mark the correct answer.

I. Inspiratory Capacity (IC) = Tidal volume + Residual Volume

II. Vital Capacity (VC) = Tidal Volume (TV) + Inspiratory Reserve Volume (IRV) + Expiratory Reserve Volume (ERV)

III. Residual Volume (RV) = Vital Capacity (VC) – Inspiratory Reserve Volume (IRV)

IV. Tidal Volume (TV) = Inspiratory Capacity (IC) – Inspiratory Reserve Volume (IRV)

Codes **[NCERT Exemplar]**

(a) I-Incorrect, II-Incorrect, III-Incorrect, IV-Correct
(b) I-Incorrect, II-Correct, III-Incorrect, IV-Correct
(c) I-Correct, II-Correct, III-Incorrect, IV-Correct
(d) I-Correct, II-Incorrect, III-Correct, IV-Incorrect

64 Ventilation rate may be expressed as

(a) Inspiratory reserve volume × Frequency of respiration
(b) Expiratory reserve volume × Frequency of respiration
(c) Tidal volume × Frequency of respiration
(d) Vital capacity × Frequency of respiration

65 Dead space air is the

(a) amount of air remaining in the alveoli
(b) amount of air left behind in lungs at the end of deep expiration
(c) amount of air taken in and out
(d) air left in the bronchial tree

66 In breathing movements, air volume can be estimated by **[NCERT Exemplar]**

(a) Stethoscope (b) Hygrometer
(c) Sphigmomanometer (d) Spirometer

Regulation of Breathing

67 Respiration mechanism is controlled by

(a) central nervous system
(b) autonomic nervous system
(c) sympathetic nervous system
(d) parasympathetic nervous system

68 Inspiratory centre is located in

(a) cerebellum (b) cerebrum
(c) hypothalamus (d) medulla oblongata

69 The impulse for voluntary forced breathing starts in

(a) medulla (b) vagus
(c) cerebral hemisphere (d) spinal cord

70 Respiratory process is regulated by certain specialised centres in the brain. One of the following listed centres can reduce the inspiratory duration upon stimulation. **[NCERT Exemplar]**

(a) Medullary inspiratory (b) Pneumotaxic centre
(c) Apneustic centre (d) Chemosensitive centre

71 Hering-Breuer reflex is related to the

(a) effect of pH on respiratory centre
(b) effect of CO_2 on respiratory centre
(c) effect of nerves on respiratory centre
(d) effect of temperature on respiratory centre

72 Which of the following conditions is responsible for increase in ventilation rate of lungs?

(a) Increase of CO_2 content in inhaled air
(b) Increase of CO_2 content in exhaled air
(c) Decrease of O_2 content in inhaled air
(d) Decrease of O_2 content in exhaled air

73 If CO_2 concentration increases in blood, then breathing will

(a) increase (b) decrease
(c) stop (d) remain unchanged

74 When blood CO_2 level rises

(a) only the rate of breathing decreases
(b) respiratory acidosis may occur
(c) peripheral pressure receptors respond
(d) Both the rate and depth of breathing decreases

75 After taking a long deep breath, we do not respire for some seconds due to

(a) more CO_2 in blood (b) more O_2 in blood
(c) less CO_2 in blood (d) less O_2 in blood

76 Consider the following statements.

I. Blood O_2 concentration does not have direct effect on respiratory centre.

II. Increased metabolic activities due to increased temperature accelerate breathing rate.

Choose the correct option.

(a) Statement I is true, but II is false
(b) Statement I is false, but II is true
(c) Both statements I and II are true
(d) Both statements I and II are false

77 A person with high fever may be breathing faster than normal. This faster breathing may be due to

(a) additional requirement of oxygen for germs
(b) high temperature of the body
(c) mental worry of patient
(d) loss of appetite

78 Apnoea is

(a) the absence of breathing
(b) decreased ventilation
(c) increased ventilation
(d) laboured breathing

79 Choose the incorrectly matched pair.

(a) Eupnea - Decreased breathing rate
(b) Dyspnea - Difficult breathing
(c) Tachypnea - Rapid shallow breathing
(d) Hypercapnia - Excess CO_2 in body

Exchange of Gases and Transport of Gases in Blood

80 Diffusion membrane is made up of which layer?
(a) Thin squamous epithelium of alveoli
(b) Basement membrane
(c) Endothelium of alveolar capillaries
(d) All of the above

81 For proper transport of O_2 and CO_2 blood should be
(a) slightly acidic (b) strongly acidic
(c) strongly alkaline (d) slightly alkaline

82 pO_2 in arterial capillaries of alveoli is
(a) 95 mm Hg (b) 104 mm Hg
(c) 40 mm Hg (d) 110 mm Hg

83 pCO_2 in alveolar air is
(a) 40 mm Hg (b) 45 mm Hg
(c) 10 mm Hg (d) 105 mm Hg

84 What will be the pO_2 and pCO_2 in the atmospheric air as compared to those in the alveolar air? **[NCERT]**
(a) pO_2 lesser, pCO_2 higher
(b) pO_2 higher, pCO_2 lesser
(c) pO_2 higher, pCO_2 higher
(d) pO_2 lesser, pCO_2 lesser

85 What is the partial pressure of oxygen and carbon dioxide in atmospheric air?
(a) pO_2 159 mm Hg, pCO_2 0.3 mm Hg
(b) pO_2 104 mm Hg, pCO_2 40 mm Hg
(c) pO_2 40 mm Hg, pCO_2 45 mm Hg
(d) pO_2 95 mm Hg, pCO_2 40 mm Hg

86 The table shows the percentage composition of four samples of air.

Which sample has been breathed out by a person?

	Oxygen	Carbon dioxide	Water vapours
(a)	16	0.3	Saturated
(b)	16	4	Saturated
(c)	21	0.03	Trace
(d)	21	3	Trace

87 The diagram shows a section of an alveolus and a capillary in a lung.

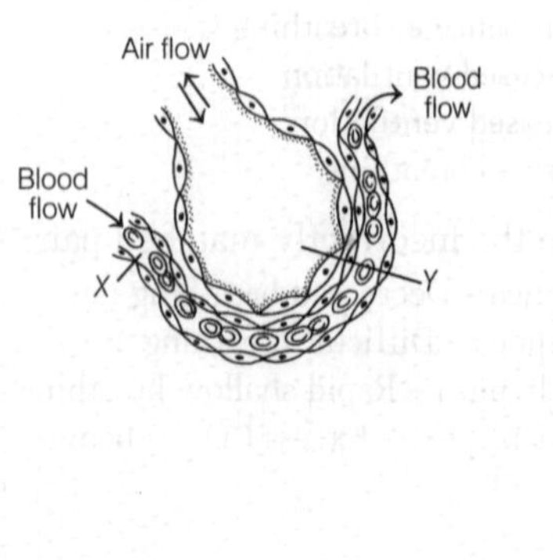

What are the concentrations of oxygen at X and Y.

	X	Y
(a)	Low	High
(b)	Low	Low
(c)	High	Low
(d)	High	High

88 Diffusion of gases along the respiratory surface occurs because
(a) pCO_2 is more in alveoli than blood
(b) pO_2 is more in alveoli than blood
(c) pCO_2 is more in blood than in tissues
(d) pO_2 is more in blood than in tissues

89 What is incorrect about oxygen binding with haemoglobin?
(a) The bond between oxygen and Hb is very loose bond
(b) Oxygen binds with the globin part of Hb
(c) Hb and oxygen are readily reversible combinations
(d) None of the above

90 The binding of oxygen with haemoglobin is primarily related to which of the following factors?
(a) Partial pressure of CO_2
(b) Partial pressure of O_2
(c) Hydrogen ion concentration
(d) Temperature

91 Combination of haemoglobin with oxygen in lungs can be promoted by
(a) increasing carbon dioxide concentration in blood
(b) increasing oxygen concentration in blood
(c) decreasing oxygen concentration in blood
(d) introducing carbon monoxide in blood

92 Body tissues obtain oxygen from oxyhaemoglobin because of its dissociation caused by
(a) low oxygen concentration
(b) high carbon dioxide concentration
(c) low carbon dioxide concentration
(d) low oxygen and high carbon dioxide concentration

93 Mark the incorrect option in context to O_2 binding to Hb. **[NCERT Exemplar]**
(a) Higher pH (b) Lower temperature
(c) Lower pCO_2 (d) Higher pO_2

94 The oxygen-haemoglobin dissociation curve will show a right shift in case of **[NCERT Exemplar]**
(a) high pCO_2 (b) high pO_2
(c) low pCO_2 (d) less H^+ concentration

95 It is known that exposure to carbon monoxide is harmful to animals because **[NCERT Exemplar]**
(a) it reduces CO_2 transport
(b) it reduces O_2 transport
(c) it increases CO_2 transport
(d) it increases O_2 transport

96 In contrast to adult Hb, the foetal Hb contains
(a) 2 α chains (b) 2 β chains
(c) 2 γ chains (d) Both (a) and (c)

97 In which form CO_2 is carried in blood?
(a) Sodium bicarbonate (b) Sodium carbonate
(c) Potassium carbonate (d) Magnesium carbonate

98 Carbonic anhydrase is found in high concentration in
(a) leucocytes (b) blood plasma
(c) erythrocytes (d) lymphocytes

99 CO_2 dissociates from carbaminohaemoglobin when
(a) pCO_2 is high and pO_2 is low **[NCERT Exemplar]**
(b) pO_2 is high and pCO_2 is low
(c) pCO_2 and pO_2 are equal
(d) None of the above

100 During transport of CO_2, blood does not become acid due to
(a) neutralisation of H_2CO_3 by Na_2CO_3
(b) absorption by leucocytes
(c) blood buffers
(d) non-accumulation

101 Of the following, the one which is an example of buffer system in blood is
(a) haemoglobin and oxyhaemoglobin
(b) oxygen and carbon dioxide
(c) albumin and globulin
(d) sodium bicarbonate and carbonic acid

102 In lungs, there is definite exchange of ions between RBC and plasma. Removal of CO_2 from blood involves
(a) influx of Cl^- ions into RBC
(b) influx of HCO_3^- ions into RBC
(c) efflux of Cl^- ions from RBC
(d) efflux of HCO_3^- ions from RBC

103 Select the incorrect statement.
(a) Decrease in pO_2 promotes the amount of CO_2 transport
(b) Oxyhaemoglobin acts as a strong alkali
(c) The chloride content of RBC increases in deoxygenated blood
(d) None of the above

104 Chloride shift is essential for the transport of
(a) CO_2 and O_2 (b) N_2
(c) CO_2 (d) O_2

105 During the transport of gases in human body,
(a) about 97% of O_2 is transported by RBCs in the blood
(b) about 3% of O_2 is carried in dissolved state in the plasma
(c) about 20-25% of CO_2 is transported by RBCs
(d) about 70% carbon dioxide is carried in dissolved state in plasma

Disorders of Respiratory System

106 Expiratory wheezing due to spasm of smooth muscles occur in
(a) asthma (b) eupnea
(c) emphysema (d) artherosclerosis

107 Increased asthmatic attacks in certain seasons are related to
(a) hot and humid environment
(b) eating fruits preserved in tin containers
(c) inhalation of seasonal pollen
(d) low temperature

108 Reduction in respiratory surface of the lungs due to the breakdown of partition in the alveoli is known as
(a) asphyxia (b) bronchitis
(c) asthma (d) emphysema

109 Incidence of emphysema-a respiratory disorder is high in cigarette smokers. In such cases, **[NCERT Exemplar]**
(a) the bronchioles are found damaged
(b) the alveolar walls are found damaged
(c) the plasma membrane is found damaged
(d) the respiratory muscles are found damaged

110 A stage when lung collapse, specially the alveoli is
(a) atelectasis (b) poliomyelitis
(c) asthma (d) epistaxis

111 Which of the following can cause atelectasis?
(a) Blockage of small bronchi with mucus
(b) Obstruction of a major bronchus
(c) Lack of surfactant in fluids lining the alveoli
(d) All of the above

112 Inflammation of the lung covering causing severe chest pain in
(a) emphysema (b) pleurisy (c) asphyxia (d) hypoxia

113 Find the incorrectly matched pair.
(a) Atelactasis - Surfactant deficiency
(b) Tuberculosis - Bacterial disease
(c) Sleep apnoea - Bony deformity
(d) None of the above

(B) Cumulative Exercise

(Assorted Questions of the Chapter for Advanced Level Practice)

1 Through which sequence does carbon dioxide pass as it leaves the lungs?
(a) Alveolar wall → In the alveoli → Blood → Capillary wall
(b) Blood → Capillary wall → Alveolar wall → In the alveoli
(c) Capillary wall → Blood → In the alveoli → Alveolar wall
(d) In the alveoli → Alveolar wall → Capillary wall → Blood

2 Select the incorrect statement.
(a) In human beings, oblique fissures are associated with both lungs
(b) When P_{50} value of Hb rises, its affinity to combine with O_2 decreases
(c) Air-filled sacs in lungs do not collapse even after forceful expiration because of residual volume.
(d) Snoring occurs due to involuntary contraction of diaphragm

3 Division of mammalian lungs into a very large number of tiny alveoli around alveolar ducts opening into bronchioles is
(a) an inefficient system of ventilation of alveoli through with very little residual air in the lungs
(b) an inefficient system of ventilation of alveoli resulting in very high percentage of residual air in the lungs
(c) a very efficient system of ventilation of alveoli with no residual air in the lungs
(d) an efficient system of ventilation of alveoli with little residual air in the lungs

4 Blood analysis of a patient reveals an unusually high quantity of carboxyhaemoglobin content. Which of the following conclusions is most likely to be correct? The patient has been inhaling polluted air containing usually high content of
(a) carbon disulphide (b) chloroform
(c) carbon dioxide (d) carbon monoxide

5 A student's rate and depth of breathing are measured before exercise and then after exercise.
The results are plotted on two graphs showing the volume of air breathed in and out in 12 seconds.

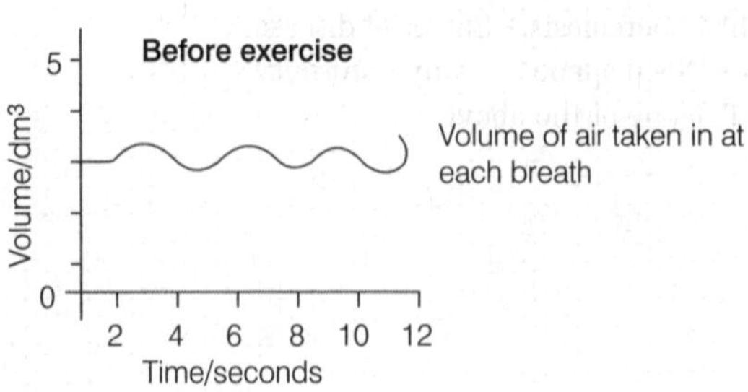

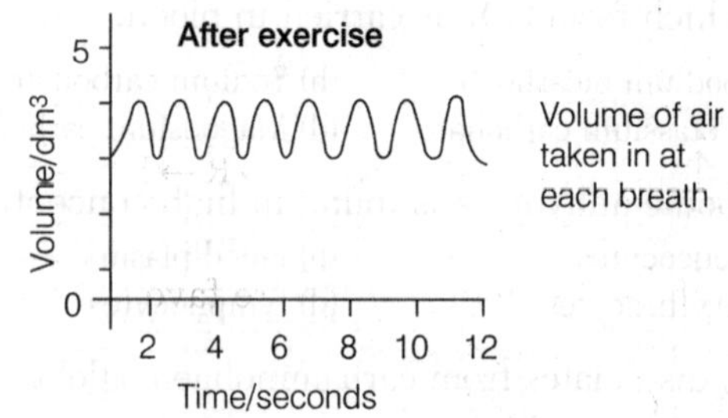

After exercise, how much air does the student take in during 12 seconds?
(a) 0.5 dm^3 (b) 1 dm^3 (c) 1.5 dm^3 (d) 8 dm^3

6 Which feature of alveoli decreases the distance over which oxygen and carbon dioxide molecules diffuse?
(a) Each alveolus has a large blood suppply
(b) Each alveolus is only 0.1-0.2 mm in diameter
(c) There are approximately 150 million alveoli in each lung
(d) The walls of the alveoli are only one cell thick

7 Oxyhaemoglobin is an unstable compound because
(a) O_2 and haemoglobin reaction depends upon partial pressure
(b) haemoglobin is a conjugated protein
(c) haemoglobin is contained within RBC
(d) one haemoglobin binds with four molecules of O_2

8 What would happen when blood is acidic?
(a) Binding of oxygen with heamoglobin increases
(b) Red blood corpuscles are formed in higher number
(c) Binding of oxygen with haemoglobin decreases
(d) There is no change in oxygen binding nor in number of RBCs

9 Arrange the following in the order of increasing volume.
I. Tidal volume
II. Residual volume
III. Expiratory reserve volume
IV. Vital capacity

Codes
(a) I < II < III < IV (b) I < III < II < IV
(c) I < IV < III < II (d) I < IV < II < III

10 Humans have to maintain the moderate respiratory rhythms to suit the demands of the body. For fulfilling that purpose, we have the
Respiratory rhythm centre in medulla = R
Pneumotaxic centre in pons = PT
Chemosensitive area in medulla = C_1
Peripheral chemoreceptions in aortic arch and Carotid artery = C_2

Select the correct path for the regulation of respiration.

(a) $C_1 \rightarrow PT \rightarrow C_3$, with $R \uparrow$ (R pointing up into PT)

(b) $PT \rightarrow C_2 \rightarrow C_1$, with $R \uparrow$ (R pointing up into PT)

(c) $PT \rightarrow R \rightarrow C_2$, with $C_1 \uparrow$ (C₁ pointing up into PT)

(d) $C_2 \rightarrow R \rightarrow PT \rightarrow C_1$

11 Which of the following factors are favourable for the formation of oxyhaemoglobin ?

I. High pO_2, low pCO_2

II. Lesser H^+ concentration, lower temperature

III. Low pO_2, high pCO_2

IV. High H^+, higher temperature

(a) Only I (b) I and II (c) II and III (d) III and IV

12 Given below graph shows an oxygen dissociation curve.

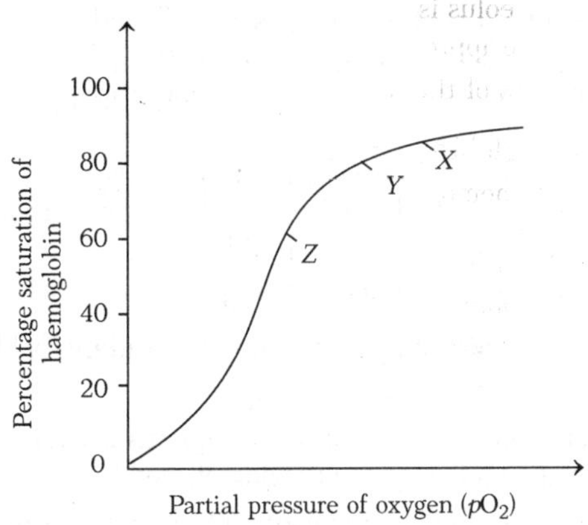

Where in the body will haemoglobin be saturated at the percentage shown at points *X*, *Y* and *Z*?

(a) *X*-Pulmonary artery, *Y*-Pulmonary vein, *Z*-Carotid artery

(b) *X*-Vena cava, *Y*-Left ventricle, *Z*-Pulmonary vein

(c) *X*-Pulmonary vein, *Y*-Systemic vein, *Z*-Systemic vein during exercise

(d) *X*-Left ventricle, *Y*-Right ventricle, *Z*-Systemic artery

13 Match the following columns.

Column I	Column II
A. Tiny air sacs in the lungs	1. Internal respiration
B. Gas diffusion between alveoli and blood in lungs	2. Haemoglobin
C. Gas exchange between environment and lungs	3. Alveoli
D. Molecule specialised for oxygen transport	4. External respiration

Codes

	A	B	C	D		A	B	C	D
(a)	3	1	4	2	(b)	1	2	3	4
(c)	4	3	2	1	(d)	2	1	4	3

14 In man and mammals, air passes from outside into the lungs through

(a) nasal cavity, larynx, pharynx, trachea, bronchi, alveoli

(b) nasal cavity, larynx, pharynx, trachea, bronchioles, alveoli

(c) nasal cavity, pharynx, larynx, trachea, bronchioles, bronchi, alveoli

(d) nasal cavity, pharynx, larynx, trachea, bronchi, bronchioles, alveoli

15 A person met with an accident and died instantly without any injury to heart, brain, stomach and kidney. One of the following is a reason for his death.

(a) Intestine got twisted (b) RBCs became coagulated

(c) Stomach stopped digestion (d) Diaphragm got punctured

16 As carbon dioxide produced in the tissues combines with water in the blood

(a) carbonic acid is formed

(b) Cl^- enters in the RBCs

(c) most of the HCO_3^- from the carbonic acid leaves the RBCs for the blood plasma

(d) All of the precedings occur

17 What are the products of anaerobic respiration in muscle?

	Alcohol	Carbon dioxide	Glucose	Lactic acid	Water
(a)	✓	✓	✓	✗	✗
(b)	✗	✓	✗	✓	✓
(c)	✓	✗	✓	✗	✓
(d)	✗	✗	✗	✓	✗

18 The table shows the percentage composition of a gas in inspired and in expired air.

Percentage composition	
Inspired air	Expired air
21.0	16.0

Which gas is mentioned in above table?

(a) Carbon dioxide (b) Nitrogen

(c) Oxygen (d) Water vapour

19 Most people require alveolar ventilation for about 4 L/min. Considering a patient suffering from damaged alveoli have tidal volume of 0.5 L and respiratory dead space of 0.3 L.

What will be the person's frequency of breathing at rest?

(a) 50 breath/min (b) 8 breath/min

(c) 66 breath/min (d) 20 breath/min

20 If O_2 concentration in tissue was almost as high as at the respiratory surface then

(a) oxyhaemoglobin would dissociate to supply to the tissue

(b) haemoglobin would combine with more O_2 at respiratory surface

(c) oxyhaemoglobin would not dissociate to supply O_2 to the tissue

(d) CO_2 will interfere the O_2 transport

21 People living at sea level have around 5 million RBCs per cubic millimetre of their blood, whereas those living at an altitude of 5400 m have around 8 million. This is because at high altitude
(a) people get pollution-free air to breath and more oxygen is available
(b) atmospheric O_2 level is less and hence, more RBCs are needed to absorb the required amount of O_2 to survive
(c) there is more UV-radiation, which enhances RBC production
(d) people eat more nutritive food, therefore more RBCs are formed

22 Artificial respiration at the rate of 10-15 times per minute is being given to a man saved from drowning. This is because
(a) the water in the respiratory passage is cleared fast at this rate
(b) it is the normal rate of breathing
(c) choking is least at this rate
(d) the lungs are ventilated best at this rate

23 Select the correct option with regard to the mode of respiration in various animals.
(a) Gills – Crustaceans (b) Skin – Insects
(c) Trachea – Annelids (d) Lungs – Arachnids

24 Granular pneumocytes secrete dipalmitoyl phosphatidyl choline. These are specialised lipids, which perform many function except,
(a) prevent oedema
(b) formation of Adam's apple
(c) prevents accumulation of lymph on alveoli
(d) preventing alveoli from collapse

25 Given in the box are certain disorders of human body.

Asthma, Cyanosis, Blue baby syndrome, Atelectasis, Jaundice, Anthracosis, Epilepsy, Asbestosis

How many disorders from above are associated with human respiratory system?
(a) Three (b) Four (c) Five (d) Six

26 What is true about haemoglobin?
(a) It is a dipeptide and present in red blood corpuscles in warm blood
(b) It is present in the dissolved state in blood plasma in earthworm
(c) It is a polypeptide in mammals and localised in red blood corpuscles
(d) It is present in dissolved state in blood plasma in scorpions

27 Mark the incorrect statement.
(a) Respiratory centres are found in medulla oblongata
(b) Near lungs Cl^- moves out of the RBC
(c) RBCs of deoxygenated blood are slightly bigger than that of oxygenated blood
(d) None of the above

28 Which one of the following statements correctly defines Bohr's effect?
(a) HbO_2 curve shifts to right with an increase in CO_2 concentration
(b) Fall in P_{50} with a decrease in pH
(c) Rise in P_{50} with an increase in pH
(d) No effect on P_{50} with a decrease in pH

29 Which one of the following statements is correct about blood constituents and transport of most accurate respiratory gases?
(a) RBCs transport oxygen, whereas plasma transports only carbon dioxide
(b) RBCs as well as WBCs transport both oxygen and carbon dioxide
(c) RBCs transport oxygen, whereas WBCs transport carbon dioxide
(d) RBCs as well as plasma transport both oxygen and carbon dioxide

30 Identify the correct statement with reference to transport of respiratory gases by blood.
(a) Haemoglobin is necessary for the transport of carbon dioxide and carbonic anhydrase for the transport of oxygen
(b) Haemoglobin is necessary for the transport of oxygen and carbonic anhydrase for the transport of carbon dioxide
(c) Only oxygen is transported by blood
(d) Only carbon dioxide is transported by blood

31 Read the following statements and identify the one which is incorrect.
(a) The upper part of larynx, i.e. glottis is covered by epiglottis during swallowing of food
(b) Apart from diaphragm and intercostal muscles, transversus abdominis and rectus abdominis also aid in respiration
(c) The pons have two centres for regulation breathing, i.e. pneumotaxic and apneustic centre
(d) Pneumotaxic centre releases neural signal, which reduces the expiration process thereby, regulating respiratory rate

32 Consider the following statements.
I. Interpulmonary pressure remains the same during inspiration and expiration.
II. Thoracic volume gets decreased during expiration.
Select the correct option.
(a) Statement I is true, but II is false
(b) Statement I is false, but II is true
(c) Both statements I and II are true
(d) Both statements I and II are false

33 Consider the following statements.
I. Band 3 proteins are a polytropic membrane protein, found on the surface of RBCs.
II. Band 3 proteins can detect metabolically active cells and supply oxygen to them.

Select the correct option.
(a) Statement I is true, but II is false
(b) Statement I is false, but II is true
(c) Both statements I and II are true
(d) Both statements I and II are false

34 The following diagram shows a section of an alveolus in a human lung.

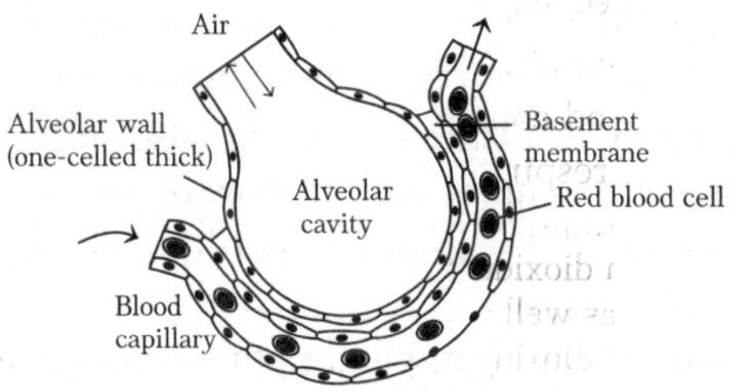

Which conditions would result in the maximum rate of diffusion of oxygen from the alveolus into the blood capillary?

	Amount of oxygen in alveolar air	Amount of oxygen in blood	Rate of blood flow
(a)	Small	Large	Fast
(b)	Small	Large	Slow
(c)	Large	Small	Fast
(d)	Large	Small	Slow

35 The diagram shows organs associated with breathing in humans.

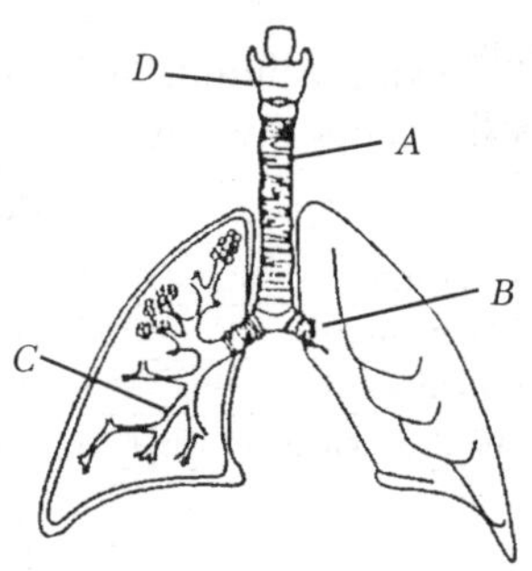

Choose the correct option for *A*, *B*, *C* and *D* in the given diagram.

	A	*B*	*C*	*D*
(a)	Bronchus	Bronchiole	Larynx	Trachea
(b)	Bronchiole	Bronchus	Larynx	Trachea
(c)	Larynx	Trachea	Bronchus	Bronchiole
(d)	Trachea	Bronchus	Bronchiole	Larynx

36 Study the following diagram and consider the statements in regard to the labels.

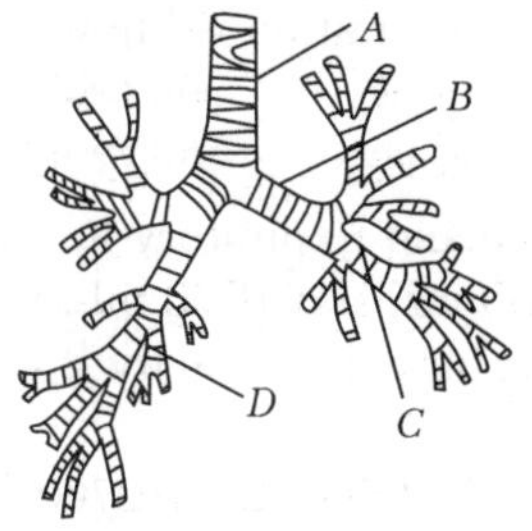

I. *A* is lined by pseudostratified ciliated columnar epithelium bearing glandular cells and cartilaginous ring.
II. *B* is the part of lung where major exchange of gases takes place.
III. *C* decreases both in diameter and thickness as it gets divided.
IV. *D* is the part of lung where exchange of gases does not take place.

Choose the correct set of statements.
(a) I, II and III (b) II, III and IV
(c) I and III (d) II and IV

37 Which of the following statements are correct?
I. The blood transports CO_2 comparatively easily due to its high solubility.
II. Approximately, 8.9% of CO_2 is transported being dissolved in the plasma of blood.
III. The CO_2 produced by the tissues diffuses passively into the bloodstream and into the red blood corpuscles and reacts with water to form H_2CO_3.
IV. The oxyhaemoglobin (HbO_2) in erythrocytes is basic in nature.
V. The chloride ions diffuse from the plasma into the erythrocytes to maintain the ionic balance.

Choose the correct option.
(a) I, III and V are correct, II and IV are incorrect
(b) II and IV are correct, I, III and V are incorrect
(c) I, II and IV are correct, III and V are incorrect
(d) III and V are correct, I, II and IV are incorrect

38 Match the following columns.

Column I	Column II
A. Carotid bodies	1. Dead space absent
B. Efficient lungs	2. Pons Varolii
C. Apneustic centre	3. Surfactant
D. Clara cells	4. Peripheral chemoreceptors

Codes

	A	B	C	D		A	B	C	D
(a)	2	3	4	1	(b)	3	2	1	4
(c)	4	1	2	3	(d)	2	3	1	4

39 An athlete runs a 1000 m race. The following changes take place in the athlete's body during the race.
1. Increased availability of oxygen to the muscles
2. Increased breathing rate
3. Increased carbon dioxide concentration in the blood
4. Increased production of carbon dioxide by the muscles

In which order do these changes occur?

First ⟶ Last

(a) 1 2 3 4 (b) 2 1 4 3
(c) 2 4 3 1 (d) 4 3 2 1

Assertion and Reason

Direction (Q.Nos. 40-47) *In each of the following questions, a statement of Assertion (A) is given followed by corresponding statement of Reason (R). Of the statements, mark the correct answer as*

(a) Both A and R are true and R is the correct explanation of A
(b) Both A and R are true, but R is not the correct explanation of A
(c) A is true, but R is false
(d) A is false, but R is true
(e) Both A and R are false

40 Assertion (A) The vital capacity is higher in athletes.
Reason (R) Vital capacity is the amount of air, which one can inhale and exhale with maximum effort.

41 Assertion (A) Tidal Volume (TV) is the volume of air inspired or expired with each normal breath.
Reason (R) Adult person contains 500 mL expired or inspired volume of air with each normal breath.

42 Assertion (A) Carbon dioxide is carried in the plasma mainly as HCO_3^- ions.
Reason (R) Enzyme carbonic anhydrase of RBC, catalyses the formation of HCO_3^- ions that enter plasma.

43 Assertion (A) Insects do not respire through body surface.
Reason (R) They have a complex system of air tubes called trachea for respiratory purpose.

44 Assertion (A) Fishes respire through gills.
Reason (R) Counter-current flow occurs in gills.

45 Assertion (A) Inspiration occurs due to muscle relaxation.
Reason (R) During inspiration, the diaphragm and external intercostal muscle contract simultaneously.

46 Assertion (A) Aquatic organisms such as protozoans respire through simple diffusion process.
Reason (R) Their outermost covering remains moist.

47 Assertion (A) Cutaneous respiration occurs in earthworm.
Reason (R) Lungs are absent in earthworm.

(C) Medical Entrances' Gallery

(Collection of questions Asked in NEET & Various Medical Entrance Exams)

1 Identify the wrong statement with reference to transport of oxygen. **[NEET 2020]**
(a) Partial pressure of CO_2 can interfere with O_2 binding with haemoglobin
(b) Higher H^+ concentration in alveoli favours the formation of oxyhaemoglobin
(c) Low pCO_2 in alveoli favours the formation of oxyhaemoglobin
(d) Binding of oxygen with haemoglobin is mainly related to partial pressure of O_2

2 Select the correct events that occur during inspiration. **[NEET 2020]**
I. Contraction of diaphragm.
II. Contraction of external intercostal muscles.
III. Pulmonary volume decreases.
IV. Intrapulmonary pressure increases.
(a) III and IV (b) I, II and IV (c) Only IV (d) I and II

3 The Total Lung Capacity (TLC) is the total volume of the accommodated in the lungs at the end of a forced inspiration. This includes **[NEET 2020]**
(a) RV, IC (Inspiratory Capacity), EC (Expiratory Capacity) and ERV
(b) RV, ERV, IC and EC
(c) RV, ERV, VC (Vital Capacity) and FRC (Functional Residual Capacity)
(d) RV (Residual Volume), ERV (Expiratory Reserve Volume), TV (Tidal Volume) and IRV (Inspiratory Reserve Volume)

4 Match the following columns and select the correct option from the codes given below.

Column I	Column II
A. Pneumotaxic centre	1. Alveoli
B. O_2 dissociation curve	2. Pons region of brain
C. Carbonic anhydrase	3. Haemoglobin
D. Primary site of exchange of gases	4. RBC

Codes **[NEET 2020]**

	A	B	C	D		A	B	C	D
(a)	1	3	2	4	(b)	2	3	4	1
(c)	3	2	4	1	(d)	4	1	3	2

5 Select the correct statement. **[NEET 2019]**
(a) Expiration occurs due to external intercostal muscles
(b) Intrapulmonary pressure is lower than the atmospheric pressure during inspiration
(c) Inspiration occurs when atmospheric pressure is less than intrapulmonary pressure
(d) Expiration is initiated due to contraction of diaphragm

6 The maximum volume of air a person can breath in after a forced expiration is known as **[NEET 2019]**
(a) expiratory capacity (b) vital capacity
(c) inspiratory capacity (d) total lung capacity

7 Tidal Volume and Expiratory Reserve Volume of an athlete is 500 mL and 1000 mL, respectively. What will be his expiratory capacity if the residual volume is 1200 mL? **[NEET 2019]**
(a) 1700 mL (b) 2200 mL (c) 2700 mL (d) 1500 mL

8 Band 3 protein in RBC has the function of **[NEET 2019]**

(a) it mediates the exchange of cellular HCO_3^- with CI^- in plasma
(b) it is a cytoplasmic protein
(c) it mediates the exchange of cations between cell and plasma
(d) Both (a) and (b)

9 Due to increasing air-borne allergens and pollutants, many people in urban areas are suffering from respiratory disorder causing wheezing due to **[NEET 2019]**

(a) inflammation of bronchi and bronchioles
(b) proliferation of fibrous tissues and damage of the alveolar walls
(c) reduction in the secretion of surfactants by pneumocytes
(d) benign growth on mucous lining of nasal cavity

10 Which one of the following options correctly represents the lung conditions in asthma and emphysema, respectively? **[NEET 2018]**

(a) Increased respiratory surface, inflammation of bronchioles
(b) Increased number of bronchioles, increased respiratory surface
(c) Inflammation of bronchioles, decreased respiratory surface
(d) Decreased respiratory surface, inflammation of bronchioles

11 Which of the following is an occupational respiratory disorder? **[NEET 2018]**

(a) Botulism (b) Silicosis
(c) Anthracis (d) Emphysema

12 Match the given items in Column I with those in Column II and select the correct option given below.

Column I	Column II
A. Tidal volume	1. 2500-3000 mL
B. Inspiratory reserve volume	2. 1100-1200 mL
C. Expiratory reserve volume	3. 500-550 mL
D. Residual volume	4. 1000-1100 mL

Codes **[NEET 2018]**

	A	B	C	D
(a)	1	4	2	3
(b)	3	1	4	2
(c)	3	2	1	4
(d)	4	3	2	1

13 Lungs are made up of air-filled sacs, the alveoli. They do not collapse even after forceful expiration, because of **[NEET 2017]**

(a) Residual Volume (RV)
(b) Inspiratory Reserve Volume (IRV)
(c) Tidal Volume (TV)
(d) Expiratory Reserve Volume (ERV)

14 Oxyhaemoglobin dissociates into oxygen and deoxyhaemoglobin at **[UP CPMT 2017, AMU 2015]**

(a) low O pressure in tissues
(b) high O pressure in tissues
(c) equal O pressure inside and outside tissues
(d) all times irrespective of O pressure

15 Oxygen dissociation curve of haemoglobin is **[JIPMER 2017, UP CPMT 2013]**

(a) sigmoid (b) hyperbolic (c) linear (d) hypobolic

16 Match the following columns.

Column I	Column II
A. pO_2 of alveoli lungs	1. 40 mm Hg
B. pO_2 of atmospheric air	2. 95 mm Hg
C. pO_2 of deoxygenated blood	3. 104 mm Hg
D. pO_2 of oxygenated blood	4. 159 mm Hg

Codes **[EAMCET 2016]**

	A	B	C	D
(a)	2	4	1	3
(b)	3	4	1	2
(c)	3	2	1	4
(d)	2	4	3	1

17 The partial pressure of oxygen in the alveoli of the lungs is **[NEET-II 2016]**

(a) equal to that in the blood
(b) more than that in blood
(c) less than that in the blood
(d) less than that of carbon dioxide

18 Lungs do not collapse between breaths and some air always remains in the lungs which can never be expelled because **[NEET-II 2016]**

(a) there is a negative pressure in the lungs
(b) there is a negative intrapleural pressure pulling at the lung walls
(c) there is a positive intrapleural pressure
(d) pressure in the lungs is higher than the atmospheric pressure

19 Name the pulmonary disease in which alveolar surface area involved in gaseous exchange is drastically reduced due to damage in alveolar wall. **[CBSE-AIPMT 2015]**

(a) Pleurisy (b) Emphysema
(c) Pneumonia (d) Asthma

20 The volume of air that will remain in the lungs after a normal expiration is called **[Kerala CEE 2015]**

(a) Vital capacity
(b) Functional residual capacity
(c) Residual volume
(d) Total lung capacity
(e) Inspiratory capacity

21 Which of the following is the most appropriate in normal circumstances? **[AMU 2015]**
(a) During inspiration, the intrapulmonary pressure is less than the atmospheric pressure
(b) During expiration, the intrapulmonary pressure is less than the atmospheric pressure
(c) During inspiration, the intrapulmonary pressure is more than the atmospheric pressure
(d) During expiration, the intrapulmonary pressure is equal to atmospheric pressure

22 The pneumotaxic and respiratory rhythm centres are respectively present in **[AMU 2015]**
(a) pons and medulla oblongata
(b) corpus callosum and pons
(c) medulla oblongata and hypothalamus
(d) diencephalon and pons

23 **Assertion** (A) In brain stem, pons is called pneumotaxic centre with the ability to moderate the function of the 'respiratory rhythm centre'.
Reason (R) Neural signals from the centre can reduce the duration of expiration, thereby alter the respiratory rate **[EAMCET 2015]**
(a) Both A and R are true and R is the correct explanation of A
(b) Both A and R are true, but R is not the correct explanation of A
(c) A is true, but R is false
(d) A is false, but R is true

24 The process of exchange of O_2 from the atmosphere with CO_2 produced by the cells is called **[J&K CET 2014]**
(a) biological respiration (b) photosynthesis
(c) biological assimilations (d) gaseous exchange

25 Inspiratory capacity of human is equal to **[UK PMT 2014]**
(a) TV + ERV (b) TV + ERV + IRV
(c) TV + IRV (d) ERV + IRV

26 Approximately 70% of carbon dioxide absorbed by the blood will be transported to the lungs **[CBSE AIPMT 2014]**
(a) as bicarbonate ions
(b) in the form of dissolved gas molecules
(c) by binding to RBC
(d) as carbaminohaemoglobin

27 The volume of air that can be breathed in by maximum forced inspiration over and above the normal inspiration is called **[WB JEE 2014]**
(a) expiratory reserve volume
(b) inspiratory reserve volume
(c) vital capacity
(d) inspiratory capacity

28 The serous membrane which covers the lungs is called
(a) pericardium (b) peritoneum **[WB JEE 2014]**
(c) perichondrium (d) pleura

29 What is the pO_2 and pCO_2 in the systemic arteries? **[Kerala CEE 2014]**
(a) pO_2 40 mm Hg, pCO_2 45 mm Hg
(b) pO_2 95 mm Hg, pCO_2 40 mm Hg
(c) pO_2 40 mm Hg, pCO_2 40 mm Hg
(d) pO_2 45 mm Hg, pCO_2 40 mm Hg
(e) pO_2 104 mm Hg, pCO_2 159 mm Hg

30 Choose the wrong statement. **[Kerala CEE 2014]**
(a) Solubility of CO_2 in blood is 20-25 times higher than that of O_2
(b) The total volume of air accommodated in the lungs at the end of a forced inspiration is called the 'vital capacity'
(c) O_2 can bind with haemoglobin in a reversible manner to form oxyhaemoglobin
(d) Every 100 mL of deoxygenated blood delivers approximately 4 mL of CO_2 to the alveoli
(e) The diffusion membrane is made of three major layers namely the thin squamous epithelium of alveoli, the endothelium of alveolar capillaries and the basement substance in between them

31 To generate pressure gradients to facilitate expiration and inspiration, the human body uses the intercostal muscles and **[Kerala CEE 2014]**
(a) alveolar sac
(b) bronchi
(c) primary, secondary and tertiary bronchioles
(d) diaphragm
(e) windpipe

32 The figure shows a diagrammatic view of human respiratory system with labels *A*, *B*, *C* and *D*. Select the option. Which option gives correct identification and main function and/or characteristic?

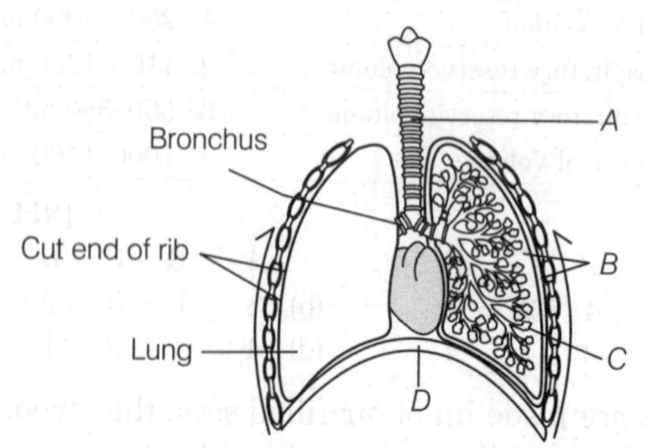

[NEET 2013]
(a) *A*-Trachea—long tube supported by complete cartilaginous rings for conducting inspired air
(b) *B*-Pleural membrane—surrounds ribs on both sides to provide cushion against rubbing
(c) *C*-Alveoli—thin-walled vascular bag-like structures for exchange of gases
(d) *D*-Lower end of lungs—diaphragm pulls it down during inspiration

33 Haemoglobin is having the maximum affinity with [UP CPMT 2013]
(a) CO_2 (b) CO
(c) O_2 (d) NH_3

34 During inspiration, the diaphragm [AFMC 2013]
(a) expands
(b) shows no change
(c) contracts and flattens
(d) relaxes to become dome-shaped

35 The oxygen toxicity is related with [Manipal 2013]
(a) blood poisoning
(b) collapse of alveolar walls
(c) Both (a) and (b)
(d) failure of ventilation of lungs

36 Skin is an accessory organ of respiration in [WB JEE 2013]
(a) human (b) frog
(c) rabbit (d) lizard

37 When the oxygen supply to the tissue is inadequate, the condition is [WB JEE 2013]
(a) dyspnea (b) hypoxia (c) asphyxia (d) apnea

38 Large volume of air a person can expire after a forceful inspiration is called [Odisha JEE 2013]
(a) tidal volume (b) vital capacity
(c) IRV (d) ERV

39 People who have migrated from the planes to an area adjoining Rohtang Pass about six months back [CBSE AIPMT 2012]
(a) have more RBCs and their haemoglobin has a lower binding affinity to O_2
(b) are not physically fit to play games like football
(c) suffer from altitude sickness with symptoms like nausea, fatigue, etc
(d) have the usual RBC count, but their haemoglobin has very high binding affinity to O_2

40 Which one of the following is the correct statement for respiration in humans? [CBSE AIPMT 2012]
(a) Cigarette smoking may lead to inflammation of bronchi
(b) Neural signals from pneumotaxic centre in pons region of brain can increase the duration of inspiration
(c) Workers in grinding and stone breaking industries may suffer from lung fibrosis
(d) About 90% of carbon dioxide (CO_2) is carried by haemoglobin as carbaminohaemoglobin

41 After forceful inspiration, the amount of air that can be breathed out by maximum forced expiration is equal to [WB JEE 2012]
(a) Inspiratory Reserve Volume (IRV) + Expiratory Reserve Volume (ERV) + Tidal Volume (TV) + Residual Volume (RV)
(b) IRV + RV + ERV
(c) IRV + TV + ERV
(d) TV + RV + ERV

42 Amount of oxygen supplied by 100 mL arterial blood while passing through the tissues is [WB JEE 2012]
(a) 0.4-0.6 mL (b) 4-6 mL
(c) 14-15 mL (d) 19-20 mL

43 The volume of 'anatomical dead space' air is normally [WB JEE 2012]
(a) 230 mL (b) 210 mL
(c) 190 mL (d) 150 mL

44 Choose the right sequential phenomena among the following during the delivery of O_2 from blood to tissue. [WB JEE 2012]
I. Absorption of CO_2 by the blood.
II. Reaction of absorbed CO_2 with H_2O to from H_2CO_3 within RBC and its conversion into H^+ and HCO_3^- ions.
III. Reaction of absorbed CO_2 with H_2O in plasma to form H_2CO_3 and its conversion into H^+ and HCO_3^- ions.
IV. Combination of H^+ with haem portion of HbO_2 to release O_2.
V. Combination of HCO_3^- with haem portion HbO_2 to form reduced haemoglobin and release of O_2. [WB JEE 2011]
(a) I, II and V (b) I, III and IV
(c) I, II and V (d) I, III and V

45 Two friends are eating together on a dining table. One of them suddenly starts coughing while swallowing some food. This coughing would have been due to improper movement of [CBSE AIPMT 2011]
(a) diaphragm (b) neck
(c) tongue (d) epiglottis

46 A large proportion of oxygen is left unused in the human blood even after its uptake by the body tissues. This O_2 [CBSE AIPMT 2011]
(a) raises the pCO_2 of blood to 75 mm of Hg
(b) is enough to keep oxyhaemoglobin
(c) helps in releasing more O_2 to the epithelial tissues
(d) acts as a reserve during muscular exercises

47 The figure given below shows a small part of human lung where exchange of gas takes place. In which one of the options given below, the one part *A*, *B*, *C* or *D* is correctly identified along with its function. [CBSE AIPMT 2011]

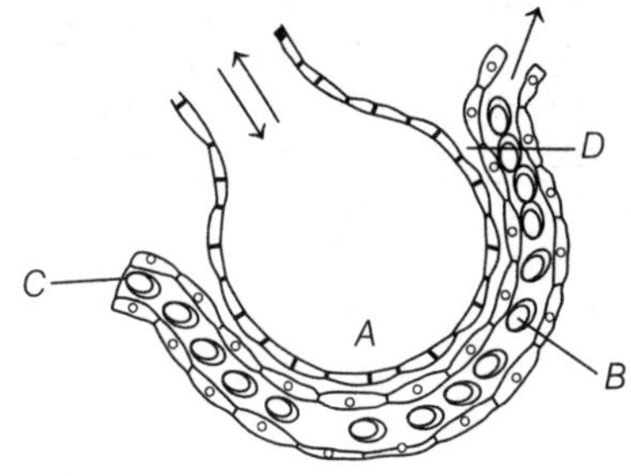

(a) A–Alveolar cavity — Main site of exchange of respiratory gases
(b) D–Capillary wall — Exchange of O_2 and CO_2 takes place here
(c) B–Red blood cell — Transport of mainly haemoglobin
(d) C–Arterial capillary — Passes oxygen to tissues

48 The factor which does not affect the rate of alveolar diffusion is **[Kerala CEE 2011]**
(a) solubility of gases (b) thickness of the membrane
(c) pressure gradient (d) concentration gradient
(e) reactivity of the gases

49 Haemoglobin of the human blood forms a stable complex compound with which of the following gas leading to death? **[MHT CET 2011]**
(a) Oxygen (b) Carbon dioxide
(c) Carbon monoxide (d) Nitrogen

50 If a large number of people are enclosed in a room, then **[MHT CET 2011]**
(a) oxygen decreases and carbon dioxide increases
(b) oxygen increases and carbon dioxide decreases
(c) Both oxygen and carbon dioxide decrease
(d) Both oxygen and carbon dioxide increases

51 The expiratory reserve volume will be **[UP CPMT 2011]**
(a) 1000 mL (b) 2000 mL
(c) 4000 mL (d) 5000 mL

52 **Assertion** (A) CO_2 transport occurs very fast through RBCs.
Reason (R) Enzyme carbonic anhydrase is absent in blood plasma. **[Gujarat CET 2011]**
(a) Both A and R are true and R is the correct explanation of A
(b) Both A and R are true, but R is not the correct explanation of A
(c) A is true, but R is false
(d) A is false, but R is true

53 Hiccups can be best described as **[DUMET 2011]**
(a) forceful sudden expiration
(b) jerky incomplete inspiration
(c) vibration of the soft palate during breathing
(d) sign of indigestion

54 Among all, CO_2 in expired air is about **[J&K CET 2011]**
(a) 0.04 % (b) 0.03 %
(c) 4.5 % (d) 2.1 %

ANSWERS

Check Point 24.1

1. (d)	2. (a)	3. (d)	4. (d)	5. (d)	6. (d)

Check Point 24.2

1. (a)	2. (c)	3. (c)	4. (b)	5. (b)	6. (c)

Check Point 24.3

1. (a)	2. (b)	3. (b)	4. (a)	5. (a)	6. (a)

(A) Topical Exercise

1. (d)	2. (a)	3. (d)	4. (c)	5. (d)	6. (b)	7. (a)	8. (b)	9. (c)	10. (d)
11. (b)	12. (d)	13. (b)	14. (a)	15. (c)	16. (a)	17. (b)	18. (a)	19. (c)	20. (a)
21. (b)	22. (c)	23. (d)	24. (c)	25. (a)	26. (c)	27. (d)	28. (b)	29. (a)	30. (b)
31. (c)	32. (b)	33. (c)	34. (c)	35. (b)	36. (b)	37. (d)	38. (c)	39. (a)	40. (d)
41. (a)	42. (c)	43. (c)	44. (d)	45. (a)	46. (b)	47. (c)	48. (b)	49. (a)	50. (b)
51. (a)	52. (d)	53. (c)	54. (c)	55. (b)	56. (d)	57. (a)	58. (b)	59. (c)	60. (c)
61. (a)	62. (b)	63. (b)	64. (c)	65. (d)	66. (d)	67. (a)	68. (d)	69. (a)	70. (b)
71. (c)	72. (a)	73. (a)	74. (b)	75. (c)	76. (c)	77. (b)	78. (a)	79. (a)	80. (d)
81. (d)	82. (a)	83. (a)	84. (b)	85. (a)	86. (b)	87. (a)	88. (b)	89. (b)	90. (b)
91. (b)	92. (d)	93. (a)	94. (a)	95. (b)	96. (d)	97. (a)	98. (c)	99. (b)	100. (c)
101. (d)	102. (a)	103. (b)	104. (c)	105. (d)	106. (a)	107. (c)	108. (d)	109. (b)	110. (a)
111. (c)	112. (b)	113. (d)							

(B) Cumulative Exercise

1. (b)	2. (d)	3. (d)	4. (d)	5. (d)	6. (d)	7. (a)	8. (c)	9. (b)	10. (c)
11. (b)	12. (b)	13. (a)	14. (d)	15. (d)	16. (d)	17. (d)	18. (c)	19. (d)	20. (c)
21. (b)	22. (d)	23. (a)	24. (b)	25. (c)	26. (c)	27. (d)	28. (a)	29. (d)	30. (b)
31. (d)	32. (b)	33. (c)	34. (c)	35. (d)	36. (c)	37. (a)	38. (c)	39. (d)	40. (b)
41. (b)	42. (a)	43. (b)	44. (a)	45. (d)	46. (a)	47. (b)			

(C) Medical Entrances' Gallery

1. (b)	2. (d)	3. (d)	4. (b)	5. (b)	6. (b)	7. (d)	8. (a)	9. (a)	10. (c)
11. (b)	12. (b)	13. (a)	14. (a)	15. (a)	16. (b)	17. (b)	18. (b)	19. (b)	20. (b)
21. (a)	22. (a)	23. (c)	24. (d)	25. (c)	26. (a)	27. (b)	28. (d)	29. (b)	30. (b)
31. (d)	32. (c)	33. (b)	34. (c)	35. (d)	36. (b)	37. (b)	38. (c)	39. (a)	40. (c)
41. (c)	42. (d)	43. (d)	44. (c)	45. (d)	46. (d)	47. (a)	48. (e)	49. (c)	50. (a)
51. (a)	52. (c)	53. (b)	54. (c)						

Explanations

(A) Topical Exercise

9 (*c*) Option (c) is incorrectly matched pair and can be corrected as

The respiratory structure in scorpion is book lungs.

Rest options are correctly matched pairs of given animals and their respiratory structures.

16 (*a*) The nasopharynx is lined by pseudostratified columnar ciliated epithelium with Goblet cells. These cells help to moisten the epithelial surface so that the ciliated cells can function. Its secretion, i.e. mucus also helps to trap the finest dust particles before they are carried all the way into alveoli. Thus, option (a) is correct.

20 (*a*) The laryngeal skeleton is formed of nine cartilages. Out of these, three are single namely, thyroid, cricoid and epiglottis. The rest six are paired, thus consisting of three types. These are arytenoid, corniculate and cuneiform.

21 (*b*) Option (b) is incorrectly matched and can be corrected as

Thyroid cartilage is most prominent in males. Cuneiform cartilage connects epiglottis with arytenoid cartilage in both males and females.

Rest options are correctly matched pairs.

30 (*b*) Option (b) represents incorrectly matched pairs and can be corrected as

Lingual tonsils are a part of oropharynx and trachea is air conducting tube with non-collapsible walls due to C-shaped hyaline cartilage.

Rest pairs are correctly matched.

35 (*b*) The trachea of mammals and insects has non-collapsible wall due to the presence of incomplete C-shaped rings.

39 (*a*) Statement in option (a) is incorrect and can be corrected as

Each terminal bronchiole gives rise respiratory bronchioles.

Rest other statements are correct regarding respiratory system.

42 (*c*) Diffusion of gases is a physical phenomenon that takes place between the tissue and blood vessels and does not occur during breathing whereas bringing the air to body temperature, its cleaning and warming occurs during the process of breathing. Thus, option (c) is correct.

44 (*d*) The process of breathing is effected by the pressure gradient in lungs with respect to atmosphere. The puncture in chest cavity would not allow the pressure gradient to develop and thus, the breath would cease.

51 (*a*) Option (a) represents incorrectly matched pair and can be corrected as

During inspiration, the diaphragm straightens so that the thoracic cavity, increases in volume.

Rest options are correctly matched pairs.

74 (*b*) Increase CO_2 concentration in blood lowers its pH due to which the blood become too acidic. This condition is known as respiratory acidosis.

79 (*a*) Option (a) represents incorrectly matched pair and can be corrected as

Eupnea refers to the normal breathing rate. The term used for decreased breathing rate is hypopnea.

Rest options are correctly matched pairs.

86 (*b*) The sample containing maximum percentage composition of CO_2 would be breathed out. Thus, option (b) is correct.

87 (*a*) The concentration of O_2 would be low in blood capillary (*X*) and high in alveolus (*Y*), due to which gaseous exchange occurs through simple diffusion, from the region of high concentration to lower concentration.

89 (*b*) Statement in option (b) is incorrect and can be corrected as

Oxygen binds with the haem part of haemoglobin. One molecule of haemoglobin contains four haem/ferrous/iron atoms. Each haem binds with one oxygen molecule.

Rest statements are correct about oxygen binding with haemoglobin.

103 (*b*) Statement in option (b) is incorrect. It can be corrected as Oxyhaemoglobin acts as strong acid. As more oxyhaemoglobin is formed, more H^+ ions are released, which increases the acidity of blood.
Rest statements are correct.

(B) Cumulative Exercise

2 (*d*) Statement in option (d) is incorrect and can be corrected as
Snoring occurs due to the partial blockage of upper respiratory tract. Involuntary contraction of diaphragm followed by sudden closure of glottis cause hiccups.
Rest statements are correct.

5 (*d*) After exercise, eight breaths are taken in 12 seconds and in each breath 1 dm^3, air is taken in. So total 8 dm^3 air is taken (during 12 seconds).

6 (*d*) Gas exchange surfaces are usually one cell thick epithelial tissues as the walls of the alveoli in lungs, where thin epithelium helps in diffusion process of gases.

12 (*b*) The haemoglobin is most saturated with O_2 in pulmonary vein (*Z*) as it carries oxygenated blood from lungs to left atrium.
While moving from pulmonary vein to left atrium and then to left ventricle (*Y*), the percentage saturation of blood with O_2 decreases slightly.
Vena cava (X) carry deoxygenated blood from body to right atrium thus, Hb is less saturated with O_2 in it.

17 (*d*) In muscles during anaerobic respiration, pyruvic acid is converted to lactic acid and lesser ATP is produced. Alcohol and CO_2 are produced in yeast.
Thus, option (d) is correct.

18 (*c*) Percentage composition shown in the table only refers to oxygen. No other gas has this compositon in air.

19 (*d*) $$\text{Number of breath} = \frac{\text{Alveolar ventilation}}{\text{Tidal volume} - \text{Dead space (mL)}}$$
$$= \frac{4000 \text{ mL/min}}{500 \text{ mL} - 300 \text{ mL}} = \frac{4000}{200}$$
$$= 20 \text{ breath/min}$$

22 (*d*) Artificial respiration is given at higher rate, i.e. about 10-15 times per minute to ensure that lungs are properly and efficiently ventilated and the life of the person can be saved.

28 (*a*) Bohr's effect states that due to increase in concentration of CO_2, HbO_2 dissociation curve shifts to right. Thus, deoxygenation of oxyhaemoglobin is directly proportional to pCO_2 of blood, CO_2 of tissue fluid and alveoli does not exert Bohr's effect.

31 (*d*) Statement in option (d) is incorrect and can be corrected as
Pneumotaxic centre reduces the inspiration so as to control the respiratory rate.
Rest statements are correct.

32 (*b*) Statement I is false, but II is true. Incorrect statement can be corrected as
Interpulmonary pressure does not remain the same during inspiration and expiration. It decreases then atmospheric pressure during inspiration, thus become negative, while during expiration, interpulmonary pressure become positive.

33 (*c*) Both statements I and II are correct.
Band 3 proteins are found on the surface of RBCs and they mediate the exchange of cellular HCO_3^- ions with Cl^- ions during chloride shift. Band 3 proteins can differentiate metabolically active cells from inactive cells and supply oxygen to former only.

37 (*a*) Statements I, III and V are correct whereas II and IV are incorrect. Incorrect statements can be corrected as

- About 20-24% of CO_2 is carried out by haemoglobin. About 7% of CO_2 is transported dissolved in plasma of blood.
- The oxyhaemoglobin in erythrocytes is acidic.

40 (*b*) Both A and R are true, but R is not the correct explanation of A.
The vital capacity is higher in athletes, mountain dwellers than in plain dwellers, because they require more energy, i.e. more oxygen through respiration. Vital capacity is the amount of air, which one can inhale and exhale with maximum effort.

41 (*b*) Both A and R are true, but R is not the correct explanation of A.
The amount of air inspired or expired with each normal breath is termed as Tidal Volume (TV). The tidal volume is 500 mL in an adult person. It constitutes about 150 mL of dead space volume and 350 mL of alveolar volume.

42 (*a*) Both A and R are true and R is the correct explanation of A.
Cells of RBCs contain a zinc-containing enzyme called carbonic anhydrase. This enzyme accelerates the binding of CO_2 molecules to the blood RBCs about 5000 times.
Due to this, about 70% of carbon dioxide received by blood from the tissues, enters the RBCs where it reacts with water to form carbonic acid (H_2CO_3). Almost as rapidly as formed, all carbonic acid of RBCs dissociates into H^+ and HCO_3^-. Then H^+ ions combine with haemoglobin to maintain the pH of the blood and thus HCO_3^- diffuse from RBCs into the plasma.

43 (*b*) Both A and R are true, but R is not the correct explanation of A.
Insects do not respire through body surface, because their body possesses a thick and impermeable integument, which minimises loss of water.
Insects have a complex system of air tubes called trachea, which help in direct transfer of air to various body tissues.

44 (*a*) Both A and R are true and R is the correct explanation of A.
Fishes possess internal gills. Gills are thin filamentous structures which are richly supplied with blood vessels. These are arranged in a way, so that the blood flows in opposite direction to the flow of water across the gills.
This arrangement is called counter-current flow and it helps in the greatest possible extraction of oxygen from water.

45 (*d*) A is false, but R is true because

Inspiration is an active process, which results due to muscular contraction.

During inspiration, the diaphragm and external muscle contract simultaneously. This moves the lateral thoracic walls outward and upward.

46 (*a*) Both A and R are true and R is the correct explanation of A.

The cell membrane of aquatic organisms such as protozoans remains moist. It serves as the respiratory surface. Oxygen and carbon dioxide simply diffuse across this thin layer.

47 (*b*) Both A and R are true, but R is not the correct explanation of A.

Earthworm does not have lungs or special respiratory organs. Respiration (gaseous exchange) in earthworm takes place through the skin (body wall). This type of respiration is known as cutaneous respiration.

Secretion of gland cells of the epidermis and coelomic fluids keeps the skin (body wall) moist which facilitates in cutaneous respiration.

(C) Medical Entrances' Gallery

1 (*b*) Statement in option (b) is wrong with reference to transport of oxygen and can be corrected as

In alveoli, high pO_2, low pCO_2, low H^+ concentration and lower temperature are the factors that favour the formation of oxyhaemoglobin.

Rest statements are correct with reference to transport of oxygen.

2 (*d*) Events I and II are correct during inspiration, i.e. the contraction of diaphragm occurs which pulls it downward, while the external intercostal muscles contract and lifts up the ribs and sternum. This increses the size of the thoracic cavity and decreases the pressure inside. As a result, air rushes in and fills the lungs.

Events III and IV are incorrect and can be corrected as

During inspiration, the volume of pulmonary or thoracic cavity increases. This causes a similar increase in pulmonary volume. An increase in pulmonary volume decreases the intrapulmonary pressure to less than the atmospheric pressure which forces the air from outside to move into the lungs.

3 (*d*) The Total Lung Capacity (TLC) is the total volume of air accommodated in the lungs at the end of a forced inspiration. This includes Residual Volume (RV), Expiratory Reserve Volume (ERV), Tidal Volume (TV) and Inspiratory Reserve Volume (IRV).

TLC is also equals to vital capacity of residual volume.

Thus, option (d) is correct.

4 (*b*) A-2, B-3, C-4, D-1

5 (*b*) Statement in option (b) is correct as

Intrapulmonary pressure is lower than the atmospheric pressure during inspiration.

Other statements are incorrect and can be corrected as

- Inspiration occurs due to external intercostal muscles.
- Inspiration occurs when atmospheric pressure is more than intrapulmonary pressure.
- Inspiration is initiated due to the contraction of diaphragm.

6 (*b*) Vital Capacity (VC) is the maximum volume of air a person can breath in after a forced expiration. It is about 3500-4500 mL and includes ERV, TV and IRV. It is also defined as the maximum volume of air a person can breath out after a forced inspiration.

7 (*d*) The Expiratory Capacity of athlete will be 1500 mL.

It can be calculated as

Given, Tidal Volume (TV) = 500 mL

Expiratory Reserve Volume (ERV) = 1000 mL

Expiratory Capacity = TV+ERV

$= 500 + 1000 = 1500$ mL

8 (*a*) Band 3 protein is a typical polytropic membrane protein that mediates the exchange of cellular bicarbonate ions (HCO_3^-) with the chloride ion (Cl^-) in plasma, which has been known as the 'chloride shift.' Owing to this, RBC can discriminate the metabolically active cells from inactive cells and deliver oxygen particularly to metabolically active tissue that produce carbon dixoide.

12 (*b*) A-3, B-1, C-4, D-2

13 (*a*) In lungs, even after the forceful expiration, some of the volume of air remains. This volume is termed as Residual Volume (RV). Due to this, lungs do not collapse. RV is about 1100 mL to 1200 mL.

14 (*a*) Oxyhaemoglobin dissociates into oxygen and deoxyhaemoglobin at low O_2 pressure in tissues. However, at high O_2 pressure in tissues, the haemoglobin binds to oxygen forming oxyhaemoglobin and transports O_2 instead of CO_2.

15 (*a*) Oxygen dissociation curve of haemoglobin is sigmoid (S-shaped). It gives the relationship between the saturation of haemoglobin and oxygen tension. It is obtained by plotting the per cent saturation of Hb against partial pressure of O_2 at 38°C and pH 7.4. Dissociation of oxyhaemoglobin can be promoted by rise in the body temperature and low pH and high pCO_2, etc.

17 (*b*) The partial pressure of oxygen (pO_2) in alveoli of lungs is 104 mm Hg, which is more than that of blood in the blood capillaries of lung alveoli (40 mm Hg). This difference allows the passive diffusion of O_2 from air-filled in the lungs to the blood vessels of lung alveoli.

18 (*b*) Lungs do not collapse between breaths and some air always remains in the lungs, which can be never be expelled, because there is a negative intrapleural pressure pulling at the lung walls. Alveoli are basic functional unit of lungs. The outer alveolar wall surface have cells, which secrete DPPC or lipid surfactant. The surfactant expands the alveoli, due to which the negative pressure inside the alveoli increases. This prevents the alveoli from collapsing.

23 (*c*) A is true, but R is false. R can be corrected as Neural signals from pneumotaxic centre can reduce the rate of inspiration.

26 (*a*) The largest fraction of carbon dioxide, i.e. about 70% is converted to bicarbonate (HCO_3^-) ions in the presence of enzyme carbonic anhydrase and transported to the lungs.

$$CO_2 + H_2O \underset{\text{anhydrase}}{\overset{\text{Carbonic}}{\rightleftharpoons}} \underset{\text{(Carbonic acid)}}{H_2CO_3} \underset{\text{anhydrase}}{\overset{\text{Carbonic}}{\rightleftharpoons}} \underset{\text{(Bicarbonate ion)}}{HCO_3^-} + H^+$$

29 (*b*) pO_2 in oxygenated blood (95 mm Hg) is greater than pO_2 in body cells (40 mmHg) and pCO_2 in oxygenated blood (40 mm Hg) is lesser than pCO_2 in body cells (45 mm Hg) during exchange of gases between blood and tissue. Therefore, in systematic arteries, pO_2 is 95 mm Hg and pCO_2 is 40 mm Hg.

30 (*b*) Statement in option (b) is wrong and can be corrected as The total volume of air accomodated in the lungs at the end of a forced inspiration is called the Total Lung Capacity (TLC).

31 (*d*) During inspiration, the diaphragm and specialised set of intercostal muscles (external and internal) contract. This moves the lateral thoracic wall outward and upward, thus increasing the capacity of thoracic cavity and generate the pressure gradient between the lungs and atmosphere. Expiration is a passive process.

32 (*c*) Options (c) is correct identification and main function and its characteristic.

C-Alveoli are thin-walled vascular bag-like structures for exchange of gases.

Rest other options are not correct identification and main functions and their characteristics. These can be corrected as

A-Trachea or windpipe is an air conducting tube through which transport of gases takes place. *B*-Pleural membrane is double-layered, which reduces friction on the lung surface. *D*-Diaphragm is involved in the inspiration and expiration process of breathing.

33 (*b*) Haemoglobin (Hb) have 250 times more affinity for CO as compared to O_2. Hb forms a cherry-red compound with CO called carboxyhaemoglobin.

34 (*c*) Diaphragm is a muscular structure present between abdominal and thoracic cavity in mammals.
In normal condition, diaphragm is dome-shaped. During inspiration, it becomes flat due to its contraction and during expiration, it becomes relaxed.

35 (*d*) Oxygen toxicity develops when 100% pure oxygen is breathed for a prolonged period. If we inhale pure oxygen, ventilation is stopped. Atmospheric air contains only 21% of oxygen.

39 (*a*) As a person moves up a hill, the pO_2 and total atmospheric pressure decreases. Decrease in pO_2 due to increasing altitude stimulates the JG cells of kidney to secrete erythropoietin hormone thus, increasing the number of RBCs (polycythemia) to compensate the supply of O_2. At higher altitude, haemoglobin has lower binding affinity for O_2, because the primary factor responsible for binding is pO_2, which decreases at higher altitude.

42 (*d*) 100 mL arterial blood carries about 19-20 mL of oxygen, but it delivers 4-6 mL of oxygen at tissue level in normal conditions.

45 (*d*) The epiglottis is a flap which is made up of elastic cartilage tissue covered with a mucous membrane, attached to the entrance of the larynx. It prevents the entry of food into the larynx and directs it to the oesophagus. Due to improper movement of epiglottis, one may suddenly start coughing, while swallowing some food.

46 (*d*) Our tissues are able to utilise only 25% of O_2 carried by arterial blood. The venous blood is still 75% saturated with O_2. This O_2 acts as a reserve during muscular exercise.

48 (*e*) Alveoli (thin, irregular-walled and vascularised bag-like structure at the end of bronchiole) are the primary sites of exchange of gases. O_2 and CO_2 are exchanged in these sites by simple diffusion, mainly based on pressure/concentration gradient.

Solubility of the gases as well as thickness of the membranes involved in diffusion are also some important factors that can affect the rate of diffusion. Reactivity of the gases does not affect the rate of alveolar diffusion.

Thus, option (e) is correct.

49 (*c*) 98.5% of O_2 is transported by blood with the help of haemoglobin. The molecule of haemoglobin has 250 times more affinity with CO as compare to O_2 and thus, prevents O_2 transport which leads to death.

52 (*c*) A is true, but R is false because

Transport of CO_2 by blood is much easier than that of oxygen due to high solubility of CO_2 in water. About 70% of CO_2 is transported through RBCs. CO_2 enters the RBCs, where it reacts with water to form carbonic acid in the presence of carbonic anhydrase.

Most of the carbonic acid rapidly dissociates to form bicarbonate and hydrogen ions.

54 (*c*) Amount of CO_2 in expired air is 4.5%. The air we inspire contains about 0.05% CO_2. The air we expire contains about 4% CO_2. In other word, exhaled air contains about 100 times the concentration of CO_2 than inhaled air does.

CHAPTER 25

Body Fluid and Circulation

All living cells require nutrients, oxygen and other essential materials. These substances take part in various metabolic processes of the cell. Various toxic substances are also produced during these processes. Accumulation of these toxic substances is hazardous for the cell. Hence, these need to be eliminated.

Therefore, an efficient mechanism and medium is required for the **to** and **fro** movements of these substances from the cells.

In humans and other higher organisms, this functioning is taken over by two special body fluids called **blood** and **lymph**. The intercellular and intracellular tissue and cellular fluids also help in this functioning.

BODY FLUIDS

As stated above, body fluids are the medium of transportation within the body. As per the most generalised categorisation, these may be categorised into intracellular and extracellular categories.

The **intracellular fluids** are present inside the cell. These contain larger amounts of potassium ions, phosphate ions and proteins.

Extracellular fluids are present outside the cell. These include blood, lymph, Cerebrospinal Fluid (CSF), etc.

Blood

It is a specialised connective tissue consisting of a fluid matrix, plasma and a cellular portion called **formed elements.** The formed elements include the cells or corpuscles namely, **erythrocytes** (RBCs), **leucocytes** (WBCs) and **thrombocytes** (platelets).

The structural and compositional details about the blood are already discussed in chapter, Animal Tissues. In this chapter, we will be focusing mainly on blood groups and its coagulation.

Inside

Blood groups

- There are certain molecules present on the surfaces of all blood cells in the body of an organism. These can be recognised as foreign, if introduced in the body of another organism. These molecules are known as **antigens**.
- In general terms, antigens are the substances which has the ability to extract immune response from our body.
- So just like others, these antigens of blood can also induce immune response by secreting a class of proteins called **antibodies.** These antibodies have the ability to bind in a specific fashion with antigens.
- Depending upon the nature of antigens present on the membrane of RBCs, various types of blood grouping has been done. These are discussed below

1. ABO blood groups

In 1901, **Karl Landsteiner** for the first time discovered a system of blood groups. It was based on two surface antigens on the RBCs. These were named **antigen-A** and **antigen-B** in human beings. This system of blood groups is commonly called ABO blood grouping system.

A brief description of this system is given in table below

Table 25.1 Types of blood groups

Phenotypes or Blood group names with their percentage in human population	Antigens on RBC membrane	Antibodies in plasma	Can receive blood from	Can donate blood to
A (40%)	A	Anti-B	A, O	A, AB
B (10%)	B	Anti-A	B, O	B, AB
AB (4%)	A and B antigens	No	A, B, AB, O **(Universal recipient)**	AB only
O (46%)	No	Anti-A and Anti-B	O only	A, B, AB, O **(Universal donor)**

Thus among the four blood groups, AB is universal recipient, while O is universal donor. The AB blood group was reported by **de Castello** and **Steini** (1902). Person with 'AB' group develop tolerance to both of these antigens and thus, do not produce either anti-A or anti-B antibodies.

2. Rhesus (Rh) blood group

It was discovered by **Landsteiner** and **Wiener** in the blood of rhesus monkeys. Later, it was also observed on the surface of RBCs of majority of human beings.

Rh positive and Rh negative

The individuals with **Rh antigen** are called **Rh positive** (Rh +ve) and in whom this antigen is not present are called **Rh negative** (Rh –ve). Always remember, Rh^+ is dominant to Rh^-.

Rh incompatibility

- **During pregnancy** it is seen that **when father's blood is Rh^+ and mother's blood is Rh^-**. Rh^+ being a dominant character is expressed in the foetus and causes a serious problem. The first child of Rh^- mother will not suffer with any problem, but blood of foetus (Rh^+) stimulates the formation of anti-Rh^+ factors in the mother's blood.
  So, in the subsequent pregnancies, the anti-Rh antibodies in the mother's blood destroy the foetal RBCs. It result in **Haemolytic Disease of the Newborn** (HDN) or **erythroblastosis foetalis** or **rhesus baby syndrome.**
- In order to prevent HDN, Rh immunoglobulin (RhoGAM) is injected to Rh^- mother after the birth of Rh^+ baby.
- **During blood transfusion** The first transfusion between Rh^+ and Rh^- blood causes no harm but Rh^- person develops anti-Rh antibodies in his blood. In the second transfusion of Rh^+ blood to Rh^- blood, the anti-Rh antibodies in the latter's blood destroy the RBCs of the donor.

Coagulation of blood

Blood exhibits coagulation or clotting in response to an injury or trauma. This is a generalised mechanism present in the body of higher animals to prevent excessive loss of blood from the body.

Clot structure

The clot or coagulum when formed is mainly consists of a network of threads called **fibrins.** In these threads, the dead and damaged elements of blood are trapped.

Mechanism of clot formation

The pathway or mechanism of blood clotting are of two types, i.e. extrinsic pathway and intrinsic pathway. Both the pathways differ in the formation of thromboplastin.

In extrinsic pathway thromboplastin is produced from injured tissues while in intrinsic pathway it is produced with the help of platelets.

Both of these can be summarised as follows

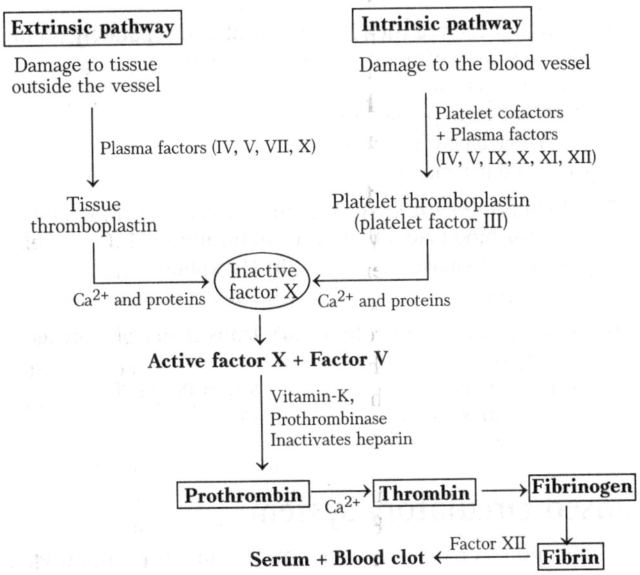

- Thromboplastin is also called **prothrombin activator complex.**
- Thromboplastin combined with calcium and certain protein helps in the formation of an enzyme prothrombinase (which inactivates heparin) that converts the inactive plasma protein, **prothrombin** into its active form **thrombin**.
- Thrombin acts as a **proteolytic enzyme** to convert **fibrinogen** molecule (produced from the liver in the presence of vitamin-K) to from insoluble fibrin monomer.
- These fibrin monomers polymerise to long, sticky fibres called fibrins, in which dead and damaged elements of blood are trapped. This finally leads to the formation of a clot or coagulum. It is a dark reddish-brown **scum** formed over the surface of injury.
- Within 2-5 minutes after clot formation, a yellow fluid called **serum** oozes out.
- The process of blood coagulation proceeds with the help of various specialised proteins called **factors.**
- The coagulation factors or clotting factors used in both the pathways above are tabulated below

Table 25.2 Clotting factors their specific characteristics and roles

Clotting factors	Synonyms	Characteristics
I	**Fibrinogen**	• Glycoprotein, synthesised in liver. Soluble in plasma, converted to fibrin on formation of clot.
II	**Prothrombin**	• Glycoprotein, synthesised in liver by vitamin-K.
III	**Thromboplastin** or **tissue factor**	• Lipoprotein, secreted in inactive form called **prothromboplastin,** which gets activated by proconvertin of plasma tissues.
IV	**Calcium ions**	• Required at all the major steps of blood clotting, i.e. for the formation of intrinsic and extrinsic thromboplastin and for the conversion of prothrombin to thrombin.
V	**Proaccelerin or labile factor**	• Glycoprotein, heat labile, synthesised in liver and absent in serum.
VI	**Accelerin** (Hypothetical factor)	• Hypothetical activation product of proaccelerin.
VII	**Serum Prothrombin Accelerator** (SPA) or stable factor or autoprothrombin or proconvertin	• Synthesised in liver by vitamin-K, associated with prothrombin and accelerates tissue thromboplastin formation from damaged tissues.
VIII	**Anti-haemophilic factor-A** or platelet cofactor or Antihaemophilic Globulin (AHG)	• Glycoprotein, synthesised in liver, required for prothrombin activator formation from blood constituents. Its deficiency causes **haemophilia-A.**
IX	**Christmas factor, anti-prothrombin II** or platelet cofactor II or Plasma Thromboplastin Component (PTC) or Antihaemophilic factor-B.	• Glycoprotein, synthesised in liver by vitamin-K. Its deficiency causes **haemophilia-B.**
X	**Stuart- Prower factor**	• Glycoprotein, synthesised in liver by vitamin-K. Its deficiency causes nose bleeding or epistaxis.
XI	**Plasma Thromboplastin Antecedent** (PTA) or Antihaemophilic factor-C	• Glycoprotein, required for stage I of intrinsic pathway, synthesised in liver, deficiency causes **haemophilia-C.**
XII	**Hageman factor** or **surface factor**	• Glycoprotein, present in both plasma and serum, required for the formation of prothrombin activator complex. Deficiency results in delayed blood clotting.
XIII	**Fibrin stabilising factor**	• Glycoprotein, causes polymerisation of soluble fibrinogen to insoluble fibrin. Deficiency causes haemorrhagic state.

Anticoagulants

- These are substances which have the property of preventing the coagulation of blood. The examples of these include **heparin, hirudin, lamphredin** and **coumarin**, etc. Among these coumarin is of plant origin and is also known as **dicumarol**. It simply acts as vitamin-K.
- The citrates of sodium and oxalates of sodium and ammonium, EDTA (Ethylene Diamine Tetra Acetate), etc., are used in laboratories and blood banks to prevent blood samples from clotting. These substances being acidic trap calcium of blood to form insoluble compounds. Thus, calcium which is required in all the three stages of both the pathways becomes not available for clotting.

Another body fluid is lymph.The details of this body fluid and related system are given at the end of this chapter.

CHECK POINT 25.1

1. Blood groups are
   (a) surface antigens (b) foreign antigens
   (c) cell surface antigens (d) All are correct
2. An Rh^- individual receives Rh^+ blood. The recipient becomes
   (a) sterile (b) dead
   (c) no reaction (d) iso-immuanised
3. Rh (–ve) person donated blood to Rh(+ve) person for second time. Then
   (a) Rh (+ve) blood starts reacting to Rh (–ve) blood
   (b) Rh (+ve) person will die
   (c) Rh (–ve) person will die
   (d) Nothing happens to Rh (+ve) person
4. A drop of each of the following is placed separately on four slides. Which of them will not coagulate?
   (a) Blood serum
   (b) Blood from pulmonary artery
   (c) Whole blood from pulmonary vein
   (d) Blood plasma
5. Which proteolytic enzyme induces lysis of fibrin which prevents blood clots that occur naturally?
   (a)Thrombokinase (b) Thrombin
   (c) Plasmin (d) Platelet cofactor VIII
6. Clotting factor whose deficiency leads to delayed blood clotting is?
   (a) Hageman factor (b) Stuart Prower factor
   (c) Christmas factor (d) PTA

CIRCULATORY SYSTEM

Simple organisms like sponges and coelenterates circulate water of their surroundings through their body cavities. The complex organisms including humans have special body fluids to circulate these substances. These fluids are flown through well defined vessels or spaces.

The system of these vessels or spaces which is primarily meant for the circulation of substances along with body fluids is called **circulatory system.**

Types of circulatory system

The two basic types of circulatory system that have evolved in animals are

(i) Open circulatory system (in many invertebrates)
(ii) Closed circulatory system (in vertebrates)

Open circulatory system

In this type of circulatory system, the blood pumped by the heart is passed through a vessel into the spaces or body cavities known as **sinuses.** From there the blood is collected by large veins, which open into the heart.

This type of systems can have following two conditions

(i) When vessels are completely absent, e.g. arthropods (like prawn, insects, etc.). Here, the blood flows in lacunae or open spaces called **blood sinuses** or **haemocoel.**
(ii) When blood is flown in larger blood vessels, e.g. in most molluscs like snails, clams, oyster, etc. Here, only at tissue level, blood vessels are absent and sinuses are visible.

Closed circulatory system

In this type of circulatory system, the blood pumped by the heart is always circulated inside the vessels (i.e. it is never present in large spaces or sinuses). From evolutionary point of view, this pattern is considered to be most advantageous as it supplies blood to the **deepest tissues** of the body. For example, in annelids (like *Pheretima*), echinoderms, some molluscs (squids) and all vertebrates.

Table 25.3 Differences between open and closed circulatory systems

Open circulatory system	Closed circulatory system
Blood remains in direct contact with tissues.	Blood never remain in contact with tissues.
Capillaries absent.	Capillaries present.
It is usually low pressure systems.	It is usually high pressure systems.
Blood distribution in different organs occurs without any control.	Blood distribution in different organs is well in control and properly regulated.
Blood returns to the heart slowly.	Blood returns to the heart rapidly.

In simple words we can say that, it is the presence or absence of capillaries at tissue level, which differentiates between these two systems.

The advantage of closed circulatory system is that it increases the efficiency of circulation and regulates blood flow. Infact, blood flow in closed systems forms well defined **circulation circuits.**

All such circuits of vertebrates are given below

Circulation circuits

In vertebrates, these are of following types

Single circuit

- It is found in fishes. The blood passes only once through the heart in each complete circuit of the body. Thus, it is referred to as single circuit or **single circulation**.
- The heart of fishes is **two-chambered** with an atrium and a ventricle. Two pouches or sacs namely sinus venosus and conus arteriosus are present in this type of heart.
- The **sinus venosus** is a distinct sac, which is specialised to receive venous blood. It opens into the auricle.
- The **truncus** or **conus arteriosus** is another similar sac into which ventricles open for the distribution of blood. It pumps out deoxygenated blood, which is first oxygenated by the gills.
- This oxygenated blood is then supplied to the body parts, which utilise the dissolved O_2 in it.
- The deoxygenated blood is then again returned to the heart to complete the circuit.
- The outline of a complete single circuit is given below

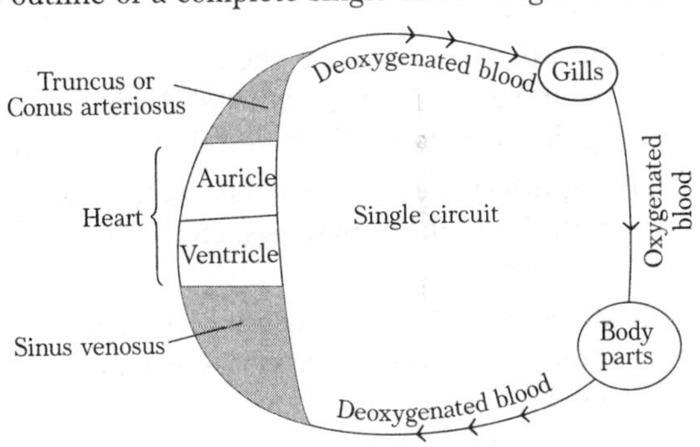

Fig 25.1 Representation of outline circulation in a fish

Double circuit

As the name suggest, the double circuit contains two distinct loops or circuits which operate through a **3** or **4 chambered heart**. Word double circuit clearly indicates that the blood follow two routes here, one for oxygenated blood and the other for deoxygenated blood. In animals, the double circuit is seen in its incomplete and complete forms.

1. Incomplete or Mixed double circuit

It is found in amphibians and reptiles (except crocodile). Both of these possess a **three-chambered** heart with two auricles and one ventricle.

(i) **In Amphibians** The left auricle receives oxygenated blood from the gills/lungs/skin.

- Deoxygenated blood from body parts goes to the right auricle.
- Both the blood forms are then mixed up in the single ventricle that pumps out mixed blood.
- The outline of this circuit is given below

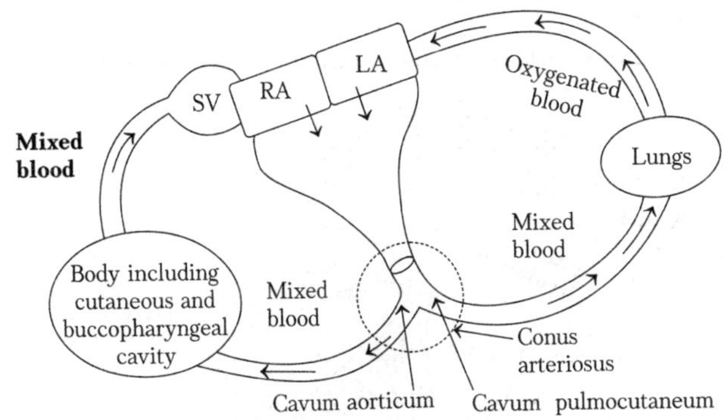

Fig 25.2 Representation of outline circulation in an amphibian

(ii) **In Reptiles** The mixed or incomplete circuit of reptiles is much more complicated than that of amphibian as they have three-chambered heart with incompletely divided ventricle. Its important features are

- Sinus venosus is present, but it is less functional.
- The conus arteriosus is divided into right systemic, left systemic and pulmonary arches.
- A connection tube called **foramen of panizzae** is present in some reptiles. It connects the left and right systemic arches.
- The SA node is present in right auricle.
- The outline of this circuit is given below

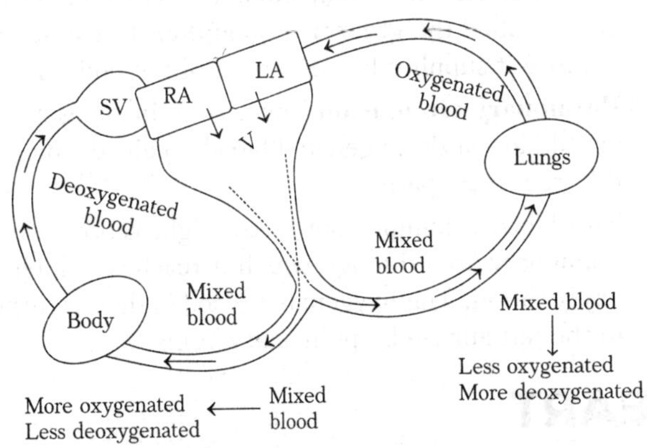

Fig 25.3 Representation of outline circulation in lower reptiles

2. Complete double circuit

- This type of circuit is found in crocodiles, birds and mammals.
- Their heart is a **four-chambered** with two atria and two ventricles. Here, the oxygenated and deoxygenated blood are completely separated.

- No sinus venosus and truncus arteriosus present, but two pulsatile pacemakers, i.e. SA node and AV node are present.
- The outline of this complete circuit is given below

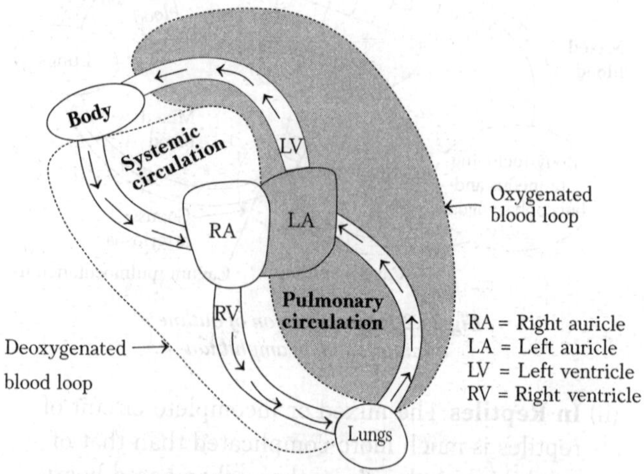

Fig 25.4 Complete double circuit circulation plan (outline in diagram representation)

- The two circuits are involved in following circulation
  (i) **Systemic circulation** This circuit involves the circulation of oxygenated blood between heart and body organs. Body organs utilise its oxygen. Deoxygenated blood from these body organs is then returned to the right auricle. Here, blood is received through two large veins called **superior** and **inferior vena cava**.
  Thus, one circuit is completed. This complete circuit is also called the **greater** or **peripheral circulation**, because it supplies blood to most of the body tissues.
  (ii) **Pulmonary circulation** This circuit involves, the circulation of deoxygenated blood received from different body parts.
  This blood is pumped out by the right ventricle to the pulmonary arch through which it reaches to lungs for oxygenation. The oxygenated blood is then returned to the left auricle by pulmonary veins.

HEART

It is the main pumping organ of all circulatory systems. Vertebrate heart is a highly muscular and chambered structure. However, a great deal of variability is seen in the hearts of invertebrates.

Types of hearts

Hearts seen in different animals can be categorised into following types

(i) **On the basis of types of blood**

(a) **Venous heart** It receives deoxygenated blood only, e.g. fishes.

(b) **Arteriovenous heart** It receive deoxygenated blood from body and oxygenated blood from lungs or gills, e.g. amphibians, reptiles, birds, mammals.

(ii) **On the basis of impulse origin**

(a) **Neurogenic heart** In such hearts impulse of contraction originates in nerves or nervous tissue, e.g. annelids, most arthropods, etc.

(b) **Myogenic heart** In such hearts impulse of contraction originates in specialised pulsatile muscular tissue of vertebrates and molluscs.

(iii) **On the basis of structure**

(a) **Tubular**, i.e. having tube like structure. Here, muscles responsible for impulse generation are external, e.g. in heart of cockroach, external alary muscles are responsible for impulse generation in this heart.

(b) **Pulsatile**, i.e. also tube like but here muscles for impulse generation are situated inside the heart, e.g. heart or dorsal blood vessel of earthworm.

(c) **Ampullary**, i.e. small hearts situated below appendages like wings, legs, etc. These are found in insects.

(d) **Chambered**, i.e. heart has auricles and ventricles as their well defined chambers, e.g. heart of vertebrates.

Human Heart

William Harvey (1628) was the first person who discovered the functioning of circulatory system in human.

Position

It is situated in the space of thoracic cavity called **mediastinum**. It is an anatomical region, present between two lungs and extending from sternum to the vertebral column. It is positioned as slightly tilted to the left side of the body.

Dimensions

The average size of adult human heart is about 12 cm long and 9 cm wide at its broadest point. It gets fully developed between the age of 17-20 years. Its weight is about 0.45 % of body weight in males, i.e. 280-340 gm and about 0.40% in females, i.e. 230-280 gm.

Coverings and surrounding cavity

- Externally, human heart is covered by a double-walled membranous covering called **pericardium** (peri-around, cardium-heart).
- This covering is made up of an inner visceral and an outer parietal membrane. Between these two membranes lie a narrow space, the pericardial cavity.
- It is filled with a watery **pericardial fluid.** It is secreted by the visceral peritoneum. It protects the heart from shocks and mechanical injuries and allows free movements of the heart.

External structure

- On removal of the pericardium, a four-chambered heart with a distinct transverse groove is observed. This groove is known as **atrioventricular groove** or **coronary sulcus.** It divides the heart into a smaller anterior atrial part and a posterior ventricular part.
- Two grooves that mark the boundary between the right and left ventricles are **anterior interventricular** and **posterior interventricular sulcus.**
- Two major arterial trunks, i.e. **pulmonary** and **aortic arch,** along with two major venous trunks, i.e. **superior** and **inferior vena cava** are also seen in intimate association with heart.

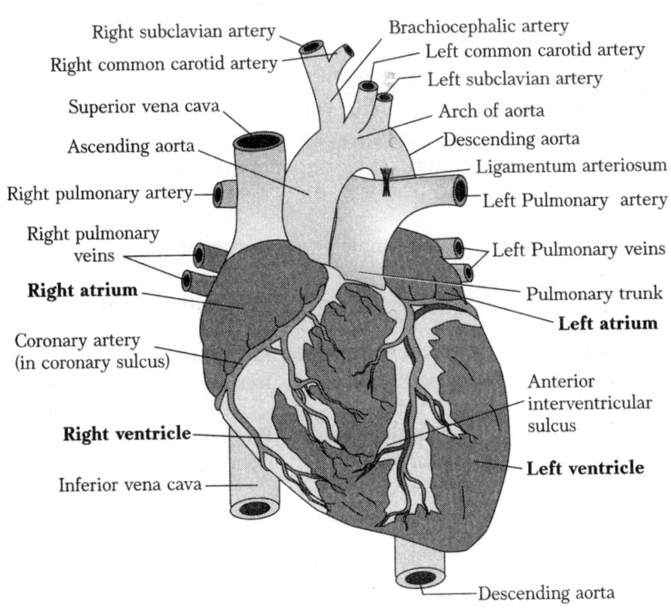

Fig 25.5 External structure of human heart

Histology

Heart is made up of cardiac muscles. The wall of heart is made up of following three layers from outside to inside

(i) **Epicardium** or **Visceral pericardium** It is the outermost layer. It is made up of simple squamous epithelium.

(ii) **Myocardium** It is the thickest middle layer. It is made up of cardiac muscles, which are striated but involuntary.

(iii) **Endocardium** It is the innermost layer. It is again made up of simple squamous epithelium.

Internal structure of heart chambers

The four-chambered human heart is divided into two auricles or atria and two ventricles.

Auricles or atria

- This part is vertically divisible into right and left auricles. Out of these, the left auricle is smaller than the right auricle.
- Each auricle extends behind into a swollen flap, i.e. the auricular appendix. It slightly covers the ventricle of each side.
- Both right and left auricles are separated completely by a thin septum called **interatrial septum** or **interauricular septum.** It prevents mixing of oxygenated and deoxygenated blood.
- An oval depression, called **fossa ovalis** is present on its posterior side. It is remnant of **foramen ovale** present in foetal stage which closes at birth.
- The **internal lining** of the auricular wall forms a network of low muscular ridges called **musculi pectinati.**
- The two auricles are separated from two ventricles by an **auriculoventricular septum.**

Ventricles

- This part of heart is broad. In particular, it is about 4-5 mm more thicker and more muscular than auricles.
- An **interventricular septum** separates the right and left ventricles.
- The walls of the ventricles are internally raised into a number of thick, muscular, column-shaped projections called **columnar carnae** or **trabeculae carnae**
- A few large muscular elevations called **papillary muscles** or **musculi papillares** are also present in

ventricles. These are two in left ventricle and three in right ventricle.

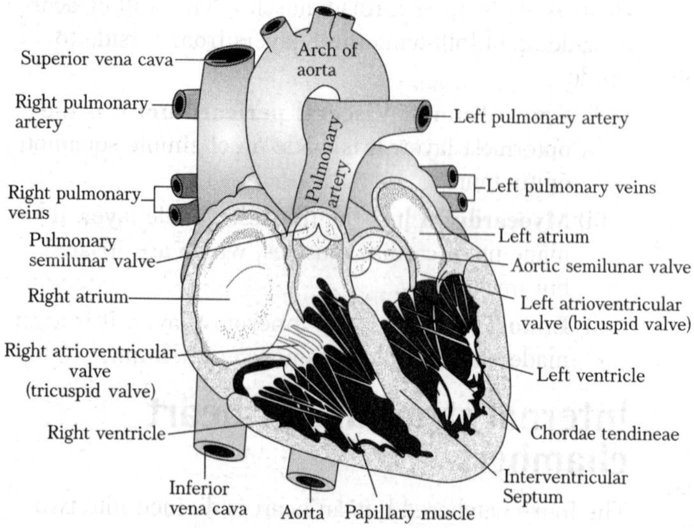

Fig 25.6 Internal structure of human heart

- All the major openings inside the heart are guarded by valves to prevent back flow of blood.
- Fine tendinous cords called **chordae tendineae** are found attached to the ventricular surface of these valves on one side and to the papillary muscles on the other side. These structures prevent inversion of valves.
- The right ventricle pumps deoxygenated blood to lungs and left ventricle pumps oxygenated blood throughout the body.

Major blood vessels associated with heart

The chambers of heart are directly connected with the following major blood vessels

(i) **Systemic aorta or Systemic arch** It is originated from left ventricle. It distributes oxygenated blood to various body parts except lungs.

(ii) **Pulmonary arch or Pulmonary trunk** It is originated from right ventricle and gets divided into two pulmonary arteries. These carry deoxygenated blood to lungs.

(iii) **Superior vena cava or Precaval** It brings deoxygenated blood from head and upper parts of the body into the right auricle.

(iv) **Inferior vena cava or Postcaval** It brings deoxygenated blood from lower parts of the body into the right auricle.

(v) **Pulmonary veins** These are four in number, i.e. two from each lung. These bring oxygenated blood from lungs to left auricle.

(vi) **Coronary veins or Coronary sinus** These return deoxygenated blood from heart wall into the right auricle.

Note • *During foetal life, lungs are non-functional hence, blood of pulmonary aorta comes into systemic aorta through a small duct called **ductus botalli** or **ductus arteriosus**. After birth, this structure gets blocked by the deposition of elastin fibres forming **ligamentum arteriosum**.*

• ***Blue-baby syndrome** is a condition which arises due to the mixing of deoxygenated and oxygenated blood in heart. It is usually seen either due to the non-closure of foramen ovale or ductus arteriosus. These conditions are also called **patent foramen ovale** and **patent ductus arteriosus**, respectively.*

Heart valves

These are specialised structures which guard all the major internal openings of heart. Their function is to prevent the back flow of blood or to maintain unidirectional flow of blood.

Following table contains the types of heart valves with their location and action

Heart valves	Location	Action
Tricuspid valve	Between right atrium and right ventricle	During ventricular contraction, it prevents back flow of blood from right ventricle into right atrium.
Bicuspid or **Mitral valve**	Between left atrium and left ventricle	During ventricular contraction, it prevents back flow of blood from left ventricle into left atrium.
Pulmonary or **Semilunar valve**	At the initial point of pulmonary artery	During ventricular relaxation, it prevents back flow of blood from pulmonary trunk into right ventricle.
Aortic valve (semilunar)	At the initial point of aorta	During ventricular relaxation, it prevents blood back flow from aorta into left ventricle.
Eustachian valve	On the opening of superior vena cava	Prevents back flow of blood from right atrium.
Thebasian or **Coronary valve**	Over the opening of coronary sinus	Prevents back flow of blood from right atrium.

Upper structure of first 4 major valves of human heart is shown below

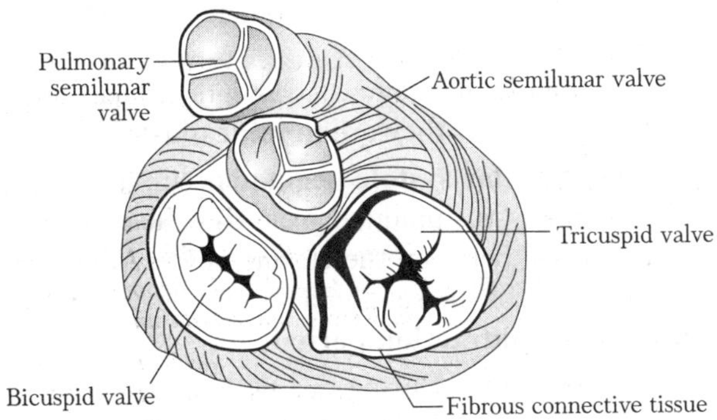

Fig 25.7 A view of heart from above showing different valves

Failure of a heart valve to open fully is known as **stenosis** (narrowing) and failure of a valve to close completely is termed as **insufficiency** or **incompetence**.

Impulse conduction system

In human heart, small groups of specialised neuromuscular cells are present within the myocardium. These initiate and conduct impulses, due to which a coordinated and synchronised contraction of the heart muscles is seen.

The components of this conducting system are as follows

(i) Sinoatrial node or SA node

- It is also called **node of Keith** and **Flack.**
- It is a mass of self-excitatory specialised cells, present in the wall of the right atrium. This mass is positioned just inferior and lateral to the opening of superior vena cava.
- Functionally, it is called as the **pacemaker of heart.** This is so because the initiation of heartbeat impulses occurs within this node.
- It initiates the impulses more rapidly than any other neuromuscular cell of heart. Its initiation rate is 70-80 impulses or beats per minute.

(ii) Atrioventricular node or AV node

- It is also called **node of Twara** and **Aschoff.**
- It is also a small mass of self-excitatory neuromuscular tissue situated in the wall of atrial septum near the atrioventricular valves.
- Normally, it works on stimulation by SA node, but it is capable of initiating its own impulses too.
- However, its initiation rate is low as compared to SA node, i.e. 40-60 impulses per minute. Functionally, it is called as the **pacesetter of heart.**

(iii) Bundle of HIS or Atrioventricular bundle

These are the mass of specialised fibres originating from AV node. At the upper end of ventricular septum, it is divided into left and right bundle branches.

(iv) Purkinje fibres

- These are the fine fibres of AV bundle in the ventricular myocardium.
- Both AV bundle and Purkinje fibres convey impulse of contraction from AV node to the apex of myocardium and brings on ventricular contraction.
- All these components of conducting system are shown in the following figure

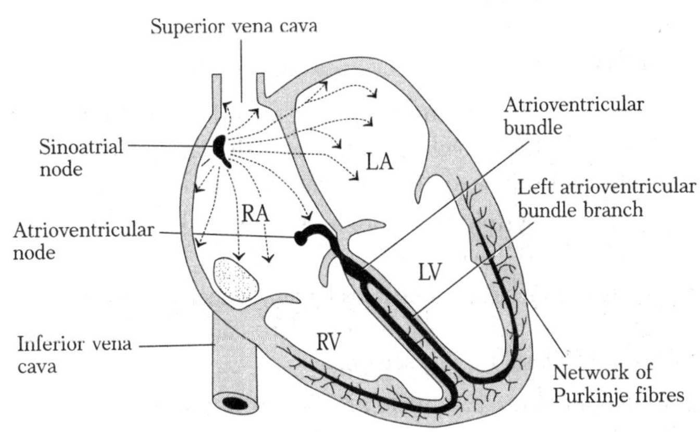

Fig. 25.8 Heart's conducting system

- ***Artificial pacemaker*** is an electric device first developed by ***Greatbach*** and ***Chardark*** in 1960. It is implanted into the body in case of any myogenic disfunctioning. It has a pulse generator having lithium halide batteries with life of about 10 years. These are covered with a biocompatible plastic cover with a fine metallic string. This string basically functions as muscles stimulating electrode. This device is placed below the skin under right clavicle and the string is passed to the right atrium *via* anterior vena cava.
- ***Ventricular escape*** is a condition when atrial impulses fail to pass into ventricles for a few seconds to few hours.

Functioning of human heart : Beat and pulse

- The heart regularly and rhythmically contracts and relaxes at short intervals throughout life without pause and fatigue. This inherent activity of heart is called its **heartbeat**.
- Each heartbeat has two phases, i.e.
  - **contraction phase** called **systole**
  - **relaxation phase** called **diastole**
- The rate of throbbing felt in arteries due to pumping action of heart (left ventricle) is called **pulse** or **heart rate**.
- The number of pulse or pulse rate is equal to number of heartbeats. It can be felt in the radial artery near the wrist, in neck and ankles. Pulse rate is about 72 times per minute in men and about 80 times per minute in women.

Note
- ***Tachycardia*** *is a term applied to rapid heart or pulse rate. It is used when the heart rate crosses over 100 times per minute.*
- ***Bradycardia*** *is the term indicating a slow heart or pulse rate. It is used when the heart rate dips rate under 50 times per minute.*

Cardiac cycle

It refers to the repeating pattern of **systole** and **diastole** of the heart. Each cycle of atrial, ventricular systole and diastole is of 0.8 second. This time period is divided as

- In auricles, systole lasts only for 0.1 sec and diastole remains for 0.7 sec.
- On the other hand, ventricles have a longer systole, i.e. of 0.3 sec due to their highly muscular walls. The diastole in ventricles lasts for 0.5 sec only.

The events of cardiac cycle are shown in the figure below

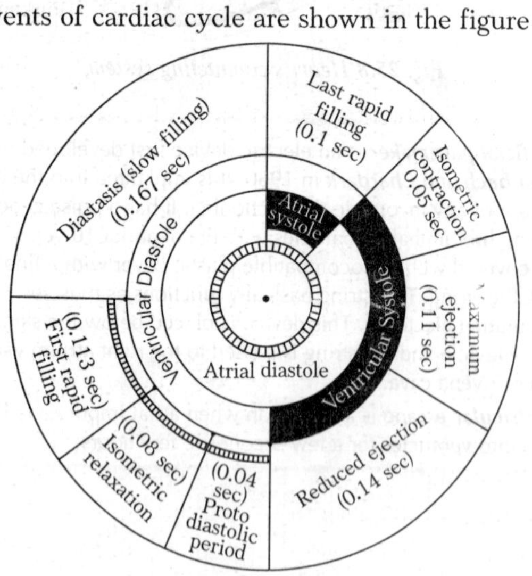

***Fig. 25.9** Cardiac cycle*

From the above figure, it is very clear that systole and diastole phases of all the chambers follow overlapping pattern. The main steps and characteristics of human cardiac cycle are listed below

(i) To begin with, all the four chambers of heart are in a relaxed state, i.e. they are in **joint diastole.**

(ii) The **ventricular systole** begin to occur. The AV valves close, thus producing the **first heart sound.** The interval between the closing of the AV valves and opening of the semilunar valves, is called the **isometric contraction period.**
It lasts for 0.05 s. During this period, ventricles contract as closed cavities and intraventricular pressure steeply rise.

(iii) It is followed by ejection period which is divided into **maximum ejection** of 0.11 sec and **reduced ejection** of 0.14 sec. During ejection period, blood is pumped out of the ventricles. The volume of blood ejected during each ventricular contraction is called **stroke volume.**

(iv) At the beginning of **ventricular diastole**, the semilunar valves close and produce the **second heart sound.** There is a brief interval between the beginning of diastole and the closure of the semilunar valves. It is known as the **protodiastolic period** of 0.04 s. Second sound occurs actually after this period.

(v) As the ventricles relax, intraventricular pressure falls below that of atria and AV valves open. This is called **isometric relaxation** of 0.08 s. As a result of this, atrial blood rushes into the ventricles. This rushing of blood produces the **third sound**. This indicates the beginning of ventricular filling.

(vi) The first part of filling is very rapid, being known as the **first rapid filling phase** of 0.113 s. The intermediate part of filling is very slow and is known as **diastasis** or **slow inflow phase**. Although, this is the longest phase of 0.167 s, yet the amount of filling is minimum in it. During this phase blood directly enters from vena cava to ventricles.

(vii) Last rapid filling phase of 0.1 s is responsible for the last part of ventricular filling. This phase overlaps with **arterial systole**. Infact, the rapid flow of blood during this phase is due to increased arterial pressure only. Due to rapid rush of blood, another sound is produced, i.e. the so called **fourth sound of heart**. Here, ventricular diastole ends and systole commences again. In this way, the cycle continues.

All the above written events along with the presences, blood volumes and heart sounds are summarised in the following figure

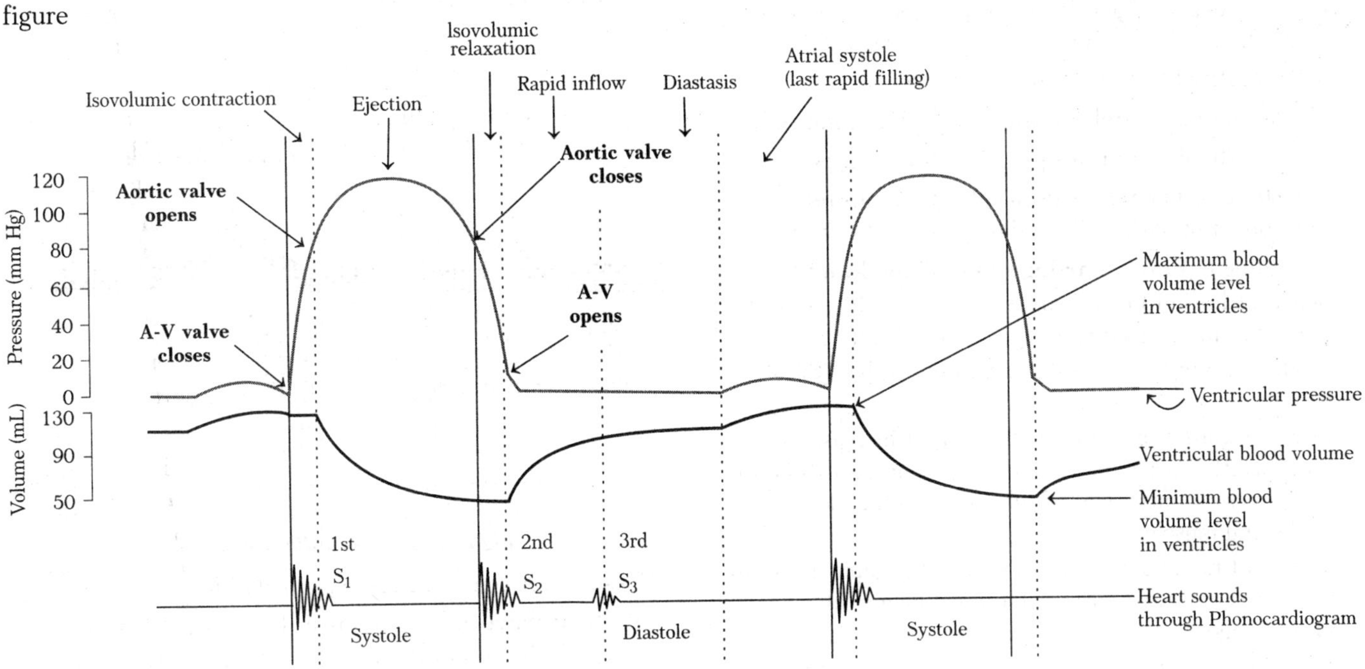

Fig 25.10 Stages of cardiac cycle

Heart sounds

The beating of heart produces sounds which are heard by stethoscope. There are four heart sounds designated as S_1, S_2, S_3 and S_4.

- First sound (**lub** $=S_1$) is a slow pitch sound. It is created by the closure of atrioventricular valves, i.e. tricuspid and bicuspid at the beginning of ventricular systole.
- Second sound (**dup** $=S_2$) is a highly pitched and sharper sound. It is created by the closure of semilunar valves.
- The S_3 and S_4 may not be audible and are usually faint. These sounds are associated with the rapid ventricular filling (S_3) and atrial systole (S_4).

Note *__Heart Murmer__ is an abnormal hissing sound. It is produced due to defective or damaged heart valves that lead to the backflow of blood either from ventricles to auricles or from aorta to ventricles.*

Functioning of heart at muscular level

Human heart is myogenic in nature. Thus, impulse for contraction and relaxation of cardiac muscles is originated within these muscles only. However, regulation of impulse is done by nerves. This complete functioning can be studied under following steps

(i) **Origin of impulse** or **heartbeat** The impulse of contraction or relaxation of heart chambers is originated at SA node.

(ii) **Conduction of impulse** or **heartbeat** The AV node picks up this impulse from SA node. From here, it is conducted to the complete heart with the help of bundles of His and Purkinje fibres.

(iii) **Regulation of impulse** or **heartbeat** Normal activities of heart are auto regulated by nodal tissues. However, a special neural centre in medulla oblongata regulates heartbeats through following two units

(a) **Cardio accelerator centre** A pair of sympathetic nerves innervates the SA node to increase the rate of heartbeat. This increase is due to nor-adrenaline secretion of sympathetic nerves.

(b) **Cardio inhibitory centre** This centre sends impulses to the SA node through cardiac branch of vagus nerve to reduce the rate of heartbeat. This reduction is due to acetylcholine secretion of parasympathetic (vagus) nerves.

Besides, the adrenaline hormone and thyroxine hormone secreted by adrenal medulla and thyroid, respectively, also increase heartbeat.

Note
- *During a cardiac cycle, once an action potential is initiated, there is a period of time that a new action potential cannot be initiated. This is termed as the __Effective Refractory Period__ (ERP) of the tissue. The period is also termed as the __Absolute Refractory Period__ (ARP), it occurs due to calcium ion influx.*
- ***Excess of*** K^+ *reduces the heart rate and strength of contraction.*
- ***Excess of*** Na^+ *results in depression of cardiac function.*
- ***Excess of*** Ca^+ *results in spastic contraction of the heart. Its reduction causes same effects as high level of* K^+.

Cardiac output

- It is the volume of blood ejected from the left or right ventricle into the aorta (pulmonary trunk) per minute.
- It is calculated by the following formula

  Cardiac output = Stroke Volume (SV) × Heart Rate (HR)

  = 70 mL × 75 beats per min = 5250 mL per min
- The stroke volume depends mainly on venous return, i.e. the amount of blood delivered to the heart by the veins.
- According to **Frank Starling's Law of the heart** (or the Starling's law), if more blood is delivered to the heart by the veins, more blood is pumped by the heart.
- Cardiac output is not constant, it varies with physical activity.
- During vigorous exercise, cardiac output increases as much as 21 L per min. This increase is primarily due to an increase in heart rate which increases the blood flow to skeletal muscles.
- The heart rate, however can increase only upto a certain maximum value, which is determined mainly by a person's age.
- It is fastest at birth, moderately fast in youth, average in adulthood and below average in old age.
- The difference between the maximum and resting cardiac output is called **cardiac reserve.**

Electrocardiogram (ECG)

- The cardiac impulse is transmitted in the form of electrical potential over the cardiac chambers in a specific wave sequence.
- These electrical changes can be sensed and recorded *via* a **galvanometer**. This recording is known as **Electrocardiogram** (ECG).
- The machine by which the electrocardiogram is recorded is called as **electrocardiograph. Einthoven** invented it in 1903, that is, why he is considered **Father of Electrocardiography.**
- An electrocardiogram represents the electric current moving through the heart during a heartbeat.
- It shows five consecutive waves, i.e. P, Q, R, S and T as shown in following figure

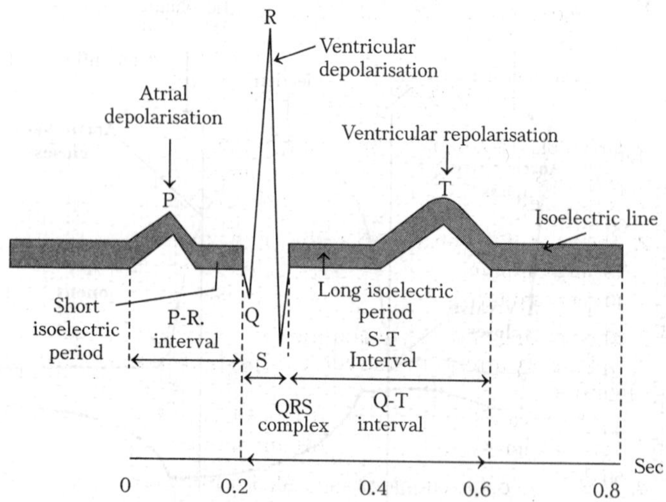

Fig 25.11 Electrocardiagram (ECG)

- These waves can be better explained as follows
  - (i) The **P-wave** is a small upward wave that indicates the depolarisation of the atria which results to atrial contraction. It is caused by the **activation of SA node** and spread of impulse through contractile fibres.
  - (ii) The **QRS wave** begins after a fraction of second of the P-wave. It begins as a small downward deflection (Q) and continues as large upright (R) and triangular wave. It end as downward wave (S) at its base. It represents ventricular depolarisation which results in ventricular contraction. It marks the beginning of systole.
  - (iii) The T-wave indicates ventricular repolarisation which results in ventricular relaxation. It is dome-shaped. The systole ends at the end of T-wave. There are two isoelectric periods
    - (a) **The shorter one,** between P and Q.
    - (b) **The longer one,** between S and T.
- Any abnormality in the working heart alters the wave pattern of ECG. Thus, ECG is of great diagnostic value in cardiac diseases.

CHECK POINT 25.2

1. 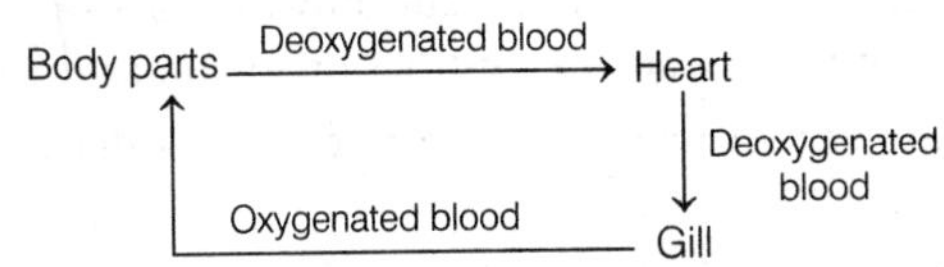

Given diagram depicts the circulation in
(a) fishes (b) mammals
(c) reptiles (d) amphibians

2. The layer of the heart responsible for its rhythmicity, is the
(a) myocardium (b) endocardium
(c) pericardium (d) epicardium

3. The opening of the right and the left ventricles into the pulmonary artery and the aorta, respectively are provided with the
(a) bicuspid valves (b) tricuspid valves
(c) semilumar valves (d) mitral valves

4. Mass of nodal excitable tissue called Atrio Ventricular Node (AVN) is present in
(a) right upper corner of right atrium
(b) right lower corner of left atrium
(c) left upper corner of righ atrium
(d) left lower corner of right atrium

5. If due to some injury, chordae tendineae of the tricuspid valve of human heart is partially non-functional, what will be the immediate effect?
(a) The flow of blood into the aorta will be slow down
(b) The pacemaker will stop working
(c) The blood will tend to flow back into the left atrium
(d) The flow of blood into the pulmonary artery will be reduced

6. What is correct about joint diastole?
(a) Semilunar valves are open
(b) Semilunar valves are partially closed
(c) Bicuspid and tricuspid valves are open
(d) Bicuspid and tricuspid valves are closed

7. While listening to heart sounds, you can best hear sound at the
(a) base of the heart
(b) apex of the heart
(c) aortic area
(d) second intercostal space to the right of the sternum

8. The beginning of ventricular systolc is represented in ECG through
(a) P and Q-wave (b) QRS-wave
(c) P and S-wave (d) S and T-wave

BLOOD VESSELS AND THEIR SYSTEM

Besides heart, blood vessels form the major portion of a closed circulatory system. The study of blood vessels is called **angiology**. The blood vessels of a closed circulatory system are of following three types

1. Arteries

- These are elastic vessels that transport blood away from the heart.
- The largest artery of the human body is **aorta.**
- It originates from the heart and branches out into smaller arteries.
- Their smallest branches are called **arterioles**, which branched off further into capillaries.

2. Veins

- These are less elastic vessels that transport blood towards the heart. These are formed by the joining of smaller vessels called **venules.**
- These are connected to the venous ends of capillaries and receive blood from the arteries *via* the capillaries.

Internal structure of arteries and veins

The wall of arteries and veins consists of the following layers from inside to outside

(i) **Tunica interna** or **intima** It has inner endothelium of long, thin cells of squamous epithelium.

These rest on a basement membrane along with an outer elastic membrane of yellow elastic tissue.

(ii) **Tunica media** It has circular smooth muscles and elastic fibres.

(iii) **Tunica externa** or **tunica adventitia** It has loose connective tissues with longitudinally arranged white and yellow fibres.

3. Capillaries

- These are extremely small vessels, located as a network within the tissues of the body. These are the main blood transport connection between the arteries and the veins.
- These were discovered by **Malpighi** in 1661.
- Capillary walls are thin and composed of a single layer of endothelium only. These walls help in exchanging oxygen, carbon dioxide, nutrients and wastes.

Table 25.4 Differences between arteries and veins

Characters	Arteries	Veins
Wall	Thick, more elastic, non-collapsible	Thin, less elastic, collapsible
• Tunica externa	Less developed, so less strong	More developed, so more strong
• Tunica media	More muscular and has many elastic fibres	Less muscular and only a few elastic fibres
• Tunica interna	Endothelial cells more elongated. Elastic membrane more developed	Endothelial cells less flat. Elastic membrane less developed.
Lumen	Narrow	Wider
Position	Deep seated except wrist, neck, etc.	Superficial
Valves	Without valves	With semilunar valves to prevent back flow, except vena cava
Direction of blood flow	From heart to body organs	From body organs to heart
Blood pressure	More, generally 120/80 mm of Hg	Less, generally 0 mm of Hg.
Speed of blood	Fast	Slow
Amount of blood	15% at any given time	64% at any given time
Distensibility	Less	More
Elastic-lamina	Present	Absent
After death	Becomes empty	Contain blood

Blood pressure

- The pressure exerted by the flow of blood on the elastic wall of arteries is called blood pressure.
- It is greater during systole than during the diastole.
- In a normal person, the systolic pressure is 120 mm Hg and diastolic pressure is 80 mm Hg.
- Blood pressure changes with age and health of the person.
- It is measured by **sphygmomanometer**. This apparatus was invented by **Samuel Ritter Von Basch** in 1881.
- **Pulse pressure** is the difference between systolic and diastolic pressure. Its average value is 40 mm Hg.

High blood pressure (Hypertension)

- Persistent increase in blood pressure above 140 mm of Hg (systolic) and 90 mm of Hg (diastolic) is termed as hypertension or high blood pressure. It is a major health problem.
- It is caused by nervous tension and emotional stress (fear, worry, anxiety, anger, excitement, physical stress, excessive joy) that leads to contraction of the arteries.
- If left untreated, it may cause heart failure, kidney diseases and stroke.

Low blood pressure (Hypotension)

- It is the term used for decrease in the arterial blood pressure. This happens when systolic pressure falls below 110 mm of Hg and diastolic pressure falls below 70 mm of Hg.
- It may result from vasodilation of arterioles, loss of blood in haemorrhage, lower pumping action of heart, nervous disorders, valvular defects, anaemia and deficient diet.

Arterial system

- The human arterial system includes aorta, arterioles and metarterioles.
- A **metarteriole** is a vessel that emerges from an arteriole and supplies blood to a group of 10-100 capillaries that constitute a **capillary bed.**
- The two main arterial vessels of human arterial system are **pulmonary arch** (from right ventricle) and **systemic aorta** (from left ventricle).
- The pulmonary arch does not have further major branching. It carries the blood to lungs for oxygenation.
- The **systemic aorta** is the major distributing vessel of human body. It gives of various major branches.

Branches from ascending aorta

These are as follows

(i) **Coronary artery** It supplies blood to the heart muscles.

(ii) **Innominate artery** or **right brachiocephalic artery** It bifurcates further into right subclavian and right common carotid artery.

(iii) **Left common carotid artery** It divides into external and internal carotid arteries. Carotids supply oxygenated blood to neck, face, mouth, eyes, scalp and brain.

(iv) **Left subclavian** It arises independently from the aorta. Subclavians carry oxygenated blood to forelimbs, chest and spinal cord.

Branches from descending aorta

The descending aorta is also called **dorsal aorta** due to its position. It is further divided into

(i) **Thoracic aorta** It gives out **oesophageal** atery (to oesophagus), **intercostal** artery (in between the ribs) and **bronchial** artery (to lungs).

(ii) **Abdominal aorta** It supplies blood to the following as paired and unpaired arteries

- Paired arteries are **phrenic** (to diaphragm), **renal** (to kidney) and **gonadial** (to gonads).
- Unpaired arteries are **coeliac** (to stomach, pancreas, spleen and liver), **anterior mesenteric** (to small intestine) and **posterior mesentric** (to large intestine).

Venous system

It consists of pulmonary veins, coronary sinus, portal system and vena cava. The two major veins of humans body are superior and inferior vena cavae.

Individually, these two are supplied with following veins

Superior vena cava

- It is formed by the fusion of two **brachiocephalic veins**.
- Each of these veins receives deoxygenated blood from a **jugular** vein, **subclavian** vein and **thoracic** vein.
- The jugular vein collects blood from brain, the subclavian vein collects from arms and thoracic vein from chest region.

Inferior vena cava

- It is formed by the union of two common **iliac** veins. These veins collect blood from pelvis and lower limbs.
- On its way to the heart, it receives blood from the **genital** (gonads), **lumbar** (muscles of back), **renal** (from kidneys) **supra-renal, hepatic** veins and **phrenic** veins.
- Finally, inferior vena cava opens into right atrium.
- A summary of all these major arteries and veins along with the organ with which these are associated is given in the table below

Supplying Artery	Organ	Collecting vein
Carotid	Head (Brain)	Jugular vein
Subclavian	Arms	Subclavian vein
Intercostal artery	Intercostal muscles (Rib)	Intercostal vein (Azygous vein)
Phrenic actery	Diaphragm	Phrenic vein
Hepatic artery	Liver	Hepatic vein
Gastric	Stomach	Superior mesentric
Splenic	Spleen	Spenic vein
Superior mesentric artery	Small intestine	Superior mesentric
Inferior mesentric artery	Colon	Inferior mesentric vein
Renal artery	Kidneys	Renal vein
Gonadial artery	Gonads	Gonadial vein
External iliac artery	Legs	External iliac vein
Internal iliac artery (hypogastric)	Urinary bladder	Internal iliac vein

Lienogastric artery

Hepatic portal vein

Some special blood vessels are as follows

- Thin blood vessels and capillaries that supply blood to the walls of various blood vessels themselves are collectively called as **vasa vasorum.**
- In certain exposed parts of body, such as lips, pinnae, tongue, fingertips, eyelids, etc. certain arterioles and venules forms direct connections in addition to their normal connections through capillary network. These direct vascular connections are called **arterio-venus anastomoses.**
- Leaky venous valves can cause veins to become dilated and twisted in appearance, a condition called **varicose veins.**

Portal systems

A specialised vascular arrangement in which blood from the capillaries of one organ is transported to the capillaries of another organ is called portal system. It is of following types

1. Hepatic portal system

- It is a special vascular connection existing between the different parts of alimentary canal of digestive system and liver in all chordates including humans.
- This system carries venous blood from digestive tract, spleen, pancreas and gall bladder to the liver.
- Human hepatic portal vein is formed by the union of following six veins
  (i) **Splenic vein** from spleen and pancreas.
  (ii) **Superior mesenteric** from colon, caecum and small intestine.
  (iii) **Inferior mesenteric** from rectal wall and anal region.
  (iv) **Cysteic vein** from gall bladder.
  (v) **Right gastric vein** from stomach.
  (vi) **Left gastric vein** from stomach and pancreas.

Significance

This system helps in the extraction of nutrients and other metabolites by breaking them up into single-celled thick capillaries and sinusoids.

2. Hypophysial portal system

- It is a specialised portal system formed by a vein from hypothalamus, which breaks up into capillaries in anterior lobe of pituitary gland. The vein is called hypophysial portal vein.
- This system is present in higher vertebrates like amphibians, reptiles, birds and mammals including humans.

Significance

- Hypothalamus produces a number of hormones for controlling endocrine activity of anterior lobe of pituitary or adenohypophysis.
- These hormones are poured into the blood in hypothalamus, which are then drained by hypophysial portal vein into the adenohypophysis.

3. Renal portal system

- It is present in well-developed form in fishes and amphibians. However, it is reduced in reptiles and birds and is absent in mammals including humans.
- It supplies blood from posterior region of the body to the kidneys by renal portal veins.

Significance

It helps to remove waste products before sending the blood to the heart *via* renal veins and postcavals.

LYMPH

- It is a colourless or yellowish fluid present in the lymph vessels.
- In real sense, it is the component of blood left in tissue fluid during normal circulation of blood. It amounts about 10% of total blood volume.
- The composition of lymph is same as that of blood plasma, but lymph has lower protein contents.
- As compared to the tissue fluid, the lymph contains very small amount of nutrients and oxygen, but contains abundant carbon dioxide and other metabolic wastes.
- Some amoeboid-shaped white blood corpuscles may also be seen in lymph.

Human lymphatic system

Lymphatic system of humans comprises of lymphatic capillaries, lymphatic vessels, lymphatic ducts and lymphatic nodes. Their main function is to collect lymph from various organs and return it back to the blood. Details of this complete system are given below

1. Lymphatic capillaries

- These capillaries are thin-walled, small and are composed of a single layer of endothelial cells.
- They lie close to the blood capillaries, but differ from them to the extent that they end blindly.
- Their major function is to collect lymph from various organs.
- The lymphatic capillaries of intestinal regions, which absorb the digested fats are called **lacteals**.

2. Lymphatic vessels

- The lymphatic capillaries unite to form larger lymphatic vessels.
- They are composed of an outer coat of fibrous tissue, middle coat of muscular tissue and an inner lining of endothelial cells.
- The lymphatic vessels have numerous valves to prevent the backflow of lymph.

3. Lymphatic ducts

- Lymph vessels of left and right side of the body combines to form a thoracic duct or left lymphatic duct and right lymphatic duct, respectively.
- The thoracic duct begins at the **cisternae chyli** present at the level of the first and second lumbar vertebrae.
- The thoracic duct discharges its lymph into the left subclavian vein, while the right lymphatic duct discharges its lymph into the right subclavian vein. Thus, the left component of blood reaches back.

4. Lymphatic nodes

- These are oval or bean-shaped structures located along the length of lymphatic vessels.
- They are about 1-25 mm long. Structurally these are covered by a capsule of dense connective tissue.
- Lymph filters through the lymph nodes. They contain phagocytic white blood corpuscles and macrophage, which eat harmful microorganisms and foreign particles from the lymph.
- Lymph nodes are connected to following two small vessels
  (i) **Afferent lymphatics** are the vessels entering the lymph node carrying lymph from interstitial space.
  (ii) **Efferent lymphatics** are vessels leaving the lymph node and draining the lymph in vein or thoracic duct.

Lymph movements

The lymph flows slowly and moves from lymphatic vessels, lymphatic ducts to venous system. Blocking of lymph flow causes **edema** (swelling). Muscular movements are helpful in its flow.

Lymphoid organs

The organs which secrete lymph are called lymphoid organs.

Besides the lymph nodes, tonsils, thymus glands, Peyer's patches, liver and spleen are the other lymphoid organs that secrete lymph.

ABNORMAL CONDITIONS RELATED TO CIRCULATORY SYSTEM

Arteriosclerosis

- It refers to the 'hardening of arteries and arterioles' due to thickening of the fibrous tissue and consequent loss of elasticity of their walls. It causes hypertension or high blood pressure.
- Typically, it occurs in elderly persons. It can be due to following two reasons
  (i) When calcium gets deposited in the tunica media of arteries.
  (ii) When the lumen of arterioles becomes narrow due to the deposition of cholesterol.

Atherosclerosis

- It refers to irregular thickening of arterial walls and narrowing of their lumen. This happens due to deposition of **yellow plaques** or **atheromas** in their tunica intima and inner part of tunica media.
- Atheromas consist of cholesterol and other lipid materials. They are formed from **Low Density Lipoproteins** or LDL.
- LDL can pass through endothelium and gets precipitated inside the artery.
- Atherosclerosis mostly develops in the aorta, coronary arteries and cerebral arteries.
- When atherosclerosis occurs in the cerebral arteries, less blood is delivered to the brain. This can lead to **Cerebro Vascular Accident** (CVA), commonly referred to as a **stroke**.
- As an atherosclerotic plaque develops and enlarges, it progressively obstructs blood flow. It is a leading contributor of heart and kidney failures as well.

CARDIOVASCULAR DISEASES

There are several cardiovascular diseases that occur due to the malfunctioning of heart and its associated conducting vessels. The two most important and common diseases related to circulatory system are given below

Coronary Artery Disease (CAD)

- It is also known as **ischemic heart disease.** It is caused by atherosclerosis, of the blood vessels that supply blood to the heart.
- The factors responsible for CAD are hypercholesterolemia (high cholesterol), smoking, hypertension, hyperglycemia, high levels of lipoproteins, lack of exercise, stress, diet low in antioxidants, obesity and diet rich in saturated fats.
- Its symptoms include chest pain, nausea, perspiration and anxiety.
- It can be **detected** through abnormal ECG.
  Preventive measures include reducing cholesterol levels in food, avoiding a sedentary lifestyle and smoking.
- The **treatments** include **coronary artery bypass grafting** or **angioplasty**. In this, breaking of arterial blockage by balloon catheter is done and **Coronary Stent-Implantation** (CAD) is one of the most common form of heart disease and the leading cause of **heart attacks** or **myocardial infarction** and **angina**.

(i) Myocardial Infarction (MI)

- It is popularly known as **heart attack**. It results from the death of a part of heart muscle following cessation of blood supply to it.
- The heart attack is characterised by severe pain in the heart or near by organ, breathlessness, restlessness, nausea and vomiting.
- A heart attack need not to be fatal, especially if one receives medical attention and treatment to deal with the blockage soon after the attack.

(ii) Angina pectoris

- It means 'pain in the chest'. It results from reduced blood supply to the heart, i.e. ischemia.
- Angina is caused by atherosclerosis.
- The typical pain of angina occurs in the chest, but it can often radiate to the left arm, shoulder or jaw.
- The pain due to angina is a warning signal of heart attack.

Heart failure

- It is also referred to as **Congestive Heart Failure** (CHF). It occurs when the heart is unable to pump blood sufficiently in order to meet the body's needs.
- Various causes of heart failure include coronary artery disease, high blood pressure, diabetes or obesity, stress, etc.
- It mostly occurs due to left ventricular dysfunction.
- The failure of adequate output by the left ventricle is marked by lungs congestion and edema, i.e. buildup of fluid in the feet, ankles and legs.
- The failure of adequate output by the right ventricle is marked by the venous enlargement, hepatic enlargement and edema.
- A complete heart failure usually leads to **cardiac arrest**.

CHECK POINT 25.3

1. In veins, valves are present to check backward flow of blood flowing at
   (a) atmospheric pressure
   (b) high pressure
   (c) low pressure
   (d) All of the above
2. Which one is present in tunica media?
   (a) Collagen fibres and smooth muscles
   (b) Yellow fibres and smooth mucles
   (c) Yellow fibres and striated muscles
   (d) Squamous epithelium and striated muscles
3. Which artery supply blood to the heart muscle?
   (a) Coronary artery
   (b) Inferior vena cava
   (c) Superior vena cava
   (d) Pulmonary artery
4. Which vascular connections are present between organs of alimentary canal and liver?
   (a) Hepatic portal system
   (b) Renal portal system
   (c) Venous system
   (d) Hypophyseal portal system
5. When brain starts receiving lesser blood volume due to atheromas, the condition progress towards
   (a) stenosis (b) cerebral failure
   (c) stroke (d) angina
6. Which disease is caused by the atherosclerosis?
   (a) Heart attack
   (b) Heart failure
   (c) Angina pectoris
   (d) Hypertension

EXAM POINTS

- Person with AB blood group is universal recipient due to the absence of antigens A and B on RBC surface. **[CBSE AIPMT 2014]**
- Blood group O is used for immediate blood transfusions. **[CBSE AIPMT 2012]**
- Erythroblastosis foetalis is a condition in which the first Rh^+ child of Rh^- mother would not suffer but blood of foetus (Rh^+) stimulates the formation anti-Rh^+ factors in the mother's blood. In the subsequent pregnancies, the anti-Rh antibodies in the mother's blood destroy the foetal RBCs and results in haemolytic disease. **[NEET 2020]**
- Factor X is the interlinking factor at the activation steps of intrinsic and extrinsic pathway of blood clotting. **[JIPMER 2017]**
- According to Best and Taylor's theory, platelets do not play any role in blood clotting. **[KCET 2012]**
- Fibrinogen is a plasma protein that is involved in blood coagulation. **[CBSE AIPMT 2011]**
- Serum is blood without corpuscles and fibrinogen. **[RPMT 2011]**
- Thrombocyte is the blood cells, whose reduction in number causes clotting disorder, leading to excessive loss of blood from the body. **[NEET 2016]**
- Fine tendinous cords called chordae tendineae is found in heart of mammals. These are attached to the ventricular surface of the valves. **[NEET 2019]**
- Foramen ovale is a small hole located in the septum, which is the wall between the two upper chambers of the heart. **[JIPMER 2017]**
- The first heart sound occurs due to the sudden closure of AV valves. **[AIIMS 2017]**
- Blood pressure is maximum during systole of the left ventricle. **[CBSE AIPMT 2016]**
- Blood pressure in the pulmonary artery is more than that in the pulmonary vein. **[NEET 2016]**
- Tricuspid valve is present between right atrium and ventricle. Bicuspid valve is present between left atrium and ventricle. Semilunar valve is present between right ventricle and pulmonary artery. **[NEET 2018]**
- SA node is known as the pacemaker because it has the highest rate of depolarisation. **[NEET 2019]**
- AV node is a specialised nodal tissue embedded in the lower corner of the right atrium, allowing the atria to empty completely. **[NEET 2019]**
- AV node is known as pacesetter because it serves as an electrical relay station. It spreads the electrical current sent by SA node to whole heart. **[UK PMT 2015]**
- Parasympathetic neural signals are associated with decrease in cardiac output. **[NEET 2020]**
- The heart rate can be calculated by dividing cardiac output with stroke volume. **[NEET 2019]**
- Number of heartbeats per minutes decreases if the stroke volume of heart increases. **[EAMCET 2015]**
- Parasympathetic neural signals decreases the cardiac output. **[NEET 2020]**
- Adrenal medullary hormones can increase cardiac output. **[Kerala CEE 2014]**
- P-wave represents atrial depolarisation. **[J&K CET 2014]**
- The QRS-complex represents depolarisation of ventricles in a standard ECG. **[NEET 2020]**
- The T-wave in one ECG represents return of the ventricles from excited state. **[KCET 2016]**
- The study of blood vessels is termed as **angiology**. **[MHT CET 2015]**
- Hepatic vein normally carry largest amount of urea. **[NEET 2016]**
- The hepatic portal vein drain blood from intestine to liver. **[NEET 2017]**
- At rest, the largest volume of blood is found in the veins. **[UK PMT 2015]**

Chapter Exercises

(A) Topical Exercise

Body Fluids

1 Select the correct statement.
(a) Intracellular fluids contain abundant potassium ions
(b) Like CSF, blood is extracellular fluid
(c) ABO grouping system was proposed by Landsteiner
(d) All of the above

2 Which of the following options describes all the components of human blood group?
(a) A and B blood group (b) AB and O blood group
(c) Rh and ABO blood group (d) Rh and AB blood group

3 Grouping of ABO blood is based on the
(a) presence or absence of surface proteins on RBCs
(b) surface lipids present on the RBCs
(c) only presence of surface proteins at WBCs
(d) presence or absence of surface antigens on both RBCs and WBCs

4 Consider the following statements.
I. The recepient's serum should not contain antigen similar to the RBCs of the donor.
II. The recepient's serum should not contain the antibodies against the RBCs of the donor.

Choose the correct option.
(a) Statement I is true, but II is false
(b) Statement I is false, but II is true
(c) Both statements I and II are true
(d) Both statements I and II are false

5 During blood transfusion, if blood groups of recepient and donor are not correctly matched, it will result in clumping of RBCs due to
(a) antigen-antibody reaction
(b) antigen-antigen reaction
(c) antibody-antibody reaction
(d) None of the above

6 Consider the following statements.
I. Rh grouping in humans is based on the presence or absence of rhesus antigen on RBCs surface.
II. Nearly 80% of humans possess Rh antigen and thus called Rh positive.

Choose the correct option.
(a) Statement I is true, but II is false
(b) Statement I is false, but II is true
(c) Both statements I and II are true
(d) Both statements I and II are false

7 What will happen if a Rh^- person's blood is exposed to the Rh^+ person?
(a) Antibody formation takes place in both Rh^- and Rh^+ person
(b) Negative and positive Rh antigen cancel out each other
(c) Antibody will be produced in Rh^- person's blood
(d) Antibody will be produced in Rh^+ person's blood

8 When is the mixture of both phospholipids and tissue factor found in plasma aiding blood coagulation released in humans?
(a) During hypertension
(b) By the traumatised cell at the place of injury
(c) In the condition of erythroblastosis foetalis
(d) During anaemia

9 Choose the incorrectly matched pairs.
(a) Factor I - Fibrinogen
(b) Factor IV - Calcium ions
(c) Factor V - Antihaemophilic factor
(d) Factor XII - Hageman factor

10 Haemophilia A and B are caused due to the deficiency of factors
(a) VIII and IX (b) X only
(c) X and XI (d) XII only

11 The pale fluid that bathes the tissue of an organism and which is considered as an important carrier for the transport of nutrients and hormone is called
(a) tissue fluid (b) lymph
(c) blood (d) Both (a) and (b)

Circulatory System

12 Identify the organisms with open circulatory system.
(a) Prawn, oyster, snail
(b) Cockroach, squids, clams
(c) *Pheretima*, frog, oyster
(d) Starfish, *Balanoglossus*, snails

13 Identify the correct statement about closed circulatory system.
(a) Blood flows in the open tissue space, the sinuses
(b) Blood is in direct contact with the tissue cells
(c) Blood pressure is high in closed network of blood vessels
(d) Blood pressure is low in closed network of blood vessels

14 In a system, in which blood does not normally fill body cavities, blood is enclosed by
(a) tubes (b) ducts
(c) valves (d) vessels

15 In the evolution of animals, a heart to pump the blood is found for the first time in
(a) earthworm (b) roundworms
(c) ribbon worm (d) flat worms

16 In frog heart, there are involuntary striated muscle tissue which consist of fibres called
(a) Purkinje fibres (b) myonemes
(c) telodendria (d) columnae carnae

17 Choose the incorrectly matched pairs.
(a) Single circuit circulation - Pisces
(b) SA node in amphibians - Sinus venosus
(c) Foramen of Panizzae - Reptiles
(d) None of the above

18 In mammals, oxygenated blood flows from
(a) left auricle to left ventricle during auricular systole
(b) right auricle to right ventricle during ventricular systole
(c) right ventricle to aorta during ventricular systole
(d) pulmonary vein to left auricle during auricular systole

19 Systemic heart refers to
(a) the heart that contracts under stimulation from nervous system
(b) left auricle and left ventricle in higher vertebrates
(c) entire heart in lower vertebrates
(d) the two ventricles together in humans

20 Systemic circulation is
(a) Left ventricle $\xrightarrow[\text{blood}]{\text{Deoxygenated}}$ Tissues $\xrightarrow[\text{blood}]{\text{Oxygenated}}$ Right ventricle
(b) Left ventricle $\xrightarrow[\text{blood}]{\text{Oxygenated}}$ Tissues $\xrightarrow[\text{blood}]{\text{Deoxygenated}}$ Left auricle
(c) Left ventricle $\xrightarrow[\text{blood}]{\text{Deoxygenated}}$ Tissues $\xrightarrow[\text{blood}]{\text{Oxygenated}}$ Right auricle
(d) Left ventricle $\xrightarrow[\text{blood}]{\text{Oxygenated}}$ Tissues $\xrightarrow[\text{blood}]{\text{Deoxygenated}}$ Right auricle

21 Which one of the following statements is correct with reference to the circulation of blood in a mammal?
(a) Left auricle receives oxygenated blood from the lungs
(b) Pulmonary artery returns oxygenated blood from the lungs to the left auricle
(c) Pulmonary vein carries venous blood from right auricle to lungs
(d) Venous blood is returned to the left auricle

22 In order for the blood to flow from right ventricle to left ventricle in mammalian heart, it must flow through
(a) Right ventricle, Pulmonary arteries, Lungs, Pulmonary veins, Left atrium
(b) Right ventricle, Pulmonary veins, Lungs, Pulmonary arteries, Left atrium
(c) Right ventricle, Right atrium, Lungs, Pulmonary veins, Left atrium
(d) Right ventricle, Systemic aorta, Lungs, Pulmonary veins, Left atrium

23 Choose the schematic diagram, which properly represents pulmonary circulation in humans.
(a) Right ventricle $\xrightarrow[\text{blood}]{\text{Deoxygenated}}$ Lungs $\xrightarrow[\text{blood}]{\text{Oxygenated}}$ Left auricle
(b) Right ventricle $\xrightarrow[\text{blood}]{\text{Oxygenated}}$ Lungs $\xrightarrow[\text{blood}]{\text{Deoxygenated}}$ Left auricle
(c) Right ventricle $\xrightarrow[\text{blood}]{\text{Oxygenated}}$ Lungs $\xrightarrow[\text{blood}]{\text{Deoxygenated}}$ Left ventricle
(d) Right auricle $\xrightarrow[\text{blood}]{\text{Deoxygenated}}$ Lungs $\xrightarrow[\text{blood}]{\text{Oxygenated}}$ Right ventricle

24 Identify whether the given statements are true or false for double circulation.
I. It checks the mixing of oxygenated and deoxygenated blood.
II. It carries only oxygenated blood.
Choose the correct option accordingly.
(a) Statement I is true, but II is false
(b) Statement I is false, but II is true
(c) Both statements I and II are true
(d) Both statements I and II are false

Heart

25 Which of the following statement is incorrect?
(a) Heart is endodermal in origin
(b) Human heart is situated in between the two lungs slightly tilted to left
(c) Heart is protected by double-walled membranous pericardium
(d) Human heart has two atria and two ventricles

26 Consider the following statements.
I. Activities of human heart are regulated extrinsically.
II. Human heart is neurogenic.
Choose the correct option.
(a) Statement I is true, but II is false
(b) Statement I is false, but II is true
(c) Both statements I and II are true
(d) Both statements I and II are false

27 The pericardial sac containing the heart and the pericardial fluid help in
(a) protecting the heart from friction, shocks and keeps it moist
(b) pumping the blood
(c) receiving the blood from various parts of the body
(d) None of the above

28 Select the incorrectly matched pair.
(a) Epicardium - Cuboidal epithelium
(b) Pulsatile heart - Earthworm
(c) Endocardium - Squamous epithelium
(d) Tubular heart - Cockroach

29 Interatrial septum is
(a) thin and muscular wall between two atria
(b) thick wall between two atria
(c) fibrous opening of atria
(d) muscular flap near the opening of atria

30 An oval depression called fossa ovalis is seen on
(a) interatrial septum
(b) interventricular septum
(c) right auriculoventricular septum
(d) left auriculoventricular septum

31 Which of the following chambers of the heart has the thickest muscular wall?
(a) Left atrium (b) Right atrium
(c) Right ventricle (d) Left ventricle

32 Chordae tendineae are found in
(a) left atrium of heart
(b) ventricles of heart
(c) right atrium of heart
(d) throughout the heart

33 Identify the correct statement for human heart.
(a) Volume of both the atria is greater than the volume of both the ventricles
(b) Volume of both the ventricle is greater than the volume of both the atria
(c) Interventricular septum separates the right and the left atria
(d) Atrioventricular septum is not found in human heart

34 Right auricle of mammalian heart receives blood from
(a) sinus venosus
(b) pulmonary veins
(c) precavals
(d) pre and postcavals

35 The blood pumped by right ventricle and left ventricle enters into
(a) aorta and vena cava
(b) pulmonary artery and aorta
(c) pulmonary vein and pulmonary artery
(d) pulmonary artery and pulmonary vein

36 Nature of valve in the heart is
(a) membranous
(b) muscular
(c) tendinous
(d) ligamentous

37 Match the following columns.

Column I	Column II
A. Tricuspid valve	1. On the opening of superior vena cava
B. Pulmonary valve	2. On the entrance of aorta
C. Eustachian valve	3. Between right arium and right ventricle
D. Coronary valve	4. Entrance of pulmonary artery
	5. Over the opening of coronary sinus

Codes

	A	B	C	D		A	B	C	D
(a)	3	5	4	1	(b)	3	4	1	5
(c)	4	2	1	5	(d)	4	3	1	5

38 The valves which allow blood to flow from the ventricles into the arteries and not in the opposite direction are
(a) bicuspid and tricuspid valves
(b) AV valve and semilunar valve
(c) aortic valve and mitral valve
(d) semilunar valve

39 The atrioventricular valve of the heart is prevented from turning inside out by tough strands of connective tissue is called as
(a) tendinous cords (b) tricuspid
(c) pocket valve (d) mitral valve

40 The muscle which attach the cusps of the atrioventricular valves occur in
(a) ventricles (b) auricles
(c) atrioventricular valves (d) pulmonary valves

41 Identify the incorrect statement.
(a) Thebasian valve is found over the opening of coronary sinus
(b) AV node is also called node of Keith and Flack
(c) Eustachian valve is present over the opening of superior vena cava
(d) Stenosis is the condition of failure of heart valve to open completely

42 The innervation of heart is primarily meant for
(a) initiation of heartbeat (b) regulation of heartbeat
(c) activation of pacemaker (d) release of acetylcholine

43 In higher vertebrates, SA node helps in
(a) conduction of blood
(b) initiation of heartbeat
(c) opening of tricuspid valve
(d) opening of bicuspid valve

44 Since, it is the sinoauricular node, which initiates the impulses in the heart of mammal is called
(a) cholinergic (b) adrenergic
(c) neurogenic (d) myogenic

45 Impulse originating from sinu-atrial node are transmitted to the
(a) atrioventricular node (b) bundle of His
(c) pacemaker (d) Purkinje system

46 The given diagram shows the human heart.

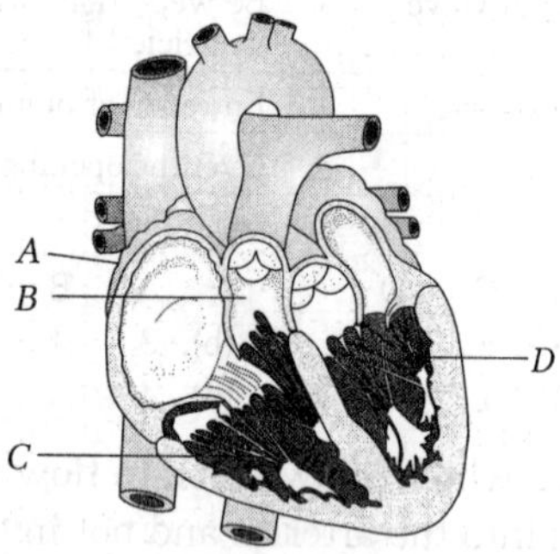

From which site initiation of heartbeat impulse is generated?
(a) *D* (b) *C* (c) *B* (d) *A*

47 To which of the following a specialised muscle bundle connecting the atrial and ventricular chambers of the heart passes stimulus of contraction?
(a) AV node (b) SA node
(c) Purkinje fibre (d) Atrium

48 The following figure shows a human heart. Which labelled structure represents the bundle of His?

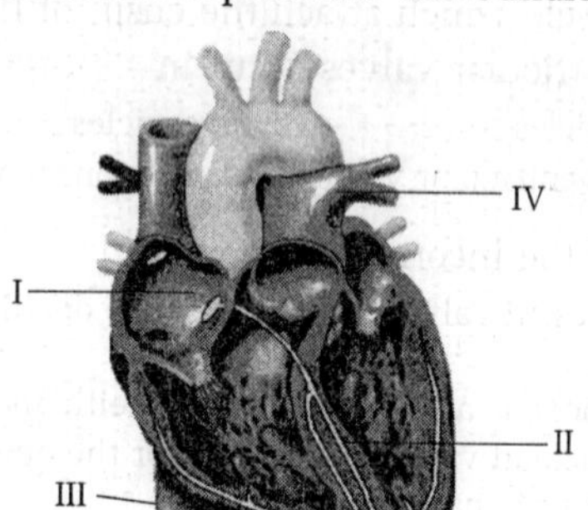

(a) IV (b) III
(c) II (d) I

49 The nerve-like modified muscle in the right auricle is known as
(a) lymph node (b) atrioventricular node
(c) pacemaker (d) bulbus arteriosus

50 Match the following columns.

Column I	Column II
A. Artificial pacemaker	1. Right atrium
B. Coronary sinus	2. Greatbach and Chardark
C. Ductus arteriosus	3. Non-functional lungs
D. Columnar carnae	4. Ventricles

Codes

	A	B	C	D		A	B	C	D
(a)	2	3	1	4	(b)	3	1	4	2
(c)	4	3	2	1	(d)	2	1	3	4

51 As blood collects in the upper chambers, the SA node sends out an electrical signal that causes the atria to contract. These activity is/are affected by
(a) carbon dioxide (b) oxygen
(c) vagus nerve (d) All of these

52 In humans, blood passes from the postcaval to the diastolic right atrium of heart due to
(a) stimulation of the sinoauricular node
(b) pressure difference between the postcaval and atrium
(c) pushing open of the venous valves
(d) suction pull

53 In the given diagram of cardiac cycle, identify *A*, *B* and *C*.

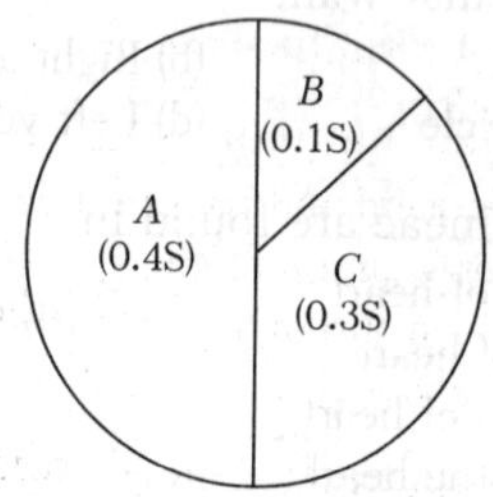

	A	*B*	*C*
(a)	Ventricular diastole	Atrial systole	Joint diastole
(b)	Joint systole	Atrial diastole	Ventricle systole
(c)	Joint diastole	Atrial systole	Ventricular systole
(d)	Atrial systole	Atrial diastole	Ventricular systole

54 At the beginning of systole, when the ventricular pressure exceeds atrial pressure. It results into
(a) opening of semilunar valve
(b) closing of semilunar valve
(c) onset of auricular svstole
(d) sudden closure of AV valves

55 Which among the following is correct during each cardiac cycle? **[NCERT Exemplar]**
(a) The volume of blood pumped out by the right and left ventricles is same
(b) The volume of blood pumped out by the right and left ventricles is different
(c) The volume of blood received by each atrium is different
(d) The volume of blood received by the aorta and pulmonary artery is different

56 The aortic closure sound and the pulmonic closure sound is associated with the closure of **[NCERT Exemplar]**
(a) tricuspid valve (b) semilunar valve
(c) bicuspid valve (d) tricuspid and bicuspid valve

57 Absolute refractory period of heart is during
(a) contraction when the heart is in non-responding period
(b) expansion
(c) negative charge
(d) positive charge

58 When heart is in relaxed state so, it is getting filled by
(a) mixed blood (b) venous blood
(c) oxygenated blood (d) deoxygenated blood

59 What would be the cardiac output of a person having 72 heartbeats per minute and a stroke volume of 50 mL? **[NCERT Exemplar]**
(a) 360 mL (b) 3600 mL
(c) 7200 mL (d) 5000 mL

60 The heart sound 'dup' is produced when
(a) tricuspid valve is opened
(b) mitral valve is opened
(c) mitral valve is closed
(d) semilunar valves at the base of aorta get closed

61 The heart sound 'murmur' is heard during
(a) closing of bicuspid and tricuspid valves
(b) closing of aortic and pulmonary semilunar valves
(c) leaking of blood through one of the valves
(d) None of the above

62 Which one of the following cardiac effects can be observed if the potassium concentration is increased two to three times of the normal value?
(a) Weakness of heart (b) Abnormal rhythm
(c) Death (d) All of these

63 Cardiac activity could be moderated by the autonomous neural system. Choose the correct answer. **[NCERT Exemplar]**
(a) The parasympathetic system stimulates heart rate and stroke volume
(b) The sympathetic system stimulates heart rate and stroke volume
(c) The parasympathetic system decreases the heart rate but increases stroke volume
(d) The sympathetic system decreases the heart rate and stroke volume

64 Neural centre in medulla oblongata can moderate the cardiac function through
(a) ANS (Autonomic Nervous System)
(b) sympathetic nervous system
(c) parasympathetic nervous system
(d) somatic nervous system

65 Neural signals through the sympathetic nerves (ANS) can increase the rate of heartbeat by
(a) increasing heart output
(b) increasing the strength of ventricular contraction
(c) Both (a) and (b)
(d) increasing the contraction of atrium

66 Given below is the diagrammatic representation of standard Electrocardiogram (ECG). Identify the figure which is the correct demonstration of ECG.

(a) 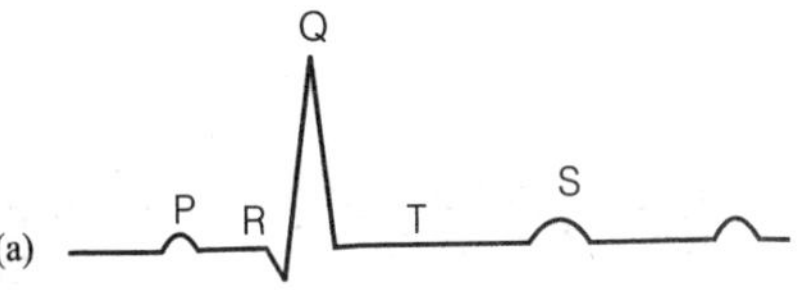

(b) 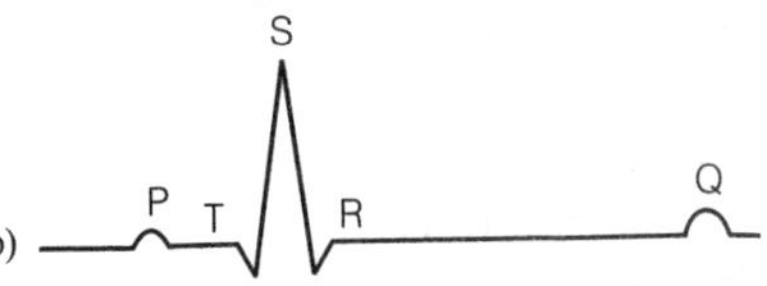

(c) 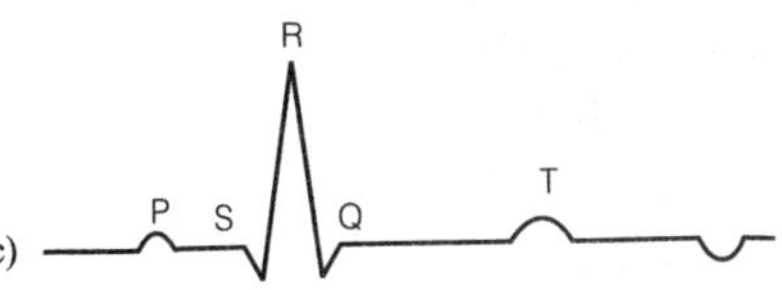

(d) 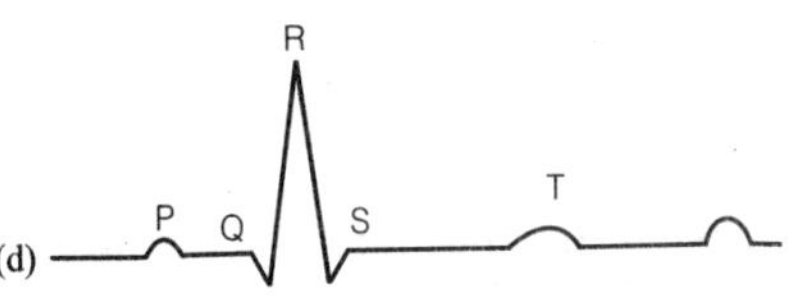

67 ECG depicts the depolarisation and repolarisation processes during the cardiac cycle. In the ECG of a normal healthy individual, one of the following waves is not represented. **[NCERT Exemplar]**
(a) Depolarisation of atria
(b) Repolarisation of atria
(c) Depolarisation of ventricles
(d) Repolarisation of ventricle

68 Which of the following correctly explains a phase/event in cardiac cycle in a standard electrocardiogram? **[NCERT Exemplar]**
(a) QRS complex indicates atrial contraction
(b) QRS complex indicates ventricular contraction
(c) Time between S and T represent atrial systole
(d) P-wave indicates beginning of ventricular contraction

69 The given figure is the ECG of a normal human. Which one of its components is correctly interpreted below?

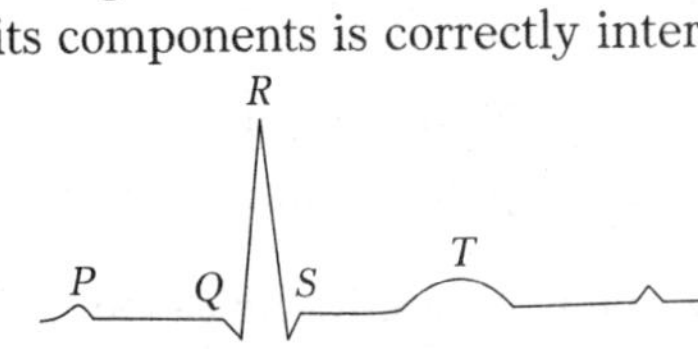

(a) Complex QRS-one complete pulse
(b) Peak T-initiation of total cardiac contraction
(c) Peak P and peak R together-systolic and diastolic blood pressures
(d) Peak P-initiation of left atrial contraction only

70 ECG is least effective in detecting abnormalities in
(a) cardiac rhythm
(b) coronary blood flow
(c) cardiac contractility
(d) atrioventricular conduction

Blood Vessels and Their System

71 Identify *A*, *B* and *C* in the given diagram.

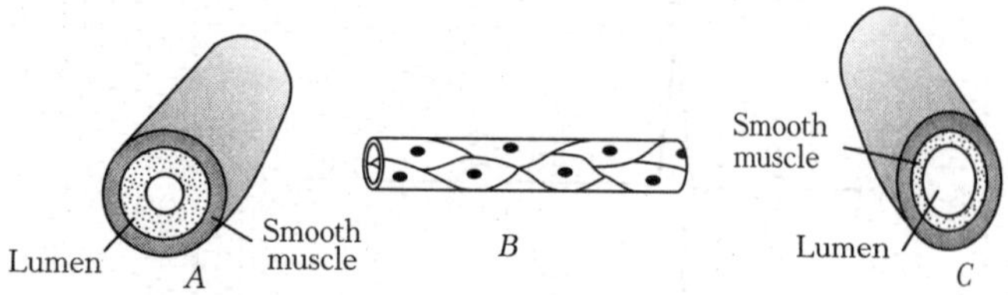

Choose the correct option.
(a) *A*-Artery, *B*-Capillary, *C*-Vein
(b) *A*-Artery, *B*-Vein, *C*-Capillary
(c) *A*-Vein, *B*-Artery, *C*-Capillary
(d) *A*-Capillary, *B*-Artery, *C*-Vein

72 Arteries are thick-walled, because
(a) they are without valves
(b) they are to pump blood
(c) the blood flows with jerks in them
(d) they have oxygenated blood

73 With the increasing distance from heart the elasticity as well as magnitude of the muscular layer of arteries would
(a) decrease (b) remain constant
(c) slightly decrease (d) increase

74 Which one of the following is different from others in the absence of muscular coat?
(a) Veins (b) Arteries
(c) Capillaries (d) Arterioles

75 Which one of the following has elastic wall?
(a) Ventral aorta (b) Dorsal aorta
(c) Descending aorta (d) Ascending aorta

76 Carotid artery carries
(a) oxygentaed blood to the brain
(b) oxygenated blood to anterior region of body or to brain
(c) oxygenated blood to the posterior region of body or to brain
(d) oxygenated blood to heart

77 Select the incorrect statements.
(a) Arteries are deep seated whereas veins are superficial in position
(b) Phrenic and renal arteries are paired
(c) The jugular vein collect blood from the chest region
(d) None of the above

78 Consider the following statements.
I. Vasa vasorum supplies blood to walls of blood vessels.
II. Coeliac arteries are unpaired arteries.
Choose the correct option.
(a) Statement I is true, but II is false
(b) Statement I is false, but II is true
(c) Both statements I and II are true
(d) Both statements I and II are false

79 In the below given diagram, which blood vessel represents vena cava?

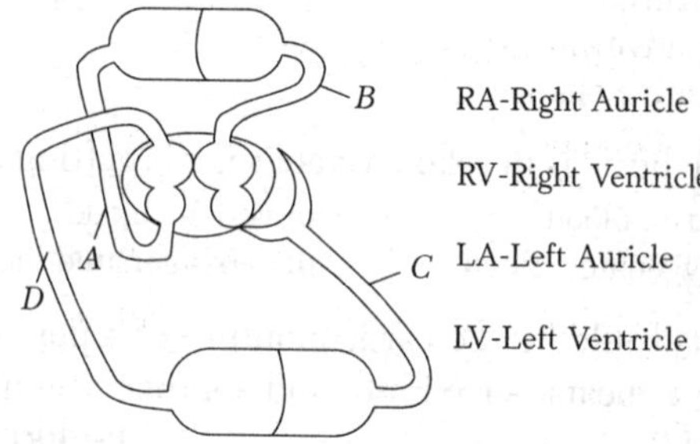

(a) C (b) D (c) A (d) B

80 Which one of the following veins break up into capillaries?
(a) Renal vein (b) Hepatic portal vein
(c) Pelvic vein (d) Pulmonary vein

81 The renal portal system is made up of
(a) femoral, renal portal veins
(b) sciatic, renal portal veins
(c) renal portal veins
(d) femoral, sciatic, renal portal veins

Human Lymphatic System

82 Identify the correct statement.
(a) Lymphatic capillaries are free and blind at one end
(b) Lacteals are fat absorbing lymphatic capillaries in intestinal regions
(c) Any blockage in the flow of lymph can cause edema
(d) All of the above

83 Which of the following is first to receive lymphatic duct from legs?
(a) Left subclavian vein
(b) Right subclavian vein
(c) Right lymphatic duct
(d) Thoracic lymphatic duct

84 The principal function of the lymph node in the man is
(a) destruction of old RBCs
(b) destruction of old WBCs
(c) collection and destruction of pathogens in the blood
(d) production of WBCs

85 The gland that filter substances that travel through the lymphatic fluid, is responsible for the formation of
(a) hormones
(b) lymphs
(c) antigens
(d) antibodies

86 Lymphoid tissue is found in
(a) thymus (b) tonsils
(c) lymph nodes (d) All of these

Abnormal Conditions Related to Circulatory System and Cardio Vascular Diseases

87 Match the following columns.

Column I (Cardiac abnormality)		Column II (Features)	
A.	Tachycardia	1.	Increased heart rate
B.	Bradycardia	2.	Irregular heartbeat
C.	Arrythmia	3.	Decreased heart rate
D.	Arteriosclerosis	4.	Hardening or loss of elasticity of arteries

Codes

	A	B	C	D
(a)	1	2	3	4
(b)	4	3	2	1
(c)	1	3	2	4
(d)	1	4	3	2

88 Atherosclerosis is caused by deposition of
(a) calcium
(b) fat and cholesterol
(c) deposition of fibrous tissues
(d) All of the above

89 Angiogram of a procedure in coronary blood vessel depicts

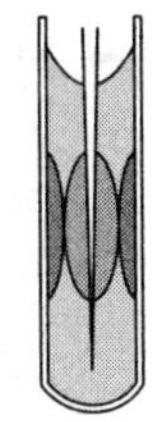

(a) removal of cancerous growth
(b) opening of defective valves
(c) cracking of a plaque blocking coronary artery
(d) removal of parasite blocking coronary vein

90 Which tube in the given figure of a heart correctly represents the result of a successful coronary bypass operation?

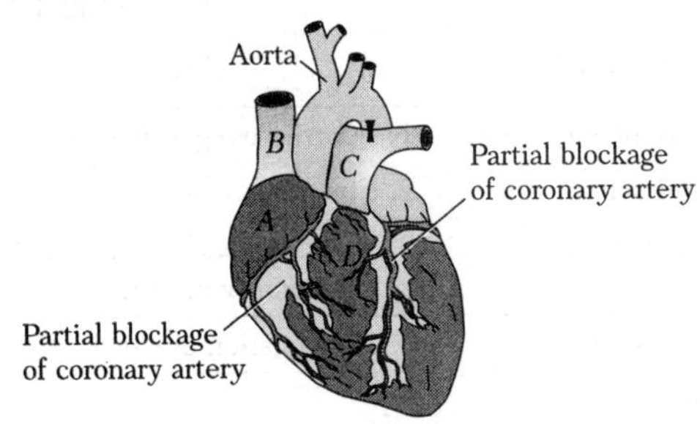

(a) *C* (b) *B* (c) *D* (d) *A*

91 Angina occurs due to the
(a) benign growth in heart
(b) decreased supply of oxygen to heart muscle
(c) the deposition of carbohydrates in artery
(d) increased blood pressure

(B) Cumulative Exercise

(Assorted Questions of the Chapter for Advanced Level Practice)

1 Which of the following statement is incorrect?
(a) The pulmonary trunk with left and right arteries is part of pulmonary circulation
(b) The aorta is a part of systemic circulation
(c) The aorta receives blood from left ventricle
(d) Pulmonary vein collects deoxygenated blood from body and pour it into right atrium

2 In the given diagram, identify *A-F*.

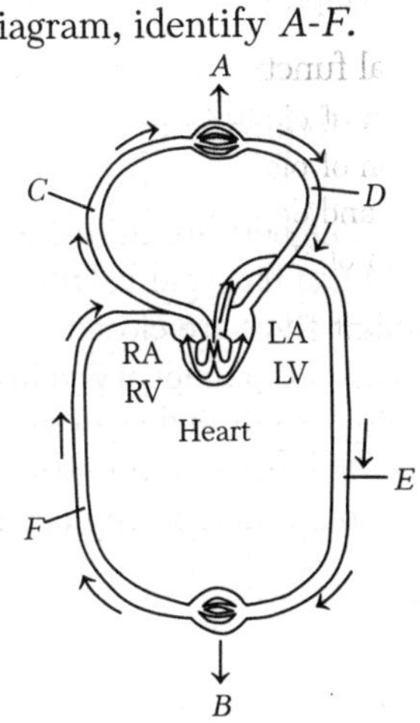

Choose the correct option.
(a) *A*-Lungs, *B*-Body parts, *C*-Pulmonary vein, *D*-Pulmonary artery, *E*-Dorsal aorta, *F*-Vena cava
(b) *A*-Lungs, *B*-Body parts, *C*-Pulmonary artery, *D*-Pulmonary vein, *E*-Dorsal aorta, *F*-Vena cava
(c) *A*-Lungs, *B*-Body parts, *C*-Pulmonary artery, *D*-Pulmonary vein, *E*-Vena cava, *F*-Dorsal aorta
(d) *A*-Body parts, *B*-Lungs, *C*-Pulmonary artery, *D*-Pulmonary vein, *E*-Vena cava, *F*-Dorsal aorta

3 Which one of the following is a correctly matched pair.

(a)	Lub	–	Sharp closure of AV valves at the beginning of ventricular systole
(b)	Dup	–	Sudden opening of semilunar valves at the beginning of ventricular diastole
(c)	Pulsation of the radial artery	–	Valves in the blood vessels
(d)	Initiation of the heartbeat	–	Purkinje fibres

4 Systolic pressure is higher than diastolic pressure because

(a) arteries are contracting during systole
(b) blood is pumped with a pressure in the arteries by the heart during systole, but not during diastole
(c) arteries resist during systole only
(d) volume of blood is higher in systole than that of diastole in the heart

5 Match the following columns. **[NCERT Exemplar]**

Column I	Column II
A. Lymphatic system	1. Carries oxygenated blood
B. Pulmonary vein	2. Immune response
C. Thrombocytes	3. To drain back the tissue fluid to the circulatory system
D. Lymphocytes	4. Coagulation of blood

Codes

	A	B	C	D		A	B	C	D
(a)	2	1	3	4	(b)	3	1	4	2
(c)	3	1	2	4	(d)	2	1	3	4

6 Arrange the following events in chronological order during the beginning of cardiac cycle.

I. SA node generates action potential
II. Atrial systole
III. Joint diastole
IV. Ventricular diastole
V. Ventricular systole
VI. Atrial diastole

Choose the correct option.

(a) III → II → I → IV → V → VI
(b) I → III → II → IV → V → VI
(c) III → I → II → V → VI → IV
(d) II → IV → I → III → V → VI

7 The diagram below shows how things get to and from the liver. They are labelled as *A*, *B*, *C*, *D*, *E* and *F*. Which one of the following labellings is the correct?

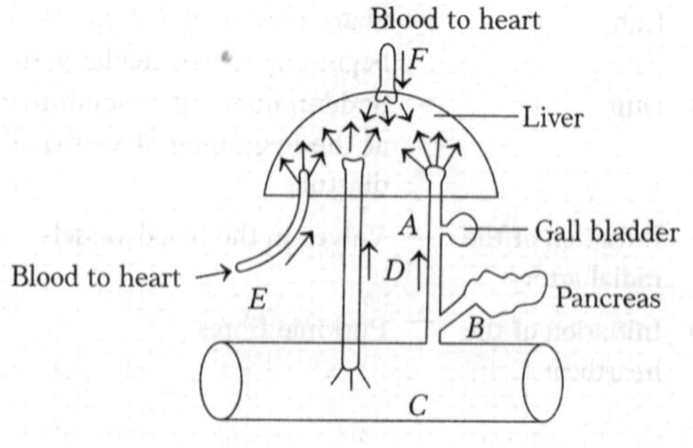

(a) *A* is the hepatic portal vein and *E* is the hepatic vein
(b) *C* is the intestine and *F* is the hepatic portal vein
(c) *D* is the hepatic portal vein and *F* is the hepatic vein
(d) *B* is pancreatic artery and *E* is the hepatic artery

8 Which of the following statements is/are incorrect about the lymph.

I. Lymph is a red coloured fluid containing large proteins and formed elements.
II. It is also called tissue fluid which leaks out through capillaries into tissue spaces.
III. It contains specialised lymphocytes which are responsible for the immunity of the body.
IV. Lymph is an important carrier for nutrients and hormones.
V. Fats are absorbed through the lymph in the lacteals present in the intestinal villi.

Choose the correct option.

(a) Only I
(b) III and IV
(c) II and III
(d) Only IV

9 Match the following columns.

Column I (Heart vessels)	Column II (Functions)
A. Superior vena cava	1. Carries oxygenated blood
B. Inferior vena cava	2. Carries deoxygenated blood
C. Pulmonary artery	3. Brings deoxygenated blood from lower part of body to right atrium
D. Pulmonary vein	4. Brings deoxygenated blood from upper part of body to right atrium

Codes

	A	B	C	D
(a)	1	2	3	4
(b)	4	3	2	1
(c)	4	2	3	1
(d)	1	4	3	2

10 Which of the following events occurs during joint diastole?

I. All four chambers are in relaxed state.
II. Tricuspid and bicuspid valves are open.
III. Semilunar valves are closed.
IV. Blood from the pulmonary veins and vena cava flows into the left and right ventricles, respectively through the left and right atria.

Choose the option containing correct choices.

(a) Only I
(b) Only III
(c) II and IV
(d) I, II, III and IV

11 In the following table of human ABO blood groups, fill up the blanks (i), (ii), (iii) and (iv) from the options given below

Blood groups	Antigens on RBCs	Antibody in plasma	Donor groups
A	A	Anti-B	A,O
B	B	Anti-A	B,O
AB	AB	(ii)	A, B, AB, O
O	(i)	(iii)	(iv)

	(i)	(ii)	(iii)	(iv)
(a)	Nil	Nil	Nil	O
(b)	Nil	Nil	Anti-A,B	AB
(c)	Nil	Anti-A,B	Nil	O
(d)	Nil	Nil	Anti-A,B	O

12 Match the following columns.

	Column I		Column II
A.	P-wave	1.	Atrial depolarisation
B.	First heart sound	2.	Ventricular depolarisation
C.	Second heart sound	3.	Ventricular repolarisation
D.	QRS complex	4.	Closure of the AV valves at the onset of systole
E.	T-wave	5.	Closure of the semilunar valves at the onset of diastole

Codes

	A	B	C	D	E
(a)	1	4	5	2	3
(b)	1	2	3	4	5
(c)	5	4	3	2	1
(d)	3	5	1	2	4

13 A portal system is a system in which
(a) a vein starts from an organ and ends up in heart
(b) an artery breaks up in an organ and restarts by the union of its capillaries
(c) the blood from the heart is brought into the kidney before it is poured into posterior vena cava
(d) a vein breaks up in an organ into capillaries and restarts by their union as a new vein in the same organ

14 In a closed circulatory system, the
I. blood is found in vessels.
II. arterial musculature and precapillary sphincter regulate flow of blood to the organs.
III. circulating fluid baths internal organs directly.
IV. the flow of blood is bidirectional.

Choose the option containing correct statements.
(a) I and II
(b) II and III
(c) III and IV
(d) Only IV

15 Choose the incorrect match.
(a) Tubular heart – Cockroach
(b) Pulsatile heart – Earthworm
(c) Ampullary heart – Aves
(d) Chambered heart – Fishes

16 Haemolytic disorder of newborn occurs when an Rh negative mother undergoes second pregnancy and father is Rh positive. This can be avoided by
(a) injection of Rh positive blood into the Rh negative mother
(b) injection of Rh antigen in mothers immediately after birth
(c) injection of anti Rh antibodies during pregnancy or immediately after birth
(d) None of the above

17 In which one of the following pairs, the two items mean one and the same thing?
(a) Malleus – Anvil
(b) SA node – Pacemaker
(c) Leucocytes – Lymphocytes
(d) Haemophilia – Blood cancer

18 Identify the correct matches with respect to etectrocardiography.
(a) Absolute refractory period – Na^+ ion influx
(b) Reduced heart rate and strength of contraction – K^+ excess
(c) Depression of cardiac function – Ca^+excess
(d) Spastic contraction – P excess

19 During the process of blood coagulation, vitamin-K assist in the
(a) formation of prothrombin
(b) conversion of prothrombin to thrombin
(c) conversion of fibrinogen to fibrin
(d) formation of thromboplastin

20 Identify the components labelled (*A-D*) in the given flowchart of the blood clotting process.

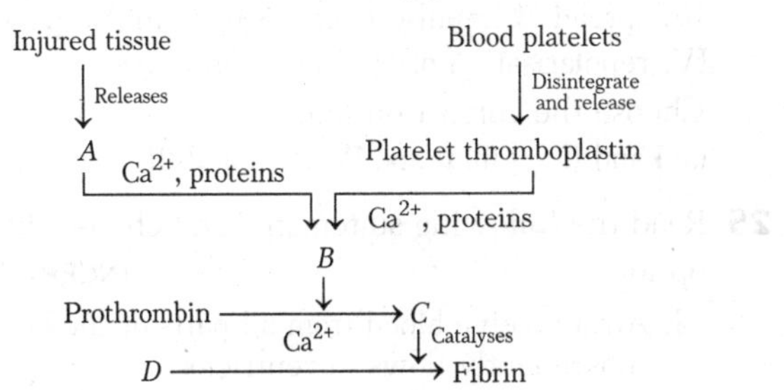

	A	B	C	D
(a)	Thrombo-plastin	Prothrom-binase	Thrombin	Fibrinogen
(b)	Fibrinogen	Thrombin	Prothrombinase	Thromboplastin
(c)	Prothrom-binase	Fibrinogen	Thromboplastin	Thrombin
(d)	Thrombin	Thrombo-plastin	Fibrinogen	Prothrombinase

21 Choose the correct proportion with respect to the distribution of blood in the body of man.

(a) 5% to heart muscles, 15% to brain, 25% to liver, 25% to kidney, 15% to bones, 15% to other organs
(b) 20% to heart muscles, 10% to brain, 10% to liver, 25% to kidney, 10% to bones, 25% to other organs
(c) 10% to heart muscles, 10% to brain, 10% to liver, 40% to kidney, 15% to bones, 15% to other organs
(d) 5% to heart muscles, 20% to brain, 20% to liver, 15% to kidney, 5% to bones, 35% to other organs

22 In the human heart, the

I. left auricle has opening of pulmonary veins and receives oxygenated blood from lungs.
II. right auricle opens into the right ventricle and the auriculoventricular aperture is guarded by a bicuspid valve.
III. valves are attached to the wall of ventricles by means of chordae tendineae.
IV. opening of left auricle into left ventricle is guarded by a tricuspid valve.

Choose the option containing correct statements.

(a) I, II and III (b) I and III
(c) II and IV (d) Only IV

23 Which of the following statement(s) is/are correct?

I. Bundle of His originates from the SA node.
II. SA node is situated in the wall of right auricle.
III. Auricular contraction stimulates the AV node.
IV. AV node sets up the cardiac impulses 72 times per minute.

Choose the correct option.

(a) Only I (b) II and III
(c) III and IV (d) II and IV

24 P-wave of ECG indicates

I. activation of SA node.
II. depolarisation of ventricle muscles.
III. spread of impulse from SA node to AV node.
IV. repolarisation of atria and ventricles.

Choose the correct option.

(a) I and II (b) II and IV (c) I and III (d) I, II and III

25 Read the following statements and choose the correct option. **(NCERT Exemplar)**

I. Atria receive blood from all parts of the body which subsequently flows to ventricles.
II. Action potential generated at SA node passes from atria to ventricles.

(a) Action mentioned in statement I is dependent on action mentioned in statement II
(b) Action mentioned in statement II is dependent on action mentioned in statement I
(c) Action mentioned in statement I and II are independent of each other
(d) Action mentioned in statement I and II are synchronous

26 ...*A*... which originates from left ventricle, distributes oxygenated blood to various body parts except ...*B*.... ...*C*... are four in number and they bring oxygenated blood from lungs to left ...*D*... .
Fill in the blanks by selecting the correct option with respect to major blood vessels of the heart.

(a) *A*-Pulmonary vein *B*- lungs
C-Coronary vein *D*- ventricle
(b) *A*- Systemic aorta *B*- lungs
C- Pulmonary vein *D*- auricle
(c) *A*- Postcaval *B*- lungs
C- Precaval *D*- auricle
(d) *A*-Precaval *B*- lungs
C-Postcaval *D*- ventricle

27 A red blood cell, entering the right side of the heart passes by or through the following structures.

1. Atrioventricular valves 2. Semilunar valves
3. Right atrium 4. Right ventricle
5. SAN

Which of the following options represents the correct sequence?

(a) $2 \rightarrow 3 \rightarrow 1 \rightarrow 4 \rightarrow 5$ (b) $3 \rightarrow 1 \rightarrow 5 \rightarrow 2 \rightarrow 4$
(c) $3 \rightarrow 5 \rightarrow 1 \rightarrow 2 \rightarrow 4$ (d) $5 \rightarrow 3 \rightarrow 1 \rightarrow 4 \rightarrow 2$

28 Given below are four statements (I-IV) regarding human blood circulatory system.

I. Arteries are thick-walled and have narrow lumen as compared to veins.
II. Angina is acute chest pain when the blood circulation to the brain is reduced.
III. Persons with blood group AB can donate blood to any person with any blood group under ABO system.
IV. Calcium ions play a very important role in blood clotting.

Which two of the above statements are correct?

(a) I and IV (b) I and II
(c) II and III (d) III and IV

29 After birth, ...*A*... gets blocked by the deposition of ...*B*... fibres and thus, forming ligamentum arteriosum. Identify *A* and *B*.

	A	*B*
(a)	ductus arteriosus	connective
(b)	Eustachian valve	elastin
(c)	trabeculae valve	connective
(d)	ductus arteriosus	elastin

30 Systolic pressure is higher than diastolic pressure because

(a) arteries are contracting during systole
(b) blood is pumped with a pressure in the arteries by the heart during systole, but not during diastole
(c) arteries resist during systole only
(d) volume of blood is higher in systole than that of diastole in the heart

31 Figure shows blood circulation in humans with labels *A* to *D*. Select the option which gives correct identification of label and functions of the part.

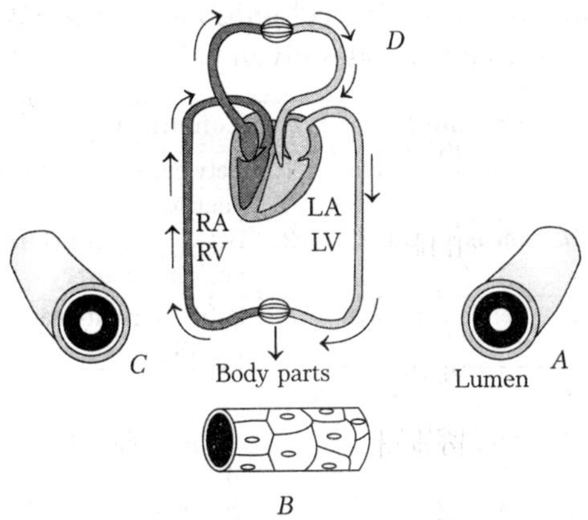

(a) *B* - Capillary — Thin without muscle layers and wall two cell thick
(b) *C* - Vein — Thin-walled and blood flows in jerks/spurts
(c) *D* - Pulmonary vein — Takes oxygenated blood to heart
(d) *A* - Artery — Thick-walled and blood flows evenly

32 Consider the following statements.

I. Four-chambered heart of bird is superior to crocodile heart.
II. Due to decreased elasticity of arteries, heart pump blood forcefully in older people.
III. Coronary circulation comprises vessels that serves heart.
IV. Thrombin activates the factors XI, V and VIII during blood clotting cascade.

Choose the correct option.

(a) I and II (b) I, II, III and IV
(c) II, III and IV (d) I, III and IV

33 The given figure shows an angiogram of the coronary blood vessle. Which one of the following statements correctly describes, what is being done?

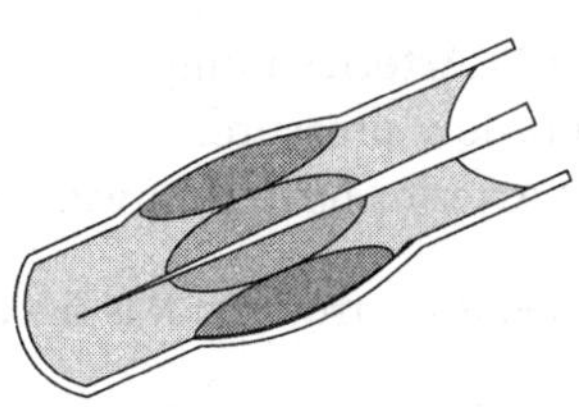

(a) It is a coronary artery which has a cancerous growth that is being removed
(b) It is a coronary artery which is blocked by a plaque and the same is being cracked
(c) It is a coronary vein in which the defective valves are being opened
(d) It is a coronary vein blocked by a parasite (blood fluke) that is being removed

Assertion and Reason

Direction (Q. Nos. 34-37) *In each of the following questions, a statement of Assertion (A) is given followed by the corresponding statement of Reason (R). Of the statements, mark the correct answer as*

(a) Both A and R are true and R is the correct explanation of A
(b) Both A and R are true, but R is not the correct explanation of A
(c) A is true, but R is false
(d) A is false, but R is true
(e) Both A and R are false

34 **Assertion** (A) Pulmonary circulation includes flow of oxygenated blood from right ventricle to the lungs.

Reason (R) The main purpose of this circulation is purification of blood.

35 **Assertion** (A) Systemic blood circulation occurs in human beings.

Reason (R) The main purpose of this circulation is to carry oxygen and nutrients to the body tissues and to remove CO_2 and other wastes from the tissues.

36 **Assertion** (A) Electrocardiogram is record of electrical activity of the heart which shows certain waves called P, Q, R, S and T waves.

Reason (R) It gives important information concerning the spread of excitation to the different parts of heart and it is of value in the diagnosis of cases of abnormal cardiac rhythm and myocardial damage.

37 **Assertion** (A) Heart pacemaker is a life-saving device when the normal heart rate of 72-80 drops to 30-40 due to disease or some other cause.

Reason (R) The pacemaker electrically stimulates the contractile heart walls.

(C) Medical Entrances' Gallery

(Collection of questions Asked in NEET & Various Medical Entrance Exams)

1 Which of the following is associated with decrease in cardiac output? **[NEET 2020]**

(a) Sympathetic nerves
(b) Parasympathetic neural signals
(c) Pneumotaxic centre
(d) Adrenal medullary hormones

2 Which of the following conditions causes erythroblastosis foetalis? **[NEET 2020]**

(a) Mother Rh+ve and foetus Rh–ve
(b) Mother Rh–ve and foetus Rh+ve
(c) Both mother and foetus Rh–ve
(d) Both mother and foetus Rh+ve

3 The QRS complex in a standard ECG represents **[NEET 2020]**

(a) depolarisation of auricles
(b) depolarisation of ventricles
(c) repolarisation of ventricles
(d) rapolarisation of auricles

4 What would be the heart rate of a person if the cardiac output is 5 L, blood volume in the ventricles at the end of diastole is 100 mL and at the end of ventricular systole is 50 mL? **[NEET 2019]**

(a) 75 beats per minute
(b) 100 beats per minute
(c) 125 beats per minute
(d) 50 beats per minute

5 All the components of the nodal tissue are autoexcitable. Why does the SA node act as the normal pacemaker? **[NEET 2019]**

(a) SA node has the lowest rate of depolarisation
(b) SA node is the only component to generate the threshold potential
(c) Only SA node can convey the action potential to the other components
(d) SA node has the highest rate of depolarisation

6 A specialised nodal tissue embedded in the lower corner of the right atrium, close to atrioventricular septum, delays the spreading of impulses to heart apex for about 0.1 sec. The delay allows **[NEET 2019]**

(a) blood to enter aorta
(b) the ventricles to empty completely
(c) blood to enter pulmonary arteries
(d) the atria to empty completely

7 Chordae tendineae is found in **[NEET 2019]**

(a) heart of cockroach
(b) heart of amphibians
(c) heart of reptiles
(d) heart of mammals

8 Match the following columns and choose the correct option from the codes given below. **[NEET 2018]**

	Column I		Column II
A.	Tricuspid valve	1.	Between left atrium and left ventricle
B.	Bicuspid valve	2.	Between right ventricle and pulmonary artery
C.	Semilunar valve	3.	Between right atrium and right ventricle

Codes

	A	B	C		A	B	C
(a)	1	2	3	(b)	1	3	2
(c)	3	1	2	(d)	2	1	3

9 Intrinsic and extrinsic pathways of blood clotting are interlinked at the activation steps of which of the following factors? **[JIPMER 2017]**

(a) Factor IX (b) Factor IV
(c) Factor X (d) Factor XIII

10 The first heart sound occurs due to the **[AIIMS 2017]**

(a) opening of semilunar valve
(b) closing of semilunar valve
(c) onset of auricular systole
(d) sudden closure of AV valves

11 Foramen ovale **[JIPMER 2017]**

(a) connects the two atria in the foetal heart
(b) is a condition in which the heart valves do not completely close
(c) is a shallow depression in the interventricular septum
(d) is a connection between the pulmonary trunk and the aorta in the foetus

12 The hepatic portal vein drains blood to liver from **[NEET 2017]**

(a) heart (b) stomach (c) kidneys (d) intestine

13 **Assertion** (A) Arteries pump blood away from the heart at very low pressure.
Reason (R) Arteries have thin wall with large lumen. **[AIIMS 2017]**

(a) Both A and R are true and R is the correct explanation of A
(b) Both A and R are true, but R is not the correct explanation of A
(c) A is true, but R is false
(d) Both A and R are false

14 Name the blood cells, whose reduction in number can cause clotting disorder, leading to excessive loss of blood from the body. **[NEET-II 2016]**

(a) Erythrocytes (b) Leucocytes
(c) Neutrophils (d) Thrombocytes

15 In mammals, which blood vessel would normally carry largest amount of urea? **[NEET-I 2016]**
(a) Dorsal aorta (b) Hepatic vein
(c) Hepatic portal vein (d) Renal vein

16 Blood pressure in the pulmonary artery is **[NEET-I 2016]**
(a) more than that in the carotid
(b) more than that in the pulmonary vein
(c) less than that in the vena cava
(d) same as that in the aorta

17 Blood pressure in the mammalian aorta is maximum during **[CBSE AIPMT 2016]**
(a) systole of the left atrium
(b) diastole of the right ventricle
(c) systole of the left ventricle
(d) diastole of the right atrium

18 At rest, the largest volume of blood is in the **[UK PMT 2015]**
(a) arteries (b) capillaries (c) veins (d) arterioles

19 In normal ECG, the component that signifies duration of one cardiac cycle is **[EAMCET 2015]**
(a) Q-T interval (b) P-R interval
(c) S-T segment (d) R-R interval

20 The study of blood vessels is termed as **[MHT CET 2015]**
(a) angiology (b) cardiology
(c) haematology (d) histology

21 Heaviness with severe chest pain which may disappear with rest indicates **[MHT CET 2015]**
(a) angina pectoris (b) atherosclerosis
(c) arteriosclerosis (d) hyperthyroidism

22 What will be the number of heartbeats per minute, if the stroke volume of heart increases, with total volume of heart increases, with total volume of blood remains same? **[EAMCET 2015]**
(a) Remains constant (b) Decreases
(c) Increases (d) Becomes erratic

23 The T-wave in one ECG represents **[KCET 2015]**
(a) return of the ventricles from excited state
(b) beginning of systole
(c) electrical excitation of atria
(d) depolarisation of ventricle

24 Which is not an important function of vertebrate circulatory system? **[UK PMT 2015]**
(a) Transport of nutrients and respiratory gases
(b) Regulation of body by circulating antibodies
(c) Protection of body by circulating antibodies
(d) Removal of waste products for excretion from the body

25 The site of exchange of wastes, nutrients, gases and hormones between the blood and body is **[UK PMT 2015]**
(a) arteries (b) arterioles
(c) capillary (d) veins

26 Which is pacesetter of the human heart? **[UK PMT 2015]**
(a) SA node (b) AV node
(c) Purkinje fibres (d) Bundle of His

27 Person with blood group AB is considered as universal recipient because he has **[CBSE AIPMT 2014]**
(a) Both A and B antigens on RBC, but no antibodies
(b) Both A and B antibodies in the plasma
(c) No antigen on RBC and no antibodies in the plasma
(d) Both A and B antigens in the plasma, but no antibodies

28 Thrombokinase is associated with **[Kerala CEE 2014]**
(a) elimination of urea and other excretory products from the body
(b) production of erythrocytes from the bone marrow
(c) pulmonary and systemic circulation
(d) cardiac cycle and its regulation
(e) enzymatic reactions in coagulations of blood

29 Which option is correct for the formation of intrinsic factor-X activator complex for blood coagulation? **[Gujarat CET 2014]**
(a) Activated Christmas factor + AHG + Phospholipid + Ca^{2+}
(b) Inactivated Christmas factor + AHG + Phospholipid + Ca^{2+}
(c) Convertin + AHG + Ca^{2+} + FSF
(d) Phospholipid – Protein complex + Proconvertin

30 Identify the correct statement regarding cardiac activity. **[Kerala CEE 2014]**
(a) Normal activities of the human heart are regulated intrinsically, hence it is neurogenic
(b) A special neural centre in the medulla oblongata can moderate the cardiac function through the CNS
(c) Parasympathetic neural signals increase the rate of heartbeat
(d) Adrenal medullary hormones can increase cardiac output
(e) The end of a T-wave marks the end of diastole

31 The depolarisation of the atria is represented by **[J&K CET 2014]**
(a) P-wave (b) Q-wave
(c) T-wave (d) QRS-complex

32 Which diagram is correct for the circulation of blood through human heart? **[Gujarat CET 2014]**

(a)
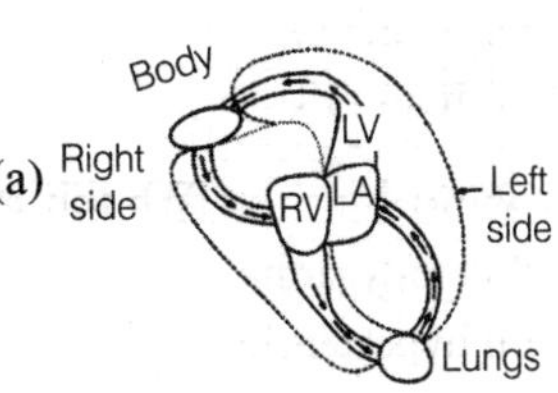

(b)
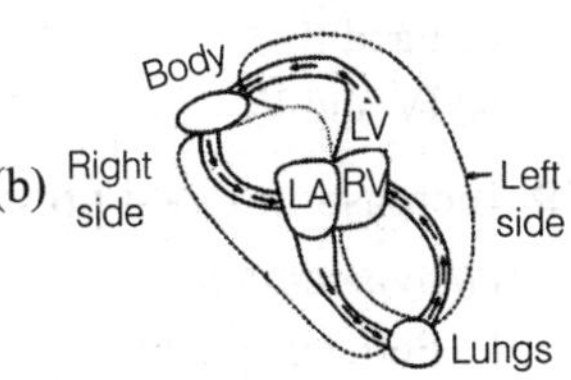

(c)
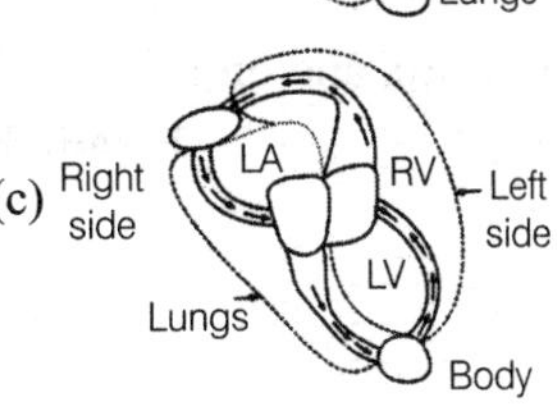

(d)
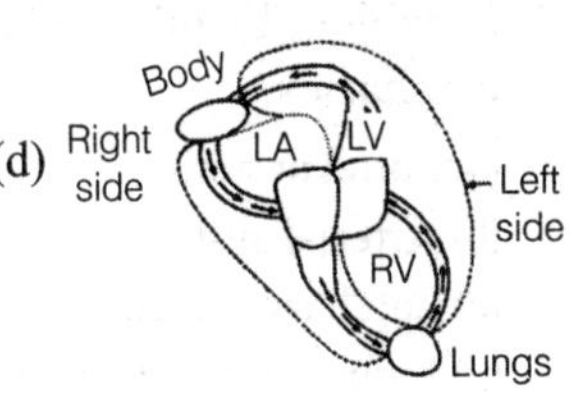

33 Blood leaving the liver and going towards heart is rich in **[UP CPMT 2013]**

(a) bile (b) urea
(c) ammonia (d) oxygen

34 The diagram given here is the standard ECG of a normal person. The P-wave represents the **[NEET 2013]**

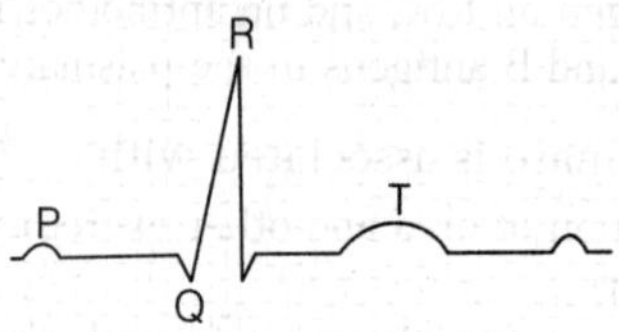

(a) contraction of both the atria
(b) initiation of the ventricular contraction
(c) beginning of the systole
(d) end of systole

35 If the systolic pressure is 120 mm Hg and diastolic pressure is 80 mm Hg, the pulse pressure is **[KCET 2013]**

(a) $120 - 80 = 40$ mm Hg (b) $120 \div 80 = 1.5$ mm Hg
(c) $120 \times 80 = 9600$ mm Hg (d) $120 + 80 = 220$ mm Hg

36 Which of the following matches correctly? **[AIIMS 2013]**

(a) Inferior vena cava – Receives deoxygenated blood from the head and body
(b) Superior vena cava – Receives deoxygenated blood from the lower body and organs
(c) Pulmonary artery – Carries deoxygenated blood to the lungs
(d) Hepatic artery – Carries deoxygenated blood to the gut

37. Arteries that carry deoxygenated blood are **[UP CPMT 2013]**

(a) pulmonary arteries (b) carotid arteries
(c) coronary arteries (d) phrenic arteries

38 The following are the branches of dorsal aorta. **[UP CPMT 2013]**

I. Intercostal II. Phrenic
III. Coeliac IV. Anterior mesenteric
V. Posterior mesenteric

(a) I and II (b) III and IV
(c) IV and V (d) II and III

39 Increase in blood pressure leads to **[WB JEE 2013]**

(a) hypotension (b) bradycardia
(c) Both (a) and (b) (d) hypertension

40 In which, blood circulation starts and ends in capillaries? **[AFMC 2013]**

(a) Portal system
(b) Arterial system
(c) Capillary system
(d) Lymphatic system

41 Which one of the following human organs is often called the 'graveyard' of RBCs? **[CBSE AIPMT 2012]**

(a) Gall bladder (b) Kidney
(c) Spleen (d) Liver

42 According to Best and Taylor's theory, which of the following does not play any role in blood clotting? **[KCET 2012]**

(a) Calcium ions (b) Prothrombin
(c) Fibrinogen (d) Platelets

43 The volume of blood that enters into the aorta with each ventricular systole is called **[KCET 2012]**

(a) vital capacity (b) cardiac cycle
(c) stroke volume (d) cardiac output

44 The frequency of heartbeat in our body is maintained by **[KCET 2012]**

(a) chordae tendineae (b) AV node
(c) SA node (d) node of Ranvier

45 After a certain road accident patient with unknown blood group needed immediate blood transfusion. One of his doctor friend offered his blood at once. What was the blood group of the donor? **[CBSE AIPMT 2012]**

(a) Blood group B
(b) Blood group AB
(c) Blood group O
(d) Blood group A

46 The function of a vein is **[AIIMS 2012]**

(a) conduction of food
(b) conduction of water and minerals
(c) conduction of hormones
(d) to collect blood from different body parts

47 Which of the following factors has little effect on blood flow in arteries? **[AIIMS 2012]**

(a) Heartbeat
(b) Blood pressure
(c) Skeletal muscle contraction
(d) Total cross-sectional area of vessels

48 The blood vessel which supplies oxygenated blood to cardiac tissue is **[WB JEE 2012]**

(a) coronary artery (b) coronary vein
(c) coronary sinus (d) pulmonary vein

49 Compared to blood, our lymph has **[AMU 2012]**

(a) no plasma (b) plasma without proteins
(c) more WBC and no RBC (d) more RBC and less WBC

50 Which one of the following plasma proteins is involved in the coagulation of blood? **[CBSE AIPMT 2011]**

(a) Serum amylase (b) A globulin
(c) Fibrinogen (d) An albumin

51 Which one of the following human cells does not contain mitochondria? **[WB JEE 2011]**

(a) Nerve cell (b) Red blood cell
(c) Liver cell (d) White blood cell

52 Serum is **[RPMT 2011]**

(a) blood without fibrinogen
(b) lymph without corpuscles
(c) blood without corpuscles and fibrinogen
(d) lymph

53 'Bundle of His' is a part of which one of the following organs in humans? **[CBSE AIPMT 2011]**

(a) Heart (b) Kidney (c) Pancreas (d) Brain

54 Which one of the following statements is correct regarding blood pressure? **[CBSE AIPMT 2011]**

(a) 100/55 mm Hg is considered an ideal blood pressure
(b) 105/50 mm Hg makes one very active
(c) 190/110 mm Hg may harm vital organs like brain and kidney
(d) 130/90 mm Hg is considered high and requires treatment

55 Pacemaker is **[MHT CET 2011]**

(a) instrument for measuring heartbeat
(b) instrument for measuring pulse rate
(c) AV node that provides impulse for heartbeat
(d) sino-auricular node that provides impulse for heartbeat

56 Which of the following is the correct pathway for propagation of cardiac impulse? **[WB JEE 2011]**

(a) SA node → AV node → Bundle of His → Purkinje fibres
(b) AV node → Bundle of His → SA node → Purkinje fibres
(c) SA node → Purkinje fibres → AV node → Bundle of His
(d) Purkinje fibres → AV node → SA node → Bundle of His

57 In a healthy adult man, the normal diastolic pressure is **[RPMT 2011]**

(a) 90 mm Hg (b) 120 mm Hg
(c) 80 mm Hg (d) 100 mm Hg

58 Arteries are best defined as the vessels which **[CBSE AIPMT 2011]**

(a) carry blood away from the heart to different organs
(b) breakup into capillaries which reunite to form a vein
(c) carry blood from one visceral organ to another visceral organ
(d) supply oxygenated blood to the different organs

59 Which blood vessels carry blood from different parts of your body to the heart? **[MHT CET 2011]**

(a) Capillaries (b) Arteries
(c) Veins (d) All of these

60 The important function of lymph is to **[MHT CET 2011]**

(a) transport oxygen to the brain
(b) transport carbon dioxide to the lungs
(c) return RBCs to the lymph nodes
(d) return interstitial fluid to the blood

61 Congestion of these lungs is one of the main symptoms in **[Kerala CEE 2011]**

(a) hypotension (b) coronary artery disease
(c) angina (d) heart failure
(e) atherosclerosis

ANSWERS

Check Point 25.1

1. (d)	2. (d)	3. (d)	4. (a)	5. (c)	6. (a)

Check Point 25.2

1. (a)	2. (a)	3. (c)	4. (d)	5. (d)	6. (c)	7. (b)	8. (b)

Check Point 25.3

1. (c)	2. (b)	3. (a)	4. (a)	5. (c)	6. (c)

(A) Topical Exercise

1. (d)	2. (c)	3. (a)	4. (b)	5. (a)	6. (c)	7. (d)	8. (b)	9. (c)	10. (a)
11. (b)	12. (a)	13. (c)	14. (d)	15. (a)	16. (a)	17. (d)	18. (a)	19. (b)	20. (d)
21. (a)	22. (a)	23. (a)	24. (a)	25. (a)	26. (d)	27. (a)	28. (a)	29. (a)	30. (a)
31. (d)	32. (b)	33. (b)	34. (d)	35. (b)	36. (a)	37. (b)	38. (a)	39. (a)	40. (a)
41. (b)	42. (b)	43. (b)	44. (d)	45. (a)	46. (d)	47. (c)	48. (c)	49. (c)	50. (d)
51. (d)	52. (b)	53. (c)	54. (d)	55. (a)	56. (b)	57. (a)	58. (b)	59. (b)	60. (d)
61. (c)	62. (d)	63. (b)	64. (a)	65. (c)	66. (d)	67. (b)	68. (b)	69. (a)	70. (c)
71. (a)	72. (c)	73. (a)	74. (c)	75. (b)	76. (b)	77. (c)	78. (c)	79. (b)	80. (b)
81. (d)	82. (d)	83. (d)	84. (c)	85. (d)	86. (d)	87. (c)	88. (d)	89. (c)	90. (a)
91. (b)									

(B) Cumulative Exercise

1. (d)	2. (b)	3. (a)	4. (b)	5. (b)	6. (c)	7. (c)	8. (a)	9. (b)	10. (d)
11. (d)	12. (a)	13. (d)	14. (a)	15. (c)	16. (c)	17. (b)	18. (b)	19. (a)	20. (a)
21. (d)	22. (b)	23. (d)	24. (c)	25. (b)	26. (b)	27. (d)	28. (a)	29. (d)	30. (b)
31. (c)	32. (b)	33. (b)	34. (d)	35. (a)	36. (a)	37. (a)			

(C) Medical Entrances' Gallery

1. (b)	2. (b)	3. (b)	4. (b)	5. (d)	6. (d)	7. (d)	8. (c)	9. (c)	10. (d)
11. (a)	12. (d)	13. (d)	14. (d)	15. (b)	16. (b)	17. (c)	18. (c)	19. (d)	20. (a)
21. (a)	22. (b)	23. (a)	24. (d)	25. (c)	26. (b)	27. (a)	28. (e)	29. (a)	30. (d)
31. (a)	32. (a)	33. (b)	34. (a)	35. (a)	36. (c)	37. (a)	38. (b)	39. (d)	40. (a)
41. (c)	42. (d)	43. (c)	44. (c)	45. (c)	46. (d)	47. (c)	48. (a)	49. (c)	50. (c)
51. (b)	52. (c)	53. (a)	54. (c)	55. (d)	56. (a)	57. (c)	58. (a)	59. (c)	60. (d)
61. (d)									

Explanations

(A) Topical Exercise

4 (*b*) Statement I is false, but statement II is true because
The recepient's serum should not contain the antibodies against the RBCs of the donor. Otherwise, the immune response is elicted.

24 (*a*) Statement I is true, but II is false.
Double circulation is the passage of the blood twice in the heart through the separate pathways, so as to complete one cardiac cycle. It consists of two pathways
(i) Pulmonary pathway (ii) Systemic pathway
Hence, it prevents the mixing of oxygenated and deoxygenated blood.
Second statement is false, its correct form would be
It carries both oxygenated and deoxygenated blood.

25 (*a*) Statement in option (a) is incorrect and can be corrected as
Heart is mesodermal in origin not endodermal.
Rest options are correct.

26 (*d*) Both statements I and II are incorrect and can be corrected as
Normal activities of human heart are regulated intrinsically, i.e. autoregulated by specialised muscle (nodal tissue). Hence, the heart is called myogenic.

28 (*a*) Option (a) represents incorrectly matched pair and can be corrected as
Epicardium is composed of simple squamous epithelium.

29 (*a*) Human heart is four-chambered, with a thin, muscular wall called the interatrial septum separating the right and the left atria. There is also present a thick-wall, i.e. the interventricular septum, which separates the left and the right ventricles.

41 (*b*) Statement in option (b) is incorrect and can be corrected as
SA node is also called node of Keith and Flack.

47 (*c*) The bundle of His passes stimulus of contraction through Purkinje fibres. This happens as the impulse from the SA node causes the atria to contract, pushing the blood through the open valves into the ventricles.
The electrical signal arrives at the AV node which is located between the two atria. From here, it travels through the bundle of His, divides into the left and right bundle branches and through the Purkinje fibres.

65 (*c*) Neural signals through the sympathetic nerves can increase the rate of heartbeat, the strength of ventricular contraction and thereby, the cardiac output. On the other hand, parasympathetic neural signals decreases the rate of heartbeat.

70 (*c*) ECG records the conduction of the action potential through the heart. The ECG is unable to detect any changes in the ability of heart to develop force and thus, ECG is least effective in detecting abnormalities in cardiac contractility.

72 (*c*) Arteries carry oxygenated blood far from the heart to the different body parts. Blood emerges from the heart under high pressure, the blood flow with jerks in them. In order to withstand their pressure, arteries have thick walls.

77 (*c*) Statement in option (c) is incorrect and can be corrected as
Jugular vein collect deoxygenated blood from the brain.

88 (*d*) Coronary artery disease, often referred to as atherosclerosis, affects the vessels that supply blood to the heart muscle. It is caused by deposition of calcium, fat, cholesterol and fibrous tissues, which make the lumen of arteries narrower.

91 (*b*) Angina, also called angina pectoris is a symptom showing acute chest pain. It occurs due to decreased supply of oxygen to the heart muscle. Angina can occur in men and women of any age, but it is more common among the middle-aged and elderly.

(B) Cumulative Exercise

1 (*d*) Statement in option (d) is incorrect and can be corrected as
The pulmonary veins carry oxygenated (not deoxygenated) blood from the lungs into the left atrium, where it is returned to systemic circulation. Rest statements are correct.

3 (*a*) Only option (a) is correctly matched pair, options (b), (c) and (d) are incorrect and can be corrected as

- At the end of ventricular systole, semilunar valves close, producing the second heart sound the 'dup'.
- Pulsation of the radial artery is due to rythmic contraction and relaxation in the aorta.
- Initiation of heartbeat occurs in SA node.

7 (*c*) The hepatic portal vein (*D*) carry blood from GIT, pancreas and spleen to liver. Hepatic vein (*F*) drain deoxygenated blood from liver into inferior vena cava.

8 (*a*) All the statements are correct except I. Incorrect statement can be corrected as
Lymph a colourless fluid containing specialised lymphocytes which are responsible for the immune responses of the body.

13 (*d*) A portal system is a system in which a vein breaks up in an organ into capillaries and restarts by their union as a new vein in the same organ. Such system is helpful in transporting products from one region to another in higher concentration.

14 (*a*) Statements I and II are correct whereas III and IV are incorrect. Incorrect statements can be corrected as
The circulating fluid is the blood which is contained inside blood vessels and it circulate unidirectionally.
The body cavity, haemocoel is not found.

15 (*c*) Option (c) is incorrect match and can be corrected as
Ampullary heart is present in insects not in Aves.

17 (*b*) Only option (b) is correctly matched pair, options (a), (c) and (d) are incorrect and can be corrected as

- Malleus and anurl are two different bones present in the middle ear which helps in amplification of sound.
- Leucocytes are also known as WBC. It helps in the defensive mechanism of the body. Lymphocytes are the type of agranular leucocytes which help in the production of antibodies.
- Haemophilia is an inherited bleeding disorder where the blood does not clot properly. It is caused when blood does not have enough clotting factor. Blood cancer is caused due to the abnormal multiplication of blood cells.

22 (*b*) Statements I and III are correct whereas II and IV are incorrect. These can be corrected as

- Right auricle opens into the right ventricle and the auriculoventricular aperture is guarded by a tricuspid valve.
- Opening of left auricle into left ventricle is guarded by a bicuspid valve.

23 (*d*) Statements II and IV are correct whereas I and III are incorrect. These can be corrected as

- Bundle of His originates from the AV node
- The auricular contraction is stimulated by the nerve impulse generated by SA node.

28 (*a*) Statements I and IV are correct whereas statements II and III are incorrect. Incorrect statements can be corrected as

- Angina is acute chest pain caused by reduced blood flow to the heart.
- Persons with blood group AB can donate blood to only AB. It can receive blood from all blood types.

34 (*d*) A is false, but R is true. A can be corrected as
Pulmonary circulation includes flow of deoxygenated blood to the lungs and the return of oxygenated blood from the lungs to the left auricle of heart. The main purpose of this circulation is to purify deoxygenated blood to oxygenated blood.

35 (*a*) Both A and R are true and R is the correct explanation of A.
Systemic circulation means the flow of blood from the left ventricle of heart to all parts of body and return of blood to right auricle of heart.
It includes all the oxygenated blood that leaves the left ventricle through the aorta and the deoxygenated blood that returns to the right auricle through the superior vena cava and inferior vena cava. The main purpose of this circulation is to carry oxygen and nutrients to body tissues and to remove CO_2 and other wastes from the tissues.

36 (*a*) Both A and R are true and R is the correct explanation of A.
ECG consitsts of P-wave, QRS wave and T-wave. ECG gives important information concerning the spread of excitation to the different parts of heart and in the diagnosis of abnormal cardiac rhythm and myocardial damage.

37 (*a*) Both A and R are true and R is the correct explanation of A.
The heart pacemaker is an electronic device, which regularly sends small amount of electrical charge for maintaining the rhythmicity of the heart. This device is implanted subcutaneously in the upper thoracic region, having a connection with the heart. This device is used when the heart rate of 72-80 drops significantly due to disease or some other condition.

(C) Medical Entrances' Gallery

1 (*b*) Parasympathetic neural signals (a component of autonomic nervous system) decreases the rate of heartbeat and increases the conduction of action potential and thereby, the cardiac output.

2. (*b*) If mother is Rh –ve and foetus is Rh + ve then there can occur a condition called erythroblastosis foetalis. It is a special case of Rh incompatibility in which during the second pregnancy of Rh – ve mother carrying Rh + ve foetus, the Rh antibodies from mother (Rh – ve) can leak into the blood of the foetus (Rh + ve) and destroy the foetal RBCs.

3 (*b*) Option (b) is correct and can be explained as
In ECG, there occurs five consecutive waves P, Q, R, S and T.
P wave represents depolarisation of atria and leads to contraction of both atria. QRS complex represents depolarisation of ventricles which leads to initiation of ventricular contraction. T-wave represents return of ventricles from excited to normal state.

4 (*b*) As per the given data, the heart rate of the person would be 100 beats per minute. It can be calculated as follows

Given, Cardiac output = 5L (5000 mL)

Blood volume in ventricles at the end of diastole = 100 mL

Blood volume at the end of ventricular systole = 50 mL

So, Stroke volume = 100 − 50 = 50 mL

Cardiac output = Stroke volume × Heart rate, i.e.

5000 mL = 50 mL × Heart rate

Therefore, Heart rate = 100 beats/min.

5 (*d*) The nodal musculature has the ability to generate action potentials without any external stimuli, i.e. it is autoexcitable. However, the number of action potentials that could be generated in a minute vary at different parts of the nodal system. The SAN (Sino-Atrial Node) can generate the maximum number of action potentials, i.e. 70-75 min, i.e. the highest rate of depolarisation and is responsible for initiating and maintaining the rhythmic contractile activity of the heart. Therefore, it is called pacemaker.

6 (*d*) Atrio-Ventricular (AV) node present in the lower corner of the right atrium, delays the spreading of impulses to heart ventricles for about 0.1 second. This pause allows the atria to empty completely into the ventricles before the ventricles pump out the blood.

7 (*d*) Chordae tendineae are found in heart muscles of mammals. They are group of tough, tendinous strands found in the heart of mammals. They are also called as 'heart strings' due to their resemblance with small pieces of string. They play an important role in holding the atrioventricular valves in place while heart is pumping blood.

9 (*c*) Both intrinsic and extrinsic pathways of blood clotting are interlinked at the activation steps of factor X.

10 (*d*) The atrioventricular (AV) valve closes due to sharp rise of intraventricular pressure. The closing of this valve produces first heart sound that is termed as 'lubb'.

11 (*a*) Foramen ovale is an opening in the interatrial septum of the foetal heart through which both the atria communicate with each other.

12 (*d*) In the hepatic portal system, the hepatic vein takes blood from intestine to the liver. In this way, it takes all the nutrients absorbed from intestine to the liver where screening and storing of nutrition takes place.

13 (*d*) Both A and R are false and these can be corrected as Arteries have a thick wall containing many smooth muscles, elastin fibres and a narrow lumen. Arteries distribute blood away from the heart to all parts of the body at a high pressure.

14 (*d*) A reduction in the number of thrombocytes or blood platelets can lead to clotting disorders which will result in excessive loss of blood from the body.

15 (*b*) Urea is synthesised in liver. So, maximum amount of urea is present in hepatic vein. In renal vein, minimum urea will be present, because it carry blood filtered by the kidneys.

16 (*b*) Blood pressure in the pulmonary artery is more than that in the pulmonary vein because arteries have higher blood pressure than vein. Also the lumen of artery is narrow.

Blood pressure in different blood vessels is represented below

Artery > Arteriole > Capillary > Venule > Vein (vena cava)

17 (*c*) Blood pressure in the mammalian aorta is maximum during systole of the left ventricle because the heart shrinks and pushes blood into the aorta. It drives blood through the aortic valve to the body and organs except lungs. As the pressure in the aorta travels down into the distributing arteries, the characteristic changes in diastolic and systolic pressures are observed. There is a very small decline in arterial pressure due to the low resistance of large distributing arteries.

18 (*c*) At rest, the largest volume of blood is in the veins because the walls of the veins are thinner than the walls of the arteries. The thinner walls of veins can expand more than the thicker walls of arteries. Thus, it can hold a larger volume of blood than arteries.

19 (*d*) In normal ECG, the component that signifies duration of one cardiac cycle is R-R interval. Q-T interval denotes duration of electric systole (ventricles). P-R interval denotes the time during which SA node transfers impulse to Purkinje fibres. S-T segment denotes ventricular depolarisation. Thus, correct option would be (d) R-R interval.

20 (*a*) The study of blood vessels is termed as angiology. Cardiology is the study and treatment of disorders of the heart and the blood vessels. Haematology studies the morphology of blood and blood forming tissues.

Histology is the branch of biology which studies the microscopic anatomy of biological tissues.

21 (*a*) Angina pectoris is a type of severe chest pain with heaviness, which is caused by reduced blood flow to the heart and with rest, it may disappear.

22 (*b*) Number of heartbeats per minute is decreased if the stroke volume of heart increases, with total volume of increase in heart, where total volume of blood remains same.

23 (*a*) The T-wave in one ECG represents ventricular repolarisation, i.e. return of the ventricles from excited state.

24 (*d*) The function of vertebrate circulatory system is to bring oxygen to tissues and to carry waste products, i.e. carbon dioxide.

They also have many functions such as it provides nutrients to different organ of the body. The blood and the lymph contains antibodies and immune cells which protect the body from infection. Removal of waste products for excretion from the body is performed by the kidney.

25 (*c*) The site of exchange of wastes, nutrients, gases and hormones between the blood and body occurs in the capillaries. They are the tiny vessels that branch out from arterioles to form networks around body cells.

26 (*b*) AV node is known as pacesetter of the the heart, it regulates the rhythm of the heart rate. The delay in conduction of impulse in this node changes the duration of contraction between the atria and the ventricles. The cardiac impulse is initiated in the sinoatrial node or natural packmaker.

27 (*a*) Individuals with AB blood group have both antigen A and antigen B on their RBCs, but no antibodies for either of the antigens in their plasma. Thus, they can receive blood of individuals with any blood group, i.e. A, B, AB or O and thus, called universal recipients.

28 (*e*) Thrombokinase is also known as factor-X or prothrombokinase or thromboplastin. It is an enzyme, which catalyses the breakdown of prothrombin into an active protein called thrombin associated with enzymatic reactions in the coagulation of blood.

29 (*a*) The formation of intrinsic factor-X activator complex for blood coagulation involves factors such as, Activated Christmas factor + AHG (Antihaemophilic Globulin) (Factor-VIII) + Phospholipid + Ca^{2+}. Blood coagulation is the complex mechanism of the formation of a clot at the site of trauma or injury. The clotting cascade occurs through two separate pathways, i.e. intrinsic and extrinsic pathways. The intrinsic pathway is slower than the extrinsic pathway.

30 (*d*) Out of all the given statements, the only correct statement is that adrenal medullary hormones can increase the cardiac output. Rest all other statements are inappropriate and can be corrected as

- Normal activities of the human heart are regulated intrinsically, i.e. autoregulated by specialised muscles (nodal tissue). Hence, it is called myogenic.
- A special neural centre in the medulla oblongata can moderate the cardiac function through the Autonomous Nervous System (ANS).
- Parasympathetic neural signals decrease the rate of heartbeat.
- The end of T-wave marks the end of ventricular systole.

31 (*a*) The P-wave is a small upward wave that represents electrical excitation or the atrial depolarisation or atrial contraction in the ECG. It is caused by the activity of SA node.

32 (*a*) Out of the diagrams given in the question, option (a) is showing the correct circulation of blood through human heart. In humans, blood passes twice through the heart to complete one cycle of blood supply to the body.

The impure blood from right side of the body passes to right atrium (RA), through vena cava and is further received by right ventricle (RV) which pumps it to pulmonary artery taking it to the lungs for purification. Pure blood is then carried through pulmonary veins into the Left Atrium (LA) from where it enters Left Ventricle (LV) which in turn distributes this blood to the whole body and organs.

33 (*b*) Blood leaving the liver and going towards the heart would be rich in urea. This is because liver is the site of the process of deamination, i.e. removal of nitrogen from the molecule.

34 (*a*) In ECG, P-wave represents depolarisation of atria, which leads to the contraction of both atria. T-wave represents the return of ventricles from excited to normal state. The QRS complex represents the depolarisation of the ventricles, which initiates ventricular contraction and the contraction starts shortly after Q-wave and marks the beginning of systole.

35 (*a*) Pulse pressure is the difference between systolic and diastolic blood pressure or the change in blood pressure seen during a contraction of the heart. Thus,

$$\text{Pulse pressure } (z) = \text{Systolic pressure} - \text{Diastolic pressure}$$

$$\therefore \quad 120 \text{ mm Hg} - 80 \text{ mm Hg} = 40 \text{ mm Hg}.$$

36 (*c*) Out of all the given pairs, the correctly matched pair is (c) stating that

Pulmonary artery carries deoxygenated blood to the lungs.

Rest of the options are incorrect matches and can be corrected as

- Inferior vena cava receives deoxygenated blood from the lower body and organs.
- Superior vena cava receives deoxygenated blood from the head and body.
- Hepatic artery carries oxygenated blood to the liver.

37 (*a*) Pulmonary arteries carry deoxygenated blood from the heart towards the lungs, for its oxygenation. Carotid arteries supply blood to the brain, neck and face. Coronary arteries supply blood to the heart. Phrenic arteries supply blood to the diaphragm.

38 (*b*) Dorsal aorta comprise of a set of arteries which supply blood to glands of digestive system. These are

(i) Coeliac artery-Unpaired, divides into three branches
  (a) Left gastric artery-To stomach.
  (b) Common hepatic artery-To pylorus, pancreas, gall bladder, liver, cystic duct, hepatic ducts, etc.
  (c) Splenic artery-To pancreas, stomach and spleen.

(ii) Anterior or Superior mesenteric-Unpaired, supplies blood to various parts of small intestine (except superior part of duodenum, colon and caecum). Its subbranches are
  (a) Pancreo (duodenal) artery-To pancreas and duodenum.
  (b) Jejunal artery-To jejunum.
  (c) Ilial artery-To ileum and jejunum.
  (d) Ileocrelic artery-To ileum and colon.

39 (*d*) Increase in blood pressure leads to hypertension, also called high blood pressure. It is dangerous because it makes the heart work harder to pump blood to the body.

40 (*a*) Portal system is a part of venous circulation, which is present between two groups of capillaries, i.e. it starts in capillaries and ends in capillaries.

41 (*c*) Spleen is often referred to as graveyard of RBCs. RBCs are formed in the red bone marrow in the adults. They have an average lifespan of 120 days after which they are destroyed in the spleen.

42 (*d*) Best and Taylor's theory explains that four factors are involved in blood clotting, i.e. prothrombin, thromboplastin, calcium ions and fibrinogen, but not platelets.

At the site of injury, the damaged tissue releases thromboplastin which in the presence of calcium ions converts inactive prothrombin into active thrombin.

The thrombin converts soluble fibrinogen into insoluble fibrin threads that overlaps each other to form a network trapping the blood cells forming blood clot.

43 (*c*) During a cardiac cycle, each ventricle pumps out approximately 70 mL of blood which is called the stroke volume. Thus, the volume of blood that enters into the aorta with each ventricular systole is called stroke volume.

44 (*c*) The Sino-Atrial (SA node) node is known as the pacemaker of heart because each wave of excitation begins here and it acts as the stimulus for the next wave of excitation. In this way, the frequency of heartbeat in our body is maintained by SA node.

45 (*c*) Donor's blood group is 'O'. Individuals with blood group O possess both antibodies in their plasma, but have no antigens in their RBCs. Therefore, they can donate blood to different individuals having different blood groups and hence called universal donors.

46 (*d*) The function of a vein is to collect blood from different parts of the body and carry it back to the heart. All veins carry deoxygenated blood back to the heart except pulmonary vein and umbilical vein.

47 (*c*) Out of the given options, skeletal muscle contraction had the least effect on blood flow in arteries. Arteries carry blood from the heart to different parts of thr body. The beating of heart depends on activity of the cardiac muscles (not skeletal muscles). It enhances heartbeat and in turn, increases blood pressure and blood flow which is regulated by the cross-sectional area of blood vessels.

48 (*a*) Coronary arteries, i.e. both left and right are the blood vessels which supply the oxygen rich blood to heart muscles (cardiac muscles).

49 (*c*) Lymph has more WBC and no RBC compared to blood. It is a clear, colourless fluid composed of lymph plasma and lymph corpuscles. It has fewer blood protein, less calcium and phosphorus concentration and high glucose concentration.

50 (*c*) Fibrinogen (factor-I) is a soluble plasma glycoprotein, synthesised by the liver. It is converted by thrombin into fibrin during blood coagulation. Fibrin is then cross-linked by factor-XIII to form a clot.

51 (*b*) Red blood cells of mammals are round, biconcave and without nucleus, mitochondria, Golgi body, ER, centrosomes, etc. These cell organelles are lost during development of the cells (reticulocyte stage).

52 (c) Serum is blood without fibrinogen and blood corpuscles.

- Blood without RBCs is lymph.
- Blood without all cells is plasma.
- Plasma without clotting factors is serum.

53 (*a*) The bundle of His, known as AV bundle (Atrio-Ventricular bundle) present in the heart are specialised for electrical conduction. These are muscle fibres of the heart named after the Swiss cardiologist Wilhelm His Jr, who discovered them in 1893.

54 (*c*) Only option (c) is correct regarding blood pressure. High blood pressure (190/110 mm Hg) may lead to heart diseases and also can affect vital organs like brain and kidney. Hypertension is the term referring to blood pressure that is higher than normal (120/80). In this measurement, 120 mm Hg (millimeter of mercury pressure) is the systolic or pumping pressure and 80 mm Hg is the diastolic or resting pressure.

55 (*d*) Pacemaker is Sino-Auricular (SA node) node that provides impulse for heartbeat. The SA node consists of a cluster of cells that are situated in the upper part of the wall of the right atrium. It initiates atrial contraction and spreads the contraction waves across the atrial walls through muscle fibres at regular intervals.

56 (*a*) When the cardiac muscles are stimulated by a specific stimulus, these get excited and initiate the waves (depolarisation) of electric potential called cardiac impulse. The SA node initiates atrial contraction.

This contraction wave is picked up by the AV node. The cardiac impulse is transferred to the ventricles through the Bundle of His and Purkinje fibres. Cardiac impulse is propagated through the following pathway

SA node → AV node → Bundle of His → Purkinje fibres.

57 (*c*) Blood pressure means the arterial blood pressure. Normal systolic BP in healthy adult man is 120 mm Hg, while the diastolic blood pressure is 80 mm Hg.

58 (*a*) Arteries are blood vessels that carry blood away from the heart and towards different organs. They generally contain oxygenated blood (except pulmonary artery). The blood flows in an artery under alternate increased pressure and with jerks.

59 (*c*) Veins carry deoxygenated blood from different parts of the body back to the heart. These have thin walls and valves to prevent the back flow of blood.

60 (*d*) The important function of lymph is to return interstitial fluid to the blood. The lymph acts as a middle man between the blood and the tissue cells as it passes on food and oxygen from blood to tissue cells and hands over excretory wastes, hormones and CO_2 from the body cells to blood.

61 (*d*) Heart failure refers to the state of the heart when it is not pumping blood effectively enough to meet the needs of the body. It is sometimes called congestive heart failure as congestion of the lungs is one of its main symptoms.

CHAPTER 26

Excretory Products and Their Elimination

Animals produce different types of nitrogenous wastes as a result of various cellular activities taking place in their body. These waste materials, if allowed to accumulate in the body, may become toxic. Therefore, these must be removed from the body.

The process of removal of these metabolic waste materials from the body is called **excretion**. Another important process taking place in the animal body is **osmoregulation**. It is the process of maintaining osmotic and ionic concentrations of body fluids.

Excretion and osmoregulation together maintain a steady state or homeostasis in the body. It is an essential requirement for the normal life processes to take place.

In this chapter, we will discuss about the basic concept of osmoregulation, along with detailed study of type of excretory products and human excretory system.

Inside

CONCEPT OF OSMOREGULATION IN ANIMALS

In animals, osmoregulation is brought about by osmoreceptors which can detect changes in osmotic pressure. On the basis of osmoregulation, animals are either osmoconformers or osmoregulators.

(i) **Osmoconformers** These animals cannot actively control the osmotic concentration of their body fluids. Instead, they change the osmolarity of body fluids according to that of the surrounding medium. Thus, they can tolerate a wide range of osmotic environments.

For example, all marine invertebrates, some freshwater invertebrates and hagfish (vertebrate).

(ii) **Osmoregulators** These animals maintain an internal osmolarity different from their surrounding medium in which they inhabit in following manner

- In **hypotonic medium**, they eliminate excess water from their body.
- In **hypertonic medium**, they continuously take in water to compensate for water loss.

Due to this, the osmoregulators have to spend energy so to move water in or out of body. For example, most vertebrates except hagfish.

TYPES OF EXCRETORY PRODUCTS

Animals excrete different types of waste products depending upon their habitat, feeding habits and metabolism.

These are as follows

(i) **Ammonia** It is the first metabolic waste of protein metabolism.
- Ammonia is produced in liver by the process of deamination.
- It is a highly toxic excretory product and it is eliminated in the form of NH_4^+ ion. It requires more water for its dissolution.
- The process of excreting ammonia is called **ammonotelism**.
- The animals performing this process are called **ammonotelic animals.**
- These animals are the normal inhabitant of freshwater and have less significant kidneys, e.g. aquatic amphibians, many bony fishes, aquatic insects, etc.

(ii) **Urea** It is produced in liver from ammonia and CO_2. It is less toxic and less soluble than ammonia.
- More urea is excreted from body when we take more proteins in diet.
- Normal blood urea level in human body is 18-38mg/100mL of blood.
- The process of excreting urea is called **ureotelism.**
- The animals performing this process are called **ureotelic animals.**
- These animals inhabit both terrestrial and aquatic area, e.g. elasmobranch fishes (shark and their relatives), adult amphibians and mammals.

(iii) **Uric acid** It is least toxic and almost insoluble in water. It is also produced in liver.
- It is excreted in semi-solid form and is helpful in conservation of water in body.
- The process of excreting uric acid is called **uricotelism**.
- The animals performing this process are called **uricotelic animals.**
- These animals inhabit terrestrial areas, e.g. land reptiles, land gastropods, most insects, birds, etc.

(iv) **Trimethylamine oxide** Few animals instead of excreting ammonia, convert it into Trimethylamine (TMA) and excrete the oxidised form of TMA, i.e. Trimethylamine Oxide (TMO).

These animals are marine molluscs, crustaceans, fishes (lung fishes, teleost), etc.

(v) **Guanine and xanthines** These are the byproducts of nucleotide metabolism. These are excreted in semi-solid form in spiders and penguins.

(vi) **Creatine** It is produced from creatine phosphates in muscle cells. Creatine excretion can be seen in foetus, pregnant and lactating women.

(vii) **Creatinine** It is produced as a result of creatine metabolism in muscles. An increased level of creatinine is indicative of kidney damage.

(viii) **Hippuric acid** It appears as excretory product only when benzoic acid is present in diet. It is present in traces in human urine. In birds, it combines with ornithine and changes into ornithuric acid for excretion.

(ix) **Allantoin** This got its name from the fact that it is excreted through the extraembryonic membrane allantois in birds and reptiles. Allantoin is formed by the oxidation of uric acid.

(x) **Amino acid** Certain animals like *Unio*, *Limnaea* and *Asterias* excrete amino acids. These are called **aminotelic animals** and the process is called **aminotelism.**

> **Dual excretion**
>
> Certain animals exhibit dual excretion, i.e. they excrete two types of excretory products. Animals like earthworm, snail, lung fishes, *Xenopus*, *Proteus*, crocodile, etc., show ammonotelic as well as ureotelic excretion depending upon the water content around them.

EXCRETORY ORGANS IN DIFFERENT ANIMALS

Excretory organs in different animals are listed below

Table 26.1 Excretory organs in different animal groups

Animal groups	Excretory organs/Surfaces
Protozoa	Plasmalemma or Body surface (e.g. *Amoeba*)
Porifera	General body surface (e.g. *Sycon*)
Coelenterata	General body surface (e.g. *Hydra*)
Platyhelminthes	Protonephridia or Flame cells (e.g. *Planaria*)
Aschelminthes	H-shaped renette cells (e.g. *Ascaris*)
Annelida	Nephridia/Metanephridia—Oligochaetes (e.g. earthworm)
	Protonephridia and Metanephridia—Polychaetes (e.g. *Nereis*)

Animal groups	Excretory organs/Surfaces
	Ciliated body—*Hirudinaria* (e.g. leech)
Arthropoda	Antennary or Green glands—Crustaceans (e.g. prawns)
	Coxal glands—Arachnids (e.g. spiders) and *Limulus*
	Malpighian tubules—Insecta (e.g. honeybee), Chilopods (scolopendra or centipede) and Diplopods (e.g. julus or millipede)
Mollusca	Metanephridia—General
	Organ of Bojanus (kidneys), Keber's organ (e.g. *Unio*)
Echinodermata	Dermal branchiae and tubefeet (e.g. starfish)
Hemichordata	Proboscis gland (e.g. *Balanoglossus*)
	Single glomerulus (e.g. Acornworm)
Chordata	Neural glands—Urochordata (e.g. *Herdmania*)
	Pharyngeal and Hatschek's nephridia with solenocytes—Cephalochordata (e.g. Amphioxus)
	Kidneys—Vertebrata (e.g. fishes, amphibians, reptiles, birds and mammals)

Different types of kidneys in vertebrates

Different types of kidneys in vertebrates are as follows

(i) **Archinephric or Holonephric kidney** It extends through the entire length of coelom. This hypothetical kidney is found in the larvae of certain cyclostome, e.g. *Myxine*. It is supposed to give rise to all other kidney types. Glomeruli may be present in posterior tubules and it is without capsule.

(ii) **Pronephric or Head kidney** It appears dorsal to the anterior end of coelom. It is associated with pronephric tubules having ciliated funnels or nephrostome. In pronephric kidney, glomerulus is external and naked.

It is functional only in embryonic and larval stages and soon gets replaced by the next stage, e.g. tadpole of frog and cyclostomes.

(iii) **Mesonephric or Middle kidney** It develops from the middle part of the intermediate mesoderm. It is functional in both larva and adults of most fishes and amphibians. In amniotes, it is functional in embryonic stage and gets replaced by metanephric kidney in adults.

(iv) **Opisthonephric or Tail kidney** The functional nephrons in this kidney belongs to the posterior region of coelom, which are displaced from their position. It is found in sharks and caecilians.

(v) **Metanephric kidney** In this, functional posterior nephrons are displaced from their original position to anteriolateral position. The nephrons of this kidney are highly differentiated, which forms loop of Henle in mammals. It is functional in all amniotes like reptiles, birds and mammals.

EXCRETORY SYSTEM OF HUMAN

The excretory system of human consists of two kidneys, two ureters, one urinary bladder and one urethra. The detailed account of these structures are given below

Kidneys

- Human kidneys are mesodermal in origin.
- Kidneys are situated in the abdominal cavity at the levels of last thoracic and third lumbar vertebra.
- Human kidneys are present outside the coelomic cavity and only their ventral surface is covered by **visceral peritoneum**. This arrangement is called **retroperitoneal.**
- Last two pairs of ribs, i.e. floating ribs protect the kidneys from backside.

External structure

- Human kidneys are bean-shaped, reddish-brown, paired structures.
- An average sized kidney is about 10-12 cm in length, 5-7 cm in width and 3-4 cm in thickness.
- Their weight is about 150 gm in males and about 135 gm in females.
- Usually, the right kidney is smaller than the left one and positioned a bit lower as compared to the left kidney. Such a difference is seen because most of the portion of right side of the body is occupied by liver.
- Towards the centre of the inner concave surface of the kidney, a notch called as **hilus renalis** or **hilum** is present. It is the region through which the blood vessels and nerves pass.
- Hilum leads to a funnel-shaped space called **renal pelvis.** The projection of renal pelvis are called **calyces.**
- Mammalian kidney is surrounded by three coverings from inside to outside. These are as follows
  (i) **Renal capsule** It is the innermost tough covering of fibrous connective tissues with few elastic fibres and few muscles.
  (ii) **Adipose capsule** It is the middle cover that possesses adipose tissue and acts as shock absorber.
  (iii) **Renal fascia** It is the outermost fibrous cover that link kidneys with the abdominal wall.

Internal structure

- The logitudinal section of kidney represents two zones, i.e. an outer cortex and an inner medulla.
  (i) **Renal cortex** Outer dark coloured part, appear granular in appearance.
  (ii) **Ranal medulla** Inner light coloured part, striated in appearance.
- The medulla is further divided into few (8-12 in human) pyramid-shaped masses projecting into the calyces (sing. calyx). These are called as **medullary pyramids.**
- The extended region of cortex in between the medullary or renal pyramids form renal columns called **columns of Bertini.**
- Each medullary pyramid consists of millions of tubular structures called **nephron** arranged in a radiating fashion.
- Each medullary (renal) pyramid have a broad base towards the cortex and a narrow end called **renal papilla** towards the pelvis.
- Each renal papilla opens into a two to three **minor calyces,** which join to form a **major calyx.**
- The major calyx opens into **renal pelvis,** which further leads into ureter.
- The area having apex or papilla with minor calyx is called **area cribrosa.**
- All these structures are shown in the diagram given below

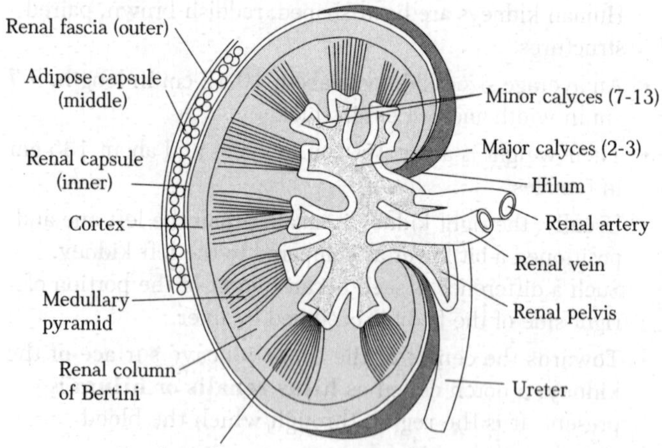

Fig 26.1 Human kidney (Posterior view)

Nephrons

A human kidney contains about one million thin, long, much convoluted tubular units. These units are called **uriniferous tubules** or nephrons. Nephrons are structural and functional unit of kidney.

A nephron consists of Malpighian body and a system of tubules, i.e. uriniferous tubules.

Malpighian body (Renal corpuscle)

- It was first observed and described by **Marcello Malpighi** in 1966.
- It is found in the cortex part of the kidney.
- It is about 200 microns in diameter.
- It performs the function of ultrafiltration of blood during urine formation.
- It further consists of following parts
  (i) **Bowman's** (Glomerular) **capsule** It is a closed, cup-shaped, double-walled structrue containing a network of thin capillaries called **glomerulus.**
  - It consists of two layers
    - **Parietal layer** (outer layer) consists of single squamous epithelium.
    - **Visceral layer** (inner layer) consists of extensively modified cells called **podocytes** or **glomerular epithelial cells.**
  - A podocyte has a number of interdigitated evagination called **pedicels** or **feet.**
  - The pedicels rest over basement membrane. They enclose slit pores or filtration slits.
  - **Capsular space or Lumen** is the space between the parietal and visceral layer of Bowman's capsule.

  (ii) **Glomerulus** It is the capillary tuft present inside the Bowman's capsule.
  - These capillaries are formed when **afferent arteriole** breaks up into about 20-50 fine tubes.
  - The afferent arteriole is short and wide.
  - All the capillaries reunite and form **efferent arteriole,** which passes out from the glomerulus.
  - The efferent arteriole is narrow and long. This arrangement facilitates the process of ultrafiltration in this region.
  - A detailed diagram representing all the above described structures is given below

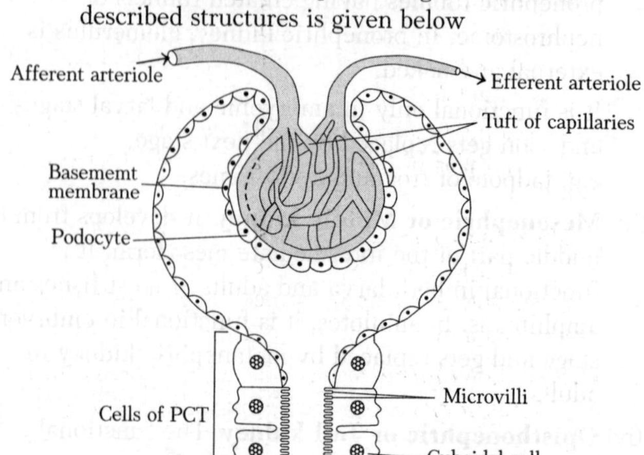

Fig. 26.2 Bowman's capsule and glomerulus (Malpighian body/renal capsule)

Uriniferous tubules

These are mainly associated with the process of reabsorption and secretion. Both the processes are normally associated with the wall of the tubules.

The tubules consist of the following four main regions

(i) **Proximal Convoluted Tubule** (PCT) **or Pars convoluta** It consists of numerous tubular coiled invaginations in the cortex region.

- It is lined by a single layer of brush-bordered cuboidal epithelial cells with long microvilli. These cells help to increase the surface area for reabsorption.
- The PCT is surrounded by peritubular vessels.
- PCT helps to maintain the pH and ionic balance of the body fluids. Nearly all essential nutrients, 70-80% of electrolytes and water are reabsorbed by this segment.
- The cells contain abundant mitochondria to perform active absorption and secretion.

(ii) **Henle's loop or Pars recta** It is a U-shaped tube which descends into renal medulla. It plays a significant role in maintaining high osmolarity of medullary intestinal fluid.

Pars recta is divisible into following three regions

- **Descending limb** This limb is lined by flat cells, i.e. simple squamous epithelium. It is permeable to water and impermeable to electrolytes. Due to this, the filtrate moving down through the limb becomes concentrated.
- **Ascending limb** It is made up of two parts
  - First, is thin ascending limb lined by squamous epithelium. It allows passive diffusion of some solutes like Na^+, Cl^-, etc.
  - Second, is thick ascending limb lined by cuboidal epithelium having microvilli and mitochondria. This segment helps in active secretion of NaCl in the medulla.
- Ascending limb is impermeable to water, but allows the transport of electrolytes and makes the filtrate dilute.
- **Main loop** This region is also lined by the flat cells, but their length is highly decreased. It is highly permeable to water and concentrate the filtrate to its maximum.

(iii) **Distal Convoluted Tubule** (DCT) It is situated in the cortex region of kidney and is about 4.5-5.5 mm long.

- It is lined by cuboidal epithelium.
- The diameter of the DCT lumen is more as compared to the diameter of PCT.
- DCT have sparse microvilli and fewer mitochondria. It signifies that active absorption and secretion does not occur in this segment.
- Conditional reabsorption of Na^+, water and HCO_3^- takes place in this segment.

(iv) **Collecting duct** The DCT continues to form a short, straight collecting tubule or duct.

- It is about 20 mm long and lined by cuboidal epithelial cells having sparse microvilli.
- The collecting ducts of different nephrons join to form **duct of Bellini**, which further open into renal papilla.

A diagram showing all the parts of nephron is given below

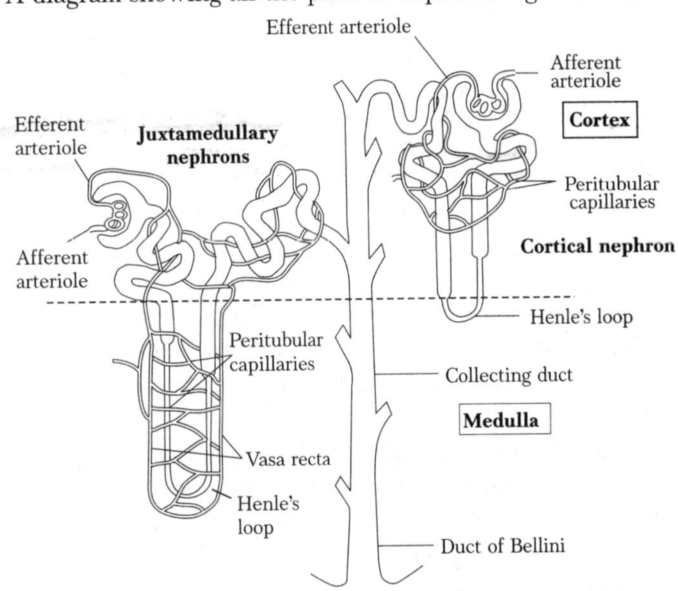

Fig 26.3 Structure of a nephron

Types of nephron

Depending upon the length of Henle's loop and location of glomeruli, nephrons are of two types as follows

(i) **Cortical nephrons** (80-85%) These possess a shorter Henle's loop. Their 1/3rd part is found in medulla while rest 2/3rd is embedded in cortex. Its glomerulus is present in the outer cortex. Vasa recta is highly reduced or absent in them.

(ii) **Juxtamedullary nephrons** (15-20%) The Henle's loop present in these nephrons is longer. Their 2/3rd part is found in medulla and only 1/3rd part is located in the cortex. Its glomerulus is found in the inner cortex. **Vasa recta** is well-developed in these nephrons and it surrounds the Henle's loop.

Blood supply to kidneys

- The blood supply to kidneys is maintained through the renal artery and renal vein.
- The outline of this circulation is shown below

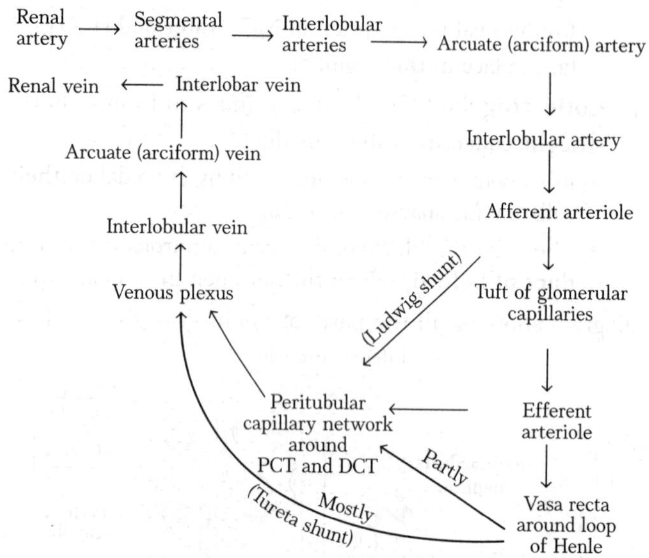

Fig 26.4 Schematic representation of blood circulation in kidney

- During the blood circulation in the kidneys, two shunts or bypass can be seen under abnormal or emergency situations.
- These are Ludwig shunt and Tureta shunt as described below
  (i) **Ludwig shunt** The afferent vessels in the corticomedullary region without forming the glomerular tuft may sometimes join directly with the medullary peritubular capillaries *via* this shunt.
  (ii) **Tureta shunt** The efferent vessels become vasa recta after a short straight run. These then partly join with the peritubular network and completely join the renal vein *via* this shunt.

Ureters

- These are a pair of muscular tubes that convey urine from kidneys to the urinary bladder. These are about 25-30 cm long with a diameter of about 3 mm.
- Ureters descend from kidney along the abdominal wall and then bend obliquely to open into the urinary bladder.
- The wall of ureter is composed of three layers, i.e. **external adventitia, middle muscular** and **inner mucosa.** These walls undergo peristalsis to propel the urine from kidneys to urinary bladder.

Urinary bladder

- Ureters open into pear-shaped structure called urinary bladder. It is the reservoir of urine.
- It is present in the pelvic cavity. The size and position of this organ varies according to the amount of urine contained in it.
- The bladder wall is composed of three layers
  (i) an outermost adventitia containing loose connective tissue.
  (ii) Middle layer of Detrusor muscle
  (iii) innermost mucosa containing transitional epithelium or urothelium towards lumen.
- Bladder has three parts, i.e. apex, fundus or body and neck.
- The body of urinary bladder has a triangular area called **trigone**. It has openings of ureters and an internal urethral orifice.
- Neck region of urinary bladder possesses two sphincters, i.e. involuntary internal sphincter and voluntary external sphincter. It then leads into urethra.

Urethra

- It is present only in mammals. The urethra is a canal extending from the neck of bladder to open outside the body through external urethral orifice.
- The length of urethra is different in males and females.
- In **males,** it is about 19-20 cm long and consists of three parts, i.e. **prostatic urethra, membranous urethra** and **spongiose** or **penile urethra.**
- In **females,** urethra is about 4 cm long and is present behind the pubic symphysis.

Route of urine flow

Glomerulus (Filtrate) → Bowman's capsule → Proximal convoluted tubule
↓
Descending limb of Henle's loop → Ascending limb of Henle's loop → Descending convoluted tubule
↓
Collecting tubule → Collecting duct (Urine) → Duct of Bellini → Papillary duct
↓
Minor calyx → Major calyx → Ureter → Urinary bladder → Urethra

CHECK POINT 26.1

1. Choose the odd one out.
   (a) Lizard
   (b) Birds
   (c) Shark
   (d) None of the above
2. Reptiles, terrestrial insects, land snails and birds
   (a) excrete ammonia directly
   (b) are ureotelic
   (c) are ammonotelic
   (d) excrete nitrogenous wastes with a minimum loss of water
3. An advantage of excreting nitrogenous wastes in the form of uric acid is that
   (a) uric acid can be excreted in almost solid form
   (b) the formation of uric acid requires extensive energy
   (c) uric acid is the first metabolic breakdown product of acids
   (d) uric acid can be excreted through the lungs
4. The African lung fishes, which are often found in small stagnant pools of freshwater, produce urea as a nitrogenous waste. What is the advantage of this adaptation?
   (a) Urea takes less energy to synthesise than ammonia
   (b) Small stagnant pools do not provide enough water to dilute the toxic ammonia
   (c) Urea forms as insoluble precipitate
   (d) The highly toxic urea makes the pool uninhabitable to potential competitors
5. The main function of pyramids within the kidney is to
   (a) contain collecting tubules of kidney
   (b) direct the urine to flow in ureter
   (c) support the openings of collecting canals
   (d) store fats and protein
6. Select the mismatched pair.
   (a) Cortical nephron – Short and coiled
   (b) Podocytes – Filtration slits
   (c) DCT – Surrounded by Henle's loop
   (d) PCT – Brush bordered epithelium

PHYSIOLOGY OF EXCRETION

The excretory physiology in humans can be categorised into two phases *viz* urea synthesis in liver and urine formation in kidneys.

Synthesis of urea

Urea is formed in the liver by two chemical processes as follows

1. Deamination

It is also known as **oxidative deamination.** During this reaction, an amino acid is converted into its corresponding keto acid by the removal of the amine functional group as ammonia.
This ammonia eventually goes into the urea cycle.

2. Ornithine cycle/Urea cycle

- It is the process of formation of urea, which takes place in the liver.
- It was described by **Krebs** and **Henseleit** (1932) hence, it is also known as **Krebs-Henseleit cycle.**
- The enzymes involved in this cycle are found in mitochondria.
- The diagram showing all the steps of urea cycle is given below

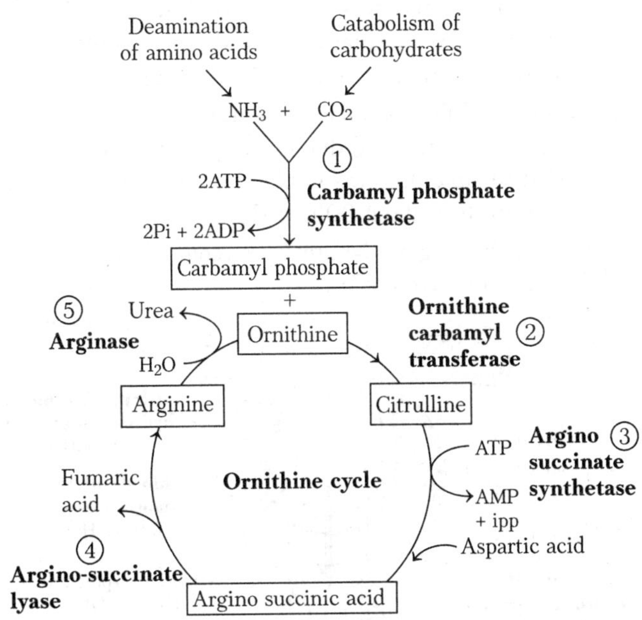

Fig 26.5 Ornithine cycle

In the above diagram,

- Enzymes are given in bold letters.
- Energy consumption occurs in two steps, i.e. 1 and 3.
- Urea formed in step 5 is the biochemical aspect of excretion.

Mechanism of urine formation in kidney

The process of formation of urine is called **uropoiesis**. It takes place in following three steps

1. Glomerular filtration or Ultrafiltration

- When the blood containing water and solutes, enters into the afferent arteriole and glomerulus, the process of ultrafiltration takes place.
- It results in the movement of water and small molecules from the glomerulus into the Bowman's capsule.
- Glomerular filtration is a result of **Net Filtration Pressure** (NFP) which includes three pressures. One of these pressures promotes filtration and other two opposes filtration. These are
  (i) **Glomerular Hydrostatic Pressure** (GHP) It is the pressure inside the glomerular capillaries. It is about 55 mm Hg. It forces water and solutes into the blood plasma through the filtration membrane.
  (ii) **Capsular Hydrostatic Pressure** (CHP) It is the hydrostatic pressure exerted against the filtration membrane. This pressure is exerted by the fluid present in the capsular space and renal tubule. CHP opposes filtration. It is about 15 mm Hg.
  (iii) **Blood Colloid Osmotic Pressure** (BCOP) This pressure is generated due to the presence of non-diffusible proteins in blood plasma. It also opposes filtration. It is about 30 mm Hg.
- A diagram showing all the pressures required in ultrafiltration is given below

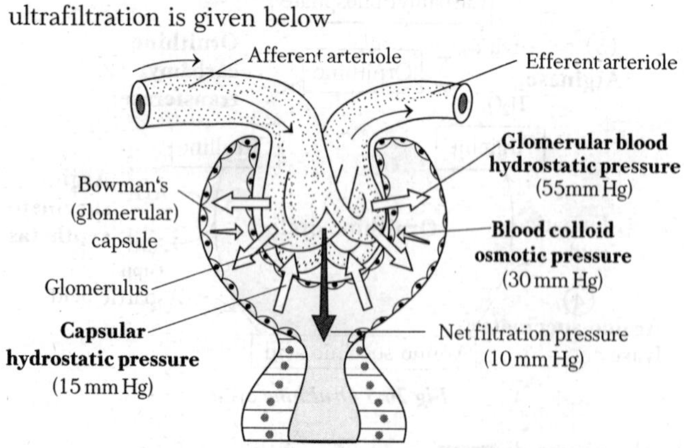

Fig 26.6 Ultrafiltration in Malpighian body

- The Net Effective Filtration Pressure (NFP or EFP) is the total pressure that promotes filtration. It is calculated as follows

$$\text{NFP or EFP} = \text{GHP} - [\text{CHP} + \text{BCOP}]$$
$$= 55 \text{ mm Hg} - [15 \text{ mm Hg} + 30 \text{ mm Hg}]$$
$$= 10 \text{ mm Hg}$$

- Thus, a pressure of only 10 mm Hg causes blood plasma to filter from glomerulus into the capsular space.
- During ultrafiltration, the blood passes through endothelial fenestrations, basement membrane and filtration slits of podocytes, so as to enter the lumen of Bowman's capsule.
- During this passage, various electrolytes (like Na^+, K^+, Ca^{2+}, Mg^{2+}, PO_4^{3-}), glucose, amino acids, urea, creatinine, etc., are filtered out. It is known as **nephric** or **glomerular filtration.**
- The glomerular filtrate therefore, essentially contains all the constituents of blood except the blood cells, proteins, certain drugs, pigments, dyes, etc.
- The amount of filtrate formed by the kidneys per minute is called **Glomerular Filtration Rate** (GFR). GFR of healthy individual is ~125 mL/minute, i.e. 180 L/day.

2. Selective or Tubular reabsorption

- The volume of the filtrate formed per day is about 180 L/day. But only 1.5 L of urine is released per day. This suggests that nearly 99% of the filtrate has to be reabsorbed by the renal tubules.
- Reabsorption involves the absorption of useful substances from glomerular filtrate. These substances are transferred to interstitial fluid and peritubular capillaries.
- The tubular epithelial cells in different segments of nephron perform this reabsorption by both active and passive mechanism.
- Reabsorption occurs in all the parts of uriniferous tubules, i.e. PCT, Henle's loop, DCT and CT.

Reabsorption in PCT

- From the Bowman's capsule, the glomerular filtrate enters the PCT after ultrafiltration.
- About 65% of glomerular filtrate is normally reabsorbed in PCT before reaching the loop of Henle.
- The filtrate remains isotonic to blood in PCT.
- The reabsorption of various substances takes place in following ways
  (i) Na^+, Ca^{2+} and K^+ are reabsorbed through the **active transport.**
  (ii) Glucose and amino acids are reabsorbed through **passive transport.**
  (iii) Reabsorption of water takes place through **osmosis.**
  (iv) Cl^-, urea and other solutes are reabsorbed through **simple diffusion.**

Reabsorption in the loop of Henle

- Reabsorption is minimum in Henle's loop. However, this region helps to maintain high osmolarity of medullary tissue fluid.
- Reabsorption in descending and ascending limbs of Henle's loop occur in following ways
  (i) The **descending limb** of loop of Henle is permeable to water but almost impermeable to electrolytes. So, here water is reabsorbed passively. This concentrates the filtrate and makes it hypertonic as it moves down.
  (ii) The **ascending limb** is impermeable to water but allow transport of electrolytes actively and passively.
- Therefore, as the concentrated filtrate passes upwards, it gets diluted due to the passage of electrolytes into the medullary fluid.
- Thus, the filtrate becomes hypotonic to the blood plasma in this region.

Reabsorption in DCT

- In DCT, Na^+ is actively reabsorbed from the filtrate due to the presence of hormone aldosterone.
- DCT is also capable of reabsorbing HCO_3^- and Cl^- ions passively.
- In the presence of ADH or vasopressin, it helps in the reabsorption of water.

Reabsorption in collecting ducts

- Water is reabsorbed extensively from the filtrate in the presence of ADH.
- The filtrate becomes more hypertonic in this region.
- A part of urea diffuses out of the bottom part of collecting duct into the medulla of kidney. It makes the medulla hyperosmotic.

3. Tubular secretion or Augmentation

- The phenomenon of secretion of metabolic wastes by tubular cells into the filtrate is known as **tubular secretion.**
- Tubular secretion is opposite to tubular reabsorption. It is an active process occurring in PCT, DCT and collecting duct.
- **In PCT**, selective secretion of H^+, NH_3^+, K^+, creatinine, uric acids, drugs, hippuric acid, etc., occurs.
- **In DCT**, secretion of H^+, K^+ occurs.
- **Collecting duct** plays a major role in the maintenance of pH and ionic balance by selective secretion of H^+ and K^+ ions.
- Tubular secretion is an important step in urine formation as it helps in the maintenance of ionic and acid-base balance of body fluid.
- In animals like marine fishes and desert amphibians, urine is formed mainly by the tubular secretion of urea, creatinine and mineral ions.
- The detailed diagram representing all the three steps of urine formation is given below

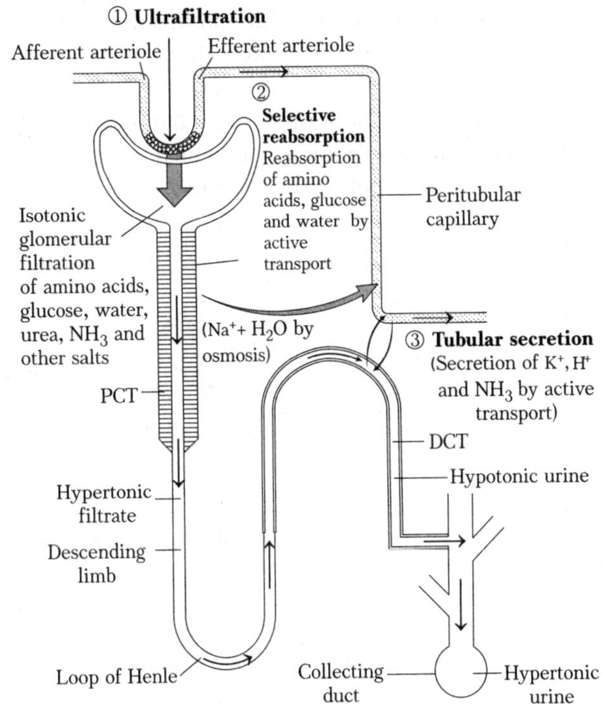

Fig 26.7 The process of urine formation

- During the process of urine formation, the concentration of filtrate is greatly affected by tubular reabsorption and secretion.
- Apart from these two processes, counter-current mechanism plays a major role in producing hypertonic urine.

Counter-current mechanism of urine concentration

- Mammals have the ability to produce concentrated urine to conserve water in the body.
- The capability of concentrating urine is largely related to the length of the loop of Henle. Vasa recta also helps in concentrating urine.
- The flow of filtrate in the two limbs of Henle's loop is in opposite directions and thus, it forms a counter-current.
- The flow of blood through the limbs of vasa recta also occur in a counter-current pattern.
- The proximity between the Henle's loop and vasa recta, as well as counter-current in them help in maintaining an increased osmolarity towards the inner medullary interstitium.

- It is from 300 $mOsmL^{-1}$ in the cortex to about 1200 $mOsmL^{-1}$ in the inner medulla.
- The gradient is mainly caused by two solutes, *viz*, NaCl and urea.
- NaCl is transported by the ascending limb of loop of Henle, which is exchanged with the descending branch of vasa recta. NaCl is returned to the interstitium by the ascending portion of vasa recta.
- Similarly, small amount of urea enters the thin segment of the ascending limb of Henle's loop, which is transported back to the interstitium by the collecting tubule.
- This transport of substances facilitated by the special arrangement of Henle's loop and vasa recta is called the **counter-current mechanism.**
- This mechanism helps to maintain a concentration gradient in the medullary interstitium. It helps in easy passage of water from the collecting tubule thereby, concentrating the filtrate, i.e. urine.
- Human kidney can produce urine nearly four times more concentrated than the initial filtrate formed (1200 mOsm/L).
- The animals with longer loop of Henle thus, produces more concentrated or hypertonic urine.

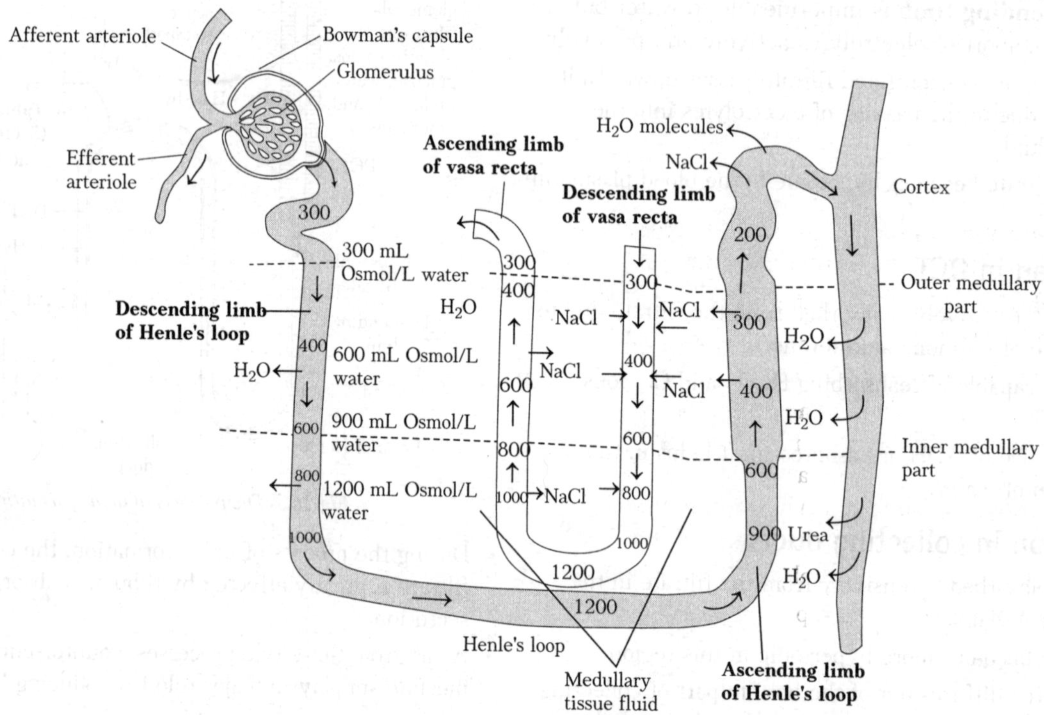

Fig 26.8 Diagrammatic representation of counter-current mechanism in vasa recta and Henle's loop

CHECK POINT 26.2

1. If the diameter of the afferent renal arteriole is decreased and that of efferent renal arteriole is increased, the ultrafiltration will
   (a) be faster
   (b) be slower
   (c) not take place
   (d) take place with the same speed
2. The moving fluid or filtrate is called urine in all of the following ducts except
   (a) renal pelvis (b) major calyx
   (c) collecting duct (d) loop of Henle
3. In a normal kidney, a substance X is completely filtered, neither secreted nor reabsorbed. The substance X is
   (a) insulin (b) glucose (c) arginine (d) inulin
4. Tubular reabsorption of water is minimum in
   (a) collecting duct
   (b) Henle's loop
   (c) proximal convoluted tubule
   (d) distal convoluted tubule
5. If a fructan compound obtained from the roots of *Dahlia* is injected into human blood, then it is subjected to
   (a) tubular secretion (b) tubular reabsorption
   (c) selective secretion (d) ultrafiltration
6. The substances that are efficiently absorbed in nephron are called high threshold substances. All of the following are high threshold substances except
   (a) glucose (b) amino acid
   (c) NaCl (d) urea

Regulation of kidney functions

The regulation of kidney functions can be summarised through the following mechanisms

Myogenic autoregulation

- A rise in blood pressure should normally increase blood flow through glomeruli.
- However, stretching of the vascular wall increases passage of Ca^{2+} ions from the extracellular fluid into the cells thus, resulting in their contraction.
- It checks overstretching of vascular walls and raises vascular resistance so that rate of blood flow and GFR are brought down to normal.

Control by Juxtaglomerular Apparatus (JGA)

- Juxtaglomerular apparatus is a complex structure that consists of three types of cells
  - (i) **Juxtaglomerular cells** The swollen smooth muscle cells of both afferent and efferent arterioles.
  - (ii) **Macula densa** The dense epithelial cells of DCT.
  - (iii) **Lacis cells** Cells forming lace-like network in between macula densa and arteriole.
- JGA works through **RAAS**, i.e. **Renin-Angiotensin-Aldosterone System**. This system normally operates at low blood pressure in the afferent arteriole of glomerulus, which is caused by the fall in Glomerular Filtration Rate (GFR).
- At that time, an enzyme renin is released from JG cells.
- The diagram showing structural aspect of JGA is given below

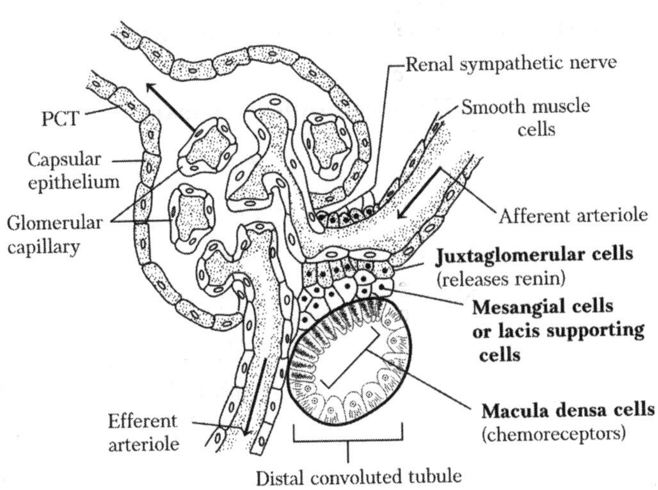

Fig 26.9 Juxtaglomerular apparatus

Mechanism

- In the blood, renin initiates a chemical reaction that converts the plasma protein **angiotensinogen** (produced by liver) into **angiotensin-I** and further to **angiotensin-II.**
- This angiotensin-II is a powerful vasoconstrictor. It increases the blood pressure by causing the arterioles to constrict. It also increases the blood volume in two ways
  - (i) By increasing water and NaCl reabsorption in PCT.
  - (ii) By stimulating adrenal glands to secrete aldosterone. It is a hormone that induces the DCT to reabsorb more Na^+ and water.
- This leads to an increase in blood volume and pressure.
- Steps involved in RAAS are shown in the diagram given below

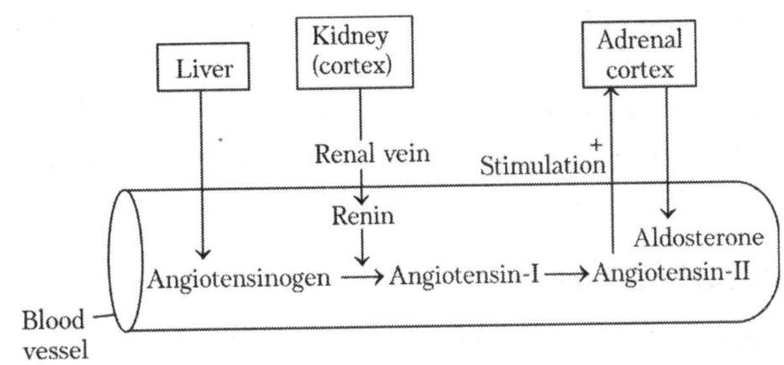

Fig 26.10 Summary of the Renin-Angiotensin-Aldosterone System (RAAS)

Control by Antidiuretic Hormone (ADH)

- Antidiuretic Hormone (ADH) is also called **vasopressin.** It is secreted by neurohypophysis of pituitary gland and produced by the hypothalamus of brain.
- The release of this hormone occurs when osmoreceptors in hypothalamus detect any change in blood volume, body fluid volume and ionic concentration.
- An excessive loss of fluid from the body can activate these receptors to release ADH and increase the reabsorption of water in DCT and collecting duct. It also promotes thirst.
- Excess of water in the body fluids can even switch off the release of this hormone to complete the feedback.
- As the release of this hormone gets lowered, the permeability of the cells of the DCT and collecting duct for water decreases.
- This decreases the reabsorption of water into the nephron. However, active reabsorption of Na^+ from the filtrate continues in these regions of the nephron.
- More filtration and less reabsorption of H_2O and normal reabsorption of Na^+ produce abundant urine and this brings down the volume of body fluids to normal.

Control by Atrial Natriuretic Factor (ANF)

- It works opposite to RAAS. It causes vasodilation when there is high blood volume and blood pressure in the body.
- Under such conditions, the walls of atria of heart produce peptide hormone (ANF) which inhibits renin and ADH secretion.
- This leads to decreased NaCl and water reabsorption. It causes normal blood pressure in the body.
- Thus, ADH, RAAS and ANF provide an elaborate system to checks, balance and regulate the kidney functioning. They control body fluid osmolarity, salt concentration, blood pressure and blood volume.

Neural control

- Blood vessels of the kidney are innervated by nerve fibres of the sympathetic nervous system.
- When activated, the nerve fibres bring about the constriction of renal arteries and cause decrease in renal blood flow as well as glomerular filtration rate.

Micturition

- It occurs as a result of contraction and relaxation of urinary bladder wall and urethral sphincter.
- Urine formed by the nephron is ultimately carried to the urinary bladder, where it is stored till a voluntary signal is given by the Central Nervous System (CNS).
- This signal is initiated by the stretching of urinary bladder as it gets filled with urine.
- These signals or impulses are then carried by the sensory neurons to the spinal cord and brain. These produce a sensation of fullness, around at 500 mL of urine.

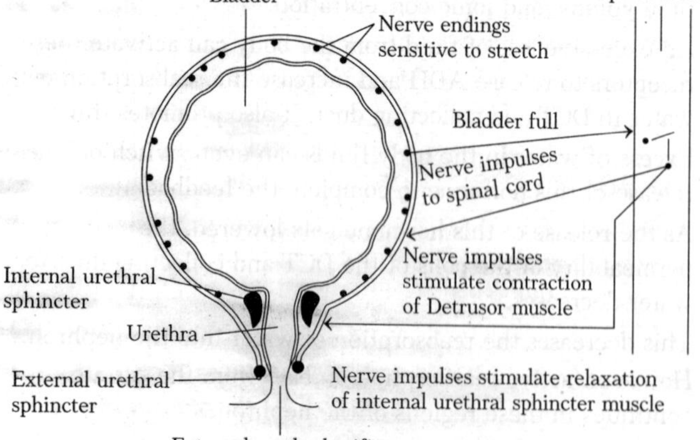

***Fig 26.11** Diagram of simple reflex control of micturition when conscious cannot override the reflex action*

- The CNS passes the motor messages to initiate the contraction of smooth muscles of the bladder and simultaneously relaxation of the urethral sphicter muscles.
- This causes the release of urine and this process is called as **micturition.** The neural mechanism causing micturition is called **micturition reflex.**

Characteristics of human urine

An adult human excretes about 1-1.5 litres of urine per day. It normally depends upon water intake, diet, temperature, humidity, mental state, etc.

The various characteristics of human urine are as follows

(i) **Colour** It is a pale-yellow or amber coloured watery fluid. This colour is given by a pigment **urochrome.** The colour of urine may vary under certain pathological conditions.

(ii) **pH** Normally urine is slightly acidic with a pH around 6.0. Intake of more fruits and proteins in diet can make urine acidic, while vegetable intake increases the alkalinity of urine. Its specific gravity is 1.02.

(iii) **Turbidity** Freshly voided urine is transparent. But alkaline urine on standing may become cloudy due to the excessive accumulation of mucus, nucleoprotein, epithelial cells and pus cells, etc., in it.

(iv) **Odour** Normal urine in fresh form is faintly aromatic. On standing, urine gives a characteristic ammonia like smell due to the presence of urinode substances and conversion of urea into ammonia by microbial action.

(v) **Composition** The composition of human urine is given below

Table 26.2 Urine constituents in man (in %)

Constituent	%
Water	96%
Urea	2%
Uric acid	0.2%
NH_3	0.25%
Creatinine	0.5%
Hippuric acid	0.025%
Salt	1%

Abnormal constituents in urine

Sometimes, unusual constituents are found in urine due to an underlying condition. These are as follows mentioned below

- **Protein** (albumin) Due to injury in renal tract, e.g. glomerulonephritis.
- **Bile salts** During jaundice.
- **Glucose** During diabetes mellitus.
- **Ketone bodies** (acetoacetic acid, β-hydroxybutyric acid) During diabetes mellitus, prolonged fasting, etc.
- **Creatinine** Hyperthyroidism, starvation, etc.

ACCESSORY EXCRETORY ORGANS

Except kidneys, some other accessory excretory organs are also present in human body. These are as follows

(i) **Skin** Human skin has two types of glands, i.e. sudoriferous (sweat) glands and sebaceous (oil) glands.

- **Sudoriferous glands** They secrete an aqueous fluid called **sweat**. Sweat consists of NaCl, water, urea, lactic acids, some amino acids and glucose.
- **Sebaceous glands** They secrete an oily or wax-like substance called **sebum**. Sebum consists of some lipids like waxes, sterols, certain hydrocarbons and fatty acids.

(ii) **Lungs** Carbon dioxide and water are metabolic wastes produced during oxidation of food in the cells. Lungs help to remove this carbon dioxide (18L/hour) and some water (400mL/day) in the form of vapour during breathing.

(iii) **Liver** The liver cells help to withdraw excess amino acids from the blood. These amino acids then gets converted into urea which is eliminated through kidneys.

Liver cells also degrade the haemoglobin of worn out RBC into bile pigments, i.e. **bilirubin** and **biliverdin**. These pigments pass into the alimentary canal with the bile for elimination in the faeces.

(iv) **Intestine** The epithelial cells of the intestine (colon) excrete certain inorganic ions like iron, magnesium and calcium. These ions are then eliminated with the faeces.

(v) **Saliva** Heavy metals and drugs are excreted through the saliva.

(vi) **Gills** These are the organs to remove CO_2 in aquatic animals. They also excrete salts in many bony fishes.

DISORDERS OF EXCRETORY SYSTEM

Disorders of excretory system are as follows

1. Renal stones

- Excessive hormonal imbalance, excess uric acid or oxalate formation, metabolic disturbance, etc., leads to the formation of renal stones or renal calculi.
- These are the insoluble masses of crytallised salts like oxalate.
- The stones are present in the kidneys or urinary tract and cause severe colic pain.
- They are removed through surgery (nephrolithotomy) or broken by laser.

2. Nephritis (Bright's disease)

- It is inflammation of any part of the kidney.
- It occurs due to congenital kidney defect or by an allergic reaction to the toxins of bacteria such as *Streptococcus*.
- Depending upon the part affected, this disease can be categorised as follows

  (i) **Pyelonephritis** Inflammation of renal pelvis and medullary tissue.

  (ii) **Glomerulonephritis** Inflammation of the glomeruli.
- Their symptoms include back pain, frequent and painful micturition.

3. Renal Failure (RF) or Kidney failure

- It is the partial or total inability of the kidneys to carryout excretory and salt water regulatory functions. It leads to **uremia**.
- It is caused due to tubular injury, bacterial toxins, inflammation of glomeruli, arterial or venous obstruction, fluid and electrolyte depletion, drug reaction, haemorrhage, etc.

4. Diabetes Insipidus (DI)

- It is a condition characterised by excessive thirst and frequent micturition.
- The urine excreted is severely diluted.
- There are several different types of DI, each with a different cause as follows

  (i) The most common type in human is **central DI**, caused by the deficiency of Arginine Vasopressin (AVP), also known as **Anti Diuretic Hormone** (ADH).

  (ii) The second common type of DI is **nephrogenic diabetes insipidus,** which is caused by an insensitivity of the kidney to ADH.

ARTIFICIAL KIDNEY

Haemodialyser is also referred to as artificial kidney. It is a machine that is used to filter blood of a damaged kidney. This process is called **haemodialysis.**

It is carried out in following manner

- Haemodialyser is a cellophane tube suspended in a salt water solution with the same composition as the normal blood plasma, but without urea.
- Blood of the patient is pumped from one of the arteries (often radial artery) into the cellophane tube after cooling it to 0°C and mixing with an anticoagulant (heparin).
- Pores of the cellophane tube allow the waste substances such as urea, uric acid, creatinine, excess salts and excess H^+ ions to diffuse from the blood into the surrounding solution.
- Diffusion of other substances like glucose, amino acids and electrolytes is prevented due to the presence of same concentration of these substances in the dialysis fluid.
- Plasma proteins remain in the blood as the pores of cellophane are too small to permit their passage.
- The purified blood is then returned to patient body through renal vein.
- The purified blood is warmed to body temperature and mixed with antiheparin to restore it to normal clotting power.
- Such a process of separating small solutes from macromolecular colloids with the help of a selectively permeable membrane is called **dialysis.**

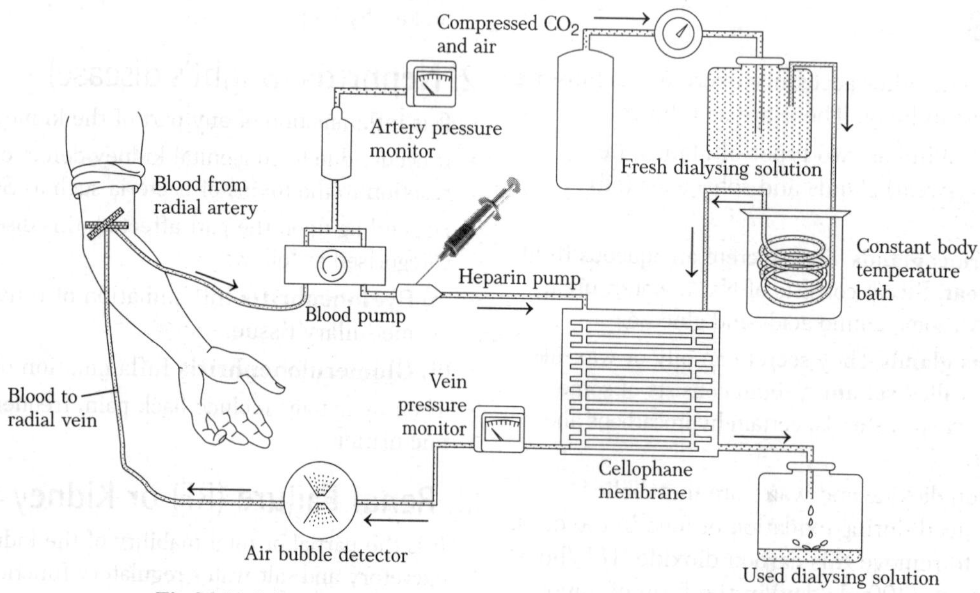

Fig 26.12 Schematic diagram to show the working of an artificial kidney

Peritoneal Dialysis (PD) It is a widely accepted method of treating end stage renal dialysis.

It uses a thin membrane called peritoneum which lines the abdomen to perform dialysis treatment.

Kidney transplantation

- It is an organ transplant of a healthy kidney into a patient with end stage renal disease.
- Kidney transplantation is classified as **cadaveric** (deceased donor) or living donor kidney transplant.
- Living donor kidney transplants are further classified as genetically related (living related) or non-related (living non-related) transplants.
- The most preferable among this is genetically related transplant as it to minimise the chances of rejection by immune system of the host.
- Before kidney transplant operations, patients are dialysed to avoid fluid overload and hyperkalemia after operation.

Terms related to kidney dysfunction

- **Oligouria** Less production of urea or urine.
- **Anuria** Failure of kidney to form urine.
- **Polyuria** or **Diuresis** Excessive production of urine.
- **Dysuria** Condition of painful micturition.
- **Cystitis** Inflammation of urinary bladder due to some bacterial infections.
- **Glycosuria** Excretion of glucose through urine. This condition is seen during diabetes mellitus.
- **Haematuria** The presence of blood in urine.
- **Proteinuria** The presence of blood protein in urine.
- **Albuminuria** The presence of albumin in urine.
- **Ketonuria** The presence of abnormally high ketone bodies in urine.
- **Haemoglobinuria** The presence of blood or blood cells in urine.
- **Uremia** The presence of excessive amount of urea in the blood. It results from decreased excretion of urea through the kidneys.

CHECK POINT 26.3

1. After blood flow is artificially reduced in one kidney, the kidney would secrete more amount of
   (a) atrial natriuretic peptide
   (b) antidiuretic hormone
   (c) renin
   (d) angiotensinogen

2. Consumption of alcoholic beverage will lead to
   (a) less volume of urination than normal
   (b) greater volume of urination than normal
   (c) no effect on urination
   (d) alcohol has no significant

3. If a person drinks very small amount of water, then after prolonged and strenuous exercise the body fluid will have
   (a) Antidiuretic hormone (b) Renin
   (c) Atrial natriuretic factor (d) Both (a) and (b)

4. If a person passes a very dilute urine but his blood glucose level is normal and experiences excessive thirst, then it may be due to
   (a) fall of blood sugar level
   (b) decrease in vasopressin release from posterior pituitary
   (c) excessive release of insulin
   (d) excessive release of glucagon

5. Which of the following is not seen in body if the blood volume is low?
   (a) More aldosterone is released by adrenal cortex
   (b) Increased tubular reabsorption of sodium
   (c) Increased secretion of Atrial Natriuretic Factor (ANF) from heart wall
   (d) Increase release of angiotensin-II

6. If the stretch receptor of urinary badder are surgically removed, then
   (a) micturition will not be affected
   (b) micturition will continue normally
   (c) urine will normally collect in urinary bladder
   (d) signals for micturition will be sent to CNS from bladder normally

EXAM POINTS

- Insects, birds and lizards are uricotelic animals. **[JIPMER 2019]**
- Mammals, terrestrial amphibians, elasmobranch fishes are ureotelic animals. **[MHT CET 2015]**
- Flame cells are present in Platyhelminthes. **[JIPMER 2019]**
- Grasshopper have Malpighian tubules, *Balanoglossus* have proboscis gland, earthworm have nephridia and Amphioxus have protonephridia. **[NEET 2019, Kerala CEE 2015]**
- Protonephridia are present in Amphioxus. **[NEET 2019]**
- Podocytes are filtration slits in Bowman's capsule. **[NEET 2019]**
- Malpighian tubules are excretory organs in insects that excrete uric acid. **[AIIMS 2019]**
- Filtration of blood is the function of Bowman's capsule and glomerulus. **[AIIMS 2019]**
- PCT helps in reabsorption of ion, water and organic nutrients. **[NEET 2019, 2016]**
- The ascending limb of loop of Henle is impermeable to water. **[NEET 2017]**
- Ascending limb of Henle's loop helps in reabsorption of salts only. **[NEET 2019]**
- Descending limb of Henle's loop helps in reabsorption of water only. **[NEET 2019]**
- Ultrafiltration occurs in Malpighian corpuscles. **[NEET 2018]**
- Concentration of urine occurs in Henle's loop. **[NEET 2018]**
- DCT helps in reabsorption of water only. **[NEET 2019]**
- Transport of urine occurs through ureter. **[NEET 2018]**
- Storage of urine occurs in urinary bladder. **[NEET 2018]**
- The ornithine cycle removes CO_2 and NH_3 from the blood in liver. **[JIPMER 2017]**
- Reduction in GFR activates JG cells to release renin. **[NEET 2020]**
- Maintaining hyperosmolarity towards inner medullary interstitium in the kidneys is responsible for the formation of concentrated urine. **[NEET 2019]**
- The increase in osmolarity from outer to inner medullary interstitium is maintained due to close proximity between Henle's loop and vasa recta and counter-current mechanism. **[NEET 2020]**
- Tubular reabsorption and secretion are two processes of urine formation taking place all along the renal tubule and collecting duct. **[AIIMS 2017]**
- Glomerular nephritis is inflammation of glomeruli. **[NEET 2018]**
- Counter-current pattern of blood flow is observed in vasa recta. **[NEET 2020]**
- Angiotensin-II is a powerful vaso-constrictor. **[NEET 2020]**
- ANF causes vosodilation and decreases in blood pressure. **[NEET 2020]**
- A decrease in blood pressure/volume do not cause the release of ANF but it causes the release of renin, aldosterone and ADH. **[NEET 2017]**
- Gout is a accumulation of uric acid in joints. **[NEET 2018]**
- Glycosuria is the presence of glucose in urine. **[NEET 2018]**
- Reabsorption of Na^+ and water from renal tubules due to aldosterone help in prevention of diuresis. **[NEET 2020]**
- Presence of ketonuria and glycosuria in the urine indicates diabetes mellitus. **[NEET 2020]**

Chapter Exercises

(A) Topical Exercise

Excretory Products and Excretory Organs

1 The process of maintaining osmotic and ionic concentrations of body fluids is known as
(a) excretion (b) osmoregulation
(c) homeostasis (d) Both (a) and (b)

2 Homeostasis is
(a) process of maintaining osmotic and ionic concentration of body fluids
(b) maintenance of a constant favourable internal environment despite the fluctuations in outer environment of the body/cell
(c) act of maintaining a steady state in the body
(d) Both (b) and (c)

3 Nitrogenous waste products are eliminated mainly as
(a) urea in tadpole and uric acid in adult frog
(b) urea in adult frog and ammonia in tadpole
(c) urea in tadpole as well as in adult frog
(d) urea in tadpole and ammonia in adult frog

4 Identify the statement, which is incorrect about ammonotelism?
(a) Ammonia is a highly toxic excretory product
(b) Ammonotelic animals live in freshwater
(c) Ammonia is excreted as ammonium ions
(d) Ammonotelic animals are elasmobranch fishes

5 Almost all the aquatic animals excrete ammonia as the nitrogenous waste product. Which of the following statements is not in an agreement with this situation?
(a) Ammonia is easily soluble in water
(b) Ammonia is released from the body in a gaseous state
(c) Ammonia is highly toxic and needs to be eliminated as and when formed
(d) Ammonia gets converted into a less toxic form called urea

6 Uricotelism is found in
(a) mammals and birds
(b) fishes and freshwater protozoans
(c) birds, reptiles and insects
(d) frogs and toads

7 Which of the following is not uricotelic?
(a) Garden lizard (b) Tiger
(c) Crow (d) Eagle

8 In which pair, both the organisms are ammonotelic?
(a) Salamander and tadpole (b) Frog and man
(c) Bony fish and toad (d) Frog and toad

9 Select the set of animals which consists exclusively of animals showing dual secretion (ammonotelic as well as ureotelic).
(a) Spiders and birds (b) Earthworm and snails
(c) Fishes and mammals (d) Gastropods and insects

10 Consider the following statements.
I. Ammonia is the most toxic waste matter.
II. Uric acid is the least toxic waste matter.
Choose the correct option.
(a) Statement I is true, but II is false
(b) Statement I is false, but II is true
(c) Both statements I and II are true
(d) Both statements I and II are false

11 Which of the following pairs is incorrect? **[NCERT Exemplar]**
(a) Uricotelic – Birds
(b) Ureotelic – Insects
(c) Ammonotelic – Tadpole
(d) Ureotelic – Elephant

12 Few animals instead of excreting ammonia, convert it into Trimethylamine (TMA).
How many of such animals do you count that follow above procedure?

Marine molluscs, Crustaceans, Fishes, Birds, Humans, Gastropods

(a) One (b) Three (c) Five (d) Four

13 Organisms belonging to which phylum excrete through their body surface?
(a) Arthropoda (b) Mollusca
(c) Aschelminthes (d) Protozoa

14 Match the following columns.

Excretory structures/organs	Animals
A. Protonephridia	1. Prawn
B. Nephridia	2. Cockroach
C. Malpighian tubules	3. Earthworm
D. Green gland or antennal gland	4. Flatworms

Codes **[NCERT Exemplar]**

	A	B	C	D		A	B	C	D
(a)	4	3	2	1	(b)	2	3	1	2
(c)	4	3	1	2	(d)	1	3	2	4

15 Which of the following is a correct match with respect to different types of kidney in living organisms?

(a) Holonephric kidney – Sharks
(b) Metanephric kidney – Birds
(c) Mesonephric kidney – *Myxine*
(d) Tail kidney – Reptiles

16 A terrestrial animal must be able to

(a) excrete large amounts of water in urine
(b) conserve water
(c) actively pump out salts through the skin
(d) excrete large amounts of salts in urine

17 Consider the following statements.

I. Flame cells are excretory structures in flatworms.
II. Dermal branchiae are the excretory organs of Hemichordata.

Choose the correct option.

(a) Statement I is true, but II is false
(b) Statement I is false, but II is true
(c) Both statements I and II are true
(d) Both statements I and II are false

18 Which one of the statement is false?

(a) Nephrons perform excretion through filtration, reabsorption and secretion
(b) Nephridia are accessory excretory organs in *Unio*
(c) Tapeworm have excretory flame cells
(d) Coxal glands are excretory organs in arachnids

Human Excretory System

19 The position of kidneys is

(a) interperitoneal
(b) retroperitoneal
(c) intraperitoneal
(d) None of the above

20 The retroperitoneal kidney is

(a) kidney of fish
(b) kidney covered by peritoneum on ventral side
(c) kidney covered by peritoneum on dorsal side
(d) kidney uncovered by peritoneum on either side

21 Parietal layer of Bowman's capsule is lined by

(a) ciliated cuboidal epithelium
(b) squamous epithelium
(c) non-ciliated cuboidal epithelium
(d) non-ciliated columnar epithelium

22 Podocytes are the cells, which are present on

(a) neck of nephron
(b) the wall of Bowman's capsule
(c) outer wall of loop of Henle
(d) wall of glomerular capillaries

23 Identify *A*, *B*, *C*, *D* and *E* in the human urinary system shown below.

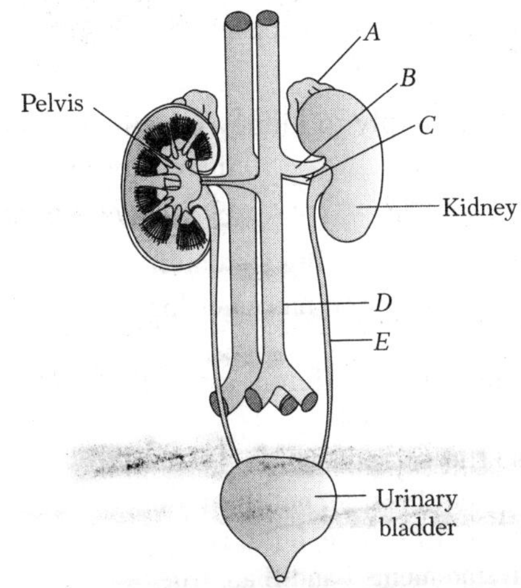

	A	B	C	D	E
(a)	Adrenal gland	Renal artery	Renal artery	Dorsal aorta	Ureter
(b)	Adrenal gland	Renal vein	Renal artery	Ureter	Urethra
(c)	Adrenal gland	Renal artery	Renal vein	Dorsal aorta	Ureter
(d)	Adrenal gland	Renal artery	Renal vein	Dorsal aorta	Urethra

24 The four structures listed are part of the human excretory system.

I. Bladder
II. Kidney
III. Ureter
IV. Urethra

In which order does a molecule of urea pass through these structures?

	First ⟶ Last			
(a)	I	II	III	IV
(b)	I	IV	III	II
(c)	II	I	III	IV
(d)	II	III	I	IV

25 Consider the following statements.

I. Renal capsule is the innermost tough covering of fibres connective tissue.
II. Renal fascia is the middle cover invading adipose tissue.

Choose the correct option.

(a) Statement I is true, but II is false
(b) Statement I is false, but II is true
(c) Both statements I and II are true
(d) Both statements I and II are false

26 In juxtamedullary nephrons,
(a) vasa recta is prominent
(b) loop of Henle is long
(c) loop of Henle runs deep into the medulla
(d) All of the above

27 The blood supply to kidney is maintained through
(a) renal artery (b) renal vein
(c) hepatic artery (d) Both (a) and (b)

28 Consider the following statements.
I. Cortical nephron has no or highly reduced vasa recta.
II. In glomerulus, efferent arteriole is wider than afferent arteriole.
Choose the correct option.
(a) Statement I is true, but II is false
(b) Statement I is false, but II is true
(c) Both statements I and II are true
(d) Both statements I and II are false

29 Glomerular area of adrenal cortex is responsible for
(a) water and electrolyte balance
(b) carbohydrate metabolism
(c) steroid and hormone secretion
(d) blood pressure

30 Which one of the following statement is correct?
(a) Renal corpuscle and glomerulus constitute Malpighian corpuscle
(b) Bowman's capsule and glomerulus together constitute renal corpuscle
(c) Bowman's capsule and Malpighian tubules constitute the glomerulus
(d) Malpighian corpuscles and glomerulus constitute the Bowman's capsule

31 Identify the correct statements with respect to human excretory system.
(a) The two kidneys lie at the level of ovaries
(b) In mammalian kidney, renal pyramids are seen in the medulla
(c) The bunch of capillaries present in the Bowman's capsule is called pacinian corpuscle
(d) Glomerulus belong to uriniferous tubules

32 Read the following statements.
I. Each nephron has two parts, i.e. glomerulus and renal tubule.
II. Vasa recta are absent or highly reduced in juxtamedullary nephrons.
Choose the correct option.
(a) Statement I is true, but II is false
(b) Statement I is false, but II is true
(c) Both statements I and II are true
(d) Both statements I and II are false

33 Consider the following statements.
I. The smallest functional unit of kidney is the nephron.
II. Urinary bladder is present in snakes and crocodiles.
Choose the correct option.
(a) Statement I is true, but II is false
(b) Statement I is false, but II is true
(c) Both statements I and II are true
(d) Both statements I and II are false

34 The region of the nephron found in the renal medulla is
(a) Malpighian corpuscle
(b) Proximal Convoluted Tubule (PCT)
(c) Distal Convoluted Tubule (DCT)
(d) Henle's loop

35 What is incorrect about kidney?
(a) Peripheral cortex and central medulla
(b) Blood enters into glomerulus through efferent arterioles
(c) Malpighian corpuscle occurs in cortex
(d) Concave part of kidney is called hilus

36 Which term is used both for a part of kidney and a part of skeleton in the mammals?
(a) Centrum
(b) Pelvis
(c) Cortex
(d) Medulla

37 The rupture of urinary bladder is prevented by
(a) pseudostratified epithelium
(b) stratified columnar epithelium
(c) stratified cuboidal epithelium
(d) transitional epithelium

38 Brush-border is characteristic of
(a) neck of nephron
(b) collecting tube
(c) proximal convoluted tubule
(d) All of the above

39 Longest loop of Henle is found in
(a) kangaroo rat (b) opossum
(c) rhesus monkey (d) All of these

40 Which one of the four parts mentioned below does not constitute a part of a single uriniferous tubule?
(a) Bowman's capsule
(b) Distal convoluted tubule
(c) Loop of Henle
(d) Collecting duct

41 Vital morphological and physiological units of mammalian kidney are
(a) ureters (b) seminiferous tubules
(c) uriniferous tubules (d) nephridia

Physiology of Excretion

42 For the formation of urea, which of the following is required with ammonia?
(a) CO_2, arginase and water
(b) O_2, CO_2 and arginase
(c) Aspartate, CO_2 and water
(d) Aspartate, O_2 and CO_2

43 The presence of arginase confirms that the
(a) urea cycle is operating
(b) urea cycle may be operating
(c) arginine is being converted into ornithine
(d) arginine is being converted into citrulline

44 Consider the following statements.
I. Urea cycle enzymes are located in mitochondria.
II. For the process of urea biosynthesis, uric acid is the starting point.

Choose the correct option.
(a) Statement I is true, but II is false
(b) Statement I is false, but II is true
(c) Both statements I and II are true
(d) Both statements I and II are false

45 The principal nitrogenous excretory compound in humans is synthesised in
(a) kidneys but eliminated mostly through liver
(b) kidneys as well as eliminated by kidneys
(c) liver and also eliminated by the same through bile
(d) the liver, but eliminated mostly through kidneys

46 I. Ludwig shunt
II. Trueta shunt

The given process are a part of
(a) ornithine cycle
(b) urea cycle
(c) RAAS
(d) Both (a) and (b)

47 Choose the correct match pair.
(a) All essential nutrients, 70–80% of electrolyses and water–Proximal convoluted tubules
(b) Conditional absorption of Na^+, water and HCO_3^- - Proximal convoluted tubules
(c) Maintenance of high osmolarity of the medullary intestinal fluid - Collecting Duct
(d) Passive reabsorption of water - Ascending limb

48 In kidneys, urine is produced by three processes
(a) dialysis, ultrafiltration and tubular secretion
(b) ultrafiltration, dialysis and tubular secretion
(c) ultrafiltration, tubular reabsorption and tubular secretion
(d) dialysis, tubular reabsorption and tubular secretion

49 When does glomerular filtration occur in Bowman's capsule?
(a) When hydrostatic pressure of blood in the glomerulus is 70 mm Hg and net filtrate pressure is −25 mm Hg
(b) When hydrostatic pressure of blood in the glomerulus is 70 mm Hg and net filtrate pressure is −35 mm Hg
(c) When hydrostatic pressure of blood in the glomerulus is 70 mm Hg and net filtrate pressure is 25 mm Hg
(d) When hydrostatic pressure of blood in the glomerulus is 70 mm Hg and net filtrate pressure is −70 mm Hg

50 Maintenance of body potassium levels is primarily by tubular
(a) absorption in PCT
(b) secretion in DCT and cortical collecting duct
(c) absorption in DCT
(d) secretion in PCT

51 The renal fluid, isotonic to the cortical fluid and blood are found in
(a) the collecting duct and ascending limb
(b) the distal convoluted tubule and ascending limb
(c) the Proximal Convoluted Tubule (PCT) and Distal Convoluted Tubule (DCT)
(d) the ascending limb and descending limb

52 During urine formation, which of the following processes create high osmotic pressure in the uriniferous tubule?
(a) Active Na^+ absorption, followed by absorption of Cl^-
(b) Active Cl^- absorption, followed by absorption of Na^+
(c) Active secretion of Na^+ into efferent arteriole followed by absorption of Cl^- into efferent renal arteriole
(d) Active secretion of Cl^- and absorption of Na^+ into efferent renal arteriole

53 If Henle's loop were absent from mammalian nephron, which of the following is to be expected?
(a) The urine will be more concentrated
(b) The urine will be more dilute
(c) There will be no urine formation
(d) There will be hardly any change in the quality and quantity of urine formed

54 Normal level of urea in blood plasma is
(a) 80-100 mg/100 mL blood
(b) 18-38 mg/100 mL blood
(c) 30-40 mg/100 mL blood
(d) 1-10 mg/100 mL blood

55 A person who is on a long hunger strike and is surviving only on water, will have
(a) more sodium in his urine
(b) less amino acids in his urine
(c) more glucose in his blood
(d) less urea in his urine

56 Consider the following statements with respect to urine.

I. Hypertonic to blood and isotonic to medullary fluid.
II. Isotonic to blood and hypertonic to medullary fluid.

Choose the correct option.

(a) Statement I is true, but II is false
(b) Statement I is false, but II is true
(c) Both statements I and II are true
(d) Both statements I and II are false

57 What will be the Glomerular Filtration Pressure (GFP) if Glomerular Hydrostatic Pressure (GHP) is 60 mm Hg, Blood Colloidal Osmotic Pressure (BCOP) is 30 mm Hg and Capsular Hydrostatic Pressure (CHP) is 20 mm Hg?

(a) 50 mm Hg (b) 30 mm Hg
(c) 70 mm Hg (d) 10 mm Hg

58 Consider the following statements.

I. When the urine moves through the descending limb, it becomes hypertonic and as it passes through the ascending limb of Henle's loop, it becomes hypotonic.
II. The descending limb is permeable to sodium ions, while the ascending limb is impermeable to sodium ions.

Choose the correct option.

(a) Statement I is true, but II is false
(b) Statement I is false, but II is true
(c) Both statements I and II are true
(d) Both statements I and II are false

59 The processes involved in urine formation are

I. tubular secretion II. glomerular filtration
III. tubular reabsorption IV. tubular filtration

Choose the correct option.

(a) Only III (b) I and II (c) III and IV (d) I, II and III

60 In nephron, water absorption is maximum in

(a) proximal convoluted tubule (b) descending loop of Henle
(c) ascending loop of Henle (d) distal convoluted tubule

61 Urine is concentrated to its maximum in loop of Henle in

(a) descending limb
(b) thick ascending limb
(c) hairpin band between descending and ascending limbs
(d) area between ascending limb and distal convoluted tubule

62 Tubular reabsorption occurs in

I. proximal convoluted tubule
II. collecting ducts
III. loop of Henle
IV. Bowman's capsule

Choose the correct option.

(a) I and IV (b) II and III
(c) I and II (d) I, II and III

63 A large quantity of fluid is filtered everyday by the nephrons in the kidneys. Only about 1% of it is excreted as urine. The remaining 99% of the filtrate

(a) gets collected in the renal pelvis
(b) is lost as sweat
(c) is stored in the urinary bladder
(d) is reabsorbed into the blood

64 The counter-current mechanism operates in

(a) ascending and descending limbs
(b) in ascending limb of Henle's loop only
(c) in descending limb of Henle's loop only
(d) loop of Henle and vasa recta

65 Reabsorption of K^+ ions in the tubules of nephrons occurs by the process of

(a) osmosis (b) diffusion
(c) filtration (d) active transport

66 What causes the liquid part of blood to filter out from the glomerulus into the renal tubule?

(a) Osmosis (b) High (hydrostatic) pressure
(c) Diapedesis (d) Dialysis

67 Glomerular filtrate is

(a) blood minus blood corpuscles and plasma protein
(b) blood minus corpuscles
(c) mixture of water, ammonia and corpuscles
(d) urine

68 If one litre of water is introduced in human blood, then

(a) BMR increases
(b) RBC collapses and urine production increases
(c) RBC collapses and urine production decreases
(d) BMR decreases

69 Capillary pressure in the glomeruli

(a) is lower than pressure in the efferent arterioles
(b) rises when the afferent arterioles constrict
(c) is higher than other capillaries in the body
(d) is reduced by about 10% when arterial pressure falls 10% below the normal level

70 Which feature enables the mammalian kidney to concentrate urine in the medullary region?

(a) Maintaining a high osmotic pressure in the tissues between the tubules
(b) Rapid removal of sodium ion from the medullary tissues
(c) Rapid flow of blood through the medulla
(d) High oxidative metabolism of medullary cells

71 We can produce a concentrated/dilute urine. This is facilitated by a special mechanism. Identify the mechanism. **[NCERT Exemplar]**

(a) Reabsorption from PCT
(b) Reabsorption from collecting duct
(c) Reabsorption/Secretion in DCT
(d) Counter-current mechanism in Henle's loop/vasa recta

72 Hypertonic urine secretion in mammals depends upon the
(a) width of Bowman's capsule
(b) length of loop of Henle
(c) length of collecting duct
(d) length of blood capillaries

73 Which blood vessel takes blood away from kidney?
(a) Renal portal vein (b) Renal vein
(c) Afferent arteriole (d) Efferent arteriole

Regulation of Kidney Function

74 Juxtaglomerular apparatus is made up of
(a) juxtaglomerular cell, macula densa and Lacis cell
(b) juxtaglomerular cell, Purkinje cell and chief cell
(c) juxtaglomerular cell, Lacis cell and myoepithelial cell
(d) juxtaglomerular cell, macula densa and Argentaffin cell

75 Renin is released from
(a) juxtaglomerular apparatus (b) cortical nephron
(c) collecting duct (d) pelvis

76 Which of the following statements is correct? **[NCERT Exemplar]**
(a) ADH prevents conversion of angiotensinogen in blood to angiotensin
(b) Aldosterone facilitates water reabsorption
(c) ANF enhances sodium reabsorption
(d) Renin causes vasodilation

77 Which one of the following is also known as antidiuretic hormone? **[NCERT Exemplar]**
(a) Oxytocin (b) Vasopressin
(c) Adrenaline (d) Calcitonin

78 Absorption of water in DCT is controlled by
(a) ACTH (b) ADH (c) LH (d) oxytocin

79 Vasopressin released from the neurohypophysis is responsible for
(a) facultative reabsorption of water through Henle's loop
(b) obligatory reabsorption of water through Bowman's capsule
(c) facultative reabsorption of water through DCT
(d) obligatory reabsorption of water through PCT

80 Why do we pass more urine during winter and wet seasons?
(a) Increased ADH secretion
(b) Increased activity of kidneys
(c) Decreased water absorption by nephrons
(d) Reduced sweating

81 The macula densa is a part of the
(a) proximal convoluted tubule
(b) afferent arteriole
(c) distal convoluted tubule
(d) efferent arteriole

82 Identify the incorrect match with respect to human urine.
(a) pH (slighlty acidic) – 6
(b) Clarity – Fresh urine is transparent
(c) Odour – Due to urinode
(d) Glucose – Present in normal urine

83 Consider the following statements.
I. Micturition is the process of passing out urine.
II. The amount of urine output per day by normal human being is 1-1.8L.

Choose the correct option.
(a) Statement I is true, but II is false
(b) Statement I is false, but II is true
(c) Both statements I and II are true
(d) Both statements I and II are false

84 Human urine is invariably acidic because
(a) the blood entering the kidney is acidic
(b) kidneys selectively filter out the acidic substances of blood into urine
(c) kidneys secrete acids to keep urine acidic
(d) urine is made acidic in urinary bladder

85 The urine of man under normal conditions does not contain glucose because
(a) glucose in the glomerular filtrate is converted into glycogen
(b) glucose in the glomerular filtrate is absorbed in the uriniferous tubules
(c) glucose of the blood is not filtered in the glomerulus
(d) the normal blood sugar is fructose

86 The pigments formed by the degradation of RBCs are
(a) bile
(b) haemoglobin
(c) bilirubin and urochrome
(d) bilirubin and biliverdin

87 Which of the following is responsible for the excretion of dilute urine?
(a) More secretion of insulin
(b) Less secretion of vasopressin
(c) More secretion of aldosterone
(d) Less secretion of glucagon

88 Choose the incorrect match pair.
(a) Angiotensinogen - Liver
(b) ADH - Pituitary gland
(c) ANF - Wall of atria
(d) Aldosterone - Medulla of adrenal gland

89 Choose the incorrect match pair.
(a) ADH - Reabsorption of water
(b) Renin - Degradation of angiotensinogen
(c) Micturition - Urethra reflex
(d) Specific gravity of urine - 1.010 to 1.015

Acessory Excretory Organs and Excretory Disorders

90 How many words correlate with accessory excretory organs?

Skin, Lungs, Liver, Pancreas, Intestine, Mouth

(a) Four (b) Three (c) Five (d) Two

91 A large quantity of one of the following is removed from our body by lungs. **[NCERT Exemplar]**

(a) CO_2 only (b) H_2O only
(c) CO_2 and H_2O (d) ammonia

92 The main cause of diabetes insipidus is

(a) deficiency of ADH
(b) increase in amount of ADH
(c) eating excess sugar
(d) None of the above

93 Diuresis is a specific pathological condition, which leads to

(a) increased volume of urine excretion
(b) decreased volume of urine excretion
(c) increased glucose excretion
(d) decreased electrolyte concentration

94 Consider the following statements.

I. Presence of excess amount of urea in blood in uremia.
II. Presence of blood cell in urine is ketonuria.

Choose the correct option.

(a) Statement I is true, but II is false
(b) Statement I is false, but II is true
(c) Both statements I and II are true
(d) Both statements I and II are false

95 Consider the following statements.

I. In Bright's disease (nephritis) occurs painful urination.
II. In isosthenuria urine has osmolarity similar to that of plasma.

Choose the correct option.

(a) Statement I is true, but II is false
(b) Statement I is false, but II is true
(c) Both statements I and II are true
(d) Both statements I and II are false

96 Excessive thirst leading to increased consumption of water is

(a) polyurea (b) glycemia
(c) polyphagia (d) polydipsia

97 What will happen, if one kidney of a person is removed?

(a) He will still survive and remain normal
(b) He will die due to blood poisoning
(c) Urea will go on accumulating in blood
(d) Urination will stop

98 Dialysing unit (artificial kidney) contains a fluid which is almost same as plasma except that it has **[NCERT Exemplar]**

(a) high glucose
(b) high urea
(c) no urea
(d) high uric acid

99 Which of the following is correct with reference to haemodialysis?

(a) Absorbs and resends excess of ions
(b) The dialysis unit has a coiled cellophane tube
(c) Blood is pumped back through a suitable artery after haemodialysis
(d) Nitrogenous wastes are removed by active process

(B) Cumulative Exercise

(Assorted Questions of the Chapter for Advanced Level Practice)

1 Which one of the following statements is incorrect? **[NCERT Exemplar]**

(a) The medullary zone of kidney is divided into a few conical masses called medullary pyramids projecting into the calyces
(b) Inside the kidney, the cortical region extends in between the medullary pyramids as renal pelvis
(c) Glomerulus along with Bowman's capsule is called the renal corpuscle
(d) Renal corpuscle, Proximal Convoluted Tubule (PCT) and Distal Convoluted Tubule (DCT) of the nephron are situated in the cortical region of kidney

2 Which one of the following statements related to the excretion by the human kidneys is correct?

(a) Descending limb of loop of Henle is impermeable to water
(b) Distal convoluted tubule is incapable of reabsorbing HCO_3
(c) Nearly 99% of the glomerular filtrate is reabsorbed by the renal tubules
(d) Ascending limb of loop of Henle is impermeable to electrolytes

3 Find the incorrect statement regarding mechanism of urine formation in man.

(a) The glomerular filtration rate is about 125 mL per minute
(b) The ultrafiltration is opposed by the colloidal osmotic pressure of plasma
(c) Aldosterone induces greater reabsorption of sodium
(d) The counter-current system contributes in diluting the urine

4 Match the following columns.

Column I	Column II
A. Proximal convoluted tubule	1. Formation of concentrated urine
B. Distal convoluted tubule	2. Filtration of blood
C Henle's loop	3. Reabsorption of 70-80% of electrolytes
D. Counter-current mechanisms	4. Ionic balance
E. Renal corpuscle	5. Maintenance of concentration gradient in medulla

Codes **[NCERT Exemplar]**

	A	B	C	D	E
(a)	3	5	4	2	1
(b)	3	4	1	5	2
(c)	1	3	2	5	4
(d)	3	1	4	5	2

5 Select the correct combinations.

I. Dysuria — Painful urination
II. Pyuria — WBCs or pus in the urine
III. Fructosuria — Presence of glucose in urine
IV. Albuminuria — Absence of ketone bodies in urine

Choose the correct option.

(a) Only I (b) II and III
(c) I and II (d) None of these

6 During haemodialysis process,

I. blood is drained from a convenient artery and anticoagulant is added (heparin).
II. removal of nitrogenous waste from blood.
III. blood is passed through a coiled porous cellophane membrane of tube bathing in dialysis fluid.
IV. blood is mixed with antiheparin and passed into vein.

Arrange the steps.

(a) I → II → III → IV (b) IV → III → II → I
(c) I → III → II → IV (d) I → IV → II → III

7 Match the following columns.

Column I	Column II
A. Uremia	1. Excess of protein level in urine
B. Haematuria	2. The presence of high ketone bodies in urine
C. Ketonuria	3. The presence of blood cells in urine
D. Glycosuria	4. The presence of glucose in urine
E. Proteinuria	5. The presence of excess urea in urine

Codes

	A	B	C	D	E
(a)	5	3	2	4	1
(b)	4	5	3	2	1
(c)	5	3	4	2	1
(d)	3	5	2	1	4

8 Human urine

I. is transparent but basic and on standing become cloudy due to excessive perspiration.
II. is pale yellow in colour due to pigment urochrome.
III. has specific gravity between 4.01 to 4.05.
IV. on standing give ammonical odour due to the conversion of urea into ammonia.

Choose the correct option.

(a) I, II and III (b) II and IV
(c) I, II and IV (d) Only IV

9 The characteristic(s) that is/are shared by urea, uric acid and ammonia is/are

I. They are nitrogenous wastes.
II. They all need very large amount of water for excretion.
III. They are all equally toxic.
IV. They are produced in the kidneys.

Choose the correct option.

(a) I and III (b) I and IV (c) I, III and IV (d) Only I

10 While urine formation progress, which of the following processes takes place in the region labelled as *A*, *B*, *C* and *D* in the given diagram?

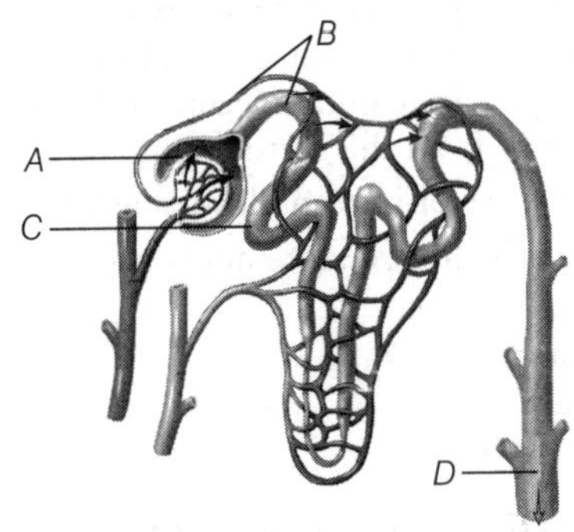

(a) *A*-Collection of urine *B*-Secretion
C-Reabsorption *D*-Pressure filtration
(b) *A*-Pressure filtration *B*-Reabsorption
C-Secretion *D*-Collection of urine
(c) *A*-Pressure filtration *B*-Secretion
C-Reabsorption *D*-Collection of urine
(d) *A*-Reabsorption *B*-Secretion
C-Pressure filtration *D*-Collection of urine

11 Find the correct pathway of blood in the kidney.

I. Peritubular capillary and vasa recta
II. Renal vein
III. Efferent arteriole
IV. Renal artery
V. Glomerulus
VI. Afferent arteriole

(a) I → II → IV → III → VI → V
(b) IV → VI → V → I → II → III
(c) IV → V → VI → I → III → II
(d) IV → VI → V → III → I → II

12 Read the following statements with respect to human kidney.

I. Juxta Glomerular Apparatus (JGA) is a special sensitive region formed by cellular modifications in the distal convoluted tubule and afferent arteriole at the location of their contact.

II. About 1.5 L of filtrate are formed with the release of 180 L of urine per day by human kidneys.

III. Reabsorption is a non-selective process.

IV. The tubular epithelial cells in different segments of nephron perform reabsorption only by active process.

V. Tubular secretion helps in the maintenance of ionic and acid-base balance of body fluids.

Which of the above statements are correct?

(a) I, II and III (b) II and IV
(c) I and V (d) IV and V

13 Read the following statements.

I. Sharks and rays retain high concentration of TMAO (Trimethylamine Oxide) and urea to minimise water loss from the body.

II. Each human kidney possesses two billion neurons.

III. Cortical nephrons of human kidney control the volume of plasma under stress.

IV. Most of the vertebrates including hagfish, sharks and rays can maintain a constant internal osmolarity different from the surrounding medium.

V. Osmolarity of human blood is about 300 $mOsmL^{-1}$

Which of the above statements are correct?

(a) I and V
(b) II, III and IV
(c) IV and V
(d) II, IV and V

14 Find the correct sequence of steps during micturition.

I. The CNS passes on motor messages to initiate the contraction of smooth muscles of the urinary bladder.

II. Simultaneous relaxation of the urethral sphincter causes release of urine.

III. Stretching of urinary bladder as it gets filled with urine.

IV. Stretch receptors on the walls of the bladder sends signals to the CNS.

(a) IV $\rightarrow$ III $\rightarrow$ II $\rightarrow$ I
(b) I $\rightarrow$ II $\rightarrow$ III $\rightarrow$ IV
(c) III $\rightarrow$ IV $\rightarrow$ I $\rightarrow$ II
(d) II $\rightarrow$ III $\rightarrow$ IV $\rightarrow$ I

15 Identify *A, B, C, D* and *E* with respect to reabsorption and secretion at different parts of nephron.

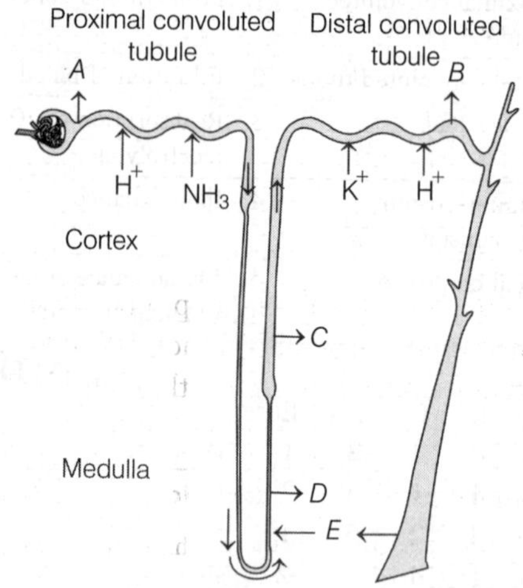

	A	*B*	*C*	*D*	*E*
(a)	HCO_3^-	Urea	NaCl	HCO_3^-	H_2O
(b)	Urea	HCO_3^-	Urea	NH_3	H_2O
(c)	HCO_3^-	HCO_3^-	NaCl	NaCl	Urea
(d)	NaCl	Urea	HCO_3^-	HCO_3^-	NH_3

16 Find out which of the following statements are true (T)/false (T) and choose the correct option with respect to the collecting duct of nephron.

I. Maintains pH and ionic balance of blood.

II. Performs selective secretion of H^+ and K^+.

III. This segment allows passage of small amounts of urea into the medullary interstitium to keep up the osmolarity.

IV. Large amounts of water can be reabsorbed to produce a concentrate urine.

	I	II	III	IV
(a)	F	F	F	F
(b)	T	T	T	F
(c)	T	F	T	T
(d)	T	T	T	T

17 Find out which of the following statements are true (T)/false (F) and choose the correct option.

I. Adult human kidney measures 10-12 cm in length, 5-7 cm in width, 2-3 cm in thickness with an average weight of 120-170 g.

II. Blood vessels, nerves and urethra enter kidney through the hilum.

III. Inner to the hilum of kidney is a broad funnel-shaped space called the renal pelvis with projections called calyces.
IV. Renal capsule is a soft layer made of perinephric adipose tissue only.

	I	II	III	IV
(a)	T	T	T	T
(b)	F	F	T	T
(c)	T	F	T	F
(d)	T	T	F	F

18 Which one of the following options gives the correct categorisation of six animals according to the type of nitrogenous wastes (*A*, *B*, *C*), they give out?

	A	*B*	*C*
	Ammonotelic	**Ureotelic**	**Uricotelic**
(a)	Pigeon, humans	Aquatic amphibians and lizards	Cockroach and frog
(b)	Frog, lizards	Aquatic amphibians and humans	Cockroach and pigeon
(c)	Aquatic amphibians	Frog and humans	Pigeon, lizards and cockroach
(d)	Aquatic amphibians	Cockroach and humans	Frog and pigeon

19 Refer the following diagram and identify the parts of a kidney indicated.

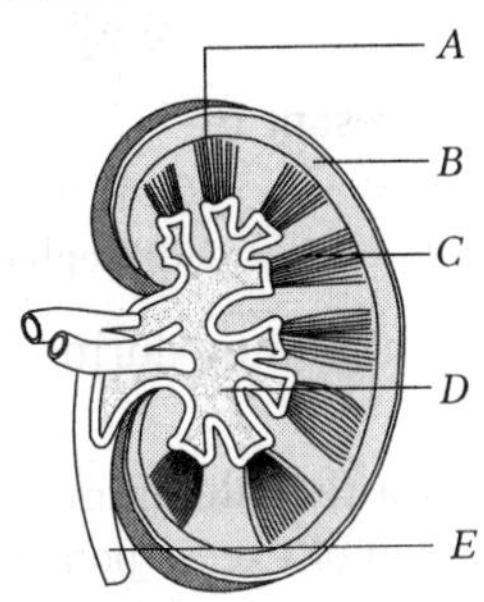

	A	*B*	*C*	*D*	*E*
(a)	Cortex	Nephron	Pelvis	Medulla	Ureter
(b)	Cortex	Medulla	Nephron	Pelvis	Ureter
(c)	Nephron	Cortex	Medulla	Ureter	Pelvis
(d)	Nephron	Cortex	Medulla	Pelvis	Ureter

20 Study the given structure and match *A*, *B*, *C*, *D*, *E*, *F* and *G* with correct option.

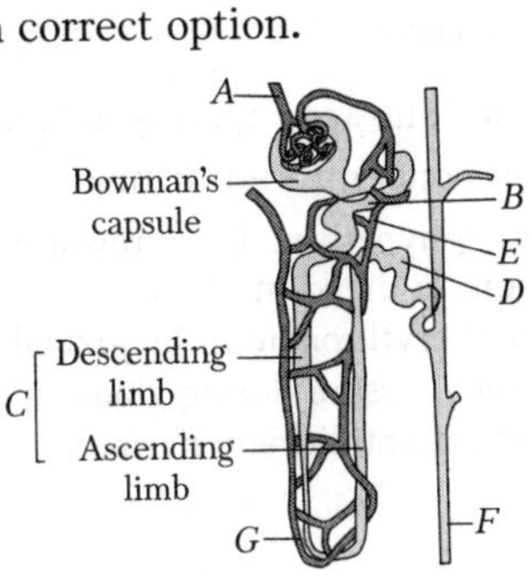

(a) *A*-Afferent arteriole, *B*-Proximal convoluted tubule, *C*-Henle's loop, *D*-Distal convoluted tubule, *E*-Peritubular capillaries, *F*-Collecting duct, *G*-Vasa recta
(b) *A*-Efferent arteriole, *B* -PCT, *C*-Henle's loop, *D* -DCT, *E*-Peritubular capillaries, *F*-Collecting duct, *G* -Vasa recta
(c) *A*-Afferent arteriole, *B*-Peritubular capillaries, *C*-Henle's loop, *D*-DCT, *E*-PCT, *F*-Collecting duct, *G*-Vasa recta
(d) *A*-Afferent arteriole, *B*-Henle's loop, *C*-Collecting duct, *D*-PCT, *E*-DCT, *F*-Peritubular capillaries, *G*-Vasa recta

21 In the skeletal form of ornithine cycle given below, some intermediate products are indicated by alphabets. Choose the answer in which these alphabets are properly matched with the name of corresponding products.

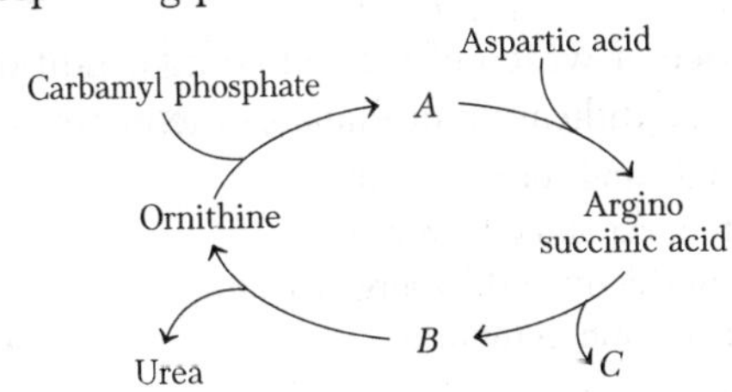

(a) *A*- Citrulline, *B* - Fumaric acid, *C* - Arginine
(b) *A*- Citrulline, *B* - Arginine, *C* - Fumaric acid
(c) *A*- Arginine, *B* - Succinic acid, *C* - Fumaric acid
(d) *A*- Citrulline, *B* - Arginine, *C* - Succinic acid

22 In reabsorption process in humans, ...*A*... and ...*B*... are reabsorbed through active transport. Glucose and amino acids are reabsorbed through ...*C*... transport. Simple diffusion is used for ...*D*... .
Fill in the blanks with appropriate options.

(a) *A*—Cl^-, *B*—water, *C*—osmosis, *D*—urea
(b) *A*—Na^+, *B*—K^+, *C*—passive, *D*—Cl^-
(c) *A*—K^+, *B*—Cl^-, *C*—active, *D*— Na^+
(d) *A*—urea, *B*—Cl^-, *C*—osmosis, *D*—K^+

Assertion and Reason

Direction (Q. Nos. 23-30) *In each of the following questions, a statement of Assertion (A) is given followed by the corresponding statement of Reason (R). Of the statements, mark the correct answer as*

(a) Both A and R are true and R is the correct explanation of A
(b) Both A and R are true, but R is not the correct explanation of A
(c) A is true, but R is false
(d) A is false, but R is true
(e) Both A and R are false

23 **Assertion** (A) Ureotelism occurs when there is excess availability of water.
Reason (R) Elasmobranches are ureotelic.

24 Assertion (A) Major portion of urinary bladder is formed of Detrusor muscle.
Reason (R) Detrusor muscles perform the function of expelling a substance.

25 Assertion (A) Land reptiles are uricotelic animals.
Reason (R) They are found in dry conditions.

26 Assertion (A) Human kidney is retroperitoneal in position.
Reason (R) It lies outside the peritoneal cavity.

27 Assertion (A) The length of Henle's loop is proportional to the concentration of urine.
Reason (R) It is opposite to concentration of urine.

28 Assertion (A) Mammals, living in deserts contain more concentrated urine.
Reason (R) They contain very long loop of Henle in their nephrons.

29 Assertion (A) Process of maintaining a constant internal environment is known as homeostasis.
Reason (R) Kidneys are excretory and homeostatic organs.

30 Assertion (A) Urea is a less toxic excretory substance comparatively to uric acid.
Reason (R) Birds and insects are ureotelic animals.

(C) Medical Entrances' Gallery

(Collection of questions Asked in NEET & Various Medical Entrance Exams)

1 Presence of which of the following conditions in urine are indicative of diabetes mellitus? **[NEET 2020]**
(a) Uremia and renal calculi
(b) Ketonuria and glycosuria
(c) Renal calculi and hyperglycaemia
(d) Uremia and ketonuria

2 Which of the following would help in prevention of diuresis? **[NEET 2020]**
(a) Reabsorption of Na^+ and water from renal tubules due to aldosterone
(b) Atrial natriuretic factor causes vasoconstriction
(c) Decrease in the secretion of renin by JG cells
(d) More water reabsorption due to undersecretion of ADH

3 Select the correct statement. **[NEET 2020]**
(a) Atrial natriuretic factor increases the blood pressure
(b) Angiotensin-II is a powerful vasodilator
(c) Counter-current pattern of blood flow is not observed in vasa recta
(d) Reduction in Glomerular Filtration Rate (GFR) activates JG cells to release renin

4 The increase in osmolarity from outer to inner medullary interstitium is maintained due to
I. close proximity between Henle's loop and vasa recta.
II. counter-current mechanism.
III. selective secretion of HCO_3^- and hydrogen ions of PCT.
IV. higher blood pressure in glomerular capillaries. **[NEET 2020]**
(a) Only II (b) III and IV (c) I, II and III (d) I and II

5 Assertion (A) Malpighian tubules are excretory organs in most of the insects.
Reason (R) These help in excretion of urea and creatinine. **[AIIMS 2019]**
(a) Both A and R are true and R is the correct explanation of A
(b) Both A and R are true, but R is not the correct explanation of A
(c) A is true, but R is false
(d) Both A and R are false

6 Which of the following is uricotelic? **[JIPMER 2019, CBSE AIPMT 2011]**
(a) Insects (b) Birds
(c) Lizards (d) All of these

7 Flame cells are present in **[JIPMER 2019]**
(a) Aschelminthes (b) Platyhelminthes
(c) Annelida (d) Cephalochordata

8 Use of an artificial kidney during haemodialysis may result in **[NEET 2019]**
I. nitrogenous waste build up in the body.
II. non-elimination of excess potassium ions.
III. reduced absorption of calcium ions from gastrointestinal tract.
IV. reduced RBC production.
Which of the following options is/are the most appropriate?
(a) II and III are correct
(b) III and IV are correct
(c) I and IV are correct
(d) I and II are correct

9 Which of the following factors is responsible for the formation of concentrated urine? **[NEET 2019]**
(a) Maintaining hyperosmolarity towards inner medullary interstitium in the kidneys
(b) Secretion of erythropoietin by juxtaglomerular complex
(c) Hydrostatic pressure during glomerular filtration
(d) Low levels of antidiuretic hormone

10 Match the following parts of a nephron (in Column I) with their functions (in Column II).

Column I	Column II
A. Descending limb of Henle's loop	1. Reabsorption of salts only
B. Proximal convoluted tubule	2. Reabsorption of water only
C. Ascending limb of Henle's loop	3. Conditional reabsorption of sodium ion and water
D. Distal convoluted tubule	4. Reabsorption of ion, water and organic nutrients

Codes [NEET 2019]

	A	B	C	D
(a)	1	3	2	4
(b)	2	4	1	3
(c)	1	4	2	3
(d)	4	1	3	2

11 Match the items in Column I with those in Column II.

Column I	Column II
A. Podocytes	1. Crystallised oxalates
B. Protonephridia	2. Annelids
C. Nephridia	3. Amphioxus
D. Renal calculi	4. Filtration slits

Codes [NEET 2019]

	A	B	C	D
(a)	3	4	2	1
(b)	3	2	4	1
(c)	4	3	2	1
(d)	4	2	3	1

12 What is the function of Bowman's capsule and glomerulus? [AIIMS 2019]

(a) Filtration of blood
(b) Reabsorption of ions from blood
(c) Reabsorption of hormones from blood
(d) Reabsorption of water from blood

13 Match the items given in Column I with those in Column II and select the correct option from the codes given below.

Column I	Column II
A. Ultrafiltration	1. Henle's loop
B. Concentration of urine	2. Ureter
C. Transport of urine	3. Urinary bladder
D. Storage of urine	4. Malpighian corpuscle
	5. Proximal convoluted tubule

Codes [NEET 2018]

	A	B	C	D
(a)	5	4	1	2
(b)	4	1	2	3
(c)	4	5	2	3
(d)	5	4	1	3

14 Match the items given in Column I with those in Column II and select the correct option given below.

Column I	Column II
A. Glycosuria	1. Accumulation of uric acid in joints
B. Gout	2. Mass of crystallised salts within the kidney
C. Renal calculi	3. Inflammation in glomeruli
D. Glomerular nephritis	4. Presence of glucose in urine

Codes [NEET 2018]

	A	B	C	D
(a)	2	3	1	4
(b)	1	2	3	4
(c)	3	1	4	2
(d)	4	1	2	3

15 A decrease in blood pressure/volume will not cause the release of [NEET 2017]

(a) renin
(b) atrial natriuretic factor
(c) aldosterone
(d) ADH

16 Which of the following statements is correct? [NEET 2017]

(a) The ascending limb of loop of Henle is impermeable to water
(b) The descending limb of loop of Henle is impermeable to water
(c) The ascending limb of loop of Henle is permeable to water
(d) The descending limb of loop of Henle is permeable to electrolytes

17 Which of the following processes of urine formation takes place all along the renal tubule and collecting duct? [AIIMS 2017]

(a) Ultrafiltration and tubular reabsorption
(b) Ultrafiltration and tubular secretion
(c) Tubular reabsorption and secretion
(d) Anti-current mechanism and reabsorption

18 The ornithine cycle removes two waste products from the blood in liver. These products are [JIPMER 2017]

(a) CO_2 and urea
(b) ammonia and urea
(c) CO_2 and ammonia
(d) ammonia and uric acid

19 The part of nephron involved in active reabsorption of sodium is [NEET-II 2016]

(a) distal convoluted tubule
(b) proximal convoluted tubule
(c) Bowman's capsule
(d) descending limb of Henle's loop

20 Which are not ureotelic? [MHT CET 2015]

(a) Mammals
(b) Terrestrial amphibians
(c) Aquatic insects
(d) Birds/snake

21 Arginosuccinase is **[EAMCET 2015, Chhattisgarh 2015]**
(a) hydrolase (b) ligase
(c) lyase (d) oxido-reductase

22 The increase in blood flow to heart stimulates secretion of **[MHT CET 2015]**
(a) renin
(b) oxytocin
(c) antidiuretic hormone
(d) atrial natriuretic factor

23 Choose the incorrect statement regarding urine formation. **[Kerala CEE 2015]**
(a) Filtration is non-selective process performed by glomerulus
(b) Glomerular capillary blood pressure causes filtration of blood through three layers
(c) GFR in a healthy individual is approximately 125 mL/min
(d) A fall in GFR activates JG cells to release renin
(e) Ascending limb of Henle's loop is permeable to water but allows transport of electrolytes actively or passively

24 Vasa recta refers to **[Kerala CEE 2015]**
(a) rectum region of insects
(b) blood capillaries in invertebrates
(c) a fine blood capillary network of afferent arteriole
(d) a fine capillary network which runs parallel to Henle's loop
(e) juxtaglomerular complex of nephrons

25 Find the incorrectly matched pair of animal and its excretory structure. **[Kerala CEE 2015]**
(a) *Balanoglossus* – Proboscis gland
(b) Earthworm – Nephridia
(c) Grasshopper – Malpighian tubules
(d) Prawn – Flame cells
(e) Amphioxus – Protonephridia

26 The condition in which kidneys fail to conserve water leading to water loss and dehydration due to impaired ADH synthesis or release is **[Kerala CEE 2015]**
(a) Graves' disease (b) Addison's disease
(c) Diabetes insipidus (d) Cretinism
(e) Acromegaly

27 Which one of the following components of urine in a healthy human does not differ much in concentration from that of blood plasma? **[WB JEE 2015]**
(a) NH_4^+ (b) K^+ (c) Na^+ (d) SO_4^{2-}

28 Human urine is usually acidic because **[CBSE AIPMT 2015]**
(a) sodium transporter exchanges one hydrogen ion for each sodium ion, in peritubular capillaries
(b) excreted plasma proteins are acidic
(c) potassium and sodium exchange generates acidity
(d) hydrogen ions are actively secreted into the filtrate

29 Which of the following statement is false?
(a) Urea is more toxic than ammonia **[UK PMT 2015]**
(b) Ammonia is converted to urea in liver
(c) Ammonia is produced in the body cells by the metabolism of proteins
(d) Fluid collected in Bowman's capsule is called glomerular fluid

30 Which determines the ability of a mammal to concentrate its urine? **[UK PMT 2015]**
(a) Number of nephrons
(b) Length of proximal convoluted tubules
(c) Length of collecting ducts
(d) Size of glomerulus

31 Concentration of urine is controlled by **[J&K CET 2015]**
(a) ACTH (b) MSH (c) ADH (d) Oxytocin

32 Choose the correct on regarding urinary excretion. **[EAMCET 2015]**
(a) Urinary excretion – Glomerular filtration - Tubular reabsorption + Tubular secretion
(b) Urinary excretion – Tubular reabsorption + Glomerular filtration - Tubular secretion
(c) Urinary excretion – Tubular secretion + Tubular reabsorption
(d) Urinary excretion – Tubular secretion - Glomerular filtration

33 Which of the following diseases shows the blockage of kidney tubules and causes severe back pain? **[Gujarat CET 2015]**
(a) Renal calculi (b) Uremia
(c) Kidney failure (d) Nephritis

34 Which segment of renal tubule is permeable to water but nearly impermeable to salts? **[Gujarat CET 2014]**
(a) Descending limb of Henle's loop
(b) Proximal convoluted tubule
(c) Ascending limb of Henle's loop
(d) Distal convoluted tubule

35 Which of the following causes an increase in sodium reabsorption in the distal convoluted tubule?
(a) Increase in aldosterone levels **[CBSE AIPMT 2014]**
(b) Increase in antidiuretic hormone levels
(c) Decrease in aldosterone levels
(d) Decrease in antidiuretic hormone levels

36 The correct match is **[MHT CET 2014]**

I. DCT – Secretion of H^+ and K^+ ions
II. Henle's loop – Reabsorption of glucose, water and Na^+ ions
III. Podocytes – Attached to parietal layer of Bowman's capsule
IV. JGA – Rise in glomerular blood pressure activates it to release renin

(a) Only III (b) Only II (c) Only I (d) Only IV

37 Identify the correct statement regarding urine formation. **[Kerala CEE 2014]**

(a) Counter-current mechanism works around the glomerulus and PCT
(b) To prevent diuresis, ADH facilitates water reabsorption from the later parts of the tubules
(c) Maximum absorption of electrolytes occurs in Henle's loop
(d) A decrease in blood pressure can increase the glomerular filtration rate
(e) The collecting duct is impermeable to water and thus helps in diluting the urine

38 Proximal convoluted tubule of nephron is responsible for **[WB JEE 2014]**

(a) filtration of blood
(b) maintenance of glomerular filtration rate
(c) selective reabsorption of glucose, amino acid, NaCl and water
(d) reabsorption of salts only

39 The accumulation of urea in the blood due to malfunctioning of kidneys is referred as **[J&K CET 2014, Kerala CEE 2011]**

(a) uremia (b) renal calculi
(c) edema (d) glomerulonephritis

40 Which are the components of ornithine cycle? **[UK PMT 2014]**

(a) Ornithine, citrulline and alanine
(b) Ornithine, citrulline and arginine
(c) Ornithine, alanine and fumaric acid
(d) Ornithine, citrulline and fumaric acid

41 Which of the following causes decreases in blood pressure? **[UK PMT 2014]**

(a) Renin (b) Angiotensin
(c) ANF (d) None of these

42 Figure shows human urinary system with structures labelled *A* to *D*. Select option which correctly identifies them and gives their characteristics and/or function. **[NEET 2013]**

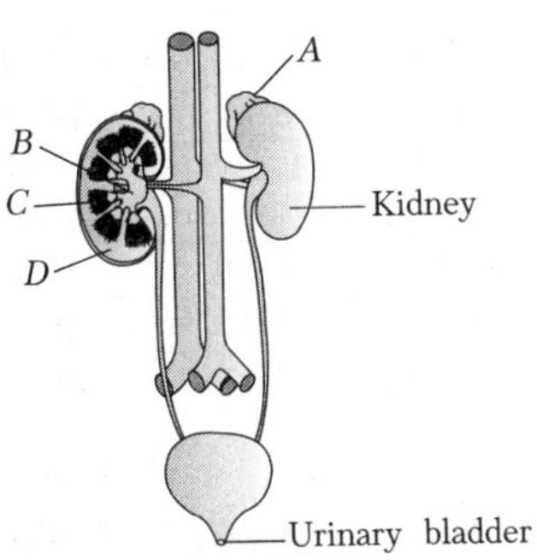

(a) *B*–Pelvis–Broad funnel-shaped space inner to hilum, directly connected to loop of Henle
(b) *C*–Medulla–Inner zone of kidney and contains complex nephrons
(c) *D*–Cortex–Outer part of kidney and does not contain any part of nephrons
(d) *A*–Adrenal glands–Located at the anterior part of kidney, secretes catecholamines which stimulate glycogen breakdown

43 Which of the following activates JG cells to release renin? **[Kerala CEE 2013]**

(a) Increase in glomerular filtration rate
(b) Passage of urea into medullary interstitium
(c) Atrial natriuretic factor
(d) Fall in glomerular filtration rate
(e) Absorption of HCO_3^- in PCT

44 In a dialysing unit, the dialysing fluid has a similar composition as the plasma except **[Kerala CEE 2013]**

(a) proteins
(b) electrolytes
(c) hormones
(d) oxygen
(e) nitrogenous wastes

45 The maximum amount of electrolytes and water (70-80%) from the glomerular filtrate is reabsorbed in which part of the nephron? **[CBSE AIPMT 2012]**

(a) Ascending limb of loop of Henle
(b) Distal convoluted tubule
(c) Proximal convoluted tubule
(d) Descending limb of loop of Henle

46 Which one of the following correctly explains the function of a specific part of a human nephron? **[CBSE AIPMT 2011]**

(a) Henle's loop	–	Most reabsorption of the major substances from the glomerular filtrate
(b) Distal convoluted tubule	–	Reabsorption of ions into the surrounding blood capillaries
(c) Afferent arteriole	–	Carries the blood away from the glomerulus towards renal vein
(d) Podocytes	–	Create minute spaces (slit pores) for the filtration of blood into the Bowman's capsule

47 Which one of the following statement is correct with respect to kidney function regulation? **[CBSE AIPMT 2011]**

(a) Exposure to cold temperature stimulates ADH release
(b) An increase in glomerular blood flow stimulates formation of angiotensin-II
(c) During summer, when body loses lot of water by evaporation, the release of ADH is suppressed
(d) When someone drinks lot of water ADH release is stopped

48 Choose the correct statement. **[Kerala CEE 2011]**
(a) The juxtamedullary nephrons have reduced Henle's loop
(b) Vasa recta is well-developed in cortical nephrons
(c) The PCT and DCT are situated in the medulla of the kidney
(d) The glomerulus encloses the Bowman's capsule
(e) The ascending limb of the Henle's loop extends as the DCT

49 The waste products produced in man which need to be excreted are **[MHT CET 2011]**
(a) carbon dioxide (b) urea and salts
(c) excess of water (d) All of these

50 The human kidney **[MHT CET 2011]**
(a) is responsible for the storage of nutrients such as glycogen
(b) concentrates the urine by actively transporting water out of the filtrate
(c) produces more dilute urine when the collecting ducts become less permeable to water
(d) responds to antidiuretic hormone by increasing urine output

51 Physiologically urea is produced by the action of an enzyme **[UP CPMT 2011]**
(a) arginase (b) urease
(c) uricase (d) None of these

52 Glucose and amino acids are reabsorbed in the **[WB JEE 2011]**
(a) proximal tubule (b) distal tubule
(c) collecting duct (d) loop of Henle

53 In 24 h, total glomerular filtrate formed in the human kidney is approximately **[WB JEE 2011]**
(a) 1.7 L (b) 7 L
(c) 17 L (d) 170 L

ANSWERS

Check point 26.1

1. (c)	2. (d)	3. (a)	4. (a)	5. (b)	6. (c)

Check point 26.2

1. (b)	2. (d)	3. (d)	4. (b)	5. (d)	6. (d)

Check point 26.3

1. (c)	2. (b)	3. (d)	4. (b)	5. (c)	6. (c)

(A) Topical Exercise

1. (b)	2. (d)	3. (b)	4. (d)	5. (d)	6. (c)	7. (b)	8. (c)	9. (b)	10. (c)
11. (b)	12. (b)	13. (d)	14. (a)	15. (b)	16. (b)	17. (a)	18. (b)	19. (b)	20. (b)
21. (b)	22. (b)	23. (c)	24. (d)	25. (a)	26. (d)	27. (d)	28. (a)	29. (a)	30. (b)
31. (b)	32. (a)	33. (a)	34. (d)	35. (b)	36. (b)	37. (d)	38. (c)	39. (a)	40. (d)
41. (c)	42. (a)	43. (c)	44. (a)	45. (d)	46. (d)	47. (a)	48. (c)	49. (b)	50. (b)
51. (c)	52. (a)	53. (b)	54. (b)	55. (d)	56. (a)	57. (d)	58. (a)	59. (d)	60. (b)
61. (c)	62. (d)	63. (d)	64. (d)	65. (d)	66. (b)	67. (a)	68. (b)	69. (c)	70. (a)
71. (d)	72. (b)	73. (b)	74. (a)	75. (a)	76. (b)	77. (b)	78. (b)	79. (c)	80. (d)
81. (c)	82. (d)	83. (c)	84. (b)	85. (b)	86. (d)	87. (b)	88. (d)	89. (a)	90. (a)
91. (c)	92. (a)	93. (a)	94. (a)	95. (c)	96. (d)	97. (a)	98. (c)	99. (b)	

(B) Cumulative Exercise

1. (b)	2. (c)	3. (d)	4. (b)	5. (c)	6. (c)	7. (a)	8. (b)	9. (d)	10. (b)
11. (d)	12. (c)	13. (a)	14. (c)	15. (c)	16. (d)	17. (c)	18. (c)	19. (d)	20. (a)
21. (b)	22. (b)	23. (d)	24. (a)	25. (a)	26. (c)	27. (c)	28. (a)	29. (b)	30. (d)

(C) Medical Entrances' Gallery

1. (b)	2. (a)	3. (d)	4. (d)	5. (c)	6. (d)	7. (b)	8. (b)	9. (a)	10. (b)
11. (c)	12. (a)	13. (b)	14. (d)	15. (b)	16. (a)	17. (c)	18. (c)	19. (b)	20. (d)
21. (c)	22. (d)	23. (e)	24. (d)	25. (d)	26. (c)	27. (d)	28. (d)	29. (a)	30. (c)
31. (c)	32. (a)	33. (a)	34. (a)	35. (a)	36. (c)	37. (b)	38. (c)	39. (a)	40. (b)
41. (c)	42. (d)	43. (d)	44. (e)	45. (c)	46. (d)	47. (d)	48. (e)	49. (d)	50. (c)
51. (a)	52. (a)	53. (d)							

Explanations

(A) Topical Exercise

4 (*d*) Statement in option (d) is incorrect and can be corrected as

Ammonotelic animals are aquatic amphibians, aquatic insects and some bony fishes, while elasmobranch fishes are ureotelic.

Rest statements are correct about ammonotelism.

5 (*d*) Statement in option (d) is incorrect for aquatic animals and can be corrected as

Ammonia is converted into less toxic form called urea in ureotelic animals, which can live in both aquatic and terrestrial area. Rest statements are correct.

17 (*a*) Statement I is true, but II is false. Incorrect statement can be correctd as

Dermal branchiae and tube feet are the excretory organs of Echinodermata.

18 (*b*) Statement in option (b) is false and can be corrected as

Nephridia are excretory organs of phylum-Annelida.

Rest statements are true.

25 (*a*) Statement I is true, but II is false. Incorrect statement can be corrected as

Renal fascia is the outermost fibrous cover linked with the abdominal wall.

28 (*a*) Statement I is true, but II is false. Incorrect statement can be corrected as

In glomerulus, afferent arteriole bring blood it has larger diameter than efferent arteriole.

31 (*b*) Statement in option (b) is correct, in mammalian kidney renal pyramids are seen in the medulla. Other statements are incorrect and can be corrected as

- The human kidneys are located on the either side of the spine at the lowest level of the ribcage.
- The bunch of capillares present in the Bowman's capsule is called glomerulus.
- Glomerulus does not belong to uriniferous tubules.

32 (*a*) Statement I is true, but II is false. Incorrect statement can be corrected as

Vasa recta are well-developed in juxtamedullary nephrons.

33 (*a*) Statement I is true, but II is false. Incorrect statement can be corrected as

The urinary bladder is absent in crocodiles, snakes and in most birds because the urine of these is semisolid consisting mostly of uric acid.

35 (*b*) Option (b) is incorrect and can be corrected as

Blood enters into glomerulus through afferent arteriole.

Rest options are correct about kidney.

44 (a) Statement I is true, but II is false. Incorrect statement can be corrected as

The starting point is ammonia which produced in the body during the amino acid catabolism is converted into less toxic urea by urea biosynthesis.

47 (*a*) Option (a) is the correct match pair.

About 70% to 80% of electrolytes and water is reabsorbed by PCT. It also helps in maintaing pH and ionic balance of our body.

Other options are not correct match pairs and can be corrected as

- DCT (Distal Convoluted Tubule) - Conditional absorption of Na^+, water and HCO_3^-
- Henle's loop - Maintenance of high osmolarity of the medullary tissue fluid
- Descending limb - Passive reabsorption of water

57 (*d*) Glomerular Hydrostatic Pressure (GHP) = 60 mmHg

Blood Colloidal Osmotic Pressure (BCOP) = 30 mmHg

Capsular Hydrostatic Pressure (CHP) = 20 mm Hg

Glomerular Filtration Pressure (GFP) = GHP – (BCOP + CHP)

$\Rightarrow \quad 60 - (30 + 20)$

$\Rightarrow \quad 60 - 50 = 10$ mmHg

Hence, option (d) is correct.

58 (*a*) Statement I is true, but II is false. Incorrect statement can be corrected as

The descending limb is permeable to water, but impermeable to sodium ions, while the ascending limb is impermeable to water, but permeable to sodium ions which make the filtrate dilute.

82 (*d*) Option (d) is incorrect match with respect to human urine and can be corrected as

Glucose is not normally found in urine. If glucose is found in urine it may be a sign of diabetes.

Rest options are correctly matched with respect to human urine.

88 (*d*) Option (d) is an incorrect match pair and can be corrected as

The zona glomerulosa, the outermost layer of the adrenal cortex is responsible for reproducing mineralocorticoids such as aldosterone.

Rest options are correctly matched pairs.

89 (*a*) Matched pair in option (a) is incorrect and can be corrected as

ADH - Active reabsorption of Na^+

Rest match pairs are correct.

94 (*a*) Statement I is true, but II is false. Incorrect statement can be corrected as

Presence of blood or blood cell in urine is called haemoglobinuria, ketonuria is a condition of presence of abnormally high ketone bodies urine.

99 (*b*) Option (b) is correct with reference to haemodialysis. Haemodialysis is done in the patients whose kidneys become non-functional. It is the process of diffusion across a semipermeable membrane to remove unwasted substances from the blood while adding derivable components. The dialysis unit consists of a coiled cellophone tube kept it the dialysing fluid. The membrane of the tube is permeable to urea, uric acid, creatinine and mineral ion.

(B) Cumulative Exercise

1 (*b*) Statement in option (b) is incorrect and can be corrected as Inside the kidney, the cortical region extends in between the medullary pyramids as renal columns are called column of Bertini.

Rest of the statements are correct.

2 (*c*) Statement in option (c) is only correct whereas other statements are incorrect. Incorrect statements can be corrected as

- Descending limb of loop of Henle is permeable to water but impermeable to sodium.
- DCT can conditionally absorb Na^+ ions.
- Ascending limb of loop of Henle is permeable to electrolytes.

3 (*d*) Statement in option (d) is incorrect and can be corrected as The counter-current system is used to concentrate urine in the kidneys.

Rest statements are correct regarding mechanism of urine formation in man.

5 (*c*) I and II are correct combinations, while II and III are not correct combinations. Incorrect combinations can be corrected as

- Glycosuria – Presence of glucose in urine
- Albuminuria – Presence of albumin in urine

12 (*c*) Statements I and V are correct, while statements II, III and IV are incorrect. Incorrect statements can be corrected as

- About 180 L filtrate is formed per/day but only 1.5 L of urine is released per day by kidneys.
- Reabsorption is a selective process where, certain molecules after being filtered out of the capillaries along with urea and water are reabsorbed.
- The tubular epithelial cells in different segments of nephron perform reabsorption by both active and passive process.

13 (*a*) Statements I and V are correct, while II, III and IV are incorrect. Incorrect statements can be corrected as

- Each human kidney possesses about 1 million nephrons approximately.
- Renal system controls the volume of plasma under stress.
- Vertebrates can maintain a constant internal osmolarity different from the surrounding medium except hagfish.

17 (*c*) Statements I and III are true, while statements II and IV are false. Incorrect statements can be corrected as

- Urethra, blood vessels, lymph vessels and nerves enter and leave at the renal hilum.
- The renal capsule is a tough fibrous layer surrounding the kidney and covered in a layer of perirenal fat known as the apipose capsule of kidney.

23 (*d*) A is false, but R is true. A can be corrected as Ureotelism occurs when there is less availability of water. Ureotelic animals inhabit terrestrial as well as aquatic areas, e.g. elasmobranch.

24 (*a*) Both A and R are true and R is the correct explanation of A.

Detrusor muscles are contractile muscles and they expel the excretory material from urinary bladder by peristaltic waves.

25 (*a*) Both A and R are true and R is the correct explanation of A.

Uricotelic organisms excrete uric acid so, as to conserve water in their body. They inhabit dry areas and excrete non-toxic uric acid, e.g. land reptiles.

26 (*c*) A is true, but R is false because

Human kidney is retroperitoneal in position and it lies inside the peritoneal cavity.

27 (*c*) A is true, but R is false.

There is a correlation between the length of loop of Henle and the concentration of the urine produced. The longer, the length of Henle's loop, the greater, the concentration gradient and more concentrated the urine will be.

28 (*a*) Both A and R are true and R is the correct explanation of A.

Functional unit of kidney is known as nephron. Nephrons contain a tubular region between proximal and distal convoluted tubule, the loop of Henle.

The length of loop of Henle is proportional to the concentration of urine. The mammals (chordates) living in desert contain longer loop of Henle in their nephrons. Due to this, these animals produce more concentrated urine.

29 (*b*) Both A and R are true, but R is not the correct explanation of A.

Homeostasis maintains the stability of the cell enivronment and this way it provides the organism with a degree of independence of the environment in order to achieve stability. Mammals have two kidneys which perform excretory functions and maintain urea level in blood.

30 (*d*) A is false, but R is true because

Highly toxic excretory substance is ammonia. Urea is less toxic than ammonia and uric acid is the least toxic excretory substance released by animals.

Birds and insects are ureotelic animals as their excretory substance is uric acid.

(C) Medical Entrances' Gallery

1 (*b*) Presence of glucose (glycosuria) and ketone bodies (ketonuria) in urine are indicative of diabetes mellitus. In diabetes mellitus, the body produces excess ketones as an indication that it is using an alternative source of energy. It is seen more commonly in type I diabetes mellitus. Presence of glucose indicates Type II diabetes. In some cases, insulin cannot transport blood sugar into the body's cell effectively. This can also cause blood sugar to be passed out in urine.

2 (*a*) Adrenal cortex secretes mineralocorticoids, i.e. aldosterone, which increases the reabsorption of Na^+ and water from renal tubules thereby preventing diuresis.

3 (*d*) Statement in option (d) is correct.

Other statements are incorrect and can be corrected as

- Atrial Natriuretic Factor (ANF) causes vasodilation and thereby decreases the blood pressure.
- Angiotensin-II is a powerful vasoconstrictor, it increases the GFR.
- Counter-current pattern of blood flow is observed in vasa recta.

4 (*d*) The close proximity between the Henle's loop and vasa recta as well as the counter-current in them help in maintaining an increasing osmolarity towards the inner medullary interstitium, i.e. from 300 $mOsmolL^{-1}$ in the cortex (outside) to about 1200 $mOsmolL^{-1}$ in the inner medulla (inside).

Thus, option (d) is correct.

5 (*c*) A is true, but R is false because

In most of the members of phylum—Arthropoda, excretion takes place through Malpighian tubules to maintain osmolarity, but they excrete their nitrogenous wastes as uric acid, which minimises the loss of water and are thus called uricotelic.

6 (*d*) All given animals are uricotelic organisms. These organisms excrete uric acid as a nitrogenous waste product. Such organisms are mostly found in dry climatic conditions where there is a need to conserve water in their bodies.

7 (*b*) Flame cells present in Platyhelminthes are excretory organs of these animals. Aschelminthes have Renette cells as excretory organs.

Annelids and cephalochordates have nepheridia as excretory organs.

8 (*b*) III and IV are the most appropriate in given statement.

Use of an artificial kidney during haemodialysis may result in reduced RBC production and reduced absorption of calcium ions from gastrointestinal tract. The former would occur due to the low level of erythropoietin hormone secreted by juxtaglomerular cells.

The later would be caused due to the elimination of calcium ions along with phosphate ions during dialysis.

I and II are not appropriate in given statement because

Dialysis helps in the removal of nitrogenous waste and potassium ions from the body.

9 (*a*) Factor in option (a) is responsible for the formation of concentrated urine.

Maintaining hyperosmolarity towards inner medullary interstitium in the kidneys is the factor responsible for the formation of concentrate urine because it provides concentration gradient necessary for water reabsorption in renal tubules.

Rest factors are not responsible for the formation of concentrated urine and can be corrected as

- Erythropoietin secretion by juxtaglomerular complex is related to red blood cell production.
- Hydrostatic pressure during glomerular filtration is related to amount of filtrate formed by kidneys.
- High levels of antidiuretic hormone and not low levels produce urine that is more concentrated.

12 (*a*) The function of Bowman's capsule and glomerulus is filtration of blood. It is known as glomerular filtration or ultrafiltration. Reabsorption of ions, hormones and water does not take place in it.

15 (*b*) A decrease in blood pressure/volume will not cause the release of Atrial Natriuretic Factor (ANF). It is produced by atria of heart during increased blood pressure/volume. It can cause vasodilation and thereby, decrease the blood pressure. On the other hand, renin, aldosterone and ADH are released during decreased blood pressure.

16 (*a*) Statement in option (a) is correct. Other statements are incorrect and can be corrected as

The ascending limb of loop of Henle is impermeable to water and permeable to electrolytes, i.e. K^+, Cl^- and Na^+ and partially permeable to urea. Due to this, sodium, potassium, calcium, magnesium and chloride are reabsorbed here, thus, making the filtrate hypotonic to blood plasma. Descending limb of Henle's loop on the other hand is permeable to water and impermeable to electrolytes. Hence, the filtrate becomes concentrated while moving down through this limb.

17 (*c*) Tubular reabsorption and tubular secretion take place all along the renal tubule and collecting duct, while ultrafiltration takes place in renal corpuscle. Counter-current mechanism helps in the concentration of urine.

18 (*c*) The ornithine cycle removes CO_2 (derived from carbohydrates catabolism) and ammonia (derived from deamination of amino acids) from the blood in liver. CO_2 and ammonia are joined to form carbamyl phosphate in the presence of carbamyl phosphate synthetase and 2 ATP molecules. Later on, in ornithine cycle, urea is formed which is transported to kidneys for excretion in urine.

19 (*b*) Proximal convoluted tubule is involved in active reabsorption of sodium. The majority (about 70%) of sodium is reabsorbed here, into the cytosol of the epithelial cells of the nephron.

21 (*c*) Arginosuccinase is a lyase as it breaks argininosuccinic acid into fumaric acid and arginic acid.

22 (*d*) Increase in blood flow to heart stimulates secretion of atrial natriuretic factor. It acts when blood volume and pressure is high. It causes vasodilation of walls.

23 (*e*) Statement in option (e) is incorrect regarding urine formation and can be corrected as

Ascending limb of Henle's loop is impermeable to water and it does not allow transport of electrolytes. It makes the hypertonic urine in descending limb to be hypotonic which again becomes isotonic in distal convoluted tubule.

Rest statement are correct regarding urine formation.

25 (*d*) Option (d) is incorrectly matched pair of animal and its excretory structure and can be corrected as

In prawns, excretion occurs by the general body surface and flame cells are excretory structures in Platyhelminthes.

Rest options are correctly matched pairs.

26 (*c*) Diabetes insipidus is a condition characterised by excessive thirst and excretion of dilute urine with the reduction of fluid intake. ADH maintains the amount of electrolytes.

29 (*a*) Statement in option (a) is false and can be corrected as Ammonia is the highly toxic excretory product.
Rest statements are true.

30 (*c*) The longer the loop of Henle (collecting duct), the greater is the ability of an animal to form hypertonic urine.

33 (*a*) Renal calculi is the condition in which stones are formed in kidney by the precipitation of uric acid or oxalate. This causes a severe back pain.

34 (*a*) Out of the options given, descending limb of Henle's loop is permeable to water, i.e. water gets reabsorbed here due to increasing osmolarity of interstitial fluid, but it is impermeable to salts, i.e. sodium and electrolytes. While, the other given options can be explained as

- **Proximal Convoluted Tubule** (PCT) It reabsorbs nearly all essential nutrients, electrolytes and water.
- **Ascending limb of loop of Henle** It is impermeable to water but permeable to electrolytes like K^+, Cl^-, Na^+ and partially permeable to urea.
- **Distal Convoluted Tubule** (DCT) It is responsible for reabsorption of water, sodium and secretion of hydrogen and potassium.

46 (*d*) Option (d) explains the function of a specific part of a human nephron.

Podocytes or visceral epithelial cells are the cells in Bowman's capsule of the kidney that wrap around the capillaries of glomerulus. They create minute pores (slit pores) for the filtration of blood into the Bowman's capsule.

47 (*d*) Statement in option (d) is correct with respect to kidney function regulation.

When someone drinks a lot of water which is not required by body, the osmolarity of the blood decreases. This decreases in osmolarity and inhibits the release of ADH.

48 (*e*) Statement in option (e) is correct.
Other statements are incorrect and can be corrected as

The ascending limb of Henle's loop continues as another highly coiled tubular region called Distal Convoluted Tubule (DCT). The juxtamedullary nephrons have long Henle's loop. Vasa recta is absent or highly reduced in cortical nephrons.
Bowman's capsule encloses the glomerulus.

The Malpighian corpuscles, Proximal Convoluted Tubule (PCT) and Distal Convoluted Tubule (DCT) are situated in the cortical region of the kidney, whereas the Henle's loop is in the medulla.

50 (*c*) From the DCT, the filtrate enters the collecting tubule, where further reabsorption of water takes place. Now, the filtrate becomes more concentrated which makes the urine hypertonic.

When the collecting duct becomes less permeable to water it produces more dilute urine.

51 (*a*) Urea cycle takes place in liver cells with the hydrolytic enzyme arginase. Arginase splits arginine into urea and ornithine with the elimination of a water molecule.

52 (*a*) Proximal convoluted tubule of nephron is a pivotal site for reabsorption of glucose, amino acids, Na^+, K^+ by active transport. Here, 80% of water is reabsorbed by passive transport.

53 (*d*) The blood pressure in glomerular capillaries becomes very high, so that there is continuous process of ultrafiltration. Glomerular Filtrate (GF) contains a large amount of water and other dissolved substances such as urea, uric acid, creatinine, amino acids, glucose, sodium, potassium, vitamins, etc. About 120 mL GF is produced per minute, i.e. approximately 170 L filtrate is produced in 24 hours.

CHAPTER 27

Movement and Locomotion

Movement is a change in posture or position of any body part of a living organism. When movement results in change of place or location, it is termed as **locomotion**. Animals show both movement and locomotion, while plants show movement only. Locomotion is rare in them. The type of movements in animals usually varies and depends upon the body part involved in it.

In this chapter, we will study different aspects and processes related to movement and locomotion in animals.

Types of movements

Living beings show following two basic types of movements

Non-muscular movements

These movements do not involve any kind of muscular tissue in them. They are further of following types

(i) **Amoeboid or Pseudopodial movement** It occurs with the help of pseudopodia, which are formed by cytoplasm. It is found in protozoans, leucocytes and macrophages for capturing food and to change their shape. Cytoskeletal elements like microfilaments also exhibit amoeboid movement.

(ii) **Ciliary movement** This type of movement occurs in protozoans and in the internal tubular organs of higher animals. These structures are lined by ciliated epithelium. For example, ciliated movement in trachea helps us in removing waste substances inhaled along with air. Likewise, cilia also facilitate the passage of ova through the female reproductive tract.

(iii) **Flagellar movement** This type of movement occurs in protozoans like *Trypanosoma, Euglena*, etc., along with sponges and coelenteron. In sponges, flagella of choanocyte cells maintain a regular current of water through the canal system. In coelenterates like *Hydra*, certain gastrodermal cells circulate fluid by regular beating of flagella. In case of human beings, the sperms swim in water or in female genital tract with the help of flagellar movements only.

Muscular movement

It is the basic type of movement used in majority of vertebrates including humans. It takes place through muscle fibres of muscular tissue. These fibres have the ability to exert force on body parts by changing their length *via* cycles of contraction and relaxation.

Inside

MUSCULAR SYSTEM

Muscle is a specialised tissue of mesodermal origin. In humans, muscles constitute about 40-50% of body weight and these are approximately 639 (317 paired and 5 unpaired) in number. Some important human muscles are shown in the diagrams below

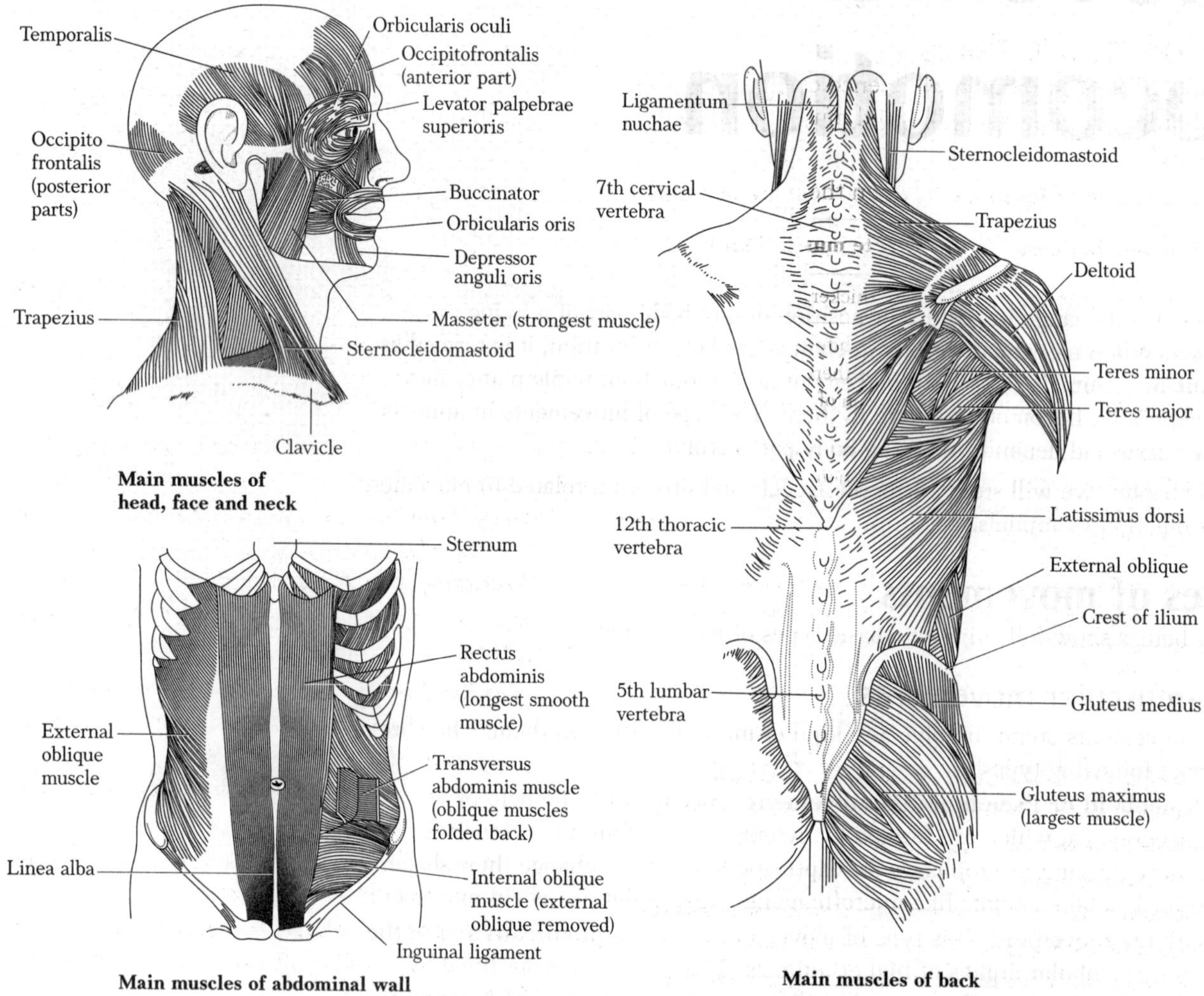

Fig. 27.1 Some important muscles in human

Muscular tissue is made up of specialised cells called **myocytes** that remain bound together. They possess various unique properties like excitability, contractibility, extensibility and elasticity, etc.

Types of muscles

Based on their location, structure and function, muscles are of following types

(i) Skeletal or striped or striated or voluntary muscles.

(ii) Visceral or smooth or unstriped or non-striated or involuntary muscles.

(iii) Cardiac muscles

Among these, skeletal muscles are found abundantly in human body. Subsequently, we will discuss the detailed account of skeletal muscles only.

Skeletal muscles

These are composed of elongated, narrow, cylindrical fibres called **muscle fibres**. These fibres are unbranched and their ends are blunt.

Types of muscle fibres

Depending upon the presence or absence of myoglobin, muscle fibres are of two types, i.e. red muscle fibre and white muscle fibre.

The characteristic differences between them are as follows

Red muscle fibres	White muscle fibres
These are thinner fibres.	They are thicker.
These are dark red in colour due to the accumulation of more myoglobin.	These are white or pale yellow in colour as there is less or no accumulation of myoglobin.
These contract slowly, but stay contracted for longer duration. Hence, they are also known as slow muscle fibres.	These contract rapidly, but for shorter duration. Hence, are called fast muscle fibres.
These are found in muscle required to undergo prolonged contraction, e.g. extensor muscle on back of human body.	These are found in muscles that contract faster, e.g. muscles of the eyeball.

Structure of skeletal muscle fibres

- Each skeletal muscle is syncytial or multinucleated with characteristic cross striations.
- The plasma membrane covering the fibre is called **sarcolemma** whereas their cytoplasm is called **sarcoplasm**.
- Sarcoplasm of each muscle fibre consists of long, proteinaceous fibrils called **myofibrils**. The mitochondria and SER present in sarcoplasm are called **sarcosomes** and **sarcoplasmic reticulum**, respectively.
- In the sarcoplasm of each muscle fibre, glycogen and myoglobin are found abundantly.
- Each skeletal muscle has three layers of connective tissue. These layers enclose the muscles and compartmentalise the muscle fibres.
- **Epimysium** is the outer sheath of dense, irregular connective tissue. The transverse section of epimysium contains muscle bundles or fasciculi.
- **Perimysium** is the middle layer of connective tissue. It surrounds the fascicles (bundles of skeletal muscle fibres). Each fasciculus contains a large number of muscle fibres.
- **Endomysium** surrounds each muscle fibre or myofibril within the fascicle. It is the innermost thin connective tissue layer of collagen and reticular fibres.

The diagram showing all the three layers is given below

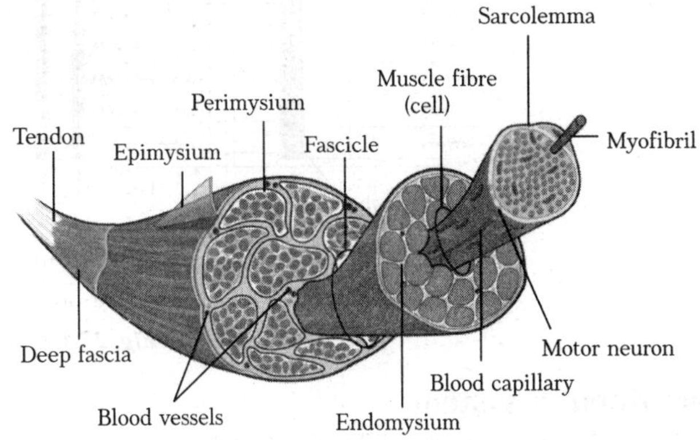

Fig. 27.2 The three connective tissue layers

- The myofibrils of skeletal muscles contain alternating dark and light bands.
  (i) **Dark band** The dark band is double refractive to polarised light that is why, it is also called **Anisotropic** or **A-band.**
  Various structures associated with A-band are as follows
    - **H-band** At the centre of A-band (dark), a comparatively light band or zone is present. It is called H-band or Henson's zone.
    - **M-line** A dark M-line passes through the centre of H-band.
    - **O-zone** The rest of the dark portion of A-band is called O-zone (overlapping zone).

  (ii) **Light band or Q-band** It is mono-refractive to polarised light. Thus, it is called as **Isotropic** or **I-band** or **J-band.**
    - The light I-band consists of a dark line which passes through its centre. This line is called as **Z-line** (Zwischenscheibe line) or **Dobie's line** or **Krause membrane.**
    - The portion between two successive Z-lines is called **sarcomere.** It is the contractile unit of skeletal muscles.
    - The length of a sarcomere is about 2 to 3 μm.
    - It comprises of a single A-band and half of each adjacent I-band.
    - On both the sides of Z-line, the N-line is present. It appears as dark, thin and transverse structure.

All the above described structures are shown in the diagram below (in outline view)

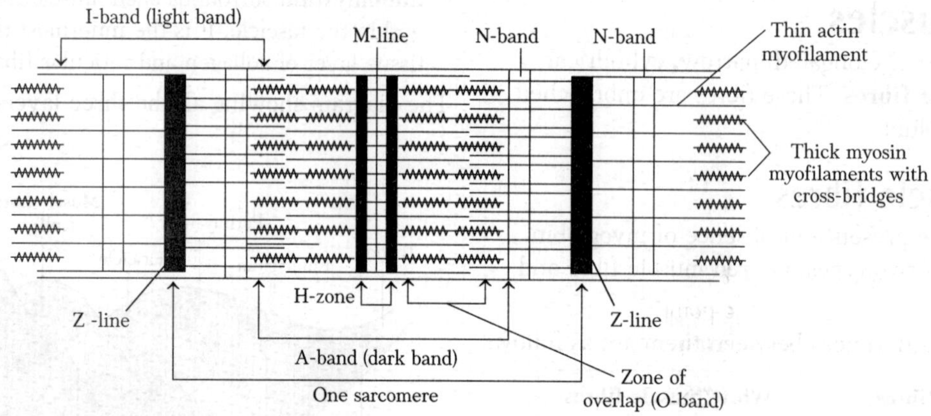

Fig. 27.3 Components of a sarcomere

Sarcotubular system

- Myofibrils are surrounded by a network of sarcoplasmic reticulum and T-tubules.
- The **T-tubules** are transverse invaginations of sarcolemma.
- The arrangement of sarcoplasmic reticulum and T-tubules forms a **triad.** It is because this network consists of a single transverse tubule lined by two enlarged cisternae of sarcoplasmic reticulum.
- The T-tubule and sarcoplasmic reticulum together forms the **sarcotubular system**. It helps to perform muscle contraction and relaxation.
- T-tubules are involved in rapid transmission of action potential from cell membrane to muscle fibres.
- On the other hand, sarcoplasmic reticulum aids in Ca^{2+} movement.

Molecular structure of skeletal muscle fibres

- The myofibrils are composed of bundles of parallel protein microfilaments called **myofilaments.** These myofilaments are made up from four types of proteins that can be categorised in two types.
  - (i) **Contractile proteins** These include myosin and actin because they are responsible for shortening of muscle fibre.
  - (ii) **Regulatory proteins** These include tropomyosin and troponin because they act like a switch to determine the contraction.
- The key structural proteins present in muscles are **titin** (connectin), **myomesin** and **dystrophin.**
- Myofilaments can be thick or thin depending upon their diameter. The thick myofilaments have diameters of about 150 Å whereas the thin myofilaments have diameter of about 70 Å.

Thick myofilaments

- The thick myofilament mainly consists of polymerised myosin protein. It contributes 55% of muscle protein by weight. It is confined to A-band only.
- Each myosin molecule is composed of six polypeptide chains, i.e. two identical heavy chains and four light chains.
- The **two heavy chains** wrap spirally around each other to form a double helix. However, one end of each of these chains is folded as a globular protein mass called **myosin head.** Thus, there are two free heads lying side by side each at one end of the double helix. This condition is called **bilamellate** condition. The elongated part of the coiled helix is called **tail.**

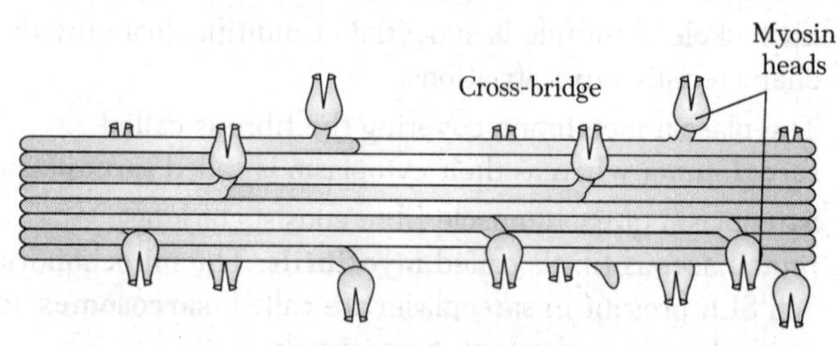

Fig 27.4 The representation of myosin molecules within the myosin filaments

- The four light chains are also included in the head region as shown in the fig (27.5).
- A small neck region is also found between these two regions. Thus, each myosin molecule appears as a tadpole with bilamellate head, a neck and a long tail.

Fractional differentiation

- This can be done with the help of enzyme trypsin.
- Myosin molecule is splitted by trypsin into two fragments, namely **HMM**, i.e. **Heavy meromyosin** and **LMM**, i.e. **Light meromyosin.**

- The word **mero** here signifies part of HMM on further cleavage gives two subfragments namely
  (i) **S-1 (globular subfragment)** Two in number each S-1 fragment contains one actin binding site and one ATPase site. Both of these sites are helpful in cross-bridge formation during muscle contraction.
  (ii) **S-2 (rod-shaped subfragment)** One in number, the S-2 fragment and LMM do not have any such functional specialisation.
  In outline view, these molecules appear as shown in diagram given below

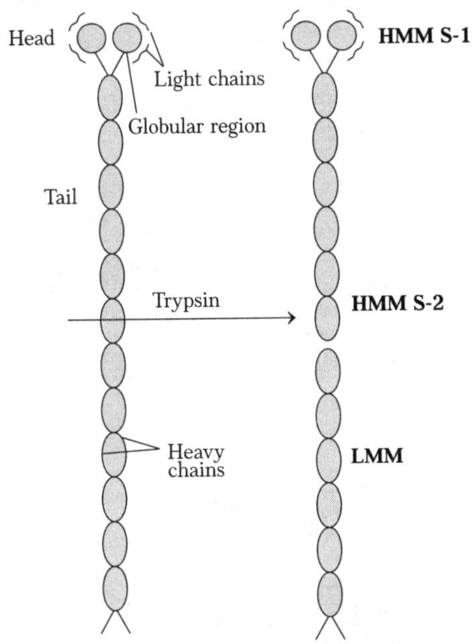

Fig. 27.5 Diagram of a myosin molecule and its enzymatic cleavage

Thin myofilaments

A thin myofilament occurs in I-band and extends into A-band. It is composed of following three different proteins

(i) **Actin** It is a globulin protein having low molecular weight.
- It occurs in two forms, the monomeric **G-actin (globular)** and the polymeric **F-actin (fibrous)**.
- G-actin polymerises to the fibrous form F-actin in the presence of Mg^{2+}.
- **Function** G-actin molecules contain myosin binding sites over them.These sites are helpful in cross-bridge formation during contraction of a muscle.

(ii) **Tropomyosin** It is a double-stranded, fibrous molecule.
- The two filaments of tropomyosin run closely along the entire length of actin.
- **Function** At resting state, tropomyosin prevents the contraction of muscle fibres. It prevents the exposure of myosin binding site of actin. Hence, the formation of cross-bridges between myosin and actin is prevented.

(iii) **Troponin** It is a complex of three polypeptides, i.e. trimeric protein.
- It is found at regular intervals on tropomyosin.
- **Function** It is responsible for keeping tropomyosin in place when muscle is relaxed.
- The three components of troponin have different binding properties, i.e.
  - **Troponin-I** It helps in binding troponin components together, i.e. troponin-T and C along with binding of troponin to F-actin.
  - **Troponin-T** It binds to tropomyosin.
  - **Troponin-C** It has affinity for calcium. So, it binds to calcium whenever available. This strong affinity of the troponin for calcium ions is believed to initiate the contraction process.

All the components of thin myofilaments are shown in the diagram given below

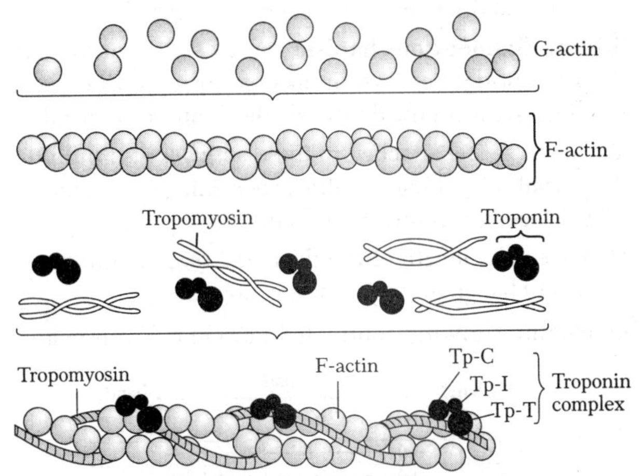

Fig. 27.6 Representation of the thin myofilament

Mechanism of muscle contraction

The mechanism of muscle contraction is explained by **sliding filament theory**. This theory was given by two groups of workers namely **Andrew Huxley** and **Ralph Niedergerke, Hugh Huxley** and **Jean Hanson** in 1954.

The detailed process of muscle contraction is explained in stepwise manner below

- The process of muscle contraction begins, when a neural signal reaches the neuromuscular junction from CNS *via* motor neurons. **Neuromuscular junction** is the junction between the nerve fibre and muscle fibre.
- The impulses then reach the terminal ends of axons of neuromuscular junction. Due to this, the vesicles having

neurotransmitter, i.e. acetylcholine fuses with the axon membranes. Thus, acetylcholine is released.

- This acetylcholine travels through **synaptic cleft** (region between axon membrane and motor end plate) and an action potential is generated within the sarcolemma.
- This action potential is transmitted to the T-tubules and then to sarcoplasmic reticulum. This relay is aided by a protein called **dystrophin**.
- Sarcoplasmic reticulum upon receiving the action potential release Ca^{2+} ions into the sarcoplasm, around the thick and thin filaments. Ca^{2+} binds with troponin-C and changes its shape. It moves the troponin-tropomyosin complex away from the myosin binding sites on actin and thus, these sites get exposed.
- These events are accompanied by the ATPase activity of globular head of myosin. It starts hydrolysing ATP molecule. The energy released is utilised by myosin cross-bridges to bind with actin.

$$\text{ATP} \xrightarrow[Ca^{2+},\ Mg^{2+}]{\text{Myosin ATPase}} \text{ADP} + \text{Pi} + \text{Energy}$$

- After the myosin-actin, cross-bridges are formed the contraction of muscle fibre begins by the sliding of thin filaments (actin) over thick filaments (myosin).
- **Sliding filament mechanism** During sliding movement, the cross-bridges of myosin undergo rotation. Due to this rotation, actin is pulled towards the centre of A-band.
- Thus, the Z-line attached to these actins is also **pulled**. As a result of pulling of Z-line, shortening of sarcomere, takes place, i.e. contraction occurs.
- However, the length of the thick and thin myofilaments does not change during muscle contractions.

The diagram showing contracted muscle is given below

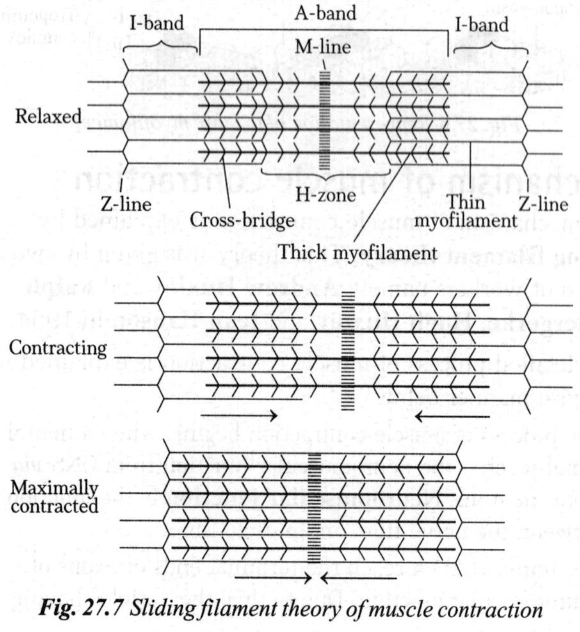

Fig. 27.7 Sliding filament theory of muscle contraction

- The myosin further releases ADP + Pi and becomes relaxed. Again, ATP binds to myosin head and cross-bridges are broken between myosin and actin.
- ATP is again hydrolysed and this cycle of formation and breakage of cross-bridges is repeated, thus causing sliding of filaments.
- The process continues till calcium ions are pumped back into sarcoplasmic reticulum.

All the above mentioned steps are shown in diagram given below

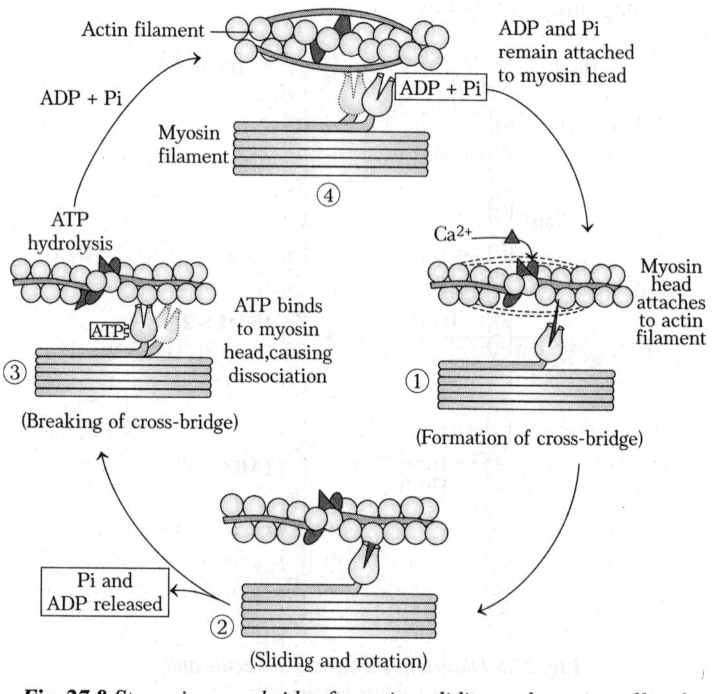

Fig. 27.8 Stages in cross-bridge formation, sliding and rotation of head and breaking of cross-bridge

Mechanism of muscle relaxation

The steps involved in this process are as follows

- It is believed that relaxation may begin due to an enzyme called **acetylcholinesterase.**
- This enzyme causes the breakdown of acetylcholine present in the synaptic cleft into acetate and choline. As a result of this, generation of action potential is retarded.
- This causes calcium ions (Ca^{2+}) to return quickly to the sarcoplasmic reticulum by active transport with the help of ATP.
- In the sarcoplasmic reticulum, Ca^{2+} binds to a calcium-binding protein called **calsequestrin.**
- This promotes troponin to again cover the myosin binding sites on actin.
- The myosin-actin cross-bridge formation stops. This in turn, prevents the sliding of filaments, thus prevent muscle contraction.

Energy release during muscle contraction

The ultimate source of energy for the process of muscle contraction is **glycogen**. It is stored in the muscles along with glucose.

Following points are important regarding the energy release

- Carbohydrates are broken down in the muscle both aerobically and anaerobically.
- In aerobic breakdown, oxygen is made available by oxymyoglobin and also from blood.
- In anaerobic breakdown, glucose is broken down through anaerobic glycolysis. This forms pyruvate which is then converted into lactic acid along with two molecules of ATP.
- ATP is an immediate source of energy for muscle contraction.
- The muscle also contains reserve energy currency in the form of Creatine Phosphate (CTP) or phosphocreatine. It gives its energy and phosphate to ADP, so as to produce ATP. It is formed in relaxed muscles by using ATP which is produced by the oxidation of carbohydrates.

Functional features of skeletal muscles

Skeletal muscles have following functional features

- They show contraction only when stimulated, i.e. they do not have the potential of automatic contractions just like found in cardiac and visceral muscles.
- A skeletal muscle may show **tonic contractions**. Some other types of contractions are treppe, fibrillation and convulsions. All of these are discussed after mechanism of contraction.

Tonic contractions

- Muscle contraction is said to be **isometric**, when the muscle does not shorten during contraction. In this system, the muscle contracts against a force transducer without decreasing the muscle length.
- The muscle contraction is called **isotonic**, when the muscle shorten while the tension on the muscle remains constant. In this system, the muscle shorten against a fixed load.
- A **twitch contraction** is a quick, jerky contraction in response to a single stimulus.
- A **tetanic contraction** (tetanus) is produced by a series of stimuli bombarding the muscles in rapid succession.

- Skeletal muscles contract with varying degrees of strength at different times.
- Skeletal muscles produce movements by pulling the bones.
- A contracting muscle applies a pulling force on a bone (lever) at the point of the muscle's attachment to the bone. This causes the bone to move about its joint (fulcrum).
- Skeletal muscles almost always act in groups rather than singly. Most movements are produced by the coordinated action of several muscles.
  Some of the muscles in the group contract, while others relax.

Types of skeletal muscles

Skeletal muscles have been classified on the basis of their functions.

Table 27.1 Functional classification of skeletal muscles

Types of skeletal muscles	Functions	Examples
Flexor	Muscles which bend one part of the body over the other.	Biceps bending forearm towards upper arm.
Extensor	Muscles which extend or straighten the limbs. It is antagonistic to flexor.	Triceps extending forearm.
Abductor	Muscles which pull a limb away from the body's median line.	Deltoids of shoulder.
Adductor	Muscles which bring a limb towards the median line of the body's. It is antagonistic to abductor.	Latissimus dorsi, which draws the whole forelimb towards the body and is antagonistic to deltoids.
Depressor	Muscles which lower some part.	Depressor anguli oris of lower jaw (similarly pectoralis major is the depressor muscle for the wings of birds).
Elevator	Antagonistic to depressors, as they raise a body part.	Masseter, which lifts the lower jaw (similarly pectoralis minor is the elevator muscle for the wings of birds).
Pronator	The muscle that turns the palm downward or backward.	Pronator rotates forearm to turn the palm downward or backward.
Supinator	Antagonistic to pronator, i.e. turns the palm upward or forward.	Supinator contracts forearm, to make palm face upward or forward.
Sphincter	Decreases the size of an opening and close it.	Pyloric sphincter of alimentary canal.
Dilator	The muscles around the openings, which increase their size and open them, antagonistic to sphincters.	Iris in the eyes.
Rotator	Associated with rotatory movements of a body part.	Pyriformis, which raises and rotates the thigh.

Contraction in smooth muscles

- Smooth muscles contract slowly and the contraction lasts for a longer period.
- It is regulated by acetylcholine, nor-epinephrine local chemical changes (e.g. pH, CO_2 and O_2 level, etc.) and action potentials from ANS.
- Smooth muscles do not contain troponin. Therefore, the regulator protein, calmodulin binds to Ca^{2+} ions to begin the contraction.

Contraction in cardiac muscles

- Cardiac muscles contain all the components similar to that of skeletal muscles, e.g. actin, myosin, Z-band, etc.
- However, gap junctions are found in cardiac muscles and not in skeletal muscles.
- These gap junctions allow the action potential to transmit from one muscle fibre to another. Thus, if a single muscle fibre is stimulated all the fibres become stimulated.
- These muscles also have a long refractory period. They can use lactic acid produed by skeletal muscles to make ATP.

Specialised muscle phenomena

Certain specialised phenomena associated with muscles are as follows

All-or-None Law

- It is also called **Bowditch's law**. This principle states that the response of a muscle or nerve to a stimulus is not proportionate to the intensity of stimulus.
  Instead, a minimum (threshold) intensity is required for muscle contraction.
  Above this intensity, a muscle contracts always with its maximum strength.
- In other words, a stimulus either causes the muscle to contract or the muscle do not contract at all.
- In case, the stimulus causes the muscle to contract, then it (muscle) always contracts to its maximum strength, i.e. all (maximum) or none.
- This rule is the characteristic of cardiac and involuntary muscle. However, skeletal muscles do not follow this rule.

Oxygen debt

- It is the extra oxygen required by the body muscles during relaxation or recovery period of a muscle.
- The events related to oxygen debt in chronological order are as follows
  - During strenuous exercise, the requirement of oxygen far exceeds its availability through breathing. Thus, other sources including oxygen from oxymyoglobin, dephosphorylation of creatine phosphate, etc., are used.
  - After their exhaustion, the muscles begin to respire anaerobically along with aerobic respiration. This leads to anaerobic contraction which produces lactic acid.
  - During recovery, extra oxygen is required for which deep breathing continues and it aids in
    (i) Regeneration of oxymyoglobin
    (ii) Oxidation of accumulated lactic acid
    (iii) Restoration of depleted ATP
    (iv) Restoration of creatine phosphate.
- Oxygen debt decreases with the regular exercise. It is because regular exercise increases oxymyoglobin content in the muscle.

Muscle fatigue

- It is the decrease in the force of contraction of a muscle after prolonged stimulation.
- The process begins with the accumulation of lactic acid within the muscle. This accumulation leads to muscle fatigue.
- Normally, pain is experienced in the fatigued muscle. The site of muscle fatigue is the neuromuscular junction.

Cori's cycle

- This is a cyclic process discovered by **Cori** and **Cori.**
- It involves the formation of lactic acid in the muscles and regeneration of glycogen from it in the liver.
- It occurs in order to reduce accumulation of lactic acid in muscles and to maintain continued supply of glucose to them.

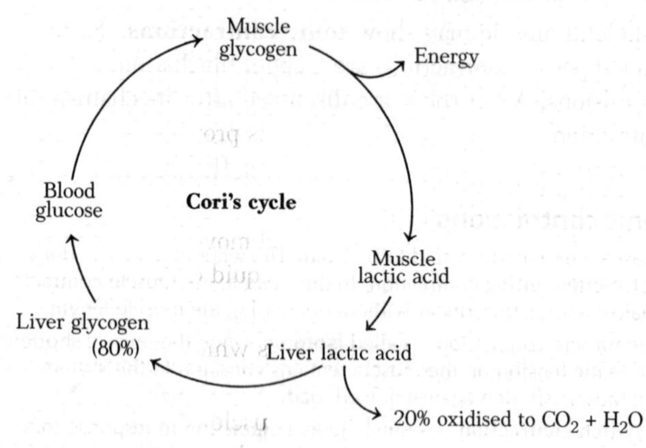

Fig 27.9 Cori's cycle

Rigor mortis

- Just few hours after death, muscles stiffen and become hard. This condition is called rigor mortis.
- It first appears in lower jaw and then appears in all body muscles.
- It occurs due to permanent or irreversible contraction between actin and myosin, which in turn occurs due to exhaustion of ATP from the blood.

CHECK POINT 27.1

1. Select the correct option.
   (a) HMM = Tail + Short arm
   (b) LMM = Tail + Head
   (c) HMM = Head + Short arm
   (d) LMM = Head
2. A protein having both structural and enzymatic properties is
   (a) actin (b) myosin
   (c) tropomyosin (d) troponin
3. The calcium-binding protein in sarcoplasmic reticulum is
   (a) tropomyosin (b) calmodulin
   (c) calsequestrin (d) phosphocreatine
4. Which of the following refers to cross arm in the myosin (thick) filament?
   (a) Light Meromyosin (LMM)
   (b) Head and short arm projecting outwards at regular distance and angle from each other
   (c) Trimers of troponin
   (d) Tropomyosin and troponin
5. Which of the following is correct set with respect to antagonistic muscles?
   (a) Protractor and retractor
   (b) Elevator and depressor
   (c) Abductor and rotator
   (d) Supinator and dilator

SKELETAL SYSTEM

- The hard, supportive or protective elements of animal body forms the skeleton. The study of skeleton is called **osteology** or **Skeletology**.
- The skeletal system may vary from simple spicules of sponges to the complex framework of arthropods and vertebrates.
- Design of the animal skeleton depends on its mode of life.
- Skeleton system can be categorised into two types as follows

Hydrostatic skeleton

- In this type of skeleton, the support is provided to the body by a watery fluid. It is found in *Hydra*, flatworms, roundworms and annelids.
- These animals control their form and movement by using muscles to change the shape of the liquid or fluid-filled compartments.
- The fluid presses against the muscles which contract against the fluid pressure.
- The combined effect of the liquid muscle contraction and fluid pressure serves to maintain the shape of the animal.

Rigid skeleton

This type of skeleton is formed of supportive structures. It can be of following two types

(i) **Exoskeleton** It includes hard parts present on the external surface of the body like scales, feathers, hairs, claws, nails, hoofs, horns, etc. It is rigid, protective and supporting framework present both in vertebrates and invertebrates.

(ii) **Endoskeleton** It includes hard parts present inside the body like spicules (e.g. sponges), calcareous endoskeleton (corals and molluscs), ossicles in invertebrates (e.g. echinoderms), cartilages (e.g. shark) and bones (e.g. humans) in vertebrates.

Table 27.2 Differences between exoskeleton and endoskeleton

Exoskeleton	Endoskeleton
Covers the body from outside.	Occurs inside the body.
Produced by ectoderm.	Produced by mesoderm.
Consists of non-living materials.	Consists of living tissues.
Muscles are attached to its inner surface.	Muscles are attached to its outer surface.
May restrict the growth of the body.	Does not restrict the growth of the body.
May be moulted periodically during growth.	Never gets moulted.

Functions of skeletal system

The skeletal system serves a number of functions in vertebrates as follows

- It forms a rigid framework, which gives support to the body against gravity.
- It encloses vital organs of the body to protect them from injury.
- It furnishes a hard surface for the attachment of muscles to make them more effective in movement.
- It helps in the regulation of mineral ion concentration of Ca^{2+}, PO_4^{2-}, etc.
- In vertebrates, it is the major site of blood production.

SKELETAL SYSTEM IN HUMAN

Human skeleton consists of **300 bones** at the time of birth. Out of these, some bones get fused during adulthood and finally form **206 bones.** On the basis of their position, the human endoskeleton is categorised as **axial skeleton** and **appendicular skeleton**. The diagram showing all the components of human skeleton is given below

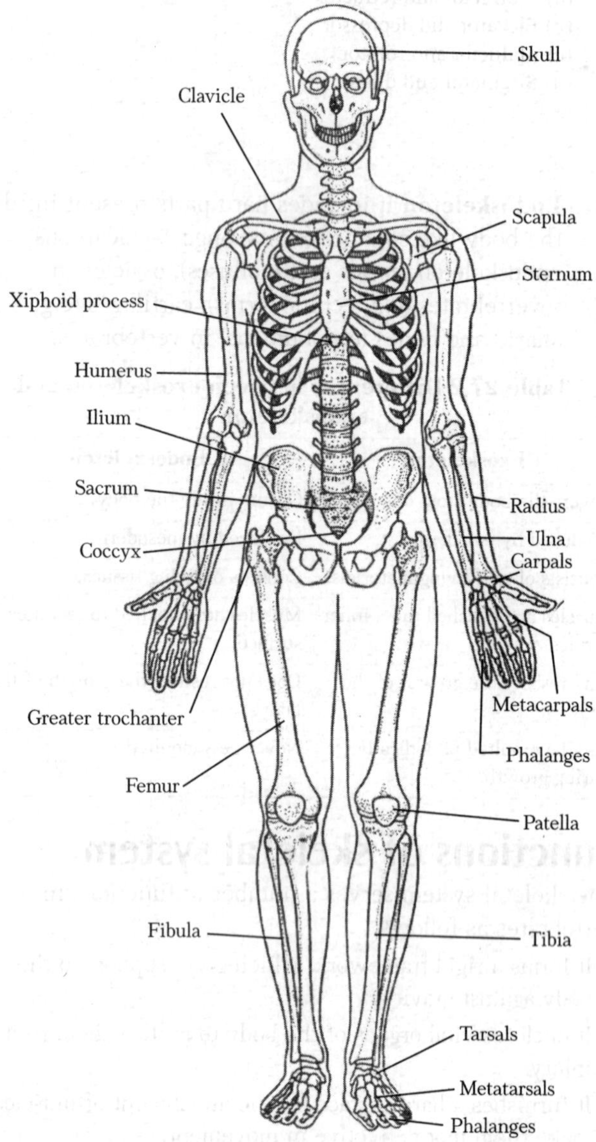

Fig 27.10 The human skeleton system anterior view

Table 27. 3 An outline of bones number in axial and appendicular skeleton

Axial skeleton (80)			Appendicular skeleton (126)		
Cranium		8	**Pectoral girdle**		
Face		14	Clavicle		2
Hyoid		1	Scapula		2
Ear ossicles	(3 × 2)	6			**4**
Vertebral column		26	**Forelimbs**		
Sternum		1			
Ribs	(12 × 2)	24	Humerus		2
		80	Ulna		2
			Radius		2
			Carpals	(8 × 2)	16
			Metacarpals	(5 × 2)	10
			Phalanges	(14 × 2)	28
					60
(i) **Cranial bones**			**Pelvic girdle**		
Frontal		1	Coxal, hip or pelvic		2
Parietals		2			**2**
Temporals		2	**Hindlimbs**		
Occipital		1	Femur		2
Sphenoid		1	Tibia		2
Ethmoid		1	Fibula		2
		8	Patella		2
			Tarsals	(7 × 2)	14
			Metatarsals	(5 × 2)	10
			Phalanges	(14 × 2)	28
					60
(ii) **Facial bones**			Total number of bones in adult human		
Nasals	Skull	2	80 + 4 + 60 + 2 + 60 = **206**		
Maxillae	Skull	2	Digital formula of both limbs		
Zygomatic bones		2	2, 3, 3, 3, 3		
Mandible		1			
Lacrimal bones		2			
Palatines		2			
Inferior nasals		2			
Vomer		1			
		14			
(iii) **Vertebral column**					
Cervical		7			
Thoracic		12			
Lumbar		5			
Sacrum		1 (5)			
Coccyx		1 (4)			
		26 (33)			

I. Axial skeleton

It is present along the middle longitudinal axis of the body. It includes skull, vertebral column, ribs and sternum.

These four components together constitute **80 bones.**

Skull

- The skeleton of head is called skull. It consists of **29 bones** and encloses brain inside it.
- Base of the skull has a large opening called **foramen magnum**. Through this opening, the brain communicates with the spinal cord.
- On both the sides of foramen, two occipital condyles are present due to which human skull is called **dicondylic.**
- Skull includes following parts
  (i) **Cranium** It is a hollow and nearly rounded bony structure. It is composed of eight bones that provide protection to the brain. The upper jaw is fused with cranium while the lower jaw remains connected with the cranium by muscles.

  The various bones of human skull are represented in the diagram given below

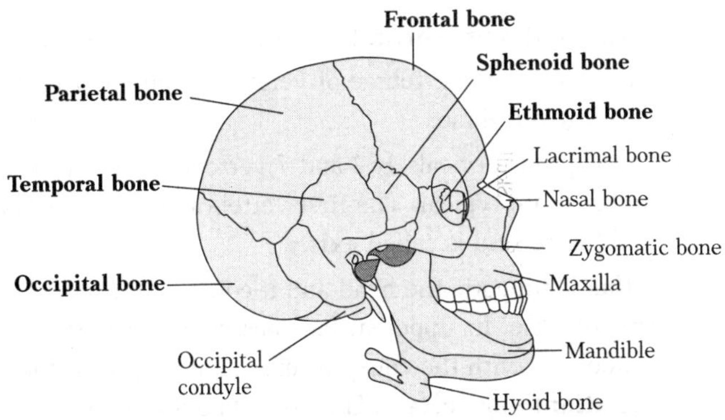

Fig. 27.11 Lateral view of human skull

The location of different bones in cranium is given in table below

Table 27.4 Locations of cranial bones in human

Cranial bones	Locations
Frontal bone	Anterior part of cranium; form forehead.
Parietal bone	Greater portion of sides and roof of cranial cavity.
Temporal bone	Inferior lateral aspect of cranium and part of cranial floor.
Occipital bone	Posterior part and most of the base of cranium. Possess foramen magnum in the inferior part through which medulla oblongata connects with spinal cord.
Sphenoid bone	Lies at the middle part of the base of skull, holds all the cranial bones together. Possess saddle-shaped structure, sella turcica to enclose pituitary gland.
Ethmoid bone	Sponge-like in appearance, located on the midline in the anterior part of cranial floor. The cribriform plate forms the roof of nasal cavity through which olfactory nerve passes to brain. Perpendicular plate forms the superior portion of nasal septum.

(ii) **Ear ossicles** There are six ear ossicles (two malleus, two incus and two stapes) in the skull. Stapes is the smallest bone of human body.

(iii) **Hyoid bone** This is a U-shaped bone situated in the wall of the upper part of the throat, just above the larynx. It is not a bone of the skull proper.

Hyoid acts as a point of attachment for tongue and serves as the floor of mouth. It does not articulate with any other bone.

(iv) **Facial bones** These constitute the frontal part of skull, i.e. the facial region. Total **14 bones** form the skeleton of the face. Thirteen of the facial bones are connected to form the bulk of the face. On the other hand, single mandible (lower jaw) forms a freely movable joint with the rest of the skull.

Table 27.5 Locations of facial bones in human

Facial bones	Locations
Nasal bone	Bridge of nose.
Maxilla	Upper jaw, articulates with every bone of face except mandible.
Zygomatic bone	Prominence of cheek.
Lacrimal bone	It is a thin bone that houses lacrimal sac.
Palatine	Form posterior part of hard palate, forms floor and lateral wall of nasal cavity.
Inferior nasal conchae	Forms part of inferior lateral wall of nasal cavity.
Vomer	It is a triangular bone that forms the inferior portion of nasal septum.
Mandible	Forms the lower jaw, largest and strongest facial bone.

Note • *Skull of mammals is **tropibasic** where the eyes are not situated much apart.*
- *Also, the brain and eyes are present at different planes in the skull.*
- *The **platybasic** skull is seen in amphibians, where eyes are situated apart with a large cranial area in between, i.e. both of these lie in same plane.*
- *Reptiles and Aves have **monocondylic** skull.*
- ***Splanchnocranium** is a part of axial skeleton that contributes to the production and encasing of jaws, gill and their related structures.*

Vertebral column (Backbone)

It forms the medial and vertical axis of the body. It is also known as **spine, spinal column** or **backbone**. It is made up of 33 serially arranged units called **vertebrae**, each of which interlocks with the bones immediately above and below.

General structure of vertebrae

A typical vertebra consists of various parts as shown in the diagram below

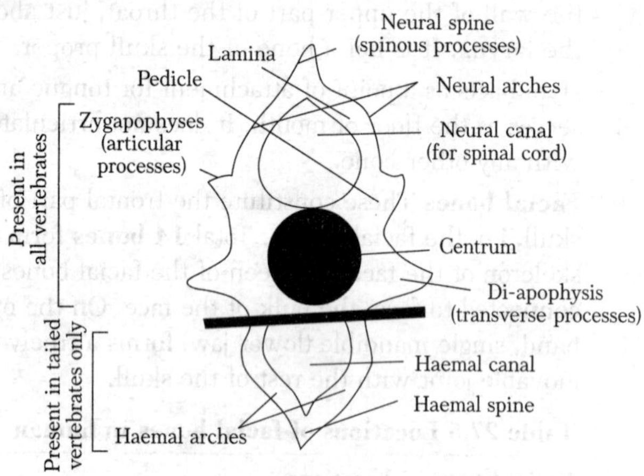

Fig. 27.12 Structure of vertebrae

Three most important parts of a vertebrae are as follows

(i) **Centrum or Body** It is a large, disc-like anterior flattened portion of the vertebrae. It contains the vestiges of notochord. It is the main identifiable part of a vertebra which is differently shaped in different animal groups as shown below

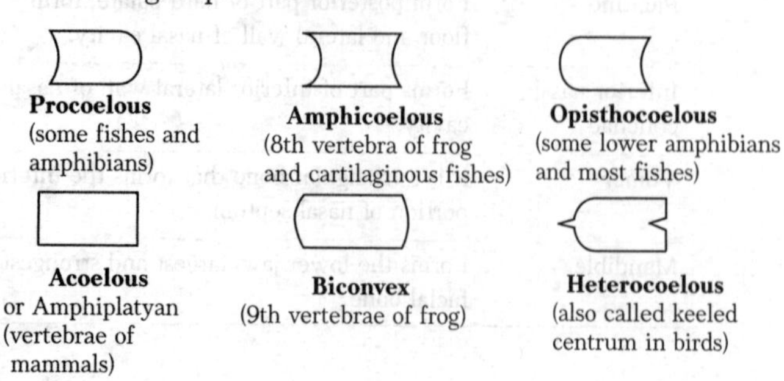

Centrum is absent in the atlas vertebrae of human beings.

(ii) **Vertebral/Neural arch** It is the posterior portion of the centrum. It is composed of pedicles (stalk-like supports) and two laminae. It encloses the spinal cord and vertebral foramen.

(iii) **Supportive processes** Each vertebra has several projections or outgrowth of tissue from the body, called as **processes**. These are supportive in function. These include a **spinous process**, four **articular processes** (two superiors and two inferiors) and two **transverse processes.**

The function of each process is as follows
- The spinous process serves as an attachment point for muscles.
- The transverse processes also serve as muscle and ligaments attachment sites.
- The articular process helps in the articulation of one vertebra with one another.

Types of vertebrae

The human vertebral column contains following five groups of vertebrae

(i) **Cervical vertebrae** (7)
- The first seven vertebrae of vertebral column are called cervical vertebrae.
- Among the cervicals 1, 2 and 7 vertebrae are atypical and 3, 6 are typical. The first vertebrae is called **atlas** and the second is called **axis.**
- **Atlas** It supports the head and it consists of a complete ring of bone. Its upper surface possesses facets that articulates with the condyles of occipital bone. It forms the **atlanto-occipital joint** (condyloid joint) with occipital bone.
- **Axis** It acts as a pivot on which atlas turns to allow movement of head.
  From the body of axis, a process of bone arises called as **odontoid peg** or **process.** It articulates with the back of the anterior arch of the atlas. The atlas moves around the odontoid peg of axis and forms a pivot joint.
- **Seventh cervical vertebra** It is the first vertebra with an undivided spinous process. It forms a prominent projection that can be observed at the lower part of the back of neck. Due to this, it is also known as **vertebra prominens.**

- The diagram showing atlas, axis and seventh cervical vertebrae is given below

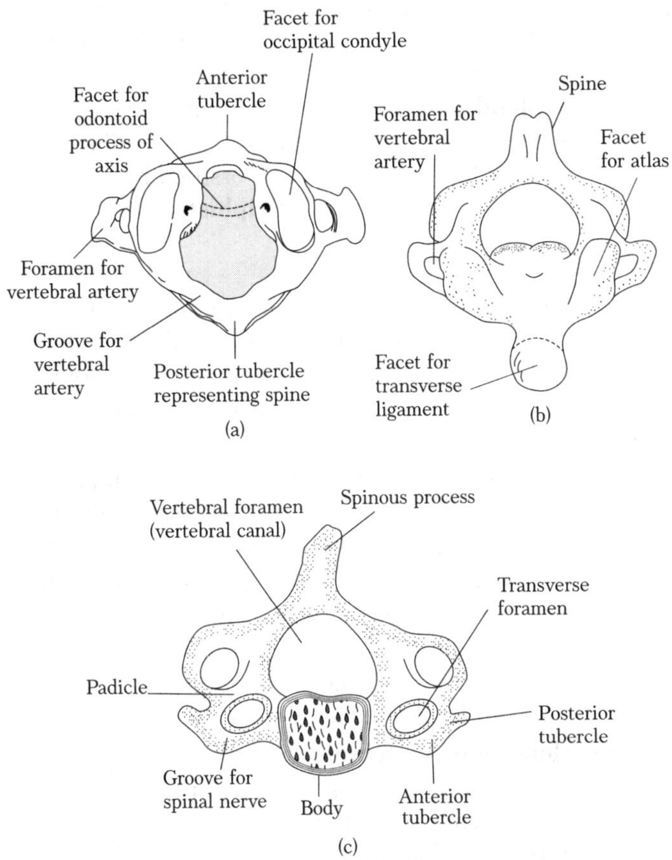

Fig. 27.13 (a) The atlas viewed from above, (b) The axis viewed from above and behind, (c) 7th cervical vertebrae

(ii) **Thoracic vertebrae** (12)

- Among these, 2-8 vertebrae are typical whereas 1, 9-12 are atypical.
- These are located in the chest region and are larger than the cervical vertebrae.

The diagram showing top and front view of thoracic vertebrae is given below

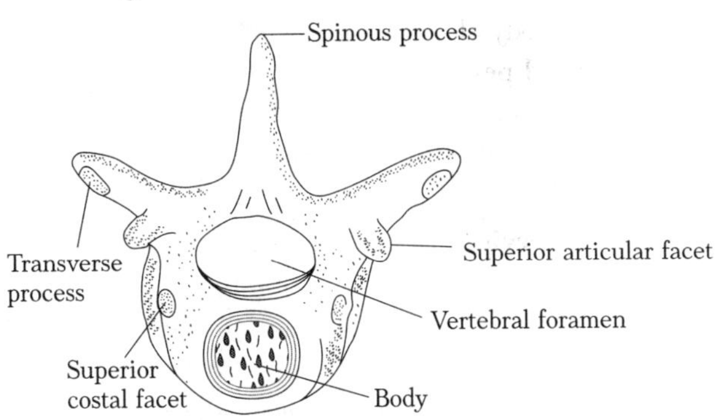

Fig. 27.14 Cervical vertebrae

- They increase in size as they extend downwards. These 12 vertebrae form the dorsal side of ribcage to protect heart and lungs.
- **Processes** Their transverse processes are thick and strong to provide additional space for the attachment of thoracic muscles and to support the ribs. Their spinous processes are long and directed downwards.

(iii) **Lumbar vertebrae (5)**

- These are the **largest** and **strongest vertebrae** of human body.
- They are found in abdominal region and bear the weight of the trunk.
- The fifth lumbar vertebra articulates with sacrum at lumbo-sacral joint.
- **Processes** Their spinous processes are broad and hatchet-shaped. Their transverse processes are long and slender. Some additional processes are also seen in the vertebrae of this region, i.e. **metapophyses, anapophyses** and **hypophyses**. All of these provide additional surface for the attachment of abdominal muscles.

The diagram showing the front view of lumbar vertebrae is given below

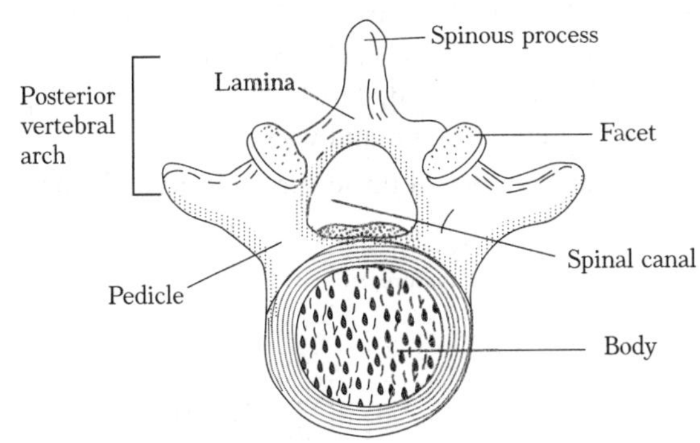

Fig. 27.15 Lumbar vertebrae

(iv) **Sacrum** (5)

- The five sacral vertebrae are fused in adult to form a single structure called sacrum.
- It lies between the innominate bones of the pelvic girdle and articulates with them through right and left sacro-iliac joint.
- All the edges of its transverse processes, **Sacral foramina** are present for the passage of nerves.
- In humans, the female sacrum is wider and shorter than male sacrum.

(v) **Coccyx** (4)

- The four coccygeal vertebrae are fused to form a curved triangular bone called coccyx.
- It articulates with sacrum from above. In humans, it is considered as a vestigial tail bone.

- The diagram showing anterior and posterior view of sacrum and coccyx vertebrae is given below

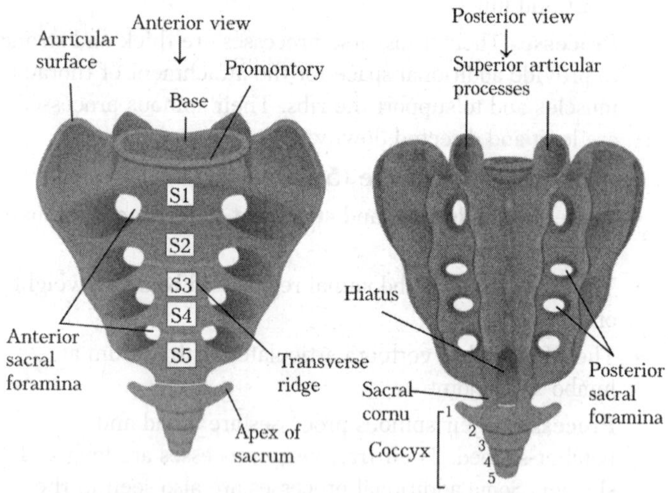

Fig. 27.16 Anterior and posterior view of sacrum and coccyx vertebrae

Curves of vertebral column

- The four antero-posterior curves in vertebral column can be broadly categorised as follows
  - (i) **Convex forward type** Cervical curve, lumbar curve.
  - (ii) **Convex backward type** Thoracic curve, pelvic curve.
- During human growth, cervical curve develops when infant raises head to look around. The lumbar curve develops when baby begin to crawl, stand and walk.
- The diagram showing all these curves is given below

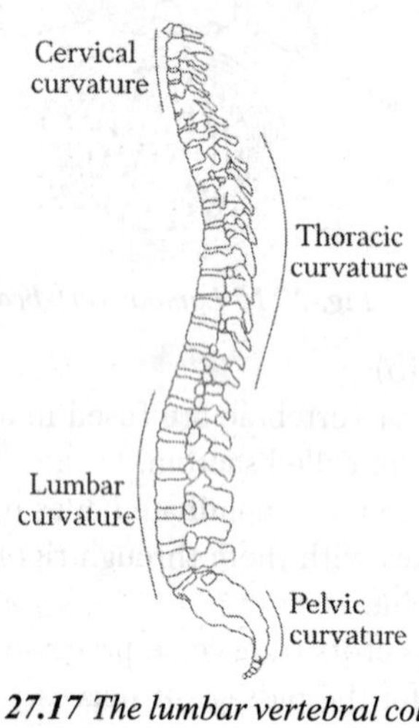

Fig. 27.17 The lumbar vertebral column

Sternum (Breast bone)

It is a flat bone in front of the chest with seven small bony segments. It is divided into the following parts

(i) **Manubrium** (Prosternum) It is the broad upper portion which is trapezoid in shape. It is fused with the body of the sternum, forming with it a slight angle called **sternal angle**.

(ii) **Body** (Merosternum) It is flat, elongated and the largest part of the sternum. It articulates with the manubrium superiorly and xiphoid process inferiorly. The lateral edges have articular facets, i.e. cartilage lined depressions.

(iii) **Xiphoid process** (Metasternum) It is the smallest and the lowest part of sternum. It is cartilaginous in early life, but usually gets ossified by adulthood. It is also called **xiphisternum**.

Ribs

They are thin flat bones which are dorsally connected to the vertebral column and ventrally to the sternum. In mammals, the number of ribs is equal to the number of the thoracic vertebrae. Thus, there are 12 pairs of ribs which belong to following categories

(i) **True or vertebrosternal ribs** First seven pairs of ribs are called true ribs. Their anterior ends are attached directly to the sternum by means of small pieces of hyaline cartilage called **costal cartilages.**

(ii) **False or vertebro chondrae ribs** The VIII, IX and X pairs of the ribs are called false ribs. These are articulated with the costal cartilage of the VII rib thus, these are attached indirectly to the sternum.

(iii) **Floating or vertebral ribs** The last two pairs of ribs are called floating ribs or **gorilla ribs**. Their anterior ends are not attached either with the sternum or the cartilage of another rib. The floating ribs protect the kidneys.

- All the ribs have two articulation surfaces on the dorsal end and hence, they are **bicephalic.**These double headed thoracic ribs are the pecularity of birds and mammals.
  The diagram showing different type of ribs alongwith sternum is given below

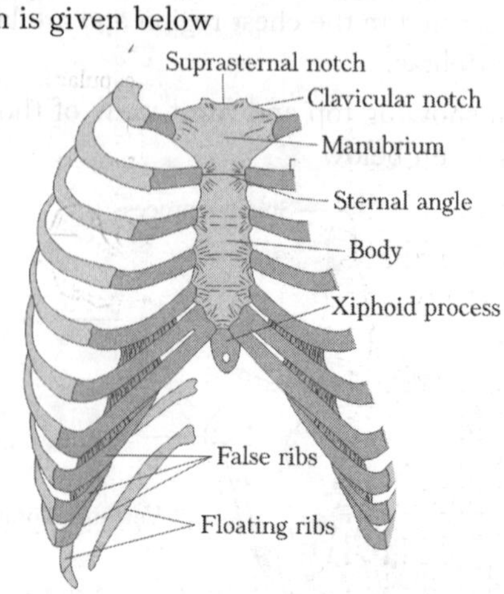

***Fig. 27.18** The sternum and ribs*

Note *Birds contain an additional **uncinate process** in their ribs for the attachment of flight muscles.*

II. Appendicular skeleton

The appendicular skeleton includes **forelimbs, hindlimbs, pectoral girdle** and **pelvic girdle**. These skeletal parts support the lateral outgrowth of the body. In humans, **126 bones** forms the appendicular skeleton.

Pectoral girdle

It is simply called **shoulder girdle**. It is meant for the attachment of the arm muscles and forelimb bones. It contains two separate right and left halves. Each half is composed of two bones, i.e. scapula and clavicle.

(i) **Scapula or Shoulder blade**

- It is a large, flat and triangular bone that forms the posterior part of shoulder girdle.
- It overlaps the posterior part of the transverse processes of second to seventh thoracic vertebra on its side.
- The two scapular bones are easily recognisable on our back as shoulder blades.
- Scapula has **acromion** and **coracoid processes**, which provide additional space for the attachment of shoulder muscles.
- A cup-like **glenoid cavity** is present just below the coracoid process into which the head of humerus bone of upper arm is articulated to form humero-scapular joint.

The diagram showing anterior view of scapula is given below

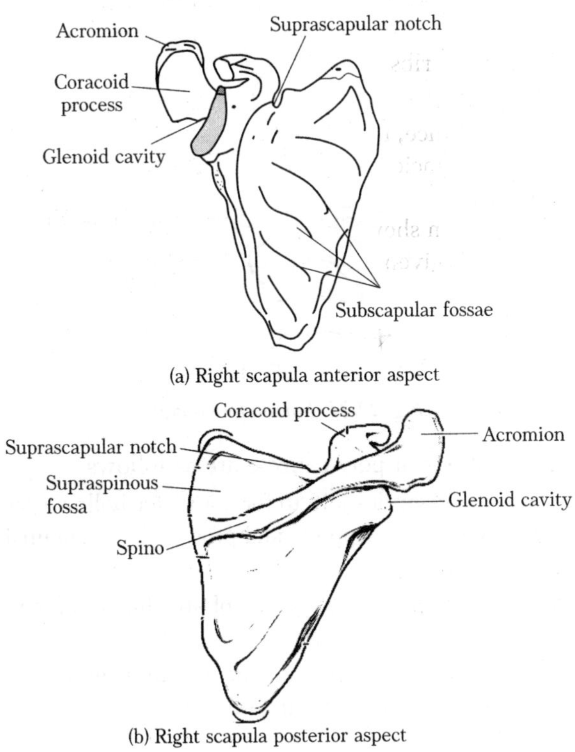

Fig. 27.19 Scapula

(ii) **Clavicle or Collar bone**

- This is a long, narrow, S-shaped bone that forms the anterior part of shoulder girdle.
- It is connected with the acromion process of scapula at acromial extremity and with the sternum at sternal extremity.

Bones of forelimb

Each forelimb consists of 30 bones. These include

(i) **Humerus**

- It is the **longest** and the **largest bone.**
- Its head fits into glenoid cavity of corresponding scapula. It has a shaft and two extremities.
- Its proximal extremity is marked by head, neck, greater and lesser tubercles and intertubecular sulcus.
- Its distal extremity has medial and lateral supraepicondylar ridges.
- It also has three depressions *viz* **coronoid, radial** and **olecranon fossae.**
- It has two processes, i.e. **capitulum** that articulates with the radius bone of the lower arm and **trochlea** that articulates with the ulna of the lower arm.

(ii) **Ulna**

- It is medial bone of forearm and longer than radius.
- It is present toward the little finger.
- It has a shaft and two extremities.
- The upper extremity is strong, thick and participates in the formation of elbow joint.
- The upper end also possesses trochlear notch and radial notch.
- The lower extremity is small and two eminences arise from it.
- It articulates with the humerus through its **olecranon process.**

(iii) **Radius**

- It is lateral bone of forearm and shorter than ulna.
- It is present toward the thumb.
- Its upper or proximal end has a head, a neck and a radial tuberosity.
- The head articulates with the capitulum of humerus. On the opposite end, it articulates with ulna and wrist bones.
- **Colles fracture** is the breaking of the lower end of radius. This may lead to deformity of wrist and hand.

(iv) **Carpals**

- Each wrist is composed of **8 small bones** called carpals. These are arranged in two rows as
  - **Proximal row** Scaphoid, Lunate, Triquetrum and Pisiform.
  - **Distal row** Trapezium, Trapezoid, Capitate and Hamate. Out of these, pisiform is a sesamoid bone and capitate is the largest carpal.

(v) **Metacarpals**

- These are **5 in number**. Each bone has a shaft and two extremities.
- One extremity articulates with carpals and it is known as carpal extremity. This region possesses carpo-metacarpal joint.
- The distal extremity called head, articulates with phalanges.

(vi) **Phalanges**

- Finger bones called phalanges, are two in our thumb or pollex and three in each of the four fingers. Thus, the digital formula is **2, 3, 3, 3, 3.**
- These are also long bones having a shaft and two extremities.
- Phalanges are attached to metacarpals through condyloid joint.

The detailed diagram showing bones of forelimbs is given below

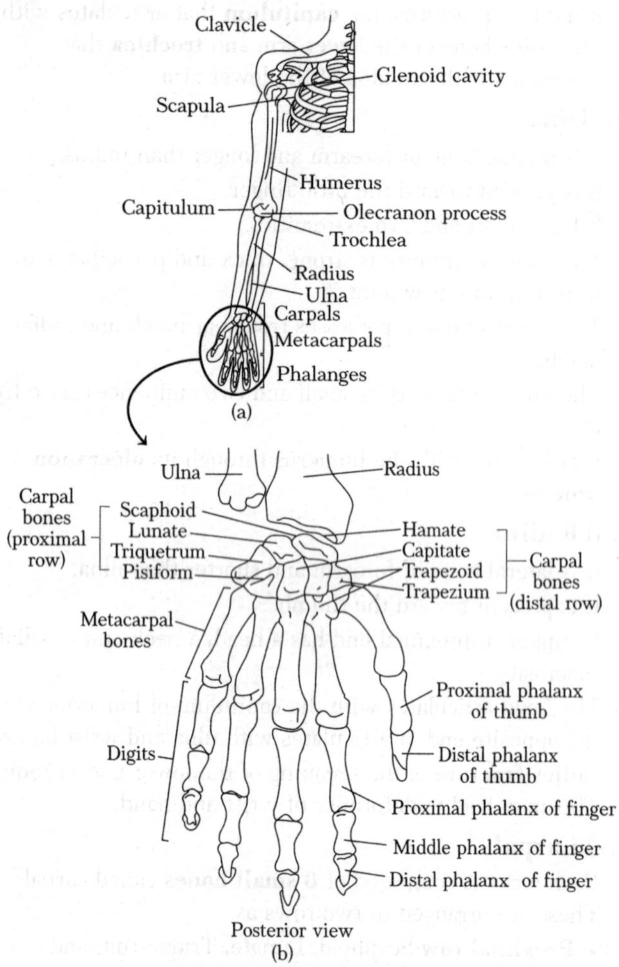

Fig. 27.20 (a) Bones of forelimb, (b) Bones of right wrist and hand

Pelvic girdle

It is also called **hip girdle.** In humans, each half of pelvic girdle is known as **coxal** or **innominate bone** (hip bone).

Specialised features of pelvic girdle are given below

- The two halves of the pelvic girdle meet ventrally to form the **pubic symphysis.** It contains fibrous cartilage.
- Each coxal bone is formed of three bones, i.e. **ilium, ischium** and **pubis.** These three bones participate equally in the formation of **acetabulum** (cup-shaped hollow cavity) to which the **thigh bone or femur** articulates.
- Ischium is the thickest and strongest bone.
- The tuberosity of ischium provides attachment site to hamstring muscles.
- The spine of ischium represents the lowest part of sciatic notch.
- While sitting, whole weight of body is exerted on ischium only.
- Presence of **obturator foramen** is a peculiar feature of mammlian pelvic girdle.
- It is a large oval foramen below the acetabulum. It remains bouded by pubis and ischium. At the upper part of obturator foramen, obturator vessels and nerve pass from pelvis into the thigh region.

A diagram showing innominate bone is given below

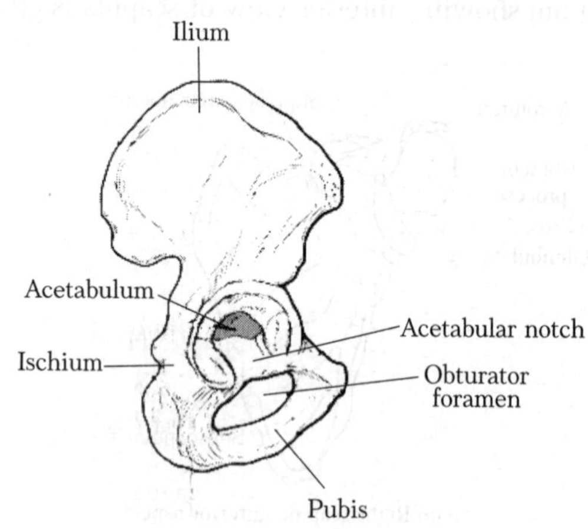

Fig. 27.21 Innominate bones

Various functions of pelvic girdle are as follows

- The two pelvic girdles join to form a wider hollow space called **pelvis.** They supports and protect the abdominal visceral structures.
- Pelvic girdle functions as the site of attachment of leg muscles and hindlimb bones.
- Pelvis in females has larger diameter than male to accommodate the foetus during pregnancy.

Bones of hindlimb

Each hindlimb has **30 bones** which form thigh, shank, knee and foot as discussed below

(i) **Femur**

- The single, longest and largest bone of thigh is called femur.
- Its one end articulates with acetabulum to form hip joint, while the other end articulates with tibia, but not fibula.
- **Projections** It shows three projections at its proximal end, i.e. head, greater trochanter and lesser trochanter.
- The greater trochanter provides attachment of gluteal muscles.
- The III trochanter is not defined in the femur of human beings.
- The lesser trochanter appears as raised region.
- **Condyles** The distal end of femur contains two condyles *viz* lateral and medial condyles. These are separated by the intercondylar fossa in between.

(ii) **Patella**

- The knee cap is formed of a triangular shaped sesamoid bone called patella.
- The sesamoid bones are bones formed by ossification of a tendon.
- Patella is the largest one among the sesamoid bones.

(iii) **Tibia and Fibula**

- The two bones present in the shank region are tibia and fibula.
- Tibia is longer, thicker and lies more medially. It is the main weight bearing bone.
- The tibia articulates with three bones *viz* femur, fibula and talus.
- On the other hand, fibula is shorter and thin. It is located more laterally.
- The lower extremity of fibula possesses malleolar fossa to which strong ligaments of ankle joint are attached.

(iv) **Bones of foot**

- The foot comprises of 7 short bones of the ankle, called **tarsals** and 5 bones of the sole, called **metatarsals**. Also, there are 19 bones forming the toe, called **phalanges**.

- The diagram showing all the bones of hindlimb is given below

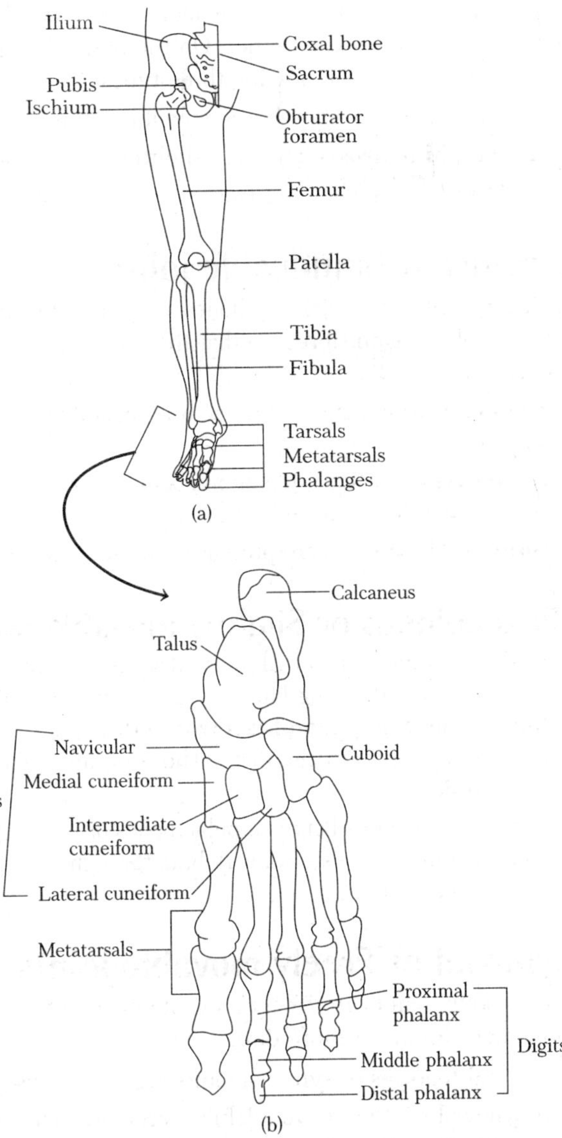

Fig. 27.22 Bone : (a) Right pelvic girdle and lower limb bones (Frontal view), (b) Bones of foot

CHECK POINT 27.2

1. Ribs are called bicephalic because
   (a) they are attached both with sternum and vertebrae
   (b) they have two articulation facets on dorsal side
   (c) they have two articulation facets on ventral side
   (d) they are attached with sternum with two cartilages

2. Mark the odd one with respect to the functions of vertebral column.
   (a) Attachment to large muscles of back
   (b) Protection of spinal cord
   (c) Attachment to ribs
   (d) Attachment to sternum

3. Choose the mismatch with respect to total number of bones in humans.
   (a) Tarsals – 14 (b) Carpals – 16
   (c) Ear ossicles – 3 (d) Thoracic vertebrae – 12

4. The number of bones present in the skeleton that runs along the middle longitudinal axis of the body is
   (a) 87 (b) 90 (c) 80 (d) 78

5. Mark the mismatch.
   (a) Upper jaw–Maxilla (b) Lower jaw–Mandible
   (c) Atlas–Olecranon process (d) Scapula–Acromion process

JOINTS

These are the structural arrangement where two bones articulate, i.e. meet each other. On the basis of degree of motion, joints are of following three types

1. **Synarthrosis** They are immovable joints.
2. **Amphiarthrosis** They are slightly movable joints.
3. **Diarthrosis** They are freely movable joints.

Fibrous or Immovable joints

In these joints, the adjoining bones are joined by strong bundles of collagen fibres, so that the bones cannot move freely, e.g.

- **Gomophoses** Articulation of teeth in mandible or maxillae.
- **Schindyleses** Articulation of one bone into the slit of other, e.g. ethmoid bone in vomer.
- **Sutures** These joints are present in the skull bones.

Cartilaginous or Slightly movable joints

- In these joints, a pad of fibro-cartilage is present between two adjoining bones. This allows very slight or limited movement (**amphiarthrosis**) through compression of cartilage discs. These are also called **symphysis**.
- For example, joints between pubic bones of pelvic girdles and the joints between the vertebrae through intervertebral discs.

Synovial or Freely movable joints

The synovial joints are freely movable joints which allow movement in one or more directions.

Structural features of synovial joints are as follows

- A synovial joint has a well-defined structure. The entire joint is enclosed within a capsule called **synovial capsule**.
- The membrane lining the capsule is called **synovial membrane**.
- It is composed of secretory epithelial cells that secrete sticky and lubricating **synovial fluid** or **synovia**.

Types of synovial joints

Synovial joints are further classified as follows

Ball and socket joint

- In this, a ball-like head of one bone fits into the socket or the concavity of the other. It is a multiaxial joint.
- It can move in all possible directions (**enarthroses**), e.g.
  (i) **Shoulder joint** Between the head of humerus and glenoid cavity of pectoral girdle.
  (ii) **Hip joint** Between the head of femur and acetabulum of pelvic girdle.

Hinge joint

- In these joints, a protuberance or condyle at the end of one bone fits into the corresponding depression of the other.
- A hinge joint allows movement in one plane only (ginglymi) hence, it is a monoaxial joint.
- It is similar to hinges of a door. For example, ankle knee, elbow joints and joints between the phalanges of digits.

Pivotal or Rotatory joint

- It is also a monoaxial joint. The primary movement at the pivotal joint is that of rotation.
- In this joint, the articulating end of one bone is fixed while that of the other can freely rotate over it.
- For example, joint between head of the radius and capitulum of humerus, between radius and ulna just below the elbow and the joint between atlas and axis of mammals.

Saddle joint

- It is so named because one bone of the joint possesses a saddle-like shape while the other bone is just like the back of a horse.
- These joints are similar to ball and socket joints, but in these, both ball and socket are poorly developed. Therefore, movement can take place in two directions hence, it is a biaxial joint.
- For example, joints between the metacarpal of thumb and carpals.

Gliding joint

- It is the simplest synovial joint. The articulating bones can slide upon one another at this joint. It is also a monoaxial joint.
- For example, the joint between radio-ulna and carpals, between zygapophyses of succesive vertebrae, between sternum and clavicle.

Ellipsoid or Angular or Condyloid joint

- Here, one bone is movable on another bone in two planes, i.e. back and forth and side to side. It is also a biaxial joint.
- For example, joint between skull and first vertebra, metacarpophalangeal joints.

The diagram showing different synovial joints is given below

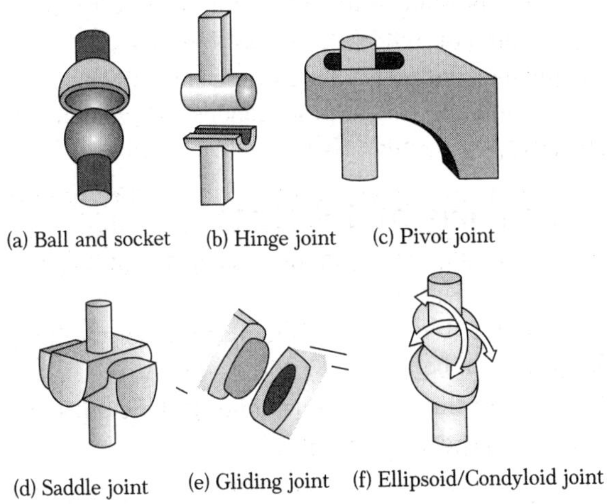

Fig 27.23 Types of synovial joint

LEVER SYSTEM IN HUMANS

In producing movement, bones act as levers and joints function as fulcrum.

Lever may be defined as rigid rod that moves around fixed point called **fulcrum** (F). Lever is acted at two different points by two different forces *viz*

- **Effort** (E) which causes movement
- **Resistance** or **Load** (R) which opposes movement.
- In human body, the effort is the amount of force exerted by muscular movements. The resistance (load) is typically the weight of the body part that is moved.

Types of lever system

Just like normal lever systems, human body also possesses all the three kinds of lever systems as follows

First class lever

- In this type, fulcrum is between effort and resistance. Head resting on the vertebral column is an example of first class lever.
- When the head is raised, the facial portion of the skull is the resistance, the joint between the atlas and occipital bone (atlantoccipital joint) is the fulcrum and the contraction of the muscles of the back is the effort.

Second class lever

- In this, resistance is exerted between the fulcrum and the effort.
- The lever operates like a wheel barrow, e.g. raising the body on the toes.
- The body weight is the resistance, the ball of the foot is the fulcrum.
- The contraction of the calf muscles to pull the heel upward is the effort.

Third class lever

- These are the most common levers in human body.
- In this type of lever system, effort is exerted between fulcrum and resistance.
- For example, flexing the forearm at elbow in which the weight of the forearm is the resistance, the contraction of the biceps is the effort and the elbow joint is the fulcrum.

The diagram showing different classes of levers is given below

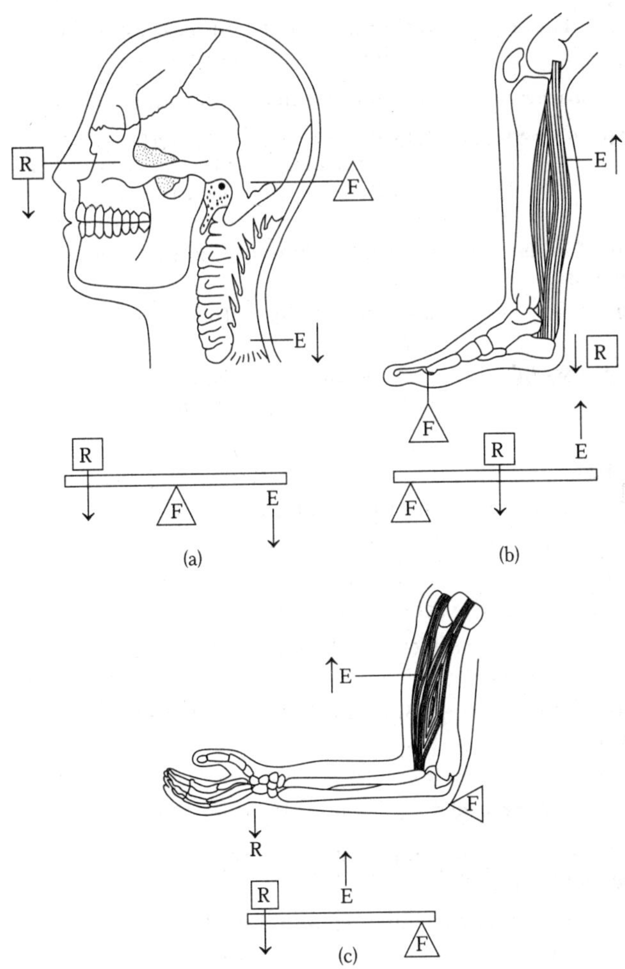

Fig. 27.24 *Classes of levers: (a) First-class lever, (b) Second-class lever and (c) Third-class lever*

DISORDERS OF BONES, MUSCLES AND JOINTS

The disorders related to bones, muscles and joints are as follows

Arthritis

- It refers to the inflammation of the joints. It is a common disease in old age people. Its main symptoms are pain and stiffness in the joints.
- It has two main forms, i.e. **degenerative** or **osteoarthritis** and **rheumatoid arthritis.**
- Osteoarthritis is common in aged persons.
- Rheumatoid arthritis usually starts at the age of 20-40 years, but can begin at any other age. Rest and exercise under medical advice may give relief.

Table 27.6 Differences between osteoarthritis and rheumatoid arthritis

Characters	Osteoarthritis	Rheumatoid arthritis
Type of disorder	Degenerative (the articular cartilage erodes)	Inflammatory, characterised by IgM (rheumatoid factor) and formation of abnormal granules (pannus) at joints.
Site of initial damage	Articular cartilage	Synovial membrane
Age	Late middle age or in old persons	Third decade/Any age
Joint affected	Large weight bearing, often single	Small joints of hands and feet, multiple

Gout

- It is an inherited disorder of purine metabolism.
- It occurs especially in men. In this, body forms excess amount of uric acid and the crystals of sodium urate are deposited in the synovial joints.
- Gout generally affects the great toe and its occurrence is related to diet.
- Persons suffering from gout should avoid paneer, meat and other protein products.

Osteoporosis

- It is the reduction in bone tissue mass. It causes weakness in skeletal strength.
- It results from excessive reabsorption of calcium and phosphorus from the bones. It mostly occurs in post-menopausal women and elderly men.
- It may also result from defective intestinal calcium absorption, smoking, excessive drinking and taking hydrocortisone.
- The symptoms of osteoporosis are pain in the bones.
- Major **causative factors** of osteoporosis are imbalance of hormones like calcitonin of thyroid, parathormone of parathyroid and sex hormones (oestrogen).
- The **preventive measures** include supplementary calcium and exercise in post-menopausal women and oestrogen replacement therapy.

Osteomalacia and Rickets

- When osteomalacia occurs in childhood, it is called rickets.
- It is characterised by insufficient amount of calcium and phosphorus in bones.
- The cause of this disease is vitamin-D deficiency or some inherited defects.

Bursitis

The bursae of joints often become inflammed due to physical **injury** or by constant pressure to the same joint over a long period of time. This condition is known as bursitis.

Dislocation

- It usually involves damage to the ligaments surrounding the joints.
- Most dislocations result from falls, blows or extreme exertion.
- Most dislocations are seen in the joints of thumb, fingers, knee or shoulder.
- Symptoms include swelling, pain and loss of motion.

Sprain and Strain

- The twisting of joint without dislocation is called **sprain**.
- Such an injury causes damage to the ligaments. It also damages tendons, muscles, blood vessels and nerves.
- **Strain** is a less severe stretching due to twisting of a joint. Muscles and tendons may be stretched and become somewhat painful, but only minor damage is done to the tissues of the joint.

Myasthenia gravis

- It is an **autoimmune neuromuscular disorder** leading to fluctuating muscle weakness and fatiguability.
- It is caused by circulating antibodies that block acetylcholine receptors at the post-synaptic neuromuscular junctions. Thus, it inhibits the excitatory effects of acetylcholine on neuromuscular junctions.
- It is medically treated with cholinesterase inhibitors.

Tetany

- It is a symptom characterised by muscle cramps, rapid spasms (wild contractions) or tremors.
- These repetitive actions of the muscles happens when our muscle contracts uncontrollably.
- Tetany may occur in any muscle in our body, such as those in face, fingers or calves. The muscle cramping associated with tetany can be long lasting and painful.
- A common cause of tetany is very low levels of calcium ions in the blood, i.e. hypocalcemia.

Muscular dystrophy

- It is a group genetic disorders characterised by damage and progressive skeletal muscle weakness overtime.
- It occurs due to lack of a protein called **dystrophin**, which is necessary for normal function of muscle.
- The absence of this protein cause problems with walking, swallowing and muscle contraction.
- Muscular dystrophy can occur at any age, but most diagnoses occur in childhood.

Bone fractures

- It is complete or partial break in a bone. It may occur due to trauma, over use or diseases that weaken the bones.
- Bone fracture may be of following kinds
  (i) **Simple or Closed fracture** Breaking of bone into two parts, which remain near to each other but are unattached.
  (ii) **Compound or Open fracture** Breaking of bone into two or more parts with some protruding out of skin, i.e. damage is done to more than one tissue, that is why, the name compound.
  (iii) **Green stick fracture** Simple crack without breaking the bone into two pieces. It is also called as **hairline fracture**.
  (iv) **Comminuted fracture** Breaking of bone into more than two pieces. Some of the bone fragments may loose connection with blood circulation.
  (v) **Exclusive fracture** A chip or small piece of bone breaks, but remains attached to the ligament.

CHECK POINT 27.3

1. Nodding of head occurs due to the joint between
(a) atlas and axis
(b) atlas and occipital condyles
(c) clavicle and scapula
(d) cervical and thoracic vertebrae

2. Which of the following movements in mammalian skeleton represent the leverage of the third order?
(a) Biceps muscle flexing arm at elbow
(b) Triceps muscle extending arm at elbow
(c) Gastrocnemius muscle raising weight of body on toes
(d) Movement of the head of femur in the acetabulum of pelvic girdle

3. An autoimmune disorder in which inflammation of synovial membrane, secretion of abnormal granules and erosion of articular cartilage occur, is
(a) osteoarthritis (b) Myasthenia gravis
(c) osteoporosis (d) rheumatoid arthritis

4. In Myasthenia gravis, ... is affected.
(a) sarcoplasmic reticulum
(b) sarcolemma
(c) neuromuscular junction
(d) T-tubules

5. Tetanus differs from tetany in having
(a) low calcium in body fluid
(b) low calcium in sarcoplasm
(c) rapid spasm in muscles due to low calcium in body fluids
(d) sustained contraction in muscles due to high calcium in sarcoplasm

6. In an accident, a person bones were broken into more than two pieces. The doctor identifies it as
(a) green stick fracture
(b) compound fracture
(c) comminuted fracture
(d) simple fracture

EXAM POINTS

- Basic unit of muscle contraction is sarcomere. These are multi protein complexes composed of three different filament systems. **[UP CPMT 2010, Punjab PMET 2012]**
- Red fibres are dark red coloured thin fibres which contract slowly, but stay contracted for longer time, they have plenty of mitochondria and these are aerobic. **[MP PMT 2012]**
- The white fibre contains less amount of myoglobin and mitochondria. **[Gujarat CET 2013]**
- Portions of myofibril between two Z-lines is called sarcomere. The dark bands are Anisotropic or A-band and the light bands are Isotropic or I-band. **[WB JEE 2014]**
- The H-zone in the skelelal muscle fibre is formed due to the central gap between actin filaments extending through myosin filaments in the A-band. **[NEET 2013]**
- Actin binding sites are located on meromyosin. **[Kerala CEE 2011]**
- The globular head of myosin contains ATPase enzyme. **[JIPMER 2017]**
- The contraction of the muscle continues in sliding filament theory till Ca^{2+} is present in sarcoplasm. **[AIIMS 2017]**
- Sliding filament theory can be best explained as actin and myosin filaments do not shorten, but rather slide pass each other. **[JIPMER 2017]**
- Acetylcholine is released when the neural signal reaches the motor end plate. **[Kerala CEE 2011]**
- Repeated activation of the muscles can lead to lactic acid accumulation. **[Kerala CEE 2011]**
- The correct order of some stage of muscle contraction from beginning to end process is
  Stimuli $\rightarrow$ Neurotransmitter $\rightarrow$ Secretion $\rightarrow$ Excitation of T-system $\rightarrow$ Release of Ca^{2+} $\rightarrow$ Cross-bridge formation $\rightarrow$ Sliding of actin filaments $\rightarrow$ H-band diminishes.
- Out of 12 pairs of ribs in humans, only 7 pairs are true ribs. True ribs are attached dorsally to vertebral column and ventrally to the sternum. **[NEET 2017]**
- Floating ribs do not connect with sternum. **[NEET 2020]**
- Acromion process is found in scapula. **[NEET 2020]**
- Scapula is located between the second and seventh ribs. **[NEET 2020]**
- Glenoid cavity is present in the head of humerus. **[NEET 2020]**
- Calcium is required for skeletal muscle contraction. **[AIIMS 2019]**
- Calcium is important in skeletal muscle contraction because it binds to troponin to remove the masking of active sites on actin for myosin. **[NEET 2019, 18]**
- There are seven pairs of vertebro-sternal, three pairs of vertebro-chondral and two pairs of vertebral ribs. **[NEET 2019]**
- Cervical vertebrae differ from other vertebrae in having transverse foramen. **[AIIMS 2019]**
- Atlas is 1 st cervical vertebrae. **[JIPMER 2019]**
- Gliding joint is between the carpals. **[NEET 2019]**
- Hinge joint is found between the humerus and ulna. **[NEET 2019]**
- Pivot joint is found between the atlas and axis. **[NEET 2019]**
- Saddle joint is found between the carpal and metacarpal of thumb. **[NEET 2019]**
- Ball and socket joint is found between humerus and pectoral girdle. **[AIIMS 2018]**
- Pivot joint between atlas and axis is a type of synovial joint. **[NEET 2017]**
- Osteoporosis is an age related disease of skeletal system due to decreased level of oestrogen in the body. **[NEET 2016, 20]**
- Gout is inflammation of joints due to the accumulation of uric acid crystals. **[NEET 2020, EAMCET 2016]**
- Muscular dystrophy is progressive degeneration of skeletal muscle due to certain genetic disorders. **[NEET 2020, EAMCET 2016]**
- Tetany is a state of prolonged contraction due to low Ca^{2+} ions in the blood. **[NEET 2020, EAMCET 2016]**
- Myasthenia gravis is an autoimmune disorder that affects the neuromuscular junctions. **[EAMCET 2016]**

Chapter Exercises

(A) Topical Exercise

Types of Movements

1 Flagellar movement is exhibited by
(a) macrophages (b) protozoans
(c) *Euglena* (d) leucocytes

2 Macrophages and leucocytes exhibit **[NCERT Exemplar]**
(a) ciliary movement (b) flagellar movement
(c) amoeboid movement (d) gliding movement

3 Ciliary movement helps in
(a) removing waste substances inhaled along with air
(b) movement of leucocytes and microfilaments
(c) contraction and relaxation of muscles
(d) All of the above

4 Choose the correct statement (s) for flagellar movements.
(a) They are found in choanocytes of sponges
(b) Human sperm also shows this movement
(c) They help in movement in *Paramecium*
(d) Both (a) and (b)

5 Sperms swim in female tract with the help of
(a) pseudopodial movement (b) ciliary movement
(c) flagellar movement (d) muscular movement

6 Which one of the following movements in man are directly concerned with locomotion?
(a) Bending of arm at elbow
(b) Rotation of head of femur in acetabulum
(c) Peristaltic movements
(d) Contraction of the heart

Muscular System

7 How many muscles are present in human body?
(a) 639 (b) 640 (c) 635 (d) 750

8 The longest smooth muscle is
(a) Rectus abdominis (b) Masseter
(c) Gluteus maximus (d) None of these

9 Which one of the following statements is incorrect? **[NCERT Exemplar]**
(a) Heart muscles are striated and involuntary
(b) The muscles of hands and legs are striated and voluntary
(c) The muscles located in the inner walls of alimentary canal are striated and involuntary
(d) Muscles located in the reproductive tracts are unstriated and involuntary

10 Which one of the following is correct for visceral muscles?
(a) Their activities are not under the voluntary control of nervous system
(b) These assist transportation of gametes through the genital tract
(c) These are located in the inner walls of hollow visceral organs of the body
(d) More than one option is correct

11 Each organised skeletal muscle in our body is made up of a number of .... held together by a common collagenous connective tissue called ....
(a) fascicles, sarcolemma **[NCERT Exemplar]**
(b) muscle bundles, sarcolemma
(c) muscle bundles, fascia
(d) fascicles, sarcomere

12 The plasma membrane covering of the muscle fibre is called as
(a) sarcolemma (b) sarcoplasmic reticulum
(c) sarcoplasma (d) Both (a) and (b)

13 The cytoplasmic segment of striated muscle fibre is termed
(a) metamere (b) neuromere
(c) sarcoplasm (d) sarcomere

14 The muscle fibres that contract slowly are
(a) red muscle fibres (b) white muscle fibres
(c) Both (a) and (b) (d) None of these

15 The difference between the red and white muscle fibres in skeletal muscle can be based on all except
(a) diameter of muscle fibre
(b) amount of myoglobin
(c) presence of sarcomere
(d) amount of sarcoplasmic reticulum

16 A skeletal muscle can show
(a) isotonic contraction (b) isometric contraction
(c) twitch contraction (d) All of these

17 Myoglobin of skeletal muscles
(a) is present in sarcoplasm (b) stores carbon dioxide
(c) stores oxygen (d) Both (a) and (c)

18 Each thin filament of muscle fibre is made up of
(a) two pairs of F (filamentous) actins helically wound to each other
(b) two F (filamentous) actins
(c) two meromyosins
(d) two meromyosins and F-actins

19 Identify the marked structures in the given figure.

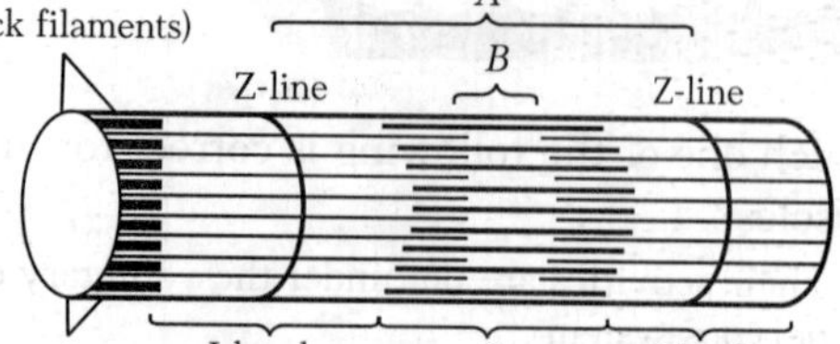

(a) A-Sarcomere, B-H-zone, C-A-band, D-Actin
(b) A-H-zone, B-A-band, C-Actin, D-Sarcomere
(c) A-Actin, B-H-zone, C-A-band, D-Sarcomere
(d) A-Sarcomere, B-A-band, C-H-zone, D-Actin

20 Skeletal muscles fibre has light and dark bands. Which is correct match of protein with its light refractive property and colour?

	Proteins	Colours	Properties
(a)	Myosin	Light	Anisotropic
(b)	Actin	Dark	Anisotropic
(c)	Myosin	Dark	Isotropic
(d)	Actin	Light	Isotropic

21 Which of the following is incorrect with respect to muscle fibre of skeletal muscles?
(a) The thin filaments (actin) are firmly attached to the Z-line
(b) The thick filaments (myosin) are held together in the middle of A-band by a thin fibrous membrane called M-line
(c) The A and I-bands are arranged alternately throughout the length of the myofibrils
(d) H-zone shows overlapping of the thin (actin) and thick (myosin) filaments

22 Choose the incorrect statement with respect to skeletal muscles.
(a) Muscle fibre is a syncytium
(b) Sarcoplasm is usually uninucleate
(c) Sarcoplasmic reticulum of muscle fibres is the storehouse of calcium ions
(d) Each myofibril has alternate dark and light bands on it

23 The proteins which in association with actin forms the thin filaments of muscles is/are
(a) tropomyosin (b) troponin
(c) Both (a) and (b) (d) myosin

24 A contractile protein in which ATPase of the muscle is located. **[NCERT Exemplar]**
(a) Actinin
(b) Troponin
(c) Myosin
(d) Actin

25 Find out the correct sequence of muscle structures/ components present one with the other.
(a) Muscle fibre → Muscle bundle → Myofilament → Myofibril
(b) Muscle bundle → Muscle fibre → Myofilament → Myofibril
(c) Muscle bundle → Muscle fibre → Myofibril → Myofilament
(d) Muscle fibre → Myofilament → Muscle bundle → Myofibril

26 The proteins present in myofilaments are
I. globulin II. actin
III. haemoglobin IV. tropomyosin
Choose correct option with correct statement (s).
(a) I and III (b) II and IV
(c) III and IV (d) Only IV

27 Consider the following statements.
I. Each meromyosin has two parts, a globular head with a short arm and a tail, the former being called the Heavy Mero-Myosin (HMM) and the latter the Light Mero-Myosin (LMM).
II. The globular head is an active ATPase enzyme and has binding sites for ATP and active sites for actin.

Choose the correct option.
(a) Statement I is true, but II is false
(b) Statement I is false, but II is true
(c) Both statements I and II are true
(d) Both statements I and II are false

28 In a myofilament, each of the thin filament
(a) is associated with two filamentous tropomyosin protein which run close to the F-actins throughout its length
(b) is made of Light Mero-Myosin (LMM)
(c) Both (a) and (b)
(d) is associated with filamentous troponin protein which is continuous throughout the length of tropomyosin

29 Given below is diagrammatic representation of an actin filament (thin filament) with labelled parts A, B and C. Which of the following are correct matches with respect to labelled parts and its description?

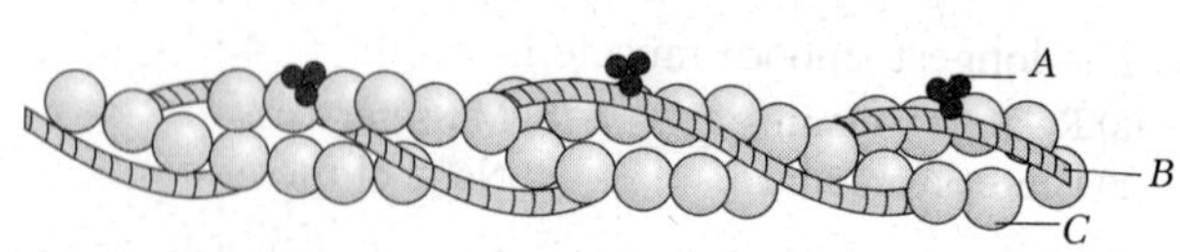

(a) A-A subunit of it masks the active binding sites for myosin on the actin filaments
(b) B-Polymer of meromyosins
(c) C-Polymer of monomeric F-actins
(d) All of the above

Muscle Contraction

30 Motor unit is a
(a) neuron
(b) muscle fibre
(c) motor neuron with muscle fibre
(d) All of the above

31 Neuromuscular junction is a junction between
(a) two neurons and muscles
(b) sensory neurons and muscles
(c) motor neurons and sarcolemma of muscles
(d) sensory neurons and sarcolemma of muscles

32 Which muscle band remains unchanged during the contraction and relaxation of the skeletal muscle?
(a) I (b) H (c) A (d) E

33 During contraction of skeletal muscle, which of the following does not occur?
(a) The length of isotropic band shortens
(b) The length of anisotropic band shortens
(c) The distance between two successive Z-lines decreases
(d) The length of Henson-zone shortens

34 Carefully observe the diagram and identify the states of sarcomere in *A*, *B* and *C*.

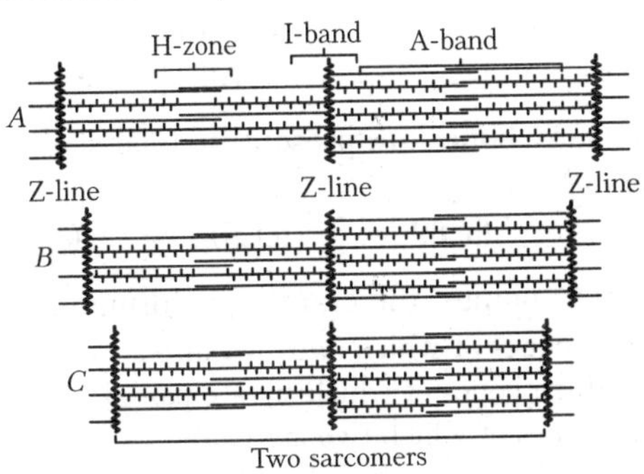

Choose the correct option identifying the state of sarcomeres.
(a) Contracting, Relaxed, Maximally contracted
(b) Relaxed, Contracting, Maximally contracted
(c) Maximally contracted, Contracting, Relaxed
(d) Relaxed, Maximally contracted, Contracting

35 During muscle contraction, ATP provides energy for
(a) cross-bridge detachment
(b) action potential genertion
(c) Ca^{2+} ions release
(d) cross-bridges attachment of myosin on actin

36 In the resting muscle fibre, tropomyosin partially covers
(a) calcium binding sites on troponin
(b) actin binding sites on myosin
(c) myosin binding sites on actin
(d) calcium binding sites on actin

37 Which of the following is related to hypertrophy of muscle?
(a) Muscles become thin and weak due to excessive work
(b) Muscles become thick and strong due to excessive work
(c) Muscles become thin and weak due to no work
(d) Muscles become thick and strong due to no work

38 During muscle contraction,
(a) chemical energy is changed into electrical energy
(b) chemical energy is charged into mechanical energy
(c) chemical energy is changed into physical energy
(d) mechanical energy is changed into chemical energy

39 Action potential in the sarcolemma of muscles is generated by
(a) neuroinhibitors (b) acetylcholine
(c) methylcholine (d) ethylcholine

40 Which of the following is important for muscle contraction and nerve impulse transmission?
(a) Ca^{2+} ions (b) Ca^{2+} and Mg^{2+} ions
(c) Mg^{2+} ions (d) Fe^{2+} ions

41 The sliding filament theory of muscle contraction was proposed by
(a) A Pullman and B Pullman
(b) Hugh Huxley and Jean Hanson
(c) R Hill and D Arnon
(d) None of the above

42 A single isolated contraction of the muscle fibre is
(a) contracture (b) tetanus
(c) twitch (d) fatigue

43 If a muscle undergoes rapid contraction and relaxation the sarcoplasmic reticulum extension
(a) requires constant plugging in and out of Ca^{2+}
(b) requires rapid synthesis of myosin
(c) does not require energy
(d) All of the above

44 Relaxation of the muscle takes place due to
I. pumping of Ca^{2+} ions in sarcoplasmic reticulum
II. presence of ATP
III. conformational changes in troponin and masking the actin filament.

Choose the option with correct statement (s).
(a) I and III (b) I and II
(c) II and III (d) I, II and III

45 The generation of excitation contraction coupling involve all of the following events except?
(a) generation of end plate potential
(b) release of calcium from troponin
(c) formation of cross linkages between actin and myosin
(d) hydrolysis of ATP to ADP

46 Muscle relaxes when
(a) myosin can attach to actin
(b) myosin can attach to tropomyosin
(c) myosin cannot attach to actin
(d) myosin cannot attach to tropomyosin

47 Apparatus used in recording muscle contraction is
(a) Kymograph (b) Electroencephalograph
(c) Electrocardiograph (d) All of these

48 Cross arms of the myosin monomer consists of
(a) outward projection of G-actin filament
(b) outward projection of the head region of meromyosin
(c) outward projection of the tail region of meromyosin
(d) Both (b) and (c)

49 Rigidity that develops in the muscle after death is known as
(a) rigor mortis (b) tetanus
(c) twitch (d) treppe

50 Which of the following is an example of isotonic contraction?
(a) Pulling of heavy object (b) Pushing of a heavy rock
(c) Simple bending of arm (d) Both (a) and (b)

51 All-or-none principle of muscle contraction is also called as
(a) Bowditch's law (b) Henry's law
(c) Johanson law (d) Mendel's law

52 A continued state of contraction caused by rapid succession of many stimuli is
(a) tetanus (b) twitch
(c) contracture (d) fatigue

53 Muscle tone can be referred to as a
(a) mild state of tetanus
(b) complete state of tetanus
(c) muscle fatigue
(d) relaxed muscle fibre

54 During isometric contraction, the tension in the muscle fibres
(a) increases (b) decreases
(c) remain same (d) do not play any role

55 If a muscle is not used for a long period, it causes muscle
(a) hypertrophy (b) tonus
(c) atrophy (d) rigor mortis

56 Consider the following statements.
I. The reduction in the force of contraction of a muscle after prolonged stimulation causes muscle fatigue.
II. A substance that accumulates in a fatigued muscle is lactic acid.

Choose the correct option.
(a) Statement I is true, but II is false
(b) Statement I is false, but II is true
(c) Both statements I and II are true
(d) Both statements I and II are false

57 Consider the following statements.
I. The muscle around the openings of iris are dilator.
II. Muscle which bend the joint is twitch.
III. Adductor muscle is antagonist to abductor.
Choose the correct option.
(a) I and II are correct
(b) II and III are correct
(c) II is correct
(d) I and III are correct

58 Select the correct matched pair.
(a) Abductor – Triceps extending forearm
(b) Adductor – Biceps bending forearm towards upper arm
(c) Elevator – Latissimus dorsi
(d) Rotator – Pyriformis that raises thigh

Skeletal System

59 The examples of exoskeletal system in vertebrates is/are
(a) nails (b) horns
(c) hoofs (d) All of these

60 How many bones are present in axial skeleton of humans?
(a) 80 (b) 89 (c) 20 (d) 26

61 In human beings, the cranium is formed by
(a) eight bones of which two are paired
(b) fourteen bones of which six are paired
(c) ten bones of which two are paired
(d) twelve bones of which four are paired

62 Vertebral formula for human beings is
(a) $C_5T_{12}L_7S_5C_{3-5} = 33-35$
(b) $C_7T_{12}\ L_5S_5C_4 = 33-35$
(c) $C_5T_{10}\ L_5S_5C_{3-5} = 33$
(d) $C_7T_{10}\ L_5S_5C_{3-5} = 33$

63 Which of the following is not a function of the skeletal system?
(a) Production of blood cells
(b) Storage of minerals
(c) Storage of carbohydrates
(d) Protection of vital organs

64 Which of the following bones serves as a point of attachment for tongue?
(a) Hyoid bone (b) Ear ossicles
(c) Cranium (d) Foramen magnum

65 The study of skeletal systems is known as
(a) Osteology (b) Myology
(c) Skeletology (d) Both (a) and (c)

66 Intervertebral disc is found in the vertebral column of **[NCERT Exemplar]**
(a) birds (b) reptiles
(c) mammals (d) amphibians

67 Cervical vertebrae are located in
(a) thoracic region (b) abdominal region
(c) neck region (d) lumbar region

68 In human beings, the second cervical vertebra helps in rotatory movements of head through knob-like process called
(a) pre-zygapophysis
(b) post-zygapophysis
(c) odontoid process
(d) metaphysis

69 Cranium of man is made up of
(a) 8 bones (b) 12 bones
(c) 10 bones (d) 16 bones

70 The vertebrae, which bear the whole weight of the skull is
(a) axis (b) sacral
(c) cervical (d) atlas

71 The only movable bone in the skull is
(a) maxilla (b) frontoparietal
(c) mandible (d) nasal

72 Centrum of man is
(a) procoelous (b) amphicoelus
(c) acoelous (d) opisthocoelous

73 The last two pairs of ribs are named floating ribs, because
(a) their sternal parts are attached to the sternum directly
(b) their sternal parts are attached on the seventh pair of ribs
(c) their sternal parts remain free and do not even reach the sternum
(d) None of the above

74 Axis vertebra of a mammal differs from atlas in the
(a) absence of centrum
(b) presence of supportive process
(c) presence of central canal
(d) presence of odontoid process

75 The eighth and ninth ribs are known as false ribs because their external portions are attached to
(a) xiphisternum
(b) costal cartilage of seventh rib
(c) they have no costa
(d) None of the above

76 Which one of the following is showing the correct sequential order of vertebrae in the vertebral column of human beings? **[NCERT Exemplar]**
(a) Cervical – Lumbar – Thoracic – Sacral – Coccygeal
(b) Cervical – Thoracic – Sacral – Lumbar – Coccygeal
(c) Cervical – Sacral – Thoracic – Lumbar – Coccygeal
(d) Cervical – Thoracic – Lumbar – Sacral – Coccygeal

77 Consider the following statements.
I. In mammals, the largest vertebra is lumbar.
II. The number of lumbar vertebrae in human vertebral column is five.

Choose the correct option.
(a) Statement I is true, but II is false
(b) Statement I is false, but II is true
(c) Both statements I and II are true
(d) Both statements I and II are false

78 In man, the thoracic basket is composed of
(a) ribs and thoracic vertebrae
(b) ribs and sternum
(c) ribs, sternum and lumbar vertebrae
(d) ribs, sternum and thoracic vertebrae

79 Choose the correct matched pair.
(a) Opisthocoelous – Lower amphibians and most fishes
(b) Heterocoelous – Mammals
(c) Amphiplatyan – Birds
(d) Amphicoelous – Frog vertebrae

80 Consider the following statements.
I. Vertebral column is connected to the pelvic girdle in sacral region.
II. The pectoral, pelvic girdle and the bones of limb form appendicular skeleton.

Choose the correct option.
(a) Statement I is true, but II is false
(b) Statement I is false, but II is true
(c) Both statements I and II are true
(d) Both statements I and II are false

81 Which of the following is incorrect with respect to bones of skull?
(a) Cranium is formed of 8 bones
(b) Mandible is the only unpaired bone of face
(c) Two maxillary bones fuse to form upper jaw
(d) Occipital bone is present at the base of skull and has two condyles to articulate with vertebral column

82 Which of the following is correct with respect to axial skeleton?
(a) Axial skeleton has bones distributed along the main axis of the body
(b) The skull, vertebral column, sternum and ribs constitute axial skeleton
(c) It comprises 80 bones
(d) All of the above are correct

83 Which of the following is incorrect with respect to vertebrae in humans?
(a) Cervical vertebrae are present in neck
(b) Lumbar vertebrae are the strongest and heaviest
(c) Sacrum is called tail bone
(d) Thoracic vertebrae serve for attachment with ribs

84 Ribs are attached to **[NCERT Exemplar]**
(a) scapula (b) sternum
(c) clavicle (d) ilium

85 Which one is an odd pair?
(a) Femur – Humerus (b) Tibia – Radius
(c) Axis – Atlas (d) Tarsal – Carpals

86 Pectoral girdle constitutes
(a) scapula and clavicle (b) radius and ulna
(c) ilium and ischium (d) maxilla and mandible

87 Transverse process of sacral vertebrae are attached to which part of pelvic girdle?
(a) Mastoid process (b) Ilium
(c) Ischium (d) Pubis

88 The vertebra in which centrum is absent and transverse process are present, is known as
(a) lumbar vertebrae (b) anterior thoracic
(c) axis vertebrae (d) atlas vertebrae

89 The longest part of sternum is
(a) merosternum (b) xiphisternum
(c) prosternum (d) metasternum

90 How many sacral vertebrae fuse to form a single structure called sacrum in adults?
(a) Five (b) Four
(c) Three (d) Six

91 Procoelous type of centrum or body is found in the vertebrae of
(a) fishes and amphibians (b) frogs
(c) mammals (d) birds

92 Surface for attachment of tongue is
(a) palatine (b) sphenoid
(c) pterygoid (d) hyoid apparatus

93 Ventral wall of pubis has a small bone called
(a) coracoid (b) cotyloid
(c) pubic symphysis (d) femur

94 Posterior terminal part of the vertebral column in man and other tailless apes is known as
(a) coccyx (b) filum terminale
(c) telson (d) urostyle

95 The number of occipital condyles in man is/are
(a) one (b) two
(c) three (d) four

96 In humans, each half of pelvic girdle is known a
(a) innominate (b) condyles
(c) pubic symphysis (d) projections

97 The number of bones in half of the lower jaw of man is
(a) one (b) four (c) six (d) eight

98 Consider the following statements.
I. In man, vertebral column has 33 bones which are organised as 28 bones.
II. Pelvic girdle is made up of two fused bones only.
Choose the correct option.
(a) Statement I is true, but II is false
(b) Statement I is false, but II is true
(c) Both statements I and II are true
(d) Both statements I and II are false

99 Coccygeal bone is formed by the fusion of ...... bones in man.
(a) three vertebrae (b) six vertebrae
(c) five vertebrae (d) four vertebrae

100 Match the following columns.

	Column I (Skeletons)		Column II (Numbers of bones)
A.	Sternum	1.	14
B.	Ribs	2.	1
C.	Pelvis	3.	24
D.	Face	4.	3

Codes

	A	B	C	D		A	B	C	D
(a)	2	3	4	1	(b)	2	4	1	3
(c)	1	3	4	2	(d)	4	1	2	3

101 Find out the correct order of number of bones in the parts of skull such as cranial bone, facial bone, hyoid bone and middle ear bone, respectively
(a) 14, 8, 1 and 6 (b) 6, 8, 14 and 1
(c) 14, 8, 6 and 1 (d) 8, 14, 1 and 6

102 Function of pre and post-zygapophysis is to
(a) allow the maximum bending of vertebrae
(b) prevent undue bending of adjacent vertebrae
(c) prevent displacement of the adjacent vertebrae
(d) Both (a) and (b)

103 Which one of the following statements is true? **[NCERT Exemplar]**
(a) Head of humerus bone articulates with acetabulum of pectoral girdle
(b) Head of humerus bone articulates with glenoid cavity of pectoral girdle
(c) Head of humerus bone articulates with a cavity called acetabulum of pelvic girdle
(d) Head of humerus bone articulates with a glenoid cavity of pelvic girdle

104 Out of the 12 pairs of ribs, true ribs are formed by
(a) VIII, IX and X pairs of ribs
(b) last two pairs of ribs
(c) first VII pairs of ribs
(d) first V pairs of ribs

105 Match the following columns.

	Column I		Column II
A.	Tarsal	1.	14
B.	Phalanges	2.	1
C.	Metatarsal	3.	7
D.	Femur	4.	5

Codes

	A	B	C	D		A	B	C	D
(a)	3	1	4	2	(b)	1	2	3	4
(c)	2	3	4	1	(d)	4	1	3	2

106 Appendicular skeleton includes bones of
(a) forelimbs (b) hindlimbs
(c) pectoral and pelvic girdle (d) All of these

107 Consider the following statements.
I. The large, flat, triangular bone present in two halves of shoulder girdle is scapula.
II. The collar bone is also known as clavicle.
Choose the correct option.
(a) Statement I is true, but II is false
(b) Statement I is false, but II is true
(c) Both statements I and II are true
(d) Both statements I and II are false

108 Ulna articulates with humerus through
(a) transverse process (b) coracoid process
(c) acromion process (d) olecranon process

109 Innominate or hip bone is formed by the fusion of how many bones?
(a) Two (b) Three (c) Four (d) Five

110 The cup-shaped structure of pelvic girdle, the acetabulum in man, is formed by
(a) ilium, ischium and pubis
(b) ilium, ischium and cotyloid
(c) ilium and ischium
(d) ilium and cotyloid

111 Deltoid ridge is found in which one of the following bones?
(a) Radius (b) Tibia
(c) Femur (d) Humerus

112 The head of the rib, which articulates with the transverse process of thoracic vertebrae is called
(a) capitulum (b) tuberculum
(c) centrum (d) shaft

113 Which one of the following bones is not the part of axial skeleton?
(a) Hyoid bone (b) Sacrum
(c) Sphenoid bone (d) Clavicle

114 Foramen magnum and occipital condyles are found in
(a) frontoparietal bone (b) occipital bone
(c) porotic bones (d) squamosal bone

115 The superior and middle conchae are bony structures of which bone?
(a) Palatine bone (b) Nasal bone
(c) Ethmoid bone (d) Maxilla

116 A facial bone that is not paired, is the
(a) maxilla (b) lacrimal bone
(c) vomer (d) nasal bone

117 An acromion process is characteristically found in the
(a) pelvic girdle of mammals
(b) skull of frog
(c) pectoral girdle of mammals
(d) sperm of mammals

118 Longest bone in lower arm is
(a) ulna (b) radius
(c) tibia (d) femur

119 Match the following columns.

	Column I		Column II
A.	True ribs	1.	11th and 12th pair
B.	False ribs	2.	Four fused pieces
C.	Floating ribs	3.	First seven pairs
D.	Coccyx	4.	8th, 9th and 10th pair

Codes

	A	B	C	D
(a)	3	1	4	2
(b)	3	4	2	1
(c)	3	2	1	4
(d)	3	4	1	2

Joints, Lever System and Bones Disorders in Human

120 The gliding joints are important for gliding movements. One example of such a joint is between the
(a) zygapophysis of adjacent vertebrae
(b) humerus and the glenoid cavity
(c) occipital condyle and odontoid process
(d) femur and tibiofibula

121 Which one of the following pair is incorrect? **[NCERT Exemplar]**
(a) Hinge joint - Between humerus and pectoral girdle
(b) Pivot joint - Between atlas, axis and occipital condyle
(c) Gliding joint - Between the carpals
(d) Saddle joint - Between carpal and metacarpals of thumb

122 Match the options given in Column I with those of Column II.

Column I	Column II
A. Carpal	1. Talus
B. Tarsal	2. Scaphoid
C. Cranium	3. Frontal
D. Face	4. Zygomatic

Codes

	A	B	C	D
(a)	2	1	3	4
(b)	1	2	3	4
(c)	4	3	2	1
(d)	1	4	3	2

123 Which one of the following pair is correctly matched?
(a) Cartilaginous joint - Skull bones
(b) Hinge joint - Between vertebrae
(c) Fibrous joint - Between phalanges
(d) Gliding joint - Between zygapophysis of the successive vertebrae

124 Which one of the following is an example of immovable/fibrous joint?
(a) Sutures (b) Gomophoses
(c) Both (a) and (b) (d) None of these

125 Joints (with respect to bones) are points of
(a) contact between bones and cartilage
(b) contact between bones
(c) Both (a) and (b)
(d) crack in the bones

126 Refer to the following Venn diagram and choose the correct option for 'X'.

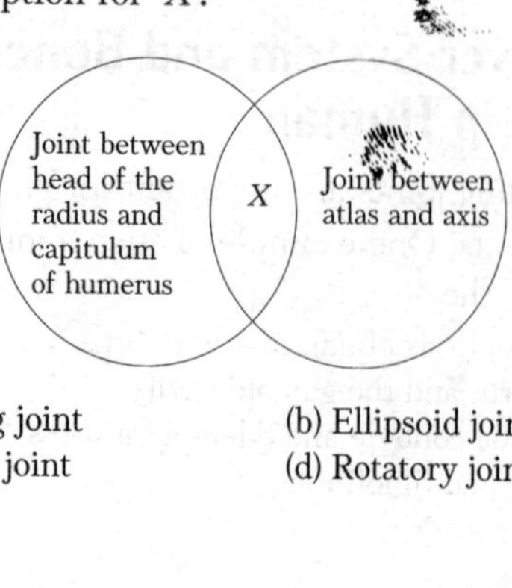

(a) Gliding joint (b) Ellipsoid joint
(c) Saddle joint (d) Rotatory joint

127 Choose the incorrect matched pair.
(a) Synarthrosis – Movable joints
(b) Amphiarthrosis – Slightly movable joints
(c) Shoulder joint – Between humerus and glenoid cavity
(d) Gliding joint – Monoaxial joint

128 The most common levers in the body is
(a) first class (b) second class
(c) third class (d) fourth class

129 In producing movement, joints act as
(a) effort (b) fulcrum (c) resistance (d) load

130 Choose the correct match.
(a) First class lever - Raising the body on the toes
(b) Second class lever - Flexing the forearm at the elbow
(c) Third class lever - Head resting on the vertebral column
(d) None of the above

131 A disease caused by the inflammation of joints is
(a) glaucoma (b) arthritis
(c) hernia (d) Horner's syndrome

132 Severe twisting of joint without dislocation is called
(a) sprain (b) dislocation
(c) strain (d) fracture

133 Gout is a disease that affects the joints and leads to arthritis. It is associated with an abnormality of
(a) pyrimidine metabolism (b) purine metabolism
(c) fat metabolism (d) protein metabolism

134 Which one of the following is not a disorder of bone? **[NCERT Exemplar]**
(a) Arthritis (b) Osteoporosis
(c) Rickets (d) Atherosclerosis

135 Progressive degeneration of skeletal muscle, mostly due to genetic disorder occurs in
(a) Myasthenia gravis (b) muscular dystrophy
(c) tetany (d) osteoporosis

136 Select the incorrectly matched pair.
(a) Green stick fracture – Hairline fracture
(b) Exclusive fracture – Breaking bone into two or more parts
(c) Comminuted fracture – Breaking bone in more than two pieces
(d) Compound fracture – More than one tissue is damaged

137 What is the type of movable joint present between the atlas and axis? **[NCERT Exemplar]**
(a) Pivot (b) Saddle (c) Hinge (d) Gliding

138 In old age, stiffness of joints is due to the
(a) hardening of bones (b) inefficiency of muscles
(c) decrease in synovial fluid (d) enlargement of bones

(B) Cumulative Exercise

(Assorted Questions of the Chapter for Advanced Level Practice)

1 Which of the following is an incorrect statement?

(a) Myasthenia gravis is an autoimmune disease affecting neuromuscular junction, leading to fatigue, weakening and paralysis of skeletal muscle
(b) Hypoparathyroidism can cause tetany where patient has rapid spasm of muscles to due to low calcium levels in body fluids
(c) Patients receiving steroids for treatment of kidney transplant or autoimmune disorders are at risk of developing osteoporosis
(d) Fibrous, hinge and pivot joints are different types of synovial joints

2 Which of the following is/are correct for human skeleton?

I. The axial and appendicular skeleton of adult man consist of 80 and 126 bones, respectively.
II. The number of bones in the skull of man is 28.
III. Number of bones in the human cranium is 8.
IV. The number of bones in vertebral column is 7.

Choose the correct option.

(a) I, II and IV (b) I and III
(c) II and III (d) Only IV

3 Ball and socket joints

I. are multiaxial, e.g. shoulder joints.
II. permit movement around many axes.
III. allow rotation around its own longitudinal axis only.
IV. produce side to side and up and down movements.

Choose the correct option.

(a) I and II (b) II, III and IV
(c) III and IV (d) I and IV

4 Given below is a diagram of the bones of the left human hindlimb as seen from front. It has certain mistakes in labelling. Two of the incorrectly labelled bones are

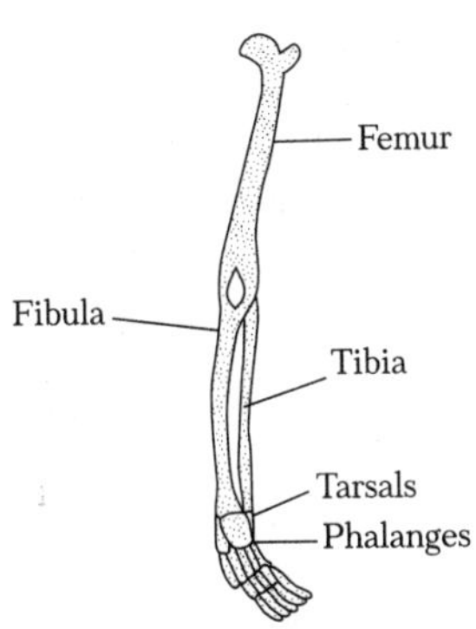

(a) tibia and tarsals
(b) femur and fibula
(c) fibula and phalanges
(d) tarsals and femur

5 Identify the types of joints in given figures *A*, *B*, *C* and *D*.

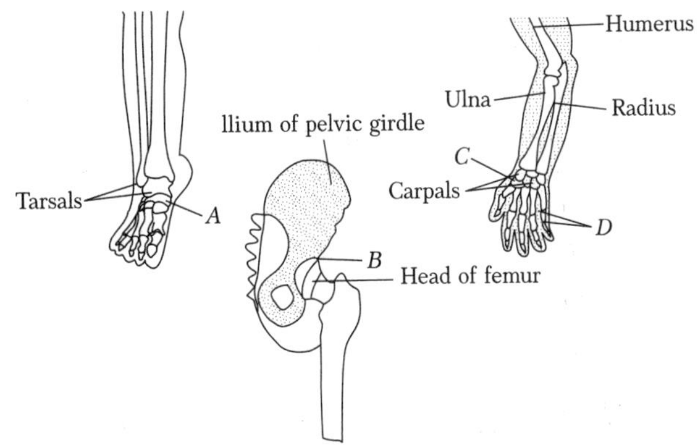

Choose the correct option.

	A	*B*	*C*	*D*
(a)	Pivot	Saddle	Gliding	Hinge
(b)	Gliding	Ball and socket	Saddle	Condyloid
(c)	Hinge	Gliding	Pivot	Saddle
(d)	Condyloid	Pivot	Saddle	Ball and socket

6 Choose the correct statement.

(a) The two condyles present at the distal end of femur are separated by intercondylar fossa
(b) The shank region consists of tarsals and metatarsals
(c) Almost 17 bones form toe
(d) Patella is the largest bone among ethmoid bones

7 Choose the incorrect statement.

(a) Innominate bone is the hip bone
(b) The two halves of pectoral girdle meet dorsally and form pelvic symphysis
(c) Pubic symphysis contains fibrous cartilage
(d) Obturator foramen is found in birds and fishes

8 Which of the following is correct regarding the bones of skull?

(a) Hyoid is not a bone of skull proper
(b) Skeleton of face is formed by 10 bones
(c) Zygomatic bones are freely movable bones
(d) The lower jaw is fused with cranium, while the upper jaw is connected with cranium by muscles

9 A cricket player is fast chasing a ball in the field. Which one of the following groups of bones are directly contributing in this movement?

(a) Femur, malleus, tibia, metatarsals
(b) Pelvis, ulna, patella, tarsals
(c) Sternum, femur, tibia, fibula
(d) Tarsals, femur, metatarsals, tibia

10 Consider the diagram given below.

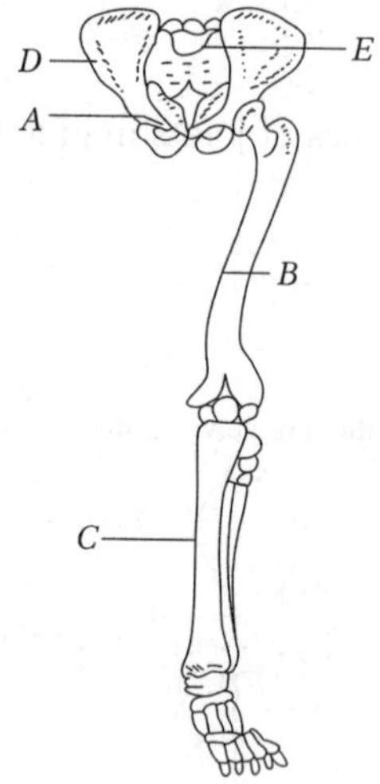

Parts labelled as *A*, *B*, *C*, *D* and *E*, respectively indicate

(a) Ilium, Femur, Tibia, Pubis and Sacrum
(b) Pubis, Tibia, Femur, Ilium and Sacrum
(c) Ilium, Femur, Tibia, Pubis and Sacrum
(d) Pubis, Femur, Tibia, Ilium and Sacrum

11 How many of these bones are a part of appendicular skeleton?

Radius, Ethmoid, Temporals, Parietal, Patella metatarsals, Carpals, Phalanges, Coxal, Hip

(a) Six (b) Seven
(c) Four (d) Eight

12 How many of these characteristics are true for scapula?

Cup-like glenoid cavity, Acromion process, Coracoid process, Flat and triangular, Narrow and S-shaped, Collar bone, Olecranon process, Capitulum process

(a) Two (b) Six
(c) Four (d) Five

13 Which of the following statements is correct?

(a) Capitulum process articulates with ulna of lower arm
(b) Trochlea articulates with radius of lower arm
(c) Ulna articulates with humerus through olecranon process
(d) Pelvis in males has larger diameter

14 Skeletal muscles are

I. attached to the bones by tendons and help in the movement of the parts of skeleton.
II. under control of conscious mind and can be moved at will.
III. exhibit transverse strips.
IV. innervated by involuntary nervous system.

Choose the option with the correct statements.

(a) I, II and III (b) II and III
(c) I and IV (d) II and IV

15 Bundles of muscle fibres are grouped as fascicule, which

I. is held together and enclosed by collagen fibres and by connective tissue.
II. is surrounded by tough external layer, called fascia.
III. contains an abundance of glycogen.
IV. is developed by repeated divisions of myoblasts.

Choose the option with the correct statements.

(a) I and II (b) III and IV
(c) II, III and IV (d) I and III

16 Red muscle fibres

I. perform slow and sustained contractions for long periods without fatigue.
II. depend mainly on anaerobic glycolysis.
III. carryout aerobic contraction without accumulating much lactic acid.
IV. perform faster contraction rate.

Choose the correct option containing correct statements from the codes given below.

(a) I and II (b) II and III
(c) III and IV (d) I and III

17 Which of the statements given below are true with reference to muscle contraction.

I. Neurotransmitter acetylcholine is released as nerve impulse reaches the end of axon and binds to receptor sites of motor end plate.
II. Action potential passes from motor end plate over sarcolemma and then into the T-tubules and sarcoplasmic reticulum.
III. Calcium ions are released into the sarcoplasm bind to troponin, causing a change in its shape and position.
IV. Myosin cross-bridges are able to bind to exposed active sites of actin.

Choose the correct option.

(a) I, II and III (b) II and IV
(c) I and IV (d) I, II, III and IV

18 Out of the following match pairs choose the incorrect one.

(a) Clavicle – Anterior region of thorax
(b) Sacrum – Lies between innominate bones of pectoral girdle
(c) Thoracic vertebrae – Are 12 in number
(d) Hyoid – Above the larynx in throat

19 Consider the diagram showing muscle structure below.

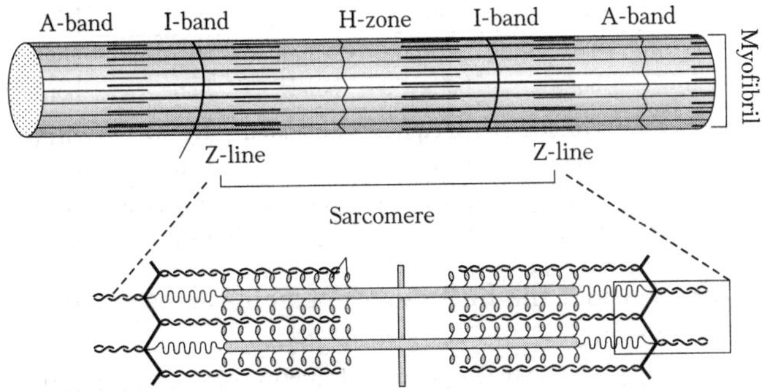

Identify the events that occur during skeletal muscle contraction.

I. I-band shortens
II. A-band shortens
III. H-zone shortens
IV. Sarcomere contracts
V. ATP changes to ADP and Pi

Choose the option with incorrect event(s).

(a) Only I (b) Only III (c) IV and V (d) Only II

20 Read the statements regarding muscle proteins.

I. Actin is a thin filament and is made up of two F-actins.
II. The complex protein, tropomyosin is distributed at regular intervals on the troponin.
III. Myosin is a thick filament which is also a polymerised protein.
IV. The globular head of meromyosin consists of Light Mero-Myosin (LMM).

Choose the correct option containing correct statements from the codes given below.

(a) I, II and III (b) I, II and IV
(c) I and III (d) II and IV

21 Match the following columns.

	Column I		Column II
A.	Fast muscle fibres	1.	Myoglobin
B.	Slow muscle fibres	2.	Lactic acid
C.	Actin filament	3.	Contractile unit
D.	Sarcomere	4.	I-band

Codes **[NCERT Exemplar]**

	A	B	C	D
(a)	1	2	4	3
(b)	2	1	3	4
(c)	2	1	4	3
(d)	3	2	4	1

22 Match the following columns.

	Column I		Column II
A.	Sternum	1.	Synovial fluid
B.	Glenoid cavity	2.	Vertebrae
C.	Freely movable joint	3.	Pectoral girdle
D.	Cartilaginous joint	4.	Flat bones

[NCERT Exemplar]

Codes

	A	B	C	D
(a)	2	1	3	4
(b)	4	3	1	2
(c)	2	1	4	3
(d)	4	1	2	3

23 Consider the diagram of thin myofilament.

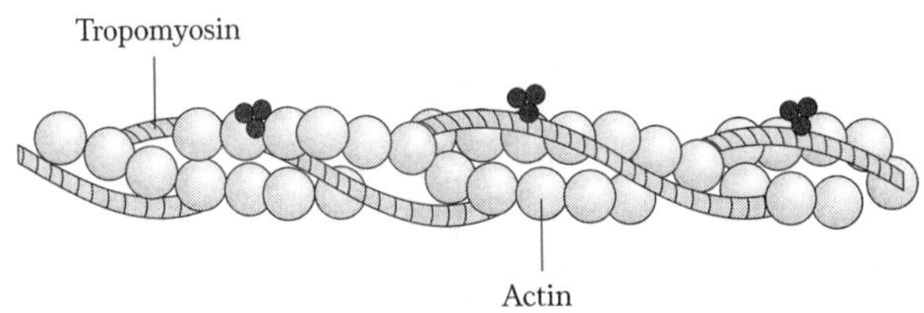

Now, read carefully the statements about molecular arrangement of actin and myosin in the filament.

I. Each actin (thin filament) is made up of 2 F-(filamentous) actins.
II. F-actin is the polymer of G-(globular) actin.
III. Two-F-actins are twisted into a helix.
IV. Two strands of tropomyosin run along the entire length of F-actin.

Choose the correct option from the codes given below.

(a) I and II (b) III and IV
(c) I and IV (d) I, II, III and IV

24 Identify the incorrect statement.

(a) Deposition of uric acid crystals in joints causes gout
(b) In patients with Myasthenia gravis, antibodies are formed against acetylcholine
(c) Ellipsoid joint is present between metacarpals and phalanges
(d) Sagittal suture is present between two parietal bones

25 Consider the following statements.

I. Joints are points of contact between bones or between bones and cartilages.
II. Force generated by muscles is used to carryout movement through joints, where muscles act as fulcrum.
III. Fibrous joints do not allow any movement.
IV. In cartilaginous joints, the bones involved are joined together with the help of dense fibrous connective tissue.
V. Synovial joints are characterised by the presence of a fluid-filled synovial cavity between the articulating surfaces of the two bones.

Which of the above statements are incorrect?

(a) I and II
(b) II and III
(c) Only IV
(d) III, IV and V

26 In the given diagram, which joints are found at the marked sites.

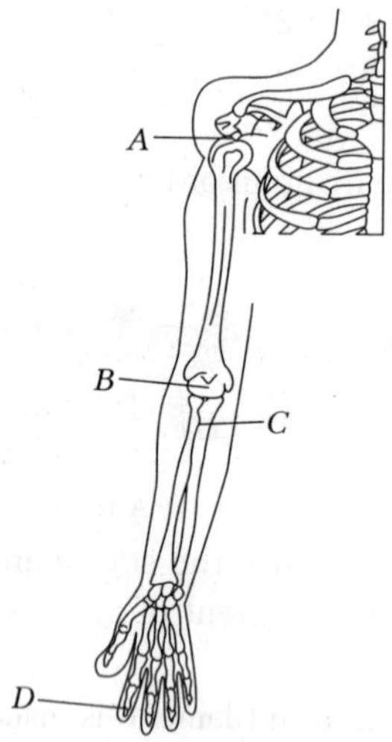

	A	*B*	*C*	*D*
(a)	Hinge	Ball and socket	Hinge	Condyloid
(b)	Ball and socket	Hinge	Pivot	Hinge
(c)	Pivot	Hinge	Ball and socket	Hinge
(d)	Ball and socket	Pivot	Hinge	Hinge

27 Which one of the following is the correct description of a certain part of a normal human skeleton?

(a) Parietal bone and the temporal bone of the skull are joined by fibrous joint

(b) First vertebra is axis, which articulates with the occipital condyles

(c) The 9th and 10th pairs of ribs are called the floating ribs

(d) Glenoid cavity is a depression to which the thigh bone articulates

28 Identify the human skull bones labelled as *A*, *B*, *C*, D and *E*.

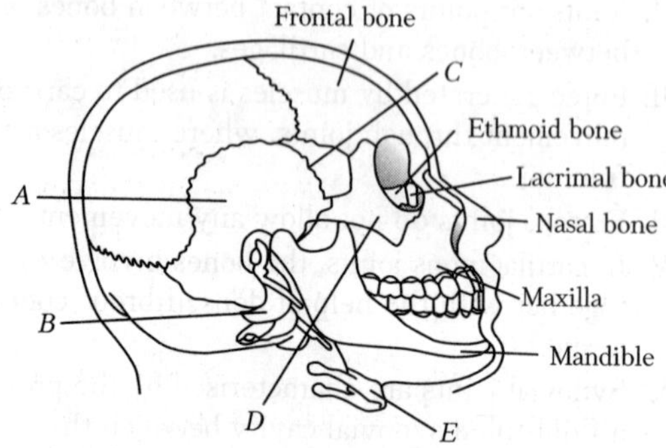

	A	*B*	*C*	*D*	*E*
(a)	Temporal bone	Occipital bone	Zygomatic bone	Sphenoid bone	Hyoid bone
(b)	Temporal bone	Occipital bone	Sphenoid bone	Zygomatic bone	Hyoid bone
(c)	Parietal bone	Occipital bone	Sphenoid bone	Hyoid bone	Zygomatic bone
(d)	Parietal bone	Occipital bone	Sphenoid bone	Zygomatic bone	Hyoid bone

29 Consider the following statements about lever system.

I. First class lever have fulcrum placed between the resistance and point of effort.

II. Second class lever have resistance placed between the fulcrum and point of effort.

III. Third class lever have effort placed in between the fulcrum and resistance.

Choose the correct option.

(a) Statements I and II are incorrect

(b) Statements II and IIII are incorrect

(c) All statements are correct

(d) All statements are incorrect

30 Identify the parts labelled as *A* to *E* in the given figure of a vertebral column and select the correct option.

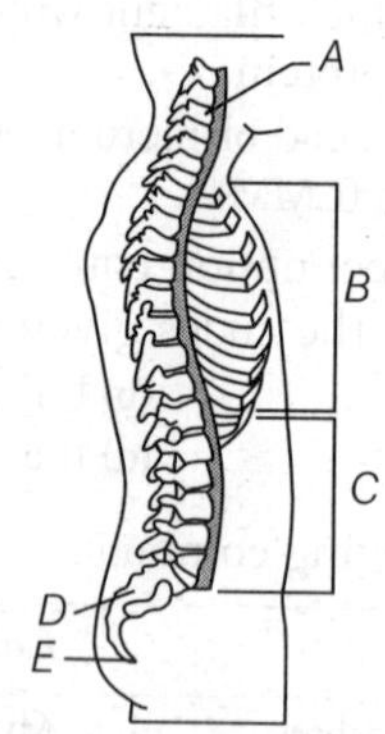

	A	*B*	*C*	*D*	*E*
(a)	Thoracic vertebra	Cervical vertebra	Lumbar vertebra	Sacrum	Coccyx
(b)	Thoracic vertebra	Cervical vertebra	Lumbar vertebra	Coccyx	Sacrum
(c)	Lumbar vertebra	Thoracic vertebra	Cervical vertebra	Coccyx	Sacrum
(d)	Cervical vertebra	Thoracic vertebra	Lumbar vertebra	Sacrum	Coccyx

31 Match the following columns and choose the correct option from the codes given below.

	Column I		Column II
A.	Boxer's muscle	1.	Serratus anterior
B.	Myalgia	2.	Inflammation of bone
C.	Myositis	3.	Pain in muscle
D.	Tailor muscle	4.	Sartorius

Codes **[NCERT Exemplar]**

	A	B	C	D		A	B	C	D
(a)	1	2	3	4	(b)	1	3	2	4
(c)	4	3	2	1	(d)	2	3	4	1

32 Read the following statements and pick the correct one.

(a) In platybasic skull, eyes are not situated much apart
(b) In tropibasic skull, eyes and brain are present in same plane
(c) Splanchnocranium helps in encasing of jaws and gills
(d) Mammals have monocondylic

33 Which of the following statements is/are correct/incorrect?

I. A-bands of the muscles are dark and contain myosin.
II. I-bands are the light bands and contain actin.
III. During muscle contraction, the A-band contracts.
IV. The part between the two Z-lines is called as sarcomere.
V. The central part of thin filament is not overlapped by thick filament is called H-zone.

Choose the correct option.

(a) I, II and III are correct, while IV and V are incorrect
(b) I, III and V are correct, while II and IV are incorrect
(c) I and II are correct, while III, IV and V are incorrect
(d) I, II and IV are correct, while III and V are incorrect

34 In the pelvic girdle of man, *A*, *B*, *C*, *D* and *E* respectively represent.

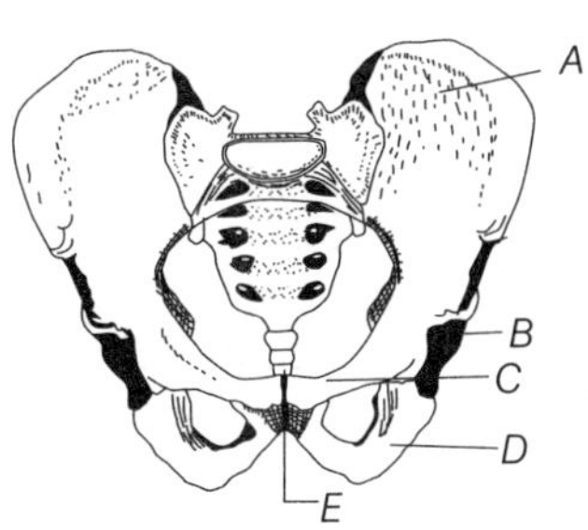

(a) *A*-Pubis, *B*-Acetabulum, *C*-Ilium, *D*-Ischium, *E*-Pubic symphysis
(b) A-Ilium, *B*-Acetabulum, *C*-Pubis, *D*-Ischium, *E*-Pubic symphysis
(c) A-Ischium, *B*-Acetabulum, *C*-Pubis, *D*-Ilium, *E*-Pubic symphysis
(d) A-Ilium, *B*-Pubis, *C*-Acetabulum, *D*-Pubic symphysis, *E*-Ischium

35 Arrange the following bones according to their decreasing length.

(a) Humerus > Ulna > Radius > Metacarpals > Carpals
(b) Humerus > Ulna > Radius > Carpals > Metacarpals
(c) Humerus > Radius > Ulna > Metacarpals > Carpals
(d) Humerus > Ulna > Metacarpals > Radius > Carpals

36 Arrange the following bones according to their decreasing length.

(a) Femur < Tibia < Fibula < Metatarsals < Tarsals
(b) Femur < Fibula< Tibia < Metatarsals < Tarsals
(c) Tarsals < Metatarsals < Tibia < Fibula < Femur
(d) Tarsals < Metatarsals < Fibula < Tibia < Femur

37 Identify *A* to *E* in the given diagram.

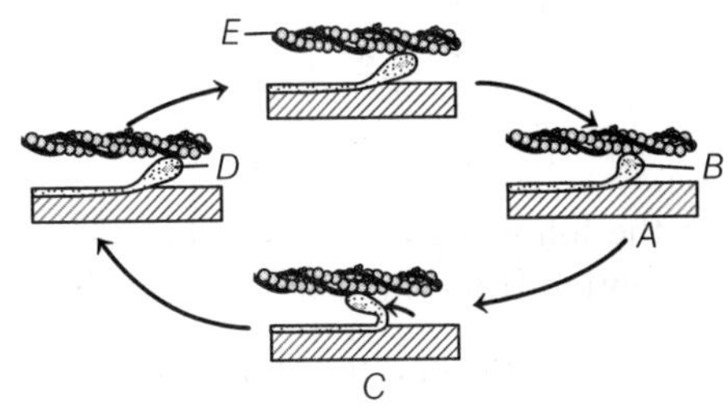

(a) *A*-Cross-bridge formation, *B*-Myosin head, *C*-Sliding, *D*-ATP, *E*-Actin filament
(b) *A*-Sliding, *B*-Actin head, *C*-Cross-bridge formation, *D*-ADP, *E*-Myosin filament
(c) *A*-Cross-bridge formation, *B*-Actin head, *C*-Sliding, *D*-ATP, *E*-Myosin filament
(d) A-Sliding, *B*-Myosin filament, *C*-Cross-bridge formation, *D*-ADP, *E*-Actin head

38 Read the following statements.

I. Summation refers to addition of continuous stimulus given to muscle before complete relaxation.
II. Rigor mortis disappears fifteen to twenty four hours after death.
III. Cori and Cori cycle occurs in the muscles.
IV. Muscle tone is a sort of mild tetanus.

	I	II	III	IV
(a)	False	False	False	True
(b)	True	True	True	True
(c)	False	False	True	True
(d)	False	True	True	False

39 Match the following columns and choose the correct option from the codes given below.

	Column I		Column II
A.	Face	1.	2 bones
B.	Ear	2.	12 pairs
C.	Ribs	3.	14 bones
D.	Pelvic girdle	4.	6 bones

Codes **[NCERT Exemplar]**

	A	B	C	D		A	B	C	D
(a)	1	2	3	4	(b)	4	2	3	1
(c)	3	4	2	1	(d)	1	3	4	2

40 Match the following columns and choose the correct option from the codes given below.

Column I	Column II
A. Arthritis	1. Damage to ligaments surrounding the joints
B. Gout	2. Inflammation of joints
C. Dislocation	3. Inherited disorder of purine metabolism
D. Myasthenia gravis	4. Autoimmune neuromuscular disorder

Codes

	A	B	C	D
(a)	1	2	3	4
(b)	2	3	1	4
(c)	1	3	2	4
(d)	4	2	1	3

41 Match the following columns and choose the correct option from the codes given below.

Column I	Column II	Column II
A. Synarthrosis joints	1. Freely movable	(i) Between vertebrae
B. Amphiarthrosis joints	2. Immovable	(ii) Between radio-ulna and carpals
C. Diarthrosis joints	3. Slightly movable	(iii) Between the Ist pair of ribs and the breast bone

Codes

	A	B	C
(a)	2-(iii)	3-(i)	1-(ii)
(b)	2-(i)	3-(ii)	1-(iii)
(c)	3-(iii)	2-(ii)	1-(i)
(d)	1-(ii)	2-(i)	3-(iii)

Assertion and Reason

Direction (Q. Nos. 42-47) *In each of the following questions, a statement of Assertion (A) is given followed by the corresponding statement of Reason (R). Of the statements, mark the correct answer as*

(a) Both A and R are true and R is the correct explanation of A
(b) Both A and R are true, but R is not the correct explanation of A
(c) A is true, but R is false
(d) A is false, but R is true
(e) Both A and R are false

42 **Assertion** (A) Movement of body parts serves to change the body posture.

Reason (R) Body parts move in relation to body axis.

43 **Assertion** (A) There are similarities between the locomotion of unicellular organisms and multicellular animals.

Reason (R) Ciliary, flagellar and amoeboid movements occur in unicellular organisms.

44 **Assertion** (A) Skeleton helps in blood cell formation.

Reason (R) Blood flows through the skeleton.

45 **Assertion** (A) Triceps is said to be an extensor muscle for elbow joint.

Reason (R) Triceps relax during extension of forearm at the elbow joint.

46 **Assertion** (A) Muscle as a whole does not obey All-or-None law.

Reason (R) Each muscle fibre contracts maximally whenever it contracts.

47 **Assertion** (A) Biceps and triceps are called antagonistic muscles.

Reason (R) This is due to the fact that they contract and relax together.

(C) Medical Entrances' Gallery

(Collection of question asked in NEET & various Medical Entrance Exam)

1. Match the following columns and select the correct option from the codes given below.

	Column I		Column II
A.	Gout	1.	Decreased levels of oestrogen
B.	Osteoporosis	2.	Low Ca^{2+} ions in the blood
C.	Tetany	3.	Accumulation of uric acid crystals
D.	Muscular dystrophy	4.	Autoimmune disorder

Codes **[NEET 2020, EAMCET 2016]**

	A	B	C	D		A	B	C	D
(a)	2	1	3	4	(b)	3	1	2	4
(c)	4	3	1	2	(d)	1	2	3	4

2 Match the following columns and select the correct option.

	Column I		Column II
A.	Floating ribs	1.	Located between second and 7th ribs
B.	Acromion	2.	Head of the humerus
C.	Scapula	3.	Clavicle
D.	Glenoid cavity	4.	Do not connect with the sternum

Codes **[NEET 2020]**

	A	B	C	D		A	B	C	D
(a)	1	3	2	4	(b)	3	2	4	1
(c)	4	3	1	2	(d)	2	4	1	3

3 Which of the following diseases is an autoimmune disorder? **[NEET 2019, EAMCET 2016]**

(a) Myasthenia gravis (b) Arthritis
(c) Osteoporosis (d) Gout

4 **Assertion** (A) Calcium required for skeletal muscle contraction.
Reason (R) Calcium influx releases acetylcholine at neuromuscular junction. **[AIIMS 2019, NEET 2018]**

(a) Both A and R are true and R is the correct explanation of A
(b) Both A and R are true, but R is not the correct explanation of A
(c) A is true, but R is false
(d) Both A and R are false

5 Match the following joints (in Column I) with the bones involved (in Column II).

	Column I		Column II
A.	Gliding joint	1.	Between carpal and metacarpal of thumb
B.	Hinge joint	2.	Between atlas and axis
C.	Pivot joint	3.	Between the carpals
D.	Saddle joint	4.	Between humerus and ulna

[NEET 2019, CBSE AIPMT 2014]

Codes

	A	B	C	D		A	B	C	D
(a)	3	4	2	1	(b)	4	2	3	1
(c)	4	1	2	3	(d)	1	3	2	4

6 Select the correct statement. **[CBSE AIPMT, NEET 2019, 12]**

(a) 11th and 12th pairs of ribs are connected to the sternum with the help of hyaline cartilage
(b) Each rib is a flat thin bone and all the ribs are connected dorsally to the thoracic vertebrae and ventrally to the sternum
(c) There are seven pairs of vertebrosternal, three pairs of vertebrochondral and two pairs of vertebral ribs
(d) 8th, 9th and 10th pairs of ribs articulate directly with the sternum

7 Cervical vertebrae differ from other vertebrae in having **[AIIMS 2019]**

(a) spinous process (b) centrum
(c) transverse process (d) transverse foramen

8 Atlas is **[JIPMER 2019]**

(a) 1st cervical vertebrae
(b) 2nd cervical vertebrae
(c) 1st thoracic vertebrae
(d) 2nd lumbar vertebrae

9 Match the following columns.

	Column I		Column II
A.	Hinge	1.	Atlas and axis
B.	Saddle	2.	Frontal and parietal
C.	Pivot	3.	Carpal and metacarpal of humans
D.	Ball and socket	4.	Knee
		5.	Humerus and pectoral girdle

Codes **[AIIMS 2018, NEET 2017]**

	A	B	C	D		A	B	C	D
(a)	2	5	3	1	(b)	3	2	5	4
(c)	4	3	1	5	(d)	1	4	2	3

10 The contraction of the muscle continues in sliding filament theory **[AIIMS 2017]**

(a) till ATP binds to myosin head
(b) till ADP binds to myosin head
(c) till Ca^{2+} is present in sarcoplasm
(d) till polymerisation of myosin head is going on

11 The globular head of myosin contains **[JIPMER 2017]**

(a) calcium ions in large quantities
(b) troponin
(c) ATPase enzyme
(d) ATP

12 Sliding filament theory can be best explained as [JIPMER 2017]

(a) when myofilaments slide pass each other actin filaments shorten, while myosin filaments do not shorten
(b) actin myosin filaments shorten and slide pass each other
(c) actin and myosin filaments do not shorten, but rather slide pass each other
(d) when myofilaments slide pass each other myosin filaments shorten, while actin filaments do not shorten

13 Out of 'X' pairs of ribs in humans only 'Y' pairs are true ribs. Select the option that correctly represents values of X and Y and provide their explanation. [NEET 2017]

(a)	$X = 12, Y = 7$	True ribs are attached dorsally to vertebral column and ventrally to the sternum.
(b)	$X = 12, Y = 5$	True ribs are attached dorsally to vertebral column and sternum on the two ends.
(c)	$X = 24, Y = 7$	True ribs are dorsally attached to vertebral column, but are free on ventral side.
(d)	$X = 24, Y = 12$	True ribs are dorsally attached to vertebral column, but are free on ventral side.

14 Osteoporosis, an age-related disease of skeletal system, may occur due to [NEET-II 2016]

(a) immune disorder affecting neuromuscular junction leading to fatigue
(b) high concentration of Ca^{2+} and Na^{+}
(c) decreased level of oestrogen
(d) accumulation of uric acid leading to inflammation of joints

15 Contraction cycle of muscle showing four parts of muscle twitch is given. Find out the correct sequence.

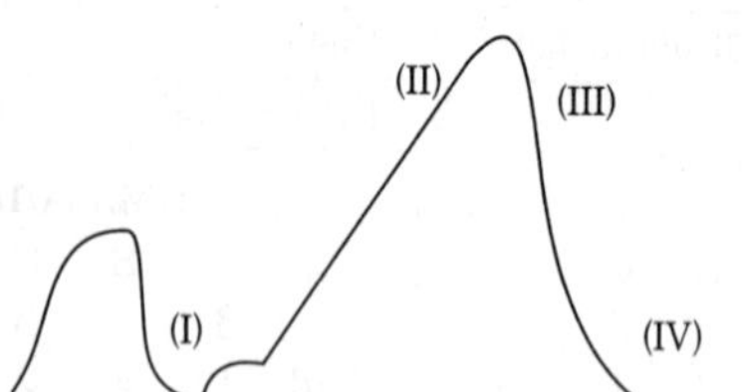

[Chhattisgarh PMT 2015]

(a) I-Latent period, II-Relaxation period, III-Refractory period, IV-Contraction
(b) I-Contraction, II-Latent period, III-Relaxation period, IV-Refractory period
(c) I-Latent period, II-Contraction, III-Relaxation period, IV-Refractory period
(d) I-Refractory period, II-Relaxation period, III-Latent period, IV-Contraction

16 Which one of the following joints will not allow movement? [CBSE AIPMT 2015]

(a) Fibrous joint
(b) Cartilaginous joint
(c) Synovial joint
(d) Ball and socket joint

17 Which is not a function of skeletal system? [CBSE AIPMT 2015]

(a) Production of erythrocytes
(b) Storage of minerals
(c) Production of body heat
(d) Locomotion

18 Choose the incorrectly matched pair. [WB JEE 2014]

(a) Portion of myofibril between two Z-lines—Sarcomere
(b) Isotropic band—Actin
(c) Anisotropic band—Myosin
(d) Central part of I-band—M-line

19 Which option is correct for the region labelled as *A*, *B*, *C* and *D* in the given diagram?

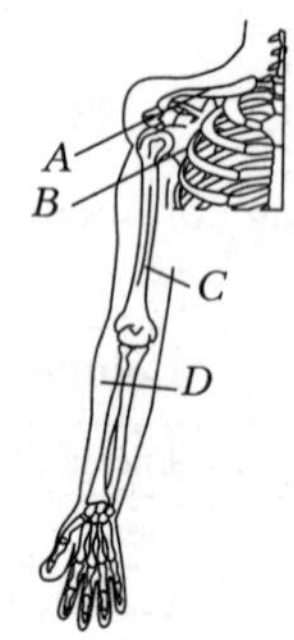

[Gujarat CET 2014]

	A	B	C	D
(a)	Scapula	Clavicle	Humerus	Ulna
(b)	Clavicle	Scapula	Humerus	Radius
(c)	Clavicle	Ulna	Radius	Humerus
(d)	Clavicle	Glenoid cavity	Radius	Ulna

20 Match the following columns.

Column I	Column II
A. Zygomatic bone	1. Keystone bone of cranium
B. Lacrimal bone	2. Cheek bone of cranium
C. Parietal bone	3. Smallest bone of face
D. Sphenoid bone	4. Roof of cranium
	5. Floor of cranium

Codes [EAMCET 2014]

	A	B	C	D		A	B	C	D
(a)	1	3	5	2	(b)	2	4	5	1
(c)	2	4	1	3	(d)	2	3	4	5

21 Which of the following type of muscles is/are not fatigued soon? [Gujarat CET 2013]

(a) Cardiac muscle
(b) Smooth muscle
(c) Both (a) and (b)
(d) Voluntary muscle

22 Which is correct statement for white muscle fibre? [Gujarat CET 2013]

(a) It contains low amount of haemoglobin and mitochondria
(b) It contains higher amount of myoglobin
(c) It contains low amount of myoglobin and mitochondria
(d) It possesses only actin type of protein

23 Select the correct statement about muscular disorder.
(a) Accumulation of urea and creatine in joints **[MP CET 2013]**
(b) An overdose of vitamin-D causes osteoporosis
(c) Rapid contraction of skeletal muscles causes dystrophy
(d) Failure of neuromuscular transmission in Myasthenia gravis can prevent normal swallowing

24 The H-zone in the skeletal muscle fibre is due to the **[NEET 2013]**
(a) central gap between myosin filaments in the A-band
(b) central gap between actin filaments extending through myosin filaments in the A-band
(c) extension of myosin filaments in the central portion of the A-band
(d) absence of myofibrils in the central portion of A-band

25 Select the correct statement with respect to locomotion in humans. **[NEET 2013]**
(a) Accumulation of uric acid crystals in joints causes their inflammation
(b) The vertebral column has 10 thoracic vertebrae
(c) The joint between adjacent vertebrae is a fibrous joint
(d) The decreased level of progesterone causes osteoporosis in old people

26 The characteristics and an example of a synovial joint in humans is **[NEET 2013]**

	Characteristics	Examples
(a)	Fluid-filled between two joints, provides cushion	Skull bones
(b)	Fluid-filled synovial cavity between two bones	Joint between atlas and axis
(c)	Lymph-filled between two bones, limited movement	Gliding, joint between carpals
(d)	Fluid cartilage between two bones, limited movements	Knee joint

27 Match the following columns.

	Column I		Column II
A.	Mastoid process	1.	Premaxilla
B.	Acromion process	2.	Axis
C.	Olecranon process	3.	Scapula
D.	Odontoid process	4.	Ulna
		5.	Periotic bone

Codes **[EAMCET 2012]**

	A	B	C	D		A	B	C	D
(a)	4	5	3	2	(b)	4	2	1	5
(c)	5	3	4	2	(d)	5	2	4	1

28 Radius is a bone found in **[UP CPMT 2012]**
(a) arms (b) legs
(c) pelvic girdle (d) None of these

29 Choose the incorrect option regarding a normal human.
I. The skull is dicondylic.
II. Metacarpals are five in numbers.
III. Patella is a cup-shaped bone covering the knee dorsally.
IV. Scapula is a large triangular flat bone, situated on the ventral side of the thorax.
V. The pelvic girdle has two coxal bones.

Identify the correct statement(s). **[Kerala CEE 2012]**
(a) I and V (b) I and II
(c) II and V (d) III and IV
(e) Only III

30 Which is correctly categorised? **[CBSE AIPMT 2012]**
(a) Troponin and myosin — Complex proteins in striated muscles
(b) Calcitonin and thymosin — Thyroid hormones
(c) Pepsin and prolactin — Digestive enzymes secreted in stomach
(d) Secretin and rhodopsin — Polypeptide hormones

31 Basic unit of muscle contraction is **[UP CPMT 2010, Punjab PMET 2012]**
(a) collagen (b) sarcomere
(c) bands (d) myofibrils

32 Which one is a mismatch? **[CBSE AIPMT 2011]**
(a) Sternum and ribs — Axial skeleton
(b) Clavicle and glenoid cavity — Pelvic girdle
(c) Humerus and ulna — Appendicular skeleton
(d) Malleus and stapes — Ear ossicles

33 Which is not true for red fibres? **[MP PMT 2012]**
(a) Muscles contain a red coloured oxygen storing pigment
(b) Muscles contain plenty of mitochondria
(c) They are also called aerobic muscles
(d) Amount of sarcoplasmic reticulum is high

34 Statements about the mechanism of muscle contraction are given below.
I. Acetylcholine is released when the neural signal reaches the motor end plate.
II. Muscle contraction is initiated by a signal sent by CNS *via* a sensory neuron.
III. During muscle contraction, isotropic band gets elongated.
IV. Repeated activation of the muscles can lead to lactic acid accumulation.

Identify the correct statements. **[Kerala CEE 2011]**
(a) I and IV (b) I and III
(c) II and III (d) I, II and III
(e) I and II

35 Actin binding sites are located on **[Kerala CEE 2011]**
(a) troponin (b) tropomyosin
(c) meromyosin (d) Both (b) and (c)
(e) Both (a) and (b)

36 Scapula is a large triangular flat bone situated in the dorsal part of the thorax between the **[Kerala CEE 2011]**
(a) second and fifth ribs (b) second and seventh ribs
(c) third and sixth ribs (d) third and eighth ribs
(e) third and five ribs

37 The coxal of the pelvic girdle is formed by the fusion of **[Kerala CEE 2011]**

(a) ilium, ischium and pubis
(b) scapula and clavicle
(c) ilium and scapula
(d) ilium, scapula and ischium
(e) clavicle and pubis

38 The joint of radio-ulna with the upper arm is **[HPMT 2008, RPMT 2011]**

(a) hinge joint (b) pivot joint
(c) socket joint (d) None of these

39 The clavicle articulates with .... of scapula. **[DUMET 2011]**

(a) acromion process (b) glenoid cavity
(c) acetabulum cavity (d) ball and socket joint

40 The sensation of fatigue in the muscles after prolonged strenuous physical work, is caused by **[MHT CET 2011]**

(a) a decrease in the supply of oxygen
(b) the depletion of glucose
(c) the accumulation of lactic acid
(d) All of the above

41 Which of the following option shows correct order of some stages of muscle contraction from the beginning to the end of the process? **[Gujarat CET 2011]**

(a) Stimuli → Neurotransmitter secretion → Release of Ca^{2+} → Cross-bridge formation → Excitation of T-system → Sliding of actin filaments
(b) Stimuli → Neurotransmitter secretion → Excitation of T-system → Release of Ca^{2+} → Cross-bridge formation → Sliding of actin filaments → H band diminishes
(c) Stimuli → Excitation of T-system → Neurotransmitter secretion → Cross-bridge formation → Sliding of actin filaments → H-band diminishes
(d) Stimuli → Neurotransmitter secretion → Cross-bridge formation → Excitation of T-system → Sliding of actin filaments

42 What is the location of troponin in the process of muscle contraction? **[Gujarat CET 2011]**

(a) Attached to myosin filament
(b) Attached to tropomyosin
(c) Attached to myosin cross-bridge
(d) Attached to T-tubule

ANSWERS

Check Point 27.1

1. (c)	2. (b)	3. (c)	4. (b)	5. (b)

Check Point 27.2

1. (b)	2. (d)	3. (c)	4. (c)	5. (c)

Check Point 27.3

1. (b)	2. (a)	3. (d)	4. (c)	5. (d)	6. (c)

(A) Topical Exercise

1. (c)	2. (c)	3. (a)	4. (d)	5. (c)	6. (b)	7. (a)	8. (a)	9. (c)	10. (d)
11. (c)	12. (a)	13. (c)	14. (a)	15. (c)	16. (d)	17. (d)	18. (b)	19. (a)	20. (d)
21. (d)	22. (b)	23. (c)	24. (c)	25. (c)	26. (b)	27. (c)	28. (a)	29. (a)	30. (c)
31. (c)	32. (c)	33. (b)	34. (b)	35. (d)	36. (c)	37. (b)	38. (b)	39. (b)	40. (b)
41. (b)	42. (c)	43. (a)	44. (d)	45. (b)	46. (c)	47. (a)	48. (b)	49. (a)	50. (c)
51. (a)	52. (a)	53. (a)	54. (c)	55. (c)	56. (c)	57. (d)	58. (d)	59. (d)	60. (a)
61. (a)	62. (b)	63. (c)	64. (a)	65. (d)	66. (c)	67. (c)	68. (c)	69. (a)	70. (d)
71. (c)	72. (c)	73. (c)	74. (d)	75. (b)	76. (d)	77. (c)	78. (d)	79. (a)	80. (c)
81. (b)	82. (d)	83. (c)	84. (b)	85. (c)	86. (a)	87. (b)	88. (d)	89. (a)	90. (a)
91. (a)	92. (d)	93. (c)	94. (a)	95. (b)	96. (a)	97. (a)	98. (b)	99. (d)	100. (a)
101. (d)	102. (c)	103. (b)	104. (c)	105. (a)	106. (d)	107. (c)	108. (d)	109. (b)	110. (a)
111. (d)	112. (b)	113. (d)	114. (b)	115. (c)	116. (c)	117. (c)	118. (a)	119. (d)	120. (a)
121. (a)	122. (a)	123. (d)	124. (c)	125. (c)	126. (d)	127. (a)	128. (c)	129. (b)	130. (d)
131. (b)	132. (a)	133. (b)	134. (d)	135. (b)	136. (b)	137. (a)	138. (c)		

(B) Cumulative Exercise

1. (d)	2. (b)	3. (a)	4. (c)	5. (b)	6. (a)	7. (b)	8. (a)	9. (d)	10. (d)
11. (b)	12. (c)	13. (c)	14. (a)	15. (a)	16. (d)	17. (d)	18. (b)	19. (d)	20. (a)
21. (c)	22. (b)	23. (d)	24. (b)	25. (c)	26. (b)	27. (a)	28. (b)	29. (c)	30. (d)
31. (b)	32. (c)	33. (d)	34. (b)	35. (a)	36. (d)	37. (a)	38. (b)	39. (c)	40. (b)
41. (a)	42. (b)	43. (b)	44. (c)	45. (c)	46. (c)	47. (c)			

(C) Medical Entrances' Gallery

1. (b)	2. (c)	3. (a)	4. (c)	5. (a)	6. (c)	7. (d)	8. (a)	9. (c)	10. (c)
11. (c)	12. (c)	13. (a)	14. (c)	15. (c)	16. (a)	17. (c)	18. (d)	19. (b)	20. (d)
21. (c)	22. (c)	23. (d)	24. (b)	25. (a)	26. (b)	27. (c)	28. (a)	29. (d)	30. (a)
31. (b)	32. (b)	33. (d)	34. (a)	35. (c)	36. (b)	37. (a)	38. (a)	39. (a)	40. (d)
41. (b)	42. (b)								

Explanations

(A) Topical Exercise

9 (*c*) Statement in option (c) is incorrect and can be corrected as
The muscles located in the inner wall of alimentary canal are non-striated and involuntary muscles.

21 (*d*) Statement in option (d) is incorrect and can be corrected as
The central part of thick filament is called H-zone, which is not overlapped by thin filament.

22 (*b*) Statement in option (b) is incorrect and can be corrected as
Each muscle fibre contains multinucleated sarcoplasm.

57 (*d*) Statements I and III are correct, while statement II is incorrect. Incorrect statement can be corrected as
Muscles which bend the joints is flexor.

58 (*d*) Option (d) is correctly matched pair.
Rotator - Pyriformis that raises thigh
Other options are incorrectly matched pairs and can be corrected as

- Abductor - Deltoids of shoulder
- Adductor - Latissimus dorsi
- Elevator - Masseter which lifts the lower Jaw

83 (*c*) Statement in option (c) is incorrect and can be corrected as
In humans, 4 coccygeal vertebrae are fused to form a curved triangular bone called coccyx and this bone is considered as vestigial tail bone.

98 (*b*) Statement I is false, but II is true. Incorrect statements can be corrected as
In human, the vertebral column is formed by 33 bones which are organised as 26 bones.

(B) Cumulative Exercise

1 (*d*) All the given statements are true except statement given in option (d), Incorrect statement can be corrected as
Synovial joints are joints having space between the adjoining bones, whereas fibrous joints are held together by fibrous connective tissue. There is no cavity and no space present between bones. Hence, hinge and pivot joints are example of synovial joint.

2 (*b*) Statements I and III are correct for human skeleton.
Statements II and IV are incorrect and can be corrected as
Human skull consists of 29 bones and vertebral column consists of 33 bones.

6 (*a*) Statement in option (a) is correct, while other statements are incorrect. Incorrect statements can be corrected as

- The shank region consists of tibia and fibula.
- Almost 19 bones form toe.
- Patella is the largest among sesamoid.

7 (*b*) Statement in option (b) is incorrect and can be corrected as
The two halves of pectoral girdle fuse in mid-ventral line.
Rest of the statements are correct.

14 (*a*) Statements I, II and III are correct about skeletal muscle, while statement IV is incorrect. Incorrect statement can be corrected as
Skeletal muscles contraction is voluntary and under the regulation of the somatic nervous system.

16 (*d*) Statements I and III are correct, while statements II and IV are incorrect. Incorrect statements can be corrected as
Red muscle fibres depend on aerobic glycolysis and they perform slow contraction for longer duration hence are called slow muscle fibre.

20 (*a*) Statements I, II and III are correct regarding muscle protein except statement IV. Incorrect statement can be corrected as

The globular head of meromyosin consists of Heavy Mero-Myosin (HMM).

25 (*c*) Statement IV is incorrect and can be corrected as

In cartilaginous joints, the bones involved are joined together with the help of cartilage.

Rest other statements are correct.

33 (*d*) Statements I, II and IV are correct, while III and V are incorrect. Incorrect statements can be corrected as

- During muscle contraction, A-band remains same, while I-band contracts.
- The central part of the thick filament is not overlapped by thin filament is called the H-zone.

42 (*b*) Both A and R are true, but R is not the correct explanation of A.

Movement of body parts serves many purposes like movement of limbs, appendages, head and trunk. This is to change the body posture and to maintain equilibrium against gravity. Limb movements are also required for carrying out locomotion. Movement of eyeballs and pinna of ear helps to collect information from the external environment.

43 (*b*) Both A and R are true, but R is not the correct explanation of A.

The movement found in unicellular organisms are ciliary, flagellar and amoeboid types. In multicellular animals also, phagocytes migrate through the tissues by amoeboid movements. Ciliary movements are seen in cells lining the upper respiratory tract, Fallopian tubes and vasa efferentia of testes, etc. Mammalian sperms move into the female reproductive tract by the flagellar movements.

44 (*c*) A is true, but R is false.

Bone marrow is a soft and semisolid fatty neurovascular tissue, present inside the bones. It carries the tissue for the formation of blood cells. Both the erythrocytes and leucocytes are produced in the red bone marrow. Blood is a connective tissue, whose continual flow in the blood vessels maintain a fairly constant environment for the body cells.

45 (*c*) A is true, but R is false.

Movements are produced at joints by the contraction of skeletal muscles inserted into articulating bones. The contraction of an extensor muscle extends a joint, hence causing extension of the limb. Contraction of triceps brings about the extension of forearm at the elbow joint hence, it is said to be the extensor muscle for elbow joint.

46 (*c*) A is true, but R is false.

Muscle as a whole does not obey All-or-None law, although individual muscle fibre constituting the muscle obey it. This is due to the fact that the strength of the threshold stimulus varies from muscle fibre to muscle fibre. A weak stimulus can stimulate only a few fibres, which have low threshold, but a stronger stimulus stimulates a larger number of fibres by stimulating even those with high thresholds.

47 (*c*) A is true, but R is false.

Antagonistic muscles are those muscles, which contract to produce opposite movements at the same joint. When a muscle contracts to produce a movement, its antagonist must relax to allow that movement to take place. Biceps and triceps are called antagonistic muscles. Because during flexion at the elbow, biceps contract and triceps relax; during extension at the same joint, triceps contract and biceps relax.

(C) Medical Entrances' Gallery

3 (*a*) Myasthenia gravis is a chronic autoimmune neuromuscular disorder that causes weakness, fatigue and paralysis in the skeletal muscles that are involved in breathing and moving various parts of the body including the arms and legs.

4 (*c*) A is true, but R is false.

Calcium ions are required for skeletal muscle contraction. Sarcoplasmic reticulum releases Ca^{2+} on receiving stimulus on motor end plate. These ions bind with troponin units of actin filament and expose the active sites to myosin for cross-bridge formation at the expense of ATP. This is sliding filament theory of muscle contraction.

6 (*c*) The statement in option (c) is correct.

There are seven pairs of vertebrosternal, three pairs of vertebrochondral and two pairs of vertebral ribs is correct.

Rest statements are incorrect and can be corrected as

- Vertebrochondral ribs (8th, 9th, 10th pairs) are false ribs which are not attached to sternum directly. They are attached to the seventh rib with the help of hyaline cartilage.
- Vertebrosternal ribs are first seven pairs of true ribs which are attached dorsally to thoracic vertebrae and ventrally to the sternum.
- Vertebral ribs are the last two pairs of floating ribs which are attached to vertebrae dorsally and are not attached ventrally.

7 (*d*) The cervical vertebrae differ from the other vertebrae in having transverse foramen. In vertebrates, cervical vertebrae are the vertebrae of the neck, immediately below the skull. In humans, they are the smallest of the true vertebrae.

8 (*a*) Atlas is the first cervical vertebrae of the spine. Along with axis, it forms the joint connecting the skull and spine.

10 (*c*) Sliding of actin and myosin filaments continues till the Ca^{2+} ions are present in the sarcoplasm. When the Ca^{2+}ions are pumped back, the actin filaments get masked again. This causes the return of *Z*-lines back to their original position, i.e. relaxation of muscle fibre takes place.

11 (*c*) Globular head of myosin contains an ATPase enzyme that along with Ca^{2+} and Mg^{2+} ions catalyses the breakdown of ATP during muscle contraction. It also has a binding site for ATP and active sites for actin.

12 (*c*) According to sliding filament theory, the sarcomere shorten but the length of actin (thin) and myosin (thick) myofilaments do not change. The myosin forms cross bridge with actin and thus, they slide pass each other.

13 (*a*) In the ribcage, the true ribs are those which are attached to the sternum in the front and vertebral column at back. These are 7 in numbers. Although, there are total 12 ribs in the ribcage. The 11th and 12th ribs are attached to the vertebral column and keep floating in the thoracic cavity, so are called floating ribs.

14 (*c*) Osteoporosis is caused by decreased level of oestrogen. It causes both the early and late forms of osteoporosis in post-menopausal women. Osteoporosis is thinning or weakening of bones which makes them fragile and more likely to break. Women have low oestrogen level when they are transitioning through menopause.

Hence, option (c) is correct.

18 (*d*) Option (d) is incorrectly matched pair and can be corrected as

I-band is considered as a light band and contains a protein actin. It has a dark membrane called Z-line (not M-line) which is an elastic fibre. It is also called Z-disc or Dobie's line. The part of the myofibril between two successive Z-lines is called sarcomere (functional unit of contraction).

Rest options are correctly matched pairs.

26 (*b*) Option (b) is the characteristic and an example of a synovial joint in humans.

Synovial joints are characterised by the presence of a fluid-filled synovial cavity between the articulating surfaces of the two bones. Such an arrangement allows considerable movement. These joints help in locomotion and many other movements. Ball and socket joint (between humerus and pectoral girdle), hinge joint (knee joint), pivot joint (between atlas and axis), gliding joint (between the carpals) and saddle joint (between carpal and metacarpal of thumb) are some examples.

28 (*a*) Each arm consists of one humerus, one radius, one ulna, 8 carpal bones, 5 metacarpal bones and 5 digits (14 phalanges).

On the other hand, pelvic girdle is formed of two innominate bones. Each innominate bone consists of ilium, pubis and ischium. The bones found in legs are femur, tibia, fibula, patella, tarsal bones, metatarsal bones and 5 digits.

31 (*b*) A sarcomere is the basic unit of a muscle's cross-striated myofibril. Sarcomeres are multi-protein complexes composed of three different filament systems.

37 (*a*) Pelvic girdle consists of two coxal bones. Each coxal bone is formed by the fusion of three bones, ilium, ischium and pubis.

38 (*a*) Hinge joint produces an angular opening and closing motion like that of a hinged door, e.g. knee, elbow and interphalangeal joints.

So, the joint of radio-ulna with the upper arm is hinge joint.

39 (*a*) Each half of pectoral girdle consists of suprascapula, scapula, coracoid, precoracoid, epicoracoid and paraglenoid cartilage. Clavicle is a slender rod, separated from the coracoid by a wide gap called coracoid foramen. Posteriorly scapula forms the upper half of a deep cup-like depression the glenoid cavity for articulation with the humerus bone of forelimbs. So, the clavicle articulates with acromion process.

40 (*d*) The reduction in force of contraction of a muscle after prolonged stimulation is called muscle fatigue. This happens due to accumulation of lactic acid. It is produced by glycolysis in the absence of O_2.

41 (*b*) The contraction of skeletal muscle includes ultrastructural and biochemical events. Ultrastructural events lead with stimuli excitation of T-system followed by cross bridge formation (myosin and actin filaments involved) and the cross-bridge breakage. The biochemical events explain that stimuli leads to neurotransmitter secretion, excitation of T-system and release of Ca^{2+}.

CHAPTER 28

Neural Control and Coordination

Multicellular animals have structurally complex body organisation. Due to this, a need to have a coordination system arises. This system must possess the ability to coordinate between various metabolic and homeostatic activities and different organ system of the body. In vertebrates, such a coordination is achieved by nervous and endocrine systems present in their body. Both these systems coordinate and integrate all such activities in complementation with each other. In this chapter, we will focus on nervous system and its various components.

Neural or Nervous system

- The system of specialised excitable tissue which precieve and respond to various kinds of stimuli is called neural system or nervous system.
- The basic unit of this system is **neuron** or a **nerve cell**. Each neuron communicates with thousands of other similar or dissimilar neurons in a complex information-processing circuits.
- The simplest nervous system is present in lower invertebrates. A true neuron or ganglion cells first appeared in coelenterates like, *Hydra*. In them neurons are linked to one another in random manner to form a **nerve net** or **plexus**.
- This neural system got more and more advanced from lower to complex animal groups. The most developed neural system is present in humans.
- The human brain contains about 10^{11} (100 billion) nerve cells.

Inside

HUMAN NEURAL OR NERVOUS SYSTEM

Humans have highly integrated neural or nervous system. It can be divided into following principal parts

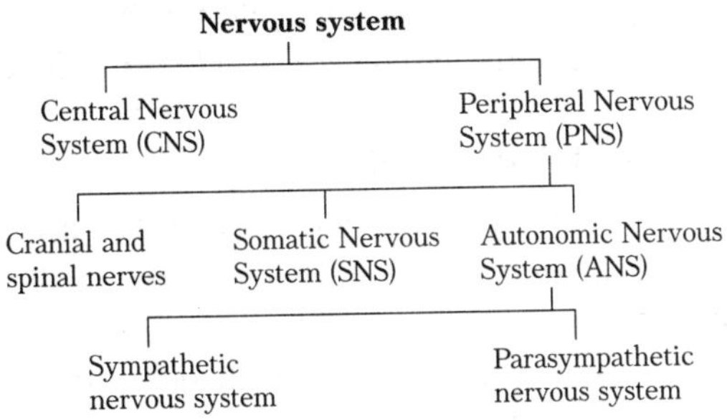

1. Central Nervous System (CNS)

It is present at the central axis of the body. It is developed from the dorsal tubular nerve cord. The anteriormost region of this nerve cord is swollen to form **brain**. Rest part is present as cord only and called **spinal cord.**

2. Peripheral Nervous System (PNS)

It consists of nerves and ganglion outside the brain and spinal cord. It basically serves to connect CNS with rest body parts. Its major units are **cranial** and **spinal nerves**.

PNS is further divided into following two parts

(i) **Somatic Nervous System** (SNS) It is also called **voluntary nervous system**. This segment of PNS is associated with voluntary control of body movements *via* skeletal muscles.

(ii) **Autonomic Nervous System** (ANS) It is also called **vegetative nervous system.** This segment of PNS is associated with involuntary control of body movements *via* smooth muscles and glands. Thus, it influences the functioning of internal organs. This system further has following two divisions

(a) **Sympathetic division**
- Thoraco-lumbar out flow
- Mobilises body system during emergency conditions

(b) **Parasympathetic division**
- Cranio-sacral out flow
- Conserves energy and promotes non-emergency systems

Embryological development of nervous system

The nervous system in humans develop from a section of the embryonic ectoderm called the **neural plate**. The edges of the neural plate elevates form the **neural folds**. These neural folds then curve upward and fuse to form the neural tube which eventually become the central nervous system. This process is called **neurulation.** The neural plate also forms the neural crest, cells of which later migrate to different parts of the body and become most of the cells in the PNS and ANS.

All these developments are shown in figure below

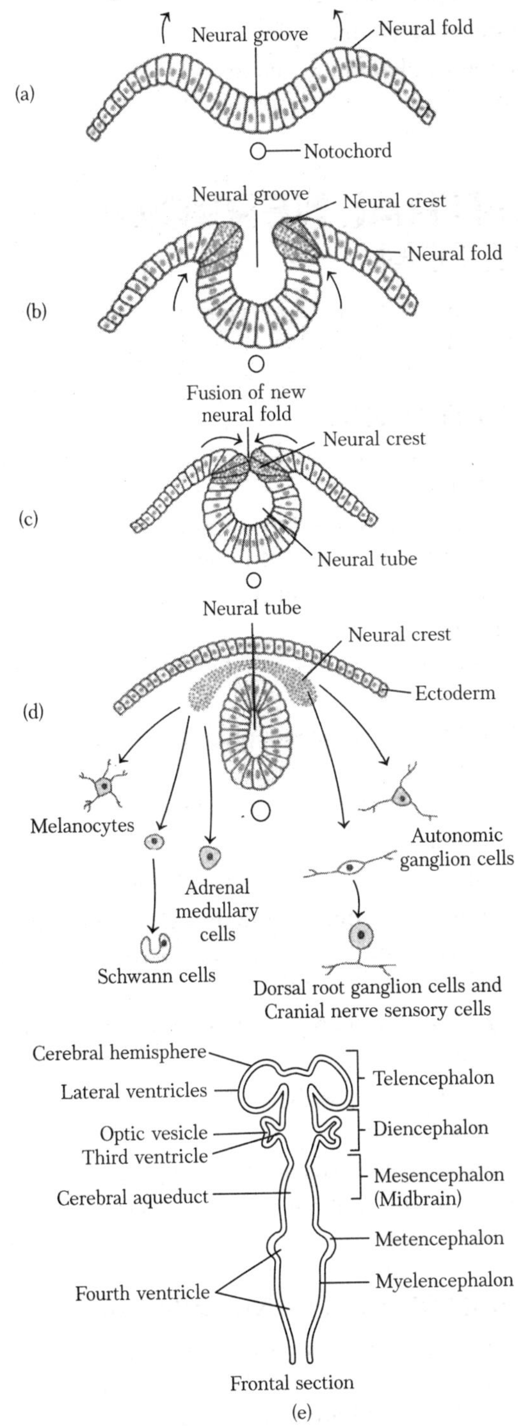

***Fig 28.1** (a-d)Embryological development of CNS, (e) Brain development (frontal section)*

Around the fourth week of development, the neural tube thicken to form brain (from anterior part) and spinal cord (from caudal part).

Before the fusion of neural folds, three vesicles are recognised in the neural tube, i.e.

- **prosencephalon** (forebrain)
- **mesencephalon** (midbrain)
- **rhombencephalon** (hindbrain)
  Each of these vesicles form the forebrain, midbrain and hindbrain, respectively.

CENTRAL NERVOUS SYSTEM

The main parts of central nervous system are **brain** and **spinal cord**.

Human brain

- **Position** It is situated in a well protected cranium or bony skull.
- This skull is composed of two sets of bones, i.e. cranial (8 in number) and facial (14 in number) bones.
- The study of all aspects of the brain is called **encephalology**.
- The brain can be divided into following three main parts
  (i) **Forebrain** or **Prosencephalon** It is further divided into
    - Olfactory lobe or Rhiencephalon
    - Cerebrum
    - Diencephalon or Thalmencephalon

  (ii) **Midbrain** or **Mesencephalon** It has four optic lobes or corpora quadrigemina along with crura cerebri as nerve bands.

  (iii) **Hindbrain** or **Rhombencephalon** It is further divided into
    - Cerebellum or Metencephalon
    - Pons Varolii
    - Medulla oblongata or Myelencephalon

  All these structures of brain are well protected by outer coverings or meninges.

Coverings of brain or Meninges or Menix

These are membranous outer coverings of brain which extend to spinal cord as well. These are made up of connective tissue and are three in number. The meninges in human brain in chronological order from outside to inside are as follows

(i) **Dura mater** Thick, tough and collagenous fibrous layer.

(ii) **Arachnoid mater** Comparatively soft but non-vascularised layer, also appears as **'spider web'**.

(iii) **Pia mater** Highly vascularised and very soft layer.

Spaces between meninges

Following spaces or gaps are identified between meninges

- The space between dura and arachnoid mater is called **sub-dural space.**
- The space between pia mater and arachnoid mater is called **sub-arachnoid space.**
- In between the dura mater and cranium, one more space is present, which is called **epidural space.**

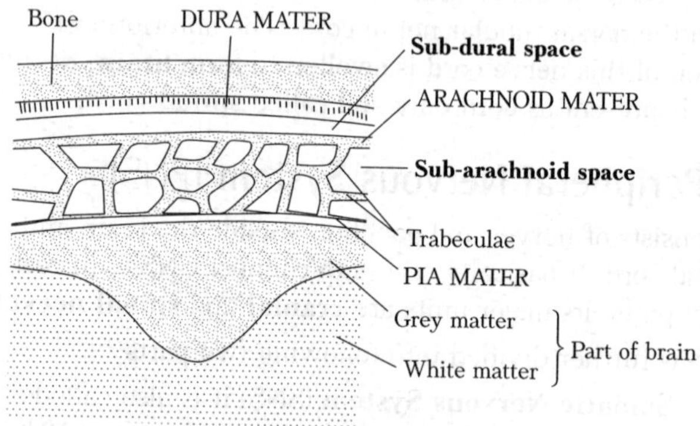

Fig 28.2 Diagram showing meninges, grey matter and white matter of brain

- This epidural space is absent in the mammalian skull, however it is seen in the spinal cord region.
- Sub-dural and sub-arachnoid spaces are filled with Cerebrospinal Fluid (CSF).

Cerebrospinal Fluid (CSF)

- It is a colourless slightly alkaline fluid similar to the plasma. Its volume is about 80-150 mL and pH is about 7.33 in an adult.
- It is secreted by modified ependymal cells (ciliated columnar epithelium cells) of brain.
- These cells form specialised plexuses from which CSF is secreted. These plexus are called **choroid plexus.**
- The main functions of CSF are as follows
  - It absorbs shock or jerk to protect brain and spinal cord.
  - It serves as medium for exchange of nutrients and waste products between blood and nervous tissue.
  - It serves as endocrine medium for the transport of hormones.

> **White and grey matter**
> The CNS has two types of tissues namely grey matter and white matter. The grey matter contains cell bodies, dendrites and axon terminals of non-myelinated neurons. On the other hand, the white matter contains cell bodies, dendrites and axon terminals of myelinated neurons.

1. Forebrain or Prosencephalon

It is the largest part of the human brain and forms most of its anterior region. It is further divided as follows

(i) Olfactory lobe or Rhiencephalon

- These are paired structures, anterioventral in position.
- Each structure is differentiated into an anterior, swollen part called **olfactory bulb** and posterior, narrow part called **olfactory tract**.
- Both of these are present as extensions of limbic system of brain which ends in olfactory area of temporal lobe of cerebral hemisphere.
- All of these are functionally related to smell.

(ii) Cerebrum

- It is the most developed part of human brain.
- A deep cleft divides the cerebrum longitudinally into two halves namely, the left and right cerebral hemispheres.

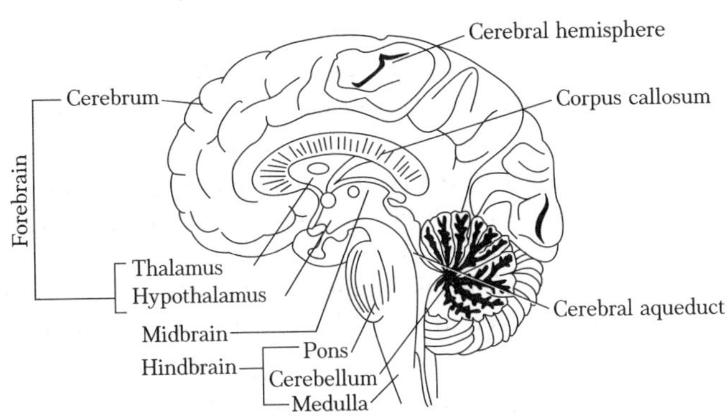

Fig. 28.3 Diagram showing sagital section of the human brain

- Both of these hemispheres are connected by a long curved nerve fibres band called **corpus callosum**. It is the largest commissure of human brain and is a characteristic of mammals only.
- The anterior curve of corpus callosum is called **genu**, the middle part is called **fornix**, the posterior curve is called **splenium**.
- Further, these cerebral hemisphere contain folds that form ridges called **gyri** along with small and large grooves called **sulci** and **fissures**, respectively.

Major fissures

Each cerebral hemisphere is further divided into five lobes by following three deep sulci or fissures.

These are

(a) **Central sulcus** or **rolandic sulcus** which demarcates frontal lobe from the parietal lobe.

(b) **Parieto-occipital sulcus** which separates the parietal and occipital lobe.

(c) **Lateral or sylvian sulcus** which demarcates the temporal lobe from the frontal and parietal lobes.

Since, these three sulci are not complete, the lobes are also not clearly demarcated from each other.

Major lobes

Five different lobes present in the cerebrum are as follows

(a) **Frontal lobe** It is the largest lobe present towards forehead and mainly contains prefrontal area, premotor area and motor area.

- This lobe is the centre of knowledge, creativity, wisdom, reasoning, judgement, logic, intelligence, ideas and dreams.
- Its motor area controls voluntary motor activities or movement of muscles. It also comprises of **Broca's area/motor speech** area which translates the thoughts into speech.
- The premotor area is involved in planning and organising movements and actions.
- The prefrontal or association area establish coordination between sensation and movements.

(b) **Parietal lobe** It is present in the middle and has somato-sensory areas for speech, i.e. **Brodmann's area**. It is also the centre for perception of pain, pressure, taste, touch and temperature.

(c) **Insular lobe** This lobe is hidden, as it is folded deep within the lateral or sylvian fissure. Its accessibility is restricted. Functionally, it is thought to be involved in sensory and motor visceral functions as well as taste perception.

(d) **Occipital lobe** It is present in the posterior most region of the brain. It is the area for vision and visual interpretations.

(e) **Temporal lobe** It is present close to ear. It has olfactory area for smell and auditory area for hearing. It also has **Wernicke's area** for understanding of spoken and written words. The cerebrum, thus shows all three types of areas, i.e. sensory, motor and association area.

These lobes and their functional areas are shown in the figures below

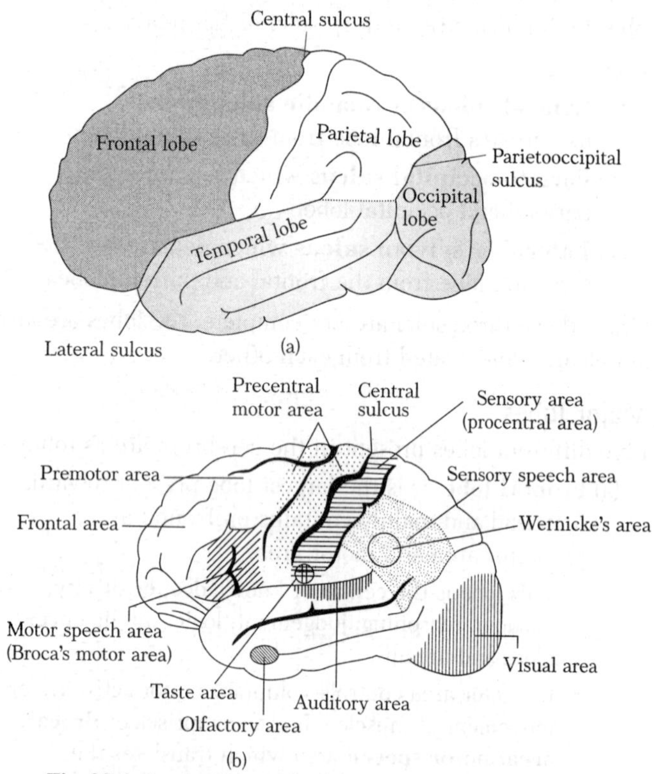

Fig. 28.4 Cerebral hemisphere : (a) The lobes and sulci of the cerebrum, (b) The cerebrum showing the functional areas

General structure

On the basis of structure, the cerebrum of human brain has following three parts

(a) **Cerebral cortex** In human, it forms about 80% of the total brain mass. It is the portion of grey matter. The white matter portion below it is called **sub cortex.** Its major features are given below

- The cortex is made up of two main kinds of neural cells called **stellate cells** and **pyramidal cells.**
- The roof of cortex is called **neopallium** in mammals and pallium in other vertebrates.
- The intermediate layer of cerebral cortex contains specialised, flask-shaped and one of the largest neurons called **Purkinje cells.**
- The presence of sulci and gyri in the cortex region increases its surface area.
- The sensory, motor and integration signals are processed in certain regions of the cerebral cortex.

(b) **Basal nuclei or Basal ganglion** These are a collection of scattered masses or bulges of grey matter, which are submerged into the white matter of cerebral hemispheres.

- The structures generally included in the basal ganglia are
  - **corpus striatum,** the largest nucleus in the cerebrum.
  - **subthalamic nuclei** of diencephalon.
  - **substantia nigra** of midbrain.
- It is the collection of three nuclei named as **caudate nucleus, putamen** and **globus pallidus.**
- The major function of basal nuclei is to control and regulate motor activities.
- Substantia nigra is associated with the secretion of dopamine and loss of this secretory activity is associated with **Parkinson's disease.**

(c) **Limbic system** The medial border of temporal lobe is called limbic system. It is a loop of cortical structures surrounding the corpus callosum and thalamus. The major components of Limbic system are

- **Hippocampus** Located inside the temporal lobe and its shape resembles the sea horse. It plays a major role in converting short term memory to long term memory. Any damage to this portion may cause **Alzheimer's disease.** Normal functioning of hippocampus require acetylcholine.
- **Amygdala** It is almond-shaped and is associated with the normal emotions.
- **Septal nuclei** or **septum** is associated with the sexual emotions like sexual arousal, etc.
- **Mammillary bodies** or **corpus mammillary** are associated with the olfactory reflexes.
- **Cingulate gyrus** is an arched convolution in the limbic system. It lies next to the corpus callosum and is separated from it by the sulcus of the corpus callosum. It helps to regulate emotions and pain. It is also involved in predicting and avoiding negative consequences.

All these components of limbic system are shown in the figure below

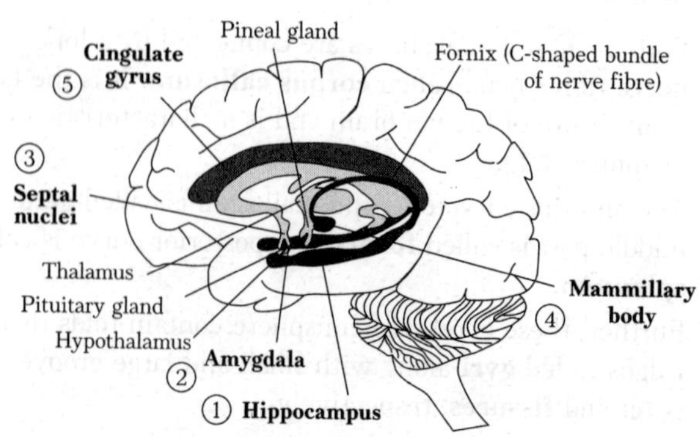

Fig. 28.5 The limbic system (1-5)

Limbic system is also called **'emotional brain'** due to its intense connections with the emotional functioning.

Thus, specialised functions like intelligence, learning skills, memory, speech, etc., all are attributed to cerebrum.

Table 28.1 Location and functions of important areas in cerebrum

Cerebral area	Location	Functions
Premotor area	Frontal lobe	Control involuntary movements of muscles and ANS
Motor area	Frontal lobe	Controls voluntary movements of specific muscles
Broca's area	Frontal lobe	Motor speech area
Somesthetic area	Parietal lobe	Perception of sensations like pain, touch and temperature
Gustatory area	Parietal lobe	Sense of taste
Auditory area	Temporal lobe	Hearing
Olfactory area	Temporal lobe	Sense of smell
Wernicke's area	Temporal lobe	Understanding speech
Visula area	Occipital lobe	Sensation of light

(iii) Diencephalon or Thalamencephalon

It is the posterioventral part of the forebrain, which is covered by cerebrum. It consists of following three major parts

(a) **Thalamus** It contains an oval mass of grey matter that underlies each cerebral hemisphere.

- Thalamus is a centre for sensory and motor signalling and is connected with the limbic system.
- All types of sensory input and other information from lower part of brain and spinal cord pass through synapses in the thalamus. It is thus, referred to as the **'relay centre'** for the cerebral cortex.
- Its connectivity with limbic system indicates that it is also related to the emotional and memory functions.

(b) **Hypothalamus** It forms the base of thalamus.

- It contains neurosecretory cells that secrete hypothalamic hormones.
- It is related to hypophysis cerebri or pituitary gland to form the posterior region of pituitary.
- It is connected with synchronisation and controlling of visceral activities.
- Along with limbic system, hypothalamus controls sexual behaviour, emotions and memory.

(c) **Epithalamus** It is basically the roof of diencephalon.

- Alongwith overlying pia mater, it forms the anterior choroid plexus. This plexus is required for the exchange of various nutrient, gases and other chemical substances between CSF and blood.
- In humans, epithalamus is also associated with pineal gland or **Epiphysis cerebri** with the help of a pineal stalk.
- This gland is related to the biological clock or biological rhythm, (i.e. circadian rhythm) set up.

2. Midbrain or Mesencephalon

The midbrain is located between the thalamus/ hypothalamus of the forebrain and pons of the hindbrain.

The dorsal portion of the midbrain consists of four round swellings called **colliculus** or **optic lobe**. These four lobes are collectively called **corpora quadrigemina** and defined in two pairs as follows

- 1 pair of **superior colliculus** associated with vision.
- 1 pair of **inferior colliculus** associated with auditory function.
- The both sides of midbrain also contains two thick nerve bands called **cerebral peduncle** or **crura cerebri** or **crus cerebrum.** These nerve bands connect forebrain with hindbrain.
- Crura cerebri help to control muscle tone and modify some muscle activities.
- A canal called **cerebral aqueduct** passes through the midbrain.

In humans, pons, medulla and midbrain portions form the **brain stem.**

3. Hindbrain or Rhombencephalon

It is the posterior region of the brain. It comprises of following parts

(i) Cerebellum or Metencephalon

- It is the second largest part of the brain and also called as **small brain** or **little cerebrum.**
- It consists of three lobes,
  - a central **vermis** and
  - two lateral **cerebellar hemispheres.**
- The central core of each cerebellar hemisphere is formed by a characteristic arrangement of white matter into the grey matter in a tree-like appearance. This arrangement is called **arbor vitae** or **tree of life.**
- Like cerebrum, cerebellum is covered by cerebellar cortex. It consist of grey matter arranged in a series of parallel ridges called **folia.**
- The middle layer of cortex contain flask shaped **Purkinje's neurons.**

- Cerebellum is connected to midbrain, pons and medulla oblongata with the help of three pairs of nerve fibre called **cerebellar peduncles.**
- Cerebellum is the centre for co-ordination of muscular movements. It helps in maintaining balance, equilibrium and orientation of the body.

(ii) Pons Varolii

It forms the floor of brain stem. It serves as neuronal link between cerebral cortex and cerebellum. It consists of

- pontine nuclei
- pneumotaxic area
- apneustic area
- Pontine nuclei are sites at which signals for voluntary movements are relayed into the cerebellum.
- Both pneumotaxic and apneustic area control respiration.

(iii) Medulla oblongata

- It is the triangular part of brain whose roof is associated with overlying pia mater to form **posterior choroid plexus.**
- The medulla oblongata is connected to the spinal cord.
- It contains centres which control respiration, cardiovascular reflexes and gastric secretions.

Note *Most of the motor and sensory nerve tracts crossover to other side in medulla. Therefore, the right half of cerebrum controls the left side of body and vice versa.*

Generalised functions of brain

Brain functions to control

- voluntary movements
- body balance
- functions of vital involuntary organs like heart, lungs, etc.
- thermoregulation
- hunger
- thirst
- circadian rhythms of our body
- activities of endocrine glands and human behaviour

Cavities or Ventricles of brain

The cavities of brain are called ventricles. Human brain contains following four ventricles

Table 28.2 Cerebral ventricles and their locations

Ventricles	Also known as	Location
Ventricle I	Paracoel or Lateral ventricle	Right cerebral hemisphere
Ventricle II	Paracoel or Lateral ventricle	Left cerebral hemisphere
Ventricle III	Diocoel	Diencephalon
Ventricle IV	Metacoel	Between brain stem and cerebellum

- Ventricles I, II and III are connected by **foramen of Monro.**
- Ventricles III and IV are connected by **cerebral aqueduct** or **aqueduct of sylvius** or **iter.**
- Roof of ventricle IV contains a pair of lateral apertures called **foramen of Luschka** and a median aperture called **foramen of Magendie.**
- The ventricle IV continues with the central canal of spinal cord.
- The entire ventricular system is lined by specialised epithelium of brain called **ependymal epithelium.**

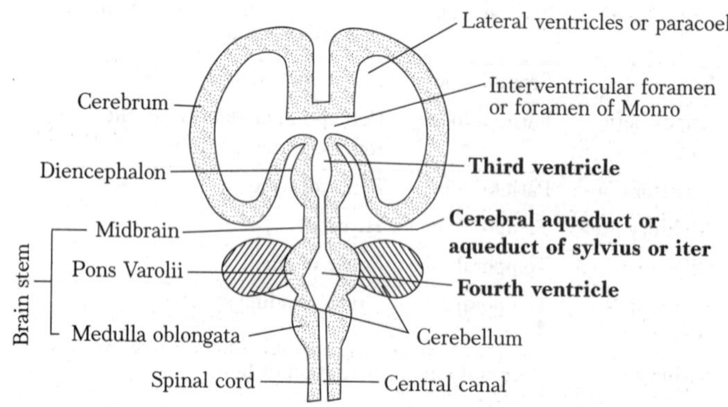

Fig. 28.6 LS of brain showing its ventricles and brain stem

Electroencephalogram (EEG)

The EEG is the pictorial graph of electrical potentials produced during different activities of brain cells. It is recorded through instruments like electroencephalograph or cathode ray oscilloscope. **Hans Berger** was the first to record EEG in 1929. The EEG can easily be seen as following rhythmic waves namely alpha, beta, theta and delta waves.

- **Alpha waves** are visible in the EEG of a normal awakened person and their frequency is 10-12 Hz.
- **Beta waves** appear during intense mental activity. Their frequency is 15-30 Hz.
- **Theta waves** appear during emotional stress like frustration. Their frequency is 5-8 Hz.
- **Delta waves** appear during sleep. Their frequency is 1-5 Hz.

Spinal cord

- The spinal cord in an adult human is about 42-45 cm long and 2 cm thick elongated cylindrical structure in the mid thoracic region.
- It is continuous with the medulla oblongata through foramen magnum of the skull.
- It lies inside the neural canal of vertebral column.
- It begins at the atlas (first cervical vertebra) and tapers to a point called **conus medullaris** in the first or second lumbar vertebra.
- It ends in a narrow, non-nervous and tapering structure called **filum terminale** in the coccyx region.

- At the posterior end of spinal cord, a collection of roots of spinal nerves appear just like horse-tail. This is called **cauda equina.**
- The spinal cord possess two enlargements *viz*, branchial enlargement (from 4th cervical to 1st thoracic vertebrae) and lumbar enlargement (from 9th thoracic to 12th thoracic vertebrae).

Internal structure

In transverse section of spinal cord following structural details are visible

- It possesses an anterior and a posterior median fissure running along its length.
- The grey matter of spinal cord is shaped like the **letter H** or a **butterfly** and is surrounded by white matter. This condition is exactly opposite to the one found in brain (see arbor vitae).
- The grey matter on each side of the spinal cord is subdivided into regions called **horns**. Among these, the one closer to the front cord are called **anterior** or **ventral grey horn** or root while those closer to the back cord are called **posterior** or **dorsal grey horn** or root.
- Due to these horns, the white matter is divided into four regions, i.e. one anterior, one posterior and 2 lateral funiculi.
- Each dorsal root has a ganglion called **dorsal root ganglion.**
- The dorsal root is sensory and ventral root is motor in function.
- All these structures in the TS of spinal cord are shown below

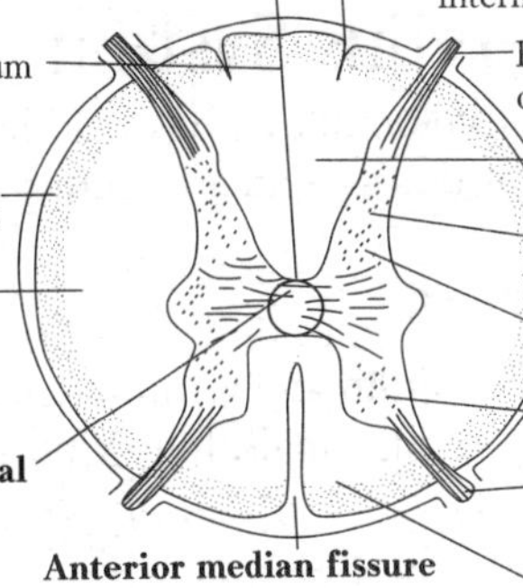

Fig. 28.7 *TS of spinal cord*

Functions

- The spinal cord conducts impulses to and from the brain.
- The dorsal funiculus has ascending nerve tract for conducting sensory impulses towards the brain.
- Lateral and ventral funiculi conduct motor impulses from the brain to spinal cord.
- It controls most of the reflex activities.

CHECK POINT 28.1

1. The central nervous system is absent in animals that have
(a) bilateral symmetry (b) a closed circulatory system
(c) a complete gut (d) radial symmetry

2. How many of the components given in the box belong to forebrain (F), midbrain (M) and hindbrain (H).

Corpora quadrigemina, Thalamus, Medulla, Hypothalamus, cerebrum, Cerebellum, Pons, Cerebral aqueduct

(a) F = 3, M = 3, H = 2 (b) F = 3, M = 2, H = 3
(c) F = 3, M = 1, H = 4 (d) F = 4, M = 2, H = 2

3. In the split brain operation for epilepsy, the nerve tract (commissure) that allows communications between two hemispheres is cut, this nerve tract is
(a) fissure of Rolando (b) corpus callosum
(c) fornix (d) hippocampal gyrus

4. After a car accident, Rahul had a lot of difficulty planning his day. This is most likely the result of damage to which lobe?
(a) Temporal (b) Parietal (c) Frontal (d) Occipital

5. Calculation, contemplation and cognition are human activities associated with increased activity in the
(a) cerebellum (b) cerebrum
(c) hypothalamus (d) epithalamus

6. Which of the given statement is not associated with midbrain?
(a) Located between thalamus of the forebrain and pons of the hindbrain
(b) The ventral portion of the midbrain is composed of mainly four, round swellings known as corpora quadrigemina
(c) The caval called cerebral aqueduct passes through the midbrain
(d) Hindbrain and midbrain constitute the brain stem

7. The cell bodies of neurons bringing afferent information into the spinal cord are present in
(a) grey matter of spinal cord
(b) white matter of spinal cord
(c) ventral root ganglia
(d) dorsal root ganglia

PERIPHERAL NERVOUS SYSTEM (PNS)

All the nerves and ganglion present outside the CNS come under the preview of PNS. It helps to transmit informations to and from the CNS. It consists of cranial and spinal nerves which constitute the **somatic nervous system.** Another part of PNS is called **Autonomic Nervous System** (ANS).

Somatic Nervous System

It is the regulator of voluntary activities of our body. Its regulatory function is associated with skeletal muscles. Two major components of this system are cranial and spinal nerves.

1. Cranial nerves

These originate in the brain and terminate mostly in the organs of the head and upper body in all amniotes, i.e. reptiles, birds and mammals. In mammals, following 12 pairs of cranial nerves are present

Table 28.3 Cranial nerves present in humans

Cranial nerves No.	Name	Origin	Distribution	Nature	Functions
I	Olfactory	Olfactory lobe	Olfactory epithelium	Sensory	Sense of smell
II	Optic	Optic chiasma	Eye retina	Sensory	Sense of vision
III	Occulomotor	Crus cerebrum	Four muscles of eyeball, iris, ciliary body	Motor	Movement of eyeball, iris, lens, eyelid
IV	Trochlear/Pathetic (smallest)	Midbrain	Superior oblique muscles of eye	Motor	Movement of eyeball
V	Trigeminal (largest, also called dentist's nerve)	Pons Varolii		Mixed	Supply to concerning part
	V$_1$-Ophthalmic		Eye, eyelids, conjunctiva	Sensory	
	V$_1$-Ophthalmic		Upper jaw, upper lip	Sensory	
	V$_3$-Mandibular		Lower jaw, lip, tongue, external ear	Motor	
VI	Abducens	Pons	External rectus muscles of eye	Motor	Movement of eyeball
VII	Facial (bear geniculate ganglion)	Pons		Mixed	Taste, facial expression, saliva secretion
	VII$_1$-Palatine		Palate	Sensory	
	VII$_2$ -Chorda tympani		Tongue, salivary gland, taste buds	Sensory	
	VII$_3$-Hyomandibular		Lower jaw, pinna, neck, hyoid	Mixed	
VIII	Auditory (acoustic)	Pons		Sensory	Hearing and equilibrium
	VIII$_1$-Vestibular		Internal ear	Sensory	
	VIII$_2$ -Cochlear		Cochlea	Sensory	
IX	Glossopharyngeal	Medulla		Mixed	Taste and saliva secretion
	IX$_1$-Lingual		Tongue, pharynx	Mixed	
	IX$_2$-Pharyngeal		Pharynx, salivary gland	Mixed	
X	Vagus	Medulla		Mixed	Visceral sensations and visceral movements like breathing, cardiac slowing, gastric and pancreatic secretion, gastrointestinal movements
	X$_1$-Superior laryngeal		Laryngeal muscles	Motor	
	X$_2$-Recurrent laryngeal		All muscles of larynx	Motor	
	X$_3$-Cardiac		Cardiac muscles	Motor	
	X$_4$-Pneumogastric		Lungs, oesophagus, stomach, ileum	Motor	
	X$_5$-Depressor		Diaphragm	Mixed	
XI	Accessory spinal	Medulla	Pharynx, larynx, neck, shoulder	Motor	Movement of pharynx, larynx, neck and shoulder
XII	Hypoglossal	Medulla	Tongue, neck	Motor	Movement of tongue

Note *In human, smallest cranial nerve is trochlear, but in all other animals, abducens is the smallest one.*

Largest cranial nerve in human is trigeminal, but in all other animals, vagus is the largest one.

2. Spinal nerves

- The spinal nerves originate from the spinal cord and extend to different parts of the body, below the head.
- There are **31 pairs** of spinal nerves in mammals.
- These are classified into five groups.
- These 5 groups along with the plexus associated with them are shown in the figure below

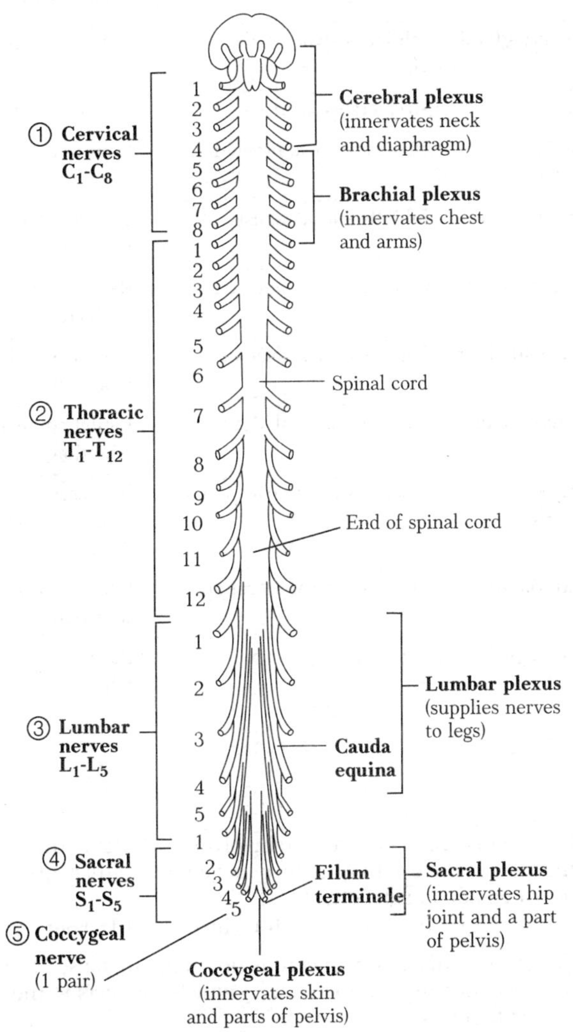

Fig. 28.8 Spinal nerves in human

- All of the spinal nerves are of mixed type, i.e. they contain axons of both sensory and motor neurons.
- Each spinal nerve is formed inside the neural canal of vertebral column by following two roots
  1. **posterior** or **dorsal sensory root** and
  2. **anterior** or **ventral root.**
- Posterior root receives the sensory nerve from the dorsal root ganglion. Cell bodies of sensory neurons are located in this ganglion.
- In contrast, the anterior/ventral root gives out the motor nerve.
- In this way, the dorsal sensory and the ventral motor nerves together form the mixed spinal nerve.
- It emerges out from both sides of the spinal cord through the intervertebral foramen.
- Each spinal nerve is further divided into four branches
  - (i) **Ramus dorsalis** This branch innervates the muscles and skin of the dorsal side.
  - (ii) **Ramus ventralis** This branch supplies muscles and skin of ventral and lateral sides. It also supplies to the upper and lower limbs.
  - (iii) **Ramus communicans** It joins the ganglionated sympathetic nerves of autonomic nervous system.
  - (iv) **Meningeal branch** It supplies to the vertebrae.

Autonomic or Visceral nervous system

- This nervous system regulates the internal environment of the body.
- It basically controls smooth and cardiac muscles and the organs of digestive, cardiovascular, excretory and endocrine system. This control is generally involuntary in nature.
- This system is originated from the lower part of brain and spinal cord.
- Its constitution involves following three components
  1. Autonomous ganglion
  2. Pre-ganglionic neuron
  3. Post-ganglionic neuron
- As the name suggest, the pre-ganglionic neuron connects CNS with the ganglion. Like wise, post-ganglonic fibre connects effector organ with ganglion.

The autonomic nervous system consists of following two antagonistic subdivisions

Sympathetic nervous system

- It consist of two chains of 21 ganglia (3 cervical, 12 thoracic, 5 lumbar, 1 sacral). These extend from the upper cervical part of sacrum, i.e. sympathetic chains.
- Pre-ganglionic fibres are the axons of neurons in spinal cord. These occur only in thoracic and lumbar region and thus form **thoraco-lumbar outflow**.
- Post-ganglionic fibres arises from sympathetic ganglia and terminates in viscera.
- It is related to the arousal and energy generation, i.e. the **fight** or **flight** responses.

Parasympathetic nervous system

- It corresponds to pre-ganglionic parasympathetic fibres arising from cranial and sacral region and constitutes **cranio-sacral outflow.**
- The ganglion in this system are situated away from the CNS. Hence, here pre-ganglionic fibres are large as compared to post-ganglionic fibres.
- The ganglion within sympathetic and parasympathetic nerves contain acetylcholine neurotransmitter only.
- However in sympathetic system, the terminal synapse has adrenaline or nor-adrenalin. On the other hand, in parasympathetic system, the terminal synapse contain acetylcholine, just as in the ganglion.
- Adrenaline and nor-adrenaline produce opposite effects to that of acetylcholine. Therefore, these two system works antagonistic to each other.
- This generally causes opposite response that promotes **calming** and **return to self-maintenance functions.**

Functions of both of these systems are summerised and compared in the table below

Table 28.4 Differences between sympathetic and parasympathetic nervous system

Organs	Sympathetic nervous system	Parasympathetic nervous system
Blood vessels	Vasoconstriction in general and vasodilation (brain, heart, lungs and skeletal muscles)	Vasodilation of coronary vessel
Iris muscle	Dilates pupil	Constricts pupil
Lacrimal gland	Increases lacrimal glands secretion	Inhibits lacrimal glands secretion
Digestive glands	Inhibits salivary + digestive glands	Stimulates them
Heart	Accelerates heartbeat	Retards heartbeat
Bronchial muscles	Dilates trachea, bronchi, lungs	Constricts them
Digestive tract	Inhibits gut peristalsis	Stimulates gut peristalsis
Anal muscles	Contracts anal sphincter	Relaxes anal sphincter
Urinary bladder	Relaxes urinary bladder	Contracts and relaxes both
Adrenal gland	Increases adrenal secretion	Inhibit adrenal secretion
Islets of Langerhans	Increases semen ejaculation	Increases external genitalia and sex desire
Sweat glands	Increases sweat secretion	Inhibits sweat secretion
Arrector pilli	Contracts arrector pili	Relaxes arrector pili

CHECK POINT 28.2

1. How many of the given cranial nerves are exclusively sensory(S) and exclusively motor(M)?

Facial, glossopharyngeal, vagus, trochlear, optic, olfactory

(a) S = 2, M = 1 (b) S = 3, M = 3
(c) S = 1, M = 3 (d) S = 2, M = 4

2. If the vagus nerves were cut, which of the following would be true?
(a) The diaphragm would by paralysed
(b) Adrenalin secretion by adrenal gland would cease
(c) Hearing will be impared
(d) Cardiac movements will be affected

3. Bell's Palsy is an affliction of one of the cranial nerves. Typically a person with Bell's Palsy has a difficult time tasting food. Based on that bit of information, which cranial nerve is affected?
(a) Facial (b) Vagus (c) Trigeminal (d) Abducen

4. How many cranial nerves are involved with the extrinsic muscles of the eye to cause eye movement (either up or down or side to side, etc.)?
(a) One (b) Two (c) Three (d) Four

5. The tail end of the spinal cord is not 'hanging loose'. It is actually anchored *via* the ...., which connects to the coccyx bone.
(a) cauda equina (b) conus medullaris
(c) filum terminale (d) corpus callosum

6. How many plexus groups off the spinal cord?
(a) Three (b) Four
(c) Five (d) Six

7. Which of the following spinal cord region is not involved with a plexus formation?
(a) Cervical (b) Thoracic
(c) Lumbar (d) Sacral

PHYSIOLOGY OF NERVE CONDUCTION

Irritability is the basic function of living beings which involves 'responses to stimulus'. The basic function of irritability is performed by specialised excitable muscle and nerve cells in animals. We have already discussed the structure of these cells in animal tissues.

Stimulus and Impulse

- A **stimulus** is any disturbance in surroundings (internal or external) that travels as an impulse in our nervous system.
- An **impulse** is defined as 'a sum of all, i.e. mechanical, chemical and electrical disturbances created by a stimulus'.
- In a nerve cell, dendrites and cytons are impulse receiving parts while axons conduct impulses away from a neuron. That is why, axons are considered as nerve fibres.
- The process of impulse conduction through a nerve fibre is essentially an **electrochemical** process in which impulses travel at a maximum speed of about 130 m/second. It is achieved through changes in membrane potentials of nerve fibre membranes.

Membrane potentials

According to the membrane potential theory, the process of nerve impulse conduction involves two main phases

(i) **Resting potential**, i.e. potential of membrane while axon is at rest.

(ii) **Action potential**, i.e. potential of membrane while axon is in action.

The complete process depends upon following three factors

- Permeability of axon membrane or axolemma
- Osmotic equilibrium
- Electrical equivalence between the axoplasm which is present within the axons and extracellular fluid (ECF) which is present outside the axons.

Fundamentals of membrane potential

Before going in much detail about functioning of nerve fibres, it is better to understand the basic physics of membrane potential.

Cause

ECF contains a large amount of NaCl and bicarbonates along with nutrients, O_2 and CO_2. The neural membrane or axolemma is impermeable to the complex proteins. But it is selectively permeable to simple organic and inorganic molecules. In resting condition, following permeability and concentration features are visible in connection with axolemma

- It is moderately permeable to Na^+ ions, but quite more (50-100 times more) permeable to K^+ and Cl^- ions.
- Concentration of Na^+ and Cl^- ions in ECF is about 10 and 14 times more as compared to axon cytoplasm.
- Similarly, the concentration of K^+ ions in axon cytoplasm is about 30 times more than in ECF.
- The result of above concentration differences is a continuous influx of Na^+ and Cl^- ions into the axon cytoplasm from ECF.
- This is accompanied by a continuous outflux of K^+ ions from axon cytoplasm into the ECF.
- This movement or diffusion of different ions is responsible for the development of a certain potential across axolemma. Diffusion of ions described above takes place through specialised protein of axolemma called **'Leak channels'**.

Nernst potential and calculation formulae

- The potential level across the membrane that prevent net diffusion of an ion in either direction through that membrane is called **Nernst potential of that particular ion.**
- Its magnitude can be determined by the ratio of ion concentration of the two sides of the membrane.
- The following equation called **Nernst equation** is used to calculate the Nernst potential for any univalent ion at normal body temperature of 37°C.

EMF (milli volts) = ± 61 log (concentration inside/concentration outside)

- When using this formula, it is assumed that the potential outside the membrane always remains zero.

Calculation of resting membrane potential

- The resting membrane potential is defined as the potential of a membrane at rest. For large nerve fibres when they are not transmitting any nerve signal, its value is about – 90 mV.
- Different contributors in the establishment of such potential are as follows

  (i) **Contribution of K^+ diffusion potential** The contribution of K^+ ions to such potential due to their diffusion is –94 mV (approximately).

  (ii) **Contribution of Na^+ diffusion potential** The contribution of Na^+ ions to such potential due to their diffusion is + 61 mV (approximately).

(iii) **Contribution of Na^+-K^+ pump** The contribution of this pump amounts to –4 mV.

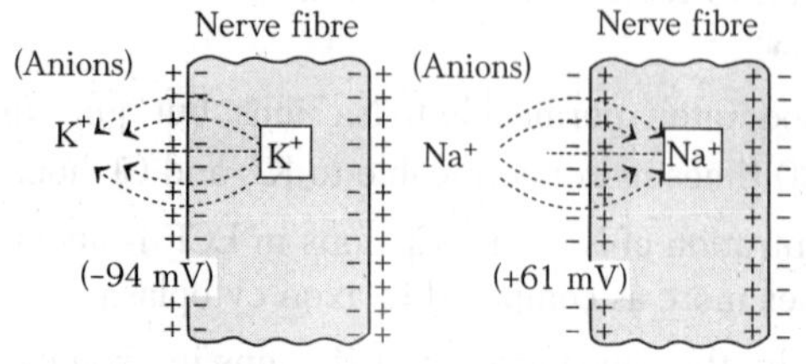

Fig. 28.9 Membrane potential in nerve fibres

- The overall value to membrane potentials due to diffusion of Na^+ and K^+ ions at rest comes out to be –86 mV.
- It is calculated with the help of a modified Nernst equation called **Goldmam-Hodgkins-Katz equation.**
- Addition of Na^+-K^+ pump's contribution makes it to –90 mV. In smaller nerve and muscle fibres, the potential at rest ranges form –40 to –90 mV.

Na^+-K^+ pump

It is a transport protein which involves active pumping of Na^+ ions outward and simultaneous K^+ pumping inward through the cell membrane.

This pump is present in all cells of the body and is responsible for the maintenance of Na^+ and K^+ levels across the membranes.

The pump functions with the help of a carrier protein which is a complex of two separate globular proteins

- A large one (α) with a molecular weight of about 100000 mass unit.
- Smaller one (β) with a molecular weight of about 55000 mass unit.

The larger one has following three specific features being important for the functioning of pump.

- It has 3 receptor sites for binding Na^+ on the portion of protein which protrudes interior (inside) of the cell.
- It has two receptor sites for K^+ towards the outside of cell.
- The inside portion of this protein adjacent to or near to this Na^+ binding has ATPase activity. This pump is electrogenic in nature. In each revolution of pump, 3 Na^+ ions are pumped outside for every $2K^+$ ions moved inside. It simply means that in each revolution of pump, a deficit of (+) ve ion inside the cell occurs. This causes unit negative potential inside, resulting in a cumulative potential of – 4 mV.

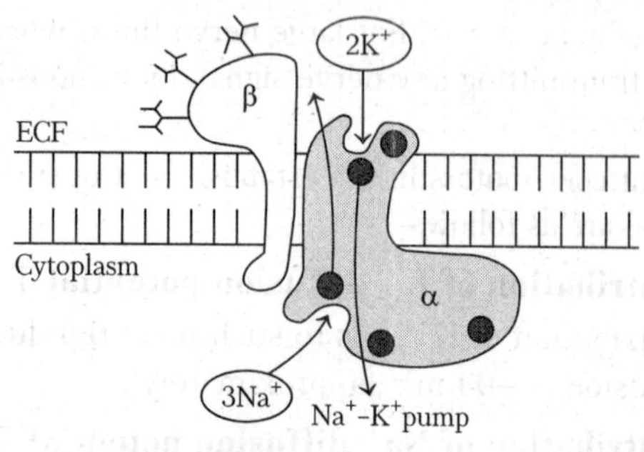

Fig. 28.10 Na^+ – K^+ pump

The action potential of nerves

The potential changes during the transmission of a nerve impulse are called **action potential.**

The successive stages of action potential are as follows

Resting stage (Polarised stage)

The membrane is said to be **polarised** during resting stage because of the presence of a - ve potential inside.

Depolarisation stage

- It is the stage when the normal –90 mV polarised state is lost, with potential rising rapidly in the + ve direction.
- This phenomenon occurs due to a sudden change in the permeability of axolemma.
- During this process, the axolemma becomes extremely permeable to Na^+ ions, allowing their tremendous inflow inside.
- This causes a sudden change in potential inside form – ve to + ve.
- The maximal change in membrane potential that occurs in axolemma is called **spike potential.**

Repolarisation stage

- After depolarisation stage, a rapid expulsion of K^+ ions to the exterior restablishes the normal - ve resting membrane potential. This is called repolarisation of the membrane.
- The phenomenon occurs within a 10000 of the second after depolarisation of the membrane.
- All the 3 phenomenon described above are shown diagrammatically in the figure below

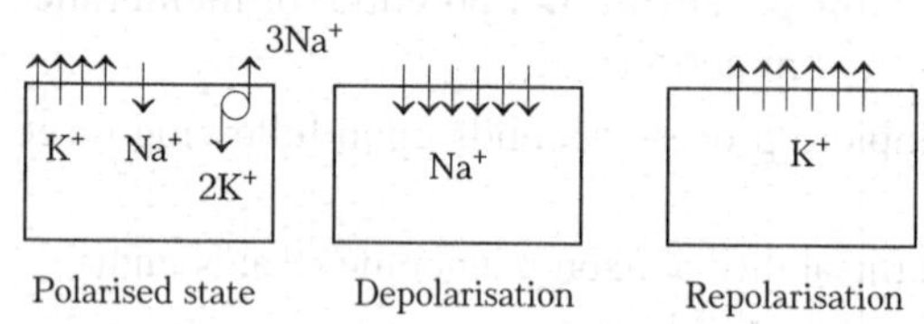

- These changes can be further understood with the help of membrane potential *vs* time graph given below

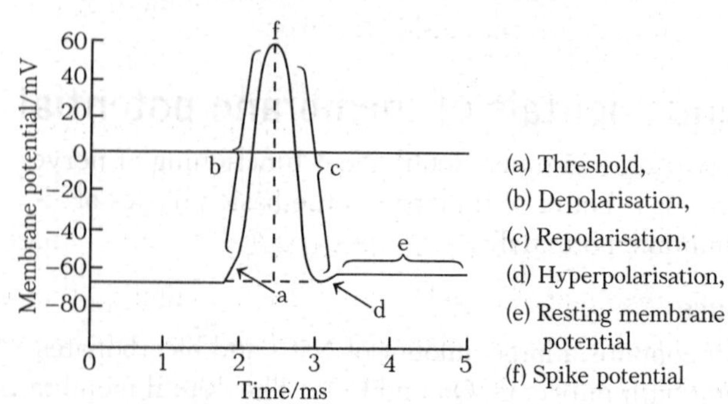

Fig. 28.11 A nerve conduction graph

Mechanism behind action potential

The essential role in both depolarisation and repolarisation is played by the voltage gated Na^+ and K^+ channels. These two voltage gated channels are present within axolemma, in addition to Na^+-K^+ pump and Na^+-K^+ leak channels.

Na^+ channels

- The voltage gated Na^+ channels has 2 gates namely, **activation** and **inactivation** gate.
- The opening and closure of these gates depends upon the potential value as shown below

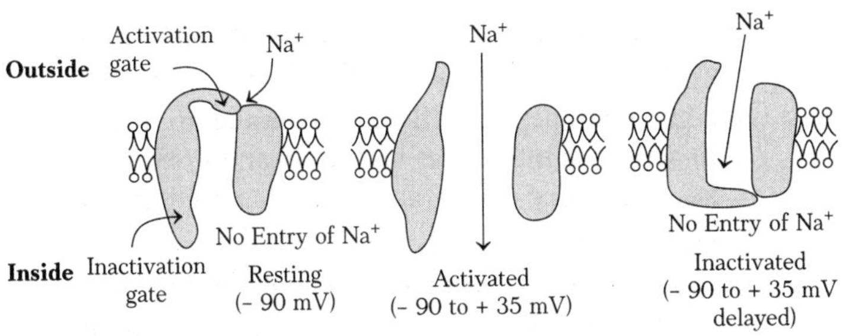

Fig. 28.12 Activation and inactivation gate

K^+ channels

- The voltage gated K^+ channels have only one gate.
- During resting state, the gate of K^+ channel is closed but when the potential rises, the voltage change causes a slow conformational opening of the gate as shown below

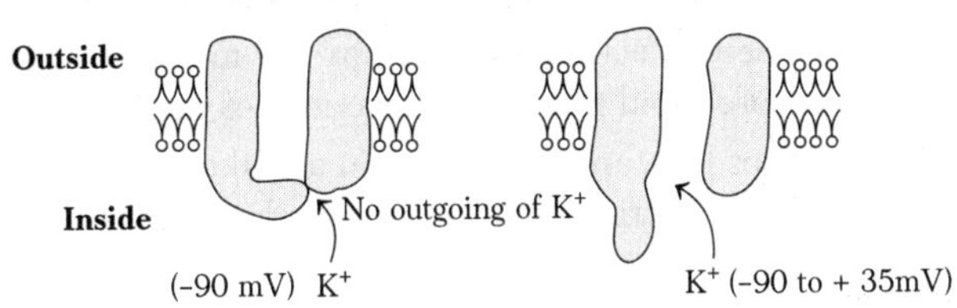

Fig. 28.13 Voltage gated K^+ channel

- This allows the increased K^+ ions diffusion outward through the channel.
- However, K^+ ion channels remain open for a bit longer period, so that the membrane potential becomes more negative than – 90 mV. It is called **hyperpolarisation.**
- The value of potential during this state is called **positive over potential**. It takes about 1-5 milli seconds for repolarisation.

Conduction of nerve impulse

The phenomenon of impulse conduction is summerised below

- Disturbance is caused in resting membrane potential or polarised state due to any stimulus of threshold value.
- This lead to the opening of voltage gated Na^+ channels. It causes depolarisation due to sudden influx of Na^+ ions.
- The membrane potential which sets in a wave of depolarisation is called **action potential.**
- It is followed by **repolarisation** which occurs once the spike potential is reached.
- The repolarisation occurs due to immediate outflux of K^+ to control the charge balancing. Outflux of K^+ ions occurs due to opening of voltage gated K^+ channels.
- Repolarisation stops after precautionary hyperpolarisation.
- Resting potential is gradually maintained afterwards with the help of leak channels and $Na^+ - K^-$ pump.
- In a myelinated nerve fibres, these changes are visible only at nodes of Ranivier. Hence, it seems like the impulse is jumping form one node to other. This phenomenon of jumping of impulse is called **saltatory conduction.**
- The speed of transmission may increase up to 50 times in saltatory conduction.

Synapse

- When an action potential reaches the terminus of an axon, it is transmitted from one neuron to another through specialised junctions called **synapses**.
- A synapse is a gap between the membranes of two nerve fibres. These membranes are called **presynaptic**, i.e. before the synapse and **postsynaptic**, i.e. after the synapse membranes.
- In a synapse, impulse reaches from presynaptic neuron and transferred to postsynaptic neuron.

Types of synapse

There are following two types of synapses

(i) **Electrical synapse** In these synapse, impulse passes directly *via* gap junctions. Thus, here gap between pre and postsynaptic membranes is about 10 nm. There features are as follows

- Each gap junction contains tubular protein structures called **connexons.**
- These structures are helpful in connecting the axoplasm of two neurons and provide a path for ionic current flow.
- The gap junctions are common in CNS, smooth muscles, cardiac muscles and developing embryo.

Advantages of electrical synapse are as follows

- Faster communication due to direct channels.
- Increased synchronisation and coordination in group of neurons.
- Two way transmission possible.

(ii) **Chemical synapse** These are most common synapses in human body.

Features of these synapse are given below

- In these synapse, the gap between pre and post synaptic membranes is about 10-20 nm. This gap is called **synaptic cleft.**
- In chemical synapse, the end of presynaptic fibre becomes bulbous expansion. This is called **synaptic knob**. It contains numerous tiny vesicles called **synaptic vesicles.**
- These vesicles are of about 50 nm size and contain approximately 10000 neurotransmitter molecules.
- The postsynaptic membrane contains receptors to receive these neurotransmitter.
- The complete arrangement of chemical synapse is shown below

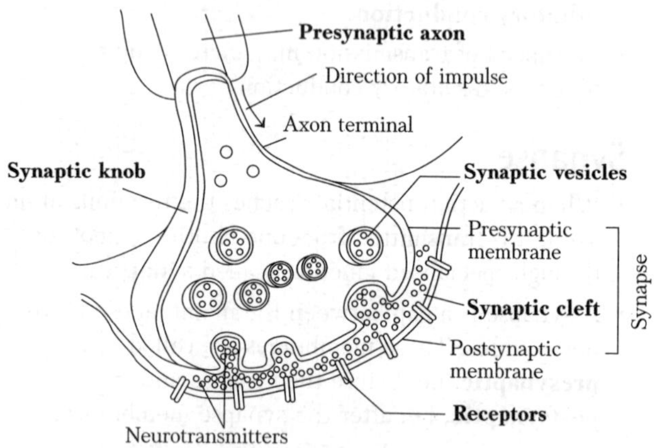

Fig. 28.14 Diagram showing axon terminal and synapse

Mechanism of transmission

The stepwise approach of transmission across the chemical synapse is given below

Step 1 Opening of voltage gated Ca^{++} channels once the wave of depolarisation reaches the presynaptic membrane.

Step 2 Diffusion of Ca^{2+} ions into synaptic cleft followed by the release of neurotransmitter due to the bursting of synaptic vesicles.

Step 3 Receiving and binding of neurotransmitter by the receptors of postsynaptic fibres.

Step 4 Development of depolarisation wave in postsynaptic fibre as per the message received by receptors *via* neurotransmitters.

Neurotransmitters

These are chemical messengers secreted by the axon terminals for transmitting impulses to the next neuron. There are various kinds of neurotransmitters which produce different effects on postsynaptic fibres.

Their names and effects are listed below

(i) **Acetylcholine** It is the most common neurotransmitter. The nerve fibres that release acetylcholine are called **cholinergic**. It has excitatory effect on the skeletal muscles and inhibitory effect on heart muscles.

(ii) **Nor-epinephrin**e or **nor-adrenaline** The nerve fibres producing this neurotransmitter are called **adrenergic**. It has excitatory or inhibitory effects.

(iii) **Glycine, dopamine** and **Gamma Amino Butyric Acid** (GABA) are inhibitory transmitters.

(iv) **Glutamate** is excitatory in function.

(v) **Serotonin** inhibits pain pathways of spinal cord. It generally controls mood and induces sleep.

(vi) Some neuropeptides like **P substance** are excitatory in action. It influences both CNS and PNS. It is a key excitatory neurotransmitter that mediates our perception of pain.

(vii) Another such neuropeptide is **endorphine.** It acts as a natural analgesic, i.e. decreases pain reception.

Note *Sometimes due to delay in the release of neurotransmitter, delay is seen in transmitting information from pre to postsynaptic neuron at synapse. This is called* ***synaptic delay.***

Reflex action

- The process of involuntary response to the peripheral nervous stimulation, that requires the involvement of a CNS part (mostly spinal cord) is called **reflex action.**
- It is an essential action exhibited by the motor nerves and usually more rapid than other reactions of the body.
- The parts of PNS and CNS form an arc like structure to perform reflex action. This arc is called **reflex arc.**

Reflex arc

It is the nerve pathway involved in a reflex action.

It has following five components

(i) **Receptor** It is any cell/tissue/organ, which receives an external or internal stimulus, e.g. skin, eye, ear.

(ii) **Sensory/Afferent nerve fibres** These take the sensory impulses generated by the receptor to the central nervous system.

(iii) **Part of central nervous system** It may be spinal cord or brain or any ganglion.

(iv) **Motor/Efferent nerve fibres** These carry the motor impulse generated in the CNS to the specific effectors.

(v) **Effector organ** It may be organ/muscle/gland which on being activated by a motor impulse, help to deal with the stimulus.

With the participation of all the above written components, the functioning of an arc is given below

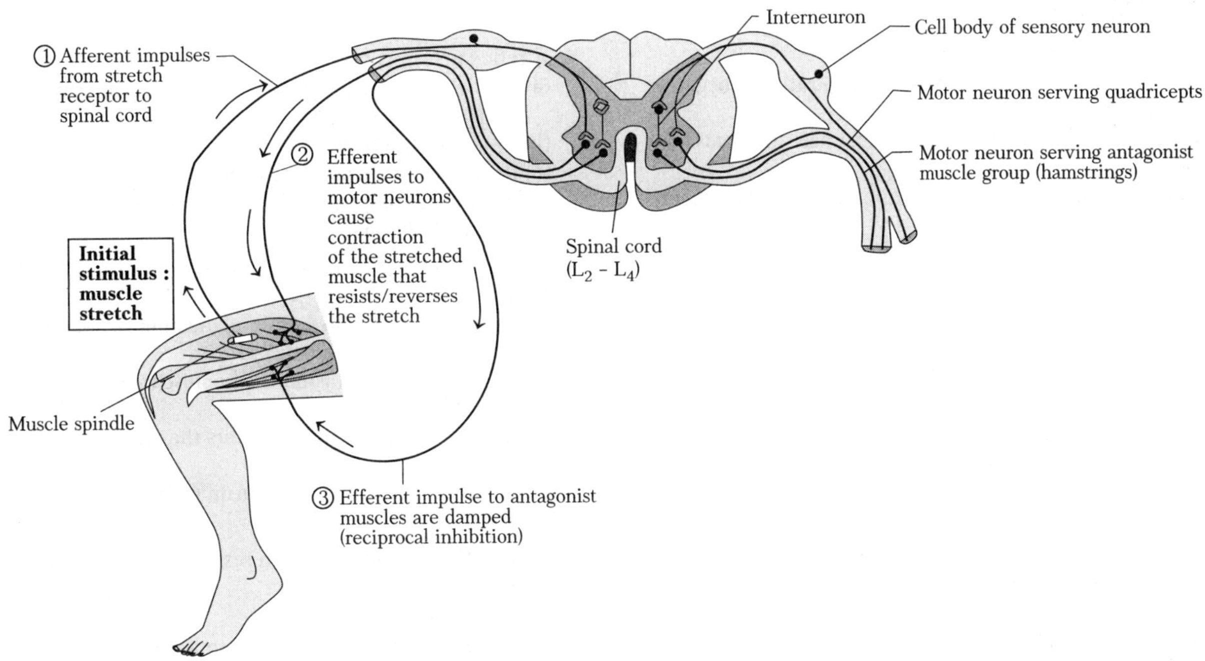

Fig. 28.15 Diagrammatic presentation of reflex action

Types of reflexes

1. Based on the part of CNS involved, reflexes are of following types
   (i) **Spinal reflexes** These reflexes are controlled by the spinal cord. Nature-wise, these are quick on response, e.g. withdrawl of leg while stepping on something hot or pointed.
   (ii) **Cranial reflexes** These reflexes are controlled by brain. Inspite of brain participation, these does not involve thinking process. These are slow in action, e.g. watering of mouth on sight or smell of food.
2. Based on the number of synapses, reflexes are of following types
   (i) **Monosynaptic** In these, there is a direct synapse found between sensory and motor neurons, e.g. knee jerk reflex.
   (ii) **Polysynaptic** In these, there are one or more small neurons present in between the sensory and motor neurons, e.g. cycling, swimming, etc.
3. Based on whether they are present from birth or acquired through experience in life, reflexes are of following types
   (i) **Unconditioned reflexes** These are present in an organism right from birth. These are also called as **simple** or **inborn** or **inherited reflexes**.
   These reflexes are specific, predictable and have survival value.
   For example,
   - withdrawl of hand or foot on being pricked
   - sweating
   - closing of eyes on sudden contact of an object
   - coughing during swallowing
   - peristalsis
   - breathing
   - jerking of knee when hit below knee cap, etc.

   (ii) **Conditional reflexes** These are not present at birth and develop later in life through learning, experience or regular association of an indifferent stimulus without affecting unconditioned stimulus. These are also called **acquired reflexes.**
   These reflexes tend to change, disappear or reappear. They actually reduce burden on brain. For example, driving a vehicle, playing a musical instrument, writing, reading, knitting and typing without looking, etc.
   These were first studied by **E Pavlov** on salivation in dog at the sight and sound of bell.

Importance of reflex action

- It controls a number of body activities.
- The response to harmful stimulus is fast, accurate and useful.
- Relieves the brain from too much work load.

CHECK POINT 28.3

1. As a nerve impulse passes along an axon
   (a) the membrane potential changes from positive to negative and then back again
   (b) sodium ions flow out through ion channels and potassium ions flow in
   (c) sodium channels open as the membrane potential become less negative
   (d) the sodium potassium pump moves sodium ions into the cell

2. Which of the following is responsible for resting membrane potential?
   (a) Greater permeability of membrane for K^+
   (b) Greater permeability of membrane for Na^+
   (c) Greater permeability of negatively charged protein of neuron across the membrane
   (d) Greater permeability of positively charged proteins of neuron across the membrane

3. The action potentials temporarily disrupts the polarisation of a neuron resulting in
   (a) depolarisation
   (b) repolarisation
   (c) closing of voltage gated Na^+ channels
   (d) –120 mV potential

4. A reflex action is not
   (a) an involuntary and nearly instantaneous movement in response to a stimulus
   (b) by conscious effort
   (c) the process of response to PNS
   (d) a pathway comprising atleast two afferent neurons and two efferent neurons

5. During a reflex action, the afferent neuron receives signal from
   (a) a sensory organ and transmits the impulse *via* dorsal root ganglion into the spinal cord
   (b) an effector and transmit the impulse *via* motor and plate into spinal cord
   (c) a motor end plate and transmits the impulse *via* an effector into the spinal cord
   (d) dorsal root ganglion and transmits the impulse *via* a sensory organ

6. Choose the odd one out with respect to reflex action.
   (a) Knee jerk
   (b) Salivation
   (c) Quick withdrawl of hands when it comes in contact with a hot object
   (d) Sweating in hot environment

SENSORY ORGANS

The sensory organs are specialised receptors or sense organs which detect all the types of changes in their surrounding environment relevant to them. They send appropriate signals to the central nervous system where these signals are processed and analysed.

Classification of sensory organs

These are classified generally on the basis of kinds of stimulus, functions and locations.

Based on their position, receptors are of following three types

(i) **Exteroceptors** These are external sense organs located near the body surface.

(ii) **Propioreceptors** These are simple receptors in joints, skeletal muscles, ligament, tendon, etc.

(iii) **Visceroreceptors** These are found within viscera. They receive stimuli that originate within the body itself, e.g. free nerve endings.

Based on kinds of stimulus, the sensory organs fall into following four categories

(i) **Mechanoreceptors** These respond to mechanical stimuli.

They are further classified as follows

- **Tangoreceptors** Sensitive to touch like Merkel's corpuscles (in epidermis of skin), Meissner's corpuscles (in dermis of skin of finger tips, lips and nipples), etc.
- **Proprioreceptors** Sensitive to position and movement of parts of body like Golgi-Mazzoni organs.
- **Rheoreceptors** Sensitive to pressure and water currents like lateral line organs in fishes and amphibian tadpoles.
- **Phonoreceptors** Sensitive to sound like organ of Corti of ear.
- **Statoreceptors** Sensitive to change in equilibrium like cristae and maculae of ear.
- **Baroreceptors** Sensitive to blood pressure change to maintain homeostasis through vasodilation or vasoconstriction in body, e.g. sensory nerve endings in walls of atrium, vena cava, etc.
- **Analgesic receptors** Sensitive to pain like free nerve endings.

(ii) **Chemoreceptors** These are sensitive to chemical molecules like water, glucose, oxygen, carbon dioxide and amino acids. The stimulus molecule binds to a specific site on the membrane to change its permeability, e.g. gustato-receptors, olfactory-receptors.

(iii) **Electromagnetic receptors** Sensitive to intensity and wavelength of light, image formation, electricity and magnetism, e.g. rods and cones.

(iv) **Thermoreceptors** Sensitive to heat or cold and helps to regulate body temperature by signaling hypothalamus, i.e. the body's thermostat.

These are further of two types

- **Frigidoreceptors** Sensitive to low temperature, e.g. Krause's corpuscles.
- **Caloreceptors** Sensitive to high temperature, e.g. Ruffini's corpuscles.

Human eye

These are paired organ of sight. These are situated in the deep protective bony cavities of skull called **orbits** or **eye sockets**. Eyes are supplied with optic nerve.

Structure

The wall of the eyeball consists of following three layers

External layer or Sclerotic

- This layer maintains the shape of eye and protects it.
- Composition-wise, it contains a dense connective tissue called the **sclera**.
- The anterior portion of this layer is called **cornea**. This portion does not have blood vessels and is refractory in function.
- Cornea is covered by a transparent membranous covering called **conjunctiva**. It provides protection and lubrication to the cornea.

Middle layer or Choroid

- This layer supplies the nutrients to outer retina. It also helps in maintaining the temperature and volume of eye.
- It contains many blood vessels and thus appears bluish in colour. In the anterior part, it becomes thickers to form the **ciliary body**.
- Some **suspensory ligaments** are found attached to the ciliary body. These ligaments are in turn, attached to the capsule surrounding the biconvex crystalline eye lens.
- The capsule and ligaments, together with the ciliary body, hold the lens in place and control its focal length.
- The ciliary body itself continues forward to form a pigmented and opaque structure called the **iris**. It is the visible coloured portion of the eye.
- The central apperture within the iris is called **pupil.** It is a central black hole like structure. Pupil helps in controlling the quantity of light entering the eye.

Lens, ciliary body and suspensory ligament divides the eye ball into following two chambers

(i) **Aqueous chamber** It is the space between the cornea and the lens.

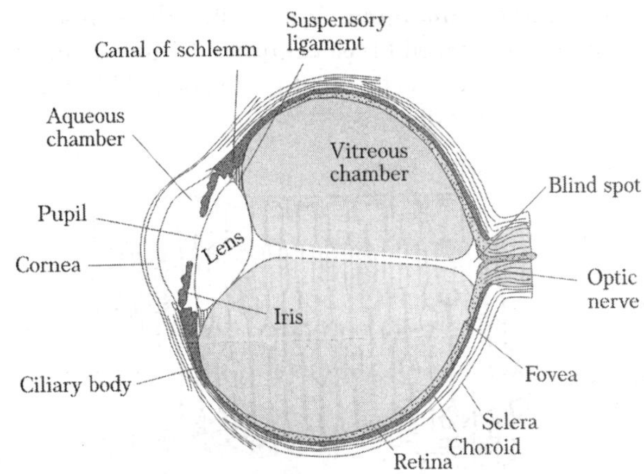

Fig. 28.16 Diagram showing parts of an eye

It contains a thin watery fluid, secreted by the veins of ciliary body called **aqueous humor.** This fluid nourishes the lens and cornea. Excess of aqueous humor is drained into the blood by **canal of Schlemm.**

(ii) **Vitreous chamber** It is the space between the lens and the retina. It is filled with a transparent gel called **vitreous humor**. It regulates eye shape and pressure to keep eye lens in position.

Inner layer or Retina

It is the innermost, delicate, non-vascular light sensitive layer. It consists of following three types of cells

(i) **Photoreceptor cells** These are further of two types, i.e. **rod cells** and **cone cells.**

- Both of these contain the light sensitive pigments called the **photopigments.**
- The daylight and coloured vision are the functions of **cones.**
- These cells contain erythropsin, iodopsin and cyanopsin containing photopsin protein.
- The twilight or scotopic or dim light vision is function of the rods.
- These cells contain a protein called **rhodopsin** or **visual purple**, which contains a **derivative of vitamin-A.**

(ii) **Bipolar cells** These are short sensory bipolar neural cells which transmits visual input from rods and cones towards ganglion cells.

(iii) **Ganglion cells** These constitute the innermost layer forming cells of the retina.

- The nerve fibres from the basal end of ganglion cells collectively form the **optic nerve.**
- It leaves the eye from a point called **blind spot.** There are no rod and cone cells in this region.
- An area, lateral to the blind spot is called **yellow area** or **macula lutea.** At its centre, there is a depression called **fovea centralis.** It has maximum density of cone cells and it is the point of formation of sharpest vision.

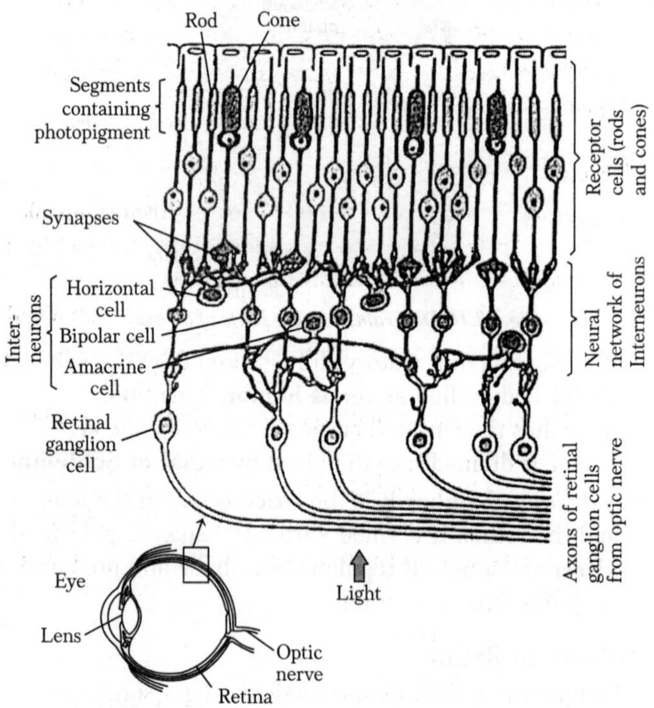

***Fig. 28.17** Schematic diagram to show the layer of the retina and main structures therein*

Accomodation Power of the lens makes fine adjustments to bring a sharp focus on retina. The ability of the lens by which the light ray from far and near objects is focused on the retina is called accomodation power of the lens.

In bright light, circular muscles of iris contract, radial muscles relaxes, pupil become smaller and thus, less light enter the eye. In dim light, *vice versa* process occurs. In total dark, the size of pupil increases. The normal eye can accomodate light from objects about 25 cm.

Accessory organs of the eye

The eyeball is associated with certain other structures which help in the functioning of the eye. These are as follows

(i) **Eyebrows** It is a skin covering over the supraorbital process of frontal bone. Eyebrows contain hairs which protect anterior aspect of eyeball from sweat and dust.

(ii) **Eyelids/Palpebrae** These are two movable folds of the skin situated above and below, in front of each eye. On their free edges, some short curved hairs are present. These hairs are called **eyelashes.**

The inner eyelids contain a transparent mucous membrane called **conjunctiva** that provides protection to the eyeball and keeps it moist.

(iii) **Glands** There are four glands associated with the eye. They are

- **Meibomian glands** sebaceous glands which help in lubrication.
- **Glands of moll** sweat glands at the edge of eyelids.
- **Glands of zeis** sebaceous glands near the follicles of eye lashes.
- **Lacrimal glands tear glands** Tears secreted by lacrimal gland helps to keep the cornea moist and transparent. Tears also contain lysozymes to prevent microbial infection.

Mechanism of vision

The mechanism of vision in stepwise manner is shown below

1. Light rays in visible wavelength focussed on the retina
↓
2. Eye lens transmit the light impulse on the photopigment of rods and cones for their activation
↓
3. Light induces dissociation of the retinal pigment, opsin and result in structural changes in opsin
↓
4. Membrane permeability of rods and cones changes
↓
5. Potential difference generated in the photoreceptor cells
↓
6. It generates action potential in the ganglionic cells through the bipolar cells
↓
7. Action potential are transmitted by the optic nerve to the visual cortex of brain
↓
8. Image formed on the retina is recognised based on earlier memory and experience

Disorders of eye

(i) **Myopia** It is also called **short-sightedness.** The person with this defect can see nearby objects clearly, but not far away objects. It occurs due to increase in convexity of lens or anterio-posterior elongation of the eyeball due to which image being formed in front of the retina. It can be corrected by using concave glasses.

(ii) **Hypermetropia** It is also called **far** or **long-sightedness**. The person with this defect can see distant objects clearly but closer objects are not clear. In this, the image is formed beyond the retina due to shortening of eyeball or lens being flattened. It can be corrected by wearing convex or convergent lenses.

Sometimes in old age, this defect occurs due to reduction of the flexibility of lens, then it is called **presbyopia**.

(iii) **Cataract** In this disease, the lens of eye becomes opaque thereby, reducing its visibility. It can be treated by replacing opaque eye lens with artificial intraocular lens or by LASER treatment.

(iv) **Glaucoma** It occurs due to an increase in intraocular pressure in aqueous chamber because of blockage in the canal of Schlemm. This closure exerts pressure on optic nerve, causing its damage. It may leads to permanent blindness (kala motia) if untreated.

(v) **Astigmatism** In this, the curvature of cornea becomes irregular and image is not clearly formed. This defect can be corrected by using cylindrical glasses.

Human ear

Ears are a pair of **statoacoustic** organs, i.e. meant for both balancing and hearing.

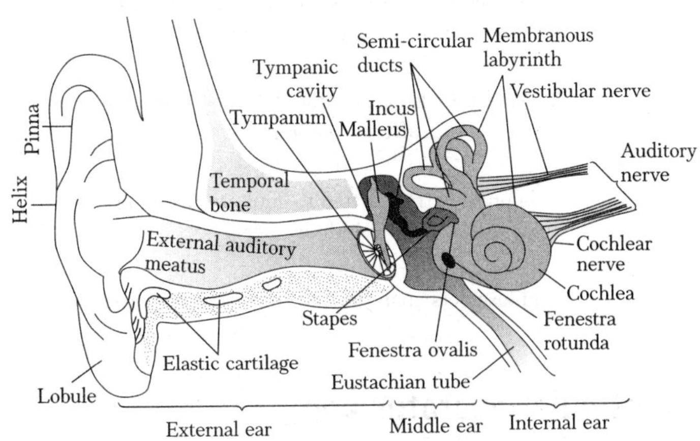

Fig. 28.18 Human ear showing external, middle and internal ear

Anatomically, ear can be divided into three parts, i.e. external ear, middle ear and inner ear.

External ear

It consists of ear pinna, auditory meatus or canal and tympanic membrane.

- The **pinna** is present in most mammals as a projecting elastic cartilage covered with skin. It has 2 parts namely, **helix** and **lobule**. It is sensitive as well as effective in collecting sound waves into the auditory canal.
- The tympanic membrane is also called the **ear drum** or **tympanum.** It is made up of connective tissue.
- The auditory canal carries the sound waves towards the eardrum.
- There are very fine (small) hairs and wax-secretory sebaceous (oil) glands in the skin of pinna and meatus.
- The wax secreting sebaceous glands produce an oily secretion called the **ear wax**.
- The fine hairs and ear wax, both prevent the foreign particles from entering the ear.

Middle ear

- The middle ear is represented by an air-filled tympanic cavity. This cavity communicates with the pharynx by a passage called **Eustachian tube**.
- The middle ear contains three bones or ossicles attached to one another in a chain-like fashion.
- These are
  - Hammer shaped **malleus** attached to the tympanic membrane
  - **Incus** is the middle bone
  - **Stapes** is attached to the oval window of cochlea.

  These ossicles with their sites of attachment are shown below

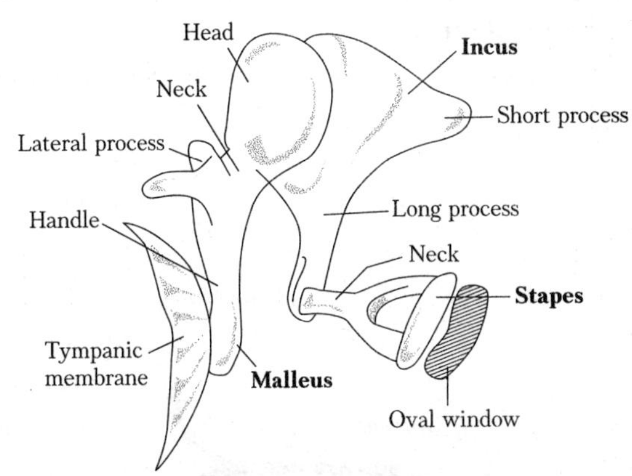

Fig 28.19 The auditory ossicles

- These three ossicles increases the efficiency of transmission of sound waves into the inner ear.
- The tympanic cavity has following two openings
  (i) An oval window called the **Fenestra ovalis** (upper)
  (ii) The round window or **Fenestra rotunda** (lower). Each of these are covered by a membrane. These openings act like a door to the inner ear.

Inner ear

The inner ear is also called the **labyrinth.** It is a fluid-filled chamber within the temporal bone of the skull. The labyrinth consists of two parts

(i) Bony labyrinth

- The bony labyrinth is present as 3 perpendicular channels or canals. Inside these canals, membranous labyrinth lies.
- These canals are filled with a fluid called **endolymph.**
- The space between the membranous labyrinth and bony labyrinth is filled by a watery fluid, **perilymph.**

(ii) Membranous labyrinth

Functionally the membranous labyrinth is divided into cochlea and vestibular apparatus.

- **Cochlea** The coiled portion of the labyrinth is called cochlea. It is the main hearing organ of the inner ear.
- The cochlea has two large canals *viz*,
  - An upper vestibular canal or **scala vestibuli**
  - A lower tympanic canal or **scala tympani.**
- Both of these canals are separated by a small cochlear duct called **scala media**. This duct contains the main organs of hearing, i.e. organ of Corti.
- The vestibular and tympanic canals contain perilymph and the cochlear duct is filled with endolymph.
- **Organ of Corti** It is the main organ of hearing, situated inside scala media.
- The scala media is partitioned from the scala vestibuli by **Reissner's membrane** and from the scala tympani by **basilar membrane.**
- Organ of Corti is located on the basilar membrane as shown below

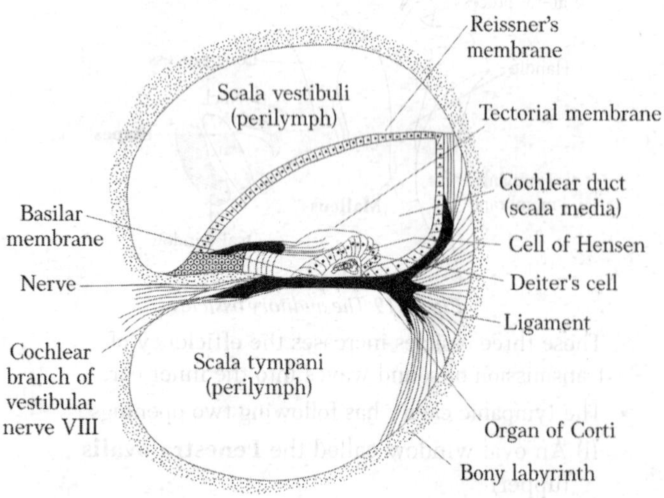

Fig 28.20 T S of cochlea

- It is a sensory structure which contains the mechanoreceptors of the ear as **hair cells.**
- The hair cells are present in rows on the internal side of the organ.
- The basal end of the hair cell is in close contact with the afferent nerve fibres, i.e. VIII cranial nerve.
- A large number of processes called **stereo cilia** are projected from the apical part of each hair cell. Above the rows of hair cells, a thin elastic membrane called **tectorial membrane** is present.

Mechanism of hearing

Stepwise mechanism of hearing is given below

1. External ear receives sound waves which are directed towards the eardrum
↓
2. When the waves strike the tympanic membrane, the alternate compression and decompression of the air cause the membrane to vibrate
↓
3. Vibrations are transmitted through the ear ossicles (mallus → incus → stapes) to oval window
↓
4. The movement of the oval window sets up wave in the perilymph of scala vestibuli which are transferred as vibration of endolymph of scala media
↓
5. The waves in the endolymph induces a ripple in the basilar membrane
↓
6. Basilar movements bend the hair cells and press them against the tectorial membrane
↓
7. Nerve impulse is generated in the associated afferent neurons
↓
8. Impulse is transmitted to auditory region of brain *via* auditory nerve
↓
9. Impulse gets analysed and sound is recognised

Vestibular aparatus and Mechanism of Balance

Besider the cochlea, the inner ear also has the vesibular apparatus which is composed of **three semi-circular canals.** Each of these have a utriculo-saccular region with **otolith organ**. These organs are specialised for balancing.

Semicircular canals

- All three semi-circular canals lie in different planes at right angle to each other. These canals are filled with endolymph.
- The base of each of canal has an ampulla in which there is a sensory spot called **crista**.
- The cristae help in maintaining equilibrium. The vestibula has two sensory spots, i.e. macula of saccule and utricle.
- The **macula** consist of hair cells and supporting cells. Tips of the hair cells project into a thick gelatinous sheath or otolithic membrane.
- Within this membrane, minute particle otocliths or otoconia are secreted. These are made of $CaCO_3$ and protein.
- The macula and crista are the receptors sensitive to the position of the head with respect to gravity.

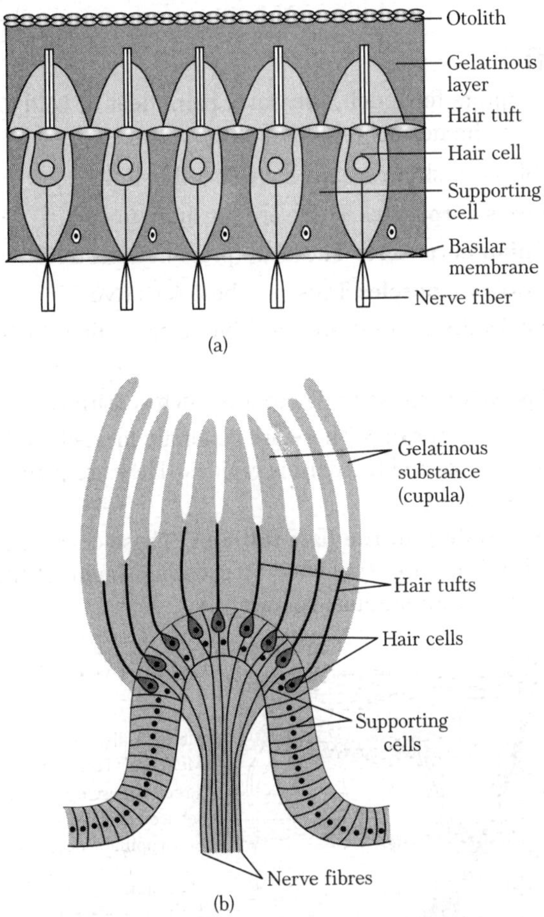

Fig. 28.21 (a) Structure of macula, (b) Structure of crista

- The three semicircular canals are arranged such a way that the movement in any plane can be detected by these cells and the balance and posture of the body is maintained.
- Receptors for dynamic balance lies in the cristae of ampullae while for static/linear balance, these are present in the maculae of utriculus and sacculus.
- Their arrangement along with cochlea is shown below

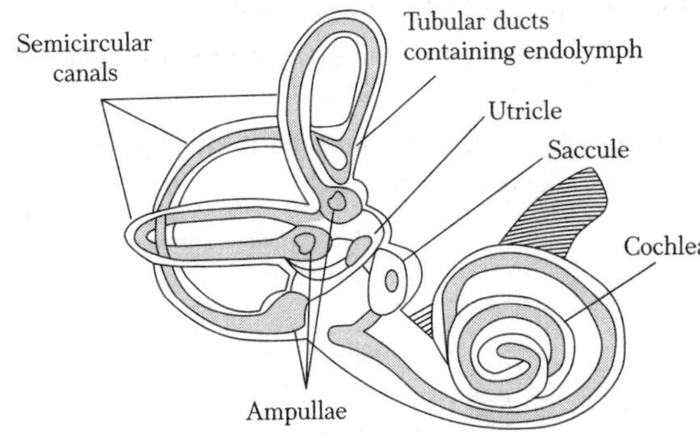

Fig. 28.22 The vestibular system : Semicircular canals and otolith organs

Note *In case of frog, external ear is not seen, only the middle and internal ears are seen. The middle ear of frog contains a single bone called **columella auris**, which can be compared with the stapes of mammals. Along with this bone, a cartilaginous palate called **stapedial plate** is also seen within the middle ear.*

Ear related disorders

(i) **Meniere's disease** It is caused by an abnormality in the composition or amount of fluid in the inner ear.

(ii) **Eustachitis** It occurs due to the inflammation of Eustachian tube.

(iii) **Otalgia** It is defined as ear pain.

(iv) **Otitis media** It is an acute infection in the middle ear.

(v) **Myringitis** Inflammation of tympanic membrane.

(vi) **Vertigo** Type of dizziness involving feeling of motion even when the person is stationary.

Nose

It is the organ of olfaction. It is well-developed in vertebrates and mammals. The upper nasal chamber is lined by epithelium called **olfactory epithelium** or **Schneiderian membrane**. This membrane is sensitive to smell.

Various features of this membrane are as follows

- It consists of sensory or **olfactory receptor cell** (sensory cell) along with **supporting cells**.
- The sensory cells have olfactory cilia at their free margin and nerve fibres at the other end, which joins to form the olfactory nerve.
- Below the olfactory epithelium a connective tissue layer having **glands of Bowman** is present. Secretions of this gland keeps the surface moist.
- Odoriferous particles coming with the inspired air get dissolved in mucus and stimulate the sensory cells.

- Supporting cells are columnar cells that lies in contact with olfactory receptor cells to support them.
- Basal cells are small cells that give rise to new olfactory receptor cells.
- The positioning of olfactory bulb with olfactory tract in human nose is given below

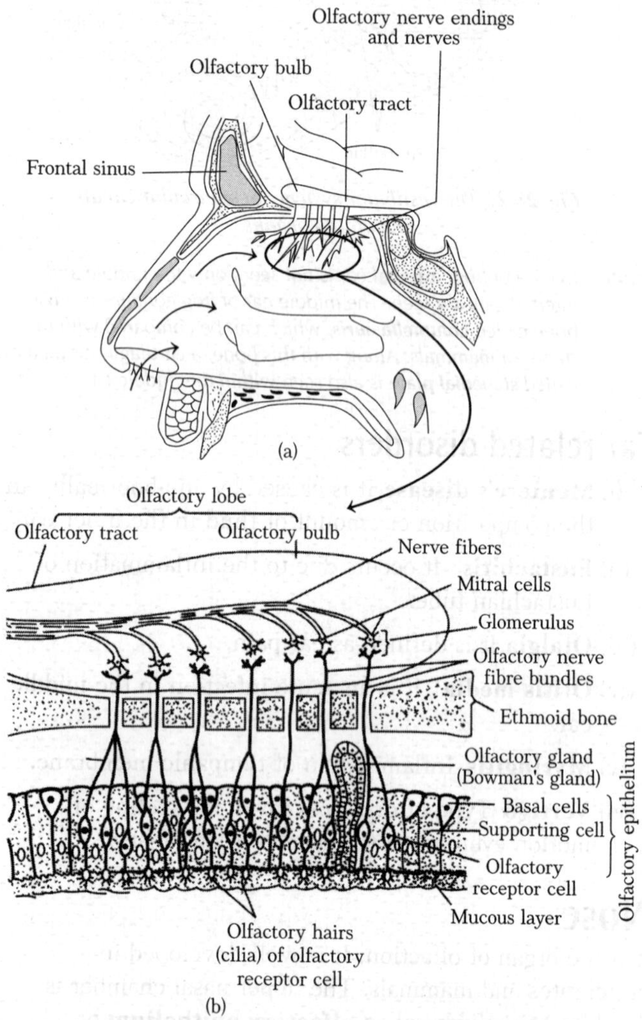

Fig. 28.23 (a) The internal structure of nose, (b) Olfactory epithelium, ethmoid bone and olfactory lobe

Tongue

The taste buds on tongue contain gustatory or taste receptors. Each taste bud comprises following three cells

(i) **Gustatory receptor cells** They possess microvilli at their free ends. These microvilli project into taste pore and have special receptor sites to detect chemicals in food. Cranial nerves VII, IX X innervates gustatory receptor cells.

(ii) **Supporting cells** They lies in close contact with gustatory receptor cells to support them. They also have microvilli, but lack nerve endings.

(ii) **Basal cells** These are present at the periphery of taste bud. They produce supporting cells and gustatory receptor cells.

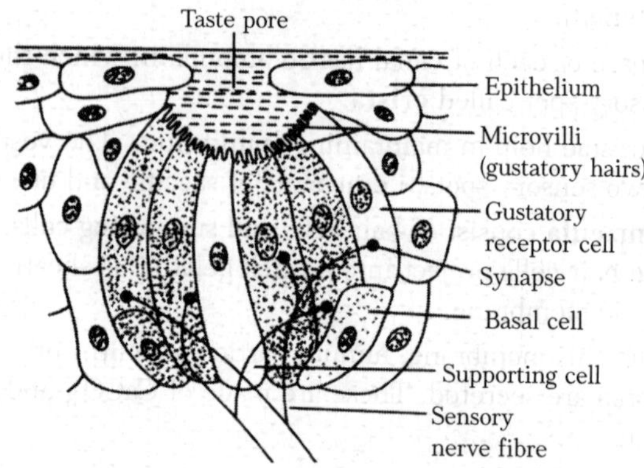

Fig. 28.24 Vertical section through a taste bud

Skin

The receptors for touch, pressure, pain, heat or cold are located in the dermis region of skin.

Some of these skin receptors are listed below

- **Krause's corpuscles** These are sensitive to cold.
- **Pacinian corpuscles** These are pressure sensitive.
- **Ruffini's corpuscles** These are heat sensitive.
- **Merkel's discs** These are sensitive to continuous tactile stimulation.
- **Meissner's corpuscles** These are touch sensitive.
- **Free nerve endings** These are generally located both in the epidermis and the dermis regions. They are pain and touch sensitive.
- **Nerve endings of the hair follicles** These cover the hair follicles. They are stimulated by each movement of the hair, even with minimal contact.

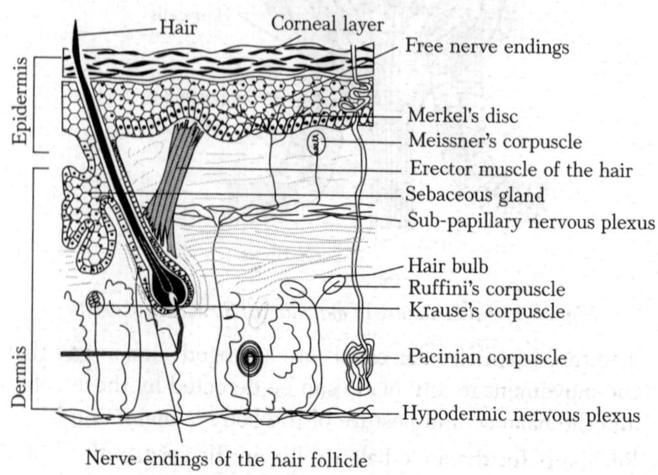

Fig. 28.25 The skin : Structure and receptors

CHECK POINT 28.4

1. The choroid layer is thin over the posterior two-third of the eyeball, but it becomes thick in the anterior part to form the
   (a) iris (b) ciliary body
   (c) pupil (d) suspensory ligament

2. Cornea and lens of eye are
   (a) sensitive and richly supplied by blood vessels
   (b) transparent and diverge light rays on retina and image formation
   (c) transparent and help in image formation on retina
   (d) sensitive and richly supplied by nerves

3. A person wears convex lenses for proper vision. Without glasses the image of the object is produced
   (a) on blind spot (b) on yellow spot
   (c) in front of retina (d) behind the retina

4. The sound vibrations are passed through the middle ear to oval window on to the fluid of the cochlea, where they generate waves in the lymph. This induce a ripple in the
   (a) basilar membrane (b) tectorial membrane
   (c) reissner's membrane (d) tympanic membrane

5. The cochlea is an organ of auditory transduction that contains
   (a) air and cells that produce wax
   (b) fluid and cells that can undergo mechano-sensory transduction
   (c) air and small bones that vibrate in response to sound waves
   (d) fluid with stacks of chemosensory cells

6. Taste and smells are similar in that
   (a) Both types of stimuli must be dissolved in a body fluid before they can be detected
   (b) Any given stimulus for one system evokes a response from the other system
   (c) Both types of stimuli are proteins
   (d) Both types of stimuli are present in thousands of different chemicals

EXAM POINTS

- Corpus callosum is a band of fibres connecting left and right cerebral hemispheres. **[NEET 2018]**
- Limbic system consists of five major components, namely hippocampus, amygdala, septal nuclei, mammillary bodies and basal ganglia. It control emotional behaviour, food habits and sexual behaviours. **[NEET 2018]**
- The cavity of diencephalon is known as third ventricle. **[NEET 2017]**
- Hypothalamus is the part of the brain responsible for thermoregulation, water balance and control of hormone function. **[NEET 2019, JIPMER 2019]**
- The pneumotaxic centre and rhythm center are respectively, present in pons and medulla oblongata. **[AMU 2015, Kerala CEE 2011]**
- Hearing is controlled by temporal lobe of cerebrum. **[Manipal 2013]**
- Sympathetic nervous system increases heart rate and releaxes bronchi. **[EAMCET 2014]**
- The parasympathetic neural signal affect the working of heart by reducing both heart rate and cardiac output. **[CBSE AIPMT 2014]**
- Oculomotor, abducens, trochlear are the group of cranial nerve which control eyeball movement. **[MHT CET 2015]**
- Trochlear is the smallest cranial nerve. **[NEET 2013]**
- Vagus nerve is mixed Xth cranial nerve. **[WB JEE 2013]**
- Medulla oblongata controls respiration and cardiovascular reflexes. **[NEET 2018]**
- Receptor sites for neurotransmitters are present on postsynaptic membrane. **[NEET 2017]**
- Arrival of an impulse at the axon terminal stimulates the release of neurotransmitters in synaptic cleft. **[NEET 2017]**
- Neurotransmitters are responsible for the opening of ion channels. **[NEET 2017]**
- An action potential in an axon does not move backward because the segment behind it is in a refractory phase. **[NEET 2019]**
- Depolarisation of hair cells of cochlea results in the opening of the mechanically gated potassium-ion channels. **[NEET 2019]**
- In the knee-jerk reflex, stimulus is the stretching of muscle and response is its contraction. **[NEET 2019]**
- Caloreceptors are heat receptors. **[NEET 2017]**
- Interoreceptor help in detecting hunger. **[NEET 2017]**
- Nociceptors are sensory nerve cells that respond to potentially damaging chemical or mechanical stimuli and send them to brain and spinal cord. **[NEET 2016]**
- Meissner's corpuscles are tango receptors. **[NEET 2016]**
- Receptors always produce graded potentials. **[NEET 2016]**
- Rods and cones are photoreceptor cells. **[NEET 2020]**
- Blind spot is the place in the eye where photoreceptor cells are absent. **[NEET 2020]**
- Fovea is the place in the eye where cones are densely packed. **[NEET 2020]**
- Iris is the visible coloured portion of eye. **[NEET 2020]**
- Photosensitive compound in human eye is made up of opsin and retinal. **[NEET 2016]**
- Sensory neuron in the retina of eye are rods and cones. **[NEET 2016]**
- Photoreceptors in the human eye are depolarised during darkness and become hyperpolarised in response to the light stimulus. **[NEET 2016]**
- The transparent lens in the human eye is held in its place by ligaments attached to the ciliary body. **[NEET 2018]**
- Cornea consists of dense matrix of collagen and is the most sensitive portion of the eye. **[NEET 2019]**
- Rods are very sensitive and contribute to night or scotopic vision. **[NEET 2019]**
- Purplish red pigment rhodopsis is derivative of vitamin-A.
- Organ of Corti is located on the basilar membrane. **[NEET 2020]**
- Glucoma is the abnormal high pressure of liquid in eye. **[NEET 2017]**
- Cochlea is the coiled part of the labyrinth. **[NEET 2020]**
- Eustachian tube connects middle ear cavity and pharynx. **[NEET 2020, 17]**
- Vestibular apparatus do not play any role in hearing. **[CBSE AIPMT 2012]**
- Stapes is attached to the oval window. **[NEET 2020]**
- Crista ampullaris and macula are responsible for the maintenance of balance of body and posture. **[NEET 2019]**

Chapter Exercises

(A) Topical Exercise

Nervous System

1 The system responsible for coordinating various activities of different body parts is
(a) excretory system
(b) circulatory system
(c) nervous system
(d) None of the above

2 Select the correct statement.
(a) Neurons regulate endocrine activity but not *vice-versa*
(b) Endocrine glands regulate neural activity but not *vice-versa*
(c) Endocrine glands regulate neural activity and nervous system regulates endocrine glands
(d) Neither hormones control neural activity nor the neurons control the endocrine activity

3 The diagram given below is the functional organisation of the human nervous system. Identify *A*, *B*, *C*, *D* and *E* in the figure.

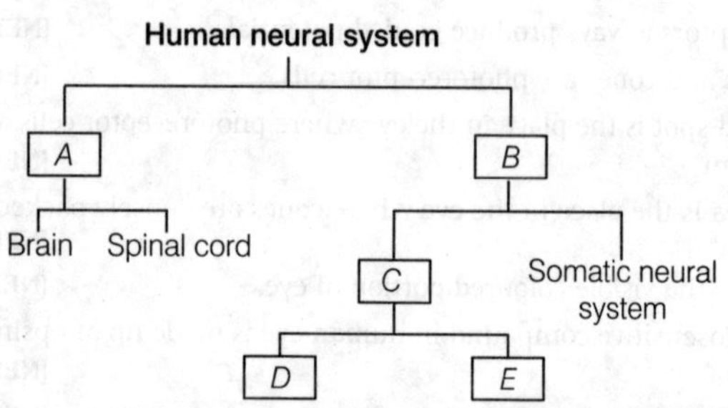

(a) *A*-PNS, *B*-CNS, *C*-ANS, *D*-Sympathetic nervous system, *E*-Parasympathetic nervous system
(b) *A*-ANS, *B*-CNS, *C*-PNS, *D*-Sympathetic nervous system, *E*-Parasympathetic nervous system
(c) *A*-CNS, *B*-PNS, *C*-ANS, *D*-Sympathetic nervous system, *E*-Parasympathetic nervous system
(d) *A*-ANS, *B*-PNS, *C*-CNS, *D*-Sympathetic nervous system, E-Parasympathetic nervous system

4 In the 4th week after conception, the neural tube develops three swellings. These are
(a) Telencephalon, Prosencephalon, Mesencephalon
(b) Rhombencephalon, Prosencephalon, Mesencephalon
(c) Metencephalon, Myelencephalon, Prosencephalon
(d) Prosencephalon, Diencephalon, Mesencephalon

5 Consider the following statements.
I. Neural system is better organised in insects due to the presence of brain, ganglia and neural tissues.
II. Nerual organisation is very simple in lower invertebrates.

Choose the correct option.
(a) Statement I is true, but II is false
(b) Statement I is false, but II is true
(c) Both statements I and II are true
(d) Both statements I and II are false

Brain

6 Select the correct sequence of meninges from inside to outside.
(a) Dura mater → Arachnoid mater → Pia mater
(b) Arachnoid mater → Dura mater → Pia mater
(c) Pia mater → Dura mater → Arachnoid mater
(d) Pia mater → Arachnoid mater → Dura mater

7 Sub-arachnoid space is
(a) between dura and arachnoid
(b) between dura mater and cranium
(c) between pia meter and arachnoid mater
(d) between pia meter and grey matter

8 The function of cerebrospinal fluid is to
(a) absorb shock or jerk to protect brain and spinal cord
(b) serve as a medium for exchange of nutrients and waste
(c) serve as endocrine medium for transport of hormones
(d) All of the above

9 The white matter of the central nervous system is always
(a) deep to the grey matter
(b) unmyelinated
(c) arranged into tracts
(d) composed of sensory fibres only

10 In the central nervous system, myelinated fibres form the ......, while the non-myelinated fibre cells form the ....
(a) grey matter, white matter
(b) white matter, grey matter
(c) ependymal cells, neurosecretory cells
(d) neurosecretory cells, ependymal cells

11 The forebrain is divided into
(a) olfactory lobe, cerebrum, diencephalon
(b) olfactory lobe, thalamus, cerebrum, pons
(c) thalamus, hypothalamus, cerebrum, medulla
(d) pons, diencephalon, cerebellum

12 The corpus callosum is a
(a) source of hypothalamic hormones
(b) neural pathway that connects the right and left hemispheres
(c) part of the neocortex
(d) structure in the cerebellum

13 Consider the following statements.
I. Wernicke's area is for understanding spoken and written words.
II. Broca's area is not present in temporal lobe.
Choose the correct option.
(a) Statement I is true, but II is false
(b) Statement I is false, but II is true
(c) Both statements I and II are true
(d) Both statements I and II are false

14 Match the following columns.

	Column I (Brain area)		Column II (Functions)
A.	Brodmann's area	1.	Perception of pain, pressure and taste
B.	Broca's area	2.	Movement of lips and tongue
C.	Wernicke's area	3.	Understanding of speech

Codes

	A	B	C		A	B	C
(a)	1	2	3	(b)	3	2	1
(c)	2	3	1	(d)	1	3	2

15 Which lobe-function pairing is incorrect?
(a) Frontal lobe — Sensory interpretation
(b) Parietal lobe — Reasoning
(c) Occipital lobe — Vision
(d) Temporal lobe — Memory

16 The given figure is that of the human brain.

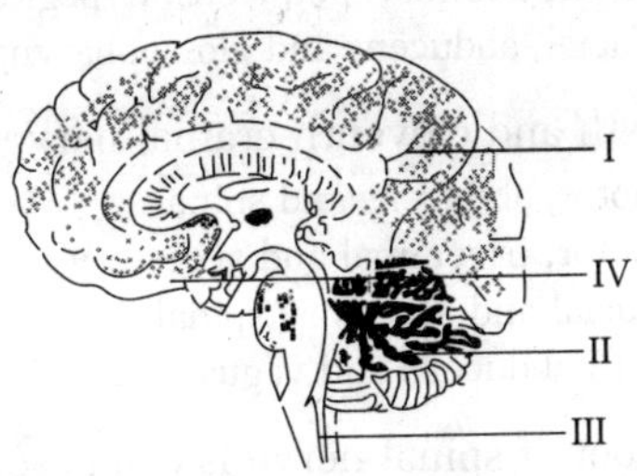

Identify the part that controls intelligence and memory is labelled.
(a) IV (b) III
(c) II (d) I

17 The basal nuclei form all of the following except the
(a) putamen (b) caudate nucleus
(c) globus pallidus (d) infundibulum

18 Parkinson's disease (characterised by tremors and progressive rigidity of limbs) is caused by degeneration of brain neurons that are involved in movement control and make use of neurotransmitter
(a) acetylcholine (b) nor-epinephrine
(c) dopamine (d) GABA

19 Parkinson's disease and other motor disorders are attributed to dysfunction or trauma of the
(a) pons (b) basal nuclei
(c) parietal lobe (d) thalamus

20 What is the work of commissural tract?
(a) Connects different parts of cerebral hemisphere
(b) Connects different areas of two cerebral hemisphere
(c) Connects cerebral cortex with grey matter of other parts of brain
(d) Connects cerebral cortex with white matter of other parts of brain

21 An area which is associated with strong emotions is
(a) cerebral cortex (b) cerebellum
(c) limbic system (d) medulla

22 Alzheimer's disease is caused by damage of
(a) amygdala (b) septum
(c) cingulate gyrus (d) hippocampus

23 The centres for the control of voluntary activities are present in the
(a) thalamus and cerebral cortex of the brain
(b) thalamus and medulla oblongata
(c) cerebral cortex and medulla oblongata
(d) None of the above

24 Diencephalon consists of three major parts of brain. They are
(a) thalamus, hippocampus, epithalamus
(b) thalamus, hypothalamus, hippocampus
(c) hypothalamus, hippocampus, epithalamus
(d) epithalamus, thalamus, hypothalamus

25 Consider the following statements.
I. Corpora quadrigemina acts as the centre for audio and vision.
II. The corpora quadrigemina, composed of the superior and inferior colliculus, is located in the mesencephalon.
Choose the correct option.
(a) Statement I is true, but II is false
(b) Statement I is false, but II is true
(c) Both statements I and II are true
(d) Both statements I and II are false

26 Consider the given diagram and identify '*A*'.

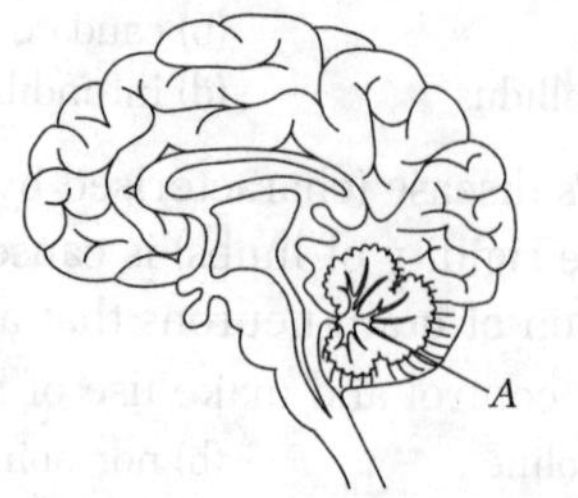

(a) Medulla oblongata (b) Cerebellum
(c) Midbrain (d) Pons

27 Which two structures of the brain control respiration?
(a) Pons and hypothalamus
(b) Cerebrum and hypothalamus
(c) Pons and medulla oblongata
(d) Hypothalamus and pituitary gland

28 Cerebellum governs the activities of
(a) cardiovascular reflex
(b) emotions and memory
(c) balancing and coordination of muscles
(d) endocrine gland

29 A patient with symptoms of tremor, halting speech and an irregular gait may have experienced trauma to the
(a) cerebrum (b) pons
(c) cerebellum (d) thalamus

30 Instruments used to record EEG is
(a) electroencephalograph (b) cathode ray oscilloscope
(c) Both (a) and (b) (d) None of these

31 Choose the incorrect statement.
(a) Ventricles of brain are lined by connective tissue
(b) Foramen of monro connects paracoel and diocoel
(c) Third ventricle of brain is situated in the roof of diencephalon
(d) Spinal cord passes out of the skull through foramen magnum

32 Choose the correct option.
(a) Delta waves – Sleeping person
(b) Theta waves – Severe emotional stress
(c) Beta waves – Intense mental activity
(d) All of the above

Spinal Cord

33 Consider the following statements.
I. Spinal cord has a central canal filled with cerebrospinal fluid.
II. In an adult human, it is about 42-45 long and 2 cm thick.

Choose the correct option.
(a) Statement I is true, but II is false
(b) Statement I is false, but II is true
(c) Both statements I and II are true
(d) Both statements I and II are false

34 Which one of the following statements is correct?
(a) Grey matter surrounds the white matter in the spinal cord
(b) Grey matter surrounds the white matter in the brain
(c) White matter surrounds the grey matter in the spinal cord
(d) White matter surrounds the grey matter in the brain

35 White matter of spinal cord contains
(a) nerve axons
(b) cell bodies
(c) clusters of neurons and cell bodies
(d) motor neurons

36 Which of the following is a function of spinal cord?
(a) Acts as minor coordinating centre
(b) Connects peripheral neural system to the brain
(c) Acts as the link between the brain and nerves that stretch throughout the body
(d) All of the above

37 Cytons of both central and autonomic sensory fibres occur in
(a) spinal ganglia (b) dorsal part of spinal cord
(c) ventral part of spinal cord (d) autonomic ganglia

38 What effect will be seen if dorsal nerve of spinal cord is broken?
(a) no impulse is transmitted
(b) impulse is transmitted but slowly
(c) impulse is transmitted fast
(d) no effect on impulse

PNS

39 II, VII, VIII and IX cranial nerves are
(a) optic, facial, auditory, glossopharyngeal
(b) optic, auditory, facial, hypoglossal
(c) oculomotor, auditory, abducens, hypoglossal
(d) optic, facial, abducens and glossopharyngeal

40 Third, sixth and eleventh cranial nerves are
(a) oculomotor, abducens and spinal
(b) oculomotor, trigeminal and accessory
(c) optic, facial and accessory spinal
(d) trochlear, abducens and vagus

41 Ventral root of spinal nerve is composed of somatic
(a) motor and visceral sensory fibres
(b) sensory and visceral sensory fibres
(c) motor and visceral motor fibres
(d) sensory and visceral motor fibres

42 Which of the following nerves are of mixed nature?

I. Trigeminal II. Hypoglossal
III. Glossopharyngeal IV. Abducens

Choose the correct option from the codes given below.

(a) I and III (b) II and IV
(c) III and IV (d) None of these

43 Identify the organ/innervation mismatch.

(a) Glossopharyngeal nerve - Tongue
(b) Optic nerve - Eye muscles
(c) Facial nerve - Olfactory epithelium
(d) Cochlear nerve - Cochlea

44 A patient with a contusion over the parotid region, the facial muscles on one side of the face are paralysed, one eye cannot be shut and the corner of the mouth droops. Which cranial nerve is damaged?

(a) Abducens nerve
(b) Facial nerve
(c) Glossopharyngeal nerve
(d) Accessory nerve

45 The rectus eye muscle capable of causing the eye ball to turn laterally in a horizontal plane is innervated by which cranial nerve?

(a) Optic nerve (b) Abducens nerve
(c) Facial nerve (d) Oculomotor nerve

46 Match the following columns.

Column I	Column II
A. Cervical nerves	1. 5 pairs
B. Thoracic nerves	2. 1 pair
C. Lumbar nerves	3. 12 pairs
D. Coccygeal nerves	4. 8 pairs

Codes

	A	B	C	D
(a)	4	3	1	2
(b)	1	3	4	2
(c)	3	2	4	1
(d)	1	4	3	2

47 Which of the following cranial nerve joins ganglionated sympathetic nerves of autonomic nervous system?

(a) Ramus dorsalis (b) Ramus ventralis
(c) Ramus communicans (d) All of these

ANS

48 The function of our visceral organs is controlled by

(a) sympathetic and somatic neural system
(b) sympathetic and parasympathetic neural system
(c) central and somatic nervous system
(d) None of the above

49 Which of the following is not a function of the ANS?

(a) Innervation of all visceral organs
(b) Transmission of sensory and motor impulses
(c) Regulation and control of vital activities
(d) Conscious control of motor activities

50 Which of the following is not the characteristic of human sympathetic system?

(a) Prepares the person for 'fight or flight'
(b) The presence of ganglia near the spinal cord
(c) Increases heart rate
(d) It is a craniosacral division of the autonomic nervous system

51 The main functions of sympathetic nervous system are

(a) contraction of skin, blood vessels and sudden increase of blood pressure
(b) contraction of muscles, secretion of sweat glands and rapid coagulation of blood
(c) dilation of bronchi, contraction of heart and sudden decrease in the number of RBC in the blood
(d) All of the above

52 The cell bodies of the pre-ganglionic neurons of the sympathetic division are located within the

(a) cervical and sacral regions of the spinal cord
(b) white matter of the spinal cord
(c) grey matter of the spinal cord
(d) brain and sacral region

53 Which of the following is not an effect of the sympathetic nervous system?

(a) Dilation of pupil
(b) Inhibition of peristalsis
(c) Elevation of blood pressure
(d) Stimulation for saliva secretion

54 Parasympathetic fibres arise from which set of cranial nerves?

(a) III, V, IX and X (b) IV, V, IX and X
(c) III, VII, IX and X (d) V, IX, X and XII

55 Which of the following is not an effect produced by parasympathetic stimulation?

(a) Pupil dilation
(b) Increased stomach activity
(c) Decreased stomach activity
(d) Constriction of bronchi

Conduction of Nerve Impulse

56 What would you call potential difference in an axon at rest and an axon in action?

(a) Depolarisation and action potential respectively
(b) Resting potential and action potential respectively
(c) Action potential and resting potential respectively
(d) Repolarisation and depolarisation respectively

57 Chemicals which are released at the synaptic junction are called [NCERT Exemplar]
(a) hormones (b) neurotransmitters
(c) cerebrospinal fluid (d) lymph

58 The sequence of events in a nerve conduction is
(a) Threshold → Depolarisation → Repolarisation → Hyperpolarisation
(b) Hyperpolarisation → Repolarisation → Depolarisation
(c) Repolarisation → Threshold → Depolarisation → Hyperpolarisation
(d) Resting potential → Repolarisation → Depolarisation

59 A synaptic cleft is
(a) the space present inside the presynaptic neuron
(b) the space present inside the postsynaptic neuron
(c) the space separating the pre and postsynaptic neurons
(d) the space present inside the synaptic vesicles

60 When an action potential reaches a synaptic terminal then
(a) Ca^{2+} ion channels open (b) Na^+ ion channels open
(c) K^+ ion channels open (d) Fe^{2+} ion channels open

61 Synaptic delay is the time taken between the
(a) actual reception of a stimulus and its perception
(b) reception of a stimulus and the resultant sensory reaction
(c) release of a neurotransmitter from one neuron and stimulation of the next neuron
(d) conduction of nerve impulse across a neuron

62 Nerve impulses are transmitted across a synapse from the
(a) cell body of one neuron to the axon terminal of another
(b) dendrites of one neuron to the axon terminal of another
(c) axon terminal of one neuron to the dendrites of another
(d) dendrites of one to cell body of another

63 Node of Ranvier occurs where
(a) nerve is covered with myelin sheath
(b) neurilemma is discontinuous
(c) neurilemma and myelin sheath are discontinuous
(d) myelin sheath is discontinuous

64 Consider the following statements.
I. The imbalance in concentration of sodium and potassium ions and proteins generates the resting potential.
II. To maintain unequal distribution of sodium and potassium, neurons use electrical energy.

Choose the correct option.
(a) Statement I is true, but II is false
(b) Statement I is false, but II is true
(c) Both statements I and II are true
(d) Both statements I and II are false

65 If a nerve is stimulated in the middle, the nerve impulse in the form of electrical charges are conducted to the
(a) left side (b) right side
(c) Both (a) and (b) (d) Either (a) or (b)

66 Once the nerve has been stimulated, it cannot be stimulated again for sometime, this period is called the
(a) refractory period (b) period of excitability
(c) period of summation (d) None of these

67 Consider the following statements.
I. Increased permeability of Na^+ ions due to the rise in stimulus is short-lived.
II. The resting membrane potential of membrane at the site of excitation is restored by Na^+ and Cl^- ions.

Choose the correct option.
(a) Statement I is true, but II is false
(b) Statement I is false, but II is true
(c) Both statements I and II are true
(d) Both statements I and II are false

68 During recovery, a nerve fibre becomes
(a) + vely charged on outside and – vely charged on inside
(b) + vely charged on both outside and inside
(c) – vely charged on outside and + vely charged on inside
(d) – vely charged on both outside and inside

69 Resting membrane potential is maintained by [NCERT Exemplar]
(a) hormones (b) neurotransmitters
(c) ion pumps (d) None of these

70 In a resting nerve fibre, the concentration of K^+ ions is
(a) more in the ECF than axoplasm
(b) more in axoplasm than ECF
(c) equal in ECF and axoplasm
(d) None of the above

71 Anaesthetics reduce pain by blocking nerve conduction due to blocking of
(a) neurotransmitter receptors
(b) Na^+ channels
(c) K^+ channels
(d) All of the above

72 Consider the diagram of synapse. Identify the parts labelled as *A*, *B*, *C* and *D*.

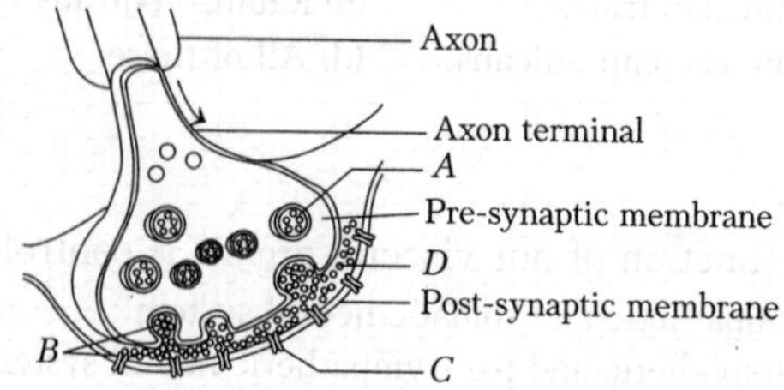

(a) A-Synaptic vesicles, B-Neurotransmitter, C-Receptor, D-Synaptic cleft
(b) A-Neurotransmitter, B-Receptor, C-Synaptic cleft, D-Synaptic vesicles
(c) A-Synaptic cleft, B-Neurotransmitter, C-Receptor, D-Synaptic vesicles
(d) A-Axon, B-Neurotransmitter, C-Receptor, D-Synaptic vesicles

73 Common feature amongst acetylcholine, nor-adrenaline and serotonin is
(a) all are anticoagulants
(b) they lower blood pressure
(c) they raise heartbeat
(d) all are neurotransmitters

74 Saltatory conduction is superior to uninterrupted conduction because of
(a) less energy required
(b) more speed
(c) less Na^+/K^+ pump
(d) All of these

75 During conduction of an impulse on the inside of plasma membrane of a neuron, electrical potential changes from
(a) – ve to + ve and remains + ve
(b) – ve to + ve and then to – ve
(c) + ve to + ve and then remains – ve
(d) + ve to – ve and then to + ve

76 Which diagram shows the distribution of Na^+ and K^+ ions in a section of non-myelinated axon which is at resting potential?

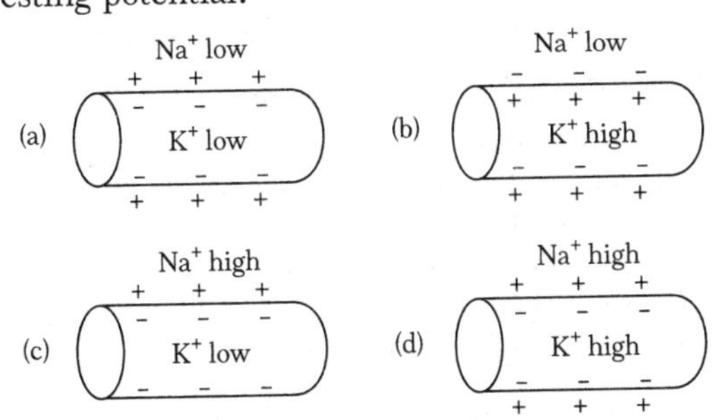

Reflex Action

77 A reflex action
(a) is an involuntary and nearly instantaneous movement in response to a stimulus
(b) involves the process of response to a peripheral nervous system
(c) requires the involvement of a part of the central nervous system
(d) All of the above

78 The knee-jerk reflex
(a) is a conditioned reflex
(b) is a monosynaptic reflex
(c) has its reflex centre in the spinal cord
(d) is mediated by a three neurons reflex arc

79 Which of the following is not involved in knee-jerk reflex? **[NCERT Exemplar]**
(a) Muscle spindle
(b) Motor neuron
(c) Brain
(d) Interneurons

80 The reflex action is
(a) blinking of eyelids
(b) swallowing of food bolus
(c) sneezing and coughing
(d) All of the above

81 An example of conditioned reflex is
(a) withdrawl of hand on touching a hot surface
(b) sneezing during cold
(c) running indoor on arrival of rain
(d) salivation in dog on seeing bread

82 Arrange the following steps in sequence with respect to reflex action and reflex arc.
I. Effector
II. Sensory of afferent neurons
III. Receptor
IV. Response
V. Stimulus
VI. Motor or Efferent neurons

Choose the correct option from the codes given below.

Codes
(a) IV → V → I → II → VI → III
(b) V → III → II → I → IV → VI
(c) V → III → I → II → IV → VI
(d) V → III → II → VI → I → IV

83 The correct route of reflex arc is
(a) effectors, grey matter, motor fibres, sensory fibres and receptors
(b) receptors, sensory fibres, grey matter and motor fibres
(c) receptors, sensory fibres, grey matter, motor fibres and effectors
(d) sensory fibres, grey matter, motor fibres, receptors and effectors

Sensory Receptors

84 The tangoreceptors are
(a) sensitive to touch
(b) present in the epidermis of skin like Merkel's corpuscles
(c) sensitive to heat
(d) Both (a) and (b)

85 Rheoreceptors are well-developed in
(a) fishes and amphibian todpoles
(b) human
(c) arthropods
(d) dogs

86 The receptors which are sensitive to blood pressure change maintain homeostasis through vasodilation or vasoconstriction are

(a) phonoreceptors (b) statoreceptors
(c) baroreceptors (d) thermoreceptors

87 Analgesic receptors are sensitive to

(a) taste buds of tongue (b) pain
(c) blood pressure change (d) sound

88 'Organ of Jacobson' helps in

(a) touch (b) vision
(c) smell (d) hear

Eye

89 Our paired eyes are located in sockets of skull called

(a) orbicules (b) orbit
(c) ocelle (d) ostiole

90 Human eyeball consists of three layers and it encloses **[NCERT Exemplar]**

(a) lens, iris, optic nerve
(b) lens, aqueous humor and viterous humor
(c) cornea, lens, iris
(d) cornea, lens, optic nerve

91 Vascular coat of eyeball is made up of

(a) sclerotic, choroid and retina
(b) optic, ciliary and iris
(c) sclerotic, cornea and conjunctiva
(d) choroid, ciliary body and iris

92 The conjunctiva

(a) covers cornea
(b) is high vascularised
(c) is transparent mucous membrane
(d) All of the above

93 Ciliary body is

(a) located at the junction of iris and choroid
(b) responsible for change in focal length of eye lens
(c) composed of epithelial tissue
(d) Both (a) and (b)

94 If the light source in front of an eye becomes bright suddenly then,

(a) focus of lens will change
(b) retinal blood supply is cut
(c) vitreous humor becomes fluid
(d) pupil will contract

95 Aqueous chamber

(a) space between the lens and the retina
(b) space between the cornea and the lens
(c) contains a thin watery fluid
(d) contains a transparent gel

96 Aqueous humor produced by the ciliary body is secreted into the posterior chamber and enters the anterior chamber through the

(a) pupil (b) scleral venous sinus
(c) vitreous body (d) suspensory ligament

97 Chamber containing vitreous humor is present

(a) in front of lens (b) behind the lens
(c) between lens and iris (d) between iris and cornea

98 Retina is the innermost layer of eye having

(a) bipolar cells and ganglion cells
(b) sensory photoreceptors cell called rods and cones
(c) point from where the optic nerve leaves the eye called blind spot
(d) All of the above

99 The direction of light striking the retina will be

(a) photosensory cells → bipolar neurons → ganglionic cells → sensory nerves
(b) sensory nerves → bipolar neurons → ganglionic cells → photosensory cells
(c) sensory nerves → ganglionic cells → bipolar neurons → photosensory cells
(d) photosensory cells → ganglionic cells → bipolar neurons → sensory nerves

100 In the central region of the retina, there is a yellowish spot, the macula lutea, with a depression in its centre that produces the sharpest vision. This depression is called

(a) optic disc (b) rods and cones
(c) vitreous body (d) fovea centralis

101 Choose the incorrect statement for the layers an adult human eye.

(a) The middle layer is called choroid which possess many blood vessels and looks deep red
(b) Choroid layer forms the ciliary body
(c) Ciliary body forms iris
(d) Retina lies inner to choroid

102 Mark the vitamin present in rhodopsin. **[NCERT Exemplar]**

(a) Vitamin-A (b) Vitamin-B
(c) Vitamin-C (d) Vitamin-D

103 Choose the incorrect matched pair about human eye.

(a) Iris – Radial and circular muscle
(b) Cornea – Abundant blood supply
(c) Vision – Binocular
(d) Lens – Biconvex

104 Choose the correct statement.

(a) Rhodopsin helps in seeing in dim light
(b) Colour blindness is due to defect in cones
(c) Human eye is most sensitive to red colour
(d) All of the above

105 The glands related to lubrication and tear production are
(a) Meibomian gland and lacrimal gland, respectively
(b) glands of Moll and glands of Zeis, respectively
(c) glands of Moll and Meiobomian glands, respectively
(d) glands of Zeis and lacrimal glands, respectively

106 In dark adaptation,
(a) only cones are involved
(b) only rods are involved
(c) Both (a) and (b)
(d) Neither rods nor cones are involved

107 Match the following columns.

	Column I (Eye defects)		Column II (Correction methods)
A.	Hypermetropia	1.	Convex lens
B.	Myopia	2.	Cylindrical lens
C.	Cataract	3.	Surgery
D.	Astigmatism	4.	Concave lens

Codes

	A	B	C	D
(a)	1	2	3	4
(b)	1	4	3	2
(c)	4	3	2	1
(d)	2	1	4	3

108 Which one is an incorrect match?
(a) Myopia — Biconvex lens
(b) Olfactory — Smell
(c) Algesic receptor — Pain
(d) Organ of Corti cells — Sensory and supporting

109 In myopia or short-sightedness,
(a) image is formed slightly in front of retina because eyeball is longer
(b) eyeball is normal but image is formed over blind spot
(c) eyeball is normal but image is formed slightly behind the retina due to faulty lens
(d) curvature of cornea becomes irregular

110 In hypermetropia, the image is formed
(a) before retina and is corrected by convex lens
(b) behind retina and is corrected by convex lens
(c) before retina and is corrected by concave lens
(d) behind retina and is corrected by concave lens

Ear

111 The external ear consists of
(a) pinna and external auditory meatus
(b) cochlea and pinna
(c) Eustachian tube and pinna
(d) All of the above

112 Eustachian canal connects
(a) middle ear with external ear
(b) middle ear with internal ear
(c) external ear with internal ear
(d) pharynx with middle ear

113 Inflammation of tympanic membrane is
(a) otalgia (b) vertigo
(c) myringitis (d) eustachitis

114 Wax gland present in the ear canal is called **[NCERT Exemplar]**
(a) sweat gland
(b) prostate gland
(c) Cowper's gland
(d) sebaceous gland/ceruminous gland

115 The organ of Corti is present in
(a) middle ear (b) Eustachian tube
(c) cochlea (d) external ear

116 Part of ear concerned with hearing is
(a) Reissner's membrane and tectorial membrane
(b) basilar membrane and tectorial membrane
(c) Reissner's membrane and basilar membrane
(d) ampulla

117 The complex system of the inner ear associated with maintenance of body balance is
(a) cochlea
(b) Reissner's membrane
(c) vestibular apparatus
(d) basilar membrane

118 Sound is transmitted from middle ear to internal ear due to
(a) vibration of tympanum (b) vibration of stapes
(c) striking of malleus (d) All of these

119 Consider the following statements.
I. The hair cells on internal side of organ of Corti are in close contact with afferent nerve fibres and possess stereocilia.
II. Saccule and utricle are the otoliths which possess macula.

Choose the correct option.
(a) Statement I is true, but II is false
(b) Statement I is false, but II is true
(c) Both statements I and II are true
(d) Both statements I and II are false

120 Choose the correct match.
(a) Eardrum — Tympanic membrane
(b) Cochlea — Receptor organ for hearing
(c) Ear fluid — Perilymph and endolymph
(d) All of the above

121 In the internal ear of humans,
(a) three semicircular canals lie in different plane at right angle to each other and suspended in perilymph
(b) crista ampullaries is formed at the base of semicircular canals as a projecting swollen ridge
(c) saccule and utricle contain projecting ridge, macula which helps in the maintenance of body balance
(d) All of the above

122 Olfactory epithelium comprises of
(a) olfactory cells + supporting cells
(b) olfactory bulb + olfactory tract
(c) olfactory nerve endings + nerves
(d) ethmoid bone

123 Identify the correct sequence of an auditory mechanoreceptor organ in human ear.
(a) Pinna-Cochlea-Tympanic membrane canal-Malleus-Stapes-Incus-Auditory nerve
(b) Pinna-Tympanic membrane-Auditory canal-Incus-Malleus- Stapes-Cochlea-Auditory nerve
(c) Pinna-Malleus-Incus-Stapes-Auditory canal-Tympanic membrane-Cochlea-Auditory nerve
(d) Pinna-Auditory canal-Tympanic membrane-Malleus-Incus-Stapes-Cochlea-Auditory nerve

124 Choose the correct statement.
(a) Largest ear ossicle is malleus
(b) Structure which is indirect contact with oval window is stapes
(c) Structure which is indirect contact with tympanic membrane is malleus
(d) All of the above

125 Choose the incorrect match.
(a) Meniere's disease - Abnormality is ear fluid
(b) Vertigo - Dizziness
(c) Otalgia - Ear pain
(d) Otitis media - Inflammation of tympanic membrane

Skin

126 Consider the following statements.
I. Pacinian corpuscles are encapsulated pressure receptors.
II. Meissner's corpuscles are located in skin and perceive gentle pressure.

Choose the correct option.
(a) Statement I is true, but II is false
(b) Statement I is false, but II is true
(c) Both statements I and II are true
(d) Both statements I and II are false

127 The receptors present on skin which are sensitive to cold are
(a) Ruffini's corpuscle
(b) Meissner's corpuscle
(c) Krause's corpuscle
(d) free nerve endings

128 Proprioceptors are those, which give the sense of
(a) chemicals
(b) temperature
(c) taste
(d) changes in the internal environment of the body

129 Organs of Ruffini are the receptors of
(a) cold (b) pressure
(c) heat (d) touch

130 Vibrissae are associated with the function of
(a) thermoregulation (b) gustation
(c) tactile perception (d) reproduction

131 Receptors for initial contact and movement of object over the skin are
(a) Pacinian corpuscles (b) hair end organs
(c) Merkel's discs (d) Ruffini's corpuscles

(B) Cumulative Exercise

(Assorted Questions of the Chapter for Advanced Level Practice)

1 An action potential in the nerve fibre is produced when positive and negative charges on the outside and the inside of the axon membrane are reversed, because
(a) more potassium ions enter the axon as compared to sodium ions leaving it
(b) more sodium ions enter the axon as compared to potassium ions leaving it
(c) all potassium ions leave the axon
(d) all sodium ions enter the axon

2 During the transmission of nerve impulse through a nerve fibre, the potential on the inner side of the plasma membrane has which type of electric charge?
(a) First negative then positive and again back to negative
(b) First positive then negative and continue to be negative
(c) First negative then positive and continue to be positive
(d) First positive then negative and again back to positive

3 Match the following columns.

Column I	Column II
A. Fovea	1. Provides opening for entry of light
B. Iris	2. Transduces RGB light
C. Pupil	3. Transmits information of CNS
D. Lens	4. Controls amount of light entering
E. Optic nerve	5. Focuses light on the retina

Codes

	A	B	C	D	E		A	B	C	D	E
(a)	4	1	5	2	3	(b)	2	4	1	5	3
(c)	3	1	2	4	5	(d)	4	3	2	5	1

4 Which one of the following is the correct difference between rod cells and cone cells of our retina?

Features	Rod cells	Cone cells
(a) Visual acuity	High	Low
(b) Visual pigment contained	Iodopsin	Rhodopsin
(c) Overall function	Vision in poor light	Colour vision and detailed vision in bright light
(d) Distribution	More concentrated in centre of retina	Evenly distributed all over retina

5 Match the following columns.

Column I	Column II
A. Frontal lobes	1. Responsible for understanding, spoken or written language
B. Occipital lobes	2. Contain primary visual cortex
C. Parietal lobes	3. Control voluntary contractions of skeletal muscles
D. Pimary motor areas	4. Contain somatic sensory areas
E. Somatic sensory areas	5. Receive afferent messages from skin and body surfaces

Codes

	A	B	C	D	E		A	B	C	D	E
(a)	1	2	4	3	5	(b)	1	2	3	4	5
(c)	5	4	3	2	1	(d)	4	3	1	2	5

6 Which of the following statement(s) is/are correct?

I. Two transmitter substances in the nervous system are dopamine and acetylcholine.

II. Motor neurons convey information from receptors in the periphery to the CNS.

III. Somatic motor nerves innervate skeletal muscles and autonomic nerves innervate smooth muscle, cardiac muscle and glands.

Chosse the correct option from the codes given below.

Codes

(a) I and II (b) II and III
(c) Only III (d) I and III

7 Match the following columns.

Column I	Column II
A. Epilespy	1. Degeneration of neurons on the cerebral cortex
B. Alzheimer's disease	2. Irregular electrical discharge in the neurons
C. Parkinson's disease	3. Decreased production of acetylcholine
D. Huntington's chorea	4. Death of dopamine releasing neurons
	5. Formation of blood clots in the brain

Codes

	A	B	C	D		A	B	C	D
(a)	5	4	3	1	(b)	2	3	1	4
(c)	2	3	4	1	(d)	2	4	3	1

8 Read the following statements.

I. Alongwith hypothalamus, limbic lobe/system is involved in the regulation of sexual behaviour, expression of emotional reactions and motivation.

II. Hypothalamus lies at the base of thalamus.

III. Hypothalamus contains a number of centres which control body temperature, urge for eating and drinking.

IV. It contains several groups of neurosecretory cells, which secrete hypothalamic hormones.

V. The inner parts of cerebral hemispheres and a group of associated structures like amygdala, hippocampus, etc., forms limbic system.

How many of the above statements are correct?

(a) Five (b) Four (c) Two (d) Three

9 Find out which of the following statements are true (T)/false (F) and choose the correct option.

I. The medulla oblongata contains centres which control respiration, cardiovascular reflexes and gastric secretions.

II. Pons consists of fibre tracts that interconnect different regions of the brain.

III. Cerebellum has very convoluted surface to provide the additional space for many more neurons.

IV. The medulla of brain is connected to the spinal cord.

Options	I	II	III	IV
(a)	T	T	T	F
(b)	T	T	T	T
(c)	F	T	T	T
(d)	T	F	T	F

10 Which of the following is incorrect?

(a) Brain depends on blood for the supply of oxygen and glucose

(b) Satiety centre of brain is present in hypothalamus

(c) Cerebellum of brain is associated with maintaining posture, body equilibrium orientation controlling and regulating tone

(d) Right cerebral hemisphere is smaller than left one

11 Read the following statements.

I. The arbor vitae is cerebellar white matter and has tree-like appearance.
II. Purkinje cells are flask-shaped complex neurons located in the cerebellar cortex of brain.
III. Vermis is worm-like central portion of cerebrum.
VI. One-third of the cranial nerves originates from medulla.
V. The central canal of spinal cord connects with the ventricles of the brain and is filled with CSF.

Which of the above statement (s) is/are incorrect?

(a) Only III (b) I, II and IV
(c) IV and V (d) II, III and IV

12 Acquired reflex is

I. first learned and then causes reflex action
II. unlearned, inborn reflex
III. controlled by spinal cord and brain
IV. monosynaptic reflex as only synapse is found

Choose the correct option from the codes given below.

(a) Only I (b) Only II
(c) I, II and III (d) I and III

13 During a reflex action, the afferent neuron receives signal from

(a) a sensory organ and transmits the impulse *via* dorsal root ganglion into the spinal cord
(b) an effector and transmits the impulse *via* motor end plate into spinal cord
(c) a motor end plate and transmits the impulse *via* an effector into the spinal cord
(d) dorsal root ganglion and transmits the impulse *via* a sensory organ

14 Read the following statements.

I. The rods contain a purpulish-red protein called rhodopsin or visual purple.
II. Rhodopsin/visual purple is a derivative of vitamin-A.
III. Cones respond to blue, red and green lights.
VI. When three different types of cones are stimulated equally then a sensation of blue, red and green light is produced in sequence.
V. Cones are photoreceptor cells, rods are not.

Which of the above statements are incorrect?

(a) IV and V (b) I, II and IV
(c) III, IV and V (d) II and III

15 The transduction of sound waves into action potentials takes place.

(a) When hair cells are bent against the tectorial membrane, causing them to depolarise and release neurotransmitter that stimulates sensory neurons
(b) Within the tectorial membrane as it is stimulated by the hair cells
(c) Within the middle ear as the vibrations are amplified by the ear ossicles
(d) As the basilar membrane vibrates at different frequencies in response to the varying volume of sounds

16 Find out which of the following statements are true (T)/false (F) and choose the correct option.

I. It is difficult to differentiate between green and red objects in dark as cones function properly in the light.
II. The innermost layer of our eye is retina.
III. Rods and cones of eye are actually the modified bipolar neurons.
IV. Yellow spot, blind spot and fovea are present on cornea.

Options	I	II	III	IV
(a)	T	T	T	T
(b)	F	F	T	F
(c)	T	T	F	F
(d)	T	T	T	F

17 Identify the incorrectly matched pair.

(a) Bony labyrinth – Surrounded by perilymph, filled with endolymph
(b) Scala media – Space within cochlea, filled with endolymph
(c) Scala vestibuli – Found at the phase of cochlea, ends at oval window
(d) Scala tympani – Terminate at round window which open to middle ear

18 Consider the following statements about the sympathetic division of the ANS.

I. All its neurons release nor-epinephrine as their primary neurotransmitter substance.
II. All the cell bodies of its post-ganglionic neurons lie in or near the organ innervated.
III. The cell bodies of its pre-ganglionic neurons lie in the thoracic and lumbar spinal cord.

Choose the option containing correct statements.

(a) Only I (b) Only II
(c) Only III (d) I and II

19 Consider the following statements about the parasympathetic division of the ANS.

I. All its neurons release acetylcholine as their primary neurotransmitter substance.
II. The cell bodies of its post-ganglionic neurons lie in or near the organ innervated.
III. The cell bodies of its pre-ganglionic neuons lie in the cervical and sacral spinal cord.

Choose the option containing correct statements.

(a) I, II and III (b) II and III
(c) I and II (d) None of these

20 Study the following figure.

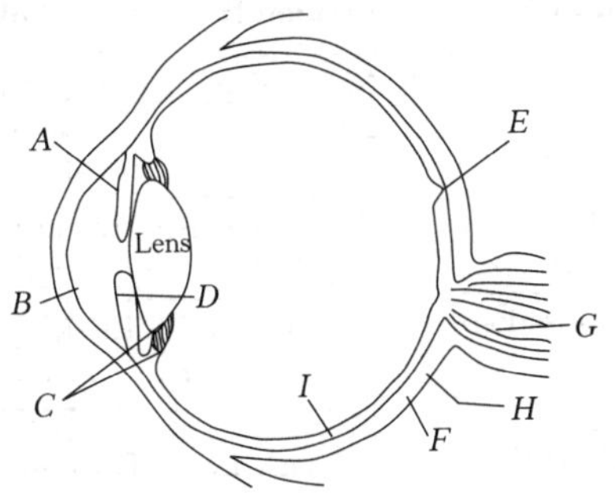

I. Carries nerve signals to the brain.
II. Regulates the size of the pupil to let more or less light into the eye.
III. Changes the shape of the lens.
IV. Photoreceptors are highly concentrated at this centre of focus.

The correct match of the above functions with parts of the eye indicated by letters is

	I	II	III	IV
(a)	*G*	*D*	*C*	*I*
(b)	*A*	*C*	*E*	*G*
(c)	*I*	*G*	*F*	*C*
(d)	*B*	*D*	*F*	*H*

21 The correct sequence of sensory processing is
(a) Sensory adaptation → Stimulus reception → Sensory transduction → Sensory perception
(b) Stimulus reception → Sensory transduction → Sensory perception → Sensory adaptation
(c) Sensory perception → Stimulus reception → Sensory transduction → Sensory adaptation
(d) Sensory perception → Sensory transduction → Stimulus reception → Sensory adaptation

22 An investigator places an isolated neuron in a calcium-free medium gives the neuron a suprathreshold stimulus and then performs an assay to test whether neurotransmitter is released into the medium. Which of the following outcomes would you predict?
(a) No neurotransmitter is detected since influx of calcium into the synaptic knob is required for neurotransmitter release
(b) No neurotransmitter is detected since influx of calcium is required in order for the neuron to conduct an action potential
(c) Neurotransmitter is detected since calcium is not required for action potential conduction and the initial stimulus was suprathreshold
(d) We cannot predict the outcome without knowing whether the neuron was myelinated or not

23 The given diagram shows the parasympathetic innervation of smooth muscle and glands of the stomach.

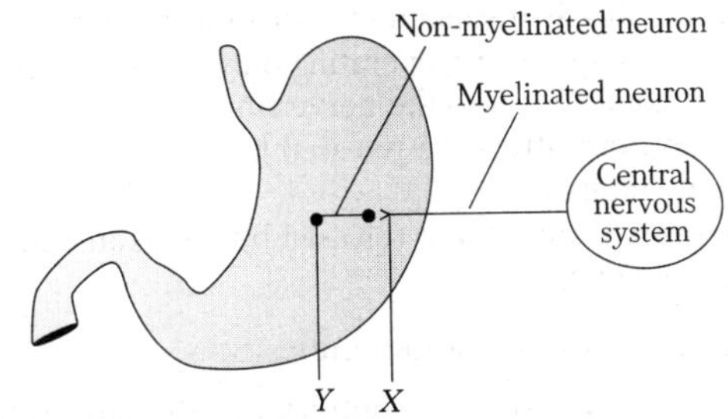

Select the right option in which the chemical transmitters are released at the synapses *X* and *Y* are correctly identified.
(a) *X*- Nor-adrenaline, *Y*- Calcium ions
(b) *X*- Acetylcholine, *Y*- Nor-adrenaline
(c) *X*- Acetylcholine, *Y*- Calcium ions
(d) *X*- Acetylcholine, *Y*- Acetylcholine

24 Identify *A*, *B*, *C*, *D* and *E* with respect to transmission of nerve impulse

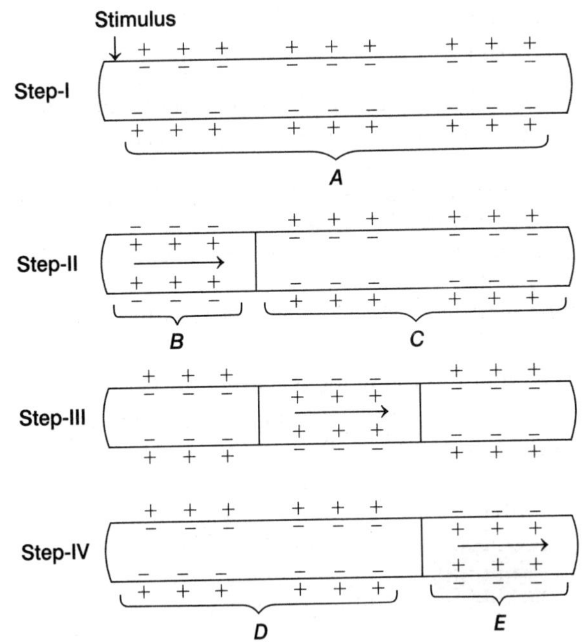

	A	*B*	*C*	*D*	*E*
(a)	Depolarised membrane	Depolarised region	Polarised region	Repolarised region	Depolarised region
(b)	Polarised membrane	Polarised region	Depolarised region	Repolarised region	Depolarised region
(c)	Polarised membrane	Depolarised region	Polarised region	Depolarised region	Polarised region
(d)	Polarised membrane	Depolarised region	Polarised region	Repolarised region	Depolarised region

25 Unidirectional transmission of a nerve impulse through nerve fibre is due to the fact that

(a) nerve fibre is insulated by a medullary sheath
(b) sodium pump starts operating only at the cyton and then continues into the nerve fibre
(c) neurotransmitters are released by dendrites and not by axon endings
(d) neurotransmitters are released by the axon endings and not by dendrites

26 Read the following statements.

I. Olfactory, optic and auditory are the pure sensory cranial nerves.
II. Trigeminal is the longest cranial nerve.
III. The vagus or pneumogastric nerve is a mixed nerve that supplies the pharynx, larynx, lungs, heart, oesophagus, stomach and most of abdominal viscera.
IV. Trigeminal nerve supplies most of the muscles of face and is thus responsible for facial expressions.

Which of the above statements are correct?

(a) I and II
(b) II and III
(c) I and III
(d) II and IV

27 Which of the following is an incorrect statement?

(a) Choroid plexus is linked with secretion of cerebrospinal fluid
(b) Comprehension of speech occurs in left temporal lobe in the Wernicke's area
(c) A learning disorder characterised by difficult reading is called dyslexia
(d) Somaesthetic area of parietal lobe controls voluntary movements

28 Which of the following is correct for the diagram I and II with respect to impulse conduction through an axon at points A and B.

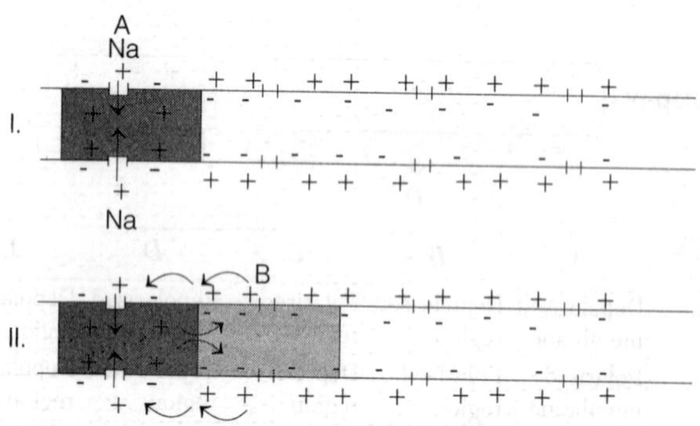

(a) **Diagram-I** When stimulus is applied at point A on the polarised membrane, the membrane at site-A undergo active transport of Na^+
(b) **Diagram-II** Axonic membrane throughout the axon is in polarised state
(c) **Diagram-I** An actional potential is initiating at site-A
(d) **Diagram-II** Current is flowing on the outer surface from site-A to site-B

29 Match the following columns.

Column I	Column II	Column III
A. Resting membrane potential	1. Na^+ channel gets open	I. Na^+ and K^+ pumps are responsible for it
B. Action potential	2. Na^+ channel is closed	II. Last for very short time
C. Depolarisation	3. Na^+ ions are more on outer side	III. K^+ ions move on outer side of membrane
D. Repolarisation	4. Na^+ ions are more on inner side of membrane	IV. Positive charge on inner side of membrane

	A	B	C	D
(a)	2-IV	1-III	3-I	4-II
(b)	3-I	4-II	1-IV	2-III
(c)	4-II	3-I	1-II	2-IV
(d)	4-I	3-II	2-III	1-III

30 Medulla oblongata controls

I. breathing and blood pressure
II. gut peristalsis and gland secretion
III. vasodilation and vasoconstriction
IV. laughing and micturition

Codes

(a) Only I
(b) II and III
(c) I, II and III
(d) Only IV

31 The given diagram represent reflex action shows knee-jerk reflex.

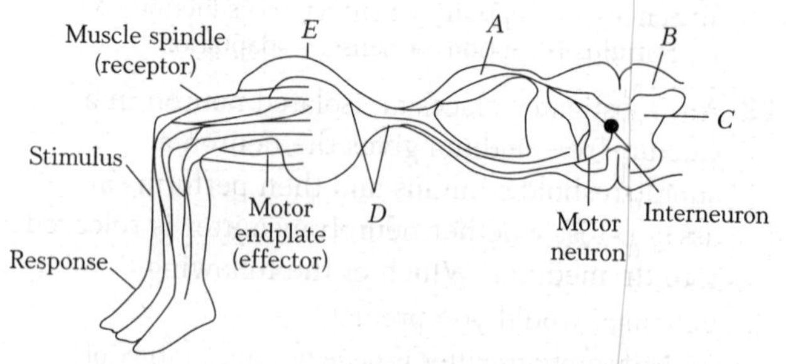

In which of the following options correct words for all the five blanks (A-E) are illustrated?

	A	B	C	D	E
(a)	Ventral root ganglion	White matter	Grey matter	Efferent pathway	Afferent pathway
(b)	Dorsal root ganglion	Grey matter	White matter	Efferent pathway	Afferent pathway
(c)	Dorsal root ganglion	White matter	Grey matter	Efferent pathway	Afferent pathway
(d)	Dorsal root ganglion	White matter	Grey matter	Afferent pathway	Efferent pathway

32 Following are the steps of mechanism of vision in random order.

I. Neural impulses are analysed and image formed on retina is recognised by visual cortex.
II. Membrane permeability changes.
III. Ganglion cells are excited.
IV. Bipolar cells are depolarised.
V. Action potential (impulse) is transmitted by optic nerves to visual cortex.
VI. Potential differences are generated in the photoreceptor cells.
VII. Light energy causes a change in shape of rhodopsin, leading to the dissociation of retinal (an aldehyde of vitamin-A) from opsin (a protein).
VIII. Structure of opsin is changed.

Choose the correct sequence.

(a) I, II, III, IV, V, VI, VII, VIII
(b) VIII, VII, VI, V, IV, III, II, I
(c) I, IV, III, II, VII, VIII, VI, V
(d) VII, VIII, II, VI, IV, III, V, I

33 The given diagram is the lateral view of the human brain, parts are indicated by alphabets. Select the answer in which these alphabets have been correctly paired with the parts which they indicate.

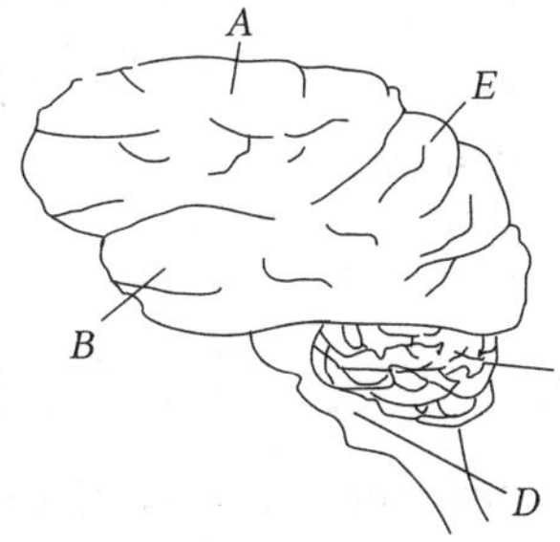

(a) *A*-Frontal lobe, *B*-Temporal lobe, *C*-Cerebellum, *D*-Medulla oblongata, *E*-Parietal lobe
(b) *A*-Temporal lobe, *B*-Parietal lobe, *C*-Cerebrum, *D*-Medulla oblongata, *E*-Frontal lobe
(c) *A*-Frontal lobe, *B*-Temporal lobe, *C*-Cerebrum, *D*-Medulla oblongata, *E*-Occipital lobe
(d) *A*-Temporal lobe, *B*-Parietal lobe, *C*-Cerebellum, *D*-Medulla oblongata, *E*-Frontal lobe

34 Na^+–K^+ pump is found in membranes of many cells, like nerve cells. It works against electrochemical gradient and involves an integral protein ATPase. For each molecule of ATP used

(a) three ions of Na^+ are pumped out and two K^+ are taken in
(b) three ions of Na^+ are taken in and two K^+ are pumped out
(c) two ions of Na^+ are thrown out and three K^+ are absorbed
(d) three ions of K^+ are absorbed and three Na^+ are pumped out

35 Match the following columns.

Column I	Column II
A. Cerebrum	1. Controls the pituitary
B. Cerebellum	2. Controls vision and hearing
C. Hypothalamus	3. Controls the rate of heartbeat
D. Midbrain	4. Seat of intelligence
	5. Maintains body posture

Codes

	A	B	C	D
(a)	5	4	2	1
(b)	4	5	2	1
(c)	5	4	1	2
(d)	4	5	1	2

36 A patient schedule 'for surgery confides in his nurse the night' before that he is 'terribly scared'. Which of the following indicate(s) increased sympathetic activity in this patient?

(a) Patient complains that his mouth feels dry
(b) Patient's gown is moist with perspiration
(c) Patient appears pale
(d) All of the above

37 Regarding cerebrospinal fluid, which of the following is a false statement?

(a) It has a specific gravity of 1.007 and buoys the brain
(b) It maintains a volume of 140 to 200 mL and a fluid pressure of 10 mm Hg
(c) It moves metabolic wastes away from the cells of nervous tissue
(d) It is produced in the choroid plexuses and drains into the cerebral arterial circle

38 See the following and identify *A* to *H*.

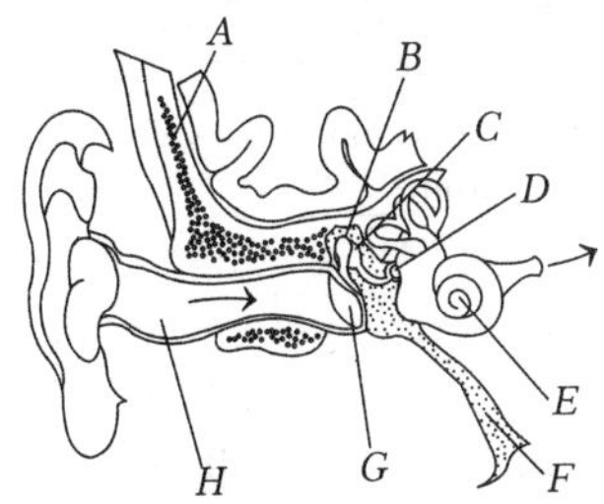

(a) *A*-Temporal bone, *B*-Malleus, *C*-Incus, *D*-Cochlea, *E*-Stapes, *F*-Eustachian tube, *G*-Tympanic membrane, *H*-External auditory canal
(b) *A*-Temporal bone, *B*-Incus, *C*-Malleus, *D*-Stapes, *E*-Cochlea, *F*-Eustachian tube, *G*-Tympanic membrane, *H*-External auditory canal
(c) *A*-Tympanic membrane, *B*-Malleus, *C*-Incus, *D*-Stapes, *E*-Cochlea, *F*-Eustachian tube, *G*-Temporal bone, *H*-External auditory canal
(d) *A*-Temporal bone, *B*-Malleus, *C*-Incus, *D*-Stapes, *E*-Cochlea, *F*-Eustachian tube, *G*-Tympanic membrane, *H*-External auditory canal

39 Which is the proper sequence of visual sensory transmission from stimulation of photoreceptors located on the medial side of the retina?

(a) Optic nerve, lateral geniculate body, optic-radiation, optic tract, cerebral cortex
(b) Optic nerve, optic chiasma, lateral geniculate body, optic tract, cerebral cortex, optic radiation
(c) Optic nerve, optic chiasma, optic tract, lateral geniculate body, optic radiation, cerebral cortex
(d) Optic nerve, optic tract, lateral geniculate body, optic radiation, cerebral cortex

40 Given below is a diagrammatic cross-section of a single loop of human cochlea.

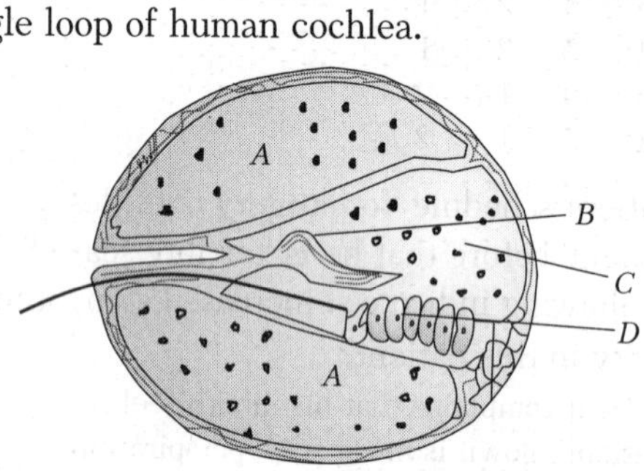

Which one of the following options correctly represents the names of three different parts?

(a) *D*-Sensory hair cells, *A*-Endolymph, *B*-Tectorial membrane
(b) *A*-Perilymph, *B*-Tectorial membrane, *C*-Endolymph
(c) *B*-Tectorial membrane, *C*-Perilymph, *D*-Secretory cells
(d) *C*-Endolymph, *D*-Sensory hair cells, *A*-Serum

41 Match the following columns.

Column I	Column II
A. Synaptic vesicles	1. Resting potential
B. Electrical potential difference across the resting plasma membrane	2. Action potential
C. Generation of a nerve impulse	3. Neurotransmitter
D. Granular bodies found in the cell body of a neuron	4. Nissl's granules

Codes

	A	B	C	D		A	B	C	D
(a)	3	1	2	4	(b)	2	3	1	4
(c)	4	3	2	1	(d)	1	4	3	2

42 Which of the following statements are correct regarding electrical synapses?

I. Pre and post-synaptic membrane neurons are in very close proximity at electric synapse.
II. Electric currents are involved in the transmission of impulses.
III. Transmission of an impulse across electrical synapses is very similar to impulse conduction along a single axon.
IV. Impulse transmission is always faster in electric synapse than that across a chemical synapse.
V. Electrical synapses are rare in our system.

Choose the correct option.

(a) I, II, III and IV
(b) I, III, IV and V
(c) I, II and IV
(d) I, II, III, IV and V

43 Which of the following statements is/are correct about resting state?

I. Neuron is not conducting any impulse.
II. Plasma membrane is electrically positive outside and negative inside.
III. The nerve fibre is stimulated and conducts mechanical or electrical impulse.
IV. Plasma membrane is negative outside and positive inside.

Choose the correct option.

(a) III and IV
(b) I and IV
(c) II and III
(d) I and II

44 Which of the following statement(s) is/are incorrect regarding synapse?

I. Synapse is a junction between axon endings of one nerve fibre and dendrite of the other.
II. At a synapse, the membrane of axon and dendrite are not in physical contact with each other.
III. Transmission of nerve impulse through synapse is only one way.
IV. Neurotransmitter is always released from axon endings and not by dendrites.

Select the correct option.

(a) Only IV
(b) Only I
(c) I, II and III
(d) None of these

45 Consider the statements as True/False regarding when a neuron is at rest and no impulse is conducting.

I. The axoplasm inside the axon contains high concentration of K^+ and negatively charged proteins.
II. The axoplasm inside the axon contains low concentration of Na^+.
III. The fluid outside the axon contains a low concentration of K^+.
IV. The fluid outside the axon contains a low concentration of Na^+ and negatively charged proteins.

The correct option is

	I	II	III	IV
(a)	True	False	False	True
(b)	True	True	False	False
(c)	True	True	True	False
(d)	False	True	False	False

46 Arrange the following events in a correct order that lead to the formation of an auditory impulse in human ears from the steps given below.

I. Vibration is transferred from the malleus to the incus and then to stapes.
II. Basilar membrane moves up and down.
III. Nerve impulse is transmitted by cochlear nerve to auditory cortex of brain for impulse analysis and recognition.
IV. Sound waves pass through ear canal.
V. Stereocilia of hair cells of organ of Corti rub against tectorial membrane.
VI. Sound waves cause eardrum to vibrate.
VII. Nerve impulse is generated.
VIII. Vibrations move from fluid of vestibular canal to the fluid of tymapanic canal.
IX. Membrane at oval window vibrates.

Codes
(a) IV, VI, I, IX, VIII, II, V, VII, III
(b) I, II, III, IV, V, VI, VII, VIII, IX
(c) IX, VIII, VII, VI, V, IV, III, II, I
(d) IV, VI, I, VIII, IX, II, V, VII, III

Assertion and Reason

Direction (Q.Nos.47-55) *In each of the following questions, a statement of Assertion (A) is given followed by corresponding statement of Reason (R). Of the statements, mark the correct answer as*

(a) Both A and R are true and R is the correct explanation of A
(b) Both A and R are true, but R is not the correct explanation of A
(c) A is true, but R is false
(d) A is false, but R is true
(e) Both A and R are false

47 Assertion (A) Nerves arising from the spinal cord are said to be sensory nerves.

Reason (R) A nerve confining only sensory nerve fibres is called a sensory nerve.

48 Assertion (A) After hearing a sound, nerve impulse passes from neurons to the brain.

Reason (R) The neurons which pass nerve impulses from the body organ to the brain are called afferent neurons.

49 Assertion (A) In man, only peripheral nervous system is present.

Reason (R) The peripheral nervous system includes nerves coursing between the central nervous system and different parts of the body.

50 Assertion (A) Medulla oblongata causes reflex actions like vomiting, coughing and sneezing.

Reason (R) It has many nerve cells, which control these autonomic reflexes.

51 Assertion (A) The pre-synaptic neuron transmits an impulse (action potential) across the synaptic cleft to the post-synaptic neuron.

Reason (R) For the transmission of impulses at synapses, chemicals called neurotransmitters are responsible.

52 Assertion (A) Neuroglial cells are known as the packing cells of brain.

Reason (R) A type of neuroglial cells forms the myelin sheath around axon.

53 Assertion (A) The axoplasm inside the axon contains high concentration of K^+ and negatively charged proteins.

Reason (R) The axonal membrane is impermeable to Na^+ and negatively charged proteins.

54 Assertion (A) The cerebral cortex is referred to as the white matter of the brain.

Reason (R) It is due to the high concentration of neuron cell bodies in it.

55 Assertion (A) Iris muscles show both dilation as well as constriction of pupil.

Reason (R) It is due to sympathetic and parasympathetic nervous system which have antagonistic functions.

(C) Medical Entrances' Gallery

(Collection of Questions Asked in NEET & in Various Medical Entrance Exams)

1 Match the following columns and select the correct option.

Column I	Column II
A. Organ of Corti	1. Connects middle ear and pharynx
B. Cochlea	2. Coiled part of the labyrinth
C. Eustachian tube	3. Attached to the oval window
D. Stapes	4. Located on the basilar membrane

[(NEET 2020]

Codes

	A	B	C	D		A	B	C	D
(a)	3	1	4	2	(b)	4	2	1	3
(c)	1	2	4	3	(d)	2	3	1	4

2 Match the following columns and select the correct option from the codes given below.

Column I	Column II
A. Rods and cones	1. Absence of photoreceptor cells
B. Blind spot	2. Cones are densely packed
C. Fovea	3. Photoreceptor cells
D. Iris	4. Visible coloured portion of eye

[(NEET 2020]

Codes

	A	B	C	D		A	B	C	D
(a)	3	1	2	4	(b)	2	3	1	4
(c)	3	4	2	1	(d)	2	4	3	1

3 Which part of the brain is responsible for thermoregulation? [NEET 2020, 19, 13, JIPMER 2019, NEET 2018]

(a) Hypothalamus (b) Corpus callosum
(c) Medulla oblongata (d) Cerebrum

4 Which of the following is not considered as a secondary messenger? [JIMPER 2019]

(a) Acetylcholine (b) cAMP
(c) Ca^{2+} (d) Diacylglycerol

5 Which of the following receptors are specifically responsible for maintenance of the balance of body and posture? [NEET 2019]

(a) Basilar membrane and otoliths
(b) Hair cells and organ of Corti
(c) Tectorial membrane and macula
(d) Crista ampullaris and macula

6 Which of the following statements is not correct? [NEET 2019]

(a) An action potential in an axon does not move backward because the segment behind is in a refractory phase
(b) Depolarisation of hair cells of cochlea results in the opening of the mechanically gated potassium-ion channels
(c) Rods are very sensitive and contribute to daylight vision
(d) In the knee-jerk reflex, stimulus is the stretching of muscle and response in its contraction

7 Which of the following statements is correct? [NEET 2019]

(a) Cornea consists of dense connective tissue of elastin and can repair itself
(b) Cornea is convex, transparent layer which is highly vascularised
(c) Cornea consists of dense matrix of collagen and is the most sensitive portion of the eye
(d) Cornea is an external, transparent and protective proteinaceous covering of the eyeball

8 Which of the following structures or regions is incorrectly paired with its function? [NEET 2018]

(a) Hypothalamus	–	Production of releasing hormones and regulation of temperature, hunger and thirst.
(b) Limbic system	–	Consists of fibre tracts that interconnect different regions of brain, controls movement.
(c) Medulla oblongata	–	Controls respiration and cardiovascular reflexes.
(d) Corpus callosum	–	Band of fibres connecting left and right cerebral hemispheres.

9 The transparent lens in the human eye is held in its place by [NEET 2018]

(a) smooth muscles attached to the iris
(b) ligaments attached to the iris
(c) ligaments attached to the ciliary body
(d) smooth muscles attached to the ciliary body

10 Which one is an incorrect match? [NEET 2017]

(a) Glucoma–Abnormal high pressure on liquid of eye
(b) Eustachian tube–Connects middle ear cavity with pharynx
(c) Caloreceptor–Heat
(d) Interoceptor–Touch

11 Receptor sites for neurotransmitters are present on [NEET 2017]

(a) membranes of synaptic vesicles
(b) pre-synaptic membrane
(c) tips of axons
(d) post-synaptic membrane

12 The cavity of diencephalon is known as **[NEET 2017]**
(a) first ventricle (b) second ventricle
(c) third ventricle (d) fourth ventricle

13 **Assertion** (A) Arrival of an impulse at the axon terminal stimulates the release of neurotransmitters in synaptic cleft.
Reason (R) These neurotransmitters are responsible for the opening of ion channels. **[AIIMS 2017]**
(a) Both A and R are true and R is the correct explanation of A
(b) Both A and R are true, but R is not the correct explanation of A
(c) A is true, but R is false
(d) Both A and R are false

14 Choose the correct statement. **[NEET-II 2016]**
(a) Nociceptors respond to changes in pressure
(b) Meissner's corpuscles are thermoreceptors
(c) Photoreceptors in the human eye are depolarised during darkness and become hyperpolarised in response to the light stimulus
(d) Receptors do not produce graded potentials

15 Photosensitive compound in human eye is made up of **[NEET-II 2016]**
(a) opsin and retinal
(b) opsin and retinol
(c) transducin and retinene
(d) guanosine and retinol

16 Destruction of the anterior horn cells of the spinal cord would result in loss of **[CBSE AIPMT 2015]**
(a) sensory impulses (b) voluntary motor impulses
(c) commissural impulses (d) integrating impulses

17 Which one of the following regions of the brain is incorrectly paired with its function? **[CBSE AIPMT 2015]**
(a) Medulla oblongata – Homeostatic control
(b) Cerebellum – Language comprehension
(c) Corpus callosum – Communication between the left and right cerebral cortices
(d) Cerebrum – Calculation and contemplation

18 A gymnast is able to balance his body upside down even in the total darkness because of **[CBSE AIPMT 2015]**
(a) cochlea (b) vestibular apparatus
(c) tectorial membrane (d) organ of Corti

19 The pneumotaxic centre and rhythm centre are respectively present in **[AMU 2015, Kerala CEE 2011]**
(a) pons and medulla oblongata
(b) corpus callosum and pons
(c) medulla oblongata and hypothalamus
(d) diencephalon and pons

20 Read the following statements and find the correct ones. **[EAMCET 2015]**
I. Midbrain, pons Varolii and medulla oblongata are together called brain stem.
II. Paracoels are connected to diocoel through iter.
III. Heartbeat, respiration, swallowing, cough related centres lie in medulla oblongata.
IV. Cerebellum is also called gyroscope of the body.
(a) I, II and III (b) I, III and IV
(c) II, III and IV (d) I, II and IV

21 The bundle of axons in the central nervous system is known as **[WB JEE 2015]**
(a) nerve (b) ganglion
(c) tract (d) neuron

22 Which of the following statements is false? **[Uttarakhand PMT 2015]**
(a) Hindbrain contains medulla, pons and cerebellum
(b) Thalamus is an important sensory relay structure
(c) Cerebral cortex in humans is folded into convolutions
(d) Hypothalamus has an important role in recognising and experiencing

23 Which group of cranial nerves control eyeball movements? **[MHT CET 2015]**
(a) Optic, abducens, pathetic
(b) Optic, oculomotor, trochlear
(c) Oculomotor, abducens, auditory
(d) Oculomotor, abducens, trochlear

24 **Assertion** (A) Nerve fibre can become excited through touch, smell, pressure and chemical changes and there is a change in polarity.
Reason (R) It is called active potential. **[Gujarat CET 2015]**
(a) Both A and R are true and R is the correct explanation of A
(b) Both A and R are true, but R is not the correct explanation of A
(c) A is true, but R is false
(d) A is false, but R is true

25 In EEG, the waves which are low in frequency but with high amplitude are **[EAMCET 2014]**
(a) theta waves (b) delta waves
(c) beta waves (d) alpha waves

26 Choose the functions of sympathetic nervous system. **[EAMCET 2014]**
(a) Constricts bronchi and pupil of eye
(b) Increases heart rate, relaxes bronchi
(c) Decreases heart rate, increases peristalsis
(d) Dilates blood vessels, stimulates salivary secretions

27 Depolarisation of nerve membrane takes place through influx of ...... ions. **[MHT CET 2014]**
(a) calcium (b) potassium
(c) sodium (d) magnesium

28 Synapse is .... **[MHT CET 2014]**
(a) crossing over between non-homologous chromosomes
(b) pairing of homologous chromosomes
(c) junction between axon and dendrite of two different neurons
(d) zig-zag junction in cardiac muscle fibres

29 Which is wrong regarding conduction of nerve impulse? **[Kerala CEE 2014]**
(a) In a resting neuron, the axonal membrane is more permeable to K^+ ions and nearly impermeable to Na^+ ions
(b) Fluid outside the axon has a high concentration of Na^+ and low concentration of K^+, in a resting neuron
(c) Ionic gradients are maintained by Na-K pumps across the resting membrane which transports 3 Na^+ ions outwards for 2 K^+ ions into the cell
(d) Resting potential is the electrical potential difference across the resting membrane
(e) A neuron is polarised only when the outer surface of axonal membrane possesses a negative charge and inner surface is positively charged

30 How do parasympathetic neural signals affect the working of the heart? **[CBSE AIPMT 2014]**
(a) Heart rate increases without affecting cardiac output
(b) Both heart rate and cardiac output increase
(c) Heart rate decreases but cardiac output increases
(d) Reduce in both heart rate and cardiac output

31 Injury localised to hypothalamus would most likely disrupt **[CBSE AIPMT 2014]**
(a) coordination during locomotion
(b) executive functions, such as decision making
(c) regulation of body temperature
(d) short term memory

32 A diagram showing axon terminal and synapse is given. Identify correctly at least two of *A-D*. **[NEET 2013]**

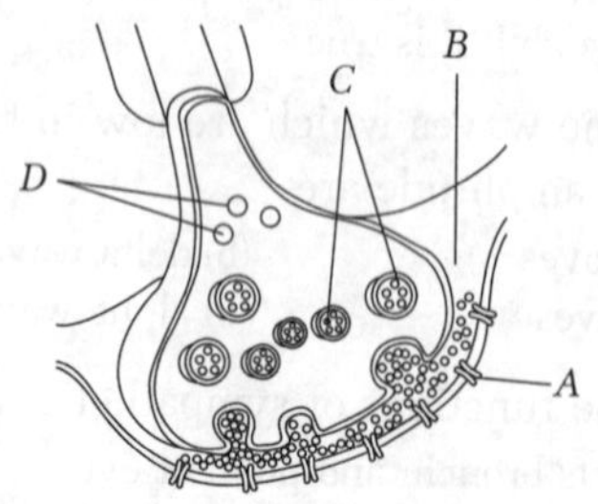

(a) *A*-Receptor, *C*-Synaptic vesicles
(b) *B*-Synaptic connection, *D* -K^+
(c) *A*-Neurotransmitter, *B* -Synaptic cleft
(d) *C*-Neurotransmitter, *D* -Ca^{2+}

33 Parts *A, B, C* and *D* of the human eyes are shown in the diagram. Select the option, which gives correct identification along with its functions/ characteristics.

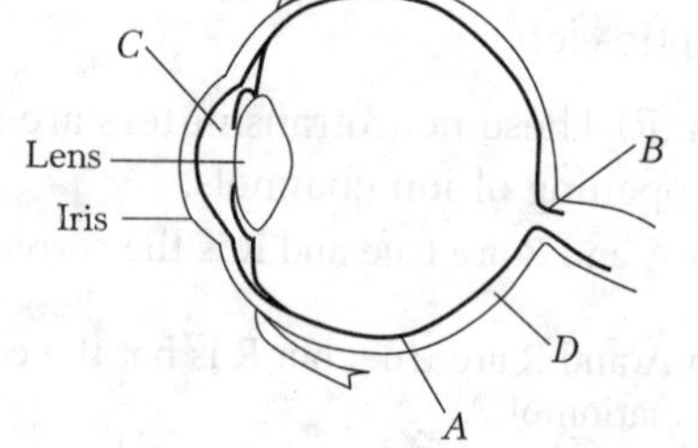

[NEET 2013]
(a) *A*-Retina; contains photoreceptors, rods and cones
(b) *B*-Blind spot; has only a few rods and cones
(c) *C*-Aqueous chamber; reflects the light, which does not pass through the lens
(d) *D*-Choroid; its anterior part forms ciliary body

34 Which of the following is the smallest cranial nerve? **[NEET 2013]**
(a) Abducens (b) Optic
(c) Trochlear (d) Facial

35 Hearing is controlled by **[Manipal 2013]**
(a) cerebellum (b) diencephalon
(c) frontal lobe of cerebrum (d) temporal lobe of cerebrum

36 The following is the scheme showing the path of reflex arc. Identify the different labellings *A, B, C, D, E* and *F* in the reflex arc.

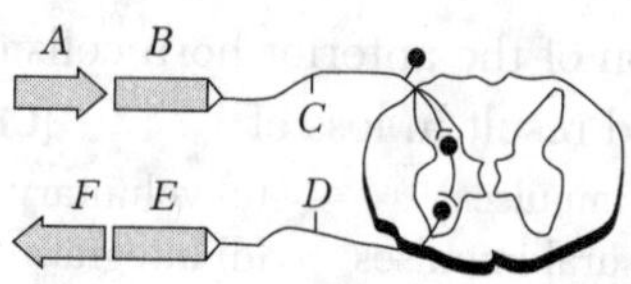

[Karnataka CET 2013, AFMC 2013]
(a) *A*-Stimulus, *B*-Effector, *C*-Motor nerve, *D*-Sensory nerve, *E*-Receptor, *F*-Response
(b) *A*-Stimulus, *B*-Receptor, *C*-Motor nerve, *D*-Sensory nerve, *E*-Effector, *F*-Response
(c) *A*-Stimulus, *B*-Effector, *C*-Sensory nerve, *D*-Motor nerve, *E*-Receptor, *F*-Response
(d) *A*-Stimulus, *B*-Receptor, *C*-Sensory nerve, *D*-Motor nerve, *E*-Effector, *F*-Response

37 The part of the brain where the centre for hunger and thirst is located on **[WB JEE 2013]**
(a) cerebrum (b) hypothalamus
(c) cerebellum (d) medulla oblongata

38 The reflex arc, which is made up of two neurons is known as **[WB JEE 2013]**
(a) monosynaptic reflex arc (b) disynaptic reflex arc
(c) polysynaptic reflex arc (d) asynaptic reflex arc

39 The cornea is very important component of the human eye. The main function of the cornea is to [AIIMS 2013]
(a) bend the light before it reaches the lens
(b) provides structural support to the eye
(c) contains concentrated amount of cone cells in the correct orientation
(d) changes the shape of the lens to enable the image to be focused on the retina

40 Adrenaline is equivalent to which neurotransmitter? [Odisha JEE 2013]
(a) Epinephrine (b) Acetylcholine
(c) Dopamine (d) GABA

41 Sensory neurons in the retina of eye are [Odisha JEE 2012]
(a) rods (b) cones
(c) ciliary body (d) Both (a) and (b)

42 The human hindbrain comprises three parts, one of which is [CBSE AIPMT 2012]
(a) spinal cord (b) corpus callosum
(c) cerebellum (d) hypothalamus

43 Which part of the human ear plays no role in hearing as such, but is otherwise very much required? [CBSE AIPMT 2012]
(a) Eustachian tube
(b) Organ of Corti
(c) Vestibular apparatus
(d) Ear ossicles

44 A person entering an empty room suddenly finds a snake right infront on opening the door. Which one of the following is likely to happen in his neurohormonal control system? [CBSE AIPMT 2012]
(a) Sympathetic nervous system is activated releasing epinephrine and nor-epinephrine from adrenal medulla
(b) Neurotransmitters diffuse rapidly across the cleft and transmit a nerve impulse
(c) Hypothalamus activates the parasympathetic division of brain
(d) Sympathetic nervous system is activated releasing epinephrine and nor-epinephrine from adrenal cortex

45 Dark adaptation in human eye involves [AMU 2012]
(a) conversion of 11 *cis*-retinene to *trans*-retinene
(b) conversion of *trans*-retinene into 11 *cis*-retinene
(c) decomposition of rhodopsin into retinene
(d) decomposition of rhodopsin to scotopsin

46 The optic lobes in humans are represented by the corpora [AMU 2012]
(a) bigemina (b) arenacea
(c) allata (d) quadrigemina

47 Pigment iodopsin is contained in [Manipal 2012]
(a) rod cells (b) cone cells
(c) amacrine cells (d) horizontal cells

48 Vagus nerve is a [WB JEE 2012]
(a) mixed Xth spinal nerve
(b) mixed XIth cranial nerve
(c) mixed Xth thoracic nerve
(d) mixed Xth cranial nerve

49 The number of spinal nerves in human is [WB JEE 2012]
(a) 10 pairs (b) 12 pairs
(c) 43 pairs (d) 31 pairs

50 Third and fourth ventricles of the brains are connected by [WB JEE 2012]
(a) Aqueduct of Sylvius (b) Foramen of Monro
(c) Foramen of Magnum (d) corpus callosum

51 During synaptic transmission of nerve impulse, neurotransmitter (P) is released from synaptic vesicles by the action of ions (Q). Choose the correct P and Q. [WB JEE 2012]
(a) P-acetylcholine, Q–Ca^{2+}
(b) P-acetylcholine, Q – Na^+
(c) P-GABA, Q–Na^+
(d) P-cholinesterase, Q – Ca^{2+}

52 The purplish red pigment rhodopsin contained in the rods type of photorecptor cell of the human eyes is a derivative of [CBSE AIPMT 2011]
(a) vitamin-C (b) vitamin-D
(c) vitamin-A (d) vitamin-B

53 When a neuron is in resting state, i.e. not conducting any impulse, the axonal membrane is [CBSE AIPMT 2011]
(a) equally permeable to both Na^+ and K^+ ions
(b) impermeable to both Na^+ and K^+ ions
(c) comparatively more permeable to K^+ ions and nearly impermeable to Na^+ ions
(d) comparatively more permeable to Na^+ ions and nearly impermeable to K^+ ions

54 The electrical potential difference between outside and inside of a nerve axon before excitation is known as [Kerala CEE 2011]
(a) resting potential (b) action potential
(c) spike potential (d) reaction potential
(e) activation potential

55 Which of the following statement(s) is/are incorrect?
I. Synaptic cleft of neurons secretes adrenaline.
II. Myelinated nerve fibres are enveloped with Schwann cells, which form a myelin sheath around the axon.
III. Non-myelinated nerve fibre is enclosed by a Schwann cell that does not form myelin sheath.
IV. Spinal cord and cranial nerves are made up of non-myelinated nerve fibres.
Of the four statements, [Kerala CEE 2011]

(a) I, II are correct, but III and IV are incorrect
(b) I, II and III are correct, but IV is incorrect
(c) III and IV are correct, but I and II are incorrect
(d) I and IV are correct, but II and III are incorrect
(e) II and III are correct, but I and IV are incorrect

56 The order of the three layers of cells in the retina of human eye from inside to outside is **[Kerala CEE 2011]**
(a) bipoloar cells, photoreceptor cells, ganglion cells
(b) ganglion cells, rods, cones
(c) ganglion cells, bipolar cells, photoreceptor cells
(d) photoreceptor cells, ganglion cells, bipolar cells
(e) ganglion cells, photoreceptor cells, bipolar cells

57 The tract of nerve fibres which connect the cerebral hemisphere is **[Kerala CEE 2011]**
(a) corpus luteum (b) corpus callosum
(c) corpora quadrigemina (d) cerebral aqueduct
(e) foramen magnum

58 Medulla oblongata originates from **[UP CPMT 2011]**
(a) ectoderm (b) mesoderm
(c) endoderm (d) ectomesoderm

59 Respiratory control centre is **[UP CPMT 2011]**
(a) cerebellum (b) medulla oblongata
(c) spinal cord (d) cerebrum

60 During the conduction of nerve impulse, the repolarisation occurs with the **[UP CPMT 2011]**
(a) influx of K^+ ions (b) influx of Na^+ ions
(c) efflux of K^+ ions (d) efflux of Mg^{2+} ions

61 In the axon of motor nerve fibre, the nerve impulse travels **[UP CPMT 2011]**
(a) towards cell body (b) away from cell body
(c) away from synapse (d) in both directions

62 Memory is the responsibility of **[MHT CET 2011]**
(a) grey matter (b) white matter
(c) cerebrum (d) cerebellum

63 What are the two types of nervous system cells? **[MHT CET 2011]**
(a) Alveoli and veins (b) Alveoli and bronchioles
(c) Neurons and nephrons (d) Neurons and glia

64 In eye donation, which one of the following parts of donor's eye is utilised? **[MHT CET 2011]**
(a) retina (b) cornea
(c) lens (d) iris

65 Rods and cones are present in **[MHT CET 2011]**
(a) iris (b) cornea
(c) sclerotic (d) retina

66 Damage to hearing is caused by sound which exceed **[MHT CET 2011]**
(a) 70 decibels (b) 100 decibels
(c) 110 decibels (d) 120 decibels

67 Gustatoreceptors are **[MHT CET 2011]**
(a) rod cells of eyes
(b) taste buds of tongue
(c) epithelium of skin
(d) cone cells of eye

68 Vomiting centre is located in the **[DUMET 2011]**
(a) medulla oblongata
(b) stomach and sometimes in duodenum
(c) GI tract
(d) hypothalamus

69 The amount of CSF in the cranial cavity is **[WB JEE 2011]**
(a) 500 mL (b) 140 mL
(c) 1 L (d) 1.5 mL

70 Movement of tongue muscle is controlled by **[WB JEE 2011]**
(a) facial nerve (b) trigeminal nerve
(c) hypoglossal nerve (d) vagus nerve

71 Which function will be lost due to damage of occipital lobe? **[WB JEE 2011]**
(a) Hearing (b) Speech
(c) Vision (d) Memory

72 Which centre is stimulated during increase in body temperature? **[WB JEE 2011]**
(a) Anterior hypothalamus (b) Posterior hypothalamus
(c) Limbic system (d) Red nucleus

73 Which converts short time memory into long time remembrance? **[Gujarat CET 2011]**
(a) Reticular system (b) Hippocampus
(c) Thalamus (d) Medulla oblongata

74 Which is a bridge between nervous system and endocrine system? **[Gujarat CET 2011]**
(a) Thalamus (b) Hypothalamus
(c) Limbic system (d) Parietal lobe

ANSWERS

Check Point 28.1

1. (d) 2. (b) 3. (b) 4. (c) 5. (b) 6. (b) 7. (d)

Check Point 28.2

1. (a) 2. (d) 3. (a) 4. (c) 5. (c) 6. (c) 7. (b)

Check Point 28.3

1. (c) 2. (a) 3. (a) 4. (d) 5. (a) 6. (d)

Check Point 28.4

1. (b) 2. (c) 3. (d) 4. (a) 5. (b) 6. (a)

(A) Topical Exercise

1. (c) 2. (a) 3. (c) 4. (b) 5. (c) 6. (d) 7. (c) 8. (d) 9. (d) 10. (b)
11. (a) 12. (b) 13. (c) 14. (a) 15. (b) 16. (d) 17. (d) 18. (c) 19. (b) 20. (a)
21. (c) 22. (d) 23. (a) 24. (d) 25. (c) 26. (b) 27. (c) 28. (c) 29. (c) 30. (c)
31. (a) 32. (d) 33. (c) 34. (c) 35. (a) 36. (d) 37. (b) 38. (b) 39. (a) 40. (a)
41. (c) 42. (a) 43. (c) 44. (b) 45. (b) 46. (a) 47. (c) 48. (b) 49. (d) 50. (d)
51. (d) 52. (c) 53. (d) 54. (c) 55. (a) 56. (b) 57. (b) 58. (a) 59. (c) 60. (a)
61. (c) 62. (c) 63. (d) 64. (a) 65. (d) 66. (a) 67. (a) 68. (a) 69. (c) 70. (b)
71. (d) 72. (a) 73. (b) 74. (b) 75. (b) 76. (d) 77. (d) 78. (b) 79. (c) 80. (d)
81. (d) 82. (b) 83. (c) 84. (d) 85. (a) 86. (c) 87. (b) 88. (c) 89. (b) 90. (b)
91. (d) 92. (d) 93. (d) 94. (d) 95. (b) 96. (b) 97. (b) 98. (d) 99. (a) 100. (d)
101. (a) 102. (a) 103. (b) 104. (d) 105. (a) 106. (b) 107. (b) 108. (a) 109. (a) 110. (b)
111. (a) 112. (d) 113. (c) 114. (d) 115. (c) 116. (b) 117. (c) 118. (b) 119. (c) 120. (d)
121. (d) 122. (a) 123. (d) 124. (d) 125. (d) 126. (c) 127. (c) 128. (d) 129. (c) 130. (c)
131. (b)

(B) Cumulative Exercise

1. (b) 2. (a) 3. (b) 4. (c) 5. (a) 6. (c) 7. (c) 8. (a) 9. (b) 10. (d)
11. (a) 12. (d) 13. (a) 14. (a) 15. (a) 16. (d) 17. (a) 18. (c) 19. (c) 20. (a)
21. (b) 22. (a) 23. (d) 24. (d) 25. (d) 26. (c) 27. (d) 28. (c) 29. (b) 30. (c)
31. (c) 32. (d) 33. (a) 34. (a) 35. (d) 36. (d) 37. (b) 38. (d) 39. (c) 40. (b)
41. (a) 42. (d) 43. (d) 44. (d) 45. (c) 46. (d) 47. (d) 48. (b) 49. (d) 50. (a)
51. (a) 52. (a) 53. (b) 54. (d) 55. (a)

(C) Medical Entrances' Gallery

1. (b) 2. (a) 3. (a) 4. (a) 5. (d) 6. (c) 7. (c) 8. (b) 9. (c) 10. (d)
11. (d) 12. (c) 13. (b) 14. (c) 15. (a) 16. (b) 17. (b) 18. (b) 19. (a) 20. (b)
21. (c) 22. (d) 23. (d) 24. (b) 25. (b) 26. (b) 27. (c) 28. (c) 29. (e) 30. (d)
31. (c) 32. (a) 33. (a) 34. (c) 35. (d) 36. (d) 37. (b) 38. (a) 39. (a) 40. (a)
41. (d) 42. (c) 43. (c) 44. (a) 45. (b) 46. (d) 47. (b) 48. (d) 49. (d) 50. (a)
51. (a) 52. (c) 53. (c) 54. (a) 55. (e) 56. (c) 57. (b) 58. (a) 59. (b) 60. (c)
61. (b) 62. (c) 63. (d) 64. (b) 65. (d) 66. (d) 67. (b) 68. (a) 69. (b) 70. (c)
71. (c) 72. (a) 73. (b) 74. (b)

Explanations

(A) Topical Exercise

15 (*b*) Match pair in option (b) is incorrect and can be corrected as
Parietal lobe is the centre for perceptions of pain, pressure, taste, etc., while frontal lobe is the centre of knowledge, reasoning, etc.

31 (*a*) Statement in option (a) is incorrect and can be corrected as

The ventricles of brain are lined by a single layer of ciliated squamous or columnar ependymal cells.

Rest of the statements are correct.

43 (*c*) Match pair in option (c) is incorrect and can be corrected as

Olfactory nerve innervates the olfactory epithelium.

49 (*d*) Option (d) is not a function of ANS and can be corrected as

The conscious control of motor activities are controlled by somatic nervous system.

Rest options are functions of the ANS.

53 (*d*) Option (d) is not an effect of the sympathetic nervous system and can be corrected as

The stimulation of saliva secretion is mediated by parasympathetic nervous system.

Rest options are effect of the sympathetic nervous system.

64 (*a*) Statement I is true, but II is false. Incorrect statement can be corrected as

To maintain unequal distribution of sodium and potassium ions, the neurons use chemical energy in the form of ATP.

67 (*a*) Statement I is true, but II is false. Incorrect statement can be corrected as

The resting membrane potential of membrane at the site of excitation is restored by Na^+ and K^+ ions.

101 (*a*) The incorrect statement in option is (a) and can be corrected as

The middle layer is called choroid which possesses many blood vessels and appears bluish in colour.

103 (*b*) Option (b) is incorrect pair and can be corrected as The anterior portion of eye is called cornea. This portion does not have blood vessels and is refractory in nature.

Rest options are correct matched pairs about human eye.

108 (*a*) Option (a) is incorrectly matched and can be corrected as

Myopia is also called short-sightedness and can be corrected by using concave glasses.

Rest options are correctly matched.

125 (*d*) Option (d) is incorrectly matched and can be corrected as

Otitis media is the acute infection in the middle ear.

Rest options are correctly matched.

(B) Cumulative Exercise

6 (*c*) Statement III is correct wheras I and II are incorrect. Incorrect statements can be corrected as

- There are more than 40 neurotransmitters are present in nervous system, which include acetylcholine, dopamine, GABA, etc.
- Sensory neurons convey information from receptors in the periphery to the CNS.

10 (*d*) Statement in option (d) is incorrect and can be corrected as Right cerebral hemisphere is slightly larger than the left one . Rest statements are correct.

14 (*a*) Statements IV and V are incorrect and can be corrected as

- When three different types of cones are stimulated equally then a sensation of white is produced.
- Both rods and cones are photoreceptor cells. The daylight and coloured vision are the function of cones while twilight or scotopic vision is function of rods.

19 (*c*) Statements I and II are correct, while III is incorrect. Incorrect statement can be corrected as
The cell bodies of parasympathetic pre-ganglionic neurons are located in the brain stem and in sacral spinal cord.

26 (*c*) Statements I and III are correct, while II and IV are incorrect. Incorrect statements can be corrected as

- Vagus nerve is the longest cranial nerve.
  Trigeminal nerve is the 5th cranial nerve and is responsible for sensation in the face and motor function such as bitting and chewing. It is the most complex cranial nerve.

43 (*d*) Statements I and II are correct, while III and IV are incorrect. Incorrect statements can be corrected as
When a nerve cell is stimulated ion channels in the plasma membrane open and Na^+ instantly diffuse down its concentration gradient into the cell. These cations over-ride the negative charges in the intercellular fluid, so the inside of the membrane briefly become positives. The charge is called as depolarisation of the membrane.

47 (*d*) A is false, but R is true. A can be corrected as
Sensory nerve fibres conduct nerve impulses from the peripheral tissues to the Central Nervous System (CNS) to evoke sensations like touch, pain, heat, cold, taste, vision and hearing. If a nerve contains only sensory fibres, it is called a sensory nerve. If a nerve contains only motor fibres, it is called a motor nerve. Its sole role is to conduct nerve impulses from the CNS to some muscles or glands to control their activities. Nerves arising from the spinal cord are neither called sensory nerves nor motor nerves, instead they are called mixed nerves because these nerves carry simultaneously both sensory and motor nerve fibres and serve both sensory and motor functions.

48 (*b*) Both A and R are true, but R is not the correct explanation of A. Activity of sense organs is related with two types of neurons (afferent and efferent). The neuron, which passes nerve impulses from the body organ to the brain, is called afferent neuron. The neuron, which passes nerve impulses from the brain to the body organs, is called efferent neuron.

49 (*d*) A is false, but R is true. A can be corrected as

The human nervous system comprises of central nervous system and peripheral nervous system.

Reasoning is a feature of frontal lobe of cerebrum. The nervous system of human comprises of central nervous system and peripheral nervous system.

50 (*a*) Both A and R are true and R is the correct explanation of A.

Medulla oblongata consists of accumulation of nerve cells which act as vital centres of many autonomic reflexes like vomiting, coughing and sneezing.

51 (*a*) Both A and R are true and R is the correct explanation of A.

The pre-synaptic neuron transmits an impulse (action potential) acrosses the synaptic cleft to the post-synaptic neuron. This occurs *via* a sequence of events occurring at the synapsis and due to the presence of neurotransmitters. As the impulse comes at the pre-synaptic membrane, the neurotransmitters get attached to neuroreceptors on the pre-synaptic membrane and thereby changing its potential and opening the ion channels.

This allows Ca^{2+} ions to get into the pre-synaptic lobe which causes the release of neurotransmitters into the synaptic cleft. These then moves towards the post-synaptic neuronal membrane. Here, again the neurotransmitters bind to the complementary receptors on the post-synaptic membrane.

The increase in the concentration causes the ion channels of post-synaptic neuron to open allowing Na^+ ions to diffuse into the post-synaptic neuron thereby creating potential difference and generation of action potential in post-synaptic neuron.

52 (*a*) Both A and R are true and R is the correct explanation of A.

The neuroglial cells are of three types, i.e. astrocytes, oligodendrocytes and microglia. Out of these, oligodendrocytes are the cells which form the myelin sheath around axon, thus are also called the packing cells of the brain. The other cells like astrocytes are responsible for the separation of two neurons by providing insulation and microglia are phagocytic cells. Nearly 50% of all brain cells are neuroglia.

53 (*b*) Both A and R are true, but R is not the correct explanation of A.

The axoplasm inside the axon contains high concentration of K^+ and negatively charged proteins because the voltage gated Na^+/K^+ channels are closed at resting state and axonal membrane is impermeable to Na^+ ions and negatively charged proteins.

54 (*d*) A is false, but R is true. A can be corrected as

The layer of cells which covers the cerebral hemispheres is called cerebral cortex and is thrown into prominent folds. It is referred to as the grey matter. It is called grey matter due to its greyish appearance. Neuron cell bodies are concentrated here which give it the greyish appearance.

55 (*a*) Both A and R are true and R is the correct explanation of A.

Autonomic nervous system includes sympathetic nervous system and parasympathetic nervous system. These two systems have antagonistic functions. Iris muscles, under sympathetic nervous system dilates pupil, while under parasympathetic nervous system constricts pupil.

(C) Medical Entrances' Gallery

3 (*a*) Hypothalamus is the thermoregulatory centre in the brain and it maintains the constant body temperature of 37°C. The hypothalamus contains a number of centres, which control body temperature.

Other can be explained as

- Corpus callosum is the thick band of nerve fibres that divide the cerebrum into left and right hemispheres.
- Medulla oblongata is the component of hindbrain. It receives and integrates signals from spinal cord and sends them to cerebellum.
- Cerebrum is the large part of the brain and consists of two hemispheres.

4 (*a*) Acetylcholine is not considered as a secondary messanger. It is a common neurotransmitter present in the neuromuscular junctions, voluntary neural synapses, synapses of pre-ganglionic nerve fibres, synapses of post-ganglionic parasympathetic nerve fibres. Cholinergic nerve fibres release acetylcholine. It has excitatory effect to the skeletal muscles and excitatory or inhibitory effect at the other sites.

5 (*d*) The inner ear contains crista ampullaris and macula as the specific receptors of the vestibular apparatus responsible for maintenance of balance of the body and posture.

6 (*c*) Statement in option (c) is not correct and can be corrected as

Rods and cones are photoreceptor cells in our eyes. The rod cells contain a purple pigment rhodopsin that is useful in night vision or scotopic vision. Daylight (photopic) vision and colour vision are the functions of cones.

Rest statements are correct.

7 (*c*) Statement in option (c) that cornea consists of dense matrix of collagen and is the most sensitive portion of the eye is correct.

Rest statements are incorrect and can be corrected as

The outer layer of the wall of eyeball, sclera consists of a dense connective tissue containing mainly collagen and some elastic fibres. Cornea is convex, transparent layer, which is non-vascularised. The cornea is the clear part of eye's protective covering.

8 (*b*) Option (b) is incorrectly paired with its function and can be corrected as

Limbic system consists of five major components, namely hippocampus, amygdala, septal nuclei, mammillary bodies and basal ganglia. It controls the emotional behaviour, food habits and sex behaviour of an organism. It is not involved in controlling movements.

The rest three options are correctly paired with their functions.

9 (*c*) The lens in the human eye is held in its place by the suspensory ligaments attached to the ciliary body.

The function of other components are as follows

- The smooth muscles attached to the ciliary body helps to control the shape of lens.
- Smooth muscle of iris helps in regulating the diameter of pupil.
- Pectinate ligament attached to iris is involved in the drainage of aqueous humor because it contains space between the fibres.

10 (*d*) Option (d) is an incorrect match and can be corrected as Interoceptor sense hunger, while tangoreceptor sense touch. The rest three options are correctly matched.

11 (*d*) The receptor sites for neurotransmitters are present on post-synaptic membrane. Binding of neurotransmitters to these receptors helps in the propagation of nerve impulse.

12 (*c*) The cavity of diencephalon is known as third ventricle. It is situated between the cerebral hemisphere (above) and midbrain (below). Third ventricle or diocoel is connected to the each lateral ventricle of cerebral hemisphere by foramen of Monro.

13 (*b*) Both A and R are ture, but R is not the correct explanation of A.

When an impulse reaches axon terminal, the synaptic vesicles move towards plasma membrane to fuse and release their neurotransmitters in synaptic cleft.
These neurotransmitters bind to specific receptors on post-synaptic membrane which open ion channels thus, allowing the entry of ions to generate a new potential.

14 (*c*) Statement in option (c) is correct.
The photosensitive compounds (rhodopsin) in the human eye is composed of opsin (a protein) and retinal (an aldehyde of vitamin-A, i.e. retinol). They are present in the rod cells (photoreceptors). Light induces dissociation of retinal from opsin thus changes the structure of opsin. This creates potential differences in the photoreceptors and they become hyperpolarised.
However during darkness, rhodopsin is resynthesised from opsin and retinal to restore the dark vision and photoreceptors are depolarised.

16 (*b*) Destruction of the anterior horn cells of the spinal cord would result in loss of voluntary motor impulses. It is because the anterior horn cells (also called anterior grey column), which is the front column of grey matter in the spinal cord contains motor neurons that affect the axial muscles.

17 (*b*) Option (d) is incorrectly paired and can be corrected as Language comprehension includes understanding of speech and written words. It requires input of sensory information (from hearing and vision), processing of the information in different centres of cerebral cortex and the coordination of motor output language comprehension is a function performed by temporal lobe of cerebrum.

Rest options are correctly paired.

18 (*b*) Vestibular apparatus of membranous labyrinth is an intricate series of interconnected and fluid-filled canals, which regulates balance and spatial orientation in the acceleration of vision. Thus, the movement of body will be regulated only by vestibular apparatus.

The otolith organs (utricle and saccule) regulate linear acceleration and semicircular canals sense rotational acceleration in various directions.

The cochlea, tectorial membrane and organ of Corti are responsible for transduction of sound waves.

24 (*b*) Both A and R are true, but R is not the correct explanation of A.

Nerve fibre can become excited through various stimulus like touch, smell, pressure and chemical change and there is a change in polarity. This change in polarity is called as action potential or active potential.

38 (*a*) Reflex arc is the arrangement of neurons in the pathway that always passes through central nervous system. The axon of one neuron ends on the dendrites of next neuron. Such a junction is called synapse. Monosynaptic reflex arc has only two neurons, i.e. sensory and motor which form one synapse.

40 (*a*) Adrenaline is equivalent to the neurotransmitter epinephrine as it is secreted in stress conditions in response to sympathetic stimulation and potentiates the conditions needed for fight or flight.

44 (*a*) Epinephrine and nor-epinephrine are secreted by adrenal medulla (under the control of sympathetic nervous system) in response to stress of any kind or during emergency situations. These are also called emergency hormones or hormones of flight, fight and fright (triple F hormone).

46 (*d*) The midbrain or mesencephalon consists of optic lobes and crura cerebri. In frog, there are two optic lobes (corpora bigemina) which are hollow, but in man, there are four solid optic lobes lying behind the diencephalon. They collectively constitute the corpora quadrigemina.

50 (*a*) The connecting passage between III ventricle (diocoel), IV ventricle (optocoels, i.e. cavities of optic lobes) is called aqueduct of sylvius.

51 (*a*) Ca^{2+} (Q) is responsible for the release of neurotransmitter, acetylcholine (P) from axon terminal by rupturing the synaptic vessicle.

52 (*c*) There are two types of photoreceptor cells of retina, namely rods and cones. The rods contain a purplish-red protein called the rhodopsin or visual purple, which contains a derivative of vitamin-A.

53 (*c*) Neurons are excitable cells because their membranes are in a polarised state. Different types of selectively permeable channels are present on the neural membrane. When a neuron is not conducting any impulse, i.e. resting, the axonal membrane is comparatively more permeable to potassium ion (K^+) and nearly impermeable to sodium ion (Na^+).

60 (*c*) When a stimulus is applied, sodium-potassium pump gets open. Sodium ions rush inside and potassium ions rush outside. This results in depolarisation (action potential). After a period of action potential, sodium-potassium pump operates (influx of $Na^+$and efflux of K^+ and axon will get resting potential by repolarisation.

61 (*b*) Nerve impulse is a wave of depolarisation of the membrane of nerve cell. The nerve impulse travels along a neuron or across a synapse. In the axon of motor nerve fibre, the nerve impulse travels away from the cell body.

62 (*c*) Cerebrum is formed of one pair largest sized lobes called cerebral hemisphere. These form 80% weight of brain. cerebral hemisphere controls all the voluntary activities of body. It is seat of memory, will, intelligency, reasoning and learning. So, memory is responsibility of cerebrum.

CHAPTER 29

Chemical Coordination and Integration

We all know human body consists of two independent integrating systems, i.e. nervous system and endocrine system.

The nervous system functions to maintain the coordination among various organ systems through nerve stimulation. But, its response lasts for the shorter period of time. Secondly, the nerve fibres (or neurons) responsible for neural coordination do not innervate all the cells of the body. Hence, another system having long lasting response and coverage of all the cells of the body is required. This system is called **endocrine system** and the branch of biology that deals with its study is known as **endocrinology**. It includes certain glands, principally ductless, which secretes hormones or chemical messengers directly into the body fluid.

In this chapter, we will discuss various glands of endocrine system, their functions and metabolic reactions regulated by them.

Before going in detail about chemical coordination and integration, it is better to have a brief idea about the glands of animals.

Inside

GLANDS

These are defined as "Any cell or tissue or organ which secretes certain useful chemical compound." In animals following 3 types of glands are visible

(i) **Exocrine glands** The glands which have ducts for discharging their secretion are called exocrine glands. These glands are also called **ducted glands**, e.g. liver, sweat glands, sebaceous glands, gastric glands and some intestinal glands.

(ii) **Endocrine glands** The glands which lack ducts and discharge their secretions into surrounding blood directly are called endocrine glands. These glands are also called **ductless glands**, e.g. pituitary, thyroid, parathyroid, adrenal, pineal body and thymus gland, etc.

(iii) **Heterocrine glands** The glands which contain both excocrine and endocrine tissues are called heterocrine glands. Some portions of these glands are ducted while others are ductless. That is why, these glands are also called **mixed glands**, e.g. pancreas and gonads.

HORMONES (Greek -'to excite')

- In 1902, **WH Bayliss** and **EH Starling** discovered the first ever hormone, i.e. secretin from intestinal mucosa.
- The term 'Hormone' was coined by **Starling** in 1905. These are defined as non-nutrient chemicals which act as intercellular messengers and are produced in trace amounts.
- Chemically, hormone is a highly active organic substance released from one part of the body directly into the blood. It is transported to far located target tissue or organs to initiate a particular metabolic change.
- Hormones are secreted by endocrine glands. In addition to endocrine glands, these can also be produced by some other organs like gastrointestinal tract, liver, kidney and heart, etc.

Types of hormones

Hormones can be classified on the following basis

On the basis of function

These are as follows

(i) **Synergistic hormone** When two or more hormone complement the function of each other. Hence, both of them are needed for full expression of hormone effect. For example, insulin and growth hormone/ thyroxine have synergistic effect on body growth. Similarly, oestrogen, progesterone, prolactin and oxytocin have synergistic effect on the physical growth of mammary gland.

(ii) **Antagonistic hormone** When two hormones oppose the actions of each other. For example,
- Insulin and glucagon
- Calcitonin and parathyroid hormone
- MSH and melatonin.

On the basis of their biochemical nature

These are as follows

(i) **Iodothyronines** These are derived from iodinated amino acid with diphenyl ether ring system, e.g. thyroxine hormone secreted from thyroid gland.

(ii) **Amines** These are amino acid derivatives, e.g. serotonin or tyrosine, epinephrine and nor-epinephrine. These all are derived from amino acid, tryptophan.

(iii) **Peptides** Hypothalamic hormones, such as TRH, GRH, GIH, MSH; hormones of posterior lobes of pituitary such as ADH, oxytocin, ACTH of anterior pituitary and calcitonin of thyroid gland belong to this category.

(iv) **Proteins** Hormones of pancreas like insulin and glucagon; hormones of gastrointestinal tract like PTH or parathyroid hormone; female hormones like relaxin, hCG are proteinaceous in nature.

(v) **Steroids** These are derived from cholesterol. For example, hormones of adrenal cortex like aldosterone, cortisol, sex corticoids and hormones of gonads like, testosterone, oestrogen, progesterone.

Special characteristics of hormones

- They have low molecular weight.
- They are soluble in water and blood.
- Some of them are soluble in fat as well. So, they easily diffuse in tissues.
- They can act in low concentration.
- They have no cumulative effect.
- They are non-antigenic and non-species specific.
- They are destroyed after use, thus do not get stored in the body except thyroxine.
- They do not participate directly in metabolic activities of target cells, but affect and control such activities.

Hormones of invertebrates

- Most of the invertebrates have specialised neurosecretory cells which secrete neuropeptides or neurohormones for chemical coordination.
- These help to regulate growth, maturation and reproduction processes in them.
- However, a well defined and developed endocrine system was first seen in phylum-Arthropoda.

Neurohormones, their source and functions in major invertebrate groups are given below

Invertebrates	Source	Neurohormone/Hormone and their function
Cnidarians	Nerve cells	Growth promoting substances
Platyhelminthes	Neurosecretory cells	Neuropeptides which promote regeneration and reproduction
Nemerteans	Dorsal and ventral pair of ganglion	Neuropeptides which control gonadal development and regulates water balance
Nematodes	Neurosecretory cells	Neuropeptides which control ecdysis

Invertebrates	Source	Neurohormone/Hormone and their function
Molluscs	Neurosecretory cells	Neuropeptides which regulate heart rate, kidney function and metabolism
Annelids	Gonads and other endocrine glands	Juvenile hormones (polychaetes), osmoregulatory hormones and hyperglycemic hormones (oligochaetes)
Arthropods	Brain has corpora allata and corpora cardiaca and prothoracic gland	Produces moulting hormone (ecdysone) and juvenile hormone (neotenin), respectively
Echinoderms	Gonads	Maturation inducing hormone

The endocrine system of all vertebrates is some what similar with minor adaptive differences.

HUMAN ENDOCRINE SYSTEM

The endocrine system consists of various endocrine glands, which are widely separated from each other with no direct anatomical links. The endocrine glands present in our body are shown below

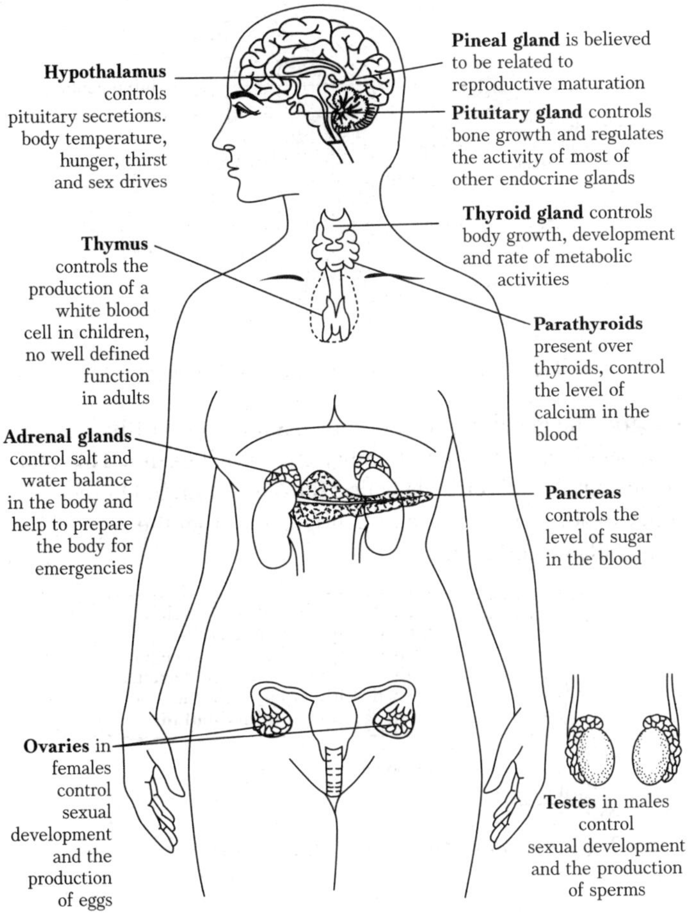

Fig. 29.1 Human endocrine system

Detailed description of endocrine glands are as follows

Hypothalamus

- It is the major link between nervous and endocrine system.
- **Position** The basal part of forebrain in diencephalon represents hypothalamus.
- **Origin** It originates from embryonic ectoderm.
- **Structure** It is made up of neurosecretory cells called **hypothalamic nuclei**. These produce neurohormones that regulate the synthesis and secretion of pituitary hormones.

Hypothalamic hormones

These are of following two types

(i) **Releasing or Trophic hormones** These stimulate the secretion of pituitary hormones. For example, a hypothalamic hormone called **Gonadotropin Releasing Hormone** (GnRH) stimulates the synthesis and release of gonadotropins from the pituitary gland.

(ii) **Inhibiting hormones** These inhibit the secretion of pituitary hormones. For example, somatostatin from the hypothalamus inhibits the release of growth hormone from the pituitary.

Origin and circulation

- All the hypothalamic hormones originate from hypothalamic neurons.
- In neurons, they pass through axons and are released from their nerve endings. These hormones then reach the anterior pituitary region through **hypophysial portal circulation.**
- The posterior pituitary is directly regulated by hypothalamus through their neurosecretory neurons. These neurons secrete precursors of oxytocin (**by paraventricular nuclei**) and vasopressin (**by supra optic nuclei**).

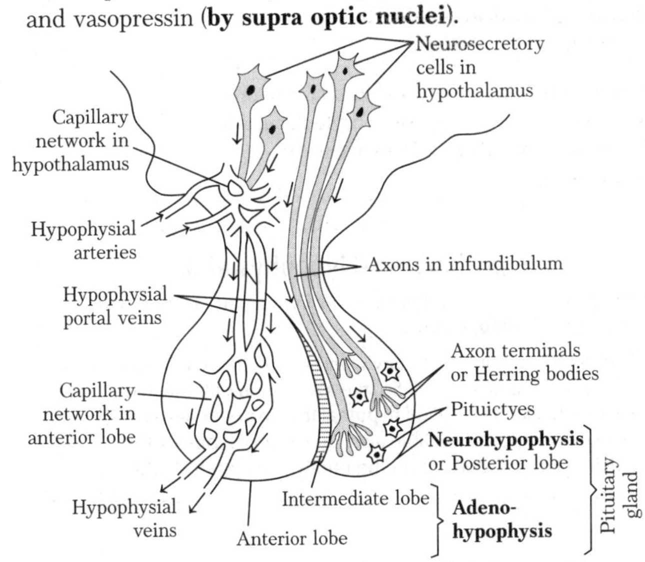

Fig. 29.2 Diagrammatic representation of pituitary and its relationship with hypothalamus

- Besides oxytocin and vasopressin precursors, hypothalamus also produces several short chain polypeptide hormones called **hypophysiotropic factors.**
- Various hypothalamic hormones, their pituitary response and target organs are given in the table below

Hypothalamic hormones	Response of pituitary	Target organ
Thyrotropin-Releasing Hormone (TRH)	Secretes Thyroid Stimulating Hormone (TSH)	Thyroid
Adrenocorticotropin Releasing Hormone (ARH)	Secretes Adrenocorticotropin Hormone (ACTH)	Adrenal cortex
Gonadotropin Hormone-Releasing Hormone (GnH-RH)	Secretes Luteinizing Hormone (LH) and Follicle Stimulating Hormone (FSH)	Ovary/Testis
Growth Hormone-Releasing Hormone (GH-RH) or **Somatocrinin**	Secretes Growth Hormone (GH) or Somatotropic Hormone (STH)	–
Growth Hormone Release Inhibiting Hormone (GH-IH) or **Somatostatin** (SRH)	Inhibits secretion of Growth Hormone (GH)	Most tissues
Prolactin-Releasing Hormone (P-RH)	Secretes Luteotropic Hormone (LTH) or Prolactin Hormone (PH)	Mammary glands
Prolactin-Inhibiting Hormone (P-IH) or **Dopamine**	Inhibits secretion of prolactin	
Melanocyte Stimulating Hormone-Releasing Hormone (MSH-RH)	Secretes Melanocyte Stimulating Hormone (MSH)	Skin pigment cells
Melanocyte Stimulating Hormone-Inhibiting Hormone (MSH-IH)	Inhibits secretion of Melanocyte Stimulating Hormone (MSH)	–

Hypothalamo-Hypophysial axis

The posterior lobe of pituitary arises from a part of embryonic brain called **hypophysis.**

Further, the action of pituitary gland is directly or indirectly under the hypothalamic control. So, the association of these two endocrine glands is called **hypothalamo-hypophysial axis** (as visible from the above table also). It is responsible for the existing overall control on the endocrine system of the body.

PITUITARY GLAND

(Hypophysis cerebri)

- This gland was previously known as **master gland due** to its contribution in regulating other glands of hormonal system.
- **Position** It is the smallest and most protected gland of the body. It lies in the bony cavity called **hypophysial fossa** or **sella turcica** of the sphenoid bone.
- This gland is attached to hypothalamus by a stalk called **infundibulum** as shown below

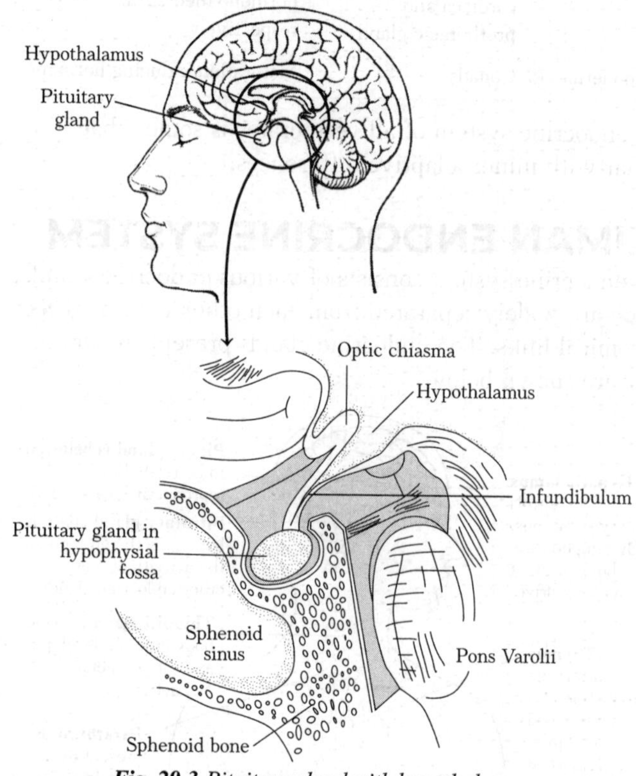

Fig. 29.3 Pituitary gland with hypothalamus

- **Origin** The gland has dual origins. Its anterior part is originated from the roof of embryonic buccal cavity. This roof comes out as a diverticula called **Rathke's pouch.** On the other hand, its posterior part is originated from the hypothalamus of brain.

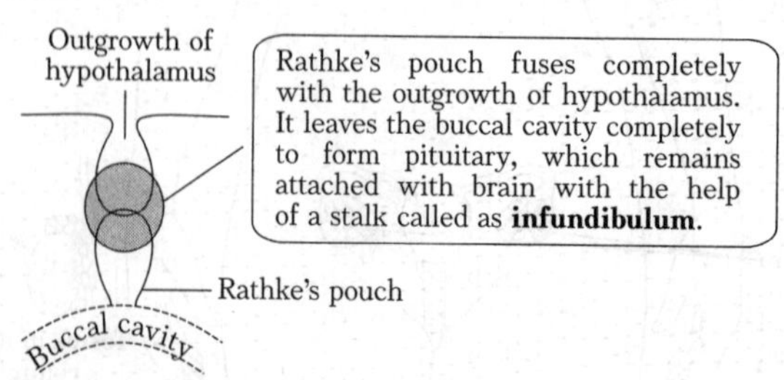

Fig. 29.4 Rathke's pouch in adenohypophysis

- **Structure** Pituitary gland is a small pink-coloured gland with a diameter of about 1-1.5 cm and weight 0.5 gm.
  - Morphologically, pituitary has three parts, i.e. **Pars dorsalis** or anterior lobe, **Pars intermedia** or intermediate lobe and **Pars nervosa** or posterior lobe.
  - It is functionally divided into two parts, i.e. adenohypophysis and neurohypophysis.

Adenohypophysis

Anterior and intermediate lobes of pituitary gland are collectively known as adenohypophysis. It is highly cellular and vascular region of pituitary. Thus, adenohypophysis is composed of following parts

Pars dorsalis

It consists of two types of cells, *viz.*, **chromophobes** and **chromophils**. These cells are further of two types, i.e. acidophils and basophils.

Hormones of pars dorsalis

The hormones secreted by pars dorsalis region of adenohypophysis are called **trophic hormones**. These are as follows

(i) GH or STH (Growth or Somatotropin Hormone)

- It is the only adenohypophysis hormone which is linked directly to body whereas other adenohypophysis hormones mostly control other glands.
- **Structure** It is a globulin protein made up of 188 amino acids secreted by specialised acidophilic cells.
- **Function** of this hormone are as follows
  - It stimulates body growth by increasing protein, fat and carbohydrate metabolism.
  - It promote the synthesis and secretion of Insulin-like Growth Factors (IGFs) or somatomedin by liver cell, skeletal muscles, bones, cartilage and other tissues.
  - GH stimulates hepatocytes to release glucose in blood by acting upon insulin-like growth factor-I (IGF-1). Thus, GH could have diabetogenic effect and indirectly, it acts as insulin antagonist.
- **Disorders** Diseases caused by abnormal secretion of GH are as follows

  Hyposecretion of this hormone causes
  - **Dwarfism** (midgets) or **pituitary nanism** during the skeletal growth period. This dwarfism occurs before puberty and often associated with the term **ateliosis**, means failure to achieve perfection.
  - **Acromicria** If the deficiency of GH occurs after puberty, then it causes acromicria. It is a condition in which the person has normal body and intelligence but small hands, feet and face.

  Hypersecretion of this hormone causes
  - **Gigantism** Hypersecretion of GH before puberty, i.e. before the ossification of epiphyseal plates, can cause gigantism. This causes exaggerated and prolonged growth in long bones along with the enlargement of internal organs.
  - **Acromegaly** If hypersecretion occurs after puberty then the individual suffers from acromegaly. It is characterised by enlarged hands, feet, jaws, nose and ears in a normal person. Sometimes, a person develops hump also due to bending of vertebral column. This is called **kyphosis**.

(ii) Gonadotropic hormones

These are hormones that stimulate gonadal activity and are produced by basophilic cells. These include

(a) **FSH** (Follicle Stimulating Hormone)
- **Structure** A glycoprotein with α and β-polypeptide units. α-chain contains 96 amino acids while β-chain contains 111 amino acids.
- **Function** It is a hormone when released in females helps in growth of ovarian follicles till ovulation. Whereas when released in males, it helps in the development of seminiferous tubules and maintenance of spermatogenesis.

(b) **LH** (Luteinizing Hormone) This hormone is also called **Interstitial Cell Stimulating Hormone** (ICSH) in males.
- **Structure** LH is a heterodimeric glycoprotein. Just like FSH, it also contains 2 polypeptide chains, i.e. α and β. Its α-chain contains 92 amino acids in humans and 96 amino acids in all other vertebrates. The β subunit of LH contains 120 amino acids.
- **Function** It stimulates the development of interstitial cells or Leydig cells in testes and is also concerned with androgens (Testosterone) in males. In females, it affects growing follicles in ovary by stimulating the release of progesterone and oestrogen.

(iii) PRL (Prolactin)

- It is a non-tropic hormone produced by acidophilic cells. It was previously called **Luteotropic Hormone** (LH).
- **Structure** It contains 198 amino acids.
- **Function** This hormone stimulates the mammary glands for lactation or milk production. That is why, it is also called **lactogenic** or **mammotrophin hormone**. It also has a direct effect on the breasts immediately after parturition.

(iv) ACTH (Adrenocorticotropic Hormone)

- It is secreted by the basophilic or corticotropic cells of anterior pituitary.
- **Structure** It is a 39 amino acids polypeptide.

- **Function** It stimulates the secretion of hormones from adrenal cortex, e.g. cortisol.
  It also increases the concentration of cholesterol and steroids within the gland. That is why, it is also given as non-disease specific.
- **Disorders** Diseases caused by the abnormal secretion of ACTH are
  Deficiency or **Hyposecretion** of this hormone can cause
  - **Acute psoriasis** It is a condition in which skin cells build up and form scales and itchy, dry patches.
  - **Dermatitis** It is skin inflammation caused by the hyposecretion of ACTH.

  Hypersecretion of this hormone can cause
  - **Cushing's disease** is characterised by obesity of trunk, but not limbs.
  - Its other symptoms include skin pigmentation, excessive hair, demineralisation of bones and loss of function.

(v) TSH (Thyroid Stimulating Hormone)

- It is a glycoproteinaceous hormone secreted by special basophilic cells.
- **Structure** Just like other glycoproteinaceous hormones, it also has α and β-chains.
- **Function** This hormone promotes growth and proper functioning of thyroid gland by stimulating synthesis and secretion of **thyroxine.**

(vi) LPH (Lipotropin or Lipolytic Hormone)

It is a polypeptide hormone secreted particularly by the corticotrophic cells.

- **Structure** This hormone is in the form of a complex with I, II and β-lipotropins. It is made up of 90 amino acids.
- **Function** It stimulates fatty acid liberation (lipolysis) from adipose tissue.
- Its hyposecretion causes obesity and hypersecretion causes thinness.

Pars intermedia

- This part is developed from the upper surface of Rathke's pouch. It contains prismatic cells having granules.
- This region produces a single hormone only known as **Melanocyte Stimulating Hormone** (MSH) or **Intermedia MSH.**
- **Function** It stimulates synthesis and dispersal of melanin pigment in skin of mammals, fishes, amphibians and some reptiles.
- In amphibians and reptiles, it causes change in colour of skin by a phenomenon known as **metachrosis.**

> **Atropy or Degeneration of anterior lobe**
>
> **Simmond's disease** or **Pituitary myxedema** is caused due to the atrophy or degeneration of anterior lobe of pituitary gland. In this disease, the skin of face becomes dry and wrinkled and shows premature ageing.

Neurohypophysis

- This lobe originates from the infundibulum. It forms the **posterior lobe** of pituitary gland and also called **pars nervosa.**
- Its cells are called pituicytes which are large fatty neuroglial supporting cells. These cells do not produce or secrete hormones.
- Posterior pituitary hormones are directly secreted from the neurosecretory cells of hypothalamus, i.e. **paraventricular nucleus** (secretes oxytocin) and **supraoptic nucleus** (secretes vasopressin).
- The axon of these neurons have swollen terminal ends called **'herring bodies'**.
- Hypothalamic hormones are stored in these bodies and are released on nerve stimulation.

Hormones of neurohypophysis

(i) Oxytocin

- It is secreted into the blood when the hypothalamic neurons are stimulated either due to the distension of uterus by the full term foetus or due to the sucking of breast by the infant.
- **Structure** It is a small peptide hormone containing 9 amino acids, i.e. a nanopeptide.
- **Function** Oxytocin acts on the smooth muscles of uterus and mammary glands. At the time of labour, oxytocin initiates vigorous uterine contraction to induce childbirth. Thus, it is called **birth hormone.**
  It also causes contraction of mammary glands, leading to milk ejection and thus, it is also called **milk-ejection hormone.** It is often injected in cows for excess milk production.

Note ***Pitocin** is a synthetic oxytocin which is often given to induce labor.*

(ii) Vasopressin or Pitressin or ADH (Antidiuretic Hormone)

- It is released from the posterior lobe when the blood osmotic pressure rises due to the loss of water from the body.
- **Structure** It is also a nanopeptide.
- **Function** It causes contraction of blood vessels and raises the blood pressure. Vasopressin is called ADH because it aids in reabsorption of water from glomerular filtrate in the nephrons of kidney. As the result of this, concentrated urine is produced.
- **Disorder** caused by abnormal secretion of ADH include
  Diabetes insipidus It is a disease caused by the hyposecretion of ADH. It leads to a condition of reduced water absorption by kidneys. Thus, the urine becomes diluted (diuresis), without sugar or glucose. Consequently elimination of a large volume of very dilute (hypotonic) urine occurs.

CHECK POINT 29.1

1. The ductless glands
   (a) produce non-nutrient intracellular messengers
   (b) produce non-nutrient intercellular messengers
   (c) produce nutrient intracellular messengers
   (d) produce nutrient intercellular messengers
2. Which of the following hormone is not present in the portal veins linking hypothalamus with anterior lobe of pituitary gland?
   (a) Somatostatin
   (b) Somatotropin
   (c) Gonadotropin releasing hormone
   (d) Prolactin inhibitory hormone
3. Which of the following hormone is secreted by hypothalamus and inhibits the release of prolactin?
   (a) Prolactin releasing hormone
   (b) Dopamine
   (c) Gonadotropin releasing hormone
   (d) Cortisol releasing hormone
4. Which of the following pituitary hormone is predominantly under inhibitory control of hypothalamus?
   (a) LH (b) TSH (c) ACTH (d) Prolactin
5. Which hormone secreted from anterior pituitary has a direct action on body rather than having action on any other gland?
   (a) Prolactin (b) FSH
   (c) ACTH (d) GH
6. The hormone which is regularly used by the milkmen to get more milk from their cows and buffaloes is
   (a) vasopressin (b) somatostatin
   (c) FSH (d) oxytocin
7. The symptoms of a disease include excretion of large amount of urine, thirst and dehydration. It is caused by
   (a) hyposecretion of vasopressin
   (b) hypersecretion of vasopressin
   (c) hyposecretion of ACTH
   (d) hypersecretion of ACTH

THYROID GLAND

- **Position** Thyroid gland is situated in the base of neck. It is present between the trachea and larynx at the level of fifth, sixth and seventh cervical and first thoracic vertebra.
- **Origin** It is the largest endocrine gland that develops from the endoderm of embryo.
- **Structure** This gland is larger in females as compared to males. It is bilobed and H-shaped. Its two lobes are connected to each other by a non-glandular flap called **isthmus**.

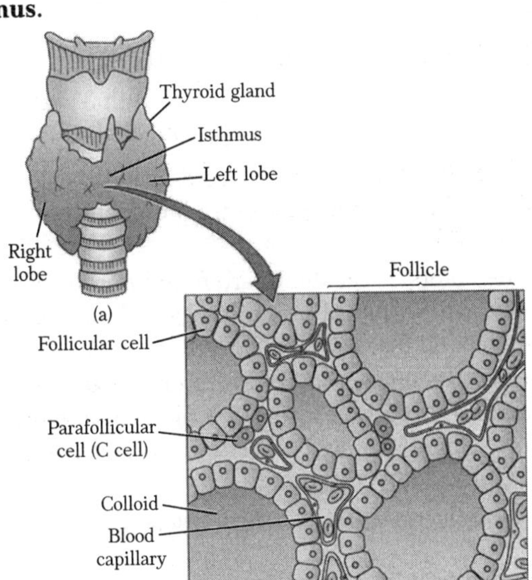

Fig. 29.5 Thyroid gland: (a) Anterior parts, (b) Internal organisation

- This gland is composed of **follicles** and **stromal tissues.**

Follicles

- The follicles are considered as the functional units of thyroid.
- These basically consist of cuboidal epithelial cells.
- An iodised colloidal substance called thyroglobulin is filled in the cavity of these follicles which synthesise hormones.
- The two main hormones secreted by these cells are
  - **Tetraiodothyronine** or **thyroxin** (T_4).
  - **Triiodothyronine** or T_3.

Stromal tissue

- This is the basic ground tissue that acts as a cementing material. It contains blood and lymph vessels, along with other ground tissue substances.
- It also contains small specialised cells called **parafollicular** or **C-cells.** These cells synthesise and secrete a hormone called **thyrocalcitonin.**

Hormones of thyroid gland

Various hormones secreted by thyroid gland are discussed below

1. Thyroxine (T_4) and triiodothyronine (T_3)

- The hormones T_4 and T_3 are named so, because they contain four and three atoms of iodine, respectively.
- T_3 is secreted in smaller amount but, it is more potent than T_4. The T_4 hormone changes to T_3 hormone on reaching the tissues.
- T_3 and T_4 hormones have similar effects, hence they are collectively known as **Thyroid Hormone** (TH).

Functions of thyroid hormone

- They control the metabolism of protein, carbohydrates and fats and hence regulate the growth of the body tissues. T_4in particular, is lipolytic and catabolic in nature.
- These hormones regulate the Basal Metabolic Rate (BMR) in the body by enhancing heat production or calorigenic effect. That is why, these are required to maintain the **"tempo of life"**.
- They help in maintaining water and electrolyte balance in human body.
- They support erythropoiesis, i.e. red blood cells formation.
- Regulation of menstrual cycle in females is attributed to these hormones.
- They enhance some actions of neurotransmitters, like adrenaline and nor-adrenaline.

Disorders of thyroid hormones

Diseases caused by abnormal secretions of thyroid hormones are as follows

Hyperthyroidism (Thyrotoxicosis)

It is seen when the body tissues get exposed to excessive levels of T_3 and T_4. It results into following conditions

(i) **Exophthalmic goitre** (Grave's disease) In this enlargement of thyroid gland takes place. It leads to protrusion of the eyeballs, increased Basal Metabolic Rate (BMR) and weight loss. It often occurs in females.

(ii) **Toxic nodular goitre** (Plummer's disease) In this type of hyperthyroidism, there occurs an increased body metabolic rate.

It may also lead to an autoimmune disorder in which the body produces antibodies that mimic the action of TSH. Function of these antibodies is not regulated by normal feedback control.

Hypothyroidism

The insufficient secretion of thyroid hormones causes hypothyroidism.

It results in following conditions

(i) **Cretinism** It is seen in children. Hypothyroidism during pregnancy causes defective development and maturation of the growing body.

- The children thus suffering from cretinism show stunted growth, mental retardation, low intelligence quotient, abnormal skin, deaf mutism, etc. They show pot-belly, pigeon like chest and a protruding tongue.
- Other features include neonatal jaundice, lethargy, respiratory problems, constipation and reduced sexual growth.

(ii) **Myxoedema** (Gull's disease) This condition is common in adults and usually affects females than males.

- In this, the deficiency of thyroid hormones occurs after the completion of normal mental and physical development of body.
- It results in an abnormally low fat and protein metabolism. Mental and physical processes become slower and skin becomes coarse. Also, the hair lacks lustre, becomes brittle and tends to fall out.
- In human female (adult) hypothyroidism may causes irregular menstrual cycle.

(iii) **Simple goitre** In this condition, there is enlargement of thyroid gland.

- It is caused due to the deficiency of iodine in diet. It may lead to myxoedema sometimes.
- When most of the people of a locality show the symptoms of this disease then it is called **endemic goitre**. Persons who eat sea food, never shows the symptoms of this goitre.

Hashimoto's thyroiditis/Hashimoto disease

It is an autoimmune disorder characterised by the production of antibodies in response to thyroid antigens. It is also known as **suicide of thyroid**. It affects more women than men. It is one of the most common cause of hypothyroidism.

2. Calcitonin or Thyrocalcitonin (TCT)

- **Structure** It is proteinaceous in nature with 32 amino acids.
- **Functions** of this hormone are as follows
  - It is a **hypocalcaemic** (i.e. decreases blood calcium levels) and **hyperphosphatemic** (i.e. increases blood phosphate levels) hormone.
  - The receptors of calcitonin are found in osteoclasts, kidney and regions of brain. It acts on bone and kidneys to reduce the blood calcium level.
  - Calcitonin reduces the reabsorption of calcium from bones and inhibits reabsorption of calcium by the renal tubules. Its effects are opposite to that of parathormone (PTH) of parathyroid gland.

Note
- *Gudernatsch (1912), first discovered the effect of thyroxine in tadpole.*
- *Kendall (1914) was the first to isolate thyroxine crystals.*
- *Harrington and Barger (1927), gave the molecular structure of thyroxine.*

> **Neoteny and paedogenesis**
>
> Two developmental processes in animals are also associated with the thyroid gland. These are
> - **Neoteny** A process of retention of larval characters in adults or development of adult characters in larva.
> - **Paedogenesis** Process in which immature stages start reproducing. A child reproducing in such a manner is called **paedogenic child**.

PARATHYROID GLAND

- **Position** It is present as four small pouches, each two of which are embedded in the posterior surface of each lobe of thyroid gland.

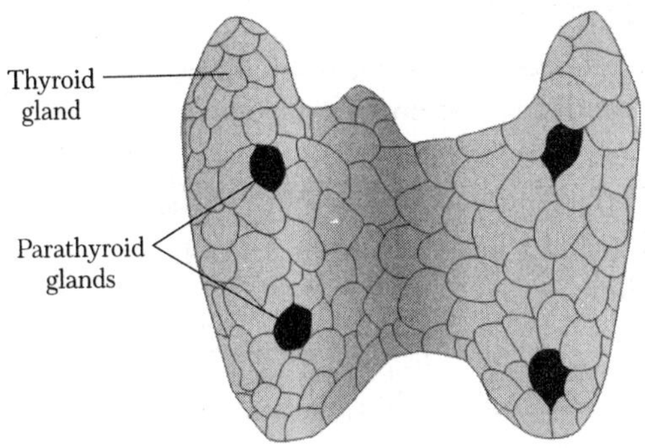

Fig. 29.6 Posterior part of thyroid gland showing parathyroid glands

- **Origin** This gland develops from the endoderm of the embryo.
- **Structure** This gland is developed as epithelial buds from third and fourth pairs of pharyngeal pouches. The two types of cells present in parathyroid gland are
  (i) **Chief cells** or **Principal cells** These cells contain prominent Golgi apparatus, endoplasmic reticulum and secretory granules. They are involved in the synthesis and secretion of hormones.
  (ii) **Oxyphil cells** or **Eosinophils** These cells contain oxyphil granules and large number of mitochondria in cytoplasm.
  Their function is unknown, although few scientists consider them as degenerated chief cells.

Hormone of parathyroid gland

- The hormone secreted by parathyroid gland is **Parathormone** (PTH) or **Collip's hormone**.
- It is a linear polypeptide with about 84 amino acids.
- It is a hypercalcaemic and hypophosphatemic hormone. Its secretion level is regulated by circulating level of Ca^{2+} in blood.
- PTH acts directly on bone to increase bone dissolution and demineralisation. Thus, it mobilises Ca^{2+} ions into the blood.
- Complete calcium homeostasis is thus, under the control of PTH and calcitonin as shown below

Calcium homeostasis

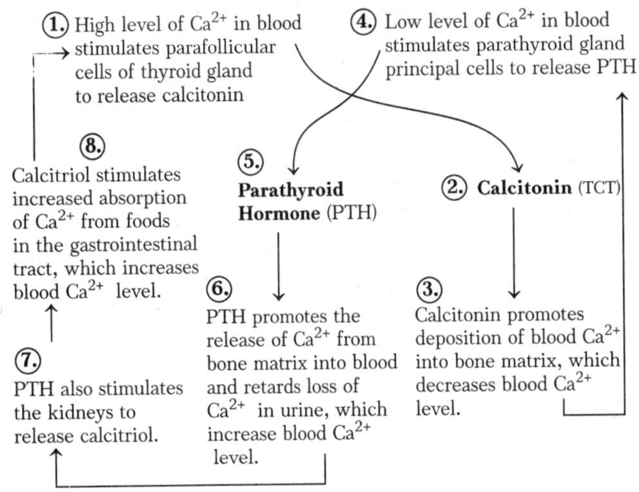

Disorders related to PTH

Disorders caused by abnormal secretion of PTH are as follows

(i) Hyperparathyroidism

Excess secretion of parathyroid hormone can result in following conditions

- Excess parathormone (PTH) release, usually occurs due to benign tumours of parathyroid gland. It causes reabsorption of calcium from bones and raises the blood calcium level (**Hypercalcaemia**).
- The effects may lead to the formation of renal calculi or stones.
  - The condition may further gets complicated by pyelonephritis, i.e. inflammation of kidney and renal failure.
  - Other associated features include muscle weakness, general fatigue, calcification of soft tissue, osteoporosis, etc.
- Hyperparathyroidism also result in **osteitis fibrosa cystica**, i.e. the condition in which normal bone is replaced by cyst and fibrous tissue.

(ii) Hypoparathyroidism

Insufficient secretion of parathyroid hormone can result in following conditions

- PTH deficiency causes **abnormally low level of ionised calcium** in the blood. This reduces absorption of calcium from the small intestine as well as reabsorption from the bones.
- Hypoparathyroidism may also lead to **parathyroid tetany**. It is a disease characterised by increased excitability of nerves and muscles, causing cramps, contraction of larynx, face, hands and feet.

ADRENAL GLANDS

- **Position** There are two adrenal glands, situated on the upper side of each kidney. That is why, the name **supra renal** is also given to them. These are enclosed within the fibrous capsule of kidney called **renal fascia.**
- **Origin** These glands develop from two separate embryological tissues, i.e. ectoderm (medulla region) and mesoderm (cortex region).
- **Structure** These glands are about 4 cm long and 3 cm thick. Internally, the gland is differentiated into two parts, i.e. **adrenal cortex** and **adrenal medulla.**

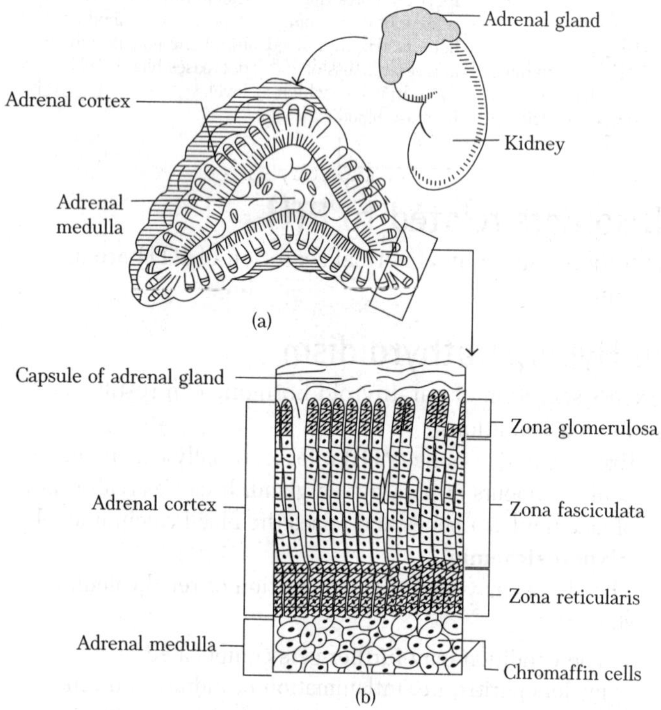

***Fig. 29.7** (a) Adrenal gland, (b) VS of adrenal gland*

Adrenal cortex

It is the external part of the adrenal gland. It is divided into following three layers

(i) Zona glomerulosa, the outermost layer
(ii) Zona fasciculata, the middle layer
(iii) Zona reticularis, the innermost layer

Hormones of adrenal cortex

- The adrenal cortex secretes more than 20 steroid hormones. These hormones are commonly called **corticoids.**
- The basic constituent of all these hormones is cholesterol.
- These hormones are grouped into following three major categories as follows

Mineralocorticoids

- These hormones are released from **zona glomerulosa.** Aldosterone is the principle mineralocorticoid which is also called **salt retaining hormone.**
- **Function** It functions to regulate the electrolyte balance in the body fluid, specially sodium (Na^+). The amount of aldosterone secreted in the blood is influenced by the sodium level in blood. If there is a fall in the blood sodium level, more aldosterone is secreted and more sodium is reabsorbed.

Glucocorticoids

- These are released from **zona fasciculata** and include **cortisol, cortisone** and **corticosterone.**
- Among these, cortisol (also called hydrocortisone) accounts for 95% of the glucocorticoid activity in the body.
- The secretion of glucocorticoids is controlled by ACTH as shown below

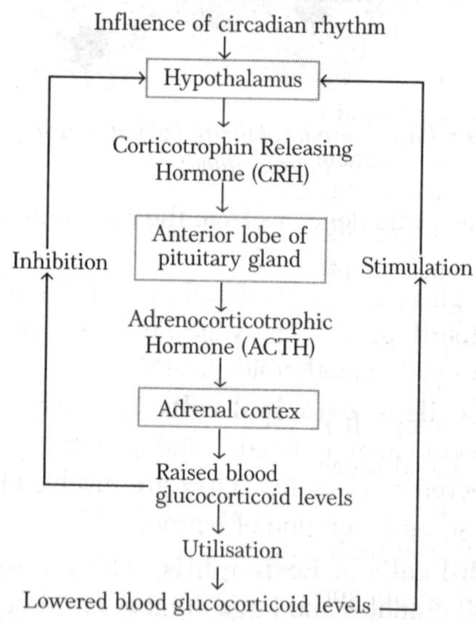

***Fig. 29.8** Diagrammatic representation showing regulation of glucocorticoid secretion*

- **Function** As the name suggests, glucocorticoids are involved in carbohydrate metabolism, as it helps to convert glycogen into glucose, i.e. **gluconeogenesis.** These are also involved in protein and fat metabolic breakdown, i.e. **proteolysis** and **lipolysis**, respectively.

> **Cortisol : The life saving glucocorticoid**
>
> Cortisol is the most potent glucocorticoid. It maintains cardiovascular system as well as the kidney functions. It produces anti-inflammatory responses and helps to suppress the immune system. Therefore, it is given during transplantation surgery to avoid tissue rejection by immune system of the recipient. It can also stimulate the production of RBCs. Cortisol is often called **life-saving hormone**, as it serves to maintain the body in living condition.

Sexcorticoids

- These are released from **zona reticularis.**
- These are sex hormones, i.e. **androgens** and **oestrogens** and are secreted as **DHEA** (Dehydroxy Epiandrosterone).
- These hormones are produced in small amount and believed to have little significance.
- These hormones help in the development of secondary sexual characters in males and females.

Disorders of adrenal cortex

Hyposecretion and hypersecretion of cortical hormone can be studied under following heads

Hypercorticism

The over secretion of adrenal corticoids may lead to following conditions

(i) **Aldosteronism** or **Conn's disease** It occurs due to over secretion of cortisol or aldosterone in the body. It is characterised by

- Increased sodium and decreased potassium level in blood plasma
- High blood pressure
- Increased blood volume
- It is also characterised by excessive breakdown of body fat and this fat is then deposited in unusual areas of body like face, upper abdomen, etc. Thus, the clinal symptoms are **moon face** and **buffalo hump**.

(ii) **Cushing's syndrome** It is caused by hypersecretion of cortisol, especially due to some sort of tumour in adrenal cortex. It is characterised by

- high blood sugar
- obesity
- wasting of limb muscles
- high sodium and potassium level in blood
- excessive hair growth in males and masculinisation in females

(iii) **Adrenal virilism** It occurs due to an excessive secretion of sex corticoids which produces male like secondary sexual characters such as

- beard and moustache
- low pitch voice in women

The main sexcorticoids responsible for this disease are androgens.

Hypocorticism

- Insufficient secretion of this hormone can cause **Addison's disease.** It is the common name given to hypocorticism.
- The condition may seen either due to the hyposecretion of adrenal corticoids or decrease in aldosterone and cortisol level or destruction of adrenal cortex by diseases like tuberculosis.
- It is characterised by
  - dark bronze colour pigmentation of skin
  - low blood sugar levels
  - excretion of sodium in urine
  - vomiting
  - weakness
  - low blood pressure and diarrhoea

Adrenal medulla

- The adrenal medulla is completely surrounded by the cortex. It consists of specialised cells called **chromaffin cells.** These are modified postganglionic cells of sympathetic nervous system.
- These chromaffin cells stimulate medulla to produce two hormones, i.e.
  - **Adrenaline** (epinephrine)
  - **Nor-adrenaline** (nor-epinephrine)

  On structural basis, these are commonly called **catecholamines.**
- These hormones are called **emergency hormone** as they are associated with initiating fight, fright and flight responses in the body under stress.
- Adrenaline and nor-adrenaline are rapidly secreted in response to stress in particular to restore the balanced condition.
- Effects of adrenaline and nor-adrenaline hormones on various body organs and processes are given below

Human system	Adrenaline	Nor-adrenaline
Heart rate	Increases	Decrease
Cardiac output	Increases	Slightly increases
Blood pressure	Raised in general	No change
• Systolic	Raised	Raised
• Diastolic	No change	Raised
• Mean arterial pressure	Raised	Raised
Skin vessels	Dilates some, constricts other	Constricts all

Human system	Adrenaline	Nor-adrenaline
Respiration	Stimulated	Stimulated
Kidney	Renal vessel constrictor	Renal vessel constrictor
CNS	Increases mental anxiety	No effect
Oxygen consumption	Increase	No effect
Blood sugar	Increase	Slight increase
Eosinophil count	Increase	No effect
Intestinal movement	Inhibited	Inhibited

Note • Adrenal gland is also known as '**4S gland**,' where 4S stands for sugar metabolism, salt retaining, sex hormone and stress hormone.

• **Pheochromocytomas** (Tumor of Adrenal Medulla) It occurs due to tumours of chromaffin cells. This causes hypersecretion of adrenaline.

PANCREAS

- **Position** It is present on the backside of stomach in abdominal cavity at the curve of duodenum.
- **Origin** It is endodermal in origin.
- **Structure** The gland is about 15 cm in length, 2.5 cm in width and 85 gms in weight. It is a **heterocrine gland** which have both exocrine as well as endocrine portions.
  - Its **exocrine part** is seen as **acini** which secrete about 500-1000 mL of pancreatic juice everyday.
  - The **endocrine part** (1-2% of gland) consists of cells that form **islets of Langerhans.**
  - The islets of Langerhans contain following 4 types of cells
    - **α-cells**
    - **β-cells**
    - **δ-cells**
    - **F-cell or PP cells**

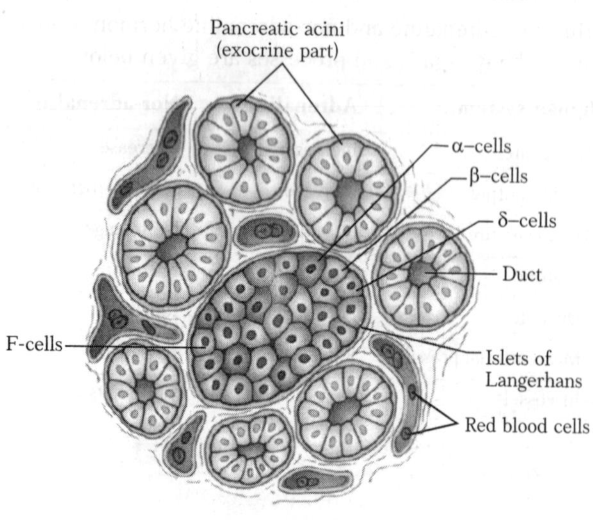

Fig. 29.9 Pancreas

Hormones of pancreas

Hormones produced by endocrine part of pancreas are as follows

1. Insulin

It is a peptide hormone secreted by β-cells of islets. Insulin acts mainly on hepatocytes and adipocytes.

Structure

- Insulin is derived from a 74 amino acid prohormone molecule called **proinsulin.** It is a protein containing 51 amino acids in 2 polypeptide chains.
- The first one, α or A chain contains 21 amino acids and the other, β or B chain contains 30 amino acids.
- These chains are linked with disulphide linkages.

Functions

- Insulin plays a major role in the regulation of glucose homeostasis.
- It stimulates the conversion of glucose to glycogen or glycogenesis thus, causing the blood sugar level to fall. It also promotes protein synthesis, fat synthesis and inhibits synthesis of ketone bodies.
- Secretion of insulin is stimulated by the increase in blood glucose, amino acid levels and gastrointestinal hormones.

Disorders : Diabetes mellitus

Hyposecretion of insulin results in the presence of excess glucose in blood. This condition is known as **hyperglycaemia.** Prolonged hyperglycaemia leads to a complex condition called **diabetes mellitus.**

Types of diabetes mellitus

(i) **Type I** (T1D) It is called Insulin Dependent Diabetes Mellitus (IDDM). This condition is also known as **juvenile onset diabetes** because most commonly it is seen in people having less than 20 year of age. It is an autoimmune disorder in which immune system destroys β-cells.

(ii) **Type II** It is Non-Insulin Dependent Diabetes Mellitus (NIDDM). It is also known as **maturity onset diabetes** because it occurs later in life. It does not arises from shortage of insulin. Instead in it, the target cells became less sensitive to insulin. Obesity may be the cause of it.

Symptoms/Effects of diabetes

The symptoms or effects of diabetes mellitus are as follows

- **Hyperglycaemia** Raised blood glucose level.
- **Glycosuria** It refers to excess of sugars in urine.

- **Polyuria** It refers to production of excess urine. Polyuria leads to hypovolemia, i.e. decreased volume of circulating blood in the body, extreme thirst or **polydipsia**, etc.
- **Weight loss** In diabetes, cells fail to metabolise glucose, resulting in weight loss due to gluconeogenesis and catabolism of excess body fat.
- **Ketoacidosis** In diabetes, the amount of available oxaloacetic acid is reduced because the glucose metabolism is reduced. As a result, excess acetyl coenzyme lowers the pH of the body fluids and cause ketoacidosis, i.e. presence of acids in blood.
- **Diabetic coma** It is a combined effect of ketoacidosis, dehydration and hyperglycaemia in patient. Due to this, patient become unconscious and may even die.

Treatment of diabetes

- Diabetes may be kept under control by the administration of insulin hormone and hypoglycaemics such as **tolbutamide, orinase** and **sulphonylurease.**
- The patient should be fed sugar-free diet in both the cases.
- Insulin cannot be taken orally as it is a proteinaceous hormone and breaks down in the digestive tract. Thus, it is administrated through injections only.

2. Glucagon

It is also a peptide hormone secreted by α-cells.

Structure

Glucagon is a 24 amino acid containing polypeptide. It is generated by the cleavage in proglucagon.

Function

It plays an important role in maintaining the normal glucose level in the blood. Functionally, it is antagonistic to insulin.

It acts mainly on hepatocytes (liver cell) and stimulates

- glycogenolysis, i.e. breakdown of glycogen.
- gluconeogenesis, i.e. synthesis of glucose molecules from fat and protein.
- It also reduces the uptake of sugar molecules from blood.

The above conditions increase the blood sugar level and thus, glucagon is also known as **hyperglycaemic hormone**.

3. Somatostatin

It is released by δ (delta) cells of islet of Langerhans.

Structure

It is a polypeptide hormone with 14 amino acid residues and a disulfide (S—S) bond.

Function

It plays an important role in the regulation of secretions of islets of Langerhans cells. Its effect is to inhibit the secretion of both glucagon and insulin.

4. Pancreatic Polypeptide (PP)

As the name suggests, it is a polypeptide hormone secreted by F-cells or PP-cells of pancreas. This hormone helps in reabsorption of food in the intestine.

CHECK POINT 29.2

1. Which of the following is the main difference between pituitary dwarfs and thyroid dwarfs?
(a) Pituitary dwarfs have dark complexion
(b) Pituitary dwarfs have mental retardation
(c) Thyroid dwarfs have mental retardation
(d) Thyroid dwarfs have high IQ

2. Which of the following disease leads to an autoimmune disorder in which person produces antibodies that mimic the action of TSH which is not regulated by normal negative feedback controls?
(a) Plummer's disease (b) Cretinism
(c) Grave's disease (d) Osteitis fibrosa cystica

3. If the secretion of TSH is stopped from pituitary gland, then which of the following condition is likely to happen?
(a) Irregular menstruation in adult women
(b) Reduction in size of thyroid gland
(c) Increased body temperature
(d) Weight loss

4. A hormone whose deficiency causes an autoimmune disorder, which affects more women than men?
(a) Parathyroid hormone (b) Thyroxine
(c) GH (d) Insulin

5. The hyposecretion of which hormones leads to loss of sodium and water through urine and lowering of blood pressure?
(a) Gonadotropins
(b) Thyroid hormones
(c) Adrenal cortex hormones
(d) Adrenal medulla hormones

6. Which of the following is not a function of catecholamines?
(a) Piloerection
(b) Increases sweat secretion
(c) Elevate blood glucose levels by promoting glycogen breakdown in liver
(d) Decrease the rate and force of muscular contraction

THYMUS GLAND

- **Position** It is a bilobed structure present in between pericardium and upper part of sternum.
- **Origin** This gland is **endodermal** in origin. It develops from the epithelium of the outer part of third gill pouch or epithelium of gill cleft.
- **Structure** From outside it is covered by a capsule of loose connective tissue.
  - It possesses a dense perished **cortex** and a loose central **medulla** on the inner side.
  - The outer cortex consists of densely packed thymocyte or cells of T-lymphocytes lineage.
  - The medulla consists of epithelial cells, few lymphocytes and macrophages. The nest of epithelial cells of thymus is called **Hassall's corpus.**

Hormone of thymus gland

- Thymus gland is well-developed in newborn child and it reduces continuously from adulthood to old age.
- The peptide hormone secreted by this gland is **thymosin.**
- This hormone promotes proliferation and maturation of T-lymphocytes responsible for **cell-mediated immunity.**
- It also promotes the production of antibodies to provide humoral immunity.

PINEAL GLAND

- **Position** The pineal gland (epiphysis) is located on the dorsal side of forebrain.
- **Origin** It develops from the **ectoderm** of the embryo.
- **Structure** It is a stalked, small, rounded, reddish-brown gland which resembles a pine cone. Its weight is about 150 mg and length is about 10 mm. It consists of pineal cells and supporting gland cells.

Hormones of pineal gland

Following three hormones are secreted by pineal glands

(i) **Melatonin** It is a derivative of amino acid tryptophan. It targets melanophore cells of skin. Various functions of melatonin are listed below

- Melatonin plays very important role in the regulation of a biological clock.
- It has antagonistic action to Melanocyte Stimulating Hormone (MSH) secreted by the pars intermedia of pituitary.
- It controls **sleep-wake cycle**, body temperature, etc. Increased production of melatonin occurs during darkness. Under intense light, retinal neurons gets stimulated and the production of melatonin is interrupted. It acts as biological clock.
- In addition, melatonin also influences metabolism, pigmentation on skin, menstrual cycle as well as defence capability.
- Melatonin hormone delays puberty in humans by opposing the action of LH and FSH and hence, affects the menstrual cycle and working of gonads.

(ii) **Serotonin** It is a biogenic amine hormone (5-Hydroxy tryptamine) which acts as a vasoconstrictor to increase blood pressure.

(iii) **Adrenoglomerulotropin** It stimulates the zona glomerulosa of adrenal cortex to secrete aldosterone.

Brain sand or Acervuli

In man, the pineal gland starts calcifying at the time of puberty. Such calcium deposits are called brain sand or Acervuli. Their presence may indicate increased secretory activity of pineal gland.

GONADS

- These are sex glands that develop from the **mesoderm** of the embryo.
- The main function of the gonads is to produce gametes.
- These are heterocrine or mixed glands, i.e. they have both exocrine and endocrine parts in function.
- The sex gland of males are called **testes** while that of females are called **ovaries.**

Testis

- **Position** In man, a pair of testis is present in the scrotal sac outside the abdomen.
- **Structure** A testis is composed of seminiferous tubules and interstitial tissue. Testes produce sperms as well as sex hormones.

Hormones of Testis

- Leydig cells or interstitial cells which are present in between the seminiferous tubules produce a group of hormones called **androgens.**
- **Testosterone** is the main androgen. These androgens are steroid hormones produced under the control of ICSH (LH) of anterior pituitary.
- **Dihydrosterone** is the active form of testosterone. It starts producing at the age of 10-13 years, i.e. at the beginning of puberty. At this time, testosterone initiates development of male secondary sex organs, i.e. seminal vesicles, prostate gland, scrotum and penis external male sex characters, like beard, moustaches, masculine voice, body hair, growth in body tissue, broadening of shoulders, aggressiveness, growth of muscles, increased sebaceous gland activity, etc.
- Androgens also play a major stimulatory role in the process of spermatogenesis, i.e. **sperm formation.**

Ovary

- **Position** It is a paired structure located in the abdomen in females. It is the primary female sex organ which produces one ovum during each menstrual cycle.
- **Structure** It is composed of ovarian follicles and stromal tissues. After ovulation, the ruptured follicle is converted to a hormone-releasing structure called **corpus luteum**.

Hormones of ovaries

The two important female sex hormones secreted by ovaries are **oestrogen** and **progesterone**. These are steroid hormones. In addition to these, some other female sex hormones are relaxin, hCG, HPL and inhibin/actin.

Details of ovarian hormones are as follows

(i) **Oestrogen** It is secreted by the ovarian follicles and influenced by FSH. It is the main female sex hormone. It initiates the development of female secondary sex organs like
- duct system of mammary glands
- accessory female sex characters like breasts, high pitch voice
- development of female body pattern and onset of menstrual cycle.

(ii) **Progesterone** It is secreted mainly by the corpus luteum under the influence of LH of anterior pituitary.
- It causes temporary changes in endometrial lining of uterus, i.e. secretory phase of menstrual cycle for receiving egg.
- This hormone basically supports pregnancy.
- Progesterone also acts on the mammary glands. It stimulates the formation of alveoli and milk secretion through mammary glands. Alveoli are sac-like structures that store milk in mammary glands.

(iii) **Relaxin** It is secreted by corpus luteum during the end of gestation period. It helps in
- loosening of pelvic ligaments
- softening and relaxing of uterus
- eases the process of parturition.

(iv) **Human Chorionic Gonadotropin** (hCG) It is secreted by the placenta for maintaining corpus luteum. Its presence in urine indicates pregnancy.

(v) **Human Somatomammotropin** (hCS) or **Human Placental Lactogenic** (HPL) **Hormone**. It is produced by the placenta and prepares mammary glands to secrete milk.

(vi) **Inhibin/Activin** It is produced by corpus luteum, placenta and testes. It can inhibit or activate gonadotropic activity of pituitary gland and hypophysis.

Disorders related to sex hormones

Abnormal functioning of sex hormones, i.e. testosterone, oestrogen and progesterone leads to following disorders in human males and females.

(i) **Eunuchoidism** It is a hormonal disorder due to deficient secretion of **testosterone** in males.

In this case
- the secondary male sex organs, such as prostate gland, seminal vesicle and penis are underdeveloped and non-functional.
- the external male sex characters like beard, moustache and masculine voice also fails to develop. Even sperm formation fails to occurs in it.

(ii) **Precocious puberty in males** It occurs in boys due to excessive production of testosterone before the age of 10 years. It causes
- early enlargement of penis
- faster body growth
- development of pubic and axillary hair
- mascularisation and then stunting

(iii) **Precocious puberty in females** In girls, it occurs before 9 years due to the excess production of oestrogen. It causes early development of breasts and pubic hair, but ovarian cycle does not occur.

(iv) **Gynaecomastia** It is the development of breast tissue in males. Gynaecomastia occurs mainly due to the disturbance in oestrogen and testosterone ratio.

HORMONES OF HEART, KIDNEY AND GASTROINTESTINAL (GI) TRACT

We have already discussed about the hormones released by major endocrine glands. However, as mentioned earlier some hormones are also secreted by such tissues and organs which are not well organised endocrine glands. Description of such hormones given below

Hormone of heart

- It secretes a peptide hormone with 28 amino acids.
- This hormone is called as **Atrial Natriuretic Factor** (ANF).
- It is released during hypertension. It decreases blood pressure by the dilation of blood vessels.

Hormones of kidney

- The juxtaglomerular cells of kidney produce a peptide hormone called as **erythropoietin**. It stimulates erythropoiesis, i.e. formation of RBC.
- In addition to this, kidney also produces **1, 25 dihydroxycholecalciferol**. It is a steroid hormone synthesised by the Proximal Convoluted Tubule (PCT) of kidney.
- It helps in forming vitamin-D and renin.
- Kidney also controls aldosterone secretion through the Renin-Angiotensin Aldosterone System (RAAS) pathway.

Hormones of gastrointestinal tract

- The hormones secreted by gastrointestinal tract includes enterogastrone, duocrinin, cholecystokinin, etc.
- The functions, secretion and the regulation of these hormones are already discussed in chapter 'digestion and absorption'.

HUMAN LOCAL HORMONES

A number of hormones are secreted by the cells which act locally on their neighbouring cells or on the same cell. They may or may not pass into the blood stream. These are widely dispersed in the body and are called local hormones or **tissue hormones** or **non-endocrine hormones**.

Description of important local hormones is given below

(i) **Histamine** This hormone is synthesised by the mast cells in the tissues and basophils in the blood. It is released as a part of the inflammatory process. It is also responsible for allergic reaction. It works through increasing capillary permeability and dilation.

(ii) **Prostaglandins** This is a group of substances produced from fatty acids by most tissues cells. These are believed to act inside cells, probably by modifying cellular responses to other stimuli.

(iii) **Leukotrienes** These are released from mast cells. They also assist in promoting allergic responses and inflammatory reactions. They attract neutrophils and eosinophils to the sites of inflammation. They are present in high concentrations in joints in case of rheumatoid arthritis.

(iv) **Thromboxanes** These are synthesised by the blood platelets. They cause vasoconstriction and platelet aggregation, thus contributing to the process of blood coagulation.

(v) **Pheromones** These are chemicals used for communication amongst the organisms of the same species like in insects.

(vi) **Cytokines** They are polypeptide hormones which help the defence cells of our body to communicate with each other.

They play very important role in the immune response of human body. The important cytokines related to immune responses are **interleukins** and **interferons.**

Note
- In addition to these hormones, some others are **acetylcholine** and **kinins** which also act as local hormones.
- The gaseous hormone nitric oxide is also a local hormone. It acts as signalling molecule of cardiovascular system that regulates blood pressure and blood flow. It is recently used in male potency drug called **viagra.**

MECHANISM OF HORMONE ACTION

All hormones produce their specific effects on target tissues/cells by binding to the specific receptors called **hormone receptors.** In target tissues, basically two types of receptors are present.

(i) **Membrane-bound extracellular receptors** These are hormone receptors present on the cell membrane of the target cells.

(ii) **Intracellular receptors** or **Nuclear receptor** These are receptors present inside the target cells.

The functioning of hormones depends upon their binding with these receptors.

On the basis of the chemical nature of hormones, their mechanism of action can be divided into following types

Peptide hormone action

- This mechanism was discovered by Nobel Prize winner, EW Sutherland in 1950.
- The hormones that are derivatives of amino acids, polypeptides or proteins are large molecules. These are collectively called **peptide hormones.**
- These hormones are lipid insoluble, hence they cannot directly enter the cell through it cell membrane.
- Thus, they act on the cell surface by binding to their respective receptors and forming hormone-receptor complex. This activates a biochemical pathway, leading to a physiological response.

- The complete process involves following steps
  - Hormone which is called **first** or **primary messenger** attaches to the cell surface receptor protein. This protein is present on the outer surface of plasma membrane of the target cell. As the result, hormone-receptor complex is formed.
  - This complex activates the enzyme, adenylate cyclase.
  - Adenylate cyclase catalyses the conversion of ATP to cyclic AMP (cAMP) on the inner surface of plasma membrane.
  - cAMP serves as the **'secondary messenger'** that conveys the information to the inside of the target cells. This activates appropriate enzyme system of cell. Thus, it accelerate the biochemical reactions in the cell.
  - cAMP has a very short existence as it is rapidly degraded by the cAMP phosphodiesterase.

  Apart from cAMP, Ca^{2+} and IP_3 also act as secondary messengers.

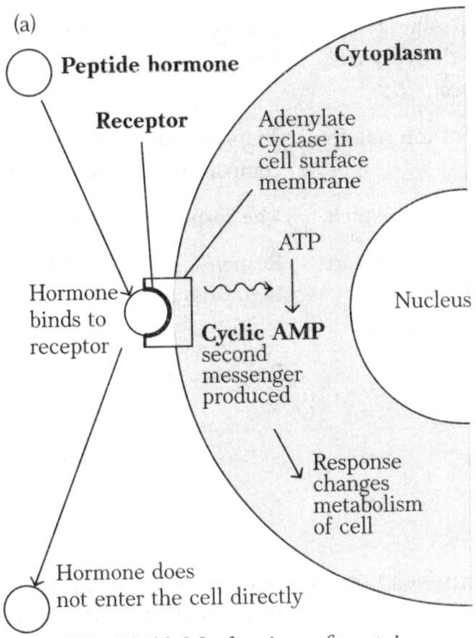

Fig. 29.10 Mechanism of protein hormone action

Steroid hormone action

- Steroid hormones are lipid soluble hormones. Thus, they can easily diffuse across the membranes and reach their nucleus to directly influence the gene action.
- Step-wise approach of complete process is given below
  - After entering into the cell, the hormone binds to the intracellular receptor forming a hormone-receptor complex.
  - This complex may inturn bind to the transcription factors. It further bind to the DNA and the particular gene is activated and transcribed.
  - Their transcription leads to the synthesis of a specific protein that can influence the metabolism of recipient cell.

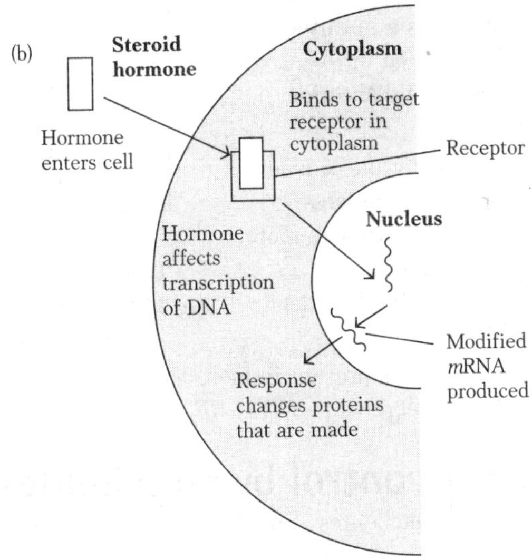

Fig 29.11 Mechanism of steroid hormone action

- Thus, the peptide hormones activate the existing enzymes in the cell, whereas the steroid hormones bring about the synthesis of new enzymes.
- Steroid hormones, usually act on a slower rate than peptides hormones. They also have a more sustained effect on metabolism as compared to peptide hormones.

REGULATION OF HORMONE ACTION

- Both hypoactivity and hyperactivity of an endocrine gland produces structural or functional abnormalities.
- Hence, the normal functioning of endocrine glands and the level of hormones in the body needs to be regulated. This is possible by **feedback mechanisms**.
- These feedback mechanism work on a simple principle that a hormone will be synthesised only when it is needed.
- Thus, feedback mechanism may be positive or negative and may operate in following ways

Feedback control by hormones

- The hyposecretion of a hormone is sometimes dependent upon the hormones secreted by other glands.
- For example, hypothalamus is stimulated by some external stimulus and produces releasing hormones.
- This stimulates anterior pituitary to secrete a specific hormone like Thyroid Stimulating Hormone (TSH), etc.

- TSH inturn stimulates thyroid gland to produce hormone thyroxine.
- If the level of thyroxine in blood is high, it exerts a negative feedback effect on hypothalamus and anterior pituitary.
- If the level of thyroxine is less than normal, it stimulates hypothalamus to produce more. Thyroid Stimulating Hormone-Releasing Factors (TSH-RF).

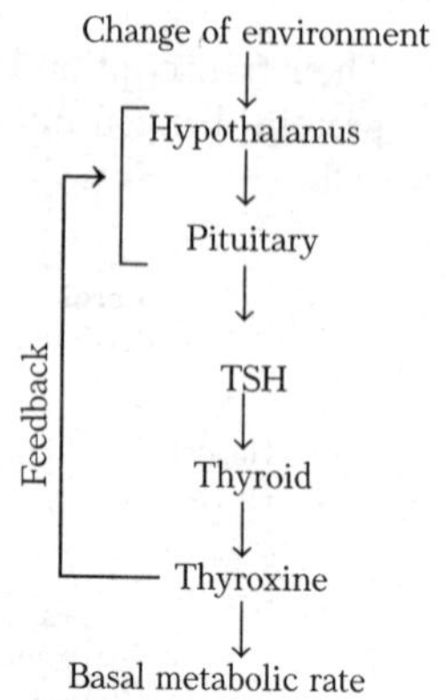

Fig 29.12 Feedback control by hormones

- But, if thyroxine is present in more amount, the hypothalamus secretes less TSH-RF.

Feedback control by metabolites

- The levels of metabolites also affect the secretion of certain hormones.
- For example, after a meal, glucose level of blood rises which stimulates the secretion of insulin to act on it. With the fall in glucose level in blood, a decrease in insulin level takes place.
- In this way, insulin maintains blood-glucose homeostasis.

Feedback control by nervous system

- Nervous system also plays important role in feedback control. For example, an emotional stress stimulates the sympathetic nervous system.
- Sympathetic nerves of adrenal gland inturn, stimulate adrenal medulla to produce adrenaline hormone. This leads to increase in blood pressure, heartbeat and rate of respiration.
- When emotional stress is over, sympathetic nerve signals the adrenal medulla to stop the release of adrenaline.
- Thus, the signals of nervous system also controls the secretion of hormones and help in maintaining homeostasis or internal environment of body.
- Besides helping in regulation of secretions, there are marked difference between nervous and endocrine systems.

These differences are summarised below

Nervous system	Endocrine system (Chemical system)
It receives and sends the information in the form of electrical impulses.	It receives and sends the information in the form of chemicals called hormones.
The system has direct connection with every part under its control.	The system is not directly connected with parts under its control.
Transmission of information is rapid.	Transmission of information is comparatively slower.
The response is very quick.	The response is comparatively slow.
Response is limited only to those cells which are innervated by neurons.	Response is produced by all the cells of target tissue.
Response is for short duration.	Response is for longer duration.

CHECK POINT 29.3

1. Which of the following is incorrect statement about melatonin?

(a) It is released from a gland attached to the roof of fourth ventricle of brain
(b) It inhibits growth of sex organs before puberty
(c) It promote synthesis and release of gonadotropins
(d) It coordinate day and night rhythms of body by influencing hypothalamus

2. Hormones artificially taken by women to avoid unwanted pregnancy are

(a) cortisol and progesterone (b) oestrogen and progesterone
(c) cortisol and oestrogen (d) progesterone and androgens

3. Steroidal hormones and iodothyronines can easily pass through the plasma membrane to bind with intracellular receptors because

(a) they are aqua-soluble
(b) they cannot bind with membrane receptors
(c) they are lipophilic
(d) they generate secondary messengers

4. The hormones bound with the cell membrane of the target cell generate ...*A*... upon interaction with the receptor. Which of the following option cannot be used to fill in the blank?

(a) cyclic AMP (b) calcium
(c) potassium (d) IP_3

5. Which of the following hormone does not bind with extracellular receptor protein to stimulate hormone action?

(a) Catecholamines
(b) Vasopressin
(c) Insulin
(d) Testosterone

6. What will be the reaction according to feedback control mechanism if the level of thyroxine is less than normal?

(a) It will produce more TSH-RF
(b) It will produce less TSH-RF
(c) No effect seen
(d) None of the above

EXAM POINTS

- Non-nutrient chemicals produced by the body in trace amount that act as intercellular messenger are known as hormones. **[JIPMER 2017]**
- Inhibin is composed of glycoproteins. **[JIPMER 2018]**
- Bradykinin is a peptide hormone. **[JIPMER 2019]**
- Epinephrine is an amino acid derivative hormone. **[NEET 2018]**
- Hypothalamus is the coordinator between nervous and endocrine system. **[MHT CET 2015]**
- GnRH, a hypothalamic hormone, needed in reproduction acts on anterior pituitary gland and stimulates secretion of LH and FSH. **[NEET 2017]**
- Function of somatostatin is to inhibit the release of growth hormone from pituitary. **[Kerala CEE 2015]**
- Hormones stored and released from neurohypophysis are oxytocin and vasopressin. **[NEET 2020]**
- Oxytocin and ADH are produced by hypothalamus and released from posterior pituitary. **[AIIMS 2019]**
- Pituitary gland releases oxytocin hormone which helps in childbirth. **[AIIMS 2019]**
- Ectopically transplanted pituitary secretes prolactin hormone abundantly. **[Chhattisgarh PMT 2015]**
- Epiphyseal plates close after adolescence therefore, hypersecretion of growth hormones in adults does not cause further increase in height. **[NEET 2017]**
- Deficiency of iodine in diet result of stunted growth, mental retardation/low intelligence quotient and abnormal skin. **[NEET 2013]**
- Grave's disease is caused due to the hypersecretion of thyroid gland. **[NEET 2018]**
- Fall in blood Ca^{2+} levels will stimulate parathyroid gland to release parathyroid hormone. **[NEET 2019]**
- Oestrogen and parathyroid hormone play a significant role in osteoporosis. **[NEET 2018]**
- Phenylalanine–N–methyl transferase enzyme is responsible for the conversion of nor-epinephrine to epinephrine. **[WB JEE 2015]**
- Fight or flight reactions cause activation of the adrenal medulla, leading to increased secretion of epinephrine and nor-epinephrine. **[CBSE AIPMT 2014]**
- Glucocorticoids shows anti-allergic and anti-inflammatory effect. **[Gujarat CET 2015]**
- Insulin causes rapid movement of glucose from blood to hepatocytes and adipocytes resulting in hypoglycaemia. **[UK PMT 2014]**
- Type-I diabetes is caused the destruction of β-cells of islets of Langerhans. For its treatment, insulin is given intravenously. **[AIIMS 2019]**
- Pineal gland in human, plays an important role in the regulation of rhythm of the body. **[J&K CET 2014]**
- Artificial light, extended work time and reduced sleep time disrupt the activity of pineal gland. **[NEET 2019]**
- Serotonin and melatonin are hormones secreted by pineal body. **[KCET 2015]**
- The diurnal rhythms are regulated by melatonin. **[MHT CET 2014]**
- The Leydig cells found in the human body are the secretory source of androgens. **[Manipal 2015]**
- Placenta produces several hormones like hCG, hPL, oestrogen and progesterone. **[NEET 2016]**
- Urine of pregnant woman contains hCG. **[JIPMER 2018]**
- Atrial Natriuretic Factor (ANF) decreases blood pressure. **[MHT CET 2015]**
- Steroid hormone influence the cellular activities by binding to DNA and forming a gene-hormone complex. **[NEET 2019]**
- Steroid hormones alter the activity of genes. **[UK PMT 2015]**

Chapter Exercises

(A) Topical Exercise

Introduction, Hypothalamus and Pituitary Gland

1 Chemical messenger secreting ductless structures are
(a) holocrine glands (b) exocrine glands
(c) endocrine glands (d) tubular glands

2 Secretin was the first hormone to be discovered. It is present in the
(a) intestinal mucosa (b) hypophysial axis
(c) heart (d) liver

3 Consider the following statements.
I. Oestrogen and prolactin have synergistic effect on mammary gland growth.
II. Insulin and glucagon oppose each other's action, thus these are antagonistic hormones.
Choose the correct option.
(a) Statement I is true, but II is false
(b) Statement I is false, but II is true
(c) Both statements I and II are true
(d) Both statements I and II are false

4 Given below are the different classes of hormones and their examples.
I. Amines - Tyrosine
II. Peptide - Thyroxine
III. Steroid - Progesterone
IV. Proteinaceous - Relaxin
Choose the incorrectly matched pairs.
(a) I and III (b) II, III and IV
(c) Only II (d) None of these

5 Choose the odd one out with respect to source of hormones in listed invertebrates.
(a) Molluscs (b) Nematodes
(c) Platyhelminthes (d) Echinoderms

6 Hypothalamic GnRH (Gonadotropin Releasing Hormone) stimulates the
(a) pancreas to release the gonadotropin
(b) pituitary for synthesise and release of gonadotropin
(c) testis is to release the gonadotropin
(d) hypothalamus to release the gonadotropin

7 The abbreviation TSH stands for
(a) Thyroxine Stimulate Hormone
(b) Thymin Stimulating Hormone
(c) Thyroxine Secreting Hormone
(d) Thyroid Stimulating Hormone

8 Transport of hormones occurs through axons from
(a) hypothalamus – adenohypophysis
(b) adenohypophysis – target organ/tissue
(c) hypothalamus – neurohypophysis
(d) neurohypophysis – target organ/tissue

9 Pituitary gland lies in the ....
(a) sella turcica (b) buccal cavity
(c) diencephalon (d) None of these

10 Functionally, the adenohypophysis of pituitary gland includes
(a) anterior lobe (b) posterior lobe
(c) intermediate lobe (d) Both (a) and (c)

11 Consider the following statements.
I. Pituitary gland is also called hypophysis cerebri.
II. Pituitary gland regulates the activity of other endocrine glands.
Choose the correct option.
(a) Statement I is true, but II is false
(b) Statement I is false, but II is true
(c) Both statements I and II are true
(d) Both statements I and II are false

12 Mammalian prolactin is secreted by
(a) adenohypophysis (b) neurohypophysis
(c) adrenal cortex (d) adrenal medulla

13 Which of the following hormones is not secreted by anterior pituitary? **[NCERT Exemplar]**
(a) Growth hormone
(b) Follicle stimulating hormone
(c) Oxytocin
(d) Adrenocorticotropic hormone

14 A globular protein made up of 188 amino acids and secreted by anterior pituitary is
(a) growth hormone (b) ACTH
(c) prolactin (d) progesterone

15 Gigantism and acromegaly are due to
(a) hypothyroidism
(b) hyperthyroidism
(c) hypopituitarism
(d) hyperpituitarism

16 Degeneration of anterior lobe of pituitary gland causes
(a) Addison's disease (b) Cushing's disease
(c) Dwarfism (d) Simmond's disease

17 The secretion of ACTH from the pituitary stimulates the release of

(a) aldosterone from the adrenal medulla
(b) cortisol from the adrenal cortex
(c) epinephrine from the adrenal medulla
(d) renin from the kidney

18 Select the incorrect statement.

(a) Pituitary gland is found in the cavity of sphenoid bone
(b) Pars intermedia develop from lower surface of Rathke's pouch and it does not produce any hormone
(c) Hormones secreted by pars distalis are called trophic hormones
(d) Pars nervosa originates from the infundibulum

19 Hormones of posterior lobes of pituitary gland are

(a) peptides
(b) steroids
(c) iodothyronines
(d) amines

20 What is true about neurohypophysis?

(a) Store hormones produced by adenohypophysis
(b) Functionless in humans
(c) Stores and releases neurohormones secreted by hypothalamus
(d) Secretes its own hormone

21 Identify the hormones, labelled as *A*, *B* and *C* in the following figure.

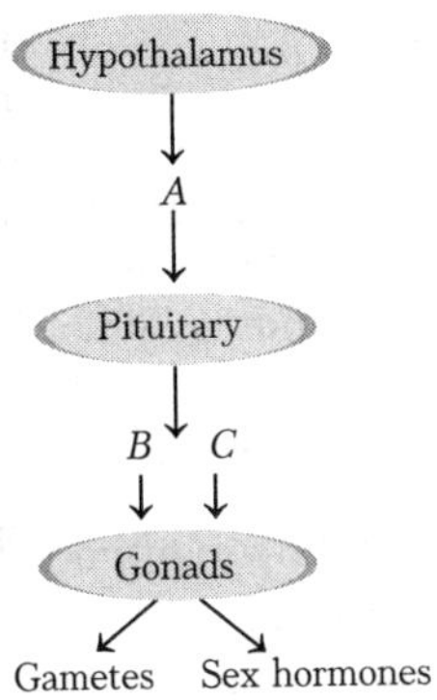

(a) *A*-GH, *B*-FSH, *C*-LH
(b) *A*-GnRH, *B*-FSH, *C*-LH
(c) *A*-GH, *B*-GnRH, *C*-PRT
(d) *A*-GnRH, *B*-GH, *C*-PRT

22 A small peptide hormone released by posterior lobe of pituitary is concerned with

(a) metabolism of carbohydrates
(b) stimulation of thyroid
(c) secondary sexual characters
(d) contraction of uterus

23 If a person takes large amount of water, the amount of ADH in blood will

(a) increase
(b) decrease
(c) first increase then decrease
(d) will remain unchanged

24 Which of the following hormones originates from the supraoptic and paraventricular nuclei of the hypothalamus?

(a) Prolactin
(b) Oestrogen
(c) Antidiuretic hormone
(d) Luteinizing hormone

25 Function of ADH is

(a) regulation of water level in blood
(b) stimulate growth of tissues
(c) maintain progesterone production
(d) manage glucose level in blood

26 A scientist develops a new drug and finds that when it is administered in humans there is a substantial increase in the volume of urine produced. When ADH/vasopressin are admistered at the same time, the volume of urine returns to normal. Which hypothesis best fits these observations? The new drug

(a) blocks the receptors for ADH on the collecting tubules of the kidney
(b) blocks the release of ADH from the pituitary
(c) mimics the action of ADH
(d) decreases blood pressure

27 Match the following columns.

Column I	Column II
A. Pitressin	1. Oxytocin
B. Pitocin	2. Metachrosis
C. MSH	3. Antidiuretic hormone

Codes

	A	B	C		A	B	C
(a)	1	2	3	(b)	1	3	2
(c)	2	3	1	(d)	3	1	2

Thyroid Gland, Parathyroid Gland, Adrenal Gland and Pancreas

28 Thyroid is

(a) a bone in thorax
(b) a waste material produced in intestine
(c) an endocrine gland located at the base of the neck
(d) an endocrine gland located near the kidneys

29 Consider the following statements.

I. The stromal tissue of thyroid gland is composed of squamous epithelial cells.
II. Thyroid hormones regulate BMR and maintain electrolyte balance in body.

Choose the correct option.

(a) Statement I is true, but II is false
(b) Statement I is false, but II is true
(c) Both statements I and II are true
(d) Both statements I and II are false

30 Which gland plays key role in metamorphosis of frog?
(a) Adrenal (b) Thyroid (c) Thymus (d) Pancreas

31 Functions of thyroid hormone are
I. fat and protein metabolism.
II. maintains the BMR.
III. maintains blood calcium levels.
IV. thermoregulation.
Choose the correct option.
(a) I and III (b) II and IV
(c) I, II and IV (d) I, III and IV

32 A man has IQ equivalent to that of a boy 5 years old, this is due to the deficiency of which hormone?
(a) Thyroxine (b) Adrenaline
(c) Aldosterone (d) Somatotropin

33 Which of the following is correct labelling of the given figure?

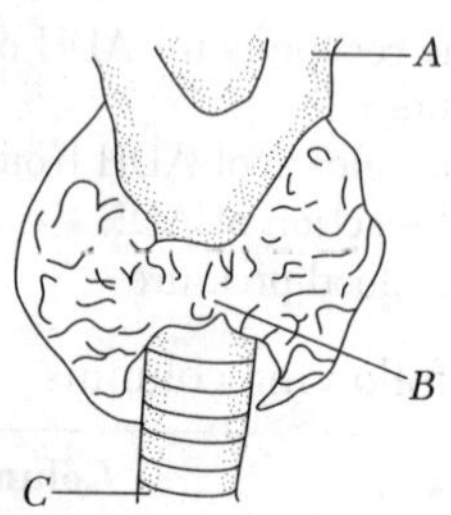

(a) *A*-Thyroid cartilage, *B*-Thyroid, *C*-Trachea
(b) *A*-Thyroid, *B*-Isthmus, *C*-Trachea
(c) *A*-Trachea, *B*-Isthmus, *C*-Thyroid
(d) *A*-Trachea, *B*-Thyroid, *C*-Isthmus

34 Which of the following diseases results from endocrine disorder?
(a) Pneumonia (b) Typhoid
(c) Diabetes (d) Jaundice

35 Which of the following conditions is not linked to deficiency of thyroid hormone? **[NCERT Exemplar]**
(a) Cretinism (b) Goitre
(c) Myxoedema (d) Exophthalmia

36 In Plummer's disease, ...
(a) BMR is increased
(b) BMR is decreased
(c) calcium level decreases
(d) protrusion of eyeballs takes place

37 Immune disease in which, body destroys the ill-functioning thyroid is
(a) Simmond's disease (b) cretinism
(c) Hashimoto's disease (d) myxoedema

38 The person showing low BMR, possibly will be suffering from
(a) myxoedema (b) cretinism
(c) Hashimoto's disease (d) Graves' disease

39 A health disorder that results from the deficiency of thyroxine in adults is characterised by
I. a low metabolic rate.
II. increase in body weight.
III. tendency to retain water in tissues.
(a) hypothyroidism (b) simple goitre
(c) myxoedema (d) cretinism

40 Calcitonin is a thyroid hormone which
(a) elevates potassium level in blood
(b) lowers calcium level in blood
(c) elevates calcium level in blood
(d) has no effect on calcium

41 Who gave the molecular structure of thyroxine hormone?
(a) Gudernatsch
(b) Kendall
(c) WB Canon
(d) Harington and Barger

42 Gull's disease is related to deficient working of
(a) thyroid (b) parathyroid
(c) adrenal cortex (d) gonads

43 The cells that secrete the parathyroid hormone are known as
(a) Chief cells (b) Herring's cells
(c) Goblet cells (d) Both (a) and (b)

44 A person is having problems with calcium and phosphorus metabolism in his body. Which one of the following glands may not be functioning properly?
(a) Parathyroid (b) Parotid (c) Pancreas (d) Thyroid

45 Which of the following two hormones have antagonistic effects?
(a) Parathormone and calcitonin
(b) FSH and LH
(c) Oestrogen and progesterone
(d) ADH and melatonin

46 Select the incorrect statement.
(a) The hormones of parathyroid gland are called Collip's hormones
(b) Parathyroid gland develops from fifth pharyngeal pouch, as epithelial outgrowth
(c) Oxyphil cells are considered as degenerated chief cells
(d) Parathormone is a linear polypeptide of about 84 amino acids

47 Which one of the following pairs of organs includes only the endocrine glands?
(a) Parathyroid and adrenal
(b) Pancreas and parathyroid
(c) Thymus and testes
(d) Adrenal and ovary

48 Identify the incorrect statement.

(a) Zona reticularis is the outermost region of adrenal cortex
(b) Adrenal cortex releases about 20 steroid hormones
(c) Cortisol is released by the zona fasciculata region of adrenal gland
(d) Adrenal glands are found enclosed within renal fascia

49 The steroid responsible for balance of water and electrolytes in our body is **[NCERT Exemplar]**

(a) insulin (b) melatonin
(c) testosterone (d) aldosterone

50 Cortisol is secreted from **[NCERT Exemplar]**

(a) pancreas (b) thyroid (c) adrenal (d) thymus

51 Which of the following hormones produces anti-inflammatory reactions, suppresses immune response, stimulates RBC production and is also involved in maintaining cardiovascular system and kidney functions?

(a) Aldosterone (b) Epinephrine
(c) Cortisol (d) Nor-epinephrine

52 Androgens are sexcorticoids secreted from

(a) zona glomerulus (b) zona fasciculata
(c) zona reticularis (d) None of these

53 Hypersecretion of cortisol (corticoids) produces a disease known as

(a) cushing's syndrome (b) haemophilia
(c) anaemia (d) mental retardation

54 Conn's syndrome is characterised by

(a) muscular weakness (b) hypertension
(c) retention of sodium (d) All of these

55 The excess of sex corticoids causes a disease in females

(a) adrenal virilism (b) aldosteronism
(c) Addison's disease (d) All of these

56 Cushing's syndrome and myxoedema are associated with these glands respectively

(a) thyroid, adrenal (b) adrenal, thyroid
(c) parathyroid, thyroid (d) adrenal, pituitary

57 Which of the following is not under direct control of pituitary gland with respect to the regulation of its secretory function?

(a) Adrenal cortex (b) Adrenal medulla
(c) Thyroid (d) Testis

58 Match the following columns.

Column I	Column II
A. Pineal	1. Epinephrine
B. Thyroid	2. Melatonin
C. Ovary	3. Oestrogen
D. Adrenal medulla	4. Tetraiodothyronine

Codes **[NCERT Exemplar]**

	A	B	C	D		A	B	C	D
(a)	4	2	1	3	(b)	2	4	1	3
(c)	3	2	1	4	(d)	2	4	3	1

59 Mary is about to face an interview. But during the first five minutes before the interview she experiences sweating, increased rate of heartbeat, respiration, etc. Which hormone is responsible for her restlessness? **[NCERT Exemplar]**

(a) Oestrogen and progesterone
(b) Oxytocin and vasopressin
(c) Adrenaline and nor-adrenaline
(d) Insulin and glucagon

60 The hormone which may increase the passage of air through the lungs

(a) adrenaline (b) prolactin
(c) ACTH (d) thyroxine

61 The functioning of adrenal medulla gland is similar to those of nerves because adrenal medulla

(a) and nervous system are derived from embryonic mesoderm
(b) and neurons secrete similar chemicals such as adrenaline and nor-adrenaline
(c) does not secrete any hormone
(d) is made up of nervous tissue

62 Which option is correct?

(a) Glucagon and insulin are secreted by adrenal gland
(b) Glucagon and insulin are exocrine secretions
(c) Glucagon and insulin both target hepatocytes
(d) None of the above

63 Glucagon produced by α-cells of islets of Langerhans

(a) converts glucose to glycogen
(b) converts glycogen to glucose
(c) decreases the concentration of glucose in blood
(d) None of the above

64 Which hormone possesses anti-insulin effect?

(a) Glucagon
(b) Calcitonin
(c) Oxytocin
(d) Aldosterone

65 A person has poorly functioning β-cells in the islets of Langerhans.

What will be the concentration of glucose, insulin and glucagon in the blood after that person has eaten a meal rich in carbohydrates?

	Glucose	Insulin	Glucagon
(a)	High	Low	High
(b)	High	Low	Low
(c)	Low	High	High
(d)	Low	High	Low

66 Source of somatostatin is the same as that of
(a) insulin and glucagon
(b) vasopressin and oxytocin
(c) thyroxine and calcitonin
(d) somatotropin and prolactin

67 Match the following columns.

Column I	Column II
A. Epinephrine	1. Stimulates muscle growth
B. Testosterone	2. Decreases blood pressure
C. Glucagon	3. Breakdown of liver glycogen
D. Atrial natriuretic factor	4. Increases heartbeat

Codes [NCERT Exemplar]

	A	B	C	D
(a)	2	1	3	4
(b)	4	1	3	2
(c)	1	2	3	4
(d)	1	4	2	3

68 A symptom of diabetes mellitus is
(a) glyconemia (b) polydypsia
(c) weight gain (d) hypoglycemia

69 Identify the incorrectly matched pair.
(a) Type-II diabetes – Autoimmune disorder
(b) Exocrine pancreas – Acini
(c) Pancreatic polypeptide – F-cells of pancreas
(d) Somatostatin – Inhibits glucagon and insulin

Thymus Gland, Pineal Gland, Gonads and Hormones of Other Organs

70 A gland which is well-developed in newborn child and reduces continuously from adulthood to oldage is
(a) thyroid gland (b) parathyroid gland
(c) thymus (d) gonads

71 Thymosin is responsible for
(a) raising the blood sugar level
(b) raising the blood calcium level
(c) maturation of T-lymphocytes
(d) decrease in blood RBCs

72 Which of the following does not play any role in calcium balance in the human body? [NCERT Exemplar]
(a) Vitamin-D (b) Parathyroid hormone
(c) Thyrocalcitonin (d) Thymosin

73 The function of pineal body is to
(a) lighten the skin colour
(b) control sexual behaviour
(c) regulate the period of puberty
(d) All of the above

74 A hormone responsible for normal sleep-wake cycle is
(a) epinephrine (b) gastrin
(c) melatonin (d) insulin

75 Leydig cells produce a group of hormones called
(a) androgens (b) oestrogens
(c) aldosterone (d) gonadotropins

76 Deficiency of testosterone secretion may cause
(a) acromegaly (b) eunuchoidism
(c) diabetes (d) cretinism

77 Which part of ovary in mammals acts as an endocrine gland after ovulation?
(a) Graafian follicle (b) Stroma
(c) Germinal epithelium (d) Vitelline membrane

78 The hormone which is not produced at the end of gestation period only is
(a) relaxin (b) oxytocin
(c) oestrogen (d) prolactin

79 Secretion of large quantities of oestrogen causes
(a) growth of Fallopian tube
(b) growth of mammary gland
(c) enlargement of female external genitalia
(d) All of the above

80 Corpus luteum secretes a hormone called [NCERT Exemplar]
(a) prolactin (b) progesterone
(c) aldosterone (d) testosterone

81 Progesterone is a/an
(a) enzyme for digesting proteins
(b) hormone to initiate uterine contraction
(c) amino acid which may be cause of alkaptonuria
(d) hormone concerned with retention and maintenance of pregnancy

82 Placenta present in mammals acts as an endocrine tissue and produces
(a) human chorionic gonadotropin
(b) oestrogen
(c) progesterone
(d) testosterone

83 A decrease in the level of progesterone causes
(a) growth and dilation of myometrium
(b) growth of endometrium
(c) constriction of uterine blood vessels leading to sloughing of uterine epithelium
(d) release of ovum from the ovary

84 Which ovarian hormone is proteinaceous?
(a) Oestradiol
(b) Progesterone
(c) Relaxin
(d) Human chorionic gonadotropin

85 If a pregnant woman is having prolonged labour pains then, to aid parturition, it is advisable to administer a hormone that can
(a) activate the smooth muscles
(b) increase the metabolic rate
(c) tighten the pelvic ligament
(d) stimulate the ovary

86 Consider the following statements.
I. Inhibin secreted by corpus luteum can inhibit gonadotropic activity of pituitary gland.
II. Disturbance in oestrogen and testosterone ratio in males causes gynaecomastia.
Choose the correct option.
(a) Statement I is true, but II is false
(b) Statement I is false, but II is true
(c) Both statements I and II are true
(d) Both statements I and II are false

87 A peptide hormone having 28 amino acids and secreted by heart is
(a) atrial natriuretic factor (b) erythropoietin
(c) oestrogen (d) duocrinin

88 Choose the incorrectly matched pair.
(a) GIT hormone - Duocrinin
(b) Leukotrienes - Released by platelets
(c) Cytokines - Interferons
(d) Nitric oxide - Viagra

89 Pheromone is
(a) a product of endocrine gland
(b) used for animal communication
(c) messenger RNA
(d) always protein

90 Which of the following organs in mammals does not consist of a central 'medullary' region surrounded by a cortical region. **[NCERT Exemplar]**
(a) Ovary (b) Adrenal (c) Liver (d) Kidney

Mechanism and Regulation of Hormone Action

91 Hormones are called chemical signals that stimulate specific target tissues. Which is the correct location of receptors of these hormones? **(NCERT Exemplar)**
(a) Extracellular matrix
(b) Blood
(c) Plasma membrane
(d) Nucleus

92 Intracellular receptors are mostly
(a) cytoplasmic receptors only
(b) membrane receptors
(c) nuclear receptors
(d) ER receptors

93 Why do some hormones need to trigger a 'second messenger' to activate a target cell?
(a) The hormone needs activation of ATP
(b) The first messenger cannot cross a plasma membrane
(c) There are no specific cell surface receptors for hormones
(d) The first messenger is not a water soluble molecule

94 In the mechanism of action of a protein hormone, one of the second messenger is **[NCERT Exemplar]**
(a) cyclic AMP (b) insulin
(c) T_3 (d) gastrin

95 Steroid hormones easily pass through the plasma membrane by simple diffusion because they
(a) are water soluble
(b) contain carbon and hydrogen
(c) enter through pores
(d) are lipid soluble

96 A steroid hormone typically alters the activity of its target cells by
(a) making holes in the members of the target cells
(b) entering the cell and altering gene expression
(c) making holes in the lysosome of target cells
(d) activation of cAMP

97 Which one is responsible for the synthesis of new enzymes?
(a) Peptide hormone (b) Steroid hormone
(c) Primary messenger (d) Secondary messenger

98 If receptor molecule is removed from target organ for hormone action, the target organ will
(a) continue to respond, but requires higher concentration of hormone
(b) continue to respond, but in opposite way
(c) continue to respond without any difference
(d) not respond to hormone

99 Feedback control may occur by means of
(a) hormones (b) metabolites
(c) nerves (d) All of these

100 Which glands are indicated as A, B and C in the given figure?

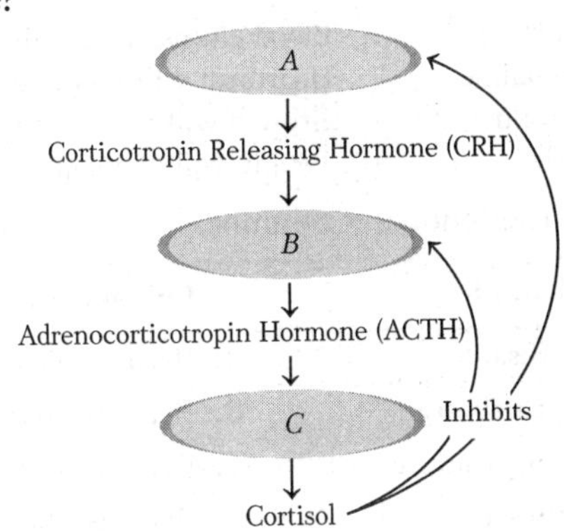

(a) *A*–Hypothalamus, *B*–Anterior pituitary, *C*–Adrenal cortex
(b) *A*–Hypothalamus, *B*–Anterior pituitary, *C*–Posterior pituitary
(c) *A*–Pituitary, *B*–Thyroid, C–Pineal
(d) *A*–Pituitary, *B*–Thyroid, C–Parathyroid

101 In a positive feedback system, where hormone Y alters the amount of protein X.
(a) A decrease in X always causes a decrease in Y
(b) A decrease in Y always produces in increase in X
(c) An increase in X always produces a decrease in Y
(d) An increase in Y always produces an increase in X

102 Two organs secrete substances which affect the body
organ 1 ⟶ product 1
organ 2 ⟶ product 2

Negative feedback control of product 2 would be achieved if
(a) product 1 counteracts product 2
(b) product 1 reinforces the effect of product 2
(c) product 2 inhibits organ 1 and product 1 stimulates organ 2
(d) product 2 stimulates organ 1 and product 1 stimulates organ 2

103 The reason that aldosterone affects only a small number of cells in the body is that
(a) only its target cells get exposed to aldosterone
(b) it is unable to enter the non-target cell
(c) only its target cells contain aldosterone receptors
(d) non-target cells destroy aldosterone before it can produce any effect

104 Through negative feedback, a hormone may shut off the secretion of an anterior pituitary hormone by
(a) stimulating the release of a (hypothalamic) releasing hormone
(b) inhibiting the release of a (hypothalamic) inhibiting hormone
(c) inhibiting the release of a (hypothalamic) releasing hormone
(d) All of the above

(B) Cumulative Exercise

(Assorted Questions of the Chapter for Advanced Level Practice)

1 In the homeostatic control of blood sugar level, which organs function as modulator and effector, respectively?
(a) Liver and islets of Langerhans
(b) Hypothalamus and liver
(c) Hypothalamus and islets of Langerhans
(d) Islets of Langerhans and hypothalamus

2 Pick out the incorrect statement.
(a) Vasopressin is an antidiuretic hormone
(b) Sex hormones are protein in nature
(c) LH and ICSH are the same hormones
(d) Glucagon is a catabolic hormone

3 In case the islets of Langerhans stop functioning, which hormone will be in short supply and what will be its effect?
(a) Insulin – Blood glucose level will rise
(b) Adrenaline – Heartbeat will increase
(c) Thyroxine – Growth will be retarded
(d) Cortin – Tetany will develop

4 Match the following columns.

Column I	Column II
A. Dwarfism	1. Hyposecretion of thyroxine
B. Grave's disease	2. Hypersecretion of thyroxine
C. Cushing's disease	3. Hyposecretion of GH
D. Cretinism	4. Hypersecretion of ACTH

Codes

	A	B	C	D
(a)	3	2	4	1
(b)	4	3	2	1
(c)	2	1	3	4
(d)	1	2	3	4

5 Which of the following statements is not false?
(a) Hormone produced by thyroid control protein and fat metabolism
(b) Hormone produced by ovary promotes the uterine contraction
(c) Hormone produced by small intestine stimulates heart
(d) Hormone produced by adrenal cortex stimulates heart beat

6 Which of the following are concerned with thymus gland?
I. It is well-developed in newborns, but reduces from adulthood to old age after sexual maturity.
II. It is located in the posterior portion of the root of the third ventricle.
III. It secretes hormone thymosin, which promotes immune competence in young T-lymphocytes.
IV. It functions as biological clock.
Choose the correct option.
(a) I, II and III
(b) II, III and IV
(c) II and IV
(d) I and III

7 Match the following columns.

Column I	Column II
A. Oestrogen	1. Gonads
B. FSH and LH	2. Anterior pituitary
C. Androgens	3. Testis
D. Gonadotropin releasing hormone	4. Ovary

Codes

	A	B	C	D
(a)	1	2	3	4
(b)	4	1	3	2
(c)	2	3	4	1
(d)	3	4	1	2

8 If you were outside for a long time on a hot, dry day, without anything to drink, which of the following would happen?

(a) The osmotic pressure of your blood would decrease
(b) The production of thyroxine by your thyroid gland would increase
(c) The secretion of ADH from your pituitary gland would increase
(d) The reabsorption of fluid from your kidney tubules would decrease

9 In a hormone control system, hormone *A* inhibits the secretion of substance *B*. If this system works by positive feedback, which of the following would be correct?

(a) Levels of *B* will stabilise at a moderate level
(b) Substance *B* stimulates the production of hormone A
(c) Secretion of substance *B* should stop completely
(d) level of hormone *A* should not be affected by the levels of substance *B*

10 Suppose you were working in a diagnostic laboratory and had measured high levels of glucagon and low levels of insulin in a blood sample from a patient. What would be the most likely explanation?

(a) The patient had consumed a large soft drink on the way to the lab
(b) The patient was suffering from diabetes
(c) The patient had not eaten anything for several hours
(d) You had made an error in one or both of the measurements

11 A scientist develops a drug that she think might inhibit the release of Gonadotropin-Releasing Hormone (GnRH) from the hypothalamus. Some volunteer female subjects take the drug. Which of the following results obtained from the volunteers would support the scientists hypothesis?

(a) Increased levels of oestrogen in the blood
(b) Earlier onset of menstruation
(c) Follicle development
(d) Decreased levels of Luteinizing Hormone (LH) in the blood

12 In the given figure, what is indicated as '*X*'?

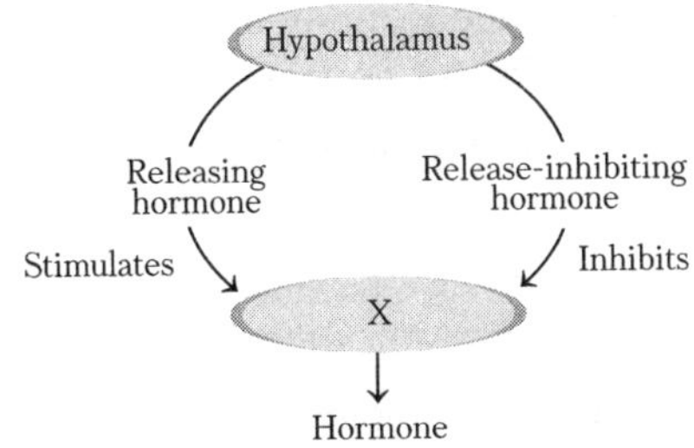

(a) Thyroid (b) Parathyroid
(c) Anterior pituitary (d) Gonads

13 Find out the correct labelling of the given figure.

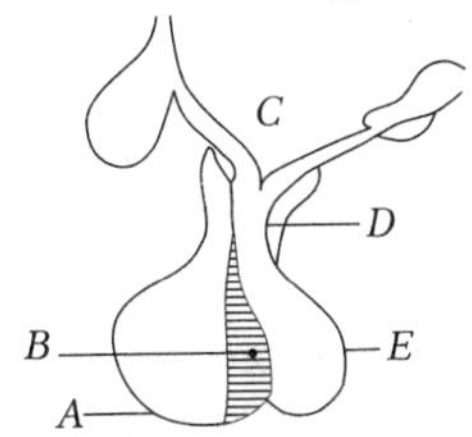

(a) *A*-Adenohypophysis, *B*-Pars intermedia, *C*-Hypothalamus, *D*-Infundibulum, *E*-Neurohypophysis
(b) *A*-Neurohypophysis, *B*-Pars intermedia, *C*-Hypothalamus, *D*-Infundibulum, *E*-Adenohypophysis
(c) *A*-Adenohypophysis, *B*-Infundibulum, *C*-Hypothalamus, *D*-Pars intermedia, *E*-Neurohypophysis
(d) *A*-Neurohypophysis, *B*-Infundibulum, *C*-Hypothalamus, *D*-Pars intermedia, *E*-Adenohypophysis

14 Some atheletes take steroids in an attempt to enhance their physical performance. This can lead to decreased sperm production and even sterility. What is the most likely explanation for this effect?

(a) Increased stimulation of anterior pituitary to produce LH
(b) Excessive growth of testicular tissue
(c) Suppression of the natural production of thyroxine due to the destruction of thyroid tissue
(d) Interference in the proper negative feedback control of testosterone

15 Match the following columns.

Column I	Column II
A. Tetany	1. Hyposecretion of parathyroid
B. Diabetes mellitus	2. Hyposecretion of insulin
C. Diabetes insipidus	3. Hyposecretion of ADH
D. Acromegaly	4. Hypersecretion of somatostatin

Codes

	A	B	C	D		A	B	C	D
(a)	1	2	3	4	(b)	3	1	4	2
(c)	4	3	2	1	(d)	2	4	1	3

16 The diagram represents the control of thyroxine secretion. P, Q and R represent organs and S, T and U represent different hormones.

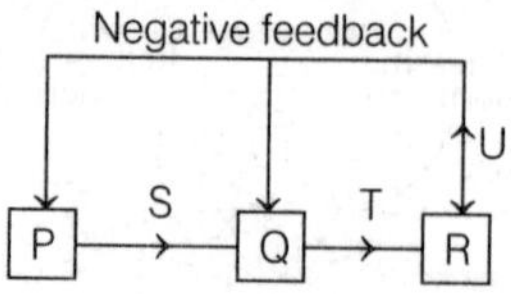

Which of the following gives the correct combination of names?

	P	Q	R	S	T	U
(a)	Anterior pituitary	—	—	—	Thyroid stimulating hormone	Thyroxine
(b)	Hypothalamus	Anterior pituitary	—	—	—	Thyroxine
(c)	Posterior pituitary	—	Thyroid gland	Thyroid stimulating hormone	—	—
(d)	Thyroid gland	Posterior pituitary	—	Thyroxine	—	—

17 Which of the following statements is false about diabetes?
(a) It is the result of deficiency of insulin
(b) Blood has excess of glucose
(c) Cells fail to pick up glucose from blood
(d) More sugar is burnt in the body of a diabetic than of a normal person

18 Which one of the following is the correctly matched pair of the organ and the hormone it secretes?
(a) Thyroid — Epinephrine
(b) Alpha cells of pancreas — Glucagon
(c) Anterior pituitary — Adrenaline
(d) Stomach epithelium — Secretin

19 Choose the true statements about a person with type-I (insulin-dependent) diabetes mellitus.
(a) There is little or no insulin secretion
(b) Dietary treatment may not be sufficient
(c) There is hyperglycaemia
(d) All of the above

20 An overdose of intravenous insulin may lead to the death of an individual due to an
(a) excessive increase of blood glucose
(b) excessive decrease of blood glucose
(c) inhibition of glucagon secretion
(d) overproduction of histamine

21 Match the following columns.

Column I	Column II
A. Calcitonin	1. Diabetes mellitus
B. Gonadotropin	2. Rickets
C. Erythropoietin	3. Diabetes insipidus
D. Insulin	4. Formation of erythrocytes
E. ADH	5. Infertility

Codes

	A	B	C	D	E
(a)	3	1	4	2	5
(b)	3	2	1	5	4
(c)	4	3	2	1	5
(d)	2	5	4	1	3

22 Match the source gland with its respective hormone and function.

	Source glands	Hormones	Functions
(a)	Anterior pituitary	Oxytocin	Contraction of uterus muscles during childbirth
(b)	Posterior pituitary	Vasopressin	Stimulates reabsorption of water in the distal tubules in nephron
(c)	Corpus luteum	Oestrogen	Supports pregnancy
(d)	Thyroid	Thyroxine	Regulates blood calcium level

23 What is/are the function of oxytocin hormone?
I. Ejection of milk from mammary glands.
II. Contraction of uterus during childbirth.
III. Maintenance of corpus luteum
IV. Stimulates reproductive system.
Choose the correct option.
(a) Only IV (b) I and II
(c) III and IV (d) II and IV

24 Find the incorrect match.
(a) β-cells of pancreas - Insulin
(b) Zone fasciculata of adrenal cortex - Cortisol
(c) Follicular cells of thyroid gland - Thyrocalcitonin
(d) Juxtaglomerular cells of kidney - Erythropoietin

25 How many of these are peptide hormones?

Erythropoietin, Gastrin, Secretin, CCK, Insulin, Glucagon, Thymosin, Parathyroid hormone, ANF, Gastroinhibitory peptide

(a) Eight (b) Seven
(c) Nine (d) Ten

26 Select the total number of hormones from the following which bind to intracellular receptors.

Cortisol, Testosterone, T_3, Glucagon, Oxytocin, FSH, Progesterone, ICSH, Oestrogen, GH

(a) Four (b) Five (c) Six (d) Seven

27 A condition characterised with increase in the excitability of nerves and muscles leading to sustained contraction of the muscles of larynx, face, hand and feet is due to the
(a) hyperactivity of thyroid
(b) hyperactivity of parathyroid
(c) hypoactivity of thyroid
(d) hypoactivity of parathyroid

28 Which of the following hormones are secreted by hypothalamus?

I. Thyrotropin releasing hormone.
II. Gonadotropin releasing hormone.
III. Prolactin inhibiting hormone.
IV. Follicle stimulating releasing hormone.

Choose the correct option.

(a) I and II (b) Only IV
(c) II and III (d) I, II, III and IV

29 In an accident, the anterior pituitary of a four-year-old boy was severely damaged, but the boy survived. What is likely to happen?

(a) High levels of thyroxine will be released
(b) Spermatogenesis will be stimulated
(c) The boy will not grow much in height
(d) The growth of mammary glands will be stimulated

30 Given below are the hormones and the number of monomeric units.

I. Luteotropic hormone - 19 amino acids
II. Adrenocorticotrophic hormone - 118 amino acids
III. Prolactin - 198 amino acids
IV. Follicle stimulating hormone - 204 amino acids

Choose the incorrectly matched pairs.

(a) I and II (b) II and IV (c) II and III (d) I and IV

31 Which one of the following four glands is correctly matched with the accompanying description?

(a) Thyroid – Hyperactivity in young children causes cretinism
(b) Thymus – Starts undergoing atrophy after puberty
(c) Parathyroid – Secretes parathormone, which promotes the movement of calcium ions from the blood into bones during calcification
(d) Pancreas – Delta cells of the islets of Langerhans secrete a hormone, which stimulates glycogenesis in liver

32 Match the following columns.

Column I (Hormones)	Column II (Target tissues/organs)
A. Thyroid stimulating hormone	1. Mammary glands
B. Adrenocorticotropic hormone	2. Thyroid gland
C. Prolactin	3. Bones
D. Growth hormone	4. Adrenal cortex

Codes

	A	B	C	D		A	B	C	D
(a)	1	2	3	4	(b)	4	1	2	3
(c)	3	2	1	4	(d)	2	4	1	3

33 A hormone X activates the *c*AMP second messenger system in its target cells. The greatest response by a cell would come from

(a) injecting a molecule of hormone X into the cytoplasm of the cell
(b) applying a molecule of *c*AMP to the extracellular fluid surrounding the cell
(c) injecting one molecule of *c*AMP into the cytoplasm of the cell
(d) applying a molecule of hormone X to the extracellular fluid surrounding the cell

34 When a steroid hormone and a peptide hormone exert similar effects on a population of target cells, then

(a) the steroid hormones affect the activity of certain proteins within the cell, whereas peptide hormone directly affects the processing of *m*RNA
(b) the steroid hormone affects only the release of proteins from the target cell, whereas peptide hormone affects only the synthesis of proteins that remain in the target cell
(c) the steroid and peptide hormones must bind to the same receptor protein
(d) the steroid hormone affects the synthesis of effector protein, whereas peptide hormone activates the effector proteins already present in the cell

35 Read the following statements.

I. Oxytocin is called milk ejecting hormone.
II. Epinephrine is methylated nor-epinephrine.
III. The endocrine portion of pancreas consists of islets of Langerhans (98-99%), whereas exocrine portion of pancreas consists of pancreatic acini (1-2%).
IV. Thyroid regulates BMR.
V. Testosterone exerts a feedback inhibitory effect on LH (ICSH) secretion.

How many of the above statements are correct ?

(a) Four (b) Three
(c) Two (d) One

36 Research has shown that an extract of the ginseng root, when fed to bullfrog tadpoles, prevents metamorphosis. The pituitary glands of the affected tadpoles show an increase in the number of cells which produce Thyroid Stimulating Hormone (TSH). Which statement about ginseng extract explain these results?

(a) It stimulates TSH production, which in turn stimulates thyroxine production
(b) It stimulates thyroxine production, which in turn stimulates TSH production
(c) It inhibits TSH production and thyroxine production increases by negative feedback
(d) It inhibits thyroxine production and TSH production increases by negative feedback

37 Tobacco smoke stimulates adrenaline production. The graph shows a person's pulse rate before, during and after smoking a cigarette.

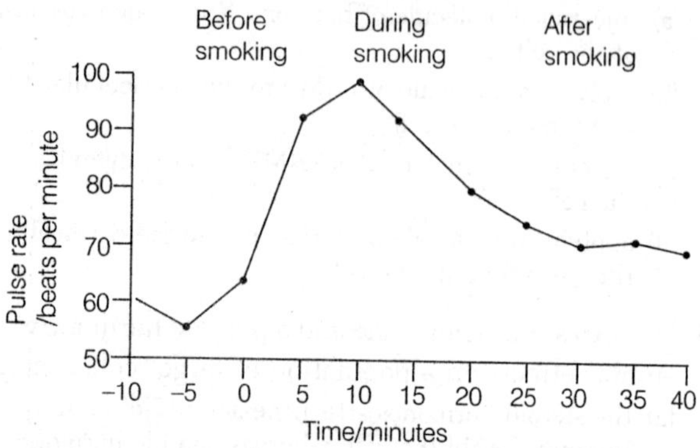

Using only the information given, which of the following is a valid interpretation of the graph?

(a) Adrenaline causes muscles to relax, so reducing their blood requirement
(b) Basal metabolic rate increases during smoking
(c) Heart rate increases as more adrenaline is produced
(d) Nicotine stimulates or increase the heart rate

38 Which of the following responses below would occur in a human body as a direct result of eating a school lunch?

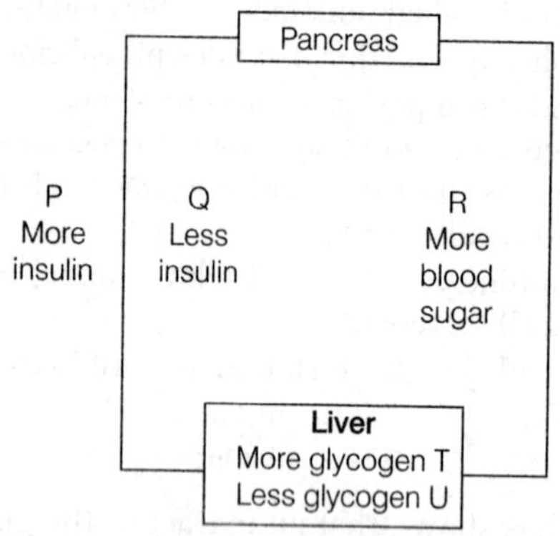

(a) R, P and T
(b) S, P and T
(c) R, Q and T
(d) R, P and U

39 Which of the following statements best describes the outcome of the negative feedback loop in a homeostatic mechanism?

(a) Glucagon is released by the pancreas is response to a drop in blood glucose concentration below the normal level
(b) The liver cells breakdown glycogen to glucose in response to glucagon
(c) Blood glucose level increases to the normal level in response to the release to glucagon
(d) Release of glucagon is stopped in response to blood glucose concentration returning to the normal level

40 Which of the following gives an accurate comparison between intracellular receptors and cell surface receptors?

	Intracellular receptors	Cell surface receptors
(a)	Their ligands can either be water-soluble or lipid-soluble because they do not need to bind with the cell membrane	Their ligands must be lipid-soluble because they need to bind with the cell membrane
(b)	All are single-pass transmembrane proteins.	All are multi-pass transmembrane proteins.
(c)	They may bind to DNA and act as gene regulators.	They may catalyse the phosphorylation of intracellular proteins.
(d)	All of the above	

41 Hypoparathyroidism causes

I. reduction in absorption of calcium
II. parathyroid tetany
III. calcification of soft tissues
IV. muscle weakness

Choose the correct option.

(a) I and II
(b) III and IV
(c) I and III
(d) All of the above

42 There are three endocrine glands involved in carbohydrate metabolism, mark the correct set.

(a) Pancreas, neurohypophysis and adrenal
(b) Pancreas, pituitary and thyroid
(c) Pancreas, pituitary and liver
(d) Pancreas, adenohypophysis and adrenal

43 Read the following statements.

I. The releasing hormones of hypothalamus always inhibits the secretion of pituitary hormones.
II. GnRH stimulates pituitary synthesis and releases of gonadotropins.
III. Somatostatin is produced in hypothalamus and inhibits the release of growth hormone from pituitary gland.
IV. Somatostatin is also called Growth Hormone-Inhibiting Hormone (GHIH).
V. Hypothalamus is connected to the anterior lobe of the pituitary gland by hypophyseal portal veins and to the posterior lobe of pituitary gland by axons of neurosecretory cells.

Which of the above statements are correct ?

(a) II, III, IV and V
(b) I, II and III
(c) I, IV and V
(d) Only I

44 Choose the correct statement.

(a) Organised endocrine glands include pineal gland, pituitary gland, pancreas, liver, ovary and testis with primary endocrine functions
(b) In nervous system, the response is for long duration whereas in endocrine system, it is for short duration
(c) Invertebrates possess very simple endocrine system whereas a large number of chemicals acts as hormones in vertebrates
(d) Hormones are produced in trace amounts by ducted glands only

45 Hormones that promote homeostasis

(a) utilise receptors that bind any hormone
(b) are not found as members of antagonistic signalling mechanisms
(c) usually operate as part of a negative feedback system
(d) initiate signal transduction in target cell without binding to receptors

Assertion and Reason

Direction (Q.Nos.46-52) *In each of the following questions, a statement of Assertion (A) is given followed by corresponding statement of Reason (R). Of the statements, mark the correct answer as*

(a) Both A and R are true and R is the correct explanation of A
(b) Both A and R are true, but R is not the correct explanation of A
(c) A is true, but R is false
(d) A is false, but R is true
(e) Both A and R are false

46 **Assertion** (A) Hypoglycaemia disorder is due to the imbalance of glucagon and insulin hormones.

Reason (R) In hypoglycaemia, the level of insulin decreases and glucagon increases.

47 **Assertion** (A) The name master gland of body is given to pituitary gland.

Reason (R) Hypothalamus exerts control over the master gland.

48 **Assertion** (A) Pitocin or oxytocin is also called as 'birth hormone'.

Reason (R) It helps in contraction of uterine muscles.

49 **Assertion** (A) Osteoporosis is related to parathyroid glands.

Reason (R) It results in softening and bending of bones.

50 **Assertion** (A) The regulation of RBC production in accomplished by FSH.

Reason (R) Erythropoietin hormone circulates in red bone marrow, where it increases stem cell mitosis and speed up the development of RBCs.

51 **Assertion** (A) Diabetes insipidus is marked by excessive urination and too much thirst for water.

Reason (R) Antidiuretic Hormone (ADH) is secreted by the posterior lobe of pituitary gland.

52 **Assertion** (A) Collip's hormone is secreted by the chief cells of parathyroid gland.

Reason (R) It regulates calcium and phosphate level between blood and tissues.

(C) Medical Entrances' Gallery

(Collection of Questions Asked in NEET & in Various Medical Entrance Exams)

1 Match the following columns and select the correct option.

Column I	Column II
A. Pituitary gland	1. Graves' disease
B. Thyroid gland	2. Diabetes mellitus
C. Adrenal gland	3. Diabetes insipidus
D. Pancreas	4. Addison's disease

Codes [NEET 2020]

	A	B	C	D
(a)	3	2	1	4
(b)	3	1	4	2
(c)	2	1	4	2
(d)	4	3	1	2

2 Match the following columns and select the correct option from the codes given below.

Column I	Column II
A. Pituitary hormone	1. Steroid
B. Epinephrine	2. Neuropeptides
C. Endorphins	3. Peptides, proteins
D. Cortisol	4. Biogenic amines

Codes [NEET 2020]

	A	B	C	D
(a)	4	1	2	3
(b)	3	4	2	1
(c)	4	3	1	2
(d)	3	4	1	2

3 Hormones stored and released from neurohypophysis are [NEET 2020]
(a) thyroid stimulating hormone and oxytocin
(b) oxytocin and vasopressin
(c) Follicle stimulating hormone and Luteinizing hormone
(d) prolactin and vasopressin

4 Match the following hormones with the respective diseases.

	Column I		Column II
A.	Insulin	1.	Addison's disease
B.	Thyroxin	2.	Diabetes insipidus
C.	Corticoids	3.	Acromegaly
D.	Growth hormone	4.	Goitre
		5.	Diabetes mellitus

Codes [NEET 2019]

	A	B	C	D
(a)	2	4	3	1
(b)	5	4	1	3
(c)	2	4	1	3
(d)	5	1	2	3

5 How does steroid hormone influence the cellular activities? [NEET 2019]
(a) Binding to DNA and forming a gene-hormone complex
(b) Activating cyclic AMP located on the cell membrane
(c) Using aquaporin channels as second messenger
(d) Changing the permeability of the cell membrane

6 Select the correct option from the following.

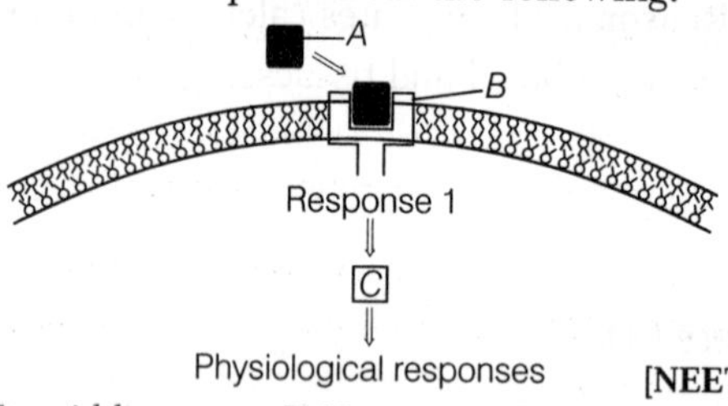

[NEET 2019]
(a) *A*-Steroid hormone, *B*-Hormone-receptor complex, *C*-Protein
(b) *A*-Protein hormone, *B*-Receptor, *C*-Cyclic AMP
(c) *A*-Steroid hormone, *B*-Receptor, *C*-Second messenger
(d) *A* - Protein hormone, *B*-Cyclic AMP, *C*-Homone-receptor complex

7 Which of the following conditions will stimulate parathyroid gland to release parathyroid hormone?
(a) Fall in active vitamin-D levels [NEET 2019]
(b) Fall in blood Ca^{+2} levels
(c) Fall in bone Ca^{+2} levels
(d) Rise in blood Ca^{+2} levels

8 Artificial light, extended work time and reduced sleep time disrupt the activity of [NEET 2019]
(a) thymus gland
(b) pineal gland
(c) adrenal gland
(d) posterior pituitary gland

9 Oxytocin and ADH are produced by hypothalamus and released from [AIIMS 2019]
(a) anterior pituitary
(b) posterior pituitary
(c) pineal gland
(d) thymus

10 **Assertion** (A) Type-I diabetes is caused by the destruction of β-cells of islets of Langerhans.

Reason (R) Insulin can be taken as pills. [AIIMS 2019]
(a) Both A and R are true and R is the correct explanation of A
(b) Both A and R are true, but R is not the correct explanation of A
(c) A is true, but R is false
(d) Both A and R are false

11 **Assertion** (A) Pituitary gland releases a hormone which is helpful in childbirth.

Reason (R) Pituitary gland releases vasopressin and antidiuretic hormone which helps in childbirth. [AIIMS 2019]
(a) Both A and R are true and R is the correct explanation of A
(b) Both A and R are true, but R is not the correct explanation of A
(c) A is true, but R is false
(d) Both A and R are false

12 Bradykinin is a [JIPMER 2019]
(a) antibody
(b) hormone
(c) enzyme
(d) receptor

13 Which of the following hormones can play a significant role in osteoporosis? [NEET 2018]
(a) Oestrogen and parathyroid hormone
(b) Progesterone and aldosterone
(c) Aldosterone and prolactin
(d) Parathyroid hormone and prolactin

14 Which of the following is an amino acid derived hormone? [NEET 2018]

Or The 'amino acid derivative' among the following hormones is [AIIMS 2018, Kerala CEE 2011]
(a) oestradiol
(b) ecdysone
(c) epinephrine
(d) oestriol

15 **Assertion** (A) Peptide and polypeptide hormones directly pass across the lipid bilayer of plasma membrane.

Reason (R) Oxytocin hormone can pass across the plasma membrane. [AIIMS 2018]
(a) Both A and R are true and R is the correct explanation of A
(b) Both A and R are true, but R is not the correct explanation of A
(c) A is true, but R is false
(d) Both A and R are false

16 Which of the following pituitary hormones works indirectly? [JIPMER 2018]
(a) MSH (b) TSH (c) GH (d) Oxytocin

17 Inhibin is composed of [JIPMER 2018]
(a) glycoprotein (b) lipoprotein
(c) steroid (d) amino acid derivative

18 Hypersecretion of growth hormone in adults does not cause further increase in height because [NEET 2017]
(a) growth hormone becomes inactive in adults
(b) epiphyseal plates close after adolescence
(c) bones lose their sensitivity to growth hormone in adults
(d) muscle fibres do not grow in size after birth

19 GnRH, a hypothalamic hormone, needed in reproduction, acts on [NEET 2017]
(a) anterior pituitary gland and stimulates secretion of LH and oxytocin
(b) anterior pituitary gland and stimulates secretion of LH and FSH
(c) posterior pituitary gland and stimulates secretion of oxytocin and FSH
(d) posterior pituitary gland and stimulates secretion of LH and relaxin

20 Which of the following statements is correct in relation to the endocrine system? [JIPMER 2017, NEET 2013]
(a) Adenohypophysis is under direct neural regulation of the hypothalamus
(b) Organs in the body like gastrointestinal tract, heart, kidney and liver do not produce any hormones
(c) Non-nutrient chemicals produced by the body in trace amount that act as intercellular messenger are known as hormones
(d) Releasing and inhibitory hormones are produced by the pituitary gland

21 Graves' disease is caused due to the [NEET-II 2016]
(a) hyposecretion of thyroid gland
(b) hypersecretion of thyroid gland
(c) hyposecretion of adrenal gland
(d) hypersecretion of adrenal gland

22 Name a peptide hormone which acts mainly on hepatocytes, adipocytes and enhances cellular glucose uptake and utilisation. [NEET-II 2016]
(a) insulin (b) glucagon
(c) secretin (d) gastrin

23 Changes in GnRH pulse frequency in females is controlled by circulating levels of [NEET-I 2016]
(a) oestrogen and inhibin
(b) progesterone only
(c) progesterone and inhibin
(d) oestrogen and progesterone

24 Match and find the correct combination.

Column I	Column II
A. Addison's disease	1. Hypothyroidism in adult
B. Cushing's syndrome	2. Hypersecretion of somatotropin in adult
C. Cretinism	3. Hyposecretion of glucocorticoids
D. Myxoedema	4. Hypothyroidism during pregnancy
E. Acromegaly	5. Overproduction of glucocorticoids

Codes [EAMCET 2015]

	A	B	C	D	E
(a)	3	5	4	1	2
(b)	5	4	3	2	1
(c)	2	3	4	1	5
(d)	3	4	5	1	2

25 **Assertion** (A) Hormones interacting with cell surface receptors do not enter the target, but they generate second messenger.
Reason (R) Glucocorticoids bind to intracellular receptors and regulate gene expression. [EAMCET 2015]
(a) Both A and R are true and R is the correct explanation of A
(b) Both A and R are true, but R is not the correct explanation of *A*
(c) A is true, but R is false
(d) A is false, but R is true

26 Select the mismatched pair from the following. [Karnataka CET 2015]
(a) Glucagon – Glycogenolysis
(b) Prolactin – Milk production
(c) Insulin – Gluconeogenesis
(d) Oxytocin – Contraction of uterine muscles

27 Which hormone stimulates the expulsion of milk during feeding of baby? [Chhattisgarh PMT 2015]
(a) Oxytocin (b) Relaxin
(c) Prolactin (d) Progesterone

28 Secretion of which hormone is not pituitary dependent? [AMU 2015]
(a) Triiodothyronine (b) Testosterone
(c) Glucocorticoids (d) Parathyroid hormone

29 The following hormone is not a steroid. [AMU 2015]
(a) Testosterone
(b) Progesterone
(c) Corticosterone
(d) Adrenocorticotropic hormone

30 Coordinator between nervous and endocrine system is [MHT CET 2015]
(a) thalamus (b) hypothalamus
(c) epithalamus (d) colliculus

31 Atrial Natriuretic Factor (ANF) decreases **[MHT CET 2015]**

(a) blood pressure (b) secretion of renin
(c) Na^+ secretion (d) vasodilation

32 Serotonin and melatonin are hormones secreted by **[Karnataka CET 2015]**

(a) pancreas (b) pineal body
(c) pituitary gland (d) thymus

33 Tetraiodothyronine refers to

(a) T_3 (b) Thyroxine
(c) TSH (d) TRH

34 Find out the correct layers in TS adrenal gland from outside to inside **[Chhattisgarh PMT 2015]**

(a) Medulla-Zona reticularis-Zona fasciculata-Zona glomerulosa
(b) Zona glomerulosa-Medulla-Zona reticularis-Zona fasciculata
(c) Zona glomerulosa-Zona fasciculata-Zona reticularis-Medulla
(d) Zona reticularis-Zona fasciculata- Zona glomerulosa-Medulla

35 Function of somatostatin is to **[Kerala CEE 2015]**

(a) stimulates pituitary synthesis and release gonadotropins
(b) inhibits the release of gonadotropins from pituitary
(c) stimulates pituitary and promotes secretion of growth hormone
(d) inhibits the release of growth hormone from pituitary
(e) stimulates the secretion of thyrotropin from thyroid

36 Ectopically transplanted pituitary secretes following hormone abundantly. **[Chhattisgarh PMT 2015]**

(a) Oxytocin (b) Vasopressin
(c) Prolactin (d) Inhibin

37 Pars distalis region of pituitary does not produce these hormones

I. Melanocyte stimulating hormone
II. Vasopressin
III. Prolactin
IV. Growth hormone

Codes **[Kerala CEE 2015]**

(a) Only III (b) I and IV (c) II and IV (d) II and III
(e) I and II

38 Which is an inhibitory hormone of GH? **[Gujarat CET 2015]**

(a) Insulin (b) Somatostatin
(c) Parathormone (d) Testosterone

39 Which is not involved in sugar metabolism? **[CBSE AIPMT 2015]**

(a) Cortisone (b) Aldosterone
(c) Insulin (d) Glucagon

40 ......show/shows anti-allergic and anti-inflammatory effect. **[Gujarat CET 2015]**

(a) Mineralocorticoids (b) Sex corticoids
(c) Glucocorticoids (d) Nor-adrenaline

41 The Leydig cells found in the human body are the secretory source of **[Manipal 2015]**

(a) glucagon (b) androgens
(c) progesterone (d) intestinal mucus

42 Which one of the following secretes glucagon? **[WB JEE 2015]**

(a) β- cells of islets of Langerhans
(b) α- cells of islets of Langerhans
(c) Acidophilic cells of adenohypophysis
(d) Basophilic cells of adenohypophysis

43 Which one of the following is called intraspecific chemical messenger? **[WB JEE 2015]**

(a) Pheromones (b) Prostaglandins
(c) Corticotropin (d) Catecholamines

44 Which enzyme is responsible for conversion of nor-epinephrine to epinephrine? **[WB JEE 2015]**

(a) Catecholamine-O-methyl transferase
(b) Phenylalanine-N-methyl transferase
(c) DOPA carboxylase
(d) Monoamine oxidase

45 Steroid hormones **[UK PMT 2015]**

(a) work *via* second messenger
(b) trigger rapid, short term response in cells
(c) alter the activity of genes
(d) initiate open channels in plasma membrane

46 Identify the hormone with its correct matching of source and function. **[CBSE AIPMT 2015]**

(a) Oxytocin — Posterior pituitary, growth and maintenance of mammary glands
(b) Melatonin — Pineal gland, regulates the normal rhythm of sleep-wake cycle
(c) Progesterone — Corpus luteum, stimulation of growth and activities of female secondary sex organs
(d) Atrial natriuretic factor — Ventricular wall increases the blood pressure

47 Fight or flight reactions cause activation of **[CBSE AIPMT 2014]**

(a) the parathyroid glands, leading to increased metabolic rate
(b) the kidney, leading to suppression of renin angiotensin aldosterone pathway
(c) the adrenal medulla, leading to increased secretion of epinephrine and nor-epinephrine
(d) the pancreas leading to a reduction in the blood sugar levels

48 The diurnal rhythms are regulated by **[MHT CET 2014]**
(a) adrenaline (b) melatonin
(c) serotonin (d) vasopressin

49 Match the hormone secreted by various endocrine structures and choose the correct option.

Column I		Column II
A. Hypothalamus	1.	Melanocyte Stimulating Hormone (MSH)
B. Pars intermedia	2.	Aldosterone
C. Pineal gland	3.	Gonadotropin Releasing Hormone (GnRH)
D. Adrenal medulla	4.	Melatonin
E. Adrenal cortex	5.	Catecholamines

Codes **[Kerala CEE 2014]**

	A	B	C	D	E
(a)	5	1	4	2	3
(b)	5	4	1	2	3
(c)	2	4	1	3	5
(d)	3	1	4	2	5
(e)	3	1	4	5	2

50 Which of the following statements is incorrect? **[Kerala CEE 2014]**
(a) Sella turcica is a bony cavity where the pituitary gland is located
(b) Parathyroid hormone decreases the Ca^{2+} levels in blood
(c) Thymosins play a major role in T-cell differentiation
(d) The middle layer of adrenal cortex is zona fasciculata
(e) Insulin stimulates glycogenesis

51 Which of the following hormones is a derivative of fatty acid? **[WB JEE 2014]**
(a) Gastrin
(b) Thyroxine
(c) Oestrogen
(d) Prostaglandins

52 Which of the following pituitary hormones is secreted without the involvement of a Releasing Hormone (RH)? **[WB JEE 2014]**
(a) Thyroid Stimulating Hormone (TSH)
(b) Follicle Stimulating Hormone (FSH)
(c) Oxytocin
(d) Prolactin

53 An endocrine gland in human, which plays an important role in the regulation of rhythm of the body is **[J&K CET 2014]**
(a) adrenal gland
(b) pineal gland
(c) thymus gland
(d) thyroid gland

54 Insulin causes rapid movement of glucose from blood to hepatocytes and adipocytes resulting in **[UK PMT 2014]**
(a) hyperglycaemia (b) hypoglycaemia
(c) diabetes (d) None of these

55 Identify the hormones, *A*, *B* and *C* that are labelled in the given flowchart. **[Karnataka CET 2014]**

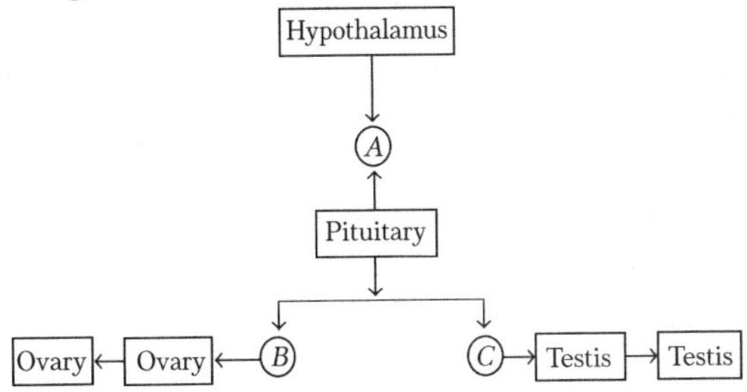

	A	*B*	*C*
(a)	GnRH	ICSH	FSH
(b)	GH	FSH	LH
(c)	GnRH	PRL	ICSH
(d)	GnRH	FSH	LH

56 Which one of the following hormones also produces anti-inflammatory reactions in man and suppresses the immune response in addition to its primary functions? **[Karnataka CET 2014]**
(a) Cortisol (b) Thymosin
(c) Thyrocalcitonin (d) Erythropoietin

57 Pituitary gland is located in *A*, which is *B* of *C* bone?

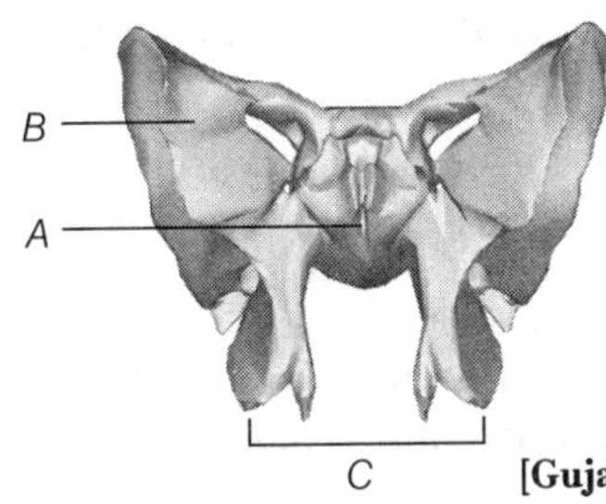

[Gujarat CET 2014]

(a) *A*-Rathke's pouch
B-Depression
C-Nasal
(b) *A*-Sella turcica
B-Raised surface
C-Ethmoid
(c) *A*-Sella turcica
B-Depression
C-Sphenoid
(d) *A*-Rathke's pouch
B-Depression
C-Sphenoid

58 A pregnant female delivers a baby who suffers from stunted growth, mental retardation/low intelligence quotient and abnormal skin. This is the result of **[NEET 2013]**
(a) low secretion of growth hormone
(b) cancer of the thyroid gland
(c) oversecretion of pars distalis
(d) deficiency of iodine in diet

59 Select the answer which correctly matches the endocrine gland with the hormone it secretes and its function/ deficiency symptom. **[NEET 2013]**

	Enodocrine glands	Hormones	Functions/Deficiency symptoms
(a)	Posterior pituitary	Growth Hormone (GH)	Oversecretion stimulates abnormal growth
(b)	Thyroid gland	Thyroxine	Lack of iodine in diet results in goitre
(c)	Corpus luteum	Testosterone	Stimulates spermatogenesis
(d)	Anterior pituitary	Oxytocin	Stimulates uterus contraction during childbirth

60 Match the following columns.

Column I	Column II
A. Adrenaline	1. Anger, fear and danger
B. Oestrogen	2. Attracting partners through sense of smell
C. Insulin	3. Females
D. Pheromones	4. Glucose

Codes **[AIIMS 2013]**

	A	B	C	D		A	B	C	D
(a)	4	1	3	2	(b)	1	3	2	4
(c)	1	3	4	2	(d)	3	1	2	4

61 'Mammalian thymus' is mainly concerned with **[BCECE 2012]**
(a) regulation of body temperature
(b) regulation of body growth
(c) immunological functions
(d) secretion of thyrotropin

62 Which one of the following pairs of hormones are the examples of those that can easily pass through the cell membrane of the target cell and bind to a receptor inside it (mostly in the nucleus)? **[CBSE AIPMT 2012]**
(a) Insulin and glucagon
(b) Thyroxine and insulin
(c) Somatostatin and oxytocin
(d) Cortisol and testosterone

63 The hormone responsible for 'fight and flight' response is **[UP CPMT 2007; WB JEE 2012]**
(a) adrenaline (b) thyroxine (c) ADH (d) oxytocin

64 Given below is an incomplete table about certain hormones, their source glands and one major effect of each on the body in humans. Identify the correct option for the three blanks *A*, *B* and *C*. **[CBSE AIPMT 2011]**

Glands	Secretions	Effects on body
A	Oestrogen	Maintenance of secondary sexual characters
Alpha cells of islets of Langerhans	*B*	Raises blood sugar level
Anterior pituitary	*C*	Oversecretion leads to gigantism

	A	*B*	*C*
(a)	Placenta	Insulin	Vasopressin
(b)	Ovary	Insulin	Calcitonin
(c)	Placenta	Glucagon	Calcitonin
(d)	Ovary	Glucagon	Growth hormone

65 Match the following columns.

	Column I	Column II
A.	ANF	1. Regulates blood calcium levels
B.	MSH	2. Decreases blood pressure
C.	GIP	3. Pigmentation
D.	TCT	4. Inhibits gastric secretion

Codes **[Kerala CEE 2011]**

	A	B	C	D		A	B	C	D
(a)	4	1	2	3	(b)	2	1	4	3
(c)	4	1	3	2	(d)	3	2	4	1
(e)	2	3	4	1					

66 Which of the following hormones of the human body regulates the blood calcium and phosphate? **[MHT CET 2011]**
(a) Glucagon (b) Growth hormone
(c) Parathyroid hormone (d) Thyroxine

67 The gland which performs both endocrine and exocrine functions is **[MHT CET 2011]**
(a) adrenal (b) thyroid
(c) pancreas (d) pituitary

68 Which is not a symptom of exophthalmic goitre? **[MHT CET 2011]**
(a) Degenerating sex organs
(b) Protrusion of eyeball
(c) Frightened look to the patient
(d) None of the above

69. Which gland secretes the most kinds of hormones? **[MHT CET 2011]**
(a) Adrenal (b) Hypothalamus
(c) Pituitary (d) Thyroid

70 The releasing hormones are produced by **[UP CPMT 2011]**
(a) testis (b) pancreas
(c) pituitary (d) hypothalamus

71 Somatostatin **[DUMET 2011]**
(a) stimulates glucagon release while inhibits insulin release
(b) stimulates the release of insulin and glucagon
(c) inhibits the release of insulin and glucagon
(d) inhibits glucagon release while stimulates insulin release

72 Insufficient quantities of antidiuretic hormone in blood lead to **[DUMET 2011]**
(a) diabetes mellitus (b) glycosuria
(c) diabetes insipidus (d) uremia

73 Thymosin hormone is secreted by **[DUMET 2011]**
(a) thyroid gland (b) parathyroid gland
(c) thymus gland (d) hypothalamus

74 Which one of the following is not a second messenger in hormone action? **[WB JEE 2011]**
(a) Calcium (b) Sodium (c) *c*AMP (d) GMP

75 The cause of cretinism is **[WB JEE 2011]**
(a) hypothyroidism (b) hypoparathyroidism
(c) hyperthyroidism (d) hyperparathyroidism

76 Which of the following is a mineralocorticoid? **[WB JEE 2011]**
(a) Testosterone (b) Progesterone
(c) Adrenaline (d) Aldosterone

77 Calcium level decreases in the blood due to the hyposecretion of **[WB JEE 2011]**
(a) parathyroid hormone (b) calcitonin
(c) thyroxine (d) adrenaline

78 Tetany is caused by **[WB JEE 2011]**
(a) hyperparathyroidism (b) hypoparathyroidism
(c) hyperthyroidism (d) hypothyroidism

79 Which of the following is a gastrointestinal hormone? **[WB JEE 2011]**
(a) Prolactin (b) Enterogastrone
(c) GH (d) FSH

80 Due to this, swelling around eyes,large and popping eyeballs are observed in an individual who has **[Gujarat CET 2011]**
(a) less secretion of thyroxine in adult
(b) excessive secretion of thyroxine
(c) excessive secretion of calcitonin
(d) less secretion of thyroxine right from birth

81 Which hormone regulates cell division, protein synthesis and growth of the bone? **[Gujarat CET 2011]**
(a) Prolactin (b) Somatotropic hormone
(c) TSH (d) MSH

82 Due to deficiency of which hormone, bones become weak in females? **[Gujarat CET 2011]**
(a) ACTH (b) TSH (c) Progesterone (d) Oestrogen

83 Which one is not a placental hormone? **[J&K CET 2011]**
(a) hCG (b) hCS (c) Progesterone (d) Melatonin

84 ADH deficiency shows which of the following condition? **[J&K CET 2011]**
(a) Polydipsia (b) Polyuria
(c) Both (a) and (b) (d) Glucosuria

ANSWERS

Check Point 29.1

1. (b)	2. (b)	3. (b)	4. (d)	5. (d)	6. (d)	7. (a)

Check Point 29.2

1. (c)	2. (a)	3. (a)	4. (b)	5. (c)	6. (d)

Check Point 29.3

1. (a)	2. (b)	3. (c)	4. (c)	5. (d)	6. (a)

(A) Topical Exercise

1. (c)	2. (a)	3. (c)	4. (c)	5. (d)	6. (b)	7. (d)	8. (c)	9. (a)	10. (d)
11. (c)	12. (a)	13. (c)	14. (a)	15. (d)	16. (d)	17. (b)	18. (b)	19. (a)	20. (c)
21. (b)	22. (d)	23. (b)	24. (c)	25. (a)	26. (b)	27. (d)	28. (c)	29. (b)	30. (b)
31. (c)	32. (a)	33. (a)	34. (c)	35. (d)	36. (a)	37. (c)	38. (b)	39. (c)	40. (b)
41. (d)	42. (a)	43. (a)	44. (a)	45. (a)	46. (b)	47. (a)	48. (a)	49. (d)	50. (c)
51. (c)	52. (c)	53. (a)	54. (d)	55. (a)	56. (b)	57. (b)	58. (d)	59. (c)	60. (a)
61. (b)	62. (c)	63. (b)	64. (a)	65. (b)	66. (b)	67. (b)	68. (b)	69. (a)	70. (c)
71. (c)	72. (d)	73. (d)	74. (c)	75. (a)	76. (b)	77. (a)	78. (c)	79. (d)	80. (b)
81. (d)	82. (a)	83. (c)	84. (c)	85. (a)	86. (c)	87. (a)	88. (b)	89. (b)	90. (c)
91. (c)	92. (c)	93. (b)	94. (a)	95. (d)	96. (b)	97. (b)	98. (d)	99. (d)	100. (a)
101. (d)	102. (c)	103. (c)	104. (c)						

(B) Cumulative Exercise

1. (c)	2. (b)	3. (a)	4. (a)	5. (a)	6. (d)	7. (b)	8. (c)	9. (c)	10. (c)
11. (d)	12. (c)	13. (a)	14. (d)	15. (a)	16. (b)	17. (d)	18. (b)	19. (d)	20. (b)
21. (d)	22. (b)	23. (b)	24. (c)	25. (d)	26. (a)	27. (d)	28. (d)	29. (c)	30. (a)
31. (b)	32. (d)	33. (d)	34. (d)	35. (a)	36. (d)	37. (c)	38. (a)	39. (d)	40. (c)
41. (a)	42. (d)	43. (a)	44. (c)	45. (c)	46. (c)	47. (b)	48. (a)	49. (b)	50. (d)
51. (b)	52. (b)								

(C) Medical Entrances' Gallary

1. (b)	2. (b)	3. (b)	4. (b)	5. (a)	6. (b)	7. (b)	8. (b)	9. (b)	10. (c)
11. (c)	12. (b)	13. (a)	14. (c)	15. (d)	16. (b)	17. (a)	18. (b)	19. (b)	20. (c)
21. (b)	22. (a)	23. (d)	24. (a)	25. (b)	26. (c)	27. (a)	28. (d)	29. (d)	30. (b)
31. (a)	32. (b)	33. (b)	34. (c)	35. (d)	36. (c)	37. (e)	38. (b)	39. (b)	40. (c)
41. (b)	42. (b)	43. (a)	44. (b)	45. (c)	46. (b)	47. (c)	48. (b)	49. (e)	50. (b)
51. (d)	52. (c)	53. (b)	54. (b)	55. (d)	56. (a)	57. (c)	58. (d)	59. (b)	60. (c)
61. (c)	62. (d)	63. (a)	64. (d)	65. (e)	66. (c)	67. (c)	68. (a)	69. (c)	70. (d)
71. (c)	72. (c)	73. (c)	74. (b)	75. (a)	76. (d)	77. (a)	78. (b)	79. (b)	80. (b)
81. (b)	82. (d)	83. (d)	84. (c)						

Explanations

(A) Topical Exercise

4 (*c*) Only II is incorrectly matched pair and can be corrected as Thyroxine is derived from iodinated amino acid, i.e. it is a iodothyronine. An example of peptide hormone is calcitonin.

18 (*b*) Statement in option (b) is incorrect and can be corrected as Pars intermedia develops from the upper surface of Rathke's pouch and it produces Melanocyte Stimulating Hormone or MSH.

29 (*b*) Statement I is false, but II is true. Incorrect statement can be corrected as
The stromal tissue of thyroid glands is composed of loose connective tissue.

46 (*b*) Statement in option (b) is incorrect and can be corrected as Parathyroid gland develops from the third and fourth pairs of pharyngeal pouch as epithelial buds.

48 (*a*) Statement in option (a) is incorrect and can be corrected as
Zona reticularis is the innermost region of adrenal cortex.

69 (*a*) Option (a) represents incorrectly matched pair and can be corrected as
Type-II diabetes is caused due to less sensitivity of target cells to insulin. Type-I diabetes is an autoimmune disorder in which β-cells are destroyed.

88 (*b*) Option (b) represents incorrectly matched pair and can be corrected as
Leukotrienes are local hormones released by the mast cells as allergic response.

(B) Cumulative Exericse

2 (*b*) Statement in option (b) is incorrect and can be corrected as Sex hormones like testosterone, are steroid in nature and these are derived from cholesterol.

5 (*a*) Except statements in option (a), other three are false. Incorrect statements can be corrected as
Hormone produced by ovary, e.g. relaxin, relax the uterus whereas uterine contractions are promoted by oxytocin released by posterior pituitary. Small intestine does not secrete any hormone that affects heart functioning. Heartbeat is affected by the hormones of adrenal medulla.

10 (*c*) Glucagon is secreted to maintain glucose levels in bloodstream when fasting, to raise glucose level. Thus, in given condition, glucagon level would be high while that of insulin would be low in patient's blood.

16 (*b*) P is hypothalamus, Q is anterior pituitary gland and R is thyroid gland. S is thyrotropin releasing hormone, T is thyroid stimulating hormone and U is thyroxine. Thus, (b) is the only option that is correct.

17 (*d*) Statement in option (d) is incorrect about diabetes. It can be corrected as
Diabetes results in increased levels of sugar in blood because less sugar is burnt in diabetic patients than a normal person.

18 (*b*) Option (b) is correctly matched pair whereas other three are incorrectly matched pairs. These can be corrected as
Thyroid secretes T_3 and T_4 hormones, whereas epinephrine is secreted by adrenal medulla. Anterior pituitary secretes several hormones like GH, FSH, etc., and adrenaline is secreted by adrenal medulla. Secretin is secreted by duodenum.

24 (*c*) Option (c) shows incorrect match and can be corrected as
Thyrocalcitonin is produced by the parafollicular cells of thyroid gland. Follicular cells produce T_3 and T_4.

26 (*a*) All the steroid hormones bind to intracellular receptors. Among listed options, four hormones are steroid hormones. These are cortisol, testosterone, progesterone, oestrogen.

30 (*a*) Points I and II are incorrectly matched pairs because Luteotropic hormone is composed of about 204 amino acids and ACTH comprises 39 amino acids.

31 (*b*) Option (b) represents correctly matched pair. Other options are incorrectly matched pairs and can be corrected as
Cretinism is caused due to hyposecretion of thyroid hormones in children. Parathormone promotes the calcium ion movement from bones into the blood. Glycogenesis in liver is promoted by insulin hormone that is secreted by β-cells of Langerhans of pancreas.

35 (*a*) Among the given statements, four (I, II, IV and V) are correct whereas III is incorrect. Incorrect statement can be corrected as
The endocrine portion of pancreas constitutes only 1-2% whereas the rest is occupied by exocrine cells.

37 (*c*) From the graph, heart rate increases during smoking, when there is increased adrenaline production. Hence, it can be concluded that as more adrenaline is produced, the heart rate increases. The other conclusions cannot be made on the basis of provided information.

38 (*a*) Food consumed will be digested into monosaccharides, mostly glucose and absorbed into the blood. A rise in blood glucose concentration stimulates the secretion of insulin. When bound to receptors on the surface membrane of cells throughout the body, insulin stimulates an increase in the rate of conversion of glucose to glycogen in liver and muscle cells.

39 (*d*) Negative feedback loop refers to a type of control circuit where any fluctuation from the set point activates the control system resulting in a series of process which ultimately returns conditions towards their optimum level.

Glucagon is secreted in response to hypoglycaemia. However, when the blood glucose concentration reaches the normal level, the secretion of glucagon is inhibited by negative feedback loop.

40 (*c*) Option (c) is correct whereas other are incorrect. Intracellular receptors are always lipid soluble as they are required to cross the cell membrane. In contrast, cell surface receptors are not lipid soluble. Intracellular receptors are multi-pass transmembrane proteins whereas cell surface receptors are single-pass transmembrane proteins.

43 (*a*) All given statements are correct except I. Incorrect statement can be corrected as

The releasing hormones of hypothalamus promotes the secretion of anterior pituitary gland.

46 (*c*) A is true, but R is false because

Secretion of glucagon hormone increases the blood glucose level. Secretion of insulin hormone lowers the blood glucose level. Hypoglycaemia disorder is due to the imbalance of glucagon and insulin hormones. In this, blood glucose level decreases due to the hypersecretion of insulin and hyposecretion of glucagon.

47 (*b*) Both A and R are true, but R is not the correct explanation of A. The name master endocrine gland of the body is given to the pituitary gland. It secretes a number of hormones such as TSH, ACTH, etc., which regulates the working of other endocrine glands. This gland itself works under the control of the hormones secreted by hypothalamus of the brain.

48 (*a*) Both A and R are true and R is the correct explanation of A. Pitocin (oxytocin) is also called as birth hormone because it promotes the contraction of the uterine muscles during childbirth.

49 (*b*) Both A and R are true, but R is not the correct explanation of A. Excess secretion of parathyroid hormone causes softening and bending of bones. Some of the bone substances are replaced by the cavities that are filled with fibrous tissue. This condition leads to osteoporosis or osteitis cystica. Such type of bones are highly prone to fractures.

50 (*d*) A is false, but R is true. Assertion can be corrected as Regulation of RBCs production is accomplished by hormone erythropoietin, which is secreted by kidney. This hormone increases stem cell mitosis and speed up RBCs development in red bone marrow.

51 (*b*) Both A and R are true, but R is not the correct explanation of A. Antidiuretic Hormone (ADH) or vasopressin is secreted by the posterior lobe (neurohypophysis) of pituitary gland. Hyposecretion of ADH results in diabetes insipidus, which is characterised by micturating dilute urine several times a day and excessive thirst.

52 (*b*) Both A and R are true, but R is not the correct explanation of A. Parathormone or parathyroid hormone is secreted by the chief cells of parathyroid gland. This is also known as Collip's hormone. This hormone maintains calcium and phosphate balance between the tissues and blood.

(C) Medical Entrances' Gallery

3 (*b*) Neurohypophysis (pass nervosa) is also known as posterior pituitary which stores and releases two hormones called oxytocin and vasopressin or ADH. These are actually synthesised by the hypothalamus and are transported axonally to neurohypophysis.

5 (*a*) Steroid hormones are able to bind to DNA and form a gene-hormone complex. These hormones can easily cross the lipid bilayer of the cell and do not require secondary messengers for the same. Steroid hormones bind to intracellular receptors in the nucleus to form hormone receptor complex which in turn interact with the genome.

6 (*b*) In the given mechanism of hormone action, the correct labels are

A-Protein hormone, *B*-Receptor, *C*-Cyclic AMP

Hormones which interact with membrane bound receptors do not enter the target cell. These bind with membrane proteins (receptors) and generate second messengers, which in turn regulate cellular metabolism, e.g. FSH on ovarian cell membrane.

7 (*b*) The fall in blood Ca^{+2} levels stimulate parathyroid gland to release parathyroid hormone. This hormone exerts its effects on bones and kidneys. When calcium levels are low, parathyroid hormone is released by the parathyroid glands into the blood and causes the bones to release calcium and increase levels in the bloodstream.

8 (*b*) Artificial light, extended work time and reduced sleep time disrupt the activity of pineal gland. It is a small pea-shaped gland in the brain. It produces melatonin, which helps to maintain circadian rhythm and regulates reproductive hormones.

10 (*c*) A is true, but R is false because

Type-I diabetes is a result of destruction of β (beta) cell of islets of Langerhans, due to which there is no production of insulin hormone. Insulin cannot be taken as pills because it is a polypeptide chain, which would breakdown in the digestive process. It is taken in the form of intravenous injection to control blood sugar level.

11 (*c*) A is true, but R is false because

Pituitary gland releases a hormone named oxytocin from its posterior lobe which causes uterine contractions during childbirth. The vasopressin released by the same lobe of pituitary gland does not play any role in childbirth. It regulates osmotic concentration of body fluids, in particular water reabsorption by kidneys.

12 (*b*) Bradykinin is a peptide hormone of kinin group. It functions as a potent endothelium dependent vasodilator, mild diuretic and an inflammatory mediator. It may cause lowering of blood pressure. It also causes contraction of non-vascular smooth muscle in the bronchus and gut and increases vascular permeability.

13 (*a*) Osteoporosis is caused due to the deficiency of oestrogen and excessive activity of parathormones. Oestrogen helps to promote the activity of osteoblast (helps in the formation of bone cells) and inhibits osteoclast activity (destruct the bones). On the other hand, parathormone promotes the mobilisation of calcium from bones into blood and hence, causes demineralisation.

The other listed hormones also contribute to osteoporosis but their effects are insignificant, e.g. low level of progesterone and aldosterone causes bone loss, whereas raised level of prolactin has been linked with osteoporosis.

14 (*c*) Among the given, epinephrine is an amino acid derived hormone. It is a catecholamine which is produced in the chromaffin cells of adrenal medulla from the amino acid, tyrosine. On the other hand, oestradiol and oestriol are steroid hormones that are involved in the regulation of oestrous and menstrual cycles.
Ecdysone is also a steroid hormone that controls moulting in insects.

15 (*d*) Both A and R are false and can be corrected as

Peptide and polypeptide hormones such as oxytocin and vasopressin are water soluble hormones and are lipid insoluble. Therefore, they cannot directly pass across the lipid bilayer of plasma membrane.

16 (*b*) Among the given hormones, TSH (Thyroid Stimulating Hormone) secreted by anterior lobe of pituitary gland, works indirectly by controlling the growth and activity of the thyroid gland. It influences the uptake of iodine and the synthesis of hormones like thyroxine and triiodothyronine by the thyroid gland.

Rest of the hormones, i.e. MSH (Melanocyte Stimulating Hormone), GH (Growth Hormone) and oxytocin works directly on their target organ. They do not stimulate any other gland.

17 (*a*) Inhibins are dimeric glycoproteins composed of an α-subunit and either a β_A (inhibin *A*) or a β_B subunit (inhibin *B*). They are mainly produced in gonads and provide negative regulation of FSH secretion.

18 (*b*) Chronic hypersecretion of Growth Hormone (GH) leads to gigantism or acromegaly depending on the age of the individual.

If its hypersecretion occurs before the ossification of epiphyseal plates, it causes exaggerated and prolonged growth in long bones resulting in gigantism.

The hypersecretion of GH after puberty leads to acromegaly. No increase in height occurs because of the ossified epiphyseal plate.

19 (*b*) GnRH is a hypothalamic hormone. It stimulates the anterior lobe of pituitary gland to secrete LH and FSH, which in turn stimulates the gonadal activity.

20 (*c*) The statement in option (c) is correct in relation to the endocrine system. Other options are incorrect. The correct information regarding the incorrect statements is as follows

Endocrine cells are present in different parts of various organs and secrete specific hormones. The gastrointestinal tract secretes gastrin, secretin, GIP.

Atrial wall of our heart secretes a peptide hormone called ANF (Atrial Natriuretic Factor). Releasing and inhibitory hormones are produced by hypothalamus. Adenohypophysis is not directly under the neural control of hypothalamus.

21 (*b*) Graves' disease or hyperthyroidism is caused due to the secretion of excessive amount of thyroid hormone by the thyroid gland, i.e. hyperthyroidism.

22 (*a*) Insulin is the peptide hormone which enhances the uptake of glucose molecules by liver cells (hepatocytes) and fat cells (adipocytes) for its cellular utilisation.

Such, an activity of insulin brings down the level of glucose in the blood.

23 (*d*) High levels of oestrogen and progesterone give negative feedback to hypothalamus for the release of GnRH, thus inhibiting the gonadotropin release.

25 (*b*) Both A and R are true, but R is not the correct explanation of A. Some hormones that are insoluble in lipids do not affect a physiological process directly, instead they activate secondary messenger that convey the information to the target cells.

On the other hand, glucocorticoids are steroid hormones that can directly affect the gene expression.

26 (*c*) Option (c) represents mismatched pair and can be corrected as Gluconeogenesis is the synthesis of glucose molecule. It is mainly controlled by glucagon. Insulin stimulates the conversion of glucose to glycogen, i.e. glycogenesis.

29 (*d*) Adrenocorticotropic hormone is a polypeptide, which promotes the release of glucocorticoids and androgens from adrenal cortex. Rest all given hormones are steroids.

33 (*b*) Thyroxine or tetraiodothyronine (T_4) is a hormone produced by follicular cells of the thyroid gland.

36 (*c*) Ectopic transplantation of pituitary is done in kidney capsule, anterior eye chamber or third cerebral ventricle. It secretes prolactin (PRL) abundantly.

37 (*e*) MSH is produced by pars intermedia and vasopressin is produced by pars nervosa.

39 (*b*) Aldosterone is not involved in sugar metabolism. It is a steroid hormone (mineralocorticoid) produced by the outer section (zona glomerulosa) of the adrenal cortex in the adrenal gland.

41 (*b*) Leydig cells are present in the testes of mammals. These produce androgens (testosterone). These hormones control the development of secondary sexual characters as well as gametogenesis (spermatogenesis) in males.

43 (*a*) Pheromones are intraspecific chemical messenger that trigger a social response in members of the same species. These are chemicals capable of acting outside the body of the secreting individual to impact the behaviour of the receiving individual.

44 (*b*) Phenylalanine-N-Methyl Transferase (PNMT) is an enzyme, found in the adrenal medulla that converts nor-epinephrine (nor-adrenaline) to epinephrine (adrenaline). PNMT catalyses the transfer of a methyl group from SAM (S-adenosyl-L-methionine) to nor-epinephrine thus, converting it into epinephrine.

Appendix

Major discoveries in physiology and anatomy

Year	Name	Contribution
1570-1657	William Harvey	Father of circulatory system, who discovered circulation of blood in animals.
1628-1694	Marcello Malpighi	Discovered Malpighian capsule in kidney, Malpighian layer in skin, Malpighian tubules in insects, blood capillaries and described plant anatomy for the first time.
1627-1705	John Ray	Coined the term 'genera'. Gave the concept of species.
1727	Stephan Hales	Father of Plant Physiology.
1769-1832	George Cuvier	Laid foundation of comparative Anatomy and Palaeontology in 1796.
1792-1876	Karl von Baer	Discovered ovum in mammals. Known as Father of Embryology. Proposed biogenetic law.
1840	JE Purkinje	Discovered sweat glands in 1833.
1842	Mayer	Proposed that light energy convert into chemical energy during photosynthesis.
1842	William Bowman	Histological structure of nephron.
1860	Louis Pasteur	Microbiology of lactic acid fermentation and role of yeast in alcoholic fermentation.
1878	Kuhne	Coined the term enzyme.
1880	Sandstrom	Discovered parathyroid glands.
1882	Robert Koch	Methods for the study of bacteria in pure culture and discovery of cause of tuberculosis.
1882	Elie Metchnikoff	Phagocytosis.
1884	Robert Koch	Koch's postulates.
1884	Eduard Strasburger	Described fertilisation in plants.
1885	Louis Pasteur	Rabies vaccine.
1890	Emil von Behring and Kitasato Shibasaburo	Diphtheria antitoxin.
1893	His	Described bundle of His in atrio ventricular node.

Year	Name	Contribution
1896	Eijkman	Discovered that disease beri-beri is caused by the deficiency of vitamin-B_1.
1897	Ronald Ross	Discovered oocytes of *Plasmodium* in stomach wall of female *Anopheles*.
1897	Buchner	Discovered enzyme.
1898	Nawaschin	Discovered double fertilisation and triple fusion in angiosperms.
1900	Karl Landsteiner	Discovered A, B and O blood groups and Rh factor.
1901	Emil von Behring	Developed antitoxin for tetanus.
1902	Alfred von Decastello and Sturli	Discovered AB blood group.
1904	Blackeslee	Discovered heterothallism in *Mucor*.
1914	EC Kendall	Isolated thyroxine (T_3).
1922	Banting and Best	Isolated insulin from pancreas of animals.
1926	Went	Carried out Avena curvature test.
1931	Van Neil	Suggested the importance of light in photolysis during photosynthesis.
1937	HA Krebs	Proposed Krebs' cycle or citric acid cycle.
1941	Ruben and Kamen	Proved that the O_2 released in photosynthesis comes from water.
1946	Lederberg and Tatum	Discovered sexual reproduction in bacteria.
1954	HE Huxley and AF Huxley	Proposed sliding filament theory of muscular contraction.
1954	D Arnon	Described photophosphorylation in chloroplast.
1957	Emerson and Arnold	Photosynthetic yield can be increased with brief flashes of light.
1959	Rodney Porter	Immunoglobulin structure.
1961	M Calvin	Described CO_2 fixation in photosynthesis.
1967	Hatch and Slack	Proposed dicarboxylic acid cycle (Hatch and Slack cycle).

Planes of the body

Different sections of the body are termed as anatomical planes (flat surfaces) by the medical professionals. These planes are imaginary lines vertical or horizontal, which are drawn through an upright body. The terms are used to describe a specific body part.

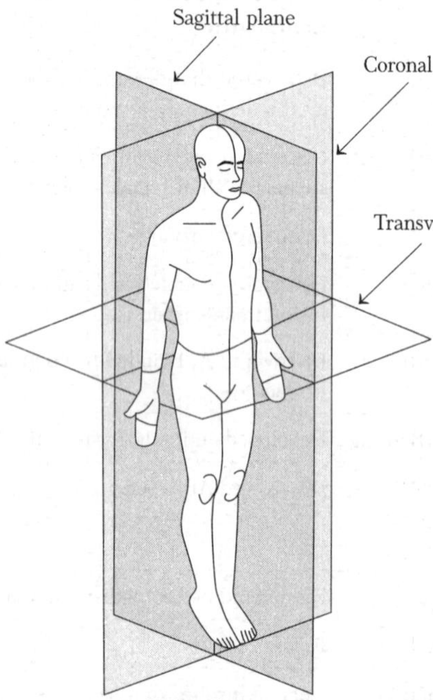

Body planes

- **Coronal plane** Coronary or frontal plane is a vertical plane running from side to side. It divides the whole body or any of its parts into anterior and posterior portions.
- **Sagittal plane** Sagittal or lateral plane is a vertical plane running from front to back. It divides the body or any of its parts into right and left sides. Median plane is a sagittal plane that runs through the midline of body.
- **Transverse plane** It is a horizontal plane. It divides the body or any of its parts into upper and lower parts.

Size of viruses, cells and organisms

Size of cells ranges from minute bacteria to large, unfertilised eggs laid by birds. The relative size ranges of some living things are shown in the following figure. In Science, we use the metric system for measuring. Here are some measurements and conversions that will aid your understanding of Biology.

1 metre =100 cm = 1,000 mm = 1,000,000 μm = 1,000,000,000 nm

1 centimetre (cm) = 1/100 metre = 10 mm

1 millimetre (mm) = 1/1000 metre = 1/10 cm

1 micrometre (μm) = 1/1,000,000 metre = 1/10,000 cm

1 nanometre (nm) = 1/1,000,000,000 metre = 1/10,000,000 cm

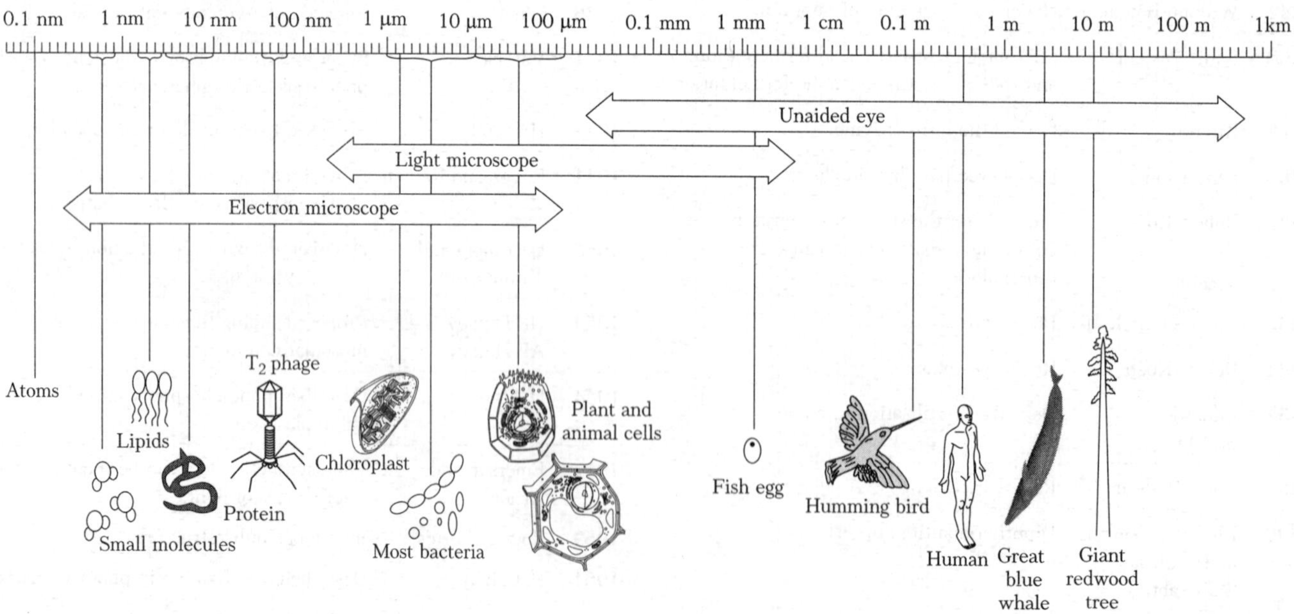

Some important parts of female flower

Types of carpel

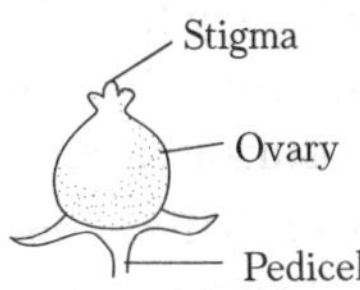 **Astylocarpellous** (without a stipe and a style)	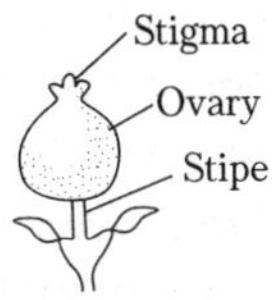 **Astylocarpepodic** (with a stipe and without a style)	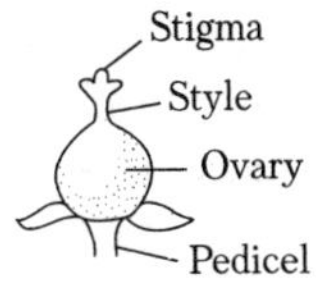 **Stylocarpellous** (Carpel without a stipe and with a style)	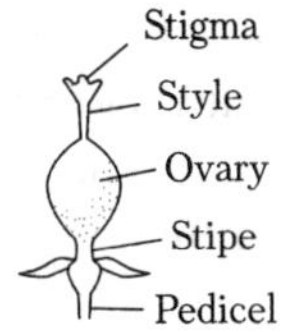 **Stylocarpepodic** (with a stipe and a style)

Types of styles

 **Geniculate** (style bent abruptly)	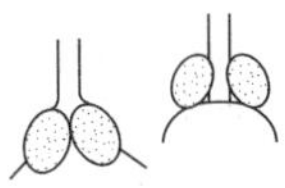 **Gynobasic** (style is attached at the base of the ovary in central depression)	 **Heterostylous** (style of different shape or length)	 **Homostylous** (style of same shape and length)	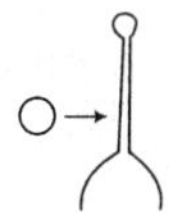 **Terete** (cylindrical and elongate)	 **Fimbriate** (presence of fimbrion style)

Types of stigma

 **Capitate** (like a cap or head)	 **Clavate** (club-shaped)	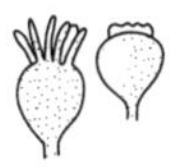 **Crested** (tuft or ridge like)	 **Decurrent** (long and extending downward)	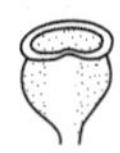 **Diffuse** (spread over a wide surface)	
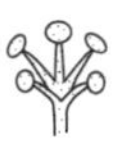 **Discoid** (like a disc)	 **Fimbriate** (fringed)	 **Lineate** (in the form of small lines)	 **Lobed** (divided into some lobes)	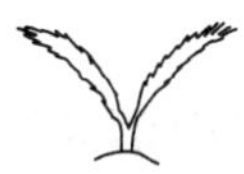 **Plumose** (like a feather)	 **Terete** (cylindrical and elongate)

Electrocardiogram (ECG)

The electric changes of the cardiac chambers follow a specific sequence. These changes are recorded by a specific device called electrocardiograph and the recorded graph is called ECG. It is represented as PQRST, where P stands for depolarisation of atria, QRS depolarisation of ventricles and T is repolarisation of ventricles. The defects in cardiac function or structure are recorded in ECG.

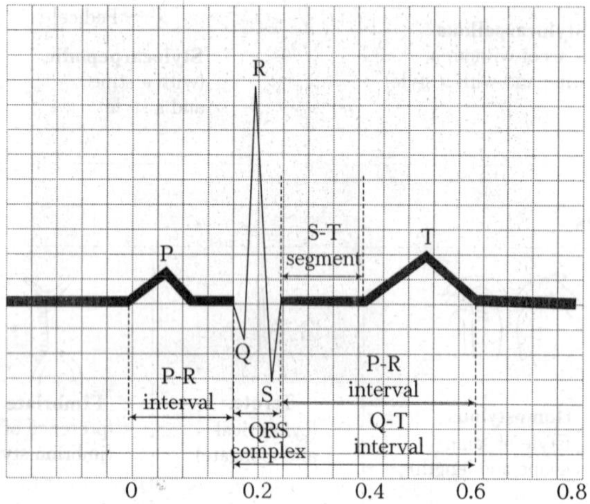

(a) Normal ECG of a single heartbeat

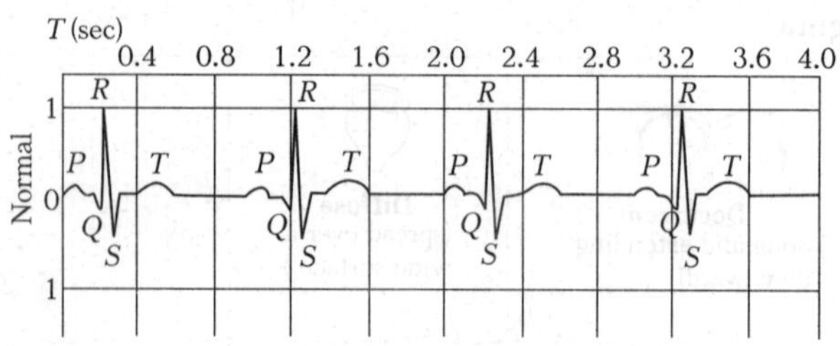

(b) Normal electrocardiogram

Electroencephalogram (EEG)

It is an electrically operated instrument having electrodes which when attached to scalp, give information about the electric waves emerging from different parts of the brain. Electroencephalogram is a recording of electrical potentials originated from different parts of the brain in the form of waves. Hans Berger (1924) was the first to record Electroencephalogram (EEG).

Types of waves

An EEG of a normal individual shows four kinds of waves—beta, alpha, theta and delta waves. The frequency of waves is measured in Hertz (Hz). 1 Hz = 1 cycle.

Various types of waves are given below

(a) **Beta waves** These are smaller and fastest waves which appear from frontal and parietal region of the brain when the person is alert and concentrating. At this time, nervous system is very active. These wave occur at a frequency of 14-30 Hz.

(b) **Alpha waves** These rhythmic waves occur at a frequency of approximately 8-13 cycles per second. They occur in all normal individuals when the person is relaxed or when awake with the eyes closed. They disappear entirely during sleep.

(c) **Theta waves** These waves appear when the person is under drowsiness. These are very large and slow waves of frequency about 4-7 Hz. They normally occur in children and in adult during emotional stress or certain brain disorders. They are abnormal in awake adults.

(d) **Delta waves** These are even slower and larger waves that occur in deep sound sleep. They have a frequency of 1-5 Hz.

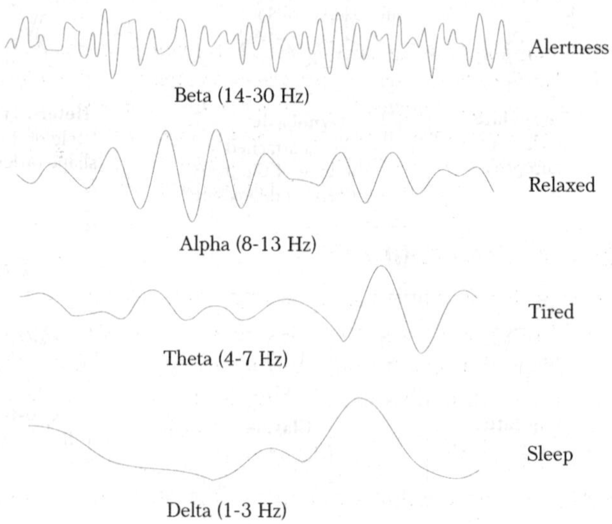

Electroencephalogram

Ultrasound imaging (Sonography) technique

It is a non-invasive technique which uses ultrasound waves for producing imaging of internal body parts.

Principle Crystal of certain substances become excited and start vibrating when electricity passes through them. Lead zirconate used as transducer convert one form of energy into other. Transducer generation of ultrasound waves by applying electrical voltage is called **Piezo-electric effect.**

As the waves travel through body tissues, they are reflected back at any point, where there is a change in tissue density. The reflected echoes are received by an electronic apparatus that determines the intensity level of the echoes and the position of the tissue giving rise to the echoes. The image thus formed can be displayed in a static form, or through the use of rapid multiple sound scans. The transducer also acts as a receiver, converting these echoes into electrical signals, which are processed and displayed on a screen of monitor to give a two dimensional image. A gel is used during ultrasound between transducer and body part to provide air free medium.

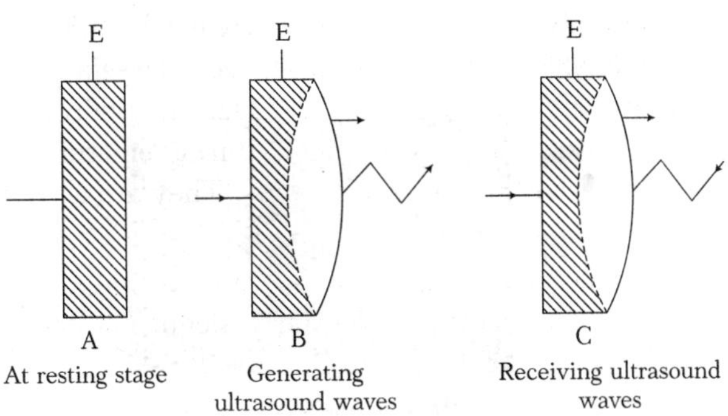

***Diagram showing** generation and reception of ultrasound by Piezo-electric effect : A–Transducer at resting stage; B–When voltage is applied transducer resonates and produces ultrasonic sound waves; C–When reflected ultrasonic sound pulse strikes the transducer, it resonates and a voltage is generated (converse Piezo-electric effect).*

Uses of ultrasound

Ultrasound imaging is useful in diagnosing the diseases of the brain, kidney stones, gallstones, cirrhosis, intestinal obstruction, Fallopian tubes, uterus, etc. It has wide applications in medicine and is especially useful in obstetrics. This technique offers no known risk to the baby and is often performed at about 16-18 weeks to reveal multiple pregnancy and foetal abnormalities like anencephaly and spina bifida.

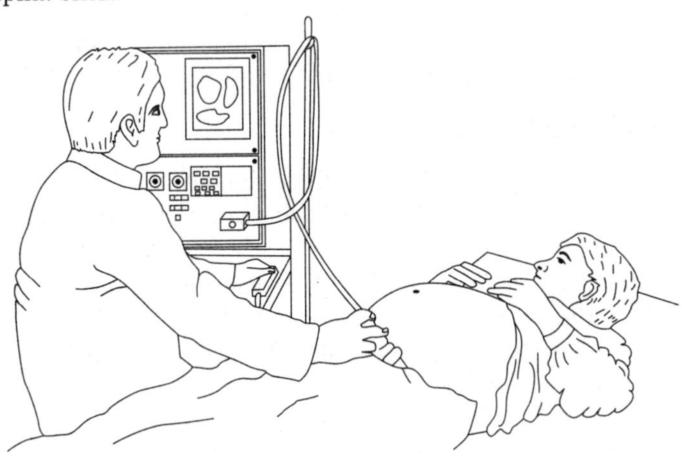

An ultrasound machine : An operator carrying out ultrasound scanning of a patient : the transducer is placed on the abdomen and the image is displayed on the screen of the monitor

X-ray imaging

It is one of the oldest and the most common medical method of imaging which began with the accidental discovery of X-rays by Wilhelm Konrad Roentgen, a German physicist, in 1895. X-rays are highly penetrating electromagnetic radiation with a wavelength in the range of 10 nanometres to 100 picometres.

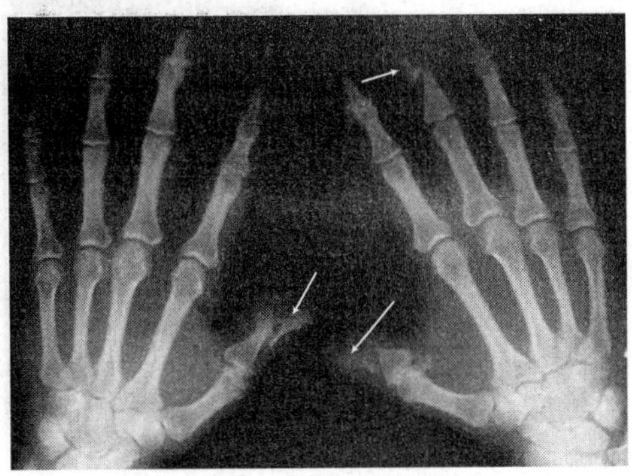

***X-ray of** hand showing resorption of terminal phallanges*

X-rays when directed on a part of the body such as chest, the rays are absorbed more by dense structures such as the ribs or heart muscles than by less dense structures such as the skin or lungs. This causes shadows of variable intensity to be cast on a photographic film. X-rays cause no sensation when passed through the body tissues. Large or frequent radiation doses may damage the skin and internal organs and may cause cancer in later life. The study of X-rays for the detection and treatment of disease is called **radiology**. X-ray imaging is the simplest form and commonly employed for diagnosing diseases of the heart, lungs and detection of bone and joint injuries.

Magnetic Resonance Imaging (MRI)

It is a diagnostic technique that provides high-quality cross-sectioned or three-dimensional images of organs and structures without using X-rays or other radiation. The technique exploits the natural behaviour of the protons (nuclei) of hydrogen atoms, when they are subjected to very strong magnetic field and radiowaves.

The most abundant source of protons in the body are hydrogen atoms in water molecules. An MRI scan reflects differences in the water content of the tissues. A newer application of MRI known as **Nuclear Magnetic Resonance** (NMR) spectroscopy relies on the detection of other chemical elements such as phosphorus and calcium. The patient lies down surrounded by massive electromagnets and is exposed to short bursts of powerful magnetic field and waves, which stimulate protons (hydrogen nuclei) in the patient's tissues to emit radio signals.

The signals are detected and analysed by a computer to create an image of a 'slice' of the patient's body.

NMR images show great sensitivity in differentiating between normal tissues and diseased or damaged ones.

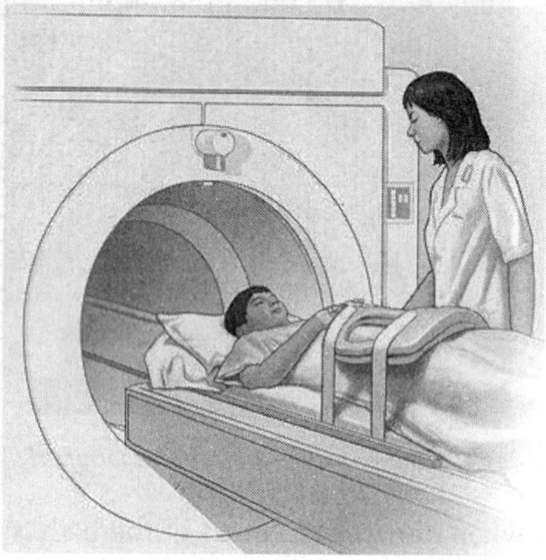

A patient being placed for MRI

This imaging technique helps to provide images of the brain, heart, liver, kidneys, spleen, pancreas, breast and other organs. MRI provides relatively high-contrast, variable-toned images that can show tumours, blood starved tissues and neural plaques resulting from multiple sclerosis. The Nobel Prize for medicine for 2003 was awarded to Paul Lauterher and Peter Mansfield for modifying MRI technique to give 3D pictures.

Uses of MRI

(i) In the examination of brain and spinal cord because white and grey matter can be differentiated.

(ii) In the examination of cancerous bone, pelvic organs, urinary bladder, joints, vertebral column and also cancerous growth in liver and kidney.

(iii) It helps in the identification of Parkinson's disease and recognition of changes that occur in the brain during this disease.

(iv) MRI also help in measuring blood flow and can assess heart diseases and strokes.

(v) MRI is being used more and more in study of joints and musculo-skeletal disorders, joint injuries and slip disc in vertebral column.

Limitation of MRI

(i) Since MRI makes the use of magnets, it cannot be used for people with artificial pacemakers or metal implants, etc.

(ii) Though radiowaves are harmless, it is still not recommended for pregnant women.

(iii) It does not show calcification which is an important pathological indicator of a disease process.

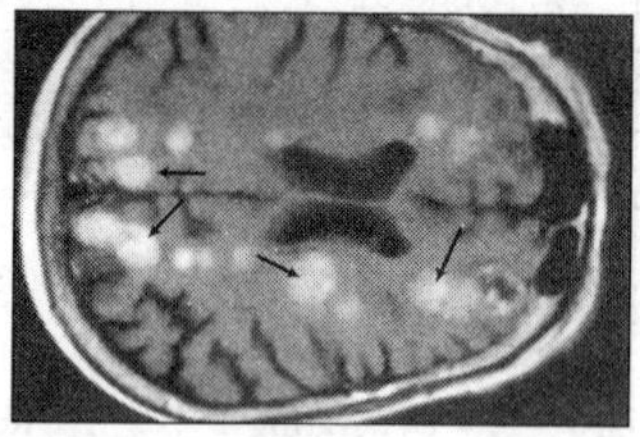

***MRI showing** brain tumours*

Positron Emission Tomography (PET)

It is a computerised imaging technique, first developed by Louis Sokoloff of USA in 1985. It is a diagnostic technique based on detection of positrons (positively charged electrons) emitted by radio isotopes such as ^{11}C, ^{13}N, ^{15}O and ^{18}F generated by the cyclotron. In PET, a chemical compound labelled with a short-lived, positron-emitting radionuclide of carbon, oxygen, nitrogen or fluorine is injected into the body.

The activity of such a radiopharmaceutical is quantitatively measured throughout the target organs by means of photomultiplier scintillator detectors. As the radionuclide decays, positrons are annihilated by electrons, giving rise to gamma rays that are detected simultaneously by the photomultiplier scintillator combinations positioned on opposite sides of the patient.

Uses of PET

(i) PET scanning provides three-dimensional images that reflects the metabolic and chemical activity of tissues being studied.

(ii) The technique is particularly valuable for the measurement of regional cerebral blood volume, blood flow, metabolic rates of glucose and oxygen in humans.

(iii) PET scanning is used for detecting tumours, for locating the origin of epileptic activity within brain and for examining brain function in various mental illnesses.

(iv) PET scanning is used to locate colour processing centres in human visual cortex.

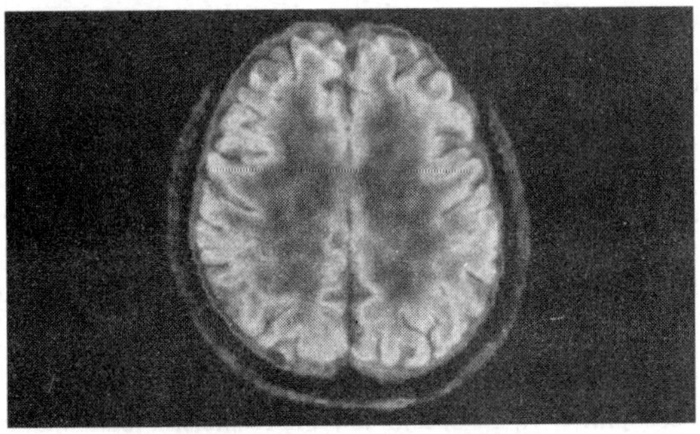

***A PET** scan image of brain*

Computerised Axial Tomography Scanning (CT scan)

CT scan was developed by sir Godfrey Hounsfield of Britain in 1967 (Nobel Prize in 1979). This technique combines the use of X-rays with computer technology to produce two or three-dimensional clear cross-sectioned image of an area. Computed tomographic scanning is also known as CAT (Computed Axial Tomography). In this technique, X-ray's beams move 360° around and pass through the thin sections of the body. Repetition of this process cover the section in all angles and then all figures are arranged by computer to give 3D view.

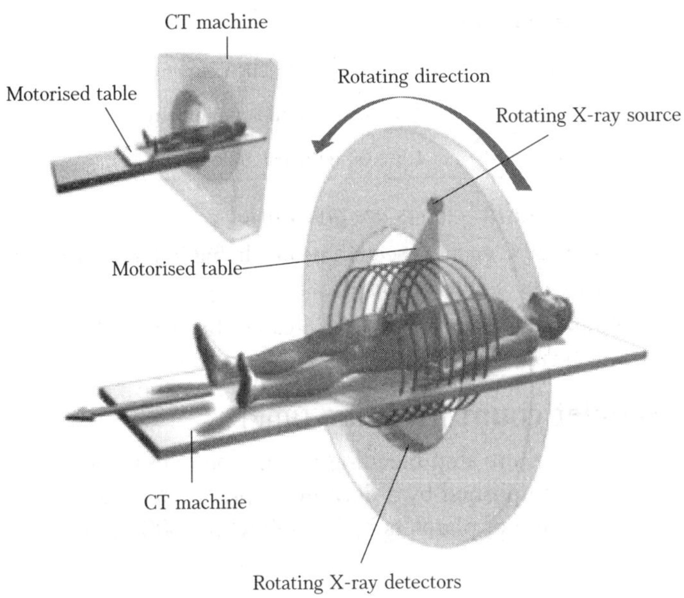

***CT scan** procedure being done on a patient*

Uses of CT scan

(i) CT scanning provides clear and more detailed information than X-rays.

(ii) It helps in the diagnosis of diseases of brain, spinal cord, chest and abdomen.

(iii) It is useful in detecting tumours and monitoring the extent of their spread to surrounding tissues and organs.

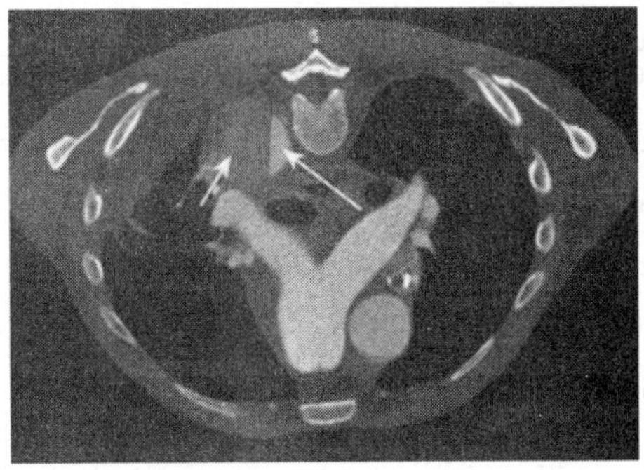

***CT scan** angiography (axial) showing aortic dissections*

Haematology

It is a branch of medical science that deals with the study of formed elements of blood (RBCs, WBCs, blood platelets), their precursors, their products and their disorders. Capillary or venous blood can be used for laboratory test.

Capillary blood is obtained from the tip of a finger while venous blood is taken from the upper part of the arm. Routine blood test is also called as the **Complete Blood Count** (CBC). Obtained blood is mixed with the anticoagulants which could be oxalates, EDTA or heparin.

Haematocrit

The Haematocrit (HT) or Packed Cell Volume (PCV) is the proportion of blood volume that is occupied by the red blood cells in plasma.

A significant drop in the value of PCV means anaemia and overhydration while a higher value means polycythemia and dehydration. Normal range of HT for male is 40-54% (average 47). Normal range of HT for female is 38-46% (average 42).

White Blood Cell Count or TLC

Leucocytes or white blood cells include granulocytes (acidophils, basophils and neutrophils) and agranulocytes (lymphocytes and monocytes) which have independent morphological, functional and staining properties. White blood cell count is an enumeration of leucocyte count which is determined by haemocytometer.

Abnormal white blood cell count and associated diseases

Leucocytosis (WBCs more than normal)	**Leucopenia** (WBCs less than normal)
Appendicitis	Influenza
Pneumonia	Typhoid
Leukemia	Hepatitis
Tonsilitis	Myxedema
Meningitis	Agranulocytosis
Diphtheria	Starvation

Normal value of Total Leucocyte Count (TLC) is given below

Adult	–	$7.5 \pm 3.5 \times 10^3/mm^3$
Full term pregnant woman	–	$15 \pm 3 \times 10^3/mm^3$
Children (4-7 yr)	–	$11 \pm 5 \times 10^3/mm^3$
Newborn	–	$18 \pm 8 \times 10^3/mm^3$
Infants (1 yr)	–	$12 \pm 6 \times 10^3\ mm^3$
Children (8-12 yr)	–	$9 \pm 4.5 \times 10^3/mm^2$

Differential Leucocyte Count (DLC)

The differential white blood cell count is the determination of the different types of white blood cells in blood.

Value of differential leucocyte count in blood

Types of leucocytes	% White blood cell count
Neutrophils	55%
Eosinophils	1-3%
Basophils	0-1%
Monocytes	2-6%
Lymphocytes	20-35%

Clinical application of Differential Leucocyte Count (DLC)

Leucocyte count	Clinical applications
High neutrophil	Bacterial infection, burn and stress.
Low neutrophil	Radiation, certain drugs and vitamin-B_{12} deficiency.
High eosinophil	Allergic reaction, parasitic infection and adrenal insufficiency.
Low eosinophil	Certain drugs, stress and cushing syndrome.
Less basophil	Pregnancy, ovulation, stress and hyperthyroidism.
More basophil	Certain allergic responses, leukemia, cancer and hypothyroidism.
High lymphocyte	Viral infections, immune diseases and leukemia.
Low lymphocyte	Prolonged severe, illness, high steroid level and immuno suppression.
High monocytes	Certain viral fungal infections, TB, some leukemia and chronic diseases.
Low monocytes	Generally does not occur.

Red Blood Cell (RBC) count

It is an enumerator of red cells or is an erythrocyte count which denotes red cells present in one cubic millimetre of blood. A haemocytometer helps in counting their number.

Normal values of RBC counts

Age group	RBC counts
Infants	5-6.5 millions/cubic milimetre
Women	4-5.5 millions/cubic millimetre
Men	4.5-6.00 millions/cubic millimetre

Variation in RBC counts are given below

(i) High RBC count is seen in people living in high altitude or patients of polycythemia.

(ii) Low RBC counts are found in cases of anaemia and leukemia.

Platelet count (Bleeding time)

Platelets help to stop bleeding when blood vessel is cut and is accomplished by a definite sequence of events. Normal value of platelets is 250000 to 500000/cu mm.

The bleeding time or coagulation time is the time required for a small cut to stop bleeding which is normally between 2-7 min. Bleeding time is prolonged when platelet count is below 10000/cu mm.

High value of platelet count occurs in case of cigarette smoking, excess production by bone marrow. Low values of platelets are found in case of immune thrombocytopenia, acute blood loss, drug effects, etc. Low platelets can also occur from clumping of the platelets. Venous blood is preferred for platelet count.

Determination of ESR
(Erythrocytic sedimentation rate)

ESR is the rate of growth at which the erythrocytes or RBCs sediment, when they are allowed to settle out from plasma. The values of sedimentation rate varies with their method of determination.

Heparin is the most suitable anticoagulant to determine ESR by Wintrobe's method. Higher ESR is observed in case of nephritis, rheumatic fever and arthritis, cancer, syphilis. Lower ESR is observed in increased allumin and sickle-shaped RBCs and pneumonia.

Normal values of ESR measured by various methods

Method	**Normal values**	
	Men (mm/hr)	**Women** (mm/hr)
Wintrobe	0-10	0-20
Westergren	0-15	0-20
Culter	0-8	0-10

Blood glucose

Normally, insulin produced by the pancreas enables the body's cells to absorb sugar or glucose from the blood. Deficiency of insulin causes glucose accumulation in the blood and urine in diabetes mellitus causing symptoms like excessive urination and thirst.

Treatment or control of glucose levels in blood is carried out by the administration of insulin. Thus, blood glucose level acts as an indicator for such disorders like hyperglycemia (decrease in blood glucose level).

Normal value of sugar is given below

(i) Whole blood sugar (fasting) 90-120 mg/100 mL. It indicates 20-30 mg/dL non-glucose fraction.
(ii) Plasma/serum sugar (fasting) 70-10 mg/dL.
(iii) Plasma/sugar post prandial sugar 100-140 mg/dL.
(iv) Whole blood post prandial sugar 140-180 mg/dL.

Blood urea

The waste products, i.e. urea and creatinine are the most frequently measured chemicals to indicate kidney disorder. There are two methods for their estimation Nesstlets reaction and Berthelots reaction. In both cases, urea is broken down by enzyme urease to liberate ammonia.

Normal value of blood urea is 30-40 mg/dL.

Higher value of blood urea may cause Loss of body fluids due to injury, haemorrhages, reduced blood flow to kidneys, increased protein catabolism, chronic renal disease or renal failure.

Lower value of blood urea may cause Liver problem, malnutrition and overhydration.

Artificial heart valves

The valves present in the heart chambers control the directional flow of blood. The artificial heart valves are inserted into the patient's heart as part of an open heart surgery called **heart valve replacement**. Artificial heart valves may be either mechanical or made of human or animal tissues.

Mechanical valves are made from special biocompatible plastics, metal alloys and ceramics.

Tissue valves are taken from cadavers of pigs or made from the pericardium of animals. Mechanical valves develop tendency of clotting of blood, so the patient must have regular medication of anticoagulants. Tissue valves do not require anticoagulants, but they tend to calcify, particularly in young people.

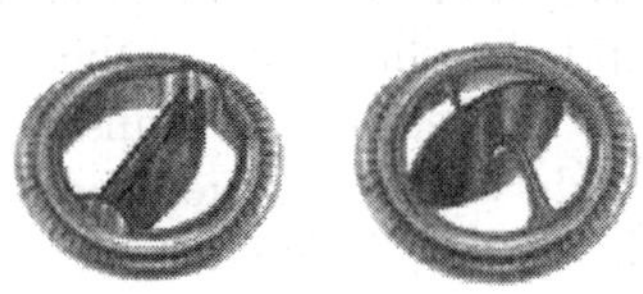

Mechanical valves

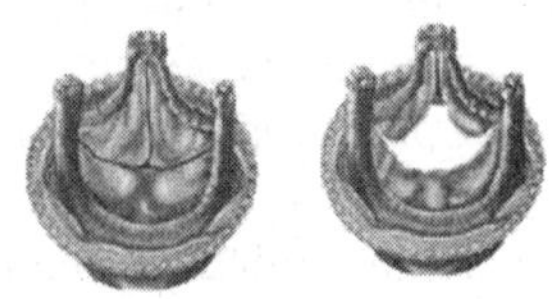

Biological valves

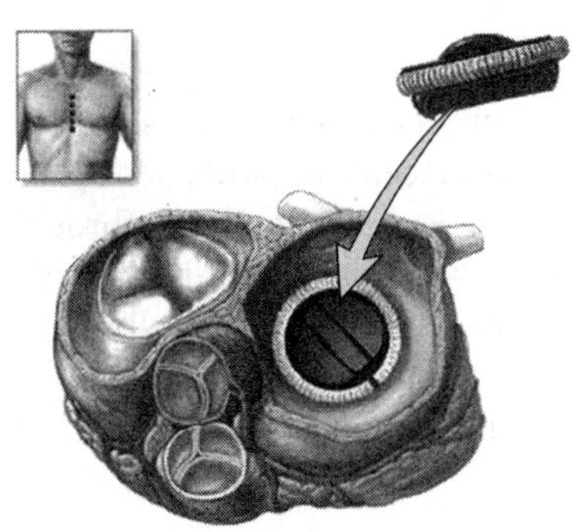

Mechanical valve implanting in heart

Vascular grafts

The tubular arteries sometimes get damaged or their walls may get thickened with advancing age (atherosclerosis) or due to certain vascular diseases. These abnormalities result in the narrowing of channel to such an extent that blood supply to the organ may become insufficient. Vascular grafts are prosthetic tubes that serves as artificial replacements for damaged blood vessels. A very commonly used material for vascular graft is Polyethylene Terephthalate (PET). Decron and teflon are also used in the making of artificial arteries.

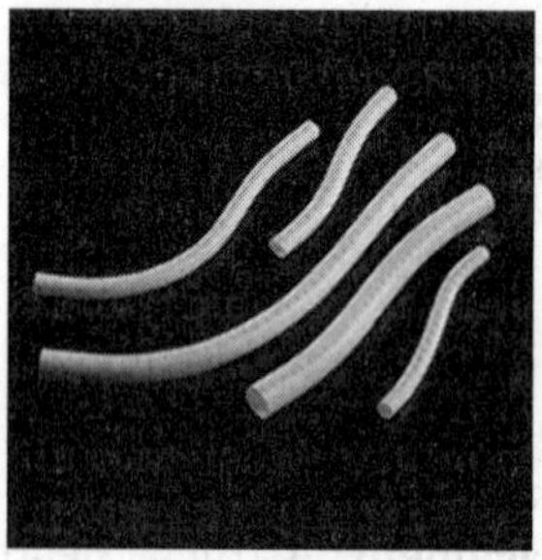

Vascular grafts

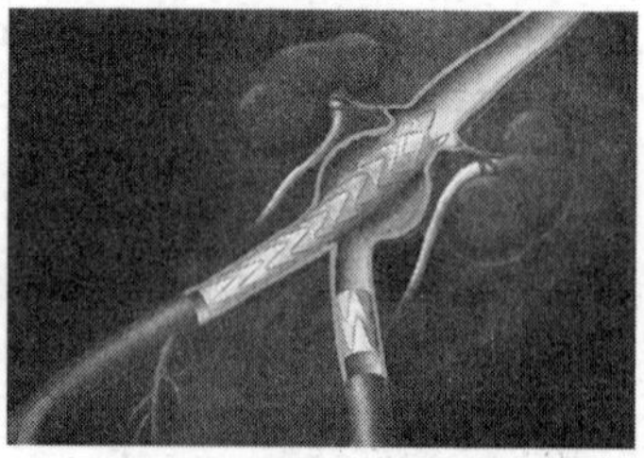

Vascular grafts implanted in blood vessels

Pacemaker

It is an electric device, first developed by Greatbach and Chardack (1960). It produces rhythmic electrical impulses that take over the regulation of the heartbeat in patients with certain types of heart disease. It is implanted in the heart of patient so that the heartbeat is generated by them.

It consists of a pulse generator and an electrode. The pulse generator is a hermetically sealed box which contains lithium halide cells to provide power for over 10 yr. The electrode is a fine metallic spring ensheathed in a thin layer of biocompatible plastic. Its special tip remains in contact with the interior of right ventricle. In temporary pacing, a miniature electrode attached to fine wires is introduced into the heart through a vein usually in the arm.

The pacing device, an electric generator, remains outside the body and produces regular pulses of electric charge to maintain the heartbeat. In permanent pacing, the electrode may again be passed into the heart through a vein or it may be surgically implanted on the heart's surface. In either case, the electrode is generally located in the right ventricle. The electric generator is placed just beneath the skin, usually in a surgically created pocket below the collarbone.

An artificial pacemaker is implanted when a person's sinu-atrial node is not functioning properly. Two basic types of pacemakers are fixed-rate and demand type. Fixed-rate pacemaker discharges impulses at a steady rate, irrespective of the heart's activity. Demand pacemaker discharges impulses only when the heart-rate slows or a beat is missed. A normal heart-rate and beat suppresses the pacemaker.

Intra aortic balloon pump

An intra aortic balloon pump is a device that increases blood flow to the heart muscle and decreases the heart's blood work load through a process called **counter pulsation.** Counter pulsation was introduced as a strategy by which the blood flow to the heart muscle could be increased during the diastole (less flow work), so the heart would not have to work so hard during the systole. The increased blood flow would provide the heart with more oxygen to improve circulation and strengthens the cardio pulmonary system.

Intra aortic balloon pump procedure was developed during the 1970s to treat atherosclerotic heart disease. This balloon catheterisation is technically called **PTCA**, i.e. Percutaneous Transluminal Coronary Angioplasty. A wire with a balloon on the tip is inserted into an artery in the leg and threaded through the aorta into the coronary artery. When the balloon reaches the atherosclerotic area, it is inflated with helium. The plaque is compressed and normal blood flow is re-established. It is estimated that about one in six coronary bypass operations can be replaced by this less dangerous procedure. PTCA was first applied by Gruentzigin in 1977.

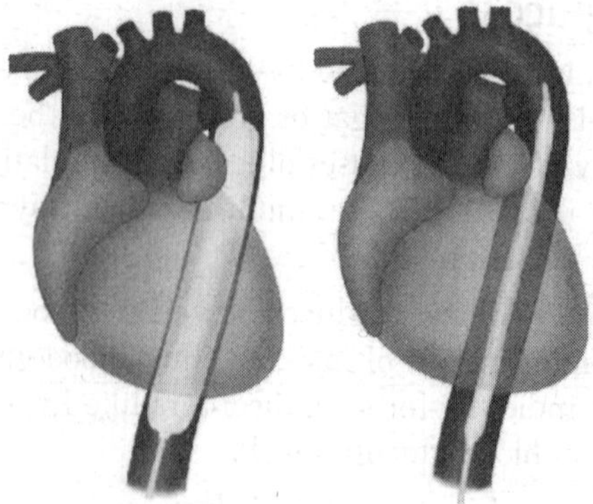

Figure showing *intr aortic balloon pump implanted in aorta*

Important plant products

Common name with plant used	Botanical name
A. Food yielding plants	
(a) **Cereals**	
1. Wheat (seed)	*Triticum aestivum*
2. Rice (seed)	*Oryza sativa*
3. Maize (seed)	*Zea mays*
(b) **Millets**	
1. Bajra (pearl millet) (seed)	*Pennisetum typhoideum*
2. Jawar (great millet) (seed)	*Sorghum vulgare*
3. Ragi (finger millet) (seed)	*Eleusine coracana*
(c) **Legumes**	
1. Matar (garden pea) (seed)	*Pisum sativum*
2. Chana (Bengal gram= chick pea) (seed)	*Cicer arietinum*
3. Arhar (red gram= pigeon pea) (seed)	*Cajanus cajan*
4. Mung (green gram) (seed)	*Phaseolus aureus*
5. Urd (black gram) (seed)	*Phaseolus mungo*
6. Soybean (seed)	*Glycine max*
7. Mungphali (ground nut = peanut) (seed)	*Arachis hypogea*
8. Lobia (cowpea) (seed)	*Vigna sinensis*
9. Masur (lentil) (seed)	*Lens culinaris*
(d) **Nuts**	
1. Almonds (badam) (seed)	*Prunus amygdalus*
2. Green almond (pista) (seed)	*Pistacia vera*
3. Cashew nut (kaju) (kernels)	*Anacardium occidentale*
4. English walnut (akhrot) (kernels)	*Juglans regia*
B. Spices and condiments	
1. Red pepper (chillies)	*Capsicum* sp.
2. Black pepper (kali mirch = black pearl)	*Piper nigrum*
3. Turmeric (haldi)	*Curcuma domestica*
4. Cumin (zira)	*Cuminum cyminum*
5. Coriander (dhania)	*Coriandrum sativum*
6. Clove (laung)	*Syzygium aromaticum*
7. Saffron (kesar)	*Crocus sativus*
8. Cardamom (chhoti ilaichi)	*Elettaria cardamomum*
9. Bengal cardamom (badi ilaichi)	*Amomum aromaticum*
10. Asafoetida (hing)	*Ferula assafoetida*

Common name with plant used	Botanical name
C. Edible oil	
1. Mungphali (ground nut=peanut)	*Arachis hypogea*
2. (a) Rape (b) Mustard	*Brassica; napus, B.campestris*
3. Til (sesame)	*Sesamum indicum*
4. Coconut	*Cocos nucifera*
5. Cotton seed	*Gossypium* sp.
D. Timber yielding plants	
1. Sisham (wood)	*Dalbergia sisso*
2. Rosewood (wood)	*D. latifolia*
3. Teak (sagaun) (wood)	*Tectona grandis*
4. Sal (wood)	*Shorea robusta*
5. Mulberry (wood)	*Morus alba*
6. Walnut (akhrot) (wood)	*Juglans regia*
7. White willow (wood)	*Solix alba*
E. Medicinal plants	
1. Sarpgandha (root)	*Rouwolfia serpentina*
2. Opium (afeem) (latex from unripe fruit capsule)	*Papaver somniferum*
3. Quinine (bark)	*Cinchona officinalis*
4. Belladona (dried leaves and roots)	*Atropa belladona*
5. Datura (fruit juice)	*Datura stramonium*
6. Amla (fruit)	*Emblica officinalis*
7. Kuchla (seed)	*Strychnos nux-vomica*
8. Isabgol (seed husk)	*Plantago ovata*
9. Liquorice (mulaithi roots)	*Glycorrhiza glabra*
10. Santonin (flowers)	*Artemesia cina*
11. Yam (tubers)	*Dioscorea* sp.
12. Digitalis (leaves)	*Digitalis purpurea*
13. Sadabahar (leaves)	*Catharanthus roseus*
F. Sugar yielding plants	
1. Sugarcane (stem)	*Saccharum officinarum*
2. Chukander (beet sugar) (root)	*Beta vulgaris*
G. Fibre yielding plants	
1. Kapas (cotton) (epidermal fibre of seed)	*Gossypium* sp.
2.Alsi (flax) (pericycle fibre)	*Linum usitatissimum*
3. Jute	*Corchorus capsularis; C.olitorius*
4. Sunn hemp (sunn) (stem fibre)	*Crotolaria juncea*
5. Coir (nariyal) (fruit wall)	*Cocos nucifera*
6. Hemp (bhang) (bark)	*Cannabis sativa*

Medical LASER

Laser is a class of devices that produces an intense beam of light of a very pure single colour. This light beam may be intense enough to vaporise the hardest and most heat-resistance materials. It is a source of intense radiation of the visible, ultraviolet or infrared portions of the light spectrum. Lasers are used in surgery to divide or weld tissues or to destroy tissue cells, hence they are known as **'light knife'**. Laser angioplasty is the opening of a blocked artery by delivering laser energy to the site through a fibre optic probe.

Uses of LASER

(i) The device is used to treat congenital cataract which occurs due to intrauterine injection. Laser beam is used to cut and breakup the lens. The injured lens tissue is then absorbed by the body.

(ii) Lasers are now used in many kinds of surgical operations.

(iii) They are widely used in Ophthalmology, especially for operations on the retina, like the repair of retinal tears and the destruction of small retinal tumours.

(iv) Lasers are also used for sealing of bleeding arteries in peptic ulcers, the destruction of abnormal cells in various parts of the body such as the cervix and treatment of endometrosis.

Oxygenators

Oxygenator is a device used in the oxygenation of blood during open heart operations. The first open heart surgery was performed by Walton Lillehei (USA) in 1952 by the use of technique called heart lung bypass.

Oxygenators are of four types

(i) **Bubble oxygenators** These are more traumatic to blood elements and are used for shorter procedures. They are least expensive and commonly used.

(ii) **Membrane oxygenators** These are more suitable for long operations and for the operation on infants. Here, the oxygenator blood does not come in direct contact with oxygen mixture, thus bubbles and foam are not formed.

(iii) **Film oxygenators** These are of several types. A feature common to them is that the blood is spread out in a thin film. Thus, is accomplished on plates, metal screens or rotating discs or spirals. An oxygen mixture flows over the thin layer of blood. These oxygenators are slightly less traumatic to the red cells, otherwise RBC membrane is so delicate and sensitive to artificial structures like oxygenator membrane that they damage and rupture easily. Film oxygenators have the disadvantage of being difficult to clean.

(iv) **Liquid-liquid oxygenators** These oxygenators use fluoridised organic compounds as a working fluid. These readily dissolve oxygen and CO_2, which then diffuse from the blood. Although, the blood is in direct contact with the liquid dissolve, it is chemically too different from the blood constituents to react with them.

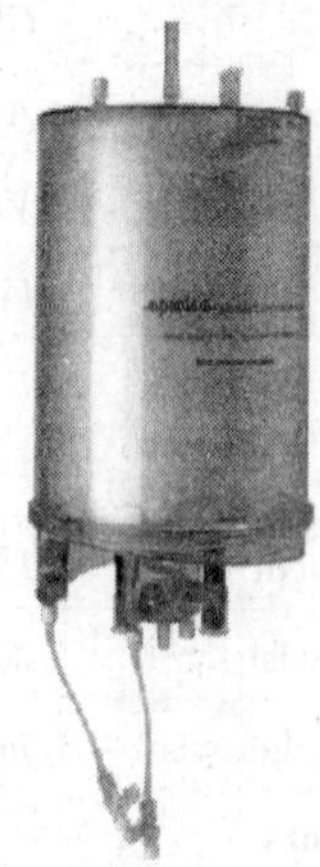

Oxygenator

MULTI-COLOURED PICTURES & DIAGRAMS

Biology as a subject always requires visualisation of concept for better understanding. Thus, some Diagrams & Pictures are given below with coloured visual effects to improved your grafting.

In May, 1986 International Committee of Virus Nomenclature named this virus as Human Immune Deficiency Virus (HIV). Its genome consists of single stranded RNA of two similar ribonucleotide chains. After entering in the body, this virus destroys the main cells of immune system, i.e. T-lymphocytes. As a result, immunity against pathogenic microbes decreases.

Structure of HIV

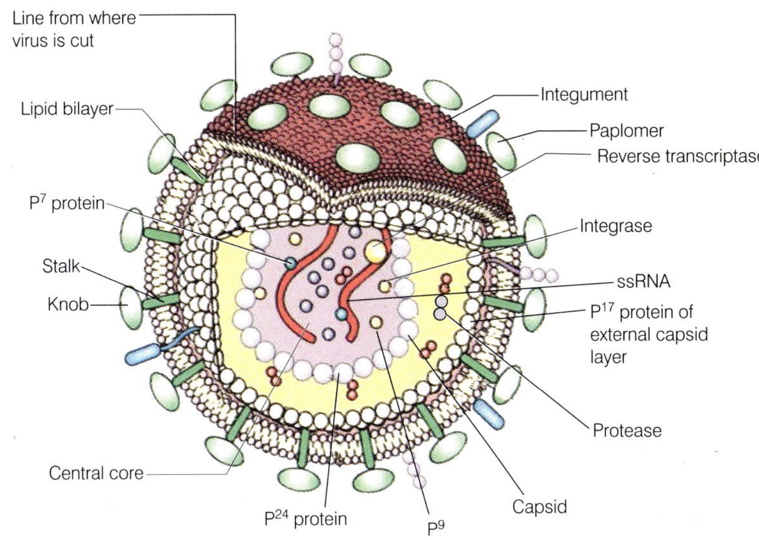

Structure of Bacterial Cell

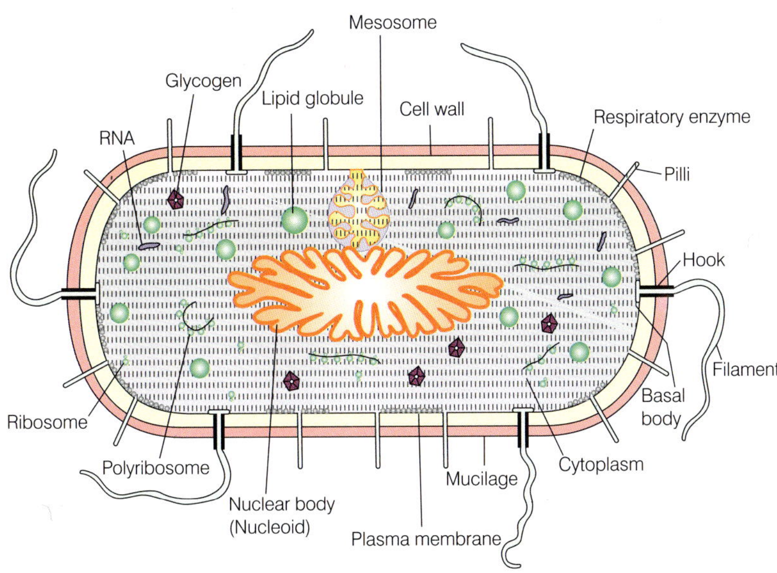

Bacterial cell is of prokaryotic type. Cell wall is clear, hard and chemically made up of peptidoglycan. Mesosomes formed by the folds of plasma membrane are helpful in respiration. Incomplete nucleus of bacterial cell is known as incipient nucleus or nucleoid. It lacks well developed cell organelles. 70 S ribosomes are formed.

Milk and Permanent Teeth

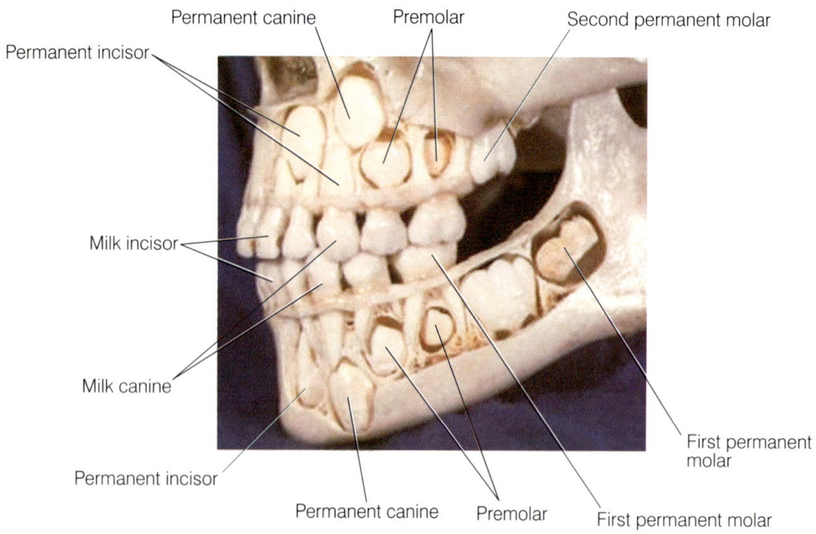

In milk teeth, there are 8 incisors, 4 canines and 8 molars. Usually, they are found between the age of 6 months to 3 years. At the age of 3 years, they are 20 in numbers, because children lack 8 premolars and 4 molars. In the figure, dentition of a 6 year old child has been shown. In it, milk and permanent, both types of teeth are visible.

Fold, Villi and Microvilli

To increase the digestion and absorption of digested materials, the absorptive area of small intestine increases many folds due to the presence of specialised folding and finger-like projections. As a result, in small intestine, the absorption of food occurs through three structures:

(i) By plicae circulares
(ii) By finger-like villi
(iii) At cellular level by finger-like microvilli

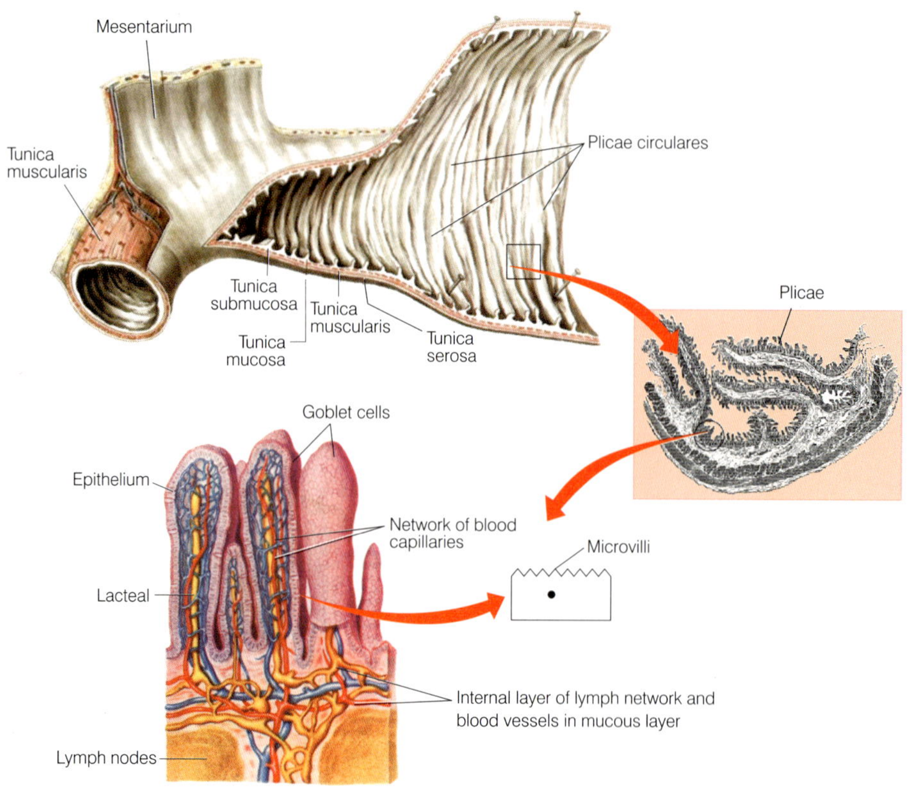

Basic functional unit of lungs are alveoli. In humans, they are about 300 million. Air is filled in their lumen. Blood capillaries form a network on the connective tissue present between walls of two adjacent alveoli. These blood capillaries are the terminal branches of pulmonary artery and pulmonary vein. Gas exchange occurs between alveolar among pulmonary of alveolar wall and blood present in capillaries.

Ultrastructure of Alveoli

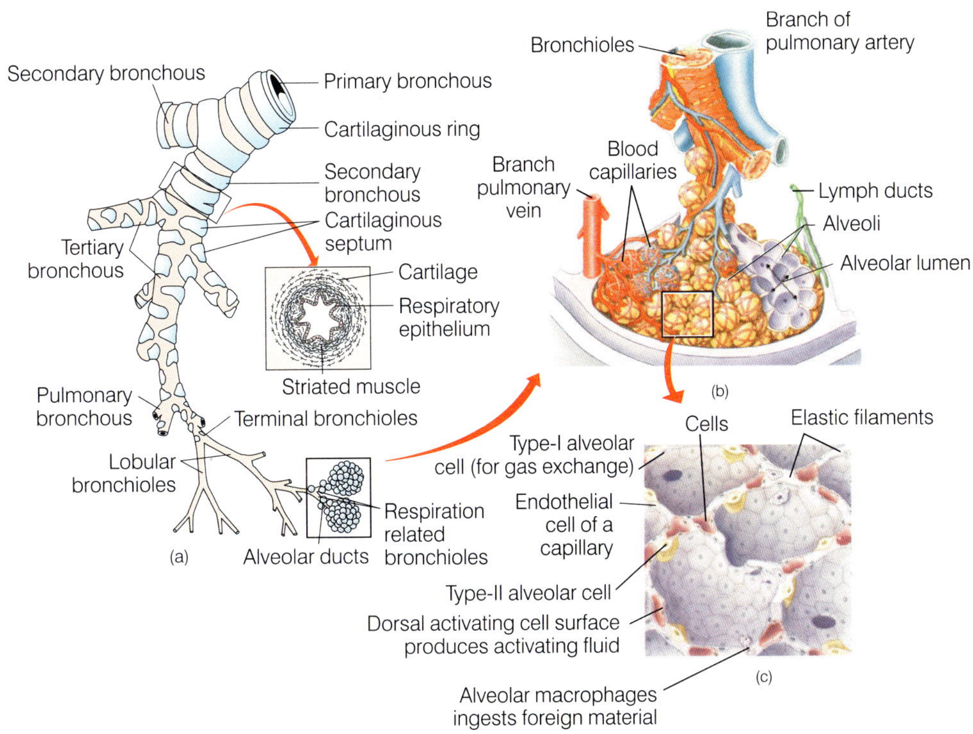

Human Ear

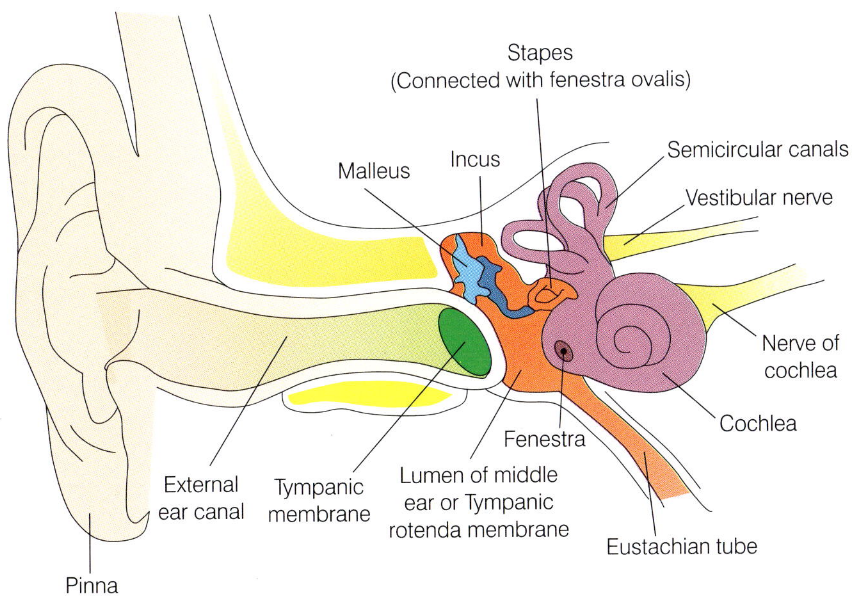

Structure of ear is most complex in mammals. Except receiving sound waves, utriculus, sacculus and three semicircular canals of ear are helpful in maintaining the balance of body. There are three parts of human ear external ear, middle ear and internal ear. Among the ossicles, malleus, incus and stapes, stapes is the smallest bone of the body.

How Heart Attack Occurs?

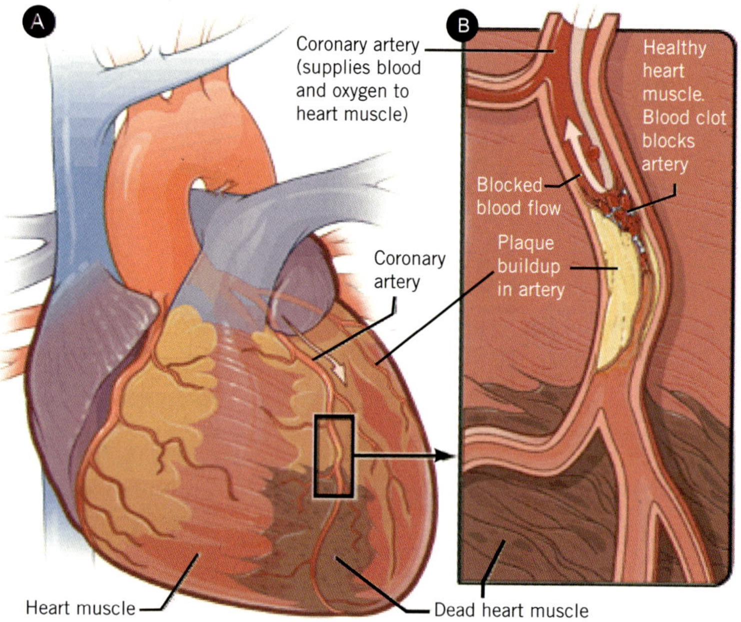

Most heart attacks happen when a clot in the coronary artery blocks the supply of blood and oxygen to the heart. A blockage that is not treated within a few hours causes the affected heart muscle to die.

Lichens

Lichens are self supporting composites of a fungus (whitish) and a photosynthesis alga (reddish) as shown in figure. The fungus gives structural support and protection to the algal cells which by photosynthesis produce carbohydrates. The fungus utilises some of these to obtain energy for growth and reproduction. This relationship enables both components to live in extreme environments.

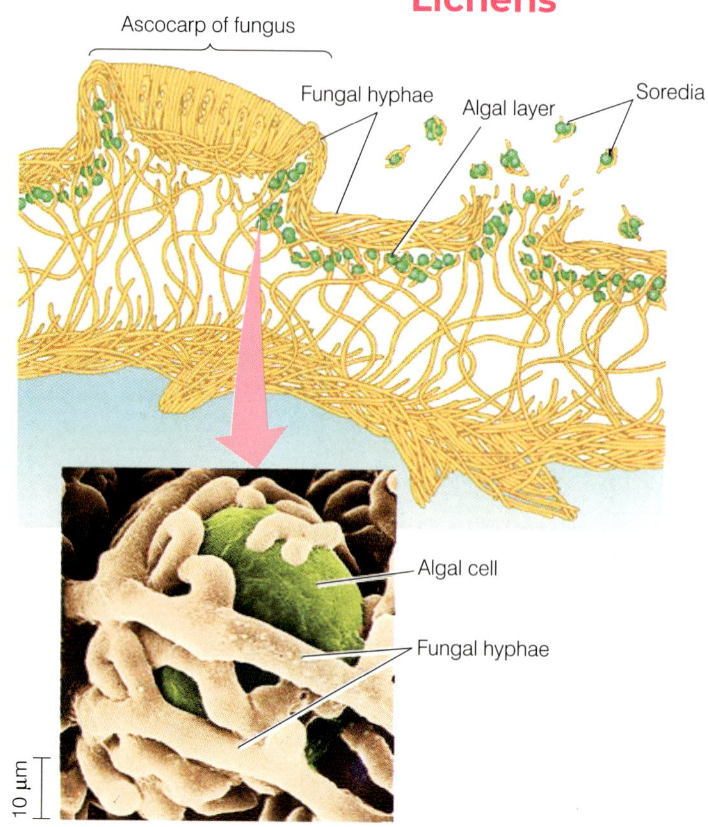

Ultrastructure of Yeast Cell (By Simple Microscope)

Saccharomyces is a less developed member of Ascomycetes. It lacks mycelium. Its cell wall consists of 6-8% protein, 8.5-13.5% lipid, 1-2% chitin and mostly two types of polysaccharides namely glucan (30-35%) and mannon (30%).

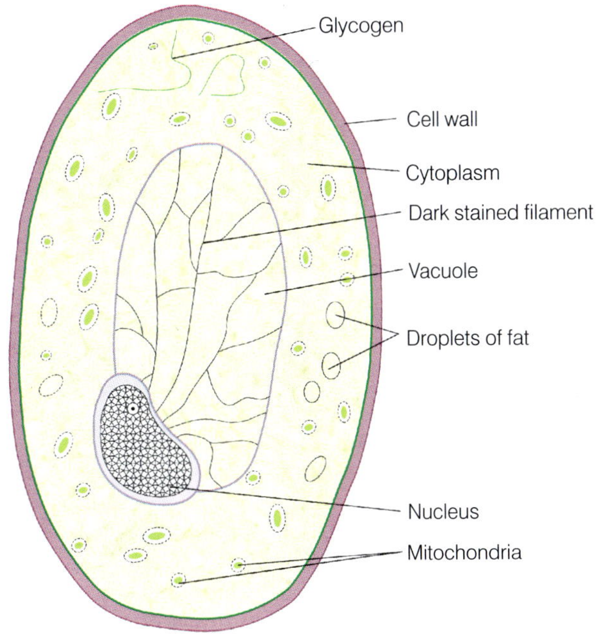

Ultrastructure of Yeast Cell (By Electron Microscope)

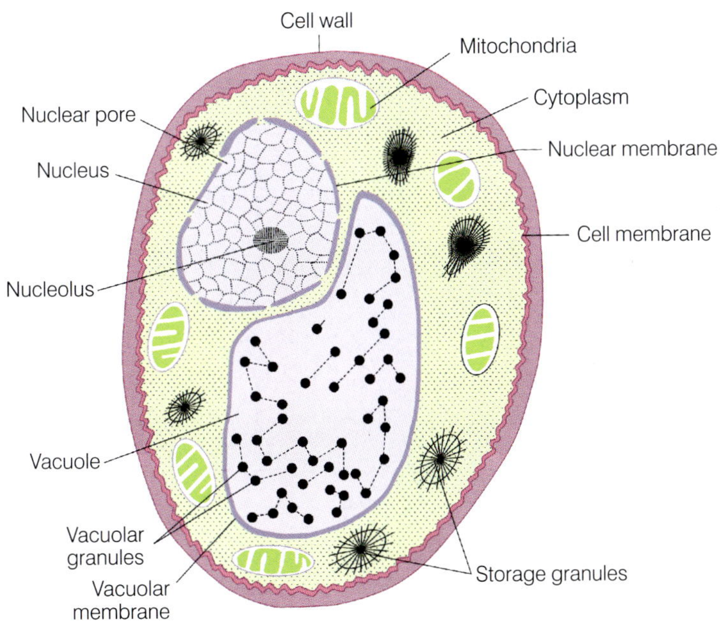

In the centre of yeast cell, there is a vacuole. Nucleus lies near the vacuole. According to the electron microscopic studies of Agar and Dougles and Carty and Naylar, nucleus has its own nuclear membrane which is different from vacuolar membrane.

Stem, Black or Cereal Rusts

It is caused by the fungus *Puccinia graminis*. It is an epidemic of stem rust on wheat is caused by race Ug99. Ug99 or TTKS, is a race of black stem rust (*Puccinia graminis tritici*). It is virulent to the great majority of wheat varieties. The stem rust fungus attacks the parts of the plant which are above ground. Spores that land on green wheat plants form pustules that invade the outer layers of the stalk. Elliptical blisters or pustules called **uredia** develop on the site of infection on the stem or leaf. Infected plants produce fewer tillers and set fewer seeds. In most of the cases, due to severe infection the plant may die.

Active Transport Through Na^+ — K^+ Pump

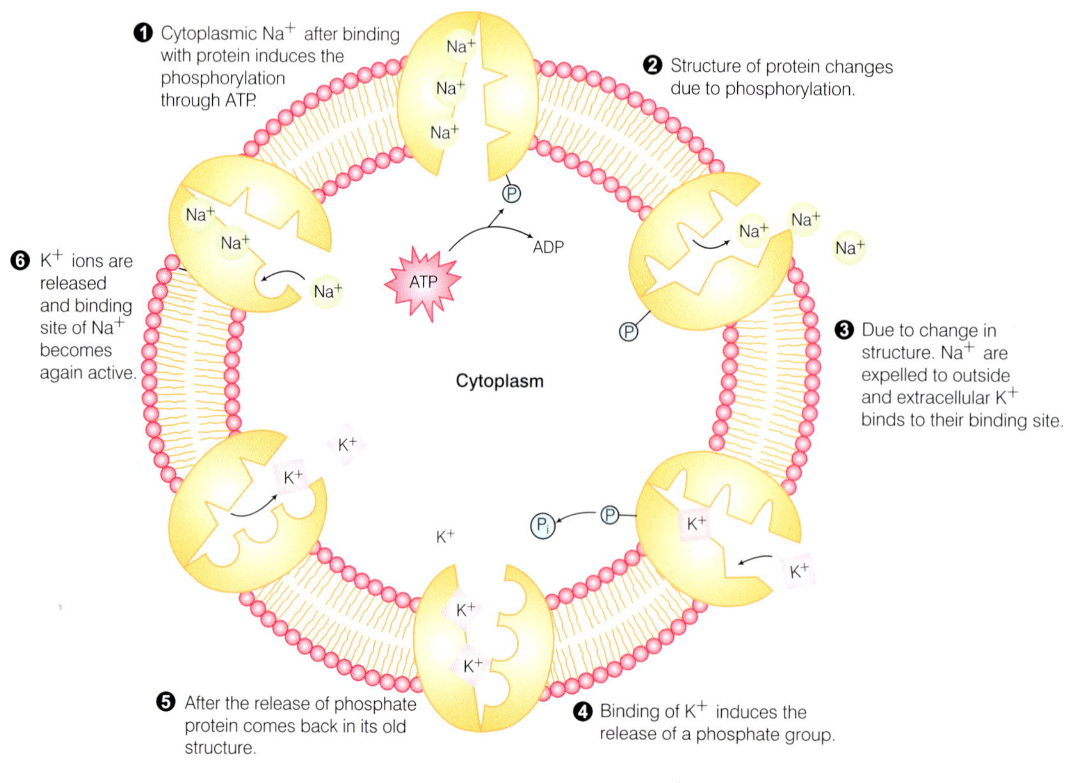

Path of Water Movement

(a) Symplast pathway through plasmodesmata
(b) Apoplast pathway through cell wall

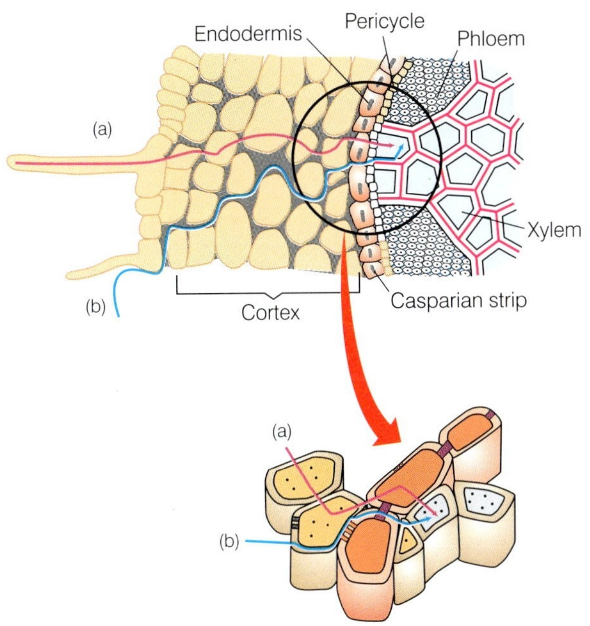

Facilitated Diffusion

(a) Uniport (b) Symport (c) Antiport

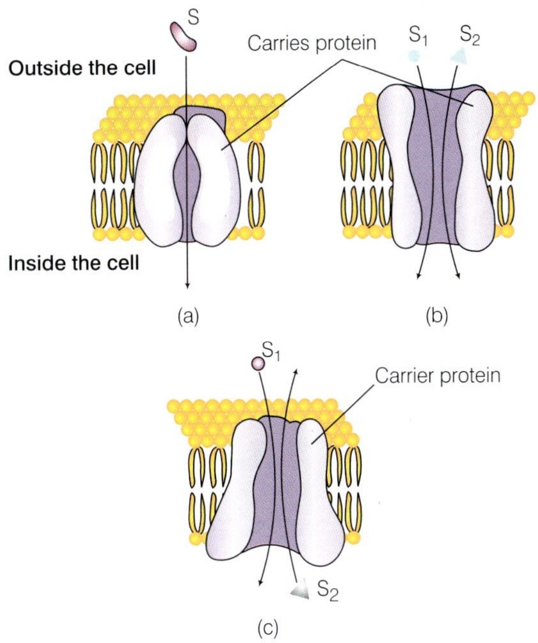

Canal System

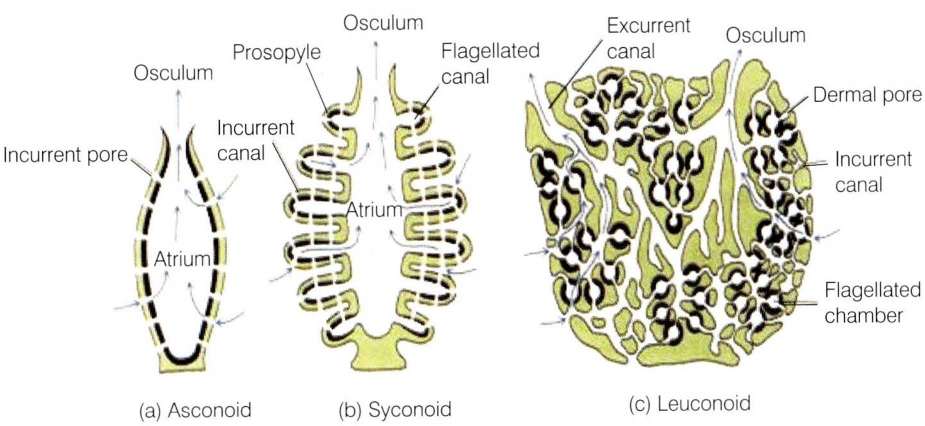

Canal system is characteristics of sponges. By contrast beating of flagella of choanocyte cells, a water current is produced which has a fixed direction. All exchange between sponges body and external medium are maintained by means of this current. The canal system helps in respiration, circulation, excretion, nutrition, reproduction, etc. It can be *Ascon*, *Sycon* and *Leucon* type.

Secondary Growth

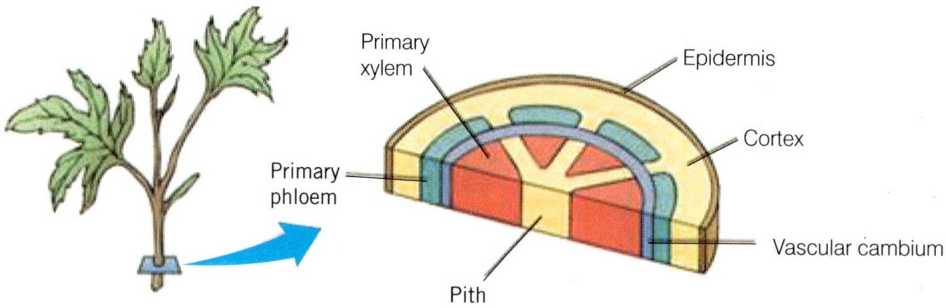

(a) At the onset of secondary growth vascular cambium arises in the parenchyma between the vascular bundles (i.e. in the pith rays), forming a cylinder of meristematic tissue (blue circle in cross section).

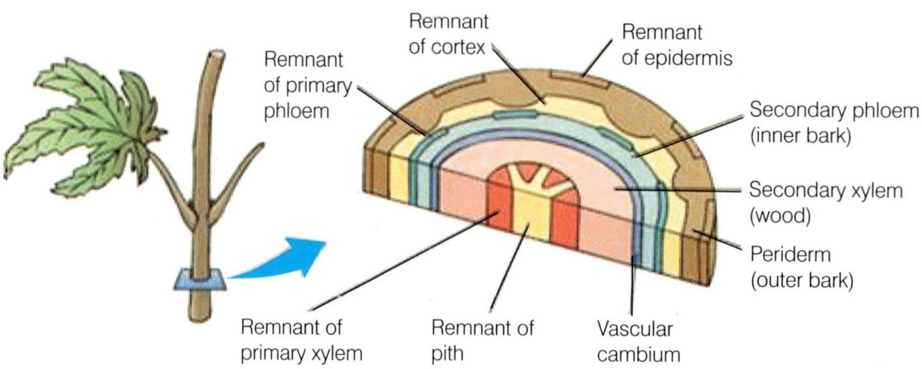

(b) Vascular cambium begins to divide, forming secondary xylem on the inside and secondary phloem on the outside.

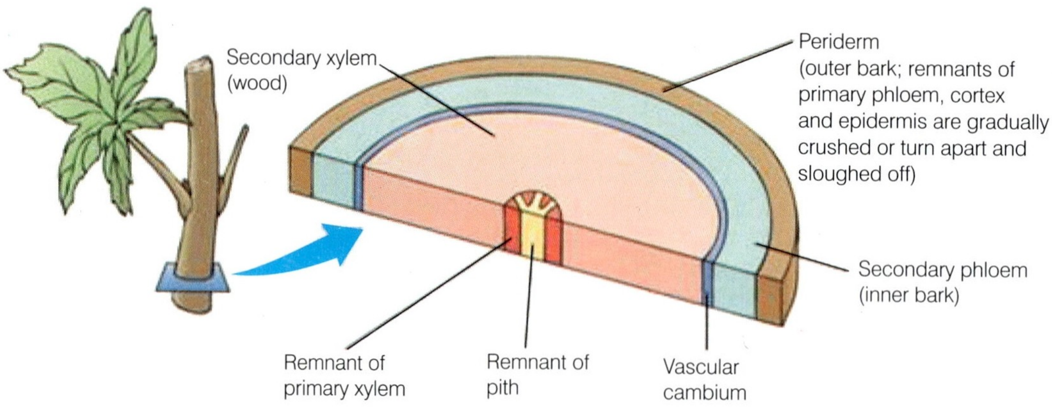

(c) A young woody stem. Vascular cambium produces significantly more secondary xylem than secondary phloem.

Secondary growth can be defined as growth that increases the girth (diameter) and thickness of the plant as a result of activities of primary and secondary lateral meristem. It occurs in perennial gymnosperms and dicots and provide strength to bear the increasing weight of the aerial growth.